INFINITE SERIES

$$\sum_{n=0}^{\infty} r^n = \frac{1}{1-r} \quad \text{(geometric series)}$$

$$\cos x = \sum_{n=0}^{\infty} \frac{(-1)^n}{(2n)!} x^{2n}$$

$$e^x = \sum_{n=0}^{\infty} \frac{x^n}{n!}$$

$$f(x) = \sum_{n=0}^{\infty} \frac{f^{(n)}(a)}{n!} (x-a)^n \quad \text{(Taylor series)}$$

$$\sin x = \sum_{n=0}^{\infty} \frac{(-1)^n}{(2n+1)!} x^{2n+1}$$

$$R_n(x) = \frac{f^{(n+1)}(c)}{(n+1)!} (x-a)^{n+1} \quad \text{(nth Taylor remainder)}$$

GEOMETRIC FORMULAS

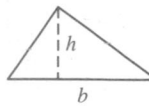

area: $\frac{1}{2}bh$

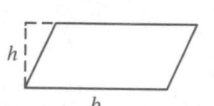

area: bh

area: πr^2

circumference: $2\pi r$

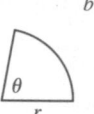

area: $\frac{\theta}{2}r^2$ (θ measured in radians)

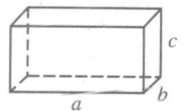

volume: abc
surface area: $2ab + 2ac + 2bc$

volume: $\frac{4}{3}\pi r^3$
surface area: $4\pi r^2$

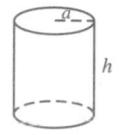

volume: $\pi a^2 h$
surface area: $2\pi a h$

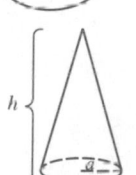

volume: $\frac{1}{3}\pi a^2 h$
surface area: $\pi a \sqrt{a^2 + h^2}$

$$|PQ| = \sqrt{(x_2 - x_1)^2 + (y_2 - y_1)^2} \quad \text{(distance formula)}$$

$y - y_1 = m(x - x_1)$ (point-slope equation of a line)

$y = mx + b$ (slope-intercept equation of a line)

$$y - y_1 = \frac{y_2 - y_1}{x_2 - x_1}(x - x_1) \quad \text{(two-point equation of a line)}$$

MEASUREMENT OF ANGLES

$$\frac{\theta_r}{2\pi} = \frac{\theta_d}{360}$$

θ_r is the angle measured in radians θ_d is the angle measured in degrees

VOLUME OF TRIGONOMETRIC FUNCTIONS FOR STANDARD ANGLES

θ (degrees)	θ (radians)	$\sin \theta$	$\cos \theta$	$\tan \theta$
0°	0	0	1	0
30°	$\pi/6$	1/2	$\sqrt{3}/2$	$\sqrt{3}/3$
45°	$\pi/4$	$\sqrt{2}/2$	$\sqrt{2}/2$	1
60°	$\pi/3$	$\sqrt{3}/2$	1/2	$\sqrt{3}$
90°	$\pi/2$	1	0	undefined

CALCULUS FIFTH EDITION

Saunders College Publishing

A Harcourt Brace Jovanovich College Publishers

Fort Worth   Philadelphia   San Diego   New York
Orlando   Austin   San Antonio   Toronto
Montreal   London   Sydney   Tokyo

CALCULUS

FIFTH EDITION

Stanley I. Grossman

University of Montana and University College London

To the Memory of
BENJAMIN GROSSMAN, *1906–1986*

Text Typeface: Times Roman
Compositor: SYNTAX INTERNATIONAL PTE LTD
Acquisitions Editor: Jay Ricci
Developmental Editor: Richard Koreto
Project Editor: Jennifer Johnson
Copy Editor: Karl Yambert
Art Director: Louise Sandy-Karkontli
Text Designer: Linda Miller
Cover Designer: Lawrence Didona
Text Artwork: SYNTAX INTERNATIONAL/ACADEMY ARTWORKS
Director of EDP: Tim Frelick
Production Manager: Lynne Bush
Marketing Manager: Monica Wilson

Cover credit: METAPHASE 3 © 1986 Barbara Kasten

Printed in the United States of America

CALCULUS

ISBN 0-03-096420-2

Library of Congress Catalog Card Number: 92-17664

2345   048   987654321

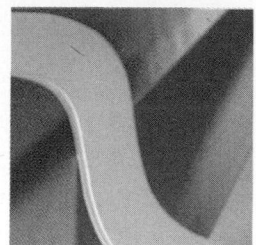

PREFACE

*One learns by doing the thing; for though you think
you know it, you have no certainty until you try.*
                                        SOPHOCLES

The study of calculus has been of central importance to scientists throughout most of recorded history. Our present understanding of the subject owes much to Archimedes of Syracuse (287–212 B.C.), who developed what were, for his time, incredibly ingenious techniques for calculating the areas enclosed by a great variety of curves. Archimedes' work led directly to the modern concept of the integral. Much later, toward the end of the seventeenth century, Sir Isaac Newton and Gottfried Leibniz independently showed how to calculate instantaneous rates of change and slopes to curves. This development inspired an enormous variety of mathematical techniques and theorems that could be used to solve problems in a number of diverse and often unrelated fields.

Many of these techniques and theorems are considered to be part of "the calculus." No textbook could discuss all the myriad results in calculus discovered over the past several hundred years. Fortunately, a weeding-out process has occurred, and there is now fairly widespread agreement which topics properly should be included in a three-semester (or four- to five-quarter) introduction to the subject. This book includes all the standard topics. It also includes a number of unique features which were written to accomplish three goals:

- to make calculus easier for students to learn through the use of examples, exercises, graphs, and devices inserted to take the mystery out of difficult ideas in the theory of differentiation and integration,

- to help students to appreciate the beauty of calculus through the use of applications in a wide variety of fields, and

- to make calculus interesting by discussing the historical development of the subject. Knowing, for example, the complicated tricks that had to be employed in order to find the equation of a tangent line (before Newton and Leibniz discovered the derivative) should help the student to appreciate the significance of Newton's and Leibniz's discovery (see pages 105–111).

In the fourth edition I made a number of changes to make the book more "user-friendly." The biggest change was to organize the problem sets into a more usable form. In this new, fifth edition, I incorporated the fourth edition changes and added several new features that, I believe, have helped further to achieve the goals outlined above. Here are some details:

EXAMPLES

The book contains approximately 800 examples. Each example includes *all* the algebraic steps needed to complete the solution. In many instances, explanations are highlighted in color to make a step easier to follow.

EXERCISES

The text includes over 7000 exercises—including both drill and applied problems. More difficult problems are marked with an asterisk (*) and a few especially difficult ones are marked with a double asterisk (**). The exercises provide the most important learning tool in any undergraduate mathematics textbook. I stress to my students that no matter how well they think they understand my lectures or the textbook, they do not really know the material until they have worked problems. A vast difference exists between understanding someone else's solution and solving a new problem by yourself. Learning mathematics without doing problems is about as easy as learning to ski without going to the slopes.

Each exercise set starts with "self-quiz" problems that are either True-False or multiple choice. These test the student's understanding of the material in the section he or she has just read. Answers to these problems are given at the end of the problem set. The problems are then divided into groups: drill, applications, calculator problems, show/prove/disprove, and challenge.

CHAPTER REVIEW EXERCISES

At the end of each chapter, I have provided a collection of review exercises. Any student who can do these exercises can feel confident that he or she understands the material in the chapter.

REVIEW CHAPTER

Chapter 0 includes a general discussion of several topics that are commonly taught in an intermediate algebra–college algebra course, including absolute value and inequalities, circles and lines, and an introduction to functions. This chapter also includes some detailed discussion of graphing techniques, including a unique section on shifting the graphs of functions—an extremely useful procedure.

INTUITION VERSUS RIGOR

Intuition rather than rigor is stressed in the early parts of the book. For example, in the introduction to the limit in Chapter 1, the student's intuition is appealed to in the initial discussion of limits, and the concept is introduced by means of examples and detailed tables. I believe that the student can appreciate the "$\epsilon - \delta$" or "neighborhood" approach to limits only after he or she has developed some feeling for the limit. However, I also feel strongly that standard definitions must be included, since mathematics depends on rigorous, unambiguous proofs. To resolve this apparent contradiction, I have put much of the one-variable theory in separate optional sections (Sections 1.8 and 4.8, and Appendix 3). These sections can be included at any stage, as the instructor sees fit, or omitted without loss of continuity if time is a problem. By the time students reach infinite series, they should have developed enough mathematical sophistication to understand and appreciate some of the subtleties of mathematical proof. Thus, beginning with the discussion of convergence of sequences and series in Chapter 10, I have included ϵ's and δ's whenever necessary.

EARLY TRIGONOMETRY

The derivatives of all six trigonometric functions are computed in Section 2.5. Equally important, all six are then used in applications throughout the rest of the book—not just $\sin x$ and $\cos x$. See, for example, the application to blood flow in Example 3.5.7 (on page 221) that makes use of the derivatives of $\csc x$ and $\cot x$.

TRIGONOMETRY REVIEW

High school trigonometry is, of course, a prerequisite for calculus. Many students, however, come to calculus without having seen trigonometry for several years. Much has been forgotten. To fill this gap and to free the instructor from the necessity of taking class time to review trigonometry, I have included a four-section appendix (Appendix A1), which contains all the precalculus trigonometry a student needs.

PEDAGOGY

The book is written to be understood by students. Details are always included and a student is never told that something is "obvious." In this edition there are several features that I believe will make calculus clearer. Here are five of them.

- The discussion of the chain rule in Section 2.3 has been extensively rewritten. Discussion of "inside" and "outside" functions makes the rule—which gives many students difficulty—much easier to use. The chain rule is used to differentiate made-up functions (Example 6 on page 150) to illustrate in yet another way how it is applied. This idea is carried into the problem set (see Problems 25–44 on pages 153–154).

- In Section 4.4 areas are approximated using rectangles. This is common. But then, in Examples 4 and 5 on pages 285 and 288, areas are approximated using the mean value theorem. With every detail included, students see proofs in two particular and very concrete cases of the fundamental theorem of calculus. When they get to the FTC in Section 4.6, it will be an old friend.

- Students often get bogged down in messy computations of integrals. In the self-quiz problems on pages 485, 490, 510, 517 and 532, students are asked to discard blatantly wrong integrals in order to choose the correct one. This helps them to develop a sense of what an integral should look like, without doing many calculations.

- There are two approaches to exponential and logarithmic functions. In Section 6.2 a^x is defined for x rational as it was done in high school or a precalculus course in college. Then $\log_a x$ is defined as the inverse of a^x and e is defined as a limit. At the end of the section, the derivatives of $\ln x$ and e^x are computed. In Section 6.3 $\ln x$ is defined as an integral; e^x is defined in Section 6.4 and a^x and $\log_a x$ are defined in Section 6.5. The two approaches are proved equivalent on page 427. A lecturer can use either approach. However, the "high school" approach in Section 6.2 is less abstract than the integral definition and will be more comfortable to many students. If both approaches are covered, then students will see how, in calculus, some very different and important ideas can lead to the same place.

- The gradient is introduced in Section 13.5 as a natural extension of the derivative of a function of one variable. Then basic ideas like tangent planes and directional derivatives are discussed in terms of the geometry of the gradient. This approach makes some basic ideas in multivariable calculus both easier to understand and easier to generalize to dimensions greater than two or three.

GRAPHS

There are 729 figures in the book—about 50% more than in the fourth edition. Graphs are used in the problem sets to develop a better understanding of central concepts. Here are some examples:

- In Section 0.4 students are asked to look at a graph to decide whether it is the graph of a function.

- In Section 0.5 students are asked to shift given graphs vertically and horizontally and reflect them around the x- and y-axes. This gives students practice with sketching graphs before they have seen a derivative.

- In Section 1.2 students are asked to evaluate limits by examining the graphs of functions.

- In Section 1.5 students are asked questions about the derivatives of functions by looking at their graphs.

- In Section 1.7 students are asked to determine points of discontinuity and non-differentiability by looking at graphs.

- In Section 4.4 students are asked to use a calculator to approximate the area under a sketched curve.

- In Section 6.1 students are asked to sketch the graph of an inverse function by reflecting a given graph about the line $y = x$.

- In Chapter 8 students are asked to match equations of ellipses and parabolas with their graphs.

- In Section 13.1 students are asked to match twelve functions of two variables with twelve three-dimensional surfaces.
- There is greater emphasis on the geometry of space. New features include a description of how to draw a plane (on pages 770–771) and the geometric interpretation of the scalar triple product (on page 762).

CALCULATORS, COMPUTERS, AND NUMERICAL ANALYSIS

Calculators and computers are playing an ever increasing role in the study of calculus and the fifth edition of *Calculus* reflects that fact. There are computer exercises at the ends of most chapters. These require the use of a reasonably sophisticated symbolic software package. Moreover, the use of the calculator is integrated throughout the text. However, I stress that no software is needed to study calculus from this text nor is any computer software package referred to in the text. I think it is meaningless to compute derivatives using MATHEMATICA, for example, before a student understands what a derivative is. There are manuals that describe how to use MAPLE, DERIVE, and MATHEMATICA that are available to users of this book. But if they are used, they should be used after a student has mastered the basic concepts.

In past editions a calculator was used in many examples and exercises. The symbol ▦ was (and still is) used to indicate the need for a calculator. In this edition the calculator is used much more extensively as a pedagogical tool in addition to its use for carrying out involved computations. Here are some examples:

- The limit of $\dfrac{\sin \theta}{\theta}$ as $\theta \to 0$ is found on a calculator (in Section 1.2) long before it is formally derived.
- In Sections 1.2 and 1.4 students are asked to use their calculators to guess at more difficult limits (Problems 59–92 in Section 1.2). In Section 1.2 and 1.4 they are asked to approximate limits involving $\ln x$ and e^x long before these functions are defined. For example, in Example 10 on page 80, $\lim\limits_{x \to \infty} \left(1 + \dfrac{1}{x}\right)^x$ is approximated and in Problems 27–35 on page 82, students are asked to approximate related limits. When students do get to these functions in Sections 6.2–6.4, they will have had practice working with them on a calculator.
- In Example 9 on page 92 (in Section 1.5) students are shown how to use a calculator to approximate, from the definition, the derivative of $\sin x$ at $x = \pi/6$. The precise derivative is computed in Section 2.5. This idea is reinforced in problems 25–38 on page 96.
- In Example 5 in Section 3.5 (pages 218–220) the bisection method is used to approximate the endpoints of an interval over which time is to be minimized to show that time is minimized when Snell's law applies. This is the only book that includes this difficult, but important, check.
- Coverage of Newton's method has been expanded. There is extensive discussion and a partial proof of a quadratic convergence theorem in

Section 3.6 (with the proof—using the mean value theorem—in Section 3.8). In Section 3.9 I apply Newton's methods to a cubic with three real zeros to obtain a fractal. While several calculus books provide pretty pictures of Mandelbrot sets and the like, this is the only calculus book that shows students how to generate their own fractals, and encourages them to do so in the problem set (of Section 3.9).

- A calculator is used to show that the area under one arch of the curve $y = \sin x$ is approximately 2, two sections before the fundamental theorem of calculus is proved.

- Students are shown how to obtain e^x on a calculator on page 423. This not-so-obvious lesson is needed by many students in their first calculus course.

- In Section 7.9 (numerical integration) there is much greater emphasis than is common on the use of error bounds to approximate an integral.

- Likewise, the coverage of approximation using Taylor polynomials is much more extensive; see, for example, page 627.

- A calculator is used to approximate the sum of a convergent series on page 670.

- There is more coverage given to error analysis for alternating series. See the improved error found on page 684 and the approximation of a cosine integral on page 702.

- Newton's method for functions of two variables is given in the new Section 13.12.

- **USE OF THE GRAPHING CALCULATOR** Many students now have access to calculators that can, if used properly, provide accurate graphs of a great number of functions. In Appendix A6 I show students how to use their calculators effectively to draw graphs of functions, sketch conic sections and polar graphs, find zeros of polynomials, and solve other types of algebraic problems. Example 14 on page A-56 shows the limitations of such calculators by discussing a polynomial whose graph *cannot* accurately be sketched on a calculator. The appendix is written to be used with any calculator now available.

APPLICATIONS

The fifth edition of *Calculus* contains all the physical applications that have become common in calculus textbooks. There are a number of less common applications as well. Here are some of them:

- The problem sets have many interesting applications such as luminosity and magnitude of a star (p. 159), Weber-Fechner laws (pp. 406 and 438), symbiosis (p. 470), and partial differential equations (Sections 13.3 and 13.4).

- Minimizing resistance in blood flow (Example 8 on page 221).

- In economics, the Lorenz curve—an application of the mean value theorem (Example 7 on page 244).

- Calculating escape velocity (Example 8 on page 362).

- The ellipse, parabola, and hyperbola in the real world (pp. 546ff, 555ff, 566ff).
- Regression Lines—Section 13.10.

HISTORICAL NOTES AND BIOGRAPHICAL SKETCHES

Calculus becomes more interesting if you know something about the people who developed it and the problems they had to overcome. *Calculus, Fifth Edition* has a well-integrated approach to history. The history comes in two ways: First, there are 23 biographical sketches of important mathematicians scattered throughout the text. This is slightly expanded from the fourth edition. New to this edition are ten essays describing the development of an important topic in calculus. These are from one to six pages in length. By seeing, for example, the extraordinarily complex tricks Descartes employed in the seventeenth century in order to find the slope of a tangent line to a curve, a student should be impressed by the beauty and simplicity of the derivative.

Here is a list of these historical discussions:

Historical Essays

1. Historical Introduction to Problems of Antiquity That Were Solved by the Invention of Calculus (Section 1.1)
2. Solutions to the Tangent Line Problem in the Early Seventeenth Century (pp. 105–111)
3. Introduction to the Area Problem (Section 4.1)
4. Development of the Integral in the Seventeenth Century (pp. 314–317)
5. Computing Arc Length in the Seventeenth Century (pp. 352–354)
6. Compound Interest in the Seventeenth Century (p. 449)
7. Johann Bernoulli and L'Hôpital's Rule (pp. 597–598)
8. The Struggle to Understand Infinite Sums (pp. 663–667)
9. Josiah Willard Gibbs and the Origins of Vector Analysis (pp. 763–764)
10. Distinguishing Between Ordinary and Partial Derivatives (pp. 862–863)

Biographical Sketches

1. Descartes (p. 14)
2. Euler (p. 34)
3. Leibniz (p. 51)
4. Newton (pp. 52 and 53) (includes an alternative biography written by Stephen Hawking)
5. Cauchy (p. 64)
6. Fermat (p. 93)
7. Barrow (p. 109)
8. Archimedes (p. 266)
9. Riemann (p. 306)
10. Cavalieri (p. 314)
11. Pascal (p. 316)
12. Wallis (p. 317)
13. Sir Christopher Wren (p. 353)

14. Pappus of Alexandria (p. 375)
15. Jakob Bernoulli (p. 449)
16. Johann Bernoulli (p. 597)
17. L'Hôpital (p. 598)
18. Lagrange (p. 624)
19. Maclauren (p. 713)
20. Gibbs (p. 764)
21. Weierstrass (p. 901)
22. Gauss (p. 1021)
23. Jacobi (p. 1029)

ADDITIONAL TOPICS

Calculus, Fifth Edition contains a number of interesting items gleaned from journals. These provide further insight into commonly taught topics. Here are a few of them:

- Using a Graph to Integrate an Inverse Function (p. 484)—an alternative to integration by parts
- Fixed point iteration—in a section on sequences (pp. 648–650)
- Parametric integration—an application of double integrals; using this unusual technique one can easily show, for example, that

$$\int_0^\infty \frac{\sin x}{x}\, dx = \frac{\pi}{2} \text{ (p. 936)}$$

- A relatively easy proof of L'Hôpital's rule in the ∞/∞ case (p. 608)
- Examples of Hardy to illustrate that an alternating series may diverge if the terms are not monotone, even if the terms approach zero; and that the comparison test cannot be applied to an alternating series (pp. 686 and 687)
- Necessary and sufficient conditions in order that a differential equation be separable (p. 1037)

ANSWERS

The answers to most odd-numbered exercises appear at the back of the book. In addition, a *Student's Solution Manual* containing detailed solutions to all odd-numbered problems is available, as is an *Instructor's Solutions Manual* containing detailed solutions to the even-numbered problems. Both manuals were prepared by Richard Lane at the University of Montana.

NUMBERING IN THE TEXT

Numbering in the book is fairly standard. Examples, problems, theorems, and equations are numbered consecutively within each section, starting with 1. Reference to an example, problem, theorem, or equation outside the section in which it appears is by chapter, section, and number. Thus, what is labeled simply as Example 4 in Section 2.3 is referred to as Example 2.3.4 outside the section. In addition, in many cases cross-referenced page numbers are included to make it easy to find an important reference.

ACCURACY

The success of a calculus book depends to a large extent on its accuracy. A few badly placed typographical errors, particularly in the answers to problems can turn a good teaching tool into a source of confusion and frustration. Consequently, we have gone to extraordinary lengths to ensure that the answers at the back of the book are correct.

Step 1. An answer key was provided by Richard Lane who both reorganized the problem sets and wrote the solutions manuals.

Step 2. Four mathematics instructors solved every odd-numbered problem.

Step 3. I compared their answers with the answers provided by Mr. Lane. I reworked those problems where there was a discrepancy and prepared a revised draft of answers for the publisher.

Step 4. Three graduate students at University College London solved the odd-numbered problems. They compared their answers with the revised draft and found a small number of answers that were incorrect. These were corrected.

Step 5. Joseph Crepeau at the University of Montana and Charles Stone at DeKalb College in Clarkstown, Georgia proofread the answers in the students' solutions manual and compared them with the typeset answers at the back of the book. A small number of errors were found and these were corrected.

Step 6. The page proofs for these answers were proofread more times than I care to remember.

The result is an answer section that is as clean as is within human ability to compile. If you do find an error in an answer or in the text, then please send it to the publisher or to me. It will be corrected in the next printing.

TO THE STUDENT: HOW TO STUDY CALCULUS

In October 1987, a conference was held at the National Academy of Sciences in Washington, D.C. The title of the conference was "Calculus for a New Century," but a more accurate title would have been "The Crisis in Calculus Instruction." The "crisis" is that in 1987 (and in 1992) 40–50% of all students who began a calculus course did not complete it successfully. This means that these students either dropped the course or failed it.

There were many proposed solutions to this problem. The most obvious ones were to teach more effectively, to use better textbooks, and to enforce stricter prerequisites before allowing students to enroll in calculus. The prerequisite for your calculus course is a reasonable knowledge of high school algebra and trigonometry. But the dropout rate of students who meet this prerequisite is far too high. Evidently, adequate preparation for calculus is not sufficient.

In the 1970s, Uri Treisman, a mathematician at the University of California at Berkeley began an extensive study to determine why students did poorly in calculus. Contrary to popular belief, he found that variables such as family wealth or social background gave little indication of a student's likely success in calculus.

What Professor Treisman did discover was that students who did homework in groups were far more likely to do well than were students who worked alone. He observed students who worked in groups of three or more. One student would get an answer that was wrong; a second student would find the error and correct it. This process was repeated continually with the result that virtually all the students in the group understood how to approach the problem under discussion correctly. More generally, students who work together often pick up lots of useful ideas and "tricks" from each other to help them do well in the course.

For students who worked alone, misconceptions went unchallenged. This led to a downward spiral of frustration and self-doubt. Loners were more likely to lose interest in the course and thereby become deflected from achieving their educational and career goals.

It is generally believed that it is easier to learn mathematics from scratch, rather than first having to unlearn bad techniques. It's simpler to teach typing or skiing to someone who has never typed or skied before, than to someone who has begun to practice either activity incorrectly. The same rule applies to learning calculus (or any other mental activity). It is truly better to do it right the first time, and the best way to do that is to have someone find and correct your errors *before* they turn into bad habits.

In the late 1980s Professor Treisman did an experiment. He and several of his colleagues in universities, state colleges, and community colleges across the country set up experimental sections of calculus. In these sections, students were required to check each other's homework before it could be submitted. The homework was graded and this had the effect of underscoring its importance.

The results were dramatic. While the dropout or failure rate in the ordinary calculus sections at Berkeley and elsewhere remained at the 40–50% level, the resulting rate in the "experimental" sections fell to below 10%.

Earlier in this preface I urged you to work problems because you cannot learn mathematics without doing so. Let me now add to that suggestion:

DO YOUR CALCULUS STUDYING AND HOMEWORK IN GROUPS.

If you do so, you are much more likely to succeed.

To learn more about Professor Treisman's findings, read his article in the November, 1992, issue of the *College Mathematics Journal.*

SUPPLEMENTS

Student:

1. **Student Solutions Manual**—Detailed solutions to every odd-numbered exercise in the text.

2. **Study Guide with Selected Solutions**—Contains worked-out solutions to selected odd-numbered problems with additional step-by-step explanations as well as chapter summary outlines, and additional practice problems with solutions.
3. **Calculus and DERIVE Manual**
4. **Calculus and MATHEMATICA Manual**
5. **Calculus and MAPLE Manual**

Instructors:

6. **Instructors Solutions Manual**—Worked-out solutions to all exercises.
7. **Instructors Resource Manual**—Includes sample course syllabi, lecture outlines with supplements referenced, suggested cooperative assignments and suggestions for how they can be integrated into the course, additional applications of calculus, Reference Guides for integrating Maple, Derive, Mathematica Manuals into the course, and Teaching Advice for TAs and inexperienced teachers.
8. **Testbank**—Over 2,000 items referenced by chapter and section.
9. **Computerized Testbank** (IBM 3.5″ & 5.25″ & MacIntosh)
10. **Transparency Masters**—125 included in the Instructor's Resource Manual.

ACKNOWLEDGMENTS

I am grateful to Richard Lane at the University of Montana both for the superb job he did on the students' and instructors' manuals and for the very useful advice he gave me on improving the problem sets.

Leon Gerber at St. John's University in New York prepared solutions manuals for earlier editions. His extremely accurate work helped a great deal in the preparation of answers and manuals for this edition.

I am grateful to Jerry Johnson and Benny Evans at Oklahoma State University for the outstanding job they did in preparing the computer exercises that appear at the end of many of the chapters in this book.

A few of the problems and examples in this text first appeared in *Mathematics for the Biological Sciences* (New York: Macmillan, 1974) written by James E. Turner and myself. I am grateful to Professor Turner for permission to use this material.

I am grateful to the National Council of Teachers of Mathematics (NCTM) for permission to use material written by John M. H. Olmstead that appeared on pages 448–450 of their book *Historical Topics for the Mathematics Classroom*. This material was the source for my historical note on ordinary and partial derivatives that appears on pages 862–863.

I wish to thank Bantam Books, a division of Bantam Doubleday, Dell Publishing Group, Inc. for permission to use two excerpts from their wonderful book *A Brief History of Time* by Stephen Hawking (New York: 1988): the alternative biography of Newton on page 52 and the graphical depictions of models of the expansion or contraction of the universe on pages 104 and 105.

Many people made suggestions that helped me make significant improvements to the accuracy and pedagogy in this book. I am especially grateful to the following reviewers for their invaluable help in preparing this fifth edition:

Kevin T. Andrews,
Oakland University;

James McKinney,
Cal State Polytechnic;

Joseph Crepeau,
University of Montana;

Eldon L. Miller,
The University of Mississippi;

John S. Cross,
The University of Northern Iowa;

James M. Osborn,
Georgia Tech;

Leon Gerber,
St. John's University;

Robert T. Ramsay,
North Carolina State Univ.;

Jerry Gold,
Cal State—Northridge;

Thomas R. Shemanske,
Darthmouth College;

Melvin Hausner,
New York University;

Charles Stone,
DeKalb College; and

Vuryl Klassen,
Cal State—Fullerton;

John A. Suvak,
Memorial University of
Newfoundland.

Michael Laidacker,
Lamar University;

I wrote a great deal of this book while I was a research associate at University College London. I am grateful to the Mathematics Department at UCL for providing office facilities, mathematical suggestions, and, especially, friendship during my annual visits there.

I very much appreciate the editorial and production staff at HBJ for the care and skill they brought to this project: Karl Yambert, the manuscript editor; Linda Miller, the designer; Louise Sandy-Karkontli, the art editor; Lynne Bush, the production manager; and, finally, Jennifer Johnson, my production editor who guided the book through galleys and page proofs with consummate skill. The production team turned what is often a tedious process into a pleasurable one. Thanks.

Finally, I wish to thank my editor Jay Ricci, my production editor, Richard Koreto and my supplements editor, Alison Muñoz who provided the help necessary to make it all come together.

Stanley I. Grossman

CONTENTS

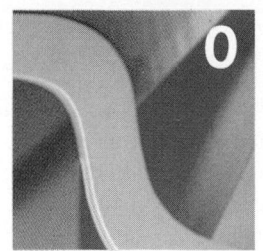

0 REVIEW OF SOME TOPICS IN ALGEBRA

In this chapter we present basic information that you will need for your study of calculus. We begin by discussing the **real-number system**. Real numbers are discussed in high-school mathematics, and we assume that you are familiar with the usual techniques of addition, subtraction, multiplication, division, determining roots, and using exponents.

The real numbers fall into several categories. The **positive integers** (sometimes called the **natural numbers**) are the numbers of counting: 1, 2, 3, 4, 5, . . . (the three dots indicate that the string of numbers goes on indefinitely). The **integers** consist of the positive integers, the negative integers, and the number 0. It is often convenient to represent the integers on a **number line**, as indicated in Figure 1.

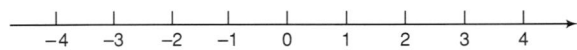

FIGURE 1 *THE NUMBER LINE*

A **rational number** is a real number that can be written as the quotient of two integers, where the integer in the denominator is not zero:

$$r = \frac{m}{n} \qquad \text{where } n \neq 0. \tag{1}$$

Every integer n is also a rational number since $n = n/1$.

EXAMPLE 1
SOME RATIONAL NUMBERS

The following are rational numbers:

    (a) $\frac{1}{2}$ **(b)** $-\frac{3}{4}$ **(c)** -5 **(d)** $0 = \frac{0}{1}$ **(e)** $-\frac{127}{105}$ **(f)** $\frac{4521}{7132}$

Any terminating decimal is a rational number, as the following example suggests.

EXAMPLE 2
A TERMINATING DECIMAL IS RATIONAL

$r = 0.721$ is a rational number, since $r = \frac{721}{1000}$.

All rational numbers can be represented in an infinite number of ways. For example,

$$\frac{1}{2} = \frac{2}{4} = \frac{4}{8} = \frac{3}{6} = \frac{125}{250} = \cdots.$$

Usually, however, a rational number is written as m/n, where m and n have no common factors. That is, we will write the rational number in *lowest terms.*

Any real number that is not rational is called **irrational**. Examples of irrational numbers are $\pi = 3.14159265\ldots$ and $\sqrt{2} = 1.41421356\ldots$. (A proof that $\sqrt{2}$ is irrational is suggested in Problem 53 at the end of this section.)

Every rational number can be written as a **repeating decimal**, whereas no irrational number can be written in this way. For example, $\frac{1}{3} = 0.33333\ldots$, $\frac{3}{11} = 0.272727\ldots$, and $\frac{2}{7} = 0.285714285714\ldots$ are examples of rational numbers written as repeating decimals. We will prove that a rational number can be written in this way in Chapter 10 in our discussion of infinite series.

The following theorem is sometimes useful:

Between any two real numbers, there is a rational number and an irrational number. $\qquad$ **(2)**

Rational and irrational numbers can be represented on the number line used to depict integers (see Figure 2). Conversely, every point on the number line corresponds to one real number.

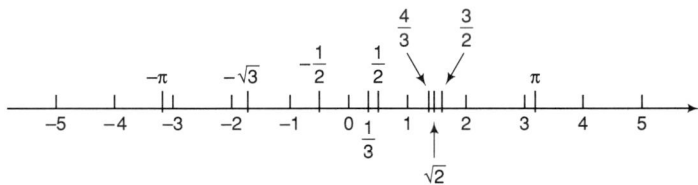

FIGURE 2 NUMBERS ON THE NUMBER LINE

The number line can be used to give us a sense of order. We put the number a to the right of the number b if a is greater than b. We then write this inequality as

$$a > b. \qquad \textbf{(3)}$$

Similarly, if $b > a$, then a is to the left of b, and we write the inequality as

$$a < b. \qquad \textbf{(4)}$$

We use the notation

$$a \leq b \qquad \textbf{(5)}$$

to indicate that a is less than or equal to b; that is, $a < b$ or $a = b$. Finally, we write

$$a \geq b \qquad \textbf{(6)}$$

to indicate that a is greater than or equal to b.

We will often be interested in **sets** of numbers.[†] A set of numbers is any *well-defined* collection of numbers. For example, the integers and rational numbers are each sets of numbers. However, the collection of "large numbers" does not constitute a set, since it is not well defined; that is, there is no universal agreement about whether a given number is or is not in this collection.

[†] When we use the word *number* in this book, we shall mean *real* number, unless otherwise stated.

$\mathbb{R}$
$\mathbb{R}^+$

NOTATION: The set of all real numbers is denoted by $\mathbb{R}$. The set of all nonnegative real numbers is denoted by $\mathbb{R}^+$.

The numbers in a set are called **members**, or **elements**, of that set. If x is an element of the set A, we write $x \in A$ and read this notation as "x is an element of A" or "x belongs to A."

The elements of a set can be written in a bracket notation. For example, the set A of integers between $\frac{1}{2}$ and $\frac{11}{2}$ can be written as

$$A = \left\{ x \in \mathbb{R}: x \text{ is an integer and } \frac{1}{2} < x < \frac{11}{2} \right\}.$$

This notation is read as "A is the set of real numbers x such that x is an integer and x is between $\frac{1}{2}$ and $\frac{11}{2}$." Alternatively, we can write

$$A = \{1, 2, 3, 4, 5\}.$$

The **empty set**, denoted by $\varnothing$, is the set containing no elements.

INTERVALS

Certain types of sets of real numbers, called *intervals*, are of great importance in calculus.

DEFINITION **OPEN AND CLOSED INTERVALS**

(i) The **open interval** (a, b) is the set of real numbers between a and b, *not including* the numbers a and b. We have

$$(a, b) = \{x: a < x < b\}.$$

Note that $a \notin (a, b)$ and $b \notin (a, b)$. The numbers a and b are called **endpoints** of the interval.

(ii) The **closed interval** $[a, b]$ is the set of numbers between a and b, *including* the numbers a and b. We have

$$[a, b] = \{x: a \leq x \leq b\}.$$

Note that $a \in [a, b]$ and $b \in [a, b]$. As before, the numbers a and b are called **endpoints** of the interval. ■

EXAMPLE 3
TWO INTERVALS

(a) $(-1, 5) = \{x: -1 < x < 5\}$ **(b)** $[0, 8] = \{x: 0 \leq x \leq 8\}$

Sometimes we will need to include one endpoint but not the other.

DEFINITION **HALF-OPEN INTERVALS**

(iii) The **half-open interval** $[a, b)$ is given by

$$[a, b) = \{x: a \leq x < b\}.$$

We include the endpoint $x = a$ but not the endpoint $x = b$. That is, $a \in [a, b)$ but $b \notin [a, b)$.

(iv) The half-open interval $(a, b]$ is given by

$$(a, b] = \{x : a < x \le b\}.$$

We have $a \notin (a, b]$ but $b \in (a, b]$. Here b is included but a is not. ■

The symbols ∞ and $-\infty$, denoting infinity and minus infinity, respectively, are *not* real numbers and do not obey the usual laws of algebra, but they can be used for notational convenience.

Intervals may be infinite. We define

INFINITE INTERVALS

$[a, \infty) = \{x : x \ge a\}$

$(a, \infty) = \{x : x > a\}$

$(-\infty, a] = \{x : x \le a\}$

$(-\infty, a) = \{x : x < a\}$

$(-\infty, \infty) = \mathbb{R}.$

We can also write

$$[a, \infty) = \{x : a \le x < \infty\} \qquad \text{and} \qquad (-\infty, a) = \{x : -\infty < x < a\}.$$

Inequalities play a central part in calculus. The following arithmetic facts will prove to be very useful. Let a, b, and c be real numbers.

THREE ARITHMETIC FACTS

If $a < b$, then $a + c < b + c$.   **(7)**

If $a < b$ and $c > 0$, then $ac < bc$.   **(8)**

If $a < b$ and $c < 0$, then $ac > bc$.   **(9)**

The rules (7), (8), and (9) can be restated as follows:

THREE ARITHMETIC FACTS WRITTEN OUT

Adding a positive or negative number to both sides of an inequality does not change the inequality.

Multiplying both sides of an inequality by a *positive* number does not change the inequality.

Multiplying both sides of an inequality by a *negative* number reverses the inequality.

EXAMPLE 4
SOLVING TWO INEQUALITIES

Solve the inequalities $-3 < (7 - 2x)/3 \le 4$.

SOLUTION: In this problem we are asked to solve *two* inequalities:

$$\frac{7 - 2x}{3} \le 4 \quad \text{and} \quad \frac{7 - 2x}{3} > -3.$$

It is necessary to find the set of numbers for which the inequalities are satisfied. This set is called the **solution set** of the inequalities. We start with

$$-3 < \frac{7 - 2x}{3} \le 4$$

$$-9 < 7 - 2x \le 12 \qquad \text{We multiplied by 3.}$$

$$-16 < -2x \le 5 \qquad \text{We subtracted 7 (added } -7)$$

$$8 > x \ge -\frac{5}{2} \quad \text{or} \quad x \in \left[-\frac{5}{2}, 8 \right) \qquad \begin{array}{l}\text{We multiplied by } -\frac{1}{2} \text{ (which} \\ \text{reversed the inequalities).}\end{array}$$

Thus, the solution set is the half-open interval $\left[-\frac{5}{2}, 8 \right)$. Note that each of the steps in the computation served to simplify the term containing x.

ABSOLUTE VALUE

DEFINITION

The **absolute value** of a number a is the positive distance from that number to zero and is written $|a|$. See Figure 3. ■

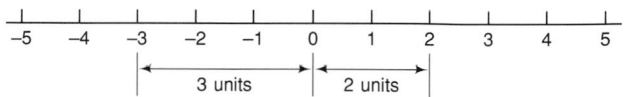

FIGURE 3 THE ABSOLUTE VALUE OF A NUMBER IS THE POSITIVE DISTANCE FROM THAT NUMBER TO 0.

Thus 2 is 2 units from zero, so that $|2| = 2$. The number -3 is 3 units from zero, so that $|-3| = 3$.

We have

ALTERNATIVE DEFINITION OF ABSOLUTE VALUE
$|a| = a \quad$ if $\quad a \ge 0;$
$|a| = -a \quad$ if $\quad a < 0.$

The absolute value of a number is a nonnegative number. Note that, for example, $|5| = 5$ and $|-5| = -(-5) = 5$, so that numbers that are negatives of one another have the same absolute value. Another way to calculate absolute value is to observe that

$$(-x)^2 = x^2 = |x|^2 \quad \text{or} \quad |x| = \sqrt{x^2}$$

where, we emphasize, the *positive* square root is taken. Thus, for example, $|-3| = \sqrt{(-3)^2} = \sqrt{9} = 3$, $|3| = \sqrt{(3)^2} = \sqrt{9} = 3$, and so on.

For all real numbers a and b, the following facts can be proven:

$$|-a| = |a| \tag{10}$$

$$|ab| = |a|\,|b| \tag{11}$$

$$|a + b| \le |a| + |b| \quad \text{Triangle inequality} \tag{12}$$

$$|a - b| = \text{the distance from } a \text{ to } b \tag{13}$$

In the solution of inequalities by using absolute values, the following property will be very useful:

SOLVING THE INEQUALITY $|x| < a$

If $a > 0$, then $|x| < a$ is equivalent to $-a < x < a$.

This holds because $|x| < a$ means that the distance from x to 0 is less than a. See Figure 4. This indicates that for any x in the open interval $(-a, a)$, the distance from x to 0 is less than a.

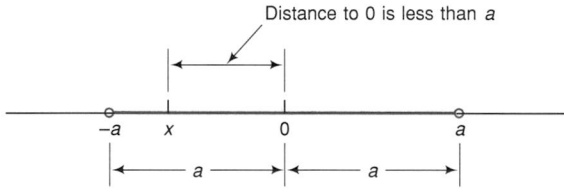

FIGURE 4   SKETCH OF THE SOLUTION SET OF THE INEQUALITY $|x| < a$

EXAMPLE 5
AN ABSOLUTE-VALUE
INEQUALITY

Solve the inequality $|x| \le 3$

SOLUTION: $|x| \le 3$ indicates that the distance from x to 0 is less than or equal to 3. Thus $-3 \le x \le 3$, and the solution set is the closed interval $[-3, 3]$. See Figure 5.

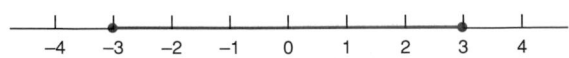

FIGURE 5   SKETCH OF THE SOLUTION SET OF THE INEQUALITY $|x| \le 3$

EXAMPLE 6
AN ABSOLUTE-VALUE
INEQUALITY

Solve the inequality $|x - 4| < 5$.

SOLUTION: The distance from $x - 4$ to 0 is less than 5, so $-5 < x - 4 < 5$, and adding 4 to each term, we see that $-1 < x < 9$. Thus the solution set is the open interval $(-1, 9)$. See Figure 6. Another way to think of this set is as the set of all x such that x is within 5 units of 4.

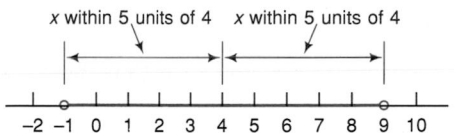

FIGURE 6 SKETCH OF THE SOLUTION SET OF THE INEQUALITY $|x - 4| < 5$

REMARK: The quantity $|b - a|$ is the (*positive*) *distance* between a and b. Thus, for example the distance between 3 and -2 is $|-2 - 3| = 5$. This is sketched in Figure 7.

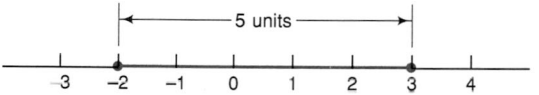

FIGURE 7 THE DISTANCE BETWEEN -2 AND 3 (OR 3 AND -2) IS 5 UNITS

Another useful property of absolute value is given next.

SOLVING THE INEQUALITY $|x| > a$

If $a > 0$, then $|x| > a$ is equivalent to $x > a$ or $x < -a$.　　　　(14)

Fact (14) is illustrated in Figure 8.

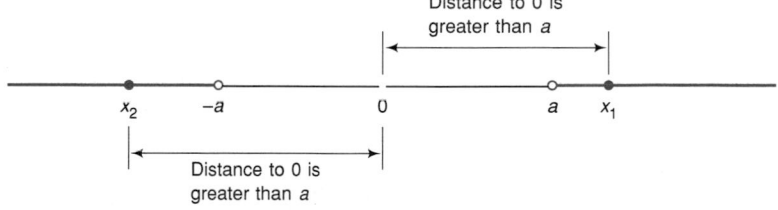

FIGURE 8 SKETCH OF THE SOLUTION SET OF THE INEQUALITY $|x| > a$

EXAMPLE 7
AN ABSOLUTE-VALUE
INEQUALITY

Solve the inequality $|x + 2| \geq 8$.

SOLUTION: The distance from $x + 2$ to 0 is greater than or equal to 8, so either $x + 2 \geq 8$ or $x + 2 \leq -8$. Hence either $x \geq 6$ or $x \leq -10$. The solution set is $\{x: x \in (-\infty, -10] \text{ or } x \in [6, \infty)\}$. See Figure 9.

FIGURE 9 SKETCH OF THE SOLUTION SET OF THE INEQUALITY $|x| \geq 10$

■ PROBLEMS 0.1

■ SELF-QUIZ

I. Fill in each blank with $<$, $=$, or $>$ so that the resulting statement is true.
  a. -3 _____ 2
  b. 4 _____ -6
  c. $\frac{3}{11}$ _____ $\frac{19}{7}$
  d. $1 + \frac{4}{11}$ _____ $3 - \frac{22}{7}$
  e. $\pi/3$ _____ 1
  f. $-\frac{22}{7}$ _____ $-\pi$
  g. $\sqrt{8}$ _____ 3.1
  h. $\sqrt{2}\sqrt{3}$ _____ $\sqrt{2} + \sqrt{3}$

II. If $3 < |x|$, then _____ [Fill in the blank with correct choice(s).]
  a. $x^2 < 9$ b. $-3 < x < 3$
  c. $3 < x < -3$ d. $x < -3$ or $3 < x$

III. If $x < y$ and _____, then $1/y < 1/x$. [Fill in the blank with correct choice(s).]
  a. $0 < y$ b. $x < 0$
  c. $0 < x$ d. $y < 0$

IV. If x lies between 3 and 5 but is not equal to either one, then _____.
  a. $x = 4$ b. $x \leq 4$
  c. $3 < x$ d. $x < 5$
  e. $3 < x$ and $x < 5$ f. $3 < x < 5$
  g. $3 > x > 5$ h. $5 < x < 3$

V. If $|x - 2| + |x - 5| = 3$, then _____.
  a. $x = 3$ b. $2 > x > 5$
  c. $2 \leq x$ and $x \leq 5$ d. $4 = x$

VI. Answer True or False.
  a. If $x < 5$, then $x^2 < 25$.
  b. If $x < -5$, then $x^2 < 25$.
  c. If $-5 < y$, then $25 < y^2$.
  d. If $x < y$, then $x^2 < y^2$.
  e. If $0 \leq x < y$, then $x^2 < y^2$.
  f. If $x < y \leq 0$, then $x^2 < y^2$.
  g. If $x^2 < y^2$, then $x < y$.
  h. If $x^2 < y^2$, then $|x| < |y|$.

■ DRILL

1. Illustrate the following collection of inequalities by marking appropriate points on a (single) number line.
  a. $2 < 3$ b. $\frac{1}{3} < \frac{1}{2}$
  c. $\frac{4}{3} < \sqrt{2} < \frac{3}{2}$ d. $-3 < 1$
  e. $-\sqrt{3} < -1 < 0$ f. $-4 < -\pi < -3$

2. List all members in the intersection of the two sets $\{-9, -6, -3, 0, 2, 4, 6, 8\}$ and $(-5, 6] = \{x : -5 < x \leq 6\}$.

In Problems 3–38, solve the inequality and graph the solution set on the number line.

3. $3x - 5 < 7$ 4. $1 - 2x < -7$
5. $3.72x > 4.06x$ 6. $1 + 2x < 3 - x$
7. $1 \leq 2x + 2 \leq 4$ 8. $-4 < (4 - 2x)/5 \leq 2$
9. $|x| < 1$ 10. $|x| \leq 4$
11. $|x| \geq 5$ 12. $|x| \leq 0$
13. $|x| \geq 0$ 14. $|x| < 3$
15. $|x| \leq -1$ 16. $|x| > -1$
17. $|x - 2| < 1$ 18. $|x + 3| \leq 4$
19. $|x + 3| \geq 4$ 20. $|x + 6| > 3$
21. $|2x + 4| < 3$ 22. $|-x + 2| < 3$
23. $|5 - x| \geq 1$ 24. $|2 - x| > 0$
25. $-4 + |-3x - 4| > 2$ 26. $|3x + 4| - 2 \geq 4$
27. $|6 - 4x| \geq |x - 2|$
    [*Hint:* Consider separate cases: $6 - 4x > 0$ and $x - 2 > 0$, etc.]
28. $\left|\dfrac{8 - 3x}{2}\right| \leq 3$ 29. $\left|\dfrac{3x + 17}{4}\right| > 9$
30. $|ax + b| < c, a > 0, c > 0$
31. $|ax + b| \geq c, a < 0, c > 0$
*32. $x \leq |x|$ 33. $1/x > 3$
34. $1/(x - 2) \leq -2$ 35. $x > \dfrac{1}{x}$
36. $x(x - 2) \leq 0$ 37. $x(x - 2) > -1$
38. $1/(x - 2) > 2/(x + 3)$
    [*Hint:* Keep track of whether you are multiplying by a positive or negative number.]

CONTINUED

■ **APPLICATIONS**

39. Use the absolute-value function to translate each of the following statements into a single inequality.
    a. $x \notin (-3, 3)$
    b. x is (strictly) between -2 and 4
    c. $x \in (-4, 10)$
    d. $x \notin [5, 11]$
    e. $x \in (-\infty, 2] \cup [9, \infty)$
    f. x is closer to 5 than to 0

40. Solve the following inequalities and graph the solution sets:
    a. $|2 - x| + |2 + x| \leq 10$
    b. $|2 - x| + |2 + x| > 6$
    c. $|2 - x| + |2 + x| \leq 4$
    d. $|2 - x| + |2 + x| \leq 3.99$

■ **SHOW/PROVE/DISPROVE**

41. Show that $(s + t + |s - t|)/2$ equals the maximum of $\{s, t\}$.

42. Prove or disprove that

$$\frac{s + t - |s - t|}{2}$$

equals the minimum of $\{s, t\}$ for any real numbers s and t.

43. Suppose p_0, q_0, p_1, q_1 are positive real numbers such that $p_0/q_0 < p_1/q_1$. Show that this implies that

$$\frac{p_0}{q_0} < \frac{p_0 + p_1}{q_0 + q_1} < \frac{p_1}{q_1}.$$

44. Show that $|xy| = |x| \, |y|$ for all real numbers x and y. [*Hint:* $|w| = \sqrt{w^2}$ for all w.] [*Alternative Hint:* Deal with each of four cases separately:
    a. $x \geq 0$ and $y \geq 0$,   b. $x \geq 0$ and $y < 0$,
    c. $x < 0$ and $y \geq 0$,   d. $x < 0$ and $y < 0$.]

45. Show that $|x + y| \leq |x| + |y|$ for all real numbers x and y by showing the following:
    a. if x and y have the same sign, then $|x + y| = |x| + |y|$.
    b. if x and y have opposite signs, then $|x + y| < |x| + |y|$.
    (This problem asks you to prove the *triangle inequality* (12).)

46. Show that $|x + y| \leq |x| + |y|$ for all real numbers x and y after first showing the following:
    a. If $a \geq 0$ and $b \geq 0$, then $a \leq b$ if and only if $a^2 \leq b^2$.
    b. $|x + y|^2 \leq (|x| + |y|)^2$.

47. Show that $||x| - |y|| \leq |x - y|$ for all real numbers x and y. [*Hint:* Write $x = (x - y) + y$ and apply the triangle inequality.]

48. a. Prove that $|a - b| \leq |a - w| + |w - b|$ for all real numbers a, b, and w.
    b. Describe those situations in which the preceding "less than or equal" statement is actually an equality.

49. An estimate for a particular number w is 1.3. Suppose this estimate is accurate to one decimal place; that is, $|w - 1.3| \leq 0.05$. Observe that $1.3^2 = 1.69$, which rounds off to 1.7. Is the estimate 1.7 for w^2 also accurate to one decimal place? Describe the shortest interval that is guaranteed to contain w^2.

50. The sides of a rectangular piece of paper (a page of this book, for example) are measured. Suppose that we measure to one-decimal-place accuracy and find that the rectangle is 16.1 cm by 23.4 cm; 16.1 cm times 23.4 cm equals 376.74 cm², which rounds off to 376.7 cm². Does 376.7 cm² estimate the true area of the rectangle with one-decimal-place accuracy? Explain.

■ **CHALLENGE**

51. a. Suppose that a and b are positive; prove that

$$\sqrt{ab} \leq \frac{a + b}{2}.$$

    (This result is referred to as the **arithmetic-geometric inequality**.)
    b. Use the inequality of part (a) to prove that among all rectangles with an area of 225 cm², the one with the shortest perimeter is a square.
    c. Use the inequality of part (a) to prove that among all rectangles with a perimeter of 300 cm, the one with the largest area is a square.

52. Show that $ac + bd \leq \sqrt{a^2 + b^2} \sqrt{c^2 + d^2}$ for any real numbers a, b, c, and d.

*53. Prove that $\sqrt{2}$ is irrational. [*Hint:* Suppose $\sqrt{2}$ were rational; then there would exist integers m and n having no common factors such that $\sqrt{2} = m/n$. Show that this result implies that $2n^2 = m^2$, and explain why this indicates that both m and n must be even. Finally, explain why this combination of results proves that $\sqrt{2}$ must be irrational.]

*54. Suppose a and b are positive integers. Prove that $\sqrt{2}$ falls between a/b and $(a + 2b)/(a + b)$.

*55. Discuss the following *false "proof"* that $-1 = 2$. Suppose a is a solution to $a = 1 + a^2$. Clearly, a cannot be 0; hence, $1 = (1/a) + a$. Use this expression to substitute for 1 in the first conditional equation: $a = [(1/a) + a] + a^2$. Therefore, $0 = (1/a) + a^2$ and $-(1/a) = a^2$; hence, $a^3 = -1$. We now infer that $a = \sqrt[3]{-1} = -1$. Substituting this value into our original equation, we have $-1 = 1 + (-1)^2 = 1 + 1 = 2$. [*Hint:* $(x^2 - x + 1)(x + 1) = x^3 + 1$ is true for all real numbers x.]

■ ANSWERS TO SELF-QUIZ

I. a. <   b. >   c. <   d. >
    e. >   f. <   g. <   h. <
II. d   III. c or d
IV. c, d, e, f   V. c
VI. a. False, e.g., let $x = -6$
     b. False, e.g., let $x = -7$

c. False, e.g., let $y = 0$
d. False, e.g., $-10 < -9$
e. True
f. False, e.g., $-10 < -9$
g. False, e.g., $(-3)^2 < (-5)^2$
h. True

0.2 THE CARTESIAN PLANE

To this point we have been concerned with single numbers. We will now discuss properties of pairs of numbers. An **ordered pair** of numbers consists of two numbers; one is called the **first component** or **element** and the other is called the **second component**. Ordered pairs are written in the form

$$(a, b) \tag{1}$$

where a and b are real numbers. Two ordered pairs are equal if their first components are equal and their second components are equal. Note that $(1, 0)$ and $(1, 1)$ are *different* ordered pairs since their second components are different. Note too that $(1, 2)$ and $(2, 1)$ are different ordered pairs.

DEFINITION   **CARTESIAN PLANE**

The set of all ordered pairs of real numbers is called the **Cartesian plane**[†] and is denoted by $\mathbb{R}^2$. Hence,

$$\mathbb{R}^2 = \{(x, y): x \in \mathbb{R} \text{ and } y \in \mathbb{R}\}.$$

Any element (x, y) of the Cartesian plane is called a **point** in the plane. ■

Any number in $\mathbb{R}$ can be represented graphically on a number line. Analogously, any point (x, y) can be represented graphically as a point in the Cartesian plane. We call the first number in the ordered pair the **x-coordinate** and the second number the **y-coordinate**. Following the technique invented

---

[†] See the biographical sketch of Descartes on page 14.

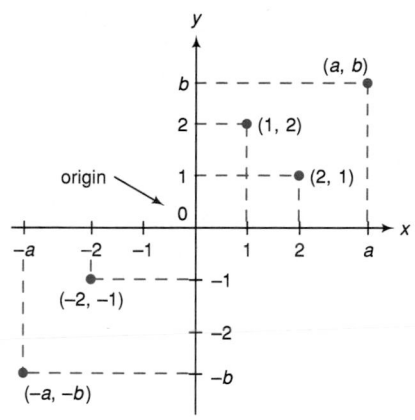

FIGURE 1
THE CARTESIAN PLANE

Positive x-values lie to the right of the y-axis.
Negative x-values lie to the left of the y-axis.
Positive y-values lie above the x-axis.
Negative y-values lie below the x-axis.

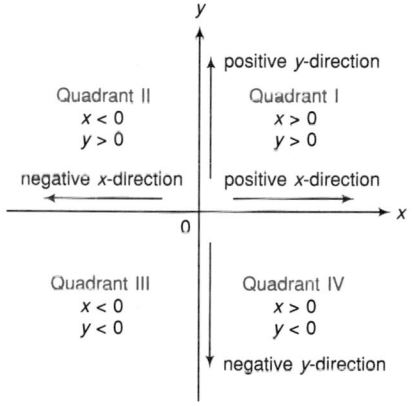

FIGURE 2
THE FOUR QUADRANTS IN THE CARTESIAN PLANE

by Descartes, we draw a horizontal line called the **x-axis** and a vertical line perpendicular to it called the **y-axis**. The point at which the two lines meet is called the **origin** and is labeled $(0, 0)$ or 0. With this orientation for the point denoted by (a, b), a is the x-coordinate and b is the y-coordinate. Along the x-axis, positive distances are measured to the right of the origin, and along the y-axis, positive distances are measured in the upward direction. This idea is illustrated in Figure 1. The arrows indicate the direction of increasing x and y. A typical point (a, b) is drawn in Figure 1, where $a > 0$ and $b > 0$. We simply measure a units in the positive direction along the x-axis and b units in the positive direction along the y-axis to arrive at the point (a, b). Note that, as in Figure 1, the distinct ordered pairs $(1, 2)$ and $(2, 1)$ represent different points in the plane.

⊗ **WARNING:** Do not become confused by the fact that (a, b) can denote either a point in the plane or an open interval. The meaning should always be clear from the context. ⊗

NOTATION: We will usually refer to the Cartesian plane as the **xy-plane**.

A glance at Figure 2 indicates that the x- and y-axes divide the xy-plane into four regions. These regions are called **quadrants** and are numbered as in the figure. In Figure 2 we see, for example, that $(1, 3)$ is in the first quadrant, $(-1, 3)$ is in the second quadrant, $(-1, -3)$ is in the third quadrant, and $(3, -1)$ is in the fourth quadrant. Points on the x- or y-axis do not belong to any quadrant.

We will often need to calculate the distance between two points in the xy-plane. Consider the two points (x_1, y_1) and (x_2, y_2), as in Figure 3. The distance d between (x_1, y_1) and (x_2, y_2) is the length of the line segment PQ. Since PQR is a right triangle, we have, by the Pythagorean theorem,

$$\overline{PQ}^2 = \overline{PR}^2 + \overline{QR}^2.^{\dagger}$$

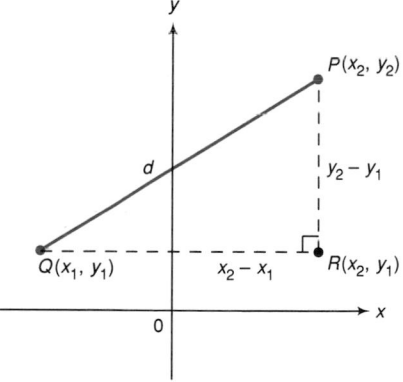

FIGURE 3 PROOF OF THE DISTANCE FORMULA $d^2 = (x_2 - x_1)^2 + (y_2 - y_1)^2$

But $\overline{PR} = |y_2 - y_1|$ and $\overline{QR} = |x_2 - x_1|$, so that $d^2 = \overline{PQ}^2 = \overline{PR}^2 + \overline{QR}^2$, or

$$d^2 = (x_2 - x_1)^2 + (y_2 - y_1)^2.$$

---

† The symbol $\overline{PQ}$ denotes the distance between the points P and Q.

Therefore, we have the distance formula:

$$d = \sqrt{(x_2 - x_1)^2 + (y_2 - y_1)^2}.$$ (2)

EXAMPLE 1
THE DISTANCE BETWEEN TWO
POINTS

Find the distance between the points $(2, 5)$ and $(-3, 7)$.

SOLUTION: Let $(x_1, y_1) = (2, 5)$ and $(x_2, y_2) = (-3, 7)$. Then, from (2), we have

$$d = \sqrt{((-3) - 2)^2 + (7 - 5)^2} = \sqrt{(-5)^2 + 2^2} = \sqrt{29}.$$

Once we have defined a coordinate system with two components, we can discuss **equations in two variables**.

EXAMPLE 2
EQUATIONS IN TWO VARIABLES

The following are equations in the two variables x and y:

(a) $y = 3x + 2$ (b) $x^2 + y^2 = 4$

DEFINITION   **GRAPH OF AN EQUATION**

The **graph** of an equation in two variables is the set of points in the xy-plane whose coordinates satisfy the equation. ∎

In Section 0.3, we will discuss equations whose graphs are straight lines. In Section 0.4, we will discuss more general equations and graphs. Now we consider the equation of a circle.

A **circle** is defined as the set of all points in a plane at a given distance from a given point. The given point is called the **center** of the circle, and the common distance from the center is called the **radius**.

EXAMPLE 3
THE UNIT CIRCLE

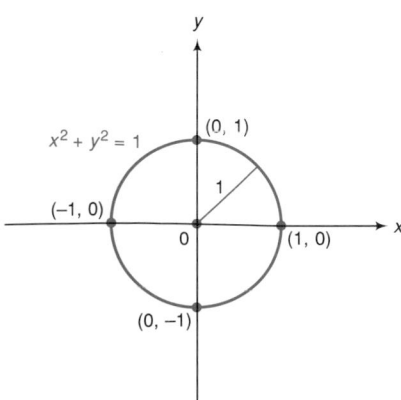

Find the equation of the circle centered at the origin with radius 1.

SOLUTION: If (x, y) is any point on the circle, then the distance from (x, y) to $(0, 0)$ is 1. From (2) we have

$$\sqrt{(x - 0)^2 + (y - 0)^2} = 1,$$

or squaring both sides, we obtain

EQUATION OF THE UNIT CIRCLE

$$x^2 + y^2 = 1.$$ (3)

This circle is sketched in Figure 4. It is called the **unit circle** and was of central importance in your study of the trigonometric functions. (See Appendix 1.2.)

FIGURE 4
THE UNIT CIRCLE

The equation of the circle with center at (h, k) and radius r is given by

> **EQUATION OF THE CIRCLE CENTERED AT (h, k) WITH RADIUS r**
>
> $$(x - h)^2 + (y - k)^2 = r^2. \tag{4}$$

EXAMPLE 4
FINDING THE EQUATION OF A CIRCLE GIVEN ITS CENTER AND RADIUS

Find the equation of the circle centered at $(2, -5)$ with radius 4.

SOLUTION: Here $(h, k) = (2, -5)$ and $r = 4$ so (4) yields

$$(x - 2)^2 + (y - (-5))^2 = 4^2 \text{ or } (x - 2)^2 + (y + 5)^2 = 16.$$

The circle is sketched in Figure 5.

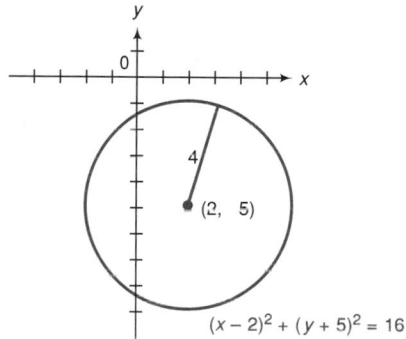

FIGURE 5 THE CIRCLE OF RADIUS 4 CENTERED AT $(2, -5)$

EXAMPLE 5
FINDING THE CENTER AND RADIUS OF A CIRCLE FROM ITS EQUATION

Show that the equation $x^2 - 6x + y^2 + 2y - 17 = 0$ represents a circle, and find its center and radius.

SOLUTION: We use the technique of completing the squares. We have

$$x^2 - 6x + y^2 + 2y - 17 = (x^2 - 6x + 9) + (y^2 + 2y + 1) - 9 - 1 - 17$$
$$= (x - 3)^2 + (y + 1)^2 - 27 = 0.$$

This result implies that

$$(x - 3)^2 + (y + 1)^2 = 27,$$

which is the equation of a circle with center at $(3, -1)$ and radius $\sqrt{27}$.

We close this section by noting that the graphs of quadratic equations in two variables—that is, equations having the form

$$ax^2 + bxy + cy^2 + dx + ey + f = 0 \tag{5}$$

—are called **conic sections**. The equation of a circle is a special case of equation (5). Another special case is discussed in Example 0.4.5 (see page 26). We will discuss conic sections in Chapter 8.

THE CARTESIAN PLANE IS NAMED after the great French mathematician and philosopher René Descartes. Born near the city of Tours in 1596, Descartes received his education first at the Jesuit school at La Flèche and later at Poitier, where he studied law. He had delicate health and, while still in school, developed the habit of spending the greater part of each morning in bed. Later, he considered these morning hours the most productive period of the day.

At the age of 16, Descartes left school and moved to Paris, where he began his study of mathematics. In 1618, he joined the army of Maurice, Prince of Nassau. He also served with Duke Maximillian I of Bavaria and with the French army at the siege of La Rochelle.

Descartes was not a professional soldier, however, and his periods of military service were broken by periods of travel and study in various European cities. After leaving the army for good, he resettled in Paris to continue his mathematical studies and then moved to Holland, where he lived for 24 years.

Much stimulated by the scientists and philosophers he met in France, Holland, and elsewhere, Descartes later became known as the "father of modern philosophy."

RENÉ DESCARTES
1596–1650

His statement "Cogito ergo sum" ("I think, therefore I am") played a central role in his philosophical writings.

Descartes's program for philosophical research was enunciated in his famous *Discours de la méthode pour bien conduire sa raison et chercher la vérité dans les sciences* (A Discourse on the Method of Rightly Conducting the Reason and Seeking Truth in the Sciences) published in 1637. This work was accompanied by three appendixes: *La dioptrique* (in which the law of refraction—discovered by Snell—was first published), *Les météores* (which contained the first accurate explanation of the rainbow), and *La géométrie*. *La géométrie*, the third and most famous appendix, took up about a hundred pages of the *Discours*. One of the major achievements of *La géométrie* was that it connected figures of geometry with the equations of algebra. The work established Descartes as the founder of analytic geometry.

In 1649 Descartes was invited to Sweden by Queen Christina. He accepted, reluctantly, but was unable to survive the harsh, Scandinavian winter. He died in Stockholm in early 1650.

■ PROBLEMS 0.2

■ SELF-QUIZ

I. The distance between $(2, 7)$ and $(-2, 10)$ is the square root of _____.
   a. $(2 - 7)^2 + (-2 - 10)^2$
   b. $(2^2 + 7^2)((-2)^2 + 10^2)$
   c. $(2 - (-2))^2 + (7 - 10)^2$
   d. $(2 - 2)^2 + (7 - 10)^2$

II. The set of points satisfying the equation

$$(x - 4)^2 + (y + 13)^2 = 36$$

is a circle with _____.
   a. center at $(-4, -13)$; radius $= 36$
   b. center at $(4, -13)$; radius $= 6$
   c. center at $(-4, 13)$; radius $= 6$
   d. center at $(4, -13)$; radius $= 36$

III. Several triangles are described below by listing their vertices (corner points). Which of these triangles are equilateral (all sides have the same length)?
   a. $(5, 5)$, $(0, 0)$, $(-5, 5)$
   b. $(-7, -7)$, $(7, -7)$, $(-7, 7)$
   c. $(3, 3)$, $(3, -3)$, $(3 + 3\sqrt{3}, 0)$

IV. Give the coordinates of (at least) five points with integer coordinates that are exactly 13 units distant from the point $(7, 11)$. [*Hint:* $5^2 + 12^2 = 13^2$.]

■ DRILL

1. Determine the quadrant in which each of the following points lies; then draw a single figure locating all of these points in the xy-plane.
   a. $(3, -2)$
   b. $(-3, 2)$
   c. $(-\sqrt{2}, -\sqrt{2})$
   d. $(\sqrt{2}, -\sqrt{2})$
   e. $(-\frac{7}{3}, -5)$
   f. $(-5, -\frac{7}{3})$

CONTINUED

In Problems 2–5, compute the distance between the two points.

2. $(1, 3), (4, 7)$
3. $(-7, 2), (4, 3)$
4. $(a, b), (0, 0)$
5. $(a, b), (b, a)$
6. Compute the perimeter of the triangle whose corner points are $(2, 3)$, $(\frac{73}{5}, -\frac{1}{5})$, and $(5, 7)$.

In Problems 7–10, find an equation of the circle with given center and radius; then sketch its graph.

7. $(1, 1), r = \sqrt{2}$
8. $(1, -1), r = 2$
9. $(-1, 4), r = 5$
10. $(3, -2), r = 4$
11. Find an equation of the circle centered at $(7, 3)$ which passes through $(-1, 7)$.
12. Sketch the graph of the set of points (x, y) such that

$$y = 3 + \sqrt{25 - (x - 4)^2}.$$

■ APPLICATIONS

13. Is the point $(3, 3)$ *inside*, *on*, or *outside* the circle of radius 9 centered at $(7, -5)$? Explain.
14. Does the graph of $(x - 3)^2 + (y - 2)^2 = 4$ *lie totally inside*, *lie totally outside*, or *overlap* the graph of $(x - 1)^2 + (y + 1)^2 = 49$? Explain.
15. For each of the following pairs of points, use the result of Problem 20 to find the midpoint of the line segment joining them.
    a. $(2, 5)$ and $(5, 12)$
    b. $(-3, 7)$ and $(4, -2)$
16. The points $(3, 1), (0, -3), (-12, 1), (-9, 5)$ are, in clockwise order, vertices of a quadrilateral (a four-sided figure). Use the result of Problem 20 to compute the coordinates for the midpoint of each diagonal of that figure.
17. Find an equation for the circle that has $(8, -4)$ and $(2, 6)$ as endpoints of a diameter. [*Hint:* The result of Problem 20 shows how to compute the coordinates of the center of this circle.]
*18. Find an equation for the (unique) circle that passes through the points $(0, -2)$, $(6, -12)$, and $(-2, -4)$.

■ SHOW/PROVE/DISPROVE

19. Show that the distance formula (2) yields $|y_1 - y_2|$ as the distance between (a, y_1) and (a, y_2).
20. Let L denote the line segment joining the points $P_1 = (x_1, y_1)$ and $P_2 = (x_2, y_2)$. Show that the midpoint of L is the point

$$P_3 = \left(\frac{x_1 + x_2}{2}, \frac{y_1 + y_2}{2}\right).$$

[*Hint:* (a) First show that $\overline{P_1 P_2} = \overline{P_1 P_3} + \overline{P_3 P_2}$ (thus, P_3 is somewhere on the line segment between P_1 and P_2); (b) then show $\overline{P_1 P_3} = \overline{P_3 P_2}$.]
21. Show that $x^2 - 6x + y^2 + 4y - 12 = 0$ is the equation of a circle; find the center and radius of that circle.
22. Show that $x^2 + ax + y^2 + by + c = 0$ is the equation of a circle if and only if $a^2 + b^2 - 4c > 0$.
23. Prove that the diagonals of a rectangle have equal lengths. [*Hint:* There is no loss in generality if we assume the rectangle is placed with one vertex at the origin, one side along the positive x-axis, and another side on the positive y-axis.]
24. Find a simple equation satisfied by all points in the plane that are twice as far from $(0, 0)$ as from $(12, 0)$. Show that this set of points is a circle.

■ CHALLENGE

25. Let $P = ([1 - \lambda]x_0 + \lambda x_1, [1 - \lambda]y_0 + \lambda y_1)$. Show that if λ is a real number between 0 and 1, then P lies on the straight-line segment between $P_0 = (x_0, y_0)$ and $P_1 = (x_1, y_1)$.

■ ANSWERS TO SELF-QUIZ

I. c
II. b
III. only c
IV. $(-5, 6), (-5, 16), (19, 6), (19, 16), (2, -1), (2, 23),$ $(12, -1), (12, 23)$

0.3 LINES

Lines will be very important to us in the study of calculus. Two distinct points (x_1, y_1) and (x_2, y_2) determine a line. The *slope* of a line is a measure of the relative rate of change of the x- and y-coordinates of points on the line as we move along the line. The following summarizes the facts we will need about lines.

FACTS ABOUT STRAIGHT LINES

(i) The **slope** m of a nonvertical line passing through the points (x_1, y_1) and (x_2, y_2) is given by (if $x_2 \neq x_1$):

$$m = \frac{y_2 - y_1}{x_2 - x_1} = \frac{\Delta y}{\Delta x}.$$

(ii) If $x_2 - x_1 = 0$ and $y_2 \neq y_1$, then the line is vertical and the slope is said to be **undefined.**[†]

(iii) The **y-intercept** of a line that is not vertical is equal to b where $(0, b)$ are the coordinates of the point at which the line crosses the y-axis.

(iv) An equation of a line can be written in many different ways. Here are three of them: Let (x_1, y_1) be a point on the line, let m be its slope, and let b denote the y-intercept.

   (a) Point-slope equation of a line: $y - y_1 = m(x - x_1)$
   (b) Slope-intercept equation of a line: $y = mx + b$
   (c) Standard equation of a line: $cx + dy = e$; either c or d (or both) is nonzero

   In the standard equation with $d \neq 0$, the slope is $m = -c/d$.

(v) Two lines are parallel if and only if they have the same slope, or if they are both vertical.

(vi) If m_1 is the slope of line L_1, and m_2 is the slope of line L_2, with $m_1 \neq 0$, then L_1 and L_2 are perpendicular, if and only if $m_2 = -1/m_1$. That is, perpendicular nonvertical lines have slopes that are negative reciprocals of one another.

(vii) Lines parallel to the x-axis (horizontal lines) have a slope of zero. An equation of a horizontal line is $y = a$.

(viii) Lines parallel to the y-axis (vertical lines) have an undefined slope. An equation of a vertical line is $x = a$.

(ix) If $m > 0$, the graph of the line will point upward as we move from left to right.

(x) If $m < 0$, the graph of the line will point downward as we move from left to right.

These last two facts are illustrated in Figure 1.

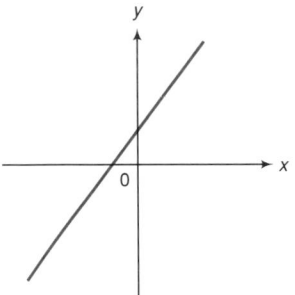

(a) Positive slope: $m > 0$.

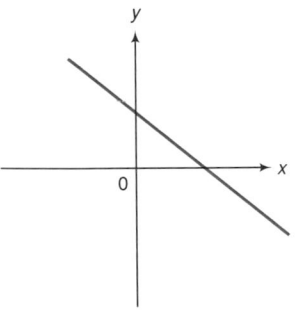

(b) Negative slope: $m < 0$.

FIGURE 1
POSITIVE AND NEGATIVE SLOPES

EXAMPLE 1
FINDING THE SLOPES OF LINES

Find the slopes of the lines containing the given pairs of points. Then sketch these lines.

[†] In some textbooks, a vertical line is said to have "an infinite slope."

(a) $(2, 3), (-1, 4)$ **(b)** $(1, -3), (4, 0)$

(c) $(2, 6), (-1, 6)$ **(d)** $(3, 1), (3, 5)$

SOLUTION:

(a) $m = \dfrac{\Delta y}{\Delta x} = \dfrac{4 - 3}{-1 - 2} = \dfrac{1}{-3} = -\dfrac{1}{3}$

(b) $m = \dfrac{\Delta y}{\Delta x} = \dfrac{0 - (-3)}{4 - 1} = \dfrac{3}{3} = 1$

(c) $m = \dfrac{6 - 6}{-1 - 2} = \dfrac{0}{-3} = 0$

That is, as the x-coordinate changes, the y-coordinate doesn't vary. This line is parallel to the x-axis.

(d) Here the slope is undefined since the line is parallel to the y-axis (the x-coordinates of both points have the same value, 3).

The lines are sketched in Figure 2.

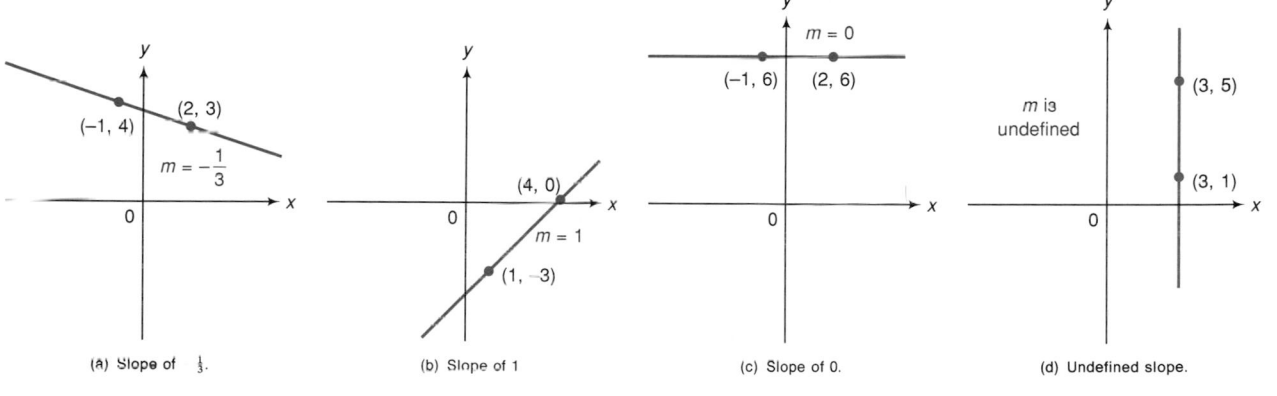

(a) Slope of $\frac{1}{3}$.   (b) Slope of 1.   (c) Slope of 0.   (d) Undefined slope.

FIGURE 2   LINES WITH VARIOUS SLOPES

EXAMPLE 2

FINDING AN EQUATION OF A LINE WHEN TWO POINTS ON IT ARE GIVEN

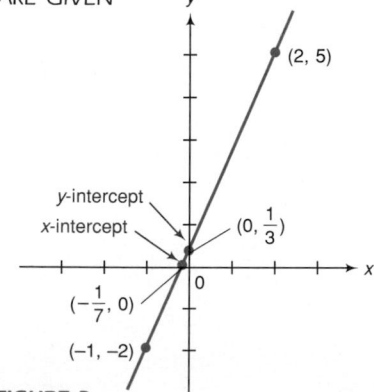

FIGURE 3

$y = \frac{7}{3}x + \frac{1}{3}$

Find the slope-intercept equation of the line passing through $(-1, -2)$ and $(2, 5)$.

SOLUTION: We first find the point-slope form:

$$m = \frac{5 - (-2)}{2 - (-1)} = \frac{7}{3}, \text{ so}$$

$$\frac{y - 5}{x - 2} = \frac{7}{3}$$

$$y - 5 = \frac{7}{3}(x - 2) = \frac{7}{3}x - \frac{14}{3} \qquad \text{Point-slope form}$$

$$y = \frac{7}{3}x - \frac{14}{3} + 5 = \frac{7}{3}x + \frac{1}{3}. \qquad \text{Slope-intercept form}$$

The line is sketched in Figure 3.

EXAMPLE 3
FINDING AN EQUATION OF A
LINE PARALLEL TO A GIVEN LINE
AND PASSING THROUGH A
GIVEN POINT

Find the slope-intercept equation of the line passing through the point
$(2, 3)$ and parallel to the line whose equation is $y = -3x + 5$.

SOLUTION: Parallel lines have the same slope. The slope of the line
$y = -3x + 5$ is -3 since the line is given in its slope-intercept form.
Then a point-slope equation of the line is

$$y - 3 = -3(x - 2)$$
$$y - 3 = -3x + 6$$
$$y = -3x + 9.$$

The line is sketched in Figure 4.

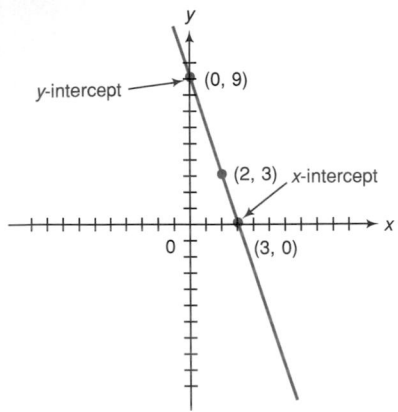

FIGURE 4
$y = -3x + 9$

EXAMPLE 4
FINDING AN EQUATION OF
A LINE PERPENDICULAR TO A
GIVEN LINE AND PASSING
THROUGH A GIVEN POINT

Find the slope-intercept equation of the line passing through $(-1, 3)$
that is perpendicular to the line $2x + 3y = 4$.

SOLUTION: From $2x + 3y = 4$, we obtain

$$3y = -2x + 4,$$

or

$$y = -\frac{2}{3}x + \frac{4}{3}.$$

Thus, the slope of the line $2x + 3y = 4$ is $-\frac{2}{3}$, so the line we seek has
the slope $-1/-\frac{2}{3} = \frac{3}{2}$. Hence, a point-slope equation of the line passing
through $(-1, 3)$ with slope $\frac{3}{2}$ is

$$y - 3 = \frac{3}{2}(x + 1)$$

$$y - 3 = \frac{3}{2}x + \frac{3}{2}$$

$$y = \frac{3}{2}x + \frac{3}{2} + 3$$

$$= \frac{3}{2}x + \frac{9}{2}.$$

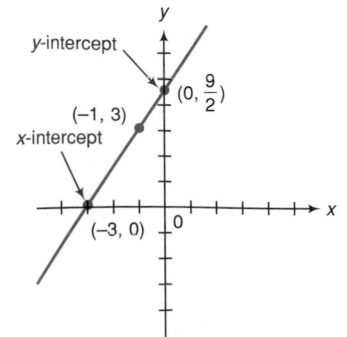

FIGURE 5
$y = \frac{3}{2}x + \frac{9}{2}$

The line is sketched in Figure 5.

NOTE: The easiest way to draw a straight line is first to find two points on the line and then connect them. Unless the line passes through the origin, the two easiest points to find are the *y*-intercept (obtained by setting *x* equal to zero) and the *x*-intercept (obtained by setting *y* equal to zero).

If the line *L* is not parallel to the *x*-axis, then its **angle of inclination** θ is the positive angle between $0°$ and $180°$ that it makes with the positive *x*-axis. If *L* is horizontal (parallel to the *x*-axis), then its angle of inclination is $0°$.

NOTE: We emphasize that for every line *L*,

$$0° \leq \theta < 180°$$

or, in radians,[†]

$$0 \leq \theta < \pi.$$

Some angles of inclination are illustrated in Figure 6.

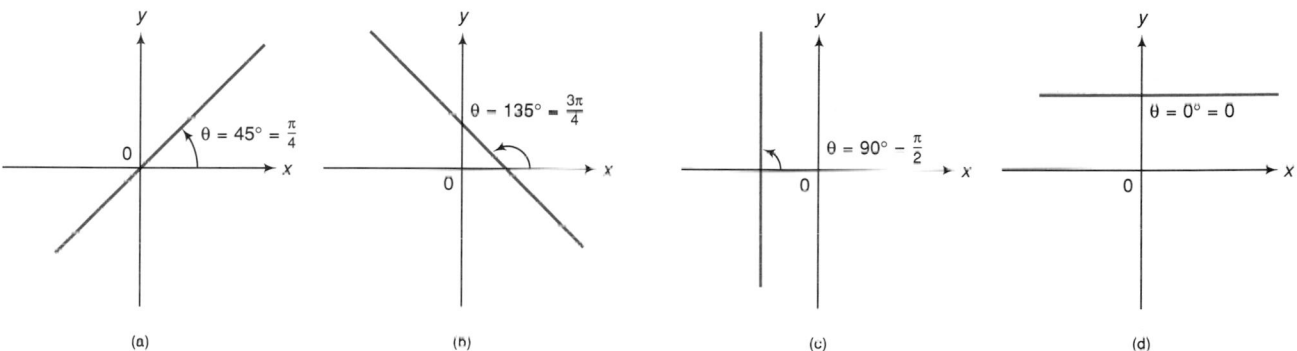

FIGURE 6   FOUR ANGLES OF INCLINATION

In many places in this text, we will use notions from trigonometry. A review of trigonometry appears in Appendix A1. This review contains all the trigonometric facts you will need in your calculus course. In the rest of this text, we will assume you are familiar with this material.

We now tie together the notions of slope and angle of inclination.

THEOREM 1

If *m* is the slope of a nonvertical line *L* and θ its angle of inclination, then

$$m = \tan \theta. \tag{1}$$

PROOF:

(i) If *L* is parallel to the *x*-axis, then $m = 0$, $\theta = 0°$, and $\tan \theta = \tan 0° = 0 = m$.

[†] A complete discussion of radian measure is given in Appendix A1.1

(ii) If $m > 0$, then $0° < \theta < 90°$ and from Figure 7a we immediately see that

$$\tan \theta = \frac{\Delta y}{\Delta x} = m.$$

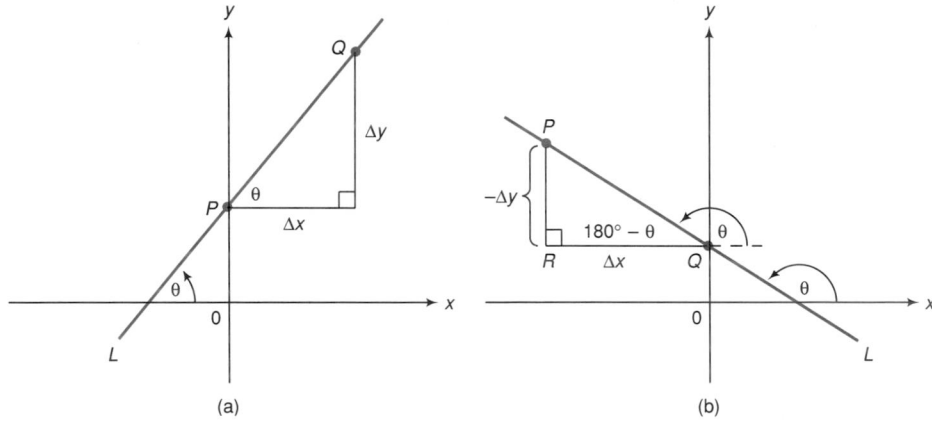

FIGURE 7 THE SLOPE OF A LINE EQUALS THE TANGENT OF THE LINE'S ANGLE OF INCLINATION.

(iii) If $m < 0$, then $90° < \theta < 180°$, and as we move from P to Q in Figure 7b, y decreases and $\Delta y < 0$. Thus the length of the line segment joining P to R is $-\Delta y$, and we have

$$\tan(180° - \theta) = \tan(\pi - \theta) = \frac{-\Delta y}{\Delta x} = -m.$$

↗ $180° = \pi$ radians

But

See identities (vi) and (vii)
Appendix A1.2, Table 2

$$\tan(\pi - \theta) = \frac{\sin(\pi - \theta)}{\cos(\pi - \theta)} = \frac{\sin \theta}{-\cos \theta} = -\tan \theta.$$

Thus $-m = -\tan \theta$ or $m = \tan \theta$, and the proof is complete. ∎

NOTE: If L is parallel to the y-axis, then m is undefined. But then $\theta = 90° = \pi/2$, and $\tan \theta$ is also undefined.

REMARK: Lines with large positive or negative slopes have angles of inclination near $90°$ ($\pi/2$ radians ≈ 1.5708[†]). These are lines that appear very "steep." Lines with small slopes have angles of inclination near $0°$ or $180°$ (near 0 or π radians).

Angles of inclination can be used to provide straightforward proofs that parallel lines have the same slope and that perpendicular lines have slopes that are negative reciprocals of one another (see Problems 61 and 62).

---

† The symbol $\approx$ stands for "is approximately equal to."

■ PROBLEMS 0.3

■ SELF-QUIZ

I. The straight line passing through $(-2, 7)$ and $(7, -2)$ _____.
   a. has positive slope   b. is vertical
   c. has negative slope   d. has zero slope

II. The straight line passing through $(5, 0)$ and $(0, 2)$ satisfies the equation _____.
   a. $y = ((0 - 2)/(5 - 0))x + 2$
   b. $(y - 2)(5 - 0) = (0 - 2)(x - 0)$
   c. $(y - 0)(0 - 5) = (2 - 0)(x - 5)$
   d. $x/5 + y/2 = 1$
   e. $2x + 5y = 10$ f. all of the above

III. Perpendicular to the graph of $2x + 5y = 10$, there is a unique line through the origin, $(0, 0)$; this line satisfies the equation _____.
   a. $5x + 2y = 0$ b. $y/x = -\frac{2}{5}$
   c. $5x = 2y$ d. $5x - 2y = 10$
   e. $y = (-\frac{5}{2})x$ f. $y = (-\frac{1}{5})x$

IV. Find a simple equation satisfied by all points $P = (x, y)$ having the property that the line through P and $(-5, 2)$ is perpendicular to the line through P and $(8, 20)$.

■ DRILL

In Problems 1–8, compute the slope of the line passing through the two given points; sketch the line.

1. $(1, 6), (2, 4)$ 2. $(-3, 4), (7, 9)$
3. $(-1, -2), (-3, -4)$ 4. $(-6, 5), (7, -2)$
5. $(1, 7), (-4, 7)$ 6. $(2, -3), (5, -3)$
7. $(-2, 4), (-2, 6)$ 8. $(a, 0), (0, a), a < 0$

In Problems 9–18, you are given information either a point and a slope (m) or two points—sufficient to determine a straight line. For each problem, write equations for that line both in a point-slope form and in the slope-intercept form; then sketch the graph of the line in the xy-plane.

9. $(4, -7), m = 0$
10. $(4, -7), m$ undefined
11. $(7, -3), m = -\frac{4}{3}$ 12. $(-5, 1), m = \frac{3}{7}$
13. $(1, 2), (3, 6)$ 14. $(-2, 3), (4, -1)$
15. $(-2, -4), (3, 7)$ 16. $(3, -\frac{1}{2}), (\frac{1}{3}, 0)$
17. $(a, b), m = 0$ 18. $(a, b), (c, d), a = c$

19. Find a standard equation of the line parallel to the line $2x + 5y = 6$ that passes through the point $(-1, 1)$.

20. Find a standard equation of the line parallel to the line $5x - 7y = 3$ that passes through the point $(2, 5)$.

21. Find a standard equation of the line perpendicular to the line $x + 3y = 7$ that passes through the point $(0, 1)$.

22. Find a standard equation of the line perpendicular to the line $ax + by = c$ that passes through the point (x_0, y_0). Assume that $a \neq 0$ and $b \neq 0$.

In Problems 23–26, find the point of intersection (if there is one) of the two lines.

23. $y - 2x = 4; 4x - 2y = 6$
24. $4x - 6y = 10; 6x - 9y = 15$
25. $3x + 4y = 5; 6x - 7y = 8$
26. $3x + y = 4; y - 5x = 2$

In Problems 27–32, find the slope of a line having the given angle of inclination.

27. $30°$ 28. $150°$ 29. $45°$
30. $135°$ 31. $120°$ 32. $60°$

In Problems 33–38, find the angle of inclination for a line having the given slope.

33. 1 34. $\sqrt{3}$ 35. $-1/\sqrt{3}$
36. -1 ▦37. 2 ▦38. -0.5

■ APPLICATIONS

In Problems 39–42, determine whether the three points lie on a single straight line (if they do, the points are said to be **collinear**).

39. $(0, 2), (1, 1), (5, -3)$
40. $(7, 13), (2, 7), (-2, 3)$
41. $(2, 1), (3, 3), (-5, -13)$
42. $(1, 1), (1, -3), (1, 27)$

In Problems 43–48, two lines are determined by the given pairs of points. Determine whether those lines are parallel, perpendicular, or neither.

43. $(1, 8), (2, 9); (1, 2), (0, 1)$
44. $(3, 1), (3, 7); (2, 4), (-1, 4)$
45. $(0, 2), (-2, 0); (0, 3), (3, 0)$
46. $(1, -2), (2, 4); (4, 1), (-8, 2)$

CONTINUED

47. $(5, 2), (1, 7); (2, 5), (7, 1)$
48. $(0, 5), (2, -1); (0, 0), (-1, 3)$

It is a geometric fact that the line T tangent to a circle at a given point is also perpendicular to the radial line R at that point (see the accompanying figure). Use this fact in Problems 49–52 to find the slope-intercept equation for the line tangent to the given circle at the specified point.

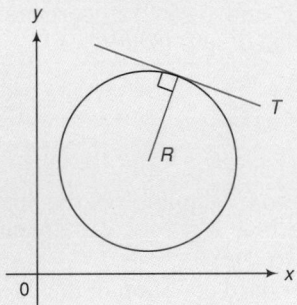

49. $x^2 + y^2 = 1; (1/\sqrt{2}, 1/\sqrt{2})$
50. $(x - 1)^2 + (y + 1)^2 = 4; (1, 1)$

51. $(x - 3)^2 + (y + 2)^2 = 5; (4, 0)$
52. $(x + 2)^2 + (y - 3)^2 = 9; (0, 3 + \sqrt{5})$

Let L be a line and let L_P denote the line perpendicular to L that passes through a given point P. Then the *distance from L to P* is defined to be the distance between P and the point of intersection of L and L_P. In Problems 53–56, find the distance between the given line and point.

53. $2x + 3y = -1; (0, 0)$
54. $3x + y = 7; (1, 2)$
55. $6y + 3x = 3; (8, -1)$
56. $2y - 5x = -2; (5, -3)$
57. Suppose $a > 0$ and $a + h > 0$. Verify that the straight line through $(a, \sqrt{a})$ and $(a + h, \sqrt{a + h})$ has slope $1/(\sqrt{a + h} + \sqrt{a})$. [*Hint:* $(\sqrt{B} - \sqrt{A})(\sqrt{B} + \sqrt{A}) = B - A$ if $A, B > 0$.]
58. Use the result of Problem 67 to calculate the areas of the triangles with the given vertices.
   a. $(2, 1), (0, 4), (3, -6)$
   b. $(4, 2), (-1, -5), (7, 3)$

■ SHOW/PROVE/DISPROVE

59. Suppose (a, b) is a point on the circle $x^2 + y^2 = r^2$. Show that the line tangent to the circle at (a, b) satisfies the equation $ax + by = r^2$.
60. Write an equation for a line that goes through the point, $(c, 0)$ and is tangent to the circle $x^2 + y^2 = 1$. Assume $c > 1$.
61. Show that two lines are parallel if and only if they have the same slope. [*Hint:* Use similar triangles.] *Note:* "Have the same slope" means either that both slopes exist and they are equal numerically or that both slopes are undefined.
62. Show that two lines (with defined slopes) are perpendicular if and only if their slopes are negative reciprocals by carrying out the following steps.

(References are to the accompanying figure and its notation.)
a. Explain why L_1 and L_2 are perpendicular if and only if $\overline{AC}^2 + \overline{CB}^2 = \overline{AB}^2$. [*Hint:* Look at the law of cosines (Problem A1.4.11).]
b. Show that $\overline{AC}^2 + \overline{CB}^2 = \overline{AB}^2$ if and only if $c^2 + d^2 = ac + bd$.
c. Show that $m_1 = -1/m_2$ if and only if
$$c^2 + d^2 = ac + bd.$$

63. Show that the distance from the point (x_0, y_0) to the line with standard equation $ax + by + c = 0$ is $|ax_0 + by_0 + c|/\sqrt{a^2 + b^2}$.
*64. The lines $x + 3y = -2$ and $x + 3y = 8$ are parallel. Find the distance between them. *Comments:* Does this problem make sense? Is there a unique reasonable answer? First, state a reasonable definition for the distance between two parallel lines. Second, devise a way to compute that distance in this case. [*Note:* Although computing $8 - (-2)$ involves easy arithmetic, 10 is not a reasonable answer to this problem.]
65. Show that $(x - 3)^2 + (y - 10)^2 = (x + 1)^2 + (y - 5)^2$ is an equation satisfied precisely by the

CONTINUED

points on the straight line that is the perpendicular bisector of the line segment joining (3, 10) and (−1, 5).

66. Suppose that $x^2 + y^2 + Ax + By + C = 0$ and $x^2 + y^2 + ax + by + c = 0$ are different circles

that meet at two distinct points. Show that an equation of the line through those two points of intersection is

$$(A - a)x + (B - b)y + (C - c) = 0.$$

■ CHALLENGE

**67. Show that the triangle with vertices (x_1, y_1), (x_2, y_2), and (x_3, y_3) has area

$$\frac{1}{2}\left|x_1y_2 - x_2y_1 + x_2y_3 - x_3y_2 + x_3y_1 - x_1y_3\right|.$$

68. Show that the area of the parallelogram with vertices at $(0, 0)$, (x_1, y_1), (x_2, y_2), and $(x_1 + x_2, y_1 + y_2)$ equals $\left|x_1y_2 - x_2y_1\right|$. [*Hint:* Use the result of Problem 67.]

■ ANSWERS TO SELF-QUIZ

I. c

II. $f = \{a, b, c, d, e\}$

III. c

IV. $(y - 2)/(x + 5) = -(x - 8)/(y - 20)$; the solutions to this equation lie on a circle with $(-5, 2)$ and $(8, 20)$ as endpoints of a diameter.

0.4 FUNCTIONS AND THEIR GRAPHS

Let us return to the equation of a straight line. For example, if $y = 3x + 5$, then the line can be thought of as the set of all ordered pairs (or points) (x, y) such that $y = 3x + 5$. The important fact here is that for every real number x there is a *unique* real number y such that $y = 3x + 5$ and the ordered pair (x, y) is on the line. We generalize this idea in the following definition.

DEFINITION   FUNCTION

Let X and Y be sets of real numbers. A **function** f is a rule that assigns to each number x in X a single number $f(x)$ in Y. The number $f(x)$ is called the **image** of the number x. X is called the **domain** of f, written dom f. The set of images of elements of X is called the **range** of f written range f or $f(X)$. ■

Simply put, a function is a rule that assigns to every x in the domain of f a unique number y in the range of f. We will usually write this in **functional notation** as follows:

FUNCTIONAL NOTATION

$$y = f(x) \tag{1}$$

is read, "y equals f of x."†

---

† See the biographical sketch on Leonhard Euler, page 34.

When we write $y = f(x)$, x is called the **independent variable**, and y is called the **dependent variable**.

EXAMPLE 1
THE SQUARING FUNCTION

$\mathbb{R}^+$

To each real number we assign its square. For example, 4 is assigned to 2, 9 is assigned to -3, 1.69 is assigned to 1.3, and 2 is assigned to $\sqrt{2}$. This assignment defines a function because each real number has a unique square. The domain is the set of real numbers $\mathbb{R}$. The range is the set of nonnegative real numbers, denoted by $\mathbb{R}^+$. We write this function as

$$y = f(x) = x^2.$$

To make this clearer, we write

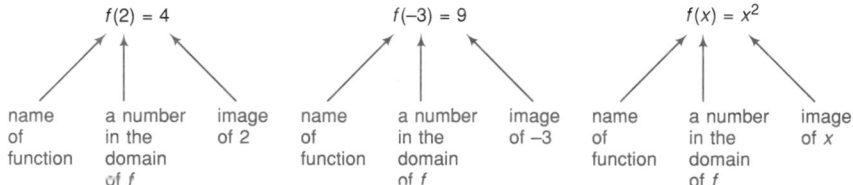

| name of function | a number in the domain of f | image of 2 | name of function | a number in the domain of f | image of -3 | name of function | a number in the domain of f | image of x |

$f(2) = 4$ $f(-3) = 9$ $f(x) = x^2$

To prove that the range of f is $\mathbb{R}^+$, we assume that $y \geq 0$. Let $x = \sqrt{y}$. Then $f(x) = x^2 = (\sqrt{y})^2 = y$. This shows that *every* nonnegative number is the image of a number in the domain of f. On the other hand, if y is in the range of f, then $y = x^2$ for some real number x. But $x^2 \geq 0$, so $y \geq 0$. That is, every number in the range of f is nonnegative. Putting these facts together, we conclude that the range of the squaring function is $\mathbb{R}^+$.

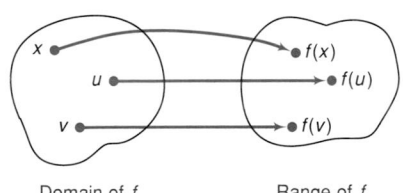

Domain of f Range of f

FIGURE 1
PICTORIAL REPRESENTATION OF A
FUNCTION

PICTORIAL REPRESENTATION OF A FUNCTION

Functions are often depicted pictorially as in Figure 1. The diagram illustrates, for example, that the function f takes a value x in the domain and assigns to it a unique value $f(x)$ in its range.

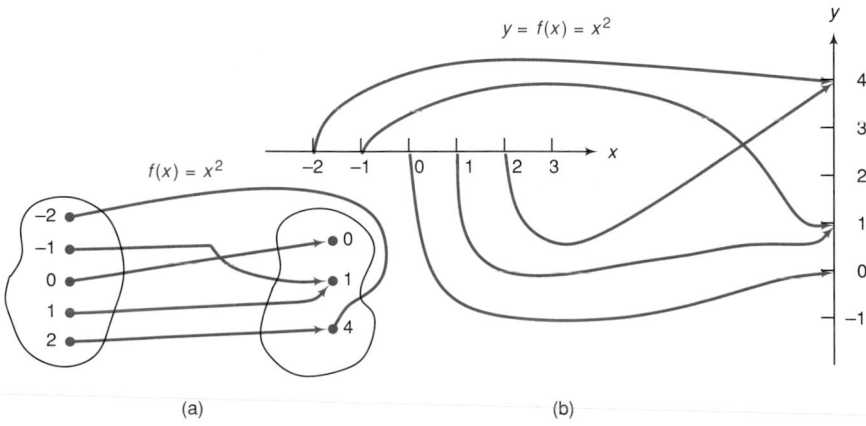

(a)   (b)

FIGURE 2 SOME VALUES TAKEN BY THE FUNCTION $y = f(x) = x^2$ (DEPICTED TWO WAYS)

In Figure 2, we depict in two different ways some of the values taken by the squaring function.

EXAMPLE 2
DETERMINING WHETHER A RULE IS A FUNCTION

Let D and R be two sets of real numbers. In Figure 3, we show four rules that assign numbers in D to numbers in R. Which of the four rules represent functions with domain D?

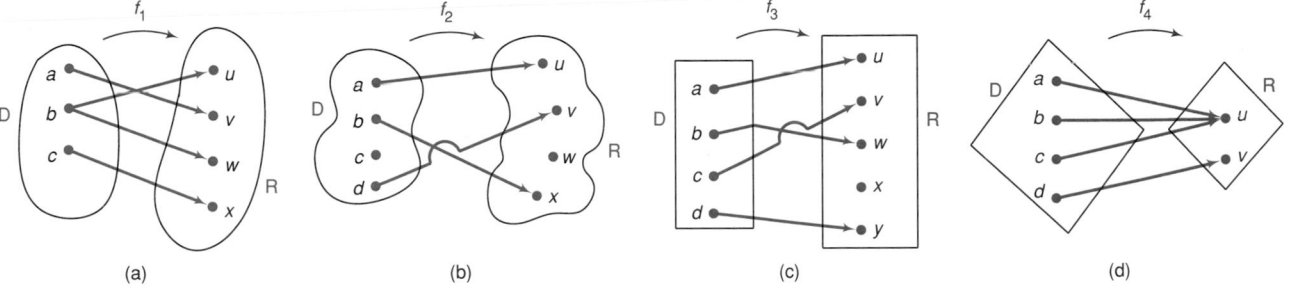

(a) (b) (c) (d)

FIGURE 3 FOUR RULES

SOLUTION:

(a) f_1 is not a function because it assigns two different members of R to one member (b) of D. That is, $f_1(b) = u$ and w. This contradicts the definition of a function. In order to be a function, f_1 must assign *exactly one* member of R to each member of D. If we delete the arrow from b to u or the arrow from b to w (but not both), then f will be a function with domain D.

(b) f_2 is not a function with domain D because f does not assign anything to c. In order to be a function with domain D, f must assign one member of R to *each* member of D.

(c) f_3 is a function because it assigns one member of R to each member of D. It is irrelevant that nothing in D is assigned to x. This may make x feel lonely, but it does not contradict the definition of a function. The range of f_3 is $\{u, v, w, y\}$, which is *not* equal to R.

(d) f_4 is a function because it assigns exactly one member of R to each member of D. The fact that u is assigned three times does not contradict the definition of a function.

To determine the domain of a function (if it is not given explicitly), we need to ask the question "For what values of x does the rule given in equation (1) make sense?" For example, if f is the function written as $f(x) = 1/x$, then since the expression $1/x$ is not defined for $x = 0$, the number 0 is not in the domain of f. However, $1/x$ is defined for any $x \neq 0$, so the domain of f is the set of all real numbers except zero. This set can be written as $\mathbb{R} - \{0\}$.

To determine the range of a function we must ask, "What values do we obtain for y as x takes on all values in dom f?" For example, if f is defined by $f(x) = x^2$, then, as we have seen, the domain of f is $\mathbb{R}$, since any real number can be squared. The range of f is the set of nonnegative real numbers, denoted by $\mathbb{R}^+$, because the square of any real number is nonnegative, and any nonnegative real number is the square of a real number.

⊗ **WARNING:** In writing $y = f(x)$, we must distinguish between the symbol f, which stands for the function rule, and the symbol $f(x)$, which is the *value* the function takes on for a given number x in the domain of f. Here $f(x)$ is a *number* in the range of f. ⊗

EXAMPLE 3
EVALUATING A QUADRATIC
FUNCTION

Let $f(x) = x^2 - 3x + 1$. Find (a) $f(2)$ and (b) $f(-5)$.

SOLUTION:

(a) Since $f(x) = x^2 - 3x + 1$, substituting 2 for x gives us

$$f(2) = 2^2 - (3)(2) + 1 = 4 - 6 + 1 = -1.$$

(b) $f(-5) = (-5)^2 - 3(-5) + 1 = 25 + 15 + 1 = 41.$

DEFINITION   **GRAPH OF A FUNCTION**

The **graph** of the function f is the set of ordered pairs $\{(x, f(x)): x \in \text{dom } f\}$. ∎

Thus a point lies on the graph of f if and only if its coordinates satisfy the equation $y = f(x)$.

EXAMPLE 4
A STRAIGHT-LINE FUNCTION

The equation $y = f(x) = 3x + 5$ can be thought of as a function, since for every real number x there is a unique real number y that is equal to $3x + 5$. Here the domain of f is $\mathbb{R}$. To calculate the range of f, we note that if y is any real number, then there is a unique number x such that $y = 3x + 5$. To see this, we simply solve for x: $x = (y - 5)/3$. Thus every real number y is in the range of f. In this example the function f is called a **straight-line function**. In general, straight-line functions have the form $y = f(x) = ax + b$, where a and b are real numbers.

EXAMPLE 5
THE GRAPH OF $f(x) = x^2$ IS A
PARABOLA THAT IS SYMMETRIC
ABOUT THE y-AXIS

Let $y = f(x) = x^2$. As we saw in Example 1, this rule constitutes a function whose domain is $\mathbb{R}$ since each real number has a unique square. We also see, as mentioned above, that the range of f equals $\{x: x \geq 0\} = \mathbb{R}^+$, the set of nonnegative real numbers, since the square of any real number is nonnegative. The graph of this function is obtained by plotting all points of the form $(x, y) = (x, x^2)$.[†] First, we note that because $f(x) = x^2$, $f(x) = f(-x)$ since $(-x)^2 = x^2$. Thus it is only necessary to calculate

---

[†] Of course, we can't plot *all* points (there are an infinite number of them). Rather, we plot some sample points and assume they can be connected to obtain the sketch of the graph.

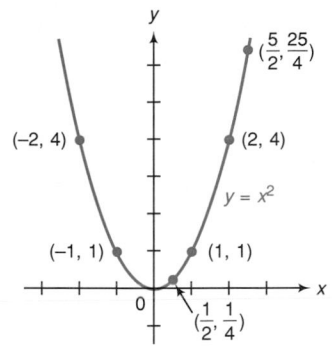

FIGURE 4
GRAPH OF $f(x) = x^2$

A parabola.

TABLE 1

x	0	$\frac{1}{2}$	1	$\frac{3}{2}$	2	$\frac{5}{2}$	3	4	5
$f(x) = x^2$	0	$\frac{1}{4}$	1	$\frac{9}{4}$	4	$\frac{25}{4}$	9	16	25

$f(x)$ for $x \geq 0$. For every $x > 0$ the number $-x$ gives the same value of y. In this situation we say that the function is **symmetric** about the y-axis. Some values for $f(x)$ are shown in Table 1. This function, which is graphed in Figure 4, is called a **parabola**. It is a second example of a conic section (the first was a circle).

REMARK: We repeat that care must be taken to distinguish between the symbols f and $f(x)$. In this example f is the *rule* that assigns to each real number the square of that number. The symbol $f(x)$ stands for the number x^2.

The function given by $y = f(x)$ is **symmetric about the y-axis** if $f(x) = f(-x)$. A function f that satisfies $f(-x) = f(x)$ is also called an **even function**.

At this point, there are essentially three reasons for restricting the domain of a function:

THREE REASONS FOR RESTRICTING THE DOMAIN OF A FUNCTION

(i) You cannot have zero in a denominator.

(ii) You cannot take the square root (or fourth root, or sixth root, etc.) of a negative number.

(iii) The domain is restricted by the nature of the applied problem under consideration. For example, if $C(q)$ denotes the cost of buying q barrels of oil, then, necessarily, $q \geq 0$ because it is impossible to buy negative quantities.

We will see a fourth kind of restriction when we discuss logarithmic functions in Sections 6.2 and 6.3.

EXAMPLE 6
THE DOMAIN AND RANGE OF A SQUARE-ROOT FUNCTION

Let $f(x) = \sqrt{2x - 6}$. Find the largest possible domain for f. Find its range.

SOLUTION: $f(x)$ is defined as long as $2x - 6 \geq 0$ or $2x \geq 6$ or $x \geq 3$. Thus,

$$\text{dom } f = \{x: x \geq 3\} = [3, \infty).$$

Let y be a nonnegative real number (i.e., $y \geq 0$). Then we can find an x such that $\sqrt{2x - 6} = y$. To do so, we solve:

$$\sqrt{2x - 6} = y$$
$$2x - 6 = y^2$$
$$2x = y^2 + 6$$
$$x = \frac{y^2}{2} + 3$$

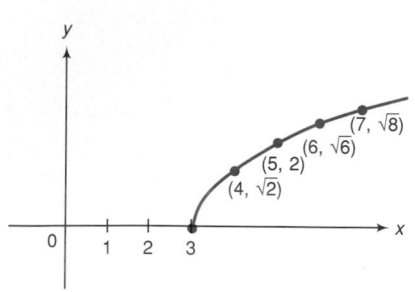

For example, if $y = 4$, then $x = \frac{4^2}{2} + 3 = \frac{16}{2} + 3 = 11$.

CHECK: $f(11) = \sqrt{2(11) - 6} = \sqrt{22 - 6} = \sqrt{16} = 4$.

Thus the range of f is $\mathbb{R}^+$. We provide a graph of this function in Figure 5.

FIGURE 5
GRAPH OF $f(x) = \sqrt{2x - 6}$

$f(x)$ is not defined if $x < 3$.

EXAMPLE 7
THE DOMAIN AND RANGE OF A
FUNCTION THAT IS NOT DEFINED
AT TWO VALUES

Consider the rule $v = g(u) = 1/(u^2 - 1)$. It is clear that $u^2 - 1 = 0$ only when $u = \pm 1$. For a fixed number $u \neq \pm 1$, there is a unique number v that is equal to $1/(u^2 - 1)$. Thus, this rule is a function, and

$$\text{dom } g = \{u \in \mathbb{R} : u \neq \pm 1\} = \mathbb{R} - \{1, -1\}.$$

Here we have used the letters u, v, and g instead of x, y, and f to illustrate the fact that there is nothing special about any particular set of letters. Often, we will use the letter t to denote time and the letter s to denote distance. The definition of a function is independent of the letters used to denote the function and variables.

We now calculate the range of the function. We note that as u becomes large, $1/(u^2 - 1)$ becomes very small. If $u > 1$ but is close to 1, then $1/(u^2 - 1)$ is very large in the positive direction. For example, $1[(1.0001)^2 - 1] \approx 5000$. If $u < 1$ and u is close to 1, then $u^2 - 1 < 0$ and $1/(u^2 - 1)$ is very large in the negative direction. For example, $1/[(0.9999)^2 - 1] \approx -5000$. Since $u^2 - 1$ can take on any positive real value, so can $1/(u^2 - 1)$. Also, $1/(u^2 - 1)$ is never equal to 0. On the other hand, if $-1 < u < 1$, then $u^2 - 1 < 0$. But $u^2 - 1 \geq -1$, so $1/(u^2 - 1) \leq -1$. Thus range $g = \{v : v \notin (-1, 0]\} = \{v : v \leq -1 \text{ or } v > 0\}$.

There is another way to determine the range. Suppose that v is in the range of g. Then there is a number u ($\neq \pm 1$) such that

$$v = g(u) = \frac{1}{u^2 - 1}.$$

We now solve for u in terms of v:

$$u^2 v - v = 1$$
$$u^2 v = 1 + v$$
$$u^2 = \frac{1 + v}{v}$$
$$u = \pm \sqrt{\frac{1 + v}{v}}.$$

Then v is in the range of g if $\sqrt{\dfrac{1+v}{v}}$ is a real number. That is, if $v \neq 0$ and $\dfrac{1+v}{v} \geq 0$. If $v > 0$, then $\dfrac{1+v}{v} > 0$, which proves that every positive real number is in the range. If $v < 0$, then

$$\frac{1+v}{v} \geq 0$$

$$1 + v \leq 0 \qquad \text{multiplying an inequality by a negative number reverses the direction of the inequality}$$

$$v \leq -1$$

Thus, as before,

range $g = \{v \colon v \leq -1 \text{ or } v > 0\}$.

We will not sketch the graph of this function. Methods for doing so will be given in Chapter 3.

EXAMPLE 8
AN EQUATION THAT DOES NOT
DEFINE A FUNCTION

Consider the graph of the unit circle $x^2 + y^2 = 1$ given in Figure 6. It is clear that for every real number x in the open interval $(-1, 1)$, there are two values of y given by $y = \pm\sqrt{1-x^2}$. Hence, the equation does not define a function. We can obtain two separate functions by defining

$$y_1 = f_1(x) - \sqrt{1-x^2} \qquad \text{and} \qquad y_2 = f_2(x) = -\sqrt{1-x^2}.$$

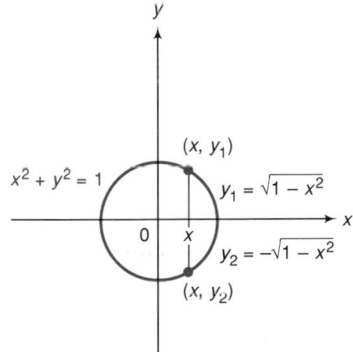

FIGURE 6 FOR EVERY x IN $(-1, 1)$, THERE ARE TWO y's SUCH THAT $x^2 + y^2 = 1$.

Then dom $f_1 = $ dom $f_2 = [-1, 1]$, range $f_1 = [0, 1]$, and range $f_2 = [-1, 0]$.

EXAMPLE 9
AN EQUATION THAT DOES
DEFINE A FUNCTION

Let

$$x^3 + y^3 = 4.$$

We can solve for y:

$$y^3 = 4 - x^3$$
$$y = \sqrt[3]{4 - x^3}.$$

If $4 - x^3 \geq 0$, then $y \geq 0$; and if $4 - x^3 < 0$, then $y < 0$. Thus, for every x there is a unique y such that $x^3 + y^3 = 4$. This defines a function with domain and range equal to $\mathbb{R}$.

THE VERTICAL-LINES TEST

In Examples 8 and 9 we saw that an equation in two variables may or may not define a function. There is a graphical test to determine this. Consider the graph of $x^2 + y^2 = 4$ in Figure 7.

We draw the vertical line $x = c$ with c in the interval $(-2, 2)$. This line intersects the circle at *two* points. This means that for every c in $(-2, 2)$ there are two y's such that $c^2 + y^2 = 4$. Thus we cannot write y as a function of x.

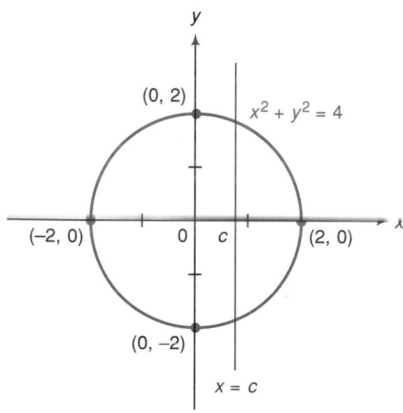

FIGURE 7
ILLUSTRATION OF THE VERTICAL-LINES TEST

VERTICAL-LINES TEST

A set of points in the Cartesian plane is the graph of a function if every vertical line in the plane intersects the set of points in *at most* one point.

EXAMPLE 10
A FUNCTION DEFINED IN PIECES

Let $f(x) = \begin{cases} 1, & x \geq 0 \\ 2, & x < 0. \end{cases}$

A graph of this function is given in Figure 8. We have dom $f = \mathbb{R}$ and range $f = \{1, 2\}$.

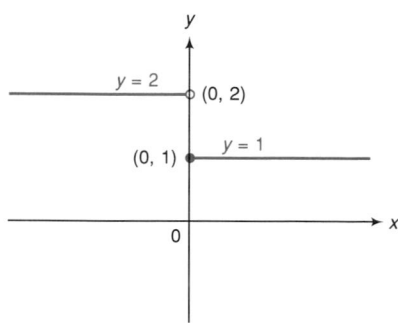

FIGURE 8   GRAPH OF $f(x) = \begin{cases} 1, x \geq 0 \\ 2, x < 0 \end{cases}$

REMARK: Example 10 gives an example of a function defined in "pieces." It is perfectly legitimate to do so as long as for each x in the domain of f there is a unique y in the range. Note that any vertical line intersects the graph in exactly one point.

DEFINITION SUM, DIFFERENCE, PRODUCT, AND QUOTIENT FUNCTIONS

Let f and g be two functions.

(i) The **sum** $f + g$ is defined by

$$(f + g)(x) = f(x) + g(x). \tag{2}$$

(ii) The **difference** $f - g$ is defined by

$$(f - g)(x) = f(x) - g(x). \tag{3}$$

(iii) The **product** $f \cdot g$ is defined by

$$(f \cdot g)(x) = f(x)g(x). \tag{4}$$

(iv) The **quotient** f/g is defined by

$$\left(\frac{f}{g}\right)(x) = \frac{f(x)}{g(x)}. \tag{5}$$

Furthermore, $f + g$, $f - g$, and $f \cdot g$ are defined for all x for which both f and g are defined. Finally, f/g is defined whenever both f and g are defined and $g(x) \neq 0$ (so that we do not divide by zero). ∎

EXAMPLE 11
THE SUM, DIFFERENCE, PRODUCT, AND QUOTIENT OF TWO FUNCTIONS

Let $f(x) = \sqrt{x + 1}$ and $g(x) = \sqrt{4 - x^2}$. Since dom $f = [-1, \infty)$ and dom $g = [-2, 2]$, we have $\text{dom}(f + g) = \text{dom}(f - g) = \text{dom}(f \cdot g) =$

$$\{x: x \geq -1 \text{ and } -2 \leq x \leq 2\} = [-1, 2]$$

and

$$\text{dom}\left(\frac{f}{g}\right) = \{x: x \geq -1, -2 \leq x \leq 2 \text{ and } x \neq -2 \text{ or } 2\} = [-1, 2).$$

We must exclude $x = \pm 2$ because $\sqrt{4 - x^2} = 0$ when $x = 2$ or -2.
The functions are

$$(f + g)(x) = \sqrt{x + 1} + \sqrt{4 - x^2}$$
$$(f - g)(x) = \sqrt{x + 1} - \sqrt{4 - x^2}$$
$$(f \cdot g)(x) = \sqrt{x + 1} \cdot \sqrt{4 - x^2}$$
$$\left(\frac{f}{g}\right)(x) = \frac{\sqrt{x + 1}}{\sqrt{4 - x^2}}.$$

NOTE: $\sqrt{x + 1} \cdot \sqrt{4 - x^2} \neq \sqrt{(x + 1)(4 - x^2)}$ because they have different domains. $\sqrt{(x + 1)(4 - x^2)}$ is defined for $x < -2$ but neither $\sqrt{x + 1}$ nor $\sqrt{4 - x^2}$ is defined for $x < -2$.

DEFINITION **COMPOSITE FUNCTION**

If f and g are functions, then the **composite function** $f \circ g$ is defined by

$$(f \circ g)(x) = f(g(x))$$

and

$$\mathrm{dom}(f \circ g) = \{x \colon x \in \mathrm{dom}\, g \text{ and } g(x) \in \mathrm{dom}\, f\}.$$

That is, $(f \circ g)(x)$ is defined for every x such that $g(x)$ and $f(g(x))$ are defined. ∎

EXAMPLE 12
THE COMPOSITION OF TWO
FUNCTIONS

Let $f(x) = \sqrt{x}$ and $g(x) = x^2 + 1$. Then,

$$(f \circ g)(x) = f(g(x)) = f(x^2 + 1) = \sqrt{x^2 + 1}$$

and

$$(g \circ f)(x) = g(f(x)) = g(\sqrt{x}) = (\sqrt{x})^2 + 1 = x + 1.$$

Note, for example, that $(f \circ g)(3) = \sqrt{10}$, while $(g \circ f)(3) = 4$. Now $\mathrm{dom}\, f = \mathbb{R}^+$, $\mathrm{dom}\, g = \mathbb{R}$, and we have

$$\mathrm{dom}(f \circ g) = \{x \colon g(x) = x^2 + 1 \in \mathrm{dom}\, f\}.$$

But since $x^2 + 1 > 0$, $x^2 + 1 \in \mathrm{dom}\, f$ for every real x, so $\mathrm{dom}(f \circ g) = \mathbb{R}$. On the other hand, $\mathrm{dom}(g \circ f) = \mathbb{R}^+$ since f is only defined for $x \geq 0$.

⊗ **WARNING:** As Example 12 suggests, it is *not* true in general that $(f \circ g)(x) = (g \circ f)(x)$. ⊗

EXAMPLE 13
THE COMPOSITION OF TWO
FUNCTIONS

Let $f(x) = 3x - 4$ and $g(x) = x^3$. Then,

$$(f \circ g)(x) = f(g(x)) = f(x^3) = 3x^3 - 4$$

and

$$(g \circ f)(x) = g(f(x)) = g(3x - 4) = (3x - 4)^3.$$

Here $\mathrm{dom}(f \circ g) = \mathrm{dom}(g \circ f) = \mathbb{R}$. Note that the functions $f \circ g$ and $g \circ f$ are, as in Example 12, quite different.

EXAMPLE 14
TWO INVERSE FUNCTIONS

Let $f(x) = 4x + 8$ and $g(x) = \frac{1}{4}x - 2$. Then,

$$(f \circ g)(x) = f(g(x)) = f\left(\frac{1}{4}x - 2\right) = 4\left(\frac{1}{4}x - 2\right) + 8$$

$$= x - 8 + 8 = x.$$

Similarly,

$$(g \circ f)(x) = g(f(x)) = g(4x + 8) = \frac{1}{4}(4x + 8) - 2 = x + 2 - 2 = x.$$

Thus,

$$(f \circ g)(x) = (g \circ f)(x) = x.$$

When these last equations hold, we say that f and g are *inverse functions*. Inverse functions are discussed in detail in Section 6.1.

DECOMPOSITION OF FUNCTIONS

In calculus, it is often necessary to write a function as the composition of two or more functions. We will do this, for example, when we study the chain rule in Section 2.3.

EXAMPLE 15
WRITING A FUNCTION AS THE COMPOSITION OF TWO FUNCTIONS

Write $f(x) = (5x + 2)^3$ as the composition of two functions.

SOLUTION: We want to write $f(x) = (h \circ g)(x) = h(g(x))$. The last thing we do is cube, so we set $h(x) = x^3$. The thing cubed is $5x + 2$, so we set $g(x) = 5x + 2$. Then,

$$(h \circ g)(x) = h(g(x)) = [g(x)]^3 = (5x + 2)^3 = f(x).$$

EXAMPLE 16
WRITING A FUNCTION AS THE COMPOSITION OF THREE FUNCTIONS

Write $\dfrac{1}{\sqrt{x^2 + 1}}$ as the composition of three functions.

SOLUTION: The last thing we do is take the reciprocal (1 divided by $\sqrt{x^2 + 1}$), so we set $h(x) = \dfrac{1}{x}$. The function $\sqrt{x^2 + 1}$ is the composition of $f(x) = x^2 + 1$ and $g(x) = \sqrt{x}$, with $g(x)$ performed second. Thus,

$$(h \circ g \circ f)(x) = h(g(f(x))) = \frac{1}{g(f(x))} = \frac{1}{\sqrt{f(x)}} = \frac{1}{\sqrt{x^2 + 1}}.$$

REMARK: There frequently are many ways to write a function as the composition of two or more functions. For example, if $u(x) = 5x$, $v(x) = x + 2$, and $w(x) = x^3$, then

$$(w \circ v \circ u)(x) = w(v(u(x))) = w(v(5x))$$
$$= w(5x + 2) = (5x + 2)^3.$$

The decompositions given in Examples 15 and 16 are the easiest ones to use, but they are not unique.

THE FUNCTIONAL NOTATION $y = f(x)$ was first used by the great Swiss mathematician Leonhard Euler (pronounced "oiler") in the *Commentarii Academia Petropolitanae* (*Petersburg Commentaries*), published in 1734–1735.

Euler was born in Basel, Switzerland. His father was a clergyman who hoped that Leonhard would follow him into the ministry. The father was adept at mathematics, however, and, together with Johann Bernoulli, instructed young Leonhard in that subject. Euler also studied theology, astronomy, physics, medicine, and several Eastern languages.

In 1727, Euler applied to and was accepted for a chair of medicine and physiology at the St. Petersburg Academy. The day Euler arrived in Russia, however, Catherine I—founder of the academy—died, and the academy was plunged into turmoil. By 1730, Euler was pursuing his mathematical career from the chair of natural philosophy. Accepting an invitation from Frederick the Great, Euler went to Berlin in 1741 to head the Prussian Academy. Twenty-five years later, he returned to St. Petersburg, where he died in 1783 at the age of 76.

The most prolific writer in the history of mathematics, Euler found new results in virtually every branch of pure and applied mathematics. Although German was his native language, he wrote mostly in Latin and occasionally in French. His amazing productivity did not decline even when he became totally blind in 1766. During his lifetime, Euler published 530 books and papers. When he died, he left so many unpublished manuscripts that the St. Petersburg Academy was still publishing his work in its *Proceedings* almost half a century later. Euler's work en-

LEONHARD EULER
·
1707–1783

riched such diverse areas as hydraulics, celestial mechanics, lunar theory, and the theory of music, as well as mathematics.

Euler had a phenomenal memory. As a young man he memorized the entire *Aeneid* by Virgil (in Latin) and many years later could still recite the entire work. He was able to solve astonishingly complex mathematical problems in his head and is said to have solved, again in his head, problems in astronomy that stymied Newton. The French academician François Arago once commented that Euler could calculate without effort "just as men breathe, as eagles sustain themselves in the air."

Euler wrote in a mathematical language that is largely in use today. Among many symbols first used by him are:

$f(x)$	for functional notation
e	for the base of the natural logarithm (see Section 6.3)
$\sum$	for the summation sign (see Section 4.3)
i	to denote $\sqrt{-1}$

Euler's textbooks were models of clarity. His texts included the *Introductio in analysin infinitorum* (1748), his *Institutiones calculi differentialis* (1755), and the three-volume *Institutiones calculi integralis* (1768–74). These and others of his works served as models for many of today's mathematics textbooks.

Some say that Euler did for mathematical analysis what Euclid did for geometry. It is no wonder that so many later mathematicians expressed their debt to him.

■ PROBLEMS 0.4

■ SELF-QUIZ

I. If a set of points (x, y) corresponds to "y is a function of x", then _____.
 a. The graph can be drawn without lifting pencil from paper
 b. No vertical line cuts the graph in more than one place
 c. No horizontal line cuts the graph more than once
 d. All of the above

II. Let $f(x) = (x - 2)/(x + 1)$. The largest possible domain for f is _____.
 a. $\{x: x \neq 2\}$ b. $\{x: x \neq -1 \text{ or } 2\}$
 c. $\{x: x \neq 1 \text{ or } -2\}$ d. $\{x: x \neq -1\}$

III. Let $g(x) = 7 - (x - 3)^2$ with domain equal to $(-\infty, \infty)$. The range of g is _____.
 a. $(-\infty, 7]$ b. $(-\infty, 7)$
 c. $(-\infty, \infty)$ d. $[3, 7]$

CONTINUED

IV. If $f(x) = 1 - 3x$ and $g(x) = (x + 1)^2$, then $(f \circ g)(-2) = $ _____.
   a. -2 b. 64   c. 16   d. -36

V. If $f(x) = x + 3$ and $g(x) = 3x + 1$, then _____ $= 3x + 10$.
   a. $f(x) + g(x)$ b. $(g - f)(x)$
   c. $(f \cdot g)(x/2)$ d. $f(g(x))$
   e. $(f \circ g)(x)$ f. $(g \circ f)(x)$

VI. Suppose that f and g are functions with the same domain. Describe the graphs of each of the following equations:
   a. $(y - f(x))(y - g(x)) = 0$
   b. $|y - f(x)| + |y - g(x)| = 0$
   c. $(y - f(x))^2 + (y - g(x))^2 = 0$

1-60 every 3rd

■ **DRILL**

In Problems 1–10, evaluate the given function at the given values.

1. $f(x) = x^2$; $f(0)$, $f(5)$, $f(v)$, and $f(2 + h)$

2. $f(x) = x^3$; $f(0)$, $f(2)$, $f(-2)$, $f(10)$, $f\left(\dfrac{1}{y}\right)$, $f(u^2)$, and $f(1 + h)$

3. $f(x) = \sqrt{x}$; $f(0)$, $f(1)$, $f(9)$, $f(u^2)$, $f(z^4)$, and $f(4 + h)$

4. $f(x) = \dfrac{1}{x}$; $f(2)$, $f(-3)$, $f\left(\dfrac{1}{t}\right)$, $f(t^5)$, and $f(x + h)$

5. $f(x) = x^4$; $f(0)$, $f(2)$, $f(-2)$, $f(\sqrt{5})$, $f(s^{1/5})$, and $f(s - 1)$

6. $g(t) = t/(t - 2)$; $g(0)$, $g(1)$, $g(-1)$, $g(3)$, $g(v^2)$, and $g(v + 5)$

7. $g(t) = \sqrt{t + 1}$; $g(0)$, $g(-1)$, $g(3)$, $g(7)$, $g(n^3 - 1)$, and $g\left(\dfrac{1}{w}\right)$

8. $h(z) = \sqrt[3]{z}$; $h(0)$, $h(8)$, $h(-\frac{1}{27})$, $h(1000)$, $h(x^{300})$, and $h(-8p)$

9. $h(z) = 1 + z + z^2$; $h(0)$, $h(2)$, $h(\frac{1}{3})$, $h(-\frac{1}{2})$, $h(z^5)$, and $h\left(\dfrac{1}{u^2}\right)$

10. $h(z) = z^3 + 2z^2 - 3z + 5$; $h(0)$, $h(1)$, $h(-1)$, $h(2)$, $h(t^{1/4})$, and $h(t + 1)$

In Problems 11–18, which of the rules depicted in the graphs are functions with each domain equal to D? If the rule is not a function, explain why it is not.

11.

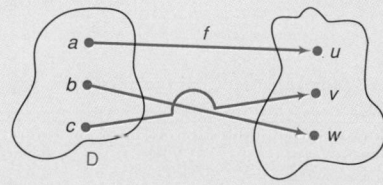

12.

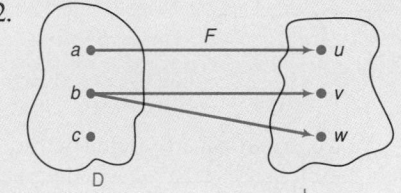

13.

14.

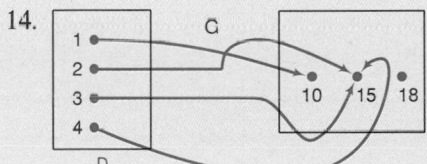

15.

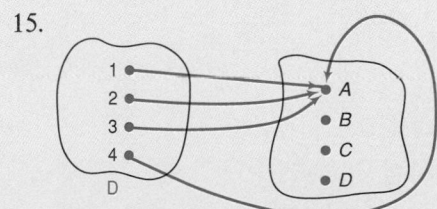

16.

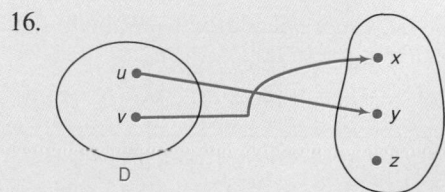

CONTINUED

17.

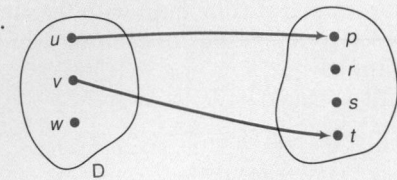

18.

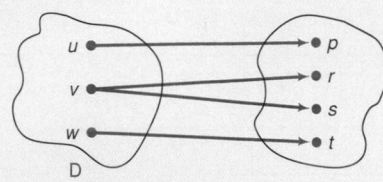

In Problems 19–28, an equation involving x and y is given. For each problem, determine whether y is a function of x with domain $= (-\infty, \infty) = \mathbb{R}$.

19. $(x - 6) = 2(y - 3)$ 20. $x/y = 2$
21. $x^2 - 3y = 4$ 22. $x - 3y^2 = 4$
23. $y = |x| - 4$ 24. $|y + 4| = x$
25. $x^2 + y^2 = 4$ 26. $|x| + |y| = 2$
27. $y^3 - x = 0$ 28. $y^4 - x = 0$

In Problems 29–40, find the largest possible domain for the given function and find the corresponding range.

29. $s = g(t) = 4t - 5$
30. $y = f(x) = 2x - 3$
31. $v = h(u) = 1/(u + 1)$
32. $y = f(x) = 1/|2 - x|$
33. $y = g(x) = 3 + 1/(1 + x^2)$
34. $y = h(x) = 3 - 1/(1 - x^2)$
35. $s = \sqrt{t^3 - 1}$
36. $y = \sqrt{(x - 1)(x - 2)}$
37. $s = \begin{cases} t & \text{if } t \geq 0 \\ -t & \text{if } t < 0. \end{cases}$

38. $y = \begin{cases} x & \text{if } x \geq 1 \\ 1 & \text{if } x < 1. \end{cases}$
39. $y = (2x + 3)/(3x + 4)$
40. $y = x^2/(x^2 + 3)$

In Problems 41–44, two functions f and g are given. Obtain simplified expressions for $f + g$, $f - g$, $f \cdot g$, and f/g. Determine the largest possible domain for each of these new functions.

41. $f(x) = 2x - 5$; $g(x) = -4x$
42. $f(x) = x^2$; $g(x) = x + 1$
43. $f(x) = \sqrt{x + 2}$; $g(x) = \sqrt{2 - x}$
44. $f(x) = x/(x + 1)$; $g(x) = (x - 1)/x$

In Problems 45–52, find $f \circ g$ and $g \circ f$, and then determine the domain of each composite function.

45. $f(x) = x + 1$; $g(x) = 2x$
46. $f(x) = x^2$; $g(x) = 2x + 3$
47. $f(x) = 2 - \sqrt{x}$; $g(x) = (x + 1)^2$
48. $f(x) = \sqrt{x + 1}$; $g(x) = x^4$
49. $f(x) = x/(x + 2)$; $g(x) = (x - 1)/x$
50. $f(x) = |x|$; $g(x) = -x$
51. $f(x) = |x|/x$; $g(x) = x^2$
52. $f(x) = \begin{cases} 2x & \text{if } x < 0 \\ x & \text{if } x \geq 0 \end{cases}$; $g(x) = \begin{cases} 5x & \text{if } x < 0 \\ -3x & \text{if } x \geq 0 \end{cases}$

In Problems 53–60, find functions f and g such that $f \circ g = h$ where h is given.

53. $h(x) = (x - 1/x)^2$
54. $h(x) = 3(x - 5)^2 + 1$
55. $h(x) = \sqrt{(x - 3)(x + 2)}$
56. $h(x) = (x + 3)^2 + 2(x + 3) - 10$
57. $h(x) = (1 - x^2)/(1 + x^2)$
58. $h(x) = -3/x^2 + 9/x^4 - 27/x^6$
59. $h(x) = (x - 1/x) - 1/(x - 1/x)$
*60. $h(x) = (8 - x\sqrt{x})/(2 - \sqrt{x})$

CALCULATOR PROBLEMS

In Problems 61–64, use a calculator to evaluate the given function at the given values.

61. $f(x) = 1.25x^2 - 3.74x + 14.38$; $f(2.34)$, $f(-1.89)$, $f(10.6)$
62. $g(t) = t^3 - 0.74t^2 + 0.756t + 1.302$; $g(0.18)$, $g(3.95)$, $g(-11.62)$
63. $h(z) = \dfrac{z + 3}{z^2 - 4}$; $h(38.2)$, $h(57.9)$, $h(238.4)$
64. $f(x) = \dfrac{x - 1.6}{x + 3.4} + \dfrac{x^2 + 5.8}{6.2 - x^2}$; $f(5.8)$, $f(-23.4)$

CONTINUED

■ APPLICATIONS

65. Consider the set of all rectangles whose perimeter is 50 cm. For any one rectangle, once its width W is measured, it is possible to do some computations that yield the area of the rectangle. Verify this assertion by producing an explicit expression for area A as a function of width W. Find the domain and the range of your function.

66. Alice, on vacation in Canada, found that she got a 25% premium on her U.S. money. When she returned, she discovered that there was a 25% discount on converting her Canadian money back into U.S. currency. Describe each conversion function. Show that one is not the inverse of the other; that is, show that after converting both ways, Alice lost money.

67. A baseball diamond is a square 90 ft long on each side. Casey runs a constant 30 ft per sec whether he hits a ground ball or a home run. Today in his first time at bat, he hit a home run. Write an expression that measures his line-of-sight distance from second base as a function of the time t, in seconds, after he left home plate.

68. Let $f(x)$ be the fifth decimal place in the decimal expansion of x. For example $f(\frac{1}{32}) = f(0.03125) = 5$, $f(\frac{7}{4}) = f(1.75000) = 0$, $f(-78.098765) = 6$, $f(\pi) = 9$, and so on. Find the domain and range of f.

69. Let $f(x) = 2\sqrt{x}$, $g(x) = 3\sqrt{x}$, and $h(x) = 6x$. Explain why $f \cdot g$ and h are different functions.

70. Explain why the equation $y^n - x = 0$ allows us to write y as a function of x if n is an odd integer but does not allow us to do so when n is an even integer. [*Hint:* First solve Problems 27 and 28.]

71. Let $f(x) = 2x + 14$ and $g(x) = 0.5x - 7$. Show that $(f \circ g)(x) = x = (g \circ f)(x)$.

72. For $f(x) = 3x + 2$, find a function g such that $(f \circ g)(x) = x = (g \circ f)(x)$.

73. For $f(x) = ax + b$, find a function g such that $(f \circ g)(x) = x = (g \circ f)(x)$. (Assume that $a \neq 0$.)

74. For $f(x) = \sqrt[3]{x} - 1$, find a function g such that $(f \circ g)(x) = x = (g \circ f)(x)$.

■ SHOW/PROVE/DISPROVE

*75. Suppose f is a function with domain $[0, \infty)$ and range $[5, \infty)$ such that if $0 \leq w < x$, then $f(w) < f(x)$. Show that there is a function g with domain $[5, \infty)$ and range $[0, \infty)$ such that
   a. $f(g(s)) = s$ for all $s \in [5, \infty)$

   b. $g(f(t)) = t$ for all $t \in [0, \infty)$

76. Let f and g be the following straight-line functions: $f(x) = ax + b$ and $g(x) = cx + d$. Find conditions on a, b, c, and d such that $f \circ g = g \circ f$.

■ CHALLENGE

*77. If T is a subset of the domain of function f, define $f(T)$ to be $\{f(t): t \in T\}$, a subset of the range of f. Show that the following relations are true for any subsets A and B of dom f:
   a. $f(A \cup B) = f(A) \cup f(B)$
   b. $f(A \cap B) \subseteq f(A) \cap f(B)$

*78. Each of the following functions satisfies an identity of the form $(f \circ f)(x) = x$ or $(g \circ g \circ g)(x) = x$ or $(h \circ h \circ h \circ h)(x) = x$, and so on. For each function, discover what type (order) of identity is appropriate.
   a. $A(x) = \sqrt[3]{1 - x^3}$ b. $B(x) = \sqrt[7]{1 - x^7}$
   c. $C(x) = 1 - 1/x$, dom $C = \mathbb{R} - \{0, 1\}$
   d. $D(x) = \dfrac{1}{1 - x}$, dom $D = \mathbb{R} - \{0, 1\}$
   e. $E(x) = \dfrac{x + 1}{x - 1}$, dom $E = \mathbb{R} - \{1\}$

   f. $F(x) = \dfrac{x - 1}{x + 1}$, dom $F = \mathbb{R} - \{-1, 0, 1\}$
   g. $G(x) = \dfrac{4x - 1}{4x + 2}$, dom $G = \mathbb{R} - \{-\frac{1}{2}, 0, \frac{1}{4}, \frac{1}{2}, 1\}$

*79. Let $R(x) = 1/x$ and $S(x) = 1 - x$. Also let I be the **identity function**; that is, $I(x) = x$. For this problem, suppose each function has domain $\mathbb{R} - \{0, 1\}$. Note that $R \circ R = I = S \circ S$.
   a. Let $T = R \circ S$. Show that $T \circ T = S \circ R$ and $T \circ T \circ T = I$.
   b. Let $U = R \circ S \circ R$. Show that $U = S \circ R \circ S$ and $U \circ U = I$.
   c. Verify that $T \circ R = R \circ T \circ T = T \circ T \circ S = S \circ T$.
   d. Show that the set of six functions $\{I, R, S, T, T \circ T, U\}$ is closed with respect

CONTINUED

to the operation of composition; that is, if F and G belong to the set, then so does $F \circ G$.

e. Show that for each function F in the set $\{I, R, S, T, T \circ T, U\}$ there is a unique func-

tion G in the set such that $F \circ G = I = G \circ F$. *Remark:* You have now shown that this collection of six functions has the algebraic structure of a **group**.

■ ANSWERS TO SELF-QUIZ

I. b   II. d   III. a

IV. a   V. f

VI. a. union of the graph of $y = f(x)$ with the graph of $y = g(x)$

b. intersection of the graph of $y = f(x)$ with the graph of $y = g(x)$

c. intersection of the graph of $y = f(x)$ with the graph of $y = g(x)$

0.5 SHIFTING THE GRAPHS OF FUNCTIONS

While more advanced methods are needed to obtain the graphs of most functions (without plotting a large number of points), there are some techniques that make it a relatively simple matter to sketch certain functions based on known graphs.

EXAMPLE 1
SHIFTING A GRAPH UP AND DOWN

Sketch the graphs of $y = x^2 + 1$ and $y = x^2 - 2$.

SOLUTION: In Figure 1(a), we have used the graph of $y = x^2$ obtained in Figure 0.4.4. To graph $y = x^2 + 1$ in Figure 1(b), we add 1 unit to every y value obtained in Figure 1(a); that is, we shift the graphs of $y = x^2$ up 1 unit. Analogously, for Figure 1(c) we shift the graph of $y = x^2$ down 2 units to obtain the graph of $y = x^2 - 2$.

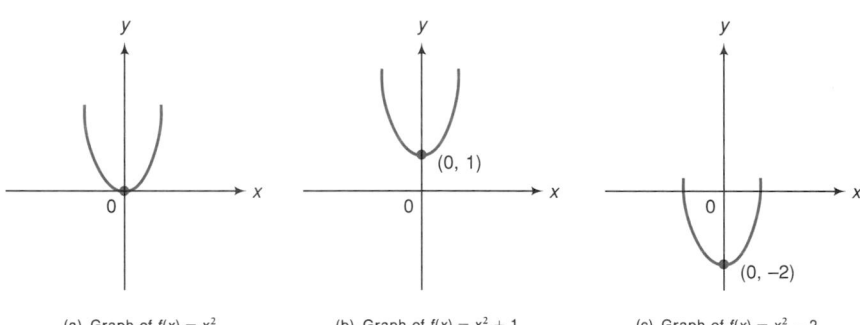

(a) Graph of $f(x) = x^2$.   (b) Graph of $f(x) = x^2 + 1$.   (c) Graph of $f(x) = x^2 - 2$.

FIGURE 1   SHIFTING THE GRAPH OF $f(x) = x^2$ UP AND DOWN

The results of Example 1 can be generalized.

VERTICAL SHIFTS OF GRAPHS

Let $y = f(x)$ and let $c > 0$.

(i) To obtain the graph of $y = f(x) + c$, shift the graph of $y = f(x)$ up c units.

(ii) To obtain the graph of $y = f(x) - c$, shift the graph of $y = f(x)$ down c units.

EXAMPLE 2
SHIFTING A GRAPH RIGHT AND LEFT

TABLE 1
VALUES TAKEN BY $(x - 1)^2$ ARE THE VALUES OF x^2 TAKEN ONE UNIT LATER

x	x^2	$(x - 1)^2$
-5	25	36
-4	16	25
-3	9	16
-2	4	9
-1	1	4
0	0	1
1	1	0
2	4	1
3	9	4
4	16	9

Sketch the graphs of $y = (x - 1)^2$ and $y = (x + 2)^2$.

SOLUTION: Let us compare the functions x^2 and $(x - 1)^2$. For example, for the function $y = x^2$, $y = 0$ when $x = 0$, and for the function $y = (x - 1)^2$, $y = 0$ when $x = 1$. Similarly, $y = 4$ when $x = -2$ if $y = x^2$, and $y = 4$ when $x = -1$ if $y = (x - 1)^2$. By continuing in this manner, you can see that y values in the graph of $y = (x - 1)^2$ are the same as y values in the graph of $y = x^2$, except that they occur 1 unit to the right on the x-axis. Some representative values are given in Table 1. Thus, we find that the graph of $y = (x - 1)^2$ is the graph of $y = x^2$ *shifted 1 unit to the right*. This is indicated in Figure 2(b).

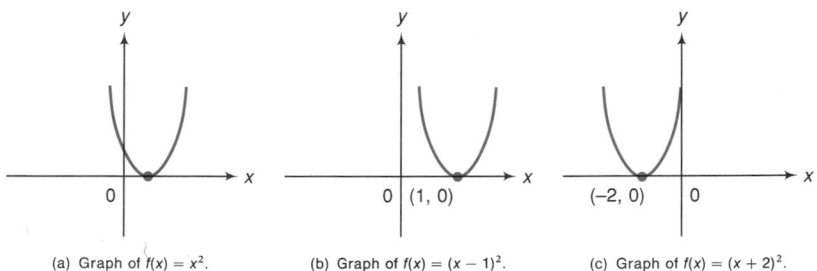

(a) Graph of $f(x) = x^2$.   (b) Graph of $f(x) = (x - 1)^2$.   (c) Graph of $f(x) = (x + 2)^2$.

FIGURE 2 SHIFTING THE GRAPH OF $f(x) = x^2$ RIGHT AND LEFT

Similarly, in Figure 2(c) we find that the graph of $y = (x + 2)^2$ is the graph of $y = x^2$ *shifted 2 units to the left*. Some values are given in Table 2.

TABLE 2
VALUES TAKEN BY $(x + 2)^2$ ARE THE VALUES OF x^2 TAKEN TWO UNITS EARLIER

x	x^2	$(x + 2)^2$
-5	25	9
-4	16	4
-3	9	1
-2	4	0
-1	1	1
0	0	4
1	1	9
2	4	16
3	9	25
4	16	36

HORIZONTAL SHIFTS OF GRAPHS

Let $y = f(x)$ and let $c > 0$.

(iii) To obtain the graph of $y = f(x - c)$, shift the graph of $y = f(x)$ c units to the right.

(iv) To obtain the graph of $y = f(x + c)$, shift the graph of $y = f(x)$ c units to the left.

⊗ **WARNING:** Be careful. $f(x + c) \neq f(x) + c$. For example, if $f(x) = x^2$, the graph of $f(x + 1) = (x + 1)^2 = x^2 + 2x + 1$ is the graph of x^2 shifted one unit to the left, but the graph of $f(x) + 1 = x^2 + 1$ is the graph of x^2 shifted 1 unit up. The graphs are not the same. ⊗

EXAMPLE 3
REFLECTING A GRAPH ABOUT THE x-AXIS

Sketch the graph of $y = -x^2$.

SOLUTION: To obtain the graph of $y = -x^2$ from the graph of $y = x^2$, note that each y value is replaced by its negative so that the graph of $y = -x^2$ is the graph of $y = x^2$ *reflected through the x-axis* (that is, turned upside down). See Figure 3.

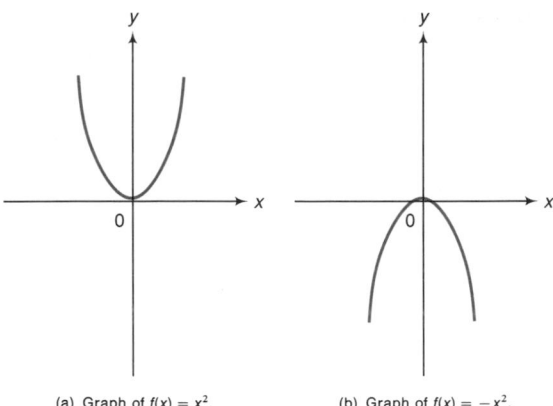

(a) Graph of $f(x) = x^2$.   (b) Graph of $f(x) = -x^2$.

FIGURE 3   REFLECTING THE GRAPH OF $f(x) = x^2$ ABOUT THE x-AXIS

EXAMPLE 4
REFLECTING A GRAPH ABOUT
THE y-AXIS

Sketch the graph of $g(x) = \sqrt{-x}$.

SOLUTION: $\sqrt{-x}$ is defined only when $x \leq 0$ (so $-x \geq 0$). In Table 3, we provide sample values of $\sqrt{-x}$ for $x \leq 0$.

TABLE 3

x	0	-0.5	-1	-2	-3	-4	-5	-10	-15	-20	-25
$g(x) = \sqrt{-x}$	0	0.707	1	1.414	1.732	2	2.236	3.162	3.873	4.472	5

These are the same values taken by $f(x) = \sqrt{x}$ for $x \geq 0$. We see that $f(x) = \sqrt{x}$ and $g(x) = \sqrt{-x}$ take the same y values. $\sqrt{x}$ takes these values for $x \geq 0$ and $\sqrt{-x}$ takes them for $x \leq 0$. The two graphs are given in Figure 4. The graph of $g(x) = \sqrt{-x}$ is *the reflection about the y-axis* of the graph of $f(x) = \sqrt{x}$.

Note that $g(x) = \sqrt{-x} = f(-x)$.

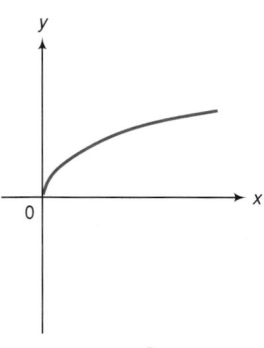

(a) Graph of $f(x) = \sqrt{x}$.

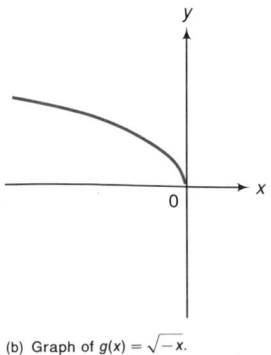

(b) Graph of $g(x) = \sqrt{-x}$.

FIGURE 4
REFLECTING THE GRAPH OF $f(x) = \sqrt{x}$
ABOUT THE y-AXIS

REFLECTION ABOUT THE x-AXIS AND THE y-AXIS

The graph of $y = -f(x)$ is obtained by reflecting the graph of $y = f(x)$ about the x-axis.

The graph of $y = f(-x)$ is obtained by reflecting the graph of $y = f(x)$ about the y-axis.

⊗ **WARNING:** Don't confuse $f(-x)$ and $-f(x)$. They are usually *not* the same. For example, if $f = x^2$, then $f(-2) = (-2)^2 = 4$, but $-f(2) = -(2^2) = -4$. If $f(x) = 2x + 3$, then $f(-5) = 2(-5) + 3 = -10 + 3 = -7$, but $-f(5) = -(2 \cdot 5 + 3) = -13$. ⊗

We can use our shifting results to sketch any quadratic function.

EXAMPLE 5
SHIFTING AND REFLECTING A GRAPH

Sketch the graph of $f(x) = -x^2 + 4x - 9$.

SOLUTION:

$$\begin{aligned}
-x^2 + 4x - 9 &= -(x^2 - 4x + 9) && \text{Factor out } -1. \\
&= -(x^2 - 4x + 2^2 - 2^2 + 9) && \text{Complete the square by} \\
& && \text{adding and subtracting } 2^2. \\
&= -[(x^2 - 4x + 4) + 5] \\
&= -[(x - 2)^2 + 5] && \text{Factor.}
\end{aligned}$$

Starting with the graph of $y = x^2$, we obtain the graph of $y = -x^2 + 4x - 9$ in three steps, as in Figure 5.

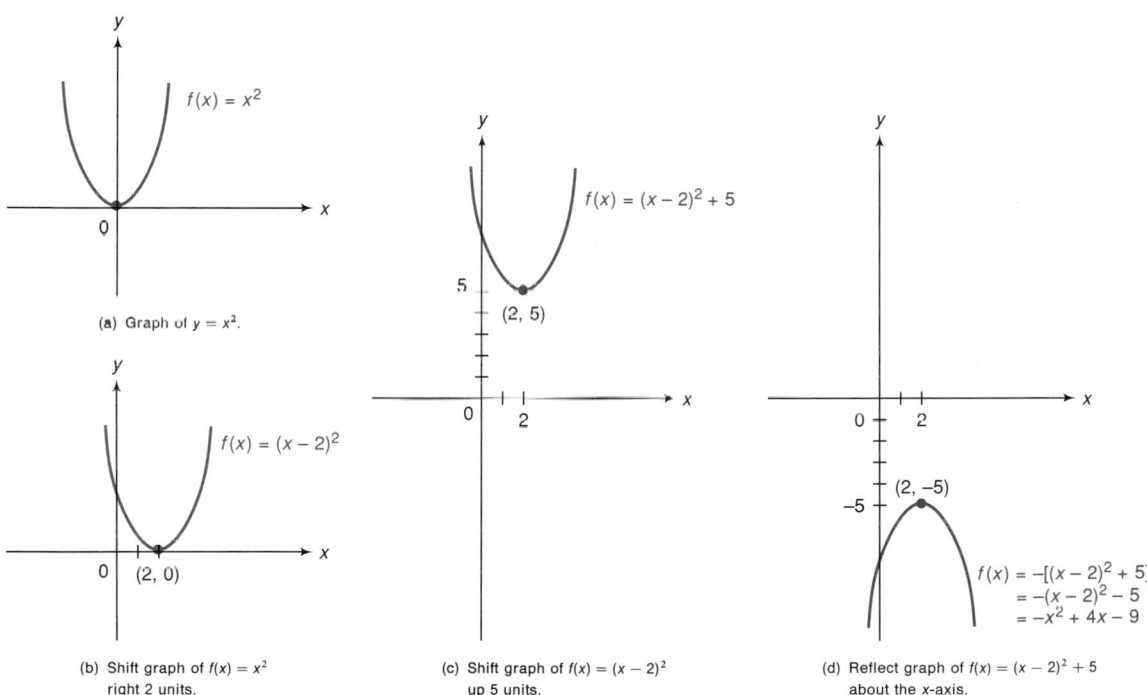

(a) Graph of $y = x^2$.

(b) Shift graph of $f(x) = x^2$ right 2 units.

(c) Shift graph of $f(x) = (x - 2)^2$ up 5 units.

(d) Reflect graph of $f(x) = (x - 2)^2 + 5$ about the x-axis.

FIGURE 5 SHIFTING AND REFLECTING THE GRAPH OF $f(x) = x^2$ TO OBTAIN THE GRAPH OF $f(x) = -x^2 + 4x - 9$

EXAMPLE 6
REFLECTING AND SHIFTING A GRAPH

The graph of a certain function $f(x)$ is given in Figure 6(a). Sketch the graph of $-f(3 - x)$.

SOLUTION: We do this in three steps:

(1) Reflect through the y-axis to obtain the graph of $f(-x)$ (Figure 6(b)).

(2) Shift to the right 3 units to obtain the graph of $f(-(x - 3)) = f(3 - x)$ (Figure 6(c)).

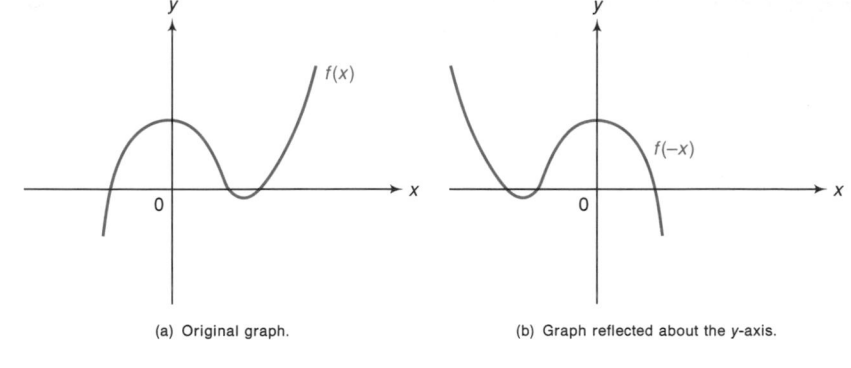

(a) Original graph. (b) Graph reflected about the y-axis.

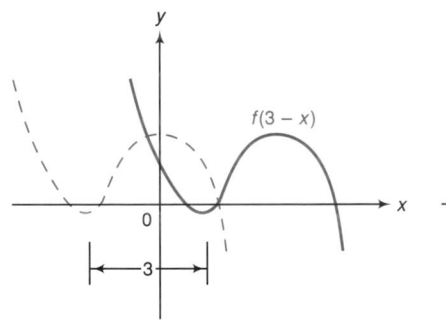 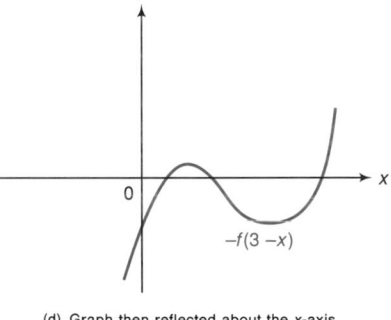

(c) Graph then shifted 3 units to the right. (d) Graph then reflected about the x-axis.

FIGURE 6 DERIVING THE GRAPH OF $-f(3 - x)$ FROM THE GRAPH OF $f(x)$

(3) Reflect through the x-axis to obtain the graph of $y = -f(3 - x)$ (Figure 6(d)).

EVEN AND ODD FUNCTIONS

On page 27, we defined an even function as a function f that satisfies $f(-x) = f(x)$. We restate that definition, and give another one here.

DEFINITION   **EVEN AND ODD FUNCTIONS**

A function $f(x)$ is an **even function** if

$$f(-x) = f(x)$$

or is an **odd function** if

$$f(-x) = -f(x). \quad \blacksquare$$

EXAMPLE 7
AN EVEN FUNCTION AND AN
ODD FUNCTION

$f(x) = x^2$ is even because $(-x)^2 = x^2$.

$f(x) = x^3$ is odd because $(-x)^3 = (-1)^3 x^3 = -x^3$.

EXAMPLE 8
DETERMINING WHETHER A
FUNCTION IS EVEN OR ODD

Determine whether each function is even, odd, or neither.

(a) $\sqrt[3]{x}$ (b) $\sqrt{x}$ (c) $x^4 + x^5$

(d) $x^4 - x^6$ (e) $\sin x$ (f) $\cos x$

SOLUTION:

(a) $\sqrt[3]{-x} = \sqrt[3]{(-1)x} = \sqrt[3]{-1}\sqrt[3]{x} = (-1)\sqrt[3]{x} = -\sqrt[3]{x}$, so $\sqrt[3]{x}$ is odd.

(b) $\sqrt{-x}$ is not defined if $x > 0$, so $\sqrt{x}$ is neither even nor odd.

(c) If $f(x) = x^4 + x^5$, then $f(-x) = (-x)^4 + (-x)^5 = x^4 - x^5$. But $-f(x) = -x^4 - x^5$. Thus $f(-x) \neq f(x)$ and $f(-x) \neq -f(x)$, so $f(x) = x^4 + x^5$ is neither even nor odd.

(d) If $f(x) = x^4 - x^6$, then $f(-x) = (-x)^4 - (-x)^6 = x^4 - x^6 = f(x)$, so $f(x) = x^4 - x^6$ is even.

(e) From identity (iii) in Appendix 1.2, $\sin(-x) = -\sin x$, so $\sin x$ is an odd function.

(f) From identity (ii) in Appendix 1.2, $\cos(-x) = \cos x$, so $\cos x$ is an even function.

Knowing that a function is even or odd helps us to draw its graph. As we saw on page 27:

The graph of an even function is symmetric about the y-axis.

Moreover:

The graph of an odd function is symmetric about the origin.

EXAMPLE 9
REFLECTING A GRAPH ABOUT
THE ORIGIN

The graph of $y = x^3$ for $x \geq 0$ is given in Figure 7. Complete the graph for $x < 0$.

SOLUTION: Since $f(x) = x^3$ is an odd function, the graph for $x \leq 0$ is the reflection about the origin of the graph for $x \geq 0$. To reflect about the origin, we first reflect about the x-axis and then reflect about the y-axis. We do this in Figure 8.

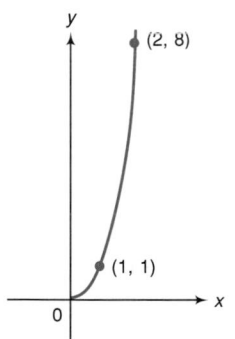

FIGURE 7
GRAPH OF $f(x) = x^3$, $x \geq 0$

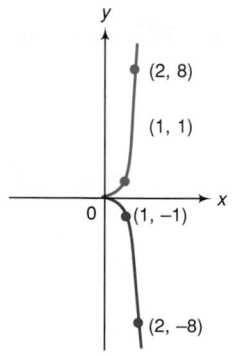

(a) Graph of $f(x) = x^3$, $x \geq 0$, reflected about the x-axis.

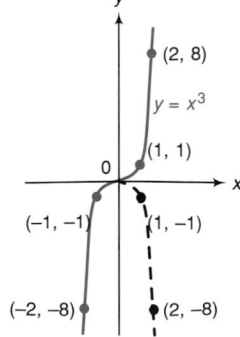

(b) The graph of $f(x) = x^3$ for $x \leq 0$ is the graph of $f(x) = x^3$, $x \geq 0$, reflected about the origin.

FIGURE 8 THE GRAPH OF $f(x) = x^3$, $x \geq 0$, IS REFLECTED ABOUT THE ORIGIN TO OBTAIN THE GRAPH OF x^3

We summarize one important technique used in Example 9.

> **TO REFLECT A GRAPH ABOUT THE ORIGIN**
>
> **(i)** Reflect it about the x-axis, and then
> **(ii)** Reflect the new graph about the y-axis.
>
> Steps (i) and (ii) can be done in the reverse order as well.

EXAMPLE 10
REFLECTING A GRAPH ABOUT
THE y-AXIS

The graph of an even function, for $x \geq 0$, is given in Figure 9. Complete the graph.

SOLUTION: Since f is even, we may obtain its graph, for $x \leq 0$, by reflecting the graph for $x \geq 0$ about the y-axis. This is done in Figure 10.

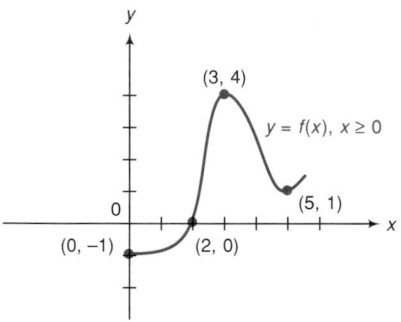

FIGURE 9
GRAPH OF AN EVEN FUNCTION FOR $x \geq 0$

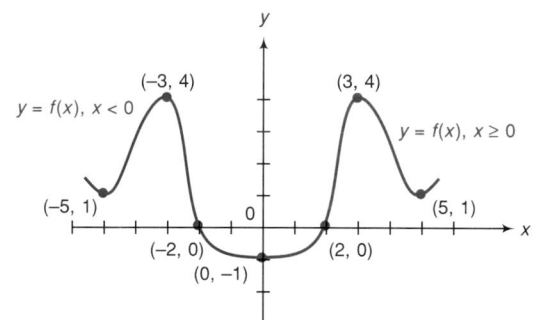

FIGURE 10
GRAPH OF $f(x)$

■ PROBLEMS 0.5

1–35, every 3rd

■ SELF-QUIZ

Manipulate the graph of $y = 3x + 7$ in the indicated fashion. Write an equation satisfied by the resulting graph.

  I. Shift 2 units to the right.

  II. Shift 5 units downward.
 III. Shift 2 units to the right and 5 units downward.
 IV. Shift 4 units to the left and 6 units upward
   V. Reflect in the x-axis.
 VI. Reflect in the y-axis.

■ DRILL

1. The graph of $f(x) = x^3$ is given in Figure 8(b). Sketch the graphs of the following functions:
   a. $g(x) = -x^3$ b. $h(x) = (x - 2)^3$
   c. $G(x) = 8 + x^3$ d. $H(x) = 5 + (4 - x)^3$

2. The graph of $f(x) = 1/x$ is given in Figure 11. Sketch the graphs of the following functions:
   a. $g(x) = -1/x$ b. $h(x) = 1/(x + 3)$
   c. $G(x) = 2 - 1/x$ *d. $H(x) = x/(x - 1)$

CONTINUED

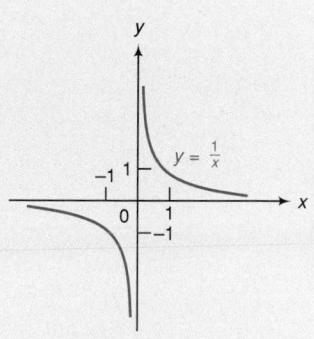

FIGURE 11

In Problems 3–8, the quadratic expression in terms of x can be rewritten in the form $b \pm (x - a)^2$. For each problem, begin by completing the square, then sketch the graph. [Each graph is related to the parabola $y = x^2$.]

3. $y = x^2 - 4x + 7$ 4. $y = x^2 + 8x + 2$
5. $y = x^2 + 3x + 4$
6. $y = -x^2 + 2x - 3$
   [*Hint:* Write $-x^2 + 2x - 3 = -(x^2 - 2x + 3)$.]
7. $y = x(8 - x)$
8. $y = 5 - (x - 3)(x - 1)$

In Problems 9–17, the graph of a function f is given. For each problem, sketch the graphs of the following equations:

a. $y = f(x - 2)$ b. $y = f(x + 3)$
c. $y = -f(x)$ d. $y = f(-x)$
e. $y = f(2 - x) + 3$

9.

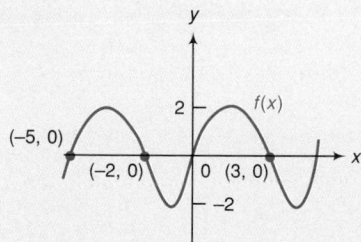

10.

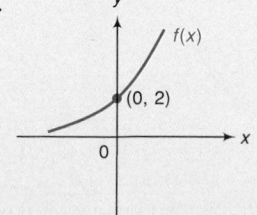

11.

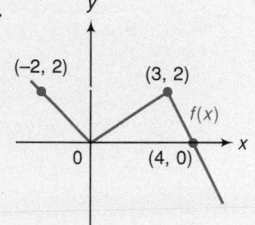

12.

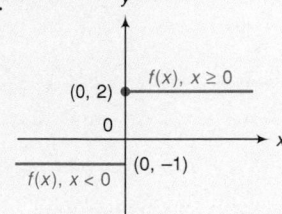

13.

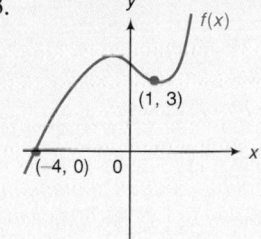

14.

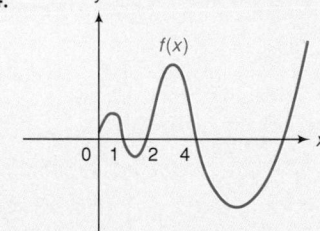

15.

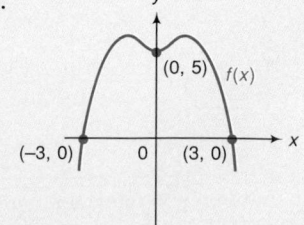

CONTINUED

16.

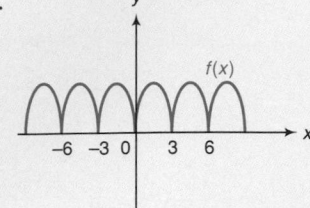

17.

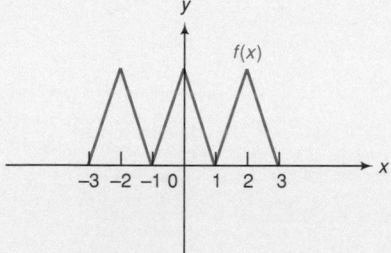

In Problems 18–27, determine whether the given function is even, odd, or neither.

18. $f(x) = x^2 + 5$

19. $f(x) = x^4 + x^2$

20. $f(x) = x^3 + x^2 + 2$

21. $f(x) = \dfrac{1}{x^2}$

22. $f(x) = \dfrac{1}{x^2 + 1}$

23. $f(x) = x^2 + x$

24. $f(x) = x^3 - x$

25. $f(x) = \dfrac{1}{x}$

26. $f(x) = \dfrac{1}{x + 1}$

27. $f(x) = \dfrac{x}{x^3 + 1}$

In Problems 28–35, the graph of an even or odd function is given for $x \geq 0$. Complete the graph of the function for $x < 0$.

28.

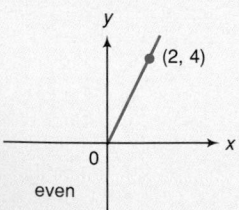

even

29.

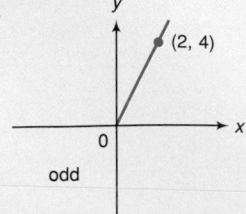

odd

30.

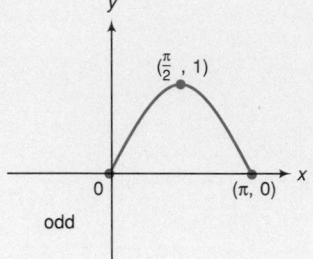

odd

31.

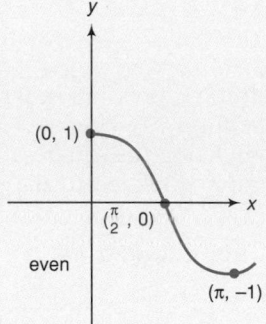

even

32.

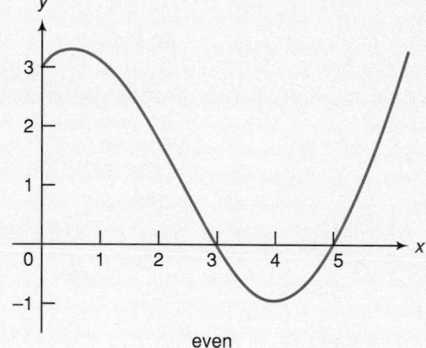

even

33.

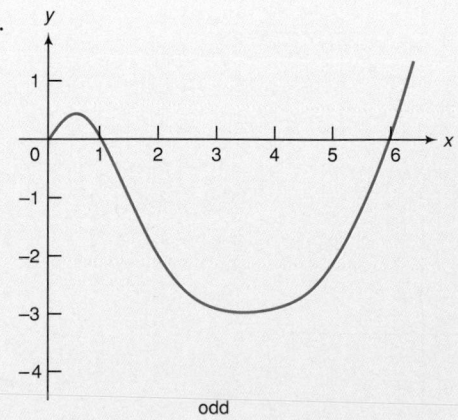

odd

CONTINUED

34.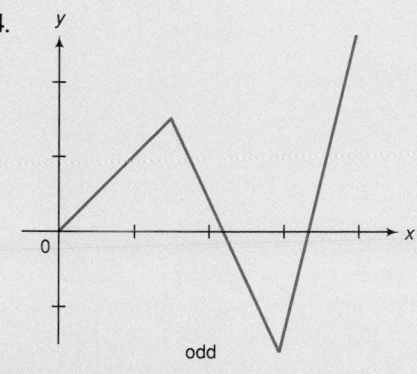

odd

35.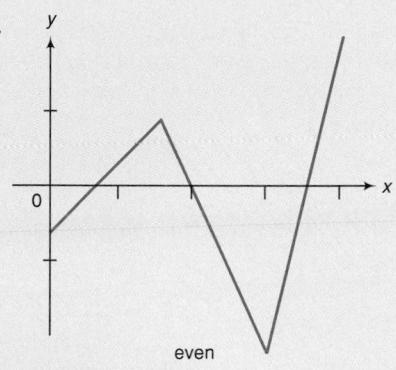

even

■ ANSWERS TO SELF-QUIZ

I. $y = 3(x - 2) + 7$

II. $y = 3x + 2$

III. $y = 3(x - 2) + 2$

IV. $y = 3(x + 4) + 13$

V. $y = -3x - 7$

VI. $y = -3x + 7$

COMPUTER EXERCISES

1. For each of the following functions, use electronic aid to sketch the graph in order to help you guess the zeros. Verify your guesses by direct evaluation. Use the zeros and the graph to solve the given inequality.

   a. $f(x) = 2x^3 - 7x^2 - 17x + 10$.
   $2x^3 - 7x^2 + 10 \leq 17x$

   b. $f(x) = 2x^3 - 77x^2 - 122x + 80$.
   $2x^3 - 77x^2 + 80 \leq 122x$

   c. $f(x) = 2x^4 - x^3 - 24x^2 - 20x + 16$.
   $2x^4 + 16 > x^3 + 24x^2 + 20x$

   d. $f(x) = |x| + 4 - |2x - 1|$.
   $|x| + 4 > |2x - 1|$

   e. $f(x) = |x + 3| + |3x| - 9$. $|x + 3| + |3x| \leq 9$

   f. $f(x) = |3x - 10| + |x| - 2$.
   $|3x - 10| + |x| < 2$

   g. $f(x) = |x - 1| + |x + 2| - |x + 1| - 2$.
   $|x - 1| + |x + 2| \leq |x + 1| + 2$

2. Use electronic aid to sketch the graphs of each of the following functions. Use a computer algebra system to find the zeros, or root approximation software to approximate the zeros. Use the zeros and the graph to solve the given inequality.

   a. $f(x) = x^6 - 10x^4 + 31x^2 - 30$.
   $x^6 + 31x^2 > 10x^4 + 30$

   b. $f(x) = x^5 - 3x^3 - 2x^2 + 6$.
   $x^5 + 6 > 3x^3 + 2x^2$

   c. $f(x) = x^4 - x^2 - 2x - 1$.
   $x^4 < x^2 + 2x + 1$

   d. $f(x) = x^4 + 7x^3 + 8x^2 + 7x + 1$. $f(x) < 0$

   e. $f(x) = |x^2 - x - 1| - 2x$.
   $|x^2 - x - 1| < 2x$

3. Find the domain of $\sqrt{x^4 + 4x^3 + x^2 - 8x - 3}$.

4. Find (or approximate) the points of intersection of the line through $(3, 5)$ with slope $\frac{1}{3}$ with the circle centered at $(1, 2)$ with radius 3.

5. Find (or approximate) the point on the circle centered at $(1, 2)$ with radius 3 that is nearest to the point $(15, 1)$.

6. $f(x) = x^3 + x^2 + x + 1$, and g is a function such that $f \circ g(x) = x = g \circ f(x)$. Find (or approximate) $g(2)$. [*Hint:* First explain why $g(2)$ must satisfy a certain cubic equation.]

7. A ball 30 inches in diameter is made of light wood having specific gravity 0.104, and will float in a pool of pure water. How much of its diameter will be beneath the surface of the water? *Discussion:* The physical principle that governs floating objects is called **Archimedes' law**: The weight of water that is displaced by a floating object is equal to the weight of that object. To say that wood has a **specific gravity** of 0.104 means that it is 0.104 times as dense as pure water. Therefore,

CONTINUED

since water weighs 62.4 lbs. per cubic foot, the wood in the ball weighs 6.4896 lbs. per cubic foot.

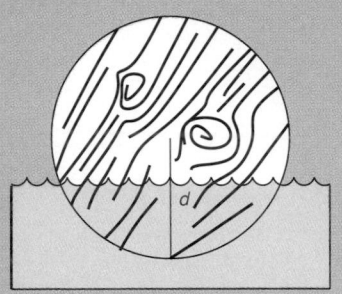

The volume of the submerged portion of a ball (as pictured) of radius r is $\pi d^2\left(r - \dfrac{d}{3}\right)$.

8. A ball 30 inches in diameter is made of light wood having specific gravity 0.208, and will float in a pool of pure water. How much of its diameter will be below the surface of the water?

**9. You must build a hollow aluminum sphere one inch in thickness that will float in water. What should its diameter be? [*Note:* This problem has some ambiguities that you must resolve before you can solve it. Think about it carefully.]

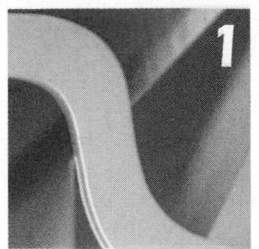

LIMITS AND DERIVATIVES

The greatest mathematical creation of antiquity was Euclidean geometry. This monumental work was not matched in importance until the discovery of calculus almost two thousand years later.

Calculus was invented independently in England by Sir Isaac Newton and in Germany by Gottfried Wilhelm Leibniz[†] in the last quarter of the seventeenth century.[‡] Though it was to some extent the answer to problems already tackled by the Greeks, calculus was created primarily to treat the major scientific problems of the seventeenth century.

There were four major types of problems. The first was, given the formula for the distance a body covers as a function of time, find the body's velocity and acceleration at any instant; and conversely, given the formula describing the acceleration of a body as a function of time, find the velocity and the distance traveled. This problem arose in the study of motion, and the difficulty it posed was that the velocities and the accelerations of concern to seventeenth-century scientists varied from instant to instant.

Let us look at this problem more closely. If an object is moving with constant velocity, v, then the distance, s, covered in t units of time is given by the **distance function**

$$s = vt. \tag{1}$$

Often, however, velocity is not constant. For example, if a ball is thrown into the air, its velocity decreases as it ascends and then increases after it begins to fall back to earth. Nevertheless, if the motion is in a straight line, we can still use formula (1) to compute the average velocity, v_{av}, which is defined by

$$\text{average velocity} = \frac{\text{total distance traveled}}{\text{elapsed time}} \quad \text{or} \quad v_{av} = \frac{s}{t}. \tag{2}$$

We can compute average velocity over very small time periods by using formula (2). However, in calculating an instantaneous velocity, for example, one cannot, as one can in the case of average velocity, divide the distance traveled by the time of travel, because at a given instant both the distance traveled and time are zero, and $\frac{0}{0}$ is meaningless. There are two problems here. First, what precisely do we mean by "instantaneous velocity," and second, how do we compute it? Both of these problems can be solved using calculus. The inverse problem of finding the distance covered, knowing the

---

[†] See the accompanying biographical sketches of Leibniz and Newton.

[‡] Much of the historical material in this section is quoted from Morris Kline's excellent book, *Mathematical Thought From Ancient to Modern Times* (New York: Oxford University Press, 1972), pp. 342–343.

formula for velocity, involves the corresponding difficulty; one cannot multiply the velocity at any one instant by the time of travel to obtain the distance traveled, because the velocity varies from instant to instant.

In the next section we shall define a limit, and in Section 1.6 we shall define and compute instantaneous velocities. This will be our first application of the concept of the **derivative**, the concept discovered by Newton and Leibniz. In Section 4.2, we shall show how to find the total distance traveled when the velocity function is known.

The second type of problem in the seventeenth century was to find the tangent to a curve. Interest in this problem stemmed from more than one source; it was a problem of pure geometry, and it was of great importance for scientific applications. Optics was one of the major scientific pursuits of the seventeenth century; the design of lenses was of direct interest to Fermat, Descartes, Huygens, and Newton. To study the passage of light through a lens, one needed to know the angle at which the ray strikes the lens in order to apply the law of refraction. The significant angle is that between the ray and the normal to the curve (Figure 1), the normal being the line perpendicular to the tangent. Hence the problem was to find either the normal or the tangent. Another scientific problem involving the tangent to a curve arose in the study of motion. The direction of motion of a moving body at any point of its path is the direction of the tangent to the path.

Actually, even the very meaning of "tangent" was open. For the conic sections, the definition of a tangent as a line touching a curve at only one point and lying on one side of the curve sufficed; this definition was used by the Greeks. But it was inadequate for the more complicated curves already in use in the seventeenth century.

The Greeks knew how to find the tangent line to a circle: it is always perpendicular to the radial line. In Section 1.5 we shall see how to find the tangent line to an arbitrary curve by computing a derivative. In so doing, we will give a definition of a tangent line that applies to a wide variety of curves.

The third problem of seventeenth-century mathematics was that of finding the maximum or minimum value of a function. When a cannonball is shot from a cannon, the distance it will travel horizontally—the range—depends on the angle at which the cannon is inclined to the ground. One "practical" problem was to find the angle that would maximize the range. Early in the seventeenth century, Galileo determined that (in a vacuum) the maximum range is obtained for an angle of fire of 45°; he also obtained the maximum heights reached by projectiles fired at various angles to the ground. The study of the motion of the planets also involved maxima and minima problems, such as finding the greatest and least distances of a planet from the sun. We shall discuss maximum and minimum problems in Section 3.5.

The fourth problem was finding the lengths of curves, such as the distance covered by a planet in a given period of time; the areas bounded by curves; volumes bounded by surfaces; centers of gravity of bodies; and the gravitational attraction that an *extended* body, a planet for example, exerts on another body. In the third century B.C., Archimedes of Syracuse discovered a method, called the *method of exhaustion*, to find some areas and volumes. Despite the fact that Greek mathematicians used the method for relatively simple areas and volumes, they had to apply much ingenuity, because the method lacked generality. Nor did they often come up with numerical ans-

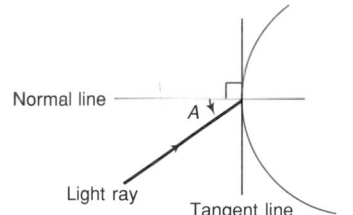

Normal line

A

Light ray

Tangent line

FIGURE 1
THE ANGLE A IS THE ANGLE BETWEEN THE LIGHT RAY AND THE LINE NORMAL TO THE LENS

wers. Interest in finding lengths, areas, volumes, and centers of gravity was revived when the work of Archimedes became known in Europe. The method of exhaustion was first modified gradually and then radically by the invention of calculus. We shall discuss the method of exhaustion in Section 4.4, and shall use calculus to compute areas and volumes in Chapters 4 and 5.

This book, then, is about the marvelous creation of Newton, Leibniz, and a number of great mathematicians who preceded and followed them. You will soon see that the basic idea of calculus, the derivative, is really quite simple—even though it took scientists over two thousand years to come up with it. You will see, too, some of the astonishingly large number of problems that can be solved using the ideas of calculus.

As Leibniz and Newton are considered to be the co-inventors of calculus, we provide biographical sketches of both.

GOTTFRIED WILHELM LEIBNIZ
1646–1716

BORN IN LEIPZIG, GERMANY, Gottfried Wilhelm Leibniz entered the university there at the age of fifteen and received the bachelor's degree before his eighteenth birthday. Truly a Renaissance man, Leibniz taught himself Latin and Greek when he was still a child. At the university he studied law, philosophy, theology, and mathematics. His studies were wide-ranging, and he is considered by many to be the last scholar to have amassed universal knowledge. Leibniz was prepared for the degree of doctor of laws before he was twenty, but the university refused to grant the degree because of his relative youth.

When his law degree was refused, Leibniz moved to Nuremberg. There he wrote a highly regarded essay on teaching law by the historical method and almost immediately thereafter was offered a professorship of law. Leibniz refused this appointment, however, and instead was commissioned by the Elector of Mainz to rewrite some statutes. From this time onward, Leibniz worked in the diplomatic service. One of the diplomats Leibniz served was an Elector of Hanover who was a great-grandson of James I of England and who became King George I in 1714.

In the diplomatic service Leibniz traveled widely. In Paris in 1672, he met the great Dutch physicist and mathematician Christiaan Huygens (1629–1695) and persuaded Huygens to tutor him in mathematics. The next year Leibniz had already begun his work on calculus. He had discovered the fundamental theorem of calculus, knew how to differentiate a wide variety of functions, and had developed much of his notation. On October 29, 1675, he first used the modern integral sign, the elongated S that stood for the Latin word *summa* (sum). Shortly thereafter he was writing derivatives and integrals in much the same way they are written today.

Leibniz's work for the Elector of Hanover gave him much leisure time to pursue subjects that intrigued him. His writing in philosophy made him one of the leading philosophers of his time, and his gift for languages won him fame as a Sanscrit scholar. He devoted a great deal of energy to a scheme for reuniting the Catholic and Protestant churches—although this was less successful than many of his other endeavors. In 1700 he founded the Berlin Academy of Science.

Many of Leibniz's mathematical papers appeared in the journal *Acta eruditorum*, which he cofounded in 1682. This journal contained his work on calculus and led to the bitter controversy with Newton over who first discovered calculus. It seems that Newton made his discoveries first. In fact, Leibniz may have seen some of Newton's work in 1673, although he was not yet sufficiently knowledgeable about analysis and geometry to understand fully what Newton had written. Nevertheless, this possibility remains one of the sources of the great controversy. In any event, Leibniz was the first to publish the important results in the calculus and was the first to use the notation that has now become standard.

In addition to everything else, Leibniz was a scientist, and he and Huygens developed the idea of kinetic energy. Unfortunately, Leibniz's work as a physicist was greatly overshadowed by the physics of Newton, so that his contributions in that area are now largely forgotten.

In 1714 Leibniz was in Hanover while his employer, the Elector of Hanover, became King George I of England. Thereafter, Leibniz was largely ignored, and when he died in 1716, his funeral was attended only by his secretary.

ISAAC NEWTON WAS BORN IN THE small English town of Woolsthorpe on Christmas Day 1642, the year of Galileo's death. His father, a farmer, had died before Isaac was born. His mother remarried when he was three and, thereafter, Isaac was raised by his grandmother. As a boy, Newton showed great cleverness and inventiveness—designing a water clock and a toy gristmill, among other things. One of his uncles, a Cambridge graduate, took an interest in the boy's education, and as a result, Newton entered Trinity College, Cambridge, in 1661. His primary interest at that time was chemistry.

Newton's interest in mathematics began with his discovery of two of the great mathematics books of his day: Euclid's *Elements* and Descartes's *La géométrie*. He also became aware of the work of the great scientists who preceded him, including Galileo and Fermat.

SIR ISAAC NEWTON
1642–1727

Many of Newton's discoveries governed physics until the discoveries of Einstein early in this century. In 1679, Newton used a new measurement of the radius of the earth, together with an analysis of the earth's motion, to formulate his universal law of gravitational attraction. Although he made many other discoveries at that time, he communicated them to no one for five years. In 1684, Edmund Halley (after whom Halley's comet is named) visited Cambridge to discuss his theories of planetary motion with Newton. The conversations with Halley stimulated Newton's interest in celestial mechanics and led him to work out many of the laws that govern the motion of bodies subject to the forces of gravitation. The result of this work was the 1687 publication of Newton's masterpiece, *Philosophiae naturalis principia mathematica* (known as the *Principia*). It was received with great acclaim throughout Europe.

By the end of 1664, Newton seems to have mastered all the mathematical knowledge of the time and had begun adding substantially to it. In 1665, he began his study of the rates of change, or *fluxions*, of quantities, such as distances or temperatures that varied continuously. The result of this study was what today we call *differential calculus*.

Newton disliked controversy so much that he delayed the publication of many of his findings for years. An unfortunate result of one of these delays was a conflict with Leibniz over who first discovered calculus. Leibniz made similar discoveries at about the same time as Newton, and to this day there is no universal agreement as to who discovered what first. The conflict stirred up so much ill will that English mathematicians (supporters of Newton) and continental mathematicians (supporters of Leibniz) had virtually no communication for more than a hundred years. English mathematics suffered greatly as a result.

Newton is considered by many to be the greatest mathematician the world has ever produced. He was the greatest "applied" mathematician, determined by his ability to discover a physical property and analyze it in mathematical terms. Leibniz once said, "Taking mathematics from the beginning of the world to the time when Newton lived, what he did was much the better half." The great English poet Alexander Pope wrote,

Nature and Nature's law lay hid in night;
God said, 'Let Newton be,' and all was light.

Newton, by contrast, was modest about his accomplishments. Late in life he wrote, "If I have seen farther than Descartes, it is because I have stood on the shoulders of giants." All who study mathematics today are standing on Isaac Newton's shoulders.

Unlike mathematics, history is often imprecise. In his best selling book *A Brief History of Time* (New York: Bantam Books, 1988), the brilliant British mathematician and physicist Stephen Hawking provides an alternative biography of Newton. We reproduce it here.[†]

Sir Isaac Newton   1642–1727

Isaac Newton was not a pleasant man. His relations with other academics were notorious, with most of his later life spent embroiled in heated disputes. Following publication of *Principia Mathematica*—surely the most influential book ever written in physics—Newton had risen rapidly into public prominence. He was appointed president of the Royal Society and became the first scientist ever to be knighted.

Newton soon clashed with the Astronomer Royal, John Flamsteed, who had earlier provided Newton with much needed data for *Principia*, but was now withholding information that Newton wanted. Newton would not take no for an answer; he had himself appointed to the governing body of the Royal Observatory and then tried to force immediate publication of the data. Eventually he arranged for Flamsteed's work to be seized and prepared for publication by Flamsteed's mortal enemy, Edmond Halley. But Flamsteed took the case to court and, in the nick of time, won a court order preventing distribution of the stolen work. Newton was incensed and sought his revenge by systematically deleting all references to Flamsteed in later editions of *Principia*.

A more serious dispute arose with the German philosopher Gottfried Leibniz. Both Leibniz and Newton had independently developed a branch of mathematics called calculus, which underlies most of modern physics. Although we now know that Newton discovered calculus years before Leibniz, he published his work much later. A major row ensued over who had been first, with scientists vigorously defending both contenders. It is remarkable, however, that most of the articles appearing in defense of Newton were originally written by his own hand—and only published in the name of friends! As the row grew, Leibniz made the mistake of appealing to the Royal Society to resolve the dispute. Newton, as president, appointed an "impartial" committee to investigate, coincidentally consisting entirely of Newton's friends! But that was not all: Newton then wrote the committee's report himself and had the Royal Society publish it, officially accusing Leibniz of plagiarism. Still unsatisfied, he then wrote an anonymous review of the report in the Royal Society's own periodical. Following the death of Leibniz, Newton is reported to have declared that he had taken great satisfaction in "breaking Leibniz's heart."

During the period of these two disputes, Newton had already left Cambridge and academe. He had been active in anti-Catholic politics at Cambridge, and later in Parliament, and was rewarded eventually with the lucrative post of Warden of the Royal Mint. Here he used his talents for deviousness and vitriol in a more socially acceptable way, successfully conducting a major campaign against counterfeiting, even sending several men to their death on the gallows.

1.2 THE CALCULATION OF LIMITS

The notation of a limit, which we discuss extensively in this chapter, plays a central role in calculus and in much of modern mathematics. However, although mathematics dates back over three thousand years, limits were not really understood until the monumental work of the great French mathematician Augustin-Louis Cauchy[†] (1789–1857) in the nineteenth century.

In this section we define a limit and show how some limits can be calculated.

[†] See the biographical sketch of Cauchy on page 64.

EXAMPLE 1
FINDING A LIMIT

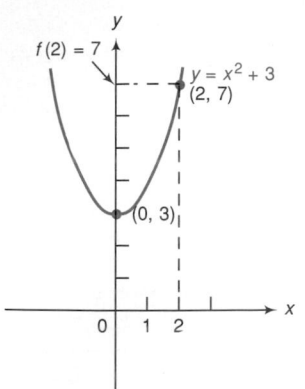

FIGURE 1
GRAPH OF $y = x^2 + 3$

TABLE 1

x	$f(x) = x^2 + 3$
2.05	7.2025
2.01	7.0401
2.001	7.004001
2.0001	7.00040001
1.95	6.8025
1.99	6.9601
1.999	6.996001
1.9999	6.99960001

We begin by looking at the function

$$y = f(x) = x^2 + 3. \tag{1}$$

This function is graphed in Figure 1 (see Section 0.5). What happens to $f(x)$ as x gets close to the value $x = 2$? This may seem to you to be a silly question. After all, isn't it obvious? If $x = 2$, then $y = f(2) = 2^2 + 3 = 4 + 3 = 7$. So, as x gets close to 2, $f(x)$ gets close to 7. This is true; if all limits were so easily calculated, there would be no need to study them. But not all limits are so simple. We will discuss this "obvious" example a bit further and then go on to problems that are not so obvious.

We ask the question again: What happens to $f(x)$ as x gets close to the value $x = 2$? Suppose that we didn't already know the answer. In that case it would be appropriate to compute a few values and see what happens as x gets closer and closer to 2. Look at Table 1, keeping in mind that x can get close to 2 from the right of 2 and from the left of 2 along the x-axis. It appears, not surprisingly, that as x gets close to 2, $f(x) = x^2 + 3$ gets close to 7.

In mathematical symbols, we write

$$\lim_{x \to 2}(x^2 + 3) = 7.$$

This notation is read, "The limit as x approaches 2 (or tends to 2) of $x^2 + 3$ is equal to 7."

NOTE: In order to guess at this limit, we did *not* have to evaluate $x^2 + 3$ at $x = 2$.

EXAMPLE 2
FINDING A LIMIT

TABLE 2

x	$f(x) = \dfrac{x^2 + 3x + 2}{13x + 8}$
4.1	0.5075040783
4.01	0.5007500416
4.001	0.5000750004
4.0001	0.5000075000
3.9	0.4925042589
3.99	0.4992500418
3.999	0.4999250004
3.9999	0.4999925000

Consider the function

$$f(x) = \frac{x^2 + 3x + 2}{13x + 8}.$$

To see what happens to this function as x approaches 4, look at Table 2. It seems that $f(x)$ approaches $0.5 = \frac{1}{2}$ as x approaches 4. This result is reasonable since

$$f(4) = \frac{4^2 + 3 \cdot 4 + 2}{13 \cdot 4 + 8} = \frac{16 + 12 + 2}{52 + 8} = \frac{30}{60} = \frac{1}{2}.$$

We write the limit we have discovered as

$$\lim_{x \to 4} \frac{x^2 + 3x + 2}{13x + 8} = \frac{1}{2}.$$

EXAMPLE 3
COMPUTING A LIMIT AT A POINT WHERE THE FUNCTION IS NOT DEFINED

Consider the function

$$f(x) = \frac{x^2 - 1}{x - 1}. \tag{2}$$

What happens to $f(x)$ as x approaches 1? We cannot substitute the value $x = 1$ because $f(1)$ is not defined. In Table 3 we provide values of $\dfrac{x^2 - 1}{x - 1}$ for x near 1.

TABLE 3

x	$\dfrac{x^2 - 1}{x - 1}$	x	$\dfrac{x^2 - 1}{x - 1}$
2	3	0	1
1.5	2.5	0.5	1.5
1.1	2.1	0.9	1.9
1.01	2.01	0.99	1.99
1.001	2.001	0.999	1.999
1.00001	2.00001	0.99999	1.99999

It seems that $f(x)$ approaches 2 as x approaches 1. This makes sense because, for $x \neq 1$,

$$\frac{x^2 - 1}{x - 1} = \frac{(x - 1)(x + 1)}{x - 1} = x + 1$$

approaches 2 as x approaches 1. We write

$$\lim_{x \to 1} \frac{x^2 - 1}{x - 1} = 2.$$

It is important to note that for $f(x) = (x^2 - 1)/(x - 1)$ it is still not permissible to set $x = 1$ because this would imply division by zero. However, we now know what happens to this function as x approaches 1. We can see why it is important that we are not required to evaluate $f(x)$ at $x = 1$ when we calculate the limit as x approaches 1.

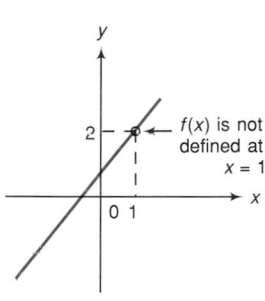

(a) Graph of $f(x) = \dfrac{x^2 - 1}{x - 1}$, $x \neq 1$.

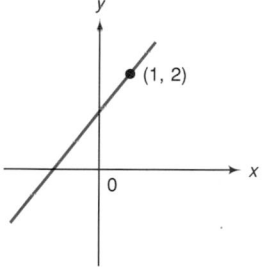

(b) Graph of $g(x) = x + 1$.

FIGURE 2

THE FUNCTIONS $f(x) = \dfrac{x^2 - 1}{x - 1}$ AND $g(x) = x + 1$ ARE NOT THE SAME BECAUSE $g(x)$ IS DEFINED WHEN $x = 1$ BUT $f(x)$ IS NOT.

Before giving further examples, we will give a more formal definition of a limit. The definition given below is meant to appeal to your intuition. It is *not* a precise mathematical definition. In this section and the ones that follow, we hope that you will begin to be comfortable with the notion of limits and will acquire some facility in calculating them. Later, at the end of this chapter (in Section 1.8), we will give a formal, mathematically precise definition.

DEFINITION **INFORMAL DEFINITION OF A LIMIT**

Let L be a real number, and suppose that $f(x)$ is defined on an open interval containing x_0 but not necessarily at x_0 itself. We say that the **limit** as x approaches x_0 of $f(x)$ is L, written

$$\lim_{x \to x_0} f(x) = L,\tag{3}$$

if, whenever x gets close to x_0 from either side with $x \neq x_0$, $f(x)$ gets close to L. ■

Here we insist that f be defined on an open interval (see page 3) containing the number x_0 except possibly at x_0 itself. This ensures us that f is defined on both sides of x_0 (see Figure 3). It is important that $f(x)$ get close to L when x gets close to x_0 from either side. In Table 1, for example, $x^2 + 3$ gets close to 7 when x gets close to 2 from the right and the left.

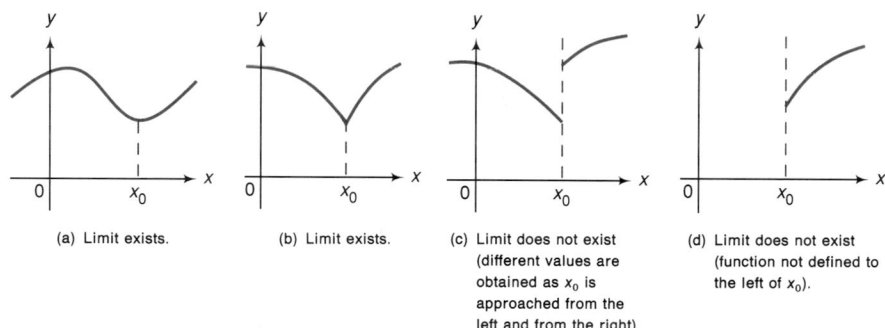

(a) Limit exists.

(b) Limit exists.

(c) Limit does not exist (different values are obtained as x_0 is approached from the left and from the right).

(d) Limit does not exist (function not defined to the left of x_0).

FIGURE 3 EXISTENCE AND NONEXISTENCE OF A LIMIT

We should emphasize that while we do not actually need to know what $f(x_0)$ is [in fact, $f(x_0)$ need not even exist], it is nevertheless often very helpful to know $f(x_0)$ in the actual computation of $\lim_{x \to x_0} f(x_0)$, since it frequently happens that $\lim_{x \to x_0} f(x)$ indeed equals $f(x_0)$. However, we again emphasize that this is *not always* the case. In Example 3 we showed that $\lim_{x \to x_0} f(x) = 2$, even though $f(1)$ did not exist.

EXAMPLE 4

COMPUTING A LIMIT THAT GIVES US THE SLOPE OF THE TANGENT LINE TO A CURVE

Calculate

$$\lim_{\Delta x \to 0} f(\Delta x) = \lim_{\Delta x \to 0} \frac{\sqrt{4 + \Delta x} - 2}{\Delta x},\tag{4}$$

where Δx denotes a real number [not the number Δ times the number x].

SOLUTION: Note that if we substitute $\Delta x = 0$ in (4), then

$$f(0) = \frac{\sqrt{4 + 0} - 2}{0} = \frac{2 - 2}{0} = \frac{0}{0},$$

TABLE 4

Δx	$f(\Delta x) = \dfrac{\sqrt{4 + \Delta x} - 2}{\Delta x}$
1	0.2360679775
0.5	0.2426406871
0.1	0.2484567313
0.01	0.2498439448
0.001	0.2499843740
0.0001	0.2499984200
-1	0.2679491924
-0.5	0.2583426132
-0.1	0.2515823419
-0.01	0.2501564457
-0.001	0.2500156290
-0.0001	0.2500015900

which is an undefined expression. In Table 4 we tabulate values of $f(\Delta x)$ for Δx near 0. From the table it appears that

$$\lim_{\Delta x \to 0} \frac{\sqrt{4 + \Delta x} - 2}{\Delta x} = 0.25.$$

We can derive this result algebraically:

Multiply and divide by $\sqrt{4 + \Delta x} + 2$

$$\lim_{\Delta x \to 0} \frac{\sqrt{4 + \Delta x} - 2}{\Delta x} = \lim_{\Delta x \to 0} \frac{(\sqrt{4 + \Delta x} - 2)(\sqrt{4 + \Delta x} + 2)}{\Delta x(\sqrt{4 + \Delta x} + 2)}$$

$(a - b)(a + b) = a^2 - b^2$

$$= \lim_{\Delta x \to 0} \frac{(\sqrt{4 + \Delta x})^2 - 2^2}{\Delta x(\sqrt{4 + \Delta x} + 2)}$$

$$= \lim_{\Delta x \to 0} \frac{(4 + \Delta x) - 4}{\Delta x(\sqrt{4 + \Delta x} + 2)} = \lim_{\Delta x \to 0} \frac{\Delta x}{\Delta x(\sqrt{4 + \Delta x} + 2)}.$$

As before, if $\Delta x \neq 0$, this last limit is equal to

$$\lim_{\Delta x \to 0} \frac{1}{\sqrt{4 + \Delta x} + 2} = \frac{1}{\sqrt{4 + 0} + 2} = \frac{1}{4} = 0.25.$$

The technique of multiplying and dividing by a nonzero number to simplify an algebraic expression is one we will use often.

GEOMETRIC INTERPRETATION OF THE RESULT OF EXAMPLE 4

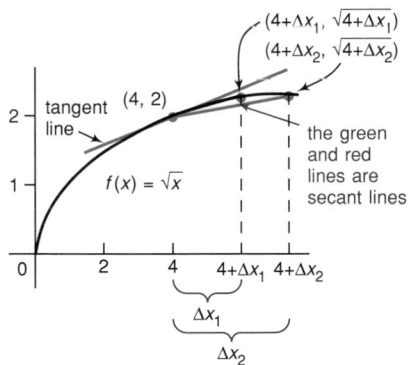

FIGURE 4
GRAPH OF $f(x) = \sqrt{x}$, WITH SELECTED TANGENT AND SECANT LINES.

In Figure 4 we draw the graph of $f(x) = \sqrt{x}$. The line joining the points $(4, 2)$ and $(4 + \Delta x, \sqrt{4 + \Delta x})$ is called a **secant line** to the curve. Two such secant lines for two different values of Δx are drawn in the figure. It is not difficult to see that as $\Delta x \to 0$ (so that the point $(4 + \Delta x, \sqrt{4 + \Delta x})$ moves along the curve toward the point $(4, 2)$), the secant lines approach the line tangent to the curve at the point $(4, 2)$. But

$$\text{slope of secant line} = \frac{\Delta y}{\Delta x} = \frac{\sqrt{4 + \Delta x} - 2}{\Delta x}$$

and

$$\text{slope of tangent line} = \lim_{\Delta x \to 0} (\text{slope of secant line})$$

$$= \lim_{\Delta x \to 0} \frac{\sqrt{4 + \Delta x} - 2}{\Delta x} = \frac{1}{4}.$$

The technique of computing a limit in order to find the slope of a line tangent to a given curve is the central technique in calculus. We will develop this idea in great detail beginning in Section 1.5.

EXAMPLE 5
COMPUTING A LIMIT

Calculate $\lim_{x \to 0} |x|$.

SOLUTION: From Section 0.1 (see page 5), we have

$$|x| = \begin{cases} x, & x \geq 0 \\ -x, & x \leq 0. \end{cases}$$

If $x > 0$, then $|x| = x$, which tends to 0 as $x \to 0$ from the right of 0. If $x < 0$, then $|x| = -x$, which again tends to 0 as $x \to 0$ from the left of 0. Then, since we get the same answer when we approach 0 from the left and from the right, we have

$$\lim_{x \to 0} |x| = 0.$$

This result is pictured in Figure 5.

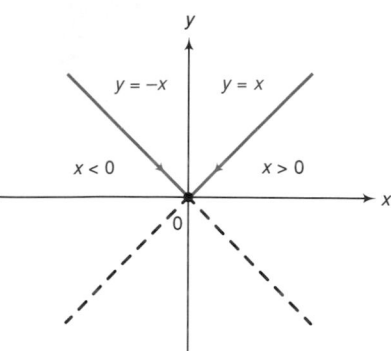

FIGURE 5
GRAPH OF $y = |x|$

The next example illustrates the fact that not all functions have limits at all points.

EXAMPLE 6
A LIMIT THAT DOES NOT EXIST

Calculate $\lim_{x \to 0} \dfrac{|x|}{x}$.

SOLUTION: If $x > 0$, then $|x| = x$, so that $|x|/x = x/x = 1$. On the other hand, if $x < 0$, then $|x| = -x$, so that $|x|/x = -x/x = -1$. Note that $|x|/x$ is not defined at $x = 0$. The graph of $|x|/x$ is sketched in Figure 6. In sum, we have

$$\frac{|x|}{x} = \begin{cases} 1, & x > 0 \\ -1, & x < 0. \end{cases}$$

From Figure 6 we conclude that $f(x) = |x|/x$ has *no* limit as $x \to 0$; for if $x > 0$, then $f(x)$ remains constant at the value 1, while for $x < 0$, $f(x)$ remains constant at the value -1. Since the value of the limit has to be the same no matter from which direction we approach the value 0, we are left to conclude that there is no limit at 0. Of course, for any other value of x there is a limit. For example, $\lim_{x \to 2} |x|/x = 1$, since near $x = 2$, $|x| = x$ and $|x|/x = 1$. Similarly, $\lim_{x \to -2} |x|/x = -1$.

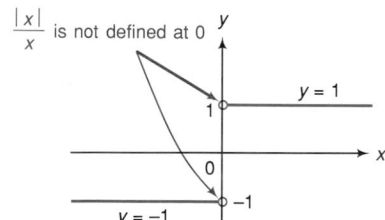

$\dfrac{|x|}{x}$ is not defined at 0

FIGURE 6

GRAPH OF $y = \dfrac{|x|}{x}$

EXAMPLE 7
COMPUTING A LIMIT USING A CALCULATOR

Compute $\lim_{\theta \to 0} \dfrac{\sin \theta}{\theta}$ where θ is measured in radians.

SOLUTION: This limit is very important in calculus. If you need to review the **trigonometric functions** (and radian measure), look at Appendix 1 before going further. Since $\sin 0 = 0$, the function $\dfrac{\sin \theta}{\theta}$ is undefined at $\theta = 0$ and takes the form $\dfrac{0}{0}$ there. Thus, inserting the value 0 will do us no good. Rather, we compute, as in Table 5, some values of $\dfrac{\sin \theta}{\theta}$ for smaller and smaller values of θ. These calculations are done with the calculator set to radian mode.

TABLE 5 VALUES OF $\dfrac{\sin \theta}{\theta}$

θ (radians)	$\sin \theta$	$\dfrac{\sin \theta}{\theta}$
1	0.8414709848	0.8414709848
0.5	0.4794255386	0.9588510772
0.1	0.09983341665	0.9983341665
0.01	0.009999833334	0.9999833334
0.001	0.0009999998333	0.9999998333
0.0001	0.00009999999983	0.9999999983

Note that $\sin(-\theta) = -\sin \theta$. Therefore, if $\theta = -0.0001$, then $\sin(-0.0001) = -\sin(0.0001)$, so $\dfrac{\sin(-0.0001)}{-0.0001} = \dfrac{-0.00009999999983}{-0.0001}$ $= 0.9999999983$. More generally, if $\theta > 0$, then $\dfrac{\sin(-\theta)}{-\theta} = \dfrac{-\sin \theta}{-\theta} = \dfrac{\sin \theta}{\theta}$. This explains why we need to consider only positive values for θ in Table 1. From the table it appears that

$$\lim_{\theta \to 0} \frac{\sin \theta}{\theta} = 1.$$

We will prove this result in Section 2.5.

⊗ **WARNING:** Calculators do many things, but they have limits (no pun intended). For example, suppose we wish to calculate

$$\lim_{x \to 0} \frac{\sin x - x}{x^3}.$$

Using a rule discussed in Section 9.1, it is not difficult to show that

$$\lim_{x \to 0} \frac{\sin x - x}{x^3} = -\frac{1}{6} \approx -0.166667.$$

However, approximating this limit on a calculator can present difficulties. If we use a calculator displaying 9 decimal places and we insert $x = 0.01$, we obtain

$$\frac{\sin 0.01 - 0.01}{(0.01)^3} \approx \frac{0.009999833 - 0.01}{0.000001} = -0.1666658$$

which is close to the limit. However, if we insert $x = 0.00001$, we obtain

$$\frac{\sin 0.00001 - 0.00001}{(0.00001)^3} = \frac{0.00001 - 0.00001}{10^{-15}} = 0.$$

The problem is that $\sin 0.00001 - 0.00001 \approx -1.67 \times 10^{-16}$ and this number is treated as zero by a calculator that carries 12 or 13 significant figures internally.

It is always a good idea to be at least a bit suspicious of results obtained on a calculator—especially when the numbers you obtain are very large or very small (i.e., close to zero). ⊗

EXAMPLE 8

AN OSCILLATING FUNCTION
THAT HAS NO LIMIT AT ZERO

Compute $\lim\limits_{x \to 0} \sin \dfrac{1}{x}$ if it exists.

SOLUTION: As $x \to 0$, $\dfrac{1}{x}$ becomes arbitrarily large if $x > 0$ and arbitrarily small (large in the negative direction) if $x < 0$. Consider the values

$$x = \frac{2}{\pi}, \frac{2}{5\pi}, \frac{2}{9\pi}, \frac{2}{13\pi}, \ldots.$$

Then

$$\frac{1}{x} = \frac{\pi}{2}, \frac{5\pi}{2}, \frac{9\pi}{2}, \frac{13\pi}{2}, \ldots$$

and

$$\sin \frac{1}{x} = 1 \text{ for each value.}$$

If $x = \dfrac{1}{n\pi}$, where $n = 1, 2, 3, \ldots$, then $\dfrac{1}{x} = n\pi$ and $\sin \dfrac{1}{x} = \sin n\pi = 0$.

Also, if $x = \dfrac{2}{3\pi}, \dfrac{2}{7\pi}, \dfrac{2}{11\pi}, \dfrac{2}{15\pi}, \ldots$, then $\dfrac{1}{x} = \dfrac{3\pi}{2}, \dfrac{7\pi}{2}, \dfrac{11\pi}{2}, \ldots$, so $\sin \dfrac{1}{x} = -1$ at each value. We summarize this and other facts in Table 6.

TABLE 6 SOME VALUES FOR $\sin \dfrac{1}{x}$

x	$\dfrac{2}{\pi}$	$\dfrac{2}{2\pi}$	$\dfrac{2}{3\pi}$	$\dfrac{2}{4\pi}$	$\dfrac{2}{5\pi}$	$\dfrac{2}{6\pi}$	$\dfrac{2}{7\pi}$	$\dfrac{2}{8\pi}$	$\dfrac{2}{9\pi}$	$\dfrac{2}{10\pi}$	$\dfrac{2}{11\pi}$	$\to 0$
$\dfrac{1}{x}$	$\dfrac{\pi}{2}$	π	$\dfrac{3\pi}{2}$	2π	$\dfrac{5\pi}{2}$	3π	$\dfrac{7\pi}{2}$	4π	$\dfrac{9\pi}{2}$	5π	$\dfrac{11\pi}{2}$	becomes arbitrarily large
$\sin \dfrac{1}{x}$	1	0	-1	0	1	0	-1	0	1	0	-1	does not approach a fixed value

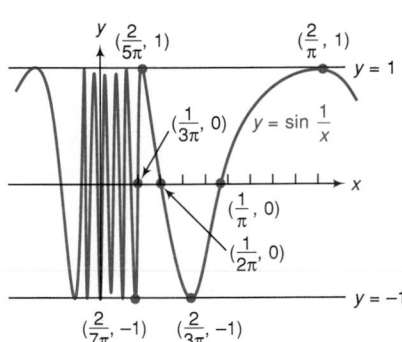

FIGURE 7

$\lim\limits_{x\to 0} \sin\dfrac{1}{x}$ DOES NOT EXIST

Evidently, $\sin\dfrac{1}{x}$ takes on the values 1, 0, and -1 infinitely often as x approaches 0. This means that $\sin\dfrac{1}{x}$ does not get close (and stay close) to any one number as $x\to 0$, so the limit does not exist. In fact, $\sin\dfrac{1}{x}$ takes on every value in the interval $[-1, 1]$ infinitely often as $x\to 0$. This oscillatory behavior is illustrated in Figure 7.

EXAMPLE 9

A FUNCTION THAT JUMPS MAY STILL HAVE A LIMIT

Let $f(x) = \begin{cases} 2x, & x\neq 1 \\ 5, & x=1 \end{cases}$. Compute $\lim\limits_{x\to 1} f(x)$, if it exists.

SOLUTION: We sketch the function in Figure 8. As x gets close to 1, *without being set equal to 1*, $y = 2x$ gets close to 2. Thus

$$\lim_{x\to 1} f(x) = 2.$$

The fact that $f(1) = 5$ is completely irrelevant. We stress that the value of f at 1, if it exists at all, is completely irrelevant in the calculation of the limit.

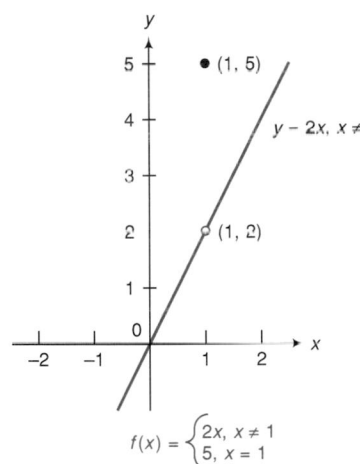

FIGURE 8

$\lim_{x\to 1} f(x) = 2$

EXAMPLE 10

A FUNCTION DOES NOT HAVE A LIMIT BECAUSE IT IS NOT DEFINED OVER A RANGE OF VALUES

Consider the function

$$f(x) = \sqrt{x}.$$

It seems as if $\lim_{x\to 0} \sqrt{x} = 0$, but that is *not* the case since $f(x) = \sqrt{x}$ is *not even defined* for $x < 0$ (we cannot take the square root of a negative number). Therefore as x approaches 0 from the left, $\sqrt{x}$ is not defined,

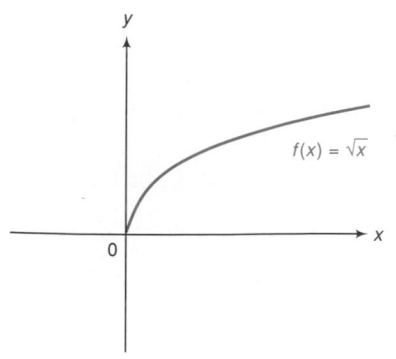

FIGURE 9
$\lim_{x \to 0} \sqrt{x}$ DOES NOT EXIST BUT
$\lim_{x \to 0^+} \sqrt{x} = 0$

so $\lim_{x \to 0} \sqrt{x}$ *does not exist.* Note also that there is *no* open interval containing 0 in which $f(x)$ is defined. The graph of $f(x) = \sqrt{x}$ is sketched in Figure 9.

The result of Example 10 leads to a new definition. Since $\sqrt{x}$ approaches 0 as x approaches 0 from the right, or positive, side, we write

$$\lim_{x \to 0^+} \sqrt{x} = 0.$$

DEFINITION   **ONE-SIDED LIMITS**

Let L be a real number.

(i) Suppose that $f(x)$ is defined near x_0 for $x > x_0$ and that as x gets close to x_0 (with $x > x_0$), $f(x)$ gets close to L. Then we say that L is the **right-hand limit** of $f(x)$ as x approaches x_0 and we write

$$\lim_{x \to x_0^+} f(x) = L. \tag{5}$$

(ii) Suppose that $f(x)$ is defined near x_0 for $x < x_0$ and that as x gets close to x_0 (with $x < x_0$), $f(x)$ gets close to L. Then we say that L is the **left-hand limit** of $f(x)$ as x approaches x_0, and we write

$$\lim_{x \to x_0^-} f(x) = L. \ \blacksquare \tag{6}$$

As before, we stress that these definitions are informal and are intended only to appeal to your intuition. We will give more precise definitions in Section 1.8.

EXAMPLE 11
TWO ONE-SIDED LIMITS

In Example 6 we discussed the function

$$f(x) = \frac{|x|}{x} = \begin{cases} 1, & x > 0 \\ -1, & x < 0. \end{cases}$$

It follows that

$$\lim_{x \to 0^+} \frac{|x|}{x} = 1 \quad \text{and} \quad \lim_{x \to 0^-} \frac{|x|}{x} = -1.$$

EXAMPLE 12
ONE-SIDED LIMITS FOR THE
GREATEST-INTEGER FUNCTION

The **greatest-integer function** is defined by

$$f(x) = [x]$$

where $[x]$ is the greatest integer smaller than or equal to x. Thus $[3] = 3$, $[\frac{1}{2}] = 0$, $[2.16] = 2$, $[-5.6] = -6$, and so on. A graph of this function is given in Figure 10.

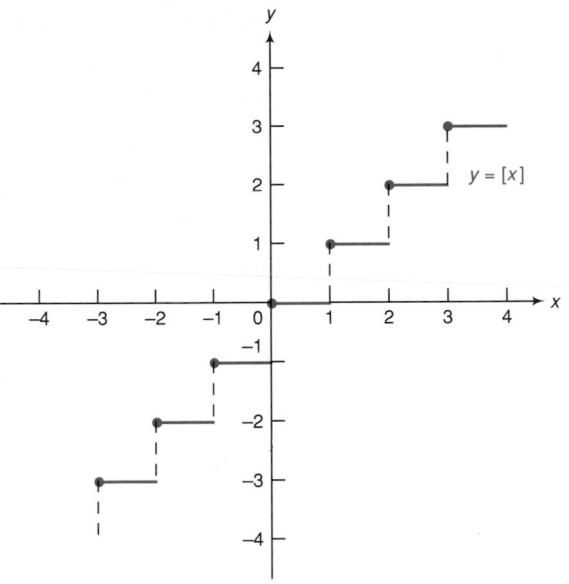

FIGURE 10 GRAPH OF THE GREATEST-INTEGER FUNCTION $y = [x]$

From the graph it is evident that

$$\lim_{x \to 2^-} [x] = 1 \qquad \text{and} \qquad \lim_{x \to 2^+} [x] = 2.$$

Moreover, for any integer n,

$$\lim_{x \to n^-} [x] = n - 1 \qquad \text{and} \qquad \lim_{x \to n^+} [x] = n$$

In Example 12 we saw that if n is an integer, then $\lim_{x \to n}[x]$ does not exist, since we get different values when $x \to n$ from the left and from the right. For the same reason, $\lim_{x \to 0}|x|/x$ does not exist. In general, we have the following theorem, whose proof is given in Section 1.8 (see page 128).

THEOREM 1

$\lim_{x \to x_0} f(x) = L$ exists if and only if the following hold:

(i) $\lim_{x \to x_0^+} f(x)$ exists.
(ii) $\lim_{x \to x_0^-} f(x)$ exists.
(iii) $\lim_{x \to x_0^+} f(x) = \lim_{x \to x_0^-} f(x) = L.$ ∎

That is, the limit exists if and only if the right- and left-hand limits exist and are equal.

AUGUSTIN-LOUIS CAUCHY IS CONsidered to be the most outstanding mathematical analyst of the first half of the nineteenth century. He was born in Paris in 1789 and received his early education from his father. In secondary school he excelled at classical studies. Entering the Ecole Polytechnique in 1805, Cauchy greatly impressed two of the greatest French mathematicians of the time: Joseph Lagrange (1736–1813) and Simon Laplace (1749–1827). Although Cauchy studied to be a civil engineer, in 1816 he was persuaded by Lagrange and Laplace to accept a professorship of mathematics at the Ecole Polytechnique.

Cauchy made many contributions to calculus. In his 1829 textbook *Leçons sur le calcul différential*, he gave the first reasonably clear definition of a limit:

When the successive values attributed to a variable approach indefinitely a fixed value so as to end by differing from it by as little as one wishes, this last is called the limit of all the others.

Cauchy was the first to define the derivative as the limit of the difference quotient

$$\frac{\Delta y}{\Delta x} = \frac{f(x + \Delta x) - f(x)}{\Delta x}.$$

He was also responsible for the modern definition of the definite integral as the limit of a sum (we shall discuss this in Chapter 4).

Cauchy wrote extensively in both pure and applied mathematics. Only Euler wrote more. Cauchy contributed to many areas including real and complex function theory, determinants, probability theory, geometry, wave propagation theory, and infinite series. In the study of calculus, his name appears in the *Cauchy-Schwarz inequality*, the *Cauchy mean-value theorem* (Section 9.1), and the *Cauchy root test for convergence of an infinite series* (Section 10.6).

Cauchy is credited with setting a new standard of rigor in mathematical publication. After Cauchy, it was much more difficult to publish a paper based on intuition; a strict adherence to formal proof was demanded.

The sheer volume of Cauchy's publication was overwhelming. When the French Academy of Sciences began publishing its journal *Comptes Rendus* in 1835, Cauchy sent his work there to be published. Soon the printing bill for Cauchy's work alone became so large that the Academy placed a limit of four pages on each published paper. This rule is still in force today.

There are some unpleasant stories told of Cauchy. One of the most tragic had to do with the Norwegian mathematician Niels Henrik Abel (1802–1829). In 1826 Abel, who had already published some brilliant results, went to Paris in search of an academic position. He approached Cauchy and gave him an important paper he had just completed. Cauchy misplaced it. Abel wrote to a friend, "Every beginner has a great deal of difficulty in getting noticed here. I have just finished an extensive treatise on a certain class of transcendental functions... but Mr. Cauchy scarely deigned to glance at it." While waiting for a suitable position, Abel lived in an unheated apartment in Paris. In 1829, he died of tuberculosis. Ironically, a letter offering Abel a professorship of mathematics at the University of Berlin arrived two days after his death.

Cauchy was a courageous defender of academic freedom. In 1843, he published a sharply worded letter in defense of freedom of conscience. This letter was partially responsible for the abolition of the oath of allegiance that Cauchy had so stubbornly refused to sign.

Cauchy died in 1857 at the age of 68 of a bronchial ailment. His last words were spoken to the Archbishop of Paris: "Men pass away, but their deeds abide."

EXAMPLE 13

A FUNCTION FOR WHICH THE LIMIT DOES NOT EXIST BUT BOTH ONE-SIDED LIMITS EXIST AT 0

Let

$$f(x) = \begin{cases} x + 1, & x > 0 \\ x - 1, & x < 0. \end{cases}$$

This function is graphed in Figure 11. We immediately see that

$$\lim_{x \to 0^+} f(x) = \lim_{x \to 0^+} (x + 1) = 1$$

and

$$\lim_{x \to 0^-} f(x) = \lim_{x \to 0} (x - 1) = -1.$$

Since these limits are different, $\lim_{x \to 0} f(x)$ does not exist.

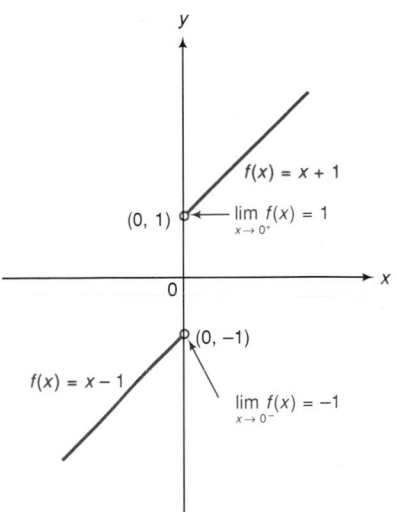

FIGURE 11   GRAPH OF $f(x) = \begin{cases} x+1, & x>0 \\ x-1, & x<0. \end{cases}$

$\lim_{x \to 0} f(x)$ *does not exist.*

■ SELF-QUIZ

I. Suppose f is a function whose domain includes $[-3, 3]$ and such that $\lim_{x \to 2} f(x) = -4$. Answer True or False to each of the following assertions:
a. $\lim_{x \to 1} f(x+1) = -4$
b. $\lim_{x \to 2/3} f(3x) = -4$
c. $\lim_{x \to -2} f(-x) = -4$
d. $\lim_{x \to 2} f(3x-4) = -4$
e. $\lim_{x \to 1} f(1+1/x) = -4$
f. $\lim_{x \to 7} f(x^2 - 8x + 9) = -4$

II. Suppose f is a function such that $\lim_{x \to 3} f(x)$ exists and equals 5. Answer True or False to each of the following assertions:
a. $f(x)$ is near 5 for any x.
b. If x is near 3, then $f(x)$ is near 5.
c. If $f(x)$ is near 5, then x must be near 3.
d. If $2.9 < x < 3$, then $|f(x) - 5| < 0.1$.
e. There is some positive number h such that if $3 - h < x < 3$, then $|f(x) - 5| < 0.1$.

III. Suppose f is a function such that $\lim_{x \to 7} f(x)$ exists and such that $f(x) > 4$ for all x. Answer

True or False to each of the following assertions:
a. $\lim_{x \to 7} f(x) > 4$
b. $\lim_{x \to 7} f(x) \geq 4$
c. $\lim_{x \to 7} f(x) > 3.995$
d. $\lim_{x \to 7} f(x) \neq 4$

IV. Find a function A such that $\lim_{x \to 2} A(x) = 7$ but $A(x) < 7$ for all x.

V. Find functions B and b such that $\lim_{x \to 3} B(x) = \lim_{t \to 3} b(t)$ but $B(w) = 2b(w)$ for all $w \neq 3$.

VI. Merlin strode into calculus class without fanfare and handed the participants the function f, where $f(x) = 7x - 3$ and domain $f = [0, 5)$. Merlin said he would close his eyes and cover his ears while, in turn, each person in the class chose a number s from the domain $[0, 5)$ and then redefined the value $f(s)$ of the function there. When the class had done so, Merlin was tapped on the shoulder. He opened his eyes and ears and said, "Your modification of f is a new function, let's call it g. I don't know what you have done to f, so I can't draw a correct graph of the function g, but I do know that $\lim_{x \to 2} g(x) = 11$." Was Merlin right? Explain.

■ DRILL

1. a. Draw the graph of the function $f(x) = x + 7$.
   b. Calculate $f(x)$ for $x = 1, 1.5, 1.9,$ and 1.99.
   c. Calculate $f(x)$ for $x = 3, 2.5, 2.1,$ and 2.01.
   d. Compute $\lim_{x \to 2}(x + 7)$.

CONTINUED

2. a. Draw the graph of the function $f(x) = (7 - x)/2$.
   b. Calculate $f(x)$ for $x = 1, 1.5, 1.9,$ and 1.99.
   c. Calculate $f(x)$ for $x = 3, 2.5, 2.1,$ and 2.01.
   d. Compute $\lim_{x \to 2}(7 - x)/2$.
3. a. Draw the graph of the function $f(x) = x^2 - 4$.
   b. Calculate $f(x)$ for $x = 0, 0.5, 0.9,$ and 0.99.
   c. Calculate $f(x)$ for $x = 2, 1.5, 1.1,$ and 1.01.
   d. Compute $\lim_{x \to 1}(x^2 - 4)$.
4. a. Draw the graph of the function $f(x) = x^2 - 3x + 4$.
   b. Calculate $f(x)$ for $x = -1.5, -1.1, -1.01,$ and -1.001.
   c. Calculate $f(x)$ for $x = -0.5, -0.9, -0.99,$ and -0.999.
   d. Compute $\lim_{x \to -1}(x^2 - 3x + 4)$.
5. Let $f(x) = (x - 1)(x - 2)/(x - 1)$.
   a. Explain why $f(x)$ is not defined for $x = 1$.
   b. Find another function that is equal to f for all $x \neq 1$ and that *is* defined for $x = 1$.
   c. Compute $\lim_{x \to 1}(x - 1)(x - 2)/(x - 1)$.
6. Let $f(x) = (x^3 - 8)/(x - 2)$.
   a. Explain why $f(x)$ is not defined for $x = 2$.
   b. Find another function that is equal to f for all $x \neq 2$ and that *is* defined for $x = 2$.
   c. Compute $\lim_{x \to 2}(x^3 - 8)/(x - 2)$.
7. Explain why $\lim_{x \to -1}\sqrt{x + 1}$ does not exist.
8. Does $\lim_{x \to -1}\sqrt{x^2 + 1}$ exist? If so, compute it.

In Problems 9–38, compute each limit, if it exists; explain why there is no limit if it does not exist.

9. $\lim_{x \to 5}(x^2 - 6)$
10. $\lim_{x \to 2}(x^4 - 9)$
11. $\lim_{x \to 0} 1/(x^5 + 6x + 2)$
12. $\lim_{x \to 0}(-x^5 + 17x^3 + 2x)$
13. $\lim_{x \to -1}(x + 1)^2/(x + 1)$
14. $\lim_{x \to 1}(x^4 - x)/(x^3 - 1)$
15. $\lim_{x \to 4}\sqrt{25 - x^2}$
16. $\lim_{x \to 4}\sqrt{x^2 - 25}$
17. $\lim_{x \to -4}\sqrt{x + 4}$
18. $\lim_{x \to 5}\sqrt{x^2 - 25}$

19. $\lim_{x \to 2}\sqrt[3]{x^3 - 8}$
20. $\lim_{x \to 2}\sqrt[3]{x^4 - 16}$
21. $\lim_{x \to 3}(x^2 - 4x + 3)/(x - 3)$ [*Hint:* Divide.]
22. $\lim_{x \to -2}(x^2 + 6x + 8)/(x + 2)$
23. $\lim_{x \to 1}(\sqrt{x} - 1)/(x - 1)$ [*Hint:* $a^2 - b^2 = (a + b)(a - b)$.]
*24. $\lim_{x \to 2}(1 - \sqrt{x/2})/(2 - x)$
25. $\lim_{x \to 2^+}\sqrt{x - 2}$
26. $\lim_{x \to -2^-}\sqrt{x + 2}$
27. $\lim_{x \to -1^-}\sqrt{-1 - x}$
28. $\lim_{x \to 1^-}\dfrac{|x - 1|}{x - 1}$
29. $\lim_{x \to -2^+}\dfrac{3x|x + 2|}{x + 2}$
30. $\lim_{x \to 1^-}\sqrt{(x - 1)(x - 2)}$
31. $\lim_{x \to (3/2)^+}[x]$
32. $\lim_{x \to 0^+}\sqrt[3]{x}$
33. $\lim_{x \to -2^+}\sqrt{x + 2}$
34. $\lim_{x \to -1^+}\sqrt{-1 - x}$
35. $\lim_{x \to 1^+}\dfrac{|x - 1|}{x - 1}$
36. $\lim_{x \to 1^+}\sqrt{(x - 1)(x - 2)}$
37. $\lim_{x \to 1^-}\sqrt{(x - 1)(x - 2)}$
38. $\lim_{x \to 2^-}[x]$

In Problems 39–50, let $f(x) = \begin{cases} x, & x \neq 3 \\ 2, & x = 3 \end{cases}$

$g(x) = \begin{cases} x, & x < 1 \\ x^2, & x \geq 1 \end{cases}$, and $h(x) = \begin{cases} 2x + 3, & x \leq -1 \\ 3 - x, & x > -1 \end{cases}$.

Compute each limit, if it exists.

39. $\lim_{x \to 3^-} f(x)$
40. $\lim_{x \to 3^+} f(x)$
41. $\lim_{x \to 3} f(x)$
42. $\lim_{x \to 1^-} g(x)$
43. $\lim_{x \to 1^+} g(x)$
44. $\lim_{x \to 1} g(x)$
45. $\lim_{x \to -1^-} h(x)$
46. $\lim_{x \to -1^+} h(x)$
47. $\lim_{x \to -1} h(x)$
*48. $\lim_{x \to 2} g(x)h(x)$
*49. $\lim_{x \to 3}\dfrac{f(x)}{g(x)}$
*50. $\lim_{x \to -2}\dfrac{f(x) + g(x)}{h(x)}$

■ **APPLICATIONS**

51. a. Graph the function $f(x) = |x - 3|$.
    b. Calculate $\lim_{x \to 3}|x - 3|$.
52. a. Graph the function $f(x) = |3x - 4|$.
    b. Calculate $\lim_{x \to 4/3}|3x - 4|$.
53. a. Graph the function $f(x) = |x + 3|/(x + 3)$.
    b. Explain why $\lim_{x \to -3}|x + 3|/(x + 3)$ does not exist.

    c. Calculate $\lim_{x \to 5}|x + 3|/(x + 3)$ and $\lim_{x \to -4}|x + 3|/(x + 3)$.
54. a. Graph the function $f(x) = |7 - 2x|/(7 - 2x)$.
    b. Explain why $\lim_{x \to 3.5}|7 - 2x|/(7 - 2x)$ does not exist.
    c. Calculate $\lim_{x \to 3}|7 - 2x|/(7 - 2x)$ and $\lim_{x \to 4}|7 - 2x|/(7 - 2x)$.

CONTINUED

55. a. Graph the curve $y = 5 - x^2$.
    b. Draw (on your graph) the straight line joining the points $(-3, -4)$ and $(-2, 1)$.
    c. Draw the straight line joining the points $(-3, -4)$ and $(-3.5, -7.25)$.
    d. For any real number $h \neq 0$, what is represented by the quotient

$$\frac{[5 - (-3 + h)^2] - (-4)}{h}?$$

    e. Calculate

$$\lim_{h \to 0} \frac{[5 - (-3 + h)^2] - (-4)}{h}.$$

    f. What is the slope of the line tangent to the curve $y = 5 - x^2$ at the point $(-3, -4)$?

56. a. Graph the curve $y = x^2 + 3$.
    b. Draw (on your graph) the straight line joining the points $(1, 4)$ and $(0, 3)$.
    c. Draw the straight line joining the points $(1, 4)$ and $(1.5, 5.25)$.
    d. For any real number $h \neq 0$, what is represented by the quotient

$$\frac{[(1 + h)^2 + 3] - 4}{h}?$$

    e. Calculate

$$\lim_{h \to 0} \frac{[(1 + h)^2 + 3] - 4}{h}.$$

    f. What is the slope of the line tangent to the curve $y = x^2 + 3$ at the point $(1, 4)$?

CALCULATOR EXERCISES

57. Let $f(x) = \sqrt{x^3 + 12}/(x + 8)$.
    a. Calculate $f(x)$ for $x = -1, -1.5, -1.9, -1.99,$ and -1.999.
    b. Calculate $f(x)$ for $x = -3, -2.5, -2.1, -2.01,$ and -2.001.
    c. Estimate $\lim_{x \to -2} \sqrt{x^3 + 12}/(x + 8)$.
    d. Calculate $f(-2)$ and compare it with your estimate in part (c).

58. Let $f(x) = (x^3 - 6x + 2)/(x^2 + x + 9)$.
    a. Calculate $f(x)$ for $x = 3, 2.5, 2.1, 2.01,$ and 2.001.
    b. Calculate $f(x)$ for $x = 1, 1.5, 1.9, 1.99,$ and 1.999.
    c. Estimate $\lim_{x \to 2}(x^3 - 6x + 2)/(x^2 + x + 9)$.
    d. Calculate $f(2)$ and compare it with your estimate in part (c).

In Problems 59–92, make an educated guess at the value of the limit by using your calculator to see what happens as x gets closer and closer to the limiting value. In problems involving trigonometric functions, make certain that the calculator is in radian mode.

59. $\lim_{x \to 1^+} \frac{x^2 - 1}{\sqrt{x - 1}}$

60. $\lim_{x \to 3} \frac{3 - x}{3 - \sqrt{3x}}$

61. $\lim_{x \to 4} \frac{4 - x}{4 - 2\sqrt{x}}$

62. $\lim_{h \to 0} \frac{\sqrt{9 + h} - 3}{h}$

63. $\lim_{h \to 0} \frac{\sqrt{0.25 - h} - 0.5}{h}$

64. $\lim_{h \to 0} \frac{\sqrt[3]{-1 + h} + 1}{h}$

65. $\lim_{u \to 4} \frac{\sqrt{u} - 2}{u - 4}$

66. $\lim_{\Delta x \to 0} \frac{\sqrt{9 + \Delta x} - 3}{\Delta x}$

67. $\lim_{\Delta x \to 0} \frac{\sqrt{0.25 - \Delta x} - 0.5}{\Delta x}$

68. $\lim_{\Delta x \to 0} \frac{\sqrt[3]{-1 + \Delta x} + 1}{\Delta x}$

69. $\lim_{\Delta x \to 0} \frac{\sqrt[3]{8 + \Delta x} - 2}{\Delta x}$

70. $\lim_{x \to 4} \frac{\sqrt{x} - 2}{x - 4}$

71. $\lim_{x \to 25} \frac{\sqrt{x} - 5}{x - 25}$

72. $\lim_{x \to 30} \frac{x^2 - 30^2}{x - 30}$

73. $\lim_{x \to 250} \frac{x^2 - 250^2}{x - 250}$

74. $\lim_{x \to 2} \frac{x^3 - 8}{x - 2}$

75. $\lim_{x \to 5} \frac{x^3 - 125}{x - 5}$

76. $\lim_{x \to -3} \frac{x^3 + 27}{x + 3}$

77. $\lim_{x \to 2} \frac{x^5 - 32}{x - 2}$

78. $\lim_{x \to 1} \frac{x^7 - 1}{x - 1}$

79. $\lim_{x \to 1} \frac{x^5 - 1}{x^3 - 1}$

80. $\lim_{x \to 3} \frac{x^4 - 81}{x^3 - 27}$

81. $\lim_{x \to -2} \frac{x^5 + 32}{x^2 - 4}$

82. $\lim_{x \to 2} \frac{x^3 - 4x^2 + 5x - 2}{x^2 - 7x + 10}$

83. $\lim_{x \to 0} \frac{\sin 2x}{x}$

84. $\lim_{x \to 0} \frac{\sin x^2}{4x^2}$

CONTINUED

85. $\lim\limits_{x\to 0} \dfrac{1-\cos x}{x}$

86. $\lim\limits_{x\to 0} x \sin \dfrac{1}{x}$

87. $\lim\limits_{x\to 0} \dfrac{\tan 3x}{5x}$

*88. $\lim\limits_{x\to 0} x^x$
[*Hint:* Use the power button $\boxed{y^x}$ or $\boxed{x^y}$.]

*89. $\lim\limits_{x\to 0}(1 + 3x)^{1/x}$

*90. $\lim\limits_{x\to 0} \dfrac{5^x - 1}{x}$ [To 5 decimal places]

*91. $\lim\limits_{x\to 0} \dfrac{3^x - 1}{2^x - 1}$ [To 5 decimal places]

92. $\lim\limits_{x\to 0} \dfrac{1 - 10^x}{1 - 100^x}$

The natural logarithmic function, $y = \ln x$, is discussed in Sections 6.2 and 6.3. Your calculator has a button $\boxed{\ln}$ or $\boxed{\log_e x}$. For example, pressing $\boxed{2}$ $\boxed{\ln}$ yields 0.69314718, and pressing 10.5 followed by $\boxed{\ln}$ gives 2.351375257. Use the $\boxed{\ln}$ button to guess at the limits in Problems 93 and 94.†

93. $\lim\limits_{x\to 0} \dfrac{\ln(1 + x)}{x}$

94. $\lim\limits_{x\to 0^+} x \ln x$

■ SHOW/PROVE/DISPROVE

95. Let

$$f(x) = \begin{cases} x^2 & \text{if } x < 1 \\ x^3 & \text{if } x \ge 1. \end{cases}$$

Show that $\lim\limits_{x\to 1} f(x) = 1$ by showing that the right- and left-hand limits exist and are equal.

96. Let

$$f(x) = \begin{cases} 0 & \text{if } x < 0 \\ x^2 & \text{if } 0 \le x < 2 \\ 4 & \text{if } x \ge 2. \end{cases}$$

Calculate $\lim\limits_{x\to 0^-} f(x)$, $\lim\limits_{x\to 0^+} f(x)$, $\lim\limits_{x\to 2^-} f(x)$, and $\lim\limits_{x\to 2^+} f(x)$.

97. Let

$$f(x) = \begin{cases} 0 & \text{if } x \le 2 \\ x - 2 & \text{if } x > 2. \end{cases}$$

Find $\lim\limits_{x\to 2^-} f(x)$ and $\lim\limits_{x\to 2^+} f(x)$.

98. Let

$$f(x) = \begin{cases} |x| & \text{if } x \le 2 \\ [x] & \text{if } x > 2. \end{cases}$$

Show that $\lim\limits_{x\to 2} f(x)$ exists and compute its value.

In Problems 99–110, use the graphs to compute each limit, if it exists.

99. $\lim\limits_{x\to 2^-} f(x)$

100. $\lim\limits_{x\to 2^+} f(x)$

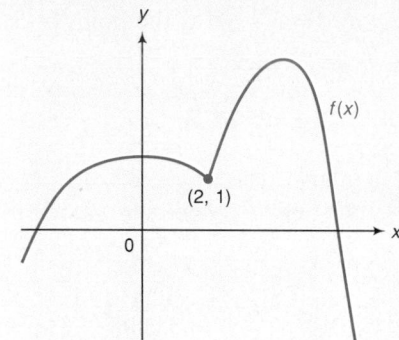

GRAPH FOR PROBLEMS 99–100.

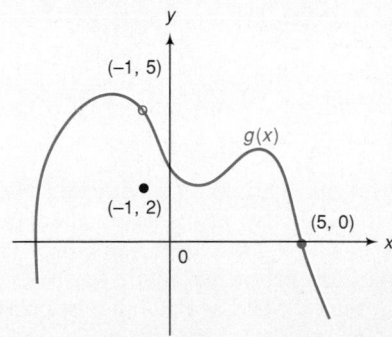

GRAPH FOR PROBLEMS 101–104.

101. $\lim\limits_{x\to -1^-} g(x)$

102. $\lim\limits_{x\to -1^+} g(x)$

103. $\lim\limits_{x\to -1} g(x)$

104. $\lim\limits_{x\to 5} g(x)$

---

† On some calculators the $\boxed{\ln}$ key is pressed first.

CONTINUED

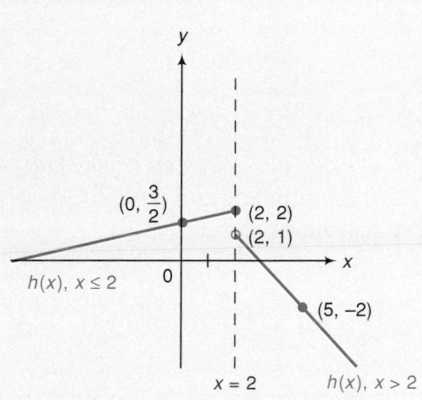

GRAPH FOR PROBLEMS 105–110.

105. $\lim_{x \to 0} h(x)$
106. $\lim_{x \to 5^-} h(x)$
107. $\lim_{x \to 5^+} h(x)$
108. $\lim_{x \to 2^-} h(x)$
109. $\lim_{x \to 2^+} h(x)$
110. $\lim_{x \to 2} h(x)$

■ **CHALLENGE**

*111. Compute $\lim_{x \to 0} [-x^2]$ (if it exists) where $[\ldots]$ denotes the *greatest-integer function* of Example 12.

*112. Compute $\lim_{x \to 0} x \cdot [1/x]$ (if it exists).

*113. For each point (x, y) on the curve $y = x^2$, let $A(x)$ be the area of the triangle with vertices $(0, 0)$, $(1, 0)$, and (x, y); let $B(x)$ be the area of the triangle with vertices $(0, 0)$, $(0, 1)$, and (x, y). Find $\lim_{x \to 0} A(x)/B(x)$, if it exists.

114. Let $f(x, y) = (2x - 5y)/(3x + 7y)$.
   a. Find $\lim_{x \to 0}(\lim_{y \to 0} f(x, y))$.
   b. Find $\lim_{y \to 0}(\lim_{x \to 0} f(x, y))$.

■ **ANSWERS TO SELF-QUIZ**

I. All are True.

II. Assertions b and c are True.

III. Assertion b is True; therefore, so is assertion c.

IV. One of many possible solutions is to let

$$A(x) = \begin{cases} 7 - (x - 2)^2 & \text{if } x \neq 2 \\ 0 & \text{if } x = 2. \end{cases}$$

V. Let $b(t) = t - 3$ and $B(x) = 2(x - 3)$. [Many other solutions exist; all have

$$\lim_{t \to 3} b(t) = 0 = \lim_{x \to 3} B(x).]$$

VI. The new function, g, can differ from the old one, f, at only a finite number of points (since each person in the class made a change at only a single point). There must be some (small) interval surrounding 2 where f is unchanged (except possibly at 2 itself); g and f agree on this set, hence $\lim_{x \to 2} g(x) = \lim_{x \to 2} f(x) = 11$.

1.3 SOME LIMIT THEOREMS

In this section we state several theorems that will make our calculations of a number of limits a great deal easier. The proofs of these theorems depend on the material in Section 1.8 and are a bit technical. For that reason they are put in an appendix (see Appendix 3).

In Example 1.2.1 we concluded that

$$\lim_{x \to 2}(x^2 + 3) = 2^2 + 3 = 7.$$

That is, the limit of $f(x) = x^2 + 3$ as x tends to 2 is equal to $f(x)$ evaluated at $x = 2$ [i.e., $f(2)$]. However, as we remarked earlier, this process of evaluation [i.e., substituting the value x_0 into $f(x)$ to find a limit as $x \to x_0$] will not always work since $f(x)$ may not even be defined at x_0. Nevertheless, it is true that if f is a polynomial, then it is always possible to calculate the limit by evaluation.

THEOREM 1

Let $p(x) = c_0 + c_1x + c_2x^2 + c_3x^3 + \cdots + c_nx^n$ be a *polynomial*, where $c_0, c_1, c_2, c_3, \ldots, c_n$ are real numbers and n is a fixed nonnegative integer. Then, for every real number x_0,

$$\lim_{x \to x_0} p(x) = p(x_0)$$
$$= c_0 + c_1x_0 + c_2x_0^2 + c_3x_0^3 + \cdots + c_nx_0^n. \quad \blacksquare \quad (1)$$

EXAMPLE 1
LIMIT OF A POLYNOMIAL

Calculate $\lim_{x \to 3}(x^3 - 2x + 6)$.

SOLUTION: $x^3 - 2x + 6$ is a polynomial. Hence,

$$\lim_{x \to 3}(x^3 - 2x + 6) = 3^3 - 2 \cdot 3 + 6 = 27 - 6 + 6 = 27.$$

NOTE: Theorem 1 and all the other theorems in this section hold for one-sided limits as well.

EXAMPLE 2
LIMIT OF A CONSTANT FUNCTION

Calculate $\lim_{x \to x_0} 4$ for any real number x_0.

SOLUTION: $f(x) = 4$ is a polynomial (of degree 0). Hence, $\lim_{x \to x_0} 4 = P(x_0) = 4$. This result simply states that

the limit of a constant function is that constant.

In Figure 1, we can see that as x gets close to x_0, $f(x)$ gets close to (is equal to) 4.

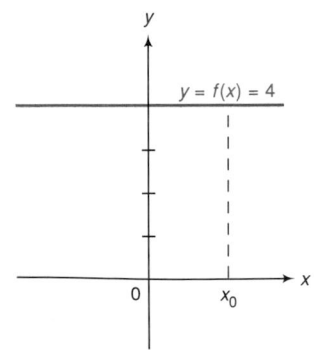

FIGURE 1
GRAPH OF $f(x) = 4$ SHOWING THAT $\lim_{x \to x_0} 4 = 4$ FOR EVERY REAL NUMBER x_0

THEOREM 2   LIMITS OF CONSTANT MULTIPLES, SUMS, PRODUCTS, AND QUOTIENTS OF FUNCTIONS

Let c be a real number, and suppose that both $\lim_{x \to x_0} f(x)$ and $\lim_{x \to x_0} g(x)$ exist. Then $\lim_{x \to x_0} cf(x)$, $\lim_{x \to x_0}[f(x) + g(x)]$, and $\lim_{x \to x_0} f(x)g(x)$ all exist. Moreover,

(i) $\lim_{x \to x_0} cf(x) = c \lim_{x \to x_0} f(x)$
   The limit of a constant times a function is equal to the product of that constant and the limit of the function.

(ii) $\lim_{x \to x_0} [f(x) + g(x)] = \lim_{x \to x_0} f(x) + \lim_{x \to x_0} g(x)$
   The limit of the sum of two functions is equal to the sum of their limits.

(iii) $\lim_{x \to x_0} [f(x)g(x)] = \left[\lim_{x \to x_0} f(x)\right] \cdot \left[\lim_{x \to x_0} g(x)\right]$
   The limit of the product of two functions is equal to the product of their limits.

(iv) If $\lim_{x \to x_0} g(x) \neq 0$, then $\lim_{x \to x_0} f(x)/g(x)$ exists and

$$\lim_{x \to x_0} \frac{f(x)}{g(x)} = \frac{\lim_{x \to x_0} f(x)}{\lim_{x \to x_0} g(x)}$$

The limit of the quotient of two functions is equal to the quotient of their limits provided that the limit in the denominator function is not zero. ■

EXAMPLE 3
LIMIT OF THE SUM AND CONSTANT MULTIPLES OF FUNCTIONS

Since $\lim_{x \to 1} \dfrac{x^2 - 1}{x - 1} = 2$ (Example 3 on page 55), we have

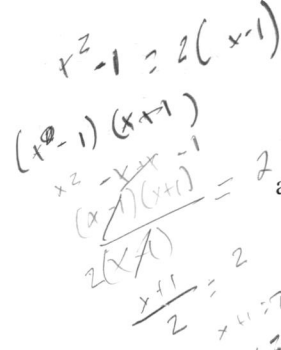

Part (i)

$$\lim_{x \to 1} 3\left(\frac{x^2 - 1}{x - 1}\right) = 3 \lim_{x \to 1} \frac{x^2 - 1}{x - 1} = 3 \cdot 2 = 6$$

and

Theorem 1

$$\lim_{x \to 1} \left(\frac{x^2 - 1}{x - 1} + 4x^3 + 3\right) = \lim_{x \to 1} \frac{x^2 - 1}{x - 1} + \lim_{x \to 1}(4x^3 + 3) = 2 + 7 = 9.$$

EXAMPLE 4
LIMIT OF A POLYNOMIAL

Calculate $\lim_{x \to -1}(x^2 - 3)^{10}$.

SOLUTION: $\lim_{x \to -1}(x^2 - 3) = (-1)^2 - 3 = -2$. We now may apply Theorem 2 (iii) several times (nine times to be exact):

$$\lim_{x \to -1}(x^2 - 3)^{10} = \underbrace{\left[\lim_{x \to -1}(x^2 - 3)\right]\left[\lim_{x \to -1}(x^2 - 3)\right] \cdots \left[\lim_{x \to -1}(x^2 - 3)\right]}_{10 \text{ terms}}$$

$$= \underbrace{(-2)(-2) \cdots (-2)}_{10 \text{ factors}} = (-2)^{10} = 1024.$$

The last example can be generalized.

COROLLARY **TO THEOREM 2**

If $\lim_{x \to x_0} f(x)$ exists and n is a positive integer, then $\lim_{x \to x_0}[f(x)]^n$ exists and

$$\lim_{x \to x_0} [f(x)]^n = \left[\lim_{x \to x_0} f(x)\right]^n. \quad \blacksquare \tag{2}$$

⊗ **WARNING:** Do not use the *limit theorem for quotients* (iv) if the limit in the denominator is zero. ⊗

Incorrect	Correct

$$\lim_{x \to 1} \frac{\sqrt{x} - 1}{x - 1} = \frac{\lim_{x \to 1}(\sqrt{x} - 1)}{\lim_{x \to 1}(x - 1)}$$

$$= \frac{0}{0}$$

which is undefined.

$$\lim_{x \to 1} \frac{\sqrt{x} - 1}{x - 1} = \lim_{x \to 1} \frac{\sqrt{x} - 1}{(\sqrt{x} - 1)(\sqrt{x} + 1)}$$

$$= \lim_{x \to 1} \frac{1}{\sqrt{x} + 1}$$

$$= \frac{\lim_{x \to 1} 1}{\lim_{x \to 1}(\sqrt{x} + 1)}$$

$$= \frac{1}{1 + 1} = \frac{1}{2}.$$

REMARK: Here and elsewhere in this section (including the problem set), we use the following result:

If $\lim_{x \to x_0} f(x) = L > 0$, then $\lim_{x \to x_0} \sqrt{f(x)} = \sqrt{L}$.

We will prove this theorem in Section 1.7 (Example 12 on page 117).

EXAMPLE 5
LIMIT OF A RATIONAL FUNCTION

Calculate $\lim_{x \to -2}(x^3 - 3x + 6)/(-x^2 + 15)$.

SOLUTION:

$$\lim_{x \to -2} (x^3 - 3x + 6) = (-2)^3 - 3(-2) + 6 = -8 + 6 + 6 = 4$$

and

$$\lim_{x \to -2} (-x^2 + 15) = -(-2)^2 + 15 = -4 + 15 = 11 \neq 0.$$

Thus,

$$\lim_{x \to -2} \frac{x^3 - 3x + 6}{-x^2 + 15} = \frac{\lim_{x \to -2} (x^3 - 3x + 6)}{\lim_{x \to -2} (-x^2 + 15)} = \frac{4}{11}.$$

DEFINITION **RATIONAL FUNCTION**

A **rational function** $r(x)$ is a function that can be written as the quotient of two polynomials; that is,

$$r(x) = \frac{p(x)}{q(x)} \tag{3}$$

where $p(x)$ and $q(x)$ are both polynomials. ■

For example, the function

$$r(x) = \frac{x^3 - 3x + 6}{-x^2 + 15}$$

given in Example 5 is a rational function.

COROLLARY **TO THEOREMS 1 AND 2**

Let $r(x) = p(x)/q(x)$ be a rational function with $q(x_0) \neq 0$. Then,

$$\lim_{x \to x_0} r(x) = \lim_{x \to x_0} \frac{p(x)}{q(x)} = \frac{p(x_0)}{q(x_0)} = r(x_0). \quad \blacksquare \qquad (4)$$

EXAMPLE 6
LIMIT OF A RATIONAL FUNCTION

Calculate $\lim_{x \to 4}(x^3 - x^2 - 3)/(x^2 - 3x + 5)$.

SOLUTION: Here $q(x) = x^2 - 3x + 5$ and $q(4) = 16 - 12 + 5 = 9 \neq 0$.
Therefore,

$$\lim_{x \to 4} \frac{x^3 - x^2 - 3}{x^2 - 3x + 5} = \frac{4^3 - 4^2 - 3}{4^2 - 3 \cdot 4 + 5} = \frac{64 - 16 - 3}{16 - 12 + 5} = \frac{45}{9} = 5.$$

■ PROBLEMS 1.3

■ SELF-QUIZ

I. Suppose $\lim_{x \to 2} f(x) = -5$ and $\lim_{x \to 2} g(x) = 3$.
Then $\lim_{x \to 2}(f(x) \cdot g(x))$ exists and it equals
_____.

 a. $f(2) \cdot g(2)$ b. $f(-5) \cdot g(3)$
 c. $2 \cdot 2 = 4$ d. $-5 \cdot 3 = -15$

II. Suppose g is a function defined on the interval
$[0, 5]$ such that $\lim_{x \to 2} g(x) = 1$. Then

$$\lim_{x \to 2}[g(x)]^3 \;\underline{\hspace{2cm}}.$$

 a. $= 2^3 = 8$ b. $= 2^{-1 \cdot 3} = 2^{-3} = \frac{1}{8}$
 c. $= (-1)^3 = -1$ d. $= 3^{-1} = \frac{1}{3}$
 e. does not exist because -1 is not in $[0, 5]$

III. True–False:
 a. If $\lim_{x \to 2}[f(x)/g(x)]$ exists, then that limit
 equals $f(2)/g(2)$.
 b. If $\lim_{x \to 2} f(x) = f(2)$ and $\lim_{x \to 2} = g(2)$, then
 $\lim_{x \to 2}[f(x)/g(x)] = f(2)/g(2)$.
 c. If $\lim_{x \to 2} f(x) = f(2)$ and $\lim_{x \to 2} = g(2) \neq 0$,
 then $\lim_{x \to 2}[f(x)/g(x)] = f(2)/g(2)$.

■ DRILL

In Problems 1–16, use the limit theorems to help
you calculate the requested limits.

1. $\lim_{x \to 3}(x^2 - 2x - 1)$
2. $\lim_{x \to -2}(-x^3 - x^2 - x - 1)$
3. $\lim_{x \to 1} 3$
4. $\lim_{x \to -1}(x^{49} + 1)$
5. $\lim_{x \to 1} 8(x^{101} + 2)$
6. $\lim_{x \to 2}(x^3 - 4x^2 + 5x - 3)(x^2 + 17x - 4)$
7. $\lim_{x \to 2}(x^2 - 1)^5$
8. $\lim_{x \to 4}(x^2 - x - 10)^7$
9. $\lim_{x \to 3} 1/(x^3 - 8)$
10. $\lim_{x \to 0} 3/(x^5 + 3x^2 + 3)$

11. $\lim_{x \to 1} \dfrac{x^8 + x^6 + x^4 + x^2 + 1}{x^7 + x^5 + x^3 + x}$

12. $\lim_{x \to -4} \dfrac{x^3 - x^2 - x + 1}{x^2 + 3}$

13. $\lim_{x \to 2} \dfrac{x^2 - x - 12}{x^2 - 5x + 4}$

14. $\lim_{x \to -1}(x^9 + 2)^{53}$

15. $\lim_{x \to 0} \dfrac{x^{28} - 17x^{14} + x^2 - 3}{x^{51} + x^{31} - 23x^2 + 2}$

16. $\lim_{x \to 0} \dfrac{x^{81} - x^{41} + 3}{23x^4 - 8x^7 + 5}$

■ APPLICATIONS

Problem 28 states a theoretical result concerning
limits and square roots. Use that result and the limit
theorems in this section to work Problems 17–26.

17. $\lim_{x \to 5} 3\sqrt{x - 1}$

18. $\lim_{x \to 3} 5\sqrt{x^2 + 7}$

CONTINUED

19. $\lim_{x \to -2} -4\sqrt{x + 3}$

20. $\lim_{x \to -2}(1 + x + x^2 + x^3 + \sqrt{x^2 - 3})$

21. $\lim_{x \to 5}(\sqrt{x - 1} + \sqrt{x^2 - 9})$

22. $\lim_{x \to 5}(\sqrt{x - 1})(\sqrt{x^2 - 9})$

23. $\lim_{x \to 0} \dfrac{\sqrt{x + 1}}{\sqrt{x^2 - 3x + 4}}$

24. $\lim_{x \to -2} \dfrac{\sqrt{x^2 - 3}}{1 + x + x^2 + x^3}$

25. $\lim_{x \to 5} \sqrt{\dfrac{x - 1}{x^2 - 9}}$

26. $\lim_{x \to 0} \dfrac{3\sqrt{x + 1}}{5\sqrt{x^2 - 3x + 4}}$

27. Find functions f and g such that $\lim_{x \to 7} f(x) = 0$ but $\lim_{x \to 7}[f(x) \cdot g(x)] \neq 0$.

■ SHOW/PROVE/DISPROVE

28. Let $p(x)$ be a polynomial function. Use the result on page 72 to prove that $\lim_{x \to x_0} \sqrt{p(x)} = \sqrt{p(x_0)}$ if $p(x_0) > 0$.

29. Prove or disprove: If $f(x) < 7$ for all x and if $\lim_{x \to 3} f(x)$ exists, then $\lim_{x \to 3} f(x) < 7$.

■ CHALLENGE

*30. Suppose that function f satisfies the inequalities $(x + 3)^2 + 41 \leq f(x)$ and $f(x) \leq |x + 3| + 41$ for all x in the interval $(-4, -2)$. Find $\lim_{x \to -3} f(x)$ and justify your result. [*Hint:* Sketch the graphs of $y = (x + 3)^2 + 41$ and $y = |x + 3| + 41$ over a short open interval containing -3.]

■ ANSWERS TO SELF-QUIZ

I. d   II. c

III. a. False, for example, let $f(x) = x^2 - 4$ and $g(x) = x - 2$.

  b. False [$g(2)$ may equal 0]

  c. True

1.4 INFINITE LIMITS AND LIMITS AT INFINITY

Consider the problem of calculating

$$\lim_{x \to 0} \frac{1}{x^2}.$$

Values of $1/x^2$ for x "near" 0 are given in Table 1. We see that as x gets closer and closer to 0, $f(x) = 1/x^2$ gets larger and larger. In fact, $1/x^2$ grows *without bound* as x approaches 0 from either side. The graph of the function

TABLE 1

x	x^2	$\dfrac{1}{x^2}$	x	x^2	$\dfrac{1}{x^2}$
1	1	1	-1	1	1
0.5	0.25	4	-0.5	0.25	4
0.1	0.01	100	-0.1	0.01	100
0.01	0.0001	10,000	-0.01	0.0001	10,000
0.001	0.000001	1,000,000	-0.001	0.000001	1,000,000
0.0001	0.00000001	100,000,000	-0.0001	0.00000001	100,000,000

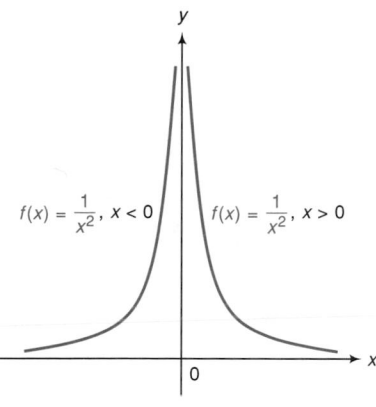

FIGURE 1

GRAPH OF $y = \dfrac{1}{x^2}$

$$\lim_{x \to 0} \frac{1}{x^2} = \infty$$

$f(x) = 1/x^2$ is given in Figure 1. In this situation we say that $f(x)$ *tends to infinity as x approaches zero*, and we write

$$\lim_{x \to 0} \frac{1}{x^2} = \infty.$$

We emphasize that ∞ is the symbol for this behavior. It is *not* a new number. In general, we have the following definition.

DEFINITION **INFINITE LIMIT**

(i) If $f(x)$ grows without bound in the positive direction as x gets close to the number x_0 from either side, then we say that $f(x)$ **tends to infinity as x approaches x_0**, and we write

$$\lim_{x \to x_0} f(x) = \infty. \tag{1}$$

(ii) If $f(x)$ grows without bound in the negative direction as x gets close to the number x_0 from either side, then we say that $f(x)$ **tends to minus infinity as x approaches x_0**, and we write

$$\lim_{x \to x_0} f(x) = -\infty. \quad\blacksquare \tag{2}$$

EXAMPLE 1
AN INFINITE LIMIT

Since $1/x^2$ grows without bound as $x \to 0$, $-1/x^2$ grows without bound in the negative direction as $x \to 0$. Thus,

$$\lim_{x \to 0} -\frac{1}{x^2} = -\infty.$$

EXAMPLE 2
AN INFINITE LIMIT

Calculate $\lim_{x \to 0} 1/(x^2 + x^3)$.

SOLUTION: With the aid of a calculator we find values for $1/(x^2 + x^3)$ for x near 0 (see Table 2). We see that as x gets close to 0, $1/(x^2 + x^3)$ grows without bound. Now $x^3 + x^2 = x^2(1 + x)$ and this is positive for x near 0. Thus, we have

$$\lim_{x \to 0} \frac{1}{x^2 + x^3} = \infty.$$

TABLE 2

x	$\dfrac{1}{x^2 + x^3}$
0.5	2.66667
0.1	90.90909
0.01	9,900.99
0.001	999,000.999
-0.5	8
-0.1	111.1111
-0.01	10,101.01
-0.001	1,001,001

To calculate this limit without making a table, we first divide the numerator and denominator by x^2. This division can be done since in the calculation of the limit $x \neq 0$, so $x^2 \neq 0$. We then have

$$\lim_{x \to 0} \frac{1}{x^2 + x^3} = \lim_{x \to 0} \frac{1/x^2}{(x^2 + x^3)/x^2} = \lim_{x \to 0} \frac{1/x^2}{1 + x}.$$

Now $\lim_{x \to 0}(1/x^2) = \infty$ and $\lim_{x \to 0}(1 + x) = 1$. Thus the numerator grows without bound while the denominator approaches 1, implying that $1/(x^2 + x^3)$ does tend to infinity. This example illustrates how a difficult calculation can be greatly simplified by a few algebraic manipulations.

EXAMPLE 3
A LIMIT THAT SEEMS TO BE
INFINITE BUT, IN FACT, DOES
NOT EXIST

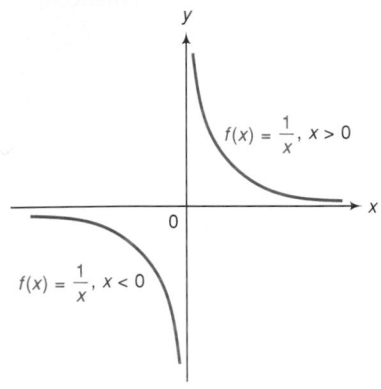

FIGURE 2

GRAPH OF $y = \dfrac{1}{x}$

$\lim\limits_{x \to 0^+} \dfrac{1}{x} = \infty$ *and* $\lim\limits_{x \to 0^-} \dfrac{1}{x} = -\infty$

TABLE 3

x	$f(x) = \dfrac{1}{x}$
1	1
10	0.1
100	0.01
1,000	0.001
10,000	0.0001

TABLE 4

x	$\dfrac{1}{x}$
-1	-1
-10	-0.1
-100	-0.01
$-1,000$	-0.001
$-10,000$	-0.0001

Calculate $\lim_{x \to 0} 1/x$.

SOLUTION: If $x > 0$, then as x gets close to 0, $1/x$ grows without bound in the positive direction. But if $x < 0$, then as x gets close to 0, $1/x$ grows without bound in the negative direction (see Figure 2). Since the behavior of $1/x$ depends on the way in which x approaches 0 (i.e., from the right or from the left), we must conclude that $1/x$ *has no limit* as $x \to 0$, or equivalently, that $\lim_{x \to 0} 1/x$ *does not exist*.

However, if we extend the definition of one-sided limits (see page 62) to infinite limits, we easily see from Figure 2 that

$$\lim_{x \to 0^+} \frac{1}{x} = \infty \qquad \text{and} \qquad \lim_{x \to 0^-} \frac{1}{x} = -\infty.$$

To this point we have considered limits as $x \to x_0$, where x_0 is a real number. But in many important applications it is necessary to determine what happens to $f(x)$ as x becomes very large. For example, we may ask what happens to the function $f(x) = 1/x$ as x becomes large. To illustrate, we again make use of a table (Table 3). It is evident that as x gets large, $1/x$ approaches zero. We then write

$$\lim_{x \to \infty} \frac{1}{x} = 0.$$

Similarly, if x grows large in the negative direction, then $1/x$ approaches 0 (see Table 4); that is,

$$\lim_{x \to -\infty} \frac{1}{x} = 0.$$

Look again at Figure 2, which illustrates the behavior of $1/x$ as $x \to \pm\infty$. These examples suggest the following definition.

DEFINITION **LIMITS AT INFINITY**

(i) The **limit as x approaches infinity** of $f(x)$ is L, written

$$\lim_{x \to \infty} f(x) = L, \tag{3}$$

if $f(x)$ is defined for all large values of x and if $f(x)$ gets close to L as x increases without bound.

(ii) The **limit as x approaches minus infinity** of $f(x)$ is L, written

$$\lim_{x \to -\infty} f(x) = L, \tag{4}$$

if $f(x)$ is defined for all values of x that are large in the negative direction and $f(x)$ gets close to L as x increases without bound in the negative direction. ∎

We emphasize that the definitions in this section, like the ones in Section 1.2, are *not* precise mathematical statements but are only intended to appeal to your intuition. Precise mathematical definitions will appear in Section 1.8.

The limit theorems of Section 1.3 apply in the same way when $x \to \infty$ or $x \to -\infty$ as they do when $x \to x_0$, where x_0 is a finite number.

EXAMPLE 4
A LIMIT AT INFINITY

Calculate $\lim_{x \to \infty} 1/x^2$.

SOLUTION: We can calculate this limit in one of three ways. First, we could construct a table of values, as in Table 5. It seems from this table that

TABLE 5

x	x^2	$\dfrac{1}{x^2}$
1	1	1
10	100	0.01
100	10,000	0.0001
1,000	1,000,000	0.000001

$$\lim_{x \to \infty} \frac{1}{x^2} = 0.$$

Second, we could use the corollary to Theorem 1.3.2. Since $\lim_{x \to \infty} 1/x = 0$, we have

$$\lim_{x \to \infty} \frac{1}{x^2} = \lim_{x \to \infty} \left(\frac{1}{x}\right)^2 = \left(\lim_{x \to \infty} \frac{1}{x}\right)^2 = 0^2 = 0.$$

Third, we note that for $x > 1$, $x^2 > x$, so that $0 < 1/x^2 < 1/x$. Then since $1/x \to 0$ as $x \to \infty$, $1/x^2$, which is between 0 and $1/x$, must also approach 0 as $x \to \infty$. This situation is illustrated in Figure 3.

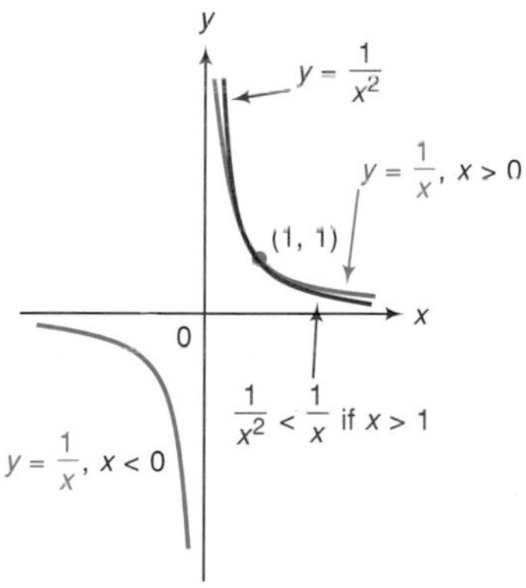

FIGURE 3
GRAPHS OF $y = \dfrac{1}{x}$ AND $y = \dfrac{1}{x^2}$

The third method used above is an example of the squeezing theorem, whose proof is given in Appendix 3.

THEOREM 1 **SQUEEZING THEOREM**

Let x_0 be a real number, ∞, or $-\infty$.

(i) Suppose the following:
    (a) $f(x) \le g(x) \le h(x)$ for all x near x_0.
    (b) $\lim_{x \to x_0} f(x) = \lim_{x \to x_0} h(x) = L$.
    Then,

$$\lim_{x \to x_0} g(x) = L.$$

(ii) Suppose the following:
    (a) $f(x) \le g(x)$ for all x near x_0.
    (b) $\lim_{x \to x_0} f(x) = \infty$.
    Then,

$$\lim_{x \to x_0} g(x) = \infty. \quad \blacksquare$$

NOTE: If $x_0 = \infty$, then the phrase "for all x near x_0" means "for all very large x."

This theorem states that if $g(x)$ is "squeezed" between $f(x)$ and $h(x)$ near x_0, and if $f(x)$ and $h(x)$ have the same limit L, then $g(x)$ must also have the limit L. In Example 4,

$$f(x) = 0, \qquad g(x) = \frac{1}{x^2}, \qquad \text{and} \qquad h(x) = \frac{1}{x}.$$

If $x > 1$, then $x^2 > x$ and $0 < 1/x^2 < 1/x$, so

$$f(x) \le g(x) \le h(x).$$

But $\lim_{x \to \infty} f(x) = \lim_{x \to \infty} 0 = 0$, and $\lim_{x \to \infty} h(x) = \lim_{x \to \infty} 1/x = 0$. Thus $L = 0$, and from the squeezing theorem

$$\lim_{x \to \infty} \frac{1}{x^2} = 0.$$

EXAMPLE 5
A LIMIT AT INFINITY

Calculate $\lim_{x \to \infty} 1/(x^3 + 8)$.

SOLUTION: For $x > 0$, $x^3 + 8 > x$, and therefore $0 < 1/(x^3 + 8) < 1/x$. Then by the squeezing theorem, since $\lim_{x \to \infty} 1/x = 0$ and $\lim_{x \to \infty} 0 = 0$, we have

$$\lim_{x \to \infty} \frac{1}{x^3 + 8} = 0.$$

EXAMPLE 6
A LIMIT AT INFINITY OF A
RATIONAL FUNCTION

Calculate

$$\lim_{x \to \infty} \frac{2x^2 - 2x + 3}{x^2 + 4x + 4}. \tag{5}$$

SOLUTION: There is an easy way to find this limit, but first we will use a calculator to see if we can find a pattern. From Table 6 we might be led to the conclusion that the limit is 2. To calculate this limit more easily, we could try to use the limit theorem for quotients [Theorem 1.3.2 (iv)]. But

$$\lim_{x \to \infty} 2x^2 - 2x + 3 = \infty, \qquad \lim_{x \to \infty} x^2 + 4x + 4 = \infty,$$

and ∞/∞ is undefined.

TABLE 6

x	$2x^2 - 2x + 3$	$x^2 + 4x + 4$	$\dfrac{2x^2 - 2x + 3}{x^2 + 4x + 4}$
1	3	9	0.33333
10	183	144	1.27083
100	19,803	10,404	1.90340
1,000	1,998,003	1,004,004	1.99003
10,000	199,980,003	100,040,004	1.99900

Fortunately, we may simplify this problem by dividing the numerator and denominator of (5) by x^2. We have

$$\lim_{x \to \infty} \frac{2x^2 - 2x + 3}{x^2 + 4x + 4} = \lim_{x \to \infty} \frac{(2x^2 - 2x + 3)/x^2}{(x^2 + 4x + 4)/x^2}$$

$$= \lim_{x \to \infty} \frac{\dfrac{2x^2}{x^2} - \dfrac{2x}{x^2} + \dfrac{3}{x^2}}{\dfrac{x^2}{x^2} + \dfrac{4x}{x^2} + \dfrac{4}{x^2}} = \lim_{x \to \infty} \frac{2 - \dfrac{2}{x} + \dfrac{3}{x^2}}{1 + \dfrac{4}{x} + \dfrac{4}{x^2}}.$$

But $2/x$, $3/x^2$, $4/x$, and $4/x^2$ all approach 0 as $x \to \infty$. Therefore,

$$\lim_{x \to \infty} \frac{2 - \dfrac{2}{x} + \dfrac{3}{x^2}}{1 + \dfrac{4}{x} + \dfrac{4}{x^2}} = \frac{2 - 0 + 0}{1 + 0 + 0} = 2.$$

NOTE: We made use of several limit theorems here. Can you name which ones?

In general, the limits as $x \to \pm\infty$ of rational expressions, like (5), can be found by first dividing both numerator and denominator by the highest power of x that appears in the denominator, and then calculating the limit as $x \to \infty$ (or $-\infty$) of both numerator and denominator.

EXAMPLE 7
A LIMIT AT INFINITY OF A
RATIONAL FUNCTION

Calculate $\lim_{x \to \infty}(x^3 + 3x + 6)/(x^5 + 2x^2 + 9)$.

SOLUTION: Divide by x^5:

$$\lim_{x \to \infty} \frac{x^3 + 3x + 6}{x^5 + 2x^2 + 9} = \frac{\lim_{x \to \infty}\left(\dfrac{1}{x^2} + \dfrac{3}{x^4} + \dfrac{6}{x^5}\right)}{\lim_{x \to \infty}\left(1 + \dfrac{2}{x^3} + \dfrac{9}{x^5}\right)} = \frac{0 + 0 + 0}{1 + 0 + 0} = 0.$$

EXAMPLE 8
A LIMIT AT INFINITY OF A
RATIONAL FUNCTION

Calculate $\lim_{x \to \infty}(x^3 + 2x + 3)/(5x^2 + 1)$.

SOLUTION: Divide by x^2:

$$\lim_{x \to \infty} \frac{x^3 + 2x + 3}{5x^2 + 1} = \lim_{x \to \infty} \frac{x + \dfrac{2}{x} + \dfrac{3}{x^2}}{5 + \dfrac{1}{x^2}}.$$

Here the numerator approaches ∞ and the denominator approaches 5. Thus

$$\lim_{x \to \infty} \frac{x^3 + 2x + 3}{5x^2 + 1} = \infty.$$

> **LIMITS OF RATIONAL FUNCTIONS**
>
> If $r(x) = p(x)/q(x)$ and the degree of the polynomial $p(x)$ is greater than the degree of $q(x)$, then $\lim_{x \to \infty} p(x)/q(x) = +\infty$ or $-\infty$.
>
> If the degree of $q(x)$ is greater than the degree of $p(x)$, then $\lim_{x \to \infty} p(x)/q(x) = 0$.
>
> Finally, if the degree of $p(x)$ is equal to the degree of $q(x)$, then $\lim_{x \to \infty} p(x)/q(x) = c_n/d_n$, where c_n is the coefficient of the highest power of x in $p(x)$ and d_n is the coefficient of the highest power of x in $q(x)$. (See Problem 40.)

EXAMPLE 9
FINDING LIMITS FROM A GRAPH

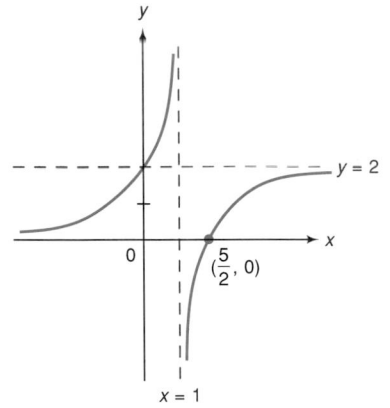

FIGURE 4
GRAPH OF $y = f(x)$

The graph of a function f is given in Figure 4. From the graph, determine

(a) $\lim_{x \to 1^+} f(x)$ (b) $\lim_{x \to 1^-} f(x)$ (c) $\lim_{x \to \infty} f(x)$

(d) $\lim_{x \to -\infty} f(x)$ (e) $\lim_{x \to 0} f(x)$ (f) $\lim_{x \to 5/2} f(x)$

SOLUTION:

(a) As x gets close to 1 from the right, $f(x)$ appears to decrease without bound. Thus

$$\lim_{x \to 1^+} f(x) = -\infty.$$

(b) Similarly,

$$\lim_{x \to 1^-} f(x) = \infty.$$

(c) As x grows, $f(x)$ gets closer and closer to the line $y = 2$. So

$$\lim_{x \to \infty} f(x) = 2.$$

(d) As x grows in the negative direction, $f(x)$ approaches the x-axis (the line $y = 0$). Therefore

$$\lim_{x \to -\infty} f(x) = 0.$$

(e), (f) From the graph it appears that

$$\lim_{x \to 0} f(x) = f(0) = 2 \quad \text{and} \quad \lim_{x \to 5/2} f(x) = f\left(\frac{5}{2}\right) = 0.$$

EXAMPLE 10
APPROXIMATING A LIMIT AT
INFINITY USING A CALCULATOR

Use a calculator to estimate $\lim_{x \to \infty} \left(1 + \dfrac{1}{x}\right)^x$.

SOLUTION: Values for $\left(1 + \dfrac{1}{x}\right)^x$ are given for ten integers x in Table 7.

It appears that $\left(1 + \dfrac{1}{x}\right)^x$ approaches a finite number between 2.7 and

TABLE 7 VALUES FOR $\left(1 + \dfrac{1}{x}\right)^x$

x	$\dfrac{1}{x}$	$1 + \dfrac{1}{x}$	$\left(1 + \dfrac{1}{x}\right)^x$
1	1	2	2
2	0.5	1.5	2.25
5	0.2	1.2	2.48832
10	0.1	1.1	2.59374246
100	0.01	1.01	2.704813829
1000	0.001	1.001	2.716923932
10,000	0.0001	1.0001	2.718145927
100,000	0.00001	1.00001	2.718268237
1,000,000	0.000001	1.000001	2.718280469
10,000,000,000	10^{-10}	$1 + 10^{-10}$	2.718281828

2.8. This number is denoted by e and is very important in calculus. We will discuss e in great detail in Section 6.3. To ten significant figures, $e = 2.718281828$. It is *not* a repeating decimal.

■ PROBLEMS 1.4

■ SELF-QUIZ

I. This section contains a discussion of the fact that $\lim_{x \to 0} 1/x^2 = \infty$. This problem asks you to do a few calculations to support those claims.
   a. How close to 0 should x be in order to guarantee that $1/x^2 > 1,000,000$?
   b. How small in absolute value must x be chosen in order that we know that $1/x^2 > 10^7$?
   c. State a condition relating x and 0 which will imply that $1/x^2$ is larger than 10^8.

II. This problem asks you to discover a few facts which can be generalized into a proof that

$$\lim_{x \to \infty} \frac{1}{\sqrt{x}} = 0.$$

   a. How large must x be in order that $1/\sqrt{x} < 0.01$?
   b. How large should x be to guarantee $1/\sqrt{x} < 0.001$?
   c. Find a large number N such that if $x > N$, then $1/\sqrt{x} < 0.0001$.

III. True–False:
   a. If $\lim_{x \to 2} f(x) = \infty$, then $\lim_{x \to 2}[1/f(x)] = 0$.
   b. If $\lim_{x \to 2} f(x) = 0$, then $\lim_{x \to 2}[1/f(x)] = \infty$.

■ DRILL

In Problems 1–26, use the limit theorems to help you calculate the requested limits where they exist and to help you identify cases where the limits do not exist.

1. $\lim_{x \to 0} 1/x^4$
2. $\lim_{x \to 0} 1/x^5$
3. $\lim_{x \to 5} 1/(x - 5)^3$
4. $\lim_{x \to 5} 1/(x - 5)^2$
5. $\lim_{x \to \pi} \pi/(x - \pi)^6$
6. $\lim_{x \to -3}(x - 4)/(x + 3)^2$
7. $\lim_{x \to 0}(x + x^2)/(x^2 + x^3)$
8. $\lim_{x \to 0}(1 - \sqrt{x})/x$
9. $\lim_{x \to -\infty} x/(1 + x)$
10. $\lim_{x \to \infty} 1/(1 - \sqrt{x})$
11. $\lim_{x \to \infty} 2x/(3x^3 + 4)$
12. $\lim_{x \to -\infty}(2x + 3)/(3x + 2)$

CONTINUED

13. $\lim_{x\to\infty}(5x - x^2)/(3x + x^2)$

14. $\lim_{x\to\infty}(1 + \sqrt{x})/(1 - \sqrt{x})$

15. $\lim_{x\to\infty}\dfrac{2x^2 + 3x + 5}{3x^2 - x + 2}$

16. $\lim_{x\to\infty}(4x^4 + 1)/(1 + 5x^4)$

17. $\lim_{x\to-\infty}\dfrac{x^5 - 3x + 4}{7x^6 + 8x^4 + 2}$

18. $\lim_{x\to-\infty}(x^8 - 2x^5 + 3)/(5x^4 + 3x + 1)$

19. $\lim_{x\to\infty}\dfrac{3x^{5/3} + 2\sqrt{x} - 3}{7x^{5/3} - 3x + 6}$

20. $\lim_{x\to\infty}(7x^{1/7} - 1)/(4x^{1/7} - x^{1/9})$

21. $\lim_{x\to1^-}1/(x^2 - 1)$

22. $\lim_{x\to1^+}1/(x^2 - 1)$

23. $\lim_{x\to1}1/(x^2 - 1)$

24. $\lim_{x\to\infty}(\sqrt{x+1})/(\sqrt{x+3})$

*25. $\lim_{x\to\infty}(x - \sqrt{x^2 + 2x})$

*26. $\lim_{x\to\infty}(\sqrt{x+1} - \sqrt{x})$

▥ CALCULATOR PROBLEMS

27. Let $f(x) = (3x^5 - 3x^3 + 17x + 2)/(x^5 + 4x^2 + 16)$.
   a. Show that $\lim_{x\to-\infty}f(x) = 3$.
   b. For $x < -2$, how small must x be (i.e., far away from zero in the negative direction) in order to guarantee that $f(x)$ is between 2.999 and 3.001?
   c. Explain why we had to stipulate $x < -2$ in part (b).

In Problems 28–31, (a) approximate the given limit and (b) give the limit in terms of $e = \lim_{x\to\infty}\left(1 + \dfrac{1}{x}\right)^x$.

28. $\lim_{x\to\infty}\left(1 + \dfrac{1}{x}\right)^{2x}$

29. $\lim_{x\to\infty}\left(1 - \dfrac{1}{x}\right)^x$

30. $\lim_{x\to\infty}\left(1 + \dfrac{1}{2x}\right)^x$

31. $\lim_{x\to\infty}(1 + x)^{1/x}$

Problems 32–35 involve the function e^x. We will discuss this function in Sections 6.2 and 6.4. In the meantime, you can get e^x on your calculator in one of two ways (depending, of course, on the calculator): (i) Use the $\boxed{e^x}$ button; (ii) enter x and then press $\boxed{\text{INV}}$ $\boxed{\ln}$ or $\boxed{\text{2}^{nd}\text{ F}}$ $\boxed{\ln}$ or $\boxed{\text{Shift}}$ $\boxed{\ln}$.[†] For example $\boxed{2}$ $\boxed{\text{2}^{nd}\text{ F}}$ $\boxed{\ln}$ yields 7.389056099 ($= e^2$) and entering -0.5 followed by $\boxed{\text{INV}}$ $\boxed{\ln}$ yields 0.606530659 ($= e^{-1/2}$). Now use your calculator to guess at each limit.

32. $\lim_{x\to\infty}xe^{-x}$

33. $\lim_{x\to\infty}\dfrac{e^{(1/10)x}}{x^4}$

34. $\lim_{x\to\infty}x^5e^{-(1/2)x}$

35. $\lim_{x\to-\infty}x^2e^x$

■ SHOW/PROVE/DISPROVE

36. Work Problem 1.3.30 on page 74 by applying the squeezing theorem.

37. Show that $\lim_{x\to0}-1/(x + x^2)$ does not exist.

38. Choose a constant c. Show that
$$\lim_{x\to\infty}(\sqrt{x(x + c)} - x) = \frac{c}{2}.$$

39. Prove or disprove: For any function f whose domain is $(-\infty, \infty)$, either
$$\lim_{x\to\infty}f(x) = \lim_{t\to0}f(1/t)$$
or both of those limits do not exist.

*40. Consider the rational function $p(x)/q(x)$ where
$$p(x) = c_0 + c_1x + \cdots + c_mx^m, \quad c_m \neq 0;$$
$$q(x) = d_0 + d_1x + \cdots + d_nx^n, \quad d_n \neq 0.$$
   a. Show that if $m < n$, then $\lim_{x\to\infty}p(x)/q(x) = 0$. [Hint: Divide numerator and denominator by x^n.]
   b. Show that if $m > n$, then $\lim_{x\to\infty}p(x)/q(x) = \pm\infty$. [That means $\lim_{x\to\infty}|p(x)/q(x)| = \infty$.]
   c. Show that if $m = n$, then $\lim_{x\to\infty}p(x)/q(x) = c_m/d_n$.

41. Using the results of Problems 32–35, give a reasonable explanation (not a "proof") that $\lim_{x\to\infty}p(x)e^{-x} = 0$, where $p(x)$ is a polynomial.

[†] On some calculators the $\boxed{\ln}$ key or the $\boxed{e^x}$ key is pressed first.

■ ANSWERS TO SELF-QUIZ

I. a. $0 < |x| < \frac{1}{1000} = 10^{-3}$
   b. $0 < |x| < 1/\sqrt{10^7} = 10^{-3.5}$
   c. $0 < |x| < 0.0001 = 10^{-4}$

II. a. $x > (1/0.01)^2 = 10^4$

b. $x > (1/0.001)^3 = 10^6$
c. Pick $N \geq 10^8 = (1/0.0001)^2$.

III. a. True
   b. False; for example, let $f(x) = x - 2$. Then $\lim_{x \to 2}[1/f(x)]$ does not exist.

1.5 TANGENT LINES AND DERIVATIVES

In this section we introduce a fundamental idea in calculus, that of the derivative. We begin by describing some of the geometric properties of tangent lines. Recall from Section 0.3 (see Figure 1 on p. 16) that if the straight line $y = mx + b$ has a positive slope ($m > 0$), then y is an *increasing* function of x, and as a point moves along the line from left to right, it also moves *upward*. If the line has a negative slope ($m < 0$), then y is a *decreasing* function of x, and as a point moves along the line from left to right, it also moves *downward*.[†]

The central concept in the study of calculus is the concept of a *derivative*. Intuitively, the derivative of a function f at the value x_0 is the slope of the line tangent to the graph of f at the point $(x_0, f(x_0))$. We use the notation $f'(x_0)$ to denote the derivative of f at x_0.

We can make this idea clearer with a picture. Consider the graph of the function f drawn in Figure 1. For each point on this graph there is a unique tangent line. Each of these tangent lines has a slope. At the point $(x_2, f(x_2))$ in Figure 1, for example, the slope of the tangent line is negative, so the derivative of f at x_2, denoted by $f'(x_2)$, is negative. Similarly, the slope of the tangent line at $(x_4, f(x_4))$ is positive, so $f'(x_4) > 0$. Since f has a unique tangent line at each point, f has a derivative at each point.

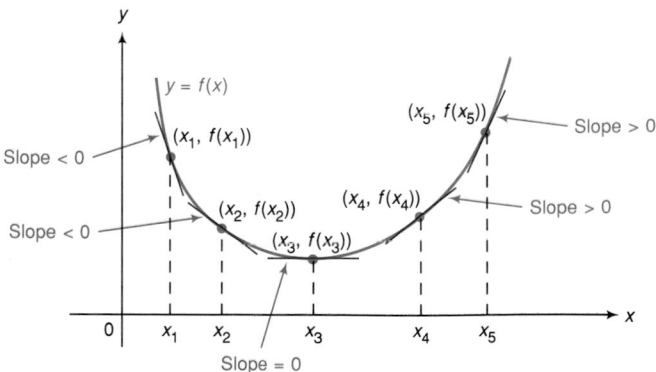

FIGURE 1   GRAPH OF $f(x)$ SHOWING SEVERAL TANGENT LINES

But what exactly is a tangent line? If you were asked to give a definition, you might say something like "it's a line that touches a curve at exactly one point and does not cross (intersect) the curve." According to the Chambers 20th Century Dictionary the first meaning of the word "tangent" is "touching

---

[†] We shall formally define increasing and decreasing functions in Section 3.2. Here we rely on your intuition.

without intersecting" (used as an adjective). However, once we have given a formal mathematical definition of a tangent line, we will see that a tangent line may not only intersect a curve at the point of tangency, it may also intersect the curve at more than one point. Three possibilities are illustrated in Figure 2.

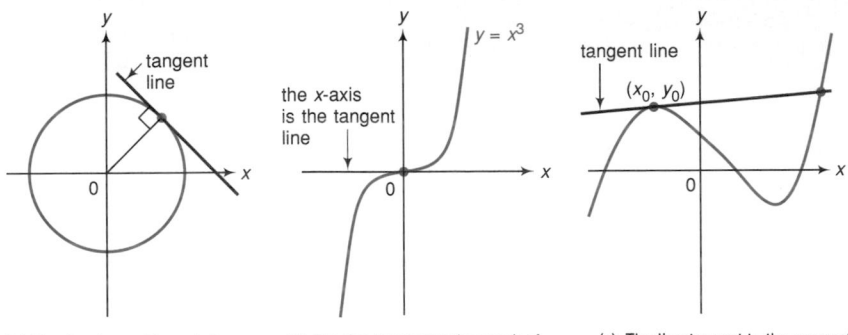

(a) The line tangent to a circle at a point touches the circle at only that one point and does not cross the circle.

(b) The line tangent to the graph of $f(x) = x^3$ at 0 crosses the curve at the point of tangency.

(c) The line tangent to the curve at (x_0, y_0) touches the curve at two points.

FIGURE 2   THREE POSSIBILITIES FOR TANGENT LINES

We need to define precisely what we mean by a tangent line. To do so, we begin with our intuitive sense of what a tangent line is, and derive a method for calculating its slope.

The method we give below is essentially the method of Newton[†] and Leibniz[‡] that resolved the tangent problem posed so long ago by the Greek mathematicians. Actually, this method was first used by the great French mathematician Pierre de Fermat (see the biographical sketch on p. 93) in his *Method of Finding Maxima and Minima* published in 1629. However, Fermat did not explain his procedure satisfactorily. It was left to Newton and Leibniz to explain the method and apply it to the calculation of tangent lines.

Let us consider the function $y = f(x)$, a part of whose graph is given in Figure 3. We calculate the slope of the line tangent to the curve at each

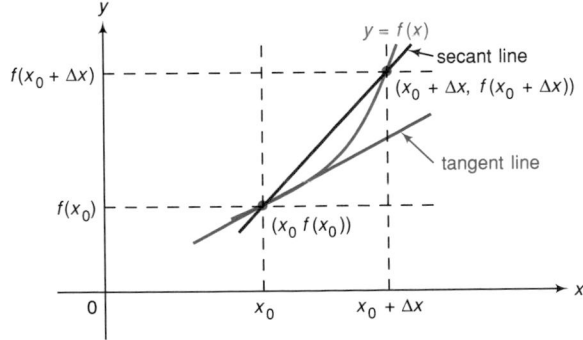

FIGURE 3   THE SECANT LINE JOINING TWO POINTS ON A CURVE AND THE TANGENT LINE AT THE POINT $(x_0, f(x_0))$

[†] *Mathematical Principles of Natural Philosophy* (*Principia*), published in 1687.

[‡] *A New Method for Maxima and Minima, and Also for Tangents, Which is not Obstructed by Irrational Quantities*, published in 1684.

point of the curve at which there is a unique tangent line. Let $(x_0, f(x_0))$ be such a point. From now on we will assume that f is defined near x_0.

If Δx is a small number (positive or negative), then $x_0 + \Delta x$ will be close to x_0. In moving from x_0 to $x_0 + \Delta x$, the values of f will move from $f(x_0)$ to $f(x_0 + \Delta x)$. Now look at the straight line in Figure 3, called a **secant line**, joining the points $(x_0, f(x_0))$ and $(x_0 + \Delta x, f(x_0 + \Delta x))$. What is its slope? If we define $\Delta y = f(x_0 + \Delta x) - f(x_0)$ and if we use m_S to denote the slope of such a secant line, we have, from Section 0.3,

$$m_S = \frac{\text{change in } y}{\text{change in } x} = \frac{f(x_0 + \Delta x) - f(x_0)}{(x_0 + \Delta x) - x_0}$$

$$= \frac{f(x_0 + \Delta x) - f(x_0)}{\Delta x} = \frac{\Delta y}{\Delta x}. \tag{1}$$

What does this equation have to do with the slope of the tangent line? The answer is suggested in Figure 4, where we see that as Δx gets closer and closer to zero, the secant line gets closer and closer to the tangent line. Put another way, as Δx approaches zero, the slope of the secant line approaches the slope of the tangent line. Thus

$$m_T = \text{slope of tangent line} = \lim_{\Delta x \to 0} m_S = \lim_{\Delta x \to 0} \frac{\Delta y}{\Delta x}$$

$$= \lim_{\Delta x \to 0} \frac{f(x_0 + \Delta x) - f(x_0)}{\Delta x} \tag{2}$$

This enables us to give a precise definition of the tangent line to a curve.

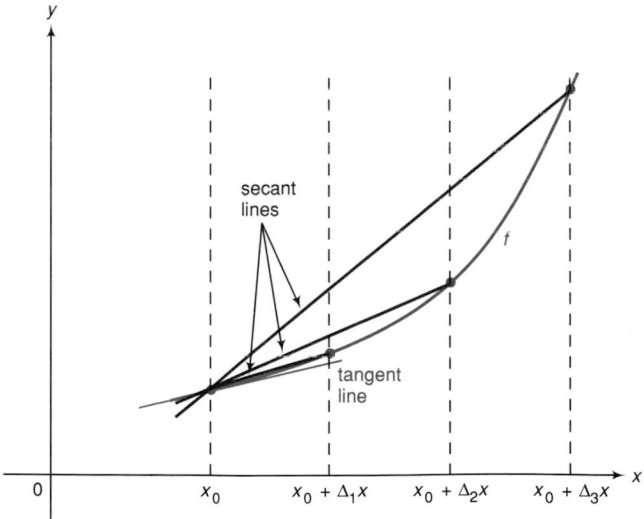

FIGURE 4   SECANT LINES APPROACH THE TANGENT LINE AS $\Delta x \to 0$

DEFINITION   **TANGENT LINE**

If the limit in (2) exists and is finite, we say that the graph of the function f has a **tangent line** at the point $(x_0, f(x_0))$. The tangent line is the line passing through the point $(x_0, f(x_0))$ with slope $f'(x_0)$.

One equation of this line is

$$y - f(x_0) = m_T(x - x_0).\quad\blacksquare$$

REMARK: According to this definition, a graph *cannot* have more than one tangent line at the point $(x_0, f(x_0))$ since this would imply that the limit in (2) had more than one value, thereby contradicting the fact that if the limit in (2) exists, it must be unique.

EXAMPLE 1
TANGENT LINE TO A
STRAIGHT-LINE FUNCTION

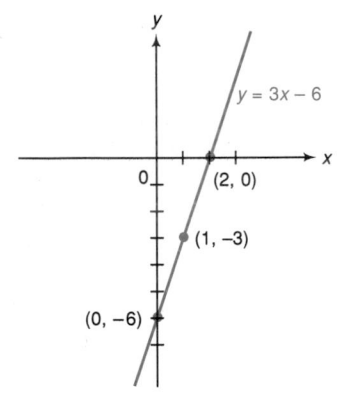

FIGURE 5
THE TANGENT LINE TO A STRAIGHT LINE IS
THE LINE ITSELF.

The slope of the line $y = 3x - 6$ is 3.

Find the slope of the tangent line to the graph of $y = 3x - 6$ at the point $(1, -3)$.

SOLUTION: The graph of the function $f(x) = 3x - 6$ is a straight line, so the tangent line at every point on the graph is the line itself. This line has slope 3. We use formula (2) to compute the slope, not because it is necessary, but because it is nice to see that this new formula will give us the correct answer in the one case we already know how to handle:

$$m_T = \lim_{\Delta x \to 0} \frac{f(x_0 + \Delta x) - f(x_0)}{\Delta x} \overset{x_0 = 1}{=} \lim_{\Delta x \to 0} \frac{[3(1 + \Delta x) - 6] - [3 \cdot 1 - 6]}{\Delta x}$$

$$= \lim_{\Delta x \to 0} \frac{3 + 3\Delta x - 6 - (-3)}{\Delta x} = \lim_{\Delta x \to 0} \frac{3\Delta x}{\Delta x} = \lim_{\Delta x \to 0} 3 = 3,$$

as expected.

EXAMPLE 2
TANGENT LINE TO A CUBIC
FUNCTION

Find the slope of the tangent line to the graph of $y = x^3$ at the point $(0, 0)$.

SOLUTION: Now we do not know the answer in advance. Using (2), we have

$$m_T = \lim_{\Delta x \to 0} \frac{(x_0 + \Delta x)^3 - x_0{}^3}{\Delta x} \overset{x_0 = 0}{=} \lim_{\Delta x \to 0} \frac{(0 + \Delta x)^3 - 0^3}{\Delta x}$$

$$= \lim_{\Delta x \to 0} \frac{(\Delta x)^3}{\Delta x} = \lim_{\Delta x \to 0} (\Delta x)^2 = 0.$$

Thus the tangent line to the graph of $y = x^3$ has slope 0 and so is horizontal at $(0, 0)$. Since it passes through the origin, the tangent line is the x-axis. This is illustrated in Figure 2(b).

We return to Figure 1.

Remember the definition of a function. It is a rule that assigns a unique real number to every number in its domain. For the function graphed in Figure 1 we have assumed that there is a unique tangent line at every point.

Thus we have a new function f', called the **derivative** of f, that assigns to each number x_0 a new number $f'(x_0)$. We emphasize that

$$f'(x_0) = \text{slope of tangent line to the graph of } f \text{ at the point } (x_0, f(x_0)).$$

The remainder of this chapter and most of the next one will be concerned with the calculation of derivatives.

We now give a formal definition of the derivative.

DEFINITION   **DERIVATIVE AT A POINT**

Let f be defined on an open interval containing the point x_0, and suppose that

$$\lim_{\Delta x \to 0} \frac{f(x_0 + \Delta x) - f(x_0)}{\Delta x}$$

exists and is a real number (i.e., it is finite). Then f is said to be **differentiable** at x_0 and the **derivative** of f at x_0, denoted by $f'(x_0)$, is given by

$$f'(x_0) = \lim_{\Delta x \to 0} \frac{\Delta y}{\Delta x} = \lim_{\Delta x \to 0} \frac{f(x_0 + \Delta x) - f(x_0)}{\Delta x}. \quad \blacksquare \qquad (3)$$

In (3), let $t = x_0 + \Delta x$. Then $t - x_0 = \Delta x$, and $\Delta x \to 0$ is equivalent to $t \to x_0$. Hence (3) can be written as follows:

DEFINITION   **ALTERNATIVE DEFINITION OF THE DERIVATIVE**

Let f be defined in an open interval containing x_0 and suppose that

$$f'(x_0) = \lim_{t \to x_0} \frac{f(t) - f(x_0)}{t - x_0} \qquad (3')$$

exists and is a real number. Then f is said to be **differentiable** at x_0 and the **derivative** of f at x_0 is given by the limit in (3'). $\quad \blacksquare$

This alternative form for the derivative at x_0 is sometimes more convenient to use in computations.

This definition tells us what we mean by the derivative of a function at a point. We next define the derivative function.

DEFINITION   **THE DERIVATIVE FUNCTION**

The **derivative** f' of the function f is the **function** defined as follows:

(i) $\text{dom } f' = \left\{ x: \lim_{\Delta x \to 0} \dfrac{f(x + \Delta x) - f(x)}{\Delta x} \text{ exists and is a real number} \right\}.$

(ii) For every x in $\text{dom } f'$

$$f'(x) = \lim_{\Delta x \to 0} \frac{f(x + \Delta x) - f(x)}{\Delta x}. \qquad (4)$$

That is, f' is the function, defined at every x for which the limit in (4) exists and is finite, that assigns to every x in its domain the derivative $f'(x)$. $\quad \blacksquare$

REMARK: It follows from the definition of the derivative at a point x that $f'(x)$ can exist only if $f(x)$ is defined. Thus

$$\text{dom } f' \text{ is contained in dom } f. \tag{5}$$

We emphasize that the derivative of a function is another function. The value of the derivative at a given number x_0 is the limit obtained in (4).

Note that these definitions do not say anything about tangent lines. Simply put, f is differentiable at x if the limit in (4) exists.

EXAMPLE 3
COMPUTING THE DERIVATIVE OF A STRAIGHT-LINE FUNCTION

Let $y = f(x) = mx + b$. Calculate $f'(x)$.

SOLUTION: To solve this problem and the ones that follow, we simply use formula (4). For $f(x) = mx + b$,

$$f(x + \Delta x) = m(x + \Delta x) + b.$$

Then

$$f'(x) = \lim_{\Delta x \to 0} \frac{f(x + \Delta x) - f(x)}{\Delta x} = \lim_{\Delta x \to 0} \frac{[m(x + \Delta x) + b] - (mx + b)}{\Delta x}$$

$$= \lim_{\Delta x \to 0} \frac{mx + m\,\Delta x + b - mx - b}{\Delta x} = \lim_{\Delta x \to 0} \frac{m\,\Delta x}{\Delta x} = \lim_{\Delta x \to 0} m = m.$$

This answer is not surprising. It simply says that the slope of the line $y = mx + b$ is equal to the constant function $f'(x) = m$.

Before giving further examples, we introduce the additional symbols dy/dx and df/dx to denote the derivative for $y = f(x)$.

The symbol dy/dx is read, "**the derivative of y with respect to x.**" We emphasize that dy/dx *is not a fraction*. At this point the symbols dy and dx have no meaning of their own. (We will define these symbols in Section 4.7.)

There are other notations for the derivative. We will often use the symbol y' or $y'(x)$ in place of f' or $f'(x)$. Thus for $y = f(x)$, we may denote the derivative in seven different ways[†]:

SEVEN WAYS TO WRITE A DERIVATIVE

$$y'(x) = f'(x) = \frac{df}{dx} = \frac{dy}{dx} = \lim_{\Delta x \to 0} \frac{\Delta y}{\Delta x} = \lim_{\Delta x \to 0} \frac{f(x + \Delta x) - f(x)}{\Delta x}$$

$$= \lim_{t \to x} \frac{f(t) - f(x)}{t - x}.$$

EXAMPLE 4
COMPUTING THE DERIVATIVE OF A QUADRATIC FUNCTION

Calculate the derivative of the function $y = x^2$. What is the equation of the line tangent to the graph of $y = x^2$ at the point $(3, 9)$?

[†] Newton (in England) and Leibniz (in Germany) independently discovered (in the 1670s) the equation for the slope of the tangent line. Newton used the symbol $\dot{y}$ (read "y dot") and Leibniz used the symbol dy/dx to indicate the derivative.

SOLUTION: For $y = f(x) = x^2$, $f(x + \Delta x) = (x + \Delta x)^2$. Then

$$\frac{dy}{dx} = \lim_{\Delta x \to 0} \frac{f(x + \Delta x) - f(x)}{\Delta x} = \lim_{\Delta x \to 0} \frac{(x + \Delta x)^2 - x^2}{\Delta x}$$

$$= \lim_{\Delta x \to 0} \frac{x^2 + 2x\,\Delta x + \Delta x^2 - x^2}{\Delta x} = \lim_{\Delta x \to 0} \frac{2x\,\Delta x + (\Delta x)^2}{\Delta x}$$

$$= \lim_{\Delta x \to 0} \frac{\Delta x(2x + \Delta x)}{\Delta x} = \lim_{\Delta x \to 0} (2x + \Delta x) = 2x.$$

At every point of the form $(x, f(x)) = (x, x^2)$, the slope of the line tangent to the curve is $2x$. For $x = 3$, $2x = 6$. Therefore the slope of the tangent line at the point $(3, 9)$ is 6; that is, $f'(3) = 6$.

We can now find the equation of the tangent line since it passes through the point $(3, 9)$ and has the slope 6. We have

$$\frac{y - 9}{x - 3} = 6 \quad \text{or} \quad y = 6x - 9.$$

using point slope

Knowing the derivative of $y = x^2$ helps us to graph the curve. First, we note that x^2 is always positive. If $x < 0$, then $y' = dy/dx = 2x < 0$, so the tangent lines have negative slopes. For $x = 0$, $dy/dx = 2 \cdot 0 = 0$, so the tangent line is horizontal. For $x > 0$, $dy/dx = 2x > 0$, and the tangent lines have positive slopes. Using this information, we obtain the graph of Figure 6.

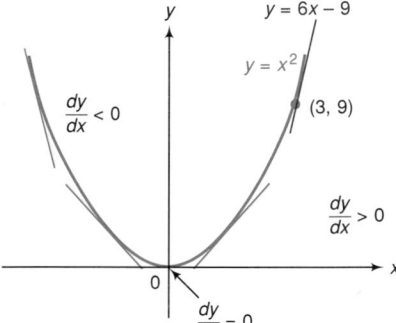

FIGURE 6
TANGENT LINES TO THE GRAPH OF $y = x^2$

At $(3, 9)$ the slope is 6.

EXAMPLE 5
COMPUTING THE DERIVATIVE
OF A SQUARE-ROOT FUNCTION

Find the derivative of $y = \sqrt{x}$, and calculate the slope of the tangent line at the point $(4, 2)$.

SOLUTION: We have

$$f'(x) = \lim_{\Delta x \to 0} \frac{f(x + \Delta x) - f(x)}{\Delta x} = \lim_{\Delta x \to 0} \frac{\sqrt{x + \Delta x} - \sqrt{x}}{\Delta x}$$

$$= \lim_{\Delta x \to 0} \frac{(\sqrt{x + \Delta x} - \sqrt{x})(\sqrt{x + \Delta x} + \sqrt{x})}{\Delta x(\sqrt{x + \Delta x} + \sqrt{x})}$$

We multiplied numerator and denominator by $\sqrt{x + \Delta x} + \sqrt{x}$.

$$= \lim_{\Delta x \to 0} \frac{(\sqrt{x + \Delta x})^2 - (\sqrt{x})^2}{\Delta x(\sqrt{x + \Delta x} + \sqrt{x})}$$

Since $(a - b)(a + b) = a^2 - b^2$.
Here $a = \sqrt{x + \Delta x}$ and $b = \sqrt{x}$.

$$= \lim_{\Delta x \to 0} \frac{(x + \Delta x) - x}{\Delta x(\sqrt{x + \Delta x} + \sqrt{x})}$$

$$= \lim_{\Delta x \to 0} \frac{\Delta x}{\Delta x(\sqrt{x + \Delta x} + \sqrt{x})}$$

$$= \lim_{\Delta x \to 0} \frac{1}{\sqrt{x + \Delta x} + \sqrt{x}} = \frac{1}{\sqrt{x} + \sqrt{x}} = \frac{1}{2\sqrt{x}}. \tag{6}$$

Thus

$$f'(x) = \frac{d}{dx}\sqrt{x} = \frac{1}{2\sqrt{x}}.$$

and

$$f'(4) = \frac{1}{2\sqrt{4}} = \frac{1}{2 \cdot 2} = \frac{1}{4} = \text{slope of tangent line at } (4, 2).$$

The equation of the tangent line is given by

$$\frac{y - 2}{x - 4} = \frac{1}{4} \qquad \text{or} \qquad y = \frac{1}{4}x + 1.$$

Recall that in Example 1.2.4 we showed, numerically, that

$$\lim_{h \to 0} \frac{\sqrt{4 + h} - \sqrt{4}}{h} = \frac{1}{4}.$$

Note that although the function $f(x) = \sqrt{x}$ is defined for $x \geq 0$, its derivative $1/(2\sqrt{x})$ is only defined for $x > 0$, since $1/(2\sqrt{0})$ is undefined.

We can again make use of the derivative to graph the curve. First note that $\sqrt{x} > 0$ if $x > 0$ and $\sqrt{x}$ is not defined if $x < 0$. Look at the derivative $f'(x) = 1/(2\sqrt{x})$. Since $\sqrt{x} > 0$, $1/(2\sqrt{x}) > 0$, so all tangents to the curve have a positive slope and the function is increasing. But we also have

$$\lim_{x \to \infty} f'(x) = \lim_{x \to \infty} \frac{1}{2\sqrt{x}} = 0.$$

This equation tells us that as x increases, the derivative approaches 0, or equivalently, the tangent lines approach the horizontal (a slope of zero). Thus we conclude that for large values of x the curve $y = \sqrt{x}$ is nearly flat. On the other hand, since $\lim_{x \to 0^+}[1/2\sqrt{x})] = \infty$, the tangent lines to the curve near $x = 0$ are nearly vertical.[†] We combine all this information in Figure 7.

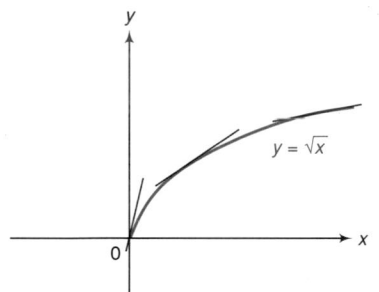

FIGURE 7
GRAPH OF $y = \sqrt{x}$

The tangent lines become steeper and steeper as x approaches 0.

EXAMPLE 6
COMPUTING THE DERIVATIVE OF
A CUBIC FUNCTION

Use the alternative definition of the derivative (on p. 87) to compute the derivative of $y = x^3$.

SOLUTION: We need the algebraic formula

$$a^3 - b^3 = (a - b)(a^2 + ab + b^2).$$

Then, using formula (3′) with $f(x) = x^3$,

$$\frac{dy}{dx} = \lim_{t \to x} \frac{f(t) - f(x)}{t - x} = \lim_{t \to x} \frac{t^3 - x^3}{t - x} = \lim_{t \to x} \frac{(t - x)(t^2 + tx + x^2)}{(t - x)}$$

$$= \lim_{t \to x}(t^2 + tx + x^2) = x^2 + x \cdot x + x^2 = 3x^2.$$

Thus the derivative of x^3 is $3x^2$. In particular, $f'(0) = 0$, as we found in Example 2. Note that $f'(x) = 3x^2 > 0$ if $x \neq 0$, which indicates that all tangent lines to the curve (except the tangent line at 0) have positive slopes. The graph of $f(x) = x^3$ is given in Figure 2(b).

[†] Lines with very large slopes are nearly vertical.

EXAMPLE 7
THE DERIVATIVE OF THE
ABSOLUTE-VALUE FUNCTION
DOES NOT EXIST AT $x = 0$

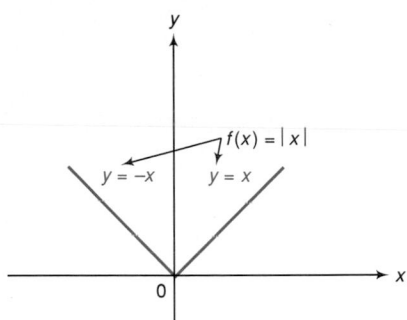

FIGURE 8
GRAPH OF $y = |x|$

There is no tangent line at the origin.

Consider the function $y = |x|$. Since

$$|x| = \begin{cases} x, & x \geq 0 \\ -x, & x \leq 0, \end{cases}$$

we obtain the graph in Figure 8. To see if the graph of f has a tangent line at the point $(0, 0)$, we calculate

$$f'(0) = \lim_{\Delta x \to 0} \frac{f(0 + \Delta x) - f(0)}{\Delta x} = \lim_{\Delta x \to 0} \frac{|0 + \Delta x| - |0|}{\Delta x} = \lim_{\Delta x \to 0} \frac{|\Delta x|}{\Delta x}.$$

But as we saw in Example 1.2.6, on page 58, this limit does not exist, so $|x|$ does not have a tangent line at $(0, 0)$. On the other hand, if $x \neq 0$, then the derivative does exist (see Problem 46).

EXAMPLE 8
THE DERIVATIVE OF A FUNCTION
MAY NOT EXIST AT A POINT
AT WHICH IT IS DEFINED

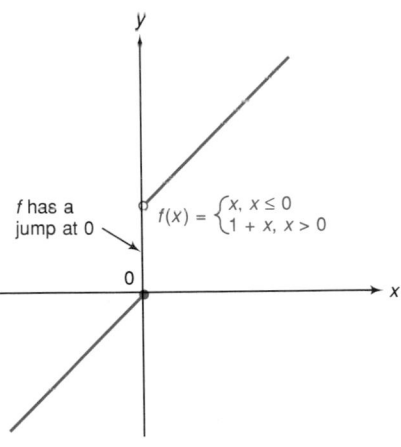

FIGURE 9
GRAPH OF $f(x) = \begin{cases} x, & x \leq 0 \\ 1 + x, & x > 0 \end{cases}$

The derivative does not exist at the origin even though $f(0)$ exists and is equal to 0.

Let

$$f(x) = \begin{cases} x, & x \leq 0 \\ 1 + x, & x > 0. \end{cases}$$

Compute $f'(0)$, if it exists.

SOLUTION: The graph of f is given in Figure 9. We have

$$f'(0) = \lim_{\Delta x \to 0} \frac{f(0 + \Delta x) - f(0)}{\Delta x} \overset{\text{Since } f(0) = 0}{=} \lim_{\Delta x \to 0} \frac{f(\Delta x)}{\Delta x}.$$

Now

$$f(\Delta x) = \begin{cases} \Delta x, & \Delta x \leq 0 \\ 1 + \Delta x, & \Delta x > 0. \end{cases}$$

So

$$\frac{f(\Delta x)}{\Delta x} = \begin{cases} 1, & \Delta x \leq 0 \\ \dfrac{1 + \Delta x}{\Delta x}, & \Delta x > 0. \end{cases}$$

But

$$\frac{1 + \Delta x}{\Delta x} = 1 + \frac{1}{\Delta x} \quad \text{and} \quad \lim_{\Delta x \to 0^+} \frac{1}{\Delta x} = \infty.$$

Thus,

$$\lim_{\Delta x \to 0^+} \frac{f(\Delta x)}{\Delta x} = \infty$$

and $f'(0)$ does not exist. [Remember, f is differentiable at x_0 if the limit in (3) exists and is *finite*.]

Almost every function we encounter in this book will be differentiable at every point in its domain. However, there are three commonly encountered situations in which a function will fail to be differentiable at a point.

CONDITIONS FOR NONDIFFERENTIABILITY

A function f is not differentiable at a point $x_0 \in \text{dom } f$ if one of the following situations holds:

(i) f has a vertical tangent at x_0 (see Figure 7; $\sqrt{x}$ has a vertical tangent at 0).

(ii) The graph of f comes to a point at x_0 (see Figure 8).

(iii) The graph of f jumps at x_0 (see Figure 9).

EXAMPLE 9
APPROXIMATING A DERIVATIVE

Use a calculator to approximate the derivative of $\sin x$ at $x = \dfrac{\pi}{6}$.

SOLUTION:

$$\frac{d}{dx} \sin x = \lim_{\Delta x \to 0} \frac{\sin(x + \Delta x) - \sin x}{\Delta x} \overset{x = \frac{\pi}{6}}{=} \lim_{\Delta x \to 0} \frac{\sin\left(\frac{\pi}{6} + \Delta x\right) - \sin \frac{\pi}{6}}{\Delta x}$$

$$\overset{\sin\left(\frac{\pi}{6}\right) = \frac{1}{2}}{=} \lim_{\Delta x \to 0} \frac{\sin\left(\frac{\pi}{6} + \Delta x\right) - 0.5}{\Delta x}$$

$$\approx \lim_{\Delta x \to 0} \frac{\sin(0.523598775 + \Delta x) - 0.5}{\Delta x}$$

In Table 1 we compute the last quotient for smaller and smaller positive values of Δx, with the calculator set to radian mode.

TABLE 1   APPROXIMATING THE DERIVATIVE OF $\sin x$ AT $x = \dfrac{\pi}{6}$

Δx	$0.523598775 + \Delta x$	$\sin(0.523598775 + \Delta x)$	$\dfrac{\sin(0.523598775 + \Delta x) - 0.5}{\Delta x}$
0.1	0.623598775	0.583960357	0.839603571
0.01	0.533598775	0.508635109	0.863510939
0.001	0.524598775	0.500865774	0.86577474
0.0001	0.523698775	0.500086599	0.86599523

In Section 2.5 we will show that if $f(x) = \sin x$, then $f'(x) = \cos x$. Thus $f'\left(\dfrac{\pi}{6}\right) = \cos \dfrac{\pi}{6} = \dfrac{\sqrt{3}}{2} \approx 0.866025$. Note that our final (blue) approximation is accurate to within 0.00003.

PIERRE DE FERMAT WAS BORN IN the French city of Toulouse in approximately 1601. (There were few accurate birth records in the early seventeenth century.) His father was a leather merchant, and what early education Fermat received, he received at home. When Fermat was 30 years old he began working for the Toulouse parliament, first as a lawyer and then as a councillor. Although these jobs were not easy, he still found time to study mathematics on his own. Fermat did have a great deal of correspondence with many of the leading mathematicians of seventeenth-century Europe, and perhaps this compensated for his lack of formal mathematical training.

Called the "prince of amateurs" in mathematics, Fermat made contributions to many branches of mathematics, including analytic geometry, infinitesimal analysis, and number theory. One of his discoveries was the formula for finding tangents to the curve, $y = x^n$. As we shall see in Section 2.4, the slope of such a tangent line is nx^{n-1}. For this and other results, Fermat is considered by many to be the discoverer of differential calculus, although his work did not approach the generality of Newton or Leibniz.

Fermat's most important mathematical work was in number theory. One of the major works of Greek mathematics was the *Arithmetica* of Diophantus of Alexandria. In 1621 this work was reintroduced to European mathematicians through the new edition of Claude Gaspard de Bachet (1591–1639). The *Arithmetica* discussed many topics in the theory of numbers, including divisibility, magic squares, and prime numbers. Fermat soon became fascinated with prime numbers and conjectured and sometimes proved facts about them. He conjectured, for example, that all numbers of the form $2^{2^n} + 1$, now called *Fermat numbers*, are always prime. Euler, about a hundred years later, showed this conjecture to be false since $2^{2^5} + 1$ is *not* prime ($2^{2^5} + 1 = 2^{32} + 1 = 4,294,967,297 = 641 \times 6,700,417$). Many of Fermat's other conjectures, however, are now known to be true.

Fermat is best known for his conjecture:

There are no positive integers a, b, c, and n such that $a^n + b^n = c^n$, where $n > 2$.

This conjecture is known as "Fermat's last theorem." Fermat wrote in the margin of his copy of Bachet's *Arithmetica*, "To divide a cube into two cubes, a fourth power, or in general any power whatever into two powers of the same denomination above the second is impossible, and I have assuredly found an admirable proof of this, but the margin is too narrow to contain it." Whether or not Fermat really had a proof of his conjecture is debatable, but as stated above, many of his other assertions have now been proved. In any case, Fermat's marginal remark has frustrated mathematicians for more than three centuries, as no one has been able either to prove the conjecture or to find a counterexample.

Before World War I, the German mathematician Paul Wolfskehl offered a prize of 100,000 marks for the first proof of Fermat's conjecture. Many professional and amateur mathematicians sought the prize in vain. More recently, the conjecture has been verified for large values of n on a computer. But still no general proof has been found.

To this day, mathematics departments around the country (and the world) receive unsolicited "proofs." Some disappointed writers of incorrect "proofs" have accused the mathematical community of a conspiracy to suppress the truth (although the reasons for this alleged conspiracy are not clear). In any event, there have been more incorrect "proofs" published for Fermat's last theorem than for any other mathematical conjecture.

Despite his many contributions to mathematics, Fermat published very little. Instead, he communicated many of his results to his friend Marin (Mersenne (1588–1648). As a result, he did not receive credit for many of his discoveries.

Before leaving this section, we give one more definition that we will use often in the rest of the book.

> **DEFINITION**   **DIFFERENTIABILITY ON AN OPEN INTERVAL**
>
> The function f is **differentiable on the open interval** (a, b) if $f'(x)$ exists for every x in (a, b). ∎

EXAMPLE 10
DIFFERENTIABILITY ON AN
OPEN INTERVAL

From Example 5 we see that $f(x) = \sqrt{x}$ is differentiable on any interval of the form $(0, b)$, where $b > 0$.

EXAMPLE 11
DIFFERENTIABILITY ON $\mathbb{R}$

From Example 4 we see that $f(x) = x^2$ is differentiable on $(-\infty, \infty)$.

■ PROBLEMS 1.5

■ SELF-QUIZ

In Problems I–VIII, the graph of a function is given and several points are marked. At each point, decide whether the function is differentiable there, and if so, state whether the derivative appears to be negative, positive, or zero.

I.

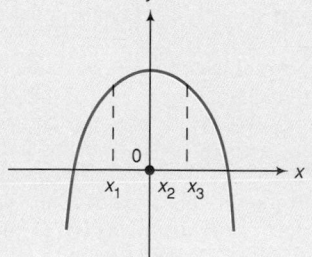

II.

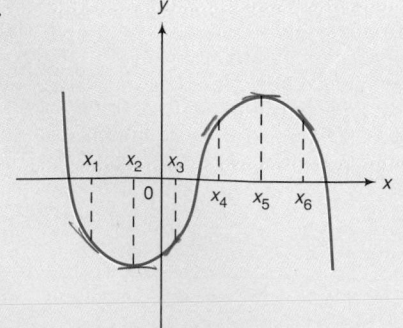

III.

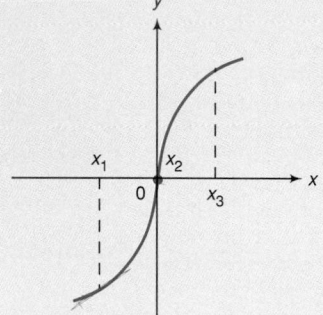

IV.

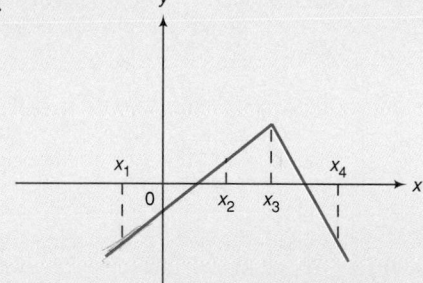

V.

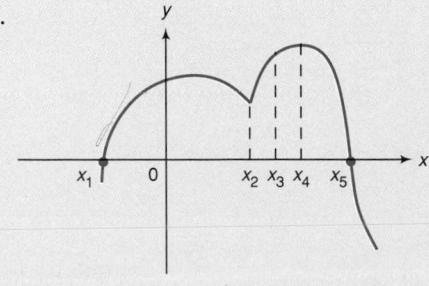

VI.

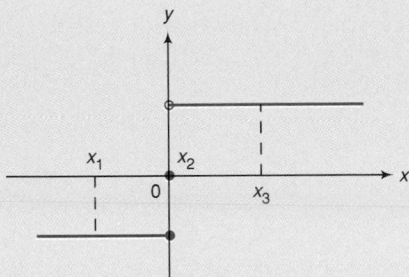

VII.

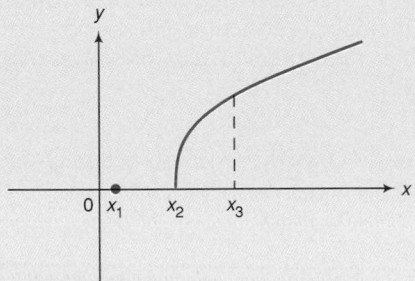

VIII.

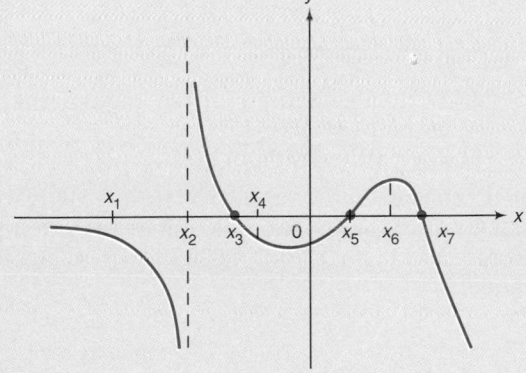

IX. Sketch the graphs of three functions, each of whose derivative function always has the value zero. In what way(s) are your three graphs similar?

X. Sketch the graphs of three functions each of whose derivative function always has value -0.75. In what way(s) are your three graphs similar?

XI. If f is a differentiable function and if $f'(2) = 5$, then

a. $\dfrac{f(2) - f(1)}{2 - 1} = 5$ b. $\dfrac{f(2) - f(1.9)}{2 - 1.9} = 5$

c. $f(1.9)$ is less than $f(2)$

d. f is increasing in an open interval surrounding $x = 2$

XII. Suppose f is a differentiable function. Consider the line that is tangent to the graph of $y = f(x)$ at the point $(a, f(a))$.

a. Under what conditions will this line intersect the x-axis?

b. In case it does meet the x-axis, locate the point of intersection (that is, compute the x-coordinate of that intersection point).

■ DRILL

In Problems 1–14, compute the derivative of the given function and find an equation for the line tangent to the curve at the specified point.

1. $f(x) = 4$; $(3, 4)$
2. $f(x) = x - 2$; $(7, 5)$
3. $f(x) = -4x + 6$; $(3, -6)$
4. $g(x) = 15(x - 3) + 40$; $(3, 40)$
5. $f(x) = 2(x - 3)^2$; $(4, 2)$
6. $g(x) = 4x^2 - 9$; $(-1, -5)$
7. $f(x) = -x^2 + 3x + 5$; $(0, 5)$
8. $g(x) = -x^2 + 3x + 5$; $(1, 7)$
9. $f(x) = x^3$; $(2, 8)$
10. $g(x) = x^3 + x^2 + x + 1$; $(1, 4)$
11. $f(x) = -\sqrt{x + 3}$; $(6, -3)$
12. $g(x) = \sqrt{2x}$; $(8, 4)$
13. $f(x) = x^4$; $(2, 16)$
14. $g(x) = 1/\sqrt{x}$; $(4, \frac{1}{2})$
15. Verify directly from the definition that if $y = mx + b$, then $dy/dx = m$.
16. Verify directly, using the definition (3) or (3′), that for any constants a, b, and c,

$$\frac{d}{dx}(ax^2 + bx + c) = 2ax + b.$$

CONTINUED

17. Verify that for any constant a,

$$\frac{d}{dx}(ax^3) = 3ax^2.$$

18. Calculate the derivative with respect to x of $ax^3 + bx^2 + cx + d$, where a, b, c, and d are arbitrary constants.

■ APPLICATIONS

19. Let $f(x) = x$ and $g(x) = x^2$. For what values of x are the tangents to these curves parallel?

20. Let $f(x) = x^2$ and $g(x) = x^3$. For what values of x are the tangents to these curves parallel?

21. Let $f(x) = ax^2$ and $g(x) = 6x - 5$. Consider the tangent to each curve at the point where that curve crosses the vertical line $x = 2$. For what value of a are these two tangents parallel?

22. Let $f(x) = 1/\sqrt{x}$. For what value(s) of x is the tangent to this curve parallel to the line $x + 8y = 10$?

23. Let $f(x) = x[-x^2]$, where $[\cdots]$ denotes the greatest integer function. Compute $f'(0)$. [*Note:* You will obtain a wrong answer if you simplify the difference quotient $[f(0 + \Delta x) - f(0)]/\Delta x$ and then take the shortcut of setting Δx equal to zero.]

24. Let $f(x) = 1/(1 + |x|)$.
   a. Find $f'(x)$ at those points where f is differentiable.
   b. Sketch the graph of $y = f(x)$.

▦ CALCULATOR PROBLEMS

25. Consider the function $f(x) = -3x^2$, especially near the point $(2, -12)$.
   a. Calculate $f(2 + \Delta x)$ for $\Delta x = -0.5$, $\Delta x = 0.1$, $\Delta x = -0.01$, and $\Delta x = 0.001$.
   b. Calculate $[f(2 + \Delta x) - f(2)]/\Delta x$ for the values of Δx in part (a) and guess the value of $f'(2)$.
   c. From the definition, equation (3) or (3′), compute a general formula for $f'(x)$; use that to compute $f'(2)$; compare this result with your guess in part (b).
   d. Write an equation for the line tangent to the graph of $f(x)$ at $(2, -12)$.
   e. Sketch the graph of $f(x)$, include information gleaned from considering $f'(x)$.

26. Consider the function $g(x) = 2/x$, especially near the point $(1, 2)$.
   a. Calculate $g(1 + \Delta x)$ for $\Delta x = 0.5$, $\Delta x = -0.1$, $\Delta x = 0.01$, and $\Delta x = -0.001$.
   b. Calculate $[g(1 + \Delta x) - g(1)]/\Delta x$ for the values of Δx in part (a) and guess the value of $g'(1)$.
   c. From (3) or (3′), compute a general formula for $g'(x)$; use that to compute $g'(1)$; compare this result with your guess in part (b).

   d. Write an equation for the line tangent to the graph of $g(x)$ at $(1, 2)$.
   e. Sketch the graph of $g(x)$.

In Problems 27–30, approximate the derivative of the given trigonometric function at the given point. Make certain that your calculator is set to radian mode.

27. $\cos x$ at $x = \dfrac{\pi}{4}$

28. $\sin x$ at $x = \dfrac{2\pi}{3}$

29. $\tan x$ at $x = \dfrac{\pi}{4}$

30. $\sin 5x$ at $x = \dfrac{\pi}{5}$

In Problems 31–34, use the $\ln x$ button on your calculator to approximate the derivative of $\ln x$ at the given value of x.

31. $x = 1$

32. $x = 4$

33. $x = 10$

34. $x = 100$

In Problems 35–38, approximate the derivative of e^x at each value. See page 82 for a description of how to get e^x on your calculator.

35. $x = 0$

36. $x = 1$

37. $x = -1$

38. $x = 2$

CONTINUED

■ SHOW/PROVE/DISPROVE

39. Let $f(x) = x \cdot |x|$. Show that $f'(x) = 2|x|$ for all x.

*40. a. Suppose that $|f(x)| \leq x^2$ for all x. Show that f is differentiable at 0.
   b. Suppose that $g(0) = 0$ and $|g(x)| \geq \sqrt{|x|}$ for all x. Show that g cannot be differentiable at 0.

Even and odd functions are defined on page 42.

41. a. Show that the function $f(x) = x^n$ is an even function if n is an even integer and it is an odd function if n is an odd integer.
   b. Show that $g(x) = |x|$ is an even function.
   c. Show that $h(x) = x \cdot |x|$ is an odd function.

*42. Show that every function with domain $(-\infty, \infty)$ can be written as the sum of an even function and an odd function.

43. Suppose that f is an even function which is differentiable everywhere. Show that f' is an odd function.

44. Suppose that g is an odd function which is differentiable everywhere. Prove or disprove that g' is an even function.

A function F is a **periodic** function if and only if there is a positive number w such that $F(x + w) = F(x)$ for all x. [The smallest such w is called the **period** of F.]

45. Suppose that f is a periodic function which is differentiable everywhere. Show that f' is also periodic (with the same period).

46. Show that $\dfrac{d}{dx}|x| = \begin{cases} 1, & x > 0 \\ -1, & x < 0. \end{cases}$

*47. Using the results of Problems 31–34, and any other computations you deem appropriate, make an educated guess at a formula for the derivative of $\ln x$ for $x > 0$.

*48. Using the results of Problems 35–38, make an educated guess at a formula for the derivative of e^x.

■ CHALLENGE

**49. Consider

$$\lim_{\Delta x \to 0^+} \frac{f(x + \Delta x) - f(x - \Delta x)}{2\,\Delta x}$$

as an alternate definition of $f'(x)$. Is this alternative definition equivalent to the definition of a derivative at a point? Prove equivalence or produce explicit counterexamples.

■ ANSWERS TO SELF-QUIZ

I. $f'(x_1) > 0$, $f'(x_2) = 0$, $f'(x_3) < 0$

II. $f'(x_1) < 0$, $f'(x_2) = 0$, $f'(x_3) > 0$, $f'(x_4) > 0$, $f'(x_5) = 0$, $f'(x_6) < 0$

III. $f'(x_1) > 0$, $f'(x_2)$ is undefined, $f'(x_3) > 0$

IV. $f'(x_1) = f'(x_2) > 0$, $f'(x_3)$ is undefined, $f'(x_4) < 0$

V. $f'(x_1) > 0$, $f'(x_2)$ is undefined, $f'(x_3) > 0$, $f'(x_4) = 0$, $f'(x_5) < 0$

VI. $f'(x_1) = f'(x_3) = 0$, $f'(x_2)$ is undefined

VII. $f'(x_1)$ and $f'(x_2)$ are undefined, $f'(x_3) > 0$

VIII. $f'(x_1) < 0$, $f'(x_2)$ is undefined, $f'(x_3) < 0$, $f'(x_4) < 0$, $f'(x_5) > 0$, $f'(x_6) = 0$, $f'(x_7) < 0$

IX. Graph three constant functions; their graphs are horizontal lines.

X. Graph three parallel lines, each with slope equal to $-\frac{3}{4}$.

XI. d

XII. a. $f'(a) \neq 0$ b. $a - [f(a)/f'(a)]$

1.6 THE DERIVATIVE AS A RATE OF CHANGE

We now show how the derivative can be used to compute instantaneous velocity and other rates of change. To begin the discussion, we again consider the line given by the equation

$$y = mx + b, \tag{1}$$

where m is the slope and b is the y-intercept. Let us examine the slope more closely. It is defined as the change in y divided by the change in x:

$$m = \frac{\Delta y}{\Delta x}. \tag{2}$$

Implicit in this definition is the understanding that no matter which two points are chosen on the line, we obtain the same value for this ratio. Thus the slope of a straight line could instead be referred to as *the rate of change of y with respect to x*. It tells us how many units y changes for every one unit that x changes. In fact, instead of following Euclid in defining a straight line as the shortest distance between two points, we could define a straight line as a curve whose rate of change is constant.

EXAMPLE 1
THE MEANING OF THE SLOPE OF
A STRAIGHT LINE

Let $y = 3x - 5$. Then in moving from the point $(1, -2)$ to the point $(2, 1)$ along the line, we see that as x has changed (increased) one unit, y has increased three units, corresponding to the slope $m = 3$. See Figure 1.

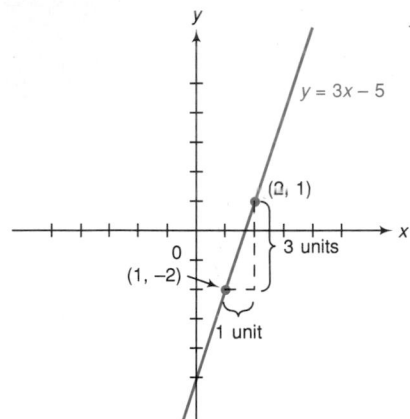

FIGURE 1
GRAPH OF $f(x) = 3x - 5$

The slope of 3 means that y increases 3 units for every 1 unit increase in x.

We would like to be able to calculate rates of change for functions that are not straight-line functions. Suppose that an object is dropped from rest from a given height. The distance s the object has dropped after t seconds (ignoring air resistance) is given by the formula

$$s = \frac{1}{2} gt^2, \tag{3}$$

where $g \approx 9.8$ m/sec$^2 \approx 32$ ft/sec^2 is the acceleration due to gravity. We now ask: What is the velocity of the object after 2 seconds? To answer this question, we first note that velocity is a rate of change. Whether measured in meters per second, feet per second, or miles per hour, velocity is the ratio of change in distance (meters, feet, miles) to the change in time (seconds, hours).

TABLE 1

t	t^2	$s = \frac{1}{2}gt^2 = 4.9t^2$ (m)	Δs = Change in s Since Last Measurement (m)	Average Velocity Over Last 0.2 sec, $\Delta s/\Delta t$ (m/sec)
0.	0.	0.	—	—
0.2	0.04	0.196	0.196	0.98
0.4	0.16	0.784	0.588	2.94
0.6	0.36	1.764	0.98	4.9
0.8	0.64	3.136	1.372	6.86
1.0	1.0	4.9	1.764	8.82
1.2	1.44	7.056	2.156	10.78
1.4	1.96	9.604	2.548	12.74
1.6	2.56	12.544	2.94	14.7
1.8	3.24	15.876	3.332	16.66
2.0	4.0	19.6	3.724	18.62
2.2	4.84	23.716	4.116	20.58
2.4	5.76	28.224	4.508	22.54
2.6	6.76	33.124	4.9	24.5
2.8	7.84	38.416	5.292	26.46
3.0	9.0	44.1	5.684	28.42

Table 1 gives (in column 3) the distance the object has fallen (in meters) after t seconds, tabulated every 0.2 second. Column 4 shows the distance the object has fallen in the previously elapsed 0.2 second. Column 5 indicates the average velocity over each 0.2-second interval of time. We have

Average Velocity

$$\text{average velocity} = \frac{\text{distance fallen}}{\text{elapsed time}} = \frac{\text{change in distance}}{\text{change in time}} = \frac{\Delta s}{\Delta t} \qquad (4)$$

where in this problem $\Delta t = 0.2$ sec. From the table, we see that the average velocity of the object is constantly increasing (which is certainly not surprising). What is the exact velocity at $t = 2$? That is, instead of an average velocity we want to know the velocity after precisely 2 seconds. Table 1 will help answer this question. When $t = 2$, the object is falling faster than at any previous time. Let $v(t)$ denote the velocity at time t. Since 18.62 m/sec is the *average* velocity for the time $t = 1.8$ sec to $t = 2$ sec, we have

$$(\text{velocity at } t = 2) = v(2) > 18.62 \text{ m/sec.} \qquad (5)$$

Similarly, at $t = 2$, the object is falling more slowly than when $t > 2$. Therefore,

$$v(2) < 20.58 \text{ m/sec,} \qquad (6)$$

which is the average velocity between the time $t = 2$ sec and $t = 2.2$ sec. Combining (4) and (5) we obtain the estimate

$$18.62 < v(2) < 20.58 \text{ m/sec.} \qquad (7)$$

To narrow this down, we choose a smaller value for Δt. If $\Delta t = 0.05$ sec, we calculate the average velocity, $\Delta s/\Delta t$, for t between 1.95 and 2.0 sec and for t between 2.0 and 2.05 sec. Then $v(2)$ will lie between these two values (see Table 2). We then have the new estimate

$$19.355 < v(2) < 19.845 \text{ m/sec.} \qquad (8)$$

TABLE 2

t	t^2	$s = 4.9t^2$ (m)	Δs (m)	$\dfrac{\Delta s}{\Delta t} = \dfrac{\Delta s}{0.05}$ (m/sec)
1.95	3.8025	18.63225	—	—
2.0	4.0	19.6	0.96775	19.355
2.05	4.2025	20.59225	0.99225	19.845

We continue this process by choosing an even smaller value for Δt, say $\Delta t = 0.001$. Then we calculate the values in Table 3. Therefore,

$$19.5951 < v(2) < 19.6049 \text{ m/sec.} \qquad (9)$$

TABLE 3

t	t^2	$s = 4.9t^2$ (m)	Δs (m)	$\dfrac{\Delta s}{\Delta t} = \dfrac{\Delta s}{0.001}$ (m/sec)
1.999	3.996001	19.5804049	—	—
2.0	4.0	19.6	0.0195951	19.5951
2.001	4.004001	19.6196049	0.0196049	19.6049

As Δt becomes smaller and smaller, we see that the average velocity $\Delta s/\Delta t$ gets closer and closer to the actual velocity at the value $t = 2$. This suggests the following definition.

DEFINITION **INSTANTANEOUS VELOCITY**

Let $s(t)$ denote the distance traveled by a moving object in t seconds. Then the **instantaneous velocity** after t seconds, denoted by ds/dt or $s'(t)$, is given by

$$s'(t) = \frac{ds}{dt} = \lim_{\Delta t \to 0} \frac{\Delta s}{\Delta t} = \lim_{\Delta t \to 0} \frac{s(t + \Delta t) - s(t)}{\Delta t}. \quad \blacksquare \tag{10}$$

NOTE: The quantity $[s(t + \Delta t) - s(t)]/\Delta t$ is the average velocity of the object between times t and $t + \Delta t$.

In our example, $s(t) = 4.9t^2$ and $s(t + \Delta t) = 4.9(t + \Delta t)^2$. Thus,

$$\frac{ds}{dt} = \lim_{\Delta t \to 0} \frac{4.9(t + \Delta t)^2 - 4.9t^2}{\Delta t} = 4.9 \lim_{\Delta t \to 0} \frac{(t + \Delta t)^2 - t^2}{\Delta t}$$

$$= 4.9 \lim_{\Delta t \to 0} \frac{t^2 + 2t\,\Delta t + (\Delta t)^2 - t^2}{\Delta t} = 4.9 \lim_{\Delta t \to 0} \frac{\Delta t(2t + \Delta t)}{\Delta t}$$

$$= 4.9 \lim_{\Delta t \to 0} (2t + \Delta t) = (4.9)(2t) = 9.8t.$$

After 2 seconds have elapsed, the velocity is $9.8(2) = 19.6$ m/sec, which agrees with our estimates (7), (8), and (9).

As you have probably noticed, the velocity ds/dt given by (10) is the derivative of s with respect to t. Although our previous definitions of derivative involved the variables x and y, there is no change in this concept when we insert t in place of x and s in place of y. Thus we can think of a derivative as a velocity or, more generally, as a rate of change. After Newton discovered the derivative, he used the word *fluxion* instead of velocity in his discussion of a moving object. (In more technical terminology, a moving particle is a particle "in flux.")

Before giving further examples, we prove an algebraic equation that is very helpful when computing certain derivatives.

THEOREM 1 **A USEFUL ALGEBRAIC FORMULA**

Let a and b be real numbers, and let n be a positive integer. Then

$$a^n - b^n = (a - b)(a^{n-1} + a^{n-2}b + a^{n-3}b^2 + \cdots + ab^{n-2} + b^{n-1})$$

$$\tag{11}$$

PROOF:

$$(a - b)(a^{n-1} + a^{n-2}b + a^{n-3}b^2 + \cdots + ab^{n-2} + b^{n-1})$$

$$= a(a^{n-1} + a^{n-2}b + \cdots + ab^{n-2} + b^{n-1})$$

$$- b(a^{n-1} + a^{n-2}b + \cdots + ab^{n-2} + b^{n-1})$$

$$= a^n + a^{n-1}b + \cdots + a^2b^{n-2} + ab^{n-1} - a^{n-1}b - a^{n-2}b^2$$

$$- \cdots - ab^{n-1} - b^n = a^n - b^n$$

as all terms but a^n and $-b^n$ cancel. $\quad \blacksquare$

EXAMPLE 2
VERIFICATION OF THE
FORMULA

You should verify that

$$a^2 - b^2 = (a - b)(a + b) \tag{12}$$

$$a^3 - b^3 = (a - b)(a^2 + ab + b^2) \tag{13}$$

and

$$a^4 - b^4 = (a - b)(a^3 + a^2b + ab^2 + b^3). \tag{14}$$

EXAMPLE 3
COMPUTING INSTANTANEOUS
POPULATION GROWTH

Let $P(t)$ denote the population of a colony of bacteria after t hours. If $P(t) = 100 + t^4$, how fast is the population growing after 3 hours?

SOLUTION: This problem is again a "rate of change" problem, even though it does not make sense to speak about "velocity." We are asking for the "instantaneous rate of growth" of the population when $t = 3$ hr. Using the alternative definition of the derivative [equation (3′) on page 87], we have

$$P'(t) = \frac{dP}{dt} = \lim_{x \to t} \frac{\Delta P}{\Delta t} = \lim_{x \to t} \frac{P(x) - P(t)}{x - t} = \lim_{x \to t} \frac{x^4 - t^4}{x - t}.$$

equation (11) with x and
t replacing a and b
↓
$$= \lim_{x \to t}(x^3 + x^2t + xt^2 + t^3) = t^3 + t^2 \cdot t + t \cdot t^2 + t^3 = 4t^3$$

When $t = 3$ hr, $dP/dt = 4 \cdot 3^3 = 4 \cdot 27 = 108$, and the population is growing at a rate of 108 individuals per hour. (We know that it is growing because $dP/dt > 0$. It would be declining if dP/dt were negative.)

EXAMPLE 4
COMPUTING DENSITY

The mass of a 4-m-long, nonuniform metal beam varies with the distance along the beam measured from the left end (see Figure 2). The mass μ is given by the formula

$$\mu(x) = x^{3/2} \text{ kilograms.}$$

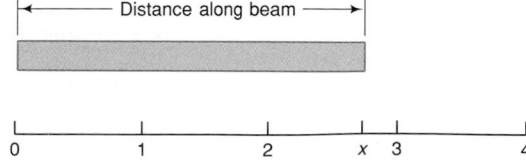

FIGURE 2 DISTANCE ALONG A BEAM

This equation tells us how much of the mass of the beam is contained in that part of the beam from the left end to the point x units from the left end. For example, at the end $x = 4$, $\mu = 4^{3/2} = 8$ kg, so the entire beam has a mass of 8 kg. On the other hand, when $x = 2$, $\mu = 2^{3/2} = 2\sqrt{2} \approx 2.83$ kg, so the left-hand half of the beam carries only approximately 35% of the mass of the beam ($2\sqrt{2}/8 \approx 0.35$). What is the density ρ of the beam at $x = 1$, $x = 2$, $x = 3$, and $x = 4$ m?

SOLUTION: Density = mass per unit of length, which is expressed in kilograms per meter (kg/m). The density is changing as we move from left to right along the beam. Since the right side carries more of the mass, the density *increases* as we move from left to right. Between $x = 0$ and $x = 2$, the "average" density is $(2\sqrt{2} \text{ kg})/(2 \text{ m}) = \sqrt{2}$ kg/m, while along the entire length of the beam (i.e., between $x = 0$ and $x = 4$), the average density is $(8 \text{ kg})/(4 \text{ m}) = 2$ kg/m.

The problem asks for the "instantaneous" density at $x = 1, 2, 3,$ and 4 or, equivalently, for the instantaneous rate of change of mass per unit of length. We therefore have (again using the alternative definition of the derivative and letting $\rho(x)$ denote the density at x),

$$\rho(x) = \frac{d\mu}{dx} = \lim_{t \to x} \frac{\Delta\mu}{\Delta x} = \lim_{t \to x} \frac{\mu(t) - \mu(x)}{t - x} = \lim_{t \to x} \frac{t^{3/2} - x^{3/2}}{t - x}$$

$$= \lim_{t \to x} \frac{(t^{1/2})^3 - (x^{1/2})^3}{(t^{1/2})^2 - (x^{1/2})^2}$$

equations (12) and (13)
$\downarrow$

$$= \lim_{t \to x} \frac{(t^{1/2} - x^{1/2})(t^{2/2} + t^{1/2}x^{1/2} + x^{2/2})}{(t^{1/2} - x^{1/2})(t^{1/2} + x^{1/2})} = \lim_{t \to x} \frac{t + t^{1/2}x^{1/2} + x}{t^{1/2} + x^{1/2}}$$

$$= \frac{x + x^{1/2}x^{1/2} + x}{x^{1/2} + x^{1/2}} = \frac{3x}{2x^{1/2}} = \frac{3}{2}x^{1/2}.$$

We see that the density ρ is given by the formula

$$\rho(x) = \frac{3}{2}\sqrt{x} \text{ kg/m.}$$

Therefore, $\rho(1) = \frac{3}{2} = 1.5$, $\rho(2) = \frac{3}{2}\sqrt{2} = \frac{3}{\sqrt{2}} \approx 2.12$, $\rho(3) = \frac{3\sqrt{3}}{2} \approx 2.6$,

and $\rho(4) = \frac{3\sqrt{4}}{2} = 3$ (all in kilograms per meter). We can clearly see how the density increases as we move from left to right.

The result of Example 4 is worth mentioning again:

The density function is the derivative of the mass function.

EXAMPLE 5
COMPUTING MARGINAL COST

A manufacturer buys large quantities of a certain machine replacement part. She finds that her cost depends on the number of cases bought at the same time, and the cost per unit decreases as the number of cases bought increases. She determines that a reasonable model for this situation is given by the formula

$$C(q) = 100 + 5q - 0.01q^2, \qquad 0 \le q \le 250 \qquad \text{(15)}$$

where q is the number of cases bought (up to 250 cases) and $C(q)$, measured in dollars, is the *total* cost of purchasing q cases. The 100 in (15) is a *fixed* cost, which does not depend on the number of cases bought. What are the incremental and marginal costs for various levels of purchase?

SOLUTION: **Incremental cost** is the cost per additional unit at a given level of purchase. **Marginal cost** is the *rate of change* of the cost with respect to the number of units purchased. These two concepts are different. This difference is like the difference between the average velocity of a falling object over a fixed period of time and the instantaneous velocity of the object at a fixed moment in time. For example, if the manufacturer buys 25 cases, the incremental cost is the cost for one *more* case, that is, the 26th case. This cost is not a constant. We can see this by calculating that 1 case costs $100 + 5 \cdot 1 - 0.01 = \$104.99$ and two cases cost $100 + 5 \cdot 2 - (0.01)4 = \$109.96$. Thus the incremental cost of buying the second case is $C(2) - C(1) = \$4.97$. On the other hand, for 100 cases it costs

$$100 + 5 \cdot 100 - (0.01)(100)^2 = 600 - 100 = \$500,$$

and it costs

$$100 + 5 \cdot 101 - (0.01)(101)^2 = 100 + 505 - (0.01)(10{,}201)$$
$$= 605 - 102.01 = \$502.99$$

to buy 101 cases. The incremental cost is now $C(101) - C(100) = \$2.99$. The 101st is cheaper than the 2nd.

Next, we compute the marginal cost:

$$\text{marginal cost} = \frac{dC}{dq} = \lim_{\Delta q \to 0} \frac{\Delta C}{\Delta q} = \lim_{\Delta q \to 0} \frac{C(q + \Delta q) - C(q)}{\Delta q}$$
$$= \lim_{\Delta q \to 0} \frac{[100 + 5(q + \Delta q) - 0.01(q + \Delta q)^2] - [100 + 5q - 0.01q^2]}{\Delta q}$$
$$= \lim_{\Delta q \to 0} \frac{5\,\Delta q - 0.01(2q\,\Delta q + \Delta q^2)}{\Delta q} = 5 - 0.02q.$$

Thus at $q = 10$ cases, the marginal cost is $5 - 0.2 = \$4.80$, while at $q = 100$ cases, the marginal cost is $5 - 2 = \$3$. This result confirms the manufacturer's statement that "the more she buys, the cheaper it gets."

EXAMPLE 6
THE BIG BANG AND, MAYBE, THE BIG CRUNCH

One of the most exciting topics in modern physics is the origin and future of the universe. Most physicists now agree that our universe began with what is called the "big bang." The big bang refers to a time when the universe was infinitely small and infinitely dense. In the ensuing cataclysmic explosion, gases were hurled uniformly (or almost uniformly) in all directions. Some of the more irregular gas formations cooled and condensed into the stars and planets that form our universe. There are at least three models that attempt to describe what happens after the big bang. In all three the universe is initially uniformly expanding. More precisely, the distance between any two nearby galaxies increases at the same rate as time increases. What happens next?

Model 1 Eventually, the rate of expansion (slope) is so small that gravitational forces cause the universe to cease expanding and to begin to contract. The rate of contraction (negative slope) increases, leading eventually to the "big crunch," which, in a real sense, represents the end of time. This model is depicted in Figure 3.

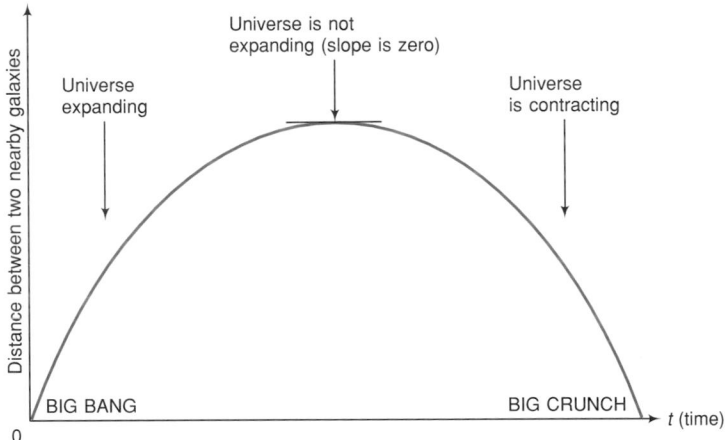

FIGURE 3 EXPANSION FOLLOWED BY CONTRACTION LEADS TO THE BIG CRUNCH

Model 2 The expansion rate is so great that gravity cannot stop it, although it can slow it down a bit. In this model, the rate of expansion approaches a constant. This means that, eventually, the expansion curve resembles a straight line. The slope of the line is the (eventual) constant rate of expansion. This model is illustrated in Figure 4.

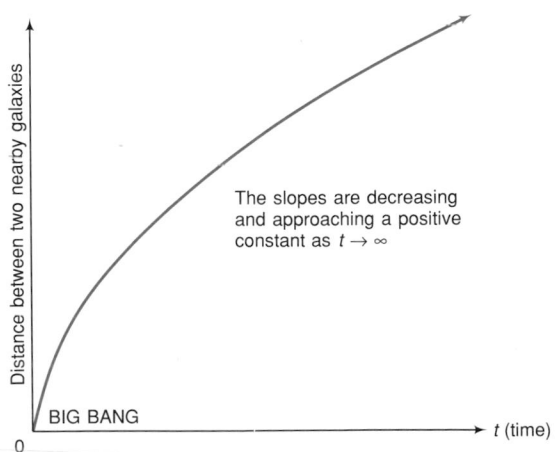

FIGURE 4 RAPID EXPANSION BECOMES CONSTANT EXPANSION, EVENTUALLY

Model 3   The expansion rate is just great enough to avoid contraction. The universe continues to expand but at a rate that approaches zero. That is, the expansion curve approaches a horizontal line (slope of zero). This is illustrated in Figure 5.

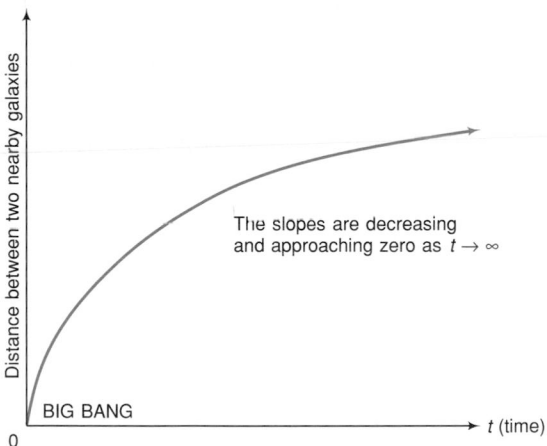

FIGURE 5   THE RATE OF EXPANSION APPROACHES ZERO

For more details on this and many other related topics, read *A Brief History of Time* by Stephen W. Hawking (New York: Bantam, 1988).

HISTORICAL NOTE: SOLUTIONS TO THE TANGENT-LINE PROBLEM IN THE EARLY SEVENTEENTH CENTURY

As we observed in the introduction to this chapter, the problem of finding the tangent line to a curve was one of the four major problems facing seventeenth-century mathematicians. The problem was solved by Newton and Leibniz at the end of the century. However, before these solutions were published, many of the important mathematicians of the time worked on the tangent-line problem and several found solutions—at least in some special cases.

Perhaps the two most brilliant mathematicians in the first half of the seventeenth century were Pierre de Fermat and René Descartes. We discussed them on pages 93 and 14. Both published solutions to the tangent-line problem for general curves. The two methods were ingenious, and very different from one another.

Fermat's method was formulated in the late 1620s and published in 1637.[†] It is closer to the process discovered by Newton and Leibniz, although Fermat did not have any clear idea of the notion of a limit. Here is what he did.

In Figure 6 we have drawn a curve. To make things clearer, we have drawn in the *x*- and *y*-axes that seventeenth-century mathematicians did not

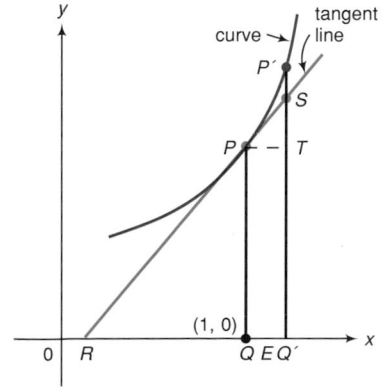

FIGURE 6
PR IS PART OF THE TANGENT LINE TO THE CURVE AT *P*.

We seek the length $\overline{RQ}$.

† In *Methodus ad Disquirendam Maximam et Minimam* (Method of finding maxima and minima), *Oeuvres* (works), 1, 133–79; 3, 121–56.

use until after the notion was developed by Descartes. But keep in mind that axes were not needed for Fermat to make his method work.

We draw the tangent line at P and extend it until it hits the x-axis at R. If we know the length of RQ, denoted by $\overline{RQ}$, then we can determine the x-coordinate of R (the y-coordinate is 0).

Fermat argued as follows: choose a point P' a small distance along the curve from P. The horizontal distance from P to P' is denoted by E (what we call Δx). Using high-school geometry, it is not difficult to show that

triangle PQR is similar to triangle PST.

Then

$$\frac{\overline{RQ}}{\overline{PQ}} = \frac{\overline{PT}}{\overline{ST}} = \frac{E}{\overline{ST}}.$$

Fermat now claimed that since E is small, $\overline{ST} \approx \overline{P'T}$. Thus

$$\frac{\overline{RQ}}{\overline{PQ}} \approx \frac{E}{\overline{P'T}} \qquad \text{or} \qquad \overline{RQ} \approx \frac{E\overline{PQ}}{\overline{P'T}}. \tag{16}$$

Using functional notation (which wasn't available to Fermat), we may write

$$\overline{PQ} = f(Q) \qquad \text{and}$$
$$\overline{P'T} = \overline{P'Q'} - \overline{Q'T} = f(Q') - f(Q) = f(Q + E) - f(Q).$$

Then (16) becomes

$$\overline{RQ} \approx \frac{Ef(Q)}{f(Q + E) - f(Q)} \tag{17}$$

Now, without worrying about the legitimacy of doing so, Fermat set $E = 0$ in (17) to obtain his answer. Let us illustrate the technique with a simple example.

EXAMPLE OF FERMAT'S METHOD

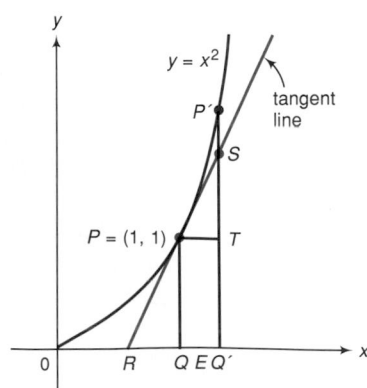

FIGURE 7
GRAPH OF $y = x^2$ SHOWING TANGENT LINE AT THE POINT (1, 1)

Find the slope of the line tangent to the curve $y = x^2$ at the point (1, 1).

SOLUTION: The situation is depicted in Figure 7. We have, from (17),

$$\overline{RQ} \approx \frac{Ef(1)}{f(1 + E) - f(1)} \overset{f(Q) = Q^2}{=} \frac{E}{(1 + E)^2 - 1}$$

$$= \frac{E}{(1 + 2E + E^2) - 1} = \frac{E}{2E + E^2} = \frac{E}{E(2 + E)}$$

$$= \frac{1}{2 + E} \overset{\text{set } E = 0}{=} \frac{1}{2}.$$

Since $\overline{OQ} = 1$ and $\overline{RQ} = \frac{1}{2}$, $\overline{OR} = \frac{1}{2}$ and the coordinates of the point R are $(\frac{1}{2}, 0)$. Then

slope of tangent line = slope of line joining $(\frac{1}{2}, 0)$ and $(1, 1)$

$$= \frac{1 - 0}{1 - \frac{1}{2}} = 2.$$

Of course, we knew this already because, from Example 4 on page 88, the slope of the tangent line is $2x$, which equals 2 when $x = 1$. However, keep in mind that Fermat found this method about half a century before the discoveries of Newton and Leibniz. And his technique certainly influenced the work of the codiscoverers of calculus.

Descartes called the tangent-line problem not only "the most useful and general problem I know but even that I have ever desired to know in geometry."[†] He found, in the 1630s, a geometric method[‡] for finding tangent lines to curves that are not circles. Descartes' method was, roughly, as follows:

(i) Choose a point on the curve.

(ii) Find a circle that is tangent to the curve at the chosen point, so that the circle and curve have the same tangent line at that point.

(iii) Find the slope of the tangent line to the circle (and also to the curve). This is easy to do because, as the Greeks knew, the tangent to a circle at a point is perpendicular to the radial line at that point.

We will illustrate Descartes's method with an example.

EXAMPLE OF DESCARTES'S METHOD

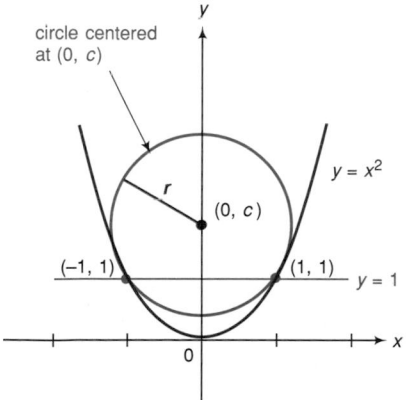

circle centered at (0, c)

$y = x^2$

(0, c)

(−1, 1)   (1, 1)

$y = 1$

0

FIGURE 8
THE CIRCLE CENTERED AT $(0, c)$ INTERSECTS THE PARABOLA $y = x^2$ AT EXACTLY TWO POINTS $(−1, 1)$ AND $(1, 1)$

Find the slope of the line tangent to the curve $y = x^2$ at the point $(1, 1)$.

SOLUTION: We begin by drawing a circle of radius r centered at $(0, c)$, where both r and c are initially unknown. The equation of this circle is

$$x^2 + (y - c)^2 = r^2. \tag{18}$$

Since the circle is to pass through $(1, 1)$, we set $x = y = 1$ in (18) to obtain

$$1 + (1 - c)^2 = r^2$$
$$1 + 1 - 2c + c^2 = r^2$$
$$r^2 = c^2 - 2c + 2$$

Thus (18) becomes

$$x^2 + (y - c)^2 = c^2 - 2c + 2$$
$$x^2 + y^2 - 2cy + c^2 = c^2 - 2c + 2$$
$$x^2 + y^2 - 2cy + 2c - 2 = 0. \tag{19}$$

[†] Descartes, *Oeuvres* (Works), Vol. VI, page 413.
[‡] Published in Book II of *La Géométrie*.

Now any point on the parabola satisfies $y = x^2$. Thus, if we insert y for x^2 in (19), we will obtain the y-coordinate of any point on both the circle and the parabola:

$$y + y^2 - 2cy + 2c - 2 = 0$$
$$y^2 + (1 - 2c)y + (2c - 2) = 0. \tag{20}$$

In order that the circle be tangent to the parabola at $(1, 1)$, the circle must intersect the parabola at exactly $y = 1$ and at no other y-value. This means that we must choose c such that the quadratic equation (20) has a unique solution. Recall that the solutions of the quadratic equation $ax^2 + bx + c = 0$ are

$$x = \frac{-b \pm \sqrt{b^2 - 4ac}}{2a}.$$

There is a unique solution $\left(y = -\dfrac{b}{2a} \right)$ if and only if the discriminant $b^2 - 4ac$ is zero. The discriminant of (20) is

$$(1 - 2c)^2 - 4(1)(2c - 2) = 0$$
$$1 - 4c + 4c^2 - 8c + 8 = 0$$
$$4c^2 - 12c + 9 = 0$$
$$(2c - 3)^2 = 0$$
$$2c - 3 = 0$$
$$c = \frac{3}{2}.$$

Thus the equation of the desired circle (sketched in Figure 8) is

$$x^2 + \left(y - \frac{3}{2} \right)^2 = \frac{5}{4}. \qquad c^2 - 2c + 2 = \left(\frac{3}{2} \right)^2 - 3 + 2 = \frac{5}{4}$$

We don't really need this, however. All we need is c. We have

slope of radial line from $(0, c)$ to $(1, 1)$

$\qquad$ = slope of line joining $(0, \frac{3}{2})$ and $(1, 1)$

$$= \frac{\frac{3}{2} - 1}{0 - 1} = -\frac{1}{2}.$$

Thus

slope of tangent line to circle and parabola

$\qquad$ = negative reciprocal of slope of radial line

$$= \frac{-1}{-\frac{1}{2}} = 2.$$

In comparing the very different methods of Fermat and Descartes, it is tempting to favor the first because it is much closer to the one we use today.

However, in the early seventeenth century the notion of a limit did not yet exist. Descartes's technique was much more precise and was much more fully grounded in the geometric techniques in use at the time. In either case, looking at early attempts to resolve the tangent-line problem should help you appreciate just how fundamental and useful were the discoveries of Newton and Leibniz.

One other mathematician worthy of mention here is Isaac Barrow (1630–1677). Barrow was educated at Cambridge and was considered one of the best Greek scholars of his day. He was the first holder of the Lucasian chair at Cambridge, a position he held from 1664 until he resigned in 1669 to become chaplain to King Charles II. One of Barrow's students at Cambridge was Isaac Newton. When Barrow resigned his post, he recommended his former student and Newton became the second holder of the Lucasian chair (the present holder is Stephen W. Hawking).

Barrow developed a technique which came closest to solving the tangent-line problem—before the problem was finally solved by Newton and Leibniz. His principal result in this area appears in his *Lectiones opticae et geometricae* (optical and geometrical lectures) published in 1669.

To illustrate Barrow's method, we draw a curve and its tangent line at P (see Figure 9). Let Q be a point on the curve close to P. Then the triangles PST and PQR are almost similar. The triangle PQR is now called a **differential triangle**. Earlier, it was called a **Barrow triangle** (but it was not so named by Barrow). Barrow then argued that as triangle PQR becomes infinitely small, we have

$$\frac{\overline{QR}}{\overline{PR}} = \frac{\overline{PS}}{\overline{TS}}. \tag{21}$$

ISAAC BARROW
1630–1677

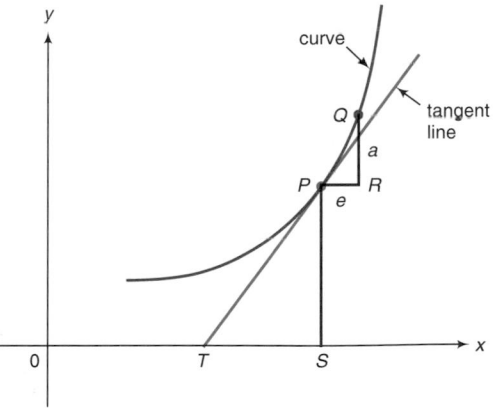

FIGURE 9 BARROW'S METHOD FOR FINDING THE SLOPE OF THE TANGENT LINE

He designated the length $\overline{QR}$ as a and $\overline{PR}$ as e. Then (21) reads

$$\frac{\overline{PS}}{\overline{TS}} = \frac{a}{e}.$$

But $\dfrac{\overline{PS}}{\overline{TS}}$ equals the slope of the tangent line, so if we can compute the ratio $\dfrac{a}{e}$, our problem is solved. Suppose that the coordinates of P are (x, y). Then the coordinates of Q are $(x + e, y + a)$. In order to calculate $\dfrac{a}{e}$, we do three things:

(i) Insert $x + e$ and $y + a$ into the equation of the curve.
(ii) Throw out squares and higher powers of e and a.
(iii) Solve for $\dfrac{a}{e}$.

EXAMPLE OF BARROW'S METHOD

Find the slope of the tangent line to the curve $y = x^2$ at the point $(1, 1)$.

SOLUTION: Substituting $y + a$ for y and $x + e$ for x in the equation $y = x^2$, we obtain

$y + a = (x + e)^2 = x^2 + 2xe + e^2$

$1 + a = 1 + 2e + e^2 \qquad (x, y) = (1, 1)$

$a = 2e + e^2 \qquad$ Simplify.

$a = 2e \qquad$ Throw out the e^2 term.

$\dfrac{a}{e} = 2. \qquad$ This is the required slope.

ANOTHER EXAMPLE OF BARROW'S METHOD

Find the slope of the tangent line to the curve $y = x^3$ at the point $(2, 8)$.

SOLUTION: Substituting $y + a$ for y and $x + e$ for x in the equation $y = x^3$, we have

$y + a = (x + e)^3 = x^3 + 3x^2e + 3xe^2 + e^3$

$8 + a = 8 + 12e + 6e^2 + e^3 \qquad (x, y) = (2, 8)$

$a = 12e + 6e^2 + e^3 \qquad$ Simplify.

$a = 12e \qquad$ Throw out the e^2 and e^3 terms.

$\dfrac{a}{e} = 12. \qquad$ This is the required slope.

In Example 6 in Section 1.5 we found that the derivative of x^3 is $3x^2$. Note that $3x^2 = 12$ when $x = 2$.

Barrow used his method to compute the tangents to a wide variety of curves. His technique of discarding the higher-order terms is equivalent to the more modern process of taking a limit. The ratio $\dfrac{a}{e}$ is now written $\dfrac{dy}{dx}$. It is likely that Newton knew of Barrow's method as early as 1664—or 23 years before the publication of the *Principia*. Fundamental discoveries are rarely made in a vacuum.

■ PROBLEMS 1.6

■ SELF-QUIZ

I. An object moves along the number line in such a way that $x(t) = t^2$ is its distance from the origin at t seconds after it started moving. The average velocity during the one-fifth second interval ending at $t = 3$ seconds is given by _____.

a. $\dfrac{3^2 - 2.8^2}{0.2} = 5.8$ b. $\dfrac{3.2^2 - 3^2}{0.2} = 6.2$

c. $\dfrac{3.1^2 - 2.9^2}{0.2} = 6$ d. $2 \cdot 3 = 6$

II. Consider a colony of dental bacteria; t hours after tooth brushing and flossing, the population size

is $P(t) = 500 + t^2$. The instantaneous rate of growth of this colony at time $t = 3$ hours is _____.

a. $\dfrac{(500 + 3^2) - (500 + 2.8^2)}{0.2} = 5.8$

b. $\dfrac{(500 + 3.1^2) - (500 + 3^2)}{0.1} = 6.1$

c. $\dfrac{(500 + 3^2) - (500 + 2.9^2)}{0.1} = 5.9$

d. $2 \cdot 3 = 6$

■ DRILL

In Problems 1–6, distance (s) is given as a function of time (t). Find the instantaneous velocity at the indicated time.

1. $s = 1 + t + t^2$; $t = 4$
2. $s = 100t - 5t^2$; $t = 6$
3. $s = t^3 - t^2 + 3$; $t = 5$
4. $s = t^4 - t^3 + t^2 - t + 5$; $t = 3$
5. $s = 1 + \sqrt{2t}$; $t = 8$
6. $s = (1 + t)^2$; $t = 2.5$

■ APPLICATIONS

7. Fuel in a rocket burns for 3.5 minutes. In the first t seconds, the rocket reaches a height of $70t^2$ feet above the earth (for any t from 0 to 210 seconds). What is the velocity of the rocket (in feet per second) after 3 seconds? After 10 seconds?
8. If an object has dropped a distance of $\frac{1}{2}gt^2 = 4.9t^2$ meters after t seconds, what is its velocity after exactly 3 seconds?
9. For the bacteria population of Example 3, how fast is the population growing after 5 hours? After one day? Comment on whether this model seems realistic for large values of t.
10. A colony of bacteria is dying out. It starts with an initial population of 10,000 organisms, and after t days the population $P(t)$ is $10,000 - 2.5t^2$.
    a. How fast is the population declining after one week?
    b. After how many days will the colony become extinct?

c. The domain was not mentioned with the formula for the function $P(t)$. What is the largest meaningful domain for the population function?
11. The volume of an expanding spherical cell is proportional to the cube of the radius of the cell. Let r denote radius and V denote volume, then

$$V = \frac{4}{3}\pi r^3.$$

What is the rate of growth of the volume with respect to the radius when the radius is 10 μm? (1 μm = 1 micrometer = (1/1,000,000) m.)
12. The surface area, S, of a spherical cell is proportional to the radius, r, of the cell: $S = 4\pi r^2$. What is the rate of growth of the surface area as a function of the radius when $r = 10$ μm? 20 μm?
13. a. Combine the formulas of Problems 11 and 12

CONTINUED

to obtain an expression for the volume of a spherical cell as a function of its surface area.

*b. How fast is the volume increasing when the surface area is $100 \ \mu m^2$? (Think about the units you use in declaring your answer.)

14. The manufacturer in Example 5 finds that her cost function for another machine part is given by $C(q) = 100q + 55$. Compute her marginal cost function for this part.

15. Suppose the cost function of Example 5 is replaced by $C(q) = 200 + 6q - 0.01q^2 + 0.01q^3$ (with the same domain for q).

   a. Find the marginal cost function.

b. Discuss whether the manufacturer should be advised to buy in large quantities.

*16. In Example 5, the actual difference between the cost of buying 101 units and buying 100 units is calculated by $C(101) - C(100) = \$2.99$. On the other hand, at $q = 100$, $dC/dq = \$3.00$. Explain this apparent discrepancy of one cent.

In Problems 17 and 18, compute the density when $x = 3$ meters.

17. $\mu(x) = \sqrt{1 + x}$ kilograms
18. $\mu(x) = (5 + 2x)^2$ kilograms

■ ANSWERS TO SELF-QUIZ

I. a   II. d

1.7 CONTINUITY

The concept of continuity is one of the central notions in mathematics. Intuitively, a function is continuous at a point if it is defined at that point and if its graph moves unbroken through that point. Figure 1 shows the graphs of six functions, three of which are continuous at x_0 and three of which are not.

There are several equivalent definitions of continuity. The one we give here depends explicitly on the theory of limits developed earlier in this chapter.

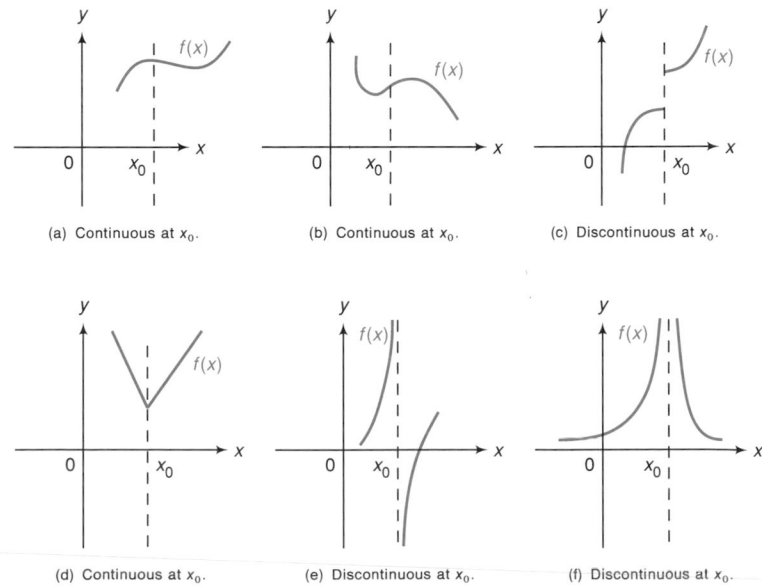

(a) Continuous at x_0.

(b) Continuous at x_0.

(c) Discontinuous at x_0.

(d) Continuous at x_0.

(e) Discontinuous at x_0.

(f) Discontinuous at x_0.

FIGURE 1   CONTINUOUS AND DISCONTINOUS FUNCTIONS

> **DEFINITION** **CONTINUITY**
>
> Let $f(x)$ be defined for every x in an interval containing the number x_0. Then
>
> $$f \text{ is } \textbf{continuous at } x_0 \text{ if } \lim_{x \to x_0} f(x) = f(x_0). \tag{1}$$
>
> This means that the following three conditions hold:
>
> **(i)** $f(x_0)$ exists (that is, x_0 is in the domain of f).
> **(ii)** $\lim_{x \to x_0} f(x)$ exists.
> **(iii)** $\lim_{x \to x_0} f(x) = f(x_0)$. ■ $\qquad\qquad\qquad$ (2)

Condition (iii) tells us that if a function f is continuous at x_0, then we can calculate $\lim_{x \to x_0} f(x)$ by evaluation. This is only one of the reasons continuous functions are so important. In the next few chapters, we will see that a large majority of the functions we encounter in applications are indeed continuous.

We give an alternative definition of continuity in the problem set of Section 1.8.

EXAMPLE 1
A CONTINUOUS MONOMIAL

Let $f(x) = x^2$. Then for any real number x_0,

$$\lim_{x \to x_0} f(x) = \lim_{x \to x_0} x^2 = x_0{}^2 = f(x_0),$$

so f is continuous at every real number.

EXAMPLE 2
CONTINUITY OF A POLYNOMIAL

Let $p(x) = c_0 + c_1 x + c_2 x^2 + c_3 x^3 + \cdots + c_n x^n$ be a polynomial. By Theorem 1, on page 70,

$$\lim_{x \to x_0} p(x) = p(x_0) \tag{3}$$

for every real number x_0. Therefore,

every polynomial is continuous at every real number.

Note that this statement also shows that any constant function is continuous.

EXAMPLE 3
CONTINUITY OF A RATIONAL
FUNCTION

Let $r(x) = p(x)/q(x)$ be a rational function [$p(x)$ and $q(x)$ are polynomials]. From the corollary [equation (4)] on page 73 we have, if $q(x_0) \neq 0$ then $\lim_{x \to x_0} r(x) = p(x_0)/q(x_0) = r(x_0)$, so

any rational function is continuous at all points x_0 at which the denominator, $q(x_0)$, is nonzero.

EXAMPLE 4
A RATIONAL FUNCTION THAT IS CONTINUOUS EXCEPT AT TWO POINTS

Let

$$f(x) = \frac{x^5 + 3x^3 - 4x^2 + 5x - 2}{x^2 - 5x + 6}.$$

Here f is a rational function and therefore is continuous at any x for which the denominator, $x^2 - 5x + 6$, is not zero. Since $x^2 - 5x + 6 = (x - 3)(x - 2) = 0$ only when $x = 2$ or 3, f is continuous at all real numbers except at these two.

DEFINITION **DISCONTINUOUS FUNCTION**

If the function f is not continuous at x_0, then f is said to be **discontinuous** at x_0. Note that f is discontinuous at x_0 if one (or more) of the three conditions given in the definition for *continuity* fails to hold. ∎

CONDITIONS FOR DISCONTINUITY

f is discontinuous at x_0 under any of the following conditions:

(i) $f(x_0)$ does not exist ($x_0 \notin \operatorname{dom} f$).
(ii) $\lim_{x \to x_0} f(x)$ does not exist.
(iii) $\lim_{x \to x_0} f(x)$ exists but is not equal to $f(x_0)$.

REMARK: There are really two somewhat conflicting definitions of the phrase "f is discontinuous at x_0." The first definition is the one given above. The second definition requires f to be defined, but not continuous, at x_0. For example, according to this second definition, the function $1/x$ is neither continuous nor discontinuous at 0, since it is not defined for $x = 0$. Analogously, the function $\sqrt{x}$ is not discontinuous at -10 (or any other negative number) because it is not defined there. In this book we will use the first definition exclusively. Thus for us the function $1/x$ is discontinuous at 0 and the function $\sqrt{x}$ is discontinuous at -10.[†]

As the examples above suggest, most commonly encountered functions are continuous. In this book, all the functions we meet will be continuous at every point except in one of the three cases discussed below.

Discontinuity Case 1 *We cannot divide by zero.* This case is exemplified by Example 4. Also, $1/x$ is discontinuous at 0.

Discontinuity Case 2 *The function is not defined over a range of values.*

[†] A more complete discussion of the meaning of the term "discontinuous" can be found in R. C. Buck and E. F. Buck, *Advanced Calculus*, 3rd ed. (New York: McGraw-Hill, 1978), p. 107.

EXAMPLE 5
DISCONTUITY OVER A RANGE OF
VALUES

$f(x) = \sqrt{x}$ is not continuous for $x < 0$ because the square-root function is not defined for negative values.

Discontinuity Case 3 *The function may be discontinuous if it is defined in pieces.*

EXAMPLE 6
A FUNCTION DEFINED IN PIECES
THAT HAS A JUMP DISCONTINUITY

A tomato wholesaler finds that the price of newly harvested tomatoes is 16¢ per pound if he purchases fewer than 100 pounds each day. However, if he purchases at least 100 pounds daily, the price drops to 14¢ per pound. Find the total cost function.

SOLUTION: Let q denote the number of pounds bought daily and C denote the cost; then,

$$C(q) = \begin{cases} 0.16q, & \text{if } 0 \le q < 100 \\ 0.14q, & \text{if } q \ge 100. \end{cases}$$

This function is sketched in Figure 2. It is discontinuous at $q = 100$. Note also that $C(q)$ is not continuous for $q < 0$ because it is not defined for $q < 0$. The discontinuity at $q = 100$ is called a **jump discontinuity**. This term is used when the function "jumps" from one *finite* value to another at a point.

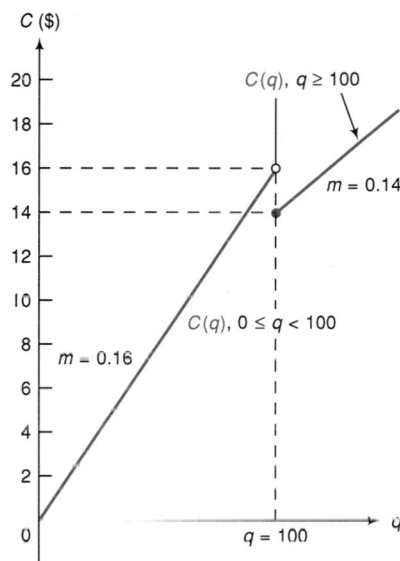

FIGURE 2
$C(q)$ HAS A JUMP DISCONTINUITY AT
$q = 100$

In Figure 2 we can see that

$$\lim_{q \to 100^-} C(q) = 16 \quad \text{while} \quad \lim_{q \to 100^+} C(q) = 14.$$

This suggests the following formal definition of a jump discontinuity.

DEFINITION   **JUMP DISCONTINUITY**

Suppose that f is defined in an open interval containing x_0, but not necessarily at x_0 itself. Suppose further that

$$\lim_{x \to x_0^-} f(x) = L_1 \quad \text{and} \quad \lim_{x \to x_0^+} f(x) = L_2$$

where L_1 and L_2 are real numbers but $L_1 \ne L_2$. Then f is said to have a **jump discontinuity** at x_0. ∎

EXAMPLE 7
A FUNCTION DEFINED IN PIECES
THAT IS CONTINUOUS
EVERYWHERE

Let

$$f(x) = \begin{cases} x^2, & x \le 1 \\ x^3, & x > 1. \end{cases}$$

In Problem 95 on page 68, you were asked to show that $\lim_{x \to 1} f(x) = 1$. Thus f is continuous at 1. In fact, since $f(x)$ is equal to either x^2 or x^3 for $x \ne 1$, f is continuous at every real number. This function provides an example of a function defined in pieces that is everywhere continuous.

EXAMPLE 8
DISCONTINUITY OF AN
OSCILLATORY FUNCTION

Let
$$f(x) = \begin{cases} \sin \dfrac{1}{x}, & x \neq 0 \\ 0, & x = 0. \end{cases}$$

In Example 8 in Section 1.2 (page 60) we showed that $\lim_{x \to 0} \sin \dfrac{1}{x}$ does not exist. Therefore, even though f is defined at 0, it is not continuous at 0.

DEFINITION   **CONTINUITY OVER AN OPEN INTERVAL**

A function f is **continuous over** (or **in**) **the open interval** (a, b) if f is continuous at every point in that interval (a may be $-\infty$ and/or b may be $+\infty$). ∎

DEFINITION   **CONTINUITY OVER A CLOSED INTERVAL**

The function f is **continuous in the closed interval** $[a, b]$ if the following conditions hold:

(i) f is continuous at every x in the open interval (a, b).

(ii) $f(a)$ and $f(b)$ both exist.

(iii) $\lim_{x \to a^+} f(x) = f(a)$ and $\lim_{x \to b^-} f(x) = f(b)$. ∎

REMARK: It may be that f is only defined in the interval $[a, b]$. Then "f is continuous at every point in $[a, b]$" means the usual thing for x in (a, b). However, by "continuity at a" we mean that $\lim_{x \to a^+} f(x) = f(a)$, and by "continuity at b" we mean that $\lim_{x \to b^-} f(x) = f(b)$. Sometimes this type of continuity at the points a and b is referred to, respectively, as **continuity from the left** and **continuity from the right**.

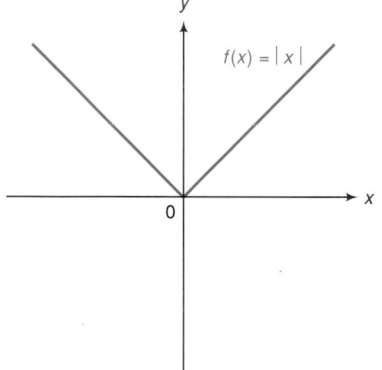

FIGURE 3
GRAPH OF $f(x) = |x|$

EXAMPLE 9
A FUNCTION CONTINUOUS OVER
$(-\infty, \infty)$

Let $f(x) = |x|$. Since for $x > 0$, $f(x) = x$, and for $x < 0$, $f(x) = -x$, f is continuous at every $x \neq 0$. If $x = 0$, then $\lim_{x \to 0} f(x) = \lim_{x \to 0} |x| = 0 = f(0)$, so f is continuous at 0 also (see Figure 3). Thus f is continuous over $(-\infty, \infty)$.

EXAMPLE 10
A FUNCTION CONTINUOUS OVER
$[0, \infty)$

Let $f(x) = \sqrt{x}$. Then since $\lim_{x \to 0^+} f(x) = 0 = f(0)$, f is continuous in $[0, \infty)$. Note that $\lim_{x \to 0} f(x)$ does not exist, since $\sqrt{x}$ is not defined for negative values of x. This example illustrates the importance of defining continuity at the endpoints of a closed interval in terms of one-sided limits.

We close this section by citing several theorems that will be very useful in the remainder of the book. The first theorem follows directly from the definition of continuity and the limit theorems of Section 1.3.

THEOREM 1 **CONSTANT MULTIPLES, SUMS, PRODUCT AND QUOTIENT OF CONTINUOUS FUNCTIONS**

Let the functions f and g be continuous at x_0, and let c be a constant. Then the following functions are continuous at x_0:

(i) cf **(ii)** $f + g$ **(iii)** $f \cdot g$

If, in addition, $g(x_0) \neq 0$, then

(iv) f/g is continuous at x_0

To prove part (iv), we have, if $g(x_0) \neq 0$,

From limit Theorem 2 (iv) on page 71

$$\lim_{x \to x_0} \left(\frac{f}{g}\right)(x) = \lim_{x \to x_0} \frac{f(x)}{g(x)} = \frac{\lim_{x \to x_0} f(x)}{\lim_{x \to x_0} g(x)} = \frac{f(x_0)}{g(x_0)} = \left(\frac{f}{g}\right)(x_0)$$

so f/g is continuous at x_0. ∎

The following theorem, whose proof is given in Appendix 3, is very useful for the calculation of limits.

THEOREM 2 **CONTINUITY THEOREM**

If f is continuous at a and if $\lim_{x \to x_0} g(x) = a$, then $\lim_{x \to x_0} f(g(x)) = f(a)$. In other words,

$$\lim_{x \to x_0} (f \circ g)(x) = \lim_{x \to x_0} f(g(x)) = f\left(\lim_{x \to x_0} g(x)\right). \quad ∎$$

EXAMPLE 11
ILLUSTRATION OF CONTINUITY THEOREM

Calculate $\lim_{x \to 3}[(x^2 - 9)/(x - 3)]^3$.

SOLUTION: Let $f(x) = x^3$ and $g(x) = (x^2 - 9)/(x - 3)$. Then $[(x^2 - 9)/(x - 3)]^3 = f(g(x))$. Since $\lim_{x \to 3} g(x) = 6$, $\lim_{x \to 3} f(g(x)) = 6^3 = 216$. Here we have used the fact that f is continuous.

EXAMPLE 12
APPLICATION OF CONTINUITY THEOREM

In Section 1.3 we made use of the following fact:

$$\text{If } \lim_{x \to x_0} g(x) = L > 0, \text{ then } \lim_{x \to x_0} \sqrt{g(x)} = \sqrt{L}.$$

This follows directly from the continuity theorem and the fact that $f(x) = \sqrt{x}$ is continuous at L if $L > 0$.

Next, we show that every differentiable function is continuous. That is, a function f is continuous at any point x_0 at which f is differentiable.

Let f be differentiable at x_0. Then f is continuous there.

This result appeals to our intuition. For if a function is differentiable at x_0, then its graph has a tangent line at the point $(x_0, f(x_0))$, and the curve seems to move "smoothly" through that point.

PROOF: We need to show that $\lim_{x \to x_0} f(x) = f(x_0)$, which is the same as showing that

$$\lim_{x \to x_0} [f(x) - f(x_0)] = 0. \text{ (Why?)}$$

Then,

$$\lim_{x \to x_0} [f(x) - f(x_0)] = \lim_{x \to x_0} \frac{[f(x) - f(x_0)](x - x_0)}{(x - x_0)}$$

$$= \lim_{x \to x_0} \frac{f(x) - f(x_0)}{x - x_0} \lim_{x \to x_0} (x - x_0)$$

$$= f'(x_0) \cdot 0 = 0.$$

Here we have used the alternative definition of the derivative [equation (3') on page 87] and the assumption that $f'(x_0)$ exists. ■

Continuity Does Not Imply Differentiability

REMARK: The converse of this theorem is not true. That is, a function that is continuous at a point is *not* necessarily differentiable at that point. For example, we saw in Example 9 that $|x|$ is continuous at 0. However, according to Example 7 on page 91, $|x|$ is *not* differentiable at 0.

Two very useful properties of continuous functions that we will need in Chapter 3 are given in the theorems that follow. The proof of Theorem 4 is beyond the scope of this text.[†] The proof of Theorem 5 is given in Appendix 3.

Before starting our results, we give some important definitions.

DEFINITION   **LEAST UPPER BOUND, GREATEST LOWER BOUND, COMPLETENESS AXIOM**

If a set of numbers is bounded above,[‡] then the **least upper bound** of the set is the smallest upper bound. If the set is bounded below, then the **greatest lower bound** of the set is the largest lower bound. For example, 1 is the least upper bound of the interval $(0, 1)$, while 0 is the greatest lower bound of the interval.

It is an axiom of real numbers, called the **completeness axiom,** that a nonempty set of numbers bounded above has a least upper bound and a set of numbers bounded below has a greatest lower bound. ■

[†] See, for example, R. C. Buck and E. F. Buck, *Advanced Calculus*, 3rd ed. (New York: McGraw-Hill, 1978), p. 91.
[‡] A nonempty set S of real numbers is *bounded above* if there exists a number K such that if $x \in S$, then $x \leq K$. It is *bounded below* if there exists a number k such that if $x \in S$, then $x \geq k$.

> ### THEOREM 4   UPPER AND LOWER BOUND THEOREM
>
> If f is continuous on the closed, bounded interval $[a, b]$, then f is bounded above and below in that interval. That is, there exist numbers m and M such that
>
> $$m \leq f(x) \leq M \qquad \text{for every } x \text{ in } [a, b].$$
>
> Moreover, if m is the greatest lower bound for f on $[a, b]$ and M is the least upper bound for f on $[a, b]$, then there exist numbers x_1 and x_2 in $[a, b]$ such that $f(x_1) = m$ and $f(x_2) = M$. ∎

REMARK 1: Theorem 4 is not true, in general, on an open interval (a, b). For example, $f(x) = 1/x$ is continuous on $(0, 1)$, but as $x \to 0^+$, $f(x) \to \infty$, so f is not bounded above on that interval. As another example, $f(x) = x$ has the least upper bound 1 on the interval $(0, 1)$ but there is no x in $(0, 1)$ such that $f(x) = 1$.

REMARK 2: Theorem 4 is also not true, in general, if f is not continuous in $[a, b]$. For example, $f(x) = 1/x$ is not bounded on the interval $[-1, 1]$ (it is not continuous at 0).

Theorem 4 is illustrated in Figure 4 for three different functions.

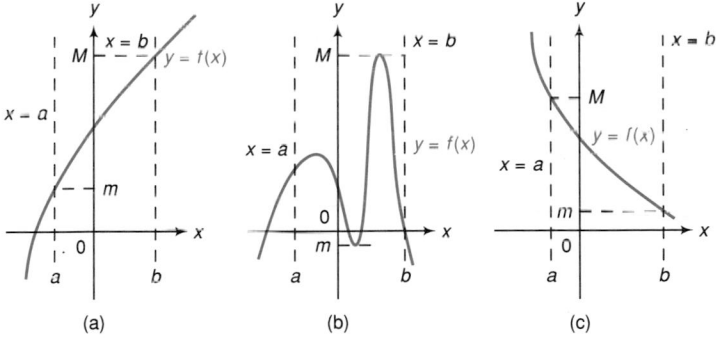

(a)   (b)   (c)

FIGURE 4   ILLUSTRATION OF UPPER AND LOWER BOUND THEOREM

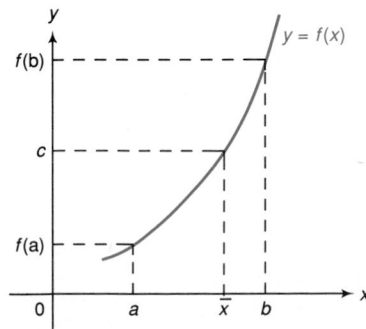

FIGURE 5
ILLUSTRATION OF INTERMEDIATE-VALUE THEOREM

> ### THEOREM 5   INTERMEDIATE-VALUE THEOREM
>
> Let f be continuous on $[a, b]$. Then if c is any number between $f(a)$ and $f(b)$, there is a number $\bar{x}$ in (a, b) such that $f(\bar{x}) = c$. ∎

This theorem has a geometrical interpretation given by Figure 5. The intermediate-value theorem could be restated: If the continuous function f takes on the values $f(a)$ and $f(b)$, then it takes on every value in between. This means that every horizontal line between $f(a)$ and $f(b)$ intersects the graph of f in a least one point. The intermediate-value theorem can be used to prove many interesting results. It is useful, for example, in finding roots of a polynomial.

EXAMPLE 13
USING THE INTERMEDIATE-VALUE
THEOREM

Show that there is a root of $P(x) = x^3 + x^2 + x - 1$ in the interval $[0, 1]$.

SOLUTION: $P(x)$ is continuous (why?). Since $P(0) = -1$ and $P(1) = 2$, there must be a number $\bar{x}$ in $(0, 1)$ such that $P(\bar{x}) = 0$ (since 0 is between -1 and 2). We can do even better. We have

$$P\left(\frac{1}{2}\right) = \frac{1}{8} + \frac{1}{4} + \frac{1}{2} - 1 = -\frac{1}{8} \quad \text{and}$$

$$P\left(\frac{3}{4}\right) = \frac{27}{64} + \frac{9}{16} + \frac{3}{4} - 1 = \frac{47}{64}.$$

Therefore there is a root between $\frac{1}{2}$ and $\frac{3}{4}$. Using a calculator, we can narrow the root down further. Since $P(0.543) = -0.002047993$ and $P(0.544) = 0.000925184$, we have located a root between 0.543 and 0.544. This method of "estimating" a root of a polynomial is very useful when used in conjunction with **Newton's method** (Section 3.6). We first use the intermediate-value theorem to get "close" to the root (within one decimal place, say) and then use Newton's method to find the root quickly to as many decimal places of accuracy as needed. Doing so, we find one root of $x^3 + x^2 + x - 1$ is $\bar{x} = 0.5436890127$, correct to 10 decimal places.

PROBLEMS 1.7

SELF-QUIZ

I. If the function f is continuous on the interval $(-2, 5)$, then _____.
  a. f is differentiable on $(-2, 5)$
  b. f is increasing throughout the interval $(-2, 5)$
  c. the graph of f neither increases nor decreases on the interval $(-2, 5)$
  d. the graph of f can be drawn without lifting pencil from paper

II. True–False: If the function f has a derivative whose values are always between -2.72 and -3.14 on the interval $(-2, 10)$, then there is some point in that interval where f is discontinuous.

III. True–False: If g is differentiable on the interval $(-2, 10)$, then g is continuous at every point of that interval.

IV. Suppose the function f is continuous on the interval $[3, 7]$; also suppose $f(3) = -2$ and $f(7) = 10$. The intermediate value theorem implies that _____.
  a. $f(5) = 4$ [*Note:* 5 is midway between 3 and 7; 4 is midway between -2 and 10.]
  b. $-2 \le f(x) \le 10$ for all $x \in (3, 7)$
  c. $-2 \le f(5) \le 10$
  d. there is some x_0 between 3 and 7 such that $f(x_0) = 4$
  e. there is some x_1 between 3 and 7 such that $f(x_1) = 0$

DRILL

In Problems 1–10, find the largest possible open interval or collection of open intervals on which the given function is continuous.

1. $f(x) = x^{17} - 3x^{15} + 2$
2. $g(x) = 1/(x + 2)$
3. $f(x) = |x + 2|/(x + 2)$
4. $g(x) = 1/(x - 10)^{15}$
5. $f(x) = -17x/(x^2 - 1)$
6. $g(x) = 2x/(x^3 - 8)$
7. $f(x) = x^{1/3}$
8. $g(x) = x^{1/4}$
9. $f(x) = x - [x]$
10. $g(x) = x[x]$

CONTINUED

In Problems 11–20, determine whether the function whose graph is drawn is continuous at x_0. Answer the same question for the derivative of f at x_0.

11.

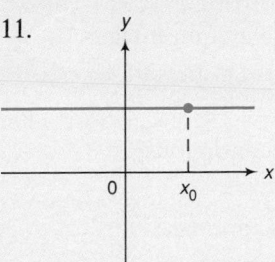

12.

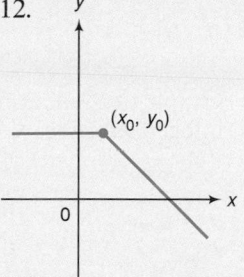

16.

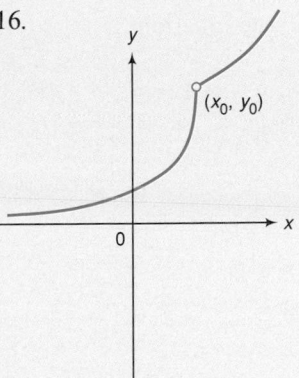

13.

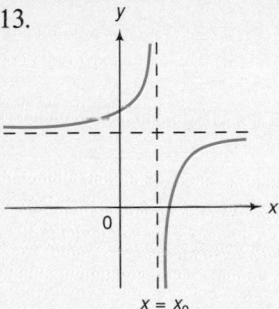

14.

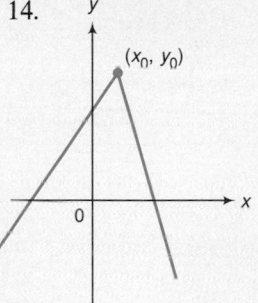

17.

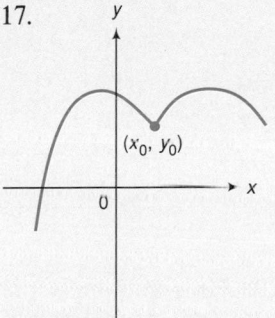

18.

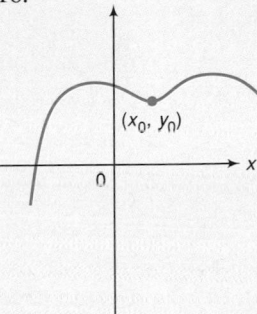

15.

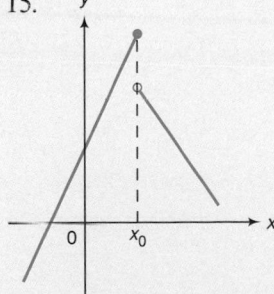

19.

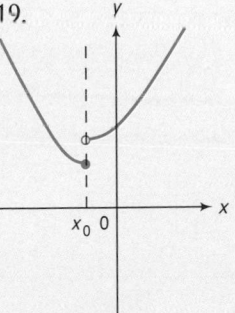

20.

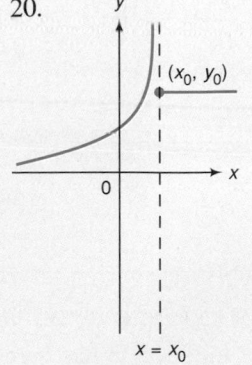

APPLICATIONS

21. A car accelerates from 0 to 80 km/hr in 30 seconds. Explain why there is some time between 0 and 30 seconds when the car is traveling exactly 50 km/hr.

22. Show that there is a number whose square equals 3. [*Hint:* Let $f(x) = x^2$ and apply the *intermediate-value theorem* over an appropriately chosen interval.]

23. Show that the function

$$f(x) = \begin{cases} \dfrac{x^3 - 1}{x - 1} & \text{if } x \neq 1 \\ 3 & \text{if } x = 1 \end{cases}$$

is continuous on $(-\infty, \infty)$.

CONTINUED

24. For what value of α is the function

$$f(x) = \begin{cases} \dfrac{x^4 - 1}{x - 1} & \text{if } x \neq 1 \\ \alpha & \text{if } x = 1 \end{cases}$$

continuous at $x = 1$?

25. Let

$$f(x) = \begin{cases} 0 & \text{if } x < 0 \\ x^2 & \text{if } 0 \leq x < 2 \\ 4 & \text{if } 2 \leq x. \end{cases}$$

Graph the function. Show that f is continuous on $(-\infty, \infty)$.

26. Let

$$f(x) = \begin{cases} x & \text{if } x < 0 \\ x^2 & \text{if } 0 \leq x < 1 \\ x^3 & \text{if } 1 \leq x. \end{cases}$$

Graph the function. Show that f is continuous on $(-\infty, \infty)$.

27. Let

$$f(x) = \begin{cases} x & \text{if } x \text{ is not an integer} \\ x^2 & \text{if } x \text{ is an integer.} \end{cases}$$

Graph the function for $-3 \leq x \leq 3$. For what integer values of x is f continuous?

*28. Let

$$f(x) = \begin{cases} 1 & \text{if } x \text{ is rational} \\ 0 & \text{if } x \text{ is irrational.} \end{cases}$$

Explain why f is not continuous at any point.

29. Use Theorem 2 to calculate $\lim_{x \to 2}(\sqrt{x - 1})/(x + 1)$.

*30. Use Theorem 2 to calculate $\lim_{x \to -1}(3x^3 + 8x^2 - 9x + 2)^{3/4}$. [Hint: $u^{3/4} = \sqrt{\sqrt{u^3}}$.]

CALCULATOR PROBLEMS

In Problems 31–40, use a calculator and apply the intermediate-value theorem to find an interval of length L in which the given function has a zero.

31. $f(x) = x^3 + 4x + 7$; $L = 0.5$
32. $f(x) = x^5 + x^3 - 8$; $L = 0.5$
33. $f(x) = 2x^5 + 5x^3 + 8x + 9$; $L = 0.5$
34. $f(x) = 14x^3 + 8x - 30$; $L = 0.5$
35. $f(x) = \dfrac{3x^3 + 2x - 20}{\sqrt{3x + 8}}$; $L = 0.5$
36. $f(x) = \sin 3x - \cos 4x$ in $\left[0, \dfrac{\pi}{2}\right]$; $L = 0.1$
37. $f(x) = \sin x - \dfrac{x}{2}$ in $(0, \pi)$; $L = 0.1$
38. $f(x) = \cos x - x$ in $\left[0, \dfrac{\pi}{2}\right]$; $L = 0.1$
39. $f(x) = x - \tan x$ in $\left(\dfrac{\pi}{2}, \dfrac{3\pi}{2}\right)$; $L = 0.1$
*40. $f(x) = \ln x - \sin x$, $x > 1$; $L = 0.1$ [Hint: Use the $\boxed{\ln}$ button.]

SHOW/PROVE/DISPROVE

The function f has a **removable discontinuity** at x_0 if and only if $\lim_{x \to x_0} f(x)$ exists and is finite, but either f is not defined at x_0 or $\lim_{x \to x_0} f(x) \neq f(x_0)$.

41. Let $f(x) = (x^2 - 1)/(x - 1)$.
    a. Explain why f is not continuous at 1.
    b. Compute $\lim_{x \to 1} f(x)$.
    c. Show that f has a removable discontinuity at $x_0 = 1$ and find a continuous function g that is equal to f for $x \neq 1$.
42. For what values of α would the function $f(x) = (x^2 - 4)/(x - \alpha)$ have a removable discontinuity at $x = \alpha$?

43. For what values of α would the function $f(x) = (x^3 - 6x^2 + 11x - 6)/(x - \alpha)$ have a removable discontinuity at $x = \alpha$?

The function f is **piecewise continuous** in $[a, b]$ if and only if f is continuous at every point in $[a, b]$ except for a finite number of points at which f has a jump or removable discontinuity. [A jump discontinuity is defined on page 115; a removable discontinuity is defined just before Problem 41.]

44. Show that the function

$$f(x) = \begin{cases} x & \text{if } x < 0 \\ 2x & \text{if } 0 \le x \le 1 \\ 3x^3 & \text{if } 1 < x \end{cases}$$

is piecewise continuous over any interval.
45. Is the function $f(x) = (x+1)/(x-2)$ piecewise continuous over the interval $[-3, 3]$? Over the interval $[-1, 1]$?

In Problems 46–51, show that the given function is piecewise continuous over the specified interval; identify those functions which are continuous throughout.

46. $f(x) = 1 + [x]; \; [-2, 2]$
47. $g(x) = x - [x]; \; [-2, 2]$
48. $f(x) = (x^2 - 1)/(x + 1); \; [-2, 2]$
49. $f(x) = \begin{cases} x + x^2 & \text{if } x \le -1; \\ x^3 & \text{if } x > -1; \end{cases} [-2, 0]$
50. $f(x) = \begin{cases} x & \text{if } x \text{ is an integer;} \\ \frac{1}{2} & \text{if } x \text{ is not an integer;} \end{cases} [-1, 3]$
51. $f(x) = \begin{cases} 2x^2 + 4 & \text{if } x \le 0; \\ (x - 2)^2 & \text{if } x > 0; \end{cases} [-5, 5]$

Discuss continuity, or the lack of it, for the functions in Problems 52–55.

52. $[x] + |x - [x]|$
53. $[x] + (x - [x])^2$
54. $[x] + \sqrt{|x - [x]|}$
55. $[x] + [-x] + 1$
56. Find a function f that is not continuous, but such that $|f|$ is continuous.
57. Give an example of a function defined in $[0, 1]$, bounded in $[0, 1]$ and continuous for $0 < x \le 1$, but discontinuous at $x = 0$.
58. Find the maximum, M, and the minimum, m, for the function $f(x) = |x + 3|$ on the interval $[-5, 7]$.
59. Let

$$f(x) = \begin{cases} x + 3 & \text{if } x \le 1 \\ 2x + 5 & \text{if } x > 1 \end{cases} \text{ and}$$

$$g(x) = \begin{cases} x^2 + 6 & \text{if } x \le 1 \\ 5x^3 - 1 & \text{if } x > 1. \end{cases}$$

a. What is the function $f(x)g(x)$?
b. Show that $\lim_{x \to 1} f(x)$ does not exist.
c. Show that $\lim_{x \to 1} g(x)$ does not exist.
d. Show that $\lim_{x \to 1} f(x)g(x)$ does exist.
60. Find two functions f and g such that $\lim_{x \to 5} f(x)$ and $\lim_{x \to 5} g(x)$ do not exist, but $\lim_{x \to 5}[f(x)/g(x)]$ does exist.

CHALLENGE

61. Let $P(x) = x^3 + cx + d$. Prove that if $c > 0$, there is exactly one real zero of P.
62. Find the maximum, M, and the minimum, m, for the function $f(x) = (x^3 + 3x^2 + 2x + 1)/(x + 1)$ on the interval $[0, 1]$.
*63. Suppose that f is continuous at x_0 and $f(x_0) > 0$. Prove that there is a number $\delta > 0$ such that if $x_0 - \delta < x < x_0 + \delta$, then $f(x) > 0$. [*Hint:* Draw a sketch.]
*64. Suppose that f is continuous on the closed interval A. In addition, suppose that B is a closed interval such that $B \subset A$. Some of the following assertions are true, others are false. Identify and

prove each of the true assertions; produce a counterexample for each of the false statements.
a. $\min_B f \le \min_A f$
b. $\min_B f \ge \min_A f$
c. $\min_B f \le \max_A f$
d. $\min_B f \ge \max_A f$
e. $\max_B f \le \min_A f$
f. $\max_B f \ge \min_A f$
g. $\max_B f \le \max_A f$
h. $\max_B f \ge \max_A f$
**65. Suppose that f is continuous in $[a, b]$ with $f'(a) < 0$ and $f'(b) > 0$. Prove there is a point c in (a, b) such that either $f'(c) = 0$ or $f'(c)$ does not exist.

ANSWERS TO SELF-QUIZ

I. d   II. False   III. True   IV. d, e

1.8 THE THEORY OF LIMITS (OPTIONAL)

The value of mathematics lies in its precision. The precision applies not only to its calculations but to the definitions of its terms and the proofs of its results. In mathematics you are not asked to believe something because it seems to be true or because some expert tells you it is true. Rather, you are asked to believe it only after seeing it proved true. On page 56, we provided an intuitive definition of a limit, and in Section 1.3 we stated, but did not prove, a number of very useful limit theorems. At this point you should be comfortable with the notion of a limit. The intuitive definition given in Section 1.2 is correct. The only problem is that the words "close to" are not precise. What do we mean by "close"? We now answer that question.

The definition of a limit we give next is a precise restatement of what you already know—namely, that as x gets close to x_0, $f(x)$ gets close to L. Before giving this definition, we recall (from page 3) that an **open interval** (a, b) is the set of all points x such that $a < x < b$. A **neighborhood** of the point x_0 is defined as an open interval that contains that point.[†] Finally, the letters ϵ (epsilon) and δ (delta) used in this section are Greek letters that almost always represent small quantities.

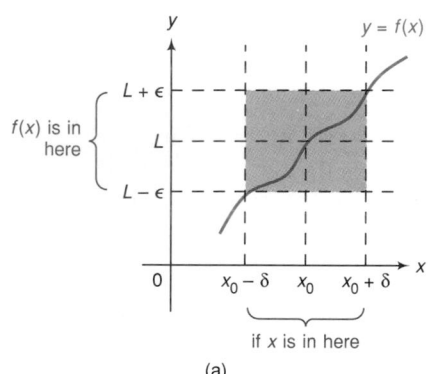

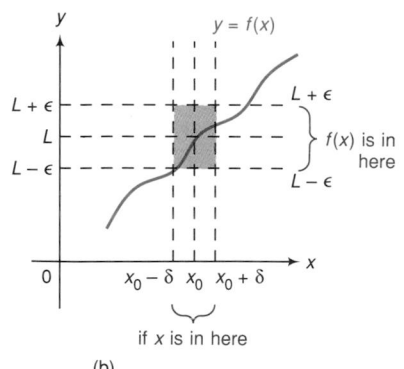

FIGURE 1
$f(x)$ IS WITHIN ϵ UNITS OF L IF x IS WITHIN δ UNITS OF x_0.

DEFINITION   **FORMAL DEFINITION OF A LIMIT**

Suppose that $f(x)$ is defined in a neighborhood of the point x_0 (a finite number) except possibly at the point x_0 itself. Then,

$$\lim_{x \to x_0} f(x) = L$$

if for every $\epsilon > 0$ there is a $\delta > 0$, such that

$$\text{if } 0 < |x - x_0| < \delta, \qquad \text{then } |f(x) - L| < \epsilon.[‡] \quad \blacksquare \qquad \textbf{(1)}$$

We must exclude $|x - x_0| = 0$ (i.e., we don't want $x = x_0$) since $f(x)$ may not be defined at x_0 (and, if it is, its value is irrelevant).

This formal definition is difficult to understand at first reading. It is worthwhile at this point to draw a picture of what is happening. We note that $0 < |x - x_0| < \delta$ means (from page 6) that $x_0 - \delta < x < x_0 + \delta, x \neq x_0$. Similarly, $|f(x) - L| < \epsilon$ means that $L - \epsilon < f(x) < L + \epsilon$. Figure 1 illustrates that no matter how small ϵ is chosen, δ can be made small enough so that $f(x)$ is within ϵ of L.

The best way to illustrate this definition further is to show by example that our intuitive notion of a limit coincides with the formal definition of a limit.

EXAMPLE 1
PROVING THAT A LIMIT EXISTS

Show that $\lim_{x \to 2} 3x = 6$.

SOLUTION: Here $x_0 = 2, L = 6,$ and $f(x) = 3x$. Let $\epsilon > 0$ be given. Then we must show that there is a $\delta > 0$ such that if $0 < |x - 2| < \delta$, then $|3x - 6| < \epsilon.[§]$ There are several ways to show this. The most

---

[†] We can also define **a neighborhood of** ∞ to be an open interval of the form (N, ∞).

[‡] Here ϵ measures how close $f(x)$ is to L, and δ measures how close x is to x_0.

[§] Note that since the function $f(x) = 3x$ is defined for all real numbers, it is certainly defined in a neighborhood of $x = 2$.

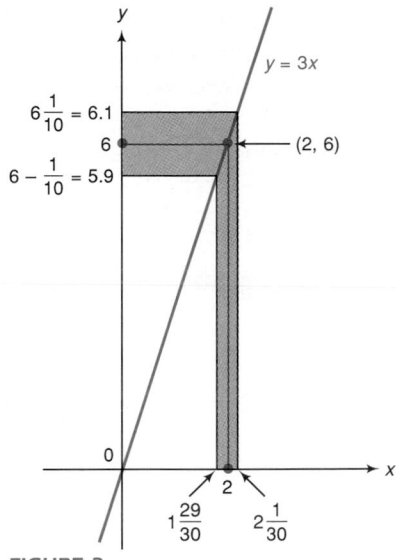

FIGURE 2
y IS WITHIN $\frac{1}{10}$ OF 6 IF x IS WITHIN $\frac{1}{30}$ OF 2

direct way is to start out with what we want and work backward. We want $|3x - 6| < \epsilon$. Thus,

$$-\epsilon < 3x - 6 < \epsilon$$

$$-\frac{\epsilon}{3} < x - 2 < \frac{\epsilon}{3}, \qquad \text{We divided by 3.}$$

which means that

$$|x - 2| < \frac{\epsilon}{3}.$$

Hence if $|x - 2| < \epsilon/3$, then $|3x - 6| < \epsilon$.

Now we are done. All we need to do is to choose $\delta = \epsilon/3$. Then, as we have seen, if $|x - 2| < \delta$, then $|3x - 6| < \epsilon$. Thus $\lim_{x \to 2} 3x = 6$. For example, if $\epsilon = 0.1 = \frac{1}{10}$, then the choice $\delta = \frac{1}{30}$ will work. This choice is illustrated in Figure 2. It is important to note that this value of δ is *not* unique. Any value of $\delta < \frac{1}{30}$ will work. This is not surprising since it merely states that if we start even closer to 2 than $\frac{1}{30}$ of a unit away, $3x$ will still remain within $\frac{1}{10}$ of a unit from 6.

Example 1 illustrates the technique for finding a δ that works. We start with the assumption $L - \epsilon < f(x) < L + \epsilon$ and work backward, using elementary algebraic steps, until we obtain an expression of the form $c < x < d$. Then if δ is chosen so that $x_0 - \delta > c$ and $x_0 + \delta < d$, then $|x - x_0| < \delta$ will imply that $c < x < d$. We can often show that $|f(x) - L| < \epsilon$ by reversing the steps that led to $c < x < d$.

EXAMPLE 2
PROVING THAT A LIMIT EXISTS

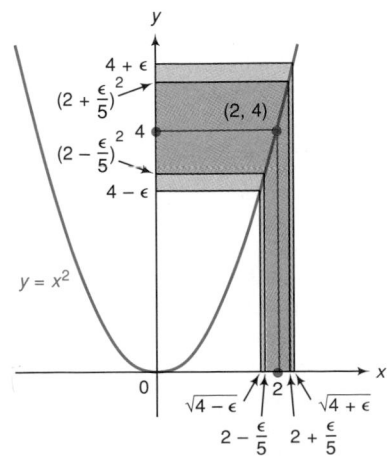

FIGURE 3
y IS WITHIN ϵ UNITS OF 4 IF x IS WITHIN $\delta = \frac{\epsilon}{5}$ UNITS OF 2

Show that $\lim_{x \to 2} x^2 = 4$.

SOLUTION: This example is a bit more involved than Example 1. Let $\epsilon > 0$ be given. We must find a $\delta > 0$ such that

$$-\epsilon < x^2 - 4 < \epsilon$$

whenever

$$-\delta < x - 2 < \delta.$$

We want to show that x^2 is near 4 whenever x is near 2. That is, $|x^2 - 4|$ is small whenever $|x - 2|$ is small. Since x is near 2, we may assume that $1 < x < 3$, so that

$$3 < x + 2 < 5 \qquad \text{and} \qquad |x + 2| < 5.$$

Then

$$|x^2 - 4| = |x + 2||x - 2| < 5|x - 2|. \tag{2}$$

Now choose $\delta = \epsilon/5$. Then if $|x - 2| < \delta$,

From (2)

$$|x^2 - 4| < 5|x - 2| < 5\delta = 5\frac{\epsilon}{5} = \epsilon.$$

This is what we wanted to show (see Figure 3).

The calculation of limits from the definition can be difficult. Fortunately, as we have seen, there are several theorems that make the calculation of limits (at least, in some cases) a relatively simple process. We restate and prove the basic limit theorems in Appendix 3.

We now discuss infinite limits and limits at infinity.

DEFINITION **FORMAL DEFINITION OF INFINITE LIMITS**

(i) $\lim_{x \to x_0} f(x) = \infty$ if for every $N > 0$ there is a $\delta > 0$ such that $f(x) > N$ for all x such that $0 < |x - x_0| < \delta$.

(ii) $\lim_{x \to x_0} f(x) = -\infty$ if for every $N > 0$ there is a $\delta > 0$ such that $f(x) < -N$ for all x such that $0 < |x - x_0| < \delta$. ■

EXAMPLE 3
PROVING THAT AN INFINITE
LIMIT EXISTS

Show that $\lim_{x \to 0}(1/x^2) = \infty$.

SOLUTION: Let $N > 0$ be chosen and choose $\delta = 1/\sqrt{N}$. If $0 \le |x - 0| < \delta = 1/\sqrt{N}$, then $-1/\sqrt{N} < x < 1/\sqrt{N}$. But if $|x| < 1/\sqrt{N}$, then $x^2 < (1/\sqrt{N})^2 = 1/N$, so that $1/x^2 > N$. See Figure 4.

NOTE: How did we know that choosing $\delta = 1/\sqrt{N}$ would work? The secret is to start with the inequality you want to end up with and work backward, as in Examples 1, 2, and 3. The inequality $1/x^2 > N$ leads naturally to the inequalities $Nx^2 < 1$, $x^2 < 1/N$, and $|x| < 1/\sqrt{N}$. Now the choice $\delta = 1/\sqrt{N}$ is obvious.

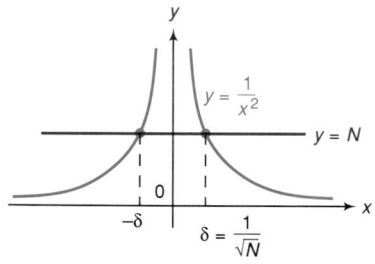

FIGURE 4
$y > N$ IF $-\delta < x < \delta$ $(|x| < \delta)$

DEFINITION **FORMAL DEFINITION OF LIMIT AT INFINITY**

(i) $\lim_{x \to \infty} f(x) = L$ if for every $\epsilon > 0$ there is an $N > 0$ such that if $x > N$, then $|f(x) - L| < \epsilon$.

(ii) $\lim_{x \to -\infty} f(x) = L$ if for every $\epsilon > 0$ there is an $N > 0$ such that if $x < -N$, then $|f(x) - L| < \epsilon$. ■

EXAMPLE 4
PROVING THAT A LIMIT AT
INFINITY EXISTS

Show that $\lim_{x \to -\infty} 1/x = 0$.

SOLUTION: Let $\epsilon > 0$ be given. Then choose $N = 1/\epsilon$. If $x < -N$, then $1/x > -1/N$ and $-\epsilon = -1/N < 1/x < 0$, so $|1/x - 0| < \epsilon$.

We next give a second example of a function that remains bounded but has no limit as $x \to 0$. [The first was $y = \sin(1/x)$—see page 60]

EXAMPLE 5
A LIMIT THAT DOES NOT EXIST

Let $f(x)$ be defined by

$$f(x) = \begin{cases} 1, & \text{if } x = \dfrac{1}{2^n} \text{ for } n = 1, 2, 3, \ldots \\ x, & \text{otherwise.} \end{cases}$$

A sketch of this function is given in Figure 5.

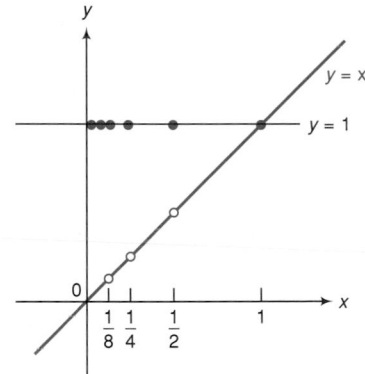

FIGURE 5
GRAPH OF $f(x) =$

$\begin{cases} 1, \text{ IF } x = \dfrac{1}{2^n}, \ n = 1, 2, 3, \ldots \\ x, \text{ OTHERWISE} \end{cases}$

$\lim_{x \to 0} f(x)$ does not exist.

Since $f(x) = x$ for all values of $x \neq 1/2^n$, it may seem that $f(x) \to 0$ as $x \to 0$. We now show that $\lim_{x \to 0} f(x)$ does not exist. First we show that $\lim_{x \to 0} f(x) \neq 0$. To this end we pick $\epsilon > 0$ such that $\epsilon < 1$. For every $\delta > 0$ there is an n such that $x = 1/2^n < \delta$ (since $1/2^n \to 0$ as $n \to \infty$). Hence, no matter how small we choose δ, there will be an x with $0 < x < \delta$ such that $|f(x) - 0| = 1 > \epsilon$. Thus $f(x)$ does not tend to 0 as $x \to 0$. But clearly $f(x)$ cannot have any other limit as $x \to 0$ since $|f(x)| < \epsilon$ if $|x| < \epsilon$ except at the points $1/2^n$. Therefore $f(x)$ has no limit as $x \to 0$. However, if $x_0 > 0$, then it is not difficult to show that $\lim_{x \to x_0} f(x) = x_0$ (see Problem 14).

We will, in a moment, turn to the theory of one-sided limits. However, before doing so, we stress that with the material already discussed in this section, it is possible to prove the limit theorems stated in Section 1.3. These proofs are given in Appendix 3.

DEFINITION   **FORMAL DEFINITION OF ONE-SIDED LIMITS**

(i) Let f be defined on the open interval (x_0, c) for some number $c > x_0$. Suppose that for every $\epsilon > 0$ there is a $\delta > 0$ such that $x_0 < x < x_0 + \delta$ implies that $|f(x) - L| < \epsilon$. Then L is called the **right-hand limit** of $f(x)$ as $x \to x_0$ and is denoted by

$$\lim_{x \to x_0^+} f(x) = L.$$

(ii) Let f be defined on the open interval (d, x_0) for some number $d < x_0$. If for every $\epsilon > 0$ there is a $\delta > 0$ such that $x_0 - \delta < x < x_0$ implies that $|f(x) - L| < \epsilon$, then L is called the **left-hand limit** of $f(x)$ as $x \to x_0$ and is denoted by

$$\lim_{x \to x_0^-} f(x) = L. \quad \blacksquare$$

These two definitions are illustrated in Figure 6. Note that the functions sketched in the figure do not have a limit as $x \to x_0$.

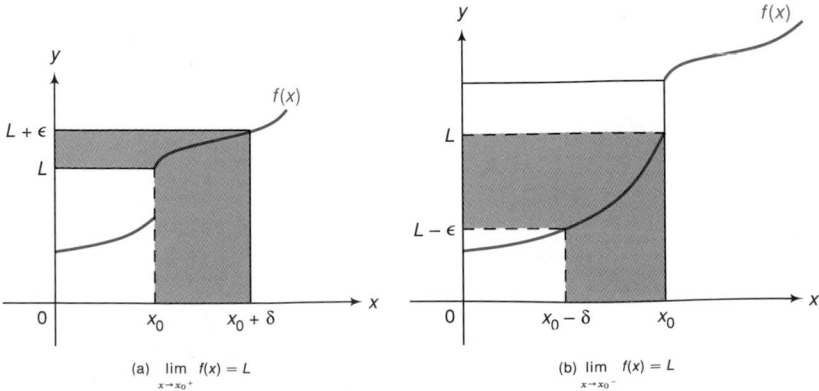

(a) $\lim_{x \to x_0^+} f(x) = L$

(b) $\lim_{x \to x_0^-} f(x) = L$

FIGURE 6   IF $x_0 < x < x_0 + \delta$, THEN $|f(x) - L| < \epsilon$.

IF $x_0 - \delta < x < x_0$, THEN $|f(x) - L| < \epsilon$.

EXAMPLE 6
PROVING THAT A ONE-SIDED LIMIT EXISTS

Prove that $\lim_{x \to 0^+} \sqrt{x} = 0$.

SOLUTION: Let $\epsilon > 0$ be given. Suppose that $|f(x) - L| = |\sqrt{x} - 0| = \sqrt{x} < \epsilon$. Then $x < \epsilon^2$. This leads to the obvious choice $\delta = \epsilon^2$. Then, if $0 < x < \delta$, $0 < \sqrt{x} < \sqrt{\delta} = \epsilon$, and so $\lim_{x \to 0^+} \sqrt{x} = 0$.

For infinite limits the definitions are similar.

DEFINITION **FORMAL DEFINITION OF INFINITE ONE-SIDED LIMITS**

(i) $\lim_{x \to x_0^+} f(x) = \infty$ means that for every $N > 0$ there is a $\delta > 0$ such that if $x_0 < x < x_0 + \delta$, then $f(x) > N$.

(ii) $\lim_{x \to x_0^-} f(x) = \infty$ if for every $N > 0$ there is a $\delta > 0$ such that if $x_0 - \delta < x < x_0$, then $f(x) > N$.

(iii) $\lim_{x \to x_0^+} f(x) = -\infty$ if for every $N > 0$ there is a $\delta > 0$ such that if $x_0 < x < x_0 + \delta$, then $f(x) < -N$.

(iv) $\lim_{x \to x_0^-} f(x) = -\infty$ if for every $N > 0$ there is a $\delta > 0$ such that if $x_0 - \delta < x < x_0$, then $f(x) < -N$. ■

EXAMPLE 7
PROVING THAT TWO INFINITE ONE-SIDED LIMITS EXIST

Prove that $\lim_{x \to 0^+} 1/x = \infty$ and $\lim_{x \to 0^-} 1/x = -\infty$.

SOLUTION: Let $N > 0$ be given. Then if $\delta = 1/N$, $0 < x < \delta$ implies that $x < 1/N$, or $1/x > N$, which shows that $\lim_{x \to 0^+} 1/x = \infty$. Similarly, if $-\delta < x < 0$, then with $\delta = 1/N$, $x > -\delta = -1/N$ implies that $1/x < -N$ and $\lim_{x \to 0^-} 1/x = -\infty$.

We now show that $\lim_{x \to x_0} f(x)$ exists if and only if the right- and left-hand limits exist and are equal. We prove this statement in the case in which the limits are finite. The infinite case is left as a problem (see Problem 21).

THEOREM 1 **(THEOREM 1.2.1 ON PAGE 63)**

$\lim_{x \to x_0} f(x) = L$ if and only if $\lim_{x \to x_0^+} f(x) = L$ and $\lim_{x \to x_0^-} f(x) = L$.

PROOF: First we show that if $\lim_{x \to x_0} f(x) = L$, then $\lim_{x \to x_0^+} f(x) = L$ and $\lim_{x \to x_0^-} f(x) = L$. To show the right-hand limit, we let $\epsilon > 0$ be given. Then there is a $\delta > 0$ such that if $0 < |x - x_0| < \delta$, then $|f(x) - L| < \epsilon$. In particular, if $x_0 < x < x_0 + \delta$, then $0 < x - x_0 < \delta$, $0 < |x - x_0| < \delta$, and $|f(x) - L| < \epsilon$. Therefore $\lim_{x \to x_0^+} f(x) = L$. The proof for the left-hand limit is similar.

Conversely, suppose that both the right- and left-hand limits exist and are equal to L. For a given $\epsilon > 0$ we choose δ_1 such that if $0 < x - x_0 < \delta_1$, then $|f(x) - L| < \epsilon$. We choose δ_2 such that if $-\delta_2 < x - x_0 < 0$, then $|f(x) - L| < \epsilon$. Let $\delta = \min\{\delta_1, \delta_2\}$. Then, if $|x - x_0| < \delta$, we have $0 < x - x_0 < \delta \le \delta_1$ and $-\delta_2 \le -\delta < x - x_0 < 0$, so that $|f(x) - L| < \epsilon$ whether $x - x_0$ is positive or negative. Hence, $\lim_{x \to x_0} f(x) = L$, and the theorem is proved. ∎

EXAMPLE 8
PROVING THAT A LIMIT EXISTS

Let

$$f(x) = \begin{cases} x, & x \le 1 \\ x^2, & x > 1. \end{cases}$$

Show that $\lim_{x \to 1} f(x) = 1$.

SOLUTION: We compute

$$\lim_{x \to 1^-} f(x) = \lim_{x \to 1^-} x = 1 \quad \text{and} \quad \lim_{x \to 1^+} f(x) = \lim_{x \to 1^+} x^2 = 1^2 = 1$$

Because $\lim_{x \to 1^-} f(x) = \lim_{x \to 1^+} f(x) = 1$, Theorem 1 tells us that $\lim_{x \to 1} f(x) = 1$.

■ PROBLEMS 1.8

■ SELF-QUIZ

I. If $0 < |x - 7| <$ _____, then $|(3x - 1) - 20| < 0.01$.
   a. 0.01   b. 0.005
   c. 0.01/8   d. 0.01/3

II. If $0 < |x - 6| <$ _____, then $|x^2 - 36| < 0.04$.
   a. 0.04   b. 0.004   c. 0.04/13
   d. $\min\{\sqrt{36.04} - 6, 6 - \sqrt{35.96}\}$

III. $\lim_{x \to a} f(x) = L$ means _____.
   a. $f(x)$ is close to L provided x is close enough to a
   b. $f(x)$ is close to $f(a)$ provided x is close enough to a
   c. x is close to a provided $f(x)$ is close enough to L
   d. $f(a) = L$

■ DRILL

In Problems 1–8, verify the declared limit directly from the definition given for *limit* on page 124. In particular, for (a) $\epsilon = 0.1$ and (b) $\epsilon = 0.01$, what values of δ will ensure that $|f(x) - L| < \epsilon$ if $0 < |x - x_0| < \delta$? Find δ as a function of ϵ.

1. $\lim_{x \to 1} 7x = 7$
2. $\lim_{x \to 4} (4x - 6) = 10$
3. $\lim_{x \to -2} (5x + 1) = -9$
4. $\lim_{x \to -2} x^2 = 4$
5. $\lim_{x \to 3} (x^2 - 6) = 3$
6. $\lim_{x \to 1} (5 - 2x^2) = 3$
7. $\lim_{x \to -1} (1 + x + x^2) = 1$
*8. $\lim_{x \to 2} x^3 = 8$
9. a. Verify from the definition of *limit at infinity* that

$$\lim_{x \to \infty} \frac{1}{\sqrt{x/10}} = 0.$$

   b. If $\epsilon = 0.01$, how large must N be so that if $N < x$, then $1/\sqrt{x/10} < \epsilon$?
10. a. Verify from the definition of an *infinite limit* that $\lim_{x \to 0} 1/x^4 = \infty$.
    b. How small must δ be chosen so that if $0 < |x| < \delta$, then $1/x^4 > 100,000,000$?
11. a. Verify from the definition of *limit of infinity* that $\lim_{x \to -\infty} |1/(x + 3)| = 0$.
    b. If $\epsilon = 0.01$, how large must N be chosen so that if $x < -N$, then $1/|x + 3| < \epsilon$?
12. a. Show that $\lim_{x \to \infty} [3 + (4/x^3)] = 3$.
    b. For $\epsilon = 0.001$, how large must N be chosen so that if $x > N$, then $|[3 + (4/x^3)] - 3| < \epsilon$?

CONTINUED

■ APPLICATIONS

13. Let

$$f(x) = \begin{cases} x^2 & \text{if } x < 4 \\ \sqrt{x} & \text{if } x \geq 4. \end{cases}$$

Prove, using the definition of *one-sided limits*, that $\lim_{x \to 4^-} f(x) = 16$ and $\lim_{x \to 4^+} f(x) = 2$.

14. Let

$$f(x) = \begin{cases} 1 & \text{if } x = 1/2^n \text{ for some integer } n \geq 1 \\ x & \text{otherwise.} \end{cases}$$

Show that if $x_0 \neq 0$, then $\lim_{x \to x_0} f(x) = x_0$.

*15. Let

$$f(x) = \begin{cases} 1 & \text{if } x \text{ is rational} \\ 0 & \text{if } x \text{ is irrational.} \end{cases}$$

Show that $\lim_{x \to x_0} f(x)$ does not exist for any real number x_0. [*Hint:* Pick an $\epsilon < 1$. Then show that

for any $\delta > 0$ there is an x in $(x_0 - \delta, x_0 + \delta)$ such that $|f(x) - f(x_0)| > \epsilon$.]

16. Let

$$f(x) = \begin{cases} 0 & \text{if } x < 0 \\ x & \text{if } x \geq 0. \end{cases}$$

Prove that $\lim_{x \to 0} f(x) = 0$.

17. Show, using Theorem 1, that $\lim_{x \to 2} f(x) = 8$ if

$$f(x) = \begin{cases} 3x + 2 & \text{if } x \leq 2 \\ 2x^2 & \text{if } x > 2. \end{cases}$$

18. Find a number c such that if

$$f(x) = \begin{cases} 3x & \text{if } x \leq -2 \\ cx^2 & \text{if } x > -2, \end{cases}$$

then $\lim_{x \to -2} f(x)$ exists.

■ SHOW/PROVE/DISPROVE

19. Let $f(x) = [x]$. Show that $\lim_{x \to x_0} f(x)$ exists if and only if x_0 is not an integer.

20. a. Show that if r is any positive integer, then $\lim_{x \to \infty} a/x^r = 0$ for any real number a.
    b. Prove part (a) when r is only assumed to be a positive rational number.
    c. Show that $\lim_{x \to \infty} a/(b + x^r) = 0$ for any positive rational number r and any real numbers a and b.

21. Prove that $\lim_{x \to a} f(x) = \infty$ if and only if $\lim_{x \to a^-} f(x) = \infty$ and $\lim_{x \to a^+} f(x) = \infty$.

22. Prove that

$$\lim_{x \to \infty} g(x) = \lim_{t \to 0^+} g(1/t)$$

if either one of those two limits exist.

■ CHALLENGE

Consider the following alternative (epsilon–delta) definition of continuity.

The function f is **continuous at** x_0 if and only if

(i) f is defined in a neighborhood of x_0,

(ii) for every $\epsilon > 0$, there is a $\delta > 0$ such that $|x - x_0| < \delta$ implies $|f(x) - f(x_0)| < \epsilon$.

In Problems 23–30, a function f is given and numbers x_0, ϵ are specified. Find a number δ such that $|f(x) - f(x_0)| < \epsilon$ whenever $|x - x_0| < \delta$.

23. $f(x) = 2x + 3$; $x_0 = 1$, $\epsilon = 0.01$

24. $f(x) = 3x - 5$; $x_0 = 2$, $\epsilon = 0.1$

25. $f(x) = x^2$; $x_0 = 0$, $\epsilon = 0.1$

26. $f(x) = x^2$; $x_0 = 3$, $\epsilon = 0.1$

27. $f(x) = 1/x$; $x_0 = 2$, $\epsilon = 0.1$

28. $f(x) = 1/x$; $x_0 = 2$, $\epsilon = 0.01$

29. $f(x) = 1/x$; $x_0 = 1$, $\epsilon = 0.1$

30. $f(x) = 1/x$; $x_0 = 1$, $\epsilon = 0.01$

31. Using limits as defined in this section, show that the definition of continuity in Section 1.7 is equivalent to the alternative definition of continuity given above. (Show that if f is continuous according to one definition, then it is continuous according to the other and vice versa.)

■ ANSWERS TO SELF-QUIZ

I. c, d   II. c, d   III. a

■ PROBLEMS CHAPTER 1 REVIEW

■ DRILL

1. Tabulate values of $f(x) = x^2 - 3x + 6$ for $x = 1$, 1.5, 1.9, 1.99; then tabulate the values of $f(x)$ for $x = 3, 2.5, 2.1, 2.01$. What does your table tell you about $\lim_{x \to 2}(x^2 - 3x + 6)$?

2. Tabulate values of $g(x) = (x + 10)x + 8$ for $x = -4, -2, -3.5, -2.5, -3.1, -2.9, -3.01$, and -2.99. What does your numerical table suggest to you about the value of $\lim_{x \to -3} g(x)$?

3. Calculate the following:
   a. $\lim_{x \to 1}(x^3 - 3x + 2)$
   b. $\lim_{x \to 5}(-x^3 + 17)$
   c. $\lim_{x \to 3} \dfrac{x^4 - 2x + 1}{x^3 + 3x - 5}$
   d. $\lim_{x \to -1} \dfrac{x^3 + x^2 + x + 1}{x^4 + x^3 + x^2 + x + 1}$

4. Calculate the following:
   a. $\lim_{x \to 0}|x + 2|$ b. $\lim_{x \to 1}|x - 3|$
   c. $\lim_{x \to -3}|x + 4|$ d. $\lim_{x \to 1}\dfrac{|x|}{x}$

5. Calculate the following:
   a. $\lim_{x \to 3}\dfrac{(x - 3)(x - 4)}{x - 3}$ b. $\lim_{x \to 5}\dfrac{x^2 - 6x + 5}{x - 5}$

6. Do the following limits exist? If not, explain why; if so, calculate them.
   a. $\lim_{x \to 1}\sqrt{x - 1}$ b. $\lim_{x \to 2}\sqrt{x - 1}$
   c. $\lim_{x \to -3}\dfrac{|x + 2|}{x + 2}$ d. $\lim_{x \to -2}\dfrac{|x + 2|}{x + 2}$
   e. $\lim_{x \to 1}\sqrt[3]{x - 1}$ f. $\lim_{x \to 1}\sqrt[4]{x - 1}$

7. Calculate the following:
   a. $\lim_{x \to 1} 23\sqrt{x - 17}$
   b. $\lim_{x \to -1}(1 - x + x^2 - x^3 + x^4)$
   c. $\lim_{x \to 4}\dfrac{x^2 + 9}{x^2 - 9}$ d. $\lim_{x \to -1} 5x^{250}$
   e. $\lim_{x \to -1} 6x^{251}$ f. $\lim_{x \to 3}(x^2 + x - 8)^5$
   g. $\lim_{x \to 0}\dfrac{x^8 - 7x^6 + x^3 - x^2 + 3}{x^{23} - 2x + 9}$
   h. $\lim_{x \to 0}\dfrac{ax^2 + bx + c}{dx^2 + ex + f}$
   (a, b, c, d, e, f are all nonzero real numbers.)

8. Calculate each of the following limits, or show that such a limit does not exist.
   a. $\lim_{x \to 1^+}\sqrt{x - 1}$ b. $\lim_{x \to 1^-}\sqrt{x - 1}$

   c. $\lim_{x \to 1^-}\dfrac{1}{x - 1}$ d. $\lim_{x \to 1^+}\dfrac{1}{x - 1}$
   e. $\lim_{x \to -2^+}\dfrac{x - 2}{x + 2}$ f. $\lim_{x \to -2^-}\dfrac{x - 2}{x + 2}$

9. Explain why $\lim_{x \to 0}(1/x^3)$ does not exist, but $\lim_{x \to 0}(1/x^4)$ does exist.

10. Calculate the following:
    a. $\lim_{x \to 2}\dfrac{x - 3}{(x - 2)^2}$ b. $\lim_{x \to -1}\dfrac{x + 10}{(x + 1)^{10}}$

11. Calculate the following:
    a. $\lim_{x \to \infty}\dfrac{1}{x^3}$ b. $\lim_{x \to \infty}\dfrac{1}{\sqrt{x + 2}}$
    c. $\lim_{x \to \infty}\dfrac{\sqrt{x}}{x^2 + 3}$ d. $\lim_{x \to \infty}\dfrac{x^3 + 6x^2 + 4x + 2}{3x^3 - 9x^2 + 11}$
    e. $\lim_{x \to \infty}\dfrac{3x^5 - 6x^2 + 3}{x^7 - 2}$
    f. $\lim_{x \to \infty}\dfrac{x^7 - 9}{30x^5 + x^4 + 161}$

12. If $x > 0$, how large must x be in order that $1/\sqrt[3]{x} < 0.01$?

13. For each labeled point of the graphs in Figure 1, decide whether the value of the derivative appears to be positive, negative, or zero.

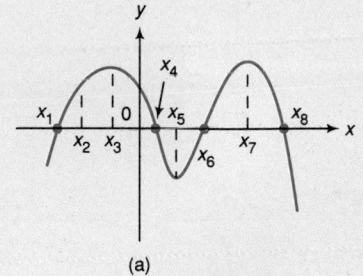

(a)

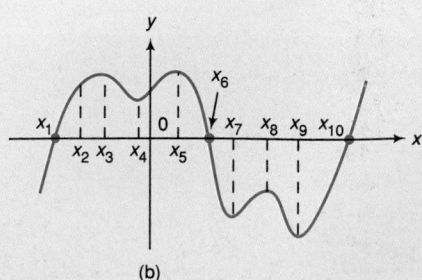

(b)

FIGURE 1

CONTINUED

14. Find an equation for the line tangent to the given curve at the specified point.
    a. $y = x^2 - 3$; $(2, 1)$
    b. $y = 7x^3 - 8$; $(1, -1)$
    c. $y = \sqrt{x + 1}$; $(3, 2)$
    d. $y = 1/(x + 1)$; $(0, 1)$

15. Explain why the curve $y = |x + 1|$ does not have a derivative at the point $x = -1$.

16. The distance a particle travels is given by $s(t) = t^3 + t^2 + 6$ kilometers after t hours. How fast is it traveling (in kilometers per hour) after 2 hours?

17. The volume of a sphere of radius r is $\frac{4}{3}\pi r^3$. How fast is the volume changing (with respect to changes in r) when the radius r is equal to 2 feet?

18. A slaughterhouse purchases cattle from a ranch at a cost of $C(q) = 200 + 8q - 0.02q^2$ dollars where q is the number of head of cattle bought at one time up to a maximum of 150. What is the house's marginal cost as a function of q?

19. At what values of x does the function graphed in Figure 2 appear not to have a derivative?

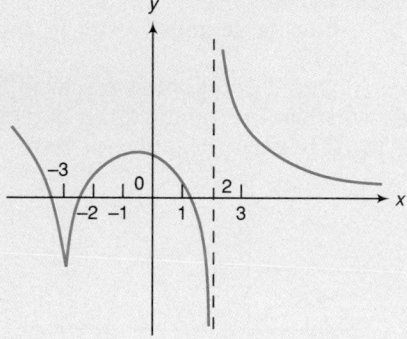

FIGURE 2

20. Show that if
    $$f(x) = \begin{cases} 2x + 3 & \text{if } x \leq -2 \\ x^2 - 5 & \text{if } x > -2 \end{cases}$$
    then $\lim_{x \to -2} f(x)$ exists. Calculate that limit.

21. Let
    $$f(x) = \begin{cases} x^2 + 3 & \text{if } x \leq 3 \\ x^3 + \alpha & \text{if } x > 3. \end{cases}$$
    For what value of α does $\lim_{x \to 3} f(x)$ exist? Calculate this limit.

In Problems 22–31, find the largest open interval in which the given function is continuous; indicate any removable discontinuities.

22. $f(x) = 2\sqrt{x}$
23. $f(x) = 3\sqrt[3]{x}$
24. $f(x) = 1/(x - 6)$
25. $f(x) = 1/(x^2 - 6)$
26. $f(x) = x/(x^2 - 4)$
27. $f(x) = |x + 2|$
28. $f(x) = \dfrac{|x + 3|}{x + 3}$
29. $f(x) = \dfrac{x^2 - 9}{x + 3}$
30. $f(x) = \dfrac{x + 3}{x^2 - 9}$
31. $f(x) = \dfrac{x^3 + 1}{x + 1}$

The last six problems rely on the theoretical material in Section 1.8. In Problems 32–36 below, use the ϵ-δ definition of the limit to verify the declared limits.

32. $\lim_{x \to -1} 4x = -4$
33. $\lim_{x \to 3}(x^2 + 2) = 11$
34. $\lim_{x \to \infty}\left(\dfrac{1}{x} + \dfrac{1}{x^2}\right) = 0$
35. $\lim_{x \to \infty}\dfrac{1}{\sqrt[3]{x + 3}} = 0$
36. $\lim_{x \to -3}|x + 4| = 1$
37. Let
    $$f(x) = \begin{cases} 1 & \text{if } x \text{ is an integer} \\ 2 & \text{if } x \text{ is not an integer.} \end{cases}$$
    Show that $\lim_{x \to \infty} f(x)$ does not exist.

COMPUTER EXERCISES

The purpose of Problems 1–6 is to estimate $\lim_{x \to x_0} f(x)$ based on its behavior near x_0. You will need a calculator or computer to evaluate functions and also to draw graphs.

1. Use five negative values and five positive values within the interval $[-0.05, 0.05]$ to estimate the value of $\lim_{x \to 0}(1 + x)^{1/x}$ to two decimal places.

Using a graphing calculator or computer software, generate a graph of the function and sketch it. Does this graph substantiate your estimate of the limit?

2. Use five negative values and five positive values within the interval $[-0.05, 0.05]$ to estimate the value of $\lim_{x \to 0}(1 + 2x)^{1/x}$ to two decimal places.

Using a graphing calculator or computer software, generate a graph of the function and sketch it. Does this graph substantiate your estimate of the limit?

3. a. Use your birth date to make a number b as follows. $b =$ "month.day year." For example, if you were born on February 7, 1971, your birthday number is $b = 2.771$.

   b. Make a table of values to estimate the value of $\lim\limits_{x \to 1} \dfrac{x^b - 1}{x - 1}$ to two decimal places. Using a graphing calculator or computer software, generate a graph of the function and sketch it. Does this graph substantiate your estimate of the limit?

4. Make a table of values to estimate the value of $\lim\limits_{x \to 0} \sin\left(\dfrac{1}{x}\right)$. Using a graphing calculator or computer software, generate a graph of the function and sketch it. Does this graph substantiate your estimate of the limit?

5. Use ten values within the interval $[0.95, 1.05]$ to estimate the value of $\lim\limits_{x \to 1} \dfrac{2^x - 2}{x - 1}$ to two decimal places. Using a graphing calculator or computer software, generate a graph of the function and sketch it. Does this graph substantiate your estimate of the limit?

6. Use ten values within the interval $[0.95, 1.05]$ to estimate the value of $\lim\limits_{x \to 1} \dfrac{3^x - 3}{x - 1}$ to two decimal places. Using a graphing calculator or computer software, generate a graph of the function and

sketch it. Does this graph substantiate your estimate of the limit?

7. a. Using the $\boxed{\ln}$ key on your calculator or the ln x function in your computer software, write a program to compute
   $$d(x) = \frac{\ln(x + 0.00001) - \ln x}{0.00001}$$
   for positive values of the number x.

   b. Using a graphing calculator or computer software, plot values of $d(x)$ for $0.1 \leq x \leq 10$ in increments of 0.1.

   c. On the same screen or printout, generate a graph of the function $\dfrac{1}{x}$ for $x > 0$ and sketch it.

   d. Using the graphs obtained in parts (b) and (c), make an educated guess at the derivative of the function $f(x) = \ln x$.

8. a. Using the $\boxed{e^x}$ or the $\boxed{\text{INV}}\,\boxed{\ln}$ key (or keys) on your calculator, or the function e^x in your computer software, write a program to compute
   $$e(x) = \frac{e^{x + 0.00001} - e^x}{0.00001}.$$

   b. Using a graphing calculator or computer software, plot values of $e(x)$ for $-4 \leq x \leq 4$ in increments of 0.1.

   c. On the same screen or printout, generate a graph of the function e^x for $-4 \leq x \leq 4$ and sketch it.

   d. Using the graphs obtained in parts (b) and (c), make an educated guess at the derivative of the function $f(x) = e^x$.

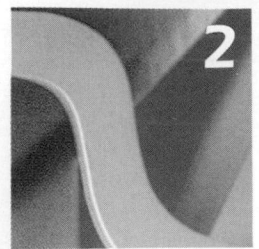

2 | MORE ABOUT DERIVATIVES

In the previous chapter, we introduced the concepts of the limit and the derivative but found that the calculation of derivatives could be extremely difficult. An even more annoying problem was the seeming necessity to come up with a special technique (like multiplying and dividing by some quantity) each time we took the limit in the process of computing a derivative.

In this chapter, we continue the discussion of properties of the derivatives of functions. We will be principally concerned with simplifying the process of differentiation so that it will no longer be necessary to deal with complicated limits.

In many of the sections of this chapter we will be deriving formulas for calculating derivatives. Formulas are usually not very exciting, and we ask you to bear with us here. By the time you have completed the chapter, you will find that differentiation is not nearly as complicated as it now seems. The work involved in memorizing the appropriate formulas will pay dividends in the chapters to come.

2.1 SOME DIFFERENTIATION FORMULAS

We begin by deriving a formula for calculating the derivative of $y = f(x) = x^n$, where n is a positive integer. First, let us look for a pattern. We have already calculated the following derivatives:

(i) $\dfrac{d}{dx} x = 1$ (since $y = x = 1 \cdot x + 0$ is the equation of a straight line with slope 1)

(ii) $\dfrac{d}{dx} x^2 = 2x$ (Example 4 on page 88)

(iii) $\dfrac{d}{dx} x^3 = 3x^2$ (Example 6 on page 90)

Do you see a pattern? The answer is given in Theorem 1.

THEOREM 1   **DERIVATIVE OF x^n**

If n is a positive integer, then x^n is differentiable and

$$\frac{d}{dx} x^n = nx^{n-1}. \tag{1}$$

PROOF: Using the alternative definition of the derivative on page 87 and Theorem 1 on page 100, we have, if $f(x) = x^n$,

$$\frac{d}{dx} x^n = f'(x) = \lim_{t \to x} \frac{f(t) - f(x)}{t - x} = \lim_{t \to x} \frac{t^n - x^n}{t - x}$$

remember from page 100:
$$a^n - b^n = (a - b)(a^{n-1} + a^{n-2}b + a^{n-3}b^2 + \cdots + ab^{n-2} + b^{n-1})$$

$$= \lim_{t \to x} \underbrace{(t^{n-1} + t^{n-2}x + t^{n-3}x^2 + \cdots + tx^{n-2} + x^{n-1})}_{n \text{ terms}}$$

$$= \underbrace{x^{n-1} + x^{n-2}x + x^{n-3}x^2 + \cdots + x \cdot x^{n-2} + x^{n-1}}_{\substack{n \text{ terms} \\ \text{all equal to } x^{n-1}}}$$

$$= nx^{n-1}. \quad \blacksquare$$

EXAMPLE 1
DERIVATIVE OF x^n

Let $f(x) = x^4$. Then $f'(x) = 4x^3$.

EXAMPLE 2
DERIVATIVE OF x^n

Let $f(x) = x^{17}$. Then $f'(x) = 17x^{16}$.

We can compute derivatives of more complicated functions by deriving some more general formulas. We first give a theorem that states the obvious fact that a constant function does not change.

THEOREM 2   **DERIVATIVE OF A CONSTANT FUNCTION**

Let $f(x) = c$, a constant. Then $f'(x) = 0$.

NOTE: This result is evident from looking at the graph of a constant function—a horizontal line with a slope of zero (see Figure 1).

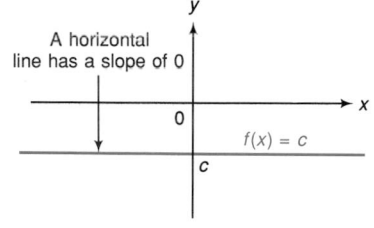

A horizontal line has a slope of 0

FIGURE 1
THE DERIVATIVE OF A CONSTANT FUNCTION = THE SLOPE OF A HORIZONTAL LINE = 0

PROOF:

$$\frac{c + \Delta x - c}{\Delta x} = \frac{c - c}{\Delta x} = 0$$

$$f'(x) = \lim_{\Delta x \to 0} \frac{f(x + \Delta x) - f(x)}{\Delta x} = \lim_{\Delta x \to 0} \frac{c - c}{\Delta x}$$

$$= \lim_{\Delta x \to 0} \frac{0}{\Delta x} = \lim_{\Delta x \to 0} 0 = 0. \quad \blacksquare$$

The converse of this theorem is also true—namely, that if $f' = 0$ on an interval, then f is a constant function on that interval. The proof of this theorem is more difficult and will be deferred until Section 3.7.

THEOREM 3   **DERIVATIVES OF CONSTANT MULTIPLES AND SUMS OF FUNCTIONS**

Let c be a constant. If f and g are differentiable, then cf and $f + g$ are also differentiable, and

(i) $\dfrac{d}{dx} cf = c \dfrac{df}{dx}.$

The derivative of a constant times a function equals the constant times the derivative of the function.

(ii) $\dfrac{d}{dx}(f + g) = \dfrac{df}{dx} + \dfrac{dg}{dx}$ The derivative of the sum equals the sum of the derivatives.

We can write these two rules in compact form:

$$(cf)' = cf'$$

$$(f + g)' = f' + g'.$$

PROOF:

$$\frac{d}{dx}(f + g) = \lim_{\Delta x \to 0} \frac{(f + g)(x + \Delta x) - (f + g)(x)}{\Delta x}$$

$$= \lim_{\Delta x \to 0} \frac{f(x + \Delta x) + g(x + \Delta x) - f(x) - g(x)}{\Delta x}$$

$$= \lim_{\Delta x \to 0} \left\{ \left[\frac{f(x + \Delta x) - f(x)}{\Delta x} \right] + \left[\frac{g(x + \Delta x) - g(x)}{\Delta x} \right] \right\}$$

By Theorem 1.3.2(ii), on page 70

$$\overset{\downarrow}{=} \lim_{\Delta x \to 0} \frac{f(x + \Delta x) - f(x)}{\Delta x} + \lim_{\Delta x \to 0} \frac{g(x + \Delta x) - g(x)}{\Delta x}$$

$$= \frac{df}{dx} + \frac{dg}{dx}$$

The proof of part (i) is left to the reader. See Problem 37. ∎

EXAMPLE 3
USING THE DIFFERENTIATION
RULES

Compute $\dfrac{d}{dx}(41x^3)$.

SOLUTION:

$$\frac{d}{dx}(41x^3) \overset{\text{Theorem 3(i)}}{=} 41 \frac{d}{dx} x^3 \overset{\text{Theorem 1}}{=} 41(3x^2) = 123x^2.$$

EXAMPLE 4
USING THE DIFFERENTIATION
RULES

Let $f(x) = 4x^3 + 3\sqrt{x}$. Find $f'(x)$.

SOLUTION:

$$\frac{d}{dx}(4x^3 + 3\sqrt{x}) \overset{\text{By Theorem 3}}{=} 4\frac{d}{dx}x^3 + 3\frac{d}{dx}\sqrt{x} \overset{\text{Example 5 on page 89}}{=} 4 \cdot 3x^2 + \frac{3}{2\sqrt{x}}$$

$$= 12x^2 + \frac{3}{2\sqrt{x}}.$$

We have shown that the derivative of the sum of two functions is the sum of the derivatives of the two functions. It is not too difficult to show that this fact applies to more than two functions. The theorem that states this result is given next. The proof is left as an exercise (see Problem 39).

THEOREM 4   **THE DERIVATIVE OF THE SUM OF *n* FUNCTIONS**

Let $f_1, f_2, \ldots, f_n$ be n differentiable functions. Then $f_1 + f_2 + \cdots + f_n$ is differentiable, and

$$(f_1 + f_2 + \cdots + f_n)' = f'_1 + f'_2 + \cdots + f'_n. \quad \blacksquare$$ (2)

EXAMPLE 5

Let $f(x) = 1 + x + x^2 + x^3$. Calculate df/dx.

SOLUTION:

$$\frac{df}{dx} = \frac{d}{dx} 1 + \frac{d}{dx}(x) + \frac{d}{dx}(x^2) + \frac{d}{dx}(x^3) = 0 + 1 + 2x + 3x^2$$

$$= 1 + 2x + 3x^2$$

■ PROBLEMS 2.1

■ SELF-QUIZ

I. True–False:
   a. A constant function has a constant derivative.
   b. If a function has a constant derivative, then the function is a constant.

II. The derivative of $5x^4$ is _____.
   a. x^5 b. $20x$
   c. $4x^3$ d. $20x^3$

III. The graph of $y = -x^2$ has tangent lines with positive slope on _____.

   a. $(-\infty, 0]$ b. $(-\infty, 0)$
   c. $(0, \infty)$ d. $(-\infty, \infty)$

IV. The line tangent to the graph of $y = x^2$ at the point $(-3, 9)$ has equation _____.
   a. $y = -6(x - 3) + 9$
   b. $y = 6(x + 3) - 9$
   c. $y = -6(x + 3) + 9$
   d. $y = -6x + 9$

■ DRILL

In Problems 1–12, calculate the derivative of the given function.

1. $f(x) = 3^4$
2. $g(x) = x^5$
3. $h(x) = x^2 + (a + b)x + ab$
4. $f(t) = t^{10} - t^3$
5. $g(t) = 1 - t + t^4 - t^7$
6. $h(t) = 3t^2 + 19t + 2$
7. $F(w) = 27w^6 - (3w^5 - 4w)$
8. $F(x) = x^5 - \sqrt{x}$
9. $F(z) = -3z^{12} + 12z^3$
10. $G(x) = x^5 - 3\sqrt{x}$
11. $G(s) = 3s^8 - 8s^6 - 7s^4 + 2s^2 + 3$
12. $G(t) = t^{100} + 100t^{10} + 10$

In Problems 13–20, find the line that is tangent to the given curve at the given point.

13. $y = x^4$; $(1, 1)$
14. $y = x^4$; $(-2, 16)$
15. $y = 2x^7 - x^6 - x^3$; $(1, 0)$
16. $y = 3x^5 - 3x^3 + 1$; $(-1, 1)$
17. $y = 1 + x + x^2 + x^3 + x^4 + x^5$; $(0, 1)$

CONTINUED

18. $y = 1 - x + 2x^2 - 3x^3 + 4x^4$; (1, 3)
19. $y = x^6 - 6\sqrt{x}$; (1, −5)
20. $y = 5x^6 - x^4 + 2x^3$; (1, 6)

A **normal line**[†] to a curve at a point is a line that is perpendicular to the tangent line[‡] at that point. In

Problems 21–24, find the normal line to the given curve at the specified point.

21. $y = x^3$; (1, 1)
22. $y = x^7 - 6x^5$; (1, −5)
23. $y = x^6 - 6\sqrt{x}$; (1, −5)
24. $y = 2x^4 - 3x^3 + 2x + 1$; (2, 13)

■ **APPLICATIONS**

25. Find the points on the graph of $y = 2x^3 + 3x^2 - 6x + 1$ where the curve has a horizontal tangent.
26. Suppose that when an airplane takes off (starting from rest), the distance (in ft) it travels during the first few seconds is given by the formula $s = 1 + 4t + 6t^2$. How fast (in ft/sec) is the plane traveling after 10 sec? After 20 sec?
27. A petri dish contains two colonies of bacteria. The population of the first colony is given by $P_1(t) = 1000 + 50t - 20\sqrt{t}$, and the population of the second is given by $P_2(t) = 2000 + 30t^2 - 80t$, where t is measured in hours.
    a. Find a function that represents the total population of the two species.
    b. What is the instantaneous growth rate of the total population?
    c. How fast is the total population growing after 4 hr? After 6 hr?
*28. For the model of Problem 27, how fast is the first population growing when the second population is growing at the rate of 160 bacteria/hr?
29. Verify that the rate of change of the area of a circle with respect to its radius is equal to its circumference.
*30. A growing grapefruit with a diameter of $2k$ inches has a skin that is $k/12$ inches thick (the

skin is included in the diameter of the grapefruit). What is the rate of growth of the volume of the skin (per unit growth in the radius) when the radius of the grapefruit is 3 inches?
31. Where, if ever, is the graph of $y = \sqrt{x}$ parallel to the line $(\frac{1}{8})x - 8y = 1$?
*32. Let $f(x) = x^2$ and $g(x) = (\frac{1}{3})x^3$. Each vertical line, $x =$ constant, meets the graph of f and the graph of g.
    a. On what vertical lines do the graphs of f and g have parallel tangents?
    b. On what horizontal lines do they have parallel tangents?
*33. Find the equation of the two lines tangent to the curve $y = x^2 + 3$ that pass through the point (1, 0). [*Hint:* Any point on the curve has the coordinates $(a, a^2 + 3)$. If L is a tangent line at the point $(a, a^2 + 3)$, then the slope of L is $2a$.]
*34. a. Verify that the normal line to $y = x^2$ at (3, 9) passes through the point (−3, 10).
    b. Are there any other points on $y = x^2$ such that the normal line there also will pass through (−3, 10)? Justify your answer.
*35. Let $f(x) = (x - 1)^2 + 3 = x^2 - 2x + 4$ and $g(x) = -f(-x) = -x^2 - 2x - 4$. Find each line that is tangent to both of the graphs $y = f(x)$ and $y = g(x)$. [*Note:* A rough sketch indicates that there is at least one such line.]
*36. Let $f(x) = x^2$ and $g(x) = -x^2 + 10x - 17$. Find each line that is tangent to both of the graphs $y = f(x)$ and $y = g(x)$.

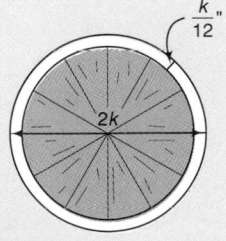

$\frac{k}{12}$"

$2k$

[†] From the Latin *norma*, meaning "square," the carpenter's square. Until the 1830s, the English word "normal" meant standing at right angles to the ground.
[‡] Recall that the slopes of two perpendicular lines are negative reciprocals of one another.

CONTINUED

SHOW/PROVE/DISPROVE

37. Suppose f is a differentiable function and c is an arbitrary real number. Prove, using the definition of the derivative, that the function $c \cdot f$ is differentiable and

$$\frac{d}{dx}(c \cdot f) = c \cdot \left(\frac{d}{dx} f\right).$$

38. Show that if $y = x^n$, where n is an integer greater than 1, then the tangent line to the curve at the point $(0, 0)$ is the x-axis.

39. a. Prove Theorem 4 when $n = 3$ and when $n = 4$.

*b. Use mathematical induction (see Appendix 2) to prove Theorem 4 for general n.

*40. Consider the graph of $y = ax^2 + bx + c$, $a \neq 0$.
   a. If a and c have the same sign ($a \cdot c > 0$), then there are exactly two tangents to the graphs which pass through the origin. Prove this statement.
   b. If a and c have opposite signs ($a \cdot c < 0$), then no tangent to the graph passes through the origin. Prove this statement.
   c. Discuss the remaining case, $c = 0$.

ANSWERS TO SELF-QUIZ

I. a. True (If $f(x) = c$, then $f'(x) = 0$ and 0 is a constant.)
   b. False (For example, if $g(x) = -3x$, then $g'(x) = -3$.)

II. d

III. b

IV. c

2.2 THE PRODUCT AND QUOTIENT RULES

In this section we develop some additional rules to simplify the calculation of derivatives. To see why additional rules are needed, consider the problem of calculating the derivatives of

$$f(x) = \sqrt{x}(x^4 + 3) \qquad \text{or} \qquad g(x) = \frac{x^4 + 3}{\sqrt{x}}.$$

To carry out the calculations from the definition would be very tedious. However, we will shortly see that these calculations can be made rather simple.

Let f and g be two differentiable functions of x. What is the derivative of the product fg? It is easy to be led astray here. The limit theorem 1.3.2 (iii) states that

$$\lim_{x \to x_0} f(x)g(x) = \lim_{x \to x_0} f(x) \lim_{x \to x_0} g(x).$$

However, the derivative of the product is *not* equal to the product of the derivatives. That is,

$$\frac{d}{dx} fg \text{ does not equal } \frac{df}{dx} \cdot \frac{dg}{dx}.$$

Originally, Leibniz, the codiscoverer of the derivative, thought that they were equal. But an easy example shows that this is false. Let $f(x) = x$ and

$g(x) = x^2$. Then,

$$\frac{df}{dx} = 1, \frac{dg}{dx} = 2x, \text{ but } \frac{d(fg)}{dx} = \frac{d(x^3)}{dx} = 3x^2 \neq \frac{df}{dx} \cdot \frac{dg}{dx}.$$

The correct formula, discovered after many false steps by both **Leibniz** and **Newton**, is given below.

THEOREM 1   **PRODUCT RULE**

Let f and g be differentiable. Then fg is differentiable, and

$$\frac{d}{dx} fg(x) = \frac{df}{dx} g(x) + f(x) \frac{dg}{dx} \tag{1}$$

or

$$(fg)' = f'g + fg' \quad \blacksquare \tag{1'}$$

Verbally, the product rule says that *the derivative of the product of two functions is equal to the first times the derivative of the second plus the second times the derivative of the first.*

PROOF:

$$\frac{d}{dx} fg = \lim_{\Delta x \to 0} \frac{f(x + \Delta x)g(x + \Delta x) - f(x)g(x)}{\Delta x}.$$

This equation doesn't look very much like the derivative of anything. We can go further if we can separate the functions f and g as in the proof of Theorem 3 (ii) on page 136 (the theorem on the derivative of a sum). We can accomplish this by adding and subtracting the term $f(x)g(x + \Delta x)$ in the numerator. As we will see, this technique makes everything come out nicely. We have (with the additional terms in blue)

$$\frac{d}{dx} fg$$

$$\overset{\text{These are the additional terms.}}{= \lim_{\Delta x \to 0} \frac{f(x + \Delta x)g(x + \Delta x) - \overbrace{f(x)g(x + \Delta x) + f(x)g(x + \Delta x)} - f(x)g(x)}{\Delta x}}$$

$$= \lim_{\Delta x \to 0} \frac{g(x + \Delta x)[f(x + \Delta x) - f(x)]}{\Delta x} + \lim_{\Delta x \to 0} \frac{f(x)[g(x + \Delta x) - g(x)]}{\Delta x}$$

$$= \lim_{\Delta x \to 0} g(x + \Delta x) \lim_{\Delta x \to 0} \frac{f(x + \Delta x) - f(x)}{\Delta x}$$

$$+ \lim_{\Delta x \to 0} f(x) \lim_{\Delta x \to 0} \frac{g(x + \Delta x) - g(x)}{\Delta x}$$

$$= g(x) \frac{df}{dx} + f(x) \frac{dg}{dx}.$$

Here we have used the fact that $\lim_{\Delta x \to 0} g(x + \Delta x) = g(x)$. This follows from the fact that g is differentiable, so that g is continuous by Theorem 1.7.3. As $\Delta x \to 0$, $x + \Delta x \to x$, so by the definition of continuity, $\lim_{x + \Delta x \to x} g(x + \Delta x) = g(x)$. ■

EXAMPLE 1
USING THE PRODUCT RULE

Let $h(x) = \sqrt{x}(x^4 + 3)$. Calculate dh/dx.

SOLUTION: $h(x) = f(x)g(x)$, where $f(x) = \sqrt{x}$ and $g(x) = x^4 + 3$. Then

$$\frac{dh}{dx} = f\frac{dg}{dx} + g\frac{df}{dx} = \sqrt{x}(4x^3) + (x^4 + 3)\frac{1}{2\sqrt{x}}.$$

This is the correct answer, but we will use some algebra to simplify the result:

$$\frac{dh}{dx} = \frac{2\sqrt{x}\sqrt{x}(4x^3) + x^4 + 3}{2\sqrt{x}} = \frac{2x(4x^3) + x^4 + 3}{2\sqrt{x}} = \frac{9x^4 + 3}{2\sqrt{x}}.$$

Having now discussed the product of two functions, we turn to their quotient.

THEOREM 2   **QUOTIENT RULE**

Let f and g be differentiable at x. Then, if $g(x) \neq 0$, $\dfrac{f}{g}$ is differentiable at x and

$$\left(\frac{f}{g}\right)' = \frac{d}{dx}\left(\frac{f}{g}\right) = \frac{g(x)(df/dx) - f(x)(dg/dx)}{g^2(x)}$$

$$= \frac{gf' - fg'}{g^2}. \quad \blacksquare \tag{2}$$

The quotient rule states that *the derivative of the quotient is equal to the denominator times the derivative of the numerator minus the numerator times the derivative of the denominator all over the denominator squared.*

PROOF:

$$\frac{d}{dx}\left(\frac{f}{g}\right) = \lim_{\Delta x \to 0} \frac{\dfrac{f(x + \Delta x)}{g(x + \Delta x)} - \dfrac{f(x)}{g(x)}}{\Delta x}$$

$$= \lim_{\Delta x \to 0} \frac{f(x + \Delta x)g(x) - f(x)g(x + \Delta x)}{\Delta x\, g(x + \Delta x)g(x)}.$$

As in the proof of the product rule, we can separate f and g by adding and subtracting an appropriate term. In this case, we add and subtract $f(x)g(x)$. We continue, with the new terms in blue,

$$\overbrace{\text{These terms sum to zero.}}$$

$$= \lim_{\Delta x \to 0} \frac{f(x + \Delta x)g(x) - f(x)g(x) + f(x)g(x) - f(x)g(x + \Delta x)}{\Delta x\, g(x + \Delta x)g(x)}$$

$$= \lim_{\Delta x \to 0} \frac{\left\{ g(x)\left(\dfrac{f(x + \Delta x) - f(x)}{\Delta x} \right) - f(x)\left(\dfrac{g(x + \Delta x) - g(x)}{\Delta x} \right) \right\}}{g(x + \Delta x)g(x)}$$

$$= \frac{\left\{ g(x) \displaystyle\lim_{\Delta x \to 0} \left(\dfrac{f(x + \Delta x) - f(x)}{\Delta x} \right) - f(x) \displaystyle\lim_{\Delta x \to 0} \left(\dfrac{g(x + \Delta x) - g(x)}{\Delta x} \right) \right\}}{g(x) \displaystyle\lim_{\Delta x \to 0}\, g(x + \Delta x)}$$

$$= \frac{g(x)f'(x) - f(x)g'(x)}{[g(x)]^2}.$$

Again we used the fact that g is continuous (being differentiable), so that $\lim_{\Delta x \to 0} g(x + \Delta x) = g(x)$. ∎

EXAMPLE 2
USING THE QUOTIENT RULE

Compute the derivative of $\dfrac{x + 1}{x^2 - 2}$.

SOLUTION:

$$\frac{d}{dx}\frac{x + 1}{x^2 - 2} = \frac{(x^2 - 2)\overset{\substack{\text{derivative}\\ \text{of}\\ \text{numerator}}}{\dfrac{d}{dx}(x + 1)} - \overset{\substack{\text{numerator}}}{(x + 1)}\overset{\substack{\text{derivative}\\ \text{of}\\ \text{denominator}}}{\dfrac{d}{dx}(x^2 - 2)}}{\underset{\substack{\text{denominator}\\ \text{squared}}}{(x^2 - 2)^2}}$$

$$= \frac{(x^2 - 2)(1) - (x + 1)(2x)}{(x^2 - 2)^2}$$

$$= \frac{(x^2 - 2) - (2x^2 + 2x)}{(x^2 - 2)^2} = \frac{-x^2 - 2x - 2}{(x^2 - 2)^2}.$$

In the last section, we proved that

$$\frac{d}{dx} x^n = n x^{n-1}$$

if n is a positive integer. With the aid of the quotient rule, we can extend this result to the case in which n is a negative integer. [*Note:* When $n = 0$, $x^0 = 1$, which we already know has a derivative of 0.]

THEOREM 3   **THE DERIVATIVE OF x^{-n}**

Let $y = x^{-n}$, where n is a positive integer (so that $-n$ is a negative integer). Then,

$$\frac{dy}{dx} = \frac{d}{dx}(x^{-n}) = -nx^{-n-1}.$$ (3)

PROOF: $x^{-n} = 1/x^n$. Then we use the quotient rule (2) to obtain

$$\frac{dy}{dx} = \frac{d}{dx}\left(\frac{1}{x^n}\right) = \frac{x^n \dfrac{d}{dx}(1) - (1)\dfrac{d}{dx}x^n}{(x^n)^2} = \frac{x^n \cdot 0 - nx^{n-1}}{x^{2n}}$$

$$= \frac{-nx^{n-1}}{x^{2n}} = -nx^{(n-1)-2n} = -nx^{-n-1}. \quad \blacksquare$$

NOTE: We now know that $d(x^n)/dx = nx^{n-1}$ holds for any integer n.

EXAMPLE 3
THE DERIVATIVE OF x^{-n}

Let $y = x^{-7}$. Calculate dy/dx.

SOLUTION: Using (3), we obtain

$$\frac{dy}{dx} = -7x^{-8}.$$

EXAMPLE 4
THE DERIVATIVE OF x^{-n}

Let $y = 1/x$. Calculate dy/dx.

SOLUTION: $1/x = x^{-1}$, so that

$$\frac{d}{dx}\left(\frac{1}{x}\right) = \frac{d}{dx}x^{-1} = -1 \cdot x^{-2} = \frac{-1}{x^2}.$$

EXAMPLE 5
GRAVITATIONAL FORCE

Newton's law of universal gravitation states that *the gravitational force between any two particles having masses* m_1 *and* m_2 *separated by a distance* r *is an attraction acting along the line joining the particles and has the magnitude*

$$F = G\frac{m_1 m_2}{r^2},$$ (4)

where G is a universal constant having the same value for all pairs of particles[†] (see Figure 1).

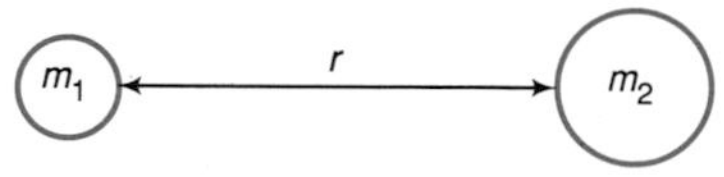

Two asteroids are approaching each other. The first has a mass of 1000 kg and the second a mass of 3000 kg.

FIGURE 1
TWO MASSES SEPARATED BY A DISTANCE r

(a) What is the gravitational force between the two asteroids when they are 10 km apart?

(b) How is this force changing at that distance?

[†] In metric units $G = 6.673 \times 10^{-11}$ N · m²/kg² (N = newton), where 1 newton is the force that will accelerate a 1-kg mass at the rate of 1 m/sec² (1 N = 1 kg-m/sec²).

SOLUTION:

(a) $r = 10$ km $= 10,000$ m. Then,

$$F = G\frac{m_1 m_2}{r^2} = G\frac{(1000)(3000)}{10,000^2} = 0.03\ G \text{ newton.}$$

(b) The rate of change of F when $r = 10,000$ m is $F'(10,000)$. But,

$$\frac{dF}{dr} = \frac{d}{dr}\,Gm_1 m_2 r^{-2} = Gm_1 m_2\,\frac{d}{dr}\,r^{-2}$$

$$= Gm_1 m_2(-2r^{-3}) = \frac{-2Gm_1 m_2}{r^3}$$

$$= \frac{(-2)(1000)(3000)G}{(10,000)^3} = -0.000006G$$

$$= -6 \times 10^{-6}G \text{ newton per meter}$$

when $r = 10,000$. Here the minus sign indicates that as r increases, the force F decreases, and vice versa. But since the asteroids are approaching one another, r is decreasing, and therefore F is increasing at a rate of $6 \times 10^{-6}G$ newton for every 1 m decrease in r.

■ PROBLEMS 2.2

■ SELF-QUIZ

I. Verify that applying the product rule to various expressions equal to x^7 yields the same result. Show that x^7, $x \cdot x^6$, $x^2 \cdot x^5$, and $x^3 \cdot x^4$ have the same derivative. (Differentiate each expression as it is shown, then multiply and simplify.)

II. Verify that x^5, x^6/x, x^7/x^2, and x^8/x^3 have the same derivative. (Apply the quotient rule to each expression as it is shown, then simplify.)

III. A beginner's mistake that is not unusual is to compute the derivative of a product as if it equaled the product of the individual derivatives. This problem should convince you that such is rarely the case. From the following functions, pick all pairs that satisfy the equation $(f \cdot g)' = f' \cdot g'$. [For instance, $f(x) = 7$ and $g(x) = 11$ works.]

  a. $a(x) = x$ b. $b(x) = 1/x$
  c. $c(x) = x + 1$ d. $d(x) = 1/x - 1$

IV. Now, let's investigate a common error in computing the derivative of a quotient. Let $g(x) = x$; for which of the following choices for function f is it true that $(f/g)' = f'/g'$?

  a. $f(x) = 1/(1 - x)$ b. $f(x) = x/(1 + x)$
  c. $f(x) = x \cdot (1 - x)$ d. $f(x) = x/(x - 1)$

■ DRILL

In Problems 1–22, find the derivative of the given function. Use Theorems 1 and 2 as appropriate; where necessary, use the fact that $(d/dx)(\sqrt{x}) = 1/(2\sqrt{x})$.

1. $f(x) = (x^2 - 9) \cdot (x - 5)$
2. $g(t) = (9 - t^2) \cdot (5 - t)$
3. $f(t) = 1/t^6$
4. $g(x) = 5/(7x^3)$

5. $f(x) = 2x \cdot (x^2 + 1)$
6. $g(t) = 2(t^2 + 1)/t$
7. $g(t) = t^{-100}$
8. $f(x) = 1/(x^5 + 3x)$
9. $F(x) = (1 + x + x^5) \cdot (2 - x + x^6)$
10. $G(x) = (1 + x + x^5)/(2 - x + x^6)$
11. $A(x) = 1/\sqrt{x}$
12. $B(x) = x^{3/2} = x \cdot \sqrt{x}$

CONTINUED

13. $g(t) = t^3 \cdot (1 + \sqrt{t})$ 14. $h(z) = z^3/(1 + \sqrt{z})$
15. $f(t) = (1 + \sqrt{t})/(1 - \sqrt{t})$
16. $g(t) = (1 + \sqrt{t}) \cdot (1 - \sqrt{t})$
17. $p(v) = (v^3 - \sqrt{v}) \cdot (v^2 + 2\sqrt{v})$
18. $q(v) = (v^3 - \sqrt{v})/(v^2 + 2\sqrt{v})$
19. $C(x) = x^{5/2} = x^2 \cdot \sqrt{x}$
20. $D(x) = x^{7/2} = x^3 \cdot \sqrt{x}$
21. $g(t) = \dfrac{1}{\sqrt{t}} \cdot \left(\dfrac{1}{t^4 + 2} \right)$

22. $f(x) = \dfrac{\sqrt{x} - (2/\sqrt{x})}{3x^3 + 4}$

In Problems 23–26, use the result of Problem 40(b) to calculate the derivative of the given function.

23. $v(x) = (1 - x^{-2}) \cdot (1 + x) \cdot (1 - x)$
24. $v(x) = x \cdot (1 + \sqrt{x}) \cdot (1 - \sqrt{x})$
25. $v(x) = (x^2 + 1) \cdot (x^3 + 2) \cdot (x^4 + 3)$
26. $v(x) = x^{-2} \cdot (2 - 3\sqrt{x}) \cdot (1 + x^3)$

■ APPLICATIONS

In Problems 27–30, find an equation for the line tangent to the given curve at the specified point.

27. $f(x) = 4x(x^5 + 1)$; $(1, 8)$
28. $g(t) = t^2/(1 + \sqrt{t})$; $(4, \frac{16}{3})$
29. $h(u) = (1 + \sqrt{u})/u^2$; $(1, 2)$
30. $p(v) = (1 + 3\sqrt{v})(1 - \sqrt{v})$; $(1, 0)$

In Problems 31–34, find an equation for the line normal to the given curve at the specified point.

31. $f(x) = 1/\sqrt{x}$; $(1, 1)$
32. $g(x) = 1/x^7$; $(1, 1)$
33. $h(x) = (1 - \sqrt{x})/(1 + 3\sqrt{x})$; $(1, 0)$
34. $p(x) = (x^3 + 2)/(x^2 + 2)$; $(0, 1)$
35. According to **Poiseuille's law**,[†] the resistance R of a blood vessel of length L and radius r is given by $R = \alpha L/r^4$, where α is a constant of proportionality determined by the viscosity of blood.

Assuming that the length of the vessel is kept constant while the radius increases, how fast is the resistance decreasing (as a function of the radius) when $r = 0.2$ mm?
*36. The mass of the earth is 5.983×10^{24} kg. Suppose a meteorite with a mass of 10,000 kg is moving toward a collision with the earth.
  a. When the meteorite is 100 km from the earth, what is the force (in newtons) of gravitational attraction between the meteorite and the earth?
  b. At that distance, how fast is this force increasing (in N/m) as the meteorite continues on its collision course?
37. Find the derivative of $f \cdot g/(f + g)$, where f and g are differentiable functions.
*38. Find $(d/dv)(v^{n/2})$ where n is a positive integer. [Hint: See Problems 12, 19, and 20.]

■ SHOW/PROVE/DISPROVE

39. Use the product rule to prove that if f is differentiable, then $d/dx([f(x)]^2) = 2f(x)f'(x)$.
40. Let $v(x) = f(x) \cdot g(x) \cdot h(x)$ where f, g, and h are differentiable functions.
  a. Using the product rule, show that

$$\frac{dv}{dx} = f \cdot \frac{d(g \cdot h)}{dx} + (g \cdot h) \cdot \frac{df}{dx}.$$

  b. Use part (a) to show that $v' = f'gh + fg'h + fgh'$.

41. Show that if f is differentiable, then

$$\left[\frac{f(x)}{x^n} \right]' = \frac{xf'(x) - nf(x)}{x^{n+1}}.$$

42. Suppose that $P(x) = (x - a)(x - b)(x - c)$. Prove that

$$\frac{P'(x)}{P(x)} = \frac{1}{x - a} + \frac{1}{x - b} + \frac{1}{x - c}.$$

[†] Jean Louis Poiseuille (1799–1869) was a French physiologist.

CONTINUED

■ **CHALLENGE**

43. If the functions f and g have positive values, then their product, $f \cdot g$, can be interpreted as computing the area of a rectangle. Examine the accompanying figure and discuss how it is related to the proof in this section of the product rule (Theorem 1). Indicate what adjustments, to the figure or to the proof, would need to be made if one function is increasing and the other is decreasing.

*44. Use the product rule and mathematical induction to prove that $\dfrac{d}{dx} x^n = nx^{n-1}$ where n is a positive integer. $\left[Hint: \dfrac{d}{dx} x^n = \dfrac{d}{dx} x \cdot x^{n-1}. \right]$

■ **ANSWERS TO SELF-QUIZ**

I. $7x^6 = 1 \cdot x^6 + x \cdot 6x^5 = 2x \cdot x^5 + x^2 \cdot 5x^4$
$= 3x^2 \cdot x^4 + x^3 \cdot 4x^3$

II. $5x^4 = (x \cdot 6x^5 - x^6 \cdot 1)/x^2$
$= (x^2 \cdot 7x^6 - x^7 \cdot 2x)/x^4$
$= (x^3 \cdot 8x^7 - x^8 \cdot 3x^2)/x^6$

III. $\{b, c\}$ is the only pair for which the "rule" turns out to be true.

IV. d

2.3 THE DERIVATIVE OF COMPOSITE FUNCTIONS:
The Chain Rule

In this section we derive a result that greatly increases the number of functions whose derivatives can be easily calculated. The idea behind the result is illustrated below.

Suppose that $y = f(u)$ is a function of u and $u = g(x)$ is a function of x.[†] Then $du/dx = g'(x)$ is the rate of change of u with respect to x, while $dy/du = f'(u)$ is the rate of change of y with respect to u. We now may ask, what is the rate of change of y with respect to x? That is, what is dy/dx?

As an illustration of this idea, suppose that a sports car on a highway is moving twice as fast as a heavy truck, while the truck is moving three times faster than a bicycle. How much faster than the bicycle is the car going? We have

$$\text{speed of car} = 2 \times \text{speed of truck}$$

$$\text{speed of truck} = 3 \times \text{speed of bicycle,}$$

so,

$$\text{speed of car} = 2 \times \text{speed of truck} = 2 \times (3 \times \text{speed of bicycle})$$

$$= (2 \times 3) \times \text{speed of bicycle} = 6 \times \text{speed of bicycle.}$$

If we play fast and loose with our $\dfrac{dy}{dx}$ notation for derivatives, we may write

$$\frac{d(\text{car})}{d(\text{bicycle})} = \frac{d(\text{car})}{d(\text{truck})} \cdot \frac{d(\text{truck})}{d(\text{bicycle})} = 2 \times 3 = 6, \tag{1}$$

[†] So that $y(x) = (f \circ g)(x)$, where $f \circ g$ denotes the composition of f and g. See Section 0.4.

where $\dfrac{d(\text{car})}{d(\text{truck})}$ represents the rate of change of the car with respect to the truck, and so on. We conclude, as was fairly obvious, that the car is moving six times faster than the bicycle.

We now generalize the idea behind this example. Let $y(x) = f(g(x))$. That is, y is the composite function $f \circ g$. It will be very useful to be able to express y' in terms of f' and g'. As will be seen shortly, we can now show that

$$(f \circ g)'(x) = y'(x) = f'(g(x)) \cdot g'(x). \tag{2}$$

In most of the functions encountered in applications, $g'(x)$ and $f'(g(x))$ exist for almost all real numbers x in the domain of $f \circ g$.

If we write $u = g(x)$, then $f'(g(x)) = f'(u) = dy/du$, $g'(x) = du/dx$, and we can write equation (2) in the form

$$\frac{dy}{dx} = \frac{dy}{du}\frac{du}{dx}. \tag{3}$$

The result given by equation (2) or (3) is called the *chain rule*. Before stating the chain rule formally, we illustrate its use with two examples.

EXAMPLE 1
USE OF THE CHAIN RULE WHEN BOTH FUNCTIONS ARE STRAIGHT-LINE FUNCTIONS

Let $u = 3x - 6$ and let $y = 7u + 10$. Then $du/dx = 3$ and $dy/du = 7$, so that

$$\frac{dy}{dx} = \frac{dy}{du}\frac{du}{dx} = 7 \cdot 3 = 21.$$

To calculate dy/dx directly, we compute $y = 7u + 10 = 7(3x - 6) + 10 = 21x - 32$. Therefore,

$$\frac{dy}{dx} = 21 = \frac{dy}{du}\frac{du}{dx}.$$

If $y = f(g(x))$, we will call f the **outside function** and g the **inside function**.

EXAMPLE 2
DETERMINING INSIDE AND OUTSIDE FUNCTIONS

Determine the inside and outside functions in each case:

    (a) $y = \sqrt{x + x^2}$ **(b)** $y = \sin 2x$ **(c)** $y = (1 + x^2)^{10}$

SOLUTION:

    (a) We obtain $\sqrt{x + x^2}$ as follows:
        Step 1: compute $x + x^2$
        Step 2: take the square root
        Thus the inside function is $x + x^2$ and the outside function is $\sqrt{x}$.
    (b) To obtain $\sin 2x$:
        Step 1: compute $2x$
        Step 2: take the sine of $2x$
        The inside function is $2x$ and the outside function is $\sin x$.

(c) To obtain $(1 + x^2)^{10}$:
  Step 1: compute $1 + x^2$
  Step 2: raise the result to the tenth power.
The inside function is $1 + x^2$ and the outside function is x^{10}.

Using the idea of inside and outside functions, we can restate the chain rule (2) as follows:

ALTERNATIVE STATEMENT OF THE CHAIN RULE

The derivative of the composite function $(f \circ g)(x)$ equals the derivative of the outside function evaluated at $g(x)$ times the derivative of the inside function evaluated at x:

$$\frac{d}{dx} \text{outside(inside}(x)) = [\text{outside}'(\text{inside}(x))] \cdot [\text{inside}'(x)] \qquad (3')$$

EXAMPLE 3
USING THE CHAIN RULE WHEN THE OUTSIDE FUNCTION IS A SQUARE-ROOT FUNCTION

Use the chain rule to compute $\dfrac{d}{dx}\sqrt{x + x^2}$.

SOLUTION: If we define $u = x + x^2$ ($=$ the inside function), then $y = \sqrt{x + x^2} = \sqrt{u}$ (the outside function). Hence, by the chain rule (3),

$$\frac{d}{du}\sqrt{u} = \frac{1}{2\sqrt{u}}$$

$$\frac{dy}{dx} = \frac{dy}{du}\frac{du}{dx} = \frac{1}{2\sqrt{u}}(1 + 2x) = \frac{1}{2\sqrt{x + x^2}}(1 + 2x) = \frac{1 + 2x}{2\sqrt{x + x^2}}.$$

Note that the only other way to calculate this derivative would be to use the original definition (on page 87), which in this example is very difficult.

Now let $u = g(x)$ and $y = f(g(x)) = f(u)$. We assume that for every point x_0 such that $g(x_0)$ is defined, $f(g(x_0))$ is also defined, so it makes sense to talk about the function $f(g(x))$ at x_0.

THEOREM 1 **CHAIN RULE**

Let g and f be differentiable functions such that the assumptions above hold. Then with $u = g(x)$, the composite function $y = (f \circ g)(x) = f(g(x)) = f(u)$ is a differentiable function of x, and

$$\frac{dy}{dx} = \frac{d}{dx}(f \circ g)(x) = f'(g(x))g'(x) = \frac{dy}{du}\frac{du}{dx}. \qquad \blacksquare \qquad (4)$$

We give a proof of the chain rule at the end of the section.

EXAMPLE 4
USING THE CHAIN RULE WHEN THE OUTSIDE FUNCTION IS A POWER FUNCTION

Let $y = (1 + x^2)^{10}$. Compute $\dfrac{dy}{dx}$.

SOLUTION: If $u = 1 + x^2$ (= the inside function), then $f(u) = u^{10}$ is the outside function and

$$\frac{dy}{dx} = \frac{dy}{du}\frac{du}{dx} = \left(\frac{d}{du}u^{10}\right)\left(\frac{d}{dx}(1+x^2)\right)$$

$$\frac{d}{du}u^{10} = 10u^9$$

$$= 10u^9(2x) = 10(1+x^2)^9(2x) = 20x(1+x^2)^9.$$

We generalize Example 4.

THEOREM 2 **POWER RULE**

Let $u(x)$ be a differentiable function of x. Then if n is an integer,

$$\frac{d}{dx}u^n = nu^{n-1}\frac{du}{dx}.$$ (5)

Alternatively, we may write

$$\frac{d}{dx}[g(x)]^n = n[g(x)]^{n-1}g'(x).$$ (6)

PROOF: If $f(u) = u^n$, then from the chain rule,

$$\frac{d}{dx}u^n = \frac{df}{du}\frac{du}{dx} = nu^{n-1}\frac{du}{dx}. \quad \blacksquare$$

The use of the chain rule may involve other variables and more complicated expressions requiring several of the rules of differentiation. However, it will usually be clear how to proceed.

EXAMPLE 5
USING THE POWER RULE AND THE QUOTIENT RULE IN THE SAME PROBLEM

Let $s(t) = \left(\dfrac{t^2}{t-1}\right)^4$. Find $\dfrac{ds}{dt}$.

SOLUTION: Because s is the fourth power of a quotient, we apply the power rule before the quotient rule. Thus

$$\frac{ds}{dt} = 4\left(\frac{t^2}{t-1}\right)^3\frac{d}{dt}\left(\frac{t^2}{t-1}\right) = 4\left(\frac{t^2}{t-1}\right)^3\left[\frac{(t-1)\dfrac{d}{dt}t^2 - t^2\dfrac{d}{dt}(t-1)}{(t-1)^2}\right]$$

$$= 4\left(\frac{t^2}{t-1}\right)^3 \frac{(t-1)(2t) - t^2(1)}{(t-1)^2}.$$

$$= 4\left(\frac{t^2}{t-1}\right)^3 \left[\frac{t^2 - 2t}{(t-1)^2}\right] = \frac{4t^8 - 8t^7}{(t-1)^5}.$$

Many students have difficulty with the chain rule. The problem seems to lie in recognizing the inside and outside functions before applying the appropriate formulas. Here is an example that may help you to apply the chain rule in a more complicated setting.

EXAMPLE 6
APPLYING THE CHAIN RULE TO
TWO MADE-UP FUNCTIONS

Suppose that Graph and Slope are two differentiable functions and

$$\frac{d}{dx}\,\text{Graph}(x) = \text{Slope}(x).$$

Determine

(a) $\dfrac{d}{dx}\,\text{Graph}(2x)$ **(b)** $\dfrac{d}{dx}\,\text{Graph}(3 - 2x^2)$ **(c)** $\dfrac{d}{dx}\,\text{Graph}^4(x)$

(d) $\dfrac{d}{dx}\,\sqrt{\text{Graph}(x)}$ **(e)** $\dfrac{d}{dx}\,\text{Graph}^5(x^3)$ **(f)** $\dfrac{d}{dx}\,\dfrac{1}{\text{Graph}^2(x)}$

(g) $\dfrac{d}{dx}\,x^2\text{Graph}(x)$

SOLUTION: In (a) and (b), the outside function is Graph. The inside function varies.

(a) The inside function is $u = 2x$ and $\dfrac{d}{dx}\,2x = 2$, so

$$\frac{d}{du}\,\text{Graph}(u) = \text{Slope}(u)$$

$$\frac{d}{dx}\,\text{Graph}(2x) = \frac{d}{du}\,\text{Graph}(u)\,\frac{d}{dx}\,2x = \text{Slope}(u)(2)$$

$$= 2\,\text{Slope}(2x).$$

(b) Now the inside function is $u = 3 - 2x^2$, so

$$\frac{d}{dx}\,\text{Graph}(3 - 2x^2) = \frac{d}{du}\,\text{Graph}(u)\,\frac{d}{dx}\,(3 - 2x^2)$$

$$= \text{Slope}(u)(-4x) = -4x\,\text{Slope}(3 - 2x^2).$$

(c) Here we use the power rule with $u = \text{Graph}(x)$:

$$\frac{d}{dx}\text{Graph}(x) = \text{Slope}(x)$$

$$\frac{d}{dx}\text{Graph}^4(x) = \frac{d}{dx}u^4 = 4u^3\frac{du}{dx} = 4\,\text{Graph}^3(x)\,\text{Slope}(x).$$

(d) Let $u = \text{Graph}(x)$ (the inside function now). Then $\sqrt{\text{Graph}(x)} = \sqrt{u}$ (the outside function) and, as in Example 3,

$$\frac{d}{dx}\sqrt{\text{Graph}(x)} = \frac{d}{dx}\sqrt{u} = \frac{1}{2\sqrt{u}}\frac{du}{dx}$$

$$= \frac{1}{2\sqrt{\text{Graph}(x)}}\cdot\text{Slope}(x) = \frac{\text{Slope}(x)}{2\sqrt{\text{Graph}(x)}}.$$

(e) Here we use the chain rule twice. First, if $u = x^3$, then $\text{Graph}(x^3) = \text{Graph}(u)$ and

$$\frac{d}{dx}\text{Graph}(x^3) = \frac{d}{dx}\text{Graph}(u) = \frac{d}{du}\text{Graph}(u)\frac{du}{dx}$$

$$= \text{Slope}(x^3)\cdot 3x^2 = 3x^2\,\text{Slope}(x^3).$$

Now, let $v = \text{Graph}(x^3)$. Then $\text{Graph}^5(x^3) = v^5$ and, by the power rule,

$$\frac{d}{dx}\text{Graph}^5(x^3) = \frac{d}{dx}v^5 = 5v^4\frac{dv}{dx} = 5v^4[3x^2\,\text{Slope}(x^3)]$$

$$= 5\,\text{Graph}^4(x^3)[3x^2\,\text{Slope}(x^3)]$$

$$= 15x^2\,\text{Graph}^4(x^3)\,\text{Slope}(x^3).$$

(f) $\dfrac{1}{\text{Graph}^2(x)} = \text{Graph}^{-2}(x)$, so, using the power rule with $u = \text{Graph}(x)$ and $n = -2$,

$$\frac{d}{du}u^{-2} = -2u^{-3}$$

$$\frac{d}{dx}\text{Graph}^{-2}(x) = \frac{d}{dx}u^{-2} = -2u^{-3}\frac{du}{dx}$$

$$= -2[\text{Graph}^{-3}(x)]\,\text{Slope}(x)$$

$$= \frac{-2\,\text{Slope}(x)}{\text{Graph}^3(x)}.$$

(g) Using the product rule (the chain rule is unnecessary here),

$$\frac{d}{dx}x^2\text{Graph}(x) = x^2\frac{d}{dx}\text{Graph}(x) + \text{Graph}(x)\frac{d}{dx}x^2$$

$$= x^2\,\text{Slope}(x) + 2x\,\text{Graph}(x).$$

We close this section with a proof of the chain rule.

PROOF OF THE CHAIN RULE

Let g and f be differentiable functions such that for every point x_0 at which $g(x)$ is defined, $f(g(x_0))$ is also defined. Then with $u = g(x)$, the composite function $y = (f \circ g)(x) = f(g(x)) = f(u)$ is a differentiable function of x, and

$$\frac{dy}{dx} = \frac{d}{dx}(f \circ g)(x) = \frac{d}{dx}f(g(x)) = f'(g(x))g'(x) = \frac{df}{du}\frac{du}{dx}.$$

PROOF: By the definition of the derivative,

$$f'(x) = \frac{dy}{dx} = \lim_{\Delta x \to 0} \frac{\Delta y}{\Delta x}. \tag{7}$$

This implies that if Δx is small, then $\Delta y/\Delta x$ is close to dy/dx. For the number x fixed, we define

$$\epsilon(\Delta x) = \frac{dy}{dx} - \frac{\Delta y}{\Delta x} = f'(x) - \frac{\Delta y}{\Delta x}. \tag{8}$$

Then from (7),

$$\lim_{\Delta x \to 0} \epsilon(\Delta x) = \lim_{\Delta x \to 0}\left(\frac{dy}{dx} - \frac{\Delta y}{\Delta x}\right) = 0.$$

Multiplying both ends of (8) by Δx and rearranging terms, we obtain

$$\Delta y = f'(x)\Delta x - \epsilon(\Delta x) \cdot \Delta x, \tag{9}$$

where $\lim_{\Delta x \to 0} \epsilon(\Delta x) = 0$.
Now,

$$\frac{d}{dx}f(g(x)) = \lim_{\Delta x \to 0} \frac{f(g(x + \Delta x)) - f(g(x))}{\Delta x},$$

if this limit exists. Since f can be written as a function of u, we can rewrite (9) as

$$\Delta f = f'(u)\Delta u - \epsilon(\Delta u)\Delta u, \tag{10}$$

where $\epsilon(\Delta u) \to 0$ as $\Delta u \to 0$; $\Delta u = g(x + \Delta x) - g(x)$ so $g(x + \Delta x) = g(x) + \Delta u = u + \Delta u$, and $f'(u) = f'(g(x))$. Then using (10), we obtain

$$f(g(x + \Delta x)) - f(g(x)) = f(u + \Delta u) - f(u) = \Delta f$$
$$= f'(u)\Delta u - \epsilon(\Delta u)\Delta u, \tag{11}$$

where $\epsilon(\Delta u) \to 0$ as $\Delta u \to 0$. Dividing (11) by Δx, we obtain

$$\frac{f(g(x + \Delta x)) - f(g(x))}{\Delta x} = f'(u)\frac{\Delta u}{\Delta x} - \epsilon(\Delta u)\frac{\Delta u}{\Delta x}.$$

Finally, we take limits as $\Delta x \to 0$. Since $u = g(x)$ and g is differentiable, $\lim_{\Delta x \to 0} \Delta u/\Delta x = du/dx = g'(x)$. Also, $\Delta u = g(x + \Delta x) - g(x)$. Since g is differentiable at x, it is continuous there, and $\lim_{\Delta x \to 0} g(x + \Delta x) = g(x)$. Or alternatively, $\lim_{\Delta x \to 0} \Delta u = \lim_{\Delta x \to 0}[g(x + \Delta x) - g(x)] = 0$.

Thus, as $\Delta x \to 0$, $\Delta u \to 0$ and $\lim_{\Delta x \to 0} \epsilon(\Delta u) = 0$. Hence

$$\lim_{\Delta x \to 0} \epsilon(\Delta u) \cdot \frac{\Delta u}{\Delta x} = \lim_{\Delta x \to 0} \epsilon(\Delta u) \lim_{\Delta x \to 0} \frac{\Delta u}{\Delta x} = 0 \cdot \frac{du}{dx} = 0.$$

Putting all this information together, we obtain

$$\lim_{\Delta x \to 0} \frac{f(g(x + \Delta x)) - f(g(x))}{\Delta x} = \lim_{\Delta x \to 0} f'(u) \frac{\Delta u}{\Delta x} - \lim_{\Delta x \to 0} \epsilon(\Delta u) \frac{\Delta u}{\Delta x}$$

$$= f'(g(x))g'(x) - 0,$$

and the chain rule is proved. ∎

■ PROBLEMS 2.3

■ SELF-QUIZ

I. Practice with the power rule and the chain rule by verifying that x^{10}, $(x^5)^2$, and $(x^2)^5$ have equal derivatives.

II. $\dfrac{d}{dx}(-5x)^3 = $ _____.

    a. $3(-5x)^2$ b. $3x^2(-5)$
    c. $3(-5x)^2 - 5$ d. $3(-5x)^2(-5)$

III. $\dfrac{d}{dx}(-5x^2)^3 = $ _____.

    a. $3(-5x^2)^2 \cdot (-5)(2x)$

    b. $3(-5x^2)^2 - 5(2x)$
    c. $(-5x^2)^3 \cdot (-5)(2x)$
    d. $-5 \cdot 3x^2 \cdot 2x$

IV. Suppose $h = f \circ g$ where f and g are differentiable functions. Answer True or False to each of the following assertions.

    a. If $f(x) > 0$ and $g(x) > 0$ for all x, then $h'(x) > 0$ for all x.
    b. If $f'(x) > 0$ and $g'(x) < 0$ for all x, then $h'(x) > 0$ for all x.
    c. If $f'(x) < 0$ and $g'(x) < 0$ for all x, then $h'(x) > 0$ for all x.

■ DRILL

In Problems 1–24, use the chain rule (and any other useful theorems) to find the derivative of the given function. (Do remember how to differentiate $\sqrt{x}$.)

1. $f(x) = (x + 1)^3$
2. $f(x) = (x^2 - 1)^2$
3. $f(x) = (1 + x^6)^6$
4. $f(x) = (x^2 - 4x + 1)^5$
5. $f(x) = (x^2 - x^3)^4$
6. $f(x) = (1 - x + x^2 - x^3 + x^4)^2$
7. $f(x) = (1 - x^2 + x^5)^3$
8. $f(x) = (\sqrt{x} + 2)^4$
9. $f(x) = (\sqrt{x} - x)^3$
10. $f(x) = 1/(\sqrt{x} - 3)^4$
11. $h(y) = (y^2 + 3)^{-4}$
12. $g(u) = 5/(u^3 + u + 1)$
13. $s(t) = [(t + 1)/(t - 1)]^3$
14. $s(t) = \left(\dfrac{t^4 + 1}{t^4 - 1}\right)^{1/2}$
15. $f(x) = (x^2 + 2)^5(x^4 + 3)^3$
16. $f(x) = (x^4 + 1)^{1/2}(x^3 + 3)^4$

17. $s(t) = \dfrac{\sqrt{t^2 + 1}}{(t + 2)^4}$

18. $g(u) = \dfrac{(u^2 + 1)^3(u^2 - 1)^2}{\sqrt{u - 2}}$

19. $g(x) = \dfrac{(x^2 + 1)^2(x^3 + 2)^3}{(x^4 + 3)^{1/2}}$

20. $f(x) = \sqrt{x + \sqrt{1 + \sqrt{x}}}$
    [*Hint:* Use the chain rule twice.]

21. $g(x) = \sqrt{x^2 + \sqrt{1 + x^2}}$
22. $h(t) = (t^2 + t^{-2})^{-1}$
23. $f(x) = [(1 + x)^{-1} + (1 - x)^{-1}]^{-1}$
24. $p(s) = [(1 - s^{-1})^{-1} + 4]^{-1}$

In Problems 25–28, Homer and Virgil are two differentiable functions having the property that $\dfrac{d}{dx}$ Homer$(x) = $ Virgil(x). Find each derivative.

25. $\dfrac{d}{dx}$ Homer$(2x + 1)$ 26. $\dfrac{d}{dx}$ Homer(x^6)

CONTINUED

27. $\dfrac{d}{dx}\,\text{Homer}^6(x)$ 28. $\dfrac{d}{dx}\,\text{Homer}^4\!\left(\dfrac{1}{x}\right)$

In Problems 29–35, L is a differentiable function with $\dfrac{d}{dx}\,L(x) = \dfrac{1}{x}$. Compute each derivative.

29. $\dfrac{d}{dx}\,L(3x)$ $\dfrac{1}{x}(3x)(3)$ 30. $\dfrac{d}{dx}\,L(x^3)$

31. $\dfrac{d}{dx}\,\sqrt{L(7x)}$ 32. $\dfrac{d}{dx}\,L^{-2}(x^5)$

33. $\dfrac{d}{dx}\,L^{12}\!\left(\dfrac{x+1}{x-1}\right)$ 34. $\dfrac{d}{dx}\,L^5(x\sqrt{1+x})$

35. $\dfrac{d}{dx}\,\dfrac{L(x)}{x}$ 36. $\dfrac{d}{dx}\,L(1+x^2)$

In Problems 37–44, E is a differentiable function which is its own derivative. That is, $\dfrac{d}{dx}\,E(x) = E(x)$. Find each derivative.

37. $\dfrac{d}{dx}\,E(4x)$ 38. $\dfrac{d}{dx}\,E(-x)$

39. $\dfrac{d}{dx}\,E(x^3 + 2x + 3)$ 40. $\dfrac{d}{dx}\,E\!\left(\dfrac{1}{x}\right)$

41. $\dfrac{d}{dx}\,xE^{1/2}(x+1)$ 42. $\dfrac{d}{dx}\,E^2(x)$

43. $\dfrac{d}{dx}\,E(L(x))$, where L is defined above.

44. $\dfrac{d}{dx}\,L(E(x))$

■ APPLICATIONS

45. A missile travels along the path $y = 6(x-3)^3 + 3x$. When $x = 1$, the missile flies off this path tangentially (i.e., along the tangent line). Where is the missile when $x = 4$? [*Hint:* Find an equation for the line tangent to the path at $x = 1$.]

46. The mass μ of the left x m of a 4-m long metal rod is given by $\mu = 4(1 + \sqrt{x})^3$ kg. What is the density of the rod (in kg/m) when $x = 1$ m?

47. Verify the result of Problem 51 in the following cases.
    a. $g(x) = 5x;\ f(x) = (\tfrac{1}{5})x$
    b. $g(x) = 17x - 8;\ f(x) = (\tfrac{1}{17})x + (\tfrac{8}{17})$
    c. $g(x) = \sqrt{x};\ f(x) = x^2$
    d. $g(x) = x^2$ on $(-\infty, 0];\ f(x) = -\sqrt{x}$

48. Differentiate $f(x) \cdot [g(x)]^{-1}$ by using the product rule and the chain rule. Verify that your result equals the derivative of the quotient $f(x)/g(x)$.

■ SHOW/PROVE/DISPROVE

49. Suppose that $p(x)$ is a polynomial which is divisible by $(x - 17)^2$, that is, there is a polynomial q such that $p(x) = (x - 17)^2 \cdot q(x)$. Prove that $p'(x)$ is divisible by $x - 17$.

*50. Show that if the polynomial $p(x)$ is divisible by $(ax + b)^n$, where n is an integer larger than 1, then $p'(x)$ is divisible by $(ax + b)^{n-1}$.

*51. Let f and g be differentiable functions with $f(g(x)) = x$ for all x in the domain of g. Show that

$$f'(g(x)) = \frac{1}{g'(x)}.$$

This formula is called the **differentiation rule for inverse functions**. Section 6.1 discusses this topic.

52. It is possible to obtain the product rule by using the chain rule, the sum rule, and the fact that $(d/dx)(x^2) = 2x$. Write out such a proof. [*Hint:* Use the fact that $2fg = (f + g)^2 - f^2 - g^2$.]

■ CHALLENGE

**53. Suppose that f is differentiable on $(0, \infty)$ and that $f(A + B) = f(A) + f(B)$ for any positive real numbers A and B. Prove that, for all $x > 0$,

$$f'(x) = \frac{f'(1)}{x}.$$

■ ANSWERS TO SELF-QUIZ

I. $10x^9 = 2(x^5)(5x^4) = 5(x^2)^4(2x)$.

II. d   III. a

IV. a. False; e.g., consider $f(x) = 1 + x^2$ and $g(x) = 2 + x^2$, then $h(x) = 1 + (2 + x^2)^2 = x^4 + 4x^2 + 5$ and $h'(x) = 8x + 4x^3 < 0$ if $x < 0$.

b. False; e.g., consider $f(x) = 2x$ and $g(x) = -3x$; then $h'(x) = -6$.

c. True; $h'(x) = f'(g(x)) \cdot g'(x)$ is the product of two negative values.

2.4 THE DERIVATIVE OF A POWER FUNCTION

In this section we derive a general formula for finding the derivative of $y = x^r$, where r is a rational number (remember that a rational number is a number of the form $r = m/n$, where m and n are integers and $n \neq 0$). Such a function is called a **power function**.

If r is a positive or negative integer, then we have shown that

$$\frac{d}{dx} x^r = rx^{r-1}.$$ (1)

We also know that (see page 90)

$$\frac{d}{dx} x^{1/2} = \frac{1}{2}x^{-1/2}$$

We will show that formula (1) holds for *all* rational numbers r, and we will suggest an extension of that result to all real numbers. This result will be obtained in several steps.

THEOREM 1

Let n be a positive integer. Then the function $f(x) = x^{1/n}$ is differentiable, and

$$\frac{d}{dx} x^{1/n} = \frac{1}{n} x^{(1/n)-1}. \quad \blacksquare$$ (2)

Equation (2) is valid whenever $x^{1/n - 1}$ is defined.

PROOF: Let $f(x) = x^{1/n}$. Using the alternative definition of the derivative on page 87 and Theorem 1.6.1 on page 100, we have

$$\frac{d}{dx} x^{1/n} = f'(x) = \lim_{t \to x} \frac{t^{1/n} - x^{1/n}}{t - x} = \frac{t^{1/n} - x^{1/n}}{(t^{1/n})^n - (x^{1/n})^n}$$

Theorem 1.6.1 with
$a = t^{1/n}$ and $b = x^{1/n}$

$$= \lim_{t \to x} \frac{t^{1/n} - x^{1/n}}{(t^{1/n} - x^{1/n})(t^{(n-1)/n} + t^{(n-2)/n}x^{1/n} + t^{(n-3)/n}x^{2/n} + \cdots + t^{1/n}x^{(n-2)/n} + x^{(n-1)/n})}$$

$$= \lim_{t \to x} \frac{1}{t^{(n-1)/n} + t^{(n-2)/n}x^{1/n} + \cdots + t^{1/n}x^{(n-2)/n} + x^{(n-1)/n}}$$

$$= \frac{1}{x^{(n-1)/n} + x^{(n-2)/n}x^{1/n} + x^{(n-3)/n}x^{2/n} + \cdots + x^{1/n}x^{(n-2)/n} + x^{(n-1)/n}}.$$ (3)

The denominator contains n terms, each of which is equal to $x^{(n-1)/n}$ (since $x^{(n-2)/n}x^{1/n} = x^{(n-2)/n+(1/n)} = x^{(n-1)/n}$, and so on). Therefore, (3) is equal to

$$\frac{1}{nx^{(n-1)/n}} = \frac{1}{nx^{1-(1/n)}} = \frac{1}{n} \cdot \frac{1}{x^{1-(1/n)}} = \frac{1}{n}x^{(1/n)-1}$$

and the theorem is proved. ■

REMARK: A simpler derivation of equation (2) is possible if we assume that $x^{1/n}$ is differentiable. This simpler derivation is suggested in Problem 28. Also an easy proof can be given once we have discussed the derivatives of inverse functions in Section 6.1.

EXAMPLE 1
DIFFERENTIATING A FUNCTION
HAVING THE FORM $x^{1/n}$

Let $y = x^{1/5}$. Calculate dy/dx.

SOLUTION:

$$\frac{dy}{dx} = \frac{1}{5}x^{(1/5)-1} = \frac{1}{5}x^{-4/5}.$$

EXAMPLE 2
USING THE CHAIN RULE FOR A
FUNCTION HAVING THE FORM $u^{1/n}$

Let $y = f(x) = (x^3 + 3)^{1/4}$. Find dy/dx.

SOLUTION: We define $u = x^3 + 3$ (the inside function). Then $y = u^{1/4}$ (the outside function), so that by the chain rule,

$$\frac{dy}{dx} = \frac{dy}{du}\frac{du}{dx} = \frac{1}{4}u^{-3/4}(3x^2) = \frac{1}{4}(x^3 + 3)^{-3/4}(3x^2).$$

We now extend Theorem 1 to rational numbers.

THEOREM 2

Let $y = x^r$, where $r = m/n$ is a rational number (m and n are integers and $n \neq 0$). Then x^r is differentiable, and

$$\frac{dy}{dx} = \frac{d}{dx}x^r = rx^{r-1}. \quad ■ \tag{4}$$

Equation (4) is valid whenever x^{r-1} is defined.

PROOF:

$$\frac{dy}{dx} = \frac{d}{dx}x^r = \frac{d}{dx}x^{m/n} = \frac{d}{dx}(x^{1/n})^m.$$

Let $u = x^{1/n}$. Then $y = u^m$, and

Chain rule

$$\frac{dy}{dx} \overset{\downarrow}{=} \frac{dy}{du}\frac{du}{dx} = mu^{m-1} \cdot \frac{1}{n}x^{(1/n)-1} = m(x^{1/n})^{m-1} \cdot \frac{1}{n}x^{(1-n)/n}$$

$$= \frac{m}{n}(x^{1/n})^{m-1}(x^{1/n})^{1-n} = \frac{m}{n}(x^{1/n})^{m-1+1-n} = \frac{m}{n}x^{(m-n)/n}$$

$$= \frac{m}{n}x^{(m/n)-1} = rx^{r-1}$$

and the theorem is proved. ∎

EXAMPLE 3
USING THE POWER RULE

Let $y = x^{2/3}$. Calculate dy/dx.

SOLUTION:

$$\frac{dy}{dx} = \frac{2}{3}x^{(2/3)-1} = \frac{2}{3}x^{-1/3} = \frac{2}{3x^{1/3}}.$$

EXAMPLE 4
USING THE POWER RULE AND
THE CHAIN RULE

Let $s = (t^3 + 2t + 3)^{11/9}$. Find ds/dt.

SOLUTION: By now you should recognize that the chain rule is called for in this problem. Let $u = t^3 + 2t + 3$ (the inside function). Then $s = u^{11/9}$ (the outside function), so that

$$\frac{ds}{dt} = \frac{ds}{du}\frac{du}{dt} = \frac{11}{9}u^{2/9}(3t^2 + 2) = \frac{11}{9}(t^3 + 2t + 3)^{2/9}(3t^2 + 2).$$

We can extend Example 4 and the power rule given on page 149.

THEOREM 3   **POWER RULE**

Let u be a differentiable function of x and let r be a rational number, then

$$\frac{d}{dx}u^r = ru^{r-1}\frac{du}{dx}. \tag{5}$$

The proof follows directly from (4) and the chain rule. ∎

It is true that the formula

$$\frac{d}{dx}x^\alpha = \alpha x^{\alpha-1} \tag{6}$$

holds when α is any real number. We cannot prove this formula here because we don't even know what x^α means if α is not a rational number. Once we have defined the exponential function (in Section 6.5), a definition of x^α and a proof of (6) will be possible (see Problem 6.5.52). However, we can get an idea of what (6) means if α is irrational by noting that if α is rational, (6) has already been proven; and if α is irrational, then it can be approximated as closely as desired by a rational number (e.g., the irrational number π can be approximated by 3.14, 3.141, 3.1415, 3.14159, etc.).

EXAMPLE 5
THE DERIVATIVE OF x^α WHERE
α IS IRRATIONAL

Let $y = x^\pi$. Calculate dy/dx.

SOLUTION: Using formula (6), we have $dy/dx = \pi x^{\pi - 1}$.

At this point we see that it is possible to differentiate a wide variety of functions. To aid you, we give in Table 1 a summary of the differentiation rules we have so far discussed. In the notation of the table, c stands for an arbitrary constant, and u and v denote differentiable functions.

TABLE 1

Function $y = f(x)$	Name	Its derivative $\dfrac{dy}{dx}$
I. c	constant function	$\dfrac{d}{dx} c = 0$
II. $cu(x)$	a constant times a function	$\dfrac{d}{dx} cu(x) = c\dfrac{du}{dx}$
III. $u(x) + v(x)$	sum of two functions	$\dfrac{d}{dx}(u + v) = \dfrac{du}{dx} + \dfrac{dv}{dx}$
IV. x^r, r a real number	power function	$\dfrac{d}{dx} x^r = rx^{r-1}$
V. $u(x) \cdot v(x)$	product of two functions	$\dfrac{d}{dx} uv = u(x)\dfrac{dv}{dx} + v(x)\dfrac{du}{dx}$ (the product rule)
VI. $\dfrac{u(x)}{v(x)}$, $v(x) \neq 0$	quotient of two functions	$\dfrac{d}{dx}\left[\dfrac{u(x)}{v(x)}\right] = \dfrac{v(x)\dfrac{du}{dx} - u(x)\dfrac{dv}{dx}}{[v(x)]^2}$ (the quotient rule)
VII. $u^r(x)$	a function raised to a power	$\dfrac{d}{dx} u^r(x) = ru^{r-1}(x)\dfrac{du}{dx}$ (the power rule)
VIII. $f(g(x))$	composite function	$\dfrac{d}{dx} f(g(x)) = f'(g(x))g'(x)$ (the chain rule)

■ PROBLEMS 2.4

■ SELF-QUIZ

I. True–False:
   a. Theorem 1 says that $(d/dx)2^x = 2^{x-1}$.
   b. Theorem 1 does not go far enough but Theorem 2 does say that $(d/dx)2^x = 2^{x-1}$.
   c. Neither Theorem 1 nor Theorem 2 say anything about $(d/dx)2^x$.
   d. Theorem 2 says that $(d/dx)x^{1.414} = 1.414x^{0.414}$.
   e. Theorem 2 says that $(d/dx)x^{\sqrt{2}} = \sqrt{2}x^{\sqrt{2}-1}$.

II. Items similar to the following appear in the table on the preceding page in which various differentiation rules are summarized. For each part, your job is to find an expression whose derivative is the item listed here. (For instance, if $F' - 2GG'$ appeared, then $F - G^2$ would be a correct answer.)
   a. $F' + G' = ($_____$)'$
   b. $(F'G - FG')/G^2 = ($_____$)'$
   c. $F'G + FG' = ($_____$)'$
   d. $($_____$)' = F'G^{-1} - FG^{-2}G'$
   e. $($_____$)' = 10F^4F'$
   f. $($_____$)' = (F' \circ g) \cdot h$ where $g(x) = 3x^2 - 1$ and $h(x) = 6x$

■ DRILL

In Problems 1–20, find the derivative of the given function.

1. $f(x) = x^{2/5} + 2x^{1/3}$
2. $f(x) = 5x^{-2/9}$
3. $f(x) = (x^2 + 1)^{5/3}$
4. $f(x) = (x^2 - 1)^{-2/3}$
5. $f(x) = (x^3 + 3)^{1/10}$
6. $f(x) = (x^5 + 2x + 1)^{3/2}$
7. $s(t) = (t^{10} - 2)^{4/7}$
8. $s(t) = (t^3 + t^2 + 1)^{5/8}$
9. $g(u) = \sqrt[3]{u^3 + 3u + 1}$
10. $g(u) = \left(\dfrac{u + 2}{u - 1}\right)^{7/2}$

11. $h(z) = \left(\dfrac{z^2 - 1}{z^2 + 1}\right)^{-7/17}$
12. $h(z) = (z^5 - 1)^{3/4}(z^3 + 2)^{8/9}$
13. $v(r) = (r^2 + 1)^{5/6}(r - 1)^{1/2}$
14. $v(r) = (r + 1)^{1/3}(r - 1)^{1/4}(r + 2)^{1/5}$
15. $f(x) = 3x^{\sqrt{2}}$
16. $f(x) = \sqrt{3}x^{\sqrt{3}}$
17. $s(t) = (t^2 + 1)^{\sqrt{3}}$
18. $s(t) = (t^4 - \pi)^{\pi^2}$
19. $g(u) = (u^{\sqrt{2}} - u^{\sqrt{3}})^4$
20. $g(u) = \dfrac{u^{\sqrt{5}} - 3}{u^{\sqrt{5}} + 1}$

■ APPLICATIONS

In Problems 21–24, find equations for the tangent and normal lines to the given curve at the specified point.

21. $y = x^{1/2} + x^{3/2}$; $(1, 2)$
22. $y = \dfrac{(x + 2)^3}{(x - 1)^{1/3}}$; $(0, -8)$
23. $y = (x - 1)^{\sqrt{2}}$; $(1, 0)$
24. $y = (x + 11x^2 + 3x^3 + x^4)^{5/4}$; $(1, 32)$
25. In the kinetic theory of gases, the **root mean square** (rms) speed of gas molecules is related to the average or typical speed of the molecules comprising the gas and is given by

$$v_{\text{rms}} = \sqrt{\frac{3P}{\rho}},$$

where P is the pressure (measured in atmospheres, atm) of the gas and ρ is its density (measured in kilograms per cubic meter).

   a. Calculate the v_{rms} of hydrogen molecules at $0°C$ and 1 atm of pressure. (At that temperature, the density of hydrogen is 8.99×10^{-2} kg/m^3.)
   b. With P fixed, calculate the instantaneous rate of change of v_{rms} with respect to ρ. (It is known that density ρ is inversely proportional to the temperature T with P fixed, so ρ will increase as the gas cools and ρ will decrease as the gas is heated.)
   c. Calculate the instantaneous rate of change of v_{rms} as a function of ρ (with P fixed) for the values of P and ρ given in part (a).
26. In astronomy, the **luminosity** of a star is the star's total energy output. Loosely speaking, a star's luminosity is a measure of how bright the star would appear at the surface of the star. The **mass-luminosity relation** gives the approximate luminosity of a star as a function of its mass. It has been found experimentally that,

CONTINUED

approximately,

$$\frac{L}{L_0} = \left(\frac{M}{M_0}\right)^r,$$

where L and M are the luminosity and mass of the star and where L_0 and M_0 are the luminosity and mass of our sun; the exponent r depends on the mass of the star according to the following table.[†]

Mass ratio, $\dfrac{M}{M_0}$	r
1.0–1.4	4.75
1.4–1.7	4.28
1.7–2.5	4.15
2.5–5	3.95
5–10	3.38
10–20	2.80
20–50	2.30
50–100	1.90

a. In this model, how is the luminosity changing as a function of mass when the mass is 2 solar masses?
b. How is it changing at a mass of 8 solar masses?
c. How is it changing at a mass of 30 solar masses?
*d. Write L as a function of M. For what values of M does dL/dM not exist? How would you suggest revising the model to avoid discontinuities in this derivative?

■ SHOW/PROVE/DISPROVE

*27. Prove that a line tangent to a circle is perpendicular to a radial line by using the following steps:
   a. Assume the circle is centered at the origin and show that its equation, for $y \geq 0$, can be written as $y = \sqrt{r^2 - x^2}$ where r is its radius.
   b. Find the equation of the line tangent to the circle at a point (x_0, y_0) on the circle.
   c. Find the equation of the line segment joining $(0, 0)$ to (x_0, y_0).

28. Let n be a positive integer; consider the function $y(x) = x^{1/n}$. This problem shows one way to ob-

tain a differentiation rule for this function provided we assume it is known to be differentiable.
   a. Show that $x = [y(x)]^n$.
   b. Using the chain rule, differentiate both sides of the equation in part (a) with respect to x.
   c. Using the result of part (b), show that $dy/dx = (1/n)[y(x)]^{1-n}$.
   d. Using part (c), show that

   $$\frac{d}{dx} x^{1/n} = \frac{1}{n} x^{1/n-1}.$$

[†] These data are based on stellar models computed by D. Ezer and A. Cameron in their paper "Early and main sequence evolution of stars in the range 0.5 to 100 solar masses," *Canadian Journal of Physics*, 45 (1967), 3429–3460. Note that r is not a function: To make it into one, use the higher value at the endpoints of interval. For example, if $M/M_0 = 1.7$, set $r = 4.28$ (not 4.15).

2.5 THE DERIVATIVES OF THE TRIGONOMETRIC FUNCTIONS

In this section we will compute the derivatives of sin x, cos x, and related functions. The material in this section depends on a knowledge of basic trigonometry. The background material for this section can be found in the first three sections of Appendix 1. Also, we emphasize that in this section, and for the remainder of this text, we assume, unless otherwise stated, that x (or θ) is measured in radians and that dom(sin x) = dom(cos x) = $\mathbb{R}$.

To compute the derivative of sin x, we need to know $\lim_{\theta \to 0} \frac{\sin \theta}{\theta}$. In Example 7 on page 58 we tabulated values of $\frac{\sin \theta}{\theta}$ for small θ that suggest that this limit is 1. We now prove this.

THEOREM 1

$$\lim_{\theta \to 0} \frac{\sin \theta}{\theta} = 1 \quad \blacksquare \tag{1}$$

Note that the limit in (1) cannot be calculated by evaluation since $(\sin \theta)/\theta$ is not defined at 0.

REMARK: Equation (1) is *false* if θ is measured in degrees rather than radians (see Problem 39).

PROOF: We prove the theorem first in the case $0 < \theta < \pi/2$. That is, we show that $\lim_{\theta \to 0^+} \frac{\sin \theta}{\theta} = 1$. Consider Figure 1. From the figure we see that

$$\text{area of } \triangle 0AC < \text{area of sector } 0AC < \text{area of } \triangle 0BC \tag{2}$$

To calculate the area of a sector, we use the fact that if $\theta = 2\pi$ radians, then the area of the sector is the area of the entire unit circle, which is π. If an angle of 2π radians corresponds to a sector area of π, then θ radians corresponds to a sector area of $\theta/2$. Thus,

$$\text{area of sector } 0AC = \frac{\theta}{2}.$$

Also,

$$\sin \theta = \overline{AD}^{\dagger} \qquad \text{Since } A \text{ is a point on the unit circle (see Appendix 1.2).}$$

$$\tan \theta = \frac{\overline{BC}}{\overline{0C}} = \overline{BC}. \qquad \text{Since } \overline{0C} = 1 \text{ (the unit circle has radius 1).}$$

Thus,

$$\text{area of } \triangle 0AC = \frac{1}{2}(\text{base} \times \text{height}) = \frac{1}{2}\overline{0C} \times \overline{AD} = \frac{1}{2}\sin \theta,$$

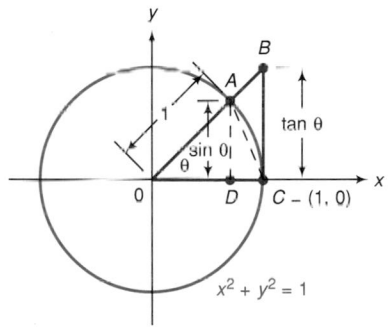

FIGURE 1
$\frac{1}{2}\sin \theta < \frac{1}{2}\theta < \frac{1}{2}\tan \theta$

---

† Recall that $\overline{AD}$ denotes the length of the line segment joining A to D.

and

$$\text{area of } \triangle 0BC = \frac{1}{2}\overline{0C} \times \overline{BC} = \frac{1}{2}\tan\theta.$$

Thus, we obtain, from (2),

$$\frac{1}{2}\sin\theta < \frac{1}{2}\theta < \frac{1}{2}\tan\theta. \tag{3}$$

For $\theta \neq 0$, we multiply the inequalities in (3) by 2 and divide by $\sin\theta$, which is positive, since $0 < \theta < \frac{\pi}{2}$, to obtain

$$1 < \frac{\theta}{\sin\theta} < \frac{1}{\cos\theta}$$

oɪ

$$\cos\theta < \frac{\sin\theta}{\theta} < 1, \qquad 0 < \theta < \frac{\pi}{2}. \qquad \begin{array}{l}\text{Since if } 0 < a < b, \\ \text{then } \dfrac{1}{a} > \dfrac{1}{b}.\end{array} \tag{4}$$

Now

$$\lim_{\theta\to 0^+} 1 = 1 \qquad \text{and} \qquad \lim_{\theta\to 0^+}\cos\theta = 1. \qquad \begin{array}{l}\text{(See Figure 2 in} \\ \text{Appendix 1.2.)}\end{array}$$

So the squeezing theorem tells us that

$$\lim_{\theta\to 0^+}\frac{\sin\theta}{\theta} = 1.$$

The limit $\lim_{\theta\to 0^-}[(\sin\theta)/\theta] = 1$ follows from what we have already proved and the fact that

$$\frac{\sin\theta}{\theta} = \frac{\sin(-\theta)}{-\theta}.$$

Thus, the proof is complete. Actually, the proof is intuitively reasonable since, in Figure 1, as $\theta \to 0^+$, the length of the line AD and the length of the arc AC (denoted by $\widehat{AC}$) get closer and closer together. Then the theorem follows, since $\overline{AD} = \sin\theta$ and $\widehat{AC} = \theta$. ∎

We can use Theorem 1 to compute other limits.

EXAMPLE 1
CALCULATING A LIMIT INVOLVING SINES

Calculate $\lim_{\theta\to 0}[(\sin 2\theta)/\theta]$.

SOLUTION:

$$\lim_{\theta\to 0}\frac{\sin 2\theta}{\theta} = \lim_{\theta\to 0}\frac{2\sin 2\theta}{2\theta} = 2\lim_{\theta\to 0}\frac{\sin 2\theta}{2\theta}.$$

But as $\theta \to 0$, $2\theta \to 0$, so that

$$2\lim_{\theta\to 0}\frac{\sin 2\theta}{2\theta} = 2\lim_{2\theta\to 0}\frac{\sin 2\theta}{2\theta} = 2\cdot 1 = 2.$$

EXAMPLE 2
A LIMIT INVOLVING $\cos \theta$

Show that

$$\lim_{\theta \to 0} \frac{\cos \theta - 1}{\theta} = 0. \tag{5}$$

SOLUTION:

$$\lim_{\theta \to 0} \frac{\cos \theta - 1}{\theta} = \lim_{\theta \to 0} \frac{(\cos \theta - 1)(\cos \theta + 1)}{\theta(\cos \theta + 1)} = \lim_{\theta \to 0} \frac{\cos^2 \theta - 1}{\theta(\cos \theta + 1)}$$

$$= \lim_{\theta \to 0} \frac{-\sin^2 \theta}{\theta(\cos \theta + 1)} \qquad \text{Since } \sin^2 \theta + \cos^2 \theta = 1.$$

$$= -\lim_{\theta \to 0} \frac{\sin \theta}{\theta} \lim_{\theta \to 0} \frac{\sin \theta}{(\cos \theta + 1)}$$

$$= -1 \cdot \frac{\displaystyle\lim_{\theta \to 0} \sin \theta}{\displaystyle\lim_{\theta \to 0} (\cos \theta + 1)} = -1 \cdot \frac{0}{2} = 0.$$

THEOREM 2   **DERIVATIVE OF $\sin x$**

$$\frac{d}{dx} \sin x = \cos x \tag{6}$$

PROOF:

$$\frac{d}{dx} \sin x = \lim_{\Delta x \to 0} \frac{\sin(x + \Delta x) - \sin x}{\Delta x}$$

Since $\sin(x + y) = \sin x \cos y + \cos x \sin y$
[see (xiii), Appendix 1.2, Table 2]

$$= \lim_{\Delta x \to 0} \frac{\sin x \cos \Delta x + \cos x \sin \Delta x - \sin x}{\Delta x} .$$

Theorem 1.3.2

$$= \lim_{\Delta x \to 0} \sin x \left(\frac{\cos \Delta x - 1}{\Delta x} \right) + \lim_{\Delta x \to 0} \cos x \cdot \frac{\sin \Delta x}{\Delta x}$$

Theorem 1.3.2

$$= \sin x \lim_{\Delta x \to 0} \frac{\cos \Delta x - 1}{\Delta x} + \cos x \lim_{\Delta x \to 0} \frac{\sin \Delta x}{\Delta x}$$

from (5)   from (1)

$$= \sin x \cdot 0 + \cos x \cdot 1 = \cos x. \quad \blacksquare$$

COROLLARY

The function $\sin x$ is continuous at every real number x.

PROOF: Since $\sin x$ is differentiable, it is continuous, by Theorem 1.7.3. $\blacksquare$

EXAMPLE 3
USING THE CHAIN RULE WITH
THE SINE FUNCTION

Compute

(a) $\dfrac{d}{dx} \sin x^2$ and   **(b)** $\dfrac{d}{dx} \sin^2 x.$

SOLUTION:

(a) If $u = x^2$, then $\sin x^2 = \sin u$. Thus, using the informal terminology discussed on page 147, x^2 is the inside function, $\sin x$
is the outside function, and

$$\frac{d}{dx} \sin x^2 = \frac{d}{dx} \sin u = \left(\frac{d}{du} \sin u\right) \overset{\overset{\displaystyle \text{from (6),}}{\overset{\displaystyle \frac{d}{du} \sin u = \cos u}{\downarrow}}}{\frac{du}{dx}} = \cos u \frac{du}{dx}$$

$$= (\cos x^2)(2x) = 2x \cos x^2.$$

derivative   derivative of
of inside   outside function
function

(b) If $u = \sin x$, then $\sin^2 x = (\sin x)^2 = u^2$, so now $\sin x$ is the inside
function (the thing being squared) and the outside function is
the squaring function x^2, so, using the power rule [equation (5)
on page 157],

$$\frac{d}{dx} \sin^2 x = \frac{d}{dx} u^2 = \left(\frac{d}{du} u^2\right)\frac{du}{dx} = 2u \frac{du}{dx}$$

$$= (2 \sin x)(\cos x).$$

derivative of   derivative of
outside function   inside function

The result of Example 3(a) can be easily generalized.

THEOREM 3

If $u(x)$ is a differentiable function of x, then

$$\frac{d}{dx} \sin u = \cos u \frac{du}{dx}. \tag{7}$$

PROOF: Let $f(u) = \sin u$. Then from the chain rule,

$$\frac{d}{dx} f(u) = \frac{df}{du}\frac{du}{dx} = \cos u \frac{du}{dx}. \quad \blacksquare$$

We now turn to the derivative of $\cos x$.

THEOREM 4	**DERIVATIVE OF cos x**

$$\frac{d}{dx}\cos x = -\sin x. \tag{8}$$

PROOF: From identity (xi) in Appendix 1.2, Table 2, $\cos x = \sin[(\pi/2) - x]$, so that

$$\frac{d}{dx}\cos x = \frac{d}{dx}\sin\left(\frac{\pi}{2} - x\right).$$

If $u = (\pi/2) - x$, then $du/dx = -1$, and

Identity (x), Appendix 1.2, Table 2

$$\frac{d}{dx}\sin\left(\frac{\pi}{2} - x\right) = -\cos\left(\frac{\pi}{2} - x\right) = -\sin x. \quad \blacksquare$$

REMARK: As with $\sin x$, $\cos x$ is evidently continuous at every real number x.

The following result can be proven by using the chain rule, as in Theorem 3.

THEOREM 5

If u is a differentiable function of x, then

$$\frac{d}{dx}\cos u = -\sin u \frac{du}{dx}. \quad \blacksquare \tag{9}$$

EXAMPLE 4
DERIVATIVE OF A COSINE
FUNCTION

Compute $(d/dx)(\cos\sqrt{x})$.

SOLUTION: Let $u = \sqrt{x}$ (the inside function). Then $du/dx = 1/2\sqrt{x}$, and, using (9), we have

$$\frac{d}{dx}\cos\sqrt{x} = \frac{d}{dx}\cos u = -\sin u \frac{du}{dx} = (-\sin\sqrt{x})\left(\frac{1}{2\sqrt{x}}\right)$$

$$= \frac{-\sin\sqrt{x}}{2\sqrt{x}}.$$

EXAMPLE 5
USING THE PRODUCT RULE AND
THE RULE FOR DIFFERENTIATING
A SINE FUNCTION

Compute $(d/dx)(x\sin 2x)$.

SOLUTION: Using the product rule, we obtain

$$\frac{d}{dx}x\sin 2x = x\frac{d}{dx}\sin 2x + 1\sin 2x$$

from (7)

$$= x\cos 2x\frac{d}{dx}(2x) + \sin 2x = 2x\cos 2x + \sin 2x.$$

We now turn to the derivatives of the other four trigonometric functions. (These functions are discussed in some detail in the review in Appendix 1.3.) Recall that

$$\tan x = \frac{\sin x}{\cos x}, \quad \cot x = \frac{\cos x}{\sin x}, \quad \sec x = \frac{1}{\cos x}, \quad \text{and} \quad \csc x = \frac{1}{\sin x}.$$

THEOREM 6 **DERIVATIVES OF FOUR TRIGONOMETRIC FUNCTIONS**

$$\frac{d}{dx} \tan x = \sec^2 x \tag{10}$$

$$\frac{d}{dx} \cot x = -\csc^2 x \tag{11}$$

$$\frac{d}{dx} \sec x = \sec x \tan x \tag{12}$$

$$\frac{d}{dx} \csc x = -\csc x \cot x \tag{13}$$

PROOF: We will prove formulas (10) and (12) and leave formulas (11) and (13) as exercises (see Problems 37 and 38).
Formula (10):

$$\frac{d}{dx} \tan x = \frac{d}{dx} \frac{\sin x}{\cos x} \overset{\text{Quotient rule}}{=} \frac{\cos x (d/dx)(\sin x) - \sin x (d/dx)(\cos x)}{\cos^2 x}$$

$$= \frac{\cos x (\cos x) - \sin x (-\sin x)}{\cos^2 x}$$

$$= \frac{\cos^2 x + \sin^2 x}{\cos^2 x} = \frac{1}{\cos^2 x}$$

$$= \left(\frac{1}{\cos x} \right)^2 = \sec^2 x.$$

Formula (12):

$$\frac{d}{dx} \sec x = \frac{d}{dx} \left(\frac{1}{\cos x} \right) = \frac{d}{dx} (\cos x)^{-1} \overset{\text{Power rule}}{=} (-1)(\cos x)^{-2} \frac{d}{dx} \cos x$$

$$= -\frac{1}{\cos^2 x} (-\sin x) = \frac{\sin x}{\cos^2 x}$$

$$= \frac{1}{\cos x} \cdot \frac{\sin x}{\cos x} = (\sec x)(\tan x). \quad \blacksquare$$

EXAMPLE 6
DERIVATIVE OF A TANGENT
FUNCTION

Compute

$$\frac{d}{dx}\sqrt{\tan x}.$$

SOLUTION:

Power rule

$$\frac{d}{dx}\sqrt{\tan x} = \frac{d}{dx}(\tan x)^{1/2} \overset{\downarrow}{=} \frac{1}{2}(\tan x)^{-1/2}\frac{d}{dx}\tan x$$

From (10)

$$\overset{\downarrow}{=} \frac{1}{2\sqrt{\tan x}}(\sec^2 x) = \frac{\sec^2 x}{2\sqrt{\tan x}}.$$

■ PROBLEMS 2.5

■ SELF-QUIZ

I. $\dfrac{d}{dx}\sin x = \cos x$ if _____.

    a. x is measured in radians
    b. x is measured in degrees
    c. x is measured either in radians or in degrees
    d. neither (a) nor (b) is correct

II. $\dfrac{d}{d\theta}\cos\theta = \sin\theta$ if _____.

    a. θ is measured in degrees
    b. θ is measured in radians
    c. both (a) and (b) are correct
    d. neither (a) nor (b) is correct

III. Compute

$$\lim_{x\to 0}\frac{\sin^2 x}{x^2}.$$

[Remember that $\sin^2 x$ is standard shorthand notation for $(\sin x)^2$.]

IV. $\dfrac{d}{dx}\sec x = $ _____.

    a. $\sec^2 x$ b. $-\csc^2 x$
    c. $\sec x \tan x$ d. $-\csc x \cot x$

V. $\sec^2\theta = \dfrac{d}{d\theta}$ _____.

    a. $\tan x$ b. $\tan\theta$
    c. $\sec\theta\tan\theta$ d. $\cot\theta$

VI. What is the simplest explanation for the fact that $\sec^2 x$ and $\tan^2 x$ have the same derivative with respect to x?

■ DRILL

In Problems 1–24, compute the derivative of the given function.

1. $y = \sin 3x$
2. $y = \cos(x - 3)$
3. $y = x\cos x$
4. $y = \sin x/x$
5. $y = \sqrt{\sin x}$
6. $y = (\cos x)^{1/3}$
7. $y = \sin^2 x$
8. $y = 1 - \cos^2 x$
9. $y = 2\sin x\cos x$
10. $y = \sin^2 x - \cos^2 x$
11. $y = \sin^2 x + \cos^2 x$
12. $y = (\sec x - \tan x)(\sec x + \tan x)$
13. $y = \sin x/\tan x$
14. $y = \sec x \tan x$
15. $y = (1 + \sec x)/(1 - \sec x)$
16. $y = (\sin x + \cos x)/(\tan x)$
17. $y = \sin(x^3 - 2x + 6)$
18. $y = \sin^2 x^3$
19. $y = \tan^2 x$
20. $y = \cot^2\sqrt{x}$
21. $y = \sec^3 x$
22. $y = \sqrt{\sin x + \tan x}$
23. $y = \csc x^2$
24. $y = \csc(1/x)$

In Problems 25–36, calculate the indicated limits.

25. $\lim\limits_{x\to 0} \dfrac{\sin(\frac{1}{2})x}{x}$

26. $\lim\limits_{x\to 0} \dfrac{3x}{\sin 2x}$

27. $\lim\limits_{x\to 0} \dfrac{\sin^2 x}{x}$

28. $\lim\limits_{x\to 0} \dfrac{\sin^2 3x}{4x}$

29. $\lim\limits_{x\to 0} \dfrac{\sin^2 4x}{x^2}$

30. $\lim\limits_{x\to 0} \dfrac{4x^7}{\sin^7 2x}$

[*Hint:* First find $\lim_{x\to 0}(\sin 4x)/x$.]

31. $\lim\limits_{x\to 0} \dfrac{\sin 3x}{\sin 4x}$

32. $\lim\limits_{x\to 0} \dfrac{\sin ax}{\sin bx}, \; ab \neq 0$

33. $\lim\limits_{x\to 0} \dfrac{\cos 2x - 1}{x}$

34. $\lim\limits_{x\to 0} \dfrac{\tan 3x}{5x}$

35. $\lim\limits_{x\to 0} \dfrac{\sin^2 (2x)}{(7x)^2}$

36. $\lim\limits_{x\to 0} \dfrac{\cos x^2 - 1}{x^4}$

■ APPLICATIONS

37. Verify that
$$\frac{d}{dx} \cot x = \frac{d}{dx}\left(\frac{\cos x}{\sin x}\right) = -\csc^2 x.$$

38. Verify that
$$\frac{d}{dx} \csc x = \frac{d}{dx}\left(\frac{1}{\sin x}\right) = -(\csc x)(\cot x).$$

For Problems 39 and 40, sine x is the sine of an angle that measures x in degrees, and cosine x is the cosine of that angle.

*39. If x is measured in degrees, compute $\lim_{x\to 0^\circ}(\text{sine } x)/x$.

*40. If x is measured in degrees, compute $\lim_{x\to 0^\circ}(\text{cosine } x - 1)/x$.

■ SHOW/PROVE/DISPROVE

*41. Find the derivative of $\tan x$ directly from the limit definition.

a. Use the fact that $\lim_{x\to 0}(\sin x)/x = 1$ to find $\lim_{x\to 0}(\tan x)/x$.

b. Use the fact that
$$\tan(A + B) = \frac{\tan A + \tan B}{1 - (\tan A)(\tan B)}$$

to show that
$$\frac{d}{dx} \tan x = \lim_{\Delta x \to 0} \frac{\tan(x + \Delta x) - \tan x}{\Delta x}$$
$$= 1 + \tan^2 x = \sec^2 x.$$

42. Show the following:
a. $(\sin x)/x$ is continuous at every real number except 0.
b. $(\sin x)/x$ has a removable discontinuity at 0.

■ ANSWERS TO SELF-QUIZ

I. a

II. d

III. $1 \left[\dfrac{\sin^2 x}{x^2} \right] = \left(\dfrac{\sin x}{x}\right)^2$

IV. c

V. b

VI. Since $\tan^2 x + 1 = \sec^2 x$, $\sec^2 x$ and $\tan^2 x$ differ by a constant (1); hence they must have the same derivative.

■ 2.6 IMPLICIT DIFFERENTIATION

*I*n most of the problems we have encountered, the variable y was given *explicitly* as a function of the variable x. For example, for each of the functions

$$y = 3x + 6, \qquad y = x^2, \qquad y = \sqrt{x + 3},$$
$$y = 1 + 2x + 4x^3 \qquad \text{and} \qquad y = (1 + 8x)^{3/2},$$

the variable y appears alone on the left-hand side. Thus we may say, "You give me an x and I'll tell you the value of $y = f(x)$." One exception to this rule was given in Section 0.2 where the variables x and y were given *implicitly* in the equation of the circle centered at $(0, 0)$ with radius r:

$$x^2 + y^2 = r^2.$$ (1)

Here x and y are not given separately. In general, we say that x and y are given *implicitly* if neither one is expressed as an explicit function of the other.

NOTE: This is *not* to say that one variable *cannot* be solved explicitly in terms of the other.

EXAMPLE 1
IMPLICIT FUNCTIONS

The following are examples in which the variables x and y are given implicitly.

(a) $x^3 + y^3 = 6xy^4$ **(b)** $(2x^{3/2} + y^{5/3})^{17} - 6y^5 = 2$

(c) $2xy(x + y)^{4/3} = 6x^{17/9}$ **(d)** $\dfrac{x + y}{\sqrt{x^2 - y^2}} = 16y^5$

(e) $xy = 1$ (Here it is easy to solve for one variable in terms of the other.)

For the example of the circle, it was possible to solve equation (1) explicitly for y in order to calculate dy/dx. However, it is very difficult or impossible to do the same thing for the functions (a)–(d) defined implicitly in Example 1 (try it!). Another difficulty is that some of the equations in Example 1 might include more than one implicit function. Nevertheless, the derivative dy/dx *may* exist. Can we calculate it?

To illustrate the answer to this question, let us again return to equation (1):

$$x^2 + y^2 = r^2.$$

If $y > 0$, then $y = \sqrt{r^2 - x^2}$, and

$$\overset{\text{Power rule}}{\frac{dy}{dx} = \frac{d}{dx}(r^2 - x^2)^{1/2} \overset{\downarrow}{=} \frac{1}{2}(r^2 - x^2)^{-1/2}\frac{d}{dx}(r^2 - x^2) = \frac{-2x}{2\sqrt{r^2 - x^2}}}$$

or

$$\frac{dy}{dx} = -\frac{x}{\sqrt{r^2 - x^2}} = -\frac{x}{y}.$$

We now calculate this derivative another way. Assuming that y can be written as a function of x, we can write $y^2 = [f(x)]^2$ for some function $f(x)$ (which we assume to be unknown[†]). Then by the chain rule,

---

[†] In this case we know that $f(x) = \sqrt{r^2 - x^2}$ or $-\sqrt{r^2 - x^2}$, but ordinarily (as for four of the functions in Example 1) $f(x)$ will *really* be unknown.

$$\frac{d}{dx}(y^2) = \frac{d}{dx}[f(x)]^2 = 2f(x) \cdot f'(x) = 2y\frac{dy}{dx}. \tag{2}$$

Now taking the derivatives of both sides of (1) with respect to x and using (2), we obtain

$$2x + 2y\frac{dy}{dx} = \frac{d}{dx}r^2 = 0.^\dagger \tag{3}$$

We can now solve for dy/dx in (3):

$$\frac{dy}{dx} = -\frac{x}{y}. \tag{4}$$

If we do not know y as a function of x, then this is as far as we can go. However, since in this case we may choose $y = \sqrt{r^2 - x^2}$, we may write equation (4) as

$$\frac{dy}{dx} = \frac{-x}{\sqrt{r^2 - x^2}}.$$

NOTE: We should keep in mind that what makes this technique work is that we are *assuming* that y is a differentiable function of x. Thus we may calculate

$$\frac{d}{dx}(y^2) = 2y\frac{dy}{dx}$$

as above.

The method we have used in the calculation above is called **implicit differentiation**, and it is the only way to calculate derivatives when it is impossible to solve for one variable in terms of the other.

In order to use implicit differentiation, it is necessary to assume that y is a differentiable function of x or that x is a differentiable function of y, without actually having a formula for the function.

If we do not make this assumption, then we cannot write $\frac{dy}{dx}\left(\text{or }\frac{dx}{dy}\right)$ because we cannot be certain that $\frac{dy}{dx}$ exists. If we assume that $\frac{dy}{dx}\left(\text{or }\frac{dx}{dy}\right)$ exists, we can find it by differentiating and then solving for it algebraically. We illustrate this method with a number of examples.

---

† We used the fact that both sides of an equation can always be differentiated provided that the derivatives of the functions involved exist.

EXAMPLE 2

FINDING $\frac{dy}{dx}$ BY IMPLICIT

DIFFERENTIATION

Suppose that

$$x^2 + x^3 = y + y^4. \qquad \text{(5)}$$

Find dy/dx.

SOLUTION: By the chain rule,

$$\frac{d}{dx} y^4 = 4y^3 \frac{d}{dx} y = 4y^3 \frac{dy}{dx}.$$

Thus, we may differentiate both sides of (5) with respect to x to obtain

$$\frac{d}{dx} x^2 + \frac{d}{dx} x^3 = \frac{d}{dx} y + \frac{d}{dx} y^4$$

or

$$2x + 3x^2 = \frac{dy}{dx} + 4y^3 \frac{dy}{dx} = (1 + 4y^3) \frac{dy}{dx}.$$

Then,

$$\frac{dy}{dx} = \frac{2x + 3x^2}{1 + 4y^3}.$$

At the point $(-1, 0)$, for example,[†]

$$\frac{dy}{dx} = \frac{2(-1) + 3 \cdot 1}{1 + 0} = 1,$$

and the equation of the tangent line at that point is

$$y = x + 1.$$

In this example there was no special reason to calculate dy/dx. We may have just as well been asked to calculate dx/dy.[†] We again use the chain rule to find that

$$\frac{d}{dy} x^2 = 2x \frac{dx}{dy} \qquad \text{and} \qquad \frac{d}{dy} x^3 = 3x^2 \frac{dx}{dy}.$$

Then differentiating both sides of (5) with respect to y yields

$$\frac{d}{dy} x^2 + \frac{d}{dy} x^3 = \frac{d}{dy} y + \frac{d}{dy} y^4$$

or

$$2x \frac{dx}{dy} + 3x^2 \frac{dx}{dy} = 1 + 4y^3 \qquad \text{and} \qquad \frac{dx}{dy} = \frac{1 + 4y^3}{2x + 3x^2}.$$

Note that $dx/dy = 1/(dy/dx)$. Although we will not prove this result here, it is true that under certain hypotheses $dy/dx = 1/(dx/dy)$ (see Problem 2.3.51 and Section 6.1).

[†] You should verify that $(-1, 0)$ is a point on the curve.

[‡] Assuming, of course, that x is a differentiable function of y.

EXAMPLE 3

SEEMING TO COMPUTE $\dfrac{dy}{dx}$
WHEN IT DOES NOT EXIST

Consider the equation

$$x^2 + y^2 + 4 = 0. \tag{6}$$

If we differentiate implicitly with respect to x, we obtain

$$2x + 2y\frac{dy}{dx} = 0 \quad \text{or} \quad \frac{dy}{dx} = -\frac{x}{y}.$$

But this answer is meaningless because there are no real numbers x and y that satisfy equation (6), so there is no real-valued function implicitly given by (6). What is the meaning of the derivative of a nonexistent function? This example illustrates that implicit differentiation only makes sense when y is a differentiable function of x.

EXAMPLE 4
COMPUTING THE EQUATION OF
A TANGENT LINE BY IMPLICIT
DIFFERENTIATION

Compute the slope of the line tangent to the curve $\sin xy = x$ at the point $(\frac{1}{2}, \pi/3)$.

SOLUTION: We have

$$\frac{d}{dx}\sin xy = \cos xy\left(\frac{d}{dx}xy\right) = \cos xy\left(x\frac{dy}{dx} + y\right)$$

and $(d/dx)x = 1$. Thus,

$$\cos xy\left(x\frac{dy}{dx} + y\right) = 1,$$

so that

$$x\frac{dy}{dx} + y = \frac{1}{\cos xy} \qquad x\frac{dy}{dx} = \frac{1}{\cos xy} - y$$

and

$$\frac{dy}{dx} = \frac{1}{x}\left(\frac{1}{\cos xy} - y\right).$$

At $(\frac{1}{2}, \pi/3)$, $xy = \pi/6$, $\cos xy = \sqrt{3}/2$, and $1/(\cos xy) = 2/\sqrt{3}$, so that

$$\frac{dy}{dx}\bigg|_{(1/2,\,\pi/3)} = 2\left(\frac{2}{\sqrt{3}} - \frac{\pi}{3}\right) = \frac{4}{\sqrt{3}} - \frac{2\pi}{3}.$$

We return briefly to the circle $x^2 + y^2 = r^2$. For $y > 0$ we calculated $dy/dx = -x/y$, which is zero when $x = 0$ (and $y = r$). Thus for $x = 0$, the tangent line has slope zero and is horizontal (see Figure 1). Now let us consider x as a function of y (for $x > 0$) and differentiate implicitly with respect to y to obtain

$$2x\frac{dx}{dy} + 2y\frac{dy}{dy} = \frac{d}{dy}r^2 = 0 \qquad \text{and} \qquad \frac{dx}{dy} = -\frac{y}{x}$$

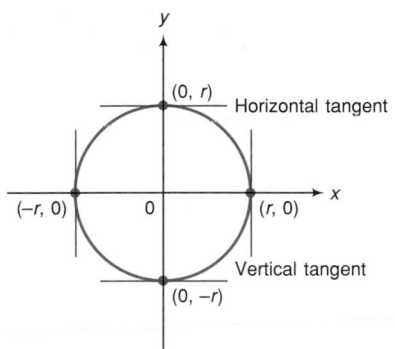

FIGURE 1
HORIZONTAL AND VERTICAL TANGENTS
TO A CIRCLE

which is zero at the point $(r, 0)$. If $dx/dy = 0$ at a point, then the tangent line to the curve at that point is parallel to the y-axis. In this case we say that the graph of the function has a *vertical tangent* at the point. This follows from the same reasoning that shows that the tangent line is parallel to the x-axis at any point at which $dy/dx = 0$. In our example, we see that the tangent line to the curve $x^2 + y^2 = r^2$ at the point $(0, r)$ is the line $x = r$. This line is vertical, as depicted in Figure 1.

We generalize this example with the following rules:

HORIZONTAL AND VERTICAL TANGENTS

(i) If $dy/dx = 0$ at the point (x_0, y_0), then the graph of $y = f(x)$ has a **horizontal tangent** at that point, given by the straight line $y = y_0$.

(ii) If $dx/dy = 0$ at the point (x_0, y_0), then the graph of $y = f(x)$ has a **vertical tangent** at that point, given by the straight line $x = x_0$.

NOTE: In both of these cases, $dy/dx \neq 1/(dx/dy)$ since we cannot divide by zero.

■ PROBLEMS 2.6

■ SELF-QUIZ

I. If $xy = 1$, then $dy/dx =$ _____.
  a. $-y/x$ b. $-y + x$
  c. $1/x$ d. y/x

II. If $y^2 = x$, then $dy/dx =$ _____.
  a. $2/y$ b. $1/(2y)$
  c. $1 - 2y$ d. $2y/x$

III. If $x^2 + y^2 = 36$, then $dx/dy =$ _____.
  a. $2x + 2y$ b. $-y/x$
  c. $-x/y$ d. $-\sqrt{y/x}$

IV. If $y + y^3 = x^2$, then $dy/dx =$ _____.
  a. $(1 + y^2)/(2x)$ b. $2x - y - 3y^2$
  c. $x^2/(3y^2)$ d. $2x/(1 + 3y^2)$

V. True–False: If we compute dy/dx and dx/dy by using implicit differentiation, then the results are negative reciprocals of one another.

■ DRILL

In Problems 1–10, find dy/dx by implicit differentiation.

1. $x^3 + y^3 = 3$
2. $x^3 + y^3 = xy$
3. $\sqrt{x} + \sqrt{y} = 2$
4. $1/x + 1/y = 1$
5. $x^{-7/8} + y^{-7/8} = \frac{7}{8}$
6. $x^{3/4} + y^{3/4} = 2$
7. $(3xy + 1)^5 = x^2$
8. $x^2 - \sqrt{xy} + y^2 = 6$
9. $(4x^2y^2)^{1/5} = 1$
10. $1/x^2 - 1/y^2 = x + y$

In Problems 11–24, find dy/dx. (Implicit differentiation is the technique to use for most of these, but in some of these problems, some preliminary algebra will greatly simplify your work.)

11. $(x + y)^{1/2} = (x^2 + y)^{1/3}$

12. $xy + x^2y^2 + x^3y^3 = 2$
13. $(x + y)(x - y) = 7$
14. $(x + y)/(x - y) = 2$
15. $\sin xy = 2$ [*Hint:* Think.]
16. $\cos(x^2 + y^2) = 2x$
17. $(\sin x)(\cos y) = y$
18. $\sin x/\cos y = \sin(x - y)$
19. $xy + x^2y^2 = x^5$
20. $x^2y^3 + x^3y^2 = xy$
21. $1/\sqrt{x^2 + y^2} = 4$
22. $\sqrt{xy^2 + yx^2} = 0$
23. $(x^2 + y^2)/(x^2 - y^2) = 4$
24. $(2xy + 1)/(3xy - 1) = 2$

CONTINUED

■ APPLICATIONS

In Problems 25–32, find the points where the given curve has a vertical tangent. Also, find the points where it has a horizontal tangent.

25. $\sqrt{x} + \sqrt{y} = 1$ 26. $1/x + 1/y = 1$
27. $xy = 1$
28. $(x - 3)^2 + (y - 4)^2 = 9$

29. $(x/a)^2 - (y/b)^2 = 1$ 30. $(x/a)^2 + (y/b)^2 = 1$
31. $y = 1/\sin x$ 32. $y \cos x = 3$
33. Find the equation of the line tangent to the curve $x^5 + y^5 = 2$ at the point $(1, 1)$.
34. Find the equation of the line tangent to the curve $x^2 - y^2 = 16$ at the point $(5, 3)$. Find the equation of the tangent line at $(-5, 3)$.

■ SHOW/PROVE/DISPROVE

35. Use implicit differentiation and the power rule for integer exponents to establish the power rule for rational exponents. Prove that if $f(x) = x^{p/q}$, where p and q are integers, then $f'(x) = x^{(p/q) - 1}$. [Hint: If $y = x^{14/3}$, then $y^3 = x^{14}$.]
36. Consider y to be a function of x implicitly specified by $\sqrt{y/x} + 4\sqrt{x/y} = \sqrt{18}$. Show that $dy/dx = y/x$.
37. Prove that the curves implicitly given by $3x - 2y + x^3 - x^2 y = 0$ and $x^2 - 2x + y^2 - 3y = 0$ are perpendicular at the origin (that is, their tangents there are perpendicular).
38. Pick a real number m.
    a. Show there are at most two points on the curve $10x^2 + (y/5)^2 = 1$ where the tangent line has the slope m.
    b. Show there are at most two points on $x^2 - y^2 = 37$ where the tangent line has slope equal to m.

*39. Suppose you are handed a function L such that $L'(x) = 1/x$ for all nonzero x. Also suppose that you have a function E with the property that $L \circ E = $ the identity function; that is, $L(E(y)) - y$ for all real y. Prove that $E' = E$.

*40. Suppose that $[y(x)]^2 = 137/\sqrt{\pi} - x^2 + 2ax$. Using implicit differentiation, show that each normal to the curve passes through the point $(a, 0)$.

■ CHALLENGE

*41. Rework the preceding problem using only algebraic tools and geometric information. Then write a brief essay comparing and contrasting this approach with the one using derivatives.

■ ANSWERS TO SELF-QUIZ

I. a.   II. b.   III. b   IV. d
V. False (omit "negative")

2.7 HIGHER-ORDER DERIVATIVES

Let $s(t)$ represent the distance a falling object has dropped after t units of time have elapsed. Then, as we saw in Section 1.6, the derivative ds/dt evaluated at a time t may be interpreted as the instantaneous velocity of the object at that time. The velocity is the rate of change of position with respect to time. By definition, **acceleration** is the rate of change of velocity with respect to time. Thus acceleration can be thought of as the derivative of the derivative, or, more simply, as the *second derivative* of the function representing position. If, for example, $s = \frac{1}{2}gt^2$ represents the position of a falling object (as in Section 1.6), then

$$\frac{ds}{dt} = gt \quad \text{and} \quad \frac{d}{dt}\left(\frac{ds}{dt}\right) = g,$$

which is the acceleration due to gravity.

We now generalize these ideas. Let $y = f(x)$ be a differentiable function. Then the derivative $y' = dy/dx = f'$ is also a function of x. This new function of x, f', may or may not be a differentiable function. If it is, we call the derivative of f' the **second derivative** of f (i.e., the derivative of the derivative) and denote it by f''. There are other commonly used notations as well. We will write

Second Derivative

$$f'' = (f')' = \left(\frac{dy}{dx}\right)' = \frac{d}{dx}\left(\frac{dy}{dx}\right) = \frac{d^2y}{dx^2} = (y')' = y''$$

The notations

$$\frac{d^2y}{dx^2} = y'' = f''(x)$$

will be used interchangeably in this book to denote the second derivative.

Similarly, if f'' exists, it might or might not be differentiable. If it is, then the derivative of f'' is called the **third derivative** of f and is denoted by $f'''(x)$. Alternative notations are

Third Derivative

$$f'''(x) - \frac{d^3y}{dx^3} = y'''.$$

We can continue this definition indefinitely as long as each successive derivative is differentiable. After the third derivative, we avoid a clumsy succession of primes by using numerals to denote higher derivatives: $f^{(4)}$, $f^{(5)}$, $f^{(6)}$, ... Alternative notations for the successive derivatives are

$$f^{(4)}(x) = \frac{d^4y}{dx^4} = y^{(4)} \quad f^{(5)}(x) = \frac{d^5y}{dx^5} = y^{(5)} \quad \cdots \quad f^{(n)}(x) = \frac{d^ny}{dx^n} = y^{(n)}.$$

We emphasize that *each higher-order derivative of $y = f(x)$ is a new function of x* (if it exists).

EXAMPLE 1
COMPUTING THE SECOND, THIRD, AND FOURTH DERIVATIVES OF A FUNCTION

Let $y = f(x) = x^3$. Then,

$$\frac{dy}{dx} = f'(x) = 3x^2.$$

The second derivative is simply the derivative of the first derivative, so

$$\frac{d^2y}{dx^2} = f''(x) = \frac{d}{dx}\,3x^2 = 6x.$$

Similarly, the third derivative is the derivative of the second derivative, and we have

$$\frac{d^3y}{dx^3} = f'''(x) = \frac{d}{dx}\,6x = 6.$$

Finally,

$$\frac{d^4y}{dx^4} = f^{(4)}(x) = \frac{d}{dx} 6 = 0.$$

Note that for $k \geq 4$, $f^{(k)}(x) = 0$ since the derivative of the zero function is zero.

EXAMPLE 2
COMPUTING ALL HIGHER
DERIVATIVES OF A FUNCTION

Let $y = f(x) = 1/x$. Then,

$$\frac{dy}{dx} = f'(x) = -\frac{1}{x^2}, \qquad \frac{d^2y}{dx^2} = f''(x) = \frac{2}{x^3}, \qquad \frac{d^3y}{dx^3} = f'''(x) = -\frac{6}{x^4},$$

and so on. In general, $d^ny/dx^n = f^{(n)}(x) = (-1)^n n!/x^{n+1}$ (see Problem 33). The symbol $n!$, which is read "n factorial," is defined (for the nonnegative integers) by

$$n! = 1 \cdot 2 \cdot 3 \cdot 4 \cdots \cdot n \qquad \text{for} \qquad n = 1, 2, 3, \ldots, \qquad \text{(1)}$$

and

$$0! = 1.$$

For example, $3! = 1 \cdot 2 \cdot 3 = 6$, and $7! = 1 \cdot 2 \cdot 3 \cdot 4 \cdot 5 \cdot 6 \cdot 7 = 5040$.

EXAMPLE 3
HIGHER DERIVATIVES OF sin x

Let $y = f(x) = \sin x$. Then $dy/dx = f'(x) = \cos x$, $d^2y/dx^2 = f''(x) = -\sin x$, $d^3y/dx^3 = f'''(x) = -\cos x$, and $d^4y/dx^4 = f^{(4)}(x) = \sin x$. That is, sin x is equal to its fourth derivative and the negative of its second derivative.

Although we have now defined derivatives of all orders, most of our major applications will involve first and/or second derivatives. Second derivatives are important for several reasons. In Section 3.3, for example, we will show that the sign of the second derivative of a function tells us something about the shape of the graph of that function. Perhaps an even more interesting application is the following: From elementary Newtonian physics, we may represent a force acting on a particle as the mass of the particle times its acceleration, $F = ma$. Now as we have seen, acceleration is merely the second derivative of the position function. Thus Newton's law states that $F = m \, d^2s/dt^2$. We can then use Newton's laws of motion to derive equations whose solutions tell us how the particle moves.

■ PROBLEMS 2.7

■ SELF-QUIZ

I. True–False:
a. d^2y/dy^2 and $(dy/dx)^2$ mean the same thing.
b. $y^{(2)}$ and y^2 mean the same thing.
c. y'' and $y^{(2)}$ mean the same thing.
d. If $y'' = 0$, then y is constant.
e. If $y'' = 0$, then y' is constant.

f. If y is constant, then $y'' = 0$.
g. If $y = 5 \cos x$, then $y'' = -y$ and $y^{(4)} = y$.
h. If $y = \sin 5x$, then $y'' = -y$ and $y^{(4)} = y$.
i. If $y = \sin 5x$, then $y'' = -25y$ and $y^{(4)} = 5^4y$.
j. If $y = 2 \sin 5x - 7 \cos 5x$, then $y'' = -25y$ and $y^{(4)} = 5^4y$.

CONTINUED

II. Rearrange the following list of functions so that those having equal second derivatives are grouped together.
a. $A(x) = -5$
b. $B(x) = (x + 5)(x - 3)$
c. $C(x) = 2x - 3$
d. $D(x) = x^2 - 3x + 2$
e. $E(x) = (x + 7)x - 5$
f. $F(x) = \frac{1}{2}(x - 7) + 15$

■ **DRILL**

In Problems 1–20, find d^2y/dx^2 and d^3y/dx^3.

1. $y = 3$
2. $y = 17x + 1$
3. $y = 4x^2$
4. $y = 9x^2$
5. $y = \sqrt{x}$
6. $y = 1/\sqrt{x}$
7. $y = 1/x^2$
8. $y = x^r$ (r is a real number)
9. $y = ax^2 + bx + c$
10. $y = a_0 + a_1x + a_2x^2 + a_3x^3$
11. $y = 1/(x + 1)^5$
12. $y = (1 + x)^{2/3}$
13. $y = \sqrt{1 - x^2}$
14. $y = (x^2 + 1)^{1/2}$
15. $y = \sin(x^2)$
16. $y = \cos\sqrt{x}$
17. $y = \tan x$
18. $y = \cot x$
19. $y = \csc x$
20. $y = \sec x$

■ **APPLICATIONS**

21. A rocket is shot upward in the earth's gravitational field so that its velocity at any time t is given by $v = 50t$. What is its acceleration?

22. A particle moves along a line so that its position along the line at time t is given by $s = 2t^3 - 4t^2 + 2t + 3$. The initial position is the position corresponding to $t = 0$.
a. What is its initial position?
b. What is its initial velocity?
c. What is its initial acceleration?
d. Show that the particle is initially decelerating.
e. For what value of t does the particle stop decelerating and begin accelerating?

23. A particle moves along the number line in such a way that its position at time t, $t \geq 0$, is $x(t) = (t - 4)(t^2 - 11t + 4) = t^3 - 15t^2 + 48t - 16$.
a. For what values of t (one or several intervals of time) is the particle moving to the right?
b. For what values of t is the particle speeding up?

c. For what values of t is the particle moving to the right and speeding up?
d. For what values of t is the particle moving to the left and slowing down?

24. Repeat the preceding problem using $x(t) = \cos t$ as the position function and restricting t to the interval $[0, 9\pi]$.

25. Let $f(x) = 1/(x^2 - 49)$. Find relatively simple expressions for $f'(x)$, $f''(x)$, and $f'''(x)$. [Hint: First consider $1/(x - 7) - 1/(x + 7)$.]

*26. Suppose that $x^2 + [y(x)]^2 = r^2$. Using implicit differentiation, find simple expressions for y'' and y'''.

27. Suppose that u and v are functions of x. Find formulas for $(uv)''$ and $(uv)'''$.

28. Suppose that $y = f(g(x))$.
a. Assuming that the appropriate derivatives of f and g do exist, find formulas for y'' and y'''.
b. Verify that your formulas work correctly in the case $f(t) = t^3$ and $g(x) = (x + 5)^2$.

■ **SHOW/PROVE/DISPROVE**

*29. Use mathematical induction (see Appendix 2) to prove the following statement about the nth derivative of x^n:

$$\frac{d^n}{dx^n} x^n = n!$$

$$= n \cdot (n - 1) \cdot (n - 2) \cdots 3 \cdot 2 \cdot 1.$$

30. Suppose that $p(x) = a_0 + a_1x + a_2x^2 + \cdots + a_{n-1}x^{n-1} + a_nx^n$. Apply the preceding problem to show that

$$\frac{d^np}{dx^n} = n!a_n \quad \text{and} \quad \frac{d^{n+1}p}{dx^{n+1}} = 0.$$

CONTINUED

*31. Suppose that k is a positive integer. Show, by mathematical induction, that

$$\frac{d^n \sin x}{dx^n} = \begin{cases} \sin x & \text{if } n = 4k \\ \cos x & \text{if } n = 4k + 1 \\ -\sin x & \text{if } n = 4k + 2 \\ -\cos x & \text{if } n = 4k + 3. \end{cases}$$

*32. Derive a formula analogous to the one in the preceding problem for the nth derivative of $\cos x$.

*33. Let $y = 1/x$. Show, using mathematical induction, that

$$\frac{d^n y}{dx^n} = \frac{(-1)^n n!}{x^{n+1}}.$$

*34. Prove or disprove that $\text{dom } f = \text{dom } f' = \text{dom } f''$ for any function f which is continuous on $(-\infty, \infty)$.

■ CHALLENGE

*35. Derive a formula for the nth derivative of a product. (Your formula will generalize the result of Problem 27.) [*Hint:* Compare terms in an expansion of $(a + b)^n$ with those in $(u \cdot v)^{(n)}$.]

*36. Derive a formula for the 4th derivative of a composite function. (This will generalize the result of Problem 28.)

■ ANSWERS TO SELF-QUIZ

I. a. False   b. False   c. True
   d. False   e. True   f. True
   g. True   h. False   i. True
   j. True

II. Functions A, C, F have second derivatives equal to zero; functions B, D, E have second derivatives equal to two.

■ PROBLEMS   CHAPTER 2 REVIEW

In Problems 1–24, calculate dy/dx.

1. $y = 3x + 4$

2. $y = 3x^2 - 6x + 2$

3. $x + y + xy = 3$

4. $y = \dfrac{x + 1}{x - 2}$

5. $y = (3x^2 + 1)\sqrt{x + 1}$

6. $y = x^5 - \sqrt{x}$

7. $x^{3/4} + y^{3/4} = 4$

8. $y = (x^2 - 1)^{1/2}(x^4 - 2)^{1/3}$

9. $y = \sqrt[3]{\dfrac{x + 2}{x - 3}}$

10. $xy^2 + yx^2 = xy$

11. $y = \dfrac{x^2 - 2x + 3}{x^3 + 5x - 6}$

12. $y = \dfrac{3}{x - \sqrt{x^3 - 2}}$

13. $y = (x^3 - 3)^{5/6}(x^3 + 4)^{6/7}$

14. $y = \dfrac{1 + x + x^5}{3 - x^2 + x^7}$

15. $x + \sqrt{x + y} = y^{1/3}$

16. $y = \dfrac{1 - \sqrt[3]{x}}{1 + \sqrt[3]{x}}$

17. $y = (1 + x^3)^{\sqrt{2}}$

18. $(1 + x^2 + y^2)^{3/2} = 6$

19. $y = \dfrac{(x^3 - 1)^{2/3}}{(x^5 + 3)^{3/4}}$

20. $y = \sqrt{1 - \sqrt{1 - \sqrt{x}}}$

21. $y = \sin x^3$

22. $y = \cos(1/x)$

23. $y = \tan x \sec x$

24. $y = \sqrt{1 + \cos x}$

In Problems 25–30, find equations for the tangent line and the normal line to the given curve at the specified point.

25. $y = x^3 - x + 4$; $(1, 4)$

26. $y = \dfrac{\sqrt{x^2 + 9}}{x^2 - 6}$; $(4, \frac{1}{2})$

27. $y = (x^2 - 4)^2(\sqrt{x} + 3)^{1/2}$; $(1, 18)$

28. $xy^2 - yx^2 = 0$; $(1, 1)$

29. $y = \cos \sqrt{x}$; $(\pi^2/9, \frac{1}{2})$

30. $y = \tan x$; $(\pi/4, 1)$

In Problems 31–38, calculate the second and third derivatives of the given function.

31. $f(x) = x^7 - 7x^6 + x^3 + 3$

32. $f(x) = \sqrt{1 + x}$

33. $f(x) = 1/(1 + x)$

34. $f(x) = \dfrac{x + 1}{x - 1}$

35. $f(x) = \dfrac{x^2 - 3}{x^2 + 5}$

36. $f(x) = \sqrt{1 + \sqrt{x}}$

37. $f(x) = \dfrac{\cos x}{x}$

38. $f(x) = \sqrt{\sin x}$

COMPUTER EXERCISES

1. Use a computer algebra system to view the first 10 derivatives of tan x. Formulate a conjecture about the general form of the nth derivative of tan x. (For example, the nth derivative of a polynomial of degree k is a polynomial of degree $k - n$ if $k \geq n$.) Provide a proof of your conjecture.

2. Use a computer algebra system to view the first 10 derivatives of sec x. Formulate a conjecture about the general form of the nth derivative of sec x. Provide a proof of your assertion.

3. If $f(x) = \dfrac{1}{1 + x^2}$, use a computer algebra system to calculate $f^{(n)}(0)$ for $n = 1, 2, 3, \ldots, 10$. In general

what is $f^{(n)}(0)$? $\left[\textit{Hint:} \text{ You will need to separate your answer into even and odd cases. It might help to look at } \dfrac{f^{(n)}(0)}{n!}. \right]$

4. A point moves along the real line so that its position at time t is given by $p(t) = -\dfrac{t^5}{5} + t^4 + t^3 + 2t^2 - t$. Find (or approximate) the time intervals during which the point is moving to the right.

5. Find 6-digit approximations to all points on the graph of $f(x) = \sin x + x^2$ where the tangent line is horizontal.

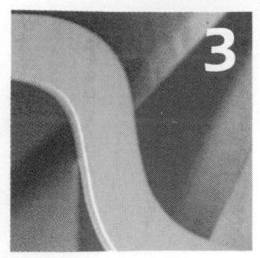

3 APPLICATIONS OF THE DERIVATIVE

We have seen, beginning in Section 1.6, that the derivative can be interpreted as a rate of change. There are many problems involving two or more variables in which it is necessary to calculate the rate of change of one or more of these variables with respect to time. After giving an example, we will suggest a procedure for handling problems of this type.

EXAMPLE 1
A ROPE AND PULLEY PROBLEM

A rope is attached to a pulley mounted on a 15-ft tower. The end of the rope is attached to a heavily loaded cart (see Figure 1). A worker can pull in rope at a rate of 2 ft/sec. How fast is the cart approaching the tower when it is 8 ft from the tower?

SOLUTION: We let s denote the horizontal distance of the cart from the tower and l the length of rope from the top of the tower to the cart, as in Figure 1. Since the speed of the cart is ds/dt, the question asks us to determine ds/dt when $s = 8$ ft. We are told that $dl/dt = -2$ (dl/dt is negative because the length represented by l is decreasing as the worker pulls in the rope). To calculate ds/dt, we must first find a relationship between s and l. From the Pythagorean theorem, we immediately obtain

$$15^2 + s^2 = l^2. \tag{1}$$

We now differentiate (1) implicitly with respect to t to find that

$$\frac{d}{dt}(15^2) + \frac{d}{dt}(s^2) = \frac{d}{dt}(l^2),$$

or

$$0 + 2s\frac{ds}{dt} = 2l\frac{dl}{dt} \quad \text{so} \quad \frac{ds}{dt} = \frac{l}{s}\frac{dl}{dt}. \tag{2}$$

When $s = 8$, $l^2 = 15^2 + 8^2 = 225 + 64 = 289$, and $l = 17$. Then inserting $s = 8$, $l = 17$, and $dl/dt = -2$ into equation (2) gives us

$$\frac{ds}{dt} = \frac{17}{8}(-2) = -\frac{17}{4} = -4\frac{1}{4} \text{ ft/sec.}$$

Thus the cart is approaching the tower at a rate of $4\frac{1}{4}$ ft/sec.

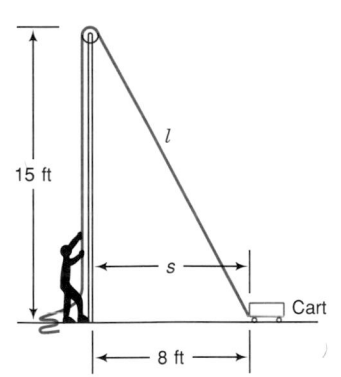

FIGURE 1
THE CART MOVES TOWARD THE TOWER AS THE ROPE IS PULLED DOWN.

15 ft

l

s

Cart

8 ft

TABLE 1

s	$\dfrac{ds}{dt} = -2\dfrac{\sqrt{s^2 + 15^2}}{s}$
8	-4.25 ft/sec
5	-6.32 ft/sec
3	-10.20 ft/sec
1	-30.03 ft/sec
0.5	-60.03 ft/sec
0.1	-300.07 ft/sec
0.01	-3000.00 ft/sec

REMARK: From equation (2) we find that

$$\frac{ds}{dt} = \frac{l}{s}\frac{dl}{dt} = -2\frac{l}{s} = -\frac{2\sqrt{s^2 + 15^2}}{s}$$

as $s \to 0^+$, $l \to 15$—the height of the tower—so $\dfrac{ds}{dt} \to -\infty$. That is, the cart moves faster and faster as it approaches the tower. In Table 1 we compute $\dfrac{ds}{dt}$ for different values of s.

⊗ **WARNING:** In solving this problem, we obtained the equation $15^2 + s^2 = l^2$ and differentiated implicitly with respect to t. We then inserted values for s and l into the resulting equation to solve for $\dfrac{ds}{dt}$. This is correct.

It is a mistake to insert the value for l or s before differentiating. If we first insert the value for l, for example, we obtain

$$15^2 + s^2 = 17^2.$$

When we differentiate now we get

$$0 + 2s\frac{ds}{dt} = 0$$

because the derivative of a constant is zero. This yields $\dfrac{ds}{dt} = 0$ which is, of course, wrong.

Another common error is to assign a letter to the length of the tower. If we assign the letter h, say, we obtain

$$h^2 + s^2 = l^2.$$

Then, differentiating with respect to t yields

$$2h\frac{dh}{dt} + 2s\frac{ds}{dt} = 2l\frac{dl}{dt}.$$

This is technically correct as long as you recognize that $\dfrac{dh}{dt} = 0$ because $h = 15$ which is a constant. However, if you fail to set $\dfrac{dh}{dt} = 0$, it is impossible to solve for $\dfrac{ds}{dt}$.

In sum, (1) remember to differentiate before inserting specific values into the equation, and (2) don't turn constants into variables. ⊗

The solution to Example 1 suggests that the following steps be taken to solve a problem involving the rates of change of related variables:

TAKE THESE STEPS TO SOLVE A RELATED-RATES PROBLEM

(i) If feasible, draw a picture of what is going on.

(ii) Determine the important variables in the problem and find an equation relating them. If the variables are related geometrically, the equation relating them often uses geometric results like similar triangles or the Pythagorean theorem.

(iii) Differentiate the equation obtained in (ii) with respect to t.

(iv) Solve for the derivative sought.

(v) Evaluate that derivative by substituting given and calculated values of the variables in the problem.

(vi) Interpret your answer in terms of the question posed in the problem.

(vii) Don't make either of the errors given in the warning that follows Example 1.

EXAMPLE 2
COMPUTING RATE OF CHANGE OF HEIGHT WHEN VOLUME IS INCREASING

An oil-storage tank is built in the form of an inverted right-circular cone with a height of 6 m and a base radius of 2 m (see Figure 2). Oil is pumped into the tank at a rate of 2 liters (L)/min = 0.002 m³/min (since 1 m³ = 1000 L). How fast is the level of the oil rising when the tank is filled to a height of 3 m?

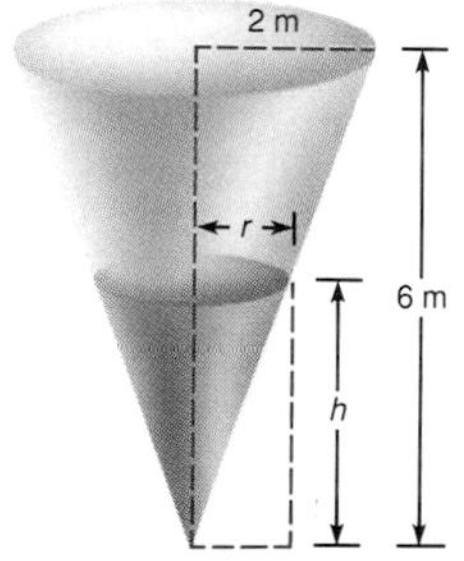

FIGURE 2
THE HEIGHT INCREASES AS OIL IS PUMPED INTO THE TANK.

SOLUTION: In mathematical terms, we are asked to calculate dh/dt when $h = 3$ m, where h denotes the height of the oil at a given time and r the radius of the cone of oil (see Figure 2). The volume of a right circular cone is $V = \frac{1}{3}\pi r^2 h$. From the data given in the problem, we see from Figure 2 (using similar triangles) that $h/r = \frac{6}{2}$, or $r = h/3$. Then,

$$V = \frac{1}{3}\pi\left(\frac{h}{3}\right)^2 h = \frac{1}{27}\pi h^3 = \text{volume of oil at height } h.$$

Differentiating with respect to t and using the fact (given to us) that $dV/dt = 0.002$, we obtain

$$\frac{dV}{dt} = \frac{1}{9}\pi h^2 \frac{dh}{dt} = 0.002, \qquad \text{or} \qquad \frac{dh}{dt} = \frac{9(0.002)}{\pi h^2} = \frac{0.018}{\pi h^2} \text{ m/min.}$$

Then for $h = 3$,

$$\frac{dh}{dt} = \frac{0.018}{\pi \cdot 9} = \frac{0.002}{\pi} \approx 6.37 \times 10^{-4} \text{ m/min,}$$

which is the rate at which the height of the oil is increasing.

EXAMPLE 3
AN ADIABATIC PROCESS

In chemistry, an **adiabatic** process is one in which there is no gain or loss of heat. During an adiabatic process, the pressure P and volume V of certain gases such as hydrogen or oxygen in a container are related by the formula

$$PV^{1.4} = \text{constant}. \tag{3}$$

At a certain time, the volume of hydrogen in a closed container is 4 m³ and the pressure is 0.75 kg/m². Suppose that the volume is increasing at a rate of $\frac{1}{2}$ m³/sec. How fast is the pressure decreasing?

SOLUTION: We use the product rule to differentiate (3) with respect to t.

$$P \frac{d}{dt}(V^{1.4}) + \frac{dP}{dt} \cdot V^{1.4} = 0, \qquad P(1.4)V^{0.4}\frac{dV}{dt} + \frac{dP}{dt} \cdot V^{1.4} = 0,$$

or

$$\frac{dP}{dt} = -\frac{1.4\,PV^{0.4}}{V^{1.4}}\frac{dV}{dt} = -1.4\,\frac{P}{V} \cdot \frac{dV}{dt}.$$

Using the data given in the problem, we find that

$$\frac{dP}{dt} = -\frac{(1.4)(0.75)\ \text{kg/m}^2}{4\ \text{m}^3} \cdot (0.5)\ \text{m}^3/\text{sec} = -0.13125\ \text{kg/m}^2/\text{sec}.$$

Thus, the pressure is decreasing at a rate of 0.13125 kg/m²/sec.

EXAMPLE 4
COMPUTING THE RATE OF
CHANGE OF AN ANGLE

A ladder 8 m long leans against a wall 4 m high. The lower end of the ladder is pulled away from the wall at a rate of 2 m/sec. How fast is the angle between the top of the ladder and the wall changing when the angle is $60° = \pi/3$ radians?

SOLUTION: We refer to Figure 3. We are asked to calculate $d\theta/dt$ when $\theta = \pi/3$. From the figure we see that $\sin \theta = x/8$, where x is the distance from the foot of the ladder to the wall. We are told that $dx/dt = 2$. Then using the chain rule, we have

$$\cos \theta \frac{d\theta}{dt} = \frac{d}{dt} \sin \theta = \frac{d}{dt}\left(\frac{x}{8}\right) = \frac{1}{8}\frac{dx}{dt} = \frac{2}{8} = \frac{1}{4}.$$

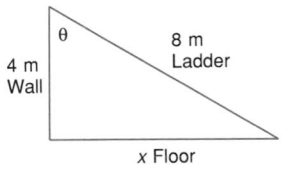

FIGURE 3
THE ANGLE θ INCREASES AS THE LADDER
SLIDES DOWN THE WALL.

Thus $d\theta/dt = 1/(4 \cos \theta)$, and when $\theta = \pi/3$, $\cos \theta = \frac{1}{2}$, so that $d\theta/dt = 1/(4 \cdot \frac{1}{2}) = \frac{1}{2}$ radian/sec, which is the rate at which the angle is increasing.

■ PROBLEMS 3.1

■ SELF-QUIZ

I. If $x + y$ is constant, then _____ .
   a. dx/dt and dy/dt have the same sign
   b. dx/dt and dy/dt have opposite signs

II. If $x \cdot y = 36$, then _____ .
   a. $dx/dt \cdot dy/dt \geq 0$
   b. $dx/dt \cdot dy/dt \leq 0$

CONTINUED

III. If $2x + 3y = 6$ and $dx/dt = -1$, then
_____.
   a. $dy/dt = -\frac{2}{3}$ b. $dy/dt = 0$
   c. $dy/dt > 0$ d. $dy/dt = \frac{3}{2}$

IV. If _____ and $dx/dt = -1$, then $dy/dt > 0$.
   a. $x - y = -5$ b. $y = 5x - 2$
   c. $5y - x = 14$ d. $x/5 + y/14 = 1$

V. Consider a rectangle with perimeter fixed at 8 m but with adjustable width W and height H. Suppose width W increases at the rate of 1 m/min, then (letting the variable t denote time) _____.

   a. $dH/dt = 0$
   b. height H does not change
   c. $dH/dt < 0$
   d. $dH/dt > 0$

VI. Consider a rectangle with area fixed at 12 m² but with adjustable width W and height H. If its width W decreases at the rate of 1 m/min, then _____. [*Hint:* Dimensions of a rectangle are positive or zero.]
   a. $dH/dt = 0$
   b. height H does not change
   c. $dH/dt < 0$
   d. $dH/dt > 0$

■ DRILL

1. Suppose $xy = 6$ and $dx/dt = 5$. Compute dy/dt when $x = 3$.
2. Suppose $x/y = 2$ and $dx/dt = 4$. Compute dy/dt when $x = 2$.
3. A 10-ft ladder is leaning against the side of a house. As the foot of the ladder is pulled away from the house, the top of the ladder slides down along the side of the house (see Figure 4). Suppose the foot of the ladder is pulled away at a rate of 2 ft/sec. How fast is the ladder top sliding down when the foot is 8 ft from the house?

4. A sailor standing on the edge of a dock 15 ft above the lake surface is pulling in her boat by means of a line attached to the boat's bow. She pulls in the line at the rate of 5 ft/min. How fast is the boat approaching the foot of the dock when the boat is 20 ft away?

5. A light is affixed to the top of a 12-ft-tall lamppost. A 6-ft-tall man walks away from the lamppost at a rate of 5 ft/sec. How fast is the length of his shadow increasing when he is 5 ft away? [*Hint:* One piece of information given here is irrelevant.]

6. An airplane at a height of 1000 m is flying horizontally at a velocity of 500 km/hr and passes directly over a police observer. How fast is the plane receding from the observer when it is 1500 m away (along a direct line of sight) from the observer?

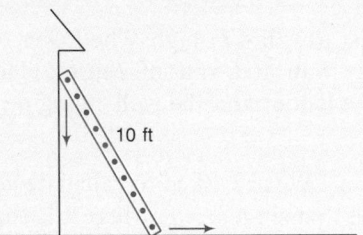

10 ft

FIGURE 4

■ APPLICATIONS

7. Joe is a baseball player whose top running speed is 25 ft/sec. He has reached second base and is now trying to steal third base. The catcher (who is 90 ft from third base) throws the ball toward the third baseman when Joe is 30 ft from third base. If the ball is thrown with speed 120 ft/sec, then what is the rate of change of the distance between the ball and Joe at the instant the ball is thrown? (See Figure 5.)
8. Two roads intersect at right angles. A car traveling 80 km/hr reaches the intersection half an hour before a bus that is traveling on the other

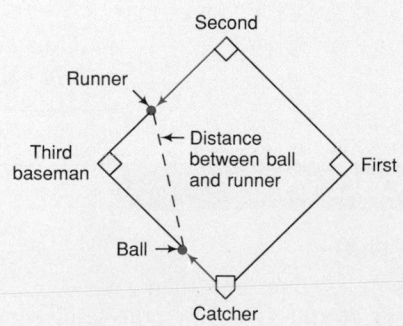

Second

Runner

Third
baseman

Distance
between ball
and runner

First

Ball

Catcher

FIGURE 5

road at 60 km/hr. How fast is the distance between the car and the bus increasing 1 hr after the bus reaches the intersection? (See Figure 6.)

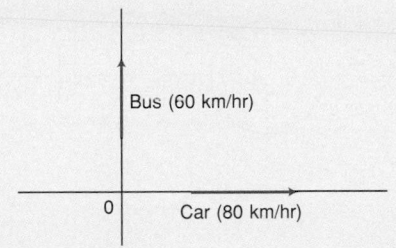

FIGURE 6

9. At 2 p.m. on a certain day, ship A is 100 km due north of ship B. At that moment, ship A begins to sail due east at a rate of 15 km/hr while ship B sails due north at a rate of 20 km/hr. How fast is the distance between the two ships changing at 5 p.m.? Is that distance increasing or decreasing?

10. Two ships, I and II, meet and then sail off in different directions, keeping an angle of 60° between their paths (see Figure 7). Ship I travels at 30 km/hr and ship II travels at 40 km/hr. How fast is the distance between them changing when ship I is 9 km from the meeting point and ship II is 12 km from the meeting point? [*Hint:* Use the law of cosines; see Problem 11 of Appendix 1.4.]

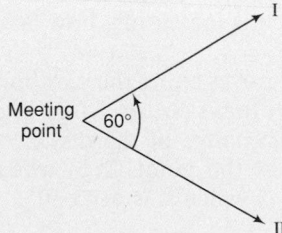

FIGURE 7

11. A cylindrical water tank 6 m high with a radius of 2 m is being filled at the rate of 10 liters/min (see Figure 8). How fast is the water rising when the water level is at a height of 0.5 m? [*Note:* 1 m³ = 1000 liters.]

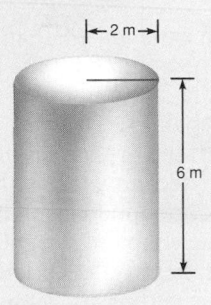

FIGURE 8

12. Sand is dropped onto a conical pile at a rate of 15 m³/min (see Figure 9). Suppose the height of the pile is always equal to its diameter. How fast is the height increasing when the pile is 7 m high?

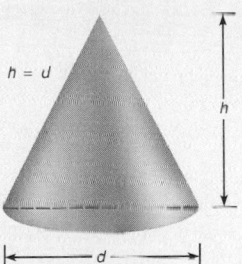

FIGURE 9

13. A rock is thrown into a pool of water. A circular wave leaves the point of impact and travels so that its radius increases at a rate of 25 cm/sec. How fast is the circumference of the wave increasing when the radius of the wave is 1 m?

14. Bacteria grow in circular colonies. The radius of one colony is increasing at the rate of 4 mm/day. On Wednesday, the radius of the colony is 1 mm. How fast is the area of the colony changing one week (i.e., seven days) later?

15. A spherical mothball is dissolving at a rate of 8π cm³/hr. How fast is the radius of the mothball decreasing when the radius is 3 cm? [*Hint:* The volume of a sphere of radius r is $(\frac{4}{3})\pi r^3$.]

16. The body of a snowman is in the shape of a sphere and is melting at a rate of 2 ft³/hr. How fast is the radius changing when the body is 3 ft in diameter (assuming the body stays spherical)?

17. When helium expands adiabatically, its pressure is related to its volume by the formula $PV^{5/3} = $ constant. At a certain time, the volume of a helium-filled balloon is 18 m³ and the pressure is 0.3 kg/m². If the pressure is increasing at a rate of 0.01 kg/m²/sec, how fast is the volume changing? Is the volume increasing or decreasing?

18. A pill is in the shape of a right circular cylinder with a hemisphere on each end (see Figure 10). The height (excluding its hemispherical ends) of the cylinder is half its radius. What is the rate of change of the volume of the pill with respect to the radius of the cylinder when the radius changes?

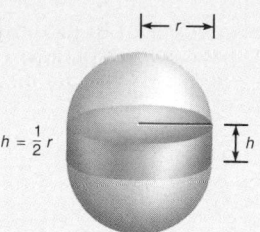

FIGURE 10

19. A circular wheel with a radius of 13 cm is rotating in the counterclockwise direction at a constant rate of 10 revolutions per minute (rpm) (see Figure 11). How fast are x and y changing when $x = 12$ and $y = 5$? [*Hint:* Express everything in terms of θ, the angle between a fixed radius of the circle and the positive x-axis.]

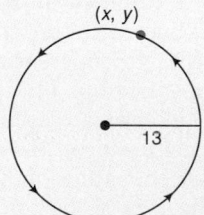

FIGURE 11

20. A prison guard tower has a light that is 300 m from a straight wall (see Figure 12). Its light revolves at a rate of 5 rpm. How fast is the light beam moving along the wall at a point 300 m along the wall (measured from the point on the wall that is closest to the tower)?

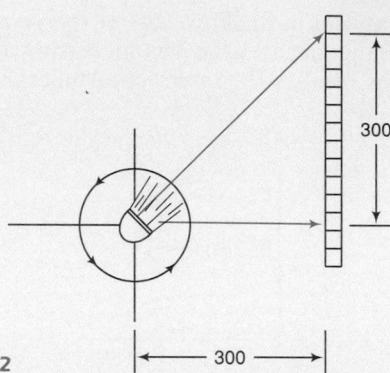

FIGURE 12

*21. In soil mechanics, the vertical stress σ_z in a soil at a point z feet deep and located r feet (horizontally) from a concentrated surface load of P pounds is given by

$$\sigma_z = \frac{3P}{2\pi z^2}\left[1 + \left(\frac{r}{z}\right)^2\right]^{2.5}.$$

a. Calculate the vertical stress in a soil at a point 2 ft deep and located 4 ft horizontally from a concentrated surface load of 10,000 lbs.

b. If that load is moved horizontally toward the point at a rate of 1 ft/hr, how is the vertical stress changing when the load is 3 ft from the point?

22. The amount of water flowing across a dam with a V-shaped notch has been determined to be $Q = 2.505h^{2.47}\tan(\theta/2)$, where Q is the volume of flow measured in m³/sec, h is the height (in meters) of the water measured from the bottom point of the notch, and θ is the angle of the notch. Assuming that h is kept constant at 2 m while the angle of the notch is increasing, how fast is Q increasing when $\theta = 40°$?

*23. A particle moves along the parabola $y = x^2$ such that dx/dt always equals 3 (see Figure 13). Find the rate of change of the distance between the particle and the point $(2, 5)$ when the particle is at $(-1, 1)$; when it is at $(3, 9)$.

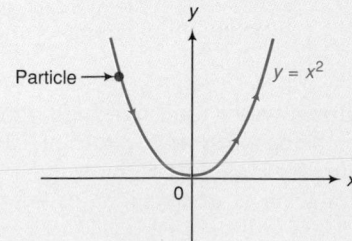

FIGURE 13

CONTINUED

24. A bright projection lamp shines from ground level on a vertical screen 25 m away. A thin man 2 m tall jogs in front of the lamp toward the screen at a constant rate of 3 m/sec (see Figure 14). When the man is 15 m from the screen, how fast is the height of his shadow on the screen changing?

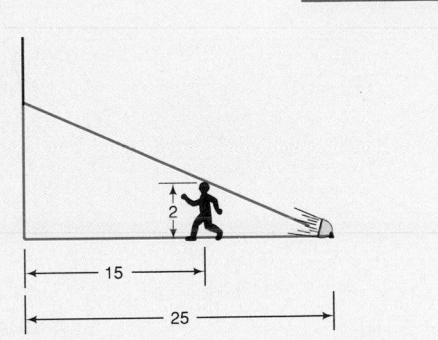

FIGURE 14

■ CHALLENGE

**25. A slide projector shines a bright light on a vertical screen 25 m away. A sphere 2 m in diameter is steadily moved in the center of the beam of light toward the screen at a constant rate of 3 m/sec. When the center of the sphere is 15 m from the screen, how fast is the height of its shadow on the screen changing?

*26. In Example 1, we found that the cart approaches the tower at a speed of $\frac{2}{s}$ ft/sec, where s is the

distance from the cart to the tower. But as $s \to 0^+$, $\frac{2}{s} \to \infty$. This means that the speed of the cart becomes infinitely large as the cart approaches the tower. But, according to relativity theory, nothing can travel faster than 186,000 miles/sec ($\approx 300{,}000$ km/sec)—the speed of light. Does this example prove relativity theory wrong, or is there some other explanation?

■ ANSWERS TO SELF-QUIZ

I. b

II. b (since $xy' + yx' = 0$ so $x'v' = -\frac{y}{x}x'^2 \le 0$

since $\frac{y}{x} > 0$)

III. c
IV. d
V. c (since $2W + 2H = 8$ and $0 < dW/dt$)
VI. d (since $WH = 12$, $0 \le W$, $0 \le H$, and $dW/dt < 0$)

3.2 ELEMENTARY CURVE SKETCHING I: Increasing and Decreasing Functions and the First-Derivative Test

In Section 1.5 we saw how we could get information about the derivative of a function by looking at the graph of that function and observing where the function is increasing (the tangent line has a positive slope) and decreasing (the tangent line has a negative slope). In Examples 1.5.4 and 1.5.5 we "reversed" the process and showed how information about the derivative of some simple functions could help us to graph those functions. Now that we know how to calculate quite a few derivatives, we will begin the important task of learning how to draw the graphs of a much wider class of functions. In this section we will make use of a theorem about increasing and decreasing functions. This theorem will enable us to get a fairly good picture about the appearance of the graph of the function in question. Later, in Section 3.3, we will use additional facts about derivatives to get even more information about these graphs.

Consider the function whose graph is depicted in Figure 1. In the interval (x_0, x_1) the function increases. In (x_1, x_2) it decreases. From x_2 to x_3 the function again increases, and so on. Moreover, when $x = x_1, x_2, x_3, x_4,$ and x_5, the tangent line is horizontal so that the derivative is zero at those points. To be more precise, we have the following definitions.

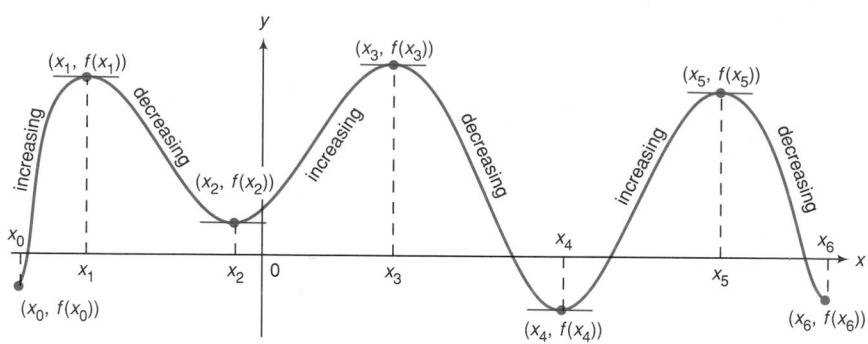

FIGURE 1 INTERVALS OVER WHICH A FUNCTION INCREASES AND DECREASES

DEFINITION   **INCREASING AND DECREASING FUNCTIONS**

The function f defined on an interval $[a, b]$ is said to be **increasing** on that interval if whenever $a \leq x_1 < x_2 \leq b$, we have

$$f(x_1) < f(x_2). \tag{1}$$

It is said to be **decreasing** on that interval if whenever $a \leq x_1 < x_2 \leq b$, we have

$$f(x_1) > f(x_2). \quad \blacksquare \tag{2}$$

DEFINITION   **CRITICAL POINT**

Let f be defined and continuous on the interval $[a, b]$[†] and let x_0 be in (a, b). Then the number x_0 is called a **critical point** of f if $f'(x_0) = 0$ *or* if $f'(x_0)$ does not exist. $\blacksquare$

REMARK 1:   Note that by this definition, a critical point x_0 is *not* an endpoint of the interval (a, b).

REMARK 2:   Sometimes the point $(x_0, f(x_0))$ is called a critical point.

EXAMPLE 1
FINDING CRITICAL POINTS BY LOOKING AT THE GRAPH

In Figure 1 the critical points are x_1, x_2, x_3, x_4, and x_5 because f' evaluated at these values is zero.

EXAMPLE 2
A CRITICAL POINT OF THE ABSOLUTE-VALUE FUNCTION

Let $f(x) = |x|$. Then 0 is a critical point since $f'(0)$ does not exist. (See Example 1.5.7.)

The critical point x_0 is sometimes called a **critical number**.

In Section 1.5, we inferred that a function was increasing when its derivative was positive and decreasing when its derivative was negative. The proof of the following theorem is given in Section 3.7 (p. 243).

[†] See page 116 for the definition of continuity on the closed interval $[a, b]$.

| THEOREM 1 | INCREASING AND DECREASING THEOREM |

Let f be differentiable for $a < x < b$ and continuous for $a \leq x \leq b$.

(i) If $f'(x) > 0$ for every x in (a, b), f is increasing on $[a, b]$.

(ii) If $f'(x) < 0$ for every x in (a, b), f is decreasing on $[a, b]$. ∎

EXAMPLE 3
INCREASING AND DECREASING
INTERVALS FOR A QUADRATIC
FUNCTION

For what values of x is the function $f(x) = x^2 - 2x + 4$ increasing and decreasing? Sketch the curve.

SOLUTION: $f'(x) = 2x - 2 = 2(x - 1)$. If $x > 1$, $x - 1 > 0$ and $2(x - 1) > 0$. If $x < 1$, $x - 1 < 0$ and $2(x - 1) < 0$. Thus for $x < 1$, f is decreasing, and for $x > 1$, f is increasing. When $x = 1$, $f'(x) = 0$, so that 1 is a critical point. These results are summarized in Table 1. We can use this

TABLE 1

	$f'(x)$	$f(x)$ is
$-\infty < x < 1$	$-$	decreasing
$x = 1$	0	(critical point)
$1 < x < \infty$	$+$	increasing

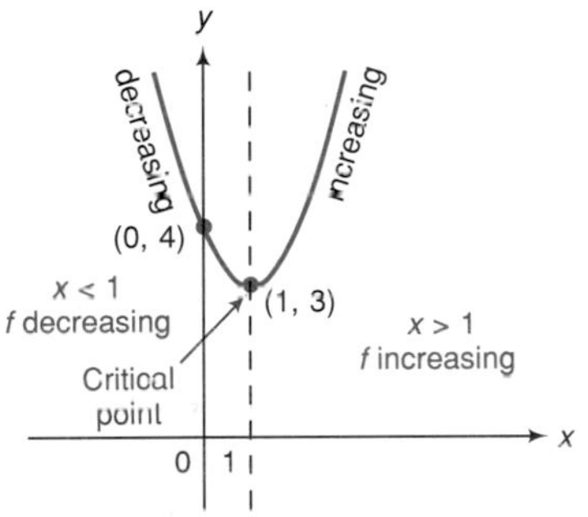

FIGURE 2
GRAPH OF $f(x) = x^2 - 2x + 4$

The function increases on $[1, \infty)$ and decreases on $(-\infty, 1]$.

information to obtain a rough sketch of the function. When $x = 1$, $f(x) = 3$. Table 1 tells us that $f(x)$ decreases until $x = 1$ and increases thereafter. Also, the y-intercept is found by setting $x = 0$ and is at the point $(0, 4)$. We put this information together in Figure 2. We notice that $f(x)$ takes its *minimum value* when $x = 1$. We will shortly see how maximum and minimum values of a function can be found before drawing a sketch of the curve by analyzing critical points more closely.

EXAMPLE 4
INCREASING AND DECREASING
INTERVALS FOR A CUBIC
FUNCTION

Let $y = x^3 + 3x^2 - 9x - 10$. For what values of x is this function increasing and decreasing? Sketch the curve.

SOLUTION: $dy/dx = 3x^2 + 6x - 9 = 3(x^2 + 2x - 3) = 3(x + 3)(x - 1)$. Now look at Table 2. The critical points are the numbers -3 and 1. At

TABLE 2

	$x + 3$	$x - 1$	$\dfrac{dy}{dx} = 3(x + 3)(x - 1)$	f is
$x < -3$	$-$	$-$	$+$	increasing
$x = -3$	0	-4	0	(critical point)
$-3 < x < 1$	$+$	$-$	$-$	decreasing
$x = 1$	4	0	0	(critical point)
$1 < x$	$+$	$+$	$+$	increasing

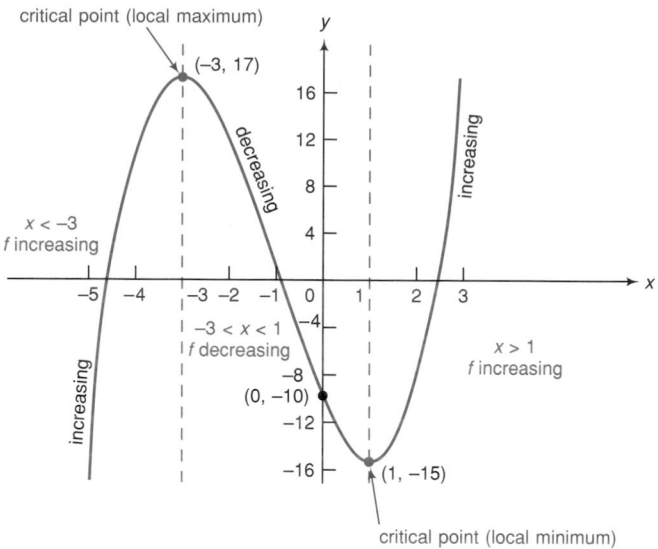

FIGURE 3   GRAPH OF $f(x) = x^3 + 3x^2 - 9x - 10$

The function increases on $(-\infty, -3]$ and $[1, \infty)$ and decreases on $[-3, 1]$.

$x = -3$, $y = 17$; at $x = 1$, $y = -15$. Setting $x = 0$, we obtain the y-intercept $y = -10$. Hence, the curve passes through the point $(0, -10)$. Using this information, we obtain the graph shown in Figure 3. We notice from the curve that the point $(-3, 17)$ is a maximum point in the sense that "near" $x = -3$, y takes its largest value at $x = -3$. However, there is no *global* (or *absolute*) maximum value for the function, since as x increases beyond the value 1, y increases without bound. For example, if $x = 10$, $y = 10^3 + 3 \cdot 10^2 - 9 \cdot 10 - 10 = 1200$, which is much bigger than 17. In this setting, the point $(-3, 17)$ is called a *local maximum* (or *relative maximum*) in the sense that the function achieves its maximum value there for points *near* $(-3, 17)$. Similarly, we call the point $(1, -15)$ a *local minimum* (or *relative minimum*).

DEFINITION   **MAXIMA AND MINIMA**

(i) The function f has a **local maximum** at x_0 if there is an open interval (c, d) containing x_0 such that $f(x_0) \geq f(x)$ for every x in (c, d).

(ii) The function f has a **local minimum** at x_0 if there is an open interval (c, d) containing x_0 such that $f(x_0) \leq f(x)$ for every x in (c, d).

(iii) The function f has a **global maximum** at x_0 if $f(x_0) \geq f(x)$ for every x in the domain of f.

(iv) The function f has a **global minimum** at x_0 if $f(x_0) \leq f(x)$ for every x in the domain of f. ■

REMARK: If f has a local maximum or minimum at x_0, then x_0 is called an **extremum**.

NOTE 1:

(i) In Example 4, f has a local, but not a global, maximum at $x = -3$. It also has a local, but not global, minimum at $x = 1$. These results are evident from Figure 3.

(ii) In Example 3, f has a local *and* global minimum at $x = 1$. There is no local or global maximum.

NOTE 2: We should point out that the interval (c, d) in the definition above could be very small.

In Figure 4, $f(x)$ has a local minimum at x_0 and a local maximum at x_1 even though these are minimum and maximum values of f over very small intervals. This result indicates why calculus methods are superior to other methods for curve sketching. Small "wiggles" such as the two in Figure 4 could easily be missed if we were to try to sketch the curve, say, by plotting points.

As we saw in Examples 3 and 4, whenever we had a local maximum or minimum, we also had a critical point. This is not a coincidence.

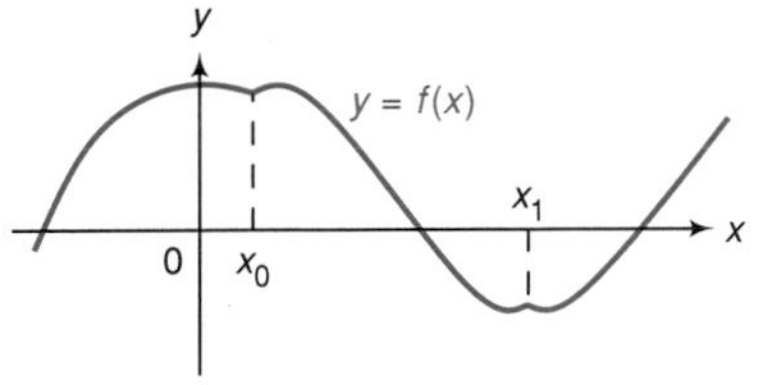

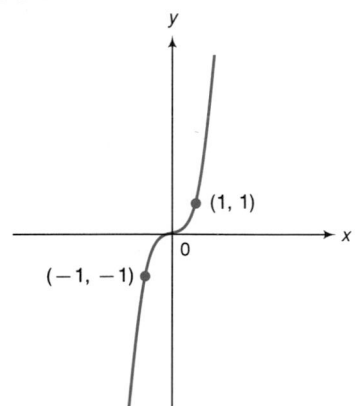

FIGURE 4
$f(x)$ HAS A LOCAL MINIMUM AT x_0 AND A LOCAL MAXIMUM AT x_1, EVEN THOUGH THE INTERVALS OVER WHICH $f(x)$ IS MINIMIZED AT x_0 AND MAXIMIZED AT x_1 ARE VERY SMALL.

THEOREM 2 **AN EXTREMUM IS A CRITICAL POINT**

Let f have a local maximum or minimum at x_0. Then x_0 is a critical point of f. ∎

The proof is given at the end of the section.

REMARK: The converse of this theorem is not true, as the next example shows.

EXAMPLE 5
A CRITICAL POINT AT WHICH THE FUNCTION HAS NEITHER A LOCAL MAXIMUM NOR A LOCAL MINIMUM

Let $y = f(x) = x^3$. Then $f'(x) = 3x^2$, which is always positive except at the critical point $x = 0$. If $x < 0$, then $f(x) < 0$; if $x > 0$, then $f(x) > 0$. So the function increases from negative values to zero to positive values at $x = 0$ (see Figure 5) and has neither a local maximum nor a local minimum there. Thus, as this example shows, a critical point may be neither a local maximum nor a local minimum.

Examples 3, 4, and 5 taken together illustrate the fact that at a critical point a function may have a local maximum, a local minimum, or neither. There are two ways to determine when a critical point is a local maximum or minimum. The first of these, the **first-derivative test**, is given in the next theorem. The second is given in the next section.

FIGURE 5
GRAPH OF $f(x) = x^3$

The function has neither a local maximum nor a local minimum.

THEOREM 3 **FIRST-DERIVATIVE TEST**

Let x_0 be a critical point of a function f with x_0 in an open interval (a, b). Suppose that f is continuous for $a \leq x \leq b$ and differentiable for $a < x < b$, except possibly at x_0 itself.

(i) If $f'(x) > 0$ for $a < x < x_0$ and $f'(x) < 0$ for $x_0 < x < b$, f has a local maximum at x_0.

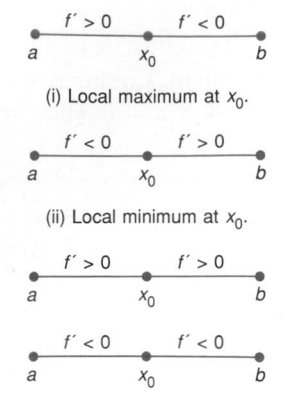

FIGURE 6
ILLUSTRATION OF FIRST-DERIVATIVE TEST

(ii) If $f'(x) < 0$ for $a < x < x_0$ and $f'(x) > 0$ for $x_0 < x < b$, f has a local minimum at x_0.

(iii) If $f'(x) < 0$ for $a < x < b$ or $f'(x) > 0$ for $a < x < b$ (except possibly at x_0 itself), f has neither a local maximum nor a local minimum at x_0. ∎

Statement (i) says that if f increases to the left of x_0 and decreases to the right of x_0, then f has a local maximum at x_0. Statement (ii) says that if f decreases to the left of x_0 and increases to the right of x_0, then f has a local minimum at x_0.

The first-derivative test is illustrated in Figure 6.

PROOF:

(i) The conditions given imply, by Theorem 1, that f is increasing on $[a, x_0]$ and decreasing on $[x_0, b]$. Thus if $a < x < x_0$, $f(x) < f(x_0)$; if $x_0 < x < b$, then $f(x_0) > f(x)$. Thus, by the definition of a local maximum f has a local maximum at x_0.

(ii) The proof of (ii) is virtually identical to the proof of (i) and is left to the reader.

(iii) Assume that $f'(x) > 0$ in (a, b) (the proof for $f'(x) < 0$ is nearly identical). Then f is increasing in (a, b). Let x be in (a, b). If $x < x_0$, then $f(x) < f(x_0)$; if $x > x_0$, then $f(x) > f(x_0)$. Thus, f cannot have either a local maximum or a local minimum at x_0. ∎

This theorem is illustrated again in Figure 7, in which we have drawn three typical tangent lines in each of four cases. We see that f has a local maximum at the critical point x_0 if the curve lies below the tangent lines near that point. Similarly, f has a local minimum if f lies above the tangent lines near that point. Finally, if neither of these "pictures" is valid near x_0, then x_0 has neither a local maximum nor a local minimum.

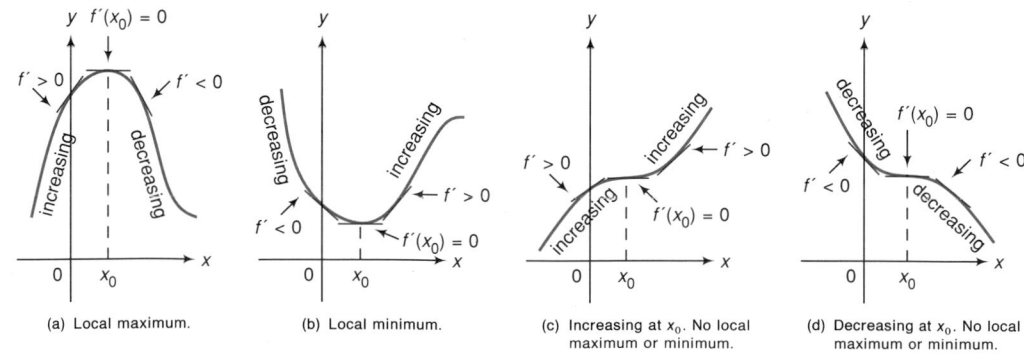

(a) Local maximum. (b) Local minimum. (c) Increasing at x_0. No local maximum or minimum. (d) Decreasing at x_0. No local maximum or minimum.

FIGURE 7 USING THE FIRST-DERIVATIVE TEST

EXAMPLE 3 (Continued)
A QUADRATIC WITH A MINIMUM

Here $f(x) = x^2 - 2x + 4$ and $f'(x) = 2x - 2 = 2(x - 1)$. Since $f'(x) < 0$ for $x < 1$, $f'(x) > 0$ for $x > 1$, and $f'(1) = 0$, f has a local minimum at $x = 1$ (which is also a global minimum).

EXAMPLE 4 (Continued)
A CUBIC WITH A LOCAL MINIMUM
AND A LOCAL MAXIMUM

Here $f'(x) = 3x^2 - 6x - 9 = 3(x + 3)(x - 1)$. We can use Table 2 to describe the nature of the critical points $x = -3$ and $x = 1$. In $(-4, -2)$, for example, $f'(x) > 0$ for $x < -3$ and $f'(x) < 0$ for $x > -3$. Thus, f has a local maximum at $x = -3$. Analogously, in $(0, 2)$, $f'(x) < 0$ for $x < 1$ and $f'(x) > 0$ for $x > 1$. Thus f has a local minimum at $x = 1$.

REMARK: In this example there was nothing special about the intervals $(-4, -2)$ and $(0, 2)$. Any intervals for which the signs of the derivatives went from positive to negative or negative to positive would work.

EXAMPLE 5 (Continued)
A CUBIC WITH NO EXTREMA

Here $f'(x) = 3x^2$, which is positive except at $x = 0$. Thus f has neither a local maximum nor a local minimum at 0.

EXAMPLE 6
SKETCHING THE GRAPH OF A
FOURTH-DEGREE POLYNOMIAL
FUNCTION

Sketch the curve $y = x^4 - 8x^2$.

SOLUTION: $dy/dx = 4x^3 - 16x = 4x(x^2 - 4) = 4x(x + 2)(x - 2)$. The critical points are $x = 0 \, (y = 0)$, $x = -2 \, (y = -16)$, and $x = 2 \, (y = -16)$. The behavior of the function is given in Table 3. When $x = 0$, $y = 0$, so the curve passes through the origin. Moreover, we can find the x-intercepts—the points on the curve and on the x-axis—by setting $y = 0$ to obtain $0 = x^4 - 8x^2 = x^2(x^2 - 8)$. This expression is equal to zero when $x = 0$ or when $x = \pm\sqrt{8}$. Using all this information, we obtain the curve drawn in Figure 8.

TABLE 3

	x	$x + 2$	$x - 2$	$f'(x) =$ $4x(x + 2)(x - 2)$	f is
$x < -2$	−	−	−	−	decreasing
$x = -2$	−	0	−	0	(critical point—local minimum)
$-2 < x < 0$	−	+	−	+	increasing
$x = 0$	0	+	−	0	(critical point—local maximum)
$0 < x < 2$	+	+	−	−	decreasing
$x = 2$	+	+	0	0	(critical point—local minimum)
$2 < x$	+	+	+	+	increasing

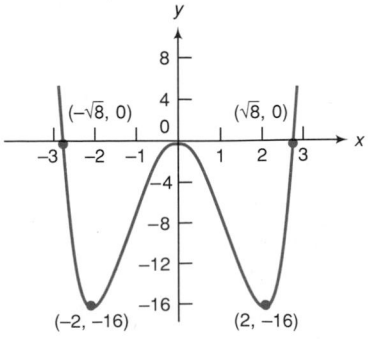

FIGURE 8   GRAPH OF $f(x) = x^4 - 8x^2$

EXAMPLE 7
SKETCHING A SINE CURVE

Sketch the curve $y = \sin x$.

SOLUTION: Since $\sin x$ is periodic of period 2π, if we can sketch $\sin x$ for $0 \le x \le 2\pi$, we can obtain the graph of $\sin x$ by extending our sketch. Now $dy/dx = \cos x$. Moreover, $\cos x$ is positive in the first and fourth quadrants, negative in the second and third quadrants, and equal to zero at $\pi/2$ and $3\pi/2$. This information is summarized in Table 4. In

TABLE 4

	$f'(x) = \cos x$	$f(x) = \sin x$ is
$0 < x < \pi/2$	+	increasing
$x = \pi/2$	0	(critical point—local maximum)
$\pi/2 < x < 3\pi/2$	−	decreasing
$x = 3\pi/2$	0	(critical point—local minimum)
$3\pi/2 < x < 2\pi$	+	increasing

addition, $\sin x = 0$ when $x = 0$, π, and 2π, $\sin \dfrac{\pi}{2} = 1$ and $\sin \dfrac{3\pi}{2} = -1$.

This is all the information we need to obtain the graph of the curve. The sketch in the interval $[0, 2\pi]$ is given in Figure 9a and the extended sketch is given in Figure 9b.

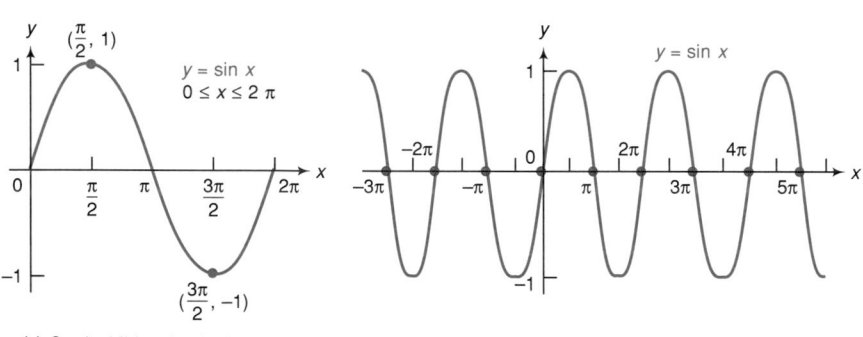

(a) Graph of $f(x) = \sin x$ for $0 \le x \le 2\pi$.

(b) Graph of $f(x) = \sin x$.

FIGURE 9 GRAPH OF $\sin x$

We close this section with a proof of Theorem 2.

PROOF OF THEOREM 2: If f has a Local Maximum or Local Minimum at x_0, Then x_0 Is a Critical Point Let f have a local maximum or local minimum at x_0. If $f'(x_0)$ does not exist, x_0 is a critical point. Otherwise, we may assume that f has a local maximum or local minimum at x_0 and that $f'(x_0)$ exists. Then it is necessary to show that $f'(x_0) = 0$. By the definition of the derivative on page 87,

$$f'(x_0) = \lim_{\Delta x \to 0} \frac{f(x_0 + \Delta x) - f(x_0)}{\Delta x}.$$

Equivalently,

$$\frac{f(x_0 + \Delta x) - f(x_0)}{\Delta x} \to f'(x_0) \text{ as } \Delta x \to 0$$

or

$$\frac{f(x_0 + \Delta x) - f(x_0)}{\Delta x} - f'(x_0) \to 0 \text{ as } \Delta x \to 0. \tag{3}$$

We define

$$\epsilon(\Delta x) = \frac{f(x_0 + \Delta x) - f(x_0)}{\Delta x} - f'(x_0). \tag{4}$$

Then, from (3),

$$\epsilon(\Delta x) \to 0 \text{ as } \Delta x \to 0$$

and we may write, from (4),

$$\frac{f(x_0 + \Delta x) - f(x_0)}{\Delta x} = f'(x_0) + \epsilon(\Delta x). \tag{5}$$

We now assume that f has a local maximum at x_0. A similar proof works when f has a local minimum at x_0. If Δx is sufficiently small, then $f(x_0 + \Delta x) \le f(x_0)$ [this is what is meant by a local maximum], so

$$f(x_0 + \Delta x) - f(x_0) \le 0. \tag{6}$$

If $\Delta x > 0$, then

$$\underset{\substack{\uparrow \\ \text{a negative over} \\ \text{a positive}}}{\frac{f(x_0 + \Delta x) - f(x_0)}{\Delta x}} \le 0$$

so, from (5),

$$f'(x_0) + \epsilon(\Delta x) \le 0. \tag{7}$$

Since $\epsilon(\Delta x) \to 0$ as $\Delta x \to 0$, (7) implies that

$$f'(x_0) \le 0. \tag{8}$$

Similarly, if $\Delta x < 0$, then

$$\underset{\substack{\uparrow \\ \text{a negative over} \\ \text{a negative}}}{\frac{f(x_0 + \Delta x) - f(x_0)}{\Delta x}} \ge 0$$

so

$$f'(x_0) + \epsilon(\Delta x) \ge 0$$

which implies that

$$f'(x_0) \ge 0. \tag{9}$$

From (8) and (9) we see that $f'(x_0) = 0$. This completes the proof. ∎

■ PROBLEMS 3.2

■ SELF-QUIZ

I. Suppose f is a differentiable function such that $f'(x) < -3$ for all x in the interval $(2, 5)$; then _____.

 a. f has a critical point at the middle of the interval $(2, 5)$
 b. every point of the interval $(2, 5)$ is a critical point for f
 c. f increases throughout the interval $(2, 5)$
 d. f decreases throughout the interval $(2, 5)$

II. Suppose $f(x) = x - 3$. Answer True or False for each of the following.
 a. f is increasing on the interval $(0, 3)$.
 b. f is increasing on the interval $(3, 25)$.
 c. f is increasing on the interval $(3, \infty)$.
 d. f is increasing everywhere.

III. Suppose $g'(x) = 3 - x$. Answer True or False for each of the following.
 a. g is decreasing on the interval $(0, 3)$.
 b. g is decreasing on the interval $(3, 25)$.
 c. g is decreasing on the interval $(-\infty, 3)$.
 d. g is decreasing everywhere.

IV. Let $f(x) = x^2 - 6x + 2$; the point $(1, -3)$ is

_____ for the function f.
 a. the global maximum
 b. the global minimum
 c. a critical point but not an extremum
 d. not a critical point

V. Let $g(x) = (x - 2)^3 + 5$; the point $(2, 5)$ is _____ for the function g.
 a. the global maximum
 b. the global minimum
 c. a critical point but not an extremum
 d. not a critical point

VI. Suppose f and g are differentiable functions which increase on the interval $[a, b]$. Answer True or False for each of the following.
 a. $f + g$ is increasing on $[a, b]$.
 b. $f \cdot g$ is increasing on $[a, b]$.
 c. If we also suppose f and g are always positive, then $f \cdot g$ is increasing on $[a, b]$.
 d. $f \circ g$ is increasing on $[a, b]$.
 e. If we also suppose the range of g is a subset of $[a, b]$, then $f \circ g$ is increasing on $[a, b]$.

■ DRILL

In Problems 1–18:

a. Find the intervals over which the given function is increasing.
b. Find the intervals over which the given function is decreasing.
c. Find all critical points.
d. Find the y-intercept (and the x-intercepts if convenient).
e. Locate all local maxima and minima by using the first derivative test.
f. Sketch the curve.

1. $y = x^2 + x - 30$
2. $y = -x^2 + 2x - 4$
3. $y = 4x^2 - 8$
4. $y = -\frac{3}{2}x^2 + 4x - 7$
5. $y = 2x^2 + 4x + 6$
6. $y = x^4 - 16$
7. $y = x^3 + x$
8. $y = x^3 - 12x + 10$
9. $y = x^3 - x^2$
10. $y = x^3 - 3x^2 - 45x + 25$
11. $y = x^4 - 18x^2$
12. $y = x^4 - 4x^3 + 4x^2 + 1$
13. $y = |2x + 7|$
14. $y = |x - 2| + |x - 5|$
15. $y = (x + 1)^{1/3}$
16. $y = (x - 2)^{2/3}$
17. $y = \sin 4x$
18. $y = \cos(x/2)$

■ APPLICATIONS

19. Let $f(x) = \sqrt{x + 1}$.
 a. Show that f is increasing wherever it is defined.
 b. Show that there are no points on the graph for which $f'(x) = 0$.
 c. Find the x- and y-intercepts.
 d. Sketch the curve. What happens as x gets close to -1?

20. Answer the items of the preceding problem for the decreasing function $g(x) = 1/\sqrt{1 + x}$.
21. Consider the function $f(x) = x/(1 + x.)$
 a. Show that this function is always increasing except at $x = -1$ where it is not defined.
 b. What happens as $x \to \pm\infty$?
 c. What happens as $x \to -1$?
 d. Show that $y > 1$ if $x < -1$.

CONTINUED

e. Show that $y < 1$ if $x > -1$.

f. Sketch the graph of $y = f(x)$.

22. Show that

$$f(x) = \frac{1}{x+1} + \frac{1}{x-1}$$

is decreasing on the interval $(-1, 1)$.

In Problems 23 and 24, draw the graph of a function having the given properties.

23. $f(-3) = f(1) = 0$; $f'(x) > 0$ for $x < -1$; $f'(-1) = 0$; $f'(x) < 0$ for $x > -1$.

24. $f(0) = f(3) = f(6) = 0$; $f'(\frac{3}{2}) = f'(\frac{7}{2}) = 0$; $f'(x) < 0$ if $x < \frac{3}{2}$ or $x > \frac{7}{2}$; $f'(x) > 0$ for $\frac{3}{2} < x < \frac{7}{2}$.

■ SHOW/PROVE/DISPROVE

25. Let $D(x) = \sqrt{(x-4)^2 + (3x-1)^2}$ and $S(x) = (x-4)^2 + (3x-1)^2$. Prove the equivalence of the following statements.

a. The global minimum of D occurs at $x = c$.

b. The global minimum of S occurs at $x = c$.

[Note the practical use of this theoretical result: S is easier to differentiate.]

26. Prove each of the *true* parts of Self-Quiz Problem VI. Provide a counterexample for each part that is *false*.

*27. Suppose that $ax^2 + bx + c \geq 0$ for each real number x. Show that $b^2 \leq 4ac$.

*28. Let $f(x) = ax^3 + bx^2 + cx + d$. Prove that f has a local extremum somewhere in $(-\infty, \infty)$ if and only if $3ac < b^2$.

■ CHALLENGE

*29. Suppose $f(x) = (x+1)^p \cdot (x-1)^q$, where p and q are integers greater than or equal to 2.

a. Show that f' is zero if $x = -1$, 1, or $(p-q)/(p+q)$.

b. Describe the local extrema of f. [*Note:* There are four different cases to analyze, depending on whether p is odd or even and whether q is odd or even.]

■ ANSWERS TO SELF-QUIZ

I. d

II. a. True   b. True
    c. True   d. True

III. a. False   b. True
     c. False   d. False

IV. d   V. c

VI. a. True   b. False   c. True
    d. False   e. True

3.3 ELEMENTARY CURVE SKETCHING II:
Concavity and the Second-Derivative Test

In this section we continue the discussion of curve sketching begun in Section 3.2. Consider the graph of $f(x) = x^2 - 2x + 4$ given in Figure 1a (see Figure 3.2.2). For $x < 1$, $f'(x) < 0$; at $x = 1$, $f'(x) = 0$; and for $x > 1$, $f'(x) > 0$. Thus, the derivative function f' *increases* continuously from negative to positive values. When this situation occurs, the function is said to be *concave up*. Functions that are concave up have the shape of the curve in Figure 1a or 1b.

Now consider the function $y = -x^2$ whose graph is given in Figure 2a. For $x < 0$, $f'(x) > 0$; at $x = 0$, $f'(x) = 0$; and for $x > 0$, $f'(x) < 0$. Thus, the derivative decreases continuously from positive to negative values, and in this case the function is called *concave down*. Functions that are concave down have the shape of the curve in Figure 2a or 2b. The more precise definition of this concept follows.

3 APPLICATIONS OF THE DERIVATIVE

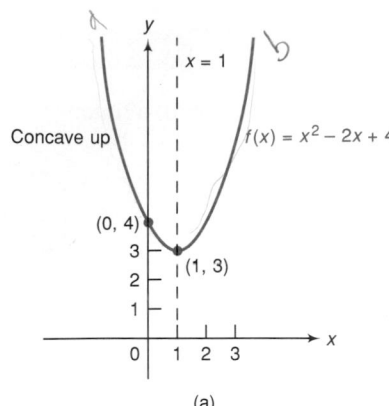

(a)

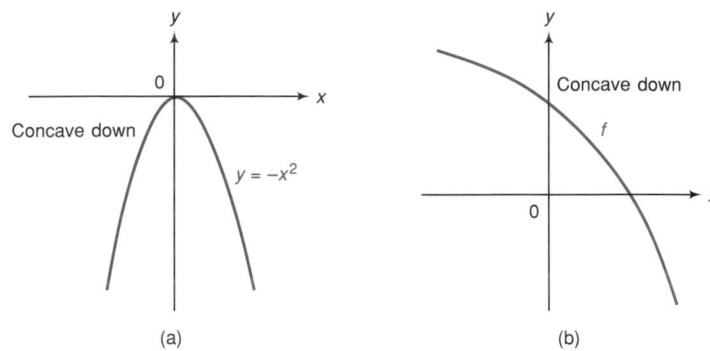

(a)   (b)

FIGURE 2 GRAPHS OF TWO CONCAVE DOWN FUNCTIONS

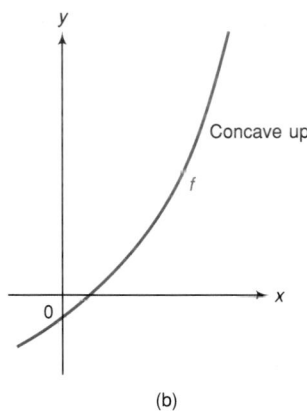

(b)

FIGURE 1
GRAPHS OF TWO FUNCTIONS THAT ARE
CONCAVE UP

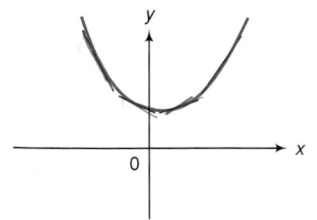

(a) Concave up: graph lies above tangent lines.

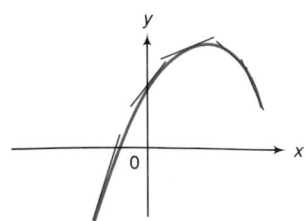

(b) Concave down: graph lies below
tangent lines.

FIGURE 3
ILLUSTRATION OF CONCAVITY

DEFINITION   **CONCAVITY**

Let f be differentiable in the interval (a, b).

  (i) f is **concave up** in (a, b) if over that interval f' is increasing.

  (ii) f is **concave down** in (a, b) if over that interval f' is decreasing. ■

REMARK: In this definition it is essential that f' exist in (a, b). For example, the function $f(x) = |x|$ has a derivative that goes from negative values (-1) to positive values $(+1)$; yet it has "no concavity" in any interval containing 0.

DEFINITION   **POINT OF INFLECTION**

The point $(x_0, f(x_0))$ is a **point of inflection** of f if f changes its direction of concavity at x_0. That is, it is a point of inflection if there is an interval (c, d) containing x_0 such that one of the following holds:

  (i) f is concave down in (c, x_0) and concave up in (x_0, d).

  (ii) f is concave up in (c, x_0) and concave down in (x_0, d). ■

REMARK: As we will see, f may have a point of inflection at $(x_0, f(x_0))$ in one of the following two cases:

  (i) $f''(x_0) = 0$.

  (ii) $f''(x_0)$ does not exist.

Another way to think of concavity is suggested by Figure 3. Here we have drawn in some tangent lines. In Figure 3a, f is concave up and all points on the curve lie *above* the tangent lines. In 3b, f is concave down and all points on the curve lie *below* the tangent lines. In some texts these facts are given as the *definition* of concavity.

Before looking at other examples, let us make some observations. The derivative of f' is f''. If $f'' > 0$, then f' has a positive derivative, and so we can conclude that f' is increasing and f is concave up. If $f'' < 0$, then f' is decreasing and f is concave down. Thus if f'' exists, *we can determine the direction of concavity by simply looking at the sign of the second derivative.* We have therefore proved the following theorem.

| THEOREM 1 | THE SECOND DERIVATIVE AND CONCAVITY |

Suppose that $f''(x)$ exists on the interval (a, b).

(i) If $f''(x) > 0$ in (a, b), f is concave up in (a, b).

(ii) If $f''(x) < 0$ in (a, b), f is concave down in (a, b). ∎

EXAMPLE 1
GRAPHING A CUBIC HAVING ONE POINT OF INFLECTION WHERE THE SECOND DERIVATIVE IS ZERO

Let $f(x) = x^3 - 2x$. Then $f'(x) = 3x^2 - 2$ and $f''(x) = 6x$. We see that $f''(x) < 0$ if $x < 0$ and $f''(x) > 0$ if $x > 0$. Thus, $(0, 0)$ is a point of inflection. Note that $f''(0) = 0$. The function is sketched in Figure 4.

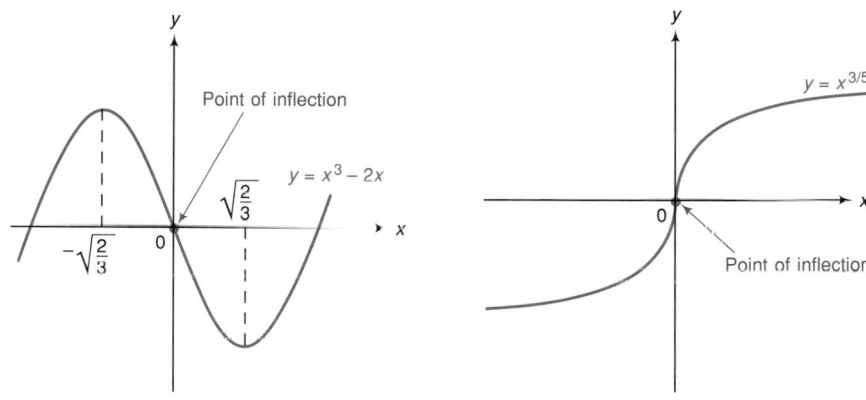

FIGURE 4
GRAPH OF $f(x) = x^3 - 2x$

FIGURE 5
GRAPH OF $f(x) = x^{3/5}$

EXAMPLE 2
A POINT OF INFLECTION WHERE THE SECOND DERIVATIVE DOES NOT EXIST

Let $f(x) = x^{3/5}$. Then $f'(x) = \frac{3}{5}x^{-2/5}$ and $f''(x) = -\frac{6}{25}x^{-7/5}$. Here $f''(x) < 0$ for $x > 0$ and $f''(x) > 0$ for $x < 0$. Thus, $(0, 0)$ is a point of inflection. Note that both $f'(x)$ and $f''(x)$ become infinite as $x \to 0$; that is, neither $f'(0)$ nor $f''(0)$ exists. The function is sketched in Figure 5.

Now suppose that f has a point of inflection at $(x_0, f(x_0))$. That means that at x_0, f' goes from increasing to decreasing (Figure 6a) or from decreasing to increasing (Figure 6b). In the first case, f' has a local maximum at x_0, and in the second case f' has a local minimum at x_0. In either case, by Theorem 3.2.2, $(f')' = f'' = 0$ at x_0. This implies the following result.

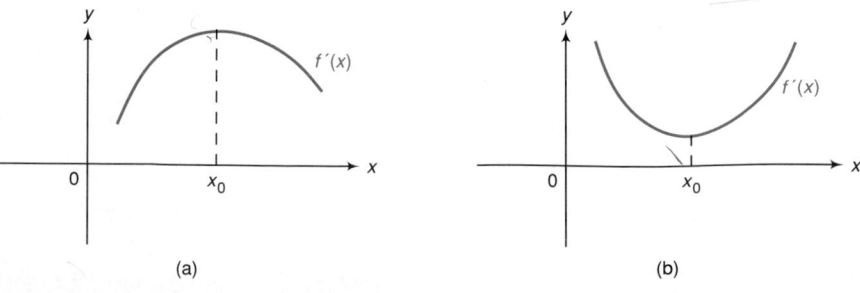

(a)

(b)

FIGURE 6 GRAPHS OF THE DERIVATIVE FUNCTION f'

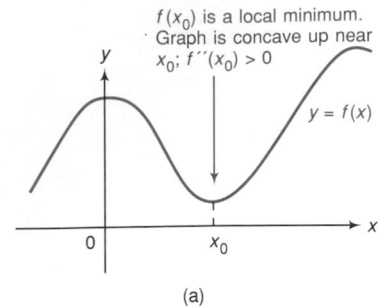

$f(x_0)$ is a local minimum. Graph is concave up near x_0; $f''(x_0) > 0$

$y = f(x)$

(a)

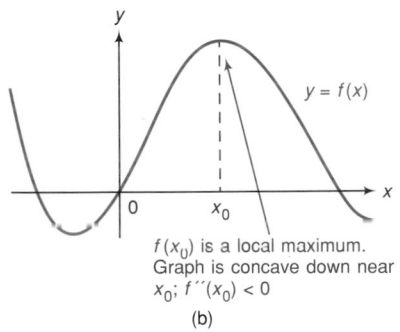

$y = f(x)$

$f(x_0)$ is a local maximum. Graph is concave down near x_0; $f''(x_0) < 0$

(b)

FIGURE 7
ILLUSTRATION OF SECOND-DERIVATIVE TEST

THEOREM 2   FINDING POSSIBLE POINTS OF INFLECTION

At a point of inflection $(x_0, f(x_0))$, if $f''(x_0)$ exists, then $f''(x_0) = 0$. ■

NOTE: The converse of this theorem is not true in general. That is, if $f''(x_0) = 0$, then x_0 is not necessarily a point of inflection. For example, if $f(x) = x^4$, then $f''(x) = 12x^2 > 0$ for $x \neq 0$. Thus f does not change concavity at 0, so 0 is *not* a point of inflection even through $f''(0) = 0$. (See Problem 33.)

Finally, suppose that $f'(x_0) = 0$. If $f''(x_0) > 0$, then the curve is concave up at x_0. A glance at Figure 7a suggests that f has a local minimum at x_0. Similarly, if $f'(x_0) = 0$ and $f''(x_0) < 0$, then f has a local maximum at x_0. It is not difficult to prove these statements (see Problems 31 and 32). In sum, we have the following theorem.

THEOREM 3   SECOND-DERIVATIVE TEST FOR A LOCAL MAXIMUM OR MINIMUM

Let f be differentiable on an open interval containing x_0 and suppose that $f''(x_0)$ exists.

(i) If $f'(x_0) = 0$ and $f''(x_0) > 0$, f has a local minimum at x_0.

(ii) If $f'(x_0) = 0$ and $f''(x_0) < 0$, f has a local maximum at x_0. ■

REMARK: If $f'(x_0) = 0$ and $f''(x_0) = 0$, then f may have a local maximum, a local minimum, or neither at x_0. For example, $y = x^3$ has neither at $x = 0$, while $y = x^4$ has a minimum at 0, and $y = -x^4$ has a maximum at 0.

EXAMPLE 3
GRAPHING A CUBIC USING THE FIRST AND SECOND DERIVATIVES

Let $f(x) = 2x^3 - 3x^2 - 12x + 5$. Sketch the curve.

SOLUTION: $f'(x) = 6x^2 - 6x - 12 = 6(x^2 - x - 2) = 6(x - 2)(x + 1)$, and $f''(x) = 12x - 6 = 6(2x - 1)$. The curve is

concave down if $x < \frac{1}{2}$ and   concave up if $x > \frac{1}{2}$.

The point $x = \frac{1}{2}(y = -\frac{3}{2})$ is a point of inflection. The critical points are $x = 2$ $(y = -15)$ and $x = -1$ $(y = 12)$. When $x = 2$, $f'' = 6(4 - 1) = 18 > 0$, so that f has a local minimum at $(2, -15)$. When $x = -1$, $f'' = -18 < 0$, so that f has a local maximum at $(-1, 12)$. From an examination of f' we see that f increases up to $x = -1$, decreases for $-1 < x < 2$, and increases thereafter. When $x = 0$, $y = 5$, so the y-intercept is the point $(0, 5)$. We present some of this information in Table 1.

TABLE 1

Critical Point x_0	$f''(x_0) = 6(2x^2 - 1)$	Sign of $f''(x_0)$	At x_0 f Has a
2	18	+	local minimum
-1	-18	-	local maximum

Putting this information all together, we obtain the curve in Figure 8. From this sketch we can see that the graph crosses the x-axis at three

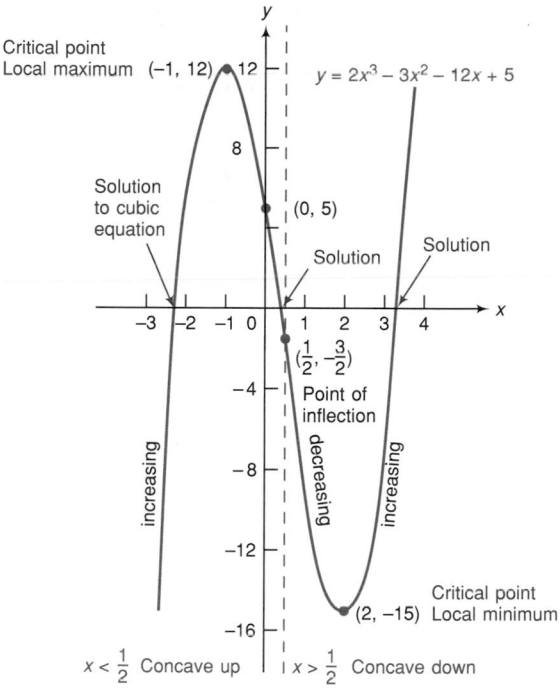

FIGURE 8 GRAPH OF $f(x) = 2x^3 - 3x^2 - 12x + 5$

places (the three x-intercepts). This tells us that the cubic equation

$$2x^3 - 3x^2 - 12x + 5 = 0$$

has three real solutions. Moreover, we see that two of these solutions are positive and one is negative. This information, which comes without additional work, can often be very useful.

EXAMPLE 4
GRAPHING A FOURTH-DEGREE
POLYNOMIAL FUNCTION

Graph the curve $y = x^4 - 4x^3 + 4x^2 + 1$.

SOLUTION: We have

$$\frac{dy}{dx} = 4x^3 - 12x^2 + 8x = 4x(x^2 - 3x + 2) = 4x(x - 1)(x - 2),$$

and

$$\frac{d^2y}{dx^2} = 12x^2 - 24x + 8 = 4(3x^2 - 6x + 2).$$

The second derivative is zero when $3x^2 - 6x + 2 = 0$, or

$$x = \frac{6 \pm \sqrt{36 - 24}}{6} = \frac{6 \pm \sqrt{12}}{6} = \frac{3 \pm \sqrt{3}}{3} = 1 \pm \frac{\sqrt{3}}{3} = 1 \pm \frac{1}{\sqrt{3}}.$$

If $x < 1 - (1/\sqrt{3})$, then $d^2y/dx^2 > 0$. If $x > 1 + (1/\sqrt{3})$, then $d^2y/dx^2 > 0$ also. If $1 - (1/\sqrt{3}) < x < 1 + (1/\sqrt{3})$, then $d^2y/dx^2 < 0$. To see this, simply plug in some sample points (try the points $x = 0$, $x = 2$, and

$x = 1$). Hence, the curve is

$$\text{concave up for} \quad x < 1 - \frac{1}{\sqrt{3}},$$

$$\text{concave down for} \quad 1 - \frac{1}{\sqrt{3}} < x < 1 + \frac{1}{\sqrt{3}},$$

$$\text{and concave up for} \quad x > 1 + \frac{1}{\sqrt{3}}.$$

The points of inflection are the points $x = 1 - (1/\sqrt{3})$ $(y = \frac{13}{9})$ and $x = 1 + (1/\sqrt{3})$ $(y = \frac{13}{9})$. The critical points are $x = 0$ $(y = 1)$, $x = 1$ $(y = 2)$, and $x = 2$ $(y = 1)$. When $x = 0$, $d^2y/dx^2 = 8$, so there is a local minimum at $(0, 1)$. Similarly, there is a local maximum at $(1, 2)$ and a local minimum at $(2, 1)$. This information is summarized in Table 2. Putting all this information together, we obtain the graph in Figure 9. Since the graph *never* crosses the x-axis, we see immediately that the quartic equation

$$x^4 - 4x^3 + 4x^2 + 1 = 0$$

has *no* real solutions.

TABLE 2

Critical Point x_0	$f''(x_0) = 4(3x^2 - 6x + 2)$	Sign of $f''(x_0)$	At x_0 f Has a
0	8	+	local minimum
1	−4	−	local maximum
2	8	+	local minimum

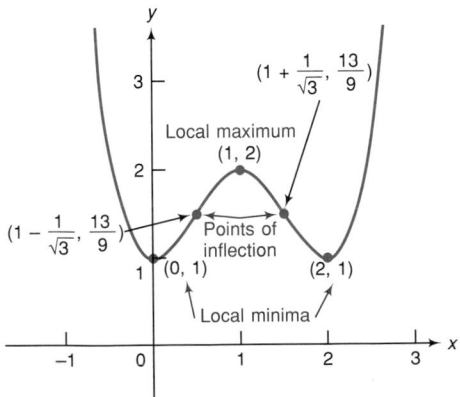

FIGURE 9   GRAPH OF $f(x) = x^4 - 4x^3 + 4x^2 + 1$

The next example shows that in some cases the first-derivative test is more useful than the second-derivative test.

EXAMPLE 5
THE FIRST-DERIVATIVE TEST CAN
SHOW A MINIMUM WHEN THE
SECOND-DERIVATIVE TEST FAILS

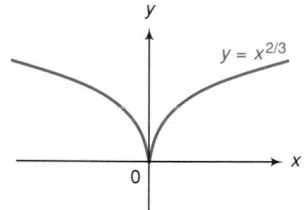

FIGURE 10
GRAPH OF $f(x) = x^{2/3}$

Sketch the curve $y = f(x) = x^{2/3}$.

SOLUTION: Since $f'(x) = \frac{2}{3}x^{-1/3} = 2/(3x^{1/3})$, the only critical point is at $x = 0$, since $f'(0)$ is not defined although $f(0)$ is defined. Also, $f''(0)$ is not defined [since $f'(0)$ isn't defined], so the second derivative test won't work. However, since $f'(x) < 0$ for $x < 0$ and $f'(x) > 0$ for $x > 0$, we see that 0 is a local minimum. (It is, in fact, a global minimum.) Also, when $x \neq 0$, $f''(x) = -2/9x^{4/3} < 0$, so the curve is concave down over any interval not containing 0. It is sketched in Figure 10.

NOTE: The graph of $x^{2/3}$ is said to have a **cusp** at $x = 0$.

■ PROBLEMS 3.3

■ SELF-QUIZ

I. Suppose $f(x) = 2 - (x - 10)^2$. The graph of $y = f(x)$ _____.
   a. is concave up over the interval (0, 20)
   b. is concave down over the interval (0, 20)
   c. has a point of inflection within the interval (0, 20)
   d. f is constant over the interval (0, 20)

II. Suppose $g'(x) = 2 - (x - 10)^2$. The graph of $y = g(x)$ _____.
   a. is concave up over the interval (9, 11)
   b. is concave down over the interval (9, 11)
   c. has a point of inflection within the interval (9, 11)
   d. g is constant over the interval (0, 20)

III. Suppose $h''(x) = 2 - (x - 10)^2$. The graph of $y = h(x)$ _____.
   a. is concave up over the interval (9, 11)
   b. is concave down over the interval (9, 11)
   c. has a point of inflection within the interval (9, 11)
   d. h is constant over the interval (0, 20)

IV. Let $F(x)$ be a polynomial function of x such that $F(-1) = 7$, $F'(-1) = 0$, and $F''(-1) = -18$. The point $(-1, 7)$ is a _____ for the graph of $y = F(x)$.
   a. local minimum

   b. local maximum
   c. point of inflection
   d. none of the above

V. Let $G(x) = 2x^3 - 3x^2 - 12x$; then $G(\frac{1}{2}) = -\frac{13}{2}$, $G'(\frac{1}{2}) = -\frac{27}{2}$, and $G''(\frac{1}{2}) = 0$. The point $(\frac{1}{2}, -\frac{13}{2})$ is a _____ for the graph of $y = G(x)$.
   a. local minimum
   b. local maximum
   c. point of inflection
   d. none of the above

VI. Let $H(x) = x^5$; then $H(0) = 0$, $H'(0) = 0$, and $H''(0) = 0$. The point $(0, 0)$ is a _____ for the graph of $y = x^5$. [*Hint:* You may want to obtain a bit more information than is given in the first sentence.]
   a. local minimum
   b. local maximum
   c. point of inflection
   d. none of the above

VII. The graph of $y = x^2 + bx + c$ has _____ point(s) of inflection.
   a. no
   b. one
   c. two
   d. an infinite number of

■ DRILL

In Problems 1–22, follow the procedures discussed in this section to sketch the graph of the given function.

1. $y = x^2 - x + 30$

2. $y = x^3 - 12x + 10$
3. $y = x^3 - 3x^2 - 45x + 25$
4. $y = 4x^3 - 3x + 2$
5. $y = 2x^3 - 9x^2 + 12x - 3$

CONTINUED

6. $y = \frac{1}{3}x^3 + \frac{1}{2}x^2 - 2x - \frac{2}{3}$
7. $y = \sin x \cos x$
*8. $y = \cos^2 x$
9. $y = (x + 2)^2 \cdot (x - 2)^2$
10. $y = 2x(x + 4)^3$
*11. $y = |x| + |x - 1| + |2 - x|$
*12. $y = |x^2 - 3| + |x^2 - 1|$

13. $y = \dfrac{x^2 - 1}{x^2 + 1}$
15. $y = 3x^5 - 5x^3$
17. $y = x\sqrt{1 + x}$
19. $y = x - \sqrt[3]{x}$
21. $y = x^{2/3}(x - 3)$

*14. $y = \dfrac{x^2 + 3}{x^2 - 6x + 10}$
*16. $y = (x + 3)^4(x - 4)^3$
18. $y = (x + 1)\sqrt{x - 1}$
*20. $y = \sqrt[3]{x^2 - 1}$
*22. $y = x^{3/2}(x - 1)^{8/3}$

■ APPLICATIONS

*23. A function f satisfies the **differential equation** $f'(x) = g(x)$ with **initial value** $f(0) = f_0$. The graph of $y = g(x)$ is sketched in Figure 11 and three choices for f_0 are shown. Draw a rough sketch of the function $f(x)$ for each of the three possible values for f_0. In your sketch, indicate where the function f is increasing and where it decreases, maxima and minima, concavity, and points of inflection.

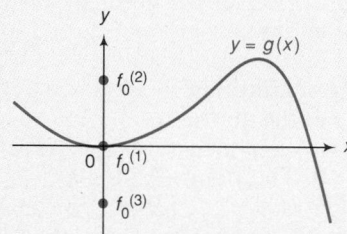

FIGURE 11

*24. Verify that $f(x) = x \cdot |x|$ has its unique point of inflection at $x = 0$; also show that $f''(0) = 0$. [*Note:* This result provides an example of a point

of inflection at x_0 where $f''(x_0)$ does not exist but, also, $|f''(x_0)|$ does not become infinite as x approaches x_0.]

25. Verify that the cubic equation $x^3 + x^2 + 5x - 15 = 0$ has exactly one real root. Is it positive or negative?

26. How many real roots does each of the following cubic equations have?
a. $x^3 + 1 = 0$
b. $x^3 + x^2 + x + 1 = 0$
c. $x^3 - x^2 + x - 1 = 0$
d. $4x^3 - 3x^2 + 2x - 1 = 0$

27. Verify that the equation $x^{11} + x^3 + x + 3 = 0$ has exactly one real root. [*Hint:* Consider its graph.]

28. Verify that the quartic equation $x^4 + x^3 - 5x^2 - 6 = 0$ has exactly two real roots. Find open intervals that contain each root.

29. Explain why the cubic equation $x^3 + cx^2 + bx + a = 0$ must have at least one real root. [*Hint:* Consider its graph.]

30. Explain why the quintic equation $a_0 + a_1x + a_2x^2 + a_3x^3 + a_4x^4 + a_5x^5 = 0$ with real coefficients and with $a_5 \neq 0$ must have at least one real root.

■ SHOW/PROVE/DISPROVE

31. Prove case (ii) of Theorem 3: Suppose f is differentiable on an open interval containing x_0, $f'(x_0) = 0$ and $f''(x_0)$ exists and is negative; then f has a local maximum at x_0. [*Hint:* Show that near x_0, $f'(x) > 0$ for $x < x_0$ and $f'(x) < 0$ for $x_0 < x$.]

32. Prove case (i) of Theorem 3: Show that if $f'(x_0) =$

0 and $f''(x_0) > 0$, then f has a local minimum at x_0.

33. Prove that if $f''(x_0) = 0$ and $f'''(x_0) \neq 0$, then x_0 is a point of inflection for f.

34. Prove or disprove: If functions f and g are both concave up at $x = c$, then the product function $f \cdot g$ is also concave up there.

■ **CHALLENGE**

CONTINUED

*35. Suppose p and q satisfy the relations $p > 1, q > 1$, $1/p + 1/q = 1$. Prove that if $u, v \geq 0$, then

$$u \cdot v \leq \frac{u^p}{p} + \frac{v^q}{q}.$$

**36. Let n be a positive integer greater than 1. Show that the curve

$$\frac{1 + y}{1 - y} = \left[\frac{1 + x}{1 - x} \right]^n$$

has precisely three points of inflection.

■ **ANSWERS TO SELF-QUIZ**

I. b	II. c	III. a	VII. a (always concave up) since $y'' = 2$
IV. b	V. c	VI. c	

3.4 ELEMENTARY CURVE SKETCHING III: Asymptotes

We complete our discussion of curve sketching by describing what happens either as $x \to \pm\infty$ or when $f(x) \to \pm\infty$ as x approaches a finite value. The material in this section uses the material from Section 1.4.

DEFINITION **HORIZONTAL ASYMPTOTE**

Suppose that $\lim_{x \to \infty} f(x) = c$ or $\lim_{x \to -\infty} f(x) = c$. Then the horizontal line $y = c$ is called a **horizontal asymptote** of the graph of f. ■

DEFINITION **VERTICAL ASYMPTOTE**

Let a be a real number. If

$$\lim_{x \to a^+} f(x) = \pm\infty \qquad \text{or} \qquad \lim_{x \to a^-} f(x) = \pm\infty,$$

then the vertical line $x = a$ is called a **vertical asymptote** of the graph of f. Pictures of horizontal and vertical asymptotes are given in Figure 1. ■

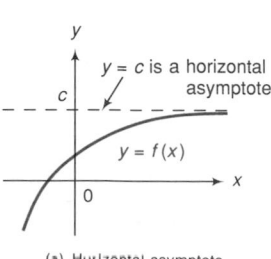

(a) Horizontal asymptote.

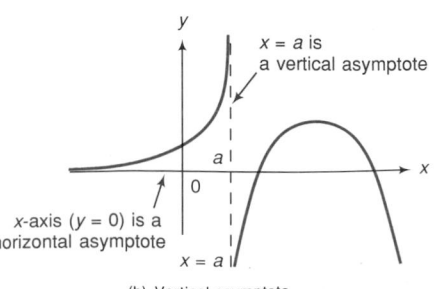

(b) Vertical asymptote.

FIGURE 1 ILLUSTRATION OF ASYMPTOTES

EXAMPLE 1
A GRAPH WITH A VERTICAL AND
A HORIZONTAL ASYMPTOTE

Let $f(x) = x/(1 + x)$. Graph the function.

SOLUTION: Before doing any calculations, we first note that we will have problems at $x = -1$ since the function is not defined there. We observe the following:

If $x < -1$, then $f(x) > 1$ (a negative over a negative, with the numerator more negative than the denominator).

If $-1 < x < 0$, then $f(x) < 0$.

If $x = 0$, then $f(x) = 0$.

If $x > 0$, then $0 < f(x) < 1$ (denominator larger than the numerator).

Moreover,

$$\lim_{x \to -1^+} \frac{x}{1 + x} = -\infty \qquad \text{The numerator is negative and the denominator is positive.}$$

and

$$\lim_{x \to -1^-} \frac{x}{1 + x} = \infty \qquad \text{The numerator and denominator are both negative.}$$

Thus, the line $x = -1$ is a vertical asymptote. Also,

$$\lim_{x \to \infty} \frac{x}{1 + x} = \lim_{x \to \infty} \frac{1}{(1/x) + 1} = 1$$

and

$$\lim_{x \to -\infty} \frac{x}{1 + x} = \lim_{x \to -\infty} \frac{1}{(1/x) + 1} = 1.$$

Therefore, as $x \to \pm\infty$, $f(x) \to 1$. Thus, the line $y = 1$ is a horizontal asymptote for the function $x/(1 + x)$. Now

$$f'(x) = \frac{d}{dx}\left(\frac{x}{1 + x}\right) = \frac{(1 + x)1 - x}{(1 + x)^2} = \frac{1}{(1 + x)^2}.$$

Therefore, f is always increasing (except at $x = -1$, where it is not defined). Also, f has *no* critical points (-1 is *not* a critical point because f is not defined at -1). We next compute

$$f''(x) = \frac{d}{dx}(1 + x)^{-2} = -\frac{2}{(1 + x)^3}.$$

If $x < -1$, then $1 + x < 0$, $(1 + x)^3 < 0$, and $-2/(1 + x)^3 > 0$. If $x > -1$, then $1 + x > 0$, $(1 + x)^3 > 0$, and $-2/(1 + x)^3 < 0$. Hence, f is concave up if $x < -1$ and concave down if $x > -1$.

We compute:

$$\lim_{x \to \pm\infty} f'(x) = \lim_{x \to \pm\infty} \frac{1}{(1 + x)^2} = 0,$$

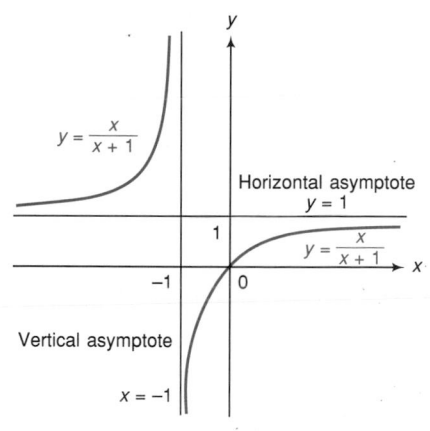

$y = \dfrac{x}{x+1}$

Horizontal asymptote
$y = 1$

$y = \dfrac{x}{x+1}$

1

−1   0

Vertical asymptote

$x = -1$

FIGURE 2

GRAPH OF $y = \dfrac{x}{x+1}$

The line $x = -1$ is a vertical asymptote.
The line $y = 1$ is a horizontal asymptote.

so the tangent lines become flat (horizontal) as $x \to \infty$. Finally,

$$\lim_{x \to -1} f'(x) = \lim_{x \to -1} \frac{1}{(1+x)^2} = \infty,$$

so the tangent lines become vertical as $x \to -1$ (from either side).
We summarize:

(i) $f(x)$ is increasing except at -1 where it is not defined.
(ii) The graph is concave up if $x < -1$ and concave down if $x > -1$.
(iii) The graph approaches the line $y = 1$ and "flattens out" as $x \to \pm\infty$.
(iv) The graph becomes more and more vertical as $x \to -1$ from either side.

With all this information, the graph must be as in Figure 2.

HELPFUL HINT: If $f(x)$ is defined in a neighborhood of x_0 but not at x_0 itself, then the graph of f is likely to have a vertical asymptote at x_0. (One exception occurs if f is piecewise continuous in a neighborhood of x_0.) In general, if $f(x) = \dfrac{g(x)}{h(x)}$ and $h(x_0) = 0$ but $g(x_0) \neq 0$, then the graph of f has a vertical asymptote at x_0.

EXAMPLE 2
THE GRAPH OF THE SECANT
FUNCTION

Sketch the curve $y = \sec x = 1/\cos x$.

SOLUTION:

$$\frac{dy}{dx} = \sec x \tan x = \frac{1}{\cos x} \frac{\sin x}{\cos x} = \frac{\sin x}{\cos^2 x}.$$

Since $\sin 0 = \sin \pi = \sin 2\pi = 0$, we see that $x = 0$, $x = \pi$, and $x = 2\pi$ are critical points in the interval $[0, 2\pi]$. This is the only interval we need to consider because $1/\cos x$, like $\cos x$, is periodic of period 2π. Moreover, since $\cos(\pi/2) = \cos(3\pi/2) = 0$, the lines $x = \pi/2$ and $x = 3\pi/2$ are vertical asymptotes.
Next, we compute

$$\frac{d^2y}{dx^2} = \sec x \sec^2 x + (\sec x \tan x) \tan x = \sec x(\sec^2 x + \tan^2 x).$$

Since $\sec^2 x + \tan^2 x$ is always positive, the sign of $\sec x = 1/\cos x$ gives us the sign of the second derivative. We summarize this information in Table 1.
Observing that $\cos x > 0$ for $0 < x < \pi/2$ and $3\pi/2 < x < 5\pi/2$, and $\cos x < 0$ for $\pi/2 < x < 3\pi/2$, we obtain the sketch in Figure 3. Note that there are no points of inflection at $x = \pi/2$ and $x = 3\pi/2$ because f is not defined at these numbers. Also, there are no horizontal asymptotes because $\lim_{x \to \infty} \sec x$ and $\lim_{x \to -\infty} \sec x$ do not exist.

TABLE 1

$$f(x) = \sec x = \frac{1}{\cos x} \qquad f'(x) = \frac{\sin x}{\cos^2 x} \qquad f''(x) = \sec x(\sec^2 x + \tan^2 x)$$

Interval or Point	f'	f	Interval or Point	f''	f
0	0	local minimum	$0 \le x < \dfrac{\pi}{2}$	+	concave up
$\left.\begin{array}{l}0 < x < \pi/2 \\ \pi/2 < x < \pi\end{array}\right\}$	+	increasing	$x = \dfrac{\pi}{2}$	doesn't exist	vertical asymptote
$x = \pi$	0	local maximum	$\dfrac{\pi}{2} < x < \dfrac{3\pi}{2}$	−	concave down
$\left.\begin{array}{l}\pi < x < 3\pi/2 \\ 3\pi/2 < x < 2\pi\end{array}\right\}$	−	decreasing	$x = \dfrac{3\pi}{2}$	doesn't exist	vertical asymptote
$x = 2\pi$	0	local minimum	$\dfrac{3\pi}{2} < x < \dfrac{5\pi}{2}$	+	concave up

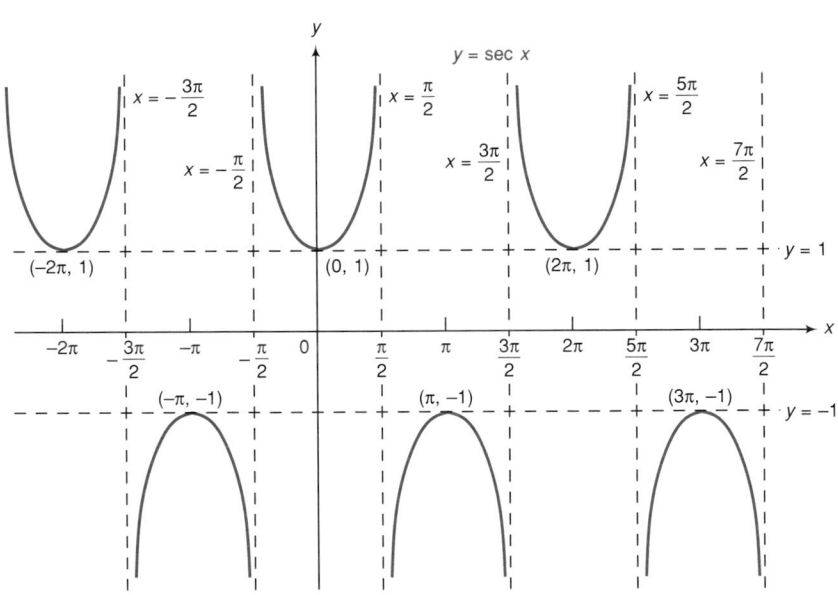

FIGURE 3 GRAPH OF $f(x) = \sec x$

The line $x = \dfrac{n\pi}{2}$ is a vertical asymptote when n is an odd integer.

REMARK: sec x is an even function (see page 42) because $\sec(-x) = \dfrac{1}{\cos(-x)} = \dfrac{1}{\cos x} = \sec x$. Thus, the graph of sec x is symmetric about the y-axis. We could have saved some work by sketching the graph for $x > 0$ and then reflecting it about the y-axis.

OTHER ASYMPTOTES

An **asymptote** to a curve is a line that the curve approaches as x approaches some fixed number, $+\infty$, or $-\infty$. More precisely, $y = mx + b$ is an asymptote to the graph of f if

$$\lim_{x \to \infty} [f(x) - (mx + b)] = 0 \qquad \text{or} \qquad \lim_{x \to -\infty} [f(x) - (mx + b)] = 0 \quad \text{(1)}$$

If $m \neq 0$, the line $y = mx + b$ is called an **oblique asymptote** to the curve.

REMARK: Vertical and horizontal asymptotes are special cases of asymptotes.

EXAMPLE 3
A GRAPH WITH AN OBLIQUE ASYMPTOTE

Sketch the graph of $f(x) = x + \dfrac{1}{x}$.

SOLUTION: We first observe that $f(0)$ is not defined. Moreover,

$$\frac{1}{x} > 0 \quad \text{if } x > 0 \qquad\qquad \frac{1}{x} < 0 \quad \text{if } x < 0$$

$$\lim_{x \to 0^+} \left(x + \frac{1}{x} \right) = \infty \qquad \text{and} \qquad \lim_{x \to 0^-} \left(x + \frac{1}{x} \right) = -\infty.$$

So f has a vertical asymptote at 0. Now

$$f'(x) = \frac{d}{dx} x + \frac{d}{dx} x^{-1} = 1 - x^{-2} = 1 - \frac{1}{x^2} = \frac{x^2}{x^2} - \frac{1}{x^2} = \frac{x^2 - 1}{x^2}.$$

First, we notice that the denominator is always positive (for $x \neq 0$). The derivative is not defined for $x = 0$. Thus $f'(x) < 0$ if $x^2 < 1$, and $f'(x) > 0$ if $x^2 > 1$; $f'(1) = f'(-1) = 0$, so 1 and -1 are critical points. Note that 0 is *not* a critical point of f because f is not defined at 0. Next we compute

$$f''(x) = \frac{d}{dx} (1 - x^{-2}) = 2x^{-3} = \frac{2}{x^3}.$$

Thus the graph is concave down if $x < 0$ and concave up if $x > 0$. Moreover, f has a local maximum at $x = -1$ $(y = -2)$ and a local minimum at $x = 1$ $(y = 2)$. Since $\dfrac{1}{x} \to 0$ as $x \to \pm \infty$,

$$\lim_{x \to \infty} [f(x) - x] = \lim_{x \to \infty} \left[\left(x + \frac{1}{x} \right) - x \right] = \lim_{x \to \infty} \frac{1}{x} = 0,$$

so the line $y = x$ is an oblique asymptote to the graph. Finally, $f(x) > 0$ if $x > 0$ and $f(x) < 0$ if $x < 0$. Putting this all together, we obtain the graph in Figure 4.

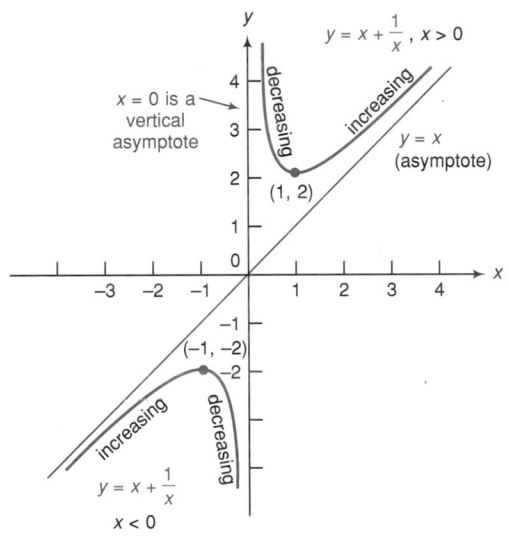

FIGURE 4   GRAPH OF $y = x + \dfrac{1}{x}$

The line $y = x$ is an oblique asymptote.

SYMMETRY ABOUT THE y-AXIS AND THE ORIGIN

In Section 0.5 we saw that the graph of the function f is symmetric about the y-axis (i.e., f is an even function) if $f(-x) = f(x)$. If

$$f(-x) = -f(x), \tag{2}$$

then f is an odd function and we saw that the graph of f is symmetric about the origin. If the graph of f is symmetric about the origin, then we can draw it by performing the following steps:

(i) Sketch the graph for $x > 0$.

(ii) Reflect the graph about both the x- and y-axes to obtain the graph for $x < 0$.

NOTE: If $f(x) = x + \dfrac{1}{x}$, then $f(-x) = -x - \dfrac{1}{x} = -\left(x + \dfrac{1}{x}\right) = -f(x)$, so this function is symmetric about the origin. In Figure 4, the graph of $f(x)$ for $x < 0$ could be obtained from the graph of $f(x)$ for $x > 0$ by reflecting about the x- and y-axes.

We now review the kinds of information that are available to help us sketch curves.

TO SKETCH THE GRAPH OF $y = f(x)$, FOLLOW THESE STEPS

(i) Check whether or not $f(-x) = f(x)$. If so, then f is symmetric about the y-axis and it is only necessary to obtain the graph for $x > 0$ and reflect it about the y-axis.

(ii) Check whether or not $f(-x) = -f(x)$. If so, then f is symmetric about the origin. Obtain the graph for $x > 0$ and reflect it about the x- and y-axes.

(iii) Calculate the derivative $dy/dx = f'$ and determine where the curve is increasing and decreasing. Find all points at which $f'(x) = 0$ or $f'(x)$ does not exist.

(iv) Calculate the second derivative $f'' = d^2y/dx^2$ and determine where the curve is concave up and concave down.

(v) Determine local maxima and minima by using either the first or the second derivative test.

(vi) Find all points of inflection.

(vii) Determine the y-intercept and (if possible) the x-intercept(s).

(viii) Determine $\lim_{x \to \infty} f(x)$ and $\lim_{x \to -\infty} f(x)$. If either of these is finite, then there are horizontal asymptotes.

(ix) If f is not defined at x_0, determine $\lim_{x \to x_0^+} f(x)$ and $\lim_{x \to x_0^-} f(x)$. Look for vertical asymptotes. These are the lines of the form $x = x_0$, where $\lim_{x \to x_0^+} f(x) = \infty$ (or $-\infty$) or $\lim_{x \to x_0^-} f(x) = \infty$ (or $-\infty$) and $f(x_0)$ is undefined.

(x) Look for oblique asymptotes. If $\lim_{x \to \infty}[f(x) - (mx + b)] = 0$ or $\lim_{x \to -\infty}[f(x) - (mx + b)] = 0$, then $y = mx + b$ is an oblique asymptote.

REMARK: In Appendix A6 we show how graphs can be obtained on a graphing calculator.

■ PROBLEMS 3.4

■ SELF-QUIZ

I. Lines _____ and _____ are asymptotes to the graph of $y = (9 - x^2)/x$.
   a. $y = 9$ b. $y = -x$ c. $y = 3x$
   d. $x = 0$ e. $x = 3$ f. $x = y$

II. The line _____ is an asymptote to the graph of $y = 1/(x + 1) + 2$.
   a. $y = 2$ b. $y = -2$
   c. $x = -1$ d. $0 = 2 + 1/(x + 1)$

III. The line $y = 7$ is an asymptote to the graph of _____.
   a. $y = 7 + x$ b. $y = 7x$
   c. $y = 7/x$ d. $y = 7 - 1/x$

IV. The line $x = 0$ (the y-axis) is an asymptote to the graph of _____.

   a. $xy = 1$ b. $-\frac{1}{2} = 1/(xy)$

   c. $y = 5/x$ d. $y = 3 - 2/x^2$

V. The graph of $y = 2x - 1/x$ is symmetric about the _____.

   a. x-axis   b. y-axis

   c. origin   d. line $x = \frac{1}{2}$

■ DRILL

In Problems 1–16, follow the steps outlined in this section to sketch the graph of the given function.

1. $y = \dfrac{1}{x - 3}$

2. $y = \dfrac{1}{|3 - x|}$

3. $y = 1 + \dfrac{1}{x^2}$

4. $y = \dfrac{x^2 - 1}{x^2}$

5. $y = \dfrac{1}{(x - 1)(x - 2)}$

6. $y = \dfrac{x - 1}{x - 2}$

7. $y = \dfrac{1}{x^2 - 1}$

8. $y = \dfrac{x}{x^2 - 1}$

9. $y = \dfrac{x^2 + 1}{x^2 - 1}$

10. $y = \dfrac{x^2 - 9}{x^2 - 3x + 2}$

11. $y = \csc x$

12. $y = \cot x$

13. $y = \dfrac{x + 1}{\sqrt{x - 1}}$

14. $y = \dfrac{x^2 + x + 7}{\sqrt{2x + 1}}$

15. $f(x) = \dfrac{2x^2 + 4}{x + 1}$

16. $f(x) = \dfrac{x^3 + 1}{2x^2}$

■ APPLICATIONS

17. Demonstrate graphically that there are an infinite number of solutions to the equation $x = \tan x$.

18. Verify that the solutions of $x = -\tan x$ are critical points of the curve $y = x \sin x$.

19. Discuss any symmetries of the graph of $y = x \sin x$.

20. Use the results of Problems 17, 18, and 19 together with whatever other information you think useful to graph $y = x \sin x$.

■ SHOW/PROVE/DISPROVE

21. Prove or Disprove: If the line $y = x$ is an asymptote for the graph of the function f, then there is no x_0 such that $x_0 = f(x_0)$.

■ ANSWERS TO SELF-QUIZ

I. b and d

II. a (horizontal asymptote) and c (vertical asymptote)

III. d   IV. a, b, c, d   V. c

3.5 APPLICATIONS OF MAXIMA AND MINIMA

In Section 3.2 we computed local maxima and minima. In this section, we compute the maximum and minimum values taken by a function over an interval (which could be finite or infinite).

Consider the closed interval $[a, b]$. If f is continuous on $[a, b]$, then by the upper and lower bound theorem on page 119, there are numbers x_1 and x_2 in $[a, b]$ such that $f(x_1) = m$ and $f(x_2) = M$, where m and M are, respectively, the minimum and maximum values taken by f over $[a, b]$. How do we find these values? If x_1 is in the open interval (a, b), then by Theorem 2 on page 191, x_1 is a critical point. This fact suggests the following procedure.

> **PROCEDURE FOR FINDING MAXIMUM AND MINIMUM VALUES OVER A CLOSED INTERVAL**
>
> For $f(x)$ continuous on the closed, finite interval $[a, b]$, to find the maximum and minimum values of f over that interval, proceed as follows:
>
> **(i)** Evaluate f at all critical points in (a, b).
> **(ii)** Evaluate $f(a)$ and $f(b)$.
> **(a)** $f_{max}[a, b]$ = the largest of the values calculated in (i) and (ii).
> **(b)** $f_{min}[a, b]$ = the smallest of the values calculated in (i) and (ii).

EXAMPLE 1
FINDING THE MAXIMUM AND MINIMUM VALUES OF A FUNCTION

Find the maximum and minimum values of $f(x) = 2x^3 - 3x^2 - 12x + 5$ (a) in the interval $[0, 4]$, and (b) for $x \geq 0$.

SOLUTION:

(a) $f'(x) = 6x^2 - 6x - 12 = 6(x^2 - x - 2) = 6(x - 2)(x + 1)$. The only place in the interval $[0, 4]$ where $f'(x) = 0$ is at $x = 2$. At $x = 2$, $f(x) = -15$ and this is the only critical point in $[0, 4]$. We also calculate $f(0) = 5$ and $f(4) = 37$. Therefore $f_{max}[0, 4] = 37$ and $f_{min}[0, 4] = -15$. This curve is sketched in Figure 1.

(b) We first note that now we are considering the infinite interval $[0, \infty)$, so our upper and lower bound theorem does not apply, and we do not know, in advance, that f has a maximum or minimum on $[0, \infty)$.

From Figure 1 we see that for $x > 0$, f takes the minimum value -15 at $x = 2$. However,

$$\lim_{x \to \infty} (2x^3 - 3x^2 - 12x + 5) = \infty,$$

so f has no maximum value on $[0, \infty)$.

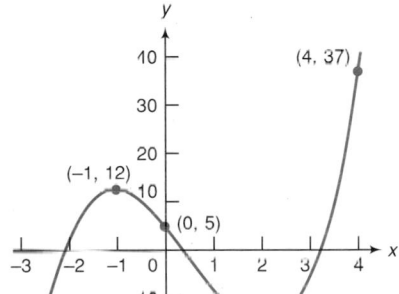

FIGURE 1
GRAPH OF $f(x) = 2x^3 - 3x^2 - 12x + 5$

EXAMPLE 2
MAXIMIZING AREA

Suppose that a farmer has 1000 yd of fence that he wishes to use to fence off a rectangular plot along the straight bank of a river (see Figure 2). What are the dimensions of the maximum area he can enclose?

SOLUTION: Since one side of the rectangular plot is taken up by the river, the farmer can use the fence for the other three sides. From the figure, the area of the plot is

$$A = xy, \tag{1}$$

where y is the length of the side parallel to the river. Without other information it would be impossible to solve this problem since the area A is a function of the *two* variables x and y. However, we are also told, since the farmer will obviously use all the available fence, that

$$2x + y = 1000, \tag{2}$$

or

$$y = 1000 - 2x, \quad \text{for } x \text{ in } [0, 500]. \tag{3}$$

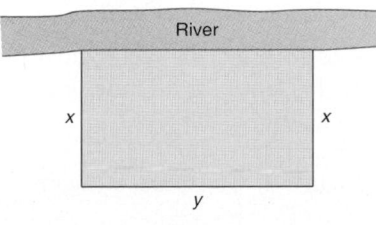

FIGURE 2
A RECTANGULAR FENCE ALONG A RIVER

Then

$$A = x(1000 - 2x) = 1000x - 2x^2. \qquad (4)$$

To find the maximum value for A, we set dA/dx equal to zero:

$$\frac{dA}{dx} = 1000 - 4x = 0$$

when $x = 250$. Also,

$$\frac{d^2A}{dx^2} = -4 < 0,$$

so that A is a maximum when $x = 250$. This is depicted in Figure 3.

When $x = 250$ yd, then $y = 500$ yd and $A = 125{,}000$ yd^2, which is the maximum area that the farmer can enclose. We note that, as is evident from Figure 3, A is positive when $0 \leq x \leq 500$ and at the end-points $x = 0$ and $x = 500$ the area is zero, so the local maximum we have obtained is a maximum over the entire interval $[0, 500]$.

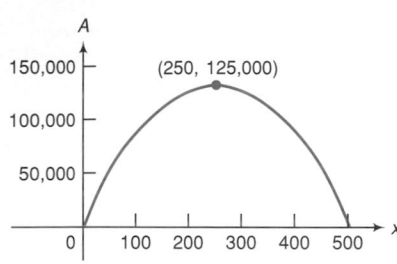

FIGURE 3
GRAPH OF $A = 1000x - 2x^2$, $0 \leq x \leq 500$

The maximum of 125,000 occurs at $x = 250$.

Example 2 suggests that the following steps be taken to solve a maximum or minimum problem.

PROCEDURE FOR SOLVING AN APPLIED MAXIMUM OR MINIMUM PROBLEM

(i) Draw a picture if it makes sense to do so.

(ii) Write all the information in the problem in mathematical terms, giving a letter name to each variable.

(iii) Determine the variable that is to be maximized or minimized and write this variable as a function of the other variables in the problem.

(iv) Using information given in the problem, eliminate all variables except one, so that the function to be maximized or minimized is written in terms of *one* of the variables of the problem.

(v) Determine the interval over which this one variable can be defined.

(vi) Follow the steps on page 213 to maximize or minimize the function over this interval (which may be infinite).

EXAMPLE 2 (Continued)
MAXIMIZING AREA

Suppose that the farmer in Example 2 wishes to build his rectangular plot away from the river (so that he must use his fence for all four sides of the rectangle). How large an area can he enclose in this case?

SOLUTION: The situation is now as in Figure 4. The area is again given by $A = xy$, but now

$$2x + 2y = 1000, \qquad (5)$$

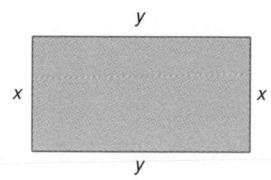

FIGURE 4
A RECTANGULAR FENCE

or

$$y = 500 - x. \tag{6}$$

Hence the problem is to maximize

$$A = x(500 - x) = 500x - x^2, \qquad \text{for } x \text{ in } [0, 500]. \tag{7}$$

Now

$$\frac{dA}{dx} = 500 - 2x = 0$$

when $x = 250$. Also, $d^2A/dx^2 = -2$, so that a local maximum is achieved when $x = y = 250$ yd and the answer is $A = 62{,}500$ yd^2. (Note that the plot is, in this case, a square.) At 0 and 500, $A = 0$, so that $x = 250$ is indeed the maximum.

REMARK 1: The reasoning in Example 2 can be used to prove the following: *For a given perimeter, the rectangle containing the greatest area is a square.*

REMARK 2: If we do not require that the plot be rectangular, then we can enclose an even greater area. Although the proof of this fact is beyond the scope of this book, it can be shown that *for a given perimeter, the geometric shape with the largest area is a circle.* For example, if the 1000 yd of fence in Example 2 are formed in the shape of a circle, then the circle has a circumference of $2\pi r = 1000$, so that the radius of the circle is $r = 1000/2\pi$ yd. Then $A = \pi r^2 = \pi(1{,}000{,}000/4\pi^2) = 1{,}000{,}000/4\pi \approx 79{,}577$ yd^2.

EXAMPLE 3
MINIMIZING CONSTRUCTION COST

A cylindrical barrel is to be constructed to hold 32π m^3 of liquid. The cost per square meter of constructing the side of the barrel is half the cost per square meter of constructing the top and bottom. What are the dimensions of the barrel that costs the least to construct?

SOLUTION: Consider Figure 5. Let h be the height of the barrel and let r be the radius of the top and bottom. Then the volume of the barrel[†] is given by $V = \pi r^2 h = 32\pi$ m^3.

Assuming that the cost of constructing the material on the side is k cents per square meter, then the cost of the top and bottom is $2k$ cents per square meter. The area of the top or the bottom is πr^2 square meters, while that of the side is $2\pi rh$ square meters (since the circumference of the side is $2\pi r$). Thus, the total cost is given by

$$C = 2\pi r^2(2k) + 2\pi rhk = k(4\pi r^2 + 2\pi rh).$$

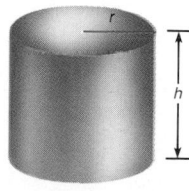

FIGURE 5
A CYLINDRICAL BARREL WITH RADIUS r AND HEIGHT h

To write C as a function of one variable only (which we must do in order to solve the problem), we use $32\pi = \pi r^2 h$ to obtain $h = 32\pi/\pi r^2 = 32/r^2$, so that

$$C = k\left(4\pi r^2 + 2\pi r \cdot \frac{32}{r^2}\right) = k\left(4\pi r^2 + \frac{64\pi}{r}\right) \qquad \text{for } r > 0. \tag{8}$$

[†] Recall that the volume of a right circular cylinder with radius r and height h is $\pi r^2 h$.

Then, since k is a constant,

$$\frac{dC}{dr} = k\left(8\pi r - \frac{64\pi}{r^2}\right).$$

Setting this expression equal to zero, we obtain

$$8\pi r = \frac{64\pi}{r^2}, \qquad 8r^3 = 64, \qquad r^3 = 8, \qquad \text{and} \qquad r = 2 \text{ m}.$$

In addition, $d^2C/dr^2 = k[8\pi + (128\pi/r^3)]$, which is greater than 0 when $r = 2$. Hence, there is a local minimum when $r = 2$ m. When $r = 2$, $h = \frac{32}{4} = 8$ m. Note that this local minimum is a true minimum since the only endpoint of the interval occurs at $r = 0$, which makes no practical sense since in that case the barrel can hold nothing. Also, from (8) it is apparent that as $r \to \infty$ or $r \to 0^+$, $C = C(r) \to \infty$.

EXAMPLE 4
SNELL'S LAW—MINIMIZING
THE TRAVEL TIME OF LIGHT

According to **Fermat's principle**,[†] *when light travels through various media, it takes the path that requires the least time.* To illustrate this principle, we suppose that light is traveling from a point beneath the surface of water to a point in the air, above the surface. In each medium (water and air), according to Fermat's principle, the light ray will travel in a straight line. We assume that the light emanates from a point P one unit of distance below the water surface to a point Q one unit above the surface and two units away in horizontal distance; see Figure 6. In Figure 6 the path from P to Q is not a straight line because the velocity of light in air is not equal to the velocity of light in water. At what point S on the surface of the water between points R and U (in the figure) will the light ray pass in order that it reach Q in minimum time?

SOLUTION: If the distance from R to S is denoted by x, then the distance from S to U is $2 - x$, so that

$$\overline{PS} = \sqrt{1 + x^2} \qquad \text{and} \qquad \overline{SQ} = \sqrt{1 + (2 - x)^2}.$$

If v_1 and v_2 represent the velocities of the light ray in water and air, respectively, then the total time for the ray to go from P to Q is given by

$$T = \frac{\sqrt{1 + x^2}}{v_1} + \frac{\sqrt{1 + (2 - x)^2}}{v_2}.$$

Then

$$\frac{dT}{dx} = \frac{x}{v_1\sqrt{1 + x^2}} - \frac{(2 - x)}{v_2\sqrt{1 + (2 - x)^2}}, \tag{9}$$

and

$$\frac{d^2T}{dx^2} = \frac{1}{v_1(1 + x^2)^{3/2}} + \frac{1}{v_2(1 + (2 - x)^2)^{3/2}}.$$

[†] Pierre de Fermat (1601–1665) was a celebrated French mathematician who helped develop a wide variety of topics of modern mathematics. See his biography on page 93.

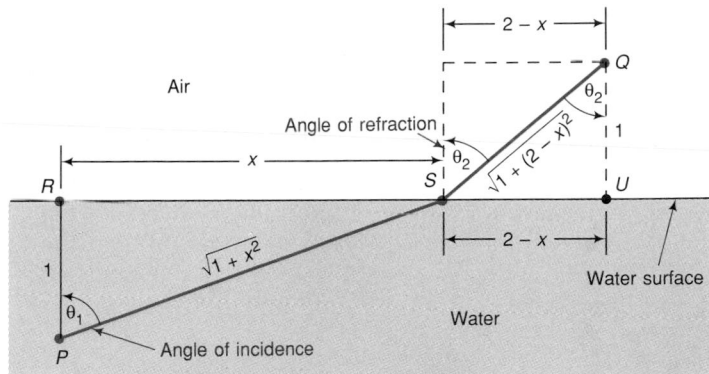

FIGURE 6 THE PATH OF A LIGHT RAY FROM A POINT IN THE WATER TO A POINT IN THE AIR

θ_1 is the angle of incidence and θ_2 is the angle of refraction.

Since $d^2T/dx^2 > 0$, the critical points will all be minima. Setting $dT/dx = 0$ in (9), we obtain

$$\frac{x}{v_1\sqrt{1 + x^2}} - \frac{2 - x}{v_2\sqrt{1 + (2 - x)^2}},$$

or

$$\frac{v_1}{v_2} = \frac{x/\sqrt{1 + x^2}}{(2 - x)/\sqrt{1 + (2 - x)^2}}. \tag{10}$$

But again from Figure 6, if θ_1 and θ_2 denote the angles at which the ray leaves the source and approaches the object, respectively, then

$$\frac{x}{\sqrt{1 + x^2}} = \sin\theta_1 \quad \text{and} \quad \frac{2 - x}{\sqrt{1 + (2 - x)^2}} = \sin\theta_2.$$

Then from (10), the minimum time is achieved when

$$\frac{v_1}{v_2} = \frac{\sin\theta_1}{\sin\theta_2}. \tag{11}$$

The number $\eta_{12} = \dfrac{v_1}{v_2}$ is called the **index of refraction**.

The angles θ_1 and θ_2 are called the angles of **incidence** and **refraction**, respectively. Equation (11) is known as **Snell's law,**[†] and it states that, according to Fermat's principle, the ratio of the sines of the angles of incidence and refraction is equal to the ratio of the velocities of light in the two media. Both are equal to the index of refraction. Note that if $v_1 = v_2$, then $\theta_1 = \theta_2$, so that the light ray will move in a straight line.

In this example we did not check the endpoints $x = 0$ and $x = 2$ because it is too complicated to do so without having specific values for v_1 and v_2. We show how to check the endpoints in the next example.

[†] Willebrord Snell (1591–1626) was a Dutch physicist.

EXAMPLE 5
CHECKING THE ENDPOINTS IN A
LIGHT REFRACTION PROBLEM:
CONTINUATION OF EXAMPLE 4

The velocity of light in air is approximately 300,000 km/sec.[†] The velocity of light in water is approximately 225,400 km/sec.

(a) Compute the index of refraction.
(b) Verify that time is minimized when x satisfies equation (10). That is, assuming that x is measured in feet, find the x that satisfies (10), find the total time the light ray travels when it refracts at x, and show that the total time is greater at $x = 0$ and $x = 2$.

SOLUTION: This is really a problem in algebra, not calculus. However, as you have already seen, it is sometimes necessary to use a great deal of algebra to solve calculus problems. The two topics are not independent.
We first compute

(a) $\eta_{12} = \dfrac{v_1}{v_2} = \dfrac{\text{velocity of light in water}}{\text{velocity of light in air}}$

$= \dfrac{225{,}400 \text{ km/sec}}{300{,}000 \text{ km/sec}} \approx 0.75133.$

(b) Then equation (10) becomes

$$\dfrac{\dfrac{x}{\sqrt{1 + x^2}}}{\dfrac{2 - x}{\sqrt{1 + (2 - x)^2}}} = 0.75133$$

or, squaring both sides and simplifying,

$$\dfrac{x^2}{1 + x^2} = (0.75133)^2 \dfrac{(2 - x)^2}{1 + (2 - x)^2}$$

$$p(x) = x^2[1 + (2 - x)^2] - 0.5645(1 + x^2)(2 - x)^2 = 0.$$

We use the symbol $p(x)$ to denote the polynomial above. Zeros of $p(x)$ are critical points of the problem. Now $p(0) \approx -4(0.5645) = -2.258 < 0$ and $p(2) = 4 > 0$, so, as we already knew, p has a zero between 0 and 2. Also,

$$p(1) = 2 - 0.5645(2) = 0.871 > 0$$

so the zero is between 0 and 1. To compute this zero we introduce a procedure called the **bisection method**. This method works as follows:

[†] More precisely, the speed of light in a vacuum is 299,792.4 km/sec. In air at 1 atmosphere of pressure and a temperature of 20°C, the speed is 299,702.5 km/sec. For our computations here and in the problems, the value of 300,000 km/sec or 186,000 miles/sec is an acceptable approximation.

> **THE BISECTION METHOD FOR FINDING A ZERO OF A FUNCTION $p(x)$**
>
> **(i)** Suppose that $p(a) < 0$ and $p(b) > 0$ (or vice versa).
>
> **(ii)** Evaluate p at the midpoint $\dfrac{a+b}{2}$.
>
> **(iii)** If $p\left(\dfrac{a+b}{2}\right) > 0$, then there is a zero in $\left(a, \dfrac{a+b}{2}\right)$.
>
> **(iv)** If $p\left(\dfrac{a+b}{2}\right) < 0$, then there is a zero in $\left(\dfrac{a+b}{2}, b\right)$.
>
> **(v)** Continue to cut the interval containing a zero in half until the length of the interval is less than the accuracy you seek.

In Table 1, we apply the bisection method to our problem.

TABLE 1 VALUES FOR $p(x) = x^2[1 + (2-x)^2] - 0.5645(1 + x^2)(2-x)^2$

$x^\dagger$	$p(x)$	Interval That Contains a Zero of $p(x)$
0	-1.129	
1	0.871	$(0, 1)$
0.5	-0.76953	$(0.5, 1)$
0.75	0.06323	$(0.5, 0.75)$
0.625	-0.35501	$(0.625, 0.75)$
0.6875	-0.14519	$(0.6875, 0.75)$
0.71875	-0.04070	$(0.71875, 0.75)$
0.73439	0.01135	$(0.71875, 0.73439)$
0.72659	-0.01462	$(0.72659, 0.73439)$
0.73049	-0.00163	$(0.73049, 0.73439)$
0.73244	0.00057	$(0.73049, 0.73244)$
0.73147	0.00163	$(0.73049, 0.73147)$
0.73098	-0.000000475	$(0.73098, 0.73147)$

$\dagger$ Each new x is obtained by taking the average of the last two x's at which $p(x)$ has opposite signs. Each new x is rounded to 5 decimal places.

To three decimal places, T has a local minimum at $x = 0.731$. Now

$$
\begin{aligned}
T(x) &= \frac{\sqrt{1+x^2}}{v_1} + \frac{\sqrt{1+(2-x)^2}}{v_2} \\
&= \frac{\sqrt{1+0.731^2}}{225{,}400} + \frac{\sqrt{1+(2-0.731)^2}}{300{,}000} \\
&\approx 0.00001088 \approx 1.09 \times 10^{-5} \text{ seconds.}
\end{aligned}
$$

Finally, we check the endpoints:

$$
T(0) = \frac{1}{v_1} + \frac{\sqrt{5}}{v_2} = \frac{1}{225{,}400} + \frac{\sqrt{5}}{300{,}000} \approx 1.19 \times 10^{-5} \text{ seconds}
$$

and

$$
T(2) = \frac{\sqrt{5}}{v_1} + \frac{1}{v_2} = \frac{\sqrt{5}}{225{,}400} + \frac{1}{300{,}000} \approx 1.33 \times 10^{-5} \text{ seconds.}
$$

Evidently, the light ray reaches Q in minimum time when Snell's law is satisfied.

To complete this problem, we compute the angles of incidence and refraction. We have

$$\sin \theta_1 = \frac{x}{\sqrt{1 + x^2}} \approx \frac{0.731}{\sqrt{1 + 0.731^2}} \approx 0.5901$$

and, using the $\sin^{-1}$ button on a calculator (and setting the calculator to degree mode),

$$\theta_1 \approx \sin^{-1} 0.5901 \approx 36.2° = \text{angle of incidence}.$$

Similarly,

$$\sin \theta_2 = \frac{2 - 0.731}{\sqrt{1 + (2 - 0.731)^2}} \approx 0.7854$$

and

$$\theta_2 \approx \sin^{-1} 0.7854 \approx 51.8° = \text{angle of refraction}.$$

Note, as a final check, that

$$\frac{\sin \theta_1}{\sin \theta_2} \approx \frac{0.5901}{0.7854} \approx 0.7513 = \frac{v_1}{v_2} = \eta_{12}.$$

REMARK: The computations in the last example were horrendous. Unfortunately, computations in real life are often horrendous. In Section 3.6 we will discuss Newton's method, which will enable us to find the zero of $p(x)$, for example, relatively quickly (see Problem 3.6.14 on page 238). However there is still a lot of work to be done. Only in carefully picked textbook examples do examples work out cleanly with a minimum of work.

In the examples we have given so far, the maximum or minimum did not occur at an endpoint. The next example shows that they can.

EXAMPLE 6

MINIMIZING THE TOTAL AREA ENCLOSING TWO SHAPES

A wire 20 cm long is cut into two pieces. One piece is bent in the shape of an equilateral triangle, and the other is bent in the shape of a circle. How should the wire be cut to maximize the total area enclosed by the shapes? How should it be cut to minimize the total area?

SOLUTION: Let x denote the length of wire used for the equilateral triangle. Then the length of each side is $x/3$, and the circumference of the circle is $20 - x$ (see Figure 7). Using the Pythagorean theorem, we find that the height of the triangle is $\sqrt{3}x/6$, so that

$$\text{area of the triangle} = \frac{1}{2}(\text{base}) \times (\text{height}) = \frac{1}{2}\left(\frac{x}{3}\right)\left(\sqrt{3}\,\frac{x}{6}\right) = \sqrt{3}\,\frac{x^2}{36}.$$

Moreover, since the circumference of the circle is $2\pi r = 20 - x$, we find that the radius of the circle is $(20 - x)/2\pi$, and its area is

$$\pi r^2 = \pi\left(\frac{20 - x}{2\pi}\right)^2 = \frac{(20 - x)^2}{4\pi}.$$

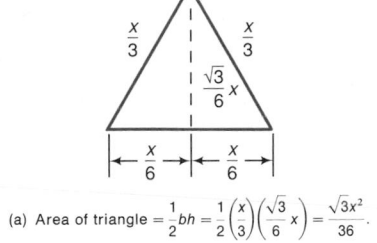

(a) Area of triangle $= \frac{1}{2}bh = \frac{1}{2}\left(\frac{x}{3}\right)\left(\frac{\sqrt{3}}{6}x\right) = \frac{\sqrt{3}x^2}{36}$.

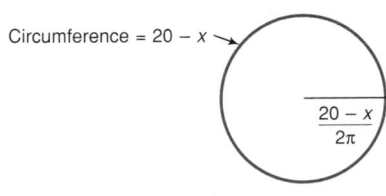

(b) Area of circle $= \pi r^2 = \pi\left(\frac{20-x}{2\pi}\right)^2 = \frac{(20-x)^2}{4\pi}$.

FIGURE 7
AREAS OF A TRIANGLE AND A CIRCLE

Thus, the total area (as a function of x) is

$$A(x) = \frac{\sqrt{3}}{36}x^2 + \frac{(20-x)^2}{4\pi}, \qquad 0 \le x \le 20.$$

Differentiating, we obtain

$$\frac{dA}{dx} = \frac{\sqrt{3}}{18}x - \frac{(20-x)}{2\pi} = \left(\frac{\sqrt{3}}{18} + \frac{1}{2\pi}\right)x - \frac{10}{\pi}.$$

This expression is equal to zero when

$$x = \frac{10/\pi}{(\sqrt{3}/18) + (1/2\pi)} \approx 12.46 \text{ cm.}$$

Also, $d^2A/dx^2 = (\sqrt{3}/18) + (1/2\pi) > 0$, so there is a local minimum at $x \approx 12.46$. For that value of x, $A \approx 11.99$ cm^2. Checking the endpoints of the interval $[0, 20]$, we find that $A(0) = 400/4\pi \approx 31.83$ cm^2, and $A(20) = (\sqrt{3}/36)(400) \approx 19.25$ cm^2. Thus, the maximum area is enclosed when we do not cut the wire at all and use it to form a circle. If the wire must be cut, then the function in this problem has no maximum. The local minimum at $x = 12.46$ is indeed the minimum over the entire interval.

EXAMPLE 7
MINIMIZING THE RESISTANCE IN BLOOD FLOW

Blood is transported in the body by means of arteries, veins, arterioles, and capillaries. This procedure is ideally carried out in such a way as to minimize the energy required[†] to transport the blood from the heart to the organs and back again. Consider the two blood vessels depicted in Figure 8, where $r_1 > r_2$. There are many ways to try to compute the minimum energy required to pump blood from the point P_1 in the main vessel to the point P_3 in the smaller vessel. We will try to find the minimum total resistance of the blood along the path $P_1P_2P_3$.

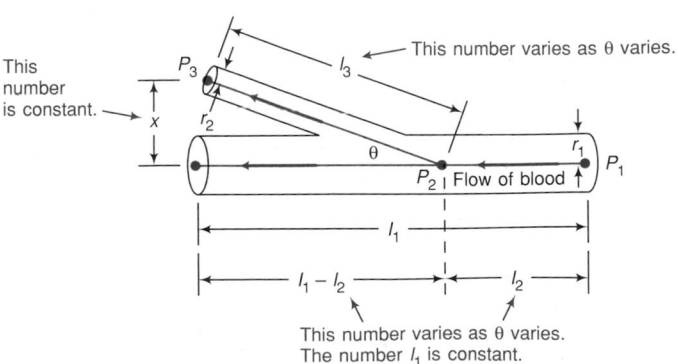

FIGURE 8   OPTIMAL BLOOD FLOW

[†] Actually, the item to be minimized is the work (energy) required to maintain the system as well as pump the blood. For a more complete discussion of this topic, consult the book by R. Rosen, *Optimality Principles in Biology* (London: Butterworth, 1967).

SOLUTION: According to Poiseuille's law (see Problem 2.2.35, page 145), the resistance is given by

$$R = \frac{\alpha l}{r^4},$$

where l is the length of the vessel, r is its radius, and α is a constant of proportionality. The problem, then, is to find the "optimal" branching angle θ at which R is a minimum. According to Figure 8, $\sin \theta = x/l_3$, so $l_3 = x/\sin \theta = x \csc \theta$. Also, $x/(l_1 - l_2) = \tan \theta$, so $l_1 - l_2 = x/(\tan \theta) = x \cot \theta$ and $l_2 = l_1 - x \cot \theta$. We now calculate the total resistance along the path $P_1 P_2 P_3$.

$$R_{1,2} = \frac{\alpha l_2}{r_1{}^4} \qquad \text{Resistance from } P_1 \text{ to } P_2$$

$$R_{2,3} = \frac{\alpha l_3}{r_2{}^4} \qquad \text{Resistance from } P_2 \text{ to } P_3$$

Then $R = R_{1,2} + R_{2,3} = \alpha[(l_2/r_1{}^4) + (l_3/r_2{}^4)]$. But $l_3 = x \csc \theta$ and $l_2 = l_1 - x \cot \theta$, so that

$$R = \alpha\left(\frac{l_1 - x \cot \theta}{r_1{}^4} + \frac{x \csc \theta}{r_2{}^4}\right).$$

We now simplify matters by using the fact that x is fixed, since x depends on P_1, P_3, and l_1, which are given in the problem, but not on P_2, l_2, l_3, or θ. (P_2, l_2, and l_3 all vary as θ varies, but P_1, P_3, and l_1 do not.) Since R is a function of θ, we can find the minimum value of R by calculating $dR/d\theta$ and setting it to zero. But,

$$\frac{dR}{d\theta} = \alpha\left(\frac{x \csc^2 \theta}{r_1{}^4} - \frac{x \csc \theta \cot \theta}{r_2{}^4}\right)$$

$$= \alpha\left(\frac{x}{\sin^2 \theta \, r_1{}^4} - \frac{x \cos \theta}{\sin^2 \theta \, r_2{}^4}\right) = \frac{\alpha x}{\sin^2 \theta}\left(\frac{1}{r_1{}^4} - \frac{\cos \theta}{r_2{}^4}\right).$$

Now $\alpha x/(\sin^2 \theta) \neq 0$, so $dR/d\theta = 0$ when

$$1/r_1{}^4 = (\cos \theta)/r_2{}^4,$$

or when

$$\cos \theta = \left(\frac{r_2}{r_1}\right)^4.$$

That this value is indeed a minimum can be verified by employing the first-derivative test. Thus we can calculate the optimum branching angle by merely considering the ratio of the radii of the blood vessels. In the cited reference by Rosen, evidence is given that indicates that branching angles of blood vessels do, in many cases, obey the optimization rule we have just derived.

■ PROBLEMS 3.5

■ SELF-QUIZ

I. True–False:
   a. If c is a critical point for function f on the interval $[a, b]$, then $f'(c) = 0$.
   b. If c is a critical point for function f on the interval $[a, b]$ and c is not an endpoint of $[a, b]$, then $f'(c) = 0$.
   c. If c is a critical point for function f on the interval $[a, b]$, if c is not an endpoint of $[a, b]$ and if $f'(c)$ exists, then $f'(c) = 0$.
   d. If $f'(c) = 0$, then $f(c)$ is the maximum value of f on the interval $[a, b]$.
   e. If $f'(c) = 0$ and if c is not an endpoint of $[a, b]$, then $f(c)$ is the maximum value of f on the interval $[a, b]$.
   f. If $f'(c) = 0$, then $f(c)$ is a local maximum value of f.
   g. If $f'(c) = 0$, then $f(c)$ is a local minimum value of f.
   h. If $f'(c) = 0$, then $f(c)$ is a local extreme value of f.
   i. If $f'(c) = 0$, then c is a critical point of f.
   j. If $f'(c)$ does not exist, then c is a critical point of f.

II. Let $f(x) = x^2 - 6x + 7$; then $f(3)$ is the _____ value of the function f on the interval $[0, 10]$.
   a. minimum
   b. maximum
   c. middle
   d. none of the above

III. Let $g(x) = x^3/3 - 4x$; then _____ is the maximum value of the function g on the interval $[-1, 4]$.
   a. $g(-1)$
   b. $g(2)$
   c. $g(\sqrt{12})$
   d. $g(4)$

IV. Let $g(x) = x^3/3 - 4x$; then _____ is the minimum value of the function g on the interval $[-1, 4]$.
   a. $g(-1)$
   b. $g(2)$
   c. $g(\sqrt{12})$
   d. $g(4)$

V. Let $g(x) = x^3/3 - 4x$; then _____ is the minimum value of the function g on the interval $[-5, 4]$.
   a. $g(-5)$
   b. $g(-2)$
   c. $g(2)$
   d. $g(4)$

■ DRILL

In Problems 1–18, find the minimum and maximum values for the given function over the specified interval.

1. $f(x) = x^3 - 12x + 10$; $[-10, 10]$
2. $f(x) = 4x^3 - 3x + 2$; $[-5, 5]$
3. $f(x) = (x + 1)^{1/3}$; $[-2, 7]$
4. $f(x) = (x - 2)^{2/3}$; $[-14, 3]$
5. $f(x) = x^3 - 3x^2 - 45x + 25$; $[-5, 5]$
6. $f(x) = x^3/3 + x^2/2 - 2x - \frac{2}{3}$; $[-3, 3]$
7. $f(x) = (x - 1)^{1/5}$; $[-31, 33]$
8. $f(x) = \sqrt{x} + 1/\sqrt{x}$; $[\frac{1}{2}, \frac{3}{2}]$
9. $f(x) = (x - 1)/(x - 2)$; $[-3, 1]$
10. $f(x) = x/(1 + x)$; $[0, \infty)$
11. $f(x) = |x^2 - 2x|$; $[-2, 3]$
12. $f(x) = |x^3 - 12x|$; $[-3, 3]$
13. $f(x) = \sin x + \cos x$; $[0, 2\pi]$
14. $f(x) = \sin x^2$; $[0, \pi]$
15. $f(x) = \cos\left(\dfrac{1 - 2x}{3}\right)$; $[-2, 2]$
16. $f(x) = \sin\left(\dfrac{x}{1 + x}\right)$; $(-1, 1]$

*17. $f(x) = \begin{cases} x^2 & \text{if } x \le 1 \\ x^3 & \text{if } x > 1 \end{cases}$; $[-4, 2]$

*18. $f(x) = \begin{cases} x - 3 & \text{if } x < 1 \\ 2x + 4 & \text{if } 1 \le x \le 3 \\ 3x - 7 & \text{if } 3 < x \end{cases}$; $[-5, 5]$

Example 2 locates the maximum of $x \cdot y$ subject to the constraint that $2x + y = 1000$. The constraint was "solved" to express y as function of x and that expression was substituted into the function of interest so that $A = x \cdot y = x \cdot (1000 - 2x)$ is explicitly written as a function of x. Implicit differentiation provides an alternative approach to some of the algebra in locating critical points. Differentiating the constraint with respect to x yields the equation $2 + y' = 0$. Differentiating the area function and then setting A' to zero yields the equation $1 \cdot y + x \cdot y' = 0$. This pair of equations imply $y' = -2$ and $y = -(x)(-2) = 2x$; substituting $y = 2x$ into the constraint yields $1000 = 2x + y = 4x$ and thus $250 = x$. Furthermore, $A'' = (y + xy')' = y' + 1 \cdot y' + x \cdot y'' = 2y' = -4$ (remember we already know $y' = -2$ and that implies $y'' = 0$); since $A'' < 0$, the critical point at $x = 250$ is a local maximum.

CONTINUED

Work Problems 19–24 by using this "implicit" technique.

19. Rework Example 2 (continued) on page 214.
20. Rework Example 3.
21. A rectangle has a perimeter of 300 m. What length and width will maximize its area?

22. A rectangle has an area of 500 m². What length and width will minimize its perimeter?
23. Find the point on the straight line $3x - 4y - 12 = 0$ that is closest to the point $(1, 2)$.
24. Find the point on the circle $x^2 + y^2 = 1$ that is closest to the point $(2, 5)$.

■ **APPLICATIONS**

25. A ball is thrown upward from ground level (Figure 9). After t seconds, its height (in feet) above the ground is $48t - 16t^2$.
   a. What is its initial vertical velocity?
   b. After how many seconds will the ball reach its maximum height?
   c. How high will the ball go?

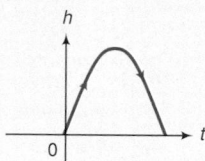

FIGURE 9

26. A farmer wishes to set aside one acre of her land for corn and wheat. To keep out the cows, she encloses a rectangular field with a fence costing 50 cents per linear foot. In addition, a fence running down the middle of the field is needed; such a fence costs 1 dollar per foot (Figure 10). Given that 1 acre = 43,560 ft², what dimensions should the field have to minimize her total cost?

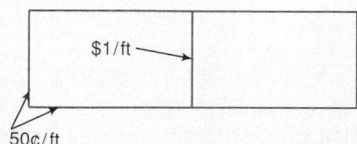

FIGURE 10

27. A farmer wishes to divide 20 acres of land along a (straight) river into 6 smaller plots by using one fence parallel to the river and 7 fences perpendicular to it (Figure 11). Verify that the total amount of fencing is minimized if the sum of the lengths of the 7 cross fences equals the length of the one fence parallel to the river.

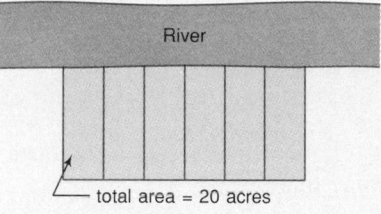

FIGURE 11

*28. A farmer has 200 m of fencing. He wishes to construct a given number of rectangular pens by running two fences from east to west and the necessary number of fences from north to south. If the total area of the pens is to be a maximum, what is the total length of the east-west fences and the total length of the north-south fences?

29. A cylindrical tin can is to hold 50 cm³ of tomato juice (Figure 12). How should the can be constructed in order to minimize the amount of material needed in its construction?

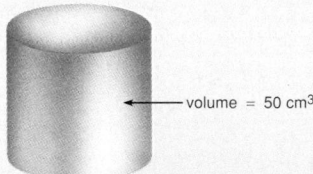

FIGURE 12

30. Find the dimensions of the right circular cylinder of largest volume that can be inscribed in a sphere of radius 10 (Figure 13).

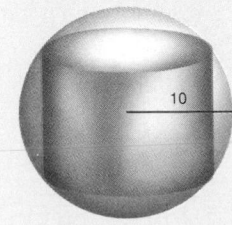

FIGURE 13

CONTINUED

31. A wire 35 cm long is cut into two pieces. One piece is bent in the shape of a square, and the other is bent in the shape of a circle (Figure 14). How should the wire be cut to maximize the total area enclosed by the pieces? How should the wire be cut to minimize the total enclosed area?

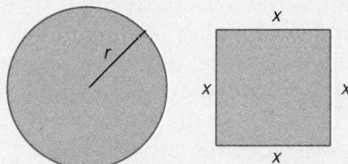

FIGURE 14

32. Answer the preceding problem if the two pieces are to be formed in the shape of a circle and an equilateral triangle (Figure 15).

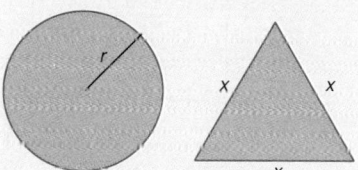

FIGURE 15

33. A Norman window is constructed from a rectangular sheet of glass surmounted by a semicircular sheet of glass. (Stained glass windows are often of this type.) The light that enters through a window is proportional to the area of the window (Figure 16). What are the dimensions of the Norman window having a perimeter of 30 ft that admits the most light?

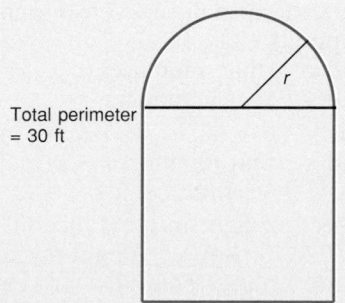

Total perimeter = 30 ft

FIGURE 16

34. Rework the preceding problem with the additional conditions that the rectangular part of the window is tinted while the circular part is clear and that the tinted glass admits light at only half the rate of the clear glass.

35. A woman is on a lake in a canoe 1 km from the closest point P of a straight shore line; she wishes to get to a point Q, 10 km along the shore from P. To do so, she paddles to a point R between P and Q and then walks the remaining distance to Q (Figure 17). She can paddle 3 km/hr and walk 5 km/hr. How should she pick the point R so that she gets to Q as quickly as possible?

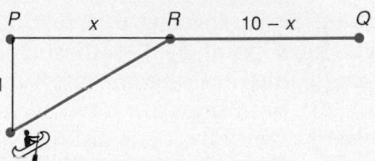

FIGURE 17

36. An oil pipeline will be built to connect points P and Q which are on opposite banks of a river. The river is straight and is $\frac{1}{2}$ km wide; the point Q is 5 km downstream from a point directly across from point P (Figure 18). The tubing used in crossing the river costs 50% more than the tubing that can be used on dry land. Describe a path for this pipeline that will minimize the total cost.

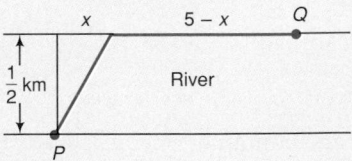

FIGURE 18

37. Two saplings, 6 and 8 ft tall, are planted 10 ft apart. So that the saplings do not bend, poles the height of the trees are pounded in, then attached to each tree, and a rope tied to the top of each pole is then fixed to the ground (between the two trees) after being pulled taut (Figure 19). How close to the taller tree will the rope be fixed if the total length of the rope is to be minimized?

CONTINUED

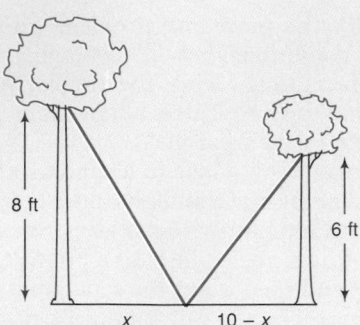

FIGURE 19

38. A submarine is traveling due east and heading straight for a point P. A battleship is traveling due south and heading for the same point P (Figure 20). Both ships are traveling at a velocity of 30 km/hr. Initially, their distances from P are 210 km for the submarine and 150 km for the battleship. The range of the submarine's torpedoes is 3 km. How close will the two vessels come to each other? Does the submarine have a chance to torpedo the battleship?

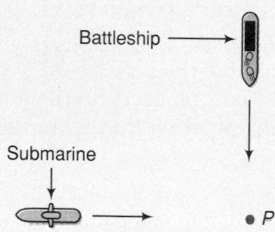

FIGURE 20

39. A storage container is constructed in the shape of a V from a rectangular sheet of stainless steel, 6 m long and 2 m wide, by bending the sheet through its middle (Figure 21). What angle θ

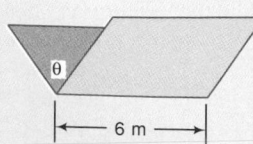

FIGURE 21

should be left between the sides of the container in order to maximize the volume? (Assume that triangular pieces, provided free of charge, are welded onto each end of the bent sheet.)

*40. A light is hung over the center of a square table 4 m². The intensity of the light hitting a point P on the table is directly proportional to the sine of the angle the path of the light makes with the table and is inversely proportional to the distance between the point and the light (Figure 22). How high above the table should the light be placed in order to maximize the light intensity at the corners of the table?

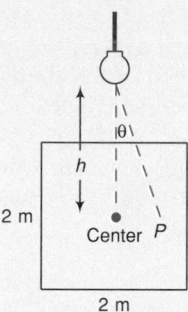

FIGURE 22

41. Refer to Example 7.
   a. Verify that if two blood vessels have equal radii, then blood resistance is minimized when the branching angle between them is zero.
   b. Find the branching angle that minimizes resistance if one blood vessel has twice the diameter of the other one.

*42. Suppose that a_1, a_2, a_3, and a_4 are four separate estimates of some single physical quantity (for example, the altitude of a volcanic cone just after an eruption).
   a. Find an x that minimizes $(a_1 - x)^2 + (a_2 - x)^2$.
   b. Find a y that minimizes
   $$\sqrt{(a_1 - y)^2 + (a_2 - y)^2 + (a_3 - y)^2}.$$
   c. Find a z that minimizes
   $$\sqrt{(a_1 - z)^2 + (a_2 - z)^2 + (a_3 - z)^2 + (a_4 - z)^2}.$$

43. The speed of light in ethyl alcohol is approximately 220,400 km/sec. A light ray leaves a point in air and strikes a point in a jar of ethyl alcohol with an angle of refraction of 38°. To the nearest tenth of a degree, find the angle of incidence.

CONTINUED

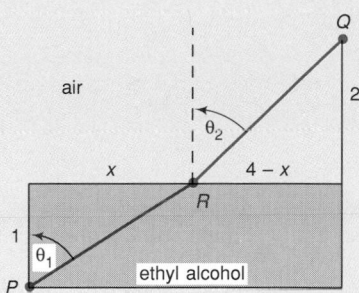

FIGURE 23

*44. Find the value of x in Figure 23 such that the time it takes a light ray at P to reach the point Q is minimized. What is the minimum time? What is the maximum time it would take if we could control the value of x (with $0 \leq x \leq 4$)? [*Hint:* Check the endpoints.] Also find the angles of incidence and refraction.

■ **SHOW/PROVE/DISPROVE**

45. The most important function of the human cough is to increase the velocity of air going out of the windpipe (*trachea*). Let R_0 denote the "rest radius" of the trachea—that is, the radius when you are relaxed and not coughing. (Measure distances in cm.) Let R be the contracted radius of the trachea during a cough ($R < R_0$); and let V be the average velocity of the air in the trachea when it is contracted to R cm. Under some fairly reasonable assumptions regarding the flow of air near the tracheal wall (we assume it is very slow) and the "perfect" elasticity of the wall, we can model the velocity of air flow during a cough by the equation[†]

$$V = \alpha(R_0 - R)/R^2 \text{ cm/sec},$$

where α is a constant depending on the length of the tracheal wall.

If you are coughing efficiently, your tracheal wall will contract in such a way as to maximize the velocity of air going out of the trachea. Show that V is maximized when the trachea is contracted by one-third of its original radius (down to $R = \frac{2}{3}R_0$).[‡]

46. Show that the minimal distance from the point (x_0, y_0) to a point on the line with standard equation $Ax + By + C = 0$ is

$$\frac{|Ax_0 + By_0 + C|}{\sqrt{A^2 + B^2}}.$$

This result appeared earlier in Problem 0.3.63 on p. 22. The tactic at that place in the text was to work with perpendicular lines. Now, your job is to construct a proof using some differential calculus tools. [*Hint:* Some calculations will be simplified if you use the fact that distance is minimized when distance-squared is minimized.]

**47. Suppose that p is a polynomial function of degree n such that $p(x) \geq 0$ for all x. Prove that

$$p(x) + p'(x) + p''(x) + \cdots + p^{(n)}(x) \geq 0$$

for all x.

*48. Pick a real number α in the interval $(1, \infty)$. On the interval $[-1, \infty)$ consider the function F where $F(x) = (1 + x)^\alpha - (1 + \alpha x)$.
a. Show that F has a unique critical point.
b. Show that $F(x) > 0$ unless $x = 0$.
c. Prove that $(1 + x)^\alpha \geq (1 + \alpha x)$ if $x \geq -1$ and $\alpha \geq 1$.

■ **CHALLENGE**

*49. State a general result that includes the solutions to the parts of Problem 42. Prove your result.

[†] This equation and a detailed description of this model appear in Philip Tuchinsky's paper "*The Human Cough*," UMAP Project, Education Development Center, Newton, Mass. (1978).

[‡] This result has been confirmed, approximately, by X-ray photographs taken during actual coughs.

▦ 3.6 NEWTON'S METHOD FOR SOLVING EQUATIONS

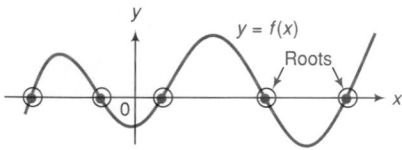

FIGURE 1
THE ROOTS, OR ZEROS, OF A FUNCTION

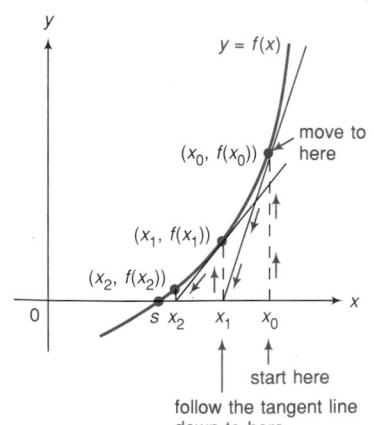

FIGURE 2
HOW TO GET THE NEXT ITERATE USING NEWTON'S METHOD

In this section, we look at a very different kind of application of the derivative. Consider the equation

$$f(x) = 0, \tag{1}$$

where f is assumed to be differentiable in some interval $[a, b]$. It is often important to calculate the *roots* of equation (1)—that is, the values of x that satisfy the equation. For example, if $f(x)$ is a polynomial of degree 5, say, then the roots of $f(x)$ could be as in Figure 1. (We have already used graphs to determine the number of roots of certain polynomials; see pages 201 and 202.)

The intermediate-value theorem (page 119) gives us a crude way to approximate the roots of $f(x) = 0$. Suppose that f is continuous, $f(a) < 0$ and $f(b) > 0$, then there is at least one number c in (a, b) such that $f(c) = 0$. If $f'(x) > 0$ in (a, b) (or $f'(x) < 0$ in (a, b)), then f is increasing or decreasing, and so there can be only one solution to $f(x) = 0$ in (a, b). We can find this root by trial and error. The bisection method described on page 219 is a more efficient method; but there is a much better way to approximate the roots.

In the seventeenth century, Newton discovered a method for estimating a solution, or root, by defining a sequence of numbers that become successively closer and closer to the root sought. His method is best illustrated graphically. Let $y = f(x)$ as in Figure 2. A number x_0 is chosen arbitrarily. We then locate the point $(x_0, f(x_0))$ on the graph and draw the tangent line to the curve at that point. Next, we follow the tangent line down until it hits the x-axis. The point of intersection of the tangent line and the x-axis is called x_1. We then repeat the process to arrive at the next point, x_2. On the graph we have labeled the solution to $f(x) = 0$ as s. That is, $f(s) = 0$. For our graph at least, it seems as if the points $x_0, x_1, x_2, \ldots$ are approaching the point $x = s$. In fact, this happens for quite a few functions, and the rate of approach to the solution is quite rapid.

Having briefly looked at a graphical representation of Newton's method, let us next develop a formula for giving us x_1 from x_0, x_2 from x_1, and so on. The slope of the tangent line at the point $(x_0, f(x_0))$ is $f'(x_0)$. Two points on this line are $(x_1, 0)$ and $(x_0, f(x_0))$. Therefore,

$$\frac{0 - f(x_0)}{x_1 - x_0} = f'(x_0). \tag{2}$$

Solving (2) for x_1 gives us

$$x_1 = x_0 - \frac{f(x_0)}{f'(x_0)}. \tag{3}$$

Similarly,

$$x_2 = x_1 - \frac{f(x_1)}{f'(x_1)}. \tag{4}$$

In general, we obtain

NEWTON'S FORMULA

$$x_{n+1} = x_n - \frac{f(x_n)}{f'(x_n)}. \tag{5}$$

This last step tells us how to obtain the $(n + 1)$st point if the nth point is given, as long as $f'(x_n) \neq 0$ so that (5) is defined. Thus, if we start with a given value x_0, we can obtain $x_1, x_2, x_3, x_4, \ldots$. The numbers $x_0, x_1, x_2, \ldots$ are called **iterates**. Thus x_0 is the **zeroth iterate** or **starting value**, x_1 is the **first iterate**, x_2 is the **second iterate**, and so on.

The formula (5) is called **Newton's formula**. The set of iterates $x_0, x_1, x_2, x_3, \ldots$ is called a **sequence**.[†] If the numbers in the sequence get closer and closer to a certain number s as n gets larger and larger, then we say that the sequence **converges** to s, and we write

$$\lim_{n \to \infty} x_n = s. \tag{6}$$

Before stating the theorem that guarantees that the sequence defined by (5) converges to a solution of $f(x) = 0$, we give two simple examples.

EXAMPLE 1
NEWTON'S FORMULA FOR
CALCULATING SQUARE ROOTS

Let $r > 1$. Formulate a rule for calculating the square root of r.

SOLUTION: We must find an x such that $x = \sqrt{r}$, or $x^2 = r$, or $x^2 - r = 0$. Let $f(x) = x^2 - r$. Then if $f(s) = 0$, s will be a square root of r. [$f(s) = 0$ means that $s^2 - r = 0$, or $s^2 = r$.] By Newton's formula, since $f'(x) = 2x$, we obtain the sequence $x_0, x_1, x_2, \ldots$, where x_0 is arbitrary and

$$x_{n+1} = x_n - \frac{f(x_n)}{f'(x_n)} = x_n - \frac{(x_n^2 - r)}{2x_n} = \frac{2x_n^2 - x_n^2 + r}{2x_n}$$

$$= \frac{1}{2}\left(x_n + \frac{r}{x_n}\right).^{‡} \tag{7}$$

NOTE: In applying Newton's method, the starting value x_0 cannot always be chosen arbitrarily. However, for many problems, you can use the results of Theorem 2 on page 231 to obtain an interval $[a, b]$ such that the sequence of iterates will converge to a unique solution to $f(x) = 0$ in the interval $[a, b]$, if you start at any x_0 in $[a, b]$.

[†] We will discuss sequences extensively in Chapter 10.

[‡] Note that this formula requires only the basic arithmetic operations of addition, subtraction, multiplication, and division. These are the operations most easily performed in a calculator, which is why many calculators use this formula, or a simple modification of it, to find square roots.

EXAMPLE 2

CALCULATING $\sqrt{2}$ BY NEWTON'S METHOD

Calculate $\sqrt{2}$ by Newton's method.

SOLUTION: In formula (7), $r = 2$, so that

$$x_{n+1} = \frac{1}{2}\left(x_n + \frac{2}{x_n}\right). \tag{8}$$

Using a calculator, we obtain the sequence in Table 1, starting with $x_0 = 1$. We can see here the remarkable accuracy of Newton's method. An answer correct to nine decimal places was obtained after only four steps! We were limited in accuracy only by the fact that our calculator could display only 10 digits.

TABLE 1 SEQUENCE OF ITERATES FOR CALCULATING $\sqrt{2}$

n	x_n	$\dfrac{2}{x_n}$	$x_n + \dfrac{2}{x_n}$	$x_{n+1} = \dfrac{1}{2}\left(x_n + \dfrac{2}{x_n}\right)$
0	1.0	2.0	3.0	1.5
1	1.5	1.333333333	2.833333333	1.416666667
2	1.416666667	1.411764706	2.828431373	1.414215686
3	1.414215686	1.414211438	2.828427125	1.414213562
4	1.414213562	1.414213562	2.828427125	1.414213562

When using Newton's method, three questions naturally arise.

(1) Is there a unique solution to $f(x) = 0$?
(2) Does the sequence of Newton iterates converge to it?
(3) If so, how fast?

We shall answer all three questions in this section. The first one is the easiest.

THEOREM 1 **EXISTENCE OF A UNIQUE ZERO IN [a, b]**

Suppose that f' is continuous in $[a, b]$ and

(i) $f(a)$ and $f(b)$ have different signs.
(ii) $f'(x) \neq 0$ on $[a, b]$.

Then the equation $f(x) = 0$ has a unique solution, s, on $[a, b]$.

PROOF: As we stated earlier in this section, the intermediate value theorem guarantees the existence of at least one solution. Also, suppose $f'(x) \neq 0$ and f' is continuous in $[a, b]$. Then $f'(x) > 0$ on $[a, b]$ or $f'(x) < 0$ on $[a, b]$ because if f' took positive and negative values, then there would be a point, c, in (a, b) where $f'(c) = 0$, again by the intermediate value theorem. Thus f is increasing or decreasing in $[a, b]$, and so the root must be unique. ∎

EXAMPLE 3
NEWTON'S METHOD FOR
CALCULATING A kth ROOT

Formulate a rule for calculating the kth root of a given number r.

SOLUTION: We must find an x such that $x = r^{1/k}$, or $x^k = r$, or $f(x) = x^k - r = 0$. Then $f'(x) = kx^{k-1}$, and

$$x_{n+1} = x_n - \frac{x_n^k - r}{kx_n^{k-1}} = x_n - \frac{1}{k}\frac{x_n^k}{x_n^{k-1}} + \frac{r}{kx_n^{k-1}}$$

$$= \left(1 - \frac{1}{k}\right)x_n + \frac{r}{kx_n^{k-1}}. \tag{9}$$

EXAMPLE 4
CALCULATING $\sqrt[3]{17}$ BY
NEWTON'S METHOD

Calculate $\sqrt[3]{17}$.

SOLUTION: By (9), with $k = 3$ and $r = 17$, we have $x_{n+1} = \frac{2}{3}x_n + (17/3x_n^2)$. Since $\sqrt[3]{17}$ is between 2 and 3, we choose $x_0 = 2$ (3 would do just as well). If $[a, b] = [2, 3]$, then since

$$f(x) = x^3 - 17, \quad f(2) = -9, \quad f(3) = 10,$$

and (i) is satisfied. Also, $f'(x) = 3x^2$ and $|f'(x)| \neq 0$ on $[2, 3]$. Thus, there is a unique solution. We now apply Newton's method. Values of x_n are tabulated in Table 2. The last number is correct to nine decimal places. Again, the rapid convergence of Newton's method is illustrated.

TABLE 2 SEQUENCE OF ITERATES FOR CALCULATING $\sqrt[3]{17}$

n	x_n	$\frac{2}{3}x_n$	x_n^2	$\dfrac{17}{3x_n^2}$	$x_{n+1} = \frac{2}{3}x_n + \dfrac{17}{3x_n^2}$
0	2.0	1.333333333	4.0	1.416666667	2.75
1	2.75	1.833333333	7.5625	0.7493112948	2.582644628
2	2.582644628	1.721763085	6.670053275	0.8495684267	2.571331512
3	2.571331512	1.714221008	6.611745745	0.8570605836	2.571281592
4	2.571281592	1.714187728	6.611489023	0.8570938629	2.571281591

We now give a remarkable theorem that answers questions (2) and (3). The proof is difficult. A partial proof is given in Section 3.8.

We assume that f, f', and f'' are continuous in $[a, b]$. By the upper and lower bound theorem on page 119, there are numbers m and M such that $|f'(x)| \geq m$ on $[a, b]$ and $|f''(x)| \leq M$ on $[a, b]$. Moreover, if $f'(x) \neq 0$ in $[a, b]$, then $m > 0$.

THEOREM 2 **QUADRATIC CONVERGENCE THEOREM FOR NEWTON'S METHOD**

Let f, f', and f'' be continuous on $[a, b]$. Suppose that

(i) $f(a)$ and $f(b)$ have different signs.

(ii) $|f'(x)| \geq m > 0$.

(iii) $|f''(x)| \leq M$.

(iv) $\dfrac{M}{m} \leq 4$.

(v) $x_0 \in [a, b]$.

Then,

(a) The sequence of Newton iterates $x_0, x_1, x_2, \ldots$ converges to the unique solution, s, of $f(x) = 0$ in $[a, b]$.

(b) The convergence is **quadratic**. That is, each iterate is accurate to twice as many decimal places as the one that precedes it. ∎

EXAMPLE 5
VERIFYING THE CONDITIONS OF THE CONVERGENCE THEOREM AND ILLUSTRATING QUADRATIC CONVERGENCE

In Example 4, $f(x) = x^3 - 17$, $|f'(x)| = 3x^2 \geq 3 \cdot 2^2 = 12$ on $[2, 3]$, and $|f''(x)| = 6x \leq 18$ on $[2, 3]$. Thus $m = 12$, $M = 18$, and $\dfrac{M}{m} = \dfrac{3}{2} < 4$. Thus Newton's method converges quadratically to $\sqrt[3]{17} = 2.571281591$. Now, from Table 2,

$$\begin{aligned}|x_1 - \sqrt[3]{17}| &= |2.75 - 2.571281591| \\ &= 0.178718409\end{aligned}$$

Correct to 0 decimal places

$$\begin{aligned}|x_2 - \sqrt[3]{17}| &= |2.582644628 - 2.571281591| \\ &= 0.011363037\end{aligned}$$

Correct to 1 decimal place (almost 2)

$$\begin{aligned}|x_3 - \sqrt[3]{17}| &= |2.571331512 - 2.571281591| \\ &= 0.000049921\end{aligned}$$

Correct to 4 decimal places

$$\begin{aligned}|x_3 - \sqrt[3]{17}| &= |2.571281592 - 2.571281591| \\ &= 0.000000001\end{aligned}$$

Correct to 8 decimal places

This illustrates nicely the quadratic convergence of Newton's method.

EXAMPLE 6
CALCULATING THE ZERO OF A CUBIC POLYNOMIAL

Find the real roots of $p(x) = x^3 + x^2 + 7x - 3$.

SOLUTION: We differentiate to find that $p'(x) = 3x^2 + 2x + 7$. This polynomial has no real roots (explain why) and, since $p'(0) = 7$, we conclude that $p'(x) > 0$ so $p(x)$ is an increasing function and has exactly one real root. It is graphed in Figure 3. Now

$$p(0) = -3 \qquad \text{and} \qquad p(1) = 6$$

so the root is in the interval $[0, 1]$. We compute

$$|p'(x)| = 3x^2 + 2x + 7 \geq 7 = m \text{ on } [0, 1],$$

and

$$|p''(x)| = 6x + 2 \leq 8 = M \text{ on } [0, 1].$$

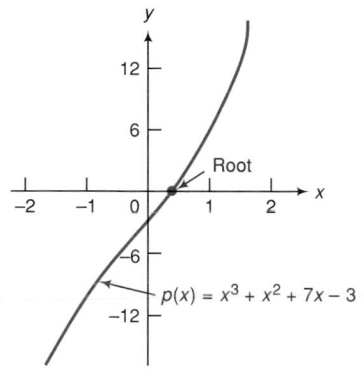

FIGURE 3
GRAPH OF $p(x) = x^3 + x^2 + 7x - 3$
SHOWING A UNIQUE ZERO IN $(0, 1)$

Since

$$\frac{M}{m} = \frac{8}{7} < 4,$$

we conclude that Newton's method converges quadratically to the unique root of $p(x) = 0$. We have

$$
\begin{aligned}
x_{n+1} = x_n - \frac{p(x_n)}{p'(x_n)} &= x_n - \frac{x_n^3 + x_n^2 + 7x_n - 3}{3x_n^2 + 2x_n + 7} \\
&= \frac{x_n(3x_n^2 + 2x_n + 7) - (x_n^3 + x_n^2 + 7x_n - 3)}{3x_n^2 + 2x_n + 7} \\
&= \frac{2x_n^3 + x_n^2 + 3}{3x_n^2 + 2x_n + 7}.
\end{aligned}
$$

If we choose $x_0 = 0$, we obtain the results in Table 3. The root is $s = 0.3970992165$, correct to 10 decimal places.

TABLE 3 SEQUENCE OF ITERATES FOR SOLVING $x^3 + x^2 + 7x - 3 = 0$

n	x_n	$2x_n^3 + x_n^2 + 3$	$3x_n^2 + 2x_n + 7$	$x_{n+1} = \dfrac{2x_n^3 + x_n^2 + 3}{3x_n^2 + 2x_n + 7}$
0	0	3.0	7.0	0.4285714286
1	0.4285714286	3.341107872	8.408163265	0.3973647712
2	0.3973647712	3.283385572	8.268425827	0.3970992352
3	0.3970992352	3.282923214	8.267261878	0.3970992165
4	0.3970992165	3.282923182	8.267261796	0.3970992165

EXAMPLE 7
FINDING THE POINT OF
INTERSECTION OF A SINE CURVE
AND A STRAIGHT LINE BY
NEWTON'S METHOD

Find all roots (if any) of the equation $\sin x = x/2$ in the interval $(0, 2\pi]$.

SOLUTION: The graphs of $y = \sin x$ and $y = x/2$ are given in Figure 4. From these it is evident that there is exactly one real root of the equation $\sin x = x/2$ in $(0, 2\pi]$, and it lies in the interval $[\pi/2, \pi]$.

Setting $f(x) = \sin x - x/2$, we have $f'(x) = \cos x - \frac{1}{2}$, and Newton's method provides the rule

$$x_{n+1} = x_n - \frac{f(x_n)}{f'(x_n)} = x_n - \frac{\sin x_n - x_n/2}{\cos x_n - \frac{1}{2}} = x_n - \frac{2 \sin x_n - x_n}{2 \cos x_n - 1}.$$

Since we know that the root is in $[\pi/2, \pi]$, we can choose this interval to be our interval. We now verify four of the conditions in Theorem 2.

(i) $f(\pi/2) = 1 - \pi/4 > 0$, and $f(\pi) = 0 - \pi/2 < 0$.

(ii) $f'(x) = \cos x - \frac{1}{2}$. $\cos x$ decreases from 0 to -1 in $\left[\dfrac{\pi}{2}, \pi\right]$, so

the smallest value for $|f'(x)| = \left|\cos \dfrac{\pi}{2} - \dfrac{1}{2}\right| = \left|0 - \dfrac{1}{2}\right| = \dfrac{1}{2} = m.$

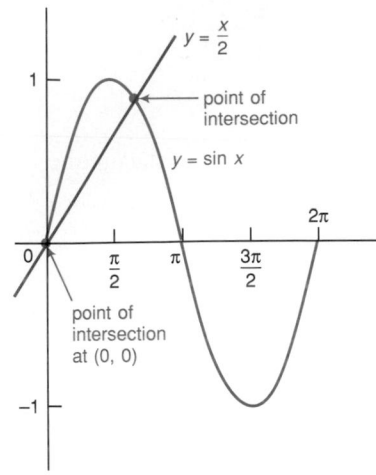

FIGURE 4

GRAPHS OF $f(x) = \dfrac{x}{2}$ AND $f(x) = \sin x$

SHOWING ONE POINT OF INTERSECTION

IN THE INTERVAL $\left(\dfrac{\pi}{2}, \pi\right)$

(iii) $f''(x) = -\sin x$ so $|f''(x)| \leq 1 = M$.

(iv) $\dfrac{M}{m} = \dfrac{1}{\frac{1}{2}} = 2 < 4$.

Thus, the Newton iterates will converge for any $x_0 \in [\pi/2, \pi]$.

We start with $x_0 = 2$ and carry out the iteration given in Table 4. We find that the unique solution (correct to 10 significant figures) is $x_n = 1.895494267$ radians ($\approx 108.6°$).

TABLE 4 SEQUENCE OF ITERATES FOR SOLVING $\sin x - \dfrac{x}{2} = 0$

n	x_n	$\sin x_n$	$2 \sin x_n - x_n$	$\cos x_n$
0	2	0.9092974268	−0.1814051463	−0.4161468365
1	1.900995594	0.9459777536	−0.0090400868	−0.3242315372
2	1.895511645	0.9477415894	−0.0000284662	−0.3190389940
3	1.895494267	0.9477471335	−0.0000000001	−0.3190225241

n	$2 \cos x_n - 1$	$\dfrac{2 \sin x_n - x_n}{2 \cos x_n - 1}$	$x_{n+1} = x_n - \dfrac{2 \sin x_n - x_n}{2 \cos x_n - 1}$
0	−1.832293673	0.0990044058	1.900995594
1	−1.648463074	0.0054839486	1.895511645
2	−1.638077988	0.0000173778	1.895494267
3	−1.638045048	−0.00000000003	1.895494267

REMARK: In performing these calculations yourself, make sure that the calculator is switched to radian mode.

Let us make some remarks about Newton's method. First, if x_n is a solution to $f(x) = 0$, then

$$x_{n+1} = x_n - \frac{f(x_n)}{f'(x_n)} = x_n - 0 = x_n.$$

This explains why, in Tables 1, 3, and 4, the value of x_n did not change after the desired accuracy was reached. Second, it is not necessary to check the conditions of Theorems 1 and 2 every time Newton's method is used. However, failure to do so could result in one or both of the following problems:

(i) The sequence generated by (5) could grow without bound or oscillate, never getting close to a solution.

(ii) The sequence could converge to a solution, but there could be *other* solutions in the interval $[a, b]$ that would be missed. This problem can be avoided if a graph is drawn that gives an idea about where (approximately) the roots are located.

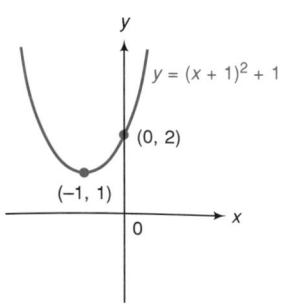

FIGURE 5

GRAPH OF $f(x) = x^2 + 2x + 2 = (x + 1)^2 + 1$

The function is never equal to zero.

Newton's method can be used to solve quickly a wide variety of problems. However, there are circumstances under which it can fail or provide unexpected results. We illustrate here some of the things that can happen if we apply Newton's method without first checking the conditions of the convergence theorem.

EXAMPLE 8
FAILURE OF NEWTON'S METHOD
BECAUSE THERE IS NO ZERO

Find a zero of $p(x) = x^2 + 2x + 2$ by Newton's method.

SOLUTION: Completing the square, we see that

$$p(x) = x^2 + 2x + 2 = (x + 1)^2 - 1 + 2 = (x + 1)^2 + 1 \geq 1$$

so $p(x)$ is never equal to 0. If we blindly apply Newton's method starting with $x_0 = 0$, we obtain the results in Table 5. Here

$$f(x_n) = x_n^2 + 2x_n + 2 \text{ and } f'(x_n) = 2x_n + 2.$$

TABLE 5 SEQUENCE OF ITERATES FOR SOLVING $x^2 + 2x + 2 = 0$
STARTING AT $x_0 = 0$

n	x_n	$x_n^2 - 2$	$2x_n + 2$	$x_{n+1} = \dfrac{x_n^2 - 2}{2x_n + 2}$
0	0	-2	2	-1
1	-1	-1	0	undefined

Then

$$x_{n+1} = x_n - \frac{x_n^2 + 2x_n + 2}{2x_n + 2} = \frac{2x_n^2 + 2x_n - x_n^2 - 2x_n - 2}{2x_n + 2}$$

$$= \frac{x_n^2 - 2}{2x_n + 2}.$$

We stop after two iterations because we cannot divide by 0. Suppose we start somewhere else. For $x_0 = 10$, we obtain the results in Table 6.

We stop after 9 iterations because we aren't getting anywhere. The values oscillate back and forth. You can see what is happening geometrically by looking at Figure 6. If we get close to the value where $p'(x) = 0$

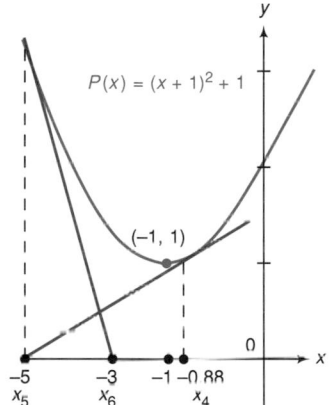

FIGURE 6
STARTING AT THE FOURTH ITERATE, NEWTON'S METHOD BRINGS US UP TO THE CURVE AND THEN DOWN THE CLOSE TO HORIZONTAL TANGENT LINE TO BRING US TO THE NEXT ITERATE $x_5 \approx -5$.

TABLE 6 SEQUENCE OF ITERATES FOR SOLVING $x^2 + 2x + 2 = 0$ STARTING AT $x_0 = 10$

n	x_n	$x_n^2 - 2$	$2x_n + 2$	$x_{n+1} = \dfrac{x_n^2 - 2}{2x_n + 2}$
0	10	98	22	4.454545455
1	4.454545455	17.84297521	10.90909091	1.635606061
2	1.635606061	0.675207185	5.271212121	0.128093343
3	0.128093343	-1.983584896	2.256242884	-0.879153973
4	-0.879153973	-1.227088292	0.241692053	-5.077073378
5	-5.077073378	23.77667408	-8.154146756	-2.915899701
6	-2.915899701	6.502471064	-3.831799401	-1.696975855
7	-1.696975855	0.879727051	-1.393951709	-0.631102961
8	-0.631102961	-1.601709053	0.737794077	-2.170943222

($x = -1$), then the next iterate "jumps" sharply to the left of -1. The next iterate will be closer to -1. When we get close to -1 again, the following iterate will similarly jump. It is apparent that the iterates will not converge.

EXAMPLE 9
SUCCESS OF NEWTON'S METHOD
EVEN WHEN THE HYPOTHESES
OF THE CONVERGENCE
THEOREM ARE NOT SATISFIED

Find a zero of $p(x) = x^2 - 4x + 4$ by Newton's method.

SOLUTION: $p(x) = x^2 - 4x + 4 = (x - 2)^2$.

Evidently $p(2) = 0$ and $p(x) > 0$ for $x \neq 2$. Thus we cannot find an interval $[a, b]$ such that $p(a)p(b) < 0$ (see Figure 7). Moreover, $p'(x) = 2x - 4$ so $p'(2) = 0$ and we cannot find a positive lower bound (m) for $p'(x)$ in any interval containing 2. Thus the convergence theorem (or even Theorem 1) does not apply. Nevertheless, the Newton iterates converge to the unique zero 2 if we start at any real value x_0. Let us see why this is true.

$$p(x) = x^2 - 4x + 4 \qquad p'(x) = 2x - 4$$

so

$$x_{n+1} = x_n - \frac{x_n^2 - 4x_n + 4}{2x_n - 4} = \frac{2x_n^2 - 4x_n - x_n^2 + 4x_n - 4}{2x_n - 4}$$

$$= \frac{x_n^2 - 4}{2x_n - 4} = \frac{(x_n - 2)(x_n + 2)}{2(x_n - 2)} = \frac{x_n + 2}{2} = \frac{x_n}{2} + 1, \qquad x_n \neq 2.$$

That is, as long as $x_n \neq 2$ (so we are not dividing by 0), we have

$$x_{n+1} = \frac{x_n}{2} + 1.$$

Starting with an arbitrary x_0, we obtain

$$x_0 = x_0$$

$$x_1 = \frac{x_0}{2} + 1$$

$$x_2 = \frac{x_1}{2} + 1 = \frac{x_0}{4} + \frac{1}{2} + 1$$

$$x_3 = \frac{x_2}{2} + 1 = \frac{x_0}{8} + \frac{1}{4} + \frac{1}{2} + 1$$

$$x_4 = \frac{x_3}{2} + 1 = \frac{x_0}{16} + \frac{1}{8} + \frac{1}{4} + \frac{1}{2} + 1$$

$$x_5 = \frac{x_4}{2} + 1 = \frac{x_0}{32} + \frac{1}{16} + \frac{1}{8} + \frac{1}{4} + \frac{1}{2} + 1$$

$$\vdots$$

$$x_n = \frac{x_{n-1}}{2} + 1 = \frac{x_0}{2^n} + \frac{1}{2^{n-1}} + \frac{1}{2^{n-2}} + \cdots + \frac{1}{4} + \frac{1}{2} + 1. \qquad (10)$$

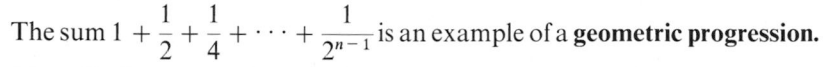

The sum $1 + \dfrac{1}{2} + \dfrac{1}{4} + \cdots + \dfrac{1}{2^{n-1}}$ is an example of a **geometric progression.**

We will discuss geometric progressions in Section 10.3. There, and in Example 4 in Appendix 2, we prove that for any real number $r \neq 1$,

$$1 + r + r^2 + \cdots + r^n = \frac{1 - r^{n+1}}{1 - r}. \tag{11}$$

Inserting $r = \frac{1}{2}$ and using $n - 1$ in place of n in (11), equation (10) becomes

$$x_n = \frac{x_0}{2^n} + \frac{1 - \left(\dfrac{1}{2}\right)^n}{1 - \dfrac{1}{2}} = \frac{x_0}{2^n} + 2\left[1 - \left(\frac{1}{2}\right)^n\right] = 2 + \frac{x_0}{2^n} - \frac{1}{2^{n-1}}.$$

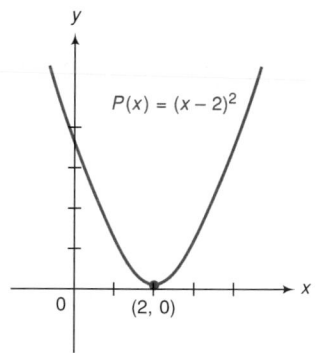

FIGURE 7
$p(x) = x^2 - 4x + 4 = (x - 2)^2$ IS 0 WHEN $x - 2$ AND IS POSITIVE OTHERWISE $p'(0) = 0$ AS WELL.

Now, as $n \to \infty$, both $\dfrac{x_0}{2^n}$ and $\dfrac{1}{2^{n-1}} \to 0$. Thus $x_n \to 2$ as $n \to \infty$. That is, the sequence of Newton iterates converges to the unique solution of $x^2 - 4x + 4 = 0$ for any starting value x_0, even though the convergence theorem does not hold (see Figure 7).

We observe, however, that in this case convergence is not quadratic. To see this, let $\epsilon_n = |x_n - 2|$ denote the error in the nth iterate. We find that

$$\epsilon_n = \frac{x_0}{2^n} - \frac{1}{2^{n-1}} = \frac{x_0 - 2}{2^n} \qquad \text{and} \qquad \epsilon_{n+1} = \frac{x_0 - 2}{2^{n+1}} = \frac{1}{2}\epsilon_n.$$

An iterative variable is said to **converge linearly** if the error is reduced by a constant factor at each iteration. Thus we see here that while Newton's method does converge, it converges only linearly.

Newton's method is one of the most successful numerical techniques ever discovered. Even if you apply it without checking the hypotheses of a convergence theorem, you are likely to get the correct answer very quickly. However there are circumstances under which Newton's method can yield some very surprising results. In Section 3.9 we show how, when a polynomial has more than one zero, applying Newton's method can lead to the phenomenon known as *chaos*.

■ PROBLEMS 3.6

■ SELF-QUIZ

I. Several tactics for approximating a solution to the equation $f(x) = 0$ are outlined below. Which is the tactic used by *Newton's method*?
  a. Find an interval $[a, b]$ such that $f(a)$ and $f(b)$ have opposite signs. Consider the midpoint of that interval, $c = (a + b)/2$. If $f(c) = 0$, then stop. Otherwise, consider the values of f at the endpoints of the subinterval $[a, c]$ and at the endpoints of $[c, b]$: f must change sign on one of those half-length subintervals. Pick that subinterval and repeat this bisection process on it, and stop the iterations when the endpoints of

CONTINUED

the current interval are close enough to each other.

b. Find an interval $[a, b]$ such that $f(a)$ and $f(b)$ have opposite signs. Connect the points $(a, f(a))$, and $(b, f(b))$ with a straight-line segment. Let c be the place where that line meets the t-axis; this c must fall somewhere between a and b. If $f(c)$ is zero, we're done. Otherwise, f must change sign either on $[a, c]$ or $[c, b]$. Pick that subinterval and continue this process,

and stop the iterations when the current interval is short enough.

c. Pick some number a. Construct the line tangent to $y = f(x)$ at the point $(a, f(a))$; let c be the place where that line meets the x-axis. If $f(c)$ turns out to be 0, we are finished. Otherwise, continue by constructing the line tangent at $(c, f(c))$, . . . ; stop when successive choices of c are close to each other.

■ DRILL

1. Find an interval over which Theorem 2 applies to the calculation of $\sqrt{90}$ by solving $x^2 - 90 = 0$.
2. Using Newton's method, find a formula for finding reciprocals without dividing (the reciprocal of x is $1/x$).

▦ CALCULATOR PROBLEMS

In Problems 3–14, repeat the iterative process until successive approximations differ by less than 10^{-8}.

3. Use the result of Problem 1 to choose a good initial approximation for $\sqrt{90}$. Continue using Newton's method.
4. Use the formula you obtained for Problem 2 to approximate $\frac{1}{7}$ and $\frac{1}{81}$.
5. Approximate $\sqrt[4]{25}$ by using Newton's method.
6. Approximate $\sqrt[5]{10}$ by using Newton's method.
7. Use Newton's method to approximate the roots of $x^2 - 7x + 5 = 0$. Compare your results with the values obtained by using the quadratic formula. [*Hint:* It will be necessary to do two separate calculations, using initial guesses in two distinct intervals.]
8. Approximate all solutions of the equation $x^3 - 6x^2 - 15x + 4 = 0$. [*Hint:* Draw a sketch and guess the roots to the nearest integer. Use each

of these guesses as your initial approximation x_0 to begin the Newton's method process.]

9. Approximate all solutions of the equation $x^3 - 8x^2 + 2x - 15 = 0$.
10. Approximate all solutions of the equation $x^3 + 3x^2 - 24x - 40 = 0$.
11. a. Show graphically that there is a unique solution to $\cos x = x$ in the interval $[0, \pi/2]$.
    b. Use Newton's method to approximate it.
*12. Use Newton's method to approximate the unique solution in the interval $(0, 3\pi/2)$ to the equation $x = \tan x$.
*13. Find the unique solution to $\sec x = 3 - x$ in the interval $(0, \pi/2)$.
*14. Find the unique solution to

$$x^2[(1 + (2 - x)^2] - 0.5645(1 + x^2)(2 - x)^2 = 0$$

in the interval $(0, 1)$.

■ SHOW/PROVE/DISPROVE

15. Show that there is no interval $[a, b]$ for which Theorem 2 applies when $f(x) = x^2 + 5x + 7$. Explain why this must be the case. Try Newton's method with $x_0 = 0$. What happens?

16. The graph of $y = x^{1/3}$ passes through the origin. Pick a nonzero initial approximation x_0. Prove or disprove that the Newton's method iteration will converge to 0.

17. Let $f(x) = x^4 - 8x^2 - 17$. A quick sketch indicates the graph of f crosses the x-axis at two places. Examine the sequence produced by Newton's method if we start with $x_0 = 1$. Does either theorem of this section apply?

18. Try to use Newton's method to find a solution to $x = \sin x$ on $[\pi/4, \pi]$. What happens? Which of the hypotheses of Theorem 2 are violated? Why is the method doomed to fail?

CONTINUED

3.7 THE MEAN-VALUE THEOREM

This chapter discusses applications. In this section we state and prove one of the most important and applicable theorems of mathematics, the *mean-value theorem*. We can use the theorem to prove some important facts about differentiation. We can also use the theorem to prove results that are useful in graphing a wide variety of functions. The mean-value theorem has many other applications as well. For example, in Chapter 4, our introductory chapter on integral calculus, the mean-value theorem will be used to prove another very important theorem—the fundamental theorem of calculus.

Before stating and proving the mean-value theorem, we need one preliminary result.

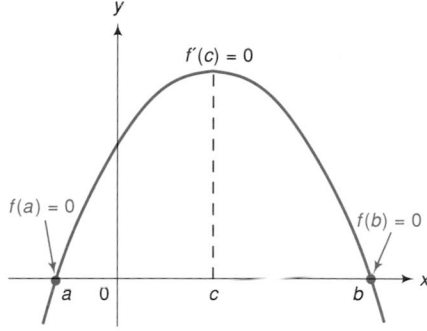

(a) There is one c in (a, b) such that $f'(c) = 0$.

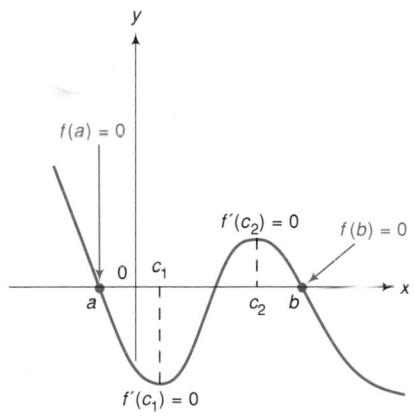

(b) There are two c's in (a, b) such that $f'(c) = 0$.

FIGURE 1
ILLUSTRATION OF ROLLE'S THEOREM

THEOREM 1 **ROLLE'S THEOREM**

Let f be continuous on the closed interval $[a, b]$ and differentiable on the open interval (a, b). If $f(a) = f(b) = 0$, then there exists at least one number c in (a, b) at which

$$f'(c) = 0. \quad ■ \tag{1}$$

REMARK: The situation described in Rolle's theorem is depicted in Figure 1. The proof of the theorem involves describing the situation in the figure.

PROOF OF ROLLE'S THEOREM If $f(x) = 0$ for all x in $[a, b]$, then $f'(x) = 0$ in (a, b) and any c in (a, b) will work. If $f(x)$ is not the zero function, then f must become positive or negative in (a, b). Suppose f takes on some positive values in (a, b), as in Figure 1. (Negative values can be handled in the same way.) By the upper and lower bound theorem (page 119) there is a number x_1 in $[a, b]$ such that $f(x_1) = M$ is the maximum value of $f(x)$ on $[a, b]$. $M > 0$, so x_1 is not equal to a or b, since $f(a) = f(b) = 0$ by hypothesis. Thus, f has a local maximum at x_1, so by Theorem 3.2.2 on page 191, x_1 is a critical point of f. Since $f'(x_1)$ exists by assumption, we have $f'(x_1) = 0$ by the definition of a critical point. ■

EXAMPLE 1
ILLUSTRATION OF
ROLLE'S THEOREM

Let $f(x) = (x - 1)(x - 2)$. Then since $f(1) = f(2) = 0$, there is a point c in $(1, 2)$ such that $f'(c) = 0$. To find this point, we must differentiate: $f'(x) = (x - 1) + (x - 2) = 2x - 3 = 0$ when $x = \frac{3}{2}$. This is illustrated in Figure 2.

---

† Named after the French mathematician Michel Rolle (1652–1719).

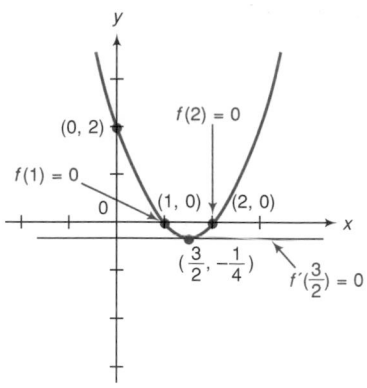

FIGURE 2 ILLUSTRATION OF ROLLE'S THEOREM FOR $f(x) = (x - 1)(x - 2)$

EXAMPLE 2
ROLLE'S THEOREM SHOWS THAT
AN OBJECT THROWN UP WILL BE
STATIONARY AT ONE INSTANT

An interesting physical illustration of Rolle's theorem is given by the following situation: Suppose an object (a ball or rock) is thrown from ground level into the air. Let $s(t)$ denote the height the object reaches at time t. Clearly, $s(0) = 0$ and $s(T) = 0$ at some future time T. (The object hits the ground at time T.) We can then conclude, from Rolle's theorem, that there is a time t_1 such that $s'(t_1) = 0$. That is, there is a time at which the velocity of the object is zero—a physical fact confirmed by experience.

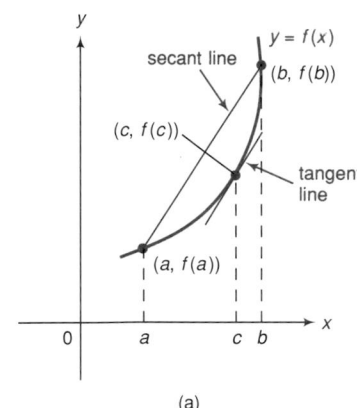

(a)

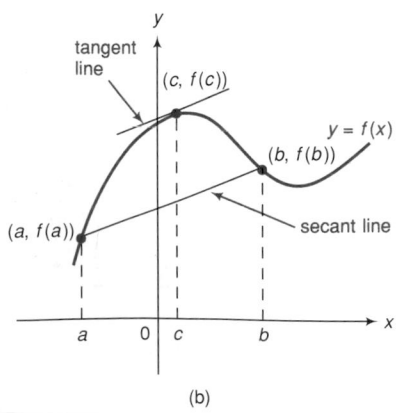

(b)

FIGURE 3
ILLUSTRATION OF MEAN VALUE THEOREM

The secant and tangent lines are parallel.

We are now ready to state the mean-value theorem.

THEOREM 2   **MEAN-VALUE THEOREM**

Let f be continuous on the closed interval $[a, b]$ and differentiable on the open interval (a, b). Then there exists at least one number c in (a, b) such that

$$f'(c) = \frac{f(b) - f(a)}{b - a}. \quad \blacksquare \tag{2}$$

Before proving the mean-value theorem, we describe what it says geometrically. Look at Figure 3. We can see that the expression $[f(b) - f(a)]/[b - a]$ represents the slope of the secant line joining the points $(a, f(a))$ and $(b, f(b))$. Then since $f'(x)$ is the slope of the tangent line to the curve, the mean-value theorem states that there is always a number c between a and b such that the slope of the tangent line at the point $(c, f(c))$ is the same as the slope of the secant line; that is, the secant and tangent lines are parallel.

PROOF OF THE MEAN-VALUE THEOREM: We   define   the function

$$g(x) = f(x) - \left[\frac{f(b) - f(a)}{b - a} (x - a) + f(a) \right].$$

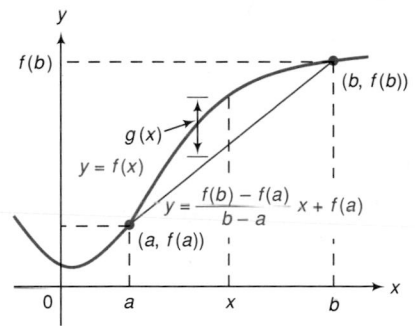

FIGURE 4
GRAPH OF THE STRAIGHT-LINE FUNCTION
$\dfrac{f(b) - f(a)}{b - a} x + f(a)$

The expression

$$y = \frac{f(b) - f(a)}{b - a}(x - a) + f(a)$$

is the equation of a straight line (the secant line discussed above) that passes through the points $(a, f(a))$ and $(b, f(b))$ since, when $x = a$, $y = f(a)$, and when $x = b$, $y = [f(b) - f(a)] + f(a) = f(b)$. See Figure 4. The function $g(x)$ represents the vertical difference between the curve $f(x)$ and the secant line. We first observe that since a linear function is everywhere differentiable (being a first-degree polynomial) and since $f(x)$ is continuous on $[a, b]$ and differentiable on (a, b), $g(x)$ is also continuous on $[a, b]$ and differentiable on (a, b). Moreover,

$$g(a) = f(a) - \left[\frac{f(b) - f(a)}{b - a}(a - a) + f(a)\right] = f(a) - f(a) = 0,$$

and

$$g(b) = f(b) - \left[\frac{f(b) - f(a)}{b - a}(b - a) + f(a)\right]$$
$$= f(b) - [f(b) - f(a) + f(a)] = 0.$$

Thus we may use Rolle's theorem and conclude that there is a number c in (a, b) such that

$$g'(c) = 0.$$

But $g'(x) = f'(x) - [f(b) - f(a)]/(b - a)$, so that since $g'(c) = 0$,

$$g'(c) = f'(c) - \left[\frac{f(b) - f(a)}{b - a}\right] = 0.$$

That is,

$$f'(c) = \frac{f(b) - f(a)}{b - a},$$

and the proof of the mean-value theorem is complete. ■

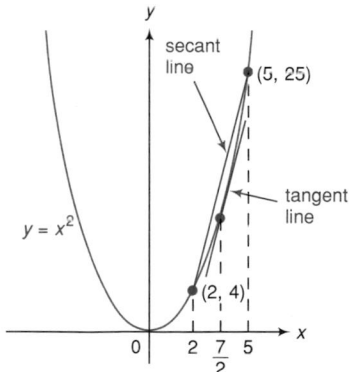

FIGURE 5
ILLUSTRATION OF MEAN-VALUE THEOREM FOR THE FUNCTION $f(x) = x^2$ IN THE INTERVAL [2, 5]

The secant and tangent lines are parallel.

EXAMPLE 3
ILLUSTRATION OF THE
MEAN-VALUE THEOREM

Let $f(x) = x^2$, $a = 2$, and $b = 5$. Then,

$$\frac{f(b) - f(a)}{b - a} = \frac{25 - 4}{3} = \frac{21}{3} = 7.$$

By the mean-value theorem, there is a number c in $(2, 5)$ such that $f'(c) = 7$. Since $f'(x) = 2x$, we have $c = \frac{7}{2}$. (See Figure 5.)

EXAMPLE 4
ILLUSTRATION OF
MEAN-VALUE THEOREM

Let $f(x) = \sin x$, $a = 0$, and $b = \pi/2$. Then

$$\frac{\sin b - \sin a}{b - a} = \frac{1}{\pi/2} = \frac{2}{\pi}.$$

There is a number c in $(0, \pi/2)$ such that $f'(c) = \cos c = 2/\pi \approx 0.6366$, and $c \approx 0.8807$. See Figure 6.

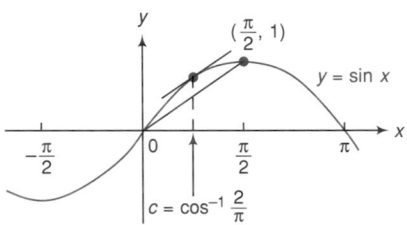

FIGURE 6 ILLUSTRATION OF MEAN-VALUE THEOREM FOR THE FUNCTION $f(x) = \sin x$ IN THE INTERVAL $\left[0, \dfrac{\pi}{2}\right]$

The secant and tangent lines are parallel.

EXAMPLE 5
ILLUSTRATION OF THE
MEAN-VALUE THEOREM FOR A
FUNCTION WHOSE DERIVATIVE
DOES NOT EXIST AT a AND b

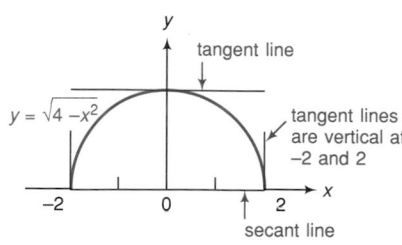

FIGURE 7
ILLUSTRATION OF MEAN-VALUE THEOREM
FOR THE FUNCTION $f(x) = \sqrt{4 - x^2}$, $-2 \le x \le 2$

The secant and tangent lines are parallel.

Let $f(x) = \sqrt{4 - x^2}$, $-2 \le x \le 2$. That is, the graph of f is the upper half of the circle $x^2 + y^2 = 4$. f is continuous in $[-2, 2]$. Moreover, $f'(x) = \dfrac{-2x}{\sqrt{4 - x^2}}$, which is continuous in $(-2, 2)$ but does not exist at 2 or -2. But, it does not matter. We only need f' to exist in $(-2, 2)$ in order to apply the mean-value theorem. We have

$$f(2) - f(-2) = \sqrt{4 - (-2)^2} - \sqrt{4 - 2^2} = 0 - 0 = 0.$$

Thus, there is a c in $(-2, 2)$ such that $f'(c) = 0$. That is,

$$\frac{-2c}{\sqrt{4 - c^2}} = 0 \qquad \text{or} \qquad c = 0.$$

The secant line joining $(-2, 0)$ and $(2, 0)$ is the line segment on the x-axis between -2 and 2. Thus the secant line is horizontal, as is the tangent line at $x = 0$. See Figure 7.

EXAMPLE 6
AN EXAMPLE WHERE THE
MEAN-VALUE THEOREM DOES
NOT APPLY

Let $f(x) = |x|$. Then f is continuous on $[-1, 1]$ but f is not differentiable on $(-1, 1)$ because $f'(0)$ does not exist (see Example 7 on page 91). Now

$$f'(x) = \frac{d}{dx}|x| = \begin{cases} 1, & x > 0 \\ -1, & x < 0. \end{cases}$$

If we try to apply the mean-value theorem, we must find a number c in $(-1, 1)$ such that

$$f(1) - f(-1) = f'(c)(1 - (-1))$$

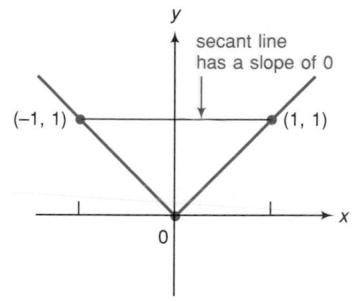

FIGURE 8
THE MEAN-VALUE THEOREM DOES NOT
APPLY TO $f(x) = |x|$ IN $[-1, 1]$ BECAUSE $f'(0)$
DOES NOT EXIST.

*No tangent line to the graph is parallel
to the secant line.*

or

$$|1| - |-1| = 2f'(c)$$
$$0 = 2f'(c)$$
$$f'(c) = 0.$$

But $f'(x)$ is never equal to 0. Thus the conclusion of the mean-value theorem is false. This happens because one of its hypotheses does not hold. See Figure 8.

We now prove, using the mean-value theorem, a theorem stated in Section 2.1.

THEOREM 3 **A FUNCTION WITH A ZERO DERIVATIVE IS CONSTANT**

If f is differentiable on (α, β) and if $f'(x) = 0$ for every x, then $f(x) = c$, a constant function on (α, β).

   PROOF: Pick any two numbers a and b in (α, β), with $a \neq b$. Then there is a number c such that

$$\frac{f(b) - f(a)}{b - a} = f'(c) = 0.$$

Since $b \neq a$, this can only occur if $f(b) - f(a) = 0$, which implies that $f(b) = f(a)$. Since this result is true for any numbers a and b, f is a constant function. ∎

We have proved:

If the derivative of f is zero in (a, b) then f is constant in (a, b).

We now prove the increasing and decreasing theorem, first stated in Section 3.2 (page 189).

THEOREM 4 **INCREASING AND DECREASING THEOREM**

Let f be differentiable for $a < x < b$ and continuous for $a \leq x \leq b$.

 (i) If $f'(x) > 0$ for every x in (a, b), then f is increasing on $[a, b]$.
 (ii) If $f'(x) < 0$ for every x in (a, b), then f is decreasing on $[a, b]$.

   PROOF:

 (i) Suppose that $a \leq x_1 < x_2 \leq b$. We must show that $f(x_2) > f(x_1)$. But since f satisfies the hypothesis of the mean-value theorem on the interval $[x_1, x_2]$, we have

$$\frac{f(x_2) - f(x_1)}{x_2 - x_1} = f'(c),$$

where $x_1 < c < x_2$. Thus,

$$f(x_2) - f(x_1) = f'(c)(x_2 - x_1). \tag{3}$$

But $f'(c) > 0$ by hypothesis and $x_2 > x_1$, so that from (3) we see that

$$f(x_2) - f(x_1) > 0 \qquad \text{or} \qquad f(x_2) > f(x_1).$$

(ii) The proof of this part is almost identical to the proof of part (i) except that now $f'(c) < 0$, so that $f(x_2) - f(x_1) < 0$ and $f(x_2) < f(x_1)$. ∎

EXAMPLE 7
THE LORENZ CURVE

TABLE 1

Fraction of People	Fraction of Income
0.0	0.0
0.2	0.05
0.4	0.16
0.6	0.33
0.8	0.55
1.0	1.00

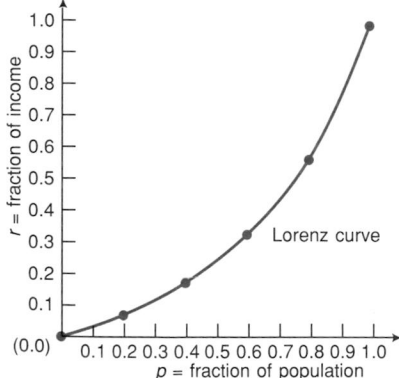

FIGURE 9
THE LORENZ CURVE

We show how the mean-value theorem can be used in a practical problem. Consider Table 1, which shows how total income was distributed among Americans in 1955.[†] In the table the ordering is from smaller incomes to greater incomes. In Figure 9 we have drawn a smooth curve through the points given in Table 1. The curve given in Figure 9 is called a **Lorenz curve**.

Consider the part of the curve between the points $(0.2, 0.05)$ and $(0.4, 0.16)$. We see that the fraction of people between 0.2 and 0.4 (20%) receives $0.11 = 11\%$ of the income. Thus this part of the population receives less than a fair share (20% of the people should, ideally, receive 20% of the income). The ratio $0.11/0.20 = 0.55$ is a measure of fairness and is easily seen to be equal to the slope of the secant line joining the points $(0.2, 0.05)$ and $(0.4, 0.16)$. Notice that things are better for that 20% of the population between 0.6 and 0.8 because they get $0.22 = 22\%$ $(0.55 - 0.33)$ of the income, with a fairness ratio of $0.22/0.20 = 1.1$. If the Lorenz curve is given by $r = f(p)$, then the slope dr/dp is the "instantaneous" fairness ratio—obtained by letting the percentage of the population (Δp) tend to zero.

If there is a point on the Lorenz curve with a slope equal to 1, then the p value of that point is called the **equal-share coefficient** (ESC) and is denoted by ϵ. If $\epsilon = 0.55$, for example, it means that 55% of the population have less than an equal share of the income and 45% have more than an equal share.

It is reasonable to assume that the Lorenz curve is continuous and differentiable. It is also reasonable to assume that the slope of the Lorenz curve is increasing. That is, as we move to the right, equal proportions of the population receive greater proportions of the income (this assumption coincides with the information given in the table). If $r = f(p) = p$, then the slope of the curve is 1, and everyone has an equal share. If $f(p) \neq p$, then by the assumption just stated, $f'(p)$ is an increasing function (as p increases, the fractions of income increase). Thus if there is one value ϵ for which $f'(\epsilon) = 1$, then there is no other value. But $f(0) = 0$, $f(1) = 1$, and by the mean-value theorem,

$$1 = \frac{f(1) - f(0)}{1 - 0} = f'(c), \qquad \text{where } 0 < c < 1.$$

[†] This table was obtained from P. A. Samuelson, *Economics, An Introductory Analysis,* 4th ed. (New York: McGraw-Hill, 1958), page 69.

Thus if $\epsilon = c$, ϵ is the ESC. That is, *every Lorenz curve has an equal-share coefficient.*[†]

NOTE: In analyzing these data, we took a discrete set of points, drew a curve through them, and assumed that the curve was differentiable. This technique is often used to analyze data because it enables us to make use of the power of calculus. In creating this differentiable model, it is necessary at some point to look at the theoretical results obtained and compare them with observed results. If the agreement is a good one, then our model is fine. If not, then modifications must be made.

EXAMPLE 8
COMPUTING AN EQUAL-SHARE
COEFFICIENT

Let $r = p^{3/2}$. Then $dr/dp = \frac{3}{2}p^{1/2}$, which is equal to 1 when $p^{1/2} = \frac{2}{3}$, or $p = \frac{4}{9} \approx 0.44$. Thus $\epsilon = 0.44$, which means that 44% of the population receive less than an equal share.

■ PROBLEMS 3.7

■ SELF-QUIZ

Answer True or False for each of the following.

I. If $f(x) = -3$ on the interval $(2, 6)$, then $f'(x) = 0$ on that interval.

II. If $f(x) = 7$ on the interval $(-5, 25)$, then $f'(x) = 0$ at many points x chosen from that interval.

III. If f is a constant function on the interval $(1, 9)$, then f is differentiable on that interval and the value of its derivative equals zero throughout that interval.

IV. If the function f is differentiable on the interval $(2, 6)$ and if $f'(4) = 0$, then $f(x) = -3$ throughout that interval.

V. If the function f is differentiable on the interval $(2, 6)$ and if $f'(4) = 0$, then f must be a constant function on that interval.

VI. Suppose the function f is differentiable on the interval $(2, 6)$; also suppose $f'(2)$, $f'(3)$, $f'(4)$, $f'(5)$, and $f'(6)$ all equal zero. Then f must be a constant function on the interval $(2, 6)$.

VII. Suppose the function f is differentiable on the interval $(2, 6)$; also suppose $f'(x) = 0$ at many points in that interval. Then f must be a constant function on the interval $(2, 6)$.

VIII. Suppose the function f is differentiable on the interval $(2, 6)$; also suppose $f'(x) = 0$ at every point in that interval. Then f must be a constant function on the interval $(2, 6)$.

IX. Suppose the function f is differentiable on the interval $(2, 6)$; also suppose $f'(x) = 0$ for all x in the interval $(3, 5)$. Then f must be a constant function on the interval $(3, 5)$.

X. Suppose f' exists and equals zero throughout the interval $(2, 6)$; also suppose $f(3) = 25$. Then $f(x) = 25$ for all x in $(2, 6)$.

XI. Suppose f' exists and equals zero throughout the interval $(2, 6)$; also suppose $f(25) = 3$. Then $f(x) = 3$ for all x in $(2, 6)$.

XII. Consider the function $f(x) = 1/x$ on the interval $(-1, 2)$. There is at least one choice of c in the interval $(-1, 2)$ such that

$$f'(c) = \frac{f(2) - f(-1)}{2 - (-1)}.$$

[†] For a much more complete discussion of Lorenz curves and many additional examples, consult Harry M. Schey's paper, "The distribution of resources," UMAP project, Educational Development Center, Newton, Mass. (1978).

CONTINUED

■ DRILL

In Problems 1–12, find a number c that satisfies the conclusion of the mean value theorem for the given function and interval.

1. $f(x) = 1 + 2x^2$; $(-1, 1)$
2. $f(x) = 1/x$; $(1, 4)$
3. $f(x) = \sqrt{x}$; $(1, 4)$
4. $f(x) = \sqrt[3]{x}$; $(-8, -1)$
5. $f(x) = x^3$; $(1, 2)$
6. $f(x) = x^3$; $(-2, 2)$
7. $f(x) = (x + 3)(x - 1)(x - 5)$; $(-3, 1)$
8. $f(x) = x^3 - 3x^2 + 1$; $(0, 3)$
9. $f(x) = \cos x$; $(0, \pi/2)$
10. $f(x) = \tan x$; $(0, \pi/4)$
11. $f(x) = \sin 2x$; $(0, \pi/12)$
12. $f(x) = \cos(2x) - 2\cos^2 x$; $(-\pi/2, \pi/4)$

■ APPLICATIONS

13. A Lorenz curve is given by $r = p^n$. Find an expression for the ESC (equal-share coefficient), ϵ_n, in terms of n.
14. Let $r = \frac{1}{2}p^{3/2}(1 + p)$. Compute the ESC to four decimal places.
15. Two police cars using radar are 7 miles apart on an interstate highway. The first one clocks a car going 65 miles an hour. 6 minutes later the second police car clocks the same car going 55 miles per hour. Assuming that the speed limit is 65 mph, show that the car must have been speeding at least one instant as it drove from one police car to the next. You may assume that the function that gives distance as a function of time is differentiable.
16. Use the result of Problem 27 to show that the only root of $x^5 - 5x = 0$ in the interval $(-1, 1)$ is $x = 0$.
17. Show that the polynomial $p(x) = 3x^4 - 20x^3 + 36x^2 - 1$ has exactly one zero in the interval $[0, 2]$.
18. Consider the parabola $y = f(x) = \alpha x^2 + \beta x + \gamma$, $\alpha \neq 0$; the hypotheses of the mean-value theorem are satisfied for this f on any interval (a, b). Verify that the conclusion of that theorem, equation (2), is satisfied if we choose $c = (a + b)/2$, the midpoint of the interval.
19. Let $f(x) = |x - 1| - 2$.
    a. Verify that $f(3) = f(-1) = 0$.
    b. Verify that there is no number c in $(-1, 3)$ such that $f'(c) = 0$.
    c. Explain why the result of parts (a) and (b) does not contradict Rolle's theorem.
20. Verify that there is no number c in the interval $[-1, 1]$ that satisfies the mean-value theorem for the function $f(x) = x^{2/3}$.

■ SHOW/PROVE/DISPROVE

21. Prove that if the conditions of the mean-value theorem hold and if $f(a) = f(b)$, there is a point c in (a, b) such that $f'(c) = 0$.
22. Suppose $f''(x)$ exists for every x; show that between any two points at which $f'(x) = 0$, there is a point at which f'' is 0.
23. Use the mean-value theorem to show that if f is differentiable and $f'(x) = c$, a nonzero constant, then there is a constant d such that $f(x) = cx + d$ for every x.
*24. Use Rolle's theorem to show that a quadratic polynomial has at most two real roots. [Hint: Show that if there are three real roots, then there are two real roots to the equation $P'(x) = 0$ and one root to the equation $P''(x) = 0$, which is impossible.]
*25. Show that a cubic polynomial has at most three real roots. [Hint: Follow steps similar to those outlined for the preceding problem.]

A function is **one-to-one** if and only if $f(x) = f(y)$ implies $x = y$.

26. Use the result of Problem 21 to show that if f is differentiable and $f'(x) \neq 0$ for every real x, then f is one-to-one.
27. Let $p(x) = c_0 + c_1 x + c_2 x^2 + \cdots + c_n x^n$. Suppose that $p'(a) = p'(b) = 0$ and $p'(x) \neq 0$ for all x in (a, b); prove there is at most one number r in (a, b) such that $p(r) = 0$.
*28. Let $p(x) = x^n + ax + b$, $n \geq 0$. Show that
    a. if n is an even integer, then p cannot have more than two real zeros;
    b. if n is an odd integer, then p cannot have more than three real zeros.
*29. Use mathematical induction (Appendix 2) to show that an nth degree polynomial has at most n real zeros. (This generalizes Problems 24 and 25.)

*30. Use the mean-value theorem to show that if f is differentiable and $f'(x)$ is a linear function, then f is a quadratic polynomial.

31. Discuss the following false proof of Problem 30. Suppose $f'(x) = mx + b$. Apply the mean-value theorem to f on the interval $[0, w]$. It implies there is some number w_1 in that interval such that

$$mw_1 + b = f'(w_1) = \frac{f(w) - f(0)}{w - 0}.$$

Therefore, $f(w) = (mw_1 + b) \cdot w + f(0)$, which is quadratic in w if $w_1 = w$.

32. Use the mean-value theorem to show that if f is differentiable and $f'(x)$ is a polynomial of degree n, then $f(x)$ is a polynomial of degree $n + 1$. (This generalizes Problem 23 and 30.)

33. Prove **Bernoulli's inequality**: If $\alpha > 1$ and $x > 0$, then $(1 + x)^\alpha > 1 + \alpha x$. [*Hint:* Define $f(x) = (1 + x)^\alpha - (1 + \alpha x)$. First show that $f(0) = 0$ and

then apply the mean-value theorem on the arbitrary interval $[0, b]$.]

34. Suppose that f has n continuous derivatives on $[a, b]$ and $f = 0$ at $n + 1$ or more points in $[a, b]$. Show that there is at least one number c in (a, b) such that $f^{(n)}(c) = 0$.

A function f is called a **contraction** on $[a, b]$ if and only if there exists some number λ such that $\lambda < 1$ and

$$|f(x) - f(w)| \le \lambda |x - w|$$

for any x and w in $[a, b]$.

35. Show that $f(x) = 1/x$ is a contraction on any interval of the form (a, ∞), where $a > 1$.

36. Show that $f(x) = \sin x$ is not a contraction on $[0, \pi]$.

37. For what values of b is the function $f(x) = x^2 + 1$ a contraction on $[0, b]$?

38. Show that if f is differentiable and $|f'(x)| \le \lambda < 1$ for every x in $[a, b]$, then f is a contraction on $[a, b]$.

■ CHALLENGE

*39. Suppose that C and S are two functions such that $C' = -S$ and $S' = C$.
   a. Prove that $C^2 + S^2$ is constant.
   b. Prove that $[C(x) - \cos x]^2 + [S(x) - \sin x]^2$ is also constant.
   c. Suppose $C(0) = 1$ and $S(0) = 0$. Prove that $C(x) = \cos x$ and $S(x) = \sin x$ for all x.

*40. Pick a real number c. Characterize the set of all polynomial functions P that have the properties
   a. $P(1) = c$.
   b. $P(w + x) = P(w) + P(x)$ for all real numbers w and x.

*41. Let f be differentiable with $f'(x) \le 2$ for every real number x. Show that there is at most one real number $s \ge 1$ such that $s^2 = f(s)$.

*42. Let f be differentiable with $f'(x) \le 3$ for every real number x. Show that there is at most one real number $t \ge 1$ such that $t^3 = f(t)$.

*43. Let f be differentiable on $[a, b]$. If $f'(a)$ and $f'(b)$ have opposite signs, show that there is at least one number c in (a, b) at which $f'(c) = 0$. (Do *not* assume that f' is continuous.)

44. **Intermediate-Value Property for Derivatives**: Use the result of the preceding problem to show that if f is differentiable on $[a, b]$ and d is any number between $f'(a)$ and $f'(b)$, then there is a number x_0 in (a, b) such that $f'(x_0) = d$. (Do *not* assume that f' is continuous.)

■ ANSWERS TO SELF-QUIZ

I. True	II. True	III. True	VIII. True	IX. True
IV. False	V. False	VI. False	X. True	XI. False
VII. False (many $\ne$ all)			XII. False [$f'(0)$ does not exist]	

3.8 AN APPLICATION OF THE MEAN-VALUE THEOREM: THE LINEARIZATION OF A FUNCTION

In this section we show how a function can be approximated by a linear function. In what follows, we assume that f, f', and f'' exist in some open interval containing x_0.

DEFINITION   **THE LINEARIZATION OF f AT x_0**

The function

$$P_1(x) = f(x_0) + f'(x_0)(x - x_0) \tag{1}$$

is called the **linearization of f at x_0**. The notation $P_1(x)$ indicates that the linearization of f is a polynomial of degree 1. ∎

EXAMPLE 1
THE LINEARIZATION OF $\sin x$

Find the linearization of $\sin x$ at 0.

SOLUTION: $f(0) = \sin 0 = 0$ and $f'(0) = \cos 0 = 1$ so, from (1),

$$P_1(x) = x$$

is the linearization of $\sin x$ at 0.

EXAMPLE 2
THE LINEARIZATION OF $\sqrt{x}$

Find the linearization of $\sqrt{x}$ at 1.

SOLUTION: $f(1) = \sqrt{1} = 1$ and $f'(1) = \dfrac{1}{2\sqrt{1}} = \dfrac{1}{2}$, so

$$P_1(x) = 1 + \frac{1}{2}(x - 1) = \frac{1}{2} + \frac{1}{2}x$$

is the linearization of $\sqrt{x}$ at 1.

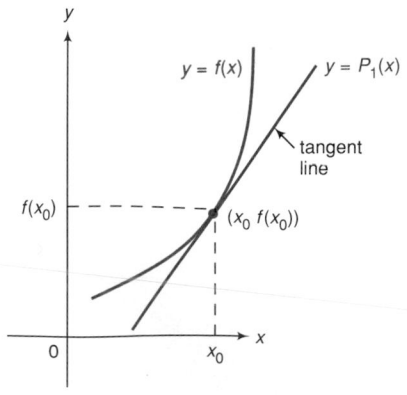

FIGURE 1
THE LINEARIZATION OF f AT x_0 IS THE FUNCTION WHOSE GRAPH IS TANGENT TO THE GRAPH OF f AT $(x_0, f(x_0))$.

THE GEOMETRIC INTERPRETATION OF THE LINEARIZATION OF A FUNCTION

The tangent line to the graph of f at the point $(x_0, f(x_0))$ has the slope $f'(x_0)$. The equation of the line is

$$\frac{y - f(x_0)}{x - x_0} = f'(x_0)$$

or

$$y = f(x_0) + f'(x_0)(x - x_0).$$

We see that the linearization of a function is the straight-line function whose graph is tangent to the graph of f at the point $(x_0, f(x_0))$. We illustrate this in Figure 1.

THE ERROR IN THE LINEAR APPROXIMATION

The linearization of a function is a linear function that approximates the given function. The question is, how good is the approximation? As we see in Figure 2, the error is the difference between $f(x)$ and $P_1(x)$. It seems, at least from the graph, that the error gets smaller as $x \to x_0$. The next theorem shows that this is true by giving a precise value for the error.

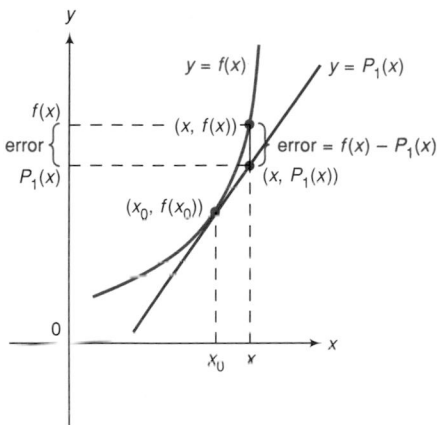

FIGURE 2 THE ERROR IN LINEAR APPROXIMATION IS THE DIFFERENCE BETWEEN $f(x)$ AND $P_1(x)$.

THEOREM 1 **ERROR OF THE LINEAR APPROXIMATION**

Suppose that $f''(x)$ exists in a neighborhood (a, b) of x_0 and that $x \in (a, b)$. Let $P_1(x)$ denote the linearization of f at x_0. Then there is a number c between x_0 and x ($x_0 < c < x$ or $x < c < x_0$) such that

$$f(x) = P_1(x) + \frac{1}{2} f''(c)(x - x_0)^2. \tag{2}$$

The term $\frac{1}{2} f''(c)(x - x_0)^2$ is the error obtained when $f(x)$ is approximated by $P_1(x)$. The value of c depends on the choice of x.

PROOF: Treat $x \neq x_0$ as a fixed number and define

$$g(t) = f(t) + f'(t)(x - t) + \frac{f(x) - P_1(x)}{(x - x_0)^2} (x - t)^2 - f(x).$$

Step 1

$$g(x_0) = \underbrace{f(x_0) + f'(x_0)(x - x_0)}_{P_1(x)} + \frac{f(x) - P_1(x)}{(x - x_0)^2} (x - x_0)^2 - f(x)$$

$$= P_1(x) + f(x) - P_1(x) - f(x) = 0$$

Step 2

$$g(x) = f(x) + f'(x)(x - x) + \frac{f(x) - P_1(x)}{(x - x_0)^2}(x - x)^2 - f(x)$$

$$= f(x) - f(x) = 0$$

Step 3

Use Rolle's theorem in the interval $[x_0, x]$ or $[x, x_0]$. Clearly g is differentiable because it is the sum of products of differentiable functions. Thus, there is a c in (x_0, x) or (x, x_0) such that $g'(c) = 0$. But remembering that x is constant (so $dx/dt = 0$),

Product rule to find $\dfrac{d}{dt} f'(t)(x - t)$

$$g'(t) = f'(t) + \overbrace{f''(t)(x - t) + f'(t)(-1)} + \frac{f(x) - P_1(x)}{(x - x_0)^2} 2(x - t)(-1)$$

$$\frac{d}{dt}(x - t) = -1$$

and

$$0 = g'(c) = f'(c) + f''(c)(x - c) - f'(c) - 2\frac{f(x) - P_1(x)}{(x - x_0)^2}(x - c)$$

$$2\frac{f(x) - P_1(x)}{(x - x_0)^2}(x - c) = f''(c)(x - c)$$

$$f(x) - P_1(x) = \frac{1}{2}(x - x_0)^2 f''(c),$$

which is what we wanted to prove. ■

The value of Theorem 1 seems limited because we don't know what the number c is (just as in the statement of the mean-value theorem). However, we can say much more if f'' is *bounded* in $[x_0, x]$ or $[x, x_0]$. Suppose f'' is continuous in $[x_0, x]$ or $[x, x_0]$. Then, according to the upper and lower bound theorem (Theorem 1.7.4 on page 119), there is a number M such that

$$|f''(t)| \le M \qquad \text{for } t \in [x_0, x] \text{ or } [x, x_0]. \tag{3}$$

COROLLARY TO THEOREM 1	**MAXIMUM VALUE FOR THE ERROR IN LINEAR APPROXIMATION**

If f'' is continuous in $[x_0, x]$ or $[x, x_0]$, then

$$|f(x) - P_1(x)| \le \frac{M}{2}(x - x_0)^2 \tag{4}$$

where M is as given in (3).

This result follows directly from (2) and (3). ■

EXAMPLE 3
THE ERROR IN THE LINEARIZATION
OF sin x

Estimate sin 0.1 using the linear approximation for sin x at 0.

SOLUTION: From Example 1, we have $P_1(x) = x$. Thus,

$$\sin 0.1 \approx P_1(0.1) = 0.1.$$

How good is our approximation? Since $f(x) = \sin x$, $f'(x) = \cos x$, and $f''(x) = -\sin x$, it follows that $|f''(x)| = |\sin x| \le 1$. Hence, $M \le 1$ in (4), $x_0 = 0$, and

$$|f(x) - P_1(x)| = |\sin x - x| \le \frac{1}{2}x^2.$$

If $x = 0.1$, then $\frac{1}{2}x^2 = \frac{1}{2}(0.01) = 0.005$.

Therefore, we know that sin 0.1 $\approx$ 0.1 with a maximum error of 0.005. To 10 decimal places, sin 0.1 = 0.0998334166, so our actual error is $0.1 - 0.0998334166 \approx 0.00016658$, which is a great deal less than 0.005. The reason our maximum error is so much larger than the actual error is that we used the value $M = 1$ rather than the true value sin c with $0 < c < 0.1$. In the interval $[0, 0.1]$ the maximum value of sin x is sin 0.1, which as we have seen, is much less than 1.

REMARK: Since $0 < c < 0.1$, we have $0 < c < \dfrac{\pi}{6}$, so $\sin c < \sin \dfrac{\pi}{6} = \dfrac{1}{2}$. Thus, we could use the value $M = \frac{1}{2}$ in (4) to get a smaller maximum error.

EXAMPLE 4
THE ERROR IN THE LINEARIZATION
OF $\sqrt{x}$

Estimate $\sqrt{0.95}$ using the linearization of $\sqrt{x}$ at 1.

SOLUTION: From Example 2, $P_1(x) = \frac{1}{2} + \frac{1}{2}x$. Thus,

$$\sqrt{0.95} \approx \frac{1}{2} + \frac{1}{2}(0.95) = 0.5 + 0.475 = 0.975.$$

If $f(x) = \sqrt{x}$, then

$$f'(x) = \frac{1}{2\sqrt{x}} = \frac{1}{2}x^{-1/2} \qquad \text{and} \qquad f''(x) = -\frac{1}{4}x^{-3/2} = -\frac{1}{4x^{3/2}}.$$

How large can this get? As x decreases, $\dfrac{1}{x^{3/2}}$ increases. The largest number whose square root we can easily obtain without a calculator and that is less than 0.95 is 0.81 ($\sqrt{0.81} = 0.9$). Thus,

$$\left| -\frac{1}{4x^{3/2}} \right| \le \frac{1}{4(0.81)^{3/2}} = \frac{1}{4(0.9)^3} = \frac{1}{4(0.729)} \approx 0.34 \ge M,$$

and

$$\left| \sqrt{x} - P_1(x) \right| \le \frac{0.34}{2}(x - 1)^2 = 0.17(x - 1)^2.$$

With $x = 0.95$, we have

$$\left|\sqrt{0.95} - P_1(0.95)\right| \leq 0.17(-0.05)^2 = 0.000425.$$

Therefore, $\sqrt{0.95} \approx 0.975$ with a maximum error of 0.000425. The actual value of $\sqrt{0.95}$ is 0.974679434, so the actual error is $0.975 - 0.974679434 = 0.000320566$. This is fairly close to the maximum error computed above.

In Sections 9.4 and 9.5 we will show how to approximate a function by a polynomial of any degree.

PARTIAL PROOF OF THE CONVERGENCE THEOREM FOR NEWTON'S METHOD

Theorem 2 can be used to prove, under certain restrictions, the convergence theorem for Newton's method stated on page 231. We restate it here:

THEOREM 2   **CONVERGENCE THEOREM FOR NEWTON'S METHOD**

Let f, f', and f'' be continuous on $[a, b]$. Suppose that

(i) $f(a)$ and $f(b)$ have different signs.
(ii) $|f'(x)| \geq m > 0$.
(iii) $|f''(x)| \leq M$.
(iv) $\dfrac{M}{m} \leq 4$.
(v) $x_0 \in [a, b]$.

Then,

(a) The sequence of Newton iterates $x_0, x_1, x_2, \ldots$ converges to the unique solution, s, of $f(x) = 0$ in $[a, b]$.
(b) The convergence is quadratic. That is, each iterate is accurate to twice as many decimal places as the one that precedes it.

PROOF: According to Theorem 3.6.1 on page 230 and conditions (i) and (ii), there is a unique number s in (a, b) such that $f(s) = 0$. Let ϵ_n denote the error of the nth iterate. That is,

$$\epsilon_n = |x_n - s|. \tag{5}$$

Suppose that x_n is accurate to k decimal places. Then,

$$\epsilon_n \leq \frac{1}{2} 10^{-k}. \tag{6}$$

We show that

$$\epsilon_{n+1} \leq \frac{1}{2} 10^{-2k}.$$

That is, x_{n+1} is accurate to $2k$ decimal places. We have

$$\epsilon_{n+1} = |s - x_{n+1}| = \left| s - \left(x_n - \frac{f(x_n)}{f'(x_n)} \right) \right| = \left| s - x_n + \frac{f(x_n)}{f'(x_n)} \right|. \quad \text{(7)}$$

The denominator in the last expression is nonzero by hypothesis (ii). From Theorem 1, the theorem on the error in linear approximation,

$$f(x) = f(x_0) + f'(x_0)(x - x_0) + \frac{1}{2}f''(c)(x - x_0)^2, \quad \text{(8)}$$

where c is between x_0 and x. The assumptions of Theorem 1 hold by assumption on $[a, b]$. If we substitute s for x and x_n for x_0 in (8), and note that $f(s) = 0$, we obtain

$$0 = f(s) = f(x_n) + f'(x_n)(s - x_n) + \frac{1}{2}f''(c_n)(s - x_n)^2, \quad \text{(9)}$$

where c_n is between x_n and s. Dividing both sides of (9) by $f'(x_n) \neq 0$ and rearranging terms, we obtain

$$0 = \frac{f(x_n)}{f'(x_n)} + s - x_n + \frac{1}{2}\frac{f''(c_n)}{f'(x_n)}(s - x_n)^2$$

$$s - x_n + \frac{f(x_n)}{f'(x_n)} = -\frac{1}{2}\frac{f''(c_n)}{f'(x_n)}(s - x_n)^2$$

$$\left| s - x_n + \frac{f(x_n)}{f'(x_n)} \right| = \frac{1}{2}\left| \frac{f''(c_n)}{f'(x_n)} \right| |s - x_n|^2$$

and, using (5) and (7),

$$\epsilon_{n+1} = \frac{1}{2}\left| \frac{f''(c_n)}{f'(x_n)} \right| \epsilon_n^2. \quad \text{(10)}$$

But, using hypotheses (ii), (iii), and (iv), $|f''(c_n)| \leq M$, $|f'(x_n)| \geq m$, and $\frac{M}{m} \leq 4$, so (10) becomes

$$\epsilon_{n+1} \leq \frac{1}{2}\frac{M}{m}\epsilon_n^2 \leq \frac{1}{2}4\epsilon_n^2 = 2\epsilon_n^2. \quad \text{(11)}$$

Finally, from (6) we conclude that

$$\epsilon_{n+1} \leq 2\left(\frac{1}{2}10^{-k} \right)^2 = 2 \cdot \frac{1}{4} \cdot 10^{-2k} = \frac{1}{2}10^{-2k},$$

which is what we wanted to prove.

This is not a proof of convergence for two reasons. First, we have not given a formal mathematical definition of convergence. This will have to wait until Section 10.1. Second, in the proof we assume that some x_n is accurate to at least one decimal place. This requires proof. However, this is not a serious problem because of (11). Suppose, for example, that $\epsilon_n < \frac{1}{3}$ for some n. Then, from (11),

$$\epsilon_{n+1} \le 2\epsilon_n^{\ 2} \le \frac{2}{9}$$

$$\epsilon_{n+2} \le 2\epsilon_{n+1}^{\ 2} \le \frac{8}{81} < 0.1$$

$$\epsilon_{n+3} \le 2\epsilon_{n+2}^{\ 2} < 2(0.01) = 0.02 < \frac{1}{2}10^{-1}$$

Thus, x_{n+3} is accurate to at least one decimal place and the accuracy doubles from here on. So, in practical terms, if we can start with an x_0 within $\frac{1}{3}$ of a unit from s, then x_3 will be accurate to at least one decimal place, x_4 to at least 2, x_5 to at least 4, and x_6 to at least 8. Of course, the convergence holds for any choice of x_0 in $[a, b]$. We just haven't proved it if $|x_0 - s| \ge \frac{1}{3}$. ■

■ PROBLEMS 3.8

■ SELF-QUIZ

I. The linearization of $\cos x$ at 0 is
   a. $-\sin x$ b. $1 - x$
   c. 1   d. $-x$
   e. none of these

II. The linearization of $\sin x$ at $\dfrac{\pi}{2}$ is

   a. $1 + \left(x - \dfrac{\pi}{2}\right)$ b. $1 + x$

   c. $1 - x$ d. 1
   e. -1

III. The linearization of $\dfrac{1}{x}$ at 0 is

   a. 1   b. $1 - x$
   c. $-x$ d. does not exist
   e. -1

■ DRILL

In Problems 1–10, (a) find the linearization of the given function at the specified point. (b) Use that linearization to approximate the requested quantity. (c) Apply the corollary to Theorem 1 to obtain an upper bound on the error of your approximation.

1. x^6 at 1; $(1.1)^6$ 2. x^{-4} at 1; $(1.1)^{-4}$
3. $\sqrt{x}$ at 4; $\sqrt{4.2}$

4. $\sqrt[3]{x}$ at 1; $\sqrt[3]{1.1}$
5. $(1 + x)/(1 - x)$ at 0; $1.05/0.95$
6. $\sqrt{1 + x^2}$ at 0; $\sqrt{1.04}$
7. $\cos x$ at 0; $\cos 0.1$
8. $\tan x$ at 0; $\tan 0.1$
9. $\sin x$ at π; $\sin(\pi - 0.05)$
10. $\cos x$ at π; $\cos(\pi - 0.1)$

■ SHOW/PROVE/DISPROVE

11. Let $f(x) = ax + b$. Show that the linearization of f is f itself.
12. Suppose that $L(x)$ is a function that is defined and differentiable for $x > 0$ and that satisfies

    (i) $L(1) = 0$ and (ii) $L'(x) = \dfrac{1}{x}$.

a. What is the linearization of L at 1?
b. Use the linearization of L at 1 to approximate $L(1.1)$ and $L(0.8)$.
c. What is the maximum error of each approximation in (b)?

CONTINUED

■ CHALLENGE

**13. Quadratic Approximation: Suppose that f, f', f'', and f''' are continuous in $[a, b]$ and let x_0 be in (a, b). Show that for every x in $[a, b]$ there is a number c between x_0 and x such that

$$f(x) = f(x_0) + f'(x_0)(x - x_0)$$
$$+ f''(x_0)\frac{(x - x_0)^2}{2} + f'''(c)\frac{(x - x_0)^3}{6}.$$

The polynomial

$$P_2(x) = f(x_0) + f'(x_0)(x - x_0) + f''(x_0)\frac{(x - x_0)^2}{2}$$

is called the **quadratic approximation** to f.

14. Under the assumptions of Problem 13, show that there is a positive number K such that, for every x in $[a, b]$,

$$\left| f(x) - P_2(x) \right| < \frac{K}{6}(x - x_0)^3.$$

In Problems 15–24, (a) find the quadratic approximation of the given function at the specified point. (b) Use that approximation to estimate the requested quantity. (c) Use the result of Problem 14 to obtain an upper bound on the error of your estimate. (d) Compare your estimates with the approximations obtained in Problems 1–10.

15. x^6 at 1; $(1.1)^6$
16. x^{-4} at 1; $(1.1)^{-4}$
17. $\sqrt{x}$ at 4; $\sqrt{4.2}$
18. $\sqrt[3]{x}$ at 1; $\sqrt[3]{1.1}$
19. $(1 + x)/(1 - x)$ at 0; $1.05/0.95$
20. $\sqrt{1 + x^2}$ at 0; $\sqrt{1.04}$
21. $\cos x$ at 0; $\cos 0.1$
22. $\tan x$ at 0; $\tan 0.1$
23. $\sin x$ at π; $\sin(\pi - 0.05)$
24. $\cos x$ at π; $\cos(\pi - 0.1)$

■ ANSWERS TO SELF-QUIZ

I. c   II. d

III. d $\left(\dfrac{1}{x} \text{ is not defined at } 0 \right)$

▦ 3.9 NEWTON'S METHOD AND CHAOS (OPTIONAL)

In this section we illustrate one of the things that can happen when Newton's method is applied to a polynomial that has several real zeros.

EXAMPLE 1
NEWTON'S METHOD MAY LEAD TO CHAOS WHEN THERE IS MORE THAN ONE ZERO

Find all zeros of $p(x) = x^3 - x$ by Newton's method.

SOLUTION:

$$p(x) = x^3 - x = x(x^2 - 1) = x(x + 1)(x - 1)$$

so the three zeros of $p(x)$ are -1, 0, and 1. The polynomial is graphed in Figure 1. We compute

$$p'(x) = 3x^2 - 1$$

$$p''(x) = 6x$$

We see that p' has zeros when $3x^2 = 1$ or when $x = \pm\dfrac{1}{\sqrt{3}} \approx \pm 0.577$.

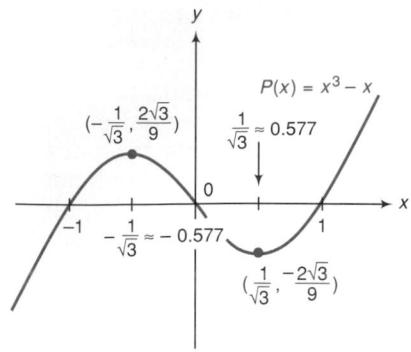

FIGURE 1
$p(x) = x^3 - x$ HAS ZEROS AT $-1, 0,$ AND 1.

TABLE 1 SOLVING $x^3 - x = 0$
USING NEWTON'S METHOD
STARTING AT $x_0 = 5$

n	x_n	$x_{n+1} = \dfrac{2x_n^3}{3x_n^2 - 1}$
0	5	
1	3.378378378	
2	2.320008885	
3	1.648781228	
4	1.252803154	
5	1.060412151	
6	1.004798914	
7	1.000034162	
8	1.000000002	
9	1.000000000	

We can find intervals containing each of the zeros over which the hypotheses of the convergence theorem (Theorem 2 on page 231) hold. For example, consider the interval $[0.8, 1.5]$.

$$p(0.8) = 0.8^3 - 0.8 = -0.288$$

and

$$p(1.5) = 1.5^3 - 1.5 = 1.875.$$

The function $3x^2 - 1$ is increasing for $x > 0$ (explain why), so

minimum value of $3x^2 - 1$ in $[0.8, 1.5]$ occurs at $x = 0.8$

or

$$m = 3(0.8)^2 - 1 = 0.92.$$

Also, $6x$ is increasing, so

maximum value of $6x$ in $[0.8, 1.5] = 6(1.5) = 9 = M.$

But $\dfrac{M}{m} = \dfrac{9}{0.92} \approx 9.8 > 4,$ so the hypotheses are not satisfied. We can bring down the value of $\dfrac{M}{m}$ by making the interval smaller. If $[a, b] = [0.95, 1.1]$, for example, then the convergence theorem applies:

$$p(0.95) = -0.092625 \qquad \text{and} \qquad p(1.1) = 0.231.$$

Also,

$$m = 3(0.95)^2 - 1 = 1.7075, \quad M = 6(1.1) = 6.6,$$

and

$$\frac{M}{m} = \frac{6.6}{1.7075} \approx 3.8653.$$

Thus we know that the sequence of Newton iterates converges to 1, the unique zero in $[0.95, 1.1]$ if x_0 is chosen in that small interval.

This is not very useful. So let us instead simply apply Newton's method and see what happens:

$$x_{n+1} = x_n - \frac{p(x_n)}{p'(x_n)} = x_n - \frac{x_n^3 - x_n}{3x_n^2 - 1} = \frac{3x_n^3 - x_n - x_n^3 + x_n}{3x_n^2 - 1}$$

or

$$x_{n+1} = \frac{2x_n^3}{3x_n^2 - 1}. \tag{1}$$

We begin our exploration by starting to the right of 1, say $x_0 = 5$. The results are given in Table 1. Thus, starting at $x_0 = 5$, the Newton iterates converge to the closest zero. In fact (see Problem 1), it is possible to show that if $x_0 > 1$, then the sequence of Newton iterates converges to 1.

Things get more interesting for $0 < x_0 < 1$. For example, if $x_0 = 0.5$, then

$$x_1 = \frac{2(0.5)^3}{3(0.5)^2 - 1} = \frac{2(\frac{1}{8})}{3(\frac{1}{4}) - 1} = \frac{\frac{1}{4}}{-\frac{1}{4}} = -1.$$

Thus, starting halfway between the zeros 0 and 1, we converge in one iteration to the third zero at -1! The reason this can happen is that 0.5 is close to a value where the derivative is zero $\left(\text{at } \frac{1}{\sqrt{3}} \approx 0.577 \right)$ and when we divide by something close to zero, we obtain relatively large numbers.

But things gets stranger. In Tables 2, 3, and 4 we give the result of Newton's method starting at $x_0 = 0.45$, 0.445, and 0.4475, respectively.

TABLE 2 NEWTON'S METHOD APPLIED TO $x^3 - x = 0$, STARTING AT $x_0 = 0.450$

n	$x_n \left[x_{n+1} = \dfrac{2x_n^3}{3x_n^2 - 1} \right]$
0	0.45
1	-0.46433121
2	0.566899062
3	-10.15637817
4	-6.79286978
5	-4.561531938
6	-3.090531005
7	-2.134858348
8	-1.535544948
9	-1.192242576
10	-1.038317582
11	-1.002021756
12	-1.000006102
13	-1.000000000

TABLE 3 NEWTON'S METHOD APPLIED TO $x^3 - x = 0$, STARTING AT $x_0 = 0.445$

n	$x_n \left[x_{n+1} = \dfrac{2x_n^3}{3x_n^2 - 1} \right]$
0	0.445
1	-0.434174416
2	0.376751651
3	-0.186273694
4	0.014428526
5	-0.000006011
6	0.000000000

TABLE 4 NEWTON'S METHOD APPLIED TO $x^3 - x = 0$, STARTING AT $x_0 = 0.4475$

n	$x_n \left[x_{n+1} = \dfrac{2x_n^3}{3x_n^2 - 1} \right]$
0	0.4457
1	-0.448936158
2	0.457700266
3	-0.516152422
4	1.369893017
5	1.110518467
6	1.01457274
7	1.000308075
8	1.000000142
9	1.000000000

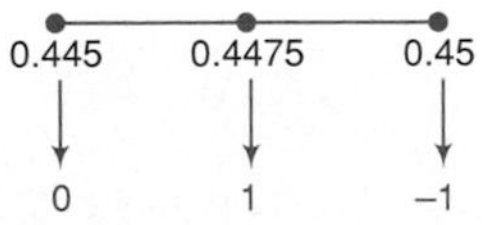

FIGURE 2
SUMMARY OF CONVERGENCE

As summarized in Figure 2:

starting at 0.4500, we converge to -1.
starting at 0.4450, we converge to 0.
starting at 0.4475, we converge to 1.

To explore this behavior in more detail, we programmed a computer to perform the Newton iterations for values of x_0 in the interval $[0, 1]$. The actual program, in TRUE BASIC, is given at the end of the section. The computer was attached to a color screen. Its instructions were to print a vertical yellow line if the sequence converged to 0, a red line if it converged to 1, a blue line if it converged to -1, and a black line if it didn't converge (there were no black lines). The results are given in Figure 3.

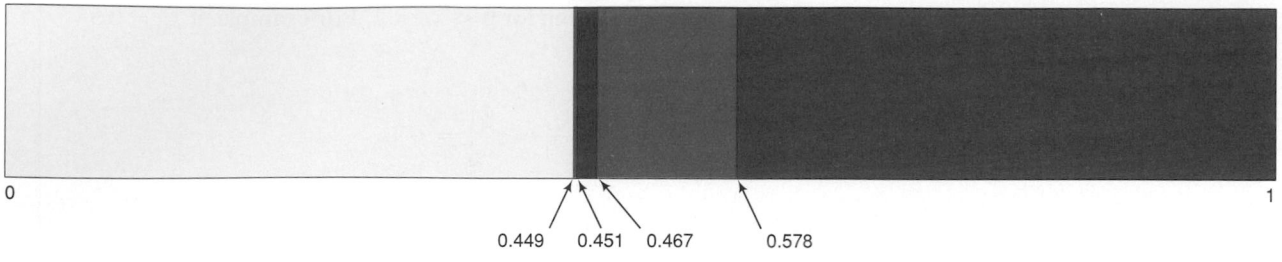

FIGURE 3   RESULTS OF ITERATIONS FOR $0 \leq x_0 \leq 1$

Note that if $0 \leq x_0 < \approx 0.449$, the iterates converge to 0. When $\approx 0.578 < x_0 < 1$, the iterates converge to 1. In between, the behavior is more erratic. To see better what is going on, we print the same kind of picture for $0.4 \leq x_0 \leq 0.5$ in Figure 4. Remarkably, except for the band of blue instead of red on the right end, the picture here is very similar to the picture in Figure 3.

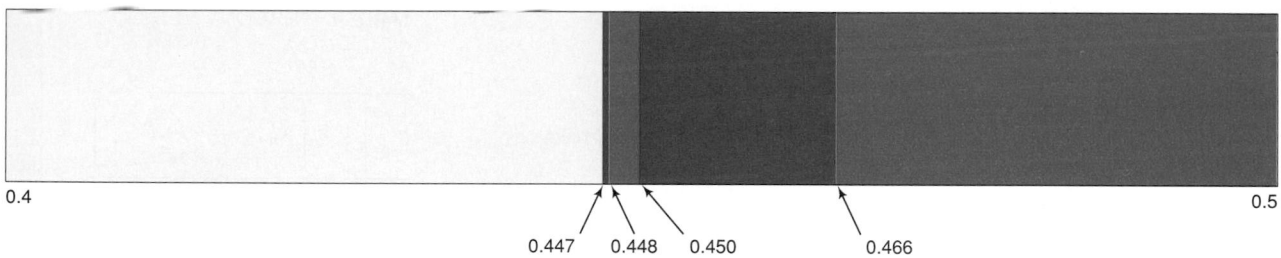

FIGURE 4   RESULTS OF ITERATIONS FOR $0.4 \leq x_0 \leq 0.5$

We again blow up the interesting strip in the middle. In Figure 5 we give the results for $0.445 \leq x_0 \leq 0.455$, a width of length 0.01. The picture is again similar to the picture for the interval $[0, 1]$. This is getting interesting. In Figures 6, 7, and 8, we reproduce the picture for x_0 in the intervals $[0.447, 0.448]$, $[0.4472, 0.4473]$, and $[0.44721, 0.447225]$, respectively. In each case we see the same pattern: wide bands of yellow, blue, or red with narrow strips of red and blue in between.[†]

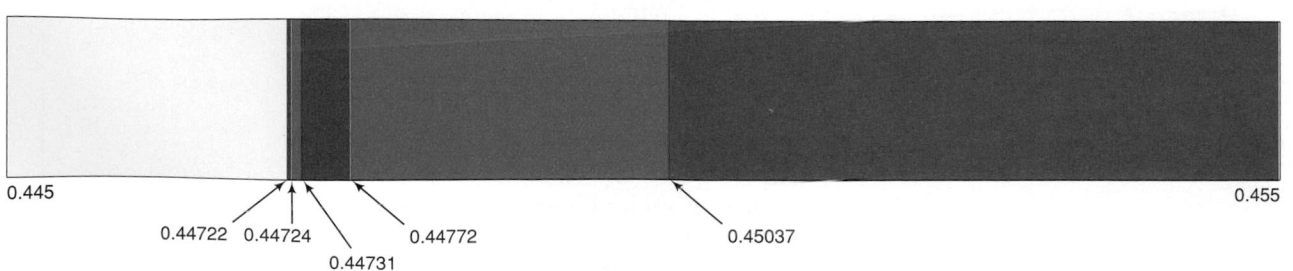

FIGURE 5   RESULTS OF ITERATIONS FOR $0.445 \leq x_0 \leq 0.455$

[†] The pictures in Figures 3–8 were printed with a resolution of 512 pixels per line. This accounts for some slight discrepancies in the threshold numbers.

0.447   0.448

0.447215   0.44723 0.4473   0.4477

0.447217

FIGURE 6   RESULTS OF ITERATIONS FOR $0.447 \leq x_0 \leq 0.448$

0.4472   0.4473

0.4472132   0.447216   0.4472275   0.447296

0.447214

FIGURE 7   RESULTS OF ITERATIONS FOR $0.4472 \leq x_0 \leq 0.4473$

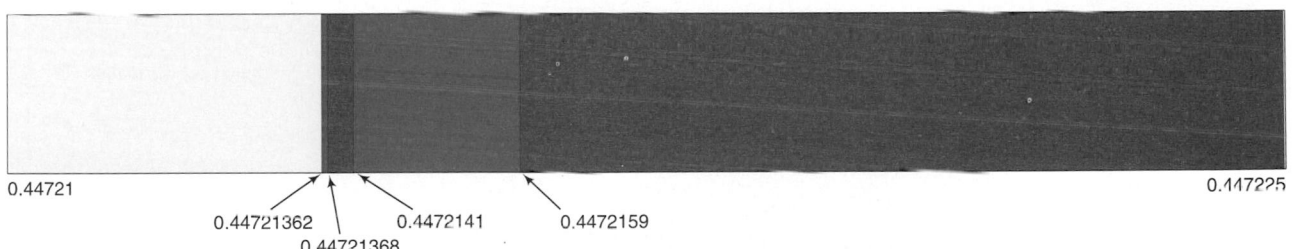

0.44721   0.447225

0.44721362   0.4472141   0.4472159

0.44721368

FIGURE 8   RESULTS OF ITERATIONS FOR $0.44721 \leq x_0 \leq 0.447225$

This is an example of a **fractal**. Loosely speaking, a fractal is the boundary of a pattern that replicates itself as you magnify smaller and smaller pieces of it. Recently, there have been many books published that seek to explain fractals,[†] and the more general notion of chaos. We will say no more about these topics here. Nor will we attempt to explain why a fractal appears in this innocent looking problem. However, to bring you back to the subject at hand:

We can avoid chaotic and other types of unwished-for behavior if we verify the hypotheses of the convergence theorem before applying Newton's method.

REMARK: The behavior of the Newton iterates for the function $P(x) = x^3 - x$ for values of x_0 in the interval $(-1, 0)$ is explored in Problems 3–5.

[†] See, for example, the interesting book *Fractals Everywhere* by Michael Barnsley (Orlando: Academic Press, 1988).

For those of you who would like to reproduce these pictures, we give below the TRUE BASIC program that accomplishes this task.

TRUE BASIC PROGRAM TO PERFORM THE ITERATION $x_{n+1} = \dfrac{2x_n{}^3}{3x_n{}^2 - 1}$ AND DRAW FRACTAL SETS

```
! JNEMT draws fractal sets using Newton's
! method for x^3-x


INPUT prompt "enter window values: x-lo, x-hi=":xn,xm
```
Enter max and min values of x_0.

```
OPEN #1: screen 0,1,0,1
SET WINDOW xn, xm, 0, 1
WINDOW #1
LET dx=(xm-xn)/512
FOR mx=xn to xm step dx
```
This gives 512 equally spaced starting values for x_0, for $x_n \le x_0 \le x_m$.

```
        LET x=mx
        FOR i=1 to 20
            LET x1=2*x*x*x/(3*x*x-1)
```
This performs the iteration $x_{n+1} = 2x_n{}^3/(3x_n{}^2 - 1)$

```
            IF abs(x-x1)<.000000001 THEN EXIT FOR
```
We stop when two successive iterates are less than 10^{-9} apart. x_1 is the name of the new iterate.

```
            LET x=x1
        NEXT i
        SET COLOR "black"
        IF abs(x1+1)<.00000001 then SET COLOR "blue"
        IF abs(x1)<.00000001 then SET COLOR "yellow"
        IF abs(x1-1)<.00000001 then SET COLOR "red"
        PLOT mx,0;mx,1

NEXT mx

END
```

■ PROBLEMS 3.9

■ APPLYING NEWTON'S METHOD TO
THE POLYNOMIAL $p(x) = x^3 - x$

*1. Let $p(x) = x^3 - x$. As we showed in Example 1, the Newton iterates are given by

$$x_{n+1} = \frac{2x_n{}^3}{3x_n{}^2 - 1}.$$

a. Prove that if $x_n > 1$, then $x_{n+1} > 1$. [Hint: Show, using calculus, that the minimum value of $f(x) = \dfrac{2x^3}{3x^2 - 1}$ is 1 and this is taken when $x = 1$.]

CONTINUED

b. Show that the Newton iterates converge to 1 for any x_0 in $[0.99, 1.2]$.

c. Show that if $x_n > \dfrac{3}{\sqrt{7}} \approx 1.134$, then $x_{n+1} < 0.9x_n$.

d. Explain why, if $x_0 > 1.2$, eventually some x_n will be < 1.2.

e. Using (b) and (d), prove that the Newton iterates converge to 1 for any $x_0 > 1$.

*2. In Problem 1, show that the Newton iterates converge to -1 for any $x_0 < -1$. [*Hint:* $x^3 - x$ is an odd function.]

3. Let $p(x) = x^3 - x$, as in Example 1. Show that starting with $x_0 = -0.5$, the Newton iterates converge to 1 in one step.

4. Show that starting with $x_0 = -0.45$ in Example 1, we obtain the negatives of the values in Table 2 and that the Newton iterates converge to 1.

5. Suppose in Example 1 that $x_0 \in (-1, 0)$. Explain why each Newton iterate is the negative of the corresponding iterate obtained by starting with $-x_0$ [which is in $(0, 1)$].

■ APPLYING NEWTON'S METHOD TO THE POLYNOMIAL $p(x) = x^3 - 6x^2 + 11x - 6$

6. Let $p(x) = (x - 1)(x - 2)(x - 3) = x^3 - 6x^2 + 11x - 6$.

a. Write a formula for the Newton iterates.

b. Write the iterates for $x_0 = 8$ and $x_0 = 5$.

*7. Prove that the Newton iterates converge to 3 for any $x_0 > 3$. Show that the Newton iterates converge to 1 in one step for $x_0 = 2.5$.

8. Show that the Newton iterates converge to 1 when $x_0 = 2.45$.

9. Show that the Newton iterates converge to 2 when $x_0 = 2.445$.

10. Show that the Newton iterates converge to 3 when $x_0 = 2.4475$.

■ APPLYING NEWTON'S METHOD TO THE POLYNOMIAL $p(x) = 3x^4 - 4x^3 - 12x^2 + 3$

11. Sketch the graph of $f(x)$ and show that it has four real zeros.

*12. Estimate each zero to four decimal places using Newton's Method.

**13. If s_4 denotes the largest zero, prove that the sequence of Newton iterates converges to s_4 if $x_0 > s_4$.

**14. If s_3 denotes the second-largest zero, show that there are values for x_0 in (s_3, s_4) starting at which the sequence of Newton iterates will converge to each of the four zeros of f.

■ PROBLEMS   CHAPTER 3 REVIEW

1. A rope is attached to a pulley mounted on top of a 5-m tower (Figure 1). One end of the rope is attached to a heavy mass. If the rope is pulled in at the rate of 1.5 m/sec, how fast is the mass approaching the tower when it is 3 m from the base of the tower?

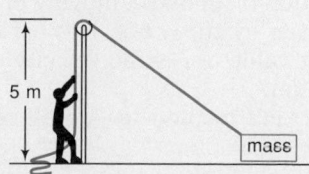

FIGURE 1

2. A storage tank is shaped as an inverted right circular cone with radius 6 ft and height 14 ft (Figure 2). If water is pumped into the tank at a rate of 20 gal/min, how fast is the water level rising when it has reached a height of 5 ft? [*Note:* 1 gallon is approximately 0.1337 ft³.]

CONTINUED

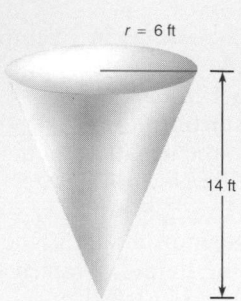

r = 6 ft

14 ft

FIGURE 2

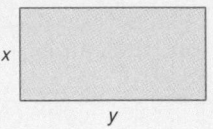

x

y

FIGURE 4

In Problems 3–18, graph the given curves by following the process summarized at the end of Section 3.4.

3. $y = x^2 - 3x - 4$
4. $y = x^3 - 3x^2 - 9x + 25$
5. $y = \sqrt[3]{x}$ 6. $y = x^{2/3}$
7. $y = |x^2 - 4|$ 8. $y = |x - 4|^3$
9. $y = (x - 7)/(x + 2)$ 10. $y = 1/(x^2 - 4)$
11. $y = x(x - 1)(x - 2)(x - 3)$
12. $y = -x(x - 1)^3$
13. $y = \sin((x + 1)/2)$
14. $y = \sin^2(x/3)$
15. $y = \cot(\pi - x)$
16. $y = \csc(\pi x)$
17. $y = (x + 2)(x - 2)/x$
18. $y = x/[(x + 1)(x - 1)]$
19. Find the maximum and minimum values for the following functions over the specified intervals.
    a. $f(x) = 2x^3 + 9x^2 - 24x + 3;\ [-2, 5]$
    b. $g(x) = (x - 2)^{1/3};\ [0, 4]$
    c. $h(x) = (x + 1)/(x^2 - 4);\ [-1, 1]$
    d. $F(x) = (x - 1)^2 + (2x + 1)^2;\ (-\infty, 0]$
20. What is the area of the largest rectangle that can be inscribed in a circle of radius 10 (see Figure 3)?

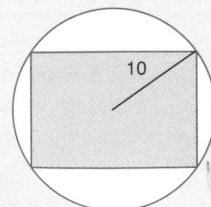

10

FIGURE 3

21. What is the maximum rectangular area that can be enclosed with 800 m of wire fencing (Figure 4)?

22. An isosceles triangle has an area of 20 cm² (Figure 5). What side lengths will minimize the perimeter of such a triangle?

x   x

y

FIGURE 5

23. A cylindrical barrel is to be constructed to hold 128π ft³ of liquid (Figure 6). The cost per square foot of constructing the side of the barrel is three times that of constructing the top and bottom. What are the dimensions of the barrel that costs the least to construct?

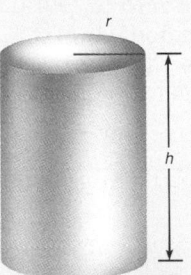

r

h

FIGURE 6

24. The position of an object moving on the number line is given by $s(t) = 6 + 4t - 6t^2$ for all $t \geq 0$. For what value of t is the velocity of the object a maximum?

25. Use Newton's method to find all roots of $x^3 - 2x^2 + 5x - 8 = 0$.

26. Approximate $\sqrt[6]{135}$ to four decimal places using Newton's method.

CONTINUED

In Problems 27–30, find some number c that satisfies the conclusion of the mean value theorem for the given function and interval.

27. $f(x) = x^5$; $[0, 1]$

28. $f(x) = x/(x + 1)$; $[0, 2]$

29. $f(x) = x^{1/5}$; $[-32, -1]$

30. $f(x) = \sin x$; $[0, \pi]$

In Problems 31–34, (a) find the linearization of the given function at the specified point, (b) use that linearization to approximate the given quantity, and (c) obtain an upper bound on the error of your approximation.

31. $\sin x$ at 0; $\sin 0.15$

32. $x^{1/3}$ at 1; $(0.9)^{1/3}$

33. $\dfrac{1}{x}$ at 2; $\dfrac{1}{2.05}$

34. $\tan x$ at $\dfrac{\pi}{4}$; $\tan 0.75$

COMPUTER EXERCISES

In Problems 1–4, you will use Newton's method to approximate roots. You will need a computer algebra system to calculate the roots to the high degree of accuracy requested. Neither a scientific calculator nor floating point software can produce the required accuracy.

1. a. Use your computer algebra system to obtain a 30-digit decimal approximation to $\sqrt{2}$.
   b. Begin Newton's method at $x_0 = 1$ and, with your computer algebra system, apply it until you get 30-decimal-place accuracy. How many iterations must you use?

2. a. Use your computer algebra system to obtain a 40-digit decimal approximation to $\sqrt[3]{25}$.
   b. Begin Newton's method at $x_0 = 2$ and, with your computer algebra system, apply it until you get 40-decimal-place accuracy. How many iterations must you use?

3. a. Use your computer algebra system to obtain a 30-digit decimal approximation to $\sqrt{703}$.
   b. Begin Newton's method at $x_0 = 20$ and, with your computer algebra system, apply it until you get 30-decimal-place accuracy. How many iterations must you use?

4. $f(x) = \dfrac{x}{10} - \sin x - \dfrac{x^2}{3} + 1$.
   a. Find the minimum of $|f'(x)|$ on $[0, 1]$.
   b. Find the maximum of $|f''(x)|$ on $[0, 1]$.
   c. Verify that the convergence theorem for Newton's method holds for $f(x)$ on $[0, 1]$ if $x_0 = \frac{1}{2}$.
   d. Find the unique zero of $f(x)$ in $[0, 1]$ to 30-decimal place accuracy.

5. Food for thought: Assume $x > 0$ and $a > 0$. Suppose x^2 is a "good" approximation to a^2. Does this mean that x is a "good" approximation to a? In particular, suppose $|x^2 - a| < \epsilon$. What can you say about $|x - \sqrt{a}|$ when (1) $a = 2$ and (2) when $a = 0.034$?

6. More food for thought: If you knew a polynomial with integer coefficients that had π as a zero you could apply Newton's method to approximate π. It is a deep fact that there is no such polynomial. Can you imagine how your computer algebra system approximates π?

7. Find the point on the curve $y = \dfrac{1}{x}$ that is nearest to the point $(1, 0)$. (It will be necessary to approximate the solution.)

8. A firm must build a cylindrical storage tank that holds 10,000 gallons of water. The top and sides may be made of inexpensive sheet iron costing $1.25 per square foot, but because the tank must sit on corrosive soil, the bottom must be made of a special alloy costing $2.75 per square foot. The contractor orders the top, bottom and cylindrical sides from a supplier, but she must weld the base and top to the prefabricated sides. With all her overhead included, it costs $4.50 per foot to do the welding. What should the dimensions be to insure the lowest cost? [*Hint:* There are approximately 7.5 gallons of water in a cubic foot.]

9. A welding firm wants to bid on a contract to build 1000 open metal boxes (no tops). The base of each box is to be square and is made of a special alloy costing $6 per square foot. The four sides are made of a cheaper metal costing $3 per square foot. The contractor must assemble the box by welding the eight seams together. Welding costs $4.50 per foot including all overhead. Assuming that the boxes are to hold 1,029

CONTINUED

cubic feet, what should the dimensions be to insure the lowest cost? If the contractor wants to make a profit of 17%, what should he bid?

10. A pipeline must be laid from town A to town B on the other side of a lake as shown in Figure 1. The cost of laying the pipeline under water is twice as much per mile as it is on land. How should the pipeline be laid in order to minimize the cost? Assuming the cost on land is $896,500 per mile, how much will the whole project cost? Write an explanation that the city commission members can understand.

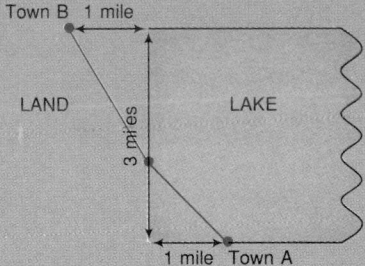

FIGURE 1

11. In Problem 10, assume the cost of laying pipe under water is K times as expensive as the cost of laying pipe on land. (K is a positive number.)
    a. How does the minimum cost pipeline change as K increases?
    b. Show that no matter how large K is, the minimum cost is never achieved by laying the underwater pipeline in a due westerly course.

12. Consider the polynomial $p(x) = x^4 - 5x^2 + 4 = (x^2 - 1)(x^2 - 4)$ which has the four zeros ± 1, ± 2.
    a. Write the sequence of Newton iterates.
    b. Write a BASIC or TRUE BASIC program that prints a vertical blue line at x_0 if the Newton sequence converges to -1 starting at x_0, a red line for convergence to 1, a yellow line for convergence to -2, and a green line for convergence to 2. [*Hint:* This can be accomplished by modifying the program on page 260. It is first necessary to change the function that follows "LET X1=".]
    c. Display on a color monitor the results of implementing your program for x in the following intervals:

$$-2 \leq x \leq 2$$
$$1.1 \leq x \leq 1.9$$
$$1.4 \leq x \leq 1.6$$
$$1.53 \leq x \leq 1.55$$
$$1.543 \leq x \leq 1.545$$
$$1.5432 \leq x \leq 1.5433$$
$$1.54325 \leq x \leq 1.54326$$

*13. Use Newton's method to find, to 20-decimal-place accuracy, all zeros of the polynomial $p(x) = x^5 - 4x^4 - 6x^3 + 8x^2 - 5x + 3$.

*14. Do the same for the polynomial $p(x) = 2x^6 + 3x^5 - 12x^4 + x^3 - x - 8$.

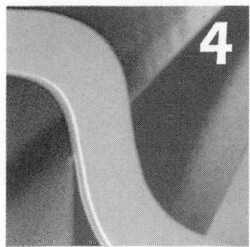

4 THE INTEGRAL

Modern calculus has its origins in two mathematical problems of antiquity. The first of these, the problem of finding the line tangent to a given curve, was, as we have noted, not solved until the seventeenth century. Its solution (by Newton and Leibniz) gave rise to what is known as *differential calculus*. The second of these problems was to find the area enclosed by a given curve. The solution of this problem led to what is now termed *integral calculus*.

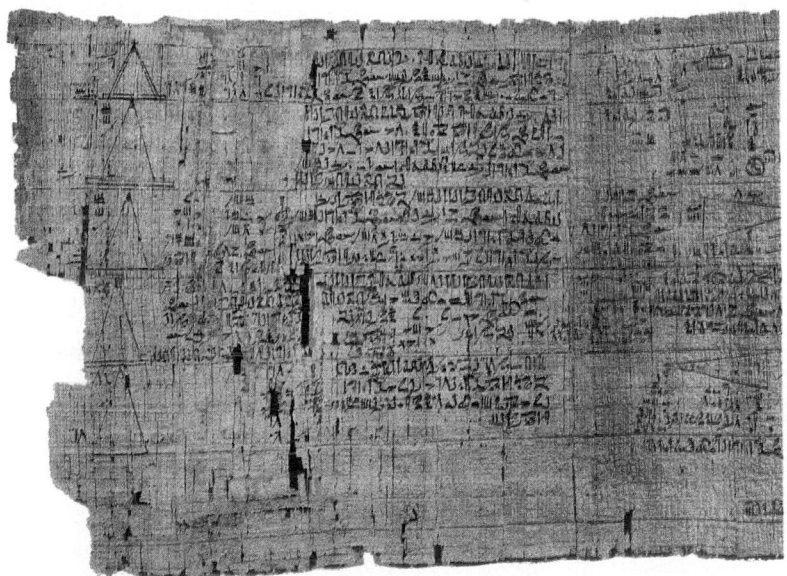

FIGURE 1   THE RHIND PAPYRUS (FRAGMENT)
Courtesy of the British Museum.

We do not know how long scientists have been concerned with finding the area bounded by a curve. In 1858, Henry Rhind, an Egyptologist from Scotland, discovered fragments of a papyrus manuscript written in approximately 1650 B.C., which came to be known as the *Rhind papyrus*.[†] The Rhind papyrus (see Figure 1) contains 85 problems and was written by the Egyptian scribe Ahmes, who wrote that he copied the problems from an

---

[†] For an interesting discussion of the Rhind papyrus and other similar finds, see C. B. Boyer, *A History of Mathematics* (New York: Wiley, 1968).

Throughout most of the Greek era, Alexandria was the center of mathematical activity. The greatest mathematician of antiquity, however, and perhaps the greatest the world has ever known, was born and died in the Roman town of Syracuse. The son of an astronomer, Archimedes of Syracuse may have been related to King Hieron of Syracuse. When he died in 212 B.C., records indicate that Archimedes was 75 years old, so the likely year of his birth was 287 B.C.

How do we measure greatness? In the case of Archimedes it is easy. The breadth and depth of his discoveries is astonishing. While he did pioneering theoretical work in a great number of scientific areas, he also invented and improved upon many useful mechanical devices.

Archimedes is best known by two stories that are told about him. One of them is related in Plutarch's life of the Roman general Marcellus. During the second Punic War, Syracuse was besieged by Romans led by Marcellus. Archimedes helped to lead the defense of the city by designing catapults with adjustable ranges, cranes that could literally lift enemy ships out of the water, and movable poles that could drop boulders on ships that came within range of the city walls. These devices all made use of the principle of the lever, and Archimedes is said to have claimed that if he were given a lever long enough and a fulcrum on which to rest it, he could move the earth. The story of the siege of Syracuse had an unhappy ending. After Marcellus failed to take Syracuse by a frontal siege, the city fell after a circuitous attack. A Roman soldier was sent by Marcellus to bring Archimedes to him. According to one version of the story, Archimedes had drawn a diagram in the sand and asked the soldier to stand away from it. This angered the soldier, who killed Archimedes with his spear.

The second story about Archimedes concerns the gold crown of King Hieron. Hieron suspected a dishonest goldsmith of filling part of his crown with cheaper silver. The king asked Archimedes for help. Archimedes pondered the problem in his bathtub and hit upon the principle of buoyancy. Absentmindedly, Archimedes leaped out of his bath and ran naked down the street shouting, "*Eureka*" ("I have found it").

The principle of buoyancy was described in Archimedes' treatise *On Floating Bodies*:

> Any solid lighter than a fluid will, if placed in a fluid, be so far immersed that the weight of the solid will be equal to the weight of the fluid displaced.

ARCHIMEDES OF
SYRACUSE
287–212 B.C.

Archimedes wrote another treatise that could properly be called the first book in mathematical physics: *On the Equilibrium of Planes*. It is concerned with centers of gravity of the triangle and the trapezoid.

At least eight other of Archimedes' treatises have survived. Each work is an model of mathematical rigor, precision, and originality. In his work *On Spirals*, Archimedes computed the area swept out by a spiral. In his work *On Conoids and Spheroids*, he computed the area of an ellipse. In his treatise *On the Sphere and the Cylinder*, Archimedes proved that if a sphere is inscribed in a cylinder the height of which is equal to its diameter, then the ratio of the volumes of the sphere and cylinder is equal to the ratio of their surface areas (3 to 2). He was so pleased with this result that he had a sketch of a sphere inscribed in a cylinder carved on his tombstone.

For the student of calculus, perhaps the most interesting of Archimedes' works is *Quadrature of the Parabola*. In his work, Archimedes described a method for estimating the area under a parabola by computing the area under a sum of rectangles. This method, called the *method of exhaustion*, forms the basis for the study of integral calculus. We will describe this method in Section 4.4.

One of the most exciting discoveries of modern times was made in 1906. The Danish scholar J. C. Heiberg had read that at Constantinople (now Istanbul) there was a mathematical palimpsest (a parchment on which some of the writing had been washed off with new writing appearing in its place). The palimpsest that Heiberg found had been, fortunately, washed off poorly, and he was able to recognize it as a treatise by Archimedes titled, simply, *Method*. By using photographs, Heiberg was able to read most of the 185 leaves of parchment that made up the treatise. The parchment contained a tenth-century copy of the text by Archimedes. *Method* is written in the form of a letter to Archimedes' friend Eratosthenes. In it, Archimedes described the process by which he had discovered many of his theorems. It is a remarkable document because in all his other works he only presented finished, polished results with no indication of how he obtained them. *Method* is a fitting tribute to the genius of this most remarkable of mathematicians.

earlier manuscript. In Problem 50 of his document, Ahmes assumed that the area of a circular field with a diameter of nine units was the same as the area of a square with a side of eight units. If we compare this statement with the correct formula for the area of a circle, we find that

$$A = \pi r^2 = \pi \left(\frac{9}{2}\right)^2 = \text{(according to Ahmes) } 8^2,$$

or

$$\pi \approx \frac{64}{(4.5)^2} = \frac{64}{20.25} \approx 3.16.$$

Thus, we see that *before 1650* B.C. the Egyptians could calculate the area of a circle from the formula

$$A = 3.16r^2.$$

Since $\pi \approx 3.1416$, we see that this is remarkably close, considering that the Egyptian formula dates back nearly four thousand years!

In ancient Greece, there was much interest in obtaining methods for calculating the areas bounded by curves other than circles and rectangles. The problem was solved for a wide variety of curves by Archimedes of Syracuse (287–212 B.C.), who is considered by many to be the greatest mathematician who ever lived. Archimedes used what he called the *method of exhaustion* to calculate an area A bounded by a parabola. We discuss this method in Section 4.4.

Before discussing the method for computing areas, we will define a new function, called the *antiderivative*. An antiderivative of a function f is a new function F having the property that $dF/dx = f$. We will discuss antiderivatives in the next section.

It is natural to ask, "What do antiderivatives have to do with the computation of area?" The answer is given by the fundamental theorem of calculus (stated on page 309). It is this beautiful result that ties together the great work of Archimedes almost twenty-three hundred years ago and the much more recent work of Newton and Leibniz. With this theorem, we will see that the two ancient problems are really very closely related.

4.2 ANTIDERIVATIVES

In Chapters 1 and 2, we discussed the problem of finding the derivative of a given function. We now consider the problem in reverse.

DEFINITION **ANTIDERIVATIVE**

Let f be defined on $[a, b]$. If there exists a function $y = F(x)$ such that F is continuous on $[a, b]$, differentiable on (a, b), and the derivative of F is f for every x in (a, b); that is, if

$$F'(x) = \frac{dF}{dx} = f(x), \qquad x \text{ in } (a, b),$$

then F is called an **antiderivative** of f on the interval $[a, b]$. That is,

> **ANTIDERIVATIVE**
>
> F is an antiderivative of f if $F'(x) = f(x)$.

EXAMPLE 1
FINDING AN ANTIDERIVATIVE

Find an antiderivative for $f(x) = 3x^2$.

SOLUTION: Since $\dfrac{d(x^3)}{dx} = 3x^2$, we see that x^3 is an antiderivative for $3x^2$. But the derivative of any constant is zero, so that $x^3 + 3$ is also an antiderivative of $3x^2$, as is $x^3 - 5$, $x^3 + 2\pi$, ... In fact, for any constant C, $x^3 + C$ is an antiderivative of $3x^2$. To see this, note that

$$\frac{d}{dx}(x^3 + C) = \frac{d}{dx}(x^3) + \frac{d}{dx}(C) = 3x^2 + 0 = 3x^2.$$

In Example 1, we saw that $3x^2$ has an infinite number of antiderivatives. If F is an antiderivative of f, then so is $F + C$ for every constant C, since

$$\frac{d}{dx}(F + C) = \frac{dF}{dx} + \frac{dC}{dx} = f + 0 = f.$$

Of course, you may ask, how do we know we have found all the antiderivatives of f? For example, are there any other functions F such that $F'(x) = 3x^2$, other than functions of the form $F(x) = x^3 + C$? The answer is *no* because of the following theorem.

THEOREM 1

If F and G are two differentiable functions that have the same derivative in (a, b), then they differ by a constant on (a, b). That is, if $F'(x) = G'(x)$ for every x in (a, b), then for all x in (a, b),

$$F(x) - G(x) = C \qquad \text{for some constant } C.$$

PROOF: Let $H(x) = F(x) - G(x)$. Then $H'(x) = F'(x) - G'(x) = 0$. Therefore, by Theorem 3.7.3, page 243, $H(x) = F(x) - G(x) = C$, and the theorem is proved. ■

DEFINITION   **INDEFINITE INTEGRAL**

This theorem allows us to define the **indefinite integral** of f as $F(x) + C$, where F is some antiderivative of f and C is an arbitrary constant. That is, the indefinite integral of f is the set of antiderivatives for f. ■

For the remainder of this section we will speak about "the integral," leaving out the word "indefinite," and will look for the most general antiderivative in our calculations.

NOTATION: The symbol that we use to indicate an integral is $\int$, a large German *s*, first used by Leibniz in the seventeenth century.[†]

INDEFINITE INTEGRAL NOTATION

Let F be an antiderivative for f; then the indefinite integral of f is written

$$\int f(x)\,dx = F(x) + C. \tag{1}$$

This notation is read "the integral of $f(x)$ with respect to x is $F(x) + C$." If such a function F exists, that is, if f has an antiderivative, then f is said to be **integrable**, and the process of calculating an integral is called **integration**. The variable x is called the **variable of integration**, and the function f is called the **integrand**.

Computing antiderivatives is often difficult or impossible. However, the antiderivatives of some elementary functions are easily obtained by the differentiation theorems of Chapter 2. Some of these are given below.

THEOREM 2

Let k be a constant. The functions k, $x^r (r \neq -1)$, $\sin x$, and $\cos x$ are integrable, and

$$\int k\,dx = kx + C \tag{2}$$

$$\int x^r\,dx = \frac{x^{r+1}}{r+1} + C \tag{3}$$

$$\int \sin x\,dx = -\cos x + C \tag{4}$$

$$\int \cos x\,dx = \sin x + C \tag{5}$$

PROOF:

$$\frac{d}{dx}\left(\frac{x^{r+1}}{r+1} + C\right) = \frac{(r+1)x^r}{r+1} + 0 = x^r$$

$$\frac{d}{dx}(-\cos x + C) = -\frac{d}{dx}\cos x + 0 = -(-\sin x) = \sin x$$

$$\frac{d}{dx}(\sin x + C) = \cos x. \quad\blacksquare$$

[†] The $\int$ was used for s in English books of the seventeenth and eighteenth centuries, too. At the end of a word s was used, and within the word $\int$ was used, as in *hou∫es* (houses).

EXAMPLE 2
THE INTEGRAL OF x^9

Calculate $\int x^9 \, dx$.

SOLUTION: $r = 9$, so

$$\int x^9 \, dx = \frac{x^{9+1}}{9+1} + C = \frac{x^{10}}{10} + C.$$

EXAMPLE 3
THE INTEGRAL OF $t^{-1/2}$

Calculate $\int (1/\sqrt{t}) \, dt$.

SOLUTION: The only difference between using x and t as the variable of integration is that t instead of x is the variable used in the antiderivative function. Then since $1/\sqrt{t} = t^{-1/2}$,

$$\int \frac{1}{\sqrt{t}} \, dt = \int t^{-1/2} \, dt = \frac{t^{(-1/2)+1}}{(-\frac{1}{2})+1} + C = \frac{t^{1/2}}{\frac{1}{2}} + C = 2\sqrt{t} + C.$$

In the last two examples, it is easy to *check the answer by differentiating.* This check should always be done since it is always easier to differentiate than to integrate. *Thus, differentiation provides a method for verifying the results of calculating an antiderivative.*

As we have already mentioned, Theorem 2 does not hold if $r = -1$. The case $r = -1$ is in some sense more interesting than the cases we have so far discussed. We will analyze this case in great detail in Chapter 6.

The following theorem follows easily from the analogous results for derivatives.

THEOREM 3 **INTEGRALS OF A CONSTANT TIMES A FUNCTION AND THE SUM OF TWO FUNCTIONS**

If f and g are integrable[†] and if k is any constant, then kf and $f + g$ are integrable. Moreover,

$$\int kf(x) \, dx = k \int f(x) \, dx \tag{6}$$

and

$$\int [f(x) + g(x)] \, dx = \int f(x) \, dx + \int g(x) \, dx. \tag{7}$$

PROOF: Let $F = \int f$ and $G = \int g$. Then,

$$\frac{d}{dx} kF = k \frac{d}{dx} F = kf \qquad \text{and}$$

$$\frac{d}{dx} (F + G) = \frac{d}{dx} F + \frac{d}{dx} G = f + g. \quad \blacksquare$$

[†] We remind you that f is integrable if f has an antiderivative.

REMARK: This theorem can be extended to more than two functions. For example, it is easy to show that $\int (f + g + h) = \int f + \int g + \int h$. The conditions (6) and (7) are called **linearity** conditions for the integral.

EXAMPLE 4
THE INTEGRAL OF THE SUM OF
TWO FUNCTIONS

Calculate $\int [(3/x^2) + 6x^2]\, dx$.

SOLUTION:

From (7)   From (6)

$$\int \left(\frac{3}{x^2} + 6x^2 \right) dx \stackrel{?}{=} \int \frac{3}{x^2}\, dx + \int 6x^2\, dx \stackrel{?}{=} 3 \int x^{-2}\, dx + 6 \int x^2\, dx$$

From (3)

$$= \frac{3x^{-2+1}}{-2+1} + \frac{6x^{2+1}}{2+1} + C = -3x^{-1} + 2x^3 + C$$

$$= -\frac{3}{x} + 2x^3 + C.$$

Check

$$\frac{d}{dx}\left(-\frac{3}{x} + 2x^3 + C \right) = -3(-x^{-2}) + 2 \cdot 3x^2 = \frac{3}{x^2} + 6x^2$$

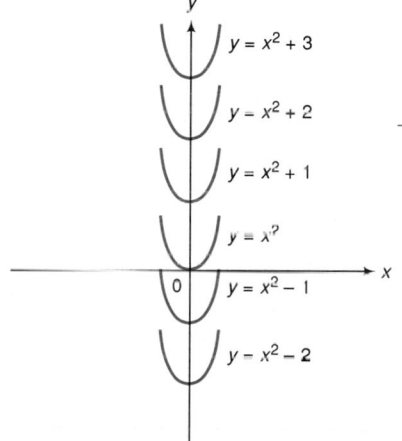

FIGURE 1
GRAPHS OF $y = x^2 + C$ FOR SIX VALUES OF C

Let us look more closely at the general integral of a function. Let $f(x) = 2x$. Then $\int f = x^2 + C$. For every value of C we get a different integral. But these integrals are very similar geometrically. For example, the curve $y = x^2 + 1$ is obtained by "shifting" the curve $y = x^2$ up one unit. The curve $y = x^2 + C$ is obtained by shifting the curve $y = x^2$ up C or down $|C|$ units (up if $C > 0$ and down if $C < 0$; see Section 0.5). Some of these curves are plotted in Figure 1. These curves never intersect. To prove this, suppose that (x_0, y_0) is a point on both curves $y = x^2 + A$ and $y = x^2 + B$. Then $y_0 = x_0^2 + A = x_0^2 + B$, which implies that $A = B$. That is, if the two curves have a point in common, then the two curves are the same. Thus if we specify one point through which the integral passes, then we know *the* antiderivative.

EXAMPLE 5
FINDING A PARTICULAR
ANTIDERIVATIVE

Find $\int 2x\, dx$ that passes through the point $(2, 7)$.

SOLUTION: $y = \int 2x\, dx = x^2 + C$. But when $x = 2$, $y = 7$, so that $7 = 2^2 + C = 4 + C$, or $C = 3$. The solution is the function $x^2 + 3$.

We now ask: What functions have antiderivatives? A partial answer is that all continuous functions have antiderivatives. That is, every continuous function is integrable. We prove the following result in Section 4.8; a graphical indication of the proof is given in Section 4.6.

THEOREM 4   EXISTENCE THEOREM FOR ANTIDERIVATIVES

If f is continuous on $[a, b]$, then f is integrable on $[a, b]$. That is, there exists a function F such that $F'(x) = f(x)$ at every point x in $[a, b]$. [Here a and b may be $-\infty$ and ∞, respectively, in which case we refer to the interval $(-\infty, \infty)$.] ∎

REMARK: This theorem does *not* say that functions that are not continuous are not integrable.

However, a serious problem remains. Consider the function $y = (1 + x^6)^{1/3}$. This function is certainly continuous (and even differentiable) everywhere. But it is impossible to find a function $F(x)$, written in terms of functions we recognize, that has the property that $F'(x) = (1 + x^6)^{1/3}$. [Try $(1 + x^6)^{4/3}/(\frac{4}{3})$ and see what happens.] This does not mean that the integral does not exist. It means that there is no way to express this integral in terms of simple functions. In fact, it is perhaps surprising that most continuous functions have antiderivatives that cannot be represented in terms of recognizable functions. We will devote most of Chapter 7 to a discussion of how to find integrals of a wide variety of functions. For the rest we will explore (also in Chapter 7) numerical methods for estimating definite integrals as closely as we wish. [Definite integrals are defined in Section 4.5.]

In Section 1.6, we gave several examples of the applicability of the derivative. We close this section by looking at some of these examples from the "reversed" point of view, that of the applicability of the integral.

EXAMPLE 6
A DISTANCE FUNCTION AS THE
ANTIDERIVATIVE OF A VELOCITY
FUNCTION

A ball is dropped from rest from a certain height. Its velocity after t seconds due to the earth's gravitational field is given by

$$v = 32t,$$

where v is measured in feet per second. How far has the ball fallen after t seconds?

SOLUTION: If $s = s(t)$ represents distance, then

$$\frac{ds}{dt} = v, \quad \text{or} \quad s(t) = \int v(t)\, dt = \int 32t\, dt = 16t^2 + C.$$

But $0 = s(0) = 16 \cdot 0^2 + C$, which implies that $C = 0$. Thus $s(t) = 16t^2$. For example, after 3 sec the ball has fallen $16 \cdot 3^2 = 16(9) = 144$ ft.

We can remember the following rule:

The velocity function is the derivative of the distance function, and the distance function is an antiderivative of the velocity function.

In Table 1, we summarize the relationships between pairs of derivative and integral functions.

TABLE 1

The Derivative[†]			The Integral		
Function	Its Derivative Function	In Symbols	Function	Its Antiderivative Function	In Symbols
Distance	Velocity	$v = \dfrac{ds}{dt}$	Velocity	Distance	$s = \displaystyle\int v(t)\,dt$
Velocity	Acceleration	$a = \dfrac{dv}{dt}$	Acceleration	Velocity	$v = \displaystyle\int a(t)\,dt$
Population	Instantaneous Growth	$\dfrac{dP}{dt}$	Instantaneous Growth	Population	$P = \displaystyle\int \dfrac{dP}{dt}\,dt$
Mass	Density	$\rho = \dfrac{d\mu}{dx}$	Density	Mass	$\mu = \displaystyle\int \rho(x)\,dx$
Cost	Marginal Cost	$MC = \dfrac{dC}{dq}$	Marginal Cost	Cost	$C(q) = \displaystyle\int \dfrac{dC}{dq}\,dq$
Revenue	Marginal Revenue	$MR = \dfrac{dR}{dq}$	Marginal Revenue	Revenue	$R(q) = \displaystyle\int \dfrac{dR}{dq}\,dq$

[†] See Section 1.6.

EXAMPLE 7
A MASS FUNCTION AS THE ANTIDERIVATIVE OF A DENSITY FUNCTION

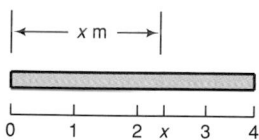

FIGURE 2
A NONUNIFORM BEAM

The density ρ of a 4-m nonuniform metal beam, measured in kilograms per meter, is given by

$$\rho(x) - 2\sqrt{x},$$

where x is the distance along the beam measured from the left end (see Figure 2).

(a) What is the mass of the beam up to the point x units from the left?

(b) What is the total mass of the beam?

SOLUTION:

(a) Since density is the rate of change of the mass μ (see Example 1.6.4 on page 107), we have

$$\rho = \frac{d\mu}{dx}, \qquad \text{or} \qquad \mu(x) = \int \rho(x)\,dx. \tag{8}$$

Then,

$$\mu(x) = \int 2\sqrt{x}\,dx = 2\int x^{1/2}\,dx = 2\,\frac{x^{3/2}}{\frac{3}{2}} + C = \frac{4}{3}x^{3/2} + C.$$

But $\mu(0) = 0$ since there is no mass zero units from the left side of the beam. Hence, $C = 0$, and

$$\mu(x) = \tfrac{4}{3}x^{3/2} \text{ kg.}$$

(b) When $x = 4$ m, $\mu(x) = \tfrac{4}{3}(4)^{3/2} = \tfrac{4}{3}(8) = \tfrac{32}{3}$ kg, which is the total mass of the beam.

EXAMPLE 8
A REVENUE FUNCTION IS THE
ANTIDERIVATIVE OF A MARGINAL
REVENUE FUNCTION

The marginal revenue that a manufacturer receives for his goods is given by

$$MR = 100 - 0.03q, \qquad 0 \le q \le 10{,}000. \tag{9}$$

Find the total revenue function.

SOLUTION: Marginal revenue is similar to marginal cost. (See Example 1.6.5 on p. 108.) Equation (9) tells us that the manufacturer receives less and less per unit the more units he sells. Marginal revenue is the rate of change of money received per unit change in the number of units sold. That is, if R denotes total revenue,

$$\frac{dR}{dq} = MR = 100 - 0.03q.$$

Then,

$$R(q) = \int MR = \int (100 - 0.03q)\,dq = 100q - 0.03\frac{q^2}{2} + C.$$

But the manufacturer certainly receives nothing if he sells nothing. Therefore, $R(0) = 0 = C$, and

$$R(q) = 100q - 0.03\frac{q^2}{2} \qquad \text{for } 0 \le q \le 10{,}000.$$

■ PROBLEMS 4.2

■ SELF-QUIZ

I. $\int 1\,dx = $ _____.

  a. C b. x
  c. $x + C$ d. Cx

II. $\int x\,dx = $ _____.

  a. 1 b. $x^2 + C$
  c. $x^2/2$ d. $x^2/2 + C$

III. $\int \sin x\,dx = $ _____.

  a. $\cos x + C$ b. $-\cos x + C$
  c. $(\sin x)^2/2 + C$ d. $(\sin x)(-\cos x) + C$

IV. $\int \cos x\,dx = $ _____.

  a. $\sin x + C$ b. $-\sin x + C$
  c. $\sin(x^2/2) + C$ d. $(\cos x)(-\sin x) + C$

V. $\int 3x^2\,dx = $ _____.

  a. $6x + C$ b. $x^3/3 + C$
  c. $9x^3 + C$ d. $x^3 + C$

VI. $\int t^2\,dt = $ _____.

  a. $t^3 + C$ b. $t^3/3 + C$
  c. $t^2x + C$ d. $3t + C$

VII. If $f'(x) = 3x^2$ and $f(2) = 15$, then $f(x) = $ _____.

  a. $3x^2 + 2$ b. $x^3 + C$
  c. $x^3 + 2$ d. $x^3 + 7$

■ DRILL

In Problems 1–16, find the indefinite integral.

1. $\int -5\,dx$

2. $\int a\,dx$, where a is a constant

3. $\int 10x\,dx$

4. $\int (ax + b)\,dx$ where a and b are constants

5. $\int (1 + 2x + 3x^2 + 4x^3)\,dx$

CONTINUED

6. $\int (1 - x + x^2 - x^3 + x^4)\,dx$

7. $\int x^{-5}\,dx$

8. $\int \dfrac{12}{x^3}\,dx$

9. $\int 7x^{-1/2}\,dx$

10. $\int \left(\sqrt{x} + \dfrac{1}{\sqrt{x}}\right)dx$

11. $\int \left(\dfrac{-17}{x^{13/17}} + \dfrac{3}{x^{4/9}}\right)dx$

12. $\int \dfrac{x^3 + x^2 - x}{x^{3/2}}\,dx$

[*Hint:* Divide first.]

13. $\int (2 \sin x - 3 \cos x)\,dx$

14. $\int \frac{1}{4}(\cos x - 5 \sin x)\,dx$

15. $\int (4x + 7 \sin x)\,dx$

16. $\int \left(1 - \dfrac{x^3}{3} - \cos x\right)dx$

In Problems 17–22, the derivative of a function and one point on its graph are given. Find the function.

17. $dy/dx = 2x(x + 1);\ (2, 0)$

18. $dy/dx = -3 + x^2 + x^3;\ (1, 5)$

19. $\dfrac{dy}{dx} = \sqrt[3]{x} + x - \dfrac{1}{3\sqrt[3]{x}};\ (-1, -8)$

20. $y' = 13x^{15/18} - 3;\ (1, 14)$

21. $y' = \cos x;\ (\pi/6, 4)$

22. $y' = \sin x - \cos x;\ (\pi/4, 3/\sqrt{2})$

■ **APPLICATIONS**

23. A particle moves along the number line with the constant acceleration of 5.8 m/sec². It starts with an initial position of 25 m and an initial velocity of 0.2 m/sec. Find the equation of motion for this particle.

24. The acceleration of a particle moving along the x-axis is given by $a(t) = 13\sqrt{t}$ m/min². Suppose the particle starts at position 100 m with an initial velocity of 25 m/min. Find the equation of motion for this particle.

25. A bullet is shot from ground level straight up into the air at an initial velocity of 2000 ft/sec. Write a function that tells us the height of the bullet after t seconds, up until the time the bullet hits the earth. (Assume the only force acting on the bullet is gravitational attraction which imparts the constant acceleration $g = 32$ ft/sec² downward.)

26. The instantaneous rate of change of a particular population is $50t^2 - 100t^{3/2}$, measured in individuals per year, and the initial population is 25,000.

a. What is the population after t years?
b. What is the population after 25 years?

27. The density of a 10-m beam x meters from one end is given by $\rho(x) = 3x - x^{3/2} + 2x^2$ kg/m. What is the total mass (in kilograms) of the beam?

28. The density of a 64-ft beam is given by $\rho(x) = 3x^{5/6}$, measured in lbs/ft. What is the total weight (in pounds) of the beam?

29. The marginal cost function for a certain manufacturer is $MC = 20 + 25q - 0.02q^2$, and it costs $2000 to produce 10 units of a particular product. How much does it cost to produce 100 units? 500 units?

30. The marginal revenue function for a particular product is given by $MR = 10 + 3q - 2\sqrt{q}$. How much money will the manufacturer receive if she sells 50 units? 100 units?

31. In Example 8, at what level of production will revenue begin to decrease as additional items are produced? (This level is often called the **point of diminishing returns**.)

*32. Find the indefinite integral of $f(x) = |x|$.

■ **SHOW/PROVE/DISPROVE**

*33. Consider the function

$$F(x) = \int \frac{1}{x}\,dx.$$

We have not yet learned how to integrate $1/x = x^{-1}$, but we can still obtain some information

about F. Assuming that $F(1) = 0$, use information about the derivatives of F to graph the curve for $x > 0$. [*Hint:* Show that F is always increasing, is concave down, and has a vertical tangent as x approaches 0 from the right.]

■ ANSWERS TO SELF-QUIZ
I. c   II. d   III. b   IV. a
V. d   VI. b   VII. d

4.3 THE $\sum$ NOTATION

In the next section we will see that one way to approximate the area under a curve involves writing a number of sums. There is a simple notation that will be very convenient for us. Consider the sum

$$a_1 + a_2 + a_3 + \cdots + a_n. \tag{1}$$

This sum is written[†]

WRITING A SUM USING SIGMA NOTATION

$$a_1 + a_2 + a_3 + \cdots a_n = \sum_{k=1}^{n} a_k \tag{2}$$

which is read, "the sum of the terms a_k as k goes from 1 to n." In this context $\sum$ is called the **summation sign**, and k is called the **index of summation**.

EXAMPLE 1
EVALUATING A SUM OF INTEGERS

Calculate $\sum_{k=1}^{4} k$.

SOLUTION: Here $a_k = k$, so that

$$\sum_{k=1}^{4} k = 1 + 2 + 3 + 4 = 10.$$

EXAMPLE 2
EVALUATING A SUM OF SQUARES

Calculate $\sum_{k=1}^{5} k^2$.

SOLUTION: Here $a_k = k^2$, and

$$\sum_{k=1}^{5} k^2 = 1^2 + 2^2 + 3^2 + 4^2 + 5^2 = 55.$$

EXAMPLE 3
USING THE SUMMATION
NOTATION

Write the sum $S_8 = 1 - 2 + 3 - 4 + 5 - 6 + 7 - 8$ by using the summation sign.

SOLUTION: Since $1 = (-1)^2, -2 = (-1)^3 \cdot 2, 3 = (-1)^4 \cdot 3, \ldots,$ we have

$$S_8 = \sum_{k=1}^{8} (-1)^{k+1} k.$$

[†] The Greek letter Σ (sigma) was first used to denote a sum by the Swiss mathematician Leonhard Euler (1707–1783). A short biography of Euler is given on page 34.

EXAMPLE 4
USING THE SUMMATION
NOTATION

Write the following sum by using the summation sign:

$$S = \left(\frac{1}{8}\right)^2 \frac{1}{8} + \left(\frac{2}{8}\right)^2 \frac{1}{8} + \cdots + \left(\frac{7}{8}\right)^2 \frac{1}{8} + \left(\frac{8}{8}\right)^2 \frac{1}{8}.$$

SOLUTION: We have

$$S = \sum_{k=1}^{8} \left(\frac{k}{8}\right)^2 \frac{1}{8} = \sum_{k=1}^{8} \left(\frac{1}{8^3}\right) k^2 = \frac{1}{8^3} \sum_{k=1}^{8} k^2.$$

In Example 4 we used the following fact:

$$\sum_{k=1}^{n} ca_k = c \sum_{k=1}^{n} a_k,$$

where c is a constant. To see why this is true, observe that

$$\sum_{k=1}^{n} ca_k = ca_1 + ca_2 + ca_3 + \cdots + ca_n$$

$$= c(a_1 + a_2 + a_3 + \cdots + a_n) = c \sum_{k=1}^{n} a_k.$$

EXAMPLE 5
USING THE SUMMATION
NOTATION

In the next section, we will encounter the sum

$$S = f(x_1)\Delta x_1 + f(x_2)\Delta x_2 + \cdots + f(x_n)\Delta x_n \qquad (3)$$

which can be written as

$$S = \sum_{k-1}^{n} f(x_k)\Delta x_k. \qquad (4)$$

NOTE: We can change the index of summation without changing the sum. For example,

$$\sum_{k=1}^{5} k^2 = \sum_{j=1}^{5} j^2 = \sum_{m=1}^{5} m^2 = 1^2 + 2^2 + 3^2 + 4^2 + 5^2 = 55.$$

■ PROBLEMS 4.3

■ SELF-QUIZ

I. $\sum_{j=1}^{5} j =$ _____ .
  a. $1 + 5$ b. $1 + 2 + 3 + 4 + 5$
  c. $(1 + 5)/2$ d. $(1 + 2 + 3 + 4 + 5)/5$

II. $\sum_{j=1}^{5} 2j =$ _____ .
  a. $2 + 4 + 6 + 8 + 10$ b. $2 + 10$
  c. $(1 + 2 + 3 + 4 + 5)^2$ d. $2 \cdot (1 + 5)$

III. $\sum_{k=1}^{5} (2k - 1) =$ _____ .
  a. $1 + 2 + 3 + 4 + 5 + 6 + 7 + 8 + 9$
  b. $(2 + 4 + 6 + 8 + 10) - 1$
  c. $2 \cdot (0 + 1 + 2 + 3 + 4)$
  d. $1 + 3 + 5 + 7 + 9$

IV. $\sum_{j=0}^{4} (2j + 1) =$ _____ .
  a. $1 + 3 + 5 + 7 + 9$
  b. $0 + 1 + 2 + 3 + 4$
  c. $(0 + 2 + 4) + 1$
  d. $2 \cdot (0 + 1 + 2 + 3 + 4) + 1$

V. $\sum_{k=1}^{5} 1 =$ _____ .
  a. $1 + 1 + 1 + 1 + 1$
  b. $5 + 4 + 3 + 2 + 1$
  c. 1
  d. $(1 + 1 + 1 + 1 + 1)/5$

VI. $\sum_{j=0}^{3} 5 =$ _____ .
  a. $0 + 1 + 2 + 3$ b. $0 + 5 + 10 + 15$
  c. $5 + 5 + 5$ d. $5 + 5 + 5 + 5$

CONTINUED

■ DRILL

In Problems 1–6, evaluate the given sum.

1. $\sum_{j=1}^{5} \dfrac{j}{2}$

2. $\sum_{j=1}^{5} j^2$

3. $\sum_{k=0}^{5} (-2)^k$

4. $\sum_{k=0}^{4} -3$

5. $\sum_{j=5}^{9} \dfrac{j}{j+1}$

6. $\sum_{k=0}^{3} a_k x^k$

In Problems 7–20, reexpress each sum using the $\sum$ notation.

7. $1 + 2 + 4 + 8 + 16$
   [*Hint:* $2^0 = 1$.]

8. $1 - 3 + 9 - 27 + 81 - 243$

9. $\dfrac{2}{3} + \dfrac{3}{4} + \dfrac{4}{5} + \dfrac{5}{6} + \dfrac{6}{7} + \dfrac{7}{8}$

10. $1/1^2 - 1/2^2 + 1/3^2 - 1/4^2 + 1/5^2 - 1/6^2 + 1/7^2$

*11. $1 \cdot 3 + 3 \cdot 5 + 5 \cdot 7 + 7 \cdot 9 + 9 \cdot 11 + 11 \cdot 13 + 13 \cdot 15 + 15 \cdot 17$

*12. $2^2 \cdot 4 + 3^2 \cdot 6 + 4^2 \cdot 8 + 5^2 \cdot 10 + 6^2 \cdot 12 + 7^2 \cdot 14$

13. $1 - 2x + 4x^2 - 8x^3 + 16x^4 - 32x^5$
    [*Hint:* $1 = x^0,\ x = x^1$.]

14. $a_0 + a_1 x + a_2 x^2 + a_3 x^3 + a_4 x^4$

15. $1 + x^3 + x^6 + x^9 + x^{12} + x^{15} + x^{18} + x^{21}$

16. $x^5 - x^{10} + x^{15} - x^{20} + x^{25} - x^{30} + x^{35}$

17. $\left(\dfrac{1}{5}\right)^2 + \left(\dfrac{2}{5}\right)^2 + \left(\dfrac{3}{5}\right)^2 + \left(\dfrac{4}{5}\right)^2 + \left(\dfrac{5}{5}\right)^2$

18. $\left(7 + \dfrac{0}{5}\right)^2 + \left(7 + \dfrac{1}{5}\right)^2 + \left(7 + \dfrac{2}{5}\right)^2 + \left(7 + \dfrac{3}{5}\right)^2 + \left(7 + \dfrac{4}{5}\right)^2$

19. $0.2 \sin(0.1) + 0.2 \sin(0.3) + 0.2 \sin(0.5) + 0.2 \sin(0.7) + 0.2 \sin(0.9)$

20. $\left(\dfrac{1}{20}\right)\sqrt{\dfrac{0}{20}} + \left(\dfrac{1}{20}\right)\sqrt{\dfrac{1}{20}} + \left(\dfrac{1}{20}\right)\sqrt{\dfrac{2}{20}} + \cdots + \left(\dfrac{1}{20}\right)\sqrt{\dfrac{18}{20}} + \left(\dfrac{1}{20}\right)\sqrt{\dfrac{19}{20}}$

Observe that there is some flexibility in the $\sum$ notation. For instance, $\sum_{i=3}^{8} i$, $\sum_{j=3}^{8} j$, $\sum_{k=5}^{10} (k-2)$, and $\sum_{L=0}^{5} (L+3)$ each equals $3 + 4 + 5 + 6 + 7 + 8$.

In each of the Problems 21–24, you are given three expressions; two yield the same sum but one is different. Identify the one that does not equal the other two.

21. a. $\sum_{i=0}^{4} 1$ b. $\sum_{j=1}^{5} 1$ c. $\sum_{k=5}^{10} 1$

22. a. $\sum_{k=0}^{7} (2k+1)$ b. $\sum_{j=1}^{15} j$
    c. $\sum_{i=2}^{9} (2i-3)$

23. a. $\sum_{k=1}^{7} k^2$ b. $\sum_{j=0}^{6} (7-j)^2$
    c. $\sum_{i=1}^{7} (7-i)^2$

24. a. $\left(\sum_{k=7}^{11} k\right)^4$ b. $\sum_{m=-11}^{-7} m^4$
    c. $\sum_{p=7}^{11} p^4$

■ SHOW/PROVE/DISPROVE

*25. Suppose that g is some function whose domain includes the integers.
   a. **(Telescoping sum)** Prove that
      $\sum_{k=1}^{n} [g(k) - g(k-1)] = g(n) - g(0)$.
   b. If we let $g(k) = \frac{1}{2}k(k+1)$, show that $g(k) - g(k-1) = k$ and part (a) yields $\sum_{k=1}^{n} k = n(n+1)/2$.

*26. a. Let $G(k) = \frac{1}{3}k(k+1)(k+2)$ and show that $G(k) - G(k-1) = k(k+1)$.
   b. Find $\sum_{k=1}^{n} k(k+1)$ by using part (a) and the result of Problem 25(a).
   c. Show that $\sum_{k=1}^{n} k^2 = n(n+1)(2n+1)/6$ by using the results of part (b) and Problem 25(b).

*27. Find a short formula with which to compute $\sum_{k=1}^{n} k^3$. [*Hint:* First look at a telescoping sum with $g(k) = \frac{1}{4}k(k+1)(k+2)(k+3)$ and then use previously obtained formulas for $\sum_{k=1}^{n} k$ and $\sum_{k=1}^{n} k^2$.]

*28. Find a short formula for computing the sum of the first n odd integers. Show that your formula is correct for all n. [*Hint:* Try a telescoping sum with $g(k) = k^2$.]

*29. Find a short formula by which to compute

$$\frac{1}{3} + \frac{1}{8} + \frac{1}{15} + \frac{1}{24} + \cdots + \frac{1}{n^2 - 1}$$

$$= \frac{1}{1 \cdot 3} + \frac{1}{2 \cdot 4} + \frac{1}{3 \cdot 5} + \frac{1}{4 \cdot 6} + \cdots$$

$$+ \frac{1}{(n-1) \cdot (n+1)}.$$

CONTINUED

*30. Find a short formula for computing

$$\frac{1}{2} + \frac{1}{6} + \frac{1}{12} + \frac{1}{20} + \cdots + \frac{1}{n^2 - n}$$

$$= \frac{1}{1 \cdot 2} + \frac{1}{2 \cdot 3} + \frac{1}{3 \cdot 4} + \frac{1}{4 \cdot 5} + \cdots$$

$$+ \frac{1}{(n-1) \cdot n}.$$

$$\left[Hint: \frac{1}{k \cdot (k+1)} = \frac{1}{k} - \frac{1}{k+1}. \right]$$

31. Show that $\sum_{i=1}^{n} ka_i = k \sum_{i=1}^{n} a_i$ for any constant k.

32. Show that $\sum_{i=1}^{n} (a_i + b_i) = \sum_{i=1}^{n} a_i + \sum_{i=1}^{n} b_i$.

■ ANSWERS TO SELF-QUIZ

I. b   II. a   III. d
IV. a   V. a   VI. d

4.4 APPROXIMATIONS TO AREA

In this section, we describe a method first used by Archimedes for computing the area under a curve. We estimate the area as the sum of areas of rectangles and then show how the estimate can be improved. Finally, we take a limit of these estimates, and it is this limit that is equal to the area sought.

EXAMPLE 1
COMPUTING THE AREA UNDER A CURVE BY ARCHIMEDES' METHOD

Compute the area bounded by the curve $y = x^2$, the x-axis, and the line $x = 1$. (See Figure 1.)

SOLUTION: We solve this problem using Archimedes' method. The area sought is approximated by rectangles under the curve and over the curve (see Figure 2). Since the equation of the parabola is $y = x^2$, the height of each rectangle (which is the y value on the curve) is the square of the x value. If we let s denote the sum of the areas of the rectangles in Figure 2a and S the sum of the areas of the rectangles in Figure 2b, then

$$s < A < S. \tag{1}$$

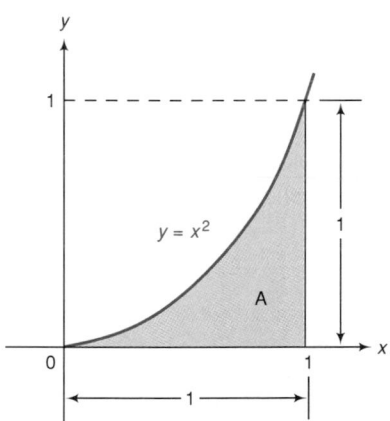

FIGURE 1
AREA UNDER THE CURVE $y = x^2$ AND ABOVE THE x-AXIS FOR $0 \le x \le 1$

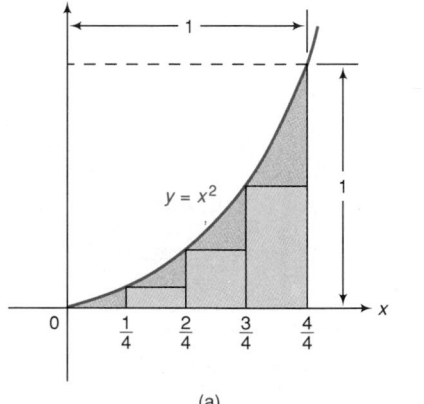

(a)

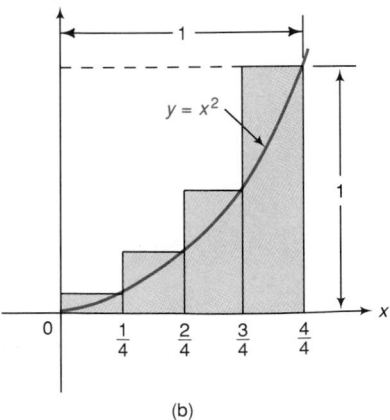

(b)

FIGURE 2 AREA UNDER CURVE APPROXIMATED BY TOTAL AREA OF RECTANGLES

Since Archimedes knew how to calculate the area of a rectangle (by multiplying its base by its height), he was able to calculate s and S exactly, thereby obtaining an estimate for the area A. We have

Height of first rectangle in Figure 2(a)   Length of subinterval   Height of second rectangle in Figure 2(a)   Height of third rectangle in Figure 2(a)

$$s = \left(\frac{1}{4}\right)^2 \frac{1}{4} + \left(\frac{2}{4}\right)^2 \frac{1}{4} + \left(\frac{3}{4}\right)^2 \frac{1}{4}$$

$$= \frac{1}{4}\left[\left(\frac{1}{4}\right)^2 + \left(\frac{2}{4}\right)^2 + \left(\frac{3}{4}\right)^2\right]$$

$$= \frac{1}{4 \cdot 4^2}(1^2 + 2^2 + 3^2) = \frac{1^2 + 2^2 + 3^2}{4^3} = \frac{14}{64} = \frac{7}{32}.$$

Using the summation notation of the last section, we may write

$$s = \sum_{k=1}^{3} \left(\frac{k}{4}\right)^2 \frac{1}{4} = \frac{7}{32}.$$

Similarly,

Height of first rectangle in Figure 2(b)   Height of second rectangle in Figure 2(b)   Height of third rectangle in Figure 2(b)   Height of fourth rectangle in Figure 2(b)

$$S = \left(\frac{1}{4}\right)^2 \frac{1}{4} + \left(\frac{2}{4}\right)^2 \frac{1}{4} + \left(\frac{3}{4}\right)^2 \frac{1}{4} + \left(\frac{4}{4}\right)^2 \frac{1}{4}$$

$$= \frac{1^2 + 2^2 + 3^2 + 4^2}{4^3} = \frac{30}{64} = \frac{15}{32},$$

and we may write

$$S = \sum_{k=1}^{4} \left(\frac{k}{4}\right)^2 \frac{1}{4} = \frac{15}{32}.$$

Thus from (1), with $s = \frac{7}{32}$ and $S = \frac{15}{32}$, we obtain

$$0.22 \approx \frac{7}{32} < A < \frac{15}{32} \approx 0.47. \tag{2}$$

It was clear to Archimedes that this estimate could be improved by increasing the number of rectangles so that the error in the estimate becomes smaller. By doubling the number of rectangles, we obtain eight rectangles with endpoints $0, \frac{1}{8}, \frac{2}{8}, \frac{3}{8}, \frac{4}{8}, \frac{5}{8}, \frac{6}{8}$, and $\frac{7}{8}$. The height of each rectangle is $\left(\frac{k}{8}\right)^2$ for some k. We then obtain the new estimates

$$s = \left(\frac{1}{8}\right)^2 \frac{1}{8} + \left(\frac{2}{8}\right)^2 \frac{1}{8} + \cdots + \left(\frac{7}{8}\right)^2 \frac{1}{8}$$

$$= \frac{1^2 + 2^2 + 3^2 + 4^2 + 5^2 + 6^2 + 7^2}{8^3} = \frac{140}{512} = \frac{35}{128},$$

and

$$s = \sum_{k=1}^{7} \left(\frac{k}{8}\right)^2 \frac{1}{8} = \frac{35}{128}.$$

Similarly, we obtain

$$S = \sum_{k=1}^{8} \left(\frac{k}{8}\right)^2 \frac{1}{8} = \frac{204}{512} = \frac{51}{128}.$$

Hence, using $s = \frac{35}{128}$ and $S = \frac{51}{128}$ in (1), we have

$$0.27 \approx \frac{35}{128} < A < \frac{51}{128} \approx 0.40. \tag{3}$$

We can continue the process of increasing the number of subintervals to get more and more accurate approximations for A.

Using his method of exhaustion, Archimedes was able to show that with the outer rectangles, the assumption $A > \frac{1}{3}$ led to a contradiction. Similarly, using the inner rectangles, he showed that $A < \frac{1}{3}$ led to a contradiction. From this result, he concluded that $A = \frac{1}{3}$.

We now prove this. Suppose that the interval $[0, 1]$ is divided into n subintervals with endpoints

$$0, \frac{1}{n}, \frac{2}{n}, \ldots, \frac{n}{n} = 1.$$

Consider the subinterval $[x_{k-1}, x_k]$ (see Figure 3). The height of the smaller rectangle is $\left(\frac{k-1}{n}\right)^2$ and the height of the larger one is $\left(\frac{k}{n}\right)^2$. The width of each rectangle is $\frac{1}{n}$. Thus,

$$\text{area of smaller rectangle} = \left(\frac{k-1}{n}\right)^2 \cdot \frac{1}{n} = \frac{(k-1)^2}{n^3}$$

$$\text{area of larger rectangle} = \left(\frac{k}{n}\right)^2 \cdot \frac{1}{n} = \frac{k^2}{n^3}.$$

If s_n denotes the sum of the areas of the smaller rectangles and S_n denotes the sum of the areas of the larger rectangles, we have

$$s_n < A < S_n$$

$$\overset{\text{area of } n\text{th small rectangle}}{\underset{}{}} \qquad \overset{\text{area of } n\text{th large rectangle}}{\underset{}{}}$$

$$\frac{0^2}{n^3} + \frac{1^2}{n^3} + \frac{2^2}{n^3} + \cdots + \frac{(n-1)^2}{n^3} < A < \frac{1^2}{n^3} + \frac{2^2}{n^3} + \frac{3^2}{n^3} + \cdots + \frac{n^2}{n^3}$$

$$\frac{1}{n^3} \sum_{k=1}^{n-1} k^2 < A < \frac{1}{n^3} \sum_{k=1}^{n} k^2. \tag{4}$$

In Appendix 2 (Example 3) we prove, using mathematical induction, that the sum of the first n squares is given by the formula

SUM OF THE FIRST n SQUARES

$$1^2 + 2^2 + \cdots + n^2 = \sum_{k=1}^{n} k^2 = \frac{n(n+1)(2n+1)}{6}. \tag{5}$$

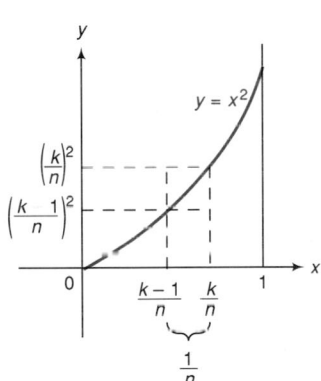

FIGURE 3
THE SMALLER AND LARGER RECTANGLES OVER ONE SUBINTERVAL

(Or see Problem 4.3.26). You should check this formula by verifying it for a few values of n. Substituting (5) into (4), we obtain (first replace n by $n - 1$ in (5))

$$\frac{(n-1)n(2n-1)}{6n^3} < A < \frac{n(n+1)(2n+1)}{6n^3}.$$

Now

$$\lim_{n \to \infty} \frac{(n-1)n(2n-1)}{6n^3} = \frac{1}{3} = \lim_{n \to \infty} \frac{n(n+1)(2n+1)}{6n^3}. \qquad \text{Check this.}$$

Thus, by the squeezing theorem (see page 77),

$$A = \lim_{n \to \infty} A = \frac{1}{3}.$$

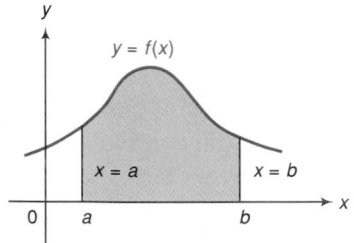

FIGURE 4
AREA UNDER THE CURVE $y = f(x)$ AND ABOVE THE x-AXIS FOR $a \leq x \leq b$

We now generalize the procedure used by Archimedes. Let $y = f(x)$ be a function that is positive and continuous on the interval $[a, b]$ (see Figure 4). We will show how the area bounded by this curve, the x-axis, and the lines $x = a$ and $x = b$ can be approximated. The method we will give here is very similar to Archimedes' method of exhaustion.

As before, we begin by dividing the interval $[a, b]$ into a number of smaller subintervals of *equal length*. Such a division is called a **regular partition**. We label the **partition points** $x_0, x_1, x_2, \ldots, x_n$, where $x_0 = a$ and $x_n = b$, and it is assumed that

$$a = x_0 < x_1 < x_2 < \cdots < x_n = b.$$

This partition is illustrated in Figure 5. Since there are n subintervals, each of the same length, and the length of the entire interval is $b - a$, we see that each subinterval has length $(b - a)/n$. We denote this quantity by Δx. We have

$$\Delta x = \frac{b - a}{n}. \tag{6}$$

Since the length of the ith subinterval is $x_i - x_{i-1}$, we also have

---

THE LENGTH OF A SUBINTERVAL

$$\Delta x = x_i - x_{i-1} = \frac{b - a}{n} \qquad \text{for } i = 1, 2, \ldots, n. \tag{7}$$

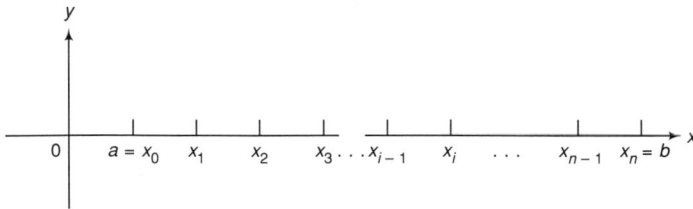

FIGURE 5 A REGULAR PARTITION OF THE INTERVAL $[a, b]$

REMARK: From (6) or (7) it follows that

$$\lim_{n \to \infty} \Delta x = 0.$$

EXAMPLE 2
REGULAR PARTITIONS OF [0, 1]

In Figure 6 we have sketched four partitions of the interval $[0, 1]$. In (i) $\Delta x = \frac{1}{2} = (1 - 0)/2$; in (ii) $\Delta x = \frac{1}{4}$; in (iii) $\Delta x = \frac{1}{6}$; and in (iv) $\Delta x = \frac{1}{10}$.

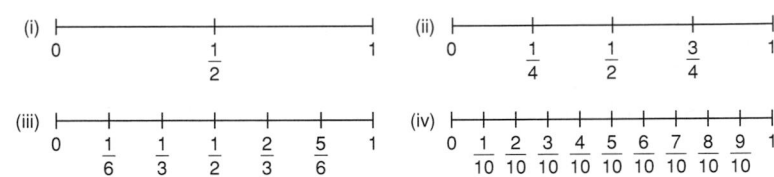

FIGURE 6 FOUR REGULAR PARTITIONS OF [0, 1]

NOTATION: We denote the area in Figure 4 by A_a^b. This is the area bounded by the curve $y = f(x)$, the x-axis, and the lines $x = a$ and $x = b$.

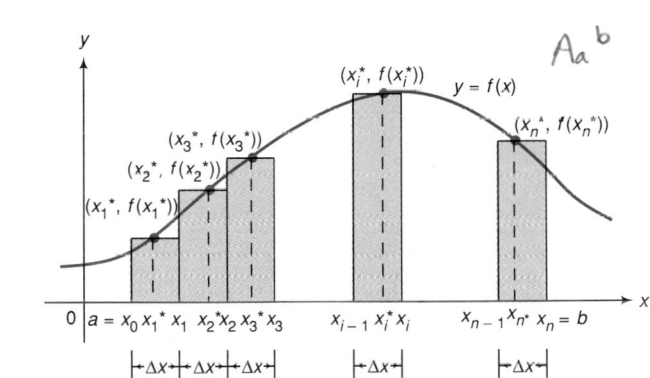

FIGURE 7 THE AREA UNDER APPROXIMATING RECTANGLES

The points at which the function is evaluated are not necessarily endpoints.

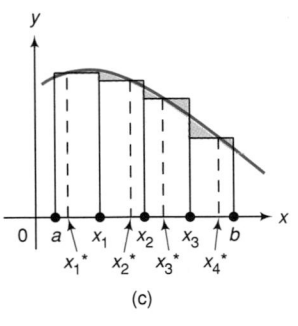

FIGURE 8
THE DIFFERENCE BETWEEN THE AREA UNDER THE CURVE AND THE AREA UNDER THE RECTANGLES DECREASES AS THE WIDTH OF THE RECTANGLES DECREASES (AS THE NUMBER OF RECTANGLES INCREASES).

Suppose that the interval $[a, b]$ is partitioned into n subintervals of equal length. We will approximate A_a^b by drawing rectangles whose total area is "close" to the actual area (see Figure 7). We first choose a point *arbitrarily* in each interval $[x_{i-1}, x_i]$ and label it x_i^*. This technique gives us n points $x_1^*, x_2^*, \ldots, x_n^*$. Next, we locate the points $(x_i^*, f(x_i^*))$ on the curve for $i = 1, 2, \ldots, n$. The numbers $f(x_i^*)$ give us the heights of our n rectangles. The base of each rectangle has length $x_i - x_{i-1} = \Delta x$. This is illustrated in Figure 7. The area A_i of the ith rectangle is the height of the rectangle times its length:

$$A_i = f(x_i^*) \Delta x. \tag{8}$$

Let us take a closer look at the region enclosed by these rectangles. In Figure 8, the shaded regions depict the differences between the region whose area we wish to calculate and the region enclosed by the rectangles. We see

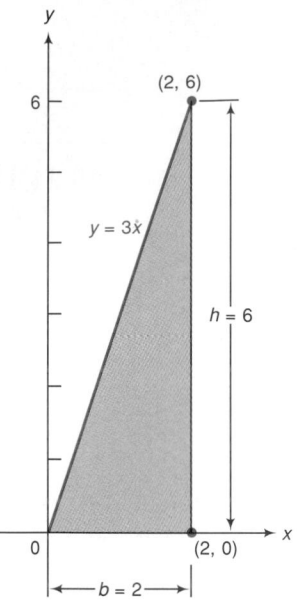

FIGURE 9
AREA UNDER THE LINE $y = 3x$ AND ABOVE THE x-AXIS FOR $0 \leq x \leq 2$

that as each rectangle becomes "thinner and thinner," the area of the regions enclosed by the rectangles seems to get closer and closer to the area of the regions under the curve. But the length of the base of each rectangle is Δx, and so if Δx is reasonably small, we have the approximation

$$A_a^b \approx A_1 + A_2 + \cdots + A_i + \cdots + A_n = \sum_{i=1}^{n} A_i,$$

or using (8), we have

$$A_a^b \approx f(x_1{}^*)\,\Delta x + f(x_2{}^*)\,\Delta x + \cdots + f(x_n{}^*)\,\Delta x,$$

and

AN APPROXIMATION TO THE AREA UNDER A CURVE

$$A_a^b \approx \sum_{i=1}^{n} f(x_i{}^*)\,\Delta x. \tag{9}$$

Earlier in this section, we saw an example of this process at work. There we chose $x_i{}^*$ to be either the right or the left endpoint of the subinterval. Our approximation technique will work for *any* choice of $x_i{}^*$ in $[x_{i-1}, x_i]$. However, it is often easier to deal with endpoints.

EXAMPLE 3
APPROXIMATING THE AREA UNDER A LINE

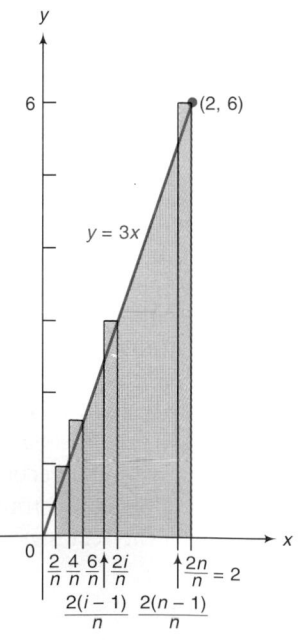

FIGURE 10
RECTANGLES FORMED WHEN THE FUNCTION IS EVALUATED AT THE RIGHT-HAND ENDPOINT

Approximate the area bounded by the curve $y = 3x$ and the x-axis from $x = 0$ to $x = 2$.

SOLUTION: The area sought is depicted in Figure 9. Since the area of a triangle is $\frac{1}{2}bh$, where b denotes its base and h its height, we can immediately calculate that $A_0^2 = \frac{1}{2}(2)(6) = 6$. We will show how our approximations lead to the same answer. To do so, we partition $[0, 2]$ into n equal subintervals, each having length $(b - a)/n = 2/n$. The partition points are then

$$0 = \frac{0}{n} < \frac{2}{n} < \frac{4}{n} < \cdots < \frac{2(i-1)}{n} < \frac{2i}{n} < \cdots < \frac{2(n-1)}{n} < \frac{2n}{n} = 2.$$

If we choose $x_i{}^* = 2i/n$, the right-hand endpoint of each subinterval, we obtain the situation depicted in Figure 10. Here $f(x_i{}^*) = f(2i/n) = 3 \cdot 2i/n = 6i/n$. Then (9) becomes

$$A_0^2 \approx \sum_{i=1}^{n} \overset{f(x_i{}^*)}{\overbrace{\quad}} \overset{\Delta x}{\overbrace{\quad}} f(x_i{}^*)\,\Delta x = \sum_{i=1}^{n} \left(\frac{6i}{n}\right)\frac{2}{n} = \frac{12}{n^2}\sum_{i=1}^{n} i. \tag{10}$$

Now the formula for the sum of the first n integers (proved in Appendix 2 and in Problem 4.3.25) is

---

THE SUM OF THE FIRST n POSITIVE INTEGERS

$$\sum_{i=1}^{n} i = 1 + 2 + 3 + \cdots + (n-1) + n = \frac{n(n+1)}{2}. \tag{11}$$

This formula should also be checked by verifying it for some sample values of n. Using (11) in (10), we obtain

$$A_0{}^2 \approx \frac{12}{n^2}\left[\frac{n(n+1)}{2}\right] = \frac{6(n+1)}{n} = 6\left(1 + \frac{1}{n}\right).$$

It is apparent that these approximations tend to the limit 6. Indeed, we have

Limit Theorem 2(i) on page 70

$$\lim_{n\to\infty} 6\left(1 + \frac{1}{n}\right) \overset{\downarrow}{=} 6\lim_{n\to\infty}\left(1 + \frac{1}{n}\right) = 6 \cdot 1 = 6.$$

Evidently, the method of exhaustion requires a great deal of computation. Rather than give further examples here, we wait until Section 4.6. There we prove a theorem that makes the computation of area no more difficult than the computation of an antiderivative.

However, we can show you what you can expect by evaluating an integral by making an interesting observation.

EXAMPLE 4
APPROXIMATING AN AREA USING THE MEAN-VALUE THEOREM

Compute the area bounded by the curve $y = x^2$, the x-axis, and the line $x = 1$.

SOLUTION: The area sought is sketched in Figure 1. We solved this problem in Example 1. Now we solve it a different way. Before proceeding, we remind you of the statement of the mean-value theorem (p. 240).

THEOREM   **MEAN-VALUE THEOREM**

If f is continuous on $[a, b]$ and differentiable on (a, b), then there is a number c in (a, b) such that

$$f(b) - f(a) = f'(c)(b - a). \quad \blacksquare \qquad (12)$$

Let $0 = x_0 < x_1 < x_2 < \cdots < x_{n-1} < x_n = 1$ be a regular partition of $[0, 1]$. Then, from (9),

$$A_0{}^1 \approx f(x_1{}^*)\Delta x + f(x_2{}^*)\Delta x + \cdots + f(x_n{}^*)\Delta x \qquad (13)$$

where each $x_i{}^*$ is an arbitrarily chosen point in the subinterval $[x_{i-1}, x_i]$. Since $f(x) = x^2$, (13) becomes

$$A_0{}^1 \approx x_1{}^{*2}\Delta x + x_2{}^{*2}\Delta x + \cdots + x_n{}^{*2}\Delta x. \qquad (14)$$

We can evaluate the sum in (14) for a particular choice of the $x_i{}^*$'s by making use of the mean-value theorem.

Let $F(x) = \dfrac{x^3}{3}$. Then F is continuous and differentiable over any interval and $F'(x) = x^2$ (that's why we chose it). By the mean-value theorem, there is a number c_i in each subinterval (x_{i-1}, x_i) such that

$$F'(x) = x^2$$
$$\downarrow$$
$$F(x_i) - F(x_{i-1}) = F'(c_i)(x_i - x_{i-1}) = c_i{}^2\Delta x. \qquad (15)$$

We now substitute c_i for $x_i{}^*$ in (14) to obtain

$$A_0{}^1 \approx c_1{}^2\,\Delta x + c_2{}^2\,\Delta x + \cdots + c_{n-1}{}^2\,\Delta x + c_n{}^2\,\Delta x$$

from (15)

$$\overset{\downarrow}{=} [F(x_1) - F(x_0)] + [F(x_2) - F(x_1)] + \cdots$$
$$+ [F(x_{n-1}) - F(x_{n-2})] + [F(x_n) - F(x_{n-1})]$$

$$x_0 = 0 \qquad F(x) = \frac{x^3}{3}$$
$$x_n = 1$$

all terms
except two cancel
$$= F(x_n) - F(x_0) = F(1) - F(0) = \frac{1^3}{3} - \frac{0^3}{3} = \frac{1}{3}.$$

To see more clearly how terms cancel, let us look at a particular sum, say with $n = 6$. Then the partition is

$$0 < \frac{1}{6} < \frac{2}{6} < \frac{3}{6} < \frac{4}{6} < \frac{5}{6} < 1$$
$$\uparrow \quad \uparrow \quad \uparrow \quad \uparrow \quad \uparrow \quad \uparrow \quad \uparrow$$
$$x_0 \quad x_1 \quad x_2 \quad x_3 \quad x_4 \quad x_5 \quad x_6$$

By the mean-value theorem, there exist numbers c_1, c_2, c_3, c_4, c_5, and c_6 such that

$$F\left(\frac{1}{6}\right) - F(0) = F'(c_1)\left(\frac{1}{6} - 0\right) = c_1{}^2\left(\frac{1}{6}\right) \qquad c_1 \in \left(0, \frac{1}{6}\right)$$

$$F\left(\frac{2}{6}\right) - F\left(\frac{1}{6}\right) = F'(c_2)\left(\frac{2}{6} - \frac{1}{6}\right) = c_2{}^2\left(\frac{1}{6}\right) \qquad c_2 \in \left(\frac{1}{6}, \frac{2}{6}\right)$$

$$F\left(\frac{3}{6}\right) - F\left(\frac{2}{6}\right) = F'(c_3)\left(\frac{3}{6} - \frac{2}{6}\right) = c_3{}^2\left(\frac{1}{6}\right) \qquad c_3 \in \left(\frac{2}{6}, \frac{3}{6}\right)$$

$$F\left(\frac{4}{6}\right) - F\left(\frac{3}{6}\right) = F'(c_4)\left(\frac{4}{6} - \frac{3}{6}\right) = c_4{}^2\left(\frac{1}{6}\right) \qquad c_4 \in \left(\frac{3}{6}, \frac{4}{6}\right)$$

$$F\left(\frac{5}{6}\right) - F\left(\frac{4}{6}\right) = F'(c_5)\left(\frac{5}{6} - \frac{4}{6}\right) = c_5{}^2\left(\frac{1}{6}\right) \qquad c_5 \in \left(\frac{4}{6}, \frac{5}{6}\right)$$

$$F(1) - F\left(\frac{5}{6}\right) = F'(c_6)\left(1 - \frac{5}{6}\right) = c_6{}^2\left(\frac{1}{6}\right) \qquad c_6 \in \left(\frac{5}{6}, 1\right)$$

Then (14) becomes, with c_i in place of $x_i{}^*$ and $\frac{1}{6}$ in place of Δx,

$$A_0{}^1 \approx c_1{}^2\left(\frac{1}{6}\right) + c_2{}^2\left(\frac{1}{6}\right) + c_3{}^2\left(\frac{1}{6}\right) + c_4{}^2\left(\frac{1}{6}\right) + c_5{}^2\left(\frac{1}{6}\right) + c_6{}^2\left(\frac{1}{6}\right)$$

$$= \left[F\left(\frac{1}{6}\right) - F(0)\right] + \left[F\left(\frac{2}{6}\right) - F\left(\frac{1}{6}\right)\right] + \left[F\left(\frac{3}{6}\right) - F\left(\frac{2}{6}\right)\right] + \left[F\left(\frac{4}{6}\right) - F\left(\frac{3}{6}\right)\right] + \left[F\left(\frac{5}{6}\right) - F\left(\frac{4}{6}\right)\right] + \left[F(1) - F\left(\frac{5}{6}\right)\right]$$

$$= F(1) - F(0) \qquad \text{since each of the pairs of numbers joined by arrows sum to 0.}$$

We have shown that no matter how many partition points are chosen, the sum in (14) is equal to $\frac{1}{3}$ for a particular choice of the x_i^*'s. This suggests that the area is $\frac{1}{3}$. We shall make this idea more precise in the next section.

EXAMPLE 5

APPROXIMATING AN AREA IN TWO WAYS: BY ADDING UP THE AREAS OF RECTANGLES AND BY USING THE MEAN-VALUE THEOREM

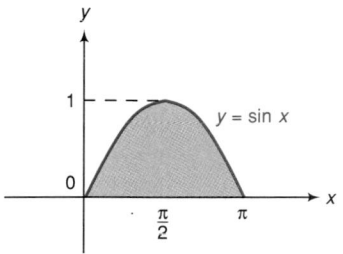

FIGURE 11
AREA UNDER THE CURVE $y = \sin x$ FOR $0 \le x \le \pi$

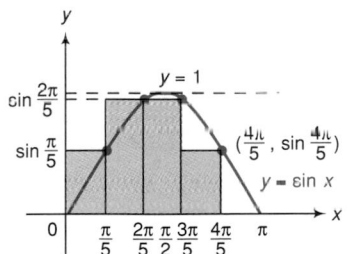

FIGURE 12
PARTITIONING $[0, \pi]$ INTO 5 SUBINTER-VALS AND EVALUATING $\sin x$ AT EACH RIGHT-HAND ENDPOINT LEADS TO 4 RECTANGLES

Approximate the area bounded by the graph of $y = \sin x$ and the x-axis for $0 \le x \le \pi$.

SOLUTION: The area we seek is the shaded area in Figure 11. We will approximate it in two ways.

Method 1: Adding up the Areas of Rectangles
We begin by partitioning the interval $[0, \pi]$.

First Approximation ($n = 5$)

Then $\Delta x = \dfrac{\pi - 0}{5} = \dfrac{\pi}{5}$ and the partition points are

$$x_0 = 0 < \frac{\pi}{5} < \frac{2\pi}{5} < \frac{3\pi}{5} < \frac{4\pi}{5} < \pi = x_5.$$

If we use the right-hand endpoint of each subinterval, we get four rectangles, as in Figure 12. Notice that the two middle rectangles have the same height because $\sin \dfrac{2\pi}{5} = \sin \dfrac{3\pi}{5}$ (since $\sin(\pi - \theta) = \sin \theta$). Likewise, the first and fourth rectangles have the same height because $\sin \dfrac{\pi}{5} = \sin \dfrac{4\pi}{5}$ (for the same reason). The last "rectangle" has a height of 0 because $\sin \pi = 0$. We compute, using (9),

$$A_0^{\pi} \approx \left(\sin \frac{\pi}{5} \right) \frac{\pi}{5} + \left(\sin \frac{2\pi}{5} \right) \frac{\pi}{5} + \left(\sin \frac{3\pi}{5} \right) \frac{\pi}{5}$$

$$+ \left(\sin \frac{4\pi}{5} \right) \frac{\pi}{5} + (\sin \pi) \frac{\pi}{5}$$

$$\approx \frac{\pi}{5} [0.5878 + 0.9511 + 0.9511 + 0.5879 + 0]$$

$$\approx 0.6283[3.0778] = 1.9338.$$

This is our first approximation to the area.

Second Approximation ($n = 10$)

Then $\Delta x = \dfrac{\pi - 0}{10} = \dfrac{\pi}{10}$ and the partition points are

$$0 < \frac{\pi}{10} < \frac{2\pi}{10} < \frac{3\pi}{10} < \frac{4\pi}{10} < \frac{5\pi}{10} < \frac{6\pi}{10} < \frac{7\pi}{10} < \frac{8\pi}{10} < \frac{9\pi}{10} < \pi$$

or

$$0 < \frac{\pi}{10} < \frac{\pi}{5} < \frac{3\pi}{10} < \frac{2\pi}{5} < \frac{\pi}{2} < \frac{3\pi}{5} < \frac{7\pi}{10} < \frac{4\pi}{5} < \frac{9\pi}{10} < \pi.$$

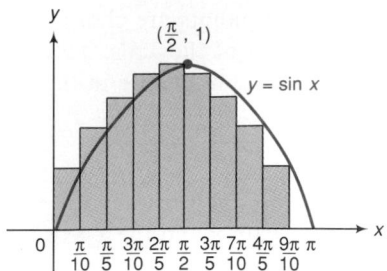

FIGURE 13
PARTITIONING $[0, \pi]$ INTO 10 SUBINTERVALS
AND EVALUATING $\sin x$ AT EACH RIGHT-
HAND ENDPOINT LEADS TO 9 RECTANGLES

The result is sketched in Figure 13. We then obtain

$$A_0{}^\pi \approx \left(\sin\frac{\pi}{10}\right)\frac{\pi}{10} + \left(\sin\frac{\pi}{5}\right)\frac{\pi}{10} + \left(\sin\frac{3\pi}{10}\right)\frac{\pi}{10} + \left(\sin\frac{2\pi}{5}\right)\frac{\pi}{10}$$

$$+ \left(\sin\frac{\pi}{2}\right)\frac{\pi}{10} + \left(\sin\frac{3\pi}{5}\right)\frac{\pi}{10} + \left(\sin\frac{7\pi}{10}\right)\frac{\pi}{10} + \left(\sin\frac{4\pi}{5}\right)\frac{\pi}{10}$$

$$+ \left(\sin\frac{9\pi}{5}\right)\frac{\pi}{10} + (\sin\pi)\frac{\pi}{10}$$

$$\approx \frac{\pi}{10}\left[0.3090 + 0.5878 + 0.8090 + 0.9511 + 1 + 0.9511\right.$$

$$\left. + 0.8090 + 0.5878 + 0.3090 + 0\right]$$

$$\approx (0.31416)(6.3138) \approx 1.9835.$$

This is our second approximation to the area.

Third Approximation ($n = 50$)

Then $\Delta x = \dfrac{\pi}{50}$ and

$$A_0{}^\pi \approx \frac{\pi}{50}\left(\sin\frac{\pi}{50} + \sin\frac{2n}{50} + \sin\frac{3\pi}{50} + \cdots + \sin\frac{49\pi}{50} + \sin\pi\right).$$

We won't write out all the numbers, but the result is

$$A_0{}^\pi \approx (0.062831853)(31.82051595) = 1.999341981.$$

It looks as if the approximations are approaching 2. The area is indeed equal to 2—a fact we will be able to prove in Section 4.6. Rather than take more approximations here, we will use our other method to give us the same answer with much less work.

Method 2: Using the Mean-Value Theorem

Suppose we partition the interval $[0, \pi]$ into n subintervals. Then $\Delta x = \dfrac{\pi - 0}{n} = \dfrac{\pi}{n}$ and the partition points are

$$x_0 = 0 < \frac{\pi}{n} < \frac{2\pi}{n} < \frac{3\pi}{n} < \cdots < \frac{(n-1)\pi}{n} < \frac{n\pi}{n} = \pi = x_n.$$

We have already seen examples of this partitioning for $n = 5$, 10, and 50. A typical subinterval has the form

$$[x_{i-1}, x_i] = \left[\frac{i-1}{n}\pi, \frac{i\pi}{n}\right].$$

Let $F(x) = -\cos x$ (an antiderivative of $\sin x$) so $F'(x) = \sin x$. Then F is everywhere continuous and differentiable, so F satisfies the hypotheses of the mean-value theorem. This means that in each interval

$\left(\dfrac{i-1}{n}\pi, \dfrac{i\pi}{n}\right)$ there is a member c_i such that

$$F\left(\frac{i\pi}{n}\right) - F\left(\frac{i-1}{n}\pi\right) = F'(c_i)\left(\frac{i\pi}{n} - \frac{i-1}{n}\pi\right) = F'(c_i)\frac{\pi}{n}$$

$$= \sin c_i\left(\frac{\pi}{n}\right). \qquad\qquad \textbf{(16)}$$

From (9), we have

$$A_0{}^\pi \approx (\sin c_1)\frac{\pi}{n} + (\sin c_2)\frac{\pi}{n} + \cdots + (\sin c_{n-1})\frac{\pi}{n} + (\sin c_n)\frac{\pi}{n}$$

from (16)

$$= \left[F\left(\frac{\pi}{n}\right) - F(0)\right] + \left[F\left(\frac{2\pi}{n}\right) - F\left(\frac{\pi}{n}\right)\right] + \cdots$$

$$+ \left[F\left(\frac{n-1}{n}\pi\right) - F\left(\frac{n-2}{n}\pi\right)\right] + \left[F(\pi) - F\left(\frac{n-1}{n}\pi\right)\right].$$

As in Example 4, all terms except two cancel and

$$A_0{}^\pi \approx F(\pi) - F(0) = -\cos\pi - (-\cos 0) = -(-1) - (-1)$$
$$= 1 + 1 = 2.$$

To illustrate this cancellation, we observe that for $n = 5$

$$A_0{}^\pi \approx (\sin c_1)\frac{\pi}{5} + (\sin c_2)\frac{\pi}{5} + (\sin c_3)\frac{\pi}{5} + (\sin c_4)\frac{\pi}{5} + (\sin c_5)\frac{\pi}{5}$$

where $c_1 \in \left(0, \dfrac{\pi}{5}\right)$, $c_2 \in \left(\dfrac{\pi}{5}, \dfrac{2\pi}{5}\right)$, $c_3 \in \left(\dfrac{2\pi}{5}, \dfrac{3\pi}{5}\right)$, $c_4 \in \left(\dfrac{3\pi}{5}, \dfrac{4\pi}{5}\right)$, and $c_5 \in \left(\dfrac{4\pi}{5}, \pi\right)$, and, for example,

$$F\left(\frac{3\pi}{5}\right) - F\left(\frac{2\pi}{5}\right) = F'(c_3)\left(\frac{3\pi}{5} - \frac{2\pi}{5}\right) = (\sin c_3)\frac{\pi}{5}.$$

Thus

$$A_0{}^\pi \approx \left[F\left(\frac{\pi}{5}\right) - F(0)\right] + \left[F\left(\frac{2\pi}{5}\right) - F\left(\frac{\pi}{5}\right)\right] + \left[F\left(\frac{3\pi}{5}\right) - F\left(\frac{2\pi}{5}\right)\right] + \left[F\left(\frac{4\pi}{5}\right) - F\left(\frac{3\pi}{5}\right)\right] + \left[F(\pi) - F\left(\frac{4\pi}{5}\right)\right]$$

$$= F(\pi) - F(0) = 2, \text{ as before.}$$

In Examples 1, 4, and 5, we saw a truly remarkable phenomenon. We can approximate the area under a curve by summing the areas of a large number of rectangles. Or we can approximate (in fact, get the exact value of) the area by evaluating an antiderivative of the function at just two values and then subtracting one from the other. The connection between these two very different processes is given in the **fundamental theorem of calculus**, discussed in Section 4.6.

■ PROBLEMS 4.4

■ SELF-QUIZ

I. The area of the region bounded by the x-axis, the graph of $y = x^2$, and the vertical line $x = 1$ is *not* approximated by _____.

a. $(\frac{1}{3})(\frac{0}{3})^2 + (\frac{1}{3})(\frac{1}{3})^2 + (\frac{1}{3})(\frac{2}{3})^2$

b. $(\frac{1}{3})(\frac{1}{3})^2 + (\frac{1}{3})(\frac{2}{3})^2 + (\frac{1}{3})(\frac{3}{3})^2$

c. $(\frac{1}{3})(\frac{1}{6})^2 + (\frac{1}{3})(\frac{3}{6})^2 + (\frac{1}{3})(\frac{5}{6})^2$

d. $(\frac{0}{3})^2 + (\frac{1}{3})^2 + (\frac{2}{3})^2$

II. The region bounded by the x-axis, the graph of $y = x^2$, and the vertical line $x = 2$ has an area that is approximated by _____.

a. $(\frac{2}{4})(\frac{1}{2}) + (\frac{2}{4})(\frac{2}{2}) + (\frac{2}{4})(\frac{3}{2}) + (\frac{2}{4})(\frac{4}{2})$

b. $(\frac{2}{4})(\frac{1}{2})^2 + (\frac{2}{4})(\frac{2}{2})^2 + (\frac{2}{4})(\frac{3}{2})^2 + (\frac{2}{4})(\frac{4}{2})^2$

c. $(\frac{1}{2})^2 + (\frac{2}{2})^2 + (\frac{3}{2})^2 + (\frac{4}{2})^2$

d. $(\frac{2}{4})^2(\frac{1}{2}) + (\frac{2}{4})^2(\frac{2}{2}) + (\frac{2}{4})^2(\frac{3}{2}) + (\frac{2}{4})^2(\frac{4}{2})$

III. The area of the region bounded by the x-axis, the graph of $y =$ _____, and the vertical lines $x = 1$ and $x = 3$ is approximated by $(\frac{2}{5})(\frac{5}{7}) + (\frac{2}{5})(\frac{5}{9}) + (\frac{2}{5})(\frac{5}{11}) + (\frac{2}{5})(\frac{5}{13}) + (\frac{2}{5})(\frac{5}{15})$.

a. $\sin x$ b. x

c. $1/x$ d. $5/x$

■ DRILL

Problems 1–4 refer to the function $f(x) = x^2$ and the interval $[0, 1]$.

1. Calculate s and S, where the interval $[0, 1]$ is divided into 16 smaller subintervals each having length $\frac{1}{16}$.

2. Calculate s and S, where the interval $[0, 1]$ is divided into 32 equal length subintervals.

3. Verify that if the interval $[0, 1]$ is divided into n pieces each having length $1/n$, then

$$s = \frac{1}{n^3}\left[0^2 + 1^2 + \cdots + (n-1)^2\right]$$

and

$$S = \frac{1}{n^3}\left[1^2 + 2^2 + \cdots + n^2\right].$$

*4. Without using the explicit formulas in Problem 3, explain why the underapproximation s and the overapproximation S for the area of the region $\{(x, y): 0 \le x \le 1, 0 \le y \le x^2\}$ differ by $1/n$.

In Problems 5–8, approximate the area bounded by the given straight line, the x-axis, and the lines $x = a$ and $x = b$, where a and b are the endpoints of the given interval. Compute three approximations with the interval $[a, b]$ divided into two, four, and eight subintervals of equal length. Then calculate the actual area using the fact that a triangle with base length b and height h has area $\frac{1}{2}bh$.

5. $y = 7x$; $[0, 4]$
6. $y = 4x$; $[1, 2]$
   [*Hint:* To calculate the exact area of this trapezoidal region, divide the region into a triangle on top of a rectangle.]

7. $y = 3x + 2$; $[0, 3]$
8. $y = 7x - 3$; $[1, 4]$

In Problems 9–20, a curve is given in the form $y = f(x)$, a closed interval is given in the form $[a, b]$, and an integer is specified in the form $\{n\}$. Your task is to approximate the area of the region bounded by the given curve, the x-axis, and the lines $x = a$ and $x = b$.

a. Divide the interval into a number (given in braces, $\{n\}$) of subintervals of equal length, choose a convenient point in each subinterval, and compute the total area of the collection of small rectangles whose tops meet the curve at those convenient points.

b. Repeat the process by doubling the number of subintervals.

c. Obtain a general formula for the area enclosed by n approximating rectangles, where n is an arbitrary positive integer.

d. Find the exact area by taking the limit of your formula obtained in part (c) as $n \to \infty$.

The following facts will be useful. (Each can be proved by mathematical induction; see Appendix 2 or Problems 4.3.25–27.)

n terms

$$(i) \quad \sum_{j=1}^{n} 1 = \overbrace{1 + 1 + \cdots + 1}^{} = n$$

$$(ii) \quad \sum_{j=1}^{n} j = 1 + 2 + \cdots + n = \frac{n(n+1)}{2}$$

$$(iii) \quad \sum_{j=1}^{n} j^2 = 1^2 + 2^2 + \cdots + n^2 = \frac{n(n+1)(2n+1)}{6}$$

$$(iv) \quad \sum_{j=1}^{n} j^3 = 1^3 + 2^3 + \cdots + n^3 = \left[\frac{n(n+1)}{2}\right]^2$$

CONTINUED

9. $y = \frac{1}{2}x^2$; $[0, 2]$; $\{4\}$
10. $y = 2x^2$; $[-1, 1]$; $\{3\}$
11. $y = 1 - x^2$; $[0, 1]$; $\{8\}$
12. $y = (1 - x)^2$; $[0, 1]$; $\{6\}$
13. $y = x^2$; $[1, 2]$; $\{8\}$ 14. $y = 17x^2$; $[0, 8]$; $\{5\}$
15. $y = x^3$; $[0, 5]$; $\{4\}$ 16. $y = x^3$; $[1, 2]$; $\{4\}$
17. $y = x + x^2$; $[0, 1]$; $\{6\}$
18. $y = 3x + x^3$; $[0, 1]$; $\{5\}$
19. $y = 1 + 2x + 3x^2$; $[0, 1]$; $\{4\}$
20. $y = 1 + x + x^2 + x^3$; $[0, 1]$; $\{4\}$

21. Consider the problem of calculating the area bounded by the curve $y = x^2$, the x-axis, and the lines $x = a$ and $x = b$.

a. Divide the interval $[a, b]$ into n subintervals of equal length; that is, $\Delta x = (b - a)/n$. What are the endpoints of each subinterval?
b. Letting $x_i{}^*$ be the right-hand endpoint of each subinterval, calculate $f(x_i{}^*) = (x_i{}^*)^2$.
c. Use formulas (ii) and (iii) to approximate $A_a{}^b$ for 4, 8, 16, and 32 subintervals.
*d. Obtain a formula for the total area under the curve on the interval $[a, b]$.

*22. Following the steps of the preceding problem and using formula (iv) above, obtain a formula for the area bounded by the curve $y = x^3$, the x-axis, and the lines $x = a$ and $x = b$ where $0 < a < b$.

🖩 **CALCULATOR PROBLEMS**

In Problems 23–30, approximate the area under each curve by partitioning the given interval into 10 subintervals, evaluating the function at the right-hand endpoint, and then computing the sum:

$$A_a{}^b \approx f(x_1)\,\Delta x + f(x_2)\,\Delta x + \cdots + f(x_{10})\,\Delta x.$$

Repeat the process using 20 subintervals.

23. $f(x) = x^4$, $0 \le x \le 2$

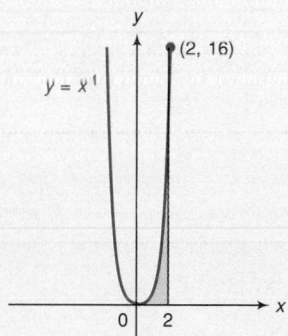

24. $f(x) = \cos x$, $0 \le x \le \dfrac{\pi}{4}$

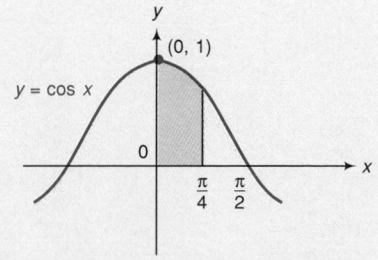

25. $f(x) = 1 + 2x^2$, $1 \le x \le 2$

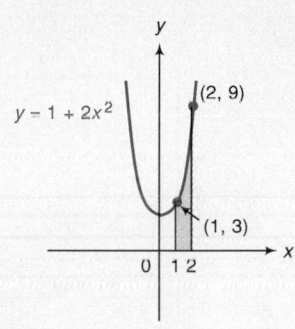

26. $f(x) = \sqrt{x}$, $0 \le x \le 1$

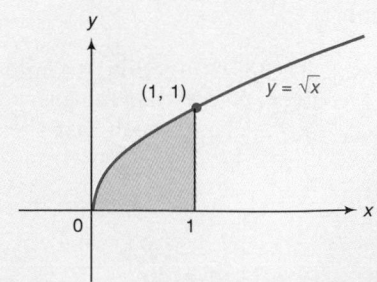

CONTINUED

27. $f(x) = \sqrt{x}$, $1 \leq x \leq 9$

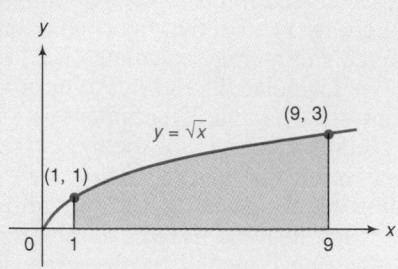

29. $y = \tan x$, $0 \leq x \leq \dfrac{\pi}{3}$

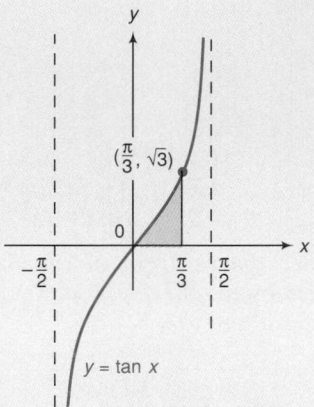

28. $y = \dfrac{1}{x^2}$, $\dfrac{1}{2} \leq x \leq 1$

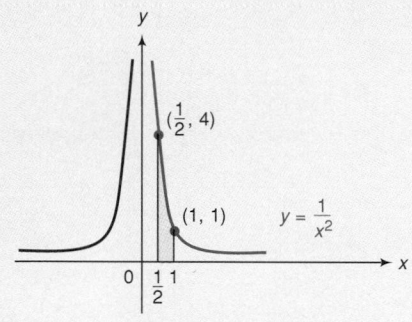

30. $f(x) = \dfrac{1}{x}$, $1 \leq x \leq 2.718281828$

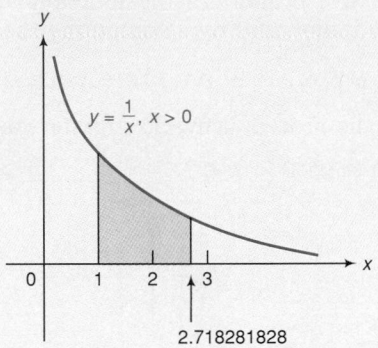

■ SHOW/PROVE/DISPROVE

Suppose that $f(x) \geq 0$ for $a \leq x \leq b$. Then, according to equation (9), the area under the graph of f for $a \leq x \leq b$ is approximated by

$$f(x_1{}^*)\,\Delta x + f(x_2{}^*)\,\Delta x + \cdots + f(x_n{}^*)\,\Delta x \qquad \textbf{(18)}$$

where $x_1{}^*, x_2{}^*, \ldots, x_n{}^*$ and Δx are as defined in this section.

In Problems 31–38, a function, an interval, and a value A are given. Show that for every n there are numbers $x_1{}^*, x_2{}^*, \ldots, x_n{}^*$ such that the sum in (18) is exactly equal to A.

31. $f(x) = x^4$; $[0, 1]$; $A = \dfrac{1}{5}$

32. $f(x) = x^3$; $[1, 3]$; $A = 20$

33. $f(x) = \sqrt{x}$; $[0, 1]$; $A = \dfrac{2}{3}$

34. $f(x) = \sqrt{x}$; $[1, 9]$; $A = \dfrac{52}{3}$

35. $f(x) = \sin x$; $\left[\dfrac{\pi}{6}, \dfrac{\pi}{3}\right]$; $A = \dfrac{\sqrt{3} - 1}{2}$

36. $f(x) = \cos x$; $\left[\dfrac{\pi}{4}, \dfrac{\pi}{2}\right]$; $A = 1 - \dfrac{\sqrt{2}}{2}$

37. $f(x) = \dfrac{1}{x^2}$; $[1, 2]$; $A = \dfrac{1}{2}$

38. $f(x) = 1 + x + x^2 + x^3$; $[0, 1]$;
$$A = 1 + \dfrac{1}{2} + \dfrac{1}{3} + \dfrac{1}{4} = \dfrac{25}{12}$$

Problems 39–46 are like Problems 31–38 except that *you* must find the number A.

39. $f(x) = x^5$; $[0, 1]$
40. $f(x) = x^{1/3}$; $[0, 1]$
41. $f(x) = x^3 - x^2$; $[1, 2]$
42. $f(x) = 4x^3 - 5$; $[2, 3]$

43. $f(x) = \dfrac{1}{x^2}$; $[1, 3]$

44. $f(x) = \cos x$; $\left[0, \dfrac{\pi}{6}\right]$

45. $f(x) = 2 \sin x$; $\left[0, \dfrac{\pi}{4}\right]$

*46. $f(x) = \sin x + \cos x$; $\left[0, \dfrac{\pi}{2}\right]$

*47. Let

$$S_n = \frac{1}{n} \sum_{k=1}^{n} \sqrt{1 - \left(\frac{k}{n}\right)^2}.$$

a. Describe a region whose area is approximated by S_n.
b. By using general information about the area of the region described in part (a), find $\lim_{n \to \infty} S_n$.

■ CHALLENGE

*48. Suppose that $f(x) \geq 0$ for $a \leq x \leq b$ and that F is an antiderivative for f. Let $a = x_0 < x_1 < x_2 < \cdots < x_n = b$ be a regular partition of $[a, b]$. Show that there are numbers $c_1, c_2, \ldots, c_n$ where c_i is in the ith subinterval such that

$$f(c_1)\,\Delta x + f(c_2)\,\Delta x + \cdots + f(c_n)\,\Delta x = F(b) - F(a).$$

When you have done this, you will have proved one case of the fundamental theorem of calculus.

■ ANSWERS TO SELF-QUIZ

I. d   II. b   III. c

4.5 THE DEFINITE INTEGRAL

In Section 4.4, we computed the area of the region bounded by the graph of a function, the x-axis, and the lines $x = a$ and $x = b$. We did so in three cases: $f(x) = x^2$ (Examples 1 and 4), $f(x) = 3x$ (Example 3), and $f(x) = \sin x$ (Example 5). Our basic procedure was to write the sum

$$\sum_{i=1}^{n} f(x_i^*)\,\Delta x \tag{1}$$

and then see what happened as $n \to \infty$ ($\Delta x \to 0$). In all the examples three conditions held:

  (i) f is continuous on $[a, b]$.
  (ii) $f(x) \geq 0$ for $x \in [a, b]$.
  (iii) All intervals in the partition have equal length. Specifically, $\Delta x = (b - a)/n$ is the length of each subinterval.

In this section, we take the limit of sums like the sum in (1) to define the definite integral. In doing so, we will relax all three of the assumptions listed above. That is, we will allow the following possibilities:

  (i) f may be discontinuous at some points in $[a, b]$; but f remains bounded on $[a, b]$.

(ii) $f(x)$ may be negative for some (or even all) points in $[a, b]$.

(iii) The subintervals $[x_{i-1}, x_i]$ might have different lengths.

DEFINITION **PARTITION**

A **partition** P of $[a, b]$ is a division of $[a, b]$ into a number of smaller subintervals. The **partition points** are labeled $x_0, x_1, x_2, \ldots, x_n$, where $x_0 = a$ and $x_n = b$. It is assumed that

$$x_0 < x_1 < x_2 < \cdots < x_n.$$

We define

$$\Delta x_i = x_i - x_{i-1}$$

so that

$$\Delta x_1 = x_1 - x_0, \Delta x_2 = x_2 - x_1, \ldots, \Delta x_n = x_n - x_{n-1},$$

and

$$\sum_{i=1}^{n} \Delta x_i = b - a = \text{total length of the interval.}$$

This is illustrated in Figure 1. ■

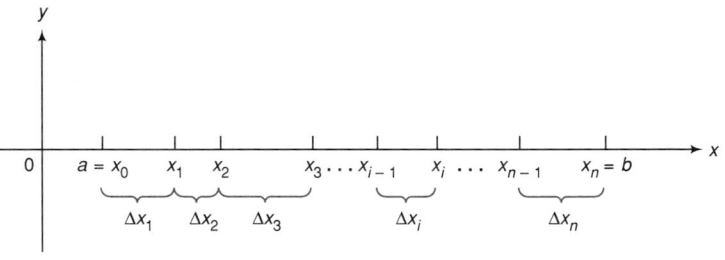

FIGURE 1 A PARTITION OF $[a, b]$

It is not regular because the lengths of the subintervals are not equal.

EXAMPLE 1
TWO PARTITIONS OF $[0, 1]$

Two partitions of the interval $[0, 1]$ are sketched in Figure 2.

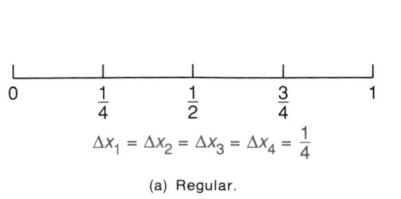

(a) Regular.

$\Delta x_1 = \frac{1}{10}$
$\Delta x_2 = \frac{1}{10}$
$\Delta x_3 = \frac{2}{5}$
$\Delta x_4 = \frac{3}{10}$
$\Delta x_5 = \frac{1}{10}$

(b) Not regular.

FIGURE 2 TWO PARTITIONS OF $[0, 1]$

As in Section 4.4, a partition is called **regular** if the partition points are equally spaced. That is,

---

REGULAR PARTITION

the partition is regular if $\Delta x_i = \dfrac{b-a}{n}$ for $i = 1, 2, \ldots, n$.   **(2)**

In Figure 2, partition (a) is regular.

In Section 4.4, we examined what happened as $n \to \infty$. Since $\Delta x = (b-a)/n$, $n \to \infty$ implies that $\Delta x \to 0$. If the partition is not regular, we need a new way to describe the fact that the lengths of the subintervals are getting small. We do so by writing

$$\max \Delta x_i \to 0, \tag{3}$$

where $\max \Delta x_i$ is the length of the largest subinterval in the partition.

EXAMPLE 2
FINDING $\max \Delta x_i$ FOR TWO PARTITIONS

In Figure 2, we have the following:

(a) $\max \Delta x_i = \frac{1}{4}$

(b) $\max \Delta x_i = \frac{2}{5}$

NOTE: If P is regular, then

$$\max \Delta x_i = \Delta x = \frac{b-a}{n}. \tag{4}$$

We can now form sums similar to (1) for any function f defined (but not necessarily continuous or nonnegative) on $[a, b]$.

Step 1. Partition the interval $[a, b]$ by

$$a = x_0 < x_1 < x_2 < \cdots < x_n = b.$$

Step 2. Choose a number x_i^* in each subinterval $[x_{i-1}, x_i]$.

Step 3. Form the sum

A RIEMANN SUM

$$f(x_1^*)\Delta x_1 + f(x_2^*)\Delta x_2 + \cdots + f(x_n^*)\Delta x_n = \sum_{i=1}^{n} f(x_i^*)\Delta x_i. \tag{5}$$

The sum (5) is called a **Riemann sum**.[†]

For every partition of the interval $[a, b]$ and for every choice of the numbers x_i^*, the Riemann sum (5) is a *real number*. In certain circumstances,

---

[†] See the biographical sketch on page 306.

that number represents an approximation to the area under a curve. So it is natural to ask what happens to the Riemann sum when the lengths of the subintervals get small—that is, what happens to $\sum_{i=1}^{n} f(x_i^*)\Delta x_i$ when max $\Delta x_i \to 0$? We are taking a limit, but it is a special kind of limit.

Let L be a real number. Then we can define the limit of Riemann sums in the following intuitive way.

DEFINITION   **INTUITIVE DEFINITION OF LIMIT OF RIEMANN SUMS**

$$\lim_{\max \Delta x_i \to 0} \sum_{i=1}^{n} f(x_i^*)\Delta x_i = L$$

if $\sum_{i=1}^{n} f(x_i^*)\Delta x_i$ is as close as we wish to L when max Δx_i is small, no matter how the numbers $x_1^*, x_2^*, \ldots, x_n^*$ are chosen.   ■

We will make this definition more mathematically precise at the end of this section.

We now define one of the most important concepts in our study of calculus.

DEFINITION   **THE DEFINITE INTEGRAL**

Let f be defined on $[a, b]$ with $a < b$. Then the **definite integral** of f over the interval $[a, b]$, written $\int_a^b f(x)\, dx$, is given by

$$\int_a^b f(x)\, dx = \lim_{\max \Delta x_i \to 0} \sum_{i=1}^{n} f(x_i^*)\Delta x_i \tag{6}$$

whenever the limit in (6) exists.

The process of calculating an integral is called **integration**, and the numbers a and b are called the **lower** and **upper limits of integration**, respectively. If the limit in (6) exists, then f is said to be **integrable on the interval $[a, b]$**. The function f in (6) is called the **integrand**.   ■

The variable x in (6) is called a **dummy variable** since it could be replaced by *any other* variable without changing the value of the integral. This is true because the definite integral is a number. We could, instead, subdivide $[a, b]$ by

$$a = t_0 < t_1 < t_2 < \cdots < t_{n-1} < t_n = b.$$

That is, we are simply renaming the same numbers chosen before. Then we could choose t_i^* in $[t_{i-1}, t_i]$ and (6) would become

$$\int_a^b f(t)\, dt = \lim_{\max \Delta t_i \to 0} \sum_{i=1}^{n} f(t_i^*)\Delta t_i. \tag{7}$$

This expression is, of course, the same definition as in (6). We can also substitute any other variable for t. Thus,

$$\int_a^b f(x)\, dx = \int_a^b f(t)\, dt = \int_a^b f(z)\, dz = \int_a^b f(\text{dummy})\, d(\text{dummy}). \tag{8}$$

REMARK: There is a great difference between the definite integral and the indefinite integral, or antiderivative. *The definite integral is a number, while the indefinite integral is a set of functions.* Nevertheless, we use the words "integral," "integrable," and "integration" in both cases. There should not be any confusion between the two uses of the words, since the two integrals stand for such different things.

We can tell the difference between a definite and indefinite integral because definite integrals always have lower and upper limits $[\int_a^b f(x)\,dx]$ while indefinite integrals do not. We will show the connection between the two integrals in Section 4.6.

Following this definition, a basic question remains: What functions are integrable? After all, the limit in (6) is usually difficult to obtain (as you could imagine after working through the problems of the previous section), and so it would be nice to know *before* doing any calculations that a given integral does exist. The following theorem, whose proof is beyond the scope of this book,[†] tells us that a great number of the functions we have already discussed are integrable.

THEOREM 1 **EXISTENCE OF DEFINITE INTEGRALS**

Let f be continuous on the interval $[a, b]$. Then f is integrable on $[a, b]$. That is, $\int_a^b f(x)\,dx$ exists. ■

EXAMPLE 3
EIGHT DEFINITE INTEGRALS

The following integrals all exist, since the functions being integrated are all continuous on the interval of integration.

(a) $\int_1^7 x^7\,dx$ **(b)** $\int_1^3 \sqrt{x}\,dx$

(c) $\int_{-10}^{10} |x|\,dx$ **(d)** $\int_0^3 \frac{1}{\sqrt{(x+1)(x+2)}}\,dx$

(e) $\int_3^5 \frac{1}{x}\,dx$ **(f)** $\int_2^4 t^{3/5}\,dt$

(g) $\int_0^{1,000,000} s^{100}\,ds$ **(h)** $\int_{-3}^0 \frac{1}{z-1}\,dz$

There are an infinite number of ways to partition an interval, and there are infinitely many numbers in the interval $[x_{i-1}, x_i]$. However, if f is continuous on $[a, b]$, then we know that f is integrable on $[a, b]$, which means that the limit in (6) exists. But if the limit exists, $\sum_{i=1}^n f(x_i^*)\,\Delta x_i$ is close to L for max Δx_i small, no matter how $[a, b]$ is partitioned or how the x_i^*s are chosen. So if f is continuous, we can make things simpler by using a regular partition of $[a, b]$, and if we wish, we can compute Riemann sums by using the right- or left-hand endpoints of each subinterval, as in Section 4.4. According to equation (4), for a regular partition max $\Delta x_i \to 0$ means $\Delta x \to 0$, or $n \to \infty$, where n is the number of subintervals. Thus, using regular partitions, we have the following alternative definition of the definite integral.

[†] For a proof, consult any book in advanced calculus, such as R. C. Buck and E. F. Buck, *Advanced Calculus*, 3rd ed. (New York: McGraw-Hill, 1978), p. 176.

DEFINITION   **ALTERNATIVE DEFINITION OF THE DEFINITE INTEGRAL**

Let f be continuous on $[a, b]$ with $a < b$. Then,

> **THE DEFINITE INTEGRAL DEFINED USING REGULAR PARTITIONS**
>
> $$\int_a^b f(x)\,dx = \lim_{\Delta x \to 0} \sum_{i=1}^{n} f(x_i^*)\,\Delta x = \lim_{n \to \infty} \sum_{i=1}^{n} f(x_i^*)\,\frac{b-a}{n}. \quad \text{(9)}$$

EXAMPLE 4
THE VALUES OF THREE DEFINITE
INTEGRALS

From Examples 4.4.1, 4.4.3, and 4.4.5 we see that

$$\int_0^1 x^2\,dx = \frac{1}{3}, \qquad \int_0^2 3x\,dx = 6, \qquad \text{and} \qquad \int_0^\pi \sin x\,dx = 2.$$

The following theorem shows how to compute a particularly simple definite interval.

THEOREM 2   **THE INTEGRAL OF A CONSTANT FUNCTION**

Let c be a constant. Then,

$$\int_a^b c\,dx = c(b - a). \quad \text{(10)}$$

In particular, for $c = 1$ we have

$$\int_a^b 1\,dx = \int_a^b dx = b - a. \quad \blacksquare \quad \text{(11)}$$

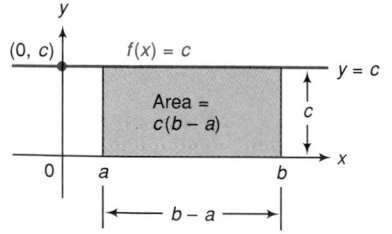

FIGURE 3
AREA UNDER A CONSTANT FUNCTION
EQUALS THE CONSTANT TIMES THE
LENGTH OF THE INTERVAL

This theorem makes sense geometrically if c is a positive constant. Look at Figure 3. Then $c(b - a)$ is equal to the area of the rectangle under the curve $y = c$ from $x = a$ to $x = b$. The proof of the theorem is left as an exercise (see Problem 39).

To this point, we have required that a be less than b in the definition of the integral. The cases $a = b$ and $a > b$ are defined next.

DEFINITION

For any real number a,

$$\int_a^a f(x)\,dx = 0. \quad \blacksquare \quad \text{(12)}$$

Note that if $\int_a^a f(x)\,dx$ represents area, then this definition states that the area of a region with a width of zero is zero. (The interval $[a, a]$ has a width of zero.) Note that it is not necessary to prove formula (12) because it is a *definition*.

> **DEFINITION**   **REVERSING THE DIRECTION OF INTEGRATION**
>
> If $a < b$, and $\int_a^b f(x)\,dx$ exists, then
>
> $$\int_b^a f(x)\,dx = -\int_a^b f(x)\,dx. \quad \blacksquare$$
> (13)

This definition says that reversing the direction of integration, that is, from b to a instead of from a to b, changes the sign of the integral, thus enabling us to define the definite integral when $a > b$.

EXAMPLE 5
REVERSING THE DIRECTION
OF INTEGRATION

Calculate $\int_1^0 x^2\,dx$.

SOLUTION:

$$\int_1^0 x^2\,dx = -\int_0^1 x^2\,dx \overset{\text{Example 4}}{=} -\frac{1}{3}$$

The definite integral does not always represent the area under a curve. The following example illustrates this fact.

EXAMPLE 6
A NEGATIVE DEFINITE INTEGRAL

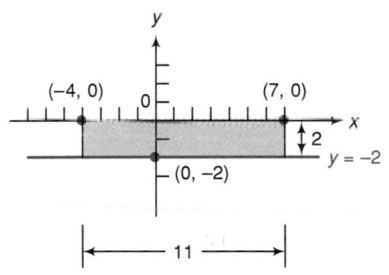

FIGURE 4
THE INTEGRAL OF A CONSTANT FUNCTION EQUALS THE CONSTANT TIMES THE LENGTH OF THE INTERVAL.

Calculate $\int_{-4}^7 (-2)\,dx$.

SOLUTION: $b - a = 7 - (-4) = 11$, so that

$$\int_{-4}^7 (-2)\,dx \overset{\text{From Theorem 2}}{=} -2[7 - (-4)] = (-2)11 = -22.$$

Here the integral is negative. To see why, look at Figure 4. The area enclosed in the figure is equal to 22 square units. However, we see that *the integral treats areas below the x-axis as negative.* Therefore we cannot calculate an area under the curve $y = f(x)$ by simple integration unless $f(x) \geq 0$ on $[a, b]$. To handle the more general case, we define area in terms of the definite integral.

> **DEFINITION**   **AREA**
>
> The **area**[†] bounded by the function $y = f(x)$, the x-axis, and the lines $x = a$ and $x = b$ (for $a < b$) is denoted by $A_a{}^b$ and is defined by the formula
>
> $$A_a{}^b = \int_a^b |f(x)|\,dx. \quad \blacksquare$$
> (14)

[†] If $f(x)$ is negative for some values of x in $[a, b]$, then $A_a{}^b$ is sometimes called the **net area** of f over $[a, b]$.

NOTE: If $a < b$ and if $f(x)$ is nonnegative for $a \leq x \leq b$, then $|f(x)| = f(x)$, and in this case $A_a^b = \int_a^b f(x)\,dx$.

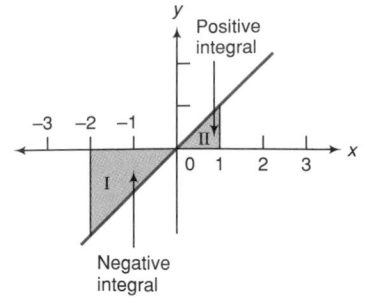

FIGURE 5
AREA BETWEEN GRAPH OF $y = x$ AND x-AXIS FOR $-2 \leq x \leq 1$

⊗ **WARNING:**   A fairly common error made by students when they first face the problem of calculating areas by integration is to employ the following *incorrect* reasoning: "If I get a negative answer when calculating an area, then all I need to do is to take the absolute value of my answer to make it right." To see why this reasoning is faulty, consider the problem of computing the area bounded by the line $y = x$ and the x-axis for x between -2 and 1. This area is drawn in Figure 5. It is easy to see that the area of triangle I is 2 $[A = \frac{1}{2}bh = \frac{1}{2}(2)(2)]$ and the area of triangle II is $\frac{1}{2}$. Thus, the total area is $\frac{5}{2}$. However, it is not difficult to show that

$$\int_{-2}^{1} x\,dx = -\frac{3}{2}.$$

This is the answer we would get if we forgot to take the absolute value in $\int_{-1}^{1} |x|\,dx$. Changing the $-\frac{3}{2}$ to $\frac{3}{2}$ will not give us the correct answer. ⊗

We shall compute a number of areas under and between curves after we have discussed the fundamental theorem of calculus in Section 4.6.

We now state three theorems that can be very useful for calculating integrals. They are proved at the end of the section.

THEOREM 3

If f is continuous on $[a, b]$ and if $a < c < b$, then f is integrable on $[a, c]$ and on $[c, b]$, and

$$\int_a^b f = \int_a^c f + \int_c^b f. \quad \blacksquare \tag{15}$$

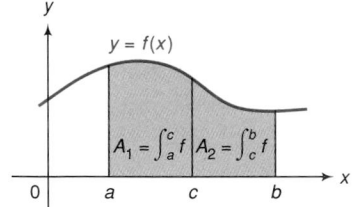

FIGURE 6
THE AREA UNDER THE CURVE FROM a TO b EQUALS THE AREA FROM a TO c PLUS THE AREA FROM c TO b.

Theorem 3 is obvious if $\int_a^b f$ represents the area under a curve. In Figure 6, the area under the curve is given by $\int_a^b f$, which is equal to $A_1 + A_2 = \int_a^c f + \int_c^b f$.

THEOREM 4   **MULTIPLICATION BY A CONSTANT**

If f is integrable on $[a, b]$ and if k is any constant, then kf is integrable on $[a, b]$, and

$$\int_a^b kf = k \int_a^b f. \quad \blacksquare \tag{16}$$

THEOREM 5   **INTEGRAL OF THE SUM OF TWO FUNCTIONS**

If the functions f and g are both integrable on $[a, b]$, then $f + g$ is integrable on $[a, b]$, and

$$\int_a^b (f + g) = \int_a^b f + \int_a^b g.$$

That is, *the integral of the sum is the sum of the integrals.*   ■

REMARK: Theorem 5 can be extended to the integral of a finite sum of integrable functions. Conditions (16) and (17) are called **linearity** conditions for the definite integral.

The next two theorems give us a way to compare integrals without tedious calculations.

THEOREM 6 **THE INTEGRAL OF A NONNEGATIVE FUNCTION IS NONNEGATIVE**

If f is integrable on $[a, b]$ and $f \geq 0$ there, then

$$\int_a^b f(x)\,dx \geq 0.$$

PROOF: This result follows since every term in the limit (6) defining the integral is nonnegative. ∎

THEOREM 7 **COMPARISON THEOREM**

Let f and g be integrable on $[a, b]$ and suppose that for every x in $[a, b]$,

$$f(x) \leq g(x). \tag{18}$$

Then,

$$\int_a^b f(x)\,dx \leq \int_a^b g(x)\,dx. \tag{19}$$

PROOF: Define $h(x) = g(x) - f(x)$. Then by (18), $h(x) \geq 0$ on $[a, b]$. By Theorem 6, $\int_a^b h(x)\,dx \geq 0$. But

Theorem 5
↓

$$0 \leq \int_a^b h(x)\,dx = \int_a^b \lfloor g(x) - f(x) \rfloor\,dx = \int_a^b g(x)\,dx + \int_a^b \left[-f(x) \right]\,dx$$

Theorem 4
↓

$$= \int_a^b g(x)\,dx - \int_a^b f(x)\,dx.$$

Therefore, $\int_a^b f(x)\,dx \leq \int_a^b g(x)\,dx$. ∎

EXAMPLE 7
USING THE COMPARISON
THEOREM

On the interval $[0, 1]$, $x^3 \leq x^2$. This inequality tells us (from Theorem 7) that

$$\int_0^1 x^3\,dx \leq \int_0^1 x^2\,dx.$$

This is indeed the case since $\int_0^1 x^2\,dx = \frac{1}{3}$ and, as we shall show in the next section (in Example 4.6.2), $\int_0^1 x^3\,dx = \frac{1}{4}$.

EXAMPLE 8
USING THE COMPARISON
THEOREM

Since $x^5 \leq x^7$ on $[2, 5]$, we have

$$\int_2^5 x^5\,dx \leq \int_2^5 x^7\,dx.$$

The following theorem is sometimes extremely useful for estimating integrals when calculations are difficult.

THEOREM 8 **UPPER AND LOWER BOUND THEOREM FOR INTEGRALS**

Suppose that on $[a, b]$

$$m \leq f(x) \leq M \tag{20}$$

for all x in $[a, b]$. That is, m and M are lower and upper bounds, respectively, for the function f on the interval $[a, b]$. Then if f is integrable on $[a, b]$,

$$m(b - a) \leq \int_a^b f(x)\, dx \leq M(b - a). \tag{21}$$

PROOF: If we define g by $g(x) = M$, then $f(x) \leq g(x)$, and by the comparison theorem,

Theorem 2

$$\int_a^b f(x)\, dx \leq \int_a^b M\, dx = M(b - a).$$

Similarly,

$$m(b - a) = \int_a^b m\, dx \leq \int_a^b f(x)\, dx,$$

and the theorem is proved. ∎

This result is illustrated in Figure 7.

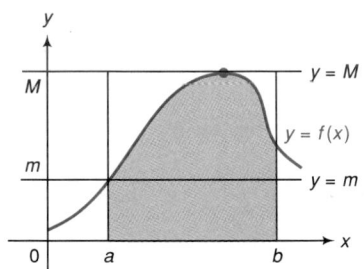

FIGURE 7
GRAPHICAL SKETCH OF THE PROOF OF THE UPPER AND LOWER BOUND THEOREM

REMARK: According to Theorem 1.7.4, if f is continuous on $[a, b]$, then f takes minimum and maximum values in that interval, so the numbers m and M do exist. The function $1/\sqrt[3]{x}$, for example, is not continuous and does not have such bounds in the interval $[-1, 1]$.

EXAMPLE 9
FINDING UPPER AND LOWER BOUNDS FOR AN INTEGRAL

Estimate $\int_0^{\pi/6} \sin^5 x\, dx$.

SOLUTION: The function $\sin x$ is increasing on $[0, \pi/6]$, so

$$0 = \sin 0 \leq \sin x \leq \sin \frac{\pi}{6} = \frac{1}{2}$$

and

$$0 \leq \sin^5 x \leq \left(\frac{1}{2}\right)^5 = \frac{1}{32} \qquad \text{for} \qquad 0 \leq x \leq \frac{\pi}{6}.$$

Thus,

$$m = 0 \qquad \text{and} \qquad M = \frac{1}{32},$$

and

$$0 \leq \int_0^{\pi/6} \sin^5 x\, dx \leq \frac{1}{32}\left(\frac{\pi}{6} - 0\right) = \frac{\pi}{192} \approx 0.01636.$$

We now show how definite integrals can arise in an application having nothing at all to do with area.

Suppose that a particle is moving with velocity $v = v(t) \geq 0$. If $v(t) = c$, a constant, then we can calculate the distance traveled between the initial time $t = t_0$ and the final time $t = t_f$ by

$$\text{distance} = \text{velocity} \times \text{elapsed time} = v \cdot (t_f - t_0). \tag{22}$$

However, if the velocity is changing (i.e., if the particle is accelerating or decelerating), then formula (22) simply will not work. Still, we can calculate the distance traveled by a method identical to the one we used to calculate areas. The basic idea is simple: Even though $v(t)$ is changing, over a very small period of time it will be almost constant.

We begin by dividing the time interval $[t_0, t_f]$ into n equal subintervals of length $(t_f - t_0)/n$ (see Figure 8). Let $t_i{}^*$ denote a time in the interval $[t_{i-1}, t_i]$. If n is large enough (i.e., if the lengths of the intervals are small), then the velocity of the particle in the time period $[t_{i-1}, t_i]$ is almost constant, so that it can be approximated by $v(t_i{}^*)$ in $[t_{i-1}, t_i]$. Let $\Delta t = t_i - t_{i-1}$. If s_i denotes the distance traveled by the particle between the time $t = t_{i-1}$ and the time $t = t_i$, then from (22),

$$s_i \approx v(t_i{}^*)\Delta t. \tag{23}$$

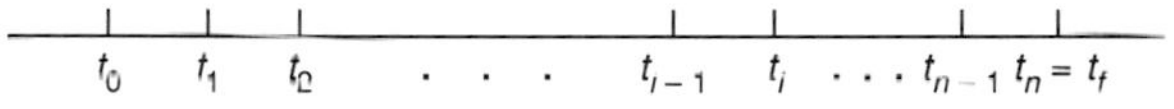

FIGURE 8   REGULAR PARTITION OF THE INTERVAL $[t_0, t_f]$

The total distance traveled by the particle over the entire time interval $[t_0, t_f]$ is then given (approximately) by

$$s = s_1 + s_2 + \cdots + s_n \approx v(t_1{}^*)\Delta t + v(t_2{}^*)\Delta t + \cdots + v(t_n{}^*)\Delta t$$

$$= \sum_{i=1}^{n} v(t_i{}^*)\Delta t. \tag{24}$$

This should look familiar. Formula (24) is only an approximation because $v(t)$ is not really constant over $[t_{i-1}, t_i]$. However, as the length of the interval becomes smaller and smaller, the approximation gets better and better. Therefore, we can conclude that

$$s = \lim_{\Delta t \to 0} \sum_{i=1}^{n} v(t_i{}^*)\Delta t = \int_{t_0}^{t_f} v(t)\, dt. \tag{25}$$

Thus, we obtain the distance traveled by "adding up" a great number of small distances, just as we obtained the area under a curve by "adding up" a great number of areas of "thin" rectangles.

There is an interesting interpretation of Theorem 3 in the use of the definite integral in formula (25). If $v(t)$ represents the velocity of a moving object and $v(t) \geq 0$ for $t \in [a, b]$, then

$$\int_a^b v(t)\, dt = \text{distance traveled from time } a \text{ to time } b,$$

$$\int_a^c v(t)\, dt = \text{distance traveled from time } a \text{ to time } c,$$

and

$$\int_c^b v(t)\,dt = \text{distance traveled from time } c \text{ to time } b.$$

It is not difficult to see why it should be true that if $a < c < b$,

$$\int_a^b v(t)\,dt = \int_a^c v(t)\,dt + \int_c^b v(t)\,dt.$$

In Theorem 1, we stated that all continuous functions are integrable. Many other kinds of functions are integrable as well. On page 122 we defined a *piecewise continuous* function as a function that is continuous at every point in $[a, b]$ except for a finite number of points at which f has a jump discontinuity (a jump discontinuity is a *finite jump*). Thus, for example, the greatest integer function $[x]$ is piecewise continuous on every closed, bounded interval. The following result is true:

If f is piecewise continuous on $[a, b]$, then $\int_a^b f(x)\,dx$ exists.   **(26)**

We now prove Theorems 3, 4, and 5. Each proof relies on the fact that, if f is integrable, then

$$\lim_{\max \Delta x_i \to 0} \sum_{i=1}^n f(x_i^*)\,\Delta x_i = \int_a^b f(x)\,dx \tag{27}$$

exists and is the same no matter how $[a, b]$ is partitioned and no matter which points x_i^* are chosen in $[x_{i-1}, x_i]$.

PROOF OF THEOREM 3:

By Theorem 1, all three integrals exist. Let $a = x_0 < x_1 < x_2 < \cdots < x_n = b$ be a partition of $[a, b]$. By (27)

$$\int_a^b f = \lim_{\max \Delta x_i \to 0} \sum_{i=1}^n f(x_i^*)\,\Delta x_i$$

and the limit is (by definition) the same no matter how $[a, b]$ is partitioned. Therefore, by adding it if necessary, assume that c is a partition point in each (nonregular) partition which we now write as $a = x_0 < x_1 < \cdots < x_k = c < x_{k+1} < \cdots < x_n$. Then

$$\int_a^b f = \lim_{\max \Delta x_i \to 0} \sum_{i=1}^n f(x_i^*)\,\Delta x_i$$

$$= \lim_{\max \Delta x_i \to 0} \left[\sum_{i=1}^k f(x_i^*)\,\Delta x_i + \sum_{i=k+1}^n f(x_i^*)\,\Delta x_i \right]$$

The limit of the sum
= the sum of the limits

$$\downarrow$$

$$= \lim_{\max \Delta x_i \to 0} \sum_{i=1}^k f(x_i^*)\,\Delta x_i + \lim_{\max \Delta x_i \to 0} \sum_{i=k+1}^n f(x_i^*)\,\Delta x_i$$

$$= \int_a^c f + \int_c^b f. \quad \blacksquare$$

PROOF OF THEOREM 4:

$$(27) \longrightarrow \int_a^b kf(x)\,dx = \lim_{\max \Delta x_i \to 0} \sum_{i=1}^{n} kf(x_i^*)\,\Delta x_i$$

Problem 31 on page 279 $\longrightarrow = \displaystyle\lim_{\max \Delta x_i \to 0} k \sum_{i=1}^{n} f(x_i^*)\,\Delta x_i$

limit of a constant times a
function equals the constant $\longrightarrow = k \displaystyle\lim_{\max \Delta x_i \to 0} \sum_{i=1}^{n} f(x_i^*)\,\Delta x_i$
times the limit of the function

$$(27) \longrightarrow = k \int_a^b f(x)\,dx. \quad \blacksquare$$

PROOF OF THEOREM 5:

$$(27) \longrightarrow \int_a^b [f(x) + g(x)]\,dx = \lim_{\max \Delta x_i \to 0} \sum_{i=1}^{n} [f(x_i^*) + g(x_i^*)]\,\Delta x_i$$

Problem 32 on page 279 $\longrightarrow = \displaystyle\lim_{\max \Delta x_i \to 0} \left[\sum_{i=1}^{n} f(x_i^*)\,\Delta x_i + \sum_{i=1}^{n} g(x_i^*)\,\Delta x_i \right]$

Limit of the sum
= the sum of the limits $\longrightarrow = \displaystyle\lim_{\max \Delta x_i \to 0} \sum_{i=1}^{n} f(x_i^*)\,\Delta x_i$

$$+ \lim_{\max \Delta x_i \to 0} \sum_{i=1}^{n} g(x_i^*)\,\Delta x_i$$

$$(27) \longrightarrow = \int_a^b f(x)\,dx + \int_a^b g(x)\,dx. \quad \blacksquare$$

We close this section with a formal definition of the limit of Riemann sums. This definition is somewhat similar to the $\epsilon - \delta$ definition of a limit given in Section 1.8.

DEFINITION **FORMAL DEFINITION OF THE LIMIT OF RIEMANN SUMS**

Let L be a real number. Then,

$$\lim_{\max \Delta x_i \to 0} \sum_{i=1}^{n} f(x_i^*)\,\Delta x_i = L$$

if for every $\epsilon > 0$ there is a number $\delta > 0$ such that for every partition P with $\max \Delta x_i < \delta$,

$$\left| \sum_{i=1}^{n} f(x_i^*)\,\Delta x_i - L \right| < \epsilon.$$

That is, if we take any partition of $[a, b]$, with each subinterval of length less than δ, then any Riemann sum $\sum_{i=1}^{n} f(x_i^*)\,\Delta x_i$, with $x_i^* \in [x_{i-1}, x_i]$, is within ϵ units of L. $\quad \blacksquare$

GEORG FRIEDRICH RIEMANN 1826–1866

EORG FRIEDRICH RIEMANN WAS born in 1826 in the German village of Hanover. His father was a Lutheran pastor. Throughout his life, Riemann was exceedingly shy and in frail health. Although his family was by no means wealthy, Riemann was, nevertheless, able to get a good education, first at the University of Berlin and then at the University of Göttingen. His work at Göttingen culminated with a brilliant thesis in the area of functions of a complex variable (a complex variable is a variable of the form $x + iy$, where x and y are real numbers and $i = \sqrt{-1}$). At Göttingen, he worked under the greatest mathematician of the nineteenth century, Karl Friedrich Gauss (1777–1855).

In 1854, Riemann was appointed *Privatdozent* (official but unpaid lecturer) at the University of Göttingen. According to the custom of the day, he was asked to give a probationary lecture. The result was the greatest paper of comparable size ever presented in the history of mathematics. The title of the lecture was "Über die Hypothesen welche der Geometrie zu Grunde liegen" ("On the Hypotheses which Lie at the Foundation of Geometry"). Rather than discussing a specific example, it urged a global view of geometry that revolutionized the study of that subject. After this lecture, and perhaps for the only time in

a career that spanned approximately sixty years, Gauss paid compliments to the work of someone else.

Riemann's great paper was presented in 1854 but was not published until 1867. One of the central ideas in the paper was that geometry could be discussed in general curved spaces rather than only in the sphere (which is one specific curved space). Albert Einstein made use of this idea in his general theory of relativity.

Riemann made important contributions to many other areas of mathematics and theoretical physics. Most important for us, he clarified the concept of the definite integral. It is this definition, now known as the *Riemann integral*, that is the basis for the material in this chapter and much of the rest of the book.

In 1859, Riemann became a full professor at the University of Göttingen, having become an assistant professor only two years earlier. The chair that he occupied was previously held by another great German mathematician, Peter Gustav Lejeune Dirichlet (1805–1859). Dirichlet, in turn, had succeeded Gauss. Riemann's career was cut short in 1866, in Italy, where he had gone to seek a cure for tuberculosis. He died at the age of 40.

■ PROBLEMS 4.5

■ SELF-QUIZ

I. The area between the x-axis and the graph of $y = 1 + x$ on the interval $[2, 5]$ is computed by _____.

   a. $\int_2^5 (1 + x)\, dx$ b. $1 + \int_2^5 x\, dx$

   c. $\int_0^{(1+x)} 1\, dx$ d. $\int_0^{(1+x)} (5 - 2)\, dx$

II. $\int_2^5 4\, dx =$ _____.

   a. $4x + C$ b. $\int 4\, dx$

   c. $4 \cdot 5$ d. $4 \cdot (5 - 2)$

III. The area between the x-axis and the graph of $y = 2x$ on the interval $[-1, 3]$ can be computed by _____.

   a. $\int_{-1}^3 2x\, dx$

   b. $\left| \int_{-1}^3 2x\, dx \right|$

   c. $\int_{|-1|}^3 2x\, dx$

   d. $\int_{-1}^3 |2x|\, dx$

IV. $\int_1^4 x^2\,dx + \int_4^6 x^2\,dx = $ _____.

  a. $\int_1^6 x^2\,dx$ b. $\int_1^4 (x^2 - x^2)\,dx$

  c. $\int_1^4 (x^2 + x^2)\,dx$ d. $\int_1^6 (x^2 + x^2)\,dx$

V. $\int_2^7 \cos x\,dx - \int_2^4 \cos x\,dx = $ _____.

  a. $\int_2^7 [(7-2)\cos x - (4-2)\cos x]\,dx$

  b. $\int_4^7 \cos x\,dx$

  c. $\int_5^7 \cos x\,dx$

  d. $(\cos 7 - \cos 2) - (\cos 4 - \cos 2)$

VI. Suppose the function g is integrable on $[2, 7]$ and $g(x) \le 50$ for all x; then $\int_2^7 g(x)\,dx$ _____.

  a. $= 50 \cdot (7 - 2)$ b. $\ne 50 \cdot (7 - 2)$

  c. $\ge 50 \cdot (7 - 2)$ d. $\le 50 \cdot (7 - 2)$

VII. True–False: (Assume functions f and g are continuous on $(-\infty, \infty)$.)

  a. $\int_a^b f = -\int_b^a f$ for all a, b, f.

  b. $\int_a^b f + \int_a^b g = \int_a^b (f + g)$ for all a, b, f, g.

  c. $\int_a^b f + \int_b^c f = \int_a^c f$ for all a, b, c, f.

  d. $\int_a^b f(x)\,dx = \int_a^b f(t)\,dt$ for all a, b, f.

  e. If $f(x) \le M$ on $[a, b]$, then $\int_a^b f \le M \cdot (b - a)$.

■ DRILL

In Problems 1–10, find (reasonable) lower and upper bounds for the given integrals. (Do not calculate exact values for any of these definite integrals.)

1. $\int_1^4 4\sqrt{x}\,dx$ 2. $\int_1^8 7x^{1/3}\,dx$

3. $\int_{-1}^1 x^{10}\,dx$ 4. $\int_2^3 (x^2 + x^3)\,dx$

5. $\int_1^{100} (1/x)\,dx$ 6. $\int_1^9 (1/\sqrt{x})\,dx$

7. $\int_0^{\pi/2} \sin x\,dx$ 8. $\int_0^{\pi/3} \cos x\,dx$

9. $\int_{-5}^{-4} x^3\,dx$ 10. $\int_{-5}^{-4} x^4\,dx$

In Problems 11–30, calculate the given definite integral by using the definitions, theorems, and examples of this section.

11. $\int_0^4 7x\,dx$ 12. $\int_1^2 4x\,dx$

13. $\int_2^5 (3x + 2)\,dx$ 14. $\int_1^4 (7x - 3)\,dx$

*15. $\int_0^5 [x]\,dx$

    [Hint: Remember that $[x]$ is the greatest integer function; its graph is piecewise horizontal.]

*16. $\int_{-4}^4 [x]\,dx$ 17. $\int_0^2 x^2\,dx$

18. $\int_{-1}^1 x^2\,dx$ 19. $\int_{-1}^1 x^3\,dx$

20. $\int_0^5 x^3\,dx$ 21. $\int_0^1 (1 - t^2)\,dt$

22. $\int_0^1 (1 - t^3)\,dt$

23. $\int_0^1 (1 + x + x^2 + x^3)\,dx$

24. $\int_1^0 (x^3 - x^2)\,dx$

25. $\int_1^0 t\,dt$

26. $\int_5^{-2} (3t - 7)\,dt$

27. $\int_{-1}^1 |x|\,dx$

    [Hint: Draw a picture.]

*28. $\int_{-3}^2 |x + 1|\,dx$

*29. $\int_a^b x^2\,dx$

    [Hint: See Problem 4.4.21.]

*30. $\int_a^b x^3\,dx$

    [Hint: See Problem 4.4.22.]

■ APPLICATIONS

31. Explain why $\int_{23}^{47} x^{17}\,dx < \int_{23}^{47} x^{55/3}\,dx$.

32. Explain why $\int_0^1 \sqrt{x}\,dx < \int_0^1 \sqrt[3]{x}\,dx$.

33. Which is larger: $\int_1^2 x\,dx$ or $\int_1^2 \sqrt{x}\,dx$?

34. Which is larger: $\int_{-1}^0 x^{1/3}\,dx$ or $\int_{-1}^0 x^{1/5}\,dx$?

*35. Show that $\int_0^1 \sqrt{1 + x^3}\,dx$ lies between 1 and $\frac{5}{4}$. [Hint: Obtain a close upper bound for the definite integral of $\sqrt{1 + x^3}$ over each of the intervals $[0, \frac{1}{2}]$ and $[\frac{1}{2}, 1]$; then combine those bounds.]

CONTINUED

36. Without doing any calculations, explain why the following are true.

a. $\int_0^{2\pi} \sin x \, dx = \int_0^{2\pi} \cos x \, dx = 0$.

b. $\int_{-\pi/2}^{\pi/2} \sin x \, dx = \int_0^{\pi} \cos x \, dx = 0$.

37. A ball is dropped from a height of 400 ft. Its velocity after t seconds is given by the formula $v = 32t$ ft/sec.

a. How fast is the ball dropping after 4 sec?

b. How far has the ball dropped after 4 sec?

c. After how many seconds will the ball hit the ground?

38. A bullet is shot straight into the air with an initial velocity of 500 m/sec. Its velocity after t sec have elapsed is $v(t) = (500 - 9.81t)$ m/sec.

a. After how many seconds will the bullet begin to fall?

b. How high will the bullet go?

■ SHOW/PROVE/DISPROVE

39. Prove Theorem 2. [Hint: $f(x_i^*) = c$ for any $x_i^* \in [a, b]$.]

*40. Suppose f is integrable on $[a, c]$ and on $[c, b]$ where $a < c < b$. Prove that those conditions imply f is integrable on $[a, b]$ and $\int_a^b f = \int_a^c f + \int_c^b f$. [Note: This result is different from Theorem 3.]

41. Suppose a, b, and c are three distinct real numbers and f is a function such that the three definite integrals $\int_a^b f$, $\int_a^c f$, and $\int_c^b f$ exist. Show that

$$\int_a^b f = \int_a^c f + \int_c^b f$$

where it is not required that $a < c < b$. [Hint: There are six possible orderings for a, b, c: $a < c < b, a < b < c, b < a < c$, etc. You can test each of these using Theorem 3 and equation (13).]

*42. Suppose f is continuous on $[a, b]$. Show that

$$\left| \int_a^b f(x) \, dx \right| \le \int_a^b |f(x)| \, dx.$$

[Hint: Show from the definition that

$$\int_a^b f(x) \, dx \le \int_a^b |f(x)| \, dx \qquad \text{and}$$

$$-\int_a^b f(x) \, dx \le \int_a^b |f(x)| \, dx.]$$

43. Prove or Disprove:

a. If $\int_a^b f = \int_b^a f$, then f is equal to 0 throughout the interval $[a, b]$.

b. If f is continuous on $[a, b]$ and $\int_a^b f = \int_b^a f$, then f is equal to 0 throughout the interval $[a, b]$.

c. If f is continuous on $[a, b]$ and $\int_a^b |f| = \int_b^a |f|$, then f is equal to 0 throughout the interval $[a, b]$.

*44. Show that

$$\int_0^n [x] \, dx = \frac{(n-1) \cdot n}{2}$$

if n is a positive integer.

■ CHALLENGE

*45. Obtain a general expression for $\int_a^b [x] \, dx$ (you may suppose that $0 \le a < b$). Prove that your formula is correct.

*46. Suppose that f is continuous on $[a, b]$. We define a new function g by $g(x) = f(a + b - x)$. Note that g is also continuous on $[a, b]$. Prove that $\int_a^b f = \int_a^b g$.

*47. Let $f(x) = 1$ if x is a rational number and $f(x) = 0$ if x is irrational. [This function appeared in Problem 1.7.28.]

a. Explain why, for any partition of the interval $[0, 1]$, it is possible to choose points t_i^* in $[x_{i-1}, x_i]$ such that $f(t_i^*) = 1$.

b. Explain why it is also possible to choose points w_i^* in $[x_{i-1}, x_i]$ such that $f(w_i^*) = 0$.

c. Show that $\lim_{\Delta x \to 0} [f(x_1^*) \Delta x + f(x_2^*) \Delta x + \cdots + f(x_n^*) \Delta x]$ depends on the choice of the points x_n^*.

d. Explain why $\int_0^1 f$ does not exist.

■ ANSWERS TO SELF-QUIZ

I. a II. d III. d VII. All are True
IV. a V. b VI. d

4.6 THE FUNDAMENTAL THEOREM OF CALCULUS

In Chapter 1, we saw how the tangent-line problem, formulated long ago by Greek mathematicians and not solved until the seventeenth century, gave rise to the modern theory of derivatives. In Sections 4.4 and 4.5, we saw how the area problem whose origins are lost in antiquity gave rise to the idea behind the definite integral. Finally, in Section 4.2, we introduced the indefinite integral, which was seemingly unrelated to the definite integral and was defined as the "inverse" operation to differentiation.

In this section, we show how these three operations on functions (the derivative and the two integrals) are intimately related. The remarkable theorem linking these operations is called the **fundamental theorem of calculus**.

THEOREM 1 **FUNDAMENTAL THEOREM OF CALCULUS**

Let f be continuous on $[a, b]$. If F is any antiderivative of f on $[a, b]$, then

$$\int_a^b f(t) \, dt = F(b) - F(a). \quad \blacksquare \tag{1}$$

This theorem simply asserts that we may calculate a definite integral by evaluating any antiderivative at the endpoints of the interval of integration and then subtracting. We already proved the theorem in the special cases $\int_0^1 x^2 \, dx$ and $\int_0^\pi \sin x \, dx$ (Examples 4.3.4 and 4.3.5). We provide the proof for all cases now. Before looking at the proof, however, you should look over those two examples. The general proof is really not any different than those specific ones.

PROOF OF THE FUNDAMENTAL THEOREM OF CALCULUS: Since f is continuous on $[a, b]$, we know from Theorem 4.5.1 that $\int_a^b f(x) \, dx$ exists. Let F be an antiderivative for f. By the definition of the antiderivative, F is continuous on $[a, b]$ and differentiable on (a, b). Let

$$a = x_0 < x_1 < x_2 < \cdots < x_n = b$$

be a regular partition of $[a, b]$. Consider the subinterval $[x_{i-1}, x_i]$. By the mean value theorem,

$$F(x_i) - F(x_{i-1}) = F'(x_i^*)(x_i - x_{i-1}) = F'(x_i^*) \Delta x, \tag{2}$$

where $x_{i-1} < x_i^* < x_i$. Thus,

$$\sum_{i=1}^n [F(x_i) - F(x_{i-1})] = \sum_{i=1}^n F'(x_i^*) \Delta x. \tag{3}$$

But,

$$\sum_{i=1}^n [F(x_i) - F(x_{i-1})]$$
$$= [F(x_1) - F(x_0)] + [F(x_2) - F(x_1)] + [F(x_3) - F(x_2)] + \cdots + [F(x_{n-1}) - F(x_{n-2})] + [F(x_n) - F(x_{n-1})]$$
$$= F(x_n) - F(x_0) = F(b) - F(a),$$

since all terms except the first and the last "cancel." Thus from (3),

$$F(b) - F(a) = \sum_{i=1}^{n} F'(x_i^*) \Delta x,$$

and taking the limit as $\Delta x \to 0$ of both sides, we obtain

$$\lim_{\Delta x \to 0} \left[F(b) - F(a) \right] = \lim_{\Delta x \to 0} \sum_{i=1}^{n} F'(x_i^*) \Delta x. \tag{4}$$

Since F is an antiderivative of f, we have

$$F'(x_i^*) = f(x_i^*), \tag{5}$$

and inserting (5) into (4) gives us

$$F(b) - F(a) = \lim_{\Delta x \to 0} \sum_{i=1}^{n} f(x_i^*) \Delta x. \tag{6}$$

But now we are done, since $\int_a^b f(x)\,dx$ exists and is equal to the limit in the right-hand side of (6), and this limit is independent of the way x_i^*'s are chosen[†]. Hence from (6),

$$F(b) - F(a) = \int_a^b f(x)\,dx. \quad \blacksquare$$

EXAMPLE 1
USING THE FUNDAMENTAL
THEOREM

Calculate $\int_0^1 x^2\,dx$.

SOLUTION: We have seen that $x^3/3$ is an antiderivative for x^2. Thus,

$$\int_0^1 x^2\,dx = \left(\frac{x^3}{3} \text{ evaluated at } x = 1 \right) - \left(\frac{x^3}{3} \text{ evaluated at } x = 0 \right)$$

$$= \frac{1}{3} - 0 = \frac{1}{3}.$$

There is a simple notation we will use to avoid writing the words "evaluated at" each time.

NOTATION:

$$F(x)\Big|_a^b = F(b) - F(a).$$

In Example 1, we could have written

$$\int_0^1 x^2\,dx = \frac{x^3}{3}\Big|_0^1 = \frac{1}{3} - 0 = \frac{1}{3}.$$

REMARK: It doesn't make any difference which antiderivative we choose to evaluate the definite integral. For example, if C is any constant, then

$$\int_0^1 x^2\,dx = \left(\frac{x^3}{3} + C \right)\Big|_0^1 = \left(\frac{1}{3} + C \right) - (0 + C) = \frac{1}{3} + C - C = \frac{1}{3}.$$

[†] Remember that this last statement is part of the definition of the definite integral.

The constants will always cancel in this manner. Thus we will use the easiest antiderivative in our evaluation of $\int_a^b f$, which will almost always be the one in which $C = 0$.

EXAMPLE 2
USING THE FUNDAMENTAL THEOREM

Compute **(a)** $\int_0^1 x^3\, dx$, **(b)** $\int_0^3 x^3\, dx$, and **(c)** $\int_1^3 x^3\, dx$.

SOLUTION: $\dfrac{x^4}{4}$ is an antiderivative for x^3. So,

(a) $\displaystyle\int_0^1 x^3\, dx = \frac{x^4}{4}\Big|_0^1 = \frac{1}{4} - 0 = \frac{1}{4}$ $F(1) - F(0)$

(b) $\displaystyle\int_0^3 x^3\, dx = \frac{x^4}{4}\Big|_0^3 = \frac{3^4}{4} - 0 = \frac{81}{4}$

(c) $\displaystyle\int_1^3 x^3\, dx = \frac{x^4}{4}\Big|_1^3 = \frac{3^4}{4} - \frac{1}{4} = \frac{81}{4} - \frac{1}{4} = 20.$

These examples illustrate Theorem 4.5.3 ($\int_a^c f = \int_a^b f + \int_b^c f$). Note that

$$\int_0^3 x^3\, dx = \int_0^1 x^3\, dx + \int_1^3 x^3\, dx = \frac{1}{4} + \frac{80}{4} = \frac{81}{4}.$$

EXAMPLE 3
USING THE FUNDAMENTAL THEOREM

Calculate $\int_1^2 (3x^4 - x^5)\, dx$.

SOLUTION: $\int (3x^4 - x^5)\, dx = 3x^5/5 - x^6/6 + C$. Thus,

$$\int_1^2 (3x^4 - x^5)\, dx = \left(\frac{3x^5}{5} - \frac{x^6}{6}\right)\Big|_1^2 = \left[\frac{3(2)^5}{5} - \frac{2^6}{6}\right] - \left[\frac{3(1)^5}{5} - \frac{1^6}{6}\right]$$

$$= \left(\frac{96}{5} - \frac{64}{6}\right) - \left(\frac{3}{5} - \frac{1}{6}\right)$$

$$= \left(\frac{576}{30} - \frac{320}{30}\right) - \left(\frac{18}{30} - \frac{5}{30}\right) = \frac{243}{30} = \frac{81}{10}.$$

⊗ **WARNING:** It is easy to lose track of minus signs when performing these calculations. Often the minus signs will cancel each other. Be careful! ⊗

EXAMPLE 4
USING THE FUNDAMENTAL THEOREM

Calculate

$$\int_1^4 \left(\frac{3}{\sqrt{s}} - 5\sqrt{s}\right) ds.$$

SOLUTION:

$$\int_1^4 \left(\frac{3}{\sqrt{s}} - 5\sqrt{s}\right) ds = \int_1^4 (3s^{-1/2} - 5s^{1/2})\, ds = \left(6\sqrt{s} - \frac{10}{3} s^{3/2}\right)\Big|_1^4$$

$$= \left[6\sqrt{4} - \frac{10}{3}(4)^{3/2}\right] - \left[6\sqrt{1} - \frac{10}{3}(1)^{3/2}\right]$$

$$= 6(2) - \frac{10}{3}(8) - 6 + \frac{10}{3} = 6 - \frac{70}{3} = -\frac{52}{3}.$$

EXAMPLE 5
THE DEFINITE INTEGRAL OF A
POWER FUNCTION

Calculate $\int_a^b x^r \, dx$, where $r \neq -1$ is a real number.

SOLUTION:

$$\int x^r \, dx = \frac{x^{r+1}}{r+1} + C,$$

so that

$$\int_a^b x^r \, dx = \frac{x^{r+1}}{r+1} \bigg|_a^b = \frac{1}{r+1}(b^{r+1} - a^{r+1}). \tag{7}$$

EXAMPLE 6
USING THE FUNDAMENTAL
THEOREM

Calculate $\int_0^{\pi/2} \sin x \, dx$

SOLUTION:

$$\int_0^{\pi/2} \sin x \, dx = -\cos x \bigg|_0^{\pi/2} = -\cos \frac{\pi}{2} - (-\cos 0) = -0 - (-1) = 1.$$

EXAMPLE 7
USING THE FUNDAMENTAL
THEOREM TO COMPUTE AREA

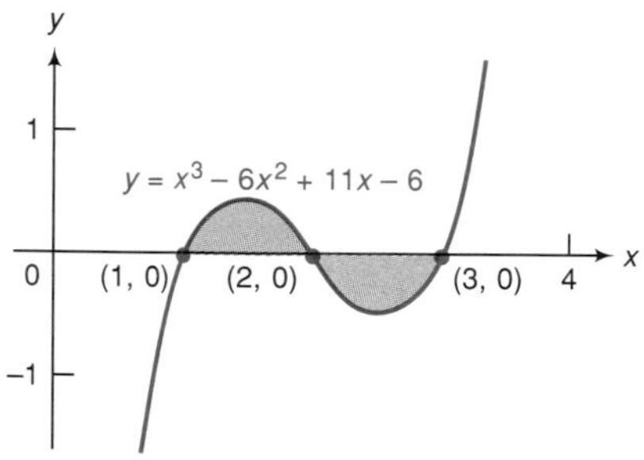

FIGURE 1
AREA BETWEEN GRAPH OF $f(x) = x^3 - 6x^2 + 11 - 6$ AND x-AXIS FOR $1 \leq x \leq 3$

Calculate the area bounded by the curve $y = x^3 - 6x^2 + 11x - 6$ and the x-axis.

SOLUTION: We have $y = x^3 - 6x^2 + 11x - 6 = (x - 1)(x - 2)(x - 3)$. The curve is graphed in Figure 1. The desired area is the shaded part of the graph. We know that

$$A = \int_1^3 |x^3 - 6x^2 + 11x - 6| \, dx \qquad \begin{array}{l} x^3 - 6x^2 + 11x - 6 \text{ is positive} \\ \text{in } (1, 2) \text{ and negative in } (2, 3). \end{array}$$

$$= \int_1^2 (x^3 - 6x^2 + 11x - 6) \, dx + \int_2^3 -(x^3 - 6x^2 + 11x - 6) \, dx$$

$$= \left(\frac{x^4}{4} - 2x^3 + \frac{11x^2}{2} - 6x \right) \bigg|_1^2 - \left(\frac{x^4}{4} - 2x^3 + \frac{11x^2}{2} - 6x \right) \bigg|_2^3$$

$$= \left(4 - 16 + 22 - 12 - \frac{1}{4} + 2 - \frac{11}{2} + 6 \right)$$

$$- \left(\frac{81}{4} - 54 + \frac{99}{2} - 18 - 4 + 16 - 22 + 12 \right) = \frac{1}{2}.$$

Note that

$$\int_1^3 (x^3 - 6x^2 + 11x - 6) \, dx = \left(\frac{x^4}{4} - 2x^3 + \frac{11x^2}{2} - 6x \right) \bigg|_1^3 = 0.$$

This example illustrates why, in the process of calculating area, care must be taken so that "positive" areas and "negative" areas don't cancel each other.

In the statement of the fundamental theorem of calculus, we left one important question unanswered. We know that if f has an antiderivative F, then $\int_a^b f(x)\,dx = F(b) - F(a)$. But how do we know that every continuous function has an antiderivative? Theorem 2 below answers that question.

We assume that the function f is continuous on $[a, b]$ (piecewise continuity would do just as well here). Then by Theorem 4.5.1, $\int_a^b f(t)\,dt$ exists. But we have even more than that. If x is any number in $[a, b]$, then f is certainly continuous on the smaller interval $[a, x]$, and so again by Theorem 4.5.1, $\int_a^x f(t)\,dt$ exists. For every value of x in $[a, b]$, this integral is a real number. Now we define a new function G by

$$G(x) = \int_a^x f(t)\,dt.$$

We emphasize that $G(x)$ is the *function* that assigns to every number x in $[a, b]$ the value of the definite integral of f over the interval $[a, x]$. The theorem below tells us that $G(x)$ is really an antiderivative for the function f over the interval $[a, b]$. Its proof is difficult and will be delayed until Section 4.8. However, we will indicate by a graph why the theorem is plausible. The following important result is sometimes called the **second fundamental theorem of calculus**.

THEOREM 2   **SECOND FUNDAMENTAL THEOREM OF CALCULUS**

If f is continuous on $[a, b]$, then the function $G(x) = \int_a^x f(t)\,dt$ is continuous on $[a, b]$, differentiable on (a, b), and for every x in (a, b),

$$G'(x) = f(x).$$

That is, G is an antiderivative of f on the interval $[a, b]$.

Graphical Indication of Proof: Consider Figure 2, paying particular attention to the various areas under the curve $f(x)$. By definition of the derivative,

$$G'(x) = \lim_{\Delta x \to 0} \frac{G(x + \Delta x) - G(x)}{\Delta x}$$

Since $G(x) = \int_a^x f(t)\,dt$ = area under the curve $f(x)$ between a and x, and assuming that $f(x) > 0$ on (a, b), we have

$$G(x + \Delta x) = \text{area between } a \text{ and } x + \Delta x$$

$$G(x) = \text{area between } a \text{ and } x$$

$$G(x + \Delta x) - G(x) = \text{area between } x \text{ and } x + \Delta x.$$

This last area is denoted by $A_{\Delta x}$ in Figure 2. Now if Δx is small, then x and $x + \Delta x$ are close, and since f is continuous, $f(x + \Delta x)$ is close to $f(x)$. That is, area $A_{\Delta x}$ is approximately equal to the area of the rectangle with height $f(x)$ and base Δx. We therefore see that

$$G(x + \Delta x) - G(x) = A_{\Delta x} \approx f(x)\,\Delta x.$$

Then for Δx small,

$$\frac{G(x + \Delta x) - G(x)}{\Delta x} \approx \frac{f(x)\,\Delta x}{\Delta x} = f(x).$$

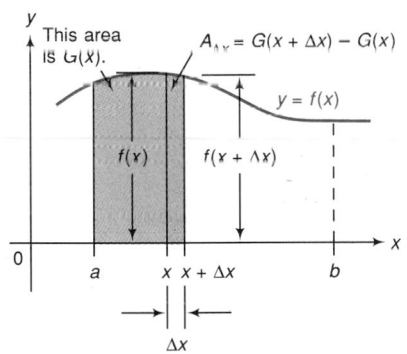

FIGURE 2
THE AREA $G(x)$ IS THE AREA UNDER THE CURVE BETWEEN a AND x. THE AREA $A_{\Delta x}$ IS THE AREA UNDER THE CURVE BETWEEN x AND $x + \Delta x$.

Since as $\Delta x \to 0$, this approximation gets better and better, we may assert that

$$G'(x) = \lim_{\Delta x \to 0} \frac{G(x + \Delta x) - G(x)}{\Delta x} = f(x),$$

which indicates that the derivative of G is f, as we wanted to show.   ∎

BONAVENTURA
CAVALIERI
—·—
1598–1647

HISTORICAL NOTE: THE DEVELOPMENT OF THE INTEGRAL IN THE SEVENTEENTH CENTURY

Archimedes' method of exhaustion provided a powerful tool for approximating area. After this monumental work, little was done to solve the area problem until the seventeenth century. Europeans didn't even know about Archimedes' work until a translation of a tenth-century copy of his manuscripts was found in Constantinople (now Istanbul) around 1450. It was not printed until 1540.

One of the first European mathematicians to work on the area problem was the Italian Bonaventura Cavalieri (1598–1647). Cavalieri, who had been a student of Galileo, served as a professor of mathematics at the University of Bologna (the first university in Europe) from 1629 until his death in 1647.

Cavalieri's greatest contribution was his work *Geometria Indivisibilibus Continuorum Nova quadam Ratione Promota* (Geometry advanced by a thus far unknown method, indivisibles of continua), published in 1635. It is not clear exactly what he meant by *indivisibles*, but it seems that he intended, for example, that an indivisible part of a plane figure is a chord (line segment) of that figure and an indivisible of a solid is a planar section (the part cut by a plane) of the solid. Using this notion, Cavalieri was able to formulate a number of geometrical principles that can be proved using modern calculus. Here is one of them:

> Suppose that two planar regions are included between two parallel lines. Suppose further that whenever both objects are cut by a line parallel to the two given lines, the ratio of the lengths of the line segments cut off by the regions is always the same. Then the areas of the two regions are always in this ratio.

For example, consider the following curves:

the circle of radius a centered at $(0, 0)$: $x^2 + y^2 = a^2$

the ellipse that intersects the axes at $(\pm a, 0)$, $(0, \pm b)$: $\dfrac{x^2}{a^2} + \dfrac{y^2}{b^2} = 1$, $a > b$.

These are drawn in Figure 3. Each curve lies between the parallel lines $x = -a$ and $x = a$. For each curve, we solve for y:

circle: $y = \pm\sqrt{a^2 - x^2}$ ellipse: $y = \pm\dfrac{b}{a}\sqrt{a^2 - x^2}$.

Suppose both curves are cut by the line $x = c$. Then

length of line segment in the circle $= 2\sqrt{a^2 - c^2}$

length of line segment in the ellipse $= 2\dfrac{b}{a}\sqrt{a^2 - c^2}$,

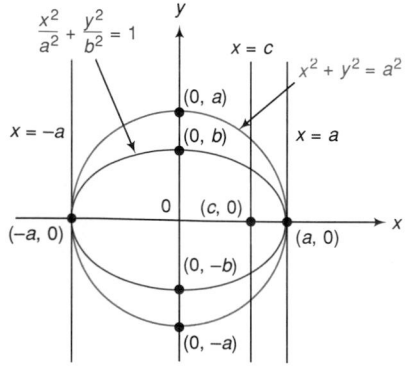

FIGURE 3
THE CURVES $x^2 + y^2 = a^2$ (CIRCLE) AND $\dfrac{x^2}{a^2} + \dfrac{y^2}{b^2} = 1$ (ELLIPSE).

and

$$\text{ratio of lengths of line segments} = \frac{b}{a}.$$

From Cavalieri's principle, we conclude that

$$\frac{\text{area of ellipse}}{\text{area of circle}} = \frac{b}{a}$$

and

$$\text{area of ellipse} = \frac{b}{a} \times \text{area of circle} = \frac{b}{a} \times \pi a^2 = \pi ab.$$

We can obtain the same result using calculus once we have developed further integration techniques.

Cavalieri was able to show, in 1639, that (in modern notation)

$$\int_0^b x^n \, dx = \frac{b^{n+1}}{n+1} \qquad \text{for any positive integer } n.$$

Cavalieri's work was harshly criticized by other mathematicians of the seventeenth century because the notion of indivisibles as small pieces of much larger objects was too vague and it sometimes led to contradictory results. Nevertheless, many used his ideas to solve problems that, until then, had remained a mystery. Integrals of such functions as x^r, $\sin x$, $\cos x$, and $\sin^2 x$ were found using Cavalieri's techniques.

Gilles Persone de Roberval (1602–1675) was a French mathematician who did much to extend the work on indivisibles. Roberval was less naive than Cavalieri and described a technique that is closer to the one we use today.

EXAMPLE OF ROBERVAL'S METHOD

Compute the area under $y = x^2$ for $0 \leq x \leq b$.

SOLUTION: We draw the curve in Figure 4 and divide the interval $[0, b]$ into n equal pieces, each having length $h = \frac{b}{n}$. Then, as we have done before, we compute the sum of areas of rectangles using right-hand endpoints:

$$A \approx h \cdot h^2 + h(2h)^2 + h(3h)^2 + \cdots + h(nh)^2$$
$$= h^3(1^2 + 2^2 + 3^2 + \cdots + n^2).$$

Fermat and others had obtained formulas for the kth powers of the first n positive integers. By the mid-1600s it was known that

$$1 + 2 + \cdots + n = \frac{n^2}{2} + \frac{n}{2}$$

$$1^2 + 2^2 + \cdots + n^2 = \frac{n^3}{3} + \frac{n^2}{2} + \frac{n}{6} \tag{8}$$

$$1^3 + 2^3 + \cdots + n^3 = \frac{n^4}{4} + \frac{n^3}{2} + \frac{n^2}{4}$$

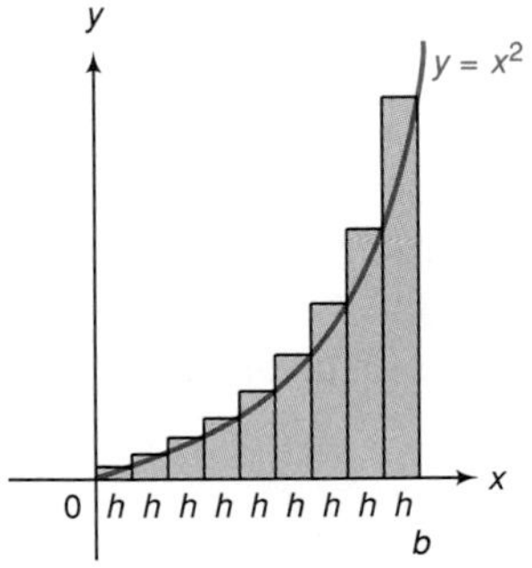

FIGURE 4
CALCULATING THE AREA UNDER $y = x^2$ FOR $0 \leq x \leq b$

The interval is divided into n subintervals, each of length $h = \frac{b}{n}$.

and, in general,

$$1^k + 2^k + \cdots + n^k = \frac{n^{k+1}}{k+1} + \text{terms of lower degree}.$$

In particular, Roberval knew the formula (8), so he could conclude that

$$A \approx h^3\left(\frac{n^3}{3} + \frac{n^2}{2} + \frac{n}{6}\right) = \frac{n^3}{6}\left(2n^3 + 3n^2 + 6\right).$$

Roberval argued, and this was his most significant contribution to this problem, that for n large, $2n^3$ is much larger than $3n^2$ or n. Therefore, he discarded those lower-degree terms to conclude that

$$h = \frac{b}{n}$$

$$A \approx h^3\left(\frac{2n^3}{6}\right) \;\downarrow\; = \frac{1}{3}\left(\frac{b}{n}\right)^3 n^3 = \frac{b^3}{3}.$$

What is significant here is not the result. Fermat and others knew around 1635 that $\int_0^b x^r \, dx = \frac{b^{r+1}}{r+1}$ whenever r is a rational number not equal to -1. In throwing out the terms $\frac{3n^2}{6}$ and $\frac{n}{6}$, Roberval was essentially using a limiting process. His most important work in this area appeared in his *Traité des indivisibles*, which was, however, not published until 1693, 18 years after his death and a few years after the work of Newton and Leibniz had become known.

Roberval claimed to be the first to make important discoveries for which others (like Cavalieri) received credit. His problem was that he often delayed publishing his results. There is an amusing reason for this. Roberval was appointed to a prestigious professorship in 1634 at the Collège Royale. According to the rules in force at the time, the position automatically became vacant every three years. A contest involving solving mathematical problems was held—with the chair going to the winner. The questions for the contest were set by the outgoing professor. Roberval withheld his best discoveries and used them to formulate questions for the contest. He could answer them; others could not. In this way he held the professorship for forty years. Who said life was fair?

Blaise Pascal (1623–1662) was a brilliant French mathematician and philosopher who continued the work on indivisibles. He used the more modern term *infinitesimals*. In his *Traité des sinus du quart de cercle*, published in 1659, the picture in Figure 5 appears:

Pascal claimed that

$$\frac{AD}{DI} = \frac{EE'}{RR'} = \frac{EE'}{EK}$$

and when the interval RI is sufficiently small, the arc length could be substituted for the tangent. He made use of facts like these to determine the sum of the sines of the x-values and, in this way, to compute the area under a sine curve.

BLAISE PASCAL
1623–1662

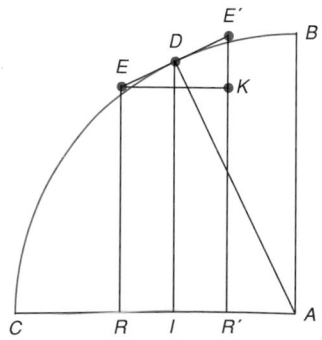

FIGURE 5
PASCAL'S TECHNIQUE FOR COMPUTING AREA UNDER A CURVE

J̲OHN W̲ALLIS
1616–1703

Pascal was more interested in geometry than in arithmetic formulas. This was unfortunate for him because, had he wished to find formulas for tangent lines, he almost certainly would have come upon the difference quotient that was the fundamental discovery in calculus made later by Newton and Leibniz. In fact, Leibniz made use of the figure above to develop his notion of a derivative. In a letter to James Bernoulli, written in 1703, Leibniz wrote that sometimes Pascal "seemed to have a bandage over his eyes."[†]

John Wallis (1616–1703) was a prolific mathematician who held a professorship of geometry at Oxford from 1649 until his death 54 years later. He took up the methods of Cavalieri and Roberval and was the first to explain with any precision what was meant by the formula $\int_0^1 x^n \, dx = \dfrac{1}{n+1}$ when n is negative or n is a rational number that is not an integer.

Wallis used his techniques to obtain an interesting expression for π. The area of the circle $x^2 + y^2 = 1$ is π. Thus the area of that quarter of the circle that lies in the first quadrant is $\dfrac{\pi}{4}$. This means that

$$\frac{\pi}{4} = \int_0^1 \sqrt{1 - x^2} \, dx.$$

Wallis did not know how to compute this integral. Instead, he computed

$$\int_0^1 (1 - x^2)^0 \, dx = 1, \quad \int_0^1 (1 - x^2)^1 \, dx = \frac{2}{3}, \quad \int_0^1 (1 - x^2)^2 \, dx = \frac{8}{15}, \text{ and so on.}$$

He then used interpolation and some very complicated reasoning to find what we now call a **Wallis product**:

$$\frac{\pi}{2} = \frac{2 \cdot 2 \cdot 4 \cdot 4 \cdot 6 \cdot 6 \cdot 8 \cdot 8 \cdot \cdots}{1 \cdot 3 \cdot 3 \cdot 5 \cdot 5 \cdot 7 \cdot 7 \cdot 9 \cdot 9 \cdots}.$$

You are asked to prove this formula in Problem 55 in Section 7.3.

Finally, we mention Isaac Barrow for the second time (see p. 109). Besides working on the tangent-line problem, Barrow also did significant work on the area problem and is considered to be the first to recognize the connection between the two. The fundamental theorem of calculus seems to be stated and proved (in a very different form than in this section) in his *Lectiones opticae et geometricae* (Optical and geometrical lectures), published in 1669.

■ **PROBLEMS 4.6**

■ **SELF-QUIZ**

I. $\int_2^5 6x \, dx =$ _____.

   a. $6 \cdot (5 - 2)$ b. $3 \cdot 2^2 - 3 \cdot 5^2$
   c. $12 \cdot 5^2 - 12 \cdot 5^2$ d. $3 \cdot 5^2 - 3 \cdot 2^2$

II. $\int_0^{\pi/2} \sin x \, dx =$ _____.

   a. $\cos(\pi/2) - \cos 0$

   b. $\sin(\pi/2) - \sin 0$
   c. $\cos 0 - \cos(\pi/2)$
   d. $-\cos(\pi/2) - (-\cos 0)$

III. $\int_{-1}^{7}$ _____ $dx = 6 \cdot 7^3 - 6 \cdot (-1)^3$.

   a. $\frac{3}{2}x^4$ b. $\frac{3}{2}x^4 + C$
   c. $18x^2$ d. $2x^2$

[†] Leibniz, *The Early Mathematical Manuscripts*, pp. 15–16, published in Chicago in 1920.

CONTINUED

IV. $\int_2^8$ _____ $= (8 - \pi)^5 - (2 - \pi)^5$.

   a. $5(x - \pi)^4$ b. $5(x - \pi)^4 + C$

   c. $(x - \pi)^6/6$ d. $\pi^5 x^4$

V. $\int_2^5 (-1/x^2)\, dx =$ _____ .

   a. $(-3)(\frac{1}{125}) - (-3)(\frac{1}{8})$

   b. $\frac{1}{125} - \frac{1}{8}$

   c. $(-\frac{1}{3})(1/5^3) - (-\frac{1}{3})(1/2^3)$

   d. $\frac{1}{5} - \frac{1}{2}$

VI. Let $f(x) = -1/x^2$ and $F(x) = 1/x$. Answer True or False to each of the following:

   a. $F'(x) = f(x)$

   b. $\int f = F$

   c. $\int f = F + \text{constant}$

   d. $\dfrac{d}{dx} \int_1^x f(t)\, dt = f(x)$

VII. $\dfrac{d}{dx} \int_3^x \left[2 + \left(\dfrac{d}{dt} \right)(\cos t) \right] dt =$ _____ .

   a. $2 - \sin x$ b. $2 - \sin t$

   c. $-\sin x$ d. 0

▪ DRILL

In Problems 1–20, calculate the given definite integral.

1. $\int_{-1}^2 x^4\, dx$

2. $\int_2^5 3s^3\, ds$

3. $\int_1^9 \frac{1}{2} \sqrt{t}\, dt$

4. $\int_1^3 (5 + 3x + x^2)\, dx$

5. $\int_1^8 \left(\frac{1}{\sqrt[3]{x}} + 7\sqrt[3]{x} \right) dx$

6. $\int_2^4 \left(\frac{1}{z^3} - \frac{1}{z^2} \right) dz$

7. $\int_0^{\pi/4} 2 \sin x\, dx$

8. $\int_0^{\pi/6} (\cos x - \sin x)\, dx$

9. $\int_{-1}^1 (p^9 - p^{17})\, dp$

10. $\int_{-A}^A x^{2n+1}\, dx$ where n is a positive integer and A is a real number.

11. $\int_2^3 (s - 1)(s + 2)\, ds$

   [Hint: Multiply.]

12. $\int_{-2}^2 (t^2 - 4)(t^5 + 6)\, dt$

13. $\int_0^1 (\sqrt{x} - x)^2\, dx$

14. $\int_0^1 (y^{3/2} - y^{2/3})(y^{4/3} - y^{3/4})\, dy$

15. $\int_2^4 \frac{6 + 7x + x^2}{1 + x}\, dx$

   [Hint: Divide.]

16. $\int_0^1 \frac{(t + 1)(t - 2)^3}{t^2 - t - 2}\, dt$

17. $\int_9^{16} \frac{s + 1}{\sqrt{s}}\, ds$

18. $\int_1^4 \frac{t^2 + 2t + 5}{t^{3/2}}\, dt$

19. $\int_{-5}^3 |x|\, dx$

20. $\int_{-5}^3 |x - 2|\, dx$

In Problems 21–32, find the area bounded by the given curve and the given lines. [Hint: Sketch the region.]

21. $y = 9 - x^2$; x-axis

22. $y = x^2 - 4$; x-axis

23. $y = x^2 - 6x + 5$; x-axis

24. $y = 34 + 15x - x^2$; x-axis

25. $y = (x - a)(x - b)$, $a < b$; x-axis

26. $y = x^2 + 2x - 3$; $x = 1$, $x = 3$, x-axis

27. $y = x^3 + 2x^2 - x - 2$; x-axis

28. $y = (x + 5)(x - 1)(x - 2)$; $x = 0$, $x = 3$, x-axis

29. $y = x^4$; $x = -2$, $x = 4$, x-axis

30. $y = (x^2 - 1)(x^2 - 4)$; x-axis

31. $y = \cos x - \sin x$; $x = 0$, $x = \pi$, x-axis

32. $y = 2 \sin x$; $x = -\pi$, $x = \pi$, x-axis

In Problems 33–38, use Theorem 2 to calculate the derivative $F'(x)$; then evaluate $F'(x_0)$ for the specified x_0.

33. $F(x) = \int_3^x \frac{1}{1 + t^3}\, dt$; $x_0 = 2$

34. $F(x) = \int_{-1}^x \frac{s^2}{s^2 + 5}\, ds$; $x_0 = 3$

35. $F(x) = \int_0^x \sqrt{\frac{u - 1}{u + 1}}\, du$; $x_0 = 1$

36. $F(x) = \int_3^x \left(\frac{d}{dt} \cos t^2 \right) dt$; $x_0 = 4$

*37. $F(x) = \int_1^{1 + 3x} \frac{1}{t}\, dt$; $x_0 = 1$

   [Hint: Use the chain rule.]

*38. $F(x) = \int_1^{x^2} \frac{1}{t}\, dt$; $x_0 = 2$

CONTINUED

■ APPLICATIONS

39. Compute the following integrals and think about their similarities.

   a. $\int_{-1}^{1} |x|\,dx$ b. $\int_{0}^{2} |t - 1|\,dt$

   c. $\int_{1}^{3} |u - 2|\,du$ d. $\int_{-5}^{-3} |s + 4|\,ds$

40. Suppose that a particle is constrained to move only along a straight line with velocity $v(t)$ at time t. Does $\int_{a}^{b} v(t)\,dt$ equal the total distance traveled during the time interval $[a, b]$, or does it equal the net distance traveled (i.e., the distance traveled in the positive direction minus the distance traveled in the negative direction)? Write an integral that computes the other distance.

41. A charged particle enters a linear accelerator. Its velocity increases during a $\frac{1}{100}$-sec period with a constant acceleration from an initial velocity of 500 m/sec to a velocity of 10,500 m/sec.

   a. Compute the acceleration.

   b. How far does the particle travel in this initial $\frac{1}{100}$ sec?

42. A ball is thrown down from a tower with an initial speed of 25 ft/sec. The tower is 500 ft high (Figure 6).

   a. How fast is the ball traveling after 3 sec?
   [*Note:* Acceleration due to gravity $\approx$ 32 ft/sec².]

   b. How far does the ball travel in the first 3 sec?

   c. How long does it take for the ball to hit the ground?

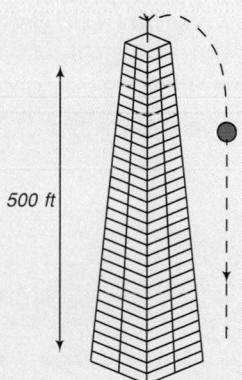

500 ft

FIGURE 6

*43. A small collection of insects is placed into an experimental environment. After t weeks have elapsed, the population of insects is found to be increasing at a rate of $3000/\sqrt{t}$ individuals per week. Assuming that to be true, how many individuals were added to the population between the 9th and the 25th week? Is there more than one way to interpret "between"? Discuss.

44. The velocity of a moving particle after t sec is given by $v(t) = t^{3/2} + 16t + 1$, where v is measured in m/sec. How far does the particle travel between the times $t = 4$ and $t = 9$ sec?

45. The marginal revenue a manufacturer receives is given by $MR = 2 - 0.02q + 0.003q^2$ dollars per additional unit sold. How much additional money does the manufacturer receive if he increases sales from 50 to 100 units?

46. The marginal cost to produce a certain tool is given by $MC = 2 - 0.003q + 0.00005q^2$ dollars. How much does it cost to increase production from 100 to 200 tools?

*47. Find all functions f such that f' is continuous on $[a, b]$ and $d/dx \int_{a}^{x} f(t)\,dt = \int_{a}^{x} f'(t)\,dt$.

*48. Find two nonconstant functions f and g such that $\int_{0}^{1} (f \cdot g) = (\int_{0}^{1} f) \cdot (\int_{0}^{1} g)$. (The real purpose of this exercise is to have you discover that this relation is unlikely to be true for a randomly chosen pair of functions.)

The arithmetic mean of a collection of n values $\{v_1, v_2, v_3, \ldots, v_n\}$ is $(v_1 + v_2 + \cdots + v_n)/n$. That is, the average value is computed by summing the values and dividing by n, the count of the number of values. By analogy with this notion, if the function f is continuous over the interval $[a, b]$, we can define the **average value of f over $[a, b]$** to be

$$\text{average value} = \frac{1}{b - a} \int_{a}^{b} f(x)\,dx.$$

In Problems 49–56, compute the average value of the given function over the given interval.

49. $f(x) = C$; $[a, b]$
50. $f(x) = 3x + 5$; $[1, 2]$
51. $f(x) = 1/x^2$; $[1, 3]$
52. $f(x) = x^2$; $[0, 2]$
53. $f(x) = x^3$; $[0, 2]$
54. $f(x) = x^r$, $r \neq -1$; $[0, 2]$
55. $f(x) = x^2 - 2x + 5$; $[-2, 2]$
56. $f(x) = (x - a)(b - x)$; $[a, b]$

*57. A ball was dropped from a height of 400 ft. Assuming the initial velocity was zero, what was its average velocity on the way to the ground?

CONTINUED

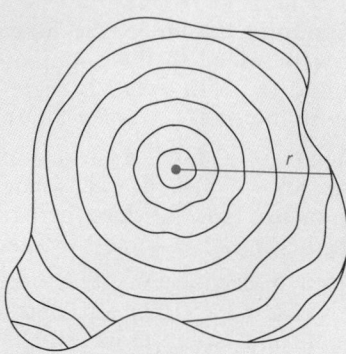

FIGURE 7

58. Consider a circular wave which spreads in such a way that its radius r increases at the rate of 2 ft/sec (Figure 7). Suppose this wave was created by dropping a pebble into a still pond. What is the average area of this circle during the 5-sec period after the wave started?

59. If the marginal cost of a given product is $50 - q/20$ dollars, what is the average cost per unit of production if 200 units are produced?

60. One day the air temperature t hours after noon was found to be $60 + 40t - t^2/3$ degrees (Fahrenheit). What was the average temperature between noon and 5 P.M.?

■ SHOW/PROVE/DISPROVE

Recall from page 42 that a function is said to be **even** if and only if $f(-x) = f(x)$ for every real number x. [For example, 1, x^2, x^4, $|x|$, $\cos x$, and $1/(1 + x^2)$ are all even functions.]

61. Show that if f is an integrable even function, then $\int_{-a}^{a} f = 2 \int_{0}^{a} f$ for every real number a. Explain this fact geometrically.

As on page 42, a function is said to be **odd** if and only if $f(-x) = -f(x)$ for every real number x. [For example, x, x^3, $\sin x$, and $1/(x + x^3)$ are all odd functions.]

62. Show that if f is an integrable odd function, then $\int_{-a}^{a} f = 0$ for every real number a. Explain this fact geometrically.

*63. Suppose that $p > 0$. Prove that

$$\frac{p}{p + 1} < \int_{0}^{1} \frac{1}{1 + x^p} \, dx < 1.$$

[Hint: First show that $1 - x^p < 1/(1 + x^p) < 1$.]

64. Watch carefully. Recall that

$$\frac{d}{dx}\left(\frac{-1}{x}\right) = (-1) \cdot \frac{d}{dx} x^{-1}$$

$$= (-1)(-1)x^{-2} = \frac{1}{x^2}.$$

Therefore,

$$\int_{-3}^{3} \frac{1}{x^2} \, dx = \frac{-1}{x}\Big|_{-3}^{3} = \frac{-1}{3} - \frac{-1}{-3} = \frac{-2}{3}?$$

But x^2 cannot be negative and neither can $1/x^2$. What went wrong to produce a negative value for the definite integral? [Hint: Sketch $y = 1/x^2$ and reread the statement of the fundamental theorem of calculus.]

■ ANSWERS TO SELF-QUIZ

I. d	II. d = c	III. c		VI. a. True	b. False	c. True d. True
IV. a		V. d		VII. a		

4.7 INTEGRATION BY SUBSTITUTION AND DIFFERENTIALS

$\mathbf{A}$s we saw in the previous section, the problem of finding a definite or indefinite integral is solved if we can find one antiderivative for the function f. However, as we have already mentioned, this task can often be very difficult. So far, the only functions we know how to integrate are of the form x^r (where $r \neq -1$), $\sin x$, and $\cos x$.

There are many other functions for which antiderivatives can be found, but learning to recognize them takes a lot of experience. We shall devote an entire chapter (Chapter 7) to the problem of recognizing the types of functions for which integrals can be expressed in terms of certain elementary functions. In this section, we enlarge the class of functions whose integrals can be immediately determined.

Before continuing, we need to discuss the notion of a differential. In using the chain rule, we obtained the formula

$$\frac{dy}{dx} = \frac{dy}{du}\frac{du}{dx}.$$
(1)

It seems as if in the right-hand side of (1), we have "canceled" the terms du to obtain the left-hand side. But the expressions dx, dy, and du have been given no meaning by themselves and are only part of a larger expression like dy/dx, which stands for "the derivative of y with respect to x." However, it is often convenient to treat the expressions dx, dy, and the like as separate entities. In fact, we did this when we wrote $\int_a^b f(x)\,dx$.

In Section 3.8 we talked about the linearization of a function. If f is differentiable, then the linearization of f at x_0 is given by

$$P_1(x) = f(x_0) + f'(x_0)(x - x_0) = f(x_0) + f'(x_0)\Delta x.$$

We saw that if x is close to x_0 (so that Δx is small), then $P_1(x)$ is a good approximation to $f(x)$ [see formula (4) on page 250]. We may write, for Δx small,

$$f(x) \approx f(x_0) + f'(x_0)\Delta x.$$
(2)

If $\Delta y = f(x) - f(x_0)$, then (2) becomes

$$\Delta y \approx f'(x_0)\Delta x.$$
(3)

We used the approximation (2) in a number of examples in Section 3.8.

The following definitions are motivated by the approximation (3).

DEFINITION   THE DIFFERENTIAL

Let $y = f(x)$, where f is a differentiable function, and let Δx be any nonzero real number.

(i) The **differential dx** is given by $dx = \Delta x$.

(ii) The **differential dy** is given by $dy = f'(x)\,dx$.   ■

Note that dx does not have to be a small number in accordance with this definition. In fact, dx can take on any nonzero value between $-\infty$ and ∞.

The first thing to notice about these definitions is that (since $\Delta x = dx$ is assumed to be nonzero)

$$\frac{dy}{dx} = \frac{f'(x)\,dx}{dx} = f'(x)$$

which is certainly not surprising since dy was chosen so that the equation above would be satisfied. We should also note that the definitions here are artificial in the sense that they have been created so that we can manipulate the symbols dx and dy. It must be emphasized that formulas like the chain rule (1) are true not because differentials can automatically be canceled but because they were proven true before we even had such things as differentials around.

EXAMPLE 1
CALCULATING A DIFFERENTIAL

Let $y = x^2$. Calculate the differential dy.

SOLUTION: Since $f'(x) = 2x$, we have $dy = 2x\,dx$.

The differentiation formulas given in Table 2.4.1 on page 158 can be extended to differential formulas, as shown in Table 1.

TABLE 1   DIFFERENTIATION FORMULAS IN DIFFERENTIAL NOTATION

$y = f(x)$	dy
I. c	$dc = 0$
II. $cu(x)$	$d(cu) = c\,du$
III. $u(x) + v(x)$	$d(u + v) = du + dv$
IV. x^r, r real	$d(x^r) = rx^{r-1}\,dx$
V. $u(x) \cdot v(x)$	$d(uv) = u\,dv + v\,du$
VI. $\dfrac{u(x)}{v(x)}$, $v(x) \neq 0$	$d\left(\dfrac{u}{v}\right) = \dfrac{v\,du - u\,dv}{v^2}$
VII. $u^r(x)$	$d[u^r(x)] = ru^{r-1}(x)u'(x)\,dx$
VIII. $f(g(x))$	$d[f(g(x))] = f'(g(x))g'(x)\,dx$
IX. $\sin u(x)$	$d[\sin u(x)] = [\cos u(x)]u'(x)\,dx$
X. $\cos u(x)$	$d[\cos u(x)] = [-\sin u(x)]u'(x)\,dx$

EXAMPLE 2
CALCULATING A DIFFERENTIAL USING THE CHAIN RULE

Let $y = \sin 2x$. Calculate the differential dy.

SOLUTION:

$$f'(x) = \frac{d}{dx}\sin 2x = (\cos 2x)(2) = 2\cos 2x$$

so

$$dy = f'(x)\,dx = 2\cos 2x\,dx.$$

EXAMPLE 3
CALCULATING A DIFFERENTIAL USING THE POWER RULE

Let $y = (1 + x^3)^{10}$. Calculate the differential dy.

SOLUTION:

$$f'(x) = \frac{d}{dx}(1 + x^3)^{10} = 10(1 + x^3)^9 \frac{d}{dx}x^3$$

$$= 10(1 + x^3)^9(3x^2) = 30x^2(1 + x^3)^9.$$

Thus

$$dy = f'(x)\,dx = 30x^2(1 + x^3)^9\,dx.$$

INTEGRATION BY SUBSTITUTION

Consider the formula

$$\int u^r \, du = \frac{u^{r+1}}{r+1} + C, \qquad \text{if } r \neq -1. \tag{4}$$

This is formula (3) on page 269 (with u in place of x). We can use this formula to prove the following basic result.

THEOREM 1   **INTEGRATING THE DERIVATIVE OF THE POWER OF A FUNCTION**

If $u = g(x)$ is a differentiable function of x and $du = g'(x) \, dx$ is its differential, then if $r \neq -1$,

$$\int [g(x)]^r g'(x) \, dx = \int u^r \, du.$$

PROOF: Applying the chain rule to the function $u^{r+1}/(r+1)$, we have

$$\frac{d}{dx}\left(\frac{u^{r+1}}{r+1}\right) = \frac{d}{dx}\left\{\frac{[g(x)]^{r+1}}{r+1}\right\} = [g(x)]^r g'(x).$$

Thus,

$$\int [g(x)]^r g'(x) \, dx = \frac{u^{r+1}}{r+1} + C = \int u^r \, du \quad \blacksquare$$

SUBSTITUTION TECHNIQUE FOR COMPUTING $\int f(x) \, dx$

   (i) Make a substitution $u = g(x)$ so that the integral can be expressed in the form $u^r \, du$ (if possible).
   (ii) Calculate the differential $du = g'(x) \, dx$.
   (iii) Write $\int f(x) \, dx$ as $\int u^r \, du$ for some number $r \neq -1$.
   (iv) Integrate using formula (4).
   (v) Substitute $g(x)$ for u to obtain the answer in terms of x.

⊗ **WARNING:**   In making a substitution, the *entire* expression must be converted from one variable to another.   ⊗

EXAMPLE 4
INTEGRATING THE SQUARE
ROOT OF A FUNCTION

Calculate $\int \sqrt{3+x} \, dx$.

SOLUTION: Let $u = g(x) = 3 + x$. Then $du = dx$, and

$$\int \overbrace{\sqrt{3+x}}^{\sqrt{u}} \overbrace{dx}^{du} = \int \sqrt{u} \, du = \int u^{1/2} \, du = \frac{2}{3} u^{3/2} + C$$

$$= \frac{2}{3}(3+x)^{3/2} + C.$$

EXAMPLE 5
INTEGRATING THE DERIVATIVE OF
THE POWER OF A FUNCTION

Calculate $\int_0^1 (x^3 - 1)^{11/5} 3x^2\, dx$.

SOLUTION: Let $u = g(x) = x^3 - 1$. Then $du = 3x^2\, dx$, and

$$\int \overbrace{(x^3 - 1)^{11/5}}^{u^{11/5}} \overbrace{3x^2\, dx}^{du} = \int u^{11/5}\, du = \frac{5}{16} u^{16/5} + C$$

$$= \frac{5}{16} (x^3 - 1)^{16/5} + C.$$

Then

$$\int_0^1 (x^3 - 1)^{11/5} 3x^2\, dx = \frac{5}{16}(x^3 - 1)^{16/5}\Big|_0^1$$

$$= \frac{5}{16}\left[0 - (-1)^{16/5}\right] = -\frac{5}{16}.$$

There is a quicker way to obtain this value:

If $x = 0$, then $u = x^3 - 1 = 0^3 - 1 = -1$.

If $x = 1$, then $u = x^3 - 1 = 1^3 - 1 = 0$.

Thus

$$\int_0^1 (x^3 - 1)^{11/5} 3x^2\, dx = \int_{u=-1}^0 u^{11/5}\, du = \frac{5}{16} u^{16/5}\Big|_{-1}^0$$

$$= \frac{5}{16}\left[0^{16/5} - (-1)^{16/5}\right] = -\frac{5}{16}.$$

The process of changing the integrand, the differential, *and* the limits of integration when computing a definite integral by making a substitution is discussed more fully in Section 7.4.

In the two examples above, the integrand was already in the form $\int u^r\, du$ for the appropriate value of r. Sometimes it is not, but we can salvage the problem by multiplying and dividing by an appropriate constant.

EXAMPLE 6
INTEGRATING A FUNCTION THAT
IS THE DERIVATIVE OF THE
POWER OF A FUNCTION WHEN IT
IS MULTIPLIED BY A CONSTANT

Calculate $\int x\sqrt[3]{1 + x^2}\, dx$.

SOLUTION: Let $u = 1 + x^2$. Then $du = 2x\, dx$. All we have is $x\, dx$. We therefore multiply and divide by 2 to obtain

$$\int x\sqrt[3]{1 + x^2}\, dx = \frac{1}{2} \int \overbrace{\sqrt[3]{1 + x^2}}^{\sqrt[3]{u}}\ \overbrace{2x\, dx}^{du} = \frac{1}{2} \int u^{1/3}\, du$$

$$= \frac{1}{2} \cdot \frac{3}{4} u^{4/3} + C = \frac{3}{8}(1 + x^2)^{4/3} + C.$$

We were allowed to multiply and divide by the constant 2 because of the theorem that states that $\int kf = k \int f$ for any constant k.

⊗ **WARNING:** *We cannot multiply and divide in the same way by functions that are not constants.* If we try, we get incorrect answers such as

$$\int \sqrt{1 + x^2}\, dx = \frac{1}{2x} \int \sqrt{1 + x^2}\, 2x\, dx \qquad \text{We multiplied and divided by } 2x.$$

$$= \frac{1}{2x} \int u^{1/2}\, du \qquad \text{Where } u = 1 + x^2.$$

$$= \frac{1}{2x} \cdot \frac{2}{3} u^{3/2} + C = \frac{1}{3x}(1 + x^2)^{3/2} + C.$$

But the derivative of $(1 + x^2)^{3/2}/3x$ is not even close to $\sqrt{1 + x^2}$. (Check!) It is possible to calculate $\int \sqrt{1 + x^2}\, dx$, but we will have to wait until Chapter 7 to see how to do so. ⊗

EXAMPLE 7
INTEGRATING A FUNCTION THAT IS THE DERIVATIVE OF THE POWER OF A FUNCTION WHEN IT IS MULTIPLIED BY A CONSTANT

Calculate $\int [1 + (1/t)]^5/t^2\, dt$.

SOLUTION: Let $u = 1 + (1/t)$. Then $du = -(1/t^2)\, dt$. We then multiply and divide by -1 to obtain

$$\int \frac{[1 + (1/t)]^5}{t^2}\, dt = -\int \overbrace{\left(1 + \frac{1}{t}\right)^5}^{u^5} \overbrace{\left(-\frac{1}{t^2}\right)}^{du} dt = -\int u^5\, du$$

$$= -\frac{u^6}{6} + C = -\frac{[1 + (1/t)]^6}{6} + C.$$

We know that

$$\int \sin u\, du = -\cos u + C$$

and

$$\int \cos u\, du = \sin u + C.$$

These formulas follow immediately from the facts that

$$\frac{d}{du}(-\cos u) = \sin u \qquad \text{and} \qquad \frac{d}{du} \sin u = \cos u.$$

Thus, we can prove the next theorem.

THEOREM 2 **INTEGRATING SINE AND COSINE FUNCTIONS**

If $u = g(x)$ is a differentiable function of x and $du = g'(x)\, dx$ is its differential, then

$$\int \sin[g(x)]g'(x)\, dx = \int \sin u\, du = -\cos u + C$$

$$= -\cos[g(x)] + C \tag{5}$$

and

$$\int \cos[g(x)]g'(x)\,dx = \int \cos u\,du = \sin u + C$$
$$= \sin[g(x)] + C. \quad \blacksquare \qquad (6)$$

EXAMPLE 8
INTEGRATING A COSINE
FUNCTION

Compute $\int \cos 3x\,dx$.

SOLUTION: If $u = 3x$, then $du = 3\,dx$, so multiplying and dividing by 3, we obtain

$$\int \cos 3x\,dx = \frac{1}{3}\int \overbrace{(\cos 3x)}^{\cos u}\overbrace{3\,dx}^{du} = \frac{1}{3}\int \cos u\,du$$
$$= \frac{1}{3}\sin u + C = \frac{1}{3}\sin 3x + C.$$

■ PROBLEMS 4.7

■ SELF-QUIZ

I. If $u = ax^2 + bx + c$, then _____.
   a. $du = 2ax + b$
   b. $du = 2a\,dx + b$
   c. $du = (2ax + b)\,dx$
   d. $dx = (2au + b)\,du$

II. $\int \sin 5x\,dx =$ _____.

   a. $\int \sin w\,dw$ with $w = 5x$

   b. $\int \sin v\left(\frac{1}{5}\,dv\right)$ with $v = 5x$

   c. $\int \sin u(5\,du)$ with $u = 5x$

   d. $\int \sin(t/5)(5\,dt)$ with $5t = x$

III. $\int 6x(x^2 - 1)^2\,dx =$ _____.

   a. $\int 3t^2\,dt$ with $t = x^2 - 1$

   b. $\int u^2\,du$ with $u = x^2 - 1$

   c. $\int 2v\,dv$ with $v = x^3$

   d. $\int 6w^5\,dw$ with $w^2 - 1 = x$

IV. $\int$ _____ $dx = (ax^2 - a^2)^3 + C.$
   a. $3(ax^2 - a^2)^2$

   b. $2ax(ax^2 - a^2)^3$
   c. $3(ax^2 - a^2)^2 2ax$
   d. $3(ax^2 - a^2)^2(2ax - 2a)$

V. $\int 3a(ax + b)^2\,dx =$ _____.

   a. $a(ax + b)^3 + C$
   b. $(ax + b)^3 + C$
   c. $(ax + b)^3/b + C$
   d. $3(ax + b)^3/(2a) + C$

VI. $\int x(1 + x^2)\,dx =$ _____.

   a. $(1 + x^2)^2/4 + C$
   b. $(1 + x^2)^2/2 + C$
   c. $(1 + x^2)^2 + C$
   d. $x^2(1 + x^2)^2/4 + C$

VII. $\int x \sin x^2\,dx =$ _____.

   a. $\sin x^2 + 2x^2 \cos x^2 + C$
   b. $\frac{1}{2}\cos x^2 + C$
   c. $-\cos x^2 + C$
   d. $-\frac{1}{2}\cos x^2 + C$

VIII. $\int -2(\sin x)(\cos x)\,dx =$ _____.

   a. $(\cos x)^2 + C$
   b. $-(\sin x)^2 + C$
   c. $(\cos x + \sin x)(\cos x - \sin x)/2 + C$
   d. all of the above

■ DRILL

In Problems 1–10, find the differential dy.

1. $y = 3x + 2$ 2. $y = -5 - 7x$
3. $y = x^4$ 4. $y = (-5 - 7x)^4$
5. $y = (1 + x^2)^4$ 6. $y = 1/(x^2 + 1)$
7. $y = \sqrt{1 + x^2}$ 8. $y = \tan x$
9. $y = \cos \sqrt{x}$ 10. $y = \sqrt[3]{1 + x^3}$

In Problems 11–42, carry out the indicated integration by making an appropriate substitution $u = g(x)$.

11. $\int (1 + x^2)^5 2x \, dx$

12. $\int_0^1 (1 + x)^4 \, dx$

13. $\int_0^2 (3 - x)^3 \, dx$

14. $\int_3^4 \sqrt{5 - t} \, dt$

15. $\int_0^1 (2 - x^4) 4x^3 \, dx$

16. $\int \frac{2x}{\sqrt{5 + x^2}} \, dx$

17. $\int_0^7 \sqrt{9 + x} \, dx$

18. $\int \sqrt{10 + 3x} \, dx$

19. $\int_0^1 \sqrt{10 - 9x} \, dx$

20. $\int_0^2 x^2 \sqrt{1 + x^3} \, dx$

21. $\int x^3 \sqrt[5]{1 + 3x^4} \, dx$

22. $\int_0^2 \frac{t}{(1 + 2t^2)^{5/2}} \, dt$

23. $\int (s^4 + 1) \sqrt{s^5 + 5s} \, ds$

24. $\int_{-2}^2 \sqrt{1 + |s|} \, ds$

25. $\int_1^2 \frac{t + 3t^2}{\sqrt{t^2 + 2t^3}} \, dt$

26. $\int_0^1 (3w - 2)^{99} \, dw$

27. $\int \frac{dx}{\sqrt{x}(1 + \sqrt{x})^3}$

28. $\int_1^2 \frac{w + 1}{\sqrt{w^2 + 2w - 1}} \, dw$

29. $\int \frac{\left[1 + \dfrac{1}{v^2}\right]^{5/3}}{v^3} \, dv$

30. $\int_0^3 \left(\frac{x}{3} - 1\right)^{77} \, dx$

31. $\int (ax + b)\sqrt{ax^2 + 2bx + c} \, dx$

32. $\int (ax^2 + bx + c)\sqrt{2ax^3 + 3bx^2 + 6cx + d} \, dx$

33. $\int_0^\alpha t\sqrt{t^2 + \alpha^2} \, dt, \alpha > 0$

34. $\int_0^\alpha t^n \sqrt{\alpha^2 + t^{n+1}} \, dt, \alpha > 0, n \geq 0$

35. $\int \frac{s^{n-1}}{\sqrt{a + bs^n}} \, ds$

36. $\int_{-\alpha}^\alpha p^2 \sqrt{\alpha^3 - p^3} \, dp, \alpha > 0$

37. $\int_{-\alpha}^\alpha p^5 \sqrt{\alpha^6 - p^6} \, dp, \alpha > 0$

38. $\int_{-\alpha}^\alpha p^{7/3}(\alpha^{10/3} - p^{10/3})^{19/7} \, dp$

39. $\int_0^{\pi/4} \sin 2x \, dx$

40. $\int x^2 \cos x^3 \, dx$

41. $\int \frac{\sin \sqrt{x}}{\sqrt{x}} \, dx$

42. $\int_0^{\sqrt{\pi}} x \cos(\pi + x^2) \, dx$

■ APPLICATIONS

In Problems 43–48, find the area of the region bounded by the given curve, lines, and axes.

43. $y = \sqrt{x + 2}$; x-axis, y-axis, $x = 7$

44. $y = x\sqrt{x^2 + 7}$; x-axis, y-axis, $x = 3$

45. $y = x^2 \sqrt{\dfrac{x^3}{3} + 1}$; x-axis, y-axis, $x = 3$

46. $y = \dfrac{x + 1}{x^3} = \dfrac{1 + (1/x)}{x^2}$; x-axis, $x = \dfrac{1}{3}$, $x = \dfrac{1}{2}$

47. $y = \sin \dfrac{x}{3}$; x-axis, $x = 0$, $x = \pi$

48. $y = x \cos x^2$; x-axis, $x = 0$, $x = \dfrac{1}{2}\sqrt{\pi}$

49. A particle moves with velocity $v(t) = 1/(3\sqrt{2 + t})$ m/min after t min of motion. How far does the particle travel between the times $t = 2$ and $t = 7$?

50. The density of a 19-m-long metal beam is given by $\rho(x) = 1/(3\sqrt[3]{8 + x})$, where x is measured in meters from the left end of the beam and ρ is measured in kg/m.
    a. What is the mass of the beam?
    *b. For what value of x does the interval $[0, x]$ contain exactly half the mass of the beam?

51. The marginal cost incurred in the manufacture of a certain item is given by $MC = 4/\sqrt{q + 4}$. What is the cost incurred by raising production from 60 to 77 units? [Costs are in dollars.]

52. What is the average velocity of the particle in Problem 49? [*Note:* The **average value** of a function over an interval is defined in the remarks preceding Problem 4.6.49 on page 319.]

53. What is the average density of the beam in Problem 50?

54. What is the average cost per unit to manufacture 96 units if the marginal cost function is the one given in Problem 51?

4.8 ADDITIONAL INTEGRATION THEORY (OPTIONAL)

In this chapter, we introduced the definite and indefinite integrals. In Chapter 7, we will show how a large variety of integrals can be calculated. The basic theorem underlying these calculations is that if f is continuous in $[a, b]$, then $\int_a^b f(x)\,dx$ exists. The proof of this theorem requires tools from advanced calculus that cannot be developed in an elementary calculus text. Nevertheless, there are interesting results that can be derived by using facts about continuous and differentiable functions that we already know. In this section, we will obtain two of these results, assuming only the basic integrability theorem cited above and the intermediate-value theorem, restated below.

THEOREM 1 **INTERMEDIATE-VALUE THEOREM (THEOREM 1.7.5)**

Let f be continuous on $[a, b]$. Then if d is any number between $f(a)$ and $f(b)$, there is a number c in (a, b) such that $f(c) = d$. ■

We can now state our first result.

THEOREM 2 **MEAN-VALUE THEOREM FOR INTEGRALS**

Suppose that f and g are continuous on $[a, b]$ and that $g(x)$ is never zero on (a, b). Then there exists a number c in (a, b) such that

$$\int_a^b f(x)g(x)\,dx = f(c) \int_a^b g(x)\,dx. \tag{1}$$

PROOF: Since f is continuous on $[a, b]$, there exist numbers m and M such that

$$m \le f(x) \le M, \tag{2}$$

and there are numbers x_1 and x_2 in $[a, b]$ such that $f(x_1) = m$ and $f(x_2) = M$ (from Theorem 1.7.4).[†] Since $g(x) \ne 0$ and g is continuous, we see that either $g(x) > 0$ or $g(x) < 0$ on (a, b). (Otherwise the intermediate-value theorem would be contradicted.) Suppose that $g > 0$. The case $g < 0$ is left as a problem (see Problem 12). Then we multiply (2) by $g(x)$ to obtain

$$mg(x) \le f(x)g(x) \le Mg(x),$$

which implies, by the comparison theorem (Theorem 4.5.7), that

$$m \int_a^b g(x)\,dx \le \int_a^b f(x)g(x)\,dx \le M \int_a^b g(x)\,dx. \tag{3}$$

[†] Recall from Theorem 1.7.4 that this result applies in the case in which m and M are the greatest lower bound and least upper bound, respectively, for f on $[a, b]$.

Since $g(x) > 0$, $\int_a^b g(x)\,dx > 0$, and dividing (3) by this integral yields

$$m \le \frac{\int_a^b f(x)g(x)\,dx}{\int_a^b g(x)\,dx} \le M.$$

Since $m = f(x_1)$ and $M = f(x_2)$ are values for $f(x)$, it follows that $[\int_a^b f(x)g(x)\,dx]/\int_a^b g(x)\,dx$ is a number between $f(x_1)$ and $f(x_2)$, so that, by the intermediate-value theorem, there is a number c such that

$$f(c) = \frac{\int_a^b f(x)g(x)\,dx}{\int_a^b g(x)\,dx},$$

and the theorem is proved. ∎

REMARK: We showed here that $c \in [a, b]$. In Problem 13 you are asked to show that $c \in (a, b)$.

The mean-value theorem for integrals is very useful. Some applications are suggested in the problem set. We use the theorem now to prove the second fundamental theorem of calculus.

THEOREM 3   **SECOND FUNDAMENTAL THEOREM OF CALCULUS (THEOREM 4.6.2)**

If f is continuous on $[a, b]$, then the function $G(x) = \int_a^x f(t)\,dt$ is continuous on $[a, b]$, differentiable on (a, b), and for every x in (a, b),

$$G'(x) = f(x). \tag{4}$$

That is, G is an antiderivative of f.

PROOF: Let x be in (a, b). Then

Theorem 4.5.3

$$G(x + \Delta x) = \int_a^{x+\Delta x} f(t)\,dt = \int_a^x f(t)\,dt + \int_x^{x+\Delta x} f(t)\,dt. \tag{5}$$

Thus,

$$G(x + \Delta x) - G(x) = \int_a^x f(t)\,dt + \int_x^{x+\Delta x} f(t)\,dt - \int_a^x f(t)\,dt$$
$$= \int_x^{x+\Delta x} f(t)\,dt. \tag{6}$$

From the mean-value theorem for integrals [using $g(x) = 1$], there is a number c in $(x, x + \Delta x)$ such that

$$\int_x^{x+\Delta x} f(t)1\,dt = f(c) \int_x^{x+\Delta x} 1\,dt = f(c)\,\Delta x. \tag{7}$$

Then using (7) in (6), we see that

$$G(x + \Delta x) - G(x) = f(c)\,\Delta x,$$

or for $\Delta x \ne 0$,

$$\frac{G(x + \Delta x) - G(x)}{\Delta x} = f(c). \tag{8}$$

We now take the limit on both sides of (8) to obtain

$$\lim_{\Delta x \to 0} \frac{G(x + \Delta x) - G(x)}{\Delta x} = \lim_{\Delta x \to 0} f(c), \tag{9}$$

or

$$G'(x) = \lim_{\Delta x \to 0} f(c). \tag{10}$$

But $c \in (x, x + \Delta x)$, so that as $\Delta x \to 0$, $c \to x$. Moreover, since $x \in (a, b)$, f is continuous at x, so

$$\lim_{c \to x} f(c) = f(x). \tag{11}$$

This shows that $G'(x) = f(x)$ for any x in (a, b).

Since a differentiable function is continuous, we see that G is continuous in (a, b). The only thing left to prove is the continuity of G at $x = a$ and $x = b$ (see the definition of continuity over a closed interval, page 116). That is, we must show that

$$\lim_{x \to a^+} G(x) = G(a) = \int_a^a f(t)\,dt = 0 \tag{12}$$

and

$$\lim_{x \to b^-} G(x) = G(b) = \int_a^b f(t)\,dt. \tag{13}$$

The limit in (12) is equivalent to

$$\lim_{\Delta x \to 0^+} \int_a^{a + \Delta x} f(t)\,dt = 0. \tag{14}$$

But $\int_a^{a+\Delta x} f(t)\,dt = f(c)\,\Delta x$, where $a < c < a + \Delta x$. And since f is continuous at a, $\lim_{c \to a^+} f(x) = f(a)$ and, as before, $\lim_{\Delta x \to 0^+} f(c) = f(a)$. Thus,

$$\lim_{\Delta x \to 0^+} \int_a^{a + \Delta x} f(t)\,dt = \lim_{\Delta x \to 0^+} f(c)\,\Delta x = \lim_{\Delta x \to 0^+} f(c) \lim_{\Delta x \to 0^+} \Delta x$$
$$= f(a) \cdot 0 = 0.$$

Hence, (12) holds.

To prove (13), we observe that

$$\int_a^b f(t)\,dt - \int_a^{b - \Delta x} f(t)\,dt = \int_{b - \Delta x}^b f(t)\,dt.$$

The same argument used to prove (14) shows that

$$\lim_{\Delta x \to 0^+} \int_{b - \Delta x}^b f(t)\,dt = 0.$$

Therefore,

$$\lim_{x \to b^-} G(x) = \lim_{\Delta x \to 0^+} \int_a^{b - \Delta x} f(t)\,dt = \int_a^b f(t)\,dt = G(b). \tag{15}$$

and (13) is proved. ∎

■ PROBLEMS 4.8

■ SELF-QUIZ

I. $\dfrac{d}{dx} \displaystyle\int_\pi^x u^2 \cos u\, du = $ _____.

  a. $x^2 \cos x$ b. $u^2 \cos u$
  c. $x^2 \cos x + C$ d. $x^2 \cos x - \pi^2 \cos \pi$

II. $\dfrac{d}{dt} \displaystyle\int_{-3}^t \dfrac{x}{16 + x^2}\, dx = $ _____.

  a. $-\dfrac{3}{25}$ b. $-\left(-\dfrac{3}{25}\right)$
  c. $x/(16 + x^2)$ d. $t/(16 + t^2)$

III. $\dfrac{d}{dx} \displaystyle\int_x^{10} u^2 \cos u\, du = $ _____.

  a. $x^2 \cos x$
  b. $-x^2 \cos x$
  c. $-2x \cos x + x^2 \sin x$
  d. $10^2 \cos 10 - x^2 \cos x$

IV. $\displaystyle\int_1^3 \left(\dfrac{d}{dt} \sqrt{t^2 - 1} \right) dt = $ _____.

  a. $\sqrt{8} - \sqrt{0}$
  b. $(d/dt)(\sqrt{8} - \sqrt{0}) = 0$
  c. $\sqrt{2 \cdot 3} - \sqrt{2 \cdot 1}$
  d. $3/\sqrt{8}$

■ DRILL

In Problems 1–6, apply the results of Problems 15–18 to differentiate the given function.

1. $f(x) = \displaystyle\int_0^{5x} \cos t\, dt$

2. $g(x) = \displaystyle\int_{\pi/3}^{x^7} \sin t\, dt$

3. $F(x) = \displaystyle\int_{1 + 2x}^5 \dfrac{t}{\sqrt{1 + t^2}}\, dt$

4. $G(x) = \displaystyle\int_{\sqrt{x}}^9 \sqrt{1 + t^2}\, dt$

5. $f(x) = \displaystyle\int_{\sin x}^{\cos x} t(5 - t)\, dt$

6. $g(x) = \displaystyle\int_{2x}^{x^2} \cos t\, dt$

■ APPLICATIONS

Establish the inequalities stated in Problems 7–10 by applying the mean-value theorem for integrals as stated in Theorem 2 and Problems 11–13.

7. Show that $\int_0^{2\pi} \sin t^3\, dt \le 2\pi$.

8. Show that $\int_0^{\pi/2} (\sin^8 t)(\sqrt[3]{\cos t})\, dt \le \pi/2$.

9. Suppose function f is continuous and positive. Show that $\int_0^\pi f(t) \sin t\, dt \le \int_0^\pi f(t)\, dt$.

10. Suppose function f is continuous and positive. Show that $\int_a^b f(t) \cos^6(t^2 + 1)\, dt \le \int_a^b f(t)\, dt$.

■ SHOW/PROVE/DISPROVE

11. Prove another form of the mean-value theorem for integrals: Let f be continuous on $[a, b]$. Then there exists a number c in (a, b) such that

$$\int_a^b f(t)\, dt = f(c) \cdot (b - a).$$

12. Show that Theorem 2, the mean-value theorem for integrals, does hold for the case that $g < 0$ on (a, b).

13. a. Suppose that f is continuous, $f(x) \ge 0$, and f is not identically equal to zero. Prove that $\int_a^b f(t)\, dt > 0$.

   b. Use the result of part (a) to show that in the proof of Theorem 2 the inequalities in (3) are strict, so that c is strictly between a and b [unless $f(x)$ is constant, in which case c can be any number between a and b].

*14. Let

$$f(x) = \int_x^{3x} \dfrac{1}{t}\, dt.$$

Prove that f is constant on $(0, \infty)$. [Hint: $\int_x^{3x} = \int_1^{3x} - \int_1^x$. Don't forget the chain rule.]

15. Let $G(u) = \int_a^u f(t)\,dt$, where f is continuous. Use the chain rule and Theorem 3 to prove that

$$\frac{d}{dx}\,G(u) = f(u)\cdot\frac{du}{dx}.$$

16. Use the result of Problem 15 to prove that if $b(x)$ is differentiable and f is continuous, then

$$\frac{d}{dx}\int_a^{b(x)} f(t)\,dt = f(b(x))\cdot b'(x).$$

17. Use the result of Problem 16 to prove that if $a(x)$ is differentiable and f is continuous, then

$$\frac{d}{dx}\int_{a(x)}^b f(t)\,dt = -f(a(x))\cdot a'(x).$$

18. Use the results of Problems 16 and 17 to prove that

$$\frac{d}{dx}\int_{a(x)}^{b(x)} f(t)\,dt = f(b(x))\cdot b'(x) - f(a(x))\cdot a'(x).$$

[*Hint:* Write $\int_{a(x)}^{b(x)} = \int_{a(x)}^c + \int_c^{b(x)}$.]

■ ANSWERS TO SELF-QUIZ

I. a

II. d

III. b $\left[\int_x^{10} u^2\cos u\,du = -\int_{10}^x u^2\cos u\,du\right]$

IV. a

■ PROBLEMS   CHAPTER 4 REVIEW

In Problems 1–4, calculate the derivative of the given function and evaluate it at the specified point.

1. $f(x) = \int_0^x t^3/\sqrt{t^2 + 17}\,dt$; $x_0 = 8$

2. $g(t) = \int_3^t (s^4 - 17s + 8)^{11/3}\,ds$; $t_0 = 1$

3. $F(s) = \int_{-1}^{2s} (1 + u^2)^{2001}\,du$; $s_0 = 0$

4. $G(u) = \int_{2u}^{u^2} x^2(1 + x^3)^4\,dx$; $u_0 = -1$

In Problems 5–8, find the differential dy.

5. $y = x/3$
6. $y = 1/(x + 1)$
7. $y = x^2 + 1$
8. $y = x^3 + 7$

In Problems 9–18, calculate the given definite or indefinite integral.

9. $\int x^5\,dx$

10. $\int (u + 3)^{-3}\,du$

11. $\int_{-1}^1 (x^2 - 3)(x^5 + 2)\,dx$

12. $\int_0^2 (t^3 + 3t + 5)\,dt$

13. $\int_0^{\pi/3} (\sin 2x + \cos 3x)\,dx$

14. $\int_0^1 x\sqrt{x^2 + 1}\,dx$

15. $\int t^2(t^3 + 7)^{-3/4}\,dt$

16. $\int (3ax^2 + 2bx + c)(ax^3 + bx^2 + cx + d)^{-1/9}\,dx$

17. $\int_{-3}^4 |s + 2|\,ds$

18. $\int_0^\pi |\sin(x/3)|\,dx$

In Problems 19–24, find the area of the region bounded by the given curve and lines.

19. $y = 3x - 7$; x-axis, $x = -2$, $x = 5$
20. $y = \sqrt{x + 1}$; x-axis, y-axis, $x = 15$
21. $y = 2 - x - x^2$; x-axis
22. $y = 15 + 7x - 7x^2 + x^3$; x-axis
23. $y = (x + 2)^{-2}$; x-axis, y-axis, $x = 3$
24. $y = \sin(x/3)$; x-axis, $x = 0$, $x = \pi/6$
25. A particle is moving with the velocity $v(t) = t + (t + 1)^{-1/2}$ m/sec.
    a. How far does the particle move in the first 15 sec?
    b. What is the average velocity of the particle over this initial 15-sec interval?
26. The density (measured in kg/m) of a tree is given by $\rho(h) = 50(h + 1)^{-1/2}$, where h denotes the distance (in meters) above the ground. The height of the tree is 24 m.
    a. What is the total mass of the tree?
    b. For what value of h does the first h meters of the tree contain half the total mass of the tree?
27. Show that $\int |x|\,dx = \frac{1}{2}x|x| + C$.

▣ COMPUTER EXERCISES

1. Consider the problem of calculating the area bounded by the curve $y = \sin \sqrt{x}$, the x-axis, and the lines $x = 0$ and $x = 1$.
   a. Divide the interval $[0, 1]$ into n subintervals of equal length. What are the endpoints of the ith subinterval?
   b. Let $x_i{}^*$ denote the left-hand endpoint of each subinterval. Use the corresponding Riemann sum to approximate the area for $n = 10$, 50, and 100.
   c. Let $x_i{}^*$ denote the right-hand endpoint of each subinterval. Use the corresponding Riemann sum to approximate the area for $n = 10$, 50, and 100.
   d. For any n, how does the approximation of the area using left-hand endpoints compare with the approximation using right-hand endpoints? (*Hint:* Show first that $f(x)$ is increasing on the interval $[0, 1]$.)
   e. How close to the correct answer is the average of the two values you got for $n = 100$ above? That is, what is the maximum error of your answer?

2. Consider the problem of finding the area bounded by the curve $y = x^2 + x^4$, the x-axis, and the lines $x = 0$ and $x = 1$.
   a. Divide the interval $[0, 1]$ into n subintervals of equal length. What are the endpoints of the ith subinterval?
   b. Letting x_i be the right-hand endpoint of the ith subinterval, use sigma notation to write an expression that approximates the area. (Your answer should be expressed in terms of n.)
   c. Use a computer algebra system to find a closed-form expression for your approximation above.
   d. Use your answer in (c) to find the area.

3. Use the fundamental theorem of calculus to calculate the area bounded by the curve $y = x^5 - 3x^3 - 2x^2 + 6$ and the x-axis.

4. Use the fundamental theorem of calculus to calculate $\int_0^2 |x^5 + x^4 + 2x^3 - 4x^2 - 4x - 8| \, dx$.

In Exercises 5 and 6, use a computer algebra system to find the indicated antiderivatives. Based on the answers reported by the computer (and additional examples of your own if needed) you will be asked to conjecture the form of more general antiderivatives. The following example should help make the idea clear to you.

A computer algebra system reports the values of the following integrals:

   (a) $\int x \sin x \, dx = \sin x - x \cos x + C$
   (b) $\int x^2 \sin x \, dx = 2x \sin x + (2 - x^2) \cos x + C$
   (c) $\int x^3 \sin x \, dx = (3x^2 - 6) \sin x + (6x - x^3) \cos x + C$

What is the form of $\int x^n \sin x \, dx$ if n is a positive integer?

5. Use a computer algebra system to find the following antiderivatives.

$$\int \cos^2 x \, dx, \qquad \int \cos^4 x \, dx, \qquad \int \cos^6 x \, dx$$

What is the form of $\int \cos^n x \, dx$ if n is an even integer?

6. Use a computer algebra system to find the following antiderivatives.

$$\int \sin^3 x \, dx, \qquad \int \sin^5 x \, dx, \qquad \int \sin^7 x \, dx$$

What is the form of $\int \sin^n x \, dx$ if n is an odd integer?

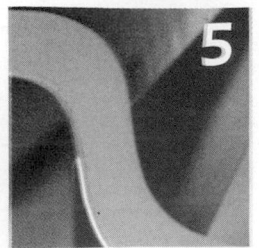

5 | # APPLICATIONS OF THE DEFINITE INTEGRAL

In this chapter, we illustrate the usefulness of the integral with a number of applications. The integral arises in an astonishingly wide variety of fields, including physics, chemistry, biology, and economics. The applications in this chapter provide a small sample of its use.

5.1 THE AREA BETWEEN TWO CURVES

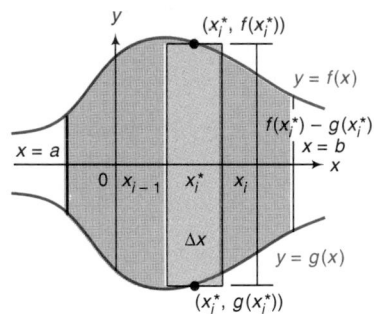

FIGURE 1
AREA BETWEEN TWO CURVES

We introduced the definite integral by calculating the area under a curve. We now show that by using similar methods, we can calculate the area between two curves.

Consider the two functions f and g as graphed in Figure 1. We will calculate the area between the curves $y = f(x)$ and $y = g(x)$ and the lines $x = a$ and $x = b$ by adding together the areas of a large number of rectangles. Note that $f(x) \geq g(x)$ for $a \leq x \leq b$. Let P be a regular partition of the interval $[a, b]$. That is, $\Delta x = \dfrac{b - a}{n}$. Then a typical rectangle (see Figure 1) has the area

$$[f(x_i^*) - g(x_i^*)]\,\Delta x,$$

where x_i^* is a point in the interval $[x_{i-1}, x_i]$ and $\Delta x = x_i - x_{i-1}$. Then proceeding exactly as we have proceeded before, we find that

$$\text{area} = \lim_{\Delta x \to 0} \sum_{i=1}^{n} [f(x_i^*) - g(x_i^*)]\,\Delta x = \int_a^b [f(x) - g(x)]\,dx. \tag{1}$$

This formula is valid as long as $f(x) \geq g(x)$ in $[a, b]$. We will deal with more general cases later in this section.

REMARK: There is no difficulty if one (or both) of the functions takes on negative values in $[a, b]$. If you are uncomfortable with these negative values, simply *shift upward*. If we define $\bar{f}(x) = f(x) + C$ and $\bar{g}(x) = g(x) + C$, where C is a positive constant, then

$$\text{area between } \bar{f} \text{ and } \bar{g} = \int_a^b (\bar{f} - \bar{g}) = \int_a^b [(f + C) - (g + C)]$$
$$= \int_a^b (f - g) = \text{area between } f \text{ and } g.$$

This is illustrated in Figure 2. The shaded areas retain the same area.

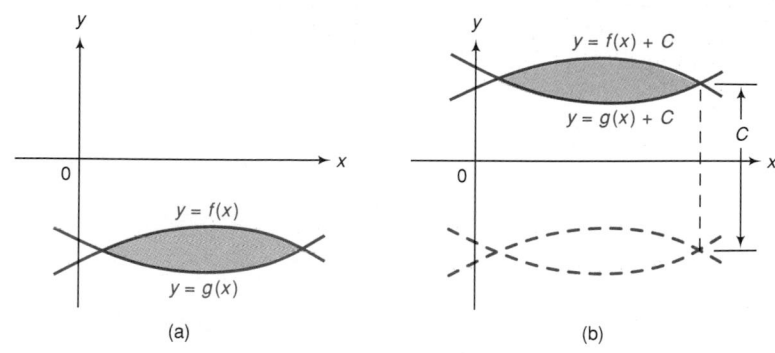

FIGURE 2 AREA BETWEEN TWO CURVES IS UNCHANGED WHEN THE CURVES ARE SHIFTED UPWARD

EXAMPLE 1
FINDING THE AREA BETWEEN
A PARABOLA AND A LINE

Find the area of the region bounded by $y = x^2$ and $y = 4x$, for x between 0 and 1.

SOLUTION: In any problem of this type, it is helpful to draw a graph. In Figure 3, the required area is shaded. We have

$$A = \int_0^1 (4x - x^2)\,dx = \left(2x^2 - \frac{x^3}{3}\right)\Big|_0^1 = 2 - \frac{1}{3} = \frac{5}{3}.$$

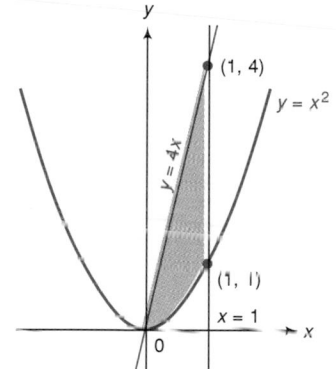

FIGURE 3
AREA BETWEEN THE CURVES $y = 4x$ AND
$y = x^2$

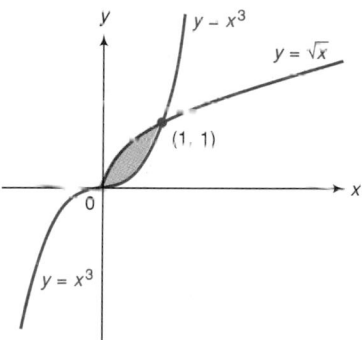

FIGURE 4 AREA BETWEEN THE CURVES $y = x^3$ AND $y = \sqrt{x}$.

EXAMPLE 2
FINDING THE AREA BETWEEN
A CUBIC FUNCTION AND A
SQUARE-ROOT FUNCTION

Find the area in the first quadrant bounded by the curves $y = x^3$ and $y = \sqrt{x}$.

SOLUTION: This problem only makes sense if the curves intersect at at least two points. (Otherwise we would have to be given other bounding lines.) See Figure 4. We need to find the points of intersection of these two curves. To find them, we set the two functions equal. If $x^3 = \sqrt{x}$, then $x^6 = x$, or $x^6 - x = x(x^5 - 1) = 0$. This occurs when $x = 0$ and $x = 1$. Then,

$$A = \int_0^1 (\sqrt{x} - x^3)\,dx = \left(\frac{2x^{3/2}}{3} - \frac{x^4}{4}\right)\Big|_0^1 = \frac{2}{3} - \frac{1}{4} = \frac{5}{12}.$$

In Examples 1 and 2, note that we have had no trouble deciding which function came first in the expression $f(x) - g(x)$. We always put the larger function first. Thus, in $[0, 1]$, $\sqrt{x} \geq x^3$, so that $\sqrt{x}$ comes first.

EXAMPLE 3
FINDING THE AREA BETWEEN
TWO PARABOLAS

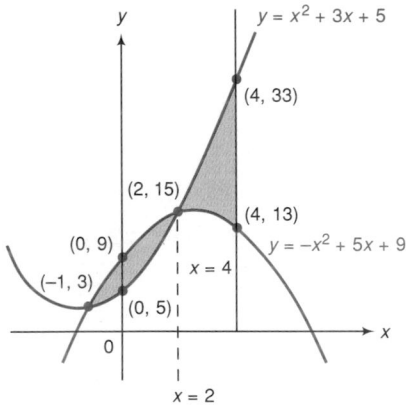

FIGURE 5
AREA BETWEEN THE CURVES
$y = x^2 + 3x + 5$ AND $y = -x^2 + 5x + 9$
FOR $-1 \leq x \leq 4$

Find the area bounded by the two curves $y = x^2 + 3x + 5$ and $y = -x^2 + 5x + 9$ for x between -1 and 4.

SOLUTION: See Figure 5. There are two areas to be added together. In the first,

$$-x^2 + 5x + 9 \geq x^2 + 3x + 5.$$

In the second,

$$x^2 + 3x + 5 \geq -x^2 + 5x + 9.$$

We, therefore, break the calculation into two parts:[†]

$$A = \int_{-1}^{2} \left[(-x^2 + 5x + 9) - (x^2 + 3x + 5) \right] dx$$

$$+ \int_{2}^{4} \left[(x^2 + 3x + 5) - (-x^2 + 5x + 9) \right] dx$$

$$= \int_{-1}^{2} (-2x^2 + 2x + 4) \, dx + \int_{2}^{4} (2x^2 - 2x - 4) \, dx$$

$$= \left(-\frac{2x^3}{3} + x^2 + 4x \right) \Big|_{-1}^{2} + \left(\frac{2x^3}{3} - x^2 - 4x \right) \Big|_{2}^{4}$$

$$= \frac{20}{3} + \frac{7}{3} + \frac{32}{3} + \frac{20}{3} = \frac{79}{3}.$$

We note that $(-x^2 + 5x + 9) - (x^2 + 3x + 5)$ in $[-1, 2]$ and $(x^2 + 3x + 5) - (-x^2 + 5x + 9)$ in $[2, 4]$ can be written as $\left| (x^2 + 3x + 5) - (-x^2 + 5x + 9) \right|$ in $[-1, 4]$. (Why?)

We can generalize the result of the last example to obtain the following rule:

THE AREA BETWEEN TWO CURVES

The area between the curves $y = f(x)$ and $y = g(x)$ between $x = a$ and $x = b$ $(a < b)$ is given by

$$A = \int_{a}^{b} |f(x) - g(x)| \, dx. \tag{2}$$

If $f(x) \leq g(x)$ on $[a, b]$, then $|f(x) - g(x)| = g(x) - f(x)$ and

$$A = \int_{a}^{b} [g(x) - f(x)] \, dx \qquad \text{if } f(x) \leq g(x). \tag{2'}$$

This rule forces the integrand to be positive, so we cannot run into the problem of adding negative areas (however, it does not make the calculation any easier).

NOTE: Reread the warning on page 300.

[†] Note that the two curves intersect when $x^2 + 3x + 5 = -x^2 + 5x + 9$ or $0 = 2x^2 - 2x - 4 = 2(x^2 - x - 2) = 2(x - 2)(x + 1)$ or when $x = 2$ and $x = -1$.

EXAMPLE 4
THE AREA BETWEEN A SINE
CURVE AND A COSINE CURVE

Calculate the area of the region in the first quadrant bounded by $y = \sin x$, $y = \cos x$, and the y-axis.

SOLUTION: The area requested is drawn in Figure 6. The curves intersect when $\sin x = \cos x$ or $\tan x = 1$, and $x = \pi/4$. Then,

$$A = \int_0^{\pi/4} (\cos x - \sin x)\, dx = (\sin x + \cos x)\Big|_0^{\pi/4} = \sqrt{2} - 1.$$

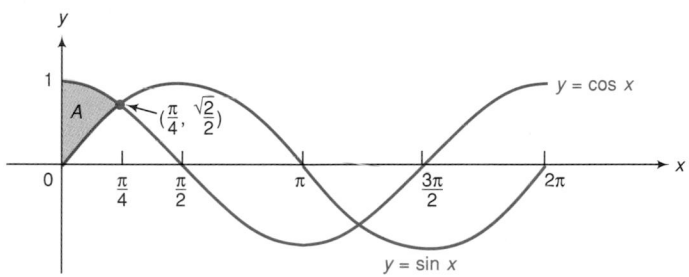

FIGURE 6 AREA BETWEEN $y = \sin x$ AND $y = \cos x$ FOR $0 \le x \le \pi/4$

EXAMPLE 5
FINDING THE AREA BETWEEN
TWO CURVES BY INTEGRATING
ALONG THE y-AXIS

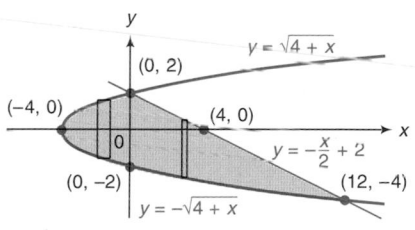

FIGURE 7
AREA BETWEEN $y^2 = 4 + x$ AND
$x + 2y = 4$, CALCULATED BY SUBDIVIDING
AN INTERVAL ALONG THE x-AXIS

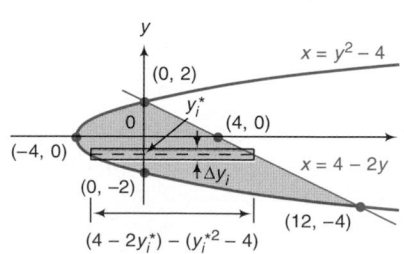

FIGURE 8
AREA BETWEEN $y^2 = 4 + x$ AND $x + 2y = 4$

The computation is easier when we begin by partitioning an interval along the y-axis.

Find the area bounded by the curves $y^2 = 4 + x$ and $x + 2y = 4$.

SOLUTION: This problem is more complicated than it might at first appear. Look at Figure 7. We can easily calculate that the two curves intersect at $(0, 2)$ and $(12, -4)$ by setting $x = y^2 - 4$ equal to $x = 4 - 2y$ to find that $y = 2$ and $y = -4$. Calculating the area is more complicated. We see that x ranges from -4 to 12. But we can't simply integrate from -4 to 12 with respect to x because the integrand changes. In $[-4, 0]$ a typical rectangle has a height of $\sqrt{4 + x_i^*} - (-\sqrt{4 + x_i^*}) = 2\sqrt{4 + x_i^*}$. But between 0 and 12, a typical rectangle has the height

$$\left(-\frac{x_i^*}{2} + 2\right) - (-\sqrt{4 + x_i^*}) = \sqrt{4 + x_i^*} - \frac{x_i^*}{2} + 2.$$

Thus,

$$A = \int_{-4}^0 2\sqrt{4 + x}\, dx + \int_0^{12} \left(\sqrt{4 + x} - \frac{x}{2} + 2\right) dx$$

$$= \frac{4}{3}(4 + x)^{3/2}\Big|_{-4}^0 + \left[\frac{2}{3}(4 + x)^{3/2} - \frac{x^2}{4} + 2x\right]\Big|_0^{12} = \frac{32}{3} + \frac{76}{3} = 36.$$

There is a much easier way to solve this problem. We are in the habit of writing every relation between x and y in terms of y as a function of x. However, in this problem it is clearly more convenient to treat x as a function of y. It is, for example, not difficult to obtain the graph of $x = y^2 - 4$. This graph is just like the graph of $y = x^2 - 4$ with the x- and y-axes interchanged. We again draw the graph, with the only change being that everything is labeled as a function of y. This graph is drawn in Figure 8. We now construct our rectangles horizontally instead of vertically by partitioning along the y-axis. The height of a typical rectangle is Δy, and the width is $(4 - 2y_i*) - (y_i*^2 - 4)$. We may, therefore, write

$$A = \int_{-4}^{2} [(4 - 2y) - (y^2 - 4)] \, dy$$

$$= \int_{-4}^{2} (8 - 2y - y^2) \, dy = \left(8y - y^2 - \frac{y^3}{3} \right) \Bigg|_{-4}^{2} = 36.$$

In general, if it is easier to treat y as the independent variable, then for $a < b$,

---

AREA BETWEEN TWO CURVES: INTEGRATING ALONG THE y-AXIS

$$A = \int_{y=a}^{y=b} |f(y) - g(y)| \, dy. \tag{3}$$

Whether to integrate with respect to x or y is a matter of judgment. In most cases an inspection of a sketch of the area being considered will indicate which is preferable. The goal is to have as few integrals as possible and to have those integrals relatively easy to compute. [It may not always be possible to achieve both these goals at the same time.]

■ PROBLEMS 5.1

■ SELF-QUIZ

I. The region bounded by the x-axis and the graph of $y = \sin x$ for x in $[-2, 5]$ has area equal to _____.

   a. $\int_{|-2|}^{|5|} |\sin x| \, dx$ b. $\int_{-2}^{5} \sin x \, dx$

   c. $\int_{-2}^{5} \sin|x| \, dx$ d. $\int_{-2}^{5} |\sin x| \, dx$

II. The region bounded by $y = x$ and $y = x^2$ for x between 2 and 5 has area equal to _____.

   a. $\int_{2}^{5} (x - x^2) \, dx$

   b. $\int_{2}^{5} (x^2 - x) \, dx$

   c. $\int_{5}^{2} |x^2 - x| \, dx$

   d. $\int_{2}^{5} |x| \, dx - \int_{2}^{5} |x^2| \, dx$

III. The region bounded by $y = x$ and $y = x^2$ for x between 0 and 5 has area equal to _____.

   a. $\int_{0}^{5} (x - x^2) \, dx$ b. $\int_{0}^{5} (x^2 - x) \, dx$

   c. $\left| \int_{0}^{5} (x - x^2) \, dx \right|$ d. $\int_{0}^{5} |x^2 - x| \, dx$

IV. $\int_{-2}^{7} |5 + x - \sin x| \, dx$ computes the area of the region bounded by the curves _____ for x between -2 and 7.

   a. $y = 5$ and $y = x - \sin x$

   b. $y = 5 - \sin x$ and $y = x$

   c. $y = 5 + x$ and $y = \sin x$

   d. $y = |5 + x|$ and $y = |\sin x|$

V. $\int_{-\pi}^{\pi} |\cos x - 4 + x^2| \, dx$ computes the area of the region bounded by the curves _____ for x between $-\pi$ and π.

   a. $y = \cos x$ and $y = 4 - x^2$

   b. $y = \sin x$ and $y = 4x - x^3/3$

   c. $y = \cos x - 4$ and $y = x^2$

   d. $y = \cos x$ and $y = -4 + x^2$

VI. Consider the upper half of the circle with radius 5 and centered at the origin. The area of this semicircle is computed by _____.

   a. $\int_{0}^{5} \sqrt{25 - x^2} \, dx$ b. $\int_{-5}^{5} \sqrt{25 - x^2} \, dx$

   c. $\int_{-5}^{5} (25 - x^2) \, dx$ d. $\int_{-5}^{5} |25 - x^2| \, dx$

CONTINUED

■ DRILL

In Problems 1–24, calculate the area of the region bounded by the given curves and lines. Where it's simple to do so, check your work by computing the area using both integration with respect to x and integration with respect to y. (See Example 5.)

1. $y = x^2$, $y = x$
2. $y = x^2$, $y = x^3$
3. $y = 3x^2 + 6x + 8$, $y = 2x^2 + 9x + 18$
4. $y = 2x^2 + 3x + 5$, $y = x^2 + 3x + 6$
5. $y = 3 - 7x + x^2$, $y = 5 - 4x - x^2$
6. $y = x^4 + x - 81$, $y = x$
7. $y = x^2$, $y = x^3$, $x = 3$
8. $y = 2x - x^2$, $y = x - 2$, $x = 3$
9. $y = x^3$, $y = x^3 + x^2 + 6x + 5$
10. $y = x^3 + x + 1$, $y = x^2 + x + 1$

11. $xy^2 = 1$, $x = 5$, $y = 5$
*12. $xy^2 = 1$, $y = 3 - 2\sqrt{x}$
13. $y = x^{1/3}$, $y = x$
14. $y = -1 + 2/x^2$, $y = x^2$, x-axis
*15. $y = \sin(\pi x/2)$, $y = x$, $x = 0$, $x = 2$
16. $y = \sin(x/2)$, $y = \cos(x/2)$, $x = 0$, $x = \pi$
17. $x + y^2 = 8$, $x + y = 2$
18. $x = y^2$, $x = 3y + 10$
19. $x = y^2$, $x^2 = 6 - 5y$
*20. $y = 2x$, $y = \sqrt{4x - 24}$, $y = 0$, $y = 10$
21. $\sqrt{x} + \sqrt{y} = 4$, $x = 0$, $y = 0$
22. $x = y^3 - y^2$, $x = 5y^2$
*23. $y = ax + b$, $y = ax^2$, y-axis, $a, b > 0$
24. $y = a^2 - x^2$, $y = x^2 - a^2$

■ APPLICATIONS

25. Find the area of the triangle with vertices at $(1, 6)$, $(2, 4)$, and $(-3, 7)$. [*Hint:* Find equations for the straight lines forming the sides, and draw a sketch.]
26. Find the area of the triangle with vertices at $(2, 0)$, $(3, 2)$, and $(6, 7)$.
27. Find the area of the triangle with sides on the lines $x - y + 6 = 0$, $2x - y - 2 = 0$, $3x - y + 4 = 0$.

28. Find the area of the triangle with sides given by $x + y = 3$, $2x - y = 4$, $6x + 3y = 3$.
29. Find a horizontal line that divides the region bounded by the x-axis and the graph of $y = a^2 - x^2$ into two parts of equal area.
30. Let $A(m)$ be the area of the region below $y = mx$ and above $y = x^2$. Find $A'(m)$.

■ SHOW/PROVE/DISPROVE

31. Consider the region between $y = a^2 - x^2$ and $y = x^2 - a^2$. Prove that any line through the origin splits this region in half (i.e., into two pieces having the same area).
*32. Pick two points $A = (a, a^2)$ and $B = (b, b^2)$ on the parabola $y = x^2$. Somewhere between A and B is a point where the tangent to the parabola is parallel to the line through A and B. (What theorem guarantees this?) Consider the parallelogram that would be traced out if we start

at point A, move along a straight line to B, drop vertically from B down to the parallel tangent line, move along this tangent line until directly beneath A, then rise vertically to A. Within this parallelogram, there is a connected region lying above the parabola and a two-part region lying below the parabola. Prove that the ratio of the areas of these two regions is constant.[†]

■ ANSWERS TO SELF-QUIZ

I. d II. b III. d
IV. c V. a VI. b

[†] This result dates back to Archimedes.

5.2 VOLUMES

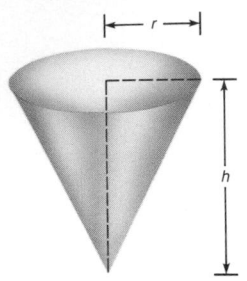

FIGURE 1
RIGHT-CIRCULAR CONE WITH HEIGHT h
AND RADIUS r

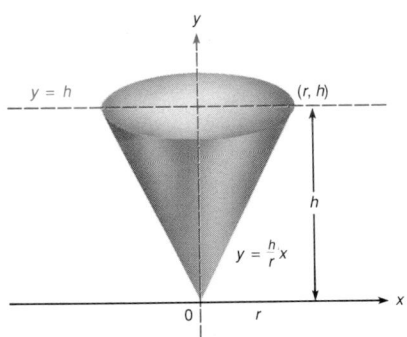

FIGURE 2
RIGHT-CIRCULAR CONE PLACED WITH ITS
VERTEX AT THE ORIGIN

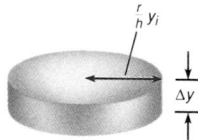

FIGURE 3
TYPICAL DISK

In this section, we show how the definite integral can be used to calculate the volumes of certain solid figures. We begin by deriving the familiar formula for the volume of a right-circular cone. Such a cone is depicted in Figure 1. To calculate its volume, we place the cone with its vertex at the origin and its central axis along the y-axis (see Figure 2). The line joining the points $(0, 0)$ and (r, h) has the equation $y = (h/r)x$, and we may think of the cone as the solid obtained by rotating the region bounded by this line, the y-axis, and the line $y = h$ around the y-axis. In this context, the cone is called a **solid of revolution**.

We will calculate the volume in several steps, using two different methods.

METHOD 1: DISK METHOD

Step 1: Partition an appropriate interval. We form an equally spaced partition of the interval along the y-axis between $y = 0$ and $y = h$ by using the partition points

$$0 = y_0 < y_1 < y_2 < \cdots < y_{n-1} < y_n = h,$$

where the width of each subinterval given by $\Delta y = y_i - y_{i-1}$ is assumed to be small. This has the effect of dividing the cone into "slices," or **disks**, parallel to the x-axis (see Figure 3).

Step 2: Approximate the volume of each disk. As drawn in Figure 4, each disk is, for Δy small, very close to a right circular cylinder with radius $x_i = (r/h)y_i$ and height Δy. It is not quite a cylinder because at the top the radius is $(r/h)y_i$, while at the bottom the radius is $(r/h)y_{i-1}$. However, if Δy is small, this difference will be small.

The volume of a cylinder with radius R and height H is $\pi R^2 H$. Thus, if ΔV_i denotes the volume of the ith disk, we have

$$\Delta V_i \approx \pi x_i^2 \, \Delta y = \pi \left(\frac{r}{h} y_i\right)^2 \Delta y = \frac{\pi r^2}{h^2} y_i^2 \, \Delta y.$$

Step 3: Add up the volumes of the disks to obtain an approximation for the volume of the cone. Then

$$V = \Delta V_1 + \Delta V_2 + \cdots + \Delta V_n$$
$$\approx \frac{\pi r^2}{h^2} (y_1^2 \, \Delta y + y_2^2 \, \Delta y + \cdots + y_n^2 \, \Delta y)$$
$$= \frac{\pi r^2}{h^2} \sum_{i=1}^{n} y_i^2 \, \Delta y. \tag{1}$$

We emphasize that the expression in (1) is an approximation to the volume. It provides the exact volume of the object drawn in Figure 5.

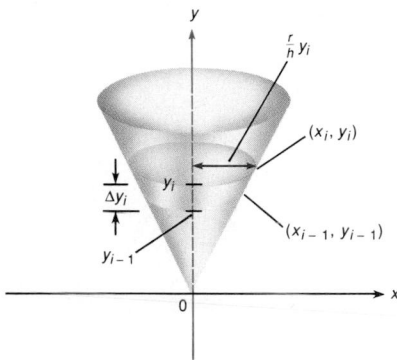

FIGURE 4
RIGHT-CIRCULAR CONE SHOWING A TYP-
ICAL DISK OBTAINED BY PARTITIONING
THE INTERNAL $[0, h]$ ALONG THE y-AXIS

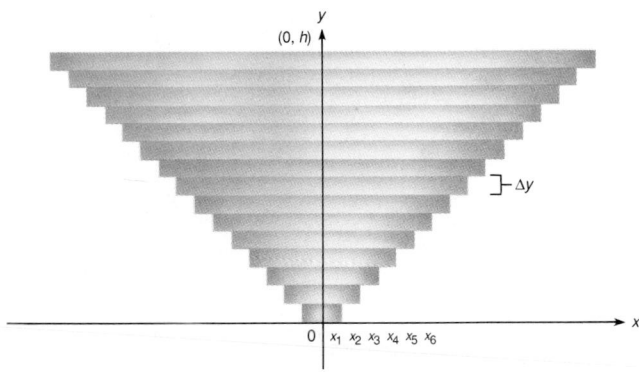

FIGURE 5 *THE SUM OF THE VOLUMES OF n DISKS APPROXIMATES THE VOLUME OF THE CONE.*

Step 4: Take the limit as $\Delta y \to 0$. The approximations in (1) get better and better as $\Delta y \to 0$. Thus, we have

$$V = \lim_{\Delta y \to 0} \frac{\pi r^2}{h^2} \sum_{i=1}^{n} y_i^2 \, \Delta y = \frac{\pi r^2}{h^2} \lim_{\Delta y \to 0} \sum_{i=1}^{n} y_i^2 \, \Delta y$$

Equation (4.5.9) on page 298

$$\overset{\downarrow}{=} \frac{\pi r^2}{h^2} \int_0^h y^2 \, dy = \frac{\pi r^2}{h^2} \left. \frac{y^3}{3} \right|_0^h = \frac{1}{3} \pi r^2 h,$$

which is the formula you might have seen earlier for the volume of a cone.

METHOD 2: SHELL METHOD

We now calculate the same volume in a different way.

Step 1: We form an equally spaced partition of the x-axis between $x = 0$ and $x = r$ by using the partition points

$$0 = x_0 < x_1 < x_2 < \cdots < x_{n-1} < x_n = r,$$

where the length of each subinterval, given by $\Delta x = x_i - x_{i-1}$, is assumed to be small. The situation is depicted in Figure 6. This has the effect of dividing the cone into what are called **cylindrical shells** (a cylindrical shell has the shape of an empty tin can with the ends removed). A typical cylindrical shell is drawn in Figure 7.

Step 2: Approximate the volume of each cylindrical shell. First, we note that a cylindrical shell is an "outer" cylinder with an "inner" cylinder removed. In Figure 6, the "shell" between x_{i-1} and x_i is not quite the same as the shell in Figure 7 because the bottom is not flat. To visualize this, note that the inner cylinder has height $h - y_{i-1}$, and the

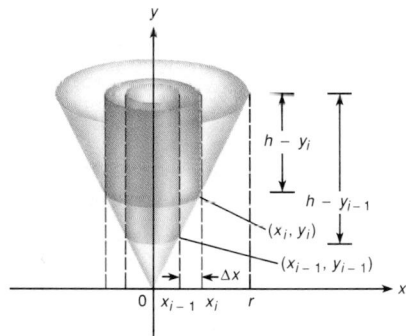

FIGURE 6
PARTITIONING ALONG THE x-AXIS LEADS
TO CYLINDRICAL SHELLS

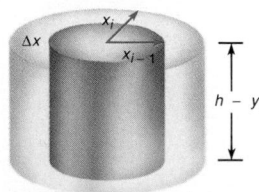

FIGURE 7
TYPICAL CYLINDRICAL SHELL

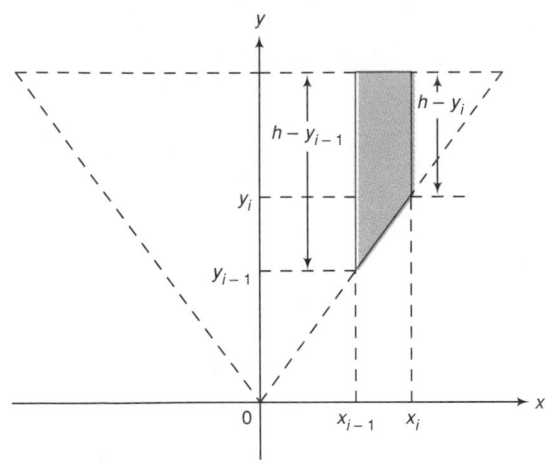

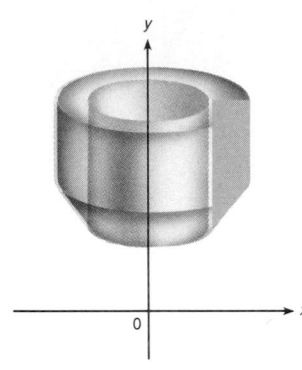

(a) If this piece is rotated about the y-axis . . .

(b) . . . this solid is obtained.

FIGURE 8 WHEN THE AREA ABOVE THE SUBINTERVAL $[x_{i-1}, x_i]$ IS ROTATED ABOUT THE y-AXIS, A SOLID THAT RESEMBLES THE SHELL OF A CYLINDER IS OBTAINED.

outer cylinder has height $h - y_i$ (see Figure 8). If Δx is small, then the difference in the volume between the cylindrical shell in Figure 7 and the shell in Figure 8b will be small. Thus, we use the volume of the cylindrical shell as an approximation to the volume of the actual shell obtained in the partition.

We have

volume of cylindrical shell
= volume of outer cylinder minus volume of inner cylinder.

The volume of a cylinder is $\pi r^2 h$, where r is its radius and h is its height. From Figure 7 we see that

volume of outer cylinder $= \pi x_i^2 (h - y_i)$

and

volume of inner cylinder $= \pi x_{i-1}^2 (h - y_i)$.

Thus, if ΔV_i denotes the volume of the ith shell (sketched in Figure 8b), then

$$\Delta V_i \approx \text{volume of } i\text{th cylindrical shell}$$
$$= \pi[x_i^2(h - y_i) - x_{i-1}^2(h - y_i)]$$
$$= \pi(x_i^2 - x_{i-1}^2)(h - y_i)$$
$$= \pi(x_i + x_{i-1})(x_i - x_{i-1})(h - y_i)$$
$$= \pi(x_i + x_{i-1})(h - y_i)\Delta x = 2\pi\left(\frac{x_i + x_{i-1}}{2}\right)(h - y_i)\Delta x$$

or

$$\Delta V_i \approx 2\pi\left(\frac{x_i + x_{i-1}}{2}\right)(h - y_i)\,\Delta x. \tag{2}$$

The right side of equation (2) gives the volume of the ith cylindrical shell exactly. Since the average of the numbers x_i and x_{i-1} is $(x_i + x_{i-1})/2$, we have the following result:

approximate volume of a cylindrical shell
$= 2\pi(\text{average radius})(\text{height})(\text{thickness}).$

Now if Δx is small, then $x_{i-1} = x_i - \Delta x \approx x_i$, so

$$\text{average radius} = \frac{x_i + x_{i-1}}{2} \approx \frac{x_i + x_i}{2} = x_i,$$

and (2) becomes

$$\Delta V_i \approx 2\pi x_i(h - y_i)\,\Delta x. \tag{3}$$

Since $y = (h/r)x$, $y_i = (h/r)x_i$, and we obtain, from (3),

$$\Delta V_i \approx 2\pi x_i\left[h - \frac{h}{r}x_i\right]\Delta x. \tag{4}$$

Step 3: Add up the volumes of the cylindrical shells to obtain an approximation to the volume of the cone.

$$V = \Delta V_1 + \Delta V_2 + \cdots + \Delta V_n = \sum_{i=1}^{n} \Delta V_i$$

$$\approx \sum_{i=1}^{n} 2\pi x_i\left(h - \frac{h}{r}x_i\right)\Delta x.$$

Step 4: Take the limit as $\Delta x \to 0$.

$$V = \lim_{\Delta x \to 0} \sum_{i=1}^{n} 2\pi x_i\left(h - \frac{h}{r}x_i\right)\Delta x$$

$$= \int_0^r 2\pi x\left(h - \frac{h}{r}x\right)dx = 2\pi h \int_0^r \left(x - \frac{x^2}{r}\right)dx$$

$$= 2\pi h\left(\frac{x^2}{2} - \frac{x^3}{3r}\right)\Big|_0^r = 2\pi h\left(\frac{r^2}{2} - \frac{r^2}{3}\right) = 2\pi h\frac{r^2}{6} = \frac{1}{3}\pi r^2 h,$$

as before.

The two methods used above can be applied in a wide variety of problems. When the curve $y = f(x)$ is revolved about an axis, we obtain a **solid of revolution** whose volume can be obtained by the **disk method** or the **cylindrical shell method**. Before giving additional examples, however, we summarize the procedure that we used here. We will use this procedure in every application in this chapter and elsewhere. We call our procedure the **integration process**.

THE INTEGRATION PROCESS FOR APPLICATIONS

Suppose that a quantity Q is defined in the interval $[a, b]$. To obtain Q, follow these steps:

(i) Break the quantity Q into a number of small pieces, denoted by ΔQ_i, so that

$$Q = \sum_{i=1}^{n} \Delta Q_i.$$

Do so by partitioning either the x-axis or the y-axis.

(ii) Select an arbitrary piece and work the problem on that piece, obtaining an approximate value:

$$\Delta Q_i \approx q(x_i{}^*)\,\Delta x \ (\text{or } \Delta Q_i \approx q(y_i{}^*)\,\Delta y).$$

(iii) Add the results for all the pieces:

$$Q \approx \sum_{i=1}^{n} q(x_i{}^*)\,\Delta x.$$

(iv) Take the limit of this sum to obtain a definite integral:

$$Q = \lim_{\Delta x \to 0} \sum_{i=1}^{n} q(x_i{}^*)\,\Delta x = \lim_{n \to \infty} \sum_{i=1}^{n} q(x_i{}^*)\,\Delta x.$$

(v) Evaluate the definite integral to obtain the answer:

$$Q = \int_{a}^{b} q(x)\,dx.$$

As we solve more and more problems by using the integration process for applications, we will obtain a number of formulas. But you should keep in mind that *it is the process that is important, not the formulas.* Once you have mastered the process, you will be able to solve an astonishingly wide variety of problems that once seemed very difficult or even impossible.

EXAMPLE 1

CALCULATING A VOLUME BY TWO DIFFERENT METHODS

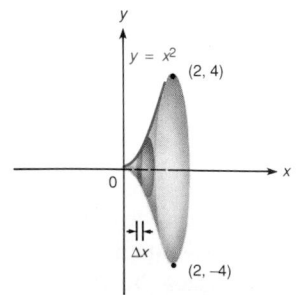

FIGURE 9
THE SOLID OBTAINED WHEN THE GRAPH OF $y = x^2$ IS REVOLVED ABOUT THE x-AXIS

The region bounded by the curve $y = x^2$, the x-axis, and the line $x = 2$ is revolved about the x-axis. Calculate the volume of the solid generated.

SOLUTION: The solid is depicted in Figure 9. By partitioning the x-axis, we obtain disks. The volume of a typical disk is given by

$$\Delta V_i \approx \pi r^2 h = \pi y_i^2\,\Delta x = \pi x_i^4\,\Delta x,$$

so that

$$V = \int_0^2 \pi x^4\,dx = \left.\frac{\pi x^5}{5}\right|_0^2 = \frac{32\pi}{5}.$$

Using cylindrical shells, we have, using Figure 10,

$$\Delta V_i \approx 2\pi y_i(2 - x_i)\,\Delta y = 2\pi y_i(2 - \sqrt{y_i})\,\Delta y,$$

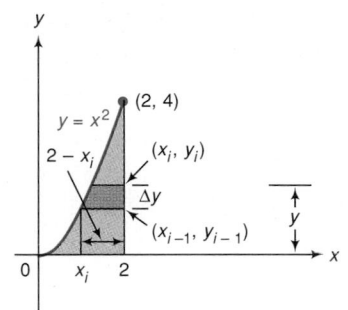

FIGURE 10
THE SHADED REGION IS ROTATED ABOUT THE x-AXIS, GENERATING A CYLINDRICAL SHELL.

so that

$$V = \int_0^4 2\pi(2 - \sqrt{y})y\,dy = 2\pi \int_0^4 (2y - y^{3/2})\,dy = 2\pi \left(y^2 - \frac{2}{5}y^{5/2} \right) \Big|_0^4$$

$$= 2\pi \left(16 - \frac{64}{5} \right) = \frac{32\pi}{5}.$$

It is, of course, not necessary to use both methods to calculate one volume. Typically, one method will be easier to use than the other in a given problem, and deciding which method to use will be a matter of judgment. The calculation of a given volume follows fairly easily once one of the axes has been partitioned and the volume of a typical disk or shell has been calculated, provided that the function involved is easy to integrate. We give two definitions that are used in the calculation of volumes.

DEFINITION **VOLUME BY DISK METHOD**

Let f be continuous and nonnegative on $[a, b]$. (See Figure 11a.) Then the **volume** of the solid generated by revolving the area below the graph of f about the x-axis between $x = a$ and $x = b$ (see Figure 11b) is defined by

$$V = \int_a^b \pi[f(x)]^2\,dx. \quad \blacksquare \qquad (5)$$

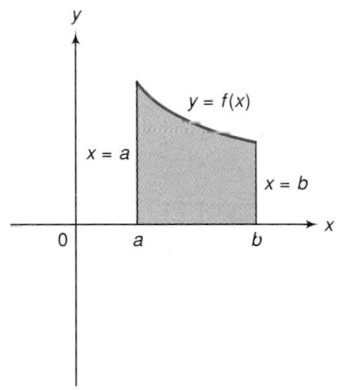

(a) This area is rotated about the x- or y-axis.

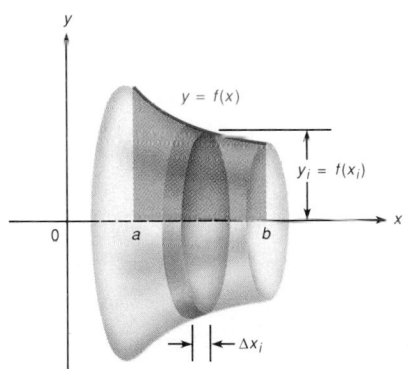

(b) This volume is obtained when the area is rotated about the x-axis. A typical disk is shown.

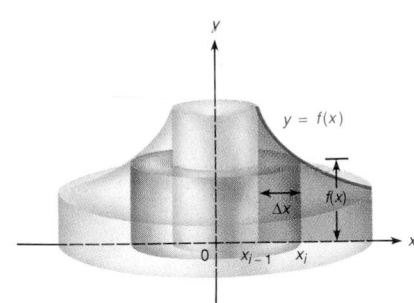

(c) This volume is obtained when the area is rotated about the y-axis. A typical cylindrical shell is shown.

FIGURE 11 THE VOLUME OBTAINED BY ROTATING AN AREA ABOUT THE x-AXIS OR y-AXIS

DEFINITION **VOLUME BY CYLINDRICAL SHELL METHOD**

If the area in Figure 11(a) is revolved about the y-axis (see Figure 11(c)) then the **volume** of the solid generated is defined by

$$V = \int_a^b 2\pi x f(x)\,dx. \quad \blacksquare \qquad (6)$$

⊗ **WARNING:** Formulas (5) and (6) are given as guidelines. They are useless if you forget how they were obtained. The important thing is the integration process, not a formula that comes as a result of using that process. ⊗

EXAMPLE 2
CALCULATING THE VOLUME
OBTAINED BY ROTATING AN AREA
ABOUT THE x-AXIS

The area bounded by the curve $y = x^2 + 2$ and the line $y = x + 8$ is rotated around the x-axis. Find the volume of the solid generated.

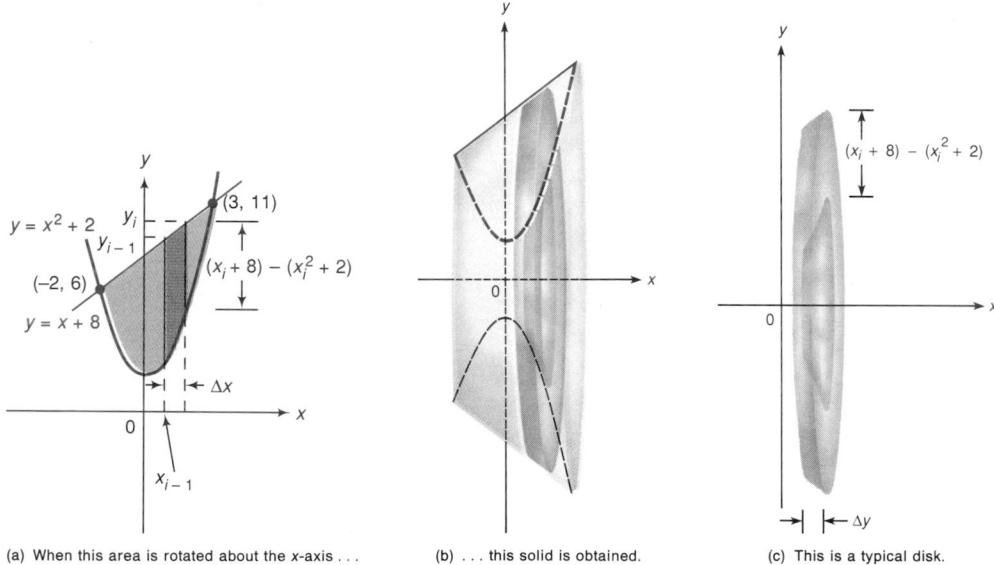

(a) When this area is rotated about the x-axis . . .   (b) . . . this solid is obtained.   (c) This is a typical disk.

FIGURE 12 ROTATING AN AREA ABOUT THE x-AXIS

SOLUTION The area is depicted in Figure 12(a), and the solid generated is depicted in Figure 12(b). The points of intersection of the two curves are found to be $(-2, 6)$ and $(3, 11)$. When the strip depicted in Figure 12(a) is rotated around the x-axis, it generates rings, one of which is sketched in Figure 12(c). The volume of this ring is the volume of the large disk (radius $x_i + 8$) minus the volume of the smaller disk (radius $x_i{}^2 + 2$). Thus,

$$\Delta V_i = \pi(x_i + 8)^2 \, \Delta x - \pi(x_i{}^2 + 2)^2 \, \Delta x,$$

and the total volume is given by

$$V = \pi \int_{-2}^{3} \left[(x + 8)^2 - (x^2 + 2)^2 \right] dx$$

$$= \pi \int_{-2}^{3} (x^2 + 16x + 64 - x^4 - 4x^2 - 4) \, dx$$

$$= \pi \int_{-2}^{3} (-x^4 - 3x^2 + 16x + 60) \, dx$$

$$= \pi \left(-\frac{x^5}{5} - x^3 + 8x^2 + 60x \right) \Bigg|_{-2}^{3} = 250\pi.$$

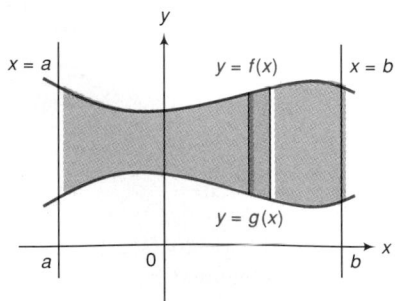

FIGURE 13
THE AREA BETWEEN TWO CURVES

The example above can be generalized to obtain the following definition (see Figure 13).

> **DEFINITION** **VOLUME OBTAINED BY ROTATING THE AREA BETWEEN TWO CURVES ABOUT THE x-AXIS**
>
> Let f and g be continuous and nonnegative on $[a, b]$ with $f(x) \geq g(x) \geq 0$ for every x in $[a, b]$. Then the **volume** generated by rotating the region bounded by the curves and lines $y = f(x)$, $y = g(x)$, $x = a$, and $x = b$ around the x-axis is defined by
>
> $$V = \pi \int_a^b \{[f(x)]^2 - [g(x)]^2\}\, dx. \quad \blacksquare \qquad (7)$$

REMARK: In (7), note that we have the difference of squares, $f^2 - g^2$, *not* the square of the difference, $(f - g)^2$.

The techniques of this section can also be used to calculate the volumes of certain solids in which parallel cross-sections all have the same basic shape. By a **cross-section** we mean the figure (in two dimensions) obtained by slicing the solid with a plane. By slicing the figure with parallel planes, we obtain a number of disks. (Think of cutting an egg into many thin, parallel slices. Each cross-section has the shape of an oval.) Each "slice" has a volume approximately equal to the area of the cross-section times its thickness. To calculate the volume of the solid, we simply "add up" the volumes of the separate slices by the now-familiar process of integration.

EXAMPLE 3
FINDING THE VOLUME OF A SOLID
BY COMPUTING THE VOLUMES
OF CROSS-SECTIONS

The base of a certain solid is the circular disk $x^2 + y^2 \leq 4$ in the xy-plane. Each plane perpendicular to the x-axis cuts the solid in an equilateral triangle. Find the volume of the solid.

SOLUTION: The solid is drawn in Figure 14a. If the solid is sliced along the x-axis, the volume of a typical slice is $A_i \Delta x$, where A_i is the area of an equilateral triangle. One side of the triangle has length

$$\sqrt{4 - x_i^2} - (-\sqrt{4 - x_i^2}) = 2\sqrt{4 - x_i^2}$$

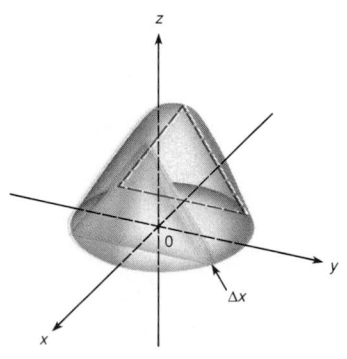

(a) Here is the solid.

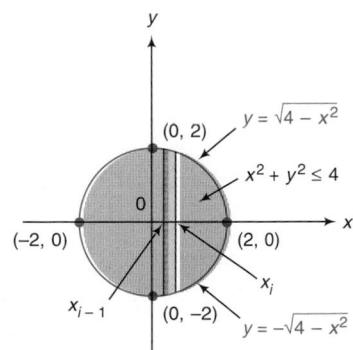

(b) Looking at the solid from the bottom.

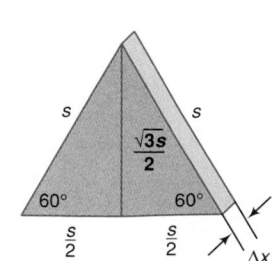

(c) The area of a cross-section is the area of an equilateral triangle.

FIGURE 14 FINDING THE VOLUME OF A SOLID WITH A CIRCULAR BASE AND TRIANGULAR CROSS-SECTION

(according to Figure 14b). Using Figure 14c, we calculate that the area of an equilateral triangle with side s is given by $A = (\sqrt{3}/4)s^2$. Thus, the area A_i above is given by

$$A_i = \frac{\sqrt{3}}{4}(4)(4 - x_i{}^2) = \sqrt{3}(4 - x_i{}^2), \quad \text{and} \quad \Delta V_i \approx \sqrt{3}(4 - x_i{}^2)\,\Delta x,$$

so that

$$V = \sqrt{3}\int_{-2}^{2}(4 - x^2)\,dx = \sqrt{3}\left(4x - \frac{x^3}{3}\right)\Bigg|_{-2}^{2} = \frac{32\sqrt{3}}{3}.$$

■ PROBLEMS 5.2

■ SELF-QUIZ

I. Consider a small rectangle, tall and skinny, with corners at $(a, 0)$, $(a + \Delta x, 0)$, $(a, 3a - 1)$, and $(a + \Delta x, 3a - 1)$. Rotating that rectangle about the x-axis generates a disk whose volume is

_____.

a. $\pi(3a - 1)^2\,\Delta x$ b. $\pi(3a - 1)^2(\Delta x)^2$
c. $(3a - 1)^2\,\Delta x$ d. $\pi(3a - 1)\,\Delta x$

II. Consider the region between the line $y = 3x - 1$ and the x-axis for x between 5 and 8. Rotate that region about the x-axis; the resulting solid has volume _____.

a. $\int_{5}^{8}\pi(3x - 1)^2\,dx$

b. $\int_{5}^{8}\pi(3x - 1)^2\,dx^2$

c. $\int_{5}^{8}(3x - 1)^2\,dx$

d. $\int_{5}^{8}\pi(3x - 1)\,dx$

III. Once again, consider our tall and skinny rectangle with corners at $(a, 0)$, $(a + \Delta x, 0)$, $(a, 3a - 1)$, $(a + \Delta x, 3a - 1)$. Rotating that rectangle about the y-axis generates a cylindrical shell whose volume is approximately _____.
a. $\pi(3a - 1)^2\,\Delta x$ b. $\pi a^2(3a - 1)\,\Delta x$
c. $2\pi a(3a - 1)\,\Delta x$ d. $\pi a(3a - 1)\,\Delta x$

IV. Consider the region between the line $y = 3x - 1$ and the x-axis for x between 5 and 8. Rotate that region about the y-axis; the resulting solid has volume _____.

a. $\int_{5}^{8}\pi(3x - 1)^2\,dx$

b. $\int_{5}^{8}\pi x^2(3x - 1)\,dx$

c. $\int_{5}^{8}2\pi x(3x - 1)\,dx$

d. $\int_{5}^{8}\pi x(3x - 1)\,dx$

V. Consider the region between the curve $y = \sqrt{x}$ and the x-axis for x between 0 and 9. Rotate that region about the x-axis; the resulting solid has volume _____.

a. $\int_{0}^{3}\pi\sqrt{x}\,dx$ b. $\int_{0}^{9}\pi(\sqrt{x})^2\,dx$

c. $\int_{0}^{9}2\pi x\sqrt{x}\,dx$ d. $\int_{0}^{\sqrt{9}}2\pi x\sqrt{x}\,dx$

VI. Consider the region between the curve $y = \sqrt{x}$ and the x-axis for x between 0 and 9. Rotate that region about the y-axis; the resulting solid has volume _____.

a. $\int_{0}^{3}\pi\sqrt{x}\,dx$ b. $\int_{0}^{9}\pi(\sqrt{x})^2\,dx$

c. $\int_{0}^{9}2\pi x\sqrt{x}\,dx$ d. $\int_{0}^{\sqrt{9}}2\pi x\sqrt{x}\,dx$

■ DRILL

In Problems 1–10, find the volume generated when the region bounded by the given curves and lines is rotated about the x-axis.

1. $y = x$, $x = 2$, $y = 0$
2. $y = 2x + 3$, $x = 1$, $x = 4$, $y = 0$
3. $y = 2x^2$, $x = 1$, $y = 0$
4. $y = x^3$, $x = 2$, $y = 1$
5. $y = \sqrt{x + 1}$, $x = 1$, $x = 5$, $y = 0$
6. $y = \sqrt{1 - x^3}$, $x = 0$, $y = 0$
7. $y = x$, $y = 2x$, $x = 2$, $x = 5$
8. $y = x^2$, $y = x^3$
*9. $x = 2 - |y - 2|$, $x = 0$
*10. $x = y - y^2$, $x = 0$

CONTINUED

In Problems 11–18, find the volume generated when the region bounded by the given curves and lines is rotated about the y-axis.

11. $x + y = 3$, $x = 0$, $y = 0$
12. $y - x = 1$, $x = 0$, $y = 5$

13. $y = x^2/2$, $x = 1$, $y = 0$
14. $y = x^2$, $y = x^3$
15. $y = \sqrt{x}$, $y = x^2$
16. $y = x^{1/3}$, $x = 1$, $y = 0$
17. $x = 1 + y^2$, $x = 2$
18. $x - 3 = (y - 3)^2$, $3x + 8y = 44$

APPLICATIONS

19. Calculate the volume of the sphere of radius r by rotating the interior of the semicircle $x = \sqrt{r^2 - y^2}$ about the y-axis.

20. The equation of an ellipse centered at the origin is given by $(x/a)^2 + (y/b)^2 = 1$. Calculate the volume of the "football" shaped solid generated by rotating this ellipse about the x-axis.

21. The circle centered at $(a, 0)$ with radius r $(r < a)$ is rotated about the y-axis. Verify that the volume of the **torus** so generated is equal to $2\pi a \cdot \pi r^2 = 2a\pi^2 r^2$.

22. Find the volume generated by rotating the triangle with vertices at $(1, 1)$, $(2, 3)$, and $(3, 2)$ about the y-axis.

23. Calculate the volume generated by rotating the area bounded by $y = \sqrt{x}$, $x = 1$, and $y = 0$ about the line $x = 4$. About the line $y = 2$.

24. What is the volume generated when the area bounded by the line $y = x$ and the curve $y = x^2$ is rotated about the line $x = 2$? About the line $y = 2$?

*25. A water tank in the shape of a sphere has a radius of 10 m. What is the mass of water in the tank when the tank is filled to a depth of 2 m? [*Note:* The density of water is approximately 1000 kg/m^3.]

*26. A right circular cylinder of radius 2 m is cut by two planes. The first is parallel to the base of the cylinder, while the second makes an angle of $45°$ with the base. The two planes intersect at a diameter of the circular cross-section of the cylinder, thereby cutting out a wedge. What is the volume of this wedge? [*Hint:* Look at Figure 15. The volume can be calculated two ways:

by taking (a) cross sections parallel to the base or (b) cross sections perpendicular to it.]

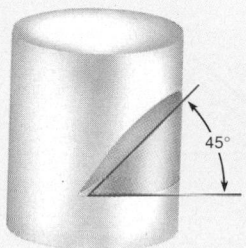

FIGURE 15

27. The base of a certain solid is a circle with radius 2, while each cross-section perpendicular to the base of the solid is a square. Find the volume of this solid.

28. The base of a certain solid is the region between the line $y = x$ and the curve $y = x^2$. Find the volume of the solid if cross sections perpendicular to the x-axis are:
   a. squares
   b. semicircles
   c. isosceles right triangles with their hypotenuses in the xy-plane.

*29. Let (a_1, b_1), (a_2, b_2), (a_3, b_3) be three distinct points lying in the first quadrant. Rotating the triangle with those points as its vertices about the x-axis generates a solid. Calculate the volume of that solid.

ANSWERS TO SELF-QUIZ

I. a   II. a   III. c
IV. c   V. b   VI. c

5.3 ARC LENGTH

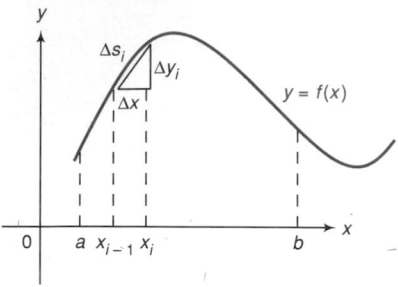

FIGURE 1
TYPICAL PLANE CURVE

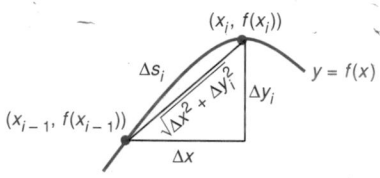

FIGURE 2
THE APPROXIMATE LENGTH OF A SMALL
PIECE OF A PLANE CURVE IS $\sqrt{\Delta x^2 + \Delta y_i^2}$.

In this section we show how to calculate the length of a curve in the xy-plane (called a **plane curve**). A typical curve is sketched in Figure 1. The problem is to calculate the length of the curve from $x = a$ to $x = b$. Such a problem arises in many interesting physical examples. For example, if an object is thrown or shot into the air, then its path as a function of time is a parabola opening downward. To calculate the length of the path traveled by the object, one must find the length of the parabola from some initial time t_0 to a final time t_f.

To find the length of the curve in Figure 1, we go through our integration process. We begin by dividing the curve into small pieces. We do so by partitioning the interval $[a, b]$ into n subintervals of equal length. We then find the approximate length Δs_i of the arc in each subinterval and add up these lengths. (This technique is, by now, very familiar.) The appearance of the curve near a typical subinterval is sketched in Figure 2. If Δx is small, and if $f(x)$ is continuous in the interval $[x_{i-1}, x_i]$, then the length of the curve between x_{i-1} and x_i, which we will denote by Δs_i, can be approximated by the straight-line segment joining the points $(x_{i-1}, f(x_{i-1}))$ and $(x_i, f(x_i))$. The length of this line segment is, by the Pythagorean theorem, given by $\sqrt{\Delta x^2 + \Delta y_i^2}$. Thus

$$\Delta s_i \approx \sqrt{\Delta x^2 + \Delta y_i^2}. \tag{1}$$

Now $\Delta y_i = f(x_i) - f(x_{i-1})$. If f is continuous on $[a, b]$ and differentiable on (a, b), then f is continuous in the smaller interval $[x_{i-1}, x_i]$ and differentiable on (x_{i-1}, x_i). By the mean-value theorem (Theorem 3.7.2), there is a number x_i^* in (x_{i-1}, x_i) such that

$$\Delta y_i = f(x_i) - f(x_{i-1}) = f'(x_i^*)(x_i - x_{i-1}) = f'(x_i^*)\,\Delta x. \tag{2}$$

Then inserting (2) into (1), we have

$$\Delta s_i \approx \sqrt{\Delta x^2 + [f'(x_i^*)]^2\,\Delta x^2}$$
$$= \sqrt{\Delta x^2(1 + [f'(x_i^*)]^2)} = \sqrt{1 + [f'(x_i^*)]^2}\,\Delta x$$

so that

$$s = \sum_{i=1}^{n} \Delta s_i \approx \sum_{i=1}^{n} \sqrt{1 + [f'(x_i^*)]^2}\,\Delta x.$$

If f' is also continuous in $[a, b]$, then $(f')^2$ is continuous in $[a, b]$, and $\sqrt{1 + [f'(x)]^2}$ is continuous in $[a, b]$. Thus $\int_a^b \sqrt{1 + [f'(x)]^2}\,dx$ exists. But,

$$\int_a^b \sqrt{1 + [f'(x)]^2}\,dx = \lim_{\Delta x \to 0} \sum_{i=1}^{n} \sqrt{1 + [f'(x_i^*)]^2}\,\Delta x.$$

This derivation is used as motivation for the following definition.

DEFINITION **LENGTH OF A CURVE (ARC LENGTH)**

Suppose that f and f' are continuous in $[a, b]$. Then the **length of the curve (arc length)** $f(x)$ between $x = a$ and $x = b$ is defined by

$$s = \int_a^b \sqrt{1 + [f'(x)]^2}\,dx. \quad\blacksquare \tag{3}$$

This integral is often written as $\int_0^s ds$, where $ds = \sqrt{1 + [f'(x)]^2}\,dx$ denotes the **differential of arc length**.

EXAMPLE 1
CALCULATING THE LENGTH OF A
LINE SEGMENT

Calculate the length of the straight-line segment $y = x$ between $x = 0$ and $x = 1$.

SOLUTION: We easily see that this distance is simply the straight-line distance between $(0, 0)$ and $(1, 1)$ and is equal to $\sqrt{2}$. However, we will also calculate it by using formula (3), mainly to illustrate for ourselves the formula's validity. Since $f'(x) = 1$, we have

$$s = \int_0^1 \sqrt{1 + 1^2}\,dx = \int_0^1 \sqrt{2}\,dx = \sqrt{2}.$$

EXAMPLE 2
CALCULATING ARC LENGTH

Calculate the length of the arc of the curve $y = 2 + x^{3/2}$ between $x = 1$ and $x = 4$.

SOLUTION: $f'(x) = \frac{3}{2}x^{1/2}$ and $[f'(x)]^2 = \frac{9}{4}x$, so

$$s = \int_1^4 \sqrt{1 + \frac{9}{4}x}\,dx = \frac{1}{2}\int_1^4 \sqrt{4 + 9x}\,dx = \frac{1}{18}\int_1^4 (4 + 9x)^{1/2} 9\,dx$$

$$= \frac{2}{54}(4 + 9x)^{3/2}\Big|_1^4 = \frac{1}{27}(40^{3/2} - 13^{3/2}).$$

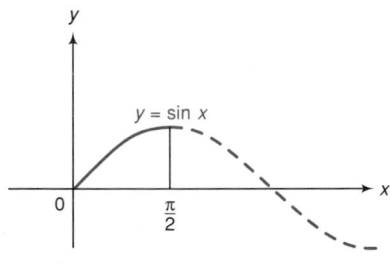

EXAMPLE 3
FINDING AN INTEGRAL THAT
REPRESENTS ARC LENGTH

Calculate the length of the loop of the sine function between $x = 0$ and $x = \pi/2$.

SOLUTION: The arc whose length is requested is sketched in Figure 3. Since $f'(x) = \cos x$, we have

$$s = \int_0^{\pi/2} \sqrt{1 + \cos^2 x}\,dx.$$

There is no way to find an antiderivative for $\sqrt{1 + \cos^2 x}$. We can approximate s by using the definition of the definite integral. We partition $\left[0, \dfrac{\pi}{2}\right]$ into 10 subintervals:

$$0 < \frac{\pi}{20} < \frac{2\pi}{20} < \frac{3\pi}{20} < \frac{4\pi}{20} < \frac{5\pi}{20} < \frac{6\pi}{20} < \frac{7\pi}{20} < \frac{8\pi}{20} < \frac{9\pi}{20} < \frac{10\pi}{20} = \frac{\pi}{2}.$$

FIGURE 3
LENGTH OF THE CURVE $y = \sin x$ FOR
$0 \le x \le \pi/2$

Then

$$s = \int_0^{\pi/2} \sqrt{1 + \cos^2 x}\,dx \approx \sum_{i=1}^n (\sqrt{1 + \cos^2 x_i{}^*})\,\Delta x$$

$$= \frac{\pi}{20}\sum_{i=1}^{10} \sqrt{1 + \cos x_i{}^{*2}}$$

where

$$x_i^* \in [x_{i-1}, x_i] = \left[\frac{(i-1)\pi}{20}, \frac{i\pi}{20}\right].$$

Choosing x_i^* to be the right-hand endpoint of each subinterval, we obtain

$$s \approx \frac{\pi}{20}\left[\sqrt{1 + \cos^2 \frac{\pi}{20}} + \sqrt{1 + \cos^2 \frac{\pi}{10}} + \sqrt{1 + \cos^2 \frac{3\pi}{20}}\right.$$
$$+ \sqrt{1 + \cos^2 \frac{\pi}{5}} + \sqrt{1 + \cos^2 \frac{\pi}{4}} + \sqrt{1 + \cos^2 \frac{3\pi}{10}}$$
$$+ \sqrt{1 + \cos^2 \frac{7\pi}{20}} + \sqrt{1 + \cos^2 \frac{2\pi}{5}} + \sqrt{1 + \cos^2 \frac{9\pi}{20}}$$
$$\left. + \sqrt{1 + \cos^2 \frac{\pi}{2}}\right]$$
$$\approx 0.15708[1.4055 + 1.3800 + 1.3394 + 1.2863 + 1.2247 + 1.1600$$
$$+ 1.0982 + 1.0467 + 1.0122 + 1] \approx 1.88.$$

In the interval $\left[0, \frac{\pi}{2}\right]$, $\cos x$ is a decreasing function, so $\sqrt{1 + \cos^2 x}$ is also a decreasing function (explain why). Thus, since we used the right-hand endpoints, the value 1.88 is an underestimate for the arc length (we used the smallest value for the function in each subinterval). If we use the left-hand endpoints, where $\sqrt{1 + \cos^2 x}$ takes its largest value in each subinterval, we obtain

$$S \approx \frac{\pi}{20}\left[\sqrt{1 + \cos^2 0} + \sqrt{1 + \cos^2 \frac{\pi}{20}} + \sqrt{1 + \cos^2 \frac{\pi}{10}}\right.$$
$$\left. + \sqrt{1 + \cos^2 \frac{3\pi}{20}} + \cdots + \sqrt{1 + \cos^2 \frac{9\pi}{20}}\right]$$
$$\approx 0.15708[12.3672] \approx 1.94.$$

Thus $1.88 < s < 1.94$. In Section 7.9 we will describe two more efficient ways to obtain this integral. To two decimal places, the value of the integral is 1.93.

We close this section by noting that the calculation of arc length is usually more difficult than the calculation of volume because it is often difficult or impossible to find an antiderivative for $\sqrt{1 + [f'(x)]^2}$. This is one of the many reasons why numerical techniques are so important.

HISTORICAL NOTE: COMPUTING ARC LENGTH IN THE SEVENTEENTH CENTURY

The problem of computing the length of a curve was one of the four principal problems that faced seventeenth-century mathematicians. The arc-length problem was different from the tangent-line and area problems because, for

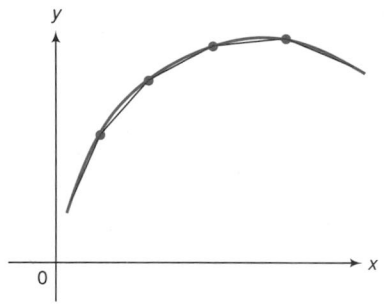

FIGURE 4
APPROXIMATING ARC LENGTH BY ADDING UP THE LENGTHS OF THE SIDES OF AN INSCRIBED POLYGON

SIR CHRISTOPHER
WREN
1632–1723

a long time, it was considered insoluble. Until the seventeenth century, mathematicians believed that a curved line could not have exactly the same length as a straight line segment. Put another way, no one believed that the length of a curve could be a finite number. Blaise Pascal once expressed admiration for the natural order which did not allow a curve to equal a line.

By the year 1660 this had changed. Several mathematicians computed the lengths of curves. The principal method employed was to inscribe polygons in a curve (see Figure 4). The greater the number of sides, the closer is the approximation to the length of the curve. One mathematician who carried out the calculations was John Wallis, discussed earlier (page 317) in connection with the area problem.

Sir Christopher Wren (1632–1723) was one of the great architects of the Western world. After the Great Fire of London in 1666, Wren played a significant role in rebuilding St. Paul's Cathedral and over fifty other churches and monuments.

However, Wren's earlier career was as a mathematician. If a tack, say, is placed on the circumference of a wheel, then as the wheel rolls forward on a road, the tack traces out a curve called a **cycloid** (see Problem 43 in Section 12.1). Wren showed, in 1658, that the length of one arch of the cycloid is equal to eight times the length of the radius of the wheel.

It is believed that if it hadn't been for the fire, Wren would now be better known as a mathematician than as an architect. However, his reputation as an architect is considerable. The oldest academic building still in use in the United States is the Wren Building at the College of William and Mary in Williamsburg, Virginia. The building was erected in 1695 and its design is attributed to Sir Christopher Wren. At his death, Wren was buried in St. Paul's Cathedral with the epitaph *Si monumentum requiris, circumspice* (If you seek a monument, look about you).

When he heard that others had computed arc length, Fermat, who dabbled in everything, set out to do the same. He used a very clever trick to compute the length of the **semicubical parabola** $y^2 = x^3$. Choose a point P on the curve with coordinates (a, b). Fermat knew how to find the equation of the tangent line at P (we discussed his method on page 106). Using modern methods, we compute

$$y = x^{3/2}$$

$$y' = \frac{3}{2}x^{1/2} = \frac{3}{2}a^{1/2} \qquad \text{when } x = a.$$

So the equation of the tangent line at (a, b) is

$$\frac{y - b}{x - a} = \frac{3}{2}a^{1/2}$$

or

$$y = \frac{3}{2}a^{1/2}(x - a) + b. \tag{4}$$

Suppose we increase a by the small amount e. Then we move along the curve from P to P'. Fermat argued that if e is small, then the arc length $\overparen{PP'}$ is approximately equal to the length $\overline{PQ}$ where Q is the point on the tangent

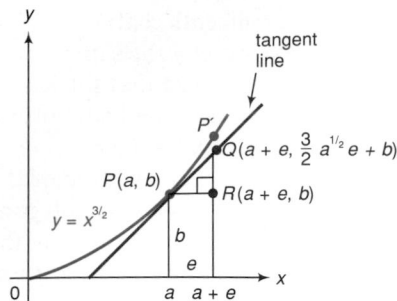

FIGURE 5
FERMAT'S METHOD FOR CALCULATING
THE LENGTH OF THE SEMICUBICAL
PARABOLA $y^2 = x^3$

line that corresponds to $x = a + e$. To find the y-coordinate of Q (we know that the x-coordinate is $a + e$), we use equation (4):

$$y\text{-coordinate of } Q = \frac{3}{2}a^{1/2}e + b. \qquad \text{Since } x - a = (a + e) - a = e$$

From Figure 5 and the Pythagorean theorem,

$$\overline{PQ}^2 = \overline{PR}^2 + \overline{RQ}^2 = e^2 + \left(\frac{3}{2}a^{1/2}e\right)^2 = e^2\left(1 + \frac{9}{4}a\right)$$

so

$$\overline{PQ} = e\sqrt{1 + \frac{9}{4}a}. \qquad\qquad \textbf{(5)}$$

Then the total arc length for $0 \le x \le 1$, say, is approximated by adding up terms of the form (5). If, in (5), we replace e by Δx and a by x_i, then we are led to the formula

$$\text{arc length} \approx \sum_{i=1}^{n} \sqrt{1 + \frac{9}{4}x_i}\,\Delta x$$

and

$$\text{arc length} = \int_0^1 \sqrt{1 + \frac{9}{4}x}\,dx \qquad = \frac{1}{27}(40^{3/2} - 13^{3/2}). \text{ See Example 2.}$$

Fermat recognized that adding up terms of the form (5) was equivalent to calculating the area under the curve $y = \sqrt{1 + \frac{9}{4}x}$ and this he was able to do by the methods described at the end of Section 4.6.

Because Fermat was able to connect the tangent and area problems so closely, as in the technique described above, he has been called by some the real inventor of calculus. The great eighteenth-century French mathematicians Lagrange, Laplace, and Fourier all believed him to be so. However this designation has been hotly disputed. Most believe that as Fermat did not recognize that the area and tangent problems are really inverse problems, he doesn't deserve to be credited with the invention of calculus. Nevertheless, he, more than anyone else, anticipated the great discoveries of Newton and Leibniz.

■ PROBLEMS 5.3

■ SELF-QUIZ

I. Consider the portion of the curve $x/3 + y/4 = 1$ for x between 0 and 3. The length of that arc is _____ .

a. $\int_0^3 \sqrt{1 - (\frac{4}{3})^2}\,dx$ b. $\int_0^3 \sqrt{1 + (-\frac{1}{3})^2}\,dx$

c. $\int_0^3 \sqrt{1 + (-\frac{3}{4})^2}\,dx$ d. $\int_0^3 \sqrt{1 + (-\frac{4}{3})^2}\,dx$

II. $\int_1^4 \sqrt{1 + x}\,dx$ computes the arc length for that part of the curve _____ which lies between $x = 1$ and $x = 4$.

a. $y = -13 + x^{3/2}$ b. $y = x^{3/2}$
c. $y = 7 + (\frac{2}{3})x^{3/2}$ d. $y = x^2$
e. $y = 25 + x^2/2$ f. $y = 25 - x^2/2$

CONTINUED

III. $\int_3^7 \sqrt{x^2 - 6x + 10}\, dx$ computes the arc length for that part of the curve _____ which lies between $x = 3$ and $x = 7$.
   a. $y = x - 3$ b. $y = -(x - 3)^2$
   c. $y = (x - 3)^2/2$ d. $y = x^2 - 6x + 10$

IV. The arc length of $y = x^2/4$ between $x = 2$ and $x = 4$ is computed by _____.
   a. $\int_2^4 \sqrt{1 + 1/x}\, dx$ b. $\int_1^4 \sqrt{1 - 1/x}\, dx$

   c. $\int_2^4 \sqrt{1 + x}\, dx$ d. $\int_2^4 \sqrt{1 + x^2/4}\, dx$

V. The arc length of $y = 2\sqrt{x}$ between $x = 1$ and $x = 4$ is computed by _____.
   a. $\int_1^4 \sqrt{1 + 1/x}\, dx$ b. $\int_2^4 \sqrt{1 + 1/x}\, dx$
   c. $\int_1^4 \sqrt{1 + x}\, dx$ d. $\int_1^4 \sqrt{1 + x^2/4}\, dx$

■ DRILL

In Problems 1–4, find a definite integral that is equal to the length of the given curve over the specified interval. Do not try to evaluate your integral.

1. $y = 2 + x^2/2$; $[0, 1]$
2. $y = \frac{1}{3}(x^{3/2} - 3\sqrt{x})$; $[1, 4]$
3. $y = \cos x$; $[0, \pi]$
4. $y = \tan x$; $[0, \pi/4]$

In Problems 5–14, calculate the length of the arc of the curve $y = f(x)$ over the specified interval.

5. $y = 3x + 5$; $[1, 5]$
6. $2x + 5y - 3 = 0$; $[0, 4]$
7. $y = x^{3/2}$; $[0, 1]$
8. $y = \frac{1}{3}(x^2 + 2)^{3/2}$; $[0, 3]$

9. $y = \frac{1}{6}\left(x^3 + \frac{3}{x}\right)$; $[1, 3]$
10. $y = \frac{x^4}{4} + \frac{1}{8x^2}$; $[1, 2]$
11. $y = \frac{1}{3}(x^{3/2} - 3\sqrt{x})$; $[1, 4]$
12. $y = \frac{5}{24}x\left(2\sqrt[5]{x} - \frac{3}{\sqrt[5]{x}}\right)$; $[1, 32]$
13. $y = \frac{2}{3}\left(a + \frac{x^2}{u^7}\right)^{3/2}$; $[0, 1]$
*14. $y = x(ax^k + bx^{-k})$, with $|k| \neq 1$ and $4ab = 1/(k^2 - 1)$; $[1, 2]$

■ APPLICATIONS

*15. Find the length of the curve $x^{2/3} + y^{2/3} = 1$.
16. Find the length of the arc of the curve $(y + 1)^2 = 4x^3$ between $x = 0$ and $x = 1$.

▥ CALCULATOR PROBLEMS

In Problems 17–20, approximate, to one decimal place, the length of the given curve over the specified interval.

17. $y = 2 + \frac{x^2}{2}$; $[0, 1]$

18. $y = \frac{1}{3}(x^{3/2} - 3\sqrt{x})$; $[1, 4]$
19. $y = \cos x$; $[0, \pi]$
*20. $y = \tan x$; $\left[0, \frac{\pi}{4}\right]$

■ SHOW/PROVE/DISPROVE

*21. a. Using the technique of this section, show that the circumference of the unit circle, $x^2 + y^2 = 1$, is 2π. [*Hint:* Show that $\pi/4$ is the length of the arc in the first quadrant between $x = 0$ and $x = 1/\sqrt{2}$.]

b. Discuss the logical difficulties in using this calculation to find the length of the circumference of the unit circle because it makes use of the fact that a circle of radius r has circumference $2\pi r$.

CONTINUED

*22. a. Show that the graph of $(x - a)^2 + (y - b)^2 = r^2$ has total arc length equal to $2\pi r$.

 b. Answer part (b) of the preceding problem as it applies here.

23. Show that the arcs considered in Self-quiz exercises IV and V have the same length. Do this in

two ways:

 a. Give a geometric argument.

 b. Find a substitution that transforms the correct answer for IV into the definite integral which is the answer for V. [*Note:* This problem can, and should, be worked without evaluating either integral.]

■ **CHALLENGE**

24. For some choices of positive constants a, b, c, computing the length of an arc of $y = c(ax^2 + b)^{3/2}$ involves an integral we know how to evaluate. For instance, Problem 13 asks you

to consider the result when $c = \frac{2}{3}$ and $a \cdot b^2 = 1$. Are there other choices that lead to "do-able" integrals? Write a brief essay summarizing what you discover.

■ **ANSWERS TO SELF-QUIZ**

 I. d   II. c   III. c

 IV. d   V. a

5.4 WORK, POWER, AND ENERGY

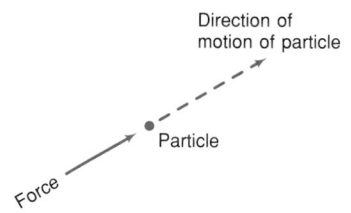

Direction of
motion of particle

Particle

Force

FIGURE 1
THE DIRECTION OF A FORCE

$\mathbf{C}$onsider a particle acted on by a force. In the simplest case, the force is constant, and the particle moves in a straight line in the direction of the force (see Figure 1). In this situation, we define the *work W done by the force on the particle* as the product of the magnitude of the force F and the distance s through which the particle travels:

THE WORK DONE BY A CONSTANT FORCE OVER A DISTANCE s

$$W = Fs. \tag{1}$$

One unit of work is the work done by a unit force in moving a body a unit distance in the direction of the force. In the metric system, the unit of work is 1 newton[†] meter (N·m), called 1 *joule* (J). In the British system, the unit of work is the foot pound[‡] (ft·lb).

EXAMPLE 1
THE WORK DONE BY A CONSTANT FORCE (IN ENGLISH UNITS)

How much work is done in lifting a 25-lb weight 5 ft off the ground?

SOLUTION: $W = Fs = 25$ lb $\times$ 5 ft $= 125$ ft·lb.

EXAMPLE 2
THE WORK DONE BY A CONSTANT FORCE (IN METRIC UNITS)

A block of mass 10.0 kg is raised 5 m off the ground. How much work is done?

SOLUTION: From Newton's second law, force = (mass) × (acceleration). The acceleration here is opposing acceleration due to gravity and

---

[†] 1 newton (N) is the force that will accelerate a 1-kg mass at the rate of 1 m/sec²; 1 N = 0.2248 lb.

[‡] 1 joule (J) = 0.7376 ft·lb, or 1 ft·lb = 1.356 J.

is therefore equal to 9.81 m/sec². We have $F = ma = (10 \text{ kg})(9.81 \text{ m/sec}^2)$
$= 98.1$ N. Therefore, $W = Fs = (98.1) \times 5$ m $= 490.5$ J.

Since we will be dealing with mass and force in many problems in this
text, we take a moment here to explain units of weight, mass, and force in
both the English and metric systems.

In the English system, we start with the unit of force, which is the *pound*.
(It is also the standard unit of weight.) Since $F = mg$ by Newton's second
law of motion, we obtain $m = F/g$. In the English system, the unit of mass
is the *slug*, and a 1-lb weight (force) has a mass of $(1/32.2)$ lb/(ft/sec²) $= 1$ slug.
Or we can equivalently define a slug as the mass of an object whose acceleration is 1 ft/sec² when it is subjected to a force of 1 lb. This explains why, in
Example 1, we did not multiply the weight of 25 lb by the acceleration to
obtain the force. The weight (in pounds) was *already* given as a force.

In the metric system, we start with the standard unit of mass, which is
the *kilogram*. In this case, to obtain the force, we must multiply the mass by
the acceleration:

F (in newtons) $= [$mass (in kilograms)$] \times g(= 9.81 \text{ m/sec}^2)$.

In sum, we have the following:

The pound is a force, but not a mass.

The kilogram is a mass but not a force.

We now return to the calculation of work. The case in which the motion
of the particle is not in a straight line in the direction of the force is more
complicated than the case we discussed in Examples 1 and 2. This case is
analyzed in a discussion of vector calculus.[†] However, we can use the integral
to handle the case in which the force is variable.

We wish to calculate how much work is done in moving from the point
$x = a$ to the point $x = b$. As usual, we partition the interval into n subintervals, a typical interval being $[x_{i-1}, x_i]$. Now if the force is given by $F(x)$
(which is assumed to be continuous), then over a very small interval the force
will be almost constant. Therefore, if x_i^* is a point in the interval $[x_{i-1}, x_i]$,
then the work done in moving from x_{i-1} to x_i is, approximately,

$$\Delta W \approx F(x_i^*)(x_i - x_{i-1}) = F(x_i^*)\Delta x. \tag{2}$$

Adding the work done over these small intervals and taking the limit as
$\Delta x \to 0$ leads to the formula

DEFINITION   **WORK DONE BY A VARIABLE FORCE OVER
AN INTERVAL**

$$W = \lim_{\Delta x \to 0} \sum_{i=1}^{n} F(x_i^*)\Delta x = \int_a^b F(x)\, dx. \quad \blacksquare \tag{3}$$

[†] We discuss work in more general settings in Section 15.2.

EXAMPLE 3
WORK DONE BY A SPRING:
HOOKE'S LAW

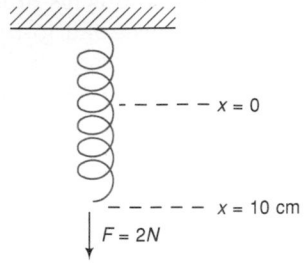

FIGURE 2
A STRETCHED SPRING

A spring is stretched 10 cm by a force of 2 N. How much work is needed to stretch it 50 cm $= \frac{1}{2}$ m (see Figure 2), assuming that the spring satisfies Hooke's law?

SOLUTION: **Hooke's law**[†] states that the force needed to stretch a spring is proportional to the amount of the spring displaced—that is,

$$F(x) = kx, \tag{4}$$

where k is a constant of proportionality, called the **spring constant**, and x is the amount displaced. From the information given, we have 2 N $= k(0.1$ m$)$, or $k = 2/0.1 = 20$. Therefore, the force is given by $F(x) = 20x$. Then,

$$W = \int_0^{1/2} F(x)\,dx = \int_0^{1/2} 20x\,dx = 10x^2\Big|_0^{1/2} = \frac{10}{4} = \frac{5}{2}\,\text{J}.$$

We should point out that Hooke's law is valid only if the spring is not stretched too far. After a certain point, the spring will become stretched out of shape, and Hooke's law will fail to hold.

EXAMPLE 4
CALCULATING THE WORK DONE
IN PUMPING OIL OUT OF A
CONICAL TANK

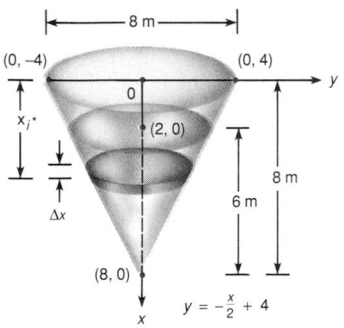

FIGURE 3
A CONICAL TANK

The work necessary to pump the oil in the shaded layer out of the tank is the product of the cross-sectional area of the layer, the thickness of the layer (Δx), the gravitational acceleration constant ($a = 9.81$ m/sec²), and the distance from the layer to the top ($4 - x_i$).

A storage tank in the shape of an inverted right-circular cone has a radius of 4 m and a height of 8 m. It is filled to a height of 6 m with olive oil (density $= 920$ kg/m³). To bottle the olive oil, the bottler must first pump it to the top of the tank. How much work is done in accomplishing this task?

SOLUTION: This problem involves a lot of things going on at once. First, we draw a sketch (see Figure 3). It helps if we put in some identifying coordinates.[‡] We place the origin 8 m up and in the center of the base of the cone. Then if the x-axis is the vertical axis and the y-axis is the horizontal axis, we can label various points on the cone as in the figure. Now the "x-axis" in the cone includes the interval $[0, 8]$, and the olive oil fills the interval $[2, 8]$. We partition that last interval in the usual way and look at a layer of oil in the ith subinterval $[x_{i-1}, x_i]$.

The work needed to get this layer of oil to the top is the force needed to overcome gravity times the distance the layer must travel. When Δx is small, this distance is approximately equal to $x_i{}^*$ for any $x_i{}^*$ in $[x_{i-1}, x_i]$. What is the force that must be overcome to lift the layer of oil? We have $F = ma$, where $a = 9.81$ m/sec². The mass m of the oil in the layer is the volume of the oil times its density ($=920$ kg/m³). The volume of the layer is the cross-sectional area times the thickness Δx. Thus, the only thing left to calculate is the cross-sectional area. This area is $\pi y_i{}^2$, where y_i is the radius of the layer. We can think of this radius as varying along the straight line from $(8, 0)$ to $(0, 4)$ (see Figure 3

[†] Named for Robert Hooke (1635–1703), who was an English philosopher, physicist, chemist, and inventor.
[‡] The coordinate system given in Figure 3 may seem a bit strange to you at first. If you are confused, turn the sketch sideways so that the x- and y-axes point in their usual directions.

again). The equation of this straight line is

$$\frac{y - 4}{x - 0} = -\frac{4}{8}, \quad \text{or} \quad y = -\frac{x}{2} + 4.$$

Then $y_i \approx -(x_i^*/2) + 4$.

Now we put everything together:

cross-sectional area of layer $= \pi\left(4 - \dfrac{x_i^*}{2}\right)^2$ square meters
($A = \pi r^2$)

volume of layer $= \pi\left(4 - \dfrac{x_i^*}{2}\right)^2 \overset{\text{Area} \quad \times \quad \text{Thickness}}{\Delta x}$ cubic meters
($V = A \times$ thickness)

mass of layer $= \overset{\text{Density} \quad \text{Volume}}{920\pi\left(4 - \dfrac{x_i^*}{2}\right)^2} \Delta x$ kilograms
($m = \mu V$)

force to overcome force of
gravity to raise layer $= \overset{a \qquad\qquad m}{(9.81)(920)\pi\left(4 - \dfrac{x_i^*}{2}\right)^2 \Delta x}$ newtons
($F = ma$)

$$= 9025.2\pi\left(4 - \frac{x_i^*}{2}\right)^2 \Delta x \text{ newtons}$$

Work to raise layer $x_i^*\, m = \overset{s}{x_i^*} 9025.2\pi\left(4 - \dfrac{x_i^*}{2}\right)^2 \underset{F}{\Delta x}$ joules.
($W = Fs$)

Therefore, the total work is given by

$$W = \int_2^8 9025.2\pi\left(4 - \frac{x}{2}\right)^2 x\, dx = 9025.2\pi \int_2^8 \left(16x - 4x^2 + \frac{x^3}{4}\right) dx$$

$$= (9025.2\pi)\left(8x^2 - \frac{4x^3}{3} + \frac{x^4}{16}\right)\Bigg|_2^8 = (9025.2\pi)(63) = 568{,}587.6\pi \text{ J.}$$

You should go through the steps in this long calculation again. The key to solving the problem is to look at a typical layer and determine, using the definition of work, what it takes to move that layer to the top.

POWER

So far, we have not considered the time spent in doing work. The work it takes to pump oil out of a tank is the same whether it is done in one minute or in one month. Sometimes, however, the rate at which work is done is more important than the actual amount of work done. The rate at which work is done is called **power**. If the work is done at a constant rate, then the power produced in t units of time is $P = W/t$. The **instantaneous power** is the rate of change of work with respect to time:

INSTANTANEOUS POWER

$$P = \frac{dW}{dt}.$$

(5)

In the metric system, the unit of power is the *watt* (W),[†] and

$$1 \text{ W} = 1 \text{ J/sec}.$$

In the British system, the unit of power is 1 ft·lb/sec. Since this unit is very small, another unit called the **horsepower** is often used. By definition

$$1 \text{ hp} = 550 \text{ ft·lb/sec} = 746 \text{ W} \approx \frac{3}{4} \text{ kilowatt (kW)}.$$

Sometimes work is expressed in terms of power × time. This convention explains the use of the term kilowatt-hour (kWh), which is the work done by something (such as twenty 50 W light bulbs) working 1 hr at a constant rate of 1 kW.

EXAMPLE 5
COMPUTING POWER WHEN WORK IS DONE AT A CONSTANT RATE

A man works at a constant rate and performs 129,600 J of work in 4 hr. Then his power output is

$$P = \frac{\text{work done}}{\text{time (in seconds)}} = \frac{129,600 \text{ J}}{4(3600) \text{ sec}} = \frac{129,600}{14,400} = 9\text{W}.$$

EXAMPLE 6
COMPUTING WORK BY INTEGRATING THE POWER FUNCTION

The instantaneous power of a machine is $50 - 3\sqrt{t + 4}$ watts. How much work does the machine do in 1 min?

SOLUTION: Since $P = dW/dt$, we have

$$W = \int_0^{60} P(t)\,dt = \int_0^{60} (50 - 3\sqrt{t + 4})\,dt$$

$$= \left[50t - 2(t + 4)^{3/2}\right]\Big|_0^{60} = 1992 \text{ J}.$$

(Here the 60 is for the 60 sec in 1 min.)

ENERGY

Let $x(t)$ represent the position of a moving particle at time t, and assume that the particle starts at the origin. Then $x'(t) = v(t)$ (velocity) and $x''(t) = a(t)$ (acceleration). If a is constant, then the velocity at time t is given by

---

[†] Named after James Watt (1736–1819), the inventor of the steam engine.

$v = \int a\, dt = at + C$. But $v(0) = v_0$, the initial velocity, so that $C = v_0$ and $v = at + v_0$, or

$$a = \frac{v - v_0}{t}. \tag{6}$$

Since velocity changes at a constant rate ($a = $ constant), the average velocity of the particle is simply the average of the initial velocity v_0 and the final velocity v. That is,

$$v_{av} = \frac{v + v_0}{2}.$$

Therefore the distance the particle travels is $x = v_{av} \cdot t$, or

$$x = \frac{v + v_0}{2}\, t. \tag{7}$$

The work done on the particle in moving from 0 to x is [using (6) and (7) and Newton's second law $F = ma$]

$$W = Fx = max = m\left(\frac{v - v_0}{t}\right)\left(\frac{v + v_0}{2}\right)t = \frac{1}{2}mv^2 - \frac{1}{2}mv_0{}^2. \tag{8}$$

The product $\frac{1}{2}mv^2$ is called the **kinetic energy** of the particle. It is denoted by the symbol K. We have

---

KINETIC ENERGY

$$K = \frac{1}{2}mv^2. \tag{9}$$

Equation (8) states the following:

The work done by the force acting on a particle is equal to the change in kinetic energy of the particle.

This result is known as the **work-energy law**.

We have only shown why the work-energy law is true when a is constant. If a is not constant, we can still illustrate it as follows: From (3), the work done in moving from x_0 to x_1 is

$$W = \int_{x_0}^{x_1} F\, dx.$$

Now,

$$a = \frac{dv}{dt} = \frac{dv}{dx}\frac{dx}{dt} \text{ (by the chain rule)} = \frac{dv}{dx}v = v\frac{dv}{dx}. \tag{10}$$

Then,

$$W = \int_{x_0}^{x_1} F\, dx = \int_{x_0}^{x_1} ma\, dx = \int_{x_0}^{x_1} mv\frac{dv}{dx}\, dx.$$

We know how to integrate $mv\,dv$ ($\int mv\,dv = mv^2/2$ since m is constant). Using a technique discussed in more detail in Section 7.4, we change the variable of integration. When $x = x_0$, $v = v(x_0) = v_0$, and when $x = x_1$, $v = v(x_1) = v_1$. Furthermore,

$$\frac{dv}{dx}\,dx = dv.$$

Therefore, we have

$$W = \int_{x_0}^{x_1} F\,dx = \int_{x_0}^{x_1} mv\,\frac{dv}{dx}\,dx = \int_{v_0}^{v_1} mv\,dv = \frac{1}{2}mv^2\Big|_{v_0}^{v_1} = \frac{1}{2}mv_1{}^2 - \frac{1}{2}mv_0{}^2$$

$$= \text{change in kinetic energy.}$$

EXAMPLE 7
EQUATING WORK AND KINETIC ENERGY

A block weighing 5 kg slides on a horizontal table with a speed of 0.3 m/sec. It is stopped by a spring with a spring constant $k = \frac{1}{4}$ placed in its path. How much is the spring compressed?

SOLUTION: We assume (for simplicity) that until the block hits the spring, its velocity is constant. Then its kinetic energy is the constant $\frac{1}{2}mv^2 = \frac{1}{2}(5)(0.3)^2 = 0.45/2$ J. The work done in compressing the spring x units is

$$W = \int_0^x kx\,dx = \frac{kx^2}{2} = \frac{x^2}{8},$$

since $k = \frac{1}{4}$. Therefore, equating work to the change in kinetic energy (from 0.45/2 to 0), we have

$$\frac{x^2}{8} = \frac{0.45}{2}, \quad \text{or} \quad x^2 = 1.8, \quad \text{and} \quad x \approx 1.34 \text{ m.}$$

EXAMPLE 8
ESCAPE VELOCITY

Neglecting air resistance, find the velocity v_0 at which a rocket at the surface of the earth would have to be fired in order to escape the earth's gravitational field.

SOLUTION: The only force the rocket must overcome is the force of gravity. Let δ denote the distance from the center of the earth to its surface ($\delta \approx 6378$ km ≈ 3963 mi). The force of gravity is, by Newton's law of gravitational attraction, inversely proportional to the square of the distance u between the rocket and the center of the earth. That is, $F = \alpha/u^2$, where α is a constant of proportionality. To find α, we observe that when $u = \delta$ (so that the rocket is at the surface of the earth), $F = mg$ (where m denotes the mass of the rocket). Then $F(\delta) = mg = \alpha/\delta^2$, or $\alpha = mg\delta^2$, and

$$F = mg\,\frac{\delta^2}{u^2}.$$

The work needed to lift the rocket from a distance of δ to a height of x is

$$W = \int_\delta^x F(u)\,du = \int_\delta^x mg\,\frac{\delta^2}{u^2}\,du = -\frac{mg\delta^2}{u}\Big|_\delta^x = mg\delta^2\left(\frac{1}{\delta} - \frac{1}{x}\right). \quad \textbf{(11)}$$

To escape the earth's gravitational pull, the rocket must have an initial velocity sufficient to thrust it to an "infinite" height (once a certain height is reached, the rocket keeps "rising" indefinitely). We therefore let $x \to \infty$ in (11) to find that the work needed to allow the rocket to escape is given by

$$W = mg\delta^2\,\frac{1}{\delta} = mg\delta. \quad \textbf{(12)}$$

Now the rocket starts at rest and then moves with a velocity of v_0. The change in kinetic energy is, therefore,

$$\Delta K = \frac{1}{2}mv_0{}^2. \quad \textbf{(13)}$$

Using the work-energy law, we equate (12) and (13) to obtain

$$v_0 = \sqrt{2g\delta}. \quad \textbf{(14)}$$

This value of v_0 is called the **escape velocity**. Note that the escape velocity is independent of the mass of the rocket. In metric units,

$$v_0 = \sqrt{2 \times 9.81 \text{ m/sec}^2 \times 6378 \times 10^3 \text{ m}}$$
$$\approx 11{,}186 \text{ m/sec} \approx 11.2 \text{ km/sec.} \approx 40{,}320 \text{ km/hr.}$$

In English units,

$$v_0 = \sqrt{2 \times 32.2 \text{ ft/sec}^2 \times 3963 \times 5280 \text{ ft}}$$
$$\approx 36{,}709 \text{ ft/sec} \approx 6.95 \text{ mi/sec.} \approx 25{,}000 \text{ miles/hr.}$$

■ PROBLEMS 5.4

■ SELF-QUIZ

I. The work done in vertically lifting a 5-kg rock a distance of 2 m is _____.
   a. (9.81)(5)(2) joules
   b. (9.81)(5) newtons
   c. (5)(2) joules
   d. (550)(9.81)(5^2) newtons

II. The work done in lifting an 8-lb weight through a vertical distance of 6 ft is _____.

   a. (8)(6)
   b. (8)(6) foot-pounds
   c. (8/32.2)(6) slug-feet
   d. (8)(6)(9.81) newtons

III. The work done by a 40 W lightbulb burning continuously for two and one-half days is _____.

   a. 40 watts

CONTINUED

b. (40)(2.5) watt-hours

c. (40)(2.5)(24) watt-hours

d. (550)(40)(2.5) horsepower

IV. Suppose the instantaneous power of a machine is $270 - 4t$ W at a time t sec after it is turned on. This machine will do _____ J of work during the first minute of operation.

a. $\int_0^1 (270 - 4t)\,dt$

b. $(270 - 4 \cdot 60) - (270 - 4 \cdot 0)$

c. $270t - 2t^2$

d. $\int_0^{60} (270 - 4t)\,dt$

V. The work done by a force acting to move a particle is _____.

a. the sum of initial and final kinetic energies of the particle

b. the product of initial and final kinetic energies of the particle

c. the maximum of initial and final kinetic energies of the particle

d. the change in kinetic energy of the particle

VI. The work done by me as I sit here quietly pondering this problem is _____.

a. 0 newtons

b. 0 joules

c. 14 words per hour

d. (14)(32.2) word-slugs

■ APPLICATIONS

1. How much work is needed to lift a 20-lb stone 6 ft off the ground?

2. How much work is needed to lift a 1.5-kg book a distance of 0.2 m (e.g., from one's lap to a convenient reading position)?

3. A force of 3 N stretches a spring 50 cm. How much work is done in stretching the spring 3 m?

4. A force of 10 N stretches a spring 25 cm. How much work is done stretching the spring 1.5 m?

5. A force of 8 N stretches the spring of Problem 3 how many meters?

6. How much force is required to stretch the spring of Problem 4 a distance of 2 m? 4 m?

7. A cylindrical tank 10 m high and 10 m in diameter is filled with water. How much work does it take to pump the water out?

8. A tank in the shape of an inverted cone has a radius of 3 m and a height of 9 m; it is filled with water. How much work is needed to pump out all of the water?

9. How much work is needed if only half of the water is pumped out of the tank in Problem 7?

*10. How much work is needed in Problem 8 if only half of the water is to be pumped out?

11. The conical tank of Problem 8 is lifted, while empty, a distance of 8 m. How much work is required to then fill the tank from a hose that runs from ground level to a hole at the vertex (bottom) of the tank?

*12. The cylindrical tank of Problem 7 is now placed 10 m above the ground. How much work is needed to fill the tank from a hose that runs from ground level to a hole halfway up the side of the tank?

13. A 10-ft metal chain weighs 1.5 lb/ft. The chain is lifted vertically from one end until the other end is 5 ft off the ground (Figure 4). How much work is done?

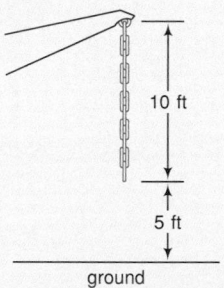

10 ft

5 ft

ground

FIGURE 4

14. A 40-ft chain that weighs 1.5 lb/ft is hanging from the top of a building (Figure 5). How much work is needed to pull 10 ft of it to the top? [*Hint:* Integrate from 30 to 40 since it is the *last* 10 ft that is being pulled to the top.]

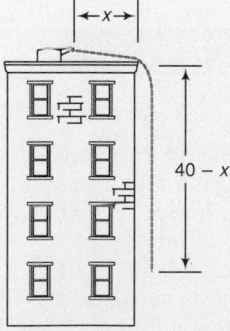

$\leftarrow x \rightarrow$

$40 - x$

FIGURE 5

CONTINUED

15. A swimming pool is 20 ft on each side and is 6 ft deep throughout (Figure 6). If it is filled with water, how much work is done by the water as it empties out of a hole in the bottom of the pool? [*Hint:* Density of water = 62.4 lb/ft^3.]

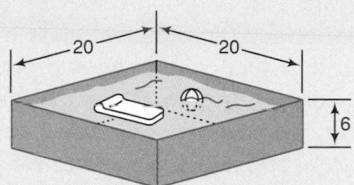

FIGURE 6

16. How much work is needed to fill the pool of the preceding problem to a depth of 2 ft?
17. How long will it take a 2-hp motor to do the work described in the preceding problem?
*18. Suppose the swimming pool of Problem 15 has a bottom that slopes down from a depth of 3 ft at the shallow end to 7 ft at the deep end (Figure 7). How much work is needed to fill this pool? How long will it take a pump with a 2-hp motor to do that work?

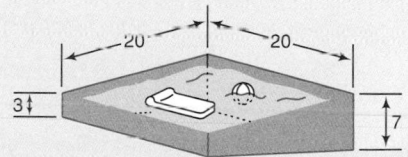

FIGURE 7

19. The instantaneous power of a particular machine t seconds after it is turned on is $P = 30 + 4t - t^2/2$ W. How much work is performed in the first 15 sec?
20. How much work is performed during the second minute of operation by a machine having instantaneous power $P = t + (30/\sqrt{1 + t})$ W?

*21. An arrow weighing 0.0322 lb is shot at full draw of $d = 16$ in from a bow with a 35-lb draw. Assume that the force of the bow (in pounds) is proportional to the draw (in inches). At what velocity does the arrow leave the bow? [*Hint:* Mass in slugs equals weight in pounds divided by acceleration of gravity.]
22. A block weighing 12 kg slides on a horizontal table with a speed of 4 m/sec. It is stopped by a spring, with a spring constant of 2, placed in its path. How much is the spring compressed?
23. The diameter of Mars is 6860 km. The acceleration due to gravity at its surface is 3.92 m/sec^2. What is the escape velocity from Mars?
*24. Our moon is approximately 380,000 km from the earth. By Newton's law of gravitational attraction (see Example 2.2.5), the attraction between two objects of masses m_1 and m_2 is Gm_1m_2/r^2, where G is a universal constant and r is the distance between the objects. The mass of the moon is 0.01228 times the mass of the earth. In the moon's revolving around the earth, a force is created which counteracts the force of gravity and keeps the moon in orbit around the earth. A cult space series on television once portrayed the moon, blown out of orbit by nuclear explosions, wandering aimlessly through space. Using the information above, calculate how much work would be required to accomplish this feat. (Express your answer in terms of M, the mass of the earth, and G. Assume that the loss of mass by the moon is negligible and that the final position of the moon is essentially an infinite distance from the earth.)
25. An object is dropped from a distance x_0 meters above the surface of the earth. Assuming the force of gravity is constant for small distances above the earth, what is the kinetic energy of the object when it hits the earth? What is its final velocity?
26. From what height would you have to drop an automobile in order for it to gain the same kinetic energy it has while driving 55 mi/hr?

■ ANSWERS TO SELF-QUIZ

I. a   II. b   III. c
IV. d   V. d   VI. b

5.5 CENTER OF MASS AND THE FIRST MOMENT

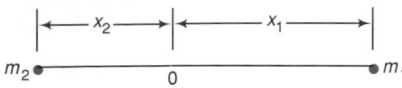

FIGURE 1
TWO MASSES

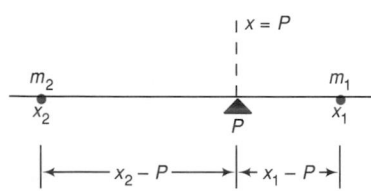

FIGURE 2
TWO MASSES ON EITHER SIDE OF A PIVOT

The distance from the pivot to the left-hand mass m_2 is treated as negative.

In the physical applications given earlier, we treated moving objects as though they were particles, having mass but not size. However, in the real world objects are made up of many particles. If a moving object is rotating or vibrating, for example, then different particles in the object may exhibit different kinds of behavior. Fortunately, we can deal with this kind of situation by representing the motion of the object as the motion of a special point, called the **center of mass** of that object. In general, we can describe the motion of an entire system if we can describe the motion of its center of mass and the motion of the system around that center of mass.

We can derive a simple equation for the center of mass by considering the motion of a seesaw. Consider a system containing two point masses m_1 and m_2, located at distances x_1 and x_2 from a reference point, which we label 0. In Figure 1 we have indicated two masses, one of which is located to the left of 0 so that its "distance" is treated as negative. We place the two masses on opposite ends of a seesaw as in Figure 2 and label the pivot point P. From experience we know that the farther away a mass is from the pivot point, the more likely the mass will "control" the motion. To put this idea more precisely, we consider the quantity $m_1(x_1 - P)$, measured in kilogram meters. By the principle of the lever, the seesaw will exactly balance if

$$m_1(x_1 - P) + m_2(x_2 - P) = 0. \tag{1}$$

The quantity on the left-hand side of (1) is called the **first moment of the system around the line $x = P$**. If $P = 0$, then the first moment is equal to $m_1 x_1 + m_2 x_2$ and is called the first moment about the y-axis (the line $x = 0$). When the first moment is zero, the seesaw will balance, and we say that the system is in **equilibrium**. Solving equation (1) for P, we obtain

$$m_1 x_1 + m_2 x_2 = m_1 P + m_2 P = (m_1 + m_2)P$$

or

$$P = \frac{m_1 x_1 + m_2 x_2}{m_1 + m_2}. \tag{2}$$

The point P is called the **center of mass** of the two masses and is denoted by $\bar{x}$. That is, from (2)

$$\bar{x} = \frac{m_1 x_1 + m_2 x_2}{m_1 + m_2}. \tag{3}$$

We have seen that the seesaw will balance if the pivot is at its center of mass. (Of course, in this derivation we ignored the mass of the board of the seesaw.)

If the system contains n particles, where $n > 2$, then we can, in an analogous manner, define the center of mass by

---

THE CENTER OF MASS OF n POINT MASSES

$$\bar{x} = \frac{m_1 x_1 + m_2 x_2 + \cdots + m_n x_n}{m_1 + m_2 + \cdots + m_n}$$

$$= \frac{\text{first moment of the system about the } y\text{-axis}}{\text{total mass of the system}} \tag{4}$$

EXAMPLE 1
CALCULATING THE CENTER OF
MASS OF FIVE POINT MASSES

Five particles of masses 2, 1, 4, 3, and 5 kg, respectively, are located on the x-axis with x values 3, −2, 4, 7, and −3 m. Calculate the center of mass of the system.

SOLUTION:

$$\bar{x} = \frac{m_1 x_1 + m_2 x_2 + m_3 x_3 + m_4 x_4 + m_5 x_5}{m_1 + m_2 + m_3 + m_4 + m_5}$$

$$= \frac{2(3) + 1(-2) + 4(4) + 3(7) + 5(-3)}{2 + 1 + 4 + 3 + 5} = \frac{26}{15} \text{ m}.$$

Now consider a rigid, horizontal rod, with variable density $\rho(x)$, that extends along the x-axis from $x = a$ to $x = b$. We can think of the rod as a system of very closely packed particles. In fact, the number of particles (atoms) is so large, and the spacing between them so small, that we can think of the body as having a continuous distribution of mass (i.e., having a continuous density function). We first calculate the first moment of the rod and begin by partitioning the interval $[a, b]$:

$$a = x_0 < x_1 < x_2 < \cdots < x_n = b.$$

If $\Delta x = x_i - x_{i-1}$ is small, the density $\rho(x)$ (which is assumed to be continuous) is almost constant in $[x_{i-1}, x_i]$. Then the mass Δm_i of the "subrod" over the subinterval $[x_{i-1}, x_i]$ is given by

Mass − density × length

$$\Delta m_i \approx \rho(x_i)\,\Delta x. \tag{5}$$

FIGURE 3
APPROXIMATING A CONTINUOUS ROD BY
n POINT MASSES

What we have done here is to approximate the continuous rod by a system of n point masses situated at the n partition points $x_1, x_2, \ldots, x_n$. (See Figure 3.) Combining equations (4) and (5), we see that the center of mass of this approximating system is given by

$$\bar{x}_{\text{approx}} = \frac{x_1 \rho(x_1)\,\Delta x + x_2 \rho(x_2)\,\Delta x + \cdots + x_n \rho(x_n)\,\Delta x}{\rho(x_1)\,\Delta x + \rho(x_2)\,\Delta x + \cdots + \rho(x_n)\,\Delta x}$$

$$= \frac{\sum_{i=1}^{n} x_i \rho(x_i)\,\Delta x}{\sum_{i=1}^{n} \rho(x_i)\,\Delta x}. \tag{6}$$

Now as the lengths of the individual sections of the partition approach zero, $\bar{x}_{\text{approx}}$ approaches the center of mass of the continuous rod. That is,

DEFINITION **THE CENTER OF MASS $\bar{x}$ OF A CONTINUOUS ROD WITH VARIABLE DENSITY $\rho(x)$**

$$\bar{x} = \lim_{\Delta x \to 0} \bar{x}_{\text{approx}} = \frac{\int_a^b x \rho(x)\,dx}{\int_a^b \rho(x)\,dx}. \qquad \blacksquare \tag{7}$$

As we saw earlier (p. 273), $\int_a^b \rho(x)\,dx$ = the mass of the rod. We define $\int_a^b x\rho(x)\,dx$ to be the **first moment of the rod around the y-axis** (the line $x = 0$). We can then rewrite (7) as

$$\bar{x} = x - \text{coordinate of the center of mass}$$

$$= \frac{\text{first moment around the } y\text{-axis}}{\text{total mass}}. \tag{8}$$

EXAMPLE 2
CALCULATING THE CENTER OF MASS OF A BEAM OF VARIABLE DENSITY

The density ρ of a 4-m nonuniform metal beam is given by $\rho(x) = 2\sqrt{x}$ kilograms per meter, where x is the distance along the beam. Find the center of mass of the beam.

SOLUTION: We place the beam along the x-axis from $x = 0$ to $x = 4$. (See Figure 4.) Then

$$\bar{x} = \frac{2\int_0^4 x\sqrt{x}\,dx}{2\int_0^4 \sqrt{x}\,dx} = \frac{\frac{4}{5}x^{5/2}\Big|_0^4}{\frac{4}{3}x^{3/2}\Big|_0^4} = \frac{\frac{4}{5}\cdot 32}{\frac{4}{3}\cdot 8} = \frac{12}{5}\text{ m}.$$

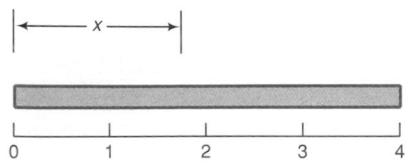

FIGURE 4
x MEASURES DISTANCE ALONG THE BEAM, STARTING AT THE LEFT END

Here the first moment is $\frac{128}{5}$ kg·m and the total mass of the beam is $\frac{32}{3}$ kg.

■ PROBLEMS 5.5

■ SELF-QUIZ

I. Consider a physical system with two point-masses on the x-axis: a 4-kg mass located 5 m to the left of the origin and a 6-kg mass located 3 m to the right of the origin. The first moment of this system about the y-axis is _____.
   a. $(4)(5) + (6)(3)$ b. $(4)(-5) + (6)(3)$
   c. $(4 + 6)(5 + 3)$ d. $(4 - 5)(6 + 3)$

II. The system described in the preceding question has a center of mass with x-coordinate equal to _____.
   a. $(4)(-5) + (6)(3)$
   b. $[(4)(-5) + (6)(3)]/[4 + 6]$
   c. $[(4)(-5) + (6)(3)]/[-5 + 3]$
   d. $[(4)(-5) + (6)(3)]/([4 + 6][-5 + 3])$

III. Each of the following physical systems consists of several masses placed upon the x-axis. Which system does not have its centroid at the origin?
   a. 4 kg at $(-1, 0)$ and 2 kg at $(2, 0)$
   b. 3 kg at $(-2, 0)$ and 2 kg at $(3, 0)$
   c. 5 g at $(-3, 0)$, 1 g at $(-1, 0)$, 8 g at $(2, 0)$
   d. 5 g at each of $(-6, 0)$, $(1, 0)$, $(2, 0)$, and $(3, 0)$
   e. 7 g at each of $(-3, 0)$, $(-2, 0)$, and $(6, 0)$
   f. 7 kg at $(-2, 0)$, 14 kg at $(0, 0)$, 2 kg at $(7, 0)$

IV. A rod of constant density lies along the x-axis between 3 and 17. The x-coordinate of its centroid is _____.
   a. 20   b. 14   c. 10
   d. unknown; we have insufficient information

■ DRILL

In Problems 1–4, find the first moment about the origin and the center of mass for the system of masses m_i located at the points P_i on the x-axis (each mass is measured in grams and each distance is measured in centimeters).

1. $m_1 = 4$, $m_2 = 6$; $P_1 = (3, 0)$, $P_2 = (5, 0)$

2. $m_1 = 3$, $m_2 = 4$, $m_3 = 7$; $P_1 = (2, 0)$, $P_2 = (-3, 0)$, $P_3 = (1, 0)$
3. $m_1 = 2$, $m_2 = 8$, $m_3 = 5$, $m_4 = 3$; $P_1 = (-6, 0)$, $P_2 = (2, 0)$, $P_3 = (20, 0)$, $P_4 = (-1, 0)$
4. $m_1 = 4$, $m_2 = 3$, $m_3 = 17$, $m_4 = 12$, $m_5 = 6$; $P_1 = (-6, 0)$, $P_2 = (8, 0)$, $P_3 = (0, 0)$, $P_4 = (-2, 0)$, $P_5 = (4, 0)$

CONTINUED

In Problems 5–10, a rod of variable density $\rho(x)$ lies along the x-axis in the specified interval $[a, b]$. Find (a) the total mass of the rod, (b) the first moment of the rod about zero, and (c) the rod's center of mass.

5. $\rho(x) = x$; $[a, b] = [0, 1]$

6. $\rho(x) = x - 1$; $[1, 2]$
7. $\rho(x) = |x|$; $[-1, 1]$
8. $\rho(x) = |x - 1|$; $[0, 2]$
9. $\rho(x) = x^{1/3}$; $[1, 8]$
10. $\rho(x) = x^3$; $[1, 2]$

■ **SHOW/PROVE/DISPROVE**

11. A rod lies along the interval $[a, b]$ on the x-axis and its varying density is given by the differentiable function $\rho(x)$. Think about splitting the rod into two pieces at $x = c$, where $a < c < b$. The first piece lies along the interval $[a, c]$, with mass m_1 and center of mass $\bar{x}_1$. The second piece lies along the interval $[c, b]$, with mass m_2 and center of mass $\bar{x}_2$. Clearly, $m_1 + m_2$ is the mass of the entire rod. Prove or disprove:

$$\frac{m_1 \bar{x}_1 + m_2 \bar{x}_2}{m_1 + m_2}$$

is the center of mass for the whole (unsplit) rod.

■ **ANSWERS TO SELF-QUIZ**

I. b
II. b

III. c (the centroid is at $(\frac{1}{3}, 0)$)
IV. c

5.6 THE CENTROID OF A PLANE REGION

We began the last section by considering the center of mass of a finite number of point masses located along the x-axis. We found that the center of mass $\bar{x}$ was given by

$$\bar{x} = \frac{\text{first moment of the system around the } y\text{-axis}}{\text{total mass}}. \tag{1}$$

We denote the first moment of the system around the y-axis by M_y, and we use the Greek letter μ (mu) to denote the total mass $(m_1 + m_2 + \cdots + m_n)$ of the system. Then we have

$$\bar{x} = \frac{M_y}{\mu}. \tag{2}$$

Now suppose, instead, that the n masses m_i are located at the points (x_i, y_i) in the xy-plane. Then we calculate the center of mass by treating separately the x- and y-coordinates of the points at which the masses are located. For the x-coordinates we simply use formula (2). We then define the first moment of the system around the x-axis (the line $y = 0$) by

$$M_x = m_1 y_1 + m_2 y_2 + \cdots + m_n y_n = \sum_{i=1}^{n} m_i y_i, \tag{3}$$

and we have

$$\bar{y} = \frac{M_x}{\mu}. \tag{4}$$

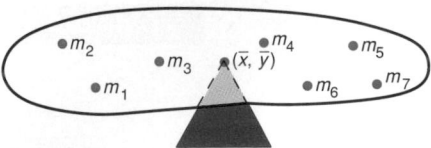

FIGURE 1
THE CENTER OF MASS IN TWO DIMEN-
SIONS CAN BE THOUGHT OF AS A BAL-
ANCING POINT.

The center of mass of the n particles is then located at the point $(\bar{x}, \bar{y})$. We can think of *the center of mass as a balancing point*. To picture this, imagine that the n masses are placed on a thin, uniform plate. Then the plate will balance on a pivot if the pivot is located at the point $(\bar{x}, \bar{y})$. This is illustrated in Figure 1.

EXAMPLE 1
COMPUTING THE CENTER OF
MASS OF THREE POINT MASSES
IN TWO DIMENSIONS

Calculate the center of mass of the system of three particles with masses $m_1 = 10$ g, $m_2 = 5$ g, and $m_3 = 8$ g located, respectively, at the points $(2, -1)$, $(3, 2)$, and $(-6, 1)$ (measured in centimeters).

SOLUTION: We calculate $M_y = 10(2) + 5(3) + 8(-6) = -13$ g·cm, $M_x = 10(-1) + 5(2) + 8(1) = 8$ g·cm, and $\mu = 10 + 5 + 8 = 23$ g. Then $\bar{x} = M_y/\mu = -\frac{13}{23}$ cm, $\bar{y} = M_x/\mu = \frac{8}{23}$ cm, and the center of mass is located at the point $(-\frac{13}{23}, \frac{8}{23})$.

If the masses in the preceding discussion all have the same mass m, then

$$M_y = mx_1 + mx_2 + \cdots + mx_n = m(x_1 + x_2 + \cdots + x_n) = m \sum_{i=1}^{n} x_i,$$

$$M_x = my_1 + my_2 + \cdots + my_n = m(y_1 + y_2 + \cdots + y_n) = m \sum_{i=1}^{n} y_i,$$

and

$$\mu = m + m + \cdots + m = nm,$$

so that

$$\bar{x} = \frac{m(x_1 + x_2 + \cdots + x_n)}{nm} = \frac{x_1 + x_2 + \cdots + x_n}{n} = \frac{\sum_{i=1}^{n} x_i}{n} \tag{5}$$

and

$$\bar{y} = \frac{m(y_1 + y_2 + \cdots + y_n)}{nm} = \frac{y_1 + y_2 + \cdots + y_n}{n} = \frac{\sum_{i=1}^{n} y_i}{n}. \tag{6}$$

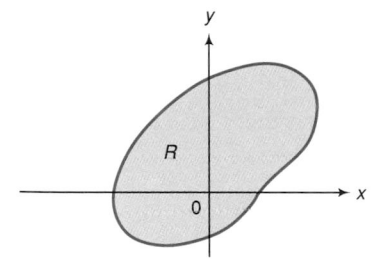

FIGURE 2
A PLANE LAMINA

We see that $\bar{x}$ is simply the arithmetic average of the x values, $\bar{y}$ is the arithmetic average of the y values, and the center of mass *does not depend* on the common mass value m. Since in this setting the actual value of the mass is irrelevant, we give the center of mass another name: the **centroid**. *The centroid is defined as the center of mass in the case of equal masses.*

Now let R denote a region in the plane, which we can think of as a thin sheet of material with *uniform* (i.e., constant) area density ρ. With this interpretation, the region is called a **plane lamina** and ρ is measured in kg/m², g/cm², lb/ft², or lb/in². (See Figure 2.) If $\rho = 1$, then the total mass of the lamina is equal to the area of the region R. We now calculate the centroid of such a lamina. We can think of the center of mass $(\bar{x}, \bar{y})$ as the pivot point at which the lamina would exactly balance. In particular, the centroid

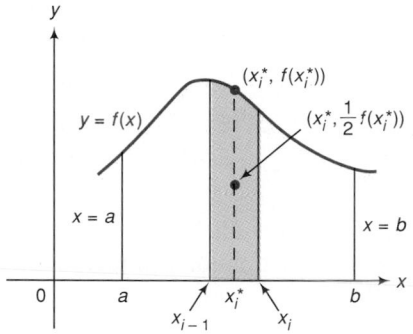

FIGURE 3
COMPUTING THE FIRST MOMENT OF A PLANE LAMINA BY SUMMING THE FIRST MOMENTS OF APPROXIMATING RECTANGLES

of a rectangle is its geometric center. We use this fact to find centroids of more complicated regions.

Suppose, first, that R is the region bounded by the graph of a continuous, nonnegative function f, the x-axis, and the lines $x = a$ and $x = b$ (see Figure 3). We partition the interval $[a, b]$ in the usual manner:

$$a = x_0 < x_1 < x_2 < \cdots < x_n = b.$$

The shaded region in Figure 3 is roughly rectangular in shape. Since density is constant, the center of mass of the rectangular lamina lies in the center of the rectangle, which is at the point $(x_i^*, \frac{1}{2}f(x_i^*))$, where $x_i^* = (x_{i-1} + x_i)/2$. The area of the rectangular lamina is, approximately, $f(x_i^*)(x_i - x_{i-1}) = f(x_i^*)\Delta x$, and the mass of the lamina is

$$\text{Mass} = \text{density} \times \text{area}$$
$$\Delta \mu_i \approx \overbrace{\rho f(x_i^*) \Delta x}.$$

The first moment of this ith sublamina around the y-axis (mass times average distance to the y-axis) is then given by

$$\Delta(M_y)_i \approx \Delta\mu_i x_i^* \approx \rho x_i^* f(x_i^*) \Delta x,$$

and the first moment of the entire lamina around the y-axis is defined as

$$M_y = \lim_{\Delta x \to 0} \sum_{i=1}^{n} \rho x_i^* f(x_i^*) \Delta x = \rho \int_a^b x f(x)\, dx. \tag{7}$$

Similarly, the first moment of the ith sublamina around the x-axis (mass times average distance to the x-axis) is given by

$$\Delta(M_x)_i \approx \overbrace{[\rho f(x_i^*) \Delta x]}^{\text{Mass}\atop\text{Density} \times \text{Area}} \cdot \overbrace{\left[\frac{1}{2}f(x_i^*)\right]}^{\text{Average distance}\atop\text{to }x\text{-axis}} = \frac{\rho}{2}[f(x_i^*)]^2 \Delta x,$$

and the first moment of the entire lamina around the x-axis is defined to be

$$M_x = \lim_{\Delta x \to 0} \sum_{i=1}^{n} \frac{\rho}{2}[f(x_i^*)]^2 \Delta x = \frac{\rho}{2}\int_a^b [f(x)]^2\, dx. \tag{8}$$

Since ρ is constant, the mass μ of the lamina is equal to ρ times the area of the region, or

$$\mu = \rho \int_a^b f(x)\, dx.$$

DEFINITION **CENTER OF MASS**

We define the **center of mass** $(\bar{x}, \bar{y})$ of the lamina by

$$x = \frac{M_y}{\mu} = \frac{\rho \int_a^b x f(x)\, dx}{\rho \int_a^b f(x)\, dx} = \frac{\int_a^b x f(x)\, dx}{\int_a^b f(x)\, dx} \tag{9}$$

and

$$\bar{y} = \frac{M_x}{\mu} = \frac{\frac{\rho}{2} \int_a^b [f(x)]^2 \, dx}{\rho \int_a^b f(x) \, dx} = \frac{\frac{1}{2} \int_a^b [f(x)]^2 \, dx}{\int_a^b f(x) \, dx}. \quad \blacksquare \tag{10}$$

Equations (9) and (10) tell us that if the lamina is uniform (i.e., if it has a constant density), then its center of mass depends only on the region and not on the density. As before, we then call the center of mass the **centroid** of the plane region R, and we can define the **first moments** of the region R around the y- and x-axes by, respectively.

FIRST MOMENTS

$$M_y = \rho \int_a^b x f(x) \, dx \quad \text{and} \quad M_x = \frac{\rho}{2} \int_a^b [f(x)]^2 \, dx \tag{11}$$

(around the y-axis) \qquad\qquad (around the x-axis)

CENTROID OF A REGION WITH CONSTANT DENSITY

$$\bar{x} = \frac{M_y}{\rho A} \quad \text{and} \quad \bar{y} = \frac{M_x}{\rho A} \tag{12}$$

---

where A denotes the area of the region.

REMARK: Since $\bar{x}$ and $\bar{y}$ do not depend on the density, we can compute $\bar{x}$ and $\bar{y}$ most easily by setting $\rho = 1$ in (11) and (12).

EXAMPLE 2
COMPUTING THE CENTROID OF A REGION BOUNDED BY A PARABOLA

Calculate the centroid of the region in the first quadrant bounded by the curve $y = x^2$, the x-axis, and the line $x = 1$.

SOLUTION: We have

Setting $\rho = 1$

$$M_y = \int_0^1 x^3 \, dx = \frac{1}{4}, \qquad M_x = \frac{1}{2} \int_0^1 x^4 \, dx = \frac{1}{10},$$

and

$$A = \int_0^1 x^2 \, dx = \frac{1}{3}.$$

Then

$$\bar{x} = \frac{M_y}{A} = \frac{\frac{1}{4}}{\frac{1}{3}} = \frac{3}{4} \quad \text{and} \quad \bar{y} = \frac{\frac{1}{10}}{\frac{1}{3}} = \frac{3}{10}.$$

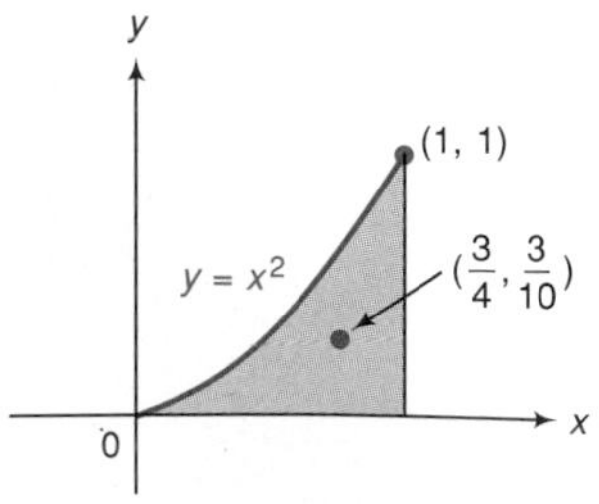

FIGURE 4
THE CENTROID OF A PARABOLIC REGION

The centroid is the point $(\frac{3}{4}, \frac{3}{10})$. This is depicted in Figure 4.

EXAMPLE 3

FINDING A CENTROID WHICH LIES OUTSIDE THE REGION UNDER CONSIDERATION

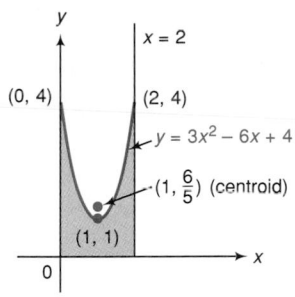

FIGURE 5
THE CENTROID OF THIS REGION LIES OUTSIDE THE REGION.

Find the centroid of the region bounded by the curve $y = 3x^2 - 6x + 4$, the x- and y-axes, and the line $x = 2$.

SOLUTION: We have

$$M_y \overset{\text{Setting } \rho = 1}{=} \int_0^2 (3x^3 - 6x^2 + 4x)\, dx = 4,$$

$$M_x = \frac{1}{2} \int_0^2 (3x^2 - 6x + 4)^2\, dx$$

$$= \frac{1}{2} \int_0^2 (9x^4 - 36x^3 + 60x^2 - 48x + 16)\, dx = \frac{24}{5},$$

and

$$A = \int_0^2 (3x^2 - 6x + 4)\, dx = 4,$$

so that $(\bar{x}, \bar{y}) = \left(\dfrac{4}{4}, \dfrac{24}{5 \cdot 4} \right) = \left(1, \dfrac{6}{5} \right)$. This is sketched in Figure 5. Note that in this case the centroid lies *outside* the region under consideration.

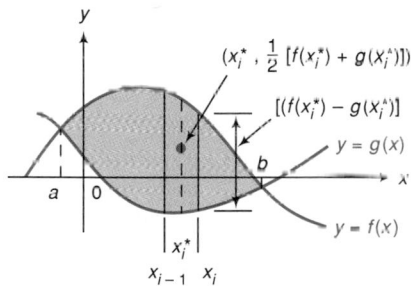

FIGURE 6
CALCULATING THE CENTROID OF A REGION BOUNDED BY TWO CURVES

We now generalize these results. Let the region R be bounded by the two curves $f(x)$ and $g(x)$, where both f and g are continuous and $f \geq g$. Then the situation is as depicted in Figure 6. If x ranges over the interval $[a, b]$, then we again partition the interval and find that the centroid of a typical rectangular subregion is, approximately, $(\bar{x}_i, \bar{y}_i) = (x_i^*, \frac{1}{2}(f(x_i^*) + g(x_i^*)))$, where $x_i^* = \frac{1}{2}[x_{i-1} + x_i]$. Then replacing the function f in equation (7) by $f - g$, we find that

---

CENTROID OF A REGION BOUNDED BY TWO CURVES

$$\bar{x} = \frac{M_y}{\rho A} \quad \text{and} \quad \bar{y} = \frac{M_x}{\rho A}, \tag{13}$$

where

$$M_y = \rho \int_a^b x[f(x) - g(x)]\, dx, \tag{14}$$

$$M_x = \frac{\rho}{2} \int_a^b [f(x) + g(x)][f(x) - g(x)]\, dx$$

$$= \frac{\rho}{2} \int_a^b \{[f(x)]^2 - [g(x)]^2\}\, dx, \tag{15}$$

and

$$A = \int_a^b [f(x) - g(x)]\, dx. \tag{16}$$

The expression for M_x follows from the fact that the area of the ith subregion is approximately equal to $\Delta x[f(x_i^*) - g(x_i^*)]$.

REMARK: Again we may set $\rho = 1$ in (13), (14), and (15) to simplify computations.

EXAMPLE 4
FINDING THE CENTROID OF
A REGION BOUNDED BY
TWO CURVES

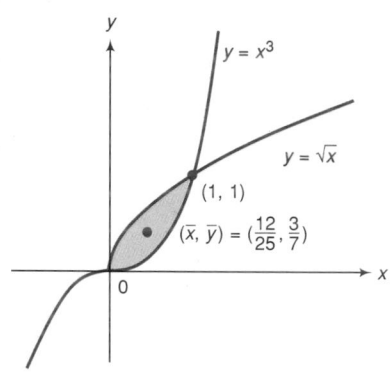

FIGURE 7
THE CENTROID OF THE REGION BETWEEN
THE GRAPHS OF $y = x^3$ AND $y = \sqrt{x}$

Find the centroid of the region bounded by the curves $y = x^3$ and $y = \sqrt{x}$.

SOLUTION: The two curves intersect at $(0, 0)$ and $(1, 1)$. Since $\sqrt{x} \geq x^3$ on $[0, 1]$, we have

Setting $\rho = 1$

$$M_y = \int_0^1 x(\sqrt{x} - x^3)\, dx = \int_0^1 (x^{3/2} - x^4)\, dx = \frac{1}{5},$$

$$M_x = \frac{1}{2} \int_0^1 (x - x^6)\, dx = \frac{5}{28},$$

and

$$A = \int_0^1 (\sqrt{x} - x^3)\, dx = \frac{5}{12},$$

so that $\bar{x} = \frac{12}{25}$ and $\bar{y} = \frac{3}{7}$. This is sketched in Figure 7.

The following theorem, whose proof is suggested in Problems 17 and 18, can be very useful in the calculation of centroids.

THEOREM 1 **SYMMETRY THEOREM FOR FINDING THE CENTROID**

   (i) If the plane region R is symmetric about the line $x = c$, then $\bar{x} = c$.
   (ii) If the plane region R is symmetric about the line $y = d$, then $\bar{y} = d$. ∎

EXAMPLE 5
FINDING THE CENTROID OF A
SEMICIRCLE

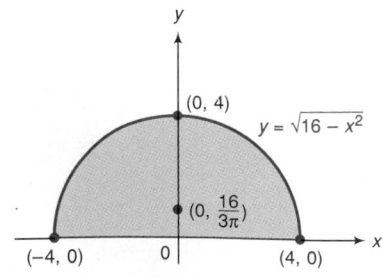

FIGURE 8
THE SEMICIRCLE $x^2 + y^2 = 16$, $y \geq 0$

The x-coordinate of the centroid is 0 because the curve is symmetric about the line $x = 0$ (the y-axis).

Calculate the centroid of the semicircle $y = \sqrt{16 - x^2}$.

SOLUTION: The semicircle (half the circle $x^2 + y^2 = 16$) is sketched in Figure 8. Since the region is symmetric about the y-axis (the line $x = 0$), we have $\bar{x} = 0$. We then calculate

$$M_x = \frac{1}{2} \int_{-4}^4 (16 - x^2)\, dx = \frac{128}{3},$$

and since $A = \frac{1}{2}\pi r^2 = 8\pi$, we find that $\bar{y} = 16/3\pi$.

COMPUTING VOLUME USING THE FIRST THEOREM OF PAPPUS

When a plane region is rotated about a line in the xy-plane, we can compute the volume generated very easily if we know the area and centroid of the region. The proof of the following theorem is given in Section 14.3.

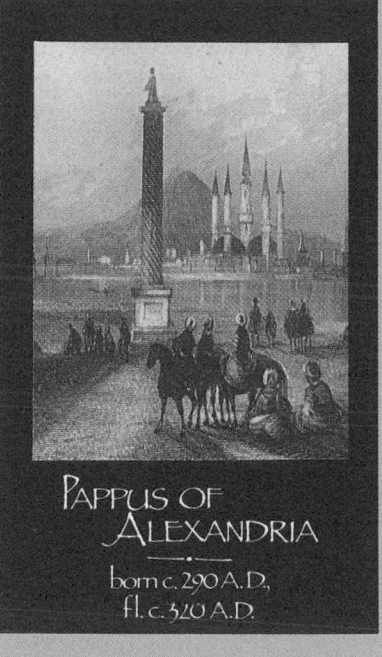

ORN DURING THE REIGN OF THE Roman emperor Diocletian (A.D. 284–305), Pappus of Alexandria lived approximately 600 years after the time of Euclid and Archimedes and devoted much of his life attempting to revitalize interest in the traditional study of Greek geometry.

Pappus's greatest work was his *Mathematical Collection*, written between A.D. 320 and 340. This work is significant for three reasons. First, in it are collected works of more than 30 different mathematicians of antiquity. We owe much of our knowledge of Greek geometry to the *Collection*. Second, it provides alternative proofs of the results of the greatest of the Greeks, including Euclid and Archimedes. Third, the *Collection* contains a variety of discoveries not found in any earlier work.

The *Collection* comprises eight books, each one containing a variety

of interesting and sometimes amusing results. In Book V, for example, Pappus showed that if two regular polygons have equal perimeters, then the one with the greater number of sides has the larger area. Pappus used this result to suggest the great wisdom of bees, as bees construct their hives using hexagonal (6-sided) cells, rather than square or triangular ones.

Book VII contains the theorem proved in this section. It is also the book that is the most important to the history of mathematics in that it contains the *Treasury of Analysis*. The *Treasury* is a collection of mathematical facts that, together with Euclid's *Elements*, claimed to contain the material the practicing mathematician in the fourth century needed to know. Although mathematicians wrote in Greek for another thousand or so years, no follower wrote a work of equal significance.

A harbor view of Alexandria, where Pappus lived in the fourth century

THEOREM 2　**FIRST THEOREM OF PAPPUS**

Suppose that the plane region R is revolved about a line L in the xy-plane that does not intersect it. Then the volume generated is equal to the product of the area of R and the length of the circumference of the circle traced by the centroid of R. ∎

EXAMPLE 6
COMPUTING THE VOLUME OF A TORUS

Calculate the volume of the doughnut shaped region called a **torus**, obtained when the circle of radius 1 centered at $(3, 0)$ is revolved about the y-axis.

SOLUTION:　The circle is sketched in Figure 9(a) and the resulting torus is sketched in Figure 9(b). The area of the circle is $\pi r^2 = \pi \cdot 1^2 = \pi$. The circle is symmetric about the line $x = 3$ and the line $y = 0$. Thus, by the symmetry theorem, the centroid is $(3, 0)$. As this point is revolved about the y-axis, a circle of radius 3 is traced out. The circumference of this circle is $2\pi r = 2\pi(3) = 6\pi$. Thus, by the first theorem of Pappus,

$$V = (\text{area of circle of radius 1}) \times (\text{circumference of circle of radius 3})$$
$$= \pi \cdot 6\pi = 6\pi^2.$$

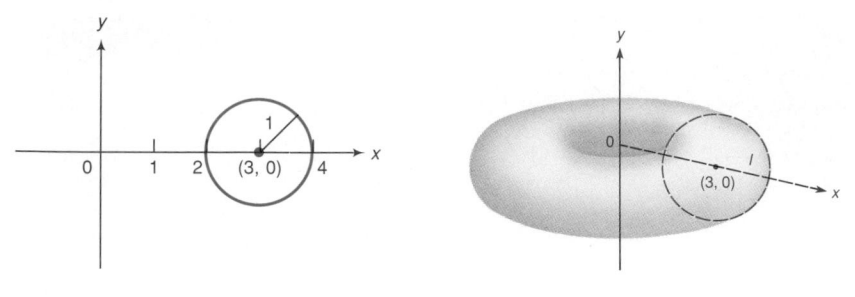

(a) Circle of radius 1 centered at (3, 0).

(b) A torus.

FIGURE 9 WHEN A CIRCLE THAT DOES NOT INTERSECT THE y-AXIS IS REVOLVED ABOUT THE y-AXIS, A TORUS IS GENERATED.

■ PROBLEMS 5.6

■ SELF-QUIZ

I. Suppose that Ω is a region of constant density with centroid $(\bar{x}, \bar{y})$. If Ω is symmetric about the x-axis, then _____.
 a. $\bar{x} = 0$ b. $\bar{y} = 0$
 c. $\bar{x} = \bar{y}$ d. $\bar{x} = \bar{y} = 0$

II. Consider the plane region bounded by the x-axis, the vertical line $x = 5$, and the line $y = 6x$. The first moment of this region about the x-axis is _____. (Assume that $\rho = 1$.)
 a. $\int_0^5 6x^2\,dx$ b. $\int_0^5 9x^2\,dx$

 c. $\int_0^5 18x^2\,dx$ d. $\int_0^5 3x^2\,dx$

III. Let Ω be the triangular region with vertices $(0, -1)$, $(1, 1)$, and $(0, 1)$. True or False: the centroid of Ω lies on the line $y = x$.

IV. Let Ω be the triangular region whose vertices are $(1, 1)$, $(1, 3)$, and $(2, 2)$. The first moment of Ω about the y-axis is _____.
 a. $\int_1^2 4(2 - x)\,dx$ b. $\int_1^2 4(4 - 2x)\,dx$
 c. $\int_1^2 x(2 - x)\,dx$ d. $\int_1^2 x(4 - x)\,dx$

■ DRILL

In Problems 1–4, find the center of mass for the system of masses m_i located at the points P_i (each mass is measured in kilograms; the x and y units are meters).

1. $m_1 = 4$, $m_2 = 6$; $P_1 = (3, 4)$, $P_2 = (-5, 3)$
2. $m_1 = 3$, $m_2 = 4$, $m_3 = 7$; $P_1 = (2, 3)$, $P_2 = (-3, -4)$, $P_3 = (1, 2)$
3. $m_1 = 2$, $m_2 = 8$, $m_3 = 5$, $m_4 = 3$; $P_1 = (-6, 2)$, $P_2 = (2, 3)$, $P_3 = (20, -5)$, $P_4 = (-1, 8)$
4. $m_1 = 4$, $m_2 = 3$, $m_3 = 17$, $m_4 = 12$, $m_5 = 6$; $P_1 = (-6, 3)$, $P_2 = (8, 3)$, $P_3 = (0, -3)$, $P_4 = (-2, 2)$, $P_5 = (4, 1)$

In Problems 5–12, find the centroid of the plane region bounded by the specified curves and lines.

5. $y = 2x + 3$, x-axis, y-axis, $x = 1$
6. $y = x^{1/3}$, x-axis, y-axis, $x = 1$
7. $x + y = 4$, x-axis, y-axis
*8. $y = \sqrt{10 + x}$, x-axis, $x = -1$, $x = 6$
9. $y = x^2$, $y = 4x$, $x = 0$, $x = 1$
10. $y = x^2 + 3x + 5$, $y = 2x^2 - x + 8$
11. $x + y^2 = 8$, $x + y = 2$
12. $x^2 = 4 + y$, $2x + y = 4$

■ APPLICATIONS

13. Use the symmetry theorem to find the centroid of the plane region bounded by the curves $y = x^2$ and $y = 2 - x^2$.

14. Use the symmetry theorem to find the centroid of the region bounded by $y = x^8$ and $y = 1$.

CONTINUED

15. Use the symmetry theorem and the first theorem of Pappus to calculate the volume of the torus generated by rotating the unit circle $(x^2 + y^2 = 1)$ about the line $y = 4 - x$.

16. Use the symmetry theorem and the first theorem of Pappus to calculate the volume of the solid generated by rotating the triangle with vertices at $(-1, 2)$, $(2, 1)$, and $(0, 4)$ about the x-axis.

■ SHOW/PROVE/DISPROVE

*17. Prove Theorem 1(i). [*Hint:* Divide the interval $[a, b]$ into $2n$ subintervals, half of which lie to the left of c and half of which lie to the right. Then treat the line $x = c$ as if it were the y-axis and show that $\bar{x} = 0$.]

*18. Prove Theorem 1(ii).

19. Consider the triangle with vertices at (x_1, y_1), (x_2, y_2), and (x_3, y_3).
    a. Show that the centroid of this triangle is at the point

$$\left(\frac{x_1 + x_2 + x_3}{3}, \frac{y_1 + y_2 + y_3}{3}\right).$$

    b. Suppose three equal weights are placed at the vertices of the triangle. Show that the center of mass for this system is the point mentioned in part (a).
    c. Show that the point mentioned in part (a) lies on each line joining a vertex of the triangle to the midpoint of the opposite side.

■ ANSWERS TO SELF-QUIZ

I. b

II. c

III. True (the centroid is $(\frac{1}{3}, \frac{1}{3})$)

IV. d

5.7 FLUID PRESSURE (OPTIONAL)

In this section we show how the integral can be used to calculate fluid pressures. By a **fluid** we mean a substance that can flow. This definition includes both liquids and gases. It is reasonable only to talk about forces applied to the *surface* of a fluid. If the fluid is at rest, then such forces can be applied only perpendicular to the surface (Figure 1(a)), since if a force were applied tangentially (i.e., from the side, as in Figure 1(b)), then layers of fluid would simply slide over one another, and the fluid would no longer be at rest, as we have assumed.

When a force is applied perpendicular to the surface of a fluid it is called a **normal force**. In this case, we define the **pressure** P to be the magnitude of the normal force per unit of area. Commonly used units of pressure are lb/ft^2, lb/in^2, N/m^2, and N/cm^2. For example, the air pressure at sea level is $14.7 \text{ lb/in}^2 = 1.013 \times 10^5 \text{ N/m}^2$.

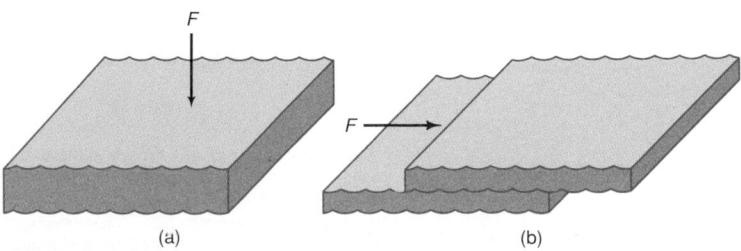

(a)   (b)

FIGURE 1 PERPENDICULAR AND TANGENTIAL FORCES ON A FLUID

Pascal's principle[†] in physics states that when an object is submerged in a fluid, then the pressure at any depth is the same in every direction (up, down, sideways). Thus in computing the pressure on a submerged object, we only need to be concerned with the fluid depth, not its direction.

Now suppose that a thin horizontal lamina is submerged at a certain depth in a fluid and we wish to calculate the total force exerted on the lamina by the column of the fluid on one side of the lamina. The force is given by

$$F = PA, \tag{1}$$

where P is the pressure and A is area. Then

$$P = \frac{F}{A} = \frac{mg}{A}, \tag{2}$$

where g is the acceleration due to gravity. Over a unit of area, the mass of a fluid is given by the product of its density ρ (in lb/in^3, lb/ft^3, kg/m^3 or g/cm^3) times the depth d of the column of fluid.

TWO FORMULAS FOR FLUID PRESSURE

If the mass is measured in slugs (1 slug = 1 lb/g), then

$$P = \rho d \quad \text{(in lb/ft}^2\text{)}. \tag{3}$$

If mass is measured in kilograms, then

$$P = \rho d g^{\ddagger} \quad \text{(in N/m}^2\text{)} \tag{4}$$

where $g = 9.81$ m/sec^2. Note that 1 kg · m/sec^2 = 1 N.

EXAMPLE 1
FLUID PRESSURE ON THE BOTTOM OF A SWIMMING POOL

A swimming pool with length 8 m, width 5 m, and depth 3 m is filled with water. What is the force exerted on the bottom of the pool by the weight of the water?

SOLUTION: The density of water (at 0°C) is, approximately,

$$\rho = 1000 \text{ kg/m}^3.$$

From (4)

$$P = \rho d g = (1000 \text{ kg/m}^3)(3 \text{ m})(9.81 \text{ m/sec}^2) = 29,430 \frac{\text{kg} \cdot \text{m/sec}^2}{\text{m}^2}$$

$$= 29,430 \text{ N/m}^2.$$

[†] Named after the brilliant French mathematician Blaise Pascal (1623–1662). Pascal is best known as one of the creators of modern probability theory in a series of letters to Pierre Fermat (see page 93).

[‡] Recall that the forces are measured in pounds (in the English system) and newtons (in the metric system). To obtain force from kilogram measurements, we need to multiply by g. Thus if water density is given in terms of pounds per cubic foot, no multiplication by g is necessary to obtain force since pounds already represent force. However, if the density is given in terms of kilograms per cubic meter, multiplication by g is necessary to obtain force.

Then using (1), we have

$$F = PA = (29{,}430 \text{ N/m}^2)(5 \times 8 \text{ m}^2) = 1{,}177{,}200 \text{ N}.$$

EXAMPLE 2
FLUID PRESSURE ON A THIN
RECTANGULAR SHEET
OF METAL

A thin rectangular sheet of metal 10 ft × 14 ft is submerged horizontally in water to a depth of 12 ft. What is the force exerted on one side of the sheet of metal by the weight of the water?

SOLUTION: The density of water at 0°C is, approximately,

$$\rho = 62.4 \text{ lb/ft}^3.$$

Using (3), we have

$$P = \rho d = (62.4 \text{ lb/ft}^3)(12 \text{ ft}) = 748.8 \text{ lb/ft}^2.$$

Then

$$F = PA = (748.8 \text{ lb/ft}^2)(10 \times 14 \text{ ft}^2) = 104{,}832 \text{ lb}.$$

EXAMPLE 3
ARCHIMEDES' PRINCIPLE

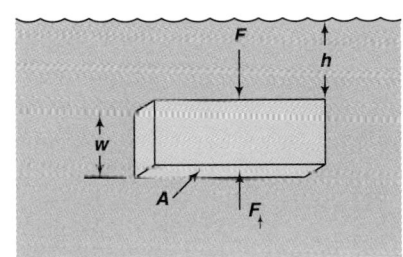

FIGURE 2
A BODY IMMERSED IN A FLUID

Archimedes' principle[†] states that a body immersed in a fluid is buoyed up by a force equal to the weight of the fluid displaced. In the case of a floating object Archimedes' principle states that a floating body must displace its own weight in water. To demonstrate Archimedes' principle, we let a block of width w and area A (see Figure 2) be submerged a distance h in a liquid of density ρ. By equation (4) the force $F_\uparrow$ on the bottom of the block points up and is given by

$$F_\uparrow = PA = \rho g(w + h)A.$$

Similarly, the force $F_\downarrow$ on the top of the block points down and is given by

$$F_\downarrow = PA = \rho g h A.$$

The net (or resultant) force is given by

$$F_\uparrow - F_\downarrow = \rho g(w + h)A - \rho g h A = \rho g w A = m_l g,$$

where m_l is the total mass of the liquid displaced by the block (since the mass of the block is its volume wA times its density ρ). Thus the block undergoes an upward force equal to the weight of the water it displaces.

In the problems discussed above, we calculated the force of the liquid on a plane surface. That is, water pressure was applied uniformly over a flat surface. However, if a body is submerged in a fluid, then the lateral pressure (pressure on the wall or side of the body) varies with the depth of the fluid so that formulas (3) or (4) cannot be used directly.

[†] It was this discovery that is reputed to have led Archimedes to jump from his bath and run home naked, shouting "Eureka" ("I have found it."). See Archimedes' biography on page 266.

EXAMPLE 4
FLUID PRESSURE ON ONE END OF
A TROUGH FILLED WITH WATER

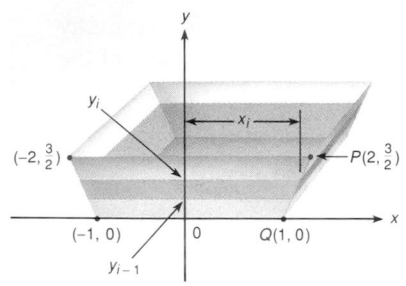

FIGURE 3
A TROUGH FILLED WITH WATER

A trough 3 m in length and filled with water has a cross-section in the shape of a trapezoid with lower base 2 m, upper base 4 m, and depth $1\frac{1}{2}$ m. Find the total force due to water pressure on one end of the trough.

SOLUTION: One end of the trough is sketched in Figure 3. To facilitate our calculations, we put in x- and y-coordinates as shown. Then we partition the interval $[0, \frac{3}{2}]$ along the y-axis. The force against the ith strip (shaded) is equal to the pressure along the strip times the area of the strip, and (since pressure is fairly constant on the strip if Δy is small) we have

$$\overset{\text{From (4)}}{\Delta F_i = P_i A_i \approx \rho d_i g A_i.}$$

But $\rho = 1000$ kg/m^3, $d_i \approx \frac{3}{2} - y_i$, $g = 9.81$ m/sec^2, and $A_i \approx 2x_i \Delta y$. Now we calculate x as a function of y on the edge of the trapezoid. The line PQ on the edge passes through the points $(2, \frac{3}{2})$ and $(1, 0)$, leading to the equation

$$\frac{y}{x-1} - \frac{-\frac{3}{2}}{-1} = \frac{3}{2}, \qquad \text{or} \qquad x = \frac{2}{3}y + 1.$$

Thus

$$\Delta F_i = \rho g\left(\frac{3}{2} - y_i\right)(2x_i)\,\Delta y = 2\rho g\left(\frac{3}{2} - y_i\right)\left(\frac{2}{3}y_i + 1\right)\Delta y$$

$$= \frac{\rho g}{3}(3 - 2y_i)(2y_i + 3)\,\Delta y = \frac{\rho g}{3}(9 - 4y_i^2)\,\Delta y.$$

Finally, we let Δy approach zero to obtain

$$F = \frac{\rho g}{3}\int_0^{3/2}(9 - 4y^2)\,dy = \frac{\rho g}{3}(9) = 3\rho g$$

$$= 3(1000)(9.81) = 29{,}430 \text{ N}.$$

EXAMPLE 5
FLUID PRESSURE ON ONE END
OF A BARREL SUBMERGED IN
SEAWATER

A cylindrical barrel whose end has a diameter of 4 ft is submerged in seawater (density approximately 64.3 lb/ft^3 at 0°C). Find the total force due to water pressure at one end if the center of the barrel is at a depth of 12 ft (see Figure 4(a)).

SOLUTION: One end of the barrel is sketched in Figure 4(b). We introduce a coordinate system that places the origin at the center of the barrel. If we now partition the "y-axis" as shown, then the force on a typical horizontal strip is given by

$$\Delta F_i = P_i A_i \approx \rho d_i A_i \qquad \text{[from (3)]}.$$

But $\rho = 64.3$ lb/ft^3, $d_i \approx 12 - y_i$, and $A_i \approx$ (length of strip) $\times (\Delta y)$. Since the equation of the circular end is $x^2 + y^2 = 4$, we have $x = \sqrt{4 - y^2}$ to the right of the y-axis and $x = -\sqrt{4 - y^2}$ to the left. Thus the length of the strip is approximately equal to $\sqrt{4 - y_i^2} - (-\sqrt{4 - y_i^2}) =$

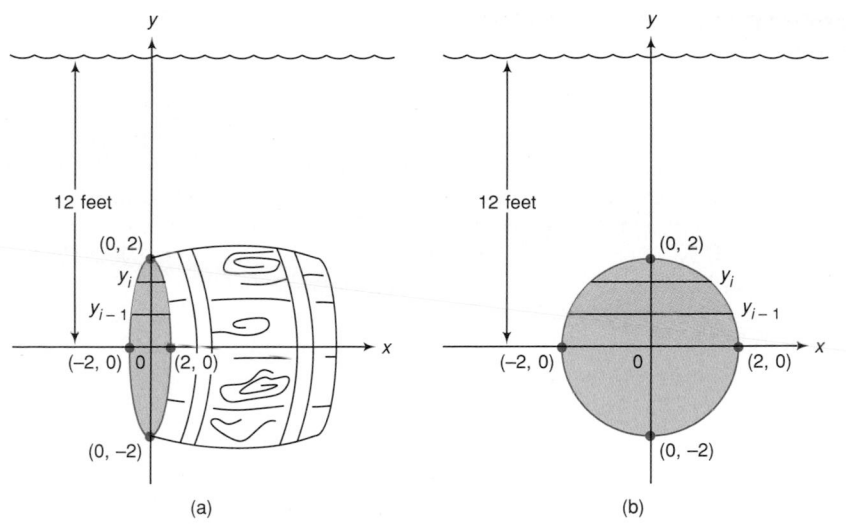

FIGURE 4 A BARREL SUBMERGED IN SEAWATER

$2\sqrt{4 - y_i^2}$, and we find that $A_i \approx 2\sqrt{4 - y_i^2}\,\Delta y$. Then

$$\Delta F_i \approx 2\rho(12 - y_l)\sqrt{4 - y_l^2}\,\Delta y$$

and

$$F = 2\rho \int_{-2}^{2} (12 - y)\sqrt{4 - y^2}\,dy$$

$$= 24\rho \int_{-2}^{2} \sqrt{4 - y^2}\,dy - 2\rho \int_{-2}^{2} y\sqrt{4 - y^2}\,dy.$$

The first integral is the area of the semicircle $x = \sqrt{4 - y^2}$ and is equal to $\frac{1}{2}\pi r^2 = \frac{1}{2}\pi(2)^2 = 2\pi$. Thus

$$F = (24\rho)(2\pi) + \frac{2\rho}{3}(4 - y^2)^{3/2}\Big|_{-2}^{2} = (48\pi)\rho + 0$$

$$= (48\pi)(64.3) = 3086.4\pi \text{ lb} \approx 9696 \text{ lb}.$$

Examples 1–5 above illustrate how to go about solving problems involving fluid pressure. We will not cite a general formula since the type of problems encountered can vary so widely. Once a convenient coordinate system is introduced, it is usually not too difficult to obtain the answer.

We conclude this section by giving the densities of some common fluids.

TABLE 1 DENSITIES OF FLUIDS

Fluid	Density (in kg/m³)	Density (in lb/ft³)
Water	1000	62.4
Sea Water	1030	64.3
Gasoline	673 (average value)	42.0 (average value)
Mercury	13,606	849
Kerosene	821	51.2
Ethyl Alcohol	810	50.5

■ PROBLEMS 5.7

■ SELF-QUIZ

I. A thin rectangular sheet of metal 10 m long and 1 m wide is submerged in water at a depth of 1 m. The force exerted on one side of the sheet of metal by the weight of the water is
a. 9,810 N   b. 98,100 N
c. 98,100 N/m²   d. 9,810 N/m²
e. 10,000 N   f. 1000 N/m²

II. A thin rectangular sheet of metal 1 ft long and 10 ft wide is submerged in water at a depth of 1 ft. The force exerted on one side of the sheet of metal by the weight of the water is
a. 62.4 lb/ft³
b. $62.4 \times 32.2 = 2021.76$ lb/ft³
c. 2021.76 lb
d. 624 lb/ft³
e. 624 lb
f. $624 \times 32.2 = 20,092.8$ lb

■ APPLICATIONS

1. A circular swimming pool has a radius of 6 m and is 2 m deep. Find the force due to water pressure on the bottom of the pool when the pool is filled with water.

2. Find the force in Problem 1 if the pool is filled with seawater ($\rho \approx 1030$ kg/m³ at 0°C).

3. A tube of mercury ($\rho = 13.6$ gm/cm³) is inverted over a beaker of mercury as shown in Figure 5. The pressures P_C and P_D at the points C and D are the same because the points are at the same height. Let P_A denote atmospheric pressure.
   a. Find a formula giving P_A as a function of h, the height of the column of mercury.
   b. For $h = 76$ cm, show that $P_A = 1.01 \times 10^5$ N/m². [Hint: Use $g = 981$ cm/sec².] This value of P_A is called **one standard atmosphere**.

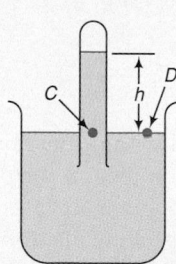

FIGURE 5

4. A rectangular swimming pool 5 m wide and 7 m long has a bottom that slopes at a constant rate, being $\frac{1}{2}$ m deep at the shallow end and 4 m deep at the deep end. When the pool is filled with water, find the force due to liquid pressure on the bottom of the pool.

5. A trough filled with water has a cross-section in the shape of an equilateral triangle with 5-ft sides (the wide end is up). What is the force due to water pressure on one end of the trough?

6. In Problem 5, find the force when the trough is filled to a depth of (a) 1 ft, (b) 2 ft, (c) 3 ft.

7. A trough filled with ethyl alcohol ($\rho = 810$ kg/m³) has a cross section in the shape of a semicircle with a radius of 1 m (the open end is up). What is the force due to liquid pressure on one end of the trough?

8. Find the force on one side of the trough in Problem 5 if the trough is 10 ft long.

9. The dam across a certain river has the shape of the parabola $y = x^2/25$. The river is 25 m wide at the top. Find the force due to water pressure on a face of the dam that is vertical.

*10. Suppose the dam across the river in Problem 9 has a face that slopes upstream from the water surface at an angle of 45°. Find the force due to liquid pressure on it. [Hint: First use Problem 9 to find the depth of the water.]

*11. Find the force in Problem 10 if the dam slopes at an angle of 30°.

12. An elliptical plate is submerged in seawater so that the center of the ellipse is at a depth of 6 m. The equation of the ellipse is $(x^2/4) + (y^2/9) = 1$. Find the force due to water pressure on the plate. [Hint: The area of a circle of radius r is πr^2.]

13. A cask with sides in the shape of the trapezoid of Example 4 is submerged in seawater so that the top of the trapezoid is level with the surface of the water. Find the force due to liquid pressure on one end of the cask.

14. The gate of a dam is in the shape of an isosceles triangle with a 6-ft base and 4-ft sides. The upper vertex of the triangle is at a depth of 22 ft. Find the force on the gate when the gate is closed.

*15. Find the force on one side of the trough in Example 4.

CONTINUED

16. Find the force on the sides of the pool (at the deeper and shallower ends) in Problem 4.

In Problems 17 and 18 a plate in the shape of the given region is submerged in water with its centroid at the given depth. Use Problem 19 to calculate the force due to liquid pressure.

17. $y = x^{1/3}$, $y = 0$, $x = 1$; $d = 15$ m.
18. $y = x^4$, $y = 0$, $x = 2$; $d = 30$ ft.

■ SHOW/PROVE/DISPROVE

19. Prove that the total force against one side of a vertical submerged region R is equal to the product of the pressure at the centroid of R times the area of R.

■ ANSWERS TO SELF-QUIZ

I. b   II. e

■ PROBLEMS   CHAPTER 5 REVIEW

In Problems 1–8, find the area of the region bounded by the given curves and lines.

1. $y = 3x - 7$; x-axis, $x = 2$, $x = 5$
2. $y = \sqrt{x + 1}$; x-axis, y-axis, $x = 15$
3. $y = 2 - x - x^2$; x-axis
4. $y = (x + 1)(x^2 - 8x + 15)$; x-axis
5. $y^2 + x = 0$, $y = 2x + 1$
6. $x^{1/3} + y^{1/3} = 1$; $x = 0$, $y = 0$
7. $y = \sin x$, $y = \cos x$; $x = 0$, $x = \pi$
8. $y = \sin 2x$, $y = \cos 3x$; $x = 0$, $x = \pi/6$

In Problems 9–16, find the volume of the solid generated by rotating the region bounded by the given curves and lines about the specified axis.

9. $y = 3x + 2$, $x = 0$, $y = 0$, $x = 2$; x-axis
10. same region as in Problem 9; y-axis
11. $y = \sqrt{2x + 1}$, $x = 0$, $y = 0$, $x = 4$; x-axis
12. same region as in Problem 11; y-axis
13. $y = x^3$, $x = 1$, $y = 0$; x-axis
14. $y = x^{1/4}$, $x = 1$, $y = 0$; y-axis
15. $y = \sqrt{2x + 1}$, $x = 0$, $y = 5$, $x = 4$; x-axis
16. same region as in Problem 15; y-axis
17. The base of a certain solid is a circle of radius 3; each vertical cross section perpendicular to a fixed diameter is a square. What is the volume of the solid?
18. Suppose the solid in Problem 17 had equilateral triangles for its perpendicular cross sections. What would be the volume of this solid?

In Problems 19–22, calculate the arc length of the specified portion of the given curve.

19. $y = 2x + 3$; $x \in [2, 7]$
20. $y = x^2$; $x \in [0, 2]$
21. $y = \frac{2}{3}x^{3/2}$; $x \in [0, 1]$
22. $y = \frac{1}{12}(3 - 4x)^{3/2}$; $x \in [-8, -5]$

23. Find a definite integral that is equal to the length of the arc of $y = x \sin x$ between $x = 0$ and $x = \pi/2$.
24. Find a definite integral that computes the arc length of that part of $y^3 = x^2$ which lies between (8, 4) and (27, 9).
25. A force of 8 N stretches a spring 30 cm. How much work is done in stretching the spring 60 cm?
26. A tank in the shape of an inverted cone has a radius of 6 ft and a height of 12 ft. It is filled with seawater (with an approximate density of 64.3 lb/ft^3). How much work is needed to empty the tank?
27. The instantaneous power of a certain machine is given by $P = 45\sqrt{t} + (t/10) + (t^2/100)$ watts, where t is the time in seconds since the machine started. How much work is done in the first 25 seconds of operation?
28. An object is dropped from a height of 500 m. What is the kinetic energy of the object when it hits the earth? What is its final velocity?

CONTINUED

29. Find the first moment around the origin and the center of mass for the system of masses $m_1 = 2$ kg, $m_2 = 5$ kg, and $m_3 = 8$ kg located along the x-axis at the points $P_1 = (4, 0)$, $P_2 = (-9, 0)$, and $P_3 = (2, 0)$, respectively.

30. Find the center of mass for the system with 2 kg at $(-2, 3)$, 7 kg at $(4, 6)$, 4 kg at $(3, -7)$, and 8 kg at $(0, -1)$.

31. Find the center of mass of a rod which lies along the x-axis in the interval $[0, 3]$ with density $\rho(x) = x^2$.

32. Find the center of mass of a rod lying along the x-axis in the interval $[1, 4]$ with density $\rho(x) = \dfrac{1}{x^3}$.

In Problems 33–36, find the centroid of the plane region bounded by the given curves.

33. $y = x^2$, $y = 0$, $x = 1$, $x = 3$
34. $y = \sqrt{1 + x}$, $y = 0$, $x = 0$, $x = 3$
35. $y = x^2 + 5x + 6$, $y = x + 3$
36. $25(x - 2)^2 + 4(y + 3)^2 = 100$

37. Find the force due to water pressure on the bottom of a rectangular swimming pool 5 m long, 4 m wide, and 2 m deep when it is filled with water.

38. Find the force due to water pressure on the bottom of the pool in Exercise 37 if the bottom slopes downward from a depth of 1 m at one end to a depth of 3 m at the other.

39. A trough filled with water is 5 m long and has a cross-section in the shape of a trapezoid 3 m wide at the bottom, $1\frac{1}{2}$ m wide at the top, and 2 m deep. What is the force due to water pressure on one end of the trough?

40. Answer the questions in Exercises 38 and 39 if the pool or trough is filled with ethyl alcohol (density $\rho = 810$ kg/m³).

41. In Exercise 39, find the force if the trough is filled with water to a depth of (a) $\frac{1}{2}$ m, (b) 1 m, (c) $1\frac{1}{2}$ m.

42. A cylindrical barrel whose end has a diameter of 2 m is submerged in seawater (density $\rho = 1030$ kg/m³). Find the total force due to water pressure at one end if the center of the barrel is at a depth of 15 m.

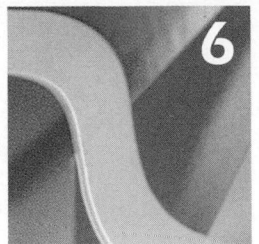

6 TRANSCENDENTAL FUNCTIONS AND THEIR INVERSES

In this chapter we discuss the calculus of three special kinds of functions: exponential, logarithmic, and trigonometric functions. The trigonometric functions are discussed in Appendix 1. You saw how to differentiate them in Section 2.5. The exponential and logarithmic functions are defined in this chapter.

Perhaps you are already familiar with exponential and logarithmic functions through some previous course. Perhaps you will be seeing these functions for the first time. In any event, we will define these functions and show why they are important by giving a large variety of applications in physics, biology, and economics.

There are two very different ways to define the exponential and logarithmic functions. The first of these starts with the basic rules of exponents with which you are all familiar. The second way starts off with a special integral. In each case we are led to the same functions. Both ways of defining these functions can give us insight into how to work with them, and both are given in this text. The first derivation is given as a review from algebra in Section 6.2. The second derivation is given in Sections 6.3–6.5.

In a sense to be made precise shortly, exponential and logarithmic functions are inverses of one another. For that reason we begin the chapter with a discussion of inverse functions.

6.1 INVERSE FUNCTIONS

Let $y = 2x + 3$. Then we can write x as a function of y: $x = (y - 3)/2$. The functions $2x + 3$ and $(x - 3)/2$ are called *inverse functions*. We have seen other examples of inverse functions. The functions x^3 and $\sqrt[3]{x}$ are inverse functions since $y = x^3$ implies that $x = \sqrt[3]{y}$. In general, we have the following definition.

DEFINITION   **INVERSE FUNCTIONS**

The functions f and g are **inverse functions** if the following conditions hold:

(i) For every x in the domain of g, $g(x)$ is in the domain of f and

$$f(g(x)) = x. \tag{1}$$

(ii) For every x in the domain of f, $f(x)$ is in the domain of g and

$$g(f(x)) = x. \tag{2}$$

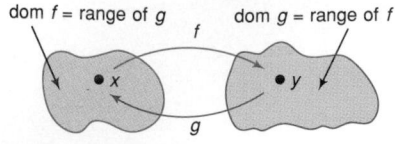

FIGURE 1
f TAKES x INTO y, AND g TAKES THAT y
BACK TO THE ORIGINAL x.

In this case, we write $f(x) = g^{-1}(x)$ or $g(x) = f^{-1}(x)$, and we say

f is the inverse of g 　　and　　 g is the inverse of f. ∎

In the notation of Section 0.4, we may write

$$(f \circ g)(x) = x \qquad \text{and} \qquad (g \circ f)(x) = x.$$ (3)

We illustrate the idea of an inverse function in Figure 1.

EXAMPLE 1
THE INVERSE OF A
STRAIGHT-LINE FUNCTION

If $f(x) = 2x + 3$ and $g(x) = (x - 3)/2$, then

$$f(g(x)) = 2\left(\frac{x - 3}{2}\right) + 3 = x - 3 + 3 = x$$

and

$$g(f(x)) = \frac{(2x + 3) - 3}{2} = x.$$

So $2x + 3$ and $(x - 3)/2$ are inverse functions.

We can illustrate this in another way. Suppose, for example, that $x = 4$. Then

$$y = f(x) = f(4) = 2 \cdot 4 + 3 = 11 \qquad \text{(4, 11) is in the graph of } f$$

$$x = g(y) = g(11) = \frac{11 - 3}{2} = 4. \qquad \text{(11, 4) is in the graph of } g$$

In Table 1 we list points in the graphs of f and g.

TABLE 1　POINTS IN THE GRAPHS OF $y = f(x) = 2x + 3$ AND $x = g(y) = \dfrac{y - 3}{2}$

x	$y = f(x) = 2x + 3$	$g(y) = \dfrac{y - 3}{2}$	Point in the Graph of f	Point in the Graph of g
-10	-17	-10	$(-10, -17)$	$(-17, -10)$
-5	-7	-5	$(-5, -7)$	$(-7, -5)$
-3	-3	-3	$(-3, -3)$	$(-3, -3)$
-1	1	-1	$(-1, 1)$	$(1, -1)$
0	3	0	$(0, 3)$	$(3, 0)$
1	5	1	$(1, 5)$	$(5, 1)$
3	9	3	$(3, 9)$	$(9, 3)$
5	13	5	$(5, 13)$	$(13, 5)$
10	23	10	$(10, 23)$	$(23, 10)$
$-\frac{3}{2}$	0	$-\frac{3}{2}$	$(-\frac{3}{2}, 0)$	$(0, -\frac{3}{2})$

We generalize the result of Example 1:

If f and g are inverse functions, then
(x, y) is in the graph of f if and only if (y, x) is in the graph of g.

EXAMPLE 2
THE INVERSE OF A CUBIC AND A
CUBE-ROOT FUNCTION

$f(x) = x^3$ and $g(x) = \sqrt[3]{x} = x^{1/3}$ are inverses because

$$f(g(x)) = f(x^{1/3}) = (x^{1/3})^3 = x$$

and

$$g(f(x)) = g(x^3) = (x^3)^{1/3} = x.$$

We do not have to worry about domains here because both x^3 and $\sqrt[3]{x}$ are defined for every real number x.

In Table 2 we list points in the graphs of f and g.

TABLE 2 POINTS IN THE GRAPHS OF $y = f(x) = x^3$ AND $x = g(y) = \sqrt[3]{y}$

x	$y = f(x) = x^3$	$g(y) = \sqrt[3]{y}$	Point in the Graph of f	Point in the Graph of g
-4	-64	-4	$(-4, -64)$	$(-64, -4)$
-3	-27	-3	$(-3, -27)$	$(-27, -3)$
-2	-8	-2	$(-2, -8)$	$(-8, -2)$
-1	-1	-1	$(-1, -1)$	$(-1, -1)$
-0.5	-0.125	-0.5	$(-0.5, -0.125)$	$(-0.125, -0.5)$
0	0	0	$(0, 0)$	$(0, 0)$
1	1	1	$(1, 1)$	$(1, 1)$
2	8	2	$(2, 8)$	$(8, 2)$
3	27	3	$(3, 27)$	$(27, 3)$
4	64	4	$(4, 64)$	$(64, 4)$

TO COMPUTE AN INVERSE FUNCTION, WE USE THE FOLLOWING PROCEDURE:

(i) Replace $f(x)$ with y.
(ii) Interchange x and y.
(iii) Solve for y in terms of x, if possible.
(iv) The resulting y is equal to $f^{-1}(x)$.

We use the phrase "if possible" in step (iii) because it is not always possible to solve explicitly for y as a *function* of x.

EXAMPLE 3
THE INVERSE OF A RATIONAL
FUNCTION

Find the inverse of $f(x) = \dfrac{3x}{1 - 2x}$.

SOLUTION: dom $f = \{x \neq \frac{1}{2}\}$ so we must avoid the value $\frac{1}{2}$. For $x \neq \frac{1}{2}$, we write:

(i) $y = \dfrac{3x}{1 - 2x}$

(ii) $x = \dfrac{3y}{1 - 2y}$ We interchanged x and y

(iii) $x(1 - 2y) = 3y$

$$x - 2xy = 3y$$

$$x = 2xy + 3y = y(2x + 3)$$

$$y = \frac{x}{2x + 3}.$$

Thus

$$g(x) = \frac{x}{2x + 3} \text{ is the inverse of } f \quad \text{and} \quad \text{dom } g = \left\{x\colon x \neq -\frac{3}{2}\right\}.$$

CHECK:

$$g(y) = \frac{y}{2y + 3} \qquad \begin{array}{l}\text{Multiply numerator}\\\text{and denominator}\\\text{by } 1 - 2x\end{array}$$

$$g(f(x)) = g\left(\frac{3x}{1 - 2x}\right) = \frac{\dfrac{3x}{1 - 2x}}{2\left(\dfrac{3x}{1 - 2x}\right) + 3} = \frac{3x}{2(3x) + 3(1 - 2x)}$$

$$= \frac{3x}{6x + 3 - 6x} = \frac{3x}{3} = x.$$

$$f(g(x)) = f\left(\frac{x}{2x + 3}\right) = \frac{3\left(\dfrac{x}{2x + 3}\right)}{1 - 2\left(\dfrac{x}{2x + 3}\right)} = \frac{3x}{(2x + 3) - 2x} = \frac{3x}{3} = x.$$

Note, for example, that

$$f(1) = -3 \quad \text{and} \quad g(-3) = 1,$$

so

$$(1, -3) \text{ is in the graph of } f \quad \text{and} \quad (-3, 1) \text{ is in the graph of } g.$$

EXAMPLE 4

A QUADRATIC FUNCTION THAT
HAS NO INVERSE

Let $y = x^2$. Then $x = \pm\sqrt{y}$. This means that each value of y comes from *two* different values of x. See Figure 2. We recall from Section 0.4 that a function $y = f(x)$ can be thought of as a rule that assigns to each x in its domain a *unique* value of y. Suppose we try to define the inverse of x^2 by taking the positive square root. Then we would like to be able to show that $f = x^2$ and $g = \sqrt{x}$ are inverses. Let $x = -2$, for example. Then

$$g(f(x)) = g(f(-2)) = \sqrt{(-2)^2} = \sqrt{4} = 2 = -(-2) = -x \neq x.$$

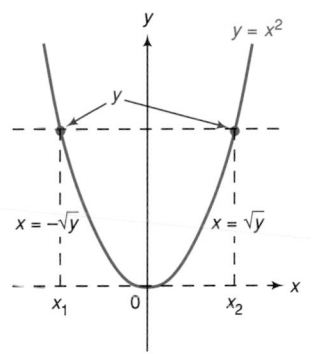

FIGURE 2
ONE y-VALUE COMES FROM TWO
DIFFERENT x-VALUES.

If x is a real number, then

$$g(f(x)) = g(x^2) = \sqrt{x^2} = |x| = \begin{cases} x, & x \geq 0 \\ -x, & x < 0 \end{cases}$$

which is not equal to x if $x < 0$. Thus, f and g are not inverses of one another.

To avoid problems like the one just encountered, we make the following definition.

DEFINITION   **ONE-TO-ONE FUNCTION**

On the interval $[a, b]$, the function $y = f(x)$ is **one-to-one**, written 1–1, if $x_1, x_2 \in [a, b]$ and $x_1 \neq x_2$ implies that $f(x_1) \neq f(x_2)$. That is, each value of $f(x)$ comes from only one value of x. ■

The definition of a 1–1 function is illustrated in Figure 3.

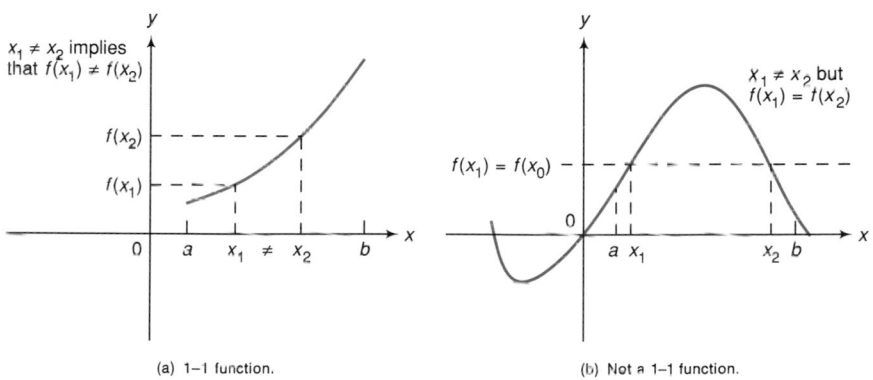

(a) 1–1 function.   (b) Not a 1–1 function.

FIGURE 3   ILLUSTRATION OF ONE-TO-ONE-NESS

This figure suggests a graphical method to determine whether a function is 1–1 or not.

HORIZONTAL-LINES TEST

If every line parallel to the x-axis intersects the graph of f in at most one point, then f is one-to-one.

NOTE: Since f is a function, the vertical-lines test on page 30 must also hold.

EXAMPLE 5
ILLUSTRATION OF
HORIZONTAL-LINES TEST

We can see, in Figure 4, that $f(x) = x^2$ is not one-to-one, but $f(x) = x^3$ and $f(x) = \dfrac{1}{x}$ are one-to-one.

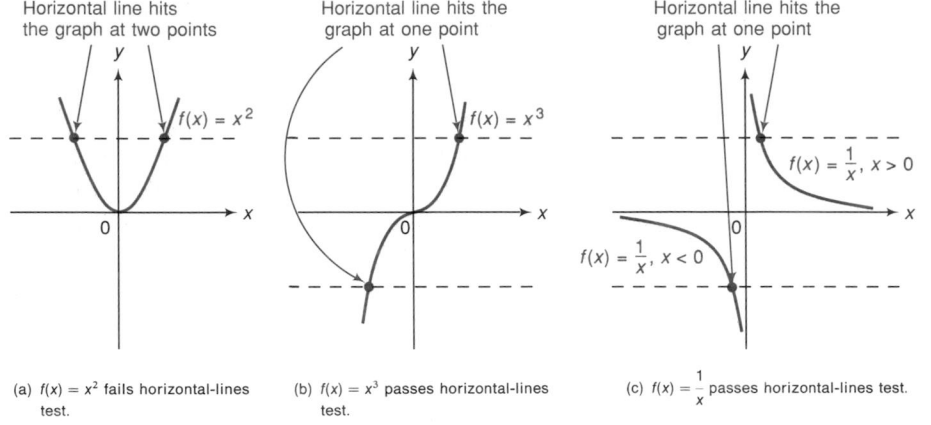

(a) $f(x) = x^2$ fails horizontal-lines test.

(b) $f(x) = x^3$ passes horizontal-lines test.

(c) $f(x) = \dfrac{1}{x}$ passes horizontal-lines test.

FIGURE 4   ILLUSTRATION OF HORIZONTAL-LINES TEST

The next two theorems help us to determine when a function is 1–1.

THEOREM 1   **AN INCREASING OR DECREASING FUNCTION IS 1–1**

If f is either an increasing or a decreasing function on $[a, b]$, then f is 1–1 on $[a, b]$.

PROOF: Suppose f is increasing and suppose that $x_1 \neq x_2$. If $x_1 > x_2$, then $f(x_1) > f(x_2)$. If $x_2 > x_1$, then $f(x_2) > f(x_1)$. In either case $f(x_1) \neq f(x_2)$, so f is 1–1.

The proof in the case when f is decreasing is similar.   ■

THEOREM 2   **A FUNCTION ON WHOSE DERIVATIVE IS NONZERO ON AN INTERVAL IS 1–1 THERE**

If f is continuous on $[a, b]$ and has a derivative on (a, b) satisfying $f'(x) \neq 0$ for every x in (a, b), then f is 1–1 on $[a, b]$.

PROOF: Suppose there are numbers x_1 and x_2 in $[a, b]$ such that $x_1 \neq x_2$, but $f(x_1) = f(x_2)$. Then, by the mean-value theorem, there is a number c in (x_1, x_2) such that

$$0 = f(x_2) - f(x_1) = f'(c)(x_2 - x_1).$$

Since $(x_2 - x_1) \neq 0$, this implies that $f'(c) = 0$, which contradicts the hypothesis that $f'(x) \neq 0$ on (a, b). Thus, if $f(x_1) = f(x_2)$, we must have $x_1 = x_2$, and f is 1–1.   ■

REMARK: There are 1–1 functions for which the condition $f'(x) \neq 0$ does not hold. For example, the function x^3 is 1–1 in $[-1, 1]$ but $f'(0) = 0$.

Why are we interested in 1–1 functions? Because if $y = f(x)$ is 1–1, then every value of y comes from a unique value of x, so that we can write x as a function of y, and this new function is the inverse of y. That is,

every 1–1 function has an inverse.

EXAMPLE 6
THE INVERSE OF A
SQUARE-ROOT FUNCTION

Let $f(x) = \sqrt{x}$. Then $f'(x) = 1/(2\sqrt{x}) \neq 0$ for $x > 0$. Thus for $x \geq 0$, f has an inverse, which we find by setting $y = \sqrt{x}$, so that $x = y^2$ and therefore $f^{-1}(x) = x^2$. Note that since dom $f^{-1} = \{x: x \geq 0\}$, f^{-1} is *not* the function discussed in Example 2 (x^2 with domain $\mathbb{R}$).

EXAMPLE 7
SHOWING THAT A FUNCTION
HAS AN INVERSE EVEN THOUGH
WE CAN'T COMPUTE IT

Let $f(x) = x^5 + x^3 + x$. Then $f'(x) = 5x^4 + 3x^2 + 1$, which is always positive, so f^{-1} exists. However, there is no way to solve explicitly the equation $y = x^5 + x^3 + x$ for x. This result does not contradict the existence of f^{-1}. It simply illustrates the limitations of algebra.

It turns out that inverses of continuous functions are continuous and that inverses of differentiable functions with nonzero derivatives are differentiable. The following theorem is proved in Appendix 3.

THEOREM 3 **CONTINUITY OF INVERSE FUNCTIONS**

Let f be continuous on $[a, b]$ and let $c = f(a)$ and $d = f(b)$. If f has an inverse $g = f^{-1}$, then g is continuous on $[c, d]$ (or $[d, c]$ if $c > d$). ∎

We now prove a theorem that shows a relationship between the derivatives of functions and the derivatives of their inverses.

THEOREM 4 **DIFFERENTIABILITY OF INVERSE FUNCTIONS**

Suppose that $y = f(x)$ is differentiable and 1–1 in some open interval containing x_0 and that $f'(x_0) \neq 0$. If $x = g(y) = f^{-1}(y)$ is an inverse function of $f(x)$ that is continuous in a neighborhood of $y_0 = f(x_0)$, then $f^{-1}(y) = g(y)$ is differentiable at y_0, and

$$g'(y_0) = \frac{1}{f'(x_0)} \qquad \text{or equivalently,} \qquad \frac{dx}{dy} = \frac{1}{dy/dx}. \qquad (4)$$

PROOF: Since

$$\lim_{\Delta x \to 0} \frac{f(x_0 + \Delta x) - f(x_0)}{\Delta x} = f'(x_0) \neq 0,$$

we have

$$\lim_{\Delta x \to 0} \frac{\Delta x}{f(x_0 + \Delta x) - f(x)} = \lim_{\Delta x \to 0} \frac{1}{\dfrac{f(x_0 + \Delta x) - f(x_0)}{\Delta x}}$$

$$\begin{array}{c} \text{Quotient Theorem} \\ \text{for limits} \\ \text{(p. 71)} \end{array} \longrightarrow = \frac{\displaystyle\lim_{\Delta x \to 0} 1}{\displaystyle\lim_{\Delta x \to 0}\left[\dfrac{f(x_0 + \Delta x) - f(x_0)}{\Delta x}\right]} = \frac{1}{f'(x_0)}.$$

(5)

Because $x_0 = g(y_0)$, and because $\Delta y = f(x_0 + \Delta x) - f(x_0) = f(x_0 + \Delta x) - y_0$, we have $y_0 + \Delta y = f(x_0 + \Delta x)$. Hence

$$g(y_0 + \Delta y) = g(f(x_0 + \Delta x)) \overset{g(f(x)) = x}{=} x_0 + \Delta x \overset{x = g(y)}{=} g(y_0) + \Delta x$$

so

$$\Delta x = g(y_0 + \Delta y) - g(y_0) \qquad \text{and}$$

$$\lim_{\Delta y \to 0} \Delta x = \lim_{\Delta y \to 0} [g(y_0 + \Delta y) - g(y_0)] \overset{\text{Since } g \text{ is continuous at } y_0}{=} 0. \tag{6}$$

Finally, since $y_0 = f(x_0)$, we find that

$$
\begin{aligned}
g'(y_0) &= \lim_{\Delta y \to 0} \frac{g(y_0 + \Delta y) - g(y_0)}{\Delta y} \\
&= \lim_{\Delta y \to 0} \frac{(x_0 + \Delta x) - x_0}{f(x_0 + \Delta x) - f(x_0)} \\
&= \lim_{\Delta y \to 0} \frac{\Delta x}{f(x_0 + \Delta x) - f(x_0)} \\
&\overset{\text{From (6)}}{=} \lim_{\Delta x \to 0} \frac{\Delta x}{f(x_0 + \Delta x) - f(x_0)} \overset{\text{From (5)}}{=} \frac{1}{f'(x_0)}.
\end{aligned}
$$

The proof of the theorem is complete. ∎

NOTE: It can be shown (see Problem 25) that if $f'(x_0) \neq 0$ and f' is continuous, then $f(x)$ is 1–1 in a neighborhood of x_0; thus the assumption that f is 1–1 is not really necessary in the statement of the theorem.

REMARK: It is tempting to prove Theorem 4 by using the chain rule. Since $g(f(x)) = x$, we have

$$\frac{d}{dx} g(f(x)) = \frac{d}{dx} x$$

$$g'(f(x))f'(x) = 1, \qquad \text{By the chain rule}$$

or

$$g'(y) = g'(f(x)) = \frac{1}{f'(x)}.$$

This proof is certainly easier, but it is incomplete. Do you see why? In using the chain rule, we assume that $g'(f(x)) = dg/dy$ exists. But this assumption is part of what we need to prove! So we must resort to the longer (but correct) proof.

EXAMPLE 8
THE DERIVATIVE OF AN INVERSE
FUNCTION

Let $y = \sqrt{x}$. Then $dy/dx = 1/(2\sqrt{x})$. The inverse function is $x = g(y) = y^2$ for $y > 0$. Then,

$$\frac{dx}{dy} = 2y = 2\sqrt{x} = \frac{1}{1/(2\sqrt{x})} = \frac{1}{dy/dx}.$$

EXAMPLE 9

CALCULATING THE DERIVATIVE OF AN INVERSE FUNCTION EVEN THOUGH WE CAN'T FIND THE INVERSE

If $y = x^5 + x^3 + x$, then as we saw in Example 7, f^{-1} exists but cannot be explicitly calculated. Nevertheless, we can compute

$$\frac{df^{-1}}{dy} = \frac{dx}{dy} = \frac{1}{dy/dx} = \frac{1}{5x^4 + 3x^2 + 1}.$$

This answer is not exactly what we want since we would like to calculate df^{-1}/dy as a function of y, not of x. Nevertheless, it *is* quite useful since, for example, if we knew a point (x_0, y_0) on the curve $x = f^{-1}(y)$, we could calculate the derivative of f^{-1} at that point. In addition, given a value of y, we can find x (by Newton's method) and, hence, we can compute df^{-1}/dy to as many decimal places of accuracy as we require.

We close this section by showing the relationship between the graph of a function f and the graph of its inverse function f^{-1}. The graphs of three functions and their inverses are sketched in Figure 5. It appears that the graphs of f and f^{-1} are symmetric about the line $y = x$. This should not be surprising.

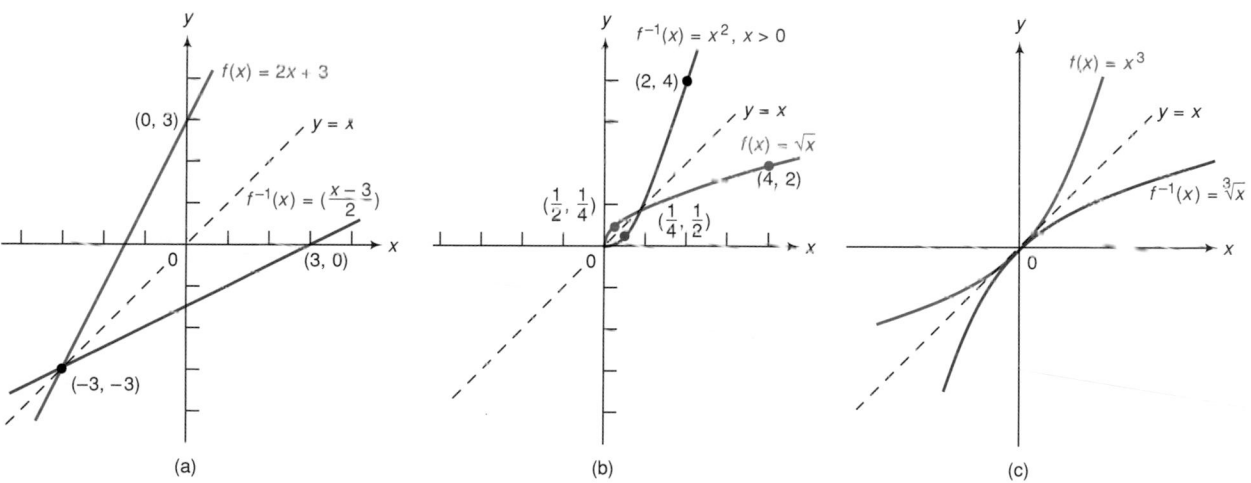

FIGURE 5 ILLUSTRATION OF THE REFLECTION PROPERTY OF INVERSE FUNCTIONS

Suppose that $y = f(x)$. If f^{-1} exists, then $x = f^{-1}(y)$. That is, (x, y) is in the graph of f if and only if (y, x) is in the graph of f^{-1}. In Figure 5(b), for example, $(\frac{1}{4}, \frac{1}{2})$, $(1, 1)$, $(4, 2)$, and $(9, 3)$ are in the graph of $f(x) = \sqrt{x}$. The points $(\frac{1}{2}, \frac{1}{4})$, $(1, 1)$, $(2, 4)$, and $(3, 9)$ are in the graph of $f^{-1}(x) = x^2$, $x \geq 0$. In Figure 5(b), if we fold the page along the line $y = x$, we find that the graphs of f and f^{-1} coincide. We conclude the following:

REFLECTION PROPERTY OF INVERSE FUNCTIONS

The graphs of f and f^{-1} are **reflections** of one another about the line

$$y = x. \tag{7}$$

This **reflection property**, as it is called, of the graphs of inverse functions enables us immediately to obtain the graph of f^{-1} once the graph of f is known.

■ PROBLEMS 6.1

■ SELF-QUIZ

I. Let $f(x) = 2x - 1$ and $g(w) = (w + 1)/2$, both with domain $(-\infty, \infty)$. Then g is the _____ function of f.
  a. reciprocal   b. opposite
  c. reflection   d. inverse

II. The function $f(x) = x^2$ is one-to-one on the interval $[0, 3]$ because _____.
  a. $f(1) = 1$
  b. each value of $f(x)$ corresponds to a unique value of x
  c. $f(x)$ equals 1 only when $x = 1$
  d. each value of x yields a single number $f(x)$

III. Answer True or False to each of the following assertions.
  a. If the function f has domain $(-3, 5)$ and is one-to-one, then every horizontal line meets the graph $y = f(x)$ exactly once.
  b. If the function f has domain $(-\infty, \infty)$ and is one-to-one, then every horizontal line meets the graph $y = f(x)$ exactly once.
  c. If the function f is one-to-one, then every horizontal line meets the graph $y = f(x)$ at most once.
  d. If the function f is one-to-one and if the range of f is $(-\infty, \infty)$, then every horizontal line meets the graph $y = f(x)$ exactly once.
  e. If every horizontal line meets the graph of $y = f(x)$ at most once, then f is one-to-one.
  f. If every horizontal line meets the graph of

$y = f(x)$ at most once, then f is an increasing function.
  g. If f is one-to-one, then f is decreasing.
  h. If f is one-to-one and if f is continuous, then f is either always increasing or always decreasing.

IV. The function $g(x) = 5 - x^2$ is not one-to-one on the interval $[-3, 3]$ because _____.
  a. $g(1) \neq 1$
  b. $g(2) = 1$ but $2 \neq 1$
  c. $g(-2) = g(2)$ and $-2 \neq 2$
  d. g is not an increasing function

V. Let $f(x) = -x$. The inverse function of f is _____.
  a. $A(x) = -x$
  b. $B(x) = x$
  c. $C(x) = 1/x$
  d. function f does not have an inverse

VI. If $F(g(x)) = x$ for all x, then the graph of $y = F(x)$ can be obtained by _____.
  a. rotating the graph of $y = g(x)$ one-quarter turn counterclockwise
  b. rotating the graph of $y = F(x)$ one-half turn clockwise
  c. reflecting the graph of $y = g(x)$ about the line $y = x$
  d. reflecting the graph of $y = g(x)$ about the y-axis

■ DRILL

In Problems 1–17, determine intervals over which the given function is 1–1, and, where it exists, find the inverse function over each interval. Then use Theorem 4 to calculate the derivative of the inverse function.

1. $f(x) = 3x + 5$ 2. $f(x) = 17 - 2x$

3. $f(x) = \dfrac{1}{x}$ 4. $f(x) = \dfrac{1}{x - 2}$

5. $f(x) = \sqrt{4x + 3}$ 6. $f(x) = (x + 1)^2$

7. $f(x) = 1 - x^3$ 8. $f(x) = \sqrt[3]{x + 1}$

9. $y = \dfrac{x + 1}{x}$ 10. $y = \dfrac{x}{x + 1}$

11. $y = (x - 2)(x - 3)$ 12. $y = x^2 + 4x + 3$

13. $y = x^3 + x$

*14. $y = x^4 + 10x^2 + 25$
  [*Hint:* First solve for x^2, then for x.]

15. $y = \cos x$ (for $x \in [0, \pi]$)

*16. $y = \sin x$ (for $x \in [0, \pi]$)

*17. $y = \tan x \left(\text{for } x \in \left(-\dfrac{\pi}{2}, \dfrac{\pi}{2} \right) \right)$

In Problems 18–22, the graph of a function over an interval $[a, b]$ is given. Sketch the graph of f^{-1} on the interval $[c, d] = [f(a), f(b)]$ or $[f(b), f(a)]$, using the reflection property.

18.

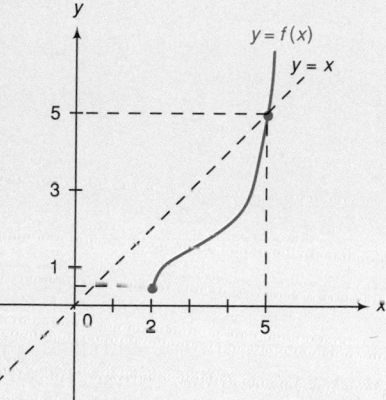

19.

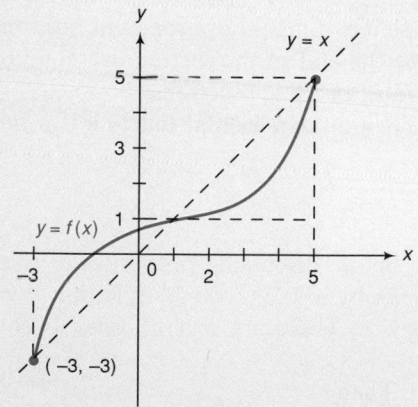

20.

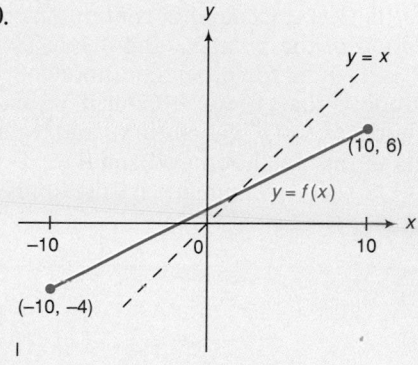

21.

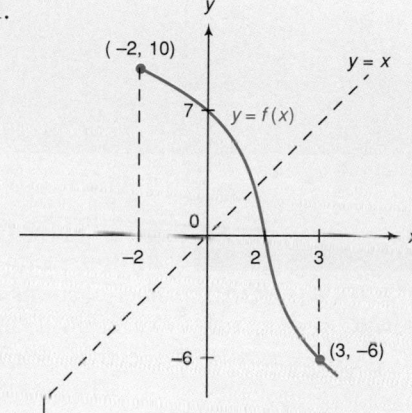

22.

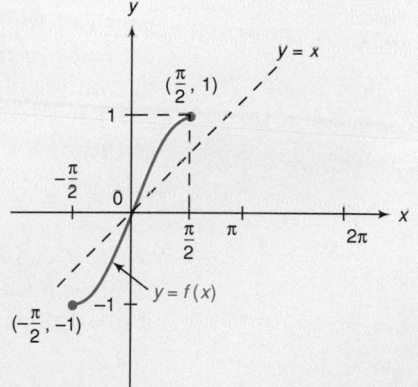

■ **SHOW/PROVE/DISPROVE**

23. Prove that

$$\frac{d}{dx} x^{1/n} = \frac{1}{n} x^{(1/n)-1}$$

if n is a positive integer. [*Hint:* Let $f(x) = x^n$. Then find f^{-1} and use Theorem 4.]

*24. Suppose that f is continuous and strictly increasing with domain $[a, b]$. Prove the following:
a. range of f is $[f(a), f(b)]$
b. f^{-1} exists and its domain is $[f(a), f(b)]$
c. f^{-1} is an increasing function
In each proof, indicate where you use the assumption that f is continuous.

■ **CHALLENGE**

*25. Prove that if $f'(x)$ exists and is continuous in a neighborhood of the point x_0 and if $f'(x_0) \neq 0$, then $f(x)$ is 1–1 in some neighborhood of x_0. [*Hint:* Suppose that $f'(x_0) > 0$. Then in some neighborhood of x_0, $f'(x) > 0$. If x_1 and x_2 are two points in this neighborhood and if $x_1 > x_2$, then $f(x_1) > f(x_2)$. A similar result holds if $f'(x_0) < 0$.]

26. **Young's inequality: Suppose that f is strictly increasing and continuous on $[0, a]$; also suppose that $f(0) = 0$. Let $g = f^{-1}$. If $0 \leq w \leq a$ and $0 \leq h \leq f(a)$, show that

$$\int_0^w f + \int_0^h g \geq w \cdot h.$$

■ **ANSWERS TO SELF-QUIZ**

I. d (The *graphs* of a pair of inverse functions are reflections of one another about the line $y = x$.)

II. b

III. a. False b. False c. True
      d. True   e. True   f. False
      g. False h. True

IV. c (Let $G(x) = 5 - x$ for an example showing that (d) is not a correct answer.)

V. a ($A(f(x)) = -(-x) = x$ and $f(A(x)) = x$ for all x.)

VI. c

6.2 REVIEW OF EXPONENTIAL AND LOGARITHMIC FUNCTIONS

In this section, we provide a brief summary of exponential and logarithmic functions as they are described in a precalculus course. In Section 6.3, we begin a discussion of these functions from a completely different perspective—starting with the fundamental theorem of calculus.

The material here is not used or needed elsewhere in the book. It appears here to remind you of the important properties of these functions and to help you appreciate how the completely different approach in Sections 6.3–6.5 leads to the same functions. At the end of the section, we compute the derivatives of the two most important of these functions.

Let a be a positive number. Then an **exponential function** is a function of the form

$$f(x) = a^x \tag{1}$$

where x can be any real number. Before discussing properties of this function, we must explain what we mean by a^x. (We will assume that $a \neq 1$, for if $a = 1$, then $a^x = 1^x = 1$ for every x.) There are several cases to consider.

Case 1. $x = n$, a positive integer. Then

$$a^x = a^n = \underbrace{a \cdot a \cdot a \cdots \cdot a}_{n \text{ times}}. \tag{2}$$

This expression is, of course, the familiar rule for raising a number to the nth power.

Case 2. $x = -n$, a negative integer. Then

$$a^x = a^{-n} = \frac{1}{a^n}. \tag{3}$$

Case 3. $x = 1/n$, where n is a positive integer. Then

$$a^x = a^{1/n} = \text{the } n\text{th root of } a. \tag{4}$$

That is, $a^{1/n} = b$ is equivalent to $a = b^n$. For example, $64^{1/6} = 2$ since $2^6 = 64$.

In the last two cases we see why it is necessary to require that $a > 0$. If $a = 0$, then $a^{-n} = 1/a^n = 1/0$ is not defined. If $a < 0$, then $a^{1/2}$, $a^{1/4}$, $a^{1/6}$, and so on are not defined [for example, look at $(-2)^{1/2} = \sqrt{-2}$, which is not a real number].

Case 4. $x = -1/n$, n is a positive integer. Then

$$a^x = a^{-1/n} = \frac{1}{a^{1/n}}. \tag{5}$$

Case 5. x is a rational number. Then $x = m/n$, where m and n are integers, and

$$a^x = a^{m/n} = (a^{1/n})^m. \tag{6}$$

These rules take care of many cases of interest. But how do we interpret expressions like $a^{\sqrt{2}}$ and a^π? To do so, we make use of a fundamental fact of mathematical analysis that any irrational number can be approximated as closely as desired by a rational number.

Case 6. x is an irrational number. Then

Definition of a^x for x irrational

$$a^x = \lim_{r \to x} a^r, \qquad r \text{ a rational number.} \tag{7}$$

This definition makes sense because it can be shown that no matter how we choose the rational numbers r that approach x, we get the same limit in (7). Once we show that, it follows that

$$a^x \text{ is continuous for all real } x.^\dagger$$

(In Section 6.5, we will define a^x in a different way. Using the alternative definition, it will be immediately clear that a^x is differentiable and is therefore continuous.)

The basic properties of exponential functions are contained in Theorem 1.

THEOREM 1   **PROPERTIES OF EXPONENTIAL FUNCTIONS**

Let $a > 0$ and let x and y be any real numbers. Then:

(i) $a^x > 0$.

(ii) $a^{-x} = \dfrac{1}{a^x}.$

---

† Proof of these facts can be found in Mary P. Dolciani *et al.*, *Modern Introductory Analysis* (Boston: Houghton Mifflin, 1967), pp. 334–337.

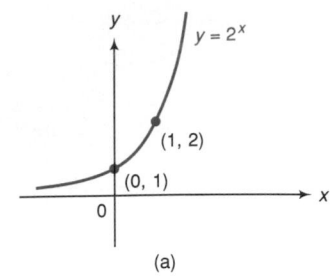

(a)

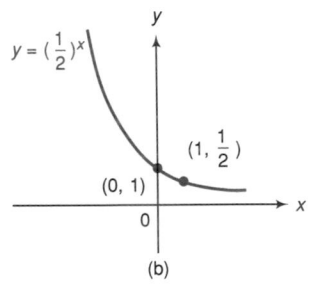

(b)

FIGURE 1

GRAPHS OF 2^x AND $\left(\dfrac{1}{2}\right)^x$

(iii) $a^{x+y} = a^x \cdot a^y$.

(iv) $a^{x-y} = a^x \cdot a^{-y} = \dfrac{a^x}{a^y}$.

(v) $a^0 = a^{x-x} = \dfrac{a^x}{a^x} = 1$.

(vi) $(a^x)^y = a^{xy}$.

(vii) If $a > 1$, then a^x is an increasing function in any interval.

(viii) If $0 < a < 1$, then a^x is a decreasing function in any interval.

(ix) a^x is continuous for every x. ∎

We will not prove this theorem here. The first six properties are familiar if x and y are integers. If x and y are rational numbers, then it is not difficult to show that properties (i)–(vi) hold by using the fact that they hold for integers. Then since (i)–(vi) are true for rational numbers, we can prove them for irrational numbers by taking limits. Properties (vii), (viii), and (ix) will follow easily once we have calculated the derivative of a^x. However, it is easy to see that these facts are true by examining the cases $a > 1$ and $a < 1$ separately and observing what happens as x grows. (Try $a = 2$ and $a = \frac{1}{2}$.) The graphs of $y = 2^x$ and $y = (\frac{1}{2})^x$ are given in Figure 1.

If $y = a^x$, we may ask the following question: Which value of x corresponds to a given y?

EXAMPLE 1
SOLVING FOR THE EXPONENT

Solve for x:

(a) $2^x = 8$ **(b)** $3^x = \frac{1}{9}$ **(c)** $(\frac{1}{2})^x = 8$

SOLUTION: (a) It is clear that $2^x = 8$ if and only if $x = 3$. Similarly, in (b) $3^x = \frac{1}{9}$ if and only if $x = -2$, and in (c), $(\frac{1}{2})^x = 8$ if and only if $x = -3$. $\left[(\frac{1}{2})^3 = \frac{1}{8}.\right]$

From Theorem 1, parts (vii) and (viii), we know that a^x is either an increasing function (if $a > 1$) or a decreasing function (if $0 < a < 1$). Thus from Theorem 6.1.1 we know that

a^x is 1–1 if $a > 0$ and $a \neq 1$.

We conclude that

a^x has an inverse if $a > 0$ and $a \neq 1$.

The inverse of a^x is called the *logarithm to the base a*.

DEFINITION   **LOGARITHM TO THE BASE *a***

If $x = a^y$, then the **logarithm to the base *a*** of x is y, written as

$y = \log_a x$. ∎

By the reflection property discussed in Section 6.1 (p. 393), the graph of $y = \log_a x$ is the reflection of the graph of $y = a^x$ about the line $y = x$. In Figure 2, typical graphs for $y = a^x$ and $y = \log_a x$ are given for $a > 1$.

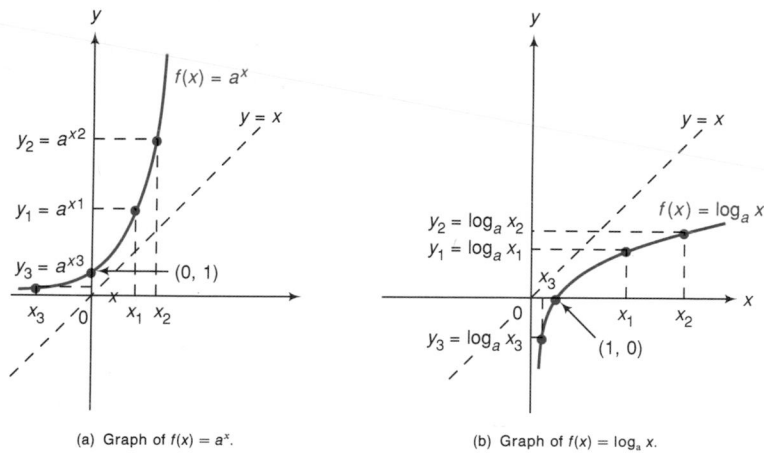

(a) Graph of $f(x) = a^x$.

(b) Graph of $f(x) = \log_a x$.

FIGURE 2 THE GRAPH OF $\log_a x$ IS OBTAINED BY REFLECTING THE GRAPH OF a^x ABOUT THE LINE $y = x$ (AND VICE VERSA).

The logarithmic function $f(x) = \log_a x$ often causes a lot of confusion when it is first encountered. It might help to think of the term $\log_a x$ as asking a question:

$\log_a x$ asks the question: to what power must we raise a in order to get x?

EXAMPLE 2
CALCULATING SIX LOGARITHMS

Compute:

(a) $\log_2 8$ (b) $\log_3 \dfrac{1}{9}$ (c) $\log_{1/2} 8$

(d) $\log_{10} 10{,}000$ (e) $\log_2 2^{3.7}$ (f) $4^{\log_4 11.5}$

SOLUTION: We calculated the first three in Example 1. We redo them here.

(a) $\log_2 8$ asks the question: to what power must we raise 2 to get 8? Since $2^3 = 8$, the answer is $\log_2 8 = 3$.

(b) $\log_3 \frac{1}{9}$ asks the question: to what power must we raise 3 to get $\frac{1}{9}$? Since $3^2 = 9$, $3^{-2} = \dfrac{1}{3^2} = \dfrac{1}{9}$. Thus the answer is -2; that is, $\log_3 \frac{1}{9} = -2$.

(c) $\log_{1/2} 8 = -3$ because $\left(\dfrac{1}{2}\right)^{-3} = \dfrac{1}{(\frac{1}{2})^3} = \dfrac{1}{\frac{1}{8}} = 8$.

(d) $\log_{10} 10{,}000 = 4$ because $10^4 = 10{,}000$.

(e) $\log_2 2^{3.7}$ asks the question: to what power must we raise 2 to get $2^{3.7}$? The question practically answers itself. If we raise 2 to the power 3.7, we get $2^{3.7}$ (read that again). Thus $\log_2 2^{3.7} = 3.7$.

(f) $\log_4 11.5$ asks the question: to what power must we raise 4 in order to get 11.5? We don't know the answer. However, when we write $4^{\log_4 11.5}$, we *are* raising 4 to that unknown power. Thus $4^{\log_4 11.5} = 11.5$. Let us put this another way: if $u = \log_4 11.5$, then $4^u = 11.5$, by definition. Thus,

$$4^{\log_4 11.5} = 4^u = 11.5.$$

We summarize the basic properties of logarithms discussed so far.

FOUR BASIC PROPERTIES OF LOGARITHMS

$y = \log_a x$ means that $x = a^y$.

$y = a^x$ means that $x = \log_a y$.

$\log_a a^x = x$ for every real number x.   See Example 2(e)

$a^{\log_a x} = x$ for every positive real number x.   See Example 2(f)

Remember that $\log_a x$ is only defined for $x > 0$ since it is impossible to raise a positive number to a power and obtain a number less than or equal to 0.

EXAMPLE 3
ILLUSTRATION OF THE FACT THAT a^x AND $\log_a x$ ARE INVERSES OF ONE ANOTHER

If $f(x) = a^x$ and $g(x) = \log_a x$, then $f(g(x)) = a^{\log_a x} = x$ and $g(f(x)) = \log_a a^x = x$, so that a^x and $\log_a x$ are indeed inverses.

NOTE: This example illustrates that $f^{-1}(f(x))$ and $f(f^{-1}(x))$ are *not* the same functions. If $f = a^x$ and $f^{-1} = \log_a x$, then $f(f^{-1}(x)) = a^{\log_a x}$ and $f^{-1}(f(x)) = \log_a a^x$. The first function is defined only when $x > 0$, and the second is defined for *all* x. The two functions have different domains and are therefore different—even though each takes the value x wherever it is defined.

Properties of the function $\log_a x$ are summarized in the next theorem.

THEOREM 2 **PROPERTIES OF $\log_a x$**

Let $a > 0$ ($a \neq 1$) and let x and y be positive.

(i) $\log_a xy = \log_a x + \log_a y$.

(ii) $\log_a(x/y) = \log_a x - \log_a y$.

(iii) $\log_a 1 = 0$.

(iv) $\log_a(1/x) = -\log_a x$.

(v) $\log_a a = 1$.

(vi) $\log_a x^y = y \log_a x$. Here y can be any real number.

(vii) $\log_a b = 1/(\log_b a)$, $b > 0$, $b \neq 1$.

(viii) If $a > 1$, $\log_a x$ is an increasing function of x for $x > 0$.

(ix) If $0 < a < 1$, $\log_a x$ is a decreasing function of x for $x > 0$.

(x) $f(x) = \log_a x$ is continuous for every $x > 0$.

PROOF:

(i) Let $u = \log_a x$ and $v = \log_a y$. Then $a^u = x$ and $a^v = y$, so $xy = a^u \cdot a^v = a^{u+v}$. By definition of the logarithm, we have $\log_a xy = u + v$, and this proves (i).

(ii) With u and v as above, $x/y = a^u/a^v = a^{u-v}$, so $\log_a(x/y) = u - v$, which proves (ii).

(iii) Since $a^0 = 1$, $0 = \log_a 1$.

(iv) Follows from (ii) and (iii).

(v) Since $a^1 = a$, $1 = \log_a a$.

(vi) Let $u = \log_a x$. Then $x = a^u$ and $x^y = (a^u)^y$. Thus $\log_a x^y = uy$, which proves (vi).

(vii) Let $u = \log_a b$ for $b > 0$. Then $a^u = b$. Now taking logarithms with respect to b, we have $\log_b a^u = \log_b b = 1$. But $\log_b a^u = u \log_b a = (\log_a b)(\log_b a) = 1$, which proves (vii).

(viii) and **(ix)** follow easily once we have calculated the derivative of $\log_a x$.

(x) $\log_a x$ is the inverse of the continuous function a^x and so is continuous by Theorem 6.1.3. ∎

EXAMPLE 4
SIMPLIFYING A LOGARITHM

Write $\log_a \dfrac{x^4 y^6}{\sqrt{x^{2/3} y^8}}$ in terms of $\log_a x$ and $\log_a y$. Assume that $x > 0$ and $y > 0$.

SOLUTION:

$$\log_a \frac{x^4 y^6}{\sqrt{x^{2/3} y^8}} = \log_a x^4 y^6 - \log_a \sqrt{x^{2/3} y^8} \qquad \text{Property ii}$$

$$= \log_a x^4 + \log_a y^6 - \log_a (x^{2/3} y^8)^{1/2}$$

$$= 4 \log_a x + 6 \log_a y - \frac{1}{2} \log_a x^{2/3} y^8 \qquad \text{Property vi}$$

$$= 4 \log_a x + 6 \log_a y - \frac{1}{2}(\log_a x^{2/3} + \log_a y^8) \qquad \text{Property i}$$

$$= 4 \log_a x + 6 \log_a y - \frac{1}{2}\left(\frac{2}{3} \log_a x + 8 \log_a y\right) \qquad \text{Property vi}$$

$$= 4 \log_a x + 6 \log_a y - \frac{1}{3} \log_a x - 4 \log_a y$$

$$= \frac{11}{3} \log_a x + 2 \log_a y$$

EXAMPLE 5
COMBINING LOGARITHMS

Write the following expression as a single logarithm:

$$\frac{1}{2} \log_a x + 4 \log_a y - 3 \log_a z$$

SOLUTION:

$$\frac{1}{2} \log_a x + 4 \log_a y - 3 \log_a z = \log_a x^{1/2} + \log_a y^4 - \log_a z^3 \qquad \text{Property vi}$$

$$= \log_a x^{1/2} y^4 - \log_a z^3 \qquad \text{Property i}$$

$$= \log_a \frac{x^{1/2} y^4}{z^3} \qquad \text{Property ii}$$

While we have discussed logarithms to any positive base, there are only two bases that are commonly used. Logarithms to base 10 are called **common logarithms** and were once, before the days of calculators, very useful for computations since our number system is based on powers of 10. For simplicity we will write $\log_{10} x$ as $\log x$, the base 10 being understood.[†]

For the remainder of this book we will be concerned with another system of logarithms, called **natural logarithms**. These are logarithms defined by using the base number e.

DEFINITION   **THE NUMBER e**

The **number e** is defined by[‡]

$$e = \lim_{n \to \infty} \left(1 + \frac{1}{n} \right)^n. \qquad \blacksquare \qquad (8)$$

Before discussing this number, we indicate that the limit in (9) exists. A proof that the limit does exist is suggested in Problems 51–56. This limit is different from the ones we encountered in Chapter 1. We can see that the exponent is becoming infinite but that $1 + (1/n)$ is approaching 1. Since 1 to any power is 1, our first guess might be that this limit is equal to 1. Table 1 illustrates what happens as n grows. The last number in the table is e, correct to ten decimal places. For convenience we will use the value 2.72 or 2.718 as an approximation to e. The number e is the second **transcendental** number we have encountered. The first was π. A transcendental number is an irrational number that cannot be written as the root of a polynomial with integer coefficients. (The irrational number $\sqrt{2}$ is not transcendental because it is a root of the polynomial equation $x^2 - 2 = 0$.)

There are equivalent definitions for e. For example,

$$e = \lim_{n \to -\infty} \left(1 - \frac{1}{n} \right)^{-n}. \qquad (9)$$

[†] On scientific calculators there is a key labeled $\boxed{\log}$ or $\boxed{\log x}$. This key refers to $\log_{10} x$.

[‡] The number e was discovered by and is named after the Swiss mathematician and physicist Leonhard Euler (1707–1783); see his biographical sketch on page 34. The number first appeared in 1728 in an unpublished manuscript.

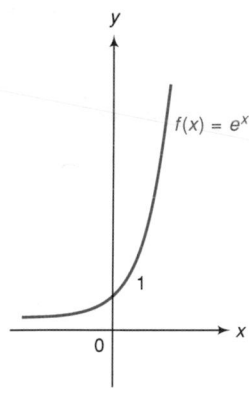

(a) Graph of $f(x) = e^x$.

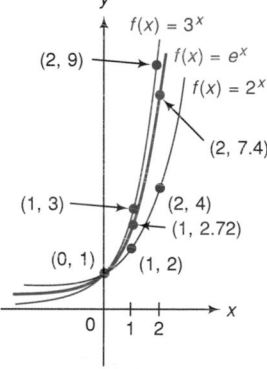

(b) Graphs of 2^x, e^x, and 3^x.

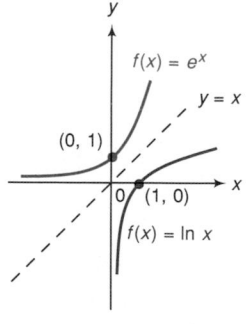

(c) Graph of $f(x) = \ln x$.

FIGURE 3
GRAPHS OF SOME EXPONENTIAL FUNCTIONS AND $\ln x$

TABLE 1 VALUES OF $\left(1 + \dfrac{1}{n}\right)^n$

n	$\dfrac{1}{n}$	$1 + \dfrac{1}{n}$	$\left(1 + \dfrac{1}{n}\right)^n$
1	1	2	2
2	0.5	1.5	2.25
5	0.2	1.2	2.48832
10	0.1	1.1	2.59374246
100	0.01	1.01	2.704813829
1000	0.001	1.001	2.716923932
10,000	0.0001	1.0001	2.718145926
100,000	0.00001	1.00001	2.718268237
1,000,000	0.000001	1.000001	2.718280469
100,000,000,000	10^{-11}	$1 + 10^{-11}$	2.7182818284

[To see this, set $n = -u$ in (8).] If $n \to \pm\infty$, then $u = 1/n \to 0$. This leads to another definition for e:

$$e = \lim_{u \to 0}(1 + u)^{1/u}. \tag{10}$$

Two other definitions for e will be given in Sections 6.3 and 10.9.

We will see in Section 6.6 why the number e is so useful and important. As we have already stated, logarithms to the base e are called *natural logarithms* and have a special notation:

THE NATURAL LOGARITHM

$\ln x = \log_e x.$

The symbol $\ln x$ should be read, "the natural logarithm of x."[†]

The graph of $y = e^x$ is given in Figure 3(a). Since $2 < e < 3$, the graph of e^x is sandwiched between the graphs of 2^x and 3^x. We have

$$2^x < e^x < 3^x \qquad \text{if } x > 0$$
$$3^x < e^x < 2^x \qquad \text{if } x < 0.$$

This is indicated in Figure 3(b). In Figure 3(c) we reflect the graph of e^x about the line $y = x$ to obtain the graph of $\ln x$.

Since we will be dealing primarily with $\ln x = \log_e x$ from now on, we list below properties of the function $y = \ln x$. These follow from the properties of logarithmic functions already discussed.

[†] All scientific calculators have ⎡ln⎤ or ⎡ln x⎤ buttons.

PROPERTIES OF THE FUNCTION $f(x) = \ln x$

(i) $y = \ln x$ is equivalent to $x = e^y$.

(ii) $y = e^x$ is equivalent to $x = \ln y$.

(iii) $e^{\ln x} = x$.

(iv) $\ln e^x = x$.

(v) $\ln e = 1$.

(vi) $\ln 1 = 0$.

(vii) $\ln xw = \ln x + \ln w$.

(viii) $\ln \dfrac{x}{w} = \ln x - \ln w$.

(ix) $\ln \dfrac{1}{x} = -\ln x$.

(x) $\ln x^r = r \ln x$.

(xi) $\ln x = \ln y$ if and only if $x = y$.

(xii) dom $\ln x = \{x : x > 0\}$.

(xiii) $\ln x$ is an increasing function.

(xiv) $\ln x$ is continuous for $x > 0$.

We now compute the derivatives of $\ln x$ and e^x.

THEOREM 3 **DERIVATIVE OF ln x**

The function $f(x) = \ln x$ is differentiable for $x > 0$ and

$$\frac{d}{dx} \ln x = \frac{1}{x}.$$

PROOF: If $y = \ln x$ and $x > 0$, then

Property (ii)
on p. 400

$$\frac{dy}{dx} = \lim_{\Delta x \to 0} \frac{\ln(x + \Delta x) - \ln x}{\Delta x} = \lim_{\Delta x \to 0} \left[\frac{1}{\Delta x} \ln\left(\frac{x + \Delta x}{x} \right) \right]$$

Multiply and
divide by x

Property (vi)
on p. 400

$$= \lim_{\Delta x \to 0} \left[\frac{x}{x\,\Delta x} \ln\left(1 + \frac{\Delta x}{x} \right) \right] = \lim_{\Delta x \to 0} \frac{1}{x} \ln\left(1 + \frac{\Delta x}{x} \right)^{x/\Delta x}.$$

We give a new name to the variable $x/\Delta x$. If $u = \Delta x/x$, then $x/\Delta x = 1/u$, and for each fixed x, $u \to 0$ as $\Delta x \to 0$, so that

$$\lim_{\Delta x \to 0} \left[\frac{1}{x} \ln\left(1 + \frac{\Delta x}{x} \right)^{x/\Delta x} \right] = \lim_{u \to 0} \left[\frac{1}{x} \ln(1 + u)^{1/u} \right]$$

$\dfrac{1}{x}$ is treated like a constant
and the limit of a constant
times a function = the
constant times the limit of
the function [see p. 70]

$$= \frac{1}{x} \lim_{u \to 0} \ln(1 + u)^{1/u}.$$

The function $\ln x$ is continuous. Thus, by the continuity theorem on page 117, we have

eq. (10)
on p. 403 $\ln e = \log_e e = 1$

$$\frac{1}{x}\lim_{u\to 0}\ln(1+u)^{1/u} = \frac{1}{x}\ln\left(\lim_{u\to 0}(1+u)^{1/u}\right) = \frac{1}{x}\ln e = \frac{1}{x}. \quad\blacksquare$$

THEOREM 4   **DERIVATIVE OF e^x**

The function $f(x) = e^x$ is differentiable for all x and

$$\frac{d}{dx}e^x = e^x.$$

That is, e^x is its own derivative

PROOF: Let $y = e^x$. Then $x = \ln y, \dfrac{dx}{dy} = \dfrac{1}{y}$, and, from equation (4) on page 391 (the differentiability of inverse functions),

$$\frac{d}{dx}e^x = \frac{dy}{dx} = \frac{1}{dx/dy} = \frac{1}{1/y} = y = e^x. \quad\blacksquare$$

■ PROBLEMS 6.2

■ SELF-QUIZ

I. $2^{\log_2 16} = $ _____.
   a. 2   b. 4   c. 16
   d. 8   e. 32

II. $\log_5 5^{125} = $ _____.
   a. 5   b. 125

   c. $\dfrac{125}{5} = 25$ d. $125 \cdot 5 = 625$

   e. 3

III. If $x = e^{4t}$, then $t = $ _____.

   a. $e^{(1/4)x}$ b. $\dfrac{\ln x}{\ln 4}$

   c. $\dfrac{\ln x}{4}$ d. $\ln\dfrac{x}{4}$

   e. none of these

IV. If $x = 5^{4u}$, then $u = $ _____.

   a. $\dfrac{1}{4}\dfrac{\ln x}{\ln 5}$ b. $\dfrac{\ln x}{\ln 4 \ln 5}$

   c. $\ln x - \ln 4 - \ln 5 = \ln x - \ln 20$

   d. $\dfrac{\ln\dfrac{x}{4}}{\ln 5}$ e. $5\ln\dfrac{x}{4}$

V. If $y = \ln 2x$, then $x = $ _____.

   a. $e^{y/2}$ b. $e^y - \ln 2$ c. $2e^y$

   d. $\dfrac{e^y}{\ln 2}$ e. $\dfrac{e^y}{2}$

VI. If $y = \ln(x + 5)$, then $x = $ _____.
   a. $e^y + 5$ b. $5e^y$ c. $e^y - \ln 5$

   d. $e^y - 5$ e. $\dfrac{e^y}{\ln 5}$

VII. The base of the natural logarithm is _____.
   a. 10   b. π c. $\mathscr{E}$
   d. e e. 3.14159   f. 2.71828

VIII. Because $\ln x$ and e^x are inverse functions, we know that _____.
   a. $e^x = 1/(\ln x)$ b. $\ln(e^x) = 1$
   c. $\ln(e^x) = x$ d. $e^{\ln x} = -x$
   e. $e^x = \ln(x^{-1})$ f. $e^x = \ln^{-1}(x)$

IX. $10^x = $ _____.
   a. $(e^{\ln 10})^x$ b. $e^{\ln(10x)}$
   c. $e^{(\ln 10)x}$ d. $x^{10}(\log e)$

X. $\log(\text{one million}) = $ _____.
   a. 6 (because $1{,}000{,}000 = 10^6$)
   b. 7 (because $1{,}000{,}000$ is written with 7 digits)
   c. $\ln 1{,}000{,}000e$
   d. 9 (because $1{,}000{,}000$ is written with 9 characters)

CONTINUED

■ DRILL

In Problems 1–32, solve for the unknown variable. (Do not use any table or calculator.)

1. $y = \log_2 4$
2. $y = \log_4 16$
3. $y = \log_{1/3} 27$
4. $y = \frac{1}{3} \log_7(\frac{1}{7})$
5. $y = \pi \log_\pi\left(\frac{1}{\pi^4}\right)$
6. $y = \log_{81} 3$
7. $y = \ln e^5$
8. $y = \ln \frac{1}{e^{3.7}}$
9. $y = \log_{1/4} 2$
10. $y = \log_a a \cdot \log_b b^2 \cdot \log_c c^3$
11. $y = \log_6 36 \log_{25} \frac{1}{5}$
12. $64 = x \log_{1/4} 64$
13. $2^{x^2} = 64$
14. $y = 1.3^{\log_{1.3} 48}$
15. $y = e^{\ln \sqrt{2}}$
16. $y = e^{\ln 14.6}$
17. $y = e^{\ln e^\pi}$
18. $\log_2 x^4 = 4$
19. $\log_x 64 = 3$
20. $\log_x 125 = 3$
21. $\log_x 32 = -5$
22. $y = \log 0.01$
23. $\log x = 10^{-23}$
24. $\log x = 3 \log 2 + 2 \log 3$
25. $\log x = 4 \log \frac{1}{2} - 3 \log \frac{1}{3}$
26. $\log x = a \log b + b \log a$
27. $\log x = \log 1 + \log 2 + \log 3 + \log 4$
28. $\log x = \log 2 - \log 3 + \log 5 - \log 7$
29. $\log x^3 = 2 \log 5 - 3 \log 2$
30. $\log \sqrt{x} = 4 \log 2 - 3 \log \frac{1}{2}$
31. $\ln x^2 - \ln x = \ln 18 - \ln 6$
32. $\log_2 x^{3/2} - \log_2 \sqrt{x} = \log_2 9 - \log_2 3$

In Problems 33–36, simplify the given expressions by using Theorem 2.

33. $\log x^3 \sqrt{x}$
34. $\log[\sqrt[3]{x^2 + 3}(x^5 - 9)^{1/8}]$
35. $\ln \frac{x^5}{(1 + x)^{18}}$
36. $\ln \sqrt[5]{\frac{(x + 1)x^2}{(x - 12)}}$.

In Problems 37–46, sketch the graph of each function. [*Hint:* Use the results of Section 0.5.]

37. $-e^x$
38. e^{-x}
39. e^{x-2}
40. $e^x + 5$
41. $3 - e^{2-x}$
42. $-\ln x$
43. $\ln(x + 2)$
44. $\ln(x - 4)$
45. $\ln(3 - x) + 1$
46. $2 \ln(4 - x)$

■ APPLICATIONS

47. The acidity of a substance is measured by the concentration of hydrogen ions $[H^+]$ in the substance. This concentration is usually measured in terms of moles per liter (mol/L). A standard way to describe this acidity is to define the pH of a substance by

$$pH = -\log[H^+].$$

Distilled water has an approximate H^+ concentration of 10^{-7} mol/L so its pH is $-\log 10^{-7} = 7$. A substance with a pH under 7 is termed an **acid**, while one with a pH above 7 is called a **base**. Determine the pH of the following substances with the indicated hydrogen ion concentrations.
   a. $[H^+] = 4.2 \times 10^{-6}$
   b. $[H^+] = 8 \times 10^{-6}$
   c. $[H^+] = 0.6 \times 10^{-7}$

48. A general psychophysical relation was established in 1834 by the physiologist Weber[†] and given a more precise phrasing later by Fechner.[‡]

By the **Weber–Fechner law**, $S = c \log(R + d)$, where S is the intensity of a sensation, R is the strength of the stimulus producing it, and c and d are constants. The Greek astronomer Ptolemy catalogued stars according to their visual brightness in six categories, or **magnitudes**. A star of the first magnitude was about $2\frac{1}{2}$ times as bright as a star of the second magnitude, which in turn was about $2\frac{1}{2}$ times as bright as a star of the third magnitude, and so on. Let b_n and b_m denote the apparent brightness of two stars having magnitudes n and m, respectively. Then modern astronomers have established the Weber–Fechner law relating the relative brightness to the difference in magnitudes as

$$(m - n) = 2.5 \log\left(\frac{b_n}{b_m}\right).$$

a. Using this formula, calculate the ratio of brightness for two stars of the second and fifth magnitudes, respectively.

[†] Ernest Weber (1796–1878) was a German physiologist.
[‡] Gustav Fechner (1801–1887) was a German physicist.

CONTINUED

b. If star A is five times as bright to the naked eye as star B, what is the difference in their magnitudes?

c. How much brighter is Sirius (magnitude 1.4) than a star of magnitude 21.5?

d. The Nova Aquilae in a 2–3-day period in June 1918 increased in brightness about 45,000 times. How many magnitudes did it rise?

*e. The bright star Castor appears to the naked eye as a single star but can be seen with the aid of a telescope to be really two stars whose magnitudes have been calculated to be 1.97 and 2.95. What is the magnitude of the two combined? [*Hint:* Brightnesses, but not magnitudes, can be added.]

49. The subjective impression of loudness can be described by a Weber–Fechner law. Let I denote the intensity of a sound. The least intense sound that can be heard is $I_0 = 10^{-12}$ watt/m^2 at a frequency of 1000 cycles/sec (this value is called the **threshold of audibility**). If L denotes the loudness of a sound, measured in decibels,[†] then $L = 10 \log(I/I_0)$.

a. If one sound has twice the intensity of another, what is the ratio of the perceived loudness of the two sounds?

b. If one sound appears to be twice as loud as another, what is the ratio of their intensities?

c. Ordinary conversation sounds six times as loud as a low whisper. What is the actual ratio of intensity of their sounds?

■ **SHOW/PROVE/DISPROVE**

50. Show, using logarithms, that $a^b = b^a$, where $a = [1 + (1/n)]^n$ and $b = [1 + (1/n)]^{n+1}$. What does this result prove in the case $n = 1$?

*51. Suppose that $b > 1$. Show that for $x > -1$ and $x \neq 0$, $f(x) > 0$, where $f(x) = (1 + x)^b - (1 + bx)$. [*Hint:* Use the mean value theorem to show that there is a number c in $[0, x]$ such that $f(x) = xb[(1 + c)^{b-1} - 1]$.]

52. Use the result of Problem 51 to show that if $b > 1$ and $x > -1$, then $(1 + x)^b > 1 + bx$.

53. Let $S_n = [1 + (1/n)]^n$ and $T_n = [1 + (1/n)]^{n+1}$. Show that $T_n > S_n$ and $T_n/S_n \to 1$ as $n \to \infty$.

*54. Use the result of Problem 52 to show that $S_{n+1}^{1/n} > S_n^{1/n}$, and conclude that S_n increases as n increases.

*55. Use the result of Problem 52 to show that $(1/T_n)^{1/n} > (1/T_{n-1})^{1/n}$, and conclude that T_n decreases as $n \to \infty$.

56. Explain why the results of Problems 53, 54 and 55 prove that $\lim_{n \to \infty}[1 + (1/n)]^n$ exists and is finite.

■ **CHALLENGE**

*57. Watch carefully. Suppose that $0 < A < B$. Because the logarithm is an increasing function, we have (a) $\log A < \log B$; then (b) $10A \cdot \log A < 10B \cdot \log B$, (c) $\log A^{10A} < \log B^{10B}$, (d) $A^{10A} < B^{10B}$. On the other hand, we run into trouble with particular choices of A and B. For instance, choose $A = \frac{1}{10}$ and $B = \frac{1}{2}$. Clearly, $0 < A < B$, but $A^{10A} = (\frac{1}{10})^1 = \frac{1}{10}$ is greater than $B^{10B} = (\frac{1}{2})^5 = \frac{1}{32}$. Where was the first false step made?

**58. Prove that $\ln x$ cannot be expressed in the form $p(x)/q(x)$, where p and q are polynomial functions.

■ **ANSWERS TO SELF-QUIZ**

I. c	II. b	III. c	VIII. c, f (It is also true that $e^{\ln x} = x$.)
IV. a	V. e	VI. d	IX. a = c
VII. d			X. a

[†] decibel (dB) = $\frac{1}{10}$ bel, named after Alexander Graham Bell (1847–1922), inventor of the telephone.

6.3 THE NATURAL LOGARITHM FUNCTION

We have seen that

$$\int t^r \, dt = \frac{t^{r+1}}{r+1} + C$$

if $r \neq -1$. In this section we see what happens when $r = -1$. To begin the discussion we remind you of the second fundamental theorem of calculus.

THEOREM 1 **SECOND FUNDAMENTAL THEOREM OF CALCULUS**

If f is continuous on $[a, b]$, then the function $G(x) = \int_a^x f(t) \, dt$ is continuous on $[a, b]$, differentiable on (a, b), and for every x in (a, b), $G'(x) = f(x)$. ■

Now let $f(t) = 1/t$. This function is continuous on any interval $[a, b]$ with $a > 0$ and $b > a$. By the second fundamental theorem of calculus, the function

$$F(x) = \int_1^x \frac{1}{t} \, dt$$

is continuous on $[a, b]$ as long as $b > a > 0$, is differentiable on (a, b), and $F'(x) = \dfrac{1}{x}$ for every $x > 0$.

We now give a formal name to this function.

DEFINITION **NATURAL LOGARITHMIC FUNCTION**

The **natural logarithm**, denoted by $\ln x$, is the function with domain $(0, \infty)$ defined by

$$\ln x = \int_1^x \frac{1}{t} \, dt \qquad \text{for all } x > 0. \quad ■ \tag{1}$$

By Theorem 1, $\ln x$ is continuous and differentiable for $x > 0$, and

DERIVATIVE OF THE NATURAL LOGARITHM FUNCTION

$$\frac{d}{dx} \ln x = \frac{1}{x}. \tag{2}$$

We emphasize that dom $\ln x = \{x \colon x > 0\}$.

Suppose that $x > 1$. Since $1/t > 0$ for $t > 0$, the function $\ln x$ gives the area under the curve $f(t) = 1/t$ between $t = 1$ and $t = x$. This is illustrated in Figure 1.

We now have a new function. As with other functions with which we are familiar, we would like to compute some of its values, find some of its properties, and draw its graph. We begin by finding an approximate value for $\ln 2$.

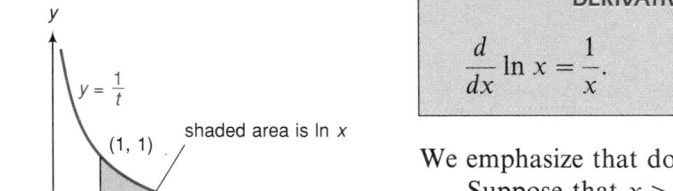

FIGURE 1

GRAPH OF $f(t) = \dfrac{1}{t}$ SHOWING $\ln x$ AS AREA UNDER THE CURVE

▦ **EXAMPLE 1**
FINDING AN APPROXIMATE VALUE FOR $\ln 2$

Approximate $\ln 2$ to 1 decimal place.

SOLUTION: According to Figure 1, $\ln 2$ is the area under the curve $y = \dfrac{1}{t}$ for $1 \le t \le 2$. We depict this area in Figure 2(a).

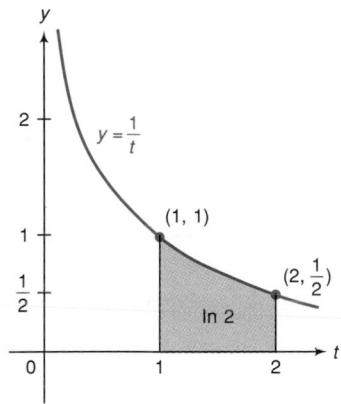

(a) Area under the curve $y = \frac{1}{t}$ for $1 \le t \le 2$.

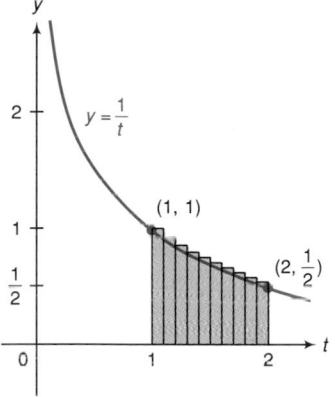

(b) Area under the curve $y = \frac{1}{t}$ for $1 \le t \le 2$ approximated by rectangles using the left-hand endpoints of each subinterval (of length 0.1).

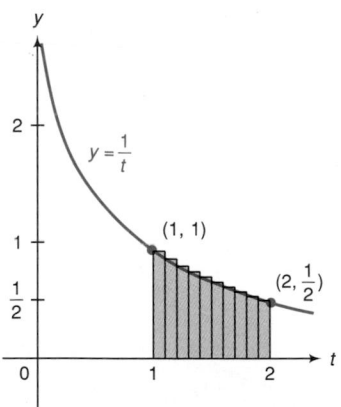

(c) Area under the curve $y = \frac{1}{t}$ approximated by rectangles using the right-hand endpoints of each subinterval (of length 0.1).

FIGURE 2
APPROXIMATING ln 2 = THE AREA UNDER
THE CURVE $y = \frac{1}{t}$ FOR $1 \le t \le 2$

To approximate this area, we partition the interval $[1, 2]$ into 10 subintervals, each having length $\Delta x = \frac{1}{10} = 0.1$ and form Riemann sums, as in Section 4.4. Now $\frac{1}{t}$ is a decreasing function, so

$$\text{value of } \frac{1}{t} \text{ at the left-hand endpoint}$$

$$= \text{maximum value of } \frac{1}{t} \text{ in each subinterval}$$

and

$$\text{value of } \frac{1}{t} \text{ at the right-hand endpoint}$$

$$= \text{minimum value of } \frac{1}{t} \text{ in each subinterval.}$$

Thus

$$s_{10} = \text{sum of areas of rectangles using right-hand endpoints} < \int_1^2 \frac{1}{t}\, dt$$

$$= \ln 2 < \text{sum of areas of rectangles using left-hand endpoints}$$

$$= S_{10}.$$

This is depicted in Figures 2(b) and 2(c). We have

$$s_{10} = f(1.1)\,\Delta x + f(1.2)\,\Delta x + f(1.3)\,\Delta x + f(1.4)\,\Delta x + f(1.5)\,\Delta x$$
$$+ f(1.6)\,\Delta x + f(1.7)\,\Delta x + f(1.8)\,\Delta x + f(1.9)\,\Delta x + f(2)\,\Delta x$$

$$f(x) = \frac{1}{x}$$

$$\downarrow$$

$$= \frac{1}{1.1}(0.1) + \frac{1}{1.2}(0.1) + \frac{1}{1.3}(0.1) + \frac{1}{1.4}(0.1) + \frac{1}{1.5}(0.1) + \frac{1}{1.6}(0.1)$$

$$\underset{f(1.1)\,\Delta x}{\uparrow \quad \uparrow}$$

$$+ \frac{1}{1.7}(0.1) + \frac{1}{1.8}(0.1) + \frac{1}{1.9}(0.1) + \frac{1}{2}(0.1)$$

$$= 0.1\left(\frac{1}{1.1} + \frac{1}{1.2} + \frac{1}{1.3} + \frac{1}{1.4} + \frac{1}{1.5} + \frac{1}{1.6} + \frac{1}{1.7} + \frac{1}{1.8} + \frac{1}{1.9} + \frac{1}{2}\right)$$

$$\approx 0.1(6.6877) \approx 0.67.$$

Analogously,

$$S_{10} = 0.1\left(\frac{1}{1} + \frac{1}{1.1} + \frac{1}{1.2} + \frac{1}{1.3} + \frac{1}{1.4} + \frac{1}{1.5}\right.$$

$$\left. + \frac{1}{1.6} + \frac{1}{1.7} + \frac{1}{1.8} + \frac{1}{1.9}\right)$$

$$\approx 0.1(7.1877) \approx 0.72.$$

Thus we have the estimate

$$0.67 < \ln 2 < 0.72$$

and, to 1 decimal place, $\ln 2 \approx 0.7$.

OBTAINING In x ON A CALCULATOR

Every scientific calculator has an $\boxed{\ln}$ or $\boxed{\ln x}$ key.

▦ **EXAMPLE 2**
COMPUTING VALUES OF
In x ON A CALCULATOR

Use a calculator to compute (a) $\ln 2$ (b) $\ln 0.02486$, (c) $\ln 2486$, (d) $\ln 10,584$, and (e) $\ln 10$.

SOLUTION:

(a) $\ln 2 = 0.6931471806$ (b) $\ln 0.02486 = -3.694495193$
(c) $\ln 2486 = 7.818430272$ (d) $\ln 10,584 = 9.267098706$
(e) $\ln 10 = 2.302585093$

REMARK: We see from part (a) above that our estimate $\ln 2 \approx 0.7$ is quite close.

PROPERTIES OF In x

Suppose that $x < 0$. Then $-x > 0$, and from the chain rule,

$$\frac{d}{dx}\ln(-x) = \frac{1}{-x}\frac{d}{dx}(-x) = \frac{-1}{-x} = \frac{1}{x}.$$

Since

$$|x| = \begin{cases} x, & x > 0 \\ -x, & x < 0, \end{cases}$$

we obtain the important formulas

$$\frac{d}{dx}\ln|x| = \frac{1}{x} \qquad \text{for } x \neq 0 \tag{3}$$

and

$$\int \frac{1}{x}\,dx = \ln|x| + C. \tag{4}$$

This definition of a logarithmic function on page 408 is very different from definitions we have seen before. But as we will see, it leads to familiar results. The next theorem gives us the properties we expect of a logarithmic function.

THEOREM 2   **PROPERTIES OF ln x**

Let x and y be positive numbers and let c be a rational number. Then,

  (i) $\ln 1 = 0$
  (ii) $\ln xy = \ln x + \ln y$
  (iii) $\ln x/y = \ln x - \ln y$
  (iv) $\ln 1/x = -\ln x$
  (v) $\ln x^c = c \ln x.$

PROOF:

(i) $\ln 1 = \int_1^1 \left(\dfrac{1}{t}\right) dt \overset{\text{equation (4.5.12)}}{=} 0.$

(ii) We have

$$\ln xy \overset{\text{Theorem 4.5.3}}{=} \int_1^{xy} \frac{1}{t}\, dt = \int_1^x \frac{1}{t}\, dt + \int_x^{xy} \frac{1}{t}\, dt$$

$$- \ln x + \int_x^{xy} \frac{1}{t}\, dt. \tag{5}$$

To evaluate $\displaystyle\int_x^{xy} \frac{1}{t}\, dt$ in (5), we make the substitution $t = xu$. Then $dt = x\,du$ and $u = t/x$. When $t = x$, $u = x/x = 1$. When $t = xy$, $u = xy/x = y$. Thus,

$$\int_x^{xy} \frac{1}{t}\, dt = \int_1^y \frac{1}{xu}\, x\, du = \int_1^y \frac{1}{u}\, du = \ln y. \tag{6}$$

Thus, inserting (6) into the last equality in (5), we have

$$\ln xy = \ln x + \ln y.$$

(iii) $\ln x = \ln\left(\dfrac{x}{y} \cdot y\right) \overset{\text{By (ii)}}{=} \ln \dfrac{x}{y} + \ln y,$

so

$$\ln \frac{x}{y} = \ln x - \ln y.$$

(iv) $\ln \dfrac{1}{x} \overset{\text{By (iii)}}{=} \ln 1 - \ln x \overset{\text{By (i)}}{=} 0 - \ln x = -\ln x.$

(v) Let $u = x^c$. Then $(d/dx)u(x) = cx^{c-1}$, and

$$\frac{d}{dx} \ln x^c = \frac{d}{dx} \ln [u(x)] \overset{\text{Chain rule}}{=} \frac{1}{u(x)} \frac{d}{dx} u(x)$$

$$= \frac{1}{x^c} \cdot cx^{c-1} = \frac{cx^{c-1}}{x^c} = \frac{c}{x}.$$

Now, let

$$f(x) = \ln x^c - c \ln x.$$

Then

$$f'(x) = \frac{c}{x} - c\left(\frac{1}{x}\right) = 0$$

so

$$f(x) = k, \text{ a constant.}$$

But

$$f(1) = \ln 1^c - c \ln 1 = \ln 1 - c \ln 1 = 0 - 0 = 0.$$

Thus,

$$k = 0, \qquad \text{so } f(x) = 0 = \ln x^c - c \ln x$$

or

$$\ln x^c = c \ln x. \quad \blacksquare$$

We now analyze the function $y = \ln x$ in order to obtain its graph. Since $(d/dx) \ln x = 1/x > 0$ if $x > 0$, $\ln x$ is an increasing function. Moreover, $(d^2/dx^2) \ln x = (d/dx)(1/x) = -1/x^2 < 0$, so that $\ln x$ is concave down. Suppose now that $x > 1$. Then from Figure 1 we see that $\ln x > 0$. If $x < 1$, then $1/x > 1$, so that $\ln (1/x) > 0$. But,

$$\ln \frac{1}{x} = -\ln x, \qquad \text{so that } \ln x < 0.$$

Thus, as we discovered earlier,

$$\ln x > 0 \qquad \text{if } x > 1$$

and

$$\ln x < 0 \qquad \text{if } 0 < x < 1.$$

From Example 1, we know that $\ln 2 \approx 0.7$ (actually, $\ln 2 \approx 0.69315$). However, the only thing we need to know about $\ln 2$ now is that $\ln 2 > 0.5 = \frac{1}{2}$.

We can use this fact to compute $\lim_{x \to \infty} \ln x$ and $\lim_{x \to 0^+} \ln x$. Let $z = 2^n$, where n is a positive integer. Then,

$$\ln z = \ln 2^n = n \ln 2 \overset{\ln 2 > \frac{1}{2}}{\geq} \frac{n}{2}. \tag{7}$$

Now let N be given and let $x = 2^{2(N+1)}$. Then, from (7),

$$\ln x \geq \frac{2(N+1)}{2} = N + 1.$$

We see that, for every number N, no matter how large, we can find a number x such that $\ln x > N$. But $\ln x$ is an increasing function, so that if $y > x$, then $\ln y > \ln x > N$. Thus, once $\ln x$ gets bigger than N, it stays bigger than N. Since N was arbitrary, we have

$$\lim_{x \to \infty} \ln x = \infty.$$

Finally, if $x \to 0^+$, then $1/x \to \infty$, so $\ln(1/x) \to \infty$. But $\ln(1/x) = -\ln x$, from which we conclude that

$$\lim_{x \to 0^+} \ln x = -\infty.$$

Let us summarize what we know about the function $\ln x$:

(i) $\ln x$ is continuous and differentiable for $x > 0$ (its domain)

(ii) $\ln x$ is an increasing function (since its derivative $\dfrac{1}{x}$ is > 0 when $x > 0$)

(iii) $\ln x \to \infty$ as $x \to \infty$

(iv) $\ln x \to -\infty$ as $x \to 0^+$.

Finally,

(v) $\dfrac{d^2}{dx^2} \ln x = \dfrac{d}{dx} \dfrac{1}{x} = -\dfrac{1}{x^2} < 0$ so the graph of $\ln x$ is concave down.

From (i), (ii), (iii), and (iv), we can also conclude that

$$\text{range } \ln x = (-\infty, \infty) = \mathbb{R}.$$

Putting this information together, we obtain the graph in Figure 3.

The function $\ln x$ is continuous. Since $\ln 1 = 0$ and $\ln x \to \infty$ as $x \to \infty$, there is, by the intermediate-value theorem (Theorem 1.7.5), a number x between 2 and 3 such that $\ln x = 1$. Because $\ln x$ is increasing, the number is unique. This is one of the most important numbers in mathematics.

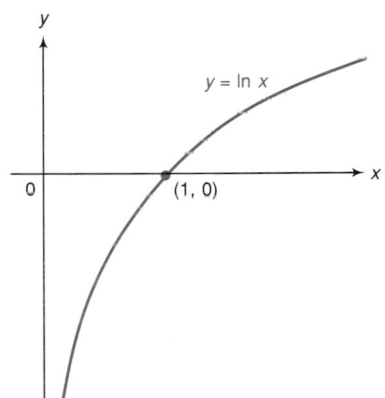

FIGURE 3
GRAPH OF $f(x) = \ln x$

DEFINITION   **THE NUMBER e**

The **number e** is defined to be the unique number having the property that $\ln x = 1$. That is,

$$\ln e = 1. \quad \blacksquare \tag{8}$$

We emphasize that e is now *defined* so that the area under the curve $1/t$ between 1 and e is equal to 1. This is sketched in Figure 4. To 10 decimal places,

$$e \approx 2.7182818284.$$

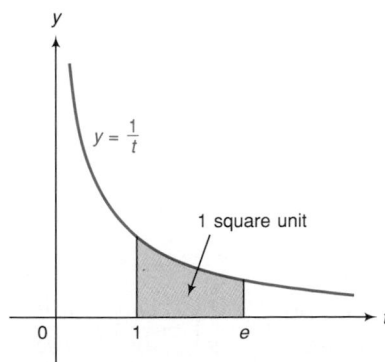

FIGURE 4

THE AREA UNDER THE CURVE $\dfrac{1}{t}$ FOR

$1 \leq t \leq e$ IS EQUAL TO 1 SQUARE UNIT

DERIVATIVES AND INTEGRALS INVOLVING LOGARITHMIC FUNCTIONS

EXAMPLE 3
THE DERIVATIVE OF A
LOGARITHMIC FUNCTION

Differentiate $y = \ln(x^3 + 3x + 1)$.

SOLUTION: Using the chain rule, we set $u = x^3 + 3x + 1$. Then,

$$y = \ln u, \quad \text{and} \quad \frac{dy}{dx} = \frac{dy}{du}\frac{du}{dx} = \frac{1}{u}(3x^2 + 3) = \frac{3x^2 + 3}{x^3 + 3x + 1}.$$

NOTE: $\ln(x^3 + 3x + 1)$ is only defined when $x^3 + 3x + 1 > 0$.

EXAMPLE 4
THE DERIVATIVE OF $\ln(\sec x + \tan x)$
IS $\sec x$

Find $(d/dx)\ln(\sec x + \tan x)$.

SOLUTION:

$$\frac{d}{dx}\ln(\sec x + \tan x) = \frac{1}{\sec x + \tan x} \cdot \frac{d}{dx}(\sec x + \tan x)$$

$$= \frac{\sec x \tan x + \sec^2 x}{\sec x + \tan x} = \frac{\sec x(\sec x + \tan x)}{\sec x + \tan x}$$

$$= \sec x.$$

NOTE: $\ln(\sec x + \tan x)$ is only defined when $\sec x + \tan x > 0$.

Generalizing these two examples, we obtain the rule

$$\frac{d}{dx}\ln u = \frac{1}{u}\frac{du}{dx}.$$

We must always remember that $\ln u$ is only defined if $u > 0$.
Now consider

$$\frac{d}{dx}\ln|u| \quad \text{for } u \neq 0.$$

If $u \neq 0$, then $|u| > 0$ and $\ln|u|$ is defined. There are two cases to consider.

Case 1 $u > 0$. Then,

$$|u| = u, \quad \text{and} \quad \frac{d}{dx}\ln|u| = \frac{d}{dx}\ln u = \frac{1}{u}\frac{du}{dx}.$$

Case 2 $u < 0$. Then,

$$|u| = -u, \quad \text{and} \quad \frac{d}{dx}\ln|u| = \frac{d}{dx}\ln(-u) = \frac{1}{-u}\frac{d(-u)}{dx} = -\frac{1}{u}\left(-\frac{du}{dx}\right) = \frac{1}{u}\frac{du}{dx}.$$

Thus, in either case we have

$$\frac{d}{dx}\ln|u| = \frac{1}{u}\frac{du}{dx}, \quad u \neq 0. \tag{9}$$

This fact is useful, for now we only have to worry about zero values of u. Negative values of u pose no problem.

Using (9), we immediately obtain

$$\int \frac{du}{u} = \ln|u| + C. \tag{10}$$

In other words, *if the numerator is the differential of the denominator, then the integral of the fraction is the natural logarithm of the absolute value of the denominator.*

We can write (10) in another way. Since $du = u'(x)\,dx$, (10) becomes

$$\int \frac{du}{u} = \int \frac{u'(x)}{u(x)}\,dx = \ln|u(x)| + C. \tag{11}$$

Here we see clearly that we are integrating a fraction where the numerator is the derivative of the denominator.

EXAMPLE 5
INTEGRATING A RATIONAL FUNCTION WHERE THE NUMERATOR IS THE DERIVATIVE OF THE DENOMINATOR

Calculate $\int [2x/(x^2 + 1)]\,dx$.

SOLUTION: Since $2x$ is the derivative of $x^2 + 1$.

$$\int \frac{2x}{x^2 + 1}\,dx = \ln|x^2 + 1| + C = \ln(x^2 + 1) + C$$

(since $x^2 + 1$ is always positive).

EXAMPLE 6
INTEGRATING A FUNCTION THAT CAN BE WRITTEN AS THE DERIVATIVE OF A FUNCTION DIVIDED BY THAT FUNCTION

Calculate

$$\int \frac{dx}{x \ln x}, \qquad x > 0.$$

SOLUTION: We write the integral as

$$\int \frac{1/x}{\ln x}\,dx.$$

Then the numerator is the derivative of the denominator, and

$$\int \frac{1/x}{\ln x}\,dx = \ln|\ln x| + C.$$

EXAMPLE 7
MULTIPLYING AND DIVIDING BY A CONSTANT IN ORDER TO SET THE NUMERATOR EQUAL TO THE DERIVATIVE OF THE DENOMINATOR

Calculate

$$\int \frac{x^3 + 1}{x^4 + 4x}\,dx.$$

SOLUTION: Here the numerator is not quite the derivative of the denominator. The derivative of the denominator is $4x^3 + 4 = 4(x^3 + 1)$. Thus we need to multiply and divide by the constant 4 to obtain

$$\int \frac{x^3 + 1}{x^4 + 4x}\,dx = \frac{1}{4} \int \frac{4x^3 + 4}{x^4 + 4x}\,dx = \frac{1}{4} \ln|x^4 + 4x| + C.$$

EXAMPLE 8
COMPUTING THE INTEGRAL OF
tan x

Calculate $\int \tan x \, dx$.

SOLUTION:

$$\int \tan x \, dx = \int \frac{\sin x}{\cos x} \, dx.$$

Let $u = \cos x$. Then $du = -\sin x \, dx$. We therefore multiply and divide by -1 to obtain

$$\int \tan x \, dx = \int \frac{\sin x}{\cos x} \, dx = -\int \frac{-\sin x}{\cos x} \, dx = -\int \frac{du}{u}$$

$$= -\ln|u| + C = -\ln|\cos x| + C.$$

LOGARITHMIC DIFFERENTIATION

Before leaving this section, we show how the use of logarithms can sometimes simplify the process of differentiation. We need only recall the rules given in Theorem 2.

EXAMPLE 9
USING LOGARITHMIC
DIFFERENTIATION

Differentiate $y = \sqrt[4]{(x^3 + 1)/x^{7/9}}$, $x > 0$.

SOLUTION: We first take the natural logarithm of both sides of the equation:

$$\ln y = \ln\left(\frac{x^3 + 1}{x^{7/9}}\right)^{1/4} = \frac{1}{4} \ln\left(\frac{x^3 + 1}{x^{7/9}}\right) = \frac{1}{4} \ln(x^3 + 1) - \frac{7}{36} \ln x.$$

Now we differentiate implicitly:

$$\frac{1}{y} \frac{dy}{dx} = \frac{3x^2}{4(x^3 + 1)} - \frac{7}{36x}$$

or

$$\frac{dy}{dx} = y\left(\frac{3x^2}{4(x^3 + 1)} - \frac{7}{36x}\right) = \left(\sqrt[4]{\frac{x^3 + 1}{x^{7/9}}}\right)\left(\frac{3x^2}{4(x^3 + 1)} - \frac{7}{36x}\right).$$

In Example 9 it was not necessary to use the quotient rule at all. The process we used to simplify the differentiation is called **logarithmic differentiation**.

Recall that, by definition,

$$\ln x = \int_1^x \frac{1}{t} \, dt \qquad \text{and} \qquad \frac{d}{dx} \ln x = \frac{1}{x}.$$

If u is a function of x, then

$$\ln u(x) = \int_1^{u(x)} \frac{1}{t} \, dt. \tag{12}$$

The expression (12) is defined as long as $u(x) > 0$. Assume that u is differentiable. By the chain rule,

$$\frac{d}{dx} \ln u(x) = \frac{d}{du} \ln u \frac{d}{dx} u(x) = \frac{1}{u} \frac{du}{dx}.$$

That is,

$$\frac{d}{dx} \int_1^{u(x)} \frac{1}{t} \, dt = \frac{u'(x)}{u(x)}.$$ (13)

EXAMPLE 10
DIFFERENTIATING A FUNCTION
DEFINED BY AN INTEGRAL

Compute $\dfrac{d}{dx} \displaystyle\int_1^{x^2} \dfrac{1}{t} \, dt$.

SOLUTION: In (13), we set $u = x^2$. Then

$$\frac{d}{dx} \int_1^{x^2} \frac{1}{t} \, dt = \frac{u'(x)}{u(x)} = \frac{2x}{x^2} = \frac{2}{x}.$$

We can verify our answer by computing the derivative in another way:

$$\int_1^{x^2} \frac{1}{t} \, dt = \ln x^2. \qquad \text{By definition}$$

But $\ln x^2 = 2 \ln x$ and $\dfrac{d}{dx} 2 \ln x = \dfrac{2}{x}$.

■ PROBLEMS 6.3

■ SELF-QUIZ

I. $\dfrac{d}{dx} \ln x^3 = $ _____.

   a. $\dfrac{1}{x^3}$ b. $\dfrac{1}{3x^2}$

   c. $\dfrac{1}{3x}$ d. $\dfrac{3}{x}$

II. $\displaystyle\int_1^{2x} \left(\dfrac{1}{t}\right) dt = $ _____.

   a. $(\ln x)^2$ b. $2 \ln x$
   c. $\ln(x^2)$ d. $\ln(2x)$

III. $\dfrac{d}{dx} \ln 2x = $ _____.

   a. $1/x$ b. e^{2x}

   c. $2/x$ d. $\dfrac{2}{x} + C$

IV. $\displaystyle\int \left(\dfrac{2}{x}\right) dx = $ _____.

   a. $\ln x^2 + C$ b. $2 \ln |x| + C$
   c. $\ln |x^5| - 3 \ln |x| + C$ d. $|\ln x|^2 + C$
V. If $x > 0$, then $\ln x = $ _____.

   a. $\displaystyle\int_x^{2x} \left(\dfrac{1}{t}\right) dt$ b. $\displaystyle\int_x^{x^2} \left(\dfrac{1}{t}\right) dt$

   c. $\displaystyle\int_{1+2}^{x+2} \left(\dfrac{1}{t}\right) dt$ d. $\displaystyle\int_{-x}^{x} \left(\dfrac{1}{t}\right) dt$

VI. If $\displaystyle\int_1^x \left(\dfrac{1}{t}\right) dt = 1$, then $x = $ _____.

   a. e b. π
   c. $\sqrt{2\pi}$ d. $1/e$

VII. If $\displaystyle\int_1^t \left(\dfrac{1}{u}\right) du = -1$, then _____.

   a. $t = e$ b. $t = 1/e$
   c. $t = 1 - e$ d. $t = -e$
   e. $t = -1/e$
   f. no value can be found for t

VIII. $\displaystyle\int \dfrac{?}{x^2 + x} \, dx = \ln|x^2 + x| + C$

   a. $2x$ b. $2x + 1$ c. $2x - 1$

   d. $x + 1$ e. $\dfrac{x^3}{3} + \dfrac{x^2}{2}$

IX. $\displaystyle\int \dfrac{?}{\sin x} \, dx = \ln|\sin x| + C$

   a. $\cos x$ b. $\cot x$ c. $-\sin x$

   d. $\dfrac{\sin^2 x}{2}$ e. $-\cos x$

CONTINUED

■ **DRILL**

In Problems 1–6, simplify the given expression by using Theorem 2.

1. $\ln|x^7| - 2\ln|x|$

2. $\ln \sqrt[3]{x} + \ln \sqrt[5]{x^2}$

3. $\ln(x^3 \cdot \sqrt{x})$

4. $\ln\left(\dfrac{x^5}{(1+x)^{18}}\right)$

5. $\ln[\sqrt[3]{x^2+3}(x^5-9)^{1/8}]$

6. $\ln \sqrt[5]{\dfrac{(x+1)x^7}{(x-12)}}$

In Problems 7–20, find the domain and compute the derivative of the given function.

7. $\ln x$

8. $\ln(x^{-1})$

9. $\ln(1 + 2x)$

10. $\ln(1 + x^5)$

11. $\ln((x-2)(x+3))$

12. $\ln|x^2 - 5x + 6|$

13. $\ln\left|\dfrac{x+1}{x}\right|$

14. $\ln\left|1 - \dfrac{1}{x}\right|$

15. $\ln(\sin x)$

16. $\ln|\cos x|$

17. $x \ln x$

18. $x \ln x - x$

19. $\ln\left|\dfrac{x+1}{x-1}\right|$

20. $\ln\left|\dfrac{1+\sin x}{1-\sin x}\right|$

In Problems 21–24, use logarithmic differentiation to obtain a (relatively) simple expression for dy/dx.

21. $y = \sqrt[5]{\dfrac{x^3 - 3}{x^2 + 1}}$

22. $y = \dfrac{x^5(x^4 - 3)}{x^8\sqrt{x^5 + 1}}$

23. $y = \sqrt{x} \cdot \sqrt[3]{x+2} \cdot \sqrt[5]{x-1}$

*24. $y = \left[\dfrac{x^3(x+3)(x^{11/9} + 2)^{1/2}}{(x^5 - 3)\sqrt{x^7 + 2}}\right]^{-9/4}$

In Problems 25–32, perform the indicated integration.

25. $\displaystyle\int \dfrac{2x}{1+x^2}\,dx$

26. $\displaystyle\int \dfrac{2}{1+x}\,dx$

27. $\displaystyle\int \dfrac{x^2}{7+x^3}\,dx$

28. $\displaystyle\int \dfrac{\ln x}{x}\,dx$

29. $\displaystyle\int \left(\dfrac{1}{x-1} + \dfrac{1}{x+1}\right)dx$

*30. $\displaystyle\int \dfrac{2}{x^2 - 1}\,dx$

[*Hint:* Modify the preceding problem slightly.]

In Problems 31–34, compute each derivative.

31. $\dfrac{d}{dx}\displaystyle\int_1^{x^4} \dfrac{1}{t}\,dt$

32. $\dfrac{d}{dx}\displaystyle\int_1^{\sin x} \dfrac{1}{t}\,dt, 0 < x < \pi$

33. $\dfrac{d}{dx}\displaystyle\int_1^{\cos^2 x} \dfrac{1}{t}\,dt, 0 < x < \dfrac{\pi}{2}$

34. $\displaystyle\int_1^{\ln x} \dfrac{1}{t}\,dt, x > 1$

▦ **CALCULATOR PROBLEMS**

Natural logarithms can be approximated on a hand calculator even if the calculator does not have an $\boxed{\ln}$ key. If $0.5 \le x \le 1.5$, then let $A = (x-1)/(x+1)$ and use

$$\ln x \approx \left[\left(\dfrac{3A^2}{5} + 1\right)\cdot\dfrac{A^2}{3} + 1\right]\cdot 2A.$$

35. Use the technique described above to approximate ln 0.8 and ln 1.2.

36. Combine the above technique with Theorem 2 to approximate ln 2. [*Hint:* Consider $\ln(\tfrac{3}{2} \cdot \tfrac{4}{3})$.]

■ **APPLICATIONS**

In Problems 37–40, sketch the given curve. Indicate maxima, minima, and points of inflection.

37. $y = \ln|x|$

38. $y = \ln(x - 5)^2$

39. $y = x \ln x - x$

40. $y = \ln(1 + x^2)$

In Problems 41–44, find the area of the region bounded by the given curves and lines.

41. $y = 3/(x - 1)$; $x = 4$, $x = 16$, $y = 0$

42. $y = x/(1 + x^2)$; $x = 0$, $x = 3$, $y = 0$

43. $y = (\ln x)/x$; $x = 2$, $x = 4$, $y = 0$

*44. $y = \ln(x + 1)$; $y = 3$, $x = 0$

45. For $x > 0$, express $\int_{-1}^{-x} (1/t)\,dt$ in terms of the natural logarithm. [*Hint:* Draw a picture.]

*46. Calculate

$$\lim_{n \to \infty} \left(\frac{1}{n+2} + \frac{1}{n+4} + \frac{1}{n+6} + \cdots + \frac{1}{3n-2} + \frac{1}{3n} \right).$$

[*Hint:* Partition the interval $[1, 3]$ into n subintervals of equal length. Then this limit represents the integral of a simple function over that interval.]

■ **CHALLENGE**

*47. Suppose that $0 < q < p$, where p and q are integers. Prove that

$$\lim_{n \to \infty} \sum_{k=1+nq}^{np} \frac{1}{k} = \ln\left(\frac{p}{q} \right).$$

**48. Suppose that L is a differentiable function such

that

$$L(u \cdot v) = L(u) + L(v) \qquad \text{for all } u, v > 0.$$

How must this function be related to the natural logarithm function $\ln$?

**49. Prove that $\ln x$ cannot be expressed in the form $p(x)/q(x)$, where p and q are polynomial functions.

■ **ANSWERS TO SELF-QUIZ**

I. d	II. d	III. a	VII. b	VIII. b	IX. a
IV. a, b, c	V. b	VI. a			

6.4 THE EXPONENTIAL FUNCTION e^x

In the last section we saw that $\ln x$ is an increasing function with range $(-\infty, \infty)$. Since it is increasing, it is 1–1 (by Theorem 6.1.1), and therefore, it has an inverse. Before writing it down, we make some observations. Let $r = m/n$ be a positive rational number and let e be the number defined on page 413 ($e \approx 2.7182818284$). Then,

$$e^{1/n} = \text{the } n\text{th root of } e,$$

$$e^{m/n} = (e^{1/n})^m,$$

and

$$e^{-m/n} = \frac{1}{e^{m/n}}.$$

Thus e^x is defined whenever x is a rational number. Moreover,

$$\underset{\text{Theorem 6.3.2(v)}}{\ln e^x} = x \ln e = x \cdot 1 = x. \qquad \text{(1)}$$

e is chosen so that $\ln e = 1$

Let $\ln^{-1} x$ denote the inverse of $\ln x$. Then, from (1) and the definition of an inverse function,

$$\ln e^x = x$$

$$\ln^{-1}(\ln e^x) = \ln^{-1} x \qquad \ln^{-1}(\ln u) = u \text{ if } u > 0$$

$$e^x = \ln^{-1} x. \qquad \text{(2)}$$

Equation (2) holds for every rational number x. It allows us to define e^x for all x. This new function agrees with the values taken by e^x when x is rational.

DEFINITION **THE EXPONENTIAL FUNCTION e^x**

The **exponential function** e^x is defined to be the inverse function of $\ln x$. That is,

$$y = e^x \text{ if and only if } x = \ln y.^\dagger \qquad \blacksquare \qquad (3)$$

NOTE: Since range of $\ln x = \mathbb{R}$, it follows that

$$\text{dom } e^x = \mathbb{R}.$$

We stress that

$$e^{\ln x} = x \text{ for } x > 0, \text{ and } \ln e^x = x \text{ for every } x. \qquad (4)$$

This is simply a restatement of the fact that if f and g are inverses of one another, then $f(g(x)) = g(f(x)) = x$.

We now prove some facts about e^x.

THEOREM 1 **PROPERTIES OF THE FUNCTION e^x**

Let x, y, and c be real numbers.

(i) $e^x > 0$ for every real number x (5)

(ii) $e^0 = 1$ (6)

(iii) $e^1 = e$ (7)

(iv) $e^{x+y} = e^x e^y$ (8)

(v) $e^{x-y} = e^x/e^y$ (9)

(vi) $e^{-x} = 1/e^x$ (10)

(vii) $e^{cx} = (e^x)^c$ (11)

(viii) $\lim_{x \to \infty} e^x = \infty$ (12)

(ix) $\lim_{x \to -\infty} e^x = 0$ (13)

(x) $\ln e^x = x$ (14)

(xi) $e^{\ln x} = x$, if $x > 0$ (15)

PROOF:

(i) Let $y = e^x$. Then $x = \ln y$, which is defined only if $y > 0$. Thus, $y = e^x > 0$.

(ii) Since $\ln 1 = 0$, $e^0 = 1$ by (3).

(iii) Since $\ln e = 1$, $e^1 = e$ by (3).

(iv) Let $u = e^x$ and $v = e^y$. Then by (3), $x = \ln u$ and $y = \ln v$. Thus,

$$x + y = \ln u + \ln v \overset{\text{Theorem 6.3.2 (ii)}}{=} \ln uv,$$

and by (3)

$$e^{x+y} = uv = e^x e^y.$$

---

† In some books, the function e^x is written $\exp(x)$.

(v) Defining u and v as in (iv), we have

Theorem 6.3.2 (iii)

$$x - y = \ln u - \ln v = \ln\left(\frac{u}{v}\right),$$

so

$$\frac{u}{v} = e^{(x-y)} \quad \text{and} \quad \frac{e^x}{e^y} = e^{x-y}.$$

(vi) $e^{-x} = e^{0-x} = e^0/e^x = 1/e^x.$

(vii) Let $u = e^x$. Then $x = \ln u$, and

Theorem 6.3.2 (v)

$$cx = c \ln u = \ln u^c,$$

so

$$u^c = e^{cx} \quad \text{and} \quad (e^x)^c = e^{cx}.$$

NOTE: At this point $(e^x)^c$ is defined only when c is rational. The proof is correct, but the statement will have meaning for all real c only after we have defined a^x for x real. This is done in the next section.

(viii) We know that $e > 2$. Thus,

$$e^n = e^{1 \cdot n} = (e^1)^n > 2^n.$$

We will show, in Theorem 2, that $\dfrac{d}{dx} e^x = e^x > 0$, so e^x is an increasing function. Thus, if $x > n$, $e^x > 2^n$, which tends to infinity as $n \to \infty$.

   Alternatively, in Problem 6.1.24(c) on page 395 you were asked to show that the inverse of a strictly increasing function is strictly increasing. But e^x is the inverse of $\ln x$, which is strictly increasing; thus, e^x is an increasing function.

(ix) Part (x) and (xi) follow from the definition of e^x. By (viii),

$$\lim_{x \to -\infty} e^x = \lim_{x \to \infty} e^{-x} = \lim_{x \to \infty} \frac{1}{e^x} = 0. \quad \blacksquare$$

We now compute the derivative of e^x.

THEOREM 2   **THE DERIVATIVE OF e^x**

The exponential function e^x is differentiable and

$$\frac{d}{dx} e^x = e^x. \tag{16}$$

That is, *the function e^x is its own derivative!*

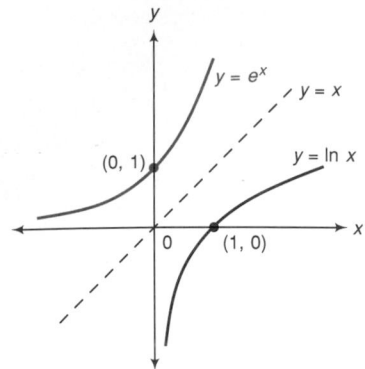

FIGURE 1
GRAPH OF $f(x) = e^x$ OBTAINED BY RE-FLECTING THE GRAPH OF ln x ABOUT THE LINE $y = x$

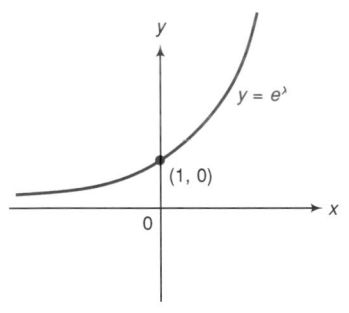

(a) Graph of $f(x) = e^x$ (exponential growth).

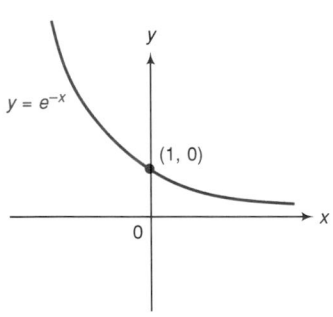

(b) Graph of $f(x) = e^{-x}$ (exponential decay).

FIGURE 2
GRAPHS SHOWING EXPONENTIAL GROWTH AND DECAY

PROOF: We use the fact that e^x is the inverse of ln x. Let $y = e^x$. Then $y > 0$, $x = \ln y$, and $dx/dy = 1/y \neq 0$. Thus, by Theorem 6.1.4, e^x is differentiable, and

$$\frac{d}{dx} e^x = \frac{dy}{dx} = \frac{1}{dx/dy} = \frac{1}{1/y} = y = e^x. \quad \blacksquare$$

Since e^x is the inverse of ln x, we obtain the graph of e^x by reflecting the graph of ln x about the line $y = x$. This is done in Figure 1.

If $f(x) = e^x$, then $f(-x) = e^{-x}$. Recall from page 40 that the graph of $f(-x)$ is the graph of $f(x)$ reflected about the y-axis. Thus the graph of e^{-x} is obtained by reflecting the graph of e^x about the y-axis (see Figure 2(b)).

The growth given by the increasing function e^x is called **exponential growth**. The behavior of the decreasing function e^{-x} (decreasing because $\frac{d}{dx} e^{-x} = -e^{-x} < 0$) is called **exponential decay**. We will say a great deal more about exponential growth and decay in Section 6.6.

On page 413 we defined the number e as the unique number such that $\int_1^e \frac{1}{t} \, dt = 1$. There is an alternative definition of e that is sometimes useful.

DEFINITION **ALTERNATIVE DEFINITION OF e**

The number e is defined by

$$e = \lim_{n \to \infty} \left(1 + \frac{1}{n}\right)^n. \quad \blacksquare \tag{17}$$

At the end of this section we will prove that the limit above exists and that it is equal to the number e as defined on page 413. This proves that the two definitions are equivalent. Values of $\left(1 + \frac{1}{n}\right)^n$ are given in Table 1. We see that, to nine decimal places, $e \approx 2.718281828$.

TABLE 1

n	$\dfrac{1}{n}$	$1 + \dfrac{1}{n}$	$\left(1 + \dfrac{1}{n}\right)^n$
1	1	2	2
2	0.5	1.5	2.25
5	0.2	1.2	2.48832
10	0.1	1.1	2.59374246
100	0.01	1.01	2.704813829
1000	0.001	1.001	2.716923932
10,000	0.0001	1.0001	2.718145927
100,000	0.00001	1.00001	2.718268237
1,000,000	0.000001	1.000001	2.718280469
10,000,000,000	10^{-10}	$1 + 10^{-10}$	2.718281828

OBTAINING VALUES OF e^x ON A CALCULATOR

Values of e^x can be obtained on a scientific calculator, usually in one of two ways: If there is an $\boxed{e^x}$ or $\boxed{\exp x}$ button, then press the number x followed by e^x or $\boxed{\exp x}$. If not, then there will be an inverse or second function button $\boxed{\text{INV}}$ or $\boxed{\text{2nd F}}$. In this case, enter x and then press $\boxed{\text{INV}}\,\boxed{\ln}$ or $\boxed{\text{2nd F}}\,\boxed{\ln}$.[†]

EXAMPLE 1
USING A CALCULATOR TO COMPUTE
VALUES OF e^x

Compute **(a)** e^2, **(b)** $e^{0.46}$, and **(c)** $e^{-3.14}$ on a calculator.

SOLUTION

(a) By pressing $\boxed{2}\,\boxed{\text{INV}}\,\boxed{\ln x}$ or $\boxed{2}\,\boxed{\text{2nd F}}\,\boxed{\ln}$, we obtain 7.389056099. We achieve the same result on a calculator with an $\boxed{e^x}$ key by pressing $\boxed{2}\,\boxed{e^x}$.

(b) $\boxed{.}\,\boxed{4}\,\boxed{6}\,\boxed{\text{INV}}\,\boxed{\ln x}$ yields 1.584073985.

(c) $\boxed{3}\,\boxed{.}\,\boxed{1}\,\boxed{4}\,\boxed{+/-}$ displays -3.14, and then pressing $\boxed{\text{INV}}$ $\boxed{\ln x}$ yields 0.043282797.

Alternatively, we could first compute $e^{3.14}$ and then use the reciprocal key $\boxed{1/x}$ to compute $e^{-3.14} = 1/e^{3.14}$.

DERIVATIVES AND INTEGRALS OF EXPONENTIAL FUNCTIONS

EXAMPLE 2
DERIVATIVE OF AN EXPONENTIAL
FUNCTION

Differentiate $y = e^{-4x}$.

SOLUTION: Let $u = -4x$. Then $\dfrac{du}{dx} = -4$, $y = e^u$, and using the chain rule we obtain

$$\frac{d}{du}e^u = e^u \quad u = -4x$$

$$\frac{dy}{dx} = \frac{dy}{du}\frac{du}{dx} = e^u(-4) = -4e^{-4x}.$$

Extending the result of Example 2, we have, for every real number a,

$$\frac{d}{dx}e^{ax} = ae^{ax}. \tag{18}$$

[†] On some calculators the keys $\boxed{e^x}$ or $\boxed{\text{INV}}\,\boxed{\ln}$ are pressed *before* entering x.

EXAMPLE 3

Differentiate $y = e^{\cos^2 x}$.

SOLUTION: If $u = \cos^2 x$, then

$$\frac{du}{dx} = 2 \cos x \frac{d}{dx} (\cos x) = -2 \cos x \sin x$$

and

$$\frac{dy}{dx} = \frac{dy}{du}\frac{du}{dx} = e^u(-2 \cos x \sin x) = -2 \cos x \sin x (e^{\cos^2 x}).$$

We now generalize Examples 2 and 3 to see that

DERIVATIVE OF AN EXPONENTIAL FUNCTION

$$\frac{d}{dx} e^u = e^u \frac{du}{dx}. \tag{19}$$

If we integrate (9), we get the rule

$$\int e^u \, du = e^u + C. \tag{20}$$

That is, *the integral of* e *raised to a power times the differential of the power is simply* e *to that power.*

Since $du = u'(x)\, dx$, we can rewrite (20) to obtain

THE INTEGRAL OF AN EXPONENTIAL FUNCTION

$$\int e^u \, du = \int e^{u(x)} u'(x) \, dx = e^{u(x)} + C. \tag{21}$$

EXAMPLE 4
THE INTEGRAL OF e RAISED TO A
POWER TIMES THE DERIVATIVE OF
THAT POWER

Calculate $\int e^{x^3} \cdot 3x^2 \, dx$.

SOLUTION: If $u = x^3$, then $du = 3x^2 \, dx$, so that

$$\int e^{x^3} \cdot 3x^2 \, dx = \int e^u \, du = e^u + C = e^{x^3} + C.$$

EXAMPLE 5
MULTIPLYING AND DIVIDING BY A
CONSTANT IN ORDER TO
INTEGRATE DIRECTLY

Calculate $\int e^{-2x} \, dx$.

SOLUTION: If $u = -2x$, then $du = -2 \, dx$, so we multiply and divide by -2 to obtain

$$\int e^{-2x} \, dx = -\frac{1}{2} \int e^{-2x}(-2) \, dx = -\frac{1}{2} \int e^u \, du = -\frac{1}{2} e^u + C$$

$$= -\frac{1}{2} e^{-2x} + C.$$

EXAMPLE 6
MULTIPLYING AND DIVIDING BY A
CONSTANT IN ORDER TO
INTEGRATE DIRECTLY

Calculate $\int (e^{\sqrt{x}}/\sqrt{x})\,dx$.

SOLUTION: If $u = \sqrt{x}$, then $du = (1/2\sqrt{x})$, and we multiply and divide by $\frac{1}{2}$ to obtain

$$\int \frac{e^{\sqrt{x}}}{\sqrt{x}}\,dx = 2 \int \frac{e^{\sqrt{x}}}{2\sqrt{x}}\,dx = 2 \int e^u\,du = 2e^u + C = 2e^{\sqrt{x}} + C.$$

EXAMPLE 7
A GRAPH INVOLVING AN
EXPONENTIAL FUNCTION

Graph the curve $y = e^{-x} \sin x$.

SOLUTION: Since $-1 \le \sin x \le 1$ and since $e^{-x} \to 0$ as $x \to \infty$, $\lim_{x \to \infty} e^{-x} \sin x = 0$. (Be careful: $\lim_{x \to \infty} \sin x$ *does not exist*, since $\sin x$ oscillates between -1 and 1 but does not approach any single number.) Also, since $e^{-x} > 0$, $-1 \le \sin x \le 1$ implies that $-e^{-x} \le e^{-x} \sin x \le e^{-x}$, so the graph of the function $e^{-x} \sin x$ stays between the graphs of the functions $-e^{-x}$ and e^{-x}. We then calculate

$$\frac{dy}{dx} = e^{-x} \cos x - e^{-x} \sin x.$$

This expression is zero when $e^{-x} \cos x = e^{-x} \sin x$, or when $\sin x = \cos x$ and $\sin x/\cos x = \tan x = 1$. Since $\tan x = 1$ when $x = \pi/4$, and since $\tan x$ is periodic of period π, the critical points are $\pi/4$, $-3\pi/4$, $5\pi/4$, $-7\pi/4$, $9\pi/4$, ... Now $d^2y/dx^2 = e^{-x} \sin x - e^{-x} \cos x - e^{-x} \cos x + e^{-x} \sin x = -2e^{-x} \cos x$, with points of inflection at $x = \pi/2$, $3\pi/2$, $5\pi/2$, ... When $x = \pi/4$, $d^2y/dx^2 < 0$; and when $x = 5\pi/4$, $\cos x < 0$ (since $5\pi/4$ is in the third quadrant), so that $d^2y/dx^2 > 0$. Hence we have a local maximum at $x = \pi/4$ and a local minimum at $x = 5\pi/4$. We now draw the curve by using the fact that $e^{-x} \sin x$ always remains between $-e^{-x}$ and e^{-x}. See Figure 3.

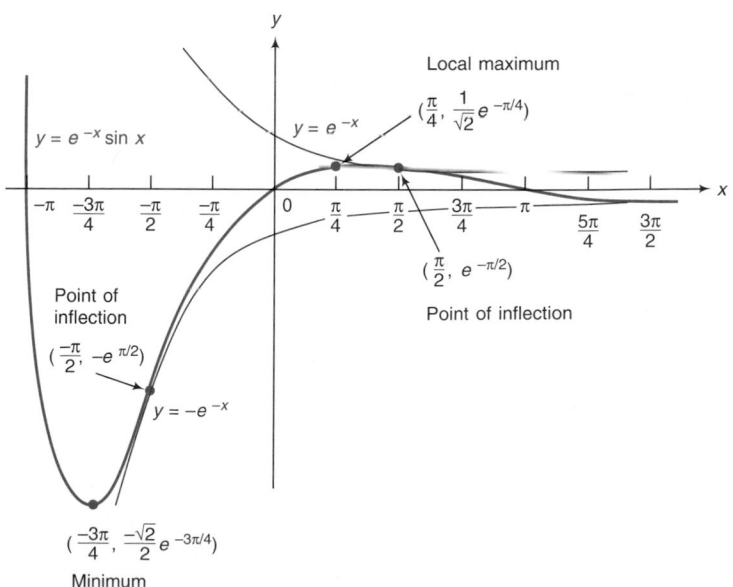

FIGURE 3   GRAPH OF $f(x) = e^{-x} \sin x$

EXAMPLE 8
THE GRAPH OF A FUNCTION
INVOLVING THE NATURAL
LOGARITHMIC FUNCTION

Sketch the curve $y = (\ln^2 x)/x$. [*Note:* By $\ln^2 x$, we mean $(\ln x)^2$.]

SOLUTION: We differentiate to obtain

$$\frac{dy}{dx} = \frac{x \cdot (d/dx)(\ln^2 x) - \ln^2 x}{x^2} = \frac{x \cdot 2(\ln x)(1/x) - \ln^2 x}{x^2}$$

$$= \frac{2 \ln x - \ln^2 x}{x^2}.$$

The function and its derivative are only defined for $x > 0$. Now $dy/dx = 0$ when $2 \ln x - \ln^2 x = (\ln x)(2 - \ln x) = 0$. This occurs when $\ln x = 0$ and when $\ln x = 2$. When $\ln x = 0$, $x = 1$. When $\ln x = 2$, $x = e^2 \approx 7.4$. If $0 < x < 1$, then $dy/dx < 0$ and y is decreasing. For $1 < x < e^2$, $dy/dx > 0$ and y is increasing. For $x > e^2$, y is again decreasing. Therefore by the first-derivative test, f has a local minimum at $x = 1$ [which gives the point $(1, 0)$] and a local maximum at $x = e^2$ [the point $(e^2, 4/e^2) \approx (7.4, 0.54)$]. To find concavity, we compute

$$\frac{d^2y}{dx^2} = \frac{x^2(d/dx)(2 \ln x - \ln^2 x) - (2 \ln x - \ln^2 x)(2x)}{x^4}$$

$$= \frac{x^2\left(\dfrac{2}{x} - \dfrac{2 \ln x}{x}\right) - 4x \ln x + 2x \ln^2 x}{x^4}$$

$$= \frac{2x - 6x \ln x + 2x \ln^2 x}{x^4} = \frac{2 - 6 \ln x + 2 \ln^2 x}{x^3}.$$

To find the points of inflection, we must solve the equation

$$2 \ln^2 x - 6 \ln x + 2 = 0$$

$$\ln^2 x - 3 \ln x + 1 = 0 \qquad \text{Divide by 2}$$

$$u^2 - 3u + 1 = 0 \qquad \text{Set } u = \ln x$$

$$\ln x = u = \frac{3 \pm \sqrt{9 - 4}}{2} = \frac{3 \pm \sqrt{5}}{2}. \qquad \text{Use the quadratic formula}$$

If $\ln x = (3 + \sqrt{5})/2 \approx 2.62$, then $x \approx e^{2.62} \approx 13.74$. If $\ln x = (3 - \sqrt{5})/2 \approx 0.38$, then $x \approx e^{0.38} \approx 1.46$. These give the two points of inflection. If $x < 1.46$, then the curve is concave up. For $1.46 < x < 13.74$ the curve is concave down, and for $x > 13.74$ the curve is again concave up. Finally, since $\ln x$ seems to increase much more slowly than x itself, it is reasonable to believe that

$$\lim_{x \to 0^+} \frac{\ln^2 x}{x} = \infty \qquad \text{and} \qquad \lim_{x \to \infty} \frac{\ln^2 x}{x} = 0.$$

We will prove these facts in Chapter 9. Putting all this information together, we obtain the curve in Figure 4.

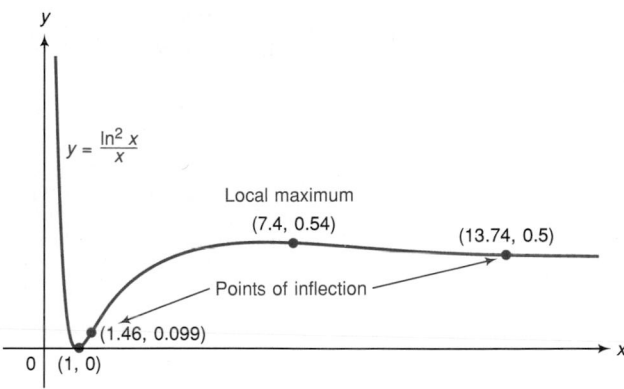

FIGURE 4 GRAPH OF $f(x) = \dfrac{\ln^2 x}{x}$

GROWTH RATES OF POWER AND EXPONENTIAL FUNCTIONS

Which grows faster: a power function or an exponential function? To get some idea, we give, in Table 2, values for x^2, x^5, x^{10}, e^x, and $\ln x$.

TABLE 2 SOME APPROXIMATE VALUES FOR x^2, x^5, x^{10}, e^x, AND $\ln x$

x	x^2	x^5	x^{10}	e^x	$\ln x$
1	1	1	1	2.71828	0
10	100	100,000	10^{10}	22,026	2.3026
50	2500	312,500,000	9.7656×10^{16}	5.1847×10^{21}	3.9120
100	10,000	10^{10}	10^{20}	2.688×10^{43}	4.6052
250	62,500	9.7656×10^{11}	9.5367×10^{23}	3.746×10^{108}	5.5215
1000	1,000,000	10^{15}	10^{30}	1.9701×10^{434}	6.9078

It seems that e^x grows much faster than the function x^{10}. In fact, the following is true:

$$\lim_{x \to \infty} \frac{e^x}{x^n} = \infty \qquad \text{for any positive integer } n. \tag{22}$$

We will prove this fact in Section 9.2 (Example 5), but Table 2 makes it seem plausible.

PROOF OF ALTERNATIVE FORM FOR e (OPTIONAL)

We close this section by showing that if e is defined as the unique number such that $\ln e = 1$, then $\displaystyle\lim_{n \to \infty} \left(1 + \frac{1}{n}\right)^n = e$.

THEOREM 3   **ALTERNATIVE FORM FOR e**

Suppose e is defined by $\ln e = \int_1^e \frac{1}{t}\,dt = 1$. Then

$$e = \lim_{n \to \infty} \left(1 + \frac{1}{n}\right)^n. \tag{23}$$

PROOF: Suppose that $t \in \left[1, 1 + \frac{1}{n}\right]$. Then

$$\frac{1}{1 + \frac{1}{n}} \le \frac{1}{t} \le 1 \tag{24}$$

and, by the comparison theorem on page 301,

$$\int_1^{1 + 1/n} \frac{1}{1 + \frac{1}{n}}\,dt \le \int_1^{1 + 1/n} \frac{1}{t}\,dt \le \int_1^{1 + 1/n} 1\,dt.$$

This is a constant   This is the definition of $\ln\left(1 + \frac{1}{n}\right)$ This is a constant

Then

$$\frac{1}{1 + \frac{1}{n}}\left[\left(1 + \frac{1}{n}\right) - 1\right] \le \ln\left(1 + \frac{1}{n}\right) \le \left(1 + \frac{1}{n}\right) - 1 = \frac{1}{n}.$$

$$= \frac{1}{n}$$

Observe that

$$\frac{1}{1 + \frac{1}{n}} \cdot \frac{1}{n} = \frac{1}{n\left(1 + \frac{1}{n}\right)} = \frac{1}{n + 1} \tag{25}$$

so we have

$$\frac{1}{n + 1} \le \ln\left(1 + \frac{1}{n}\right) \le \frac{1}{n}. \tag{26}$$

From (26), we obtain

$$(n + 1)\ln\left(1 + \frac{1}{n}\right) \ge 1 \qquad \text{and} \qquad n\ln\left(1 + \frac{1}{n}\right) \le 1.$$

Since $\ln x^r = r \ln x$, we see that

This is the definition of e that we are assuming

$$(n + 1)\ln\left(1 + \frac{1}{n}\right) = \ln\left(1 + \frac{1}{n}\right)^{n+1} \ge 1 = \ln e. \tag{27}$$

Since $\ln x$ is an increasing function, we conclude, from (27), that

$$\left(1 + \frac{1}{n}\right)^{n+1} \geq e. \tag{28}$$

Since $\left(1 + \frac{1}{n}\right)^{n+1} = \left(1 + \frac{1}{n}\right)^n \left(1 + \frac{1}{n}\right)$, we can divide both sides of (28) by $1 + \frac{1}{n}$ to obtain

$$\frac{e}{1 + \frac{1}{n}} \leq \left(1 + \frac{1}{n}\right)^n. \tag{29}$$

Similarly,

$$n \ln\left(1 + \frac{1}{n}\right) = \ln\left(1 + \frac{1}{n}\right)^n \leq 1 = \ln e,$$

so

$$\left(1 + \frac{1}{n}\right)^n \leq e. \tag{30}$$

Combining (29) and (30), we obtain

$$\frac{e}{1 + \frac{1}{n}} \leq \left(1 + \frac{1}{n}\right)^n \leq e \qquad \text{for } n = 1, 2, 3, \ldots . \tag{31}$$

Since $1 + \frac{1}{n} \to 1$ as $n \to \infty$, $\dfrac{e}{1 + \frac{1}{n}} \to \dfrac{e}{1} = e$ as $n \to \infty$. Then if we take limits as $n \to \infty$ in (31), the result follows from the squeezing theorem (see page 77). ■

REMARK 1: A slightly different proof is suggested in Problems 53–58.

REMARK 2: Using (28) and (30), we obtain

$$\left(1 + \frac{1}{n}\right)^n \leq e \leq \left(1 + \frac{1}{n}\right)^{n+1} \qquad \text{for } n = 1, 2, 3, \ldots . \tag{32}$$

The inequalities (32) give us estimates on the size of e. For example, for $n = 10$ we obtain

$$\left(1 + \frac{1}{10}\right)^{10} \leq e \leq \left(1 + \frac{1}{10}\right)^{11},$$

or

$$2.59374246 \leq e \leq 2.853116706.$$

For $n = 1000$ we obtain

$$\left(1 + \frac{1}{1000}\right)^{1000} \leq e \leq \left(1 + \frac{1}{1000}\right)^{1001},$$

or

$$2.716923932 \leq e \leq 2.719640856.$$

■ PROBLEMS 6.4

■ SELF-QUIZ

I. The base of the natural logarithms is
_____.

   a. 10   b. π c. $\mathscr{E}$
   d. e e. 3.14159   f. 2.71828

II. Because $\ln x$ and e^x are inverse functions, we know that _____.
   a. $e^x = 1/(\ln x)$ b. $\ln(e^x) = 1$
   c. $\ln(e^x) = x$ d. $e^{\ln x} = -x$
   e. $e^x = \ln(x^{-1})$ f. $e^x = \ln^{-1}(x)$

III. Let $u = e^x$, then $\dfrac{d}{dx} u^2 = $ _____.

   a. $2u = 2e^x$ b. $u = e^x$
   c. $2uu' = 2u^2 = 2e^{2x}$ d. $u^2 = e^{2x}$

IV. $\int e^x \, dx = $ _____.

   a. $C + e^x$ b. $C \cdot e^x$
   c. e^{x+C} d. $-e^{-x} + C$

V. $\dfrac{d}{dx} e^{\sin x} = $ _____.

   a. $e^{\sin x}$ b. $e^{\cos x}$

   c. $\cos x e^{\sin x}$ d. $\sin x e^{\cos x}$
   e. $\cos x$

VI. $\int e^{\ln x} \, dx = $ _____ $+ C$.

   a. $e^{\ln x}$ b. $e^{1/x}$ c. $xe^{\ln x}$

   d. x e. $\dfrac{x^2}{2}$

   f. cannot be done by the methods of this section

VII. $\int e^{3x^2}[?] \, dx = e^{3x^2} + C$.

   a. x^3 b. $3x^2$ c. $\dfrac{x^2}{2}$

   d. $6x$ e. 6

VIII. $\int e^{\tan x}[?] \, dx = e^{\tan x} + C$.

   a. $\dfrac{\tan^2 x}{2}$ b. $-\ln|\cos x|$

   c. $\tan x$ d. $\sec x \tan x$
   e. $\sec^2 x$

■ DRILL

In Problems 1–10, compute the value of x. (Do not use any numerical table or electronic calculator.)

1. $x = \ln e^5$ 2. $e^{\ln(5\pi)} = x$
3. $\ln(1/e^{2.718}) = x$ 4. $x = e^{-\ln 14.6}$
5. $\ln x = \ln 1 + \ln 2 + \ln 3 + \ln 4$
6. $\ln x = \ln 2 - \ln 3 + \ln 5 - \ln 7$
7. $\ln(x^2) - \ln x = \ln 18 - \ln 6$
8. $\ln \sqrt{x} - \ln(x^{-1/2}) = e + 10$
9. $e^x = e^2 \cdot e^{-\pi}$
10. $\ln(e^x/e^{-x}) = e^{\ln 1} + \ln(e^1)$

In Problems 11–22, compute the derivative of the given function.

11. $e^x \cdot e^{x+1}$ 12. e^{2x+1} 13. $\sqrt{e^x}$

14. $e^{\sqrt{x}}$ 15. $e^{-x} \cdot \cos x$ 16. $e^{-x^2/2}$
17. xe^x 18. $x^2 e^x$
19. $(x - 1)e^x$ 20. $(x^2 - 2x + 2)e^x$
21. $\ln(xe^x)$ 22. $\ln(1 + e^{2x})$

In Problems 23–36, compute the given integral.

23. $\int_1^{\ln 4} e^x \, dx$ 24. $\int e^{x+3} \, dx$

25. $\int e^{3x} \, dx$ 26. $\int_{-1}^2 (1/e^x) \, dx$

27. $\int e^x \cdot e^{x+1} \, dx$ 28. $\int (2x - 1)e^{x^2 - x} \, dx$

29. $\int (\sin e^x)(e^x \, dx)$ 30. $\int (\sin x)e^{\cos x} \, dx$

31. $\int (e^{1/x}/x^2) \, dx$ 32. $\int (e^x/(1 + e^x)) \, dx$

33. $\int_1^4 (e^{\sqrt{x}}/\sqrt{x})\,dx$ 34. $\int_1^{64} (e^{\sqrt[3]{x}}/x^{2/3})\,dx$ *36. $\int [1/(e^x + e^{-x} + 2)]\,dx$

*35. $\int x \cdot e^{x^2} \cdot (5 + e^{x^2})^3\,dx$

[*Hint:* Multiply numerator and denominator by e^x and then simplify.]

▦ CALCULATOR PROBLEMS

37. If x is in $\left[-\frac{1}{2}, \frac{1}{2}\right]$, then a simple fifth-degree polynomial is fairly close to e^x. We have the following approximation

$$e^x \approx \left(\left\{\left[\left\langle \frac{x}{5} + 1\right\rangle \frac{x}{4} + 1\right]\frac{x}{3} + 1\right\}\frac{x}{2} + 1\right)\frac{x}{1} + 1.$$

   a. Calculate an approximate value for $e^{0.13}$.
   b. Calculate an approximate value for $e^{-0.37}$.
   c. Compute the error $(= true - approx)$ for parts (a) and (b).

38. It can be shown that

$$\lim_{n \to \infty} \left(1 + \frac{x}{n}\right)^n = e^x.$$

If we fix $n = 5$, then $(1 + x/5)^5$ approximates e^x (it is an alternative to the approximation given in the preceding problem). Repeat parts (a)–(c) of Problem 37; compare results for these two approximation techniques.

■ APPLICATIONS

In Problems 39–44, sketch the given curve; indicate minima, maxima, and points of inflection.

39. $y = e^{2x+3}$ 40. $y = \ln(1 + e^x)$
41. $y = e^{-|x|}$ 42. $y = e^{-x^2}$
43. $y = e^{-x}\cos x$ *44. $y = e^x \ln x$

In Problems 45–48, compute the area of the region bounded by the given curves and lines

45. $y = 5 - e^x$; $x = 0$, $x = 1$, $y = 0$
46. $y = e^{-x}$; $x = 0$, $x = 1$, $y = 2$
47. $y = e^{2x} - 2x$; $x = -1$, $x = 1$, $y = 0$
48. $y = e^x$; $y = xe^{x^2}$; $x = 0$

49. What is the acceleration after 10 minutes of a particle whose equation of motion is

$$s(t) = 30 + 3t + 0.01t^2 + \ln t + e^{2t^2}?$$

(Time t is measured in minutes and distance s is measured in meters.)

50. The revenue R (in dollars) received by a manufacturer when q units of a given product are sold is given by $R(q) = 0.5qe^{-0.001q}$.
   a. What is the marginal revenue when $q = 100$ units?
   b. At what level of sales is revenue maximized?
   c. What is the maximum revenue?

■ SHOW/PROVE/DISPROVE

51. Prove that if f is continuous and positive and if $A > 0$, then

$$\int_0^A e^{-t}f(t)\,dt < \int_0^A f(t)\,dt.$$

52. Prove or Disprove: If $\beta \neq 0$, then there is at least one line which is tangent to the graph of $y = e^{\beta x}$ and which also passes through the origin.

*53. Suppose that $b > 1$ and let $f(x) = (1 + x)^b - (1 + bx)$. Show that if $x > -1$ and $x \neq 0$, then $f(x) > 0$. [*Hint:* Use the mean-value theorem to show that there is a number c in $[0, x]$ such that $f(x) = b[(1 + c)^{b-1} - 1]x$.]

54. **Bernoulli's Inequality:** Use the result of the preceding problem to show that if $b > 1$ and $x > -1$, then $(1 + x)^b \geq 1 + bx$ with equality only when $x = 0$.

Let $S_n = \left[1 + \dfrac{1}{n}\right]^n$ and $T_n = \left[1 + \dfrac{1}{n}\right]^{n+1}$.

55. Show that $T_n > S_n$ and $T_n/S_n \to 1$ as $n \to \infty$.
56. Show that

$$S_{n+1}^{1/n} > S_n^{1/n},$$

and conclude that S_n increases as n increases.

CONTINUED

57. Show that

$$\left(\frac{1}{T_n}\right)^{1/n} > \left(\frac{1}{T_{n-1}}\right)^{1/n}$$

and conclude that T_n decreases as $n \to \infty$.

*58. Show that $\lim_{n \to \infty} [1 + (1/n)]^n$ exists and is finite (the value of this limit is given in equation (17)). [*Hint:* Apply the results of the preceding three problems.]

■ **CHALLENGE**

**59. Suppose that E is a differentiable function such that $E(s + t) = E(s) \cdot E(t)$ for all real numbers s and t. How must this function be related to the exponential function e^x?

**60. Use the following steps† to prove that

$$\lim_{n \to \infty} \frac{\sqrt[n]{n!}}{n} = \frac{1}{e}$$

where $n! = n(n-1)(n-2) \cdots 3 \cdot 2 \cdot 1$.

a. Show that

$$\frac{1}{n} \ln\left(\frac{k-1}{n}\right) < \int_{(k-1)/n}^{k/n} \ln x \, dx < \frac{1}{n} \ln\left(\frac{k}{n}\right)$$

for $k = 2, 3, \ldots, n$.

b. Show that

$$\ln\left[\left(\frac{1}{n}\right)\left(\frac{2}{n}\right) \cdots \left(\frac{n-1}{n}\right)\right]^{1/n}$$

$$< \int_{1/n}^{1} \ln x \, dx < \ln\left[\left(\frac{2}{n}\right)\left(\frac{3}{n}\right) \cdots \left(\frac{n}{n}\right)\right]^{1/n}.$$

c. Show that

$$-1 + \frac{1}{n} < \ln\left[\frac{n!}{n^n}\right]^{1/n} < -1 + \frac{1}{n} + \frac{\ln n}{n}.$$

[*Hint:* You may choose to simplify some of your work here by using a fact that can be proven using the results in Section 9.2:

$$\lim_{x \to \infty} \frac{\ln x}{x} = 0.]$$

■ **ANSWERS TO SELF-QUIZ**

I. d $(2.71828 \approx e)$

II. c, f (It is also true that $e^{\ln x} = x$.)

III. c IV. a V. c

VI. e $[e^{\ln x} = x$ by definition of an inverse function$]$

VII. d VIII. e

---

† Adapted from the paper "Using Riemann sums in evaluating a familiar limit" by F. Burk, S. Goel, and D. Rodriquez in *College Mathematics Journal* 17, No. 2 (March 1986): 170–171.

6.5 THE FUNCTIONS a^x AND $\log_a x$

In this section we define logarithmic and exponential functions with bases other than e.

DEFINITION **THE FUNCTION a^x**

Let $a > 0$ and x be real numbers. Then we define

$$a^x = e^{x \ln a}. \quad ■ \tag{1}$$

The domain of a^x is $\mathbb{R}$. Since $\ln e^u = u$, it immediately follows from (1) that

$$\ln a^x = x \ln a. \tag{2}$$

REMARK: The definition of the function a^x seems artificial, but is really quite important. Recall that, up to now, the expression a^x was only defined if x was a rational number. For example, $2^{3/2} = (\sqrt{2})^3$ was defined, but $2^{\sqrt{2}}$ was not. Now, a^x is defined for every real x if $a > 0$.

THEOREM 1   **DERIVATIVE OF a^x**

For $a > 0$, a^x is everywhere differentiable and

$$\frac{d}{dx} a^x = a^x \ln a. \tag{3}$$

PROOF:

equation (6.4.19)

$$\frac{d}{dx} a^x = \frac{d}{dx} e^{x \ln a} = e^{x \ln a} \frac{d}{dx} x \ln a$$

$$= e^{x \ln a} \ln a = a^x \ln a. \quad \blacksquare$$

Remember that $\ln a$ is a constant and that the derivative of a constant times x is simply that constant.

Since $e^x > 0$ for every real number x, it follows from (1) that $a^x > 0$ for every real number x. Since $\ln a < 0$ if $0 < a < 1$ and $\ln a > 0$ if $a > 1$, it follows from (3) that a^x is a decreasing function if $0 < a < 1$ and an increasing function if $a > 1$. We can obtain other properties of a^x as well. For example,

$$e^{u + v} = e^u e^v$$

$$a^{x+y} = e^{(x+y) \ln a} = e^{x \ln a + y \ln a} = e^{x \ln a} e^{y \ln a}$$

$$= a^x a^y. \tag{4}$$

Additional properties of a^x are given in Theorem 2. The proofs follow as in the proof of (4).

THEOREM 2   **PROPERTIES OF THE FUNCTION $f(x) = a^x$**

Let $a > 0$ and let x and y be any real numbers. Then,

(i) $a^x > 0$.

(ii) $a^{-x} = \dfrac{1}{a^x}$.

(iii) $a^{x+y} = a^x \cdot a^y$.

(iv) $a^{x-y} = a^x \cdot a^{-y} = \dfrac{a^x}{a^y}$.

(v) $a^0 = e^{0 \ln a} = e^0 = 1$.

(vi) $(a^x)^y = a^{xy}$.

(vii) If $a > 1$, then a^x is an increasing function in any interval.

(viii) If $0 < a < 1$, then a^x is a decreasing function in any interval.

(ix) a^x is continuous for every x.

(x) If $a > 1$, then

$$\lim_{x \to \infty} a^x = \infty \qquad \text{and} \qquad \lim_{x \to -\infty} a^x = 0.$$

(xi) If $0 < a < 1$, then

$$\lim_{x \to \infty} a^x = 0 \qquad \text{and} \qquad \lim_{x \to -\infty} a^x = \infty. \quad \blacksquare$$

EXAMPLE 1
SKETCHING AN EXPONENTIAL
GRAPH

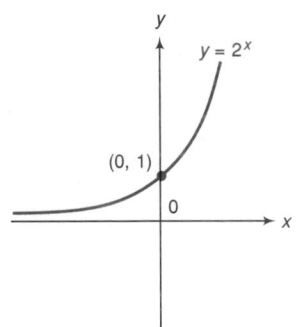

FIGURE 1
GRAPH OF $y = 2^x$

Sketch the graph of $y = 2^x$.

SOLUTION: $\dfrac{dy}{dx} = 2^x \ln 2$ and $\dfrac{d^2y}{dx^2} = 2^x \ln^2 2$. We see that 2^x is an increasing function whose graph is concave up. There are no critical points or points of inflection. Also, $\lim_{x \to \infty} 2^x = \infty$ and $\lim_{x \to -\infty} 2^x = 0$. The graph is given in Figure 1.

Using (3), we obtain the following integration formula:

THE INTEGRAL OF a^x

$$\int a^x \, dx = \frac{a^x}{\ln a} + C. \tag{5}$$

EXAMPLE 2
THE INTEGRAL OF AN
EXPONENTIAL FUNCTION

$\int 2^x \, dx = 2^x/(\ln 2) + C.$

Since a^x is increasing or decreasing, it is 1–1 and it has an inverse. The inverse of a^x is called the *logarithm to the base* a.

DEFINITION   **LOGARITHM TO THE BASE a**

Let a be a positive number with $a \neq 1$. Then the **logarithmic function with base a** is defined by

$$y = \log_a x \qquad \text{if and only if} \qquad a^y = x. \quad \blacksquare \tag{6}$$

Since $a^y > 0$, dom $\log_a x = \mathbb{R}^+$, the positive real numbers.

THEOREM 3

Let $a > 0$ with $a \neq 1$.

(i) $\log_a x = \dfrac{\ln x}{\ln a}$ First conversion formula   **(7)**

(ii) $\log_a e = \dfrac{1}{\ln a}$ Second conversion formula   **(8)**

(iii) $\dfrac{d}{dx} \log_a x = \dfrac{1}{x \ln a} = \dfrac{\log_a e}{x}$ Derivative of $\log_a x$ **(9)**

PROOF:

eq. (2)

(i) Let $y = \log_a x$. Then $x = a^y$ and $\ln x = \ln a^y \overset{\downarrow}{=} y \ln a$. Thus, $y = \ln x / \ln a$.

(ii) From (i)

$$\log_a e = \frac{\ln e}{\ln a} = \frac{1}{\ln a}.$$

from (ii)

(iii) $\dfrac{d}{dx} \log_a x = \dfrac{d}{dx} \left(\dfrac{\ln x}{\ln a} \right) = \dfrac{1}{\ln a} \dfrac{d}{dx} \ln x = \dfrac{1}{x \ln a} \overset{\downarrow}{=} \dfrac{1}{x} \log_a e.$ ∎

REMARK 1: From (i) we see that $\log_e x = \ln x$ since $\ln e = 1$.

REMARK 2: Logarithms to the base 10, $\log_{10} x$, are called **common logarithms** and are usually denoted by $\log x$, with the subscript 10 omitted.[†]

The following theorem summarizes properties of logarithmic functions. The proof, which is omitted, follows directly from Theorem 3 (i) and properties of the function $\ln x$.

THEOREM 4 **PROPERTIES OF LOGARITHMS**

Let $a > 0$ ($a \neq 1$) and let x and y be positive.

(i) $\log_a xy = \log_a x + \log_a y$.

(ii) $\log_a(x/y) = \log_a x - \log_a y$.

(iii) $\log_a 1 = 0$.

(iv) $\log_a(1/x) = -\log_a x$.

(v) $\log_a a = 1$.

(vi) $\log_a x^y = y \log_a x$. Here y can be any real number.

(vii) $\log_a b = 1/(\log_b a)$, $b > 0$, $b \neq 1$.

(viii) If $a > 1$, $\log_a x$ is an increasing function of x for $x > 0$.

(ix) If $0 < a < 1$, $\log_a x$ is a decreasing function of x for $x > 0$.

(x) $f(x) = \log_a x$ is continuous for every $x > 0$. ∎

EXAMPLE 3
COMPUTING LOGARITHMS

Compute **(a)** $\log_2 8$, **(b)** $\log 10{,}000$, **(c)** $\log_{1/3} 9$.

SOLUTION:

(a) $2^3 = 8$ so $\log_2 8 = 3$.

(b) $10{,}000 = 10^4$ so $\log 10{,}000 = \log_{10} 10{,}000 = 4$.

(c) $\left(\frac{1}{3}\right)^{-2} = 9$ so $\log_{1/3} 9 = -2$.

[†] Most scientific calculators have common logarithm function keys. These are usually labeled $\boxed{\log}$ or $\boxed{\log x}$.

EXAMPLE 4
SKETCHING THE GRAPH OF A
LOGARITHMIC FUNCTION

Sketch the graph of $y = \log_2 x$.

SOLUTION: The graph of $\log_2 x$ is obtained by reflecting the graph of $y = 2^x$ about the line $y = x$. This is done in Figure 2.

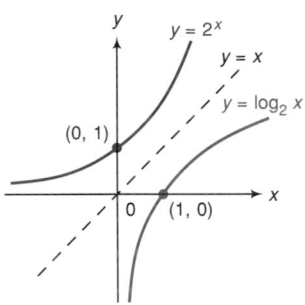

FIGURE 2   GRAPH OF $\log_2 x$. THIS IS THE GRAPH OF 2^x REFLECTED ABOUT THE LINE $y = x$.

EXAMPLE 5
COMPUTING LOGARITHMS TO
OTHER BASES ON A CALCULATOR

Compute

(a) $\log_7 125$ (b) $\log_{1/2} 10$

SOLUTION: We use equation (7):

(a) $\log_7 125 = \dfrac{\ln 125}{\ln 7} = \dfrac{4.828313737}{1.945910149} = 2.481262426$

(b) $\log_{1/2} 10 = \dfrac{\ln 10}{\ln \frac{1}{2}} = \dfrac{2.302585093}{-0.69314718} = -3.321928095$

EXAMPLE 6
USING LOGARITHMIC
DIFFERENTIATION

Differentiate $y = x^x$.

NOTE: x^x is neither a power function nor an exponential function because both the base and the exponent vary. This function is not likely to arise in applications. Nevertheless it is useful to compute its derivative—if only as an exercise in logarithmic differentiation (see page 416).

SOLUTION: If $y = x^x$, then $\ln y = x \ln x$, and

$$\frac{1}{y}\frac{dy}{dx} = x \cdot \frac{1}{x} + \ln x = 1 + \ln x,$$

so that

$$\frac{dy}{dx} = y(1 + \ln x) = x^x(1 + \ln x).$$

■ PROBLEMS 6.5

■ SELF-QUIZ

I. log (one million) = _____.
  a. 6 (because $1,000,000 = 10^6$)
  b. 7 (because 1,000,000 is written with 7 digits)
  c. ln $1,000,000e$
  d. 9 (because 1,000,000 is written with 9 characters)

II. $\log_5 125$ = _____.
  a. 25 (because $5 \cdot 25 = 125$)
  b. 3 (because $5^3 = 125$)
  c. $5 \cdot \frac{1}{125}$ (because $d/dx \ln x = 1/x$)
  d. 243 (because $3^5 = 243$)

III. 10^x = _____.
  a. $(e^{\ln 10})^x$
  b. $e^{\ln(10x)}$
  c. $e^{(\ln 10)x}$
  d. $x^{10}(\log e)$

IV. $(\log x)/(\log e)$ = _____.
  a. $\ln x$
  b. $\log(x/e)$
  c. $\log(x - e)$
  d. $(\log x)^{-\log e}$

V. $\dfrac{d}{dx} 10^x$ = _____.
  a. $10^x/(\ln 10)$
  b. $10^x \cdot \ln 10$
  c. $10^x \cdot \log e$
  d. $10^x/(\log e)$

VI. $\displaystyle\int \frac{1}{cabin}\, d\, cabin$ = _____.
  a. $\log(cabin)$
  b. $\log|cabin| + C$
  c. $\log_e|cabin| + C$
  d. $\ln|cabin| + C$

■ DRILL

In Problems 1–8, compute the value of the given expression. (Do not use any numerical table or electronic calculator.)

1. log 10,000
2. log 0.001
3. $\log_7(\frac{1}{7})$
4. $\log_3(\frac{3}{81})$
5. $\log_{1/2} 32$
6. $\log_{1/3} 27$
7. $\log_{25} 5$
8. $\log_{81} 27$

In Problems 9–18, compute the value of x. (Once again, tables and calculators are not needed.)

9. $x = \log_a a \cdot \log_b b^2 \cdot \log_c c^3$
10. $x - 10^{-\log 100}$
11. $\log x = 10^{-23}$
12. $\log_2 x^4 = 4$
13. $\log_x 64 = 3$
14. $\log_x 125 = -3$
15. $\log x = 3 \log 2 + 2 \log 3$
16. $\log x = 4 \log \frac{1}{2} - 3 \log \frac{1}{3}$
17. $\log x^3 = 2 \log 5 - 3 \log 2$
18. $\log \sqrt{x} = 4 \log 2 - 3 \log \frac{1}{2}$

In Problems 19–28, compute the derivative of the given function.

19. 5^x
20. $(\frac{1}{3})^x$
21. 2^{3x}
22. 10^{3-2x}
23. $\log(100x)$
24. $\log(x^3)$
25. $\log_\pi(3 - 2x)$
26. $\log_\pi(3x^2)$
27. $x^{\sqrt{x}}$
28. $x^{2(1-x^3)}$

In Problems 29–32, compute the given integral.

29. $\displaystyle\int (\frac{1}{2})^x dx$
30. $\displaystyle\int_1^3 2^x dx$
31. $\displaystyle\int \pi^x dx$
32. $\displaystyle\int x^\pi dx$

▦ CALCULATOR PROBLEMS

In Problems 33–42, use equation (7) and a calculator to compute each value.

33. $\log_5 40$
34. $\log_2 2583$
35. $\log_2 0.1524$
36. $\log_2 \pi$
37. $\log_\pi 2$
38. $\log_3 237.8$
39. $\log_\pi 1.285$
40. $\log_2 e$
41. $\log_{1/3} 26$
42. $\log_{0.135} 7.1812$

43. Which is larger: 999^{1000} or 1000^{999}? [*Hint:* Most pocket calculators will show you an error signal if you naively try the keypress sequence: "999 $\boxed{y^x}$ 1000." Find an equivalent comparison that can be resolved by examining things the calculator can handle.]

44. The **factorial** function can be defined for positive integer n by $1 \cdot 2 \cdot 3 \cdots (n-1) \cdot n$; it is usually denoted by $n!$. As n increases, $n!$ grows very rapidly; **Stirling's approximation** says that, when n is large,

$$n! \approx \sqrt{2\pi n}\left(\frac{n}{e}\right)^n.$$

Use the right-hand-side formula to approximate 50!, 100!, and 200!. [*Hint:* Use common logarithms and express your results in scientific notation—i.e., in the form $\alpha \times 10^k$ where $|\alpha| \in [1, 10)$ and k is an integer.]

CONTINUED

■ APPLICATIONS

In Problems 45–48, sketch the graph of the given function.

45. $y = \pi^{x-1}$ **46.** $y = (1/\pi)^{2-x}$
47. $y = \log(x/10)$ ***48.** $y = x^x$

A general psychophysical relation was established in 1834 by the physiologist Weber and given a more precise phrasing later by Fechner.[†] **The Weber–Fechner law** states that $S = c \log(R + d)$ where S is the intensity of a sensation, R is the strength of the stimulus producing it, and c and d are constants.

49. The subjective impression of loudness can be described by a Weber–Fechner law. Let l denote the intensity of a sound. The least intense sound that can be heard is $l_0 = 10^{-12}$ watt/m^2 at a frequency of 1000 cycles/sec (this value is called the *threshold of audibility*). If L denotes the loudness of a sound, measured in decibels,[‡] then $L = 10 \log(l/l_0)$ relates intensity and loudness.

a. If one sound has twice the intensity of another, what is the ratio of the perceived loudness of the two sounds?
b. If one sound appears to be twice as loud as another, what is the ratio of their intensities?
c. Ordinary conversation sounds six times as loud as a low whisper. What is the actual ratio of intensity of their sounds?

■ SHOW/PROVE/DISPROVE

50. Show, using logarithms, that if $a = [1 + (1/n)]^n$ and $b = [1 + (1/n)]^{n+1}$, then $a^b = b^a$. What does this result state for the case $n = 1$?

****51.** Prove that $2^\alpha < 1 + \alpha$ for all α in $(0, 1)$. [*Hint:* Let $f(x) = 2^x$ and show that f is concave up.]

52. Suppose β is a fixed real number and $x > 0$.
a. Show that $x^\beta = e^{\beta \ln x}$.
b. Show that $\dfrac{d}{dx} e^{\beta \ln x} = (\beta/x)e^{\beta \ln x}$. [*Hint:* Set $u = \beta \ln x$ and apply the chain rule.]
c. Using the results of parts (a) and (b), show that for any real number β, $\dfrac{d}{dx} x^\beta = \beta x^{\beta - 1}$.

***53.** Suppose that $\alpha > 0$. Define the function A by $A(x) = \alpha^x$. Prove directly from the definition of the derivative that $A'(x) = A'(0) \cdot A(x)$.

***54.** Suppose $a > 0$ ($a \neq 1$); show that the function $\log_a x$ is continuous throughout the interval $(0, \infty)$.

55. Show, without using a calculator, that $e^\pi > \pi^e$ by carrying out the following analysis of the function $f(x) = e^x/x^e$.
a. Show that $f(1) > 1$, $f(e) = 1$, and $f(10) > 1$.
b. Explain why f must reach a minimum value in the open interval $(1, 10)$.
c. Compute $f'(x)$.
d. Show that the minimum of f in $[1, 10]$ occurs at $x = e$.
e. Explain why $e^\pi > \pi^e$.
f. Show that $e^x > x^e$ for any $x \neq e$ in the interval $[1, 10]$.

■ CHALLENGE

***56.** Prove that $x^p \cdot y^q \leq px + qy$ if $x, y, p, q > 0$ and $p + q = 1$.

****57.** Suppose that x and y are positive real numbers such that $x + y = 1$. Prove that $x^x + y^y \geq \sqrt{2}$, and discuss conditions for equality.

****58.** For what values of a is the graph of $y = a^x$ tangent to the graph of $y = \log_a x$?

■ ANSWERS TO SELF-QUIZ

I. a   II. b   V. b, d   VI. c, d
III. $a = c$ IV. a

[†] Ernst Weber (1795–1878) was a German physiologist. Gustav Fechner (1801–1887) was a German physicist.
[‡] 1 decibel (dB) = $\frac{1}{10}$ bel, named after Alexander Graham Bell (1847–1922), inventor of the telephone.

6.6 DIFFERENTIAL EQUATIONS OF EXPONENTIAL GROWTH AND DECAY

In this section, we begin to illustrate the great importance of the exponential function e^x in applications. Before citing examples, we will discuss a very basic type of mathematical model.

Let $y = f(x)$ represent some physical quantity such as the volume of a substance, the population of a certain species, the mass of a decaying radioactive substance, the number of dollars invested in bonds, and so on. Then the growth of $f(x)$ is given by its derivative dy/dx. Thus, if $f(x)$ is growing at a constant rate, then $dy/dx = k$ and $y = kx + C$, that is, $y = f(x)$ is a straight-line function.

It is sometimes more interesting and more appropriate to consider the **relative rate of growth**, defined by

$$\text{relative rate of growth} = \frac{\text{actual rate of growth}}{\text{size of } f(x)} = \frac{f'(x)}{f(x)} = \frac{dy/dx}{y}. \tag{1}$$

The relative rate of growth indicates the percentage increase or decrease in f. For example, an increase of 100 individuals for a species with a population size of 500 would probably have a significant impact, being an increase of 20%. On the other hand, if the population were 1,000,000, then the addition of 100 would hardly be noticed, being an increase of only 0.01%.

In many applications, we are told that the relative rate of growth of the given physical quantity is constant. That is,

$$\frac{dy/dx}{y} = \alpha \tag{2}$$

or

$$\frac{dy}{dx} = \alpha y \tag{3}$$

where α is the constant percentage increase or decrease in the quantity.

Another way to view equation (3) is that it tells us that the function is changing at a rate proportional to itself. If the constant of proportionality α is greater than 0, the quantity is increasing; while if $\alpha < 0$, it is decreasing. Equation (3) is called a **differential equation** because it is an equation involving a derivative. Differential equations arise in a great variety of settings in the physical, biological, engineering, and social sciences. It is a vast subject, and many of you will take courses in differential equations after completing your basic calculus sequence. We will study differential equations in some detail in Chapter 16.

We return to equation (3). To solve it,[†] we carry out the following steps:

(i) Multiply both sides of equation (3) by $e^{-\alpha x}$ to obtain

$$\frac{dy}{dx} e^{-\alpha x} - \alpha y e^{-\alpha x} = 0. \tag{4}$$

(ii) Observe that

$$\frac{d}{dx}(e^{-\alpha x} y) \overset{\text{Product rule}}{=} e^{-\alpha x} \frac{dy}{dx} + y(-\alpha e^{-\alpha x}). \tag{5}$$

[†] By a solution to this differential equation, we mean a function f that is differentiable and for which $f'(x) = \alpha f(x)$.

(iii) Using (5), write (4) as

$$\frac{d}{dx}(e^{-\alpha x}y) = 0.$$

(iv) Since the derivative of $ye^{-\alpha x}$ is zero,

$$ye^{-\alpha x} = c \qquad \text{for some constant } c. \tag{6}$$

(v) Multiply both sides of (6) by $e^{\alpha x}$ to obtain

$$y = ce^{\alpha x}.$$

Thus, any solution to (3) can be written as

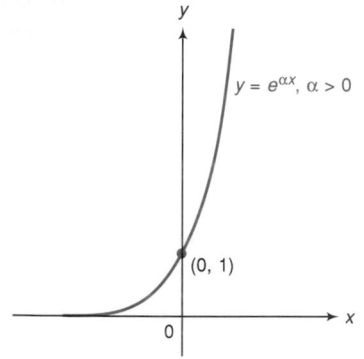

(a) Exponential growth.

> **SOLUTION TO THE DIFFERENTIAL EQUATION OF CONSTANT RELATIVE GROWTH**
>
> $$y = ce^{\alpha x} \tag{7}$$

where c can be any real number. To check this, we simply differentiate:

$$\frac{dy}{dx} = \frac{d}{dx}ce^{\alpha x} = c\frac{d}{dx}e^{\alpha x} = c\alpha e^{\alpha x} = \alpha(ce^{\alpha x}) = \alpha y,$$

so that $y = ce^{\alpha x}$ does satisfy (3). We therefore have proven the next theorem.

THEOREM 1

If α is any real number, then there are an infinite number of solutions to the differential equation $y' = \alpha y$. They take the form $y = ce^{\alpha x}$ for any real number c. ■

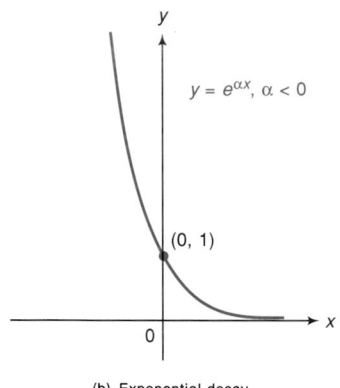

(b) Exponential decay.

FIGURE 1
CURVES SHOWING EXPONENTIAL
GROWTH AND DECAY

If $\alpha > 0$, we say that the quantity described by $f(x)$ is **growing exponentially**. If $\alpha < 0$, it is **decaying exponentially** (see Figure 1). Of course, if $\alpha = 0$, then there is no growth and $f(x)$ remains constant.

For a physical problem, we often know more than the rate of growth. We may also know one particular value of x, say $y(x_0) = y_0$. This value is called an **initial condition**, and it will give us a unique solution to the problem. The differential equation (3) together with an initial condition is called an **initial-value problem**. We will find the unique solutions to a number of initial-value problems in this section.

EXAMPLE 1
SOLVING AN INITIAL-VALUE
PROBLEM

(a) Find all solutions to $dy/dx = 3y$.
(b) Find the solution that satisfies the initial condition $y(0) = 2$.

SOLUTION:

(a) Since $\alpha = 3$, all solutions are of the form $y = ce^{3x}$.
(b) $2 = y(0) = ce^{3 \cdot 0} = c \cdot 1 = c$, so $c = 2$ and the unique solution is
$y = 2e^{3x}$.

EXAMPLE 2
GROWTH OF A BACTERIAL
POPULATION

A bacterial population is growing continuously at a rate equal to 10% of its population each day. Its initial size is 10,000 organisms. How many bacteria are present after 10 days? After 30 days?

SOLUTION: Since the percentage growth of the population is $10\% = 0.1$, we have

$$\frac{dP/dt}{P} = 0.1 \quad \text{or} \quad \frac{dP}{dt} = 0.1P.$$

Here $\alpha = 0.1$, and all solutions have the form

$$P(t) = ce^{0.1t}$$

where t is measured in days. Since $P(0) = 10,000$, we have

$$ce^{(0.1)(0)} = c = 10,000, \quad \text{and} \quad P(t) = 10,000e^{0.1t}.$$

This is the unique solution to the initial-value problem. After 10 days $P(10) = 10,000e^{(0.1)(10)} = 10,000e \approx 27,183$, and after 30 days $P(30) = 10,000e^{0.1(30)} = 10,000e^3 \approx 200,855$ bacteria.

Before giving further examples, we summarize the principal result of this section:

> **THE EQUATION OF EXPONENTIAL GROWTH OR DECAY**
>
> If $P(t)$ is growing or declining continuously at a constant relative rate (so that $P'(t)/P(t) = \pm \alpha$) with initial size $P(0)$, then $P(t)$ is given by (with $\alpha > 0$)
>
> $$P(t) = P(0)e^{\alpha t} \quad \text{or} \quad P(t) = P(0)e^{-\alpha t}. \tag{8}$$

The term $P(0)e^{\alpha t}$ represents exponential growth and the term $P(0)e^{-\alpha t}$ represents exponential decay.

EXAMPLE 3
POPULATION GROWTH IN INDIA

The population of India was estimated to be 574,220,000 in 1974 and 746,388,000 in 1984. Assume that the relative growth rate remains constant and that growth is continuous.

    (a) Estimate the population in 1994.

    (b) When will the population reach 1.5 billion?

SOLUTION:

    (a) From (8), we have

$$P(t) = P(0)e^{\alpha t}.$$

We have $\alpha > 0$ because the population is increasing.

Treat the year 1974 as year zero. Then $1984 =$ year 10. We are told that

$$P(0) = 574{,}220{,}000 \qquad \text{and} \qquad P(10) = 746{,}388{,}000.$$

Thus

$$P(t) = 574{,}220{,}000e^{\alpha t}$$

$$P(10) = 746{,}388{,}000 = 574{,}220{,}000e^{10\alpha}$$

$$e^{10\alpha} = \frac{746{,}388{,}000}{574{,}220{,}000} \approx 1.3$$

$$\ln e^{10\alpha} \approx \ln 1.3$$

$$10\alpha \approx \ln 1.3 \qquad \ln e^x = x$$

$$\alpha \approx \frac{\ln 1.3}{10}.$$

The year 1994 is year 20. Thus

$$P(20) = 574{,}220{,}000e^{20\alpha} \approx 574{,}220{,}000e^{\left(20(\ln 1.3/10)\right)}$$

$$r \ln x = \ln x^r$$

$$= 574{,}220{,}000e^{2 \ln 1.3} = 574{,}220{,}000e^{\ln 1.3^2}$$

$$e^{\ln x} = x$$

$$P(20) = (574{,}220{,}000)(1.3)^2$$

$$= (574{,}220{,}000)(1.69) \approx 970{,}432{,}000.$$

(b) We seek a number t such that $P(t) = 1{,}500{,}000{,}000$. That is,

$$P(t) = 1{,}500{,}000{,}000 = 574{,}220{,}000e^{\alpha t}$$

$$e^{\alpha t} = \frac{1{,}500{,}000{,}000}{574{,}220{,}000} \approx 2.61$$

$$\ln e^{\alpha t} = \ln 2.61$$

$$\alpha t = \ln 2.61$$

$$t = \frac{\ln 2.61}{\alpha}$$

$$= \frac{\ln 2.61}{\ln 1.3/10} = \frac{10 \ln 2.61}{\ln 1.3}$$

$$\approx 36.6 \text{ years.}$$

Then

$$\text{Year } 36.6 = 1974 + 36.6 = 2010.6.$$

We conclude that the population of India would reach 1.5 billion sometime in the year 2010 if its population growth rate continued at the same rate as it was between 1974 and 1984.

In this problem we found that

$$\alpha \approx \frac{\ln 1.3}{10} \approx 0.0262.$$

This means that the problem is growing at a rate of about 2.62% a year. We stress that all the calculations in this problem were made under the assumption that this percentage did not vary and will not vary at all from 1974 until now and into the foreseeable future. This might not be a reasonable assumption. In that case our answers are not reasonable.

In this example, we created a mathematical model to solve a problem based on certain assumptions about population growth. Solving the problem was much easier than checking the validity of our assumptions. You should keep this in mind whenever you use a mathematical model.

EXAMPLE 4
NEWTON'S LAW OF COOLING

Newton's law of cooling states that the rate of change of the temperature difference between an object and its surrounding medium is proportional to the temperature difference. If $D(t)$ denotes this temperature difference at time t and if α denotes the constant of proportionality, then we obtain

$$\frac{dD}{dt} = -\alpha D.$$

The minus sign indicates that this difference decreases. (If the object is cooler than the surrounding medium—usually air—it will warm up; if it is hotter, it will cool.) The solution to this differential equation is

$$D(t) = ce^{-\alpha t}.$$

If we denote the initial ($t = 0$) temperature difference by D_0, then

$$D(t) = D_0 e^{-\alpha t} \qquad (9)$$

is the formula for the temperature difference for any $t > 0$. Notice that for t large $e^{-\alpha t}$ is very small, so that, as we have all observed, temperature differences tend to die out rather quickly.

We can rewrite (9) in a form that is often more useful. Let $T(t)$ denote the temperature of the object at time t, let T_s denote the temperature of the surroundings (which is assumed to be constant throughout), and let T_0 denote the initial temperature of the object. Then $D(t) = T(t) - T_s$ and $D_0 = T_0 - T_s$, so (9) becomes

$$T(t) - T_s = (T_0 - T_s)e^{-\alpha t}$$

or

$$T(t) = T_s + (T_0 - T_s)e^{-\alpha t}. \qquad (10)$$

We now may ask: In terms of the constant α, how long does it take for the temperature difference to decrease to half its original value?

SOLUTION: The original value is D_0. We are therefore looking for a value of t for which $D(t) = \frac{1}{2}D_0$. That is, $\frac{1}{2}D_0 = D_0 e^{-\alpha t}$, or $e^{-\alpha t} = \frac{1}{2}$.

Taking natural logarithms, we obtain

$$-\alpha t = \ln\frac{1}{2} = -\ln 2 = -0.6931, \quad \text{and} \quad t \approx \frac{0.6931}{\alpha}.$$

Notice that this value of t does *not* depend on the initial temperature difference D_0.

EXAMPLE 5
USING NEWTON'S LAW OF COOLING

With the air temperature equal to 30°C, an object with an initial temperature of 10°C warmed to 14°C in 1 hr.

(a) What was its temperature after 2 hr?

(b) After how many hours was its temperature 25°C?

SOLUTION: Here $T_s = 30$, $T_0 = 10$ and $T(1) = 14$. From (10), we have

$$T(t) = 30 - 20e^{-\alpha t}.$$

But

$$14 = T(1) = 30 - 20e^{-\alpha}$$

$$20e^{-\alpha} = 16$$

$$e^{-\alpha} = \frac{4}{5} = 0.8.$$

Thus,

$$T(t) = 30 - 20e^{-\alpha t} = 30 - 20(e^{-\alpha})^t = 30 - 20(0.8)^t.$$

We can now answer the two questions.

(a) $T(2) = 30 - 20(0.8)^2 = 30 - 20(0.64) = 17.2°C.$

(b) We need to find t such that $T(t) = 25$. That is,

$$25 = 30 - 20(0.8)^t, \quad \text{or} \quad (0.8)^t = \frac{1}{4} \quad \text{or} \quad t\ln(0.8) = -\ln 4,$$

and

$$t = \frac{-\ln 4}{\ln(0.8)} \approx \frac{1.3863}{0.2231} \approx 6.2 \text{ hr} = 6 \text{ hr } 12 \text{ min.}$$

EXAMPLE 6
CARBON DATING

Carbon dating is a technique used by archaeologists, geologists, and others who want to estimate the ages of certain artifacts and fossils they uncover. The technique is based on certain properties of the carbon atom. In its natural state the nucleus of the carbon atom ^{12}C has 6 protons and 6 neutrons. The **isotope** carbon-14, ^{14}C, is produced through cosmic-ray bombardment of nitrogen in the atmosphere. Carbon-14 has 6 protons and 8 neutrons and is **radioactive** It decays by beta emission. That is, when an atom of ^{14}C decays, it gives up an electron to form a stable nitrogen atom ^{14}N. We make the assumption that the ratio of ^{14}C to ^{12}C in the atmosphere is constant. This assumption has been shown experimentally to be approximately valid, for although ^{14}C is being constantly lost through **radioactive decay** (as this process is often termed), new ^{14}C is constantly being produced by the cosmic bombardment of nitrogen in the upper atmosphere. Living plants and animals do not distinguish between ^{12}C and ^{14}C, so at the time of death the ratio of ^{12}C to ^{14}C in an organism is the same as the ratio in the atmosphere. However, this ratio changes after death since ^{14}C is converted to ^{14}N but no further ^{14}C is taken in.

It has been observed that ^{14}C decays at a rate proportional to its mass and that its **half-life** is approximately 5730 years.[†] That is, if a substance starts with 1 g of ^{14}C, then 5730 years later it will have $\frac{1}{2}$ g of ^{14}C, the other $\frac{1}{2}$ g having been converted to ^{14}N.

We may now pose a question typically asked by an archaeologist. A fossil is unearthed and it is determined that the amount of ^{14}C present is 40% of what it would be for a similarly sized living organism. What is the approximate age of the fossil?

SOLUTION: Let $M(t)$ denote the mass of ^{14}C present in the fossil. Then since ^{14}C decays at a rate proportional to its mass, we have

$$\frac{dM}{dt} = -\alpha M,$$

where α is the constant of proportionality. Then $M(t) = ce^{-\alpha t}$, where $c = M_0$, the initial amount of ^{14}C present. When $t = 0$, $M(0) = M_0$; when $t = 5730$ years, $M(5730) = \frac{1}{2}M_0$, since half the original amount of ^{14}C has been converted to ^{12}C. We can use this fact to solve for α since we have

$$\frac{1}{2}M_0 = M_0 e^{-\alpha \cdot 5730}, \qquad \text{or} \qquad e^{5730\alpha} = \frac{1}{2}.$$

Thus,

$$(e^{-\alpha})^{5730} = \frac{1}{2}, \quad \text{or} \quad e^{-\alpha} = \left(\frac{1}{2}\right)^{1/5730}, \quad \text{and} \quad e^{-\alpha t} = \left(\frac{1}{2}\right)^{t/5730},$$

so

$$M(t) = M_0 \left(\frac{1}{2}\right)^{t/5730}.$$

[†] This number was first determined in 1949 by the American chemist W. S. Libby, who based his calculations on the wood from sequoia trees, whose ages were determined by rings marking years of growth. Libby's method has come to be regarded as the archaeologist's absolute measuring scale. But in truth, this scale is flawed. Libby used the assumption that the atmosphere had at all times a constant amount of ^{14}C. However, the American chemist C. W. Ferguson of the University of Arizona deduced from his study of tree rings in 4000-year-old American giant trees that before 1500 B.C. the radiocarbon content of the atmosphere was considerably higher than it was later. This result implied that objects from the pre–1500 B.C. era were much older than previously believed, because Libby's "clock" allowed for a smaller amount of ^{14}C than actually was present. For example, a find dated at 1800 B.C. was in fact from 2500 B.C. This has had a considerable impact on the study of prehistoric times. For a fascinating discussion of this subject, see Gerhard Herm, *The Celts* (New York: St. Martin's Press, 1975), pages 90–92.

Now we are told that after t years (from the death of the fossilized organism to the present) $M(t) = 0.4M_0$, and we are asked to determine t. Then,

$$0.4M_0 = M_0\left(\frac{1}{2}\right)^{t/5730},$$

and taking natural logarithms (after dividing by M_0), we obtain

$$\ln 0.4 = \frac{t}{5730}\ln\left(\frac{1}{2}\right), \quad \text{or} \quad t = \frac{5730 \ln(0.4)}{\ln(\frac{1}{2})} \approx 7575 \text{ years.}$$

The carbon-dating method has been used successfully on numerous occasions. It was this technique that established that the Dead Sea scrolls were prepared and buried about two thousand years ago.

COMPOUND INTEREST

Suppose A_0 dollars are invested in an enterprise (which may be a bank, bonds, a common stock, etc.) with an annual interest rate of i. **Simple interest** is the amount earned on the $$A_0$ over a period of time. If the $$A_0$ is invested for t years, then the simple interest I is given by

$$I = A_0 ti. \tag{11}$$

EXAMPLE 7
COMPUTING SIMPLE INTEREST

If $1000 is invested for 5 years with an interest rate of 6%, then $i = 0.06$, and the simple interest earned is

$$I = (\$1000)(5)(0.06) = \$300.$$

Compound interest is interest paid on the interest previously earned as well as on the original investment. Suppose that interest is paid annually. Then if $$A_0$ is invested, the interest after one year is $$iA_0$, and the original investment is now worth $A_0 + iA_0 = A_0(1 + i)$ dollars. After two years, the compound interest paid would be $i[A_0(1 + i)]$ dollars, and the investment would then be worth

$$A_0(1 + i) + iA_0(1 + i) = A_0(1 + i)(1 + i) = A_0(1 + i)^2$$

dollars. Continuing in this fashion, we see that after t years the investment would be worth

$$A(t) = A_0(1 + i)^t \text{ dollars.} \tag{12}$$

We have used the notation $A(t)$ to denote the value of the investment after t years.

EXAMPLE 8
COMPUTING COMPOUND
INTEREST WITH ANNUAL
COMPOUNDING

If the interest in Example 7 is compounded annually, then after 5 years the investment is worth

$$A(5) = 1000(1 + 0.06)^5 = 1000(1.33823) = \$1338.23.$$

The actual interest paid is $338.23.

In practice, interest is compounded more frequently than annually. If it is paid m times a year, then in each interest period the rate of interest is i/m, and in t years there are tm pay periods. Then, according to formula (12), we obtain the following rule:

> **COMPOUND-INTEREST FORMULA: COMPOUNDING m TIMES A YEAR**
>
> The value of an investment of $\$A_0$ compounded m times a year with an annual interest rate of i after t years is
>
> $$A(t) = A_0\left(1 + \frac{i}{m}\right)^{mt}. \qquad (13)$$

EXAMPLE 9
COMPUTING COMPOUND
INTEREST WITH QUARTERLY
COMPOUNDING

If the interest in Example 7 is compounded quarterly (4 times a year), then after 5 years the investment is worth

$$A(5) = 1000\left(1 + \frac{0.06}{4}\right)^{(4)(5)} = 1000(1.015)^{20} = 1000(1.34686)$$

$$= \$1346.86.$$

The interest paid is now $346.86.

It is clear from the examples above that the interest actually paid increases as the number of interest periods increases. We are naturally led to ask: What is the value of an investment if interest is **compounded continuously**—that is, when the number of interest payments approaches infinity? Alternatively, what is the interest when the length of a payment period approaches zero?

In the time period t to $t + \Delta t$, the interest earned is $A(t + \Delta t) - A(t)$. If Δt is small, then the interest paid on $A(t)$ dollars would be [from formula (11)] approximately equal to $A(t)\,\Delta t\, i$. We say "approximately" because $A(t)\,\Delta t\, i$ represents simple interest between t and $t + \Delta t$. However, the difference between this approximation and the actual interest paid is small if Δt is small. Thus,

$$A(t + \Delta t) - A(t) \approx A(t)\,\Delta t\, i,$$

or dividing by Δt, we have

$$\frac{A(t + \Delta t) - A(t)}{\Delta t} \approx iA(t).$$

Then taking the limit as $\Delta t \to 0$, we obtain

$$\frac{dA}{dt} = iA(t),$$

or, from (8),

$$A(t) = ce^{it}.$$

But $A(0) = A_0$, which tells us that $c = A_0$, and we have

COMPOUND-INTEREST FORMULA: CONTINUOUS COMPOUNDING

$$A(t) = A_0 e^{it} \tag{14}$$

is the value of an original investment after t years with an interest rate of i compounded continuously.

EXAMPLE 10
COMPUTING COMPOUND
INTEREST WITH CONTINUOUS
COMPOUNDING

If the interest in Example 7 is compounded continuously, then

$$A(5) = 1000e^{(0.06)(5)} = 1000e^{0.3} = \$1349.86,$$

and the interest paid is $349.86. Compare this result to the results in Examples 8 and 9.

EXAMPLE 11
FUTURE VALUE OF A BOND WITH
CONTINUOUS COMPOUNDING

$5000 is invested in a bond yielding $8\frac{1}{2}\%$ annually. What will the bond be worth in 10 years if interest is compounded continuously?

SOLUTION: $A(t) = A_0 e^{it} = 5000e^{(0.085)(10)} = 5000e^{0.85} = \$11{,}698.23.$

EXAMPLE 12
COMPUTING THE DOUBLING TIME

How long does it take for an investment to double if the annual interest rate is 6% compounded continuously?

SOLUTION: We need to determine a value of t such that $A(t) = 2A_0$. That is,

$$A_0 e^{0.06t} = 2A_0, \quad \text{or} \quad e^{0.06t} = 2.$$

Taking natural logarithms, we obtain $0.06t = \ln 2 \approx 0.6931$, and $t \approx 0.6931/0.06 = 11.55$ years.

EXAMPLE 13
COMPUTING THE INTEREST RATE

What must be the interest rate in order for an investment to double in 7 years when interest is compounded continuously?

SOLUTION: We need to determine i such that $A_0 e^{7i} = 2A_0$. Then $e^{7i} = 2$, $7i = \ln 2$, and $i \approx 0.6931/7 = 0.099 = 9.9\%$.

EXAMPLE 14
COMPUTING THE EFFECTIVE
INTEREST RATE

If money is invested at 8% compounded continuously, what is the effective rate of interest?

SOLUTION: The **effective interest rate** is the actual rate of simple interest received on $1 over a 1-year period. If we start with 1 dollar, there will be $1e^{0.08} \approx 1(1.0833)$ after 1 year. Thus, the effective interest rate is about $0.0833 \approx 8\frac{1}{3}\%$.

HISTORICAL NOTE: COMPOUND INTEREST IN THE SEVENTEENTH CENTURY

JAKOB BERNOULLI
1654–1705

In 1690, the great Swiss mathematician, Jakob Bernoulli (1654–1705), proposed the following problem: "*Quaeritur, si creditor aliquis pecuniam suam foenori exponat, ea lege, ut singulis momentis pars proportionalis usarae annuae sorti annumeretur; quantum ipsi finito anno debeatur?*" ["This question is posed: if a creditor lends his money at interest with the understanding that a proportionate part of the annual interest be added to the principal periodically, how much is due to him at the end of a year?"]

Bernoulli's passage expresses clearly the concept of compound interest on a loan, and reflects a culture in which this practice is new and stands in contrast to the more familiar practice of simple annual interest. Under the older, simple-interest system, the creditor who lent $100 for a year at 10% would be due $110 at the end of the year. Under the system described in the question, assuming the *singulis momentis*[†] to be "quarterly," the principal would increase from $100 to $102.50 at the end of three months ($10 annual interest divided by 4 = $2.50, added to the principal). During the second quarter then, the annual interest would be reckoned, not on the original $100, but on $102.50, and so on.

Bernoulli partially answered his own question by deriving the compound-interest formula (13). But his real question amounted to this: as m, the number of times interest is paid each year, increases without bound, would the depositor acquire an unbounded fortune? In our notation, Bernoulli was asking

What happens to $\left(1 + \dfrac{x}{m}\right)^m$ as $m \to \infty$?

We know that the limit is e^x and that the investor would not acquire unlimited wealth. Bernoulli concluded the same thing. But he could not go further than to claim from numerical evidence that the limit for ($x = 1$) was somewhere between 2.7 and 2.8. He did not have the advantage of using the number e because the number e was not introduced by Euler until 1728, 38 years later (and 23 years after Bernoulli's death).

[†] That is, "periodically"; "from time to time"; "by periodic increments".

■ PROBLEMS 6.6

■ SELF-QUIZ

I. Suppose y depends smoothly on x (i.e., $dy/dx = y'$ exists). Also suppose that its relative rate of growth is constant. Then y satisfies a differential equation of the form _____ (where α is a nonzero constant).
  a. $yy' = \alpha$ b. $y'/y = \alpha$
  c. $y' + 1/y = \alpha$ d. $y' = \alpha/y$

II. The general solution to $y'/y = 10$ is _____.
  a. $\ln|y| = 10x + C$ b. $y = K \cdot e^{10x}$
  c. $x = K + e^{10y}$ d. $y = 5y^2 + K$

III. Suppose y is a smooth function of x such that $y'/y = -3$; then $\lim_{x \to \infty} y$ _____.
  a. $= 1$ b. $= -\frac{3}{2}$ c. $= \infty$
  d. $= 0$ e. doesn't exist
  f. can't be found from this information alone

IV. Suppose y is a smooth function of x such that $y'/y = -3$; then $\lim_{x \to 0} y$ _____.
  a. $= 1$ b. $= -\frac{3}{2}$ c. $= \infty$
  d. $= 0$ e. doesn't exist
  f. can't be determined from this information

▦ ■ APPLICATIONS

1. The growth rate of a bacteria population is proportional to its size. Initially the population is 10,000, while after 10 days its size is 25,000. What will the population size be after 20 days (that's 10 days beyond the time when the population is 25,000)? After 30 days?

2. In Problem 1, suppose instead that the population after 10 days is 6000. How large will this population be after 20 days? After 30 days?

3. The following data were collected from a bacteria population which is known to grow exponentially.

Number of Days	Number of Bacteria
5	936
10	2190
20	11,993

  a. What was the initial population?
  b. If the present growth rate were to continue, what would be the population after 60 days?

4. During one experiment, a particular bacteria population declines exponentially; the following data were collected during the first two days of the experiment:

Number of Hours	Number of Bacteria
12	5969
24	3563
48	1269

  a. What was the initial population?
  b. How many bacteria will be left after one week?
  c. When will there be no bacteria left (i.e., how soon is $P(t) < 1$)?

5. When the air temperature is 70°F, a particular object cools from 170°F to 140°F in $\frac{1}{2}$ hr. Assume Newton's law of cooling applies.
  a. What will be the object's temperature after 1 hr?
  b. When will the temperature be 90°F?

6. A hot coal (temperature 150°C) is immersed in ice salt water (temperature −10°C). After 30 seconds the temperature of the coal is 60°C. Assume the ice water temperature is maintained at −10°C.
  a. What is the temperature of the coal after 2 min?
  b. When will the temperature of the coal be 0°C?

7. A fossilized leaf contains 70% of a "normal" amount of ^{14}C. How old is the fossil?

8. Forty percent of a particular radioactive substance disappears in 100 yrs.
  a. What is its half-life?
  b. After how many years will 90% be gone?

9. A lump sum of $5000 is invested with a return of 7%/yr.
  a. What amount would be paid in simple interest during an 8-yr period?
  b. If the interest is compounded annually, what is the investment worth after 8 yr?
  c. If interest is compounded monthly, what is the investment worth after 8 yr?
  d. If the interest is compounded continuously, what is the investment worth after 8 yr?

10. a. As a gimmick to lure depositors, some banks offer 5% interest compounded continuously in comparison to their competitors who offer

$5\frac{1}{8}\%$ compounded annually. Which bank would you choose?

b. Suppose a competitor now compounds $5\frac{1}{8}\%$ semiannually. Now, which bank offers the better deal?

11. a. What must be the interest rate in order for an investment to triple in 15 yr if interest is continuously compounded?

b. If the investor changes his goals and now wants the initial amount to double in 10 yr, what should the interest rate be?

12. a. What is the most a banker should pay for a $100,000 note due in 5 yr if she can invest a like amount of money at 9% compounded annually? (This is called the **present value** of the note.)

b. Suppose the 9% were compounded continuously; recalculate the present value of that $100,000 due in 5 yr.

13. Atmospheric pressure P is a function of altitude a above sea level and satisfies the relation $dP/da = \beta P$, where β is a constant. The pressure is measured in millibars (mbar). At sea level ($a = 0$), $P(0)$ is 1013.25 mbar, which means that the atmosphere at sea level will support a column of mercury 1013.25 mm high at a standard temperature of 15°C. At an altitude of $a = 1500$ m, the pressure is 845.6 mbar.

a. What is the pressure at $a = 4000$ m?

b. What is the pressure at 10 km?

c. In California the highest and lowest points are Mount Whitney (4418 m) and Death Valley (86 m below sea level). What is the difference in their atmospheric pressures?

d. What is the atmospheric pressure atop Mount Everest (8848 m)?

e. At what elevation is the atmospheric pressure equal to 1 mbar?

†14. In a certain medical treatment, a tracer dye is injected into the pancreas to measure its function rate. A normally active pancreas will excrete 4% of the dye each minute. A physician injects 0.3 gm of the dye into a sick patient; 30 min later 0.14 gm remain. How much dye would remain if the pancreas were functioning normally?

15. Salt decomposes in water into sodium $[Na^+]$ and chloride $[Cl^-]$ ions at a rate proportional to its mass. Suppose there were initially 25 kg of salt and 15 kg after 10 hr.

a. How much salt would be left after one day?

b. After how many hours would there be less than $\frac{1}{2}$ kg of salt left?

16. X-rays are absorbed into a uniform, partially opaque body as a function not of time but of penetration distance. The rate of change of the intensity $I(x)$ of the X-ray is proportional to the intensity. Here x measures the distance of penetration. The more the X-ray penetrates, the lower the intensity is. The constant of proportionality is the density D of the medium being penetrated.

a. Formulate a differential equation describing this phenomenon.

b. Solve for $I(x)$ in terms of x, D, and the initial (surface) intensity $I(0)$.

17. The estimated world population in 1986 was 4,845,000,000. Assume that the population grows at a constant rate of 1% a year. When will the world population reach 8 billion?

18. In Problem 17, at what constant annual rate would the population grow if it reached 6 billion in the year 2000?

19. According to the official U.S. census, the population of the U.S. was 203,302,031 in 1970 and 226,549,448 in 1980. Assume that population grows at a constant relative rate.

a. Estimate the population in the year 2000.

b. When will the population reach $\frac{1}{2}$ billion?

20. The population of the U.S. was 76,212,168 in 1900 and 92,228,496 in 1910. If population had grown at a constant percentage until 1980, what would the population have been in 1970 and 1980? Compare this answer with the data given in Problem 19. To explain this discrepancy is a problem in history, not calculus.

‡21. In 1920, the consumer price index (CPI) for perishable goods was 213.4 with 1913 assigned the base rate of 100. Assuming that inflation was constant (that is, assume continuous compounding), find the annual rate of inflation of the price of perishable goods between 1913 and 1920.

---

† This and similar mathematical models in medicine are discussed in the paper by J. S. Rustagi, "Mathematical Models in Medicine," *International Journal of Mathematics Education in Science and Technology* 2 (1971): 193–203.

‡ This means that an average item costing $1.00 in 1913 would cost 2.13\frac{4}{10}$ in 1920.

22. The average annual earnings of full-time employees in the U.S. was \$4,743 in 1960 and \$7,564 in 1970. Assuming a continuous increase in earnings at a constant rate during the period 1960–70, what was the annual increase in wages?

23. The CPI was 88.7 in 1960 and 116.3 in 1970 (1967 = 100). Assuming a continuous increase in prices at a constant rate between 1960 and 1970, find the annual rate of inflation.

24. The *real* increase in earnings is defined as the percentage increase in wages minus the rate of inflation divided by the quantity 100 (%) plus the rate of inflation. That is $RI = \dfrac{w - i}{1 + i}$ where w and i are given as fractions. Using the data in Problems 22 and 23, determine the real annual increase in the average U.S. worker's earnings between 1960 and 1970.

■ CHALLENGE

*25. The president and vice president sit down for coffee. They are both served a cup of hot black coffee (at the same temperature). The president takes a container of cream and immediately adds it to her coffee, stirs it, and waits. The vice-president waits 10 minutes and then adds the same amount of cream (which has been kept cool at its original temperature) to his coffee and stirs it in. Then they both drink. Assuming the temperature of the cream is lower than that of the air, who drinks the hotter coffee? [*Hint:* Use Newton's law of cooling. It is necessary to treat each case separately and to keep track of the volumes of coffee, cream, and the coffee-cream mixture.[†]]

■ ANSWERS TO SELF-QUIZ

I. b

II. a, b (a implies b)

III. d

IV. f [The limit is $y(0)$.]

[†] This is a hoary problem that keeps on popping up, with an ever-changing pair of characters. The problem is nontrivial and has stymied many a student; do not get frustrated if you cannot solve it right away. The trick is to write everything down and to keep track of all the variables. The fact that the air is warmer than the cream is critical. Also note that guessing the correct answer is fairly easy; proving your guess is correct is where your work is apt to become nontrivial.

6.7 INTEGRATION OF TRIGONOMETRIC FUNCTIONS

In Section 2.5, we computed the derivatives of the six trigonometric functions. The following theorem results from the formulas obtained in Section 2.5 and the chain rule. Keep in mind that, as in Section 2.5, if x, θ, or u represents an angle, then x, θ, or u is measured in *radians*.

THEOREM 1

Suppose $u(x)$ is a differentiable function of x and $u(x)$ is measured in radians.

(i) $\dfrac{d}{dx} \sin u = \cos u \dfrac{du}{dx}$

(ii) $\dfrac{d}{dx} \cos u = -\sin u \dfrac{du}{dx}$

(iii) $\dfrac{d}{dx} \tan u = \sec^2 u \dfrac{du}{dx}$

TABLE 1

(i) $\int \cos u \, du = \sin u + C$

(ii) $\int \sin u \, du = -\cos u + C$

(iii) $\int \sec^2 u \, du = \tan u + C$

(iv) $\int \csc^2 u \, du = -\cot u + C$

(v) $\int \sec u \tan u \, du = \sec u + C$

(vi) $\int \csc u \cot u \, du = -\csc u + C$

(iv) $\dfrac{d}{dx} \cot u = -\csc^2 u \dfrac{du}{dx}$

(v) $\dfrac{d}{dx} \sec u = \sec u \tan u \dfrac{du}{dx}$

(vi) $\dfrac{d}{dx} \csc u = -\csc u \cot u \dfrac{du}{dx}$ ■

We can reverse the process of differentiation to calculate the integrals of certain trigonometric functions. The formulas in Table 1 follow from Theorem 1.

In this section we will integrate functions that can be written in one of the forms displayed in the table. In Chapter 7 we will see how other trigonometric functions can be integrated.

EXAMPLE 1
INTEGRATING A COSINE FUNCTION

Calculate $\int 2 \cos 2x \, dx$.

SOLUTION: If $u = 2x$, then $du = 2 \, dx$, so that

$$\int 2 \cos 2x \, dx = \int \cos u \, du = \sin u + C = \sin 2x + C.$$

EXAMPLE 2
INTEGRATING A TRIGONOMETRIC FUNCTION

Calculate $\int_0^{\pi/3} \tan^5 x \sec^2 x \, dx$.

SOLUTION: If we set $u = \tan x$, then $du = \sec^2 x \, dx$, so that

$$\int \tan^5 x \sec^2 x \, dx = \int u^5 \, du = \frac{u^6}{6} + C = \frac{\tan^6 x}{6} + C,$$

and

$$\tan \frac{\pi}{3} = \sqrt{3}$$

$$\int_0^{\pi/3} \tan^5 x \sec^2 x \, dx = \frac{\tan^6 x}{6} \Big|_0^{\pi/3} = \frac{(\sqrt{3})^6}{6} = \frac{27}{6} = \frac{9}{2}.$$

EXAMPLE 3
INTEGRATING $\sin^2 x$

Calculate $\int_0^{\pi/2} \sin^2 x \, dx$.

SOLUTION: By identity (xxi) in Appendix 1.2,

$$\int_0^{\pi/2} \sin^2 x \, dx = \int_0^{\pi/2} \frac{1 - \cos 2x}{2} \, dx = \int_0^{\pi/2} \frac{1}{2} \, dx - \int_0^{\pi/2} \frac{\cos 2x}{2} \, dx$$

$$= \frac{x}{2} \Big|_0^{\pi/2} - \frac{\sin 2x}{4} \Big|_0^{\pi/2} = \frac{\pi}{4}.$$

■ PROBLEMS 6.7

■ SELF-QUIZ

I. $\int \sin 2x \, dx =$ _____.

  a. $\sin x^2 + C$ b. $-\cos x^2 + C$
  c. $\frac{1}{2}\cos x + C$ d. $-\cos 2x + C$
  e. $\frac{1}{2}\cos 2x + C$ f. $-\frac{1}{2}\cos 2x + C$

II. $\int \cos(x/5) \, dx =$ _____.

  a. $-5\sin(x/5) + C$ b. $5\sin(x/5) + C$
  c. $\sin(x^2/10) + C$ d. $\sin(x/5) + C$

III. $\int 2\sin x \cos x \, dx =$ _____.

  a. $(\sin x)^2 + C$
  b. $C - (\cos x)^2$
  c. $(\sin x)^2 - (\cos x)^2 + C$
  d. $\frac{1}{2}(\sin x + \cos x)(\sin x - \cos x) + C$

IV. $\int \cot x \, dx =$ _____.

  a. $\log \sin x + C$ b. $\ln \sin x + C$
  c. $\ln|\sin x| + C$ d. $\ln|\csc x| + C$

V. $\int_0^{\pi/3} \sec x \cos x \, dx =$ _____.

  a. $\pi/3$ b. $\tan^2(\pi/3)$
  c. $\sec(\pi/3) - \sec 0$ d. $\cos(\pi/3) - \sec 0$

VI. $\int_1^{5\pi/3} (\sin^2 x + \cos^2 x) \, dx =$ _____.

  a. 1 b. $x + C$
  c. $5\pi x/3 - 1$ d. $5\pi/3 - 1$

VII. $\int (1/\cos x)^2 \, dx =$ _____.

  a. $C + \sec^2 x$ b. $C + \tan^2 x$
  c. $-\cot x + C$ d. $\tan x + C$

VIII. $\int \csc(3x) \cot(3x) \, dx =$ _____.

  a. $-3\cot(3x) + C$ b. $(-\frac{1}{3})\csc(3x) + C$
  c. $(\frac{1}{3})\csc(3x) + C$ d. $(-\frac{1}{3})\cot(3x) + C$

■ DRILL

In Problems 1–36, evaluate the given integral.

1. $\int \cos 2x \, dx$

2. $\int \sin(x/5) \, dx$

3. $\int_0^{\pi/12} 3\cos 2x \, dx$

4. $\int_{-\pi/8}^{\pi/8} \frac{1}{2}\sin 4x \, dx$

5. $\int e^{\cos x} \sin x \, dx$

6. $\int \sin(\cos x) \sin x \, dx$

7. $\int \frac{\cos x}{1 + \sin x} \, dx$

8. $\int \frac{\sin 2x}{3 + 5\cos 2x} \, dx$

9. $\int_0^{3\pi/4} \tan(x/3) \, dx$

10. $\int \cot(x/8) \, dx$

11. $\int \cot(2 - x) \, dx$

12. $\int x \tan x^2 \, dx$

13. $\int_0^{\pi/3} \sin 2x \cos 2x \, dx$

14. $\int_0^{\pi/10} \cos 5x \sin 5x \, dx$

15. $\int \sin^2 x \cos x \, dx$

16. $\int \sin x \cos^3 x \, dx$

17. $\int_0^{\pi/6} \sqrt{\sin x} \cos x \, dx$

18. $\int 3\sin^2(x/2) \cos(x/2) \, dx$

19. $\int (\sin 2x)^{-1/3} \cos 2x \, dx$

*20. $\int \sqrt[3]{\tan x}(\sin x)^{2/3} \, dx$

21. $\int \sec 4x \tan 4x \, dx$

22. $\int e^{-2\tan x} \sec^2 x \, dx$

23. $\int_0^{\pi/4} 2\sec x[\sec x \tan x] \, dx$

24. $\int (\sec x)^6 \tan x \, dx$

25. $\int x \cot(x^2) \csc(x^2) \, dx$

26. $\int \csc^{3/2} x \cot x \, dx$

*27. $\int [\sqrt{\csc x} \tan x]^{-1} \, dx$

28. $\int \frac{\csc \sqrt{x} \cot \sqrt{x}}{\sqrt{x}} \, dx$

29. $\int_0^{\pi/3} 2\tan(x + \pi) \sec^2(x + \pi) \, dx$

30. $\int x \csc^2(x^2) \, dx$

31. $\int \sec^2 x \cot x \, dx$
  [*Hint:* $\cot x = 1/(\tan x)$.]

32. $\int_0^{\pi/4} (\sec^2 x + \tan^2 x) \, dx$
  [*Hint:* Use the identity $\tan^2 x = \sec^2 x - 1$.]

CONTINUED

33. $\int \sin^2(x/2)\, dx$

34. $\int_0^{\pi/2} \cos^2 x\, dx$

35. $\int_{\pi/4}^{\pi/2} \dfrac{1 + \cot^2 x}{\csc^2 x}\, dx$

36. $\int x^{-1/2} \cos^2 \sqrt{x} \sec \sqrt{x}\, dx$

■ **APPLICATIONS**

37. Calculate

$$\int \sec x\, dx = \int \frac{\sec^2 x + (\sec x)(\tan x)}{\sec x + \tan x}\, dx.$$

38. Calculate $\int \csc x\, dx$. [*Hint:* First multiply and divide by $\csc x + \cot x$.]

39. Calculate $\int \sec x\, dx = \int (1/\cos x)\, dx$ after doing the following preparatory steps.
    a. Calculate $\int [\cos x/(1 + \sin x)]\, dx$.
    b. Calculate $\int [\sin x/\cos x]\, dx$.
    c. Show that
    $\cos x/(1 + \sin x) = (1 - \sin x)/\cos x$.

40. Calculate $\int \csc x\, dx = \int (1/\sin x)\, dx$ after doing the following preparatory steps.
    a. Calculate $\int [\sin x/(1 - \cos x)]\, dx$.
    b. Calculate $\int [\cos x/\sin x]\, dx$.
    c. Show that
    $\sin x/(1 - \cos x) = (1 + \cos x)/\sin x$.

In Problems 41–46, calculate the area of the region bounded by the given curves (or parts of curves) and the specified lines.

41. $y = \cot x$; x-axis, $x = \pi/4$, $x = \pi/3$
42. $y = \tan x$; y-axis, $y = 1$

43. $y = \csc^2 x$; x-axis, $x = \pi/4$, $x = \pi/2$
*44. One arch of $y = \csc^2 x$; $y = 2$
*45. One arch of $y = \cos^2(x/3)$; x-axis
*46. $y = \sin^2 x$; $y = \cos^2 x$; y-axis

47. You may have already noticed that if you differentiate something of the form $e^x \cdot f(x)$, then e^x is a factor of the derivative. That suggests we tackle computations of the form $\int e^x g(x)\, dx$ by a "trial-and-adjustment" procedure. The following steps demonstrate such a technique.

    a. Compute $\dfrac{d}{dx} e^x \sin x$.

    b. Compute $\dfrac{d}{dx} e^x \cos x$.

    c. Compute $\dfrac{d}{dx} e^x(\sin x + \cos x)$.

    d. Compute $\dfrac{d}{dx} e^x(\sin x - \cos x)$.

    e. Use the result of part (d) to find $\int e^x \sin x\, dx$.
    f. Use the result of part (c) to find $\int e^x \cos x\, dx$.

48. Apply the technique used in the preceding problem to find $\int e^{-x} \cos x\, dx$ and $\int e^{-x} \sin x\, dx$.

■ **SHOW/PROVE/DISPROVE**

49. To calculate $\int \sin x \cos x\, dx$, we can first set $u = \sin x$. Then,

$$\int \sin x \cos x\, dx = \int u\, du = \frac{1}{2} u^2 + C$$

$$= \frac{1}{2} \sin^2 x + C.$$

On the other hand, if we set $v = \cos x$, so that $dv = -\sin x\, dx$, we obtain

$$\int \sin x \cos x\, dx = -\int \cos x(-\sin x)\, dx$$

$$= -\int v\, dv = -\frac{1}{2} v^2 + C$$

$$= -\frac{1}{2} \cos^2 x + C.$$

Resolve this apparent discrepancy.

50. What is wrong with the following calculation?

$$\int_0^\pi \sec^2 x\, dx = \tan x \Big|_0^\pi$$

$$= \tan \pi - \tan 0 = 0 - 0 = 0$$

*51. Suppose a and b are positive constants, and that θ satisfies $\tan \theta = b/a$. Show that

$$\int e^{ax} \sin bx\, dx$$

$$= \frac{1}{\sqrt{a^2 + b^2}} e^{ax} \sin(bx - \theta) + C$$

CONTINUED

and

$$\int e^{ax} \cos bx\, dx$$

$$= \frac{1}{\sqrt{a^2 + b^2}}\, e^{ax} \cos(bx - \theta) + C.$$

[*Hint:* Begin by showing that $\dfrac{d}{dx}\, e^{ax} \sin bx = \sqrt{a^2 + b^2}\, e^{ax} \sin(bx + \theta)$.]

■ **ANSWERS TO SELF-QUIZ**

I. f   II. b   V. a   VI. d
III. a, b, d (Parts (a) and (b) imply (d).)   VII. d   VIII. b
IV. c (Ponder why (b) is wrong.)

6.8 THE INVERSE TRIGONOMETRIC FUNCTIONS

Let $y = \sin x$. Can we solve for x as a function of y? Using what we now know, the answer is no. Suppose that $\sin x = \frac{1}{2}$. Then x could be $\pi/6$, or it could be $5\pi/6$, or it could be $13\pi/6$, and so on. In fact, if y is any number in the interval $[-1, 1]$, then there are an *infinite* number of values of x for which $\sin x = y$. This is illustrated in Figure 1. At the circled points, $\sin x = y_0$. We can eliminate this problem by restricting x to lie in a certain interval over which $\sin x$ is 1–1. By convention, the interval $[-\pi/2, \pi/2]$ is chosen. In the interval $[-\pi/2, \pi/2]$, $\sin x$ is 1–1 (see page 389). Thus, as in Figure 1, for each value of y in $[-1, 1]$, there is a unique value x in $[-\pi/2, \pi/2]$ such that $\sin x = y$. Then from the discussion in Section 6.1, $\sin x$ has an inverse.

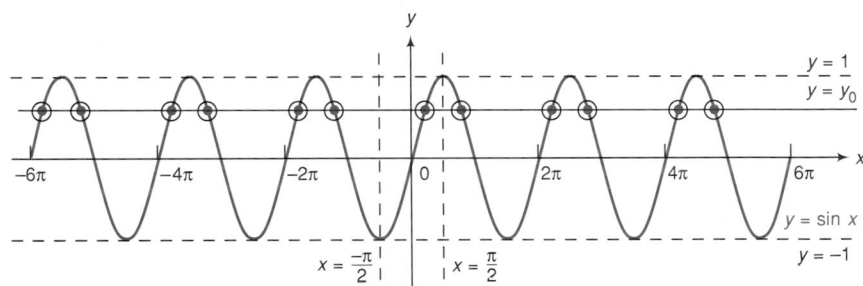

FIGURE 1 GRAPH SHOWS THAT $y = \sin x$ IS NOT 1–1 UNLESS x IS RESTRICTED TO THE INTERVAL $[-\pi/2, \pi/2]$.

DEFINITION   **THE INVERSE SINE FUNCTION**

The **inverse sine function** is the function that assigns to each number x in $[-1, 1]$ the unique number y in $[-\pi/2, \pi/2]$ such that $x = \sin y$. We write

$$y = \sin^{-1} x. \tag{1}$$

The domain of $\sin^{-1} x$ is $[-1, 1]$ and its range is $[-\pi/2, \pi/2]$. ■

NOTE: The -1 appearing in (1) does *not* mean $1/(\sin x)$, which is equal to $(\sin x)^{-1} = \csc x$.

For x in the interval $[-1, 1]$, $y = \sin^{-1} x$ is the unique number in $\left[-\dfrac{\pi}{2}, \dfrac{\pi}{2}\right]$ such that $\sin y = x$.

Another commonly used notation for the inverse sine function is

$$y = \arcsin x. \tag{2}$$

EXAMPLE 1
VALUES OF $\sin^{-1} x$

Calculate $\sin^{-1} x$ for the following:

(a) $x = 0$ (b) $x = 1$ (c) $x = \frac{1}{2}$
(d) $x = -1$ (e) $x = 2$

SOLUTION:

(a) $\sin^{-1} 0 = 0$ since $\sin 0 = 0$.
(b) $\sin^{-1} 1 = \pi/2$ since $\sin \pi/2 = 1$.
(c) $\sin^{-1} \frac{1}{2} = \pi/6$ since $\sin \pi/6 = \frac{1}{2}$.
(d) $\sin^{-1}(-1) = -\pi/2$ since $\sin(-\pi/2) = -1$.
(e) $y = \sin^{-1} 2$ is not defined because, for every real number y, $-1 \le \sin y \le 1$.

We emphasize that the function $y = \sin^{-1} x = \arcsin x$ is only defined if we restrict y to lie in $[-\pi/2, \pi/2]$. Note, for the moment, that if x is in $[-1, 1]$ and if y is in $[-\pi/2, \pi/2]$, then $\sin(\sin^{-1} x) = x$ and $\sin^{-1}(\sin y) = y$. (This result follows from the definition of the inverse sine function and is consistent with the definition of an inverse function). However, it is not true in general. For example, let $y = 2\pi$. Then $x = \sin y = 0$ and $y = \sin^{-1} x = \sin^{-1}(\sin y) = \sin^{-1}(0) = 0$, which is not equal to 2π, the original value of y.

We now differentiate $y = \sin^{-1} x$ for y in $[-\pi/2, \pi/2]$.

THEOREM 1   **DERIVATIVE OF THE INVERSE SINE FUNCTION**

In the open interval $(-1, 1)$, $y = \sin^{-1} x = \arcsin x$ is differentiable, and

$$\frac{d}{dx} \sin^{-1} x = \frac{d}{dx} \arcsin x = \frac{1}{\sqrt{1 - x^2}} \qquad -1 < x < 1. \tag{3}$$

PROOF: Let $y = \sin^{-1} x$ with $x \in (-1, 1)$. Then $x = \sin y$, and

$$\frac{dx}{dy} = \cos y.$$

Since $x \ne \pm 1$, $y \ne \pm\pi/2$, so $\cos y \ne 0$ and $dx/dy \ne 0$ for $-1 < x < 1$. Thus, from Theorem 6.1.4,

$$\frac{dy}{dx} = \frac{1}{dx/dy} = \frac{1}{\cos y}. \tag{4}$$

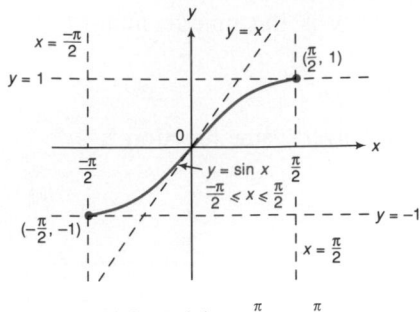

(a) Graph of sin x, $-\frac{\pi}{2} \le x \le \frac{\pi}{2}$.

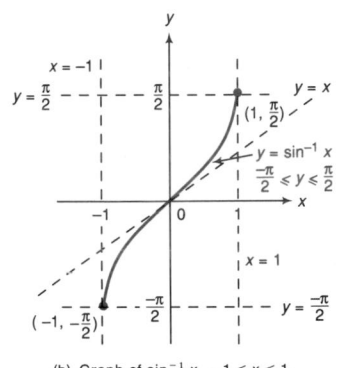

(b) Graph of $\sin^{-1} x$, $-1 \le x \le 1$.

FIGURE 2
GRAPHS OF $y = \sin x$ AND $y = \sin^{-1} x$

The graph of $\sin^{-1} x$ is the reflection about the line $y = x$ of the graph of $\sin x$.

If $x \in (-1, 1)$, then $y = \sin^{-1} x \in (-\pi/2, \pi/2)$ by the definition of the inverse sine function. In the interval $(-\pi/2, \pi/2)$, $\cos y > 0$. Thus,

$$\frac{dy}{dx} = \frac{1}{\cos y} \overset{\text{Since } \sin^2 y + \cos^2 y = 1}{=} \frac{1}{\sqrt{1 - \sin^2 y}} \overset{\text{Since } x = \sin y}{=} \frac{1}{\sqrt{1 - x^2}}. \quad \blacksquare$$

To graph $y = \sin^{-1} x$ for x in $(-1, 1)$, we first observe that $dy/dx = 1/\sqrt{1 - x^2} > 0$, so that the function is increasing. There are no critical points. In addition,

$$\frac{d^2 y}{dx^2} = \frac{d}{dx}(1 - x^2)^{-1/2} = \frac{x}{(1 - x^2)^{3/2}},$$

which is negative for $x < 0$ and positive for $x > 0$. The origin is a point of inflection. The graph is given in Figure 2(b); above it, in Figure 2(a), we place, as a frame of reference, the graph of $y = \sin x$.

Alternatively, note [see statement (6.1.7) on page 393] that the graph of $\sin^{-1} x$ is the reflection about the line $y = x$ of the graph of $\sin x$.

For the remainder of this chapter, we will use the notation $y = \sin^{-1} x$ because it expresses more clearly the fact that the function is the inverse of the sine function.

Next, consider the graph of $\cos x$ in Figure 3(a). If $y = \cos x$ and x is restricted to lie in the interval $[0, \pi]$, then for every y in $[-1, 1]$, there is a unique x such that $\cos x = y$. That is, $\cos x$ is 1–1 in the interval $[0, \pi]$.

DEFINITION   **INVERSE COSINE FUNCTION**

The **inverse cosine function** is the function that assigns to each number x in $[-1, 1]$, the unique number y in $[0, \pi]$ such that $x = \cos y$. We write $y = \cos^{-1} x$.

     The domain of $\cos^{-1} x$ is $[-1, 1]$ and its range is $[0, \pi]$. $\blacksquare$

NOTE: $\cos^{-1} x$ is also written as $\arccos x$.

For x in the interval $[-1, 1]$, $y = \cos^{-1} x$ is the unique number in $[0, \pi]$ such that $\cos y = x$.

EXAMPLE 2
VALUES OF $\cos^{-1} x$

Calculate $y = \cos^{-1} x$ for the following:

   (a) $x = 0$ **(b)** $x = 1$ **(c)** $x = \frac{1}{2}$

   (d) $x = -1$ **(e)** $x = \pi/2$

SOLUTION

   (a) $\cos^{-1} 0 = \pi/2$ since $\cos \pi/2 = 0$.

   (b) $\cos^{-1} 1 = 0$ since $\cos 0 = 1$.

   (c) $\cos^{-1} \frac{1}{2} = \pi/3$ since $\cos \pi/3 = \frac{1}{2}$.

   (d) $\cos^{-1}(-1) = \pi$ since $\cos \pi = -1$.

   (e) $y = \cos^{-1}(\pi/2)$ is not defined because $\pi/2 \approx 1.57 > 1$ and $-1 \le \cos y \le 1$.

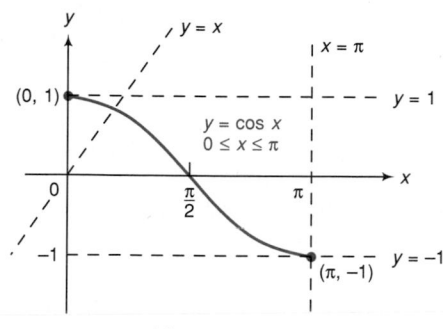

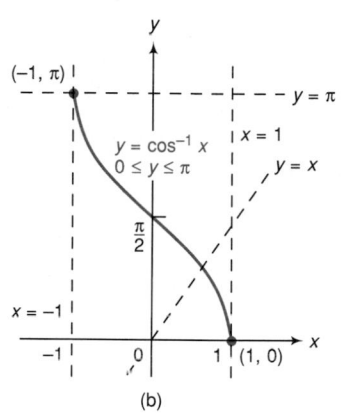

FIGURE 3
GRAPHS OF $y = \cos x$ AND $y = \cos^{-1} x$

The graph of $\cos^{-1} x$ is the reflection about the line $y = x$ of the graph of $\cos x$.

| THEOREM 2 | DERIVATIVE OF $\cos^{-1} x$ |

$\text{Cos}^{-1} x$ is differentiable for $-1 < x < 1$ and

$$\frac{d}{dx} \cos^{-1} x = -\frac{1}{\sqrt{1 - x^2}} \qquad -1 < x < 1. \qquad (5)$$

PROOF: If $y = \cos^{-1} x$, then $x = \cos y$ and $1 = (-\sin y)\, dy/dx$, or

$$\frac{dy}{dx} = -\frac{1}{\sin y} = -\frac{1}{\sqrt{1 - x^2}}$$

($\sin y = +\sqrt{1 - \cos^2 y}$ because $0 \le y \le \pi$). Note that dy/dx exists according to Theorem 6.1.4. ∎

The graphs of $y = \cos x$ and $y = \cos^{-1} x$ are given in Figure 3.

We next consider the graph of $y = \tan x$ given in Figure 4(a). The function $\tan x$ can take on values in the interval $(-\infty, \infty)$. To get a unique x for a given y, we restrict x to the interval $(-\pi/2, \pi/2)$.

| DEFINITION | INVERSE TANGENT FUNCTION |

The **inverse tangent function** is the function that assigns to each real number x the unique number y in $(-\pi/2, \pi/2)$ such that $x = \tan y$. We write $y = \tan^{-1} x = \arctan x$.

The domain of $\tan^{-1} x$ is $(-\infty, \infty)$ and the range of $\tan^{-1} x$ is $(-\pi/2, \pi/2)$. ∎

For any real number x, $y = \tan^{-1} x$ is the unique number in $\left(-\dfrac{\pi}{2}, \dfrac{\pi}{2}\right)$ such that $\tan y = x$.

EXAMPLE 3
VALUES OF $\tan^{-1} x$

Calculate $y = \tan^{-1} x$ for the following:

(a) $x = 0$ **(b)** $x = 1$ **(c)** $x = -\dfrac{1}{\sqrt{3}}$ **(d)** $x = \sqrt{3}$

SOLUTION:

(a) $\tan^{-1} 0 = 0$ since $\tan 0 = 0$.

(b) $\tan^{-1} 1 = \dfrac{\pi}{4}$ since $\tan \dfrac{\pi}{4} = 1$.

(c) $\tan^{-1}\left(-\dfrac{1}{\sqrt{3}}\right) = -\dfrac{\pi}{6}$ since $\tan\left(-\dfrac{\pi}{6}\right) = -\dfrac{1}{\sqrt{3}}$.

(d) $\tan^{-1} \sqrt{3} = \dfrac{\pi}{3}$ since $\tan \dfrac{\pi}{3} = \sqrt{3}$.

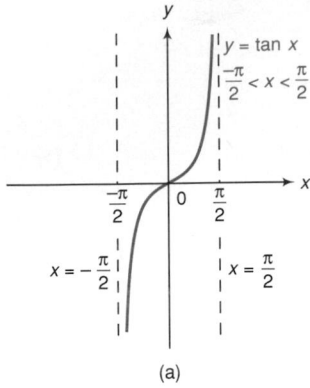

(a)

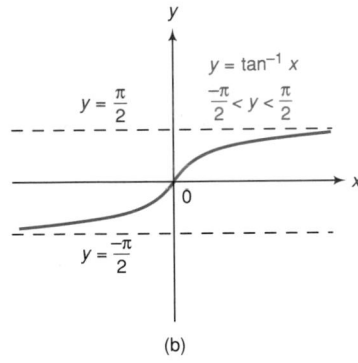

(b)

FIGURE 4
GRAPHS OF $y = \tan x$ AND $y = \tan^{-1} x$

THEOREM 3   **DERIVATIVE OF $\tan^{-1} x$**

$\tan^{-1} x$ is differentiable and

$$\frac{d}{dx} \tan^{-1} x = \frac{1}{1 + x^2} \qquad -\infty < x < \infty. \tag{6}$$

PROOF: If $y = \tan^{-1} x$, then $x = \tan y$ and $1 = (\sec^2 y)\, dy/dx$, or

See formula (1) in Appendix 1.3

$$\frac{dy}{dx} = \frac{1}{\sec^2 y} = \frac{1}{1 + \tan^2 y} = \frac{1}{1 + x^2}.$$

Note that dy/dx exists according to Theorem 6.1.4. ■

The graphs of $\tan x$ and $\tan^{-1} x$ are given in Figure 4.

There are, of course, three other inverse trigonometric functions. These arise less frequently in applications, and so we will delay their introduction until the problems.

We summarize in Table 1 the formulas that can be obtained from the results of this section. The last three rows will be derived in the problem set.

TABLE 1

$$\frac{d}{dx} \sin^{-1} u = \frac{1}{\sqrt{1 - u^2}} \frac{du}{dx} \qquad\qquad \int \frac{du}{\sqrt{1 - u^2}} = \sin^{-1} u + C$$

$$\frac{d}{dx} \cos^{-1} u = -\frac{1}{\sqrt{1 - u^2}} \frac{du}{dx} \qquad -\int \frac{du}{\sqrt{1 - u^2}} = \cos^{-1} u + C$$

$$\frac{d}{dx} \tan^{-1} u = \frac{1}{1 + u^2} \frac{du}{dx} \qquad\qquad \int \frac{du}{1 + u^2} = \tan^{-1} u + C$$

$$\frac{d}{dx} \cot^{-1} u = -\frac{1}{1 + u^2} \frac{du}{dx} \qquad -\int \frac{du}{1 + u^2} = \cot^{-1} u + C$$

$$\frac{d}{dx} \sec^{-1} u = \frac{1}{|u|\sqrt{u^2 - 1}} \frac{du}{dx} \qquad\quad \int \frac{du}{u\sqrt{u^2 - 1}} = \sec^{-1}|u| + C$$

$$\frac{d}{dx} \csc^{-1} u = -\frac{1}{|u|\sqrt{u^2 - 1}} \frac{du}{dx} \quad -\int \frac{du}{u\sqrt{u^2 - 1}} = \csc^{-1}|u| + C$$

REMARK: In Table 1 it may seem surprising that $\int du/\sqrt{1 - u^2} = \sin^{-1} u + C$ but $-\int du/\sqrt{1 - u^2} = \cos^{-1} u + C$. This result is explained by noting that $\sin^{-1} u + \cos^{-1} u$ is a constant function (see Problem 71).

EXAMPLE 4
THE DERIVATIVE OF AN INVERSE
SINE FUNCTION

Calculate dy/dx for $y = \sin^{-1} x^2$.

SOLUTION: If $u = x^2$, then $du/dx = 2x$, and using the chain rule we have

$$\frac{dy}{dx} = \frac{dy}{du} \frac{du}{dx} = \frac{1}{\sqrt{1 - u^2}} \cdot 2x = \frac{2x}{\sqrt{1 - x^4}}.$$

EXAMPLE 5
AN INTEGRAL LEADING TO AN
INVERSE TANGENT FUNCTION

Calculate $\int \dfrac{dx}{a^2 + x^2}$.

SOLUTION: We want to write the integral in the form

$$\int \frac{du}{1 + u^2}$$

since we can integrate this. To do so, we factor the a^2 term in the denominator to obtain

$$\int \frac{dx}{a^2 + x^2} = \frac{1}{a^2} \int \frac{dx}{1 + (x^2/a^2)}.$$

Let $u = x/a$; then $du = (1/a)\,dx$, so that

$$\frac{1}{a^2} \int \frac{dx}{1 + (x^2/a^2)} = \frac{1}{a} \int \frac{(1/a)\,dx}{1 + (x/a)^2} = \frac{1}{a} \int \frac{du}{1 + u^2} = \frac{1}{a} \tan^{-1} u + C$$

$$= \frac{1}{a} \tan^{-1} \frac{x}{a} + C.$$

NOTE: The formula

$$\int \frac{dx}{a^2 + x^2} = \frac{1}{a} \tan^{-1} \frac{x}{a} + C$$

is worth memorizing. It is encountered quite often in applications.

EXAMPLE 6
AN INTEGRAL LEADING TO AN
INVERSE SINE FUNCTION

Calculate $\int \dfrac{x^2}{\sqrt{1 - x^6}}\,dx$.

SOLUTION: If $u = x^3$, then $du = 3x^2\,dx$, so we multiply and divide by 3 to obtain

$$\int \frac{x^2}{\sqrt{1 - x^6}}\,dx = \frac{1}{3} \int \frac{3x^2}{\sqrt{1 - (x^3)^2}}\,dx$$

$$= \frac{1}{3} \int \frac{du}{\sqrt{1 - u^2}} = \frac{1}{3} \sin^{-1} u + C$$

$$= \frac{1}{3} \sin^{-1} x^3 + C.$$

In some applications, it is necessary to evaluate expressions like $\sin(\tan^{-1} 3)$. To do so, we draw a triangle with an angle whose tangent is 3. See Figure 5. If $\tan \theta = 3$, then with the sides as in the figure, we use the Pythagorean theorem to find that the hypotenuse must be $\sqrt{10}$. Then $\sin \theta = 3/\sqrt{10}$, $\cos \theta = 1/\sqrt{10}$, and so on.

FIGURE 5
IN THIS TRIANGLE $\tan \theta = 3$ SO $\theta = \tan^{-1} 3$.

EXAMPLE 7
COMPUTING A TRIGONOMETRIC
VALUE USING A TRIANGLE

Calculate $\tan(\cos^{-1}\frac{3}{7})$.

SOLUTION: By using the triangle in Figure 6, we see that $\tan(\cos^{-1}\frac{3}{7}) = \tan\theta = \sqrt{40}/3 = 2\sqrt{10}/3$.

Note that this problem was solved without a calculator.

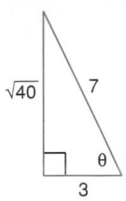

FIGURE 6

IN THIS TRIANGLE $\cos\theta = \dfrac{3}{7}$ SO $\theta = \cos^{-1}\dfrac{3}{7}$.

FINDING INVERSE TRIGONOMETRIC VALUES ON A CALCULATOR

In order to find values of $\sin^{-1} x$, for example, on a scientific calculator, press $\boxed{\text{INV}}\ \boxed{\sin}$ or $\boxed{\text{2nd F}}\ \boxed{\sin}$ [some calculators do have a $\boxed{\sin^{-1}}$ key and some have other names for the inverse key]. Make certain that the calculator is set to radians (unless you have a particular trigonometric application that calls for degrees).

EXAMPLE 8
FINDING INVERSE
TRIGONOMETRIC VALUES
ON A CALCULATOR

Compute each value on a calculator.

(a) $\sin^{-1} 0.6$ **(b)** $\cos^{-1}(-0.83)$ **(c)** $\tan^{-1} 10$

(d) $\tan^{-1}\left(-\dfrac{1}{2}\right)$ **(e)** $\sin^{-1}\dfrac{\pi}{3}$

SOLUTION:

(a) $\sin^{-1} 0.6 = 0.643501108\ (\approx 36.9°)$
(b) $\cos^{-1}(-0.83) = 2.549904011\ (\approx 146.1°)$
(c) $\tan^{-1} 10 = 1.471127674\ (\approx 84.3°)$
(d) $\tan^{-1}(-0.5) = -0.463647609\ (\approx -26.6°)$

(e) $\dfrac{\pi}{3} \approx 1.047 > 1$ so $\sin^{-1}\dfrac{\pi}{3}$ is not defined. If you try to compute it, your calculator will give you an error message.

■ PROBLEMS 6.8

■ SELF-QUIZ

I. $\sin^{-1}(\sin\theta) = \theta$ for all $\theta \in$ _____.
 a. $[0, \pi]$ b. $(-\infty, \infty)$
 c. $[-\pi/2, \pi/2]$ d. $[0, 2\pi]$
 e. $[0, 1]$ f. $[-1, 1]$

II. $\sin(\sin^{-1} x) = x$ for all $x \in$ _____.
 a. $[0, \pi]$ b. $(-\infty, \infty)$
 c. $[-\pi/2, \pi/2]$ d. $[0, 2\pi]$
 e. $[0, 1]$ f. $[-1, 1]$

III. $\sin[\sin^{-1}(\sin\theta)] = \sin\theta$ for all $\theta \in$ _____.
 a. $[0, \pi]$ b. $(-\infty, \infty)$

 c. $[-\pi/2, \pi/2]$ d. $[0, 2\pi]$
 e. $[0, 1]$ f. $[-1, 1]$

IV. The range of the arccosine function, $\cos^{-1}$, is _____.
 a. $[0, \pi]$ b. $(-\infty, \infty)$
 c. $[-\pi/2, \pi/2]$ d. $[0, 2\pi]$
 e. $[0, 1]$ f. $[-1, 1]$

V. $\tan^{-1}(\tan\theta) = \theta$ for all $\theta \in$ _____.
 a. $[0, \pi]$ b. $(-\infty, \infty)$
 c. $(-\pi/2, \pi/2)$ d. $[0, 2\pi]$
 e. $[0, 1]$ f. $[-1, 1]$

■ DRILL

In Problems 1–18, calculate the indicated value. Do not use a numerical table or a calculator.

1. $\sin^{-1}[\sqrt{3}/2]$
2. $\cos^{-1}[-\sqrt{3}/2]$
3. $\sin^{-1}[-1/2]$
4. $\cos^{-1}[-\sqrt{2}/2]$
5. $\tan^{-1}[1/\sqrt{3}]$
6. $\tan^{-1}[\sqrt{3}]$
7. $\tan^{-1}[-1/\sqrt{3}]$
8. $\tan^{-1}[-1]$
9. $\sin[\cos^{-1}(\frac{3}{5})]$
10. $\cos[\sin^{-1}(-\frac{3}{5})]$
11. $\tan[\sin^{-1}(\frac{3}{5})]$
12. $\sin[\tan^{-1}(\frac{3}{5})]$
13. $\sin[\tan^{-1}(-5)]$
*14. $\tan[\sin^{-1}(-5)]$
    [*Hint:* Think.]
15. $\sin[\cos^{-1}x]$
16. $\sin[\tan^{-1}x]$
17. $\tan[\sin^{-1}x]$
18. $\tan[\cos^{-1}x]$

In Problems 19–34, calculate the derivative of the given function.

19. $\sin^{-1}3x$
20. $\cos^{-1}(1-2x)$
21. $\cos^{-1}(x-5)$
22. $\tan^{-1}(x+7)$
23. $\tan^{-1}(x/2)$
24. $\tan^{-1}(2/x)$
25. $\sin^{-1}(\sqrt{x})$
26. $\tan^{-1}(\sqrt{x})$
*27. $\sin^{-1}(1+x^2)$
28. $\cos^{-1}(x+x^3)$
29. $\cos^{-1}(x^3+1)$
30. $\sqrt{\sin^{-1}(x)+\cos^{-1}(x)}$
31. $(\sin^{-1}x)\cdot(\sin x)$
32. $\cos^{-1}(x^{3/2})$

*33. $x\sqrt{1-x^2}+\sin^{-1}x$
*34. $-x\sqrt{1-x^2}-\cos^{-1}x$

In Problems 35–50, calculate the given integral.

35. $\int_0^2 \dfrac{1}{1+x^2}\,dx$
36. $\int \dfrac{1}{4x^2+1}\,dx$
37. $\int \dfrac{1}{x^2-2x+2}\,dx$
    [*Hint:* $x^2-2x+2 = (x-1)^2+1$.]
38. $\int \dfrac{x}{x^4+1}\,dx$
39. $\int \dfrac{e^{-x}}{4+e^{-2x}}\,dx$
40. $\int \dfrac{\sqrt{x}}{x^3+1}\,dx$
41. $\int \dfrac{\cos x}{1+\sin^2 x}\,dx$
42. $\int \dfrac{\sec x \tan x}{1+\sec^2 x}\,dx$
43. $\int_0^{1/2} \dfrac{1}{\sqrt{1-x^2}}\,dx$
44. $\int_0^{\sqrt{3}/2} \dfrac{-dx}{\sqrt{1-x^2}}$
45. $\int \dfrac{dx}{\sqrt{1-9x^2}}$
46. $\int \dfrac{-15\,dx}{\sqrt{9-25x^2}}$
47. $\int \dfrac{\sin x}{\sqrt{4-\cos^2 x}}\,dx$
48. $\int \dfrac{\sqrt{x}}{\sqrt{1-x^3}}\,dx$
49. $\int \dfrac{\cos^{-1}x}{\sqrt{1-x^2}}\,dx$
50. $\int \dfrac{\sin^{-1}(x/3)}{\sqrt{9-x^2}}\,dx$

■ APPLICATIONS

51. Find the area of the region bounded by the curve $y = 1/(1+x^2)$, the x- and y-axes, and the line $x = 1$.

52. Compute the area of the region bounded below by $y = x/(1+x^2)$, above by $y = 1/(1+x^2)$, to the left by the y-axis, and to the right by $x = 1$.

53. Compute the area between $y = 1/\sqrt{1-x^2}$, the x- and y-axes, and the line $x = \frac{1}{2}$.

54. Find the area of the region between $y = -1/\sqrt{1-x^2}$ and the line $y = -2$. [*Hint:* Sketch the curve.]

55. A man stands on top of a vertical cliff, 100 m above a lake. He watches a rowboat move directly away from the foot of the cliff at the rate of 30 m/min. How fast is the angle of depression of his line-of-sight changing when the boat is 60 m from the foot of the cliff?

56. A department store wishes to place a large advertising display near its main entrance. The display is a rectangular sign 7 ft tall that is mounted on a wall and whose bottom edge is 8 ft from the floor. The advertising manager wishes to place the entrance walkway at a distance x ft from the wall to maximize the "viewing angle" (the angle, at the viewer's eye, between sight lines to the top and bottom of the display). If we assume the eye level of an average customer is 5 ft from the floor, how far from the wall should the walkway be placed?

57. A visitor to the Kennedy Space Center in Florida watches a rocket blast off. The visitor is standing 3 km from the blast site. At a given moment the rocket makes an angle of 45° with her line of sight. The visitor estimates that this angle is changing at a rate of 20° per sec. If the rocket is traveling vertically, what is its velocity at the given moment? [*Hint:* Convert angles from degrees to radians.]

CONTINUED

58. A lighthouse containing a revolving beacon light is on a small island 3 km from the nearest point Q on a straight shoreline. The light from the spotlight moves along the shore as the beacon re-volves. At a point 1 km from Q, the light is sweeping along the shoreline at a rate of 40 km/min. How fast (in rpms) is the beacon revolving?

CALCULATOR PROBLEMS

In Problems 59–68, compute each value on a calculator. Given your answer in radians.

59. $\sin^{-1}(0.78)$

60. $\sin^{-1}\left(\dfrac{\sqrt{2}}{2}\right)$

61. $\sin^{-1}\left(-\dfrac{\sqrt{3}}{2}\right)$

62. $\sin^{-1}(\sqrt{2})$

63. $\cos^{-1}\left(\dfrac{\pi}{4}\right)$

64. $\cos^{-1}\left(-\dfrac{1}{5}\right)$

65. $\cos^{-1}(-0.93)$

66. $\tan^{-1}\left(\dfrac{3}{4}\right)$

67. $\tan^{-1}(100{,}000)$

68. $\tan^{-1}(-2)$

SHOW/PROVE/DISPROVE

69. Show that if $-1 \le x \le 1$, then $\sin(\cos^{-1} x) = \cos(\sin^{-1} x)$.

70. Show that $\lim_{x \to \infty} \tan^{-1} x = \pi/2$ and $\lim_{x \to -\infty} \tan^{-1} x = -\pi/2$.

*71. Show that $\sin^{-1} x + \cos^{-1} x = \pi/2$. [Hint: First show by differentiation that $\sin^{-1} x + \cos^{-1} x$ is constant; then compute that constant by evaluating the expression at a convenient choice of x.]

72. Investigate the function $\tan^{-1} x + \tan^{-1}(1/x)$. What is its minimum value? its maximum value? [Hint: 0 is not in the domain of this function.]

73. Prove or Disprove: If a is a nonzero constant, then
$$\int \frac{dx}{\sqrt{a^2 - x^2}} = \sin^{-1} \frac{x}{|a|} + C.$$

74. Show that
$$\frac{d}{dx}\left[x \tan^{-1} x - \frac{1}{2}\ln(1 + x^2)\right] = \tan^{-1} x.$$
(This gives us the integral of the arctangent function.)

75. Show that
$$\frac{d}{dx}\left[x \cos^{-1} x - \sqrt{1 - x^2}\right] = \cos^{-1} x.$$
(This gives us the integral of $\cos^{-1} x$.)

76. Show that
$$\frac{d}{dx}\left[x \sin^{-1} x + \sqrt{1 - x^2}\right] = \sin^{-1} x.$$
(This gives us the integral of $\sin^{-1} x$.)

*77. Let $f(x) = \sin^{-1}[(x^2 - 8)/8]$ and $g(x) = 2\sin^{-1}(x/4) - \pi/2$.
a. Verify that $f(0) = g(0)$.
b. Find the maximal domain for each of the functions.
c. Find $f'(x)$ and $g'(x)$; compare them.
d. Find all x such that $f(x) = g(x)$.

*78. Discuss the following "proof" that $\tan^{-1}[(1 + x)/(1 - x)] = \pi/4 + \tan^{-1} x$.
a. $$\frac{d}{dx}\tan^{-1}\left(\frac{1 + x}{1 - x}\right) = \frac{1}{1 + \left(\dfrac{1 + x}{1 - x}\right)^2} \cdot \frac{2}{(1 - x)^2}$$
$$= \frac{2}{(1 - x)^2 + (1 + x)^2} = \frac{1}{1 + x^2} = \frac{d}{dx}\tan^{-1} x.$$
b. $\tan^{-1}[(1 + x)/(1 - x)] - \tan^{-1} x$ is constant; evaluating this expression for $x = 0$, we get $\tan^{-1} 1 - \tan^{-1} 0 = \pi/4 - 0 = \pi/4$. [Hint: Notice that $\tan^{-1}\sqrt{3} = \pi/3$, whereas $\tan^{-1}[(1 + \sqrt{3})/(1 - \sqrt{3})] = -5\pi/12$; include this fact in your discussion.]

cot^{-1}

79. **Definition of cot^{-1}:** Consider the graph of $y = \cot x$ in Figure 1(b) in Appendix 1.3. Show that if $y = \cot x$, then for every y in $(-\infty, \infty)$ there is a unique x in $(0, \pi)$ such that $y = \cot x$. Then define the function $y = \cot^{-1} x$ by $y = \cot^{-1} x$ if $x = \cot y$, $0 < y < \pi$.

CONTINUED

80. Calculate the following:
    a. $\cot^{-1} 0$ b. $\cot^{-1}(-1)$
    c. $\cot^{-1}(-\sqrt{3})$ d. $\cot^{-1}(1/\sqrt{3})$
81. Show that
$$\frac{d}{dx} \cot^{-1} x = \frac{-1}{1+x^2}.$$
82. Graph the function $y = \cot^{-1} x$, $0 < y < \pi$.
83. Show that $\tan^{-1} x + \cot^{-1} x = \pi/2$ for all x.
84. Find the minimum of the function $\tan^{-1} x - \cot^{-1}(1/x)$.

*85. Let $A(x) = \cot^{-1} x$ and $B(x) = \tan^{-1}(1/x)$.
    a. Verify that $A(1) = B(1)$ and $A(-1) = B(-1) + \pi$.
    b. Compare $A'(x)$ with $B'(x)$.
    c. Discuss and graph the function $A(x) - B(x)$.
*86. Prove that $\cot^{-1}[(1 + x)/(1 - x)] + \cot^{-1} x$ is a constant if $x > 1$ and is a different constant for $x < 1$; then find those constants. [*Hint:* Differentiate the function and see what you get, then think about the domain of the function.]

■ **$\sec^{-1}$**

87. **Definition of $\sec^{-1}$:** Consider the graph of $y = \sec x$ in Figure 1(c) in Appendix 1.3. Show that if $y = \sec x$, then for every y in $(-\infty, -1]$ or $[1, \infty)$ there is a unique x in $(\pi/2, \pi]$ or $[0, \pi/2)$ such that $y = \sec x$. (Explain why the value $x = \pi/2$ must be excluded.) Then define the function $y = \sec^{-1} x$ by $y = \sec^{-1} x$ if $x = \sec y$, $0 \le y < \pi/2$ or $\pi/2 < y \le \pi$.[†]
88. Show that $\sec^{-1} x = \cos^{-1}(1/x)$ for $|x| \ge 1$.
89. Calculate the following:
    a. $\sec^{-1} 2$ b. $\sec^{-1}(-2)$
    c. $\sec^{-1} \sqrt{2}$ d. $\sec^{-1}(2/\sqrt{3})$
    e. $\sec^{-1} 1$ f. $\sec^{-1}(-1)$
90. a. Show that if $x \ge 1$, then
$$\frac{d}{dx} \sec^{-1} x = \frac{1}{x\sqrt{x^2-1}}.$$
    b. Show that if $x \le -1$, then
$$\frac{d}{dx} \sec^{-1} x = \frac{-1}{x\sqrt{x^2-1}}.$$

c. Conclude that[‡]
$$\frac{d}{dx} \sec^{-1} x = \frac{1}{|x|\sqrt{x^2-1}}.$$
d. What is the domain of the function
$$y' = \frac{d}{dx} \sec^{-1} x?$$

91. Graph the function $y = \sec^{-1} x$, $0 \le y < \pi/2$ and $\pi/2 < y \le \pi$.
92. Show that
$$\int \frac{dx}{x\sqrt{x^2-a^2}} = \frac{1}{|a|} \sec^{-1}\left|\frac{x}{a}\right| + C$$
$$= \frac{1}{|a|} \cos^{-1}\left|\frac{a}{x}\right| + C.$$

■ **$\csc^{-1}$**

93. **Definition of $\csc^{-1}$:** Consider the graph of $y = \csc x$ in Figure 1(d) in Appendix 1.3. Show that if $y = \csc x$, then for every y in $(-\infty, -1]$ or $[1, \infty)$ there is a unique x in $[-\pi/2, 0)$ or $(0, \pi/2]$ such that $y = \csc x$. (Explain why the value $x = 0$ must be excluded.) Then define the function $y = \csc^{-1} x$ by $y = \csc^{-1} x$ if $x = \sec y$ with $-\pi/2 \le y < 0$ or $0 < y \le \pi/2$.

[†] This problem suggests one way to define $\sec^{-1} x$. An alternative way is to restrict the range of $\sec^{-1} x$ to the intervals $[-\pi, -\pi/2)$ and $[0, \pi/2)$. Then for every y in $(-\infty, -1]$ or $[1, \infty)$, there is a unique x in $[-\pi, -\pi/2)$ or $[0, \pi/2)$ such that $y = \sec x$.

[‡] With the alternative definition of $\sec^{-1} x$ given in the previous footnote, it is not difficult to show that $\frac{d}{dx} \sec^{-1} x = 1/x\sqrt{x^2-1}$. Some mathematicians prefer this alternative definition because it is then not necessary to keep track of the absolute value in the derivative of $\sec^{-1} x$.

CONTINUED

94. Show that $\csc^{-1} x = \sin^{-1}(1/x)$ for $|x| \geq 1$.

95. Calculate the following:
    a. $\csc^{-1} 2$ b. $\csc^{-1}(-2)$
    c. $\csc^{-1} \sqrt{2}$ d. $\csc^{-1}(2/\sqrt{3})$
    e. $\csc^{-1} 1$ f. $\csc^{-1}(-1)$

96. a. Show that if $x \geq 1$, then
$$\frac{d}{dx} \csc^{-1} x = \frac{-1}{x\sqrt{x^2 - 1}}.$$

b. Show that if $x \leq -1$, then
$$\frac{d}{dx} \csc^{-1} x = \frac{1}{x\sqrt{x^2 - 1}}.$$

c. Conclude that
$$\frac{d}{dx} \csc^{-1} x = \frac{-1}{|x|\sqrt{x^2 - 1}}.$$

*97. Show that $\sec^{-1} x + \csc^{-1} x$ is constant.

■ **DRILL WITH $\cot^{-1}$, $\sec^{-1}$, $\csc^{-1}$**

98. Calculate the following:
    a. $\sin(\csc^{-1} 4)$ b. $\sec(\cot^{-1} 8)$
    c. $\cos[\sec^{-1}(-10)]$ d. $\cot[\sec^{-1}(-3)]$
    e. $\tan[\csc^{-1}(4)]$ f. $\sin[\cot^{-1}(\frac{1}{5})]$

In Problems 99–106, compute the derivative of the given function.

99. $\cot^{-1}(e^x)$ 100. $\cot^{-1}(x^2 + 5)$
101. $\sec^{-1}(4x + 2)$ 102. $x^2 \sec^{-1}(x^2)$
103. $\csc^{-1}(\sqrt{x})$ 104. $\csc^{-1}(\ln x)$
105. $\sec^{-1}(\cot x)$ 106. $\cot^{-1}(\sec x)$

In Problems 107–112, evaluate the given integral.

107. $\displaystyle\int \frac{dx}{1 + (3 - x)^2}$

108. $\displaystyle\int_0^{\pi/4} \frac{\sin x}{1 + \cos^2 x}\, dx$

109. $\displaystyle\int_{2/\sqrt{3}}^2 \frac{dx}{x\sqrt{x^2 - 1}}$

110. $\displaystyle\int_0^{\ln 2} \frac{e^x\, dx}{e^x\sqrt{e^{2x} - 1}}$

*111. $\displaystyle\int \frac{dx}{x\sqrt{x^2 - a^2}}$

*112. $\displaystyle\int \frac{dx}{x\sqrt{x - 1}}$
[*Hint:* Let $u = \sqrt{x}$.]

■ **ANSWERS TO SELF-QUIZ**

I. c   II. f   III. b
IV. a   V. c

6.9 THE HYPERBOLIC FUNCTIONS

In many physical applications (especially those involving differential equations), functions arise that are combinations of e^x and e^{-x}. In fact, this happens so often that certain of the combinations that occur most frequently have been given special names.

DEFINITION   THE HYPERBOLIC SINE AND COSINE FUNCTIONS

The **hyperbolic cosine function** is defined by
$$\cosh x = \frac{e^x + e^{-x}}{2}. \tag{1}$$

The domain of $\cosh x$ is $(-\infty, \infty)$.
    The **hyperbolic sine function** is defined by
$$\sinh x = \frac{e^x - e^{-x}}{2}. \tag{2}$$

The domain of $\sinh x$ is $(-\infty, \infty)$.[†] ■

[†] $\cosh x$ is pronounced "kosh" x and $\sinh x$ is pronounced "cinch" x.

Since $e^{-x} > 0$ for all x, we see that for every x

$$\cosh x > \sinh x. \tag{3}$$

THEOREM 1 **DERIVATIVES OF cosh x AND sinh x**

$$\frac{d}{dx} \cosh x = \sinh x \tag{4}$$

and

$$\frac{d}{dx} \sinh x = \cosh x. \tag{5}$$

PROOF:

$$\frac{d}{dx} \cosh x = \frac{d}{dx}\left(\frac{e^x + e^{-x}}{2}\right) = \frac{e^x - e^{-x}}{2} = \sinh x.$$

The proof for the derivative of sinh is just as easy. ■

It is not very difficult to draw the graphs of these functions. We immediately see that $\sinh x < 0$ if $x < 0$ and $\sinh x > 0$ if $x > 0$. If $x = 0$, $\sinh x = 0$ and $\cosh x = 1$. Therefore, $\cosh x$ is decreasing if $x < 0$, $\cosh x$ is increasing if $x > 0$, and the point $x = 0$ is a critical point for $\cosh x$. In addition,

$$\frac{d^2}{dx^2}(\cosh x) = \cosh x > 0,$$

so that $\cosh x$ is concave up and the point $(0, 1)$ is a minimum. This information is all depicted in Figure 1(b).

Since

$$\frac{d}{dx} \sinh x = \cosh x > 0,$$

$\sinh x$ is always increasing (and has no critical points). We have

$$\frac{d^2}{dx^2} \sinh x = \sinh x$$

so that $\sinh x$ is concave down for $x < 0$ and concave up for $x > 0$. At $(0, 0)$ there is a point of inflection. This figure is drawn in Figure 1(a).

Consider the expression $\cosh^2 x - \sinh^2 x$. From the definition of the hyperbolic sine and cosine functions, it is not difficult to show that

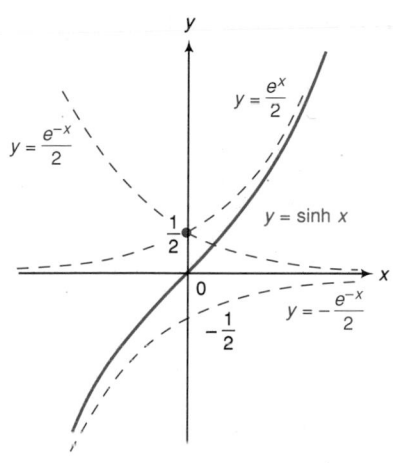

(a) $f(x) = \sinh x$

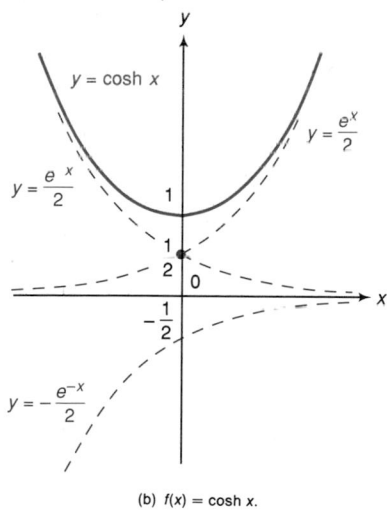

(b) $f(x) = \cosh x.$

FIGURE 1
GRAPHS OF THE HYPERBOLIC SINE AND COSINE FUNCTIONS

BASIC HYPERBOLIC FUNCTION IDENTITY
$$\cosh^2 x - \sinh^2 x = 1. \tag{6}$$

(See Problem 25.)

The trigonometric, or circular, functions $\sin \theta$ and $\cos \theta$ are defined with reference to the unit circle whose equation is $x^2 + y^2 = 1$ (see Figure 1 in

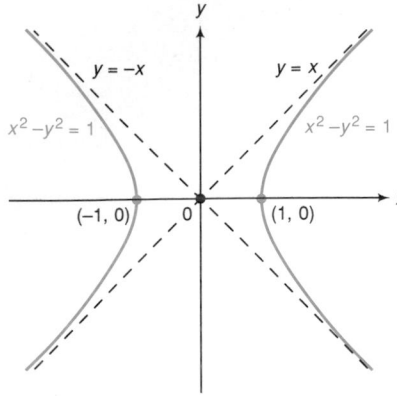

FIGURE 2
GRAPH OF HYPERBOLA $x^2 - y^2 = 1$

Appendix 1.2). That is, if (x, y) is a point on the unit circle and θ is the angle between the x-axis and the radial line joining $(0, 0)$ to (x, y), then by definition $x = \cos \theta$ and $y = \sin \theta$.

Now the equation $x^2 - y^2 = 1$ is the equation of a hyperbola, sometimes called the *unit* hyperbola (see Section 8.3), whose graph is given in Figure 2. If $x = \cosh \theta$ and $y = \sinh \theta$ for some number θ, then from (6), (x, y) is a point on the right branch of this hyperbola. This is the reason $\cosh x$ and $\sinh x$ are called *hyperbolic* functions.

From Theorem 1 we obtain the differentiation and integration formulas shown in Table 1.

TABLE 1   DERIVATIVES AND INTEGRALS OF $\cosh u$ AND $\sinh u$

$$\frac{d}{dx} \cosh u = \sinh u \frac{du}{dx} \qquad \int \sinh u \, du = \cosh u + C$$

$$\frac{d}{dx} \sinh u = \cosh u \frac{du}{dx} \qquad \int \cosh u \, du = \sinh u + C$$

EXAMPLE 1
DERIVATIVE OF A HYPERBOLIC
COSINE FUNCTION

Calculate $(d/dx)(\cosh \sqrt{x^3 - 1})$.

SOLUTION: If $u = \sqrt{x^3 - 1}$, then

$$\frac{du}{dx} = \frac{3x^2}{2\sqrt{x^3 - 1}},$$

and

$$\frac{d}{dx} \cosh \sqrt{x^3 - 1} = \sinh u \frac{du}{dx} = (\sinh \sqrt{x^3 - 1}) \frac{3x^2}{2\sqrt{x^3 - 1}}.$$

EXAMPLE 2
AN INTEGRAL INVOLVING $\sinh x$

Calculate $\int x \sinh 3x^2 \, dx$.

SOLUTION: If $u = 3x^2$, then $du = 6x \, dx$, so that

$$\int x \sinh 3x^2 \, dx = \frac{1}{6} \int \sinh 3x^2 (6x) \, dx = \frac{1}{6} \int \sinh u \, du$$

$$= \frac{1}{6} \cosh u + C = \frac{1}{6} \cosh 3x^2 + C.$$

EXAMPLE 3
A CATENARY

A **catenary** is the curve formed by a uniform flexible cable hanging from two points under the influence of its own weight. If the lowest point of the catenary is at the point $(0, a)$, then its equation (which we will not derive here) is given by

$$y = a \cosh \frac{x}{a}, \qquad a > 0.$$

This curve is graphed in Figure 3.

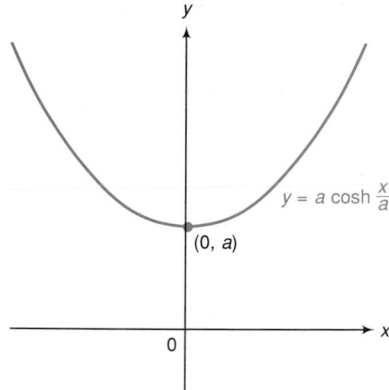

FIGURE 3
GRAPH OF $f(x) = a \cosh \dfrac{x}{a}$, $a > 0$

This is a catenary.

As you might expect, there are four other hyperbolic functions. They do not arise nearly as often in applications as the two we have already discussed.

DEFINITION   **OTHER HYPERBOLIC FUNCTIONS**

(i) $\tanh x = \dfrac{\sinh x}{\cosh x} = \dfrac{e^x - e^{-x}}{e^x + e^{-x}}$.

The domain of $\tanh x$ is $(-\infty, \infty)$.

(ii) $\coth x = \dfrac{1}{\tanh x} = \dfrac{\cosh x}{\sinh x} = \dfrac{e^x + e^{-x}}{e^x - e^{-x}}$.

The domain of $\coth x$ is all x except $x = 0$.

(iii) $\operatorname{sech} x = \dfrac{1}{\cosh x} = \dfrac{2}{e^x + e^{-x}}$.

The domain of $\operatorname{sech} x$ is $(-\infty, \infty)$.

(iv) $\operatorname{csch} x = \dfrac{1}{\sinh x} = \dfrac{2}{e^x - e^{-x}}$.

The domain of $\operatorname{csch} x$ is all x except $x = 0$. ■

The graphs of these four functions are given in Figure 4.

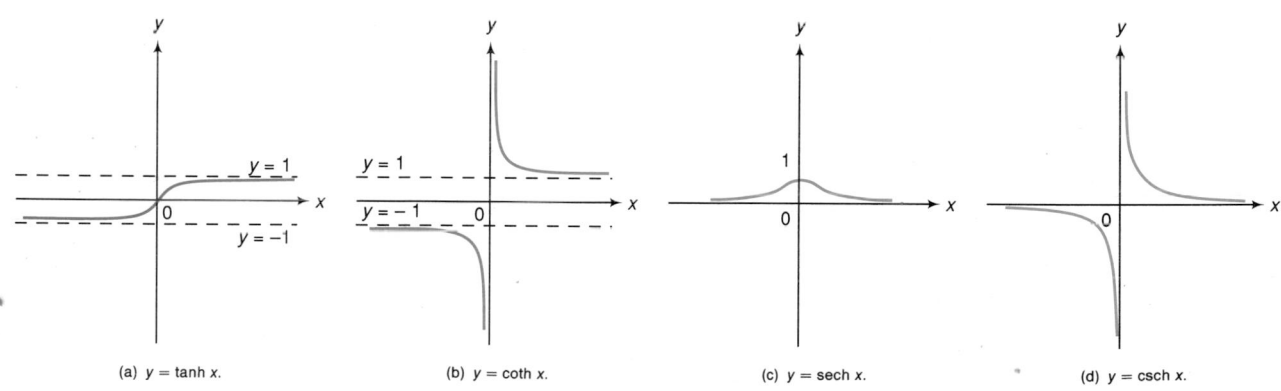

(a) $y = \tanh x$.   (b) $y = \coth x$.   (c) $y = \operatorname{sech} x$.   (d) $y = \operatorname{csch} x$.

FIGURE 4   GRAPHS OF FOUR HYPERBOLIC FUNCTIONS

Integrals and derivatives of these four hyperbolic functions are given in Table 2. The proofs of these facts are left as problems.

TABLE 2   DERIVATIVES AND INTEGRALS OF HYPERBOLIC FUNCTIONS

$$\frac{d}{dx}\tanh u = \operatorname{sech}^2 u \,\frac{du}{dx} \qquad\qquad \int \operatorname{sech}^2 u\, du = \tanh u + C$$

$$\frac{d}{dx}\coth u = -\operatorname{csch}^2 u \,\frac{du}{dx} \qquad\qquad \int \operatorname{csch}^2 u\, du = -\coth u + C$$

$$\frac{d}{dx}\operatorname{sech} u = -\operatorname{sech} u \tanh u \,\frac{du}{dx} \qquad \int \operatorname{sech} u \tanh u\, du = -\operatorname{sech} u + C$$

$$\frac{d}{dx}\operatorname{csch} u = -\operatorname{csch} u \coth u \,\frac{du}{dx} \qquad \int \operatorname{csch} u \coth u\, du = -\operatorname{csch} u + C$$

■ PROBLEMS 6.9

■ SELF-QUIZ

I. $\sinh x = $ _____ .
   a. $\sin(h \cdot x)$ b. $e^x + e^{-x}$
   c. $\frac{1}{2}(e^x - e^{-x})$ d. $\frac{1}{2}(e^x + e^{-x})$

II. $\cosh x = $ _____ .
   a. $\cos(h \cdot x)$ b. $e^x + e^{-x}$
   c. $\frac{1}{2}(e^x - e^{-x})$ d. $\frac{1}{2}(e^x + e^{-x})$

III. $\tanh x = $ _____ .
   a. $(\sinh x)/(\cosh x)$
   b. $(e^x - e^{-x})/(e^x + e^{-x})$
   c. $(\cosh x)/(\sinh x)$
   d. $(e^x + e^{-x})/(e^x - e^{-x})$

IV. $d/dx \cosh(2x) = $ _____ .
   a. $2\cosh(2x)$ b. $-2\sinh(2x)$
   c. $2\sinh(2x)$ d. $\sinh(2x)$

V. If $y = \sinh(3x)$, then $y'' = $ _____ .
   a. $9y$ b. $9y^2$
   c. $-9y$ d. $-27y^3$

■ DRILL

1. Use equation (6) to compute $\cosh x$ if $\sinh x = -\frac{2}{5}$. Then compute $\tanh x$, $\coth x$, $\text{sech } x$, and $\text{csch } x$.

2. Suppose $\text{csch } x = 5$; calculate the other five hyperbolic functions at this x.

In Problems 3–12, compute the derivative (with respect to x) of the given function.

3. $\sinh(4x + 2)$ 4. $\cosh(1/x)$
5. $\sin(\sinh x)$ 6. $\sinh(\sin x)$
7. $\tanh(1/x)$ 8. $\text{csch}(\ln x)$
9. $\tanh(\tan^{-1} x)$ 10. $\tanh(\text{csch } x)$
*11. $\sin^{-1}(\cosh x)$ *12. $x^{\cosh x}$
   [*Hint:* Be cautious.]

In Problems 13–22, perform the indicated integration.

13. $\int \sinh 2x \, dx$

14. $\int (\cosh x)(\sinh x) \, dx$

15. $\int \dfrac{\cosh 2x}{1 + \sinh 2x} \, dx$

16. $\int e^{\sinh x} \cosh x \, dx$

17. $\int \dfrac{\text{sech}^2 \sqrt{x}}{\sqrt{x}} \, dx$

18. $\int \dfrac{\text{csch}^2 x}{\sqrt[3]{\coth x}} \, dx$

19. $\int \coth x \, dx$

20. $\int \tanh x \, dx$

21. $\int \dfrac{\sinh(1/x)}{x^2} \, dx$

*22. $\int \cosh^3 x \, dx$
   [*Hint:* Use equation (6).]

■ APPLICATIONS

23. Calculate the area bounded by the catenary $y = a \cosh(x/a)$, the x-axis, and the lines $x = -a$ and $x = a$.

*24. **Symbiosis** is defined as the relationship of two or more different organisms in a close association that may be of benefit to each. If two species are living in a symbiotic relationship, then we may suppose that the rate of growth of the first population is proportional to the size of the second, and vice versa. If we let $x(t)$ and $y(t)$ be the populations of the two respective species, then there are positive constants α and β such that

$$\frac{dx}{dt} = \alpha y(t) \quad \text{and} \quad \frac{dy}{dt} = \beta x(t).$$

a. Verify that $x(t)$ satisfies

$$\frac{d^2 x}{dt^2} = \alpha \beta x(t).$$

b. Suppose ω is a positive constant such that $\omega^2 = \alpha\beta$. Verify that for any constants c_1 and c_2, $x(t) = c_1 \sinh \omega t + c_2 \cosh \omega t$ satisfies the condition asserted in part (a).

c. Suppose time is measured in months; also suppose $\alpha = 4$ and $\beta = 1$ for two populations which begin with $x(0) = 500$ and $y(0) = 1000$. What is the size of each population after one-half month?

CONTINUED

■ SHOW/PROVE/DISPROVE

25. Show directly from the definitions that $\cosh^2 x - \sinh^2 x = 1$.

26. Show that $e^x = \cosh x + \sinh x$ and $e^{-x} = \cosh x - \sinh x$.

27. Show that $\sinh(-x) = -\sinh x$ (i.e., sinh is an *odd* function).

28. Show that $\cosh(-x) = \cosh x$ (i.e., cosh is an *even* function).

29. Show that $(d/dx) \tanh x = \operatorname{sech}^2 x$. [*Hint:* Use $\tanh x = (\sinh x)/(\cosh x)$.]

30. Show that $(d/dx) \coth x = -\operatorname{csch}^2 x$.

31. Show that $(d/dx) \operatorname{sech} x = -(\operatorname{sech} x)(\tanh x)$.

32. Show that $(d/dx) \operatorname{csch} x = -(\operatorname{csch} x)(\coth x)$.

33. Use the results of Problem 26 to prove that, for any real numbers n and x,

$$(\cosh x + \sinh x)^n = \cosh nx + \sinh nx.$$

34. Use the results of Problem 26 to prove that, for any real numbers n and x,

$$(\cosh x - \sinh x)^n = \cosh nx - \sinh nx.$$

35. Show that $\sinh(x + y) = \sinh x \cdot \cosh y + \cosh x \cdot \sinh y$. [*Hint:* Write everything in terms of $e^{\pm x}$ and $e^{\pm y}$. For instance, $\sinh(x + y) = (e^{x+y} - e^{-(x+y)})/2 = (e^x e^y - e^{-x} e^{-y})/2$.]

36. Show that $\cosh(x + y) = \cosh x \cdot \cosh y + \sinh x \cdot \sinh y$.

37. Show that $\sinh 2x = 2 \sinh x \cdot \cosh x$.

38. Show that

$$\cosh 2x = \cosh^2 x + \sinh^2 x$$
$$= 2 \sinh^2 x + 1.$$

39. Use equation (6) to show both of the following identities:
   a. $\operatorname{sech}^2 x + \tanh^2 x = 1$
   b. $\coth^2 x - \operatorname{csch}^2 x = 1$

*40. Obtain general formulas expressing $\sinh(x - y)$ and $\cosh(x - y)$ in terms of hyperbolic functions of x and y.

■ ANSWERS TO SELF-QUIZ

I. c   II. d   III. a = b
IV. c   V. a

6.10 THE INVERSE HYPERBOLIC FUNCTIONS (OPTIONAL)

From Figure 6.9.1(a) we see that for every real number y there is a unique x such that $y = \sinh x$. Thus $\sinh x$ is 1–1 and therefore has an inverse function.

DEFINITION **INVERSE HYPERBOLIC SINE**

The **inverse hyperbolic sine function** is defined by $y = \sinh^{-1} x$ if and only if $x = \sinh y$. The domain of $y = \sinh^{-1} x$ is $(-\infty, \infty)$. ■

NOTE: In contrast to the definitions of the inverse trigonometric functions, with the inverse hyperbolic sine function there is no need to restrict y to a fixed interval since the function $\sinh x$ is 1–1.

From Figure 6.9.1(b) we see that for each $y \geq 1$ there are two values of x for which $y = \cosh x$: one positive, one negative. Thus we have the following definition.

DEFINITION **INVERSE HYPERBOLIC COSINE**

The **inverse hyperbolic cosine function** is defined by $y = \cosh^{-1} x$ if and only if $\cosh y = x$, $y \geq 0$. The domain of $\cosh^{-1} x$ is the interval $[1, \infty)$. ■

REMARK: Here we need to restrict y to be nonnegative so that for each x there will be a *unique y* such that $x = \cosh y$ (otherwise, there would be two values for $\cosh^{-1} x$, and $\cosh^{-1} x$ would then *not* be a function).

THEOREM 1 **DERIVATIVES OF $\sinh^{-1} x$ AND $\cosh^{-1} x$**

(i) $\dfrac{d}{dx} \sinh^{-1} x = \dfrac{1}{\sqrt{x^2 + 1}}$ (1)

(ii) $\dfrac{d}{dx} \cosh^{-1} x = \dfrac{1}{\sqrt{x^2 - 1}}$ $\qquad x > 1$ (2)

PROOF:

(i) If $y = \sinh^{-1} x$, then $x = \sinh y$, so that

$$1 = \cosh y \frac{dy}{dx} \qquad \text{and} \qquad \frac{dy}{dx} = \frac{1}{\cosh y}.$$

But $\cosh^2 y - \sinh^2 y = 1$, so that $\cosh y = \sqrt{1 + \sinh^2 y} = \sqrt{1 + x^2}$, which proves (i). Note that dy/dx exists according to Theorem 6.1.4.

(ii) If $y = \cosh^{-1} x$, then $x = \cosh y$ and

$$1 = \sinh y \frac{dy}{dx}, \qquad \text{or} \qquad \frac{dy}{dx} = \frac{1}{\sinh y}.$$

Now $\sinh y = \pm \sqrt{\cosh^2 y - 1}$. But, $\sinh y \geq 0$ because $y \geq 0$, so that

$$\sinh y = \sqrt{\cosh^2 y - 1} = \sqrt{x^2 - 1},$$

which proves (ii). ∎

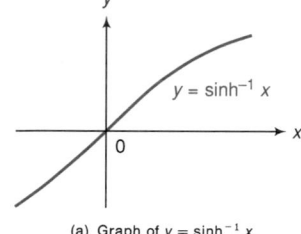

(a) Graph of $y = \sinh^{-1} x$.

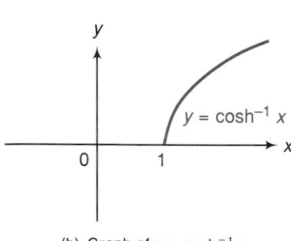

(b) Graph of $y = \cosh^{-1} x$.

FIGURE 1
GRAPHS OF $y = \sinh^{-1} x$ AND $y = \cosh^{-1} x$

The graphs of $y = \sinh^{-1} x$ and $y = \cosh^{-1} x$ are given in Figure 1.

From the graph of $y = \tanh x$ in Figure 6.9.4(a), we see that for every y in $(-1, 1)$ there is an x such that $y = \tanh x$.

DEFINITION **INVERSE HYPERBOLIC TANGENT**

The **inverse hyperbolic tangent function** is defined by $y = \tanh^{-1} x$ if and only if $x = \tanh y$. The domain of $\tanh^{-1} x$ is $(-1, 1)$. ∎

THEOREM 2 **DERIVATIVE OF $\tanh^{-1} x$**

$$\frac{d}{dx} \tanh^{-1} x = \frac{1}{1 - x^2} \qquad -1 < x < 1$$

PROOF: If $y = \tanh^{-1} x$, then $x = \tanh y$, and

$$1 = \operatorname{sech}^2 y \frac{dy}{dx}, \qquad \text{or} \qquad \frac{dy}{dx} = \frac{1}{\operatorname{sech}^2 y} = \frac{1}{1 - \tanh^2 y} = \frac{1}{1 - x^2}$$

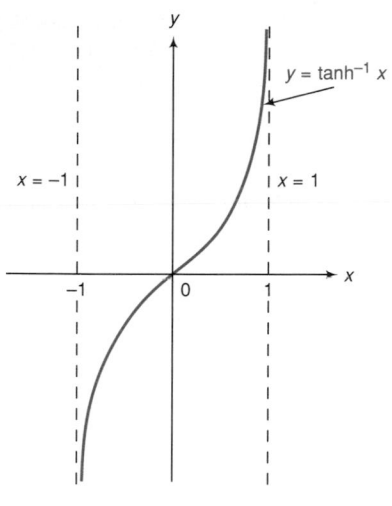

FIGURE 2
GRAPH OF $y = \tanh^{-1} x$

[see Problem 6.9.39a]. Again, dy/dx exists according to Theorem 6.1.4. ■

A sketch of $y = \tanh^{-1} x$ appears in Figure 2.

We summarize in Table 1 the rules we have discovered in this section. The derivation of the last formula is suggested in Problem 22.

TABLE 1   DERIVATIVES AND INTEGRALS OF INVERSE HYPERBOLIC FUNCTIONS

$$\frac{d}{dx} \sinh^{-1} u = \frac{1}{\sqrt{u^2 + 1}} \frac{du}{dx} \qquad \int \frac{du}{\sqrt{u^2 + 1}} = \sinh^{-1} u + C$$

$$\frac{d}{dx} \cosh^{-1} u = \frac{1}{\sqrt{u^2 - 1}} \frac{du}{dx}, u > 1 \qquad \int \frac{du}{\sqrt{u^2 - 1}} = \cosh^{-1} u + C, u > 1$$

$$\frac{d}{dx} \tanh^{-1} u = \frac{1}{1 - u^2} \frac{du}{dx}, |u| < 1 \qquad \int \frac{du}{1 - u^2} = \begin{cases} \tanh^{-1} u + C, |u| < 1 \\ \tanh^{-1}(1/u) + C, |u| > 1 \end{cases}$$

EXAMPLE 1
AN INTEGRAL INVOLVING $\sinh^{-1} x$

Compute $\displaystyle\int \frac{x^2}{\sqrt{1 + x^6}} dx$.

SOLUTION: Let $u = x^3$. Then $du = 3x^2 dx$ so that

$$\int \frac{x^2}{\sqrt{1 + x^6}} dx = \frac{1}{3} \int \frac{3x^2}{\sqrt{1 + x^6}} dx = \frac{1}{3} \int \frac{du}{\sqrt{1 + u^2}} = \frac{1}{3} \sinh^{-1} u$$

$$= \frac{1}{3} \sinh^{-1} x^3 + C.$$

There are three other inverse hyperbolic functions, but these are rarely used, so we will not discuss them here. In fact, the inverse hyperbolic functions occur rarely in applications and are usually used only as devices to help us to integrate.

■ PROBLEMS 6.10

■ SELF-QUIZ

I. Which of the hyperbolic functions cosh, sinh, tanh needs to have its domain restricted in order to obtain a 1–1 function?

II. True or False:
  a. $\cosh^{-1}(\cosh x) = x$ for all x.
  b. $\sinh^{-1}(\sinh x) = x$ for all x.
  c. $\tanh^{-1}(\tanh x) = x$ for all x.
  d. $\cosh^{-1}(\cosh x) - x$ for all $x \geq 0$.
  e. $\cosh(\cosh^{-1} x) = x$ for all x.
  f. $\cosh(\cosh^{-1} x) = x$ for all $x \geq 1$.
  g. $\sinh(\sinh^{-1} x) = x$ for all x.
  h. $\tanh(\tanh^{-1} x) = x$ if $-1 \leq x \leq 1$.

III. $d/dx \sinh^{-1}(x/2) = $ _____.
  a. $1/\sqrt{x^2 - 4}$ b. $1/\sqrt{x^2 + 1}$
  c. $-1/\sqrt{x^2 + 4}$ d. $1/\sqrt{x^2 + 4}$

IV. $d/dx \cosh^{-1}(3x) = $ _____.
  a. $1/\sqrt{x^2 - 9}$ b. $3/\sqrt{9x^2 - 1}$
  c. $3/\sqrt{x^2 + 1}$ d. $-1/\sqrt{9x^2 - 1}$

V. $d/dx \tanh^{-1}(2x - 1) = $ _____.
  a. $1/(4x - x^2)$ b. $2/(1 - 4x^2)$
  c. $1/(2x - 2x^2)$ d. $1/(4x^2 - 4x)$

CONTINUED

■ DRILL

In Problems 1–8, find the domain and calculate the derivative of the given function.

1. $\sinh^{-1}(3x + 2)$
2. $\cosh^{-1}(\sqrt{x})$
3. $\tanh^{-1}(\ln x)$
4. $\tanh^{-1}(1/x^2)$
5. $\sqrt{\sinh^{-1} x}$
6. $(\sinh^{-1} x)/(\cosh^{-1} x)$
7. $\cosh(\sinh^{-1} x)$
8. $\sinh^{-1}(\cosh x)$

In Problems 9–18, compute the given integral.

9. $\int \dfrac{x}{\sqrt{x^4 - 4}}\, dx$

10. $\int \dfrac{x}{\sqrt{x^4 + 4}}\, dx$

11. $\int \dfrac{x}{4 - x^4}\, dx$

12. $\int \dfrac{e^x}{16 - e^{2x}}\, dx$

13. $\int \dfrac{e^x}{\sqrt{1 + e^{2x}}}\, dx$

14. $\int \dfrac{e^x}{\sqrt{e^{2x} - 9}}\, dx$

15. $\int \dfrac{\cos x}{\sqrt{1 + \sin^2 x}}\, dx$

16. $\int \dfrac{\sec x \tan x}{\sqrt{4 + \sec^2 x}}\, dx$

17. $\int \dfrac{dx}{x\sqrt{\ln^2 x - 25}}$

18. $\int \dfrac{dx}{\sqrt{x + x^2}}$

[*Hint:* Complete the square.]

■ SHOW/PROVE/DISPROVE

19. Show that
$$\frac{d}{dx} \sinh^{-1} \frac{x}{a} = \frac{1}{\sqrt{x^2 + a^2}} \quad \text{for every } x.$$

20. Show that
$$\frac{d}{dx} \cosh^{-1} \frac{x}{a} = \frac{1}{\sqrt{x^2 - a^2}} \quad \text{for } \frac{x}{a} > 1.$$

21. Show that
$$\frac{d}{dx} \tanh^{-1} \frac{x}{a} = \frac{a}{a^2 - x^2} \quad \text{for } x^2 < a^2,$$
so that
$$\int \frac{du}{a^2 - u^2} = \frac{1}{a} \tanh^{-1}\left(\frac{u}{a}\right) + C.$$

22. Show that for $|x| > 1$
$$\int \frac{dx}{1 - x^2} = \tanh^{-1}\left(\frac{1}{x}\right) + C.$$

*23. Show that $\sinh^{-1} x = \ln(x + \sqrt{x^2 + 1})$. [*Hint:* Begin by confirming that these two functions have the same domain. Then differentiate each function and show their derivatives are identical; this will imply the original functions differ by a constant. Finally, show that this constant is zero by evaluating both functions at a convenient choice of x (e.g., $x = 0$).]

*24. Show that $\cosh^{-1} x = \ln(x + \sqrt{x^2 - 1})$.

*25. Show that
$$\tanh^{-1} x = \tfrac{1}{2} \ln[(1 + x)/(1 - x)].$$

*26. Show that
$$\int \frac{du}{\sqrt{u^2 + a^2}} = \ln(u + \sqrt{u^2 + a^2}) + C.$$
[The results of Problems 19 and 23 can be used to show this.]

*27. Show that if $u > a$, then
$$\int \frac{du}{\sqrt{u^2 - a^2}} = \ln(u + \sqrt{u^2 - a^2}) + C.$$
[The results of Problems 20 and 24 can be used to show this.]

28. Show that if $u^2 < a^2$, then
$$\int \frac{a\, du}{a^2 - u^2} = \frac{1}{2} \ln\left|\frac{a + u}{a - u}\right| + C.$$
[The results of Problems 21 and 25 can be used to show this.]

■ ANSWERS TO SELF-QUIZ

I. cosh

II. a. False   b. True   c. True
    d. True   e. False f. True

g. True   h. True

III. d

IV. b

V. c

■ PROBLEMS　　CHAPTER 6 REVIEW

In Problems 1–12, compute the value of x.

1. $x = \log_3 9$
2. $x = \log_{1/100} 1000$
3. $4 = \log_x(\frac{1}{16})$
4. $x = e^{\ln 17.2}$
5. $\log x = 10^{-2}$
6. $\log_x 32 = -5$
7. $x = \sin[\cos^{-1}(\frac{7}{10})]$
8. $x = \tan[\sec^{-1}(-4)]$
9. $x = \csc[\cot^{-1}(\frac{2}{3})]$
10. $x = \sinh[\cosh^{-1} 2]$
11. $x = \operatorname{sech}[\tanh^{-1}(\frac{3}{7})]$
*12. $7^x = 25 \cdot e^{2x}$

13. If $y = 3 \ln x$, what happens to y if x doubles?
14. Simplify

$$\ln\left(\left[\frac{\sqrt{x^3 - 1}\,(x^5 + 1)^{4/3}}{\sqrt{(x - 3)(x - 2)}}\right]^{-5/6}\right).$$

In Problems 15–36, compute the derivative of the given function.

15. $\ln(1 + x^2)$
16. $\ln|e^x - 5|$
17. $e^{x^2 + 2x + 1}$
18. $e^{\sqrt{x^3 - 3}}$
19. $\ln[x + \ln(x + 3)]$
20. $\dfrac{(x - 5)\sqrt[3]{x + 5}}{(x + 2)\sqrt{x + 1}}$
21. $2^{x + 5}$
*22. $x^{x + 1}$
23. $\tan^{-1}\left(\dfrac{1}{x^2}\right)$
24. $\cot\left(\dfrac{1 + x}{1 - x}\right)$
25. $\csc(\sqrt{x})$
26. $\sin^{-1}(e^x + 1)$
27. $\ln|\sec x - \tan x|$
28. $\cos^{-1}(\sqrt{x})$
29. $\tan^{-1}(\sqrt[3]{x})$
30. $\tan^{-1}[(x - 4)^2]$
31. $\sin^{-1}(e^{-x})$
32. $x^2 \cos^{-1}(1 - x)$
33. $\sinh(x^2 + 3x)$
34. $\operatorname{csch}(\sqrt[3]{x})$
35. $\tanh^{-1}(\sqrt{1 + x})$
36. $\cosh^{-1}(2/x)$

In Problems 37–54, evaluate the integral.

37. $\displaystyle\int \frac{3x}{1 + x^2}\,dx$
38. $\displaystyle\int \frac{e^{\sqrt{x}}}{\sqrt{x}}\,dx$
39. $\displaystyle\int \frac{e^{-1/x^2}}{x^3}\,dx$
40. $\displaystyle\int \frac{dx}{3x \ln x}$
41. $\displaystyle\int \frac{4^{\ln x}}{x}\,dx$
*42. $\displaystyle\int \frac{1}{x}\sqrt{\log_5 x}\,dx$
43. $\displaystyle\int_0^{5\pi/6} \tan\left(\frac{x}{5}\right)dx$
44. $\displaystyle\int \frac{dx}{\sqrt{1 - x^2}}$
45. $\displaystyle\int \frac{3x\,dx}{x^4 + 1}$
46. $\displaystyle\int \frac{e^x\,dx}{\sqrt{1 - e^{2x}}}$
47. $\displaystyle\int \sin 3x \cos 3x\,dx$
48. $\displaystyle\int x \cos(6x^2 - 2)$
49. $\displaystyle\int \coth x\,dx$
50. $\displaystyle\int x^3 \sinh x^4\,dx$
51. $\displaystyle\int \frac{\operatorname{sech}^2(\ln x)}{x}\,dx$
52. $\displaystyle\int \frac{x}{9 - x^4}\,dx$
53. $\displaystyle\int \frac{\sec^2 x}{\sqrt{\tan^2 x - 4}}\,dx$
54. $\displaystyle\int \frac{e^{4x}\,dx}{\sqrt{16 + e^{8x}}}$

In Problems 55–60, determine intervals over which the given function is 1–1, and determine the inverse function over each such interval, if possible. Then calculate the derivative of each inverse function.

55. $f(x) = \sqrt{2x - 5}$
56. $y = x^2 + 5x + 6$
57. $g(x) = \ln(3 - 4x)$
58. $F(x) = \sin(x - \pi)$
59. $G(x) = \tan(x/2)$
60. $H(x) = e^x - e^{-x}$

■ APPLICATIONS

61. The population of a certain city grows 6%/yr. If the population in 1970 was 250,000, what was the population in 1980? What will it be in the year 2000? [*Hint:* Assume the population grows continuously.]
62. The relative annual growth rate of a particular population is 15%. If the initial population is 10,000, what is the population after 5 yr? After 10 yr? How long will it take for the population to double?
63. When a cake is taken out of the oven, its temperature is 125°C. Room temperature is 23°C; the temperature of the cake is 80°C after 10 min.
    a. What temperature will the cake have after 20 min?

    b. How long will the cake take to cool to 25°C?
64. A fossil contains 35% of the normal amount of ^{14}C. What is its approximate age?
65. What is the half-life of an exponentially decaying substance that loses 20% of its mass in one week? How long will it take this substance to lose 75% of its mass? How long to lose 95%?
66. Radioactive beryllium is sometimes used to date fossils found in deep-sea sediment. The decay of beryllium satisfies the equation

$$\frac{dA}{dt} = -\alpha A, \qquad \text{where } \alpha = 1.5 \times 10^{-7}.$$

What is the half-life of beryllium?

CONTINUED

67. A sum of $10,000 is invested at an interest rate of 6% compounded quarterly. What will the investment be worth in 8 yr? If the interest were compounded continuously, what then would be the worth after 8 yr?

68. A certain government bond sells for $750 and can be redeemed for $1000 in 8 yr. Assuming continuous compounding, what is the rate of interest paid?

69. Sketch the graph of $y = \sinh(x - 1)$ and the graph of $y = 1 + \sinh^{-1} x$.

70. Find the values of $\sinh x$ and $\tanh x$ if $\cosh x = 2$.

📱 COMPUTER EXERCISES

1. Use a computer algebra system to evaluate the following integrals.

   a. $\displaystyle\int \frac{1}{(x-1)(x-2)} \, dx$

   b. $\displaystyle\int \frac{1}{(x+3)(x-4)} \, dx$

   c. $\displaystyle\int \frac{1}{(x-2)(x-5)} \, dx$

   Based on the answers obtained above, what is the form of $\displaystyle\int \frac{1}{ax^2 + bx + c} \, dx$ if $ax^2 + bx + c$ has zeros p and q? $(p \neq q.)$ Use the computer to view additional examples if necessary.

2. Use a computer algebra system to evaluate the following integrals.

   a. $\displaystyle\int xe^x \, dx$ b. $\displaystyle\int x^2 e^x \, dx$ c. $\displaystyle\int x^3 e^x \, dx$

   Based on the answers obtained above, what is the form of $\int x^n e^x \, dx$ if n is a positive integer? (Use the computer to view additional examples if necessary.)

3. Approximate all extrema of $f(x) = \ln^3 x - x$ to four decimal places and sketch the graph.

4. Approximate all extrema of $f(x) = e^{2x} - x^2 e^x$ to four decimal places and sketch the graph.

5. You have the opportunity to invest your money at a 10% yearly rate of return. How many times per year must the interest be compounded in order to ensure that your money is approximately doubled in 7 years?

6. Show that the function $f(x) = \dfrac{x^5}{5} - \dfrac{x^4}{2} - \dfrac{x^3}{3} + x^2 + 10x + 17$ has an inverse. [*Hint:* Use a computer algebra system to factor $f'(x)$ and show that it has no real zeros.]

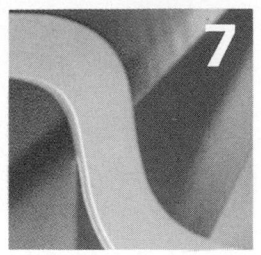

7 TECHNIQUES OF INTEGRATION

In Chapters 2 and 6, we developed methods for differentiating almost any conceivable function we could write down. However, as we have seen, the process of integration is much more difficult. There are many harmless-looking functions that are continuous (and so have antiderivatives) but for which it is impossible to express antiderivatives in terms of the functions we have so far discussed. The functions e^{x^2} and $\sin(1/\sqrt{x})$ fall into this category. We simply do not have enough functions and must do the best we can with algebraic functions (i.e., functions like $[\sqrt{x^2 - 1}/(x^4 - 2\sqrt{x})]^{3/5}$), exponential and logarithmic functions, trigonometric functions and their inverses, and the hyperbolic functions (which are really exponential functions). It is true that most continuous functions do not have integrals expressible in terms of the functions listed above.[†]

The aim of this chapter is to show how to recognize classes of functions that can be integrated in the sense of actually finding a recognizable antiderivative. Many techniques have evolved over a period of decades, and we will present the more general ones. This chapter consists primarily of a set of techniques. Each of these was discovered in response to the need to calculate a particular integral. There is no short, easy way to learn these techniques, and it will be necessary to do great numbers of problems.

After you have completed the first eight sections of this chapter, it will be apparent that there are many integrals for which none of our techniques will yield an answer. For such cases, we can still approximate the definite integral since the definite integral of any continuous function can be calculated to any number of decimal places of accuracy by a variety of numerical techniques. Some of these techniques will be discussed in Section 7.9.

7.1 REVIEW OF THE BASIC FORMULAS OF INTEGRATION

We give below the integration formulas that were developed in Chapters 4 and 6.

$$\int u^n \, du = \frac{u^{n+1}}{n+1} + C \qquad (n \neq -1) \tag{1}$$

$$\int \frac{1}{u} \, du = \ln|u| + C \tag{2}$$

$$\int a^u \, du = \frac{1}{\ln a} a^u + C \tag{3}$$

[†] Actually, there are theorems that give us ways to determine whether a function can be integrated in terms of the functions listed. For an interesting discussion of this topic, see the paper by D. G. Mead, "Integration," *American Mathematical Monthly*, 68, (1961): 152–156.

$$\int e^u \, du = e^u + C \tag{4}$$

$$\int \cos u \, du = \sin u + C \tag{5}$$

$$\int \sin u \, du = -\cos u + C \tag{6}$$

$$\int \sec^2 u \, du = \tan u + C \tag{7}$$

$$\int \csc^2 u \, du = -\cot u + C \tag{8}$$

$$\int \sec u \tan u \, du = \sec u + C \tag{9}$$

$$\int \csc u \cot u \, du = -\csc u + C \tag{10}$$

$$\int \tan u \, du = -\ln|\cos u| + C = \ln|\sec u| + C$$
(Example 8, p. 416) \tag{11}

$$\int \cot u \, du = \ln|\sin u| + C \tag{12}$$

$$\int \sec u \, du = \ln|\sec u + \tan u| + C$$
(Example 4, p. 414)[†] \tag{13}

$$\int \csc u \, du = -\ln|\csc u + \cot u| + C \qquad \text{(Problem 38, p. 455)} \tag{14}$$

$$\int \frac{du}{\sqrt{1 - u^2}} = \sin^{-1} u + C = -\cos^{-1} u + C$$
(Theorems 1 and 2, pp. 457 and 459) \tag{15}

$$\int \frac{du}{1 + u^2} = \tan^{-1} u + C \qquad \text{(Theorem 3, p. 460)}$$
$$= -\cot^{-1} u + C \qquad \text{(Problem 81, p. 465)} \tag{16}$$

$$\int \frac{du}{a^2 + u^2} = \frac{1}{a} \tan^{-1} \frac{u}{a} + C \qquad \text{(Example 5, p. 461)} \tag{17}$$

$$\int \frac{du}{u\sqrt{u^2 - 1}} = \sec^{-1}|u| + C \qquad \text{(Problem 92, p. 465)} \tag{18}$$

$$\int \sinh u \, du = \cosh u + C \tag{19}$$

$$\int \cosh u \, du = \sinh u + C \tag{20}$$

$$\int \operatorname{sech}^2 u \, du = \tanh u + C \tag{21}$$

$$\int \operatorname{csch}^2 u \, du = -\coth u + C \tag{22}$$

$$\int \operatorname{sech} u \tanh u \, du = -\operatorname{sech} u + C \tag{23}$$

[†] This integral seems to have been pulled out of a hat. For a fascinating discussion of the original derivation of $\int \sec \theta \, d\theta$, see V. Frederick Rickey and Philip M. Tuchinsky, "An application of geography to mathematics: History of the integral of the secant," *Mathematics Magazine*, 53, No. 3 (May 1980): 162–166.

$$\int \operatorname{csch} u \coth u \, du = -\operatorname{csch} u + C \tag{24}$$

$$\int \frac{du}{\sqrt{u^2+1}} = \sinh^{-1} u + C \qquad \text{(Theorem 1, p. 472)}$$

$$= \ln|u + \sqrt{1+u^2}| + C \qquad \text{(Problem 23, p. 474)} \tag{25}$$

$$\int \frac{du}{\sqrt{u^2-1}} = \cosh^{-1} u + C \qquad \text{(Theorem 1, p. 472)}$$

$$= \ln|u + \sqrt{u^2-1}| + C \qquad (u > 1) \text{ (Problem 24, p. 474)} \tag{26}$$

$$\int \frac{du}{1-u^2} = \tanh^{-1} u + C \qquad \text{(Theorem 2, p. 472)}$$

$$= \frac{1}{2} \ln\left|\frac{1+u}{1-u}\right| + C \qquad (u^2 < 1) \qquad \text{(Problem 25, p. 474)} \tag{27}$$

A much more complete table of integrals appears at the end of the book.

7.2 INTEGRATION BY PARTS

Of all the methods to be discussed in this chapter, the most powerful is called **integration by parts**. It is derived from the product rule of differentiation (written in terms of differentials):

$$d(uv) = u \, dv + v \, du. \tag{1}$$

Integrating both sides of (1), we obtain

$$uv = \int u \, dv + \int v \, du,$$

or rearranging terms,

INTEGRATION BY PARTS: INDEFINITE INTEGRAL FORM

$$\int u \, dv = uv - \int v \, du. \tag{2}$$

As we will see, the idea in using integration by parts is to rewrite an expression that is difficult to integrate in terms of an expression for which an integral is readily obtainable.

EXAMPLE 1
COMPUTING AN INDEFINITE INTEGRAL BY INTEGRATING BY PARTS

Calculate $\int xe^x \, dx$.

SOLUTION: We cannot integrate xe^x directly because the x term gets in the way. However, if we set $u = x$ and $dv = e^x \, dx$, then $du = dx$, $v = \int e^x \, dx = e^x$, and

$$\int xe^x \, dx = \int u \, dv = uv - \int v \, du = xe^x - \int e^x \, dx = xe^x - e^x + C.$$

This result can be checked by differentiation. Note how the x term "disappeared." This example is typical of the type of problem that can be solved by integration by parts.

EXAMPLE 2
COMPUTING THE INTEGRAL OF
ln x BY INTEGRATING BY PARTS

Calculate $\int \ln x \, dx$.

SOLUTION: There are two terms here, $\ln x$ and dx. If we are to integrate by parts, the only reasonable choices we have for u and dv are $u = \ln x$ and $dv = dx$. Then $du = (1/x)\,dx$, $v = x$, and

$$\overset{u}{\overbrace{\int \ln x}} \overset{dv}{\overbrace{dx}} = \int u \, dv = uv - \int v \, du = x \ln x - \int \overset{v}{x} \cdot \overset{du}{\frac{1}{x} \, dx}$$

$$= x \ln x - x + C.$$

NOTE: There are other choices for u and dv but they will not get us an answer. We could, for example, set $u = \dfrac{\ln x}{x}$ and $dv = x \, dx$. Then $du =$

$\dfrac{x\left(\dfrac{1}{x}\right) - \ln x}{x^2} \, dx = \dfrac{1 - \ln x}{x^2} \, dx$ and $v = \dfrac{x^2}{2}$ so $\int \ln x \, dx = \int \dfrac{\ln x}{x} x \, dx =$

$\int u \, dv = uv - \int v \, du = \dfrac{x \ln x}{2} - \dfrac{1}{2} \int (1 - \ln x) \, dx$. This is correct but it is not helpful because $1 - \ln x$ is no easier to integrate than $\ln x$.

In many of the integrations involving $\ln x$, we may take $u = \ln x$ so that $du = (1/x)\,dx$ and the $\ln$ term "vanishes."

Integration by parts works just as well when computing a definite integral.

INTEGRATION BY PARTS: DEFINITE INTEGRAL FORM

$$\int_a^b u \, dv = uv \Big|_a^b - \int_a^b v \, du = u(b)v(b) - u(a)v(a) - \int_a^b v \, du.$$

EXAMPLE 3
EVALUATING A DEFINITE INTEGRAL
BY INTEGRATING BY PARTS

Calculate $\int_0^1 x^3 e^{x^2} \, dx$.

SOLUTION: There are several choices for u and dv. However, most of these choices won't help because we won't be able to integrate dv. The key is to observe that $\dfrac{d}{dx} e^{x^2} = 2xe^{x^2}$ so $\int xe^{x^2} \, dx = \frac{1}{2}e^{x^2}$. That is, if we choose $dv = xe^{x^2} \, dx$, we can find v. Thus, we set $u = x^2$ and $dv = xe^{x^2} \, dx$ to obtain

$$\int_0^1 x^3 e^{x^2} \, dx = \int_0^1 \overset{u}{x^2} \cdot \overset{dv}{\overbrace{xe^{x^2} \, dx}} = \overset{u}{x^2} \overset{v}{\frac{e^{x^2}}{2}} \Big|_0^1 - \int_0^1 \overset{v}{\frac{1}{2}e^{x^2}} \cdot \overset{du}{\overbrace{2x \, dx}}$$

$$= \frac{e}{2} - \int_0^1 xe^{x^2} \, dx = \frac{e}{2} - \frac{1}{2}e^{x^2} \Big|_0^1 = \frac{e}{2} - \frac{e}{2} + \frac{1}{2} = \frac{1}{2}.$$

REMARK: We were able to complete the integration because we could integrate $v\dfrac{du}{dx} = xe^{x^2}$. If, for example, the integrand had been $x^4 e^{x^2}$, then $v = x^3$, $dv = xe^{x^2}dx$, $du = 3x^2\,dx$, $v = \frac{1}{2}e^{x^2}$, and $v\dfrac{du}{dx} = \frac{3}{2}x^2 e^{x^2}$, which we cannot integrate.

Example 3 points out two things to look for in choosing u and dv:

TWO THINGS TO LOOK FOR WHEN CHOOSING u AND dv

(i) It must be possible to evaluate $\int dv$.

(ii) $\int v\,du$ should be easier to evaluate than $\int u\,dv$.

EXAMPLE 4
INTEGRATING BY PARTS TWICE IN ORDER TO COMPUTE AN INTEGRAL

Evaluate $\int e^x \sin x\,dx$.

SOLUTION: We set $I = \int e^x \sin x\,dx$. (The reason for this will become apparent as we work the problem.) Let $u = e^x$ and $dv = \sin x\,dx$. Then $du = e^x\,dx$, $v = -\cos x$, and

$$I = -e^x \cos x + \int e^x \cos x\,dx.$$

It seems as if we have not accomplished anything, but this is not the case. We now set $u = e^x$ and $dv = \cos x\,dx$, so that $du = e^x\,dx$, $v = \sin x$, and

$$I = -e^x \cos x + e^x \sin x - \int e^x \sin x\,dx = e^x(\sin x - \cos x) - I,$$

or

$$2I = e^x(\sin x - \cos x)$$

and

$$I = \int e^x \sin x\,dx = \frac{e^x(\sin x - \cos x)}{2} + C.$$

EXAMPLE 5
FINDING A VOLUME BY INTEGRATING BY PARTS

Find the volume generated when the area under one loop of the curve $y = \sin x$ between $x = 0$ and $x = \pi$ is rotated about the y-axis.

SOLUTION: The solid is sketched in Figure 1. From equation (6) on page 345, the volume is given by

$$V = 2\pi \int_0^\pi x \sin x\,dx.$$

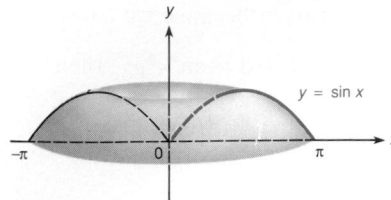

FIGURE 1
VOLUME GENERATED WHEN AREA UNDER
ONE LOOP OF $y = \sin x$ IS ROTATED ABOUT
THE y-AXIS

We can integrate this expression by parts. Let $u = x$ and $dv = \sin x \, dx$. Then $du = dx$, $v = -\cos x$, and

$$V = 2\pi \left(\overbrace{x}^{u}\overbrace{(-\cos x)}^{v}\Big|_0^\pi + \int_0^\pi \overbrace{\cos x}^{-v} \, \overbrace{dx}^{du} \right) = 2\pi \left(\pi + \sin x \Big|_0^\pi \right) = 2\pi^2.$$

EXAMPLE 6

COMPUTING THE INTEGRAL OF
$\sec^3 x$ BY INTEGRATING BY PARTS

Evaluate $\int \sec^3 x \, dx$.

SOLUTION: Let $I = \int \sec^3 x \, dx$. We know that $(d/dx)\tan x = \sec^2 x$. We set $u = \sec x$ and $dv = \sec^2 x \, dx$, so that

$$du = \sec x \tan x \, dx, \qquad v = \tan x,$$

and

$$I = \sec x \tan x - \int \sec x \tan^2 x \, dx. \tag{3}$$

Now

$$\tan^2 x = \sec^2 x - 1, \tag{4}$$

and by inserting (4) into (3), we obtain

$$I = \sec x \tan x - \int \sec x(\sec^2 x - 1) \, dx$$

$$= \sec x \tan x - \int \sec^3 x \, dx + \int \sec x \, dx$$

$$= \sec x \tan x - I + \ln|\sec x + \tan x|,$$

or

$$2I = \sec x \tan x + \ln|\sec x + \tan x|,$$

and

$$I = \int \sec^3 x \, dx = \frac{1}{2}(\sec x \tan x + \ln|\sec x + \tan x|) + C.$$

This integral comes up frequently in applications.

REDUCTION FORMULAS

We can use the technique of integration by parts to obtain **reduction formulas**. A reduction formula allows us to evaluate a whole class of integrals by reducing each integral in the class to a simpler one. This idea is best illustrated by an example.

EXAMPLE 7
A REDUCTION FORMULA
INVOLVING sin^n x

Prove that

$$\int \sin^n x \, dx = -\frac{1}{n} \sin^{n-1} x \cos x + \frac{n-1}{n} \int \sin^{n-2} x \, dx \qquad \text{(5)}$$

where $n > 1$ is a positive integer.

SOLUTION: Let $u = \sin^{n-1} x$ and $dv = \sin x \, dx$. Then

$$du = (n-1) \sin^{n-2} x \cos x \, dx, \qquad v = -\cos x,$$

and

$$I = \int \sin^n x \, dx = -\sin^{n-1} x \cos x + (n-1) \int \sin^{n-2} x \cos^2 x \, dx.$$

Using $\cos^2 x = 1 - \sin^2 x$, we obtain

$$I = -\sin^{n-1} x \cos x + (n-1) \int (\sin^{n-2} x)(1 - \sin^2 x) \, dx$$

$$= -\sin^{n-1} x \cos x + (n-1) \int \sin^{n-2} x \, dx - (n-1) \int \sin^n x \, dx$$

$$= -\sin^{n-1} x \cos x + (n-1) \int \sin^{n-2} x \, dx - (n-1)I$$

$$I + (n-1)I = nI = -\sin^{n-1} x \cos x + (n-1) \int \sin^{n-2} x \, dx$$

$$I = -\frac{1}{n} \sin^{n-1} x \cos x + \frac{n-1}{n} \int \sin^{n-2} x \, dx.$$

REMARK: As Example 7 indicates, we can integrate sin^n x if we know how to integrate sin^{n-2} x. By using the reduction formula (5) repeatedly, if necessary, we can integrate any positive integral power of sin x since we can integrate 1 and sin x (because even powers of sin x will "reduce" to 1 and odd powers will "reduce" to sin x).

EXAMPLE 8
USING THE REDUCTION
FORMULA FOR $\int$ sin^n x dx

Evaluate $\int \sin^3 x \, dx$.

SOLUTION: Using (5) with $n = 3$, we have

$$\int \sin^3 x \, dx = -\frac{1}{3} \sin^2 x \cos x + \frac{2}{3} \int \sin x \, dx$$

$$= -\frac{1}{3} \sin^2 x \cos x - \frac{2}{3} \cos x + C.$$

There are many other reduction formulas. Several are given in the table of integrals. You are asked to derive four of them in Problems 53, 54, and 55.

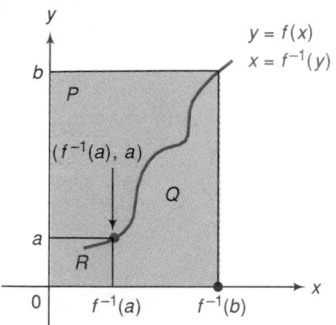

FIGURE 2
THE AREA OF $P = \int_a^b f^{-1}(y)\,dy$

THE AREA OF $Q = \int_{f^{-1}(a)}^{f^{-1}(b)} f(x)\,dx.$

THE AREA OF $R = af^{-1}(a).$
THE SUM OF THE THREE AREAS IS
$bf^{-1}(b).$

A PERSPECTIVE: ON INTEGRATING AN INVERSE FUNCTION[†]

We usually use integration by parts to compute definite integrals like $\int_1^e \ln x\,dx$ or $\int_0^1 \sin^{-1} x\,dx$. In this perspective, we present an alternative procedure.[†]

Let the function f be strictly increasing and differentiable; the case of f strictly decreasing is similar. Look at Figure 2. The area of the region marked P is the area "under" the curve $x = f^{-1}(y)$ from $y = a$ to $y = b$. That is, we compute the area by interchanging the roles of x and y in the usual computation of area under a curve. Thus

$$\text{area of } P = \int_a^b f^{-1}(y)\,dy.$$

The area of Q is computed in the usual way:

$$\text{area of } Q = \int_{f^{-1}(a)}^{f^{-1}(b)} f(x)\,dx.$$

Finally, the region marked R is a rectangle, so

$$\text{area of } R = \text{base} \times \text{height} = f^{-1}(a) \times a = af^{-1}(a).$$

Now, the region $P + Q + R$ is a larger rectangle with base $f^{-1}(b)$ and height b. Thus,

$$\text{area of } P = \int_a^b f^{-1}(y)\,dy$$
$$= \text{area of } (P + Q + R) - (\text{area of } R) - (\text{area of } Q)$$

or

$$\int_a^b f^{-1}(y)\,dy = bf^{-1}(b) - af^{-1}(a) - \int_{f^{-1}(a)}^{f^{-1}(b)} f(x)\,dx. \qquad (6)$$

EXAMPLE 9
USING THE INVERSE FUNCTION
TO INTEGRATE

Compute $\int_1^e \ln x\,dx.$

SOLUTION: Here $f^{-1}(x) = \ln x,\quad f(x) = \ln^{-1} x = e^x,\quad a = 1,\quad b = e,$
$f^{-1}(a) = \ln 1 = 0, f^{-1}(b) = \ln e = 1$ and from (6),

$$\int_1^e \ln x\,dx = bf^{-1}(b) - af^{-1}(a) - \int_{f^{-1}(a)}^{f^{-1}(b)} f(x)\,dx$$
$$= e - 0 - \int_0^1 e^x\,dx = e - e^x\Big|_0^1 = e - (e - 1) = 1.$$

EXAMPLE 10
USING THE INVERSE FUNCTION
TO INTEGRATE

Compute $\int_0^1 \sin^{-1} x\,dx.$

SOLUTION: Here $f^{-1}(x) = \sin^{-1} x, f(x) = \sin x, a = 0, b = 1, f^{-1}(0) = 0, f^{-1}(1) = \dfrac{\pi}{2}$ and

$$\int_0^1 \sin^{-1} x\,dx = \frac{\pi}{2} - \int_0^{\pi/2} \sin x\,dx = \frac{\pi}{2} + \cos x\Big|_0^{\pi/2} = \frac{\pi}{2} - 1.$$

[†] Adapted from "Integrating an inverse function" by A. K. Austin in the *Mathematical Gazette* (March, 1986): 40.

■ PROBLEMS 7.2

■ SELF-QUIZ

I. The integration technique by parts corresponds to the _____ rule for differentiation.
   a. power   b. chain
   c. product   d. quotient

II. Which of the following statements involving differentials is easily transformed into integration by parts?
   a. $d(u^k) = k \cdot u^{k-1} \cdot du$
   b. $d(g(v)) = g'(v) \cdot dv$
   c. $d(u \cdot v) = v \cdot du + u \cdot dv$
   d. $d(u/v) = (v \cdot du - u \cdot dv)/v^2$

III. Which of the following facts is most useful in finding $\int x\, e^x\, dx$?
   a. $d(x^e) = e \cdot x^{e-1} \cdot dx$
   b. $d(e^{x^2}) = e^{x^2} \cdot d(x^2) = 2xe^{x^2}\, dx$
   c. $d(x \cdot e^x) = e^x \cdot d(x) + x \cdot d(e^x)$
   d. $d\left(\dfrac{e^x}{x}\right) = \dfrac{x \cdot d(e^x) - e^x \cdot d(x)}{x^2}$

In addition to the specific facts and theorems you have been learning, you probably have acquired some general notions about patterns. Such notions can suggest the general form of an answer; they can also allow you to discard quickly an incorrect expression because it is of the wrong form. For instance, it doesn't take any messy computations to realize that

$$\frac{d}{dx}(x^7 - 13x^4 + 5x - 7) \cdot (x^3 - 3x)$$

$$\neq \cos x + 10x^9 - 21x,$$

because the derivative of a polynomial is also a polynomial and no cos x term can appear. Similarly, you might perceive immediately that

$$\frac{d}{dx}(\cos x^2 \cdot e^{-3x}) \neq -2x \sin x^2 - 3 \cos x^2 \cdot e^{-4x}$$

using the general idea that the derivative of a product $f(x) \cdot e^{g(x)}$ is also of the form of something times $e^{g(x)}$.

The following quiz problems give you a chance to develop and exercise such intuitions. Pick the right answer by seeing what remains after you have excluded the blatantly wrong choices.

IV. Pick the right answer: $\int (x + 1)e^x\, dx =$ _____ $+ C$.
   a. xe^x b. $\frac{1}{2}e^{(x+1)^2}$
   c. $(x + 1)^{e+1}$ d. e^x/x

V. Pick the right answer: $\int xe^x\, dx =$ _____ $+ C$.
   a. x^{e+1} b. $\frac{1}{2}e^{x^2}$
   c. $(e^x/x) + (e^x/x^2)$ d. $xe^x - e^x$

VI. Pick the right answer: $\int x \cos x\, dx =$ _____ $+ C$.
   a. $\frac{1}{2}x^2 + \sin x$
   b. $\frac{1}{2} \sin x^2$
   c. $x \sin x + \cos x$
   d. $\frac{1}{2}x^2 \cos x - x \sin x$

VII. Pick the right answer: $\int 25e^{3x} \cos 4x\, dx =$ _____ $+ C$.
   a. $4e^{3x} - 3 \cos 4x$
   b. $e^{3x} \cdot (3 \cos 4x + 4 \sin 4x)$
   c. $3e^{3x} \cos 4x + 4e^{4x} \cos 3x$
   d. $(3e^{3x} - 4e^{4x}) \cdot (-3 \cos 3x + 4 \sin 4x)$

■ DRILL

In Problems 1–30, evaluate the given integral.

1. $\int xe^{3x}\, dx$

2. $\int_0^1 xe^{-7x}\, dx$

3. $\int x^2 e^{-x}\, dx$
   [*Hint:* Integrate by parts twice.]

4. $\int_0^4 x^2 e^{x/4}\, dx$

5. $\int_1^2 x \ln x\, dx$

6. $\int x^7 \ln x\, dx$

7. $\int x \sinh x\, dx$

8. $\int_0^{\pi/2} x \sin x\, dx$

9. $\int_0^2 x\sqrt{1 - x/2}\, dx$

10. $\int_0^5 x\sqrt{3x + 1}\, dx$

11. $\int_0^1 x^2 \cosh 2x\, dx$

12. $\int x^2 \cos 2x\, dx$

13. $\int \cos(\ln x)\, dx$

14. $\int_1^2 \ln^2 x\, dx$

15. $\int_0^1 x^5 e^{x^3}\, dx$

16. $\int x5^x\, dx$

17. $\int e^x \cos x\, dx$

18. $\int e^{3x} \sin 2x\, dx$

19. $\int \sin^5 x\, dx$

20. $\int \sin^4 x\, dx$

21. $\int_0^{1/\sqrt{2}} \cos^{-1} x\, dx$

22. $\int \sin^{-1} x\, dx$

23. $\int_0^1 \tan^{-1} x\, dx$

24. $\int_0^{1/2} \tan^{-1} 2x\, dx$

*25. $\int x \tan^{-1} x \, dx$

$$\left[Hint: \frac{x^2}{1+x^2} = 1 - \frac{1}{1+x^2} \right]$$

*26. $\int x^2 \tan^{-1} x \, dx$

*27. $\int e^{ax} \cos bx \, dx, \; a, b \neq 0$

*28. $\int e^{ax} \sin bx \, dx, \; a, b \neq 0$

*29. $\int \sinh 2x \cosh 3x \, dx$ *30. $\int \sinh 2x \sinh 5x \, dx$

In Problems 31–32, use equation (6) to evaluate the given definite integral.

31. $\int_{-1}^{x} \cos^{-1} t \, dt$ 32. $\int_{0}^{x} \tan^{-1} t \, dt$

■ APPLICATIONS

In Problems 33–38, find the volume generated when the region bounded by the given curves and lines is rotated about the x-axis.

33. $y = \cos x; \; x = \pi/2, \; x = 0, \; y = 0$
34. $y = \tan x; \; x = \pi/4, \; y = 0$
35. $y = \ln x; \; x = 1, \; x = e, \; y = 0$
36. $y = \sin^{-1} x; \; x = 1, \; y = 0$
37. $y = \sin x, \; y = \cos x; \; 0 \leq x \leq \pi/4$
38. $y = xe^x, \; x = 1, \; y = 0$

In Problems 39–42, find the volume generated when the region bounded by the given curves and lines is rotated about the y-axis.

39. $y = \ln x; \; x = 1, \; x = 3, \; y = 0$
40. $y = e^{x^2}; \; x = 0, \; x = 1, \; y = 0$
41. $y = \sin x, \; y = \cos x; \; 0 \leq x \leq \pi/4$
42. $y = \cos(x^2); \; x = 0, \; x = \sqrt{\pi/6}, \; y = 0$

43. a. Graph the curve $y = xe^{-x}$. [Hint: Use the fact (which we will prove later) that $\lim_{x \to \infty} xe^{-x} = 0$.]
 b. Calculate the area bounded by the curve $y = xe^{-x}$, the x-axis, and the lines $x = -1$ and $x = 1$.

44. Find the length of the arc of $y = e^{-x}$ between $x = 0$ and $x = 1$.

45. Use the result of Problem 52 to evaluate $\int x^4 e^x \, dx$.

46. Use the result of Problem 53 to evaluate $\int \cos^3 x \, dx$.

47. Use the result of Problem 54 to evaluate $\int \sec^5 x \, dx$.

48. Use the result of Problem 55 to evaluate $\int x^3 \cos x \, dx$.

■ SHOW/PROVE/DISPROVE

49. Show that if $a \neq 0$, then

$$\int x^3 \sin ax \, dx = \left(\frac{3x^2}{a^2} - \frac{6}{a^4} \right) \sin ax$$
$$+ \left(\frac{6x}{a^3} - \frac{x^3}{a} \right) \cos ax + C.$$

50. Show that if $a \neq 0$, then

$$\int x^2 \sinh ax \, dx = \left(\frac{x^2}{a} + \frac{2}{a^3} \right) \cosh ax$$
$$- \left(\frac{2x}{a^2} \right) \sinh ax + C.$$

51. What is wrong with the following "reasoning"?

$$\int \cot x \, dx = \int \csc x \left(\frac{d}{dx} \sin x \right) dx$$

$$= \csc x \sin x - \int \sin x \left(\frac{d}{dx} \csc x \right) dx$$
$$= 1 + \int \cot x \, dx;$$

therefore, subtracting $\int \cot x \, dx$, we infer that $0 = 1$.

*52. Using mathematical induction (Appendix 2) and integration by parts, show that if n is a positive integer and $a \neq 0$, then

$$\int x^n e^{ax} \, dx = \frac{e^{ax}}{a} \left[x^n - \frac{nx^{n-1}}{a} \right.$$
$$\left. + \frac{n(n-1)x^{n-2}}{a^2} - \cdots + \frac{(-1)^n n!}{a^n} \right] + C,$$

where $n! = n \cdot (n-1) \cdot (n-2) \cdots 3 \cdot 2 \cdot 1$.

CONTINUED

53. Suppose $n > 1$ is an integer. Prove the reduction formula

$$\int \cos^n x \, dx = \frac{1}{n} \cos^{n-1} x \sin x$$

$$+ \frac{n-1}{n} \int \cos^{n-2} x \, dx.$$

*54. Suppose $n > 1$ is an integer; show that

$$\int \sec^n x \, dx = \frac{1}{n-1} \tan x \sec^{n-2} x$$

$$+ \frac{n-2}{n-1} \int \sec^{n-2} x \, dx.$$

55. Suppose that $n > 1$ is an integer.

a. Show that

$$\int x^n \cos x \, dx = x^n \sin x - n \int x^{n-1} \sin x \, dx.$$

b. Show that

$$\int x^n \sin x \, dx = -x^n \cos x + n \int x^{n-1} \cos x \, dx.$$

56. Prove the result summarized in Figure 2 and equation (6). [*Hint:* Substitute $y = f(x)$ into $\int_a^b f^{-1}(y) \, dy$ to obtain

$$\int_{f^{-1}(a)}^{f^{-1}(b)} f^{-1}(f(x)) f'(x) \, dx = \int_{f^{-1}(a)}^{f^{-1}(b)} x f'(x) \, dx.$$

Then integrate this right-hand expression by parts.]

■ **CHALLENGE**

57. Let $G_n(x) = \int_0^x e^{-t} t^{n-1} \, dt$.
   a. Find a recursion relation connecting G_{n+1} with G_n.
   b. Find explicit expressions for $G_1(x)$, $G_2(x)$, and $G_3(x)$.

58. For $-\pi/2 < x < \pi/2$, let $T_n(x) = \int_0^x \tan^n \theta \, d\theta$.
   a. Find explicit formulas for $T_0(x)$ and $T_1(x)$.
   *b. Find a recursion formula connecting $T_n(x)$ with $T_{n-2}(x)$. [*Hint:* $\tan^2 \theta = \sec^2 \theta - 1$.]
   c. Use your answers to parts (a) and (b) to find $T_2(x)$ and $T_3(x)$.

59. For $0 < x < \pi$, let $C_n(x) = \int_{\pi/2}^x \cot^n \theta \, d\theta$.
   a. Find explicit formulas for $C_0(x)$ and $C_1(x)$.
   *b. Find a recursion formula connecting $C_n(x)$ with $C_{n-2}(x)$. [*Hint:* $\cot^2 \theta = \csc^2 \theta - 1$.]
   c. Use your answers to parts (a) and (b) to find $C_2(x)$ and $C_3(x)$.

*60. Let $I_n(x) = \int_0^x (t^n/\sqrt{t^2 + 1}) \, dt$.

a. Prove that $I_0(x) = \ln(x + \sqrt{x^2 + 1})$.
b. Find $I_1(x)$.
c. Find a recursion relation for $I_n(x)$ of the form

$$I_n(x) = P_n(x) \cdot \sqrt{x^2 + 1} + C_n \cdot I_{n-2}(x),$$

where $P_n(x)$ is a polynomial function and C_n is a constant.

**61. Let $J_n(x) = \int (x^n/\sqrt{x^k + 1}) \, dx$, where k is a fixed positive integer. Find a recursion relation connecting J_n with J_{n-k}.

**62. Suppose that f and g are continuous on $[a, b]$, and that f has a continuous derivative that is never zero. Show that there is at least one choice of c such that $a \leq c \leq b$ and

$$\int_a^b f \cdot g = f(a) \cdot \int_a^c g + f(b) \cdot \int_c^b g.$$

■ ANSWERS TO SELF-QUIZ

I. c	II. c	III. c	IV. a
V. d	VI. c	VII. b	

7.3 INTEGRALS OF CERTAIN TRIGONOMETRIC FUNCTIONS

In this section we show how to integrate some classes of trigonometric functions. In each case, some technique is used to make the integration easier. In Table 1 below, we describe each class and the technique that is needed. In the table, m and n are positive integers. The following trigonometric identities

TABLE 1

Integral	How to Compute the Integral
case (i) $\int \sin^m x \cos^n x \, dx$, either m or n odd[†]	Break off $\sin^2 x$ factors (if m is odd) or $\cos^2 x$ factors (if n is odd). Use identity (6) to obtain integrals of the form $\int \cos^p x \sin x \, dx$ or $\int \sin^q x \cos x \, dx$.
case (ii) $\int \sin^m x \cos^n x \, dx$, both m and n even	Use identity (1), (2), or (8).
case (iii) $\int \sin mx \cos nx \, dx$	Use identity (3).
case (iv) $\int \sin mx \sin nx \, dx$	Use identity (4).
case (v) $\int \cos mx \cos nx \, dx$	Use identity (5).
case (vi) $\int \tan^n x \, dx$	Use identity (7) to write $\tan^n x = \tan^{n-2} x(\sec^2 x - 1)$. Use the fact that $\dfrac{d}{dx} \tan x = \sec^2 x$.
case (vii) $\int \sec^n x \, dx$	Use identity (7) to write $\sec^n x = \sec^2 x \sec^{n-2} x = \sec^2 x(1 + \tan^2 x)^{(n-2)/2}$ if n is even or $\sec^3 x(1 + \tan^2 x)^{(n-3)/2}$ if n is odd. Also, integration by parts may be useful (see Example 7.2.6).
case (viii) $\int \tan^m x \sec^n x \, dx$, n even	Use identity (7). Break off one $\sec^2 x$ term since $$\frac{d}{dx} \tan x = \sec^2 x.$$
case (ix) $\int \tan^m x \sec^n x \, dx$, n and m odd	Break off a term of the form $\sec x \tan x$ since $\dfrac{d}{dx} \sec x = \sec x \tan x$.
case (x) $\int \tan^m x \sec^n x \, dx$, n odd, m even	Use identity (7) to write everything in terms of $\sec x$. Then proceed as in case (vii).

[†] Here if m is an odd integer, then n need not be an integer, and vice versa.

are useful:

$$\sin^2 x = \frac{1 - \cos 2x}{2} \tag{1}$$

$$\cos^2 x = \frac{1 + \cos 2x}{2} \tag{2}$$

$$\sin mx \cos nx = \tfrac{1}{2}\{\sin[(m + n)x] + \sin[(m - n)x]\} \tag{3}$$

$$\sin mx \sin nx = \tfrac{1}{2}\{\cos[(m - n)x] - \cos[(m + n)x]\} \tag{4}$$

$$\cos mx \cos nx = \tfrac{1}{2}\{\cos[(m - n)x] + \cos[(m + n)x]\} \tag{5}$$

[See Problems 39, 40, and 41 for the derivation of formulas (3), (4), and (5).]

$$\sin^2 x + \cos^2 x = 1 \tag{6}$$

$$1 + \tan^2 x = \sec^2 x \tag{7}$$

$$\sin x \cos x = \tfrac{1}{2} \sin 2x \tag{8}$$

EXAMPLE 1
AN INTEGRAL OF THE FORM
$\int \sin^m x \cos^n x\, dx$, m ODD

Calculate $\int \sin^3 x \cos^{4/3} x\, dx$.

SOLUTION: Write $\sin^3 x = (\sin^2 x) \sin x = (1 - \cos^2 x) \sin x$ (case (i) in Table 1). Then

$$\int \sin^3 x \cos^{4/3} x\, dx = \int (1 - \cos^2 x) \sin x \cos^{4/3} x\, dx$$

$$= \int \cos^{4/3} x \sin x\, dx - \int \cos^{10/3} x \sin x\, dx$$

$$= -\frac{3}{7} \cos^{7/3} x + \frac{3}{13} \cos^{13/3} x + C.$$

EXAMPLE 2
AN INTEGRAL OF THE FORM
$\int \sin^m x \cos^n x\, dx$, m AND n BOTH EVEN

Calculate $\int_0^{\pi/2} \sin^2 x \cos^4 x\, dx$.   Case (ii)

SOLUTION:

$$\int_0^{\pi/2} \sin^2 x \cos^4 x\, dx = \int_0^{\pi/2} (\sin x \cos x)^2 \cos^2 x\, dx$$

from (2) and (8)
$$\downarrow$$
$$= \int_0^{\pi/2} \left(\frac{1}{2} \sin 2x \right)^2 \frac{1}{2} (1 + \cos 2x)\, dx$$

$$= \frac{1}{8} \int_0^{\pi/2} \sin^2 2x\, dx + \frac{1}{8} \int_0^{\pi/2} \sin^2 2x \cos 2x\, dx$$

from (1)
$$\downarrow$$
$$= \frac{1}{16} \int_0^{\pi/2} (1 - \cos 4x)\, dx + \frac{1}{8} \cdot \frac{1}{3} \cdot \frac{1}{2} \sin^3 2x \Big|_0^{\pi/2}$$

$$= \frac{1}{16} \left[x - \frac{\sin 4x}{4} \right]\Big|_0^{\pi/2} + \frac{1}{48}(\sin^3 \pi - \sin^3 0)$$

$$= \frac{1}{16} \left[\frac{\pi}{2} - \frac{\sin 2\pi}{4} \right] + \frac{1}{48}(0 - 0) = \frac{\pi}{32}.$$

EXAMPLE 3
AN INTEGRAL OF THE FORM
$\int \tan^n x\, dx$

Calculate $\int \tan^2 x\, dx$.   Case (vi)

SOLUTION: $\int \tan^2 x\, dx = \int (\sec^2 x - 1)\, dx = \tan x - x + C.$

EXAMPLE 4
AN INTEGRAL OF THE FORM
$\int \tan^m x \sec^n x\, dx$, n EVEN

Calculate $\int_0^{\pi/3} \tan^7 x \sec^4 x\, dx$.

SOLUTION: This is case (viii).

from (7)

$$\int_0^{\pi/3} \tan^7 x \sec^2 x \sec^2 x \, dx = \int_0^{\pi/3} (\tan^7 x)(1 + \tan^2 x) \sec^2 x \, dx$$

$$= \int_0^{\pi/3} \tan^7 x \sec^2 x \, dx + \int_0^{\pi/3} \tan^9 x \sec^2 x \, dx$$

$$= \left(\frac{\tan^8 x}{8} + \frac{\tan^{10} x}{10} \right) \Big|_0^{\pi/3} = \frac{\sqrt{3}^8}{8} + \frac{\sqrt{3}^{10}}{10}$$

$$= \frac{3^4}{8} + \frac{3^5}{10} = \frac{81}{8} + \frac{243}{10} = \frac{1377}{40}.$$

EXAMPLE 5
AN INTEGRAL OF THE FORM
$\int \tan^m x \sec^n x \, dx$, n ODD, m EVEN

Calculate $\int \tan^2 x \sec x \, dx$.

SOLUTION: This is case (x).

$$\int \tan^2 x \sec x \, dx = \int (\sec^2 x - 1) \sec x \, dx = \int \sec^3 x \, dx - \int \sec x \, dx$$

(From Example 7.2.6)

$$= \frac{1}{2}(\sec x \tan x + \ln|\sec x + \tan x|)$$

$$- \ln|\sec x + \tan x| + C$$

$$= \frac{1}{2}[\sec x \tan x - \ln|\sec x + \tan x|] + C.$$

■ PROBLEMS 7.3

■ SELF-QUIZ

I. By discarding the blatantly wrong answers, pick the right answer:
$\int [(1 - \sin^2 x) + (1 - \cos^2 x) \, dx$
$= \underline{\hspace{2cm}} + C.$
   a. x
   b. $\sin^3 x - \cos^3 x$
   c. $2x - \frac{1}{3}(\sin^3 x + \cos^3 x)$
   d. $2x + \frac{1}{3}(-\cos 3x + \sin 3x)$

II. Pick the right answer: $\int (\tan^2 x - \sec^2 x) \, dx = \underline{\hspace{2cm}} + C.$
   a. $\dfrac{\tan^3 x}{3} - \dfrac{\sec^3 x}{3}$ b. $-x$
   c. $-\frac{1}{3}(\ln \cos x)^3 - \tan x$ d. 0

III. Pick the right answer: $\int (\tan^3 x + \tan x) \, dx = \underline{\hspace{2cm}} + C.$
   a. $\frac{1}{2} \tan^2 x$
   b. $\frac{1}{4} \tan^4 x - \ln \cos x$
   c. $\frac{1}{3} \sec^3 x$
   d. $-\frac{1}{4}(\ln \cos x)^4 + \ln \cos x$

IV. Pick the right answer: $\int \sin^2 x \cos^3 x \, dx = \underline{\hspace{2cm}} + C.$
   a. $\frac{1}{3} \sin^3 x - \frac{1}{5} \sin^5 x$
   b. $\frac{1}{3} \sin^3 x \cos^3 x - \frac{1}{4} \sin^2 x \cos^4 x$
   c. $\frac{1}{4} \cos^4 x - \frac{1}{6} \cos^6 x$
   d. $\cos^4 x - \cos^6 x$

■ DRILL

In Problems 1–28, evaluate the given integral (use the tactics summarized in Table 1).

1. $\int_0^\pi \sin^3 x \cos^2 x \, dx$ 2. $\int \sin^3 x \sec x \, dx$

3. $\int \cos^3 x \sqrt{\sin x} \, dx$ 4. $\int \cos^3 x \sqrt{\csc x} \, dx$

5. $\int \sin^5 x \cos^2 x \, dx$ 6. $\int \dfrac{\sin^5 x}{\cos^2 x} \, dx$

CONTINUED

7. $\int \sin^2 x \cos^5 x \, dx$ 8. $\int_0^{\pi/4} \sin^2 x \tan x \, dx$

9. $\int_0^{\pi/4} \cos^2 x \, dx$ 10. $\int \sin^4 x \, dx$

11. $\int \cos^4 2x \, dx$

12. $\int_0^{\pi/2} \sin^2 3x \cos^4 3x \, dx$

13. $\int \sin^4 x \cos^2 x \, dx$

14. $\int_0^{\pi} \sin^6 x \cos^4 x \, dx$ 15. $\int_0^{\pi/8} \tan^2 2x \, dx$

16. $\int \tan^5 x \, dx$ 17. $\int_0^{\pi/4} \sec^4 x \, dx$

18. $\int_0^{\pi/3} \sec^6 x \, dx$ 19. $\int \tan^3 x \sec^4 x \, dx$

20. $\int \tan^4 x \sec^3 x \, dx$ 21. $\int_0^{\pi/6} \sec^3 2x \tan 2x \, dx$

22. $\int_0^{\pi/4} \sec^3 x \tan^3 x \, dx$

23. $\int \sin(2x) \cos(3x) \, dx$

24. $\int_0^{\pi/2} \sin(x/2) \cos(x/3) \, dx$

25. $\int \sin(\sqrt{2}x) \cos(x/\sqrt{2}) \, dx$

26. $\int_0^{\pi} \sin(2x) \cos(x/2) \, dx$

27. $\int \cos(10x) \cos(100x) \, dx$

28. $\int \sin(3x) \sin(7x) \, dx$

■ APPLICATIONS

Using the identity $1 + \cot^2 x = \csc^2 x$, we can evaluate integrals of the form $\int \cot^n x \, dx$, $\int \csc^n x \, dx$, $\int \cot^m x \csc^n x \, dx$. We need only use the method outlined in Table 1 for integrals of types (vi)–(x). Use similar techniques to evaluate the integrals in Problems 29–36.

29. $\int \cot^2 x \, dx$ 30. $\int \cot^5 x \, dx$

31. $\int_{\pi/4}^{\pi/2} \csc^4 x \, dx$ 32. $\int \csc^3 x \, dx$

33. $\int_{\pi/6}^{\pi/2} \cot^7 x \csc^4 x \, dx$ 34. $\int \cot^3 x \csc^5 x \, dx$

35. $\int \cot^2 x \csc x \, dx$ 36. $\int \cot^2 x \csc^6 x \, dx$

*37. a. Find $\int (\cos^2 x - \sin^2 x) \, dx$ without using double-angle formulas such as $\cos(2x) = \cos^2 x - \sin^2 x$. [Hint: $(c + s)(c - s) = c^2 - s^2$.]
   b. Find $\int \cos^2 x \, dx$ and $\int \sin^2 x \, dx$. [Hint: $\int \cos^2 x \, dx - \int \sin^2 x \, dx$ was found in part (a) and $\int \cos^2 x \, dx + \int \sin^2 x \, dx$ is easy.]

38. Verify the formula

$$\int \tan^n ax \, dx = \frac{\tan^{n-1} ax}{(n-1)a} - \int \tan^{n-2} ax \, dx.$$

■ SHOW/PROVE/DISPROVE

39. Show that $\sin mx \cos nx = \frac{1}{2}\{\sin[(m+n)x] + \sin[(m-n)x]\}$. [Hint: Add the identities $\sin(a+b) = \sin a \cos b + \cos a \sin b$ and $\sin(a-b) = \sin a \cos b - \cos a \sin b$.]

40. Prove identity (4).   41. Prove identity (5).

42. Show that if $m \neq \pm n$, then

$$\int \sin mx \sin nx \, dx = \frac{1}{2}\left\{\frac{\sin[(m-n)x]}{m-n} - \frac{\sin[(m+n)x]}{m+n}\right\} + C.$$

43. Show that if $m \neq \pm n$, then

$$\int \cos mx \cos nx \, dx = \frac{1}{2}\left\{\frac{\sin[(m-n)x]}{m-n} + \frac{\sin[(m+n)x]}{m+n}\right\} + C.$$

44. Show that $\int_0^{2\pi} \sin mx \cos nx \, dx = 0$ for any integers m and n.

45. Show that $\int_0^{2\pi} \cos mx \cos nx \, dx = 0$ if $m \neq \pm n$.

46. Show that $\int_0^{2\pi} \sin mx \sin nx \, dx = 0$ if $m \neq \pm n$.

47. Show that $\int_0^{\pi/2} \sin^2 x \, dx = \frac{\pi}{4} = \frac{1}{2} \cdot \frac{\pi}{2}$.

48. Show that $\int_0^{\pi/2} \sin^4 x \, dx = \frac{3}{4} \cdot \frac{1}{2} \cdot \frac{\pi}{2}$.

49. Show that $\int_0^{\pi/2} \sin^6 x \, dx = \frac{5}{6} \cdot \frac{3}{4} \cdot \frac{1}{2} \cdot \frac{\pi}{2}$.

50. By using the reduction formula of Example 7.2.7, show that

$$\int_0^{\pi/2} \sin^{2n} x \, dx = \frac{2n-1}{2n} \int_0^{\pi/2} \sin^{2n-2} x \, dx.$$

CONTINUED

*51. Apply the results of the preceding four problems to prove that

$$\int_0^{\pi/2} \sin^{2n} x \, dx$$

$$= \frac{2n-1}{2n} \cdot \frac{2n-3}{2n-2} \cdot \frac{2n-5}{2n-4} \cdot \ldots \cdot \frac{5}{6} \cdot \frac{3}{4} \cdot \frac{1}{2} \cdot \frac{\pi}{2}.$$

This is called a **Wallis product**.

*52. Show that

$$\int_0^{\pi/2} \sin^{2n+1} x \, dx$$

$$= \frac{2n}{2n+1} \cdot \frac{2n-2}{2n-1} \cdot \frac{2n-4}{2n-3} \cdot \ldots \cdot \frac{4}{5} \cdot \frac{2}{3}.$$

This is another Wallis product.

53. Show that $\left| \int_0^{\pi/2} \sin^{2n} x \, dx \right| \le \pi/2$.

*54. Show, using the reduction formula of Example 7.2.7, that

$$\lim_{n \to \infty} \frac{\int_0^{\pi/2} \sin^{2n} x \, dx}{\int_0^{\pi/2} \sin^{2n+1} x \, dx} = 1.$$

*55. Divide the two Wallis products and use the result of the preceding problem to derive the **Wallis formula**:

$$\frac{\pi}{2} = \lim_{n \to \infty} \frac{2}{1} \cdot \frac{2}{3} \cdot \frac{4}{3} \cdot \frac{4}{5} \cdot \frac{6}{5} \cdot \frac{6}{7}$$

$$\ldots \cdot \frac{2n}{2n-1} \cdot \frac{2n}{2n+1}.$$

■ CHALLENGE

**56. Compute

$$I_k = \int_0^{\pi/2} \frac{\sin^2(k\theta)}{\sin \theta} \, d\theta,$$

where k is a positive integer. [*Hint:* Compute I_1 and I_2; then compute $I_{k+1} - I_k$; guess a generalized result and use mathematical induction to prove that your guess is correct.]

■ ANSWERS TO SELF-QUIZ

I. a   II. b   III. a   IV. a

7.4 TRIGONOMETRIC SUBSTITUTIONS

In Section 4.7, we saw how a substitution could be made to change $\int f(x) \, dx$ into an integral of the form $\int u^r \, du$. There are many different kinds of substitutions that will allow us to integrate more easily. Some of these will be illustrated in this and the next section. We begin with two examples.

EXAMPLE 1
COMPUTING A DEFINITE INTEGRAL BY MAKING A TRIGONOMETRIC SUBSTITUTION

To compute $\int_0^1 \sqrt{1 - x^2} \, dx$, we make the substitution $x = \sin u$ (the reason for making this substitution will be clear shortly). In converting from an integral in x into an integral in which u is the variable of integration, we must change three things:

(i) Change *the integrand*: $\sqrt{1 - x^2} = \sqrt{1 - \sin^2 u} = \sqrt{\cos^2 u} = \cos u.$[†]

(ii) Change dx: Since $dx/du = (d/du) \sin u = \cos u$, we have

$$dx = \cos u \, du.$$

[†] When we substitute $x = \sin u$, we mean that $u = \sin^{-1} x$. By the definition of the inverse sine function on page 456, $-\pi/2 \le u \le \pi/2$. On the interval $[-\pi/2, \pi/2]$, $\cos u \ge 0$, which is why we take the positive square root of $\cos^2 u$ here.

(iii) Change *the limits of integration*: x ranges between 0 and 1. When $x = 0$, $\sin u = 0$ and $u = \sin^{-1} 0 = 0$; when $x = 1$, $\sin u = 1$ and $u = \sin^{-1} 1 = \pi/2$.

Thus,

$$\int_{x=0}^{1} \sqrt{1 - x^2}\, dx = \int_{u=0}^{\pi/2} \underbrace{\cos u}_{} \cdot \overbrace{\cos u\, du}^{dx} = \int_{0}^{\pi/2} \cos^2 u\, du$$

$$= \frac{1}{2} \int_{0}^{\pi/2} (1 + \cos 2u)\, du$$

$$= \frac{1}{2}\left(u + \frac{\sin 2u}{2}\right)\Bigg|_{0}^{\pi/2} = \frac{1}{2}\left(\frac{\pi}{2} + 0\right) = \frac{\pi}{4}.$$

NOTE: This result should not be surprising since $\int_{0}^{1} \sqrt{1 - x^2}\, dx$ is the area enclosed by the part of the unit circle lying in the first quadrant.

EXAMPLE 2
COMPUTING AN INDEFINITE
INTEGRAL BY MAKING AN
APPROPRIATE SUBSTITUTION

Compute $\displaystyle\int \frac{dx}{2 + \sqrt{x}}$.

SOLUTION: We make the substitution $u = x^{1/2}$.

(1) Change the *integrand*: $\dfrac{1}{2 + \sqrt{x}} = \dfrac{1}{2 + u}$.

(2) Change dx: $u^2 = x$ so $\dfrac{dx}{du} = 2u$ and $dx = 2u\, du$.

(3) *Integrate*:

$$\int \frac{dx}{1 + \sqrt{x}} = \int \frac{2u}{2 + u}\, du \overset{\text{Divide } 2u \text{ by } 2 + u}{=} \int \left(2 - \frac{4}{u + 2}\right) du$$

$$= 2u - 4 \ln|u + 2| + C.$$

But this answer is in terms of u, not x, so we

(4) Write the answer in terms of x. Here, $u = \sqrt{x}$ so

$$2u - 4 \ln|u + 2| + C = 2\sqrt{x} - 4 \ln(\sqrt{x} + 2) + C.$$

The techniques used in Examples 1 and 2 can be used whenever you can find an appropriate substitution (this is often the hardest part). Let us see why this is true.

Consider the definite integral

$$\int_{a}^{b} f(x)\, dx.$$

Let g be a differentiable, 1–1 function. Suppose we wish to make the substitution $x = g(u)$ in order to change the integral with respect to x to an integral with respect to u. We want to do so because with the proper choice of the function g, it will be easier to carry out the integration.

Since g is 1–1, g has an inverse function. Let

$$c = g^{-1}(a) \qquad \text{and} \qquad d = g^{-1}(b),$$

so that

$$a = g(c) \qquad \text{and} \qquad b = g(d).$$

Let F be an antiderivative for f. Then, by the chain rule,

$$\frac{d}{du} F(g(u)) = F'(g(u))g'(u) = f(g(u))g'(u).$$

Then, using the fundamental theorem of calculus twice, we obtain

$$\int_c^d f(g(u))g'(u)\,du = \int_c^d \left[\frac{d}{du} F(g(u)) \right] du$$

$$= F(g(u)) \Big|_{u=c}^{u=d} = F(g(d)) - F(g(c))$$

$$= F(b) - F(a) = \int_a^b f(x)\,dx.$$

Thus,

MAKING A SUBSTITUTION TO INTEGRATE

Let $x = g(u)$. Then

$$\int_a^b f(x)\,dx = \int_c^d f(g(u))g'(u)\,du,$$

where

$$c = g^{-1}(a) \qquad \text{and} \qquad d = g^{-1}(b). \tag{1}$$

In Example 2, we computed an antiderivative. There it was necessary to convert back to the original variable after the integration was completed. For a definite integral (as in Example 1), this is not necessary. The steps used in computing an integral by making an appropriate substitution follow.

WHEN USING A SUBSTITUTION TO COMPUTE A DEFINITE INTEGRAL $\int_a^b f(x)\,dx$

1. Change the integrand.
2. Change dx.
3. Change the limits of integration.
4. Integrate.

WHEN USING A SUBSTITUTION TO FIND AN ANTIDERIVATIVE $\int f(x)\,dx$

1. Change the integrand.
2. Change dx.
3. Integrate.
4. Write the result in terms of x.

We now turn to the question of how to find a substitution that works. For a certain class of integrals an appropriate trigonometric substitution will be effective.

INTEGRALS INVOLVING $\sqrt{a^2 - x^2}$, $\sqrt{a^2 + x^2}$, AND $\sqrt{x^2 - a^2}$

We may simplify a large number of complicated-looking integrals by making an appropriate trigonometric substitution. Our technique makes use of the following two identities, each of which we have used before:

$$\sin^2 \theta + \cos^2 \theta = 1$$

$$1 + \tan^2 \theta = \sec^2 \theta. \tag{2}$$

We use these identities in Table 1.

TABLE 1   USEFUL TRIGONOMETRIC SUBSTITUTIONS

Term	Substitution	Term Becomes	dx becomes	
$a^2 - x^2$	$x = a \sin \theta$	$a^2 - a^2 \sin^2 \theta = a^2(1 - \sin^2 \theta) = a^2 \cos^2 \theta$	$a \cos \theta \, d\theta$	(3)
$a^2 + x^2$	$x = a \tan \theta$	$a^2 + a^2 \tan^2 \theta = a^2(1 + \tan^2 \theta) = a^2 \sec^2 \theta$	$a \sec^2 \theta \, d\theta$	(4)
$x^2 - a^2$	$x = a \sec \theta$	$a^2 \sec^2 \theta - a^2 = a^2(\sec^2 \theta - 1) = a^2 \tan^2 \theta$	$a \sec \theta \tan \theta \, d\theta$	(5)

We used the substitution (3) in Example 1.

EXAMPLE 3
EVALUATING AN INDEFINITE INTEGRAL BY MAKING A SUBSTITUTION OF THE FORM $x = a \sin \theta$

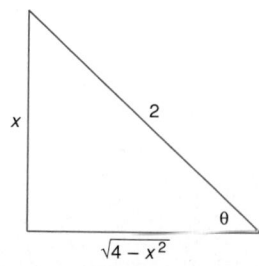

FIGURE 1

IN THIS TRIANGLE $\sin \theta = \dfrac{x}{2}$.

Compute $\displaystyle\int \frac{dx}{(4 - x^2)^{3/2}}$.

SOLUTION: $4 - x^2$ can be written in the form $a^2 - x^2$ with $a = 2$. Thus we make the substitution (from (3)) $x = 2 \sin \theta$.

(1) $(4 - x^2)^{3/2} = (4 - 4 \sin^2 \theta)^{3/2} = [4(1 - \sin^2 \theta)]^{3/2} = (4 \cos^2 \theta)^{3/2} = (2 \cos \theta)^3 = 8 \cos^3 \theta$.

(2) $dx = 2 \cos \theta \, d\theta$.

(3) $\displaystyle\int \frac{dx}{(4 - x^2)^{3/2}} = \int \frac{2 \cos \theta}{8 \cos^3 \theta} \, d\theta = \frac{1}{4} \int \frac{d\theta}{\cos^2 \theta}$

$$= \frac{1}{4} \int \sec^2 \theta = \frac{1}{4} \tan \theta + C.$$

(4) In the triangle of Figure 1 we have

$$\sin \theta = \frac{\text{opposite}}{\text{hypotenuse}} = \frac{x}{2}.$$

The adjacent side has length $\sqrt{4 - x^2}$ by the Pythagorean theorem. Then

$$\tan \theta = \frac{x}{\sqrt{4 - x^2}}$$

so

$$\int \frac{dx}{(4 - x^2)^{3/2}} = \frac{1}{4} \tan \theta + C = \frac{x}{4\sqrt{4 - x^2}} + C.$$

EXAMPLE 4
FINDING AN INDEFINITE
INTEGRAL BY MAKING THE
SUBSTITUTION $x = a \sin \theta$

Calculate $\int \dfrac{dx}{a^2 - x^2}$, $a \neq 0$.

SOLUTION: We make the substitution $x = a \sin \theta$.

(1) $\dfrac{1}{a^2 - x^2} = \dfrac{1}{a^2 - a^2 \sin^2 \theta} = \dfrac{1}{a^2(1 - \sin^2 \theta)} = \dfrac{1}{a^2 \cos^2 \theta}$

(2) $dx = a \cos \theta \, d\theta$

(3) $\displaystyle\int \frac{dx}{a^2 - x^2} = \int \frac{a \cos \theta}{a^2 \cos^2 \theta} \, d\theta = \frac{1}{a} \int \frac{d\theta}{\cos \theta}$

formula (13) on p. 478

$$= \frac{1}{a} \int \sec \theta \, d\theta = \frac{1}{a} \ln|\sec \theta + \tan \theta| + C$$

(4) In the triangle of Figure 2, we have

$$\sin \theta = \frac{\text{opposite}}{\text{hypotenuse}} = \frac{x}{a}.$$

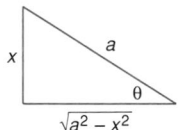

FIGURE 2

IN THIS TRIANGLE $\sin \theta = \dfrac{x}{a}$.

The adjacent side has length $\sqrt{a^2 - x^2}$ by the Pythagorean theorem. Then

$$\tan \theta = \frac{x}{\sqrt{a^2 - x^2}} \qquad \text{and}$$

$$\sec \theta = \frac{1}{\cos \theta} = \frac{1}{\sqrt{a^2 - x^2}/a} = \frac{a}{\sqrt{a^2 - x^2}}$$

so that

$$\int \frac{dx}{a^2 - x^2} = \frac{1}{a} \ln \left| \frac{x + a}{\sqrt{a^2 - x^2}} \right| = \frac{1}{a} \ln \sqrt{\frac{(x + a)^2}{(a - x)(a + x)}}$$

$\ln u^{1/2} = \frac{1}{2} \ln u$

$$= \frac{1}{a} \ln \sqrt{\frac{a + x}{a - x}} = \frac{1}{2a} \ln \left| \frac{a + x}{a - x} \right|.$$

That is,

$$\int \frac{dx}{a^2 - x^2} = \frac{1}{2a} \ln \left| \frac{a + x}{a - x} \right| + C.$$

REMARK: It is possible to compute this integral without making a trigonometric substitution. In Section 7.7 we will discuss the technique of **partial fractions**. There, we will show that

$$\frac{1}{a^2 - x^2} = \frac{1}{(a + x)(a - x)} = \frac{1}{2a}\left[\frac{1}{a + x} + \frac{1}{a - x}\right]$$

so

$$\int \frac{dx}{a^2 - x^2} = \frac{1}{2a} \int \left[\frac{1}{a + x} + \frac{1}{a - x}\right] dx$$

$$= \frac{1}{2a}\left[\ln|a + x| - \ln|a - x|\right] + C$$

$$\ln u - \ln v = \ln\frac{u}{v}$$

$$\downarrow$$

$$= \frac{1}{2a}\ln\left|\frac{a + x}{a - x}\right| + C.$$

EXAMPLE 5
FINDING AN INDEFINITE INTEGRAL BY MAKING A SUBSTITUTION OF THE FORM $x = a \sec \theta$

Compute $\int \sqrt{x^2 - 25}\, dx$, $x > 5$.

SOLUTION: We make the substitution $x = 5 \sec \theta$. Then

$$\sqrt{x^2 - 25} = \sqrt{25 \sec^2 \theta - 25} = \sqrt{25(\sec^2 \theta - 1)}$$
$$= \sqrt{25 \tan^2 \theta} = 5 \tan \theta^\dagger$$

and

$$dx = 5 \sec \theta \tan \theta\, d\theta.$$

Then

$$\int \sqrt{x^2 - 25}\, dx$$

$$= \int 5 \tan \theta(5 \sec \theta \tan \theta)\, d\theta = 25 \int \tan^2 \theta \sec \theta\, d\theta$$

$$= 25 \int (\sec^2 \theta - 1) \sec \theta\, d\theta = 25 \int (\sec^3 \theta - \sec \theta)\, d\theta$$

Example 6 on page 482
and formula (13) on page 478

$$\downarrow$$

$$= 25\left[\frac{1}{2} \sec \theta \tan \theta + \frac{1}{2}\ln|\sec \theta + \tan \theta| - \ln|\sec \theta + \tan \theta|\right]$$

$$= \frac{25}{2}\left[\sec \theta \tan \theta - \ln|\sec \theta + \tan \theta|\right] + C.$$

In the triangle in Figure 3, we see that if $\sec \theta = \dfrac{x}{5}$, then $\tan \theta = \dfrac{\sqrt{x^2 - 25}}{5}$,

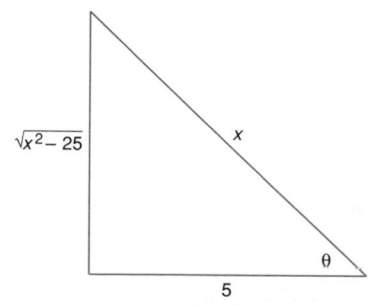

FIGURE 3

IN THIS TRIANGLE $\cos \theta = \dfrac{5}{x}$ SO $\sec \theta =$
$\dfrac{1}{\cos \theta} = \dfrac{x}{5}$.

$\dagger$ If $x > 5$, then by the definition of the inverse secant function on page 465, $\theta = \sec^{-1}\dfrac{x}{5}$ is in the interval $\left[0, \dfrac{\pi}{2}\right)$. In this interval $\tan \theta > 0$. This is why we choose the positive square root.

so

$$\int \sqrt{x^2 - 25}\, dx = \frac{25}{2}\left[\frac{x}{5} \cdot \frac{\sqrt{x^2 - 25}}{5} - \ln\left| \frac{x + \sqrt{x^2 - 25}}{5} \right| \right] + C.$$

We can simplify this somewhat. Observe that

$$\ln\left| \frac{x + \sqrt{x^2 - 25}}{5} \right| = \ln|x + \sqrt{x^2 - 25}| - \ln 5$$

and $\ln 5$ is a constant. We can incorporate this constant into the "general" constant C to obtain

$$\int \sqrt{x^2 - 25}\, dx = \frac{x\sqrt{x^2 - 25}}{2} - \frac{25}{2}\ln|x + \sqrt{x^2 - 25}| + C.$$

CHECK:

$$\frac{d}{dx} \frac{1}{2}\left[x\sqrt{x^2 - 25} - 25\ln|x + \sqrt{x^2 - 25}| \right]$$

$$= \frac{1}{2}\left[x \cdot \frac{x}{\sqrt{x^2 - 25}} + \sqrt{x^2 - 25} - \frac{25}{x + \sqrt{x^2 - 25}}\left(1 + \frac{x}{\sqrt{x^2 - 25}} \right) \right]$$

$$= \frac{1}{2}\left[\frac{x^2}{\sqrt{x^2 - 25}} + \sqrt{x^2 - 25} - \frac{25\sqrt{x^2 - 25}\left(1 + \frac{x}{\sqrt{x^2 - 25}} \right)}{\sqrt{x^2 - 25}(x + \sqrt{x^2 - 25})} \right]$$

$$= \frac{1}{2}\left[\sqrt{x^2 - 25} + \frac{x^2}{\sqrt{x^2 - 25}} - \frac{25(\sqrt{x^2 - 25} + x)}{\sqrt{x^2 - 25}(x + \sqrt{x^2 - 25})} \right]$$

$$= \frac{1}{2}\left[\sqrt{x^2 - 25} + \frac{x^2}{\sqrt{x^2 - 25}} - \frac{25}{\sqrt{x^2 - 25}} \right]$$

$$= \frac{1}{2}\left[\sqrt{x^2 - 25} + \frac{x^2 - 25}{\sqrt{x^2 - 25}} \right] = \frac{1}{2}\left[\sqrt{x^2 - 25} + \sqrt{x^2 - 25} \right]$$

$$= \frac{1}{2}\left[2\sqrt{x^2 - 25} \right] = \sqrt{x^2 - 25}.$$

EXAMPLE 6
CALCULATING A DEFINITE INTEGRAL
BY MAKING A SUBSTITUTION OF THE
FORM $x = a \tan \theta$

Compute $\int dx/(x^2\sqrt{x^2 + 5})$.

SOLUTION: This example is more complicated than others we have done, but our procedure will still work. The term $x^2 + 5$ is of the form $x^2 + a^2$ with $a = \sqrt{5}$. We thus choose $x = \sqrt{5}\tan\theta$, so that

$$dx = \sqrt{5}\sec^2\theta\, d\theta$$

and

$$x^2\sqrt{x^2 + 5} = 5\tan^2\theta\sqrt{5\tan^2\theta + 5} = 5\tan^2\theta\sqrt{5\sec^2\theta}$$
$$= 5\sqrt{5}\tan^2\theta\sec\theta.$$

We choose the positive square root of $5\sec^2\theta$ because $\theta = \tan^{-1}\dfrac{x}{\sqrt{5}}$

takes values in $\left(-\frac{\pi}{2}, \frac{\pi}{2}\right)$ (see p. 459) and $\sec \theta = \frac{1}{\cos \theta}$ is positive in $\left(-\frac{\pi}{2}, \frac{\pi}{2}\right)$. Thus

$$\int \frac{dx}{x^2 \sqrt{x^2 + 5}} = \int \frac{\sqrt{5} \sec^2 \theta \, d\theta}{5\sqrt{5} \tan^2 \theta \sec \theta} = \frac{1}{5} \int \frac{\sec \theta}{\tan^2 \theta} \, d\theta.$$

Now we can simplify this expression by writing everything in terms of sines and cosines:

$$\frac{1}{5} \int \frac{\sec \theta}{\tan^2 \theta} \, d\theta = \frac{1}{5} \int \sec \theta \cot^2 \theta \, d\theta = \frac{1}{5} \int \frac{1}{\cos \theta} \cdot \frac{\cos^2 \theta}{\sin^2 \theta} \, d\theta$$

$$= \frac{1}{5} \int \frac{\cos \theta}{\sin^2 \theta} \, d\theta = \frac{1}{5} \left(\frac{-1}{\sin \theta}\right) = -\frac{1}{5} \csc \theta + C.$$

If $x = \sqrt{5} \tan \theta$, then $\tan \theta = x/\sqrt{5}$, and from the triangle in Figure 4, $\csc \theta = \sqrt{x^2 + 5}/x$. Thus

$$\int \frac{dx}{x^2 \sqrt{x^2 + 5}} = -\frac{\sqrt{x^2 + 5}}{5x} + C.$$

NOTE: In problems this complicated the answer should be checked by differentiation.

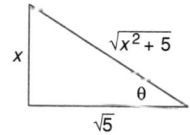

FIGURE 4

IN THIS TRIANGLE $\tan \theta = \dfrac{x}{\sqrt{5}}$.

EXAMPLE 7
CALCULATING ARC LENGTH BY MAKING A TRIGONOMETRIC SUBSTITUTION

Calculate the length of the arc of the curve $y = x^2$ between $x = 0$ and $x = 2$.

SOLUTION: Since $f'(x) = 2x$, $[f'(x)]^2 = 4x^2$ and, from equation (3) on page 350,

$$s = \int_0^2 \sqrt{1 + 4x^2} \, dx = 2 \int_0^2 \sqrt{x^2 + \frac{1}{4}} \, dx.$$

Let $x = \frac{1}{2} \tan \theta$. Then $\sqrt{x^2 + \frac{1}{4}} = \sqrt{\frac{1}{4} \tan^2 \theta + \frac{1}{4}} = \sqrt{\frac{1}{4}(\tan^2 \theta + 1)} = \sqrt{\frac{1}{4} \sec^2 \theta} = \frac{1}{2} \sec \theta$ and $dx = \frac{1}{2} \sec^2 \theta \, d\theta$. Also, $\theta = \tan^{-1} 2x$ so when $x = 0$, $\theta = 0$ and when $x = 2$, $\theta = \tan^{-1} 4$. Thus,

$$s = 2 \int_0^{\tan^{-1} 4} \left(\frac{1}{2} \sec \theta\right)\left(\frac{1}{2} \sec^2 \theta\right) d\theta = \frac{1}{2} \int_0^{\tan^{-1} 4} \sec^3 \theta \, d\theta$$

$$= \frac{1}{4}(\ln|\sec \theta + \tan \theta| + \sec \theta \tan \theta)\Big|_0^{\tan^{-1} 4}. \tag{6}$$

In the triangle in Figure 5 we labeled the sides so that $\tan \theta = 4$. Then $\sec \theta = \dfrac{\text{hypotenuse}}{\text{adjacent}} = \sqrt{17}$. That is, $\tan(\tan^{-1} 4) = 4$ and $\sec(\tan^{-1} 4) = \sqrt{17}$. Thus, from (6), we find that

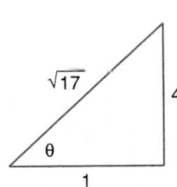

FIGURE 5

IN THIS TRIANGLE $\tan \theta = \dfrac{4}{1} = 4$.

$$s = \frac{1}{4}[\ln(\sqrt{17} + 4) + 4\sqrt{17}] \approx 4.65.$$

EXAMPLE 8
VOLUME OF A TORUS

The circular disk $(x - 2)^2 + y^2 \leq 1$ is rotated around the y-axis. Find the volume of the doughnut-shaped region generated. This solid is called a **torus**.

SOLUTION: The circular disk $(x - 2)^2 + y^2 \leq 1$ consists of the circle centered at $(2, 0)$ with radius 1 and all points interior to it [see Figure 6(a)]. In the upper part of the circle (i.e., in the first quadrant) $y = \sqrt{1 - (x - 2)^2}$, while in the lower part of the circle y is negative and is equal to $-\sqrt{1 - (x - 2)^2}$. Thus, the height of a typical cylindrical shell is equal to

$$\sqrt{1 - (x_i - 2)^2} - \left[-\sqrt{1 - (x_i - 2)^2}\right] = 2\sqrt{1 - (x_i - 2)^2}.$$

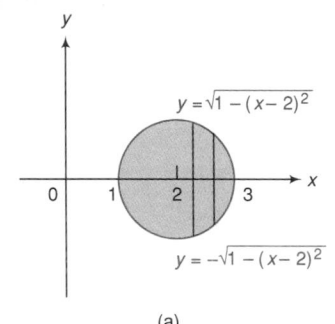

Hence,

$$\Delta V_i \approx 2\pi x_i \left[2\sqrt{1 - (x_i - 2)^2}\right]\Delta x$$

$$V \approx \sum_{i=1}^{n} 2\pi x_i \left[2\sqrt{1 - (x_i - 2)^2}\right]\Delta x.$$

So taking the limit as $\Delta x \to 0$, we have

$$V = \lim_{\Delta x \to 0} \sum_{i=1}^{n} 2\pi x_i \left[2\sqrt{1 - (x_i - 2)^2}\right]\Delta x = 4\pi \int_1^3 x\sqrt{1 - (x - 2)^2}\,dx.$$

To integrate this, we make the substitution $x - 2 = \sin\theta$. Then

$$dx = \cos\theta\,d\theta, \qquad x = 2 + \sin\theta, \qquad \text{and}$$

$$\sqrt{1 - (x - 2)^2} = \sqrt{1 - \sin^2\theta} = \cos\theta.$$

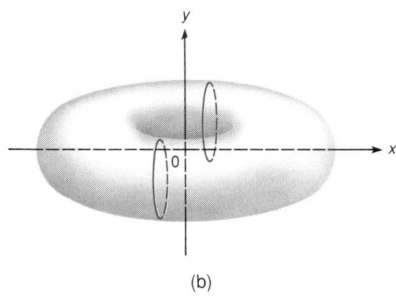

(b)

FIGURE 6
A TORUS IS OBTAINED WHEN THE AREA ENCLOSED BY A CIRCLE THAT DOES NOT INTERSECT THE y-AXIS IS ROTATED ABOUT THE y-AXIS.

When $x = 1$, $\sin\theta = 1 - 2 = -1$ and $\theta = -\pi/2$. When $x = 3$, $\sin\theta = 1$ and $\theta = \pi/2$. Thus,

$$V = 4\pi \int_{-\pi/2}^{\pi/2} (2 + \sin\theta)\cos^2\theta\,d\theta$$

$$= 8\pi \int_{-\pi/2}^{\pi/2} \cos^2\theta\,d\theta + 4\pi \int_{-\pi/2}^{\pi/2} \cos^2\theta\,\sin\theta\,d\theta$$

$$= 4\pi \int_{-\pi/2}^{\pi/2} (1 + \cos 2\theta)\,d\theta - \frac{4\pi}{3}\cos^3\theta\Big|_{-\pi/2}^{\pi/2}$$

$$= 4\pi\left(\theta + \frac{\sin 2\theta}{2}\right)\Big|_{-\pi/2}^{\pi/2} - 0 = 4\pi^2.$$

■ PROBLEMS 7.4

■ SELF-QUIZ

I. The substitution $x = \sin u$ transforms $\int_0^1 x\sqrt{1 - x^2}\,dx$ into _____.

   a. $\int_0^{\pi/2} \sin u \cos u\,du$ b. $\int_0^1 \sin u \cos^2 u\,du$

   c. $\int_0^{\pi/2} \sin u \cos^2 u\,du$ d. $\int_0^{\pi/2} x \cos u\,du$

II. The substitution $x = \cos u$ transforms $\int_0^1 x(1 - x^2)^{3/2}\,dx$ into _____.

   a. $\int_0^{\pi/2} \cos u \sin^3 u(-\sin u\,du)$

   b. $-\int_{\pi/2}^0 \cos u \sin^4 u\,du$

CONTINUED

c. $\int_0^1 x \sin^3 x \, dx$

d. $\int_{-\pi/2}^0 \sin u \cos^4 u \, du$

III. Which one of the following substitutions does **not** simplify

$$\int \frac{x \, dx}{\sqrt{25 - x^2}}?$$

a. $x = 5 \cos \theta$ b. $u = 25 - x^2$
c. $x = 5 \tan \theta$ d. $x = 5 \sin \theta$

IV. Which of the following substitutions do **not** simplify

$$\int \frac{dx}{7 + \sqrt{x}}?$$

a. $u = \sqrt{x}$ b. $x = 49 \tan^2 \theta$
c. $x = (u - 7)^2$ d. $u = 7 + \sqrt{x}$

■ **DRILL**

In Problems 1–46, evaluate the given integral by making an appropriate trigonometric substitution.

1. $\int_0^1 \frac{dx}{\sqrt{4 - x^2}}$

2. $\int \frac{dx}{\sqrt{x^2 - 25}}$

3. $\int -\frac{x}{\sqrt{9 - x^2}} \, dx$

4. $\int \frac{x}{(25 - x^2)^{3/2}} \, dx$

5. $\int_0^1 \sqrt{4 - x^2} \, dx$

6. $\int -\frac{x^2}{\sqrt{9 - x^2}} \, dx$

7. $\int_0^5 \frac{x}{\sqrt{x^2 + 25}} \, dx$

8. $\int_0^1 x^3 \sqrt{4 - x^2} \, dx$

9. $\int \frac{x}{\sqrt{x^2 - 4}} \, dx$

10. $\int \frac{\sqrt{4 - 9x^2}}{x^2} \, dx$

11. $\int \frac{dx}{(x^2 + 36)^{3/2}}$

12. $\int \frac{dx}{x^2 \sqrt{x^2 + 25}}$

13. $\int \frac{x}{36 + x^2} \, dx$

*14. $\int (x^2 + 49)^{3/2} \, dx$

15. $\int_0^1 (4 - x^2)^{3/2} \, dx$

16. $\int \sqrt{a^2 - x^2} \, dx$

17. $\int_0^2 \frac{x^2}{\sqrt{x^2 + 4}} \, dx$

18. $\int \frac{dx}{x \sqrt{a^2 - x^2}}$

19. $\int \frac{x^2}{\sqrt{x^2 - 36}} \, dx$

20. $\int x^2 \sqrt{a^2 - x^2} \, dx$

21. $\int_4^5 \frac{dx}{(x^2 - 9)^{3/2}}$

22. $\int_0^{\sqrt{3}} \frac{x^5}{1 + x^2} \, dx$

23. $\int \frac{dx}{x \sqrt{x^2 + a^2}}$

24. $\int \frac{dx}{\sqrt{x^2 + 81}}$

25. $\int \sqrt{a^2 + x^2} \, dx$

26. $\int \frac{x^2}{(x^2 + a^2)^{3/2}} \, dx$

27. $\int \frac{x^2}{x^2 + a^2} \, dx$

28. $\int \frac{dx}{x^2(a^2 + x^2)^2}$

29. $\int \frac{dx}{x^3(x^2 + a^2)}$

30. $\int_0^{1/\sqrt{2}} \frac{x^3}{2 + x^2} \, dx$

31. $\int \frac{x^3}{\sqrt{a^2 - x^2}} \, dx$

32. $\int_0^2 x^3 \sqrt{x^2 + 4} \, dx$

*33. $\int \frac{\sqrt{5x^2 + 9}}{x} \, dx$

*34. $\int_1^3 \frac{dx}{x^4 \sqrt{x^2 + 3}}$

*35. $\int \frac{dx}{x^3 \sqrt{a^2 - x^2}}$

36. $\int \frac{dx}{x \sqrt{x^2 - a^2}}$

37. $\int \frac{\sqrt{9 - 5x^2}}{x} \, dx$

38. $\int \frac{dx}{\sqrt{a^2 - b^2 x^2}}$

39. $\int \frac{\sqrt{5x^2 - 9}}{x} \, dx$

40. $\int \frac{dx}{x^3 \sqrt{x^2 - a^2}}$

41. $\int \frac{dx}{(a^2 - x^2)^{3/2}}$

42. $\int_2^4 \sqrt{4x^2 - 9} \, dx$

*43. $\int \frac{(x^2 - a^2)^{3/2}}{x^3} \, dx$

*44. $\int_4^8 \frac{x^2}{(x^2 - 4)^{3/2}} \, dx$

*45. $\int x^3 \sqrt{x^2 - 49} \, dx$

*46. $\int x^3 (x^2 - a^2)^{3/2} \, dx$

■ **APPLICATIONS**

In Problems 47–50, find the volume generated when the region bounded by the given curves and lines is rotated about the x-axis.

47. $y = \cos x; y = 0, 0 \le x \le \pi/2$

48. $y = \tan x; y = 0, 0 \le x \le \pi/4$

49. $y = \sin x, y = \cos x; 0 \le x \le \pi/4$

50. $y = \sin^{-1} x; y = 0, 0 \le x \le 1$

CONTINUED

In Problems 51–52, find the volume generated when the region bounded by the given curves and lines is rotated about the y-axis.

51. $y = \sqrt{x^4 + 1}$; $0 \le x \le 1$
52. $y = \cos(x^2)$; $y = 0$, $0 \le x \le \sqrt{\pi/6}$

53. a. Graph the curve $y = \sqrt{1 - x^2}$ for $0 \le x \le 1$.
    b. Find the area bounded by the curve and the x- and y-axes.
54. a. Graph the curve $y = \sqrt{9 + x^2}$ for $x \ge 0$.
    b. Find the area bounded by the curve, the x- and y-axes, and the line $x = 4$.

■ **SHOW/PROVE/DISPROVE**

*55. Show that if $x < -a < 0$, then

$$\int \sqrt{x^2 - a^2}\,dx = -\frac{a^2}{2}\left[\frac{x}{a}\left(\frac{-\sqrt{x^2 - a^2}}{a}\right) \right. $$
$$\left. - \ln\left|\frac{x}{a} - \frac{\sqrt{x^2 - a^2}}{a}\right|\right] + C$$
$$= \frac{a^2}{2}\left[\frac{x}{a}\frac{\sqrt{x^2 - a^2}}{a}\right.$$
$$\left. + \ln\left|\frac{x - \sqrt{x^2 - a^2}}{a}\right|\right] + C.$$

56. a. Show, using properties of logarithms, that

$$-\ln\left|\frac{x + \sqrt{x^2 - a^2}}{a}\right| = \ln\left|\frac{x - \sqrt{x^2 - a^2}}{a}\right| + C.$$

   b. Use the results of the preceding problem and of part (a) of this one to show, if $|x| > a > 0$,

$$\int \sqrt{x^2 - a^2}\,dx = \frac{a^2}{2}\left[\frac{x}{a}\frac{\sqrt{x^2 - a^2}}{a}\right.$$
$$\left. - \ln\left|\frac{x + \sqrt{x^2 - a^2}}{a}\right|\right] + C.$$

[*Hint:* First consider the case $x > a > 0$.]

57. a. Show that

$$\int \frac{dx}{\sqrt{1 + x^2}} = \ln(x + \sqrt{1 + x^2}) + C_1.$$

   b. Recall that

$$\int \frac{dx}{\sqrt{1 + x^2}} = \sinh^{-1} x + C_2. \quad \text{(See Section 6.10)}$$

   Show that $\sinh^{-1} x = \ln(x + \sqrt{1 + x^2})$.

■ **CHALLENGE**

*58. Let

$$A = \int_{-1}^{1} \frac{1}{1 + x^2}\,dx.$$

Because $1/(1 + x^2) > \frac{1}{2}$ for all $x \in (-1, 1)$, we see that $A > \int_{-1}^{1} \left(\frac{1}{2}\right)dx = 1$. [*Note:* The true value of A is $\tan^{-1}(1) - \tan^{-1}(-1) = (\pi/4) - (-\pi/4) = \pi/2 \approx 1.57$.] However, if we try the change of variable $u = 1/x$, then $x = 1/u$, $dx = (-1/u^2)\,du$, and

$$A = \int_{-1}^{1} \frac{1}{1 + x^2}\,dx$$

$$(u = 1/x)$$

$$\overset{\downarrow}{=} \int_{1/-1}^{1/1} \frac{1}{1 + (1/u)^2} \cdot \left(\frac{-1}{u^2}\right)du$$

$$= -\int_{-1}^{1} \frac{1}{1 + u^2}\,du = -A;$$

but the only number satisfying $A = -A$ is $A = 0$. What went wrong?

*59. Compute the following integral without making any substitutions:[†]

$$\int_{0}^{1} \left(\sqrt{2 - x^2} - \sqrt{2x - x^2}\right)dx.$$

[*Hint:* Find a region, bounded above and below by simple curves, whose area is computed by this integral. The first integral gives the area of a triangle plus one-eighth of a circle. The second integral gives the area of one-quarter of a different circle.]

[†] This is Problem Q370 from *Mathematics Magazine*, volume 38 (1965), pages 326 and 302.

7.5 OTHER SUBSTITUTIONS (OPTIONAL)

There are many other types of expressions that can be integrated by making an appropriate substitution. Unfortunately, many of these substitutions work only for a very small class of integrals. Even when a function can be integrated by substitution, there is usually no obvious way to find the substitution that works, and it is often necessary to resort to trial and error. In this section, we provide a few examples of substitutions that do work.

When an integrand contains a term of the form $[f(x)]^{m/n}$, then the substitution

$$u = f(x)^{1/n}$$

will often be useful.

EXAMPLE 1
COMPUTING A DEFINITE INTEGRAL BY MAKING AN APPROPRIATE SUBSTITUTION

Calculate $\int_{-1}^{0} x \sqrt[3]{x + 1} \, dx$.

SOLUTION: Let $u = (x + 1)^{1/3}$. Then $u^3 = x + 1$, or $x = u^3 - 1$, and $dx = 3u^2 \, du$. When $x = -1$, $u = 0$, and when $x = 0$, $u = 1$. Thus,

$$\int_{-1}^{0} x \sqrt[3]{x + 1} \, dx = \int_{0}^{1} (u^3 - 1)u \cdot 3u^2 \, du = \int_{0}^{1} (3u^6 - 3u^3) \, du$$

$$= \frac{3}{7}u^7 - \frac{3}{4}u^4 \Big|_{0}^{1} = -\frac{9}{28}.$$

This problem could also be solved by the method of integration by parts or by making the substitution $u = x + 1$.

Many rational functions of $\sin x$ and $\cos x$ can be integrated with the use of a special trigonometric substitution.

EXAMPLE 2
COMPUTING AN INDEFINITE INTEGRAL INVOLVING cos x BY MAKING AN APPROPRIATE SUBSTITUTION

Calculate $\int dx/(2 + \cos x)$.

SOLUTION: Let $u = \tan(x/2)$.[†] Then $x = 2 \tan^{-1} u$, and

$$dx = \frac{2}{1 + u^2} \, du. \tag{1}$$

From Figure 1,

$$\sin \frac{x}{2} = \frac{u}{\sqrt{1 + u^2}} \qquad \text{and} \qquad \cos \frac{x}{2} = \frac{1}{\sqrt{1 + u^2}}.$$

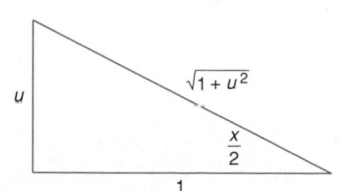

FIGURE 1

IN THIS TRIANGLE $\tan \dfrac{x}{2} = u$.

[†] The reason for making this seemingly arbitrary substitution will soon become apparent. It is certainly not obvious that it will work before we try it.

Using the double-angle formulas, $\sin 2x = 2 \sin x \cos x$ and $\cos 2x = \cos^2 x - \sin^2 x$, we obtain

$$\sin x = 2 \sin \frac{x}{2} \cos \frac{x}{2} = \frac{2u}{1 + u^2} \tag{2}$$

and

$$\cos x = \cos^2 \frac{x}{2} - \sin^2 \frac{x}{2} = \frac{1 - u^2}{1 + u^2}. \tag{3}$$

Thus, using (1) and (3), we obtain

$$\int \frac{dx}{2 + \cos x} = \int \frac{2/(1 + u^2)}{2 + [(1 - u^2)/(1 + u^2)]} \, du = \int \frac{2}{3 + u^2} \, du$$

Formula (17) on page 478 $\rightarrow$

$$= \frac{2}{\sqrt{3}} \tan^{-1} \frac{u}{\sqrt{3}} = \frac{2}{\sqrt{3}} \tan^{-1} \left[\frac{\tan(x/2)}{\sqrt{3}} \right] + C.$$

REMARK: The substitutions (1), (2), and (3) will always work to simplify a rational function of $\sin x$ and $\cos x$.

■ PROBLEMS 7.5

■ SELF-QUIZ

I. The substitution $u = 3 + x^2$ transforms

$$\int_0^1 \frac{2x^3}{(3 + x^2)^{5/2}} \, dx \text{ into } \underline{\hspace{2cm}}.$$

a. $\int_0^1 \frac{u - 3}{u^{5/2}} \, du$ b. $\int_3^4 \frac{u - 3}{u^{5/2}} \, du$

c. $\int_1^3 \frac{(u - 3)^{3/2}}{u^{5/2}} \, du$ d. $\int_4^3 \frac{(u - 3)^{3/2}}{u^{5/2}} \, du$

II. If $u = \tan(x/2)$, then $\cos x = \underline{\hspace{2cm}}$.

a. $\dfrac{2u}{1 + u^2}$ b. $\dfrac{1 + u^2}{1 - u^2}$

c. $\dfrac{2}{1 + u^2} - 1$ d. $\dfrac{1 - u^2}{1 + u^2}$

III. Which one of the following substitutions does **not** simplify

$$\int \frac{dx}{4 + x^2}?$$

a. $u = 4 + x^2$ b. $x = 2 \tan \theta$
c. $x = 2 \sinh t$ d. $\phi = \tan^{-1}(x/2)$

IV. Which (one or several) of the following substitutions do **not** simplify

$$\int \frac{dx}{\sqrt{9 + x^2}}?$$

a. $u = 9 + x^2$ b. $x = 3 \tan \theta$
c. $x = -3 \cot t$ d. $\phi = \tan^{-1}(x/3)$

■ DRILL

In Problems 1–16, calculate the given integral using an appropriate substitution.

1. $\int \dfrac{x}{(x - 1)^4} \, dx$

2. $\int \dfrac{dx}{4 + 5\sqrt{x}}$

3. $\int_0^{13} x(1 + 2x)^{2/3} \, dx$

4. $\int_0^2 \dfrac{3x^5}{(1 + x^3)^{3/2}} \, dx$

5. $\int_0^1 x^8(1 - x^3)^{1/3} \, dx$

6. $\int_0^1 x^8(1 - x^3)^{6/5} \, dx$

7. $\int x\sqrt{1 + x} \, dx$

8. $\int \dfrac{1 + x}{\sqrt{1 - x}} \, dx$

9. $\int \dfrac{3x^2 - x}{\sqrt{1 + x}} \, dx$

10. $\int \dfrac{x^3 - x}{\sqrt{x^2 - 2}} \, dx$

*11. $\displaystyle\int \frac{dx}{\sqrt{e^{2x}-1}}$

[*Hint:* Try the substitution $u = e^{-x}$.]

*12. $\displaystyle\int \frac{dx}{x^3(16-x^3)^{1/3}}$

[*Hint:* Try the substitution $u = 1/x^3$.]

13. $\displaystyle\int \frac{dx}{1+\sin x}$

14. $\displaystyle\int_0^{\pi/2} \frac{dx}{1+\sin x + \cos x}$

15. $\displaystyle\int_0^{\pi/3} \frac{dx}{3+2\cos x}$

16. $\displaystyle\int \frac{dx}{\sin x + \tan x}$

The integrals in Problems 17–26 also appeared in Section 7.4 where you were asked to handle them by means of a trigonometric substitution. For the following problems, you are to reevaluate those integrals by making a nontrigonometric substitution.

17. $\displaystyle\int \frac{x\,dx}{\sqrt{9-x^2}}$

18. $\displaystyle\int \frac{x\,dx}{\sqrt{x^2-25}}$

19. $\displaystyle\int \frac{x\,dx}{36+x^2}$

20. $\displaystyle\int_0^5 \frac{x\,dx}{\sqrt{x^2+25}}$

21. $\displaystyle\int_0^1 x^3\sqrt{4-x^2}\,dx$

22. $\displaystyle\int_0^2 x^3\sqrt{4+x^2}\,dx$

23. $\displaystyle\int \frac{x^3}{\sqrt{a^2-x^2}}\,dx$

24. $\displaystyle\int \frac{dx}{x\sqrt{x^2-a^2}}$

25. $\displaystyle\int x^3\sqrt{x^2-49}\,dx$

26. $\displaystyle\int x^3(x^2-a^2)^{3/2}\,dx$

■ **APPLICATIONS**

In Problems 27–28, find the volume generated when the region bounded by the given curves and lines is rotated about the x-axis.

27. $y = \ln x$; $y = 0$, $1 \le x \le e$

28. $y = xe^x$; $y = 0$, $0 \le x \le 1$

In Problems 29–30, find the volume generated when the region bounded by the given curves and lines is rotated about the y-axis.

29. $y = \ln x$; $y = 0$, $1 \le x \le 3$

30. $y = e^{x^2}$; $y = 0$, $0 \le x \le 1$

*31. Apply the substitution $x = \tan^{-1}[\sinh(e^u)]$ to find $\int \sec x\,dx$.

*32. Apply the substitution $x = \tan^{-1}[\sinh \theta]$ to find $\int \sec^3 x\,dx$.

†33. Use **Euler's substitution**, $t = x + \sqrt{x^2+a}$, to find $\int dx/\sqrt{x^2+5}$.

*34. Compute $\int \sqrt{\sec^2 x + A}\,dx$, $A \ge 0$. [*Hint:* Start with the substitution $u = \sec x$; two other substitutions are needed.]

■ **SHOW/PROVE/DISPROVE**

35. Prove that if f is continuous on $[a, b]$, then

$$\int_a^b f(x)\,dx = \int_a^b f(a+b-x)\,dx.$$

*36. Suppose that f is continuous on $[a, b]$ and $f(m+x) = f(m-x)$ where $m = (a+b)/2$ is the midpoint of $[a, b]$. Prove that

$$\int_a^b xf(x)\,dx = m\int_a^b f(x)\,dx.$$

■ **CHALLENGE**

*37. What goes wrong with the following procedure used to compute $\int_{-1}^7 (3+5x^2)x^2\,dx$? [*Note:* The true value of $\int_{-1}^7 (3+5x^2)x^2\,dx = \int_{-1}^7 (3x^2+5x^4)\,dx = \{7^3 + 7^5\} - \{(-1)^3 + (-1)^5\} = 17{,}152$.] Let $x^2 = t$; then $2x\,dx = dt$ and $dx = dt/2x = dt/2\sqrt{t}$. Since $t = (-1)^2 = 1$ when $x = -1$ and $t = 7^2 = 49$ when $x = 7$, we compute

$$\int_{-1}^7 (3+5x^2)x^2\,dx = \int_{(-1)^2}^{7^2} (3+5t)t\left(\frac{dt}{2\sqrt{t}}\right)$$

$$= \int_1^{49} \left(\frac{3}{2}t^{1/2} + \frac{5}{2}t^{3/2}\right)dt = (t^{3/2}+t^{5/2})\Big|_{t=1}^{49}$$

$$= (343 + 16{,}807) - (1+1) = 17{,}148.$$

38. Consider $\int x^2\sqrt{x^2-1}\,dx$. We can integrate this expression by making the substitution $x = \sec \theta$. Show that other "reasonable" substitutions will *not* work. [*Hint:* Let $u = \sqrt{x^2-1}$ and see what happens.]

CONTINUED

■ ANSWERS TO SELF-QUIZ

I. b   II. c = d   III. a   IV. a

7.6 THE INTEGRATION OF RATIONAL FUNCTIONS I: LINEAR AND QUADRATIC DENOMINATORS

In this section and the next one, we will show how, at least in theory, every rational function can be integrated. Recall that a **rational function** is a quotient of the form $p(x)/q(x)$, where p and q are both polynomials. In this section, we will deal with the case in which $q(x)$ is either a linear or a quadratic polynomial.

Case 1. $q(x)$ is a **linear function**. That is,

$$q(x) = ax + b. \tag{1}$$

We illustrate what can happen.

EXAMPLE 1
INTEGRATING A RATIONAL FUNCTION WITH A LINEAR TERM IN THE DENOMINATOR

Compute $\int dx/(ax + b)$.

SOLUTION: Let $u = ax + b$. Then $du = a\,dx$ and

$$\int \frac{dx}{ax + b} = \frac{1}{a} \int \frac{a\,dx}{ax + b} = \frac{1}{a} \int \frac{du}{u} = \frac{1}{a} \ln|u| + C,$$

or

$$\int \frac{dx}{ax + b} = \frac{1}{a} \ln|ax + b| + C. \tag{2}$$

Thus we can integrate $1/q(x)$ if $q(x) = ax + b$. To integrate $p(x)/(ax + b)$, we simply divide. If the degree of $p(x) = n$, then

$$\frac{p(x)}{ax + b} = \frac{c}{ax + b} + r(x), \tag{3}$$

where $r(x)$ is a polynomial of degree $n - 1$. Instead of proving formula (3), we give an example.

EXAMPLE 2
INTEGRATING A RATIONAL FUNCTION WITH A LINEAR TERM IN THE DENOMINATOR BY FIRST DIVIDING

Compute $\int [(x^3 - x^2 + 4x - 5)/(x - 2)]\,dx$.

SOLUTION: We divide:

$$
\begin{array}{r}
x^2 + x + 6 \\
x - 2 \overline{)\, x^3 - x^2 + 4x - 5} \\
\underline{x^3 - 2x^2} \\
x^2 + 4x \\
\underline{x^2 - 2x} \\
6x - 5 \\
\underline{6x - 12} \\
7.
\end{array}
$$

So

$$\frac{x^3 - x^2 + 4x - 5}{x - 2} = x^2 + x + 6 + \frac{7}{x - 2}$$

and

$$\int \frac{x^3 - x^2 + 4x - 5}{x - 2}\, dx = \int \left(x^2 + x + 6 + \frac{7}{x - 2} \right) dx$$

$$= \frac{x^3}{6} + \frac{x^2}{2} + 6x + 7\ln|x - 2| + C.$$

As Examples 1 and 2 illustrate, we can integrate any rational function with a linear denominator.

Case 2. $q(x)$ is a **quadratic function**. That is,

$$q(x) = px^2 + qx + r.$$

This case is a bit more complicated. We need three basic integrals. In Example 5 in Section 6.8 (page 461), we showed that

$$\int \frac{dx}{a^2 + x^2} = \frac{1}{a}\tan^{-1}\frac{x}{a} + C. \tag{4}$$

In Example 4 in Section 7.4 (on page 496), we computed

$$\int \frac{dx}{a^2 - x^2} = \frac{1}{2a}\ln\left|\frac{a + x}{a - x}\right| + C. \tag{5}$$

From (5), we obtain

$$\int \frac{dx}{x^2 - a^2} = -\int \frac{dx}{a^2 - x^2} = -\frac{1}{2a}\ln\left|\frac{a + x}{a - x}\right|$$

$$\underset{|-u| = |u|}{} \qquad \underset{-\ln u = \ln\frac{1}{u}}{}$$

$$\overset{\downarrow}{=} -\frac{1}{2a}\ln\left|\frac{x + a}{x - a}\right| \overset{\downarrow}{=} \frac{1}{2a}\ln\left|\frac{x - a}{x + a}\right|$$

or

$$\int \frac{dx}{x^2 - a^2} = \frac{1}{2a}\ln\left|\frac{x - a}{x + a}\right| + C. \tag{6}$$

I. PROCEDURE FOR COMPUTING $\int \dfrac{dx}{px^2 + qx + r}$

(i) Factor the number p from the denominator.

(ii) Complete the square.

(iii) Make an appropriate substitution to write the denominator as $a^2 - u^2$, $u^2 - a^2$, or $a^2 + u^2$.

(iv) Use formula (4), (5), or (6) to complete the integration.

EXAMPLE 3
INTEGRATING A RATIONAL
FUNCTION WITH A QUADRATIC
TERM IN THE DENOMINATOR
BY WRITING THE DENOMINATOR
IN THE FORM $u^2 - a^2$

Compute $\int dx/(x^2 + 4x - 5)$.

SOLUTION:

$$\int \frac{dx}{x^2 + 4x - 5} = \int \frac{dx}{(x + 2)^2 - 4 - 5} = \int \frac{dx}{(x + 2)^2 - 9}.$$

Let $u = x + 2$; then $du = dx$ and

$$\int \frac{dx}{x^2 + 4x - 5} = \int \frac{du}{u^2 - 9} = \int \frac{du}{u^2 - 3^2}.$$

Then, from equation (6), this expression is equal to

$$\frac{1}{2 \cdot 3} \ln \left| \frac{u - 3}{u + 3} \right| = \frac{1}{6} \ln \left| \frac{(x + 2) - 3}{(x + 2) + 3} \right| = \frac{1}{6} \ln \left| \frac{x - 1}{x + 5} \right| + C.$$

EXAMPLE 4
INTEGRATING A RATIONAL
FUNCTION WITH A QUADRATIC
TERM IN THE DENOMINATOR

Compute $\int (x + 8)/(x^2 + 6x + 12)\,dx$.

SOLUTION: We first note that

$$\frac{d}{dx}(x^2 + 6x + 12) = 2x + 6.$$

Thus, if we can put $2x + 6$ in the numerator and let $u = x^2 + 6x + 12$, we will have an expression of the form du/u, which is easily integrated. Let's see how this can be done:

$$\int \frac{x + 8}{x^2 + 6x + 12}\,dx \overset{\overset{\text{Multiply and divide by 2}}{\downarrow}}{=} \frac{1}{2} \int \frac{2x + 16}{x^2 + 6x + 12}\,dx \overset{\overset{\text{We want } 2x + 6 \text{ in the numerator}}{\downarrow}}{=} \frac{1}{2} \int \frac{2x + 6 + 10}{x^2 + 6x + 12}\,dx$$

$$= \frac{1}{2} \int \frac{2x + 6}{x^2 + 6x + 12}\,dx + \frac{1}{2} \int \frac{10}{x^2 + 6x + 12}\,dx$$

$$= \frac{1}{2} \ln|x^2 + 6x + 12| + 5 \int \frac{dx}{x^2 + 6x + 12}$$

$$= \frac{1}{2} \ln|x^2 + 6x + 12| + 5 \int \frac{dx}{(x + 3)^2 + 3}.$$

Finally,

$$\int \frac{dx}{(x + 3)^3 + 3} \overset{\overset{u = x + 3}{\underset{du = dx}{\downarrow}}}{=} \int \frac{du}{u^2 + 3} \overset{\overset{\text{from (4)}}{\downarrow}}{=} \frac{1}{\sqrt{3}} \tan^{-1} \frac{u}{\sqrt{3}} = \frac{1}{\sqrt{3}} \tan^{-1} \frac{x + 3}{\sqrt{3}}$$

and

$$\int \frac{x + 8}{x^2 + 6x + 12}\,dx = \frac{1}{2} \ln|x^2 + 6x + 12| + \frac{5}{\sqrt{3}} \tan^{-1} \frac{x + 3}{\sqrt{3}} + C.$$

II. PROCEDURE FOR COMPUTING $\int \dfrac{ax + b}{x^2 + cx + d}\,dx,\ a \ne 0$

(i) Factor out $\dfrac{a}{2}$ to obtain

$$\int \frac{ax + b}{x^2 + cx + d}\,dx = \frac{a}{2} \int \frac{2x + \dfrac{2b}{a}}{x^2 + cx + d}\,dx.$$

(ii) Add and subtract c in the numerator:

$$\int \frac{ax + b}{x^2 + cx + d}\,dx = \frac{a}{2} \int \frac{2x + c + \dfrac{2b}{a} - c}{x^2 + cx + d}\,dx$$

$$= \frac{a}{2} \int \frac{2x + c}{x^2 + cx + d}\,dx + \frac{a}{2} \int \frac{\dfrac{2b}{a} - c}{x^2 + cx + d}\,dx.$$

(iii) The first integral equals $\dfrac{a}{2}\ln|x^2 + cx + d| + k$ since the numerator is the derivative of the denominator. The second integral can be obtained using Procedure I.

REMARK: Let $n =$ degree of $p(x)$ with $n \ge 2$. If $q(x)$ is a quadratic polynomial, then we can divide to obtain

$$\frac{p(x)}{q(x)} = \frac{ax + b}{q(x)} + r(x), \tag{7}$$

where a and b are constants and $r(x)$ is a polynomial of degree $n - 2$. For example,

$$\int \frac{x^5 + 3x^4 - x^3 + 9x^2 - 4x + 12}{x^2 + 2x + 5}\,dx$$

$$= \int \left(x^3 + x^2 - 8x + 20 + \frac{-4x - 88}{x^2 + 2x + 5} \right) dx$$

$$= \int \left[x^3 + x^2 - 8x + 20 - 2\left(\frac{2x + 2}{x^2 + 2x + 5} \right) + \frac{-84}{(x + 1)^2 + 4} \right] dx$$

$$= \frac{x^4}{4} + \frac{x^3}{3} - 4x^2 + 20x - 2\ln(x^2 + 2x + 5) - 42\tan^{-1}\left(\frac{x + 1}{2} \right) + C.$$

As the preceding examples indicate, we can integrate any rational function with a linear or quadratic denominator. In the next section, we will show how other rational functions can be integrated.

■ PROBLEMS 7.6

■ SELF-QUIZ

I. By discarding the blatantly wrong answers, pick the right answer:

$$\int \frac{-2x}{1 - x^2} \, dx = \underline{\hspace{2cm}} + C.$$

  a. $\dfrac{1}{1 - x}$ b. $\dfrac{x}{1 - x}$

  c. $\dfrac{1 + x}{1 - x}$ d. $\ln|1 + x| + \ln|1 - x|$

II. Pick the right answer:

$$\int \frac{x - 1}{x + 1} \, dx = \underline{\hspace{2cm}} + C.$$

  a. $x + 2 \ln|x - 1|$

  b. $\dfrac{2}{3} \cdot \dfrac{1}{x - 1} - \dfrac{5}{3} \cdot \dfrac{1}{x + 1}$

  c. $x - 2 \ln|x + 1|$

  d. $\dfrac{1}{x - 1} + \dfrac{1}{x + 1}$

III. Pick the right answer:

$$\int \frac{1}{x^2 - 6x + 10} \, dx = \underline{\hspace{2cm}} + C.$$

  a. $\tan^{-1}(x + 3)$ b. $\tan^{-1}(x - 3)$
  c. $\tan^{-1}(x/3)$ d. $\frac{1}{2} \ln|(x - 3)^2 + 1|$

IV. Pick the right answer:

$$\int \frac{x^2 - 6x + 10}{x^2 - 6x + 8} \, dx = \underline{\hspace{2cm}} + C.$$

  a. $1 + \dfrac{1}{x - 4} - \dfrac{1}{x - 2}$

  b. $x + \ln\left|\dfrac{x - 4}{x - 2}\right|$

  c. $\ln\left|\dfrac{x - 3 - \sqrt{17}}{x - 3 + \sqrt{17}}\right|$

  d. $\sqrt{x} - \tan^{-1}\left(\dfrac{x - 3}{2\sqrt{2}}\right)$

■ DRILL

In Problems 1–24, calculate the given integral.

1. $\displaystyle\int \frac{dx}{2x - 5}$

2. $\displaystyle\int \frac{dx}{5x + 10}$

3. $\displaystyle\int \frac{dx}{3x + 11}$

4. $\displaystyle\int \frac{dx}{3x - 11}$

5. $\displaystyle\int \frac{dx}{x^2 + 9}$

6. $\displaystyle\int \frac{dx}{2x^2 + 8}$

7. $\displaystyle\int \frac{dx}{x^2 - 16}$

8. $\displaystyle\int \frac{dx}{x^2 - 12}$

9. $\displaystyle\int \frac{x - 1}{x + 1} \, dx$

10. $\displaystyle\int \frac{x + 1}{x - 1} \, dx$

11. $\displaystyle\int \frac{2x + 3}{x - 4} \, dx$

12. $\displaystyle\int \frac{x^2 + 4x + 5}{x - 2} \, dx$

13. $\displaystyle\int \frac{x^3 - x}{2x + 6} \, dx$

14. $\displaystyle\int \frac{x^4 - x^3 + 3x^2 - 2x + 7}{x + 2} \, dx$

15. $\displaystyle\int \frac{dx}{x^2 + 12}$

16. $\displaystyle\int \frac{x + 1}{x^2 + 1} \, dx$

17. $\displaystyle\int \frac{3x + 5}{x^2 + 9} \, dx$

18. $\displaystyle\int \frac{x^2 - 1}{x^2 + 1} \, dx$

19. $\displaystyle\int \frac{2x - 3}{x^2 - 4} \, dx$

20. $\displaystyle\int \frac{x^5 - 1}{x^2 - 1} \, dx$

21. $\displaystyle\int \frac{x - 1}{x^2 - 6x + 16} \, dx$

22. $\displaystyle\int \frac{x^2 - 1}{x^2 - 6x + 25} \, dx$

23. $\displaystyle\int \frac{x^3 - x^2 + 2x + 5}{x^2 + x + 1} \, dx$

24. $\displaystyle\int \frac{x^4 - 2x^3 + x - 3}{2x^2 - 8x + 10} \, dx$

■ APPLICATIONS

25. Calculate the area bounded by the curve $y = (1 + x^2)/(1 + x)^2$, the x- and y-axes, and the line $x = 1$.

26. Calculate the area bounded by the curve $y = 1/(x^2 + x)$, the y-axis, and the lines $x = 1$ and $x = 2$.

CONTINUED

In Problems 27–38, evaluate the given integral; one or several substitutions may be appropriate.

27. $\int \dfrac{\sqrt{x+2}-1}{\sqrt{x+2}+1}\, dx$

28. $\int_0^8 \dfrac{\sqrt[3]{x}-1}{\sqrt[3]{x}+1}\, dx$

29. $\int \dfrac{\sin x}{\cos^2 x + 4\cos x + 4}\, dx$

30. $\int \dfrac{dx}{e^x + e^{-x}}$

31. $\int_0^{1/\sqrt{2}} \dfrac{x^3}{2+x^2}\, dx$

32. $\int_0^{\sqrt{3}} \dfrac{x^5}{1+x^2}\, dx$

*33. $\int \dfrac{dx}{x^{1/3}+x^{1/4}}$ [*Hint:* Try $u = x^{1/12}$.]

*34. $\int \dfrac{x^{1/2}}{x^{1/3}+x^{1/4}}\, dx$

*35. $\int \dfrac{x-4}{\sqrt{x}(1+\sqrt[3]{x})}\, dx$

36. $\int \dfrac{2-\sin x}{2+\sin x}\, dx$ [*Hint:* Start with $u = \tan(x/2)$.]

*37. $\int \dfrac{\csc x}{3\csc x + 2\cot x + 2}\, dx$

*38. $\int \dfrac{dx}{10\sec x - 6}$

■ **SHOW/PROVE/DISPROVE**

39. Show that if $r > q^2/4$, then

$$\int \frac{dx}{x^2 + qx + r} = \frac{2}{\sqrt{4r - q^2}}\tan^{-1}\!\left(\frac{2x+q}{\sqrt{4r-q}}\right) + C.$$

40. a. Show that if $r < q^2/4$, then

$$\int \frac{dx}{x^2 + qx + r} = \frac{1}{\sqrt{q^2 - 4r}}\ln\left|\frac{\sqrt{q^2-4r}+2x+q}{\sqrt{q^2-4r}-2x-q}\right| + C.$$

b. What happens if $r = q^2/4$?

■ **ANSWERS TO SELF-QUIZ**

I. d   II. c   III. b   IV. b

7.7 THE INTEGRATION OF RATIONAL FUNCTIONS II: THE METHOD OF PARTIAL FRACTIONS

In the last section, we saw how to integrate $p(x)/q(x)$, where $q(x)$ is a linear or a quadratic polynomial. In this section, we show how techniques from algebra can be used to evaluate (at least theoretically) any rational function.

In your algebra course, you learned how to combine rational expressions. For example,

$$\frac{1}{x+3} + \frac{3}{x-2} = \frac{(x-2)+3(x+3)}{(x+3)(x-2)} = \frac{4x+7}{x^2+x-6}.$$

In the technique called **partial fractions**, we reverse this process.

EXAMPLE 1

USING PARTIAL FRACTIONS TO INTEGRATE A RATIONAL FUNCTION

$$\int \frac{4x+7}{x^2+x-6}\, dx = \int \left[\frac{1}{x+3} + \frac{3}{x-2}\right] dx$$
$$= \ln|x+3| + 3\ln|x-2| + C$$
$$= \ln|(x+3)(x-2)^3| + C.$$

REMARK: The sum $\dfrac{1}{x+3} + \dfrac{3}{x-2}$ is called the **partial-fraction decomposition** of the rational function $\dfrac{4x+7}{x^2+x-6}$.

This section is basically algebraic in nature. We will show how any rational function can be written in a partial-fraction decomposition where the denominator of each term is linear or quadratic. We can then use the techniques of Section 7.6 to complete the integration.

Before doing examples, we cite some results from algebra. A quadratic polynomial is called **irreducible** if it cannot be factored into linear factors. That means that the polynomial $ax^2 + bx + c$ is irreducible if the quadratic equation $ax^2 + bx + c = 0$ has no real roots—that is, if

$$b^2 - 4ac < 0. \tag{1}$$

EXAMPLE 2
AN IRREDUCIBLE QUADRATIC

The polynomial $2x^2 + 4x + 3$ is irreducible because

$$4^2 - 4(2)(3) = 16 - 24 = -8 < 0.$$

EXAMPLE 3
A QUADRATIC THAT
IS NOT IRREDUCIBLE

The polynomial $x^2 - 5x + 6$ is not irreducible because $(-5)^2 - 4(1)(6) = 25 - 24 = 1 > 0$. Note that $x^2 - 5x + 6 = (x - 3)(x - 2)$, which is a product of linear factors.

$\otimes$ **WARNING:** Keep in mind that there is a difference between linear factors of multiplicity two [such as $(x + 1)^2$] and irreducible quadratic factors (such as $x^2 + 1$). To emphasize this point, we note that

$$\int \frac{dx}{(x + 1)^2} = -\frac{1}{x + 1} + C \quad \text{and} \quad \int \frac{dx}{x^2 + 1} = \tan^{-1} x + C. \otimes$$

The following theorem from algebra will be very helpful to us. We will not prove it, but will indicate instead how it can be used.

THEOREM 1   **PARTIAL-FRACTION DECOMPOSITION THEOREM**

Let $r(x) = p(x)/q(x)$ be a rational function with degree of $p(x)$ < degree of $q(x)$. Then,

$$\frac{p(x)}{q(x)} = R_1 + R_2 + \cdots + R_n, \tag{2}$$

where each R_i is a rational function whose denominator is a positive integral power of either a linear polynomial of the form $ax + b$ or an irreducible quadratic polynomial of the form $ax^2 + bx + c$. ■

The formula (2) is called the **partial-fraction decomposition** of the function $p(x)/q(x)$.[†]

We emphasize that according to Theorem 1, we can integrate (at least theoretically) any rational function by writing it as a sum of terms we already know how to integrate. To use Theorem 1, we provide a four-step procedure for carrying out the partial-fraction decomposition of any rational function.

[†] Sometimes a higher-degree polynomial looks irreducible but is not. For example, $x^4 + 1 = 0$ has no real roots and $x^4 + 1$ seems to be irreducible. However, it can be factored: $x^4 + 1 = (x^2 + \sqrt{2}x + 1) \times (x^2 - \sqrt{2}x + 1)$. Both $x^2 + \sqrt{2}x + 1$ and $x^2 - \sqrt{2}x + 1$ *are* irreducible.

To simplify our discussion, we will assume that if the degree of $q(x)$ is n, then the coefficient of the x^n term, called the **leading coefficient**, is 1. If it is not, we can factor out the coefficient to make the leading coefficient equal to 1. For example,

$$\frac{x^3 - x^2 + 2}{4x^4 - 7x^3 + 6x^2 - 2x + 3} = \frac{1}{4}\left[\frac{x^3 - x^2 + 2}{x^4 - \frac{7}{4}x^3 + \frac{3}{2}x^2 - \frac{1}{2}x + \frac{3}{4}}\right],$$

and we can integrate the function in brackets and then multiply by $\frac{1}{4}$ at the end.

Here, then, is our procedure.

PROCEDURE FOR INTEGRATING $p(x)/q(x)$ USING PARTIAL FRACTIONS

Step 1. If degree of $p(x) \geq$ degree of $q(x)$, first divide to obtain

$$\frac{p(x)}{q(x)} = s(x) + \frac{t(x)}{q(x)}, \tag{3}$$

where s, t, and q are polynomials and degree of $t(x) <$ degree of $q(x)$.

Step 2. Factor $q(x)$ into linear and/or irreducible quadratic factors.[†] A linear factor will have the form $x - a_i$ [where a_i is a root of $q(x)$], and a quadratic factor will have the form $x^2 + bx + c$. This follows from the fact that the leading coefficient of $q(x)$ is 1. (4)

Step 3. For each factor of the form $(x - a_i)^k$, the partial-fraction decomposition (2) contains a sum of k partial fractions

$$\frac{A_1}{x - a_i} + \frac{A_2}{(x - a_i)^2} + \cdots + \frac{A_k}{(x - a_i)^k}. \tag{5}$$

In particular, if $k = 1$, then the decomposition has one term of the form

$$\frac{A}{x - a_i}. \tag{6}$$

Step 4. For each factor of the form $(x^2 + bx + c)^m$, the partial-fraction decomposition (2) contains a sum of m partial fractions

$$\frac{A_1x + B_1}{x^2 + bx + c} + \frac{A_2x + B_2}{(x^2 + bx + c)^2} + \cdots + \frac{A_mx + B_m}{(x^2 + bx + c)^m}. \tag{7}$$

In particular, if $m = 1$, then the decomposition has one term of the form

$$\frac{Ax + B}{x^2 + ax + b}. \tag{8}$$

[†] This step is the hardest step since there are no rules for factoring high-degree polynomials. In every example we give, such a factoring will be possible. But it would be impossible to factor exactly a polynomial of degree 5, say, if its coefficients were randomly chosen. Thus we say that "theoretically" an antiderivative can be found for any rational function. In practice, this step may be impossible.

We illustrate this technique with a number of examples.

EXAMPLE 4
INTEGRATION USING PARTIAL
FRACTIONS: TWO LINEAR TERMS

Compute $\int \dfrac{dx}{x^2 + 4x - 5}$.

SOLUTION: We solved this problem in Example 3 in Section 7.6. We now solve it using partial fractions. We have

$$\frac{1}{x^2 + 4x - 5} = \frac{1}{(x + 5)(x - 1)} = \frac{A}{x + 5} + \frac{B}{x - 1}.$$

Then, cross multiplying,

$$\frac{A(x - 1) + B(x + 5)}{(x + 5)(x - 1)} = \frac{1}{(x + 5)(x - 1)}$$

$$A(x - 1) + B(x + 5) = 1 \qquad \text{Equate numerators since denominators are the same.}$$

$$(A + B)x + (-A + 5B) = 1 = 0x + 1$$

$$\left. \begin{array}{r} A + B = 0 \\ -A + 5B = 1 \end{array} \right\} \qquad \text{Equate coefficients and add.}$$

$$6B = 1, \qquad B = \frac{1}{6}$$

$$A = -B = -\frac{1}{6}.$$

Thus,

$$\frac{1}{(x + 5)(x - 1)} = \frac{-\frac{1}{6}}{x + 5} + \frac{\frac{1}{6}}{x - 1}$$

and

$$\int \frac{dx}{x^2 + 4x - 5} = \frac{1}{6} \int \left(\frac{1}{x - 1} - \frac{1}{x + 5} \right) dx = \frac{1}{6}[\ln |x - 1| - \ln |x + 5|]$$

$$= \frac{1}{6} \ln \left| \frac{x - 1}{x + 5} \right| + C.$$

EXAMPLE 5
INTEGRATION USING PARTIAL
FRACTIONS: THREE LINEAR TERMS

Compute $\int dx/[(x - 1)(x - 2)(x - 3)]$.

SOLUTION: By (6),

$$\frac{1}{(x - 1)(x - 2)(x - 3)} = \frac{A}{x - 1} + \frac{B}{x - 2} + \frac{C}{x - 3}. \tag{9}$$

There are several ways to find A, B, and C. The simplest one is as follows: Multiply both sides of (9) by $(x - 1)(x - 2)(x - 3)$. Then

$$1 = A(x - 2)(x - 3) + B(x - 1)(x - 3) + C(x - 1)(x - 2). \tag{10}$$

Equation (10) holds for every value of x. Setting $x = 1$ makes the last two terms in (10) equal to 0, which gives us the equation

$$1 = A(-1)(-2), \quad \text{or} \quad A = \frac{1}{2}.$$

Similarly, setting $x = 2$ gives us

$$1 = B(1)(-1), \quad \text{or} \quad B = -1.$$

Setting $x = 3$ leads to

$$1 = C(2)(1), \quad \text{or} \quad C = \frac{1}{2}.$$

Thus,

$$\frac{1}{(x-1)(x-2)(x-3)} = \frac{\frac{1}{2}}{x-1} + \frac{-1}{x-2} + \frac{\frac{1}{2}}{x-3}, \tag{11}$$

so that

$$\int \frac{dx}{(x-1)(x-2)(x-3)} = \frac{1}{2}\ln|x-1| - \ln|x-2| + \frac{1}{2}\ln|x-3| + C.$$

We can also combine these terms to obtain

$$\ln\left|\frac{\sqrt{|(x-1)(x-3)|}}{x-2}\right| + C.$$

EXAMPLE 6
INTEGRATION USING PARTIAL FRACTIONS: A LINEAR TERM AND THE SQUARE OF A LINEAR TERM

Calculate $\int dx/[x(x-1)^2]$.

SOLUTION: From (5) we have

$$\frac{1}{x(x-1)^2} = \frac{A}{x} + \frac{B}{x-1} + \frac{C}{(x-1)^2}. \tag{12}$$

Multiplying both sides of (12) by $x(x-1)^2$, we obtain

$$1 = A(x-1)^2 + Bx(x-1) + Cx. \tag{13}$$

Setting $x = 0$ in (13) gives us $1 = A$. Setting $x = 1$ in (13) leads to $1 = C$. Finally, since $A = C = 1$, (13) can be rewritten as

$$1 = (x-1)^2 + Bx(x-1) + x,$$

and setting $x = 2$ (any number other than 0 or 1 will work), we obtain

$$1 = 1 + 2B + 2, \quad \text{or} \quad 2B = -2 \quad \text{and} \quad B = -1.$$

Thus,

$$\int \frac{dx}{x(x-1)^2} = \int \left[\frac{1}{x} - \frac{1}{x-1} + \frac{1}{(x-1)^2} \right] dx$$

$$= \ln|x| - \ln|x-1| - \frac{1}{x-1} + C$$

$$= \ln\left| \frac{x}{x-1} \right| - \frac{1}{x-1} + C.$$

EXAMPLE 7
INTEGRATION USING PARTIAL
FRACTIONS: A LINEAR TERM AND
AN IRREDUCIBLE QUADRATIC TERM

Compute $\int dx/(x^3 + x)$.

SOLUTION: From (8),

$$\frac{1}{x^3 + x} = \frac{1}{x(x^2 + 1)} = \frac{A_1}{x} + \frac{A_2 x + B_2}{x^2 + 1} \qquad (14)$$

since $x^2 + 1$ is irreducible. Multiplying both sides of (14) by $x(x^2 + 1)$ and setting $x = 0$ yields

$$1 = A_1(x^2 + 1) + (A_2 x + B_2)x \qquad \text{and} \qquad 1 = A_1.$$

Thus,

$$1 - (x^2 + 1) = A_2 x^2 + B_2 x, \qquad \text{or} \qquad -x^2 = A_2 x^2 + B_2 x.$$

Dividing by x, we have

$$-x = A_2 x + B_2,$$

and again setting $x = 0$, we obtain

$$B_2 = 0.$$

Thus $-x = A_2 x$, or $A_2 = -1$, and

$$\int \frac{dx}{x(x^2 + 1)} = \int \left[\frac{1}{x} - \frac{x}{x^2 + 1} \right] dx = \ln|x| - \frac{1}{2} \ln|x^2 + 1|$$

$$= \ln\left| \frac{x}{\sqrt{x^2 + 1}} \right| + C.$$

There is one case that we have not considered, that of repeated irreducible quadratic factors. We illustrate this case with an example.

EXAMPLE 8
INTEGRATING A RATIONAL
FUNCTION WITH THE SQUARE
OF AN IRREDUCIBLE
QUADRATIC IN THE
DENOMINATOR

Compute $\int dx/(x^2 + 2x + 10)^2$.

SOLUTION: The polynomial $x^2 + 2x + 10 = (x + 1)^2 + 9$ is irreducible. We make the substitution $(x + 1) = 3 \tan \theta$. Then,

$$dx = 3 \sec^2 \theta \, d\theta$$

and

$$(x^2 + 2x + 10)^2 = [(x + 1)^2 + 9]^2 = (9 \tan^2 \theta + 9)^2$$
$$= (9 \sec^2 \theta)^2 = 81 \sec^4 \theta.$$

Thus,

$$\int \frac{dx}{(x^2 + 2x + 10)^2} = \int \frac{3 \sec^2 \theta}{81 \sec^4 \theta} d\theta = \frac{1}{27} \int \frac{1}{\sec^2 \theta} d\theta = \frac{1}{27} \int \cos^2 \theta \, d\theta$$

$$= \frac{1}{54} \int (1 + \cos 2\theta) \, d\theta$$

$$= \frac{1}{54} \left(\theta + \frac{\sin 2\theta}{2} \right) = \frac{1}{54} (\theta + \sin \theta \cos \theta)$$

Figure 1 $\sin \theta$ $\cos \theta$

$$= \frac{1}{54} \left[\tan^{-1} \left(\frac{x+1}{3} \right) + \overbrace{\frac{x+1}{\sqrt{(x+1)^2 + 9}}}^{} \cdot \overbrace{\frac{3}{\sqrt{(x+1)^2 + 9}}}^{} \right]$$

$$= \frac{1}{54} \left[\tan^{-1} \left(\frac{x+1}{3} \right) + \frac{3(x+1)}{x^2 + 2x + 10} \right] + C.$$

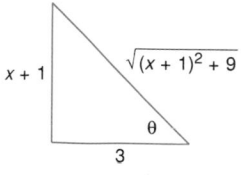

FIGURE 1

IN THIS TRIANGLE $\tan \theta = \dfrac{x+1}{3}$.

■ **SELF-QUIZ**

I. By discarding the blatantly wrong answers, pick the right answer:

$$\int \frac{x}{(x-1)^2} \, dx = \underline{\hspace{2cm}} + C.$$

a. $\dfrac{1}{x-1} + \dfrac{1}{(x-1)^2}$ b. $\frac{1}{2} \ln|x^2 - 2x + 1|$

c. $\ln|1 - x| - \dfrac{1}{x-1}$ d. $\frac{1}{2} \ln|x^2 - 1|$

II. Pick the right answer:

$$\int \frac{dx}{x^2(x+1)} = \underline{\hspace{2cm}} + C.$$

a. $\dfrac{1}{x+1} - \dfrac{1}{x} + \dfrac{1}{x^2}$ b. $\ln \left| \dfrac{x}{x+1} \right| - \dfrac{1}{x-1}$

c. $\ln \left| \dfrac{x+1}{x} \right| - \dfrac{1}{x}$ d. $\tan^{-1} \left(\dfrac{x}{x+1} \right)$

III. Pick the right answer:

$$2 \int_0^1 \frac{dx}{(x^2 + 1)^2} = \underline{\hspace{2cm}}.$$

a. $\sin \theta \cos \theta + \theta + C$

b. $\dfrac{x}{x^2 + 1} + \tan^{-1} x + C$

c. $\sqrt{3} - \pi$

d. $\dfrac{1}{2} + \dfrac{\pi}{4}$

IV. Pick the right answer:

$$\int_1^2 \frac{x^2 + 3}{x(x^2 + 1)} \, dx = \underline{\hspace{2cm}}.$$

a. $\dfrac{3}{x} - \dfrac{2x}{x^2 + 1}$ b. $3 \ln|x| - \ln|x^2 + 1|$

c. $\ln(\frac{16}{5})$ d. 0

■ **DRILL**

In Problems 1–34, evaluate the given integral. The method of partial fractions can be used in each case; in some, you may find a substitution that eases your work.

1. $\displaystyle\int \frac{dx}{(x-1)(x-4)}$

2. $\displaystyle\int \frac{dx}{x^2 - 4x - 21}$

3. $\displaystyle\int \frac{3x - 5}{(x+3)(x-7)} \, dx$

4. $\displaystyle\int \frac{2x - 3}{x^2 - 5x + 4} \, dx$

5. $\displaystyle\int \frac{dx}{(x-a)(x-b)}, \quad a \neq b$

6. $\displaystyle\int \frac{cx + d}{(x-a)(x-b)} \, dx, \quad \begin{array}{l} c \neq 0 \\ a \neq b \end{array}$

CONTINUED

7. $\displaystyle\int \frac{x^2 + 3x + 4}{x^2 + x}\, dx$

8. $\displaystyle\int \frac{x^3 - 2x^2 + 2x + 5}{x^2 - 6x - 7}\, dx$

9. $\displaystyle\int \frac{x^5}{x^3 - x}\, dx$ 10. $\displaystyle\int \frac{x^3}{x^4 - x^2}\, dx$

11. $\displaystyle\int_{-1}^{0} \frac{dx}{(x - 1)(x - 2)(x - 3)}$

12. $\displaystyle\int \frac{x + 2}{(x - 1)(x - 2)(x - 3)}\, dx$

13. $\displaystyle\int \frac{x^2 + 3x + 4}{(x - 1)(x - 2)(x - 3)}\, dx$

14. $\displaystyle\int \frac{x^3 + 3x^2 + 2x + 1}{(x - 1)(x - 2)(x - 3)}\, dx$

15. $\displaystyle\int \frac{x^2 + 2}{x(x - 1)^2(x + 1)}\, dx$

16. $\displaystyle\int_{0}^{1} \frac{dx}{(x - 2)^2(x - 3)^2}$

17. $\displaystyle\int \frac{dx}{x^2(x - 1)}$ 18. $\displaystyle\int_{1}^{2} \frac{x^4}{16x - x^3}\, dx$

19. $\displaystyle\int \frac{dx}{x^2 + x^4}$ 20. $\displaystyle\int \frac{x}{(x + 1)^2(x^2 + 1)}\, dx$

21. $\displaystyle\int \frac{dx}{x^2(x + 2)^2}$ 22. $\displaystyle\int \frac{x^2 + 1}{(x^2 - 1)^2}\, dx$

23. $\displaystyle\int \frac{x}{x^4 - 1}\, dx$ 24. $\displaystyle\int \frac{x^4 + 1}{x^5 + x^3}\, dx$

25. $\displaystyle\int \frac{dx}{(x + 2)(x^2 + 1)}$ *26. $\displaystyle\int \frac{x^2 - 2}{(x^2 + 1)^2}\, dx$

27. $\displaystyle\int \frac{4x + 3}{(x^2 + 1)(x^2 + 2)}\, dx$

*28. $\displaystyle\int \frac{x^3 + 5}{(x^2 - x + \frac{1}{2})(x^2 - 6x + 10)}\, dx$

*29. $\displaystyle\int_{1}^{32} \frac{dx}{x(1 + x^{1/5})}$ *30. $\displaystyle\int \frac{x^{2/3} + 1}{x^{2/3} - 1}\, dx$

31. $\displaystyle\int_{0}^{\pi/2} \frac{\cos x}{\sin^2 x + 2\sin x + 2}\, dx$

32. $\displaystyle\int_{0}^{\pi} \frac{\sin x}{\cos^2 x - 4\cos x + 13}\, dx$

*33. $\displaystyle\int \frac{dx}{\sin 2x + \cos 2x}$

*34. $\displaystyle\int \frac{\sin x}{\cos^2 x - 5\cos x + 6}\, dx$

Use the result of Problem 39 to evaluate the integrals given in Problems 35–38.

35. $\displaystyle\int \frac{x^2 - 1}{x^3 - 2x}\, dx$ 36. $\displaystyle\int \frac{x^6 + 1}{x^7 - x}\, dx$

37. $\displaystyle\int \frac{3x^4 + 4}{2x^5 + 5x}\, dx$ 38. $\displaystyle\int \frac{5x^{10} + 8}{-x^{11} + 2x}\, dx$

■ SHOW/PROVE/DISPROVE

39. a. Show that

$$\frac{Ax^N + B}{x(Cx^N + D)} = \frac{\alpha}{x} + \frac{\beta x^{N-1}}{Cx^N + D},$$

where $\alpha = B/D$, and $\beta = A - BC/D$.

b. Show that

$$\int \frac{Ax^N + B}{x(Cx^N + D)}\, dx$$

$$= \frac{B}{D}\ln|x| + \frac{1}{N}\left(\frac{A}{C} - \frac{B}{D}\right)\ln|Cx^N + D| + k.$$

■ CHALLENGE

40. An alternative approach to the integral of the preceding problem involves finding a real number r such that if numerator and denominator of the given fraction are each multiplied by x^r, then an easy substitution reduces the task to $\int u^{-1}\, du$. Investigate this approach; write a brief summary of your findings (include a comparison of the two approaches).

7.8 USING THE INTEGRAL TABLES

Many types of integrals are encountered in applications. For that reason, extensive tables of integrals have been made available. One of the most complete can be found in *Table of Integrals, Series and Products, Corrected and Enlarged Edition*, by I. S. Gradshteyn and I. M. Ryzhik (New York: Academic Press, 1980). It gives 150 pages of indefinite integrals and 400 pages of definite integrals. A more modest table of integrals appears at the back of this text.

In the first seven sections of this chapter, we saw how to integrate a wide variety of functions. All the integrals given in the tables can be obtained by one or more of our methods. However, a great deal of time can be saved by using the tables when one of the more common integrals is encountered. We illustrate this technique with several examples.

EXAMPLE 1
FINDING AN INTEGRAL BY USING AN INTEGRAL TABLE

Use the integral table to find $\int dx/\sqrt{5x^2 + 8}$.

SOLUTION:

Entry 22 in the table of integrals

$$\int \frac{du}{\sqrt{u^2 + a^2}} = \ln|u + \sqrt{u^2 + a^2}| + C. \tag{1}$$

If $u = \sqrt{5x^2} = \sqrt{5}x$ and $a = \sqrt{8}$, then $\sqrt{u^2 + a^2} = \sqrt{5x^2 + 8}$. Also $du = \sqrt{5}\,dx$. Thus multiplying and dividing the integrand by $\sqrt{5}$ and using (1), we obtain

$$\int \frac{dx}{\sqrt{5x^2 + 8}} = \frac{1}{\sqrt{5}} \int \frac{\sqrt{5}\,dx}{\sqrt{5x^2 + 8}} = \frac{1}{\sqrt{5}} \int \frac{du}{\sqrt{u^2 + 8}}$$

$$= \frac{1}{\sqrt{5}} \ln|u + \sqrt{u^2 + 8}|$$

$$= \frac{1}{\sqrt{5}} \ln|\sqrt{5}x + \sqrt{5x^2 + 8}| + C.$$

REMARK: In Example 1, we had to simplify the "inner" function $5x^2 + 8$ before using the table. This kind of simplification is often necessary. So, before looking for an integral in the table of integrals, simplify as much as possible the arguments (the "inner" functions) of the function in the integrand.

EXAMPLE 2
USING AN INTEGRAL TABLE

Use the integral table to find $\int dx/(4 - 3 \sin 5x)$.

SOLUTION:

Entry 107 in the table of integrals

$$\int \frac{du}{p + q \sin u} = \frac{2}{\sqrt{p^2 - q^2}} \tan^{-1}\left[\frac{p \tan(u/2) + q}{\sqrt{p^2 - q^2}}\right] + C, \qquad |p| > |q|. \tag{2}$$

Here $p = 4$, $q = -3$, $|4| > |-3|$ and $u = 5x$. Then $du = 5\,dx$, so that

$$\int \frac{dx}{4 - 3\sin 5x} = \frac{1}{5}\int \frac{5\,dx}{4 - 3\sin 5x} = \frac{1}{5}\int \frac{du}{4 - 3\sin u}$$

$$= \frac{1}{5}\frac{2}{\sqrt{16 - 9}}\tan^{-1}\left[\frac{4\tan(u/2) - 3}{\sqrt{16 - 9}}\right]$$

$$= \frac{2}{5\sqrt{7}}\tan^{-1}\left[\frac{4\tan(5x/2) - 3}{\sqrt{7}}\right] + C.$$

EXAMPLE 3
USING AN INTEGRAL TABLE

Use the integral table to find $\int xe^{-5x^2}\sin 4x^2\,dx$.

SOLUTION:

Entry 168

$$\int e^{au}\sin bu\,du = \frac{e^{au}}{a^2 + b^2}(a\sin bu - b\cos bu) + C. \tag{3}$$

If $u = x^2$, then $du = 2x\,dx$, so that with $a = -5$ and $b = 4$,

$$\int xe^{-5x^2}\sin 4x^2\,dx = \frac{1}{2}\int e^{-5x^2}\sin 4x^2(2x\,dx)$$

$$= \frac{1}{2}\int e^{-5u}\sin 4u\,du$$

$$= \frac{1}{2}\left(\frac{e^{-5u}}{25 + 16}\right)(-5\sin 4u - 4\cos 4u)$$

$$= \frac{-e^{-5x^2}}{82}(5\sin 4x^2 + 4\cos 4x^2) + C.$$

EXAMPLE 4
USING AN INTEGRAL TABLE

Use the table of integrals to find $\int \sin^4 3x\,dx$.

SOLUTION:

Entry 109 or Example 7 on page 483

$$\int \sin^n au\,du = -\frac{1}{an}\sin^{n-1} au\cos au + \frac{n-1}{n}\int \sin^{n-2} au\,du. \tag{4}$$

So, using (4) twice with $a = 3$, we have

$$\int \sin^4 3x\,dx = -\frac{1}{12}\sin^3 3x\cos 3x + \frac{3}{4}\int \sin^2 3x\,dx$$

$$\overbrace{\sin^{2-2} x = 1}$$

$$= -\frac{1}{12}\sin^3 3x\cos 3x + \frac{3}{4}\left[-\frac{1}{6}\sin 3x\cos 3x + \frac{1}{2}\int dx\right]$$

$$= -\frac{1}{12}\sin^3 3x\cos 3x - \frac{1}{8}\sin 3x\cos 3x + \frac{3}{8}x + C.$$

■ PROBLEMS 7.8

■ SELF-QUIZ

I. After the substitution $u = x + 3$, we can find

$$\int \frac{dx}{x^2 + 6x + 8}$$

by using the integral table entry for _____.

a. $\int \dfrac{du}{u^2 + a^2}$ b. $\int \dfrac{du}{u^2 - a^2}$

c. $\int \dfrac{du}{u^2 \pm a^2}$ d. $\int u^k \, du$

II. The integral table entry for $\int e^{au} \cos bu \, du$ does **not** apply directly to _____.

a. $\int e^{-3x} \cos 2x \, dx$ b. $\int e^{5x} \cos 5x \, dx$

c. $\int xe^{-x^2} \cos x^2 \, dx$ d. $\int xe^{2x} \cos x \, dx$

III. After a substitution, $\int \dfrac{x}{3 + x^4} \, dx$ can be found using the integral table entry for _____.

a. $\int \dfrac{du}{u^2 + a^2}$ b. $\int \dfrac{u}{u^2 + a^2} \, du$

c. $\int \dfrac{du}{u + a}$ d. $\int \dfrac{u^2}{u^2 + a^2} \, du$

IV. The integral table entry for

$$\int \frac{du}{u^2 + a^2}$$

does **not** apply to _____.

a. $\int \dfrac{\sin x}{1 + \cos^2 x} \, dx$ b. $\int \dfrac{\sec^2 \theta}{1 + \tan^2 \theta} \, d\theta$

c. $\int \dfrac{ds}{e^{-s} + e^s} = \int \dfrac{e^s}{1 + e^{2s}} \, ds$

d. $\int \dfrac{\tan t}{1 + \sec^2 t} \, dt$

■ DRILL

In Problems 1–34, use the table of integrals to evaluate the given integral.

1. $\int \dfrac{x}{\sqrt{16 + 3x^2}} \, dx$

2. $\int \dfrac{x}{\sqrt{16 - 3x^2}} \, dx$

3. $\int \dfrac{dx}{9 - 4x^2}$

4. $\int \dfrac{dx}{9 + 4x^2}$

5. $\int \dfrac{x}{3 + 2x} \, dx$

6. $\int \dfrac{x^2}{3 + 2x} \, dx$

7. $\int \dfrac{dx}{x(3 + 2x)}$

8. $\int \dfrac{dx}{x^2(3 + 2x)}$

9. $\int x\sqrt{3 + 2x} \, dx$

10. $\int \dfrac{\sqrt{3 + 2x}}{x} \, dx$

11. $\int x^2 e^{3x} \, dx$

12. $\int x^2 \sin 3x \, dx$

13. $\int \sin^3 x \cos^4 x \, dx$

14. $\int \cos^4 2x \, dx$

15. $\int \sin 3x \cos 4x \, dx$

16. $\int x^2 \cos 4x^3 \cos 5x^3 \, dx$

17. $\int \tan^4\left(\dfrac{x}{3}\right) dx$

18. $\int x \sec^4 x^2 \, dx$

19. $\int x \tan^{-1} x^2 \, dx$

20. $\int x^5 \tan^{-1} x^3 \, dx$

21. $\int x \cos^{-1} 2x \, dx$

*22. $\int x^3 \sin^{-1}\left(\dfrac{x}{4}\right) dx$

23. $\int \sqrt{16 - 3x^2} \, dx$

24. $\int \dfrac{dx}{\sqrt{16 + 3x^2}}$

25. $\int \dfrac{\sqrt{10 + 2x^2}}{x} \, dx$

26. $\int \dfrac{\sqrt{10 - 2x^2}}{x} \, dx$

27. $\int \dfrac{dx}{x\sqrt{2x^2 - 9}}$

28. $\int \dfrac{\sqrt{25x^2 - 4}}{3x} \, dx$

29. $\int \dfrac{dx}{(4x^2 - 16)^{3/2}}$

30. $\int \dfrac{4x^2}{\sqrt{9 + 4x^2}} \, dx$

31. $\int \dfrac{x^2}{\sqrt{2x^2 - 5}} \, dx$

*32. $\int \dfrac{dx}{x^2\sqrt{9x^2 - 1}}$

33. $\int x^2 e^{4x^3} \cos 7x^3 \, dx$

*34. $\int x\sqrt{3x - 4x^2} \, dx$

■ ANSWERS TO SELF-QUIZ

I. b   II. d   III. a   IV. d

▦ 7.9 NUMERICAL INTEGRATION

Consider the problem of evaluating

$$\int_0^1 \sqrt{1 + x^3}\, dx \qquad \text{or} \qquad \int_0^1 e^{x^2}\, dx.$$

Since both $\sqrt{1 + x^3}$ and e^{x^2} are continuous in $[0, 1]$, we know that both the definite integrals given here exist. They represent the areas under the curves $y = \sqrt{1 + x^3}$ and $y = e^{x^2}$ for x between 0 and 1. The problem is that none of the methods we have studied (or any other method for that matter) will enable us to find the antiderivative of $\sqrt{1 + x^3}$ or e^{x^2} because neither antiderivative can be expressed in terms of the functions we know.

In fact, there are a great number of continuous functions for which an antiderivative cannot be expressed in terms of functions we know. In those cases we cannot use the fundamental theorem of calculus to evaluate a definite integral. Nevertheless, it may be very important to approximate the value of such an integral. For that reason, many methods have been devised to approximate the value of a definite integral to as many decimal places as are deemed necessary. All these techniques come under the heading of **numerical integration**. We will not discuss this vast subject in great generality here. Rather, we will introduce two reasonably effective methods for estimating a definite integral: the **trapezoidal rule** and **Simpson's rule**.[†]

Consider the problem of calculating

$$\int_a^b f(x)\, dx. \tag{1}$$

We know that

$$\int_a^b f(x)\, dx = \lim_{\Delta x \to 0} \left[f(x_1{}^*)\Delta x + f(x_2{}^*)\Delta x + \cdots + f(x_n{}^*)\Delta x \right]$$

$$= \lim_{\Delta x \to 0} \sum_{i=1}^n f(x_i{}^*)\Delta x. \tag{2}$$

In other words, when the lengths of the subintervals in a partition of $[a, b]$ are small, the sum in the right-hand side of (2) gives us a crude approximation of the integral. Here the area is approximated by a sum of areas of rectangles. We saw some examples of this type of approximation in Section 4.4. We now develop a more efficient way to approximate the integral.

TRAPEZOIDAL RULE

Let f be as in Figure 1 and let us partition the interval $[a, b]$ by the equally spaced points

$$a = x_0 < x_1 < x_2 < \cdots < x_{i-1} < x_i < \cdots < x_n = b, \tag{3}$$

where $x_i - x_{i-1} = \Delta x = (b - a)/n$. In Figure 1 we have indicated that the area under the curve can be approximated by the sum of the areas of n trap-

---

[†] One reasonably elementary book that gives a more complete discussion of numerical integration is by S. D. Conte and C. De Boor, *Elementary Numerical Analysis: An Algorithmic Approach*, 3rd ed. (New York: McGraw-Hill, 1980).

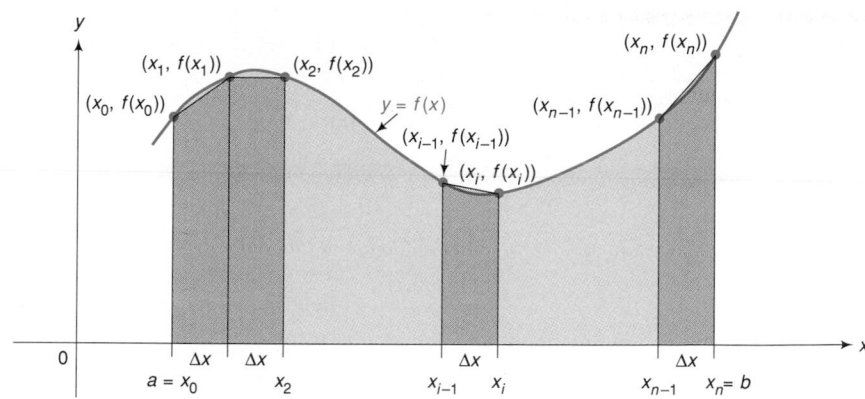

FIGURE 1 *APPROXIMATION OF A DEFINITE INTEGRAL BY ADDING UP THE AREAS UNDER TRAPEZOIDS*

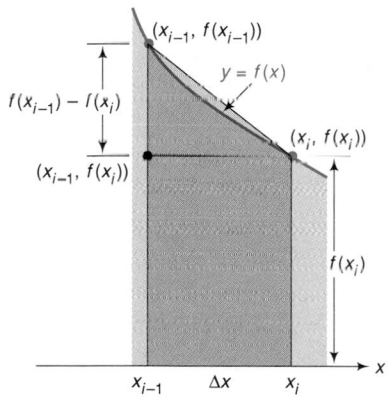

FIGURE 2
AREA UNDER A TRAPEZOID

ezoids. One typical trapezoid is sketched in Figure 2. The area of the trapezoid is the area of the rectangle plus the area of the triangle. But the area of the rectangle is $f(x_i)\,\Delta x$, and the area of the triangle is $\frac{1}{2}[f(x_{i-1}) - f(x_i)]\,\Delta x$, so that

Area of rectangle ⌐⌐ Area of triangle

$$\text{area of trapezoid} = \overbrace{f(x_i)\,\Delta x} + \overbrace{\frac{1}{2}[f(x_{i-1}) - f(x_i)]\,\Delta x}$$

$$= \frac{1}{2}[f(x_{i-1}) + f(x_i)]\,\Delta x.^{\dagger} \tag{4}$$

Then,

$$\int_a^b f(x)\,dx \approx \text{sum of the areas of the trapezoids}$$

$$= \frac{1}{2}[f(x_0) + f(x_1)]\,\Delta x + \frac{1}{2}[f(x_1) + f(x_2)]\,\Delta x + \cdots$$

$$+ \frac{1}{2}[f(x_{n-2}) + f(x_{n-1})]\,\Delta x + \frac{1}{2}[f(x_{n-1}) + f(x_n)]\,\Delta x,$$

or

$$\int_a^b f(x)\,dx \approx \frac{1}{2}\,\Delta x[f(x_0) + 2f(x_1) + 2f(x_2) + \cdots + 2f(x_{n-1}) + f(x_n)]. \tag{5}$$

The approximation formula (5) is called the **trapezoidal rule** for numerical integration. Note that since $\Delta x = (b-a)/n$, we can write (5) as

$$\int_a^b f(x)\,dx \approx \frac{b-a}{2n}\,[f(x_0) + 2f(x_1) + 2f(x_2) + \cdots + 2f(x_{n-1}) + f(x_n)]. \tag{6}$$

---

† Note that this expression is the same as the average of the area of the rectangle R_{i-1} whose height is $f(x_{i-1})$ (the left-hand endpoint) and the area of the rectangle R_i whose height is $f(x_i)$ (the right-hand endpoint). That is,

$$\frac{1}{2}[(\text{area of } R_{i-1}) + (\text{area of } R_i)] = \frac{1}{2}[f(x_{i-1})\,\Delta x + f(x_i)\,\Delta x] = \frac{1}{2}[f(x_{i-1}) + f(x_i)]\,\Delta x.$$

EXAMPLE 1

USING THE TRAPEZOIDAL RULE
TO APPROXIMATE AN INTEGRAL
WE ALREADY KNOW

Estimate $\int_1^2 (1/x)\,dx$ by using the trapezoidal rule first with $n = 5$ and then with $n = 10$.

SOLUTION:

(i) Here $n = 5$ and

$$\Delta x = \frac{b - a}{n} = \frac{2 - 1}{5} = \frac{1}{5} = 0.2.$$

Then $x_0 = 1$, $x_1 = 1.2$, $x_2 = 1.4$, $x_3 = 1.6$, $x_4 = 1.8$, and $x_5 = 2$. From (5),

$$\int_1^2 \frac{1}{x}\,dx \approx \frac{1}{2}\Delta x\big[f(x_0) + 2f(x_1) + 2f(x_2)$$
$$+ 2f(x_3) + 2f(x_4) + f(x_5)\big]$$
$$= \frac{0.2}{2}\left(\frac{1}{1} + \frac{2}{1.2} + \frac{2}{1.4} + \frac{2}{1.6} + \frac{2}{1.8} + \frac{1}{2}\right)$$
$$\approx 0.1(1 + 1.6667 + 1.4286 + 1.25 + 1.1111 + 0.5)$$
$$= 0.1(6.9564) = 0.6956.$$

(ii) Now $n = 10$ and $\Delta x = \frac{1}{10} = 0.1$, so that $x_0 = 1$, $x_1 = 1.1, \ldots$, $x_9 = 1.9$, and $x_{10} = 2$. Thus,

$$\int_1^2 \frac{1}{x}\,dx \approx \frac{1}{2}(0.1)\left[1 + \frac{2}{1.1} + \frac{2}{1.2} + \frac{2}{1.3} + \frac{2}{1.4} + \frac{2}{1.5}\right.$$
$$\left. + \frac{2}{1.6} + \frac{2}{1.7} + \frac{2}{1.8} + \frac{2}{1.9} + \frac{1}{2}\right]$$
$$\approx 0.05[1 + 1.8182 + 1.6667 + 1.5385 + 1.4286$$
$$+ 1.3333 + 1.25 + 1.1765 + 1.1111 + 1.0526 + 0.5]$$
$$= 0.05[13.8755] = 0.6938.$$

We can check our calculations by integrating:

$$\int_1^2 \frac{1}{x}\,dx = \ln x \Big|_1^2 = \ln 2 - \ln 1 = \ln 2 \approx 0.6931.$$

We can see that by increasing the number of intervals, we increase the accuracy of our answer. This, of course, is not surprising. However, we are naturally led to ask what kind of accuracy we can expect by using the trapezoidal rule. In general, there are two kinds of errors encountered when we use a numerical method to integrate. The first is the kind we have already encountered—the error obtained by approximating the curve between the points $(x_{i-1}, f(x_{i-1}))$ and $(x_i, f(x_i))$ by the straight line joining those points. Since we now consider the function at a finite or *discrete* number of points, the error incurred by this approximation is called **discretization error**. However, there is another kind of error we will always encounter. As we saw in Example 1, we rounded our calculations to four decimal places. Each such "rounding" led to an error in our calculation. The accumulated effect of this rounding is called **round-off error**. Note that as we increase the number of intervals in our calculation, we improve the accuracy of our approximation to the area under the curve. This, evidently, has the effect of reducing the discretization error. On the other hand, an increase in the number of sub-intervals leads to an increase in the number of computations, which, in turn, leads to an increase in the accumulated round-off error. In fact, there is a delicate balance between these two types of errors, and often there is an "optimal" number of intervals to be chosen so as to minimize the total error. Round-off error depends on the type of device used for the computations (pencil and paper, hand calculator, computer, etc.) and will not be discussed further here. However, we can give a formula for estimating the discretization error incurred in using the trapezoidal rule.

Let the sum in (5) be denoted by T and let $\epsilon_n{}^T$ denote the discretization error:

$$\epsilon_n{}^T = \int_a^b f(x)\,dx - T$$

when n subintervals are used.

THEOREM 1 **DISCRETIZATION ERROR OF THE TRAPEZOIDAL APPROXIMATION**

Let f be continuous on $[a, b]$ and suppose that f' and f'' exist on $[a, b]$. Then there is a number c in (a, b) such that

$$\epsilon_n{}^T = -\frac{(b - a)}{12}(\Delta x)^2 f''(c). \quad \blacksquare \tag{7}$$

The proof of this theorem is beyond the scope of this book but it can be found in most standard numerical analysis texts, including the one cited earlier in this section.

COROLLARY 1 **ERROR BOUND FOR TRAPEZOIDAL RULE**

If $|f''(x)| \leq M$ for all x in $[a, b]$, then

$$|\epsilon_n{}^T| \leq M\frac{(b - a)^3}{12n^2}. \tag{8}$$

PROOF: From (7),

$$|\epsilon_n{}^T| = \frac{b - a}{12}\Delta x^2 |f''(c)| \leq \frac{b - a}{12}\left(\frac{b - a}{n}\right)^2 M = \frac{M(b - a)^3}{12n^2} \quad \blacksquare$$

REMARK: The expression in (8) gives the maximum possible value of the error. Often, as in the next example, the actual error will be quite a bit less. You may think of the error bound as a *guarantee* that the error will not be any greater. Accurate estimates of the error are more difficult to obtain.

EXAMPLE 2
FINDING A BOUND ON THE ERROR USING THE TRAPEZOIDAL RULE

Find a bound on the discretization error incurred when estimating $\int_1^2 (1/x)\,dx$ by using the trapezoidal rule with n subintervals.

SOLUTION: $f(x) = 1/x$, $f'(x) = -1/x^2$, and $f''(x) = 2/x^3$. Hence, $f''(x)$ is bounded above by 2 for x in $[1, 2]$. Then from (8),

$$|\epsilon_n{}^T| \leq \frac{2(2 - 1)^3}{12n^2} = \frac{1}{6n^2}.$$

For example, for $n = 5$ we calculated $\int_1^2 (1/x)\,dx \approx 0.6956$. Then the actual error is

$$\epsilon_n{}^T \approx 0.6931 - 0.6956 = -0.0025.$$

This result compares with a maximum possible error of $1/6n^2 = 1/(6 \cdot 25) = \frac{1}{150} \approx 0.0067$. For $n = 10$ the actual error is

$$\epsilon_{10}{}^T \approx 0.6931 - 0.6938 = -0.0007.$$

This result compares with a maximum possible error of $1/6n^2 = \frac{1}{600} \approx 0.0017$. Hence we see, in this example at least, that the error bound in (8) is a crude estimate of the actual error. Nevertheless, even this crude bound allows us to estimate the accuracy of our calculation in the cases where we cannot check our answers by computing an antiderivative. Of course, these are the only cases of interest since we would not use a numerical technique if we could calculate the answer exactly.

EXAMPLE 3

USING THE TRAPEZOIDAL RULE TO APPROXIMATE AN INTEGRAL WE DO NOT KNOW

Use the trapezoidal rule to estimate $\int_0^2 e^{x^2}\,dx$ with a maximum error of 1.

SOLUTION: We must choose n large enough so that $|\epsilon_n{}^T| \leq 1$. For $f(x) = e^{x^2}$, we have $f'(x) = 2xe^{x^2}$ and $f''(x) = (2 + 4x^2)e^{x^2}$. Since this function is an increasing function, its maximum over the interval $[0, 2]$ occurs at 2. Then $M = f''(2) = 18e^4 \approx 983$. Hence from (8),

$$|\epsilon_n{}^T| \leq \frac{M(b - a)^3}{12n^2} \leq \frac{(983)2^3}{12n^2} \approx \frac{655}{n^2}.$$

We need $655/n^2 \leq 1$, or $n^2 \geq 655$, or $n \geq \sqrt{655}$. The smallest n that meets this requirement is $n = 26$. Hence, we use the trapezoidal rule with $n = 26$ and $\Delta x = (b - a)/n = \frac{2}{26} = \frac{1}{13}$. We have $x_0 = 0$, $x_1 = \frac{1}{13}$, $x_2 = \frac{2}{13}$, ..., $x_{25} = \frac{25}{13}$, and $x_{26} = \frac{26}{13} = 2$. Then,

$$\int_0^2 e^{x^2}\,dx \approx \frac{1}{2} \cdot \frac{1}{13}\left[e^0 + 2e^{(1/13)^2} + 2e^{(2/13)^2} + \cdots\right.$$

$$\left. + 2e^{(25/13)^2} + e^{(26/13)^2}\right]$$

$$\approx \frac{1}{26}(1 + 2.012 + 2.048 + 2.109 + 2.199 + 2.319 + 2.475$$

$$+ 2.673 + 2.921 + 3.230 + 3.614 + 4.092 + 4.689$$

$$+ 5.437 + 6.378 + 7.572 + 9.097 + 11.059 + 13.603$$

$$+ 16.933 + 21.328 + 27.184 + 35.060 + 45.756$$

$$+ 60.427 + 80.751 + 54.598)$$

$$= \frac{1}{26}(430.564) \approx 16.560.^\dagger$$

This answer is correct to within 1 unit. The next method we discuss will enable us to calculate this integral with greater accuracy and less work.

$\dagger$ Values of the function $\int_0^x e^{t^2}\,dt$ have been tabulated. To six decimal places, the correct value of $\int_0^2 e^{t^2}\,dt$ is 16.452627. Thus, our answer is actually correct to within 0.11.

SIMPSON'S RULE

We now derive a second method for estimating a definite integral. Look at the three sketches in Figure 3. In Figure 3(a), the area under the curve $y = f(x)$ over the interval $[x_i, x_{i+2}]$ is approximated by rectangles, where the height of each rectangle is the value of the function at an endpoint of an interval. In Figure 3(b), we have depicted the trapezoidal approximation to

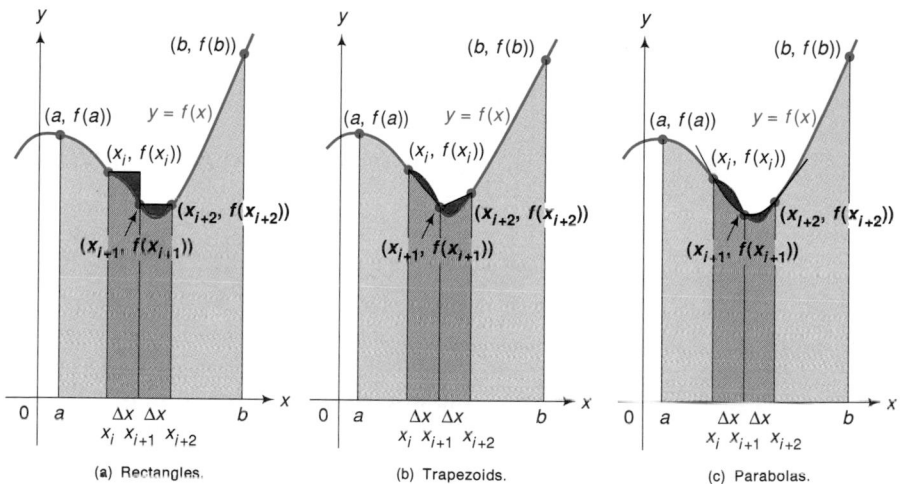

FIGURE 3   APPROXIMATING AREA USING RECTANGLES, TRAPEZOIDS, AND PARABOLAS

this area. The "top" of the first trapezoid is the straight line joining the *two* points $(x_i, f(x_i))$ and $(x_{i+1}, f(x_{i+1}))$ and the "top" of the second trapezoid is given in an analogous manner. In Figure 3(c), we are approximating the required area by drawing a figure whose "top" is the parabola passing through the *three* points $(x_i, f(x_i))$, $(x_{i+1}, f(x_{i+1}))$, and $(x_{i+2}, f(x_{i+2}))$. As we will see, this method will give us a better approximation to the area under the curve. First, we need to calculate the area depicted in Figure 3c.

THEOREM 2

The area A_{i+2} bounded by the parabola[†] passing through the points $(x_i, f(x_i))$, $(x_{i+1}, f(x_{i+1}))$, and $(x_{i+2}, f(x_{i+2}))$, the lines $x = x_i$ and $x = x_{i+2}$, and the x-axis (where $x_{i+1} - x_i = x_{i+2} - x_{i+1} = \Delta x$) is given by

$$A_{i+2} = \frac{1}{3}\,\Delta x[f(x_i) + 4f(x_{i+1}) + f(x_{i+2})]. \quad \blacksquare \tag{9}$$

The proof is left as an exercise (see Problem 37).

[†] In Figure 3(c), we assume that f is positive over the interval $[x_i, x_{i+2}]$. The method we are about to develop does *not* require that f be positive. However, the method is easier to motivate if we make this assumption.

Let n be an *even* number and suppose that the interval $[a, b]$ is divided into n subintervals of equal lengths $\Delta x = (b - a)/n$. Then from (9), we have

$$\int_a^b f(x)\, dx \approx A_2 + A_4 + A_6 + \cdots + A_n$$

$$= \frac{1}{3}\Delta x[f(x_0) + 4f(x_1) + f(x_2)]$$

$$+ \frac{1}{3}\Delta x[f(x_2) + 4f(x_3) + f(x_4)]$$

$$+ \cdots + \frac{1}{3}\Delta x[f(x_{n-2}) + 4f(x_{n-1}) + f(x_n)],$$

or

$$\int_a^b f(x)\, dx \approx \frac{1}{3}\Delta x[f(x_0) + 4f(x_1) + 2f(x_2) + 4f(x_3) + 2f(x_4)$$

$$+ \cdots + 2f(x_{n-2}) + 4f(x_{n-1}) + f(x_n)]. \tag{10}$$

The approximation in (10) is called **Simpson's rule**[†] (or the **parabolic rule**) for approximating a definite integral. From (10), we see that there is a bit more work needed to estimate an integral by using Simpson's rule than by using the trapezoidal rule with the same number of subintervals. However, the discretization error in Simpson's rule is usually a good deal less, as is suggested in the following theorem, whose proof can be found in the text cited earlier in this section.

REMARK: It is important to note that in the formula for Simpson's rule, n must be an even number.

THEOREM 3 **DISCRETIZATION ERROR OF SIMPSON'S RULE**

Let f be continuous on $[a, b]$ and suppose that f', f'', f''', and $f^{(4)}$ all exist on $[a, b]$. Then the discretization error ϵ_n^S of Simpson's rule (10), using n equally spaced subintervals of length $\Delta x = (b - a)/n$, where n is an even number, is given by

$$\epsilon_n^S = -\frac{(b - a)}{180}(\Delta x)^4 f^{(4)}(c), \tag{11}$$

where c is some number in the open interval (a, b). ∎

COROLLARY 2 **ERROR BOUND FOR SIMPSON'S RULE**

If $|f^{(4)}(x)| \leq M$ for all x in $[a, b]$, then

$$|\epsilon_n^S| \leq \frac{M(b - a)^5}{180n^4}. \tag{12}$$

[†] Named after the British mathematician Thomas Simpson (1710–1761), who published the result in his *Mathematical Dissertations on Physical and Analytical Subjects* in 1743.

PROOF:

$$\left|\epsilon_n{}^S\right| = \frac{b-a}{180}(\Delta x)^4\left|f^{(4)}(c)\right| \le \left(\frac{b-a}{180}\right)\left(\frac{b-a}{n}\right)^4 M$$

$$\le \frac{M(b-a)^5}{180n^4}. \quad \blacksquare$$

EXAMPLE 4
USING SIMPSON'S RULE TO
APPROXIMATE AN INTEGRAL WE
ALREADY KNOW

Use Simpson's rule to estimate $\int_1^2 (1/x)\,dx$ by using 10 subintervals. What is the maximum error in your estimate? Compare this error with the exact answer of ln 2.

SOLUTION: Here we have $\Delta x = \frac{1}{10}$ and $n = 10$, so that from (10)

$$\int_1^2 \frac{1}{x}\,dx$$

$$\approx \frac{1}{30}\left(\frac{1}{1} + \frac{4}{1.1} + \frac{2}{1.2} + \frac{4}{1.3} + \frac{2}{1.4} + \frac{4}{1.5} + \frac{2}{1.6} + \frac{4}{1.7} + \frac{2}{1.8} + \frac{4}{1.9} + \frac{1}{2}\right)$$

$$= \frac{1}{30}(1 + 3.636364 + 1.666667 + 3.076923 + 1.428571 + 2.666667$$

$$+ 1.25 + 2.352941 + 1.111111 + 2.105263 + 0.5)$$

$$= \frac{1}{30}(20.794507) \approx 0.693150.$$

To six decimal places, ln 2 = 0.693147, so our answer is very accurate indeed. To calculate the maximum possible error, we first need to calculate $f^{(4)}$. But $f(x) = 1/x$, $f'(x) = -1/x^2$, $f''(x) = 2/x^3$, $f'''(x) = -6/x^4$, and $f^{(4)}(x) = 24/x^5$. Over the interval $[1, 2]$, $|24/x^5| \le 24$, so that $M = 24$. Then we use formula (12) with $M = 24$ and $n = 10$ to obtain

$$\left|\epsilon_{10}{}^S\right| \le \frac{24}{(180)(10^4)} = \frac{24}{1,800,000} \approx 0.0000133.$$

Our actual error is $0.693150 - 0.693147 = 0.000003$, which is about one-fourth the maximum possible discretization error. Notice that in this example Simpson's rule gives a far more accurate answer than the trapezoidal rule using the same number of subintervals (10).

EXAMPLE 5
USING SIMPSON'S RULE TO
APPROXIMATE AN INTEGRAL WE
DO NOT KNOW

Use Simpson's rule to estimate $\int_0^2 e^{x^2}\,dx$ with a maximum error of 0.1.

SOLUTION: If $f(x) = e^{x^2}$, we have already calculated (in Example 3) that $f''(x) = (2 + 4x^2)e^{x^2}$. Then $f'''(x) = (12x + 8x^3)e^{x^2}$ and $f^{(4)}(x) = (12 + 48x^2 + 16x^4)e^{x^2}$. This is an increasing function for x in $[0, 2]$, so that $M = f^{(4)}(2) = 460e^4 \approx 25{,}115$. Since $(b - a)^5 = 2^5 = 32$, we must choose n such that

$$\left|\epsilon_n{}^S\right| \le \frac{M(b-a)^5}{180n^4} \approx \frac{(25{,}115)(32)}{180n^4} \approx \frac{4465}{n^4} < 0.1 = \frac{1}{10}.$$

We then need $n^4 > 44{,}650$, or $n > 44{,}650^{1/4} \approx 14.54$. The smallest even integer n that satisfies this inequality is $n = 16$. Thus to obtain the required accuracy, we use Simpson's rule with $n = 16$ subintervals. Then

$\Delta x = (b - a)/n = \frac{2}{16} = \frac{1}{8}$, and we have

$$\int_0^2 e^{x^2}\,dx \approx \frac{1}{3} \cdot \frac{1}{8}\left(e^0 + 4e^{(1/8)^2} + 2e^{(2/8)^2} + 4e^{(3/8)^2} + \cdots\right.$$

$$\left. + 2e^{(14/8)^2} + 4e^{(15/8)^2} + e^{(16/8)^2}\right)$$

$$\approx \frac{1}{24}(1 + 4.0630 + 2.1290 + 4.6040 + 2.5681$$

$$+ 5.9116 + 3.5101 + 8.6014 + 5.4366$$

$$+ 14.1812 + 9.5415 + 26.4940 + 18.9755$$

$$+ 56.0879 + 42.7619 + 134.5478 + 54.5982)$$

$$\approx \frac{1}{24}(395.0118) \approx 16.4588.$$

This answer is correct to within one-tenth of a unit.[†] Notice how in the calculation of $\int_0^2 e^{x^2}\,dx$, Simpson's rule give us more accuracy with fewer calculations than the trapezoidal rule does.

EXAMPLE 6

USING SIMPSON'S RULE TO APPROXIMATE AN INTEGRAL WE DO NOT KNOW

Use Simpson's rule to approximate

$$\int_0^{\pi/4} \cos x^2\,dx \qquad \text{using 10 subintervals.}$$

What is the maximum error in this approximation?

SOLUTION: We have $n = 10$ so $\Delta x = \dfrac{b - a}{n} = \dfrac{\pi/4}{10} = \dfrac{\pi}{40}$ and the partition points are

$$0 < \frac{\pi}{40} < \frac{2\pi}{40} < \frac{3\pi}{40} < \frac{4\pi}{40} < \frac{5\pi}{40} < \frac{6\pi}{40} < \frac{7\pi}{40} < \frac{8\pi}{40} < \frac{9\pi}{40} < \frac{10\pi}{40} = \frac{\pi}{4}.$$

Then (10) becomes (after making certain that the calculator is set to radian mode)

$$\int_0^{\pi/4} \cos x^2\,dx = \frac{1}{3}\frac{\pi}{40}\left[\cos 0 + 4\cos\left(\frac{\pi}{40}\right)^2 + 2\cos\left(\frac{\pi}{20}\right)^2 \right.$$

$$+ 4\cos\left(\frac{3\pi}{40}\right)^2 + 2\cos\left(\frac{\pi}{10}\right)^2 + 4\cos\left(\frac{\pi}{8}\right)^2$$

$$+ 2\cos\left(\frac{3\pi}{20}\right)^2 + 4\cos\left(\frac{7\pi}{40}\right)^2 + 2\cos\left(\frac{\pi}{5}\right)^2$$

$$\left. + 4\cos\left(\frac{9\pi}{40}\right)^2 + \cos\left(\frac{\pi}{4}\right)^2\right]$$

$$= \frac{\pi}{120}[1 + 3.999923899 + 1.999391224$$

$$+ 3.993837414 + 1.990266995$$

$$+ 3.952531153 + 1.950888965 + 3.818668704$$

$$+ 1.846159187 + 3.511003676 + 0.815704517]$$

$$= (0.026179938)(28.87837573) = 0.756034108.$$

[†] Since the correct value is 16.452627 (correct to six decimal places), our answer is really correct to within 0.007 units.

We leave it to you to verify that if $f(x) = \cos x^2$, then

$$f^{(4)}(x) = 48x^2 \sin x^2 + (16x^4 - 12) \cos x^2$$

and

$$f^{(5)}(x) = 160x^3 \cos x^2 + x(120 - 32x^4) \sin x^2.$$

Observe the following:

if $0 \leq x \leq \dfrac{\pi}{4} \approx 0.784$, then $x^4 < 1$ and $32x^4 < 32$

then

$$120 - 32x^4 > 120 - 32 = 88.$$

Also, if $0 \leq x \leq \dfrac{\pi}{4}$, then $0 \leq x^2 \leq \dfrac{\pi^2}{16} \approx 0.62 < \dfrac{\pi}{4}$, so, in $\left(0, \dfrac{\pi}{4}\right)$, $\cos x^2 > 0$ and $\sin x^2 > 0$. Thus,

$$f^{(5)}(0) = 0 \qquad \text{and} \qquad f^{(5)}(x) > 0 \qquad \text{if } 0 < x \leq \dfrac{\pi}{4}.$$

Since $f^{(5)}$ is the derivative of $f^{(4)}$, we see that $f^{(4)}$ is an increasing function in $\left(0, \dfrac{\pi}{4}\right)$. Now $f^{(4)}(0) = -12$ and

$$f^{(4)}\left(\frac{\pi}{4}\right) = 48\left(\frac{\pi}{4}\right)^2 \sin\left(\frac{\pi}{4}\right)^2 + \left[16\left(\frac{\pi}{4}\right)^4 - 12\right] \cos\left(\frac{\pi}{4}\right)^2 \approx 12.31.$$

Thus $f^{(4)}$ increases from -12 to 12.31 in $\left[0, \dfrac{\pi}{4}\right]$ and

$$M = \text{maximum value of } \left|f^{(4)}(x)\right| \text{ for } 0 \leq x \leq \frac{\pi}{4} \approx 12.31.$$

From (12), with $n = 10$, we obtain

$$\left|\epsilon_{10}{}^S\right| \leq \frac{12.31\left(\dfrac{\pi}{4}\right)^5}{180 \cdot 10^4} \approx 0.000002043.$$

This implies that

$$0.756034108 - 0.000002043 < \int_0^{\pi/4} \cos x^2 \, dx$$
$$< 0.756034108 + 0.000002043$$

or

$$0.756032064 < \int_0^{\pi/4} \cos x^2 \, dx < 0.756036151.$$

We can conclude that, to 4 decimal places,

$$\int_0^{\pi/4} \cos x^2 \, dx = 0.7560.$$

Using a technique that we discuss in Section 10.9 (see Example 9 in Section 10.9), we can show that, to 10 decimal places,

$$\int_0^{\pi/4} \cos x^2 \, dx = 0.756035278.$$

Thus the actual error in our approximation is 0.00000117, which is about 57% of the maximum possible error of 0.000002043.

There are many other methods that can be used to approximate definite integrals. For example, there are methods in which the points x_0, $x_1, \ldots, x_n$ are *not* equally spaced. One such method is called **Gaussian quadrature**. We will not discuss this very useful method here except to note that it can be found in any introductory book on numerical analysis. Finally, as we will discuss in Chapter 10, techniques using infinite series can be used to approximate certain definite integrals.

We conclude by noting that every definite integral that is known to exist can be evaluated to any number of decimal places of accuracy if one is supplied with an appropriate calculating tool. The problems at the end of this section can all be done reasonably quickly by using a scientific hand calculator. For more accuracy, it may be necessary to evaluate a function at hundreds, or even thousands, of points. This problem is a manageable one only if you have access to a high-speed computer or a programmable calculator. If, in fact, you do have such access, you should write a computer program to estimate an integral using Simpson's rule and then use it to calculate each of the integrals in the problem set to at least six decimal places of accuracy.

■ PROBLEMS 7.9

■ SELF-QUIZ

I. $\int_3^5 \sin x \, dx \approx$ _____.

   a. -8.63 b. -1.27 c. 0.08 d. 3.14

II. $\int_0^1 \sqrt{1 + x^3} \, dx \approx$ _____.

   a. -0.900 b. 0.841 c. 1.111 d. 5.073

III. $\int_0^1 \sqrt{1 - x^3} \, dx \approx$ _____.

   a. -0.900 b. 0.841 c. 1.111 d. 5.073

IV. $\int_0^1 e^{-x^2/2} \, dx \approx$ _____.

   a. -0.341 b. 0.856 c. 1.121 d. 7.389

V. $\int_0^1 e^{x^{10}} \, dx \approx$ _____.

   a. -0.341 b. 0.856 c. 1.121 d. 7.389

▦ CALCULATOR PROBLEMS

In Problems 1–10, you will get some practice with numerical methods by applying them in a situation where you have an alternative, exact way to obtain a numerical value. For each problem, do the following:

a. Approximate the given definite integral by the trapezoidal rule using the specified number of subintervals of the interval of integration.

b. Approximate the definite integral by Simpson's rule using the same subintervals (and function values) as for part (a).

c. Use Corollary 1 and inequality (8) to obtain an upper bound for the error of the trapezoidal approximation.

d. Use Corollary 2 and inequality (12) to obtain an upper bound for the error of the approximation using Simpson's rule.

CONTINUED

e. Calculate the integral exactly.

f. Compare the actual errors in your approximations with upper bounds found in parts (c) and (d).

[*Note:* If you write a computer program to do the calculations requested for these problems, it would be a good idea to validate your program by having it repeat the calculations for some examples in this section. Do that before depending on it as a tool in the following problems.]

1. $\int_0^1 x \, dx$; 4 subintervals

2. $\int_{-2}^2 x \, dx$; 6 subintervals

3. $\int_0^1 x^2 \, dx$; 4 subintervals

4. $\int_1^2 \frac{1}{x^2} \, dx$; 6 subintervals

5. $\int_0^1 e^x \, dx$; 4 subintervals

6. $\int_0^2 e^x \, dx$; 6 subintervals

7. $\int_0^{\pi/2} \cos x \, dx$; 4 subintervals

8. $\int_0^{\pi/3} \tan x \, dx$; 8 subintervals

9. $\int_1^e \ln x \, dx$; 6 subintervals

10. $\int_0^1 \frac{1}{1 + x^2} \, dx$; 10 subintervals

In Problems 11–22, approximate the given definite integral by using both the trapezoidal rule and Simpson's rule with the specified number of subintervals.

11. $\int_0^1 \sqrt{x + x^2} \, dx$; 4 subintervals

12. $\int_1^3 \frac{x^2}{\sqrt[3]{1 + x}} \, dx$; 10 subintervals

13. $\int_1^2 \frac{\sin x}{x} \, dx$; 6 subintervals

14. $\int_1^3 \frac{x}{\sin x} \, dx$; 6 subintervals

15. $\int_0^1 e^{\sqrt{x}} \, dx$; 6 subintervals

16. $\int_0^\pi \sin(x^2) \, dx$; 8 subintervals

17. $\int_{-1}^1 e^{-x^2} \, dx$; 10 subintervals

18. $\int_0^1 e^{x^3} \, dx$; 8 subintervals

19. $\int_0^1 \sqrt{1 + x^3} \, dx$; 10 subintervals

20. $\int_0^1 \frac{1}{\sqrt{1 + x^3}} \, dx$; 10 subintervals

21. $\int_0^1 \ln(1 + e^x) \, dx$; 8 subintervals

22. $\int_1^2 \sqrt{\ln x} \, dx$; 10 subintervals

In Problems 23–28, compute close upper bounds for the discretization errors made by using the trapezoidal rule and by using Simpson's rule for

23. the integral of Problem 17.
24. the integral of Problem 16.
25. the integral of Problem 19.
26. the integral of Problem 20.
27. the integral of Problem 21.
28. the integral of Problem 22.

■ **APPLICATIONS**

29. The integral

$$\frac{1}{\sqrt{2\pi}} \int_{-a}^a e^{-x^2/2} \, dx$$

is very important in probability theory. Using Simpson's rule, approximate that integral for $a = 1$ with an error less than 0.001. [*Hint:* Begin by showing that $\int_{-a}^a e^{-x^2/2} \, dx = 2 \int_0^a e^{-x^2/2} \, dx$.]

30. Approximate $\frac{1}{\sqrt{2\pi}} \int_{-5}^5 e^{-x^2/2} \, dx$ with an error less than 0.001.

*31. a. Approximate $\frac{1}{\sqrt{2\pi}} \int_{-50}^{50} e^{-x^2/2} \, dx$ with an error less than 0.1.

b. Guess what happens to $\frac{1}{\sqrt{2\pi}} \int_{-N}^N e^{-x^2/2} \, dx$ as $N \to \infty$.

32. We know that

$$\int_0^1 \frac{dx}{1 + x^2} = \tan^{-1} x \Big|_0^1 = \frac{\pi}{4}.$$

CONTINUED

Therefore,

$$\pi = 4 \int_0^1 \frac{dx}{1 + x^2}.$$

a. With Simpson's rule, how many subintervals of $[0, 1]$ does it take to approximate π with an error less than 0.0001?
b. Using this number of subintervals, give your approximation of π.

33. How many subintervals of $[1, 2]$ would it take to approximate $\ln 2$ with an error less than 10^{-10} by using the trapezoidal rule applied to the integral $\int_1^2 (1/x)\,dx$.

34. How many subintervals would it take to perform the approximation of the preceding problem by using Simpson's rule?

*35. The function

$$J_{1/2}(x) = \sqrt{\frac{2}{\pi x}}\, \sin x$$

is called the **Bessel function of order** $\frac{1}{2}$; it occurs frequently in physics and engineering. Approximate $\int_{1/2}^1 J_{1/2}(x)\,dx$ with an error less than 0.01.

■ SHOW/PROVE/DISPROVE

36. Show that Simpson's rule provides the exact answer for $\int_a^b P(x)\,dx$ if $P(x)$ is a polynomial of degree three or less.

37. Prove Theorem 2 by carrying out the following steps:
   a. Write the parabola in the form $ax^2 + bx + c$.
   b. Using the fact that $x_{i+2} = x_i + 2\,\Delta x$, show that (with a, b, and c as in part (a))

$$A_{i+2} = \frac{1}{3}\,\Delta x[a(6x_i{}^2 + 12x_i\,\Delta x + 8\,\Delta x^2)$$
$$+ b(6x_i + 6\,\Delta x) + 6c].$$

   c. Using the fact that the parabola passes through the points in Figure 3(c), show that

$$f(x_i) = ax_i{}^2 + bx_i + c$$
$$f(x_{i+1}) = a(x_i{}^2 + 2x_i\,\Delta x + \Delta x^2)$$
$$+ b(x_i + \Delta x) + c$$
$$f(x_{i+2}) = a(x_i{}^2 + 4x_i\,\Delta x + 4\,\Delta x^2)$$
$$+ b(x_i + 2\,\Delta x) + c.$$

   d. Use the results of parts (b) and (c) to show that

$$A_{i+2} = \frac{1}{3}\,\Delta x[f(x_i) + 4f(x_{i+1}) + f(x_{i+2})].$$

■ CHALLENGE

38. Show that a Simpson's rule approximation can be obtained as the weighted average of two trapezoidal rule approximations. More specifically, if T_0 is a trapezoidal rule approximation using subintervals of length h, and if T_1 is a similar trapezoidal rule approximation using subintervals of length $h/2$, then $(4 \cdot T_1 - T_0)/3$ equals the Simpson's rule approximation, which also uses subintervals of length $h/2$.[†]

■ ANSWERS TO SELF-QUIZ

I. b   II. c   III. b
IV. b   V. c

[†] To pursue these notions further, examine a book on numerical analysis for its discussions of Romberg integration, Richardson extrapolation, and adaptive quadrature.

■ PROBLEMS CHAPTER 7 REVIEW

■ SELF-QUIZ

I. By quickly discarding blatantly wrong answers, pick the right answer: $\int xe^{-x}\,dx =$ _____ $+\,C$.
   a. x^{-e+1}
   b. $-\frac{1}{2}e^{-x^2}$
   c. $-(x+1)e^{-x}$
   d. $\frac{1}{2}x^2 - e^{-x}$

II. Pick the right answer:
   $\int x\sin x\,dx =$ _____ $+\,C$.
   a. $-\frac{1}{2}\cos x^2$
   b. $\sin x - x\cos x$
   c. $\cos x + x\tan x$
   d. $\frac{1}{2}x^2 \sin x - x\cos x$

III. Pick the right answer: $\int \sin^3 x\cos^2 x\,dx =$ _____ $+\,C$.
   a. $\frac{1}{5}\cos^5 x - \frac{1}{3}\cos^3 x$
   b. $\frac{1}{4}\sin^3 x\cos^2 x - \frac{1}{3}\cos^3 x\sin^3 x$
   c. $\sin^4 t - \sin^6 t$
   d. $\frac{1}{4}\sin^4 x - \frac{1}{6}\sin^6 x$

IV. Which of the following substitutions does **not** simplify
   $$\int \frac{x}{9 + 16x^2}\,dx?$$
   a. $x = \frac{3}{4}\tan\theta$
   b. $u = 9 + 16x^2$
   c. $x = \frac{3}{4}\sec\theta$
   d. $x = \frac{3}{4}\cot\theta$

V. If $u = \tan(x/2)$, then $\sin x =$ _____.
   a. $\dfrac{2u}{1 + u^2}$
   b. $\dfrac{1 + u^2}{1 - u^2}$
   c. $\dfrac{2}{1 + u^2} - 1$
   d. $\dfrac{1 - u^2}{1 + u^2}$

VI. Pick the right answer:
   $$\int \frac{x^2 + 4x + 5}{(x+1)(x+3)}\,dx =$$ _____ $+\,C$.
   a. $1 + \dfrac{1}{x+1} - \dfrac{1}{x+3}$
   b. $x + \ln\left|\dfrac{x+1}{x+3}\right|$
   c. $e^x \cdot \dfrac{x+1}{x+3}$
   d. $\tan^{-1}\left(\dfrac{x+2}{\sqrt{3}}\right)$

VII. Pick the right answer:
   $$\int \frac{dx}{x^2(1 + x^2)} =$$ _____ $+\,C$.
   a. $\dfrac{1}{x^2} - \dfrac{1}{1 + x^2}$
   b. $\dfrac{-2}{x^3} + e^{1+x^2}$
   c. $\dfrac{-1}{x} - \tan^{-1} x$
   d. $\tan^{-1} x$

VIII. $\int_0^1 e^{-x^2}\,dx \approx$ _____.
   a. 0
   b. -3.4103
   c. 17.0058
   d. 0.7468

■ DRILL

In Problems 1–86, use one or more of the techniques of this chapter to evaluate the given integral. Do **not** use the integral tables for these problems.

1. $\int_0^1 \dfrac{x^2}{\sqrt{4 + x^2}}\,dx$

2. $\int x\sqrt{x + 2}\,dx$

3. $\int \dfrac{x^2 + 3}{(x^2 + 1)^2}\,dx$

4. $\int \dfrac{dx}{(x - 1)(x + 3)}$

5. $\int_0^{\pi/2} \cos^3 x \sin^2 x\,dx$

6. $\int_0^1 \dfrac{\sqrt{x}}{1 + \sqrt{x}}\,dx$

7. $\int xe^{-2x}\,dx$

8. $\int \sin 4x \cos 5x\,dx$

9. $\int \tan^5 x\,dx$

10. $\int_0^{\pi/4} \sec^2 x \tan^3 x\,dx$

11. $\int_0^1 \dfrac{dx}{1 + \sqrt[3]{x}}$

12. $\int_0^1 (1 + x^2)^{3/2}\,dx$

13. $\int_{\sqrt{2}}^2 \dfrac{x^2}{(x^2 - 1)^{3/2}}\,dx$

14. $\int \dfrac{dx}{\cos x - \sin x}$

15. $\int e^{2x} \sin 2x\,dx$

16. $\int_1^2 x^3 \ln x\,dx$

17. $\int \sin^3 x \cos^3 x\,dx$

18. $\int \sec^4 x \tan^2 x\,dx$

19. $\int \csc^3 x \cot^3 x\,dx$

20. $\int \dfrac{dx}{(x - 1)(x^2 + x + 1)}$

21. $\int \dfrac{x^5 - 1}{x^3 - x}\,dx$

22. $\int \dfrac{x^2}{(x + 1)^2(x - 1)^2}\,dx$

CONTINUED

23. $\int x \sinh x \, dx$

24. $\int \dfrac{dx}{x^2 - 4x + 5}$

25. $\int \dfrac{dx}{x^2 - 4x + 4}$

26. $\int \sin 5x \sin 6x \, dx$

27. $\int \cos 6x \cos 7x \, dx$

28. $\int \dfrac{1 + x + x^2}{(3 + 2x + x^2)^2} \, dx$

29. $\int \dfrac{x^2 + 2}{(x - 3)(x + 4)(x - 5)} \, dx$

30. $\int_0^3 \dfrac{x^2}{\sqrt{9 + x^2}} \, dx$

31. $\int x^3 \sqrt{x^2 - 4} \, dx$

32. $\int x \cos^{-1} x \, dx$

33. $\int \sec^4 x \, dx$

34. $\int \dfrac{x^3}{(x + 1)^4} \, dx$

35. $\int \dfrac{(x + 1)^3}{x^3} \, dx$

36. $\int_0^{\pi/4} \sin x \tan^2 x \, dx$

37. $\int_0^{\pi/2} \sin^6 x \, dx$

38. $\int \sec^4 x \tan^3 x \, dx$

39. $\int_0^1 (1 + x^2)^{-1/2} \, dx$

40. $\int \dfrac{dx}{2 - \cos x}$

41. $\int \dfrac{dx}{1 + 5 \sin x}$

42. $\int_0^1 x^3 (1 - x)^{2/3} \, dx$

43. $\int \dfrac{2x^3}{(1 + x^4)^{4/3}} \, dx$

44. $\int \dfrac{1 - x}{1 + \sqrt{x}} \, dx$

45. $\int \dfrac{x - 2x^3}{\sqrt{2 + 3x}} \, dx$

46. $\int \sin(\ln x) \, dx$

47. $\int e^{-x} \cos\left(\dfrac{x}{3}\right) dx$

48. $\int \sinh x \cosh 2x \, dx$

49. $\int \cosh x \cosh 3x \, dx$

50. $\int \dfrac{dx}{2 \sec x - 1}$

51. $\int x^2 e^{-x} \, dx$

52. $\int x^{3/2} \ln x \, dx$

53. $\int \dfrac{\sin \sqrt{x} \cos \sqrt{x}}{\sqrt{x}} \, dx$

54. $\int \dfrac{dx}{\sqrt{4x^2 - 9}}$

55. $\int \dfrac{dx}{x \sqrt{9 - 4x^2}}$

56. $\int \dfrac{x}{(2x + 3)^2} \, dx$

57. $\int \dfrac{\sqrt{x^2 - 16}}{x^2} \, dx$

58. $\int \dfrac{dx}{x \sqrt{4x - x^2}}$

59. $\int x \sin^{3/4}(2x^2) \cos(2x^2) \, dx$

60. $\int_0^{\pi/2} \cos^6 x \, dx$

61. $\int_0^{\pi/2} \sin^7 x \, dx$

62. $\int_0^{\pi/2} \cos^5 x \, dx$

63. $\int \sinh^2 3x \, dx$

64. $\int x^5 \ln 4x \, dx$

65. $\int x^2 e^{-3x^3} \sin(2x^3) \, dx$

66. $\int \csc^4 2x \cot 2x \, dx$

67. $\int \sec^4 3x \, dx$

68. $\int \operatorname{sech} 4x \, dx$

69. $\int e^{2x} \sinh 3x \, dx$

70. $\int e^{-x} \cosh 5x \, dx$

71. $\int \tanh^2 3x \, dx$

72. $\int \dfrac{dx}{1 + \cos 2x}$

73. $\int \dfrac{dx}{4 - 3 \cos 2x}$

74. $\int \dfrac{dx}{1 - \sin 2x}$

75. $\int \dfrac{dx}{1 + \sin 2x}$

76. $\int x^3 \sin 4x \, dx$

77. $\int x^3 \cos 4x \, dx$

78. $\int x(x^2 - 1)^{7/3} \, dx$

79. $\int \dfrac{x^5}{9 + x^{12}} \, dx$

80. $\int \dfrac{x^5}{9 - x^{12}} \, dx$

81. $\int x^5 \sqrt{9 + x^{12}} \, dx$

82. $\int \dfrac{dx}{x^2 + 4x + 13}$

83. $\int \dfrac{x^2 - 3}{x^3 - 1} \, dx$

84. $\int \dfrac{x^4}{(x - 1)(x^2 + 2)} \, dx$

85. $\int \dfrac{2x - 3}{(x - 1)^2} \, dx$

86. $\int \dfrac{x + 2}{(x^2 + 1)(x^2 + 4)} \, dx$

In Problems 87–106, use the table of integrals in the back of this text to simplify your work in evaluating the given integrals.

87. $\int \dfrac{dx}{16 - x^2}$

88. $\int \dfrac{dx}{\sqrt{25 - 4x^2}}$

89. $\int \dfrac{x}{\sqrt{25 - 9x^2}} \, dx$

90. $\int \dfrac{dx}{x \sqrt{4x^2 - 25}}$

91. $\int \dfrac{9x^2}{\sqrt{16 + 9x^2}} \, dx$

92. $\int \sin^4 3x \, dx$

93. $\int x^3 \cos^{-1} 3x \, dx$

94. $\int x^4 e^{-2x} \, dx$

95. $\int x^2 \tan^{-1} x^3 \, dx$

96. $\int x \sin 2x^2 \sin 3x^2 \, dx$

97. $\int \dfrac{x}{(x - 4)^3} \, dx$

98. $\int \dfrac{x^4}{(x - 4)^3} \, dx$

99. $\int \dfrac{x^3}{x^4 - x^2} \, dx$

100. $\int \dfrac{x}{x^4 - 18x^2 + 81} \, dx$

CONTINUED

101. $\displaystyle\int \frac{dx}{x^4 - 5x^2 + 4}$

103. $\displaystyle\int \frac{x^3 - 1}{x^2 + x + 1}\, dx$

104. $\displaystyle\int \frac{x + 8}{x^2 + 6x - 7}\, dx$

102. $\displaystyle\int \frac{x^3}{x^4 - 8x^2 + 16}\, dx$

*105. $\displaystyle\int_0^\pi \ln \sin \theta\, d\theta$

106. $\displaystyle\int \frac{dx}{x\sqrt{a^2 - x^2}}$

■ **APPLICATIONS**

107. Calculate the area of the region bounded by the curve $y = x^3/(x^2 + 1)^3$, the x- and y-axes, and the line $x = 1$.

108. Compute $\int dx/(1 + e^x)$ by making an appropriate substitution and then integrating via partial fractions.

109. Suppose $a < b$. Evaluate $\int_a^b \sqrt{(x - a)(b - x)}\, dx$. [*Hint:* Think about the graph of $y = \sqrt{(x - a)(b - x)}$.]

*110. Apply the substitution $x = \cot^{-1}(\sinh \theta)$ to find $\int \csc x\, dx$.

▦ **CALCULATOR PROBLEMS**

In Problems 111–116, use the trapezoidal rule (T) or Simpson's rule (S) to approximate the given integral with the specified number of subintervals.

111. $\displaystyle\int_0^1 e^{x^3}\, dx$; (T), 4 subintervals

112. $\displaystyle\int_0^1 e^{x^3}\, dx$; (S), 4 subintervals

113. $\displaystyle\int_0^1 \frac{1}{\sqrt{1 + x^4}}\, dx$; (T), 6 subintervals

114. $\displaystyle\int_0^1 \frac{1}{\sqrt{1 + x^4}}\, dx$; (S), 6 subintervals

115. $\displaystyle\int_0^{\pi/2} \cos \sqrt{x}\, dx$; (S), 6 subintervals

116. $\displaystyle\int_{\pi/6}^{\pi/2} \ln(\sin x)\, dx$; (S), 8 subintervals

117. How many subintervals are needed in Problem 111 to obtain a discretization error less than 0.01?

118. How many subintervals are needed in Problem 112 to obtain a discretization error less than 0.00001?

119. Use Simpson's rule to approximate $\int_1^2 (1/x^2)\, dx$ with an error less than 0.0001. Compare your answer with the actual value of that integral which is easily obtained by the fundamental theorem of calculus.

120. Answer the questions in the preceding problem for the integral $\int_1^2 \ln x\, dx$.

■ **ANSWERS TO SELF-QUIZ**

I. c   II. b   III. a   IV. c
V. a   VI. b   VII. c   VIII. d

🖫 **COMPUTER EXERCISES**

1. a. Show that $\displaystyle\int_0^1 \frac{1}{1 + x^2}\, dx = \frac{\pi}{4}$.

   b. If $f(x) = \dfrac{1}{1 + x^2}$, find the absolute maximum of $|f^{(4)}(x)|$ on the interval $[0, 1]$.

   c. With Simpson's rule, how many subintervals of $[0, 1]$ does it take to approximate $\displaystyle\int_0^1 \frac{1}{1 + x^2}\, dx$ with an error less than 10^{-11}?

CONTINUED

d. Obtain an approximation of $\int_0^1 \frac{1}{1+x^2}\,dx$ with error less than 10^{-11} and use your answer to obtain an approximation of π. Give an estimate of the error in your approximation of π.

2. a. Show that $\int_1^2 \frac{1}{x}\,dx = \ln 2$.

   b. If $f(x) = \frac{1}{x}$, find the absolute maximum of $|f^{(4)}(x)|$ on the interval $[1, 2]$.

   c. With Simpson's rule, how many subintervals of $[1, 2]$ would it take to approximate $\ln 2$ with an error less than 10^{-11}?

   d. Use Simpson's rule to calculate $\int_1^2 \frac{1}{x}\,dx = \ln 2$ with error less than 10^{-11}.

3. Use Simpson's rule to obtain an approximation of $\ln 100$ with an error less than 10^{-11}. You may use either of the following integrals for your approximation. Choose wisely.

   a. $\int_1^{100} \frac{1}{x}\,dx = \ln 100$

   b. $\ln 100 = -\ln \frac{1}{100} = \int_{1/100}^1 \frac{1}{x}\,dx$

4. Explain carefully what procedure you would follow to calculate $\int x^{10} e^{3x}\,dx$ by hand. Without actually doing the calculations, explain what the form of the antiderivative should be. Use a computer algebra system to calculate the integral and compare it with the expected form you proposed above.

5. Explain carefully what procedure you would follow to calculate $\int \frac{x+4}{(x-2)^2(x^2+x+1)(x+3)}\,dx$. Without actually doing the calculations, explain what the form of the antiderivative should be. Use a computer algebra system to calculate the integral and compare it with the expected form you proposed above.

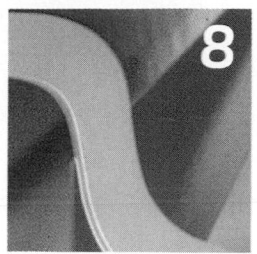

8 CONIC SECTIONS AND POLAR COORDINATES

The period from about 300 to 200 B.C. is known as the Golden Age of Greek mathematics because three of the world's greatest mathematicians lived during that period. The first of these, Euclid, invented much of the geometry that is studied in high schools today. The second, Archimedes, is considered by many to be the finest mathematician of any era. The third of the great Greek mathematicians, Apollonius of Perga, made many of the discoveries that are part of what we now call *analytic geometry*. The exact dates of Apollonius' life are not known, but he is believed to have lived from approximately 260 to 190 B.C.

Like most of the mathematicians of his day, Apollonius was an *applied* mathematician. He studied certain kinds of curves because they arose in practical ways. For example, he applied his work to the analysis of planetary motion and is considered to be the founder of Greek mathematical astronomy.

One of Apollonius' most important discoveries was that four different types of curves are obtained if a right-circular cone is cut by a plane. These curves are circles, ellipses, parabolas, and hyperbolas. Because of the way they are formed, they are called **conic sections**. In Figure 1, we illustrate how the four curves can be obtained.

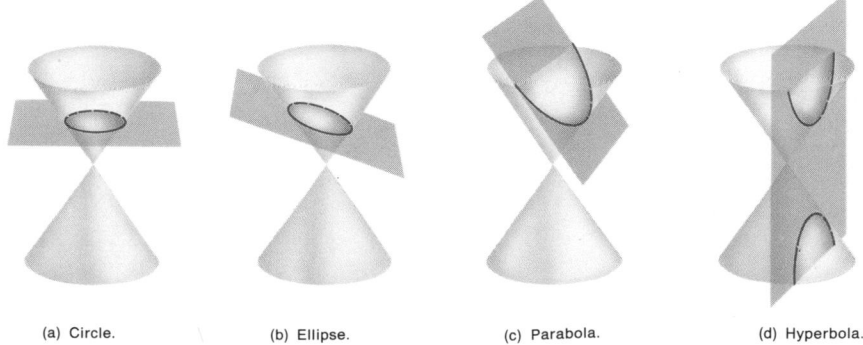

(a) Circle.   (b) Ellipse.   (c) Parabola.   (d) Hyperbola.

FIGURE 1   FOUR CROSS-SECTIONS OF A RIGHT-CIRCULAR CONE

It turns out that each of the four curves can be written in the form

$$Ax^2 + Bxy + Cy^2 + Dx + Ey + F = 0 \tag{1}$$

for certain constants A, B, C, D, E, and F.

We saw special cases of equation (1) earlier in this book. In Section 0.2 (page 13) we found that the equation of a circle centered at (a, b) with radius r is

$$(x - a)^2 + (y - b)^2 = r^2.$$

This equation can be written as

$$x^2 - 2ax + y^2 - 2by + (a^2 + b^2 - r^2) = 0,$$

which is in the form (1). In Example 5 in Section 0.4 (p. 27), we drew the graph of $f(x) = x^2$ and said that this graph was a parabola.

In the following four sections, we shall discuss conic sections in considerable detail. In Sections 8.5–8.7, we shall describe another way to represent points and functions in $\mathbb{R}^2$: the polar-coordinate system.

8.1 THE ELLIPSE AND TRANSLATION OF AXES

DEFINITION **ELLIPSE**

An **ellipse** is the set of points (x, y) such that the sum of the distances from (x, y) to two given points is fixed. Each of the two points is called a **focus** of the ellipse.[†] ∎

EXAMPLE 1
FINDING AN EQUATION OF AN ELLIPSE

Find an equation of the ellipse with foci at $(-2, 0)$ and $(2, 0)$ such that the sum of the distances to the foci is 6.

SOLUTION: As in Figure 1, let (x, y) denote a point on the ellipse. Then

[distance from (x, y) to $(2, 0)$] + [distance from (x, y) to $(-2, 0)$] = 6

or, from the distance formula (equation (2) on page 12),

$$\sqrt{(x - 2)^2 + (y - 0)^2} + \sqrt{(x + 2)^2 + (y - 0)^2} = 6$$

or

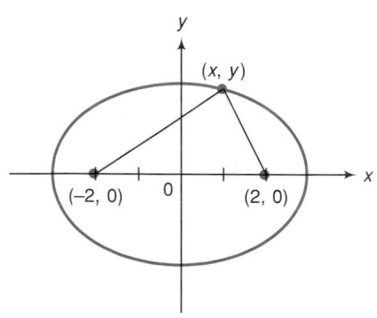

FIGURE 1
ELLIPSE WITH FOCI $(-2, 0)$ AND $(2, 0)$

$$\sqrt{(x - 2)^2 + y^2} = 6 - \sqrt{(x + 2)^2 + y^2}$$

$$(x - 2)^2 + y^2 = 36 - 12\sqrt{(x + 2)^2 + y^2} + (x + 2)^2 + y^2 \quad \text{Square both sides}$$

$$x^2 - 4x + 4 + y^2 = 36 - 12\sqrt{(x + 2)^2 + y^2} + x^2 + 4x + 4 + y^2$$

$$-4x = 36 - 12\sqrt{(x + 2)^2 + y^2} + 4x \quad \text{Subtract } x^2 + y^2 + 4 \text{ from both sides}$$

$$36 + 8x = 12\sqrt{(x + 2)^2 + y^2} \quad \text{Rearrange terms}$$

$$3 + \frac{2}{3}x = \sqrt{(x + 2)^2 + y^2} \quad \text{Divide both sides by 12}$$

$$9 + 4x + \frac{4}{9}x^2 = (x + 2)^2 + y^2 = x^2 + 4x + 4 + y^2 \quad \text{Square again}$$

$$\frac{5}{9}x^2 + y^2 = 5 \quad \text{Combine terms}$$

$$\frac{x^2}{9} + \frac{y^2}{5} = 1. \quad \text{Divide by 5} \tag{2}$$

[†] The plural of *focus* is **foci**. Thus we speak about the two foci of an ellipse.

> This is the **standard equation** of the ellipse with foci at $(-2, 0)$ and $(2, 0)$ and sum of distances equal to 6.

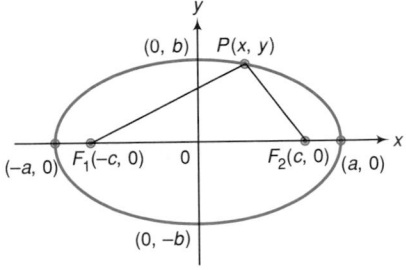

FIGURE 2
ELLIPSE WITH FOCI $(-c, 0)$ AND $(c, 0)$ SUCH THAT THE SUM OF THE DISTANCES FROM A POINT ON THE ELLIPSE TO EACH OF THE FOCI IS $2a$

Suppose now that the foci, F_1 and F_2, of an ellipse are the points $(-c, 0)$ and $(c, 0)$. (See Figure 2.) If $P = (x, y)$ is on the ellipse, then by the definition of the ellipse, the distance from P to the first focus plus the distance from P to the second focus equals $2a$; that is,

$$\overline{F_1P} + \overline{F_2P} = 2a. \tag{3}$$

Since a straight line is the shortest distance between two points, we have

$$\overline{F_1F_2} = 2c < \overline{F_1P} + \overline{F_2P} = 2a.$$

That is,

$$2a > 2c \quad \text{or} \quad a > c$$

($2c$ is the distance between the foci and $2a$ is the given sum of the distances). Then

$$\sqrt{(x + c)^2 + y^2} + \sqrt{(x - c)^2 + y^2} - 2a \qquad \text{From (3)}$$

or

$$\sqrt{(x + c)^2 + y^2} = 2a - \sqrt{(x - c)^2 + y^2}$$

$$(x + c)^2 + y^2 = 4a^2 - 4a\sqrt{(x - c)^2 + y^2} + (x - c)^2 + y^2 \qquad \begin{array}{l}\text{Square} \\ \text{both sides}\end{array}$$

$$x^2 + 2xc + c^2 + y^2 = 4a^2 - 4a\sqrt{(x - c)^2 + y^2} + x^2 - 2xc + c^2 + y^2$$

$$4xc = 4a^2 - 4a\sqrt{(x - c)^2 + y^2} \qquad \begin{array}{l}\text{Subtract } x^2 + c^2 + y^2 \\ \text{from both sides and} \\ \text{add } 2xc \text{ to both sides}\end{array}$$

$$\sqrt{(x - c)^2 + y^2} = a - \frac{c}{a}x \qquad \begin{array}{l}\text{Divide by } 4a \text{ and} \\ \text{rearrange terms}\end{array}$$

$$(x - c)^2 + y^2 = a^2 - 2cx + \frac{c^2}{a^2}x^2. \qquad \text{Square again}$$

Then

$$x^2 - 2xc + c^2 + y^2 = a^2 - 2xc + \frac{c^2}{a^2}x^2$$

$$x^2\left(1 - \frac{c^2}{a^2}\right) + y^2 = a^2 - c^2$$

or

$$x^2\left(\frac{a^2 - c^2}{a^2}\right) + y^2 = a^2 - c^2$$

and after dividing both sides by $a^2 - c^2$, which is positive since $a > c$, we have

$$\frac{x^2}{a^2} + \frac{y^2}{a^2 - c^2} = 1.$$

Finally, we define the positive number b by $b^2 = a^2 - c^2$ to obtain the standard equation of the ellipse.

The **standard equation of the ellipse** is given by

$$\frac{x^2}{a^2} + \frac{y^2}{b^2} = 1, \tag{4}$$

where $a^2 = b^2 + c^2$ and $(c, 0)$ and $(-c, 0)$ are the foci.

Note that this ellipse is symmetric about both the x- and y-axes. Since $a > b$, the line segment from $(-a, 0)$ to $(a, 0)$ is called the **major axis** and the line segment from $(0, -b)$ to $(0, b)$ is called the **minor axis**. The point $(0, 0)$, which is at the intersection of the axes, is called the **center** of the ellipse. The points $(a, 0)$ and $(-a, 0)$ are called the **vertices** of the ellipse. In general, the vertices of an ellipse are the endpoints of the major axis.

REMARK: If $a = b$ in (2), then the equation is the equation of a circle centered at $(0, 0)$ with radius a. But if $a = b$, then $c = 0$, so the two foci are both located at $(0, 0)$. That is, we can think of a circle as an ellipse in which the two foci have collapsed into one. (Look at Figure 2 and imagine the curve obtained as F_1 and F_2 approach each other. This limiting curve is a circle centered at $(0, 0)$.)

EXAMPLE 2
FINDING THE STANDARD EQUATION OF AN ELLIPSE

Find the equation of the ellipse with foci at $(-3, 0)$ and $(3, 0)$ and with $a = 5$.

SOLUTION: $c = 3$ and $a = 5$, so that $b^2 = a^2 - c^2 = 16$, and we obtain

$$\frac{x^2}{25} + \frac{y^2}{16} = 1.$$

The ellipse is sketched in Figure 3.

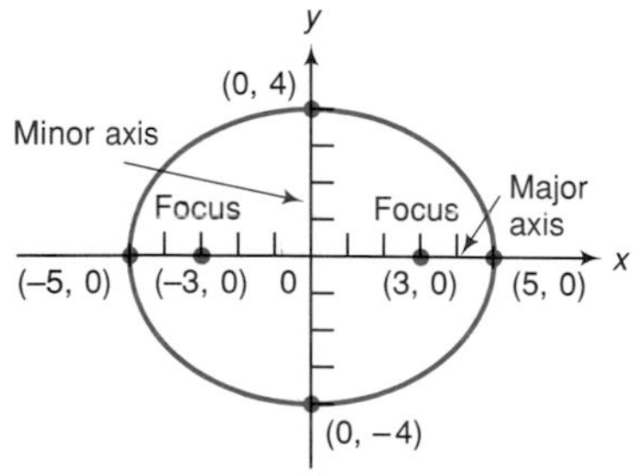

FIGURE 3
THE ELLIPSE $\dfrac{x^2}{25} + \dfrac{y^2}{16} = 1$

We can reverse the roles of x and y in the preceding discussion. Suppose that the foci are at $(0, c)$ and $(0, -c)$ on the y-axis. Then if the fixed sum of the distances is given as $2b$, we obtain, using similar reasoning,

$$\frac{x^2}{a^2} + \frac{y^2}{b^2} = 1 \qquad \text{Standard equation of an ellipse with major axis on the } y\text{-axis}$$

where $a^2 = b^2 - c^2$. Now the major axis is on the y-axis, the minor axis is on the x-axis, and the vertices are $(0, b)$ and $(0, -b)$.

EXAMPLE 3
AN ELLIPSE WITH MAJOR AXIS ON THE y-AXIS

Discuss the curve $9x^2 + 4y^2 = 36$.

SOLUTION: Dividing both sides by 36, we obtain

$$\frac{x^2}{4} + \frac{y^2}{9} = 1.$$

Here $a = 2$, $b = 3$, and the major axis is on the y-axis. Since $c^2 = 9 - 4 = 5$, the foci are at $(0, \sqrt{5})$ and $(0, -\sqrt{5})$. This curve is sketched in Figure 4.

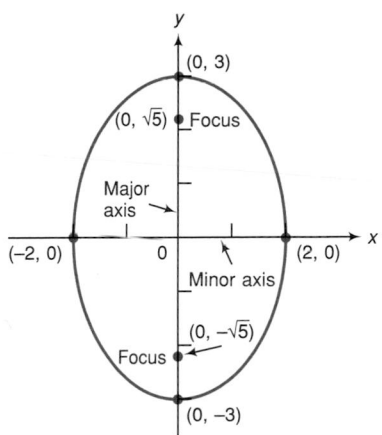

FIGURE 4 THE ELLIPSE $\dfrac{x^2}{4} + \dfrac{y^2}{9} = 1$

EXAMPLE 4
FINDING THE EQUATION OF
AN ELLIPSE WITH GIVEN FOCUS
AND AXIS

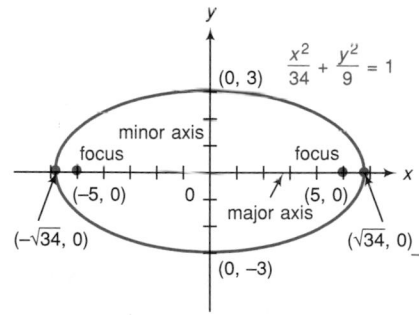

FIGURE 5
THE ELLIPSE WITH FOCI $(-5, 0)$ AND $(5, 0)$
AND MINOR AXIS EXTENDING FROM $(0, -3)$
TO $(0, 3)$

Find the equation of the ellipse with foci $(-5, 0)$ and $(5, 0)$ and whose minor axis is the line segment extending from $(0, -3)$ to $(0, 3)$.

SOLUTION: We have $c = 5$ and $b = 3$, so that

$$a^2 = b^2 + c^2 = 34$$

and the equation of the ellipse is

$$\frac{x^2}{34} + \frac{y^2}{9} = 1.$$

It is sketched in Figure 5.

The **eccentricity** e of an ellipse is defined by

$$e = \frac{c}{a} \qquad \text{if } a \geq b \tag{5}$$

and

$$e = \frac{c}{b} \qquad \text{if } b \geq a. \tag{6}$$

NOTE: If $a \geq b$, then the length of the major axis is $2a$; if $b \geq a$, the length of the major axis is $2b$. In both cases the distance between the foci is $2c$. Since $2c/2a = c/a$ and $2c/2b = c/b$, we have

ECCENTRICITY
$e = \dfrac{\text{distance between foci}}{\text{length of major axis}} = \dfrac{\text{distance between foci}}{\text{distance between vertices}}.$ **(7)**

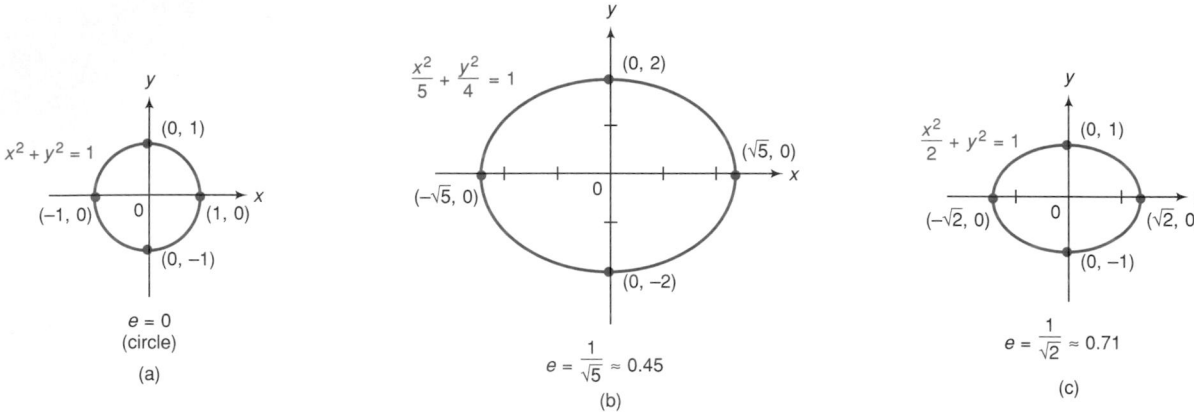

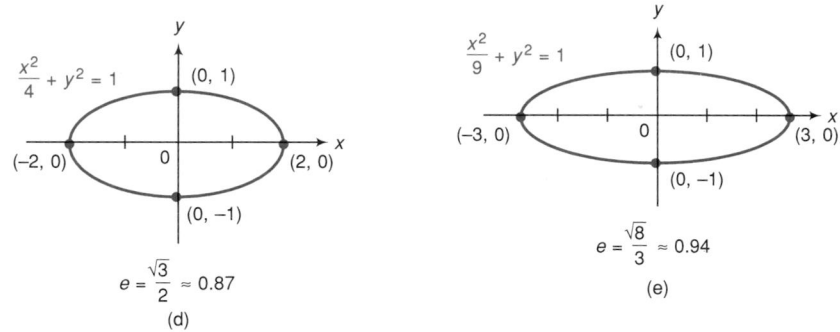

FIGURE 6   *THE LARGER THE ECCENTRICITY, THE FLATTER THE ELLIPSE*

The eccentricity of an ellipse is a measure of the shape of the ellipse and is always a number in the interval $[0, 1]$. If $e = 0$, then the ellipse is a circle, since in that case $c = 0$, so that $a^2 = b^2$ and the foci now coincide and are the center of the circle. As e approaches 1, the ellipse becomes progressively flatter and approaches the major axis: the straight-line segment from $(-a, 0)$ to $(a, 0)$ if $a > b$, and from $(0, -b)$ to $(0, b)$ if $b > a$. In general,

the larger the eccentricity, the flatter the ellipse.

In Example 2, $e = 3/5 = 0.6$. In Example 3, $c^2 = b^2 - a^2 = 5$, so that $e = \sqrt{5}/3 \approx 0.74536$. In Example 4, $e = 5/\sqrt{34} \approx 0.85749$.

In Figure 6, we draw five different ellipses showing how the ellipses get flatter as the eccentricity increases.

TRANSLATION OF AXES

We now turn to a different question. What happens if the center of the ellipse $(x^2/a^2) + (y^2/b^2) = 1$ is shifted from $(0, 0)$ to a new point (x_0, y_0)? Consider the equation

$$\frac{(x - x_0)^2}{a^2} + \frac{(y - y_0)^2}{b^2} = 1. \tag{8}$$

If we define two new variables by

$$x' = x - x_0 \qquad \text{and} \qquad y' = y - y_0,$$

then (8) becomes

$$\frac{(x')^2}{a^2} + \frac{(y')^2}{b^2} = 1.$$

This expression is the equation of an ellipse centered at the origin in the new coordinate system (x', y'). But $(x', y') = (0, 0)$ implies $(x - x_0, y - y_0) = (0, 0)$, or $x = x_0$ and $y = y_0$. That is, equation (8) is the equation of the "shifted" ellipse. We have performed what is called a **translation of axes**. That is, we moved (or **translated**) the x- and y-axes to new positions so that they intersect at the point (x_0, y_0).

EXAMPLE 5
FINDING THE EQUATION OF A TRANSLATED ELLIPSE

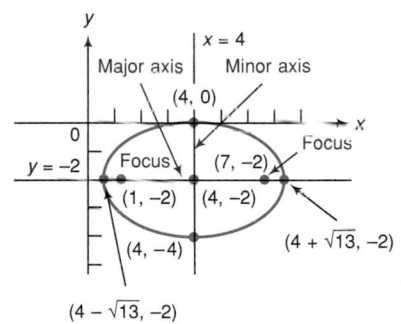

FIGURE 7
THE ELLIPSE $\dfrac{(x-4)^2}{13} + \dfrac{(y+2)^2}{4} = 1$

Find the equation of the ellipse centered at $(4, -2)$ with foci $(1, -2)$ and $(7, -2)$ and minor axis joining the points $(4, 0)$ and $(4, -4)$.

SOLUTION: Here $c = (7 - 1)/2 = 3$, $b = [0 - (-4)]/2 = 2$, and $a^2 = b^2 + c^2 = 13$. Thus, from (8) we have

$$\frac{(x - 4)^2}{13} + \frac{(y + 2)^2}{4} = 1.$$

This ellipse is sketched in Figure 7. Its major axis is on the line $y = -2$ and its minor axis is on the line $x = 4$.

EXAMPLE 6
COMPLETING THE SQUARE IN ORDER TO OBTAIN THE STANDARD EQUATION OF A TRANSLATED ELLIPSE

Discuss the curve given by

$$9x^2 + 36x + 4y^2 - 8y + 4 = 0.$$

SOLUTION: We write this expression as

$$9(x^2 + 4x) + 4(y^2 - 2y) = -4.$$

Then after completing the squares, we obtain

$$\overset{\text{Added Subtracted } 9 \cdot 4}{\overbrace{}} \qquad \overset{\text{Added Subtracted } 4 \cdot 1}{\overbrace{}}$$

$$9(x^2 + 4x + 4) - 36 + 4(y^2 - 2y + 1) - 4 = -4,$$

or

$$9(x + 2)^2 + 4(y - 1)^2 = 36,$$

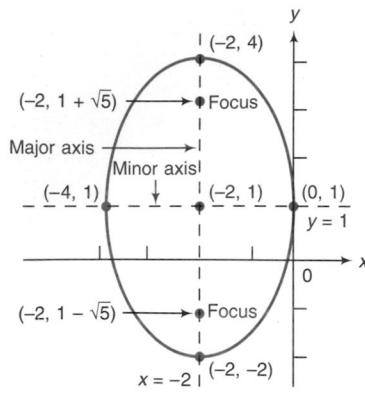

FIGURE 8
THE ELLIPSE $9x^2 + 36x + 4y^2 - 8y + 4 = 0$
OR $\dfrac{(x+2)^2}{4} + \dfrac{(y-1)^2}{9} = 1$

and, dividing both sides by 36,

$$\frac{(x + 2)^2}{4} + \frac{(y - 1)^2}{9} = 1.$$

This is the equation of an ellipse centered at $(-2, 1)$. Here $a = 2$, $b = 3$, and $c = \sqrt{b^2 - a^2} = \sqrt{5}$. The foci are therefore at $(-2, 1 - \sqrt{5})$ and $(-2, 1 + \sqrt{5})$. The major axis is on the line $x = -2$ and the minor axis is on the line $y = 1$. The ellipse is sketched in Figure 8.

HOW TO DRAW AN ELLIPSE

Pick two positive numbers a and c with $a > c$. Place two tacks $2c$ units apart as in Figure 9, and attach a string of length $2a$ to the tacks. Then proceed as in the figure to obtain the ellipse whose equation is $\dfrac{x^2}{a^2} + \dfrac{y^2}{a^2 - c^2} = 1$.

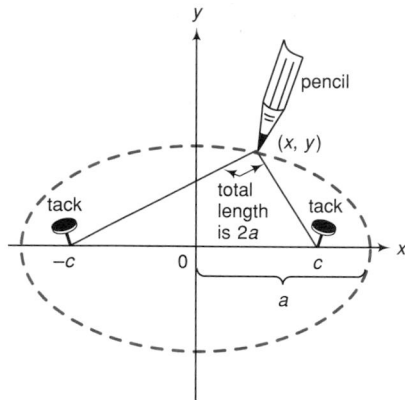

FIGURE 9
DRAWING AN ELLIPSE

THE ELLIPSE IN THE REAL WORLD

Ellipses are all around you. To see one, take a circular glass of water (or any other liquid) and tilt it. The surface of the liquid now forms an ellipse. As another common example, hold a spherical ball in front of a light so that its shadow falls on a table. If the light is not shining perpendicular to the table, the ball will cast an elliptical shadow (see Figure 10). If the light shines perpendicular to the table, the ball will cast a circular shadow.

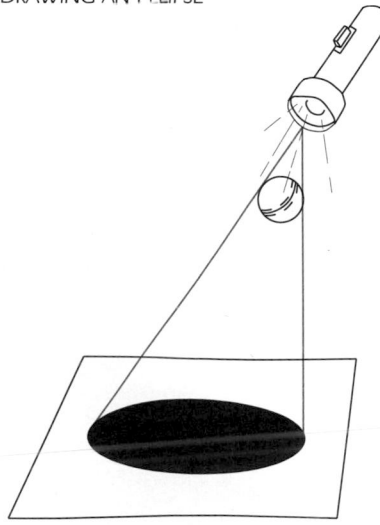

FIGURE 10
A SPHERICAL BALL CAN
CAST AN ELLIPTICAL SHADOW.

TABLE 1

Planet	Eccentricity, e
Mercury	0.2056
Venus	0.0068
Earth	0.0167
Mars	0.0934
Jupiter	0.0484
Saturn	0.0543
Uranus	0.0460
Neptune	0.0082
Pluto	0.2481

Ellipses are very important in astronomy. In 1609, the German astronomer Johannes Kepler (1571–1630) discovered that each planet follows an elliptical path with the sun at one focus. Except for Mercury and Pluto, the nearest and farthest planets from the sun, the orbits are nearly circular; that is, the eccentricities are close to 0. Table 1 on page 546 gives the eccentricities of the nine planets.

EXAMPLE 7
FINDING THE ECCENTRICITY OF THE ORBIT OF A COMET

The comet Katounek orbits the earth in a very flat elliptical path. If 1 AU (astronomical unit) denotes the distance between the earth and the sun (1 AU ≈ 93 million miles), then the length of the minor axis of Katounek's orbit is 44 AU, while the length of the major axis is 3600 AU. Find the eccentricity of Katounek's orbit.

SOLUTION: If the major axis is on the x-axis and the minor axis is on the y-axis, then

$$2a = 3600 \text{ AU}, \qquad a = 1800 \text{ AU}$$

$$2b = 44 \text{ AU}, \qquad b = 22 \text{ AU},$$

and

$$c = \sqrt{a^2 - b^2} = \sqrt{(1800)^2 - (22)^2} = \sqrt{3{,}239{,}516}$$
$$= 1799.865551.$$

Then

$$e = \frac{c}{a} = \frac{1799.865551}{1800} \approx 0.999925.$$

In Figure 11, we depict the very flat orbit of Katounek and use the orbit of Pluto as a reference.

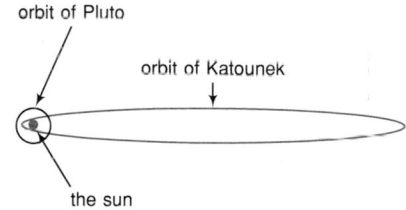

orbit of Pluto

orbit of Katounek

the sun

FIGURE 11
THE ORBIT OF KATOUNEK

THE REFLECTED-WAVE PROPERTY OF ELLIPSES AND THE WHISPERING-GALLERY EFFECT

Let P be a point on an ellipse that is not a vertex (see Figure 12). We draw the tangent line T that passes through P and draw lines joining P to each of the two foci of the ellipse. Let θ_1 denote the angle between T and PF_1 and let θ_2 denote the angle between T and PF_2. Then $\theta_1 = \theta_2$. This property is called the **reflected-wave property** of ellipses. You are asked to prove this in Problem 47.

The reflected-wave property has been used to good effect by both Renaissance and modern architects. If the upper half of the ellipse in Figure 12 is rotated around the x-axis, then it will form a dome. In any room with an elliptical domed ceiling, a sound made at one focus will be reflected to the other focus where it will be heard very clearly. This phenomenon is known as the **whispering-gallery effect**. Some of the most famous rooms that exhibit the whispering-gallery effect include St. Paul's Cathedral in London (designed

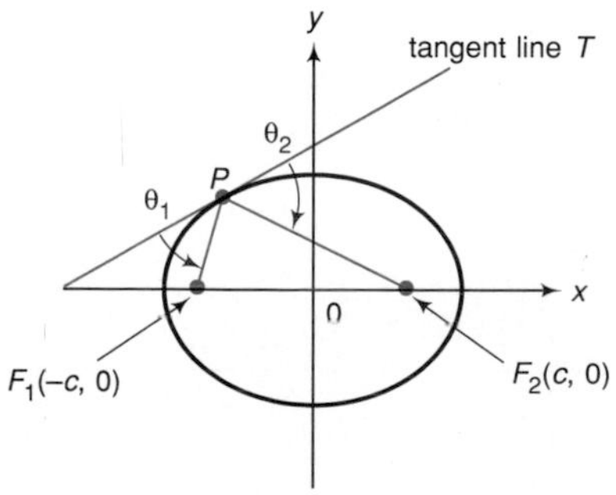

y

tangent line T

θ_2

θ_1

P

0

x

$F_1(-c, 0)$

$F_2(c, 0)$

FIGURE 12
THE ANGLES θ_1 AND θ_2 ARE EQUAL.

by the most famous of British architects, Sir Christopher Wren), the Caryatids room in the Louvre Museum in Paris, and the National Statuary Hall at the U.S. Capitol (the original House of Representatives) in Washington, D.C.[†]

■ PROBLEMS 8.1

■ SELF-QUIZ

I. Of the following, _____ is *not* an equation for an ellipse.
   a. $x^2 + y^2 = 36$
   b. $(x/3)^2 + (y/4)^2 = 1$
   c. $16x^2 + 9y = 144$
   d. $16x^2 + y^2/9 = 12$

II. Among the ellipses satisfying the following equations, the one satisfying _____ is *not* a translation of the others.
   a. $x^2/25 + y^2/16 = 1$
   b. $25(x - 1)^2 + 16(y - 3)^2 = 400$
   c. $(x/5)^2 + (y + 3)^2/16 = 1$
   d. $16(x + 7)^2 + 25(y - 8)^2 = 16 \cdot 25$

III. The ellipse with foci $(-6, 0)$ and $(6, 0)$ and vertices $(-10, 0)$ and $(10, 0)$ can be drawn by attaching the ends of a string, _____ units in length, to thumb-tacks at the foci and then tracing the outline with a pencil which moves in such a way as to keep the string taut.
   a. 24   b. 20   c. 16   d. 12

IV. The ellipse with foci $(-3, 0)$ and $(3, 0)$ and vertices $(-5, 0)$ and $(5, 0)$ has equation _____.
   a. $(x/5)^2 + (y/4)^2 = 1$
   b. $(x/3)^2 + (y/4)^2 = 1$
   c. $(x/3)^2 + (y/5)^2 = 1$
   d. $(x/4)^2 + (y/5)^2 = 1$

■ DRILL

In Problems 1–18, the equation of an ellipse is given. Locate its center, foci, vertices, and major and minor axes; then sketch it. Identify any ellipse which turns out to be a circle.

1. $\dfrac{x^2}{16} + \dfrac{y^2}{25} = 1$

2. $\dfrac{x^2}{25} + \dfrac{y^2}{16} = 1$

3. $x^2 + \dfrac{y^2}{9} = 1$

4. $\dfrac{x^2}{9} + y^2 = 1$

5. $x^2 + 4y^2 = 16$

6. $4x^2 + y^2 = 16$

7. $\dfrac{(x - 1)^2}{16} + \dfrac{(y + 3)^2}{25} = 1$

8. $\dfrac{(x + 3)^2}{25} + \dfrac{(y - 1)^2}{16} = 1$

9. $2x^2 + 2y^2 = 2$
10. $4x^2 + y^2 = 9$
11. $x^2 + 4y^2 = 9$
12. $4(x - 3)^2 + (y - 7)^2 = 9$
13. $4x^2 + 8x + y^2 + 6y = 3$
14. $x^2 + 6x + 4y^2 + 8y = 3$
15. $4x^2 + 8x + y^2 - 6y = 3$
16. $x^2 + 2x + 4y^2 + 2y = 7$

17. $3x^2 + 12x + 8y^2 - 4y = 20$
18. $2x^2 - 3x + 4y^2 + 5y = 37$

In Problems 19–24, find the eccentricity of each ellipse.

19. the ellipse of Problem 1
20. the ellipse of Problem 2
21. the ellipse of Problem 11
22. the ellipse of Problem 12
23. the ellipse of Problem 17
24. the ellipse of Problem 18

In Figure 13, eight ellipses are sketched. Match each sketch with the equations given in Problems 25–32. Find the unmarked foci and vertices of each ellipse.

25. $(x - 3)^2 + 16(y - 2)^2 = 16$
26. $4x^2 + 25y^2 = 100$
27. $25x^2 + 21y^2 = 525$
28. $9(x + 2)^2 + 25(y - 1)^2 = 225$
29. $16(x + 2)^2 + 4(y + 1)^2 = 64$
30. $13x^2 + 4y^2 = 52$
31. $9x^2 + 16y^2 = 144$
32. $9(x - 3)^2 + 4(y - 1)^2 = 36$

[†] For more details on these and other interesting applications, see the excellent article, "The Standup Conic Presents: The Ellipse and Applications" by Lee Whitt in *The UMAP Journal*, Vol. 4, No. 2, 1983, pp. 157–186.

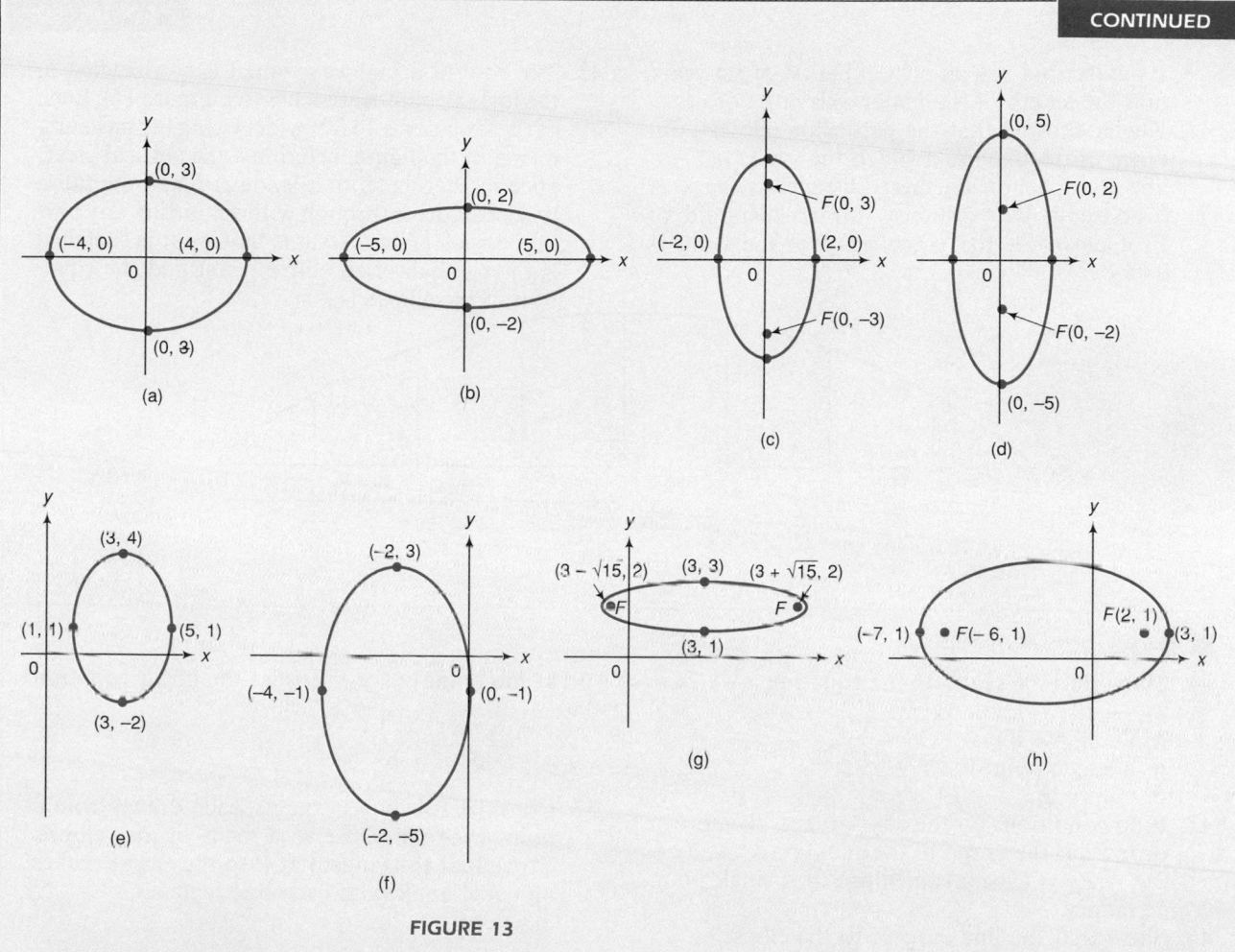

FIGURE 13

■ APPLICATIONS

33. Find an equation for the ellipse with foci $(0, 4)$ and $(0, -4)$ and vertices $(0, 5)$ and $(0, -5)$.

34. Find equations for two ellipses with centers at $(-1, 4)$ that have the same shape as the ellipse of the preceding problem and that contain the same area.

35. Find an equation for the line tangent to the ellipse $2x^2 + 3y^2 = 14$ at the point $(1, 2)$.

*36. Find two values for c such that each line $x + 3y = c$ is tangent to the ellipse $6x^2 + y^2 = 24$; also find each line's point of tangency.

37. Verify that the region enclosed by the ellipse

$$\frac{x^2}{a^2} + \frac{y^2}{b^2} = 1$$ has area equal to $|\pi ab|$.

38. Find the equation of the ellipse with vertices $(-2, 0)$ and $(2, 0)$ and eccentricity 0.8.

39. The orbit of Halley's comet is an ellipse with major axis approximately 36.2 AU (astronomical units) and minor axis approximately 9.1 AU. Find the eccentricity of the orbit.

*40. The major axis of the earth's orbit is approximately 185.5 million miles and the eccentricity of the orbit is 0.0167. Find the largest and smallest distances between the earth and the sun. (The closest and farthest positions of the planet from the sun are called the **perihelion** and **aphelion**, respectively.) [*Hint:* Remember that the sun is at one focus of the orbit.]

41. A body orbits around the sun with major axis $2a$ measured in astronomical units (AU). Let T, measured in years, denote the period of the orbit. Then, according to **Kepler's third law**,

$$T^2 = a^3.$$

CONTINUED

If an asteroid has an orbital period of 8.4 years, find the length of the major axis of its orbit.

42. Kepler showed that the perihelion (closest) distance of the planet's orbit to the sun is $a(1 - e)$ and its aphelion (farthest) distance is $a(1 + e)$. Find these two distances for the asteroid of Problem 41 if the eccentricity of the orbit is 0.058.

43. The roof of a highway tunnel is constructed in the form of elliptical arches (see Figure 14). Each of the six lanes is 14 feet wide. Using the measurements in the figure, determine the vertical clearance in each lane; that is, determine how tall a truck can drive through without hitting any part of the roof. [*Hint:* Assume that what is sketched is the top half of an ellipse. First find the equation of the ellipse.]

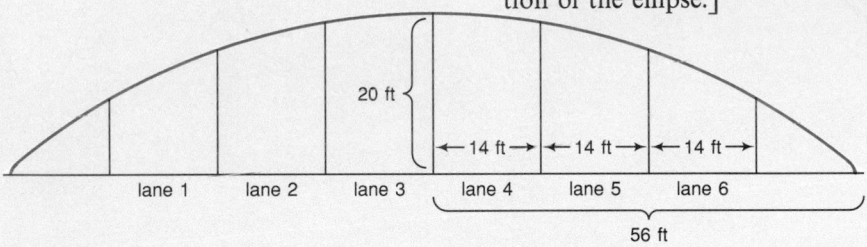

FIGURE 14

■ SHOW/PROVE/DISPROVE

44. Show that the graph of the equation $x^2 + 2x + 2y^2 + 12y = c$ is _____.
    a. an ellipse if $c > -19$
    b. a single point if $c = -19$
    c. empty if $c < -19$

*45. Find conditions on the constants A, B, and C in order that the graph of the equation $x^2 + Ax + 2y^2 + By = C$ be (a) an ellipse, (b) a single point, (c) empty.

46. Show that the line tangent to the ellipse

$$\frac{x^2}{a^2} + \frac{y^2}{b^2} = 1$$

at the point (x_0, y_0) satisfies the linear equation

$$\frac{x_0 \cdot x}{a^2} + \frac{y_0 \cdot y}{b^2} = 1.$$

47. Consider two line segments, each drawn from a point P of an ellipse to a focus of that ellipse. Prove that the tangent at P to the ellipse makes an equal angle with each line segment.

■ ANSWERS TO SELF-QUIZ

I. c   II. b   III. b   IV. a

8.2 THE PARABOLA

DEFINITION **PARABOLA**

A **parabola** is the set of points (x, y) equidistant from a fixed point and a fixed line that does not contain the fixed point. The fixed point is called the **focus**, and the fixed line is called the **directrix**. ■

EXAMPLE 1
FINDING AN EQUATION OF A PARABOLA

Find an equation of the parabola whose focus is the point $(0, 2)$ and whose directrix is the line $y = -2$.

SOLUTION: As in Figure 1(a), if (x, y) is a point on the parabola, then the distance from $(0, 2)$ to (x, y) is, from the distance formula (p. 12),

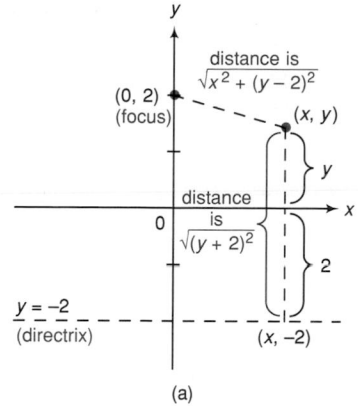

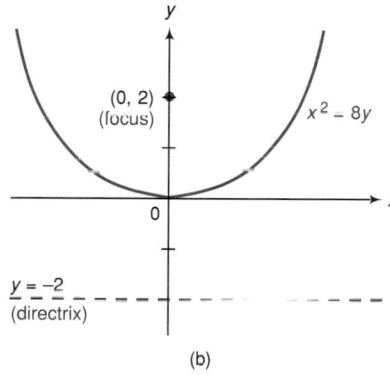

FIGURE 1
THE PARABOLA WITH FOCUS (0, 2) AND
DIRECTRIX THE LINE $y = -2$

equal to $\sqrt{(x - 0)^2 + (y - 2)^2}$. The distance between (x, y) and the line $y = -2$ is defined as the shortest distance from the point to the line. This is obtained by "dropping a perpendicular" from (x, y) to the line. Since the line $y = -2$ is horizontal, the perpendicular line will be vertical and will intersect $y = -2$ at the point $(x, -2)$. The distance between (x, y) and $(x, -2)$ is $\sqrt{(x - x)^2 + (y + 2)^2} = \sqrt{(y + 2)^2}$. Setting these two distances equal and squaring, we obtain

$$x^2 + (y - 2)^2 = (y + 2)^2$$
$$x^2 + y^2 - 4y + 4 = y^2 + 4y + 4$$
$$x^2 - 4y = 4y$$
$$x^2 = 8y.$$

This is an equation of the parabola we sought. Its **vertex** is at the origin. The parabola is sketched in Figure 1(b).

We now calculate a more general equation of a parabola. We place the axes so that the focus is the point $(0, c)$ and the directrix is the line $y = -c$. (See Figure 2.) If $P = (x, y)$ is a point on the parabola, then, as in Example 1, we obtain (with c instead of 2)

$$\sqrt{x^2 + (y - c)^2} = \sqrt{(y + c)^2}$$

or squaring,

$$x^2 + (y - c)^2 = (y + c)^2$$

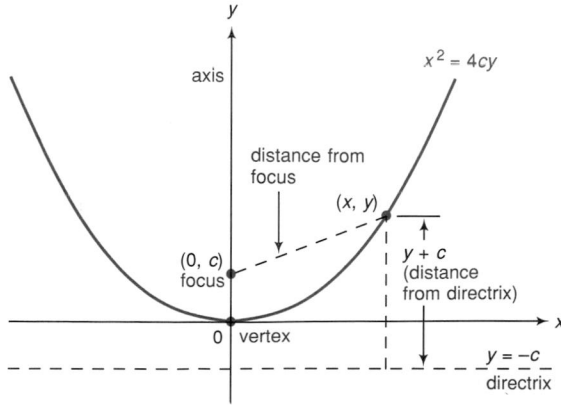

FIGURE 2   THE PARABOLA WITH FOCUS (0, c) AND DIRECTRIX THE LINE $y = -c$

which reduces to

> **STANDARD EQUATION OF A PARABOLA WITH VERTEX AT THE ORIGIN, FOCUS AT (0, c), AND DIRECTRIX THE LINE $y = -c$**
>
> $$x^2 = 4cy.$$
>
> (1)

The parabola given by (1) is symmetric about the y-axis. This line is called the **axis** of the parabola. Note that the axis contains the focus and is perpendicular to the directrix. The point at which the axis and the parabola intersect is called the **vertex**. The vertex is equidistant from the focus and the directrix.

EXAMPLE 2
SKETCHING A PARABOLA

Describe the parabola given by $x^2 = 12y$.

SOLUTION: Here, as in equation (1), $4c = 12$, so that $c = 3$, the focus is the point $(0, 3)$, and the directrix is the line $y = -3$. The axis of the parabola is the y-axis and the vertex is the origin. The curve is sketched in Figure 3.

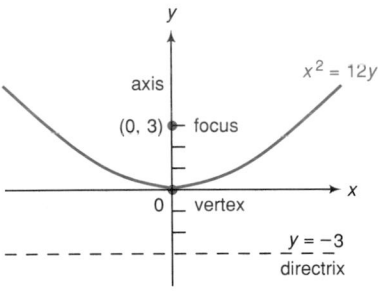

FIGURE 3   THE PARABOLA $x^2 = 12y$

EXAMPLE 3
SKETCHING A PARABOLA THAT
OPENS DOWNWARD

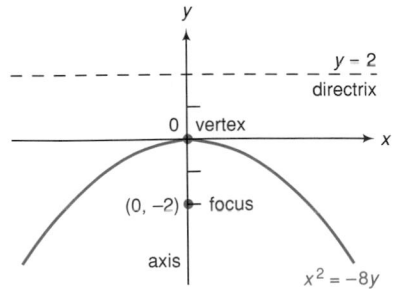

FIGURE 4
THE PARABOLA $x^2 = -8y$

Describe the parabola given by $x^2 = -8y$.

SOLUTION: Here $4c = -8$, so that $c = -2$, and the focus is $(0, -2)$, the directrix is the line $y = 2$, and the curve opens downward, as shown in Figure 4.

The parabola described by $x^2 = 4cy$ opens upward if $c > 0$ and opens downward if $c < 0$.

To verify this statement, note that

$$y = \frac{x^2}{4c} \qquad y' = \frac{x}{2c} \qquad \text{and} \qquad y'' = \frac{1}{2c}.$$

Thus the graph of $x^2 = 4cy$ is concave up if $c > 0$ and is concave down if $c < 0$.

As with the ellipse, we can exchange the role of x and y. We then obtain the following:

> **THE STANDARD EQUATION OF THE PARABOLA WITH FOCUS AT**
> **$(c, 0)$ AND DIRECTRIX THE LINE $x = -c$**
>
> $$y^2 = 4cx. \tag{2}$$
>
> If $c > 0$, the parabola opens to the right; if $c < 0$, the parabola opens to the left.

EXAMPLE 4
SKETCHING A PARABOLA THAT
OPENS TO THE RIGHT

Describe the parabola $y^2 = 16x$.

SOLUTION: Here $4c = 16$, so that $c = 4$, and the focus is (4, 0), the directrix is the line $x = -4$, the axis is the x-axis (since the parabola is symmetric about the x-axis), and the vertex is the origin. The curve is sketched in Figure 5.

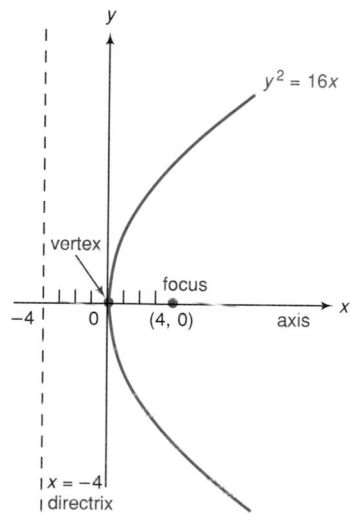

FIGURE 5
THE PARABOLA $y^2 = 16x$

TRANSLATING PARABOLAS

All the parabolas we have drawn so far have had their vertices at the origin. Other parabolas can be obtained by a simple translation of axes. The parabolas

$$(x - x_0)^2 = 4c(y - y_0) \tag{3}$$

and

$$(y - y_0)^2 = 4c(x - x_0) \tag{4}$$

have vertices at the point (x_0, y_0).

EXAMPLE 5
SKETCHING A TRANSLATED
PARABOLA OPENING TO THE LEFT

Describe the parabola $(y - 2)^2 = -8(x + 3)$.

SOLUTION: This parabola has its vertex at $(-3, 2)$. It is obtained by shifting the parabola $y^2 = -8x$ three units to the left and two units up (see Section 0.5). The focus and directrix of $y^2 = -8x$ are $(-2, 0)$ and $x = 2$. Hence, after translation, the focus and directrix of $(y - 2)^2 = -8(x + 3)$ are $(-5, 2)$ and $x = -1$. The axis of this curve is $y = 2$. The two parabolas are sketched in Figure 6.

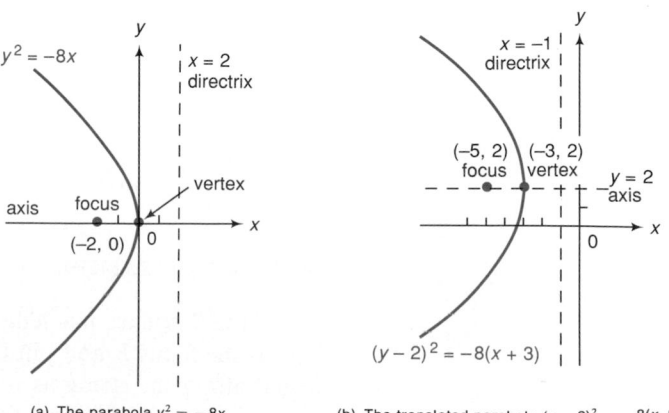

(a) The parabola $y^2 = -8x$.

(b) The translated parabola $(y - 2)^2 = -8(x + 3)$.

FIGURE 6 TRANSLATING THE PARABOLA $y^2 = -8x$ SO THAT ITS VERTEX MOVES FROM (0, 0) TO (-3, 2)

EXAMPLE 6
FINDING THE STANDARD
EQUATION OF A TRANSLATED
PARABOLA

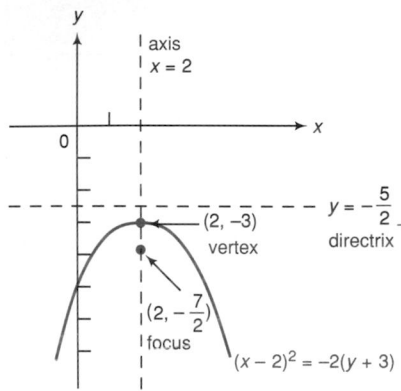

FIGURE 7
THE PARABOLA $(x - 2)^2 = -2(y + 3)$

Describe the curve $x^2 - 4x + 2y + 10 = 0$.

SOLUTION: We first complete the square:

$$x^2 - 4x + 2y + 10 = (x - 2)^2 - 4 + 2y + 10 = 0,$$

or

$$(x - 2)^2 = -2(y + 3).$$

This expression is the equation of a parabola with vertex at $(2, -3)$. Since $4c = -2$, $c = -\frac{1}{2}$, and the focus is $(2, -3 - \frac{1}{2}) = (2, -\frac{7}{2})$. The directrix is the line $y = -3 - (-\frac{1}{2}) = -\frac{5}{2}$, and the axis is the line $x = 2$. The curve is sketched in Figure 7.

HOW TO DRAW A PARABOLA

Much of what we now know about the parabola was discovered by the great German physicist, astronomer, and mathematician Johannes Kepler (1571–1630). Kepler was the first to use the term *focus* (Latin for "hearthside") in the context of this section. He constructed parabolas using a table, a piece of string, a pencil, and the seventeenth-century version of a T-square. Place a piece of paper along a wall above the edge of a horizontal table as in Figure 8. The table's edge is the directrix of our parabola.

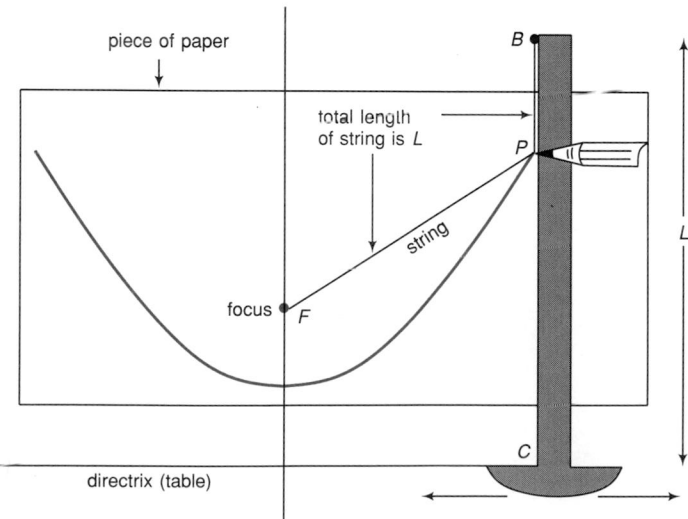

FIGURE 8 KEPLER'S METHOD FOR DRAWING A PARABOLA

If the T-square has length L, then choose a string of length L. Pin one end to the focus F and pin the other end to the top of the T-square. Slide a pencil along the string as in Figure 8, keeping the string taut. Then, as the T-square is moved along the side of the table, the pencil will trace out a parabola. The reason for this is that $\overline{FP} + \overline{PB} = \overline{CP} + \overline{PB} = L$ so that $\overline{FP} = \overline{CP}$. That is, the distance from a point P on the parabola to the focus equals the distance from the point to the directrix.

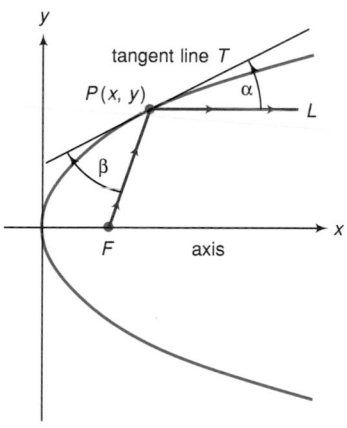

FIGURE 9
A LIGHT OR RADIO BEAM EMANATING FROM F IS REFLECTED PARALLEL TO THE AXIS ALONG THE LINE L.

THE REFLECTIVE PROPERTY OF A PARABOLA

Consider the parabola with focus F sketched in Figure 9. Let $P(x, y)$ be a point on the parabola, T the line tangent to the parabola at P, and L the line passing through P that is parallel to the axis. Finally, let α denote the angle between T and L and let β denote the angle between T and PF. Then $\alpha = \beta$. This means that a beam that starts at the focus F will be reflected off the parabola parallel to the axis. This **reflective property of the parabola**, as it is called, is useful in a wide variety of applications, as we shall soon see. You are asked to prove this in Problem 41.

The opposite effect is also seen: a beam coming in parallel to the axis will be reflected back to the focus. This can be illustrated in Figure 9 by reversing the arrows.

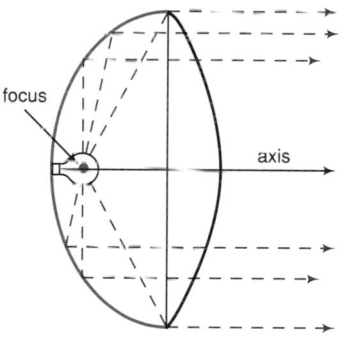

FIGURE 10
A PARABOLIC HEADLIGHT

Light rays from a bulb at the focus are reflected parallel to the axis.

THE PARABOLA IN THE REAL WORLD

Parabolas are all around us. The automobile headlight has the parabolic shape obtained by rotating a parabola around its axis. All the light emanating from a bulb placed at the focus is reflected parallel to the parabola's axis: that is, parallel to the ground. This follows from the reflective property just discussed (see Figure 10).

Parabolic reflectors can be found in communications systems, electronic surveillance systems, radar systems, and telescopes. In radar, an electromag-

FIGURE 11 THE PARABOLIC DISH OF THIS RADIO TELESCOPE REFLECTS SIGNALS TO INSTRUMENTS AT THE FOCUS.

netic beam is bounced off a target, and a parabolic reflector is used to collect and concentrate the deflected beam for signal processing. This works because the reflector reflects all returning signals back to the focus. In Figure 11, we show a parabolic reflector used on earth to track space probes.

When a projectile is shot into the air, its path takes the shape of a parabola. This famous phenomenon was discovered by Galileo (1564–1642). Renaissance scientists were fascinated by this fact. The great artist Leonardo da Vinci drew the path of exploding mortar shells toward the end of the fifteenth century (Figure 12).

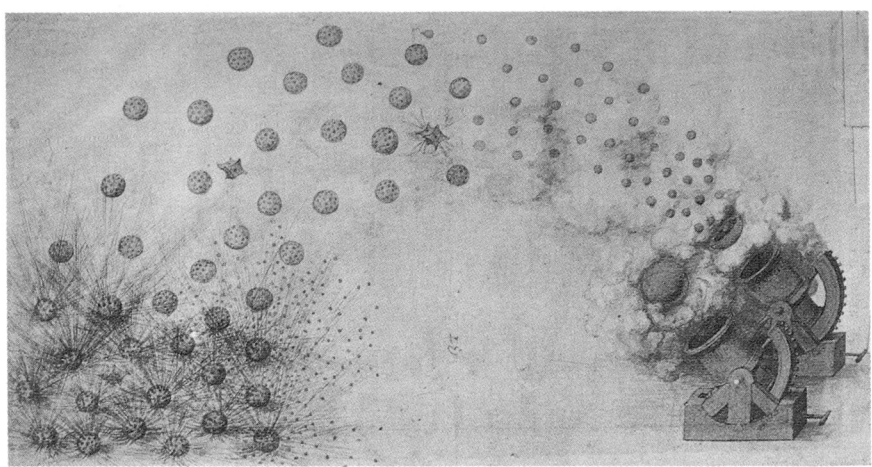

FIGURE 12 CANNON IN ACTION
The flight of exploding mortar shells, drawn by Leonardo da Vinci.

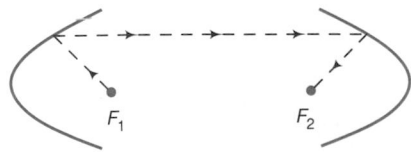

FIGURE 13
A SOUND MADE AT FOCUS F_1 WILL BE REFLECTED TWICE AND CLEARLY HEARD AT FOCUS F_2.

Sound waves can be transmitted effectively using parabolic reflectors. This is illustrated in Figure 13. One such "double reflector" device can be found at the Exploratorium in San Francisco. In this model, the reflectors have diameters of 8 feet and are placed facing each other about 50 feet apart. Someone whispering at the focus of one reflector can be clearly heard at the focus of the other reflector.

Parabolic reflectors also have been used in warfare and other unfortunate circumstances. The most famous example comes from Archimedes of Syracuse (287–212 B.C.), who seemed to have dabbled in everything. According to the Greek historian Plutarch, Syracuse was besieged by Romans led by their great general Marcellus. Archimedes helped save the city by designing, among other things, "burning" mirrors. These were parabolic mirrors capable of concentrating the rays of the sun onto attacking ships. Archimedes' clever devices also included a catapult that used the principle of the lever to hurl huge boulders. Because of Archimedes, Syracuse was able to hold out for nearly 3 years.

The Greeks were fascinated by the parabola and sometimes suggested applications that were, to put it mildly, very unpleasant. The Greek Diocles, who lived in the second century B.C., wrote a book entitled *Burning Mirrors*. In his work, Diocles suggested that if victims were to be sacrificed in front

FIGURE 14   IN THE IDEAL CASE, THE MAIN CABLE OF A SUSPENSION BRIDGE IS PARABOLIC.

of large crowds, parabolic mirrors could be used to provide a visible burning spot on the victim's body. It is not clear whether this idea was ever put into practice.

Parabolas are used in civil engineering. Bridges are built with twin parabolic cables that, ideally, will support a uniform horizontal load (see Figure 14.)†

■ PROBLEMS 8.2

■ SELF-QUIZ

I. The graph of $x/(-4) = (y/3)^2$ is a parabola opening _____.
   - a. to the right
   - b. to the left
   - c. upwards
   - d. downwards

II. The set of points $\{(a, b): 4a = -b^2\}$ is a _____.
   - a. vertical line
   - b. horizontal line
   - c. single point
   - d. parabola

III. Answer True or False to each of the following assertions about the parabola satisfying

$$y = -(x - 1)^2.$$

   - a. The focus is $(0, 0)$.
   - b. The vertex is $(0, 0)$.
   - c. The vertex is $(1, 0)$.
   - d. The focus is $(1, 4)$.
   - e. The focus is $(1, -\frac{1}{4})$.
   - f. The directrix passes through the vertex.
   - g. The directrix passes through the focus.
   - h. The directrix is perpendicular to the axis of the parabola.

---

† These and many other applications of parabolas can be found in the fascinating paper, "The Standup Conic Presents: The Parabola and Applications," by Lee Whitt in *The UMAP Journal*, Vol. 3, No. 3, 1983, pp. 285–316.

CONTINUED

■ **DRILL**

In Problems 1–18, the equation of a parabola is given. Locate its focus, directrix, axis, and vertex; then sketch it.

1. $x^2 = 16y$
2. $y^2 = 16x$
3. $x^2 = -16y$
4. $y^2 = -16x$
5. $2x^2 = 3y$
6. $2y^2 = -3x$
7. $4x^2 = -9y$
8. $7y^2 = -20x$
9. $(x - 1)^2 = -16(y + 3)$
10. $(y - 1)^2 = 16(x + 3)$
11. $x^2 + 4y = 9$
12. $(x + 1)^2 + 25y = 50$
13. $x^2 + 2x + y + 1 = 0$
14. $x + y - y^2 = 4$

15. $x^2 + 4x + y = 0$
16. $y^2 + 4y + x = 0$
17. $x^2 + 4x - y = 0$
18. $y^2 + 4y - x = 0$

In Figure 15, the graphs of eight parabolas are given. Match the graphs with the equations given in Problems 19–26.

19. $x^2 + 10x + 3y + 13 = 0$
20. $x^2 - 4x - 5y - 11 = 0$
21. $4x - y^2 = 0$
22. $x^2 - 3y = 0$
23. $2x + y^2 = 0$
24. $y^2 + 4y + x = 0$
25. $y^2 - 2x - 4y + 6 = 0$
26. $x^2 + 4y = 0$

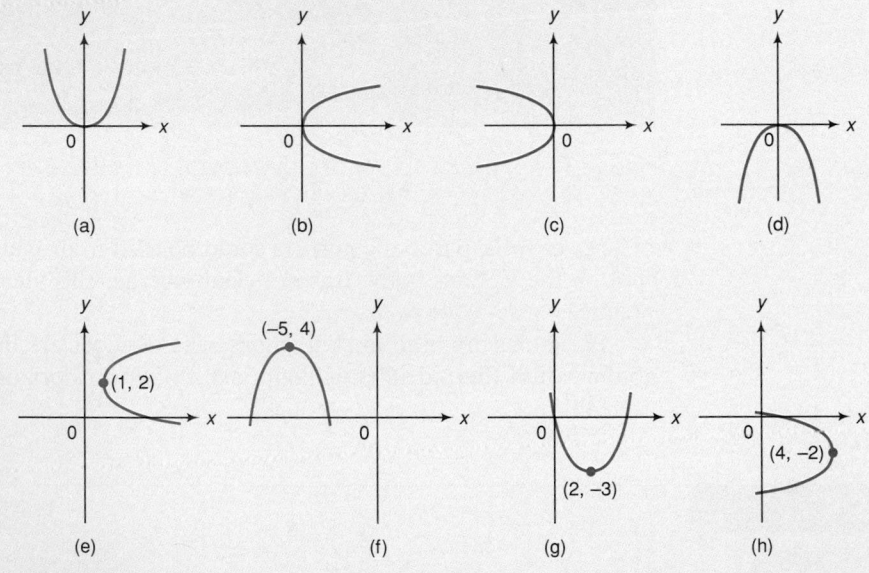

FIGURE 15

■ **APPLICATIONS**

27. Find an equation for the parabola with focus $(0, 4)$ and directrix the line $y = -4$.
28. Find an equation for the parabola with $(-3, 0)$ as its focus and the line $x = 3$ as its directrix.
29. Find an equation for the parabola obtained when the parabola of Problem 27 is shifted so that its vertex is at the point $(-2, 5)$.
30. The parabola of Problem 28 is translated so that its vertex is now at $(3, -1)$. Find its new focus and directrix.

31. Find an equation for the line tangent to the parabola $x^2 = 9y$ at the point $(3, 1)$.
*32. Find an equation for the parabola with vertex at $(1, 2)$ and directrix the line $x = y$. [*Hint:* The result of Problem 0.3.63 may be useful to you here.]
**33. An asteroid orbits the earth making a parabolic orbit around it—that is, the earth is at the focus of the parabola. The asteroid is 150,000 km from the earth at a point where the line from the as-

CONTINUED

teroid to the earth makes an angle of 30° with the axis of the parabola. How close will the asteroid come to the earth? [*Hint:* Use the result of Problem 38.]

**34. Answer the preceding problem if the angle is 57°.

35. The tops of two towers of a suspension bridge (like the one in Figure 14) are 100 feet above water level and 375 feet apart. The lowest point of the parabolic cable connecting the two towers is 40 feet above the water. How high is a point on the cable that is 60 feet (horizontally) from one of the towers?

36. The receiver of a parabolic signal receptor is at the focus, which is 2 feet from the vertex. If the receptor is placed as in Figure 16, find an equation for the cross-sectional parabola that lies in the xy-plane.

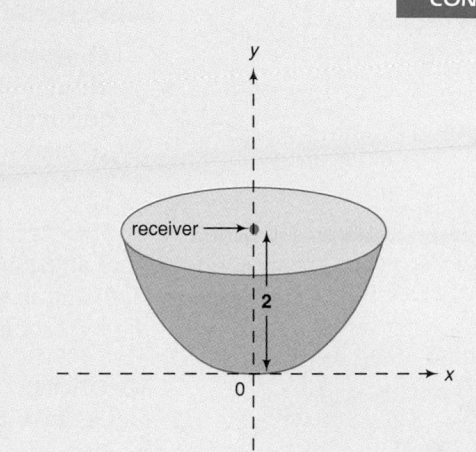

FIGURE 16

■ SHOW/PROVE/DISPROVE

37. a. Show that the line tangent to the parabola $4cy = x^2$ at the point (x_0, y_0) satisfies the equation $2c(y + y_0) - x_0 x$.
   b. Find where a line tangent to a parabola intersects the axis of that parabola. [*Hint:* Find where the tangent line in part (a) intersects the y-axis.]
   c. Describe a simple method to construct the line tangent to a given parabola at a specific point. [*Hint:* Use your result for part (b).]

38. Prove or Disprove: Among all points lying on a particular parabola, the point closest to its focus is its vertex.

*39. Let F be the focus of a parabola, let P be any point except the vertex on that parabola, and let Q be the point where the tangent at P intersects the directrix of the parabola. Prove that these three points are vertices of a right triangle.

*40. Pick a point on the directrix of a parabola; from this point, two lines can be drawn which are tangent to the parabola. Prove that the line segment connecting the two points of tangency passes through the focus of the parabola.

*41. A parabolic mirror will concentrate light at its focus. Let (x_0, y_0) be a point on a parabola. Let L_1 be the line passing through (x_0, y_0) and the focus; let L_2 be the line through (x_0, y_0) and parallel to the axis of the parabola. Show that the line tangent to the parabola at (x_0, y_0) will make an equal angle with L_1 and L_2.

*42. The **latus rectum** of a parabola is the chord passing through the focus that is perpendicular to the axis. (See Figure 17.) Show that the length of the latus rectum of a parabola is twice the distance between the focus and the directrix.

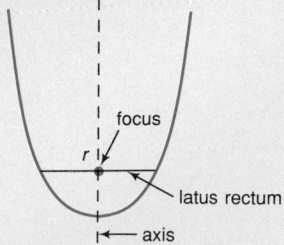

FIGURE 17

8.3 THE HYPERBOLA

DEFINITION **HYPERBOLA**

A **hyperbola** is a set of points (x, y) with the property that the positive difference between the distances from (x, y) and each of two given (distinct) points is a constant. Each of the two given points is called a **focus** of the hyperbola. ■

EXAMPLE 1
FINDING AN EQUATION OF A HYPERBOLA

Find an equation of the hyperbola whose foci are the points $(-3, 0)$ and $(3, 0)$ and in which the difference of the distances from a point (x, y) on the hyperbola to the foci is equal to 4.

SOLUTION: If P is a point on the hyperbola and F_1 and F_2 denote the foci as in Figure 1, then the difference of the distances from P to the foci is

$$\overline{PF}_1 - P\overline{F}_2 = 4.$$

Using the distance formula, we obtain

$$\sqrt{(x + 3)^2 + y^2} - \sqrt{(x - 3)^2 + y^2} = 4$$

$$\sqrt{(x + 3)^2 + y^2} = \sqrt{(x - 3)^2 + y^2} + 4$$

$$(x + 3)^2 + y^2 = (x - 3)^2 + y^2 + 8\sqrt{(x - 3)^2 + y^2} + 16$$

                                               Square both sides

$$x^2 + 6x + 9 + y^2 = x^2 - 6x + 9 + y^2 + 8\sqrt{(x - 3)^2 + y^2} + 16$$

$$6x = -6x + 8\sqrt{(x - 3)^2 + y^2} + 16$$

              Subtract $x^2 + 9 + y^2$ from both sides

$$12x - 16 = 8\sqrt{(x - 3)^2 + y^2} \qquad \text{Rearrange terms}$$

$$\frac{3}{2}x - 2 = \sqrt{(x - 3)^2 + y^2} \qquad \text{Divide by 8}$$

$$\frac{9}{4}x^2 - 6x + 4 = (x - 3)^2 + y^2 \qquad \text{Square again}$$

$$\frac{9}{4}x^2 - 6x + 4 = x^2 - 6x + 9 + y^2$$

$$\left(\frac{9}{4} - 1\right)x^2 - y^2 = 5 \qquad \text{Combine terms}$$

$$\frac{5}{4}x^2 - y^2 = 5$$

$$\frac{x^2}{4} - \frac{y^2}{5} = 1. \qquad \text{Divide by 5} \qquad\qquad \text{(1)}$$

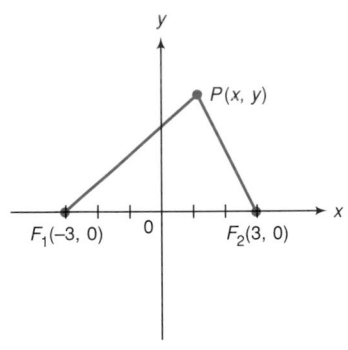

FIGURE 1
IN THE HYPERBOLA WITH FOCI $F_1 = (-3, 0)$ AND $F_2 = (3, 0)$,

$$|\overline{PF}_1 - \overline{PF}_2| = \text{A CONSTANT}.$$

In this example, the constant is 4.

Equation (1) is the **standard equation** of the hyperbola. Here we assumed that $\overline{PF}_1 > \overline{PF}_2$. If $\overline{PF}_2 > \overline{PF}_1$, we obtain the same equation. You should verify this.

To graph the hyperbola, we make several observations. First,

$$\frac{x^2}{4} = 1 + \frac{y^2}{5} \qquad \text{or} \qquad x^2 = 4 + \frac{4}{5}y^2. \qquad \text{From equation (1)}$$

Since $y^2 \geq 0$ for all real numbers y, we have $x^2 \geq 4$ so that

$$x \geq 2 \qquad \text{or} \qquad x \leq -2.$$

That is, no point on the hyperbola has an x-coordinate in the interval $(-2, 2)$. Second, if we replace x by $-x$ in (1), we obtain the same equation. This means the hyperbola has two symmetric branches about the y-axis: one for $x > 0$ and one for $x < 0$. These correspond to the two cases $\overline{PF}_1 - \overline{PF}_2 = 4$ and $\overline{PF}_2 - \overline{PF}_1 = 4$. We now solve (1) for y:

$$\frac{y^2}{5} = \frac{x^2}{4} - 1$$

$$y^2 = \frac{5}{4}x^2 - 5$$

$$y = \pm\sqrt{\frac{5}{4}x^2 - 5}.$$

Suppose $|x|$ is large, then $\frac{5}{4}x^2 - 5 \approx \frac{5}{4}x^2$. This is true because

$$\lim_{x \to \infty} \frac{\frac{5}{4}x^2 - 5}{\frac{5}{4}x^2} = 1.$$

For x large,

$$y \approx \pm\sqrt{\frac{5}{4}x^2} = \pm\frac{\sqrt{5}}{2}x.$$

The lines $y = \dfrac{\sqrt{5}}{2}x$ and $y = -\dfrac{\sqrt{5}}{2}x$ are oblique asymptotes for the hyperbola (see page 209). All together, this yields the graph in Figure 2.

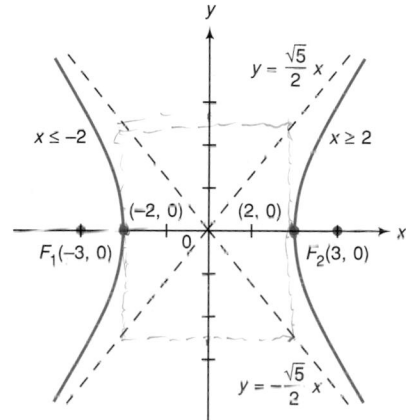

FIGURE 2

THE HYPERBOLA $\dfrac{x^2}{4} - \dfrac{y^2}{5} = 1$

To calculate the equation of a more general hyperbola, we place the axes so that the foci are the points $(c, 0)$ and $(-c, 0)$ and the difference of the distances from a point (x, y) on the hyperbola to the foci is equal to $2a > 0$. In Figure 3, we assume that $\overline{PF}_2 > \overline{PF}_1$. Then

$$\text{difference of distances} = \overline{PF}_2 - \overline{PF}_1 = 2a$$

so

$$\overline{PF}_2 = 2a + \overline{PF}_1.$$

But, as the shortest distance between two points is a straight line, we have

$$2a + \overline{PF}_1 = \overline{PF}_2 < \overline{PF}_1 + \overline{F_1F_2} = \overline{PF}_1 + 2c.$$

Therefore, $2a + \overline{PF}_1 < 2c + \overline{PF}_1$

or

$$2a < 2c \qquad \text{and} \qquad c > a.$$

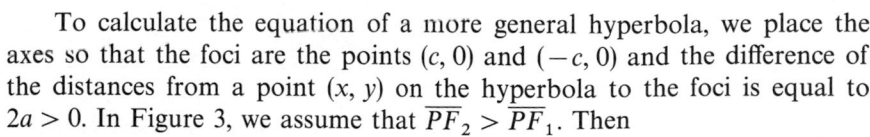

FIGURE 3

IN THE HYPERBOLA WITH FOCI $(-c, 0)$ AND $(c, 0)$, $|\overline{PF}_1 - \overline{PF}_2| = 2a$ WHERE a IS A CONSTANT

If (x, y) is a point on the hyperbola,

$$\overline{PF}_2 - \overline{PF}_1 = 2a$$

or

$$\sqrt{(x + c)^2 + y^2} - \sqrt{(x - c)^2 + y^2} = 2a$$

(assuming that $\overline{PF}_2 > \overline{PF}_1$ so that $\overline{PF}_2 - \overline{PF}_1$ gives a positive distance). Then,

$$\sqrt{(x + c)^2 + y^2} = \sqrt{(x - c)^2 + y^2} + 2a$$

$$(x + c)^2 + y^2 = (x - c)^2 + y^2 + 4a\sqrt{(x - c)^2 + y^2} + 4a^2 \qquad \text{Square}$$

$$x^2 + 2cx + c^2 + y^2 = x^2 - 2cx + c^2 + y^2 + 4a\sqrt{(x - c)^2 + y^2} + 4a^2$$

$$4cx - 4a^2 = 4a\sqrt{(x - c)^2 + y^2} \qquad \text{Simplify}$$

$$\frac{c}{a}x - a = \sqrt{(x - c)^2 + y^2} \qquad \text{Divide by } 4a$$

$$\frac{c^2}{a^2}x^2 - 2cx + a^2 = (x - c)^2 + y^2 \qquad \text{Square again}$$

$$\frac{c^2}{a^2}x^2 - 2cx + a^2 = x^2 - 2cx + c^2 + y^2$$

$$\left(\frac{c^2}{a^2} - 1\right)x^2 - y^2 = c^2 - a^2. \qquad (2)$$

Since $c > a > 0$, $c^2 - a^2 > 0$, and we can define the positive number b by

$$b^2 = c^2 - a^2.$$

Then dividing both sides of (2) by $c^2 - a^2$, we obtain, since $\frac{c^2}{a^2} - 1 = \frac{c^2 - a^2}{a^2}$,

---

STANDARD EQUATION OF A HYPERBOLA CENTERED AT THE ORIGIN WITH HORIZONTAL TRANSVERSE AXIS

$$\frac{x^2}{a^2} - \frac{y^2}{b^2} = 1. \qquad (3)$$

A similar derivation shows that if we assume that $\overline{PF}_1 > \overline{PF}_2$, we also obtain equation (3).

These two cases correspond to the right and left branches of the hyperbola sketched in Figure 4. Note that the hyperbola given by (3) is symmetric about both the x-axis and the y-axis. The **principal axis** is the line containing the foci. The **vertices** of the hyperbola are the points of intersection of the hyperbola and its principal axis. The midpoint of the line segment joining the foci is called the **center** of the hyperbola. The **transverse axis** of the hyperbola is the line segment joining the vertices. The **conjugate axis** is the line joining the points $(0, -b)$ and $(0, b)$.

As in Example 1, we can write (3) as

$$y = \pm\sqrt{\frac{b^2}{a^2}x^2 - b^2} \approx \pm\sqrt{\frac{b^2}{a^2}x^2} = \pm\frac{b}{a}x \text{ for } |x| \text{ large.}$$

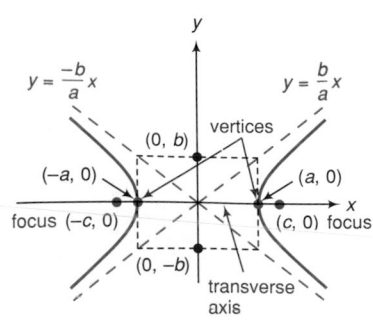

FIGURE 4

THE HYPERBOLA $\dfrac{x^2}{a^2} - \dfrac{y^2}{b^2} = 1$

Thus, the lines $y = \pm\dfrac{b}{a}x$ are oblique asymptotes to the hyperbola.

REMARK: To make it easier to sketch the hyperbola in Figure 4, we have drawn the rectangle with sides $x = \pm a$, $y = \pm b$. The lines that pass through the diagonals of this rectangle are the asymptotes of the hyperbola.

EXAMPLE 2
FINDING THE STANDARD EQUATION OF A HYPERBOLA WITH HORIZONTAL TRANSVERSE AXIS

Find the equation of the hyperbola with foci at $(4, 0)$ and $(-4, 0)$ and with $a = 3$.

SOLUTION: Here $a^2 = 9$ and $b^2 = c^2 - a^2 = 16 - 9 = 7$, so the equation is

$$\frac{x^2}{9} - \frac{y^2}{7} = 1.$$

The asymptotes are $y = \pm(b/a)x = \pm(\sqrt{7}/3)x$. (See Figure 5.)

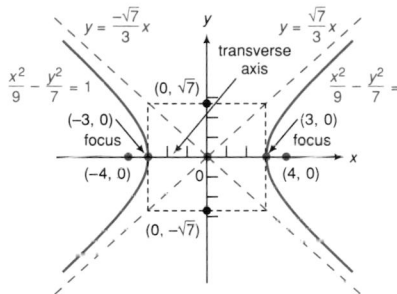

FIGURE 5

THE HYPERBOLA $\dfrac{x^2}{9} - \dfrac{y^2}{7} = 1$

As with the ellipse and parabola, the roles of x and y can be reversed.

> **STANDARD EQUATION OF A HYPERBOLA CENTERED AT THE ORIGIN WITH VERTICAL TRANSVERSE AXIS**
>
> $$\frac{y^2}{a^2} - \frac{x^2}{b^2} = 1. \qquad (4)$$

The graph of the equation (4) has its transverse axis on the y-axis. From (2) we have $c^2 = a^2 + b^2$, and the foci are $(0, c)$ and $(0, -c)$. The vertices of the hyperbola given by (4) are the points $(0, a)$ and $(0, -a)$. The asymptotes are the lines $x = \pm(b/a)y$ (or $y = \pm(a/b)x$).

EXAMPLE 3
SKETCHING A HYPERBOLA WITH VERTICAL TRANSVERSE AXIS

Discuss the curve given by $4y^2 - 9x^2 = 36$.

SOLUTION: Dividing by 36, we obtain

$$\frac{y^2}{9} - \frac{x^2}{4} = 1.$$

Hence $a = 3$, $b = 2$, and $c^2 = 9 + 4 = 13$, so that $c = \sqrt{13}$. The foci are $(0, \sqrt{13})$ and $(0, -\sqrt{13})$, and the vertices are $(0, 3)$ and $(0, -3)$. The asymptotes are the lines $x = \frac{2}{3}y$ and $x = -\frac{2}{3}y$. (See Figure 6.)

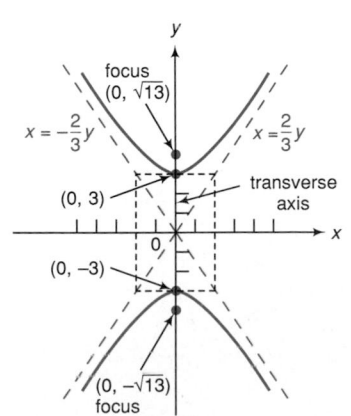

FIGURE 6

THE HYPERBOLA $\dfrac{y^2}{9} - \dfrac{x^2}{4} = 1$

TRANSLATING HYPERBOLAS

The hyperbolas we have sketched to this point are **standard hyperbolas**. These hyperbolas have their centers at the origin. Other hyperbolas can be obtained by a translation of the axes. The hyperbolas

$$\frac{(x - x_0)^2}{a^2} - \frac{(y - y_0)^2}{b^2} = 1 \quad \text{and} \quad \frac{(y - y_0)^2}{a^2} - \frac{(x - x_0)^2}{b^2} = 1$$

have centers at the point (x_0, y_0).

EXAMPLE 4
TRANSLATING A HYPERBOLA

Describe the hyperbola $(y - 2)^2/9 - (x + 3)^2/4 = 1$.

SOLUTION: This curve is the hyperbola of Figure 6 shifted three units to the left and two units up, so that its center is at $(-3, 2)$. It is sketched in Figure 7.

EXAMPLE 5
FINDING THE EQUATION OF A
TRANSLATED HYPERBOLA BY
COMPLETING THE SQUARE

Describe the curve $x^2 - 4y^2 - 4x - 8y - 9 = 0$.

SOLUTION: We have

$$(x^2 - 4x) - (4y^2 + 8y) - 9 = 0,$$

or completing the squares,

$$(x - 2)^2 - 4 - 4[(y + 1)^2 - 1] - 9 = 0,$$

or

$$(x - 2)^2 - 4(y + 1)^2 = 9,$$

and

$$\frac{(x - 2)^2}{9} - \frac{(y + 1)^2}{(\frac{3}{2})^2} = 1.$$

This is the equation of a hyperbola with center at $(2, -1)$. It is sketched in Figure 8.

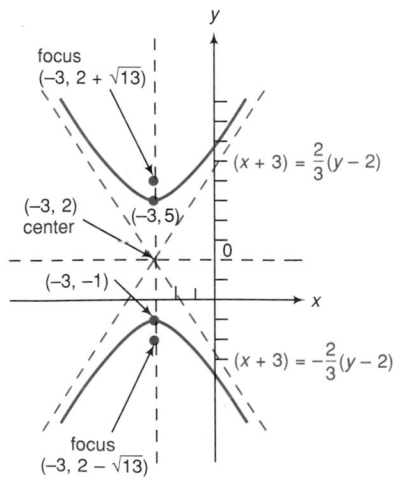

FIGURE 7

THE HYPERBOLA $\dfrac{(y - 2)^2}{9} - \dfrac{(x + 3)^2}{4} = 1$

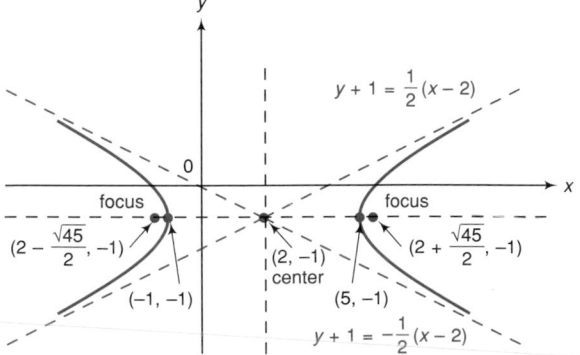

FIGURE 8   THE HYPERBOLA $\dfrac{(x - 2)^2}{9} - \dfrac{(y + 1)^2}{(\frac{3}{2})^2} = 1$

In Table 1 we summarize properties of ellipses, parabolas, and hyperbolas with centers (or vertices in the case of parabolas) at the origin.

TABLE 1 STANDARD CONIC SECTIONS

	Equation	Description
Ellipses	$\dfrac{x^2}{a^2} + \dfrac{y^2}{b^2} = 1,\ a > b$	Ellipse with major axis on x-axis; $a^2 = b^2 + c^2$; foci $(-c, 0)$ and $(c, 0)$
	$\dfrac{x^2}{a^2} + \dfrac{y^2}{b^2} = 1,\ b > a$	Ellipse with major axis on y-axis; $b^2 = a^2 + c^2$; foci $(0, c)$ and $(0, -c)$
	$\dfrac{x^2}{a^2} + \dfrac{y^2}{b^2} = 1,\ b = a$	Circle with radius $a\ (= b)$
	$\dfrac{x^2}{a^2} + \dfrac{y^2}{b^2} = 0$	Degenerate ellipse; single point $(0, 0)$
	$\dfrac{x^2}{a^2} + \dfrac{y^2}{b^2} = -1$	Degenerate ellipse; graph is empty
Parabolas	$x^2 = 4cy$	Parabola Focus: $(0, c)$ Directrix: $y = -c$ Axis: y-axis Vertex: $(0, 0)$ Curve opens upward if $c > 0$ and downward if $c < 0$
	$y^2 = 4cx$	Parabola Focus: $(c, 0)$ Directrix: $x = -c$ Axis: x-axis Vertex: $(0, 0)$ Curve opens to the right if $c > 0$ and to the left if $c < 0$
	$x^2 = 0$	Degenerate parabola; graph is y-axis (one line)
	$y^2 = 0$	Degenerate parabola; graph is x-axis (one line)
	$x^2 = 1$	Degenerate parabola; graph consists of the two lines $x = 1$ and $x = -1$
	$y^2 = 1$	Degenerate parabola; graph consists of the two lines $y = 1$ and $y = -1$
Hyperbolas	$\dfrac{x^2}{a^2} - \dfrac{y^2}{b^2} = 1$	Hyperbola with foci $(c, 0)$ and $(-c, 0)$, where $c^2 = a^2 + b^2$; transverse axis is the line segment joining $(-a, 0)$ and $(a, 0)$; center at origin; asymptotes $y = \pm(b/a)x$; curve opens to right and left
	$\dfrac{y^2}{a^2} - \dfrac{x^2}{b^2} = 1$	Hyperbola with foci $(0, c)$ and $(0, -c)$, where $c^2 = a^2 + b^2$; transverse axis is the line segment joining $(0, -a)$ and $(0, a)$; center at origin; asymptotes $x = \pm(b/a)y$ $(y = \pm(a/b)x)$; curve opens at top and bottom
	$\dfrac{x^2}{a^2} - \dfrac{y^2}{b^2} = 0$	Degenerate hyperbola; graph consists of two lines: $y = \pm(b/a)x$

THE HYPERBOLA IN THE REAL WORLD

Hyperbolas do not appear in physical constructions as often as ellipses and parabolas, but nevertheless they are useful. Hyperbolas do frequently appear as the graphs of important equations in physics, chemistry, biology, business, and economics. Examples include Ohm's law and supply and demand curves. In Einstein's theory of special relativity, an observer in an inertial reference frame sees a particle in a parallel force field follow a hyperbolic path in space time. The British physicist Ernest Rutherford (1871–1937) developed his model of the atom by measuring the hyperbolic orbits of scattered positive-charged particles.

Capillary action is the elevation of the surface of liquids in narrow spaces due to surface tension and other forces. Suppose, as in Figure 9, that two pieces of glass are joined along one pair of edges and are slightly separated along the other pair. If the glass configuration is placed vertically in a dish of colored water, then capillary action will force the water to rise in such a way as to form a hyperbola. Try it.

Hyperbolas are very useful in certain types of navigation. In order to explain why, we first do an example.

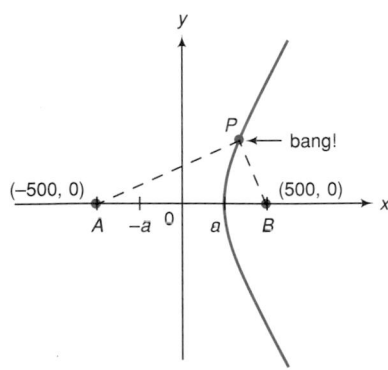

FIGURE 9
COLORED WATER IS DRAWN UP BY CAPIL-LARY ACTION.

EXAMPLE 6
DETERMINING THE SOURCE OF AN EXPLOSION

FIGURE 10
ONE BRANCH OF THE HYPERBOLA
$$\frac{x^2}{65,025} - \frac{y^2}{184,975} = 1$$

An explosion was heard on two ships 1 kilometer apart. Sailors on Ship B heard the explosion $1\frac{1}{2}$ seconds before those on Ship A. Relative to the two ships, where did the explosion occur?

SOLUTION: The speed of sound in air (at 20°C) is approximately 340 meters/sec. In $1\frac{1}{2}$ seconds the sound travelled $1\frac{1}{2} \times 340 = 510$ meters. Therefore, the explosion took place at a point 510 meters closer to Ship B than to Ship A. In Figure 10, we draw a coordinate system and place A and B on the x-axis, equidistant from the origin. Since $\overline{AB} = 1$ km $= 1000$ m, the coordinates of A and B are $(-500, 0)$ and $(500, 0)$. The explosion took place at a point P such that $\overline{PA} - \overline{PB} = 510$. Thus, P is on one branch of the hyperbola with foci $(-500, 0)$ and $(500, 0)$ such that the difference of the distances to the foci is $2a = 510$. Thus, $a = \frac{510}{2} = 255$, $c = 500$, $b^2 = c^2 - a^2 = 250,000 - 65,025 = 184,975$. The equation of this hyperbola is

$$\frac{x^2}{65,025} - \frac{y^2}{184,975} = 1$$

and the point P is on the branch of the hyperbola containing points that are closer to B than to A.

Of course, we have not located the point P precisely. However, if we have a third ship, Ship C, that hears the explosion, then we can obtain two more hyperbolas (one for Ships A and C and one for Ships B and C), and the point of the explosion is the single point at which the three hyperbolas intersect.

Because of the technique illustrated in the last example, the hyperbola is very useful in navigation—particularly in the LORAN (LOng RAnge Navigation) system. During World War II, LORAN served as a navigational aid for the strategic night bombing of Germany and for the long-range bomb-

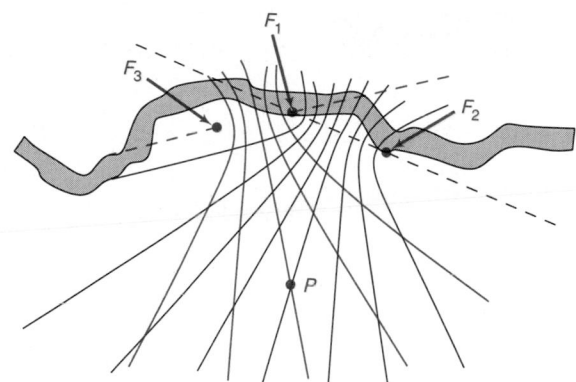

FIGURE 11 A LORAN MAP WITH HYPERBOLAS OF CONSTANT TIME DIFFERENCE

Two hyperbolas are needed to get a cross-fix at point P.

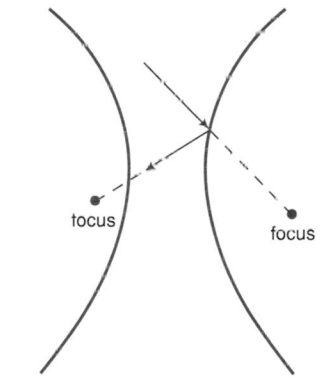

FIGURE 12
ILLUSTRATING THE REFLECTION PROPERTY
OF A HYPERBOLA

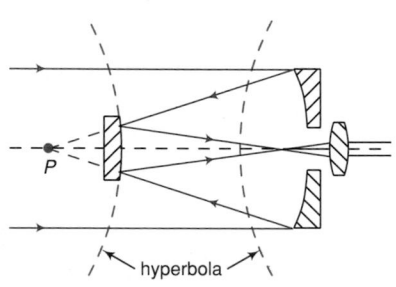

FIGURE 13
THE CASSEGRAIN TELESCOPE

ing of Japan from islands in the South Pacific. LORAN was used to draw highly accurate navigational maps, and one of these may have been used by the crew of the Enola Gay when it dropped the first atomic bomb over Hiroshima.[†]

The LORAN map in Figure 11 shows two sets of confocal hyperbolas (i.e., hyperbolas with the same foci) with the foci located at three radio broadcasting stations. As in Example 6, LORAN is based on the time difference between the reception of signals sent simultaneously from the stations in each broadcasting pair. A ship records the time difference as the signals from one pair (F_1 and F_2, say) arrive and determines its own position on one branch of a hyperbola. It uses the signals from the other pair (F_1 and F_3) to determine its position on one branch of a second hyperbola. The point of intersection of the two hyperbolas is the location of the ship.

LORAN has several advantages over other navigational systems. Radio waves are not affected by clouds or fog that do hamper celestial and visual navigation. Sunspot activity and atmospheric storms can bend radio signals so that direction finders may be inaccurate. However, these storms do not seriously affect the *velocity* of the radio signals so measurements based on time differences remain accurate.

Like the ellipse and parabola, the hyperbola has a useful **reflection property**. A light ray approaching (or leaving) one focus will be reflected toward (or away from) the other focus. This is illustrated in Figure 12, in which the direction of the arrows may be reversed. You are asked to prove this property in Problem 52.

The reflection property is exploited in the design of telescopes. In 1672, the French sculptor and astronomer Guillaume Cassegrain designed a reflecting telescope with a large parabolic mirror and a smaller hyperbolic mirror both sharing a common focus. This is illustrated in Figure 13. The same principle is used in the 200-inch Hale telescope on Mount Palomar in California. See also Problem 58.[‡]

[†] LORAN is now available in small two-seat and four-seat aircraft.

[‡] For more details of these and other interesting applications of the hyperbola, see Lee Whitt's delightful article, "The Standup Conic Presents: The Hyperbola and Applications," in *The UMAP Journal*, Vol. 5, No. 1, 1984, pp. 9–21.

■ PROBLEMS 8.3

■ SELF-QUIZ

I. The graph of $(y/3)^2 - (x/4)^2 = 1$ is a hyperbola that opens _____.
   a. to the left
   b. to the left and right
   c. upward
   d. upward and downward

II. The transverse axis of the hyperbola satisfying $x^2 - y^2 = 1$ is _____.
   a. the line segment between $(-\sqrt{2}, 0)$ and $(\sqrt{2}, 0)$
   b. the line segment between $(-1, 0)$ and $(1, 0)$
   c. the line $x = 0$
   d. the line $y = 0$

III. The vertices of the hyperbola satisfying $(x/4)^2 - (y/3)^2 = 1$ are _____.
   a. $(0, 0)$
   b. $y = 3x/4$ and $y = -3x/4$
   c. $(-5, 0)$ and $(5, 0)$
   d. $(-4, 0)$ and $(4, 0)$

IV. The asymptotes of the hyperbola satisfying $(x/4)^2 - (y/3)^2 = 1$ are _____.
   a. $y = 3x/4$ and $y = -3x/4$
   b. $y = 4x/3$ and $y = -4x/3$
   c. $y = x$ and $y = -x$
   d. the x-axis and the y-axis

■ DRILL

In Problems 1–20, the equation of a hyperbola is given. Locate its center, vertices, foci, transverse axis, and asymptotes; then sketch it.

1. $\dfrac{x^2}{16} - \dfrac{y^2}{25} = 1$

2. $\dfrac{y^2}{16} - \dfrac{x^2}{25} = 1$

3. $\dfrac{y^2}{25} - \dfrac{x^2}{16} = 1$

4. $\dfrac{x^2}{25} - \dfrac{y^2}{16} = 1$

5. $y^2 - x^2 = 1$
6. $x^2 - y^2 = 1$
7. $x^2 - 4y^2 = 9$
8. $4y^2 - x^2 = 9$
9. $y^2 - 4x^2 = 9$
10. $4x^2 - y^2 = 9$
11. $2x^2 - 3y^2 = 4$
12. $3x^2 - 2y^2 = 4$
13. $2y^2 - 3x^2 = 4$
14. $3y^2 - 2x^2 = 4$
15. $(x - 1)^2 - 4(y + 2)^2 = 4$

16. $\dfrac{(y + 3)^2}{4} - \dfrac{(x - 2)^2}{9} = 1$

17. $4x^2 + 8x - y^2 - 6y = 21$
18. $-4x^2 - 8x + y^2 - 6y = 20$
19. $2x^2 - 16x - 3y^2 + 12y = 45$
20. $2y^2 - 16y - 3x^2 + 12x = 45$

In Problems 21–30, find the standard equation of the indicated hyperbola.

21. foci: $(-4, 0)$, $(4, 0)$
    vertices: $(-3, 0)$, $(3, 0)$
22. foci: $(-1, 0)$, $(1, 0)$
    vertices: $(-\frac{1}{2}, 0)$, $(\frac{1}{2}, 0)$

23. foci: $(0, -3)$, $(0, 3)$
    vertices: $(0, -2)$, $(0, 2)$
24. foci: $(0, -6)$, $(0, 6)$
    vertices: $(0, -4)$, $(0, 4)$
25. foci: $(-1, 1)$, $(5, 1)$
    vertices: $(0, 1)$, $(4, 1)$
26. foci: $(-3, -1)$, $(-3, 5)$
    vertices: $(-3, 1)$, $(-3, 3)$
27. vertices: $(-2, 0)$, $(2, 0)$
    asymptotes: $y = \pm x$
28. vertices: $(0, -3)$, $(0, 3)$
    asymptotes: $y = \pm 2x$
29. vertices: $(1, 1)$, $(5, 1)$
    asymptotes: $y = 2x - 5$, $y = -2x + 7$
30. vertices: $(-2, -1)$, $(-2, 9)$
    asymptotes: $y = \frac{5}{2}x + 9$, $y = -\frac{5}{2}x - 1$

The **eccentricity** of a hyperbola is defined in Problem 53, and a useful computational result is stated in Problem 54. In Problems 31–40, compute the eccentricity of the specified hyperbola.

31. the hyperbola of Problem 1
32. the hyperbola of Problem 4
33. the hyperbola of Problem 7
34. the hyperbola of Problem 10
35. the hyperbola of Problem 3
36. the hyperbola of Problem 2
37. the hyperbola of Problem 15
38. the hyperbola of Problem 16
39. the hyperbola of Problem 19
40. the hyperbola of Problem 20

■ **APPLICATIONS**

41. Find an equation for the hyperbola with foci $(5, 0)$ and $(-5, 0)$ and vertices $(4, 0)$ and $(-4, 0)$.

42. Find an equation for the hyperbola with foci $(0, -5)$ and $(0, 5)$ and vertices $(0, -4)$ and $(0, 4)$.

43. Find an equation for the unique hyperbola with center at $(0, 0)$ and vertices $(-2, 0)$ and $(2, 0)$ which is asymptotic to the lines $y = \pm 3x$.

44. Shift the hyperbola of the preceding problem so that $(4, -3)$ is its center; write an equation satisfied by this translated hyperbola.

45. Find an equation for the hyperbola obtained by shifting the hyperbola of Problem 17 two units down and five units to the right.

*46. Find an equation for the hyperbola with center at $(0, 0)$ and transverse axis parallel to the x-axis which passes through the points $(1, 2)$ and $(5, 12)$.

47. Find an equation for the curve having the property that the difference of the distances from a point on the curve to the points $(1, -2)$ and $(4, 3)$ is 5.

*48. A tangent line to the hyperbola $9y^2 - 16x^2 = 25$

passes through the point $(1, 0)$. Find the point or points where this line is tangent to the hyperbola.

*49. The speed of sound in sea water (at $25°C$) is 1533 meters/second ($= 5030$ ft/sec). Submarine A heard the sound of an exploding depth charge 2 seconds before Submarine B heard the sound. The submarines are 4 kilometers apart, and the depth charge was dropped by an enemy destroyer. Find an equation for the hyperbola that contains the point at which the destroyer dropped the charge. [*Hint:* Draw a coordinate system and put A and B on the x-axis, equidistant from the origin.]

*50. In the coordinate system of Problem 49, assume that the positive x-axis points east. Submarine C, located 2 km due north of Submarine A, heard the sound of the depth charge 1 second after Submarine A heard it. Find the exact location of the destroyer at the moment it dropped the charge.

■ **SHOW/PROVE/DISPROVE**

51. Show that the line tangent to the hyperbola $(x/a)^2 - (y/b)^2 = 1$ at the point (x_0, y_0) satisfies the linear equation

$$\frac{x_0 \cdot x}{a^2} - \frac{y_0 \cdot y}{b^2} = 1.$$

52. **The reflection property of a hyperbola** Pick a point on a hyperbola. Prove that the tangent there makes equal angles with the two line segments connecting that point to each focus of the hyperbola.

53. The **eccentricity**, e, of a hyperbola is defined to be

$$e \equiv \frac{\text{distance between foci}}{\text{distance between vertices}}.$$

Show that the eccentricity of any hyperbola is greater than 1.

54. For the hyperbola satisfying $(x/a)^2 - (y/b)^2 = 1$, show that we can compute its eccentricity as follows: $e = \sqrt{a^2 + b^2}/|a| = \sqrt{1 + (b/a)^2}$.

55. Show that the graph of the curve $x^2 + 4x - 3y^2 + 6y = c$ is (a) a hyperbola if $c \neq -1$, (b) a pair of straight lines if $c = -1$. (Also find the equations of the two lines in the case $c = -1$.)

56. Find conditions relating the constants A, B, and C such that the graph of $2x^2 + Ax - 3y^2 + By = C$ is (a) a hyperbola, (b) a pair of straight lines.

■ **CHALLENGE**

*57. Prove that a hyperbola and an ellipse with the same two points as foci will intersect at right angles (i.e., their tangents at an intersection point are perpendicular).

**58. Use the reflecting properties of parabolas (see

page 555) and hyperbolas (see Problem 52 above) to design a simple telescope with two mirrors.[†] [*Hint:* Have incoming light reflect off a parabolic mirror onto another mirror shaped as one branch of a hyperbola.]

[†] This problem invites you to replicate the invention of N. Cassegrain.

8.4 SECOND-DEGREE EQUATIONS AND ROTATION OF AXES

The curves we have so far considered in this chapter have had their axes parallel to the two coordinate axes. This is not necessary. The curves sketched in Figure 1 are, respectively, an ellipse, a parabola, and a hyperbola. To obtain curves like those pictured, we must rotate the coordinate axes through an appropriate angle. How do we do so? Suppose that the x- and y-axes are rotated counterclockwise through an angle of θ with respect to the origin (see Figure 2). Let $P(x, y)$ represent a typical point in the coordinates x and y. We now seek a representation of that point in the "new" coordinates x' and y'. From Figure 2 we see that

$$x = \overline{0A} = \overline{0P} \cos(\theta + \alpha)$$

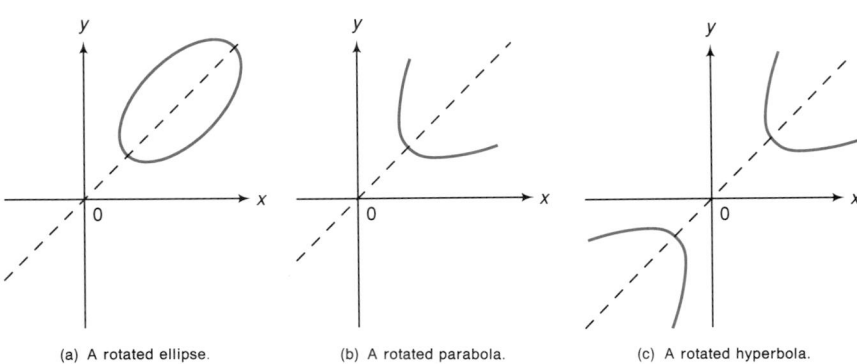

(a) A rotated ellipse.   (b) A rotated parabola.   (c) A rotated hyperbola.

FIGURE 1   THREE ROTATED CONIC SECTIONS

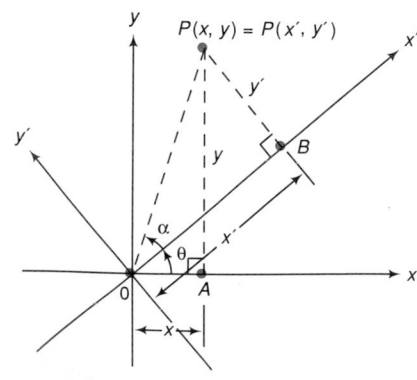

FIGURE 2
ROTATION OF AXES

When the x- and y-axes are rotated through an angle of θ, the coordinates of a point labelled (x, y) are now labelled (x′, y′). These are the coordinates with respect to the new, rotated coordinate axes.

and

$$y = \overline{AP} = \overline{0P} \sin(\theta + \alpha).$$

But

$$\cos(\theta + \alpha) = \cos \theta \cos \alpha - \sin \theta \sin \alpha,$$

and

$$\sin(\theta + \alpha) = \sin \alpha \cos \theta + \cos \alpha \sin \theta.$$

Thus,

$$x = \overline{0P}(\cos \theta \cos \alpha - \sin \theta \sin \alpha) \tag{1}$$

and

$$y = \overline{0P}(\sin \alpha \cos \theta + \cos \alpha \sin \theta). \tag{2}$$

Now from the right triangle OBP we find that

$$\sin \alpha = \frac{y'}{\overline{OP}} \qquad \text{or} \qquad \overline{OP} \sin \alpha = y'.$$

Similarly,

$$\cos \alpha = \frac{x'}{\overline{OP}} \qquad \text{or} \qquad \overline{OP} \cos \alpha = x'.$$

Substituting these last expressions into (1) and (2) yields

CONVERTING FROM (x', y') COORDINATES TO (x, y) COORDINATES

$$x = x' \cos \theta - y' \sin \theta \qquad \qquad \text{(3)}$$

$$y = x' \sin \theta + y' \cos \theta \qquad \qquad \text{(4)}$$

Equations (3) and (4) can be solved simultaneously to express the "new" coordinates x' and y' in terms of the "old" coordinates x and y. Switching from (x', y') to (x, y) is the same as rotating through an angle of $-\theta$. Since $\cos(-\theta) = \cos \theta$ and $\sin(-\theta) = -\sin \theta$, we obtain, from (3) and (4),

---

CONVERTING FROM (x, y) COORDINATES TO (x', y') COORDINATES

$$x' = x \cos \theta + y \sin \theta \qquad \qquad \text{(5)}$$

$$y' = -x \sin \theta + y \cos \theta. \qquad \qquad \text{(6)}$$

EXAMPLE 1
A ROTATED ELLIPSE

Find the equation of the curve obtained from the graph of $x^2 - xy + y^2 = 10$ by rotating the axes through an angle of $\pi/4$.

SOLUTION: Since $\cos \pi/4 = \sin \pi/4 = 1/\sqrt{2}$, we obtain, from equations (3) and (4),

$$x = \frac{x' - y'}{\sqrt{2}} \qquad \text{and} \qquad y = \frac{x' + y'}{\sqrt{2}}.$$

Substitution of these into the equation $x^2 - xy + y^2 = 10$ yields

$$\left(\frac{x' - y'}{\sqrt{2}} \right)^2 - \left(\frac{x' - y'}{\sqrt{2}} \right) \left(\frac{x' + y'}{\sqrt{2}} \right) + \left(\frac{x' + y'}{\sqrt{2}} \right)^2 = 10,$$

or

$$\frac{(x')^2 - 2x'y' + (y')^2}{2} - \left[\frac{(x')^2 - (y')^2}{2} \right] + \frac{(x')^2 + 2x'y' + (y')^2}{2} = 10,$$

which after simplification yields

$$\frac{(x')^2 + 3(y')^2}{2} = 10, \qquad \text{or} \qquad (x')^2 + 3(y')^2 = 20,$$

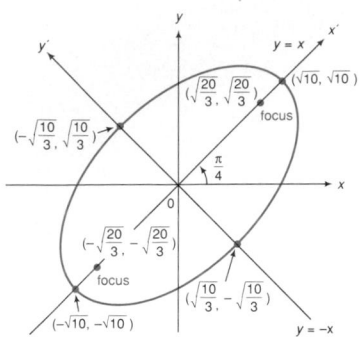

FIGURE 3
THE ELLIPSE $x^2 - xy + y^2 = 10$

and, finally,

$$\frac{(x')^2}{20} + \frac{(y')^2}{\frac{20}{3}} = 1.$$

In the new coordinates x' and y' this is the equation of an ellipse with $a = \sqrt{20}$, $b = \sqrt{\frac{20}{3}}$, $c = \sqrt{20 - \frac{20}{3}} = \sqrt{\frac{40}{3}}$, and foci $(\sqrt{\frac{40}{3}}, 0)$ and $(-\sqrt{\frac{40}{3}}, 0)$. [This ellipse can be obtained by rotating the ellipse $(x^2/20) + (y^2/\frac{20}{3}) = 1$ through an angle of $\pi/4$.] It is sketched in Figure 3. We can obtain the foci in the coordinates (x, y) by using equations (3) and (4). Since

$$x = \frac{x' - y'}{\sqrt{2}} \qquad \text{and} \qquad y = \frac{x' + y'}{\sqrt{2}},$$

we find that $(\sqrt{\frac{40}{3}}, 0)$ in the coordinates (x', y') comes from $(\sqrt{\frac{20}{3}}, \sqrt{\frac{20}{3}})$ in the coordinates (x, y) and $(-\sqrt{\frac{40}{3}}, 0)$ comes from $(-\sqrt{\frac{20}{3}}, -\sqrt{\frac{20}{3}})$. The major axis is on the line $y = x$.

EXAMPLE 2
A ROTATED HYPERBOLA

Find the equation of the curve obtained from the graph of $xy = 1$ by rotating the axes through an angle of $\pi/4$.

SOLUTION: As in Example 1, we have $x = (x' - y')/\sqrt{2}$ and $y = (x' + y')/\sqrt{2}$, so that

$$1 = xy = \left(\frac{x' - y'}{\sqrt{2}}\right)\left(\frac{x' + y'}{\sqrt{2}}\right) = \frac{(x')^2}{2} - \frac{(y')^2}{2}.$$

Thus in the new (rotated) coordinate system we obtain the hyperbola $[(x')^2/2] - [(y')^2/2] = 1$. This curve is sketched in Figure 4. We may say that the equation $xy = 1$ is really the equation of a "standard" hyperbola rotated through an angle of $\pi/4$.

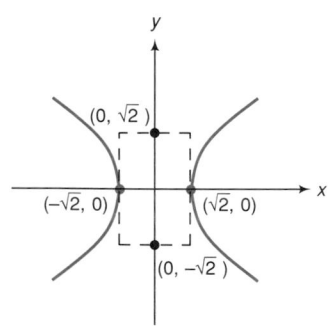

(a) The standard hyperbola $\dfrac{x^2}{2} - \dfrac{y^2}{2} = 1$.

The last two examples illustrate the fact that the equations of our three basic curves can take forms other than the standard forms given in Table 8.3.1 on page 565. It turns out that any second-degree equation $Ax^2 + Bxy + Cy^2 + Dx + Ey + F = 0$ can be written in the standard form whose graph is a circle, an ellipse, a parabola, a hyperbola, or a degenerate form such as a line, a pair of lines, a point, or an empty set of points. We will not discuss this fact further except to state the following theorem. (You are asked to prove the theorem in Problem 24.)

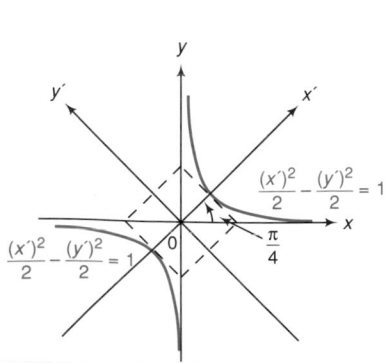

(b) The hyperbola $xy = 1$.

FIGURE 4
A ROTATED HYPERBOLA

| THEOREM 1 | DETERMINING THE TYPE OF CONIC FROM THE DISCRIMINANT |

Consider the second-degree equation

$$Ax^2 + Bxy + Cy^2 + Dx + Ey + F = 0. \tag{7}$$

(i) If $B^2 - 4AC = 0$, then (7) is the equation of a parabola, a line, or two parallel lines, or is imaginary.[†]

(ii) If $B^2 - 4AC < 0$, then (7) is the equation of a circle, an ellipse, a single point, or is imaginary.

(iii) If $B^2 - 4AC > 0$, then (7) is the equation of a hyperbola or two intersecting lines. ■

The number $B^2 - 4AC$ is called the **discriminant** of the conic section.

EXAMPLE 3
DETERMINING THE TYPE OF CURVE BY EVALUATING THE DISCRIMINANT

Determine the type of curve represented by the equation

$$16x^2 - 24xy + 9y^2 + 100x - 200y + 100 = 0. \tag{8}$$

Then write the equation in a standard form by finding an appropriate translation and rotation of axes.

SOLUTION: Here $A = 16$, $B = -24$, and $C = 9$, so that $B^2 - 4AC = (24)^2 - 4(16)(9) = 576 - 576 = 0$. Thus the equation represents a parabola (or a degenerate form of a parabola). To write it in a standard form, we first rotate the axes through an appropriate angle θ to eliminate the xy term in (8). To determine θ, we substitute

$$x = x' \cos \theta - y' \sin \theta \qquad \text{and} \qquad y = x' \sin \theta + y' \cos \theta$$

in equation (8). We then obtain

$$16(x' \cos \theta - y' \sin \theta)^2 - 24(x' \cos \theta - y' \sin \theta)(x' \sin \theta + y' \cos \theta)$$
$$+ 9(x' \sin \theta + y' \cos \theta)^2 + 100(x' \cos \theta - y' \sin \theta)$$
$$- 200(x' \sin \theta + y' \cos \theta) + 100 = 0.$$

The idea now is to choose θ so that the coefficient of the term $x'y'$ is zero. After simplification we find that this coefficient is given by

$$-24(\cos^2 \theta - \sin^2 \theta) - 14 \sin \theta \cos \theta.$$

[†] By "imaginary" we mean that the graph contains no real points; for example, $x^2 + 2xy + y^2 + 1 = 0$ satisfies $B^2 - 4AC = 0$, but there are no real values of x and y that satisfy $0 = x^2 + 2xy + y^2 + 1 = (x + y)^2 + 1$. (Explain why.)

Setting this expression equal to zero and using the fact that $\cos 2\theta = \cos^2 \theta - \sin^2 \theta$ and $\sin 2\theta = 2 \sin \theta \cos \theta$, we obtain

$$24 \cos 2\theta + 7 \sin 2\theta = 0,$$

or, dividing by $\cos 2\theta$ and rearranging terms,

$$\tan 2\theta = -\frac{24}{7}.$$

If $\tan 2\theta = -\frac{24}{7}$, then $\cos 2\theta = -\frac{7}{25}$,[†] and also

$$\cos \theta = \sqrt{\frac{1 + \cos 2\theta}{2}} = \sqrt{\frac{9}{25}} = \frac{3}{5}.$$

Furthermore, $\sin \theta = \frac{4}{5}$. At this point we do not need to know θ but only the values of $\cos \theta$ and $\sin \theta$. Next,

$$x = x' \cos \theta - y' \sin \theta = \frac{1}{5}(3x' - 4y')$$

and

$$y = x' \sin \theta + y' \cos \theta = \frac{1}{5}(4x' + 3y').$$

We substitute these expressions into (8) to obtain (after simplification)

$$25(y')^2 - 200y' - 100x' + 100 = 0.$$

To further simplify, we divide by 25 and complete the square:

$$(y')^2 - 8y' - 4x' + 4 = 0, \quad \text{or} \quad (y' - 4)^2 - 16 - 4x' + 4 = 0, \quad \text{or}$$

$$(y' - 4)^2 = 4x' + 12 = 4(x' + 3).$$

This is the equation of a parabola with vertex at $(-3, 4)$ in the new (rotated) coordinate system. To sketch the parabola, we need to calculate the angle of rotation given by $\theta = \cos^{-1} \frac{3}{5} \approx 0.927 \approx 53°$. The curve is sketched in Figure 5.

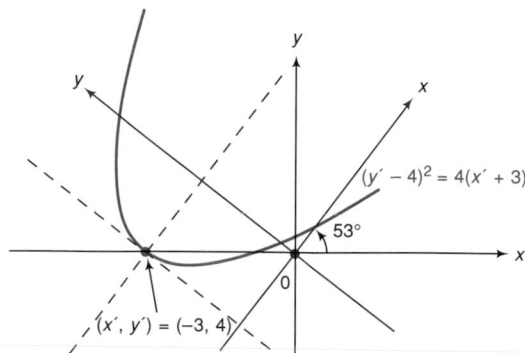

FIGURE 5 THE PARABOLA $16x^2 - 24xy + 9y^2 + 100x - 200y + 100 = 0$

[†] Alternatively, we could choose $\cos 2\theta = \frac{7}{25}$, yielding $\cos \theta = \frac{4}{5}$ and $\sin \theta = -\frac{3}{5}$. This, however, is merely a 90° rotation of the coordinate axes obtained with the choice $\cos 2\theta = -\frac{7}{25}$.

We note here that the xy term in (7) can always be eliminated by a rotation of axes through an angle θ, where θ is given by the equation

> **ROTATING THROUGH AN ANGLE θ WILL ELIMINATE THE xy TERM WHEN**
>
> $$\cot 2\theta = \frac{A - C}{B}. \tag{9}$$

The proof of this fact follows exactly as in the derivation of the angle of rotation in Example 3 and is left as an exercise. (See Problem 21.)

We conclude by pointing out that instead of rotating axes, we can rotate curves, keeping the axes fixed. Note the following important fact:

Rotating a curve through an angle θ has the effect of rotating the axes through an angle $-\theta$.

■ PROBLEMS 8.4

■ SELF-QUIZ

I. If the graph of $y = x^2$ is rotated $\pi/\sqrt{19}$, then it turns into a _____.
   a. parabola
   b. straight line
   c. ellipse
   d. hyperbola

II. The set of angles through which the graph of $y = x^2$ can be rotated to produce an ellipse is _____.
   a. empty
   b. $\{\theta: \theta = k \cdot \pi/4, k \text{ is an integer}\}$
   c. $\{\theta: -\pi/6 < \theta < \pi/8\}$
   d. $\{\pi/6, \pi/4, \pi/3, \pi/2\}$

III. If the graph of $x^2 - (y/3)^2 = 1$ is rotated $\pi/2$, the resulting curve satisfies equation _____.
   a. $(x')^2 - (y'/3)^2 = 1$
   b. $(y'/3)^2 - (x')^2 = 1$
   c. $(-y')^2 - (x'/3)^2 = 1$
   d. $9(x')^2 - (y')^2 = 1$

IV. Rotating a curve through an angle θ has the same result as rotating the coordinate axes through the angle _____.
   a. $\pi/2 + \theta$
   b. $\pi - \theta$
   c. $\pi + 2\theta$
   d. $2\pi - \theta$
   e. $-\theta$
   f. $\theta - 2\pi$

V. If the graph of $Ax^2 + Bxy + Cy^2 + Dx + Ey + F = 0$ is an ellipse, then $B^2 - 4AC$ _____.
   a. is undefined
   b. $= 0$
   c. < 0
   d. > 0
   e. $= \pi$
   f. $\to 0$ as $x \to \infty$

■ DRILL

1. Describe the curve obtained from the graph of $4x^2 - 2xy + 4y^2 = 45$ by rotating the axes through an angle of $\pi/4$.

2. Describe the curve obtained from the graph of $x^2 + 2\sqrt{3}xy - y^2 = 4$ by rotating the axes through an angle of $\pi/6$.

3. What is the equation of the line obtained from the line $2x - 3y = 6$ by rotating the axes through an angle of $\pi/6$?

4. Find the equation of the line obtained from the line $ax + by + c = 0$ by rotating the axes through an angle of θ.

5. Find the equation of a parabola obtained from the parabola $y^2 = -12x$ if the axes are rotated until the axis of the parabola coincides with the line $y = \sqrt{3}x$.

*6. Find the equation of the ellipse whose major axis is on the line $y = -x$, which is obtained by rotating the ellipse $(x/5)^2 + (y/4)^2 = 1$. [*Hint:* Find the angle through which the axes must be rotated to accomplish this.]

CONTINUED

■ APPLICATIONS

In Problems 7–20, use the results of Problems 21 and 22 to find a rotation of the coordinate axes such that the given equation has no xy term when written in the new coordinates (i.e., such that the $x'y'$ term has zero coefficient). Describe the curve and then sketch it.

7. $4x^2 + 4xy + y^2 = 9$
8. $4x^2 + 4xy - y^2 = 9$
9. $3x^2 - 2xy - 5 = 0$
10. $xy = 2$

11. $xy = a, a > 0$
12. $xy = a, a < 0$
13. $4x^2 + 4xy + y^2 + 20x - 10y = 0$
14. $x^2 + 4xy + 4y^2 - 6 = 0$
15. $2x^2 + xy + y^2 = 4$
16. $9x^2 + 6xy + y^2 + 10x - 30y = 0$
17. $3x^2 - 6xy + 5y^2 = 36$
18. $x^2 - 3xy + 4y^2 = 1$
19. $3y^2 - 4xy + 30y - 20x + 40 = 0$
20. $6x^2 + 5xy - 6y^2 + 7 = 0$

■ SHOW/PROVE/DISPROVE

21. Show that if $A \neq C$, then the xy term in the second-degree equation

$$Ax^2 + Bxy + Cy^2 + Dx + Ey + F = 0$$

will be eliminated by rotation through an angle θ where θ satisfies the relation

$$\cot 2\theta = \frac{A - C}{B}.$$

22. Show that if $A = C$ in the preceding problem,

then the xy term will be eliminated by a rotation through an angle of either $\pi/4$ or $-\pi/4$.

*23. Suppose that a rotation converts $Ax^2 + Bxy + Cy^2$ into $A'(x')^2 + B'(x'y') + C'(y')^2$.
   a. Show that $A + C = A' + C'$.
   b. Show that $B^2 - 4AC = (B')^2 - 4A'C'$.

*24. Use the results of Problems 21–23 to prove Theorem 1. [*Hint:* Rotate the axes so that $B' = 0$.]

25. Show that $xy = k$ is the equation of a hyperbola for every constant k, $k \neq 0$.

■ ANSWERS TO SELF-QUIZ

   I. a (once a parabola, always a parabola)
  II. a (rotation does not change the graph's shape)
 III. c   IV. d, e   V. c

8.5 THE POLAR-COORDINATE SYSTEM

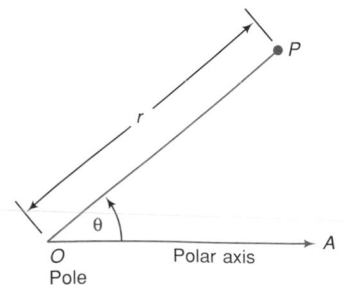

FIGURE 1
REPRESENTING A POINT IN TERMS OF r AND θ

In Section 0.2, we introduced the Cartesian plane (the xy-plane), and up to this moment we have represented points in the plane by their x- and y-coordinates. There are many other ways to represent points in the plane, the most important of which is called the **polar-coordinate system**.

We begin by choosing a fixed point, which we label O, and a ray (half line) that extends in one direction from O, which we label OA. The fixed point is called the **pole** (or **origin**), and the ray OA is called the **polar axis**. In Figure 1, the polar axis is drawn as a horizontal ray that extends indefinitely to the right (just like the positive x-axis). This representation is a matter of convention, although it would be correct to have the polar axis extend in any direction.

If $P \neq O$ is any other point in the plane, let r denote the distance between O and P, and let θ represent the angle (in radians) between OA and OP, measured counterclockwise from OA to OP. With this representation, we initially use the convention that $0 \leq \theta < 2\pi$ and $r > 0$ (except at the pole). Then every point in the plane, except the pole, can be represented by a pair

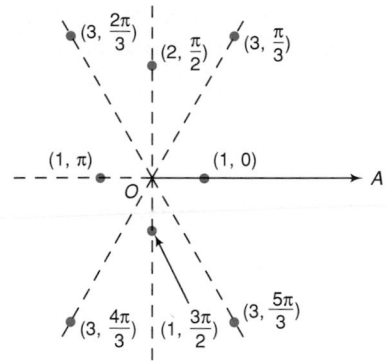

FIGURE 2
POINTS IN POLAR COORDINATES

of numbers (r, θ), where $r > 0$ and $0 \leq \theta < 2\pi$. The pole can be represented as $(0, \theta)$ for any number θ. The representation $P = (r, \theta)$ is called the **polar representation** of the point, and r and θ are called the **polar coordinates** of P. Some typical points are sketched in Figure 2.

If we are given two numbers r and θ, with $r \geq 0$ and $0 \leq \theta < 2\pi$, then we can draw the point $P = (r, \theta)$ in the plane. But what do we do if either $r < 0$ or $\theta > 2\pi$? To avoid this difficulty, we simply extend our definition. Let (r, θ) be any pair of real numbers. To locate the point $P = (r, \theta)$, we first rotate the polar axis through an angle of θ in the counterclockwise direction if θ is positive and in the clockwise direction if θ is negative. Let us call this new ray OB. Then if $r > 0$, the point P is placed on the ray OB, r units from the pole. If $r < 0$, the ray OB is extended backward through the pole and the point P is then located $|r|$ units from the pole along this extended ray.

EXAMPLE 1
PLOTTING POINTS IN POLAR
COORDINATES

Plot the following points:

(a) $\left(2, \dfrac{8\pi}{3} \right)$ **(b)** $\left(-1, \dfrac{\pi}{4} \right)$

(c) $\left(3, -22\dfrac{1}{2}\pi \right)$ **(d)** $\left(-2, -\dfrac{\pi}{2} \right)$

SOLUTION: The four points are plotted in Figure 3.

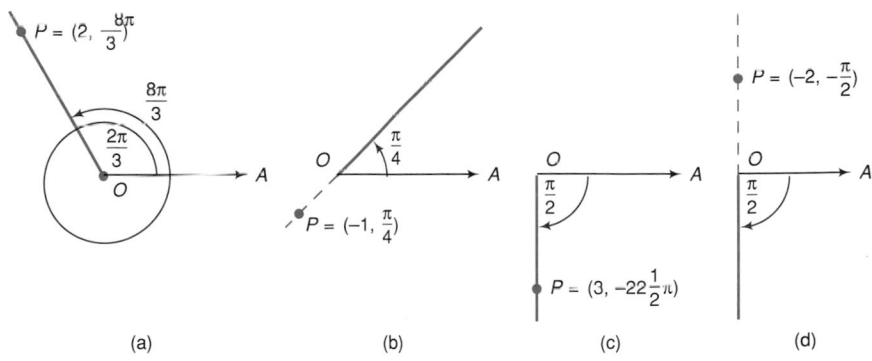

(a)   (b)   (c)   (d)

FIGURE 3 FOUR POINTS IN POLAR COORDINATES

In (a), since $8\pi/3 = 2\pi + (2\pi/3)$, the line OB makes an angle of $2\pi/3$ with OA and the point P lies 2 units along this line.

In (b) we rotate $\pi/4$ units and then extend OB backward since $r < 0$. The point P is located 1 unit along this extended line.

In (c) we rotate in a clockwise direction and find that the line OB makes an angle of $\pi/2$ with the polar axis since $22\frac{1}{2}\pi = 11(2\pi) + (\pi/2)$. The point P is located 3 units along OB.

In (d) we rotate OA in a clockwise direction $\pi/2$ radians and then extend OB backward (and therefore upward) since $r < 0$. The point P is then located 2 units along this line.

We immediately notice that if we do not restrict r and θ, then there are many (in fact, an infinite number of) representations for each point in the plane. For example, in Figure 3(d) we see that $(-2, -\pi/2) = (2, \pi/2) = (2, (\pi/2) + 2n\pi)$ for any integer n. In general,

$$P = (r, \theta) = (-r, \theta + \pi) = (r, \theta + 2n\pi) \qquad \text{for } n = \pm 1, \pm 2, \pm 3, \ldots$$

(see Problem 37). Actually, these ordered pairs are not equal. The equal sign indicates that the points they represent are the same. Therefore, with this understanding we will write $(r_1, \theta_1) = (r_2, \theta_2)$ in the rest of this chapter when the points they represent are the same, even though r_1 may not be equal to r_2, and θ_1 may not be equal to θ_2. However, to avoid difficulties, if a point $P \neq O$ is given, then we can write $P = (r, \theta)$ in polar coordinates in a unique way if we specify that $r > 0$ and $0 \leq \theta < 2\pi$. If P is the pole O, then $P = (0, \theta)$ for any real number θ, and there is no unique representation unless we specify a value for θ.

What is the relationship between polar and rectangular (Cartesian) coordinates? To find it, we place the pole at the origin in the xy-plane with the polar axis along the x-axis as in Figure 4. Let $P = (x, y) = (r, \theta)$ be a point in the plane. Then as is evident from Figure 4, we have

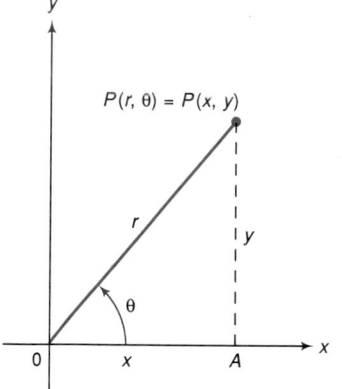

FIGURE 4

A POINT IN BOTH RECTANGULAR AND POLAR COORDINATES

CHANGING FROM POLAR TO RECTANGULAR COORDINATES

$$\cos \theta = \frac{x}{r} \qquad \text{and} \qquad \sin \theta = \frac{y}{r},$$

or

$$x = r \cos \theta \qquad \text{and} \qquad y = r \sin \theta. \tag{1}$$

CHANGING FROM RECTANGULAR TO POLAR COORDINATES

From the Pythagorean theorem we see that $x^2 + y^2 = r^2$, so that if we specify that $r > 0$, we have

$$r = \sqrt{x^2 + y^2}. \tag{2}$$

Finally, we can calculate θ from x and y by the relation

$$\tan \theta = \frac{y}{x} \qquad \text{if } x \neq 0. \tag{3}$$

Before citing examples, we note that these conversion formulas have only been illustrated for $r > 0$ and $0 \leq \theta < \pi/2$. The fact that $0 \leq \theta < \pi/2$ in Figure 4 is irrelevant since, using the circular definition of $\sin x$ and $\cos x$ in Appendix 1 (Section A1.2), we see that $x/r = \cos \theta$ and $y/r = \sin \theta$ for any real number θ.

⊗ **WARNING:** Formula (3) does *not* determine θ uniquely. The signs of x and y must be taken into account. For example, for both $(1, 1)$ and $(-1, -1)$,

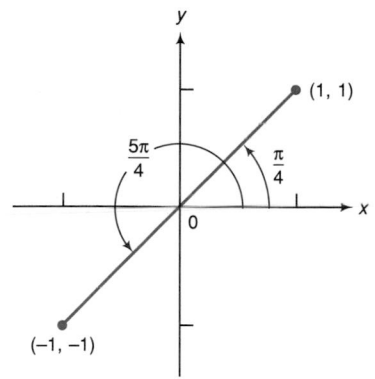

FIGURE 5

TAN $\theta = \dfrac{y}{x}$ DOES NOT DETERMINE θ UNIQUELY

The point $(1, 1)$ *makes an angle of* $\dfrac{\pi}{4}$ *with the positive x-axis while the point* $(-1, -1)$ *makes an angle of* $\dfrac{5\pi}{4}$. *In both cases* $\tan \theta = \dfrac{y}{x} = 1$.

$\tan \theta = 1$. But in the first case $\theta = \pi/4$ and in the second case $\theta = 5\pi/4$ (see Figure 5). ⊗

A good rule that works in all cases where $x \neq 0$ is given below. In formula (3),

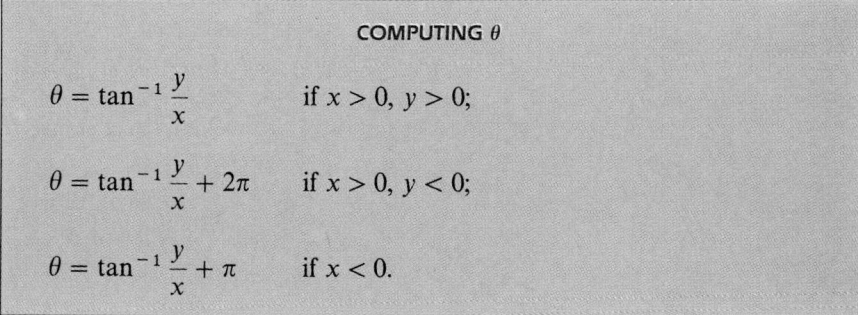

COMPUTING θ

$$\theta = \tan^{-1} \frac{y}{x} \qquad \text{if } x > 0, y > 0;$$

$$\theta = \tan^{-1} \frac{y}{x} + 2\pi \qquad \text{if } x > 0, y < 0;$$

$$\theta = \tan^{-1} \frac{y}{x} + \pi \qquad \text{if } x < 0.$$

Note that by the definition on page 459, $-\dfrac{\pi}{2} < \tan^{-1} \dfrac{y}{x} < \dfrac{\pi}{2}$.

To deal with the case $r < 0$, we note the identity

$$(-r, \theta) = (r, \theta + \pi). \tag{4}$$

The proof is left as an exercise (see Problem 37).

With these formulas we can always convert from polar to rectangular coordinates in a unique way. To convert in this way, we must specify that $r > 0$ and $0 \leq \theta < 2\pi$.

EXAMPLE 2
CONVERTING FROM POLAR TO
RECTANGULAR COORDINATES

Convert from polar to rectangular coordinates:

(a) $\left(3, \dfrac{\pi}{6}\right)$ **(b)** $\left(4, \dfrac{2\pi}{3}\right)$ **(c)** $\left(-6, \dfrac{\pi}{4}\right)$ **(d)** $(2, 0)$ **(e)** $(1, -\pi)$.

SOLUTION:

(a) $x = 3 \cos \pi/6$ and $y = 3 \sin \pi/6$, so that $(x, y) = (3\sqrt{3}/2, \frac{3}{2})$.
(b) $x = 4 \cos 2\pi/3 = -4 \cos \pi/3 = -\frac{4}{2} = -2$ and $y = 4 \sin 2\pi/3 = 4 \sin \pi/3 = 4\sqrt{3}/2 = 2\sqrt{3}$, so that $(x, y) = (-2, 2\sqrt{3})$.
(c) $(-6, \pi/4) = (6, (\pi/4) + \pi) = (6, 5\pi/4)$ [from (4)], so we have $x = 6 \cos 5\pi/4 = 6(-\sqrt{2}/2) = -3\sqrt{2}$, $y = 6 \sin 5\pi/4 = -3\sqrt{2}$, and $(x, y) = (-3\sqrt{2}, -3\sqrt{2})$.
(d) $x = 2 \cos 0 = 2$ and $y = 2 \sin 0 = 0$, so that $(x, y) = (2, 0)$.
(e) Since $(1, -\pi) = (1, -\pi + 2\pi) = (1, \pi)$, we have $x = \cos \pi = -1$, $y = \sin \pi = 0$, and $(x, y) = (-1, 0)$.

These five points are sketched in Figure 6.

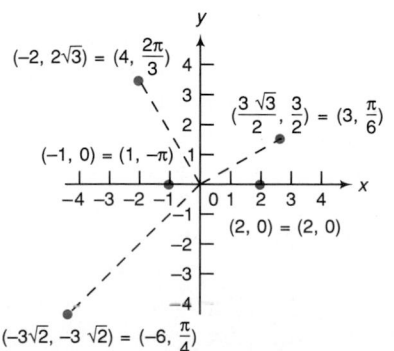

FIGURE 6
FIVE POINTS IN BOTH RECTANGULAR AND
POLAR COORDINATES

EXAMPLE 3
CONVERTING FROM RECTANGULAR
TO POLAR COORDINATES

Convert from rectangular to polar coordinates (with $r > 0$ and $0 \le \theta < 2\pi$):

(a) $(1, \sqrt{3})$ (b) $(2, -2)$ (c) $(-4\sqrt{3}, -4)$.

SOLUTION:

(a) $r = \sqrt{x^2 + y^2} = \sqrt{1^2 + \sqrt{3}^2} = \sqrt{1+3} = 2$ and $\tan \theta = \sqrt{3}/1 = \sqrt{3}$. Since $(1, \sqrt{3})$ is in the first quadrant, we have $\theta = \tan^{-1} \sqrt{3} = \pi/3$, so that $(r, \theta) = (2, \pi/3)$.

(b) $r = \sqrt{(2)^2 + (-2)^2} = \sqrt{8} = 2\sqrt{2}$ and $\tan \theta = -\frac{2}{2} = -1$. Since $(2, -2)$ is in the fourth quadrant, $\theta = \tan^{-1}(-1) + 2\pi = -\pi/4 + 2\pi = 7\pi/4$ and $(r, \theta) = (2\sqrt{2}, 7\pi/4)$.

(c) $r = \sqrt{(-4\sqrt{3})^2 + (-4)^2} = \sqrt{48 + 16} = 8$ and $\tan \theta = -4/(-4\sqrt{3}) = 1/\sqrt{3}$. Since $(-4\sqrt{3}, -4)$ is in the third quadrant, we have $\theta = \tan^{-1}(1/\sqrt{3}) + \pi = \pi/6 + \pi = 7\pi/6$ and $(r, \theta) = (8, 7\pi/6)$.

These three points are sketched in Figure 7.

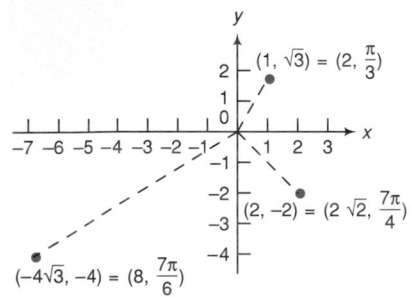

FIGURE 7
THREE POINTS IN BOTH RECTANGULAR
AND POLAR COORDINATES

■ PROBLEMS 8.5

■ SELF-QUIZ

I. Three of the following ordered pairs of numbers are polar coordinates for the same point in the plane. _____ is the odd one.
   a. $(2, 3\pi/4)$ b. $(-2, -\pi/4)$
   c. $(-2, -3\pi/4)$ d. $(2, -5\pi/4)$

II. Three of the following ordered pairs are polar coordinates for points lying on a single line through the origin. _____ is not on that line.
   a. $(1, \pi/6)$ b. $(2, 13\pi/6)$
   c. $(-3, 7\pi/6)$ d. $(-4, 5\pi/6)$

III. Three of the following ordered pairs are polar coordinates for points lying on a single circle centered at the origin. _____ is not on that circle.
   a. $(1, \pi/8)$ b. (π^e, π)
   c. $(2/(1-3), 5\pi)$ d. $(e^0, -3\pi/2)$

IV. Three of the following ordered pairs are polar coordinates for points which are moderately close to each other. _____ is not in that cluster.
   a. $(9, -5\pi/7)$ b. $(9, -7\pi/5)$
   c. $(\pi^2, 9\pi/13)$ d. $(10, 27\pi/10)$

■ DRILL

In Problems 1–18, a point is specified in polar coordinates. Find the rectangular coordinates for that point, plot the point in the xy-plane, and label the point with both coordinate representations.

1. $(3, 0)$ 2. $(-5, 0)$
3. $(4, \pi/4)$ 4. $(-4, \pi/4)$
5. $(6, 7\pi/6)$ 6. $(-3, -\pi/6)$
7. $(5, \pi)$ 8. $(5, -\pi)$
9. $(5, -\pi/3)$ 10. $(-2, 2\pi/3)$
11. $(-2, 3\pi/4)$ 12. $(-2, 5\pi/4)$
13. $(2, 7\pi/4)$ 14. $(-2, -\pi/4)$
15. $(1, 3\pi/2)$ 16. $(1, -\pi/2)$
17. $(-1, \pi/2)$ 18. $(-1, -3\pi/2)$

In Problems 19–36, a point is given in rectangular coordinates. Write the point in polar coordinates with $r \ge 0$ and $0 \le \theta < 2\pi$, then plot the point.

19. $(0, 0)$ 20. $(7, 0)$ 21. $(3, 0)$
22. $(-3, 0)$ 23. $(0, 1)$ 24. $(0, -1)$
25. $(1, 1)$ 26. $(-2, -2)$ 27. $(3, -3)$
28. $(-4, 4)$ 29. $(2, 2\sqrt{3})$ 30. $(-2, 2\sqrt{3})$
31. $(5, -5\sqrt{3})$ 32. $(-7, -7\sqrt{3})$
33. $(\sqrt{3}, 1)$ 34. $(\sqrt{3}, -1)$
35. $(-4\sqrt{3}, 4)$ 36. $(-8\sqrt{3}, -8)$

CONTINUED

■ **SHOW/PROVE/DISPROVE**

37. Show that (r, θ) and $(-r, \theta + \pi)$ are polar coordinates for the same point. [*Hint:* Draw a sketch.]

38. Show that $\sqrt{r_1{}^2 - 2r_1 r_2 \cos(\theta_2 - \theta_1) + r_2{}^2}$ is the distance between the points with polar coordinates (r_1, θ_1) and (r_2, θ_2).

■ **ANSWERS TO SELF-QUIZ**

I. c   II. d   III. b   IV. a

8.6 GRAPHING IN POLAR COORDINATES

In rectangular coordinates we define the graph of the equation $y = f(x)$, or more generally, $F(x, y) = 0$, as the set of points (x, y) whose coordinates satisfy the equation. In polar coordinates, however, we must be careful, since each point in the plane has an infinite number of representations.

DEFINITION   **GRAPH IN POLAR COORDINATES**

The **graph** of an equation written in polar coordinates r and θ consists of those points P having at least one representation $P = (r, \theta)$ whose coordinates satisfy the equation. ■

EXAMPLE 1
THE GRAPH OF A CIRCLE IN POLAR COORDINATES

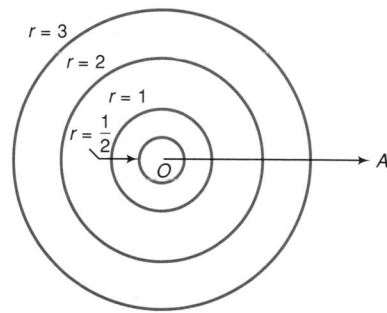

FIGURE 1
FOUR CIRCLES

Find the graph of the polar equation $r = 1$.

SOLUTION: The set of points for which $r = 1$ is simply the set of points one unit from the pole, which is the definition of the unit circle. Thus the graph of $r = 1$ is the unit circle. To see this in another way, note that $r = \sqrt{x^2 + y^2} = 1$ implies that $x^2 + y^2 = 1$. This curve is sketched in Figure 1, together with the curves $r = 2$ (the circle of radius 2), $r = 3$, and $r = \frac{1}{2}$.

REMARK: Every point on the circle $r = 1$ has a representation for which $r \neq 1$ since the point $(-1, \theta)$ is on the circle for any real number θ. This presents no problem since $(-1, \theta) = (1, \theta + \pi)$.

EXAMPLE 2
THE GRAPH OF A LINE IN POLAR COORDINATES

Sketch the curve $\theta = \pi/4$.

SOLUTION: The curve $\theta = \pi/4$ is the straight line passing through the pole making an angle of $\pi/4$ with the polar axis (see Figure 2). It extends in both directions because it contains points of the form $(-r, \pi/4) = (r, 5\pi/4)$. To see in another way that this graph is a straight line, note that $y/x = \tan \theta = \tan \pi/4 = 1$, so that $y = x$ is the rectangular representation of the line.

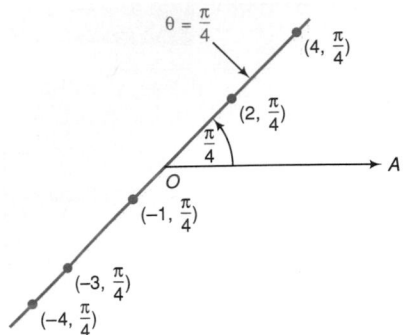

FIGURE 2

GRAPH OF THE LINE $\theta = \dfrac{\pi}{4}$

Examples 1 and 2 give us some very useful information. Previously, we sketched graphs by representing a typical point $P = (x_0, y_0)$ as the intersection of the line $x = x_0$ with the line $y = y_0$ [Figure 3(a)]. In polar coordinates we can represent the point $P = (r_0, \theta_0)$ as the intersection of the circle $r = r_0$ with the ray $\theta = \theta_0$ ($r \geq 0$) [Figure 3(b)].

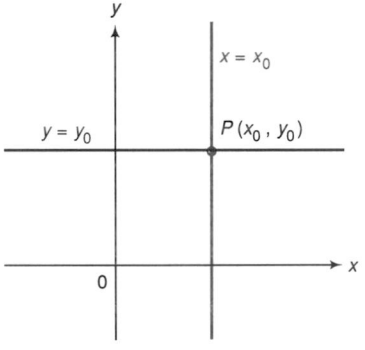

(a) The graphs of $x = x_0$, $y = y_0$ (Cartesian coordinates).

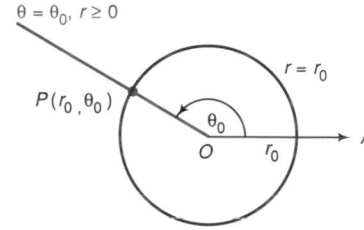

(b) The graphs of $r = r_0$, $\theta = \theta_0$ (polar coordinates).

FIGURE 3 IN CARTESIAN COORDINATES A POINT IS DETERMINED BY THE INTERSECTION OF A VERTICAL LINE AND A HORIZONTAL LINE. IN POLAR COORDINATES A POINT IS DETERMINED BY THE INTERSECTION OF A RAY THAT STARTS AT THE POLE AND A CIRCLE CENTERED AT THE POLE.

We now consider the graphs of some more general curves in polar coordinates. To aid us in obtaining sketches of these curves, we cite three rules that are often useful.

RULES OF SYMMETRY

(i) If in a polar equation θ can be replaced by $-\theta$ without changing the equation, then the polar graph is symmetric about the polar axis [see Figure 4(a)].

(ii) If θ can be replaced by $\pi - \theta$ without changing the equation, then the polar graph is symmetric about the line $\theta = \pi/2$ [see Figure 4(b)].

(iii) If r can be replaced by $-r$, or equivalently, if θ can be replaced by $\theta + \pi$, without changing the equation, then the polar graph is symmetric about the pole [see Figure 4(c)].

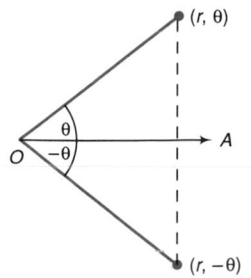

(a) Symmetry about polar axis: $f(r, \theta) = f(r, -\theta)$.

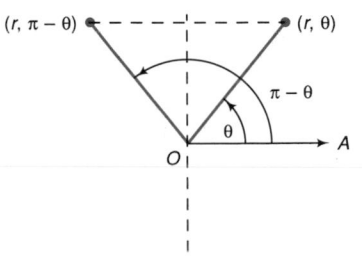

(b) Symmetry about line $\theta = \pi/2$: $f(r, \pi - \theta) = f(r, \theta)$.

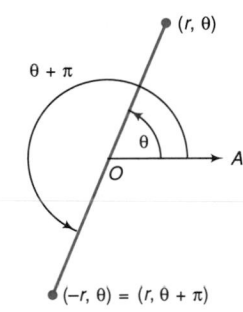

(c) Symmetry about pole: $f(-r, \theta) = f(r, \theta)$.

FIGURE 4 ILLUSTRATION OF THREE RULES OF SYMMETRY

REMARK: These rules do not give necessary conditions for symmetry. That is, there may be symmetry in a situation where none of these rules holds.

EXAMPLE 3
A CIRCLE

Sketch the curve $r = \cos \theta$.

SOLUTION: Since $\cos(-\theta) = \cos \theta$, rule (i) implies that the graph is symmetric about the polar axis, so we need only consider values of θ between 0 and π. In Table 1 we tabulate r for some typical values of θ in $[0, \pi]$. From this table we can plot the curve for $0 \le \theta < \pi$. This is done in Figure 5. Here we have used the facts that

$$\left(-\frac{1}{2}, \frac{2\pi}{3}\right) = \left(\frac{1}{2}, \frac{5\pi}{3}\right), \qquad \left(-\frac{\sqrt{2}}{2}, \frac{3\pi}{4}\right) = \left(\frac{\sqrt{2}}{2}, \frac{7\pi}{4}\right)$$

and

$$\left(-\frac{\sqrt{3}}{2}, \frac{5\pi}{6}\right) = \left(\frac{\sqrt{3}}{2}, \frac{11\pi}{6}\right).$$

TABLE 1

θ	0	$\dfrac{\pi}{6}$	$\dfrac{\pi}{4}$	$\dfrac{\pi}{3}$	$\dfrac{\pi}{2}$	$\dfrac{2\pi}{3}$	$\dfrac{3\pi}{4}$	$\dfrac{5\pi}{6}$	π
$r = \cos \theta$	1	$\dfrac{\sqrt{3}}{2}$	$\dfrac{\sqrt{2}}{2}$	$\dfrac{1}{2}$	0	$-\dfrac{1}{2}$	$-\dfrac{\sqrt{2}}{2}$	$-\dfrac{\sqrt{3}}{2}$	-1

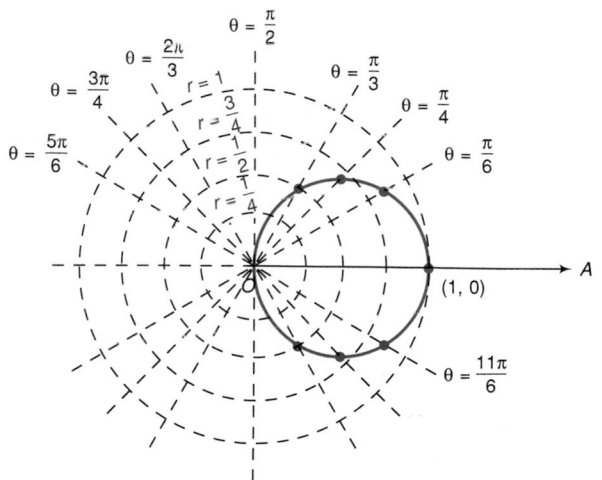

FIGURE 5 THE CIRCLE $r = \cos \theta$

From the sketch in Figure 5 it appears that the graph is a circle. To see this analytically, we convert to rectangular coordinates. Since $r = \cos \theta$, we have, for $r \ne 0$,

$$r^2 = r \cos \theta$$

($r = 0$ is just the origin, which is already known to be on the graph), or

$$x^2 + y^2 = x.$$

This equation can be written as

$$x^2 + y^2 - x = 0.$$

Then completing the square, we obtain $(x - \frac{1}{2})^2 + y^2 = \frac{1}{4}$, which is the equation of a circle centered at $(\frac{1}{2}, 0)$ with radius $\frac{1}{2}$.

CIRCLE IN POLAR COORDINATES

The graph of the equation $r = a \cos \theta + b \sin \theta$ is a circle for any real numbers a and b. (See Problems 89, 90, and 91.)

EXAMPLE 4
A CARDIOID

Sketch the curve $r = 1 + \sin \theta$.

SOLUTION: Since $\sin(\pi - \theta) = \sin \theta$, the curve is symmetric about the line $\theta = \pi/2$. We therefore need only consider values of θ in $[0, \pi/2]$ and $[3\pi/2, 2\pi]$. Typical values of the function are given in Table 2.

TABLE 2

θ	0	$\dfrac{\pi}{6}$	$\dfrac{\pi}{4}$	$\dfrac{\pi}{3}$	$\dfrac{\pi}{2}$	$\dfrac{3\pi}{2}$	$\dfrac{5\pi}{3}$	$\dfrac{7\pi}{4}$	$\dfrac{11\pi}{6}$
$r = 1 + \sin \theta$	1	$\dfrac{3}{2}$	$1 + \dfrac{\sqrt{2}}{2}$	$1 + \dfrac{\sqrt{3}}{2}$	2	0	$1 - \dfrac{\sqrt{3}}{2}$	$1 - \dfrac{\sqrt{2}}{2}$	$\dfrac{1}{2}$

We use these values in Figure 6 and then use symmetry to reflect the curve about the line $\theta = \pi/2$. The heart-shaped curve we have sketched is called a **cardioid**, from the Greek *kardia*, meaning "heart." Note that the cardioid has a corner at the pole.

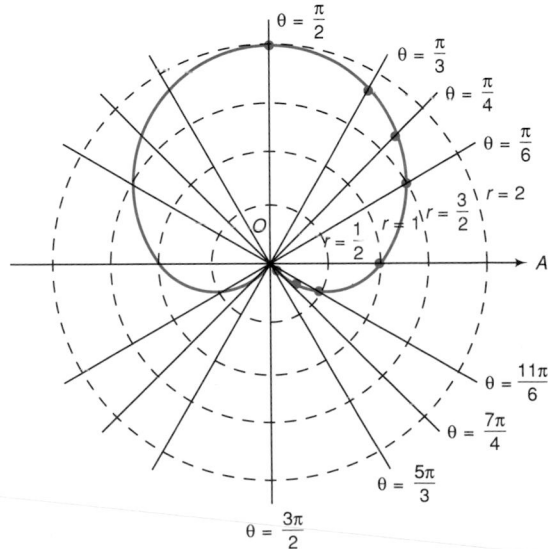

FIGURE 6   THE CARDIOID $r = 1 + \sin \theta$

EXAMPLE 5
A LIMAÇON

Sketch the curve $r = 3 - 2 \cos \theta$.

SOLUTION: The curve is symmetric about the polar axis, and we therefore calculate values of r for θ in $[0, \pi]$ (Table 3).

TABLE 3

θ	0	$\frac{\pi}{6}$	$\frac{\pi}{4}$	$\frac{\pi}{3}$	$\frac{\pi}{2}$	$\frac{2\pi}{3}$	$\frac{3\pi}{4}$	$\frac{5\pi}{6}$	π
$r = 3 - 2\cos\theta$	1	$3 - \sqrt{3}$	$3 - \sqrt{2}$	2	3	4	$3 + \sqrt{2}$	$3 + \sqrt{3}$	5

We then obtain the graph sketched in Figure 7. The curve sketched in Figure 7 is called a **limaçon**.[†]

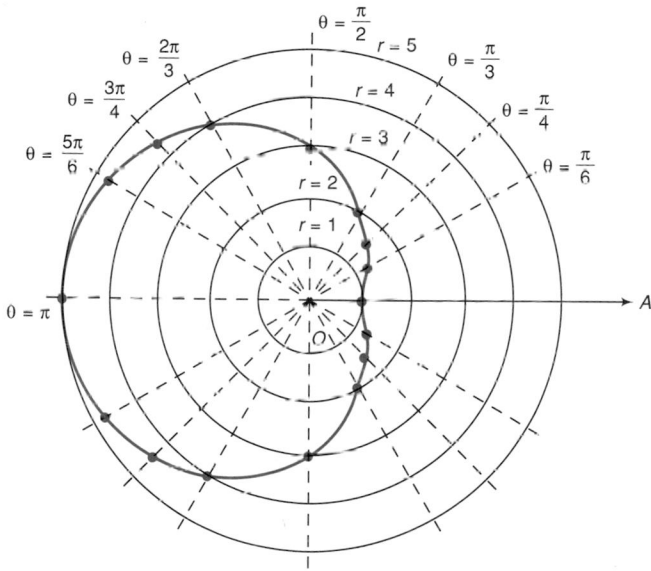

FIGURE 7 THE LIMAÇON $r = 3 - 2 \cos \theta$

EXAMPLE 6
A LIMAÇON WITH LOOP

Sketch the curve $r = 1 + 2 \cos \theta$.

SOLUTION: This curve is also symmetric about the polar axis, and we therefore tabulate the function for θ in $[0, \pi]$ (Table 4).

TABLE 4

θ	0	$\frac{\pi}{6}$	$\frac{\pi}{4}$	$\frac{\pi}{3}$	$\frac{\pi}{2}$	$\frac{2\pi}{3}$	$\frac{3\pi}{4}$	$\frac{5\pi}{6}$	π
$r = 1 + 2\cos\theta$	3	$1 + \sqrt{3}$	$1 + \sqrt{2}$	2	1	0	$1 - \sqrt{2}$	$1 - \sqrt{3}$	-1

[†] From the Latin word *limax*, meaning "snail." The curve is also referred to as *Pascal's limaçon* since it was discovered by Etienne Pascal (1588–1640), father of the famous French mathematician Blaise Pascal (1623–1662).

Using the facts that

$$\left(1 - \sqrt{2}, \frac{3\pi}{4}\right) = \left(\sqrt{2} - 1, \frac{7\pi}{4}\right), \quad \left(1 - \sqrt{3}, \frac{5\pi}{6}\right) = \left(\sqrt{3} - 1, \frac{11\pi}{6}\right),$$

and

$$(-1, \pi) = (1, 2\pi) = (1, 0),$$

together with the symmetry around the polar axis, we obtain the graph in Figure 8. This curve is also called a limaçon.

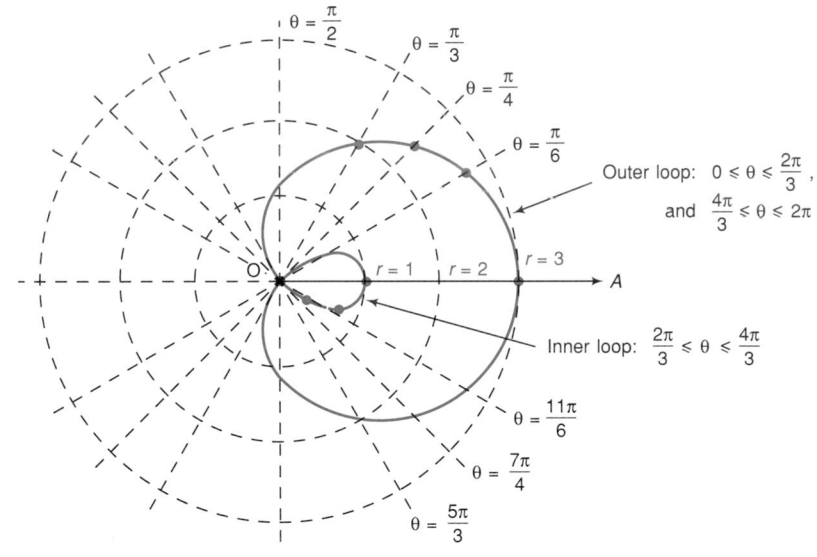

FIGURE 8 THE LIMAÇON WITH A LOOP $r = 1 + 2\cos\theta$

We can generalize the results of the last three examples.

CARDIOIDS AND LIMACONS

The graph of the equation

$$r = a \pm b\cos\theta \qquad \text{or} \qquad r = a \pm b\sin\theta, \qquad a > 0, b > 0,$$

is a cardioid or a limaçon:

 (i) If $a = b$, we obtain a cardioid as in Figure 6.
 (ii) If $b < a < 2b$, we obtain a limaçon having the appearance of the curve in Figure 7.
 (iii) If $b > a$, we obtain a limaçon with a loop as in Figure 8.
 (iv) If $a \geq 2b$ the limaçon is convex. That is, the "dent" in Figure 7 is missing.

EXAMPLE 7
A LEMNISCATE

Sketch the curve $r^2 = 4 \sin 2\theta$.

SOLUTION: Two things should be immediately evident from the equation. First, since $(-r)^2 = r^2$, the graph is symmetric about the pole. Second, since $r^2 \geq 0$, the function is only defined for values of θ such that $\sin 2\theta \geq 0$. If θ is restricted to the interval $[0, 2\pi]$, then $\sin 2\theta \geq 0$ if and only if θ is in $[0, \pi/2]$ or $[\pi, 3\pi/2]$. Then using Table 5 and the symmetry about the pole, we obtain the graph sketched in Figure 9. This curve is called a **lemniscate**.[†]

TABLE 5

θ	0	$\dfrac{\pi}{12}$	$\dfrac{\pi}{8}$	$\dfrac{\pi}{6}$	$\dfrac{\pi}{4}$	$\dfrac{\pi}{3}$	$\dfrac{5\pi}{12}$	$\dfrac{\pi}{2}$
$r = \sqrt{4 \sin 2\theta}$	0	$\sqrt{2}$	$2^{3/4}$	$\sqrt{2} \cdot 3^{1/4}$	2	$\sqrt{2} \cdot 3^{1/4}$	$\sqrt{2}$	0

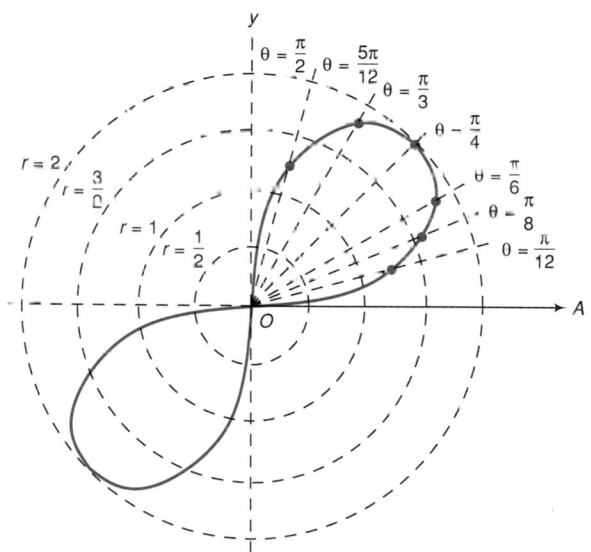

FIGURE 9 THE LEMNISCATE $r^2 = 4 \sin 2\theta$

LEMNISCATES

Any curve with the equation

$$r^2 = a \sin b\theta \qquad \text{or} \qquad r^2 = a \cos b\theta$$

is a *lemniscate*.

[†] From the Greek *lemniskos* and the Latin *lemniscus*, meaning "knotted ribbon." Actually, this curve was originally called the *lemniscate of Bernoulli*, named after the Swiss mathematician Jakob Bernoulli (1654–1705), who described the curve in the *Acta Eruditorum* published in 1694.

EXAMPLE 8
A SPIRAL OF ARCHIMEDES

Sketch the curve $r = \theta$, $\theta \geq 0$.

SOLUTION: This curve has no symmetry. In particular, none of the rules on page 582 apply. To sketch its graph, we note that as we move around the pole, θ increases as r increases. The graph is sketched in Figure 10, and the curve depicted is called the **spiral of Archimedes**.[†] It intersects the polar axis when $\theta = 2n\pi$, n an integer, and it intersects the line $\theta = \pi/2$ when $\theta = [(2n + 1)/2]\pi$.

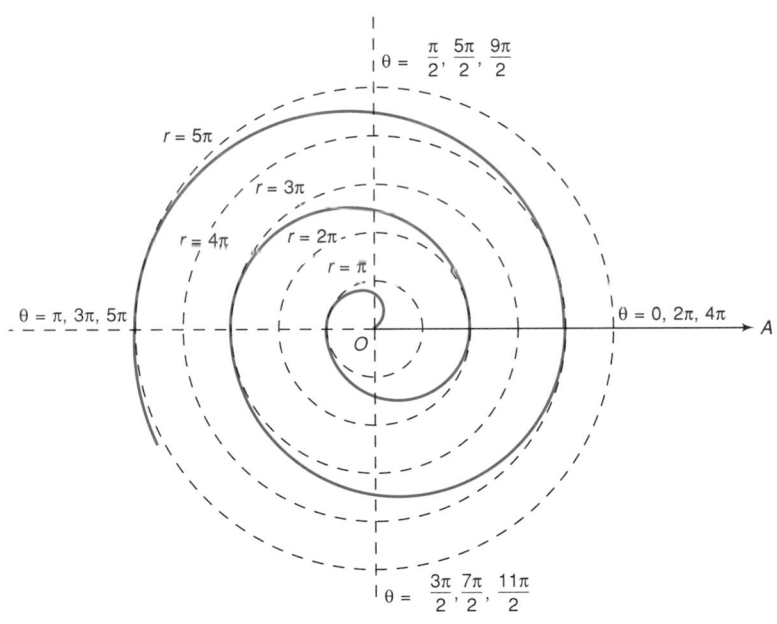

FIGURE 10 THE SPIRAL OF ARCHIMEDES $r = \theta$

■ PROBLEMS 8.6

■ SELF-QUIZ

I. Three of the following ordered pairs are polar coordinates for points that are on the graph of the polar equation $r = 2$; _____ is not on that graph. [*Note:* $(-2, \pi/2)$ and $(2, 3\pi/2)$ are polar coordinates for the same point; $(2, 3\pi/2)$ does satisfy the equation $r = 2$.]
 a. $(-2, \pi/2)$ b. $(2, 7\pi)$
 c. (π, π) d. $(-2, -2)$

II. Three of the following ordered pairs are polar coordinates for points that are on the graph of $\theta = 5\pi/3$; _____ is not on that graph.
 a. $(2, 2\pi/3)$ b. $(2, 8\pi/3)$
 c. $(7, -\pi/3)$ d. $(-7, \pi/3)$

III. Three of the following ordered pairs are polar coordinates for points that are on the graph of $r = \theta$; _____ is not on that graph.
 a. $(3\pi/2, \pi/2)$ b. $(\pi, 3\pi)$
 c. $(\pi/2, -3\pi/2)$ d. $(5, 5 + 2\pi)$

IV. The graphs of the polar equations $r = 2$ and $\theta = \pi/3$ meet at points with polar coordinates _____.
 a. $(2, 7\pi/3)$ b. $(2, -5\pi/3)$
 c. $(2, 601\pi/3)$ d. $(-2, \pi/3)$
 e. $(2, 4\pi/3)$ f. $(2, -2\pi/3)$

[†] Archimedes devoted a great deal of study to the spiral that can be described by the polar equation $r = a\theta$. He used it initially in his attempt to solve the ancient problem of trisecting the angle. Later he calculated part of its area by his method of exhaustion. He described his work on this subject in one of his most important works, entitled *On Spirals*.

■ **DRILL**

In Problems 1–54, sketch the graph of the given equation; indicate any symmetry about the polar axis, the vertical line $\theta = \pi/2$, or the pole (the origin).

1. $r = 5$
2. $r = -4$
3. $\theta = -\pi/6$
4. $\theta = 13\pi/5$
5. $r = 5 \sin \theta$
6. $r = 5 \cos \theta$
7. $r = -5 \sin \theta$
8. $r = -5 \cos \theta$
9. $r = |\cos \theta|$
10. $r = 3|\sin \theta|$
11. $r = 5 \cos \theta + 5 \sin \theta$
12. $r = 5 \cos \theta - 5 \sin \theta$
13. $r = 2 + 2 \sin \theta$
14. $r = 2 + 2 \cos \theta$
15. $r = 2 - 2 \sin \theta$
16. $r = 2 - 2 \cos \theta$
17. $r = -2 - 2 \sin \theta$
18. $r = -2 + 2 \cos \theta$
19. $r = 1 + 3 \sin \theta$
20. $r = 3 + \sin \theta$
21. $r = -2 + 4 \cos \theta$
22. $r = -4 + 2 \cos \theta$
23. $r = -3 - 4 \cos \theta$
24. $r = -4 - 3 \cos \theta$
25. $r = -3 - 4 \sin \theta$
26. $r = -4 - 3 \sin \theta$
27. $r = 4 - 3 \cos \theta$
28. $r = 3 - 4 \cos \theta$
29. $r = 4 + 3 \sin \theta$
30. $r = 3 + 4 \sin \theta$
31. $r = 3 \sin 2\theta$ [This is called a **four-leafed rose.**]
32. $r = -3 \cos 2\theta$
33. $r = -3 \sin 2\theta$
34. $r = 3 \cos 2\theta$
35. $r = 5 \sin 3\theta$ [This curve is called a **three-leafed rose.**]
36. $r = 5 \cos 3\theta$
37. $r = -5 \sin 3\theta$
38. $r = -5 \cos 3\theta$
39. $r = 2 \cos 4\theta$ [This curve is an **eight-leafed rose.**]
40. $r = 2 \sin 4\theta$
41. $r = -5\theta, \theta \geq 0$
42. $r = \theta/2, \theta \geq 0$
43. $r = e^\theta$ [This curve is called a **logarithmic spiral** since $\theta = \ln r$.]
44. $r = e^{\theta/2}$
45. $r^2 = -4 \sin 2\theta$
46. $r^2 = \cos 2\theta$
47. $r^2 = -25 \cos 2\theta$
48. $r^2 = -25 \sin 2\theta$
49. $r = \sin \theta \tan \theta$ [This curve is a **cissoid.**]
50. $r = \cos \theta \cot \theta$
51. $r = 4 + 3 \csc \theta$ [This curve is a **conchoid.**]
52. $r = 2 - 3 \sec \theta$
53. $r^2 = \theta$ [This curve is a **parabolic spiral.**]
54. $(r + 1)^2 = 3\theta$

■ **APPLICATIONS**

55. Use the results of Problems 89–91 to find the polar equation of the circle centered at $(-2, \frac{3}{2})$ with radius $\frac{5}{2}$.
56. Use the results of Problems 92–94 to graph the following.
   a. $r \sin \theta = 3$
   b. $r \cos \theta = -2$
   c. $3r \cos \theta = 8$
   d. $r(2 \sin \theta - 3 \cos \theta) = 4$
   e. $r(-5 \sin \theta + 10 \cos \theta) = 20$
   *f. $r(4 - 3 \tan \theta) = 5 \sec \theta$

Points of Intersection of Polar Curves: Because each point has more than one polar coordinate representation, it is sometimes difficult to determine where polar curves intersect. In Problems 57–76, sketch the graphs of both given curves and then locate all points they have in common.

57. $r = 1, r = \sin \theta$
58. $r = \frac{1}{2}, r = \cos \theta$
59. $r = \sin \theta, r = \cos \theta$
60. $r = \cos \theta, r = 2 \sin \theta$
61. $r = 3, \theta = \pi/4$
62. $r = -2, \theta = \pi/3$
63. $r = -2 + 2 \cos \theta, r = -4 + 2 \sin \theta$
64. $r = \sqrt{2} \sin \theta, r = \cos \theta$
65. $r = \sqrt{3} \sin \theta, r = \cos \theta$
66. $r^2 = \sin 2\theta, r^2 = \cos 2\theta$
67. $r = \frac{3}{2}, r = 3 \cos 2\theta$
68. $r = 2, r = 4 \sin 2\theta$
69. $r^2 = \cos \theta, r = \cos \theta$
70. $r^2 = \sin \theta, r = \sin \theta$
71. $r = \theta, r^2 = \theta$ for $\theta \geq 0$ and $0 \leq r \leq 2$
72. $r = 5 \sin 3\theta, r = 5 \cos 3\theta$
73. $r = \sin 2\theta, r = \cos 2\theta$
74. $r = \pi/3, r = 2\theta$
75. $r = \sin \theta, r = \sec \theta$
76. $r = |\sin \theta|, r = |\cos \theta|$

Using the results of Problems 95 and 96, we can find all points of intersection of the graphs of $r = f(\theta)$ and $r = g(\theta)$ by checking the pole and solving equations (a), (b), and (c) in Problem 95. In Problems 77–88, use this method to find all points of intersection.

77. $r = 1 + \sin \theta, r = 1 + 2 \cos \theta$
78. $r = 1 - \sin \theta, r = 1 + \sqrt{3} \cos \theta$
79. $r = \tan 2\theta, r = 1$
80. $r = 3 \tan \theta \sin \theta, r = 3 \cos \theta$
81. $r = 8(1 + \sin \theta), r = 6/(1 - \sin \theta)$
82. $r = 4(1 + \cos \theta), r = 3/(1 - \cos \theta)$
83. $r^2 = -\cos \theta, r = \sec \theta$
84. $r^2 = \sin \theta, r = \csc \theta$
85. $r = \frac{1}{2} \tan \theta, r = (1/\sqrt{3}) \sin \theta$
86. $r = \csc \theta, r = \cot \theta$
87. $r = \theta, r = 2\theta$ for $\theta \geq 0$
88. $r = 2\theta, r = 3\theta$ for $\theta \geq 0$

CONTINUED

■ SHOW/PROVE/DISPROVE

89. By translating into rectangular coordinates and completing the square, show that for any real number a, the graph of the equation $r = a \cos \theta$ is a circle in the xy-plane with center at $(a/2, 0)$ and radius $|a/2|$.

90. Show that the graph of the equation $r = b \sin \theta$ in the xy-plane is a circle with radius $|b/2|$ centered at $(0, b/2)$.

91. Show that the graph of the equation $r = a \cos \theta + b \sin \theta$ in the xy-plane is a circle centered at $(a/2, b/2)$ with radius $\sqrt{a^2 + b^2}/2$.

92. Show that the polar equation $r \cos \theta = a$ is the equation of a vertical line for any real number a.

93. Show that the polar equation $r \sin \theta = a$ is the equation of a horizontal line for any number a.

94. Show that the polar equation $r(a \cos \theta + b \sin \theta) = c$ (with a, b, c constant real numbers) is the equation of a straight line provided a and b are not both zero.

*95. Suppose the graphs of the polar equations $r = f(\theta)$ and $r = g(\theta)$ intersect at a point (R, Θ) which is not the origin. Explain why one of the following conditions must hold for R and Θ:
   a. $R = f(\Theta) = g(\Theta)$.
   b. There is some integer n such that $R = f(\Theta) = g(\Theta + 2n\pi)$.
   c. $R = f(\Theta)$ and there is some integer n such that $-R = g(\Theta + [2n + 1]\pi)$.

96. Show that if $r = f(\theta)$ and $r = g(\theta)$ intersect at the pole, then there exist numbers θ_1 and θ_2 such that $f(\theta_1) = 0 = g(\theta_2)$.

97. Suppose f is a smooth function and consider a point (r_0, θ_0) on the graph of the polar equation $r = f(\theta)$. Let β be the angle, measured counterclockwise, from the radial line $\theta = \theta_0$ to the line tangent to the graph at that point. Show that
   a. $\beta = \pi/2$ if and only if $f'(\theta_0) = 0$;
   b. if $f'(\theta_0) \neq 0$, then $\tan \beta = f(\theta_0)/f'(\theta_0) = r_0/f'(\theta_0)$.

98. Suppose f is a smooth function such that $f(\theta_0) = 0$ and $f'(\theta_0) \neq 0$. Show that the line $\theta = \theta_0$ is tangent to the graph of $r = f(\theta)$ at the pole.

■ ANSWERS TO SELF-QUIZ

I. c   II. d   III. a

IV. a, b, c, d, e, f (each is a point on the circle and on the straight line)

8.7 AREAS IN POLAR COORDINATES

In this section we derive a method for calculating the area of a region whose boundary is given in polar coordinates. We begin by recalling the formula for the area of a sector. Consider the sector bounded by the lines $\theta = \theta_1$ and $\theta = \theta_2$ and the circle $r = r_0$ (the shaded region in Figure 1). Let φ denote the angle (given in radians) between these two lines ($\varphi = \theta_2 - \theta_1$) and let A denote the area of the sector. Then since the area of the circle is πr_0^2, we have the proportion

$$\frac{\text{number of radians in sector}}{\text{number of radians in circle}} = \frac{\text{area of sector}}{\text{area of circle}},$$

or

$$\frac{\varphi}{2\pi} = \frac{A}{\pi r_0^2}, \tag{1}$$

or

$$A = \frac{1}{2}r_0^2 \varphi. \tag{2}$$

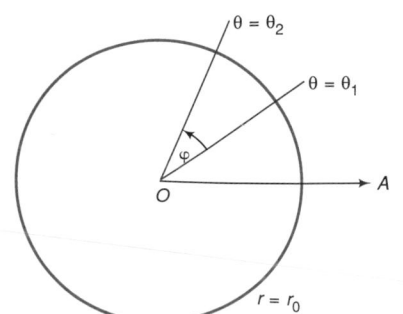

FIGURE 1
AREA OF A SECTOR

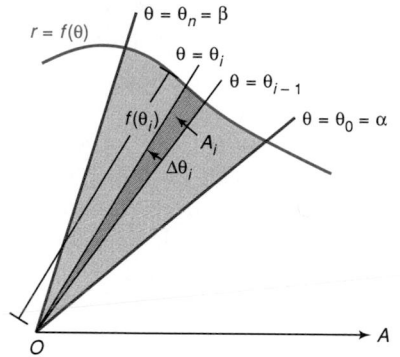

FIGURE 2
THE PLANE REGION BOUNDED BY THE GRAPH OF $f(\theta)$ AND THE RAYS $\theta = \alpha$ AND $\theta = \beta$

Now consider the plane region bounded by the continuous polar curve $r = f(\theta)$ and the lines $\theta = \alpha$ and $\theta = \beta$ (see Figure 2). To calculate the area of this region, we divide the region into n equal subregions by the rays $\alpha = \theta_0 < \theta_1 < \theta_2 < \cdots < \theta_{n-1} < \theta_n = \beta$. We use the notation $\Delta\theta$ to denote the angle between any two successive lines:

$$\Delta\theta = \theta_i - \theta_{i-1}.$$

Let r_i and R_i be the smallest and largest values of $f(\theta)$ for $\theta_{i-1} \le \theta \le \theta_i$, and let ΔA_i denote the area of the ith subregion. Then,

$$\frac{1}{2}r_i^2 \, \Delta\theta \le \Delta A_i \le \frac{1}{2}R_i^2 \, \Delta\theta.$$

Since $r = f(\theta)$ is continuous, so is $r^2 = f^2(\theta)$ and so is the function $g(\theta) = \frac{1}{2}f^2(\theta)\Delta\theta$ (for $\Delta\theta$ fixed). The function $g(\theta)$ takes on the values $\frac{1}{2}r_i^2 \, \Delta\theta$ and $\frac{1}{2}R_i^2 \, \Delta\theta$. Since ΔA_i is between these values, there is, by the intermediate-value theorem, a number θ_i^* in (θ_{i-1}, θ_i) such that $g(\theta_i^*) = \Delta A_i$. But

$$\Delta A_i = g(\theta_i^*) = \frac{1}{2}[f(\theta_i^*)]^2 \, \Delta\theta.$$

Then using a familiar argument, we have

$$A = \lim_{\Delta\theta \to 0} \sum_{i=1}^{n} \Delta A_i = \lim_{\Delta\theta \to 0} \sum_{i=1}^{n} \frac{1}{2}[f(\theta_i^*)]^2 \, \Delta\theta,$$

or

AREA OF THE PLANE REGION SKETCHED IN FIGURE 2

$$A = \frac{1}{2}\int_{\alpha}^{\beta} [f(\theta)]^2 \, d\theta. \tag{3}$$

EXAMPLE 1
THE AREA ENCLOSED BY A CARDIOID

Calculate the area of the region enclosed by the graph of the cardioid $r = 1 + \sin\theta$.

SOLUTION: The curve is symmetric about the line $\theta = \pi/2$ and is sketched in Figure 3. (See Example 8.6.4.) Thus we need only calculate the area for θ in $[-\pi/2, \pi/2]$ and then multiply it by 2. We have

$$\text{half the area} = \frac{1}{2}\int_{-\pi/2}^{\pi/2} (1 + \sin\theta)^2 \, d\theta$$

$$= \frac{1}{2}\int_{-\pi/2}^{\pi/2} (1 + 2\sin\theta + \sin^2\theta) \, d\theta$$

$$= \frac{1}{2}\theta\Big|_{-\pi/2}^{\pi/2} + \int_{-\pi/2}^{\pi/2} \sin\theta \, d\theta + \frac{1}{4}\int_{-\pi/2}^{\pi/2} (1 - \cos 2\theta) \, d\theta$$

$$= \frac{\pi}{2} + 0 + \frac{\pi}{4} = \frac{3\pi}{4},$$

and the total area $A = 2 \cdot 3\pi/4 = 3\pi/2$.

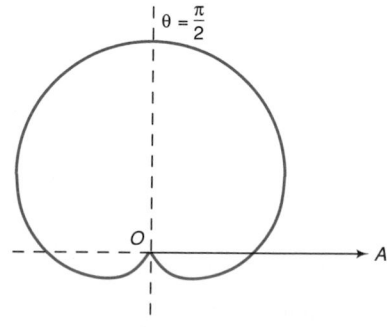

FIGURE 3
THE CARDIOID $r = 1 + \sin\theta$

EXAMPLE 2
THE AREA BETWEEN A CIRCLE
AND A LIMAÇON

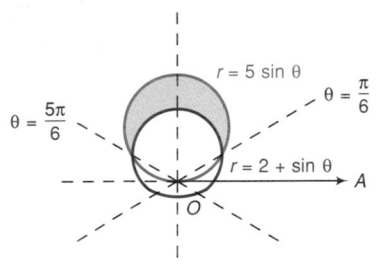

FIGURE 4
REGION BETWEEN A CIRCLE AND A
LIMAÇON

Find the area inside the circle $r = 5 \sin \theta$ and outside the limaçon $r = 2 + \sin \theta$.

SOLUTION: The region is sketched in Figure 4. The two curves intersect whenever $2 + \sin \theta = 5 \sin \theta$, or $\sin \theta = \frac{1}{2}$, so that $\theta = \pi/6$ and $\theta = 5\pi/6$. The area of the region can be calculated by calculating the area of the limaçon between $\pi/6$ and $5\pi/6$ and subtracting it from the area of the circle for θ in that interval. By symmetry, we need only calculate the area between $\theta = \pi/6$ and $\theta = \pi/2$ and then multiply it by 2. We therefore have

$$A = \int_{\pi/6}^{\pi/2} (5 \sin \theta)^2 \, d\theta - \int_{\pi/6}^{\pi/2} (2 + \sin \theta)^2 \, d\theta$$

(we have already multiplied by 2)

$$= \int_{\pi/6}^{\pi/2} (25 \sin^2 \theta - 4 - 4 \sin \theta - \sin^2 \theta) \, d\theta$$

$$= \int_{\pi/6}^{\pi/2} [12(1 - \cos 2\theta) - 4 - 4 \sin \theta] \, d\theta$$

$$= (8\theta - 6 \sin 2\theta + 4 \cos \theta)\Big|_{\pi/6}^{\pi/2} = \frac{8\pi}{3} + \sqrt{3}.$$

■ PROBLEMS 8.7

■ SELF-QUIZ

I. The area of the region $\{(r, \theta): 0 \le r \le f(\theta), \alpha \le \theta \le \beta\}$ is computed by _____.

   a. $\int_{\alpha}^{\beta} [f(\theta)]^2 \, d\theta$ b. $\frac{1}{2} \int_{\alpha}^{\beta} [f(\theta)] \, d\theta$

   c. $\frac{1}{2} \int_{\alpha}^{\beta} [f(\theta)]^2 \, d\theta$ d. $\frac{1}{2} \int_{\alpha}^{\beta} \theta^2 \, d\theta$

II. The area of the circle $\{(r, \theta): r = 5, 0 \le \theta \le 2\pi\}$ is computed by _____.

   a. $\frac{1}{2} \int_0^{\pi} 5^2 \, d\theta$ b. $\frac{1}{2} \int_0^{\pi} \theta^2 \, d\theta$

   c. $\frac{1}{2} \int_0^{2\pi} 5 \cdot \theta \, d\theta$ d. $\frac{1}{2} \int_0^{2\pi} 5^2 \, d\theta$

III. The area of the region (a circle) enclosed by $r = \sin \theta$ is _____.

   a. $\frac{1}{2} \int_0^{\pi} (\sin \theta)^2 \, d\theta$ b. $\frac{1}{2} \int_0^{2\pi} (\sin \theta)^2 \, d\theta$

   c. $\frac{1}{2} \int_0^{\pi} \sin(\theta^2) \, d\theta$ d. $\frac{1}{2} \int_{-\pi}^{\pi} (\sin \theta)^2 \, d\theta$

■ DRILL

In Problems 1–10, calculate the area of the smallest region bounded by the given curves and lines.

1. $r = \theta$; $\theta = 0$; $\theta = \pi/2$; $0 \le r \le \pi/2$
2. $r = 3\theta$; $\theta = 0$; $\theta = \pi$
3. $r = 2 \cos \theta$; $\theta = -\pi/2$; $\theta = \pi/2$
4. $r = a \sin \theta$; $\theta = 0$; $\theta = \pi$
5. $r = a \cos 5\theta$; $\theta = -\pi/10$; $\theta = \pi/10$
6. $r = a \sin 3\theta$; $\theta = 0$; $\theta = \pi/6$
7. $r = 2\theta^2$; $\theta = 0$; $\theta = \pi/2$
8. $r = e^{\theta}$; $\theta = 0$; $\theta = 2\pi/3$
9. $r = 5\theta^4$; $\theta = 0$; $\theta = \pi$
10. $r = 1/\theta$; $\theta = \pi/6$; $\theta = \pi/2$

In Problems 11–26, sketch the given curve; then compute the area of the entire region bounded by the curve. (Where appropriate, assume $a > 0$.)

11. $r = a \sin \theta$
12. $r = a \cos \theta$
13. $r = a(\cos \theta + \sin \theta)$
14. $r = a(\cos \theta - \sin \theta)$
15. $r = a(\sin \theta - \cos \theta)$
16. $r = 3 + 2 \sin \theta$
17. $r = 3 + 2 \cos \theta$
18. $r = 3 - 2 \sin \theta$
19. $r = 3 - 2 \cos \theta$
20. $r = a \cos 2\theta$
21. $r = a \sin 2\theta$
22. $r = a \sin 3\theta$
23. $r = a \cos 5\theta$
24. $r^2 = a^2 \sin 2\theta$
25. $r^2 = a^2 \cos 2\theta$
26. $r = \sqrt{\cos \theta}$

■ APPLICATIONS

In Problems 27–41, sketch the given region and find its area.

27. The smaller loop of the limaçon $r = 1 + 2 \cos \theta$
   [*Hint:* According to Example 8.6.6, the inner loop corresponds to $\theta \in [2\pi/3, 4\pi/3]$.]
28. The smaller loop of the limaçon $r = 2 - 3 \cos \theta$
29. The smaller loop of the limaçon $r = 2 + 3 \cos \theta$
30. The smaller loop of the limaçon $r = a + b \cos \theta$, $0 < a < b$
31. The region outside the smaller loop but inside the larger loop of the limaçon in Problem 27
32. The region outside the smaller loop but inside the larger loop of the limaçon in Problem 28
33. The region outside the smaller loop but inside the larger loop of the limaçon in Problem 29
34. The region bounded by the graphs of $r = 2 \sin \theta$, $r = 2 + 2 \sin \theta$, $\theta = 0$, and $\theta = \pi$
35. The region inside the cardioid $r = a(1 + \sin \theta)$ and outside the circle $r = a$ $(a > 0)$

*36. The region outside the circle $r = a \cos \theta$ and inside the lemniscate $r^2 = a^2 \sin 2\theta$
37. The region inside the two circles $r = 3 \sin \theta$ and $r = 4 \cos \theta$
38. The region enclosed by the graphs of $r = 4 \sin 2\theta$ and $r = 3 \cos \theta$
*39. The region inside the circle $r = \cos \theta$ and inside the four-leafed rose $r = \cos 2\theta$
*40. The region enclosed by one "leaf" of the rose $r = a \cos n\theta$, $(a > 0$ and n is a fixed positive integer)
*41. The region enclosed by both circles $r = a \sin \theta$ and $r = b \cos \theta$ $(a > 0$ and $b > 0)$
42. A roller coaster encloses a region bounded by the graph of the equation $r^2 = 9 \sin 2\theta$. If r is measured in meters, what is the area of this region?

■ ANSWERS TO SELF-QUIZ

I. c   II. d   III. a

■ PROBLEMS   **CHAPTER 8 REVIEW**

In Problems 1–14, identify the type of conic. If it is an ellipse (or a circle), give its foci, center, vertices, major and minor axes, and eccentricity. If it is a parabola, give its focus, directrix, axis, and vertex. If it is a hyperbola, give its foci, transverse axis, center, vertices, asymptotes, and eccentricity. Finally, sketch the curve.

1. $\dfrac{x^2}{9} + \dfrac{y^2}{16} = 1$

2. $\dfrac{x^2}{9} - \dfrac{y^2}{16} = 1$

3. $\dfrac{x^2}{9} - \dfrac{y}{16} = 0$

4. $\dfrac{y^2}{16} - \dfrac{x}{9} = 0$

5. $\dfrac{y^2}{9} - \dfrac{x^2}{16} = 1$

6. $\dfrac{x^2}{16} + \dfrac{y^2}{9} = 1$

7. $\dfrac{(x - 1)^2}{4} + \dfrac{(y + 1)^2}{9} = 1$

8. $\dfrac{(x + 2)^2}{25} - \dfrac{(y + 3)^2}{4} = 1$

9. $\dfrac{(x + 2)^2}{25} + \dfrac{(y - 5)^2}{25} = 0$

10. $x^2 + 2x + y^2 + 2y = 0$
11. $x^2 + 2x - y^2 + 2y = 0$
12. $x^2 + 2x - 2y = 0$
13. $4x^2 + 4x + 3y^2 + 24y = 0$
14. $-3x^2 + 6x + 2y^2 + 4y = 6$

15. Find the equation of an ellipse with foci $(3, 0)$ and $(-3, 0)$ and eccentricity 0.6.
16. Find the equation of the parabola with focus $(3, 0)$ and directrix the line $x = -4$.
17. Find the equation of the hyperbola with foci $(0, 3)$ and $(0, -3)$ and vertices $(0, 2)$ and $(0, -2)$.
18. Find an equation for the hyperbola centered at $(0, 0)$ with vertices $(3, 0)$ and $(-3, 0)$ that is asymptotic to the lines $y = \pm 5x$.
19. Describe the curve obtained from the graph of $2x^2 + xy + 2y^2 = 10$. [*Hint:* Rotate the axes through an angle of $\pi/4$.]
20. Find the equation of the line obtained by rotating the line $2x - 3y = 7$ through an angle of $\pi/3$.

CONTINUED

In Problems 21–26, find a rotation of coordinate axes in which the given equation written in the new coordinates has no xy term. Describe each curve and then sketch it.

21. $x^2 + 4xy + 4y^2 = 9$
22. $x^2 + 4xy - 4y^2 = 9$
23. $xy = -3$
24. $4x^2 + 3xy + y^2 = 1$
25. $-2x^2 + 3xy + 4y^2 = 5$
26. $x^2 + 2xy + y^2 = 6$

In Problems 27–32, convert from polar to rectangular coordinates.

27. $(2, 0)$
28. $(3, \pi/6)$
29. $(-7, \pi/2)$
30. $(3, 23\pi/3)$
31. $(-1, -\pi/2)$
32. $(2, 11\pi/2)$

In Problems 33–38, convert from rectangular to polar coordinates with $r \geq 0$ and $0 \leq \theta < 2\pi$.

33. $(2, 0)$
34. $(1, \sqrt{3})$
35. $(\sqrt{3}, -1)$
36. $(6, 6)$
37. $(-6, -6)$
38. $(-6, 6)$

In Problems 39–52, sketch the graph of the given polar equation. Indicate any symmetry about the polar axis, the vertical line $\theta = \pi/2$, or the pole (the origin).

39. $r = 8$
40. $\theta = \pi/3$
41. $r = 2 \cos \theta$
42. $r = 3 - 3 \sin \theta$
43. $r = 3 - 2 \sin \theta$
44. $r = 2 - 3 \sin \theta$
45. $r = 5 \cos 2\theta$
46. $r^2 = 4 \sin 2\theta$
47. $r = 3 \sin 4\theta$
48. $r = 3\theta, \ \theta \geq 0$
49. $r = e^{2\theta}$
50. $r^2 = 4\theta, r \geq 0$
51. $r \sin \theta = 4$
52. $r \cos \theta = -2$

53. Find, in rectangular coordinates, the equation of the circle $r = 3 \sin \theta + 4 \cos \theta$.
54. Find the polar equation of the circle centered at $(\frac{5}{2}, -6)$ with radius $\frac{13}{2}$.

In Problems 55–62, find all points of intersection of the graphs of the two polar equations.

55. $r = 1; r = \cos \theta$
56. $r = 2 + \cos \theta; r = 2 - \sin \theta$
57. $r^2 = \sin \theta; r = \sin \theta$
58. $r^2 = 4 \cos 2\theta; r^2 = 4 \sin 2\theta$
59. $r = \theta; r^2 = 9\theta; (\theta, r \geq 0)$
60. $r = \sin \theta; r = \sqrt{3} \cos \theta$
61. $r = \sqrt{3} \tan 2\theta; r = 1$
62. $r = 2 \cot \theta; r = 8 \cos \theta$

In Problems 63–73, find the area of the given region.

63. The region bounded by $r = \theta$, $\theta = \pi/6$, and $\theta = \pi/2$ for $\pi/6 \leq r \leq \pi/2$
64. The region bounded by $r = a \cos 4\theta$, $\theta = 0$, and $\theta = \pi/12$
65. The region enclosed by the circle $r = a(\sin \theta + \cos \theta)$ (assume $a > 0$)
66. The region enclosed by the limaçon $r = 4 + 3 \cos \theta$
67. The region enclosed by the cardioid $r = -2 - 2 \sin \theta$
68. The region enclosed by the smaller loop of the limaçon $r = 3 - 4 \cos \theta$
69. The region outside the smaller loop and inside the larger loop of the limaçon of the preceding problem
70. The region inside the cardioid $r = 4(1 + \cos \theta)$ and outside the circle $r = 4$
71. The region inside the two circles $r = 2 \sin \theta$ and $r = 2 \cos \theta$
72. The region enclosed by the lemniscate $r^2 = 36 \cos 2\theta$
73. The region enclosed by one petal of the rose $r = 4 \sin 10\theta$

COMPUTER EXERCISES

1. Find a rotation of the coordinate axes such that $3x^2 + 3xy + y^2 = 1$ has no xy term when written in the new coordinates. Describe the curve and then sketch it. (You may do this problem without electronic aid; however, you will find a computer algebra system to be a real time-saver in your substitutions and simplifications.)
2. Find the area enclosed by the spiral $r = \theta$ and the cardioid $r = 1 + \cos \theta$. (One of the intersection points cannot be found in closed form and must be approximated.)
3. Find the area inside the circle $r = 3 \cos \theta$ to the right of the conchoid $r = \sec \theta + 1$. (You may need graphics software to display the graphs. The intersection points will have to be approximated. A computer algebra system can help you with the integration.)

9 INDETERMINATE FORMS, IMPROPER INTEGRALS, AND TAYLOR'S THEOREM

9.1 THE INDETERMINATE FORM $\frac{0}{0}$ AND L'HÔPITAL'S RULE

In earlier chapters we encountered quotients of the form $f(x)/g(x)$. In Theorem 2 in Section 1.3 we stated that if the limits $\lim_{x \to x_0} f(x)$ and $\lim_{x \to x_0} g(x)$ both exist, then

$$\lim_{x \to x_0} \frac{f(x)}{g(x)} = \frac{\lim\limits_{x \to x_0} f(x)}{\lim\limits_{x \to x_0} g(x)} \tag{1}$$

provided that $\lim_{x \to x_0} g(x) \neq 0$.

However, this last condition is frequently an obstacle in important applications. For example, in Section 2.5 we showed that $\lim_{x \to 0}[(\sin x)/x] = 1$. We did so despite the fact that $\lim_{x \to 0} x = 0$ so that rule (1) did not apply.

DEFINITION   **INDETERMINATE FORM $\frac{0}{0}$**

Let f and g be two functions having the property that $\lim_{x \to x_0} f(x) = 0$ and $\lim_{x \to x_0} g(x) = 0$. Then the function f/g has **the indeterminate form $\frac{0}{0}$ at x_0.** ■

We now give a rule for finding $\lim_{x \to x_0}[f(x)/g(x)]$ in certain cases when f/g has the indeterminate form $\frac{0}{0}$ at x_0. The proof of this result will be given at the end of this section, and other indeterminate forms will be considered in Section 9.2.

THEOREM 1   **L'HÔPITAL'S† RULE FOR THE INDETERMINATE FORM $\frac{0}{0}$**

Let x_0 be a real number, $+\infty$, or $-\infty$, and let f and g be two functions that satisfy the following:

(i) f and g are differentiable at every point in an open interval containing x_0, except possibly at x_0 itself.
(ii) $\lim_{x \to x_0} f(x) = 0$ and $\lim_{x \to x_0} g(x) = 0$.
(iii) $\lim_{x \to x_0}[f'(x)/g'(x)] = L$, where L is a real number, $+\infty$, or $-\infty$.

Then

$$\lim_{x \to x_0} \frac{f(x)}{g(x)} = L.$$

---

† Named after the French mathematician, the Marquis de l'Hôpital (1661–1704). See the historical note on page 597.

That is, under the conditions of the theorem, the limit of the quotient of the two functions is equal to the limit of the quotient of their derivatives.[†] Furthermore, Theorem 1 is also true for right- and left-hand limits. ∎

EXAMPLE 1
USING L'HÔPITAL'S RULE

Calculate $\lim_{x \to 1}(x^3 - 5x^2 + 6x - 2)/(x^5 - 4x^4 + 7x^2 - 9x + 5)$.

SOLUTION: We first note that

$$\lim_{x \to 1}(x^3 - 5x^2 + 6x - 2) = 0 \qquad \text{and}$$

$$\lim_{x \to 1}(x^5 - 4x^4 + 7x^2 - 9x + 5) = 0,$$

so that the indicated limit has the form $\frac{0}{0}$. Then, applying l'Hôpital's rule, we have

$$\lim_{x \to 1} \frac{x^3 - 5x^2 + 6x - 2}{x^5 - 4x^4 + 7x^2 - 9x + 5} = \lim_{x \to 1} \frac{3x^2 - 10x + 6}{5x^4 - 16x^3 + 14x - 9}$$

$$= \frac{-1}{-6} = \frac{1}{6}.$$

EXAMPLE 2
USING L'HÔPITAL'S RULE TO
EVALUATE A ONE-SIDED LIMIT

Calculate $\lim_{x \to 0^+}[\sqrt{x}/(\sin 3\sqrt{x})]$.

SOLUTION: Since $\sqrt{x}$ is only defined for $x \geq 0$ and since we are considering only a right-hand limit, the required open interval is an interval of the form $(0, b)$. In this context the hypotheses of Theorem 1 hold, and we have

$$\lim_{x \to 0^+} \frac{\sqrt{x}}{\sin 3\sqrt{x}} = \lim_{x \to 0^+} \frac{1/2\sqrt{x}}{(\cos 3\sqrt{x})(3/2\sqrt{x})} = \lim_{x \to 0^+} \frac{1}{3 \cos 3\sqrt{x}} = \frac{1}{3}.$$

EXAMPLE 3
USING L'HÔPITAL'S RULE TO
OBTAIN AN INFINITE LIMIT

Calculate $\lim_{x \to 0}[(\sin x)/x^3]$.

SOLUTION:

$$\lim_{x \to 0} \frac{\sin x}{x^3} = \lim_{x \to 0} \frac{\cos x}{3x^2} = \infty.$$

NOTE: We cannot apply l'Hôpital's rule to the last expression because $\lim_{x \to 0} \cos x = 1 \neq 0$.

EXAMPLE 4
USING L'HÔPITAL'S RULE

Calculate $\lim_{x \to 1}(\ln x)/(x - 1)$.

SOLUTION: We note that $\lim_{x \to 1} \ln x = 0 = \lim_{x \to 1}(x - 1)$, so that l'Hôpital's rule applies. Then

$$\lim_{x \to 1} \frac{\ln x}{x - 1} = \lim_{x \to 1} \frac{1/x}{1} = \lim_{x \to 1} \frac{1}{x} = 1.$$

[†] We emphasize that we are taking the quotient of the two derivatives (f'/g'), *not* the derivative of the quotient ($(f/g)'$).

EXAMPLE 5
TAKING NATURAL LOGARITHMS
BEFORE APPLYING L'HÔPITAL'S
RULE IN ORDER TO OBTAIN
A LIMIT

Calculate $\lim_{x\to\infty}[1+(1/x)]^x$.

SOLUTION: First, note that the expression $\lim_{x\to\infty}[1+(1/x)]^x$ can be used to define the number e. We cannot apply l'Hôpital's rule directly since the indicated limit does not have the form of a quotient. Let us define $y = [1+(1/x)]^x$. Then

$$\ln y = x\ln\left(1+\frac{1}{x}\right) = \frac{\ln[1+(1/x)]}{1/x}.$$

Since $\lim_{x\to\infty}\ln[1+(1/x)] = 0$ and $\lim_{x\to\infty} 1/x = 0$, we can now apply l'Hôpital's rule to obtain

$$\lim_{x\to\infty}\ln y = \lim_{x\to\infty}\frac{\ln[1+(1/x)]}{1/x} = \lim_{x\to\infty}\frac{\dfrac{1}{1+(1/x)}\cdot\dfrac{-1}{x^2}}{-1/x^2}$$

$$= \lim_{x\to\infty}\frac{1}{1+(1/x)} = 1.$$

Hence, $\ln y \to 1$ as $x \to \infty$. Thus,

Since e^x is continuous

$$y = e^{\ln y} \to e^1 = e$$

(which is what we expected).

$\otimes$ **WARNING:** Do not try to apply l'Hôpital's rule when either the numerator or denominator of f/g has a finite, nonzero limit at x_0. For example, we easily see that

$$\lim_{x\to 0}\frac{x}{1+\sin x} = \frac{\displaystyle\lim_{x\to 0} x}{\displaystyle\lim_{x\to 0}(1+\sin x)} = \frac{0}{1} = 0.$$

But if we try to apply l'Hôpital's rule, we obtain

$$\lim_{x\to 0}\frac{x}{1+\sin x} = \lim_{x\to 0}\frac{1}{\cos x} = 1,$$

which is incorrect. $\otimes$

HISTORICAL NOTE: JOHANN BERNOULLI AND L'HÔPITAL'S RULE

Many mathematicians worked on calculus after its discovery by Newton and Leibniz towards the end of the seventeenth century. Among these were members of the Bernoulli family of Switzerland—perhaps the most distinguished family in the history of science. The Bernoulli mathematical dynasty began with two brothers: Jakob Bernoulli (1654–1705) and Johann Bernoulli (1667–1748). We encountered Jakob in our discussion of compound interest on page 449.

Johann Bernoulli was a more prolific contributor to mathematics than was his brother. He worked in a wide variety of areas, including the calculation of arc length, reflection and refraction in optics, the calculation of

JOHANN BERNOULLI
1667–1748

MARQUIS DE
L'HÔPITAL
1661–1704

area, and analytical trigonometry. Johann was a very influential teacher and helped significantly in getting the power and techniques of calculus appreciated in Europe. Among his many accomplishments was his discovery and proof, in 1694, of what is now (inaccurately) called l'Hôpital's rule.

Guillaume François Antoine de l'Hôpital (1661–1704), later known as the Marquis de l'Hôpital, was a French mathematician who was a student of Johann Bernoulli. L'Hôpital did not make any significant discoveries in calculus. His major contribution was his book, the *Analyse des infiniment petits* (the analysis of infinitely small things) published in 1696. The *Analyse* was the first calculus book ever written. However, this work was not created by l'Hôpital. Under a rather strange agreement, l'Hôpital published the lectures of his teacher, Johann Bernoulli. Bernoulli lectured in German and the *Analyse* was published in French, so, at least, we can give l'Hôpital credit for the translation. In any event, Bernoulli sent the proof of l'Hôpital's rule to l'Hôpital in 1694 and the rule was published in the *Analyse* in 1696. The name "l'Hôpital's rule" was used in subsequent books on calculus and the name stuck. However, "Bernoulli's rule" would be a more accurate (and fair) term.

PROOF OF L'HÔPITAL'S RULE (If Time Permits)

We now prove l'Hôpital's rule for the indeterminate form $\frac{0}{0}$. Before doing so, however, we need to prove another kind of mean-value theorem that will be useful here.

THEOREM 2   **CAUCHY[†] MEAN-VALUE THEOREM**

Suppose that the two functions f and g are continuous in the closed interval $[a, b]$ and differentiable in the open interval (a, b). Suppose further that $g'(x) \neq 0$ for x in (a, b). Then there exists at least one number c in (a, b) such that

$$\frac{f(b) - f(a)}{g(b) - g(a)} = \frac{f'(c)}{g'(c)}. \tag{2}$$

PROOF: Let

$$h(x) = [g(b) - g(a)][f(x) - f(a)] - [g(x) - g(a)][f(b) - f(a)].$$

Then clearly $h(a) = h(b) = 0$. Since h is continuous in $[a, b]$ and differentiable in (a, b), we can apply Rolle's theorem (see page 239) which tells us that there exists at least one number c in (a, b) such that $h'(c) = 0$. But

$$h'(c) = [g(b) - g(a)][f'(c)] - [g'(c)][f(b) - f(a)] = 0. \tag{3}$$

Then $f'(c)[g(b) - g(a)] = g'(c)[f(b) - f(a)]$. Since $g'(c) \neq 0$ by assumption, we can divide by it and by $[g(b) - g(a)]$ to obtain equation (2). ∎

[†] See the biography of Cauchy on page 64.

NOTE: It is *not* necessary to assume that $g(b) - g(a) \neq 0$ since this fact is implied by the hypotheses (see Problem 42).

We can now prove l'Hôpital's rule for the indeterminate form $\frac{0}{0}$. We show the following: If

> **(i)** f and g are differentiable at every point in an open interval containing x_0, except possibly at x_0 itself,
>
> **(ii)** $\lim_{x \to x_0} f(x) = 0$ and $\lim_{x \to x_0} g(x) = 0$,
>
> **(iii)** $\lim_{x \to x_0}[f'(x)/g'(x)] = L$, where L is a real number, $+\infty$, or $-\infty$,

then $\lim_{x \to x_0}[f(x)/g(x)] = L$.

Furthermore, the theorem is true for right- and left-hand limits as well.

REMARK: The idea of the proof is simple. The proof, however, is not so simple because of a number of small details. Here's the idea: Since $\lim_{x \to x_0} f(x) = 0 = \lim_{x \to x_0} g(x)$, we assume that $f(x_0) = g(x_0) = 0$. Then if $x > x_0$, we find from the Cauchy mean-value theorem that there is a number c in (x_0, x) such that

$$\frac{f(x)}{g(x)} = \frac{f(x) - 0}{g(x) - 0} = \frac{f(x) - f(x_0)}{g(x) - g(x_0)} = \frac{f'(c)}{g'(c)} \to L \qquad \text{as } x \to x_0,$$

since as $x \to x_0, c \to x_0$ (because c is squeezed between x_0 and x). Now to the details.

> **PROOF:** We prove the theorem first in the case in which $x \to x_0^+$, where x_0 is a real number.
>
> *Step 1.* Choose a positive number $\Delta_1 x$ such that f and g are differentiable in $(x_0, x_0 + \Delta_1 x)$ and continuous in $(x_0, x_0 + \Delta_1 x]$.
> We can do this by hypothesis (i).
>
> *Step 2.* Choose $\Delta_2 x$ such that $f'(x)/g'(x)$ exists in $(x_0, x_0 + \Delta_2 x)$.
> We can do this because $\lim_{x \to x_0^+}[f'(x)/g'(x)]$ exists.
>
> *Step 3.* Note that $g'(x) \neq 0$ for x in $(x_0, x_0 + \Delta_2 x)$.
> Since $f'(x)/g'(x)$ exists (by Step 2) in $(x_0, x_0 + \Delta_2 x)$.
>
> *Step 4.* Define $\Delta x = \min(\Delta_1 x, \Delta_2 x)$.
>
> *Step 5.* Since $\lim_{x \to x_0} f(x) = 0$, $f(x_0) = 0$ or f is not defined at x_0 or $f(x_0) = k \neq 0$. In the latter cases we simply redefine f to be zero at x_0. Similarly, either $g(x_0) = 0$ or g can be defined to be zero at x_0. Then by hypothesis (ii), both functions are continuous in the closed interval $[x_0, x_0 + \Delta x]$.
>
> *Step 6.* Apply the Cauchy mean-value theorem to f and g in the interval $(x_0, x_0 + \Delta x)$. There is a number $c_{\Delta x}$ in $(x_0, x_0 + \Delta x)$[†] such that
>
> $$\frac{f'(c_{\Delta x})}{g'(c_{\Delta x})} = \frac{f(x_0 + \Delta x) - f(x_0)}{g(x_0 + \Delta x) - g(x_0)} = \frac{f(x_0 + \Delta x)}{g(x_0 + \Delta x)}.$$

[†] The notation $c_{\Delta x}$ implies that the number depends on the choice of Δx.

Step 7. Observe that $c_{\Delta x} \to x_0^+$ as $\Delta x \to 0^+$, so

$$\lim_{x \to x_0^+} \frac{f(x)}{g(x)} = \lim_{\Delta x \to 0^+} \frac{f(x_0 + \Delta x)}{g(x_0 + \Delta x)} = \lim_{\Delta x \to 0^+} \frac{f'(c_{\Delta x})}{g'(c_{\Delta x})}$$

$$= \lim_{c_{\Delta x} \to x_0^+} \frac{f'(c_{\Delta x})}{g'(c_{\Delta x})} = \lim_{x \to x_0^+} \frac{f'(x)}{g'(x)} = L.$$

This step completes the proof in the case $x \to x_0^+$.

If $x \to x_0^-$, then we obtain the same proof by considering the interval $(x_0 - \Delta x, x_0)$. If $x \to x_0$, then $x \to x_0^-$ and $x \to x_0^+$. Finally, for the case in which $x \to \infty$, let $y = 1/x$. Then $x = 1/y$, $dx/dy = -1/y^2$, and by the chain rule,

$$f'(x) = \left[f\left(\frac{1}{y}\right) \right]' = f'\left(\frac{1}{y}\right) \frac{d}{dy}\left(\frac{1}{y}\right) = f'\left(\frac{1}{y}\right)\left(-\frac{1}{y^2}\right).$$

Thus

$$\lim_{x \to \infty} \frac{f'(x)}{g'(x)} = \lim_{y \to 0^+} \frac{f'(1/y)(-1/y^2)}{g'(1/y)(-1/y^2)} = \lim_{y \to 0^+} \frac{f'(1/y)}{g'(1/y)}$$

By the case proven above

$$= \lim_{y \to 0^+} \frac{f(1/y)}{g(1/y)} = \lim_{x \to \infty} \frac{f(x)}{g(x)},$$

and the case where $x \to \infty$ is proved. The case in which $x \to -\infty$ is handled in a similar manner. ∎

■ PROBLEMS 9.1

■ SELF-QUIZ

I. $\lim\limits_{x \to 1} \dfrac{3 - (3/x)}{5x - 5} = $ _____.

    a. 1   b. 3   c. -5 d. $\frac{3}{5}$

II. $\lim\limits_{x \to 0} \dfrac{\cos x - 1}{x} = $ _____.

    a. -1 b. 1
    c. 0   d. undefined

III. True or False: l'Hôpital's rule can be used to compute $\lim\limits_{x \to 0} \dfrac{7}{x^2 - 3}$ because $\dfrac{\lim\limits_{x \to 0} \frac{d}{dx} 7}{\lim\limits_{x \to 0} \frac{d}{dx}(x^2 - 3)}$ is a $\frac{0}{0}$ indeterminate form.

IV. Suppose $f(x) = x^2 - x + 3$ and $g(x) = 5x - 7$. Then, easy computations show that

$$\lim_{x \to 2} \frac{f(x)}{g(x)} = \frac{5}{3}, \quad \text{but} \quad \frac{\lim\limits_{x \to 2} f'(x)}{\lim\limits_{x \to 2} g'(x)} = \frac{3}{5};$$

it is also easy to see that $\frac{5}{3} \neq \frac{3}{5}$. We get the wrong numerical result using l'Hôpital's rule here because _____.

    a. $g''(x) = 0$, therefore

$$\frac{\lim f''}{\lim g''} \text{ does not exist, so}$$

$$\frac{\lim f'}{\lim g'} \text{ and } \frac{\lim f}{\lim g} \text{ don't either.}$$

    b. l'Hôpital's rule applies only to limits computed as x approaches zero.

    c. f/g is not an indeterminate form $\frac{0}{0}$ as $x \to 2$.

    d. we misremembered it: l'Hôpital's rule has an "escape clause" saying

$$\lim \frac{f}{g} = 1 \left/ \frac{\lim f'}{\lim g'} \right.$$

in special circumstances.

CONTINUED

■ **DRILL**

In Problems 1–30, calculate the given limit.

1. $\lim\limits_{x \to 0} \dfrac{x + \sin 5x}{x - \sin 5x}$

2. $\lim\limits_{x \to a} \dfrac{a - x}{a - \sqrt{ax}},\ a \neq 0$

3. $\lim\limits_{x \to 0} \dfrac{\cos 3x - 1}{\sin 5x}$

4. $\lim\limits_{x \to \pi/2} \dfrac{x - (\pi/2)}{\cos x}$

5. $\lim\limits_{x \to 0^+} \dfrac{\sin x}{\sqrt{x}}$

6. $\lim\limits_{x \to 0} \dfrac{\sin x}{x^5}$

7. $\lim\limits_{x \to 1} \dfrac{x^4 - x^3 + x^2 - 1}{x^3 - x^2 + x - 1}$

8. $\lim\limits_{x \to \infty} \dfrac{1/x}{\ln[1 + (1/x)]}$

9. $\lim\limits_{x \to x_0} \dfrac{\sqrt{x} - \sqrt{x_0}}{x - x_0}$

10. $\lim\limits_{x \to 0} \dfrac{x + \sin ax}{x - \sin ax},\ a \neq 1$

11. $\lim\limits_{x \to 0} \dfrac{x^2}{\sin^{-1} x}$

12. $\lim\limits_{x \to 0} \dfrac{\tan^{-1} x}{x}$

13. $\lim\limits_{x \to 0^+} \left(\dfrac{e^x - 1}{x^2} - \dfrac{1}{x} \right)$

*14. $\lim\limits_{x \to 0} \left(\dfrac{2}{\sin x} - \dfrac{2}{x} \right)$

15. $\lim\limits_{x \to 0} \dfrac{e^x - 1}{x(3 + x)}$

16. $\lim\limits_{x \to 0} \dfrac{3 + x - 3e^x}{x(2 + 5e^x)}$

17. $\lim\limits_{x \to \infty} x \sin(1/x)$

18. $\lim\limits_{x \to \infty} x^2 \tan(1/x)$

19. $\lim\limits_{x \to 1^-} \dfrac{x^2 - 1}{\sqrt{1 - x}}$

*20. $\lim\limits_{x \to 1^-} \dfrac{\sqrt{1 - x^3}}{\sqrt{1 - x^4}}$

21. $\lim\limits_{x \to 0^+} \dfrac{4^x - 3^x}{\sqrt{x}}$

22. $\lim\limits_{x \to \pi/2} \dfrac{\cos x}{\sin 2x}$

23. $\lim\limits_{x \to 0} (1 + x)^{2/x}$

24. $\lim\limits_{x \to 0} (x + e^x)^{1/x}$

25. $\lim\limits_{x \to -2} \dfrac{3x^3 + 16x^2 + 28x + 16}{x^5 + 4x^4 + 4x^3 + 3x^2 + 12x + 12}$

26. $\lim\limits_{x \to 3} \dfrac{x^5 - 7x^4 + 10x^3 + 18x^2 - 27x - 27}{x^4 - 16x^3 + 90x^2 - 216x + 189}$

27. $\lim\limits_{x \to 0} \dfrac{x - \sin x}{x^3}$

28. $\lim\limits_{x \to 0} \dfrac{\sin^{-1} x - x}{5x^2}$

29. $\lim\limits_{x \to 4\pi} \dfrac{(x - 4\pi)^2}{\ln \cos x}$

*30. $\lim\limits_{x \to 0} \dfrac{\sinh x - \sin x}{\sin^3 x}$

*31. Calculate $\lim\limits_{x \to 0} \left(\displaystyle\int_0^x \cos t^2\, dt \right) \Big/ \left(\displaystyle\int_0^x e^{t^2}\, dt \right)$.

[*Hint:* Use the second fundamental theorem of calculus (Theorem 4.6.2).]

32. Calculate $\lim\limits_{x \to 0^+} (\int_0^x \sin t^3\, dt)/x^4$.

33. Suppose f is a smooth function with all desired derivatives. Calculate

$$\lim_{h \to 0} \frac{f(x + h) - f(x - h)}{2h}.$$

34. Suppose f is a smooth function with all desired derivatives. Calculate

$$\lim_{h \to 0} \frac{f(x + h) - 2f(x) + f(x - h)}{h^2}.$$

■ **SHOW/PROVE/DISPROVE**

35. Show that if $x_0 > 0$ and if a and b are real numbers with $b \neq 0$, then

$$\lim_{x \to x_0} \frac{x^a - x_0^a}{x^b - x_0^b} = \frac{a}{b} x_0^{a-b}.$$

36. a. Show that for any real number $a \neq 1$ and for any positive integer n, the initial $n + 1$ powers of a, a **geometric progression**, add as follows:

$$1 + a + a^2 + \cdots + a^n = \sum_{k=0}^{n} a^k = \frac{1 - a^{n+1}}{1 - a}.$$

b. Calculate $\lim\limits_{a \to 1} \dfrac{1 - a^{n+1}}{1 - a}$ and compare this limit with the sum of the first $n + 1$ terms of the geometric progression with $a = 1$.

37. Show that $\lim\limits_{x \to 0} (1 + x)^{a/x} = e^a$ for any real number $a \neq 0$.

38. Show that

$$\lim_{x \to \infty} x^a \sin(1/x) = \begin{cases} 0 & \text{if } 0 < a < 1, \\ \infty & \text{if } 1 < a. \end{cases}$$

39. Suppose that f is a function continuous on the interval $[a, b]$. Define a new function A with domain $(a, b]$ as follows:

$$A(x) = \frac{1}{x - a} \cdot \int_a^x f(t)\, dt.$$

That is, $A(x)$ is the **average value of f on the interval** $[a, x]$. Find $\lim\limits_{x \to a^+} A(x)$.

*40. Fix two positive real numbers a and b and let $f(x) = \int_a^b t^x\, dt$. Show that $\lim\limits_{x \to -1} f(x)$ exists and obtain a simple expression for it.

CONTINUED

*41. Compute

$$\lim_{\epsilon \to 0} \frac{\int \left(\frac{1}{x - a - \epsilon} - \frac{1}{x - a} \right) dx}{\epsilon}.$$

42. Use the mean-value theorem (Theorem 3.7.2) to show that under the hypotheses of Theorem 2, $g(a) \neq g(b)$.

■ CHALLENGE

43. Find the maximum p such that

$$\lim_{x \to 0} \frac{\cos x - 1}{x^p}$$

exists and is finite.

44. Find the maximum p such that

$$\lim_{x \to 0} \frac{e^x - (1 + x)}{x^p}$$

exists and is finite.

■ ANSWERS TO SELF-QUIZ

I. d   II. c   III. False   IV. c

9.2 OTHER INDETERMINATE FORMS

There are other situations in which $\lim_{x \to x_0} [f(x)/g(x)]$ cannot be evaluated directly. One important case is defined below.

DEFINITION   **INDETERMINATE FORM** ∞/∞

Let f and g be two functions having the property that $\lim_{x \to x_0} f(x) = \pm \infty$ and $\lim_{x \to x_0} g(x) = \pm \infty$, where x_0 is real, $+\infty$, or $-\infty$. Then the function f/g has the **indeterminate form** ∞/∞ at x_0.   ■

THEOREM 1   **L'HÔPITAL'S RULE FOR THE INDETERMINATE FORM** ∞/∞

Let x_0 be a real number, $+\infty$, or $-\infty$ and let f and g be two functions that satisfy the following:

(i) f and g are differentiable at every point in a neighborhood of x_0, except possibly at x_0 itself.

(ii) $\lim_{x \to x_0} f(x) = \pm \infty$ and $\lim_{x \to x_0} g(x) = \pm \infty$.

(iii) $\lim_{x \to x_0} [f'(x)/g'(x)] = L$, where L is a real number, $+\infty$, or $-\infty$.

Then

$$\lim_{x \to x_0} \frac{f(x)}{g(x)} = L. \tag{1}$$

The result is also true for right- and left-hand limits.   ■

The proof of this theorem is more complicated than the proof of the earlier version of l'Hôpital's rule and is omitted.[†] However, an elementary proof in the case f' and g' are continuous and $g'(x) \neq 0$ is suggested in Problems 42–48.

EXAMPLE 1
AN ∞/∞ INDETERMINATE FORM

Calculate $\lim_{x \to \infty}(e^x/x)$.

SOLUTION: Since $e^x \to \infty$ as $x \to \infty$, l'Hôpital's rule applies, and we have

$$\lim_{x \to \infty} \frac{e^x}{x} = \lim_{x \to \infty} \frac{e^x}{1} = \infty.$$

EXAMPLE 2
A $-\infty/\infty$ INDETERMINATE FORM

Calculate $\lim_{x \to 0^+}(x \ln x)$.

SOLUTION: We write $x \ln x = (\ln x)/(1/x)$. Then since $\lim_{x \to 0^+} \ln x = -\infty$ and $\lim_{x \to 0^+}(1/x) = \infty$, we have

$$\lim_{x \to 0^+} (x \ln x) = \lim_{x \to 0^+} \frac{\ln x}{1/x} - \lim_{x \to 0^+} \frac{1/x}{-1/x^2} = \lim_{x \to 0^+} (-x) = 0.$$

EXAMPLE 3
A $0°$ INDETERMINATE FORM THAT BECOMES AN $\infty/-\infty$ INDETERMINATE FORM AFTER TAKING LOGARITHMS

Calculate $\lim_{x \to 0^+} x^{ax}$, $a \neq 0$.

SOLUTION: If $y = x^{ax}$, then $\ln y = ax \ln x$, and

$$\lim_{x \to 0^+} \ln y = \lim_{x \to 0^+} (ax \ln x) = \lim_{x \to 0^+} \frac{\ln x}{1/ax} = \lim_{x \to 0^+} \frac{1/x}{-1/ax^2}$$
$$= \lim_{x \to 0^+} (-ax) = 0.$$

Since $\ln y \to 0$, we have $y = x^{ax} \to 1$ as $x \to 0^+$.

EXAMPLE 4
A SITUATION WHERE L'HÔPITAL'S RULE DOES NOT APPLY

Compute $\lim_{x \to \infty} 3x/(5x + 2 \sin x)$.

SOLUTION: By dividing numerator and denominator by x, we obtain

$$\lim_{x \to \infty} \frac{3x}{5x + 2 \sin x} = \lim_{x \to \infty} \frac{3}{5 + (2 \sin x)/x} = \frac{3}{5}.$$

Also, $\lim_{x \to \infty} 3x = \lim_{x \to \infty}(5x + 2 \sin x) = \infty$. However, if we try to apply l'Hôpital's rule, we obtain

$$\lim_{x \to \infty} \frac{3x}{5x + 2 \sin x} \overset{?}{=} \lim_{x \to \infty} \frac{3}{5 + 2 \cos x},$$

which *does not exist*. This is one example of a situation in which $\lim_{x \to x_0} f(x)/g(x)$ exists, but $\lim_{x \to x_0} f'(x)/g'(x)$ does not exist. That is, the

[†] For a proof, consult E. F. Buck and R. C. Buck, *Advanced Calculus*, 3rd ed. (New York: McGraw-Hill, 1978), pp. 121–122.

nonexistence of $\lim_{x \to x_0} f'(x)/g'(x)$ does not imply the nonexistence of $\lim_{x \to x_0} f(x)/g(x)$. It illustrates the fact that in order to apply l'Hôpital's rule we must know that $\lim_{x \to x_0} f'(x)/g'(x)$ exists. Otherwise, the rule cannot be used.

EXAMPLE 5
COMPUTING $\lim_{x \to \infty} x^a e^{-bx}$
WHERE $b > 0$

Calculate $\lim_{x \to \infty} x^a e^{-bx}$ for any real number a and any positive real number b.

Case 1. $a \leq 0$. Then, for $x > 0$, $x^a \leq x^0 = 1$ and

$$0 \leq \lim_{x \to \infty} x^a e^{-bx} \leq \lim_{x \to \infty} x^0 e^{-bx} = \lim_{x \to \infty} \frac{1}{e^{bx}} = 0,$$

so, by the squeezing theorem, $\lim_{x \to \infty} x^a e^{-bx} = 0$.

Case 2. $a > 0$. Then,

Since $f(u) = u^a$ is continuous

$$\lim_{x \to \infty} \frac{x^a}{e^{bx}} = \lim_{x \to \infty} \left(\frac{x}{e^{bx/a}} \right)^a \stackrel{\downarrow}{=} \left(\lim_{x \to \infty} \frac{x}{e^{bx/a}} \right)^a = 0^a = 0$$

since

$$\lim_{x \to \infty} \frac{x}{e^{bx/a}} = \lim_{x \to \infty} \frac{\dfrac{d}{dx} x}{\dfrac{d}{dx} e^{bx/a}} = \lim_{x \to \infty} \frac{1}{\dfrac{b}{a} e^{bx/a}} = 0.$$

Therefore,

$$\lim_{x \to \infty} x^a e^{-bx} = 0 \qquad \text{if } b > 0. \tag{2}$$

This result is very interesting and useful. It tells us that the exponential function e^x grows much faster than any power function.

There are many other ways to prove this important result. One of these is suggested in Problem 36. Another is given in Problem 10.6.30.

Now let $P_n(x)$ be a polynomial of degree n. That is,

$$P_n(x) = a_n x^n + a_{n-1} x^{n-1} + a_{n-2} x^{n-2} + \cdots + a_2 x^2 + a_1 x + a_0.$$

From (2) we see that, for $k = 0, 1, 2, \ldots, n$,

$$\lim_{x \to \infty} a_k x^k e^{-bx} = a_k \lim_{x \to \infty} x^k e^{-bx} = a_k \cdot 0 = 0.$$

Thus we have the following theorem.

THEOREM 2

If $b > 0$, then

$$\lim_{x \to \infty} P_n(x) e^{-bx} = 0 \qquad \text{for all values of } n. \blacksquare \tag{3}$$

We will make use of this theorem in later chapters.

There are other indeterminate forms that can be dealt with by applying l'Hôpital's rule. For example, in Example 2 we calculated $\lim_{x \to 0^+}(x \ln x)$. Since $\lim_{x \to 0^+} x = 0$ and $\lim_{x \to 0^+} \ln x = -\infty$, this indeterminate expression is really of the form $0 \cdot \infty$. An indeterminate form of this type can often be treated by putting it into one of the forms $\frac{0}{0}$ or ∞/∞.

EXAMPLE 6
AN $\infty - \infty$ INDETERMINATE FORM

Calculate $\lim_{x \to 0^+}(\csc x - \cot x)$.

SOLUTION: We first note that

$$\lim_{x \to 0^+} \csc x = \lim_{x \to 0^+} \frac{1}{\sin x} = \infty$$

and

$$\lim_{x \to 0^+} \cot x = \lim_{x \to 0^+} \frac{\cos x}{\sin x} = \infty.$$

We now write

$$\lim_{x \to 0^+}(\csc x - \cot x) = \lim_{x \to 0^+} \frac{1}{\sin x} - \frac{\cos x}{\sin x}$$

$$= \lim_{x \to 0^+} \frac{1 - \cos x}{\sin x}$$

l'Hôpital's rule for $\frac{0}{0}$

$$\overset{\downarrow}{=} \lim_{x \to 0^+} \frac{\sin x}{\cos x} = 0.$$

EXAMPLE 7
AN ∞^0 INDETERMINATE FORM

Calculate $\lim_{x \to \infty} x^{1/x}$.

SOLUTION: This expression is of the form ∞^0. We set $y = x^{1/x}$. Then $\ln y = (1/x) \ln x$, and

$$\lim_{x \to \infty} \ln y = \lim_{x \to \infty} \frac{\ln x}{x} = \lim_{x \to \infty} \frac{1/x}{1} = 0,$$

so that $\ln y \to 0$ and $y = x^{1/x} \to 1$.

The technique used in Example 7 can often be used to evaluate limits of expressions having the form

$$f(x)^{g(x)}.$$

REMARK: In using this technique, we make use of the fact that e^x is continuous. Without that we couldn't conclude, for example, that $y \to 1$ just because $\ln y \to 0$. But because of the continuity of e^x, we may conclude that

$$\lim_{x \to \infty} y = \lim_{x \to \infty} e^{\ln y} = e^{\lim_{x \to \infty} \ln y} = e^0 = 1.$$

In Table 1 we give a list of indeterminate forms. We stress, however, that only the indeterminate forms $\frac{0}{0}$ and ∞/∞ can be evaluated directly. In all other cases it is necessary to bring the expression into the form $\frac{0}{0}$ or ∞/∞. This is done either by an algebraic manipulation (as in Example 2) or by taking logarithms (as in Example 3).

TABLE 1 INDETERMINATE FORMS

Indeterminate Form	Example
$\dfrac{0}{0}$	$\displaystyle\lim_{x\to 0}\frac{\sin x}{x}$ (Section 2.5)
$\dfrac{\infty}{\infty}$	$\displaystyle\lim_{x\to\infty}\frac{2x^2 - 2x + 3}{x^2 + 4x + 4}$ (Example 6 in Section 1.4)
$0\cdot\infty$	$\displaystyle\lim_{x\to 0^+} x\ln x$ (Example 2)
$\infty - \infty$	$\displaystyle\lim_{x\to 0^+}(\csc x - \cot x)$ (Example 6)
0^0	$\displaystyle\lim_{x\to 0^+} x^{ax}$ (Example 3)
∞^0	$\displaystyle\lim_{x\to\infty} x^{1/x}$ (Example 7)
1^∞	$\displaystyle\lim_{x\to\infty}\left(1 + \frac{1}{x}\right)^x$ (Example 5 in Section 9.1)

■ PROBLEMS 9.2

■ SELF-QUIZ

I. $\displaystyle\lim_{x\to\infty}\frac{\ln x}{x} = $ _____.

   a. 0   b. 1   c. e d. ∞

II. $\displaystyle\lim_{x\to\infty}\frac{x^2}{e^x} = $ _____.

   a. ∞ b. 2   c. 1   d. 0

III. $\displaystyle\lim_{x\to 0}\left(\frac{1}{x} - \frac{1}{\sin x}\right) = $ _____.

   a. -1 b. 0   c. 1   d. ∞

IV. $\displaystyle\lim_{x\to\infty}\frac{(5x - 1)^3}{(7 - 2x)^3} = $ _____.

   a. $5^3/7^3$ b. $(-1)^3/7^3$
   c. $5^3/(-2)^3$ d. $(-1)^3/(-2)^3$
   e. 0   f. $-\infty$

V. True–False: Every indeterminate form of type 0^0 actually converges to 1.

VI. $\displaystyle\lim_{x\to\infty}\left(\frac{1}{x}\right)^{p/\ln x} = $ _____.

   a. 0^0 b. 1   c. e
   d. p e. e^{-p} f. ∞

■ DRILL

In Problems 1–32, evaluate the given limit.

1. $\displaystyle\lim_{x\to\infty}\frac{x^3 + 3x + 4}{2x^3 - 4x + 2}$

2. $\displaystyle\lim_{x\to\infty}\frac{4x^{5/2} + 3\sqrt{x} - 10}{3x^{5/2} - 8x^{3/2} + 45x^2}$

3. $\lim\limits_{x \to \infty} \dfrac{\ln x}{\sqrt{x}}$

4. $\lim\limits_{x \to \pi/2^-} \dfrac{\sec x}{\tan x}$

5. $\lim\limits_{x \to \infty} \dfrac{\ln x}{x^a}, a > 0$

6. $\lim\limits_{x \to 0^+} \dfrac{\ln(\sin x)}{\ln(\tan x)}$

*7. $\lim\limits_{x \to 0^+} \dfrac{\ln x}{\cot x}$

8. $\lim\limits_{x \to \pi/2^-} \dfrac{\tan 2x}{\tan x}$

9. $\lim\limits_{x \to \infty} \left(x \tan \dfrac{1}{x} \right)$

10. $\lim\limits_{x \to 0^+} x^{-1/x}$

*11. $\lim\limits_{x \to \infty} \dfrac{x}{e^{\sqrt{x}}}$

*12. $\lim\limits_{x \to \infty} \dfrac{x}{e^{x^a}}, a > 0$

13. $\lim\limits_{x \to 0^+} x^{\sin x}$

14. $\lim\limits_{x \to \pi/2^+} \left(x - \dfrac{\pi}{2} \right)^{\cos x}$

15. $\lim\limits_{x \to 0^+} x \cdot \ln \sin x$

16. $\lim\limits_{x \to 0^+} (\sin x)^x$

17. $\lim\limits_{x \to \infty} \left(1 + \dfrac{1}{2x} \right)^{x^2}$

*18. $\lim\limits_{x \to \infty} \left(1 + \dfrac{1}{ax} \right)^{x^a}, a > 0$

[Hint: Consider separately the cases $a < 1$, $a = 1$, and $a > 1$.]

19. $\lim\limits_{x \to 0^+} (1 + \sinh x)^{1/x}$

20. $\lim\limits_{x \to 0^+} (1 + \sinh x)^{a/x}, a > 0$

21. $\lim\limits_{x \to 0^+} \csc 2x \sin 3x$

22. $\lim\limits_{x \to \pi/2} \sec x \cos 3x$

*23. $\lim\limits_{x \to 0^+} (\sin x)^{(\sin x) - x}$

24. $\lim\limits_{x \to \pi/2^-} (\cos x)^{\sec x}$

25. $\lim\limits_{x \to 2} x^{1/(2-x)}$

26. $\lim\limits_{x \to \infty} (1 + 5x)^{e^{-x}}$

27. $\lim\limits_{x \to \infty} \left(\cos \dfrac{1}{x} \right)^x$

28. $\lim\limits_{x \to 1} \dfrac{\tanh^{-1} x}{\tan(\pi x/2)}$

29. $\lim\limits_{x \to 1^-} \left(\dfrac{1}{1-x} - \dfrac{1}{\ln x} \right)$

30. $\lim\limits_{x \to 2^+} \left(\dfrac{x}{x-2} - \dfrac{1}{\ln(x-1)} \right)$

31. $\lim\limits_{x \to \infty} (\sqrt{x^2 + x} - x)$

32. $\lim\limits_{x \to \infty} (x^3 - \sqrt{x^6 - 5x^3 + 3})$

*33. Compute
a. $\lim\limits_{x \to 0^+} x^{-k/\ln x}, k > 0$
b. $\lim\limits_{x \to 0^+} x^{-k/\sqrt{-\ln x}}, k > 0$

*34. Let $g(x) = (x \ln x)/(x^2 - 1)$. Compute the following:
a. $\lim\limits_{x \to 0^+} g(x)$ b. $\lim\limits_{x \to 1} g(x)$

■ SHOW/PROVE/DISPROVE

35. Show that $\ln x$ grows more slowly than any positive power of x: if $\beta > 0$, then

$$\lim_{x \to \infty} \dfrac{\ln x}{x^\beta} = 0.$$

*36. Fill in the details of the following outline of an alternative proof of the result of Example 5: if $b > 0$, then $\lim\limits_{x \to \infty} x^a e^{-bx} = 0$ for any a.
a. If $a \le 0$, the limit expression is not an indeterminate form; its value is "obviously" equal to zero.
b. If $a > 0$, then x^a/e^{bx} is an indeterminate form ∞/∞; repeated application of l'Hôpital's rule eventually yields an expression of the type discussed in part (a) [times a constant].

37. a. Show that $\lim\limits_{x \to \infty} \dfrac{x + \cos x}{x + \sin x} = 1$.

b. Explain why $\lim\limits_{x \to \infty} \dfrac{1 - \sin x}{1 + \cos x}$ does not exist.

Do the results of parts (a) and (b) show that l'Hôpital's rule is not always true?

*38. Since $\lim\limits_{x \to \infty} \sin x$ does not exist (but oscillates continuously between -1 and $+1$), it is evident that $\lim\limits_{x \to \infty} e^{-\sin x}$ also does not exist. But

$$\lim_{x \to \infty} e^{-\sin x} = \lim_{x \to \infty} \dfrac{2x + \sin 2x}{(2x + \sin 2x)e^{\sin x}}.$$

Now show the following successively:
a. $2 + 2\cos 2x = 4\cos^2 x$.
[Hint: $\cos 2x = 2\cos^2 x - 1$.]

b. $\lim\limits_{x \to \infty} \dfrac{4\cos x}{(2x + 4\cos x + \sin 2x)e^{\sin x}} = 0$.

Then apply l'Hôpital's rule [using (a) and (b)] to show that:

c. $\lim\limits_{x \to \infty} \dfrac{2x + \sin 2x}{(2x + \sin 2x)e^{\sin x}} = 0$.

Thus $\lim\limits_{x \to \infty} e^{-\sin x} = 0$.
Where have we been led astray?

CONTINUED

39. Show that

$$\lim_{x \to 0} \frac{e^{-1/x^2}}{x} = 0.$$

[*Hint:* Write $\lim_{x \to 0}(e^{-1/x^2}/x) = \lim_{x \to 0}[(1/x)/e^{1/x^2}]$.]

40. Show that for every integer n,

$$\lim_{x \to 0} \frac{e^{-1/x^2}}{x^n} = 0.$$

*41. Use the result of the preceding problem to show that the function

$$f(x) = \begin{cases} e^{-1/x^2} & \text{if } x \neq 0 \\ 0 & \text{if } x = 0 \end{cases}$$

has continuous derivatives of all orders at every real number x. [*Hint:* Show that $f^{(n)}(x)$ is a sum of terms of the form $ax^{-m}e^{-1/x^2}$, where $n + 2 \leq m \leq 3n$.]

■ CHALLENGE

The following problems provide a proof of l'Hôpital's rule for the indeterminate form ∞/∞ in the case f' and g' are continuous and $g'(x)$ is never zero. The proof appears in the note "L'Hôpital's Rule via Integration" by Donald Hartig on pages 156–157 of the February 1991 issue of *The American Mathematical Monthly*.

In these problems we assume that

(i) $\lim_{x \to \infty} \dfrac{f'(x)}{g'(x)} = L$, (ii) $\lim g(x) = \infty$, (iii) f' and g' are continuous, and (iv) $g'(x) \neq 0$ for all x.

42. Show that $g'(x) > 0$ for all x. [*Hint:* Use the intermediate-value theorem.]

43. Show that for every $\epsilon > 0$, there is an $M > 0$ such that $|f'(x) - Lg'(x)| < \epsilon g'(x)$ if $x > M$. $\left[\text{*Hint:* Explain why } \left|\dfrac{f'(x)}{g'(x)} - L\right| < \epsilon \text{ if } x > M \text{ and then multiply through by } g'(x) > 0.\right]$

44. Explain why, if $b > a > M$ (where M is as in Problem 43),

$$\left|\int_a^b [f'(x) - Lg'(x)] \, dx\right|$$
$$\leq \int_a^b |f'(x) - Lg'(x)| \, dx < \int_a^b \epsilon g'(x) \, dx.$$

[*Hint:* Use the result of Problem 4.5.42 on page 308.]

45. Use the result of Problem 44 to show that, if $b > a > M$,

$$|f(b) - f(a) - L[g(b) - g(a)]| < \epsilon[g(b) - g(a)].$$

46. Show that

$$\left|\frac{f(b)}{g(b)} - \frac{f(a)}{g(b)} - L\left[1 - \frac{g(a)}{g(b)}\right]\right| < \epsilon\left[1 - \frac{g(a)}{g(b)}\right] < \epsilon.$$

[*Hint:* Divide the inequality in Problem 45 through by $g(b) > 0$ and use the fact that $g'(x) > 0$.]

47. Show that

$$\left|\frac{f(b)}{g(b)} - L\right| < \epsilon + \frac{|f(a)|}{g(b)} + |L|\frac{g(a)}{g(b)}.$$

[*Hint:* Use the equation in Problem 46 and the fact that $|x - y| \geq ||x| - |y||$ (see Problem 0.1.47 on page 9).]

48. Prove the following special case of l'Hôpital's rule for the indeterminate form ∞/∞: Let f and g have continuous derivatives with $g'(x) \neq 0$. If $\lim_{x \to \infty} g(x) = \infty$ and $\lim_{x \to \infty} f'(x)/g'(x) = L$, then $\lim_{x \to \infty} f(x)/g(x) = L$ also. [*Hint:* With a fixed, let $b \to \infty$ in the inequality in Problem 47.] Note that in this theorem it is *not* necessary to assume that $\lim_{x \to \infty} f(x) = \infty$, since this condition is not needed in the proof.

■ ANSWERS TO SELF-QUIZ

I. a   II. d   III. b   IV. c

V. False (see VI.)

VI. $e\left[\left(\dfrac{1}{x}\right)^{p/\ln x} = e^{-p} \text{ for all } p > 0, x > 0\right]$

9.3 IMPROPER INTEGRALS

In our introduction to the definite integral in Chapter 4, we cited Theorem 1 in Section 4.5, which states that $\int_a^b f(x)\,dx$ exists if $f(x)$ is continuous in the closed interval $[a, b]$. However, in many interesting applications one of two situations occurs: either (1) a or b is infinite or (2) f becomes infinite at one or more values in the interval $[a, b]$. If one of these situations occurs, we say that the integral in question is an **improper integral**. In this section, we will learn how to evaluate these two different types of improper integrals.

IMPROPER INTEGRAL, TYPE 1: INFINITE LIMIT OF INTEGRATION; $b = +\infty$, $a = -\infty$, OR BOTH

Before dealing with the general case, we give an example.

EXAMPLE 1
THE AREA UNDER A CURVE OVER AN INFINITE INTERVAL

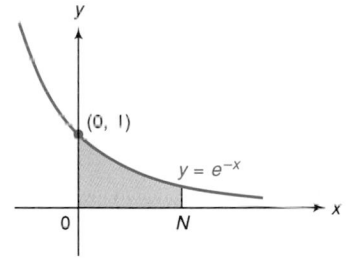

FIGURE 1
THE SHADED REGION IS THE AREA UNDER THE GRAPH OF e^{-x} FOR $0 \le x \le N$.

Calculate the area in the first quadrant under the curve $y = e^{-x}$.

SOLUTION: The region in question is sketched in Figure 1. For the first time we are asked to calculate the area of a region that stretches an infinite distance. To deal with this situation, we calculate the area from 0 to N, where N is some "large" number, and then we see what happens as $N \to \infty$. We have

$$\text{area from 0 to } N = A_0^{\ N} = \int_0^N e^{-x}\,dx = -e^{-x}\Big|_0^N = 1 - e^{-N}.$$

Then

$$\text{total area} = \lim_{N \to \infty} A_0^{\ N} = \lim_{N \to \infty}(1 - e^{-N}) = 1.$$

Thus the total area in the first quadrant under the curve $y = e^{-x}$ is 1.

Example 1 leads to the following general definitions.

DEFINITION   **IMPROPER INTEGRAL, TYPE 1: INFINITE LIMIT OF INTEGRATION**

(i) Let a be a real number and let f be a function having the property that $\int_a^N f(x)\,dx$ exists for every real number $N \ge a$. Then we define the **improper integral**

$$\int_a^\infty f(x)\,dx = \lim_{N \to \infty} \int_a^N f(x)\,dx \tag{1}$$

provided that this limit exists.

If $\int_a^\infty f(x)\,dx$ exists and is finite, we say that the improper integral is **convergent**. If the limit in (1) does not exist, or if it exists and is infinite, then we say that the improper integral is **divergent**.

(ii) If $\int_{-N}^b f(x)\,dx$ exists for every real number N such that $-N \le b$, we define

$$\int_{-\infty}^b f(x)\,dx = \lim_{N \to \infty} \int_{-N}^b f(x)\,dx \tag{2}$$

whenever the limit exists. We define the terms "convergent" and "divergent" as in (i).

(iii) If $\int_{-M}^{0} f(x)\,dx$ and $\int_{0}^{N} f(x)\,dx$ exist for every real N and M, then we define

$$\int_{-\infty}^{\infty} f(x)\,dx = \lim_{N \to \infty} \int_{0}^{N} f(x)\,dx + \lim_{M \to \infty} \int_{-M}^{0} f(x)\,dx \qquad \textbf{(3)}$$

whenever both of these limits exist. ■

EXAMPLE 2
A DIVERGENT IMPROPER INTEGRAL

Evaluate $\int_{1}^{\infty} (1/x)\,dx$.

SOLUTION: $\int_{1}^{N} (1/x)\,dx = \ln x \big|_{1}^{N} = \ln N - \ln 1 = \ln N$. But $\lim_{N \to \infty} \ln N = \infty$, so that the improper integral is divergent.

EXAMPLE 3
A DIVERGENT IMPROPER INTEGRAL

Evaluate $\int_{0}^{\infty} e^{x}\,dx$.

SOLUTION: $\int_{0}^{N} e^{x}\,dx = e^{N} - 1$, which approaches ∞ as $N \to \infty$, so this improper integral is also divergent.

EXAMPLE 4
AN OSCILLATING DIVERGENT INTEGRAL

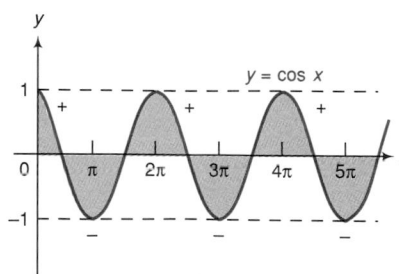

Evaluate $\int_{0}^{\infty} \cos x\,dx$.

SOLUTION $\int_{0}^{N} \cos x\,dx = \sin x \big|_{0}^{N} = \sin N$, which has no limit as $N \to \infty$, so that the improper integral $\int_{0}^{\infty} \cos x\,dx$ diverges. This makes sense graphically. In Figure 2, the integral in question is the area of the shaded region, where the areas below the x-axis are treated as negative. Thus the total area has no finite value because we must keep on adding and subtracting areas of equal size.

FIGURE 2
GRAPH OF $\cos x$

EXAMPLE 5
A CONVERGENT IMPROPER INTEGRAL

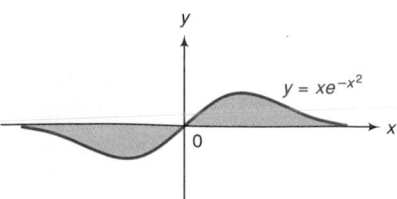

Evaluate $\int_{-\infty}^{\infty} xe^{-x^{2}}\,dx$.

SOLUTION: The function $xe^{-x^{2}}$ is sketched in Figure 3.

$$\int_{0}^{N} xe^{-x^{2}}\,dx = -\frac{1}{2}e^{-x^{2}}\Big|_{0}^{N} = \frac{1}{2}(1 - e^{-N^{2}}) \to \frac{1}{2}$$

as $N \to \infty$, and

$$\int_{-M}^{0} xe^{-x^{2}}\,dx = -\frac{1}{2}e^{-x^{2}}\Big|_{-M}^{0} = \frac{1}{2}(e^{-M^{2}} - 1) \to -\frac{1}{2}$$

as $M \to \infty$. Thus, since both limits exist,

$$\int_{-\infty}^{\infty} xe^{-x^{2}}\,dx = \int_{0}^{\infty} xe^{-x^{2}}\,dx + \int_{-\infty}^{0} xe^{-x^{2}}\,dx = \frac{1}{2} - \frac{1}{2} = 0.$$

FIGURE 3
THE GRAPH OF $xe^{-x^{2}}$ IS SYMMETRIC ABOUT THE ORIGIN

EXAMPLE 6
A DIVERGENT IMPROPER INTEGRAL

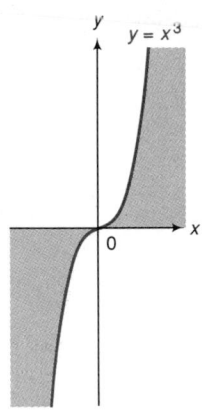

Calculate $\int_{-\infty}^{\infty} x^3\,dx$.

SOLUTION: The area to be calculated is sketched in Figure 4. Since $\lim_{N \to \infty} \int_0^N x^3\,dx = \lim_{N \to \infty}(N^4/4) = \infty$, the integral diverges. Note, however, that

$$\lim_{N \to \infty} \int_{-N}^{N} x^3\,dx = \lim_{N \to \infty} \frac{x^4}{4}\Big|_{-N}^{N} = \lim_{N \to \infty}\left(\frac{N^4}{4} - \frac{N^4}{4}\right) = \lim_{N \to \infty} 0 = 0,$$

which explains why we *do not* define $\int_{-\infty}^{\infty} f(x)\,dx$ as $\lim_{N \to \infty} \int_{-N}^{N} f(x)\,dx$. The problem here is that $\int_0^{\infty} x^3\,dx = \infty$ and $\int_{-\infty}^0 x^3\,dx = -\infty$. We simply cannot "cancel off" infinite terms. The expression $\infty - \infty$ is not defined.

FIGURE 4
THE GRAPH OF x^3 IS SYMMETRIC ABOUT THE ORIGIN

Finite positive and negative areas cancel.

EXAMPLE 7
THE INTEGRAL $\int_1^{\infty} \dfrac{1}{x^p}\,dx$

Evaluate

$$\int_1^{\infty} \frac{1}{x^p}\,dx, \qquad p > 0,\ p \neq 1.$$

SOLUTION:

$$\int_1^N \frac{1}{x^p}\,dx = \int_1^N x^{-p}\,dx = \frac{x^{-p+1}}{-p+1}\Big|_1^N = \frac{1}{p-1}(1 - N^{1-p}).$$

If $0 < p < 1$, then $N^{1-p} \to \infty$ as $N \to \infty$ and the integral diverges. If $p > 1$, then $N^{1-p} = 1/N^{p-1} \to 0$ as $N \to \infty$ and the integral converges. Combining this result with that of Example 2, we have

$$\int_1^{\infty} \frac{1}{x^p}\,dx \quad \begin{cases} \text{diverges if } 0 < p \leq 1, \\ \text{converges to } 1/(p-1) \text{ if } p > 1. \end{cases}$$

EXAMPLE 8
AN IMPROPER INTEGRAL
EVALUATED BY COMPLETING
A SQUARE

Evaluate

$$\int_{-\infty}^{\infty} \frac{dx}{x^2 - 4x + 9}.$$

SOLUTION:

$$\int_0^N \frac{dx}{x^2 - 4x + 9} = \int_0^N \frac{dx}{(x-2)^2 + 5} = \frac{1}{\sqrt{5}} \tan^{-1} \frac{x-2}{\sqrt{5}}\Big|_0^N$$

$$= \frac{1}{\sqrt{5}}\left[\tan^{-1} \frac{N-2}{\sqrt{5}} - \tan^{-1}\left(-\frac{2}{\sqrt{5}}\right)\right]$$

and $\lim_{N \to \infty} \tan^{-1}[(N - 2)/\sqrt{5}] = \pi/2$. Thus,

$$\int_0^\infty \frac{dx}{x^2 - 4x + 5} = \frac{1}{\sqrt{5}}\left[\frac{\pi}{2} - \tan^{-1}\left(-\frac{2}{\sqrt{5}}\right)\right].$$

Similarly,

$$\int_{-\infty}^0 \frac{dx}{x^2 - 4x + 5} = \frac{1}{\sqrt{5}}\left[\tan^{-1}\left(-\frac{2}{\sqrt{5}}\right) - \left(-\frac{\pi}{2}\right)\right].$$

Thus,

$$\int_{-\infty}^\infty \frac{dx}{x^2 - 4x + 5} = \int_0^\infty \frac{dx}{x^2 - 4x + 5} + \int_{-\infty}^0 \frac{dx}{x^2 - 4x + 5}$$

$$= \frac{1}{\sqrt{5}}\left(\frac{\pi}{2} + \frac{\pi}{2}\right) = \frac{\pi}{\sqrt{5}}.$$

IMPROPER INTEGRAL, TYPE 2: INTEGRAL BECOMES INFINITE IN [a, b]

We now turn to the second type of improper integral in which the integrand becomes infinite in the closed interval $[a, b]$. As before, we begin with an example.

EXAMPLE 9
EVALUATING AN IMPROPER INTEGRAL WHEN THE INTEGRAL BECOMES INFINITE AT AN ENDPOINT

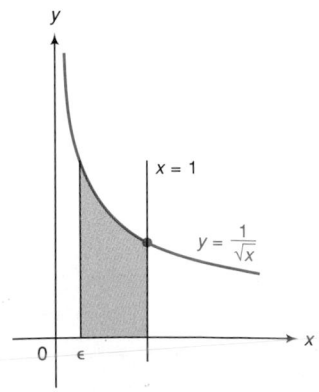

FIGURE 5
THE SHADED REGION IS THE REGION UNDER THE GRAPH OF $\dfrac{1}{\sqrt{x}}$ FOR $\epsilon \leq x \leq 1$.

Calculate the area in the first quadrant under the curve $y = 1/\sqrt{x}$ between $x = 0$ and $x = 1$.

SOLUTION: The region in question is sketched in Figure 5. As in Example 1, the region is infinite. However, if ϵ is a small, positive real number, then we can calculate $\int_\epsilon^1 (1/\sqrt{x})\,dx$ and see what happens as $\epsilon \to 0^+$. We have

$$A_\epsilon^{\ 1} = \int_\epsilon^1 \frac{1}{\sqrt{x}}\,dx = 2\sqrt{x}\,\Big|_\epsilon^1 = 2(1 - \sqrt{\epsilon}).$$

Then

$$\text{total area} = \lim_{\epsilon \to 0^+} A_\epsilon^{\ 1} = \lim_{\epsilon \to 0^+} 2(1 - \sqrt{\epsilon}) = 2.$$

Before citing the general definition, we recall that if f is continuous on $(a, b]$ but $f(x) \to +\infty$ or $-\infty$ as $x \to a^+$, then the graph of f has a **vertical asymptote** at $x = a$. That is, as $x \to a^+$, the graph of $y = f(x)$ approaches the vertical line $x = a$. In Figure 5 we see this illustrated for $a = 0$.

DEFINITION	**IMPROPER INTEGRAL, TYPE 2: INTEGRAND BECOMES INFINITE IN [a, b]**

Let a and b be finite numbers.

(i) If $\int_{a+\epsilon}^{b} f(x)\,dx$ exists for every ϵ in $(0, b - a]$ and if f has a vertical asymptote at $x = a$, then

$$\int_{a}^{b} f(x)\,dx = \lim_{\epsilon \to 0^+} \int_{a+\epsilon}^{b} f(x)\,dx \qquad (4)$$

provided that the limit exists.

(ii) If $\int_{a}^{b-\epsilon} f(x)\,dx$ exists for every ϵ in $(0, b - a]$ and if f has a vertical asymptote at $x = b$, then

$$\int_{a}^{b} f(x)\,dx = \lim_{\epsilon \to 0^+} \int_{a}^{b-\epsilon} f(x)\,dx \qquad (5)$$

provided that the limit exists.

(iii) If for c in (a, b) f has a vertical asymptote at $x = c$ and if the integrals $\int_{a}^{c-\epsilon_1} f(x)\,dx$ and $\int_{c+\epsilon_2}^{b} f(x)\,dx$ exist for ϵ_1 in $(0, c - a]$ and ϵ_2 in $(0, b - c]$, then

$$\int_{a}^{b} f(x)\,dx = \lim_{\epsilon_1 \to 0^+} \int_{a}^{c-\epsilon_1} f(x)\,dx + \lim_{\epsilon_2 \to 0^+} \int_{c+\epsilon_2}^{b} f(x)\,dx \qquad (6)$$

provided that both of these integrals exist. ■

For each of the cases (i), (ii), and (iii), the improper integral is **convergent** if the appropriate limit or limits exist and are finite. Otherwise, it is **divergent**.

The three cases of Type 2 are illustrated in Figure 6. Of course, we could have just as well drawn the curves approaching $-\infty$ instead of $+\infty$, but the basic idea is the same.

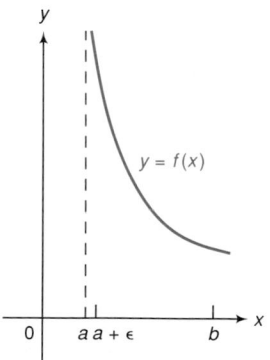
(a) Vertical asymptote at left endpoint.

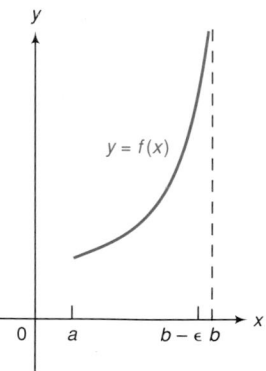
(b) Vertical asymptote at right endpoint.

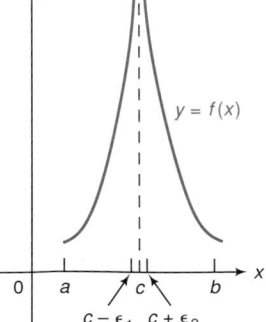
(c) Vertical asymptote in (a, b).

FIGURE 6 THREE PLACES FOR A VERTICAL ASYMPTOTE

In Example 9, we found that $\int_{0}^{1} (1/\sqrt{x})\,dx$ is a converging improper integral.

EXAMPLE 10
A DIVERGENT IMPROPER INTEGRAL

Evaluate $\int_0^1 (1/x)\,dx$.

SOLUTION: $\int_\epsilon^1 (1/x)\,dx = -\ln \epsilon$, which approaches ∞ as $\epsilon \to 0^+$, so that $\int_0^1 (1/x)\,dx$ diverges.

EXAMPLE 11

EVALUATING $\int_0^a \dfrac{1}{x^p}\,dx$

Evaluate $\int_0^a (1/x^p)\,dx$, where $p > 0$, $p \neq 1$, $a > 0$.

SOLUTION:

$$\int_\epsilon^a \left(\frac{1}{x^p}\right) dx = \int_\epsilon^a x^{-p}\,dx = \frac{x^{-p+1}}{-p+1}\bigg|_\epsilon^a = \frac{1}{1-p}(a^{1-p} - \epsilon^{1-p}).$$

If $p < 1$, then $1 - p > 0$, so $\lim_{\epsilon \to 0^+} \epsilon^{1-p} = 0$. If $p > 1$, then $p - 1 > 0$ and $\lim_{\epsilon \to 0^+} \epsilon^{1-p} = \lim_{\epsilon \to 0^+}(1/\epsilon^{p-1}) = \infty$. Thus adding in the case $p = 1$ from Example 10, we have

$$\int_0^a \frac{dx}{x^p} \quad \begin{cases} \text{diverges} & \text{if } p \geq 1, \\[2mm] = \dfrac{a^{1-p}}{1-p} & \text{if } 0 < p < 1. \end{cases} \tag{7}$$

EXAMPLE 12
AN INTEGRAL WHERE THE INTEGRAND BECOMES INFINITE AT A POINT THAT IS NOT AN ENDPOINT OF THE INTERVAL

Calculate

$$\int_0^3 \frac{dx}{(x-1)^3}.$$

SOLUTION: See Figure 7. In this problem, the integrand has a vertical asymptote at $x = 1$, so we must consider two improper integrals:

$$\int_0^3 \frac{dx}{(x-1)^3} = \int_0^1 \frac{dx}{(x-1)^3} + \int_1^3 \frac{dx}{(x-1)^3}.$$

Let $y = x - 1$. Then $dy = dx$, and

$$\int_0^3 \frac{dx}{(x-1)^3} = \int_{-1}^2 \frac{dy}{y^3} = \int_{-1}^0 \frac{dy}{y^3} + \int_0^2 \frac{dy}{y^3}.$$

Using (7), we see that the last integral diverges, since $3 > 1$. Thus $\int_0^3 (dx/(x-1)^3)$ diverges.

NOTE: If we had not noticed the discontinuity at $x = 1$, we might have blindly integrated to obtain

$$\int_0^3 \frac{dx}{(x-1)^3} = -\frac{1}{2(x-1)^2}\bigg|_0^3 = \frac{1}{2}\left(1 - \frac{1}{4}\right) = \frac{3}{8},$$

which is, of course, wrong. The reason that we obtain this finite answer is that two infinite areas have been incorrectly canceled (as illustrated in Figure 7).

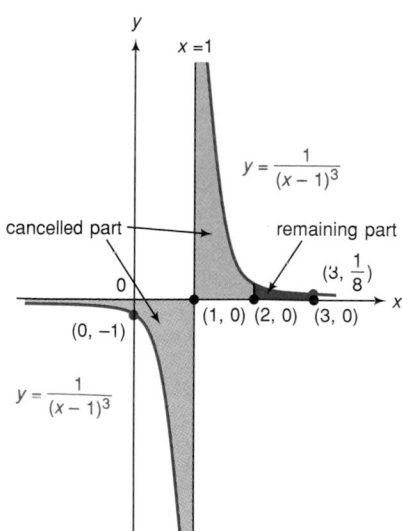

FIGURE 7

THE GRAPH OF $\dfrac{1}{(x-1)^3}$ FOR $x \geq 0$, $x \neq 1$

EXAMPLE 13
EVALUATING AN IMPROPER
INTEGRAL BY MAKING A
TRIGONOMETRIC SUBSTITUTION

Calculate

$$\int_{-1}^{1} \frac{dx}{\sqrt{1 - x^2}}.$$

SOLUTION: Here we have vertical asymptotes at both endpoints. Before integrating, we make the substitution $x = \sin \theta$. Then

$$\int_{-1}^{1} \frac{dx}{\sqrt{1 - x^2}} = \int_{-\pi/2}^{\pi/2} \frac{\cos \theta \, d\theta}{\cos \theta} = \pi.$$

Here we have transformed an improper integral into an ordinary one. This conversion is legitimate as long as the vertical asymptotes occur at the endpoints. Note that if we integrate directly, we get $\sin^{-1}(1 - \epsilon_1)$ and $\sin^{-1}(-1 + \epsilon_2)$, which tend, respectively, to $\pi/2$ and $-\pi/2$ as ϵ_1 and $\epsilon_2 \to 0^+$.

EXAMPLE 14
A CONVERGENT IMPROPER
INTEGRAL WITH A VERTICAL
ASYMPTOTE AT A POINT THAT
IS NOT AN ENDPOINT OF
THE INTERVAL

Calculate $\int_{-1}^{8} (1/\sqrt[3]{x}) \, dx$.

SOLUTION: Since we have a vertical asymptote at $x = 0$, we write this expression as the sum of two integrals:

$$\int_{-1}^{8} \frac{1}{\sqrt[3]{x}} \, dx = \int_{-1}^{0} \frac{1}{\sqrt[3]{x}} \, dx + \int_{0}^{8} \frac{1}{\sqrt[3]{x}} \, dx$$

$$= \lim_{\epsilon_1 \to 0^+} \int_{-1}^{-\epsilon_1} x^{-1/3} \, dx + \lim_{\epsilon_2 \to 0^+} \int_{\epsilon_2}^{8} x^{-1/3} \, dx$$

$$= \lim_{\epsilon_1 \to 0^+} \frac{3}{2} x^{2/3} \Big|_{-1}^{-\epsilon_1} + \lim_{\epsilon_2 \to 0^+} \frac{3}{2} x^{2/3} \Big|_{\epsilon_2}^{8}$$

$$= \lim_{\epsilon_1 \to 0^+} \frac{3}{2} [(-\epsilon_1)^{2/3} - 1] + \lim_{\epsilon_2 \to 0^+} \frac{3}{2} [4 - \epsilon_2^{3/2}]$$

$$= -\frac{3}{2} + 6 = \frac{9}{2}.$$

■ PROBLEMS 9.3

■ SELF-QUIZ

I. Which of the following is *not* an improper integral?

a. $\int_{0}^{\pi} \cot x \, dx$

b. $\int_{\pi}^{\infty} \sin e^x \, dx$

c. $\int_{0}^{\infty} \frac{\sin x}{x} \, dx$

d. $\int_{0.03}^{3.12} \left(\frac{1}{x} - 5 \cot x \right) dx$

II. $\int_{1}^{\infty} t^3 \, dt =$ _____.

a. $-\frac{1}{4}$ b. $\frac{1}{4}$
c. diverges   d. 0

III. $\int_{-\infty}^{1} t^3 \, dt =$ _____.

a. $-\frac{1}{4}$ b. $\frac{1}{4}$
c. diverges   d. 0

IV. True–False:

$$\int_{-\infty}^{\infty} t^3 \, dt = \lim_{x \to \infty} \int_{-x}^{x} t^3 \, dt.$$

CONTINUED

V. $\int_1^3 \dfrac{t-2}{(t+1)(t-3)}\, dt =$ _____.

   a. $\displaystyle\lim_{x\to 2} \int_x^3 \dfrac{t-2}{(t+1)(t-3)}\, dt$

   b. $\displaystyle\lim_{x\to 1} \int_x^3 \dfrac{t-2}{(t+1)(t-3)}\, dt$

   c. $\displaystyle\lim_{x\to 1^+} \int_x^3 \dfrac{t-2}{(t+1)(t-3)}\, dt$

   d. $\displaystyle\lim_{x\to 3^-} \int_x^3 \dfrac{t-2}{(t+1)(t-3)}\, dt$

   e. $\displaystyle\lim_{x\to 3} \int_1^x \dfrac{t-2}{(t+1)(t-3)}\, dt$

   f. $\displaystyle\lim_{x\to 3^-} \int_1^x \dfrac{t-2}{(t+1)(t-3)}\, dt$

VI. True–False: If $\int_0^\infty f(t)\, dt$ diverges, then $\lim_{N\to\infty} \left| \int_0^N f(t)\, dt \right| = \infty$.

■ DRILL

In Problems 1–46, determine whether the given improper integral converges or diverges. If it converges, calculate its value.

1. $\displaystyle\int_0^\infty \dfrac{dx}{x^3}$

2. $\displaystyle\int_0^\infty \dfrac{dx}{\sqrt{x}}$

3. $\displaystyle\int_0^1 \dfrac{dx}{x^{3/2}}$

4. $\displaystyle\int_2^3 \dfrac{dx}{\sqrt{x-2}}$

5. $\displaystyle\int_2^3 \dfrac{dx}{(x-2)^{1/3}}$

6. $\displaystyle\int_2^3 \dfrac{dx}{(x-2)^{4/3}}$

7. $\displaystyle\int_0^\infty e^{-2x}\, dx$

8. $\displaystyle\int_{-\infty}^4 e^{3x}\, dx$

9. $\displaystyle\int_{-\infty}^0 e^x\, dx$

10. $\displaystyle\int_{-\infty}^\infty e^{-0.01x}\, dx$

11. $\displaystyle\int_0^\infty x e^{-2x}\, dx$

12. $\displaystyle\int_0^\infty \dfrac{e^{-2\sqrt{x}}}{\sqrt{x}}\, dx$

13. $\displaystyle\int_{-\infty}^\infty x^3 e^{-x^4}\, dx$

14. $\displaystyle\int_{-\infty}^\infty x^2 e^{-x^3}\, dx$

15. $\displaystyle\int_{-\infty}^1 \dfrac{dx}{\sqrt{4-x}}$

16. $\displaystyle\int_{-\infty}^1 \dfrac{dx}{(4-x)^{3/2}}$

17. $\displaystyle\int_a^\infty \dfrac{dx}{b^2+x^2}$

18. $\displaystyle\int_0^\infty \dfrac{2x}{x^2+1}\, dx$

19. $\displaystyle\int_{-\infty}^\infty \dfrac{x^2}{x^2+1}\, dx$

20. $\displaystyle\int_{-\infty}^0 \dfrac{x}{(x^2+1)^{5/2}}\, dx$

21. $\displaystyle\int_0^{\pi/2} \tan x\, dx$

22. $\displaystyle\int_0^\pi \tan x\, dx$

23. $\displaystyle\int_0^{\pi/4} \csc x\, dx$

24. $\displaystyle\int_0^{\pi/2} \sec x\, dx$

25. $\displaystyle\int_0^2 \dfrac{dx}{x-1}$

26. $\displaystyle\int_0^2 \dfrac{dx}{(x-1)^{1/3}}$

27. $\displaystyle\int_{10}^\infty \dfrac{dx}{x^2-1}$

28. $\displaystyle\int_0^2 \dfrac{x}{(x^2-1)^{1/3}}\, dx$

29. $\displaystyle\int_0^\infty \sin x\, dx$

30. $\displaystyle\int_0^\infty e^{-x} \sin x\, dx$

31. $\displaystyle\int_{-\infty}^0 e^x \cos x\, dx$

32. $\displaystyle\int_0^\infty e^{-ax} \sin bx\, dx,\ a>0$

33. $\displaystyle\int_0^5 \dfrac{dx}{\sqrt{25-x^2}}$

34. $\displaystyle\int_0^\infty \dfrac{x^5}{e^{x^6}}\, dx$

35. $\displaystyle\int_{-2}^0 \dfrac{dx}{(x+2)^5}$

36. $\displaystyle\int_{-2}^0 \dfrac{dx}{(x+2)^{1/5}}$

37. $\displaystyle\int_2^4 \dfrac{dx}{(x-3)^7}$

38. $\displaystyle\int_2^4 \dfrac{dx}{(x-3)^{1/7}}$

*39. $\displaystyle\int_0^{\pi/2} \dfrac{dx}{1-\sin x}$

*40. $\displaystyle\int_0^\pi \dfrac{dx}{1-\cos x}$

41. $\displaystyle\int_5^\infty \dfrac{dx}{x^2-6x+8}$

42. $\displaystyle\int_1^\infty \dfrac{dx}{x^2-6x+8}$

43. $\displaystyle\int_1^\infty \dfrac{dx}{x \ln x}$

44. $\displaystyle\int_5^\infty \dfrac{dx}{x(\ln x)^2}$

45. $\displaystyle\int_0^1 \ln x\, dx$

*46. $\displaystyle\int_1^\infty \dfrac{dx}{1+e^x}$

■ APPLICATIONS

47. Find the volume of the solid generated when the curve $y = e^{-x}$ for $x \ge 0$ is rotated about the x-axis.

48. In Example 2, we showed that the area under the curve $y = 1/x$ for $x \ge 1$ is infinite.

a. Show that if this curve is rotated about the x-axis, the resulting solid of revolution has a finite volume.
b. Calculate that volume.

Use the comparison theorems stated in Problem 66 to do Problems 49–54.

49. Show that $\int_4^\infty dx/\sqrt{x^3 + 1 + \sin x}$ converges.
50. Show that $\int_1^\infty (\sqrt{1 + x^{1/8}}/x^{3/4})\,dx$ diverges.
51. Show that $\int_1^\infty [1/\ln(1 + x)]\,dx$ diverges.
52. Show that $\int_{-\infty}^\infty e^{-x^2}\,dx$ converges.
53. Show that $\int_1^\infty [(\sin^2 x)/x^2]\,dx$ converges.
54. Show that $\int_0^1 (dx/\sqrt{\sin x})$ converges. [*Hint:* Show that $\sin x \geq kx$ for some constant k so that $1/\sqrt{\sin x} \leq 1/\sqrt{kx}$, which has a convergent integral on $[0, 1]$.]

*55. Suppose N is a positive integer; compute
$$\int_0^1 \frac{1 - (1 - x)^N}{x}\,dx.$$

*56. Using only the fact that $\int_0^\infty [1/(1 + t^2)]\,dt$ exists (i.e., is finite), find the solution A to the equation

$$\int_0^A \frac{dt}{1 + t^2} = \frac{1}{2}\int_0^\infty \frac{dt}{1 + t^2} = \int_A^\infty \frac{dt}{1 + t^2}.$$

Suppose the function f has a domain which includes $[0, \infty)$. The **Laplace transform** of f is denoted by $\hat{f}$ and is the function with domain $[0, \infty)$ defined by

$$\hat{f}(s) = \int_0^\infty e^{-st}f(t)\,dt.$$

In Problems 57–60, find the Laplace transform of the given function. [Problems 67–68 suggest some facts for you to prove about Laplace transforms.]

57. $f(x) = 1$　　　　58. $g(x) = x$
59. $F(x) = e^x$　　　60. $G(x) = \cos x$

61. The substitution $u = x^{1/3}$ transforms the improper integral $\int_0^8 x^{-1/3}\,dx$ into a proper integral. Evaluate both integrals and comment.
62. Evaluate each of the following expressions (or show it diverges), then comment on why the two results are or are not the same.

a. $\displaystyle\int_{-\infty}^\infty \frac{1 + x}{1 + x^2}\,dx$　　b. $\displaystyle\lim_{N \to \infty} \int_{-N}^N \frac{1 + x}{1 + x^2}\,dx$

■ **SHOW/PROVE/DISPROVE**

*63. Prove each of the following:

a. $\displaystyle\int_0^1 \ln t\,dt < \frac{1}{N}\sum_{k=1}^N \ln\left(\frac{k}{N}\right) < \int_{1/N}^1 \ln t\,dt$

for all positive integers N.

b. $\displaystyle\lim_{N \to \infty} \frac{1}{N}\sum_{k=1}^N \ln\left(\frac{k}{N}\right) = \int_0^1 \ln t\,dt$.

c. $\displaystyle\lim_{N \to \infty} \frac{\sqrt[N]{N!}}{N} = \frac{1}{e}$.

[*Note:* This result is not quite as strong as **Stirling's formula**: $N! \approx \sqrt{2\pi N}(N/e)^N$ for N large.]

*64. Show that for any integer $n > 0$,

$$\int_0^\infty x^n e^{-x}\,dx = n!.$$

[*Hint:* Use mathematical induction (Appendix 2) and integration by parts.]

*65. Suppose that α is a positive real number; define the **gamma function**, $\Gamma(\alpha)$, by

$$\Gamma(\alpha) \equiv \int_0^\infty t^{\alpha - 1}e^{-t}\,dt.$$

Prove that $\Gamma(\alpha + 1) = \alpha \cdot \Gamma(\alpha)$.
*66. Let f and g be two continuous functions such that $0 \leq g(x) \leq f(x)$ if $x \geq a$. Prove the following **comparison theorems**.
a. If $\int_a^\infty f(x)\,dx$ converges, then so does $\int_a^\infty g(x)\,dx$.
b. If $\int_a^\infty g(x)\,dx$ diverges, then so does $\int_a^\infty f(x)\,dx$.

[*Hint:* For (a), show that if N is large, then $\int_0^N g(x)\,dx \leq \int_0^N f(x)\,dx \leq L$ where $L = \int_0^\infty f(x)\,dx$. For (b), show that $\int_0^N f(x)\,dx \geq \int_0^N g(x)\,dx$, which approaches ∞ as $N \to \infty$.]
67. The Laplace transform, $\hat{f}$, of a function f with domain $[0, \infty)$ is defined prior to Problems 57–60. Suppose that $\int_0^\infty |f(t)|\,dt$ converges; show that $\hat{f}(s)$ exists and is finite for every nonnegative real number s. [*Note:* The assumption that $\int_0^\infty |f(t)|\,dt$ converges is sufficient, but it is not

CONTINUED

necessary—other functions have Laplace transforms (see Problems 57–60).]

*68. Suppose f is a differentiable function with Laplace transform $\hat{f}$.

a. Let $g(x) = x \cdot f(x)$. Show that $\hat{g}(s) = -\dfrac{d}{ds} \hat{f}(s)$.

b. Let $h(x) = f'(x)$. Show that $\hat{h}(s) = s \cdot \hat{f}(s) - f(0)$.

■ APPLICATIONS TO PROBABILITY THEORY

A **probability density function** is a function f whose domain is the set of all real numbers and such that

a. $f(t) \geq 0$ for each t in $(-\infty, \infty)$,

b. $\int_{-\infty}^{\infty} f(t)\, dt = 1$.

Such functions are very useful in probability theory to make probability statements about a number which is the result of some random event: $\int_{-\infty}^{x} f(t)\, dt$ computes the probability that the result is less than or equal to the real number x.

69. Suppose a and b are real numbers such that $a < b$; show that

$$f(x) = \begin{cases} 0 & \text{if } x < a \\ 1/(b-a) & \text{if } a \leq x \leq b \\ 0 & \text{if } x > b \end{cases}$$

is a probability density function. (This function is called a **uniform** density function.)

70. Suppose $a > 0$; show that

$$f(x) = \begin{cases} 0 & \text{if } x < 0 \\ ae^{-ax} & \text{if } 0 \leq x \end{cases}$$

is a probability density function. (This function is called, unsurprisingly, an **exponential** density function. It can be used to help analyze random times spent in waiting lines.)

71. Show that

$$\int_{-\infty}^{\infty} \frac{1}{\sqrt{2\pi}} e^{-t^2/2}\, dt$$

converges. The integrand here is called the **unit normal** probability density function. Tables of the function

$$\int_{-\infty}^{x} \frac{1}{\sqrt{2\pi}} e^{-t^2/2}\, dt$$

are available in most books on probability or statistics.

The **expected value** μ (also known as the **mean**) associated with the probability density function f is defined by

$$\mu = \int_{-\infty}^{\infty} t f(t)\, dt.$$

(We can think of the expected value as an average of the possible values of a random experiment weighted by their associated probabilities.)

72. Calculate the expected value associated with the uniform probability density function given in Problem 69.

73. Calculate the expected value associated with the exponential probability density function given in Problem 70.

74. Calculate the expected value associated with the unit normal probability density function given in Problem 71.

The **variance** σ^2 associated with the probability density function f is defined by

$$\sigma^2 = \int_{-\infty}^{\infty} (t - \mu)^2 f(t)\, dt.$$

(The variance is often referred to as the "mean squared deviation.") We also define the **standard deviation**, σ, associated with f to be the square root of the variance ("root mean square").

75. Calculate the variance and standard deviation associated with the uniform probability density function given in Problem 69.

76. Calculate the variance and standard deviation associated with the exponential probability density function given in Problem 70.

*77. Show that the variance and standard deviation associated with the unit normal probability density function given in Problem 71 are finite.

■ CHALLENGE

**78. Let

$$H(r) = \int_0^\infty \frac{1 - e^{-t}}{t^r}\, dt,$$

where $1 < r < 2$. Express H in terms of the gamma function defined in Problem 65.

9.4 TAYLOR'S THEOREM AND TAYLOR POLYNOMIALS

In Section 3.8, we discussed the linearization of a function. We proved that if $f''(x)$ exists in an open interval (a, b) containing x_0 and if $x \in (a, b)$, then there is a number c between x_0 and x such that

$$f(x) = \overbrace{f(x_0) + f'(x_0)(x - x_0)}^{P_1(x)} + \frac{1}{2}f''(c)(x - x_0)^2.$$

Moreover, if $|f''(x)| \le M$ in $[x_0, x]$ or $[x, x_0]$, then

$$|f(x) - P_1(x)| \le \frac{M}{2}(x - x_0)^2.$$

In short, we showed how a differentiable function could be approximated by a polynomial of degree 1. In this section we show how a function can be approximated as closely as desired by a polynomial provided that the function possesses a sufficient number of derivatives.

We begin by reminding you of the factorial notation defined for all positive integers n:

$$n! = n(n - 1)(n - 2) \cdots 3 \cdot 2 \cdot 1.$$

That is, $n!$ is the product of the first n positive integers. For example, $3! = 3 \cdot 2 \cdot 1 = 6$ and $5! = 5 \cdot 4 \cdot 3 \cdot 2 \cdot 1 = 120$. By convention, we define $0!$ to be equal to 1.

DEFINITION   TAYLOR[†] POLYNOMIAL

Let the function f and its first n derivatives exist on the closed interval $[x_0, x_1]$. Then, for $a \in (x_0, x_1)$ and $x \in (x_0, x_1)$ the nth-degree **Taylor polynomial** of f at a is the nth-degree polynomial $P_n(x)$, given by

$$P_n(x) = f(a) + \frac{f'(a)}{1!}(x - a) + \frac{f''(a)}{2!}(x - a)^2 + \frac{f'''(a)}{3!}(x - a)^3$$

$$+ \cdots + \frac{f^{(n)}(a)}{n!}(x - a)^n$$

$$= \sum_{k=0}^{n} \frac{f^{(k)}(a)(x - a)^k}{k!}.\text{[‡]} \quad ■$$

(1)

[†] Named after the English mathematician Brook Taylor (1685–1731), who published what we now call *Taylor's formula* in *Methodus Incrementorum* in 1715. There was a considerable controversy over whether Taylor's discovery was, in fact, a plagiarism of an earlier result of the Swiss mathematician Johann Bernoulli (1667–1748). We discussed Johann Bernoulli on page 597.

[‡] In this notation we have $f^{(0)}(a) = f(a)$.

EXAMPLE 1
A TAYLOR POLYNOMIAL FOR $\sin x$ AT 0

Calculate the fifth-degree Taylor polynomial of $f(x) = \sin x$ at 0.

SOLUTION: We have $f(x) = \sin x$, $f'(x) = \cos x$, $f''(x) = -\sin x$, $f'''(x) = -\cos x$, $f^{(4)}(x) = \sin x$, and $f^{(5)}(x) = \cos x$. Then $f(0) = 0$, $f'(0) = 1$, $f''(0) = 0$, $f'''(0) = -1$, $f^{(4)}(0) = 0$, $f^{(5)}(0) = 1$, and we obtain

$$P_5(x) = f(0) + \frac{f'(0)}{1!} x + \frac{f''(0)}{2!} x^2 + \frac{f'''(0)}{3!} x^3$$

$$+ \frac{f^{(4)}(0)}{4!} x^4 + \frac{f^{(5)}(0)}{5!} x^5$$

$$= x - \frac{x^3}{3!} + \frac{x^5}{5!} = x - \frac{x^3}{6} + \frac{x^5}{120}.$$

DEFINITION   REMAINDER TERM

Let $P_n(x)$ be the nth-degree Taylor polynomial of the function f. Then the **remainder term**, denoted by $R_n(x)$, is given by

$$R_n(x) = f(x) - P_n(x). \quad \blacksquare \tag{2}$$

Why do we study Taylor polynomials? Because of the following remarkable result that tells us that a Taylor polynomial provides a good approximation to a function f. The proof is given at the end of this section.

THEOREM 1   TAYLOR'S THEOREM (TAYLOR'S FORMULA WITH REMAINDER)

Suppose that $f^{(n+1)}(x)$ exists on the closed interval $[x_0, x_1]$. Let a be in (x_0, x_1) and let x be any number in $[x_0, x_1]$. Then there is a number $c^{\dagger}$ in (a, x) or (x, a) such that

$$R_n(x) = \frac{f^{(n+1)}(c)}{(n+1)!} (x - a)^{n+1}. \quad \blacksquare \tag{3}$$

The expression in (3) is called **Lagrange's form of the remainder.**[‡] Using (3), we can write Taylor's formula as

$$f(x) = f(a) + \frac{f'(a)}{1!} (x - a) + \frac{f''(a)}{2!} (x - a)^2 + \cdots$$

$$+ \frac{f^{(n)}(a)}{n!} (x - a)^n + \frac{f^{(n+1)}(c)}{(n+1)!} (x - a)^{n+1}. \tag{4}$$

[†] c depends on x.
[‡] See the biographical sketch of Lagrange on page 624.

EXAMPLE 2
A TAYLOR POLYNOMIAL FOR sin x
AT $\pi/6$

Calculate the fifth-degree Taylor polynomial of $f(x) = \sin x$ at $\pi/6$.

SOLUTION: Using the derivatives found in Example 1, we have $f(\pi/6) = \frac{1}{2}$, $f'(\pi/6) = \sqrt{3}/2$, $f''(\pi/6) = -\frac{1}{2}$, $f'''(\pi/6) = -\sqrt{3}/2$, $f^{(4)}(\pi/6) = \frac{1}{2}$, and $f^{(5)}(\pi/6) = \sqrt{3}/2$, so that in this case

$$
\begin{aligned}
P_5(x) &= \frac{1}{2} + \frac{\sqrt{3}}{2}\left(x - \frac{\pi}{6}\right) - \frac{1}{2}\frac{[x - (\pi/6)]^2}{2!} - \frac{\sqrt{3}}{2}\frac{[x - (\pi/6)]^3}{3!} \\
&\quad + \frac{1}{2}\frac{[x - (\pi/6)]^4}{4!} + \frac{\sqrt{3}}{2}\frac{[x - (\pi/6)]^5}{5!} \\
&= \frac{1}{2} + \frac{\sqrt{3}}{2}\left(x - \frac{\pi}{6}\right) - \frac{1}{4}\left(x - \frac{\pi}{6}\right)^2 - \frac{\sqrt{3}}{12}\left(x - \frac{\pi}{6}\right)^3 \\
&\quad + \frac{1}{48}\left(x - \frac{\pi}{6}\right)^4 + \frac{\sqrt{3}}{240}\left(x - \frac{\pi}{6}\right)^5.
\end{aligned}
$$

Examples 1 and 2 illustrate that in many cases it is easiest to calculate the Taylor polynomial at 0. In this situation, we have

TAYLOR POLYNOMIAL AT 0

$$
P_n(x) = f(0) + f'(0)x + \frac{f''(0)}{2!}x^2 + \cdots + \frac{f^{(n)}(0)}{n!}x^n. \tag{5}
$$

EXAMPLE 3
A TAYLOR POLYNOMIAL FOR
e^x AT 0

Find the eighth-degree Taylor polynomial of $f(x) = e^x$ at 0.

SOLUTION: Here $f(x) = f'(x) = f''(x) = \cdots = f^{(8)}(x) = e^x$, and $e^0 = 1$, so that

$$
P_8(x) = 1 + x + \frac{x^2}{2!} + \frac{x^3}{3!} + \frac{x^4}{4!} + \frac{x^5}{5!} + \frac{x^6}{6!} + \frac{x^7}{7!} + \frac{x^8}{8!} = \sum_{k=0}^{8} \frac{x^k}{k!}.
$$

EXAMPLE 4
A TAYLOR POLYNOMIAL FOR
$1/(1 - x)$ AT 0

Find the fifth-degree Taylor polynomial of $f(x) = 1/(1 - x)$ at 0.

SOLUTION: Here,

$$
f(x) = \frac{1}{1 - x}, \qquad f'(x) = \frac{1}{(1 - x)^2}, \qquad f''(x) = \frac{2}{(1 - x)^3},
$$

$$
f'''(x) = \frac{6}{(1 - x)^4}, \qquad f^{(4)}(x) = \frac{24}{(1 - x)^5}, \qquad f^{(5)}(x) = \frac{120}{(1 - x)^6}.
$$

Thus, $f(0) = 1$, $f'(0) = 1$, $f''(0) = 2$, $f'''(0) = 6$, $f^{(4)}(0) = 24$, $f^{(5)}(x) = 120$, and

$$
\begin{aligned}
P_5(x) &= 1 + x + \frac{2x^2}{2!} + \frac{6x^3}{3!} + \frac{24x^4}{4!} + \frac{120x^5}{5!} \\
&= 1 + x + x^2 + x^3 + x^4 + x^5 = \sum_{k=0}^{5} x^k.
\end{aligned}
$$

Note that in Examples 1, 2, and 3, the given function had continuous derivatives of all orders defined for all real numbers. In Example 4, $f(x)$ is defined over intervals of the form $[-b, b]$, where $b < 1$. Thus Taylor's theorem does *not* apply, for example, in any interval containing 1. It is always necessary to check whether the hypothesis of Taylor's theorem holds over a given interval.

Before leaving this section, we observe that the nth-degree Taylor polynomial at a of a function is the polynomial that agrees with the function and each of its first n derivatives at a. This follows immediately from (1):

$$P_n(a) = f(a)$$

$$P_n'(a) = \left[f'(a) + f''(a)(x - a) + f'''(a)\frac{(x - a)^2}{2!} + \cdots \right.$$
$$\left. + \frac{f^{(n)}(a)}{(n - 1)!}(x - a)^{n-1} \right]\bigg|_{x=a} = f'(a),$$

$$P_n'' = \left[f''(a) + f'''(a)(x - a) + \cdots + \frac{f^{(n)}(a)}{(n - 2)!}(x - a)^{n-2} \right]\bigg|_{x=a} = f''(a),$$

and so on. In particular, since the $(n + 1)$st derivative of an nth-degree polynomial is zero, we find that if $Q(x)$ is a polynomial of degree n, then $P_n(x) = Q(x)$. This follows immediately from the fact that the remainder term given by (3) will be zero since $Q^{(n+1)}(c) = 0$.

EXAMPLE 5
A TAYLOR POLYNOMIAL
FOR A POLYNOMIAL

Let $Q(x) = 3x^4 + 2x^3 - 4x^2 + 5x - 8$. Compute $P_4(x)$ at 0.

SOLUTION: We have

$$Q(0) = -8,$$

$$Q'(0) = (12x^3 + 6x^2 - 8x + 5)\big|_{x=0} = 5,$$

$$Q''(0) = (36x^2 + 12x - 8)\big|_{x=0} = -8,$$

$$Q'''(0) = (72x + 12)\big|_{x=0} = 12,$$

$$Q^{(4)}(0) = 72.$$

Therefore,

$$P_4(x) = -8 + 5x - \frac{8x^2}{2!} + \frac{12x^3}{3!} + \frac{72x^4}{4!}$$
$$= -8 + 5x - 4x^2 + 2x^3 + 3x^4 = Q(x),$$

as expected.

PROOF OF TAYLOR'S THEOREM

We show that if $f^{(n+1)}(x)$ exists in $[a, b]$, then for any number $x \in [a, b]$, there is a number c in $[a, x]$ such that

$$f(x) = f(a) + \frac{f'(a)}{1!}(x-a) + \frac{f''(a)}{2!}(x-a)^2 + \cdots + \frac{f^{(n)}(a)}{n!}(x-a)^n + R_n(x),$$

where

definition of $R_n(x)$

$$f(x) - P_n(x) \stackrel{\downarrow}{=} R_n(x) = \frac{f^{(n+1)}(c)}{(n+1)!}(x-a)^{n+1}.$$

Let $x \in [a, b]$ be fixed. We define the new function $h(t)$ by

$$h(t) = f(x) - f(t) - f'(t)(x-t) - \frac{f''(t)}{2!}(x-t)^2 - \cdots$$
$$- \frac{f^{(n)}(t)}{n!}(x-t)^n - \frac{R_n(x)(x-t)^{n+1}}{(x-a)^{n+1}}. \tag{6}$$

Then, $h(x) = 0$ and

$$h(a) = f(x) - P_n(x) - R_n(x) = R_n(x) - R_n(x) = 0.$$

Since $f^{(n+1)}$ exists, $f^{(n)}$ is differentiable so that h, being a sum of products of differentiable functions, is also differentiable for t in (a, x). Remember that x is fixed so h is a function of t only.

Recall Rolle's theorem (see page 239), which states that if h is continuous on $[a, b]$ and differentiable on (a, b), and if $h(a) = h(b) = 0$, then there is at least one number c in (a, b) such that $h'(c) = 0$. We see that the conditions of Rolle's theorem hold in the interval $[a, x]$ so that there is a number c in (a, x) with $h'(c) = 0$. In Problems 30 and 31, you are asked to show that

$$h'(t) = \frac{-f^{(n+1)}(t)(x-t)^n}{n!} + \frac{(n+1)R_n(x)(x-t)^n}{(x-a)^{n+1}}.$$

Then, setting $t = c$, we obtain

$$0 = h'(c) = \frac{-f^{(n+1)}(c)(x-c)^n}{n!} + \frac{(n+1)R_n(x)(x-c)^n}{(x-a)^{n+1}}.$$

Finally, dividing the equations above through by $(x-c)^n$ and solving for $R_n(x)$, we obtain

$$R_n(x) = \frac{f^{(n+1)}(c)(x-a)^n}{(n+1)n!} = \frac{f^{(n+1)}(c)(x-a)^{n+1}}{(n+1)!}.$$

This is what we wanted to prove. ∎

REMARK: If you go over the proof, you may observe that we didn't need to assume that $x > a$; if we replace the interval $[a, x]$ with the interval $[x, a]$, then all results are still valid.

JOSEPH LOUIS LAGRANGE WAS ONE of the two greatest mathematicians of the eighteenth century—the other being Leonhard Euler (see page 34). Born in 1736 in Turin, Italy, Lagrange was the youngest of 11 children of French and Italian parents and the only one to survive to adulthood. Educated in Turin, Joseph Louis became a professor of mathematics in the military academy there when he was still quite young.

Lagrange's early publications established his reputation. When Euler left his post at the court of Frederick the Great in Berlin in 1766, he recommended that Lagrange be appointed his successor. Accepting Euler's advice, Frederick wrote to Lagrange that "the greatest king in Europe" wished to invite to his court "the greatest mathematician in Europe." Lagrange accepted and remained in Berlin for 20 years. Afterward, he accepted a post at the École Polytechnique in France.

JOSEPH LOUIS LAGRANGE

1736–1813

$1/(1 - x)$ at $x = 0$. This is the origin of the word *derivative*. The notation $f'(x), f''(x), \ldots$ was first used by Lagrange as was the form of the remainder term (3).

Lagrange is known for much else as well. Beginning in the 1750s, he invented the calculus of variations. He made significant contributions to ordinary differential equations, partial differential equations, numerical analysis, number theory, and algebra. In 1787 he published his *Mécanique analytique*, which contained the equations of motion of a dynamical system. Today these equations are known as *Lagrange's equations*.

Lagrange lived in France during the French Revolution. In 1790 he was placed on a committee to reform weights and measures and later became the head of a related committee that, in 1799, recommended the adoption of the system that we know today as the *metric system*. Despite his

Lagrange had a deep influence on nineteenth- and twentieth-century mathematics. He is perhaps best known as the first great mathematician to attempt to make calculus mathematically rigorous. His major work in this area was his 1797 paper "Théorie des fonctions analytiques contenant les principes du calcul différentiel." In this work, Lagrange tried to make calculus more logical—rather than more useful. His key idea was to represent a function $f(x)$ by a Taylor series. For example, we can write $1/(1 - x) = 1 + x + x^2 + \cdots + x^n + \cdots$ (a result that can be obtained by long division). Lagrange multiplied the coefficient of x^n by $n!$ and called the result the nth *derived function* of

work for the revolution, however, Lagrange was disgusted by its cruelties. After the great French chemist Lavoisier was guillotined, Lagrange exclaimed, "It took the mob only a moment to remove his head; a century will not suffice to reproduce it."

In his later years, Lagrange was often lonely and depressed. When he was 56, the 17-year-old daughter of his friend the astronomer P. C. Lemonier was so moved by his unhappiness that she proposed to him. The resulting marriage apparently turned out to be ideal for both.

Perhaps the greatest tribute to Lagrange was given by Napoleon Bonaparte, who said, "Lagrange is the lofty pyramid of the mathematical sciences."

■ PROBLEMS 9.4

■ SELF-QUIZ

I. The zero-degree Taylor polynomial of $f(x) = \cos x$ at 0 is _____.
  a. $0 \cdot x$ b. $f(0)$ c. 1   d. 0

II. The zero-degree Taylor polynomial of $f(x) = \cos x$ at π is _____.
  a. $0 \cdot x$ b. $1 \cdot (x - \pi)$ c. $f(0)$
  d. $f(\pi)$ e. π f. -1

III. The first-degree Taylor polynomial of $f(x) = \cos x$ at π is _____.
  a. $0 \cdot (x - \pi)$ b. $-1 + 1 \cdot x$
  c. $-1 + 0 \cdot x$ d. $-1 + 0 \cdot (x - \pi)$

IV. The second-degree Taylor polynomial of $f(x) = \cos x$ at π is _____.
  a. $-1 + 0 \cdot (x - \pi)$
  b. $-1 + 0 \cdot (x - \pi) + 1 \cdot (x - \pi)^2$
  c. $-1 + 0 \cdot x + \frac{1}{2} \cdot x^2$
  d. $-1 + 0 \cdot (x - \pi) + \frac{1}{2} \cdot (x - \pi)^2$

V. The second-degree Taylor polynomial of $f(x) = 2 - 3x$ at -5 is _____.
  a. $17 - 3 \cdot (x + 5)$
  b. $f(0) + f(1)(x - (-5)) + f(2)(x - (-5))^2/2$
  c. $17 + (-\frac{3}{1}) \cdot (x - 5) + \frac{0}{2} \cdot (x - 5)^2$

CONTINUED

d. $2 - 3x$

e. $2 - 3 \cdot (x + 5) + 4 \cdot (x + 5)^2$

f. $2 - 3 \cdot (x - 5) + 4 \cdot (x - 5)^2$

VI. The second-degree Taylor polynomial of $f(x) = 2 - 3x^2$ at -5 is _____.

a. $-73 + (-\frac{6}{2})(x + 5)^2$

b. $-73 + 30(x + 5) - 3(x + 5)^2$

c. $-73 + 30x - 3x^2$

d. $-73 + \frac{30}{1}(x - 5) + (-\frac{6}{2})(x - 5)^2$

■ **DRILL**

In Problems 1–26, a function f, a positive integer n, and a real number a are specified. Find the nth-degree Taylor polynomial of f at a. [*Note:* In some nth-degree Taylor polynomials, the coefficient of x^n is zero.]

1. $f(x) = \cos x$; $a = \pi/4$; $n = 6$

2. $f(x) = \sqrt{x}$; $a = 1$; $n = 4$

3. $f(x) = \ln x$; $a = e$; $n = 5$

4. $f(x) = \ln(1 + x)$; $a = 0$; $n = 5$

5. $f(x) = \dfrac{1}{x}$; $a = 1$; $n = 4$

6. $f(x) = \dfrac{1}{1 + x}$; $a = 0$; $n = 5$

7. $f(x) = \tan x$; $a = 0$; $n = 4$

8. $f(x) = \tan^{-1} x$; $a = 0$; $n = 6$

9. $f(x) = \tan x$; $a = \pi$; $n = 4$

10. $f(x) = \tan^{-1} x$; $a = 1$; $n = 6$

11. $f(x) = \dfrac{1}{1 + x^2}$; $a = 0$; $n = 4$

12. $f(x) = \dfrac{1}{\sqrt{x}}$; $a = 4$; $n = 3$

13. $f(x) = \sinh x$; $a = 0$; $n = 4$

14. $f(x) = \cosh x$; $a = 0$; $n = 4$

15. $f(x) = \ln \sin x$; $a = \pi/2$; $n = 3$

16. $f(x) = \ln \cos x$; $a = 0$; $n = 3$

17. $f(x) = \dfrac{1}{\sqrt{1 \ x}}$; $a = 0$; $n = 4$

18. $f(x) = \dfrac{1}{\sqrt{4 - x}}$; $a = 3$; $n = 4$

19. $f(x) = e^{\beta x}$, β real; $a = 0$; $n - 6$

20. $f(x) = \sin(\beta x)$, β real; $a = 0$; $n = 6$

21. $f(x) = \sin^{-1} x$; $a = 0$; $n = 3$

22. $f(x) = 1 + x + x^2$; $a = 0$; $n = 10$

23. $f(x) = a_0 + a_1 x + a_2 x^2 + a_3 x^3$; $a = 1$; $n = 10$

24. $f(x) = e^{x^2}$; $a = 0$; $n = 4$

25. $f(x) = \sin(x^2)$; $a = 0$; $n = 4$

26. $f(x) = \cos(x^2)$; $a = 0$; $n = 4$

■ **SHOW/PROVE/DISPROVE**

27. Show that the nth-degree Taylor polynomial of $f(x) = 1/(1 - x)$ at 0 is

$$1 + x + x^2 + \cdots + x^n = \sum_{k=0}^{n} x^k.$$

28. Show that the nth-degree Taylor polynomial of $f(x) = \ln(1 + x)$ at 0 is

$$x - \frac{x^2}{2} + \frac{x^3}{3} - \frac{x^4}{4} + \cdots + (-1)^{n-1} \frac{x^n}{n}$$

$$= \sum_{k=1}^{n} (-1)^{k-1} \frac{x^k}{k}.$$

29. Show that the $(2n + 1)$st and $(2n + 2)$nd degree Taylor polynomials for $f(x) = \sin x$ at $a = 0$ are

equal and are given by

$$x - \frac{x^3}{3!} + \frac{x^5}{5!} - \cdots + \frac{(-1)^n x^{2n+1}}{(2n + 1)!}$$

$$= \sum_{k=0}^{n} \frac{(-1)^k x^{2k+1}}{(2k + 1)!}$$

30. Show that for $1 \le k \le n$,

$$\frac{d}{dt}\left[\frac{f^{(k)}(t)}{k!} \cdot (x - t)^k\right] = \frac{-f^{(k)}(t)}{(k - 1)!} \cdot (x - t)^{k-1}$$

$$+ \frac{f^{(k+1)}(t)}{k!} \cdot (x - t)^k.$$

CONTINUED

31. Use the results of the preceding problem to show that the function h defined by equation (6) has the following derivative:

$$h'(t) = \frac{-f^{(n+1)}(t)}{n!} \cdot (x - t)^n$$
$$+ \frac{(n+1) \cdot R_n(x) \cdot (x - t)^n}{(x - a)^{n+1}}.$$

*32. Suppose the function f and its first $n - 1$ derivatives exist throughout an open interval containing a. Also suppose that $f^{(n)}(a)$ exists. Prove that $f(x) = P_n(x) + R(x)$, where P_n is the nth-degree Taylor polynomial of f centered at a and

$$\lim_{x \to a} \frac{R(x)}{(x - a)^n} = 0.$$

■ **CHALLENGE**

Suppose that f and its first $n + 1$ derivatives are continuous on the closed interval $[a, b]$. (Note that this assumption is a bit different from and slightly stronger than the one used in the preceding problem.) The following three problems constitute a proof that Taylor polynomials are unique: P_n is the only nth-degree polynomial whose remainder term satisfies (7).

33. Show that

$$\lim_{x \to a^+} \frac{R_n(x)}{(x - a)^n} = 0. \tag{7}$$

34. Suppose that $S_n(x)$ satisfies (7); show that

$$\lim_{x \to a^+} \frac{S_k(x)}{(x - a)^k} = 0 \qquad \text{for } k = 0, 1, 2, \dots, n.$$

$$\tag{8}$$

*35. **Uniqueness Theorem for Taylor Polynomials** Suppose that

$$f(x) = P_n(x) + R_n(x)$$
$$= Q_n(x) + S_n(x)$$

where $P_n(x)$ is the nth-degree Taylor polynomial, $Q_n(x)$ is another nth-degree polynomial, and $S_n(x)$ satisfies (7). Show that $P_n(x) = Q_n(x)$ for all $x \in [a, b]$. [*Hints:*

- Let $D(x) = P_n(x) - Q_n(x)$
  $$= d_0 + d_1(x - a) + \cdots + d_n(x - a)^n.$$
- Use the preceding problem to show that

$$\lim_{x \to a^+} \frac{D_k(x)}{(x - a)^k} = 0 \quad \text{for } k = 0, 1, 2, \dots, n.$$

- Conclude that $d_0 = d_1 = \cdots = d_n = 0$.]

■ **ANSWERS TO SELF-QUIZ**

I. $b = c$ II. $d = f$
III. d (c gives same answer, but for the wrong reason)
IV. d   V. $a = d$ VI. b

■ **9.5 APPROXIMATION USING TAYLOR POLYNOMIALS**

In this section we show how Taylor's formula can be used as a tool for making approximations. In many of the examples that follow, results have been obtained by making use of a hand calculator.

For convenience, we summarize the results of the preceding section.

Let f, f', f'', ..., $f^{(n+1)}$ exist in $[x_0, x_1]$. Then if $a \in (x_0, x_1)$ and $x \in [x_0, x_1]$,

$$f(x) = f(a) + f'(a)(x - a) + f''(a)\frac{(x - a)^2}{2!} + \cdots + f^{(n)}(a)\frac{(x - a)^n}{n!} + R_n(x)$$

$$= \sum_{k=0}^{n} \frac{f^{(k)}(a)(x - a)^k}{k!} + R_n(x), \tag{1}$$

where

$$R_n(x) = f^{(n+1)}(c) \frac{(x - a)^{n+1}}{(n + 1)!}, \tag{2}$$

for some number c (which depends on x) in the interval (a, x) or (x, a).

THEOREM 1 **THE MAXIMUM ERROR OF THE TAYLOR POLYNOMIAL APPROXIMATION**

If f has $n + 1$ continuous derivatives on $[a, b]$, there exists a positive number M_n such that

$$|R_n(x)| \leq M_n \frac{|x - a|^{n+1}}{(n + 1)!} \tag{3}$$

for all x in $[a, b]$. Here, M_n is an upper bound for the $(n + 1)$st derivative of f in the interval $[a, b]$.

PROOF: Since $f^{(n+1)}$ is continuous on $[a, b]$, it is bounded above and below on that interval (from Theorem 4 in Section 1.7 (page 119)). That is, there is a number M_n such that $|f^{(n+1)}(x)| \leq M_n$ for every x in $[a, b]$. Since

$$R_n(x) = f^{(n+1)}(c) \frac{(x - a)^{n+1}}{(n + 1)!},$$

with c in (a, x), we see that

$$|R_n(x)| = |f^{(n+1)}(c)| \frac{|x - a|^{n+1}}{(n + 1)!} \leq M_n \frac{|x - a|^{n+1}}{(n + 1)!},$$

and the theorem is proved. ■

We stress that (3) provides an upper bound for the error. In many cases the actual error [the difference $|f(x) - P_n(x)|$] will be considerably less than $M_n(x - a)^{n+1}/(n + 1)!$.

EXAMPLE 1
THE ERROR IN APPROXIMATING
$\sin x$ BY A TAYLOR POLYNOMIAL

In Example 1 in Section 9.4 we found that the fifth-degree Taylor polynomial of $f(x) = \sin x$ at 0 is $P_5(x) = x - (x^3/3!) + (x^5/5!)$. We then have

$$\sin x = x - \frac{x^3}{3!} + \frac{x^5}{5!} + R_5(x), \tag{4}$$

where

$$R_5(x) = \frac{f^{(6)}(c)(x - 0)^6}{6!}.$$

But $f^{(6)}(c) = \sin^{(6)}(c) = -\sin c$ and $|-\sin c| \leq 1$. Thus for x in $[0, 1]$,

$$|R_5(x)| \leq \frac{1(x - 0)^6}{6!} = \frac{x^6}{720}.$$

For example, suppose we wish to calculate $\sin(\pi/10)$. From (4)

$$\sin \frac{\pi}{10} = \frac{\pi}{10} - \frac{\pi^3}{3!10^3} + \frac{\pi^5}{5!10^5} + R_5\left(\frac{\pi}{10}\right)$$

with

$$\left| R_5\left(\frac{\pi}{10}\right) \right| \le \frac{(\pi/10)^6}{720} \approx 0.000001335.$$

We find that

$$\sin\frac{\pi}{10} \approx \frac{\pi}{10} - \frac{1}{3!}\frac{\pi^3}{10^3} + \frac{1}{5!}\frac{\pi^5}{10^5} \approx 0.3090170542.$$

The actual value of $\sin \pi/10 = \sin 18° = 0.3090169944$, correct to 10 decimal places, so our actual error is 0.0000000598, which is quite a bit less than 0.000001335. This illustrates the fact that the actual error (the value of the remainder term) is often quite a bit smaller than the theoretical upper bound on the error given by formula (3).

In this case we can explain the overly large error bound. Since $\sin^{(6)}(x) = -\sin x$ and $\sin^6(0) = -\sin 0 = 0$, the sixth-degree Taylor polynomial for $\sin x$ at 0 is

$$P_6(x) = x - \frac{x^3}{3!} + \frac{x^5}{5!} + 0 \cdot \frac{x^6}{6!} = x - \frac{x^3}{3!} + \frac{x^5}{5!}$$

so $P_5(x) = P_6(x)$ and we can use the error estimate for P_6. Since $|\sin^{(7)}(c)| = |-\cos c| \le 1$, we have

$$R_6(x) \le \frac{x^7}{7!} = \frac{x^7}{5040}.$$

If $x = \pi/10$, we obtain

$$\left| R_6\left(\frac{\pi}{10}\right) \right| \le \frac{(\pi/10)^7}{5040} \approx 0.0000000599.$$

Now we see that our estimate on the remainder term is really quite accurate.

EXAMPLE 2
APPROXIMATING AN INTEGRAL
USING A TAYLOR POLYNOMIAL

Approximate $\int_0^{1/4} e^{x^2}\, dx$ with an error less than 0.001.

SOLUTION: We solve this problem by approximating e^{x^2} by its nth-degree Taylor polynomial at 0 and then integrating this polynomial. If we do this, we find that

$$\text{error} = \left| \int_0^{1/4} e^{x^2}\, dx - \int_0^{1/4} P_n(x)\, dx \right| = \left| \int_0^{1/4} \left[e^{x^2} - P_n(x) \right] dx \right|$$

$$\le \int_0^{1/4} |R_n(x)|\, dx \le \int_0^{1/4} (\text{maximum value of } |R_n|)\, dx$$

$$= \frac{1}{4} \left[\text{maximum value of } |R_n(x)| \right]$$

$$= \frac{1}{4} \left[\frac{M_n \cdot \left(\frac{1}{4}\right)^{n+1}}{(n+1)!} \right]$$

$$= \frac{M_n}{(n+1)!} \left(\frac{1}{4}\right)^{n+2} \qquad \text{where } M_n = \text{upper bound for } f^{(n+1)} \text{ on } [0, \tfrac{1}{4}].$$

We need $\dfrac{M_n}{(n+1)!}\left(\dfrac{1}{4}\right)^{n+2} \le 0.001$. Since $f(x) = e^{x^2}$, we have

$$f'(x) = 2xe^{x^2}, \quad f''(x) = e^{x^2}(4x^2 + 2), \quad f'''(x) = e^{x^2}(8x^3 + 12x)$$

and

$$f^{(4)}(x) = e^{x^2}(16x^4 + 48x^2 + 12).$$

On $[0, \frac{1}{4}]$,

$$\left|f'''(x)\right| \le e^{(1/4)^2}\left[8\left(\frac{1}{4}\right)^3 + 12\left(\frac{1}{4}\right)\right] \approx 3.3 = M_2 \qquad \text{and}$$

$$\frac{M_2}{3!}\left(\frac{1}{4}\right)^4 \approx 0.002,$$

which is too big. On $[0, \frac{1}{4}]$,

$$\left|f^{(4)}(x)\right| \le e^{(1/4)^2}\left[16\left(\frac{1}{4}\right)^4 + 48\left(\frac{1}{4}\right)^2 + 12\right] \approx 16.03 = M_3$$

and

$$\frac{M_3}{4!}\left(\frac{1}{4}\right)^5 \approx 0.00065.$$

Thus, $\int_0^{1/4} P_3(x)\,dx$ will approximate $\int_0^{1/4} e^{x^2}\,dx$ with an error no greater than 0.00065. Now $f(0) = 1$, $f'(0) = 0$, $f''(0) = 2$, and $f'''(0) = 0$, so $P_3(x) = P_2(x) = 1 + x^2$ and

$$\int_0^{1/4} e^{x^2}\,dx \approx \int_0^{1/4}(1 + x^2)\,dx = \left(x + \frac{x^3}{3}\right)\Bigg|_0^{1/4} = \frac{1}{4} + \left(\frac{1}{4}\right)^3 \cdot \frac{1}{3}$$

$$\approx 0.255208.$$

This answer is within 0.00065 of the actual answer, so adding and subtracting 0.00065, we obtain the bounds

$$0.254558 < \int_0^{1/4} e^{x^2}\,dx < 0.255858.$$

We now consider a more general example.

THE LOGARITHM [ln(1 + x)]

From equation (11) on page 106 with $a = 1$ and $b = u$, it follows that

$$(1 - u)(1 + u + u^2 + \cdots + u^n) = 1 - u^{n+1}.$$

Thus, if $u \ne 1$, we have

$$1 + u + u^2 + \cdots + u^n = \frac{1 - u^{n+1}}{1 - u} = \frac{1}{1 - u} - \frac{u^{n+1}}{1 - u}. \tag{5}$$

The expression (5) is called the sum of a **geometric progression** (see also Section 10.3). From equation (5) we obtain the following:

$$\frac{1}{1-u} = 1 + u + u^2 + \cdots + u^n + \frac{u^{n+1}}{1-u}. \qquad (6)$$

Setting $u = -t$ in (6) gives us

$$\frac{1}{1+t} = 1 - t + t^2 - t^3 + \cdots + (-1)^n t^n + (-1)^{n+1} \frac{t^{n+1}}{1+t}. \qquad (7)$$

Integration of both sides of (7) from 0 to x yields

$$\ln(1+x) = x - \frac{x^2}{2} + \frac{x^3}{3} - \cdots + (-1)^n \frac{x^{n+1}}{n+1} + (-1)^{n+1} \int_0^x \frac{t^{n+1}}{t+1} \, dt, \qquad (8)$$

which is valid if $x > -1$ [since $\ln(1+x)$ is not defined for $x \leq -1$]. Now let

$$R_{n+1}(x) = (-1)^{n+1} \int_0^x \frac{t^{n+1}}{t+1} \, dt.$$

In Problems 31–33, you are asked to prove two things. First, for $-1 < x \leq 1$,

$$\lim_{n \to \infty} R_{n+1}(x) = 0.$$

This will ensure that the polynomial given in (8) provides an increasingly good approximation to $\ln(1+x)$ as n increases. Second, for every $n \geq 1$,

$$\lim_{x \to 0} \frac{R_{n+1}(x)}{x^{n+1}} = 0.$$

Then, according to the uniqueness theorem (Problem 35 in Section 9.4), we know that the polynomial in (8) is *the* $(n + 1)$st-degree Taylor polynomial for $\ln(1 + x)$ in the interval $-1 < x \leq 1$. We conclude that the Taylor polynomial

$$x - \frac{x^2}{2} + \frac{x^3}{3} - \cdots + (-1)^n \frac{x^{n+1}}{n+1}$$

may be used to approximate $\ln(1 + x)$ for sufficiently large n, provided that $-1 < x \leq 1$.

If $0 \leq x \leq 1$, then (see Problem 31)

$$R_{n+1}(x) \leq \frac{x^{n+2}}{n+2}. \qquad (9)$$

▦ **EXAMPLE 3**
APPROXIMATING ln 1.4 USING A
TAYLOR POLYNOMIAL

Calculate ln 1.4 with an error of less than 0.001.

SOLUTION: Here $x = 0.4$, and from (9) we need to find an n such that $(0.4)^{n+2}/(n+2) < 0.001$. We have $(0.4)^5/5 \approx 0.00205$ and $(0.4)^6/6 \approx 0.00068$, so choosing $n = 4$ ($n + 2 = 6$), we obtain the Taylor polynomial [from (8)]

$$P_5(x) = x - \frac{x^2}{2} + \frac{x^3}{3} - \frac{x^4}{4} + \frac{x^5}{5}$$

[remember, from (8), the last term in $P_{n+1}(x)$ is $(-1)^n x^{n+1}/(n+1)$], and

$$\ln 1.4 \approx P_5(0.4) = 0.4 - \frac{(0.4)^2}{2} + \frac{(0.4)^3}{3} - \frac{(0.4)^4}{4} + \frac{(0.4)^5}{5}$$

$$\approx 0.4 - 0.08 + 0.02133 - 0.0064 + 0.00205 = 0.33698.$$

The actual value of ln 1.4 is 0.33647, correct to five decimal places, so that the error is $0.33698 - 0.33647 = 0.00051$. This error is slightly less than the maximum possible error of 0.00068, and so in this case our error bound is fairly sharp. Note that the error formula (9) used in this problem is *not* the same as the error formula given by (3). To obtain the bound (3), we would have to compute the sixth derivative of $\ln(1+x)$. Surprisingly, if you use formula (3), you will obtain the same bound $[(0.4)^6/6]$. Try it.

THE ARC TANGENT ($\tan^{-1} x$)

If in equation (6) we substitute $u = -t^2$, then for any number t we obtain

$$\frac{1}{1+t^2} = 1 - t^2 + t^4 - t^6 + \cdots + (-1)^n t^{2n} + (-1)^{n+1} \frac{t^{2(n+1)}}{1+t^2}. \tag{10}$$

Integration of both sides of (10) from 0 to x yields

$$\tan^{-1} x = x - \frac{x^3}{3} + \frac{x^5}{5} - \frac{x^7}{5} + \cdots + \frac{(-1)^n x^{2n+1}}{2n+1} + (-1)^{n+1} \int_0^x \frac{t^{2(n+1)}}{1+t^2}\, dt. \tag{11}$$

This equation is valid for any real number $x \geq 0$.

In Problem 34, you are asked to show that

$$\lim_{x \to 0} \frac{R_{2n+1}(x)}{x^{2n+1}} = 0.$$

This ensures, according to Problem 35 in Section 9.4, that the polynomial in (11) is *the* Taylor polynomial for $\tan^{-1} x$ if $0 \leq x \leq 1$. In Problem 34, you are also asked to show that $R_{2n+1}(x) \to 0$ as $n \to \infty$ so the Taylor polynomial provides an increasingly good approximation to $\tan^{-1} x$ as n increases.

Equation (11) gives us a formula for calculating π.[†] Setting $x = 1$, we have

$$\frac{\pi}{4} = \tan^{-1} 1 \approx 1 - \frac{1}{3} + \frac{1}{5} - \frac{1}{7} + \frac{1}{9} - \frac{1}{11} + \cdots + \frac{(-1)^n}{2n+1}.$$

The error here (see Problem 34(b)) is bounded by $1/(2n+3)$, which approaches zero very slowly. To get an error less than 0.001, we would need $1/(2n+3) < \frac{1}{1000}$, or $2n+3 > 1000$, and $n \geq 499$. For example, $1 - \frac{1}{3} + \frac{1}{5} - \frac{1}{7} + \frac{1}{9} - \frac{1}{11} + \frac{1}{13} - \frac{1}{15} = 0.75427$, while $\pi/4$ is 0.78540. A better way to approximate π is suggested in Problems 27 and 28.

Under certain conditions, it is easy to show that the remainder term $R_n(x)$ approaches 0 as $n \to \infty$. The proof of the following theorem is left as an exercise (see Problems 35 and 36).

[†] This formula was discovered by the Scottish mathematician James Gregory (1638–1675) and was first published in 1712.

THEOREM 2

Suppose that $f^{(n)}(x)$ exists for $x \in [a, b]$ for $n = 0, 1, 2, \ldots$ and that there exists a number $M > 0$ such that $\left| f^{(n)}(x) \right| \leq M$ for $a \leq x \leq b$ and all nonnegative integers n. Then,

$$\lim_{n \to \infty} \left| R_n(x) \right| = 0. \quad \blacksquare$$

REMARK: Theorem 2 can be restated as follows: If all derivatives of f are uniformly bounded, then the remainder term approaches zero as $n \to \infty$.

■ PROBLEMS 9.5

■ SELF-QUIZ

I. $\sin x \approx$ _____ on the interval $[-0.25, 0.25]$.
  a. x b. $-x$
  c. $1 + x$ d. $1/2 - x^2$

II. $\cos x \approx$ _____ on the interval $[\pi/2 - 0.25, \pi/2 + 0.25]$.
  a. $x - \pi/2$ b. $-(x - \pi/2)$
  c. $1 - x^2/2$ d. $(x - \pi/2) + x^2/2$

III. True–False: Approximation of $f(x)$ by a first-degree Taylor polynomial is the same as approximation using the linearization of a function (as discussed in Section 3.8).

IV. Let f be a function whose fifth derivative is continuous on $[0, 6]$. Let P_4 be the fourth-degree Taylor polynomial for f at 1. Suppose $P_4(1.7)$ is a good approximation to $f(1.7)$. Then is it True or False that $P_4(3.8)$ is just as good an approximation to $f(3.8)$?

V. Let $f(x) = 2 - 7x + 5x^2$. The first-degree Taylor polynomial for f at 2 is $P_1(x) = f(2) + f'(2)(x - 2) = 8 + 13(x - 2)$. Also notice that f'' is constant: $f''(x) = 10$ for all x. Which of the following statements about the remainder is true?
  a. $f(x) - P_1(x)$ is constant
  b. $f(x) - P_1(x) = 10/2!$
  c. $\left| f(x) - P_1(x) \right| = 10|x - 2|^2$
  d. $\left| f(x) - P_1(x) \right| = 10|x - 2|^2/2!$
  e. $\left| f(x) - P_1(x) \right| \leq 10|x - 2|^2/2!$
  f. $f(x) - P_1(x) = 10|x - 2|^2/2!$

VI. Let $f(x) = \cos x$, then $P_3(x) = -(x - \pi/2) + (x - \pi/2)^3/3!$ is the third-degree Taylor polynomial for f at $\pi/2$. Because all derivatives of f are bounded by 1, we can make the following statement about error when approximating $f(x)$ by $P_3(x)$:
  a. $\left| f(x) - P_3(x) \right| \leq (x - \pi/2)^3$
  b. $\left| f(x) - P_3(x) \right| = (x - \pi/2)^4$
  c. $\left| f(x) - P_3(x) \right| \leq (x - \pi/2)^4/4!$
  d. $\left| f(x) - P_3(x) \right| \leq (x - \pi/2)^4/5!$

▦ ■ DRILL

In Problems 1–10, a function f, point a, degree n, and interval are given. Find a close bound for $\left| R_n(x) \right|$ where $R_n = f - P_n$ is the remainder from the nth-degree Taylor polynomial P_n.

1. $f(x) = \sin x$; $a = \pi/4$; $n = 6$; $[0, \pi/2]$
2. $f(x) = 1/x$; $a = 1$; $n = 4$; $[\frac{1}{2}, 2]$
3. $f(x) = \sqrt{x}$; $a = 1$; $n = 4$; $[\frac{1}{4}, 4]$
4. $f(x) = 1/\sqrt{x}$; $a = 5$; $n = 5$; $[\frac{19}{4}, \frac{21}{4}]$
5. $f(x) = e^{\beta x}$; $a = 0$; $n = 4$; $[0, 1]$
6. $f(x) = \sinh x$; $a = 0$; $n = 4$; $[0, 1]$
7. $f(x) = \tan x$; $a = 0$; $n = 4$; $[0, \pi/4]$
8. $f(x) = \ln \cos x$; $a = 0$; $n = 3$; $[0, \pi/6]$
9. $f(x) = e^{x^2}$; $a = 0$; $n = 4$; $[0, \frac{1}{3}]$
10. $f(x) = \sin(x^2)$; $a = 0$; $n = 4$; $[0, \pi/4]$

In Problems 11–16, use a Taylor polynomial of appropriate degree to approximate the given function value with the specified accuracy.

11. $\sin\left(\dfrac{\pi}{6} + 0.2\right)$; $|\text{error}| < 0.001$

12. $\tan\left(\dfrac{\pi}{4} + 0.1\right)$; $|\text{error}| < 0.01$

13. e; $|\text{error}| < 0.0001$ [*Note:* You may use $2 < e < 3$.]
14. e^{-1}; $|\text{error}| < 0.001$

CONTINUED

15. $\sin 33°$; $|\text{error}| < 0.001$ [*Hint:* Convert $33°$ to radians.]
16. $\tan^{-1} 0.5$; $|\text{error}| < 0.001$
17. Use the result of Problem 7 to approximate

$$\int_0^{\pi/4} \tan x \, dx.$$

What is the maximum error of your approximation?
18. Use the result of Problem 8 to approximate

$$\int_0^{\pi/6} \ln \cos x \, dx.$$

What is the maximum error of your approximation?
19. Use the result of Problem 9 to approximate

$$\int_0^{1/3} e^{x^2} \, dx,$$

What is the maximum error of your approximation?
20. Use the result of Problem 10 to approximate

$$\int_0^{\pi/4} \sin(x^2) \, dx.$$

What is the maximum error of your approximation?

In Problems 21–24, use Taylor polynomials to approximate each integral with the indicated maximum error.

21. $\int_0^{1/5} e^{x^3} \, dx$; 0.00001
22. $\int_0^{1/2} \sin(x^2) \, dx$; 0.0001
23. $\int_0^{\pi/6} \cos(x^2) \, dx$; 0.0001
*24. $\int_0^{1/4} x^2 e^{x^2} \, dx$; 0.001
25. Use the binomial theorem (Problem 29) to calculate
   a. 1.03^3 b. 0.97^4
   c. 1.2^4 d. 0.8^5
26. Use the general binomial expansion (Problem 30) to approximate the following numbers with four decimal place accuracy (i.e., $|\text{error}| \le 0.5 \cdot 10^{-4}$).
   a. $\sqrt{1.2}$ b. $0.9^{3/4}$ c. $1.8^{1/4}$
   d. $1.01^{-1/3}$ e. $2^{5/3}$ f. $0.4^{1/6}$
*27. Use the result of Problem 28 to approximate π to five decimal places.

■ **SHOW/PROVE/DISPROVE**

*28. Use the formula

$$\tan(A \pm B) = \frac{\tan A \pm \tan B}{1 \mp \tan A \cdot \tan B}$$

to prove that

$$\tan\left(4 \tan^{-1} \frac{1}{5} - \tan^{-1} \frac{1}{239}\right) = 1.$$

[*Hint:* First calculate $\tan[2 \tan^{-1}(\frac{1}{5})]$ and $\tan[4 \tan^{-1}(\frac{1}{5})]$. Then observe that $4 \tan^{-1}(\frac{1}{5}) - \tan^{-1}(\frac{1}{239}) = \pi/4$.]
*29. Let $f(x) = (1 + x)^n$, where n is a positive integer.
   a. Show that $f^{(n+1)}(x) = 0$ for all x.
   b. Show that

$$f(x) = 1 + nx + \frac{n(n-1)}{2!} x^2$$
$$+ \frac{n(n-1)(n-2)}{3!} x^3 + \cdots + \frac{n!}{n!} x^n.$$

[*Hint:* Obtain the nth-degree Taylor polynomial for f at 0 and then examine its remainder.] This result is called the **Binomial Theorem**.

30. Let $f(x) = (1 + x)^\alpha$, where α is any real number.
   a. Consider the nth-degree Taylor polynomial for f at 0 and show that

$$f(x) = 1 + \alpha x + \frac{\alpha(\alpha - 1)}{2!} x^2$$
$$+ \frac{\alpha(\alpha - 1)(\alpha - 2)}{3!} x^3 + \cdots$$
$$+ \frac{\alpha(\alpha - 1)(\alpha - 2) \cdots (\alpha - n + 1)}{n!} x^n$$
$$+ R_n(x).$$

   b. Show that if $|x| < 1$, then $R_n(x) \to 0$ as $n \to \infty$. This result is called the **general binomial theorem**.
31. Consider the expression (see equation (8)) which represents the error in the Taylor polynomial approximation for $\ln(1 + x)$:

$$R_n(x) = (-1)^n \int_0^x \frac{t^n}{1 + t} \, dt.$$

CONTINUED

Show that if $0 \leq x \leq 1$, then

$$|R_n(x)| \leq \int_0^x t^n \, dt = \frac{x^{n+1}}{n+1} \leq \frac{1}{n+1}.$$

[*Hint:* $1/(1+t) \leq 1$ if $0 \leq t \leq 1$.]

32. In the preceding problem, suppose that $-1 < x < 0$.
   a. Show that

   $$R_n(x) = (-1)^{n-1} \int_x^0 \frac{t^n}{1+t} \, dt.$$

   b. Show that $1 \leq \dfrac{1}{1+t} \leq \dfrac{1}{1+x}$ if $-1 < x \leq t \leq 0$.

   c. Show that

   $$|R_n(x)| \leq \int_x^0 \frac{|t|^n}{|1+t|} \, dt \leq \int_x^0 \frac{(-t)^n}{1+x} \, dt$$
   $$\leq \frac{(-x)^{n+1}}{(n+1)(1+x)}.$$

33. Using the results of the preceding two problems, show that if $-1 < x \leq 1$,
   a. $|R_n(x)| \to 0$ as $n \to \infty$.

b. $\displaystyle \lim_{x \to 0} \frac{|R_n(x)|}{x^n} = 0$.

34. Let (see equation (11))

$$R_{2n+1}(x) = (-1)^{n+1} \int_0^x \frac{t^{2(n+1)}}{1+t^2} \, dt.$$

   a. Show that $1/(1+t^2) \leq 1$ for all t.
   b. Show that $|R_{2n+1}(x)| \leq x^{2n+3}/(2n+3)$ for all x.
   c. Show that

   $$\lim_{x \to 0} \frac{|R_{2n+1}(x)|}{x^{2n+1}} = 0 \qquad \text{for all integers } n \geq 0.$$

   d. Show that

   $$\lim_{n \to \infty} |R_{2n+1}(x)| = 0 \qquad \text{if } |x| \leq 1.$$

*35. Show that for any real number x,

$$\lim_{n \to \infty} \frac{x^n}{n!} = 0.$$

36. Prove Theorem 2. [*Hint:* Use the result of the preceding problem.]

ANSWERS TO SELF-QUIZ

I. a 　　II. b 　　III. True

IV. False 　　V. d, e, f,

VI. c ($P_3 = P_4$; thus (d) is true too.)

PROBLEMS 　　**CHAPTER 9 REVIEW**

DRILL

In Problems 1–20, evaluate the given limit.

1. $\displaystyle \lim_{x \to 0} \frac{\sin 3x}{2x}$

2. $\displaystyle \lim_{x \to \infty} \frac{\ln x}{x^2}$

3. $\displaystyle \lim_{x \to 0} \frac{2x^3 + 3x + 4}{x - x^3}$

4. $\displaystyle \lim_{x \to \pi/4} \frac{\cos 2x}{(\pi/4) - x}$

5. $\displaystyle \lim_{x \to \infty} 3xe^{-x}$

6. $\displaystyle \lim_{x \to 0} \frac{1 - e^x}{x^2 + x}$

7. $\displaystyle \lim_{x \to 0^+} \frac{\sqrt{x}}{\sin x}$

8. $\displaystyle \lim_{x \to 0^+} (\sin x)^{2x}$

9. $\displaystyle \lim_{x \to \pi/4^-} \left[(\pi/4) - x \right]^{\cos 2x}$

10. $\displaystyle \lim_{x \to 0^+} x^{3x}$

11. $\displaystyle \lim_{x \to 0} \frac{\sin x}{x^3}$

12. $\displaystyle \lim_{x \to 0} \frac{x - \sin x}{2x^3}$

13. $\displaystyle \lim_{x \to \pi/2^-} \frac{\tan x}{\sec x}$

14. $\displaystyle \lim_{x \to 0^+} \frac{\ln \tan x}{\ln \sin x}$

15. $\displaystyle \lim_{x \to \infty} (x - \sqrt{x^2 + x})$

16. $\displaystyle \lim_{x \to \infty} (x^4 - \sqrt{x^8 + x^4 + 1})$

17. $\displaystyle \lim_{x \to \infty} \frac{e^x}{2^x}$

18. $\displaystyle \lim_{x \to 0} \frac{\tanh ax}{\tanh bx}$

19. $\lim_{x \to 0} x^{10} e^{-1/x^2}$

20. $\lim_{x \to 0^+} \dfrac{\int_0^x e^{t^3} dt}{\int_0^x e^{t^2} dt}$

35. $\int_0^\infty \dfrac{x^3}{x^3 + 1} dx$

36. $\int_0^e \ln x \, dx$

37. $\int_2^\infty \dfrac{dx}{(x - 1)^{1/100}}$

38. $\int_0^\infty e^{-x} \sin 2x \, dx$

39. $\int_0^\infty \dfrac{dx}{x^2 - 3x + 2}$

40. $\int_{-\infty}^0 e^{2x} \cos 3x \, dx$

In Problems 21–40, determine whether the given improper integral converges or diverges. If it converges, calculate its value.

21. $\int_0^\infty e^{-3x} dx$

22. $\int_{-\infty}^2 e^{50x} dx$

23. $\int_0^\infty x^3 e^{-7x} dx$

24. $\int_{-\infty}^\infty x^2 e^{-x^3} dx$

25. $\int_{50}^\infty \dfrac{dx}{x^2 - 4}$

26. $\int_0^1 \dfrac{dx}{\sqrt[4]{x}}$

27. $\int_0^1 \dfrac{dx}{x^4}$

28. $\int_{-\infty}^\infty \dfrac{dx}{x^3}$

29. $\int_{-\infty}^0 \dfrac{dx}{(1 - x)^{5/2}}$

30. $\int_3^4 \dfrac{dx}{(x - 3)^{6/5}}$

31. $\int_0^{\pi/2} \csc x \, dx$

32. $\int_0^4 \dfrac{dx}{x - 2}$

33. $\int_2^4 \dfrac{dx}{\sqrt{x - 2}}$

34. $\int_3^5 \dfrac{dx}{(x - 4)^{10}}$

In Problems 41–48, find the nth-degree Taylor polynomial for f at a.

41. $f(x) = e^x$; $a = 0$; $n = 3$
42. $f(x) = \ln x$; $a = 1$; $n = 4$
43. $f(x) = \sin x$; $a = \pi/6$; $n = 3$
44. $f(x) = \cos x$; $a = \pi/2$; $n = 5$
45. $f(x) = \cot x$; $a = \pi/2$; $n = 4$
46. $f(x) = \sinh x$; $a = 0$; $n = 3$
47. $f(x) = x^3 - x^2 + 2x + 3$; $a = 0$; $n = 8$
48. $f(x) = e^{-x^2}$; $a = 0$; $n = 5$

In Problems 49–52, find a close bound on $|R_n(x)|$ for the specified interval.

49. $f(x) = \cos x$; $a = \pi/6$; $n = 5$; $[0, \pi/2]$
50. $f(x) = \sqrt[3]{x}$; $a = 1$; $n = 4$; $\left[\frac{7}{8}, \frac{9}{8}\right]$
51. $f(x) = e^x$; $a = 0$; $n = 6$; $[-\ln e, \ln e]$
52. $f(x) = \cot x$; $a = \pi/2$; $n = 2$; $[\pi/4, 3\pi/4]$

APPLICATIONS

53. Find the volume generated when the region in the first quadrant bounded by $y = 1/(x + 1)^2$ is rotated about the x-axis.

54. Find the volume generated if the region in the preceding problem is rotated about the y-axis.

55. Find the volume of the solid generated when the curve $y = e^{-x}$ for $x \geq 0$ is rotated about the y-axis.

56. Find the volume of the solid generated when the curve $y = e^{3x}$ for $x \leq 0$ is rotated about the x-axis.

57. Use a Taylor polynomial of degree 4 to approximate

$$\int_0^{1/2} \cos x^2 \, dx.$$

Find the maximum error of the approximation.

*58. Use a Taylor polynomial of degree 4 to approximate

$$\int_{-\pi}^\pi e^{\cos x} \, dx.$$

Find the maximum error of the approximation.

SHOW/PROVE/DISPROVE

59. Show that $\int_0^{\pi/2} \dfrac{dx}{\sqrt{\cos x}}$ converges.

60. Show that $\int_{10}^\infty \dfrac{dx}{\ln \sqrt{2 + x}}$ diverges.

*61. Suppose N is a positive integer. Use the substitution $u = -\ln t$ and apply the result of Problem 64 in Section 9.3.
a. Show that $I_N = \int_0^1 (\ln t)^N dt$ converges.
b. Evaluate I_N.

COMPUTER EXERCISES

1. a. Show that $\int_1^\infty \dfrac{1}{x^3 + x + 1}\, dx$ exists. $\Big[$*Hint:* Compare with the integral $\int_1^\infty \dfrac{1}{x^3}\, dx.\Big]$

   b. Use a computer algebra system to approximate $\int_1^{10} \dfrac{1}{x^3 + x + 1}\, dx$, $\int_1^{100} \dfrac{1}{x^3 + x + 1}\, dx$, and $\int_1^{1000} \dfrac{1}{x^3 + x + 1}\, dx$, and use your answer to obtain an approximation to $\int_1^\infty \dfrac{1}{x^3 + x + 1}\, dx$.

   c. If $\int_1^{1000} \dfrac{1}{x^3 + x + 1}\, dx$ is used to approximate $\int_1^\infty \dfrac{1}{x^3 + x + 1}\, dx$, find a bound on the error. [*Hint:* Once again, compare your integral with $\int \dfrac{1}{x^3}\, dx$. You should express your answer as an integral and evaluate it.]

2. In the computer exercises for Chapter 7, you were asked to approximate π using Simpson's rule. If you wish to calculate many digits of π, this is an inefficient method. A method that has been used effectively to calculate many digits of π is offered by the formula $4 \tan^{-1} \frac{1}{5} - \tan^{-1} \frac{1}{239} = \pi/4$ produced in Problem 28 of Section 9.5. In Problem 34 of Section 9.5, a bound on the error term for the Taylor polynomial for the inverse tangent is given by $|R_{2n+1}(x)| \le \dfrac{x^{2n+3}}{2n+3}$.

   a. Show that the following expression gives an approximation of $4 \tan^{-1} \frac{1}{5} - \tan^{-1} \frac{1}{239}$ with error $\le \dfrac{4}{(2n+3)5^{2n+3}} + \dfrac{1}{(2n+3)239^{2n+3}}$:

   $$4\left(\frac{1}{5} - \frac{1}{3\cdot 5^3} + \frac{1}{5\cdot 5^5} - \cdots + \frac{(-1)^n}{(2n+1)5^{2n+1}} \right)$$
   $$- \left(\frac{1}{239} - \frac{1}{3\cdot 239^3} + \cdots + \frac{(-1)^n}{(2n+1)239^{2n+1}} \right).$$

   b. Obtain an approximation of π with error $\le 10^{-20}$.

   c. One can also use the fact that $\tan^{-1}(1) = \dfrac{\pi}{4}$ so that π can be approximated by $4\left(1 - \dfrac{1}{3} + \dfrac{1}{5} - \right.$

   $\left. \dfrac{1}{7} + \cdots + \dfrac{(-1)^n}{2n+1} \right)$. Discuss the advantages of using $4 \tan^{-1}\left(\dfrac{1}{5} \right) - \tan^{-1}\left(\dfrac{1}{239} \right) = \dfrac{\pi}{4}$.

3. In this exercise we will look at the Taylor polynomial for $\tan x$. In order to handle the remainder term, it is necessary first to obtain a bound for $\tan x$ near zero.

   PART I. (No electronic aid is needed for this part.)

   a. Let $f(x) = x - \sin x$. Show that f is an increasing function and $f(0) = 0$. Use these facts to derive the inequality $\sin x \le x$ for $x \ge 0$.

   b. If $0 \le x < \pi$, show that $\cos x \ge 1 - \dfrac{x^2}{2}$. [*Hint:* Since both $\sin x$ and x are nonnegative on the interval $[0, x]$, the area under $\sin x$ must be less than the area under x.]

   c. Use (a) and (b) to show that $\tan x \le \dfrac{x}{1 - \dfrac{x^2}{2}}$ for $0 \le x < \sqrt{2}$.

   PART II. Let $f(x) = \tan x$.

   a. Use a computer algebra system to verify that $f^{(12)}(x)$ is a polynomial in $\tan x$ whose coefficients are positive. What can you conclude about $f^{(11)}(x)$?

   b. Use your conclusion in (a) and the inequality in part I to show that $|R_{10}(0.1)| < 1.188 \times 10^{-13}$. (Use a computer algebra system to compute $f^{(11)}(x)$. Replace each occurrence of $\tan x$ with $\dfrac{0.1}{1 - \dfrac{(0.1)^2}{2}}$.)

   c. Use a computer algebra system to generate the tenth-order Taylor polynomial for $\tan x$ and obtain an estimate for $\tan(0.1)$ with error $\le 1.188 \times 10^{-13}$.

   REMARK: It is very difficult to obtain the general form of the nth-order Taylor polynomial for $\tan x$. Leonhard Euler, a Swiss mathematician, was the first to obtain such an expression.

4. Calculate $\int_0^4 \dfrac{1}{\sqrt{|x^2 + x - 7|}}\, dx$.

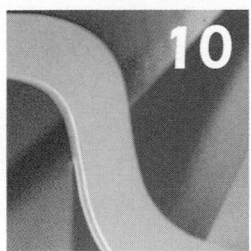

10 SEQUENCES AND SERIES

In Sections 9.4 and 9.5, we saw how a wide variety of functions could be approximated by polynomials. A polynomial in x is a *finite* sum of terms of the form $a_k x^k$. In this chapter we will discuss what we mean by an *infinite* sum. The infinite sums we describe are called *infinite series*. We will discuss the theory of infinite series and will show how it can give us a great deal of information about a wide variety of functions.

10.1 SEQUENCES OF REAL NUMBERS

According to a popular dictionary,[†] a *sequence* is "the following of one thing after another." In mathematics we could define a sequence intuitively as a succession of numbers that never terminates. The numbers in the sequence are called the *terms* of the sequence. In a sequence there is one term for each positive integer.

EXAMPLE 1
A SEQUENCE

Consider the sequence

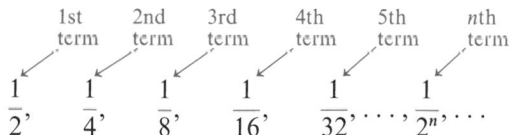

$$\frac{1}{2}, \quad \frac{1}{4}, \quad \frac{1}{8}, \quad \frac{1}{16}, \quad \frac{1}{32}, \cdots, \frac{1}{2^n}, \cdots$$

We see that there is one term for each positive integer. The terms in this sequence form an infinite set of real numbers, which we write as

$$A = \frac{1}{2}, \frac{1}{4}, \frac{1}{8}, \cdots, \frac{1}{2^n}, \cdots \tag{1}$$

That is, the set A consists of all numbers of the form $1/2^n$, where n is a positive integer. There is another way to describe this set. We define the function f by the rule $f(n) = 1/2^n$, where the domain of f is the set of positive integers. Then the set A is precisely the set of values taken by the function f.

We have the following formal definition.

DEFINITION **SEQUENCE**

A **sequence** of real numbers is a function whose domain is the set of positive integers. The values taken by the function are called **terms** of the sequence. ■

[†] *Webster's Encyclopedic Unabridged Dictionary of the English Language* (New York: Portland House, 1989).

NOTATION: We will often denote the terms of a sequence by a_n. Thus if the function given in the definition above is f, then $a_n = f(n)$. With this notation, *we can denote the set of values taken by the sequence by* $\{a_n\}$. Also, we will use n, m, and so on as integer variables and x, y, and so on as real variables.

EXAMPLE 2
SIX SEQUENCES AND THE VALUES
THEY TAKE WRITTEN OUT

Sequence	**The terms of the sequence written out**
(a) $\{a_n\} = \left\{\dfrac{1}{n}\right\}$	$1, \dfrac{1}{2}, \dfrac{1}{3}, \dfrac{1}{4}, \ldots, \dfrac{1}{n}, \ldots$
(b) $\{a_n\} = \{\sqrt{n}\}$	$1, \sqrt{2}, \sqrt{3}, \sqrt{4}, \ldots, \sqrt{n}, \ldots$
(c) $\{a_n\} = \left\{\dfrac{1}{n!}\right\}$	$1, \dfrac{1}{2}, \dfrac{1}{6}, \dfrac{1}{24}, \ldots, \dfrac{1}{n!}, \ldots$
(d) $\{a_n\} = \{\sin n\}$	$\sin 1, \sin 2, \sin 3, \sin 4, \ldots, \sin n, \ldots$
(e) $\{a_n\} = \left\{\dfrac{e^n}{n!}\right\}$	$e, \dfrac{e^2}{2}, \dfrac{e^3}{6}, \dfrac{e^4}{24}, \ldots, \dfrac{e^n}{n!}, \ldots$
(f) $\{a_n\} = \left\{\dfrac{n-1}{n}\right\}$	$0, \dfrac{1}{2}, \dfrac{2}{3}, \dfrac{3}{4}, \ldots, \dfrac{n-1}{n}, \ldots$

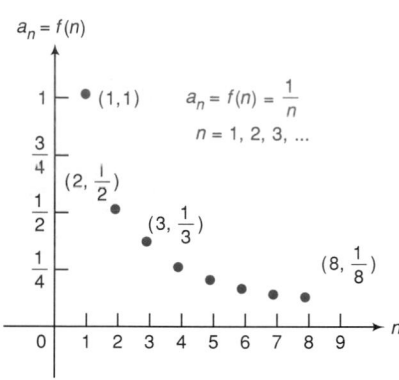

FIGURE 1

GRAPH OF $\{a_n\} = \left\{\dfrac{1}{n}\right\}$

Observe that $f(n)$ is defined only when n is a positive integer so there is one point on the graph for each positive integer.

REMARK: In Example 2 the first term in each sequence corresponds to $n = 1$. However it is sometimes more convenient to start at $n = 0$, $n = 2$, or some other integer. All the definitions and theorems in this and the next section are unchanged if we start at $n = k$ where $k \neq 1$.

Because a sequence is a function, it has a graph. In Figure 1 we draw part of the graph of the first sequence in Example 2.

EXAMPLE 3
FINDING THE GENERAL TERM OF
A SEQUENCE

Find the general term a_n of the sequence

$$-1, 1, -1, 1, -1, 1, -1, \ldots.$$

SOLUTION: We see that $a_1 = -1$, $a_2 = 1$, $a_3 = -1$, $a_4 = 1, \ldots$. Hence

$$a_n = \begin{cases} -1, & \text{if } n \text{ is odd} \\ 1, & \text{if } n \text{ is even.} \end{cases}$$

A more concise way to write this term is

$$a_n = (-1)^n.$$

It is evident that as n gets large, the numbers $1/n$ get small. We can write

$$\lim_{n \to \infty} \frac{1}{n} = 0.$$

This is also suggested by the graph in Figure 1. Similarly, it is not hard to show that as n gets large, $(n - 1)/n$ gets close to 1. We write

$$\lim_{n \to \infty} \frac{n-1}{n} = 1.$$

On the other hand, it is clear that $a_n = (-1)^n$ does not get close to any one number as n increases. It simply oscillates back and forth between the numbers $+1$ and -1.

For the remainder of this section we will be concerned with calculating the limit of a sequence as $n \to \infty$. Since a sequence is a special type of function, our formal definition of the limit of a sequence is going to be very similar to the definition of $\lim_{x \to \infty} f(x)$.

DEFINITION   **FINITE LIMIT OF A SEQUENCE**

A sequence $\{a_n\}^\dagger$ has the **limit** L if for every $\epsilon > 0$ there exists an integer $N > 0$ such that if $n \geq N$, then $|a_n - L| < \epsilon$. We write

$$\lim_{n \to \infty} a_n = L. \quad \blacksquare$$

(2)

Intuitively, this definition states that $a_n \to L$ if as n increases without bound, a_n gets arbitrarily close to L (see pages 76 and 126). We illustrate this definition in Figure 2.

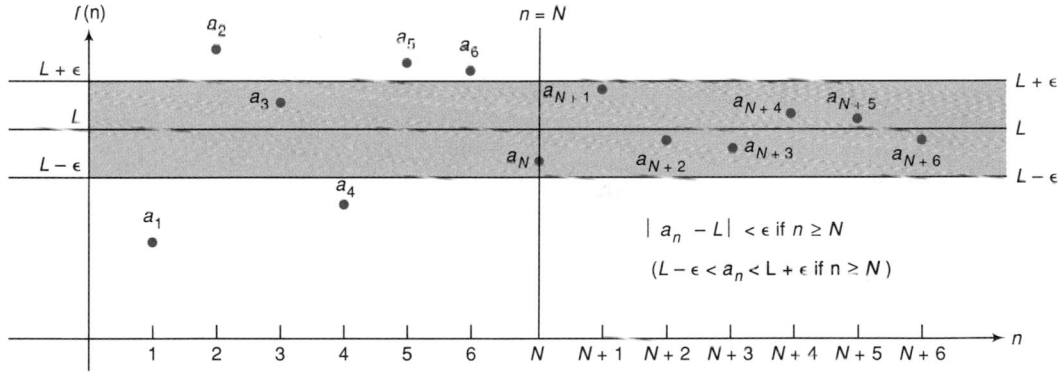

FIGURE 2   ILLUSTRATION OF THE DEFINITION OF A LIMIT

For $n > N$, the values taken by a_n lie between $L - \epsilon$ and $L + \epsilon$ (i.e. in the shaded strip).

DEFINITION   **INFINITE LIMIT OF A SEQUENCE**

The sequence $\{a_n\}$ has the limit ∞ if for every positive number M there is an integer $N > 0$ such that if $n > N$, then $a_n > M$. In this case we write

$$\lim_{n \to \infty} a_n = \infty. \quad \blacksquare$$

Intuitively, $\lim_{n \to \infty} a_n = \infty$ means that as n increases without bound, a_n also increases without bound. The expression $\lim_{n \to \infty} a_n = -\infty$ is defined in a similar way. The following theorem gives us a very useful result.

$\dagger$ To be precise, $\{a_n\}$ denotes the set of values taken by the sequence. There is a difference between the sequence, which is a function f, and the values $a_n = f(n)$ taken by this function. However, because it is more convenient to write down the values the sequence takes, we will, from now on, use the notation $\{a_n\}$ to denote a sequence.

THEOREM 1 **LIMIT OF A GEOMETRIC SEQUENCE**

Let r be a real number. Then,

$$\lim_{n \to \infty} r^n = 0 \qquad \text{if } |r| < 1$$

and

$$\lim_{n \to \infty} |r^n| = \infty \qquad \text{if } |r| > 1.$$

NOTE: The sequence $\{a_n\} = \{r^n\}$ for $r \neq 1$ is called a **geometric sequence**.

PROOF:

Case 1. $r = 0$. Then $r^n = 0$, and the sequence has the limit 0.

Case 2. $0 < |r| < 1$. For a given $\epsilon > 0$, choose N such that

$$N > \frac{\ln \epsilon}{\ln|r|}.$$

Note that since $|r| < 1$, $\ln|r| < 0$. Now if $n > N$,

$$n > \frac{\ln \epsilon}{\ln|r|} \qquad \text{and} \qquad n \ln|r| < \ln \epsilon.$$

The second inequality follows from the fact that $\ln|r|$ is negative, and multiplying both sides of an inequality by a negative number reverses the inequality. Thus,

$$\ln|r^n - 0| = \ln|r^n| = \ln|r|^n = n \ln|r| < \ln \epsilon.$$

Since $\ln|r^n - 0| < \ln \epsilon$ and $\ln x$ is an increasing function, we conclude that $|r^n - 0| < \epsilon$. Thus according to the definition of a finite limit of a sequence,

$$\lim_{n \to \infty} r^n = 0.$$

Case 3. $|r| > 1$. Let $M > 0$ be given. Choose $N > \ln M / \ln|r|$. Then if $n > N$,

$$\ln|r^n| = n \ln|r| \overset{n > N}{>} N \ln|r| > \left(\frac{\ln M}{\ln|r|} \right)(\ln|r|) = \ln M,$$

so that

$$|r^n| > M \qquad \text{if } n > N.$$

From the definition of an infinite limit of a sequence, we see that

$$\lim_{n \to \infty} |r^n| = \infty. \qquad \blacksquare$$

In Sections 1.3 and 1.8, we stated a number of limit theorems. In Section 1.4 and in Chapter 9 we saw how to evaluate limits as $x \to \infty$. All these results can be applied when n, rather than x, approaches infinity. The only difference is that as n grows, it takes on integer values. For convenience, we state without proof the major limit theorems we need for sequences. The proofs are similar to the proofs given in Appendix 3.

THEOREM 2

Suppose that $\lim_{n \to \infty} a_n$ and $\lim_{n \to \infty} b_n$ both exist and are finite.

(i) $\lim_{n \to \infty} \alpha a_n = \alpha \lim_{n \to \infty} a_n$ for any real number α. (3)

(ii) $\lim_{n \to \infty} (a_n + b_n) = \lim_{n \to \infty} a_n + \lim_{n \to \infty} b_n$. (4)

(iii) $\lim_{n \to \infty} a_n b_n = (\lim_{n \to \infty} a_n)(\lim_{n \to \infty} b_n)$. (5)

(iv) If $\lim_{n \to \infty} b_n \neq 0$, then

$$\lim_{n \to \infty} \frac{a_n}{b_n} = \frac{\lim_{n \to \infty} a_n}{\lim_{n \to \infty} b_n} \quad \blacksquare$$

(6)

THEOREM 3 **CONTINUITY THEOREM**

Suppose that L is finite and $\lim_{n \to \infty} a_n = L$. If f is continuous in an open interval containing L, then

$$\lim_{n \to \infty} f(a_n) = f\left(\lim_{n \to \infty} a_n \right) = f(L). \quad \blacksquare$$

(7)

THEOREM 4 **SQUEEZING THEOREM**

Suppose that $\lim_{n \to \infty} a_n = \lim_{n \to \infty} b_n = L$ and that $\{c_n\}$ is a sequence having the property that for $n > N$ (a positive integer), $a_n \leq c_n \leq b_n$. Then,

$$\lim_{n \to \infty} c_n = L. \quad \blacksquare$$

(8)

We now give a central definition in the theory of sequences.

DEFINITION **CONVERGENCE AND DIVERGENCE OF A SEQUENCE**

If the limit in (2) exists and if L is finite, we say that the sequence **converges** or is **convergent**. Otherwise, we say that the sequence **diverges** or is **divergent**. $\blacksquare$

EXAMPLE 4
A CONVERGENT GEOMETRIC SEQUENCE

The geometric sequence $\{1/2^n\} = \{(\frac{1}{2})^n\}$ is convergent since, by Theorem 1, $\lim_{n \to \infty} 1/2^n = \lim_{n \to \infty} (1/2)^n = 0$.

EXAMPLE 5
A DIVERGENT GEOMETRIC
SEQUENCE

The geometric sequence $\{r^n\}$ is divergent for $r > 1$ since $\lim_{n \to \infty} r^n = \infty$ if $r > 1$.

EXAMPLE 6
A SEQUENCE MAY DIVERGE EVEN
THOUGH ITS VALUES DO NOT
APPROACH INFINITY

The sequence $\{(-1)^n\}$ is divergent since the values a_n alternate between -1 and $+1$ but do not stay close to any fixed number as n becomes large.

Since we have a large body of theory and experience behind us in the calculation of ordinary limits, we would like to make use of that experience to calculate limits of sequences. The following theorem, whose proof is left as a problem (see Problem 31), is extremely useful.

THEOREM 5

Suppose that $\lim_{x \to \infty} f(x) = L$, a finite number, ∞, or $-\infty$. If f is defined for every positive integer, then the limit of the sequence $\{a_n\} = \{f(n)\}$ is also equal to L. That is, $\lim_{x \to \infty} f(x) = \lim_{n \to \infty} a_n = L$. ∎

EXAMPLE 7
A CONVERGENT SEQUENCE

Calculate $\lim_{n \to \infty} 1/n^2$.

SOLUTION: Since $\lim_{x \to \infty} 1/x^2 = 0$, we have $\lim_{n \to \infty} 1/n^2 = 0$ (by Theorem 5).

EXAMPLE 8
DETERMINING CONVERGENCE OR
DIVERGENCE BY USING
L'HÔPITAL'S RULE

Does the sequence $\{e^n/n\}$ converge or diverge?

SOLUTION: Since $\lim_{x \to \infty} e^x/x = \lim_{x \to \infty} e^x/1$ (by l'Hôpital's rule) $= \infty$, we find that the sequence diverges.

REMARK: It should be emphasized that Theorem 5 does *not* say that if $\lim_{x \to \infty} f(x)$ does not exist, then $\{a_n\} = \{f(n)\}$ diverges. For example, let

$$f(x) = \sin \pi x.$$

Then $\lim_{x \to \infty} f(x)$ does not exist, but $\lim_{n \to \infty} f(n) = \lim_{n \to \infty} \sin \pi n = 0$ since $\sin \pi n = 0$ for every integer n.

EXAMPLE 9
A SEQUENCE THAT CONVERGES
TO e

Let $\{a_n\} = \{[1 + (1/n)]^n\}$. Does this sequence converge or diverge?

SOLUTION: Since $\lim_{x \to \infty} [1 + (1/x)]^x = e$, we see that a_n converges to the limit e.

EXAMPLE 10
USING L'HÔPITAL'S RULE

Determine the convergence or divergence of the sequence $\{(\ln n)/n\}$.

SOLUTION: $\lim_{x \to \infty} [(\ln x)/x] = \lim_{x \to \infty} [(1/x)/1] = 0$ by l'Hôpital's rule, so that the sequence converges to 0.

EXAMPLE 11
LIMIT OF A SEQUENCE
DETERMINED BY A RATIONAL
FUNCTION

Let $p(x) = c_0 + c_1 x + \cdots + c_m x^m$ and $q(x) = d_0 + d_1 x + \cdots + d_r x^r$. In Problem 32 in Section 1.4 we showed that if the rational function $r(x) = p(x)/q(x)$ and if $c_m d_r \neq 0$, then

$$\lim_{x \to \infty} \frac{p(x)}{q(x)} = \begin{cases} 0, & \text{if } m < r \\ \pm \infty, & \text{if } m > r \\ \dfrac{c_m}{d_r}, & \text{if } m = r. \end{cases}$$

Thus the sequence $\{p(n)/q(n)\}$ converges to 0 if $m < r$, converges to c_m/d_r if $m = r$, and diverges if $m > r$.

EXAMPLE 12
USING THE SQUEEZING THEOREM
TO PROVE CONVERGENCE OF A
SEQUENCE

Determine the convergence or divergence of the sequence $\{\sin \alpha n / n^\beta\}$, where α is a real number and $\beta > 0$.

SOLUTION: Since $-1 \leq \sin \alpha x \leq 1$, we see that

$$-\frac{1}{x^\beta} \leq \frac{\sin \alpha x}{x^\beta} \leq \frac{1}{x^\beta} \qquad \text{for any } x > 0.$$

But $\pm \lim_{x \to \infty} 1/x^\beta = 0$, and so by the squeezing theorem (Theorem 1 in Section 1.4), $\lim_{x \to \infty} [(\sin \alpha x)/x^\beta] = 0$. Therefore, the sequence $\{(\sin \alpha n)/n^\beta\}$ converges to 0.

As in Example 12, the squeezing theorem can often be used to calculate the limit of a sequence.

■ PROBLEMS 10.1

■ SELF-QUIZ

I. The first three terms of the sequence $\{n/(n + 1)\}$, starting with $n = 1$, are _____.
 a. $a/(a + 1), b/(b + 1), c/(c + 1)$
 b. $\frac{0}{1}, \frac{1}{2}, \frac{2}{3}$
 c. $\frac{1}{2}, \frac{2}{3}, \frac{3}{4}$
 d. $\frac{2}{1}, \frac{3}{2}, \frac{4}{3}$

II. The first four terms of the sequence $\{(-n)^n\}$, starting with $n = 1$, are _____.
 a. $-1, 2, -3, 4$ b. $1, 4, 27, 256$
 c. $1, -4, 27, -256$ d. $-1, 4, -27, 256$

III. The first five terms of the sequence $\{\cos(n\pi)\}$, starting at $n = 1$, are _____.
 a. $-1, 0, 1, 0, -1$ b. $1, 0, -1, 0, 1$
 c. $0, -1, 0, 1, 0$ d. $1, -1, 1, -1, 1$
 e. $-1, 1, -1, 1, -1$
 f. $-1, 1, -1, 1, -1, \ldots$

IV. The first five terms of the sequence $\{\sin(n\pi)\}$, starting at $n = 1$, are _____.
 a. $0, 0, 0, 0, 0, \ldots$ b. $0, 0, 0, 0, 0$
 c. $-1, 0, 1, 0, -1$ d. $0, -1, 0, 1, 0$

V. The general term, a_n, of the sequence $1, -2, 3, -4, 5, -6, \ldots$ is _____.
 a. $(-n)^n$ b. n^{-n}
 c. $(-1)^n n$ d. $n(-1)^{n-1}$

VI. The general term, a_n, of the sequence $\frac{1}{2}, \frac{3}{4}, \frac{7}{8}, \frac{15}{16}, \frac{31}{32}, \ldots$ is _____.
 a. $\dfrac{2n - 1}{2n}$ b. $1 - \dfrac{1}{2^n}$
 c. $\dfrac{n}{2^n}$ d. $\dfrac{2^{n-1}}{2^n}$

VII. $\lim_{n \to \infty} (e - \pi)^n =$ _____.
 a. ∞ b. $-\infty$
 c. 0 d. does not converge

■ DRILL

In Problems 1–6, find the first five terms of the given sequence.

1. $\left\{\dfrac{1}{3^n}\right\}$ 2. $\left\{\dfrac{n+1}{n}\right\}$ 3. $\left\{1-\dfrac{1}{4^n}\right\}$

4. $\{n\cos n\}$ 5. $\left\{\sin\dfrac{n\pi}{2}\right\}$ 6. $\left\{\cos\dfrac{8\pi}{2^n}\right\}$

In Problems 7–10, find the general term, a_n, of the given sequence.

7. $\dfrac{1}{2},\dfrac{2}{3},\dfrac{3}{4},\dfrac{4}{5},\dfrac{5}{6},\cdots$ 8. $1,-\dfrac{1}{3},\dfrac{1}{9},-\dfrac{1}{27},\cdots$

9. $1, 2\cdot 5, 3\cdot 5^2, 4\cdot 5^3, 5\cdot 5^4,\ldots$

*10. $\dfrac{1}{3},\dfrac{2}{5},\dfrac{3}{7},\dfrac{4}{9},\dfrac{5}{11},\cdots$

In Problems 11–28, determine whether the given sequence is convergent or divergent. If it is convergent, find its limit.

11. $\left\{\dfrac{17}{\sqrt{n}}\right\}$ 12. $\left\{\dfrac{(-1)^n}{\sqrt{n}}\right\}$

*13. $\{\sin n\}$ 14. $\{\sin n\pi\}$

15. $\left\{\dfrac{3}{5n}\right\}$ 16. $\left\{\dfrac{(-1)^n n^3}{n^3+1}\right\}$

17. $\left\{\dfrac{n^5+3n^2+1}{n^6+4n}\right\}$ 18. $\left\{\dfrac{4n^5-3}{7n^5+n^2+2}\right\}$

19. $\left\{\left(1+\dfrac{4}{n}\right)^n\right\}$ 20. $\left\{\left(1+\dfrac{1}{4n}\right)^n\right\}$

21. $\left\{\dfrac{\sqrt{n}}{\ln n}\right\}$

22. $\{\sqrt{n+3}-\sqrt{n}\}$
[*Hint:* Multiply and divide by $\sqrt{n+3}+\sqrt{n}$.]

23. $\left\{\dfrac{2^n}{n!}\right\}$

24. $\left\{\dfrac{\beta^n}{n!}\right\}$, β is a fixed real number

25. $\left\{\dfrac{n+1}{n^{5/2}}\right\}$ 26. $\left\{\dfrac{4}{\sqrt{n^2-n+3}}\right\}$

27. $\{(-1)^n\cos n\pi\}$ 28. $\left\{\cos\left(n+\dfrac{\pi}{2}\right)\right\}$

■ APPLICATIONS

*29. Suppose that $\{a_n\}$ is a sequence such that $a_{n+1}=a_n-a_n^2$ for $n\ge 0$. Find all values of a_0 such that $\lim_{n\to\infty}a_n=0$.

**30. Suppose that $a_{n+1}=1/(2+a_n)$ for $n\ge 0$. For what choices of a_0 does the sequence $\{a_n\}$ diverge?

■ SHOW/PROVE/DISPROVE

*31. Prove Theorem 5. [*Hint:* Use the definition of limit at infinity in Section 1.8.]

32. Prove that if $|r|<1$, then the sequence $\{nr^n\}$ converges to 0.

33. Suppose that $\{a_n\}$ and $\{b_n\}$ are two sequences such that $|a_n|\le |b_n|$ for each n. Show that if $|b_n|$ converges to 0, then $|a_n|$ also converges to 0. [*Hint:* Use the squeezing theorem.]

34. Use the result of the preceding problem to show that the sequence

$$\left\{\dfrac{a\sin bn+c\cos dn}{n^{p^2}}\right\}$$

converges to 0 for any real numbers a, b, c, d, and $p\ne 0$.

35. Show that $a_n=[1+(\beta/n)]^n$ converges to e^β as $n\to\infty$.

*36. Show that if $\{a_n\}$ converges, then its limit is unique. [*Hint:* Suppose that $\lim_{n\to\infty}a_n=L$, $\lim_{n\to\infty}a_n=M$, and $L\ne M$. Then choose $\epsilon=\frac{1}{2}|L-M|$ to show that the definition of a finite limit is violated.]

*37. Prove or Disprove: If $\lim_{n\to\infty}n^p a_n=0$, then there exists an $\epsilon>0$, such that $\lim_{n\to\infty}n^{p+\epsilon}a_n$ also exists (i.e., is finite).

■ ANSWERS TO SELF-QUIZ

I. c   II. d   V. d   VI. b
III. c (but not f)   IV. b (but not a)   VII. c ($e-\pi\approx -0.42331$ satisfies $|r|<1$.)

10.2 BOUNDED AND MONOTONIC SEQUENCES

There are certain kinds of sequences that have special properties worthy of mention.

DEFINITION **BOUNDEDNESS**

(i) The sequence $\{a_n\}$ is **bounded above** if there is a number M_1 such that

$$a_n \leq M_1 \text{ for every positive integer } n. \tag{1}$$

(ii) It is **bounded below** if there is a number M_2 such that

$$M_2 \leq a_n \text{ for every positive integer } n. \tag{2}$$

(iii) It is **bounded** if there is a number $M > 0$ such that

$$|a_n| \leq M \text{ for every positive integer } n.$$

The numbers M_1, M_2, and M are called, respectively, an **upper bound**, a **lower bound**, and a **bound** for $\{a_n\}$

(iv) If the sequence is not bounded, it is called **unbounded**. ∎

REMARK: If $\{a_n\}$ is bounded above and below, then it is bounded. Simply set $M = \max\{|M_1|, |M_2|\}$.

EXAMPLE 1
A BOUNDED SEQUENCE

The sequence $\{\sin n\}$ has the upper bound of 1, the lower bound of -1, and the bound of 1 since $-1 \leq \sin n \leq 1$ for every n. Of course, any number greater than 1 is also a bound.

EXAMPLE 2
A BOUNDED SEQUENCE

The sequence $\{(-1)^n\}$ has the upper bound 1, the lower bound -1, and the bound 1.

EXAMPLE 3
AN UNBOUNDED SEQUENCE
THAT IS BOUNDED BELOW

The sequence $\{2^n\}$ is bounded below by 2 for $n \geq 1$ but has no upper bound and so is unbounded.

EXAMPLE 4
AN UNBOUNDED SEQUENCE
THAT IS NEITHER BOUNDED
ABOVE NOR BOUNDED BELOW

The sequence $\{(-1)^n 2^n\}$ is bounded neither below nor above.

It turns out that the following statement is true:

Every convergent sequence is bounded.

THEOREM 1

If the sequence $\{a_n\}$ is convergent, then it is bounded.

PROOF: Before giving the technical details, we remark that the idea behind the proof is easy. For if $\lim_{n \to \infty} a_n = L$, then a_n is close to the finite number L if n is large. Thus, for example, $|a_n| \leq |L| + 1$ if n is large enough. Since a_n is a real number for every n, the first few terms of the sequence are bounded, and these two facts give us a bound for the entire sequence.

Now to the details: Let $\epsilon = 1$. Then there is an $N > 0$ such that (according to the definition of finite limit of a sequence, in Section 10.1)

$$|a_n - L| < 1 \qquad \text{if } n \geq N. \tag{3}$$

Let

$$K = \max\{|a_1|, |a_2|, \ldots, |a_N|\}. \tag{4}$$

Since each a_n is finite, K, being the maximum of a finite number of terms, is also finite. Now let

$$M = \max\{|L| + 1, K\}. \tag{5}$$

It follows from (4) that if $n \leq N$, then $|a_n| \leq K$. If $n \geq N$, then from (3), $|a_n| < |L| + 1$; so in either case $|a_n| \leq M$, and the theorem is proved. ■

Theorem 1 is useful in another way. Since every convergent sequence is bounded, it follows that:

Every unbounded sequence is divergent.

EXAMPLE 5
THREE DIVERGENT SEQUENCES

The following sequences are divergent since they are clearly unbounded:

(a) $\{\ln \ln n\}$ (starting at $n = 2$)

(b) $\left\{ n \sin \dfrac{n\pi}{2} \right\} = \{1, 0, -3, 0, 5, 0, -7, 0, 9, \ldots\}$

(c) $\{(-\sqrt{2})^n\}$

The converse of Theorem 1 is *not* true. That is, it is not true that every bounded sequence is convergent. For example, the sequences $\{(-1)^n\}$ and $\left\{ \sin \dfrac{n\pi}{2} \right\}$ are both bounded *and* divergent. Since boundedness alone does not ensure convergence, we need some other property. We investigate this idea now.

DEFINITION **MONOTONICITY**

(i) The sequence $\{a_n\}$ is **monotone increasing** if $a_{n+1} \geq a_n$ for every $n \geq 1$.

(ii) The sequence $\{a_n\}$ is **monotone decreasing** if $a_{n+1} \leq a_n$ for every $n \geq 1$.

(iii) The sequence $\{a_n\}$ is **monotonic** if it is either monotone increasing or monotone decreasing. ■

DEFINITION   **STRICT MONOTONICITY**

(i) The sequence $\{a_n\}$ is **strictly increasing** if $a_{n+1} > a_n$ for every $n \geq 1$.

(ii) The sequence $\{a_n\}$ is **strictly decreasing** if $a_{n+1} < a_n$ for every $n \geq 1$.

(iii) The sequence $\{a_n\}$ is **strictly monotonic** if it is either strictly increasing or strictly decreasing. ∎

EXAMPLE 6
A STRICTLY DECREASING
SEQUENCE

The sequence $\{1/2^n\}$ is strictly decreasing since $1/2^n > 1/2^{n+1}$ for every n.

EXAMPLE 7
DETERMINING MONOTONICITY

Determine whether the sequence $\{2n/(3n + 2)\}$ is increasing, decreasing, or not monotonic.

SOLUTION: If we write out the first few terms of the sequence, we find that $\{2n/(3n + 2)\} = \{\frac{2}{5}, \frac{4}{8}, \frac{6}{11}, \frac{8}{14}, \frac{10}{17}, \frac{12}{20}, \ldots\}$. Since these terms are strictly increasing, we suspect that $\{2n/(3n + 2)\}$ is an increasing sequence. To check this, we try to verify that $a_n < a_{n+1}$. The easiest proof uses calculus: Let $f(x) = \dfrac{2x}{3x + 2}$. Then $f'(x) = \dfrac{4}{(3x + 2)^2}$ and f is increasing for $x \neq -\frac{2}{3}$. But $a_{n+1} = f(n + 1) > f(n) = a_n$, so $\{2n/(3n + 1)\}$ is strictly increasing.

In all the examples we have given, a divergent sequence diverges for one of two reasons: It goes to infinity (it is unbounded) or it oscillates [like $(-1)^n$, which oscillates between -1 and 1]. But if a sequence is bounded, it does not go to infinity. And if it is monotone, then it does not oscillate. Thus, the following theorem should not be surprising.

THEOREM 2   **A BOUNDED MONOTONIC SEQUENCE IS CONVERGENT**

PROOF: We will prove this theorem for the case in which the sequence $\{a_n\}$ is increasing. The proof of the other case is similar. Since $\{a_n\}$ is bounded, there is a number M such that $a_n \leq M$ for every n. Let L be the least upper bound (see page 118). Now let $\epsilon > 0$ be given. Then there is a number $N > 0$ such that $a_N > L - \epsilon$. If this were not true, then we would have $a_n < L - \epsilon$ for all $n \geq 1$. Then $L - \epsilon$ would be an upper bound for $\{a_n\}$, and since $L - \epsilon < L$, this would contradict the choice of L as the *least* upper bound. Since $\{a_n\}$ is increasing, we have, for $n \geq N$,

$$L - \epsilon < a_N \leq a_n \leq L < L + \epsilon. \tag{6}$$

But the inequalities in (6) imply that $|a_n - L| < \epsilon$ for $n \geq N$, which proves, according to the definition of convergence, that $\lim_{n \to \infty} a_n = L$. ∎

We have actually proved a stronger result. Namely, that *if the sequence* $\{a_n\}$ *is bounded above and increasing, then it converges to its least upper bound. Similarly, if* $\{a_n\}$ *is bounded below and decreasing, then it converges to its greatest lower bound.*

EXAMPLE 8
A BOUNDED MONOTONIC
SEQUENCE IS CONVERGENT

In Example 7, we saw that the sequence $\{2n/(3n + 2)\}$ is strictly increasing. Also, since $2n/(3n + 2) < 3n/(3n + 2) < 3n/3n = 1$, we see that $\{a_n\}$ is also bounded, so that by Theorem 2, $\{a_n\}$ is convergent. We find that $\lim_{n \to \infty} 2n/(3n + 2) = \frac{2}{3}$.

SEQUENCES GENERATED BY FIXED-POINT ITERATION

In our discussion of Newton's method in Sections 3.6 and 3.9, we saw a number of sequences that converged to zeros of certain functions. We recall that if f is a differentiable function, then we may define the sequence $\{x_n\}$ by

$$x_{n+1} = x_n - \frac{f(x_n)}{f'(x_n)}. \tag{7}$$

If we start at some specified initial value x_0, then we can find the terms of the sequence **recursively**, that is, one after the other. Theorem 2 on page 231 gives conditions under which the sequence of iterates generated by (7) will converge to the unique solution s of $f(x) = 0$ in an interval $[a, b]$ for any initial value x_0 in $[a, b]$.

Suppose now that F is a function and we seek a solution to the equation

$$F(x) = x. \tag{8}$$

Any number that satisfies (8) is called a **fixed point** of F. Note that if s is a zero of f, then

$$f(s) = 0 \qquad \text{and} \qquad f(s) + s = s$$

so that s is a fixed point of the function $F(x) = f(x) + x$.

EXAMPLE 9
FIXED POINTS OF THREE
FUNCTIONS

(a) 0 is a fixed point of $\sin x$ because $\sin 0 = 0$.
(b) 2 and 3 are fixed points of $x^2 - 4x + 6$ because

$$2^2 - 4 \cdot 2 + 6 = 2 \qquad \text{and} \qquad 3^2 - 4 \cdot 3 + 6 = 3.$$

(c) e is a fixed point of $f(x) = \dfrac{x}{\ln x}$ because $\dfrac{e}{\ln e} = \dfrac{e}{1} = e$.

In order to find fixed points for a function F, we generate a sequence $\{x_n\}$ according to the following algorithm: Choose x_1. Then define

$$x_{n+1} = F(x_n). \tag{9}$$

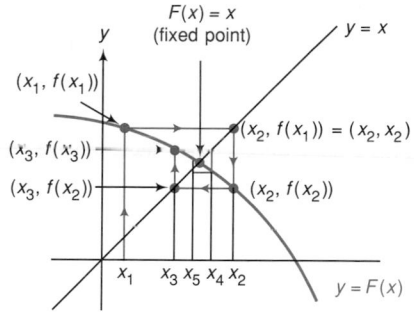

FIGURE 1
FIXED-POINT CONVERGENCE

To obtain the next value (x_{n+1}) in the sequence, start at the point $(x_n, f(x_n))$ and move horizontally until you hit the line $y = x$. The x-coordinate (or the equal y-coordinate) of the point that you hit is the value of x_{n+1}.

| THEOREM 3 | **FIXED-POINT CONVERGENCE THEOREM** |

Suppose that the function F has a fixed point u and that there is a number c such that

(i) F is continuous in $[u - c, u + c]$,
(ii) F is differentiable in $(u - c, u + c)$,
(iii) $|F'(x)| \leq M < 1$ for every x in $(u - c, u + c)$.

Then u is the only fixed point of F in $(u - c, u + c)$ and the sequence $\{x_n\}$ generated by the fixed-point iteration (9) converges to u for every choice of x_1 in $(u - c, u + c)$. ∎

An outline of the proof of this theorem is given in Problems 45–49. Note that if $f(x) = F(x) - x$, then a fixed point of F is a zero of f. By Theorem 1 in Section 3.6, if $f(u - c)f(u + c) < 0$, then f has a zero in $(u - c, u + c)$ so F has a fixed point in $(u - c, u + c)$.

The convergence of the sequence generated by (9) is illustrated in Figure 1.

▦ EXAMPLE 10
FINDING A FIXED POINT BY ITERATION

Find a fixed point of $F(x) = \dfrac{x}{\ln x}$ in the interval (2, 10).

SOLUTION: We observe that $F'(x) = \dfrac{\ln x - 1}{\ln^2 x} = \dfrac{1}{\ln x}\left(\dfrac{1}{\ln x} - 1\right)$. Since $\ln x$ is an increasing function, F' is decreasing. We have

$$F'(2) = \frac{1}{\ln 2}\left(\frac{1}{\ln 2} - 1\right) \approx 0.64 \qquad \text{and}$$

$$F'(10) = \frac{1}{\ln 10}\left(\frac{1}{\ln 10} - 1\right) = -0.25,$$

so, on (2, 10), $|F'(x)| < 0.65 < 1$. Starting at $x_1 = 5$, we obtain the sequence in Table 1.

We see that $x_5 = 2.718281828 \approx e$ (to 9 decimal places). This is no surprise since $\dfrac{e}{\ln e} = \dfrac{e}{1} = e$.

TABLE 1 SEQUENCE OF FIXED-POINT ITERATES OF $f(x) = \dfrac{x}{\ln x}$

n	x_n	$x_{n+1} = \dfrac{x_n}{\ln x_n}$
1	5	3.106674673
2	3.106674673	2.740652532
3	2.740652532	2.718372635
4	2.718372635	2.71828183
5	2.71828183	2.718281828

NOTE: This example is a special case of the following problem, which is given on page 442 of the November 1989 issue of the *College Mathematics Journal*:

Let $x_{n+1} = \dfrac{x_n}{\ln x_n}$ where $1 < x_1 < \infty$. Evaluate $\lim_{n \to \infty} x_n$.

Since $\dfrac{x}{\ln x} = x$ only when $x = e$, e is the only fixed point of $\dfrac{x}{\ln x}$. Then, using the fixed-point convergence theorem, it is not difficult to show that $\lim_{n \to \infty} x_n = e$ for every choice of x_1 in $(1, \infty)$.

EXAMPLE 11
FINDING A FIXED POINT OF A
SINE FUNCTION

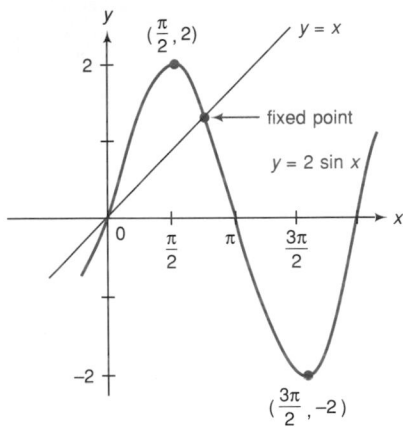

FIGURE 2
THE FUNCTION $2 \sin x$ HAS A UNIQUE
POSITIVE FIXED POINT

TABLE 2 SEQUENCE OF FIXED-POINT
ITERATES OF $F(x) = 2 \sin x$

n	x_n	$x_{n+1} = 2 \sin x_n$
1	2	1.818594854
2	1.818594854	1.938909453
3	1.938909453	1.866016016
4	1.866016016	1.913476493
5	1.913476493	1.883714959
6	1.883714959	1.902878322
7	1.902878322	1.890731275
8	1.890731275	1.898511758
9	1.898511758	1.893560345
10	1.893560345	1.896724651

Find, by iteration, the unique fixed point of $F(x) = 2 \sin x$ in the interval $\left[\dfrac{\pi}{2}, \pi \right]$.

SOLUTION: In Figure 2 we have sketched the graphs of $y = 2 \sin x$ and $y = x$. It is clear that there is a unique positive fixed point and that it lies in the interval $\left(\dfrac{\pi}{2}, \pi \right)$. (0 is also a fixed point, but it is not very interesting.) Note that $F'(x) = 2 \cos x$ and $F'(\pi) = -2$, so the condition $|F'(x)| < 1$ in $\left[\dfrac{\pi}{2}, \pi \right]$ does not hold. Nevertheless we perform the iteration and see what happens. The results are given in Table 2. We start with $x_1 = 2$ and stop after 10 iterations.

In Example 7 in Section 3.6 (page 234), we showed, using Newton's method, that the unique fixed point is 1.895494267, correct to 9 decimal places.

NOTE: Let $g(x) = 2 \sin x - x$. Then

$$g\left(\frac{\pi}{3} \right) = 2 \sin \frac{\pi}{3} - \frac{\pi}{3} = 2 \frac{\sqrt{3}}{2} - \frac{\pi}{3}$$

$$= \sqrt{3} - \frac{\pi}{3} \approx 0.68 > 0$$

and

$$g\left(\frac{2\pi}{3} \right) = 2 \sin \frac{2\pi}{3} - \frac{2\pi}{3} = \sqrt{3} - \frac{2\pi}{3} \approx -0.36 < 0.$$

Thus g has a zero in the interval $\left(\dfrac{\pi}{3}, \dfrac{2\pi}{3} \right)$, which means that $2 \sin x$ has a fixed point in $\left(\dfrac{\pi}{3}, \dfrac{2\pi}{3} \right)$. Moreover, $2 \cos \dfrac{\pi}{3} = 1$ and $2 \cos \dfrac{2\pi}{3} = -1$, so $F'(x)$ decreases from 1 to -1 and $|F'(x)| < 1$ in $\left[\dfrac{\pi}{3} + \epsilon, \dfrac{2\pi}{3} - \epsilon \right]$ where ϵ is a small number. This shows that the fixed-point iteration will converge if we choose x_0 in $\left(\dfrac{\pi}{3}, \dfrac{2\pi}{3} \right)$. The number 2 is in this interval $\left(\dfrac{2\pi}{3} \approx 2.094 \right)$, so, as we have seen, the iteration does converge if we start at $x_0 = 2$.

■ PROBLEMS 10.2

■ SELF-QUIZ

I. The sequence $\left\{ n + \left(\dfrac{1}{n} \right) \right\}$ is _____.

  a. unbounded
  b. bounded but not convergent
  c. decreasing
  d. increasing and bounded

II. The sequence $\{ 1 + (-1)^n \}$ is _____.

  a. convergent but unbounded
  b. bounded

CONTINUED

   c. decreasing
   d. increasing

III. The sequence $\left\{5 + \dfrac{(-1)^n}{n}\right\}$ is _____.

   a. unbounded and divergent
   b. convergent but unbounded
   c. bounded and convergent
   d. bounded but not convergent

IV. The sequence $\left\{\pi^2 + \dfrac{1}{n}\right\}$ is _____.

   a. unbounded but convergent

   b. increasing but not convergent
   c. decreasing and unbounded
   d. decreasing, bounded, and convergent

V. For each positive integer n, consider a regular n-sided polygon inscribed within a circle of radius 1. (Mark n equally spaced points on that circle, then join consecutive points with a line segment.) Let A_n be the area of this n-gon. The sequence $\{A_n\}$ is _____.

   a. bounded, increasing, and convergent
   b. increasing and bounded but not convergent
   c. decreasing, bounded, and convergent
   d. increasing and convergent but unbounded

■ **DRILL**

In Problems 1–12, determine whether the given sequence is bounded or unbounded. If it is bounded, find the least upper bound for $|a_n|$.

1. $\left\{\dfrac{1}{n+1}\right\}$ 2. $\{\sin n\pi\}$ 3. $\{\cos n\pi\}$

4. $\{\sqrt{n}\sin n\}$ 5. $\left\{\dfrac{2^n}{1+2^n}\right\}$ 6. $\left\{\dfrac{2^n+1}{2^n}\right\}$

7. $\left\{\dfrac{1}{n!}\right\}$ 8. $\left\{\dfrac{3^n}{n!}\right\}$ 9. $\left\{\dfrac{n^2}{n!}\right\}$

10. $\left\{\dfrac{2n}{2^n}\right\}$ 11. $\left\{\dfrac{\ln n}{n}\right\}$ *12. $\{ne^{-n}\}$

In Problems 13–28, determine whether the given sequence is monotone increasing, strictly increasing, monotone decreasing, strictly decreasing, or not monotonic.

13. $\{\sin n\pi\}$ 14. $\left\{\dfrac{3^n}{2+3^n}\right\}$ 15. $\left\{\left(\dfrac{n}{25}\right)^{1/3}\right\}$

16. $\{n + (-1)^n\sqrt{n}\}$ 17. $\left\{\dfrac{\sqrt{n+1}}{n}\right\}$

18. $\left\{\dfrac{n!}{n^n}\right\}$ *19. $\left\{\dfrac{n^n}{n!}\right\}$

*20. $\left\{\dfrac{2(n!)}{1\cdot 3\cdot 5\cdot 7\cdot\,\cdots\,(2n-1)}\right\}$

*21. $\{n + \cos n\}$ 22. $\left\{\dfrac{2^{2n}}{n!}\right\}$

23. $\left\{\dfrac{\sqrt{n}-1}{n}\right\}$ 24. $\left\{\dfrac{n-1}{n+1}\right\}$

25. $\left\{\ln\left(\dfrac{3n}{n+1}\right)\right\}$ 26. $\{\ln n - \ln(n+2)\}$

*27. $\left\{\left(1 - \dfrac{3}{n}\right)^n\right\}$ *28. $\left\{\left(1 + \dfrac{3}{n}\right)^{1/n}\right\}$

▦ **CALCULATOR PROBLEMS**

29. a. Show graphically that e^{-x} has a unique fixed point in $(0, 1)$.
   b. Find the fixed point, to 5 decimal places, by iterating starting with $x_1 = 1$.

30. a. Show graphically that $\cos x$ has a unique fixed point for $x > 0$.
   b. Find the fixed point, to 8 decimal places, by iterating starting with $x_1 = 1$. [This is easy to do; just push the ⎡cos⎤ key repeatedly, making sure that the calculator is set to radians.]

31. Find, to 5 significant figures, a fixed point for $F(x) = 5 \ln x$.

32. Let $F(x) = x^2 - 3x + 3$.
   a. Show that 3 is a fixed point for F.
   b. Starting with $x_1 = 3.01$, compute the iterates $x_2, x_3, x_4, x_5, x_6, x_7$, and x_8 and conclude that the sequence of iterates diverges even though we started within 0.01 of the fixed point. [This behavior is explained in Problem 51.]

33. Let $F(x) = x^2 + x - 2$.

CONTINUED

a. Compute the two fixed points of F.
b. Show that, starting with $x_1 = 0$, the iterates oscillate between 0 and -2 so that the sequences of iterates diverges.

34. Let $F(x) = \frac{1}{5}x^3 + \frac{1}{4}x^2 - 1$.
    a. Find an interval (a, b) over which $|F'(x)| < 1$.
    b. Find a fixed point for F in that interval (correct to 5 decimal places).

■ SHOW/PROVE/DISPROVE

35. a. Show that $\{5^n/n!\}$ is bounded. [*Hint:* See Problem 9.5.35.]
    🔲 b. Find the smallest integer M such that $5^n/n! \leq M$ for all positive integers n.

*36. a. Show that for $n > 2^{10}$, $n^{10}/n! > (n + 1)^{10}/(n + 1)!$. Use this result to infer that $\{n^{10}/n!\}$ is bounded.
    🔲 b. Find a relatively small integer M such that $n^{10}/n! \leq M$ for all positive integers n.

*37. Show that the sequence $\{(2^n + 3^n)^{1/n}\}$ is convergent.

*38. Show that $\{(a^n + b^n)^{1/n}\}$ is convergent for any positive real numbers a and b. [*Hint:* First do the preceding problem, then treat the cases $a = b$ and $a \neq b$ separately.]

*39. Show that the sequence $\{n!/n^n\}$ is bounded. [*Hint:* Show that $n!/n^n > (n + 1)!/(n + 1)^{n+1}$ for sufficiently large n.]

40. Prove that the sequence $\{n!/n^n\}$ converges. [*Hint:* Use the result of the preceding problem.]

41. a. Show that the sequence $\{(\ln n)/n\}$, $n \geq 1$ is not monotonic.
    b. Show that the sequence $\{(\ln n)/n\}$, $n > 2$ is monotonic.

42. Use Theorem 2 to show that $\{\ln n - \ln(n + 4)\}$ converges.

43. Show that the sequence of Problem 20 is convergent.

44. Show that the sequence $\{2 \cdot 5 \cdot 8 \cdot 11 \cdots (3n - 1)/3^n n!\}$ is convergent.

■ CHALLENGE

In Problems 45–49, assume that the hypotheses of Theorem 3 (the fixed-point convergence theorem) hold. In particular, assume that u is a fixed point for F.

45. Show that if v is a fixed point for F in $(u - c, u + c)$, then $v = u$. [*Hint:* Use the fact that $F(u) - F(v) = u - v$ and apply the mean-value theorem to the expression $F(u) - F(v)$.]

46. Let E_n, the error term, be defined by $E_n = x_n - u$. Show that $|E_{n+1}| \leq M|E_n|$ for $n = 1, 2, 3, \ldots$.

47. Use the result of Problem 46 to show that if x_1 is in $(u - c, u + c)$, then $x_n \in (u - c, u + c)$ for $n = 2, 3, 4, \ldots$. This shows that the iterates are all defined.

48. Use the result of Problem 46 to show that

$$|E_n| \leq M^{n-1}|E_1|.$$

49. Show that $E_n \to 0$ as $n \to \infty$ and conclude that $\{x_n\}$ converges to u.

*50. Assume that F is continuous and that the sequence of iterates given by (9) converges to a number p. Prove that p must be a fixed point of F. [*Hint:* Use the continuity theorem on page 117.]

*51. **Fixed-Point Divergence Theorem** Suppose that the hypotheses of Theorem 3 hold except that $|F'(x)| \geq k > 1$ for every x in $(u - c, u + c)$. Show that if $x_1 \neq u$, then the sequence of iterates $\{x_n\}$ given by (9) does not converge to u (it may converge to a different fixed point of F in a different interval—if there is one). [*Hint:* Show that if $x_n \in (u - c, u + c)$, then $|E_{n+1}| > |E_n|$ where E_n is defined as in Problem 46.

■ ANSWERS TO SELF-QUIZ

I. a   II. b   III. c   IV. d
V. a (It converges to π, the area of the circle.)

10.3 GEOMETRIC SERIES

Consider the sum

$$S_7 = 1 + 2 + 4 + 8 + 16 + 32 + 64 + 128.$$

This can be written as

$$S_7 = 1 + 2 + 2^2 + 2^3 + 2^4 + 2^5 + 2^6 + 2^7 = \sum_{k=0}^{7} 2^k.$$

GEOMETRIC PROGRESSION

In general, the sum of a **geometric progression** is a sum of the form

$$S_n = 1 + r + r^2 + r^3 + \cdots + r^{n-1} + r^n = \sum_{k=0}^{n} r^k, \tag{1}$$

where r is a real number and n is a fixed positive integer.

In Section 9.5 (equation (5) on page 629), we derived the following result:

THEOREM 1

If $r \neq 1$, the sum of a geometric progression (1) is given by

$$S_n = \frac{1 - r^{n+1}}{1 - r}. \quad \blacksquare \tag{2}$$

NOTE: If $r = 1$, we obtain

$$S_n = \overbrace{1 + 1 + \cdots + 1}^{n + 1 \text{ terms}} = n + 1.$$

EXAMPLE 1
COMPUTING THE SUM OF A
GEOMETRIC PROGRESSION

Calculate $S_7 = 1 + 2 + 4 + 8 + 16 + 32 + 64 + 128$, using formula (2).

SOLUTION: Here $r = 2$ and $n = 7$, so that

$$S_7 = \frac{1 - 2^8}{1 - 2} = 2^8 - 1 = 256 - 1 = 255.$$

EXAMPLE 2
COMPUTING THE SUM OF A
GEOMETRIC PROGRESSION

Calculate $\sum_{k=0}^{10} \left(\frac{1}{2}\right)^k$.

SOLUTION: Here $r = \frac{1}{2}$ and $n = 10$, so that

$$S_{10} = \frac{1 - \left(\frac{1}{2}\right)^{11}}{1 - \frac{1}{2}} = \frac{1 - \frac{1}{2048}}{\frac{1}{2}} = 2\left(\frac{2047}{2048}\right) = \frac{2047}{1024}.$$

EXAMPLE 3
COMPUTING THE SUM OF
A GEOMETRIC PROGRESSION
WITH A NEGATIVE r

Calculate

$$S_6 = 1 - \frac{2}{3} + \left(\frac{2}{3}\right)^2 - \left(\frac{2}{3}\right)^3 + \left(\frac{2}{3}\right)^4 - \left(\frac{2}{3}\right)^5 + \left(\frac{2}{3}\right)^6 = \sum_{k=0}^{6} \left(-\frac{2}{3}\right)^k.$$

SOLUTION: Here $r = -\frac{2}{3}$ and $n = 6$, so that

$$S_6 = \frac{1 - (-\frac{2}{3})^7}{1 - (-\frac{2}{3})} = \frac{1 + \frac{128}{2187}}{\frac{5}{3}} = \frac{3}{5}\left(\frac{2315}{2187}\right) = \frac{463}{729}.$$

The sum of a geometric progression is the sum of a finite number of terms. We now see what happens if the number of terms is infinite. Consider the sum

$$S = 1 + \frac{1}{2} + \frac{1}{4} + \frac{1}{8} + \frac{1}{16} + \cdots = \sum_{k=0}^{\infty} \left(\frac{1}{2}\right)^k.$$

What can such a sum mean? We will give a formal definition in a moment. For now, let us show why it is reasonable to say that $S = 2$. Let $S_n = \sum_{k=0}^{n} (\frac{1}{2})^k = 1 + \frac{1}{2} + \frac{1}{4} + \cdots + (\frac{1}{2})^n$. Then,

$$S_n = \frac{1 - (\frac{1}{2})^{n+1}}{1 - \frac{1}{2}} = 2\left[1 - \left(\frac{1}{2}\right)^{n+1}\right] = 2 - \left(\frac{1}{2}\right)^n.$$

Thus, for any n (no matter how large), $1 \leq S_n < 2$. Hence, the numbers S_n are bounded. Also, since $S_{n+1} = S_n + (\frac{1}{2})^{n+1} > S_n$, the numbers S_n are monotone increasing. Thus, the sequence $\{S_n\}$ converges. But

$$S = \lim_{n \to \infty} S_n.$$

Thus, S has a finite sum. To compute it, we note that

$$S = \lim_{n \to \infty} S_n = \lim_{n \to \infty} 2[1 - (\tfrac{1}{2})^{n+1}] = 2 \lim_{n \to \infty} [1 - (\tfrac{1}{2})^{n+1}] = 2$$

since $\lim_{n \to \infty}(\frac{1}{2})^{n+1} = 0$.

GEOMETRIC SERIES

The infinite sum $\sum_{k=0}^{\infty} (\frac{1}{2})^k$ is called a *geometric series*. In general, a **geometric series** is an infinite sum of the form

$$S = \sum_{k=0}^{\infty} r^k = 1 + r + r^2 + r^3 + \cdots. \tag{3}$$

DEFINITION **CONVERGENCE AND DIVERGENCE OF A GEOMETRIC SERIES**

Let $S_n = \sum_{k=0}^{n} r^k$. Then we say that the geometric series **converges** if $\lim_{n \to \infty} S_n$ exists and is finite. Otherwise, the series is said to **diverge**. ∎

EXAMPLE 4

THE GEOMETRIC SERIES DIVERGES WHEN $r = 1$

Let $r = 1$. Then

$$S_n = \sum_{k=0}^{n} 1^k = \sum_{k=0}^{n} 1 = \underbrace{1 + 1 + \cdots + 1}_{n+1 \text{ times}} = n + 1.$$

Since $\lim_{n \to \infty}(n + 1) = \infty$, the series $\sum_{k=0}^{\infty} 1^k$ diverges.

THEOREM 2

Let $S = \sum_{k=0}^{\infty} r^k = 1 + r + r^2 + \cdots$ be a geometric series.

(i) The series converges to

$$\frac{1}{1 - r} \qquad \text{if } |r| < 1.$$

(ii) The series diverges if $|r| \geq 1$.

PROOF:

(i) If $|r| < 1$, then $\lim_{n \to \infty} r^{n+1} = 0$. Thus,

$$S = \lim_{n \to \infty} S_n = \lim_{n \to \infty} \frac{1 - r^{n+1}}{1 - r} = \frac{1}{1 - r} \lim_{n \to \infty} (1 - r^{n+1})$$

$$= \frac{1}{1 - r}(1 - 0) = \frac{1}{1 - r}.$$

(ii) If $|r| > 1$, then $\lim_{n \to \infty}|r|^{n+1} = \infty$. Thus, $1 - r^{n+1}$ does not have a finite limit and the series diverges. Finally, if $r = 1$, then the series diverges, by Example 4, and if $r = -1$, then S_n alternates between the numbers 0 and 1, so that the series diverges. ∎

EXAMPLE 5

THE GEOMETRIC SERIES FOR $r = -\frac{2}{3}$

$$1 - \frac{2}{3} + \left(\frac{2}{3}\right)^2 - \cdots = \sum_{k=0}^{\infty} \left(-\frac{2}{3}\right)^k = \frac{1}{1 - (-\frac{2}{3})} = \frac{1}{\frac{5}{3}} = \frac{3}{5}.$$

EXAMPLE 6

THE GEOMETRIC SERIES FOR $r = \pi/4$

$$1 + \frac{\pi}{4} + \left(\frac{\pi}{4}\right)^2 + \left(\frac{\pi}{4}\right)^3 + \cdots = \sum_{k=0}^{\infty} \left(\frac{\pi}{4}\right)^k = \frac{1}{1 - (\pi/4)}$$

$$= \frac{4}{4 - \pi} \approx 4.66.$$

EXAMPLE 7

A GEOMETRIC SERIES THAT DOES NOT START WITH THE TERM 1

Compute $\left(\frac{4}{5}\right)^2 + \left(\frac{4}{5}\right)^3 + \left(\frac{4}{5}\right)^4 + \cdots = \sum_{k=2}^{\infty} \left(\frac{4}{5}\right)^k.$

SOLUTION: There are at least two ways to compute this sum.

Method 1: Factor   We factor out the first term:

$$\left(\frac{4}{5}\right)^2 + \left(\frac{4}{5}\right)^3 + \left(\frac{4}{5}\right)^4 + \cdots = \left(\frac{4}{5}\right)^2\left[1 + \frac{4}{5} + \left(\frac{4}{5}\right)^2 + \cdots\right]$$

$$= \left(\frac{4}{5}\right)^2 \sum_{k=0}^{\infty}\left(\frac{4}{5}\right)^k = \left(\frac{4}{5}\right)^2\left(\frac{1}{1 - \frac{4}{5}}\right)$$

$$= \left(\frac{4}{5}\right)^2\left(\frac{1}{\frac{1}{5}}\right) = \left(\frac{4}{5}\right)^2 \cdot 5 = \frac{16}{5}.$$

Method 2:   Add and subtract the two missing terms:

$$\left(\frac{4}{5}\right)^2 + \left(\frac{4}{5}\right)^3 + \left(\frac{4}{5}\right)^4 + \cdots = \overbrace{1 + \frac{4}{5}}^{\substack{\text{these} \\ \text{terms} \\ \text{added}}} + \left(\frac{4}{5}\right)^2 + \left(\frac{4}{5}\right)^3 + \cdots \overbrace{- 1 - \frac{4}{5}}^{\substack{\text{the same} \\ \text{two terms} \\ \text{subtracted}}}$$

$$= \frac{1}{1 - \frac{4}{5}} - 1 - \frac{4}{5} = 5 - \frac{9}{5} = \frac{16}{5}.$$

■ PROBLEMS 10.3

■ SELF-QUIZ

I. $\displaystyle\sum_{k=0}^{5}\left(\frac{1}{2}\right)^k =$ _____.

　　a. $\dfrac{1}{2^6}$　　　　　b. $\dfrac{1 - 2^6}{1 - 2}$

　　c. $\dfrac{1 - (\frac{1}{2})^5}{1 - (\frac{1}{2})}$　　　d. $\dfrac{1 - (\frac{1}{2})^6}{1 - (\frac{1}{2})}$

　　e. $\dfrac{1 - (\frac{1}{2})}{1 - (\frac{1}{2})^6}$　　　f. $\dfrac{1 - 2}{1 - 2^5}$

II. $\displaystyle\sum_{k=0}^{4} 2^k =$ _____.

　　a. $\dfrac{1}{2^5}$　　　　　b. $\dfrac{1 - 2^5}{1 - 2}$

　　c. $\dfrac{1 - (\frac{1}{2})^5}{1 - (\frac{1}{2})}$　　　d. $\dfrac{1 - (\frac{1}{2})^4}{1 - (\frac{1}{2})}$

　　e. $\dfrac{1 - (\frac{1}{2})}{1 - (\frac{1}{2})^4}$　　　f. $\dfrac{1 - 2}{1 - 2^5}$

III. $2 - \displaystyle\sum_{k=0}^{7} \frac{1}{2^k} =$ _____.

　　a. $\dfrac{1}{2}$　　　　　b. $\dfrac{1}{2^7}$

　　c. $\dfrac{1}{2^8}$　　　　　d. $\dfrac{-1}{2^8}$

IV. $\displaystyle\sum_{k=0}^{\infty}\left(\frac{1}{10}\right)^k =$ _____.

　　a. $\frac{10}{9}$　　　　　b. $\frac{9}{10}$

　　c. $\frac{10}{11}$　　　　d. $1.23456\ldots$

V. $\displaystyle\sum_{k=0}^{\infty}\left(\frac{-1}{10}\right)^k =$ _____.

　　a. $\frac{10}{9}$　　　　　b. $\frac{9}{10}$

　　c. $\frac{10}{11}$　　　　d. $1.23456\ldots$

VI. $0.9 + 0.09 + 0.009 + \cdots = \displaystyle\sum_{k=1}^{\infty} 9 \cdot \left(\frac{1}{10}\right)^k =$ _____.

　　a. $1.0 - \left(\dfrac{1}{10}\right)^k$　　　b. $1.0 - \left(\dfrac{1}{10}\right)^{k+1}$

　　c. $0.99999\ldots$　　　　d. 1.0

■ DRILL

In Problems 1–12, calculate the sum of the given geometric progression.

1. $1 + 3 + 9 + 27 + 81 + 243$

2. $1 + \dfrac{1}{4} + \dfrac{1}{16} + \cdots + \dfrac{1}{4^8}$

3. $1 - 5 + 25 - 125 + 625 - 3125$

CONTINUED

4. $0.2 + 0.2^2 + 0.2^3 + \cdots + 0.2^9$

5. $0.3^2 - 0.3^3 + 0.3^4 - 0.3^5 + 0.3^6 - 0.3^7 + 0.3^8$

6. $1 + b^3 + b^6 + b^9 + b^{12} + b^{15} + b^{18} + b^{21}$

7. $1 - \dfrac{1}{b^2} + \dfrac{1}{b^4} - \dfrac{1}{b^6} + \dfrac{1}{b^8} - \dfrac{1}{b^{10}} + \dfrac{1}{b^{12}} - \dfrac{1}{b^{14}}$

8. $\pi - \pi^3 + \pi^5 - \pi^7 + \pi^9 - \pi^{11} + \pi^{13}$

9. $1 + \sqrt{2} + 2 + 2^{3/2} + 4 + 2^{5/2} + 8 + 2^{7/2} + 16$

10. $1 - \dfrac{1}{\sqrt{3}} + \dfrac{1}{3} - \dfrac{1}{3\sqrt{3}} + \dfrac{1}{9} - \dfrac{1}{9\sqrt{3}} + \dfrac{1}{27}$

11. $-16 + 64 - 256 + 1024 - 4096$

12. $\dfrac{3}{4} - \dfrac{9}{8} + \dfrac{27}{16} - \dfrac{81}{32} + \dfrac{243}{64} - \dfrac{729}{128}$

In Problems 13–20, calculate the sum of the given geometric series.

13. $1 + \dfrac{1}{4} + \dfrac{1}{4^2} + \dfrac{1}{4^3} + \cdots$

14. $1 - \dfrac{1}{2} + \dfrac{1}{4} - \dfrac{1}{8} + \dfrac{1}{16} - \cdots$

15. $1 - \dfrac{1}{3} + \dfrac{1}{9} - \dfrac{1}{27} + \dfrac{1}{81} - \cdots$

16. $1 + \dfrac{1}{\pi} + \dfrac{1}{\pi^2} + \dfrac{1}{\pi^3} + \cdots$

17. $1 + 0.7 + 0.7^2 + 0.7^3 + \cdots$

18. $1 - 0.62 + 0.62^2 - 0.62^3 + 0.62^4 - \cdots$

19. $\dfrac{1}{4} + \dfrac{1}{16} + \dfrac{1}{64} + \cdots$

[*Hint:* Factor out the term $\frac{1}{4}$.]

20. $-\dfrac{3}{5} + \dfrac{3}{25} - \dfrac{3}{125} + \cdots$

CALCULATOR PROBLEMS

21. How large must n be in order that $\left(\frac{1}{2}\right)^n < 0.01$?

22. How large must n be in order that $0.8^n < 0.01$?

23. How large must n be in order that $0.99^n < 0.01$?

APPLICATIONS

24. A bacteria population initially contains 1000 organisms; each bacterium produces two live bacteria every 2 hrs. If none of the bacteria dies during a 12-hr growth period, then how many organisms will be alive at its end?

SHOW/PROVE/DISPROVE

25. Show that if $x > 1$, then

$$1 + \frac{1}{x} + \frac{1}{x^2} + \frac{1}{x^3} + \cdots = \frac{x}{x-1}.$$

[*Note:* You need to do two tasks here: Show that the series converges; then show that the right-hand-side formula does equal the limit of the finite sums.]

CHALLENGE

*26. **Assertion:** *The legendary Greek hero Achilles could never overtake a turtle in a footrace.*

Argument: Suppose that Achilles can run 100 times faster than the turtle; also suppose the turtle starts with a 1000 m lead. While Achilles runs his first 1000 m, the turtle runs 10 m; when Achilles has run 10 m, the turtle has run $\frac{1}{10}$ m further. By the time Achilles reaches the turtle's current position, the turtle will have moved on— thus maintaining its lead.[†]

Your problem: Resolve this paradox using tools and insights from your current studies. (If you feel more detail is required, you may assume that the turtle runs at 1 km/hr and that Achilles runs 100 km/hr.)

[†] The Greek mathematician and philosopher Zeno of Elea (ca. 495—430 B.C.) posed this paradox.

10.4 INFINITE SERIES

In Section 10.3 we defined the geometric series $\sum_{k=0}^{\infty} r^k$ and showed that if $|r| < 1$, the series converges to $1/(1 - r)$. Let us again look at what we did. If S_n denotes the sum of the first $n + 1$ terms of the geometric series, then

$$S_n = 1 + r + r^2 + \cdots + r^n = \frac{1 - r^{n+1}}{1 - r}, \qquad r \neq 1. \tag{1}$$

For each n we obtain the number S_n, and therefore we can define a new sequence $\{S_n\}$ to be the sequence of **partial sums** of the geometric series. If $|r| < 1$, then

$$\lim_{n \to \infty} S_n = \lim_{n \to \infty} \frac{1 - r^{n+1}}{1 - r} = \frac{1}{1 - r}.$$

That is, the convergence of the geometric series is implied by the convergence of the sequence of partial sums $\{S_n\}$.

We now give a more general definition of these concepts.

DEFINITION   **INFINITE SERIES**

Let $\{a_n\}$ be a sequence. Then the infinite sum

$$\sum_{k=1}^{\infty} a_k = a_1 + a_2 + a_3 + \cdots + a_n + \cdots \tag{2}$$

is called an **infinite series** (or simply, **series**). Each a_k in (2) is called a **term** of the series. The **partial sums** of the series are given by

$$S_n = \sum_{k=1}^{n} a_k.$$

The sum S_n is called the **nth partial sum** of the series. If the sequence of partial sums $\{S_n\}$ converges to L, then we say that the infinite

Convergent Series

series $\sum_{k=1}^{\infty} a_k$ **converges** to L, and we write

$$\sum_{k=1}^{\infty} a_k = L.$$

Divergent Series

Otherwise, we say that the series $\sum_{k=1}^{\infty} a_k$ **diverges**. ■

REMARK: Occasionally a series will be written with the first term other than a_1. For example, $\sum_{k=0}^{\infty} \left(\frac{1}{2}\right)^k$ and $\sum_{k=2}^{\infty} 1/(\ln k)$ are both examples of infinite series. In the second case we must start with $k = 2$ since $1/(\ln 1)$ is not defined.

EXAMPLE 1

EXPRESSING A REPEATING
DECIMAL AS A RATIONAL
NUMBER BY USING A
GEOMETRIC SERIES

Express the **repeating decimal** $0.123123123\ldots$ as a rational number (the quotient of two integers).

SOLUTION:

$$0.123123123\ldots = 0.123 + 0.000123 + 0.000000123 + \cdots$$

$$= \frac{123}{10^3} + \frac{123}{10^6} + \frac{123}{10^9} + \cdots$$

$$= \frac{123}{10^3}\left[1 + \frac{1}{10^3} + \frac{1}{(10^3)^2} + \cdots\right]$$

$$= \frac{123}{1000}\sum_{k=0}^{\infty}\left(\frac{1}{1000}\right)^k = \frac{123}{1000}\left[\frac{1}{1 - (\frac{1}{1000})}\right]$$

$$= \frac{123}{1000}\cdot\frac{1}{\frac{999}{1000}}$$

$$= \frac{123}{1000}\cdot\frac{1000}{999} = \frac{123}{999} = \frac{41}{333}.$$

As a matter of fact, any decimal number x can be thought of as a convergent infinite series, for if $x = 0.\,a_1a_2a_3\ldots a_n\ldots$, then

$$x = \frac{a_1}{10} + \frac{a_2}{100} + \frac{a_3}{1000} + \cdots + \frac{a_n}{10^n} + \cdots = \sum_{k=1}^{\infty}\frac{a_k}{10^k}.^{\dagger}$$

In general, we can use the geometric series to write any repeating decimal in the form of a fraction by using the technique of Example 1. In fact, the rational numbers are exactly those real numbers that can be written as repeating decimals. Repeating decimals include numbers like $3 = 3.00000\ldots$, $\frac{1}{4} = 0.25 = 0.25000000\ldots$, and $\frac{1}{3} = 0.3333\ldots$.

EXAMPLE 2

A TELESCOPING SERIES

Consider the infinite series $\sum_{k=1}^{\infty} 1/k(k+1)$. We write the first three partial sums:

$$S_1 = \sum_{k=1}^{1}\frac{1}{k(k+1)} = \frac{1}{1\cdot 2} = \frac{1}{2} = 1 - \frac{1}{2},$$

$$S_2 = \sum_{k=1}^{2}\frac{1}{k(k+1)} = \frac{1}{1\cdot 2} + \frac{1}{2\cdot 3} = \frac{1}{2} + \frac{1}{6} = \frac{2}{3} = 1 - \frac{1}{3},$$

$$S_3 = \sum_{k=1}^{3}\frac{1}{k(k+1)} = \frac{1}{1\cdot 2} + \frac{1}{2\cdot 3} + \frac{1}{3\cdot 4} = \frac{1}{2} + \frac{1}{6} + \frac{1}{12}$$

$$= \frac{3}{4} = 1 - \frac{1}{4}.$$

---

† Since $0 \le a_k < 10$,

$$\sum_{k=1}^{\infty}\frac{a_k}{10^k} < \sum_{k=1}^{\infty}\frac{10}{10^k} = \sum_{k=1}^{\infty}\frac{1}{10^{k-1}} = 1 + \frac{1}{10} + \left(\frac{1}{10}\right)^2 + \cdots = \frac{1}{1 - \frac{1}{10}} = \frac{10}{9}.$$

In Section 10.5 we will prove the comparison test. Once we have this test, the inequality given above implies that $\sum_{k=1}^{\infty}(a_k/10^k)$ converges.

We can use partial fractions to rewrite the general term as

$$a_k = \frac{1}{k(k+1)} = \frac{1}{k} - \frac{1}{k+1},$$

from which we can get a better view of the nth partial sum:

$$S_n = \left(\frac{1}{1} - \frac{1}{2}\right) + \left(\frac{1}{2} - \frac{1}{3}\right) + \left(\frac{1}{3} - \frac{1}{4}\right) + \cdots$$
$$+ \left(\frac{1}{n-1} - \frac{1}{n}\right) + \left(\frac{1}{n} - \frac{1}{n+1}\right)$$
$$= 1 - \frac{1}{n+1}$$

because all other terms cancel. Since

$$\lim_{n \to \infty} S_n = \lim_{n \to \infty} \left(1 - \frac{1}{n-1}\right) = 1,$$

we see that

$$\sum_{k=1}^{\infty} \frac{1}{k(k+1)} = 1.$$

When, as here, alternate terms cancel, we say that the series is a **telescoping series**.

REMARK: Often, it is not possible to calculate the exact sum of an infinite series, even if it can be shown that the series converges.

EXAMPLE 3
THE HARMONIC SERIES: AN EXAMPLE OF A DIVERGENT SERIES

Consider the series

$$\sum_{k=1}^{\infty} \frac{1}{k} = 1 + \frac{1}{2} + \frac{1}{3} + \frac{1}{4} + \cdots + \frac{1}{n} + \cdots. \tag{3}$$

This series is called the **harmonic series**. Although $a_n = 1/n \to 0$ as $n \to \infty$, it is not difficult to show that the harmonic series diverges. To see this, we write

$$\sum_{k=1}^{\infty} \frac{1}{k} = 1 + \frac{1}{2} + \underbrace{\left(\frac{1}{3} + \frac{1}{4}\right)}_{> \frac{1}{2}} + \underbrace{\left(\frac{1}{5} + \frac{1}{6} + \frac{1}{7} + \frac{1}{8}\right)}_{> \frac{1}{2}} + \underbrace{\left(\frac{1}{9} + \cdots + \frac{1}{16}\right)}_{> \frac{1}{2}} + \cdots.$$

(with labels "2 terms", "4 terms", "8 terms" above the respective groups)

Here we have written the terms in groups containing 2^n numbers. Note that $\frac{1}{3} + \frac{1}{4} > \frac{2}{4} = \frac{1}{2}$, $\frac{1}{5} + \frac{1}{6} + \frac{1}{7} + \frac{1}{8} > \frac{1}{8} + \frac{1}{8} + \frac{1}{8} + \frac{1}{8} = \frac{1}{2}$, and so on. Thus $\sum_{k=1}^{\infty} 1/k > 1 + \frac{1}{2} + \frac{1}{2} + \cdots$, and the series diverges.

⊗ **WARNING:** Example 3 clearly shows that even though the sequence $\{a_n\}$ converges to 0, the series $\sum a_n$ may, in fact, diverge. That is, if $a_n \to 0$, then $\sum_{k=1}^{\infty} a_k$ may or may not converge. Some additional test is needed to determine convergence or divergence. ⊗

It is often difficult to determine whether a series converges or diverges. For that reason, a number of techniques have been developed to make it easier to do so. We will present some easy facts here, and then we will develop additional techniques in the three sections that follow.

THEOREM 1

Let c be a constant. Suppose that $\sum_{k=1}^{\infty} a_k$ and $\sum_{k=1}^{\infty} b_k$ both converge. Then $\sum_{k=1}^{\infty} (a_k + b_k)$ and $\sum_{k=1}^{\infty} ca_k$ converge, and

(i) $\displaystyle\sum_{k=1}^{\infty} (a_k + b_k) = \sum_{k=1}^{\infty} a_k + \sum_{k=1}^{\infty} b_k,$ (4)

(ii) $\displaystyle\sum_{k=1}^{\infty} ca_k = c \sum_{k=1}^{\infty} a_k.$ ■ (5)

This theorem should not be surprising. Since the sum in a series is the limit of a sequence (the sequence of partial sums), the first part, for example, simply restates the fact that the limit of the sum is the sum of the limits. This is Theorem 10.1.2 (ii).

PROOF OF THEOREM 1:

(i) Let $S = \sum_{k=1}^{\infty} a_k$ and $T = \sum_{k=1}^{\infty} b_k$. The partial sums are given by $S_n = \sum_{k=1}^{n} a_k$ and $T_n = \sum_{k=1}^{n} b_k$. Then,

$$\sum_{k=1}^{\infty} (a_k + b_k) = \lim_{n \to \infty} \sum_{k=1}^{n} (a_k + b_k) = \lim_{n \to \infty} \left(\sum_{k=1}^{n} a_k + \sum_{k=1}^{n} b_k \right)$$

$$= \lim_{n \to \infty} (S_n + T_n) = \lim_{n \to \infty} S_n + \lim_{n \to \infty} T_n$$

$$= S + T = \sum_{k=1}^{\infty} a_k + \sum_{k=1}^{\infty} b_k.$$

(ii) $\displaystyle\sum_{k=1}^{\infty} ca_k = \lim_{n \to \infty} \sum_{k=1}^{n} ca_k = \lim_{n \to \infty} c \sum_{k=1}^{n} a_k = \lim_{n \to \infty} cS_n$

$$= c \lim_{n \to \infty} S_n = cS = c \sum_{k=1}^{\infty} a_k.$$ ■

EXAMPLE 4
A CONSTANT MULTIPLE OF THE
HARMONIC SERIES DIVERGES

Does $\sum_{k=1}^{\infty} 1/50k$ converge or diverge?

SOLUTION: We show that the series diverges by assuming that it converges to obtain a contradiction. If $\sum_{k=1}^{\infty} 1/50k$ did converge, then $50 \sum_{k=1}^{\infty} 1/50k$ would also converge by Theorem 1. But then $50 \sum_{k=1}^{\infty} 1/50k = \sum_{k=1}^{\infty} 50 \cdot 1/50k = \sum_{k=1}^{\infty} 1/k$, and this series is the harmonic series, which we know diverges. Hence $\sum_{k=1}^{\infty} 1/50k$ diverges.

Another useful test is given by the following theorem and corollary.

THEOREM 2

If $\sum_{k=1}^{\infty} a_k$ converges, then $\lim_{n \to \infty} a_n = 0$.

PROOF: Let $S = \sum_{k=1}^{\infty} a_k$. Then the partial sums S_n and S_{n-1} are given by

$$S_n = \sum_{k=1}^{n} a_k = a_1 + a_2 + \cdots + a_{n-1} + a_n$$

and

$$S_{n-1} = \sum_{k=1}^{n-1} a_k = a_1 + a_2 + \cdots + a_{n-1},$$

so that

$$S_n - S_{n-1} = a_n.$$

Then

$$\lim_{n \to \infty} a_n = \lim_{n \to \infty}(S_n - S_{n-1}) = \lim_{n \to \infty} S_n - \lim_{n \to \infty} S_{n-1}$$
$$= S - S = 0. \quad \blacksquare$$

We have already seen that the converse of this theorem is false. The convergence of $\{a_n\}$ to 0 does *not* imply that $\sum_{k=1}^{\infty} a_k$ converges. For example, the harmonic series does not converge, but the sequence $\{1/n\}$ does converge to zero.

COROLLARY **A TEST FOR DIVERGENCE**

If $\{a_n\}$ does not converge to 0, then $\sum_{k=1}^{\infty} a_k$ diverges. $\blacksquare$

EXAMPLE 5
A SERIES THAT DIVERGES BECAUSE THE PARTIAL SUMS OSCILLATE

$\sum_{k=1}^{\infty}(-1)^k$ diverges since the sequence $\{(-1)^k\}$ does not converge to zero. To see this more clearly, look at the partial sums:

$S_1 = (-1)^1 = -1.$

$S_2 = (-1)^1 + (-1)^2 = -1 + 1 = 0.$

$S_3 = (-1)^1 + (-1)^2 + (-1)^3 = -1 + 1 - 1 = -1.$

$S_4 = (-1)^1 + (-1)^2 + (-1)^3 + (-1)^4 = -1 + 1 - 1 + 1 = 0.$

We see that

$$S_n = \begin{cases} -1 & \text{if } n \text{ is odd,} \\ 0 & \text{if } n \text{ is even.} \end{cases}$$

Thus the partial sums oscillate between -1 and 0 and do not converge.

EXAMPLE 6
A DIVERGENT SERIES

$\sum_{k=1}^{\infty} k/(k+100)$ diverges since $\lim_{n \to \infty} a_n = \lim_{n \to \infty} n/(n+100) = 1 \neq 0.$

HISTORICAL NOTE: THE STRUGGLE TO UNDERSTAND INFINITE SUMS

The basic idea in the study of infinite series is that an infinite number of numbers can have a finite sum. This concept may seem natural now, but it took mathematicians over two thousand years to come to grips with it. More generally, the notion of the "infinite" was poorly understood and this lack of understanding led to some very confusing results.

Some of the early work on series was motivated by unresolved questions that had been posed by Greek mathematicians. For example, the fifth-century B.C. philosopher and mathematician Zeno (ca. 495–435 B.C.) posed four problems which came to be known as *Zeno's paradoxes*. In the second of these, Zeno argued that the legendary Greek hero Achilles could never overtake a tortoise. Suppose that the tortoise starts 100 yards ahead and that Achilles can run 10 times as fast as the tortoise. Then when Achilles has run 100 yards, the tortoise has run 10 yards, and when Achilles has covered 10 yards, the tortoise is still a yard ahead, and so on. It seems that the tortoise will stay ahead!

We can view this seeming paradox in another way which is equally contradictory of common sense. Suppose that a woman is standing a certain distance, say 10 feet, from a door (see Figure 1). Using Zeno's reasoning, we may claim that it is impossible for the woman to walk to the door. In order to reach the door, she must walk half the distance (5 feet) to the door. She then reaches point 1 on Figure 1. From point 1, 5 feet from the door, she must again walk halfway ($2\frac{1}{2}$ feet) to the door, to point 2. Continuing in this manner, no matter how close she comes to the door, she must walk halfway to the door and halfway from there and halfway from there ... and so on. Thus, no matter how close the woman gets to the door, she still has half of some remaining distance to cover. It seems that the woman will never actually reach the door. Of course, this contradicts our common sense. But where is the flaw in Zeno's reasoning?

With our knowledge of geometric series, we can reason as follows: Suppose the woman walks at a speed of 5 feet/second (≈ 3.4 miles/hour). Then

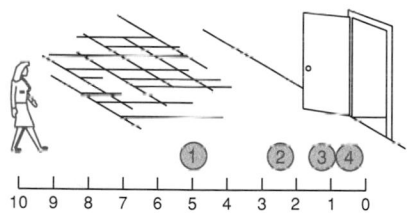

FIGURE 1

A WOMAN WALKING TOWARD A DOOR MUST COVER AN INFINITE NUMBER OF DISTANCES, EACH HALF AS LONG AS THE PRECEDING ONE.

$$t_1 = \text{time required to walk first 5 feet} = \frac{\text{distance}}{\text{velocity}} = \frac{5}{5} = 1 \text{ second}$$

$$t_2 = \text{time required to walk next } \frac{5}{2} \text{ feet} = \frac{\frac{5}{2}}{5} = \frac{1}{2} \text{ second}$$

$$t_3 = \text{time required to walk next } \frac{5}{4} \text{ feet} = \frac{\frac{5}{4}}{5} = \frac{1}{4} \text{ second}$$

$$t_4 = \text{time required to walk next } \frac{5}{8} \text{ feet} = \frac{\frac{5}{8}}{5} = \frac{1}{8} \text{ second}$$

and so on. We see that

total time to walk 10 feet $= 1 + \frac{1}{2} + \frac{1}{4} + \frac{1}{8} + \cdots = 2$ seconds.

This is not surprising as it tells us that it takes 2 seconds to walk 10 feet if she walks at a rate of 5 feet/second. This result seems hardly worth the bother, yet the paradox troubled mathematicians and philosophers for over 2000 years.

The great Archimedes used the idea of an infinite series (sort of) in order to compute area. In his *Quadrature of the Parabola*, Archimedes used the following procedure to obtain the area of a parabolic segment: He began by drawing the triangle T with the same base as the parabolic segment and its third vertex at the vertex of the parabola (see Figure 2). He chose points C and D and inscribed two triangles (denoted by T_1 and T_2 in the figure). He then inscribed four more triangles, using the sides of T_1 and T_2 as bases. One such triangle, T_3, is drawn in Figure 2. Continuing in this manner, Archimedes obtained a sequence of inscribed polygons which gave better and better approximations to the area of the parabolic segment. Using geometric arguments, Archimedes showed that

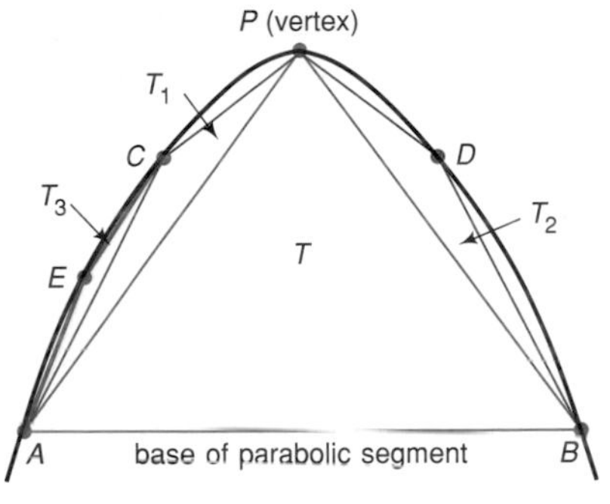

Area of nth inscribed polygon

$$= (\text{area of } T) \times \left(1 + \frac{1}{4} + \frac{1}{4^2} + \cdots + \frac{1}{4^{n-1}} \right).$$

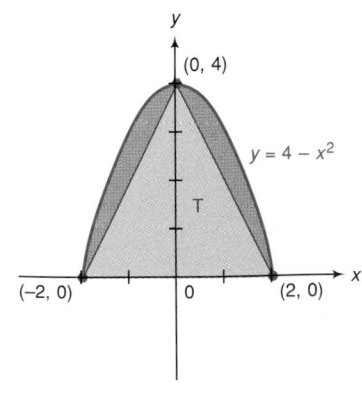

FIGURE 2
THE AREA OF THE PARABOLIC SEGMENT IS APPROXIMATED BY THE SUM OF THE AREAS OF INSCRIBED TRIANGLES.

Since $1 + \frac{1}{4} + \frac{1}{4^2} + \cdots = \frac{1}{1 - \frac{1}{4}} = \frac{4}{3}$, it follows that

$$\text{area of parabolic segment} = \frac{4}{3} \times \text{area of triangle } T.$$

However, like most Greek mathematicians before him, Archimedes avoided all reference to the infinite. Rather, he argued by contradiction that the area of the segment could be neither less than nor greater than $\frac{4}{3}$ times the area of T. It would have been simpler if he had been able to sum a geometric series.

To see that Archimedes' method works, consider the area, A, above the x-axis and under the curve $y = 4 - x^2$ (see Figure 3). T is the triangle with vertices at $(-2, 0)$, $(2, 0)$, and $(0, 4)$. We have

$$\text{area of } T = \frac{1}{2}(\text{base} \times \text{height}) = \frac{1}{2}(4 \times 4) = 8.$$

Then

$$A = \frac{4}{3} \cdot 8 = \frac{32}{3}.$$

We can verify this using an integral:

$$A = \int_{-2}^{2} (4 - x^2)\, dx = \left(4x - \frac{x^3}{3} \right) \Bigg|_{-2}^{2} = \frac{32}{3}.$$

FIGURE 3
THE AREA UNDER THE CURVE $y = 4 - x^2$ EQUALS $\frac{4}{3}$ × AREA OF T.

In thinking about infinite sums, Aristotle distinguished between two infinities: the *possible* and the *actual*. Put another way, he claimed that there was a difference between thinking about an infinite sum and actually having one. The concept of an infinite sum was so difficult that he categorically denied that such a sum could exist.

This state of ignorance continued well into the middle ages. In the thirteenth century, Petrus Hisparus, who became Pope John XXI, maintained the Aristotelian distinction between the two kinds of infinity by describing what he called the **categorematic infinity** (the realizable infinity) and the **syncategorematic infinity** (the merely conceivable infinity).

One of the most important and interesting philosophers of the fourteenth century was William of Occam (or Ockham) (*c.* 1285–*c.* 1349). Occam studied and taught at Oxford University from 1310 until 1324. In 1324 he was summoned to the papal court in Avignon, France, to answer charges of heresy. His principal crime was to deny the existence of universal truths except in language. This directly contradicted the teachings of the church. Occam is best remembered for the principle known as "Occam's razor," which can be stated as "what can be done with fewer [assumptions] is done in vain with more." A modern statement of Occam's razor, highly simplified, is "less is more." Occam's principle of economy led him into agreement with Aristotle that the categorematic infinity could never be realized. And, no, Occam was not executed for his heresy. He stayed in Avignon until 1328 when, it seemed, Pope John XXII was about to condemn his writings. At the last moment he fled to the protection of the Holy Roman Emperor Louis IV. It is believed that Occam died in the Black Death that swept through Europe in the 1340s.

During the Middle Ages, the question of the convergence of an infinite sum was largely philosophical. This changed in the seventeenth century. Galileo attempted to resolve Zeno's paradox by vague arguments that the "end of motion" (the time just before Achilles catches the tortoise) is a time of infinite speed just as the time just before an object that is thrown straight up begins its descent is a time of infinite slowness. Newton made similar arguments and is the first to have realized, but only in a vague sense, that some sort of limiting argument was necessary. In 1669 Newton wrote a treatise *De Analysi per Aequationes Numero Terminorum Infinitas* (On analysis by means of equations with an infinite number of terms) that was published in 1711. In attempting to compute a certain area, Newton was led to the integral of $\frac{1}{1 + x^2}$. He divided to obtain

$$\frac{1}{1 + x^2} = 1 - x^2 + x^4 - x^6 + \cdots \qquad \text{This is a geometric series.}$$

so

$$\int \frac{dx}{1 + x^2} = x - \frac{x^3}{3} + \frac{x^5}{5} - \frac{x^7}{7} + \cdots.$$

He then argued that this formula could be used when x was small enough. On the other hand, if x is large, he argued, then the following expansion works:

$$\frac{1}{1 + x^2} = \frac{1}{x^2 + 1} = \frac{1}{x^2\left(1 + \frac{1}{x^2}\right)} = \frac{1}{x^2}\left(1 - \frac{1}{x^2} + \frac{1}{x^4} - \frac{1}{x^6} + \cdots\right)$$

$$= \frac{1}{x^2} - \frac{1}{x^4} + \frac{1}{x^6} - \frac{1}{x^8} + \cdots$$

and

$$\int \frac{dx}{1 + x^2} = -\frac{1}{x} + \frac{1}{3x^3} - \frac{1}{5x^5} + \frac{1}{7x^7} - \cdots.$$

We know that the first series converges if $|x| < 1$ and the second converges if $|x| > 1$. However Newton had no sense of convergence. In his treatise, he wrote

> And whatever the common Analysis performs by Means of Equations of a finite Number of Terms (provided that can be done) this can always perform the same by Means of infinite Equations so that I have not made any question of giving this the name of Analysis likewise. For the reasonings in this are no less certain than in the other; nor the equations less exact; albeit we Mortals whose reasoning powers are confined within narrow limits, can neither express, nor so conceive all the Terms of these Equations, as to know exactly from thence the quantities we want.

In this rather vague statement Newton seems to suggest that term-by-term integration of finite sums can be extended to infinite sums without any loss of precision but that human beings can only conceive of the finite. This seems a reversal of Aristotle's claim that the infinite could be conceived, but not realized.

Throughout the seventeenth and eighteenth centuries, mathematicians worked with infinite series but had no clear sense of convergence and divergence. This sometimes led to some absurd problems. For example, seventeenth-century mathematicians were perplexed by the series

$$1 - 1 + 1 - 1 + 1 - \cdots.$$

In Example 5, we saw that this series diverges because the partial sums oscillate. But without the concept of divergence, one is led to problems by the simple process of placing parentheses:

$$1 - 1 + 1 - 1 + 1 - 1 + \cdots = (1 - 1) + (1 - 1) + (1 - 1) + \cdots$$
$$= 0 + 0 + 0 + \cdots = 0,$$

but

$$1 - 1 + 1 - 1 + 1 - 1 + \cdots = 1 - (1 - 1) - (1 - 1) - \cdots$$
$$= 1 - 0 - 0 - 0 - \cdots = 1.$$

Thus $0 = 1$. This paradoxical result confused mathematicians for about two centuries.

Euler (see p. 34) found the following absurd series:

$$0 = \frac{x}{1 - x} - \frac{x}{1 - x} = \frac{x}{1 - x} + \frac{x}{x - 1} = x\left(\frac{1}{1 - x}\right) + \frac{1}{1 - \dfrac{1}{x}}$$

$$= x(1 + x + x^2 + x^3 + \cdots) + 1 + \frac{1}{x} + \frac{1}{x^2} + \cdots \tag{6}$$

which results in

$$\sum_{k = -\infty}^{\infty} x^k = \cdots + \frac{1}{x^3} + \frac{1}{x^2} + \frac{1}{x} + 1 + x + x^2 + x^3 + \cdots = 0.$$

This is false for every real number x (look at $x = 1$, for instance). The problem is that the geometric series $1 + x + x^2 + \cdots$ converges only when $|x| < 1$, while the geometric series $1 + \dfrac{1}{x} + \dfrac{1}{x^2} + \cdots$ converges only when $\left|\dfrac{1}{x}\right| < 1$ or

$|x| > 1$. Thus the series (6) converges only when $|x| < 1$ and $|x| > 1$, an impossibility. This is clear to us but it was a mystery to Euler, who had no notion of convergence.

The term "convergent series" was first used by the Scottish mathematician James Gregory (1638–1675) in 1668 and the term "divergent series" was coined by Nicolaus Bernoulli (1687–1759) in 1721, but these concepts were not really understood until much later. From the middle of the eighteenth century, progress was made on developing the idea of convergence. This culminated in a formal definition of convergence given by Cauchy (see p. 64) in 1821 in his monograph *Analyse Algébrique*. Cauchy was the first to give a rigorous definition of a limit, and it was this definition that he applied to the limit of partial sums. Before Cauchy, some mathematicians gave criteria for convergence of certain series. Notable among these was the great German mathematician Carl Friedrich Gauss (1775–1855) (see p. 1021), who discussed the hypergeometric series important in probability theory.

For more details about the history of criteria for determining convergence of series, see *History of Mathematics, 3rd Edition*, by Florian Cajori (New York: Chelsea, 1980), pages 373–375.

■ PROBLEMS 10.4

■ SELF-QUIZ

I. The difference between 1.0 and the repeating decimal 0.999 . . . is _____.
   - a. 0.0
   - b. 0.001
   - c. 0.0001
   - d. 0.00 . . . 01

II. $\sum_{k=1}^{\infty} \frac{1}{k}$ _____.

   - a. converges to 2.75
   - b. converges to some number between 5 and 12.3
   - c. diverges to 13.47
   - d. diverges

III. $\dfrac{2}{3} + \dfrac{2}{8} + \dfrac{2}{15} + \dfrac{2}{24} + \dfrac{2}{35} + \cdots$

$$= \frac{2}{1 \cdot 3} + \frac{2}{2 \cdot 4} + \frac{2}{3 \cdot 5} + \frac{2}{4 \cdot 6} + \frac{2}{5 \cdot 7} + \cdots$$

$$= \sum_{k=2}^{\infty} \frac{2}{(k - 1) \cdot (k + 1)}$$

$$= \sum_{k=2}^{\infty} \left(\frac{1}{k - 1} - \frac{1}{k + 1} \right) \text{_____}.$$

   - a. $= 1 + \frac{1}{2} = \frac{3}{2}$
   - b. $= 1 - \frac{1}{2} = \frac{1}{2}$
   - c. $= 1 - \frac{1}{3} = \frac{2}{3}$
   - d. diverges

■ DRILL

In Problems 1–10, write the repeating decimal as a rational number.

1. 0.333 . . .
2. 0.6666 . . .
3. 0.353535 . . .
4. 0.282828 . . .
5. 0.717171 . . .
6. 0.214214214 . . .
7. 0.501501501 . . .
8. 0.124242424 . . .
9. 0.11362362362 . . .
10. 0.0513651365136 . . .

In Problems 11–26, a convergent infinite series is given; find its sum.

11. $\sum_{k=0}^{\infty} \frac{9}{10^k}$

12. $\sum_{k=0}^{\infty} \frac{1}{4^k}$

13. $\sum_{k=2}^{\infty} \frac{1}{2^k}$

14. $\sum_{k=1}^{\infty} \frac{3}{2^{k-1}}$

15. $\sum_{k=0}^{\infty} \left(-\frac{2}{3} \right)^k$

16. $\sum_{k=3}^{\infty} \left(\frac{2}{3} \right)^k$

17. $\sum_{k=0}^{\infty} \frac{100}{5^k}$

18. $\sum_{k=0}^{\infty} \frac{5}{100^k}$

19. $\sum_{k=2}^{\infty} \frac{1}{k(k + 1)}$

20. $\sum_{k=3}^{\infty} \frac{1}{k(k - 1)}$

21. $\sum_{k=0}^{\infty} \frac{1}{(k + 1)(k + 2)}$

22. $\displaystyle\sum_{k=-1}^{\infty} \frac{1}{(k+3)(k+4)}$

25. $\displaystyle\sum_{k=4}^{\infty} \frac{5^{k-2}}{6^{k+1}}$

23. $\displaystyle\sum_{k=2}^{\infty} \frac{2^{k+3}}{3^k}$

24. $\displaystyle\sum_{k=2}^{\infty} \frac{2^{k+4}}{3^{k-1}}$

26. $\displaystyle\sum_{k=-3}^{\infty} \frac{\sqrt{2}^k}{2^{k+3}}$

■ APPLICATIONS

In Problems 27–32, use Theorem 1 to calculate the sum of the given convergent series.

27. $\displaystyle\sum_{k=0}^{\infty} \left[\left(\frac{1}{3}\right)^k + \left(\frac{2}{3}\right)^k \right]$

28. $\displaystyle\sum_{k=0}^{\infty} \left[\frac{1}{2^k} + \frac{1}{5^k} \right]$

29. $\displaystyle\sum_{k=0}^{\infty} \left[\frac{3}{5^k} - \frac{7}{4^k} \right]$

30. $\displaystyle\sum_{k=3}^{\infty} \left[\frac{12 \cdot 2^{k+1}}{3^{k-2}} - \frac{15 \cdot 3^{k+1}}{4^{k+2}} \right]$

31. $\displaystyle\sum_{k=1}^{\infty} \left[\frac{8}{5^k} - \frac{7}{(k+3)(k+4)} \right]$

32. $\displaystyle\sum_{k=1}^{\infty} \left[\frac{1}{k(k+1)} + \frac{1}{(k+1)(k+2)} \right]$

*33. At what time between 1 P.M. and 2 P.M. is the minute hand of a clock exactly over the hour hand? [*Hint:* The minute hand moves 12 times as fast as the hour hand. Start at 1:00 P.M. When the minute hand has reached 1, the hour hand points to $1 + \frac{1}{12}$; when the minute hand has reached $1 + \frac{1}{12}$, the hour hand has reached $1 + \frac{1}{12} + \frac{1}{12} \cdot \frac{1}{12}$; etc. Now add up the geometric series.]

*34. At what time between 7 A.M. and 8 A.M. is the minute hand exactly over the hour hand?

35. A ball is dropped from a height of 8 m. Each time it hits the ground, it rebounds to a height of two-thirds the height from which it fell. Find the total distance traveled by the ball until it comes to rest (i.e., until it stops bouncing).

36. All banks in the state of Mondaho are required to maintain cash reserves of 20% for their accounts—only 80% of each deposit is available for loan to other customers. Suppose one bank receives an out-of-state deposit for $25,000 and immediately loans the maximum permissible amount to a customer; suppose that customer then deposits that amount in a second bank which, in its turn, then loans its maximum permissible amount to a third customer who deposits it in another bank.... What is the least upper bound on the increase in capitalization caused by this process triggered by the initial $25,000?

37. Use the result of Problem 42 to show that $\sum_{k=1}^{\infty} 1/(k+6)$ diverges.

38. Use the result of Problem 43 to show that $\sum_{k=1}^{\infty} \left[(3/2^k) + 2 \cdot 5^k \right]$ diverges.

■ SHOW/PROVE/DISPROVE

39. Use the geometric series to show that

$$\frac{1}{1+x} = \sum_{k=0}^{\infty} (-1)^k x^k$$

for any real number x with $|x| < 1$.

40. Prove or disprove: If $|x| < 1$, then

$$\frac{1}{1+x^2} = \sum_{k=0}^{\infty} (-1)^k x^{2k}.$$

41. Prove or disprove: For any nonzero real numbers a and b, $\sum_{k=1}^{\infty} a/(bk)$ diverges.

42. Show that if the sequences $\{a_k\}$ and $\{b_k\}$ differ only for a finite number of terms, then either both $\sum_{k=1}^{\infty} a_k$ and $\sum_{k=1}^{\infty} b_k$ converge or both diverge.

*43. Show that if $\sum_{k=1}^{\infty} a_k$ converges and $\sum_{k=1}^{\infty} b_k$ diverges, then $\sum_{k=1}^{\infty} (a_k + b_k)$ diverges. [*Hint:* Show that if $\sum_{k=1}^{\infty} (a_k + b_k)$ converges, then we get a contradiction of Theorem 1.]

44. Give an example in which $\sum_{k=1}^{\infty} a_k$ and $\sum_{k=1}^{\infty} b_k$ both diverge, but $\sum_{k=1}^{\infty} (a_k + b_k)$ converges.

CONTINUED

■ **CHALLENGE**

*45. Pick a_0 and a_1. For $n \geq 2$, compute a_n recursively so that

$$n(n - 1)a_n = (n - 1)(n - 2)a_{n-1} - (n - 3)a_{n-2}.$$

Evaluate $\sum_{n=0}^{\infty} a_n$.

■ **ANSWERS TO SELF-QUIZ**

I. a   II. d   III. a

10.5 SERIES WITH NONNEGATIVE TERMS I:
TWO COMPARISON TESTS AND THE INTEGRAL TEST

In this section and the next one, we consider series of the form $\sum_{k=1}^{\infty} a_k$, where each a_k is nonnegative. Such series are often easier to handle than others. One fact is easy to prove. The sequence $\{S_n\}$ of partial sums is a monotone increasing sequence since $S_{n+1} = S_n + a_{n+1}$ and $a_{n+1} \geq 0$ for every n. Then if $\{S_n\}$ is bounded, it is convergent by Theorem 10.2.2, and we have the following theorem:

THEOREM 1

An infinite series of nonnegative terms is convergent if and only if its sequence of partial sums is bounded. ■

EXAMPLE 1
A CONVERGENT SERIES

Show that $\sum_{k=1}^{\infty} 1/k^2$ is convergent.

SOLUTION: We group the terms as follows:

$$\sum_{k=1}^{\infty} \frac{1}{k^2} = \frac{1}{1^2} + \overbrace{\frac{1}{2^2} + \frac{1}{3^2}}^{2 \text{ terms}} + \overbrace{\frac{1}{4^2} + \frac{1}{5^2} + \frac{1}{6^2} + \frac{1}{7^2}}^{4 \text{ terms}} + \overbrace{\frac{1}{8^2} + \cdots + \frac{1}{15^2}}^{8 \text{ terms}} + \cdots$$

$$\leq 1 + \overbrace{\frac{1}{2^2} + \frac{1}{2^2}}^{2 \text{ terms}} + \overbrace{\frac{1}{4^2} + \frac{1}{4^2} + \frac{1}{4^2} + \frac{1}{4^2}}^{4 \text{ terms}} + \overbrace{\frac{1}{8^2} + \cdots + \frac{1}{8^2}}^{8 \text{ terms}} + \cdots$$

$$= 1 + \frac{2}{2^2} + \frac{4}{4^2} + \frac{8}{8^2} + \cdots = 1 + \frac{1}{2} + \frac{1}{4} + \frac{1}{8} + \cdots$$

$$= \sum_{k=0}^{\infty} \frac{1}{2^k} = 2.$$

Thus the sequence of partial sums is bounded by 2 and is therefore convergent. If we draw a picture, it is easy to see that the series is convergent. Figure 1 is self-explanatory.[†]

[†] Figure 1 was adapted from the interesting article "Convergence with pictures" by P. J. Rippon in *American Mathematical Monthly*, 93(1986): 476.

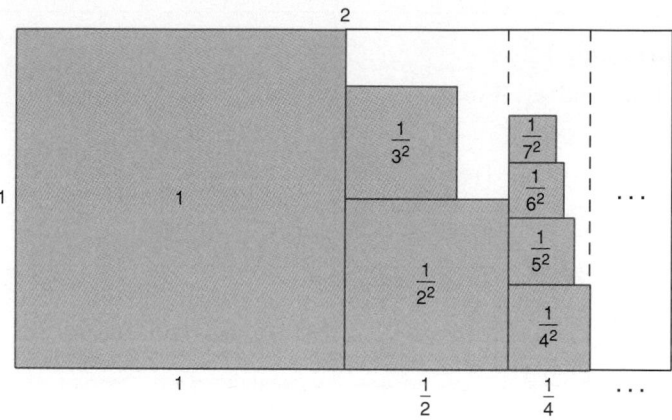

FIGURE 1   ILLUSTRATION OF FACT THAT

$$1 + \frac{1}{2^2} + \frac{1}{3^2} + \frac{1}{4^2} + \cdots < 1 + \frac{1}{2} + \frac{1}{4} + \cdots = 2$$

Since we know that the series converges, we can approximate its value by computing some partial sums on a computer.

$$S_{10} = \sum_{k=1}^{10} \frac{1}{k^2} = 1.5497677312$$

$$S_{50} = \sum_{k=1}^{50} \frac{1}{k^2} = 1.6251327336$$

$$S_{500} = \sum_{k=1}^{500} \frac{1}{k^2} = 1.6429360655$$

$$S_{10,000} = \sum_{k=1}^{10,000} \frac{1}{k^2} = 1.6448340718$$

It can be shown that $\sum_{k=1}^{\infty} \frac{1}{k^2} = \frac{\pi^2}{6} \approx 1.644934067.$[†]

⊗ **WARNING:**   You cannot use a computer to determine whether a series converges or diverges. You can, as in Example 1, get approximate values for the sum of a series once you *know* that it converges, but you cannot use a calculator to prove that it does converge. If you have trouble with this idea, try to use a calculator to determine convergence or divergence of the harmonic series (see Problem 43).   ⊗

With Theorem 1 the convergence or divergence of a series of nonnegative terms depends on whether or not its partial sums are bounded. There are several tests that can be used to determine whether or not the sequence of partial sums of a series is bounded. We will deal with these one at a time.

[†] See "A Simple Proof of the Formula $\sum_{k=1}^{\infty} k^{-2} = \pi^2/6$" in *American Mathematical Monthly*, 80(1973): 424.

THEOREM 2   COMPARISON TEST

Let $\sum_{k=1}^{\infty} a_k$ be a series with $a_k \geq 0$ for every k.

(i) If there exists a convergent series $\sum_{k=1}^{\infty} b_k$ and a number N such that $a_k \leq b_k$ for every $k \geq N$, then $\sum_{k=1}^{\infty} a_k$ converges.

(ii) If there exists a divergent series $\sum_{k=1}^{\infty} c_k$ and a number N such that $a_k \geq c_k \geq 0$ for every $k \geq N$, then $\sum_{k=1}^{\infty} a_k$ diverges.

PROOF: In either case the sum of the first N terms is finite, so we need only consider the series $\sum_{k=N+1}^{\infty} a_k$ since if this is convergent or divergent, then the addition of a finite number of terms does not affect the convergence or divergence.

(i) $\sum_{k=N+1}^{\infty} b_k$ is a nonnegative series (since $b_k \geq a_k \geq 0$ for $k > N$) and is convergent. Thus the partial sums $T_n = \sum_{k=N+1}^{n} b_k$ are bounded. If $S_n = \sum_{k=N+1}^{n} a_k$, then $S_n \leq T_n$, and so the partial sums of $\sum_{k=N+1}^{\infty} a_k$ are also bounded, implying that $\sum_{k=N+1}^{\infty} a_k$ is convergent.

(ii) Let $U_n = \sum_{k=N+1}^{n} c_k$. By Theorem 1 these partial sums are unbounded since $\sum_{k=N+1}^{\infty} c_k$ diverges. Since in this case $S_n \geq U_n$, the partial sums of $\sum_{k=N+1}^{\infty} a_k$ are also unbounded, and the series $\sum_{k=N+1}^{\infty} a_k$ diverges. ∎

REMARK: One fact mentioned in the proof of (i) is important enough to state again:

If for some positive integer N, $\sum_{k=N+1}^{\infty} a_k$ converges, then $\sum_{k=1}^{\infty} a_k$ also converges. If $\sum_{k=N+1}^{\infty} a_k$ diverges, then $\sum_{k=1}^{\infty} a_k$ diverges. That is, the addition of a finite number of terms does not affect convergence or divergence.

EXAMPLE 2
A SERIES THAT DIVERGES BECAUSE EACH OF ITS TERMS IS GREATER THAN OR EQUAL TO THE CORRESPONDING TERM OF THE HARMONIC SERIES

Determine whether $\sum_{k=1}^{\infty} 1/\sqrt{k}$ converges or diverges.

SOLUTION: Since $1/\sqrt{k} \geq 1/k$ for $k \geq 1$, and since $\sum_{k=1}^{\infty} 1/k$ diverges, we see that by the comparison test, $\sum_{k=1}^{\infty} 1/\sqrt{k}$ diverges.

EXAMPLE 3
A CONVERGENT SERIES

Determine whether $\sum_{k=1}^{\infty} 1/k!$ converges or diverges.

SOLUTION: If $k \geq 4$, $k! \geq 2^k$. To see this, note that $4! = 24$ and $2^4 = 16$. Then $5! = 5 \cdot 24$ and $2^5 = 2 \cdot 16$ and since $5 > 2$, $5! > 2^5$, and so on. Since $\sum_{k=1}^{\infty} 1/2^k$ converges, we see that $\sum_{k=1}^{\infty} 1/k!$ converges. In fact, as we will show in Section 10.9, it converges to $e - 1$. That is,

$$e = 1 + 1 + \frac{1}{2!} + \frac{1}{3!} + \frac{1}{4!} + \cdots. \tag{1}$$

This series converges very rapidly because $\dfrac{1}{n!}$ approaches 0 very rapidly.

For example,

$$10! = 3{,}628{,}800 \qquad \text{and} \qquad \frac{1}{10!} = 0.000000275573 \approx 2.76 \times 10^{-7};$$

$$20! \approx 2.43 \times 10^{18} \qquad \text{and} \qquad \frac{1}{20!} = 4.11 \times 10^{-19}.$$

We have (remembering that $0! = 1$)

$$\sum_{k=0}^{10} \frac{1}{k!} = 2.718281801 \qquad \text{and} \qquad \sum_{k=0}^{15} \frac{1}{k!} = 2.718281828.$$

THEOREM 3   **THE INTEGRAL TEST**

Let f be a function that is continuous, positive, and decreasing for all $x \geq 1$. Then the series

$$\sum_{k=1}^{\infty} f(k) = f(1) + f(2) + f(3) + \cdots + f(n) + \cdots \tag{2}$$

converges if and only if $\int_1^{\infty} f(x)\,dx$ converges, and diverges if and only if $\int_1^n f(x)\,dx \to \infty$ as $n \to \infty$.

PROOF: The idea behind this proof is fairly easy. Take a look at Figure 2. Comparing areas, we immediately see that

$$f(2) + f(3) + \cdots + f(n) \leq \int_1^n f(x)\,dx \leq f(1) + f(2) + \cdots + f(n-1).$$

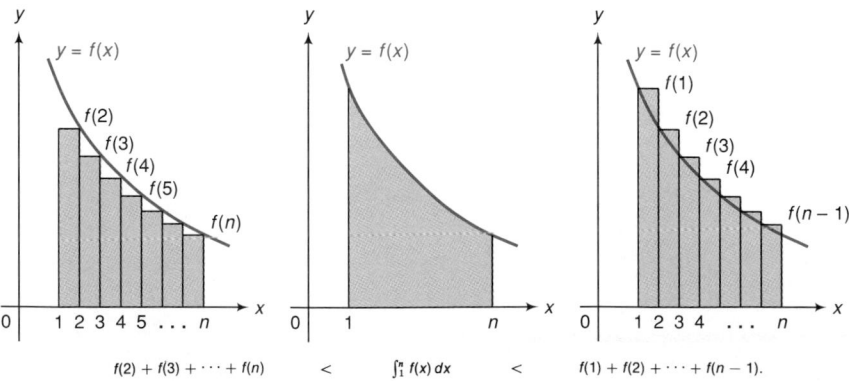

FIGURE 2   ILLUSTRATION OF THE PROOF OF THE INTEGRAL TEST

If $\lim_{n \to \infty} \int_1^n f(x)\,dx$ is finite, then the partial sums $[f(2) + f(3) + \cdots + f(n)]$ are bounded and the series converges. On the other hand, if $\lim_{n \to \infty} \int_1^n f(x)\,dx = \infty$, then the partial sums $[f(1) + f(2) + \cdots + f(n-1)]$ are unbounded and the series diverges.

Conversely, if the series diverges, then $f(2) + f(3) + \cdots + f(n) \to \infty$, so $\int_1^n f(x)\,dx \to \infty$. If it converges, then $f(1) + f(2) + \cdots + f(n-1)$ is bounded, so $\int_1^n f(x)\,dx$ converges. ∎

EXAMPLE 4
THE p SERIES

Consider the series $\sum_{k=1}^{\infty} 1/k^p$ with $p > 0$. We have already seen that this series diverges for $p = 1$ (the harmonic series) and converges for $p = 2$ (Example 1). Now let $f(x) = 1/x^p$. We saw in Example 7 on page 611 that $\int_1^{\infty} (1/x^p) \, dx$ diverges if $0 < p \leq 1$ and converges to $1/(p-1)$ if $p > 1$. Thus, by the integral test,

$$\sum_{k=1}^{\infty} \frac{1}{k^p} \quad \begin{cases} \text{diverges if } 0 < p \leq 1, \\ \text{converges if } p > 1. \end{cases}$$

The series $\sum_{k=1}^{\infty} 1/k^p$ is called the **p series**.

EXAMPLE 5
USING BOTH THE COMPARISON
AND INTEGRAL TESTS TO PROVE
CONVERGENCE

Determine whether $\sum_{k=1}^{\infty} (\ln k)/k^2$ converges or diverges.

SOLUTION: We see, using l'Hôpital's rule, that

$$\lim_{x \to \infty} \frac{\ln x}{\sqrt{x}} = \lim_{x \to \infty} \frac{1/x}{1/2\sqrt{x}} = 2 \lim_{x \to \infty} \frac{\sqrt{x}}{x} = 2 \lim_{x \to \infty} \frac{1}{\sqrt{x}} - 0,$$

so that for k sufficiently large, $\ln k \leq \sqrt{k}$. Thus,

$$\frac{\ln k}{k^2} \leq \frac{\sqrt{k}}{k^2} = \frac{1}{k^{3/2}}.$$

But $\sum_{k=1}^{\infty} 1/k^{3/2}$ converges by the result of Example 4, and therefore, by the comparison test, $\sum_{k=1}^{\infty} (\ln k)/k^2$ also converges.

NOTE: The integral test can also be used directly here since $\int_1^{\infty} (\ln x/x^2) \, dx$ can be integrated by parts with $u = \ln x$.

We now give another test that is an extension of the comparison test.

THEOREM 4   **THE LIMIT COMPARISON TEST**

Let $\sum_{k=1}^{\infty} a_k$ and $\sum_{k=1}^{\infty} b_k$ be series with positive terms. If there is a number $c > 0$ such that

$$\lim_{k \to \infty} \frac{a_k}{b_k} = c, \tag{3}$$

then either both series converge or both series diverge.

PROOF: We have $\lim_{k \to \infty} (a_k/b_k) = c > 0$. In the definition of a limit on page 639, let $\epsilon = c/2$. Then there is a number $N > 0$ such that

$$\left| \frac{a_k}{b_k} - c \right| < \frac{c}{2} \qquad \text{if } k \geq N. \tag{4}$$

Equation (4) is equivalent to

$$-\frac{c}{2} < \frac{a_k}{b_k} - c < \frac{c}{2}, \qquad \text{or} \qquad \frac{c}{2} < \frac{a_k}{b_k} < \frac{3c}{2}. \tag{5}$$

From the last inequality in (5), we obtain

$$a_k < \frac{3c}{2} b_k.$$ (6)

If $\sum b_k$ is convergent, then so is $(3c/2)\sum b_k = \sum(3c/2)b_k$. Thus from (6) and the comparison test, $\sum a_k$ is convergent. From the next-to-the-last inequality in (5), we have

$$a_k > \frac{c}{2} b_k.$$ (7)

If $\sum b_k$ is divergent, then so is $(c/2)\sum b_k = \sum(c/2)b_k$. Then using (7) and the comparison test, we find that $\sum a_k$ is also divergent. Thus if $\sum b_k$ is convergent, then $\sum a_k$ is convergent; and if $\sum b_k$ is divergent, then $\sum a_k$ is divergent. This proves the theorem. ■

EXAMPLE 6
USING THE LIMIT COMPARISON
TEST TO PROVE CONVERGENCE

Show that $\sum_{k=1}^{\infty} 1/(ak^2 + bk + c)$ is convergent, where a, b, and c are positive real numbers.

SOLUTION: We know from Example 1 that $\sum_{k=1}^{\infty} 1/k^2$ is convergent. If

$$a_k = \frac{1}{k^2} \quad \text{and} \quad b_k = \frac{1}{ak^2 + bk + c},$$

then

$$\frac{a_k}{b_k} = \frac{1/k^2}{1/(ak^2 + bk + c)} = \frac{ak^2 + bk + c}{k^2} = a + \frac{b}{k} + \frac{c}{k^2},$$

so that $\lim_{k \to \infty}(a_k/b_k) = a > 0$. Thus, by the limit comparison test,

$$\sum_{k=1}^{\infty} \frac{1}{ak^2 + bk + c}$$

is convergent.

■ **PROBLEMS 10.5**

■ **SELF-QUIZ**

I. True–False:
   a. $\sum_{k=1}^{\infty} (5/k)$ diverges because we can compare each term with corresponding, but smaller, terms of the harmonic series $\sum_{k=1}^{\infty} (1/k)$ which is already known to diverge (Example 10.4.3).
   b. If $k \geq 1$, then $k^2 \geq k$ and $1/k \geq 1/k^2$. Therefore, by comparing terms of $\sum_{k=1}^{\infty} (1/k^2)$ with the divergent harmonic series $\sum_{k=1}^{\infty} (1/k)$, we infer that $\sum_{k=1}^{\infty} (1/k^2)$ also diverges.

II. True–False:
   a. If $2 \leq k$, then $0 < k^2 - 1 < k^2$ and $1/k^2 < 1/(k^2 - 1)$. Because $\sum_{k=2}^{\infty} (1/k^2)$ is known to converge (see Example 1 and the remark following Theorem 2), our term-by-term comparison allows us to conclude that $\sum_{k=2}^{\infty} (1/[k^2 - 1])$ must also converge.
   b. Since each term of $\sum_{k=1}^{\infty} (-3/k)$ is less than a corresponding term of the convergent

CONTINUED

series $\sum_{k=2}^{\infty} (1/k^2)$ $(-3/k < 0 < 1/k^2$ for all $k \geq 2$), Theorem 2 implies that $\sum_{k=1}^{\infty} (-3/k)$ converges.

III. True–False:

a. $\sum_{k=2}^{\infty} (1/[k^2 - 1])$ converges because $\int_2^{\infty} (1/[x^2 - 1])\,dx$ converges (to $\frac{1}{2}\ln 3$).

b. $\sum_{k=1}^{\infty} (1/[7k^2 + 3\sin k + 13])$ converges because

$$\lim_{k \to \infty} \frac{1/[7k^2 + 3\sin k + 13]}{1/k^2} = \frac{1}{7}$$

and $\sum_{k=1}^{\infty} (1/k^2)$ converges.

■ DRILL

In Problems 1–34, determine the convergence or divergence of the given series.

1. $\sum_{k=1}^{\infty} (-1)^k$

2. $\sum_{k=1}^{\infty} \frac{k}{k+100}$

3. $\sum_{k=4}^{\infty} \frac{1}{5k+50}$

4. $\sum_{k=1}^{\infty} \frac{k+1}{(k+3)(k+5)}$

5. $\sum_{k=1}^{\infty} \frac{1}{k^2+1}$

6. $\sum_{k=10}^{\infty} \frac{1}{k(k-3)}$

7. $\sum_{k=1}^{\infty} \frac{1}{\sqrt{k^2+2k}}$

8. $\sum_{k=1}^{\infty} \frac{1}{\sqrt{k^3+1}}$

9. $\sum_{k=0}^{\infty} ke^{-k}$

10. $\sum_{k=3}^{\infty} k^3 e^{-k^4}$

11. $\sum_{k=2}^{\infty} \frac{4}{k\ln k}$

12. $\sum_{k=1}^{\infty} \frac{1}{k\ln(k+5)}$

13. $\sum_{k=5}^{\infty} \frac{1}{k(\ln k)^3}$

14. $\sum_{k=4}^{\infty} \frac{1}{k^2\sqrt{\ln k}}$

15. $\sum_{k=1}^{\infty} \frac{1}{(3k-1)^{3/2}}$

16. $\sum_{k=1}^{\infty} \frac{1}{\sqrt{k^2+3}}$

17. $\sum_{k=2}^{\infty} \frac{1}{k\sqrt{\ln k}}$

18. $\sum_{k=1}^{\infty} \frac{1}{50+\sqrt{k}}$

19. $\sum_{k=3}^{\infty} \left(\frac{k}{k+1}\right)^k$

20. $\sum_{k=1}^{\infty} \left(\frac{k}{k+1}\right)^{1/k}$

21. $\sum_{k=4}^{\infty} \frac{1}{k\ln\ln k}$

22. $\sum_{k=1}^{\infty} \sin\frac{1}{k}$

23. $\sum_{k=1}^{\infty} \frac{1}{(k+2)\sqrt{\ln(k+1)}}$

24. $\sum_{k=1}^{\infty} \frac{e^{1/k}}{k^2}$

25. $\sum_{k=1}^{\infty} \tan^{-1} k$

26. $\sum_{k=1}^{\infty} \frac{\tan^{-1} k}{1+k^2}$

27. $\sum_{k=1}^{\infty} \frac{\ln k}{k^3}$

28. $\sum_{k=10}^{\infty} \frac{1}{k(\ln k)(\ln\ln k)}$

29. $\sum_{k=1}^{\infty} \operatorname{sech} k$

30. $\sum_{k=1}^{\infty} \frac{1}{\cosh^2 k}$

*31. $\sum_{k=2}^{\infty} \frac{1}{\sqrt{k}\ln^{10} k}$

32. $\sum_{k=1}^{\infty} \frac{\sqrt{k}}{3k^2+2k+20}$

33. $\sum_{k=1}^{\infty} \frac{(k+1)^{7/8}}{k^3+k^2+3}$

34. $\sum_{k=1}^{\infty} \frac{k^5+2k^4+3k+7}{k^6+3k^4+2k^2+1}$

■ SHOW/PROVE/DISPROVE

35. Suppose that $\sum_{k=0}^{\infty} a_k$ is a convergent series of positive terms; prove that $\sum_{k=0}^{\infty} a_k^2$ also converges.

36. Suppose that $\sum_{k=0}^{\infty} a_k$ and $\sum_{k=0}^{\infty} b_k$ are both convergent series of positive terms; prove that $\sum_{k=0}^{\infty} a_k \cdot b_k$ also converges. [Hint: $(a+b)^2 - a^2 - b^2 = 2ab$.]

*37. Let $p(x)$ be a polynomial of degree n with positive coefficients and let $q(x)$ be a polynomial of degree $\leq n+1$ with positive coefficients. Show that $\sum_{k=1}^{\infty} p(k)/q(k)$ diverges.

*38. With $p(x)$ as in the preceding problem and with $r(x)$ a polynomial of degree $\geq n+2$ having positive coefficients, show that $\sum_{k=1}^{\infty} p(k)/r(k)$ converges.

*39. Determine whether $\sum_{n=1}^{\infty} \left(\frac{1}{n}\right)^{1+(1/n)}$ converges or diverges.

40. Let $S_n = \ln(n!) = \ln 2 + \ln 3 + \cdots + \ln n$. By calculating $\int_1^n \ln x\,dx$ and comparing areas, as in the proof of the integral test, show that for $n \geq 2$

$$\ln[(n-1)!] < n\ln n - n + 1 < \ln[n!]$$

and thus

$$(n-1)! < n^n e^{-(n-1)} < n!.$$

CONTINUED

*41. Let $S_n = \sum_{k=1}^{n} 1/k$. Show that

$$\ln(n + 1) < S_n < \ln n + 1.$$

[*Hint:* Use the inequality $1/(k + 1) \leq 1/x \leq 1/k$ when $0 < k \leq x \leq k + 1$, integrate and add (as in the proof of the integral test).]

■ CHALLENGE

42. Let $S_n = \sum_{k=1}^{n} 1/k$.
   a. Show that the sequence $\{S_n - \ln(n + 1)\}$ is increasing.
   *b. Show that this sequence is bounded by 1. [*Hint:* $S_n - \ln(n + 1) =$ the sum of the areas of "triangular" shaped regions which touch the graph of $y = 1/x$. Move those "triangles" over so that they all fit into a single rectangle with area 1.]
   c. Show that $\lim_{n \to \infty}[S_n - \ln(n + 1)]$ exists. This limit is denoted by γ; it is called the **Euler–Mascheroni constant**:

$$\gamma = \lim_{n \to \infty}\left[1 + \frac{1}{2} + \frac{1}{3} + \cdots + \frac{1}{n} - \ln(n + 1)\right]$$

$$= \lim_{n \to \infty}\left[1 + \frac{1}{2} + \frac{1}{3} + \cdots + \frac{1}{n} - \ln n\right].$$

This number arises in physical applications. To seven decimal places, $\gamma = 0.5772157.$[†]

*43. Consider a computer programmed to add terms of the harmonic series. Suppose it can add one million terms per second. Also suppose it works with perfect accuracy (i.e., no round-off or truncation errors).
   a. How many hours will it need to work before the divergent sum exceeds 25?
   b. How many days will it need to work before the divergent sum exceeds 30?
   c. How many years will it need to work before the divergent sum exceeds 35?[‡]
   [*Hint:* Use the preceding problem and the value given there for the Euler–Mascheroni constant.]

■ ANSWERS TO SELF-QUIZ

I. (a) True   (b) False
II. (a) False (It does converge, but not for the reason given.)

(b) False
III. (a) True   (b) True

10.6 SERIES WITH NONNEGATIVE TERMS II: THE RATIO AND ROOT TESTS

In this section we discuss two more tests that can be used to determine whether an infinite series converges or diverges. The first of these, the ratio test, is useful in a wide variety of applications.

THEOREM 1   THE RATIO TEST

Let $\sum_{k=1}^{\infty} a_k$ be a series with $a_k > 0$ for every k, and suppose that

$$\lim_{a_n \to \infty} \frac{a_{n+1}}{a_n} = L. \tag{1}$$

   (i) If $L < 1$, $\sum_{k=1}^{\infty} a_k$ converges.
   (ii) If $L > 1$, $\sum_{k=1}^{\infty} a_k$ diverges.
   (iii) If $L = 1$, $\sum_{k=1}^{\infty} a_k$ may converge or diverge and the ratio test is inconclusive; some other test must be used.

[†] It is an interesting fact that as of July, 1992, no one has determined whether or not γ is rational.

[‡] There are many interesting ways to use computers to help us understand calculus better; doing such a summation is not one of them.

PROOF:

(i) Pick $\epsilon > 0$ such that $L + \epsilon < 1$. By the definition of the limit in (1), there is a number $N > 0$ such that if $n \geq N$, we have

$$\frac{a_{n+1}}{a_n} < L + \epsilon.$$

Then

$$a_{n+1} < a_n(L + \epsilon), \qquad a_{n+2} < a_{n+1}(L + \epsilon) < a_n(L + \epsilon)^2,$$

and

$$a_{n+k} < a_n(L + \epsilon)^k \tag{2}$$

for each $k \geq 1$ and each $n \geq N$. In particular, for $k \geq N$ we use (2) to obtain

$$a_k = a_{(k-N)+N} \leq a_N(L + \epsilon)^{k-N}.$$

Then

$$S_n = \sum_{k=N}^{n} a_k \leq \sum_{k=N}^{n} a_N(L + \epsilon)^{k-N} = \frac{a_N}{(L + \epsilon)^N} \sum_{k=N}^{n} (L + \epsilon)^k.$$

But since $L + \epsilon < 1$, $\sum_{k=0}^{\infty} (L + \epsilon)^k = 1/[1 - (L + \epsilon)]$ (since this last sum is the sum of a geometric series). Thus,

$$S_n \leq \frac{a_N}{(L + \epsilon)^N} \cdot \frac{1}{1 - (L + \epsilon)},$$

and so the partial sums of $\sum_{k=N}^{\infty} a_k$ are bounded, implying that $\sum_{k=N}^{\infty} a_k$ converges. Thus, $\sum_{k=1}^{\infty} a_k = \sum_{k=1}^{N-1} a_k + \sum_{k=N}^{\infty} a_k$ also converges.

(ii) If $1 < L < \infty$, pick ϵ such that $L - \epsilon > 1$. Then for $n \geq N$, the same proof as before (with the inequalities reversed) shows that

$$\frac{a_{n+1}}{a_n} > L - \epsilon > 1.$$

Then $\{a_n\}$ is an increasing sequence and so does not have the limit 0. Thus $\sum_{k=1}^{\infty} a_k$ diverges. The proof in the case $L = \infty$ is suggested in Problem 35.

(iii) To illustrate (iii), we show that $L = 1$ can occur for a converging or diverging series.

(a) The harmonic series $\sum_{k=1}^{\infty} 1/k$ diverges. But

$$\lim_{n \to \infty} \frac{a_{n+1}}{a_n} = \lim_{n \to \infty} \frac{1/(n+1)}{1/n} = \lim_{n \to \infty} \frac{n}{n+1} = 1.$$

(b) The series $\sum_{k=1}^{\infty} 1/k^2$ converges. Here

$$\lim_{n \to \infty} \frac{a_{n+1}}{a_n} = \lim_{n \to \infty} \frac{1/(n+1)^2}{1/n^2} = \lim_{n \to \infty} \left(\frac{n}{n+1}\right)^2 = 1. \quad \blacksquare$$

REMARK 1: The ratio test is very useful. But in those cases where $L = 1$, we must try another test to determine whether the series converges or diverges.

REMARK 2: If the terms of the series $\sum_{k=1}^{\infty} a_k$ are nonzero, then we can apply the ratio test to the series $\sum_{k=1}^{\infty} |a_k|$, which is a series of positive terms. We then have

$$\sum_{k=1}^{\infty} |a_k| \text{ converges if } \lim_{n \to \infty} \frac{|a_{n+1}|}{|a_n|} < 1$$

and

$$\sum_{k=1}^{\infty} |a_k| \text{ diverges if } \lim_{n \to \infty} \frac{|a_{n+1}|}{|a_n|} > 1.$$

COROLLARY

(i) If $0 < \dfrac{a_{n+1}}{a_n} \le L < 1$ for $n \ge 1$, then $\sum_{k=1}^{\infty} a_k$ converges.

(ii) If $\dfrac{a_{n+1}}{a_n} > L > 1$ for $n \ge 1$, then $\sum_{k=1}^{\infty} a_k$ diverges. ∎

The proof of this result is almost identical to the proof of Theorem 1. Note that, using the corollary, we can prove convergence or divergence in cases where $\lim_{n \to \infty} \dfrac{a_{n+1}}{a_n}$ does not exist.

EXAMPLE 1
USING THE RATIO TEST TO PROVE CONVERGENCE

We have used the comparison test to show that $\sum_{k=1}^{\infty} 1/k!$ converges. Using the ratio test, we find that

$$\lim_{n \to \infty} \frac{a_{n+1}}{a_n} = \lim_{n \to \infty} \frac{1/(n+1)!}{1/n!} = \lim_{n \to \infty} \frac{n!}{(n+1)!} = \lim_{n \to \infty} \frac{1}{n+1} = 0 < 1,$$

so that the series converges.

EXAMPLE 2
USING THE RATIO TEST TO PROVE CONVERGENCE

Determine whether the series $\sum_{k=0}^{\infty} (100)^k/k!$ converges or diverges.

SOLUTION: Here

$$\lim_{n \to \infty} \frac{a_{n+1}}{a_n} = \lim_{n \to \infty} \frac{(100)^{n+1}/(n+1)!}{(100)^n/n!} = \lim_{n \to \infty} \frac{100}{n+1} = 0,$$

so that the series converges.

EXAMPLE 3
USING THE RATIO TEST TO PROVE DIVERGENCE

Determine whether the series $\sum_{k=1}^{\infty} k^k/k!$ converges or diverges.

SOLUTION:

$$\lim_{n \to \infty} \frac{a_{n+1}}{a_n} = \lim_{n \to \infty} \frac{[(n+1)^{n+1}/(n+1)!]}{n^n/n!} = \lim_{n \to \infty} \frac{(n+1)^{n+1}}{(n+1)n^n}$$

$$= \lim_{n \to \infty} \left(\frac{n+1}{n}\right)^n = \lim_{n \to \infty} \left(1 + \frac{1}{n}\right)^n = e > 1,$$

so that the series diverges.

EXAMPLE 4
A DIVERGENT SERIES FOR WHICH
THE RATIO TEST DOES NOT WORK

Determine whether the series $\sum_{k=1}^{\infty} (k + 1)/[k(k + 2)]$ converges or diverges.

SOLUTION: Here,

$$\lim_{n \to \infty} \frac{a_{n+1}}{a_n} = \lim_{n \to \infty} \frac{(n + 2)/(n + 1)(n + 3)}{(n + 1)/n(n + 2)} = \lim_{n \to \infty} \frac{n(n + 2)^2}{(n + 1)^2(n + 3)}$$

$$= \lim_{n \to \infty} \frac{n^3 + 4n^2 + 4n}{n^3 + 5n^2 + 7n + 3} = 1.$$

Thus the ratio test fails. However, $\lim_{k \to \infty}[(k + 1)/k(k + 2)]/(1/k) = 1$, so that $\sum_{k=1}^{\infty} (k + 1)/[k(k + 2)]$ diverges by the limit comparison test.

THEOREM 2 | **THE ROOT TEST**

Let $\sum_{k=1}^{\infty} a_k$ be a series with $a_k > 0$ and suppose that $\lim_{n \to \infty}(a_n)^{1/n} = R$.

(i) If $R < 1$, $\sum_{k=1}^{\infty} a_k$ converges.
(ii) If $R > 1$, $\sum_{k=1}^{\infty} a_k$ diverges.
(iii) If $R = 1$, the series either converges or diverges, and no conclusions can be drawn from this test.

The proof of this theorem is similar to the proof of the ratio test and is left as an exercise (see Problems 31–33). ∎

EXAMPLE 5
USING THE ROOT TEST TO
PROVE CONVERGENCE

Determine whether $\sum_{k=2}^{\infty} 1/(\ln k)^k$ converges or diverges.

SOLUTION: Note first that we start at $k = 2$ since $1/(\ln 1)^1$ is not defined.

$$\lim_{n \to \infty} \left[\frac{1}{(\ln n)^n} \right]^{1/n} = \lim_{n \to \infty} \frac{1}{\ln n} = 0,$$

so that the series converges.

EXAMPLE 6
USING THE ROOT TEST TO
PROVE DIVERGENCE

Determine whether the series $\sum_{k=1}^{\infty} (k^k/3^{4k+5})$ converges or diverges.

SOLUTION: $\lim_{n \to \infty}(n^n/3^{4n+5})^{1/n} = \lim_{n \to \infty}(n/3^{4+5/n}) = \infty$, since $\lim_{n \to \infty} 3^{4+5/n} = 3^4 = 81$. Thus, the series diverges.

■ PROBLEMS 10.6

■ SELF-QUIZ

I. The ratio test is inconclusive for which of the following series?

a. $\sum_{k=1}^{\infty} \frac{1}{k^3}$

b. $\sum_{k=1}^{\infty} \left(\frac{5}{13}\right)^k$

c. $\sum_{k=1}^{\infty} \frac{k}{k + 7}$

d. $\sum_{k=1}^{\infty} \frac{5^k}{k!}$

e. $\sum_{k=1}^{\infty} \frac{1 \cdot 3 \cdot 5 \cdot \cdots \cdot (2k - 1)}{4 \cdot 7 \cdot 10 \cdot \cdots \cdot (3k + 1)}$

f. $\sum_{k=1}^{\infty} 2^{-n+(-1)^n}$

c. $\sum_{k=1}^{\infty} \frac{k-27}{k}$ d. $\sum_{k=1}^{\infty} \frac{3^k}{k!}$

II. The root test is inconclusive for which of the following series?

e. $\sum_{k=1}^{\infty} 2^{-n-(-1)^n}$

a. $\sum_{k=1}^{\infty} \frac{1}{k^4}$ b. $\sum_{k=1}^{\infty} \left(\frac{11}{13}\right)^k$

■ DRILL

In Problems 1–26, determine whether the given series converges or diverges.

1. $\sum_{k=1}^{\infty} \frac{2^k}{k^2}$

2. $\sum_{k=1}^{\infty} \frac{5^k}{k^5}$

3. $\sum_{k=1}^{\infty} \frac{r^k}{k^r}, 0 < r < 1$

4. $\sum_{k=1}^{\infty} \frac{r^k}{k^r}, r > 1$

5. $\sum_{k=2}^{\infty} \frac{k!}{k^k}$

6. $\sum_{k=1}^{\infty} \frac{k^k}{(2k)!}$

7. $\sum_{k=1}^{\infty} \frac{e^k}{k^5}$

8. $\sum_{k=1}^{\infty} \frac{e^k}{k!}$

9. $\sum_{k=1}^{\infty} \frac{k^{2/3}}{10^k}$

10. $\sum_{k=1}^{\infty} \frac{3^k + k}{k! + 2}$

11. $\sum_{k=2}^{\infty} \frac{k}{(\ln k)^k}$

12. $\sum_{k=1}^{\infty} \frac{4^k}{k^3}$

13. $\sum_{k=2}^{\infty} \left(1 + \frac{1}{k}\right)^k$

14. $\sum_{k=1}^{\infty} \frac{\sqrt{k}\ln k}{k^3 + 1}$

15. $\sum_{k=1}^{\infty} \frac{3^{4k+5}}{k^k}$

16. $\sum_{k=1}^{\infty} \frac{a^{mk+b}}{k^k}, a > 1$

17. $\sum_{k=1}^{\infty} \frac{k^k}{a^{mk+b}}, a > 1$

18. $\sum_{k=1}^{\infty} \frac{k^6 5^k}{(k+1)!}$

19. $\sum_{k=1}^{\infty} \frac{k^2 k!}{(2k)!}$

20. $\sum_{k=1}^{\infty} \frac{(2k)!}{k^2 k!}$

*21. $\sum_{k=1}^{\infty} \left(\frac{k!}{k^k}\right)^k$

*22. $\sum_{k=1}^{\infty} \left(\frac{k^k}{k!}\right)^k$

23. $\sum_{k=2}^{\infty} \frac{e^k}{(\ln k)^k}$

24. $\sum_{k=1}^{\infty} \frac{(\ln k)^k}{k^2}$

25. $\sum_{k=1}^{\infty} \left(\frac{k}{3k+2}\right)^k$

26. $\sum_{k=1}^{\infty} \left(\frac{1}{2} + \frac{1}{k}\right)^k$

■ APPLICATIONS

27. Let

$$a_k = \begin{cases} 3/k^2 & \text{if } k \text{ is even,} \\ 1/k^2 & \text{if } k \text{ is odd.} \end{cases}$$

Verify that $\lim_{n \to \infty}(a_{n+1}/a_n)$ does not exist, but that $\sum_{k=1}^{\infty} a_k$ converges.

28. Construct a series of positive terms for which $\lim_{n \to \infty}(a_{n+1}/a_n)$ does not exist but for which $\sum_{k=1}^{\infty} a_k$ diverges.

■ SHOW/PROVE/DISPROVE

29. Show that $\sum_{k=0}^{\infty} x^k/k!$ converges for every real number x.

30. a. Show that for any real number a,

$$\lim_{k \to \infty}(k^a e^{-k})^{1/k} = \frac{1}{e}.$$

[*Hint:* Use l'Hôpital's rule.]
   b. Show that $\sum_{k=1}^{\infty} k^a e^{-k}$ converges.
   c. Prove that $\lim_{k \to \infty} k^a e^{-k} = 0$ for any real number a.

*31. Prove part (i) of the root test (Theorem 2). [*Hint:* If $R < 1$, choose $\epsilon > 0$ so that $R + \epsilon < 1$. Show that there is an N such that if $n \geq N$, then $a_n < (R + \epsilon)^n$. Then complete the proof by comparing $\sum a_k$ with the sum of a geometric series.]

32. Prove part (ii) of the root test. [*Hint:* Parallel the steps suggested for the preceding problem.]

33. Show that if $\sqrt[n]{a_n} \to 1$, then $\sum_{k=1}^{\infty} a_k$ may converge or diverge. [*Hint:* Consider $\sum 1/k$ and $\sum 1/k^2$.]

*34. Prove that $k!/k^k \to 0$ as $k \to \infty$.

CONTINUED

35. Prove that if $a_n > 0$ and $\lim_{n \to \infty}(a_{n+1}/a_n) = \infty$, then $\sum_{k=1}^{\infty} a_k$ diverges. [*Hint:* Show that for sufficiently large N, $a_k \geq 2^{k-N} \cdot a_N$.]

36. Prove both parts of the corollary to Theorem 1.

■ *ANSWERS TO SELF-QUIZ*

I. a (converges), c (diverges), f (converges)
II. a (converges), c (diverges)

10.7 ABSOLUTE AND CONDITIONAL CONVERGENCE: ALTERNATING SERIES

In Sections 10.5 and 10.6 all the series we dealt with had positive terms. In this section we consider special types of series that have positive and negative terms.

DEFINITION   **ABSOLUTE CONVERGENCE**

The series $\sum_{k=1}^{\infty} a_k$ is said to **converge absolutely** if the series $\sum_{k=1}^{\infty} |a_k|$ converges. ■

Absolute convergence is important for the following reason: The series $\sum |a_k|$ is a series with nonnegative terms. Therefore, all the tests we used in the last two sections can be applied to determine whether or not the series $\sum |a_k|$ converges; that is, whether or not $\sum a_k$ converges absolutely. But, as Theorem 1 below tells us, if $\sum |a_k|$ converges, then $\sum a_k$ converges also. That is, we can use the ratio, root, comparison, and integral tests to prove that a series converges even when some of its terms are negative.

EXAMPLE 1
AN ABSOLUTELY CONVERGENT SERIES

The series

$$\sum_{k=1}^{\infty} \frac{(-1)^{k+1}}{k^2} = \frac{1}{1^2} - \frac{1}{2^2} + \frac{1}{3^2} - \frac{1}{4^2} + \cdots$$

converges absolutely because $\sum_{k=1}^{\infty} |(-1)^{k+1}/k^2| = \sum_{k=1}^{\infty} 1/k^2$ converges.

EXAMPLE 2
A SERIES THAT DOES NOT CONVERGE ABSOLUTELY

The series

$$\sum_{k=1}^{\infty} \frac{(-1)^{k+1}}{k} = \frac{1}{1} - \frac{1}{2} + \frac{1}{3} - \frac{1}{4} + \frac{1}{5} + \cdots$$

does not converge absolutely because $\sum_{k=1}^{\infty} 1/k$ diverges.

The importance of absolute convergence is given in the theorem below.

THEOREM 1

If $\sum_{k=1}^{\infty} |a_k|$ converges, then $\sum_{k=1}^{\infty} a_k$ also converges. That is,

absolute convergence implies convergence. ■

REMARK: The converse of this theorem is false. That is, there are series that are convergent but not absolutely convergent. We will see examples of this phenomenon shortly.

PROOF: Since $a_k \leq |a_k|$, we have $0 \leq a_k + |a_k| \leq 2|a_k|$.

Since $\sum_{k=1}^{\infty} |a_k|$ converges, we see that $\sum_{k=1}^{\infty} (a_k + |a_k|)$ converges by the comparison test. Then since $a_k = (a_k + |a_k|) - |a_k|$, $\sum_{k=1}^{\infty} a_k$ converges because it is the sum of two convergent series. ∎

EXAMPLE 3
ABSOLUTE CONVERGENCE
IMPLIES CONVERGENCE

The series $\sum_{k=1}^{\infty} (-1)^{k+1}/k^2$ considered in Example 1 converges since it converges absolutely.

DEFINITION **ALTERNATING SERIES**

A series in which successive terms have opposite signs is called an **alternating series**. ∎

EXAMPLE 4
AN ALTERNATING SERIES

The series

$$\sum_{k=1}^{\infty} \frac{(-1)^{k+1}}{k} = 1 - \frac{1}{2} + \frac{1}{3} - \frac{1}{4} + \frac{1}{5} - \frac{1}{6} + \cdots$$

is an alternating series.

EXAMPLE 5
A SERIES WITH BOTH POSITIVE
AND NEGATIVE TERMS THAT IS
NOT AN ALTERNATING SERIES

The series $1 + \frac{1}{2} - \frac{1}{3} - \frac{1}{4} + \frac{1}{5} + \frac{1}{6} - \cdots$ is not an alternating series because two successive terms have the same sign.

Let us consider the series of Example 4:

$$S = 1 - \frac{1}{2} + \frac{1}{3} - \frac{1}{4} + \frac{1}{5} - \frac{1}{6} + \cdots.$$

Calculating successive partial sums, we find that

$$S_1 = 1, \quad S_2 = \frac{1}{2}, \quad S_3 = \frac{5}{6}, \quad S_4 = \frac{7}{12}, \quad S_5 = \frac{47}{60}, \ldots.$$

It seems that this series is not diverging to infinity (indeed, $\frac{1}{2} \leq S_n \leq 1$) and that the partial sums are getting "narrowed down." At this point it is reasonable to suspect that the series converges. But it does *not* converge absolutely (since the series of absolute values is the harmonic series), and we cannot use any of the tests of the previous section since the terms are not nonnegative. The result we need is given in the theorem below.

THEOREM 3 **ALTERNATING SERIES TEST**

Let $\{a_k\}$ be a decreasing sequence of positive numbers such that $\lim_{k \to \infty} a_k = 0$. Then the alternating series $\sum_{k=1}^{\infty} (-1)^{k+1} a_k = a_1 - a_2 + a_3 - a_4 + \cdots$ converges. ∎

PROOF: Looking at the odd-numbered partial sums of this series, we find that

$$S_{2n+1} = (a_1 - a_2) + (a_3 - a_4) + (a_5 - a_6) + \cdots$$
$$+ (a_{2n-1} - a_{2n}) + a_{2n+1}.$$

Since $\{a_k\}$ is decreasing, all the terms in parentheses are nonnegative, so that $S_{2n+1} \geq 0$ for every n. Moreover,

$$S_{2n+3} = S_{2n+1} - a_{2n+2} + a_{2n+3} = S_{2n+1} - (a_{2n+2} - a_{2n+3}),$$

and since $a_{2n+2} - a_{2n+3} \geq 0$, we have

$$S_{2n+3} \leq S_{2n+1}.$$

Hence, the sequence of odd-numbered partial sums is bounded below by 0 and is decreasing and is therefore convergent by Theorem 2 in Section 10.2. Thus S_{2n+1} converges to some limit L. Now let us consider the sequence of even-numbered partial sums. We find that $S_{2n+2} = S_{2n+1} - a_{2n+2}$ and since $a_{2n+2} \to 0$,

$$\lim_{n \to \infty} S_{2n+2} = \lim_{n \to \infty} S_{2n+1} - \lim_{n \to \infty} a_{2n+2} = L - 0 = L,$$

so that the even partial sums also converge to L. Since both the odd and even sums converge to L, we see that the partial sums converge to L, and the proof is complete. ∎

EXAMPLE 6
FOUR CONVERGENT
ALTERNATING SERIES

The following alternating series are convergent by the alternating series test:

(a) $1 - \dfrac{1}{2} + \dfrac{1}{3} - \dfrac{1}{4} + \dfrac{1}{5} - \dfrac{1}{6} + \cdots$

(b) $1 - \dfrac{1}{\sqrt{2}} + \dfrac{1}{\sqrt{3}} - \dfrac{1}{\sqrt{4}} + \dfrac{1}{\sqrt{5}} - \dfrac{1}{\sqrt{6}} + \dfrac{1}{\sqrt{7}} - \cdots$

(c) $\dfrac{1}{\ln 2} - \dfrac{1}{\ln 3} + \dfrac{1}{\ln 4} - \dfrac{1}{\ln 5} + \dfrac{1}{\ln 6} - \cdots$

(d) $1 - \dfrac{1}{2} + \dfrac{1}{2^2} - \dfrac{1}{2^3} + \dfrac{1}{2^4} - \dfrac{1}{2^5} + \dfrac{1}{2^6} - \dfrac{1}{2^7} + \cdots$

DEFINITION **CONDITIONAL CONVERGENCE**

An alternating series is said to be **conditionally convergent** if it is convergent but not absolutely convergent. ∎

In Example 6 all the series are conditionally convergent except the last one, which is absolutely convergent.

It is not difficult to estimate the sum of a convergent alternating series. We again consider the series

$$S = 1 - \dfrac{1}{2} + \dfrac{1}{3} - \dfrac{1}{4} + \dfrac{1}{5} - \cdots.$$

Suppose we wish to approximate S by its nth partial sum S_n. Then,

$$S - S_n = \pm \left(\frac{1}{n+1} - \frac{1}{n+2} + \frac{1}{n+3} - \frac{1}{n+4} + \cdots \right) = R_n.$$

But we can estimate the remainder term R_n:

$$|R_n| = \left\| \left[\frac{1}{n+1} - \left(\frac{1}{n+2} - \frac{1}{n+3} \right) - \left(\frac{1}{n+4} - \frac{1}{n+5} \right) - \cdots \right] \right\| \le \frac{1}{n+1}.$$

That is, the error is less than the first term that we left out! For example, $|S - S_{20}| \le \frac{1}{21} \approx 0.0476$.

In general, we have the following result, whose proof is left as an exercise (see Problem 43).

THEOREM 3

If $S = \sum_{k=1}^{\infty} (-1)^{k+1} a_k$ is a convergent alternating series with monotone decreasing terms, then for any n,

$$|S - S_n| \le |a_{n+1}|. \quad \blacksquare \tag{1}$$

EXAMPLE 7
APPROXIMATING A CONVERGENT
ALTERNATING SERIES

The series

$$\sum_{k=1}^{\infty} \frac{(-1)^{k+1}}{\ln(k+1)} = \frac{1}{\ln 2} - \frac{1}{\ln 3} + \frac{1}{\ln 4} - \frac{1}{\ln 5} + \cdots$$

can be approximated by S_n with an error of less than $1/\ln(n+2)$. For example, with $n = 10$, $1/\ln(n+2) = 1/\ln 12 \approx 0.4$. Hence, the sum

$$\sum_{k=1}^{\infty} \frac{(-1)^{k+1}}{\ln(k+1)} = \frac{1}{\ln 2} - \frac{1}{\ln 3} + \cdots$$

can be approximated by

$$S_{10} = \frac{1}{\ln 2} - \frac{1}{\ln 3} + \frac{1}{\ln 4} - \frac{1}{\ln 5} + \frac{1}{\ln 6} - \frac{1}{\ln 7} + \frac{1}{\ln 8}$$
$$- \frac{1}{\ln 9} + \frac{1}{\ln 10} - \frac{1}{\ln 11}$$
$$\approx 0.7197,$$

with an error of less than 0.4.

By modifying Theorem 3, we can significantly improve on the last result.

THEOREM 4

Suppose that the hypotheses of Theorem 3 hold and that, in addition, the sequence $\{|a_n - a_{n+1}|\}$ is monotone decreasing. Let $T_n =$

$S_{n-1} - (-1)^n \frac{1}{2} a_n$. Then,

$$|S - T_n| \le \frac{1}{2}|a_n - a_{n+1}|. \quad \blacksquare \tag{2}$$

This result follows from Theorem 3 and is also left as an exercise (see Problem 44).

EXAMPLE 8

IMPROVING THE APPROXIMATION
OF A CONVERGENT ALTERNATING
SERIES

We can improve the estimate in Example 7. We may approximate $\sum_{k=1}^{\infty} (-1)^{k+1}/\ln(k+1)$ by

$$\frac{1}{\ln 2} - \frac{1}{\ln 3} + \frac{1}{\ln 4} - \frac{1}{\ln 5} + \frac{1}{\ln 6} - \frac{1}{\ln 7} + \frac{1}{\ln 8} - \frac{1}{\ln 9} + \frac{1}{\ln 10} - \frac{1}{2 \ln 11}$$

$$\approx 0.9282.$$

With $n = 10$ (so that $n + 1 = 11$),

$$T_{10} = S_9 - \frac{1}{2}\left(\frac{1}{\ln 11}\right),$$

which is precisely the sum given above. Thus,

$$|S - T_{10}| < \frac{1}{2}|a_{10} - a_{11}| = \frac{1}{2}\left(\frac{1}{\ln 11} - \frac{1}{\ln 12}\right) \approx 0.0073.$$

This result is a considerable improvement.

Note that in order to justify this result, we must verify that $|a_n - a_{n+1}|$ is monotone decreasing. This fact is left as an exercise (see Problem 50).

There is one fascinating fact about an alternating series that is conditionally convergent:

By reordering the terms of a conditionally convergent alternating series, the new series of rearranged terms can be made to converge to any real number.

Let us illustrate this fact with the series

$$S = 1 - \frac{1}{2} + \frac{1}{3} - \frac{1}{4} + \frac{1}{5} - \frac{1}{6} + \cdots.$$

The odd-numbered terms sum to a divergent series:

$$1 + \frac{1}{3} + \frac{1}{5} + \frac{1}{7} + \cdots. \tag{3}$$

The even-numbered terms are likewise a divergent series:

$$-\frac{1}{2} - \frac{1}{4} - \frac{1}{6} - \cdots. \tag{4}$$

If either of these series converged, then the other one would too (by Theorem 1(i) in Section 10.4) and then the entire series would be absolutely convergent (which we know to be false). Now choose any real number, say 1.5.

Then,

(i) Choose enough terms from the series (3) so that the sum exceeds 1.5. We can do so since the series diverges.

$$1 + \frac{1}{3} + \frac{1}{5} = 1.53333\ldots.$$

(ii) Add enough negative terms from (4) so that the sum is now just under 1.5.

$$1 + \frac{1}{3} + \frac{1}{5} - \frac{1}{2} = 1.0333\ldots.$$

(iii) Add more terms from (3) until 1.5 is exceeded.

$$1 + \frac{1}{3} + \frac{1}{5} - \frac{1}{2} + \frac{1}{7} + \frac{1}{9} + \frac{1}{11} + \frac{1}{13} + \frac{1}{15} \approx 1.5218.$$

(iv) Again subtract terms from (4) until the sum is under 1.5.

$$1 + \frac{1}{3} + \frac{1}{5} - \frac{1}{2} + \frac{1}{7} + \frac{1}{9} + \frac{1}{11} + \frac{1}{13} + \frac{1}{15} - \frac{1}{4} \approx 1.2718.$$

We continue the process to "converge" to 1.5. Since the terms in each series are decreasing to 0, the amount above or below 1.5 will approach 0 and the partial sums converge.

We will indicate in Section 10.10 that without rearranging, we have

$$\sum_{k=1}^{\infty} \frac{(-1)^{k+1}}{k} = 1 - \frac{1}{2} + \frac{1}{3} - \frac{1}{4} + \frac{1}{5} - \frac{1}{6} + \cdots = \ln 2 \approx 0.693147. \tag{5}$$

REMARK: *Any* rearrangement of the terms of an *absolutely converging* series converges to the same number.

EXAMPLE 9

AN ALTERNATING SERIES MAY DIVERGE IF $\{a_n\}$ IS NOT A MONOTONE DECREASING SEQUENCE

The following series was given by the great English mathematician G. H. Hardy (1877–1947)[†] to illustrate the necessity of the "decreasing" condition in Theorem 2:

$$\frac{1}{\sqrt{2}+1} - \frac{1}{\sqrt{3}-1} + \frac{1}{\sqrt{4}+1} - \frac{1}{\sqrt{5}-1} + \cdots = \sum_{k=2}^{\infty} \frac{(-1)^k}{\sqrt{k}+(-1)^k}. \tag{6}$$

For $k \geq 2$, $\sqrt{k} \pm 1 > 0$ and $\dfrac{1}{\sqrt{k} \pm 1} \to 0$ as $k \to \infty$ so the series (6) is an alternating series with terms approaching zero. However, the terms are not decreasing. For example, since $\sqrt{5} - 1 < \sqrt{4} + 1$, $a_5 = \dfrac{1}{\sqrt{5}-1} >$ $\dfrac{1}{\sqrt{4}+1} = a_4 \left(\dfrac{1}{\sqrt{5}-1} \approx 0.81 > \dfrac{1}{\sqrt{4}+1} \approx 0.33 \right)$. We can show, using a

---

[†] See G. H. Hardy, *Pure Mathematics* (Cambridge University Press, 1967), pages 377–378.

bit of algebra, that the series (6) diverges. We compute

$$\frac{(-1)^k}{\sqrt{k}} - \frac{1}{k + (-1)^k\sqrt{k}} = \frac{(-1)^k[k + (-1)^k\sqrt{k}] - \sqrt{k}}{\sqrt{k}[k + (-1)^k\sqrt{k}]}$$

$$= \frac{(-1)^k k + (-1)^{2k}\sqrt{k} - \sqrt{k}}{\sqrt{k}[k + (-1)^k\sqrt{k}]}$$

$$\underset{(-1)^{2k} = 1}{\overset{\downarrow}{=}} \frac{(-1)^k k}{\sqrt{k}[k + (-1)^k\sqrt{k}]} \overset{\underset{\text{divide numerator}}{\text{and denominator by } k}}{\underset{\downarrow}{=}} \frac{(-1)^k}{\sqrt{k} + (-1)^k}.$$

Thus

$$\sum_{k=2}^{\infty} \frac{(-1)^k}{\sqrt{k} + (-1)^k} = \sum_{k=2}^{\infty} \frac{(-1)^k}{\sqrt{k}} - \sum_{k=2}^{\infty} \frac{1}{k + (-1)^k\sqrt{k}}.$$

The first series on the right converges by the alternating series test. The second series diverges. To see this, note that $k - \sqrt{k} > 0$ if $k \geq 2$, so the terms are positive. Moreover, $k \pm \sqrt{k} < 2k$, so $\frac{1}{k + (-1)^k} > \frac{1}{2k}$ and $\sum_{k=2}^{\infty} \frac{1}{2k}$ diverges. This implies that $\sum_{k=2}^{\infty} \frac{1}{k + (-1)^k\sqrt{k}}$ diverges by the comparison test. Thus the series (6) diverges.

EXAMPLE 10
THE COMPARISON TEST
CANNOT BE USED FOR AN
ALTERNATING SERIES

We again consider the series (6):

$$\sum_{k=2}^{\infty} \frac{(-1)^k}{\sqrt{k} + (-1)^k}.$$

Since $\sqrt{k} + 1 > \sqrt{k} - 1$, we then have $\frac{1}{\sqrt{k} + 1} < \frac{1}{\sqrt{k} - 1}$. The series $\sum_{k=2}^{\infty} \frac{(-1)^k}{\sqrt{k} - 1}$ converges by the alternating series test. Then, because $\frac{(-1)^k}{\sqrt{k} + (-1)^k} \leq \frac{(-1)^k}{\sqrt{k} - 1}$, the series (6) converges by the comparison test. But, as we saw in Example 9, this series diverges. We conclude that

the comparison test cannot be used to determine convergence or divergence of an alternating series.

Of course, the comparison test *can* be used to determine absolute convergence of an alternating series, because $\sum |a_k|$ is a series of nonnegative terms.

NOTE: Examples 9 and 10 are slight modifications of examples found in the article "Counterexamples to a Comparison Test for Alternating Series" by J. Richard Morris in *The College Mathematics Journal* 17(2), March, 1986, pages 165–166.

We close this section by providing in Table 1 a summary of the convergence tests we have discussed.

TABLE 1 TESTS OF CONVERGENCE

Test	First discussed on page	Description	Examples and Comments
Convergence test for a geometric series	655	$\sum_{k=0}^{\infty} r^k$ converges to $1/(1-r)$ if $|r| < 1$ and diverges if $|r| > 1$	$\sum_{k=0}^{\infty} (\frac{1}{2})^k$ converges to 2; $\sum_{k=0}^{\infty} 2^k$ diverges
Look at the terms of the series—the limit test	662	If $|a_k|$ does not converge to 0, then $\sum a_n$ diverges	If $a_k \to 0$, then $\sum_0^{\infty} a_k$ may converge ($\sum_{k=0}^{\infty} 1/k^2$) or it may not (the harmonic series $\sum_{k=0}^{\infty} 1/k$)
Comparison test	671	If $0 \le a_k \le b_k$ and $\sum b_k$ converges, then $\sum a_k$ converges. If $a_k \ge b_k \ge 0$ and $\sum b_k$ diverges, then $\sum a_k$ diverges	It is not necessary that $a_k \le b_k$ or $a_k \ge b_k$ for *all* k, only for $k \ge N$ for some integer N; convergence or divergence of a series is not affected by the values of the first few terms
Integral test	672	If $a_k = f(k) \ge 0$ and f is decreasing, then $\sum_{k=1}^{\infty} a_k$ converges if $\int_1^{\infty} f(x)\,dx$ converges and $\sum_{k=1}^{\infty} a_k$ diverges if $\int_1^{\infty} f(x)\,dx$ diverges	Use this test whenever $f(x)$ can easily be integrated
$\sum_{k=1}^{\infty} 1/k^{\alpha}$	673	$\sum_{k=1}^{\infty} 1/k^{\alpha}$ diverges if $0 \le \alpha \le 1$ and converges if $\alpha > 1$	
Limit comparison test	673	If $a_k > 0$, $b_k > 0$ and there is a number $c > 0$ such that $\lim_{k \to \infty} a_k/b_k = c$, then either both series converge or both series diverge	Use the limit comparison test when a series $\sum b_k$ can be found such that (a) it is known whether $\sum b_k$ converges or diverges and (b) it appears that a_k/b_k has an easily computed limit; (b) will be true, for instance, when $a_k = 1/p(k)$ and $b_k = 1/q(k)$ where $p(k)$ and $q(k)$ are polynomials.
Ratio test	676	If $a_k > 0$ and $\lim_{n \to \infty} a_{n+1}/a_n = L$, then $\sum_{k=1}^{\infty} a_k$ converges if $L < 1$ and diverges when $L > 1$	This is often the easiest test to apply; note that if $L = 1$, then the series may either converge ($\sum 1/k^2$) or diverge ($\sum 1/k$)
Root test	679	If $a_k > 0$ and $\lim_{n \to \infty} (a_n)^{1/n} = R$, then $\sum_{k=1}^{\infty} a_k$ converges if $R < 1$ and diverges if $R > 1$	If $R = 1$, the series may either converge ($\sum 1/k^2$) or diverge ($\sum 1/k$); the root test is the hardest test to apply; it is most useful when a_k is something raised to the kth power [$\sum 1/(\ln k)^k$, for example]
Alternating series test	682	$\sum (-1)^k a_k$ with $a_k \ge 0$ converges if (i) $a_k \to 0$ as $k \to \infty$ and (ii) $\{a_k\}$ is a decreasing sequence; also, $\sum (-1)^k a_k$ diverges if $\lim_{k \to \infty} a_k \ne 0$	This test can only be applied when the terms are alternately positive and negative; if there are two or more positive (or negative) terms in a row, then try another test
Absolute convergence test for a series with both positive and negative terms	681	$\sum a_k$ converges absolutely if $\sum |a_k|$ converges	To determine whether $\sum |a_k|$ converges, try any of the tests that apply to series with nonnegative terms

■ PROBLEMS 10.7

■ SELF-QUIZ

I. True–False:

$$1 + \frac{1}{2} - \frac{1}{3} - \frac{1}{4} + \frac{1}{5} + \frac{1}{6} - \frac{1}{7} + \frac{1}{8} - \frac{1}{9} + \frac{1}{10}$$

$$-\frac{1}{11} + \cdots + \frac{(-1)^n}{n} + \cdots$$

does not converge because it is not a strictly alternating series.

II. Suppose $\sum_{k=0}^{\infty} a_k$ is absolutely convergent. Which of the following must converge?

a. $\sum_{k=0}^{\infty} (-1)^k \cdot a_k$ b. $\sum_{k=0}^{\infty} \frac{a_k}{2^k}$

c. $\sum_{k=0}^{\infty} 2^k \cdot a_k$ d. $\sum_{k=0}^{\infty} e^{a_k}$

e. $\sum_{k=0}^{\infty} \frac{1}{a_k}$ f. $\sum_{k=0}^{\infty} a_k^2$

g. $\sum_{k=0}^{\infty} |a_k|$ h. $\sum_{k=0}^{\infty} \sqrt{|a_k|}$

III. Suppose that the alternating series $\sum_{k=0}^{\infty} b_k$ is conditionally convergent. Which of the following must diverge?

a. $\sum_{k=0}^{\infty} (-1)^k \cdot b_k$ b. $\sum_{k=0}^{\infty} \frac{b_k}{2^k}$

c. $\sum_{k=0}^{\infty} \frac{1 - b_k}{1 + |b_k|}$ d. $\sum_{k=0}^{\infty} e^{b_k}$

e. $\sum_{k=0}^{\infty} \frac{1}{b_k}$ f. $\sum_{k=0}^{\infty} b_k^2$

g. $\sum_{k=0}^{\infty} |b_k|$ h. $\sum_{k=0}^{\infty} \sqrt{|b_k|}$

■ DRILL

In Problems 1–30, determine whether the given series is absolutely convergent, conditionally convergent, or divergent.

1. $\sum_{k=1}^{\infty} (-1)^k$ 2. $\sum_{k=1}^{\infty} \frac{(-1)^{k+1}}{2k}$

3. $\sum_{k=0}^{\infty} \cos \frac{k\pi}{2}$ 4. $\sum_{k=0}^{\infty} \sin \frac{k\pi}{2}$

5. $\sum_{k=1}^{\infty} \frac{(-1)^k \sqrt{k}}{k+3}$ 6. $\sum_{k=1}^{\infty} \frac{(-1)^k}{k^{3/2}}$

7. $\sum_{k=2}^{\infty} \frac{(-1)^k}{k \ln k}$ 8. $\sum_{k=2}^{\infty} \frac{(-1)^k k}{\ln k}$

9. $\sum_{k=2}^{\infty} \frac{(-1)^k}{k\sqrt{\ln k}}$ 10. $\sum_{k=2}^{\infty} \frac{(-1)^k}{\sqrt[3]{\ln k}}$

11. $\sum_{k=1}^{\infty} \frac{(-1)^{k+1}}{5k - 4}$ 12. $\sum_{k=1}^{\infty} \frac{(-1)^k \ln k}{k}$

13. $\sum_{k=1}^{\infty} \frac{k!}{(-3)^k}$ 14. $\sum_{k=1}^{\infty} \frac{k^2}{(-2)^k}$

15. $\sum_{k=1}^{\infty} \frac{(-3)^k}{k^2}$ 16. $\sum_{k=1}^{\infty} \frac{(-2)^k}{k!}$

17. $\sum_{k=2}^{\infty} \frac{(-1)^k k^2}{k^3 + 1}$ 18. $\sum_{k=2}^{\infty} \frac{(-1)^{k+1}}{\sqrt{k(k-1)}}$

19. $\sum_{k=3}^{\infty} \frac{\sin(k\pi/7)}{k^3}$ 20. $\sum_{k=1}^{\infty} \frac{\cos(k\pi/6)}{k^2}$

21. $\sum_{k=1}^{\infty} \frac{(-1)^k(k+2)}{k(k+1)}$ 22. $\sum_{k=2}^{\infty} \frac{(-1)^k k(k+1)}{(k+2)^3}$

23. $\sum_{k=2}^{\infty} \frac{(-1)^k k(k+1)}{(k+2)^4}$ 24. $\sum_{k=1}^{\infty} \frac{(-1)^k k^k}{k!}$

25. $\sum_{k=1}^{\infty} \frac{(-1)^k 2^k}{k}$ 26. $\sum_{k=1}^{\infty} \frac{(-1)^{k+1}}{k!}$

27. $\sum_{k=1}^{\infty} \frac{(-1)^k k^2}{4 + k^2}$ 28. $\sum_{k=1}^{\infty} (-1)^k \left(1 + \frac{1}{k}\right)^k$

29. $\sum_{k=2}^{\infty} \frac{(-1)^k(k^2 + 3)}{k^3 + 4}$ 30. $\sum_{k=2}^{\infty} \frac{(-1)^k k^3}{k^3 + 2k^2 + k - 1}$

▦ CALCULATOR PROBLEMS

In Problems 31–36, use the result of Theorem 3 or of Theorem 4 to approximate the given sum within the specified accuracy.

31. $\sum_{k=1}^{\infty} \frac{(-1)^{k+1}}{k!}$; $|\text{error}| < 0.001$

CONTINUED

32. $\displaystyle\sum_{k=1}^{\infty} \frac{(-1)^{k+1}}{k^2}$; $|\text{error}| < 0.01$

33. $\displaystyle\sum_{k=1}^{\infty} \frac{(-1)^{k+1}}{k^4}$; $|\text{error}| < 0.0001$

34. $\displaystyle\sum_{k=2}^{\infty} \frac{(-1)^{k+1}}{k \ln k}$; $|\text{error}| < 0.05$

35. $\displaystyle\sum_{k=1}^{\infty} \frac{(-1)^{k+1}}{k^k}$; $|\text{error}| < 0.0001$

36. $\displaystyle\sum_{k=1}^{\infty} \frac{(-1)^{k+1}}{\sqrt{k}}$; $|\text{error}| < 0.1$

■ APPLICATIONS

37. Find the first 10 terms of a rearrangement of the series $\sum_{k=1}^{\infty} (-1)^{k+1}/k$ that converges to 0. [As on page 685, start with 1, then add just enough negative terms in the order $-\frac{1}{2}, -\frac{1}{4}, \ldots$ to obtain a negative sum. Then add positive terms, starting with $\frac{1}{3}$, and so on.]

38. Find the first 10 terms of a rearrangement of the series $\sum_{k=1}^{\infty} (-1)^{k+1}/k$ that converges to 0.3. [Use the same method as in Problem 37.]

39. Give an example of a sequence $\{a_n\}$ such that $\sum_{k=1}^{\infty} a_k{}^2$ converges but $\sum_{k=1}^{\infty} a_k$ diverges.

*40. Give an example of a sequence $\{a_n\}$ such that $\sum_{k=1}^{\infty} a_k$ converges but $\sum_{k=1}^{\infty} a_k{}^3$ diverges.

■ SHOW/PROVE/DISPROVE

41. Show that if $\sum_{k=1}^{\infty} a_k$ is absolutely convergent, then $\sum_{k=1}^{\infty} a_k{}^p$ is convergent for any integer $p \geq 1$.

42. Prove that if $\sum_{k=1}^{\infty} a_k$ is a convergent series of nonzero terms, then $\sum_{k=1}^{\infty} 1/|a_k|$ diverges.

*43. Prove Theorem 3. [*Hint:* Assume that the odd-numbered terms are positive. Show that the sequence $\{S_{2n}\}$ is increasing and that $S_{2n} < S_{2n+2} < S$ for all $n \geq 1$. Then show that the sequence of odd-numbered partial sums is decreasing and that $S < S_{2n+1} < S_{2n-1}$ for all $n \geq 1$. Conclude that for all $n \geq 1$,
  a. $0 < S - S_{2n} < a_{2n+1}$ and
  b. $0 < S_{2n-1} - S < -a_{2n}$.
  Use inequalities (a) and (b) to prove the theorem.]

*44. Prove Theorem 4. [*Hint:* Think about writing the original series as

$$S = \frac{1}{2}a_1 + \frac{1}{2}(a_1 - a_2) - \frac{1}{2}(a_2 - a_3)$$

$$+ \frac{1}{2}(a_3 - a_4) - \cdots$$

$$= \frac{1}{2}a_1 + \sum_{k=1}^{\infty} (-1)^{k+1} \frac{a_k - a_{k+1}}{2}.$$

Then apply Theorem 3.]

45. Explain why there is no rearrangement of the series $\sum_{k=1}^{\infty} (-1)^k/k^2$ that converges to -1.

*46. Consider the following rearrangement of the alternating harmonic series:

$$S^* = 1 - \frac{1}{2} - \frac{1}{4} + \frac{1}{3} - \frac{1}{6} - \frac{1}{8}$$

$$+ \frac{1}{5} - \frac{1}{10} - \frac{1}{12} + \cdots. \tag{7}$$

By inserting parentheses in (6), show that $S^* = \frac{1}{2}\ln 2$.[†] $\left[\textit{Hint: } \displaystyle\sum_{k=1}^{\infty} \frac{(-1)^{n+1}}{n} = \ln 2.\right]$

*47. Show that

$$1 + \frac{1}{3} - \frac{1}{2} + \frac{1}{5} + \frac{1}{7} - \frac{1}{4} + \frac{1}{9} + \frac{1}{11} - \frac{1}{6} + \cdots$$

$$= \frac{3}{2}\ln 2.$$

**48. Show that

$$1 - \frac{1}{2} - \frac{1}{4} - \frac{1}{6} + \frac{1}{3} - \frac{1}{8} - \frac{1}{10} - \frac{1}{12} + \frac{1}{5} - \cdots$$

$$= \ln 2 - \frac{1}{2}\ln 3.$$

*49. Suppose $\sum_{k=0}^{\infty} a_k$ is conditionally convergent. Prove or disprove: $\sum_{k=0}^{\infty} 2^k a_k$ must diverge.

[†] The sums in Problems 46–48 are discussed in the paper "Rearranging the Alternating Harmonic Series" by C. C. Cowen, K. R. Davidson, and R. P. Kaufman, *American Mathematical Monthly*, 87 (December 1980): 817–819.

for $0 < u < 2$. Integration then yields

$$\ln u = \int \frac{du}{u} = u - \frac{(u-1)^2}{2} + \frac{(u-1)^3}{3} - \cdots + C.$$

Since $\ln 1 = 0$, we immediately find that $C = -1$, so that

$$\ln u = \sum_{k=0}^{\infty} (-1)^k \frac{(u-1)^{k+1}}{k+1} \tag{5}$$

for $0 < u < 2$. Here we have expressed the logarithmic function defined on the interval $(0, 2)$ as a power series.

EXAMPLE 5
WRITING e^x AS A POWER SERIES

The series

$$f(x) = 1 + x + \frac{x^2}{2!} + \frac{x^3}{3!} + \cdots = \sum_{k=0}^{\infty} \frac{x^k}{k!} \tag{6}$$

converges for every real number x (i.e., $R = \infty$; see Example 4 in Section 10.8). But

$$f'(x) = \frac{d}{dx} 1 + \frac{d}{dx} x + \frac{d}{dx} \frac{x^2}{2!} + \cdots = 1 + x + \frac{x^2}{2!} + \cdots = f(x).$$

Thus f satisfies the differential equation

$$f' = f,$$

and so from the discussion in Section 6.6, we find that

$$f(x) = ce^x \tag{7}$$

for some constant c. Then substituting $x = 0$ into equations (6) and (7) yields

$$f(0) = 1 = ce^0 = c,$$

so that $f(x) = e^x$. We have obtained the important expansion that is valid for any real number x:

$$e^x = 1 + x + \frac{x^2}{2!} + \frac{x^3}{3!} + \cdots = \sum_{k=0}^{\infty} \frac{x^k}{k!}. \tag{8}$$

For example, if we substitute the value $x = 1$ into (8), we obtain partial sum approximations for $e = 1 + 1 + 1/2! + 1/3! + \cdots$ (see Table 1). The last value ($\sum_{k=0}^{8} 1/k!$) is correct to five decimal places.

TABLE 1

n	0	1	2	3	4	5	6	7	8
$S_n = \sum_{k=0}^{n} \frac{1}{k!}$	1	2	2.5	2.66667	2.70833	2.71667	2.71806	2.71825	2.71828

EXAMPLE 6
OBTAINING THE SERIES FOR e^{-x}

Substituting $x = -x$ in (8), we obtain

$$e^{-x} = 1 - x + \frac{x^2}{2!} - \frac{x^3}{3!} + \cdots = \sum_{k=0}^{\infty} (-1)^k \frac{x^k}{k!}. \tag{9}$$

Since this is an alternating series if $x > 0$, Theorem 3 in Section 10.7 tells us that the error $|S - S_n|$ in approximating e^{-x} for $x > 0$ is bounded by $|a_{n+1}| = x^{n+1}/(n+1)!.^{\dagger}$ For example, to calculate e^{-1} with an error of less than 0.0001, we must have $|S - S_n| \le 1/(n+1)! < 0.0001 = 1/10,000$, or $(n+1)! > 10,000$. If $n = 7$, $(n+1)! = 8! = 40,320$, so that $\sum_{k=0}^{7} (-1)^k/k!$ will approximate e^{-1} correct to four decimal places. We obtain

$$e^{-1} \approx 1 - 1 + \frac{1}{2!} - \frac{1}{3!} + \frac{1}{4!} - \frac{1}{5!} + \frac{1}{6!} - \frac{1}{7!} \approx 0.36786.$$

Note that $e^{-1} \approx 0.36788$, correct to five decimal places.

EXAMPLE 7
THE UNIT NORMAL DISTRIBUTION

The function $f(x) = \dfrac{1}{\sqrt{2\pi}} e^{-x^2/2}$ is very important in probability theory. Its graph is the famous "bell-shaped" curve sketched in Figure 1.

The probabilistic interpretation of this function is as follows: If a random variable is normally distributed with mean 0 and standard deviation 1, and if P_a^b denotes the probability that the random variable takes a value between a and b, then

$$P_a^b = \frac{1}{\sqrt{2\pi}} \int_a^b e^{-t^2/2} \, dt.$$

In this context $f(x) = \dfrac{1}{\sqrt{2\pi}} e^{-x^2/2}$ is called the **density function of the unit normal distribution**. In particular,

$$P_0^x = \begin{matrix} \text{probability that the random variable} \\ \text{takes a value between 0 and } x \end{matrix} = \frac{1}{\sqrt{2\pi}} \int_0^x e^{-t^2/2} \, dt.$$

Estimate P_0^1 to 5 decimal places.

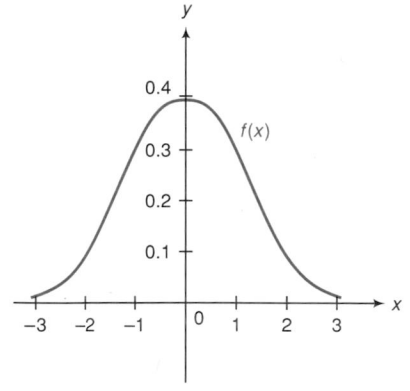

FIGURE 1

GRAPH OF THE FUNCTION $f(x) = \dfrac{1}{\sqrt{2\pi}} e^{-t^2/2}$

SOLUTION: Substituting $-\dfrac{t^2}{2}$ for x in (8), we obtain

$$e^{-t^2/2} = 1 - \frac{t^2}{2} + \frac{(t^2/2)^2}{2!} - \frac{(t^2/2)^3}{3!} + \frac{(t^2/2)^4}{4!} - \cdots$$

$$= 1 - \frac{t^2}{2} + \frac{t^4}{2^2 2!} - \frac{t^6}{2^3 3!} + \frac{t^8}{2^4 4!} - \cdots = \sum_{k=0}^{\infty} \frac{(-1)^k t^{2k}}{2^k k!}.$$

---

† We can apply Theorem 3 in Section 10.7 here because the terms in the sequence $\{x^n/n!\}$ are monotone decreasing for $n > x - 1$.

This series converges for every real number t and we can integrate it term-by-term to obtain

$$\int_0^x e^{-t^2/2}\, dt = \int_0^x \left(1 - \frac{t^2}{2} + \frac{t^4}{2^2 2!} - \frac{t^6}{2^3 3!} + \frac{t^8}{2^4 4!} - \cdots \right) dt$$

$$= \left(t - \frac{t^3}{3 \cdot 2} + \frac{t^5}{5 \cdot 2^2 \cdot 2!} - \frac{t^7}{7 \cdot 2^3 \cdot 3!} + \frac{t^9}{9 \cdot 2^4 \cdot 4!} - \cdots \right) \Big|_0^x$$

$$= x - \frac{x^3}{3 \cdot 2} + \frac{x^5}{5 \cdot 2^2 \cdot 2!} - \frac{x^7}{7 \cdot 2^3 \cdot 3!} + \frac{x^9}{9 \cdot 2^4 \cdot 4!} - \cdots$$

$$= \sum_{k=0}^{\infty} \frac{(-1)^k x^{2k+1}}{(2k+1)2^k k!}$$

and

$$\frac{1}{\sqrt{2\pi}} \int_0^x e^{-t^2/2}\, dt = \frac{1}{\sqrt{2\pi}} \sum_{k=0}^{\infty} \frac{(-1)^k x^{2k+1}}{(2k+1)2^k k!}.$$

In particular,

$$\text{set } x = 1$$
$$\frac{1}{\sqrt{2\pi}} \int_0^1 e^{-t^2/2}\, dt = \frac{1}{\sqrt{2\pi}} \sum_{k=0}^{\infty} \frac{(-1)^k}{(2k+1)2^k k!}.$$

If we approximate this integral by taking a partial sum of this alternating series, then the error is, by Theorem 3 in Section 10.7,[†] less than the absolute value of the first term omitted. In Table 2 we give the values of $\dfrac{1}{\sqrt{2\pi}(2k+1)2^k k!}$ for $k = 1, 2, 3, 4, 5,$ and 6.

We see that if we compute $\displaystyle\sum_{k=0}^{5} \frac{(-1)^k}{\sqrt{2\pi}(2k+1)2^k k!}$, then our error will be less than 0.000000665 and our answer will be correct to 5 decimal places. [Explain why stopping at $k = 4$ might not work.] We compute

$$\sum_{k=0}^{5} \frac{(-1)^k}{\sqrt{2\pi}(2k+1)2^k k!} = \frac{1}{\sqrt{2\pi}} \left(1 - \frac{1}{6} + \frac{1}{40} - \frac{1}{336} + \frac{1}{3456} - \frac{1}{42,240} \right)$$

$$= (0.39894228)(0.85562282) \approx 0.34134.$$

Tables of values of $\dfrac{1}{\sqrt{2\pi}} \displaystyle\int_0^x e^{-t^2/2}\, dt$ (or $\dfrac{1}{\sqrt{2\pi}} \displaystyle\int_{-\infty}^x e^{-t^2/2}\, dt$) can be found in virtually every book on statistics, although most of these books do not mention how these values are obtained.

TABLE 2 VALUES OF $\dfrac{1}{\sqrt{2\pi}(2k+1)2^k k!}$

k	$(2k+1)2^k k!$	$\dfrac{1}{\sqrt{2\pi}(2k+1)2^k k!}$
1	6	0.06649038
2	40	0.009973557
3	336	0.001187328
4	3456	0.000115434
5	42,240	0.000009444
6	599,040	0.000000665

EXAMPLE 8
A SERIES REPRESENTATION
OF cos x

Consider the series

$$f(x) = 1 - \frac{x^2}{2!} + \frac{x^4}{4!} - \frac{x^6}{6!} + \cdots = \sum_{k=0}^{\infty} (-1)^k \frac{x^{2k}}{(2k)!}. \tag{10}$$

[†] As in Example 6, we can apply this theorem as long as the terms are decreasing. Here $|a_{n+1}| < |a_n|$ when $2(2n+3)(n+1)/(2n+1) > x^2$. This holds for every n when $x = 1$.

It is easy to see that $R = \infty$ since the series is absolutely convergent for every x by comparison with the series (8) for e^x. [The series (8) is larger than the series (10) since it contains the terms $x^n/n!$ for n both even and odd, not just for n even, as in (10).] Differentiating, we obtain

$$f'(x) = -x + \frac{x^3}{3!} - \frac{x^5}{5!} + \frac{x^7}{7!} - \cdots = \sum_{k=0}^{\infty} (-1)^{k+1} \frac{x^{2k+1}}{(2k+1)!}. \qquad \text{(11)}$$

Since this series has a radius of convergences $R = \infty$, we can differentiate once more to obtain

$$f''(x) = -1 + \frac{x^2}{2!} - \frac{x^4}{4!} + \frac{x^6}{6!} - \cdots = \sum_{k=0}^{\infty} \frac{(-1)^{k+1} x^{2k}}{(2k)!} = -f(x).$$

Thus, we see that f satisfies the differential equation

$$f'' + f = 0. \qquad \text{(12)}$$

Moreover, from equations (10) and (11), we see that

$$f(0) = 1 \qquad \text{and} \qquad f'(0) = 0. \qquad \text{(13)}$$

It is easily seen that the function $f(x) = \cos x$ satisfies equation (12) together with the conditions (13). In fact, although we do not prove it here, it is the only function that does so.[†] Thus, we have

$$\cos x = 1 - \frac{x^2}{2!} + \frac{x^4}{4!} - \frac{x^6}{6!} + \cdots = \sum_{k=0}^{\infty} (-1)^k \frac{x^{2k}}{(2k)!}. \qquad \text{(14)}$$

Since

$$\frac{d}{dx} \cos x = -\sin x,$$

we obtain, from (14), the series (after multiplying both sides by -1)

$$\sin x = x - \frac{x^3}{3!} + \frac{x^5}{5!} - \frac{x^7}{7!} + \cdots = \sum_{k=0}^{\infty} (-1)^k \frac{x^{2k+1}}{(2k+1)!}. \qquad \text{(15)}$$

EXAMPLE 9
APPROXIMATING A COSINE
INTEGRAL

Approximate $\int_0^{\pi/4} \cos t^2 \, dt$ with an error $< 10^{-10}$.

SOLUTION: We first note that we used Simpson's rule to approximate this integral in Example 6 in Section 7.9 (p. 530).
Substituting t^2 for x in (14) yields

$$\cos t^2 = 1 - \frac{(t^2)^2}{2!} + \frac{(t^2)^4}{4!} - \frac{(t^2)^6}{6!} + \cdots = \sum_{k=0}^{\infty} \frac{(-1)^k t^{4k}}{(2k)!},$$

[†] This follows from a basic existence-uniqueness result for differential equations. For a statement and proof, see W. R. Derrick and S. I. Grossman, *Introduction to Differential Equations with Boundary Value Problems*, 3rd ed. (St. Paul: West, 1987), Appendix 3.

so

$$\int_0^x \cos t^2 \, dt = \int_0^x \left(1 - \frac{t^4}{2!} + \frac{t^8}{4!} - \frac{t^{12}}{6!} + \cdots \right) dt$$

$$= \left(t - \frac{t^5}{5 \cdot 2!} + \frac{t^9}{9 \cdot 4!} - \frac{t^{13}}{13 \cdot 6!} + \cdots \right) \Bigg|_0^x$$

$$= x - \frac{x^5}{5 \cdot 2!} + \frac{x^9}{9 \cdot 4!} - \frac{x^{13}}{13 \cdot 6!} + \cdots = \sum_{k=0}^{\infty} \frac{(-1)^k x^{4k+1}}{(4k+1)(2k)!}$$

and, in particular,

$$\int_0^{\pi/4} \cos t^2 \, dt = \frac{\pi}{4} - \frac{(\pi/4)^5}{5 \cdot 2!} + \frac{(\pi/4)^9}{9 \cdot 4!} - \frac{(\pi/4)^{13}}{13 \cdot 6!} + \cdots .$$

The error in stopping after the $(k-1)$st term is less than $\dfrac{(\pi/4)^{4k+1}}{(4k+1)(2k)!}$.
We compute

$$k = 3 \qquad \frac{(\pi/4)^{13}}{13 \cdot 6!} = 0.000004622.$$

$$k = 4 \qquad \frac{(\pi/4)^{17}}{17 \cdot 8!} = 0.000000024.$$

$$k = 5 \qquad \frac{(\pi/4)^{21}}{21 \cdot 10!} \approx 8.22 \times 10^{-11} < 10^{-10}.$$

Thus the desired accuracy is obtained by stopping at $k = 4$.

$$\int_0^{\pi/4} \cos t^2 \, dt \approx \frac{\pi}{4} - \frac{(\pi/4)^5}{5 \cdot 2!} + \frac{(\pi/4)^9}{9 \cdot 4!} - \frac{(\pi/4)^{13}}{13 \cdot 6!} + \frac{(\pi/4)^{17}}{17 \cdot 8!}$$

$$= 0.75603527754.$$

In Section 7.9, we discussed the trapezoidal rule and Simpson's rule as examples of some more general ways to calculate definite integrals that do not require the existence of a series expansion of the function being integrated. However, as shown in Examples 7 and 9, a power series provides an easy method of numerical integration when the power series representation of a function is readily obtainable.

■ PROBLEMS 10.9

■ SELF-QUIZ

I. By differentiating the power series

$$1 + x + x^2 + x^3 + \cdots = \sum_{k=0}^{\infty} x^k,$$

we obtain

$$0 + 1 + 2x + 3x^2 + \cdots = \sum_{k=0}^{\infty} k \cdot x^{k-1},$$

which, for $|x| < 1$, converges to _____.

a. $-\ln|1-x|$ b. $\dfrac{1}{1-x}$

c. $\dfrac{-1}{(1-x)^2}$ d. $\dfrac{1}{(1-x)^2}$

II. If $|x| < 1$, then

$$x + \frac{x^2}{2} + \frac{x^3}{3} + \frac{x^4}{4} + \cdots$$

$$= \sum_{j=1}^{\infty} \frac{x^j}{j}$$

$$= \sum_{k=0}^{\infty} \frac{x^{k+1}}{k+1} = \sum_{k=0}^{\infty} \left(\int_0^x t^k \, dt \right)$$

$$= \int_0^x \left(\sum_{k=0}^{\infty} t^k \right) dt = \underline{\hspace{1.5cm}}?$$

a. $-\ln|1-x|$ b. $\ln|1-x|$

c. $\dfrac{1}{1-x}$ d. $\dfrac{1}{(1-x)^2}$

III. If $|x| < 1$, then

$$x - \frac{x^2}{2} + \frac{x^3}{3} - \frac{x^4}{4} + \cdots$$

$$= \sum_{j=1}^{\infty} (-1)^{j-1} \frac{x^j}{j}$$

$$= \int_0^x \left(\sum_{k=0}^{\infty} (-1)^k t^k \right) dt = \underline{\hspace{1.5cm}}?$$

a. $\ln|1+x|$ b. $-\ln|1+x|$

c. $\dfrac{1}{1+x}$ d. $\dfrac{-1}{(1+x)^2}$

IV. If $|x| < 1$, then

$$x + \frac{x^3}{3} + \frac{x^5}{5} + \frac{x^7}{7} + \cdots$$

$$= \sum_{k=0}^{\infty} \frac{x^{2k+1}}{2k+1}$$

$$= \frac{1}{2} \left(\sum_{j=1}^{\infty} \frac{x^j}{j} + \sum_{j=1}^{\infty} (-1)^{j-1} \frac{x^j}{j} \right)$$

$$= \underline{\hspace{1.5cm}}?$$

a. $\dfrac{1}{2} \ln|1 - x^2|$ b. $\dfrac{1}{2} \ln \left| \dfrac{1-x}{1+x} \right|$

c. $\dfrac{1}{2} \ln \left| \dfrac{1+x}{1-x} \right|$ d. $\dfrac{1}{2(1-x^2)}$

■ DRILL

1. By substituting x^2 for x in equation (4), find a series expansion for $1/(1 + x^2)$ that is valid for $|x| < 1$.
2. Find a series expansion for xe^x that is valid for all real values of x.
3. Integrate the series obtained in Problem 1 to obtain a series expansion for $\tan^{-1} x$.
4. Use the result of Problem 2 to find a series expansion for $\int_0^x te^t \, dt$.

5. Expand $1/x$ as a power series of the form $\sum_{k=0}^{\infty} a_k (x-1)^k$. What is the interval of convergence of this series?
6. Find a power series expansion for

$$\int_0^x \frac{\ln(1+t)}{t} \, dt.$$

▦ CALCULATOR PROBLEMS

7. Use the result of Problem 3 to obtain an approximation of π that is correct to two decimal places (i.e., error bounded by 0.5×10^{-2}). [*Hint:* The series for $\pi/4 = \tan^{-1} 1$ converges very slowly. Instead, use one of the following facts: $\pi/6 = \tan^{-1}(1/\sqrt{3})$, $\pi/12 = \tan^{-1}(2 - \sqrt{3})$.]

8. Use the series expansion for $\ln x$ (see Example 4) to calculate the following to two decimal places of accuracy:
   a. $\ln 0.5$ b. $\ln 1.6$

In Problems 9–16, approximate the given integral within the specified accuracy.

9. $\int_0^1 e^{-t^2} \, dt$; $|\text{error}| < 0.01$

10. $\int_0^1 e^{-t^3} \, dt$; $|\text{error}| < 0.001$

11. $\int_0^{1/2} \cos t^2 \, dt$; $|\text{error}| < 0.001$

12. $\int_0^{1/2} \sin t^2 \, dt$; $|\text{error}| < 0.0001$

13. $\int_0^1 \cos \sqrt{t} \, dt$; $|\text{error}| < 0.01$

CONTINUED

14. $\int_0^1 t \sin \sqrt{t}\, dt;$ $|\text{error}| < 0.001$

15. $\int_0^1 t^2 e^{-t^2}\, dt;$ $|\text{error}| < 0.01$

   [*Hint:* The series expansion of $t^2 e^{-t^2}$ can be obtained by multiplying each term of the series expansion for e^{-t^2} by t^2.]

16. $\int_0^{1/2} \dfrac{dt}{1 + t^8};$ $|\text{error}| < 0.0001$

In Problems 17–20, approximate $\dfrac{1}{\sqrt{2\pi}} \int_0^x e^{-t^2/2}\, dt$ to 5 decimal places.

17. $x = 0.25$ 18. $x = 0.5$

19. $x = 2$ *20. $x = 50$

■ SHOW/PROVE/DISPROVE

21. Use the result of Problem 4 to show that

$$\sum_{k=0}^\infty \frac{1}{(k+2)k!} = 1.$$

*22. Define the function J_0 by

$$J_0(x) = \sum_{k=0}^\infty \frac{(-1)^k}{(k!)^2}\left(\frac{x}{2}\right)^{2k}.$$

a. What is the interval of convergence of this series?

b. Show that J_0 satisfies the differential equation

$$x^2 J''(x) + x J'(x) + x^2 J(x) = 0.$$

The function J_0 is called a **Bessel function of order zero.**[†]

■ CHALLENGE

23. Use the power series for $\ln|(1 + x)/(1 - x)|$ which was obtained in the Self-quiz (II–IV) to approximate ln 1.5, ln 0.5, and ln 2 within four decimal places. Consider that series and those for $\ln|1 \pm x|$ as computational tools. Which seems to be the best? Why?

■ ANSWERS TO SELF-QUIZ

I. d   II. a   III. a   IV. c

10.10 TAYLOR AND MACLAURIN SERIES

In the last two sections, we used the fact that within its interval of convergence, the function

$$f(x) = \sum_{k=0}^\infty a_k(x - x_0)^k$$

is differentiable and integrable. In this section, we look more closely at the coefficients a_k and show that they can be represented in terms of derivatives of the function f.

We begin with the case $x_0 = 0$ and assume that $R > 0$, so that the theorem on power series differentiation applies. We have

$$f(x) = \sum_{k=0}^\infty a_k x^k = a_0 + a_1 x + a_2 x^2 + \cdots + a_n x^n + \cdots, \tag{1}$$

[†] Named after the German physicist and mathematician Friedrich Wilhelm Bessel (1784–1846), who used the function in his study of planetary motion. The Bessel functions of various orders arise in many applications in modern physics and engineering.

and clearly,

$$f(0) = a_0 + 0 + 0 + \cdots + 0 + \cdots = a_0. \tag{2}$$

If we differentiate (1), we obtain

$$f'(x) = \sum_{k=1}^{\infty} k a_k x^{k-1} = a_1 + 2a_2 x + 3a_3 x^2 + \cdots + n a_n x^{n-1} + \cdots \tag{3}$$

and

$$f'(0) = a_1. \tag{4}$$

Continuing to differentiate, we obtain

$$f''(x) = \sum_{k=2}^{\infty} k(k-1)a_k x^{k-2}$$
$$= 2a_2 + 3 \cdot 2a_3 x + 4 \cdot 3a_4 x^2 + \cdots + n(n-1)a_n x^{n-2} + \cdots$$

and

$$f''(0) = 2a_2, \quad \text{or} \quad a_2 = \frac{f''(0)}{2} = \frac{f''(0)}{2!}. \tag{5}$$

Similarly,

$$f'''(x) = \sum_{k=3}^{\infty} k(k-1)(k-2)a_k x^{k-3},$$

so

$$f'''(0) = 3 \cdot 2a_3 \quad \text{and} \quad a_3 = \frac{f'''(0)}{3 \cdot 2} = \frac{f'''(0)}{3!}. \tag{6}$$

It is not difficult to see that this pattern continues and that for every positive integer n,

---

THE nth COEFFICIENT OF THE POWER SERIES FOR f

$$a_n = \frac{f^{(n)}(0)}{n!}. \tag{7}$$

For $n = 0$, we use the convention $0! = 1$ and $f^{(0)}(x) = f(x)$. Then formula (7) holds for every n, and we have the following:

MACLAURIN SERIES

If $f(x) = \sum_{k=0}^{\infty} a_k x^k$, then

$$f(x) = \sum_{k=0}^{\infty} \frac{f^{(k)}(0)}{k!} x^k$$

$$= f(0) + f'(0)x + f''(0)\frac{x^2}{2!} + \cdots + f^{(n)}(0)\frac{x^n}{n!} + \cdots \tag{8}$$

for every x in the interval of convergence.

In the general case, if

$$f(x) = \sum_{k=0}^{\infty} a_k(x - x_0)^k$$
$$= a_0 + a_1(x - x_0) + a_2(x - x_0)^2 + \cdots + a_n(x - x_0)^n + \cdots, \qquad (9)$$

then $f(x_0) = a_0$, and differentiating as before, we find that

$$a_n = \frac{f^{(n)}(x_0)}{n!}. \qquad (10)$$

Thus, we have the following:

TAYLOR SERIES

If $f(x) = \sum_{k=0}^{\infty} a_k(x - x_0)^k$, then

$$f(x) = \sum_{k=0}^{\infty} \frac{f^{(k)}(x_0)}{k!}(x - x_0)^k$$
$$= f(x_0) + f'(x_0)(x - x_0) + f''(x_0)\frac{(x - x_0)^2}{2!} + \cdots$$
$$+ f^{(n)}(x_0)\frac{(x - x_0)^n}{n!} + \cdots \qquad (11)$$

for every x in the interval of convergence.

DEFINITION **TAYLOR AND MACLAURIN SERIES**

The series in (11) is called the **Taylor series**[†] of the function f at x_0. The special case $x_0 = 0$ in (8) is called a **Maclaurin series**.[†] We see that the first n terms of the Taylor series of a function are simply the Taylor polynomial described in Section 9.4. ∎

⊗ **WARNING:** We have shown here that *if* $f(x) = \sum_{k=0}^{\infty} a_k(x - x_0)^k$, then f is infinitely differentiable (i.e., f has derivatives of all orders) and that the series for f is the Taylor series (or Maclaurin series if $x_0 = 0$) of f. What we have *not* shown is that if f is infinitely differentiable at x_0, then f has a Taylor series expansion at x_0. In general, this last statement is false, as we will see in Example 3. ⊗

EXAMPLE 1
THE MACLAURIN SERIES FOR e^x

Find the Maclaurin series for e^x.

SOLUTION: If $f(x) = e^x$, then $f(0) = f'(0) = \cdots = f^{(k)}(0) = 1$, and

$$e^x = \sum_{k=0}^{\infty} \frac{x^k}{k!} = 1 + x + \frac{x^2}{2!} + \frac{x^3}{3!} + \cdots + \frac{x^n}{n!} + \cdots. \qquad (12)$$

[†] The history of the Taylor series is somewhat muddied. It has been claimed that the basis for its development was found in India before 1550. (Taylor published the result in 1715.) For an interesting discussion of this controversy, see the paper by C. T. Rajagopal and T. V. Vedamurthi, "On the Hindu proof of Gregory's series," *Scripta Mathematica*, 17 (1951): 65–74.

[‡] See the biographical sketch of Maclaurin on page 713.

This series is the series we obtained in Example 5 in Section 10.9. It is important to note here that this example shows that *if* e^x has a Maclaurin series expansion, then the series must be the series (12). It does *not* show that e^x actually does have such a series expansion. To prove that the series in (12) is really equal to e^x, we differentiate, as in Example 5 in Section 10.9, and use the fact that the only continuous function that satisfies

$$f'(x) = f(x), \qquad f(0) = 1,$$

is the function e^x.

EXAMPLE 2

THE MACLAURIN SERIES FOR cos *x*

Assuming that the function $f(x) = \cos x$ can be written as a Maclaurin series, find that series.

SOLUTION: If $f(x) = \cos x$, then $f(0) = 1, f'(0) = 0, f''(0) = -1, f'''(0) = 0$, $f^{(4)}(0) = 1$, and so on, so that if

$$\cos x = \sum_{k=0}^{\infty} a_k x^k,$$

then

$$\cos x = f(0) + f'(0) + \frac{f''(0)x^2}{2!} + \frac{f'''(0)x^3}{3!} + \frac{f^{(4)}(0)x^4}{4!} + \cdots,$$

or

$$\cos x = 1 - \frac{x^2}{2!} + \frac{x^4}{4!} - \frac{x^6}{6!} + \cdots = \sum_{k=0}^{\infty} \frac{(-1)^k x^{2k}}{(2k)!}. \qquad (13)$$

This series is the series found in Example 8 in Section 10.9.

NOTE: Again, this does not prove that the equality in (13) is correct. It only shows that *if* cos *x* has a Maclaurin expansion, then the expansion must be given by (13). We will show that cos *x* has a Maclaurin series in Example 5.

EXAMPLE 3

AN INFINITELY DIFFERENTIABLE FUNCTION THAT CANNOT BE REPRESENTED BY A MACLAURIN SERIES

Let

$$f(x) = \begin{cases} e^{-1/x^2}, & \text{if } x \neq 0 \\ 0, & \text{if } x = 0. \end{cases}$$

Find a Maclaurin expansion for f if one exists.

SOLUTION: First, we note that since $\lim_{x \to 0} e^{-1/x^2} = 0$, f is continuous. Now recall from Example 5 in Section 9.2 that $\lim_{x \to \infty} x^a e^{-bx} = 0$ if $b > 0$. Let $y = 1/x^2$. Then, as $x \to 0$, $y \to \infty$. Also, $1/x^n = (1/x^2)^{n/2}$, so that $\lim_{x \to 0}(e^{-1/x^2}/x^n) = \lim_{y \to \infty} y^{n/2} e^{-y} = 0$.

Now for $x \neq 0$, $f'(x) = (2/x^3)e^{-1/x^2} \to 0$ as $x \to 0$, so that f' is continuous at 0. Similarly, $f''(x) = [(4/x^6) - (6/x^4)]e^{-1/x^2}$, which also approaches 0 as $x \to 0$ by the limit result above. In fact, *every* derivative of f is continuous and $f^{(n)}(0) = 0$ for every n. (You were asked to prove this result in Problem 47 in Section 9.2.) Thus, f is infinitely differentiable, and *if* it had a Maclaurin series that represented the function, then we would have

$$f(x) = f(0) + f'(0)x + f''(0)\frac{x^2}{2!} + \cdots.$$

But $f(0) = f'(0) = f''(0) = \cdots = 0$, so that the Maclaurin series would be the zero series. But since f is obviously not the zero function, we can only conclude that there is *no* Maclaurin series that represents f at any point other than 0.

Example 3 illustrates that infinite differentiability is not sufficient to guarantee that a given function can be represented by its Taylor series. Something more is needed.

DEFINITION   **ANALYTIC FUNCTION**

We say that a function f is **analytic** at x_0 if f can be represented by a Taylor series in some neighborhood of x_0. ■

We see that the functions e^x and $\cos x$ are analytic at 0, while the function

$$f(x) = \begin{cases} e^{-1/x^2}, & x \neq 0 \\ 0, & x = 0 \end{cases}$$

is not. A condition that guarantees analyticity of an infinitely differentiable function is given below.

THEOREM 1

Suppose that the function f has continuous derivatives of all orders in a neighborhood $N(x_0)$ of the number x_0.
Then f is analytic at x_0 if and only if

$$\lim_{n \to \infty} R_n(x) = \lim_{n \to \infty} \frac{f^{(n+1)}(c_n)}{(n+1)!}(x - x_0)^{n+1} = 0 \qquad \text{(14)}$$

for every x in $N(x_0)$ where c_n is between x_0 and x. ■

REMARK: The expression between the equal signs in (14) is simply the remainder term given by Taylor's theorem (see page 620).

PROOF: The hypotheses of Taylor's theorem apply, so that we can write, for any n,

$$f(x) = P_n(x) + R_n(x), \qquad \text{(15)}$$

where $P_n(x)$ is the nth-degree Taylor polynomial for f. To show that f is analytic, we must show that

$$\lim_{n \to \infty} P_n(x) = f(x) \qquad \text{(16)}$$

for every x in $N(x_0)$. But if x is in $N(x_0)$, we obtain, from (14) and (15),

$$\lim_{n \to \infty} P_n(x) = \lim_{n \to \infty} [f(x) - R_n(x)] = f(x) - \lim_{n \to \infty} R_n(x)$$

$$= f(x) - 0 = f(x).$$

Conversely, if f is analytic, then $f(x) = \lim_{n \to \infty} P_n(x)$ so $R_n(x) \to 0$ as $n \to \infty$. ∎

EXAMPLE 4
SHOWING THAT e^x IS ANALYTIC

If $f(x) = e^x$, then $f^{(n)}(x) = e^x$, and

$$\lim_{n \to \infty} \left| \frac{f^{(n+1)}(c_n)}{(n+1)!} (x-0)^{n+1} \right| = \lim_{n \to \infty} \frac{e^{c_n} |x|^{n+1}}{(n+1)!} \overset{\displaystyle 0 < c_n < |x|}{\leq} e^{|x|} \lim_{n \to \infty} \frac{|x|^{n+1}}{(n+1)!} \to 0,$$

since $|x|^{n+1}/(n+1)!$ is the $(n+2)$nd term in the converging power series $\sum_{k=0}^{\infty} |x|^k/k!$ and the terms in a converging power series $\to 0$ by Theorem 2 in Section 10.4. Since this result is true for any $x \in \mathbb{R}$, we may take $N = (-\infty, \infty)$ to conclude that the series (12) is valid for every real number x.

EXAMPLE 5
SHOWING THAT $\cos x$ IS ANALYTIC

Let $f(x) = \cos x$. Since all derivatives of $\cos x$ are equal to $\pm \sin x$ or $\pm \cos x$, we see that $|f^{(n+1)}(c_n)| \leq 1$. Then for $x_0 = 0$,

$$|R_n(x)| \leq |x|^{n+1}/(n+1)!$$

which $\to 0$ as $n \to \infty$, so that the series (13) is also valid for every real number x.

EXAMPLE 6
A FUNCTION THAT IS NOT ANALYTIC

It is evident that for the function in Example 3, $R_n(x) \nrightarrow 0$ if $x \neq 0$. This follows from the fact that $R_n(x) = f(x) - P_n(x) = e^{-1/x^2} - 0 = e^{-1/x^2} \neq 0$ if $x \neq 0$.

In the rest of this section we will not prove that remainder terms go to zero. However, you should be aware that unless this is done, there is no guarantee that the series you obtain by using formula (8) or (11) will be valid.

EXAMPLE 7
THE TAYLOR SERIES FOR $\ln x$ AT $x = 1$

Find the Taylor expansion for $f(x) = \ln x$ at $x = 1$.

SOLUTION: Since $f'(x) = 1/x$, $f''(x) = -1/x^2$, $f'''(x) = 2/x^3$, $f^{(4)}(x) = -6/x^4, \ldots, f^{(n)}(x) = (-1)^{n+1}(n-1)!/x^n$, we find that $f(1) = 0$, $f'(1) = 1$, $f''(1) = -1$, $f'''(1) = 2$, $f^{(4)}(1) = -6, \ldots, f^{(n)}(1) = (-1)^{n+1}(n-1)!$. Then wherever valid,

$$\ln x = \sum_{k=0}^{\infty} f^{(k)}(1) \frac{(x-1)^k}{k!}$$

$$= 0 + (x-1) - \frac{(x-1)^2}{2} + \frac{2(x-1)^3}{3!}$$

$$- \frac{3!(x-1)^4}{4!} + \frac{4!(x-1)^5}{5!} + \cdots,$$

or

$$\ln x = (x-1) - \frac{(x-1)^2}{2} + \frac{(x-1)^3}{3} - \frac{(x-1)^4}{4} + \cdots$$

$$= \sum_{k=1}^{\infty} \frac{(-1)^{k+1}(x-1)^k}{k}. \tag{17}$$

The radius of convergence of this power series is 1. Thus, the series converges to ln x for $-1 < x - 1 < 1$ or $0 < x < 2$. When $x = 2$, we obtain the series $\sum \frac{(-1)^{k+1}}{k}$, which converges by the alternating series test. This implies that the series (17) converges to $\ln(x)$ for $0 < x \le 2$. When $x = 2$, we obtain, from (17),

$$\ln 2 = 1 - \frac{1}{2} + \frac{1}{3} - \frac{1}{4} + \frac{1}{5} - \cdots = \sum_{k=1}^{\infty} \frac{(-1)^{k+1}}{k}. \tag{18}$$

EXAMPLE 8

THE TAYLOR SERIES FOR sin x AT x = π/3

Find a Taylor series for $f(x) = \sin x$ at $x = \pi/3$.

SOLUTION: Here we have $f(\pi/3) = \sqrt{3}/2$, $f'(\pi/3) = \frac{1}{2}$, $f''(\pi/3) = -\sqrt{3}/2$, $f'''(\pi/3) = -\frac{1}{2}$, and so on, so that

$$\sin x = \frac{\sqrt{3}}{2} + \frac{1}{2}\left(x - \frac{\pi}{3}\right) - \frac{\sqrt{3}}{2}\frac{[x - (\pi/3)]^2}{2!} - \frac{1}{2}\frac{[x - (\pi/3)]^3}{3!}$$
$$+ \frac{\sqrt{3}}{2}\frac{[x - (\pi/3)]^4}{4!} + \cdots.$$

The proof that this series is valid for every real number x is similar to the proof in Example 5.

We provide here a list of useful Maclaurin series:

SOME USEFUL MACLAURIN SERIES

$$e^x = \sum_{k=0}^{\infty} \frac{x^k}{k!} = 1 + x + \frac{x^2}{2!} + \frac{x^3}{3!} + \cdots \tag{19}$$

$$\cos x = \sum_{k=0}^{\infty} \frac{(-1)^k x^{2k}}{(2k)!} = 1 - \frac{x^2}{2!} + \frac{x^4}{4!} - \frac{x^6}{6!} + \cdots \tag{20}$$

$$\sin x = \sum_{k=0}^{\infty} \frac{(-1)^k x^{2k+1}}{(2k+1)!} = x - \frac{x^3}{3!} + \frac{x^5}{5!} - \frac{x^7}{7!} + \cdots \tag{21}$$

$$\cosh x = \sum_{k=0}^{\infty} \frac{x^{2k}}{(2k)!} = 1 + \frac{x^2}{2!} + \frac{x^4}{4!} + \frac{x^6}{6!} + \cdots \tag{22}$$

$$\sinh x = \sum_{k=0}^{\infty} \frac{x^{2k+1}}{(2k+1)!} = x + \frac{x^3}{3!} + \frac{x^5}{5!} + \frac{x^7}{7!} + \cdots \tag{23}$$

$$\frac{1}{1-x} = \sum_{k=0}^{\infty} x^k = 1 + x + x^2 + \cdots, \qquad |x| < 1 \tag{24}$$

$$\ln(1+x) = \sum_{k=0}^{\infty} \frac{(-1)^k x^{k+1}}{k+1} \tag{25}$$

You are asked to prove, in Problems 27 and 28, that the series (21), (22), and (23) are valid for every real number x.

BINOMIAL SERIES

We close this section by deriving another series that is quite useful. Let $f(x) = (1 + x)^r$, where r is a real number not equal to an integer. We have

$$f'(x) = r(1 + x)^{r-1},$$

$$f''(x) = r(r - 1)(1 + x)^{r-2},$$

$$f'''(x) = r(r - 1)(r - 2)(1 + x)^{r-3},$$

$$\vdots$$

$$f^{(n)}(x) = r(r - 1)(r - 2) \cdots (r - n + 1)(1 + x)^{r-n}.$$

Note that since r is not an integer, $r - n$ is never equal to 0, and all derivatives exist and are nonzero as long as $x \neq -1$. Then,

$$f(0) = 1,$$

$$f'(0) = r,$$

$$f''(0) = r(r - 1),$$

$$\vdots$$

$$f^{(n)}(0) = r(r - 1) \cdots (r - n + 1),$$

and we can write

$$(1 + x)^r = 1 + rx + \frac{r(r - 1)}{2!} x^2 + \frac{r(r - 1)(r - 2)}{3!} x^3 + \cdots$$

$$+ \frac{r(r - 1) \cdots (r - n + 1)}{n!} x^n + \cdots$$

$$= 1 + \sum_{k=1}^{\infty} \frac{r(r - 1) \cdots (r - k + 1)}{k!} x^k, \quad |x| < 1 \qquad (26)$$

The series (26) is called the **binomial series**.

EXAMPLE 9
USING THE BINOMIAL SERIES TO
APPROXIMATE AN INTEGRAL

Approximate $\int_0^{0.6} \sqrt{1 + t^5}\, dt$ with an error < 0.00001 using an appropriate binomial series.

SOLUTION: Substituting $r = \frac{1}{2}$ and $x = t^5$ in (26) yields

$$(1 + t^5)^{1/2} = 1 + \frac{t^5}{2} + \frac{\frac{1}{2}(\frac{1}{2} - 1)}{2!} t^{10} + \frac{\frac{1}{2}(\frac{1}{2} - 1)(\frac{1}{2} - 2)}{3!} t^{15} + \cdots$$

$$= 1 + \frac{t^5}{2} - \frac{1}{4 \cdot 2!} t^{10} + \frac{3}{8 \cdot 3!} t^{15} - \frac{15}{16 \cdot 4!} t^{20} + \cdots.$$

Then

$$\int_0^x (1 + t^5)^{1/2}\, dt = x + \frac{x^6}{6 \cdot 2} - \frac{x^{11}}{11 \cdot 4 \cdot 2!} + \frac{3x^{16}}{16 \cdot 8 \cdot 3!} - \frac{15x^{21}}{21 \cdot 16 \cdot 4!} + \cdots$$

and

$$\int_0^{0.6} (1 + t^5)^{1/2}\, dt = 0.6 + \frac{0.6^6}{12} - \frac{0.6^{11}}{88} + \frac{3(0.6)^{16}}{768} - \frac{15(0.6)^{21}}{8064} + \cdots.$$

After the first term, the series above is alternating. Since $\dfrac{0.6^{11}}{88} \approx 0.000041$

and $\dfrac{3(0.6)^{16}}{768} \approx 0.0000011$, we can obtain the desired accuracy by stopping

after the term $-\dfrac{0.6^{11}}{88}$. We have

$$\int_0^{0.6} (1 + t^5)^{1/2}\, dt \approx 0.6 + \frac{0.6^6}{12} - \frac{0.6^{11}}{88} \approx 0.603847.$$

ONSIDERED THE FINEST BRITISH mathematician of the generation after Newton, Colin Maclaurin was certainly one of the best mathematicians of the eighteenth century.

Born in Scotland, Maclaurin was a mathematical prodigy and entered Glasgow University at the age of 11. By the age of 19, he was a professor of mathematics in Aberdeen and later obtained a post at the University of Edinburgh.

Maclaurin is best known for the term *Maclaurin series*, which is the Taylor series in the case $x_0 = 0$. He used this series in his 1742 work, *Treatise of Fluxions*. (Maclaurin acknowledged that the series had first been used by Taylor in 1715.) The *Treatise of Fluxions* was most significant in that it presented the first logical description of Newton's method of fluxions. This work was written to defend Newton from the attacks of the powerful Bishop George Berkeley (1685–1753). Berkeley was troubled (as are many of today's calculus students) by the idea of a quotient that takes the form $\frac{0}{0}$. This, of course, is what we obtain when we take a derivative. Berkeley wrote:

COLIN MACLAURIN
1698–1746

And what are these fluxions? The velocities of evanescent increments And what are these same evanescent increments? They are neither finite quantities nor quantities infinitely small nor yet nothing. May we not call them ghosts of departed quantities?

Maclaurin answered Berkeley using geometric arguments. Later, Newton's calculus was put on an even firmer footing by the work of Lagrange in 1797 (see page 624).

Maclaurin made many other contributions to mathematics—especially in the areas of geometry and algebra. He published his *Geometria organica* when only 21 years old. His posthumous work *Treatise of Algebra*, published in 1748, contained many important results, including the well-known *Cramer's rule* for solving a system of equations (Cramer published the result in 1750).

In 1745, when "Bonnie Prince Charlie" marched against Edinburgh, Maclaurin helped defend the city. When the city fell, Maclaurin escaped, fleeing to York, where he died in 1746 at the age of 48.

■ PROBLEMS 10.10

■ SELF-QUIZ

Each of the following power series is the Maclaurin series for a well-known function. Identify each one.

I. $1 + x + x^2 + x^3 + \cdots + x^k + \cdots$

II. $1 - x + x^2 - x^3 + \cdots + (-1)^k x^k + \cdots$

III. $1 + x + \dfrac{x^2}{2!} + \dfrac{x^3}{3!} + \cdots + \dfrac{x^k}{k!} + \cdots$

CONTINUED

IV. $1 - x + \dfrac{x^2}{2!} - \dfrac{x^3}{3!} + \cdots + \dfrac{(-x)^k}{k!} + \cdots$

V. $1 - \dfrac{x^2}{2!} + \dfrac{x^4}{4!} - \dfrac{x^6}{6!} + \cdots + \dfrac{(-1)^k x^{2k}}{(2k)!} + \cdots$

VI. $x - \dfrac{x^3}{3!} + \dfrac{x^5}{5!} - \dfrac{x^7}{7!} + \cdots + \dfrac{(-1)^k x^{2k+1}}{(2k+1)!} + \cdots$

VII. $x + \dfrac{x^3}{3!} + \dfrac{x^5}{5!} + \dfrac{x^7}{7!} + \cdots + \dfrac{x^{2k+1}}{(2k+1)!} + \cdots$

VIII. $1 + \dfrac{x^2}{2!} + \dfrac{x^4}{4!} + \dfrac{x^6}{6!} + \cdots + \dfrac{x^{2k}}{(2k)!} + \cdots$

IX. $1 - x^2 + x^4 - x^6 + \cdots + (-1)^k (x^2)^k + \cdots$

X. $x - \dfrac{x^3}{3} + \dfrac{x^5}{5} - \dfrac{x^7}{7} + \cdots + \dfrac{(-1)^k x^{2k+1}}{(2k+1)} + \cdots$

■ DRILL

1. Find the Taylor series for e^x at 1.
2. Find the Maclaurin series for e^{-x}.
3. Find the Taylor series for $\cos x$ at $\pi/4$.
4. Find the Taylor series for $\sinh x$ at $\ln 2$.
5. Find the Maclaurin series for $e^{\beta x}$, β real.
6. Find the Taylor series for e^x at $x = -1$.
7. Find the Maclaurin series for xe^x.
8. Find the Maclaurin series for $x^2 e^{-x^2}$.
9. Find the Maclaurin series for $(\sin x)/x$.
10. Find the Maclaurin series for $\sin^2 x$. [Hint: $\sin^2 x = (1 - \cos 2x)/2$.]
11. Find the Taylor series for $(x - 1)\ln x$ at 1. Over what interval is this representation valid?
12. Find the first three nonzero terms of the Maclaurin series for $\tan x$. What is the interval of convergence for that Maclaurin series?
13. Find the first four terms of the Taylor series for $\csc x$ at $\pi/2$. What is the interval of convergence for that Taylor series?
14. Find the first three terms of the Maclaurin series for $\ln|\cos x|$. What is its interval of convergence? [Hint: $\int \tan x \, dx = -\ln|\cos x|$.]
15. Find the Taylor series of $\sqrt{x}$ at $x = 4$. What is its radius of convergence?
*16. Find the Maclaurin series of $\sin^{-1} x$. What is its radius of convergence? [Hint: Find the Maclaurin series for $1/\sqrt{1-x}$; then find the Maclaurin series for $1/\sqrt{1-x^2}$; then integrate.]
17. Use the Maclaurin series for $\sin x$ to obtain the Maclaurin series for $\sin x^2$.
18. Find the Maclaurin series for $\cos x^2$.
19. Find the Maclaurin series for $\tan^{-1} x$. What is its radius of convergence? [Hint: Integrate the Maclaurin series for $1/(1 + x^2)$.]
20. Find the Maclaurin series for $\ln|(1 + x)/(1 - x)|$. What is its radius of convergence? [Hint: Integrate the Maclaurin series for $1/(1 - x^2)$.]
21. Use the binomial series (equation (26)) to find a power series representation for $\sqrt[4]{1 + x}$.
22. Use the result of the preceding problem to find a power series representation for $\sqrt[4]{1 + x^3}$.

▦ CALCULATOR PROBLEMS

23. Use the result of Problem 22 to approximate $\int_0^{0.5} \sqrt[4]{1 + x^3} \, dx$ to four significant figures.
24. Use the technique used for Problems 21–23 or Example 9 to approximate $\int_0^{1/4} (1 + \sqrt{x})^{3/5} \, dx$ to four significant figures.

■ APPLICATIONS

25. The **error function** (which arises in mathematical statistics) is defined by

$$\operatorname{erf}(x) = \frac{2}{\sqrt{\pi}} \int_0^x e^{-t^2} \, dt.$$

   a. Find a Maclaurin series for $\operatorname{erf}(x)$ by integrating the Maclaurin series for e^{-x^2}.
   ▦ b. Use the series obtained in part (a) to approximate $\operatorname{erf}(1)$ and $\operatorname{erf}(0.5)$, each with an error less than 0.0001.

▦ 26. The **complementary error function** is defined by

$$\operatorname{erfc}(x) = 1 - \operatorname{erf}(x) = 1 - \frac{2}{\sqrt{\pi}} \int_0^x e^{-t^2} \, dt$$

$$= \frac{2}{\sqrt{\pi}} \int_x^\infty e^{-t^2} \, dt.$$

Find the Maclaurin series for erfc(x); use it to approximate erfc(1) and erfc(0.5) with a maximum error of 0.0001. (Note that for large values of x, erfc(x) can be approximated by integrating the last integral by parts.)

■ SHOW/PROVE/DISPROVE

27. Prove that the series (21) represents $\sin x$ for all real x.

28. a. Prove that the series (22) represents $\cosh x$ for all real x.

    b. Use the fact that $\sinh x = \dfrac{d}{dx} \cosh x$ to derive the series in (23).

29. Differentiate the Maclaurin series for $\sin x$ and show that it is equal to the Maclaurin series for $\cos x$.

30. Differentiate the Maclaurin series for $\sinh x$ and show that it is equal to the Maclaurin series for $\cosh x$.

31. Using the fact that if f has a Taylor series at x_0, then the Taylor series is given by (11), show that $1 + x + x^2 + x^3 + \cdots$ is the Taylor series for $1/(1 - x)$ at $x_0 = 0$ with interval of convergence $(-1, 1)$.

32. a. Show that the Maclaurin series for f has only even powers of x (i.e., the odd powers have coefficient zero) if and only if f is an even function (i.e., $f(-x) = f(x)$ for all x).

    b. The Maclaurin series for f has only odd powers if and only if f is an odd function (i.e., $f(-x) = -f(x)$ for all x). Prove this.

33. Show that the binomial series (equation (26)) converges if $|x| < 1$.

34. Show that for any real number r

$$1 + \frac{r}{2} + \frac{r(r-1)}{2^2 2!} + \cdots$$
$$+ \frac{r(r-1)\cdots(r-k+1)}{2^k k!} + \cdots = \left(\frac{3}{2}\right)^r.$$

*35. The **sine integral** is defined by

$$\text{Si}(x) = \int_0^x \frac{\sin t}{t}\, dt.$$

  a. Show that $\text{Si}(x)$ is defined and continuous for all real x.

  b. Show that $\lim_{x \to \infty} \text{Si}(x)$ exists and is finite.

  c. Find a Maclaurin series expansion for $\text{Si}(x)$.

  d. Approximate $\text{Si}(1)$ and $\text{Si}(0.5)$ with a maximum error of 0.0001.

■ CHALLENGE

*36. Suppose f is a well-behaved function with values tabulated at $a + k \cdot \Delta x$ where $\Delta x = 0.05$. Compare

$$\frac{f(a + \Delta x) - f(a)}{\Delta x} \quad \text{and} \quad \frac{f(a + \Delta x) - f(a - \Delta x)}{2\,\Delta x}$$

as numerical approximations to $f'(a)$.

*37. Suppose $f, f',$ and f'' are continuous in a neighborhood of a; also suppose $f'(a) = 0$. Prove the second-derivative test for maxima-minima.

*38. Prove that e is irrational. [*Hint:* If $e = p/q$, then consider the qth-degree Taylor polynomial for e^1 and its remainder.]

**39. Suppose that $|f(x)| \leq 1$ and $|f''(x)| \leq 1$ on $[-1, 1]$. Prove that, for every x in $[-1, 1]$, $|f'(x)| \leq \sqrt{2}$.

■ ANSWERS TO SELF-QUIZ

   I. $1/(1 - x)$ for $|x| < 1$

  II. $1/(1 + x)$ for $|x| < 1$

 III. e^x for all x

 IV. e^{-x} for all x

  V. $\cos x$ for all x

  VI. $\sin x$ for all x

 VII. $\sinh x$ for all x

VIII. $\cosh x$ for all x

 IX. $1/(1 + x^2)$ for $|x| < 1$

  X. $\tan^{-1} x$ for $|x| < 1$

■ PROBLEMS   CHAPTER 10 REVIEW

■ **DRILL**

1. Find the first five terms of the sequence $\{(n-2)/n\}$.
2. Find the first seven terms of the sequence $\{n^2 \sin n\}$.
3. Find the general term of the sequence $\{\frac{1}{8}, \frac{3}{16}, \frac{5}{32}, \frac{7}{64}, \ldots\}$.
4. Find the general term of the sequence $\{1, -\frac{1}{5}, \frac{1}{25}, -\frac{1}{125}, \frac{1}{625}, \ldots\}$.

In Problems 5–10, determine whether the given sequence is convergent or divergent. If it is convergent, find its limit.

5. $\left\{\dfrac{-7}{n}\right\}$ 6. $\{\cos \pi n\}$ 7. $\left\{\dfrac{\ln n}{\sqrt{n}}\right\}$

8. $\left\{\dfrac{7^n}{n!}\right\}$ 9. $\left\{\left(1-\dfrac{2}{n}\right)^n\right\}$ 10. $\left\{\dfrac{3}{\sqrt{n^2+8}-n}\right\}$

In Problems 11–18, determine whether the given sequence is bounded or unbounded and whether it is increasing, decreasing, or not monotonic. Start with $n = 1$.

11. $\{\sqrt{n} \cos n\}$ 12. $\left\{\dfrac{3}{n+2}\right\}$ 13. $\left\{\dfrac{2^n}{1+2^n}\right\}$

14. $\left\{\dfrac{n!}{n^n}\right\}$ 15. $\left\{\dfrac{\sqrt{n}+1}{n}\right\}$ 16. $\left\{\left(1-\dfrac{1}{n}\right)^{1/n}\right\}$

17. $\left\{\dfrac{n-7}{n+4}\right\}$ 18. $\{(3^n+5^n)^{1/n}\}$

In Problems 19–22, evaluate the given sum.

19. $\displaystyle\sum_{k=2}^{10} 4^k$ 20. $\displaystyle\sum_{k=1}^{\infty} \dfrac{1}{3^k}$

21. $\displaystyle\sum_{k=3}^{\infty}\left[\left(\dfrac{3}{4}\right)^k - \left(\dfrac{2}{5}\right)^k\right]$ 22. $\displaystyle\sum_{k=2}^{\infty} \dfrac{1}{k(k-1)}$

23. Write $0.797979\ldots$ as a rational number.
24. Write $14.2314231423\ldots$ as a rational number.

In Problems 25–36, determine whether the given series converges or diverges.

25. $\displaystyle\sum_{k=1}^{\infty} \dfrac{1}{k^3-5}$ 26. $\displaystyle\sum_{k=5}^{\infty} \dfrac{1}{k(k+6)}$

27. $\displaystyle\sum_{k=1}^{\infty} \dfrac{1}{\sqrt{k^3+4}}$ 28. $\displaystyle\sum_{k=2}^{\infty} \dfrac{3}{\ln k}$

29. $\displaystyle\sum_{k=4}^{\infty} \dfrac{1}{\sqrt[3]{k^3+50}}$ 30. $\displaystyle\sum_{k=1}^{\infty} \dfrac{r^k}{k^r}, \, 0 < r < 1$

31. $\displaystyle\sum_{k=2}^{\infty} \dfrac{10^k}{k^5}$ 32. $\displaystyle\sum_{k=1}^{\infty} \dfrac{k^{6/5}}{8^k}$

33. $\displaystyle\sum_{k=1}^{\infty} \dfrac{\sqrt{k} \ln(k+3)}{k^2+2}$ 34. $\displaystyle\sum_{k=1}^{\infty} \operatorname{csch} k$

35. $\displaystyle\sum_{k=2}^{\infty} \dfrac{e^{1/k}}{k^{3/2}}$

36. $\displaystyle\sum_{k=1}^{\infty} \dfrac{k(k+6)}{(k+1)(k+3)(k+5)}$

In Problems 37–48, determine whether the given alternating series is absolutely convergent, conditionally convergent, or divergent.

37. $\displaystyle\sum_{k=1}^{\infty} \dfrac{(-1)^{k+1}}{50k}$ 38. $\displaystyle\sum_{k=2}^{\infty} \dfrac{(-1)^k \sqrt{k}}{\ln k}$

39. $\displaystyle\sum_{k=2}^{\infty} \dfrac{(-1)^{k+1}}{\sqrt{k(k-1)}}$ 40. $\displaystyle\sum_{k=2}^{\infty} \dfrac{(-1)^k k^2}{k^3+1}$

41. $\displaystyle\sum_{k=2}^{\infty} \dfrac{(-1)^k k^2}{k^4+1}$

42. $\displaystyle\sum_{k=2}^{\infty} \dfrac{(-1)^k k^3}{k^3+1}$

43. $\displaystyle\sum_{k=3}^{\infty} \dfrac{(-1)^k(k+2)(k+3)}{(k+1)^3}$

44. $\displaystyle\sum_{k=2}^{\infty} \dfrac{(-1)^k 3^k}{3^{k+1}}$

45. $\displaystyle\sum_{k=1}^{\infty} \dfrac{(-1)^k k^k}{k!}$

46. $\displaystyle\sum_{k=1}^{\infty} \dfrac{(-1)^k k^4}{k^4+20k^3+17k+2}$

47. $\displaystyle\sum_{k=1}^{\infty} (-1)^k\left(1+\dfrac{1}{k}\right)^k$

48. $\displaystyle\sum_{k=1}^{\infty} \dfrac{(-1)^k k!}{k^k}$

49. Approximate $\sum_{k=1}^{\infty} (-1)^{k+1}/k^3$ with an error less than 0.001.
50. Approximate $\sum_{k=0}^{\infty} (-1)^k/k!$ with an error less than 0.0001.
51. At what time between 9 p.m. and 10 p.m. is the minute hand of a clock exactly over the hour hand?
52. Find the first 10 terms of a rearrangement of the conditionally convergent series $\sum_{k=1}^{\infty} (-1)^{k+1}/k$ that converges to 0.5.

CONTINUED

In Problems 53–62, find the radius of convergence and the interval of convergence of the given power series.

53. $\sum_{k=0}^{\infty} \dfrac{x^k}{3^k}$

54. $\sum_{k=0}^{\infty} \dfrac{(-1)^k x^k}{3^k}$

55. $\sum_{k=0}^{\infty} \dfrac{x^k}{k^2 + 2}$

56. $\sum_{k=1}^{\infty} \dfrac{x^k}{k!}$

57. $\sum_{k=2}^{\infty} \dfrac{x^k}{(2 \ln k)^k}$

58. $\sum_{k=0}^{\infty} \dfrac{(3x + 5)^k}{k!}$

59. $\sum_{k=0}^{\infty} \dfrac{(3x - 5)^k}{3^k}$

60. $\sum_{k=0}^{\infty} \left(\dfrac{k}{6}\right)^k x^k$

61. $\sum_{k=0}^{\infty} (-1)^k x^{3k}$

62. $\sum_{k=1}^{\infty} \dfrac{(\ln k)(x - 1)^k}{k + 2}$

▦ 63. Approximate $\int_0^{1/2} e^{-t^2} \, dt$ with an error less than 0.00001.

▦ 64. Approximate $\int_0^{1/2} \sin t^2 \, dt$ with an error less than 0.0001.

▦ 65. Approximate $\int_0^{1/2} t^3 e^{-t^3} \, dt$ with an error less than 0.001.

▦ 66. Approximate $\int_0^{1/2} \left[1/(t^4 + 1) \right] dt$ with an error less than 0.00001.

67. Find the Maclaurin series for $x^2 e^x$.

68. Find the Taylor series for e^x at $\ln 3$.

69. Find the Maclaurin series for $\cos^2 x$. [*Hint:* $\cos x = (1 + \cos 2x)/2.$]

70. Find the Maclaurin series for $\sin \beta x$, β real.

COMPUTER EXERCISES

1. a. Show that $\sum_{n=1}^{\infty} \dfrac{1}{n^2}$ is a convergent series.

   b. Use a computer algebra system to evaluate the following partial sums.

   (i) $\sum_{n=1}^{10} \dfrac{1}{n^2}$ (ii) $\sum_{n=1}^{100} \dfrac{1}{n^2}$ (iii) $\sum_{n=1}^{500} \dfrac{1}{n^2}$

   c. How does the integral $\int_{500}^{\infty} \dfrac{1}{x^2} \, dx$ compare with the sum $\sum_{n=501}^{\infty} \dfrac{1}{n^2}$? (See Figure 1.)

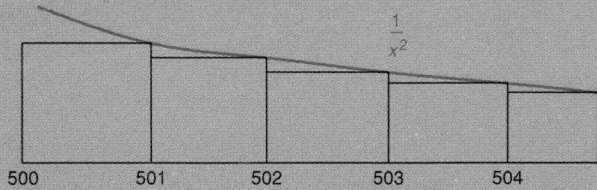

FIGURE 1

   d. If $\sum_{n=1}^{500} \dfrac{1}{n^2}$ is used to approximate $\sum_{n=1}^{\infty} \dfrac{1}{n^2}$, find a bound on the error. Express your answer as an integral and evaluate it.

2. In the eighteenth century, Leonhard Euler showed that $\sum_{n=1}^{\infty} \dfrac{1}{n^2} = \dfrac{\pi^2}{6}$. While Euler was in fact able to

evaluate $\sum_{n=1}^{\infty} \dfrac{1}{n^k}$ for every positive even integer k, he was unable to obtain an "exact value" for $\sum_{n=1}^{\infty} \dfrac{1}{n^3}$. This problem has resisted the efforts of the strongest mathematicians for two centuries and remains unsolved today. The series was only recently shown to converge to an irrational number. Find, accurate to 10 decimal places,

   a. $\sum_{n=1}^{100} \dfrac{1}{n^3}$ b. $\sum_{n=1}^{1000} \dfrac{1}{n^3}$ c. $\sum_{n=1}^{5000} \dfrac{1}{n^3}$

3. Euler's constant is defined to be $\lim_{n \to \infty} \gamma_n = \gamma$ where $\gamma_n = (1 + \frac{1}{2} + \cdots + \frac{1}{n}) - \ln(n + 1)$. (Notice that $\lim_{n \to \infty} (1 + \frac{1}{2} + \cdots + \frac{1}{n})$ is the harmonic series and so diverges and that $\lim_{n \to \infty} \ln(n + 1) = \infty$. One of the goals of this exercise is to show that the limit of the difference exists. Does this square with theorems you know concerning sums and differences of limits?)

   a. Show that Euler's constant is in fact the sum of the shaded areas in Figure 2.

   b. Use Figure 2 to show that the sequence $\{\gamma_n\}$ is increasing and bounded above. (Therefore the limit exists.)

   c. Calculate γ_n for $n = 100$, 500, and 1,000. Use your answers to obtain an approximation of γ.

   d. Write down and evaluate an integral that gives a bound on the error in using $\gamma_{1,000}$ to approximate γ. (Look back at Figure 2.)

CONTINUED

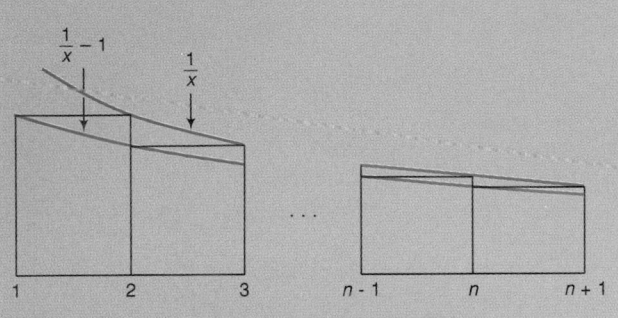

FIGURE 2

e. Use the technique you used to obtain an error estimate in (d) to find the value of n that is needed if γ_n is to approximate γ with error $\leq 10^{-10}$.

4. If $f(x)$ is a function on an open interval about the origin, the partial sums of its Maclaurin series are polynomials which can be graphed along with f. If you compare the graphs, you can see what the geometric meaning of convergence is when f is analytic. Use a computer algebra system to do the calculations and the graphing in each of the following:

   a. Graph $f(x) = e^x$ along with the first three distinct partial sums of its Maclaurin series on the same axes.

   b. By zooming into the relevant spots on the graph, estimate the largest value of a such that

      i. the first-degree Maclaurin polynomial is a "good" approximation to $f(x)$ on the inter-val $[0, a)$. (By "good" we mean an error of about 0.05.)

      ii. the second-degree Maclaurin polynomial is a "good" approximation to $f(x)$ on the interval $[0, a)$.

      iii. the third-degree Maclaurin polynomial is a "good" approximation to $f(x)$ on the interval $[0, a)$.

   c. Explain how you can use your computer algebra system to approximate these three numbers without resorting to a graph. What results do you get to two decimal places?

5. a. Graph $f(x) = \tan x$ along with the first three distinct partial sums of its Maclaurin series on the same axes.

   b. By zooming into the relevant spots on the graph, estimate the largest value of a such that

      i. the first-degree Maclaurin polynomial is a "good" approximation to $f(x)$ on the inter-val $[0, a)$. (By "good" we mean an error of about 0.05.)

      ii. the third-degree Maclaurin polynomial is a "good" approximation to $f(x)$ on the interval $[0, a)$.

      iii. the fifth-degree Maclaurin polynomial is a "good" approximation to $f(x)$ on the in-terval $[0, a)$.

   c. Explain how you can use your computer algebra system to approximate these three numbers without resorting to a graph. What results do you get to two decimal places?

11

VECTORS IN THE PLANE AND IN SPACE

With this chapter we begin a new subject in the study of calculus: the study of vectors and vector functions. This will lead us, in Chapter 13, to the study of functions of two and more variables, and it is a sharp departure from the first 10 chapters of this book.

The modern study of vectors began essentially with the work of the great Irish mathematician Sir William Rowan Hamilton (1805–1865) who worked with what he called quaternions.[†] After Hamilton's death, his work on quaternions was supplanted by the more adaptable work on vector analysis by the American mathematician and physicist Josiah Willard Gibbs (1839–1903) and the general treatment of ordered n-tuples by the German mathematician Hermann Grassmann (1809–1877).

Throughout Hamilton's life and for the remainder of the nineteenth century, there was considerable debate over the usefulness of quaternions and vectors. At the end of the century, the great British physicist Lord Kelvin wrote that quaternions, "although beautifully ingenious, have been an unmixed evil to those who have touched them in any way. . . . Vectors . . . have never been of the slightest use to any creature."

But Kelvin was wrong. Today nearly all branches of classical and modern physics are represented using the language of vectors. Vectors are also used with increasing frequency in the social and biological sciences. Quaternions, too, have recently been used in physics—in particle theory and other areas.

In this chapter, we will explore properties of vectors in the plane and in space. When going through this material, keep in mind that, like most important discoveries, vectors have been a source of great controversy—a controversy that was not resolved until well into the twentieth century.[‡]

11.1 VECTORS AND VECTOR OPERATIONS

In many applications of mathematics to the physical and biological sciences and engineering, scientists are concerned with entities that have both magnitude (length) and direction. Examples include the notions of force, velocity, acceleration, and momentum. It is frequently useful to express these quantities both geometrically and algebraically.

Let P and Q be two different points in the plane. Then the **directed line segment** from P to Q, denoted by $\overrightarrow{PQ}$, is the straight-line segment extending

---

[†] See the biographical sketch of Hamilton on p. 727.

[‡] For interesting discussions of the development of modern vector analysis, consult the book by M. J. Crowe, *A History of Vector Analysis* (Notre Dame: University of Notre Dame Press, 1967), or Morris Kline's excellent book *Mathematical Thought from Ancient to Modern Times* (New York: Oxford University Press, 1972), Chapter 32.

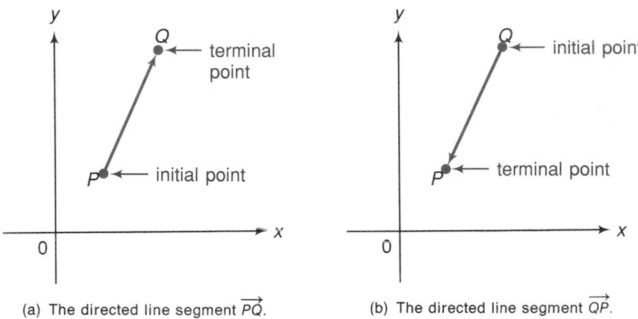

(a) The directed line segment $\overrightarrow{PQ}$. (b) The directed line segment $\overrightarrow{QP}$.

FIGURE 1 THE DIRECTED LINE SEGMENTS $\overrightarrow{PQ}$ AND $\overrightarrow{QP}$ POINT IN OPPOSITE DIRECTIONS.

from P to Q [see Figure 1(a)]. Note that the directed line segments $\overrightarrow{PQ}$ and $\overrightarrow{QP}$ are different since they point in opposite directions [Figure 1(b)].[†]

The point P in the directed line segment $\overrightarrow{PQ}$ is called the **initial point** of the segment and the point Q is called the **terminal point**. The two important properties of a directed line segment are its magnitude (length) and its direction. If two directed line segments $\overrightarrow{PQ}$ and $\overrightarrow{RS}$ have the same magnitude and direction, we say that they are **equivalent** no matter where they are located with respect to the origin. The directed line segments in Figure 2 are all equivalent.

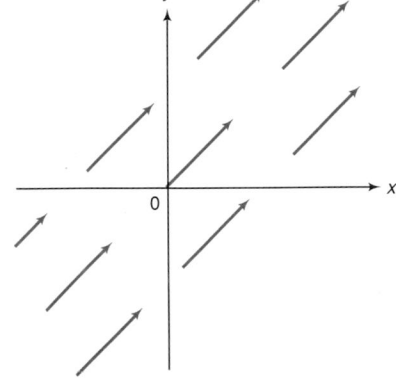

FIGURE 2
EQUIVALENT DIRECTED LINE SEGMENTS

DEFINITION **GEOMETRIC DEFINITION OF A VECTOR**

The set of all directed line segments equivalent to a given directed line segment is called a **vector**. Any directed line segment in that set is called a **representation** of the vector. ∎

REMARK: The directed line segments in Figure 2 are all representations of the same vector.

NOTATION: We will denote vectors by lowercase boldface letters such as **v, w, a, b**.

From the definition, we see that a given vector **v** can be represented in many different ways. In fact, let $\overrightarrow{PQ}$ be a representation of **v**. Then without changing magnitude or direction, we can move $\overrightarrow{PQ}$ in a parallel way so that its initial point is shifted to the origin. We then obtain the directed line segment $\overrightarrow{OR}$, which is another representation of the vector **v** (see Figure 3). It is called the **standard representation** of the vector. Now suppose that R has the Cartesian coordinates (a, b). Then we can describe the directed line segment $\overrightarrow{OR}$ by the coordinates (a, b). That is, $\overrightarrow{OR}$ is the directed line segment with initial point $(0, 0)$ and terminal point (a, b). Since one representation of a vector is as good as another, we can write the vector **v** as (a, b). In sum, we see that a vector can be thought of as a point in the xy-plane.

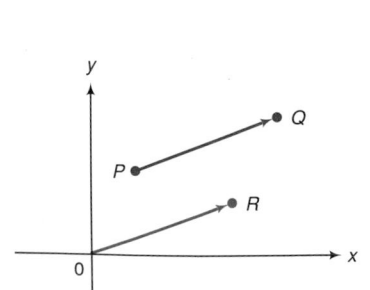

FIGURE 3
$\overrightarrow{OR}$ IS THE STANDARD REPRESENTATION OF THE DIRECTED LINE SEGMENT $\overrightarrow{PQ}$.

We can move $\overrightarrow{PQ}$ to obtain an equivalent directed line segment with its initial point at the origin. Note that $\overrightarrow{OR}$ and $\overrightarrow{PQ}$ are parallel and have the same length.

REMARK: In a number of places in this text, we shall use phrases such as "a vector lies in a plane." This is common shorthand usage. The precise statement is "a vector has a representation as a directed line segment all of

---

[†] We give a formal definition of the direction of a vector on page 722. Here we rely on your intuition.

whose points lie in the plane." In the remainder of this book, we shall not worry about this distinction.

DEFINITION   **ALGEBRAIC DEFINITION OF A VECTOR**

A **vector v** in the xy-plane is an ordered pair of real numbers (a, b). The numbers a and b are called the **components** of the vector **v**. The **zero vector** is the vector $(0, 0)$ and is denoted by **0**. Two vectors are **equal** if their corresponding components are equal. That is, $(a, b) = (c, d)$ if $a = c$ and $b = d$. ∎

DEFINITION   **SCALAR**

Since we will often have to distinguish between real numbers and vectors (which are pairs of real numbers), we will use the term **scalar**[†] to denote a real number. ∎

DEFINITION   **MAGNITUDE OF A VECTOR**

Since a vector is really a set of equivalent directed line segments, we define the **magnitude** or **length** of a vector as the length of any one of its representations.

Using the representation $\overrightarrow{OR}$, and writing the vector $\mathbf{v} = (a, b)$, we find that

$$|\mathbf{v}| = \text{magnitude of } \mathbf{v} = \sqrt{a^2 + b^2}. \quad ∎ \tag{1}$$

This follows from the Pythagorean theorem (see Figure 4). We have used the notation $|\mathbf{v}|$ to denote the magnitude of **v**. Note that $|\mathbf{v}|$ is a *scalar*.

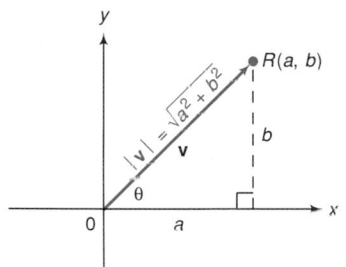

FIGURE 4
THE MAGNITUDE OF A VECTOR

EXAMPLE 1
CALCULATING THE MAGNITUDES OF SIX VECTORS

Calculate the magnitudes of the vectors **(a)** $(2, 2)$; **(b)** $(2, 2\sqrt{3})$; **(c)** $(-2\sqrt{3}, 2)$; **(d)** $(-3, -3)$; **(e)** $(6, -6)$; **(f)** $(0, 3)$.

SOLUTION:

(a) $|\mathbf{v}| = \sqrt{2^2 + 2^2} = \sqrt{8} = 2\sqrt{2}$

(b) $|\mathbf{v}| = \sqrt{2^2 + (2\sqrt{3})^2} = \sqrt{16} = 4$

(c) $|\mathbf{v}| = \sqrt{(-2\sqrt{3})^2 + 2^2} = 4$

(d) $|\mathbf{v}| = \sqrt{(-3)^2 + (-3)^2} = \sqrt{18} = 3\sqrt{2}$

(e) $|\mathbf{v}| = \sqrt{6^2 + (-6)^2} = \sqrt{72} = 6\sqrt{2}$

(f) $|\mathbf{v}| = \sqrt{0^2 + 3^2} = \sqrt{9} = 3$

[†] The term "scalar" originated with Hamilton. His definition of the quaternion included what he called a *real part* and an *imaginary part*. In his paper, "On Quaternions, or on a New System of Imaginaries in Algebra," in *Philosophical Magazine*, 3rd Ser., 25 (1844): 26–27, he wrote

The algebraically *real* part may receive . . . all values contained on the one *scale* of progression of numbers from negative to positive infinity; we shall call it therefore the *scalar part*, or simply the *scalar* of the quaternion.

Moreover, in the same paper, Hamilton went on to define the imaginary part of his quaternion as the *vector* part. Although this was not the first usage of the word *vector*, it was the first time it was used in the context of the definition of a vector given above. In fact, it is fair to say that the paper from which the quotation above was taken marks the beginning of modern vector analysis.

DEFINITION **DIRECTION OF A VECTOR**

We now define the **direction** of the nonzero vector $\mathbf{v} = (a, b)$ to be the angle θ, measured in radians, that the standard representation of the vector makes with the positive x-axis. By convention, we choose θ such that $0 \le \theta < 2\pi$.

It follows from Figure 4 that if $a \ne 0$, then

$$\tan \theta = \frac{b}{a}. \quad \blacksquare \tag{2}$$

Tan θ is periodic of period π, so for $a \ne 0$ there are always *two* numbers in $[0, 2\pi)$ such that $\tan \theta = b/a$. For example, $\tan \pi/4 = \tan 5\pi/4 = 1$. In order to determine θ uniquely, it is necessary to determine the quadrant of $\mathbf{v}$ as we will see in the next example.

REMARK 1: The zero vector has a magnitude of 0. Since the initial and terminal points coincide, we say that *the zero vector has no direction.*

REMARK 2: It follows from the definition that *parallel vectors have the same direction*. This is because two parallel vectors of the same magnitude have the same standard representation.

EXAMPLE 2
CALCULATING THE DIRECTIONS
OF SIX VECTORS

Calculate the directions of the vectors in Example 1.

SOLUTION: We depict these six vectors in Figure 5.

(a) Here $\mathbf{v}$ is in the first quadrant, and since $\tan \theta = 2/2 = 1$, $\theta = \pi/4$.

(b) Here $\theta = \tan^{-1} 2\sqrt{3}/2 = \tan^{-1} \sqrt{3} = \pi/3$.

(c) We see that $\mathbf{v}$ is in the second quadrant, and since $\tan^{-1} 2/(2\sqrt{3}) = \tan^{-1} 1/\sqrt{3} = \pi/6$, we see from the figure that $\theta = \pi - (\pi/6) = 5\pi/6$.

(d) Here $\mathbf{v}$ is in the third quadrant, and since $\tan^{-1} 1 = \pi/4$, $\theta = \pi + (\pi/4) = 5\pi/4$.

(e) Since $\mathbf{v}$ is in the fourth quadrant, and since $\tan^{-1}(-1) = -\pi/4$, $\theta = 2\pi - (\pi/4) = 7\pi/4$.

(f) We cannot use equation (2) because b/a is undefined. However, we see in Figure 5(f) that $\theta = \pi/2$.

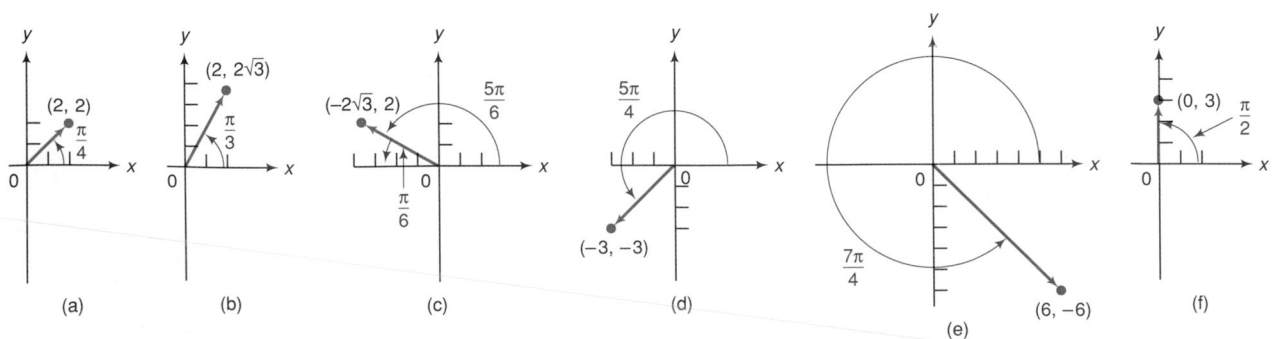

FIGURE 5 THE DIRECTIONS OF SIX VECTORS

In general, if $b > 0$, the direction of $(0, b) = \dfrac{\pi}{2}$ and the direction of $(0, -b) = \dfrac{3\pi}{2}$.

⊗ **WARNING:** Recall from the definition of the function $\tan^{-1} x$ on page 459 (Section 6.8) that $-\pi/2 < \tan^{-1} x < \pi/2$. But $0 \leq$ direction of $\mathbf{v} < 2\pi$. Thus,

$$\theta = \text{direction of } \mathbf{v} \text{ is not necessarily equal to } \tan^{-1} \dfrac{b}{a}.$$

In fact, $\theta = \tan^{-1} b/a$ only if θ is in the first quadrant. In the other cases, we compute θ as in Example 2. ⊗

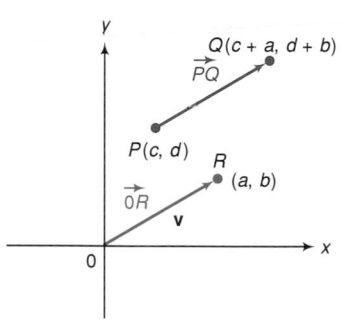

FIGURE 6
THE DIRECTED LINE SEGMENTS $\overrightarrow{PQ}$ AND $\overrightarrow{OR}$ ARE EQUIVALENT.

Let $\mathbf{v} = (a, b)$. Then as we have seen, $\mathbf{v}$ can be represented in many different ways. For example, the representation of $\mathbf{v}$ with the initial point (c, d) has the terminal point $(c + a, d + b)$. This is depicted in Figure 6. It is not difficult to show that the directed line segment $\overrightarrow{PQ}$ in Figure 6 has the same magnitude and direction as the vector $\mathbf{v}$ (see Problems 44 and 57).

EXAMPLE 3
FINDING A REPRESENTATION OF A VECTOR WHOSE INITIAL POINT IS NOT AT THE ORIGIN

Find a representation of the vector $(2, -1)$ whose initial point is the point $P = (5, -4)$.

SOLUTION: Let $Q = (5 + 2, -4 - 1) = (7, -5)$. Then $\overrightarrow{PQ}$ is a representation of the vector $(2, -1)$. This is illustrated in Figure 7.

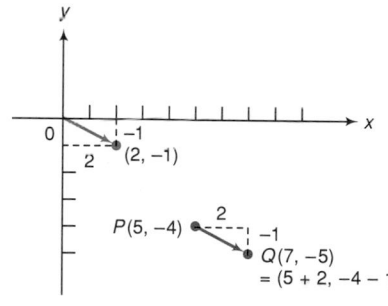

FIGURE 7
EQUIVALENT DIRECTED LINE SEGMENTS

We now turn to the question of adding vectors and multiplying them by scalars.

DEFINITION   **ADDITION AND SCALAR MULTIPLICATION OF A VECTOR**

Let $\mathbf{u} = (a_1, b_1)$ and $\mathbf{v} = (a_2, b_2)$ be two vectors in the plane and let α be a scalar. Then we define

    (i) $\mathbf{u} + \mathbf{v} = (a_1 + a_2, b_1 + b_2)$
    (ii) $\alpha\mathbf{u} = (\alpha a_1, \alpha b_1)$
    (iii) $-\mathbf{v} = (-1)\mathbf{v} = (-a_2, -b_2)$
    (iv) $\mathbf{u} - \mathbf{v} = \mathbf{u} + (-\mathbf{v}) = (a_1 - a_2, b_1 - b_2)$. ∎

We have:

> **VECTOR ADDITION AND SCALAR MULTIPLICATION**
>
> To add two vectors, we add their corresponding components. To multiply a vector by a scalar, we multiply each of its components by that scalar.

EXAMPLE 4
FINDING SUMS AND SCALAR
MULTIPLES OF VECTORS

Let $\mathbf{u} = (1, 3)$ and $\mathbf{v} = (-2, 4)$. Calculate **(a)** $\mathbf{u} + \mathbf{v}$; **(b)** $3\mathbf{u}$; **(c)** $-\mathbf{v}$; **(d)** $\mathbf{u} - \mathbf{v}$; and **(e)** $-3\mathbf{u} + 5\mathbf{v}$.

SOLUTION:

(a) $\mathbf{u} + \mathbf{v} = (1 + (-2), 3 + 4) = (-1, 7)$
(b) $3\mathbf{u} = 3(1, 3) = (3, 9)$
(c) $-\mathbf{v} = (-1)(-2, 4) = (2, -4)$
(d) $\mathbf{u} - \mathbf{v} = \mathbf{u} + (-\mathbf{v}) = (1 + 2, 3 - 4) = (3, -1)$
(e) $-3\mathbf{u} + 5\mathbf{v} = (-3, -9) + (-10, 20) = (-13, 11)$

There are interesting geometric interpretations of vector addition and scalar multiplication. First, let $\mathbf{v} = (a, b)$ and let α be any scalar. Then,

$$|\alpha\mathbf{v}| = |(\alpha a, \alpha b)| = \sqrt{\alpha^2 a^2 + \alpha^2 b^2} = |\alpha|\sqrt{a^2 + b^2} = |\alpha|\,|\mathbf{v}|.$$

That is, multiplying a vector by a nonzero scalar has the effect of multiplying the length of the vector by the absolute value of that scalar.

Moreover, if $\alpha > 0$, then $\alpha\mathbf{v}$ is in the same quadrant as $\mathbf{v}$ and, since $\tan^{-1}(\alpha b/\alpha a) = \tan^{-1}(b/a)$, the direction of $\alpha\mathbf{v}$ is the same as the direction of $\mathbf{v}$. If $\alpha < 0$, then the direction of $\alpha\mathbf{v}$ is equal to the direction of $\mathbf{v}$ plus π (which is the direction of $-\mathbf{v}$). Thus:

Scalar multiples of $\mathbf{v}$ have the same direction as $\mathbf{v}$ or $-\mathbf{v}$.

EXAMPLE 5
GEOMETRIC INTERPRETATION OF
THE SCALAR MULTIPLE OF
A VECTOR

Let $\mathbf{v} = (1, 1)$. Then $|\mathbf{v}| = \sqrt{1 + 1} = \sqrt{2}$ and $|2\mathbf{v}| = |(2, 2)| = \sqrt{2^2 + 2^2} = \sqrt{8} = 2\sqrt{2} = 2|\mathbf{v}|$. Also, $|-2\mathbf{v}| = \sqrt{(-2)^2 + (-2)^2} = 2\sqrt{2} = 2|\mathbf{v}|$. Moreover, the direction of $2\mathbf{v}$ is $\pi/4$, while the direction of $-2\mathbf{v}$ is $5\pi/4$.

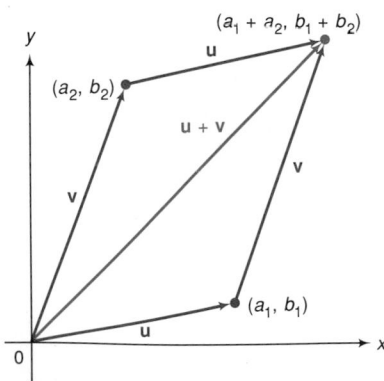

FIGURE 8
SUM OF TWO VECTORS

Now suppose we add the vectors $\mathbf{u} = (a_1, b_1)$ and $\mathbf{v} = (a_2, b_2)$, as in Figure 8. From the figure we see that the vector $\mathbf{u} + \mathbf{v} = (a_1 + a_2, b_1 + b_2)$ can be obtained by shifting the representation of the vector $\mathbf{v}$ so that its initial point coincides with the terminal point (a_1, b_1) of the vector $\mathbf{u}$. We can therefore obtain the vector $\mathbf{u} + \mathbf{v}$ by drawing a parallelogram with one vertex at the origin and sides $\mathbf{u}$ and $\mathbf{v}$. Then $\mathbf{u} + \mathbf{v}$ is the vector that points from the origin along the diagonal of the parallelogram.

NOTE: Since a straight line is the shortest distance between two points, it immediately follows from Figure 8 that

---

TRIANGLE INEQUALITY

$$|\mathbf{u} + \mathbf{v}| \le |\mathbf{u}| + |\mathbf{v}|.$$

For obvious reasons this inequality is called the **triangle inequality**.

We can also obtain a geometric representation of the vector $\mathbf{u} - \mathbf{v}$. Since $\mathbf{u} = \mathbf{u} - \mathbf{v} + \mathbf{v}$, the vector $\mathbf{u} - \mathbf{v}$ is the vector that must be added to $\mathbf{v}$ to obtain $\mathbf{u}$. This is illustrated in Figure 9.

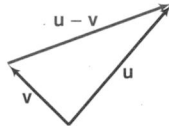

FIGURE 9
THE VECTOR **u** − **v**

The following theorem lists several properties that hold for any vectors **u**, **v**, and **w** and any scalars α and β. Since the proof is easy, we leave it as an exercise (see Problem 58). Some parts of this theorem have already been proven.

THEOREM 1   **PROPERTIES OF VECTORS**

Let **u**, **v**, and **w** be any three vectors in the plane, let α and β be scalars, and let **0** denote the zero vector.

(i) $\mathbf{u} + \mathbf{v} = \mathbf{v} + \mathbf{u}$ **(ii)** $\mathbf{u} + (\mathbf{v} + \mathbf{w}) = (\mathbf{u} + \mathbf{v}) + \mathbf{w}$

(iii) $\mathbf{v} + \mathbf{0} = \mathbf{v}$ **(iv)** $0\mathbf{v} = \mathbf{0}$ (here the 0 on the left is the scalar zero)

(v) $\alpha\mathbf{0} = \mathbf{0}$ **(vi)** $(\alpha\beta)\mathbf{v} = \alpha(\beta\mathbf{v})$

(vii) $\mathbf{v} + (-\mathbf{v}) = \mathbf{0}$ **(viii)** $1\mathbf{v} = \mathbf{v}$

(ix) $(\alpha + \beta)\mathbf{v} = \alpha\mathbf{v} + \beta\mathbf{v}$ **(x)** $\alpha(\mathbf{u} + \mathbf{v}) = \alpha\mathbf{u} + \alpha\mathbf{v}$

(xi) $|\alpha\mathbf{v}| = |\alpha|\,|\mathbf{v}|$ **(xii)** $|\mathbf{u} + \mathbf{v}| \le |\mathbf{u}| + |\mathbf{v}|$ ∎

$\mathbb{R}^2$

When a set of vectors together with a set of scalars and the operations of addition and scalar multiplication have the properties given in Theorem 1(i)–(x), we say that the vectors form a **vector space**. The set of vectors of the form (a, b), where a and b are real numbers, is denoted by $\mathbb{R}^2$. We will not discuss properties of abstract vector spaces here, except to say that all abstract vector spaces have properties very similar to the properties of the vector space $\mathbb{R}^2$.

There are two special vectors in $\mathbb{R}^2$ that allow us to represent other vectors in $\mathbb{R}^2$ in a convenient way. We will denote the vector $(1, 0)$ by the vector symbol **i** and the vector $(0, 1)$ by the vector symbol **j** (see Figure 10). If (a, b) denotes any vector in $\mathbb{R}^2$, then since $(a, b) = a(1, 0) + b(0, 1)$, we may write

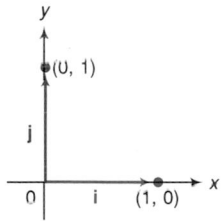

FIGURE 10
THE VECTORS **i** AND **j**

WRITING A VECTOR IN TERMS OF i AND j

$$\mathbf{v} = (a, b) = a\mathbf{i} + b\mathbf{j}. \tag{3}$$

Moreover, any vector in $\mathbb{R}^2$ can be represented in a unique way in the form $a\mathbf{i} + b\mathbf{j}$ since the representation of (a, b) as a point in the plane is unique. (Put another way, a point in the xy-plane has one and only one x-coordinate and one and only one y-coordinate.) Thus Theorem 1 holds with this new representation as well.

HISTORICAL NOTE: The symbols **i** and **j** were first used by Hamilton. He defined his quaternion as a quantity of the form $a + b\mathbf{i} + c\mathbf{j} + d\mathbf{k}$, where a was the "scalar part" and $b\mathbf{i} + c\mathbf{j} + d\mathbf{k}$ the "vector part." In Section 11.4 we will write vectors in space in the form $b\mathbf{i} + c\mathbf{j} + d\mathbf{k}$.

When the vector **v** is written in the form $\mathbf{v} = a\mathbf{i} + b\mathbf{j}$, we say that **v** *is resolved into its horizontal and vertical components*, since a is the horizontal component of **v** while b is its vertical component. The vectors **i** and **j** are called **basis vectors** for the vector space $\mathbb{R}^2$.

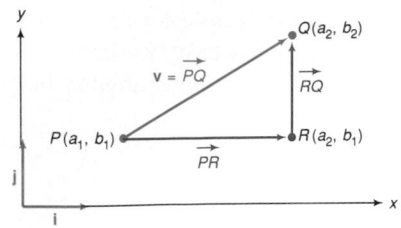

FIGURE 11
$\overrightarrow{PQ} = \overrightarrow{PR} + \overrightarrow{RQ} = (a_2 - a_1)\mathbf{i} + (b_2 - b_1)\mathbf{j}$

Now suppose that a vector $\mathbf{v}$ can be represented by the directed line segment $\overrightarrow{PQ}$, where $P = (a_1, b_1)$ and $Q = (a_2, b_2)$. (See Figure 11). If we label the point (a_2, b_1) as R, then we immediately see that

$$\mathbf{v} = \overrightarrow{PQ} = \overrightarrow{PR} + \overrightarrow{RQ}. \tag{4}$$

If $a_2 \geq a_1$, then the length of $\overrightarrow{PR}$ is $a_2 - a_1$, and since $\overrightarrow{PR}$ has the same direction as $\mathbf{i}$ (they are parallel), we can write

$$\overrightarrow{PR} = (a_2 - a_1)\mathbf{i}. \tag{5}$$

If $a_2 < a_1$, then the length of $\overrightarrow{PR}$ is $a_1 - a_2$, but then $\overrightarrow{PR}$ has the same direction as $-\mathbf{i}$ so $\overrightarrow{PR} = (a_1 - a_2)(-\mathbf{i}) = (a_2 - a_1)\mathbf{i}$ again. Similarly,

$$\overrightarrow{RQ} = (b_2 - b_1)\mathbf{j}, \tag{6}$$

and we may write [using (4), (5), and (6)]

$$\mathbf{v} = (a_2 - a_1)\mathbf{i} + (b_2 - b_1)\mathbf{j}. \tag{7}$$

This is the **resolution of v into its horizontal and vertical components**.

EXAMPLE 6
RESOLVING A VECTOR INTO ITS
HORIZONTAL AND VERTICAL
COMPONENTS

Resolve the vector represented by the directed line segment from $(-2, 3)$ to $(1, 5)$ into its horizontal and vertical components.

SOLUTION: Using (7), we have

$$\mathbf{v} = (a_2 - a_1)\mathbf{i} + (b_2 - b_1)\mathbf{j} = [1 - (-2)]\mathbf{i} + (5 - 3)\mathbf{j} = 3\mathbf{i} + 2\mathbf{j}.$$

We conclude this section by defining a kind of vector that is very useful in certain types of applications.

DEFINITION **UNIT VECTOR**

A **unit vector** $\mathbf{u}$ is a vector that has length 1. ■

EXAMPLE 7
A UNIT VECTOR

The vector $\mathbf{u} = (\frac{1}{2})\mathbf{i} + (\sqrt{3}/2)\mathbf{j}$ is a unit vector since

$$|\mathbf{u}| = \sqrt{\left(\frac{1}{2}\right)^2 + \left(\frac{\sqrt{3}}{2}\right)^2}$$

$$= \sqrt{\frac{1}{4} + \frac{3}{4}} = 1.$$

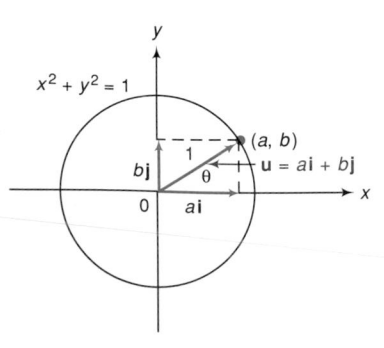

FIGURE 12
REPRESENTATION OF A UNIT VECTOR

Let $\mathbf{u} = a\mathbf{i} + b\mathbf{j}$ be a unit vector. Then $|\mathbf{u}| = \sqrt{a^2 + b^2} = 1$, so $a^2 + b^2 = 1$ and $\mathbf{u}$ is a point on the unit circle (see Figure 12). If θ is the direction of $\mathbf{u}$, then we immediately see that $a = \cos \theta$ and $b = \sin \theta$. Thus any unit vector $\mathbf{u}$ can be written in the form

$$\mathbf{u} = (\cos \theta)\mathbf{i} + (\sin \theta)\mathbf{j} \tag{8}$$

where θ is the direction of $\mathbf{u}$.

SIR WILLIAM ROWAN
HAMILTON
· · ·
1805–1865

Born in Dublin in 1805, where he spent most of his life, William Rowan Hamilton was without question Ireland's greatest mathematician. Hamilton's father (an attorney) and mother died when he was a small boy. His uncle, a linguist, took over the boy's education. By his fifth birthday, Hamilton could read English, Hebrew, Latin, and Greek. By his 13th birthday he had mastered not only the languages of continental Europe, but also Sanscrit, Chinese, Persian, Arabic, Malay, Hindi, Bengali, and several others as well. Hamilton liked to write poetry, both as a child and as an adult, and his friends included the great English poets Samuel Taylor Coleridge and William Wordsworth. Hamilton's poetry was considered so bad, however, that it is fortunate that he developed other interests—especially in mathematics.

Although he enjoyed mathematics as a young boy, Hamilton's interest was greatly enhanced by a chance meeting at the age of 15 with Zerah Colburn, the American lightning calculator. Shortly afterwards, Hamilton began to read important mathematical books of the time. In 1823, at the age of 18, he discovered an error in Simon Laplace's *Mécanique céleste* and wrote an impressive paper on the subject. A year later he entered Trinity College in Dublin.

Hamilton's university career was astonishing. At the age of 21, while still an undergraduate, he had so impressed the faculty that he was appointed Royal Astronomer of Ireland and Professor of Astronomy at the University. Shortly thereafter, he wrote what is now considered a classical work on optics. Using only mathematical theory, he predicted conical refraction in certain types of crystals. Later, this theory was confirmed by physicists. Largely because of this work, Hamilton was knighted in 1835.

Hamilton's first great purely mathematical paper appeared in 1833. In this work he described an algebraic way to manipulate pairs of real numbers. This work gives rules that are used today to add, subtract, multiply, and divide complex numbers. At first, however, Hamilton was unable to devise a multiplication for triples or *n*-tuples of numbers for $n > 2$. For 10 years he pondered this problem, and it is said that he solved it in an inspiration while walking on the Brougham Bridge in Dublin in 1843. The key was to discard the familiar commutative property of multiplication. The new objects he created were called *quaternions*, which were the precursors of what we now call vectors.

For the rest of his life, Hamilton spent most of his time developing the algebra of quaternions. He felt that they would have revolutionary significance in mathematical physics. His monumental work on this subject, *Treatise on Quaternions*, was published in 1853. Thereafter, he worked on an enlarged work, *Elements of Quaternions*. Although Hamilton died in 1865 before his *Elements* was completed, the work was published by his son in 1866.

Students of mathematics and physics know Hamilton in a variety of other contexts. In mathematical physics, for example, one encounters the Hamiltonian function, which often represents the total energy in a system and the Hamilton-Jacobi differential equations of dynamics. In matrix theory, the Cayley-Hamilton theorem states that every matrix satisfies its own characteristic equation.

Despite the great work he was doing, Hamilton's final years were a torment to him. His wife was a semi-invalid and he was plagued by alcoholism. It is therefore gratifying to point out that during these last years, the newly formed American National Academy of Sciences elected Sir William Rowan Hamilton to be its first foreign associate.

For a fascinating account of Hamilton's discoveries, read the article "Hamilton, Rodrigues, and the Quaternion Scandal—what went wrong with one of the major mathematical discoveries of the nineteenth century" in *Mathematics Magazine* 62(5) December, 1989, 291–308.

EXAMPLE 8
WRITING A UNIT VECTOR AS
$\cos \theta \mathbf{i} + \sin \theta \mathbf{j}$

The unit vector $\mathbf{u} = (\frac{1}{2})\mathbf{i} + (\sqrt{3}/2)\mathbf{j}$ of Example 7 can be written in the form (8) with $\theta = \cos^{-1}(\frac{1}{2}) = \pi/3$. Note that since $\cos \theta = \frac{1}{2}$ and $\sin \theta = \sqrt{3}/2$, θ is in the first quadrant. We need this fact to conclude that $\theta = \pi/3$. It is also true that $\cos 5\pi/3 = \frac{1}{2}$, but $5\pi/3$ is in the fourth quadrant.

Finally (see Problem 55):

Let $\mathbf{v}$ be any nonzero vector. Then $\mathbf{u} = \mathbf{v}/|\mathbf{v}|$ is the unit vector having the same direction as $\mathbf{v}$.

EXAMPLE 9
FINDING A UNIT VECTOR THAT HAS THE SAME DIRECTION AS A GIVEN VECTOR

Find the unit vector having the same direction as $\mathbf{v} = 2\mathbf{i} - 3\mathbf{j}$.

SOLUTION: Here $|\mathbf{v}| = \sqrt{4 + 9} = \sqrt{13}$, so

$$\mathbf{u} = \mathbf{v}/|\mathbf{v}| = (2/\sqrt{13})\mathbf{i} - (3/\sqrt{13})\mathbf{j}$$

is the required unit vector.

■ PROBLEMS 11.1

■ SELF-QUIZ

I. A *vector* is _____.
   a. two points in the xy-plane
   b. a line segment between two points
   c. a directed line segment from one point to another
   d. a collection of equivalent directed line segments

II. If $P = (3, -4)$ and $Q = (8, 6)$, the vector $\overrightarrow{PQ}$ has length _____.
   a. $|3| + |-4|$ b. $(3)^2 + (-4)^2$
   c. $(3 - 8)^2 + (-4 - 6)^2$
   d. $\sqrt{(8 - 3)^2 + (6 - (-4))^2}$

III. The *direction* of the vector $(4, 8)$ is _____.
   a. π b. $\tan^{-1}(8 - 4)$
   c. $(\frac{8}{4})\pi$ d. $\tan^{-1}(\frac{8}{4})$

IV. If $\mathbf{u} = (3, 4)$ and $\mathbf{v} = (5, 8)$, then $\mathbf{u} + \mathbf{v} =$ _____.
   a. $(7, 13)$ b. $(8, 12)$
   c. $(2, 4)$ d. $(15, 32)$

V. If $\mathbf{u} = (4, 3)$ then the unit vector with the same direction as $\mathbf{u}$ is _____.
   a. $(0.4, 0.3)$ b. $(0.8, 0.6)$
   c. $(\frac{4}{5}, \frac{3}{5})$ d. $(\frac{4}{7}, \frac{3}{7})$

■ DRILL

In Problems 1–6, a vector $\mathbf{v}$ and a point P are given. Find a point Q such that the directed line segment $\overrightarrow{PQ}$ is a representation of $\mathbf{v}$. Sketch $\mathbf{v}$ and $\overrightarrow{PQ}$.

1. $\mathbf{v} = (2, 5)$; $P = (1, -2)$
2. $\mathbf{v} = (5, 8)$; $P = (3, 8)$
3. $\mathbf{v} = (-3, 7)$; $P = (7, -3)$
4. $\mathbf{v} = -\mathbf{i} - 7\mathbf{j}$; $P = (0, 1)$
5. $\mathbf{v} = 5\mathbf{i} - 3\mathbf{j}$; $P = (-7, -2)$
6. $\mathbf{v} = e\mathbf{i} + \pi\mathbf{j}$; $P = (\pi, \sqrt{2})$

In Problems 7–18, find the magnitude and direction of the given vector $\mathbf{v}$.

7. $\mathbf{v} = (4, 4)$ 8. $\mathbf{v} = (-4, 4)$
9. $\mathbf{v} = (4, -4)$ 10. $\mathbf{v} = (-4, -4)$
11. $\mathbf{v} = (\sqrt{3}, 1)$ 12. $\mathbf{v} = (1, \sqrt{3})$
13. $\mathbf{v} = (-1, \sqrt{3})$ 14. $\mathbf{v} = (1, -\sqrt{3})$
15. $\mathbf{v} = (-1, -\sqrt{3})$ 16. $\mathbf{v} = (1, 2)$
17. $\mathbf{v} = (-5, 8)$ 18. $\mathbf{v} = (11, -14)$

In Problems 19–26, the vector $\mathbf{v}$ is represented by $\overrightarrow{PQ}$ where P and Q are given. Write $\mathbf{v}$ in the form $a\mathbf{i} + b\mathbf{j}$. Sketch $\overrightarrow{PQ}$ and $\mathbf{v}$.

19. $P = (1, 2)$; $Q = (1, 3)$
20. $P = (2, 4)$; $Q = (-7, 4)$
21. $P = (5, 2)$; $Q = (-1, 3)$
22. $P = (8, -2)$; $Q = (-3, -3)$
23. $P = (7, -1)$; $Q = (-2, 4)$

24. $P = (3, -6)$; $Q = (8, 0)$
25. $P = (-3, -8)$; $Q = (-8, -3)$
26. $P = (2, 4)$; $Q = (-4, -2)$
27. Let $\mathbf{u} = (2, 3)$ and $\mathbf{v} = (-5, 4)$. Compute and sketch the following vectors:
   a. $3\mathbf{u}$ b. $\mathbf{u} + \mathbf{v}$
   c. $\mathbf{v} - \mathbf{u}$ d. $2\mathbf{u} - 7\mathbf{v}$
28. Let $\mathbf{u} = 2\mathbf{i} - 3\mathbf{j}$ and $\mathbf{v} = -4\mathbf{i} + 6\mathbf{j}$. Compute and sketch the following vectors:
   a. $\mathbf{u} + \mathbf{v}$ b. $\mathbf{u} - \mathbf{v}$ c. $3\mathbf{u}$
   d. $-7\mathbf{v}$ e. $8\mathbf{u} - 3\mathbf{v}$ f. $4\mathbf{v} - 6\mathbf{u}$

In Problems 29–34, find a unit vector having the same direction as the given vector $\mathbf{v}$.

29. $\mathbf{v} = 2\mathbf{i} + 3\mathbf{j}$ 30. $\mathbf{v} = \mathbf{i} - \mathbf{j}$
31. $\mathbf{v} = (3, 4)$ 32. $\mathbf{v} = (3, -4)$
33. $\mathbf{v} = -3\mathbf{i} + 4\mathbf{j}$ 34. $\mathbf{v} = (a, a)$ $a \neq 0$

35. For $\mathbf{v} = 2\mathbf{i} - 3\mathbf{j}$, find $\sin \theta$ and $\cos \theta$.
36. For $\mathbf{v} = -3\mathbf{i} + 8\mathbf{j}$, find $\sin \theta$ and $\cos \theta$.

A vector $\mathbf{u}$ has a direction **opposite** to that of a vector $\mathbf{v}$ if and only if $|\text{direction } \mathbf{v} - \text{direction } \mathbf{u}| = \pi$. In Problems 37–42, find a unit vector $\mathbf{u}$ that has direction opposite the direction of the given vector $\mathbf{v}$.

37. $\mathbf{v} = \mathbf{i} + \mathbf{j}$ 38. $\mathbf{v} = 2\mathbf{i} - 3\mathbf{j}$
39. $\mathbf{v} = (-3, 4)$ 40. $\mathbf{v} = (-2, 3)$
41. $\mathbf{v} = -3\mathbf{i} - 4\mathbf{j}$ 42. $\mathbf{v} = (8, -3)$

CONTINUED

43. Let $\mathbf{u} = 2\mathbf{i} - 3\mathbf{j}$ and $\mathbf{v} = -\mathbf{i} + 2\mathbf{j}$. For each of the following, find a unit vector having the same direction.
   a. $\mathbf{u} + \mathbf{v}$ b. $\mathbf{u} - \mathbf{v}$
   c. $2\mathbf{u} - 3\mathbf{v}$ d. $3\mathbf{u} + 8\mathbf{v}$
44. Let $P = (c, d)$ and $Q = (c + a, d + b)$. Verify that the magnitude of $\overrightarrow{PQ}$ is $\sqrt{a^2 + b^2}$.

In Problems 45–52, find a vector $\mathbf{v}$ having the given magnitude and direction.

45. $|\mathbf{v}| = 3$; $\theta = \pi/6$
46. $|\mathbf{v}| = 8$; $\theta = \pi/3$
47. $|\mathbf{v}| = 7$; $\theta = \pi$
48. $|\mathbf{v}| = 4$; $\theta = \pi/2$
49. $|\mathbf{v}| = 1$; $\theta = \pi/4$
50. $|\mathbf{v}| = 6$; $\theta = 2\pi/3$
51. $|\mathbf{v}| = 8$; $\theta = 3\pi/2$
52. $|\mathbf{v}| = 6$; $\theta = 11\pi/6$

■ SHOW/PROVE/DISPROVE

53. Show that the vectors $\mathbf{i}$ and $\mathbf{j}$ are unit vectors.

54. Show that the vector $\dfrac{1}{\sqrt{2}}\mathbf{i} - \dfrac{1}{\sqrt{2}}\mathbf{j}$ is a unit vector.

55. Show that if $\mathbf{v} = a\mathbf{i} + b\mathbf{j} \neq \mathbf{0}$, then

$$\mathbf{u} = \frac{a}{\sqrt{a^2 + b^2}}\mathbf{i} + \frac{b}{\sqrt{a^2 + b^2}}\mathbf{j}$$

is a unit vector having the same direction as $\mathbf{v}$.

56. Show that if $\mathbf{v} = a\mathbf{i} + b\mathbf{j} \neq \mathbf{0}$, then $a/\sqrt{a^2 + b^2} = \cos\theta$ and $b/\sqrt{a^2 + b^2} = \sin\theta$, where θ is the direction of $\mathbf{v}$.

57. Show that the direction of $\overrightarrow{PQ}$ in Problem 44 is the same as the direction of the vector (a, b).

[*Hint:* If $R = (a, b)$, show that the line passing through the points P and Q is parallel to the line passing through the points 0 and R.]

58. Prove Theorem 1. [*Hint:* Use the definitions of addition and scalar multiplication of vectors.]

59. Show algebraically (i.e., strictly from the definitions of vector addition and magnitude) that for any two vectors $\mathbf{u}$ and $\mathbf{v}$,

$$|\mathbf{u} + \mathbf{v}| \leq |\mathbf{u}| + |\mathbf{v}|.$$

60. Show that if neither $\mathbf{u}$ nor $\mathbf{v}$ is the zero vector, then $|\mathbf{u} + \mathbf{v}| = |\mathbf{u}| + |\mathbf{v}|$ if and only if $\mathbf{u}$ is a positive scalar multiple of $\mathbf{v}$.

■ ANSWERS TO SELF-QUIZ

I. d   II. d   III. d
IV. b   V. b = c

11.2 THE DOT PRODUCT

In Section 11.1, we showed how a vector could be multiplied by a scalar but not how two vectors could be multiplied. Actually, there are several ways to define the product of two vectors, and in this section we will discuss one of them. We will discuss a second product operation in Section 11.6.

DEFINITION   **DOT PRODUCT**

Let $\mathbf{u} = (a_1, b_1) = a_1\mathbf{i} + b_1\mathbf{j}$ and $\mathbf{v} = (a_2, b_2) = a_2\mathbf{i} + b_2\mathbf{j}$. Then the **dot product** of $\mathbf{u}$ and $\mathbf{v}$, denoted by $\mathbf{u} \cdot \mathbf{v}$, is defined by

$$\mathbf{u} \cdot \mathbf{v} = a_1 a_2 + b_1 b_2. \quad ■ \tag{1}$$

REMARK: The dot product of two vectors is a *scalar*. For this reason, the dot product is often called the **scalar product**. It is also called the **inner product**.

EXAMPLE 1
COMPUTING A DOT PRODUCT

If $\mathbf{u} = (1, 3)$ and $\mathbf{v} = (4, -7)$, then

$$\mathbf{u} \cdot \mathbf{v} = 1(4) + 3(-7) = 4 - 21 = -17.$$

THEOREM 1 **PROPERTIES OF THE DOT PRODUCT**

For any vectors $\mathbf{u}$, $\mathbf{v}$, $\mathbf{w}$, and scalar α,

(i) $\mathbf{u} \cdot \mathbf{v} = \mathbf{v} \cdot \mathbf{u}$

(ii) $(\mathbf{u} + \mathbf{v}) \cdot \mathbf{w} = \mathbf{u} \cdot \mathbf{w} + \mathbf{v} \cdot \mathbf{w}$

(iii) $(\alpha\mathbf{u}) \cdot \mathbf{v} = \alpha(\mathbf{u} \cdot \mathbf{v})$

(iv) $\mathbf{u} \cdot \mathbf{u} \geq 0$ and $\mathbf{u} \cdot \mathbf{u} = 0$ if and only if $\mathbf{u} = \mathbf{0}$

(v) $|\mathbf{u}|^2 = \mathbf{u} \cdot \mathbf{u}$

PROOF: Let $\mathbf{u} = (u_1, u_2)$, $\mathbf{v} = (v_1, v_2)$, and $\mathbf{w} = (w_1, w_2)$.

(i) $\mathbf{u} \cdot \mathbf{v} = u_1v_1 + u_2v_2 = v_1u_1 + v_2u_2 = \mathbf{v} \cdot \mathbf{u}$

(ii) $(\mathbf{u} + \mathbf{v}) \cdot \mathbf{w} = (u_1 + v_1, u_2 + v_2) \cdot (w_1, w_2)$
$= (u_1 + v_1)w_1 + (u_2 + v_2)w_2$
$= u_1w_1 + u_2w_2 + v_1w_1 + v_2w_2 = \mathbf{u} \cdot \mathbf{w} + \mathbf{v} \cdot \mathbf{w}$

(iii) $(\alpha\mathbf{u}) \cdot \mathbf{v} = (\alpha u_1, \alpha u_2) \cdot (v_1, v_2) = \alpha u_1 v_1 + \alpha u_2 v_2 = \alpha(u_1v_1 + u_2v_2)$
$= \alpha(\mathbf{u} \cdot \mathbf{v})$

(iv) $\mathbf{u} \cdot \mathbf{u} = u_1{}^2 + u_2{}^2 \geq 0$ and $\mathbf{u} \cdot \mathbf{u} = 0$ if and only if $u_1 = u_2 = 0$.

(v) $\mathbf{u} \cdot \mathbf{u} = (u_1, u_2) \cdot (u_1, u_2) = u_1{}^2 + u_2{}^2 = |\mathbf{u}|^2$ ∎

The dot product is useful in a wide variety of applications. An interesting one follows.

DEFINITION **ANGLE BETWEEN TWO VECTORS**

Let $\mathbf{u}$ and $\mathbf{v}$ be two nonzero vectors. If $\mathbf{u}$ and $\mathbf{v}$ are not parallel, then the **angle** φ between $\mathbf{u}$ and $\mathbf{v}$ is defined to be the smallest angle between the representations of $\mathbf{u}$ and $\mathbf{v}$ that have the origin as their initial points. If $\mathbf{u} = \alpha\mathbf{v}$ for some scalar α, then we define $\varphi = 0$ if $\alpha > 0$ and $\varphi = \pi$ if $\alpha < 0$. By this definition

$$0 \leq \varphi \leq \pi. \quad ∎$$

In Figure 1 we illustrate the angles between five sets of vectors.

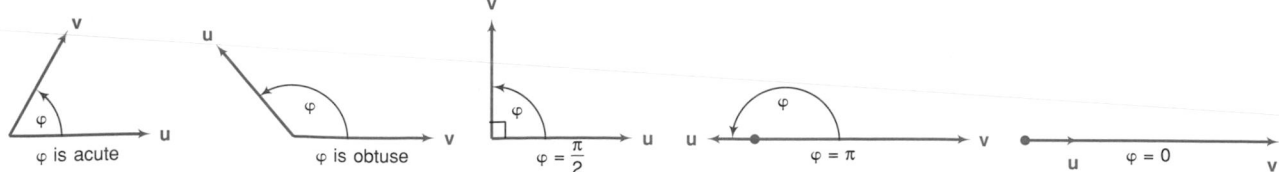

FIGURE 1 THE ANGLES BETWEEN FIVE SETS OF VECTORS

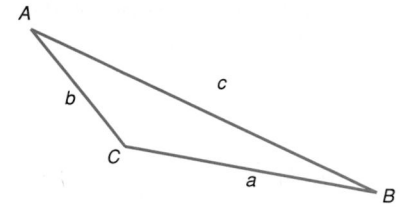

FIGURE 2
FROM THE LAW OF COSINES,
$c^2 = a^2 + b^2 - 2ab \cos C$

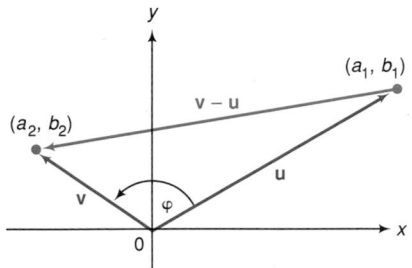

FIGURE 3
$|\mathbf{v} - \mathbf{u}|^2 = |\mathbf{u}|^2 + |\mathbf{v}|^2 - 2|\mathbf{u}||\mathbf{v}| \cos \psi$

THEOREM 2

Let $\mathbf{u}$ and $\mathbf{v}$ be two nonzero vectors. Then if φ is the angle between them,

$$\cos \varphi = \frac{\mathbf{u} \cdot \mathbf{v}}{|\mathbf{u}||\mathbf{v}|} \qquad \text{and} \qquad \varphi = \cos^{-1}\left(\frac{\mathbf{u} \cdot \mathbf{v}}{|\mathbf{u}||\mathbf{v}|}\right). \tag{2}$$

PROOF: The law of cosines (see Problem 11, Appendix 1.4) states that in the triangle of Figure 2,

$$c^2 = a^2 + b^2 - 2ab \cos C. \tag{3}$$

We now place the representations of $\mathbf{u}$ and $\mathbf{v}$ with initial points at the origin so that $\mathbf{u} = (a_1, b_1)$ and $\mathbf{v} = (a_2, b_2)$ (see Figure 3). Then from the law of cosines,

$$|\mathbf{v} - \mathbf{u}|^2 = |\mathbf{v}|^2 + |\mathbf{u}|^2 - 2|\mathbf{u}||\mathbf{v}| \cos \varphi.$$

But using Theorem 1 several times, we have

$$|\mathbf{v} - \mathbf{u}|^2 = \overset{\text{part (v)}}{(\mathbf{v} - \mathbf{u}) \cdot (\mathbf{v} - \mathbf{u})} = \mathbf{v} \cdot \mathbf{v} - 2\mathbf{u} \cdot \mathbf{v} + \mathbf{u} \cdot \mathbf{u}$$
$$= |\mathbf{v}|^2 - 2\mathbf{u} \cdot \mathbf{v} + |\mathbf{u}|^2.$$

Thus after simplification, we obtain

$$-2\mathbf{u} \cdot \mathbf{v} = -2|\mathbf{u}||\mathbf{v}| \cos \varphi,$$

from which the theorem follows. ∎

REMARK: Using Theorem 2, we could define the dot product $\mathbf{u} \cdot \mathbf{v}$ by

$$\mathbf{u} \cdot \mathbf{v} = |\mathbf{u}||\mathbf{v}| \cos \varphi \tag{4}$$

if neither $\mathbf{u}$ nor $\mathbf{v} = \mathbf{0}$ and $\mathbf{u} \cdot \mathbf{v} = 0$ if $\mathbf{u}$ or $\mathbf{v}$ is the zero vector.

EXAMPLE 2
COMPUTING THE ANGLE
BETWEEN TWO VECTORS

Find the cosine of the angle between the vectors $\mathbf{u} = 2\mathbf{i} + 3\mathbf{j}$ and $\mathbf{v} = -7\mathbf{i} + \mathbf{j}$.

SOLUTION: $\mathbf{u} \cdot \mathbf{v} = -14 + 3 = -11$, $|\mathbf{u}| = \sqrt{2^2 + 3^2} = \sqrt{13}$, and $|\mathbf{v}| = \sqrt{(-7)^2 + 1^2} = \sqrt{50}$, so

$$\cos \varphi = \frac{\mathbf{u} \cdot \mathbf{v}}{|\mathbf{u}||\mathbf{v}|} = \frac{-11}{\sqrt{13}\sqrt{50}} = \frac{-11}{\sqrt{650}} \approx -0.431455497,$$

so $\varphi = \cos^{-1}(-0.431455497) \approx 2.0169^\dagger$ ($\approx 115.6°$).

DEFINITION **PARALLEL VECTORS**

Two nonzero vectors $\mathbf{u}$ are $\mathbf{v}$ are **parallel** if the angle between them is 0 or π. ∎

---

† When doing this computation yourself, make certain that the calculator is set to radian mode.

EXAMPLE 3
TWO PARALLEL VECTORS

Show that the vectors $\mathbf{u} = (2, -3)$ and $\mathbf{v} = (-4, 6)$ are parallel.

SOLUTION:

$$\cos \varphi = \frac{\mathbf{u} \cdot \mathbf{v}}{|\mathbf{u}| |\mathbf{v}|} = \frac{-8 - 18}{\sqrt{13}\sqrt{52}} = \frac{-26}{\sqrt{13}(2\sqrt{13})} = \frac{-26}{2(13)} = -1,$$

so $\varphi = \pi$ (so that $\mathbf{u}$ and $\mathbf{v}$ have opposite directions).

THEOREM 3

If $\mathbf{u} \neq \mathbf{0}$, then $\mathbf{v} = \alpha\mathbf{u}$ for some nonzero constant α if and only if $\mathbf{u}$ and $\mathbf{v}$ are parallel.

PROOF: This follows from the last part of the definition of an angle between two vectors (see also Problem 69). ∎

DEFINITION   ORTHOGONAL VECTORS

The nonzero vectors $\mathbf{u}$ and $\mathbf{v}$ are called **orthogonal** (or **perpendicular**) if the angle between them is $\pi/2$. ∎

EXAMPLE 4
TWO ORTHOGONAL VECTORS

Show that the vectors $\mathbf{u} = 3\mathbf{i} - 4\mathbf{j}$ and $\mathbf{v} = 4\mathbf{i} + 3\mathbf{j}$ are orthogonal.

SOLUTION: $\mathbf{u} \cdot \mathbf{v} = 3 \cdot 4 - 4 \cdot 3 = 0$. This implies that $\cos \varphi = (\mathbf{u} \cdot \mathbf{v})/(|\mathbf{u}| |\mathbf{v}|) = 0$. Since φ is in the interval $[0, \pi]$, $\varphi = \pi/2$.

THEOREM 4

The nonzero vectors $\mathbf{u}$ and $\mathbf{v}$ are orthogonal if and only if $\mathbf{u} \cdot \mathbf{v} = 0$.

PROOF: This proof is not difficult and is left as an exercise (see Problem 70). ∎

REMARK: The condition $\mathbf{u} \cdot \mathbf{v} = 0$ is sometimes given as the *definition* of orthogonality.

A number of interesting problems involve the notion of the *projection* of one vector onto another. Before defining this term, we prove the following theorem.

THEOREM 5

Let $\mathbf{v}$ be a nonzero vector. Then for any other nonzero vector $\mathbf{u}$, the vector

$$\mathbf{w} = \mathbf{u} - [(\mathbf{u} \cdot \mathbf{v})/|\mathbf{v}|^2]\mathbf{v} \text{ is orthogonal to } \mathbf{v}.$$

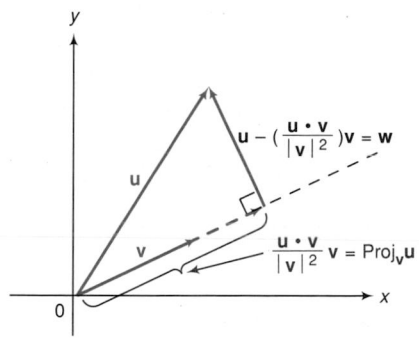

FIGURE 4
PROJECTION OF **u** ONTO **v**

PROOF:

$$\mathbf{w} \cdot \mathbf{v} = \left(\mathbf{u} - \frac{(\mathbf{u} \cdot \mathbf{v})\mathbf{v}}{|\mathbf{v}|^2} \right) \cdot \mathbf{v} = \mathbf{u} \cdot \mathbf{v} - \frac{(\mathbf{u} \cdot \mathbf{v})(\mathbf{v} \cdot \mathbf{v})}{|\mathbf{v}|^2}$$

$$= \mathbf{u} \cdot \mathbf{v} - \frac{(\mathbf{u} \cdot \mathbf{v})|\mathbf{v}|^2}{|\mathbf{v}|^2} = \mathbf{u} \cdot \mathbf{v} - \mathbf{u} \cdot \mathbf{v} = 0$$

The vectors **u**, **v**, and **w** are illustrated in Figure 4. ∎

DEFINITION **PROJECTION AND COMPONENT**

Let **u** and **v** be nonzero vectors. Then the **projection of u onto v is** a vector, denoted by Proj$_\mathbf{v}$ **u**, which is defined by

$$\text{Proj}_\mathbf{v}\, \mathbf{u} = \frac{\mathbf{u} \cdot \mathbf{v}}{\mathbf{v} \cdot \mathbf{v}} \mathbf{v} = \frac{\mathbf{u} \cdot \mathbf{v}}{|\mathbf{v}|^2} \mathbf{v} = \left(\frac{\mathbf{u} \cdot \mathbf{v}}{|\mathbf{v}|} \right) \frac{\mathbf{v}}{|\mathbf{v}|}. \tag{5}$$

The **component** of **u** in the direction **v** is $\mathbf{u} \cdot \mathbf{v}/|\mathbf{v}|$. ∎ **(6)**

Note that $\mathbf{v}/|\mathbf{v}|$ is a unit vector in the direction of **v**.

REMARK 1: From Figure 4 and the fact that $\cos \varphi = (\mathbf{u} \cdot \mathbf{v})/(|\mathbf{u}||\mathbf{v}|)$, we find that **v** and Proj$_\mathbf{v}$ **u** have

  (i) The same direction if $\mathbf{u} \cdot \mathbf{v} > 0$ and
  (ii) opposite directions if $\mathbf{u} \cdot \mathbf{v} < 0$.

REMARK 2: Proj$_\mathbf{v}$ **u** can be thought of as the "**v**-component" of the vector **u**.

REMARK 3: If **u** and **v** are orthogonal, then $\mathbf{u} \cdot \mathbf{v} = 0$, so that Proj$_\mathbf{v}$ **u** = **0**.

REMARK 4: An alternative definition of projection is: If **u** and **v** are nonzero vectors, then Proj$_\mathbf{v}$ **u** is the unique vector having the properties

  (i) Proj$_\mathbf{v}$ **u** is parallel to **v** and
  (ii) **u** − Proj$_\mathbf{v}$ **u** is orthogonal to **v**.

EXAMPLE 5
CALCULATING A PROJECTION

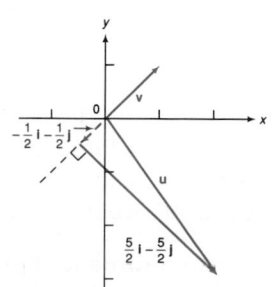

FIGURE 5
THE PROJECTION OF **u** ONTO **v** IS THE
VECTOR $-\frac{1}{2}\mathbf{i} - \frac{1}{2}\mathbf{j}$.

Let $\mathbf{u} = 2\mathbf{i} - 3\mathbf{j}$ and $\mathbf{v} = \mathbf{i} + \mathbf{j}$. Find Proj$_\mathbf{v}$ **u**.

SOLUTION: Here $(\mathbf{u} \cdot \mathbf{v})/|\mathbf{v}|^2 = -\frac{1}{2}$, so that Proj$_\mathbf{v}$ $\mathbf{u} = -\frac{1}{2}\mathbf{i} - \frac{1}{2}\mathbf{j}$. This is illustrated in Figure 5.

In earlier chapters we solved problems involving the notion of force (see page 357, for example). We started such problems with a statement like "a force of x newtons is applied to . . . ." Implicit in such a problem is the idea that a force with a certain magnitude (measured in pounds or newtons) is exerted in a certain direction. In this context force can (and should) be thought of as a vector.

If more than one force is applied to an object at the same point, then we define the **resultant** of the forces applied to the object to be the *vector sum* of these forces. We can think of the resultant as the *net* applied force.

EXAMPLE 6
COMPUTING THE RESULTANT
OF THREE FORCES

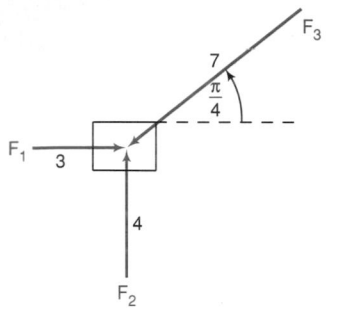

FIGURE 6
THREE FORCES ACT ON THE SAME OBJECT

A force of 3 N is applied from the left side of an object, one of 4 N is applied from the bottom, and a force of 7 N is applied from an angle of $\pi/4$ to the horizontal.[†] What is the resultant of forces applied to the object?

SOLUTION: The forces are indicated in Figure 6. We write each force as a magnitude times a unit vector in the indicated direction. For convenience we can think of the center of the object as being at the origin and that all three forces are applied at the center. Then $\mathbf{F}_1 = 3\mathbf{i}$; $\mathbf{F}_2 = 4\mathbf{j}$; and $\mathbf{F}_3 = -(7/\sqrt{2})(\mathbf{i} + \mathbf{j})$. This last vector follows from the fact that the vector $-(1/\sqrt{2})(\mathbf{i} + \mathbf{j})$ is a unit vector pointing toward the origin making an angle of $\pi/4$ with the x-axis. Then the resultant is given by

$$\mathbf{F} = \mathbf{F}_1 + \mathbf{F}_2 + \mathbf{F}_3 = \left(3 - \frac{7}{\sqrt{2}}\right)\mathbf{i} + \left(4 - \frac{7}{\sqrt{2}}\right)\mathbf{j}.$$

The magnitude of $\mathbf{F}$ is

$$|\mathbf{F}| = \sqrt{\left(3 - \frac{7}{\sqrt{2}}\right)^2 + \left(4 - \frac{7}{\sqrt{2}}\right)^2} = \sqrt{74 - \frac{98}{\sqrt{2}}} \approx 2.17 \text{ N}.$$

The direction θ can be calculated by first finding the unit vector in the direction of $\mathbf{F}$:

$$\frac{\mathbf{F}}{|\mathbf{F}|} = \frac{3 - (7/\sqrt{2})}{\sqrt{74 - (98/\sqrt{2})}}\mathbf{i} + \frac{4 - (7/\sqrt{2})}{\sqrt{74 - (98/\sqrt{2})}}\mathbf{j}$$
$$= (\cos\theta)\mathbf{i} + (\sin\theta)\mathbf{j}.$$

Then

$$\cos\theta = \frac{3 - (7/\sqrt{2})}{\sqrt{74 - (98/\sqrt{2})}} \approx -0.8990 \quad \text{and}$$

$$\sin\theta = \frac{4 - (7/\sqrt{2})}{\sqrt{74 - (98/\sqrt{2})}} \approx -0.4379.$$

This means that θ is in the third quadrant, and $\theta \approx 3.5949 \approx 206°$ (or $-154°$). This is illustrated in Figure 7.

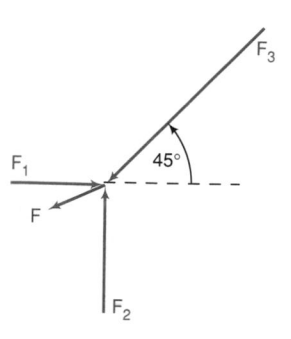

FIGURE 7
$\mathbf{F}$ IS THE RESOLVENT OF $\mathbf{F}_1$, $\mathbf{F}_2$, AND $\mathbf{F}_3$.

Recall from Section 5.4 (page 356) that the work done by a force F in moving an object a distance d is given by

$$W = Fd \tag{7}$$

where units of work are newton-meters or foot-pounds. In formula (7) it is assumed that the force is applied in the same direction as the direction of motion. However, this is not always the case. In general, we may define

$$W = (\text{component of } \mathbf{F} \text{ in the direction of motion}) \times (\text{distance moved}). \tag{8}$$

[†] Units of force are discussed on page 357.

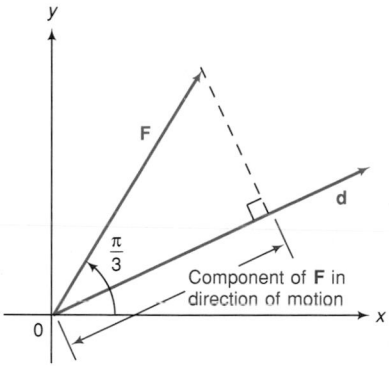

FIGURE 8

$\dfrac{\mathbf{F}\cdot\mathbf{d}}{|\mathbf{d}|}$ IS THE COMPONENT OF THE FORCE VECTOR IN THE DIRECTION OF THE DISPLACEMENT VECTOR **d**.

If the object moves from P to Q, then the distance moved is $|\overrightarrow{PQ}|$. The vector **d**, one of whose representations is $\overrightarrow{PQ}$, is called a **displacement vector**. Then from equation (6),

$$\text{component of } \mathbf{F} \text{ in direction of motion} = \frac{\mathbf{F}\cdot\mathbf{d}}{|\mathbf{d}|}. \tag{9}$$

Finally, combining (8) and (9), we obtain

$$W = \frac{\mathbf{F}\cdot\mathbf{d}}{|\mathbf{d}|}|\mathbf{d}| = \mathbf{F}\cdot\mathbf{d}. \tag{10}$$

That is, *the work done is the dot product of the force* **F** *and the displacement vector* **d**. Note that if **F** acts in the direction **d** and if φ denotes the angle (which is zero) between **F** and **d**, then $\mathbf{F}\cdot\mathbf{d} = |\mathbf{F}||\mathbf{d}|\cos\varphi = |\mathbf{F}||\mathbf{d}|\cos 0 = |\mathbf{F}||\mathbf{d}|$, which is formula (7).

EXAMPLE 7
COMPUTING WORK WHEN THE FORCE DOES NOT ACT IN THE DIRECTION OF MOTION

A force of 4 N has the direction $\pi/3$. What is the work done in moving an object from the point $(1, 2)$ to the point $(5, 4)$, where distances are measured in meters?

SOLUTION: A unit vector with direction $\pi/3$ is given by $\mathbf{u} = (\cos \pi/3)\mathbf{i} + (\sin \pi/3)\mathbf{j} = \frac{1}{2}\mathbf{i} + (\sqrt{3}/2)\mathbf{j}$. Thus $\mathbf{F} = 4\mathbf{u} = 2\mathbf{i} + 2\sqrt{3}\mathbf{j}$. The displacement vector **d** is given by $(5-1)\mathbf{i} + (4-2)\mathbf{j} = 4\mathbf{i} + 2\mathbf{j}$. Thus

$$W = \mathbf{F}\cdot\mathbf{d} = (2\mathbf{i} + 2\sqrt{3}\mathbf{j})\cdot(4\mathbf{i} + 2\mathbf{j})$$
$$= (8 + 4\sqrt{3}) \approx 14.93 \text{ N·m.}$$

The component of **F** in the direction of motion is sketched in Figure 8.

■ PROBLEMS 11.2

■ SELF-QUIZ

I. $\mathbf{i}\cdot\mathbf{j} = $ _____.
   a. 1   b. $\sqrt{(0-1)^2 + (1-0)^2}$
   c. 0   d. $\mathbf{i}+\mathbf{j}$

II. $(3, 4)\cdot(3, 2) = $ _____.
   a. $(3+3)(4+2) = 36$
   b. $(3)(3) + (4)(2) = 17$
   c. $(3-3)(2-4) = 0$
   d. $(3)(3) - (4)(2) = 1$

III. The cosine of the angle between $\mathbf{i}+\mathbf{j}$ and $\mathbf{i}-\mathbf{j}$ is _____.
   a. $0\mathbf{i} + 0\mathbf{j}$ b. 0
   c. $\sqrt{2}$ d. $1/\sqrt{2+0}$

IV. The vectors $2\mathbf{i} - 12\mathbf{j}$ and $3\mathbf{i} + (1/2)\mathbf{j}$ are _____.
   a. neither parallel nor orthogonal
   b. parallel
   c. orthogonal
   d. identical

V. $\text{Proj}_{\mathbf{w}}\mathbf{u} = $ _____.
   a. $\dfrac{\mathbf{u}\cdot\mathbf{w}}{|\mathbf{w}|}$ b. $\dfrac{\mathbf{w}}{|\mathbf{w}|}$
   c. $\dfrac{\mathbf{u}\cdot\mathbf{w}}{|\mathbf{w}|}\dfrac{\mathbf{w}}{|\mathbf{w}|}$ d. $\dfrac{\mathbf{u}\cdot\mathbf{w}}{|\mathbf{u}|}\dfrac{\mathbf{u}}{|\mathbf{u}|}$

CONTINUED

DRILL

In Problems 1–10, two vectors are given. Calculate both the dot product of the two vectors and the cosine of the angle between them.

1. $\mathbf{u} = \mathbf{i} + \mathbf{j}; \mathbf{v} = \mathbf{i} - \mathbf{j}$
2. $\mathbf{u} = 3\mathbf{i}; \mathbf{v} = -7\mathbf{j}$
3. $\mathbf{u} = -5\mathbf{i}; \mathbf{v} = 18\mathbf{j}$
4. $\mathbf{u} = \alpha\mathbf{i}; \mathbf{v} = \beta\mathbf{j}$
5. $\mathbf{u} = 2\mathbf{i} + 5\mathbf{j}; \mathbf{v} = 5\mathbf{i} + 2\mathbf{j}$
6. $\mathbf{u} = 2\mathbf{i} + 5\mathbf{j}; \mathbf{v} = 5\mathbf{i} - 3\mathbf{j}$
7. $\mathbf{u} = -3\mathbf{i} + 4\mathbf{j}; \mathbf{v} = -2\mathbf{i} - 7\mathbf{j}$
8. $\mathbf{u} = 4\mathbf{i} + 5\mathbf{j}; \mathbf{v} = 7\mathbf{i} - 4\mathbf{j}$
9. $\mathbf{u} = 11\mathbf{i} - 8\mathbf{j}; \mathbf{v} = 4\mathbf{i} - 7\mathbf{j}$
10. $\mathbf{u} = -13\mathbf{i} + 8\mathbf{j}; \mathbf{v} = 2\mathbf{i} + 11\mathbf{j}$

In Problems 11–18, determine whether the given vectors are orthogonal, parallel, or neither. Then sketch each pair of vectors.

11. $\mathbf{u} = 3\mathbf{i} + 5\mathbf{j}; \mathbf{v} = -6\mathbf{i} - 10\mathbf{j}$
12. $\mathbf{u} = 2\mathbf{i} + 3\mathbf{j}; \mathbf{v} = 6\mathbf{i} - 4\mathbf{j}$
13. $\mathbf{u} = 2\mathbf{i} + 3\mathbf{j}; \mathbf{v} = 6\mathbf{i} + 4\mathbf{j}$

14. $\mathbf{u} = 2\mathbf{i} + 3\mathbf{j}; \mathbf{v} = -6\mathbf{i} + 4\mathbf{j}$
15. $\mathbf{u} = 7\mathbf{i}; \mathbf{v} = -23\mathbf{j}$
16. $\mathbf{u} = 2\mathbf{i} - 6\mathbf{j}; \mathbf{v} = -\mathbf{i} + 3\mathbf{j}$
17. $\mathbf{u} = \mathbf{i} + \mathbf{j}; \mathbf{v} = \alpha\mathbf{i} + \alpha\mathbf{j}$
18. $\mathbf{u} = -2\mathbf{i} + 3\mathbf{j}; \mathbf{v} = -\mathbf{i} + 2\mathbf{j}$

In Problems 19–32, calculate $\text{Proj}_\mathbf{v}\, \mathbf{u}$.

19. $\mathbf{u} = 3\mathbf{i}; \mathbf{v} = \mathbf{i} + \mathbf{j}$
20. $\mathbf{u} = -5\mathbf{j}; \mathbf{v} = \mathbf{i} + \mathbf{j}$
21. $\mathbf{u} = 2\mathbf{i} + \mathbf{j}; \mathbf{v} = \mathbf{i} - 2\mathbf{j}$
22. $\mathbf{u} = 2\mathbf{i} + 3\mathbf{j}; \mathbf{v} = 4\mathbf{i} + \mathbf{j}$
23. $\mathbf{u} = \mathbf{i} + \mathbf{j}; \mathbf{v} = 2\mathbf{i} - 3\mathbf{j}$
24. $\mathbf{u} = \mathbf{i} + \mathbf{j}; \mathbf{v} = 2\mathbf{i} + 3\mathbf{j}$
25. $\mathbf{u} = 4\mathbf{i} + 5\mathbf{j}; \mathbf{v} = 2\mathbf{i} + 4\mathbf{j}$
26. $\mathbf{u} = 4\mathbf{i} + 5\mathbf{j}; \mathbf{v} = 2\mathbf{i} - 4\mathbf{j}$
27. $\mathbf{u} = -4\mathbf{i} + 5\mathbf{j}; \mathbf{v} = 2\mathbf{i} - 4\mathbf{j}$
28. $\mathbf{u} = -4\mathbf{i} - 5\mathbf{j}; \mathbf{v} = -2\mathbf{i} - 4\mathbf{j}$
29. $\mathbf{u} = \alpha\mathbf{i} + \beta\mathbf{j}; \mathbf{v} = \mathbf{i} + \mathbf{j}$
30. $\mathbf{u} = \mathbf{i} + \mathbf{j}; \mathbf{v} = \alpha\mathbf{i} + \beta\mathbf{j}$
31. $\mathbf{u} = \alpha\mathbf{i} - \beta\mathbf{j}; \mathbf{v} = \mathbf{i} + \mathbf{j}$
32. $\mathbf{u} = \alpha\mathbf{i} - \beta\mathbf{j}; \mathbf{v} = -\mathbf{i} + \mathbf{j}$

APPLICATIONS

33. Let $\mathbf{u} = 3\mathbf{i} + 4\mathbf{j}$ and $\mathbf{v} = \mathbf{i} + \alpha\mathbf{j}$.
    a. Determine α such that $\mathbf{u}$ and $\mathbf{v}$ are orthogonal.
    b. Determine α such that $\mathbf{u}$ and $\mathbf{v}$ are parallel.
    c. Determine α such that the angle between $\mathbf{u}$ and $\mathbf{v}$ is $2\pi/3$.
    d. Determine α such that the angle between $\mathbf{u}$ and $\mathbf{v}$ is $\pi/3$.
34. Let $\mathbf{u} = -2\mathbf{i} + 5\mathbf{j}$ and $\mathbf{v} = \beta\mathbf{i} - 2\mathbf{j}$.
    a. Determine β such that $\mathbf{u}$ and $\mathbf{v}$ are orthogonal.
    b. Determine β such that $\mathbf{u}$ and $\mathbf{v}$ are parallel.
    c. Determine β such that the angle between $\mathbf{u}$ and $\mathbf{v}$ is $2\pi/3$.
    d. Determine β such that the angle between $\mathbf{u}$ and $\mathbf{v}$ is $\pi/3$.
35. A triangle has vertices $(1, 3)$, $(4, -2)$, and $(-3, 6)$. Find the cosine of each of its angles.
36. A triangle has vertices (a_1, b_1), (a_2, b_2), and (a_3, b_3). Find a formula for the cosine of each of its angles.
37. Let $P = (2, 3)$, $Q = (5, 5)$, $R = (2, -3)$, and $S = (1, 2)$. Calculate $\text{Proj}_{\overrightarrow{PQ}}\, \overrightarrow{RS}$ and $\text{Proj}_{\overrightarrow{RS}}\, \overrightarrow{PQ}$.
38. Let $P = (-1, 3)$, $Q = (2, 4)$, $R = (-6, -2)$, and $S = (3, 0)$. Calculate $\text{Proj}_{\overrightarrow{PQ}}\, \overrightarrow{RS}$ and $\text{Proj}_{\overrightarrow{RS}}\, \overrightarrow{PQ}$.
39. Find the distance between $P = (2, 3)$ and the line through the points $Q = (-1, 7)$ and $R = (3, 5)$. [*Hint:* Draw a picture and use the Pythagorean theorem.]

40. Find the distance between the point $(3, 7)$ and the line along the vector $\mathbf{v} = b\mathbf{i} - a\mathbf{j}$ which passes through the origin.

In Problems 41–49, find the resultant of the forces acting at the center of an object. Then find the force vector that must be applied so that the object will remain at rest.

41. 2 N (from right), 5 N (from above)
42. 2 N (from left), 5 N (from below)
43. 3 N (from left), 5 N (from right), 3 N (from above)
44. 10 lb (from right), 8 lb (from below)
45. 5 lb (from above), 4 lb (from direction $\pi/6$)
46. 6 N (from left), 4 N (from direction $\pi/4$), 2 N (from direction $\pi/3$)
47. 2 N (from above), 3 N (from direction $3\pi/4$)
48. 5 N (from direction $\pi/3$), 5 N (from direction $2\pi/3$)
49. 7 N (from direction $\pi/6$), 7 N (from direction $\pi/3$), 14 N (from direction $5\pi/4$)

In Problems 50–58, find the work done when the force with given magnitude and direction moves an object from P to Q. All distances are measured in meters. (Note that work can be negative.)

50. $|\mathbf{F}| = 3$ N; $\theta = 0$; $P = (2, 3)$; $Q = (1, 7)$

51. $|\mathbf{F}| = 2$ N; $\theta = \pi/2$; $P = (5, 7)$; $Q = (1, 1)$

52. $|\mathbf{F}| = 6$ N; $\theta = \pi/4$; $P = (2, 3)$; $Q = (-1, 4)$

53. $|\mathbf{F}| = 4$ N; $\theta = \pi/6$; $P = (-1, 2)$; $Q = (3, 4)$

54. $|\mathbf{F}| = 7$ N; $\theta = 2\pi/3$; $P = (4, -3)$; $Q = (1, 0)$

55. $|\mathbf{F}| = 3$ N; $\theta = 3\pi/4$; $P = (2, 1)$; $Q = (1, 2)$

56. $|\mathbf{F}| = 6$ N; $\theta = \pi$; $P = (3, -8)$; $Q = (5, 10)$

57. $|\mathbf{F}| = 4$ N; θ is the direction of $2\mathbf{i} + 3\mathbf{j}$; $P = (2, 0)$; $Q = (-1, 3)$

58. $|\mathbf{F}| = 5$ N; θ is the direction of $-3\mathbf{i} + 2\mathbf{j}$; $P = (1, 3)$; $Q = (4, -6)$

59. Two tugboats are towing a barge (see Figure 9). Tugboat 1 pulls with a force of 500 N at an angle of 20° with the horizontal. Tugboat 2 pulls with a force of x newtons at an angle of 30°. The barge moves horizontally (i.e., $\theta = 0$). Find x.

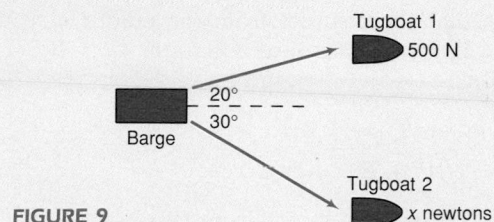

FIGURE 9

60. Answer the question of Problem 59 if the angles are 50° and 75°, respectively, and all other data remain the same.

61. In Problem 59, how much work is done by each tugboat in moving the barge a distance of 750 m?

62. In Problem 60, how much work is done by each tugboat in moving the barge a distance of 2 km?

■ **SHOW/PROVE/DISPROVE**

63. Prove or Disprove: For any nonzero real numbers α and β, the vectors $\mathbf{u} = \alpha\mathbf{i} + \beta\mathbf{j}$ and $\mathbf{v} = \beta\mathbf{i} - \alpha\mathbf{j}$ are orthogonal.

64. Let $\mathbf{u}$, $\mathbf{v}$, and $\mathbf{w}$ be three arbitrary vectors. Explain why the triple dot-product $\mathbf{u} \cdot \mathbf{v} \cdot \mathbf{w}$ is *not defined*.

65. Prove or Disprove: There is at least one value of α for which $3\mathbf{i} + 4\mathbf{j}$ and $\mathbf{i} + \alpha\mathbf{j}$ have opposite directions.

66. Prove or Disprove: There is a unique value of β for which $-2\mathbf{i} + 5\mathbf{j}$ and $\mathbf{v} = \beta\mathbf{i} - 2\mathbf{j}$ have the same direction.

67. Let $\mathbf{u} = a_1\mathbf{i} + b_1\mathbf{j}$ and $\mathbf{v} = a_2\mathbf{i} + b_2\mathbf{j}$. Give a condition on a_1, b_1, a_2, and b_2 which will ensure that $\mathbf{v}$ and $\text{Proj}_\mathbf{v} \mathbf{u}$ have the same direction.

68. In the preceding problem, give a condition which will ensure that $\mathbf{v}$ and $\text{Proj}_\mathbf{v} \mathbf{u}$ have opposite directions.

69. Prove that the nonzero vectors $\mathbf{u}$ and $\mathbf{v}$ are parallel if and only if $\mathbf{v} = \alpha\mathbf{u}$ for some nonzero constant α. [*Hint:* Show that $\cos \varphi = \pm 1$ if and only if $\mathbf{v} = \alpha\mathbf{u}$.]

70. Prove that the nonzero vectors $\mathbf{u}$ and $\mathbf{v}$ are orthogonal if and only if $\mathbf{u} \cdot \mathbf{v} = 0$.

71. Show that the vector $\mathbf{v} = a\mathbf{i} + b\mathbf{j}$ is orthogonal to the line $ax + by + c = 0$.

72. Show that the vector $\mathbf{v} = b\mathbf{i} - a\mathbf{j}$ is parallel to the line $ax + by + c = 0$.

73. **Parallelogram law:** Prove that for any vectors $\mathbf{a}$ and $\mathbf{b}$,

$$|\mathbf{a} + \mathbf{b}|^2 + |\mathbf{a} - \mathbf{b}|^2 = 2|\mathbf{a}|^2 + 2|\mathbf{b}|^2.$$

[*Hint:* $|\mathbf{u}|^2 = \mathbf{u} \cdot \mathbf{u}$ for any vector $\mathbf{u}$.]

74. **Polarization identity:** Prove that for any vectors $\mathbf{a}$ and $\mathbf{b}$,

$$\mathbf{a} \cdot \mathbf{b} = \frac{1}{4}(|\mathbf{a} + \mathbf{b}|^2 - |\mathbf{a} - \mathbf{b}|^2).$$

*75. One form of the **Cauchy-Schwarz inequality** states that for any real numbers a_1, a_2, b_1, and b_2,

$$\left| \sum_{i=1}^{2} a_i b_i \right| \le \left(\sum_{i=1}^{2} a_i^2 \right)^{1/2} \left(\sum_{i=1}^{2} b_i^2 \right)^{1/2}.$$

Use the dot product to prove this formula. Under what circumstances can the inequality be replaced by an equality? [*Hint:* Show that $|\mathbf{a} \cdot \mathbf{b}| \le |\mathbf{a}| |\mathbf{b}|$.]

76. Use the dot product and the Cauchy-Schwarz inequality to prove the **triangle inequality**:

$$|\mathbf{a} + \mathbf{b}| \le |\mathbf{a}| + |\mathbf{b}| \qquad \text{for all vectors } \mathbf{a}, \mathbf{b}.$$

77. Prove that the shortest distance between a point and a line is measured along a line through the point and perpendicular to the line.

CONTINUED

78. Prove that the distance from the point (x_0, y_0) to the line with standard equation $Ax + By + C = 0$ is

$$\frac{|Ax_0 + By_0 + C|}{\sqrt{A^2 + B^2}}.$$

[*Hint:* Let **u** be a vector from (x_0, y_0) to some point on the line and let **v** be a vector with the same direction as the line. Remember Theorem 5 and the result of the preceding problem, then consider the length of $\mathbf{u} - \text{Proj}_{\mathbf{v}}\,\mathbf{u}$. *Note:* This problem asks you to use vector methods to re-prove the result stated in Problem 63 in Section 0.3.]

■ ANSWERS TO SELF-QUIZ

I. c   II. b   III. b
IV. c   V. c

11.3 THE RECTANGULAR COORDINATE SYSTEM IN SPACE

In Section 0.2, we showed how any point in a plane can be represented as an ordered pair of real numbers. It is not surprising, then, that any point in space can be represented by an **ordered triple** of real numbers

$$(a, b, c), \tag{1}$$

where a, b, and c are real numbers.

DEFINITION   **THREE-DIMENSIONAL SPACE** $\mathbb{R}^3$

The set of ordered triples of the form (1) is called **real three-dimensional space** and is denoted by $\mathbb{R}^3$. ■

There are many ways to represent a point in $\mathbb{R}^3$. Two others will be given in Section 11.9. However, the most common representation (given in the definition above) is very similar to the representation of a point in the plane by its x- and y-coordinates. We begin, as before, by choosing a point in $\mathbb{R}^3$ and calling it the **origin**, denoted by 0. Then we draw three mutually perpendicular axes, called the **coordinate axes**, which we label the **x-axis**, the **y-axis**, and the **z-axis**. These axes can be selected in a variety of ways, but the most common selection has the x- and y-axes drawn horizontally with the z-axis vertical. On each axis we choose a positive direction and measure distance along each axis as the number of units in this positive direction measured from the origin.

A standard way of drawing these axes is depicted in Figure 1. The system is called a **right-handed system**. In the figure the arrows indicate the positive directions on the axes. The reason for this choice of terms is as follows: In a right-handed system, if you place your right hand so that your index finger points in the positive direction of the x-axis while your middle finger points in the positive direction of the y-axis, then your thumb will point in the positive direction of the z-axis. This is illustrated in Figure 2. For the remainder of this text we will follow common practice and depict the coordinate axes using a right-handed system.

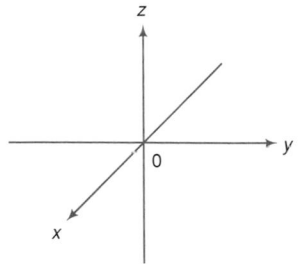

FIGURE 1
A RIGHT-HANDED SYSTEM

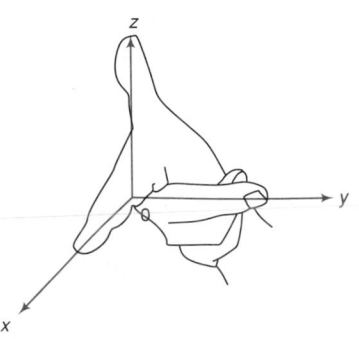

FIGURE 2
A RIGHT-HANDED SYSTEM

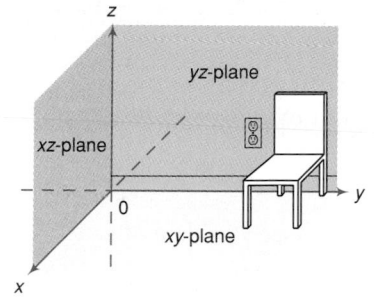

FIGURE 3
THE THREE COORDINATE PLANES

If you have trouble visualizing the placement of these axes, do the following. Face any uncluttered corner (on the floor) of the room in which you are sitting. Call the corner the origin. Then the x-axis lies along the floor, along the wall, and to your left; the y-axis lies along the floor, along the wall, and to your right; and the z-axis lies along the vertical intersection of the two perpendicular walls. This is illustrated in Figure 3.

The three axes in our system determine three **coordinate planes** that we will call the **xy-plane**, the **xz-plane**, and the **yz-plane**. The xy-plane contains the x- and y-axes and is simply the plane with which we have been dealing to this point in most of this book. The xz- and yz-planes can be thought of in a similar way.

Having built our structure of coordinate axes and planes, we can describe any point P in space in a unique way:

$$P = (x, y, z), \tag{2}$$

where the first coordinate x is the directed distance from the yz-plane to P (measured in the positive direction of the x-axis), the second coordinate y is the directed distance from the xz-plane to P (measured in the positive direction of the y-axis), and the third coordinate z is the directed distance from the xy-plane to P (measured in the positive direction of the z-axis). Thus, for example, any point in the xy-plane has z-coordinate 0; any point in the xz-plane has y-coordinate 0; and any point in the yz-plane has x-coordinate 0. Some representative points are sketched in Figure 4.

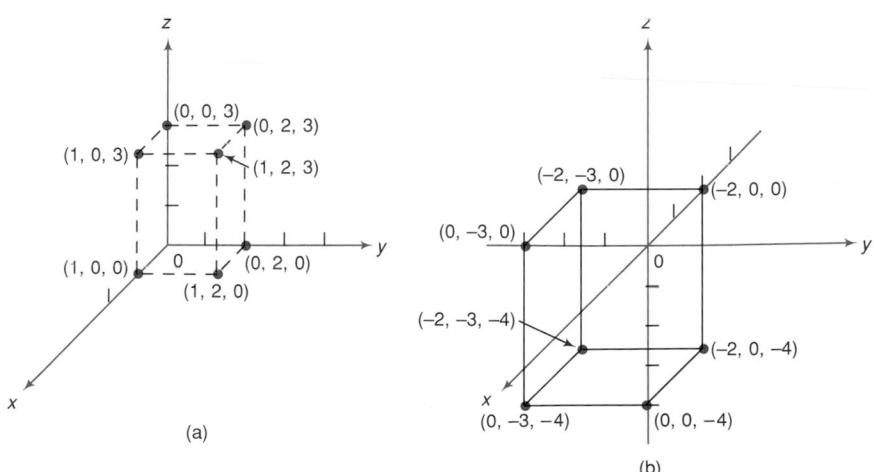

(a)

(b)

FIGURE 4   TYPICAL POINTS IN $\mathbb{R}^3$

In this system, the three coordinate planes divide $\mathbb{R}^3$ into eight **octants**, just as in $\mathbb{R}^2$ the two coordinate axes divide the plane into four quadrants. The first octant is always chosen to be the one in which the three coordinates are positive (the other seven octants are usually not numbered).

The coordinate system we have just established is often referred to as the **rectangular coordinate system**, or the **Cartesian coordinate system**. Once we are comfortable with the notion of depicting a point in this system, then we can generalize many of our ideas from the plane.

THEOREM 1

Let $P = (x_1, y_1, z_1)$ and $Q = (x_2, y_2, z_2)$ be two points in space. Then the distance $\overline{PQ}$ between P and Q is given by

$$\overline{PQ} = \sqrt{(x_1 - x_2)^2 + (y_1 - y_2)^2 + (z_1 - z_2)^2}. \quad \blacksquare \qquad (3)$$

The proof of this theorem is left as an exercise (see Problem 40).

EXAMPLE 1
CALCULATING THE DISTANCE
BETWEEN TWO POINTS IN $\mathbb{R}^3$

Calculate the distance between the points $(3, -1, 6)$ and $(-2, 3, 5)$.

SOLUTION: $\overline{PQ} = \sqrt{[3 - (-2)]^2 + (-1 - 3)^2 + (6 - 5)^2} = \sqrt{42}$.

DEFINITION GRAPH IN $\mathbb{R}^3$

The **graph** of an equation in $\mathbb{R}^3$ is the set of all points in $\mathbb{R}^3$ whose coordinates satisfy the equation. $\blacksquare$

One of our first examples of a graph in $\mathbb{R}^2$ was the graph of the unit circle $x^2 + y^2 = 1$. This example can easily be generalized.

DEFINITION SPHERE

A **sphere** is the set of points in space at a given distance from a given point. The given point is called the **center** of the sphere, and the given distance is called the **radius** of the sphere. $\blacksquare$

EXAMPLE 2
THE EQUATION OF THE UNIT
SPHERE

Suppose that the center of a sphere is the origin $(0, 0, 0)$ and the radius of the sphere is 1. Let (x, y, z) be a point on the sphere. Then from (3)

$$1 = \sqrt{(x - 0)^2 + (y - 0)^2 + (z - 0)^2}.$$

Simplifying and squaring, we obtain

$$x^2 + y^2 + z^2 = 1,$$

which is the equation of the **unit sphere**.

In general, if the center of a sphere is (a, b, c), the radius is r, and if (x, y, z) is a point on the sphere, we obtain $r = \sqrt{(x - a)^2 + (y - b)^2 + (z - c)^2}$, or

---

EQUATION OF A SPHERE CENTERED AT (a, b, c) WITH RADIUS r

$$(x - a)^2 + (y - b)^2 + (z - c)^2 = r^2. \qquad (4)$$

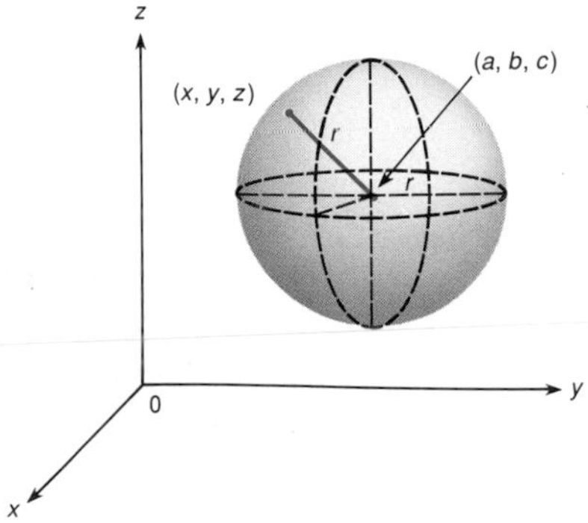

FIGURE 5
SPHERE OF RADIUS r CENTERED AT (a, b, c)

This is sketched in Figure 5.

EXAMPLE 3
FINDING THE EQUATION OF A
SPHERE

Find the equation of the sphere with center at $(1, -3, 2)$ and radius 5.

SOLUTION: From (4) we obtain

$$(x - 1)^2 + (y + 3)^2 + (z - 2)^2 = 25.$$

EXAMPLE 4
SHOWING THAT A SECOND-
DEGREE EQUATION IS THE
EQUATION OF A SPHERE

Show that

$$x^2 - 6x + y^2 + 2y + z^2 + 10z + 5 = 0 \qquad \text{(5)}$$

is the equation of a sphere, and find its center and radius.

SOLUTION: We complete the square three times:

$$
\begin{aligned}
0 &= x^2 - 6x + y^2 + 2y + z^2 + 10z + 5 \\
&= x^2 - 6x + 9 \quad 9 + y^2 + 2y + 1 - 1 + z^2 + 10z + 25 - 25 + 5 \\
&= (x - 3)^2 \quad 9 + (y + 1)^2 - 1 + (z + 5)^2 - 25 + 5
\end{aligned}
$$

or

$$(x - 3)^2 + (y + 1)^2 + (z + 5)^2 = 30,$$

which is the equation of a sphere with center $(3, -1, -5)$ and radius $\sqrt{30}$.

$\otimes$ **WARNING:** Not every second-degree equation in a form similar to (5) is the equation of a sphere. For example, if the number 5 in (5) is replaced by 40, we obtain

$$(x - 3)^2 + (y + 1)^2 + (z + 5)^2 = -5.$$

Clearly, the sum of squares cannot be negative, so there are *no* points in $\mathbb{R}^3$ that satisfy this equation. On the other hand, if we replaced the 5 by 35, we would obtain

$$(x - 3)^2 + (y + 1)^2 + (z + 5)^2 = 0.$$

This equation can hold only when $x = 3$, $y = -1$, and $z = -5$. In this case, the graph of the equation consists of the single point $(3, -1, -5)$. $\otimes$

■ PROBLEMS 11.3

■ SELF-QUIZ

I. True–False: The common practice, followed in this text, is to display the *xyz*-axes for $\mathbb{R}^3$ as a right-handed system.

II. The distance between the points $(1, 2, 3)$ and $(3, 5, -1)$ is _____.

a. $\sqrt{(1 + 2 + 3)^2 + (3 + 5 - 1)^2}$
b. $\sqrt{2^2 + 3^2 + 2^2}$
c. $\sqrt{2^2 + 3^2 + 4^2}$
d. $\sqrt{4^2 + 7^2 + 2^2}$

III. The point $(0.3, 0.5, 0.2)$ is _____ the unit sphere.
   a. tangent to   b. on
   c. inside   d. outside

IV. $(x - 3)^2 + (y + 5)^2 + z^2 = 81$ is the equation of the sphere with _____.

   a. center 81 and radius $(-3, 5, 0)$
   b. radius 81 and center $(-3, 5, 0)$
   c. radius -9 and center $(3, -5, 0)$
   d. radius 9 and center $(3, -5, 0)$

■ **DRILL**

In Problems 1–16, sketch the given point in $\mathbb{R}^3$.

1. $(1, 4, 2)$
2. $(3, -2, 1)$
3. $(-1, 5, 7)$
4. $(8, -2, 3)$
5. $(-2, 1, -2)$
6. $(1, -2, 1)$
7. $(3, 2, -5)$
8. $(-2, -3, -8)$
9. $(2, 0, 4)$
10. $(-3, -8, 0)$
11. $(0, 4, 7)$
12. $(1, 3, 0)$
13. $(3, 0, 0)$
14. $(0, 8, 0)$
15. $(0, 0, -7)$
16. $(5, 5, 5)$

In Problems 17–26, calculate the distance between the two given points.

17. $(8, 1, 6); (8, 1, 4)$
18. $(3, -4, 3); (3, 2, 5)$
19. $(3, -4, 7); (3, -4, 9)$
20. $(-2, 1, 3); (4, 1, 3)$
21. $(2, -7, 5); (8, -7, -1)$
22. $(1, 3, -2); (4, 7, -2)$
23. $(3, 1, 2); (1, 2, 3)$
24. $(5, -6, 4); (3, 11, -2)$
25. $(-1, -7, -2); (-4, 3, -5)$
26. $(8, -2, -3); (-7, -5, 1)$

■ **APPLICATIONS**

27. Find the equation of the sphere with center $(2, -1, 4)$ and radius 2.
28. Find the equation of the sphere with center $(-1, 8, -3)$ and radius $\sqrt{5}$.
29. Use the result stated in Problem 38 to verify that the points $(3, 0, 1)$, $(0, -4, 0)$ and $(6, 4, 2)$ are collinear.
30. Use the result stated in Problem 39 to find the midpoint of the line joining the points $(2, -1, 4)$ and $(5, 7, -3)$.
31. Verify that $x^2 + y^2 + z^2 - 4x - 4y + 8z + 8 = 0$ is the equation of a sphere. Find the center and radius of that sphere.
32. Verify that $x^2 + 3x + y^2 - y + z^2 + 2z - 1 = 0$ is the equation of a sphere. Find the center and radius of that sphere.

33. One sphere is said to be **concentric** to a second sphere if it has a different radius and the same center as the second sphere. Find the equation of a sphere of radius 1 concentric to the sphere given by $x^2 - 2x + y^2 - 4y + z^2 + z - 2 = 0$.
34. Find the equation of the sphere that has a diameter with endpoints $(3, 1, -2)$ and $(4, 1, 5)$. [*Hint:* First find the center and radius of the sphere.]
35. Find a number α such that the equation $x^2 - 2x + y^2 + 8y + z^2 - 5z + \alpha = 0$ has exactly one solution (x, y, z).
36. For the equation in the preceding problem, give a condition on α such that the resulting equation has no solution.

■ **SHOW/PROVE/DISPROVE**

37. For fixed constants a, b, c, and d the equation

   $$x^2 + ax + y^2 + by + z^2 + cz + d = 0 \qquad \textbf{(6)}$$

   is a second-degree equation in three variables x, y, and z. Let

   $$E = d - (a/2)^2 - (b/2)^2 - (c/2)^2.$$

   Show that equation (6)
   a. is the equation of a sphere if $E < 0$,
   b. has exactly one solution if $E = 0$,
   c. has no solutions if $E > 0$.
38. Three points P, Q, and R are **collinear** if and only if they lie on the same straight line. Show that, in $\mathbb{R}^2$, points P, Q, and R are collinear if and

only if one of the following conditions holds:

- $\overline{PR} = \overline{PQ} + \overline{QR}$ (i.e., Q is between the others)
- $\overline{PQ} = \overline{PR} + \overline{RQ}$ (i.e., R is between the others)
- $\overline{QR} = \overline{QP} + \overline{PR}$ (i.e., P is between the others)

[Here $\overline{PQ}$ denotes the length of the directed line segment $\overrightarrow{PQ}$: $\overline{PQ} = |\overrightarrow{PQ}|$.] Apply the same idea in $\mathbb{R}^3$ to show that the points $(-1, -1, -1)$, $(5, 8, 2)$, and $(-3, -4, -2)$ are collinear.

*39. Let $P = (x_1, y_1, z_1)$ and $Q = (x_2, y_2, z_2)$. Show that the midpoint of PQ is the point

$$R = \left(\frac{x_1 + x_2}{2}, \frac{y_1 + y_2}{2}, \frac{z_1 + z_2}{2} \right).$$

[*Hint:* Show that P, Q, and R are collinear and that $\overline{PR} = \overline{RQ}$.]

40. Prove Theorem 1. [*Hint:* Use the Pythagorean theorem twice. Show, with points labeled as in Figure 6, that
  a. $\overline{PQ}^2 = \overline{PR}^2 + \overline{RQ}^2$,
  b. $\overline{PR}^2 = \overline{PS}^2 + \overline{SR}^2$,
  c. $\overline{PQ}^2 = \overline{PS}^2 + \overline{SR}^2 + \overline{RQ}^2$.]

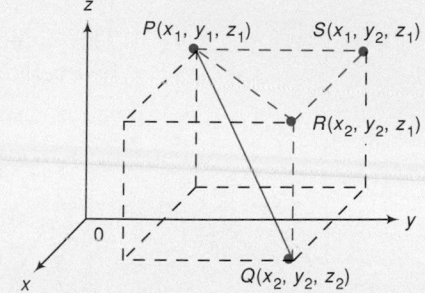

FIGURE 6

■ ANSWERS TO SELF-QUIZ

I. True II. c III. c IV. d

11.4 VECTORS IN $\mathbb{R}^3$

In Sections 11.1 and 11.2, we developed properties of vectors in the plane $\mathbb{R}^2$. Given the similarity between the coordinate systems in $\mathbb{R}^2$ and $\mathbb{R}^3$, it should come as no surprise to learn that vectors in $\mathbb{R}^2$ and $\mathbb{R}^3$ have very similar structures. In this section we will develop the notion of a vector in space. The development will closely follow the development in Sections 11.1 and 11.2 and, therefore, some of the details will be omitted.

Let P and Q be two distinct points in $\mathbb{R}^3$. Then the **directed line segment** $\overrightarrow{PQ}$ is the straight line segment that extends from P to Q. Two directed line segments are **equivalent** if they have the same magnitude and direction. A **vector** in $\mathbb{R}^3$ is the set of all directed line segments equivalent to a given line segment, and any directed line segment $\overrightarrow{PQ}$ in that set is called a **representation** of the vector.

So far, our definitions are identical. For convenience we will choose P to be the origin and label the endpoint of the vector R, so that the vector $\mathbf{v} = \overrightarrow{0R}$ can be described by the coordinates (x, y, z) of the point R. As in $\mathbb{R}^2$, two vectors are **equal** if their corresponding coordinates are equal. Then the **magnitude** of $\mathbf{v} = |\mathbf{v}| = \sqrt{x^2 + y^2 + z^2}$ (from Theorem 11.3.1).

EXAMPLE 1

CALCULATING THE MAGNITUDE OF A VECTOR IN $\mathbb{R}^3$

Let $\mathbf{v} = (1, 3, -2)$. Find $|\mathbf{v}|$.

SOLUTION: $|\mathbf{v}| = \sqrt{1^2 + 3^2 + (-2)^2} = \sqrt{14}$.

Let $\mathbf{u} = (x_1, y_1, z_1)$ and $\mathbf{v} = (x_2, y_2, z_2)$ be two vectors and let α be a real number (scalar). Then we define

---

SUM AND SCALAR MULTIPLE OF VECTORS IN $\mathbb{R}^3$

$$\mathbf{u} + \mathbf{v} = (x_1 + x_2, y_1 + y_2, z_1 + z_2)$$

and

$$\alpha\mathbf{u} = (\alpha x_1, \alpha y_1, \alpha z_1).$$

This is the same definition of vector addition and scalar multiplication we had before and is illustrated in Figure 1.

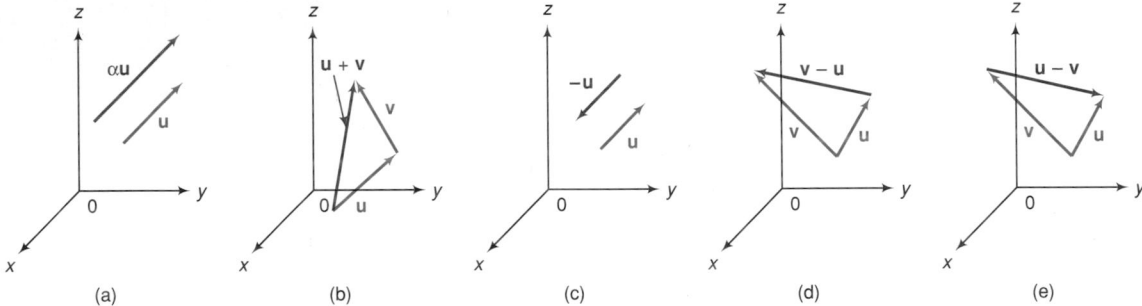

FIGURE 1 ILLUSTRATING THE VECTOR OPERATIONS $\alpha\mathbf{u}$, $\mathbf{u} + \mathbf{v}$, $-\mathbf{u}$, $\mathbf{v} - \mathbf{u}$, AND $\mathbf{u} - \mathbf{v}$

EXAMPLE 2
FINDING SUMS AND SCALAR MULTIPLES OF VECTORS

Let $\mathbf{u} = (2, 3, -1)$ and $\mathbf{v} = (-6, 2, 4)$. Find **(a)** $\mathbf{u} + \mathbf{v}$, **(b)** $3\mathbf{v}$, **(c)** $-\mathbf{u}$, and **(d)** $4\mathbf{u} - 3\mathbf{v}$.

SOLUTION:

(a) $\mathbf{u} + \mathbf{v} = (2 - 6, 3 + 2, -1 + 4) = (-4, 5, 3)$
(b) $3\mathbf{v} = (-18, 6, 12)$
(c) $-\mathbf{u} = (-2, -3, 1)$
(d) $4\mathbf{u} - 3\mathbf{v} = (8, 12, -4) - (-18, 6, 12) = (26, 6, -16)$

The following theorem extends to three dimensions the results of Theorem 11.1.1. Its proof is easy and is left as an exercise (Problem 56).

THEOREM 1   **PROPERTIES OF VECTORS IN $\mathbb{R}^3$**

Let $\mathbf{u}$, $\mathbf{v}$, and $\mathbf{w}$ be any three vectors in space, let α and β be scalars, and let $\mathbf{0}$ denote the zero vector $(0, 0, 0)$.

(i) $\mathbf{u} + \mathbf{v} = \mathbf{v} + \mathbf{u}$ **(ii)** $\mathbf{u} + (\mathbf{v} + \mathbf{w}) = (\mathbf{u} + \mathbf{v}) + \mathbf{w}$

(iii) $\mathbf{v} + \mathbf{0} = \mathbf{v}$ **(iv)** $0\mathbf{v} = \mathbf{0}$

(v) $\alpha\mathbf{0} = \mathbf{0}$ **(vi)** $(\alpha\beta)\mathbf{v} = \alpha(\beta\mathbf{v})$

(vii) $\mathbf{v} + (-\mathbf{v}) = \mathbf{0}$ **(viii)** $(1)\mathbf{v} = \mathbf{v}$

(ix) $(\alpha + \beta)\mathbf{v} = \alpha\mathbf{v} + \beta\mathbf{v}$ **(x)** $\alpha(\mathbf{u} + \mathbf{v}) = \alpha\mathbf{u} + \alpha\mathbf{v}$

(xi) $|\alpha\mathbf{v}| = |\alpha|\,|\mathbf{v}|$ **(xii)** $|\mathbf{u} + \mathbf{v}| \le |\mathbf{u}| + |\mathbf{v}|$ ■

triangle inequality

DEFINITION   UNIT VECTOR

A **unit vector u** is a vector with magnitude 1. If $\mathbf{v}$ is any nonzero vector, then $\mathbf{u} = \mathbf{v}/|\mathbf{v}|$ is a unit vector having the same direction as $\mathbf{v}$. ■

EXAMPLE 3
FINDING A UNIT VECTOR THAT
HAS THE SAME DIRECTION AS
A GIVEN VECTOR

Find a unit vector having the same direction as $\mathbf{v} = (2, 4, -3)$.

SOLUTION: Since $|\mathbf{v}| = \sqrt{2^2 + 4^2 + (-3)^2} = \sqrt{29}$, we have

$$\mathbf{u} = \left(\frac{2}{\sqrt{29}}, \frac{4}{\sqrt{29}}, \frac{-3}{\sqrt{29}} \right).$$

We can now formally define the direction of a vector in $\mathbb{R}^3$. We cannot define it to be the angle θ the vector makes with the positive x-axis, since, for example, if $0 < \theta < \pi/2$, then there are an *infinite number* of unit vectors making the angle θ with the positive x-axis, and these together form a cone (see Figure 2).

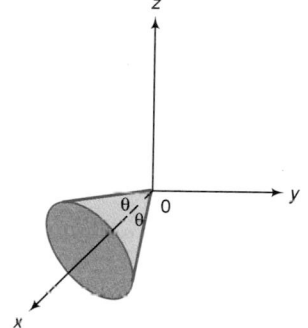

FIGURE 2
EVERY VECTOR IN THE CONE MAKES THE
ANGLE θ WITH THE POSITIVE x-AXIS.

DEFINITION   DIRECTION OF A VECTOR

The **direction** of a nonzero vector $\mathbf{v}$ in $\mathbb{R}^3$ is defined to be the unit vector $\mathbf{u} = \mathbf{v}/|\mathbf{v}|$. ■

REMARK: We could have defined the direction of a vector $\mathbf{v}$ in $\mathbb{R}^2$ in this way. For if $\mathbf{u} = \mathbf{v}/|\mathbf{v}|$, then $\mathbf{u} = (\cos \theta, \sin \theta)$, where θ is the direction of $\mathbf{v}$ (according to the $\mathbb{R}^2$ definition).

DIRECTION COSINES

It would still be useful to define the direction of a vector in terms of some angles. Let $\mathbf{v} = (x_0, y_0, z_0)$ be the vector $\overrightarrow{OP}$ depicted in Figure 3. We define α to be the angle between $\mathbf{v}$ and the positive x-axis, β the angle between $\mathbf{v}$ and the positive y-axis, and γ the angle between $\mathbf{v}$ and the positive z-axis. The angles α, β, and γ are called the **direction angles** of the vector $\mathbf{v}$. Then, from Figure 3,

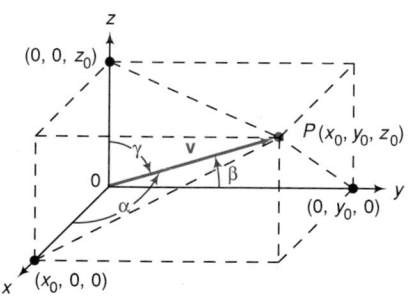

FIGURE 3
DIRECTION ANGLES

THREE DIRECTION ANGLES			
$\cos \alpha = \dfrac{x_0}{|\mathbf{v}|}$,	$\cos \beta = \dfrac{y_0}{|\mathbf{v}|}$,	$\cos \gamma = \dfrac{z_0}{|\mathbf{v}|}.$	**(1)**

If $\mathbf{v}$ is a unit vector, then $|\mathbf{v}| = 1$ and

$$\cos \alpha = x_0, \qquad \cos \beta = y_0, \qquad \cos \gamma = z_0. \tag{2}$$

By definition, each of these three angles lies between 0 and π. The cosines of these angles are called the **direction cosines** of the vector **v**.

Note from (1) that

$$\cos^2 \alpha + \cos^2 \beta + \cos^2 \gamma = \frac{x_0{}^2 + y_0{}^2 + z_0{}^2}{|\mathbf{v}|^2} = \frac{x_0{}^2 + y_0{}^2 + z_0{}^2}{x_0{}^2 + y_0{}^2 + z_0{}^2} = 1. \quad \text{(3)}$$

If α, β, and γ are any three numbers between 0 and π such that condition (3) is satisfied, then they uniquely determine a unit vector given by $\mathbf{u} = (\cos \alpha, \cos \beta, \cos \gamma)$.

REMARK: If $\mathbf{v} = (a, b, c)$ and $|\mathbf{v}| \neq 0$, then the numbers a, b, and c are called **direction numbers** of the vector **v**.

EXAMPLE 4
FINDING THE DIRECTION
COSINES OF A VECTOR IN $\mathbb{R}^3$

Find the direction cosines of the vector $\mathbf{v} = (4, -1, 6)$.

SOLUTION: The direction of **v** is $\mathbf{v}/|\mathbf{v}| = \mathbf{v}/\sqrt{53} = (4/\sqrt{53}, -1/\sqrt{53}, 6/\sqrt{53})$. Then $\cos \alpha = 4/\sqrt{53} \approx 0.5494$, $\cos \beta = -1/\sqrt{53} \approx -0.1374$, and $\cos \gamma = 6/\sqrt{53} \approx 0.8242$. From these we use a calculator to obtain $\alpha \approx 56.7° \approx 0.989$ radian, $\beta \approx 97.9° \approx 1.71$ radians, and $\gamma = 34.5° \approx 0.602$ radian.

EXAMPLE 5
FINDING A VECTOR IN $\mathbb{R}^3$
GIVEN ITS MAGNITUDE AND
DIRECTION COSINES

Find a vector **v** of magnitude 7 whose direction cosines are $1/\sqrt{6}$, $1/\sqrt{3}$, and $1/\sqrt{2}$.

NOTE: We can solve this problem because $(1/\sqrt{6})^2 + (1/\sqrt{3})^2 + (1/\sqrt{2})^2 = 1$.

SOLUTION: Let $\mathbf{u} = (1/\sqrt{6}, 1/\sqrt{3}, 1/\sqrt{2})$. Then **u** is a unit vector since $|\mathbf{u}| = 1$. Thus, the direction of **v** is given by **u**, and so

$$\mathbf{v} = |\mathbf{v}|\mathbf{u} = 7\mathbf{u} = \left(\frac{7}{\sqrt{6}}, \frac{7}{\sqrt{3}}, \frac{7}{\sqrt{2}}\right).$$

In Section 11.1, we showed how any vector in the plane can be written in terms of the basis vectors **i** and **j**. To extend this idea to $\mathbb{R}^3$, we define

THE VECTORS i, j, AND k IN $\mathbb{R}^3$
$\mathbf{i} = (1, 0, 0), \qquad \mathbf{j} = (0, 1, 0), \qquad \mathbf{k} = (0, 0, 1).$ (4)

Here **i**, **j**, and **k** are unit vectors. The vector **i** lies along the x-axis, **j** along the y-axis, and **k** along the z-axis. These vectors are sketched in Figure 4. If $\mathbf{v} = (x, y, z)$ is any vector in $\mathbb{R}^3$, then

$$\mathbf{v} = (x, y, z) = (x, 0, 0) + (0, y, 0) + (0, 0, z) = x\mathbf{i} + y\mathbf{j} + z\mathbf{k}. \quad \text{(5)}$$

That is, *any vector* **v** *in* $\mathbb{R}^3$ *can be written in a unique way in terms of the vectors* **i**, **j**, *and* **k**.

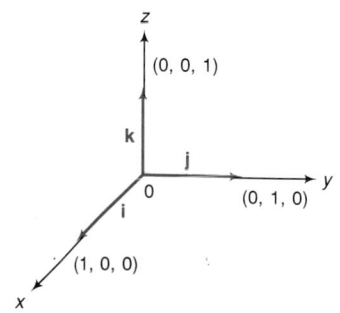

FIGURE 4
THE VECTORS **i**, **j**, AND **k**

Let $P = (a_1, b_1, c_1)$ and $Q = (a_2, b_2, c_2)$. Then as in Section 11.1, the vector $\mathbf{v} = \overrightarrow{PQ}$ can be written (see Problem 57)

$$\mathbf{v} = (a_2 - a_1)\mathbf{i} + (b_2 - b_1)\mathbf{j} + (c_2 - c_1)\mathbf{k}. \tag{6}$$

EXAMPLE 6
WRITING A VECTOR IN TERMS OF THE BASIS VECTORS **i**, **j**, AND **k**

Find a vector in space that can be represented by the directed line segment from $(2, -1, 4)$ to $(5, 1, -3)$.

$$\mathbf{v} = (5 - 2)\mathbf{i} + [1 - (-1)]\mathbf{j} + (-3 - 4)\mathbf{k} = 3\mathbf{i} + 2\mathbf{j} - 7\mathbf{k}.$$

We now turn to the notion of dot product (or scalar product) in $\mathbb{R}^3$.

DEFINITION **DOT PRODUCT IN $\mathbb{R}^3$**

If $\mathbf{u} = x_1\mathbf{i} + y_1\mathbf{j} + z_1\mathbf{k}$ and $\mathbf{v} = x_2\mathbf{i} + y_2\mathbf{j} + z_2\mathbf{k}$, then we define the **dot product** (or **scalar product** or **inner product**) by

$$\mathbf{u} \cdot \mathbf{v} = x_1 x_2 + y_1 y_2 + z_1 z_2. \quad\blacksquare \tag{7}$$

As before, the dot product of two vectors is a *scalar*. Note that $\mathbf{i} \cdot \mathbf{i} = 1$, $\mathbf{j} \cdot \mathbf{j} = 1$, $\mathbf{k} \cdot \mathbf{k} = 1$, $\mathbf{i} \cdot \mathbf{j} = 0$, $\mathbf{j} \cdot \mathbf{k} = 0$, and $\mathbf{i} \cdot \mathbf{k} = 0$.

EXAMPLE 7
COMPUTING A DOT PRODUCT IN $\mathbb{R}^3$

For $\mathbf{u} = 2\mathbf{i} - 3\mathbf{j} - 4\mathbf{k}$ and $\mathbf{v} = -3\mathbf{i} + \mathbf{j} - 2\mathbf{k}$, calculate $\mathbf{u} \cdot \mathbf{v}$.

SOLUTION: $\mathbf{u} \cdot \mathbf{v} = 2(-3) + (-3)(1) + (-4)(-2) = -1$.

THEOREM 2 **PROPERTIES OF THE DOT PRODUCT IN $\mathbb{R}^3$**

For any vectors **u**, **v**, and **w** in space, and for any scalar α, we have

(i) $\mathbf{u} \cdot \mathbf{v} = \mathbf{v} \cdot \mathbf{u}$
(ii) $(\mathbf{u} + \mathbf{v}) \cdot \mathbf{w} = \mathbf{u} \cdot \mathbf{w} + \mathbf{v} \cdot \mathbf{w}$
(iii) $(\alpha\mathbf{u}) \cdot \mathbf{v} = \alpha(\mathbf{u} \cdot \mathbf{v})$
(iv) $\mathbf{u} \cdot \mathbf{u} \geq 0$, and $\mathbf{u} \cdot \mathbf{u} = 0$ if and only if $\mathbf{u} = \mathbf{0}$
(v) $|\mathbf{u}| = \sqrt{\mathbf{u} \cdot \mathbf{u}}$.

PROOF: The proof is almost identical to the proof of Theorem 11.2.1 and is left as an exercise (see Problem 58).   ∎

THEOREM 3

If φ denotes the angle between two nonzero vectors **u** and **v**, we have

$$\cos \varphi = \frac{\mathbf{u} \cdot \mathbf{v}}{|\mathbf{u}|\,|\mathbf{v}|}. \tag{8}$$

PROOF: The proof is almost identical to the proof of Theorem 11.2.2 and is left as an exercise (Problem 59). For an interesting corollary to this theorem, see Problem 55.   ∎

EXAMPLE 8

CALCULATING THE COSINE OF
THE ANGLE BETWEEN TWO
VECTORS IN $\mathbb{R}^3$

Calculate the cosine of the angle between $\mathbf{u} = 3\mathbf{i} - \mathbf{j} + 2\mathbf{k}$ and $\mathbf{v} = 4\mathbf{i} + 3\mathbf{j} - \mathbf{k}$.

SOLUTION: $\mathbf{u} \cdot \mathbf{v} = 7$, $|\mathbf{u}| = \sqrt{14}$, and $|\mathbf{v}| = \sqrt{26}$, so that $\cos \varphi = 7/\sqrt{(14)(26)} = 7/\sqrt{364} \approx 0.3669$ and $\varphi = 68.5° \approx 1.2$ radians.

DEFINITION   **PARALLEL AND ORTHOGONAL VECTORS**

   (i) Two nonzero vectors $\mathbf{u}$ and $\mathbf{v}$ are **parallel** if the angle between them is 0 or π.

   (ii) Two nonzero vectors $\mathbf{u}$ and $\mathbf{v}$ are **orthogonal** (or **perpendicular**) if the angle between them is $\pi/2$. ■

THEOREM 4

   (i) If $\mathbf{u} \neq \mathbf{0}$, then $\mathbf{u}$ and $\mathbf{v}$ are parallel if and only if $\mathbf{v} = \alpha\mathbf{u}$ for some constant α.

   (ii) If $\mathbf{u}$ and $\mathbf{v}$ are nonzero, then $\mathbf{u}$ and $\mathbf{v}$ are orthogonal if and only if $\mathbf{u} \cdot \mathbf{v} = 0$.

     PROOF: Again the proof is not difficult and is left as an exercise (see Problem 60). ■

We now turn to the definition of the projection of one vector on another. First, we state the theorem that is the analog of Theorem 11.2.5 (and has an identical proof).

THEOREM 5

Let $\mathbf{v}$ be a nonzero vector. Then for any other vector $\mathbf{u}$,

$$\mathbf{w} = \mathbf{u} - \frac{\mathbf{u} \cdot \mathbf{v}}{|\mathbf{v}|^2} \mathbf{v}$$

is orthogonal to $\mathbf{v}$. ■

DEFINITION   **PROJECTION AND COMPONENT**

Let $\mathbf{u}$ and $\mathbf{v}$ be nonzero vectors. Then the **projection**[†] **of u onto v**, denoted by $\text{Proj}_{\mathbf{v}} \mathbf{u}$, is defined by

$$\text{Proj}_{\mathbf{v}} \mathbf{u} = \frac{\mathbf{u} \cdot \mathbf{v}}{\mathbf{v} \cdot \mathbf{v}} \mathbf{v} = \frac{\mathbf{u} \cdot \mathbf{v}}{|\mathbf{v}|^2} \mathbf{v} = \left(\frac{\mathbf{u} \cdot \mathbf{v}}{|\mathbf{v}|}\right) \frac{\mathbf{v}}{|\mathbf{v}|}. \tag{9}$$

The **component** of $\mathbf{u}$ in the direction $\mathbf{v}$ is given by $(\mathbf{u} \cdot \mathbf{v})/|\mathbf{v}|$.   **(10)** ■

[†] The projection vector in $\mathbb{R}^2$ is sketched in Figure 4 of Section 11.2. The derivation and a sketch of a projection vector in $\mathbb{R}^3$ are virtually the same.

EXAMPLE 9
CALCULATING A
PROJECTION IN $\mathbb{R}^3$

Let $\mathbf{u} = 2\mathbf{i} + 3\mathbf{j} + \mathbf{k}$ and $\mathbf{v} = \mathbf{i} + 2\mathbf{j} - 6\mathbf{k}$. Find $\text{Proj}_\mathbf{v}\, \mathbf{u}$.

SOLUTION: Here $(\mathbf{u} \cdot \mathbf{v})/|\mathbf{v}|^2 = \frac{2}{41}$, so

$$\text{Proj}_\mathbf{v}\, \mathbf{u} = \frac{2}{41}\mathbf{i} + \frac{4}{41}\mathbf{j} - \frac{12}{41}\mathbf{k}.$$

The component of $\mathbf{u}$ in the direction $\mathbf{v}$ is $(\mathbf{u} \cdot \mathbf{v})/|\mathbf{v}| = 2/\sqrt{41}$.

Note that, as in the planar case, $\text{Proj}_\mathbf{v}\, \mathbf{u}$ is a vector that has the same direction as $\mathbf{v}$ if $\mathbf{u} \cdot \mathbf{v} > 0$ and the direction opposite to that of $\mathbf{v}$ if $\mathbf{u} \cdot \mathbf{v} < 0$.

■ PROBLEMS 11.4

■ SELF-QUIZ

I. $\mathbf{j} - (4\mathbf{k} - 3\mathbf{i}) = $ _____.
   a. $(1, -4, -3)$ b. $(1, -4, 3)$
   c. $(-3, 1, -4)$ d. $(3, 1, -4)$
II. $(\mathbf{i} + 3\mathbf{k} - \mathbf{j}) \cdot (\mathbf{k} - 4\mathbf{j} + 2\mathbf{i}) = $ _____.
   a. $2 + 4 + 3 = 9$
   b. $(1 + 3 - 1)(1 - 4 + 2) = -3$
   c. $1 - 12 - 2 = -13$
   d. $2 - 4 - 3 = -5$
III. The unit vector in the same direction as $2\mathbf{i} - 2\mathbf{j} + \mathbf{k}$ is _____.

   a. $\mathbf{i} - \mathbf{j} + \mathbf{k}$ b. $\frac{1}{5}(2\mathbf{i} - 2\mathbf{j} + \mathbf{k})$
   c. $\frac{1}{3}(2\mathbf{i} - 2\mathbf{j} + \mathbf{k})$ d. $\frac{1}{3}(2\mathbf{i} + 2\mathbf{j} + \mathbf{k})$
IV. The component of $\mathbf{u}$ in the direction $\mathbf{w}$ is _____.

   a. $\dfrac{\mathbf{u} \cdot \mathbf{w}}{|\mathbf{w}|}$ b. $\dfrac{\mathbf{w}}{|\mathbf{w}|}$
   c. $\dfrac{\mathbf{u} \cdot \mathbf{w}}{|\mathbf{w}|}\dfrac{\mathbf{w}}{|\mathbf{w}|}$ d. $\dfrac{\mathbf{u} \cdot \mathbf{w}}{|\mathbf{w}|}\dfrac{\mathbf{u}}{|\mathbf{u}|}$

■ DRILL

In Problems 1–20, find the magnitude and the direction cosines of the given vector $\mathbf{v}$.

1. $\mathbf{v} = 3\mathbf{j}$ 2. $\mathbf{v} = -3\mathbf{i}$
3. $\mathbf{v} = 14\mathbf{k}$ 4. $\mathbf{v} = -8\mathbf{j}$
5. $\mathbf{v} = 4\mathbf{i} - \mathbf{j}$ 6. $\mathbf{v} = \mathbf{i} + 2\mathbf{k}$
7. $\mathbf{v} = -2\mathbf{i} + 3\mathbf{j}$ 8. $\mathbf{v} = \mathbf{i} + \mathbf{j} + \mathbf{k}$
9. $\mathbf{v} = \mathbf{i} - \mathbf{j} + \mathbf{k}$ 10. $\mathbf{v} = \mathbf{i} + \mathbf{j} - \mathbf{k}$
11. $\mathbf{v} = -\mathbf{i} + \mathbf{j} + \mathbf{k}$ 12. $\mathbf{v} = \mathbf{i} - \mathbf{j} - \mathbf{k}$
13. $\mathbf{v} = -\mathbf{i} + \mathbf{j} - \mathbf{k}$ 14. $\mathbf{v} = -\mathbf{i} - \mathbf{j} + \mathbf{k}$
15. $\mathbf{v} = -\mathbf{i} - \mathbf{j} - \mathbf{k}$ 16. $\mathbf{v} = 2\mathbf{i} + 5\mathbf{j} - 7\mathbf{k}$
17. $\mathbf{v} = -7\mathbf{i} + 2\mathbf{j} - 13\mathbf{k}$
18. $\mathbf{v} = \mathbf{i} + 7\mathbf{j} - 7\mathbf{k}$
19. $\mathbf{v} = -3\mathbf{i} - 3\mathbf{j} + 8\mathbf{k}$
20. $\mathbf{v} = -2\mathbf{i} - 3\mathbf{j} - 4\mathbf{k}$

In Problems 21–40, let $\mathbf{u} = 2\mathbf{i} - 3\mathbf{j} + 4\mathbf{k}$, $\mathbf{v} = -2\mathbf{i} - 3\mathbf{j} + 5\mathbf{k}$, $\mathbf{w} = \mathbf{i} - 7\mathbf{j} + 3\mathbf{k}$, and $\mathbf{t} = 3\mathbf{i} + 4\mathbf{j} + 5\mathbf{k}$; calculate the value of the specified expression.

21. $\mathbf{u} + \mathbf{v}$ 22. $2\mathbf{u} - 3\mathbf{v}$
23. $-18\mathbf{u}$ 24. $\mathbf{w} - \mathbf{u} - \mathbf{v}$
25. $\mathbf{t} + 3\mathbf{w} - \mathbf{v}$ 26. $2\mathbf{u} - 7\mathbf{w} + 5\mathbf{v}$
27. $2\mathbf{v} + 7\mathbf{t} - \mathbf{w}$ 28. $\mathbf{u} \cdot \mathbf{v}$
29. $|\mathbf{w}|$ 30. $\mathbf{u} \cdot \mathbf{w} - \mathbf{w} \cdot \mathbf{t}$
31. angle between $\mathbf{u}$ and $\mathbf{w}$
32. angle between $\mathbf{t}$ and $\mathbf{w}$
33. angle between $\mathbf{v}$ and $\mathbf{t}$
34. angle between $\mathbf{v} - \mathbf{t}$ and $\mathbf{v} + \mathbf{t}$
35. $\text{Proj}_\mathbf{v}\, \mathbf{u}$ 36. $\text{Proj}_\mathbf{u}\, \mathbf{v}$
37. $\text{Proj}_\mathbf{t}\, \mathbf{w}$ 38. $\text{Proj}_\mathbf{w}\, \mathbf{t}$
39. $\text{Proj}_\mathbf{w}\, \mathbf{u}$ 40. $\text{Proj}_\mathbf{w}(\mathbf{u} - \mathbf{t})$

■ APPLICATIONS

41. The three direction angles of a certain unit vector are equal and are between 0 and $\pi/2$. What is the vector?

42. Find a vector of magnitude 12 that has the same direction as the unit vector of the preceding problem.

CONTINUED

43. Let $P = (2, 1, 4)$ and $Q = (3, -2, 8)$. Find a unit vector in the direction of $\overrightarrow{PQ}$.

44. Let $P = (-3, 1, 7)$ and $Q = (8, -1, -7)$. Find a unit vector whose direction is opposite that of $\overrightarrow{PQ}$.

45. Verify that there is no unit vector whose direction angles are $\pi/6$, $\pi/3$, and $\pi/4$.

*46. Find the angle between the diagonal of a cube and the diagonal of one of its faces.

47. Let $P = (-3, 1, 7)$ and $Q = (8, 1, 7)$. Find all points R such that $\overrightarrow{PR} \perp \overrightarrow{PQ}$.

48. Let $P = (-3, 2, -5)$ and $Q = (8, 2, -5)$. Verify that the set of all points R such that $\overrightarrow{PR} \perp \overrightarrow{PQ}$ and $|\overrightarrow{PR}| = 1$ is a circle. Then find its center and radius.

49. Find the distance between the point $P = (2, 1, 3)$ and the line passing through the points $Q = (-1, 1, 2)$ and $R = (6, 0, 1)$. [*Hint:* See Problem 11.2.77 and the hint for Problem 11.2.78.]

50. Find the distance between the point $P = (1, 0, 1)$ and the line passing through the points $Q = (2, 3, -1)$ and $R = (6, 1, -3)$.

51. Verify that the points $P = (3, 5, 6)$, $Q = (1, 2, 7)$, and $R = (6, 1, 0)$ are vertices of a right triangle.

52. Verify that the points $P = (3, 2, -1)$, $Q = (4, 1, 6)$, $R = (7, -2, 3)$, and $S = (8, -3, 10)$ are vertices of a parallelogram.

53. Use the dot product to find two unit vectors orthogonal to the vectors $(1, 2, 3)$ and $(-4, 1, 5)$.

54. Find two unit vectors perpendicular to both of the vectors $(-2, 0, 4)$ and $(3, -2, -1)$.

■ SHOW/PROVE/DISPROVE

55. a. Use Theorem 3 to prove one form of the **Cauchy-Schwarz inequality**:

$$\left(\sum_{i=1}^{3} u_i v_i \right)^2 \le \left(\sum_{i=1}^{3} u_i^2 \right) \left(\sum_{i=1}^{3} v_i^2 \right) \qquad \text{(10)}$$

where the u_i's and v_i's are real numbers. (See Problem 75 in Section 11.2 for another form.)

b. Show that equality holds in (10) if and only if at least one of the vectors $\mathbf{u} = (u_1, u_2, u_3)$

and $\mathbf{v} = (v_1, v_2, v_3)$ is a scalar multiple of the other.

56. Prove Theorem 1.

57. Prove that formula (6) is correct. [*Hint:* Follow the steps leading to formula (11.1.7).]

58. Prove Theorem 2.

59. Prove Theorem 3.

60. Prove Theorem 4.

■ ANSWERS TO SELF-QUIZ

I. d   II. a   III. c   IV. a

11.5 LINES IN $\mathbb{R}^3$

In Section 0.3, we derived the equation of a line in the plane. In that section we showed that the equation of the line could be determined if we knew either (i) two points on the line or (ii) one point on the line and the direction (slope) of the line. In $\mathbb{R}^3$ our intuition tells us that the basic ideas are the same. Since two points determine a line, we should be able to calculate the equation of a line in space if we know two points on it. Alternatively, if we know one point and the direction of a vector parallel to the line, we should also be able to find its equation.

We begin with two points $P = (x_1, y_1, z_1)$ and $Q = (x_2, y_2, z_2)$ on a line L. A vector parallel to L is a vector with representation $\mathbf{v} = \overrightarrow{PQ}$, or [from formula (6) in Section 11.4],

$$\mathbf{v} = (x_2 - x_1)\mathbf{i} + (y_2 - y_1)\mathbf{j} + (z_2 - z_1)\mathbf{k}. \qquad \text{(1)}$$

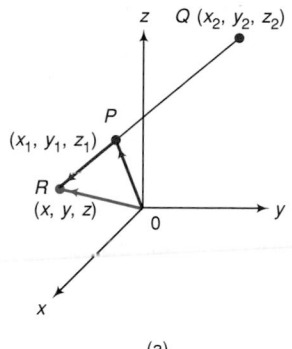

(a)

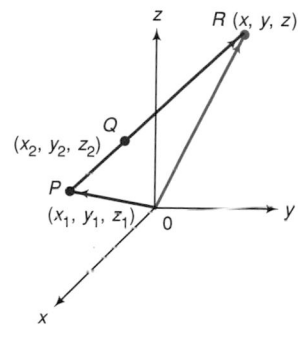

(b)

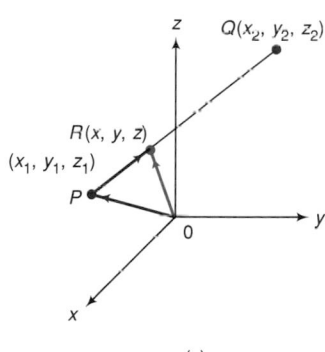

(c)

FIGURE 1
IN ALL THREE CASES $\overrightarrow{0R} = \overrightarrow{0P} + \overrightarrow{PR}$

Now let $R = (x, y, z)$ be another point on the line. Then $\overrightarrow{PR}$ is parallel to $\overrightarrow{PQ}$, which is parallel to $\mathbf{v}$, so that by Theorem 4(i) in Section 11.4,

$$\overrightarrow{PR} = t\mathbf{v} \tag{2}$$

for some real number t. Now look at Figure 1. From the figure we have (in each of the three possible cases)

$$\overrightarrow{0R} = \overrightarrow{0P} + \overrightarrow{PR}, \tag{3}$$

and combining (2) and (3),

$$\overrightarrow{PR} = \overrightarrow{0R} - \overrightarrow{0P} = t\mathbf{v}.$$

Thus,

VECTOR EQUATION OF A LINE

$$\overrightarrow{0R} = \overrightarrow{0P} + t\mathbf{v} \tag{4}$$

or

$$(x, y, z) = (x_1, y_1, z_1) + t(x_2 - x_1, y_2 - y_1, z_2 - z_1). \tag{4'}$$

Equation (4) or (4′) is called the **vector equation** of the line L. For if R is on L, then (4) is satisfied for some real number t. Conversely, if (4) is satisfied, then reversing our steps, we see that $\overrightarrow{PR}$ is parallel to $\mathbf{v}$, which means that R is on L.

NOTE: Since $\mathbf{v} = \overrightarrow{PQ} = \overrightarrow{0Q} - \overrightarrow{0P}$, (4) can be rewritten as

$$\overrightarrow{0R} = \overrightarrow{0P} + t(\overrightarrow{0Q} - \overrightarrow{0P}),$$

or

$$\overrightarrow{0R} = (1 - t)\overrightarrow{0P} + t\overrightarrow{0Q}. \tag{5}$$

The vector equation (5) is sometimes very useful.

If we write out the components of equation (4′), we obtain

$$x\mathbf{i} + y\mathbf{j} + z\mathbf{k} = x_1\mathbf{i} + y_1\mathbf{j} + z_1\mathbf{k} + t(x_2 - x_1)\mathbf{i} + t(y_2 - y_1)\mathbf{j} + t(z_2 - z_1)\mathbf{k},$$

or

PARAMETRIC EQUATIONS OF A LINE

$$x = x_1 + t(x_2 - x_1),$$
$$y = y_1 + t(y_2 - y_1), \tag{6}$$
$$z = z_1 + t(z_2 - z_1).$$

The equations (6) are called the **parametric equations** of a line.

Finally, solving for t in (6), and defining $x_2 - x_1 = a$, $y_2 - y_1 = b$, and $z_2 - z_1 = c$, we find that

---

SYMMETRIC EQUATIONS OF A LINE

$$\frac{x - x_1}{a} = \frac{y - y_1}{b} = \frac{z - z_1}{c}, \qquad \text{where } a, b, c \neq 0. \tag{7}$$

The equations (7) are called the **symmetric equations** of the line.

Here, a, b, and c are direction numbers of the vector $\mathbf{v}$. In Examples 3 and 4, we illustrate what happens if one or more of these numbers is zero.

EXAMPLE 1
DETERMINING EQUATIONS OF A LINE

Find a vector equation, parametric equations, and symmetric equations of the line L passing through the points $P = (2, -1, 6)$ and $Q = (3, 1, -2)$.

SOLUTION: First, we calculate

$$\mathbf{v} = (3 - 2)\mathbf{i} + [1 - (-1)]\mathbf{j} + (-2 - 6)\mathbf{k} = \mathbf{i} + 2\mathbf{j} - 8\mathbf{k}.$$

Then from (4), if $R = (x, y, z)$ is on the line,

$$\overrightarrow{OR} = x\mathbf{i} + y\mathbf{j} + z\mathbf{k} = \overrightarrow{OP} + t\mathbf{v} = 2\mathbf{i} - \mathbf{j} + 6\mathbf{k} + t(\mathbf{i} + 2\mathbf{j} - 8\mathbf{k}),$$

or

$$x = 2 + t, \qquad y = -1 + 2t, \qquad z = 6 - 8t. \qquad \text{Parametric equations}$$

Finally, since $a = 1$, $b = 2$, and $c = -8$, we find the symmetric equations

$$\frac{x - 2}{1} = \frac{y + 1}{2} = \frac{z - 6}{-8}.$$

REMARK: If in Example 1 we write $P = (3, 1, -2)$ and $Q = (2, -1, 6)$, then we find that $\mathbf{v} = (2 - 3)\mathbf{i} + (-1 - 1)\mathbf{j} + (6 - (-2))\mathbf{k} = -\mathbf{i} - 2\mathbf{j} + 8\mathbf{k}$ and the direction numbers are now $a = -1$, $b = -2$, and $c = 8$. A new set of parametric equations is

$$x = 3 - t, \qquad y = 1 - 2t, \qquad z = -2 + 8t$$

and the corresponding symmetric equations are

$$\frac{x - 3}{-1} = \frac{y - 1}{-2} = \frac{z + 2}{8}.$$

Note that $(2, -1, 6)$ is on this line because

$$\frac{2 - 3}{-1} = \frac{-1 - 1}{-2} = \frac{6 + 2}{8} = 1.$$

The point here is that for a line, there are an infinite number of sets of parametric or symmetric equations (two for every point on the line) and two sets of direction numbers.

EXAMPLE 2
FINDING SYMMETRIC EQUATIONS
OF A LINE

Find symmetric equations of the line passing through the point $(1, -2, 4)$ and parallel to the vector $\mathbf{v} = \mathbf{i} + \mathbf{j} - \mathbf{k}$.

SOLUTION: We simply choose $\overrightarrow{0P} = \mathbf{i} - 2\mathbf{j} + 4\mathbf{k}$ and $\mathbf{v}$ as above. Then $a = 1, b = 1, c = -1$, and we obtain

$$\frac{x-1}{1} = \frac{y+2}{1} = \frac{z-4}{-1}.$$

What happens if one of the direction numbers a, b, or c is zero?

EXAMPLE 3
FINDING SYMMETRIC EQUATIONS
OF A LINE WHEN ONE OF THE
DIRECTION NUMBERS IS ZERO

Find symmetric equations of the line containing the points $P = (3, 4, -1)$ and $Q = (-2, 4, 6)$.

SOLUTION: Here $\mathbf{v} = -5\mathbf{i} + 7\mathbf{k}$, and $a = -5, b = 0, c = 7$. Then a parametric representation of the line is

$$x = 3 - 5t, \qquad y = 4, \qquad z = -1 + 7t.$$

Solving for t, we find that

$$\frac{x-3}{-5} = \frac{z+1}{7} \qquad \text{and} \qquad y = 4.$$

The equation $y = 4$ is the equation of a plane parallel to the xz-plane, so we have obtained an equation of a line in that plane.

Now, what happens if two of the direction numbers are zero?

EXAMPLE 4
FINDING SYMMETRIC
EQUATIONS OF A LINE WHEN
TWO OF THE DIRECTION
NUMBERS ARE ZERO

Find the symmetric equations of the line passing through the points $P = (2, 3, -2)$ and $Q = (2, -1, -2)$.

SOLUTION: Here $\mathbf{v} = -4\mathbf{j}$, so that $a = 0, b = -4$, and $c = 0$. A parametric representation of the line is, by equations (6), given by

$$x = 2, \qquad y = 3 - 4t, \qquad z = -2.$$

Now $x = 2$ is the equation of a plane parallel to the yz-plane, while $z = -2$ is the equation of a plane parallel to the xy-plane. Their intersection is the line $x = 2, z = -2$, which is parallel to the y-axis. In fact, the equation $y = 3 - 4t$ says, essentially, that y can take on any value (while x and z remain fixed).

The results of Examples 1, 3, and 4 are summarized in Theorem 1.

THEOREM 1

Let L be a line passing through the point (x_1, y_1, z_1) and parallel to the vector $\mathbf{v} = a\mathbf{i} + b\mathbf{j} + c\mathbf{k}$. Then symmetric equations of the line are

as follows: **(i)** $\dfrac{x - x_1}{a} = \dfrac{y - y_1}{b} = \dfrac{z - z_1}{c}$, if a, b, and c are all nonzero.

(ii) $x = x_1, \qquad \dfrac{y - y_1}{b} = \dfrac{z - z_1}{c},$

if $a = 0$. Then the line is parallel to the yz-plane. If either b or $c = 0$, but $a \neq 0$, similar results hold.

(iii) $x = x_1, \qquad y = y_1, \qquad z = z_1 + ct,$

if a and b are 0. Then the line is parallel to the z-axis. If a and c or b and c are 0, similar results hold. ∎

⊗ **WARNING:** We repeat that the parametric or symmetric equations of a line are *not* unique. To see this, simply choose two other points on the line. ⊗

EXAMPLE 5
ILLUSTRATING THE FACT THAT SYMMETRIC EQUATIONS ARE NOT UNIQUE

In Example 1, the line contains the point $(5, 5, -18)$; set $t = 3$. Choose $P = (5, 5, -18)$ and $Q = (3, 1, -2)$. We find that $\mathbf{v} = -2\mathbf{i} - 4\mathbf{j} + 16\mathbf{k}$, so that

$$x = 5 - 2t, \qquad y = 5 - 4t, \qquad z = -18 + 16t.$$

[Note that if $t = \frac{3}{2}$, we obtain $(x, y, z) = (2, -1, 6)$.] The symmetric equations are now

$$\frac{x - 5}{-2} = \frac{y - 5}{-4} = \frac{z + 18}{16}.$$

■ PROBLEMS 11.5

■ SELF-QUIZ

I. The line through the points $(1, 2, 4)$ and $(5, 10, 15)$ satisfies the equation _____.
   a. $(x, y, z) = (1, 2, 4) + t(4, 8, 11)$
   b. $\dfrac{x - 1}{4} = \dfrac{y - 2}{8} = \dfrac{z - 4}{11}$
   c. $(x, y, z) = (5, 10, 15) + s(4, 8, 11)$
   d. $\dfrac{x - 5}{4} = \dfrac{y - 10}{8} = \dfrac{z - 15}{11}$

II. The line through the point $(7, 3, -4)$ and parallel to the vector $\mathbf{i} + 5\mathbf{j} + 2\mathbf{k}$ satisfies the equation _____.
   a. $\dfrac{x - 7}{1} = \dfrac{y - 3}{5} = \dfrac{z + 4}{2}$

   b. $(x, y, z) = (1, 5, 2) + t(7, 3, -4)$
   c. $\dfrac{x - 7}{8} = \dfrac{y - 3}{8} = \dfrac{z + 4}{-2}$
   d. $(x, y, z) = (7, 3, -4) + s(8, 8, -2)$

III. The vector equation $(x, y, z) - (3, 5, -7) = t(-1, 4, 8)$ describes _____.
   a. the line through $(-1, 4, 8)$ and parallel to $3\mathbf{i} + 5\mathbf{j} - 7\mathbf{k}$
   b. the line through $(-3, -5, 7)$ and parallel to $-\mathbf{i} + 4\mathbf{j} + 8\mathbf{k}$
   c. the line through $(3, 5, -7)$ and perpendicular to $-\mathbf{i} + 4\mathbf{j} + 8\mathbf{k}$
   d. the line through $(3, 5, -7)$ and parallel to $-\mathbf{i} + 4\mathbf{j} + 8\mathbf{k}$

CONTINUED

■ DRILL

In Problems 1–22, find a vector equation, parametric equations, and symmetric equations for the specified line.

1. the line passing through $(2, 1, 3)$ and $(1, 2, -1)$
2. the line passing through $(1, -1, 1)$ and $(-1, 1, -1)$
3. the line passing through $(1, 3, 2)$ and $(2, 4, -2)$
4. the line passing through $(-2, 4, 5)$ and $(3, 7, 2)$
5. the line passing through $(-4, 1, 3)$ and $(-4, 0, 1)$
6. the line passing through $(2, 3, -4)$ and $(2, 0, -4)$
7. the line passing through $(1, 2, 3)$ and $(3, 2, 1)$
8. the line passing through $(7, 1, 3)$ and $(-1, -2, 3)$
9. the line passing through $(1, 2, 4)$ and $(1, 2, 7)$
10. the line passing through $(-3, -1, -6)$ and $(-3, 1, 6)$
11. the line passing through $(2, 2, 1)$ and parallel to $2\mathbf{i} - \mathbf{j} - \mathbf{k}$
12. the line passing through $(-1, -6, 2)$ and parallel to $4\mathbf{i} + \mathbf{j} - 3\mathbf{k}$
13. the line passing through $(1, 0, 3)$ and parallel to $\mathbf{i} - \mathbf{j}$
14. the line passing through $(2, 1, -4)$ and parallel to $\mathbf{i} + 4\mathbf{k}$
15. the line passing through $(-1, -2, 5)$ and parallel to $-3\mathbf{j} + 4\mathbf{k}$
16. the line passing through $(-2, 3, -2)$ and parallel to $4\mathbf{k}$
17. the line passing through $(-1, -3, 1)$ and parallel to $-7\mathbf{j}$
18. the line passing through $(2, 1, 5)$ and parallel to $3\mathbf{i}$
19. the line passing through (a, b, c) and parallel to $d\mathbf{i} + e\mathbf{j}$, $d, e \neq 0$
20. the line passing through (a, b, c) and parallel to $d\mathbf{k}$, $d \neq 0$
21. the line passing through $(4, 1, -6)$ and parallel to

$$\frac{x-2}{3} = \frac{y+1}{6} = \frac{z-5}{2}$$

22. the line passing through $(3, 1, -2)$ and parallel to

$$\frac{x+1}{3} = \frac{y+3}{2} = \frac{z-2}{4}$$

■ APPLICATIONS

In the plane, two lines that are not parallel have exactly one point of intersection. In $\mathbb{R}^3$, this is not the case. For example the lines $L_1: x = 2$, $y = 3$ (parallel to the z-axis) and $L_2: x = 1$, $z = 3$ (parallel to the y-axis) are not parallel and have no points in common. It takes a bit of work to determine whether two lines in $\mathbb{R}^3$ do have a point in common (they usually do not).

23. Determine whether the lines

$$L_1: x = 1 + t, \ y = -3 + 2t; \ z = -2 - t$$

and

$$L_2: x = 17 + 3s, \ y = 4 + s, \ z = -8 - s$$

have a point of intersection. [*Hint:* If (x, y, z) is a point common to both lines, then $x = 1 + t = 17 + 3s$, $y = -3 + 2t = 4 + s$, $z = -2 - t = -8 - s$. Find numbers s and t that satisfy all three of these equations or show that no such numbers s and t can exist.]

24. Determine whether the lines

$$L_1: x = 2 - t, \ y = 1 + t, \ z = -2 - t$$

and

$$L_2: x = 1 + s, \ y = -2s, \ z = 3 + 2s$$

have a point in common.

In Problems 25–30, determine whether the given pair of lines has a point of intersection. If so, find it.

25. $L_1: x = 2 + t, \ y = -1 + 2t, \ z = 3 + 4t$;
    $L_2: x = 9 + s, \ y = -2 - s, \ z = 1 - 2s$
26. $L_1: x = 3 + 2t, \ y = 2 - t, \ z = 1 + t$;
    $L_2: x = 4 - s, \ y = -2 + 3s, \ z = 2 + 2s$
27. $L_1: \dfrac{x-4}{-3} = \dfrac{y-1}{7} = \dfrac{z+2}{-8}$;
    $L_2: \dfrac{x-5}{1} = \dfrac{y-3}{-1} = \dfrac{z-1}{2}$
28. $L_1: \dfrac{x-2}{-5} = \dfrac{y-1}{1} = \dfrac{z-3}{4}$;
    $L_2: \dfrac{x+3}{4} = \dfrac{y-2}{-1} = \dfrac{z-7}{6}$
29. $L_1: x = 4 - t, \ y = 7 + 5t, \ z = 2 - 3t$;
    $L_2: x = 1 + 2s, \ y = 6 - 2s, \ z = 10 + 3s$

CONTINUED

30. L_1: $x = 1 + t$, $y = 2 - t$, $z = 3t$;
    L_2: $x = 3s$, $y = 2 - s$, $z = 2 + s$

31. Verify that

$$L_1: \frac{x - 1}{1} = \frac{y + 3}{2} = \frac{z + 3}{3}$$

and

$$L_2: \frac{x - 3}{3} = \frac{y - 1}{6} = \frac{z - 3}{9}$$

are equations of the same straight line.

32. Verify that L_1: $x = 1 - 2t$, $y = -3 - 6t$, $z = 5 + 10t$ and L_2: $x = -5 + s$, $y = -21 + 3s$, $z = 35 - 5s$ are equations for the same line.

33. Let the line L be given in its vector form $\overrightarrow{OR} = \overrightarrow{OP} + t\mathbf{v}$. Find a number t such that $\overrightarrow{OR}$ is perpendicular to $\mathbf{v}$.

34. Apply your solution of the preceding problem to find the distance between the origin 0 and the line L which passes through the given P and is parallel to the given $\mathbf{v}$.
    a. $P = (2, 1, -4)$; $\mathbf{v} = \mathbf{i} + \mathbf{j} + \mathbf{k}$
    b. $P = (1, 2, -3)$; $\mathbf{v} = 3\mathbf{i} - \mathbf{j} - \mathbf{k}$
    c. $P = (-1, -4, 2)$; $\mathbf{v} = -\mathbf{i} + \mathbf{j} + 2\mathbf{k}$

35. Find two different lines that pass through the point $(2, -3, 1)$ and are also perpendicular to the line

$$\frac{x + 2}{4} = \frac{y - 1}{-4} = \frac{z + 2}{3}.$$

■ SHOW/PROVE/DISPROVE

36. Prove or Disprove: There are at least two different lines which pass through the point $(2, -3, 1)$ and are also perpendicular to the line

$$\frac{x + 2}{4} = \frac{y - 1}{2} = \frac{z + 2}{-1}.$$

37. Show that direction vectors of the lines

$$L_1: \frac{x - 3}{2} = \frac{y + 1}{4} = \frac{z - 2}{-1}$$

and

$$L_2: \frac{x - 3}{5} = \frac{y + 1}{-2} = \frac{z - 3}{2}$$

are orthogonal.

38. Let L_1 be given by

$$\frac{x - x_1}{a_1} = \frac{y - y_1}{b_1} = \frac{z - z_1}{c_1}, \quad a_1, b_1, c_1 \neq 0,$$

and L_2 be given by

$$\frac{x - x_2}{a_2} = \frac{y - y_2}{b_2} = \frac{z - z_2}{c_2}, \quad a_2, b_2, c_2 \neq 0.$$

Show that the direction vector of L_1 is orthogonal to the direction vector of L_2 if and only if $a_1a_2 + b_1b_2 + c_1c_2 = 0$.

39. a. Find symmetric equations for the line in the xy-plane that passes through the points $(x_1, y_1, 0)$ and $(x_2, y_2, 0)$.
    b. Show those equations can be rewritten in the form

$$y = mx + b, \qquad z = 0.$$

This shows the symmetric equations of a line in $\mathbb{R}^3$ generalize the slope-intercept equation of a line in $\mathbb{R}^2$.

*40. Show that the lines L_1: $x = x_1 + a_1t$, $y = y_1 + b_1t$, $z = z_1 + c_1t$ and L_2: $x = x_2 + a_2s$, $y = y_2 + b_2s$, $z = z_2 + c_2s$ have at least one point in common or they are parallel if and only if

$$\begin{vmatrix} a_1 & a_2 & x_1 - x_2 \\ b_1 & b_2 & y_1 - y_2 \\ c_1 & c_2 & z_1 - z_2 \end{vmatrix} = 0.$$

[*Note*: See Appendix 4 if you need to learn or review how to evaluate a 3 × 3 determinant.]

■ ANSWERS TO SELF-QUIZ

I. a, b, c, d   II. a   III. d

11.6 THE CROSS PRODUCT OF TWO VECTORS

To this point, the only product of vectors we have considered has been the dot or scalar product. We now define a new product, called the *cross product* (or *vector product*), which is defined only in $\mathbb{R}^3$.

There are several ways to define the product of two vectors. In many applications, it is useful to find a vector that is orthogonal to two given vectors. The cross product of two vectors $\mathbf{u}$ and $\mathbf{v}$ is indeed a vector that is orthogonal to both $\mathbf{u}$ and $\mathbf{v}$. How do we find such a vector?

Suppose that $\mathbf{u} = a_1\mathbf{i} + b_1\mathbf{j} + c_1\mathbf{k}$ and $\mathbf{v} = a_2\mathbf{i} + b_2\mathbf{j} + c_2\mathbf{k}$. Let $\mathbf{w} = a_3\mathbf{i} + b_3\mathbf{j} + c_3\mathbf{k}$ be a vector orthogonal to both $\mathbf{u}$ and $\mathbf{v}$. Then

$$\mathbf{u} \cdot \mathbf{w} = a_1a_3 + b_1b_3 + c_1c_3 = 0 \tag{1}$$

and

$$\mathbf{v} \cdot \mathbf{w} = a_2a_3 + b_2b_3 + c_2c_3 = 0. \tag{2}$$

Multiply equation (1) by c_2 and equation (2) by c_1 and subtract:

$$\begin{array}{l} a_1c_2a_3 + b_1c_2b_3 + c_1c_2c_3 = 0 \\ a_2c_1a_3 + b_2c_1b_3 + c_1c_2c_3 = 0 \\ \hline (a_1c_2 - a_2c_1)a_3 + (b_1c_2 - b_2c_1)b_3 = 0 \end{array} \tag{3}$$

Equation (3) has the form

$$\alpha a_3 + \beta b_3 = 0 \qquad \text{where } \alpha = a_1c_2 - a_2c_1 \text{ and } \beta = b_1c_2 - b_2c_1.$$

One solution is $a_3 = \beta$ and $b_3 = -\alpha$. Thus one solution to (3) is

$$a_3 = b_1c_2 - b_2c_1 \qquad \text{and} \qquad b_3 = -(a_1c_2 - a_2c_1) = a_2c_1 - a_1c_2.$$

Inserting these values into equation (1) yields $c_3 = a_1b_2 - b_1a_2$ (check this). Thus the vector

$$\mathbf{w} = (b_1c_2 - b_2c_1)\mathbf{i} + (c_1a_2 - a_1c_2)\mathbf{j} + (a_1b_2 - b_1a_2)\mathbf{k}$$

is a vector orthogonal to both $\mathbf{u}$ and $\mathbf{v}$.

CHECK:

$$\mathbf{u} \cdot \mathbf{w} = a_1b_1c_2 - a_1b_2c_1 + b_1c_1a_2 - b_1a_1c_2 + c_1a_1b_2 - c_1b_1a_2 = 0.$$

$$\mathbf{v} \cdot \mathbf{w} = a_2b_1c_2 - a_2b_2c_1 + b_2c_1a_2 - b_2a_1c_2 + c_2a_1b_2 - c_2b_1a_2 = 0.$$

This informal derivation motivates the following definition:

DEFINITION **CROSS PRODUCT**

Let $\mathbf{u} = a_1\mathbf{i} + b_1\mathbf{j} + c_1\mathbf{k}$ and $\mathbf{v} = a_2\mathbf{i} + b_2\mathbf{j} + c_2\mathbf{k}$. Then the **cross product (vector product)** of $\mathbf{u}$ and $\mathbf{v}$, denoted by $\mathbf{u} \times \mathbf{v}$, is a new vector defined by

$$\mathbf{u} \times \mathbf{v} = (b_1c_2 - c_1b_2)\mathbf{i} + (c_1a_2 - a_1c_2)\mathbf{j} + (a_1b_2 - b_1a_2)\mathbf{k}. \quad \blacksquare \tag{4}$$

Note that the result of the *cross product* is a *vector*, while the result of the *dot product* is a scalar.

HISTORICAL NOTE: The cross product was defined by Hamilton in one of a series of papers discussing his quaternions, which were published in *Philosophical Magazine* between the years 1844 and 1850.

EXAMPLE 1
CALCULATING THE CROSS
PRODUCT OF TWO VECTORS

Let $\mathbf{u} = \mathbf{i} - \mathbf{j} + 2\mathbf{k}$ and $\mathbf{v} = 2\mathbf{i} + 3\mathbf{j} - 4\mathbf{k}$. Calculate $\mathbf{w} = \mathbf{u} \times \mathbf{v}$.

SOLUTION: Using formula (1), we have

$$\mathbf{w} = [(-1)(-4) - (2)(3)]\mathbf{i} + [(2)(2) - (1)(-4)]\mathbf{j}$$
$$+ [(1)(3) - (-1)(2)]\mathbf{k}$$
$$= -2\mathbf{i} + 8\mathbf{j} + 5\mathbf{k}.$$

NOTE: In this example $\mathbf{u} \cdot \mathbf{w} = \mathbf{v} \cdot \mathbf{w} = 0$. That is, $\mathbf{u} \times \mathbf{v}$ is orthogonal to both $\mathbf{u}$ and $\mathbf{v}$.

Before continuing our discussion of the uses of the cross product, there is an easy way to remember how to calculate $\mathbf{u} \times \mathbf{v}$ if you are familiar with the elementary properties of 3×3 determinants. If you are not, we suggest that you turn to Appendix A4 where these properties are discussed.

THEOREM 1

$$\mathbf{u} \times \mathbf{v} = \begin{vmatrix} \mathbf{i} & \mathbf{j} & \mathbf{k} \\ a_1 & b_1 & c_1 \\ a_2 & b_2 & c_2 \end{vmatrix}^\dagger \tag{5}$$

PROOF:

$$\begin{vmatrix} \mathbf{i} & \mathbf{j} & \mathbf{k} \\ a_1 & b_1 & c_1 \\ a_2 & b_2 & c_2 \end{vmatrix} = \mathbf{i}\begin{vmatrix} b_1 & c_1 \\ b_2 & c_2 \end{vmatrix} - \mathbf{j}\begin{vmatrix} a_1 & c_1 \\ a_2 & c_2 \end{vmatrix} + \mathbf{k}\begin{vmatrix} a_1 & b_1 \\ a_2 & b_2 \end{vmatrix}$$
$$- (b_1 c_2 - c_1 b_2)\mathbf{i} + (a_2 c_1 - a_1 c_2)\mathbf{j} + (a_1 b_2 - b_1 a_2)\mathbf{k}$$

which is equal to $\mathbf{u} \times \mathbf{v}$ according to the definition of the cross product. ∎

EXAMPLE 2
USING A DETERMINANT TO
CALCULATE A CROSS PRODUCT

Calculate $\mathbf{u} \times \mathbf{v}$, where $\mathbf{u} = 2\mathbf{i} + 4\mathbf{j} - 5\mathbf{k}$ and $\mathbf{v} = -3\mathbf{i} - 2\mathbf{j} + \mathbf{k}$.

SOLUTION:

$$\mathbf{u} \times \mathbf{v} = \begin{vmatrix} \mathbf{i} & \mathbf{j} & \mathbf{k} \\ 2 & 4 & -5 \\ -3 & -2 & 1 \end{vmatrix} = (4 - 10)\mathbf{i} - (2 - 15)\mathbf{j} + (-4 + 12)\mathbf{k}$$
$$= -6\mathbf{i} + 13\mathbf{j} + 8\mathbf{k}.$$

The following theorem summarizes some properties of the cross product.

---

† The determinant is defined as a real number, not a vector. This use of the determinant notation is simply a convenient way to denote the cross product. In this notation we always expand the determinant along the first row.

THEOREM 2 **PROPERTIES OF THE CROSS PRODUCT**

Let $\mathbf{u}$, $\mathbf{v}$, and $\mathbf{w}$ be vectors in $\mathbb{R}^3$, and let α be a scalar.

(i) $\mathbf{u} \times \mathbf{0} = \mathbf{0} = \mathbf{0} \times \mathbf{u}$.

(ii) $\mathbf{u} \times \mathbf{v} = -(\mathbf{v} \times \mathbf{u})$.

(iii) $(\alpha\mathbf{u} \times \mathbf{v}) = \alpha(\mathbf{u} \times \mathbf{v})$.

(iv) $\mathbf{u} \times (\mathbf{v} + \mathbf{w}) = (\mathbf{u} \times \mathbf{v}) + (\mathbf{u} \times \mathbf{w})$.

(v) $(\mathbf{u} \times \mathbf{v}) \cdot \mathbf{w} = \mathbf{u} \cdot (\mathbf{v} \times \mathbf{w})$. (This product is called the **scalar triple product** of $\mathbf{u}$, $\mathbf{v}$, and $\mathbf{w}$.)

(vi) $\mathbf{u} \cdot (\mathbf{u} \times \mathbf{v}) = \mathbf{v} \cdot (\mathbf{u} \times \mathbf{v}) = 0$. (That is, $\mathbf{u} \times \mathbf{v}$ is orthogonal to both $\mathbf{u}$ and $\mathbf{v}$.)

(vii) If $\mathbf{u}$ and $\mathbf{v}$ are parallel, then $\mathbf{u} \times \mathbf{v} = \mathbf{0}$.

PROOF:

(i) Let $\mathbf{u} = a_1\mathbf{i} + b_1\mathbf{j} + c_1\mathbf{k}$. Then,

$$\mathbf{u} \times \mathbf{0} = \begin{vmatrix} \mathbf{i} & \mathbf{j} & \mathbf{k} \\ a_1 & b_1 & c_1 \\ 0 & 0 & 0 \end{vmatrix} = 0\mathbf{i} + 0\mathbf{j} + 0\mathbf{k} = \mathbf{0}.$$

Similarly, $\mathbf{0} \times \mathbf{u} = \mathbf{0}$.

(ii) Let $\mathbf{v} = a_2\mathbf{i} + b_2\mathbf{j} + c_2\mathbf{k}$. Then,

$$\mathbf{u} \times \mathbf{v} = \begin{vmatrix} \mathbf{i} & \mathbf{j} & \mathbf{k} \\ a_1 & b_1 & c_1 \\ a_2 & b_2 & c_2 \end{vmatrix} = -\begin{vmatrix} \mathbf{i} & \mathbf{j} & \mathbf{k} \\ a_2 & b_2 & c_2 \\ a_1 & b_1 & c_1 \end{vmatrix} = -(\mathbf{v} \times \mathbf{u}),$$

since interchanging the rows of a determinant has the effect of multiplying the determinant by -1 [see Theorem 5(iii), Appendix A4].

(iii)
$$(\alpha\mathbf{u}) \times \mathbf{v} = \begin{vmatrix} \mathbf{i} & \mathbf{j} & \mathbf{k} \\ \alpha a_1 & \alpha b_1 & \alpha c_1 \\ a_2 & b_2 & c_2 \end{vmatrix}$$

$$= \alpha\begin{vmatrix} \mathbf{i} & \mathbf{j} & \mathbf{k} \\ a_1 & b_1 & c_1 \\ a_2 & b_2 & c_2 \end{vmatrix} = \alpha(\mathbf{u} \times \mathbf{v})$$

The second equality follows from Theorem 5(iv) in Appendix A4.

(iv) Let $\mathbf{w} = a_3\mathbf{i} + b_3\mathbf{j} + c_3\mathbf{k}$. Then

$$\mathbf{u} \times (\mathbf{v} + \mathbf{w}) = \begin{vmatrix} \mathbf{i} & \mathbf{j} & \mathbf{k} \\ a_1 & b_1 & c_1 \\ a_2 + a_3 & b_2 + b_3 & c_2 + c_3 \end{vmatrix}$$

$$= \begin{vmatrix} \mathbf{i} & \mathbf{j} & \mathbf{k} \\ a_1 & b_1 & c_1 \\ a_2 & b_2 & c_2 \end{vmatrix} + \begin{vmatrix} \mathbf{i} & \mathbf{j} & \mathbf{k} \\ a_1 & b_1 & c_1 \\ a_3 & b_3 & c_3 \end{vmatrix}$$

$$= (\mathbf{u} \times \mathbf{v}) + (\mathbf{u} \times \mathbf{w}).$$

The second equality is easily verified by direct calculation.

(v) $(\mathbf{u} \times \mathbf{v}) \cdot \mathbf{w} = [(b_1c_2 - c_1b_2)\mathbf{i} + (c_1a_2 - a_1c_2)\mathbf{j} + (a_1b_2 - b_1a_2)\mathbf{k}]$
$\cdot (a_3\mathbf{i} + b_3\mathbf{j} + c_3\mathbf{k})$
$= b_1c_2a_3 - c_1b_2a_3 + c_1a_2b_3 - a_1c_2b_3$
$\quad + a_1b_2c_3 - b_1a_2c_3$
$$= \begin{vmatrix} a_1 & a_2 & a_3 \\ b_1 & b_2 & b_3 \\ c_1 & c_2 & c_3 \end{vmatrix}$$

You should show that $\mathbf{u} \cdot (\mathbf{v} \times \mathbf{w})$ is equal to the same expression (see Problem 47).

NOTE: We provide an interesting geometric interpretation of the scalar triple product on page 762. See also Problem 50.

(vi) We have already shown this. Here is an alternative proof. We know that $\mathbf{u} \cdot (\mathbf{u} \times \mathbf{v}) = (\mathbf{u} \times \mathbf{v}) \cdot \mathbf{u}$ [since the dot product is commutative—see Theorem 2(i) of Section 11.4]. But from parts (ii) and (v) of this theorem,

$$(\mathbf{u} \times \mathbf{v}) \cdot \mathbf{u} = \mathbf{u} \cdot (\mathbf{v} \times \mathbf{u}) = \mathbf{u} \cdot (-\mathbf{u} \times \mathbf{v}) = -\mathbf{u} \cdot (\mathbf{u} \times \mathbf{v}).$$

Thus $\mathbf{u} \cdot (\mathbf{u} \times \mathbf{v}) = -\mathbf{u} \cdot (\mathbf{u} \times \mathbf{v})$, which can only occur if $\mathbf{u} \cdot (\mathbf{u} \times \mathbf{v}) = 0$. A similar computation shows that $\mathbf{v} \cdot (\mathbf{u} \times \mathbf{v}) = 0$.

(vii) If $\mathbf{u}$ and $\mathbf{v}$ are parallel, then $\mathbf{v} = \alpha\mathbf{u}$ for some scalar α [from Theorem 4(i) of Section 11.4], so that

$$\mathbf{u} \times \mathbf{v} = \begin{vmatrix} \mathbf{i} & \mathbf{j} & \mathbf{k} \\ a_1 & b_1 & c_1 \\ \alpha a_1 & \alpha b_1 & \alpha c_1 \end{vmatrix} = \mathbf{0}$$

since the third row is a multiple of the second row [see Theorem 5(ii) in Appendix A4]. ∎

NOTE: We could have proved this theorem without using determinants, but the proof would have involved many more computations.

Part (vi), which motivated our definition of the cross product, is the most commonly used part of this theorem. We restate it below:

The cross product $\mathbf{u} \times \mathbf{v}$ is orthogonal to both $\mathbf{u}$ and $\mathbf{v}$.[†]

What happens when we take cross products of the basis vectors $\mathbf{i}, \mathbf{j}, \mathbf{k}$? It is not difficult to verify the following:

$$\mathbf{i} \times \mathbf{i} = \mathbf{j} \times \mathbf{j} = \mathbf{k} \times \mathbf{k} = 0,$$

$$\mathbf{i} \times \mathbf{j} = \mathbf{k}, \quad \mathbf{k} \times \mathbf{i} = \mathbf{j}, \quad \mathbf{j} \times \mathbf{k} = \mathbf{i},$$

$$\mathbf{j} \times \mathbf{i} = -\mathbf{k}, \quad \mathbf{i} \times \mathbf{k} = -\mathbf{j}, \quad \mathbf{k} \times \mathbf{j} = -\mathbf{i}.$$

[†] Technically, this statement is correct only when $\mathbf{u} \times \mathbf{v} \neq \mathbf{0}$ because orthogonality is only defined for nonzero vectors.

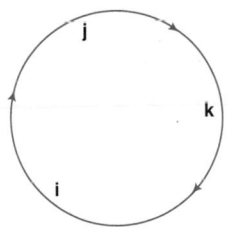

FIGURE 1
CROSS PRODUCTS OF BASIS VECTORS

The cross product of two consecutive vectors in the clockwise direction is positive, while the cross product of two consecutive vectors in the counterclockwise directions is negative.

To remember these results, consider the circle in Figure 1. The cross product of two consecutive vectors in the clockwise direction is positive, while the cross product of two consecutive vectors in the counterclockwise direction is negative. Note that the formulas above show that the cross product is *not* associative, since, for example, $\mathbf{i} \times (\mathbf{i} \times \mathbf{j}) = \mathbf{i} \times \mathbf{k} = -\mathbf{j}$ while $(\mathbf{i} \times \mathbf{i}) \times \mathbf{j} = \mathbf{0} \times \mathbf{j} = \mathbf{0}$, so that

$$\mathbf{i} \times (\mathbf{i} \times \mathbf{j}) \neq (\mathbf{i} \times \mathbf{i}) \times \mathbf{j}.$$

In general,

$$\mathbf{u} \times (\mathbf{v} \times \mathbf{w}) \neq (\mathbf{u} \times \mathbf{v}) \times \mathbf{w}.$$

This means that, in most cases, $\mathbf{u} \times \mathbf{v} \times \mathbf{w}$ is undefined because we get different answers depending on whether we multiply $\mathbf{u} \times \mathbf{v}$ or $\mathbf{v} \times \mathbf{w}$ first.

EXAMPLE 3
FINDING A LINE WHOSE
DIRECTION VECTOR IS
ORTHOGONAL TO TWO
GIVEN LINES

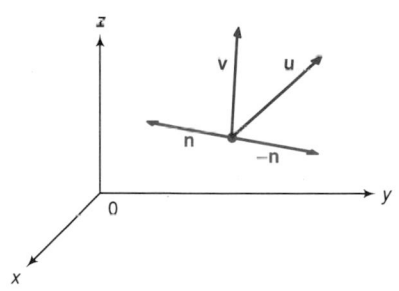

FIGURE 2
BOTH **n** AND −**n** ARE ORTHOGONAL TO
u AND **v**.

Find a line whose direction vector is orthogonal to the direction vectors of the lines $(x - 1)/3 = (y + 6)/4 = (z - 2)/-2$ and $(x + 2)/-3 = (y - 3)/4 = (z + 1)/1$ and that passes through the point $(2, -1, 1)$.

SOLUTION: The directions of these lines are

$$\mathbf{v}_1 = 3\mathbf{i} + 4\mathbf{j} - 2\mathbf{k} \qquad \text{and} \qquad \mathbf{v}_2 = -3\mathbf{i} + 4\mathbf{j} + \mathbf{k}.$$

A vector orthogonal to these vectors is

$$\mathbf{w} = \mathbf{v}_1 \times \mathbf{v}_2 = \begin{vmatrix} \mathbf{i} & \mathbf{j} & \mathbf{k} \\ 3 & 4 & -2 \\ -3 & 4 & 1 \end{vmatrix} = 12\mathbf{i} + 3\mathbf{j} + 24\mathbf{k}.$$

Then symmetric equations of a line satisfying the requested conditions are given by

$$L_1: \frac{x - 2}{12} = \frac{y + 1}{3} = \frac{z - 1}{24}.$$

NOTE: $\mathbf{w}_1 = \mathbf{v}_2 \times \mathbf{v}_1 = -(\mathbf{v}_1 \times \mathbf{v}_2)$ is also orthogonal to $\mathbf{v}_1$ and $\mathbf{v}_2$, so symmetric equations of another line are given by

$$L_2: \frac{x - 2}{-12} = \frac{y + 1}{-3} = \frac{z - 1}{-24}.$$

However, L_1 and L_2 are really the same line. (Explain why.)

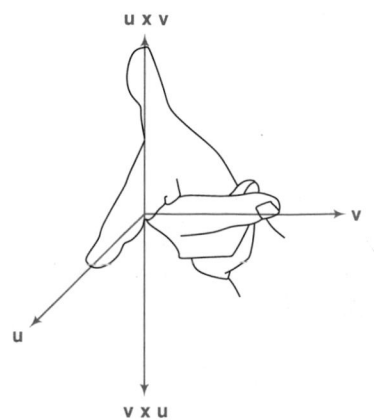

FIGURE 3
RIGHT-HAND RULE FOR THE CROSS
PRODUCT

The preceding example leads to a basic question. We know that $\mathbf{u} \times \mathbf{v}$ is a vector orthogonal to $\mathbf{u}$ and $\mathbf{v}$. But there are always *two* unit vectors orthogonal to $\mathbf{u}$ and $\mathbf{v}$ (see Figure 2). The vectors $\mathbf{n}$ and $-\mathbf{n}$ (**n** stands for *normal*, of course) are both orthogonal to $\mathbf{u}$ and $\mathbf{v}$. Which one is in the direction of $\mathbf{u} \times \mathbf{v}$? The answer is given by the **right-hand rule**. If the right hand is placed so that the index finger points in the direction of $\mathbf{u}$ while the middle finger points in the direction of $\mathbf{v}$, then the thumb points in the direction of $\mathbf{u} \times \mathbf{v}$ (see Figure 3).

We will not prove the right-hand rule here. However, we note that the right-hand rule for cross products holds if the three coordinate axes constitute a right-handed system as defined on page 738. If we instead had a left-handed system (obtained by interchanging the x- and y-axes), then the direction of $\mathbf{u} \times \mathbf{v}$ could be found by using a left-hand rule.

Having discussed the direction of the vector $\mathbf{u} \times \mathbf{v}$, we now turn to a discussion of its magnitude.

THEOREM 3

If φ is the angle between $\mathbf{u}$ and $\mathbf{v}$, then

$$|\mathbf{u} \times \mathbf{v}| = |\mathbf{u}|\,|\mathbf{v}| \sin \varphi. \tag{6}$$

PROOF: It is not difficult to show (by comparing components) that

$$|\mathbf{u} \times \mathbf{v}|^2 = |\mathbf{u}|^2|\mathbf{v}|^2 - (\mathbf{u} \cdot \mathbf{v})^2 \tag{7}$$

(see Problem 48). Then since $(\mathbf{u} \cdot \mathbf{v})^2 = |\mathbf{u}|^2|\mathbf{v}|^2 \cos^2 \varphi$ (from Theorem 3 in Section 11.4),

$$|\mathbf{u} \times \mathbf{v}|^2 = |\mathbf{u}|^2|\mathbf{v}|^2 - |\mathbf{u}|^2|\mathbf{v}|^2 \cos^2 \varphi = |\mathbf{u}|^2|\mathbf{v}|^2(1 - \cos^2 \varphi)$$
$$= |\mathbf{u}|^2|\mathbf{v}|^2 \sin^2 \varphi,$$

and the theorem follows after taking square roots of both sides. ∎

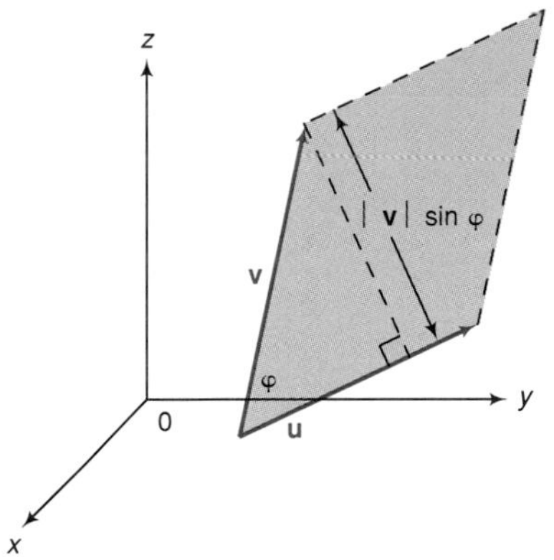

There is an interesting geometric interpretation of Theorem 3. The vectors $\mathbf{u}$ and $\mathbf{v}$ are sketched in Figure 4 and can be thought of as two adjacent sides of a parallelogram. Then from elementary geometry we see that

$$\text{area of the parallelogram} = |\mathbf{u}|\,|\mathbf{v}| \sin \varphi = |\mathbf{u} \times \mathbf{v}|. \tag{8}$$

FIGURE 4
AREA OF A PARALLELOGRAM

EXAMPLE 4
FINDING THE AREA OF A PARALLELOGRAM

Find the area of a parallelogram with vertices at $P = (1, 3, -2)$, $Q = (2, 1, 4)$, and $R = (-3, 1, 6)$.

SOLUTION: One such parallelogram is sketched in Figure 5 (there are two others). We have

$$\text{area} = |\overrightarrow{PQ} \times \overrightarrow{QR}| = |(\mathbf{i} - 2\mathbf{j} + 6\mathbf{k}) \times (-5\mathbf{i} + 2\mathbf{k})|$$

$$= \left\| \begin{array}{ccc} \mathbf{i} & \mathbf{j} & \mathbf{k} \\ 1 & -2 & 6 \\ -5 & 0 & 2 \end{array} \right\| = |-4\mathbf{i} - 32\mathbf{j} - 10\mathbf{k}|$$

$$= \sqrt{1140} \text{ square units.}$$

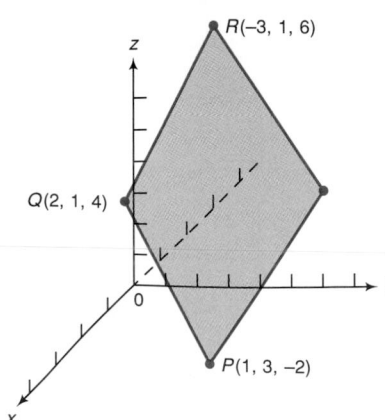

FIGURE 5
THE PARALLELOGRAM WITH VERTICES AT $(1, 3, -2)$, $(2, 1, 4)$, AND $(-3, 1, 6)$

GEOMETRIC INTERPRETATION OF THE SCALAR TRIPLE PRODUCT

Let $\mathbf{u}$, $\mathbf{v}$, and $\mathbf{w}$ be three vectors that are not in the same plane. Then they form the sides of a **parallelepiped** in space (see Figure 6). Let us compute its volume. The base of the parallelepiped is a parallelogram. Its area, from (6), is equal to $|\mathbf{u} \times \mathbf{v}|$.

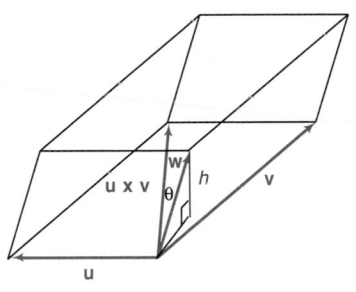

FIGURE 6
THE PARALLELEPIPED WITH SIDES **u**, **v**, AND **w**

The vector **u** × **v** is orthogonal to both **u** and **v** and is therefore orthogonal to the parallelogram determined by **u** and **v**. The height of the parallelepiped, h, is measured along a vector orthogonal to the parallelogram.

From our discussion of projections on page 735, we see that h is the absolute value of the component of **w** in the (orthogonal) direction **u** × **v**. Thus, from equation (10) on page 735,

$$h = |\text{component of } \mathbf{w} \text{ in the direction } \mathbf{u} \times \mathbf{v}| = \left| \frac{\mathbf{w} \cdot (\mathbf{u} \times \mathbf{v})}{|\mathbf{u} \times \mathbf{v}|} \right|.$$

Thus

volume of parallelepiped = area of base × height

$$= |\mathbf{u} \times \mathbf{v}| \left[\frac{|\mathbf{w} \cdot (\mathbf{u} \times \mathbf{v})|}{|\mathbf{u} \times \mathbf{v}|} \right] = |\mathbf{w} \cdot (\mathbf{u} \times \mathbf{v})|.$$

That is:

VOLUME OF A PARALLELEPIPED

The volume of the parallelepiped determined by the three vectors **u**, **v**, and **w** is equal to $|(\mathbf{u} \times \mathbf{v}) \cdot \mathbf{w}|$ = the absolute value of the scalar triple product of **u**, **v**, and **w**.

We can derive another interesting and useful fact about the scalar triple product. Suppose that neither **w** nor **u** × **v** is the zero vector. If **w** is in the plane of **u** and **v**, then **w** is perpendicular to **u** × **v**, which means that $\mathbf{w} \cdot (\mathbf{u} \times \mathbf{v}) = 0$. Conversely, if $(\mathbf{u} \times \mathbf{v}) \cdot \mathbf{w} = 0$, then **w** is perpendicular to $(\mathbf{u} \times \mathbf{v})$, so **w** is in the plane determined by **u** and **v**. We conclude that

three vectors **u**, **v**, and **w** are coplanar if and only if their scalar triple product is zero.

HISTORICAL NOTE: JOSIAH WILLARD GIBBS AND THE ORIGINS OF VECTOR ANALYSIS

As we have already noted, the study of vectors originated with Hamilton's invention of quaternions. Quaternions were developed by Hamilton and others as mathematical tools for the exploration of physical space. But the results were disappointing because quaternions proved to be too complicated for quick mastery and easy application. Fortunately, there was a solution. Quaternions contained a scalar part and a vector part and difficulties arose when these parts were treated simultaneously. Scientists soon learned that many problems could be dealt with by considering the vector part separately, and the study of vector analysis began.

This work was due principally to the American physicist Josiah Willard Gibbs (1839–1903). As a native of New Haven, Connecticut, Gibbs studied mathematics and physics at Yale University, receiving a doctorate in physics in 1863. He then studied mathematics and physics further at Paris, Berlin, and Heidelberg. In 1871 he was appointed professor of mathematical physics at Yale. He was a highly original physicist who published widely in mathematical physics. Gibbs's book *Vector Analysis* appeared in 1881 and again in 1884. In 1902 he published his *Elementary Principles of Statistical Mechanics*. Students of applied mathematics encounter the curious **Gibbs's phenomenon** of Fourier series.

Gibbs's pioneering book *Vector Analysis* was actually a small pamphlet based on Gibbs's lectures and printed for private distribution—primarily for the use of his students. Nevertheless, it created great excitement among those who were looking for an alternative to quaternions and the book soon became widely known. The material was finally turned into a standard book by E. B. Wilson. The book, *Vector Analysis* by Gibbs and Wilson, was published in 1901.

Gibbs's work is encountered by every student of elementary physics. In introductory physics, a vector in space is regarded as a directed line segment, or arrow. Gibbs gave definitions of equality, addition, and multiplication of vectors; these are essentially the definitions given in this chapter. In particular, the vector part of a quaternion was written as $a\mathbf{i} + b\mathbf{j} + c\mathbf{k}$ and this is one way we now depict vectors in $\mathbb{R}^3$.

Gibbs defined the scalar product initially only for the vectors $\mathbf{i}$, $\mathbf{j}$, $\mathbf{k}$:

$$\mathbf{i} \cdot \mathbf{i} = \mathbf{j} \cdot \mathbf{j} = \mathbf{k} \cdot \mathbf{k} = 1$$

$$\mathbf{i} \cdot \mathbf{j} = \mathbf{j} \cdot \mathbf{i} = \mathbf{i} \cdot \mathbf{k} = \mathbf{k} \cdot \mathbf{i} = \mathbf{j} \cdot \mathbf{k} = \mathbf{k} \cdot \mathbf{j} = 0.$$

The more general definition followed soon thereafter. Gibbs applied the scalar product in a problem involving force [remember, he was first a physicist]: If $\mathbf{F}$ is a force vector of magnitude $|\mathbf{F}|$ acting in the direction of the segment $\overrightarrow{OQ}$ (see Figure 7), then the effectiveness of this force in pushing an object along the segment $\overrightarrow{OP}$ (i.e., along the vector $\mathbf{u}$) is given by $\mathbf{F} \cdot \mathbf{u}$. If $|\mathbf{u}| = 1$, then $\mathbf{F} \cdot \mathbf{u}$ is the component of $\mathbf{F}$ in the direction $\mathbf{u}$. The cross product, too, has physical significance. Suppose that a force vector $\mathbf{F}$ acts at a point P in space in the direction $\overrightarrow{PQ}$ (see Figure 8). If $\mathbf{u}$ denotes the vector represented by $\overrightarrow{OP}$, then the moment of force exerted by $\mathbf{F}$ around the origin is the vector $\mathbf{u} \times \mathbf{F}$.

Both the scalar and cross products of vectors appear prominently in physical applications involving multivariable calculus. These include the famous Maxwell equations of electromagnetism.

Despite the phenomenal success of Gibbs's work, followers of Hamilton clung stubbornly to quaternions and severely criticized those who favored the much more applicable vector analysis. In 1890, the Scottish mathematician Peter Guthrie Tait (1831–1901) wrote:

> Even Prof. Willard Gibbs must be ranked as one of the retarders of quaternion progress, in virtue of his pamphlet on *Vector Analysis*, a sort of hermaphrodite monster, compounded on the notations of Hamilton and Grassman.

But vectors were not monstrous and Tait's view soon became a minority one.

In studying mathematics at the end of the twentieth century, it is important that we not lose sight of the fact that much of modern mathematics was developed to solve real-world problems. Vectors were developed by Gibbs and others to make it easier to analyze physical phenomena. In that role they have been hugely successful.

JOSIAH WILLARD GIBBS
1839–1903

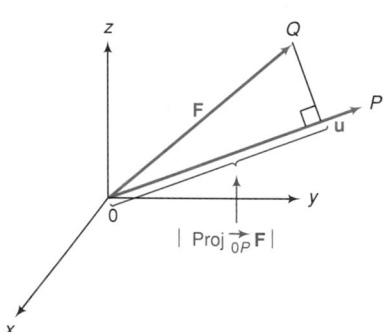

FIGURE 7
THE EFFECTIVENESS OF **F** IN THE DIRECTION OF $\overrightarrow{OP}$ IS THE COMPONENT OF **F** IN THE DIRECTION $\overrightarrow{OP}$.

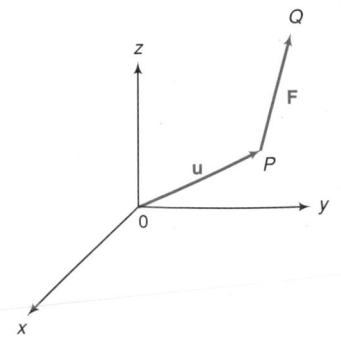

FIGURE 8
THE FORCE **F** ACTS AT A POINT P IN SPACE. THE VECTOR **u** × **F** IS THE MOMENT OF FORCE AROUND THE ORIGIN.

■ PROBLEMS 11.6

■ SELF-QUIZ

I. $\mathbf{i} \times \mathbf{k} - \mathbf{k} \times \mathbf{i} =$ _____.
   a. $\mathbf{0}$ b. $\mathbf{j}$ c. $2\mathbf{j}$ d. $-2\mathbf{j}$

II. $\mathbf{i} \cdot (\mathbf{j} \times \mathbf{k}) =$ _____.
   a. 0   b. $\mathbf{0}$
   c. 1   d. $\mathbf{i} - \mathbf{j} + \mathbf{k}$

III. $\mathbf{i} \times \mathbf{j} \times \mathbf{k}$ _____.
   a. $= 0$ b. $= \mathbf{0}$
   c. $= 1$ d. is undefined

IV. $(\mathbf{i} + \mathbf{j}) \times (\mathbf{j} + \mathbf{k}) =$ _____.
   a. 0   b. $\mathbf{0}$ c. 1   d. $\mathbf{i} - \mathbf{j} + \mathbf{k}$

V. The sine of the angle between vectors $\mathbf{u}$ and $\mathbf{w}$ is _____.

   a. $\dfrac{|\mathbf{u} \times \mathbf{w}|}{|\mathbf{u}|\,|\mathbf{w}|}$ b. $\dfrac{|\mathbf{u} \times \mathbf{w}|}{|\mathbf{u} \cdot \mathbf{w}|}$

   c. $\dfrac{|\mathbf{u} \cdot \mathbf{w}|}{|\mathbf{u} \times \mathbf{w}|}$ d. $\big|\,|\mathbf{u} \times \mathbf{w}| - |\mathbf{u} \cdot \mathbf{w}|\,\big|$

VI. $\mathbf{u} \times \mathbf{u} =$ _____.
   a. $|\mathbf{u}|^2$ b. 1   c. $\mathbf{0}$ d. 0

■ DRILL

In Problems 1–20, calculate the cross product $\mathbf{u} \times \mathbf{v}$.

1. $\mathbf{u} = \mathbf{i} - 2\mathbf{j}$; $\mathbf{v} = 3\mathbf{k}$
2. $\mathbf{u} = 3\mathbf{i} - 7\mathbf{j}$; $\mathbf{v} = \mathbf{i} + \mathbf{k}$
3. $\mathbf{u} = \mathbf{i} - \mathbf{j}$; $\mathbf{v} = \mathbf{j} + \mathbf{k}$
4. $\mathbf{u} = -7\mathbf{k}$; $\mathbf{v} = \mathbf{j} + 2\mathbf{k}$
5. $\mathbf{u} = -2\mathbf{i} + 3\mathbf{j}$; $\mathbf{v} = 7\mathbf{i} + 4\mathbf{k}$
6. $\mathbf{u} = a\mathbf{i} + b\mathbf{j}$; $\mathbf{v} = c\mathbf{i} + d\mathbf{j}$
7. $\mathbf{u} = a\mathbf{i} + b\mathbf{k}$; $\mathbf{v} = c\mathbf{i} + d\mathbf{k}$
8. $\mathbf{u} = a\mathbf{j} + b\mathbf{k}$; $\mathbf{v} = c\mathbf{i} + d\mathbf{k}$
9. $\mathbf{u} = 2\mathbf{i} - 3\mathbf{j} + \mathbf{k}$; $\mathbf{v} = \mathbf{i} + 2\mathbf{j} + \mathbf{k}$
10. $\mathbf{u} = 3\mathbf{i} - 4\mathbf{j} + 2\mathbf{k}$; $\mathbf{v} = 6\mathbf{i} - 3\mathbf{j} + 5\mathbf{k}$
11. $\mathbf{u} = -3\mathbf{i} - 2\mathbf{j} + \mathbf{k}$; $\mathbf{v} = 6\mathbf{i} + 4\mathbf{j} - 2\mathbf{k}$
12. $\mathbf{u} = \mathbf{i} + 7\mathbf{j} - 3\mathbf{k}$; $\mathbf{v} = -\mathbf{i} - 7\mathbf{j} + 3\mathbf{k}$
13. $\mathbf{u} = \mathbf{i} - 7\mathbf{j} - 3\mathbf{k}$; $\mathbf{v} = -\mathbf{i} + 7\mathbf{j} - 3\mathbf{k}$
14. $\mathbf{u} = 2\mathbf{i} - 3\mathbf{j} + 5\mathbf{k}$; $\mathbf{v} = 3\mathbf{i} - \mathbf{j} - \mathbf{k}$
15. $\mathbf{u} = 10\mathbf{i} + 7\mathbf{j} - 3\mathbf{k}$; $\mathbf{v} = -3\mathbf{i} + 4\mathbf{j} - 3\mathbf{k}$
16. $\mathbf{u} = 2\mathbf{i} + 4\mathbf{j} - 6\mathbf{k}$; $\mathbf{v} = -\mathbf{i} - \mathbf{j} + 3\mathbf{k}$
17. $\mathbf{u} = 2\mathbf{i} - \mathbf{j} + \mathbf{k}$; $\mathbf{v} = 4\mathbf{i} + 2\mathbf{j} + 2\mathbf{k}$
18. $\mathbf{u} = 3\mathbf{i} - \mathbf{j} + 8\mathbf{k}$; $\mathbf{v} = \mathbf{i} + \mathbf{j} - 4\mathbf{k}$
19. $\mathbf{u} = a\mathbf{i} + a\mathbf{j} + a\mathbf{k}$; $\mathbf{v} = b\mathbf{i} + b\mathbf{j} + b\mathbf{k}$
20. $\mathbf{u} = a\mathbf{i} + b\mathbf{j} + c\mathbf{k}$; $\mathbf{v} = a\mathbf{i} + b\mathbf{j} - c\mathbf{k}$

21. Find two unit vectors orthogonal to both $\mathbf{u} = 2\mathbf{i} - 3\mathbf{j}$ and $\mathbf{v} = 4\mathbf{j} + 3\mathbf{k}$.
22. Find two unit vectors orthogonal to both $\mathbf{u} = \mathbf{i} + \mathbf{j} + \mathbf{k}$ and $\mathbf{v} = \mathbf{i} - \mathbf{j} - \mathbf{k}$.
23. Use the cross product to find the sine of the angle φ between the vectors $\mathbf{u} = 2\mathbf{i} + \mathbf{j} - \mathbf{k}$ and $\mathbf{v} = -3\mathbf{i} - 2\mathbf{j} + 4\mathbf{k}$.
24. Use the dot product to find the cosine of the angle φ between the vectors of the preceding problem; then verify that for the sine and cosine values you have calculated, $\sin^2 \varphi + \cos^2 \varphi = 1$.

In Problems 25–28, find a line whose direction vector is orthogonal to the direction vectors of the two given lines and that passes through the specified point.

25. $\dfrac{x+2}{-3} = \dfrac{y-1}{4} = \dfrac{z}{-5}$; $\dfrac{x-3}{7} = \dfrac{y+2}{-2} = \dfrac{z-8}{3}$; $(1, -3, 2)$

26. $\dfrac{x-2}{-4} = \dfrac{y+3}{-7} = \dfrac{z+1}{3}$; $\dfrac{x+2}{3} = \dfrac{y-5}{-4} = \dfrac{z+3}{-2}$; $(-4, 7, 3)$

27. $x = 3 - 2t$, $y = 4 + 3t$, $z = -7 + 5t$; $x = -2 + 4s$, $y = 3 - 2s$, $z = 3 + s$; $(-2, 3, 4)$

28. $x = 4 + 10t$, $y = -4 - 8t$, $z = 3 + 7t$; $x = -2s$, $y = 1 + 4s$, $z = -7 - 3s$; $(4, 6, 0)$

In Problems 29–34, calculate the area of a parallelogram having the specified points as consecutive vertices.

29. $(1, -2, 3)$; $(2, 0, 1)$; $(0, 4, 0)$
30. $(-2, 1, 1)$; $(2, 2, 3)$; $(-1, -2, 4)$
31. $(-2, 1, 0)$; $(1, 4, 2)$; $(-3, 1, 5)$
32. $(7, -2, -3)$; $(-4, 1, 6)$; $(5, -2, 3)$
33. $(a, 0, 0)$; $(0, b, 0)$; $(0, 0, c)$
34. $(a, b, 0)$; $(a, 0, b)$; $(0, a, b)$

■ APPLICATIONS

35. Use equation (6) to calculate the area of the triangle with vertices at $(2, 1, -4)$, $(1, 7, 2)$, and $(3, -2, 3)$.
36. Calculate the area of the triangle with vertices at $(3, 1, 7)$, $(2, -3, 4)$, and $(7, -2, 4)$.

37. Sketch the triangle with vertices at $(1, 0, 0)$, $(0, 1, 0)$, and $(0, 0, 1)$. Then calculate its area.
*38. Sketch the triangle with vertices at $(0, 2, 2)$, $(2, 0, 2)$, and $(2, 2, 0)$. Then sketch the triangle whose vertices are the midpoints of the sides of

the first triangle. Calculate the area of each triangle.

39. Calculate the volume of the parallelepiped determined by the vectors $\mathbf{u} = 2\mathbf{i} - \mathbf{j} + \mathbf{k}$, $\mathbf{v} = 3\mathbf{i} + 2\mathbf{j} - 2\mathbf{k}$, and $\mathbf{w} = 3\mathbf{i} + 2\mathbf{j}$.

40. Calculate the volume of the parallelepiped determined by the vectors $\mathbf{u} = \mathbf{i} + \mathbf{j}$, $\mathbf{v} = \mathbf{j} - \mathbf{k}$, and $\mathbf{w} = \mathbf{k} + \mathbf{i}$.

41. Calculate the volume of the parallelepiped determined by the vectors $\mathbf{i} - \mathbf{j}$, $3\mathbf{i} + 2\mathbf{k}$, and $-7\mathbf{j} + 3\mathbf{k}$.

42. Calculate the volume of the parallelepiped determined by the vectors $\overrightarrow{PQ}$, $\overrightarrow{PR}$, and $\overrightarrow{PS}$ where $P = (2, 1, -1)$, $Q = (-3, 1, 4)$, $R = (-1, 0, 2)$, and $S = (-3, -1, 5)$.

*43. Calculate the distance between the lines

$$L_1: \frac{x - 2}{3} = \frac{y - 5}{2} = \frac{z - 1}{-1}$$

and

$$L_2: \frac{x - 4}{-4} = \frac{y - 5}{4} = \frac{z + 2}{1}.$$

[*Hint:* The distance is measured along a vector $\mathbf{v}$ that is perpendicular to both L_1 and L_2. Let P be a point on L_1 and Q a point on L_2. Then the length of the projection of $\overrightarrow{PQ}$ on $\mathbf{v}$ is the distance between the lines, measured along a

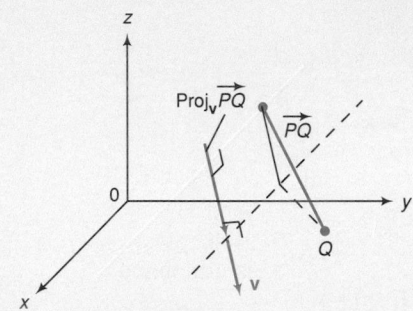

FIGURE 9

vector that is perpendicular to them both. See Figure 9.

*44. Find the distance between the lines

$$L_1: \frac{x + 2}{3} = \frac{y - 7}{-4} = \frac{z - 2}{2}$$

and

$$L_2: \frac{x - 1}{-3} = \frac{y + 2}{4} = \frac{z + 1}{1}.$$

*45. Find the distance between the lines $x = 2 - 3t$, $y = 1 + 2t$, $z = -2 - t$ and $x = 1 + 4s$, $y = -2 - s$, $z = 3 + s$.

*46. Find the distance between the lines $x = -2 - 5t$, $y = -3 - 2t$, $z = 1 + 4t$ and $x = 2 + 3s$, $y = -1 + s$, $z = 3s$.

■ SHOW/PROVE/DISPROVE

47. Show that if $\mathbf{a} = (a_1, a_2, a_3)$, $\mathbf{b} = (b_1, b_2, b_3)$, and $\mathbf{c} = (c_1, c_2, c_3)$, then

$$\mathbf{a} \cdot (\mathbf{b} \times \mathbf{c}) = \begin{vmatrix} a_1 & a_2 & a_3 \\ b_1 & b_2 & b_3 \\ c_1 & c_2 & c_3 \end{vmatrix}.$$

48. Show that $|\mathbf{a} \times \mathbf{b}|^2 = |\mathbf{a}|^2|\mathbf{b}|^2 - (\mathbf{a} \cdot \mathbf{b})^2$. [*Hint:* Write out in terms of components.]

*49. Suppose $\mathbf{a}$ and $\mathbf{b}$ are two vectors in $\mathbb{R}^3$; let $\mathbf{c} = \mathbf{a} - \mathbf{b}$. Show that the law of sines can be inferred from the fact that $\mathbf{c} \times \mathbf{c} = \mathbf{0}$ implies $\mathbf{a} \times \mathbf{c} = \mathbf{b} \times \mathbf{c}$.[†]

*50. Suppose $\mathbf{a}$, $\mathbf{b}$, and $\mathbf{c}$ are vectors in $\mathbb{R}^3$. Prove each of the following assertions.

   a. There are constants β and γ such that $\mathbf{a} \times (\mathbf{b} \times \mathbf{c}) = \beta\mathbf{b} + \gamma\mathbf{c}$.

   b. $\mathbf{a} \times (\mathbf{b} \times \mathbf{c}) = (\mathbf{a} \cdot \mathbf{c})\mathbf{b} - (\mathbf{a} \cdot \mathbf{b})\mathbf{c}$.

   c. **Jacobi identity:** $\mathbf{a} \times (\mathbf{b} \times \mathbf{c}) + \mathbf{b} \times (\mathbf{c} \times \mathbf{a}) + \mathbf{c} \times (\mathbf{a} \times \mathbf{b}) = \mathbf{0}$.

■ CHALLENGE

*51. The parallelepiped in Figure 6 has six faces. The diagonals of each face meet at a point. Show that these six "center points" are themselves vertices of a solid. Calculate the volume of this solid.

[†] Analogously, we could interpret the computations in the text's proof of Theorem 2 in Section 11.2 as showing that the law of cosines follows from "multiplying out" the relation $\mathbf{c} \cdot \mathbf{c} = (\mathbf{a} - \mathbf{b}) \cdot (\mathbf{a} - \mathbf{b})$.

■ ANSWERS TO SELF-QUIZ

I. d   II. c   IV. d V. a

III. b = zero vector [*Note:* **i × j × k** *is* defined be-   VI. c = zero vector
cause **(i × j) × k = 0 = i × (j × k)**.]

11.7 PLANES

In Section 11.5, we derived the equation of a line in space by specifying a point on the line and a vector *parallel* to this line. We can derive the equation of a plane in space by specifying a point in the plane and a vector orthogonal to every vector in the plane. This orthogonal vector is called a **normal vector** and is denoted by **N**. (See Figure 1.)

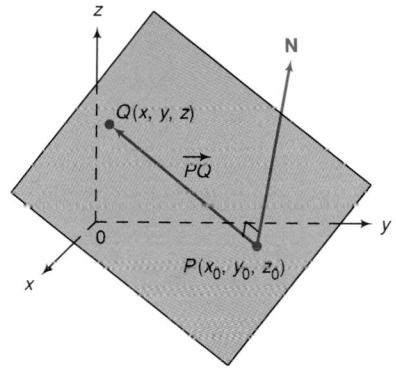

FIGURE 1
A PLANE IS DETERMINED BY A POINT AND A NORMAL VECTOR.

DEFINITION **PLANE**

Let P be a point in space and let **N** be a given nonzero vector. Then the set of all points Q for which $\overrightarrow{PQ}$ and **N** are orthogonal constitutes a **plane** in $\mathbb{R}^3$. ■

NOTATION: We will usually denote a plane by the symbol Π (a capital pi).

Let $P = (x_0, y_0, z_0)$ and $\mathbf{N} = a\mathbf{i} + b\mathbf{j} + c\mathbf{k}$. Then if $Q = (x, y, z)$,
$\overrightarrow{PQ} = (x - x_0)\mathbf{i} + (y - y_0)\mathbf{j} + (z - z_0)\mathbf{k}$.

If $\overrightarrow{PQ} \perp \mathbf{N}$, then $\overrightarrow{PQ} \cdot \mathbf{N} = 0$. But this implies that

$$a(x - x_0) + b(y - y_0) + c(z - z_0) = 0. \tag{1}$$

A more common way to write the equation of a plane is easily derived from (1):

EQUATION OF A PLANE

$$ax + by + cz = d \tag{2}$$

where

$$d = ax_0 + by_0 + cz_0 = \overrightarrow{0P} \cdot \mathbf{N}.$$

EXAMPLE 1
FINDING AN EQUATION OF THE PLANE PASSING THROUGH A GIVEN POINT WITH GIVEN NORMAL VECTOR

Find an equation of the plane Π passing through the point $(2, 5, 1)$ and normal to the vector $\mathbf{N} = \mathbf{i} - 2\mathbf{j} + 3\mathbf{k}$.

SOLUTION: From (1) we immediately obtain

$$(x - 2) - 2(y - 5) + 3(z - 1) = 0,$$

or

$$x - 2y + 3z = -5. \tag{3}$$

This plane is sketched in Figure 2.

REMARK: We show you how to draw a plane on page 770.

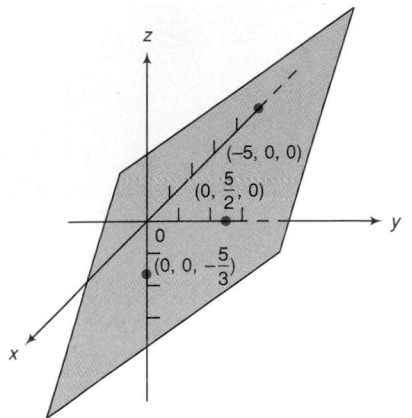

FIGURE 2
THE PLANE $x - 2y + 3z = -5$

The three coordinate planes are represented as follows:

(i) **The xy-plane** passes through the origin $(0, 0, 0)$ and any vector lying along the z-axis is normal to it. The simplest such vector is $\mathbf{k}$. Thus, from (1), we obtain

$$0(x - 0) + 0(y - 0) + 1(z - 0) = 0,$$

which yields

$$z = 0 \tag{4}$$

as the equation of the xy-plane. (This result should not be very surprising.)

REMARK: The equation $z = 0$ is really a shorthand notation for the equation of the xy-plane. The full notation is

the xy-plane $= \{(x, y, z): z = 0\}$.

The shorthand notation is fine as long as we don't lose sight of the fact that we are in $\mathbb{R}^3$.

(ii) **The xz-plane** has the equation

$$y = 0. \tag{5}$$

(iii) **The yz-plane** has the equation

$$x = 0. \tag{6}$$

Three points that are not collinear determine a plane, since they determine two nonparallel vectors that intersect at a point.

EXAMPLE 2
FINDING AN EQUATION OF
THE PLANE PASSING THROUGH
THREE GIVEN POINTS

Find an equation of the plane Π passing through the points $P = (1, 2, 1)$, $Q = (-2, 3, -1)$ and $R = (1, 0, 4)$.

SOLUTION: The vectors $\overrightarrow{PQ} = -3\mathbf{i} + \mathbf{j} - 2\mathbf{k}$ and $\overrightarrow{QR} = 3\mathbf{i} - 3\mathbf{j} + 5\mathbf{k}$ lie on the plane and are therefore orthogonal to the normal vector. Thus,

$$\mathbf{N} = \overrightarrow{PQ} \times \overrightarrow{QR} = \begin{vmatrix} \mathbf{i} & \mathbf{j} & \mathbf{k} \\ -3 & 1 & -2 \\ 3 & -3 & 5 \end{vmatrix} = -\mathbf{i} + 9\mathbf{j} + 6\mathbf{k},$$

and we obtain

$$\Pi: -(x - 1) + 9(y - 2) + 6(z - 1) = 0,$$

or

$$-x + 9y + 6z = 23.$$

Note that if we choose another point, say Q, we get the equation

$$-(x + 2) + 9(y - 3) + 6(z + 1) = 0,$$

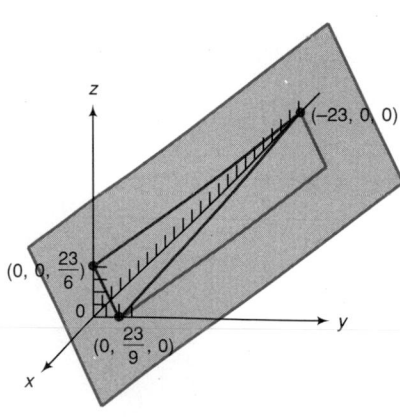

FIGURE 3
THE PLANE $-x + 9y + 6z = 23$

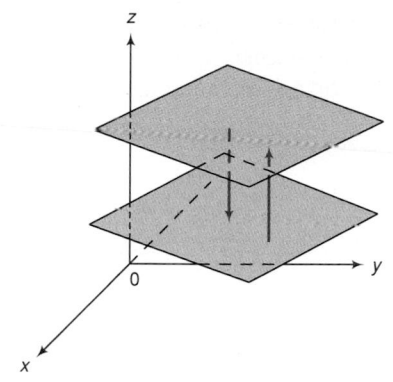

FIGURE 4
PARALLEL PLANES

which reduces to

$$-x + 9y + 6z = 23.$$

This plane is sketched in Figure 3.

DEFINITION **PARALLEL PLANES**

Two planes are **parallel** if their normal vectors are parallel. ∎

Two parallel planes are drawn in Figure 4.

EXAMPLE 3
TWO PARALLEL PLANES

The planes Π_1: $2x + 3y - z = 3$ and Π_2: $-4x - 6y + 2z = 8$ are parallel since $N_1 = 2i + 3j - k$, $N_2 = -4i - 6j + 2k$, and $N_2 = -2N_1$ (note also that $N_1 \times N_2 = 0$).

If two distinct planes are not parallel, then they intersect in a straight line.

EXAMPLE 4
FINDING POINTS OF
INTERSECTION OF PLANES

Find all points of intersection of the planes $2x - y - z = 3$ and $x + 2y + 3z = 7$.

SOLUTION: When the planes intersect, we have

$$2x - y - z = 3$$

and

$$x + 2y + 3z = 7.$$

Multiplying the first equation by 2 and adding it to the second, we obtain

$$\begin{aligned} 4x - 2y - 2z &= 6 \\ x + 2y + 3z &= 7 \\ \hline 5x + \; z &= 13 \end{aligned}$$

or $z = -5x + 13$. Then from the first equation,

$$y = 2x - z - 3 = 2x - (-5x + 13) - 3 = 7x - 16.$$

Then setting $x = t$, we obtain the parametric representation of the line of intersection:

$$x = t, \qquad y = -16 + 7t, \qquad z = 13 - 5t.$$

Note that this line is orthogonal to both normal vectors $N_1 = 2i - j - k$ and $N_2 = i + 2j + 3k$.

⊗ **WARNING:** Notice the difference in information required to find the equation of a plane and the equation of a line. For a plane, we need a point on the plane and a vector *normal* to the plane; for a line, we need a point on the line and a vector *parallel* to the line. ⊗

ANOTHER GEOMETRIC WAY TO LOOK AT A PLANE

Consider equation (2),

$$ax + by + cz = d,$$

where at least one of the numbers a, b, c is nonzero. Let $\mathbf{N} = (a, b, c)$ and $\mathbf{x} = (x, y, z)$. Then (2) can be written

$$\mathbf{N} \cdot \mathbf{x} = d, \tag{7}$$

or, since $\mathbf{N} \neq \mathbf{0}$, we can divide both sides of (7) by $|\mathbf{N}|$ to obtain

$$\frac{\mathbf{N}}{|\mathbf{N}|} \cdot \mathbf{x} = \frac{d}{|\mathbf{N}|}.$$

Let $\mathbf{v} = \dfrac{\mathbf{N}}{|\mathbf{N}|}$, a unit vector. Then, $\mathbf{v} \cdot \mathbf{x} = \dfrac{\mathbf{N} \cdot \mathbf{x}}{|\mathbf{N}|} = \dfrac{d}{|\mathbf{N}|}$ and, from equation (9) on page 748,

$$\text{Proj}_\mathbf{v}\, \mathbf{x} = \frac{\mathbf{v} \cdot \mathbf{x}}{|\mathbf{v}|^2}\, \mathbf{v} \overset{\overset{|\mathbf{v}|=1}{\downarrow}}{=} \frac{d}{|\mathbf{N}|}\, \mathbf{v}$$

and the component of $\mathbf{x}$ in the direction $\mathbf{N}$ ($=$ the direction of $\mathbf{v}$) is $\dfrac{d}{|\mathbf{N}|}$. That is,

the plane is the set of vectors $\mathbf{x}$ such that the component of $\mathbf{x}$ in the direction $\mathbf{N}$ is $\dfrac{d}{|\mathbf{N}|}$. Moreover, $\dfrac{d}{|\mathbf{N}|}$ is the distance from the origin to the plane.

This is illustrated in Figure 5. Vectors $\mathbf{x}$ such that $\dfrac{\mathbf{N}}{|\mathbf{N}|} \cdot \mathbf{x} > \dfrac{d}{|\mathbf{N}|}$ and $\dfrac{\mathbf{N}}{|\mathbf{N}|} \cdot \mathbf{x} < \dfrac{d}{|\mathbf{N}|}$ lie on opposite sides of the plane.

HOW TO DRAW A PLANE

In many places in this and subsequent chapters it will be useful to draw a plane. This is not difficult to do. We consider two cases:

Case 1: The plane is parallel to a coordinate plane If the plane is parallel to one of the coordinate planes, then the equation of the plane is one of the

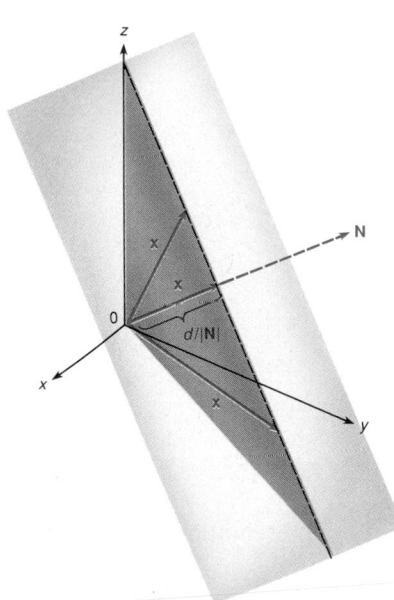

FIGURE 5
THE VECTOR **x** IS ON THE PLANE IF THE COMPONENT OF **x** IN THE DIRECTION **N**
IS EQUAL TO $\dfrac{d}{|\mathbf{N}|}$

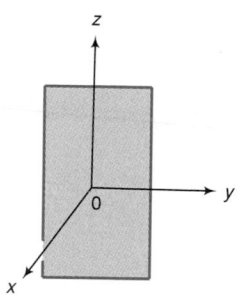

(a) The plane $x = a$ (parallel to the yz-plane).

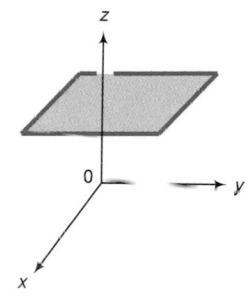

(b) The plane $y = b$ (parallel to the xy-plane).

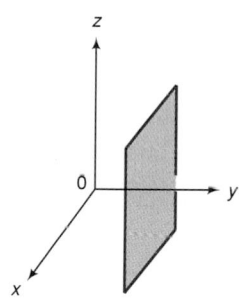

(c) The plane $z = c$ (parallel to the xy-plane).

FIGURE 6
THREE PLANES PARALLEL TO A COORDI-
NATE PLANE

following:

$$x = a \qquad \text{(parallel to the } yz\text{-plane)}$$
$$y = b \qquad \text{(parallel to the } xz\text{-plane)}$$
$$z = c \qquad \text{(parallel to the } xy\text{-plane).}$$

We draw each plane as a rectangle with sides parallel to the two other co-ordinate axes. We sketch three of these planes in Figure 6.

Case 2: The plane intersects each coordinate axis Suppose an equation of the plane is

$$ax + by + cz = d \qquad \text{with } abc \neq 0.$$

The x-intercept is the point $\left(\dfrac{d}{a}, 0, 0\right)$, the y-intercept is the point $\left(0, \dfrac{d}{b}, 0\right)$, and the z-intercept is the point $\left(0, 0, \dfrac{d}{c}\right)$.

Step 1: Plot the three intercepts.

Step 2: Join the three intercepts to form a triangle.

Step 3: By drawing two parallel lines, draw a parallelogram with one diagonal being the third side of the triangle.

Step 4: Expand the parallelogram by drawing four parallel lines.

We illustrate this process by drawing the plane $x + 2y + 3z = 6$ in Figure 7 (a second illustration appears in Figure 3). The intercepts are $(6, 0, 0)$, $(0, 3, 0)$, and $(0, 0, 2)$.

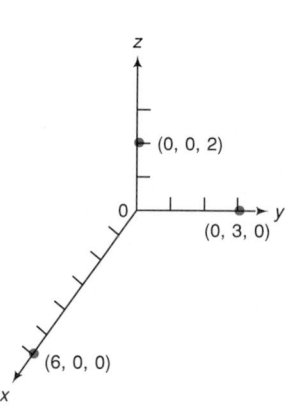

(a) Step 1: Plot the three intercepts.

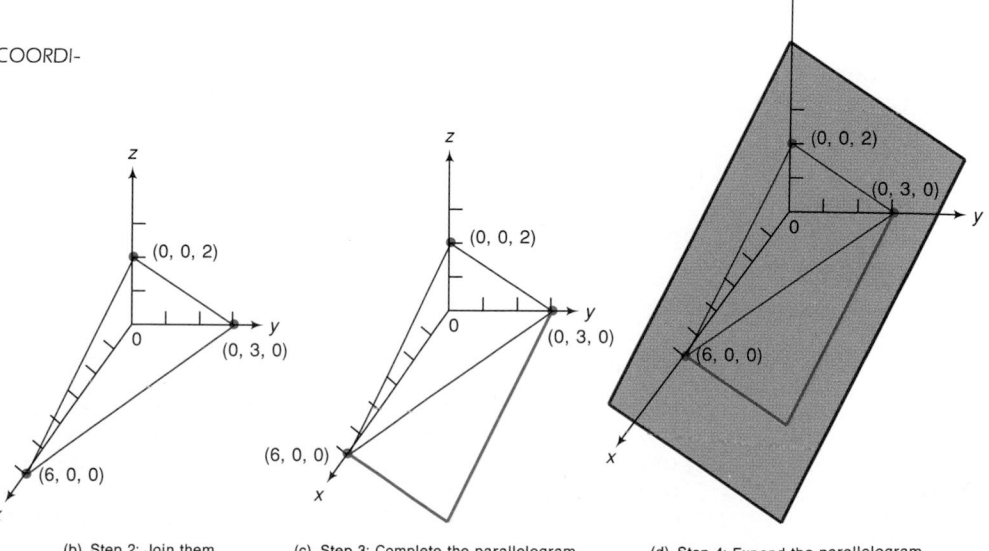

(b) Step 2: Join them.　(c) Step 3: Complete the parallelogram.　(d) Step 4: Expand the parallelogram.

FIGURE 7　DRAWING THE PLANE $x + 2y + 3z = 6$ IN FOUR STEPS

■ PROBLEMS 11.7

■ SELF-QUIZ

I. The plane passing through $(5, -4, 3)$ that is orthogonal to $\mathbf{j}$ satisfies _____.
   a. $y = -4$
   b. $(x - 5) + (z - 3) = 0$
   c. $x + y + z = 4$
   d. $5x - 4y + 3z = -4$

II. The plane passing through $(5, -4, 3)$ that is orthogonal to $\mathbf{i} + \mathbf{j} + \mathbf{k}$ satisfies _____.
   a. $y = -4$
   b. $(x - 5)/1 = (y + 4)/1 = (z - 3)/1$
   c. $x + y + z = 4$
   d. $5x - 4y + 3z = -4$

III. The vector _____ is orthogonal to the plane satisfying $2(x - 3) - 3(y + 2) + 5(z - 5) = 0$.
   a. $-3\mathbf{i} + 2\mathbf{j} - 5\mathbf{k}$
   b. $2\mathbf{i} - 3\mathbf{j} + 5\mathbf{k}$
   c. $(2 - 3)\mathbf{i} + (-3 + 2)\mathbf{j} + (5 - 5)\mathbf{k}$
   d. $(2)(-3)\mathbf{i} + (-3)(2)\mathbf{j} + (5)(-5)\mathbf{k}$

IV. The plane satisfying $6x + 18y - 12z = 17$ is _____ to the plane $-5x - 15y + 10z = 29$.
   a. identical
   b. parallel
   c. orthogonal
   d. neither parallel nor orthogonal

■ DRILL

In Problems 1–12, find an equation of the plane passing through the given point such that the specified vector $\mathbf{N}$ is normal to the plane. Sketch the plane.

1. $P = (0, 0, 0)$; $\mathbf{N} = \mathbf{i}$
2. $P = (0, 0, 0)$; $\mathbf{N} = \mathbf{j}$
3. $P = (0, 0, 0)$; $\mathbf{N} = \mathbf{k}$
4. $P = (1, 2, 3)$; $\mathbf{N} = \mathbf{i} + \mathbf{j}$
5. $P = (1, 2, 3)$; $\mathbf{N} = \mathbf{i} + \mathbf{k}$
6. $P = (1, 2, 3)$; $\mathbf{N} = \mathbf{j} + \mathbf{k}$
7. $P = (2, -1, 6)$; $\mathbf{N} = 3\mathbf{i} - \mathbf{j} + 2\mathbf{k}$
8. $P = (-4, -7, 5)$; $\mathbf{N} = -3\mathbf{i} - 4\mathbf{j} + \mathbf{k}$
9. $P = (-3, 11, 2)$; $\mathbf{N} = 4\mathbf{i} + \mathbf{j} - 7\mathbf{k}$
10. $P = (3, -2, 5)$; $\mathbf{N} = 2\mathbf{i} - 7\mathbf{j} - 8\mathbf{k}$
11. $P = (4, -7, -3)$; $\mathbf{N} = -\mathbf{i} - \mathbf{j} - \mathbf{k}$
12. $P = (8, 1, 0)$; $\mathbf{N} = -7\mathbf{i} + \mathbf{j} + 2\mathbf{k}$

In Problems 13–16, find an equation for the plane passing through the three given points. Sketch the plane.

13. $(1, 2, -4)$, $(2, 3, 7)$, and $(4, -1, 3)$
14. $(-7, 1, 0)$, $(2, -1, 3)$, and $(4, 1, 6)$
15. $(1, 0, 0)$, $(0, 1, 0)$, and $(0, 0, 1)$
16. $(2, 3, -2)$, $(4, -1, -1)$, and $(3, 1, 2)$

Two planes are said to be **orthogonal** if their normal vectors are orthogonal. In Problems 17–24, determine whether the given planes are parallel, orthogonal, coincident (i.e., the same), or none of these.

17. $\Pi_1: x + y + z = 2$; $\Pi_2: 2x + 2y + 2z = 4$
18. $\Pi_1: x - y + z = 3$; $\Pi_2: -3x + 3y - 3z = -9$
19. $\Pi_1: 2x - y + z = 3$; $\Pi_2: x + y - z = 7$
20. $\Pi_1: 2x - y + z = 3$; $\Pi_2: x + y + z = 3$
21. $\Pi_1: 3x - 2y + 7z = 4$; $\Pi_2: -2x + 4y + 2z = 16$
22. $\Pi_1: -4x + 4y - 6z = 7$; $\Pi_2: 2x - 2y + 3z = -3$
23. $\Pi_1: -4x + 4y - 6z = 6$; $\Pi_2: 2x - 2y + 3z = -3$
24. $\Pi_1: 3x - y + z = y - 2x$; $\Pi_2: 5x - 4y + 3z = 2z - 2y$

In Problems 25–28, find an equation for the set of all points common to the two given planes (i.e., describe the points of intersection).

25. $\Pi_1: x + y + z = 1$; $\Pi_2: x - y - z = -3$
26. $\Pi_1: x - y + z = 2$; $\Pi_2: 2x - 3y + 4z = 7$
27. $\Pi_1: 3x - y + 4z = 3$; $\Pi_2: -4x - 2y + 7z = 8$
28. $\Pi_1: -2x - y + 17z = 4$; $\Pi_2: 2x - y - z = -7$

The **angle between two planes** is defined to be the acute angle between their normal vectors. In Problems 29–32, calculate the angle between the two planes given in the specified problem.

29. Problem 25   30. Problem 26
31. Problem 27   32. Problem 28

In Problems 33–36, use the result of Problem 51 to determine whether the three given position vectors (i.e., with one endpoint at the origin) are coplanar. If they are coplanar, then find an equation for the plane containing them.

33. $\mathbf{u} = 2\mathbf{i} - 3\mathbf{j} + 4\mathbf{k}$; $\mathbf{v} = 7\mathbf{i} - 2\mathbf{j} + 3\mathbf{k}$; $\mathbf{w} = 9\mathbf{i} - 5\mathbf{j} + 7\mathbf{k}$
34. $\mathbf{u} = -3\mathbf{i} + \mathbf{j} + 8\mathbf{k}$; $\mathbf{v} = -2\mathbf{i} - 3\mathbf{j} + 5\mathbf{k}$; $\mathbf{w} = 2\mathbf{i} + 14\mathbf{j} - 4\mathbf{k}$
35. $\mathbf{u} = 2\mathbf{i} + \mathbf{j} - 2\mathbf{k}$; $\mathbf{v} = 2\mathbf{i} - \mathbf{j} - 2\mathbf{k}$; $\mathbf{w} = 2\mathbf{i} - \mathbf{j} + 2\mathbf{k}$

CONTINUED

36. $\mathbf{u} = 3\mathbf{i} - 2\mathbf{j} + \mathbf{k};\ \mathbf{v} = \mathbf{i} + \mathbf{j} - 5\mathbf{k};$
$\mathbf{w} = -\mathbf{i} + 5\mathbf{j} - 16\mathbf{k}$

In Problems 37–40, use the result of Problem 56 to find the distance between the given point and plane.

37. $(2, -1, 4);\ 3x - y + 7z = 2$
38. $(4, 0, 1);\ 2x - y + 8z = 3$
39. $(-7, -2, -1);\ -2x + 8z = -5$
40. $(-3, 0, 2);\ -3x + y + 5z = 0$

In Problems 41–50, draw each plane.

41. $x = 1$
42. $y = 5$
43. $z = -2$
44. $y = -3$
45. $x + y + z = 1$
46. $x - y + z = 1$
47. $x - y - z = 1$
48. $x + 2y - 4z = 4$
49. $-2x + 3y + 5z = 10$
50. $3x - y - 2z = 6$

■ SHOW/PROVE/DISPROVE

51. Three vectors $\mathbf{u}$, $\mathbf{v}$, and $\mathbf{w}$ are said to be **coplanar** if and only if they have representative directed line segments that all lie in the same plane Π. Show that if $\mathbf{u}$, $\mathbf{v}$, and $\mathbf{w}$ all have an endpoint at the origin, then they are coplanar if and only if their scalar triple product equals zero: $\mathbf{u} \cdot (\mathbf{v} \times \mathbf{w}) = 0$.

*52. Let $\mathbf{u}$ and $\mathbf{v}$ be two nonparallel vectors that lie in a particular plane Π. Show that if $\mathbf{w}$ is any other vector in Π, then there are scalars α and β such that $\mathbf{w} = \alpha\mathbf{u} + \beta\mathbf{v}$. This expression is called the **parametric representation** of the plane Π. [*Hint:* Draw a parallelogram in which $\alpha\mathbf{u}$ and $\beta\mathbf{v}$ form adjacent sides and the diagonal vector is $\mathbf{w}$.]

*53. Let $P = (x_1, y_1, z_1)$, $Q = (x_2, y_2, z_2)$, and $R = (x_3, y_3, z_3)$ be three points in $\mathbb{R}^3$ that are not col-linear. Show that the plane passing through those three points satisfies the equation

$$\begin{vmatrix} x & y & z & 1 \\ x_1 & y_1 & z_1 & 1 \\ x_2 & y_2 & z_2 & 1 \\ x_3 & y_3 & z_3 & 1 \end{vmatrix} = 0.$$

*54. Let three planes be described by the equations $a_1x + b_1y + c_1z = d_1$, $a_2x + b_2y + c_2z = d_2$, and $a_3x + b_3y + c_3z = d_3$. Show that the planes have a unique point in common if

$$\begin{vmatrix} a_1 & b_1 & c_1 \\ a_2 & b_2 & c_2 \\ a_3 & b_3 & c_3 \end{vmatrix} \neq 0.$$

■ CHALLENGE

*55. Let Π be a plane, P a point on the plane, $\mathbf{N}$ a vector normal to that plane, and Q a point not on the plane (see Figure 8). Show that the perpendicular distance D from R to the plane Π is given by

$$D = \left|\text{Proj}_{\mathbf{N}} \overrightarrow{PQ}\right| = \frac{|\overrightarrow{PQ} \cdot \mathbf{N}|}{|\mathbf{N}|}.$$

*56. Prove that the distance between the plane $ax + by + cz + d = 0$ and the point (x_0, y_0, z_0) is given by

$$D = \frac{|ax_0 + by_0 + cz_0 + d|}{\sqrt{a^2 + b^2 + c^2}}.$$

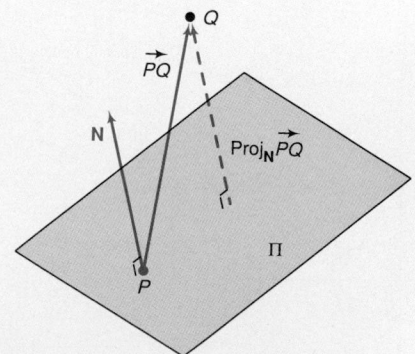

FIGURE 8

■ ANSWERS TO SELF-QUIZ

I. a   II. c   III. b   IV. b

11.8 QUADRIC SURFACES

A **surface** in space is defined as the set of points in $\mathbb{R}^3$ satisfying the equation $F(x, y, z) = 0$ for F a continuous function. For example, the equation

$$F(x, y, z) = x^2 + y^2 + z^2 - 1 = 0 \qquad (1)$$

is the equation of the unit sphere, as we saw in Section 11.3. In this section we will take a brief look at some of the most commonly encountered surfaces in $\mathbb{R}^3$. We will take a more detailed look at general surfaces in $\mathbb{R}^3$ in Chapter 13.

Having already discussed the sphere, we turn our attention to the cylinder.

DEFINITION **CYLINDER**

Let a plane curve C and a line L not in the plane of C be given. A **cylinder** is the surface generated when a line parallel to L moves around C, remaining parallel to L. The line L is called the **generatrix** of the cylinder, and the curve C is called its **directrix**. ■

EXAMPLE 1
SKETCHING A RIGHT
CIRCULAR CYLINDER

Let L be the z-axis and C the circle $x^2 + y^2 = a^2$ in the xy-plane. Sketch the cylinder.

SOLUTION: As we move a line along the circle $x^2 + y^2 = a^2$ and parallel to the z-axis, we obtain the **right circular cylinder** $x^2 + y^2 = a^2$ sketched in Figure 1.

REMARK: We can write the equation $x^2 + y^2 = 4$ as

$$x^2 + y^2 + 0 \cdot z^2 = 4.$$

This illustrates that we are talking about a cylinder in $\mathbb{R}^3$ rather than the circle $x^2 + y^2 = 4$ in $\mathbb{R}^2$.

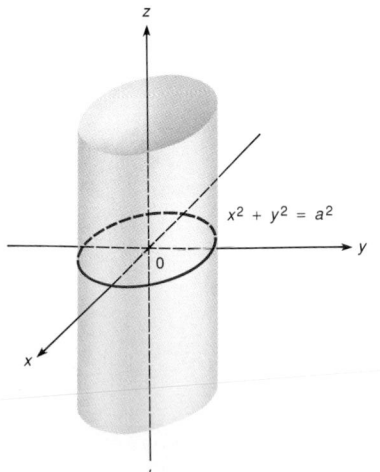

FIGURE 1
RIGHT CIRCULAR CYLINDER

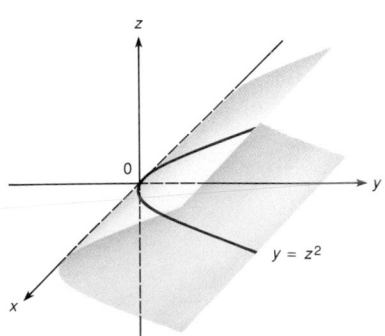

FIGURE 2 PARABOLIC CYLINDER

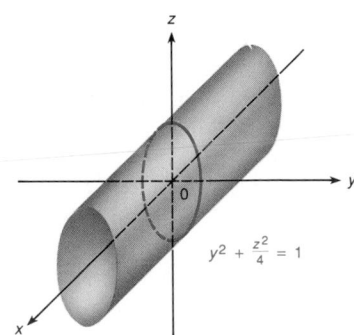

FIGURE 3 ELLIPTIC CYLINDER

EXAMPLE 2
SKETCHING A PARABOLIC
CYLINDER

Suppose L is the x-axis and C is given by $y = z^2$. Sketch the resulting cylinder.

SOLUTION: The curve $y = z^2$ is a parabola in the yz-plane. As we move along it, parallel to the x-axis, we obtain the **parabolic cylinder** sketched in Figure 2.

EXAMPLE 3
SKETCHING AN ELLIPTIC CYLINDER

Suppose L is the x-axis and C is given by $y^2 + (z^2/4) = 1$. Sketch the resulting cylinder.

SOLUTION: $y^2 + (z^2/4) = 1$ is an ellipse in the yz-plane. As we move along it parallel to the x-axis, we obtain the **elliptic cylinder** sketched in Figure 3.

EXAMPLE 4
SKETCHING A HYPERBOLIC
CYLINDER

The **hyperbolic cylinder**: $(y^2/b^2) - (x^2/a^2) = 1$. This is the equation of a hyperbola in the xy-plane. See Figure 4.

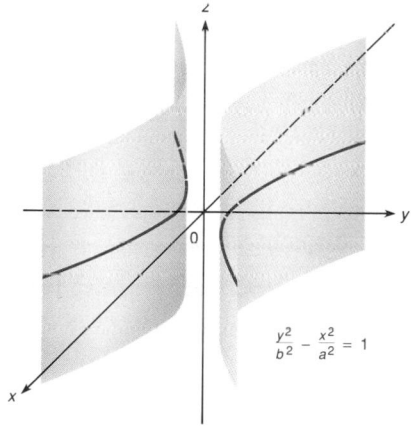

$$\frac{y^2}{b^2} - \frac{x^2}{a^2} = 1$$

FIGURE 4
HYPERBOLIC CYLINDER

As we saw from the discussion in Chapter 8, the graph of a second-degree equation in the variables x and y is a circle, parabola, ellipse, or hyperbola (or a degenerate form of one of these such as a single point or a straight line or a pair of straight lines) in the plane. In $\mathbb{R}^3$ we have the following definition.

DEFINITION **QUADRIC SURFACE**

A **quadric surface** in $\mathbb{R}^3$ is the graph of a second-degree equation in the variables x, y, and z. Such an equation takes the form

$$Ax^2 + By^2 + Cz^2 + Dxy + Exz + Fyz + Gx + Hy + Jz + K = 0. \quad \blacksquare$$

(2)

We have already seen sketches of several quadric surfaces. We list below the eleven types of nondegenerate[†] quadric surfaces. Although we will not prove this result here, any quadric surface can be written in one of these forms by a translation or rotation of the coordinate axes. Here is the list:

1. sphere   **2.** right circular cylinder
3. parabolic cylinder   **4.** elliptic cylinder
5. hyperbolic cylinder   **6.** ellipsoid
7. hyperboloid of one sheet   **8.** hyperboloid of two sheets
9. elliptic paraboloid   **10.** hyperbolic paraboloid
11. elliptic cone

[†] By **degenerate** we again mean a surface whose graph contains a finite number of points (or none) or consists of a pair of planes. We saw examples of degenerate spheres (zero or one point) in Section 11.3.

We have already seen sketches of surfaces 1, 2, 3, 4, and 5. We can best describe the remaining six surfaces by looking at **cross-sections** parallel to a given coordinate plane. Cross-sections are obtained by setting one of the variables x, y, or z equal to a constant. For example, in the unit sphere $x^2 + y^2 + z^2 = 1$, cross-sections parallel to the xy-plane are circles. To see this, let $z = c$, where $-1 < c < 1$ (this is a plane parallel to the xy-plane). Then,

$$x^2 + y^2 = 1 - c^2,$$

which is the equation of a circle in the xy-plane.

Typical graphs and specific, computer-drawn graphs of the six remaining quadric surfaces appear in Table 1. The equations given in Table 1 are examples of **standard forms**[†] of the given surfaces. We now describe the six remaining quadric surfaces.

6. **The ellipsoid**: $(x^2/a^2) + (y^2/b^2) + (z^2/c^2) = 1$. Cross-sections parallel to the xy-plane, the xz-plane, and the yz-plane are all ellipses.

NOTE: If the ellipse $(x^2/a^2) + (y^2/b^2) = 1$ is revolved about the x-axis, the resulting surface is the ellipsoid $(x^2/a^2) + (y^2/b^2) + (z^2/b^2) = 1$. What do you get if the ellipse is rotated around the y-axis? (See Problem 37.)

7. **The hyperboloid of one sheet**: $(x^2/a^2) + (y^2/b^2) - (z^2/c^2) = 1$. Cross-sections parallel to the xy-plane are ellipses. Cross-sections parallel to the xz-plane and the yz-plane are hyperbolas.

8. **The hyperboloid of two sheets**: $(z^2/c^2) - (x^2/a^2) - (y^2/b^2) = 1$. Cross-sections are the same as those of the hyperboloid of one sheet. Note that the equation implies that $|z| \geq c$ (explain why).

9. **The elliptic paraboloid**: $z = (x^2/a^2) + (y^2/b^2)$. For each positive fixed z, $(x^2/a^2) + (y^2/b^2) = z$ is the equation of an ellipse. Hence, cross-sections parallel to the xy-plane are ellipses. If x or y is fixed, then we obtain parabolas. Hence cross-sections parallel to the xz- or yz-planes are parabolas.

10. **The hyperbolic paraboloid**: $z = (x^2/a^2) - (y^2/b^2)$. For each fixed z we obtain a hyperbola parallel to the xy-plane. Hence, cross-sections parallel to the xy-plane are hyperbolas. If x or y is fixed, we obtain parabolas. Thus cross-sections parallel to the xz- and yz-planes are parabolas. The shape of the graph suggests why the hyperbolic paraboloid is often called a **saddle surface**.

11. **The elliptic cone**: $(x^2/a^2) + (y^2/b^2) = z^2$. We get one **nappe** of the cone for $z > 0$ and another for $z < 0$. Cross-sections cut by planes not passing through the origin are either parabolas, circles, ellipses, or hyperbolas. That is, the cross-sections are the **conic sections** discussed in Chapter 8. If $a = b$, we obtain the equation of a **circular cone**.

[†] A standard form is one in which the surface has its center or vertex at the origin and its axes parallel to the coordinate axes.

TABLE 1 QUADRIC SURFACES

Name of Surface and Typical Equation	Typical Graph	Computer-Drawn Graph
Ellipsoid $$\frac{x^2}{a^2} + \frac{y^2}{b^2} + \frac{z^2}{c^2} = 1$$	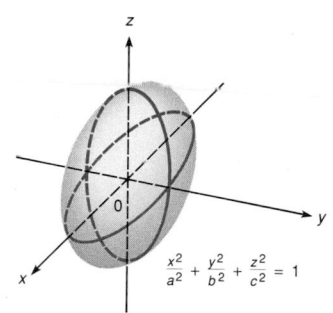 $$\frac{x^2}{a^2} + \frac{y^2}{b^2} + \frac{z^2}{c^2} = 1$$	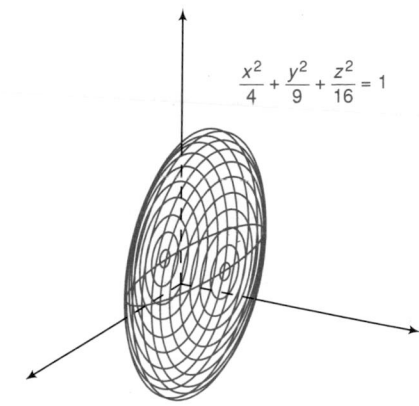 $$\frac{x^2}{4} + \frac{y^2}{9} + \frac{z^2}{16} = 1$$
Hyperboloid of one sheet $$\frac{x^2}{a^2} + \frac{y^2}{b^2} - \frac{z^2}{c^2} = 1$$	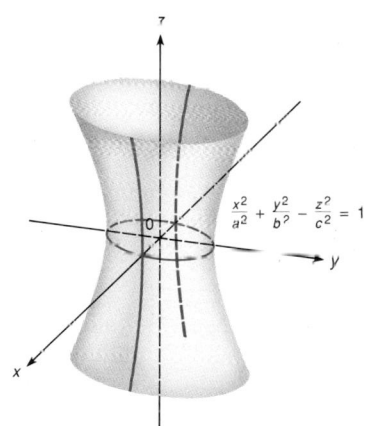 $$\frac{x^2}{a^2} + \frac{y^2}{b^2} - \frac{z^2}{c^2} = 1$$	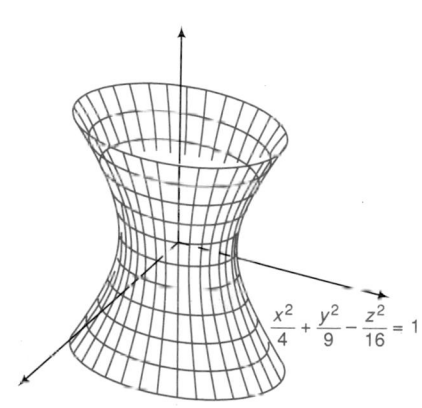 $$\frac{x^2}{4} + \frac{y^2}{9} - \frac{z^2}{16} = 1$$
Hyperboloid of two sheets $$\frac{z^2}{c^2} - \frac{x^2}{a^2} - \frac{y^2}{b^2} = 1$$	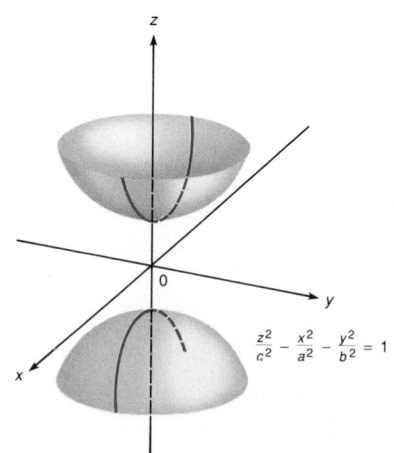 $$\frac{z^2}{c^2} - \frac{x^2}{a^2} - \frac{y^2}{b^2} = 1$$	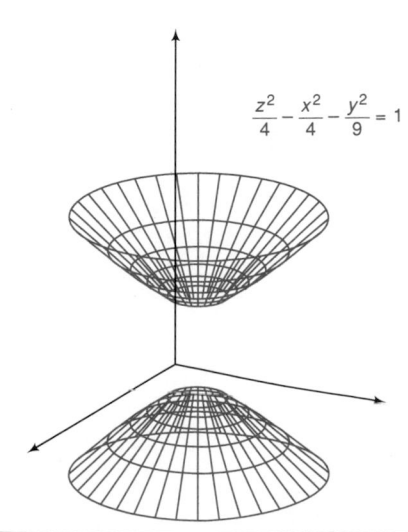 $$\frac{z^2}{4} - \frac{x^2}{4} - \frac{y^2}{9} = 1$$

(*continued*)

TABLE 1 CONTINUED

Name of Surface and Typical Equation	Typical Graph	Computer-Drawn Graph

Elliptic paraboloid

$$z = \frac{x^2}{a^2} + \frac{y^2}{b^2}$$

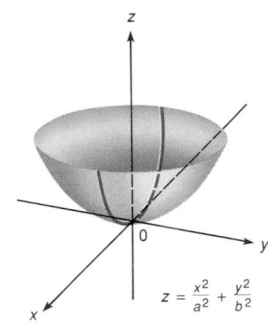

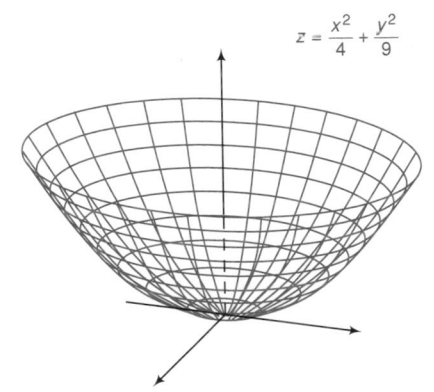

$$z = \frac{x^2}{4} + \frac{y^2}{9}$$

Hyperbolic paraboloid (or saddle surface)

$$z = \frac{x^2}{a^2} - \frac{y^2}{b^2}$$

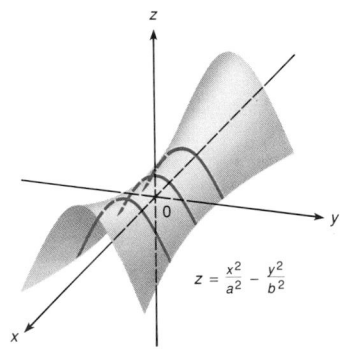

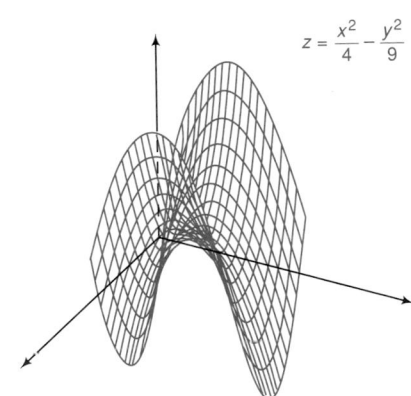

$$z = \frac{x^2}{4} - \frac{y^2}{9}$$

Elliptic cone

$$\frac{x^2}{a^2} + \frac{y^2}{b^2} = z^2$$

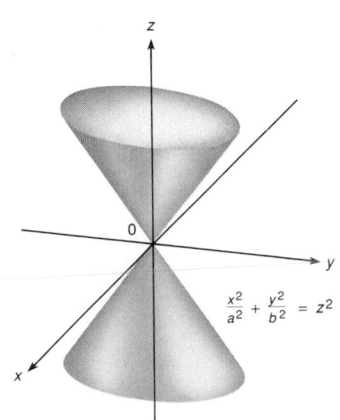

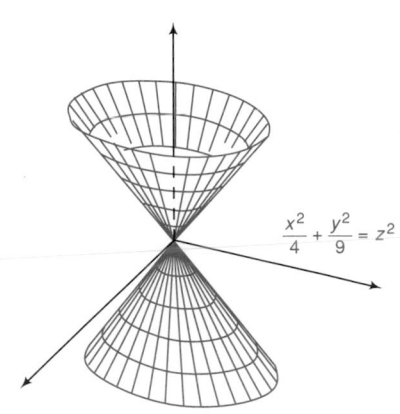

$$\frac{x^2}{4} + \frac{y^2}{9} = z^2$$

For all the quadric surfaces discussed above, we can interchange the roles of x, y, and z.

EXAMPLE 5
A HYPERBOLOID OF TWO SHEETS

Describe the surface given by $x^2 - 4y^2 - 9z^2 = 25$.

SOLUTION: Dividing by 25, we obtain

$$\frac{x^2}{25} - \frac{4y^2}{25} - \frac{9z^2}{25} = 1, \quad \text{or} \quad \frac{x^2}{25} - \frac{y^2}{(\frac{5}{2})^2} - \frac{z^2}{(\frac{5}{3})^2} = 1.$$

This is the equation of a hyperboloid of two sheets. Note here that cross-sections parallel to the yz-plane are ellipses. The surface is sketched in Figure 5.

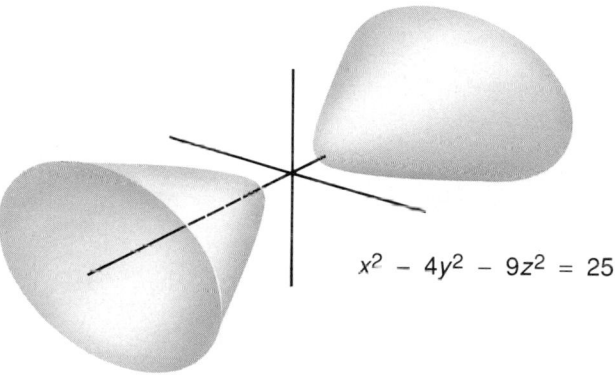

$$x^2 - 4y^2 - 9z^2 = 25$$

FIGURE 5 THE HYPERBOLOID OF TWO SHEETS $\dfrac{x^2}{25} - \dfrac{y^2}{(\frac{5}{2})^2} - \dfrac{z^2}{(\frac{5}{3})^2} = 1$

All of the surfaces discussed so far have been centered at the origin. We can describe other quadrics as well.

EXAMPLE 6
A HYPERBOLOID OF ONE SHEET

Describe the surface $x^2 + 8x - 2y^2 + 8y + z^2 = 0$.

SOLUTION: Completing the squares, we obtain

$$(x + 4)^2 - 2(y - 2)^2 + z^2 = 8,$$

or dividing by 8,

$$\frac{(x + 4)^2}{8} - \frac{(y - 2)^2}{4} + \frac{z^2}{8} = 1.$$

This surface is a hyperboloid of one sheet, which, however, is centered at $(-4, 2, 0)$ instead of the origin. Moreover, cross-sections parallel to the xz-plane are ellipses, while those parallel to the other coordinate planes are hyperbolas. The surface is sketched in Figure 6.

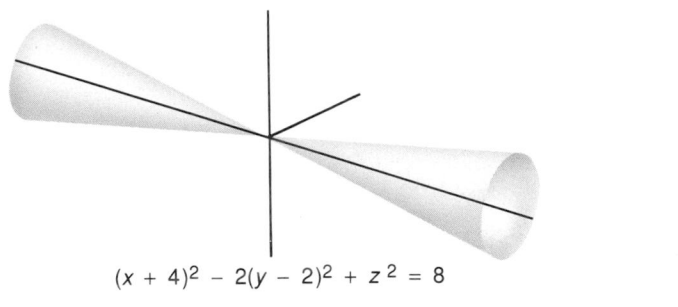

$(x + 4)^2 - 2(y - 2)^2 + z^2 = 8$

Generally, to identify a quadric surface with x, y, and z terms, completing the square will resolve the problem. If terms of the form cxy, dyz, or exz are present, also rotate the axes. We will not discuss this technique except to note it is similar to the $\mathbb{R}^2$ technique discussed in Section 8.4.

⊗ **WARNING:** Watch for degenerate surfaces. For example, no points satisfy

$$x^2 + \frac{y^2}{4} + \frac{z^2}{9} + 1 = 0. \ \otimes$$

■ PROBLEMS 11.8

■ SELF-QUIZ

I. The graph of _____ is a parabolic cylinder.
   a. $z = y^2$
   b. $y^2 + z^2 = 1$
   c. $x^2 - z^2 = 1$
   d. $(x/3)^2 + z^2 = 1$

II. The graph of _____ is an elliptic cylinder.
   a. $x = z^2$
   b. $x^2 + y^2 = 9$
   c. $y^2 - z^2 = 9$
   d. $(y/5)^2 + (z/2)^2 = 1$

III. The graph of _____ is a right circular cylinder.
   a. $z = x^2$
   b. $x^2 + z^2 = 25$
   c. $x^2 - y^2 = 25$
   d. $(x/2)^2 + (z/7)^2 = 1$

IV. The graph of _____ is a hyperbolic cylinder.
   a. $y = z^2$
   b. $y^2 + z^2 = 36$
   c. $z^2 - y^2 = 36$
   d. $(y/5)^2 + (z/6)^2 = 1$

■ DRILL

In Problems 1–8, draw a sketch of the given cylinder. Here the directrix C is given; the generatrix L is the axis of the variable missing in the equation.

1. $y = \sin x$ 2. $z = \sin y$ 3. $y = \cos z$
4. $y = \cosh x$ 5. $z = x^3$ 6. $z = |y|$
7. $|y| + |z - 5| = 1$
8. $x^2 + y^2 + 2y = 0$
   [*Hint:* Complete the square.]

In Problems 9–36, identify the quadric surface and sketch it.

9. $x^2 + y^2 = 4$
10. $y^2 + z^2 = 4$
11. $x^2 + z^2 = 4$
12. $(x/2)^2 - (y/3)^2 = 1$
13. $x^2 + 4z^2 = 1$
14. $y^2 - 2y + 4z^2 = 1$
15. $x^2 - z^2 = 1$
16. $3x^2 - 4y^2 = 4$
17. $x^2 + y^2 + z^2 = 1$
18. $x^2 + y^2 - z^2 = 1$
19. $y^2 + 2y - z^2 + x^2 = 1$

20. $x^2 + 2y^2 + 3z^2 = 4$
21. $x + 2y^2 + 3z^2 = 4$
22. $x^2 - y^2 - 3z^2 = 4$
23. $x^2 + 4x + y^2 + 6y - z^2 - 8z = 2$
24. $4x - x^2 + y^2 + z^2 = 0$
25. $5x^2 + 7y^2 + 8z^2 = 8z$
26. $x^2 - y^2 - z^2 = 1$
27. $x^2 - 2y^2 - 3z^2 = 4$
28. $z^2 - x^2 - y^2 = 2$
29. $y^2 - 3x^2 - 3z^2 = 27$
30. $x^2 + y^2 + z^2 = 2(x + y + z)$
31. $4x^2 + 4y^2 + 16z^2 = 16$
32. $4y^2 - 4x^2 + 8z^2 = 16$
33. $z + x^2 - y^2 = 0$
34. $-x^2 + 2x + y^2 - 6y = z$
35. $x + y + z = x^2 + z^2$
36. $x - y + z = y^2 - z^2$

APPLICATIONS

37. Identify and sketch the surface generated when the ellipse $(x/a)^2 + (y/b)^2 = 1$ is revolved about the y-axis.

ANSWERS TO SELF-QUIZ

In each of the four quiz problems,

- (a) is a parabolic cylinder,
- (b) is a right circular cylinder,
- (c) is a hyperbolic cylinder, and
- (d) is an elliptic cylinder.

Therefore, the answers are
I. a
II. d, b (a circle is also an ellipse)
III. b
IV. c

11.9 CYLINDRICAL AND SPHERICAL COORDINATES

In this chapter, so far, we have represented points using rectangular (Cartesian) coordinates. However, there are many ways to represent points in space. In this section, we will briefly introduce two common ways to do so. The first is the generalization of the polar coordinate system in the plane.

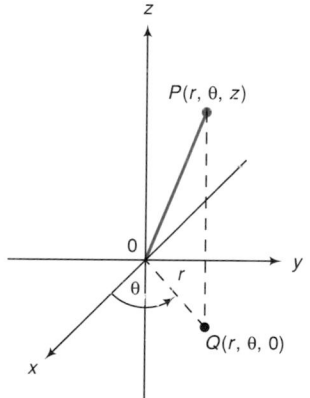

FIGURE 1
CYLINDRICAL COORDINATES

CYLINDRICAL COORDINATES

In the **cylindrical coordinate system**, a point P is given by

A POINT IN CYLINDRICAL COORDINATES

$$P = (r, \theta, z),$$ (1)

where $r \geq 0$, $0 \leq \theta < 2\pi$, r and θ are polar coordinates of the projection of P onto the xy-plane, called the **polar plane**, and z is the directed distance (measured in the positive direction) of this plane from P (see Figure 1). In this figure, $\overrightarrow{0Q}$ is the projection of $\overrightarrow{0P}$ on the xy-plane.

EXAMPLE 1
FOUR STANDARD GRAPHS IN
CYLINDRICAL COORDINATES

Discuss the graphs of the equations in cylindrical coordinates:

(a) $r = c, c > 0$ (b) $\theta = c$
(c) $z = c$ (d) $r_1 \leq r \leq r_2, \theta_1 \leq \theta \leq \theta_2, z_1 \leq z \leq z_2$

SOLUTION:

(a) If $r = c$, a constant, then θ and z can vary freely, and we obtain a right circular cylinder with radius c [see Figure 2(a)]. This is the analog of the circle whose equation in polar coordinates is $r = c$ and is the reason the system is called the *cylindrical* coordinate system.

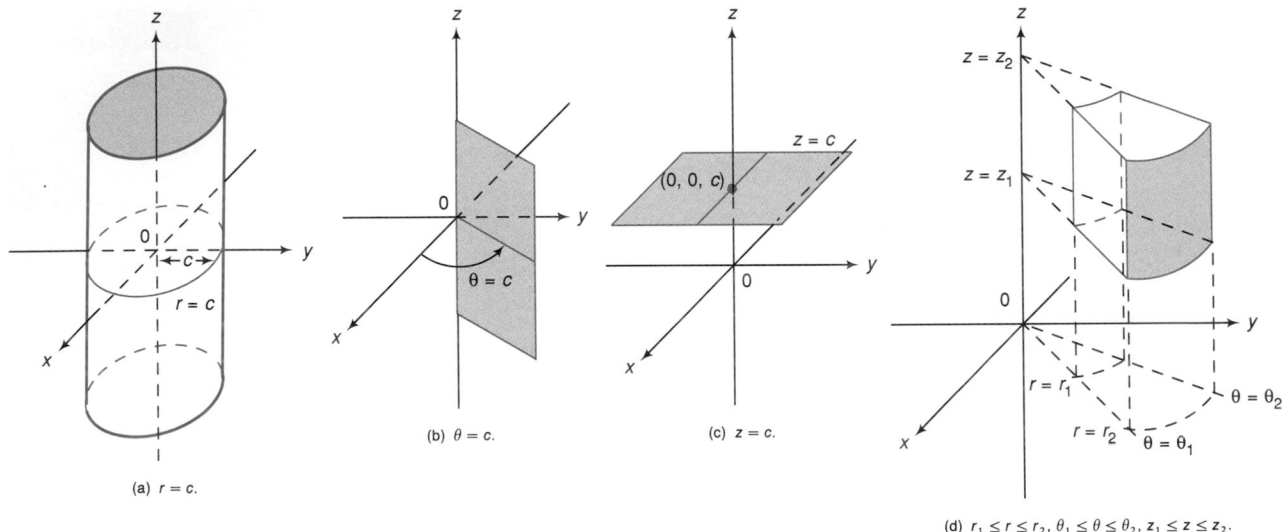

(a) $r = c$.

(b) $\theta = c$.

(c) $z = c$.

(d) $r_1 \leq r \leq r_2, \theta_1 \leq \theta \leq \theta_2, z_1 \leq z \leq z_2$.

FIGURE 2   FOUR STANDARD GRAPHS IN CYLINDRICAL COORDINATES

(b) If $\theta = c$, we obtain a half plane through the z-axis [see Figure 2(b)].

(c) If $z = c$, we obtain a plane parallel to the polar plane. This plane is sketched in Figure 2(c).

(d) $r_1 \leq r \leq r_2$ gives the region between the cylinders $r = r_1$ and $r = r_2$. $\theta_1 \leq \theta \leq \theta_2$ is the wedge-shaped region between the half planes $\theta = \theta_1$ and $\theta = \theta_2$. Finally, $z_1 \leq z \leq z_2$ is the "slice" of space between the planes $z = z_1$ and $z = z_2$. Putting these results together, we get the rectangular shaped solid in Figure 2(d).

The following formulas need no proof, because they follow from the formulas for polar coordinates in Section 8.5.

TO CHANGE FROM CYLINDRICAL TO RECTANGULAR COORDINATES

$$x = r \cos \theta, \qquad y = r \sin \theta, \qquad z = z \tag{2}$$

TO CHANGE FROM RECTANGULAR TO CYLINDRICAL COORDINATES

$$\text{If } x \neq 0, \qquad r = \sqrt{x^2 + y^2}, \qquad \tan \theta = \frac{y}{x}, \qquad z = z. \tag{3}$$

If $x = 0$, then $\theta = \dfrac{\pi}{2}$ if $y > 0$ and $\theta = \dfrac{3\pi}{2}$ if $y < 0$.

EXAMPLE 2
CONVERTING FROM CYLINDRICAL
TO RECTANGULAR COORDINATES

Convert $P = (2, \pi/3, -5)$ from cylindrical to rectangular coordinates.

SOLUTION: $x = r \cos \theta = 2 \cos(\pi/3) = 1, \quad y = r \sin \theta = 2 \sin(\pi/3) = \sqrt{3}$, and $z = -5$. Thus, in rectangular coordinates, $P = (1, \sqrt{3}, -5)$.

EXAMPLE 3
CONVERTING FROM
RECTANGULAR TO
CYLINDRICAL COORDINATES

Convert $(-3, 3, 7)$ from rectangular to cylindrical coordinates.

SOLUTION: $r = \sqrt{(-3)^2 + 3^2} = 3\sqrt{2}$, $\tan \theta = (3/-3) = -1$, so $\theta = 3\pi/4$ [since $(-3, 3)$ is in the second quadrant], and $z = 7$, so $(-3, 3, 7) = (3\sqrt{2}, 3\pi/4, 7)$ in cylindrical coordinates.

SPHERICAL COORDINATES

The second new coordinate system in space is the **spherical coordinate system**. This system is not a generalization of any system in the plane. A typical point P in space is represented as

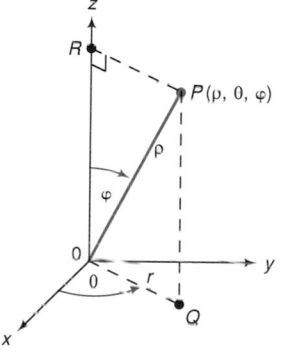

FIGURE 3
SPHERICAL COORDINATES

> ### A POINT IN SPHERICAL COORDINATES
>
> $$P = (\rho, \theta, \varphi) \tag{4}$$
>
> where $\rho \geq 0$, $0 \leq \theta < 2\pi$, $0 \leq \varphi \leq \pi$, and
>
> ρ is the (positive) distance between the point and the origin,
> θ is the same as in cylindrical coordinates, and
> φ is the angle between $\overrightarrow{0P}$ and the positive z-axis.

This system is illustrated in Figure 3.

EXAMPLE 4
FOUR STANDARD GRAPHS IN
SPHERICAL COORDINATES

Discuss the graphs of the equations in spherical coordinates:

(a) $\rho = c, c > 0$ **(b)** $\theta = c$
(c) $\varphi = c$ **(d)** $\rho_1 \leq \rho \leq \rho_2, \theta_1 \leq \theta \leq \theta_2, \varphi_1 \leq \varphi \leq \varphi_2$

SOLUTION:

(a) The set of points of which $\rho = c$ is the set of points c units from the origin. This set constitutes a sphere centered at the origin with radius c and is what gives the coordinate system its name. This surface is sketched in Figure 4(a).

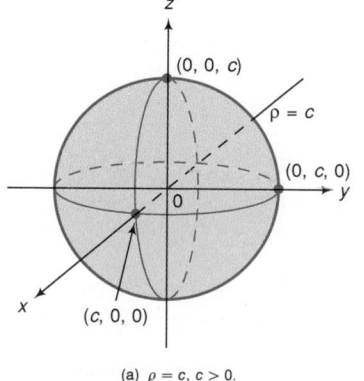

(a) $\rho = c, c > 0$.

(b) $\theta = c$.

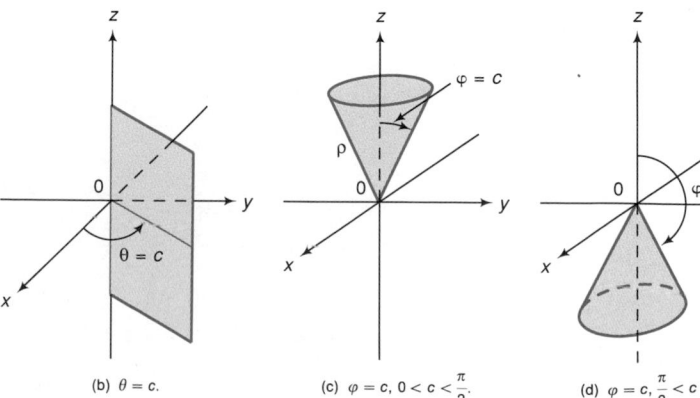

(c) $\varphi = c, 0 < c < \dfrac{\pi}{2}$.

(d) $\varphi = c, \dfrac{\pi}{2} < c < \pi$.

FIGURE 4 FOUR STANDARD GRAPHS IN SPHERICAL COORDINATES

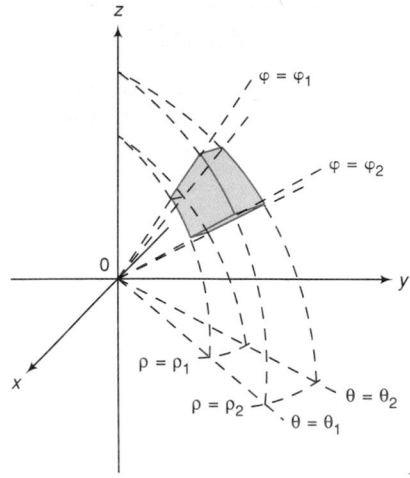

FIGURE 5
SPHERICAL WEDGE

(b) As with cylindrical coordinates, the graph is a half plane containing the z-axis. See Figure 4(b).

(c) If $0 < c < \pi/2$, then the graph of $\varphi = c$ is obtained by rotating the vector $\overrightarrow{OP}$ around the z-axis. This yields the circular cone sketched in Figure 4(c). If $\pi/2 < c < \pi$, then we obtain the cone of Figure 4(d). Finally, if $c = 0$, we obtain the positive z-axis; if $c = \pi$, we obtain the negative z-axis; and if $c = \pi/2$, we obtain the xy-plane.

(d) $\rho_1 \leq \rho \leq \rho_2$ gives the region between the sphere $\rho = \rho_1$ and the sphere $\rho = \rho_2$. $\theta_1 \leq \theta \leq \theta_2$ consists, as before, of the wedge-shaped region between the half planes $\theta = \theta_1$ and $\theta = \theta_2$. Finally, $\varphi_1 \leq \varphi \leq \varphi_2$ yields the region between the cones $\varphi = \varphi_1$ and $\varphi = \varphi_2$. Putting these results together, we obtain the solid (called a **spherical wedge**) sketched in Figure 5.

To convert from spherical to rectangular coordinates, we have, from Figure 3,

$$x = \overline{0Q} \cos \theta \qquad \text{and} \qquad y = \overline{0Q} \sin \theta.$$

But from Figure 3,

$$\overline{0Q} = \overline{PR} = \rho \sin \varphi, \qquad \text{so } x = \rho \sin \varphi \cos \theta \text{ and } y = \rho \sin \varphi \sin \theta.$$

Finally, from triangle OPR we have $z = \rho \cos \varphi$. In sum, we have the following:

TO CHANGE POINTS FROM SPHERICAL TO RECTANGULAR COORDINATES

$$x = \rho \sin \varphi \cos \theta \tag{5}$$

$$y = \rho \sin \varphi \sin \theta \tag{6}$$

$$z = \rho \cos \varphi \tag{7}$$

TO CHANGE POINTS FROM RECTANGULAR TO SPHERICAL COORDINATES

$$\rho = \sqrt{x^2 + y^2 + z^2} \tag{8}$$

$$\varphi = \cos^{-1} \frac{z}{\rho}, \qquad 0 \leq \varphi \leq \pi \tag{9}$$

$$\cos \theta = \frac{x}{\rho \sin \varphi} \qquad \text{or} \qquad \sin \theta = \frac{y}{\rho \sin \varphi}, \qquad 0 \leq \theta < 2\pi \tag{10}$$

Formulas (8), (9), and (10) follow from formulas (5), (6), and (7). First, we observe, from Figure 3, that $\overline{OP} = \sqrt{x^2 + y^2 + z^2} = \rho$, so that, from (7), $\cos \varphi = \dfrac{z}{\rho} = \dfrac{z}{\sqrt{x^2 + y^2 + z^2}}$. Knowing ρ and φ, we can then calculate θ from equations (5) and (6).

EXAMPLE 5
CONVERTING FROM SPHERICAL
TO RECTANGULAR COORDINATES

Convert $(4, \pi/6, \pi/4)$ from spherical to rectangular coordinates.

SOLUTION:

$$x = \rho \sin \varphi \cos \theta = 4 \sin \frac{\pi}{4} \cos \frac{\pi}{6} = 4 \frac{\sqrt{2}}{2} \frac{\sqrt{3}}{2} = \sqrt{6},$$

$$y = \rho \sin \varphi \sin \theta = 4 \sin \frac{\pi}{4} \sin \frac{\pi}{6} = 4 \frac{\sqrt{2}}{2} \frac{1}{2} = \sqrt{2}, \text{ and}$$

$$z = \rho \cos \varphi = 4 \cos \frac{\pi}{4} = 2\sqrt{2}.$$

Therefore, in rectangular coordinates $(4, \pi/6, \pi/4)$ is $(\sqrt{6}, \sqrt{2}, 2\sqrt{2})$.

EXAMPLE 6
CONVERTING FROM RECTANGULAR
TO SPHERICAL COORDINATES

Convert $(1, \sqrt{3}, -2)$ from rectangular to spherical coordinates.

SOLUTION: $\rho = \sqrt{x^2 + y^2 + z^2} = \sqrt{1 + 3 + 4} = \sqrt{8} = 2\sqrt{2}.$ Then from (7),

$$\cos \varphi = \frac{z}{\rho} = \frac{-2}{2\sqrt{2}} = -\frac{1}{\sqrt{2}},$$

so that $\varphi = 3\pi/4$. Finally, from (5) and (6),

$$\cos \theta = \frac{x}{\rho \sin \varphi} = \frac{1}{(2\sqrt{2})1/\sqrt{2}} = \frac{1}{2} \text{ and } \sin \theta = \frac{\sqrt{3}}{(2\sqrt{2})(1/\sqrt{2})} = \frac{\sqrt{3}}{2},$$

so that θ is in the first quadrant and $\theta = \pi/3$. Thus $(1, \sqrt{3}, -2)$ is $(2\sqrt{2}, \pi/3, 3\pi/4)$ in spherical coordinates.

We will not discuss the graphs of different kinds of surfaces given in cylindrical and spherical coordinates, because that would take us too far afield. Both coordinate systems are useful in a wide variety of physical applications. For example, points on the earth and its interior are much more easily described by using a spherical coordinate system than a rectangular system: $\{(\rho, \theta, \varphi): \rho \le a\}$, where $a =$ the radius of the earth.

PROBLEMS 11.9

SELF-QUIZ

I. True–False: Cylindrical coordinates are "just" polar coordinates (r, θ) with the Cartesian z-coordinate appended.

II. True–False: In spherical coordinates, the angles θ and φ have a constant sum.

III. True–False: In comparing cylindrical and spherical coordinates for a single point, $r^2 = \rho^2$.

IV. The _____ coordinate system is a generalization of polar coordinates.

a. solar   b. spherical
c. cylindrical   d. cubical

V. In converting from spherical coordinates to rectangular coordinates, $y = $ _____.
a. $\rho \sin \varphi$ b. $\rho \cos \varphi$
c. $\rho \sin \theta$ d. $\rho \cos \theta$
e. $\rho \sin \varphi \cos \theta$ f. $\rho \sin \varphi \sin \theta$
g. $\rho \cos \varphi \cos \theta$ h. $\rho \cos \varphi \sin \theta$

CONTINUED

■ DRILL

In Problems 1–10, convert from cylindrical to rectangular coordinates.

1. $(2, \pi/3, 5)$ 2. $(1, 0, -3)$
3. $(8, 2\pi/3, 1)$ 4. $(4, \pi/2, -7)$
5. $(3, 3\pi/4, 2)$ 6. $(10, 5\pi/3, 1)$
7. $(10, \pi, -3)$ 8. $(13, 3\pi/2, 4)$
9. $(7, 5\pi/4, 2)$ 10. $(5, 11\pi/6, -5)$

In Problems 11–20, convert from rectangular to cylindrical coordinates.

11. $(1, 0, 0)$ 12. $(0, 1, 0)$
13. $(0, 0, 1)$ 14. $(1, 1, 2)$
15. $(-1, 1, 4)$ 16. $(2, 2\sqrt{3}, -5)$
17. $(2\sqrt{3}, -2, 8)$ 18. $(2, -2\sqrt{3}, 4)$
19. $(-2\sqrt{3}, -2, 1)$ 20. $(-5, -5, -3\sqrt{2})$

In Problems 21–30, convert from spherical to rectangular coordinates.

21. $(2, 0, \pi/3)$ 22. $(4, \pi/4, \pi/4)$
23. $(6, \pi/2, \pi/3)$ 24. $(3, \pi/6, 5\pi/6)$
25. $(7, 7\pi/4, 3\pi/4)$ 26. $(3, \pi/2, \pi/2)$
27. $(4, \pi/3, 2\pi/3)$ 28. $(4, 2\pi/3, \pi/3)$
29. $(5, 11\pi/6, 5\pi/6)$ 30. $(\sqrt{3}, 5\pi/4, \pi/4)$

In Problems 31–40, convert from rectangular to spherical coordinates.

31. $(1, 1, 0)$ 32. $(1, 1, \sqrt{2})$
33. $(1, -1, \sqrt{2})$ 34. $(-1, -1, \sqrt{2})$
35. $(1, -\sqrt{3}, 2)$ 36. $(-\sqrt{3}, -1, 2)$
37. $(2, \sqrt{3}, 4)$ 38. $(-2, \sqrt{3}, -4)$
39. $(-\sqrt{3}, -2, -4)$
40. $(-1/\sqrt{2}, -\sqrt{2}/2, -\sqrt{\frac{1}{2}})$

■ APPLICATIONS

In Problems 41–54, convert the given equation into the specified coordinate system. (For instance, Problem 41 asks for a particular sphere to be described by an equation in cylindrical coordinates and by an equation in spherical coordinates.)

41. $x^2 + y^2 + z^2 = 25$, cylindrical and spherical
42. $ax + by + cz = d$, cylindrical and spherical
43. $r = 9 \sin \theta$, rectangular
44. $r^2 \sin 2\theta = z^3$, rectangular

45. $x^2 + y^2 - z^2 = 1$, cylindrical and spherical
46. $x^2 + 4y^2 + 4z^2 = 1$, cylindrical and spherical
47. $z = r^2$, rectangular
48. $\rho \cos \varphi = 1$, rectangular
49. $\rho^2 \sin \varphi \cos \varphi = 1$, rectangular
50. $\rho = \sin \theta \cos \varphi$, rectangular
51. $z = r^2 \sin 2\theta$, rectangular
52. $\rho = 2 \cot \theta$, rectangular
53. $x^2 + (y - 3)^2 + z^2 = 9$, spherical
54. $r = 6 \cos \theta$, spherical

■ SHOW/PROVE/DISPROVE

55. Use the dot product of appropriate vectors to show that $\cos \varphi = z/\rho$.

■ ANSWERS TO SELF-QUIZ

I. True   II. False   III. False
IV. c   V. f

■ PROBLEMS   CHAPTER 11 REVIEW

■ DRILL

1. Let $\mathbf{u} = (2, 1)$ and $\mathbf{v} = (-3, 4)$. Calculate the following:
   a. $5\mathbf{u}$ b. $\mathbf{u} - \mathbf{v}$
   c. $-8\mathbf{u} + 5\mathbf{v}$ d. $3\mathbf{u} + 2\mathbf{v}$

2. Let $\mathbf{u} = -4\mathbf{i} + \mathbf{j}$ and $\mathbf{v} = -3\mathbf{i} - 4\mathbf{j}$. Calculate the following:
   a. $-3\mathbf{v}$ b. $\mathbf{u} + \mathbf{v}$
   c. $4\mathbf{u} + \mathbf{v}$ d. $3\mathbf{u} - 6\mathbf{v}$

3. Let $\mathbf{u} = (1, -3, 5)$ and $\mathbf{v} = (5, 0, 2)$. Calculate the following:
   a. $3\mathbf{v}$ b. $\mathbf{v} - \mathbf{u}$
   c. $2\mathbf{u} - 2\mathbf{v}$ d. $5\mathbf{u} - \mathbf{v}$

4. Let $\mathbf{u} = \mathbf{i} - 2\mathbf{j} + 3\mathbf{k}$ and $\mathbf{v} = -3\mathbf{i} + 2\mathbf{j} + 5\mathbf{k}$. Calculate the following:
   a. $2\mathbf{v}$ b. $\mathbf{u} + \mathbf{v}$
   c. $3\mathbf{u} + 5\mathbf{v}$ d. $5\mathbf{u} - 3\mathbf{v}$

In Problems 5–14, find the magnitude and direction of the given vector.

5. $\mathbf{u} = (3, 3)$ 6. $\mathbf{u} = (2, -2\sqrt{3})$
7. $\mathbf{u} = -12\mathbf{i} - 12\mathbf{j}$ 8. $\mathbf{u} = \mathbf{i} + 4\mathbf{j}$
9. $\mathbf{w} = (\sqrt{3}, 1, 0)$ 10. $\mathbf{w} = (3, -12, -5)$
11. $\mathbf{w} = 7\mathbf{i} - 3\mathbf{j}$ 12. $\mathbf{w} = 2\mathbf{i} - \mathbf{k}$
13. $\mathbf{w} = -3\mathbf{i} + 4\mathbf{j} + 5\mathbf{k}$
14. $\mathbf{w} = \sqrt{6}\mathbf{i} + 3\mathbf{j} - \sqrt{10}\mathbf{k}$

In Problems 15–18, sketch the two given points and then calculate the distance between them.

15. $(3, 1, 2)$; $(-1, -3, -4)$
16. $(4, -1, 7)$; $(-5, 1, 3)$
17. $(-2, 4, -8)$; $(0, 0, 6)$
18. $(2, -7, 0)$; $(0, 5, -8)$

In Problems 19–26, write the vector $\mathbf{u} = a\mathbf{i} + b\mathbf{j}$ or $\mathbf{w} = \alpha\mathbf{i} + \beta\mathbf{j} + \gamma\mathbf{k}$ that is represented by $\overrightarrow{PQ}$.

19. $P = (2, 3)$; $Q = (4, 5)$
20. $P = (1, -2)$; $Q = (7, 12)$
21. $P = (-1, -6)$; $Q = (3, -4)$
22. $P = (-1, 3)$; $Q = (3, -1)$
23. $P = (2, -3, 0)$; $Q = (-2, 3, 0)$
24. $P = (0, 8, 9)$; $Q = (0, -3, 9)$
25. $P = (-3, 5, 12)$; $Q = (-1, 1, 8)$
26. $P = (4, -5, 6)$; $Q = (6, -5, 4)$

In Problems 27–36, find a unit vector whose direction is opposite to that of the given vector $\mathbf{v}$.

27. $\mathbf{v} = \mathbf{i} + \mathbf{j}$ 28. $\mathbf{v} = 5\mathbf{i} + 2\mathbf{j}$
29. $\mathbf{v} = 10\mathbf{i} - 7\mathbf{j}$ 30. $\mathbf{v} = a\mathbf{i} - a\mathbf{j}$
31. $\mathbf{v} = 3\mathbf{j} + 11\mathbf{k}$ 32. $\mathbf{v} = 4\mathbf{i} - 3\mathbf{k}$
33. $\mathbf{v} = \mathbf{i} - 2\mathbf{j} - 3\mathbf{k}$ 34. $\mathbf{v} = -4\mathbf{i} + \mathbf{j} + 6\mathbf{k}$
35. $\mathbf{v} = \overrightarrow{PQ}$ where $P = (3, -1)$ and $Q = (-4, 1)$
36. $\mathbf{v} = \overrightarrow{PQ}$ where $P = (1, -3, 0)$ and $Q = (-7, 1, -4)$

In Problems 37–44, find a vector $\mathbf{v}$ having the given magnitude and direction.

37. $|\mathbf{v}| = 2$; $\theta = \pi/3$ 38. $|\mathbf{v}| = 1$; $\theta = \pi/2$
39. $|\mathbf{v}| = 4$; $\theta = \pi$ 40. $|\mathbf{v}| = 7$; $\theta = 5\pi/6$

41. $|\mathbf{v}| = 5$; direction $= (0.6, 0.8, 0)$
42. $|\mathbf{v}| = 2$; direction $= (0, 1/2, \sqrt{3}/2)$
43. $|\mathbf{v}| = \sqrt{2}$; direction $= (1/2, -1/\sqrt{2}, -1/2)$
44. $|\mathbf{v}| = 12$; direction $= (\cos \pi/3, \cos 4\pi/3, \cos \pi/4)$

In Problems 45–52, calculate the dot product of the two given vectors and then calculate the cosine of the angle between them.

45. $\mathbf{u} = \mathbf{i} - \mathbf{j}$; $\mathbf{v} = \mathbf{i} + 2\mathbf{j}$
46. $\mathbf{u} = -4\mathbf{i}$; $\mathbf{v} = 11\mathbf{j}$
47. $\mathbf{u} = 4\mathbf{i} - 7\mathbf{j}$; $\mathbf{v} = 5\mathbf{i} + 6\mathbf{j}$
48. $\mathbf{u} = -\mathbf{i} - 2\mathbf{j}$; $\mathbf{v} = 4\mathbf{i} + 5\mathbf{j}$
49. $\mathbf{u} = \mathbf{i} + 2\mathbf{k}$; $\mathbf{v} = 2\mathbf{j} - \mathbf{k}$
50. $\mathbf{u} = -3\mathbf{i} + \mathbf{k}$; $\mathbf{v} = \mathbf{i} - \mathbf{j} + 2\mathbf{k}$
51. $\mathbf{u} = \mathbf{i} + \mathbf{j} + \mathbf{k}$; $\mathbf{v} = 2\mathbf{i} - \mathbf{j} - \mathbf{k}$
52. $\mathbf{u} = \mathbf{i} + \mathbf{j} - \mathbf{k}$; $\mathbf{v} = -\mathbf{i} - \mathbf{j} + \mathbf{k}$

In Problems 53–60, determine whether the given vectors are orthogonal, parallel, or neither; then sketch each pair of vectors.

53. $\mathbf{u} = 2\mathbf{i} - 6\mathbf{j}$; $\mathbf{v} = -\mathbf{i} + 3\mathbf{j}$
54. $\mathbf{u} = -7\mathbf{i} - 7\mathbf{j}$; $\mathbf{v} = -\mathbf{i} + \mathbf{j}$
55. $\mathbf{u} = 4\mathbf{i} - 5\mathbf{j}$; $\mathbf{v} = 5\mathbf{i} - 4\mathbf{j}$
56. $\mathbf{u} = 4\mathbf{i} + 5\mathbf{j}$; $\mathbf{v} = -5\mathbf{i} + 4\mathbf{j}$
57. $\mathbf{u} = 3\mathbf{i} - 4\mathbf{j} + 5\mathbf{k}$; $\mathbf{v} = -5\mathbf{i} + 4\mathbf{j} + 3\mathbf{k}$
58. $\mathbf{u} = 3\mathbf{i} - 4\mathbf{j} + 5\mathbf{k}$; $\mathbf{v} = -5\mathbf{i} + 3\mathbf{k}$
59. $\mathbf{u} = 2\mathbf{i} + \mathbf{j} - 3\mathbf{k}$; $\mathbf{v} = 6\mathbf{i} + 3\mathbf{j} + 9\mathbf{k}$
60. $\mathbf{u} = -3\mathbf{i} + 2\mathbf{j} + 5\mathbf{k}$; $\mathbf{v} = 6\mathbf{i} - 4\mathbf{j} - 10\mathbf{k}$

In Problems 61–68, calculate $\text{Proj}_{\mathbf{v}} \mathbf{u}$.

61. $\mathbf{u} = 14\mathbf{i}$; $\mathbf{v} = \mathbf{i} + \mathbf{j}$
62. $\mathbf{u} = 14\mathbf{i}$; $\mathbf{v} = \mathbf{i} - \mathbf{j}$
63. $\mathbf{u} = 3\mathbf{i} - 2\mathbf{j}$; $\mathbf{v} = 3\mathbf{i} + 2\mathbf{j}$
64. $\mathbf{u} = 3\mathbf{i} + 2\mathbf{j}$; $\mathbf{v} = \mathbf{i} - 3\mathbf{j}$
65. $\mathbf{u} = 2\mathbf{i} - 5\mathbf{j} + \mathbf{k}$; $\mathbf{v} = -3\mathbf{i} - 7\mathbf{j} + \mathbf{k}$
66. $\mathbf{u} = 4\mathbf{i} - 5\mathbf{j} - \mathbf{k}$; $\mathbf{v} = -3\mathbf{i} - \mathbf{j} + \mathbf{k}$
67. $\mathbf{u} = \mathbf{i} - 2\mathbf{j} + 3\mathbf{k}$; $\mathbf{v} = 2\mathbf{i} - 4\mathbf{j} + \mathbf{k}$
68. $\mathbf{u} = 2\mathbf{i} - 4\mathbf{j} + \mathbf{k}$; $\mathbf{v} = -3\mathbf{i} + 2\mathbf{j} + 5\mathbf{k}$

In Problems 69–72, calculate $\mathbf{u} \times \mathbf{v}$.

69. $\mathbf{u} = 7\mathbf{j}$; $\mathbf{v} = \mathbf{i} - \mathbf{k}$
70. $\mathbf{u} = 3\mathbf{i} - \mathbf{j}$; $\mathbf{v} = 2\mathbf{i} + 4\mathbf{k}$
71. $\mathbf{u} = 4\mathbf{i} - \mathbf{j} + 7\mathbf{k}$; $\mathbf{v} = -7\mathbf{i} + \mathbf{j} - 2\mathbf{k}$
72. $\mathbf{u} = -2\mathbf{i} + 3\mathbf{j} - 4\mathbf{k}$; $\mathbf{v} = -3\mathbf{i} + \mathbf{j} - 10\mathbf{k}$

In Problems 73–76, find a vector equation, parametric equations, and symmetric equations for the specified line.

73. containing $(3, -1, 4)$ and $(-1, 6, 2)$
74. containing $(-4, 1, 0)$ and $(3, 0, 7)$

CONTINUED

75. containing $(3, 1, 2)$ and parallel to $3\mathbf{i} - \mathbf{j} - \mathbf{k}$
76. containing $(1, 2, -3)$ and parallel to the line $(x + 1)/5 = (y - 2)/(-3) = (z - 4)/2$.

In Problems 77–80, find an equation for the plane containing the given point and having the given vector as a normal vector.

77. $P = (-1, 1, 1); \mathbf{N} = \mathbf{j} - \mathbf{k}$
78. $P = (1, 3, -2); \mathbf{N} = \mathbf{i} + \mathbf{k}$
79. $P = (1, -4, 6); \mathbf{N} = 2\mathbf{j} - 3\mathbf{k}$
80. $P = (-4, 1, 6); \mathbf{N} = 2\mathbf{i} - 3\mathbf{j} + 5\mathbf{k}$

In Problems 81–88, convert from one coordinate system to another as specified.

81. $(3, \pi/6, -1)$; cylindrical to rectangular
82. $(2, 2\pi/3, 4)$; cylindrical to rectangular
83. $(2, 2, -4)$; rectangular to cylindrical
84. $(-2, 2\sqrt{3}, -4)$; rectangular to cylindrical
85. $(3, \pi/3, \pi/4)$; spherical to rectangular
86. $(2, 7\pi/3, 2\pi/3)$; spherical to rectangular
87. $(-1, 1, -\sqrt{2})$; rectangular to spherical
88. $(2, -\sqrt{3}, 4)$; rectangular to spherical
89. Convert the equation $x^2 + y^2 + z^2 = 25$ into cylindrical and spherical coordinates.
90. Convert the equation $r^2 \cos 2\theta = z^3$ into rectangular coordinates.
91. Convert the equation $x^2 - y^2 + z^2 = 1$ into cylindrical and spherical coordinates.
92. Convert the equation $\rho \sin \varphi = 1$ into rectangular coordinates.

In Problems 93–96, draw a sketch of the specified cylinder. The directrix curve C is given; the generatrix line L is the axis of the variable missing from the equation.

93. $y = 3 - 5x$
94. $x = \cos y$
95. $y = z^2$
96. $z = \sqrt[3]{x}$

■ APPLICATIONS

97. Find two unit vectors in $\mathbb{R}^2$ that are orthogonal to $\mathbf{u} = \mathbf{i} - \mathbf{j}$.
98. Find two unit vectors in $\mathbb{R}^3$ that are orthogonal to both $\mathbf{u} = \mathbf{i} - \mathbf{j} + 3\mathbf{k}$ and $\mathbf{v} = -2\mathbf{i} - 3\mathbf{j} + 4\mathbf{k}$.
99. Let $\mathbf{u} = 2\mathbf{i} + 3\mathbf{j}$ and $\mathbf{v} = 4\mathbf{i} + \alpha\mathbf{j}$. Determine α such that
    a. $\mathbf{u}$ and $\mathbf{v}$ are parallel,
    b. $\mathbf{u}$ and $\mathbf{v}$ are orthogonal,
    c. the angle between $\mathbf{u}$ and $\mathbf{v}$ is $\pi/4$,
    d. the angle between $\mathbf{u}$ and $\mathbf{v}$ is $\pi/6$.
100. Let $\mathbf{u} = 6\mathbf{i} + \beta\mathbf{j} - 15\mathbf{k}$ and $\mathbf{v} = -2\mathbf{i} + 4\mathbf{j} + 5\mathbf{k}$. Determine β such that
    a. $\mathbf{u}$ and $\mathbf{v}$ are parallel,
    b. $\mathbf{u}$ and $\mathbf{v}$ are orthogonal,
    c. the angle between $\mathbf{u}$ and $\mathbf{v}$ is $\pi/4$,
    d. the angle between $\mathbf{u}$ and $\mathbf{v}$ is $\pi/3$.
101. Find an equation for the plane containing the points $(-2, 4, 1)$, $(3, -7, 5)$, and $(-1, -2, -1)$.
102. Find an equation for the plane containing the points $(6, -5, 1)$, $(1, -6, 5)$, and $(5, -1, 6)$.
103. Find all points of intersection of the planes Π_1: $-x + y + z = 3$ and Π_2: $-4x + 2y - 7z = 5$.
104. Find all points of intersection of the planes Π_1: $-4x + 6y + 8z = 12$ and Π_2: $2x - 3y - 4z = 5$.
105. Find the distance from the point $P = (2, 3)$ to the line with equation $(x, y) = (7, -3) + t(1, 2)$.
106. Find the distance from the point $P = (3, -1, 2)$ to the line passing through the points $Q = (-2, -1, 6)$ and $R = (0, 1, -8)$.
107. Find the distance from $(1, -2, 3)$ to the plane $2x - y - z = 6$.
108. Find the distance from the origin to the line passing through the point $(3, 1, 5)$ and having the direction $\mathbf{v} = 2\mathbf{i} - \mathbf{j} + \mathbf{k}$.
109. Find an equation for the line passing through $(-1, 2, 4)$ and orthogonal to both L_1: $(x - 1)/4 = (y + 6)/3 = z/(-2)$ and L_2: $(x + 3)/5 = y - 1 = (z + 3)/4$.
110. Find an equation for the line passing through $(-1, 2, 4)$ and parallel to the intersection of the planes Π_1: $-x + y + z = 3$ and Π_2: $-4x + 2y - 7z = 5$.
111. Do calculations to determine whether or not the points $(1, 3, 0)$, $(3, -1, -2)$, and $(-1, 7, 2)$ are collinear.
112. Calculate the area of the triangle with vertices $(2, 1, 3)$, $(-4, 1, 7)$, and $(-1, -1, 3)$.
113. Calculate the area of a parallelogram with adjacent vertices $(1, 4, -2)$, $(-3, 1, 6)$, and $(1, -2, 3)$.
114. Calculate the volume of the parallelepiped determined by the vectors $\mathbf{i} + \mathbf{j}$, $2\mathbf{i} - 3\mathbf{k}$, and $2\mathbf{j} + 7\mathbf{k}$.

In Problems 115–120, identify the quadric surface satisfying the given equation; then sketch its graph.

115. $x^2 + y^2 = 9$
116. $4x^2 + y^2 + 4z^2 - 8y = 0$

CONTINUED

117. $x^2 - \dfrac{y^2}{4} + \dfrac{z^2}{9} = 1$

118. $x^2 - \dfrac{y^2}{4} - \dfrac{z^2}{9} = 1$

119. $-9x^2 + 16y^2 - 9z^2 = 25$

120. $x = \dfrac{y^2}{4} - \dfrac{z^2}{9}$

■ **SHOW/PROVE/DISPROVE**

121. Show that the lines $L_1: x = 3 - 2t$, $y = 4 + t$, $z = -2 + 7t$ and $L_2: x = -3 + s$, $y = 2 - 4s$, $z = 1 + 6s$ have no points of intersection.

122. Show that the position vectors $\mathbf{u} = \mathbf{i} - 2\mathbf{j} + \mathbf{k}$, $\mathbf{v} = 3\mathbf{i} + 2\mathbf{j} - 3\mathbf{k}$, and $\mathbf{w} = 9\mathbf{i} - 2\mathbf{j} - 3\mathbf{k}$ are coplanar. Then find an equation for the plane containing them.

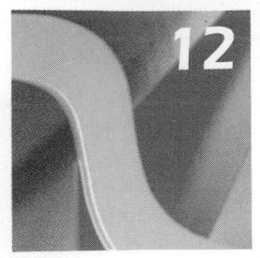

12

VECTOR FUNCTIONS, VECTOR DIFFERENTIATION, AND PARAMETRIC EQUATIONS

12.1 VECTOR FUNCTIONS AND PARAMETRIC EQUATIONS

In Chapter 11, we considered vectors that could be written as

$$\mathbf{v} = (a, b) = a\mathbf{i} + b\mathbf{j} \text{ in } \mathbb{R}^2 \text{ or}$$

$$\mathbf{v} = (a, b, c) = a\mathbf{i} + b\mathbf{j} + c\mathbf{k} \text{ in } \mathbb{R}^3. \qquad (1)$$

In this chapter, we see what happens when the numbers a, b, and c are replaced by functions $f_1(t)$, $f_2(t)$, and $f_3(t)$.

DEFINITION   **VECTOR FUNCTION IN $\mathbb{R}^2$**

Let f_1 and f_2 be functions of the real variable t. Then for all values of t for which $f_1(t)$ and $f_2(t)$ are defined, we define the **vector-valued function f** by

$$\mathbf{f}(t) = (f_1(t), f_2(t)) = f_1(t)\mathbf{i} + f_2(t)\mathbf{j}. \quad \blacksquare \qquad (2)$$

The **domain** of **f** is the intersection of the domains of f_1 and f_2. It is a set of real numbers. The **range** of **f** is a set of vectors in $\mathbb{R}^2$.

REMARK: For simplicity, we will refer to vector-valued functions as **vector functions**.

EXAMPLE 1
FINDING THE DOMAIN OF A VECTOR-VALUED FUNCTION

Let $\mathbf{f}(t) = f_1(t)\mathbf{i} + f_2(t)\mathbf{j} = (1/t)\mathbf{i} + \sqrt{t + 1}\,\mathbf{j}$. Find the domain of **f**.

SOLUTION: The domain of **f** is the set of all t for which f_1 and f_2 are defined. Since $f_1(t)$ is defined for $t \neq 0$ and $f_2(t)$ is defined for $t \geq -1$, we see that the domain of **f** is the set $\{t: t \geq -1 \text{ and } t \neq 0\}$.

DEFINITION   **PLANE CURVES AND PARAMETRIC EQUATIONS**

Let **f** be a vector function. Then for each t in the domain of **f**, the endpoint of the vector $f_1(t)\mathbf{i} + f_2(t)\mathbf{j}$ is a point (x, y) in the xy-plane, where

$$x = f_1(t) \qquad \text{and} \qquad y = f_2(t). \qquad (3)$$

Suppose that the interval $[a, b]$ is in the domain of the function **f** and that both f_1 and f_2 are continuous in $[a, b]$. Then the set of points $(f_1(t), f_2(t))$ for $a \leq t \leq b$ is called a **plane curve** C. Equation (2) is called the **vector equation** of C, while equations (3) are called the **parametric equations** or **parametric representation** of C. In this context, the variable t is called a **parameter**. $\blacksquare$

REMARK: Any plane curve can be thought of as the range of a vector function whose domain is restricted to the interval $[a, b]$.

EXAMPLE 2
PARAMETRIC EQUATIONS OF THE
UNIT CIRCLE

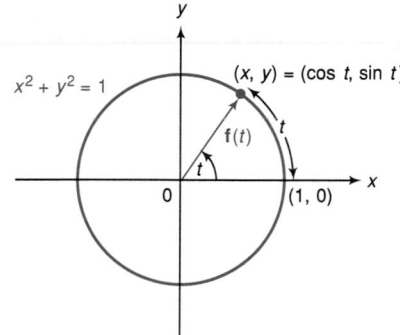

FIGURE 1
THE UNIT CIRCLE

Describe the curve given by the vector equation

$$\mathbf{f}(t) = (\cos t)\mathbf{i} + (\sin t)\mathbf{j}, \qquad 0 \le t \le 2\pi. \tag{4}$$

SOLUTION: We first see that for every t, $|\mathbf{f}(t)| = 1$ since $|\mathbf{f}(t)| = \sqrt{\cos^2 t + \sin^2 t} = 1$. Moreover, if we write the curve in its parametric representation, we find that

$$x = \cos t, \qquad y = \sin t, \tag{5}$$

and since $\cos^2 t + \sin^2 t = 1$, we have

$$x^2 + y^2 = 1,$$

which is the equation of the unit circle. This curve is sketched in Figure 1. Note that in the sketch the parameter t represents both the length of the arc from $(1, 0)$ to the endpoint of the vector and the angle (measured in radians) the vector makes with the positive x-axis. The representation $x^2 + y^2 = 1$ is called the *Cartesian equation* of the curve given by (5).

REMARK: As t increases from 0 to 2π, we move around the unit circle in the counterclockwise direction. If t were instead restricted to the range $0 \le t \le \pi$, then we would not get the entire circle. Rather, we would stop at the point $(\cos \pi, \sin \pi) = (-1, 0)$, which would give us the upper semicircle only.

On the other hand, suppose that $0 \le t \le 4\pi$. Starting at $(1, 0)$ when $t = 0$, we get back to $(1, 0)$ when $t = 2\pi$. Then, because $\cos t$ and $\sin t$ are periodic of period 2π, we simply go around the circle again. That is, the curve given by $\mathbf{f}(t) = (\cos t)\mathbf{i} + (\sin t)\mathbf{j}$, $0 \le t \le 4\pi$ is the unit circle. It appears as in Figure 1 but, in this case, it is traversed *twice*. Finally, the curve given by $\mathbf{f}(t) = (\cos t)\mathbf{i} + (\sin t)\mathbf{j}$, $0 \le t \le 2n\pi$ is the unit circle traversed n times.

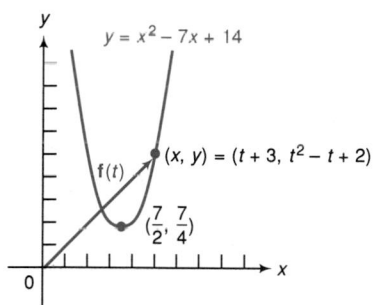

FIGURE 2
GRAPH OF THE PARABOLA $x = t + 3$,
$y = t^2 - t + 2$

DEFINITION **CARTESIAN EQUATION OF A PLANE CURVE**

A **Cartesian equation** of the curve $\mathbf{f}(t) = x(t)\mathbf{i} + y(t)\mathbf{j}$ is an equation relating the variables x and y only. ∎

EXAMPLE 3
OBTAINING A CARTESIAN
EQUATION FROM THE
PARAMETRIC EQUATIONS OF
A CURVE

Describe and sketch the curve given parametrically by $x = t + 3$, $y = t^2 - t + 2$.

SOLUTION: With problems of this type, the easiest thing to do is to write t as a function of x or y, if possible. Since $x = t + 3$, we immediately see that $t = x - 3$ and $y = t^2 - t + 2 = (x - 3)^2 - (x - 3) + 2 = x^2 - 7x + 14$. This is the Cartesian equation of the curve and is the equation of a parabola. It is sketched in Figure 2. Note that in the Cartesian equation of the parabola, the parameter t does not appear.

EXAMPLE 4
OBTAINING A CARTESIAN
EQUATION FROM THE
PARAMETRIC EQUATIONS OF
A CURVE

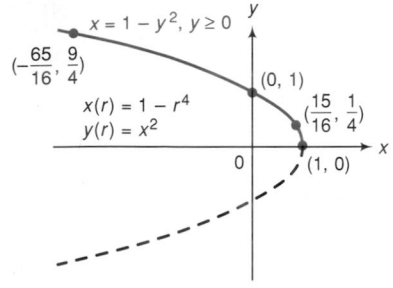

FIGURE 3
UPPER HALF OF PARABOLA $x = 1 - y^2$

Describe and sketch the curve given by the vector equation

$$\mathbf{f}(r) = (1 - r^4)\mathbf{i} + r^2\mathbf{j}. \tag{6}$$

SOLUTION: First, we note that the parameter in this problem is r instead of t. This makes absolutely no difference since, as in the case of the variable of integration, the parameter is a "dummy" variable. Now to get a feeling for the shape of this curve, we display in Table 1 values of x and y for various values of r. Plotting some of these points leads to the sketch in Figure 3. To write the Cartesian equation for this curve, we square both sides of the equation $y = r^2$ to obtain $y^2 = r^4$ and $x = 1 - r^4 = 1 - y^2$, which is the equation of the parabola sketched in Figure 3. Note that this curve is *not* the graph of the parabola $x = 1 - y^2$ since the parametric representation $y = r^2$ requires that y be nonnegative. Thus, the curve described by (6) is only the *upper half* of the parabola described by the equation $x = 1 - y^2$.

TABLE 1

r	0	$\pm\frac{1}{2}$	± 1	$\pm\frac{3}{2}$	± 2
$x = 1 - r^4$	1	$\frac{15}{16}$	0	$-\frac{65}{16}$	-15
$y = r^2$	0	$\frac{1}{4}$	1	$\frac{9}{4}$	4

Having seen how vectors in $\mathbb{R}^2$ generalize to vectors in $\mathbb{R}^3$, we can imagine how properties of vector functions are extended to $\mathbb{R}^3$.

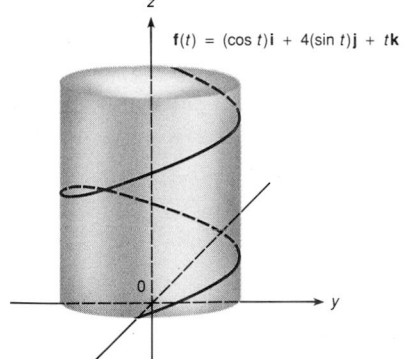

FIGURE 4
ELLIPTICAL HELIX

DEFINITION **VECTOR FUNCTION IN $\mathbb{R}^3$ AND CURVE IN SPACE**

Let f_1, f_2, and f_3 be functions of the real variable t. Then for all values of t for which $f_1(t)$, $f_2(t)$, and $f_3(t)$ are defined, we define the **vector-valued function f of a real variable** t by

$$\mathbf{f}(t) = (f_1(t), f_2(t), f_3(t)) = f_1(t)\mathbf{i} + f_2(t)\mathbf{j} + f_3(t)\mathbf{k}. \tag{7}$$

If f_1, f_2, and f_3 are continuous over an interval I, then as t varies over I, the set of points traced out by the end of the vector **f** is called a **curve in space**. ∎

EXAMPLE 5
AN ELLIPTICAL HELIX

Sketch the curve $\mathbf{f}(t) = (\cos t)\mathbf{i} + 4(\sin t)\mathbf{j} + t\mathbf{k}$.

SOLUTION: Here $x = \cos t$ and $y = 4 \sin t$, so eliminating t, we obtain $x^2 + (y^2/16) = 1$, which is the equation of an ellipse in the xy-plane. Since $z = t$ increases as t increases, the curve is a spiral that climbs up the side of an elliptical cylinder, as sketched in Figure 4. This curve is called an **elliptical helix**.

■ PROBLEMS 12.1

■ SELF-QUIZ

I. The graph of $x = 1 + t$, $y = 3 - 2t$ is the same as the graph of _____.
 a. $x + y = 4$
 b. $2x + y = 5$
 c. $y = -2x + 1$ and $x \geq 0$
 d. $|2x| + |y| = 5$

II. The graph of $x = 1 + t^2$, $y = 3 + t^2$ is the same as the graph of _____.
 a. $y = x + 2$
 b. $x = y + 2$
 c. $y = x + 2$ and $x \geq 0$
 d. $y = x + 2$ and $x \geq 1$

III. The graph of $x = \cos t$, $y = -\cos t$ is the same as the graph of _____.
 a. $x = -y$
 b. $x = -y$ and $x \geq 0$
 c. $y = -x$ and $x \leq 1$
 d. $y = -x$ and $-1 \leq x \leq 1$

IV. The unit circle $x^2 + y^2 = 1$ can be parameterized by _____.
 a. $x = \sin t$, $y = \cos t$
 b. $x = \cos(t - \pi/2)$, $y = \sin(t - \pi/2)$
 c. $x = -\cos t$, $y = \sin t$
 d. $x = \cos 2t$, $y = \sin 2t$

■ DRILL

In Problems 1–8, find the domain of the given vector-valued function.

1. $\mathbf{f}(t) = \dfrac{1}{t}\mathbf{i} + \dfrac{1}{t - 1}\mathbf{j}$ 2. $\mathbf{f}(t) = \sqrt{t}\,\mathbf{i} + \dfrac{1}{t}\mathbf{j}$

3. $\mathbf{f}(s) = \dfrac{1}{s^2 - 1}\mathbf{i} + (s^2 - 1)\mathbf{j}$

4. $\mathbf{f}(s) = e^{1/s}\mathbf{i} + e^{-1/(s+1)}\mathbf{j}$

5. $\mathbf{f}(r) = (\ln r)\mathbf{i} + \ln(1 - r)\mathbf{j} + \ln(1 + r)\mathbf{k}$

6. $\mathbf{f}(r) = (\sin r)\mathbf{i} + r\mathbf{j} - (1 + r^2)\mathbf{k}$

7. $\mathbf{f}(r) = (\sec t)\mathbf{i} + (\csc t)\mathbf{j} + (\cos 2t)\mathbf{k}$

8. $\mathbf{f}(t) = (\tan t)\mathbf{i} + (\tan 2t)\mathbf{j} + (\cot t)\mathbf{k}$

In Problems 9–28, find the Cartesian equation for each curve; then sketch the curve (in $\mathbb{R}^2$ or $\mathbb{R}^3$).

9. $\mathbf{f}(t) = t^2\mathbf{i} + 2t\mathbf{j}$

10. $\mathbf{f}(t) = (2t - 3)\mathbf{i} + t^2\mathbf{j}$ 11. $\mathbf{f}(t) = t^2\mathbf{i} + t^3\mathbf{j}$

12. $\mathbf{f}(t) = 3(\sin t)\mathbf{i} + 3(\cos t)\mathbf{j}$

13. $\mathbf{f}(t) = (2t - 1)\mathbf{i} + (4t + 3)\mathbf{j}$

14. $\mathbf{f}(t) = 2(\cosh t)\mathbf{i} + 2(\sinh t)\mathbf{j}$

15. $\mathbf{f}(t) = (t^4 + t^2 + 1)\mathbf{i} + t^2\mathbf{j}$

16. $\mathbf{f}(t) = t^2\mathbf{i} + t^8\mathbf{j}$ 17. $\mathbf{f}(t) = t^3\mathbf{i} + (t^9 - 1)\mathbf{j}$

18. $\mathbf{f}(t) = t\mathbf{i} + e^t\mathbf{j}$ 19. $\mathbf{f}(t) = e^t\mathbf{i} + t^2\mathbf{j}$

20. $\mathbf{f}(t) = (t^2 + t - 3)\mathbf{i} + \sqrt{t}\,\mathbf{j}$

*21. $\mathbf{f}(t) = e^t(\sin t)\mathbf{i} + e^t(\cos t)\mathbf{j}$
 [*Hint:* Show that $|\mathbf{f}(t)| = e^t$.]

22. $\mathbf{f}(t) = \mathbf{i} + (\tan t)\mathbf{j}$ 23. $\mathbf{f}(t) = e^t\mathbf{i} + e^{2t}\mathbf{j}$

*24. $\mathbf{f}(t) = \dfrac{6t}{1 + t^3}\mathbf{i} + \dfrac{6t^2}{1 + t^3}\mathbf{j}$

25. $\mathbf{g}(t) = 2t\mathbf{i} - 3t\mathbf{j} + t\mathbf{k}$

26. $\mathbf{g}(t) = (1 - t)\mathbf{i} + (3 + 5t)\mathbf{j} + 3t\mathbf{k}$

27. $\mathbf{g}(t) = t\mathbf{i} + (\cos t)\mathbf{j} + (\sin t)\mathbf{k}$

28. $\mathbf{g}(t) = (\cos t)\mathbf{i} - (\sin t)\mathbf{j} + (\cos t)\mathbf{k}$

In Problems 29–38, use the result of Problem 49 to find a parametric representation of the straight line that passes through the two given points.

29. $(2, 4)$; $(1, 6)$ 30. $(-3, 2)$; $(0, 4)$

31. $(3, 5)$; $(-1, -7)$ 32. $(4, 6)$; $(7, 9)$

33. $(-2, 3)$; $(4, 7)$ 34. $(-4, 0)$; $(3, -2)$

35. $(1, 3, 5)$; $(2, 4, 6)$ 36. $(1, 3, 5)$; $(-2, 4, -6)$

37. $(-3, 0, 7)$; $(-5, 3, 0)$ 38. $(7, -3, 0)$; $(5, 0, 3)$

■ APPLICATIONS

39. Verify that a vector equation for the ellipse $(x^2/a^2) + (y^2/b^2) = 1$ is

 $$\mathbf{f}(t) = a(\cos t)\mathbf{i} + b(\sin t)\mathbf{j}.$$

40. Verify that a parametric representation for the hyperbola $(x^2/a^2) - (y^2/b^2) = 1$ is

 $$x(\theta) = a \sec \theta \quad \text{and} \quad y(\theta) = b \tan \theta.$$

*41. Suppose that a wheel of radius r is rolling in a straight line without slipping. Let P be a fixed point on the wheel which is distance s from the center. As the wheel rotates, the point P traces out a curve known as a **trochoid**.† Look at Figure 5. Suppose that P moves through an angle of α radians to reach its position shown in part (b) of the figure.

---
† From the Greek word *trochos*, meaning *wheel*.

CONTINUED

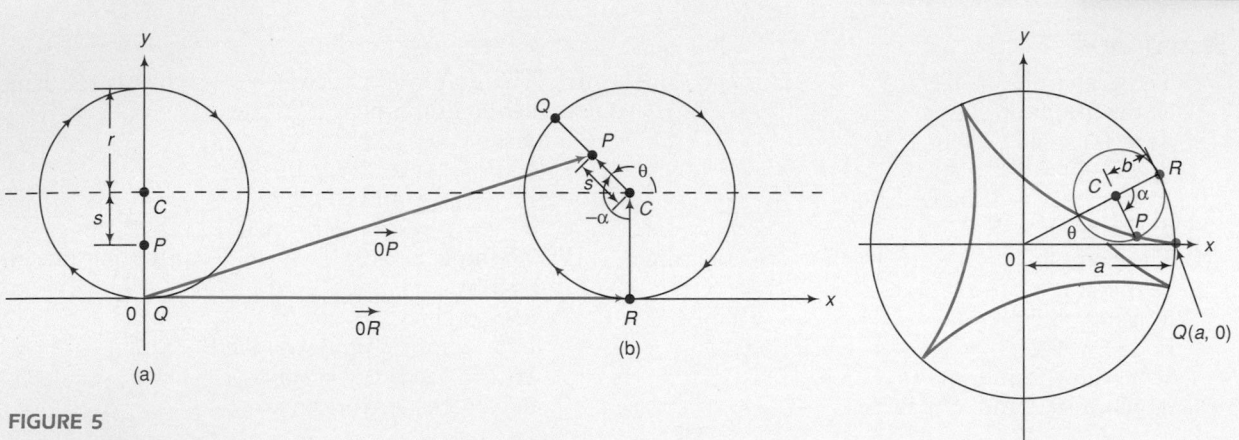

FIGURE 5

FIGURE 6

a. Show that the new position vector is
$$\overrightarrow{OP} = \overrightarrow{OR} + \overrightarrow{RC} + \overrightarrow{CP}.$$
b. Show that $\overrightarrow{OR} = \alpha r \mathbf{i}$.
c. Show that $\overrightarrow{CP} = s[(\cos \theta)\mathbf{i} + (\sin \theta)\mathbf{j}]$.
d. Show that
$$\overrightarrow{OP} = [\alpha r + s \cos \theta]\mathbf{i} + [r + s \sin \theta]\mathbf{j}.$$
e. Show that $\theta = (3\pi/2) - \alpha$.
f. Show that $\cos \theta = -\sin \alpha$ and $\sin \theta = -\cos \alpha$.
g. Show that the trochoid satisfies the parametric equations

$$x = r\alpha - s \sin \alpha \quad \text{and} \quad y = r - s \cos \alpha. \qquad \text{(8)}$$

42. A point is located 25 cm from the center of a wheel 1 m in diameter. Find a parametric representation of the trochoid traced by that point as the wheel rolls.

*43. A trochoid generated by a point P on the circumference of the wheel (i.e., $s = r$) has the special name **cycloid**.[†]
   a. Find parametric equations for the cycloid.
   b. Verify that the portion of the cycloid for $0 \le x \le 2r$ has the Cartesian equation

$$x = r \cos^{-1}\left(\frac{r-y}{r}\right) - \sqrt{2ry - y^2}.$$

44. A point is located on the circumference of a wheel 1 m in diameter. Find a parametric representa-

tion for the cycloid traced by that point as the wheel rolls.

*45. A **hypocycloid**[‡] is a curve generated by the motion of a point P on the circumference of a circle that rolls internally (without slipping) on a larger circle (see Figure 6). Assume that the radius of the larger circle is a while that of the smaller circle is b. Let θ be the angle indicated in Figure 6. Verify that the hypocycloid has parametric representation

$$x = (a - b) \cos \theta + b \cos\left[\left(\frac{a-b}{b}\right)\theta\right],$$

$$y = (a - b) \sin \theta - b \sin\left[\left(\frac{a-b}{b}\right)\theta\right].$$

[*Hint:* First show that $a\theta = b\alpha$.]

46. If $a = 4b$ in Figure 6, then the curve generated is called a **hypocycloid of four cusps**. Sketch this curve.

*47. Verify that the hypocycloid of four cusps has the Cartesian equation

$$x^{2/3} + y^{2/3} = a^{2/3}.$$

[*Hint:* Use the identities $\cos 3\theta = 4 \cos^3 \theta - 3 \cos \theta$ and $\sin 3\theta = 3 \sin \theta - 4 \sin^3 \theta$ together with the facts that $a - b = 3b$ and $(a - b)/b = 3$.]

48. Calculate the area bounded by the hypocycloid of four cusps.

[†] From the Greek words *kyklos*, meaning *circle*. The cycloid was a source of great controversy in the seventeenth century. Many of its properties were discovered by the French mathematician Gilles Personne de Roberval (1602–1675), although the curve was first discussed by Galileo. Unfortunately, Roberval, for unknown reasons, did not publish discoveries concerning the cycloid, which meant that he lost credit for most of them. The ensuing arguments over who discovered what were so bitter that the cycloid became known as the "Helen of geometers" (after Helen of Troy—the source of the intense jealousy that led to the Trojan War).

[‡] The *hypo* comes from the Greek *hupo*, meaning *under*.

■ SHOW/PROVE/DISPROVE

CONTINUED

49. Figure 7 shows three points P, Q, R on a straight line.
   a. Show that $\overrightarrow{PR} = t\overrightarrow{PQ}$ for some real number t.
   b. Show that $\overrightarrow{OR} = \overrightarrow{OP} + \overrightarrow{PR} = \overrightarrow{OP} + t\overrightarrow{PQ}$.
   c. Use the results of parts (a) and (b) to show that the line passing through the points (x_1, y_1) and (x_2, y_2) satisfies the parametric equations

   $$x = x_1 + t(x_2 - x_1) \quad \text{and}$$
   $$y = y_1 + t(y_2 - y_1).$$

50. Use the substitution $\tan \theta = \sin t$ to show that the lemniscate $r^2 = \cos 2\theta$ has the Cartesian parametric representation

   $$x = \frac{\cos t}{1 + \sin^2 t} \quad \text{and} \quad y = \frac{\sin t \cos t}{1 + \sin^2 t}.$$

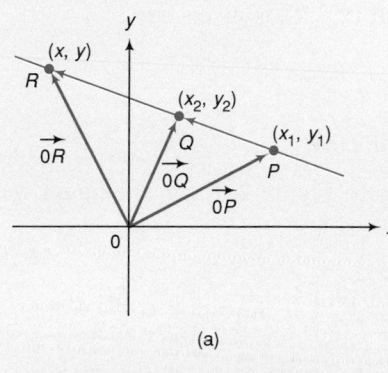

(a)

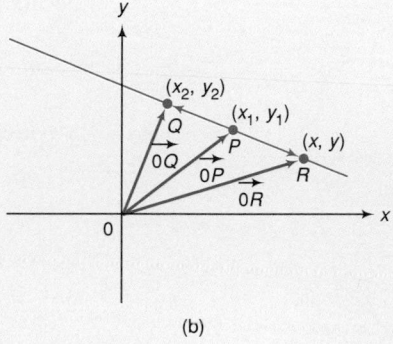

(b)

FIGURE 7

12.2 THE EQUATION OF THE TANGENT LINE TO A PLANE CURVE AND SMOOTHNESS

Suppose that $x = f_1(t)$ and $y = f_2(t)$ is the parametric representation of a curve C. We would like to be able to calculate the equation of the line tangent to the curve without having to determine the Cartesian equation of the curve. However, there are complications that might occur since the curve could intersect itself. This happens if there are two numbers $t_1 \neq t_2$ such that $f_1(t_1) = f_1(t_2)$ and $f_2(t_1) = f_2(t_2)$. Thus, there are three possibilities. At a given point the curve could have the following:

(i) a unique tangent

(ii) no tangent

(iii) two or more tangents.

In Figure 1 we sketch three plane curves.

A condition that ensures that there is at least one tangent line at each point is that

$$f_1' \text{ and } f_2' \text{ exist and } [f_1'(t)]^2 + [f_2'(t)]^2 \neq 0. \tag{1}$$

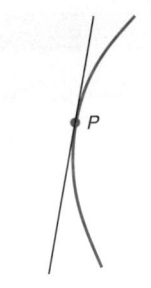

(a) Unique tangent at P.

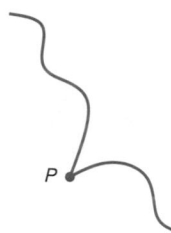

(b) No tangent at P.

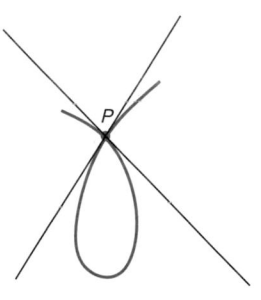

(c) Two tangents at P.

FIGURE 1
THREE PLANE CURVES

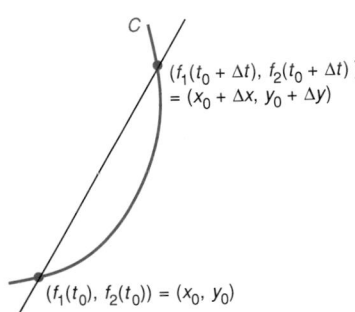

FIGURE 2
THE SECANT LINE JOINS
(x_0, y_0) AND $(x_0 + \Delta x, y_0 + \Delta y)$.

REMARK: Condition (1) simply states that the derivatives of f_1 and f_2 are not zero at the same value of t.

DEFINITION SMOOTH CURVE

The parametric curve $\mathbf{f}(t) = f_1(t)\mathbf{i} + f_2(t)\mathbf{j}$ is said to be **smooth** on an interval $I = (a, b)$ if $f'_1(t)$ and $f'_2(t)$ exist and are continuous on I and, for every t in I, $[f'_1(t)]^2 + [f'_2(t)]^2 \neq 0$. ∎

THEOREM 1

Let (x_0, y_0) be on the smooth curve C given by $x = f_1(t)$ and $y = f_2(t)$. If the curve passes through (x_0, y_0) when $t = t_0$,[†] then the slope m of the line tangent to C at (x_0, y_0) is given by

$$m = \lim_{t \to t_0} \frac{f'_2(t)}{f'_1(t)}, \tag{2}$$

provided that this limit exists.

PROOF: We refer to Figure 2. From the figure we see that

$$m = \lim_{\Delta t \to 0} \frac{f_2(t_0 + \Delta t) - f_2(t_0)}{f_1(t_0 + \Delta t) - f_1(t_0)} \overset{\text{L'Hôpital's rule}}{=} \lim_{\Delta t \to 0} \frac{f'_2(t_0 + \Delta t)}{f'_1(t_0 + \Delta t)} \overset{\text{let } t = t_0 + \Delta t}{=} \lim_{t \to t_0} \frac{f'_2(t)}{f'_1(t)}. \quad ∎$$

Suppose now that $f'_1(t_0) \neq 0$. Then, by Theorem 6.1.4 on page 391, f_1 has an inverse that is differentiable in a neighborhood of $x_0 = f_1(t_0)$ and, in this neighborhood, we may write $t = f_1^{-1}(x)$. Then $y = f_2(t) = f_2(f_1^{-1}(x))$ in this neighborhood of x_0. That is, we may write y as a function of x.

COROLLARY TO THEOREM 1

If $f'_1(t_0) \neq 0$, then the smooth curve C has a representation $y = f(x) = f_2(f_1^{-1}(x))$ in a neighborhood of $x_0 = f_1(t_0)$. Moreover, the derivative $f'(x_0)$ exists and is given by

$$\frac{dy}{dx} = f'(x_0) = \frac{f'_2(t_0)}{f'_1(t_0)}. \tag{3}$$

NOTE: This result is often written in the symbolic form

$$\frac{dy}{dx} = \frac{dy/dt}{dx/dt}.$$

PROOF: By the quotient rule for limits and the continuity of f'_1 and f'_2, we have

$$\lim_{t \to t_0} \frac{f'_2(t)}{f'_1(t)} = \frac{\lim_{t \to t_0} f'_2(t)}{\lim_{t \to t_0} f'_1(t)} = \frac{f'_2(t_0)}{f'_1(t_0)},$$

[†] The curve may pass through the point (x_0, y_0) for other values of t as well, and therefore it may have other tangent lines at that point, as in Figure 1(c).

so, by Theorem 1, $\dfrac{f_2'(t_0)}{f_1'(t_0)}$ is the slope of the line tangent to the curve at $t = t_0$. This result also follows from the chain rule and Theorem 6.1.4:

$$y(x_0) = f_2(f_1^{-1}(x_0))$$

so

$$\underset{\text{chain rule}}{\downarrow} \qquad \qquad \underset{\text{equation (6.1.4) on p. 391}}{\downarrow}$$

$$\dfrac{dy}{dx}\bigg|_{x=x_0} = f_2'(f_1^{-1}(x_0))[f_1^{-1}(x_0)]' = \dfrac{f_2'(t_0)}{f_1'(t_0)}$$

since $f_1^{-1}(x_0) = t_0$. ∎

Using the corollary, we find that the tangent line to C at t_0 is given by

$$\dfrac{y - y_0}{x - x_0} = \dfrac{f_2'(t_0)}{f_1'(t_0)},$$

or after simplification,

EQUATION OF THE TANGENT LINE TO A SMOOTH CURVE AT
$(x_0, y_0) = (f_1(t_0), f_2(t_0))$

$$f_2'(t_0)(x - x_0) - f_1'(t_0)(y - y_0) = 0. \tag{4}$$

If f is smooth, then $f_1'(t)$ and $f_2'(t)$ are never zero at the same time. If a curve has a derivative of zero at a point, then the tangent is horizontal at that point. If the derivative approaches infinity as we approach a point, then the tangent line is vertical at that point. The following facts follow from (3).

VERTICAL AND HORIZONTAL TANGENTS

(i) C has a vertical tangent line at t_0 if

$$\dfrac{dx}{dt}\bigg|_{t=t_0} = f_1'(t_0) = 0 \qquad \text{and} \qquad f_2'(t_0) \neq 0.$$

(ii) C has a horizontal tangent line at t_0 if

$$\dfrac{dy}{dt}\bigg|_{t=t_0} = f_2'(t_0) = 0 \qquad \text{and} \qquad f_1'(t_0) \neq 0.$$

EXAMPLE 1
FINDING THE TANGENT LINE TO A CURVE

Find the equation of the line tangent to the curve

$$x = 2t^3 - 15t^2 + 24t + 7, \qquad y = t^2 + t + 1 \qquad \text{at } t = 2.$$

SOLUTION: Here $f_1'(2) = (6t^2 - 30t + 24)|_{t=2} = -12$ and $f_2'(2) = (2t + 1)|_{t=2} = 5$. When $t = 2$, $x = 11$ and $y = 7$. Then using (4), we obtain

$$5(x - 11) + 12(y - 7) = 0, \qquad \text{or} \qquad 5x + 12y = 139.$$

EXAMPLE 2
FINDING HORIZONTAL AND
VERTICAL TANGENTS

Find all horizontal and vertical tangents to the curve of Example 1.

SOLUTION: There are vertical tangents when

$$f'_1(t) = 6t^2 - 30t + 24 = 0 = 6(t^2 - 5t + 4) = 6(t - 4)(t - 1),$$

or when $t = 1$ and $t = 4$, since $f'_2(t)$ is nonzero at these points. When $t = 1$, $x = 18$, and when $t = 4$, $x = -9$. Thus, the vertical tangents are the lines $x = 18$ and $x = -9$. There is a horizontal tangent when $2t + 1 = 0$, or $t = -\frac{1}{2}$. When $t = -\frac{1}{2}$, $y = \frac{3}{4}$, and the line $y = \frac{3}{4}$ is a horizontal tangent.

■ PROBLEMS 12.2

■ SELF-QUIZ

I. The smooth curve given by $x = 3\cos t$, $y = 4\sin t$ has a horizontal tangent when $t = $ _____.
   a. $\pi/4$ b. $\pi/2$
   c. $3\pi/4$ d. π

II. The smooth curve given by $x = 4\cos t$, $y = 3\sin t$ has a vertical tangent when $t = $ _____.
   a. $\pi/4$ b. $\pi/2$
   c. $3\pi/4$ d. π

III. The smooth curve given by $x = \cos 2t$, $y = \sin 2t$ has a horizontal tangent when $t = $ _____.
   a. $\pi/4$ b. $\pi/2$
   c. $3\pi/4$ d. π

IV. The smooth curve given by $x = 3\cos 2t$, $y = 4\sin 2t$ has a vertical tangent when $t = $ _____.
   a. $\pi/4$ b. $\pi/2$
   c. $3\pi/4$ d. π

■ DRILL

In Problems 1–10, find the slope of the line tangent to the given curve for the specified value of the parameter.

1. $x = t^3$; $y = t^4 - 5$; $t = -1$
2. $x = t^2$; $y = \sqrt{1 - t}$; $t = 1/2$
3. $x = \cos \theta$; $y = \sin \theta$; $\theta = \pi/4$
4. $x = \tan \theta$; $y = \sec \theta$; $\theta = \pi/4$
5. $x = \sec \theta$; $y = \tan \theta$; $\theta = \pi/4$
6. $x = \cosh t$; $y = \sinh t$; $t = 0$
7. $x = 8\cos \theta$; $y = -3\sin \theta$; $\theta = 2\pi/3$
8. $x = \cos^3 \theta$; $y = \sin^3 \theta$; $\theta = \pi/2$
9. $x = \theta$; $y = 1/\theta$; $\theta = \pi/4$
10. $x = t\cosh t$; $y = (\tanh t)/(1 + t)$; $t = 0$

In Problems 11–16, find an equation for the line tangent to the given curve for the specified value of the parameter.

11. $x = t^2 - 2$; $y = 4t$; $t = 3$
12. $x = t + 4$; $y = t^3 - t + 4$; $t = 2$

13. $x = e^{2t}$; $y = e^{-2t}$; $t = 1$
14. $x = \cos 2\theta$; $y = \sin 2\theta$; $\theta = \pi/4$
15. $x = \sqrt{1 - \sin \theta}$; $y = \sqrt{1 + \cos \theta}$; $\theta = 0$
16. $x = \cos^3 \theta$; $y = \sin^3 \theta$; $\theta = \pi/6$

In Problems 17–26, find all points [in the form (x, y)] at which the given curve has a vertical or horizontal tangent.

17. $x = t^2 - 1$; $y = 4 - t^2$
18. $x = 1/\sqrt{1 - t^2}$; $y = \sqrt{1 - t^2}$
19. $x = \sinh t$; $y = \cosh t$
20. $x = \sin \theta + \cos \theta$; $y = \sin \theta - \cos \theta$
21. $x = 2\cos \theta$; $y = 3\sin \theta$
22. $x = 2 + 7\cos \theta$; $y = 8 + 3\sin \theta$
**23. $x = \ln(1 + t^2)$; $y = \ln(1 + t^3)$
    [*Hint:* Use l'Hôpital's rule.]
24. $x = e^{3t}$; $y = e^{-5t}$
*25. $x = \sin 3\theta$; $y = \cos 5\theta$
*26. $x = \theta \sin \theta$; $y = \theta \cos \theta$

CONTINUED

■ APPLICATIONS

In Problems 27–34, use the result of Problem 39 to calculate the slope of the tangent line to the given curve for the specified value of θ.

27. $r = 5 \sin \theta$; $\theta = \pi/6$
28. $r = 5 \cos \theta + 5 \sin \theta$; $\theta = \pi/4$
29. $r = -4 + 2 \cos \theta$; $\theta = \pi/3$
30. $r = 4 + 3 \sin \theta$; $\theta = 2\pi/3$
31. $r = 3 \sin 2\theta$; $\theta = \pi/6$
32. $r = 5 \sin 3\theta$; $\theta = \pi/4$
33. $r = e^{\theta/2}$; $\theta = 0$
34. $r^2 = \cos 2\theta$; $\theta = \pi/6$

35. Find an equation for each of the two tangents to the curve $x = t^3 - 2t^2 - 3t + 11$ and $y =$ $t^2 - 2t - 5$ at the point $(11, -2)$. [*Hint:* First find the relevant values of the parameter by solving the equation $y = t^2 - 2t - 5 = -2$.]

*36. Write an equation for each of the three lines that are tangent to the curve $x = t^3 - 2t^2 - t + 3$ and $y = -t^2 + 2t - 2/t$ at the point $(1, -1)$.

37. Calculate the slope of the tangent to the cycloid given in Problem 12.1.43, and then show that the tangent is vertical when α is a multiple of 2π.

38. For what values of α does the trochoid given by equations (12.1.8) (in Problem 12.1.48) have a vertical tangent?

■ SHOW/PROVE/DISPROVE

39. Suppose a curve is given in polar coordinates by the equation $r = f(\theta)$. Show that the curve can then be given parametrically by

$$x = f(\theta) \cos \theta \quad \text{and} \quad y = f(\theta) \sin \theta.$$

■ ANSWERS TO SELF-QUIZ

I. b   II. d   III. a, c   IV. b, d

12.3 THE DIFFERENTIATION AND INTEGRATION OF A VECTOR FUNCTION

In Section 12.2, we showed how the derivative dy/dx could be calculated when x and y were given parametrically in terms of t. In this section, we will show how to calculate the derivative of a vector function. The definitions of limit, continuity, and differentiability are virtually the same in $\mathbb{R}^2$ and $\mathbb{R}^3$. For that reason, we give definitions of these basic concepts in terms of vectors in the plane. We begin with the definition of a limit.

DEFINITION **LIMIT OF A VECTOR FUNCTION**

Let $\mathbf{f}(t) = f_1(t)\mathbf{i} + f_2(t)\mathbf{j}$. Let t_0 be any real number, $+\infty$, or $-\infty$. If $\lim_{t \to t_0} f_1(t)$ and $\lim_{t \to t_0} f_2(t)$ both exist, then we define

$$\lim_{t \to t_0} \mathbf{f}(t) = \left[\lim_{t \to t_0} f_1(t) \right] \mathbf{i} + \left[\lim_{t \to t_0} f_2(t) \right] \mathbf{j}. \quad \blacksquare \tag{1}$$

That is, *the limit of a vector function is determined by the limits of its component functions.* Thus, in order to calculate the limit of a vector function, it is only necessary to calculate two ordinary limits.

EXAMPLE 1
COMPUTING A LIMIT

Compute

$$\lim_{t \to 0} \left[\frac{\sin t}{t} \mathbf{i} + 3 \cos t \mathbf{j} \right]$$

SOLUTION: From page 161 we know that $\lim\limits_{t \to 0} \dfrac{\sin t}{t} = 1$. Also,

$$\lim_{t \to 0} 3 \cos t = 3 \lim_{t \to 0} \cos t = 3 \cdot 1 = 3.$$

Thus

$$\lim_{t \to 0} \left[\frac{\sin t}{t} \mathbf{i} + 3 \cos t \mathbf{j} \right] = \left(\lim_{t \to 0} \frac{\sin t}{t} \right) \mathbf{i} + \left(\lim_{t \to 0} 3 \cos t \right) \mathbf{j}$$

$$= 1\mathbf{i} + 3\mathbf{j} = \mathbf{i} + 3\mathbf{j}.$$

DEFINITION   **CONTINUITY OF A VECTOR FUNCTION**

f is **continuous** at t_0 if the component functions f_1 and f_2 are continuous at t_0. Thus, **f** is continuous at t_0 if

$$\lim_{t \to t_0} \mathbf{f}(t) = \mathbf{f}(t_0).$$

This means that the following three conditions hold:

  (i) **f** is defined at t_0,
 (ii) $\lim\limits_{t \to t_0} \mathbf{f}(t)$ exists,
(iii) $\lim\limits_{t \to t_0} \mathbf{f}(t) = \mathbf{f}(t_0)$. ■

DEFINITION   **DERIVATIVE OF A VECTOR FUNCTION**

Let **f** be defined at t. Then **f** is **differentiable** at t if

$$\lim_{\Delta t \to 0} \frac{\mathbf{f}(t + \Delta t) - \mathbf{f}(t)}{\Delta t} \tag{2}$$

exists and is finite. The vector function **f′** defined by

$$\mathbf{f}'(t) = \frac{d\mathbf{f}}{dt} = \lim_{\Delta t \to 0} \frac{\mathbf{f}(t + \Delta t) - \mathbf{f}(t)}{\Delta t} \tag{3}$$

is called the **derivative** of **f**, and the domain of **f′** is the set of all t such that the limit in (2) exists. ■

DEFINITION   **DIFFERENTIABILITY IN AN OPEN INTERVAL**

The vector function **f** is **differentiable** on the open interval I if $\mathbf{f}'(t)$ exists for every t in I. ■

Before giving examples of the calculation of derivatives, we prove a theorem that makes this calculation no more difficult than the calculation of "ordinary" derivatives.

THEOREM 1

If $\mathbf{f}(t) = f_1(t)\mathbf{i} + f_2(t)\mathbf{j}$, then at any value t for which $f'_1(t)$ and $f'_2(t)$ exist,

$$\mathbf{f}'(t) = f'_1(t)\mathbf{i} + f'_2(t)\mathbf{j}. \tag{4}$$

That is, the derivative of a vector function is determined by the derivatives of its component functions.

PROOF:

$$\mathbf{f}'(t) = \lim_{\Delta t \to 0} \frac{\mathbf{f}(t + \Delta t) - \mathbf{f}(t)}{\Delta t}$$

$$= \lim_{\Delta t \to 0} \frac{[f_1(t + \Delta t)\mathbf{i} + f_2(t + \Delta t)\mathbf{j}] - [f_1(t)\mathbf{i} + f_2(t)\mathbf{j}]}{\Delta t}$$

$$= \lim_{\Delta t \to 0} \frac{[f_1(t + \Delta t) - f_1(t)]\mathbf{i} + [f_2(t + \Delta t) - f_2(t)]\mathbf{j}}{\Delta t}$$

$$= \lim_{\Delta t \to 0} \left[\frac{f_1(t + \Delta t) - f_1(t)}{\Delta t} \right]\mathbf{i} + \lim_{\Delta t \to 0} \left[\frac{f_2(t + \Delta t) - f_2(t)}{\Delta t} \right]\mathbf{j}$$

$$= f'_1(t)\mathbf{i} + f'_2(t)\mathbf{j}. \quad \blacksquare$$

EXAMPLE 2
DIFFERENTIATING A VECTOR FUNCTION

Let $\mathbf{f}(t) = (\cos t)\mathbf{i} + e^{2t}\mathbf{j}$. Calculate $\mathbf{f}'(t)$.

SOLUTION:

$$\mathbf{f}'(t) = \frac{d}{dt}(\cos t)\mathbf{i} + \frac{d}{dt}e^{2t}\mathbf{j} = -(\sin t)\mathbf{i} + 2e^{2t}\mathbf{j}$$

Once we know how to calculate the first derivative of $\mathbf{f}$, we can calculate higher derivatives as well.

DEFINITION   **SECOND DERIVATIVE**

If the function $\mathbf{f}'$ is differentiable at t, we define the **second derivative** of $\mathbf{f}$ to be the derivative of $\mathbf{f}'$. That is,

$$\mathbf{f}'' = (\mathbf{f}')'. \quad \blacksquare \tag{5}$$

EXAMPLE 3
FINDING A SECOND DERIVATIVE

Find the second derivative of $\mathbf{f}(t) = (\cos t)\mathbf{i} + (\sin t)\mathbf{j} + t\mathbf{k}$.

SOLUTION: $\mathbf{f}'(t) = -(\sin t)\mathbf{i} + (\cos t)\mathbf{j} + \mathbf{k}$ and
$\mathbf{f}''(t) = -(\cos t)\mathbf{i} - (\sin t)\mathbf{j} + 0\mathbf{k} = -\cos t\mathbf{i} - \sin t\mathbf{j}$.

As with real-valued functions, the following are true:

the (vector-valued) velocity function is the derivative of the position function;

the acceleration function is the derivative of the velocity function and so is the second derivative of the position function.

EXAMPLE 4
PARAMETRIC EQUATIONS OF
MOTION

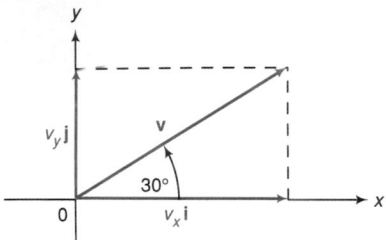

FIGURE 1
RESOLUTION OF A VELOCITY VECTOR

A cannonball shot from a cannon has an initial velocity of 600 m/sec. The muzzle of the cannon is inclined at an angle of 30°. Ignoring air resistance and the rotation of the earth, determine the path of the cannonball.

SOLUTION: If the earth's rotation is ignored, then all motion takes place in a plane perpendicular to the earth. We place the x- and y-axes so that the mouth of the cannon is at the origin (see Figure 1). The velocity vector $\mathbf{v}$ can be resolved into its vertical and horizontal components:

$$\mathbf{v} = v_x\mathbf{i} + v_y\mathbf{j}.$$

A unit vector in the direction of $\mathbf{v}$ is

$$\mathbf{u} = (\cos 30°)\mathbf{i} + (\sin 30°)\mathbf{j} = \frac{\sqrt{3}}{2}\mathbf{i} + \frac{1}{2}\mathbf{j},$$

so initially,

$$\mathbf{v} = |\mathbf{v}|\mathbf{u} = 600\mathbf{u} = 300\sqrt{3}\mathbf{i} + 300\mathbf{j}.$$

The scalar $|\mathbf{v}|$ is called the **speed** of the cannonball. Thus, initially, $v_x = 300\sqrt{3}$ m/sec (the initial speed in the horizontal direction) and $v_y = 300$ m/sec (the initial speed in the vertical direction). Now the vertical acceleration (due to gravity) is

$$a_y = -9.81 \text{ m/sec}^2 \quad \text{and} \quad v_y = \int a_y \, dt = -9.81t + C.$$

Since, initially, $v_y(0) = 300$, we have $C = 300$ and

$$v_y = -9.81t + 300.$$

Then,

$$y(t) = \int v_y \, dt = -\frac{9.81t^2}{2} + 300t + C_1,$$

and since $y(0) = 0$ (we start at the origin), we find that

$$y(t) = -\frac{9.81t^2}{2} + 300t.$$

To calculate the x-component of the position vector, we note that, ignoring air resistance, the velocity in the horizontal direction is constant;[†] that is,

$$v_x = 300\sqrt{3} \text{ m/sec.}$$

Then $x(t) = \int v_x \, dt = 300\sqrt{3}t + C_2$, and since $x(0) = 0$, we obtain

$$x = 300\sqrt{3}t.$$

Thus, the position vector describing the location of the cannonball is

$$\mathbf{s}(t) = x(t)\mathbf{i} + y(t)\mathbf{j} = 300\sqrt{3}t\mathbf{i} + \left(300t - \frac{9.81t^2}{2}\right)\mathbf{j}.$$

[†] There are no forces acting on the ball in the horizontal direction since the force of gravity acts only in the vertical direction.

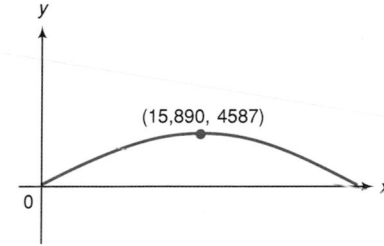

FIGURE 2
GRAPH OF THE POSITION FUNCTION
$y = \dfrac{x}{\sqrt{3}} - \dfrac{9.81}{540,000} x^2$

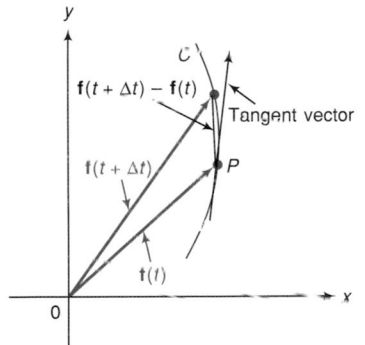

FIGURE 3
THE TANGENT VECTOR TO THE GRAPH
OF **f** AT P

To obtain the Cartesian equation of this curve, we start with

$$x = 300\sqrt{3}t \text{ so } t = x/300\sqrt{3} \text{ and}$$

$$y = 300\left(\frac{x}{300\sqrt{3}}\right) - \frac{9.81}{2}\frac{x^2}{(300\sqrt{3})^2} = \frac{x}{\sqrt{3}} - \frac{9.81}{540,000}x^2.$$

This parabola is sketched in Figure 2.

GEOMETRIC INTERPRETATION OF f′

We now seek a geometric interpretation for **f′**. As we have seen, the set of vectors $\mathbf{f}(t) = f_1(t)\mathbf{i} + f_2(t)\mathbf{j}$ for t in the domain of **f** form a curve C in the plane. We see in Figure 3 that the vector $[\mathbf{f}(t + \Delta t) - \mathbf{f}(t)]/\Delta t$ has the direction of a secant vector whose direction approaches that of the tangent vector as $\Delta t \to 0$. Hence, $\mathbf{f}'(t)$ is tangent to the graph of **f** at the point P.

DEFINITION **UNIT TANGENT VECTOR**

It is sometimes useful to calculate a **unit tangent vector** to a curve C. This vector is a tangent vector with a magnitude of 1. The unit tangent vector is usually denoted by **T** and can be calculated by the formula

$$\mathbf{T}(t) = \frac{\mathbf{f}'(t)}{|\mathbf{f}'(t)|} \tag{6}$$

for any number t, as long as $\mathbf{f}'(t) \neq \mathbf{0}$. This follows since $\mathbf{f}'(t)$ is a tangent vector and $\mathbf{f}'/|\mathbf{f}'|$ is a unit vector. ∎

EXAMPLE 5
FINDING A UNIT TANGENT
VECTOR

Find the unit tangent vector to the curve $\mathbf{f} = (\ln t)\mathbf{i} + (1/t)\mathbf{j}$ at $t = 1$.

SOLUTION: $\mathbf{f}'(t) = (1/t)\mathbf{i} - (1/t^2)\mathbf{j} = \mathbf{i} - \mathbf{j}$ when $t = 1$. Since $|\mathbf{f}'| = |\mathbf{i} - \mathbf{j}| = \sqrt{2}$, we find that

$$\mathbf{T} = \frac{1}{\sqrt{2}}\mathbf{i} - \frac{1}{\sqrt{2}}\mathbf{j}.$$

EXAMPLE 6
FINDING A UNIT TANGENT
VECTOR TO A CIRCULAR HELIX

Find the unit tangent vector to the **circular helix** $\mathbf{f}(t) = \cos t\mathbf{i} + \sin t\mathbf{j} + t\mathbf{k}$ at $t = \pi/3$.

SOLUTION: $\mathbf{f}'(t) = -\sin t\mathbf{i} + \cos t\mathbf{j} + \mathbf{k}$ and $|\mathbf{f}'(t)|^2 = \sin^2 t + \cos^2 t + 1 = 2$ so $|\mathbf{f}'(t)| = \sqrt{2}$. Thus,

$$\mathbf{T}(t) = \frac{1}{\sqrt{2}}(-\sin t\mathbf{i} + \cos t\mathbf{j} + \mathbf{k}) \quad \text{and}$$

$$\mathbf{T}\left(\frac{\pi}{3}\right) = -\frac{\sqrt{3}}{2\sqrt{2}}\mathbf{i} + \frac{1}{2\sqrt{2}}\mathbf{j} + \frac{1}{\sqrt{2}}\mathbf{k}.$$

We now turn briefly to the integration of vector functions.

DEFINITION **INTEGRAL OF A VECTOR FUNCTION**

(i) Let $\mathbf{f}(t) = f_1(t)\mathbf{i} + f_2(t)\mathbf{j}$ and suppose that the component functions f_1 and f_2 have antiderivatives. Then we define the **indefinite integral** of $\mathbf{f}$ by

$$\int \mathbf{f}(t)\,dt = \left(\int f_1(t)\,dt\right)\mathbf{i} + \left(\int f_2(t)\,dt\right)\mathbf{j} + \mathbf{C}. \tag{7}$$

where $\mathbf{C}$ is a constant vector of integration. For each vector $\mathbf{C}$ the expression (7) is an **antiderivative** of $\mathbf{f}$.

(ii) If f_1 and f_2 are integrable over the interval $[a, b]$, then we define the **definite integral** of $\mathbf{f}$ by

$$\int_a^b \mathbf{f}(t)\,dt = \left(\int_a^b f_1(t)\,dt\right)\mathbf{i} + \left(\int_a^b f_2(t)\,dt\right)\mathbf{j}. \quad \blacksquare \tag{8}$$

REMARK 1: An antiderivative of $\mathbf{f}$ is a new vector function and is *not unique* since $\int f_1(t)\,dt$ and $\int f_2(t)\,dt$ are not unique.

REMARK 2: The definite integral of $\mathbf{f}$ is a constant vector since $\int_a^b f_1(t)\,dt$ and $\int_a^b f_2(t)\,dt$ are constants.

EXAMPLE 7
INTEGRATING A VECTOR
FUNCTION

Let $\mathbf{f}(t) = (\cos t)\mathbf{i} + (\sin t)\mathbf{j}$. Calculate the following:

(a) $\displaystyle\int \mathbf{f}(t)\,dt$ **(b)** $\displaystyle\int_0^{\pi/2} \mathbf{f}(t)\,dt$

SOLUTION:

(a) $\displaystyle\int \mathbf{f}(t)\,dt = \left(\int \cos t\,dt\right)\mathbf{i} + \left(\int \sin t\,dt\right)\mathbf{j}$

$= (\sin t + C_1)\mathbf{i} + (-\cos t + C_2)\mathbf{j}$

$= (\sin t)\mathbf{i} - (\cos t)\mathbf{j} + \mathbf{C},$

where $\mathbf{C} = C_1\mathbf{i} + C_2\mathbf{j}$ is a constant vector.

(b) $\displaystyle\int_0^{\pi/2} \mathbf{f}(t)\,dt = \left(\sin t\Big|_0^{\pi/2}\right)\mathbf{i} + \left(-\cos t\Big|_0^{\pi/2}\right)\mathbf{j} = \mathbf{i} + \mathbf{j}$

■ **PROBLEMS 12.3**

■ **SELF-QUIZ**

I. If $\mathbf{f}(t) = -3\mathbf{i} + t^2\mathbf{j}$, then $\mathbf{f}'(t) = $ _____.
   a. $2t$ b. $-3\mathbf{i} + 2t\mathbf{j}$
   c. $2t\mathbf{i}$ d. $2t\mathbf{j}$

II. If $\mathbf{f}(t) = 15\mathbf{i} - 37\mathbf{j}$, then $\mathbf{f}'(t) = $ _____.
   a. 0 b. $\mathbf{0}$
   c. -22 d. $15t\mathbf{i} - 37t\mathbf{j}$

III. If $\mathbf{f}'(t) = \mathbf{j}$ and $\mathbf{f}(0) = 2\mathbf{i} + \mathbf{j}$, then
   $\mathbf{f}(t) = $ _____.

   a. $2\mathbf{i} + t\mathbf{j}$ b. $2\mathbf{i} + (1 + t)\mathbf{j}$
   c. $(1 + t)\mathbf{j}$ d. $t\mathbf{j}$

IV. If $\mathbf{f}(t) = 3\cos t\mathbf{i} + 4\sin t\mathbf{j}$, then the unit tangent
   vector when $t = \pi/4$ is _____.
   a. $0.6\mathbf{i} + 0.8\mathbf{j}$ b. $-0.8\mathbf{i} + 0.6\mathbf{j}$
   c. $-0.6\mathbf{i} + 0.8\mathbf{j}$ d. $(-3/\sqrt{2})\mathbf{i} + (4/\sqrt{2})\mathbf{j}$

CONTINUED

■ **DRILL**

In Problems 1–10, calculate the first and second derivatives of the given vector function and then determine their domains.

1. $\mathbf{f}(t) = t\mathbf{i} - t^5\mathbf{j}$　　2. $\mathbf{f}(t) = (1 + t^2)\mathbf{i} + \dfrac{2}{t}\mathbf{j}$

3. $\mathbf{f}(t) = (\sin 2t)\mathbf{i} + (\cos 3t)\mathbf{j}$

4. $\mathbf{f}(t) = \dfrac{t}{1 + t}\mathbf{i} + \dfrac{1}{\sqrt{t}}\mathbf{j}$

5. $\mathbf{f}(t) = (\ln t)\mathbf{i} + e^{3t}\mathbf{j}$
6. $\mathbf{f}(t) = e^t(\sin t)\mathbf{i} + e^t(\cos t)\mathbf{j}$
7. $\mathbf{f}(t) = (\tan t)\mathbf{i} + (\sec t)\mathbf{j}$
8. $\mathbf{f}(t) = (\tan^{-1} t)\mathbf{i} + (\sin^{-1} t)\mathbf{j}$
9. $\mathbf{f}(t) = (\ln \cos t)\mathbf{i} + (\ln \sin t)\mathbf{j}$
10. $\mathbf{f}(t) = (\cosh t)\mathbf{i} + (\sinh t)\mathbf{j}$

In Problems 11–26, calculate the unit tangent vector **T** for the specified value of t.

11. $\mathbf{f}(t) = t^2\mathbf{i} + t^3\mathbf{j}; \ t = 1$

12. $\mathbf{f}(t) = t\mathbf{i} + \dfrac{1}{t}\mathbf{j}; \ t = 1$

13. $\mathbf{f}(t) = (\cos t)\mathbf{i} + (\sin t)\mathbf{j}; \ t = 0$
14. $\mathbf{f}(t) = (\cos t)\mathbf{i} + (\sin t)\mathbf{j}; \ t = \pi/2$
15. $\mathbf{f}(t) = (\cos t)\mathbf{i} + (\sin t)\mathbf{j}; \ t = \pi/4$
16. $\mathbf{f}(t) = (\cos t)\mathbf{i} + (\sin t)\mathbf{j}; \ t = 3\pi/4$
17. $\mathbf{f}(t) = (\tan t)\mathbf{i} + (\sec t)\mathbf{j}; \ t = 0$
18. $\mathbf{f}(t) = (\ln t)\mathbf{i} + e^{2t}\mathbf{j}; \ t = 1$

19. $\mathbf{f}(t) = \dfrac{t}{t + 1}\mathbf{i} + \dfrac{t + 1}{t}\mathbf{j}; \ t = 2$

20. $\mathbf{f}(t) = \dfrac{t + 1}{t}\mathbf{i} + \dfrac{t}{t + 1}\mathbf{j}; \ t = 2$

21. $\mathbf{g}(t) = t\mathbf{i} + t^2\mathbf{j} + t^3\mathbf{k}; \ t = 1$
22. $\mathbf{g}(t) = t^3\mathbf{i} + t^5\mathbf{j} + t^7\mathbf{k}; \ t = 1$
23. $\mathbf{g}(t) = t\mathbf{i} + e^t\mathbf{j} + e^{-t}\mathbf{k}; \ t = 0$
24. $\mathbf{g}(t) = t^2\mathbf{i} + t^2\mathbf{j} + t^{5/2}\mathbf{k}; \ t = 4$
25. $\mathbf{g}(t) = 4(\cos 2t)\mathbf{i} + 9(\sin 2t)\mathbf{j} + t\mathbf{k}; \ t = \pi/4$
26. $\mathbf{g}(t) = (\cosh t)\mathbf{i} + (\sinh t)\mathbf{j} + t^2\mathbf{k}; \ t = 0$

In Problems 27–36, calculate the given indefinite or definite integral.

27. $\displaystyle\int \left[(\sin 2t)\mathbf{i} + e^t\mathbf{j} \right] dt$

28. $\displaystyle\int (t^{-1/2}\mathbf{i} + t^{1/2}\mathbf{j}) \, dt$

29. $\displaystyle\int_0^{\pi/4} \left[(\cos 2t)\mathbf{i} - (\sin 2t)\mathbf{j} \right] dt$

30. $\displaystyle\int_1^e \left[\left(\dfrac{1}{t}\right)\mathbf{i} - \left(\dfrac{3}{t}\right)\mathbf{j} \right] dt$

31. $\displaystyle\int \left[(\ln t)\mathbf{i} + te^t\mathbf{j} \right] dt$

32. $\displaystyle\int \left[(\tan t)\mathbf{i} + (\sec t)\mathbf{j} \right] dt$

33. $\displaystyle\int_0^2 (t^2\mathbf{i} - t^3\mathbf{j} + t^4\mathbf{k}) \, dt$

34. $\displaystyle\int_0^1 \left[(\sinh t)\mathbf{i} - (\cosh t)\mathbf{j} - (\tanh t)\mathbf{k} \right] dt$

35. $\displaystyle\int 2t[(\sin t^2)\mathbf{i} + (\cos t^2)\mathbf{j} + e^{t^2}\mathbf{k}] \, dt$

36. $\displaystyle\int t[(\cos t)\mathbf{i} + (\sin t)\mathbf{j} + e^t\mathbf{k}] \, dt$

■ **APPLICATIONS**

37. Suppose $\mathbf{f}'(t) = t^3\mathbf{i} - t^5\mathbf{j}$ and $\mathbf{f}(0) = 2\mathbf{i} + 5\mathbf{j}$. Find $\mathbf{f}(t)$.
38. Suppose $\mathbf{f}'(t) = (1/\sqrt{t})\mathbf{i} + \sqrt{t}\mathbf{j}$ and $\mathbf{f}(1) = -2\mathbf{i} + \mathbf{j}$. Find $\mathbf{f}(t)$.
39. Suppose $\mathbf{f}'(t) = (\cos t)\mathbf{i} + (\sin t)\mathbf{k}$ and $\mathbf{f}(\pi/2) = \mathbf{i}$. Find $\mathbf{f}(t)$.
40. Suppose $\mathbf{f}'(t) = (\cos t)\mathbf{i} - (\sin t)\mathbf{k}$ and $\mathbf{f}(0) = \mathbf{j} + \mathbf{k}$. Find $\mathbf{f}(t)$.
41. The ellipse $(x/a)^2 + (y/b)^2 = 1$ can be written parametrically as $x = a \cos \theta$, $y = b \sin \theta$. Find a unit tangent vector to this ellipse at $\theta = \pi/4$.
42. Find a unit tangent vector to the cycloid $\mathbf{f}(\alpha) = r(\alpha - \sin \alpha)\mathbf{i} + r(1 - \cos \alpha)\mathbf{j}$ for $\alpha = 0$.
*43. Find, for $\theta = \pi/6$, a unit tangent vector to the hypocycloid of Problem 12.1.45, assuming that $a = 5$ and $b = 2$.

44. A cannonball is shot upward from ground level at an angle of 45° with an initial speed of 1300 ft/sec. Find a parametric representation of the path of the cannonball; then find a Cartesian equation for this path.
45. How many feet (horizontally) does the cannonball of the preceding problem travel before it hits the ground?
46. An object is thrown down from the top of a 150 m building at an angle of 30° (below the horizontal) with an initial velocity of 100 m/sec. Determine a parametric representation of the path of the object. [*Hint:* Draw a picture.]
47. When the object in the preceding problem hits the ground, how far is it from the base of the building?

■ ANSWERS TO SELF-QUIZ

I. d II. b III. b IV. c

12.4 SOME DIFFERENTIATION FORMULAS

In Section 2.1, we gave a number of rules for differentiation. Many of these rules carry over with little change to the differentiation of a vector function. The proof of Theorem 1 is a straightforward application of the differentiation rules discussed in Chapter 2 and is left as an exercise (see Problems 31–36).

THEOREM 1

Let $\mathbf{f}$ and $\mathbf{g}$ be vector functions that are differentiable in an interval I. Let the scalar function h be differentiable in I. Finally, let α be a scalar and let $\mathbf{v}$ be a constant vector. Then, on I, we have

(i) $\mathbf{f} + \mathbf{g}$ is differentiable and

$$\frac{d}{dt}(\mathbf{f} + \mathbf{g}) = \frac{d\mathbf{f}}{dt} + \frac{d\mathbf{g}}{dt} = \mathbf{f}' + \mathbf{g}'. \tag{1}$$

(ii) $\alpha\mathbf{f}$ is differentiable and

$$\frac{d}{dt}\alpha\mathbf{f} = \alpha\frac{d\mathbf{f}}{dt} = \alpha\mathbf{f}'. \tag{2}$$

(iii) $\mathbf{v} \cdot \mathbf{f}$ is differentiable and

$$\frac{d}{dt}\mathbf{v} \cdot \mathbf{f} = \mathbf{v} \cdot \frac{d\mathbf{f}}{dt} = \mathbf{v} \cdot \mathbf{f}'. \tag{3}$$

(iv) $h\mathbf{f}$ is differentiable and

$$\frac{d}{dt}h\mathbf{f} = h\frac{d\mathbf{f}}{dt} + \frac{dh}{dt}\mathbf{f} = h\mathbf{f}' + h'\mathbf{f}. \tag{4}$$

(v) $\mathbf{f} \cdot \mathbf{g}$ is differentiable and

$$\frac{d}{dt}\mathbf{f} \cdot \mathbf{g} = \mathbf{f} \cdot \frac{d\mathbf{g}}{dt} + \frac{d\mathbf{f}}{dt} \cdot \mathbf{g} = \mathbf{f} \cdot \mathbf{g}' + \mathbf{f}' \cdot \mathbf{g}. \tag{5}$$

(vi) If $\mathbf{f}$ and $\mathbf{g}$ are differentiable vector functions in $\mathbb{R}^3$, then $\mathbf{f} \times \mathbf{g}$ is differentiable and

$$(\mathbf{f} \times \mathbf{g})' = (\mathbf{f}' \times \mathbf{g}) + (\mathbf{f} \times \mathbf{g}'). \quad ■ \tag{6}$$

EXAMPLE 1
USING THE DIFFERENTIATION FORMULAS

Let $\mathbf{f}(t) = t\mathbf{i} + t^3\mathbf{j}$, $\mathbf{g}(t) = (\cos t)\mathbf{i} + (\sin t)\mathbf{j}$ and $\mathbf{v} = 2\mathbf{i} - 3\mathbf{j}$. Calculate **(a)** $(\mathbf{f} + \mathbf{g})'$, **(b)** $(\mathbf{v} \cdot \mathbf{f})'$, and **(c)** $(\mathbf{f} \cdot \mathbf{g})'$.

SOLUTION:

(a) $(\mathbf{f} + \mathbf{g})' = \mathbf{f}' + \mathbf{g}' = (\mathbf{i} + 3t^2\mathbf{j}) + [-(\sin t)\mathbf{i} + (\cos t)\mathbf{j}]$
$= (1 - \sin t)\mathbf{i} + (3t^2 + \cos t)\mathbf{j}$

(b) $(\mathbf{v} \cdot \mathbf{f})' = \mathbf{v} \cdot \mathbf{f}' = (2\mathbf{i} - 3\mathbf{j}) \cdot (\mathbf{i} + 3t^2\mathbf{j}) = 2 - 9t^2$

(c) $(\mathbf{f} \cdot \mathbf{g})' = \mathbf{f} \cdot \mathbf{g}' + \mathbf{f}' \cdot \mathbf{g}$

$\qquad = (t\mathbf{i} + t^3\mathbf{j}) \cdot [-(\sin t)\mathbf{i} + (\cos t)\mathbf{j}]$

$\qquad \quad + (\mathbf{i} + 3t^2\mathbf{j}) \cdot [(\cos t)\mathbf{i} + (\sin t)\mathbf{j}]$

$\qquad = -t \sin t + t^3 \cos t + \cos t + 3t^2 \sin t$

$\qquad = (\cos t)(t^3 + 1) + (\sin t)(3t^2 - t)$

EXAMPLE 2
THE DERIVATIVE OF $\mathbf{f} \cdot \mathbf{f}'$ IS ZERO
IF $|\mathbf{f}(t)|$ IS CONSTANT

Let $\mathbf{f}(t) = (\cos t)\mathbf{i} + (\sin t)\mathbf{j}$. Calculate $\mathbf{f} \cdot \mathbf{f}'$.

SOLUTION:

$\mathbf{f} \cdot \mathbf{f}' = [(\cos t)\mathbf{i} + (\sin t)\mathbf{j}] \cdot [-(\sin t)\mathbf{i} + (\cos t)\mathbf{j}]$

$\qquad = -\cos t \sin t + \sin t \cos t = 0$

Example 2 can be generalized to the following interesting result.

THEOREM 2

Let $\mathbf{f}(t) = f_1\mathbf{i} + f_2\mathbf{j}$ be a differentiable vector function such that $|\mathbf{f}(t)| = \sqrt{f_1{}^2(t) + f_2{}^2(t)}$ is constant. Then

$\mathbf{f} \cdot \mathbf{f}' = 0.$

PROOF: Suppose $|\mathbf{f}(t)| = C$, a constant. Then

$\mathbf{f} \cdot \mathbf{f} = f_1{}^2 + f_2{}^2 = |\mathbf{f}|^2 = C^2,$

so

$\dfrac{d}{dt}(\mathbf{f} \cdot \mathbf{f}) = \dfrac{d}{dt} C^2 = 0.$

But

$\dfrac{d}{dt}(\mathbf{f} \cdot \mathbf{f}) = \mathbf{f} \cdot \mathbf{f}' + \mathbf{f}' \cdot \mathbf{f} = 2\mathbf{f} \cdot \mathbf{f}' = 0,$

so

$\mathbf{f} \cdot \mathbf{f}' = 0.$ ∎

NOTE: In Example 2, $|\mathbf{f}(t)| = \sqrt{\cos^2 t + \sin^2 t} = 1$ for every t.

UNIT NORMAL VECTOR TO A PLANE CURVE

There is an interesting geometric application of Theorem 2. Let $\mathbf{f}(t) = f_1(t)\mathbf{i} + f_2(t)\mathbf{j}$ be a differentiable vector function. For all t for which $\mathbf{f}'(t) \neq \mathbf{0}$, we let $\mathbf{T}(t)$ denote the unit tangent vector to the curve $\mathbf{f}(t)$. Then since $|\mathbf{T}(t)| = 1$ by the definition of a *unit* tangent vector, we have, from Theorem 2 [assuming that $\mathbf{T}'(t)$ exists],

$$\mathbf{T}(t) \cdot \mathbf{T}'(t) = 0. \tag{7}$$

That is, $\mathbf{T}'(t)$ is *orthogonal* to $\mathbf{T}(t)$.

DEFINITION **UNIT NORMAL VECTOR**

The vector $\mathbf{T}'$ which is orthogonal to the tangent line is called a **normal vector**† to the curve $\mathbf{f}$. Whenever $\mathbf{T}'(t) \neq \mathbf{0}$, we can define the **unit normal vector** to the curve $\mathbf{f}$ at t as

$$\mathbf{n}(t) = \frac{\mathbf{T}'(t)}{|\mathbf{T}'(t)|}. \quad \blacksquare \tag{8}$$

NOTE: If $\mathbf{T}'(t) = \mathbf{0}$, then we may define $\mathbf{n}(t)$ to be one of the two unit vectors that are orthogonal to $\mathbf{T}(t)$. For example, if $\mathbf{f}(t) = t\mathbf{i} + 2t\mathbf{j}$, then $\mathbf{f}'(t) = \mathbf{i} + 2\mathbf{j}$ and $\mathbf{T}(t) = \dfrac{1}{\sqrt{5}}\mathbf{i} + \dfrac{2}{\sqrt{5}}\mathbf{j}$. Two choices for $\mathbf{n}(t)$ are $\mathbf{n}_1(t) = \dfrac{2}{\sqrt{5}}\mathbf{i} - \dfrac{1}{\sqrt{5}}\mathbf{j}$ and $\mathbf{n}_2(t) = -\dfrac{2}{\sqrt{5}}\mathbf{i} + \dfrac{1}{\sqrt{5}}\mathbf{j}$.

EXAMPLE 3
FINDING A UNIT NORMAL VECTOR
TO THE UNIT CIRCLE

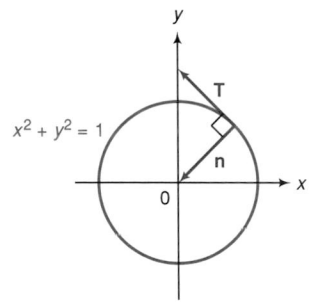

FIGURE 1
UNIT TANGENT AND UNIT NORMAL VEC-
TORS TO THE UNIT CIRCLE

Calculate a unit normal vector to the curve $\mathbf{f}(t) = (\cos t)\mathbf{i} + (\sin t)\mathbf{j}$ at $t = \pi/4$.

SOLUTION: First, we calculate $\mathbf{f}'(t) = -(\sin t)\mathbf{i} + (\cos t)\mathbf{j}$, and since $|\mathbf{f}'(t)| = 1$, we find that

$$\mathbf{T}(t) = \frac{\mathbf{f}'(t)}{|\mathbf{f}'(t)|} = -(\sin t)\mathbf{i} + (\cos t)\mathbf{j}.$$

Then

$$\mathbf{T}'(t) = -(\cos t)\mathbf{i} - (\sin t)\mathbf{j} = \mathbf{n}(t)$$

since $|\mathbf{T}'(t)| = 1$, so that at $t = \pi/4$,

$$\mathbf{n} = -\frac{1}{\sqrt{2}}\mathbf{i} - \frac{1}{\sqrt{2}}\mathbf{j}.$$

This is sketched in Figure 1 [remember that $\mathbf{f}(t) = (\cos t)\mathbf{i} + (\sin t)\mathbf{j}$ is the parametric equation of the unit circle]. The reason the vector $\mathbf{n}$ points inward is that it is the negative of the position vector at $t = \pi/4$.

EXAMPLE 4
FINDING A UNIT NORMAL VECTOR

Calculate a unit normal vector to the curve $\mathbf{f}(t) = [(t^3/3) - t]\mathbf{i} + t^2\mathbf{j}$ at $t = 3$.

SOLUTION: Here $\mathbf{f}'(t) = (t^2 - 1)\mathbf{i} + 2t\mathbf{j}$ and

$$|\mathbf{f}'(t)| = \sqrt{(t^2 - 1)^2 + 4t^2} = \sqrt{t^4 - 2t^2 + 1 + 4t^2}$$
$$= \sqrt{t^4 + 2t^2 + 1} = t^2 + 1.$$

---

† From the Latin *norma*, meaning "square," the carpenter's square. Until the 1830s the English "normal" meant standing at right angles to the ground.

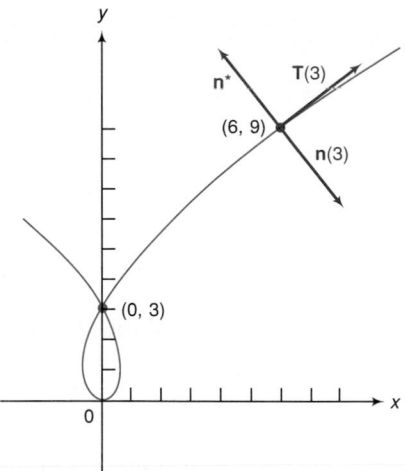

FIGURE 2

GRAPH OF $\mathbf{f}(t) = \left(\dfrac{t^3}{3} - t\right)\mathbf{i} + t^2\mathbf{j}$

there are two unit vectors orthogonal to the unit tangent vector when $t = 3$.

Thus

$$\mathbf{T}(t) = \frac{\mathbf{f}'(t)}{|\mathbf{f}'(t)|} = \frac{t^2 - 1}{t^2 + 1}\mathbf{i} + \frac{2t}{t^2 + 1}\mathbf{j}.$$

Then

$$\mathbf{T}'(t) = \frac{d}{dt}\left(\frac{t^2 - 1}{t^2 + 1}\right)\mathbf{i} + \frac{d}{dt}\left(\frac{2t}{t^2 + 1}\right)\mathbf{j} = \frac{4t}{(t^2 + 1)^2}\mathbf{i} + \frac{2 - 2t^2}{(t^2 + 1)^2}\mathbf{j}.$$

Finally,

$$|\mathbf{T}'(t)| = \left\{\left[\frac{4t}{(t^2 + 1)^2}\right]^2 + \left[\frac{2 - 2t^2}{(t^2 + 1)^2}\right]^2\right\}^{1/2}$$

$$= \frac{1}{(t^2 + 1)^2}(16t^2 + 4 - 8t^2 + 4t^4)^{1/2}$$

$$= \frac{1}{(t^2 + 1)^2}\sqrt{4t^4 + 8t^2 + 4} = \frac{2}{(t^2 + 1)^2}\sqrt{t^4 + 2t^2 + 1}$$

$$= \frac{2(t^2 + 1)}{(t^2 + 1)^2} = \frac{2}{t^2 + 1},$$

so that

$$\mathbf{n}(t) = \frac{\mathbf{T}'(t)}{|\mathbf{T}'(t)|} = \frac{t^2 + 1}{2}\left[\frac{4t}{(t^2 + 1)^2}\mathbf{i} + \frac{2 - 2t^2}{(t^2 + 1)^2}\mathbf{j}\right]$$

$$= \frac{2t}{t^2 + 1}\mathbf{i} + \frac{1 - t^2}{t^2 + 1}\mathbf{j}.$$

At $t = 3$, $\mathbf{T}(t) = \frac{4}{5}\mathbf{i} + \frac{3}{5}\mathbf{j}$ and $\mathbf{n} = \frac{3}{5}\mathbf{i} - \frac{4}{5}\mathbf{j}$. Note that $\mathbf{T}(3) \cdot \mathbf{n}(3) = 0$. This is sketched in Figure 2.

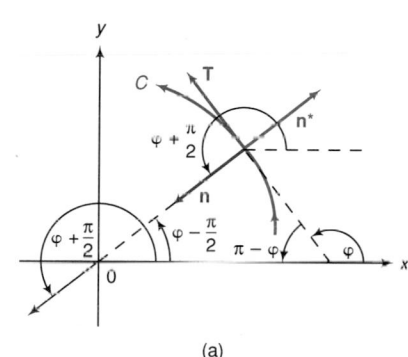

(a)

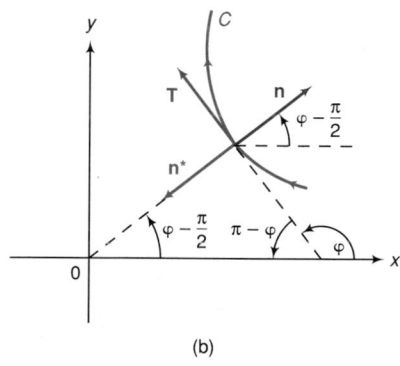

(b)

FIGURE 3
THE UNIT NORMAL VECTOR **n** POINTS IN THE DIRECTION OF THE CONCAVE SIDE OF THE CURVE

THE DIRECTION OF THE NORMAL VECTOR TO A PLANE CURVE

In Figure 2, we see that there are two vectors perpendicular to **T** at the point P: one pointing "inward" (**n**) and one pointing "outward" (**n***). In general, the situation is as shown in Figure 3. How do we know which of the two vectors is **n**? If φ is the direction of **T**, then since **T** is a unit vector,

$$\mathbf{T} = (\cos \varphi)\mathbf{i} + (\sin \varphi)\mathbf{j},$$

so that

$$\frac{d\mathbf{T}}{dt} = \frac{d\mathbf{T}}{d\varphi}\frac{d\varphi}{dt}$$

and

$$\left|\frac{d\mathbf{T}}{dt}\right| = \left|\frac{d\mathbf{T}}{d\varphi}\right|\left|\frac{d\varphi}{dt}\right| = \left|\frac{d\varphi}{dt}\right| \qquad \left(\text{since } \left|\frac{d\mathbf{T}}{d\varphi}\right| = 1\right).$$

Then

$$\mathbf{n}(t) = \frac{d\mathbf{T}/dt}{|d\mathbf{T}/dt|} = \frac{d\mathbf{T}}{d\varphi}\frac{d\varphi/dt}{|d\varphi/dt|} = \frac{d\varphi/dt}{|d\varphi/dt|}\left[-(\sin\varphi)\mathbf{i} + (\cos\varphi)\mathbf{j}\right].$$

Since $d\varphi/dt$ is a scalar, $(d\varphi/dt)/|d\varphi/dt| = \pm 1$. It is $+1$ in Figure 3(a) since φ increases as t increases so that $d\varphi/dt > 0$, while it is -1 in Figure 3(b) since φ decreases as t increases. In Figure 3(a), then,

$$\mathbf{n}(t) = -(\sin\varphi)\mathbf{i} + (\cos\varphi)\mathbf{j} = \cos[\varphi + (\pi/2)]\mathbf{i} + \sin[\varphi + (\pi/2)]\mathbf{j}.$$

That is, the direction of $\mathbf{n}$ is $\varphi + (\pi/2)$ and $\mathbf{n}$ must point inward (on the concave side of C) as in Figure 3(a). In the other case, $(d\varphi/dt)/|d\varphi/dt| = -1$, so

$$\mathbf{n}(t) = (\sin\varphi)\mathbf{i} - (\cos\varphi)\mathbf{j} = \cos[\varphi - (\pi/2)]\mathbf{i} + \sin[\varphi - (\pi/2)]\mathbf{j}.$$

That is, the direction of $\mathbf{n}$ is $\varphi - (\pi/2)$ and so $\mathbf{n}$ points as in Figure 3(b). In either case:

> The unit normal vector $\mathbf{n}$ always points in the direction of the concave side of the curve.

REMARK: It is more difficult to define a unit normal vector in space ($\mathbb{R}^3$) because there is a whole "plane-full" of vectors orthogonal to a given tangent vector $\mathbf{T}$. We will define normal vectors in $\mathbb{R}^3$ in Section 12.6, after we have discussed the notion of curvature.

■ PROBLEMS 12.4

■ SELF-QUIZ

I. $\dfrac{d}{dt}\{[(\cos 3t)\mathbf{i} + (\sin 4t)\mathbf{j}] \cdot [-4\mathbf{i} + 3\mathbf{j}]\} =$

_____.
a. $-3(\sin 3t)\mathbf{i} + 4(\cos 4t)\mathbf{j}$
b. $[-3(\sin 3t)\mathbf{i} + 4(\cos 4t)\mathbf{j}] \cdot [-4\mathbf{i} + 3\mathbf{j}]$
c. $-12\sin 3t + 12\cos 4t$
d. $12(\sin 3t + \cos 4t)$

II. If $\mathbf{f}(t) = (\cos 3t)\mathbf{i} + (\sin 3t)\mathbf{j}$, then $\mathbf{f} \cdot \mathbf{f}' =$ _____.
a. 0   b. **0**
c. 1   d. $3(\cos 3t)^2 - 3(\sin 3t)^2$

III. If $\mathbf{f}(t) = (\cos 2t)\mathbf{i} + (\sin 2t)\mathbf{j}$, then the unit tangent vector is _____.
a. $-2(\sin 2t)\mathbf{i} + 2(\cos 2t)\mathbf{j}$
b. $-2(\sin t)\mathbf{i} + 2(\cos t)\mathbf{j}$
c. $-(\sin t)\mathbf{i} + (\cos t)\mathbf{j}$
d. $-(\sin 2t)\mathbf{i} + (\cos 2t)\mathbf{j}$

IV. If $\mathbf{f}(t) = (\cos 2t)\mathbf{i} + (\sin 2t)\mathbf{j}$, then the unit normal vector when $t = \pi/6$ is _____.
a. $-\mathbf{i} - \sqrt{3}\mathbf{j}$ b. $\frac{1}{2}\mathbf{i} + (\sqrt{3}/2)\mathbf{j}$
c. $-\frac{1}{2}\mathbf{i} - (\sqrt{3}/2)\mathbf{j}$ d. $-(\sqrt{3}/2)\mathbf{i} - \frac{1}{2}\mathbf{j}$

■ DRILL

In Problems 1–8, calculate the indicated derivative.

1. $\dfrac{d}{dt}\{[2t\mathbf{i} + (\cos t)\mathbf{j}] + [(\tan t)\mathbf{i} - (\sec t)\mathbf{j}]\}$

2. $\dfrac{d}{dt}\left[(t^3\mathbf{i} - t^5\mathbf{j}) \cdot \left(\dfrac{-1}{t^3}\mathbf{i} + \dfrac{1}{t^5}\mathbf{j}\right)\right]$

3. $\dfrac{d}{dt}[(t^3\mathbf{i} - t^5\mathbf{j}) + (-t^3\mathbf{i} + t^5\mathbf{j})]$

4. $\dfrac{d}{dt}\{[(\sinh t)\mathbf{i} + (\cosh t)\mathbf{j}] \cdot [(\sin t)\mathbf{i} + (\cos t)\mathbf{j}]\}$

5. $\dfrac{d}{dt}\{[\sin^{-1} t\,\mathbf{i} + \cos^{-1} t\,\mathbf{j}] + [\cos t\mathbf{i} + \sin t\mathbf{j}]\}$

6. $\dfrac{d}{dt}\{[(\sin^{-1} t)\mathbf{i} + (\cos^{-1} t)\mathbf{j}] \cdot [2\mathbf{i} - 10\mathbf{j}]\}$

CONTINUED

7. $\dfrac{d}{dt}\{[2\mathbf{i} + t\mathbf{j} - t^2\mathbf{k}] \times [3t\mathbf{i} + 5\mathbf{k}]\}$

8. $\dfrac{d}{dt}\{[t\mathbf{i} + t^2\mathbf{j} + t^3\mathbf{k}] \times [\mathbf{i} + 2t\mathbf{j} + 3t^2\mathbf{k}]\}$

In Problems 9–16, find the unit tangent $\mathbf{T}(t)$ and the particular vector $\mathbf{T}$ when $t = t_0$.

9. $\mathbf{g}(t) = (\cos t)\mathbf{i} + (\sin t)\mathbf{j} + t\mathbf{k}; \ t = \pi/4$
10. $\mathbf{g}(t) = 2(\cos t)\mathbf{i} + 2(\sin t)\mathbf{j} + t\mathbf{k}; \ t = \pi/6$
11. $\mathbf{g}(t) = 3(\cos t)\mathbf{i} + 4(\sin t)\mathbf{j} + t\mathbf{k}; \ t = 0$
12. $\mathbf{g}(t) = 4(\cos t)\mathbf{i} + 3(\sin t)\mathbf{j} + t\mathbf{k}; \ t = \pi/2$
13. $\mathbf{g}(t) = \mathbf{i} + t\mathbf{j} + t^2\mathbf{k}; \ t = 1$
14. $\mathbf{g}(t) = t\mathbf{i} + t^2\mathbf{j} + t^3\mathbf{k}; \ t = 0$
15. $\mathbf{g}(t) = e^t(\cos 2t)\mathbf{i} + e^t(\sin 2t)\mathbf{j} + e^t\mathbf{k}; \ t = 0$
16. $\mathbf{g}(t) = e^t(\cos 2t)\mathbf{i} + e^t(\sin 2t)\mathbf{j} + e^t\mathbf{k}; \ t = \pi/4$

In Problems 17–30, find unit tangent $\mathbf{T}(t)$, unit normal $\mathbf{n}(t)$, and the particular vectors $\mathbf{T}$ and $\mathbf{n}$ when $t = t_0$. Then sketch the curve near $t = t_0$ and include the vectors $\mathbf{T}$ and $\mathbf{n}$ in your sketch.

17. $\mathbf{f}(t) = (\cos 3t)\mathbf{i} + (\sin 3t)\mathbf{j}; \ t = 0$
18. $\mathbf{f}(t) = (\cos 5t)\mathbf{i} + (\sin 5t)\mathbf{j}; \ t = \pi/2$
19. $\mathbf{f}(t) = 2(\cos 4t)\mathbf{i} + 2(\sin 4t)\mathbf{j}; \ t = \pi/4$
20. $\mathbf{f}(t) = -3(\cos 10t)\mathbf{i} - 3(\sin 10t)\mathbf{j}; \ t = \pi$
21. $\mathbf{f}(t) = 8(\cos t)\mathbf{i} + 8(\sin t)\mathbf{j}; \ t = \pi/4$
22. $\mathbf{f}(t) = 4t\mathbf{i} + 2t^2\mathbf{j}; \ t = 1$
23. $\mathbf{f}(t) = (2 + 3t)\mathbf{i} + (8 - 5t)\mathbf{j}; \ t = 3$
    [*Hint:* See the note on page 808.]
24. $\mathbf{f}(t) = (4 - 7t)\mathbf{i} + (-3 + 5t)\mathbf{j}; \ t = -5$
25. $\mathbf{f}(t) = (a + bt)\mathbf{i} + (c + dt)\mathbf{j}; \ t = t_0$
26. $\mathbf{f}(t) = 2t\mathbf{i} + (e^t + e^{-t})\mathbf{j}; \ t = 0$
*27. $\mathbf{f}(t) = (t - \cos t)\mathbf{i} + (1 - \sin t)\mathbf{j}; \ t = \pi/2$
*28. $\mathbf{f}(t) = (t - \cos t)\mathbf{i} + (1 - \sin t)\mathbf{j}; \ t = \pi$
*29. $\mathbf{f}(t) = (t - \cos t)\mathbf{i} + (1 - \sin t)\mathbf{j}; \ t = \pi/4$
*30. $\mathbf{f}(t) = (\ln \sin t)\mathbf{i} + (\ln \cos t)\mathbf{j}; \ t = \pi/6$

■ SHOW/PROVE/DISPROVE

31. Prove part (i) of Theorem 1.
32. Prove part (ii) of Theorem 1.
33. Prove part (iii) of Theorem 1.
34. Prove part (iv) of Theorem 1.
35. Prove part (v) of Theorem 1.

36. Prove part (vi) of Theorem 1.
*37. Let $\mathbf{f}(t) = a(\cos t)\mathbf{i} + a(\sin t)\mathbf{j} + t\mathbf{k}$ (a circular helix). Show that the angle between the unit tangent vector $\mathbf{T}$ and the z-axis is constant.

■ ANSWERS TO SELF-QUIZ

I. b, d II. a (scalar zero)
III. d IV. c

12.5 ARC LENGTH REVISITED

In Section 5.3, we derived a formula for the length of an arc of the curve $y = f(x)$ between $x = a$ and $x = b$ [equation (5.3.3) on page 350]. We now derive a formula for arc length in a more general setting. Let the curve C be given parametrically by

$$x = f_1(t), \qquad y = f_2(t). \tag{1}$$

We will assume in this section that f_1' and f_2' exist. Let t_0 be a fixed number, which fixes a point $P_0 = (x_0, y_0) = (f_1(t_0), f_2(t_0))$ on the curve (see Figure 1). The arrows in Figure 1 indicate the direction in which a point moves along the curve as t increases. We define the function $s(t)$ by

$$s(t) = \text{length along the curve } C \text{ from } (f_1(t_0), f_2(t_0)) \text{ to } (f_1(t), f_2(t)).$$

The following theorem allows us to calculate the length of a curve given parametrically.

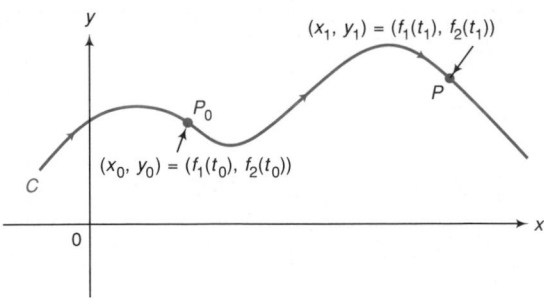

FIGURE 1 $P_0 = (f_1(t_0), f_2(t_0))$ AND $P = (f_1(t_1), f_2(t_1))$ ARE TWO POINTS ON THE CURVE C. AS t INCREASES FROM t_0 TO t_1, WE MOVE ALONG C FROM P_0 TO P.

THEOREM 1

Suppose that f'_1 and f'_2 are continuous in the interval $[t_0, t_1]$. Then $s(t)$ is a differentiable function of t for $t \in [t_0, t_1]$ and

$$\frac{ds}{dt} = \sqrt{[f'_1(t)]^2 + [f'_2(t)]^2} = \sqrt{\left(\frac{dx}{dt}\right)^2 + \left(\frac{dy}{dt}\right)^2}. \tag{2}$$

SKETCH OF PROOF: The proof of this theorem is quite difficult. However, it is possible to give an intuitive idea of what is going on. Consider an arc of the curve between t and $t + \Delta t$ (see Figure 2).

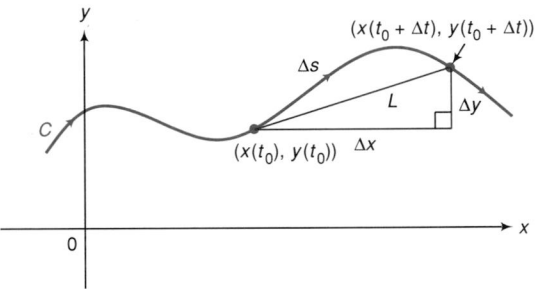

FIGURE 2   THE ARC LENGTH $\triangle s$ IS APPROXIMATELY EQUAL TO L. THE APPROXIMATION IMPROVES AS $\triangle t \to 0$.

First, we note that as $\Delta t \to 0$, the ratio of the length of the secant line L to the length of the arc Δs between the points $(x(t_0), y(t_0))$ and $(x(t_0 + \Delta t), y(t_0 + \Delta t))$ approaches 1. That is,

$$\lim_{\Delta t \to 0} \frac{\Delta s}{L} = 1. \tag{3}$$

Then since $L = \sqrt{\Delta x^2 + \Delta y^2}$, we have

$$\frac{\Delta s}{\Delta t} = \left(\frac{\Delta s}{L}\right)\left(\frac{L}{\Delta t}\right) = \left(\frac{\Delta s}{L}\right)\left(\frac{\sqrt{\Delta x^2 + \Delta y^2}}{\Delta t}\right)$$

$$= \left(\frac{\Delta s}{L}\right)\sqrt{\left(\frac{\Delta x}{\Delta t}\right)^2 + \left(\frac{\Delta y}{\Delta t}\right)^2}$$

and

$$\frac{ds}{dt} = \lim_{\Delta t \to 0} \frac{\Delta s}{\Delta t} = \lim_{\Delta t \to 0} \frac{\Delta s}{L} \cdot \lim_{\Delta t \to 0} \sqrt{\left(\frac{\Delta x}{\Delta t}\right)^2 + \left(\frac{\Delta y}{\Delta t}\right)^2}$$

$$= \sqrt{\left(\frac{dx}{dt}\right)^2 + \left(\frac{dy}{dt}\right)^2}.$$

The last step can be justified under the assumption that dx/dt and dy/dt are continuous, but its proof is beyond the scope of this book.[†] ∎

We use equation (2) to define the length of the arc from t_0 to t_1.

DEFINITION   **ARC LENGTH IN** $\mathbb{R}^2$

Suppose $\mathbf{f}$ has a continuous derivative in the interval $[t_0, b]$. Then $\mathbf{f}$ is said to be **rectifiable** in the interval $[t_0, b]$, and the **arc length** $s(t_1)$ of the curve $\mathbf{f}(t)$ in the interval $[t_0, t_1]$ is given by

$$s(t_1) = \text{length of arc from } t_0 \text{ to } t_1$$

$$= \int_{t_0}^{t_1} \left(\frac{ds}{dt}\right) dt = \int_{t_0}^{t_1} \sqrt{\left(\frac{dx}{dt}\right)^2 + \left(\frac{dy}{dt}\right)^2} \, dt. \quad ∎ \qquad (4)$$

REMARK: This formula generalizes equation (5.3.3) on page 350. See Problem 35.

Equation (4) can be rewritten in a slightly different form. Using the chain rule (applied to differentials), we have

$$\frac{ds}{dt} dt = ds, \qquad (5)$$

and so (4) becomes

$$s(t_1) = \int_{t_0}^{t_1} ds, \qquad (6)$$

where

PARAMETER OF ARC LENGTH
$ds = \sqrt{\left(\dfrac{dx}{dt}\right)^2 + \left(\dfrac{dy}{dt}\right)^2} \, dt.$ $\qquad$ (7)

In this context, the variable s is called the **parameter of arc length**. Note that while x measures distance along the horizontal axis and y measures distance along the vertical axis, s measures distance *along the curve* given parametrically by equations (1).

[†] See, for example, R. C. Buck and E. F. Buck, *Advanced Calculus*, 3rd ed. (New York: McGraw-Hill, 1978), p. 404.

EXAMPLE 1
COMPUTING THE LENGTH OF THE UNIT CIRCLE

Calculate the length of the curve $x = \cos t$, $y = \sin t$ in the interval $[0, 2\pi]$.

SOLUTION:

$$ds = \sqrt{\left(\frac{dx}{dt}\right)^2 + \left(\frac{dy}{dt}\right)^2}\, dt = \sqrt{(-\sin t)^2 + (\cos t)^2}\, dt = dt,$$

so that

$$s = \int_0^{2\pi} ds = \int_0^{2\pi} dt = 2\pi.$$

Since $x = \cos t$, $y = \sin t$ for t in $[0, 2\pi]$ is a parametric representation of the unit circle, we have verified the accuracy of formula (4) in this instance, since the circumference of the unit circle is 2π.

NOTE: As in the remark on page 791, we can parametrize the unit circle by $\mathbf{f}(t) = (\cos t)\mathbf{i} + (\sin t)\mathbf{j}$, $0 \le t \le 4\pi$. Now the circle is traversed twice and the arc length is $s = \int_0^{4\pi} ds = \int_0^{4\pi} dt = 4\pi$. This does not contradict the result above. Rather, it states the obvious fact that if you go around the circle twice, you cover twice the distance you cover if you go around only once.

EXAMPLE 2
CALCULATING ARC LENGTH

Let the curve C be given by $x = t^2$ and $y = t^3$. Calculate the length of the arc from $t = 0$ to $t = 3$.

SOLUTION: Here,

$$\sqrt{\left(\frac{dx}{dt}\right)^2 + \left(\frac{dy}{dt}\right)^2} = \sqrt{(2t)^2 + (3t^2)^2} = \sqrt{4t^2 + 9t^4} = t\sqrt{4 + 9t^2},$$

so that

$$s = \int_0^3 t\sqrt{4 + 9t^2}\, dt = \frac{1}{27}(4 + 9t^2)^{3/2}\Big|_0^3 = \frac{85^{3/2} - 8}{27}.$$

There is a more concise way to write our formula for arc length by using vector notation. Let the curve C be given by

$$\mathbf{f}(t) = f_1(t)\mathbf{i} + f_2(t)\mathbf{j}.$$

Then

$$\mathbf{f}'(t) = f_1'(t)\mathbf{i} + f_2'(t)\mathbf{j}$$

and

$$|\mathbf{f}'(t)| = \sqrt{[f_1'(t)]^2 + [f_2'(t)]^2} = \frac{ds}{dt}, \tag{8}$$

so that the length of the arc between t_0 and t_1 is given by

---

THE LENGTH OF AN ARC BETWEEN t_0 AND t_1

$$s = \int_{t_0}^{t_1} |\mathbf{f}'(t)|\, dt. \tag{9}$$

EXAMPLE 3
CALCULATING ARC LENGTH

Calculate the length of the arc of the curve $\mathbf{f}(t) = (2t - t^2)\mathbf{i} + \frac{8}{3}t^{3/2}\mathbf{j}$ between $t = 1$ and $t = 3$.

SOLUTION: $\mathbf{f}'(t) = (2 - 2t)\mathbf{i} + 4\sqrt{t}\mathbf{j}$ and

$$|\mathbf{f}'(t)| = \sqrt{4 - 8t + 4t^2 + 16t} = 2\sqrt{t^2 + 2t + 1} = 2(t + 1),$$

so

$$s = \int_1^3 2(t + 1)\,dt = (t^2 + 2t)\Big|_1^3 = 12.$$

All the results in this section hold, with very little change, in $\mathbb{R}^3$. If

$$\mathbf{f}(t) = x(t)\mathbf{i} + y(t)\mathbf{j} + z(t)\mathbf{k},$$

then

$$\mathbf{f}'(t) = \frac{dx}{dt}\mathbf{i} + \frac{dy}{dt}\mathbf{j} + \frac{dz}{dt}\mathbf{k}$$

and

$$\mathbf{f}'(t) = \sqrt{\left(\frac{dx}{dt}\right)^2 + \left(\frac{dy}{dt}\right)^2 + \left(\frac{dz}{dt}\right)^2}.$$

DEFINITION **ARC LENGTH IN $\mathbb{R}^3$**

Suppose $\mathbf{f}$ has a continuous derivative in the interval $[t_0, b]$ and suppose that for every t_1 in $[t_0, b]$, $\int_{t_0}^{t_1} |\mathbf{f}'(t)|\,dt$ exists. Then $\mathbf{f}$ is said to be **rectifiable** in the interval $[t_0, b]$, and the **arc length** of the curve $\mathbf{f}(t)$ in the interval $[t_0, t_1]$ is given by

$$s(t_1) = \int_{t_0}^{t_1} |\mathbf{f}'(t)|\,dt$$

where $t_0 \leq t_1 \leq b$. ∎

THEOREM 2

Let $\mathbf{f}$ in $\mathbb{R}^3$ have a continuous derivative and let $s(t)$ denote the length of the arc from t_0 to t, then

$$\frac{ds}{dt} = |\mathbf{f}'(t)|. \quad ∎ \tag{10}$$

EXAMPLE 4
CALCULATING ARC LENGTH IN $\mathbb{R}^3$

Let $\mathbf{f}(t) = (\cos t)\mathbf{i} + (\sin t)\mathbf{j} + t\mathbf{k}$. Find the length of the arc from $t = 0$ to $t = 4$.

SOLUTION:

$$f'(t) = -\sin t\mathbf{i} + \cos t\mathbf{j} + \mathbf{k}$$

and

$$|\mathbf{f}'(t)| = \sqrt{\sin^2 t + \cos^2 t + 1} = \sqrt{2},$$

so

$$s(4) = \int_0^4 |\mathbf{f}'(t)| \, dt = \int_0^4 \sqrt{2} \, dt = 4\sqrt{2}.$$

ARC LENGTH AS A PARAMETER

In many problems, it is convenient and more natural to use the arc length s as a parameter. We can think of the vector function $\mathbf{f}$ as the *position vector* of a particle moving in the plane or in space. Then, if $P_0 = (x_0, y_0)$ or (x_0, y_0, z_0) is a fixed point on the curve C described by the vector function $\mathbf{f}$, we may write

$$\mathbf{f}(s) = x(s)\mathbf{i} + y(s)\mathbf{j}, \qquad \text{or} \qquad \mathbf{f}(s) = x(s)\mathbf{i} + y(s)\mathbf{j} + z(s)\mathbf{k},$$

where s is the distance along the curve measured from P_0 in the direction of increasing s. In this way we can determine the x-, y-, and z-components of the position vector as we move s units *along the curve*, and we are using a parameter that is intrinsic to the curve.

EXAMPLE 5
THE UNIT CIRCLE PARAMETRIZED
BY ARC LENGTH

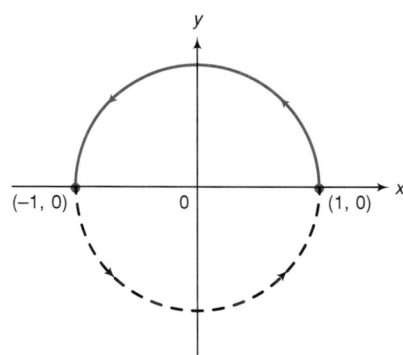

FIGURE 3
GRAPH OF $\mathbf{f}(s) = \cos s\mathbf{i} + \sin s\mathbf{j}$, $0 \le t \le \pi$

Write the vector $\mathbf{f}(t) = \cos t\mathbf{i} + \sin t\mathbf{j}$ (which describes the unit circle) with arc length as a parameter. Take $P_0 = (1, 0)$.

SOLUTION: We have $ds/dt = 1$ (from Example 1), so that since $(1, 0)$ is reached when $t = 0$, we find that

$$s = \int_0^t ds = \int_0^t \frac{ds}{du} \, du = \int_0^t du = t.$$

Thus, we may write

$$\mathbf{f}(s) = \cos s\mathbf{i} + \sin s\mathbf{j}.$$

For example, if we begin at the point $(1, 0)$ and move π units along the unit circle (which is half the unit circle), then we move to the point $(\cos \pi, \sin \pi) = (-1, 0)$. This is what we would expect. See Figure 3.

EXAMPLE 6
WRITING A CURVE WITH ARC
LENGTH AS THE PARAMETER

Let $\mathbf{f}(t) = (2t - t^2)\mathbf{i} + \frac{8}{3}t^{3/2}\mathbf{j}$, $t \ge 0$. Write this curve with arc length as a parameter.

SOLUTION: Suppose that the fixed point is $P_0 = (0, 0)$ when $t = 0$. Then from Example 3 we have

$$\frac{ds}{dt} = 2(t + 1), \qquad \text{and} \qquad s = \int_0^t \frac{ds}{du} \, du = \int_0^t 2(u + 1) \, du = t^2 + 2t.$$

This leads to the equations

$$t^2 + 2t - s = 0 \qquad \text{and} \qquad t = \frac{-2 + \sqrt{4 + 4s}}{2} = \sqrt{1 + s} - 1.$$

We took the positive square root here since it is assumed that t starts at 0 and increases. Then

$$x = 2t - t^2 = 4\sqrt{1 + s} - 4 - s$$

$$y = \tfrac{8}{3}t^{3/2} = \tfrac{8}{3}(\sqrt{1 + s} - 1)^{3/2},$$

and we obtain

$$\mathbf{f}(s) = (4\sqrt{1 + s} - 4 - s)\mathbf{i} + \tfrac{8}{3}(\sqrt{1 + s} - 1)^{3/2}\mathbf{j}.$$

As Example 6 illustrates, writing $\mathbf{f}$ explicitly with arc length as a parameter can be tedious (or, more often, impossible).

EXAMPLE 7
A CIRCULAR HELIX PARAMETRIZED BY ARC LENGTH

Write the equation of a circular helix $\mathbf{f}(t) = \cos t\,\mathbf{i} + \sin t\,\mathbf{j} + t\mathbf{k}$ with arc length as a parameter.

SOLUTION: From Example 4, $\dfrac{ds}{dt} = |\mathbf{f}'(t)| = \sqrt{2}$ so $s = \int_0^t \sqrt{2}\,du = \sqrt{2}t$.

Then $t = \dfrac{s}{\sqrt{2}}$ and we have

$$\mathbf{f}(s) = \cos\frac{s}{\sqrt{2}}\,\mathbf{i} + \sin\frac{s}{\sqrt{2}}\,\mathbf{j} + \frac{s}{\sqrt{2}}\,\mathbf{k}.$$

There is an interesting and important relationship between position vectors, tangent vectors, and normal vectors that becomes apparent when we use s as a parameter.

THEOREM 3 **THE UNIT TANGENT VECTOR PARAMETRIZED BY ARC LENGTH**

If the curve C is parametrized by $\mathbf{f}(s) = x(s)\mathbf{i} + y(s)\mathbf{j}$ or $x(s)\mathbf{i} + y(s)\mathbf{j} + z(s)\mathbf{k}$, where s is arc length and x, y, and z have continuous derivatives, then the unit tangent vector $\mathbf{T}$ is given by

$$\mathbf{T}(s) = \frac{d\mathbf{f}}{ds}. \tag{11}$$

PROOF: With *any* parametrization of C, the unit tangent vector is given by [see equation (12.3.6) on p. 803]

$$\mathbf{T}(t) = \frac{\mathbf{f}'(t)}{|\mathbf{f}'(t)|}.$$

So choosing $t = s$ yields

$$\mathbf{T}(s) = \frac{d\mathbf{f}/ds}{|d\mathbf{f}/ds|}.$$

But from equation (2) or (10), $|\mathbf{f}'(t)| = ds/dt$, so

$$\left|\frac{d\mathbf{f}}{ds}\right| = \left|\frac{d\mathbf{f}/dt}{ds/dt}\right| = \left|\frac{\mathbf{f}'(t)}{ds/dt}\right| = 1,$$

and the proof is complete. ∎

Theorem 3 is quite useful in that it provides a check of our calculation of the parametrization in terms of arc length. For if

$$\mathbf{f}(s) = x(s)\mathbf{i} + y(s)\mathbf{j} \qquad \text{or} \qquad \mathbf{f}(s) = x(s)\mathbf{i} + y(s)\mathbf{j} + z(s)\mathbf{k},$$

then

$$\mathbf{T} = \frac{d\mathbf{f}}{ds} = \frac{dx}{ds}\mathbf{i} + \frac{dy}{ds}\mathbf{j} \qquad \text{or} \qquad \mathbf{T} = \frac{dx}{ds}\mathbf{i} + \frac{dy}{ds}\mathbf{j} + \frac{dz}{ds}\mathbf{k}.$$

But $|\mathbf{T}| = 1$, so that $|\mathbf{T}|^2 = 1$, which implies that

$$\left(\frac{dx}{ds}\right)^2 + \left(\frac{dy}{ds}\right)^2 = 1 \qquad \text{or} \qquad \left(\frac{dx}{ds}\right)^2 + \left(\frac{dy}{ds}\right)^2 + \left(\frac{dz}{ds}\right)^2 = 1. \qquad (12)$$

In Problems 37 and 38, you are asked to show that equation (12) holds for the functions computed in Examples 6 and 7.

■ PROBLEMS 12.5

■ SELF-QUIZ

I. Let $\mathbf{f}(t) = (3 + t)\mathbf{i} + (2 - t)\mathbf{j} + (1 + 3t)\mathbf{k}$. The arc length between the points where $t = 1$ and $t = 4$ is _____.

   a. 3   b. $\sqrt{11}$ c. 99   d. $\sqrt{99}$

II. If we reparametrize $\mathbf{f}(t) = (-1, 7, 2) + t(3, 4, 12)$ in terms of the arc length s from the point where $t = 0$, then $\mathbf{f}(s) = $ _____.

   a. $(-1, 7, 2) + s(\frac{3}{13}, \frac{4}{13}, \frac{12}{13})$
   b. $s(\frac{3}{13}, \frac{4}{13}, \frac{12}{13})$
   c. $s(-1/\sqrt{54}, 7/\sqrt{54}, 2/\sqrt{54})$
   d. $(3, 4, 12) + s(-1/\sqrt{54}, 7/\sqrt{54}, 2/\sqrt{54})$

■ DRILL

In Problems 1–12, find the length of the given arc over the specified interval or the length of the closed curve.

1. $x = t^3$; $y = t^2$; $-1 \leq t \leq 1$ [Hint: $\sqrt{t^2} = -t$ for $t < 0$.]
2. $x = \cos 2\theta$; $y = \sin 2\theta$; $0 \leq \theta \leq \pi/2$
3. $x = t^3 + 1$; $y = 3t^2 + 2$; $0 \leq t \leq 2$
4. $x = 1 + t$; $y = (1 + t)^{3/2}$; $0 \leq t \leq 1$
5. One arc of the cycloid $x = a(\theta - \sin \theta)$, $y = a(1 - \cos \theta)$
6. The hypocycloid of four cusps $x = a \cos^3 \theta$, $y = a \sin^3 \theta$ [Hint: Calculate the length in the first quadrant and multiply by 4.]
*7. $\mathbf{f}(t) = \dfrac{1}{\sqrt{1 + t}}\mathbf{i} + \dfrac{t}{2(1 + t)}\mathbf{j}$; $0 \leq t \leq 4$

   [Hint: Substitute $u^2 = t + 2$ and integrate by partial fractions.]
8. $\mathbf{g}(t) = \frac{2}{3}t^3\mathbf{i} + (1 + t^{9/2})\mathbf{j} + (1 - t^{9/2})\mathbf{k}$; $0 \leq t \leq 2$
9. $\mathbf{f}(t) = e^t(\cos t)\mathbf{i} + e^t(\sin t)\mathbf{j}$; $0 \leq t \leq \pi/2$

10. $\mathbf{g}(t) = e^t(\cos 2t)\mathbf{i} + e^t(\sin 2t)\mathbf{j} + e^t\mathbf{k}$; $1 \leq t \leq 4$
11. $\mathbf{f}(t) = (\sin^2 t)\mathbf{i} + (\cos^2 t)\mathbf{j}$; $0 \leq t \leq \pi/2$
12. $\mathbf{g}(t) = 2(\cos 3t)\mathbf{i} + 2(\sin 3t)\mathbf{j} + t^2\mathbf{k}$; $0 \leq t \leq 10$

In Problems 13–20, use the result of Problem 36 to find the length of the given arc over the specified interval or the length of the given closed curve.

13. $r - a \sin \theta$; $0 \leq \theta \leq \pi/2$
14. $r = a \cos \theta$; $0 \leq \theta \leq \pi$
15. $r = \theta^2$; $0 \leq \theta \leq \pi$
16. $r = 6 \cos^2(\theta/2)$; $0 \leq \theta \leq \pi/2$
*17. The cardioid $r = a(1 + \sin \theta)$ $(a > 0)$ [Hint:

$$\int \sqrt{1 + \sin \theta}\, d\theta = \int \sqrt{1 + \sin \theta} \cdot \frac{\sqrt{1 - \sin \theta}}{\sqrt{1 - \sin \theta}}\, d\theta.$$

Pay attention to signs.]
*18. $r = a\theta$; $0 \leq \theta \leq 2\pi$; $a > 0$
19. $r = e^\theta$; $0 \leq \theta \leq 3$
20. $r = \sin^3(\theta/3)$; $0 \leq \theta \leq \pi/2$

   [Hint: $\sin^2(\theta/3) = \frac{1}{2}(1 - \cos(2\theta/3)).$]

CONTINUED

■ APPLICATIONS

In Problems 21–32, find parametric equations in terms of the arc length s measured from the point reached when $t = 0$ (suppose all constants a, b, c, d are positive). Verify your solution by using formula (12).

21. $\mathbf{f}(t) = 3t^2\mathbf{i} + 2t^3\mathbf{j}$
22. $\mathbf{f}(t) = t^3\mathbf{i} + t^2\mathbf{j}$
23. $\mathbf{f}(t) = (t^3 + 1)\mathbf{i} + (t^2 - 1)\mathbf{j}$
24. $\mathbf{f}(t) = (3t^2 + a)\mathbf{i} + (2t^3 + b)\mathbf{j}$
25. $\mathbf{f}(t) = 3(\cos t)\mathbf{i} + 3(\sin t)\mathbf{j}$
26. $\mathbf{f}(t) = a(\sin t)\mathbf{i} + a(\cos t)\mathbf{j}$
27. $\mathbf{f}(t) = a(\cos t)\mathbf{i} - a(\sin t)\mathbf{j}$
28. $\mathbf{f}(t) = 3(\cos t + t \sin t)\mathbf{i} + 3(\sin t - t \cos t)\mathbf{j}$
29. $\mathbf{f}(t) = (a + b \cos t)\mathbf{i} + (c + b \sin t)\mathbf{j}$
30. $\mathbf{f}(t) = ae^t(\cos t)\mathbf{i} + ae^t(\sin t)\mathbf{j}$

31. One cusp of the hypocycloid of four cusps $\mathbf{f}(t) = a(\cos^3 t)\mathbf{i} + a(\sin^3 t)\mathbf{j}$; $0 \le t \le \pi/2$.
*32. The cycloid $x = a(\theta - \sin \theta)$, $y = a(1 - \cos \theta)$.
33. The ellipse $(x/a)^2 + (y/b)^2 = 1$ has parametric representation $x = a \cos \theta$, $y = b \sin \theta$. Find an integral that computes the length of the circumference of this ellipse. [*Note:* Do not try to evaluate this **elliptic integral**. It arises in a variety of physical applications, but cannot be integrated in terms of common functions (except numerically) unless $a = b$.]
*34. A tack stuck in the front tire of a bicycle wheel has a diameter of 1 m. What is the total distance traveled by the tack if the bicycle moves forward a total of 30π m?

■ SHOW/PROVE/DISPROVE

35. This problem asks you to reconcile the arc length formula presented in equation (4) with that given in equation (5.3.3): Suppose $\mathbf{f}(t) = x(t)\mathbf{i} + y(t)\mathbf{j}$ is a smooth function of t for $t_0 \le t \le t_1$. Also suppose y can be expressed as a differentiable monotonic function of x with $x(t_0) = a < x(t_1) = b$. Show that

$$\int_a^b \sqrt{1 + \left(\frac{dy}{dx}\right)^2}\, dx = \int_{t_0}^{t_1} \sqrt{\left(\frac{dx}{dt}\right)^2 + \left(\frac{dy}{dt}\right)^2}\, dt.$$

36. Consider a smooth curve which satisfies the polar coordinate equation $r = f(\theta)$. Show that the length of the arc for $\alpha \le \theta \le \beta$ is

$$s = \int_\alpha^\beta \sqrt{r^2 + \left(\frac{dr}{d\theta}\right)^2}\, d\theta.$$

[*Hint:* Problem 12.2.39 has a useful parametrization for Cartesian coordinates.]

37. In Example 6, show that $(dx/ds)^2 + (dy/ds)^2 = 1$.
38. In Example 7, show that $(dx/ds)^2 + (dy/ds)^2 + (dz/ds)^2 = 1$.

■ ANSWERS TO SELF-QUIZ

I. d   II. a

12.6 CURVATURE AND THE ACCELERATION VECTOR (OPTIONAL)

The derivative dy/dx of a curve $y = f(x)$ measures the rate of change of the vertical component of the curve with respect to the horizontal component. As we have seen, the derivative ds/dt represents the change in the length of the arc traced out by the vector $\mathbf{f} = x(t)\mathbf{i} + y(t)\mathbf{j}$ as t increases. Another quantity of interest is the rate of change of the direction of the curve with respect to the length of the curve. That is, how much does the direction change for every one-unit change in the arc length?

We begin our discussion in $\mathbb{R}^2$.

DEFINITION   **CURVATURE IN $\mathbb{R}^2$ AND RADIUS OF CURVATURE**

(i) Let the curve C be given by the differentiable vector function $\mathbf{f}(t) = f_1(t)\mathbf{i} + f_2(t)\mathbf{j}$. Let $\varphi(t)$ denote the direction of $\mathbf{f}'(t)$. Then the **curvature** of C, denoted by $\kappa(t)$, is the absolute value of the rate of change of direction with respect to arc length; that is,

$$\kappa(t) = \left| \frac{d\varphi}{ds} \right|. \tag{1}$$

Note that $\kappa(t) \geq 0$.

(ii) The **radius of curvature** $\rho(t)$ is defined by

$$\rho(t) = \frac{1}{\kappa(t)} \qquad \text{if } \kappa(t) > 0. \quad \blacksquare \tag{2}$$

REMARK 1: The curvature is a measure of *how fast* the curve turns as we move along it.

REMARK 2: If $\kappa(t) = 0$, we say that the radius of curvature is *infinite*. To understand this idea, note that if $\kappa(t) = 0$, then the "curve" does not bend and so is a straight line (see Example 2). A straight line can be thought of as an arc of a circle with *infinite* radius.

In $\mathbb{R}^3$, we will need a definition of curvature that does not depend on the angle φ. We first prove the following theorem in $\mathbb{R}^2$.

THEOREM 1

If $\mathbf{T}(t)$ denotes the unit tangent vector to $\mathbf{f}$, then

$$\kappa(t) = \left| \frac{d\mathbf{T}}{ds} \right|. \tag{3}$$

PROOF: By the chain rule (which applies just as well to vector-valued functions),

$$\frac{d\mathbf{T}}{ds} = \frac{d\mathbf{T}}{d\varphi} \frac{d\varphi}{ds}.$$

Since φ is the direction of $\mathbf{f}'$, and therefore is also the direction of $\mathbf{T}$, we have

$$\mathbf{T} = (\cos \varphi)\mathbf{i} + (\sin \varphi)\mathbf{j},$$

so that

$$\frac{d\mathbf{T}}{d\varphi} = -(\sin \varphi)\mathbf{i} + (\cos \varphi)\mathbf{j} \qquad \text{and}$$

$$\left| \frac{d\mathbf{T}}{d\varphi} \right| = \sqrt{\sin^2 \varphi + \cos^2 \varphi} = 1.$$

Thus,

$$\left| \frac{d\mathbf{T}}{ds} \right| = \left| \frac{d\mathbf{T}}{d\varphi} \right| \left| \frac{d\varphi}{ds} \right| = 1 \left| \frac{d\varphi}{ds} \right| = \kappa(t),$$

and the theorem is proved. $\quad \blacksquare$

We now derive an easier way to calculate $\kappa(t)$.

THEOREM 2

With the curve C given as in the definition of curvature in $\mathbb{R}^2$, the curvature of C is given by the formula

$$\kappa(t) = \frac{|(dx/dt)(d^2y/dt^2) - (dy/dt)(d^2x/dt^2)|}{[(dx/dt)^2 + (dy/dt)^2]^{3/2}}$$

$$= \frac{|x'y'' - y'x''|}{[(x')^2 + (y')^2]^{3/2}} \tag{4}$$

where $x(t) = f_1(t)$ and $y(t) = f_2(t)$.

PROOF: By the chain rule,

$$\frac{d\varphi}{ds} = \frac{d\varphi}{dt}\frac{dt}{ds} = \frac{d\varphi/dt}{ds/dt}. \tag{5}$$

From equation (12.5.7) on page 813,

$$\frac{ds}{dt} = \sqrt{\left(\frac{dx}{dt}\right)^2 + \left(\frac{dy}{dt}\right)^2}. \tag{6}$$

Let φ denote the direction of $\mathbf{f}'$. Then, from equation (11.1.2) on page 722,

$$\tan \varphi = \frac{dy/dt}{dx/dt}, \tag{7}$$

or

$$\varphi = \tan^{-1}\frac{dy/dt}{dx/dt}. \tag{8}$$

Differentiating both sides of (8) with respect to t, we obtain

$$\frac{d\varphi}{dt} = \frac{1}{1 + \left(\dfrac{dy/dt}{dx/dt}\right)^2}\frac{d}{dt}\left(\frac{dy/dt}{dx/dt}\right)$$

$$= \frac{\left(\dfrac{dx}{dt}\right)^2}{\left(\dfrac{dx}{dt}\right)^2 + \left(\dfrac{dy}{dt}\right)^2} \cdot \frac{\dfrac{dx}{dt}\dfrac{d^2y}{dt^2} - \dfrac{dy}{dt}\dfrac{d^2x}{dt^2}}{\left(\dfrac{dx}{dt}\right)^2}. \tag{9}$$

Substitution of (6) and (9) into (5) completes the proof of the theorem. ∎

EXAMPLE 1

THE CURVATURE AND RADIUS OF CURVATURE OF A CIRCLE

We certainly expect that the curvature of a circle is constant and that its radius of curvature is its radius. Show that this is true.

SOLUTION: The circle of radius r centered at the origin is given parametrically by

$$x = r \cos t, \qquad y = r \sin t.$$

Then $dx/dt = -r \sin t$, $d^2x/dt^2 = -r \cos t$, $dy/dt = r \cos t$, and $d^2y/dt^2 = -r \sin t$, and from (4),

$$\kappa(t) = \frac{|r^2 \sin^2 t + r^2 \cos^2 t|}{[r^2 \sin^2 t + r^2 \cos^2 t]^{3/2}} = \frac{r^2}{r^3} = \frac{1}{r} \quad \text{and} \quad \rho(t) = \frac{1}{\kappa(t)} = r,$$

as expected.

EXAMPLE 2
THE CURVATURE OF A STRAIGHT
LINE IS ZERO

Show that for a straight line $\kappa(t) = 0$.

SOLUTION: A line can be represented parametrically [see Problem 12.1.49] by $x = x_1 + t(x_2 - x_1)$ and $y = y_1 + t(y_2 - y_1)$. Then $dx/dt = x_2 - x_1$, $d^2x/dt^2 = 0$, $dy/dt = y_2 - y_1$, and $d^2y/dt^2 = 0$. Substitution of these values into (4) immediately yields $\kappa(t) = 0$.

EXAMPLE 3
FINDING THE CURVATURE AND
RADIUS OF CURVATURE OF A
CURVE

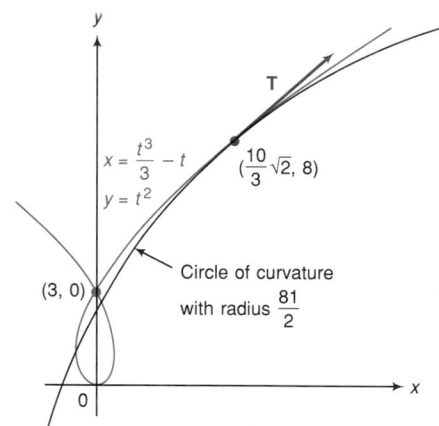

Find the curvature and radius of curvature of the curve given parametrically by $x = (t^3/3) - t$ and $y = t^2$ (see Example 12.4.4) at $t = 2\sqrt{2}$.

SOLUTION: We have $dx/dt = t^2 - 1$, $d^2x/dt^2 = 2t$, $dy/dt = 2t$, and $d^2y/dt^2 = 2$, so

$$\kappa(t) = \frac{|2(t^2 - 1) - 2t(2t)|}{[(t^2 - 1)^2 + (2t)^2]^{3/2}} = \frac{2(t^2 + 1)}{(t^4 + 2t^2 + 1)^{3/2}} = \frac{2(t^2 + 1)}{(t^2 + 1)^3}$$

$$= \frac{2}{(t^2 + 1)^2}.$$

Then $\kappa(2\sqrt{2}) = 2/81$, and the radius of curvature $\rho(2\sqrt{2}) = 1/\kappa(2\sqrt{2}) = 81/2$. A portion of this curve near $t = 2\sqrt{2}$ is sketched in Figure 1. For reference purposes, the unit tangent vector $\mathbf{T}$ at $t = 2\sqrt{2}$ is included.

NOTE: In Figure 1, the circle with radius of curvature $\rho(t)$ that lies on the concave side of C and is tangent to the curve at a point is called the **circle of curvature**, or **osculating circle**.

FIGURE 1
THE CIRCLE OF CURVATURE OR OSCU-
LATING CIRCLE AT $t = 2\sqrt{2}$ OF THE CURVE
$\mathbf{f}(t) = \left(\frac{t^3}{3} - t\right)\mathbf{i} + t^2\mathbf{j}$

CURVATURE AND NORMAL VECTORS IN $\mathbb{R}^3$

The curvature $\kappa(t)$ and the unit normal vector $\mathbf{n}(t)$ are defined differently in $\mathbb{R}^3$. They would have to be. For one thing, there is a whole "plane-full" of unit vectors orthogonal to $\mathbf{T}$. Also, curvature cannot be defined as $|d\varphi/ds|$ since there are now three angles that define the direction of C. Therefore, we use an alternative definition of curvature (see Theorem 1).

DEFINITION **CURVATURE IN $\mathbb{R}^3$**

If $\mathbf{f}$ has a continuous derivative, then the **curvature** of $\mathbf{f}$ is given by

$$\kappa(t) = \left|\frac{d\mathbf{T}}{ds}\right|. \quad \blacksquare \tag{10}$$

REMARK: By Theorem 1, equation (10) is equivalent in $\mathbb{R}^2$ to the definition $\kappa(t) = |d\varphi/ds|$, which was our $\mathbb{R}^2$ definition.

DEFINITION **PRINCIPAL UNIT NORMAL VECTOR**

For any value of t for which $\kappa(t) \neq 0$, the **principal unit normal vector** **n** is defined by

$$\mathbf{n}(t) = \frac{1}{\kappa(t)} \frac{d\mathbf{T}}{ds}. \quad \blacksquare \tag{11}$$

REMARK: Equation (11) also holds in $\mathbb{R}^2$.

Having defined a unit normal vector, we must show that it is orthogonal to the unit tangent vector in order to justify its name.

THEOREM 3

$\mathbf{n}(t) \perp \mathbf{T}(t)$.

PROOF: Since $1 = \mathbf{T} \cdot \mathbf{T}$, we differentiate both sides with respect to s to obtain

$$0 = \mathbf{T} \cdot \frac{d\mathbf{T}}{ds} + \frac{d\mathbf{T}}{ds} \cdot \mathbf{T} = 2\mathbf{T} \cdot \frac{d\mathbf{T}}{ds} = (2\kappa)\mathbf{T} \cdot \mathbf{n}. \quad \blacksquare$$

EXAMPLE 4
THE CURVATURE AND PRINCIPAL UNIT NORMAL VECTOR FOR A CIRCULAR HELIX

Find the curvature and principal unit normal vector, $\kappa(t)$ and $\mathbf{n}(t)$, for the circular helix $f(t) = \cos t\mathbf{i} + \sin t\mathbf{j} + t\mathbf{k}$ at $t = \pi/3$.

SOLUTION: From Example 12.5.7, $t = s/\sqrt{2}$. From Example 12.3.6,

$$\mathbf{T}(t) = \frac{1}{\sqrt{2}}(-\sin t\mathbf{i} + \cos t\mathbf{j} + \mathbf{k}).$$

Inserting $t = s/\sqrt{2}$ into this equation yields

$$\mathbf{T}(s) = -\frac{\sin(s/\sqrt{2})}{\sqrt{2}}\mathbf{i} + \frac{\cos(s/\sqrt{2})}{\sqrt{2}}\mathbf{j} + \frac{1}{\sqrt{2}}\mathbf{k}$$

and

$$\frac{d\mathbf{T}}{ds} = -\frac{\cos(s/\sqrt{2})}{2}\mathbf{i} - \frac{\sin(s/\sqrt{2})}{2}\mathbf{j}.$$

Hence,

$$\kappa(t) = \sqrt{\frac{\cos^2(s/\sqrt{2})}{4} + \frac{\sin^2(s/\sqrt{2})}{4}} = \frac{1}{2}$$

and

$$\mathbf{n}(t) = \frac{1}{\kappa(t)} \frac{d\mathbf{T}}{ds} = -(\cos t)\mathbf{i} - (\sin t)\mathbf{j}.$$

At $t = \pi/3$,

$$\mathbf{n}\left(\frac{\pi}{3}\right) = -\frac{1}{2}\mathbf{i} - \frac{\sqrt{3}}{2}\mathbf{j}.$$

REMARK: When it is not convenient to solve for t in terms of s, we can calculate $d\mathbf{T}/ds$ directly by the relation

$$\frac{d\mathbf{T}}{ds} = \frac{d\mathbf{T}/dt}{ds/dt}. \tag{12}$$

There is a third vector that is often useful in applications.

DEFINITION **BINORMAL VECTOR**

The **binormal vector B** to the curve $\mathbf{f}$ in $\mathbb{R}^3$ is defined by

$$\mathbf{B} = \mathbf{T} \times \mathbf{n}. \quad \blacksquare \tag{13}$$

From this definition, we see that $\mathbf{B}$ is orthogonal to both $\mathbf{T}$ and $\mathbf{n}$. Moreover, since $\mathbf{T} \perp \mathbf{n}$, the angle θ between $\mathbf{T}$ and $\mathbf{n}$ is $\pi/2$, and

$$|\mathbf{B}| = |\mathbf{T} \times \mathbf{n}| = |\mathbf{T}|\,|\mathbf{n}|\sin\theta = 1,$$

so that $\mathbf{B}$ is a unit vector. Also, $\mathbf{T}$, $\mathbf{n}$, and $\mathbf{B}$ form a right-handed system, just like the vectors $\mathbf{i}$, $\mathbf{j}$, and $\mathbf{k}$. Unlike these last three vectors, $\mathbf{T}$, $\mathbf{n}$, and $\mathbf{B}$ are derived from properties of the curve itself.

EXAMPLE 5
CALCULATING A BINORMAL
VECTOR

Find the binormal vector to the curve of Example 4 at $t = \pi/3$.

SOLUTION:

$$\mathbf{B} = \mathbf{T} \times \mathbf{n} = \begin{vmatrix} \mathbf{i} & \mathbf{j} & \mathbf{k} \\ -\dfrac{\sin t}{\sqrt{2}} & \dfrac{\cos t}{\sqrt{2}} & \dfrac{1}{\sqrt{2}} \\ -\cos t & -\sin t & 0 \end{vmatrix} = \frac{\sin t}{\sqrt{2}}\mathbf{i} - \frac{\cos t}{\sqrt{2}}\mathbf{j} + \frac{1}{\sqrt{2}}\mathbf{k}$$

When $t = \pi/3$, $\mathbf{B} = (\sqrt{3}/2\sqrt{2})\mathbf{i} - (1/2\sqrt{2})\mathbf{j} + (1/\sqrt{2})\mathbf{k}$. The vectors $\mathbf{T}$, $\mathbf{n}$, and $\mathbf{B}$ are sketched in Figure 2.

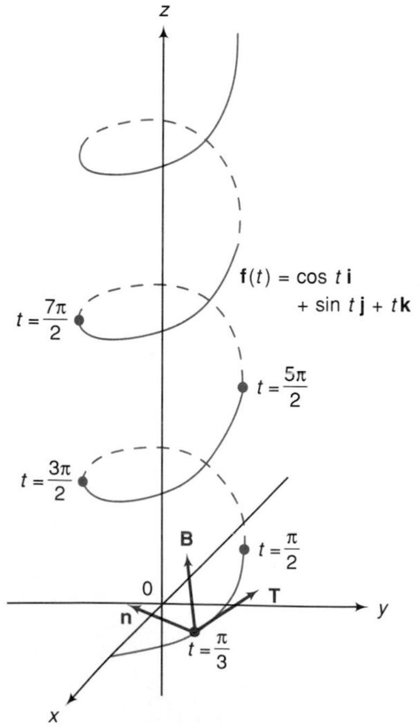

$f(t) = \cos t\,\mathbf{i} + \sin t\,\mathbf{j} + t\,\mathbf{k}$

FIGURE 2
TANGENT, NORMAL, AND BINORMAL
VECTORS FOR CIRCULAR HELIX

TANGENTIAL AND NORMAL COMPONENTS OF ACCELERATION

There is an interesting relationship between curvature and acceleration vectors. Suppose that a particle is moving along the curve C with position vector

$$\mathbf{f}(t) = x(t)\mathbf{i} + y(t)\mathbf{j}, \qquad \text{or} \qquad \mathbf{f}(t) = x(t)\mathbf{i} + y(t)\mathbf{j} + z(t)\mathbf{k}.$$

We know that velocity is the derivative of distance and acceleration is the derivative of velocity, so

velocity vector: $\quad \mathbf{v}(t) = x'(t)\mathbf{i} + y'(t)\mathbf{j}$ or

$$\mathbf{v}(t) = x'(t)\mathbf{i} + y'(t)\mathbf{j} + z'(t)\mathbf{k}.$$

acceleration vector: $\mathbf{a}(t) = \dfrac{d^2x}{dt^2}\,\mathbf{i} + \dfrac{d^2y}{dt^2}\,\mathbf{j}$ or   (14)

$$\mathbf{a}(t) = \frac{d^2x}{dt^2}\,\mathbf{i} + \frac{d^2y}{dt^2}\,\mathbf{j} + \frac{d^2z}{dt^2}\,\mathbf{k}.$$ (15)

The representation (14) resolves **a** into its horizontal and vertical components. However, there is another representation that is often more useful. Imagine yourself driving on the highway. If the car in which you are riding accelerates forward, you are pressed to the back of your seat. If it turns sharply to one side, you are thrown to the other. Both motions are due to acceleration. The second force is related to the rate at which the car turns, which is, of course, related to the curvature of the road. Thus, we would like to express the acceleration vector as a component in the direction of motion and a component somehow related to the curvature of the path. How do we do so? A glance back at Figure 1 on page 822 reveals the answer. The radial line of the circle of curvature is perpendicular to the unit tangent vector **T** since **T** is tangent to the circle of curvature, and in a circle, tangent and radial lines at a point are orthogonal. (See Example 12.4.3.) But a vector that is perpendicular to **T** has the direction of the unit normal vector **n**. Thus, the component of acceleration in the direction **n** will be a measure of the acceleration due to turning. We would like to write

$$\mathbf{a} = a_\mathrm{T}\mathbf{T} + a_\mathrm{n}\mathbf{n}$$ (16)

where a_T and a_n are, respectively, the components of **a** in the tangential and normal directions.

THEOREM 4

$$\mathbf{a}(t) = \frac{d^2s}{dt^2}\,\mathbf{T} + \left(\frac{ds}{dt}\right)^2 \kappa\mathbf{n}$$ (17)

> Chain rule   Theorem 12.5.3

PROOF: $\mathbf{v}(t) = (d\mathbf{f}/dt) \overset{\downarrow}{=} (d\mathbf{f}/ds)(ds/dt) \overset{\downarrow}{=} \mathbf{T}(ds/dt)$. Then

> Product rule

$$\mathbf{a}(t) = \frac{d\mathbf{v}}{dt} \overset{\downarrow}{=} \mathbf{T}\frac{d^2s}{dt^2} + \frac{d\mathbf{T}}{dt}\frac{ds}{dt} = \mathbf{T}\frac{d^2s}{dt^2} + \left(\frac{d\mathbf{T}}{ds}\frac{ds}{dt}\right)\left(\frac{ds}{dt}\right)$$

$$= \frac{d^2s}{dt^2}\,\mathbf{T} + \kappa\mathbf{n}\left(\frac{ds}{dt}\right)^2.$$

The last step follows from equation (11). ■

NOTE: $v = \dfrac{ds}{dt} = |\mathbf{f}'(t)|$ is called the **speed** of a moving particle. Then, since

$\kappa = \dfrac{1}{\rho}$, where ρ is the radius of curvature, we can write (16) as

WRITING a IN TERMS OF T AND n

$$\mathbf{a} = \frac{dv}{dt}\,\mathbf{T} + v^2\kappa\mathbf{n} = \frac{dv}{dt}\,\mathbf{T} + \frac{v^2}{\rho}\,\mathbf{n}.$$ (18)

Thus, the tangential component of acceleration is $a_T = dv/dt$, and the normal component of acceleration is $a_n = v^2/\rho$.

Using Theorem 4, we can prove the next theorem.

THEOREM 5

If $\mathbf{f}(t) \in \mathbb{R}^3$ and $\mathbf{f}'$ and $\mathbf{f}''$ exist,

$$\kappa = \frac{|\mathbf{f}' \times \mathbf{f}''|}{|\mathbf{f}'|^3}.$$

PROOF: $\mathbf{f}' \times \mathbf{f}'' = (\mathbf{T}\, ds/dt) \times [(d^2s/dt^2)\mathbf{T} + \kappa \mathbf{n}(ds/dt)^2]$. Now $\mathbf{T} \times \mathbf{T} = \mathbf{0}$ and $\mathbf{T} \times \mathbf{n} = \mathbf{B}$. Thus using Theorem 11.6.2 (iv) on page 759, we have

$$\mathbf{f}' \times \mathbf{f}'' = \left(\frac{ds}{dt}\right)^3 \kappa \mathbf{B}.$$

Then taking absolute values and using the fact that $|ds/dt| = |\mathbf{f}'|$ yields

$$|\mathbf{f}' \times \mathbf{f}''| = \kappa |\mathbf{f}'|^3,$$

from which the result follows. ■

Theorem 5 is useful for calculating curvature when it is not easy to write $\mathbf{f}$ in terms of the arc length parameter s.

EXAMPLE 6
CALCULATING THE CURVATURE
OF A CURVE IN $\mathbb{R}^3$

Calculate the curvature of $\mathbf{f} = t^2\mathbf{i} + t^3\mathbf{j} + t^4\mathbf{k}$ at $t = 1$.

SOLUTION: Here,

$$\mathbf{f}' = 2t\mathbf{i} + 3t^2\mathbf{j} + 4t^3\mathbf{k}$$

$$|\mathbf{f}'| = \sqrt{4t^2 + 9t^4 + 16t^6} = t\sqrt{4 + 9t^2 + 16t^4}$$

$$\mathbf{f}'' = 2\mathbf{i} + 6t\mathbf{j} + 12t^2\mathbf{k}$$

$$\mathbf{f}' \times \mathbf{f}'' = \begin{vmatrix} \mathbf{i} & \mathbf{j} & \mathbf{k} \\ 2t & 3t^2 & 4t^3 \\ 2 & 6t & 12t^2 \end{vmatrix} = 12t^4\mathbf{i} - 16t^3\mathbf{j} + 6t^2\mathbf{k}$$

$$|\mathbf{f}' \times \mathbf{f}''| = \sqrt{144t^8 + 256t^6 + 36t^4} = t^2\sqrt{144t^4 + 256t^2 + 36}.$$

Thus,

$$\kappa = \frac{\sqrt{144t^4 + 256t^2 + 36}}{t(4 + 9t^2 + 16t^4)^{3/2}}.$$

At $t = 1$, $\kappa = \sqrt{436}/29^{3/2} \approx 0.1337$.

The result given by (17) or (18) is very important in physics. If an object of constant mass m is traveling along a trajectory, then the force on the object to produce that trajectory is given by

$$\mathbf{F} = m\mathbf{a} = m\frac{dv}{dt}\mathbf{T} + mv^2\kappa\mathbf{n}. \tag{19}$$

The term $mv^2\kappa$ is the magnitude of the force necessary to keep the project from "moving off" the trajectory in the direction of $\mathbf{T}$. In the case of automobiles, the force is supplied by tire friction.

EXAMPLE 7
COMPUTING FRICTIONAL FORCE

A 1500-kg race car is driven at a speed of 150 km/hr on an unbanked circular race track of radius 250 m. What frictional force must be exerted by the tires on the road surface to keep the car from skidding?

SOLUTION: The frictional force exerted by the tires must be equal to the component of the force (due to acceleration) normal to the circular race track. That is,

$$F = mv^2\kappa = \frac{mv^2}{\rho} = (1500 \text{ kg})\frac{(150{,}000 \text{ m})^2}{(3600 \text{ sec})^2} \cdot \frac{1}{250 \text{ m}}$$

$$= 10{,}416\tfrac{2}{3}(\text{kg})(\text{m})/\text{sec}^2 = 10{,}416\tfrac{2}{3} \text{ N}.$$

EXAMPLE 8
FINDING THE SMALLEST POSSIBLE
COEFFICIENT OF FRICTION IN
ORDER THAT A CAR REMAIN
ON A ROAD

Let the car of Example 7 have the **coefficient of friction** μ. That is, the maximum frictional force that can be exerted by the car on the road surface is μmg, where mg is the **normal force** of the car on the road (the force of the car on the road due to gravity). What is the minimum value μ can take in order that the car not slide off the road?

SOLUTION: We must have $\mu mg \geq 10{,}416\tfrac{2}{3}$ N. But $\mu mg = \mu(9.81)(1500)$, so we obtain

$$\mu \geq \frac{10{,}416\tfrac{2}{3}}{(9.81)(1500)} \approx 0.71.$$

■ PROBLEMS 12.6

■ **SELF-QUIZ**

I. True–False:
   a. A parabola that opens downward has negative curvature.
   b. A circle with radius r has curvature equal to $1/r$.
   c. Every straight line has curvature equal to 1.
   d. There is exactly one straight line with curvature equal to 1.
   e. There is exactly one straight line with curvature equal to 0.
   f. $\mathbf{n}(t) \perp \mathbf{T}(t)$ for every smooth curve in $\mathbb{R}^3$.

CONTINUED

■ DRILL

In Problems 1–28, find the curvature and radius of curvature for each curve in $\mathbb{R}^2$. Sketch the unit tangent vector and the circle of curvature (the osculating circle). [*Note:* For some problems stated in Cartesian coordinates, you may find it convenient to use the result of Problem 55; for those involving polar coordinates, use the result of Problem 56.]

1. $\mathbf{f}(t) = t\mathbf{i} + t^2\mathbf{j}; t = 1$
2. $\mathbf{f}(t) = -t\mathbf{i} + t^2\mathbf{j}; t = 2$
3. $\mathbf{f}(t) = 2\cos t\mathbf{i} + 2\sin t\mathbf{j}; t = \pi/4$
4. $\mathbf{f}(t) = 2\cos t\mathbf{i} + 2\sin t\mathbf{j}; t = \pi/2$
5. $\mathbf{f}(t) = 3\sin t\mathbf{i} + 4\cos t\mathbf{j}; t = 0$
6. $\mathbf{f}(t) = 3\sin t\mathbf{i} + 4\cos t\mathbf{j}; t = \pi/2$
7. $\mathbf{f}(t) = 3\cos t\mathbf{i} + 4\sin t\mathbf{j}; t = \pi/4$
8. $\mathbf{f}(t) = 3\cos t\mathbf{i} - 4\sin t\mathbf{j}; t = \pi/3$
9. $\mathbf{f}(t) = (t - \sin t)\mathbf{i} + (1 - \cos t)\mathbf{j}; t = \pi/3$
10. $\mathbf{f}(t) = (\cos t + t\sin t)\mathbf{i} + (\sin t - t\cos t)\mathbf{j}; t = \pi/6$
11. $xy = 1; (1, 1)$ 12. $y = 3/x; (1, 3)$
13. $y = e^x; (0, 1)$ 14. $y = \ln x; (1, 0)$
15. $y = \ln x; (e, 1)$ 16. $y = e^x; (1, e)$
17. $y = \cos x; (\pi/3, \frac{1}{2})$
18. $y = \ln\cos x; (\pi/4, -\frac{1}{2}\ln 2)$
19. $y = ax^2 + bx + c; a \neq 0; (0, c)$
20. $y = \sqrt{1 - x^2}; (0, 1)$
*21. $y = \sin^{-1} x; (1, \pi/2)$
*22. $x = \cos y; (0, \pi/2)$
23. $r = 2a\cos\theta; \theta = \pi/3$
24. $r = a\sin 2\theta; \theta = \pi/8$
25. $r = a(1 + \sin\theta); \theta = \pi/2$
26. $r = a(1 - \cos\theta); \theta = \pi$
27. $r = a\theta; \theta = 1$ 28. $r = e^{a\theta}; \theta = 1$

In Problems 29–36, find the unit tangent vector $\mathbf{T}$, the curvature κ, the principal unit normal $\mathbf{n}$, and the binormal vector $\mathbf{B}$ at the specified value of t; then verify that $\mathbf{n} \cdot \mathbf{T} = 0$.

29. $\mathbf{g}(t) = \mathbf{i} + t\mathbf{j} + t^2\mathbf{k}; t = 1$
30. $\mathbf{g}(t) = t\mathbf{i} + t^2\mathbf{j} + t^3\mathbf{k}; t = 0$
31. $\mathbf{g}(t) = a(\sin t)\mathbf{i} + a(\cos t)\mathbf{j} + t\mathbf{k}; a > 0; t = \pi/4$
32. $\mathbf{g}(t) = a(\sin t)\mathbf{i} + a(\cos t)\mathbf{j} + t\mathbf{k}; a > 0; t = \pi/6$
33. $\mathbf{g}(t) = a(\cos t)\mathbf{i} + b(\sin t)\mathbf{j} + t\mathbf{k}; a > 0, b > 0; a \neq b; t = 0$
34. $\mathbf{g}(t) = a(\cos t)\mathbf{i} + b(\sin t)\mathbf{j} + t\mathbf{k}; a > 0, b > 0, a \neq b; t = \pi/2$
35. $\mathbf{g}(t) = e^t(\cos 2t)\mathbf{i} + e^t(\sin 2t)\mathbf{j} + e^t\mathbf{k}; t = 0$
36. $\mathbf{g}(t) = e^t(\cos 2t)\mathbf{i} + e^t(\sin 2t)\mathbf{j} + e^t\mathbf{k}; t = \pi/4$

■ APPLICATIONS

In Problems 37–44, find the tangential and normal components of acceleration for the given position vector.

37. $\mathbf{f}(t) = (\cos 2t)\mathbf{i} + (\sin 2t)\mathbf{j}$
38. $\mathbf{f}(t) = 2(\cos t)\mathbf{i} + 3(\sin t)\mathbf{j}$
39. $\mathbf{f}(t) = t\mathbf{i} + t^2\mathbf{j}$
40. $\mathbf{f}(t) = t^2\mathbf{i} + t^3\mathbf{j}$
41. $\mathbf{f}(t) = t\mathbf{i} + (\cos t)\mathbf{j}$
42. $\mathbf{f}(t) = (t^3 - 3t)\mathbf{i} + (t^2 - 1)\mathbf{j}$
43. $\mathbf{f}(t) = e^{-t}\mathbf{i} + e^t\mathbf{j}$
44. $\mathbf{f}(t) = (\sin t^2)\mathbf{i} + (\cos t^2)\mathbf{j}$
45. At what point on the curve $y = \ln x$ is the curvature a maximum?
46. For what value of t in the interval $[0, \pi/2]$ is the curvature of the four-cusp hypocycloid $\mathbf{f}(t) = a(\cos^3 t)\mathbf{i} + a(\sin^3 t)\mathbf{j}$ a minimum? For what value is it a maximum?
47. Suppose that the driver of the race car in Example 7 reduces her speed by a factor of M. The frictional force needed to keep the car from skidding is reduced by what factor?
48. If the race car of Example 7 is placed on a track with half the radius of the original one, how much slower would it have to be driven so as not to increase the normal component of acceleration?
*49. A truck traveling at 80 km/hr and weighing 10,000 kg is moving on an unbanked curved stretch of track. The equation of the curved section is the parabola $y = x^2 - x$ meters.
   a. What is the frictional force exerted by the wheels of the truck at the "point" $(0, 0)$ on the track?
   b. If the coefficient of friction for the truck is 2.5, what is the maximum speed it can achieve at the point $(0, 0)$ without going off the track?
*50. A child swings a rope attached to a bucket containing 3 kg of water. The pail rotates in the vertical plane in a circular path with a radius of 1 m. What is the smallest number of revolutions that must be made every minute in order that the water stay in the pail? [*Hint:* Calculate the pressure of the water on the bottom of the pail. This can be determined by first calculating the normal force of the motion. Then the water will stay in the bucket if this normal force exceeds the force due to gravity.]

The local twisting of a curve can be measured by examining $d\mathbf{B}/ds$. Problems 60–63 ask you to think about something called the **torsion** of a curve. Use the computational result of Problem 62 to work Problems 51–54.

51. Calculate the torsion of the straight line given parametrically by $f(t) = t\mathbf{i} + (1 - t)\mathbf{j} + (2 + 3t)\mathbf{k}$.

*52. Calculate the torsion of the curve $\mathbf{f}(t) = e^t(\cos t)\mathbf{i} + e^t(\sin t)\mathbf{j} + e^t\mathbf{k}$ at $t = 0$.

*53. Calculate the torsion of the circular helix $\mathbf{f}(t) = (\cos t)\mathbf{i} + (\sin t)\mathbf{j} + t\mathbf{k}$ at $t = \pi/6$.

*54. Calculate the torsion of the circular helix $\mathbf{f}(t) = a(\cos \omega t)\mathbf{i} + a(\sin \omega t)\mathbf{j} + bt\mathbf{k}$.

■ **SHOW/PROVE/DISPROVE**

*55. Suppose that a curve satisfies the smooth Cartesian equation $y = f(x)$. Show that its curvature can be computed by

$$\kappa(x) = \left|\frac{d\varphi}{dx}\right| = \frac{\left|\dfrac{d^2y}{dx^2}\right|}{\left[1 + \left(\dfrac{dy}{dx}\right)^2\right]^{3/2}}.$$

*56. Suppose that a curve satisfies the smooth polar equation $r = g(\theta)$. Show that its curvature can

be computed by

$$\kappa(\theta) = \frac{\left|r^2 + 2\left(\dfrac{dr}{d\theta}\right)^2 - r\dfrac{d^2r}{d\theta^2}\right|}{\left[r^2 + \left(\dfrac{dr}{d\theta}\right)^2\right]^{3/2}}.$$

57. Show that if a particle is moving at a constant speed, then the tangential component of acceleration is zero.

58. Show that the curvature of a straight line in space is zero.

■ **CHALLENGE**

*59. Let $\mathbf{B}$ be the binormal vector to a curve $\mathbf{f}$, and let s denote arc length. Show that if $d\mathbf{B}/ds \neq 0$, then $d\mathbf{B}/ds \perp \mathbf{B}$ and $d\mathbf{B}/ds \perp \mathbf{T}$. [*Hint:* Differentiate $\mathbf{B} \cdot \mathbf{T} = 0$ and $\mathbf{B} \cdot \mathbf{B} = 1$.]

*60. $d\mathbf{B}/ds$ must be parallel to $\mathbf{n}$ because it is orthogonal to $\mathbf{B}$ and $\mathbf{T}$ (see the preceding problem). Therefore, there must be some number τ such that

$$\frac{d\mathbf{B}}{ds} = -\tau\mathbf{n}.$$

This number τ is called the **torsion** of the curve. It is a measure of how much the curve twists. Show that

$$\frac{d\mathbf{n}}{ds} = -\kappa\mathbf{T} + \tau\mathbf{B}.$$

**61. Show that

$$\mathbf{f}'''(t) = \left[\frac{d^3s}{dt^3} - \left(\frac{ds}{dt}\right)^3 \kappa^2\right]\mathbf{T}$$
$$+ \left[3\left(\frac{d^2s}{dt^2}\right)\frac{ds}{dt}\kappa + \left(\frac{ds}{dt}\right)^2\frac{d\kappa}{dt}\right]\mathbf{n}$$
$$+ \left(\frac{ds}{dt}\right)^3 \kappa\tau\mathbf{B}.$$

**62. Use the result of the preceding problem to show that

$$\tau = \frac{\mathbf{f}''' \cdot (\mathbf{f}' \times \mathbf{f}'')}{\kappa^2\left(\dfrac{ds}{dt}\right)^6} = \frac{\mathbf{f}''' \cdot (\mathbf{f}' \times \mathbf{f}'')}{|\mathbf{f}' \times \mathbf{f}''|^2}.$$

**63. Show that any smooth curve with the property that the position vector $\mathbf{r}(t)$ lies in a fixed plane must have torsion equal to zero everywhere.

■ **ANSWERS TO SELF-QUIZ**

I. a. False   b. True   c. False
   d. False   e. False   f. True

■ PROBLEMS   CHAPTER 12 REVIEW

■ DRILL

In Problems 1–8, find a Cartesian equation for the curve and then sketch the curve in the xy-plane.

1. $\mathbf{f}(t) = t\mathbf{i} + 2t\mathbf{j}$
2. $\mathbf{f}(t) = (2t - 6)\mathbf{i} + t^2\mathbf{j}$
3. $\mathbf{f}(t) = t^2\mathbf{i} + (2t - 6)\mathbf{j}$
4. $\mathbf{f}(t) = t^2\mathbf{i} + t^4\mathbf{j}$
5. $\mathbf{f}(t) = (\cos 4t)\mathbf{i} + (\sin 4t)\mathbf{j}$
6. $\mathbf{f}(t) = 4(\sin t)\mathbf{i} + 9(\sin t)\mathbf{j}$
7. $\mathbf{f}(t) = t^6\mathbf{i} + t^2\mathbf{j}$
8. $\mathbf{f}(t) = e^t(\cos t)\mathbf{i} + e^t(\sin t)\mathbf{j}$

In Problems 9–16, find the slope of the line tangent to the given curve for the specified value of the parameter; then find all points at which the curve has a vertical or horizontal tangent.

9. $x = t^2$; $y = 6t$; $t = 1$
10. $x = t^7$; $y = t^8 - 5t$; $t = 2$
11. $x = \sin 5\theta$; $y = \cos 5\theta$; $\theta = \pi/3$
12. $x = \cos^2 \theta$; $y = -3\theta$; $\theta = \pi/4$
13. $x = \cosh t$; $y = \sinh t$; $t = 0$
14. $x = 2/\theta$; $y = -3\theta$; $\theta = 10\pi$
15. $x = 3 \cos \theta$; $y = 4 \sin \theta$; $\theta = \pi/3$
16. $x = 3 \cos \theta$; $y = -4 \sin \theta$; $\theta = 2\pi/3$

In Problems 17–22, calculate the first and second derivatives of the given vector function.

17. $\mathbf{f}(t) = 2t\mathbf{i} - t^2\mathbf{j}$
18. $\mathbf{f}(t) = (1/t^2)\mathbf{i} + \sqrt[3]{t}\mathbf{j}$
19. $\mathbf{f}(t) = (\cos 5t)\mathbf{i} + 2(\sin t)\mathbf{j}$
20. $\mathbf{f}(t) = (\tan t)\mathbf{i} + (\cot t)\mathbf{j}$
21. $\mathbf{g}(t) = t^3\mathbf{i} - t^2\mathbf{j} + t\mathbf{k}$
22. $\mathbf{g}(t) = (2 - 3t)\mathbf{i} + (e^t - 1 - t)\mathbf{j} + (e^{-t})\mathbf{k}$

In Problems 23–26, calculate the given definite or indefinite integral.

23. $\int_0^3 (t^2\mathbf{i} + t^5\mathbf{j})\, dt$
24. $\int [(\cos 3t)\mathbf{i} + (\sin 3t)\mathbf{j}]\, dt$
25. $\int_0^{\pi/2} [(\cos t)\mathbf{i} + (\sin t)\mathbf{j} + t\mathbf{k}]\, dt$
26. $\int_0^1 [e^t\mathbf{i} + (\cos 3t)\mathbf{j} - 4t\mathbf{k}]\, dt$

In Problems 27–30, calculate the derivative.

27. $\dfrac{d}{dt}\left[(2t\mathbf{i} + \sqrt{t}\mathbf{j}) \cdot (t\mathbf{i} - 3\mathbf{j})\right]$
28. $\dfrac{d}{dt}\{(2\mathbf{i} - 11\mathbf{j}) \cdot [-(\tan t)\mathbf{i} + (\sec t)\mathbf{j}]\}$
29. $\dfrac{d}{dt}\{(3\mathbf{i} + 4\mathbf{j}) \times ((\cos t)\mathbf{i} + (\sin t)\mathbf{j} + t\mathbf{k})\}$
30. $\dfrac{d}{dt}\{[(\cos t)\mathbf{i} + (\sin t)\mathbf{j} + t\mathbf{k}] \times (3\mathbf{i} + 4\mathbf{j} + 5\mathbf{k})\}$

In Problems 31–34, find the unit tangent and unit normal vectors to the given curve for the specified value of t.

31. $\mathbf{f}(t) = t^4\mathbf{i} + t^5\mathbf{j}$; $t = 1$
32. $\mathbf{f}(t) = (\cos 2t)\mathbf{i} + (\sin 2t)\mathbf{j}$; $t = \pi/6$
33. $\mathbf{f}(t) = (\ln t)\mathbf{i} + \sqrt{t}\mathbf{j}$; $t = 1$
34. $\mathbf{f}(t) = (\tan t)\mathbf{i} + (\cot t)\mathbf{j}$; $t = \pi/3$

■ APPLICATIONS

In Problems 35–40, calculate the length of the arc over the specified interval or the length of the given closed curve.

35. $x = \cos 4\theta$; $y = \sin 4\theta$; $0 \le \theta \le \pi/12$
36. $\mathbf{f}(t) = e^t \sin t\mathbf{i} + e^t \cos t\mathbf{j}$; $0 \le t \le \pi/12$
37. $r = 2(1 + \cos \theta)$
38. $r = 5 \sin \theta$
39. $x = -2t$; $y = \cos 2t$; $z = \sin 2t$; $0 \le t \le \pi/6$
40. $\mathbf{g}(t) = 3(\sin 2t)\mathbf{i} + 3(\cos 2t)\mathbf{j} + t^2\mathbf{k}$; $0 \le t \le 5$

In Problems 41–46, the position vector of a moving particle is given. For the specified value of t, calculate the velocity vector, the acceleration vector, the speed, and the acceleration scalar. Then sketch a portion of the trajectory showing the velocity and acceleration vectors.

41. $\mathbf{f}(t) = (\cos 2t)\mathbf{i} + (\sin 2t)\mathbf{j}$; $t = \pi/6$
42. $\mathbf{f}(t) = 6t\mathbf{i} + 2t^3\mathbf{j}$; $t = 1$
43. $\mathbf{f}(t) = (2^t + e^{-t})\mathbf{i} + 2t\mathbf{j}$; $t = 0$
44. $\mathbf{f}(t) = (1 - \cos t)\mathbf{i} + (t - \sin t)\mathbf{j}$; $t = \pi/2$
45. $\mathbf{f}(t) = 2(\sinh t)\mathbf{i} + 4t\mathbf{j}$; $t = 1$
46. $\mathbf{f}(t) = (3 + 5t)\mathbf{i} + (2 + 8t)\mathbf{j}$; $t = 6$

In Problems 47–50, find the tangential and normal components of acceleration for the given position vector.

47. $\mathbf{f}(t) = 2(\sin t)\mathbf{i} + 2(\cos t)\mathbf{j}$
48. $\mathbf{f}(t) = 4(\cos t)\mathbf{i} + 9(\sin t)\mathbf{j}$
49. $\mathbf{f}(t) = 3t^2\mathbf{i} + 2t^3\mathbf{j}$
50. $\mathbf{f}(t) = (\cos t^2)\mathbf{i} + (\sin t^2)\mathbf{j}$

In Problems 51–60, find the curvature and radius of curvature of the given curve at the specified point. Sketch the curve in the neighborhood of that point, the unit tangent vector there, and the circle of curvature.

51. $\mathbf{f}(t) = (\cos 2t)\mathbf{i} + (\sin 2t)\mathbf{j}$; $t = \pi/3$
52. $\mathbf{f}(t) = t^2\mathbf{i} + 2t\mathbf{j}$; $t = 2$
53. $\mathbf{f}(t) = 4(\cos t)\mathbf{i} + 9(\sin t)\mathbf{j}$; $t = \pi/4$
54. $\mathbf{f}(t) = 2t\mathbf{i} + (t^2/2)\mathbf{j}$; $t = 0$
55. $y = 2x^2$; $(0, 0)$
56. $xy = 1$; $(2, \frac{1}{2})$
57. $y = e^{-x}$; $(1, 1/e)$
58. $y = \sqrt{x}$; $(4, 2)$
59. $r = 1 + \sin\theta$; $\theta = \pi/2$
60. $r = 3\theta$; $\theta = \pi$

In Problems 61–64, find parametric equations in terms of the arc length s measured from the point

reached when $t = 0$. Verify your answer by using equation (12.5.12).

61. $\mathbf{f}(t) = 3t\mathbf{i} + 4t^{3/2}\mathbf{j}$
62. $\mathbf{f}(t) = \frac{2}{9}t^{9/2}\mathbf{i} + \frac{1}{3}t^3\mathbf{j}$
63. $\mathbf{f}(t) = 2(\cos 3t)\mathbf{i} + 2(\sin 3t)\mathbf{j}$
64. $\mathbf{f}(t) = e^t(\sin t)\mathbf{i} + e^t(\cos t)\mathbf{j}$

In Problems 65–68, find the unit tangent vector $\mathbf{T}$, the curvature κ, the principal unit normal vector $\mathbf{n}$, and the binomial vector $\mathbf{B}$ for the specified value of t,

65. $\mathbf{g}(t) = \mathbf{i} + t\mathbf{j} + \frac{1}{2}t^2\mathbf{k}$; $t = 2$
66. $\mathbf{g}(t) = t^2\mathbf{i} + t^3\mathbf{j} - t\mathbf{k}$; $t = 0$
67. $\mathbf{g}(t) = 2(\cos t)\mathbf{i} + 2(\sin t)\mathbf{j} + t\mathbf{k}$; $t = \pi/6$
68. $\mathbf{g}(t) = e^t[\mathbf{i} + (\cos t)\mathbf{j} + (\sin t)\mathbf{k}]$; $t = 0$

🖳 COMPUTER EXERCISES

1. This is a marble race. Marbles are started at the origin and allowed to roll down a decreasing path to the point $(\pi, -2)$. The challenge is that each contestant may select the path that his or her marble will follow. (Any point below the x-axis and not on the y-axis would work, but $(\pi, -2)$ turns out to simplify calculations.)

   The first contestant chooses the brachistochrone curve given by $(\pi t - \sin \pi t, \cos \pi t - 1)$, $0 \leq t \leq 1$. The second contestant chooses the straight line $(\pi t, -2t)$, $0 \leq t \leq 1$, and the third contestant chooses the parabolic arch $(\pi t, 2(t-1)^2 - 2)$, $0 \leq t \leq 1$. As the fourth contestant, you may select any path you want: just be sure that your parametric curve decreases from the origin to $(\pi, -2)$ as t moves from 0 to 1.

   With the aid of some elementary physics, it is not difficult to show that the time required for a marble to slide from the origin along the decreasing parametric curve $(x(t), y(t))$ from $t = 0$ to $t = 1$ is given by the following integral:

$$\text{time} = \frac{1}{\sqrt{g}} \int_0^1 \sqrt{\frac{(x'(t))^2 + (y'(t))^2}{-2y(t)}} \, dt,$$

where g is the constant of gravitational attraction near the surface of the earth.

   Calculate the finish times for contestants 1, 2, 3, and for yourself. (You should be able to calculate by hand the time required for contestants 1 and 2. For contestant 3, you will need a computer

algebra system to approximate the integral. You are almost certain to need similar help with your own integral.)

   Based on the outcome of the race, formulate a conjecture about a physical property of the brachistochrone curve. Can you prove your assertion? (The curve presented here is called a *cycloid* reflected through the x-axis. It is the path taken by a point on the rim of a tire as its rolls down a road.)

   Problems 2, 3, and 4 can be completed without electronic aid. However you will find a computer algebra system to be very helpful in your calculations.

2. Show that the curvature of the ellipse $(3 \cos t, \sin t)$ reaches a maximum at $(\pm 3, 0)$ and is a minimum at $(0, \pm 1)$. Calculate the curvature at these points.

3. Calculate the curvature of the elliptical helix $(3 \cos t, \sin t, t)$. Where does the curvature reach a maximum? Where does it reach a minimum?

4. Calculate the curvature of $(\cos at, \sin at, t)$. What happens to the curvature as a approaches 0? What happens to the curvature as a approaches infinity? Give a geometric explanation of your answers above. You may wish to use graphics software to look at a picture of the curve for various values of a.

13

DIFFERENTIATION OF FUNCTIONS OF TWO AND THREE VARIABLES

For most of the functions we have so far encountered in this book, we have been able to write $y = f(x)$. This means that we could write the variable y explicitly in terms of the single variable x. However, in a great variety of applications it is necessary to write the quantity of interest in terms of two or more variables. We have already encountered this situation. For example, the volume of a right circular cylinder is given by

$$V = \pi r^2 h,$$

where r is the radius of the cylinder and h is its height. That is, V is a function of the *two* variables r and h.

As a second example, the **ideal-gas law**, which relates pressure, volume, and temperature for an ideal gas, can be written as

$$PV = nRT,$$

where P is the pressure of the gas, V is the volume, T is the absolute temperature (i.e., in degrees kelvin), n is the number of moles of the gas, and R is a constant. Solving for P, we find that

$$P = \frac{nRT}{V}.$$

That is, we can write P as a function of the *three* variables n, T, and V.

As a third example, we saw in Problem 2.2.35 (page 145) that according to Poiseuille's law, the resistance R of a blood vessel of length l and radius r is given by

$$R = \frac{\alpha l}{r^4},$$

where α is a constant of proportionality. If l and r are allowed to vary, then R is written as a function of the *two* variables l and r.

As a final example, let the vector $\mathbf{v}$ be given by

$$\mathbf{v} = x\mathbf{i} + y\mathbf{j} + z\mathbf{k}.$$

Then the magnitude of $\mathbf{v}$, $|\mathbf{v}|$, is given by

$$|\mathbf{v}| = \sqrt{x^2 + y^2 + z^2}.$$

That is, the magnitude of **v** is written as a function of the *three* variables, x, y, and z.

There are many examples like the four cited above. It is probably fair to say that very few physical, biological, or economic quantities can be properly expressed in terms of one variable alone. Often we write these quantities in terms of one variable simply because functions of only one variable are the easiest functions to handle.

In this chapter, we will see how many of the operations we have studied in our discussion of the "one-variable" calculus can be extended to functions of two and three variables. We will begin by discussing the basic notions of limits and continuity and will go on to differentiation and applications of differentiation. In Chapter 14, we will discuss the integration of functions of two and three variables.

13.1 FUNCTIONS OF TWO AND THREE VARIABLES

In Chapter 12, we discussed the notion of vector-valued functions. For example, if $\mathbf{f}(t) = f_1(t)\mathbf{i} + f_2(t)\mathbf{j}$, then for every t in the domain of $\mathbf{f}$, we obtain a vector $\mathbf{f}(t)$. In defining functions of two or more variables, we obtain a somewhat reversed situation. For example, in the formula for the volume of a right circular cylinder, we have

$$V = \pi r^2 h. \tag{1}$$

Here the volume V is written as a function of the two variables r and h. Put another way, for every ordered pair of positive real numbers (r, h), there is a unique positive number V such that $V = \pi r^2 h$. To indicate the dependence of V on the variables r and h, we write $V(r, h)$. That is, V can be thought of as a function that assigns a positive real number to every ordered pair of real numbers. Thus we see what we meant when we said that the situation is "reversed." Instead of having a vector function of one variable (a scalar), we have a scalar function of an ordered pair of variables (which is a vector in $\mathbb{R}^2$). We now give a definition of a function of two variables.

DEFINITION **FUNCTION OF TWO VARIABLES**

Let D be a subset of $\mathbb{R}^2$. Then a **real-valued function of two variables** f is a rule that assigns to each point (x, y) in D a unique real number that we denote by $f(x, y)$. The set D is called the **domain** of f. The set $\{f(x, y): (x, y) \in D\}$, which is the set of values the function f takes on, is called the **range** of f. ■

EXAMPLE 1
A FUNCTION OF TWO VARIABLES

Let $D = \mathbb{R}^2$. For each point $(x, y) \in \mathbb{R}^2$, we assign the number $f(x, y) = x^2 + y^4$. Since $x^2 + y^4 \geq 0$ for every pair of real numbers (x, y), we see that the range of f is $\mathbb{R}^+$, the set of nonnegative real numbers.

When the domain D is not given, we will take the domain of f to be the largest subset of $\mathbb{R}^2$ for which the expression $f(x, y)$ makes sense.

EXAMPLE 2
FINDING THE DOMAIN AND
RANGE OF A FUNCTION OF
TWO VARIABLES

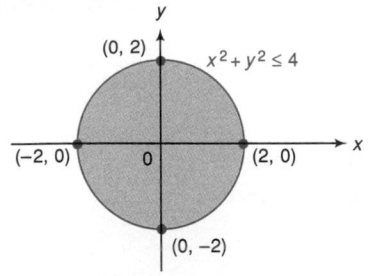

FIGURE 1
GRAPH OF THE DOMAIN OF
$f(x, y) = \sqrt{4 - x^2 - y^2}$

Find the domain and range of the function f given by $f(x, y) = \sqrt{4 - x^2 - y^2}$, and find $f(0, 1)$ and $f(1, -1)$.

SOLUTION: Clearly, f is defined when the expression under the square-root sign is nonnegative. Thus, $D = \{(x, y): x^2 + y^2 \le 4\}$. This is the disk[†] centered at the origin with radius 2. It is sketched in Figure 1. Since x^2 and y^2 are nonnegative, $4 - x^2 - y^2$ is largest when $x = y = 0$. Thus, the largest value of $\sqrt{4 - x^2 - y^2} = \sqrt{4} = 2$. Since $x^2 + y^2 \le 4$, the smallest value of $4 - x^2 - y^2$ is 0, taken when $x^2 + y^2 = 4$ (at all points on the circle $x^2 + y^2 = 4$). Thus, the range of f is the closed interval $[0, 2]$. Finally,

$$f(0, 1) = \sqrt{4 - 0^2 - 1^2} = \sqrt{3} \quad \text{and}$$
$$f(-1, 1) = \sqrt{4 - (-1)^2 - 1^2} = \sqrt{2}.$$

REMARK: We emphasize that the domain of f is a subset of $\mathbb{R}^2$, while the range is a subset of $\mathbb{R}$, the real numbers.

CAUTION: In Figure 1, we sketched the *domain* of the function. We did *not* sketch its graph. The graph of a function of two or more variables is more complicated and will be discussed shortly.

In Chapter 0, we wrote $y = f(x)$. That is, we used the letter y to denote the value of a function of one variable. Here we can use the letter z (or any other letter for that matter) to denote the value taken by f, which is now a function of two variables. We then write

$$z = f(x, y). \tag{2}$$

EXAMPLE 3
FINDING THE DOMAIN AND
RANGE OF A FUNCTION OF TWO
VARIABLES

Find the domain and range of $f(x, y) = \tan^{-1}(y/x)$.

SOLUTION: $\tan^{-1}(y/x)$ is defined as long as $x \ne 0$. Hence the domain of $f = \{(x, y): x \ne 0\}$. From the definition of $\tan^{-1} x$ in Section 6.8, we find that the range of f is the open interval $(-\pi/2, \pi/2)$.

▦ **EXAMPLE 4**
THE COBB-DOUGLAS
PRODUCTION FUNCTION

In a manufacturing process, costs are typically divided between labor costs and capital costs. The total sum spent on production is usually fixed, but often a manufacturer has some choice in allocating money between capital and labor. For example, if part of the process is automated, more money will be spent on capital and less will be spent on labor.

Suppose that L units of labor and K units of capital are used in production. How many units will be produced? Economists have determined that, in some cases, the answer is given by the **Cobb-Douglas production function**:

number of units produced $= F(L, K) = cL^a K^{1-a}$,

[†] The **disk** of radius r is the circle of radius r together with all points interior to that circle.

where c and a are constants that depend on the particular manufacturing process.

In the manufacture of a certain type of die, the Cobb-Douglas production function is given by

$$F(L, K) = 200L^{2/5}K^{3/5}. \tag{3}$$

(a) How many units are produced if 100 units of labor and 300 units of capital are used?

(b) If the number of units of labor and capital are both doubled, what is the change in the number of units produced?

SOLUTION:

(a) $F(100, 300) = 200(100)^{0.4}(300)^{0.6} \approx 38{,}668$ units

(b) We are asked to determine what happens to F if L becomes $2L$ and K becomes $2K$. We have

$$F(2L, 2K) = 200(2L)^{2/5}(2K)^{3/5} = 200 \cdot 2^{2/5}L^{2/5}2^{3/5}K^{3/5}$$
$$= 2^{2/5}2^{3/5}[200L^{2/5}K^{3/5}] = 2F(L, K).$$

$$\underbrace{2^{5/5} = 2^1 = 2}_{} \quad \underbrace{F(L, K)}_{}$$

We have shown that production is doubled if both labor and production costs are doubled.

We now turn to the definition of a function of three variables.

DEFINITION **FUNCTION OF THREE VARIABLES**

Let D be a subset of $\mathbb{R}^3$. Then a **real-valued function of three variables** f is a rule that assigns to each point (x, y, z) in D a unique real number that we denote by $f(x, y, z)$. The set D is called the **domain** of f, and the set $\{f(x, y, z): (x, y, z) \in D\}$, which is the set of values the function f takes on, is called the **range** of f. ∎

NOTE: We will often use the letter w to denote the values that a function of three variables takes. We then write

$$w = f(x, y, z).$$

EXAMPLE 5
FINDING THE DOMAIN AND
RANGE OF A FUNCTION OF
THREE VARIABLES

Let $w = f(x, y, z) = \sqrt{1 - x^2 - (y^2/4) - (z^2/9)}$. Find the domain and range of f. Calculate $f(0, 1, 1)$.

SOLUTION: $f(x, y, z)$ is defined if $1 - x^2 - (y^2/4) - (z^2/9) \geq 0$, which occurs if $x^2 + (y^2/4) + (z^2/9) \leq 1$. From Section 11.8, we see that the equation $x^2 + (y^2/4) + (z^2/9) = 1$ is the equation of an ellipsoid. Thus the domain of f is the set of points (x, y, z) in $\mathbb{R}^3$ that are on and interior to this ellipsoid (sketched in Figure 2). The range of f is the closed

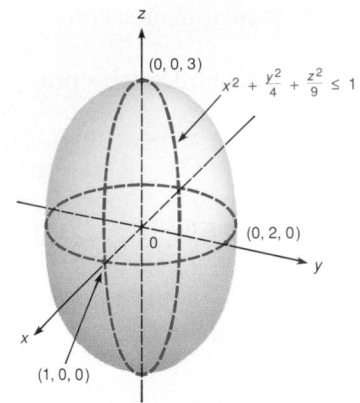

FIGURE 2
GRAPH OF THE DOMAIN OF $f(x, y, z) =$
$\sqrt{1 - x^2 - \dfrac{y^2}{4} - \dfrac{z^2}{9}}$

interval $[0, 1]$. This follows because $x^2 + (y^2/4) + (z^2/9) \geq 0$, so that $1 - x^2 - (y^2/4) - (z^2/9)$ is maximized when $x = y = z = 0$. Finally,

$$f(0, 1, 1) = \sqrt{1 - 0^2 - \frac{1^2}{4} - \frac{1^2}{9}} = \sqrt{1 - \frac{1}{4} - \frac{1}{9}} = \sqrt{1 - \frac{13}{36}} = \frac{\sqrt{23}}{6}.$$

NOTE: Figure 2 is a sketch of the domain of f, not the graph of f itself.

We will not here define a function of more than three variables, since such a definition would be very similar to the definitions in $\mathbb{R}^2$ and $\mathbb{R}^3$.

We now turn to a discussion of the graph of a function. Recall that the graph of a function of one variable f is the set of all points (x, y) in the plane such that $y = f(x)$. Using this definition as our model, we have the following definition.

DEFINITION   **GRAPH OF A FUNCTION OF TWO VARIABLES AND SURFACE**

The **graph** of a function f of two variables x and y is the set of all points (x, y, z) in $\mathbb{R}^3$ such that $z = f(x, y)$. The graph of a continuous[†] function of two variables is called a **surface** in $\mathbb{R}^3$. ∎

EXAMPLE 6
SKETCHING THE GRAPH OF A
FUNCTION OF TWO VARIABLES

Sketch the graph of the function

$$z = f(x, y) = \sqrt{1 - x^2 - y^2}. \tag{4}$$

SOLUTION: We first note that $z \geq 0$. Then squaring both sides of (4), we have $z^2 = 1 - x^2 - y^2$, or $x^2 + y^2 + z^2 = 1$. This equation is the equation of the unit sphere. However, since $z \geq 0$, the graph of f is the hemisphere sketched in Figure 3.

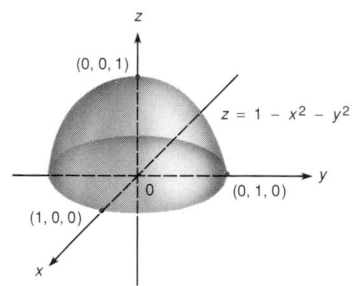

FIGURE 3
GRAPH OF THE SURFACE $z = \sqrt{1 - x^2 - y^2}$

It is often very difficult to sketch the graph of a function $z = f(x, y)$ since, except for the quadric surfaces we discussed in Section 11.8, we really do not have a vast "catalog" of surfaces to which to refer. Moreover, the techniques of curve plotting in three dimensions are tedious, to say the least, and plotting points in space will, except in the most trivial of cases, not get us very far. The best we can usually do is to describe **cross-sections** of the surface that lie in planes parallel to the coordinate planes. This will give an idea of what the surface looks like. Then if we have access to a computer with graphing capabilities, we can obtain a computer-drawn sketch of the surface. We illustrate this process with two examples.

EXAMPLE 7
SKETCHING A SURFACE AND ITS
CROSS-SECTIONS

Obtain a sketch of the surface

$$z = f(x, y) = x^3 + 3x^2 - y^2 - 9x + 2y - 10. \tag{5}$$

SOLUTION: The xz-plane has the equation $y = 0$. Planes parallel to the xz-plane have the equation $y = c$, where c is a constant. Setting $y = c$

[†] We will discuss continuity in the next section.

in (5), we have

$$z = x^3 + 3x^2 - 9x - 10 + (-c^2 + 2c)$$
$$= x^3 + 3x^2 - 9x - 10 + k,$$

where $k = -c^2 + 2c$ is a constant. We obtained a graph of the curve $y = x^3 + 3x^2 - 9x - 10$ in the xy-plane in Example 3.2.4 on page 189 (see Figure 3.2.3).

Using the material in Section 0.5, we can obtain the graph of $z = x^3 + 3x^2 - 9x - 10 + k$ by shifting the graph of $z = x^3 + 3x^2 - 9x - 10$ up or down $|k|$ units. For several values of k, cross sections lying in planes parallel to the xz-plane are given in Figure 4(a).

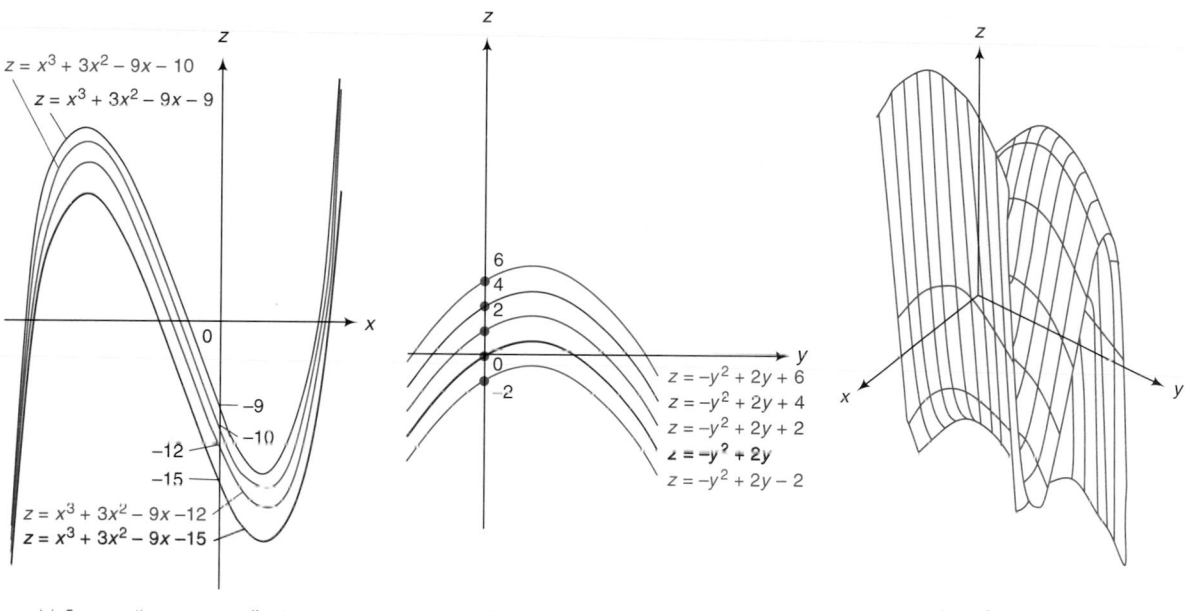

(a) Cross-sections corresponding to $y = c$.

(b) Cross-sections corresponding to $x = c$.

(c) $z = x^3 + 3x^2 - 9x - y^2 + 2y - 10$.

FIGURE 4   THE GRAPH OF $z = x^3 + 3x^2 - 9x - y^2 + 2y - 10$

Cross-sections are sketched in (a) and (b) while the surface is sketched in (c).

REMARK: The maximum value of the function $g(c) = -c^2 + 2c$ is 1, taken when $c = 1$ (you should verify this). Thus, 1 is the largest value k can assume in the parallel cross-sections, so $-10 + k \leq -9$, and the "highest" cross-section is $z = x^3 + 3x^2 - 9x - 9$.

The yz-plane has the equation $x = 0$. Planes parallel to the yz-plane have the equation $x = c$, where c is a constant. Setting $x = c$ in (5), we obtain

$$z = -y^2 + 2y + (c^3 + 3c^2 - 9c - 10) = -y^2 + 2y + k.$$

For several values of k, cross-sections parallel to the yz-plane are drawn in Figure 4(b). Note that this k can take on any real value. In Figure 4(c), we provide a computer-drawn sketch of the surface.

As we have stated, except for the standard quadric surfaces described in Section 11.8, it is extremely difficult to sketch three-dimensional graphs by

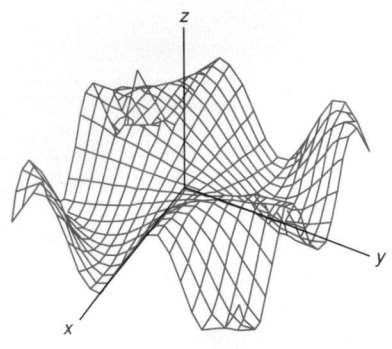

(a) Graph of surface $z = \sin xy$.

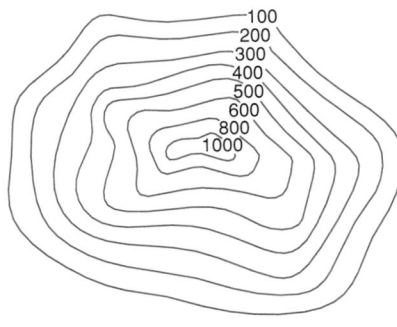

(b) Graph of surface $z = x^3 + \cos y$.

FIGURE 5
COMPUTER-DRAWN GRAPHS OF TWO
SURFACES

FIGURE 6
CONTOUR CURVES

hand. However, virtually every modern computer (and microcomputer) has graphing capabilities. In Figure 5, we provide microcomputer-drawn sketches of the graphs of two different functions of two variables.

The situation becomes much more complicated when we try to sketch the graph of a function of three variables $w = f(x, y, z)$. We would need *four dimensions* to sketch such a surface. Being human, we are limited to three dimensions, and we see that we have reached the point where our comfortable three-dimensional geometry fails us. We are *not* saying that curves and surfaces in four-dimensional space do not exist—they do—only that we are not able to sketch them.

We will not spend much time in this text discussing spaces of dimension higher than three. To give you a taste of the subject, we can define four-dimensional space $\mathbb{R}^4$ as the set of all "points" or *four-tuples* (x, y, z, w), where x, y, z, and w are real numbers. Then the graph of a function $w = f(x, y, z)$ is the set of points (x, y, z, w) in $\mathbb{R}^4$ with $w = f(x, y, z)$. *It is important to note that the only difference between a graph of a function of two or three variables is our ability to sketch one (sometimes) but not the other.* If we do not insist on being able to sketch things, then there is really no essential difference between $\mathbb{R}^3$ or $\mathbb{R}^4$ (or $\mathbb{R}^5$, $\mathbb{R}^6$, . . . , for that matter). For the remainder of this chapter we will focus on $\mathbb{R}^2$ and $\mathbb{R}^3$, keeping in mind that all our definitions apply to higher dimensions as well.

We have shown that it is usually very difficult to sketch the graph of a function of two variables. Fortunately, there is a way to represent such a function geometrically in two dimensions. The idea for what we are about to do comes from cartographers (mapmakers). Cartographers have the problem of indicating three-dimensional features (such as mountains and valleys) on a two-dimensional surface. They solve the problem by drawing a **contour map** (**topographic map**), which is a map in which points of constant elevation are joined to form curves, called **contour curves**. The closer together these contour curves are drawn, the steeper is the terrain. A portion of a typical contour map is sketched in Figure 6.

LEVEL CURVES

We can use the same idea to depict the function $z = f(x, y)$ geometrically. If z is fixed, then the equation $f(x, y) = z$ is the equation of a curve in the xy-plane, called a **level curve**.

We can think of a level curve as the projection of a cross-section lying in a plane parallel to the xy-plane. Each value of z gives us such a curve. In other words, a level curve is the projection of the intersection of the surface $z = f(x, y)$ with the plane $z = c$. This idea is best illustrated with an example.

EXAMPLE 8
SKETCHING LEVEL CURVES

Sketch some level curves of $z = x^2 + y^2$.

SOLUTION: If $z > 0$, then $z = a^2$ for some positive number $a > 0$. Hence, all level curves are circles of the form $x^2 + y^2 = a^2$. The number a^2 can be thought of as the "elevation" of points on a level curve. Some of these

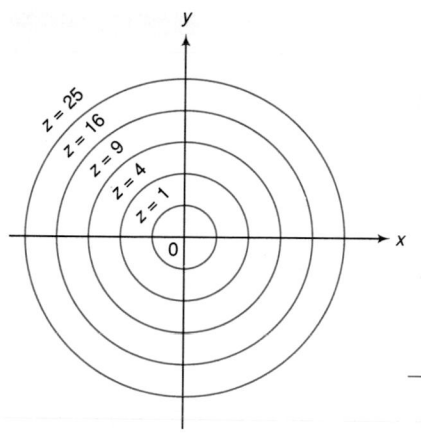

FIGURE 7
LEVEL CURVES FOR $z = x^2 + y^2$

curves are sketched in Figure 7. Each level curve encloses a projection of a "slice" of the actual graph of the function in three dimensions. In this case, each circle is the projection onto the xy-plane of a part of the surface in space. Actually, this example is especially simple because we can, without much difficulty, sketch the graph in space. From Section 11.8, the equation $z = x^2 + y^2$ is the equation of an elliptic paraboloid (actually, a circular paraboloid). It is sketched in Figure 8. In this easy case, we can see that if we slice this surface parallel to the xy-plane, we obtain the circles whose projections onto the xy-plane are the level curves. In most cases, of course, we will not be able to sketch the graph in space, so we will have to rely on our contour map sketch in $\mathbb{R}^2$.

There are some interesting applications of level curves in the sciences and economics. Three are given below.

(i) Let $T(x, y)$ denote the temperature at a point (x, y) in the xy-plane. The level curves $T(x, y) = c$ are called **isothermal curves**. All points on such a curve have the same temperature.

(ii) Let $V(x, y)$ denote the voltage (or potential) at a point in the xy-plane. The level curves $V(x, y) = c$ are called **equipotential curves**. All points on such a curve have the same voltage.

(iii) A manufacturer makes two products. Let x and y denote the number of units of the first and second products produced during a given year. If $P(x, y)$ denotes the profit the manufacturer receives each year, then the level curves $P(x, y) = c$ are **constant profit curves**. All points on such a curve yield the same profit.

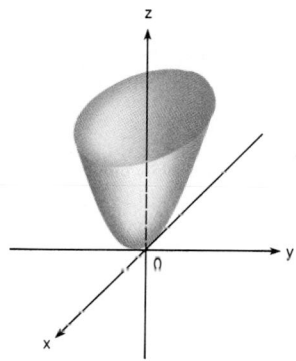

FIGURE 8
GRAPH OF CIRCULAR PARABOLOID $z = x^2 + y^2$

We conclude this section by noting that the idea behind level curves can be used to describe functions of three variables of the form $w = f(x, y, z)$. For each fixed w, the equation $f(x, y, z) = w$ is the equation of a surface in space, called a **level surface**. Since most surfaces are so difficult to draw, we will not pursue the notion of level surfaces here.

■ PROBLEMS 13.1

■ SELF-QUIZ

I. The domain of the function $f(x, y) = \sin x / \cos y$ is _____, where k denotes an integer.
   a. $\{(x, y): y \neq k\pi\}$
   b. $\{(x, y): y \neq (k + \frac{1}{2})\pi\}$
   c. $\{(x, y): x \neq k\pi\}$
   d. $\{(x, y): x \neq j\pi, y \neq (k + \frac{1}{2})\pi\}$

II. The range of the function $g(x, y) = \cos x + \sin y$ is _____.
   a. $[-1, 1]$ b. $[-2, 2]$
   c. $[-\pi, \pi]$ d. $(-\infty, \infty)$

III. The level curves of $z = x^2 + 4y^2$ are _____.
   a. circles   b. ellipses
   c. straight lines   d. diamonds

IV. The level curves of $z = |x| + 2|y|$ are _____.
   a. circles   b. ellipses
   c. straight lines   d. diamonds

V. The level surfaces of $f(x, y, z) = 3x + 2y - 5z$ are _____.
   a. spheres   b. planes
   c. straight lines   d. ellipsoids

CONTINUED

■ DRILL

In Problems 1–34, find the domain and range of the given function.

1. $f(x, y) = \sin(x + y)$ 2. $f(x, y) = e^x + e^y$

3. $f(x, y) = \sqrt{x^2 + y^2}$
4. $f(x, y) = \sqrt{1 + x + y}$
5. $f(x, y) = x/y$ 6. $f(x, y) = y/|x|$
7. $f(x, y) = \sqrt{1 - x^2 - 4y^2}$
8. $f(x, y) = \sqrt{1 - x^2 + 4y^2}$
9. $f(x, y) = \ln(1 + x^2 - y^2)$
10. $f(x, y) = (x^2 - y^2)^{-3/2}$

11. $f(x, y) = \sqrt{\dfrac{x + y}{x - y}}$ 12. $f(x, y) = \sqrt{\dfrac{x - y}{x + y}}$

13. $f(x, y) = \sin^{-1}(x + y)$
14. $f(x, y) = \cos^{-1}(x - y)$

15. $f(x, y) = \dfrac{x^2 - y^2}{x + y}$

*16. $f(x, y) = \left(\dfrac{x}{2y}\right) + \left(\dfrac{2y}{x}\right)$

17. $g(x, y, z) = x + y + z$
18. $g(x, y, z) = \sqrt{x + y + z}$
19. $g(x, y, z) = e^{xy + z}$
20. $g(x, y, z) = \sin x + \cos y + \sin z$
21. $g(x, y, z) = xyz$ 22. $g(x, y, z) = 1/(xyz)$

23. $g(x, y, z) = xy/z$ 24. $g(x, y, z) = \dfrac{x}{y + z}$

25. $g(x, y, z) = \tan^{-1}\left(\dfrac{x + z}{y}\right)$

26. $g(x, y, z) = \dfrac{e^x + e^y}{e^z}$

27. $g(x, y, z) = \sin(x + y - z)$
28. $g(x, y, z) = \sin^{-1}(x + y - z)$
29. $g(x, y, z) = \ln(x + y - z)$
30. $g(x, y, z) = \ln(x - 2y - 3z + 4)$

31. $g(x, y, z) = \dfrac{1}{\sqrt{x^2 + y^2 + z^2}}$

32. $g(x, y, z) = \dfrac{1}{\sqrt{x^2 - y^2 + z^2}}$

33. $g(x, y, z) = \dfrac{1}{\sqrt{x^2 - y^2 - z^2}}$

34. $g(x, y, z) = \sqrt{-x^2 - y^2 - z^2}$

In Problems 35–42, sketch the surface satisfying the given equation.

35. $z = x - y^2$ 36. $z = 4x^2 + 4y^2$
37. $y = x^2 + 4z^2$ 38. $x = 4z^2 - 4y^2$
39. $z = x^2 - 4y^2$ 40. $z = \sqrt{x^2 + 4y^2 + 4}$
41. $y = \sqrt{x^2 - 4z^2 + 4}$ 42. $x = \sqrt{4 - z^2 - 4y^2}$

In Problems 43–50, describe the level curves of the given function and then sketch those curves for the specified values of z.

43. $z = \sqrt{1 + x + y}$; $z = 0, 1, 5, 10$
44. $z = x/y$; $z = 1, 3, 5, -1, -3$
45. $z = \sqrt{1 - x^2 - 4y^2}$; $z = 0, \frac{1}{4}, \frac{1}{2}, 1$
46. $z = \sqrt{1 + x^2 - y}$; $z = 0, 1, 2, 5$
47. $z = \cos^{-1}(x - y)$; $z = 0, \pi/6, \pi/3, \pi/2$
48. $z = \sqrt{\dfrac{x + y}{x - y}}$; $z = 0, 1, 2, 5$
49. $z = \tan(x + y)$; $z = 0, 1, -1, \sqrt{3}$
50. $z = \tan^{-1}(x - y)$; $z = 0, \pi/6, \pi/4$

■ APPLICATIONS

51. The temperature T at any point on an object in the plane is given by $T(x, y) = 20 + x^2 + 4y^2$. Sketch the isothermal curves for $T = 50$, $T = 60$, and $T = 70$ degrees.

52. The voltage at a point (x, y) on a metal plate placed in the xy-plane is given by $V(x, y) = \sqrt{1 - 4x^2 - 9y^2}$. Sketch the equipotential curves for $V = 1.0$, $V = 0.5$, and $V = 0.25$ volts.

53. A manufacturer earns $P(x, y) = 100 + 2x^2 + 3y^2$ dollars each year for producing x and y units, respectively, of two products. Sketch the constant profit curves for $P = 100$, $P = 200$, and $P = 1000$ dollars.

54. The Cobb-Douglas production function for a given product is

$$F(L, K) = 250L^{0.7}K^{0.3}.$$

a. Compute $F(50, 80)$.
b. Show that if labor and capital costs both triple, then the total output (number of units produced) triples as well.

55. The Cobb-Douglas production function for a given product is

$$F(L, K) = 500L^{1/3}K^{2/3}.$$

a. Compute $F(250, 150)$.

CONTINUED

b. If labor costs double while capital costs are halved, what is the change in the total output? (Give this change as a percentage increase or decrease.)

c. Answer the question in part (b) if labor costs are halved while capital costs double.

56. A jeweler sells ordinary and digital watches. Each ordinary watch costs her $8, and each digital watch costs her $25. The demand functions for the two watches are

$$q_1 = 80 - 2.5p_1 + 0.8p_2 \qquad \text{for the ordinary watches}$$

and

$$q_2 = 120 + p_1 - 1.8p_2 \qquad \text{for the digital watches.}$$

a. Find the total profit as a function of the prices of the two watches.

b. What is the profit when ordinary watches are sold for $24 and digital watches are sold for $30?

57. In Problem 56, what is the profit if ordinary and digital watches are sold for $30 and $40, respectively?

■ CHALLENGE

In Problems 58–69, a function of two variables is given. Match each function with one of the graphs in Figure 9.

58. $z = x^2 + \sin y$

59. $z = \ln x + e^{-y}$

60. $z = \cos xy$

61. $z = y^2 - x^2$

62. $z = x^2 - y^3 - x + 2y - 2$

63. $z = (y - 1.5x^2)(y - 0.5x^2)$

64. $z = e^{-(x+y)}$

65. $z = \sin x \tan y, \quad -\pi/2 < y < \pi/2$

66. $z = \sec x \cos y, \quad -\pi/2 < x < \pi/2$

67. $z = x - 2y + 4$

68. $z = (2x^2 + 3y^2)e^{1-(x^2+y^2)}$

69. $z = \dfrac{1}{(x + 0.15)^2 + y^2 + 0.2}$
$$- \dfrac{1}{(x - 0.15)^2 + y^2 + 0.2}$$

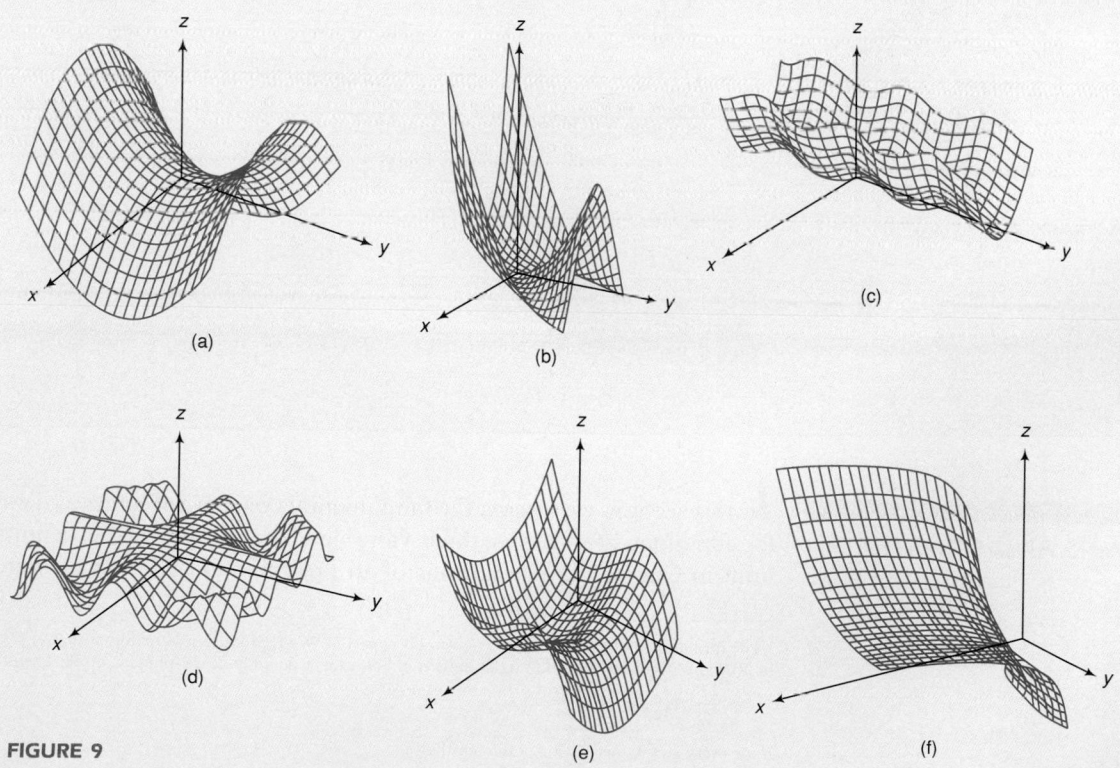

(a)   (b)   (c)

(d)   (e)   (f)

FIGURE 9

CONTINUED

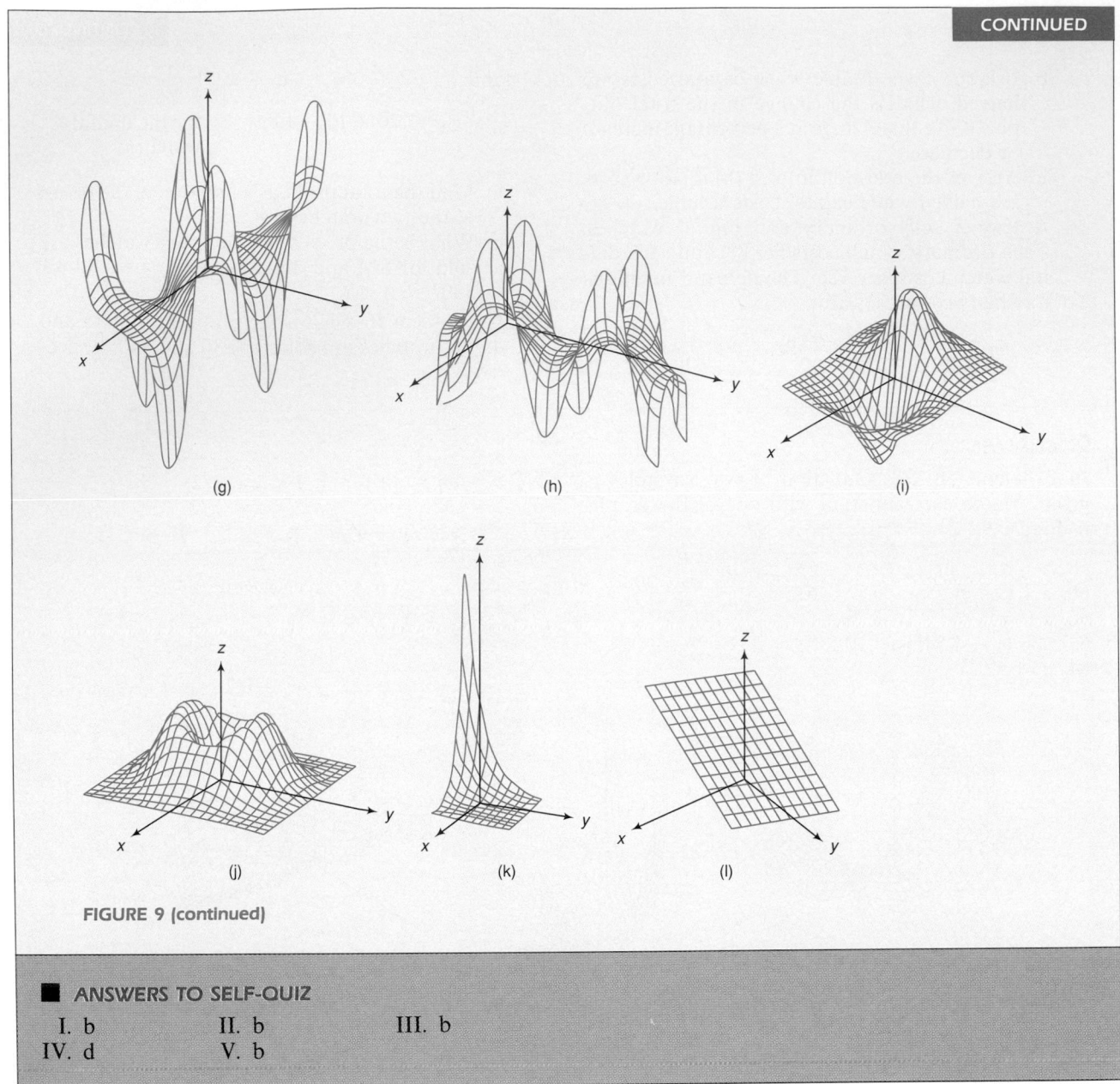

FIGURE 9 (continued)

ANSWERS TO SELF-QUIZ

I. b   II. b   III. b
IV. d   V. b

13.2 LIMITS AND CONTINUITY

In this section, we discuss the fundamental concepts of limits and continuity for functions of two and three variables. Recall that in the definition of a limit in Chapter 1,[†] the notions of an open and closed interval were funda-

---

[†] The formal definition of a limit was given on page 124. We reproduce it here: Suppose that $f(x)$ is defined in a neighborhood of the point x_0 (a finite number) except possibly at the point x_0 itself. Then

$$\lim_{x \to x_0} f(x) = L$$

if for every $\epsilon > 0$, there is a $\delta > 0$, such that if

$$0 < |x - x_0| < \delta, \qquad \text{then} \qquad |f(x) - L| < \epsilon.$$

mental. We could say, for example, that x was close to x_0 if $|x - x_0|$ was sufficiently small or, equivalently, if x was contained in a small open interval (neighborhood) centered at x_0. It is interesting to see how these ideas extend to $\mathbb{R}^2$ or $\mathbb{R}^3$.

Let (x_0, y_0) be a point in $\mathbb{R}^2$. What do we mean by the equation

$$|(x, y) - (x_0, y_0)| = r? \tag{1}$$

This is the definition of the circle of radius r centered at (x_0, y_0). Since (x, y) and (x_0, y_0) are vectors in $\mathbb{R}^2$,

Equation (11.1.1) on page 721

$$|(x, y) - (x_0, y_0)| = |(x - x_0, y - y_0)| = \sqrt{(x - x_0)^2 + (y - y_0)^2}. \tag{2}$$

Inserting (2) in (1) and squaring both sides yields

$$(x - x_0)^2 + (y - y_0)^2 = r^2, \tag{3}$$

which is the equation of the circle of radius r centered at (x_0, y_0). Then the set of points whose coordinates (x, y) satisfy the inequality

$$|(x, y) - (x_0, y_0)| < r \tag{4}$$

is the set of all points in $\mathbb{R}^2$ interior to the circle given by (3). This set is sketched in Figure 1. Similarly, the inequality

$$|(x, y) - (x_0, y_0)| \leq r \tag{5}$$

describes the set of all points interior to and on the circle given by (3). This is the set of points in Figure 1 plus the points on the circle. Setting $\mathbf{x} = (x, y)$ and $\mathbf{x}_0 = (x_0, y_0)$, we can write (4) and (5) as

$$|\mathbf{x} - \mathbf{x}_0| < r \tag{4'}$$

$$|\mathbf{x} - \mathbf{x}_0| \leq r. \tag{5'}$$

This leads to the following definitions:

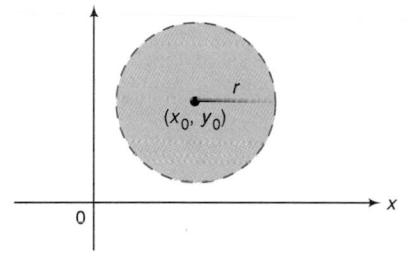

FIGURE 1
AN OPEN DISK

DEFINITION   **OPEN AND CLOSED DISKS; NEIGHBORHOODS**

(i) The **open disk** D_r centered at (x_0, y_0) with radius r is the subset of $\mathbb{R}^2$ given by

$$\{(x, y): |(x, y) - (x_0, y_0)| < r\}.$$

(ii) The **closed disk** centered at (x_0, y_0) with radius r is the subset of $\mathbb{R}^2$ given by

$$\{(x, y): |(x, y) - (x_0, y_0)| \leq r\}.$$

(iii) The **boundary** of the open or closed disk defined in (i) or (ii) is the circle

$$\{(x, y): |(x, y) - (x_0, y_0)| = r\}.$$

(iv) A **neighborhood** of a point (x_0, y_0) in $\mathbb{R}^2$ is an open disk centered at (x_0, y_0). ■

REMARK: In this definition, the words "open" and "closed" have meanings very similar to their meanings in the terms "open interval" and "closed interval." An open interval does not contain its endpoints. An open disk does not contain any point on its boundary. Similarly, a closed interval contains all its boundary points, as does a closed disk.

With these definitions, it is easy to define a limit of a function of two variables. Intuitively, we say that $f(x, y)$ approaches the limit L as (x, y) approaches (x_0, y_0) if $f(x, y)$ gets arbitrarily "close" to L as (x, y) approaches (x_0, y_0) along any **path**.[†] We define this notion below.

DEFINITION **LIMIT**

Let $f(x, y)$ be defined in a neighborhood of (x_0, y_0) but not necessarily at (x_0, y_0) itself. Then the **limit** of $f(x, y)$ as (x, y) approaches (x_0, y_0) is the real number L, written

$$\lim_{(x,y)\to(x_0,y_0)} f(x, y) = L, \tag{6}$$

if for every number $\epsilon > 0$, there is a number $\delta > 0$ such that $|f(x, y) - L| < \epsilon$ for every $(x, y) \neq (x_0, y_0)$ in the open disk centered at (x_0, y_0) with radius δ (see Figure 2). ∎

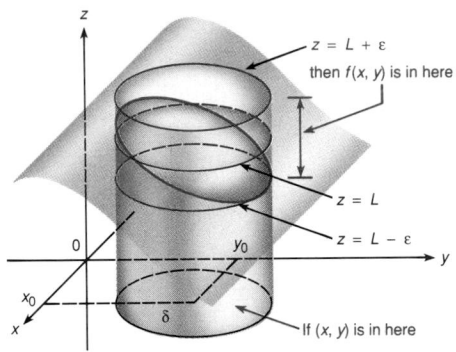

FIGURE 2 ILLUSTRATION OF THE DEFINITION OF A LIMIT IN $\mathbb{R}^2$

EXAMPLE 1
SHOWING FORMALLY THAT A LIMIT EXISTS

Show, using the definition of a limit, that $\lim_{(x,y)\to(1,2)}(3x + 2y) = 7$.

SOLUTION: Let $\epsilon > 0$ be given. We need to choose a $\delta > 0$ such that $|3x + 2y - 7| < \epsilon$ if $0 < \sqrt{(x - 1)^2 + (y - 2)^2} < \delta$. We start with

Triangle inequality on page 6

$$|3x + 2y - 7| = |3x - 3 + 2y - 4| \leq |3x - 3| + |2y - 4|$$

Explain why

$$= |3(x - 1)| + |2(y - 2)| = 3|x - 1| + 2|y - 2|. \tag{7}$$

[†] A path is another name for a curve joining (x, y) to (x_0, y_0).

Now

$$|x - 1| = \sqrt{(x-1)^2} \leq \sqrt{(x-1)^2 + (y-2)^2}$$

and

$$|y - 2| = \sqrt{(y-2)^2} \leq \sqrt{(x-1)^2 + (y-2)^2}.$$

So from (7),

$$|3x + 2y - 7| \leq 3\sqrt{(x-1)^2 + (y-2)^2} + 2\sqrt{(x-1)^2 + (y-2)^2}$$
$$= 5\sqrt{(x-1)^2 + (y-2)^2}. \tag{8}$$

Now we want $|3x + 2y - 7| < \epsilon$. We choose $\delta = \epsilon/5$. Then from (8),

$$|3x + 2y - 7| \leq 5\sqrt{(x-1)^2 + (y-2)^2} < 5\delta = 5\left(\frac{\epsilon}{5}\right) = \epsilon,$$

and the requested limit is shown.

We can prove the limit much faster if we use the Cauchy-Schwarz inequality (see Problem 11.2.75 on page 737):

$$\sum_{k=1}^{2} a_k b_k \leq \left(\sum_{k=1}^{2} a_k^2\right)^{1/2} \left(\sum_{k=1}^{2} b_k^2\right)^{1/2}$$

We have

$$|3x + 2y - 7| = |3(x-1) + 2(y-2)|.$$

Set $a_1 = 3$, $b_1 = x - 1$, $a_2 = 2$, $b_2 = y - 2$. Then we have

$$|3x + 2y - 7| = |3(x-1) + 2(y-2)| = |a_1 b_1 + a_2 b_2|$$
$$= \left|\sum_{k=1}^{2} a_k b_k\right| \leq \left(\sum_{k=1}^{2} a_k^2\right)^{1/2} \left(\sum_{k=1}^{2} b_k^2\right)^{1/2}$$
$$= \sqrt{a_1^2 + a_2^2} \sqrt{b_1^2 + b_2^2}$$
$$= \sqrt{3^2 + 2^2} \sqrt{(x-1)^2 + (y-2)^2}$$
$$= \sqrt{13} \sqrt{(x-1)^2 + (y-2)^2}.$$

Now we may choose $\delta = \epsilon/\sqrt{13}$ and the result is proved.

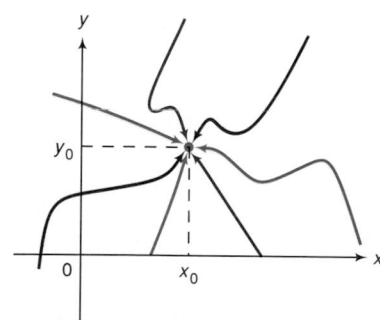

FIGURE 3
MANY PATHS LEAD TO (x_0, y_0)

In the intuitive definition of a limit which preceded the definition of a limit, we used the phrase "if $f(x, y)$ gets arbitrarily close to L as (x, y) approaches (x_0, y_0) *along any path.*" Recall that $\lim_{x \to x_0} f(x) = L$ only if $\lim_{x \to x_0^+} f(x) = \lim_{x \to x_0^-} f(x) = L$. That is, the limit exists only if we get the same value from either side of x_0. In $\mathbb{R}^2$ the situation is more complicated because (x, y) can approach (x_0, y_0) not just along two but along an *infinite number* of paths. Some of these are illustrated in Figure 3. Thus, the only way we can verify a limit is by making use of the definition or some appropriate limit theorem that can be proven directly from the definition. In the next two examples we illustrate the kinds of problems we can encounter.

EXAMPLE 2

A LIMIT DOES NOT EXIST AT A
POINT IN $\mathbb{R}^2$ IF WE GET
DIFFERENT VALUES AS WE
APPROACH THE POINT ALONG
DIFFERENT PATHS

Let $f(x, y) = (y^2 - x^2)/(y^2 + x^2)$ for $(x, y) \neq (0, 0)$. We will show that $\lim_{(x,y)\to(0,0)} f(x, y)$ does not exist. There are an infinite number of approaches to the origin. For example, if we approach along the x-axis, then $y = 0$ and, if the indicated limit exists,

$$\lim_{(x,y)\to(0,0)} \frac{y^2 - x^2}{y^2 + x^2} = \lim_{(x,y)\to(0,0)} \frac{-x^2}{x^2} = \lim_{(x,y)\to(0,0)} -1 = -1.$$

On the other hand, if we approach along the y-axis, then $x = 0$ and

$$\lim_{(x,y)\to(0,0)} \frac{y^2 - x^2}{y^2 + x^2} = \lim_{(x,y)\to(0,0)} \frac{y^2}{y^2} = \lim_{(x,y)\to(0,0)} 1 = 1.$$

Thus, we get different answers depending on how we approach the origin. To prove that the limit cannot exist, we note that we have shown that in any open disk centered at the origin, there are points at which f takes on the values $+1$ and -1. Hence, f cannot have a limit as $(x, y) \to (0, 0)$.

Example 2 leads to the following general rule:

RULE FOR NONEXISTENCE OF A LIMIT

If we get two or more different values for $\lim_{(x,y)\to(x_0,y_0)} f(x, y)$ as we approach (x_0, y_0) along different paths, or if the limit fails to exist along some path, then $\lim_{(x,y)\to(x_0,y_0)} f(x, y)$ does not exist.

EXAMPLE 3

A LIMIT THAT DOES NOT EXIST

Let $f(x, y) = xy^2/(x^2 + y^4)$. We will show that $\lim_{(x,y)\to(0,0)} f(x, y)$ does not exist. First, let us approach the origin along a straight line passing through the origin that is not the y-axis. Then $y = mx$ and

$$f(x, y) = \frac{x(m^2 x^2)}{x^2 + m^4 x^4} = \frac{m^2 x}{1 + m^2 x^2},$$

which approaches 0 as $x \to 0$. Thus, along every straight line, $f(x, y) \to 0$ as $(x, y) \to (0, 0)$. But if we approach $(0, 0)$ along the parabola $x = y^2$, then

$$f(x, y) = \frac{y^2(y^2)}{y^4 + y^4} = \frac{1}{2},$$

so that along this parabola, $f(x, y) \to \frac{1}{2}$ as $(x, y) \to (0, 0)$, and the limit does not exist.

EXAMPLE 4

A LIMIT THAT DOES EXIST

Prove that $\lim_{(x,y)\to(0,0)} [xy(y^2 - x^2)/(x^2 + y^2)] = 0$.

SOLUTION: It is easy to show that this limit is zero along any straight line passing through the origin. But as we have seen, this is not enough. We must rely on our definition. Let $\epsilon > 0$ be given. Then we must show that there is a $\delta > 0$ such that if $0 < \sqrt{x^2 + y^2} < \delta$, then

$$|f(x, y)| = \left| \frac{xy(y^2 - x^2)}{x^2 + y^2} \right| < \epsilon.$$

But $|x| \le \sqrt{x^2 + y^2}$, $|y| \le \sqrt{x^2 + y^2}$ and $|y^2 - x^2| \le x^2 + y^2$, so

$$\left| \frac{xy(y^2 - x^2)}{x^2 + y^2} \right| \le \frac{\sqrt{x^2 + y^2}\,\sqrt{x^2 + y^2}\,(x^2 + y^2)}{x^2 + y^2} = x^2 + y^2 < \delta^2.$$

Hence, if we choose $\delta = \sqrt{\epsilon}$, we will have $|f(x, y)| < \epsilon$ if $\sqrt{x^2 + y^2} < \delta$. A computer-drawn graph of this function is given in Figure 4.

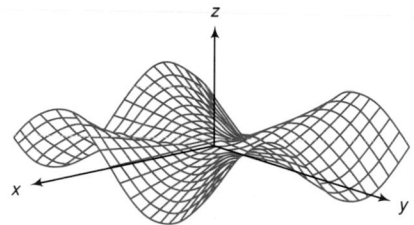

FIGURE 4   GRAPH OF $z = \begin{cases} \dfrac{xy(y^2 - x^2)}{x^2 + y^2} & \text{if } (x, y) \ne (0, 0) \\ 0 & \text{if } (x, y) = (0, 0) \end{cases}$

We can solve this problem more easily by using polar coordinates. Recall from Section 8.5 (see page 578) that, in polar coordinates,

$$x = r \cos \theta, \qquad y = r \sin \theta, \qquad \text{and} \qquad x^2 + y^2 = r^2.$$

Then $(x, y) \to (0, 0)$ is equivalent to $r \to 0$ and we have

$$\lim_{(x,y) \to (0,0)} \frac{xy(y^2 - x^2)}{x^2 + y^2} = \lim_{r \to 0} \frac{(r \cos \theta)(r \sin \theta)(r^2 \sin^2 \theta - r^2 \cos^2 \theta)}{r^2}$$

$$= \lim_{r \to 0} \frac{r^4 \sin \theta \cos \theta (\sin^2 \theta - \cos^2 \theta)}{r^2}$$

$$= \lim_{r \to 0} r^2 \sin \theta \cos \theta (\sin^2 \theta - \cos^2 \theta).$$

Now $\sin^2 \theta - \cos^2 \theta = -(\cos^2 \theta - \sin^2 \theta) = -\cos 2\theta$, so, as $|\sin \theta| \le 1$ and $|\cos \theta| \le 1$

$$|\sin \theta \cos \theta (\sin^2 \theta - \cos^2 \theta)| = |\sin \theta \cos \theta \cos 2\theta| \le 1.$$

Thus

$$-r^2 \le r^2 \sin \theta \cos \theta (\sin^2 \theta - \cos^2 \theta) \le r^2$$

and, as both r^2 and $-r^2 \to 0$ as $r \to 0$,

$$r^2 \sin \theta \cos \theta (\sin^2 \theta - \cos^2 \theta) \to 0 \qquad \text{by the squeezing theorem.}$$

NOTE: $\sin \theta \cos \theta = \frac{1}{2} \sin 2\theta$, so we can replace 1 by $\frac{1}{2}$ in the inequalities above. This, however, makes no difference in the final result.

REMARK: This example illustrates the fact that while the existence of different limits for different paths implies that no limit exists, only the definition or an appropriate limit theorem can be used to prove that a limit does exist.

As Examples 1 and 4 illustrate, it is often tedious to calculate limits from the definition. Fortunately, just as in the case of functions of a single variable, there are a number of theorems that greatly facilitate the calculation of limits. As you will recall, one of our important definitions stated that if f is continuous, then $\lim_{x \to x_0} f(x) = f(x_0)$. We now define continuity of a function of two variables.

DEFINITION   **CONTINUITY**

 (i) Let $f(x, y)$ be defined at every point (x, y) in a neighborhood of (x_0, y_0). Then f is **continuous** at (x_0, y_0) if all of the following conditions hold:
   (a) $f(x_0, y_0)$ exists [i.e., (x_0, y_0) is in the domain of f].
   (b) $\lim_{(x,y) \to (x_0,y_0)} f(x, y)$ exists.
   (c) $\lim_{(x,y) \to (x_0,y_0)} f(x, y) = f(x_0, y_0)$.
 (ii) If one or more of these three conditions fail to hold, then f is said to be **discontinuous** at (x_0, y_0).
 (iii) f is **continuous in a subset** S of $\mathbb{R}^2$ if f is continuous at every point (x, y) in S. ∎

REMARK: Condition (c) tells us that if a function f is continuous at (x_0, y_0), then we can calculate $\lim_{(x,y) \to (x_0,y_0)} f(x, y)$ by evaluation of f at (x_0, y_0).

EXAMPLE 5
A FUNCTION THAT IS
DISCONTINUOUS AT (0, 0)

Let $f(x, y) = \dfrac{xy^2}{x^2 + y^2}$. Then f is discontinuous at $(0, 0)$ because $f(0, 0)$ is not defined.

EXAMPLE 6
ANOTHER FUNCTION THAT IS
DISCONTINUOUS AT (0, 0)

Let
$$f(x, y) = \begin{cases} \dfrac{xy^2}{x^2 + y^4}, & (x, y) \neq (0, 0) \\ 0, & (x, y) = (0, 0). \end{cases}$$

Here f is defined at $(0, 0)$, but it is still discontinuous there because $\lim_{(x,y) \to (0,0)} f(x, y)$ does not exist (see Example 3).

EXAMPLE 7
A FUNCTION THAT IS
CONTINUOUS AT (0, 0)

Let
$$f(x, y) = \begin{cases} \dfrac{xy(y^2 - x^2)}{x^2 + y^2}, & (x, y) \neq (0, 0) \\ 0, & (x, y) = (0, 0). \end{cases}$$

Here f is continuous at $(0, 0)$, according to Example 4.

EXAMPLE 8
A FUNCTION THAT IS
DISCONTINUOUS AT (0, 0)

Let

$$f(x, y) = \begin{cases} \dfrac{xy(y^2 - x^2)}{x^2 + y^2}, & (x, y) \neq (0, 0) \\ 1, & (x, y) = (0, 0). \end{cases}$$

Then f is discontinuous at $(0, 0)$ because $\lim_{(x,y)\to(0,0)} f(x, y) = 0 \neq f(0, 0) = 1$, so condition (c) is violated.

Naturally, we would like to know what functions are continuous. We can answer this question if we look at the continuous functions of one variable. We start with polynomials.

DEFINITION   **POLYNOMIAL AND RATIONAL FUNCTION**

(i) A **polynomial** $p(x, y)$ in the two variables x and y is a finite sum of terms of the form $Ax^m y^n$, where m and n are nonnegative integers and A is a real number.

(ii) A **rational function** $r(x, y)$ in the two variables x and y is a function that can be written as the quotient of two polynomials in x and y: $r(x, y) = p(x, y)/q(x, y)$. ■

EXAMPLE 9
A POLYNOMIAL IN x AND y

$p(x, y) = 5x^5 y^2 + 12xy^9 - 37x^{82}y^5 + x + 4y - 6$ is a polynomial.

EXAMPLE 10
A RATIONAL FUNCTION
IN x AND y

$$r(x, y) = \frac{8x^3 y^7 - 7x^2 y^4 + xy - 2y}{1 - 3y^3 + 7x^2 y^2 + 18yx^7}$$

is a rational function.

The limit theorems we stated in Section 1.3 and proved in Appendix 3 can be extended, with minor modifications, to functions of two variables. We will not state them here. However, by using them it is not difficult to prove the following theorem about continuous functions.

THEOREM 1   **FACTS ABOUT CONTINUITY**

(i) Any polynomial p is continuous at any point in $\mathbb{R}^2$.

(ii) Any rational function $r = p/q$ is continuous at any point (x_0, y_0) for which $q(x_0, y_0) \neq 0$. It is discontinuous when $q(x_0, y_0) = 0$ because it is then not defined at (x_0, y_0).

(iii) If f and g are continuous at (x_0, y_0), then $f + g$, $f - g$, and $f \cdot g$ are continuous at (x_0, y_0).[†]

[†] $(f + g)(x, y) = f(x, y) + g(x, y)$, just as in the case of functions of one variable. Similarly, $(f \cdot g)(x, y) = f(x, y)g(x, y)$ and $(f/g)(x, y) = f(x, y)/g(x, y)$.

(iv) If f and g are continuous at (x_0, y_0) and if $g(x_0, y_0) \neq 0$, then f/g is continuous at (x_0, y_0).

(v) If f is continuous at (x_0, y_0) and if h is a function of one variable that is continuous at $f(x_0, y_0)$, then the composite function $h \circ f$, defined by $(h \circ f)(x, y) = h(f(x, y))$, is continuous at (x_0, y_0). ∎

EXAMPLE 11
THE LIMIT OF A RATIONAL FUNCTION

Calculate $\lim_{(x,y)\to(4,1)}(x^3y^2 - 4xy)/(x + 6xy^3)$.

SOLUTION: $(x^3y^2 - 4xy)/(x + 6xy^3)$ is continuous at $(4, 1)$ so we can calculate the limit by evaluation. We have

$$\lim_{(x,y)\to(4,1)} \frac{x^3y^2 - 4xy}{x + 6xy^3} = \left.\frac{x^3y^2 - 4xy}{x + 6xy^3}\right|_{(4,1)} = \frac{64 \cdot 1 - 4 \cdot 4 \cdot 1}{4 + 6 \cdot 4 \cdot 1} = \frac{48}{28} = \frac{12}{7}.$$

All the ideas in this section can be generalized to functions of three variables. A **neighborhood** of radius r of the point (x_0, y_0, z_0) in $\mathbb{R}^3$ consists of all points in $\mathbb{R}^3$ interior to the sphere

$$(x - x_0)^2 + (y - y_0)^2 + (z - z_0)^2 = r^2.$$

That is, the neighborhood (also called an **open ball**) is described by

$$\{(x, y, z): (x - x_0)^2 + (y - y_0)^2 + (z - z_0)^2 < r^2\}.$$

A typical neighborhood is sketched in Figure 5. The **closed ball** is described by

$$\{(x, y, z): (x - x_0)^2 + (y - y_0)^2 + (z - z_0)^2 \leq r^2\}.$$

Using an open ball instead of an open disk, the definition of the limit

$$\lim_{(x,y,z)\to(x_0,y_0,z_0)} f(x, y, z) = L$$

is analogous to the definition of $\lim_{(x,y)\to(x_0,y_0)} = L$ given on page 844. All other definitions, remarks, and theorems given in this section apply to functions of three (or more) variables.

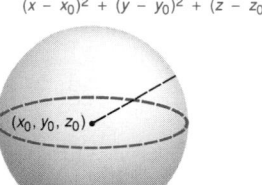

$(x - x_0)^2 + (y - y_0)^2 + (z - z_0)^2 < r^2$

(x_0, y_0, z_0)

FIGURE 5
AN OPEN BALL IN $\mathbb{R}^3$

■ PROBLEMS 13.2

■ SELF-QUIZ

I. Let $f(x, y) = (x^2 - y^2)/(x^2 + y^2)$. Then $\lim_{(x,y)\to(0,0)} f(x, y)$ _____.
  a. $= \lim_{x\to 0}(\lim_{y\to 0} f(x, y))$
  b. $= \lim_{y\to 0}(\lim_{x\to 0} f(x, y))$
  c. $= \lim_{x\to 0}(\lim_{y\to x} f(x, y))$
  d. does not exist

II. Let $g(x, y) = x^2 \cdot y/(x^4 + y^2)$. Then $\lim_{(x,y)\to(0,0)} g(x, y)$ _____.
  a. $= \lim_{x\to 0}(\lim_{y\to 0} g(x, y))$
  b. $= \lim_{y\to 0}(\lim_{x\to 0} g(x, y))$
  c. $= \lim_{x\to 0}(\lim_{y\to x} g(x, y))$
  d. does not exist

III. Let $h(x, y) = (x^2 - 3y + 5)/(1 + x^2 + y^2)$. Then $\lim_{(x,y)\to(0,0)} h(x, y)$ _____.
  a. $= \lim_{x\to 0}(\lim_{y\to 0} h(x, y))$
  b. $= \lim_{y\to 0}(\lim_{x\to 0} h(x, y))$
  c. $= \lim_{x\to 0}(\lim_{y\to x} h(x, y))$
  d. does not exist

IV. Let $F(x, y) = y/x$. Then $\lim_{(x,y)\to(0,0)} F(x, y)$ _____.
  a. $= \lim_{x\to 0}(\lim_{y\to 0} F(x, y))$
  b. $= \lim_{y\to 0}(\lim_{x\to 0} F(x, y))$
  c. $= \lim_{x\to 0}(\lim_{y\to x} F(x, y))$
  d. does not exist

■ **DRILL**

In Problems 1–4, sketch the indicated region.
1. The open disk centered at (3, 2) with radius 3.
2. The closed disk centered at $(-1, 1)$ with radius 1.
3. The open ball centered at (1, 0, 0) with radius 1.
4. The closed ball centered at (0, 1, 1) with radius 2.

In Problems 5–10, use the definition on page 844 to verify the declared limit.
5. $\lim_{(x,y)\to(1,2)}(3x + y) = 5$.
6. $\lim_{(x,y)\to(3,-1)}(x - 7y) = 10$.
7. $\lim_{(x,y)\to(5,-2)}(ax + by) = 5a - 2b$.

* 8. $\lim_{(x,y)\to(1,1)} \dfrac{x}{y} = 1$.　　9. $\lim_{(x,y)\to(0,0)} \dfrac{2x^2y}{x^2 + y^2} = 0$.

10. $\lim_{(x,y)\to(4,1)}(x^2 + 3y^2) = 19$.

In Problems 11–20, show that the indicated limit does not exist.

11. $\lim_{(x,y)\to(0,0)} \dfrac{x + y}{x - y}$
12. $\lim_{(x,y)\to(0,0)} \dfrac{xy}{x^2 - y^2}$

13. $\lim_{(x,y)\to(0,0)} \dfrac{xy}{x^2 + y^2}$
14. $\lim_{(x,y)\to(0,0)} \dfrac{xy^3}{x^4 + y^4}$

15. $\lim_{(x,y)\to(0,0)} \dfrac{xy}{x^3 + y^3}$
16. $\lim_{(x,y)\to(0,0)} \dfrac{(x^2 + y^2)^2}{x^4 + y^4}$

17. $\lim_{(x,y)\to(0,0)} \dfrac{x^2 - 2y}{y^2 + 2x}$

18. $\lim_{(x,y)\to(0,0)} \dfrac{ax^2 + by}{cy^2 + dx}$, $a, b, c, d > 0$.

19. $\lim_{(x,y,z)\to(0,0,0)} \dfrac{xy + 2xz + 3yz}{x^2 + y^2 + z^2}$

20. $\lim_{(x,y,z)\to(0,0,0)} \dfrac{xyz}{x^3 + y^3 + z^3}$

In Problems 21–30, calculate the indicated limit.

21. $\lim_{(x,y)\to(-1,2)}(xy + 4y^2x^3)$

22. $\lim_{(x,y)\to(-1,2)} \dfrac{4x^3y^2 - 2xy^5 + 7y - 1}{3xy - y^4 + 3x^3}$

23. $\lim_{(x,y)\to(-4,3)} \dfrac{1 + xy}{1 - xy}$

24. $\lim_{(x,y)\to(\pi,\pi/3)} \ln(x + y)$

25. $\lim_{(x,y)\to(1,2)} \ln(1 + e^{x+y})$

26. $\lim_{(x,y)\to(2,5)} \sinh\left(\dfrac{x + 1}{y - 2}\right)$　　27. $\lim_{(x,y)\to(1,1)} \dfrac{\sin(x - y)}{\cos(x - y)}$

28. $\lim_{(x,y)\to(2,2)} \dfrac{x^3 - 2xy + 3x^2 - 2y}{x^2y + 4y^2 - 6x^2 + 24y}$

29. $\lim_{(x,y,z)\to(1,1,3)} \dfrac{xy^2 - 4xz^2 + 5yz}{3z^2 - 8z^3y^7x^4 + 7x - y + 2}$

30. $\lim_{(x,y,z)\to(4,1,3)} \ln(x - yz + 4x^3y^5z)$

In Problems 31–44, describe the maximum region over which the given function is continuous.

31. $f(x, y) = \sqrt{x - y}$

32. $f(x, y) = \dfrac{x^3 + 4xy^6 - 7x^4}{x^2 + y^2}$

33. $f(x, y) = \dfrac{x^3 + 4xy^6 - 7x^4}{x^3 - y^3}$

*34. $f(x, y) = \dfrac{xy^3 - 17x^2y^5 + 8x^3y}{xy + 3y - 4x - 12}$

35. $f(x, y) = \ln(3x + 2y + 6)$
36. $f(x, y) = \tan^{-1}(x - y)$　　37. $f(x, y) = e^{xy+2}$

38. $f(x, y) = \dfrac{x^3 - 1 + 3y^5x^2}{1 - xy}$

39. $f(x, y) = \dfrac{x}{\sqrt{1 - (x/2)^2 - y^2}}$

40. $f(x, y, z) = e^{xy + yz - \sqrt{x}}$

41. $f(x, y, z) = \dfrac{xyz^2 + yzx^2 - 3x^3yz^5}{x - y + 2z + 4}$

42. $f(x, y, z) = y \ln(xz)$
43. $f(x, y, z) = z + \cos^{-1}(x^2 - y)$

44. $f(x, y, z) = \dfrac{1}{\sqrt{1 - x^2 - y^2 - z^2}}$

■ **APPLICATIONS**

45. Find a function $g(x)$ such that the function

$$f(x, y) = \begin{cases} \dfrac{x^2 - y^2}{x - y}, & \text{if } x \neq y \\ g(x) & \text{if } x = y \end{cases}$$

is continuous at every point in $\mathbb{R}^2$.

46. Find a number c such that the function

$$f(x, y) = \begin{cases} \dfrac{3xy}{\sqrt{x^2 + y^2}}, & \text{if } (x, y) \neq (0, 0) \\ c, & \text{if } (x, y) = (0, 0) \end{cases}$$

is continuous at the origin; show that your choice of c is unique.

CONTINUED

47. Find a number c such that the function

$$f(x, y) = \begin{cases} \dfrac{xy}{|x| + |y|}, & \text{if } (x, y) \neq (0, 0) \\ c, & \text{if } (x, y) = (0, 0) \end{cases}$$

is continuous at the origin; show that your choice of c is unique.

*48. Discuss the continuity at the origin of

$$f(x, y, z) = \begin{cases} \dfrac{yz - x^2}{x^2 + y^2 + z^2}, & \text{if } (x, y, z) \neq (0, 0, 0) \\ 0, & \text{if } (x, y, z) = (0, 0, 0). \end{cases}$$

■ SHOW/PROVE/DISPROVE

In Problems 49–52, show that the indicated limit exists and then calculate it.

49. $\displaystyle\lim_{(x,y)\to(0,0)} \frac{3xy}{\sqrt{x^2 + y^2}}$

50. $\displaystyle\lim_{(x,y)\to(0,0)} \frac{5x^2y^2}{x^4 + y^2}$

51. $\displaystyle\lim_{(x,y)\to(0,0)} \frac{x^3 + y^3}{x^2 + y^2}$

52. $\displaystyle\lim_{(x,y,z)\to(0,0,0)} \frac{yx^2 + z^3}{x^2 + y^2 + z^2}$

*53. Explain why the following reasoning is false:

$$\lim_{(x,y)\to(1,1)} \frac{x - y}{x^2 - y^2} = \lim_{(x,y)\to(1,1)} \frac{x - y}{(x - y)(x + y)}$$
$$= \lim_{(x,y)\to(1,1)} \frac{1}{x + y}$$
$$= \frac{1}{1 + 1} = \frac{1}{2}.$$

■ ANSWERS TO SELF-QUIZ

I. d $(a = 1, b = -1, c = 0)$

II. d (along the curve $y = x^2$ the limiting value is $\frac{1}{2}$ while along $y = -x^2$ the limiting value is $-\frac{1}{2}$)

III. $a = b = c = 5$

IV. d $(a = 0, b$ does not exist, $c = 1)$

13.3 PARTIAL DERIVATIVES

In this section, we show one of the ways a function of several variables can be differentiated. The idea is simple. Let $z = f(x, y)$. If we keep one of the variables, say y, fixed, then f can be treated as a function of x only and we can calculate the derivative (if it exists) of f with respect to x. This new function is called the *partial derivative of f with respect to x* and is denoted by $\partial f / \partial x$.[†] Before giving a more formal definition, we give an example.

EXAMPLE 1
COMPUTING A PARTIAL DERIVATIVE

Let $z = f(x, y) = x^2 y + \sin xy^2$. Calculate $\partial f / \partial x$.

SOLUTION: Treating y as if it were a constant, we have

$$\frac{\partial f}{\partial x} = \frac{\partial}{\partial x}(x^2 y + \sin xy^2) = \frac{\partial}{\partial x} x^2 y + \frac{\partial}{\partial x} \sin xy^2 = 2xy + y^2 \cos xy^2.$$

How should we define a partial derivative of a function f of two variables? Suppose that y is held fixed. Then only x varies. We define

$$g_y(x) = f(x, y).$$

[†] See the historical note on page 862.

g_y is a function of x only, and by the definition of the derivative of a function of one variable on page 87,

$$g_y(x) = f(x, y)$$

$$\frac{dg_y}{dx} = \lim_{\Delta x \to 0} \frac{g_y(x + \Delta x) - g_y(x)}{\Delta x} = \lim_{\Delta x \to 0} \frac{f(x + \Delta x, y) - f(x, y)}{\Delta x}.$$

This suggests the following definitions of partial derivatives.

DEFINITION   **PARTIAL DERIVATIVES IN $\mathbb{R}^2$**

Let $z = f(x, y)$.

(i) The **partial derivative of f with respect to x** is the function

$$\frac{\partial z}{\partial x} = \frac{\partial f}{\partial x} = \lim_{\Delta x \to 0} \frac{f(x + \Delta x, y) - f(x, y)}{\Delta x}. \tag{1}$$

$\partial f / \partial x$ is defined at every point (x, y) in the domain of f such that the limit (1) exists.

(ii) The **partial derivative of f with respect to y** is the function

$$\frac{\partial z}{\partial y} = \frac{\partial f}{\partial y} = \lim_{\Delta y \to 0} \frac{f(x, y + \Delta y) - f(x, y)}{\Delta y}. \tag{2}$$

$\partial f / \partial y$ is defined at every point (x, y) in the domain of f such that the limit (2) exists. ∎

REMARK 1:   This definition allows us to calculate partial derivatives in the same way we calculate ordinary derivatives by allowing only one of the variables to vary while the other one is held fixed. It also allows us to use all the formulas from one-variable calculus.

REMARK 2:   The partial derivatives $\partial f / \partial x$ and $\partial f / \partial y$ give us the rate of change of f as each of the variables x and y changes with the other one held fixed. They do *not* tell us how f changes when x and y change simultaneously. We will discuss this different topic in Section 13.5.

REMARK 3:   It should be emphasized that while the functions $\partial f / \partial x$ and $\partial f / \partial y$ are computed with one of the variables held constant, each is a function of both variables.

EXAMPLE 2
COMPUTING TWO PARTIAL
DERIVATIVES

Let $f(x, y) = \sqrt{x + y^2}$. Calculate $\partial f / \partial x$ and $\partial f / \partial y$ at the point $(2, -3)$.

SOLUTION:

$$\frac{\partial f}{\partial x} = \frac{1}{2\sqrt{x + y^2}} \frac{\partial}{\partial x}(x + y^2) = \frac{1}{2\sqrt{x + y^2}}(1 + 0) = \frac{1}{2\sqrt{x + y^2}}$$

since we are treating y as a constant. At $(2, -3)$,

$$\frac{\partial f}{\partial x} = \frac{1}{2\sqrt{2 + (-3)^2}} = \frac{1}{2\sqrt{11}}$$

$$\frac{\partial f}{\partial y} = \frac{1}{2\sqrt{x + y^2}} \frac{\partial}{\partial y}(x + y^2) = \frac{1}{2\sqrt{x + y^2}}(0 + 2y) = \frac{y}{\sqrt{x + y^2}}$$

since we are treating x as a constant. At $(2, -3)$, $\dfrac{\partial f}{\partial y} = \dfrac{-3}{\sqrt{11}}$.

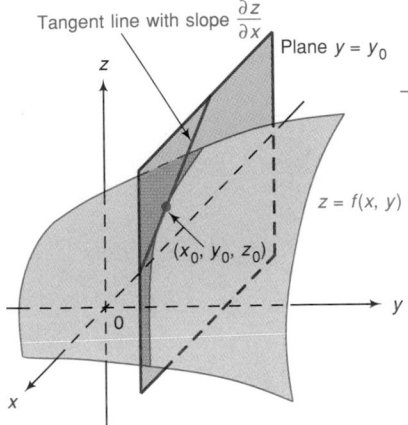

FIGURE 1

GRAPHICAL INTERPRETATION OF $\dfrac{\partial f}{\partial x}$

We now obtain a geometric interpretation of the partial derivative. Let $z = f(x, y)$. As we saw in Section 13.1, this is the equation of a surface in $\mathbb{R}^3$. To obtain $\partial z/\partial x$, we hold y fixed at some constant value y_0. The equation $y = y_0$ is a plane in space parallel to the xz-plane (whose equation is $y = 0$). Thus if y is constant, $\partial z/\partial x$ is the rate of change of f with respect to x as x changes along the curve C, which is at the intersection of the surface $z = f(x, y)$ and the plane $y = y_0$. This is illustrated in Figure 1. To be more precise, let (x_0, y_0, z_0) be a point on the surface $z = f(x, y)$. The intersection of this surface and the plane $y = y_0$ is a curve. Then $\partial z/\partial x$ evaluated at (x_0, y_0) is the slope of the tangent line to this curve at the point (x_0, y_0, z_0). Analogously, the intersection of the surface with the plane $x = x_0$ is a different curve and $\partial z/\partial y$ evaluated at (x_0, y_0) is the slope of the line tangent to the new curve at the point (x_0, y_0, z_0). (This is true because x is held fixed in order to calculate $\partial z/\partial y$.) This is illustrated in Figure 2.

There are other ways to denote partial derivatives. We will often write

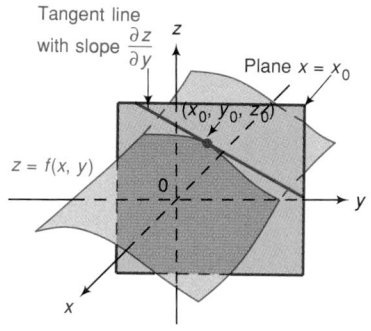

FIGURE 2

GRAPHICAL INTERPRETATION OF $\dfrac{\partial f}{\partial y}$

NOTATION FOR PARTIAL DERIVATIVES

$$f_x = \frac{\partial f}{\partial x} \qquad \text{and} \qquad f_y = \frac{\partial f}{\partial y}. \tag{3}$$

As an example of the use of this notation, we may write the result of Example 2 as $f_x(2, -3) = \dfrac{1}{2\sqrt{11}}$ and $f_y(2, -3) = -\dfrac{3}{\sqrt{11}}$. If f is a function of other variables, say s and t, then we may write $\partial f/\partial s = f_s$ and $\partial f/\partial t = f_t$.

We now turn to the question of finding partial derivatives of functions of three variables.

DEFINITION **PARTIAL DERIVATIVES IN $\mathbb{R}^3$**

Let $w = f(x, y, z)$.

(i) The **partial derivative of f with respect to** x is the function

$$\frac{\partial w}{\partial x} = \frac{\partial f}{\partial x} = f_x = \lim_{\Delta x \to 0} \frac{f(x + \Delta x, y, z) - f(x, y, z)}{\Delta x}. \tag{4}$$

f_x is defined at every point (x, y, z) in the domain of f at which the limit in (4) exists.

(ii) The **partial derivative of** f **with respect to** y is the function

$$\frac{\partial w}{\partial y} = \frac{\partial f}{\partial y} = f_y = \lim_{\Delta y \to 0} \frac{f(x, y + \Delta y, z) - f(x, y, z)}{\Delta y}. \tag{5}$$

f_y is defined at every point (x, y, z) in the domain of f at which the limit in (5) exists.

(iii) The **partial derivative of** f **with respect to** z is the function

$$\frac{\partial w}{\partial z} = \frac{\partial f}{\partial z} = f_z = \lim_{\Delta z \to 0} \frac{f(x, y, z + \Delta z) - f(x, y, z)}{\Delta z}. \tag{6}$$

f_z is defined at any point (x, y, z) in the domain of f at which the limit in (6) exists. ■

REMARK: As can be seen from this definition, to calculate $\partial f/\partial x$, we simply treat y and z as constants and then calculate an ordinary derivative.

EXAMPLE 3
CALCULATING THREE PARTIAL DERIVATIVES

Let $w = f(x, y, z) = xz + e^{y^2 z} + \sqrt{xy^2 z^3}$. Calculate $\partial w/\partial x$, $\partial w/\partial y$, and $\partial w/\partial z$.

SOLUTION: To calculate $\partial w/\partial x$, we keep y and z fixed. Then,

$$\frac{\partial w}{\partial x} = \frac{\partial f}{\partial x} = f_x = \frac{\partial}{\partial x} xz + \frac{\partial}{\partial x} e^{y^2 z} + \frac{1}{2\sqrt{xy^2 z^3}} \frac{\partial}{\partial x} (xy^2 z^3)$$

$$= z + 0 + \frac{y^2 z^3}{2\sqrt{xy^2 z^3}} = z + \frac{y^2 z^3}{2\sqrt{xy^2 z^3}}.$$

To calculate $\partial w/\partial y$, we keep x and z fixed. Then,

$$\frac{\partial w}{\partial y} = \frac{\partial f}{\partial y} = f_y = \frac{\partial}{\partial y} xz + e^{y^2 z} \frac{\partial}{\partial y} (y^2 z) + \frac{1}{2\sqrt{xy^2 z^3}} \frac{\partial}{\partial y} (xy^2 z^3)$$

$$= 0 + 2yz e^{y^2 z} + \frac{2xyz^3}{2\sqrt{xy^2 z^3}} = 2yz e^{y^2 z} + \frac{xyz^3}{\sqrt{xy^2 z^3}}.$$

To calculate $\partial w/\partial z$, we keep x and y fixed. Then,

$$\frac{\partial w}{\partial z} = \frac{\partial f}{\partial z} = f_z = x + e^{y^2 z} \frac{\partial}{\partial z} y^2 z + \frac{1}{2\sqrt{xy^2 z^3}} \frac{\partial}{\partial z} (xy^2 z^3)$$

$$= x + y^2 e^{y^2 z} + \frac{3xy^2 z^2}{2\sqrt{xy^2 z^3}}.$$

We can use our definitions to develop a definition of partial derivatives of functions of four or more variables. Although we will not give a formal definition, the general idea is to hold constant all the variables but one and then perform an "ordinary" differentiation.

In Section 1.7 we showed that a differentiable function is continuous. The situation is more complicated for a function of two variables. In fact, there are functions that are discontinuous at a point but for which all partial derivatives exist at that point.

EXAMPLE 4
PARTIAL DERIVATIVES MAY EXIST
AT A POINT OF DISCONTINUITY

Let

$$f(x, y) = \begin{cases} \dfrac{xy}{x^2 + y^2}, & (x, y) \neq (0, 0) \\ 0, & (x, y) = (0, 0). \end{cases}$$

Show that f_x and f_y exist at $(0, 0)$ but that f is not continuous there.

SOLUTION: We first show that $\lim_{(x,y)\to(0,0)} f(x, y)$ does not exist, so that f cannot be continuous at $(0, 0)$. To show that, we first let $(x, y) \to (0, 0)$ along the line $y = x$. Then,

$$\frac{xy}{x^2 + y^2} = \frac{x^2}{x^2 + x^2} = \frac{1}{2},$$

so if $\lim_{(x,y)\to(0,0)} f(x, y)$ existed, it would have to equal $\frac{1}{2}$. But if we now let $(x, y) \to (0, 0)$ along the line $y = -x$, we have

$$\frac{xy}{x^2 + y^2} = \frac{-x^2}{x^2 + x^2} = -\frac{1}{2},$$

so along this line the limit is $-\frac{1}{2}$. Hence, the limit does not exist. On the other hand, we have

$$f_x(0, 0) = \lim_{\Delta x \to 0} \frac{f(0 + \Delta x, 0) - f(0, 0)}{\Delta x} = \lim_{\Delta x \to 0} \frac{\dfrac{(0 + \Delta x) \cdot 0}{\Delta x^2 + 0^2}}{\Delta x}$$

$$= \lim_{\Delta x \to 0} \frac{0}{\Delta x} = \lim_{\Delta x \to 0} 0 = 0.$$

Similarly, $f_y(0, 0) = 0$. Hence, both f_x and f_y exist at $(0, 0)$ (and are equal to zero) even though f is not continuous there.

In Section 13.5, we will show a relationship between continuity and partial derivatives and show that a certain kind of differentiability does imply continuity.

■ PROBLEMS 13.3

■ SELF-QUIZ

I. If $f(x, y) = x^2 - y^3 + x \cos y$, then

$$\frac{\partial f(x, y)}{\partial y} = \underline{\hspace{2cm}}.$$

a. $-3y^2 - x \sin y$
b. $-3y^2 + \cos y - x \sin y$
c. $2x - 3y^2 + \cos y - x \sin y$
d. $-\sin y$

II. If $f(x, y, z) = xy - y^2z + xe^z - ye^x$, then
$$\underline{\hspace{2cm}} = x - 2yz - e^x.$$

a. $\dfrac{\partial f}{\partial x}$ b. $\dfrac{\partial f}{\partial y}$ c. $\dfrac{\partial f}{\partial z}$ d. $\dfrac{dy}{dt}$

III. If $f(x, y) = \underline{\hspace{2cm}}$, then $\dfrac{\partial f}{\partial x} = 0$.

a. $\sin^2 x + \cos^2 y$
b. $e^y + \sin(3x)$
c. $7 + \cos y - \ln|5y + 2|$
d. $(x + y)^2 - (x - y)^2$

IV. If $z = f(x, y)$ and $\dfrac{\partial z}{\partial x} = 0$, then $\underline{\hspace{2cm}}$.

a. $f(x, y) = 7 + \cos y - \ln|5y + 2|$
b. $f(x, y) = e^y - 3y + \tan(2 - y)$
c. $f(x, y)$ is a constant
d. the value of $f(x, y)$ is independent of x

CONTINUED

V. True–False: Suppose $f(x, y) = (x + \sin y)(y - \cos x)$ and $g(x) = f(x, -\pi/2)$, then

$$\frac{\partial f}{\partial x}(\pi/3, -\pi/2) = \frac{dg}{dx}(\pi/3).$$

VI. True–False: Suppose $f(x, y) = (x + \sin y)(y - \cos x)$ and $h(y) = f(\pi/3, y)$, then

$$\frac{\partial f}{\partial y}(\pi/3, -\pi/2) = \frac{dh}{dy}(-\pi/2).$$

■ **DRILL**

In Problems 1–16, calculate $\dfrac{\partial z}{\partial x}$ and $\dfrac{\partial z}{\partial y}$.

1. $z = x^2 y$
2. $z = xy^2$
3. $z = 3e^{xy^3}$
4. $z = \sin(x^2 + y^3)$
5. $z = 4x/y^5$
6. $z = e^y \tan x$
7. $z = \ln(x^3 y^5 - 2)$
8. $z = \sqrt{xy + 2y^3}$
9. $z = (x + 5y \sin x)^{4/3}$
10. $z = \sinh(2x - y)$
11. $z = \sin^{-1}(x - y)$
12. $z = \sec xy$
13. $xz + yz^3 = z$
    [*Hint:* differentiate implicitly.]
14. $\ln(x + y + z) = yz$
15. $\sin(z - x) = y/z$
16. $2xz + 3xyz^2 = e^{xy^2 z^3}$

In Problems 17–24, calculate the value of the given partial derivative at the specified point.

17. $f(x, y) = x^3 - y^4$; $f_x(1, -1)$
18. $f(x, y) = \ln(x^2 + y^4)$; $f_y(3, 1)$
19. $f(x, y) = \sin(x + y)$; $f_x(\pi/6, \pi/3)$
20. $f(x, y) = e^{\sqrt{x^2 + y}}$; $f_y(0, 4)$
21. $f(x, y) = \sinh(x - y)$; $f_x(3, 3)$
22. $f(x, y) = \sqrt[3]{x^2 y - y^2 x^5}$; $f_y(-2, 4)$
23. $f(x, y) = \dfrac{x^2 - y^2}{x^2 + y^2}$; $f_y(2, -3)$
24. $f(x, y) = \tan^{-1}(y/x)$; $f_x(4, 4)$

In Problems 25–32, calculate $\dfrac{\partial w}{\partial x}, \dfrac{\partial w}{\partial y}$, and $\dfrac{\partial w}{\partial z}$.

25. $w = \sqrt{x + y + z}$
26. $w = \sin(xyz)$
27. $w = e^{x + 2y + 3z}$
28. $w = \cosh \sqrt{x + 2y + 5z}$
29. $w = \dfrac{x + y}{z}$
30. $w = \tan^{-1}(xz/y)$
31. $w = \ln(x^3 + y^2 + z)$
32. $w = \dfrac{x^2 - y^2 + z^2}{x^2 + y^2 + z^2}$

In Problems 33–42, calculate the value of the given partial derivative at the specified point.

33. $f(x, y, z) = xyz$; $f_x(2, 3, 4)$
34. $f(x, y, z) = \sqrt{x + 2y + 3z}$; $f_y(2, -1, 3)$
35. $f(x, y, z) = \dfrac{x - y}{z}$; $f_z(-3, -1, 2)$
36. $f(x, y, z) = \sin(z^2 - y^2 + x)$; $f_y(0, 1, 0)$
37. $f(x, y, z) = \ln(x + 2y + 3z)$; $f_z(2, 2, 5)$
38. $f(x, y, z) = \tan^{-1}\left(\dfrac{xy}{z}\right)$; $f_x(1, 2, -2)$
39. $f(x, y, z) = \dfrac{y^3 - z^5}{x^2 y + z}$; $f_y(4, 0, 1)$
40. $f(x, y, z) = \sqrt{\dfrac{x + y - z}{x + y + z}}$; $f_z(1, 1, 1)$
41. $f(x, y, z) = e^{xy}(\cosh z - \sinh z)$; $f_z(2, 3, 0)$
42. $f(x, y, z) = \sqrt{x^2 + y^2 + z^2}$; $f_x(a, b, c)$

■ **APPLICATIONS**

43. Find the equation of the line tangent to the surface $z = x^3 - 4y^3$ at the point $(1, -1, 5)$ that
    a. lies in the plane $x = 1$;
    b. lies in the plane $y = -1$.
44. Find the equation of the line tangent to the surface $z = \tan^{-1}(y/x)$ at the point $(\sqrt{3}, 1, \pi/6)$ that
    a. lies in the plane $x = \sqrt{3}$;
    b. lies in the plane $y = 1$.

45. Find the equation of the line tangent to the surface $x^2 + 4y^2 + 4z^2 = 9$ at the point $(1, 1, 1)$ that lies in the plane $y = 1$.
*46. Find the equation of the line tangent to the surface $x^2 + 4y^2 + 4z^2 = 9$ at the point $(1, 1, 1)$ that lies in the plane $z = 1$.
47. The cost to a manufacturer of producing x units of product A and y units of product B is given

(in dollars) by

$$C(x, y) = \frac{50}{2 + x} + \frac{125}{(3 + y)^2}.$$

Calculate the marginal cost of each of the two products.

48. The revenue received by the manufacturer of the preceding problem is given by

$$R(x, y) = \ln(1 + 50x + 75y) + \sqrt{1 + 40x + 125y}.$$

Calculate the marginal revenue from each of the two products.

49. A **partial-differential equation** is an equation involving partial derivatives. Verify that the function $z = f(x, y) = e^{(x + \sqrt{3}y)/4} - 4x - 2y - 4 - 2\sqrt{3}$ satisfies the partial-differential equation

$$\frac{\partial z}{\partial x} + \sqrt{3}\frac{\partial z}{\partial y} - z = 4x + 2y.$$

50. Verify that the function $f(x, y) = z = e^{x + (5/4)y} + \frac{7}{2}(1 - e^{y/2})$ is a solution to the partial-differential equation $3f_x - 4f_y + 2f = 7$ which also satisfies $f(x, 0) = e^x$.

51. The **ideal-gas law** states that $P = nRT/V$. Assume that the number n of moles of an ideal gas and the temperature T of the gas are held constant at the values 10 and $20°C\,(= 293°K)$, respectively. What is the rate of change of the pressure P as a function of the volume V when the volume of the gas is 2 liters?

52. If the current annual interest rate is i, then the **present value** of an annuity which pays B dollars each year for t years is given by

$$A_0 = \frac{B}{i}\left[1 - \left(\frac{1}{1 + i}\right)^t\right].$$

a. If time period t is fixed, how does the present value of the annuity change as the rate of interest i changes?
b. How is the present value changing with respect to i if $B = \$500$ and $i = 6\%$?
c. If interest rate i is fixed, how does the present value of the annuity change as the number of years during which the payments are made, t, is increased?

53. Let C denote the oxygen consumption (per unit weight) of a fur-bearing animal. Let T denote its internal body temperature (in deg Celsius). Let t denote the outside temperature of its fur, and let w denote its weight (in kg). It has been experimentally determined that if T is considerably larger than t, then a reasonable model for the oxygen consumption of the animal is given by

$$C = \frac{5(T - t)}{2w^{2/3}}.$$

Consider a particular fur-bearing animal weighing 10 kg which has a constant internal body temperature of 23°C. If the outside temperature is dropping, how is the animal's oxygen consumption changing when the outside temperature of its fur is 5°C?

54. If a particle is falling in a fluid, then according to **Stoke's law**, the velocity of the particle is given by

$$V = \frac{2g}{9}(P - F)\frac{r^2}{\eta},$$

where g is the acceleration due to gravity, P is the density of the particle, F is the density of the fluid, r is the radius of the particle, and η is the absolute viscosity of the liquid. Calculate V_P, V_F, V_r, and V_η.

■ SHOW/PROVE/DISPROVE

55. Let $f(x, y) = g(x)h(y)$, where g and h are differentiable functions of a single variable. Show that the partial derivatives $\frac{\partial f}{\partial x}$ and $\frac{\partial f}{\partial y}$ exist and that

$$\frac{\partial f}{\partial x} = g'(x)h(y) \quad \text{and} \quad \frac{\partial f}{\partial y} = g(x)h'(y).$$

56. Suppose g, h, and k are differentiable functions of a single variable. Let $f(x, y, z) = g(x)h(y)k(z)$.

Show that $\frac{\partial f}{\partial x}$, $\frac{\partial f}{\partial y}$, and $\frac{\partial f}{\partial z}$ exist and that

$$\frac{\partial f}{\partial x} = g'(x)h(y)k(z), \qquad \frac{\partial f}{\partial y} = g(x)h'(y)k(z),$$

$$\frac{\partial f}{\partial z} = g(x)h(y)k'(z).$$

CONTINUED

In the following four problems, suppose f and g are functions such that $\dfrac{\partial f}{\partial x}$ and $\dfrac{\partial g}{\partial x}$ exist at every point (x, y) in $\mathbb{R}^2$.

57. Prove that $\dfrac{\partial}{\partial x}(f + g)$ exists and

$$\frac{\partial(f + g)}{\partial x} = \frac{\partial f}{\partial x} + \frac{\partial g}{\partial x}.$$

58. Prove that $\dfrac{\partial}{\partial x}(af)$ exists for every constant a and

$$\frac{\partial(af)}{\partial x} = a\left(\frac{\partial f}{\partial x}\right).$$

59. Prove that $\dfrac{\partial}{\partial x}(f \cdot g)$ exists and

$$\frac{\partial(f \cdot g)}{\partial x} = \frac{\partial f}{\partial x} \cdot g + f \cdot \frac{\partial g}{\partial x}.$$

60. Suppose $\dfrac{\partial g}{\partial x} \neq 0$. Prove that $\dfrac{\partial}{\partial x}(f/g)$ exists and

$$\frac{\partial(f/g)}{\partial x} = \frac{\dfrac{\partial f}{\partial x} \cdot g - f \cdot \dfrac{\partial g}{\partial x}}{g^2}.$$

61. Let

$$f(x, y) = \begin{cases} \dfrac{x + y}{x - y}, & \text{if } (x, y) \neq (0, 0) \\ 0, & \text{if } (x, y) = (0, 0). \end{cases}$$

  a. Show that f is not continuous at $(0, 0)$.
  b. Do $f_x(0, 0)$ and $f_y(0, 0)$ exist? Explain.

*62. The ideal-gas law (see Problem 51) allows each of V, T, and P to be expressed as a function of the other two quantities. Suppose that is done so we have $V = V(T, P)$, $T = T(P, V)$, and $P = P(V, T)$. Show that

$$\frac{\partial V}{\partial T} \cdot \frac{\partial T}{\partial P} \cdot \frac{\partial P}{\partial V} = -1.$$

■ ANSWERS TO SELF-QUIZ

I. a   II. b   III. c
IV. d   V. True   VI. True

13.4 HIGHER-ORDER PARTIAL DERIVATIVES

$\mathbf{W}$e have seen that if $y = f(x)$, then

$$y' = \frac{df}{dx} \qquad \text{and} \qquad y'' = \frac{d^2 f}{dx^2} = \frac{d}{dx}\left(\frac{df}{dx}\right).$$

That is, the second derivative of f is the derivative of the first derivative of f. Analogously, if $z = f(x, y)$, then we can differentiate each of the two "first" partial derivatives $\partial f/\partial x$ and $\partial f/\partial y$ with respect to both x and y to obtain four **second partial derivatives** as follows:

DEFINITION   **SECOND PARTIAL DERIVATIVES IN** $\mathbb{R}^2$

(i) Differentiate twice with respect to x:

$$\frac{\partial^2 z}{\partial x^2} = \frac{\partial^2 f}{\partial x^2} = f_{xx} = \frac{\partial}{\partial x}\left(\frac{\partial f}{\partial x}\right).$$ (1)

(ii) Differentiate first with respect to x and then with respect to y:

$$\frac{\partial^2 z}{\partial y \, \partial x} = \frac{\partial^2 f}{\partial y \, \partial x} = f_{xy} = \frac{\partial}{\partial y}\left(\frac{\partial f}{\partial x}\right). \tag{2}$$

(iii) Differentiate first with respect to y and then with respect to x:

$$\frac{\partial^2 z}{\partial x \, \partial y} = \frac{\partial^2 f}{\partial x \, \partial y} = f_{yx} = \frac{\partial}{\partial x}\left(\frac{\partial f}{\partial y}\right). \tag{3}$$

(iv) Differentiate twice with respect to y:

$$\frac{\partial^2 z}{\partial y^2} = \frac{\partial^2 f}{\partial y^2} = f_{yy} = \frac{\partial}{\partial y}\left(\frac{\partial f}{\partial y}\right). \tag{4}$$

The derivatives $\partial^2 f / \partial x \, \partial y$ and $\partial^2 f / \partial y \, \partial x$ are called the **mixed second partials**. ∎

REMARK: It is much easier to denote the second partials by f_{xx}, f_{xy}, f_{yx}, and f_{yy}. We will therefore use this notation for the remainder of this section. Note that the symbol f_{xy} indicates that we differentiate first with respect to x and then with respect to y.

EXAMPLE 1
COMPUTING SECOND PARTIAL
DERIVATIVES

Let $z = f(x, y) = x^3 y^2 - xy^5$. Calculate the four second partial derivatives.

SOLUTION: We have $f_x = 3x^2 y^2 - y^5$ and $f_y = 2x^3 y - 5xy^4$.

(a) $f_{xx} = \dfrac{\partial}{\partial x}(f_x) = 6xy^2$ **(b)** $f_{xy} = \dfrac{\partial}{\partial y}(f_x) = 6x^2 y - 5y^4$

(c) $f_{yx} = \dfrac{\partial}{\partial x}(f_y) = 6x^2 y - 5y^4$ **(d)** $f_{yy} = \dfrac{\partial}{\partial y}(f_y) = 2x^3 - 20xy^3$

In Example 1 we saw that $f_{xy} = f_{yx}$. This result is no accident, as we see by the following theorem, whose proof can be found in any intermediate calculus text.[†]

THEOREM 1   **EQUALITY OF MIXED PARTIALS[‡] IN** $\mathbb{R}^2$

Suppose that $f, f_x, f_y, f_{xy},$ and f_{yx} are all continuous at (x_0, y_0). Then

$$f_{xy}(x_0, y_0) = f_{yx}(x_0, y_0). \quad\blacksquare \tag{5}$$

The definition of second partial derivatives and the theorem on the equality of mixed partials are easily extended to functions of three variables. If $w = f(x, y, z)$, then we have the nine second partial derivatives (assuming

---

[†] See, for example, S. Grossman, *Multivariable Calculus, Linear Algebra, and Differential Equations*, 2nd ed. (San Diego: Harcourt Brace Jovanovich, 1986), p. 195.

[‡] This theorem was first stated by Euler in a 1734 paper devoted to a problem in hydrodynamics.

that they exist):

$$\frac{\partial^2 f}{\partial x^2} = f_{xx}, \qquad \frac{\partial^2 f}{\partial y\,\partial x} = f_{xy}, \qquad \frac{\partial^2 f}{\partial z\,\partial x} = f_{xz},$$

$$\frac{\partial^2 f}{\partial x\,\partial y} = f_{yx}, \qquad \frac{\partial^2 f}{\partial y^2} = f_{yy}, \qquad \frac{\partial^2 f}{\partial z\,\partial y} = f_{yz},$$

$$\frac{\partial^2 f}{\partial x\,\partial z} = f_{zx}, \qquad \frac{\partial^2 f}{\partial y\,\partial z} = f_{zy}, \qquad \frac{\partial^2 f}{\partial z^2} = f_{zz}.$$

THEOREM 2   EQUALITY OF MIXED PARTIALS IN $\mathbb{R}^3$

If f, f_x, f_y, f_z, and all six mixed partials are continuous at a point (x_0, y_0, z_0), then at that point

$$f_{xy} = f_{yx}, \qquad f_{xz} = f_{zx}, \qquad \text{and} \qquad f_{yz} = f_{zy}. \quad \blacksquare$$

EXAMPLE 2
COMPUTING SECOND PARTIAL
DERIVATIVES

Let $f(x, y, z) = xy^3 - zx^5 + x^2yz$. Calculate all nine second partial derivatives and show that all three pairs of mixed partials are equal.

SOLUTION: We have

$$f_x = y^3 - 5zx^4 + 2xyz,$$

$$f_y = 3xy^2 + x^2z,$$

and

$$f_z = -x^5 + x^2y.$$

Then

$$f_{xx} = -20zx^3 + 2yz, \qquad f_{yy} = 6xy, \qquad f_{zz} = 0,$$

$$f_{xy} = \frac{\partial}{\partial y}(y^3 - 5zx^4 + 2xyz) = 3y^2 + 2xz,$$

$$f_{yx} = \frac{\partial}{\partial x}(3xy^2 + x^2z) = 3y^2 + 2xz,$$

$$f_{xz} = \frac{\partial}{\partial z}(y^3 - 5zx^4 + 2xyz) = -5x^4 + 2xy,$$

$$f_{zx} = \frac{\partial}{\partial x}(-x^5 + x^2y) = -5x^4 + 2xy,$$

$$f_{yz} = \frac{\partial}{\partial z}(3xy^2 + x^2z) = x^2,$$

$$f_{zy} = \frac{\partial}{\partial y}(-x^5 + x^2y) = x^2.$$

We conclude this section by pointing out that we can easily define partial derivatives of orders higher than two. For example,

$$f_{zyx} = \frac{\partial^3 f}{\partial x\,\partial y\,\partial z} = \frac{\partial}{\partial x}\left(\frac{\partial^2 f}{\partial y\,\partial z}\right) = \frac{\partial}{\partial x}(f_{zy}).$$

EXAMPLE 3
COMPUTING HIGHER-ORDER
PARTIAL DERIVATIVES

Calculate f_{xxx}, f_{xzy}, f_{yxz}, and f_{yxzx} for the function of Example 2.

SOLUTION: We obtain the three third partial derivatives:

$$f_{xxx} = \frac{\partial}{\partial x}(f_{xx}) = \frac{\partial}{\partial x}(-20zx^3 + 2yz) = -60zx^2$$

$$f_{xzy} = \frac{\partial}{\partial y}(f_{xz}) = \frac{\partial}{\partial y}(-5x^4 + 2xy) = 2x$$

$$f_{yxz} = \frac{\partial}{\partial z}(f_{yx}) = \frac{\partial}{\partial z}(3y^2 + 2xz) = 2x.$$

Note that $f_{xzy} = f_{yxz}$. This again is no accident and follows from the generalization of Theorem 2 to mixed third partial derivatives. Finally, the fourth partial derivative f_{yxzx} is given by

$$f_{yxzx} = \frac{\partial}{\partial x}(f_{yxz}) = \frac{\partial}{\partial x}(2x) = 2.$$

HISTORICAL NOTE: DISTINGUISHING BETWEEN ORDINARY AND PARTIAL DERIVATIVES[†]

In the closing years of the seventeenth century it was already becoming evident, as disclosed in the writings of Isaac Newton and Gottfried Wilhelm von Leibniz, that in both the theory and application of mathematics proper attention must be given to the concept of partial derivatives, wherein differentiation of a function of several variables is performed with respect to one variable at a time. In most of the early researches in which partial derivatives appeared, no special notation was used. Thus the symbol du/dx was alternatively interpreted, according to context, to mean the ordinary or total derivative of u with respect to x or the partial derivative of u with respect to x with other independent variables being held fixed, now usually denoted $\partial u/\partial x$. The need for a distinct notation for partial derivatives, however, initiated a conflict in symbols that persisted throughout most of the eighteenth and nineteenth centuries. Of the dozens of proposals made, most died out rather quickly. Under consideration were the symbols d, D, δ, ϑ, and ∂, often in combination with subscripts or superscripts or both. For a period toward the end of the eighteenth century it seemed possible that the letters d and D might be exclusively appropriated by those working in finite differences, and the round ∂ was introduced for ordinary derivatives. The round ∂ was used in 1770 by Marquis de Condorcet (1743–1794) for partial differentials, and in 1776 by Leonhard Euler in the form $\partial^\lambda/p \cdot V$, now written $\partial^\lambda V/\partial p^\lambda$. This symbol was first used in the modern combination $\partial v/\partial x$ in 1786 by Adrien Marie Legendre (1752–1833) and the letter δ was introduced in an identical role in 1824 by William Rowan Hamilton. However, this use of the "round dee" (∂) was not generally adopted until close to the end of the ninetcenth

[†] This note is adapted from an article by John M. H. Olmsted that appears in *Historical Topics for the Mathematics Classroom* published in 1989 by the National Council of Teachers of Mathematics (NCTM). It is used here with the permission of the NCTM.

century, and Carl Gustav Jacob Jacobi, who outlined the advantages of the symbol ∂ in 1841, is often incorrectly credited with its invention.

A substantial proportion of the applications of mathematics, especially to physics and astronomy, rests on partial derivatives and in particular on solutions of partial differential equations. One of the early pioneers in this area was Daniel Bernoulli (1700–1782). In 1747, Jean Le Rond d'Alembert (1717–1783) solved the fundamental problem of the vibrating string by formulating it in terms of the differential equation

$$\frac{\partial^2 y}{\partial t^2} = a^2 \frac{\partial^2 y}{\partial x^2}$$

and expressing the solution in the form (see Problem 27)

$$u = f(x + ct) + \varphi(x - ct).$$

Others whose early work in partial differential equations had far-reaching effects in the history of mathematics and its applications are Euler, Joseph Louis Lagrange (1736–1813), Pierre Simon Laplace (1749–1827), who introduced and studied the equation (see Problem 35)

$$\frac{\partial^2 u}{\partial x^2} + \frac{\partial^2 u}{\partial y^2} + \frac{\partial^2 u}{\partial z^2} = 0,$$

and Jean Joseph Fourier (1768–1830), who exploited his celebrated series (now called *Fourier series*) in the study of the heat equation

$$\frac{\partial T}{\partial t} = \delta \frac{\partial^2 T}{\partial x^2}$$

(see Problem 28). Partial differential equations continue to be a subject of active research activity today.

■ PROBLEMS 13.4

■ SELF-QUIZ

I. The notation $\partial^2 f / \partial x \, \partial y$ is shorthand for _____.

a. $\dfrac{\partial}{\partial x}\left(\dfrac{\partial}{\partial y} f\right)$ b. $\dfrac{\partial}{\partial y}\left(\dfrac{\partial}{\partial x} f\right)$

II. The notation f_{yx} is shorthand for _____.

a. $\dfrac{\partial \cdot}{\partial x}\left(\dfrac{\partial}{\partial y} f\right)$ b. $\dfrac{\partial}{\partial y}\left(\dfrac{\partial}{\partial x} f\right)$

III. $f_{xy} = $ _____.

a. $\dfrac{\partial^2 f}{\partial y \, \partial x}$ b. $\dfrac{\partial^2 f}{\partial x \, \partial y}$

IV. $g_{zxy} = $ _____.

a. $\dfrac{\partial^3 g}{\partial y \, \partial x \, \partial z}$ b. $\dfrac{\partial}{\partial y}\left[\dfrac{\partial}{\partial x}\left(\dfrac{\partial g}{\partial z}\right)\right]$

V. True–False:

$$\frac{\partial}{\partial x}\left(\frac{\partial f}{\partial y}\right) \neq \frac{\partial}{\partial y}\left(\frac{\partial f}{\partial x}\right)$$

for all functions f with domain $\mathbb{R}^2$.

VI. True–False:

$$\frac{\partial}{\partial x}\left(\frac{\partial g}{\partial y}\right) = \frac{\partial}{\partial y}\left(\frac{\partial g}{\partial x}\right)$$

for all functions g with domain $\mathbb{R}^2$.

VII. How many third-order partial derivatives are there for a function of
a. two variables?   b. three variables?

CONTINUED

■ DRILL

In Problems 1–12, calculate the four second partial derivatives and verify that the two mixed partial derivatives are equal.

1. $f(x, y) = x^2 y$
2. $f(x, y) = xy^2$
3. $f(x, y) = 3e^{xy^3}$
4. $f(x, y) = \sin(x^2 + y^3)$
5. $f(x, y) = 4x/y^5$
6. $f(x, y) = e^y \tan x$
7. $f(x, y) = \ln(x^3 y^5 - 2)$
8. $f(x, y) = \sqrt{xy + 2y}$
9. $f(x, y) = (x + 5y \sin x)^{4/3}$
10. $f(x, y) = \sinh(2x - y)$
11. $f(x, y) = \sin^{-1}(x - y)$
12. $f(x, y) = \sec(xy)$

In Problems 13–20, calculate the nine second partial derivatives and then verify that the three pairs of mixed partials are equal.

13. $f(x, y, z) = xyz$

14. $f(x, y, z) = \cos(xyz)$
15. $f(x, y, z) = x^2 y^3 z^4$
16. $f(x, y, z) = \sin(x + 2y + z^2)$

17. $f(x, y, z) = \dfrac{x + y}{z}$

18. $f(x, y, z) = \tan^{-1}\left(\dfrac{xz}{y}\right)$

19. $f(x, y, z) = e^{3xy} \cos z$
20. $f(x, y, z) = \ln(xy + z)$

In Problems 21–26, calculate the indicated partial derivative.

21. $f(x, y) = x^2 y^3 + 2y; \ f_{xyx}$
22. $f(x, y) = \sin(2xy^4); \ f_{xyy}$
23. $f(x, y) = \ln(3x - 2y); \ f_{yxy}$
24. $f(x, y, z) = \cos(x + 2y + 3z); \ f_{zzx}$
25. $f(x, y, z) = x^2 y + y^2 z - 3\sqrt{xz}; \ f_{xyz}$
26. $f(x, y, z) = e^{xy} \sin z; \ f_{zxyx}$

■ APPLICATIONS

27. Consider a string that is stretched tightly between two fixed points 0 and L on the x-axis. The string is pulled back and released at a time $t = 0$, causing it to vibrate. One position of the string is sketched in Figure 1. Let $y(x, t)$ denote the height of the string at any time $t \geq 0$ and at any point x in the interval $[0, L]$.

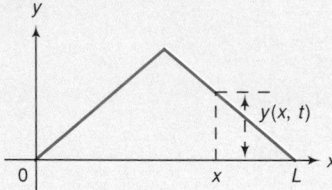

FIGURE 1

It can be shown that $y(x, t)$ satisfies the partial differential equation[†]

$$\frac{\partial^2 y}{\partial t^2} = c^2 \frac{\partial^2 y}{\partial x^2},$$

where c is a constant. (This equation is called the one-dimensional **wave equation**.) Verify that $y(x, t) = \frac{1}{2}[(x - ct)^2 + (x + ct)^2]$ is a solution to the wave equation.

28. Consider a cylindrical rod composed of a uniform heat-conducting material of length L and radius r. Assume that heat can enter and leave the rod only through its ends. Let $T(x, t)$ denote the absolute temperature (deg Kelvin) at time t at a point x units along the rod (see Figure 2).

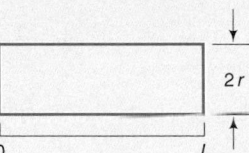

FIGURE 2

Then it can be shown[‡] that T satisfies the partial differential equation

$$\frac{\partial T}{\partial t} = \delta \frac{\partial^2 T}{\partial x^2},$$

[†] For a derivation of this wave equation, see W. Derrick and S. Grossman, *Introduction to Differential Equations With Boundary Value Problems*, 3rd ed. (St. Paul: West, 1987), Section 12.5.

[‡] See Derrick and Grossman, *Introduction to Differential Equations*, Section 12.7.

CONTINUED

where δ is a positive constant called the **diffusivity** of the rod. (This equation is called the **heat equation** or **diffusion equation**.) Verify that the function

$$T(x, t) = e^{-\alpha^2 \delta t} \sin(\alpha x)$$

satisfies the heat equation for any constant α.

29. Verify that

$$T(x, t) = \frac{1}{\sqrt{t}} e^{-x^2/(4 \delta t)}$$

satisfies the heat equation.

30. Find constants α and β such that $T(x, t) = e^{\alpha x + \beta t}$ satisfies the heat equation.

31. One of the most important partial-differential equations of mathematical physics is **Laplace's equation** in $\mathbb{R}^2$, given by

$$\frac{\partial^2 f}{\partial x^2} + \frac{\partial^2 f}{\partial y^2} = 0.$$

Laplace's equation can be used to model steady-state heat flow in a closed, bounded region in $\mathbb{R}^2$. Verify that the function $f(x, y) = x^2 - y^2$ satisfies Laplace's equation.

32. Verify that $f(x, y) = \tan^{-1}(y/x)$ satisfies Laplace's equation.

33. Verify that $f(x, y) = \ln(x^2 + y^2)$ satisfies Laplace's equation.

34. Verify that $f(x, y) = \sin x \sinh y$ satisfies Laplace's equation.

35. Laplace's equation in $\mathbb{R}^3$ is given by

$$\frac{\partial^2 f}{\partial x^2} + \frac{\partial^2 f}{\partial y^2} + \frac{\partial^2 f}{\partial z^2} = 0.$$

Verify that $f(x, y, z) = x^2 + y^2 - 2z^2$ satisfies Laplace's equation in $\mathbb{R}^3$.

■ **CHALLENGE**

*36. Show that $f_{xy}(0, 0)$ and $f_{yx}(0, 0)$ both exist but are not equal for the following function[†]

$$f(x, y) = \begin{cases} xy \dfrac{x^2 - y^2}{x^2 + y^2} & \text{if } x^2 + y^2 \neq 0, \\ 0 & \text{if } x = 0 \text{ and } y = 0. \end{cases}$$

■ **ANSWERS TO SELF-QUIZ**

I. a   II. a   III. a   IV. a = b

V. False (Equality holds in certain cases according to Theorem 1.)

VI. False (The hypotheses for Theorem 1 suggest equality does not always hold; see Problem 36 for such an exception.)

VII. a. $2^3 = 8$; for smooth functions only 4 are different.

b. $3^3 = 27$; for smooth functions only 10 are different.

[†] B. R. Gelbaum and J. M. H. Olmstead, *Counterexamples in Analysis* (San Francisco: Holden-Day, 1964), p. 120.

13.5 DIFFERENTIABILITY AND THE GRADIENT

In this section, we discuss the notion of the differentiability of a function of several variables. There are several ways to introduce this subject and the way we have chosen is designed to illustrate the great similarities between differentiation of functions of one variable and differentiation of functions of several variables.

We begin with a function of one variable,

$$y = f(x).$$

If f is differentiable, then

$$f'(x) = \frac{dy}{dx} = \lim_{\Delta x \to 0} \frac{\Delta y}{\Delta x}. \tag{1}$$

If we define the new function $\epsilon(\Delta x)^\dagger$ by

$$\epsilon(\Delta x) = \frac{\Delta y}{\Delta x} - f'(x), \tag{2}$$

we have

$$\lim_{\Delta x \to 0} \epsilon(\Delta x) = \lim_{\Delta x \to 0} \left(\frac{\Delta y}{\Delta x} - f'(x) \right) = \lim_{\Delta x \to 0} \frac{\Delta y}{\Delta x} - f'(x)$$
$$= f'(x) - f'(x) = 0. \tag{3}$$

Multiplying both sides of (2) by Δx and rearranging terms, we obtain

$$\Delta y = f'(x)\,\Delta x + \epsilon(\Delta x)\,\Delta x.$$

Note that here Δy depends on both Δx and x. Finally, since $\Delta y = f(x + \Delta x) - f(x)$, we obtain $\tag{4}$

$$f(x + \Delta x) - f(x) = f'(x)\,\Delta x + \epsilon(\Delta x)\,\Delta x.$$

Why did we do all this? We did so in order to be able to state the following alternative definition of differentiability of a function f of one variable.

ALTERNATIVE DEFINITION OF DIFFERENTIABILITY OF A FUNCTION OF ONE VARIABLE

Let f be a function of one variable. Then f is **differentiable** at a number x if there is a number $f'(x)$ and a function $g(\Delta x)$ such that

$$f(x + \Delta x) - f(x) = f'(x)\,\Delta x + g(\Delta x), \tag{5}$$

where $\lim_{\Delta x \to 0}[g(\Delta x)/\Delta x] = 0$.

We will soon show how the definition (5) can be extended to a function of two or more variables. First, we give a definition.

DEFINITION DIFFERENTIABILITY OF A FUNCTION OF TWO VARIABLES

Let f be a real-valued function of two variables that is defined in a neighborhood of a point (x, y) and such that $f_x(x, y)$ and $f_y(x, y)$ exist. Then f is **differentiable** at (x, y) if there exist functions $\epsilon_1(\Delta x, \Delta y)$

$^\dagger \epsilon$ is really a function of both x and Δx. We write $\epsilon(\Delta x)$ rather than $\epsilon(x, \Delta x)$ to simplify notation.

and $\epsilon_2(\Delta x, \Delta y)$ such that

$$f(x + \Delta x, y + \Delta y) - f(x, y)$$
$$= f_x(x, y)\,\Delta x + f_y(x, y)\,\Delta y + \epsilon_1(\Delta x, \Delta y)\,\Delta x + \epsilon_2(\Delta x, \Delta y)\,\Delta y$$

(6)

where

$$\lim_{(\Delta x, \Delta y) \to (0,0)} \epsilon_1(\Delta x, \Delta y) = 0 \qquad \text{and}$$

$$\lim_{(\Delta x, \Delta y) \to (0,0)} \epsilon_2(\Delta x, \Delta y) = 0. \quad \blacksquare$$

(7)

EXAMPLE 1
SHOWING THAT A FUNCTION IS
EVERYWHERE DIFFERENTIABLE
IN $\mathbb{R}^2$

Let $f(x, y) = xy$. Show that f is differentiable at every point (x, y) in $\mathbb{R}^2$.

SOLUTION:

$$f(x + \Delta x, y + \Delta y) - f(x, y) = (x + \Delta x)(y + \Delta y) - xy$$
$$= xy + y\,\Delta x + x\,\Delta y + \Delta x\,\Delta y - xy$$
$$= y\,\Delta x + x\,\Delta y + \Delta x\,\Delta y.$$

Now $f_x = y$ and $f_y = x$, so we have

$$f(x + \Delta x, y + \Delta y) - f(x, y) = f_x(x, y)\,\Delta x + f_y(x, y)\,\Delta y$$
$$+ \Delta y\,\Delta x + 0 \cdot \Delta y.$$

Setting $\epsilon_1(\Delta x, \Delta y) = \Delta y$ and $\epsilon_2(\Delta x, \Delta y) = 0$, we see that

$$\lim_{(\Delta x, \Delta y) \to (0,0)} \epsilon_1(\Delta x, \Delta y) = \lim_{(\Delta x, \Delta y) \to (0,0)} \epsilon_2(\Delta x, \Delta y) = 0.$$

This shows that $f(x, y) = xy$ is differentiable at every point in $\mathbb{R}^2$.

NOTE: The functions ϵ_1 and ϵ_2 are not unique. For example, we could set $\epsilon_1 = 0$ and $\epsilon_2 = \Delta x$ to show that xy is differentiable.

We now rewrite our definition of differentiability in a more compact form. Since a point (x, y) is a vector in $\mathbb{R}^2$, we write (as we have done before) $\mathbf{x} = (x, y)$. Then if $z = f(x, y)$, we can simply write

$$z = f(\mathbf{x}).$$

Similarly, if $w = f(x, y, z)$, we may write

$$w = f(\mathbf{x}),$$

where $\mathbf{x}$ is the vector (x, y, z). With this notation we may use the symbol $\Delta\mathbf{x}$ to denote the vector $(\Delta x, \Delta y)$ in $\mathbb{R}^2$ or $(\Delta x, \Delta y, \Delta z)$ in $\mathbb{R}^3$.

Next, we write

$$g(\Delta\mathbf{x}) = \epsilon_1(\Delta x, \Delta y)\,\Delta x + \epsilon_2(\Delta x, \Delta y)\,\Delta y.$$

(8)

Note that $(\Delta x, \Delta y) \to (0, 0)$ can be written in the compact form $\Delta \mathbf{x} \to \mathbf{0}$. Then if the conditions (7) hold, we see that

$$|\Delta \mathbf{x}| = \sqrt{\Delta x^2 + \Delta y^2}$$

$$\lim_{\Delta \mathbf{x} \to \mathbf{0}} \frac{|g(\Delta \mathbf{x})|}{|\Delta \mathbf{x}|} \leq \lim_{\Delta \mathbf{x} \to \mathbf{0}} |\epsilon_1(\Delta x, \Delta y)| \frac{|\Delta x|}{\sqrt{\Delta x^2 + \Delta y^2}}$$

$$+ \lim_{\Delta \mathbf{x} \to \mathbf{0}} |\epsilon_2(\Delta x, \Delta y)| \frac{|\Delta y|}{\sqrt{\Delta x^2 + \Delta y^2}}$$

$$\frac{|\Delta x|}{\sqrt{\Delta x^2 + \Delta y^2}} \leq 1$$

$$\leq \lim_{\Delta \mathbf{x} \to \mathbf{0}} |\epsilon_1(\Delta x, \Delta y)| + \lim_{\Delta \mathbf{x} \to \mathbf{0}} |\epsilon_2(\Delta x, \Delta y)| = 0 + 0 = 0.$$

Finally, we have the following important definition.

DEFINITION   **THE GRADIENT IN** $\mathbb{R}^2$

Let f be a function of two variables such that f_x and f_y exist at a point $\mathbf{x} = (x, y)$. Then the **gradient** of f at $\mathbf{x}$, denoted by $\nabla f(\mathbf{x})$, is given by

$$\nabla f(\mathbf{x}) = f_x(x, y)\mathbf{i} + f_y(x, y)\mathbf{j}. \quad \blacksquare \tag{9}$$

Note that the gradient of f is a **vector function**. That is, for every point $\mathbf{x}$ in $\mathbb{R}^2$ for which $\nabla f(\mathbf{x})$ is defined, we see that $\nabla f(\mathbf{x})$ is a vector in $\mathbb{R}^2$.

HISTORICAL NOTE: The gradient of f is denoted by ∇f, which is read "del" f. This symbol, an inverted Greek delta, was first used in the 1850s, although the name "del" first appeared in print only in 1901. The symbol ∇ is also called *nabla*. This name is used because someone once suggested to the Scottish mathematician Peter Guthrie Tait (1831–1901) that ∇ looks like an Assyrian harp, the Assyrian name of which is nabla.† Tait, incidentally, was one of the mathematicians who helped carry on Hamilton's development of the theory of quaternions and vectors in the nineteenth century.

EXAMPLE 2
COMPUTING A GRADIENT IN $\mathbb{R}^2$

In Example 1, $f(x, y) = xy$, $f_x = y$, and $f_y = x$, so that

$$\nabla f(\mathbf{x}) = \nabla f(x, y) = y\mathbf{i} + x\mathbf{j}.$$

Using this new notation, we observe that

$$\nabla f(\mathbf{x}) \cdot \Delta \mathbf{x} = (f_x \mathbf{i} + f_y \mathbf{j}) \cdot (\Delta x \mathbf{i} + \Delta y \mathbf{j}) = f_x(x, y) \Delta x + f_y(x, y) \Delta y.$$

Also,

$$f(x + \Delta x, y + \Delta y) = f(\mathbf{x} + \Delta \mathbf{x}).$$

---

† Fortunately, most (but certainly not all) of the mathematical terms currently in use have more to do with the objects they describe. We might further point out that ∇ is also called *atled*, which is delta spelled backward.

Thus we have the following definition, which is equivalent to the definition of differentiability.

DEFINITION   **DIFFERENTIABILITY IN $\mathbb{R}^2$ (DEFINITION GIVEN IN TERMS OF THE GRADIENT)**

Let f be a function of two variables that is defined in a neighborhood of a point $\mathbf{x} = (x, y)$. Let $\Delta\mathbf{x} = (\Delta x, \Delta y)$. If $f_x(x, y)$ and $f_y(x, y)$ exist, then f is **differentiable** at $\mathbf{x}$ if there is a function g such that

$$f(\mathbf{x} + \Delta\mathbf{x}) - f(\mathbf{x}) = \nabla f(\mathbf{x}) \cdot \Delta\mathbf{x} + g(\Delta\mathbf{x}) \tag{10}$$

where

$$\lim_{\Delta\mathbf{x} \to \mathbf{0}} \frac{g(\Delta\mathbf{x})}{|\Delta\mathbf{x}|} = 0. \quad \blacksquare \tag{11}$$

REMARK 1:   Although formulas (5) and (10) look very similar, there are two fundamental differences. First, f' is a scalar, while ∇f is a vector. Second, $f'(x)\,\Delta x$ is an ordinary product of real numbers, while $\nabla f(\mathbf{x}) \cdot \Delta\mathbf{x}$ is a dot product of vectors.

REMARK 2:   According to this definition, f is *not* differentiable at $\mathbf{x}$ if one or more of its partial derivatives fails to exist at $\mathbf{x}$.

There are two reasons for giving you this new definition. First, it illustrates that the gradient of a function of two variables is the natural extension of the derivative of a function of one variable. Second, the definition (10), (11) can be used to define the notion of differentiability for a function of three or more variables as well. We will say more about this subject shortly.

One important question remains: What functions are differentiable? A partial answer is given in Theorem 1. A proof is given at the end of the section.

THEOREM 1   **A FUNCTION WITH CONTINUOUS PARTIAL DERIVATIVES IS DIFFERENTIABLE**

Let f, f_x, and f_y be defined and continuous in a neighborhood of $\mathbf{x} = (x, y)$. Then f is differentiable at $\mathbf{x}$. $\blacksquare$

REMARK: We cannot omit the hypothesis that f_x and f_y are continuous in this theorem. In Example 13.3.4 on page 856, we saw an example of a function for which f_x and f_y exist at $(0, 0)$ but f itself is not continuous there. As we will see in Theorem 2, if f is differentiable at a point, then it is continuous there. Thus the function in Example 13.3.4 is not differentiable at $(0, 0)$ even though $f_x(0, 0)$ and $f_y(0, 0)$ exist.

DEFINITION   **CONTINUOUSLY DIFFERENTIABLE OR SMOOTH FUNCTION**

If the hypotheses of Theorem 1 are satisfied, then f is said to be **continuously differentiable** or **smooth** at $\mathbf{x}$. $\blacksquare$

EXAMPLE 3
SHOWING THAT A FUNCTION IS
DIFFERENTIABLE AND
COMPUTING ITS GRADIENT

Let $z = f(x, y) = \sin xy^2 + e^{x^2y^3}$. Show that f is differentiable and calculate ∇f. Find $\nabla f(1, 1)$.

SOLUTION: $\partial f/\partial x = y^2 \cos xy^2 + 2xy^3 e^{x^2y^3}$ and $\partial f/\partial y = 2xy \cos xy^2 + 3x^2y^2 e^{x^2y^3}$. Since $\partial f/\partial x$ and $\partial f/\partial y$ are continuous, f is differentiable and

$$\nabla f(x, y) = (y^2 \cos xy^2 + 2xy^3 e^{x^2y^3})\mathbf{i} + (2xy \cos xy^2 + 3x^2y^2 e^{x^2y^3})\mathbf{j}.$$

At $(1, 1)$, $\nabla f(1, 1) = (\cos 1 + 2e)\mathbf{i} + (2 \cos 1 + 3e)\mathbf{j}$.

In Section 13.3, we showed that the existence of all of its partial derivatives at a point does *not* ensure that a function is continuous at that point. However, differentiability does ensure continuity.

THEOREM 2 **A DIFFERENTIABLE FUNCTION IS CONTINUOUS**

If f is differentiable at $\mathbf{x}_0 = (x_0, y_0)$, then f is continuous at $\mathbf{x}_0$.

PROOF: We must show that $\lim_{(x,y)\to(x_0,y_0)} f(x, y) = f(x_0, y_0)$, or, equivalently, $\lim_{\mathbf{x}\to\mathbf{x}_0} f(\mathbf{x}) = f(\mathbf{x}_0)$. But if we define $\Delta\mathbf{x}$ by $\Delta\mathbf{x} = \mathbf{x} - \mathbf{x}_0$, this is the same as showing that

$$\lim_{\Delta\mathbf{x}\to 0} f(\mathbf{x}_0 + \Delta\mathbf{x}) = f(\mathbf{x}_0). \tag{12}$$

Since f is differentiable at $\mathbf{x}_0$,

$$f(\mathbf{x}_0 + \Delta\mathbf{x}) - f(\mathbf{x}_0) = \nabla f(\mathbf{x}_0) \cdot \Delta\mathbf{x} + g(\Delta\mathbf{x}). \tag{13}$$

But as $\Delta\mathbf{x} \to 0$, both terms on the right-hand side of (13) approach zero, so

$$\lim_{\Delta\mathbf{x}\to 0} \left[f(\mathbf{x}_0 + \Delta\mathbf{x}) - f(\mathbf{x}_0) \right] = 0,$$

which means that (12) holds and the theorem is proved. ∎

The converse to this theorem is false, as it is in one-variable calculus. That is, there are functions that are continuous, but not differentiable, at a given point. For example, the function

$$f(x, y) = \sqrt[3]{x} + \sqrt[3]{y}$$

is continuous at any point (x, y) in $\mathbb{R}^2$, but

$$\nabla f(x, y) = \frac{1}{3x^{2/3}} \mathbf{i} + \frac{1}{3y^{2/3}} \mathbf{j},$$

so f is not differentiable at any point (x, y) for which either x or y is zero. That is, $\nabla f(x, y)$ is not defined on the x- and y-axes. Hence f is not differentiable along these axes.

In Section 2.1 we showed that

$$(f + g)' = f' + g' \qquad \text{and} \qquad (\alpha f)' = \alpha f'.$$

That is, the derivative of the sum of two functions is the sum of the derivatives of the two functions, and the derivative of a scalar multiple of a function is the scalar times the derivative of the function. These results can be extended to the gradient vector. The proof of the following theorem is left as an exercise (see Problems 24 and 25).

THEOREM 3   **THE GRADIENT OF A SCALAR MULTIPLE AND SUM OF FUNCTIONS**

Let f and g be differentiable in a neighborhood of $\mathbf{x} = (x, y)$. Then for every scalar α, αf and $f + g$ are differentiable at $\mathbf{x}$, and

(i) $\mathbf{V}(\alpha f) = \alpha\, \mathbf{V}f$, and

(ii) $\mathbf{V}(f + g) = \mathbf{V}f + \mathbf{V}g$.   ∎

REMARK: Any rule that satisfies conditions (i) and (ii) of Theorem 3 is called a **linear mapping** or a **linear operator**. Linear operators play an extremely important role in advanced mathematics.

All the definitions and theorems in this section hold for functions of three or more variables. We give the equivalent results for functions of three variables below.

DEFINITION   **THE GRADIENT IN** $\mathbb{R}^3$

Let f be a scalar function of three variables such that f_x, f_y, and f_z exist at a point $\mathbf{x} = (x, y, z)$ Then the **gradient** of f at $\mathbf{x}$, denoted by $\mathbf{V}f(\mathbf{x})$, is given by the vector

$$\mathbf{V}f(\mathbf{x}) = f_x(x, y, z)\mathbf{i} + f_y(x, y, z)\mathbf{j} + f_z(x, y, z)\mathbf{k}. \quad ∎ \tag{14}$$

DEFINITION   **DIFFERENTIABILITY IN** $\mathbb{R}^3$

Let f be a function of three variables that is defined in a neighborhood of $\mathbf{x} = (x, y, z)$, and let $\Delta\mathbf{x} = (\Delta x, \Delta y, \Delta z)$. If $f_x(x, y, z), f_y(x, y, z)$, and $f_z(x, y, z)$ exist, then f is **differentiable** at $\mathbf{x}$ if there is a function g such that

$$f(\mathbf{x} + \Delta\mathbf{x}) - f(\mathbf{x}) = \mathbf{V}f \cdot \Delta\mathbf{x} + g(\Delta\mathbf{x})$$

where

$$\lim_{|\Delta\mathbf{x}| \to \mathbf{0}} \frac{g(\Delta\mathbf{x})}{|\Delta\mathbf{x}|} = 0.$$

Equivalently, we can write

$$f(x + \Delta x, y + \Delta y, z + \Delta z) - f(x, y, z)$$
$$= f_x(x, y, z)\,\Delta x + f_y(x, y, z)\,\Delta y + f_z(x, y, z)\,\Delta z + g(\Delta x, \Delta y, \Delta z)$$

where

$$\lim_{(\Delta x, \Delta y, \Delta z) \to (0,0,0)} \frac{g(\Delta x, \Delta y, \Delta z)}{\sqrt{\Delta x^2 + \Delta y^2 + \Delta z^2}} = 0. \quad ∎$$

THEOREM 1

If f, f_x, f_y, and f_z exist and are continuous in a neighborhood of $\mathbf{x} = (x, y, z)$, then f is differentiable at $\mathbf{x}$. ∎

THEOREM 2

Let f be a function of three variables that is differentiable at $\mathbf{x}_0$. Then f is continuous at $\mathbf{x}_0$. ∎

EXAMPLE 4
SHOWING THAT A FUNCTION OF THREE VARIABLES IS DIFFERENTIABLE AND COMPUTING ITS GRADIENT

Let $f(x, y, z) = xy^2z^3$. Show that f is differentiable at any point $\mathbf{x}_0$, calculate ∇f, and find $\nabla f(3, -1, 2)$.

SOLUTION: $\partial f/\partial x = y^2z^3$, $\partial f/\partial y = 2xyz^3$, and $\partial f/\partial z = 3xy^2z^2$. Since f, $\partial f/\partial x$, $\partial f/\partial y$, and $\partial f/\partial z$ are all continuous, we know that f is differentiable and that

$$\nabla f = y^2z^3\mathbf{i} + 2xyz^3\mathbf{j} + 3xy^2z^2\mathbf{k}$$

and

$$\nabla f(3, -1, 2) = 8\mathbf{i} - 48\mathbf{j} + 36\mathbf{k}.$$

THEOREM 3

Let f and g be differentiable in a neighborhood of $\mathbf{x} = (x, y, z)$. Then for any scalar α, αf and $f + g$ are differentiable at $\mathbf{x}$, and

(i) $\nabla(\alpha f) = \alpha \nabla f$, and
(ii) $\nabla(f + g) = \nabla f + \nabla g$. ∎

We conclude this section with a proof of Theorem 1. The proof of Theorem 1′ is similar.

PROOF OF THEOREM 1: We begin by restating the mean value theorem for a function f of one variable.

Mean-Value Theorem Let f be continuous on $[a, b]$ and differentiable on (a, b). Then there is a number c in (a, b) such that

$$f(b) - f(a) = f'(c)(b - a).$$

Now we have assumed that f, f_x, and f_y are all continuous in a neighborhood N of $\mathbf{x} = (x, y)$. Choose $\Delta \mathbf{x}$ so small that $\mathbf{x} + \Delta \mathbf{x}$ is in N. Then

$$\Delta f(\mathbf{x}) = f(x + \Delta x, y + \Delta y) - f(x, y)$$

$$\overbrace{}^{\text{This term was added and subtracted}}$$

$$= [f(x + \Delta x, y + \Delta y) - f(x + \Delta x, y)] + [f(x + \Delta x, y) - f(x, y)].$$

(15)

If $x + \Delta x$ is fixed, then $f(x + \Delta x, y)$ is a function of y that is continuous and differentiable in the interval $[y, y + \Delta y]$. Hence, by the mean-value theorem, there is a number c_2 between y and $y + \Delta y$ such that

$$f(x + \Delta x, y + \Delta y) - f(x + \Delta x, y) = f_y(x + \Delta x, c_2)[(y + \Delta y) - y]$$
$$= f_y(x + \Delta x, c_2) \Delta y. \tag{16}$$

Similarly, with y fixed, $f(x, y)$ is a function of x only, and we obtain

$$f(x + \Delta x, y) - f(x, y) = f_x(c_1, y) \Delta x, \tag{17}$$

where c_1 is between x and $x + \Delta x$. Thus using (16) and (17) in (15), we have

$$\Delta f(\mathbf{x}) = f_x(c_1, y) \Delta x + f_y(x + \Delta x, c_2) \Delta y. \tag{18}$$

Now both f_x and f_y are continuous at $\mathbf{x} = (x, y)$, so since c_1 is between x and $x + \Delta x$ and c_2 is between y and $y + \Delta y$, we obtain

$$\lim_{\Delta \mathbf{x} \to \mathbf{0}} f_x(c_1, y) = f_x(x, y) = f_x(\mathbf{x}) \tag{19}$$

and

$$\lim_{\Delta \mathbf{x} \to \mathbf{0}} f_y(x + \Delta x, c_2) = f_y(x, y) = f_y(\mathbf{x}). \tag{20}$$

Let

$$\epsilon_1(\Delta \mathbf{x}) = f_x(c_1, y) - f_x(x, y). \tag{21}$$

From (19) it follows that

$$\lim_{|\Delta \mathbf{x}| \to 0} \epsilon_1(\Delta \mathbf{x}) = 0. \tag{22}$$

Similarly, if

$$\epsilon_2(\Delta \mathbf{x}) = f_y(x + \Delta x, c_2) - f_y(x, y), \tag{23}$$

then

$$\lim_{|\Delta \mathbf{x}| \to 0} \epsilon_2(\Delta \mathbf{x}) = 0. \tag{24}$$

Now define

$$g(\Delta \mathbf{x}) = \epsilon_1(\Delta \mathbf{x}) \Delta x + \epsilon_2(\Delta \mathbf{x}) \Delta y. \tag{25}$$

From (22) and (24) it follows that

$$\lim_{|\Delta \mathbf{x}| \to 0} \frac{g(\Delta \mathbf{x})}{|\Delta \mathbf{x}|} = 0. \tag{26}$$

Finally, since

$$f_x(c_1, y) = f_x(x, y) + \epsilon_1(\Delta \mathbf{x}) \qquad \text{From (21)} \tag{27}$$

and

$$f_y(x + \Delta x, c_2) = f_y(x, y) + \epsilon_2(\Delta \mathbf{x}), \qquad \text{From (23)} \tag{28}$$

we may substitute (27) and (28) into (18) to obtain

$$\Delta f(\mathbf{x}) = f(\mathbf{x} + \Delta \mathbf{x}) - f(\mathbf{x})$$
$$= \left[f_x(\mathbf{x}) + \epsilon_1(\Delta \mathbf{x}) \right] \Delta x + \left[f_y(\mathbf{x}) + \epsilon_2(\Delta \mathbf{x}) \right] \Delta y$$
$$= f_x(\mathbf{x}) \Delta x + f_y(\mathbf{x}) \Delta y + g(\Delta \mathbf{x}) = (f_x \mathbf{i} + f_y \mathbf{j}) \cdot (\Delta \mathbf{x}) + g(\Delta \mathbf{x}),$$

where $\lim_{\Delta \mathbf{x} \to \mathbf{0}} [g(\Delta \mathbf{x})/|\Delta \mathbf{x}|] \to 0$, and the proof is (at last) complete. ∎

PROBLEMS 13.5

SELF-QUIZ

I. True–False: If $f(x, y) = \cos y$, then $\nabla f(x, y) = -\sin y$.

II. If $f(x, y) = $ _____, then $\nabla f(x, y) = \mathbf{i} - \sin y \mathbf{j}$.
 a. $y + \cos x$ b. $x + \cos y$
 c. $1 + x \cos y$ d. $1 - \sin y$

III. If $f(x, y) = y/x$, then $x^2 \nabla f(x, y) = $ _____.
 a. $x\mathbf{i} + y\mathbf{j}$ b. $x\mathbf{i} - y\mathbf{j}$
 c. $y\mathbf{i} - x\mathbf{j}$ d. $-y\mathbf{i} + x\mathbf{j}$

IV. If $f(x, y) = $ _____, then $\nabla f(x, y) = y\mathbf{i} + x\mathbf{j}$.
 a. $\frac{1}{2} y^2 + \frac{1}{2} x^2$ b. x/y
 c. xy d. y/x

V. If $f(x, y) = $ _____, then $\nabla f(x, y) = x\mathbf{i} + y\mathbf{j}$.
 a. $\frac{1}{2} y^2 + \frac{1}{2} x^2$ b. x/y
 c. xy d. y/x

VI. If $g(x, y, z) = xyz$, then $\nabla g(1, 2, 3) = $ _____.
 a. $\mathbf{0}$ b. $\mathbf{i} + 2\mathbf{j} + 3\mathbf{k}$
 c. $yz\mathbf{i} + xz\mathbf{j} + xy\mathbf{k}$ d. $6\mathbf{i} + 3\mathbf{j} + 2\mathbf{k}$

DRILL

In Problems 1–20, calculate the gradient of the given function. If a point is also specified, then evaluate the gradient at that point.

1. $f(x, y) = (x + y)^2$
2. $f(x, y) = \ln(2x - y + 1)$
3. $f(x, y) = e^{\sqrt{xy}}$; $(1, 1)$
4. $f(x, y) = \cos(x - y)$, $(\pi/2, \pi/4)$
5. $f(x, y) = \sqrt{x^2 + y^2}$
6. $f(x, y) = \tan^{-1}(y/x)$; $(3, 3)$
7. $f(x, y) = y \tan(y - x)$
8. $f(x, y) = x^2 \sinh y$
9. $f(x, y) = \sec(x + 3y)$; $(0, 1)$
10. $f(x, y) = \dfrac{x - y}{x + y}$; $(3, 1)$

11. $f(x, y) = \dfrac{x^2 - y^2}{x^2 + y^2}$

12. $f(x, y) = \dfrac{e^{x^2} - e^{-y^2}}{3y}$

13. $f(x, y, z) = \sin x \cos y \tan z$; $(\pi/6, \pi/4, \pi/3)$
14. $f(x, y, z) = x \sin y \ln z$; $(1, 0, 1)$

15. $f(x, y, z) = \dfrac{x^2 - y^2 + z^2}{3xy}$; $(1, 2, 0)$

16. $f(x, y, z) = x \ln y - z \ln x$
17. $f(x, y, z) = xy^2 + y^2 z^3$; $(2, 3, -1)$
18. $f(x, y, z) = (y - z)e^{x + 2y + 3z}$; $(-4, -1, 3)$

19. $f(x, y, z) = \dfrac{x - z}{\sqrt{1 - y^2 + x^2}}$; $(0, 0, 1)$

20. $f(x, y, z) = x \cosh z - y \sin x$

SHOW/PROVE/DISPROVE

21. Let $f(x, y) = x^2 + y^2$. Show, by directly using the definition on page 866, that f is differentiable at any point in $\mathbb{R}^2$. (Do not merely cite Theorem 1.)
22. Let $g(x, y) = x^2 y^2$. Show, by using the definition of a limit (don't use Theorem 1), that g is differentiable at any point in $\mathbb{R}^2$.

23. Let $f(x, y)$ be any polynomial function of the two variables x and y. Show that f is differentiable at any point in $\mathbb{R}^2$.
24. Prove Theorem 3 by using the definition of a limit. [*Hint* (for part (i)): If f and g satisfy equa-

CONTINUED

tion (10) and if α is constant, then

$$[\alpha f(\mathbf{x} + \Delta \mathbf{x})] - [\alpha f(\mathbf{x})]$$
$$= [\alpha \nabla f(\mathbf{x})] \cdot \Delta \mathbf{x} + [\alpha g(\Delta \mathbf{x})]$$

and

$$\lim_{\Delta \mathbf{x} \to 0} \frac{\alpha g(\Delta \mathbf{x})}{|\Delta \mathbf{x}|} = \alpha \lim_{\Delta \mathbf{x} \to 0} \frac{g(\Delta \mathbf{x})}{|\Delta \mathbf{x}|}.]$$

25. Show that if f and g are differentiable functions of three variables, then so is $f + g$ and $\nabla(f + g) = \nabla f + \nabla g$.
26. Show that if f and g are differentiable functions of three variables, then fg is also differentiable and $\nabla(fg) = (\nabla f)g + f(\nabla g)$.
*27. Show that $\nabla f = \mathbf{0}$ if and only if f is constant.
*28. Show that if $\nabla f = \nabla g$, then there is a constant c for which $f(x, y) = g(x, y) + c$. [*Hint:* Use the result of the preceding problem.]
*29. Let

$$f(x, y) = \begin{cases} (x^2 + y^2) \sin\left(\dfrac{1}{\sqrt{x^2 + y^2}}\right) & \text{if } (x, y) \neq (0, 0), \\ 0 & \text{if } (x, y) = (0, 0). \end{cases}$$

   a. Calculate $f_x(0, 0)$ and $f_y(0, 0)$.
   b. Explain why f_x and f_y are *not* continuous at $(0, 0)$.
   c. Show that f is differentiable at $(0, 0)$.
*30. Suppose that f is a differentiable function of one variable and g is a differentiable function of three variables. Show that $f \circ g$ is differentiable and $\nabla(f \circ g) = (f' \circ g) \nabla g$.

■ CHALLENGE

*31. What is the most general function f such that $\nabla f(\mathbf{x}) = \mathbf{x}$ for every $\mathbf{x}$ in $\mathbb{R}^2$?
*32. Show that the converse of Theorem 1 is not true. [*Hint:* Find a function $F(x, y)$ that is differentiable everywhere but whose partial derivatives are not continuous in a neighborhood of $(0, 0)$. You may find it useful to contemplate the following functions of a single variable:

$$f(x) = \begin{cases} x \sin\left(\dfrac{1}{x}\right) & \text{if } x \neq 0 \\ 0 & \text{if } x = 0; \end{cases}$$

$$g(x) = \begin{cases} x^2 \sin\left(\dfrac{1}{x}\right) & \text{if } x \neq 0 \\ 0 & \text{if } x = 0; \end{cases}$$

$$h(x) = \begin{cases} x^3 \sin\left(\dfrac{1}{x}\right) & \text{if } x \neq 0 \\ 0 & \text{if } x = 0. \end{cases}$$

The function f is continuous but not differentiable at 0; g is differentiable but its derivative is discontinuous at 0; h is continuously differentiable everywhere.]

■ ANSWERS TO SELF-QUIZ

I. False $(-\sin y \mathbf{j})$ II. b   V. a   VI. d
III. d   IV. c

13.6 THE CHAIN RULE

In this section, we derive several chain rules for functions of two and three variables. Let us recall the chain rule for the composition of two functions of one variable:

Let $y = f(u)$ and $u = g(x)$ and assume that f and g are differentiable. Then

$$\frac{dy}{dx} = \frac{dy}{du}\frac{du}{dx} = f'(g(x))g'(x). \tag{1}$$

If $z = f(x, y)$ is a function of two variables, then there are two versions of the chain rule.

THEOREM 1 **CHAIN RULE**

Let $z = f(x, y)$ be differentiable and suppose that $x = x(t)$ and $y = y(t)$. Assume further that dx/dt and dy/dt exist and are continuous. Then z can be written as a function of the parameter t, and

$$\frac{dz}{dt} = \frac{\partial z}{\partial x}\frac{dx}{dt} + \frac{\partial z}{\partial y}\frac{dy}{dt} = f_x\frac{dx}{dt} + f_y\frac{dy}{dt}. \tag{2}$$

We can also write this result using our gradient notation. If $\mathbf{g}(t) = x(t)\mathbf{i} + y(t)\mathbf{j}$, then $\mathbf{g}'(t) = (dx/dt)\mathbf{i} + (dy/dt)\mathbf{j}$, and (2) can be written as

$$\frac{d}{dt} f(x(t), y(t)) = (f \circ \mathbf{g})'(t) = [f(\mathbf{g}(t))]' = \nabla f \cdot \mathbf{g}'(t). \quad \blacksquare \tag{3}$$

THEOREM 2 **CHAIN RULE**

Let $z = f(x, y)$ be differentiable and suppose that x and y are functions of the two variables r and s. That is, $x = x(r, s)$ and $y = y(r, s)$. Suppose further that $\partial x/\partial r$, $\partial x/\partial s$, $\partial y/\partial r$, and $\partial y/\partial s$ all exist and are continuous. Then z can be written as a function of r and s, and

$$\frac{\partial z}{\partial r} = \frac{\partial z}{\partial x}\frac{\partial x}{\partial r} + \frac{\partial z}{\partial y}\frac{\partial y}{\partial r} \tag{4}$$

$$\frac{\partial z}{\partial s} = \frac{\partial z}{\partial x}\frac{\partial x}{\partial s} + \frac{\partial z}{\partial y}\frac{\partial y}{\partial s}. \quad \blacksquare \tag{5}$$

We will leave the proofs of these theorems until the end of this section.

EXAMPLE 1
USING THE CHAIN RULE TO
COMPUTE $\dfrac{dz}{dt}$

Let $z = f(x, y) = xy^2$. Let $x = \cos t$ and $y = \sin t$. Calculate dz/dt.

SOLUTION:

$$\frac{dz}{dt} = \frac{\partial z}{\partial x}\frac{dx}{dt} + \frac{\partial z}{\partial y}\frac{dy}{dt} = y^2(-\sin t) + 2xy(\cos t)$$

$$= (\sin^2 t)(-\sin t) + 2(\cos t)(\sin t)(\cos t)$$

$$= 2 \sin t \cos^2 t - \sin^3 t.$$

EXAMPLE 2
USING THE CHAIN RULE TO
CALCULATE TWO PARTIAL
DERIVATIVES

Let $z = f(x, y) = \sin xy^2$. Suppose that $x = r/s$ and $y = e^{r-s}$. Calculate $\partial z/\partial r$ and $\partial z/\partial s$.

SOLUTION:

$$\frac{\partial z}{\partial r} = \frac{\partial z}{\partial x}\frac{\partial x}{\partial r} + \frac{\partial z}{\partial y}\frac{\partial y}{\partial r} = (y^2 \cos xy^2)\frac{1}{s} + (2xy \cos xy^2)e^{r-s}$$

$$= \frac{e^{2(r-s)} \cos[(r/s)e^{2(r-s)}]}{s} + \frac{2r}{s}\{\cos[(r/s)e^{2(r-s)}]\}e^{2(r-s)}$$

and

$$\frac{\partial z}{\partial s} = \frac{\partial z}{\partial x}\frac{\partial x}{\partial s} + \frac{\partial z}{\partial y}\frac{\partial y}{\partial s} = (y^2 \cos xy^2)\frac{-r}{s^2} + (2xy \cos xy^2)(-e^{r-s})$$

$$= \frac{-re^{2(r-s)} \cos[(r/s)e^{2(r-s)}]}{s^2} - \frac{2r}{s}\{\cos[(r/s)e^{2(r-s)}]\}e^{2(r-s)}.$$

The chain rules given in Theorem 1 and Theorem 2 can be extended to functions of three or more variables.

THEOREM 1 **CHAIN RULE**

Let $w = f(x, y, z)$ be a differentiable function. If $x = x(t)$, $y = y(t)$, $z = z(t)$, and if dx/dt, dy/dt, and dz/dt exist and are continuous, then w may be considered as a differentiable function of t and

$$\frac{dw}{dt} = \frac{\partial w}{\partial x}\frac{dx}{dt} + \frac{\partial w}{\partial y}\frac{dy}{dt} + \frac{\partial w}{\partial z}\frac{dz}{dt}. \quad \blacksquare \tag{6}$$

THEOREM 2 **CHAIN RULE**

Let $w = f(x, y, z)$ be a differentiable function and let $x = x(r, s, t)$, $y = y(r, s, t)$, and $z = z(r, s, t)$. If all indicated partial derivatives exist and are continuous, then w may be considered as a function of r, s, and t and

$$\frac{\partial w}{\partial r} = \frac{\partial w}{\partial x}\frac{\partial x}{\partial r} + \frac{\partial w}{\partial y}\frac{\partial y}{\partial r} + \frac{\partial w}{\partial z}\frac{\partial z}{\partial r},$$

$$\frac{\partial w}{\partial s} = \frac{\partial w}{\partial x}\frac{\partial x}{\partial s} + \frac{\partial w}{\partial y}\frac{\partial y}{\partial s} + \frac{\partial w}{\partial z}\frac{\partial z}{\partial s}, \tag{7}$$

$$\frac{\partial w}{\partial t} = \frac{\partial w}{\partial x}\frac{\partial x}{\partial t} + \frac{\partial w}{\partial y}\frac{\partial y}{\partial t} + \frac{\partial w}{\partial z}\frac{\partial z}{\partial t}. \quad \blacksquare$$

We saw in Section 3.1 that the chain rule is useful to solve "related rates" types of problems. The chain rules for functions of two or more variables are useful in a similar way.

EXAMPLE 3
APPLYING THE CHAIN RULE TO THE IDEAL-GAS LAW

According to the ideal-gas law, the pressure, volume, and absolute temperature of n moles of an ideal gas are related by

$$PV = nRT,$$

where R is a constant. Suppose that the volume of an ideal gas is increasing at a rate of 10 cm³/min and the pressure is decreasing at a rate of 0.3 N/cm²/min. How is the temperature of the gas changing when the volume of 5 mol of a gas is 100 cm³ and the pressure is 2 N/cm²?

SOLUTION: We have $T = PV/nR$, where $n = 5$. Then

$$\frac{dT}{dt} = \frac{\partial T}{\partial P}\frac{dP}{dt} + \frac{\partial T}{\partial V}\frac{dV}{dt} = \frac{V}{nR}(-0.3) + \frac{P}{nR}(10)$$

$$= \frac{100}{5R}(-0.3) + \frac{2}{5R}(10) = \frac{-2}{R}\,°\text{K/min}.$$

We conclude this section with a proof of Theorem 2. Theorem 1 follows easily from Theorem 2. (Explain why.)

PROOF OF THEOREM 2: We will show that

$$\frac{\partial z}{\partial r} = \frac{\partial z}{\partial x}\frac{\partial x}{\partial r} + \frac{\partial z}{\partial y}\frac{\partial y}{\partial r}.$$

Equation (5) follows in an identical manner. Since $x = x(r, s)$ and $y = y(r, s)$, a change Δr in r will cause a change Δx in x and a change Δy in y. We may therefore write

$$\Delta x = x(r + \Delta r, s) - x(r, s)$$

and

$$\Delta y = y(r + \Delta r, s) - y(r, s).$$

Since z is differentiable, we may write (from equation (13.5.6) on page 867)

$$\Delta z = \frac{\partial z}{\partial x}\Delta x + \frac{\partial z}{\partial y}\Delta y + \epsilon_1(\Delta x, \Delta y)\,\Delta x + \epsilon_2(\Delta x, \Delta y)\,\Delta y, \qquad \textbf{(8)}$$

where

$$\lim_{(\Delta x, \Delta y)\to(0,0)} \epsilon_1(\Delta x, \Delta y) = \lim_{(\Delta x, \Delta y)\to(0,0)} \epsilon_2(\Delta x, \Delta y) = 0.$$

Then, dividing both sides of (8) by Δr and taking limits, we obtain

$$\lim_{\Delta r\to 0}\frac{\Delta z}{\Delta r} = \lim_{\Delta r\to 0}\left[\frac{\partial z}{\partial x}\frac{\Delta x}{\Delta r} + \frac{\partial z}{\partial y}\frac{\Delta y}{\Delta r} + \epsilon_1(\Delta x, \Delta y)\frac{\Delta x}{\Delta r} + \epsilon_2(\Delta x, \Delta y)\frac{\Delta y}{\Delta r}\right].$$

$$\textbf{(9)}$$

Since x and y are continuous functions of r, we have

$$\lim_{\Delta r\to 0} \Delta x = 0 \qquad \text{and} \qquad \lim_{\Delta r\to 0} \Delta y = 0,$$

so that

$$\lim_{\Delta r\to 0} \epsilon_1(\Delta x, \Delta y) = \lim_{(\Delta x, \Delta y)\to(0,0)} \epsilon_1(\Delta x, \Delta y) = 0$$

and

$$\lim_{\Delta r\to 0} \epsilon_2(\Delta x, \Delta y) = \lim_{(\Delta x, \Delta y)\to(0,0)} \epsilon_2(\Delta x, \Delta y) = 0.$$

Thus, the limits in (9) become

$$\frac{\partial z}{\partial r} = \frac{\partial z}{\partial x}\frac{\partial x}{\partial r} + \frac{\partial z}{\partial y}\frac{\partial y}{\partial r} + 0 \cdot \frac{\partial x}{\partial r} + 0 \cdot \frac{\partial y}{\partial r},$$

and the theorem is proved. ■

■ PROBLEMS 13.6

■ SELF-QUIZ

I. If $z = f(x, y) = xy$ and $x = t$ and $y = -t$, then $\frac{dz}{dt} = $ _____.
  a. $-2t\mathbf{k}$ b. $-2t$
  c. 0 d. $0\mathbf{i} + 0\mathbf{j}$

II. If $z = 3x + 4y - 5$ and $x = r \cos \theta$ and $y = r \sin \theta$, then $\frac{\partial z}{\partial \theta} = $ _____.
  a. $-3y + 4x$
  b. $-3r \sin \theta + 4r \cos \theta$
  c. $-3r \sin \theta + 4r \cos \theta - 5$
  d. $-3 \sin \theta + 4 \cos \theta$

III. Suppose $w = x^2 + y^2 + z^2$ and $x = \rho \sin \varphi \cos \theta, y = \rho \sin \varphi \sin \theta, z = \rho \cos \varphi$; then $\frac{\partial w}{\partial \rho} = $ _____.
  a. 0 b. $(\cos \theta + \sin \theta) \sin \varphi + \cos \varphi$
  c. ρ d. 2ρ

IV. Suppose $w = x - y + 3z$, $x = \sin z$, and $y = x^2$, then $\frac{\partial w}{\partial z} = $ _____.
  a. 3
  b. $\cos z - 0 + 3$
  c. $\cos z - 2 \sin z \cos z + 3$
  d. $\cos z - 2 \sin z \cos z$

■ DRILL

In Problems 1–10, use the chain rule to calculate $\frac{dz}{dt}$ or $\frac{dw}{dt}$. Check your result by writing z or w explicitly as a function of t and then differentiating this expression.

1. $z = xy$, $x = e^t$, $y = e^{2t}$
2. $z = x^2 + y^2$, $x = \cos t$, $y = \sin t$
3. $z = y/x$, $x = t^2$, $y = t^3$
4. $z = e^x \sin y$, $x = \sqrt{t}$, $y = \sqrt[3]{t}$
5. $z = \tan^{-1}(y/x)$, $x = \cos 3t$, $y = \sin 5t$
6. $z = \sinh(x - 2y)$, $x = 2t^4$, $y = t^2 + 1$
7. $w = x^2 + y^2 + z^2$, $x = \cos t$, $y = \sin t$, $z = t$
8. $w = \ln(2x - 3y + 4z)$, $x = e^t$, $y = \ln t$, $z = \cosh t$
9. $w = xy - yz + zx$, $x = e^t$, $y = e^{2t}$, $z = e^{3t}$
10. $w = (x + y)/z$, $x = t$, $y = t^2$, $z = t^3$

In Problems 11–26, use the chain rule to calculate the indicated partial derivatives.

11. $z = xy$, $x = r + s$, $y = r - s$; $\frac{\partial z}{\partial r}, \frac{\partial z}{\partial s}$

12. $z = x^2 + y^2$, $x = \cos(r + s)$, $y = \sin(r - s)$; $\frac{\partial z}{\partial r}, \frac{\partial z}{\partial s}$

13. $z = y/x$, $x = e^r$, $y = e^s$; $\frac{\partial z}{\partial r}, \frac{\partial z}{\partial s}$

14. $z = \sin(y/x)$, $x = r/s$, $y = s/r$; $\frac{\partial z}{\partial r}, \frac{\partial z}{\partial s}$

15. $z = \dfrac{e^{x+y}}{e^{x-y}}$, $x = \ln rs$, $y = \ln(r/s)$; z_r, z_s

16. $z = x^2 y^3$, $x = r - s^2$, $y = 2s + r$; z_r, z_s

17. $w = x + y + z$, $x = rs$, $y = r + s$, $z = r - s$; w_r, w_s

18. $w = xy/z$, $x = r$, $y = s$, $z = t$; $\frac{\partial w}{\partial r}, \frac{\partial w}{\partial s}, \frac{\partial w}{\partial t}$

19. $w = xy/z$, $x = r + s$, $y = t - r$, $z = s + 2t$; w_r, w_s, w_t

20. $w = \sin xyz$, $x = s^2 r$, $y = r^2 s$, $z = r - s$; w_r, w_s, w_t

21. $w = \sinh(x + 2y + 3z)$, $x = \sqrt{r + s}$, $y = \sqrt[3]{s - t}$, $z = 1/(r + t)$; w_s

22. $w = x^2 y + yz^2$, $x = rst$, $y = rs/t$, $z = 1/(rst)$; w_r, w_s, w_t

CONTINUED

23. $w = \ln(x + 2y + 3z)$, $x = r^2 + t^2$, $y = s^2 - t^2$, $z = r^2 + s^2$; $\dfrac{\partial w}{\partial r}, \dfrac{\partial w}{\partial s}, \dfrac{\partial w}{\partial t}$

24. $w = e^{xy/z}$, $x = r^2 + t^2$, $y = s^2 - t^2$, $z = r^2 + s^2$; $\dfrac{\partial w}{\partial r}, \dfrac{\partial w}{\partial s}, \dfrac{\partial w}{\partial t}$

*25. $w = r - \cos\theta + t$, $r = \sqrt{x^2 + y^2}$, $\theta = \tan^{-1}(y/x)$, $t = z$; w_x, w_y, w_z

*26. $u = xy + w^2 - z^3$, $x = t + r - q$, $y = q^2 + s^2 - t + r$, $z = (qr + st)/r^4$, $w = (r - s)/(t + q)$; u_q, u_r, u_s, u_t

■ APPLICATIONS

27. The radius of a right circular cone is increasing at a rate of 7 in/min, while its height is decreasing at a rate of 20 in/min. Is the volume of the cone increasing or decreasing when $r = 45$ in and $h = 100$ in? How fast is the volume changing then?

28. The radius of a right circular cylinder is decreasing at a rate of 12 cm/sec, while its height is increasing at a rate of 25 cm/sec. How is the volume changing when $r = 180$ cm and $h = 500$ cm? Is the volume increasing or decreasing then?

29. The volume of 10 moles of an ideal gas is decreasing at a rate of 25 cm³/min and its temperature is increasing at a rate of 1°C/min. How fast is the pressure changing when $V = 1000$ cm³ and $P = 3$ N/cm²? (Leave your answer expressed in terms of R.) Is the pressure increasing or decreasing?

30. The pressure of 8 moles of an ideal gas is decreasing at a rate of 0.4 N/cm²/min, while the temperature is decreasing at a rate of 0.5°K/min. How fast is the volume of the gas changing when $V = 1000$ cm³ and $P = 3$ N/cm²? Is the volume increasing or decreasing? [*Hint:* Use the value $R = 8.314$ J/mol °K.]

31. The angle A of a triangle ABC is increasing at a rate of 3°/sec, the side AB is increasing at a rate of 1 cm/sec, and the side AC is decreasing at a rate of 2 cm/sec. How fast is the side BC changing when $A = 30°$, $AB = 10$ cm, and $AC = 24$ cm? Is the length of BC increasing or decreasing? [*Hint:* Use the law of cosines (after converting degrees to radians).]

32. Suppose $z = f(x, y)$ is differentiable. Write the expression $\left(\dfrac{\partial z}{\partial x}\right)^2 + \left(\dfrac{\partial z}{\partial y}\right)^2$ in terms of polar coordinates r and θ.

33. Suppose $w = g(x, y, z)$ is differentiable. Let $x = r\cos\theta$, $y = r\sin\theta$, and $z = t$ (these are cylindrical coordinates). Calculate $\dfrac{\partial w}{\partial r}, \dfrac{\partial w}{\partial \theta}$, and $\dfrac{\partial w}{\partial t}$.

34. Suppose $w = g(x, y, z)$ is differentiable. Let $x = \rho\sin\varphi\cos\theta$, $y = \rho\sin\varphi\sin\theta$, and $z = \rho\cos\varphi$ (these are spherical coordinates). Calculate $\dfrac{\partial w}{\partial \rho}$, $\dfrac{\partial w}{\partial \theta}$, and $\dfrac{\partial w}{\partial \varphi}$.

■ SHOW/PROVE/DISPROVE

35. The wave equation (see Problem 13.4.27) is the partial-differential equation (PDE) $\dfrac{\partial^2 y}{\partial t^2} = c^2 \dfrac{\partial^2 y}{\partial x^2}$. Show that if f is any differentiable function, then

$$y(x, t) = \frac{1}{2}[f(x - ct) + f(x + ct)]$$

is a solution to the wave equation.

36. The function $f(x, y)$ is said to be **homogeneous of degree** n if and only if $f(tx, ty) = t^n f(x, y)$ holds for all t, x, y. Show that if f is homogeneous of degree n, then

$$x\frac{\partial f}{\partial x}(x, y) + y\frac{\partial f}{\partial y}(x, y) = nf(x, y).$$

*37. Suppose $z = F(x, y)$, $x = X(r, s)$, and $y = Y(r, s)$. Show that

$$\frac{\partial^2 z}{\partial r^2} = \frac{\partial}{\partial x}\left(\frac{\partial F}{\partial x}\frac{\partial X}{\partial r} + \frac{\partial F}{\partial y}\frac{\partial Y}{\partial r}\right)\left(\frac{\partial X}{\partial r}\right)$$
$$+ \frac{\partial}{\partial y}\left(\frac{\partial F}{\partial x}\frac{\partial X}{\partial r} + \frac{\partial F}{\partial y}\frac{\partial Y}{\partial r}\right)\left(\frac{\partial Y}{\partial r}\right).$$

CONTINUED

**38. Laplace's equation (see Problem 13.4.31) is the partial-differential equation

$$\frac{\partial^2 f}{\partial x^2} + \frac{\partial^2 f}{\partial y^2} = 0.$$

If we write (x, y) in polar coordinates $(x = r \cos \theta, \ y = r \sin \theta)$, show that Laplace's equa-

tion becomes

$$\frac{\partial^2 f}{\partial r^2} + \frac{1}{r^2} \frac{\partial^2 f}{\partial \theta^2} + \frac{1}{r} \frac{\partial f}{\partial r} = 0.$$

[*Hint:* Use the result of the preceding problem to write $\partial^2 f / \partial x^2$ and $\partial^2 f / \partial y^2$ in terms of r and θ.]

■ **CHALLENGE**

*39. Find the general solution to the partial-differential equation (PDE)

$$\frac{\partial z}{\partial y} = \frac{1}{a} \frac{\partial z}{\partial x}$$

where a is some nonzero constant. [*Hint:* Rewrite the given PDE in terms of new independent variables

$$u = y + ax \quad \text{and} \quad v = y - ax.]$$

**40. Generalize the result of Problem 35 by finding the general smooth solution to the wave equation

$$\frac{\partial^2 y}{\partial t^2} = c^2 \frac{\partial^2 y}{\partial x^2} \quad \text{and} \quad c \neq 0.$$

[*Hint:* Rewrite the given PDE in terms of new independent variables

$$u = x + ct \quad \text{and} \quad v = x - ct$$

and assume that mixed partials are equal.]

■ ANSWERS TO SELF-QUIZ

I. b   II. b = a

III. d $(2\rho \sin^2 \varphi (\cos^2 \theta + \sin^2 \theta) + 2\rho \cos^2 \varphi = 2\rho \sin^2 \varphi + 2\rho \cos^2 \psi = 2\rho)$

IV. $c \left(\dfrac{dy}{dz} = \dfrac{dy}{dx} \dfrac{dx}{dz} \right)$

13.7 TANGENT PLANES, NORMAL LINES, AND GRADIENTS

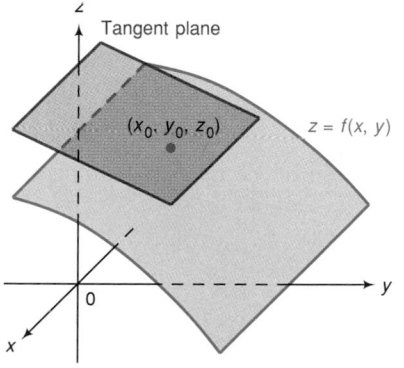

FIGURE 1
A TANGENT PLANE TO A SURFACE

Let $z = f(x, y)$ be a continuous function of two variables. As we have seen, the graph of f is a surface in $\mathbb{R}^3$. More generally, the graph of the equation $F(x, y, z) = 0$ is a surface in $\mathbb{R}^3$ if F is continuous. The surface $F(x, y, z) = 0$ is called **continuously differentiable** or **smooth** at a point (x_0, y_0, z_0) if $\partial F / \partial x$, $\partial F / \partial y$, and $\partial F / \partial z$ all exist and are continuous at (x_0, y_0, z_0). In $\mathbb{R}^2$ a differentiable curve has a unique tangent line at each point. A differentiable surface in $\mathbb{R}^3$ has a unique tangent plane at each point at which $\partial F / \partial x$, $\partial F / \partial y$, and $\partial F / \partial z$ are not all zero. We will formally define what we mean by a tangent plane to a surface after a bit, although it should be easy enough to visualize (see Figure 1). We note here that not every surface has a tangent plane at every point. For example, the cone $z = \sqrt{x^2 + y^2}$ has no tangent plane at the origin (see Figure 2).

Assume that the surface S given by $F(x, y, z) = 0$ is smooth. Let C be any curve lying on S. That is, C can be given parametrically by $\mathbf{g}(t) = x(t)\mathbf{i} + y(t)\mathbf{j} + z(t)\mathbf{k}$. (Recall from Section 12.1 (page 792) the definition of a curve in $\mathbb{R}^3$.) Then for points on the curve, $F(x, y, z)$ can be written as a function of t, and from the vector form of the chain rule [equation (13.6.3)] we have

$$F'(t) = \nabla F \cdot \mathbf{g}'(t). \tag{1}$$

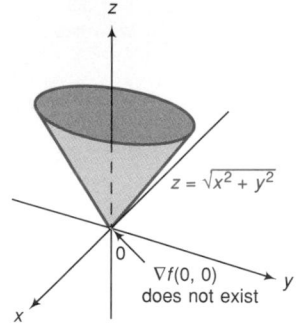

FIGURE 2
THERE IS NO TANGENT PLANE AT THE
ORIGIN.

But since $F(x(t), y(t), z(t)) = 0$ for all t [since $(x(t), y(t), z(t))$ is on S], we see that $F'(t) = 0$ for all t. But $\mathbf{g}'(t)$ is tangent to the curve C for every number t. Thus (1) implies the following:

> The gradient of F at a point $\mathbf{x}_0 = (x_0, y_0, z_0)$ on S is orthogonal to the tangent vector at $\mathbf{x}_0$ to any curve C remaining on S and passing through $\mathbf{x}_0$.

This statement is illustrated in Figure 3.

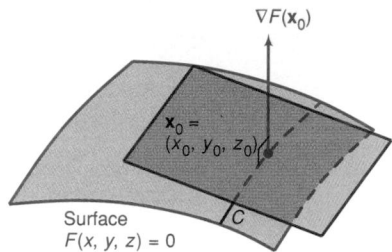

FIGURE 3   THE GRADIENT AT x_0 IS ORTHOGONAL TO ANY VECTOR THAT IS TANGENT TO THE SURFACE AT x_0.

Thus if we think of all the vectors tangent to a surface at a point $\mathbf{x}_0$ as constituting a plane, then $\boldsymbol{\nabla}F(\mathbf{x}_0)$ is a *normal* vector to that plane. This motivates the following definition.

DEFINITION   **TANGENT PLANE AND NORMAL LINE**

Let F be continuously differentiable at $\mathbf{x}_0 = (x_0, y_0, z_0)$ and let the smooth surface S be defined by $F(x, y, z) = 0$.

(i) The **tangent plane** to S at (x_0, y_0, z_0) is the plane passing through the point (x_0, y_0, z_0) with normal vector $\boldsymbol{\nabla}F(\mathbf{x}_0)$.

(ii) The **normal line** to S at $\mathbf{x}_0$ is the line passing through $\mathbf{x}_0$ having the same direction as $\boldsymbol{\nabla}F(\mathbf{x}_0)$. ∎

EXAMPLE 1
FINDING EQUATIONS OF THE
TANGENT PLANE AND NORMAL
LINE AT A POINT

Find the equation of the tangent plane and symmetric equations of the normal line to the ellipsoid $x^2 + (y^2/4) + (z^2/9) = 3$ at the point $(1, 2, 3)$.

SOLUTION: Since $F(x, y, z) = x^2 + (y^2/4) + (z^2/9) - 3 = 0$, we have

$$\boldsymbol{\nabla}F = \frac{\partial F}{\partial x}\mathbf{i} + \frac{\partial F}{\partial y}\mathbf{j} + \frac{\partial F}{\partial z}\mathbf{k} = 2x\mathbf{i} + \frac{y}{2}\mathbf{j} + \frac{2z}{9}\mathbf{k}.$$

Then $\boldsymbol{\nabla}F(1, 2, 3) = 2\mathbf{i} + \mathbf{j} + \frac{2}{3}\mathbf{k}$, and the equation of the tangent plane is

$$2(x - 1) + (y - 2) + \frac{2}{3}(z - 3) = 0 \text{ or } 2x + y + \frac{2}{3}z = 6.$$

The normal line is given by

$$\frac{x - 1}{2} = y - 2 = \frac{3}{2}(z - 3).$$

The situation is even simpler if we can write the surface in the form $z = f(x, y)$. That is, the surface is the graph of a function of two variables. Then $F(x, y, z) = z - f(x, y) = 0$, so that

$$F_x = -f_x, \qquad F_y = -f_y, \qquad \text{and} \qquad F_z = 1,$$

and the normal vector $\mathbf{N}$ to the tangent plane is

NORMAL VECTOR TO THE SURFACE $z = f(x, y)$

$$\mathbf{N} = -f_x(x_0, y_0)\mathbf{i} - f_y(x_0, y_0)\mathbf{j} + \mathbf{k}. \tag{2}$$

REMARK: One interesting consequence of this fact is that if $z = f(x, y)$ and if $\nabla f(x_0, y_0) = \mathbf{0}$, then *the tangent plane to the surface at $(x_0, y_0, f(x_0, y_0))$ is parallel to the xy-plane (that is, it is horizontal)*. This occurs because at $(x_0, y_0, f(x_0, y_0))$, $\mathbf{N} = -(\partial f/\partial x)\mathbf{i} - (\partial f/\partial y)\mathbf{j} + \mathbf{k} - -\nabla f + \mathbf{k} = \mathbf{k}$. Thus the z-axis is normal to the tangent plane.

EXAMPLE 2
FINDING EQUATIONS OF THE TANGENT PLANE AND NORMAL LINE AT A POINT

Find the tangent plane and normal line to the surface $z = x^3 y^5$ at the point $(2, 1, 8)$.

SOLUTION: $\mathbf{N} = -(\partial z/\partial x)\mathbf{i} - (\partial z/\partial y)\mathbf{j} + \mathbf{k} = -3x^2 y^5 \mathbf{i} - 5x^3 y^4 \mathbf{j} + \mathbf{k} = -12\mathbf{i} - 40\mathbf{j} + \mathbf{k}$ at $(2, 1, 8)$. Then the tangent plane is given by

$$-12(x - 2) \quad 40(y - 1) + (z - 8) = 0,$$

or

$$12x + 40y - z = 56.$$

Symmetric equations of the normal line are

$$\frac{x - 2}{12} = \frac{y - 1}{40} = \frac{z - 8}{-1}.$$

■ PROBLEMS 13.7

■ SELF-QUIZ

I. The plane tangent to the surface $x^2 + y^2 + z^2 = 1$ at the point $(1, 0, 0)$ satisfies the equation _____.

  a. $x = 1$ b. $y + z = 0$
  c. $2x + 2y + 2z = 2$ d. $2x = 2y + 2z$

II. The line normal to the surface $x^2 + y^2 + z^2 = 1$ at the point $(1, 0, 0)$ satisfies the equation _____.

  a. $(x - 1)/2 = (y - 0)/2 = (z - 0)/2$
  b. $x\mathbf{i} + y\mathbf{j} + z\mathbf{k} = \mathbf{i} + t(2\mathbf{i})$
  c. $y = 0, z = 0$
  d. $x = 1, y = z$

III. The plane tangent to the surface $x^2 + y^2 + z^2 = 2$ at the point $(0, 1, 1)$ satisfies the equation _____.

  a. $y + z = 0$ b. $2y + 2z = 4$
  c. $x = 0, y = z$ d. $x + y + z = 2$

IV. The plane tangent to the surface $z = 2x + \cos y$ at the point $(1, 0, 3)$ satisfies the equation _____.

  a. $2(x - 1) + (-1)(z - 3) = 0$
  b. $x\mathbf{i} + y\mathbf{j} + z\mathbf{k} = (\mathbf{i} + 3\mathbf{k}) + t(2\mathbf{i} - \mathbf{k})$
  c. $[(x - 1)\mathbf{i} + (y - 0)\mathbf{j} + (z - 3)\mathbf{k}] \cdot (2\mathbf{i} - \mathbf{k}) = 0$
  d. $(x - 1)/2 = (z - 3)/(-1), y = 0$

CONTINUED

  e. $[x\mathbf{i} + y\mathbf{j} + z\mathbf{k}] \cdot (\mathbf{i} + 3\mathbf{k}) = 0$
  f. $x\mathbf{i} + y\mathbf{j} + z\mathbf{k} = (2\mathbf{i} - \mathbf{k}) \cdot (\mathbf{i} + 3\mathbf{k})$
V. The line normal to the surface $2x + \cos y = z$ at the point $(1, 0, 3)$ satisfies the equation _____.

  a. $2(x - 1) + (-1)(z - 3) = 0$

  b. $x\mathbf{i} + y\mathbf{j} + z\mathbf{k} = (\mathbf{i} + 3\mathbf{k}) + t(2\mathbf{i} - \mathbf{k})$
  c. $[(x - 1)\mathbf{i} + (y - 0)\mathbf{j} + (z - 3)\mathbf{k}] \cdot (2\mathbf{i} - \mathbf{k}) = 0$
  d. $(x - 1)/2 = (z - 3)/(-1),\ y = 0$
  e. $[x\mathbf{i} + y\mathbf{j} + z\mathbf{k}] \cdot (\mathbf{i} + 3\mathbf{k}) = 0$
  f. $x\mathbf{i} + y\mathbf{j} + z\mathbf{k} = (2\mathbf{i} - \mathbf{k}) \cdot (\mathbf{i} + 3\mathbf{k})$

■ **DRILL**

In Problems 1–14, a surface is given and a point is specified. Find an equation for the plane tangent to the surface at that point and symmetric (or vector) equations for the normal line.

1. $x^2 + y^2 + z^2 = 1$; $(0, 1, 0)$
2. $x^2 - y^2 + z^2 = 1$; $(1, 1, 1)$
3. $(x/a)^2 + (y/b)^2 + (z/c)^2 = 3$; (a, b, c)
4. $(x/a)^2 + (y/b)^2 + (z/c)^2 = 3$; $(-a, b, -c)$
5. $\sqrt{x} + \sqrt{y} + \sqrt{z} = 6$; $(4, 1, 9)$
6. $ax + by + cz = d$; $(1/a, 1/b, (d - 2)/c)$
7. $xyz = 4$; $(1, 2, 2)$
8. $xy^2 - yz^2 + zx^2 = 1$; $(1, 1, 1)$
9. $4x^2 - y^2 - 5z^2 = 15$; $(3, 1, -2)$
10. $xe^y - ye^z = 1$; $(1, 0, 0)$
11. $\sin xy - 2 \cos yz = 0$; $(\pi/2, 1, \pi/3)$
12. $x^2 + y^2 + 4x + 2y + 8z = 7$; $(2, -3, -1)$
13. $e^{xyz} = 5$; $(1, 1, \ln 5)$
14. $\sqrt{\dfrac{x + y}{z - 1}} = 1$; $(1, 1, 3)$

In Problems 15–22, write an equation for the plane tangent to $z = f(\mathbf{x})$ at the point $\mathbf{x}_0$ in the form (discussed in Problem 40 below)

$$z = f(\mathbf{x}_0) + (\mathbf{x} - \mathbf{x}_0) \cdot \nabla f(\mathbf{x}_0)$$

and write a vector equation satisfied by the normal line at that point.

15. $z = xy^2$; $(1, 1, 1)$
16. $z = \ln(x - 2y)$; $(3, 1, 0)$
17. $z = \sin(2x + 5y)$; $(\pi/8, \pi/20, 1)$
18. $z = \sqrt{\dfrac{x + y}{x - y}}$; $(5, 4, 3)$
19. $z = \tan^{-1}(y/x)$; $(-2, 2, -\pi/4)$
20. $z = \sinh xy^2$; $(0, 3, 0)$
21. $z = \sec(x - y)$; $(\pi/2, \pi/6, 2)$
22. $z = e^x \cos y + e^y \cos x$; $(\pi/2, 0, e^{\pi/2})$

■ **APPLICATIONS**

In Problems 23–28, use the results of Problems 37 and 38 to find a normal vector and the equation of the line tangent to the curve at the specified point.

23. $xy = 5$; $(1, 5)$
24. $x^2 + xy + y^2 + 3x - 5y = 16$; $(1, -2)$
25. $(x + y)/(x - y) = 7$; $(4, 3)$
26. $xe^{xy} = 1$; $(1, 0)$

27. $(x/2)^2 + (y/4)^2 = 1$; $(\sqrt{2}, 2\sqrt{2})$
28. $\tan(x + y) = 1$; $(\pi/4, 0)$

*29. Find the two points of intersection of the surface $z = x^2 + y^2$ and the line $(x - 3)/1 = (y + 1)/(-1) = (z + 2)/(-2)$. At each of these intersection points, calculate the cosine of the angle between the normal line there and the given line.

■ **SHOW/PROVE/DISPROVE**

30. Show that the plane tangent to the surface $z = ax^2 + by^2$ at the point (x_0, y_0, z_0) satisfies the equation

$$\frac{z + z_0}{2} = ax_0x + by_0y.$$

31. Show that the plane tangent to the surface $ax^2 + by^2 + cz^2 = d$ at the point (x_0, y_0, z_0) satisfies the equation $ax_0x + by_0y + cz_0z = d$.

*32. Show that every line normal to the surface of a sphere passes through the center of the sphere.

CONTINUED

*33. Show that every line normal to the cone $z^2 = a(x^2 + y^2)$ intersects the z-axis.

*34. Suppose f is a differentiable function of one variable and let $z = yf(y/x)$. Show that all planes tangent to the surface described by this equation have a point in common.

35. The **angle between two surfaces** at a point of intersection is defined to be the angle between their normal lines. Show that if two surfaces $F(x, y, z) = 0$ and $G(x, y, z) = 0$ intersect at right angles at a point $\mathbf{x}_0$, then $\nabla F(\mathbf{x}_0) \cdot \nabla G(\mathbf{x}_0) = 0$.

36. Show that the sum of the squares of the x-, y-, and z-intercepts of any plane tangent to the surface $x^{2/3} + y^{2/3} + z^{2/3} = a^{2/3}$ is a constant.

37. Suppose the equation $F(x, y) = 0$ defines a curve in $\mathbb{R}^2$. Show that if F is differentiable, then $\nabla F(x, y)$ is normal to the curve at each point of the curve.

38. Use the result of the preceding problem to find an equation for the line tangent to the curve $F(x, y) = 0$ at a point (x_0, y_0).

39. Show that at any point (x, y), $y \neq 0$, the curve $x/y = a$ is orthogonal to the curve $x^2 + y^2 = r^2$ for any constants a and r.

40. Show that the plane tangent at $\mathbf{x}_0 = (x_0, y_0)$ to the smooth surface $z = F(x, y)$ satisfies the equation

$$z = F(\mathbf{x}_0) + (\mathbf{x} - \mathbf{x}_0) \cdot \nabla F(\mathbf{x}_0).$$

[*Note:* If we focus on a function of a single variable, we can describe the line tangent to the curve $y = f(x)$ at $x = x_0$ by the equation $y = f(x_0) + (x - x_0)f'(x_0)$. This problem asks you to show that the gradient is the "right" generalization of the derivative of a function of one variable.]

■ ANSWERS TO SELF-QUIZ

I. a   II. b = c   III. b

IV. a = c   V. b = d

13.8 DIRECTIONAL DERIVATIVES AND THE GRADIENT

Let us take another look at the partial derivatives $\partial f/\partial x$ and $\partial f/\partial y$ of the function $z = f(x, y)$. We have

$$\frac{\partial f}{\partial x}(x_0, y_0) = \lim_{\Delta x \to 0} \frac{f(x_0 + \Delta x, y_0) - f(x_0, y_0)}{\Delta x}. \tag{1}$$

This measures the rate of change of f as we approach the point (x_0, y_0) along a vector parallel to the x-axis [since $(x_0 + \Delta x, y_0) - (x_0, y) = (\Delta x, 0) = \Delta x \mathbf{i}$]. Similarly

$$\frac{\partial f}{\partial y}(x_0, y_0) = \lim_{\Delta x \to 0} \frac{f(x_0, y_0 + \Delta y) - f(x_0, y_0)}{\Delta y} \tag{2}$$

measures the rate of change of f as we approach the point (x_0, y_0) along a vector parallel to the y-axis.

It is frequently of interest to compute the rate of change of f as we approach (x_0, y_0) along a vector that is not parallel to one of the coordinate axes. The situation is depicted in Figure 1. Suppose that (x, y) approaches the fixed point (x_0, y_0) along the line segment joining them, and let t denote the distance between the two points. We want to determine the relative rate of change in f with respect to a change in t.

Let $\mathbf{u}$ denote a unit vector with the initial point at (x_0, y_0) and parallel to $\overrightarrow{PQ}$ (see Figure 2). Since $\mathbf{u}$ and $\overrightarrow{PQ}$ are parallel, there is, by Theorem 11.2.3,

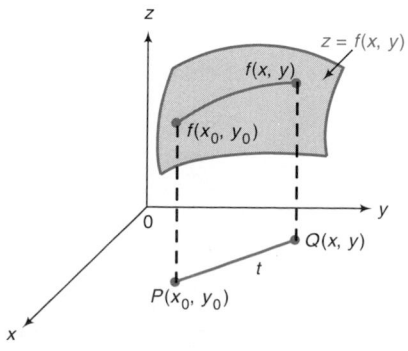

FIGURE 1

(x, y) MOVES ALONG THE VECTOR $\overrightarrow{PQ}$

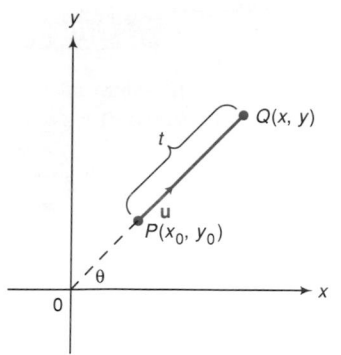

FIGURE 2
u AND $\overrightarrow{PQ}$ ARE PARALLEL.

There is a number t such that $\overrightarrow{PQ} = t\mathbf{u}$.
*Since **u** is a unit vector, $t = |\overrightarrow{PQ}|$.*

a value of t such that

$$\overrightarrow{PQ} = t\mathbf{u}. \tag{3}$$

Note that $t > 0$ if **u** and $\overrightarrow{PQ}$ have the same direction and $t < 0$ if **u** and $\overrightarrow{PQ}$ have opposite directions. Now

$$\overrightarrow{PQ} = (x - x_0)\mathbf{i} + (y - y_0)\mathbf{j}, \tag{4}$$

and since **u** is a unit vector, we have

$$\mathbf{u} = \cos\theta\mathbf{i} + \sin\theta\mathbf{j}, \tag{5}$$

where θ is the direction of **u**. Thus, inserting (4) and (5) into (3), we have

$$(x - x_0)\mathbf{i} + (y - y_0)\mathbf{j} = t\cos\theta\mathbf{i} + t\sin\theta\mathbf{j},$$

or

$$x = x_0 + t\cos\theta$$
$$y = y_0 + t\sin\theta. \tag{6}$$

The equations (6) are the parametric equations of the line passing through P and Q. Using (6), we have

$$z = f(x, y) = f(x_0 + t\cos\theta, y_0 + t\sin\theta). \tag{7}$$

Remember that θ is fixed—it is the direction of approach. Thus $(x, y) \rightarrow (x_0, y_0)$ along $\overrightarrow{PQ}$ is equivalent to $t \rightarrow 0$ in (7). Hence, to compute the instantaneous rate of change of f as $(x, y) \rightarrow (x_0, y_0)$ along the vector $\overrightarrow{PQ}$, we need to compute dz/dt. But by the chain rule,

$$\frac{dz}{dt} = \frac{\partial f}{\partial x}(x, y)\frac{dx}{dt} + \frac{\partial f}{\partial y}(x, y)\frac{dy}{dt}, \text{ or}$$

$$\frac{dz}{dt} = f_x(x, y)\cos\theta + f_y(x, y)\sin\theta, \text{ and} \tag{8}$$

$$\frac{dz}{dt} = \left[f_x(x_0 + t\cos\theta, y_0 + t\sin\theta)\right]\cos\theta$$
$$+ \left[f_y(x_0 + t\cos\theta, y_0 + t\sin\theta)\right]\sin\theta. \tag{9}$$

If we set $t = 0$ in (9), we obtain the instantaneous rate of change of f in the direction $\overrightarrow{PQ}$ at the point (x_0, y_0). That is,

$$\left.\frac{dz}{dt}\right|_{t=0} = f_x(x_0, y_0)\cos\theta + f_y(x_0, y_0)\sin\theta. \tag{10}$$

But (10) can be written [using (5)] as

$$\left.\frac{dz}{dt}\right|_{t=0} = \nabla f(x_0, y_0) \cdot \mathbf{u}. \tag{11}$$

This leads to the following definition.

DEFINITION   **DIRECTIONAL DERIVATIVE**

Let f be differentiable at a point $\mathbf{x}_0 = (x_0, y_0)$ in $\mathbb{R}^2$ and let $\mathbf{u}$ be a unit vector. Then the **directional derivative of f in the direction $\mathbf{u}$**, denoted by $f'_{\mathbf{u}}(\mathbf{x}_0)$, is given by

$$f'_{\mathbf{u}}(\mathbf{x}_0) = \nabla f(\mathbf{x}_0) \cdot \mathbf{u}. \quad \blacksquare \tag{12}$$

REMARK 1:   Note that if $\mathbf{u} = \mathbf{i}$, then $\nabla f \cdot \mathbf{u} = \partial f / \partial x$ and (12) reduces to the partial derivative $\partial f / \partial x$. Similarly, if $\mathbf{u} = \mathbf{j}$, then (12) reduces to $\partial f / \partial y$.

REMARK 2:   This definition makes sense if f is a function of three variables. Then, of course, $\mathbf{u}$ is a unit vector in $\mathbb{R}^3$.

REMARK 3:   There is another definition of the directional derivative. It is given by

$$f'_{\mathbf{u}}(\mathbf{x}_0) = \lim_{h \to 0} \frac{f(\mathbf{x}_0 + h\mathbf{u}) - f(\mathbf{x}_0)}{h}. \tag{13}$$

It can be shown that if the limit in (13) exists, it is equal to $\nabla f(\mathbf{x}_0) \cdot \mathbf{u}$ if f is differentiable.

EXAMPLE 1
COMPUTING A DIRECTIONAL
DERIVATIVE IN $\mathbb{R}^2$

Let $z = f(x, y) = xy^2$. Calculate the directional derivative of f in the direction of the vector $\mathbf{v} = 2\mathbf{i} + 3\mathbf{j}$ at the point $(4, -1)$.

SOLUTION: A unit vector in the direction $\mathbf{v}$ is $\mathbf{u} = (2/\sqrt{13})\mathbf{i} + (3/\sqrt{13})\mathbf{j}$. Also, $\nabla f = y^2\mathbf{i} + 2xy\mathbf{j}$. Thus,

$$f'_{\mathbf{u}}(x, y) = \nabla f(\mathbf{x}) \cdot \mathbf{u} = \frac{2y^2}{\sqrt{13}} + \frac{6xy}{\sqrt{13}} = \frac{2y^2 + 6xy}{\sqrt{13}}.$$

At $(4, -1)$, $f'_{\mathbf{u}}(4, -1) = -22/\sqrt{13}$.

EXAMPLE 2
COMPUTING A DIRECTIONAL
DERIVATIVE IN $\mathbb{R}^3$

Let $f(x, y, z) = x \ln y - e^{xz^3}$. Calculate the directional derivative of f in the direction of the vector $\mathbf{v} = \mathbf{i} - \mathbf{j} + 3\mathbf{k}$. Evaluate this derivative at the point $(-5, 1, -2)$.

SOLUTION: A unit vector in the direction of $\mathbf{v}$ is $\mathbf{u} = (1/\sqrt{11})\mathbf{i} - (1/\sqrt{11})\mathbf{j} + (3/\sqrt{11})\mathbf{k}$, and

$$\nabla f = (\ln y - z^3 e^{xz^3})\mathbf{i} + \frac{x}{y}\mathbf{j} - 3xz^2 e^{xz^3}\mathbf{k}.$$

Thus,

$$f'_{\mathbf{u}}(\mathbf{x}) = \nabla f(\mathbf{x}) \cdot \mathbf{u} = \frac{\ln y - z^3 e^{xz^3} - (x/y) - 9xz^2 e^{xz^3}}{\sqrt{11}}, \quad \text{and}$$

$$f'_{\mathbf{u}}(-5, 1, -2) = \frac{5 + 188e^{40}}{\sqrt{11}}.$$

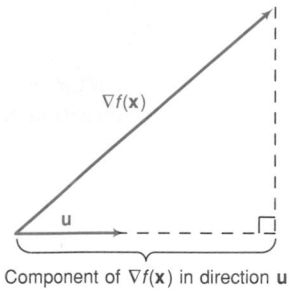

FIGURE 3
THE DIRECTIONAL DERIVATIVE OF f IN THE DIRECTION $\mathbf{u}$ IS THE COMPONENT OF ∇f IN THE DIRECTION $\mathbf{u}$.

There is an interesting geometric interpretation of the directional derivative. The projection of ∇f on $\mathbf{u}$ is given by (see page 748)

$$\text{Proj}_{\mathbf{u}} \, \nabla f = \frac{\nabla f \cdot \mathbf{u}}{|\mathbf{u}|^2} \, \mathbf{u},$$

and since $\mathbf{u}$ is a unit vector, the component of ∇f in the direction $\mathbf{u}$ is given by

$$\frac{\nabla f \cdot \mathbf{u}}{|\mathbf{u}|^2} = \nabla f \cdot \mathbf{u}.$$

Thus the *directional derivative of f in the direction $\mathbf{u}$ is the component of the gradient of f in the direction $\mathbf{u}$*. This is illustrated in Figure 3.

We now derive another remarkable property of the gradient. Recall that $\mathbf{u} \cdot \mathbf{v} = |\mathbf{u}| \, |\mathbf{v}| \cos \theta$, where θ is the smallest angle between the vectors $\mathbf{u}$ and $\mathbf{v}$. Thus, the directional derivative of f in the direction $\mathbf{u}$ can be written as

$$f'_{\mathbf{u}}(\mathbf{x}) = \nabla f(\mathbf{x}) \cdot \mathbf{u} = |\nabla f(\mathbf{x})| \, |\mathbf{u}| \cos \theta, \tag{14}$$

or since $\mathbf{u}$ is a unit vector,

$$f'_{\mathbf{u}}(\mathbf{x}) = |\nabla f(\mathbf{x})| \cos \theta.$$

Now $\cos \theta = 1$ when $\theta = 0$, which occurs when $\mathbf{u}$ has the direction of ∇f. Similarly, $\cos \theta = -1$ when $\theta = \pi$, which occurs when $\mathbf{u}$ has the direction of $-\nabla f$. Also, $\cos \theta = 0$ when $\theta = \pi/2$. Thus, since $-1 \le \cos \theta \le 1$, equation (14) implies the following important result.

THEOREM 1   **DIRECTION OF MAXIMUM RATE OF INCREASE**

Let f be differentiable. Then f increases most rapidly in the direction of its gradient and decreases most rapidly in the direction opposite to that of its gradient. It does not change in a direction perpendicular to its gradient. ∎

EXAMPLE 3
FINDING THE PATH OF STEEPEST ASCENT ON A SPHERE

Consider the sphere $x^2 + y^2 + z^2 = 1$. We can write the upper half of this sphere (i.e., the upper hemisphere) as $z = f(x, y) = \sqrt{1 - x^2 - y^2}$. Then

$$\nabla f = \frac{-x}{\sqrt{1 - x^2 - y^2}} \mathbf{i} + \frac{-y}{\sqrt{1 - x^2 - y^2}} \mathbf{j}.$$

Since $\sqrt{1 - x^2 - y^2} > 0$, we see that

direction of ∇f = direction of $-x\mathbf{i} - y\mathbf{j}$.

But $-x\mathbf{i} - y\mathbf{j}$ points from (x, y) to $(0, 0)$. Thus, if we start at a point (x, y, z) on the sphere, the path of steepest ascent (increase) is a great circle passing through the point $(0, 0, 1)$ (called the *north pole* of the sphere).

EXAMPLE 4
FINDING THE DIRECTIONS OF
MAXIMUM VOLTAGE
INCREASE AND DECREASE

The distribution of voltage on a metal plate is given by $V = 50 - x^2 - 4y^2$.

(a) At the point $(1, -2)$, in what direction does the voltage increase most rapidly?

(b) In what direction does it decrease most rapidly?

(c) What is the magnitude of this increase or decrease?

(d) In what direction does it change least?

SOLUTION: $\nabla V = V_x\mathbf{i} + V_y\mathbf{j} = -2x\mathbf{i} - 8y\mathbf{j}$. At $(1, -2)$, $\nabla V = -2\mathbf{i} + 16\mathbf{j}$.

(a) The voltage increases most rapidly as we move in the direction of $-2\mathbf{i} + 16\mathbf{j}$.

(b) It decreases most rapidly in the direction of $2\mathbf{i} - 16\mathbf{j}$.

(c) The magnitude of the increase or decrease is $\sqrt{2^2 + 16^2} = \sqrt{260}$.

(d) A unit vector perpendicular to ∇V is $(16\mathbf{i} + 2\mathbf{j})/\sqrt{260}$. The voltage does not change in this or the opposite direction.

EXAMPLE 5
FINDING THE PATH OF A PARTICLE
THAT MOVES IN THE DIRECTION
OF GREATEST VOLTAGE INCREASE

In Example 4, describe the path of a particle that starts at the point $(1, -2)$ and moves in the direction of greatest voltage increase.

SOLUTION: The path of the particle will be that of the gradient. If the particle follows the path $\mathbf{f}(t) = x(t)\mathbf{i} + y(t)\mathbf{j}$, then since the direction of the path is $\mathbf{f}'(t) = x'(t)\mathbf{i} + y'(t)\mathbf{j}$ and since this direction is also given by $\nabla V = -2x\mathbf{i} - 8y\mathbf{j}$, we must have

$$x'(t) = -2x(t) \quad \text{and} \quad y'(t) = -8y(t).$$

From Section 6.6 (see page 440), the solutions to these differential equations are

$$x(t) = c_1 e^{-2t} \quad \text{and} \quad y(t) = c_2 e^{-8t}.$$

But $x(0) = 1$ and $y(0) = -2$, so

$$x(t) = e^{-2t} \quad \text{and} \quad y(t) = -2e^{-8t}.$$

Then since $e^{-8t} = (e^{-2t})^4$, we see that the particle moves along the path

$$y = -2x^4.$$

REMARK: Technically, a direction is a unit vector, so we should choose the direction $(-2x\mathbf{i} - 8y\mathbf{j})/\sqrt{4x^2 + 64y^2}$ in our computations. But this choice would not change the final answer. A method for obtaining the answer by using unit vectors is suggested in Problem 29.

One other fact about directional derivatives and gradients should be mentioned here. Since the gradient vector is normal to the curve $f(x, y) = C$, for any constant C, we say that the directional derivative of f in the direction of the gradient is the **normal derivative** of f and is denoted by df/dn. We then have, from equation (14),

$$\text{normal derivative} = \frac{df}{dn} = |\nabla f|.$$

(15)

EXAMPLE 6
CALCULATING A NORMAL
DERIVATIVE

Let $f(x, y) = xy^2$. Calculate the normal derivative. Evaluate df/dn at the point $(3, -2)$.

SOLUTION: $\nabla f = y^2\mathbf{i} + 2xy\mathbf{j}$. Then $df/dn = |\nabla f| = \sqrt{y^4 + 4x^2y^2}$. At $(3, -2)$, $df/dn = \sqrt{16 + 144} = \sqrt{160}$.

■ PROBLEMS 13.8

■ SELF-QUIZ

I. True–False (assume f is differentiable):
   a. If $\mathbf{u} = \mathbf{i}$, then $f'_\mathbf{u} = f_x$.
   b. If $\mathbf{u} = \mathbf{k}$, then $f'_\mathbf{u} = f_z$.
   c. $f'_\mathbf{j} = f_y$.
   d. If $\mathbf{u} = -\mathbf{j}$, then $f'_\mathbf{u} = -f_y$.

II. Let $f(x, y) = 7x - 5y$ and $\mathbf{u} = \frac{3}{5}\mathbf{i} + \frac{4}{5}\mathbf{j}$. Then $f'_\mathbf{u}(2, 1)$, the directional derivative of f along $\mathbf{u}$ at $(2, 1)$, is _____.
   a. $\frac{1}{5}$ b. $\frac{1}{5}\mathbf{j}$
   c. $-\frac{1}{5}(\mathbf{i} - \mathbf{j})$ d. $\frac{22}{5}$

III. Let $g(x, y) = 7x^2 - 5y^2$ and $\mathbf{v} = 3\mathbf{i} + 4\mathbf{j}$. The directional derivative of g at $(2, 1)$ in the direction of $\mathbf{v}$ is _____.

   a. $84\mathbf{i} - 40\mathbf{j}$ b. 44
   c. $\frac{44}{5}$ d. $\frac{2}{5}$

IV. Let $f(x, y) = x^3y^4$. At the point $(1, -1)$, f increases most rapidly in the direction of _____.
   a. $3x^2y^4\mathbf{i} + 4x^3y^3\mathbf{j}$ b. $3\mathbf{i} + 4\mathbf{j}$
   c. $3\mathbf{i} - 4\mathbf{j}$ d. $\sqrt{3^2 + 4^2}$

V. If $f(x, y) = x^3y^4$, then $|f'_\mathbf{u}(1, -1)|$ is minimal for $\mathbf{u} = $ _____.
   a. $\frac{3}{5}\mathbf{i} - \frac{4}{5}\mathbf{j}$ b. $-\frac{3}{5}\mathbf{i} + \frac{4}{5}\mathbf{j}$
   c. $\frac{4}{5}\mathbf{i} + \frac{3}{5}\mathbf{j}$ d. $4\mathbf{i} - 3\mathbf{j}$

■ DRILL

In Problems 1–14, calculate the directional derivative of the given function at the specified point in the direction of the specified vector $\mathbf{v}$.

1. $f(x, y) = xy$ at $(2, 3)$; $\mathbf{v} = \mathbf{i} + 3\mathbf{j}$
2. $f(x, y) = 2x^2 - 3y^2$ at $(1, -1)$; $\mathbf{v} = -\mathbf{i} + 2\mathbf{j}$
3. $f(x, y) = \ln(x + 3y)$ at $(2, 4)$; $\mathbf{v} = \mathbf{i} + \mathbf{j}$
4. $f(x, y) = ax^2 + by^2$ at (c, d); $\mathbf{v} = \alpha\mathbf{i} + \beta\mathbf{j}$
5. $f(x, y) = \tan^{-1}(y/x)$ at $(2, 2)$; $\mathbf{v} = 3\mathbf{i} - 2\mathbf{j}$
6. $f(x, y) = \dfrac{x - y}{x + y}$ at $(4, 3)$; $\mathbf{v} = -\mathbf{i} - 2\mathbf{j}$
7. $f(x, y) = xe^y + ye^x$ at $(1, 2)$; $\mathbf{v} = \mathbf{i} + \mathbf{j}$
8. $f(x, y) = \sin(2x + 3y)$ at $(\pi/12, \pi/9)$; $\mathbf{v} = -2\mathbf{i} + 3\mathbf{j}$
9. $f(x, y, z) = xy + yz + xz$ at $(1, 1, 1)$; $\mathbf{v} = \mathbf{i} + \mathbf{j} + \mathbf{k}$
10. $f(x, y, z) = xy^3z^5$ at $(-3, -1, 2)$; $\mathbf{v} = -\mathbf{i} - 2\mathbf{j} + \mathbf{k}$
11. $f(x, y, z) = xe^{yz}$ at $(2, 0, -4)$; $\mathbf{v} = -\mathbf{i} + 2\mathbf{j} + 5\mathbf{k}$
12. $f(x, y, z) = x^2y^3 + z\sqrt{x}$ at $(1, -2, 3)$; $\mathbf{v} = 5\mathbf{j} + \mathbf{k}$
13. $f(x, y, z) = e^{-(x^2 + y^2 + z^2)}$ at $(1, 1, 1)$; $\mathbf{v} = \mathbf{i} + 3\mathbf{j} - 5\mathbf{k}$
14. $f(x, y, z) = 1/\sqrt{x^2 + y^2 + z^2}$ at $(-1, 2, 3)$; $\mathbf{v} = \mathbf{i} - \mathbf{j} + \mathbf{k}$

In Problems 15–18, find the directional derivative of the given function at the specified point in the direction from that point to Q.

15. $f(x, y) = 2x^2y$ at $(-1, 4)$; $Q = (2, -5)$
16. $f(x, y) = \dfrac{x}{y}$ at $(2, 3)$; $Q = (1, 6)$
17. $f(x, y, z) = xy^2z + x^3yz^2$ at $(1, 2, -1)$; $Q = (0, -2, 4)$
18. $f(x, y, z) = \dfrac{x + y}{z}$ at $(3, 2, -1)$; $Q = (4, 2, 5)$

In Problems 19–22, calculate the normal derivative at the given point.

19. $f(x, y) = x + 2y$ at $(1, 4)$
20. $f(x, y) = e^{x + 3y}$ at $(1, 0)$
21. $f(x, y) = \tan^{-1}(y/x)$ at $(-1, -1)$
22. $f(x, y) = \sqrt{\dfrac{x - y}{x + y}}$ at $(3, 1)$

CONTINUED

■ APPLICATIONS

23. The voltage (potential) at any point on a metal structure is given by

$$v(x, y, z) = \frac{1}{0.02 + \sqrt{x^2 + y^2 + z^2}}.$$

At the point $(1, -1, 2)$, in what direction does the voltage increase most rapidly?

24. The temperature at any point in a solid metal ball centered at the origin is given by

$$T(x, y, z) = 100e^{-(x^2+y^2+z^2)}.$$

a. Where is the ball hottest?
b. Verify that at any point (x, y, z) on the ball, the direction of greatest increase in temperature is a vector pointing toward the origin.

25. The temperature distribution of a ball centered at the origin is given by

$$T(x, y, z) = \frac{100}{1 + x^2 + y^2 + z^2}.$$

a. Where is the ball hottest?
b. Find the direction of greatest decrease of temperature at the point $(3, -1, 2)$.
c. Find the direction of greatest increase in temperature. Does this vector point toward the origin?

26. The temperature distribution on a plate is given by

$$T(x, y) = 1 - \frac{x^2}{a^2} - \frac{y^2}{b^2}.$$

Find the path of a heat-seeking particle (i.e., a particle that always moves in the direction of greatest increase in temperature) if it starts at the point (a, b).

27. Find the path of the particle in the preceding problem if it starts at the point $(-a, b)$.

28. The height of a mountain is given by $h(x, y) = 3000 - 2x^2 - y^2$, where the y-axis points east, the x-axis points north, and all distances are measured in meters. Suppose a mountain climber is at the point $(30, -20, 800)$.
a. If the climber moves in the southwest direction, will she ascend or descend?
b. In what direction should the climber move so as to ascend most rapidly?

29. We refer to Example 5.
a. Show that a unit vector in the direction of motion is given by

$$\frac{x'}{\sqrt{x'^2 + y'^2}} \mathbf{i} + \frac{y'}{\sqrt{x'^2 + y'^2}} \mathbf{j}.$$

b. Show that a unit vector having the direction of the gradient is

$$\frac{-2x}{\sqrt{4x^2 + 64y^2}} \mathbf{i} + \frac{-8y}{\sqrt{4x^2 + 64y^2}} \mathbf{j}.$$

c. By equating coordinates in parts (a) and (b), show that

$$\frac{y'(t)}{y(t)} = 4 \frac{x'(t)}{x(t)}.$$

d. Integrate both sides of the equation in part (c) to show that $\ln|y(t)| = \ln|x^4(t)| + C$.
e. Use part (d) to show that $y(t) = k[x(t)]^4$, where $k = \pm e^C$.
f. Show that $k = -2$ in part (e) by using the point $(1, -2)$ on the curve.

■ SHOW/PROVE/DISPROVE

*30. Prove that if $w = f(x, y, z)$ is differentiable at $\mathbf{x}$, then all the first partials exist at $\mathbf{x}$ and

$$\nabla f(\mathbf{x}) = f_x(\mathbf{x})\mathbf{i} + f_y(\mathbf{x})\mathbf{j} + f_z(\mathbf{x})\mathbf{k}.$$

(This result is a partial converse to Theorem 13.5.1.)

*31. Suppose f is differentiable at point $\mathbf{a}$. Prove that for any unit vector $\mathbf{u}$,

$$\lim_{h \to 0} \frac{f(\mathbf{a} + h\mathbf{u}) - f(\mathbf{a})}{h} = \nabla f(\mathbf{a}) \cdot \mathbf{u}.$$

[*Note:* Remark 3 discusses this fact.]

■ ANSWERS TO SELF-QUIZ

I. a. True   b. True
   c. True   d. True

II. a (The directional derivative is always a scalar.)
III. c   IV. c   V. c

13.9 THE TOTAL DIFFERENTIAL AND APPROXIMATION

In Section 3.8, we defined the linear approximation to a function and showed that if Δx was small, then

$$f(x + \Delta x) - f(x) = \Delta y \approx f'(x)\,\Delta x. \tag{1}$$

In Section 4.7, we defined the differential dy by

$$dy = f'(x)\,dx = f'(x)\,\Delta x \tag{2}$$

(since dx is defined to be equal to Δx). Note that in (2) it is not required that Δx be small.

We now extend these ideas to functions of two or three variables.

DEFINITION **INCREMENT AND THE TOTAL DIFFERENTIAL**

Let $f = f(\mathbf{x})$ be a function of two or three variables and let $\Delta \mathbf{x} = (\Delta x, \Delta y)$ or $(\Delta x, \Delta y, \Delta z)$.

(i) The **increment** of f, denoted by $\Delta \mathbf{f}$, is defined by

$$\Delta f = f(\mathbf{x} + \Delta \mathbf{x}) - f(\mathbf{x}). \tag{3}$$

(ii) The **total differential** of f, denoted by df, is given by

$$df = \nabla f(\mathbf{x}) \cdot \Delta \mathbf{x}. \ \blacksquare \tag{4}$$

Note that equation (4) is very similar in form to equation (2).

REMARK 1: If f is a function of two variables, then (3) and (4) become

$$\Delta f = f(x + \Delta x, y + \Delta y) - f(x, y), \tag{5}$$

and the total differential is

$$df = f_x(x, y)\,\Delta x + f_y(x, y)\,\Delta y. \tag{6}$$

REMARK 2: If f is a function of three variables, then (3) and (4) become

$$\Delta f = f(x + \Delta x, y + \Delta y, z + \Delta z) - f(x, y, z) \tag{7}$$

and

$$df = f_x(x, y, z)\,\Delta x + f_y(x, y, z)\,\Delta y + f_z(x, y, z)\,\Delta z. \tag{8}$$

REMARK 3: Note that in the definition of the total differential, it is *not* required that $|\Delta \mathbf{x}|$ be small.

From Theorems 13.5.1 and 13.5.1′ and the definition of differentiability, we see that if $|\Delta \mathbf{x}|$ is small and if f is differentiable, then

$$\Delta f \approx df. \tag{9}$$

We can use the relation (9) to approximate functions of several variables in much the same way that we used the relation (1) to approximate the values of functions of one variable.

EXAMPLE 1
USING THE TOTAL DIFFERENTIAL
TO APPROXIMATE VOLUME

The radius of a cone is measured to be 15 cm and the height of the cone is measured to be 25 cm. There is a maximum error of ± 0.02 cm in the measurement of the radius and ± 0.05 cm in the measurement of the height.

(a) What is the approximate volume of the cone?

(b) What is the maximum error in the calculation of the volume?

SOLUTION:

(a) $V = \frac{1}{3}\pi r^2 h \approx \frac{1}{3}\pi(15)^2 25 = 1875\pi$ cm^3 ≈ 5890.5 cm^3.

(b) $\nabla V = V_r \mathbf{i} + V_h \mathbf{j} = \frac{2}{3}\pi rh\mathbf{i} + \frac{1}{3}\pi r^2 \mathbf{j} = \pi(250\mathbf{i} + 75\mathbf{j})$. Then, choosing $\Delta r = 0.02$ and $\Delta h = 0.05$ to find the maximum error, we have

$$\Delta V \approx dV = \nabla V \cdot \Delta \mathbf{x} = \pi[250(0.02) + 75(0.05)]$$
$$= \pi(5 + 3.75) = 8.75\pi \approx 27.5 \text{ cm}^3.$$

Thus, the maximum error in the calculation is, approximately, 27.5 cm^3, which means that

$$5890.5 - 27.5 < V < 5890.5 + 27.5, \qquad \text{or}$$

$$5863 \text{ cm}^3 < V < 5918 \text{ cm}^3.$$

Note that an error of 27.5 cm^3 is only a *relative error* of $27.5/5890.5 \approx 0.0047 < \frac{1}{2}\%$, which is a very small relative error.

■ PROBLEMS 13.9

■ SELF-QUIZ

I. Suppose f is a differentiable function of $\mathbf{x} = (x, y)$ and consider the plane tangent to the surface $z = f(\mathbf{x})$ at the point $\mathbf{x} = \mathbf{a}$. True–False:
   a. $f(\mathbf{a}) + \nabla f(\mathbf{a}) \cdot \Delta \mathbf{x}$ is the z-value on that tangent plane which corresponds to $\mathbf{x} = \mathbf{a} + \Delta \mathbf{x}$.
   b. Approximating $f(\mathbf{a} + \Delta \mathbf{x})$ by $f(\mathbf{a}) + \nabla f(\mathbf{a}) \cdot \Delta \mathbf{x}$ is reasonable when $|\Delta \mathbf{x}|$ is small enough for the tangent plane to be close to the surface.

II. If $f(x, y) = x^2 + \sin y$, then $df(x, y)(\Delta \mathbf{x}) = $ _____.
   a. $2x\,\Delta x + \cos y\,\Delta y$
   b. $x^2\,\Delta x + \sin y\,\Delta y$
   c. $(x^3/3)\,\Delta x - \cos y\,\Delta y$
   d. $2x\mathbf{i} + \cos y\mathbf{j}$

III. We can approximate $f(x, y) = x^2 + \sin y$ in the neighborhood of $\mathbf{a} = (-1, \pi/3)$ by $f(-1, \pi/3) + df(\Delta \mathbf{x}) = 1 + \sqrt{3}/2 + df(\Delta \mathbf{x})$ where $df(\Delta \mathbf{x}) = $ _____.
   a. $-2\,\Delta x + \frac{1}{2}\Delta y$
   b. $\Delta x + \frac{1}{2}\Delta y$
   c. $-\frac{1}{3}\Delta x - (\sqrt{3}/2)\,\Delta y$
   d. $-2\mathbf{i} + (\sqrt{3}/2)\mathbf{j}$

IV. We can approximate $g(x, y) = y^2 \sin x$ in the neighborhood of $\mathbf{a} = (-\pi/4, 3)$ by $dg(\Delta \mathbf{x}) = $ _____.
   a. $\cos x\,\Delta x + 2y\,\Delta y$
   b. $(1/\sqrt{2})\,\Delta x + 6\,\Delta y$
   c. $y^2 \cos x\,\Delta x + 2y \sin x\,\Delta y$
   d. $(9/\sqrt{2})\,\Delta x + (-6/\sqrt{2})\,\Delta y$

■ DRILL

In Problems 1–10, calculate the total differential df.

1. $f(x, y) = xy^3$
2. $f(x, y) = \tan^{-1}(y/x)$
3. $f(x, y) = \sqrt{\dfrac{x - y}{x + y}}$
4. $f(x, y) = xe^y$

CONTINUED

5. $f(x, y) = \ln(2x + 3y)$
6. $f(x, y) = \sin(x - 4y)$
7. $f(x, y, z) = xy^2z^5$
8. $f(x, y, z) = xy/z$
9. $f(x, y, z) = \cosh(xy - z)$

10. $f(x, y, z) = \dfrac{x - z}{y + 3x}$

11. Let $f(x, y) = xy^2$.
    a. Calculate explicitly the difference $\Delta f - df$.
    b. Use the result of (a) to find $\Delta f - df$ at the point $(1, 2)$ where $\Delta x = -0.01$ and $\Delta y = 0.03$.
*12. Repeat the steps of the preceding problem for the function $f(x, y) = x^3y^2$.

■ APPLICATIONS

13. The radius and height of a cylinder are approximately 10 cm and 20 cm, respectively. The maximum errors in approximation are ± 0.03 cm and ± 0.07 cm.
    a. What is the approximate volume of the cylinder?
    b. What is the approximate maximum error in your calculation?
14. Two sides of a triangular piece of land were measured to be 50 m and 110 m. There was an error of at most ± 0.3 m in each measurement. The angle between the sides was measured to be $60°$ with an error of at most $\pm 1°$.
    a. Using the law of cosines, find the approximate length of the third side of the triangle.
    b. What is the approximate maximum error of your measurement? [*Hint:* Convert to radians.]
15. When three resistors are connected in parallel, the total resistance R (measured in ohms, Ω) is

given by

$$\frac{1}{R} = \frac{1}{R_1} + \frac{1}{R_2} + \frac{1}{R_3},$$

where R_1, R_2, and R_3 are the three separate resistances. Suppose $R_1 = 6 \pm 0.1 \ \Omega$, $R_2 = 8 \pm 0.03 \ \Omega$, and $R_3 = 12 \pm 0.15 \ \Omega$.
    a. Approximate R.
    b. Find an approximate value for the maximum error in your approximation of R.
16. The volume of 10 moles of an ideal gas was calculated to be 500 cm^3 at a temperature of $40°C$ ($= 313°K$). The maximum error in each measurement n, V, and T was $\frac{1}{2}\%$.
    a. Calculate the approximate pressure of the gas (in Newtons per square cm).
    b. Find the approximate maximum error in your computation. [*Hint:* Recall that according to the ideal-gas law, $PV = nRT$.]

■ ANSWERS TO SELF-QUIZ

I. a. True   b. True
II. a   III. a   IV. d

13.10 MAXIMA AND MINIMA FOR A FUNCTION OF TWO VARIABLES

In Section 3.5, we discussed methods for obtaining maximum and minimum values for a function of one variable. The first basic fact we used was that if f is continuous on a closed bounded interval $[a, b]$, then f takes on a maximum and minimum value in $[a, b]$. We then defined a critical point to be a number x at which $f'(x) = 0$ or for which $f(x)$ exists but $f'(x)$ does not. Finally, we defined local maxima and minima (which occur at critical points) and gave conditions on first and second derivatives that ensured that a critical point was a local maximum or minimum.

The theory for functions of two variables is more complicated, but some of the basic ideas are the same. The first thing we need to know is that a

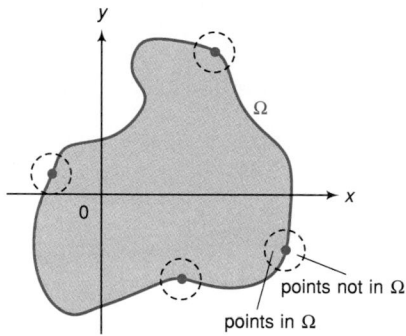

FIGURE 1
A REGION *R* SHOWING FOUR OF ITS
BOUNDARY POINTS

*R is closed because it contains all its
boundary points.*

function *has* a maximum or minimum. In order to state our basic result, we must say what we mean by a closed, bounded region in $\mathbb{R}^2$.

Let Ω be a region in the plane (we will say a great deal more about regions in $\mathbb{R}^2$ in Section 15.1). A point **P** in $\mathbb{R}^2$ is a **boundary point** of Ω if every open disk, D_r (see page 843) centered at **P** contains points that are in Ω and points that are not in Ω. The region is **closed** if Ω contains all its boundary points. A closed region is illustrated in Figure 1. Examples of closed regions in $\mathbb{R}^2$ include closed disks (see page 843), triangles and the points inside them, and rectangles and the points inside them (see Figure 2).

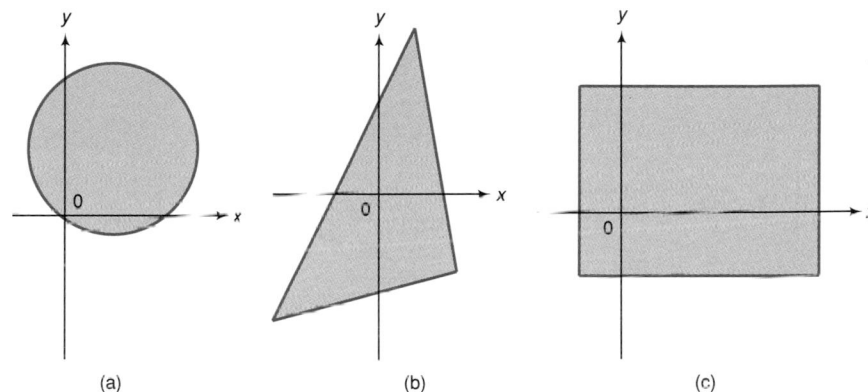

FIGURE 2 THREE CLOSED, BOUNDED REGIONS IN $\mathbb{R}^2$

A region Ω in $\mathbb{R}^2$ is **bounded** if there is an open disk that completely surrounds it. It doesn't matter how big the disk is, as long as one can be found. The regions in Figures 1 and 2 are all bounded.

A region is called **unbounded** if it is not bounded. The first quadrant, for example, is an unbounded region. The proof of the following theorem can be found in any advanced calculus book.[†]

THEOREM 1 **UPPER AND LOWER BOUND THEOREM IN** $\mathbb{R}^2$

Let f be a function of two variables that is continuous on the closed, bounded region Ω. Then there exist points $\mathbf{x}_1$ and $\mathbf{x}_2$ in Ω such that

$$m = f(\mathbf{x}_1) \leq f(\mathbf{x}) \leq f(\mathbf{x}_2) = M \tag{1}$$

for any other point **x** in Ω. The numbers m and M are called, respectively, the **global minimum** and the **global maximum** for the function f over the region Ω.[‡] ∎

We now define what we mean by local minima and maxima.

[†] See, e.g., R. C. Buck and E. F. Buck, *Advanced Calculus*, 3rd ed. (New York: McGraw-Hill, 1978), p. 91.
[‡] The first rigorous proof of this theorem was given by the German mathematician Karl Weierstrass (1815–1897). See the biographical sketch on page 901.

DEFINITION   **LOCAL MAXIMUM AND LOCAL MINIMUM**

Let f be defined in a neighborhood of a point $\mathbf{x}_0 = (x_0, y_0)$.

(i) f has a **local** or **relative maximum** at $\mathbf{x}_0$ if there is a neighborhood N_1 of $\mathbf{x}_0$ such that for every point $\mathbf{x}$ in N_1,

$$f(\mathbf{x}) \le f(\mathbf{x}_0). \tag{2}$$

(ii) f has a **local** or **relative minimum** at $\mathbf{x}_0$ if there is a neighborhood N_2 of $\mathbf{x}_0$ such that for every point $\mathbf{x}$ in N_2,

$$f(\mathbf{x}) \ge f(\mathbf{x}_0). \tag{3}$$

(iii) If f has a local maximum or minimum at $\mathbf{x}_0$, then $\mathbf{x}_0$ is called an **extreme point** of f. ∎

NOTE: Compare this definition with the definitions on page 190.

A rough sketch of a function with several local extreme points is given in Figure 3. The following theorem is a natural generalization of Theorem 3.2.2.

THEOREM 2   **THE GRADIENT IS ZERO AT AN EXTREME POINT**

Let f be differentiable at $\mathbf{x}_0$ and suppose that $\mathbf{x}_0$ is an extreme point of f. Then

$$\nabla f(\mathbf{x}_0) = \mathbf{0}. \tag{4}$$

PROOF: We must show that $f_x(x_0, y_0)$ and $f_y(x_0, y_0) = 0$. To prove that $f_x(x_0, y_0) = 0$, define a function h by

$$h(x) = f(x, y_0).$$

Then $h(x_0) = f(x_0, y_0)$ and $h'(x_0) = f_x(x_0, y_0)$. But since f has a local maximum or (minimum) at (x_0, y_0), h has a local maximum (or minimum) at x_0, so that by Theorem 3.2.2, $h'(x_0) = 0$. Thus $f_x(x_0, y_0) = 0$. In a very similar way we can show that $f_y(x_0, y_0) = 0$ by defining $h(y) = f(x_0, y)$, and the theorem is proved. ∎

DEFINITION   **CRITICAL POINT**

$\mathbf{x}_0$ is a **critical point** of f if either (a) f is differentiable at $\mathbf{x}_0$ and $\nabla f(\mathbf{x}_0) = \mathbf{0}$ or (b) $\mathbf{x}_0$ is in the domain of f but $\nabla f(\mathbf{x}_0)$ does not exist. ∎

REMARK 1:   The definitions and Theorems 1 and 2 hold for functions of three variables with very little change.

REMARK 2:   By the remark on page 883, we see that if $\mathbf{x}_0$ is a critical point of f and if $\nabla f(\mathbf{x}_0)$ exists, then the tangent plane to the surface $z = f(x, y)$ is horizontal (i.e., parallel to the xy-plane).

As in the case of a function of one variable, the fact that $\mathbf{x}_0$ is a critical point of f does not guarantee that $\mathbf{x}_0$ is an extreme point of f, as Example 3 below illustrates.

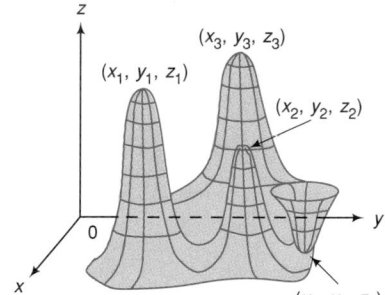

FIGURE 3
A FUNCTION WITH THREE LOCAL
MAXIMA AND ONE LOCAL MINIMUM

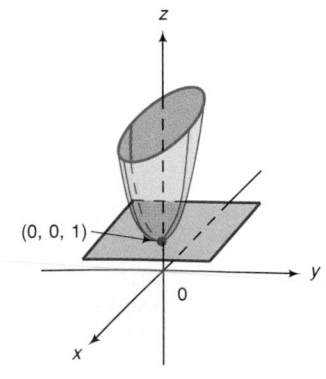

FIGURE 4
THE SURFACE $z = 1 + x^2 + 3y^2$

EXAMPLE 1
A FUNCTION WITH A LOCAL AND GLOBAL MINIMUM AT THE CRITICAL POINT (0, 0)

Let $f(x, y) = 1 + x^2 + 3y^2$. Then $\nabla f = 2x\mathbf{i} + 6y\mathbf{j}$, which is zero only when $(x, y) = (0, 0)$. Thus, $(0, 0)$ is the only critical point, and it is clearly a local (and global) minimum. This is illustrated in Figure 4.

EXAMPLE 2
A FUNCTION WITH A LOCAL AND GLOBAL MAXIMUM AT THE CRITICAL POINT (0, 0)

Let $f(x, y) = 1 - x^2 - 3y^2$. Then $\nabla f = -2x\mathbf{i} - 6y\mathbf{j}$, which is zero only at the origin. In this case $(0, 0)$ is a local (and global) maximum for f.

EXAMPLE 3
A FUNCTION FOR WHICH (0, 0) IS A CRITICAL POINT BUT IS NEITHER A LOCAL MAXIMUM NOR A LOCAL MINIMUM

Let $f(x, y) = y^2 - x^2$. Then $\nabla f = -2x\mathbf{i} + 2y\mathbf{j}$, which again is zero only at $(0, 0)$. But $(0, 0)$ is *neither* a local maximum nor a local minimum for f. To see this, we simply note that f can take positive and negative values in any neighborhood of $(0, 0)$ since $f(x, y) > 0$ if $|x| < |y|$ and $f(x, y) < 0$ if $|x| > |y|$. This is illustrated by Figure 5. The figure sketched in Figure 5 is called a **hyperbolic paraboloid**, or **saddle surface**.

We have the following definition.

> **DEFINITION** **SADDLE POINT**
>
> If $\mathbf{x}_0$ is a critical point of f but is not an extreme point of f, then $\mathbf{x}_0$ is called a **saddle point** of f. ■

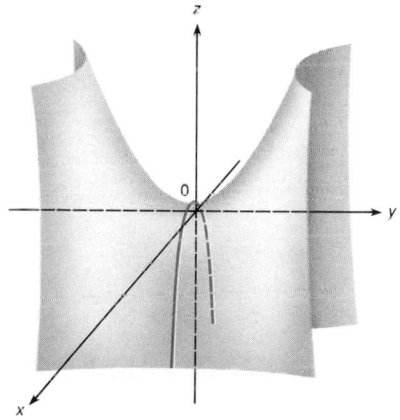

FIGURE 5 SADDLE SURFACE $z = y^2 - x^2$

The three examples above illustrate that a critical point of f may be a local maximum, a local minimum, or a saddle point.

EXAMPLE 4
A FUNCTION MAY HAVE A LOCAL MINIMUM AT A POINT WHERE IT IS NOT DIFFERENTIABLE

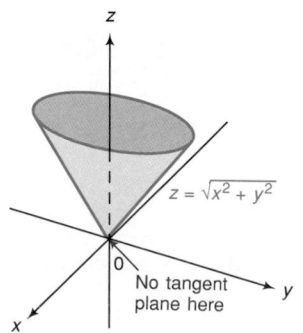

FIGURE 6
THE FUNCTION $z = \sqrt{x^2 + y^2}$ IS NOT DIFFERENTIABLE AT (0, 0) ALTHOUGH IT HAS A LOCAL (AND GLOBAL) MINIMUM THERE.

Let $f(x, y) = \sqrt{x^2 + y^2}$. Then it is obvious that f has a local (and global) minimum at $(0, 0)$. But

$$\nabla f = \frac{x}{\sqrt{x^2 + y^2}}\,\mathbf{i} + \frac{y}{\sqrt{x^2 + y^2}}\,\mathbf{j},$$

so that $\nabla f(0, 0)$ *does not exist* even though $(0, 0)$ is a critical point of f (case (b) in the definition of a critical point). We see this clearly in Figure 6. The graph of f (which is a cone) does not have a tangent plane at $(0, 0)$.

We generalize the result of Example 4: If $f(x_0, y_0)$ exists but $\nabla f(x_0, y_0)$ does not exist, then f *may* have a local maximum or minimum at (x_0, y_0).

Examples 1, 2, and 3 indicate that more is needed to determine whether a critical point is an extreme point (in most cases it will not be at all obvious). You might suspect that the answer has something to do with the signs of the second partial derivatives of f, as in the case of functions of one variable.

The following theorem gives the answer. Although we could give its proof by using results we already know, the proof is very tedious, so we will omit it. Its proof can be found (as usual) in many more advanced calculus texts.[†]

THEOREM 3 **SECOND-DERIVATIVES TEST**

Let f and all its first and second partial derivatives be continuous in a neighborhood of (x_0, y_0). Suppose that $f_x(x_0, y_0) = 0$ and $f_y(x_0, y_0) = 0$. That is, (x_0, y_0) is a critical point of f. Let

$$D(x, y) = f_{xx}(x, y)f_{yy}(x, y) - [f_{xy}(x, y)]^2 \tag{5}$$

and let D denote $D(x_0, y_0)$. D is called the **discriminant** of f.

  (i) If $D > 0$ and $f_{xx}(x_0, y_0) > 0$, then f has a local minimum at (x_0, y_0).
 (ii) If $D > 0$ and $f_{xx}(x_0, y_0) < 0$, then f has a local maximum at (x_0, y_0).
(iii) If $D < 0$, then (x_0, y_0) is a saddle point of f.
(iv) If $D = 0$, then any of the preceding is possible. ∎

REMARK 1: Using a 2×2 determinant, we can write (5) as

$$D = \begin{vmatrix} f_{xx} & f_{xy} \\ f_{yx} & f_{yy} \end{vmatrix}.$$

REMARK 2: This theorem does not provide the full answer to the questions of what are the extreme points of f in $\mathbb{R}^2$. It is still necessary to check points at which ∇f does not exist (as in Example 4).

EXAMPLE 5
USING THE SECOND-DERIVATIVES TEST TO PROVE THAT A FUNCTION HAS A LOCAL MINIMUM AT (0, 0)

Let $f(x, y) = 1 + x^2 + 3y^2$. Then as we saw in Example 1, $(0, 0)$ is the only critical point of f. But $f_{xx} = 2$, $f_{yy} = 6$, and $f_{xy} = 0$, so $D(0, 0) = 12$ and $f_{xx} > 0$, which *proves* that f has a local minimum at $(0, 0)$.

EXAMPLE 6
USING THE SECOND-DERIVATIVES TEST TO DETERMINE THE NATURE OF THE CRITICAL POINTS

Let $f(x, y) = -x^2 - y^2 + 2x + 4y + 5$. Determine the nature of the critical points of f.

SOLUTION: $\nabla f = (-2x + 2)\mathbf{i} + (-2y + 4)\mathbf{j}$. We see that $\nabla f = \mathbf{0}$ when

$$-2x + 2 = 0 \quad \text{and} \quad -2y + 4 = 0,$$

which occurs only at the point $(1, 2)$. Also $f_{xx} = -2$, $f_{yy} = -2$, and $f_{xy} = 0$, so $D = 4$ and $f_{xx} < 0$, which implies that there is a local maximum at $(1, 2)$. At $(1, 2)$, $f = 10$.

[†] See S. Grossman, *Multivariable Calculus, Linear Algebra, and Differential Equations*, 2nd ed. (San Diego: Harcourt Brace Jovanovich, 1986), p. 584.

EXAMPLE 7
USING THE SECOND-DERIVATIVES TEST TO DETERMINE THE NATURE OF THE CRITICAL POINTS

Let $f(x, y) = 2x^3 - 24xy + 16y^3$. Determine the nature of the critical points of f.

SOLUTION: We have $\nabla f(x, y) = (6x^2 - 24y)\mathbf{i} + (-24x + 48y^2)\mathbf{j}$. If $\nabla f = \mathbf{0}$, we have

$$6(x^2 - 4y) = 0 \quad \text{and} \quad -24(x - 2y^2) = 0.$$

Clearly, one critical point is $(0, 0)$. To obtain another, we must solve the simultaneous equations

$$x^2 - 4y = 0$$
$$x - 2y^2 = 0.$$

The second equation tells us that $x = 2y^2$. Substituting this expression into the first equation yields

$$4y^4 - 4y = 0,$$

with solutions $y = 0$ and $y = 1$, giving us the critical points $(0, 0)$ and $(2, 1)$. Now $f_{xx} = 12x$, $f_{yy} = 96y$, and $f_{xy} = -24$, so

$$D(x, y) = (12x)(96y) - 24^2 = 1152xy - 576.$$

Then $D(0, 0) = -576$ and $D(2, 1) = 1728$. We find that $(0, 0)$ is a saddle point, and since $f_{xx}(2, 1) = 24 > 0$, $(2, 1)$ is a local minimum.

In Problems 32–34, you are asked to show that any of the three cases can occur when $D = 0$. That is, you will show that the second-derivatives test *fails* when $D = 0$.

EXAMPLE 8
REGRESSION LINES

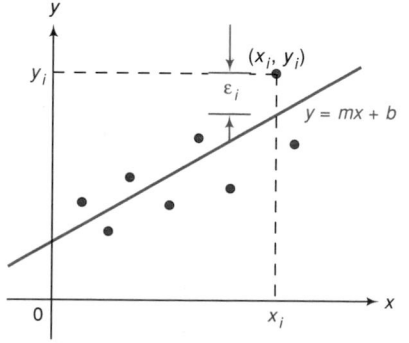

FIGURE 7
THE REGRESSION LINE $y = mx + b$

ϵ_i *is the difference between the actual y-value corresponding to $x_i (= y_i)$ and the y-value corresponding to x_i that lies on the approximating line ($= mx_i + b$).*

We can use the theory of this section in an interesting way to derive a result that is very useful for statistical analysis and, in fact, any analysis involving the use of a great deal of data. Suppose n data points (x_1, y_1), $(x_2, y_2), \ldots, (x_n, y_n)$ are collected. For example, the x's may represent average tree growth and the y's average daily temperature in a given year in a certain forest. Or x may represent a week's sales and y a week's profit for a certain business. The question arises as to whether we can "fit" these data points to a straight line. That is, is there a straight line that runs "more or less" through the points? If so, then we can write y as a linear function of x, with obvious computational advantages.

The problem is to find the "best" straight line, $y = mx + b$, passing through or near these points. Look at Figure 7. If (x_i, y_i) is one of our n points, then, on the line $y = mx + b$, corresponding to x_i we obtain $y_i = mx_i + b$. The "error," ϵ_i, between the y value of our actual point and the "approximating" value on the line is given by

$$\epsilon_i = y - mx_i - b.$$

One way to choose the approximating line is to use the line that minimizes the sum of the squares of the errors. This is called the **least-squares** criterion for choosing the line.

Now we want to choose m and b such that the function

$$f(m, b) = \epsilon_1{}^2 + \epsilon_2{}^2 + \cdots + \epsilon_n{}^2 = \sum \epsilon_i{}^2 = \sum (y_i - mx_i - b)^2$$

is a minimum. To do this, we calculate

$$\frac{\partial}{\partial m} (y_i - mx_i - b)^2 = -2x_i(y_i - mx_i - b)$$

and

$$\frac{\partial}{\partial b} (y_i - mx_i - b)^2 = -2(y_i - mx_i - b).$$

Hence,

$$\frac{\partial f}{\partial m} = -2\sum x_i(y_i - mx_i - b)$$

and

$$\frac{\partial f}{\partial b} = -2\sum (y_i - mx_i - b).$$

Setting $\partial f/\partial m = 0$ and $\partial f/\partial b = 0$ and rearranging terms, we obtain

$$\sum (x_i y_i - mx_i{}^2 - bx_i) = 0$$
$$\sum (y_i - mx_i - b) = 0.$$

This leads to the system of two equations in the unknowns m and b:

$$\left(\sum x_i{}^2\right)m + \left(\sum x_i\right)b = \sum x_i y_i \tag{6}$$

and

$$\left(\sum x_i\right)m + nb = \sum y_i. \tag{7}$$

Here we have used the fact that $\sum b = nb$. The system (6) and (7) is not hard to solve for m and b. To do so, we multiply both sides of (6) by n and both sides of (7) by $\sum x_i$ and then subtract to finally obtain

$$m = \frac{n\sum x_i y_i - \left[\sum x_i\right]\left[\sum y_i\right]}{n\sum x_i{}^2 - \left[\sum x_i\right]^2} \tag{8}$$

and

$$b = \frac{\left[\sum x_i{}^2\right]\left[\sum y_i\right] - \left[\sum x_i\right]\left[\sum x_i y_i\right]}{n\sum x_i{}^2 - \left[\sum x_i\right]^2}. \tag{9}$$

We will leave it to you to check that the numbers m and b given in (8) and (9) do indeed provide a local (and global) minimum. The line $y = mx + b$ given by (8) and (9) is called the **regression line** for the n points.

TABLE 1

i	x_i	y_i	x_i^2	$x_i y_i$
1	1	2	1	2
2	2	4	4	8
3	5	5	25	25
$\sum$	8	11	30	35

REMARK: Equations (8) and (9) make sense only if

$$n\sum x_i^2 - (\sum x_i^2) \neq 0.$$

But in fact,

$$n\sum x_i^2 - (\sum x_i)^2 \geq 0$$

and is equal to zero only when all the x_i's are equal (in which case the regression line is the vertical line $x = x_i$). We will not prove this fact.

EXAMPLE 9
FINDING A REGRESSION LINE

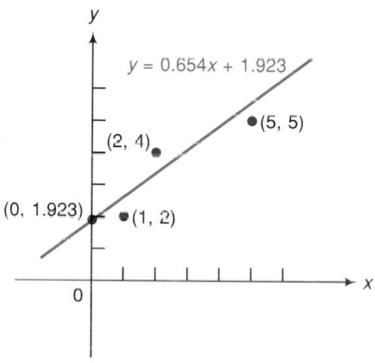

FIGURE 8
THE REGRESSION LINE FOR THE POINTS (1, 2), (2, 4), AND (5, 5)

Find the regression line for the points (1, 2), (2, 4), and (5, 5).

SOLUTION: We tabulate some appropriate values in Table 1. Then, from (8) and (9),

$$m = \frac{3(35) - (8)(11)}{3(30) - 8^2} = \frac{17}{26} \approx 0.654$$

and

$$b = \frac{(30)(11) - 8(35)}{26} = \frac{50}{26} \approx 1.923.$$

Thus the regression line is

$$y = 0.654x + 1.923$$

This is all illustrated in Figure 8.

THE FIRST RIGOROUS PROOF OF THE upper and lower bound theorem in $\mathbb{R}^2$ was given by the German mathematician Karl Weierstrass (1815–1897). Weierstrass began his academic career by studying law and finance and then became a secondary-school teacher. He did not devote himself to mathematics until the age of 40. He was able to work at mathematics full time only after being appointed to a professorship at the University of Berlin in 1864—at the age of 49. Nevertheless, he made significant contributions to many branches of mathematics.

Weierstrass was the first to prove a great number of results in the theory of functions of two or more variables. However his best-known work is his development of the theory of complex functions using power series. Many of the important results in complex function theory today bear his name.

KARL WEIERSTRASS
·—·
1815–1897

Weierstrass was a very influential teacher. Many of his mathematical discoveries were published by his students, who often got credit for his ideas. One of his most famous students was the Russian mathematician Sonya Kovalevsky (1850–1891). Kovalevsky applied Weierstrass's ideas to solve problems in mathematical physics and other areas. She was one of the first women to gain international prominence as a research mathematician.

Weierstrass developed new standards of mathematical rigor. Using "Weierstrassian analysis," most of what we today term "mathematical analysis" can be logically derived from a set of axioms that characterize the real-number system.

■ PROBLEMS 13.10

■ SELF-QUIZ

I. The second-derivatives test for extreme points of a differentiable function $f(x, y)$ includes examination of the discriminant $D = $ _____.
   a. $f_{xx}f_{xy} - f_{yy}f_{yx}$
   b. $f_{xx}{}^2 - f_{yy}{}^2$
   c. $f_{xx}f_{yx} - f_{yy}f_{xy}$
   d. $f_{xx}f_{yy} - f_{xy}f_{yx}$

II. True–False: In searching for extreme points of a function $f(x, y)$, only the various second derivatives of f are examined.

III. Let $f(x, y) = x^2 + y^2 + 4x - 2y$. Then $f_x(-2, 1) = 2(-2) + 4 = 0$, $f_y(-2, 1) = 2(1) - 2 = 0$, $f_{xx}(-2, 1) = 2 > 0$, $f_{xy}(-2, 1) = 0 = f_{yx}(-2, 1)$, $f_{yy}(-2, 1) = 2$, and $D(-2, 1) = (2)(2) - (0)(0) = 4 > 0$. Therefore, $(-2, 1)$ is _____ of the function f.

   a. a local minimum
   b. a local maximum
   c. not a critical point
   d. a saddle point

IV. Let $f(x, y) = x^3 + xy^2 + y^4$; then $f_x(1, -1) = 3 + 1 = 4 \neq 0$, $f_y(1, -1) = -2 - 4 = -6 \neq 0$, $f_{xx}(1, -1) = 6 > 0$, $f_{xy}(1, -1) = -2 = f_{yx}(1, -1)$, $f_{yy}(1, -1) = 14$, and $D(1, -1) = 6(14) - (-2)^2 = 80 > 0$. Therefore $(1, -1)$ is _____ of the function f.

   a. a local maximum
   b. a local minimum
   c. not a critical point
   d. a saddle point

■ DRILL

In Problems 1–18, find the critical points of the given function and determine their nature.

1. $f(x, y) = 7x^2 - 8xy + 3y^2 + 1$
2. $f(x, y) = x^2 + y^3 - 3xy$
3. $f(x, y) = x^2 + 3y^2 + 4x - 6y + 3$
4. $f(x, y) = x^2 + y^2 + 4xy + 6y - 3$
5. $f(x, y) = x^2 + y^2 + 4x - 2y + 3$
6. $f(x, y) = xy^2 + x^2y - 3xy$
7. $f(x, y) = x^3 + 3xy^2 + 3y^2 - 15x + 2$
8. $f(x, y) = x^3 + y^3 - 3xy$
9. $f(x, y) = \dfrac{1}{y} - \dfrac{1}{x} - 4x + y$

10. $f(x, y) = \dfrac{1}{x} + \dfrac{2}{y} + 2x + y + 1$
11. $f(x, y) = x^2 - xy + y^2 + 2x + 2y$
*12. $f(x, y) = xy + \dfrac{8}{x} + \dfrac{1}{y}$
13. $f(x, y) = (4 - x - y)xy$
*14. $f(x, y) = \sin x + \sin y + \sin(x + y)$
15. $f(x, y) = 2x^2 + y^2 + \dfrac{2}{x^2y}$
16. $f(x, y) = 4x^2 + 12xy + 9y^2 + 25$
17. $f(x, y) = x^{25} - y^{25}$
18. $f(x, y) = \tan xy$

■ APPLICATIONS

19. Find three positive numbers whose sum is 50 and such that their product is a maximum.
20. Find three positive numbers whose product is 50 and whose sum is a minimum.
21. Find three positive numbers x, y, and z whose sum is 50 such that the product xy^2z^3 is a maximum.
22. Find three numbers whose sum is 50 and the sum of whose squares is a minimum.
23. Use the methods of this section to find the minimum distance from the point $(1, -1, 2)$ to the plane $x + 2y - z = 4$. [Hint: Express the distance between $(1, -1, 2)$ and a point (x, y, z) on the plane in terms of x and y.]

24. A rectangular wooden box with an open top is to contain α cubic centimeters, where α is a given positive number. Ignoring the thickness of the wood, how is the box to be constructed so as to use the smallest amount of wood? [Hint: $V = xyz = \alpha$. Find a formula for the total area of the sides and bottom and substitute $z = \alpha/xy$ to write this formula in terms of x and y only.]
25. What is the maximum volume of an open-top rectangular box that can be built from β m^2 of wood? (Assume that any board can be cut without waste.)
*26. Find the dimensions of the rectangular box of maximum volume that can be inscribed in the

CONTINUED

ellipsoid

$$\frac{x^2}{a^2} + \frac{y^2}{b^2} + \frac{z^2}{c^2} = 1$$

such that the faces of the box are parallel to the coordinate planes.

27. A company uses two types of raw material, I and II, for its product. If it uses x units of I and y units of II, it can produce U units of the finished product where $U(x, y) = 8xy + 32x + 40y - 4x^2 - 6y^2$. Each unit of I costs \$10 and each unit of II costs \$4; each unit of the product can be sold for \$40. How can the company maximize its profits?

28. A major oil company sells both oil and natural gas. Each unit of oil costs the company \$25, and each unit of gas costs \$15. The revenue (in dollars) received from selling i units of oil and g units of gas is given by

$$R(i, g) = 60i + 50g - 0.02ig - 0.3i^2 - 0.2g^2.$$

a. How many units of each should the company sell in order to maximize profits?
b. What is the maximum profit?

29. A manufacturer is developing a new electronic garage-door opener. She hopes to build the device for \$100 and sell it for \$250. After considerable market research, it is determined that the number of units that can be sold at that price depends on the amount of money spent on advertising (a) and the amount spent on product development (d). The relationship is given by

$$N(a, d) = \text{number of units sold}$$
$$= \frac{300a}{a + 3} + \frac{160d}{d + 5}.$$

a. Write down a function that gives profit P as a function of a and d.
b. Explain why there is no expenditure on advertising and product development that will maximize profit. Is this a realistic model?

30. Find the regression line for the points $(1, 1)$, $(2, 3)$, and $(3, 6)$. Sketch the line and the points.

31. Find the regression line for the points $(-1, 3)$, $(1, 2)$, $(2, 0)$, and $(4, -2)$. Sketch the line and the points.

■ SHOW/PROVE/DISPROVE

32. Let $f(x, y) = 4x^2 - 4xy + y^2 + 5$. Show that there are an infinite number of critical points, that $D = 0$ at each one, and that each is a global (and local) minimum.

33. Let $f(x, y) = -(4x^2 - 4xy + y^2) + 5$. Show that there are an infinite number of critical points, that $D = 0$ at each one, and that each is a global (and local) maximum.

34. Let $f(x, y) = x^3 - y^3$. Show that $D = 0$ at the only critical point and that this point is a saddle point.

REMARK: Problems 32, 33, and 34 illustrate that any of the three cases can occur when $D = 0$.

35. Show that the rectangular box inscribed in a sphere that encloses the greatest volume is a cube.

*36. Show that the function

$$e^{(x^2 + y^2)/y} + \frac{11}{x} + \frac{5}{y} + \sin x^2 y$$

has a global minimum in the first quadrant.

■ ANSWERS TO SELF-QUIZ

I. d

II. False (First derivatives are examined in search for critical points.)

III. a

IV. c

13.11 CONSTRAINED MAXIMA AND MINIMA—LAGRANGE MULTIPLIERS

In the previous section, we saw how to find the maximum and minimum of a function of two variables by taking gradients and applying a second-derivative test. It often happens that there are side conditions (or **constraints**) attached to a problem. For example, we have been asked to find the shortest distance from a point (x_0, y_0) to a line $y = mx + b$. We could write this problem as follows:

Minimize the function: $z = \sqrt{(x - x_0)^2 + (y - y_0)^2}$

subject to the constraint: $y - mx - b = 0.$

As another example, suppose that a region of space containing a sphere is heated and a function $w = T(x, y, z)$ gives the temperature of every point of the region. Then if the sphere is given by $x^2 + y^2 + z^2 = r^2$, and if we wish to find the hottest point on the sphere, we have the following problem:

Maximize: $w = T(x, y, z)$

subject to the constraint: $x^2 + y^2 + z^2 - r^2 = 0.$

We now generalize these two examples. Let f and g be functions of two variables. Then we can formulate a **constrained maximization** (or **minimization**) problem as follows:

Maximize (or minimize): $z = f(x, y)$ (1)

subject to the constraint: $g(x, y) = 0.$ (2)

If f and g are functions of three variables, we have the following problem:

Maximize (or minimize): $w = f(x, y, z)$ (3)

subject to the constraint: $g(x, y, z) = 0.$ (4)

We now develop a method for dealing with problems of the type (1), (2), or (3), (4). Let C be a curve in $\mathbb{R}^2$ or $\mathbb{R}^3$ given parametrically by the differentiable function $\mathbf{F}(t)$. That is, C is given by

$\mathbf{F}(t) = x(t)\mathbf{i} + y(t)\mathbf{j}$ (in $\mathbb{R}^2$)

or

$\mathbf{F}(t) = x(t)\mathbf{i} + y(t)\mathbf{j} + z(t)\mathbf{k}$ (in $\mathbb{R}^3$).

Let $f(\mathbf{x})$ denote the function (of two or three variables) that is to be maximized.

THEOREM 1

Suppose that f is differentiable at a point $\mathbf{x}_0$ and that among all points on a differentiable curve C, f takes its maximum (or minimum) value at $\mathbf{x}_0$. Then $\nabla f(\mathbf{x}_0)$ is orthogonal to C at $\mathbf{x}_0$. That is, since $\mathbf{F}'(t)$ is tangent to C, if $\mathbf{x}_0 = \mathbf{F}(t_0)$, then

$\nabla f(\mathbf{x}_0) \cdot \mathbf{F}'(t_0) = 0.$ (5)

PROOF: For $\mathbf{x}$ on C, $\mathbf{x} = \mathbf{F}(t)$, so that the composite function $f(\mathbf{F}(t))$ has a maximum (or minimum) at t_0. Therefore its derivative at t_0 is 0. By the chain rule,

$$\frac{d}{dt} f(\mathbf{F}(t)) = \nabla f(\mathbf{F}(t)) \cdot \mathbf{F}'(t),$$

and at t_0

$$0 = \nabla f(\mathbf{F}(t_0)) \cdot \mathbf{F}'(t_0) = \nabla f(\mathbf{x}_0) \cdot \mathbf{F}'(t_0),$$

and the theorem is proved. ∎

For f as a function of two variables, the result of Theorem 1 is illustrated in Figure 1.

We can use Theorem 1 to make an interesting geometric observation.[†] Suppose that, subject to the constraint $g(x, y) = 0$, f takes its maximum (or minimum) at the point $\mathbf{x}_0 = (x_0, y_0)$. The equation $g(x, y) = 0$ determines a curve C in the xy-plane, and by Theorem 1, $\nabla f(x_0, y_0)$ is orthogonal to C at (x_0, y_0). But from Section 13.7, $\nabla g(x_0, y_0)$ is also orthogonal to C at (x_0, y_0). Thus, we see that

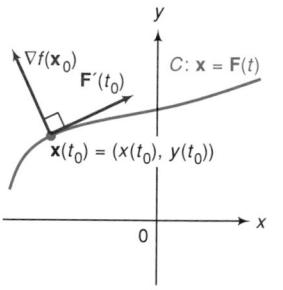

FIGURE 1
THE GRADIENT AT $\mathbf{x}_0$ IS ORTHOGONAL TO ANY TANGENT LINE AT $\mathbf{x}_0$.

LAGRANGE'S OBSERVATION

$\nabla g(x_0, y_0)$ and $\nabla f(x_0, y_0)$ are parallel.

Hence, there is a number λ such that

$$\nabla f(x_0, y_0) = \lambda \nabla g(x_0, y_0).$$

We can extend this observation to the following rule, which applies equally well to functions of three or more variables:

THE BASIC FACT ABOUT LAGRANGE MULTIPLIERS

If, subject to the constraint $g(\mathbf{x}) = 0$, f takes its maximum (or minimum) value at a point $\mathbf{x}_0$, then there is a number λ such that

$$\nabla f(\mathbf{x}_0) = \lambda \nabla g(\mathbf{x}_0). \tag{6}$$

DEFINITION   **LAGRANGE MULTIPLIER**

The number λ is called a **Lagrange multiplier**. ∎

We will illustrate the Lagrange multiplier technique with three examples.

EXAMPLE 1
USING LAGRANGE MULTIPLIERS TO FIND CLOSEST AND FARTHEST POINTS

Find the points on the sphere $x^2 + y^2 + z^2 = 1$ closest to and farthest from the point $(1, 2, 3)$.

SOLUTION: In problems like this, it is almost always easier to work with the square of the distance rather than with the distance itself. The distance from (x, y, z) to $(1, 2, 3)$ is $\sqrt{(x-1)^2 + (y-2)^2 + (z-3)^2}$. The square of the distance is $(x-1)^2 + (y-2)^2 + (z-3)^2$. You should explain why maximum and minimum values of the square of the distance

---

[†] This observation was first made by the French mathematician Joseph-Louis Lagrange (1736–1813). (See the biography of Lagrange on page 624.)

occur at the same points at which the distance is maximized and minimized.

Now, our problem is to minimize and maximize $f(x, y, z) = (x - 1)^2 + (y - 2)^2 + (z - 3)^2$ subject to $g(x, y, z) = x^2 + y^2 + z^2 - 1 = 0$. We have $\nabla f(x, y, z) = 2(x - 1)\mathbf{i} + 2(y - 2)\mathbf{j} + 2(z - 3)\mathbf{k}$, and $\nabla g(x, y, z) = 2x\mathbf{i} + 2y\mathbf{j} + 2z\mathbf{k}$. Condition (6) implies that at a maximizing point, $\nabla f = \lambda \nabla g$, so

$$2(x - 1) = 2x\lambda$$

$$2(y - 2) = 2y\lambda$$

$$2(z - 3) = 2z\lambda.$$

If $\lambda \neq 1$, then we find that

$$x - 1 = x\lambda, \quad \text{or} \quad x - x\lambda = 1, \quad \text{or} \quad x(1 - \lambda) = 1,$$

and

$$x = \frac{1}{1 - \lambda}.$$

Similarly, we obtain

$$y = \frac{2}{1 - \lambda} \quad \text{and} \quad z = \frac{3}{1 - \lambda}.$$

Then

$$1 = x^2 + y^2 + z^2 = \frac{1}{(1 - \lambda)^2} (1^2 + 2^2 + 3^2) = \frac{14}{(1 - \lambda)^2},$$

so that $(1 - \lambda)^2 = 14$, $(1 - \lambda) = \pm\sqrt{14}$, and $\lambda = 1 \pm \sqrt{14}$.

If $\lambda = 1 + \sqrt{14}$, then $(x, y, z) = (-1/\sqrt{14}, -2/\sqrt{14}, -3/\sqrt{14})$. If $\lambda = 1 - \sqrt{14}$, then $(x, y, z) = (1/\sqrt{14}, 2/\sqrt{14}, 3/\sqrt{14})$. Evaluation shows us that f is maximized at $(-1/\sqrt{14}, -2/\sqrt{14}, -3/\sqrt{14})$ and that f is minimized at $(1/\sqrt{14}, 2/\sqrt{14}, 3/\sqrt{14})$. Finally, $f_{\max} = f(-1/\sqrt{14}, -2/\sqrt{14}, -3/\sqrt{14}) \approx 22.48$ and $f_{\min} = f(1/\sqrt{14}, 2/\sqrt{14}, 3/\sqrt{14}) \approx 7.52$.

EXAMPLE 2

MAXIMIZING OUTPUT USING A COBB-DOUGLAS PRODUCTION FUNCTION

In Example 13.1.4 on page 834, we discussed the Cobb-Douglas production function. In particular, we discussed the function

$$F(L, K) = 200L^{2/5}K^{3/5},$$

which represents the number of units of a certain type of die produced if L units of labor and K units of capital are used in production. Suppose that each unit of labor costs \$400 and each unit of capital costs \$500. If \$50,000 is available for production costs, how many units of labor and capital should be used in order to maximize output? How many units will be produced?

SOLUTION: The total cost is

$$400L + 500K.$$

But the total cost is fixed at \$50,000, so we have

$$400L + 500K = 50,000$$

or

$$400L + 500K - 50,000 = 0.$$

Thus, our problem is

Maximize: $F(L, K) = 200L^{2/5}K^{3/5}$

subject to: $G(L, K) = 400L + 500K - 50,000 = 0.$

Now

$$\frac{\partial F}{\partial L} = 80L^{-3/5}K^{3/5}, \qquad \frac{\partial F}{\partial K} = 120L^{2/5}K^{-2/5},$$

$$\frac{\partial G}{\partial L} = 400, \qquad \frac{\partial G}{\partial K} = 500.$$

At a maximizing point,

$$\frac{\partial F}{\partial L} = \lambda \frac{\partial G}{\partial L} \qquad \text{and} \qquad \frac{\partial F}{\partial K} = \lambda \frac{\partial G}{\partial K},$$

so that

$$80L^{-3/5}K^{3/5} = 400\lambda \qquad \text{and} \qquad 120L^{2/5}K^{-2/5} = 500\lambda.$$

Thus

$$\lambda = \frac{80L^{-3/5}K^{3/5}}{400} = \frac{120L^{2/5}K^{-2/5}}{500}$$

and

$$\tfrac{1}{5}L^{-3/5}K^{3/5} = \tfrac{6}{25}L^{2/5}K^{-2/5}.$$

We multiply both sides by $L^{3/5}K^{2/5}$:

$$\frac{1}{5}(\overbrace{L^{-3/5}L^{3/5}}^{1}\overbrace{K^{3/5}K^{2/5}}^{K}) = \frac{6}{25}(\overbrace{L^{2/5}L^{3/5}}^{L}\overbrace{K^{-2/5}K^{2/5}}^{1}),$$

$$\tfrac{1}{5}K = \tfrac{6}{25}L,$$

$$K = \tfrac{30}{25}L = \tfrac{6}{5}L.$$

But

$$400L + 500K = 50,000$$

$$400L + 500(\tfrac{6}{5})L = 50,000$$

$$400L + 600L = 50{,}000$$

$$1000L = 50{,}000$$

$$L = 50$$

$$K = \tfrac{6}{5}L = 60.$$

Thus 50 units of labor and 60 units of capital should be used, and the maximum number of units produced is

$$F(50, 60) = 200(50)^{2/5}(60)^{3/5} \approx 11{,}156 \text{ units.}$$

We can use the result of the last example to make an interesting observation. The quantity $\partial F/\partial L$ is called the **marginal productivity of labor** and $\partial F/\partial K$ is called the **marginal productivity of capital**. At our maximizing values $L = 50$, $K = 60$, we have

$$\frac{\text{marginal productivity of labor}}{\text{marginal productivity of capital}} = \frac{\partial F/\partial L}{\partial F/\partial K} = \frac{80(50)^{-3/5}(60)^{3/5}}{120(50)^{2/5}(60)^{-2/5}}$$

$$= \frac{2}{3}\left[\frac{(60)^{3/5}(60)^{2/5}}{(50)^{2/5}(50)^{3/5}}\right] = \frac{2}{3}\cdot\frac{60}{50} = \frac{2}{3}\cdot\frac{6}{5} = \frac{4}{5}.$$

In addition,

$$\frac{\text{unit cost of labor}}{\text{unit cost of capital}} = \frac{400}{500} = \frac{4}{5}.$$

This is no coincidence! It is a general law of economics that

> when labor and capital investments are such as to maximize production, the ratio of marginal productivity of labor to marginal productivity of capital is equal to the ratio of the unit cost of labor to the unit cost of capital.

We note that we can use Lagrange multipliers in $\mathbb{R}^3$ if there are two or more constraint equations. Suppose, for example, we wish to maximize (or minimize) $w = f(x, y, z)$ subject to the constraints

$$g(x, y, z) = 0 \qquad\qquad\qquad (7)$$

and

$$h(x, y, z) = 0. \qquad\qquad\qquad (8)$$

Each of the equations (7) and (8) represents a surface in $\mathbb{R}^3$, and their intersection forms a curve in $\mathbb{R}^3$. By an argument very similar to the one we used earlier (but applied in $\mathbb{R}^3$ instead of $\mathbb{R}^2$), we find that if f is maximized (or minimized) at (x_0, y_0, z_0), then $\nabla f(x_0, y_0, z_0)$ is in the plane determined by $\nabla g(x_0, y_0, z_0)$ and $\nabla h(x_0, y_0, z_0)$. Thus, there are numbers λ and μ such that

$$\nabla f(x_0, y_0, z_0) = \lambda\,\nabla g(x_0, y_0, z_0) + \mu\,\nabla h(x_0, y_0, z_0) \qquad (9)$$

(see Problem 11.7.52). Formula (9) is the generalization of formula (6) in the case of two constraints.

EXAMPLE 3

USING LAGRANGE MULTIPLIERS IN $\mathbb{R}^3$

Find the maximum value of $w = xyz$ among all points (x, y, z) lying on the line of intersection of planes $x + y + z = 30$ and $x + y - z = 0$.

SOLUTION: Setting $f(x, y, z) = xyz$, $g(x, y, z) = x + y + z - 30$, and $h(x, y, z) = x + y - z$, we obtain

$$\nabla f = yz\mathbf{i} + xz\mathbf{j} + xy\mathbf{k} \qquad \nabla g = \mathbf{i} + \mathbf{j} + \mathbf{k} \qquad \nabla h = \mathbf{i} + \mathbf{j} - \mathbf{k},$$

and using equation (9) to obtain the maximum, we obtain the equations

$$yz = \lambda + \mu \qquad xz = \lambda + \mu \qquad xy = \lambda - \mu.$$

Multiplying the three equations by x, y, and z, respectively, we find that

$$xyz = (\lambda + \mu)x = (\lambda + \mu)y = (\lambda - \mu)z.$$

If $\lambda + \mu = 0$, then $yz = 0$ and $xyz = 0$, which is not a maximum value since xyz can be positive (for example, $x = 8$, $y = 7$, $z = 15$ is in the constraint set and $xyz = 840$). Thus we can divide the first two equations by $\lambda + \mu$ to find that $x = y$.

Since $x + y - z = 0$, we have $2x - z = 0$, or $z = 2x$. But then

$$30 = x + y + z = x + x + 2x = 4x, \qquad \text{or} \qquad x = \frac{15}{2}.$$

Then

$$y = \frac{15}{2}, \qquad z = 15,$$

and the maximum value of xyz occurs at $(\frac{15}{2}, \frac{15}{2}, 15)$ and is equal to $(\frac{15}{2})(\frac{15}{2})15 = 843\frac{3}{4}$.

We conclude this section with two observations. First, while the outlined steps make the method of Lagrange multipliers seem easy, it should be noted that solving three nonlinear equations in three unknowns or four such equations in four unknowns often entails very involved algebraic manipulations. Second, no method is given for determining whether a solution found actually yields a maximum, a minimum, or neither. Fortunately, in many practical applications the existence of a maximum or a minimum can readily be inferred from the nature of the particular problem. For a function of two variables, the following theorem is true.

THEOREM 2 **TEST TO DETERMINE WHETHER, FOR A CONSTRAINED MAX-MIN PROBLEM, (X_0, Y_0) IS A LOCAL MAXIMUM OR A LOCAL MINIMUM**

Suppose that f is a function of two variables. Define $\Delta(x_0, y_0)$ by

$$\begin{aligned}\Delta(x_0, y_0) = {} & 2g_x(x_0, y_0)g_y(x_0, y_0)[f_{xy}(x_0, y_0) - \lambda g_{xy}(x_0, y_0)] \\ & - [g_x(x_0, y_0)]^2[f_{yy}(x_0, y_0) - \lambda g_{yy}(x_0, y_0)] \\ & - [g_y(x_0, y_0)]^2[f_{xx}(x_0, y_0) - \lambda g_{xx}(x_0, y_0)].\end{aligned}$$

If $\Delta(x_0, y_0) > 0$, then f has a local maximum at (x_0, y_0) subject to the constraint $g(x, y) = 0$

If $\Delta(x_0, y_0) < 0$, then f has a local minimum at (x_0, y_0) subject to the constraint $g(x, y) = 0$

If $\Delta(x_0, y_0) = 0$, then the test is inconclusive. ∎

■ PROBLEMS 13.11

■ SELF-QUIZ

I. True–False: By using the Lagrange multiplier technique, we can look for extreme points of $F(x, y)$ subject to the constraint $3x + 2y = 18$ without needing to solve the constraint for $y = 9 - 1.5x$ and then considering $f(x) = F(x, 9 - 1.5x)$.

II. To find extreme points of xy subject to the constraint $3x + 2y = 18$, we consider the equation _____.
   a. $xy = \lambda[3x + 2y - 18]$
   b. $y + x = \lambda[3 + 2]$
   c. $y\mathbf{i} + x\mathbf{j} = \lambda[3\mathbf{i} + 2\mathbf{j}]$
   d. $x\mathbf{i} + y\mathbf{j} = \lambda[3\mathbf{i} + 2\mathbf{j}]$

III. True–False: If we change the constraint of the preceding problem to be $3x + 2y = 42$, the equation for the Lagrange multiplier λ changes also.

IV. To find extreme points of $3x + 2y$ subject to the constraint $xy = 18$, we consider the equation _____.
   a. $\lambda[xy - 18] = 3x + 2y$
   b. $\lambda[y + x] = 3 + 2$
   c. $\lambda[y\mathbf{i} + x\mathbf{j}] = 3\mathbf{i} + 2\mathbf{j}$
   d. $\lambda[x\mathbf{i} + y\mathbf{j}] = 3\mathbf{i} + 2\mathbf{j}$

■ DRILL

For Problems 1–4, use the technique of Lagrange multipliers to solve Problems 19–22 of the preceding section. Assume that $x \geq 0$, $y \geq 0$, and $z \geq$

1. Maximize xyz subject to $x + y + z = 50$.
2. Minimize $x + y + z$ subject to $xyz = 50$.
3. Maximize xy^2z^3 subject to $x + y + z = 50$.
4. Minimize $x^2 + y^2 + z^2$ subject to $x + y + z = 50$.

In Problems 5–12, use the technique of Lagrange multipliers to find the minimum distance between a point and a line or plane.

5. Point $(1, 2)$; line $2x + 3y = 5$.
6. Point $(3, -2)$; line $y = 2 - x$.
7. Point $(1, -1, 2)$; plane $x + y - z = 3$.
8. Point $(3, 0, 1)$; plane $2x - y + 4z = 5$.
9. Point $(0, 0)$; line $ax + by = d$.
10. Point $(0, 0, 0)$; plane $ax + by + cz = d$.
*11. Point (x_0, y_0); line $ax + by = d$. (Compare your result here to the formula stated in Problems 0.3.63 and 11.2.78.)
*12. Point (x_0, y_0, z_0); plane $ax + by + cz = d$. (Compare your result here to the formula stated in Problem 11.7.56.)
*13. Find the maximum and minimum values of $x^2 + y^2$ subject to the condition $x^3 + y^3 = 6xy$.
14. Find the maximum and minimum values of $2x^2 + xy + y^2 - 2y$ subject to the condition $y = 2x - 1$.
15. Find the maximum and minimum values of $x^2 + y^2 + z^2$ subject to the condition $z^2 = x^2 - 1$.

16. Find the maximum and minimum values of $x^3 + y^3 + z^3$ if (x, y, z) lies on the sphere $x^2 + y^2 + z^2 = 4$.
17. Find the maximum and minimum values of $x + y + z$ if (x, y, z) lies on the sphere $x^2 + y^2 + z^2 = 1$.
18. Find the maximum and minimum values of xyz if (x, y, z) lies on the ellipsoid $x^2 + (y^2/4) + (z^2/9) = 1$.
19. Solve the preceding problem if (x, y, z) lies on the ellipsoid $(x/a)^2 + (y/b)^2 + (z/c)^2 = 1$.
*20. Minimize the function $x^2 + y^2 + z^2$ for (x, y, z) on the planes $3x - y + z = 6$ and $x + 2y + 2z = 2$.
21. Find the minimum value of $x^3 + y^3 + z^3$ for (x, y, z) on the planes $x + y + z = 2$ and $x + y - z = 3$.
22. Find the maximum and minimum distances from the origin to a point on the ellipse $(x/a)^2 + (y/b)^2 = 1$.
23. Find the maximum and minimum distances from the origin to a point on the ellipsoid $(x/a)^2 + (y/b)^2 + (z/c)^2 = 1$.
*24. Find the maximum value of $x_1 + x_2 + x_3 + x_4$ subject to $x_1^2 + x_2^2 + x_3^2 + x_4^2 = 1$. [Hint: Use the obvious generalization of Lagrange multipliers to functions of four variables.]
*25. Find the maximum value of $x_1 + x_2 + \cdots + x_n$ subject to $x_1^2 + x_2^2 + \cdots + x_n^2 = 1$.
26. Find the maximum and minimum values of xyz subject to $x^2 + z^2 = 1$ and $x = y$.

CONTINUED

■ **APPLICATIONS**

27. Find the volume of the largest rectangular parallelepiped that can be inscribed in the ellipsoid $x^2 + 4y^2 + 9z^2 = 9$.

28. A silo is in the shape of a cylinder topped with a cone. The radius of each is 6 m and the total surface area (excluding the base) is 200 m². What are the heights of the cylinder and cone that maximize the volume enclosed by the silo?

29. The base of an open-top rectangular box costs $3/m² to construct, while the sides cost only $1/m². Find the dimensions of the box of greatest volume that can be constructed for $36.

30. A manufacturing company has three plants I, II and III, which produce x, y, and z units, respectively, of a certain product. The annual revenue from this production is given by $R(x, y, z) = 6xyz^2 - 400,000(x + y + z)$. If the company is to produce 1000 units annually, how should it allocate production among the three plants so as to maximize its profits?

31. A firm has $250,000 to spend on labor and raw materials. The output of the firm is αxy, where α is a constant and x and y are, respectively, the quantity of labor and raw materials consumed. The unit price of hiring labor is $5000, and the unit price of raw materials is $2500. Find the ratio of x to y that maximizes output.

32. The temperature of a point (x, y, z) on the unit sphere is given by $T(x, y, z) = xy + yz$. What is the hottest point on the sphere?

33. The Cobb-Douglas production function for a certain product is

$$F(L, K) = 250L^{0.7}K^{0.3}.$$

Suppose that each unit of labor costs $200, that each unit of capital costs $350, and that $25,000 is available for production costs.
a. How many units of labor and capital should be used to maximize output?
b. How many units will be produced?
c. Compute the ratio of marginal productivity of labor to marginal productivity of capital at levels of labor and capital costs that maximize output.

34. A product has the Cobb-Douglas production function

$$F(L, K) = 500L^{1/3}K^{2/3}.$$

Answer the questions of Problem 33 assuming that each unit of labor costs $1000, each unit of capital costs $1600, and a total of $250,000 is available for production.

35. A can of dog food is advertised to contain 80 units of protein. Two types of meat are used in making up the food. Each unit of liver costs 30¢, and each unit of horsemeat costs 16¢. If l units of liver and h units of horsemeat are in the can, then the number of units of protein is

$$N(l, h) = 4l^2 + 2.5h^2.$$

a. How many units of each meat should be put in a can of dog food to meet the advertised claim at minimum cost?
b. What is the minimum cost?

36. Bellingham Health Care (BHC) is a nonprofit foundation providing medical treatment to emotionally distressed children. BHC has hired you as a business consultant to aid the foundation in the development of a hiring policy that would be consistent with its overall goal of providing the most meaningful patient service possible given scarce foundation resources. In your initial analysis, you have determined that *service* can be described as a function of medical (M) and social services (S) staff input as follows:

$$\text{service} = M + 0.5S + 0.5MS - S^2.$$

BHC's staff budget for the coming year is $600,000. Annual employment costs total $15,000 for each social service staff member and $30,000 for each medical staff member.
a. Construct the function you might use to determine the optimal (service-maximizing) social service-medical employment combination.
b. Determine the optimal combination of social service and medical staff for BHC.

37. A field representative for a major pharmaceutical firm has just received the following information from a marketing research consultant who has been analyzing his recent sales performance. The consultant estimates that time spent in the two major metropolitan areas that compose his sales territory will result in monthly sales as indicated by the equation

$$\text{sales} = 500A - 20A^2 + 300B - 10B^2.$$

Here A and B represent the number of days spent in each metropolitan area respectively. Assuming that a working month is composed of twenty business days, what is the optimal number of days the salesperson should spend in each city?

CONTINUED

38. For a simple lens of focal length f, the object distance d and the image distance i are related by the formula

$$\frac{1}{d} + \frac{1}{i} = \frac{1}{f}.$$

A given lens has a focal length of 50 cm.
a. What is the minimum value of the object-image distance $d + i$?
b. For what values of d and i is this minimum achieved?

■ SHOW/PROVE/DISPROVE

39. Show that among all rectangles with the same perimeter, the square encloses the greatest area.

*40. Show that among all triangles having the same perimeter, the equilateral triangle has the greatest area. [*Hint:* If the sides have lengths a, b, and c, **Heron's formula** states the area as $\sqrt{s(s - a)(s - b)(s - c)}$ where $s = (a + b + c)/2$.]

41. Prove the **general arithmetic-geometric inequality**: If $p + q + r = 1$, then $x^p y^q z^r \le px + qy + rz$ with equality if and only if $x = y = z > 0$. Assume that p, q, r, x, y, and z are nonnegative.

■ CHALLENGE

42. The task "Find the maximal area of a rectangle inscribed within semicircle of radius r" can be rephrased as "Maximize $A(x, y) = (2x)y$ subject to the constraint that $x^2 + y^2 = r^2$" because there is no loss of generality in fixing the semicircle to be the upper half of a circle centered at the origin. You now know several ways to analyze and solve this maximization task.
   a. Think about the following fact:

   $$2xy = (x^2 + y^2) - (x - y)^2$$
   $$= r^2 - (x - y)^2 \le r^2.$$

   b. Solve the constraint for y; substitute that expression, $y = \sqrt{r^2 - x^2}$, into A; locate and analyze the critical point(s) of this function, $A(x) = (2x)\sqrt{r^2 - x^2}$, of a single variable x.

   c. Use the implicit differentiation technique: Assume the constraint determines y as a function of x and differentiate to obtain $2x + 2yy' = 0$. Therefore $y' = -x/y$ and $A' = (2)y + (2x)y' = 2y - 2x^2/y, \dots$. (Alternatively, solve the simultaneous pair of equations $2x + 2yy' = 0$ and $2y + 2xy' = 0$.)

   d. Use the technique of Lagrange multipliers: analyze the consequences of

   $$2y\mathbf{i} + 2x\mathbf{j} = \lambda[2x\mathbf{i} + 2y\mathbf{j}].$$

   Do the computations for each of these parts, then write a brief essay comparing the four techniques.

■ ANSWERS TO SELF-QUIZ

I. True   II. c   III. False   IV. c

▥ 13.12 NEWTON'S METHOD FOR FUNCTIONS OF TWO VARIABLES (OPTIONAL)

In Section 3.6, we discussed Newton's method for finding solutions of the equation $f(x) = 0$. In Theorem 3.6.2 on page 231, we gave conditions on f such that the sequence defined recursively by

$$x_{n+1} = x_n - \frac{f(x_n)}{f'(x_n)} \tag{1}$$

converges quadratically to a solution s of $f(x) = 0$ if we start at any point x_0 in a specified interval $[a, b]$.

In this section we extend Newton's method to find points (x, y) such that

$$f(x, y) = 0 \tag{2}$$

and

$$g(x, y) = 0 \qquad (3)$$

for some differentiable functions f and g. We saw a need to solve such a system in Section 13.10 in our calculation of critical points. For example, in Example 7 on page 899, we solved the system

$$6(x^2 - 4y) = 0$$

$$-24(x - 2y^2) = 0.$$

We were able to find both solutions, $(0, 0)$ and $(2, 1)$ quite easily, but in most situations the computations are likely to be much more difficult.

The most compact form of Newton's method in two (or more) variables looks very much like equation (1). However, in order to present this form, it is necessary to make use of matrices and their inverses—a topic not discussed in this book. Instead, we will describe the method without making use of matrix notation. For those of you who have studied matrices, we give the compact form at the end of the section.

NEWTON'S METHOD IN TWO VARIABLES

Let (x_0, y_0) be chosen. Generate the sequence of vectors (x_n, y_n) recursively as follows: Let

$$D(x, y) = f_x(x, y)g_y(x, y) - f_y(x, y)g_x(x, y). \qquad (4)$$

Then we define

NEWTON ITERATES IN $\mathbb{R}^2$

$$x_{n+1} = x_n - \frac{f(x_n, y_n)g_y(x_n, y_n) - f_y(x_n, y_n)g(x_n, y_n)}{D(x_n, y_n)} \qquad (5)$$

$$y_{n+1} = y_n - \frac{-f(x_n, y_n)g_x(x_n, y_n) + f_x(x_n, y_n)g(x_n, y_n)}{D(x_n, y_n)}. \qquad (6)$$

Under suitable conditions, which we do not give here, the sequence $\{(x_n, y_n)\}$ will converge to a vector (u, v) which satisfies

$$f(u, v) = g(u, v) = 0.$$

Newton's method is most likely to work when the initial vector (x_0, y_0) is close to a solution to the system (2), (3). It is not necessary that the system have only one solution.

EXAMPLE 1

USING NEWTON'S METHOD FOR A FUNCTION OF TWO VARIABLES

Use Newton's method to find a solution to the system

$$x^2 - 2x - y + \frac{1}{2} = 0$$

$$x^2 + 4y^2 - 4 = 0,$$

starting at the initial vector $(2, 0.25)$. Carry all calculations to 6 decimal places of accuracy.

SOLUTION: We first compute

$$f = x^2 - 2x - y + \frac{1}{2} \qquad f_x = 2x - 2 \qquad \text{and} \qquad f_y = -1$$

$$g = x^2 + 4y^2 - 4 \qquad g_x = 2x \qquad \text{and} \qquad g_y = 8y$$

so

$$D(x, y) = f_x g_y - f_y g_x = (2x - 2)(8y) - (-1)(2x)$$
$$= 2x - 16y + 16xy$$

$$f g_y - f_y g = \left(x^2 - 2x - y + \frac{1}{2}\right)(8y) - (-1)(x^2 + 4y^2 - 4)$$
$$= -4 + 4y + x^2 - 4y^2 - 16xy + 8x^2y$$

$$-f g_x + f_x g = -\left(x^2 - 2x - y + \frac{1}{2}\right)(2x) + (2x - 2)(x^2 + 4y^2 - 4)$$
$$= 8 - 9x + 2xy + 2x^2 - 8y^2 + 8xy^2.$$

Thus,

$$x_{n+1} = x_n - \frac{-4 + 4y_n + x_n{}^2 - 4y_n{}^2 - 16x_n y_n + 8x_n{}^2 y_n}{2x_n - 16y_n + 16x_n y_n}$$

$$y_{n+1} = y_n - \frac{8 - 9x_n + 2x_n y_n + 2x_n{}^2 - 8y_n{}^2 + 8x_n y_n{}^2}{2x_n - 16y_n + 16x_n y_n}.$$

For example, starting at $x_0 = 2$ and $y_0 = 0.25$, we obtain

$$x_1 = 2 - \frac{0.75}{8} = 1.90625$$

and

$$y_1 = 0.25 - \frac{-0.5}{8} = 0.3125.$$

We show the first four iterates in Table 1.

TABLE 1

n	x_n	y_n	x_{n+1}	y_{n+1}
0	2	0.25	1.90625	0.3125
1	1.90625	0.3125	1.900691	0.311213
2	1.900691	0.311213	1.900677	0.311219
3	1.900677	0.311219	1.900677	0.311219

We stop here since $(x_4, y_4) = (x_3, y_3)$ (to 6 decimal places). The vector $(1.900677, 0.311219)$ is correct to 6 decimal places.

CHECK:

$$x^2 - 2x - y + \frac{1}{2} = (1.900677)^2 - 2(1.900677) - 0.311219 + 0.5$$

$$= 0.000000058$$

$$x^2 + 4y^2 - 4 = (1.900677)^2 + 4(0.311219)^2 - 4 = 0.000002112.$$

The errors come from the 6-decimal-place rounding.

We see that even in a relatively simple example, implementing Newton's method in two variables can be very tedious. For that reason, it is best to carry out the computations on a computer or programmable calculator. Of course, you first have to differentiate and write out the rules (5) and (6) to begin the iteration. In Problems 3–9, you are asked to solve systems of equations on a computer.

NEWTON'S METHOD IN MATRIX FORM

We now write Newton's method in matrix form. This material should be read only if you know how to multiply matrices and compute the inverse of a matrix.

We write the vector (x, y) as $\mathbf{x}$ and define the vector function $\mathbf{F}$ by

$$\mathbf{F}(\mathbf{x}) = \begin{pmatrix} f(\mathbf{x}) \\ g(\mathbf{x}) \end{pmatrix}.$$

We define the **Jacobian matrix** J by

$$J(\mathbf{x}) = \begin{pmatrix} f_x(\mathbf{x}) & f_y(\mathbf{x}) \\ g_x(\mathbf{x}) & g_y(\mathbf{x}) \end{pmatrix}.$$

Then Newton's method becomes

> **NEWTON'S ITERATION IN MATRIX NOTATION**
>
> $$\mathbf{x}_{n+1} = \mathbf{x}_n - J^{-1}(\mathbf{x}_n)\mathbf{F}(\mathbf{x}_n). \tag{7}$$

The nice thing about this formulation is that it applies equally well to solving systems of n equations in n unknowns, where $n > 2$.

In Problem 12 you are asked to show that (7) reduces to (5) and (6) when J^{-1} is calculated and the matrix multiplication in (7) is carried out.

■ PROBLEMS 13.12

■ SELF-QUIZ

I. In order to use Newton's method to solve the system (2), (3) it is necessary that
   a. f and g be differentiable at (x_0, y_0).
   b. f_x, f_y, g_x, and g_y exist at (x_0, y_0).
   c. the system (2), (3) have a unique solution.
   d. $D(x_0, y_0) \neq 0$.

II. In order to carry out Newton's method in two variables, it is necessary to evaluate _____ functions at every step (not counting additions, subtractions, multiplications, or divisions of functions).

   a. 4   b. 5   c. 6
   d. 8   e. 10   f. 14

CONTINUED

■ DRILL

1. Consider the system

$$x^2 - y = 0$$

$$y^2 - x = 0.$$

   a. Verify that the solutions are $(0, 0)$ and $(1, 1)$.
   b. Start with $(x_0, y_0) = (\frac{1}{4}, -\frac{1}{4})$ and use Newton's method to find (x_1, y_1) and (x_2, y_2).
   c. Start with $(x_0, y_0) = (0.9, 1.25)$ and use Newton's method to find (x_1, y_1) and (x_2, y_2).

2. Consider the system

$$x^2 + y^2 - 2 = 0$$

$$x^2 - y = 0.$$

   a. Verify that the solutions are $(1, 1)$ and $(-1, 1)$.
   b. Start with $(x_0, y_0) = (0.8, 0.75)$ and use Newton's method to find (x_1, y_1) and (x_2, y_2).
   c. Start with $(x_0, y_0) = (-1.25, 0.75)$ and use Newton's method to find (x_1, y_1) and (x_2, y_2).

In Problems 3–8, a system is given.
a. Start at (x_0, y_0) and use Newton's method to find (x_1, y_1) and (x_2, y_2).
b. Write a computer program to find all the solutions to 8 decimal places of accuracy.

3. $2xy - 3 = 0$
   $x^2 - y - 2 = 0$

   $(x_0, y_0) = (1.5, 0.9)$

4. $x^2 + 4y^2 - 4 = 0$
   $x^2 - 2x - y + 1 = 0$

   (i) $(x_0, y_0) = (1.5, 0.5)$
   (ii) $(x_0, y_0) = (-0.25, 1.1)$

5. $3x^2 - 2y^2 - 1 = 0$
   $x^2 - 2x + y^2 + 2y - 8 = 0$

   (i) $(x_0, y_0) = (-1, 1)$ (ii) $(x_0, y_0) = (3, -3.4)$

6. $-x + y^2 - 2 = 0$
   $x^3 - 3x^2 + 4x - y = 0$

   (i) $(x_0, y_0) = (0.5, 1.2)$
   (ii) $(x_0, y_0) = (-0.25, -1.3)$

7. $2x^3 - 12x - y - 1 = 0$
   $3y^2 - 6y - x - 3 = 0$

   (i) $(x_0, y_0) = (2.5, 2.5)$
   (ii) $(x_0, y_0) = (2.5, -1)$
   (iii) $(x_0, y_0) = (0, 0)$
   (iv) $(x_0, y_0) = (-2.5, 2.5)$
   (v) $(x_0, y_0) = (-2.5, 0)$
   (vi) $(x_0, y_0) = (0, 2.5)$

8. $3x^2 - 2y^2 - 1 = 0$
   $x^2 - 2x + 2y - 8 = 0$

   (i) $(x_0, y_0) = (2.5, 3)$
   (ii) $(x_0, y_0) = (-1.6, 1.6)$
   (iii) $(x_0, y_0) = (5.6, -7)$
   (iv) $(x_0, y_0) = (-3, -3.6)$

9. Use Newton's method to find all nine solutions to

$$7x^3 - 10x - y - 1 = 0$$

$$8y^3 - 11y + x - 1 = 0.$$

   Use the starting points $(0, 0)$, $(1, 0)$, $(0, 1)$, $(-1, 0)$, $(0, -1)$, $(1, 1)$, $(-1, 1)$, $(1, -1)$, and $(-1, -1)$.

■ SHOW/PROVE/DISPROVE

*10. Consider the nonlinear system

$$x^2 + y^2 - 2 = 0$$

$$xy - 1 = 0.$$

   a. Verify that the solutions are $(1, 1)$ and $(-1, -1)$.
   b. What difficulties might arise if we try to use Newton's method to find the solutions?

■ CHALLENGE

(For those who know how to manipulate matrices.)

11. Show that $J^{-1}(\mathbf{x}) = \dfrac{1}{D}\begin{pmatrix} g_y & -f_y \\ -g_x & f_x \end{pmatrix}$.

12. Show that equation (7) yields equations (5) and (6) when the matrices in (7) are multiplied and added.

Consider the system

$$f(x, y, z) = 0 \tag{8}$$

$$g(x, y, z) = 0 \tag{9}$$

$$h(x, y, z) = 0. \tag{10}$$

CONTINUED

Write $\mathbf{x} = (x, y, z)$, $\mathbf{F} = \begin{pmatrix} f \\ g \\ h \end{pmatrix}$, and

$J = \begin{pmatrix} f_x & f_y & f_z \\ g_x & g_y & g_z \\ h_x & h_y & h_z \end{pmatrix}$. Then equation (7) defines

Newton's method in three variables. In Problems 13 and 14, use Newton's method to find all solutions to the given system.

**13. $x^2 - x + y^2 + z^2 - 5 = 0$
$x^2 + y^2 - y + z^2 - 4 = 0$
$x^2 + y^2 + z^2 + z - 6 = 0$
a. Start with $(x_0, y_0, z_0) = (-0.8, 0.2, 1.8)$.
b. Start with $(x_0, y_0, z_0) = (1.2, 2.2, -0.2)$.

**14. $x^2 - x + 2y^2 + yz - 10 = 0$
$5x - 6y + z = 0$
$z - x^2 - y^2 = 0$
$(x_0, y_0, z_0) = (1.1, 1.5, 3.5)$

■ ANSWERS TO SELF-QUIZ

I. b, d   II. c

■ PROBLEMS   CHAPTER 13 REVIEW

■ DRILL

In Problems 1–6, find the domain and range of the given function.

1. $f(x, y) = \sqrt{x^2 - y^2}$
2. $f(x, y) = 1/\sqrt{x^2 + y^2}$
3. $f(x, y) = \cos(x + 3y)$
4. $f(x, y, z) = \sqrt{1 - x^2 - y^2 - z^2}$
5. $f(x, y, z) = 1/\sqrt{x^2 + y^2 + z^2 - 1}$
6. $f(x, y, z) = \ln(x - y + 4z - 3)$

In Problems 7–10, describe the level curves of the given functions; then sketch these curves for the specified values of z.

7. $z = \sqrt{1 - x - y}$; $z = 0, 1, 3, 8$
8. $z = \sqrt{1 - y^2 + x}$; $z = 0, 2, 4, 7$
9. $z = \ln(x - 3y)$; $z = 0, 1, 2, 3$
10. $z = \dfrac{x^2 - y^2}{x^2 + y^2}$; $z = 1, 3, 6$

11. Sketch the open disk centered at $(-1, 2)$ with radius 4.
12. Sketch the closed ball centered at $(1, 2, 3)$ with radius 2.

13. Show that $\lim\limits_{(x,y)\to(0,0)} \dfrac{xy}{y^2 - x^2}$ does not exist.

14. Show that $\lim\limits_{(x,y)\to(0,0)} \dfrac{y^2 - 2x}{y^2 + 2x}$ does not exist.

15. Show that $\lim\limits_{(x,y)\to(0,0)} \dfrac{4xy^3}{x^2 + y^4} = 0$.

16. Show that $\lim\limits_{(x,y,z)\to(0,0,0)} \dfrac{zy^2 + x^3}{x^2 + y^2 + z^2} = 0$.

17. Calculate $\lim\limits_{(x,y)\to(1,-2)} \dfrac{1 + x^2 y}{2 - y}$.

18. Calculate $\lim\limits_{(x,y,z)\to(2,-1,1)} \dfrac{x^2 - yz^3}{1 + xyz - 2y^5}$.

19. Find the maximum region over which the function $f(x, y) = \ln(1 - 2x + 3y)$ is continuous.
20. Find the maximum region over which the function $f(x, y, z) = \ln(x - y - z + 4)$ is continuous.
21. Find the maximum region over which the function $f(x, y, z) = 1/\sqrt{1 - x^2 + y^2 - z^2}$ is continuous.
22. Find a number c such that the function

$$f(x, y) = \begin{cases} \dfrac{-2xy}{\sqrt{x^2 + y^2}} & \text{if } (x, y) \neq (0, 0) \\ c & \text{if } (x, y) = (0, 0) \end{cases}$$

is continuous at the origin.

In Problems 23–34, calculate all first partial derivatives.

23. $f(x, y) = y/x$
24. $f(x, y) = \cos(x - 3y)$

CONTINUED

25. $f(x, y) = 1/\sqrt{x^2 - y^2}$

26. $f(x, y) = \tan^{-1}\left(\dfrac{y}{1 + x}\right)$

27. $f(x, y, z) = \ln(x - y + 4z)$

28. $f(x, y, z) = 1/\sqrt{x^2 + y^2 + z^2}$

29. $f(x, y, z) = \cosh(y/x^2)$

30. $f(x, y, z) = \sec\left(\dfrac{x - y}{z}\right)$

31. $f(x, y, z) = (x^2y - y^3z^5 + x\sqrt{z})^{2/3}$

32. $f(x, y, z) = \dfrac{x^2 - y^3}{y^3 + z^4}$

33. $f(x, y, z, w) = \dfrac{x - z + w}{y + 2w - x}$

34. $f(x, y, z, w) = e^{(x - w)/(y + z)}$

In Problems 35–40, calculate all second partial derivatives and show that all pairs of mixed partials are equal.

35. $f(x, y) = xy^3$ 36. $f(x, y) = \tan^{-1}(y/x)$

37. $f(x, y) = \sqrt{x^2 - y^2}$ 38. $f(x, y) = \dfrac{x + y}{x - y}$

39. $f(x, y, z) = \ln(2 - 3x + 4y - 7z)$

40. $f(x, y, z) = 1/\sqrt{1 - x^2 - y^2 - z^2}$

41. Calculate f_{yzx} if $f(x, y, z) = x^2y^3 - zx^5$.

42. Calculate f_{zxxyz} if $f(x, y, z) = (x - y)/z$.

In Problems 43–50, calculate the gradient of the given function at the specified point.

43. $f(x, y) = x^2 - y^3$; $(1, 2)$

44. $f(x, y) = \tan^{-1}(y/x)$; $(-1, -1)$

45. $f(x, y) = \dfrac{x - y}{x + y}$; $(3, 2)$

46. $f(x, y) = \cos(x - 2y)$; $(\pi/2, \pi/6)$

47. $f(x, y, z) = xy + yz^3$; $(1, 2, -1)$

48. $f(x, y, z) = \dfrac{x - y}{3z}$; $(2, 1, 4)$

49. $f(x, y, z) = 1/\sqrt{x^2 + y^2 + z^2}$; (a, b, c)

50. $f(x, y, z) = e^{-(x^2 + y^3 + z^4)}$; $(0, -1, 1)$

In Problems 51–56, use the chain rule to calculate the indicated derivative(s).

51. $z = 2xy$, $x = \cos t$, $y = \sin t$; dz/dt

52. $z = y/x$, $x = r - s$, $y = r + s$; $\partial z/\partial s$

53. $z = xy^3$, $x = r/s$, $y = s^2/r$; $\partial z/\partial r$

54. $z = \sin(x - y)$, $x = e^{r+s}$, $y = e^{r-s}$; $\partial z/\partial s$

55. $w = xyz$, $x = rs$, $y = r/s$, $z = s^2r^3$; w_r, w_s

56. $w = x^3y + y^3z$, $x = rst$, $y = rs/t$, $z = rt/s$; w_s, w_t

In Problems 57–62, find an equation for the tangent plane and symmetric equations of the normal line to the given surface at the specified point.

57. $x^2 + y^2 + z^2 = 3$; $(1, 1, 1)$

58. $x^{1/2} + y^{3/2} + z^{1/2} = 3$; $(1, 0, 4)$

59. $3x - y + 5z = 15$; $(-1, 2, 4)$

60. $xy^2 - yz^3 = 0$; $(1, 1, 1)$

61. $xyz = 6$; $(-2, 1, -3)$

62. $\sqrt{\dfrac{x - y}{y + z}} = \dfrac{1}{2}$; $(2, 1, 3)$

In Problems 63–68, calculate the directional derivative of the given function at the specified point in the direction of the specified vector $\mathbf{v}$.

63. $f(x, y) = y/x$ at $(1, 2)$; $\mathbf{v} = \mathbf{i} - \mathbf{j}$

64. $f(x, y) = 3x^2 - 4xy$ at $(3, -1)$; $\mathbf{v} = 2\mathbf{i} + 5\mathbf{j}$

65. $f(x, y) = \tan^{-1}(y/x)$ at $(1, -1)$; $\mathbf{v} = -3\mathbf{i} + 2\mathbf{j}$

66. $f(x, y, z) = xy^2 - zy^3$ at $(1, 2, 3)$; $\mathbf{v} = \mathbf{i} - \mathbf{j} + 2\mathbf{k}$

67. $f(x, y, z) = 1/\sqrt{x^2 + y^2 + z^2}$ at $(1, -1, 2)$; $\mathbf{v} = -2\mathbf{i} + \mathbf{j} - 3\mathbf{k}$

68. $f(x, y, z) = e^{-(x + y^2 - xz)}$ at $(1, 0, -1)$; $\mathbf{v} = 2\mathbf{i} + 5\mathbf{j} + \mathbf{k}$

In Problems 69–73, calculate the total differential df.

69. $f(x, y) = x^3y^2$ 70. $f(x, y) = \cos^{-1}(y/x)$

71. $f(x, y) = \sqrt{\dfrac{x + 1}{y - 1}}$ 72. $f(x, y, z) = xy^5z^3$

73. $f(x, y, z) = \ln(x - y + 4z)$

74. Approximately how much wood is contained in the sides of a rectangular box with sides of inside measurements 1.5 m, 1.3 m, and 2 m if the thickness of the wood making up the sides is 3 cm ($= 0.03$ m)?

In Problems 75–80, determine the nature of the various critical points of the given function.

75. $f(x, y) = 6x^2 + 14y^2 - 16xy + 2$

76. $f(x, y) = x^5 - y^5$

77. $f(x, y) = \dfrac{1}{y} + \dfrac{2}{x} + 2y + x + 4$

78. $f(x, y) = 49 - 16x^2 + 24xy - 9y^2$

79. $f(x, y) = x^2 + y^2 + \dfrac{2}{xy^2}$

80. $f(x, y) = \cot xy$

CONTINUED

■ APPLICATIONS

81. Find the minimum distance from the point $(2, -1, 4)$ to the plane $x - y + 3z = 7$.
82. What is the smallest amount of wood needed to build an open-top rectangular box enclosing a volume of 25 m³?
83. What is the maximum volume of an open-top rectangular box that can be built from 10 m² of wood?

84. What are the dimensions of the rectangular parallelepiped (with faces parallel to the coordinate planes) of maximum volume that can be inscribed in the ellipsoid $x^2 + 9y^2 + 4z^2 = 36$?
85. Minimize the function $x^2 + y^2 + z^2$ for (x, y, z) on the planes $2x + y + z = 2$ and $x - y - 3z = 4$.
*86. Among all the ellipses $(x/a)^2 + (y/b)^2 = 1$ that pass through the point $(3, 5)$, which one has the smallest area?

■ SHOW/PROVE/DISPROVE

87. Show that there is no maximum or minimum value of xy^3z^2 if (x, y, z) lies on the plane $x - y + 2z = 2$.

88. Show that among all rectangles with the same area, the square has the smallest perimeter.

⚡ COMPUTER EXERCISES

1. The purpose of this exercise is to demonstrate that f_{xy} and f_{yx} need not always be equal. You may find it helpful to use a computer algebra system that will calculate partial derivatives.

$$\text{Define } f(x, y) = \begin{cases} \dfrac{xy(x^2 - y^2)}{x^2 + y^2} & \text{if } (x, y) \neq (0, 0) \\ 0 & \text{if } (x, y) = (0, 0). \end{cases}$$

   a. Calculate
      (i) $f_x(x, y)$ and $f_y(x, y)$ for $(x, y) \neq (0, 0)$.
      (ii) $f_x(0, 0)$ and $f_y(0, 0)$.
   b. Using your answers above, calculate
      (i) $f_{xy}(x, y)$ and $f_{yx}(x, y)$ for $(x, y) \neq (0, 0)$.
      (ii) $f_{xy}(0, 0)$ and $f_{yx}(0, 0)$.
   c. You should be aware of a theorem that says $f_{xy} = f_{yx}$ under certain conditions. What hypothesis of this theorem has been violated in this example? Prove your assertion.
   d. Use graphics software to view the graph of this function.

2. $f(x, y) = \frac{1}{3}x^3y + x^3 + \frac{1}{2}x^2y^2 + \frac{3}{2}x^2y - xy^2 - 4xy - 3x + y^2 + 6y$
   a. Find the critical points of $f(x, y)$.
   b. Use the discriminant to determine whether each critical point gives a local maximum, local minimum, or a saddle point.
   c. Use graphics software to view the graph of this function.

3. $f(x, y) = \dfrac{2x^3}{3} - 2x^2y + \dfrac{x^2}{2} + 2xy - x + y^2 + 2y$
   a. Find the critical points of $f(x, y)$.
   b. Use the discriminant to test each critical point and determine if it is a local maximum, local minimum, or a saddle point.
   c. Use graphics software to view the graph of this function.

4. Find the two points on the graphs of $f(x) = 3x$ and $g(x) = x^2 + 3$ that are nearest.

5. A box with no lid is to be made of special materials. The volume of the box is to be 1 cubic foot. The material for the bottom costs $1.00 per square foot. The material for the front and back panels costs $2.00 per square foot, and the material to be used on the left and right hand side panels costs $3.00 per square foot. The box is to be constructed by gluing the five panels together along their edges. It costs $1.00 per foot to do the gluing. Find the dimensions of the box of minimal cost.

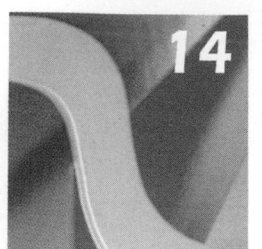

14 MULTIPLE INTEGRATION

14.1 VOLUME UNDER A SURFACE AND THE DOUBLE INTEGRAL

In our development of the definite integral in Chapter 4, we began by calculating the area under a curve $y = f(x)$ (and above the x-axis) for x in the interval $[a, b]$. We initially assumed that, on $[a, b]$, $f(x) \geq 0$. We carry out a similar development by obtaining an expression which represents a volume in $\mathbb{R}^3$.

We begin by considering an especially simple case. Let R denote the rectangle in $\mathbb{R}^2$ given by

$$R = \{(x, y): a \leq x \leq b \text{ and } c \leq y \leq d\}. \tag{1}$$

This rectangle is sketched in Figure 1. Let $z = f(x, y)$ be a continuous function that is nonnegative over R. That is, $f(x, y) \geq 0$ for every (x, y) in R. We now ask: What is the volume "under" the surface $z = f(x, y)$ and "over" the rectangle R? The volume requested is sketched in Figure 2.

We will calculate this volume in much the same way we calculated the area under a curve in Chapter 4. We begin by "partitioning" the rectangle.

Step 1. Form a **regular partition** (i.e., all subintervals have the same length) of the intervals $[a, b]$ and $[c, d]$:

$$a = x_0 < x_1 < x_2 < \cdots < x_{n-1} < x_n = b, \tag{2}$$

$$c = y_0 < y_1 < y_2 < \cdots < y_{m-1} < y_m = d. \tag{3}$$

We then define

$$\Delta x = x_i - x_{i-1} = \frac{b - a}{n}, \tag{4}$$

$$\Delta y = y_j - y_{j-1} = \frac{d - c}{m}, \tag{5}$$

and define the subrectangles R_{ij} by

$$R_{ij} = \{(x, y): x_{i-1} \leq x \leq x_i \text{ and } y_{j-1} \leq y \leq y_j\} \tag{6}$$

for $i = 1, 2, \ldots, n$ and $j = 1, 2, \ldots, m$. This is sketched in Figure 3. Note that there are nm subrectangles R_{ij} covering the rectangle R.

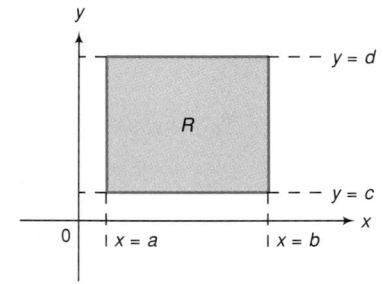

FIGURE 1
A RECTANGLE IN $\mathbb{R}^2$

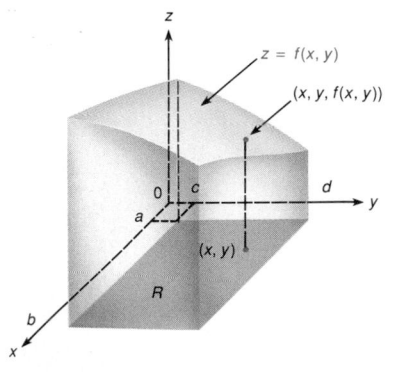

FIGURE 2
THE SHADED SURFACE LIES OVER THE RECTANGLE R.

REMARK: In defining the definite integral, we allowed for the possibility that the subintervals had different lengths (see page 294). However, we observed on page 297 that we obtained the same result by using regular partitions whenever $\int_a^b f(x)\, dx$ exists. This is true if f is continuous. Now we are assuming that $z = f(x, y)$ is continuous. Under this assumption, it is sufficient to

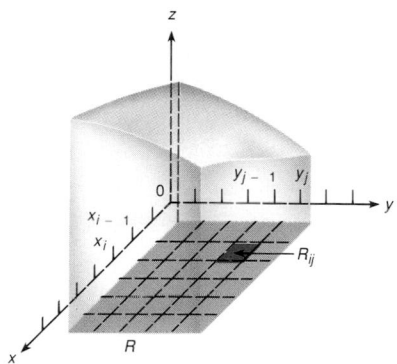

FIGURE 3
R_{ij} IS ONE OF THE nm SUBRECTANGLES THAT PARTITION R.

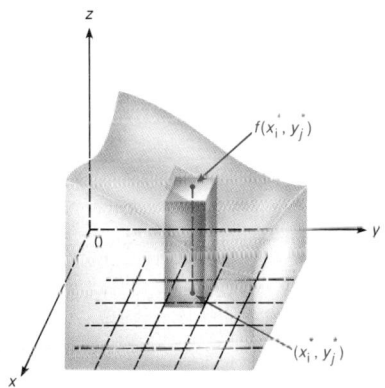

FIGURE 4
THE VOLUME OF A SMALL RECTANGULAR SOLID APPROXIMATES THE VOLUME UNDER A SMALL PART OF THE SURFACE.

consider only regular partitions. Since these are easier to deal with, we will work exclusively with regular partitions in the remainder of this book.

Step 2. Estimate the volume under the surface and over each subrectangle.

Let $(x_i{}^*, y_j{}^*)$ be a point in R_{ij}. Then the volume V_{ij} under the surface and over R_{ij} is approximated by

$$V_{ij} \approx f(x_i{}^*, y_j{}^*)\,\Delta x\,\Delta y = f(x_i{}^*, y_j{}^*)\,\Delta A, \tag{7}$$

where $\Delta A = \Delta x\,\Delta y$ is the area of R_{ij}. The expression on the right-hand side of (7) is simply the volume of the parallelepiped (three-dimensional box) with base R_{ij} and height $f(x_i{}^*, y_j{}^*)$. This volume corresponds to the approximate area $A_i \approx f(x_i{}^*)\,\Delta x_i$ that we used on page 284. Unless the graph of f is a plane parallel to the xy-plane, the expression $f(x_i{}^*, y_j{}^*)\,\Delta A$ will not in general be equal to the volume under the surface S. But if Δx and Δy are small, the approximation will be a good one. The difference between the actual V_{ij} and the approximate volume given in (7) is illustrated in Figure 4.

Step 3. Add up the approximate volumes to obtain an approximation to the total volume sought.

The total volume is

$$V = V_{11} + V_{12} + \cdots + V_{1m} + V_{21} + V_{22} + \cdots V_{2m}$$
$$+ \cdots + V_{n1} + V_{n2} + \cdots + V_{nm}. \tag{8}$$

To simplify notation, we use the summation sign $\sum$ introduced in Section 4.3. Since we are summing over two variables i and j, we need two such signs:

$$V = \sum_{i=1}^{n} \sum_{j=1}^{m} V_{ij}. \tag{9}$$

The expression in (9) is called a **double sum**. If we "write out"[†] the expression in (9), we obtain the expression in (8). Then combining (7) and (9), we have

$$V \approx \sum_{i=1}^{n} \sum_{j=1}^{m} f(x_i{}^*, y_j{}^*)\,\Delta A. \tag{10}$$

Step 4. Take a limit as both Δx and Δy approach zero.

To indicate that this is happening, we define

$$\Delta s = \sqrt{(\Delta x)^2 + (\Delta y)^2}.$$

Geometrically, Δs is the length of a diagonal of the rectangle R_{ij} whose sides have lengths Δx and Δy (see Figure 5). As $\Delta s \to 0$, the number of

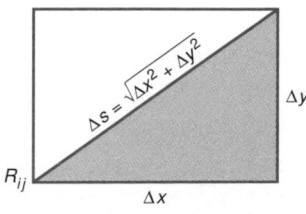

FIGURE 5
Δs IS THE LENGTH OF THE DIAGONAL OF THE SUBRECTANGLE R_{ij}.

[†] This writing out is done by summing over j first and then over i. For example,

$$\sum_{i=1}^{3} \sum_{j=1}^{4} a_{ij} = \sum_{i=1}^{3} (a_{i1} + a_{i2} + a_{i3} + a_{i4})$$
$$= a_{11} + a_{12} + a_{13} + a_{14} + a_{21} + a_{22} + a_{23} + a_{24} + a_{31} + a_{32} + a_{33} + a_{34}.$$

subrectangles R_{ij} increases without bound and the area of each R_{ij} approaches zero. This implies that the volume approximation given by (7) is getting closer and closer to the "true" volume over R_{ij}. Thus the approximation (10) gets better and better as $\Delta s \to 0$, which enables us to define

$$V = \lim_{\Delta s \to 0} \sum_{i=1}^{n} \sum_{j=1}^{m} f(x_i{}^*, y_j{}^*) \Delta A. \tag{11}$$

EXAMPLE 1
COMPUTING THE VOLUME
UNDER A PLANE AND OVER
A RECTANGLE

Calculate the volume under the plane $z = x + 2y$ and over the rectangle $R = \{(x, y): 1 \le x \le 2 \text{ and } 3 \le y \le 5\}$.

SOLUTION: The solid whose volume we wish to calculate is sketched in Figure 6.

Step 1. For simplicity, we partition each of the intervals $[1, 2]$ and $[3, 5]$ into n subintervals of equal length (i.e., $m = n$):

$$1 = x_0 < x_1 < \cdots < x_n = 2; \; x_i = 1 + \frac{i}{n}, \Delta x = \frac{1}{n}$$

$$3 = y_0 < y_1 < \cdots < y_n = 5; \; y_i = 3 + \frac{2j}{n}, \Delta y = \frac{2}{n}.$$

Step 2. Then choosing $x_i{}^* = x_i$ and $y_j{}^* = y_j$, we obtain

$$V_{ij} \approx f(x_i{}^*, y_j{}^*) \Delta A = (x_i + 2y_j) \Delta x \, \Delta y$$

$$= \left[\left(1 + \frac{i}{n} \right) + 2 \left(3 + \frac{2j}{n} \right) \right] \frac{1}{n} \cdot \frac{2}{n}$$

$$= \left(7 + \frac{i}{n} + \frac{4j}{n} \right) \frac{2}{n^2}.$$

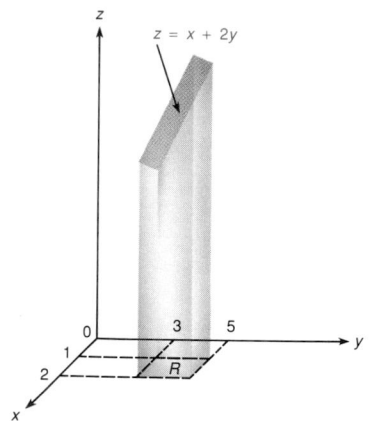

FIGURE 6
THE VOLUME OF THE SOLID UNDER A
PLANE AND OVER A RECTANGLE

Step 3.

$$V = \sum_{i=1}^{n} \sum_{j=1}^{n} V_{ij} \approx \sum_{i=1}^{n} \sum_{j=1}^{n} \left(\frac{14}{n^2} + \frac{2i}{n^3} + \frac{8j}{n^3} \right)$$

$$= \underbrace{\sum_{i=1}^{n} \sum_{j=1}^{n} \frac{14}{n^2}}_{①} + \underbrace{\sum_{i=1}^{n} \sum_{j=1}^{n} \frac{2i}{n^3}}_{②} + \underbrace{\sum_{i=1}^{n} \sum_{j=1}^{n} \frac{8j}{n^3}}_{③}.$$

It is not difficult to evaluate each of these double sums. There are n^2 terms in each sum. Since $14/n^2$ does not depend on i or j, we evaluate the sum ① by simply adding up the term $14/n^2$ a total of n^2 times. Thus,

$$\sum_{i=1}^{n} \sum_{j=1}^{n} \frac{14}{n^2} = n^2 \left(\frac{14}{n^2} \right) = 14.$$

Next, if we set $i = 1$ in ③, then we have $\sum_{j=1}^{n} 8j/n^3$. Similarly, setting $i = 2, 3, 4, \ldots, n$ in ③ yields $\sum_{j=1}^{n} 8j/n^3$. Thus in ③ we obtain the term $\left(\sum_{j=1}^{n} 8j/n^3\right)$ n times. But

$$\sum_{j=1}^{n} \frac{8j}{n^3} = \frac{8}{n^3} \sum_{j=1}^{n} j = \frac{8}{n^3}(1 + 2 + \cdots + n)$$

Equation (11) on page 284

$$\downarrow \quad = \frac{8}{n^3}\left[\frac{n(n+1)}{2}\right] = \frac{4(n+1)}{n^2}.$$

Thus,

$$\sum_{i=1}^{n} \sum_{j=1}^{n} \frac{8j}{n^3} = n\left[\sum_{j=1}^{n} \frac{8j}{n^3}\right] = n\left[\frac{4(n+1)}{n^2}\right] = \frac{4(n+1)}{n}.$$

Similarly,

$$\sum_{i=1}^{n}\left(\sum_{j=1}^{n} \frac{2i}{n^3}\right) = \sum_{i=1}^{n}\left[n\frac{2i}{n^3}\right] = \frac{2}{n^2}\sum_{i=1}^{n} i = \frac{2}{n^2}\left[\frac{n(n+1)}{2}\right] = \frac{n+1}{n}.$$

Finally, we have

$$\sum_{i=1}^{n} \sum_{j=1}^{n} V_{ij} \approx 14 + \frac{4(n+1)}{n} + \frac{n+1}{n}.$$

Step 4. As $\Delta s \to 0$, both Δx and Δy approach 0, so $n = (b - a)/\Delta x \to \infty$. Thus,

$$V = \lim_{\Delta s \to 0} \sum_{i=1}^{n} \sum_{j=1}^{n} f(x_i{}^*, y_j{}^*)\,\Delta A = \lim_{n \to \infty} \sum_{i=1}^{n} \sum_{j=1}^{n} f(x_i{}^*, y_j{}^*)\,\Delta A$$

$$= \lim_{n \to \infty}\left[14 + 4\left(\frac{n+1}{n}\right) + \frac{n+1}{n}\right] = 14 + 4 + 1 = 19.$$

The calculation we just made was very tedious. Instead of making other calculations like this one, we will define the double integral and, in Section 14.2, show how double integrals can be easily calculated.

DEFINITION   **THE DOUBLE INTEGRAL**

Let $z = f(x, y)$ and let the rectangle R be given by (1). Let $\Delta A = \Delta x \, \Delta y$. Suppose that

$$\lim_{\Delta s \to 0} \sum_{i=1}^{n} \sum_{j=1}^{m} f(x_i{}^*, y_j{}^*)\,\Delta A$$

exists and is independent of the way in which the points $(x_i{}^*, y_j{}^*)$ are chosen. Then the **double integral of f over R**, written $\iint_R f(x, y)\,dA$, is defined by

$$\iint_R f(x, y)\,dA = \lim_{\Delta s \to 0} \sum_{i=1}^{n} \sum_{j=1}^{m} f(x_i{}^*, y_j{}^*)\,\Delta A. \tag{12}$$

If the limit in (12) exists, then the function f is said to be **integrable** over R. ∎

We observe that this definition says nothing about volumes (just as the definition of the definite integral in Section 4.5 says nothing about areas). For example, if $f(x, y)$ takes on negative values in R, then the limit in (12) will not represent the volume under the surface $z = f(x, y)$. However, the limit in (12) may still exist, and in that case f will be integrable over R.

NOTE: $\iint_R f(x, y)\, dA$ is a number, not a function. This is analogous to the fact that the definite integral $\int_a^b f(x)\, dx$ is a number. We will not encounter indefinite double integrals in this book.

As we already stated, we will not calculate any other double integrals in this section but will wait until Section 14.2 to see how these calculations can be made simple. We should note, however, that the result of Example 1 can now be restated as

$$\iint_R (x + 2y)\, dA = 19,$$

where R is the rectangle $\{(x, y): 1 \leq x \leq 2 \text{ and } 3 \leq y \leq 5\}$.

What functions are integrable over a rectangle R? The following theorem is the double integral analog of Theorem 4.5.1.

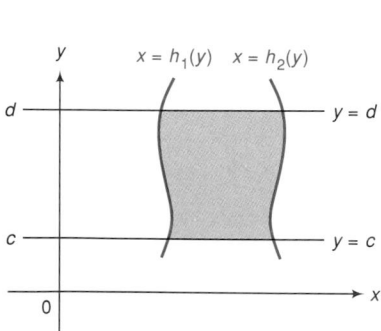

THEOREM 1 **EXISTENCE OF THE DOUBLE INTEGRAL OVER A RECTANGLE**

If f is continuous on R, then f is integrable over R. ∎

We will not give proofs of the theorems stated in this section regarding double integrals. The proofs are similar to, but more complicated than, the analogous proofs for theorems on definite integrals stated in Section 4.5. The proofs of all these theorems can be found in any standard advanced calculus text.[†]

We now turn to the question of defining double integrals over regions[‡] in $\mathbb{R}^2$ that are not rectangular. We will denote a region in $\mathbb{R}^2$ by Ω. The two types of regions in which we will be most interested are illustrated in Figure 7. In this figure $g_1, g_2, h_1,$ and h_2 denote continuous functions. A more general region Ω is sketched in Figure 8. We assume that the region is bounded. This means that there is a number M such that for every (x, y) in Ω, $|(x, y)| = \sqrt{x^2 + y^2} \leq M$. Since Ω is bounded, we can draw a rectangle R around it. Let f be defined over Ω. We then define a new function F by

$$F(x, y) = \begin{cases} f(x, y), & \text{for } (x, y) \text{ in } \Omega \\ 0, & \text{for } (x, y) \text{ in } R \text{ but not in } \Omega. \end{cases} \tag{13}$$

FIGURE 7
TWO REGIONS IN $\mathbb{R}^2$

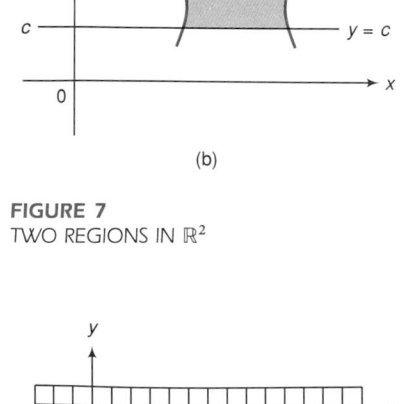

FIGURE 8
PARTITIONING THE REGION Ω BY ENCLOSING Ω IN A RECTANGLE R AND PARTITIONING R

[†] See, for example, R. C. Buck and E. F. Buck, *Advanced Calculus*, 3rd ed. (New York: McGraw-Hill, 1978), p. 175.

[‡] We will provide a formal definition of a region in Section 15.1. For our purposes in this chapter, a region is a finite union of sets in the plane that takes one of the forms in Figure 7.

DEFINITION   **INTEGRABILITY OVER A REGION IN** $\mathbb{R}^2$

Let f be defined for (x, y) in Ω and let F be defined by (13). Then we write

$$\iint\limits_{\Omega} f(x, y) \, dA = \iint\limits_{R} F(x, y) \, dA \tag{14}$$

if the integral on the right exists. In this case, we say that f is **integrable** over Ω. ∎

REMARK: If we divide R into nm subrectangles, as in Figure 8, then we can see what is happening. For each subrectangle R_{ij} that lies entirely in Ω, $F = f$, so the volume of the "parallelepiped" above R_{ij} is given by

$$V_{ij} \approx f(x_i^*, y_j^*) \, \Delta x \, \Delta y = F(x_i^*, y_j^*) \, \Delta x \, \Delta y.$$

However, if R_{ij} is in R but not in Ω, then $F = 0$, so

$$V_{ij} \approx F(x_i^*, y_j^*) \, \Delta x \, \Delta y = 0.$$

Finally, if R_{ij} is partly in Ω and partly outside of Ω, then there is no real problem since, as $\Delta s \to 0$, the sum of the volumes above these rectangles (along the boundary of Ω) will approach zero —unless the boundary of Ω is very complicated indeed. Thus, we see that the limit of the sum of the volumes of the "parallelepipeds" above R is the same as the limit of the sum of the volumes of the "parallelepipeds" above Ω. This should help explain the "reasonableness" of expression (14).

THEOREM 2   **EXISTENCE OF THE DOUBLE INTEGRAL OVER A MORE GENERAL REGION**

Let Ω be one of the regions depicted in Figure 7 where the functions g_1 and g_2 or h_1 and h_2 are continuous. Let F be defined by (13). If f is continuous over Ω, then f is integrable over Ω and its integral is given by (14). ∎

REMARK 1:   There are some regions Ω that are so complicated that there are functions continuous but not integrable over Ω. We will not concern ourselves with such regions in this book.

REMARK 2:   If f is nonnegative and integrable over Ω, then

$$\iint\limits_{\Omega} f(x, y) \, dA$$

is defined as the volume under the surface $z = f(x, y)$ and over the region Ω.

REMARK 3:   If the function $f(x, y) = 1$ is integrable over Ω, then

$$\iint\limits_{\Omega} 1 \, dA = \iint\limits_{\Omega} dA \tag{15}$$

is equal to the area of the region Ω. To see this, note that

$$V_{ij} \approx f(x_i{}^*, y_j{}^*)\,\Delta A = \Delta A,$$

so the double integral (15) is the limit of the sum of areas of rectangles in Ω.

We close this section by stating five theorems about double integrals. Each one is analogous to a theorem about definite integrals.

THEOREM 3   **DOUBLE INTEGRAL OF A CONSTANT TIMES A FUNCTION**

If f is integrable over Ω, then for any constant c, cf is integrable over Ω, and

$$\iint\limits_{\Omega} cf(x, y)\,dA = c \iint\limits_{\Omega} f(x, y)\,dA \qquad \textbf{(16)}$$

(see Theorem 4.5.4). ∎

THEOREM 4   **DOUBLE INTEGRAL OF A SUM**

If f and g are integrable over Ω, then $f + g$ is integrable over Ω, and

$$\iint\limits_{\Omega} [f(x, y) + g(x, y)]\,dA = \iint\limits_{\Omega} f(x, y)\,dA + \iint\limits_{\Omega} g(x, y)\,dA \qquad \textbf{(17)}$$

(see Theorem 4.5.5). ∎

THEOREM 5   **DOUBLE INTEGRAL OVER THE UNION OF TWO REGIONS**

If f is integrable over Ω_1 and Ω_2, where Ω_1 and Ω_2 have no points in common except perhaps those of their common boundary, then f is integrable over $\Omega = \Omega_1 \cup \Omega_2$, and

$$\iint\limits_{\Omega} f(x, y)\,dA = \iint\limits_{\Omega_1} f(x, y)\,dA + \iint\limits_{\Omega_2} f(x, y)\,dA$$

(see Theorem 4.5.3). A typical region is depicted in Figure 9. ∎

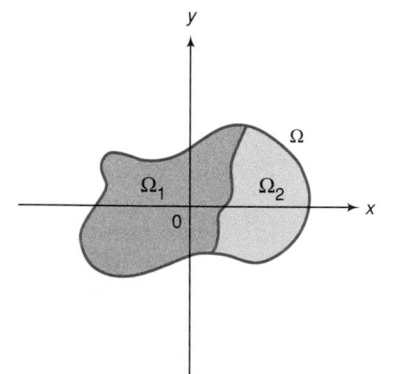

FIGURE 9
$\Omega = \Omega_1 \cup \Omega_2$

THEOREM 6   **COMPARISON THEOREM FOR DOUBLE INTEGRALS**

If f and g are integrable over Ω and $f(x, y) \le g(x, y)$ for every (x, y) in Ω, then

$$\iint\limits_{\Omega} f(x, y)\,dA \le \iint\limits_{\Omega} g(x, y)\,dA \qquad \textbf{(18)}$$

(see Theorem 4.5.7). ∎

THEOREM 7   **UPPER AND LOWER BOUND THEOREM FOR DOUBLE INTEGRALS**

Let f be integrable over Ω. Suppose that there exist constants m and M such that

$$m \le f(x, y) \le M \qquad \textbf{(19)}$$

for every (x, y) in Ω. If A_Ω denotes the area of Ω, then

$$mA_\Omega \le \iint_\Omega f(x, y)\, dA \le MA_\Omega \qquad\qquad \textbf{(20)}$$

(see Theorem 4.5.8). ∎

Theorem 7 can be useful for estimating double integrals.

EXAMPLE 2
FINDING UPPER AND LOWER
BOUNDS FOR A DOUBLE
INTEGRAL

Let Ω be the disk $\{(x, y): x^2 + y^2 \le 1\}$. Find upper and lower bounds for

$$\iint_\Omega \frac{1}{1 + x^2 + y^2}\, dA.$$

SOLUTION: Since $0 \le x^2 + y^2 \le 1$ in Ω, we easily see that

$$\frac{1}{2} < \frac{1}{1 + x^2 + y^2} \le 1.$$

Since the area of the disk is π, we have

$$\frac{\pi}{2} \le \iint_\Omega \frac{1}{1 + x^2 + y^2}\, dA \le \pi.$$

In fact, it can be shown (see Problem 14.4.24) that the value of the integral is $\pi \ln 2 \approx 0.693\pi$.

■ PROBLEMS 14.1

■ SELF-QUIZ

For each of these Self-quiz problems, let $\Omega = \{(x, y): 1 \le x \le 3, 0 \le y \le 2\}$. Also suppose f is a smooth function of (x, y).

I. If $[1, 3]$ and $[0, 2]$ are each partitioned into two subintervals of length 1, then _____ approximates $\iint_\Omega f(x, y)\, dA$.
   a. $f(0, 0) + f(0, 1) + f(1, 0) + f(1, 1)$
   b. $f(1, 0) + f(1.5, 0) + f(2, 0) + f(2.5, 0)$
   c. $f(0, 0) + f(1, 1) + f(2, 2) + f(3, 3)$
   d. $f(1.5, 0.5) + f(2.5, 0.5) + f(1.5, 1.5)$
      $+ f(2.5, 1.5)$

II. If we are told that the values of $f(x, y)$ lie be-
tween 7 and 11 when $(x, y) \in \Omega$, then we also know that $\iint_\Omega f(x, y)\, dA$ must lie between _____.
   a. 7 and 11   b. 28 and 44
   c. 30 and 50   d. -4 and 18

III. Suppose $f_x = 0$ and $f_y = 0$ throughout Ω, and also suppose $f(2, 1) = 15$; then $\iint_\Omega f(x, y)\, dA$
   _____.
   a. does not exist
   b. ≤ 57.38
   c. $= 60$
   d. lies between 30 and 45

■ DRILL

In Problems 1–8, let Ω be the rectangle $\{(x, y): 0 \le x \le 3, 1 \le y \le 2\}$. Use the technique employed in Example 1 to calculate the given double integral. Use Theorem 3 or 4 where appropriate.

1. $\displaystyle\iint_\Omega (2x + 3y)\, dA$ 2. $\displaystyle\iint_\Omega (x - y)\, dA$

3. $\displaystyle\iint_\Omega (y - x)\, dA$ 4. $\displaystyle\iint_\Omega (ax + by + c)\, dA$

CONTINUED

5. $\iint\limits_{\Omega} (x^2 + y^2)\, dA$

   [*Hint:* Use formula (2) in Appendix A.2.]

6. $\iint\limits_{\Omega} (x^2 - y^2)\, dA$

7. $\iint\limits_{\Omega} (2x^2 + 3y^2)\, dA$

8. $\iint\limits_{\Omega} (ax^2 + by^2)\, dA$

In Problems 9–14, let Ω be the rectangle $\{(x, y): -1 \le x \le 0,\ -2 \le y \le 3\}$. Calculate the given integral.

9. $\iint\limits_{\Omega} (x + y)\, dA$

10. $\iint\limits_{\Omega} (3x - y)\, dA$

11. $\iint\limits_{\Omega} (y - 2x)\, dA$

12. $\iint\limits_{\Omega} (x^2 + 2y^2)\, dA$

13. $\iint\limits_{\Omega} (y^2 - x^2)\, dA$

14. $\iint\limits_{\Omega} (3x^2 - 5y^2)\, dA$

■ **APPLICATIONS**

In Problems 15–18, suppose that $\iint_{\Omega} x\, dA = 2$ and $\iint_{\Omega} y\, dA = 7$. Compute each double integral.

15. $\iint\limits_{\Omega} (x + y)\, dA$ 16. $\iint\limits_{\Omega} (x - 2y)\, dA$

17. $\iint\limits_{\Omega} (3x + 5y)\, dA$ 18. $\iint\limits_{\Omega} (2y - 4x)\, dA$

19. Suppose that Ω_1, Ω_2, and Ω are as in Figure 9 and that $\iint_{\Omega_1} f(x, y)\, dA = 3$ and $\iint_{\Omega_2} f(x, y)\, dA = 8$. Compute $\iint_{\Omega} f(x, y)\, dA$.

20. In Problem 19 suppose that $\iint_{\Omega} f(x, y)\, dA = 12$ and $\iint_{\Omega_1} f(x, y)\, dA = 17$. Compute $\iint_{\Omega_2} f(x, y)\, dA$.

In Problems 21–25, use Theorem 7 to obtain lower and upper bounds for the given integral.

21. $\iint_{\Omega} (x^5 y^2 + xy)\, dA$, where Ω is the rectangle $\{(x, y): 0 \le x \le 1,\ 1 \le y \le 2\}$.

22. $\iint_{\Omega} e^{-(x^2 + y^2)}\, dA$, where Ω is the disk $x^2 + y^2 \le 4$.

*23. $\iint\limits_{\Omega} \left[\dfrac{x - y}{4 - x^2 - y^2} \right] dA$,

   where Ω is the disk $x^2 + y^2 \le 1$.

24. $\iint_{\Omega} \cos(\sqrt{|x|} - \sqrt{|y|})\, dA$, where Ω is the disk $x^2 + y^2 \le 1$.

25. $\iint_{\Omega} \ln(1 + x + y)\, dA$, where Ω is the region bounded by the lines $y = x$, $y = 1 - x$, and the x-axis.

*26. Let Ω be one of the regions depicted in Figure 7. Which is greater:

$$\iint\limits_{\Omega} e^{(x^2 + y^2)}\, dA \quad \text{or} \quad \iint\limits_{\Omega} (x^2 + y^2)\, dA?$$

■ **ANSWERS TO SELF-QUIZ**

I. d   II. b   III. c

14.2 THE CALCULATION OF DOUBLE INTEGRALS

In this section we derive a method for calculating $\iint_{\Omega} f(x, y)\, dx\, dy$, where Ω is one of the regions depicted in Figure 14.1.7.

   We begin, as in Section 14.1, by considering

$$\iint\limits_{R} f(x, y)\, dA, \tag{1}$$

where R is the rectangle

$$R = \{(x, y): a \le x \le b \text{ and } c \le y \le d\}. \tag{2}$$

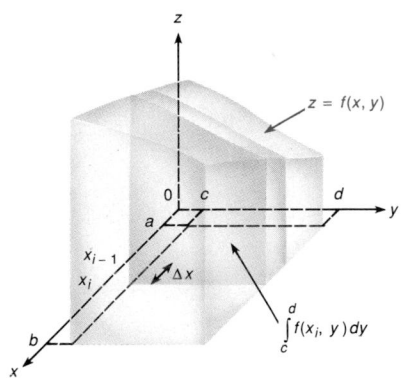

FIGURE 1
$\int_c^d f(x_i, y)\,dy$ IS THE AREA OF A CROSS-SECTION PARALLEL TO THE yz-PLANE.

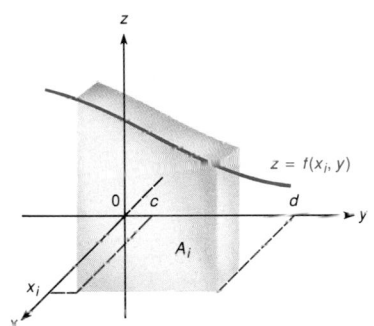

FIGURE 2
THE VOLUME OF THIS "SLICE" IS APPROXIMATED BY THE AREA OF THE FACE TIMES Δx.

If $z = f(x, y) \geq 0$ for (x, y) in R, then the double integral in (1) is the volume under the surface $z = f(x, y)$ and over the rectangle R in the xy-plane. We now calculate this volume by partitioning the x-axis, taking "slices" parallel to the yz-plane. This is illustrated in Figure 1. We can approximate the volume by adding up the volumes of the various "slices." The face of each "slice" lies in the plane $x = x_i$, and the volume of the ith slice is approximately equal to the area of its face times its thickness Δx. What is the area of the face? If x is fixed, then $z = f(x, y)$ can be thought of as a curve lying in the plane $x = x_i$. Thus, the area of the ith face is the area bounded by this curve, the xy-plane, and the planes $y = c$ and $y = d$. This area is sketched in Figure 2. If $f(x_i, y)$ is a continuous function of y, then the area of the ith face, denoted by A_i, is given by

$$A_i = \int_c^d f(x_i, y)\,dy.$$

By treating x_i as a constant, we can compute A_i as an ordinary definite integral, where the variable is y. Note, too, that $A(x) = \int_c^d f(x, y)\,dy$ is a function of x only and can therefore be integrated as in Chapter 4. Then the volume of the ith slice is approximated by

$$V_i \approx \left\{ \int_c^d f(x_i, y)\,dy \right\} \Delta x$$

so that, adding up these "subvolumes" and taking the limit as Δx approaches zero, we obtain

$$V = \int_a^b \left\{ \int_c^d f(x, y)\,dy \right\} dx = \int_a^b A(x)\,dx. \tag{3}$$

The expression in (3) is called a **repeated integral** or **iterated integral**. Since we also have

$$V = \iint_R f(x, y)\,dA,$$

we obtain

**REPEATED INTEGRAL OVER A RECTANGLE:
INTEGRATING FIRST WITH RESPECT TO y**

$$\iint_R f(x, y)\,dA = \int_a^b \left\{ \int_c^d f(x, y)\,dy \right\} dx. \tag{4}$$

REMARK 1: Usually we will write equation (4) without braces. We then have

$$\iint_R f(x, y)\,dA = \int_a^b \int_c^d f(x, y)\,dy\,dx. \tag{5}$$

REMARK 2: We should emphasize that the first integration in $\int_a^b \int_c^d f(x, y)\,dy\,dx$ is performed by treating x as a constant.

Similarly, if we instead begin by partitioning the y-axis, we find that the area of the face of a "slice" lying in the plane $y = y_i$ is given by

$$A_i = \int_a^b f(x, y_i)\,dx,$$

where now A_i is an integral in the variable x. Thus as before,

$$V = \int_c^d \left\{ \int_a^b f(x, y)\,dx \right\} dy, \tag{6}$$

and

REPEATED INTEGRAL OVER A RECTANGLE:
INTEGRATING FIRST WITH RESPECT TO x

$$\iint_R f(x, y)\,dA = \int_c^d \int_a^b f(x, y)\,dx\,dy. \tag{7}$$

EXAMPLE 1
CALCULATING A VOLUME BY
EVALUATING A REPEATED
INTEGRAL

Calculating the volume under the plane $z = x + 2y$ and over the rectangle

$$R = \{(x, y): 1 \le x \le 2 \text{ and } 3 \le y \le 5\}.$$

SOLUTION: We calculated this volume in Example 14.1.1 (see page 922). We now calculate the volume using a repeated integral. Using equation (5), we have

$$V = \iint_R (x + 2y)\,dA = \int_1^2 \left[\int_3^5 (x + 2y)\,dy \right] dx^\dagger$$

$$= \int_1^2 \left[(xy + y^2) \Big|_{y=3}^{y=5} \right] dx = \int_1^2 \left[(5x + 25) - (3x + 9) \right] dx$$

$$= \int_1^2 (2x + 16)\,dx = (x^2 + 16x) \Big|_1^2 = 19.$$

Similarly, using equation (7), we have

$$V = \int_3^5 \left[\int_1^2 (x + 2y)\,dx \right] dy = \int_3^5 \left[\left(\frac{x^2}{2} + 2yx \right) \Big|_{x=1}^{x=2} \right] dy$$

$$= \int_3^5 \left[(2 + 4y) - \left(\frac{1}{2} + 2y \right) \right] dy = \int_3^5 \left(2y + \frac{3}{2} \right) dy$$

$$= \left(y^2 + \frac{3}{2} y \right) \Big|_3^5 = 19.$$

---

† Remember, in computing the bracketed integral, we treat x as a constant.

EXAMPLE 2
CALCULATING THE VOLUME
UNDER A SURFACE AND OVER
A RECTANGLE

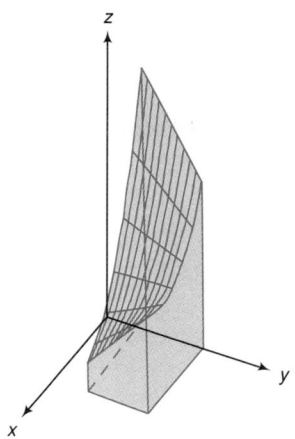

FIGURE 3
$z = xy^2 + y^3$

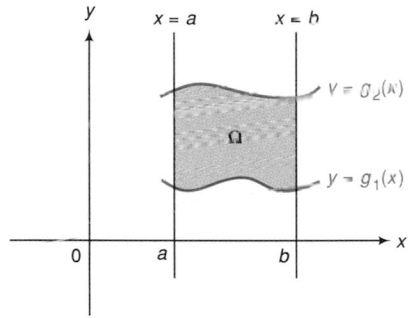

FIGURE 4
A TYPE LU REGION

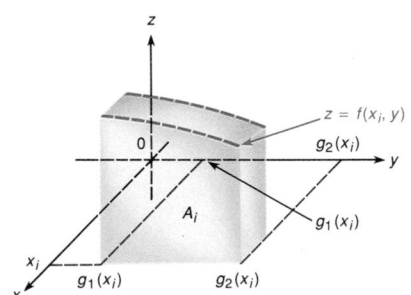

FIGURE 5
A SLICE OF VOLUME OVER AN LU REGION
OBTAINED BY PARTITIONING THE x-AXIS

Calculate the volume of the region beneath the surface $z = xy^2 + y^3$ and over the rectangle

$$R = \{(x, y): 0 \le x \le 2 \text{ and } 1 \le y \le 3\}.$$

SOLUTION: A computer-drawn sketch of this region is given in Figure 3. Using equation (5), we have

$$V = \int_0^2 \int_1^3 (xy^2 + y^3) \, dy \, dx = \int_0^2 \left[\left(\frac{xy^3}{3} + \frac{y^4}{4} \right) \Big|_1^3 \right] dx$$

$$= \int_0^2 \left[\left(9x + \frac{81}{4} \right) - \left(\frac{x}{3} + \frac{1}{4} \right) \right] dx = \int_0^2 \left(\frac{26}{3} x + 20 \right) dx$$

$$= \left(\frac{13x^2}{3} + 20x \right) \Big|_0^2 = \frac{52}{3} + 40 = \frac{172}{3}.$$

You should verify that the same answer is obtained by using equation (7).

We now extend our results to more general regions. Let

$$\Omega = \{(x, y): a \le x \le b \text{ and } g_1(x) \le y \le g_2(x)\}. \tag{8}$$

This region is sketched in Figure 4. We assume that for every x in $[a, b]$,

$$g_1(x) \le g_2(x). \tag{9}$$

This means that the *lower* and *upper* boundaries are curves. Such a region will be called a type LU region (for lower-upper). If we partition the x-axis as before, we obtain slices lying in the planes $x = x_i$, a typical one of which is sketched in Figure 5. Then

$$A_i = \int_{g_1(x_i)}^{g_2(x_i)} f(x_i, y) \, dy \qquad V_i \approx \left\{ \int_{g_1(x_i)}^{g_2(x_i)} f(x_i, y) \, dy \right\} \Delta x,$$

and

$$V = \iint_\Omega f(x, y) \, dA = \int_a^b \int_{g_1(x)}^{g_2(x)} f(x, y) \, dy \, dx. \tag{10}$$

Similarly, let

$$\Omega = \{(x, y): h_1(y) \le x \le h_2(y) \text{ and } c \le y \le d\}. \tag{11}$$

Here the *left* and *right* boundaries are curves (see Figure 6). We call this a type LR region. Then

$$V = \int_c^d \int_{h_1(y)}^{h_2(y)} f(x, y) \, dx \, dy. \tag{12}$$

We summarize these results in the following theorem.

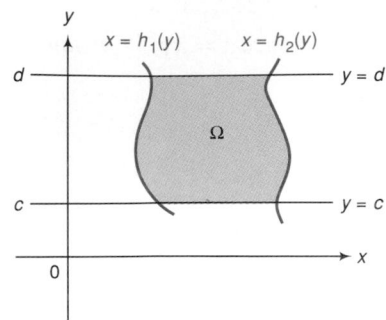

FIGURE 6
A TYPE LR REGION

THEOREM 1 **DOUBLE INTEGRALS OVER A REGION**

Let f be continuous over a region Ω given by equation (8) or (11).

(i) If Ω is a type LU region (8), where g_1 and g_2 are continuous, then

$$\iint_\Omega f(x, y)\, dA = \int_a^b \int_{g_1(x)}^{g_2(x)} f(x, y)\, dy\, dx.$$

(ii) If Ω is a type LR region (11), where h_1 and h_2 are continuous, then

$$\iint_\Omega f(x, y)\, dA = \int_c^d \int_{h_1(y)}^{h_2(y)} f(x, y)\, dx\, dy. \quad \blacksquare$$

REMARK 1: We have not actually proved this theorem here but have merely indicated why it should be so. A rigorous proof can be found in any advanced calculus text.

REMARK 2: Note that this theorem says nothing about volume. It can be used to calculate any double integral if the hypotheses of the theorem are satisfied and if each function being integrated has an antiderivative that can be written in terms of elementary functions.

REMARK 3: Many regions are of the form (8) or (11). In addition, almost all regions that arise in practical applications can be broken into a finite number of regions of the form (8) or (11), and the integration can be carried out using Theorem 14.1.5.

EXAMPLE 3
CALCULATING THE VOLUME
UNDER A SURFACE AND OVER
A TYPE LU REGION

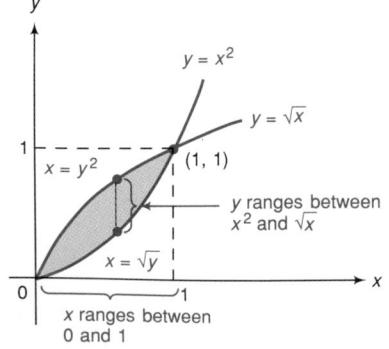

FIGURE 7
THE REGION BOUNDED BY $y = x^2$ AND
$y = \sqrt{x}$ for $0 \le x \le 1$

Find the volume of the solid under the surface $z = x^2 + y^2$ and lying above the region

$$\Omega = \{(x, y): 0 \le x \le 1 \text{ and } x^2 \le y \le \sqrt{x}\}.$$

SOLUTION: Ω is sketched in Figure 7. We see that $0 \le x \le 1$ and $x^2 \le y \le \sqrt{x}$. Then using (10), we have

$$V = \int_0^1 \int_{x^2}^{\sqrt{x}} (x^2 + y^2)\, dy\, dx = \int_0^1 \left\{ \left(x^2 y + \frac{y^3}{3} \right) \Big|_{x^2}^{\sqrt{x}} \right\} dx$$

$$= \int_0^1 \left\{ \left(x^2 \sqrt{x} + \frac{(\sqrt{x})^3}{3} \right) - \left(x^2 \cdot x^2 + \frac{(x^2)^3}{3} \right) \right\} dx$$

$$= \int_0^1 \left(x^{5/2} + \frac{x^{3/2}}{3} - x^4 - \frac{x^6}{3} \right) dx$$

$$= \left(\frac{2x^{7/2}}{7} + \frac{2x^{5/2}}{15} - \frac{x^5}{5} - \frac{x^7}{21} \right) \Big|_0^1 = \frac{2}{7} + \frac{2}{15} - \frac{1}{5} - \frac{1}{21} = \frac{18}{105}.$$

We can calculate this integral in another way. We note that x varies between the curves $x = y^2$ and $x = \sqrt{y}$. Then using (12), since $0 \le y \le 1$ and $y^2 \le x \le \sqrt{y}$, we have

$$V = \int_0^1 \int_{y^2}^{\sqrt{y}} (x^2 + y^2)\, dx\, dy,$$

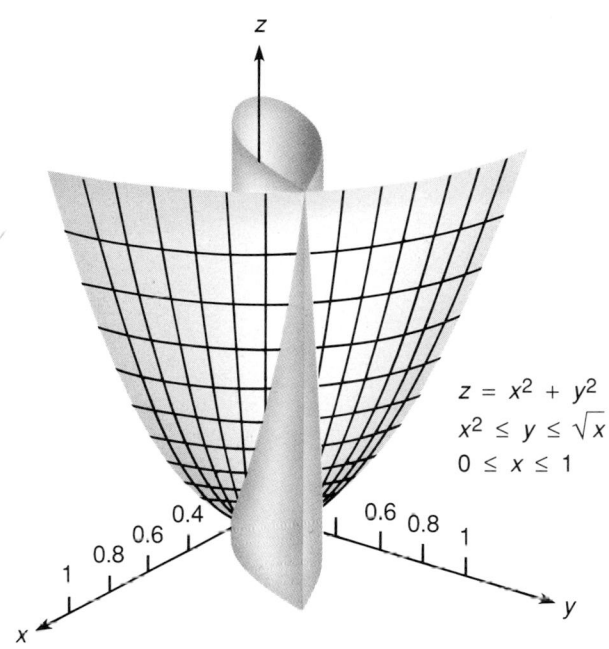

$$z = x^2 + y^2$$
$$x^2 \le y \le \sqrt{x}$$
$$0 \le x \le 1$$

FIGURE 8   COMPUTER-DRAWN SKETCH OF THE REGION UNDER THE CIRCULAR PA-
RABOLOID $z = x^2 + y^2$ AND OVER THE REGION $x^2 \le y \le \sqrt{x}, 0 \le x \le 1$

which is easily seen to be equal to $\frac{18}{105}$. The region is sketched in Figure 8.

REVERSING THE ORDER OF INTEGRATION

EXAMPLE 4

EVALUATING A DOUBLE INTEGRAL
BY REVERSING THE ORDER OF
INTEGRATION

Evaluate $\int_1^2 \int_1^{x^2} (x/y)\, dy\, dx$.

SOLUTION:

$$\int_1^2 \int_1^{x^2} \frac{x}{y}\, dy\, dx = \int_1^2 \left\{ x \ln y \Big|_1^{x^2} \right\} dx = \int_1^2 x \ln x^2\, dx = \int_1^2 2x \ln x\, dx$$

It is necessary to use integration by parts to complete the problem. Setting $u = \ln x$ and $dv = 2x\, dx$, we have $du = (1/x)\, dx$, $v = x^2$, and

$$\int_1^2 2x \ln x\, dx = x^2 \ln x \Big|_1^2 - \int_1^2 x\, dx = 4 \ln 2 - \frac{x^2}{2} \Big|_1^2 = 4 \ln 2 - \frac{3}{2}.$$

There is an easier way to calculate the repeated integral. We simply **reverse the order of integration**. The region of integration is sketched in Figure 9. If we want to integrate first with respect to x, we note that we can describe the region by

$$\Omega = \{(x, y): \sqrt{y} \le x \le 2 \text{ and } 1 \le y \le 4\}.$$

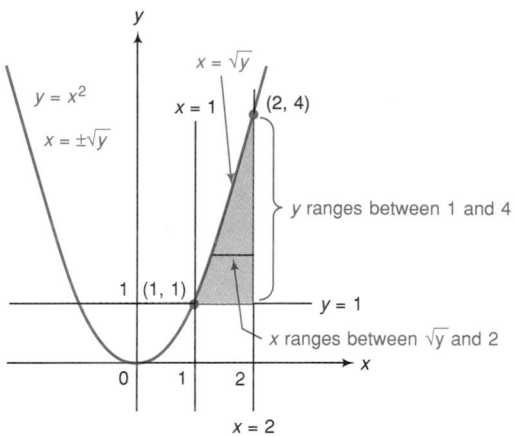

FIGURE 9   THE SHADED REGION CAN BE WRITTEN TWO WAYS: $1 \leq y \leq x^2$, $1 \leq x \leq 2$ AND $\sqrt{y} \leq x \leq 2$, $1 \leq y \leq 4$.

Then

$$\int_1^2 \int_1^{x^2} \frac{x}{y} \, dy \, dx = \iint_\Omega \frac{x}{y} \, dA = \int_1^4 \int_{\sqrt{y}}^2 \frac{x}{y} \, dx \, dy = \int_1^4 \left(\frac{x^2}{2y} \Big|_{\sqrt{y}}^2 \right) dy$$

$$= \int_1^4 \left(\frac{2}{y} - \frac{1}{2} \right) dy = \left(2 \ln y - \frac{y}{2} \right) \Big|_1^4 = 2 \ln 4 - \frac{3}{2}$$

$$= 4 \ln 2 - \frac{3}{2}.$$

Note that in this case it is easier to integrate first with respect to x.

The technique used in Example 4 suggests the following:

When changing the order of integration, first sketch the region of integration in the xy-plane.

REMARK: Why is it legitimate to reverse the order of integration? Suppose that the region Ω can be written as

$$\Omega = \{(x, y): a \leq x \leq b, g_1(x) \leq y \leq g_2(x)\}$$
$$= \{(x, y): c \leq y \leq d, h_1(y) \leq x \leq h_2(y)\}.$$

Then if f, g_1, g_2, h_1, and h_2 are continuous, we can equate the two parts of Theorem 1 to obtain

$$\iint_\Omega f(x, y) \, dA = \int_a^b \int_{g_1(x)}^{g_2(x)} f(x, y) \, dy \, dx = \int_c^d \int_{h_1(y)}^{h_2(y)} f(x, y) \, dx \, dy.$$

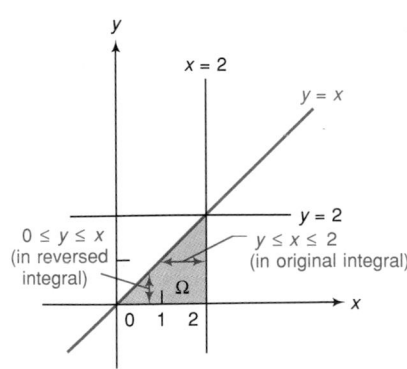

FIGURE 10
THE SHADED REGION CAN BE WRITTEN AS $y \leq x \leq 2$, $0 \leq y \leq 2$ OR $0 \leq y \leq x$, $0 \leq x \leq 2$.

EXAMPLE 5
EVALUATING A DOUBLE
INTEGRAL BY REVERSING THE
ORDER OF INTEGRATION

Compute $\int_0^2 \int_y^2 e^{x^2} \, dx \, dy$.

SOLUTION: The region of integration is sketched in Figure 10. We first observe that the double integral cannot be evaluated directly since it is impossible to find an antiderivative for e^{x^2}. Instead, we reverse the order of integration. From Figure 10 we see that Ω can be written as $0 \leq y \leq x$,

$0 \le x \le 2$, so

$$\int_0^2 \int_y^2 e^{x^2} \, dx \, dy = \iint_\Omega e^{x^2} \, dA = \int_0^2 \int_0^x e^{x^2} \, dy \, dx$$

$$= \int_0^2 \left(y e^{x^2} \Big|_{y=0}^{y=x} \right) dx = \int_0^2 x e^{x^2} \, dx$$

$$= \frac{1}{2} e^{x^2} \Big|_0^2 = \frac{1}{2} (e^4 - 1).$$

EXAMPLE 6
REVERSING THE ORDER
OF INTEGRATION

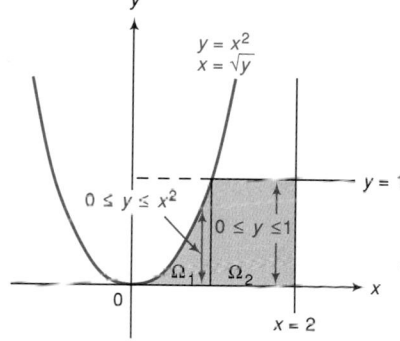

FIGURE 11
THE SHADED REGION CAN BE WRITTEN
AS $\sqrt{y} \le x \le 2$, $0 \le y \le 1$ OR AS THE
UNION OF TWO REGIONS: $\{(x, y): 0 \le y \le x^2, 0 \le x \le 1\} \cup \{(x, y): 0 \le y \le 1, 1 \le x \le 2\}$.

Reverse the order of integration in the iterated integral $\int_0^1 \int_{\sqrt{y}}^2 f(x, y) \, dx \, dy$.

SOLUTION: The region of integration is sketched in Figure 11. This region is divided into two subregions Ω_1 and Ω_2. What happens if we integrate first with respect to y? In Ω_1, $0 \le y \le x^2$. In Ω_2, $0 \le y \le 1$. Thus,

Theorem 14.1.5

$$\int_0^1 \int_{\sqrt{y}}^2 f(x, y) \, dx \, dy = \iint_\Omega f(x, y) \, dA = \iint_{\Omega_1} f(x, y) \, dA + \iint_{\Omega_2} f(x, y) \, dA$$

$$= \int_0^1 \int_0^{x^2} f(x, y) \, dy \, dx + \int_1^2 \int_0^1 f(x, y) \, dy \, dx.$$

Improper double integrals are defined in much the same way we defined improper single integrals in Section 9.3. Rather than providing a formal definition, we give an example.

EXAMPLE 7
COMPUTING THE VOLUME OF A
SOLID THAT EXTENDS OVER AN
UNBOUNDED REGION LEADS TO AN
IMPROPER DOUBLE INTEGRAL

Find the volume in the first octant bounded by the three coordinate planes and the surface $z = 1/(1 + x + 3y)^3$.

SOLUTION: The solid here extends over the unbounded region $\{(x, y): 0 \le x \le \infty$ and $0 \le y \le \infty\}$. Thus,

$$V = \int_0^\infty \int_0^\infty \frac{1}{(1 + x + 3y)^3} \, dx \, dy = \int_0^\infty \lim_{N \to \infty} \left(-\frac{1}{2(1 + x + 3y)^2} \Big|_0^N \right) dy$$

$$= \int_0^\infty \frac{1}{2(1 + 3y)^2} \, dy = \lim_{N \to \infty} \left(-\frac{1}{6(1 + 3y)} \right) \Big|_0^N = \frac{1}{6}.$$

Note that improper double integrals can be treated in the same way that we treat improper "single" integrals.

EXAMPLE 8
FINDING THE VOLUME BOUNDED
BY A PLANE AND THE THREE
COORDINATE PLANES

Find the volume of the solid bounded by the coordinate planes and the plane $2x + y + z = 2$.

SOLUTION: We have $z = 2 - 2x - y$ and this expression must be integrated over the region in the xy-plane bounded by the line $2x + y = 2$

(obtained when $z = 0$) and the x- and y-axes. See Figure 12. We therefore have

$$V = \int_0^1 \int_0^{2-2x} (2 - 2x - y)\,dy\,dx = \int_0^1 \left\{ \left(2y - 2xy - \frac{y^2}{2} \right) \Big|_0^{2-2x} \right\} dx$$

$$= \int_0^1 \left\{ 2(2 - 2x) - 2x(2 - 2x) - \frac{(2 - 2x)^2}{2} \right\} dx$$

$$= \int_0^1 (2x^2 - 4x + 2)\,dx = \left(\frac{2x^3}{3} - 2x^2 + 2x \right) \Big|_0^1 = \frac{2}{3}.$$

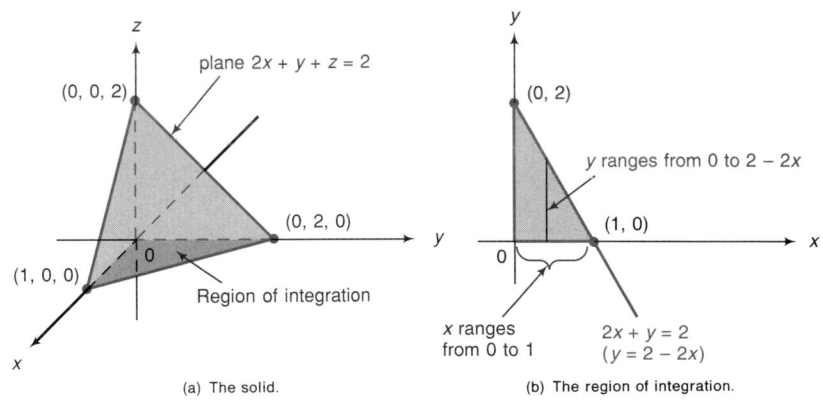

(a) The solid.

(b) The region of integration.

FIGURE 12 TO FIND THE VOLUME OF THE SOLID IN (a), WE INTEGRATE OVER THE REGION (b).

COMPUTING DIFFICULT IMPROPER INTEGRALS BY PARAMETRIC INTEGRATION[†]

There is a very clever technique that can sometimes be used to evaluate an integral that cannot otherwise be integrated using methods from calculus.

EXAMPLE 9
EVALUATING AN IMPORTANT
IMPROPER INTEGRAL[‡]

Compute $\int_0^\infty \dfrac{\sin x}{x}\,dx$.

SOLUTION: Consider $\int_0^\infty e^{-px} \sin x\,dx$. Using integration by parts, it is not difficult to show (see Example 4 in Section 7.2, page 481, or entry

---

[†] The material here was suggested by the article "Parametric Integration Techniques" by Aurel J. Zajta and Sudkir K. Goel in *Mathematics Magazine*, 62(5), December 1989, pages 318–322.

[‡] To see why the function $\dfrac{\sin x}{x}$ is important, read the article "The function $\dfrac{\sin x}{x}$," by William B. Gearhart and Harris S. Shultz in *The College Mathematics Journal*, 21(2), March 1990, pages 90–99.

168 in the table of integrals at the back of the book) that

$$\int_0^N e^{-px} \sin x \, dx = \frac{e^{-px}(-p \sin x - \cos x)}{1 + p^2}\bigg|_0^N$$

$$= \frac{1}{1 + p^2}\left[1 - e^{-pN}(-p \sin N - \cos N)\right]$$

and, for $p > 0$,

$$\int_0^\infty e^{-px} \sin x \, dx = \lim_{N \to \infty} \int_0^N e^{-px} \sin x \, dx = \frac{1}{1 + p^2}.$$

Now, let $f(p) = \int_0^\infty e^{-px} \sin x \, dx$. That is, we treat p as a parameter. Then

$$f(p) = \frac{1}{1 + p^2},$$

so

$$\int_0^\infty f(p) \, dp = \int_0^\infty \frac{1}{1 + p^2} \, dp = \lim_{N \to \infty} \int_0^N \frac{dp}{1 + p^2} = \lim_{N \to \infty} \tan^{-1} p\bigg|_0^N$$

$$= \lim_{N \to \infty} \tan^{-1} N = \frac{\pi}{2}.$$

Thus,

$$\frac{\pi}{2} = \int_0^\infty f(p) \, dp = \int_0^\infty \int_0^\infty e^{-px} \sin x \, dx \, dp$$

reverse the order
of integration

$$\downarrow$$
$$= \int_0^\infty \int_0^\infty e^{-px} \sin x \, dp \, dx^\dagger$$

$$= \int_0^\infty \left[\int_0^\infty e^{-px} \, dp\right] \sin x \, dx$$

$$= \int_0^\infty \left[\lim_{N \to \infty} \int_0^N e^{-px} \, dp\right] \sin x \, dx$$

$$= \int_0^\infty \left[\lim_{N \to \infty} \frac{-e^{-px}}{x}\bigg|_{p=0}^N\right] \sin x \, dx$$

$$= \int_0^\infty \left[\lim_{N \to \infty} \frac{1}{x}(1 - e^{-Nx})\right] \sin x \, dx$$

$$= \int_0^\infty \frac{\sin x}{x} \, dx.$$

Thus,

$$\int_0^\infty \frac{\sin x}{x} \, dx = \frac{\pi}{2}.$$

---

† We need a theorem to justify reversing the order of integration when the limits are infinite, but we won't worry about that here. The result is correct.

PROBLEMS 14.2

SELF-QUIZ

I. $\int_1^3 \int_2^4 2xy\,dy\,dx = $ _____ .

  a. $\int_1^3 12x\,dy$ b. $\int_1^3 12y\,dx$

  c. $\int_1^3 12x\,dx$ d. $\int_2^4 8x\,dy$

II. If $\Omega = \{(x, y): 1 \le x \le 3,\ 2 \le y \le 4\}$ and if f is a smooth function, then $\iint_\Omega f(x, y)\,dA = $ _____ .

  a. $\int_1^3 A(x)\,dx$ where $A(x) = \int_2^4 f(x, y)\,dy$

  b. $\int_1^3 B(y)\,dy$ where $B(y) = \int_2^4 f(x, y)\,dx$

  c. $\int_1^3 \int_2^4 f(x, y)\,dx\,dy$

  d. $\int_1^3 \int_2^4 f(x, y)\,dy\,dx$

  e. $\int_2^4 \int_1^3 f(x, y)\,dy\,dx$

  f. $\int_2^4 \int_1^3 f(x, y)\,dx\,dy$

III. The integral of $f(x, y) = x^2 y$ over the region bounded by the x-axis and the semicircle $x^2 + y^2 = 4$, $y \ge 0$ equals _____ .

  a. $\int_{-2}^2 \int_0^{\sqrt{4-x^2}} x^2 y\,dy\,dx$

  b. $\int_{-2}^2 \int_0^{\sqrt{4-x^2}} x^2 y\,dx\,dy$

  c. $\int_0^{\sqrt{4-x^2}} \int_{-2}^2 x^2 y\,dx\,dy$

  d. $\int_0^2 \int_{-\sqrt{4-y^2}}^{\sqrt{4-y^2}} x^2 y\,dx\,dy$

  e. $\int_0^2 \int_{-\sqrt{4-y^2}}^{\sqrt{4-y^2}} x^2 y\,dy\,dx$

  f. $\int_{-\sqrt{4-y^2}}^{\sqrt{4-y^2}} \int_0^2 x^2 y\,dy\,dx$

DRILL

In Problems 1–8, evaluate the given iterated integral.

1. $\int_0^1 \int_0^2 xy^2\,dx\,dy$

2. $\int_{-1}^3 \int_2^4 (x^2 - y^3)\,dy\,dx$

3. $\int_2^5 \int_0^4 e^{(x-y)}\,dx\,dy$ 4. $\int_0^1 \int_{x^2}^x x^3 y\,dy\,dx$

5. $\int_2^4 \int_{1+y}^{2+3y} (x - y^2)\,dx\,dy$

6. $\int_{\pi/4}^{\pi/3} \int_{\sin x}^{\cos x} (x + 2y)\,dy\,dx$

7. $\int_0^3 \int_{-\sqrt{9-y^2}}^{\sqrt{9-y^2}} x^2 y\,dx\,dy$ 8. $\int_1^2 \int_{y^5}^{3y^5} \frac{1}{x}\,dx\,dy$

In Problems 9–24, evaluate the given double integral by means of an appropriate iterated integral.

9. $\iint_\Omega (x^2 + y^2)\,dA$, where $\Omega = \{(x, y): 1 \le x \le 2, -1 \le y \le 1\}$.

10. $\iint_\Omega 2xy\,dA$, where $\Omega = \{(x, y): 0 \le x \le 4, 1 \le y \le 3\}$.

11. $\iint_\Omega (x - y)^2\,dA$, where $\Omega = \{(x, y): -2 \le x \le 2, 0 \le y \le 1\}$.

12. $\iint_\Omega \sin(2x + 3y)\,dA$, where $\Omega = \{(x, y): 0 \le x \le \pi/6, 0 \le y \le \pi/18\}$.

13. $\iint_\Omega xe^{(x^2+y)}\,dA$, where Ω is the region of Problem 10.

14. $\iint_\Omega (x - y^2)\,dA$, where Ω is the region in the first quadrant bounded by the x-axis, the y-axis, and the unit circle.

15. $\iint_\Omega (x^2 + y)\,dA$, where Ω is the region of Problem 14.

16. $\iint_\Omega (x^3 - y^3)\,dA$, where Ω is the region of Problem 14.

17. $\iint_\Omega (x + 2y)\,dA$, where Ω is the triangular region bounded by the lines $y = x$, $y = 1 - x$, and the y-axis.

18. $\iint_\Omega e^{(x+2y)}\,dA$, where Ω is the region of Problem 17.

19. $\iint_\Omega (x^2 + y)\,dA$, where Ω is the region in the first quadrant between the parabolas $y = x^2$ and $y = 1 - x^2$.

20. $\iint_\Omega (1/\sqrt{y})\,dA$, where Ω is the region of Problem 19.

*21. $\iint_\Omega (y/\sqrt{x^2 + y^2})\,dA$, where $\Omega = \{(x, y): 1 \le x \le y, 1 \le y \le 2\}$.

22. $\iint_\Omega [e^{-y}/(1 + x^2)]\,dA$, where Ω is the first quadrant.

23. $\iint_\Omega \sin x \cos y\,dA$ where Ω is the rectangle $\{(x, y): 0 \le x \le \pi/2, 0 \le y \le \pi\}$.

24. $\iint_\Omega \sin^2 y\,dA$ where $\Omega = \{(x, y): 0 \le x \le \cos y, 0 \le y \le \pi/6\}$.

In Problems 25–28, evaluate $\iint_\Omega x^2 y\, dA$ over the given region Ω.

25.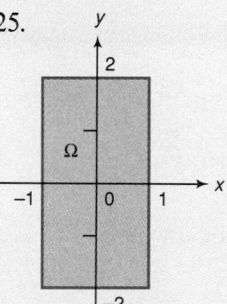

26.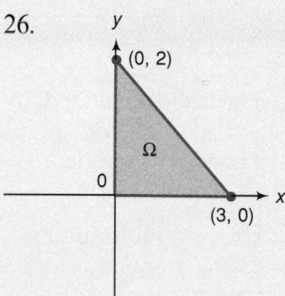

27.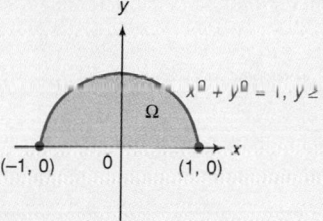

28.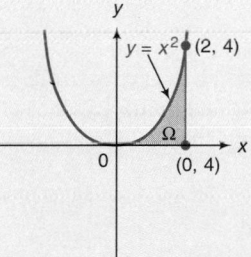

In Problems 29–32, evaluate $\iint_\Omega \dfrac{y}{x^2}\, dA$ over the given region Ω.

29.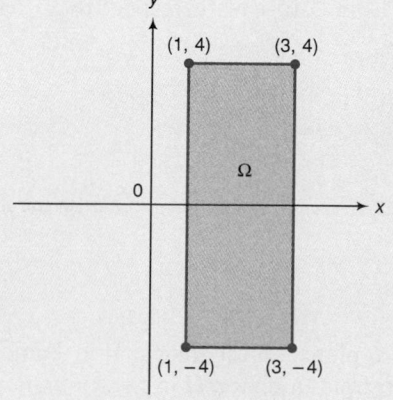

30.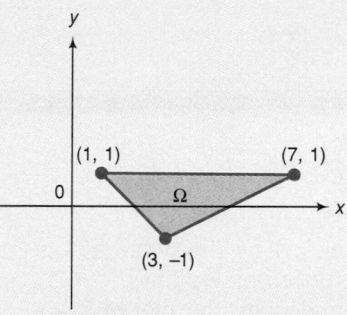

31.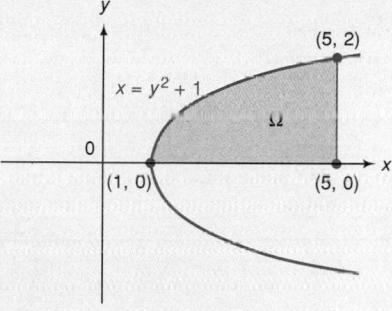

*32.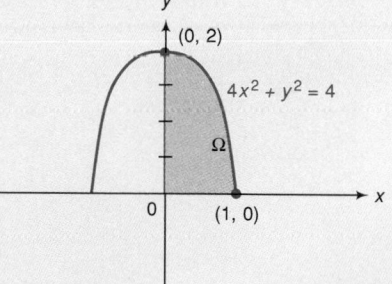

In Problems 33–42, (a) sketch the region over which the integral is taken. Then (b) change the order of integration, and (c) evaluate the given integral.

33. $\displaystyle\int_0^2 \int_{-1}^3 dx\, dy$

34. $\displaystyle\int_0^5 \int_{-5}^8 (x + y)\, dy\, dx$

35. $\displaystyle\int_2^4 \int_1^y (y^3/x^3)\, dx\, dy$

36. $\displaystyle\int_0^1 \int_0^x dy\, dx$

37. $\displaystyle\int_0^1 \int_x^1 dy\, dx$

38. $\displaystyle\int_0^{\pi/2} \int_0^{\cos y} y\, dx\, dy$

39. $\displaystyle\int_0^2 \int_0^{\sqrt{4-y^2}} (4 - x^2)^{3/2}\, dx\, dy$

40. $\displaystyle\int_0^1 \int_{\sqrt{x}}^{\sqrt[3]{x}} (1 + y^6)\, dy\, dx$

CONTINUED

41. $\int_0^1 \int_{\sqrt{y}}^1 \sqrt{3 - x^3}\, dx\, dy$

■ APPLICATIONS

In Problems 43–52, find the volume of the given solid.

43. The solid bounded by the plane $x + y + z = 3$ and the three coordinate planes.

*44. The solid bounded by the planes $x = 0$, $z = 0$, $x + 2y + z = 6$, and $x - 2y + z = 6$.

45. The solid bounded by the cylinders $x^2 + y^2 = 4$ and $y^2 + z^2 = 4$.

46. The solid bounded by the cylinder $x^2 + z^2 = 1$ and the planes $y = 0$ and $y = 2$.

*47. The ellipsoid $x^2 + 4y^2 + 9z^2 = 36$.

*48. The solid bounded above by the sphere $x^2 + y^2 + z^2 = 9$ and below by the plane $z = \sqrt{5}$.

49. The solid bounded by the surface $z = e^{-(x+y)}$ and the three coordinate planes.

*50. The solid bounded by the parabolic cylinder $x = z^2$ and the planes $y = 1$, $y = 5$, $z = 1$, and $x = 0$.

51. The solid bounded by the circular paraboloid $x = y^2 + z^2$ and the plane $x = 1$.

*52. The solid bounded by the paraboloid $y = x^2 + z^2$ and the plane $x + y = 3$.

*53. Use a double integral to find the area of each of the regions bounded by the x-axis and the curves $y = x^3 + 1$ and $y = 3 - x^2$.

54. Use a double integral to find the area in the first quadrant bounded by the curves $y = x^{1/m}$ and $y = x^{1/n}$, where m and n are positive and $n > m$.

55. Sketch the solid whose volume is given by $\int_1^3 \int_0^2 (x + 3y)\, dx\, dy$.

*56. Sketch the solid whose volume is given by $\int_0^1 \int_{x^2}^{\sqrt{x}} \sqrt{x^2 + y^2}\, dy\, dx$.

■ SHOW/PROVE/DISPROVE

57. Show that if both of the following integrals exist, then

$$\int_0^\infty \int_0^x f(x, y)\, dy\, dx = \int_0^\infty \int_y^\infty f(x, y)\, dx\, dy.$$

[*Hint:* Draw a picture.]

58. Suppose $f(x, y) = g(x)h(y)$, where g and h are continuous. Let Ω be the rectangle $\{(x, y): a \le x \le b, c \le y \le d\}$. Show that

$$\iint_\Omega f(x, y)\, dA = \left[\int_a^b g(x)\, dx \right] \cdot \left[\int_c^d h(y)\, dy \right].$$

■ CHALLENGE

*59. Evaluate $\int_0^\infty \left(\dfrac{\sin t}{t} \right)^2 dt$ using the technique of Example 9. [*Hint:* Substitute $t = \dfrac{x}{p}$ in Example 9

to obtain $\int_0^\infty \dfrac{\sin pt}{t}\, dt = \dfrac{\pi}{2}$ and then integrate both sides from 0 to q with respect to p. Then substitute $q = 2$.]

■ ANSWERS TO SELF-QUIZ

I. c   II. a = d, f   III. a, d

14.3 DENSITY, MASS, AND CENTER OF MASS (OPTIONAL)

Let $\rho(x, y)$ denote the density of a plane object (like a thin lamina, for example). Suppose that the object occupies a region Ω in the xy-plane. Then the mass of a small rectangle of sides Δx and Δy centered at the point (x, y)

is approximated by

$$\rho(x, y)\,\Delta x\,\Delta y = \rho(x, y)\,\Delta A. \tag{1}$$

MASS

If we sum these approximate masses and take a limit, then we obtain a formula for the total mass:

THE TOTAL MASS μ IS GIVEN BY

$$\mu = \iint\limits_{\Omega} \rho(x, y)\,dA. \tag{2}$$

Compare this formula with the formula for the mass of an object lying along the x-axis with density $\rho(x)$ [see equation (8) in Section 4.2 (page 273)].

In Sections 5.5 and 5.6 we showed how to calculate the first moment and center of mass of an object around the x- and y-axes. For example, we defined

$$M_y = \int_a^b x\rho(x)\,dx \tag{3}$$

to be the first moment about the y-axis when we had a system of masses distributed along the x-axis. Similarly, we calculated the x-coordinate of the center of mass of the object to be

$$\bar{x} = \frac{\int_a^b x\rho(x)\,dx}{\int_a^b \rho(x)\,dx} = \frac{\text{first moment about y-axis}}{\text{mass}} = \frac{M_y}{\mu} \tag{4}$$

[see equation (5.5.7) on page 367].

When, in Section 5.6, we calculated the centroid of a plane region, we found that it was necessary to assume that the region had a constant area density ρ. However, by using double integrals, we can get away from this restriction. Consider the plane region whose mass is given by (2). Then we define

---

FIRST MOMENT AROUND THE y-AXIS

$$M_y = \text{first moment around y-axis} = \iint\limits_{\Omega} x\rho(x, y)\,dA. \tag{5}$$

Look at Figure 1. The first moment about the y-axis of a small rectangle centered at (x, y) is given approximately by

$$x_i{}^*\rho(x_i{}^*, y_j{}^*)\,\Delta x\,\Delta y,$$

and if we add up these moments for all such "subrectangles" and take a limit, we arrive at equation (5). Finally, we define the **center of mass** of the

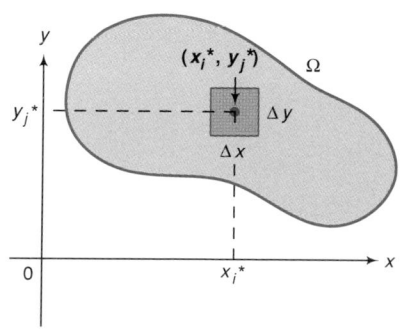

FIGURE 1
THE FIRST MOMENT ABOUT THE y-AXIS OF A SMALL SUBRECTANGLE CENTERED AT $(x_i{}^*, y_j{}^*)$ IS APPROXIMATED BY $x_i{}^*\rho(x_i{}^*, y_j{}^*)\,\Delta x\Delta y.$

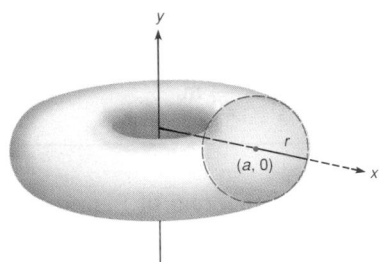

FIGURE 5
A TORUS

SOLUTION: The circle and the torus are sketched in Figure 5. The area of the circle is πr^2. The radius of the circle traced by the centroid $(a, 0)$ is a, and the circumference is $2\pi a$. Thus,

$$V = (2\pi a)\pi r^2 = 2\pi^2 a r^2.$$

■ PROBLEMS 14.3

■ SELF-QUIZ

I. If Ω is a region of constant density, then $(\iint_\Omega x\,dA)/(\iint_\Omega dA)$ computes _____.
   a. the x-coordinate of the region's center of mass
   b. the first moment of the region about the x-axis
   c. the y-coordinate of the region's center of mass
   d. the first moment of the region about the y-axis

II. Consider an object filling a region Ω and having density given by $\rho(x, y)$. The first moment of this object about the x-axis equals _____.

a. $\displaystyle\iint_\Omega \rho(x, y)\,dA$ b. $\displaystyle\iint_\Omega y\rho(x, y)\,dA$

c. $\displaystyle\iint_\Omega x\rho(x, y)\,dA$

d. $\displaystyle\left(\iint_\Omega x\rho(x, y)\,dA\right)\Big/\left(\int_\Omega \rho(x, y)\,dA\right)$

■ DRILL

In Problems 1–12, find the mass and the center of mass of an object that lies in the given region and has the given area density function.

1. $\Omega = \{(x, y): 1 \le x \le 2,\ -1 \le y \le 1\}$;
   $\rho(x, y) = x^2 + y^2$
2. $\Omega = \{(x, y): 0 \le x \le 4,\ 1 \le y \le 3\}$; $\rho(x, y) = 2xy$
3. $\Omega = \{(x, y): 0 \le x \le \pi/6,\ 0 \le y \le \pi/18\}$;
   $\rho(x, y) = \sin(2x + 3y)$
4. $\Omega = \{(x, y): -2 \le x \le 2,\ 0 \le y \le 1\}$;
   $\rho(x, y) = (x - y)^2$

5. Ω is the region of Problem 2; $\rho(x, y) = xe^{x-y}$
6. Ω is the quarter of the unit circle lying in the first quadrant; $\rho(x, y) = x + y^2$
7. Ω is the region of Problem 6; $\rho(x, y) = x^2 + y$
8. Ω is the region of Problem 6; $\rho(x, y) = x^3 + y^3$
9. Ω is the triangular region bounded by the lines $y = x$, $y = 1 - x$, and the x-axis; $\rho(x, y) = x + 2y$
10. Ω is the region of Problem 9; $\rho(x, y) = e^{x+2y}$
11. Ω is the first quadrant; $\rho(x, y) = e^{-y}/(1 + x)^3$
12. Ω is the first quadrant; $\rho(x, y) = (x + y)e^{-(x+y)}$

■ APPLICATIONS

13. Use the first theorem of Pappus to calculate the volume of the torus generated by rotating the unit circle about the line $y = 4 - x$.
14. Use the first theorem of Pappus to calculate the

volume of the "elliptical torus" generated by rotating the ellipse $(x/a)^2 + (y/b)^2 = 1$ about the line $y = 3a$. Assume that $3a > b > 0$.

■ ANSWERS TO SELF-QUIZ

I. a   II. b

14.4 DOUBLE INTEGRALS IN POLAR COORDINATES

In this section, we will see how to evaluate double integrals of functions in the form $z = f(r, \theta)$, where r and θ denote the polar coordinates of a point in the plane.

Let $z = f(r, \theta)$ and let Ω denote the "polar rectangle"

$$\theta_1 \leq \theta \leq \theta_2, \qquad r_1 \leq r \leq r_2. \tag{1}$$

This region is sketched in Figure 1. We will calculate the volume of the solid between the surface $z = f(r, \theta)$ and the region Ω. We partition Ω by small "polar rectangles" and calculate the volume over such a region. The volume of the part of the solid over the region Ω_{ij} is given, approximately, by

$$f(r_i, \theta_j) A_{ij}, \tag{2}$$

where A_{ij} is the area of Ω_{ij}. Recall from Section 8.7 (p. 591) that if $r = f(\theta)$, then the area bounded by the lines $\theta = \alpha$, $\theta = \beta$, and the curve $r = f(\theta)$ is given by

$$A = \int_\alpha^\beta \frac{1}{2} [f(\theta)]^2 \, d\theta. \tag{3}$$

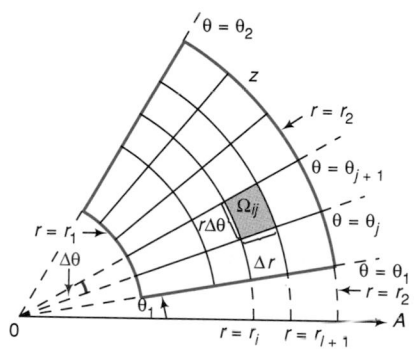

FIGURE 1
THE SHADED POLAR RECTANGLE IS BOUNDED BY THE LINES $\theta = \theta_j$ AND $\theta = \theta_{j+1}$, AND THE CIRCLES $r = r_i$ AND $r = r_{i+1}$.

Thus,

$$A_{ij} = \frac{1}{2} \int_{\theta_j}^{\theta_{j+1}} (r_{i+1}^2 - r_i^2) \, d\theta = \frac{1}{2}(r_{i+1}^2 - r_i^2)(\theta_{j+1} - \theta_j)$$

$$= \frac{1}{2}(r_{i+1} + r_i)(r_{i+1} - r_i)(\theta_{j+1} - \theta_j) = \frac{1}{2}(r_{i+1} + r_i) \, \Delta r \, \Delta \theta.$$

But if Δr is small, then $r_{i+1} \approx r_i$, and we have

$$A_{ij} \approx \frac{1}{2}(2r_i) \, \Delta r \, \Delta \theta = r_i \, \Delta r \, \Delta \theta.$$

Then

$$V_{ij} \approx f(r_i, \theta_j) A_{ij} \approx f(r_i, \theta_j) r_i \, \Delta r \, \Delta \theta,$$

so that, adding up the individual volumes and taking a limit, we obtain

---

VOLUME IN POLAR COORDINATES

$$V = \int_\Omega f(r, \theta) r \, dr \, d\theta. \tag{4}$$

NOTE: Do not forget the extra r in the formula above.

EXAMPLE 1
FINDING THE VOLUME ENCLOSED BY A SPHERE

Find the volume enclosed by the sphere $x^2 + y^2 + z^2 = a^2$.

SOLUTION: We will calculate the volume enclosed by the hemisphere $z = \sqrt{a^2 - x^2 - y^2}$ and then multiply by two. To do so, we first note that this volume is the volume of the solid under the hemisphere and above the disk $x^2 + y^2 \leq a^2$. We use polar coordinates, since in polar coordinates $x^2 + y^2 = (r \cos \theta)^2 + (r \sin \theta^2) = r^2$. On the disk, $0 \leq r \leq a$ and

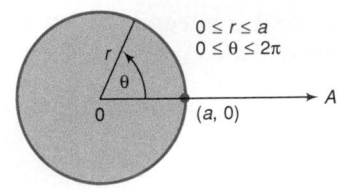

FIGURE 2
A DISK OF RADIUS r

$0 \le \theta \le 2\pi$ (see Figure 2). Then $\sqrt{a^2 - (x^2 + y^2)} = \sqrt{a^2 - r^2}$, so by (4)

$$V = \iint_{\text{disk}} \sqrt{a^2 - r^2}\, r\, dr\, d\theta = \int_0^{2\pi} \int_0^a \sqrt{a^2 - r^2}\, r\, dr\, d\theta$$

$$= \int_0^{2\pi} \left\{ -\frac{1}{3}(a^2 - r^2)^{3/2} \Big|_0^a \right\} d\theta = \int_0^{2\pi} \frac{1}{3} a^3\, d\theta = \frac{2\pi a^3}{3}.$$

Thus, the volume of the sphere is $2(2\pi a^3/3) = \frac{4}{3}\pi a^3$.

AREA FORMULA

Since area of $\Omega = \iint_\Omega dA$, we can write, using (4),

THE AREA OF A POLAR REGION

$$\iint_\Omega r\, dr\, d\theta = \text{area of } \Omega. \tag{5}$$

EXAMPLE 2
COMPUTING THE AREA ENCLOSED
BY A CARDIOID

Calculate the area enclosed by the cardioid $r = 1 + \sin \theta$.

SOLUTION: The cardioid is sketched in Figure 3. The region is described by the inequalities $0 \le r \le 1 + \sin \theta$, $0 \le \theta \le 2\pi$. Thus,

$$\text{area} = \int_0^{2\pi} \int_0^{1 + \sin \theta} r\, dr\, d\theta = \int_0^{2\pi} \left\{ \frac{r^2}{2} \Big|_0^{1 + \sin \theta} \right\} d\theta$$

$$= \frac{1}{2} \int_0^{2\pi} (1 + 2 \sin \theta + \sin^2 \theta)\, d\theta$$

$$= \frac{1}{2} \int_0^{2\pi} \left(1 + 2 \sin \theta + \frac{1 - \cos 2\theta}{2} \right) d\theta$$

$$= \frac{1}{2} \left(\frac{3\theta}{2} - 2 \cos \theta - \frac{\sin 2\theta}{4} \right) \Big|_0^{2\pi} = \frac{3\pi}{2}.$$

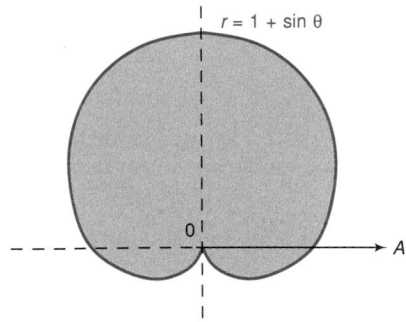

FIGURE 3
THE CARDIOID $r = 1 + \sin \theta$

As we will see, it is often very useful to write a double integral in terms of polar coordinates. Let $z = f(x, y)$ be a function defined over a region Ω. Then using polar coordinates, we can write

$$z = f(r \cos \theta, r \sin \theta),$$

and we can also describe Ω in terms of polar coordinates. The volume of the solid under f and over Ω is the same whether we use rectangular or polar coordinates. Thus, writing the volume in both rectangular and polar coordinates, we obtain the useful **change-of-variables formula:**

CONVERTING A DOUBLE INTEGRAL FROM RECTANGULAR TO POLAR COORDINATES

$$\iint_\Omega f(x, y)\, dA = \iint_\Omega f(r \cos \theta, r \sin \theta)\, r\, dr\, d\theta. \tag{6}$$

EXAMPLE 3
FINDING THE MASS OF A
SEMICIRCULAR PLANE LAMINA USING
POLAR COORDINATES

The density at any point on a semicircular plane lamina is proportional to the square of the distance from the point to the center of the circle. Find the mass of the lamina.

SOLUTION: We have $\rho(x, y) = \alpha(x^2 + y^2) = \alpha r^2$ and $\Omega = \{(r, \theta): 0 \le r \le a$ and $0 \le \theta \le \pi\}$, where a is the radius of the circle. Then

$$\mu = \int_0^\pi \int_0^a (\alpha r^2) r \, dr \, d\theta = \int_0^\pi \left\{ \frac{\alpha r^4}{4} \Big|_0^a \right\} d\theta = \frac{\alpha a^4}{4} \int_0^\pi d\theta = \frac{\alpha \pi a^4}{4}.$$

This double integral can be computed without using polar coordinates, but the computation is much more tedious. Try it!

EXAMPLE 4
COMPUTING THE VOLUME OF A
SOLID USING POLAR COORDINATES

Find the volume of the solid bounded by the xy-plane, the cylinder $x^2 + y^2 = 4$, and the paraboloid $z = 2(x^2 + y^2)$.

SOLUTION: The volume requested is the volume under the surface $z = 2(x^2 + y^2) = 2r^2$ and above the circle $x^2 + y^2 = 4$. Thus,

$$V = \int_0^{2\pi} \int_0^2 2r^2 \cdot r \, dr \, d\theta = \int_0^{2\pi} \left\{ \frac{r^4}{2} \Big|_0^2 \right\} d\theta = 16\pi.$$

The solid is sketched in Figure 4.

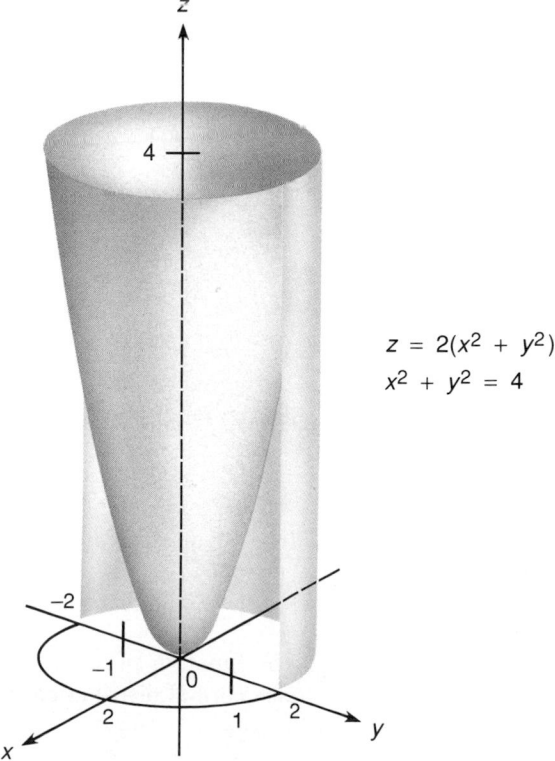

$z = 2(x^2 + y^2)$
$x^2 + y^2 = 4$

FIGURE 4   THE SOLID ABOVE THE xy-PLANE BOUNDED BY THE CYLINDER $x^2 + y^2 = 4$ AND THE PARABOLOID $z = 2(x^2 + y^2)$

EXAMPLE 5
USING POLAR COORDINATES TO
EVALUATE AN INTEGRAL IMPORTANT
IN PROBABILITY THEORY

In probability theory, one of the most important integrals is

$$\int_{-\infty}^{\infty} e^{-x^2}\, dx.$$

We now show how a combination of double integrals and polar coordinates can be used to evaluate it. Let

$$I = \int_{0}^{\infty} e^{-x^2}\, dx.$$

Then by symmetry $\int_{-\infty}^{\infty} e^{-x^2}\, dx = 2I$. Thus, we need only to evaluate I. But since any dummy variable can be used in a definite integral, we also have

$$I = \int_{0}^{\infty} e^{-y^2}\, dy.$$

Thus,

$$I^2 = \left(\int_{0}^{\infty} e^{-x^2}\, dx \right) \left(\int_{0}^{\infty} e^{-y^2}\, dy \right),$$

and from a result similar to that of Problem 14.2.58,[†]

$$I^2 = \int_{0}^{\infty} \int_{0}^{\infty} e^{-x^2} e^{-y^2}\, dx\, dy = \int_{0}^{\infty} \int_{0}^{\infty} e^{-(x^2+y^2)}\, dx\, dy = \iint\limits_{\Omega} e^{-(x^2+y^2)}\, dA,$$

where Ω denotes the first quadrant. In polar coordinates the first quadrant can be written as

$$\Omega = \left\{ (r, \theta): 0 \le r < \infty \text{ and } 0 \le \theta \le \frac{\pi}{2} \right\}.$$

Thus, since $x^2 + y^2 = r^2$, we obtain

$$I^2 = \int_{0}^{\pi/2} \int_{0}^{\infty} e^{-r^2} r\, dr\, d\theta = \int_{0}^{\pi/2} \left(\lim_{N \to \infty} \int_{0}^{N} e^{-r^2} r\, dr \right) d\theta$$

$$= \int_{0}^{\pi/2} \left(\lim_{N \to \infty} -\frac{1}{2} e^{-r^2} \Big|_{0}^{N} \right) d\theta = \frac{1}{2} \int_{0}^{\pi/2} d\theta = \frac{\pi}{4}.$$

Hence, $I^2 = \pi/4$, so $I = \sqrt{\pi}/2$, and

$$\int_{-\infty}^{\infty} e^{-x^2}\, dx = 2I = \sqrt{\pi}.$$

By making the substitution $u = x/\sqrt{2}$, it is straightforward to show that

$$\int_{-\infty}^{\infty} e^{-x^2/2}\, dx = \sqrt{2\pi}.$$

The function $\rho(x) = (1/\sqrt{2\pi}) e^{-x^2/2}$ is called the **density function for the unit normal distribution**. We have just shown that $\int_{-\infty}^{\infty} \rho(x)\, dx = 1$.

[†] Problem 14.2.58 really does not apply to improper integrals such as this one. The answer we obtain is correct, but additional theory is needed to justify this next step. The needed result is best left to a course in advanced calculus.

■ PROBLEMS 14.4

■ SELF-QUIZ

I. True–False: The area of region Ω equals the integral of the constant function 1 over the region.

II. The circle of radius A centered at the pole (the origin) has area _____.

a. $\int_0^\pi \int_0^A r\, dr\, d\theta$ b. $\int_0^A \int_0^\pi r\, dr\, d\theta$

c. $\int_0^{2\pi} \int_0^A r\, dr\, d\theta$ d. $\int_0^{2\pi} \int_0^r A\, dr\, d\theta$

III. The volume under the cone $z = r$ and above the disk $r \le a$, $z = 0$ is _____.

a. $\int_0^{2\pi} \int_0^a r^2\, dr\, d\theta$ b. $\int_0^{2\pi} \int_0^r ar\, dr\, d\theta$

c. $\int_0^{2\pi} \int_0^a (r - a)r\, dr\, d\theta$

d. $\int_{-\pi}^\pi \int_0^a (r - 0)r\, dr\, d\theta$

IV. The volume under the plane $z = x + y$ and above the quarter-circle $x^2 + y^2 \le 25$, $x \ge 0$, $y \ge 0$, $z = 0$ is _____.

a. $\int_0^\pi \int_0^{25} (r\cos\theta + r\sin\theta)r\, dr\, d\theta$

b. $\int_0^{\pi/2} \int_0^5 r(\cos\theta + \sin\theta)\, dr\, d\theta$

c. $\int_0^{\pi/4} \int_{-5}^5 (r\cos\theta + r\sin\theta)\, dr\, d\theta$

d. $\int_0^{\pi/2} \int_0^5 (r\cos\theta + r\sin\theta)r\, dr\, d\theta$

■ DRILL

In Problems 1–4, calculate the volume under the given surface that lies over the specified region in the plane $z = 0$.

1. $z = r^n$ where n is a positive integer; Ω is the circle of radius a centered at the pole.

2. $z = 3 - r$; Ω is the circle $r = 2\cos\theta$, $-\pi/2 \le \theta \le \pi/2$.

3. $z = r^2$; Ω is the cardioid $r = 4(1 - \cos\theta)$.

4. $z = r^3$; Ω is the region enclosed by the spiral of Archimedes $r = a\theta$ and the polar axis for θ between 0 and 2π.

In Problems 5–14, calculate the area of the region enclosed by the given curve or curves.

5. $r = 1 - \cos\theta$ 6. $r = 4(1 + \cos\theta)$

7. $r = 1 + 2\cos\theta$ (outer loop)

8. $r = 3 - 2\sin\theta$ 9. $r^2 = \cos 2\theta$

10. $r^2 = 4\sin 2\theta$ 11. $r = a + b\sin\theta$, $a > b > 0$

12. $r = \tan\theta$ and the line $\theta = \pi/4$

13. Outside the circle $r = 6$ and inside the cardioid $r = 4(1 + \sin\theta)$.

14. Inside the cardioid $r = 2(1 + \cos\theta)$ but outside the circle $r = 2$.

■ APPLICATIONS

15. Find the volume of the solid bounded above by the sphere $x^2 + y^2 + z^2 = 4a^2$, below by the xy-plane, and on the sides by the cylinder $x^2 + y^2 = a^2$.

16. Find the area of the region interior to the curve $(x^2 + y^2)^3 = 9y^2$.

*17. Find the volume of the solid bounded by the cone $x^2 + y^2 = z^2$ and the cylinder $x^2 + y^2 = 4y$.

*18. Find the volume of the solid bounded by the cone $z^2 = x^2 + y^2$ and the paraboloid $2z = x^2 + y^2$.

*19. Find the volume of the solid bounded by the cylinder $x^2 + y^2 = 9$ and the hyperboloid $x^2 + y^2 - z^2 = 1$.

20. Find the volume of the solid centered at the origin that is bounded above by the surface $z = e^{-(x^2 + y^2)}$ and below by the unit circle.

21. Find the centroid of the region bounded by $r = \cos\theta + 2\sin\theta$.

22. Find the centroid of the region bounded by the limaçon $r = 3 + \sin\theta$.

23. Find the centroid of the region bounded by the limaçon $r = a + b\cos\theta$, $a > b > 0$.

■ SHOW/PROVE/DISPROVE

24. Show that if Ω is the unit disk, then

$$\iint_\Omega \frac{1}{1 + x^2 + y^2}\, dA = \pi \ln 2.$$

■ ANSWERS TO SELF-QUIZ

I. True   II. c   III. a = d   IV. d

14.5 THE TRIPLE INTEGRAL

In this section we discuss the idea behind the triple integral of a function of three variables $f(x, y, z)$ over a region S in $\mathbb{R}^3$. This is really a simple extension of the double integral. For that reason we will omit a number of technical details.

We start with a parallelepiped π in $\mathbb{R}^3$, which can be written as

$$\pi = \{(x, y, z): a_1 \le x \le a_2, b_1 \le y \le b_2, c_1 \le z \le c_2\} \tag{1}$$

and is sketched in Figure 1. We construct regular partitions of the three intervals $[a_1, a_2]$, $[b_1, b_2]$, and $[c_1, c_2]$:

$$a_1 = x_0 < x_1 < \cdots < x_n = a_2$$

$$b_1 = y_0 < y_1 < \cdots < y_m = b_2$$

$$c_1 = z_0 < z_1 < \cdots < z_p = c_2,$$

to obtain nmp "boxes." The volume of a typical box B_{ijk} is given by

$$\Delta V = \Delta x \, \Delta y \, \Delta z, \tag{2}$$

where $\Delta x = x_i - x_{i-1}$, $\Delta y = y_j - y_{j-1}$, and $\Delta z = z_k - z_{k-1}$. We then form the sum

$$\sum_{i=1}^{n} \sum_{j=1}^{m} \sum_{k=1}^{p} f(x_i^*, y_j^*, z_k^*) \, \Delta V, \tag{3}$$

where (x_i^*, y_j^*, z_k^*) is in B_{ijk}. We now define

$$\Delta u = \sqrt{\Delta x^2 + \Delta y^2 + \Delta z^2}.$$

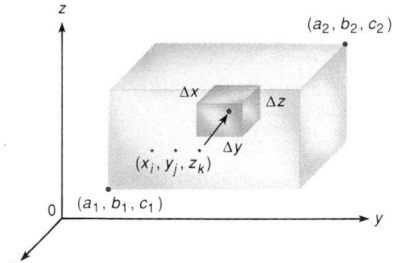

FIGURE 1
THE PARALLELEPIPED $a_1 \le x \le a_2$, $b_1 \le y \le b_2$, $c_1 \le z \le c_2$

Geometrically, Δu is the length of a diagonal of a rectangular solid with sides having lengths Δx, Δy, and Δz, respectively. We see that as $\Delta u \to 0$, Δx, Δy, and Δz approach 0, and the volume of each box tends to zero. We then take the limit as $\Delta u \to 0$.

DEFINITION   **THE TRIPLE INTEGRAL**

Let $w = f(x, y, z)$ and let the parallelepiped π be given by (1). Suppose that

$$\lim_{\Delta u \to 0} \sum_{i=1}^{n} \sum_{j=1}^{m} \sum_{k=1}^{p} f(x_i^*, y_j^*, z_k^*) \, \Delta x \, \Delta y \, \Delta z$$

exists and is independent of the way in which the points (x_i^*, y_j^*, z_k^*) are chosen. Then the **triple integral** of f over π, written $\iiint_\pi f(x, y, z) \, dV$, is defined by

$$\iiint_\pi f(x, y, z) \, dV = \lim_{\Delta u \to 0} \sum_{i=1}^{n} \sum_{j=1}^{m} \sum_{k=1}^{p} f(x_i^*, y_j^*, z_k^*) \, \Delta V. \quad \blacksquare \tag{4}$$

As with double integrals, we can write triple integrals as iterated (or repeated) integrals. If π is defined by (1), we have

---

THE TRIPLE INTEGRAL OVER A PARALLELEPIPED

$$\iiint_\pi f(x, y, z)\, dV = \int_{a_1}^{a_2} \int_{b_1}^{b_2} \int_{c_1}^{c_2} f(x, y, z)\, dz\, dy\, dx.$$ (5)

EXAMPLE 1
CALCULATING A TRIPLE INTEGRAL
OVER A PARALLELEPIPED

Evaluate $\iiint_\pi xy \cos yz\, dV$, where π is the parallelepiped

$$\left\{ (x, y, z) : 0 \le x \le 1, 0 \le y \le 1, 0 \le z \le \frac{\pi}{2} \right\}.$$

SOLUTION:

$$\iiint_\pi xy \cos yz\, dV = \int_0^1 \int_0^1 \int_0^{\pi/2} xy \cos yz\, dz\, dy\, dx$$

$$= \int_0^1 \int_0^1 \left\{ xy \cdot \frac{1}{y} \sin yz \Big|_0^{\pi/2} \right\} dy\, dx$$

$$= \int_0^1 \int_0^1 x \sin\left(\frac{\pi}{2} y \right) dy\, dx$$

$$= \int_0^1 \left\{ -\frac{2}{\pi} x \cos \frac{\pi}{2} y \Big|_0^1 \right\} dx = \int_0^1 \frac{2}{\pi} x\, dx = \frac{x^2}{\pi} \Big|_0^1$$

$$= \frac{1}{\pi}.$$

THE TRIPLE INTEGRAL OVER A MORE GENERAL REGION

We now define the triple integral over a more general region S. We assume that S is bounded. Then we can enclose S in a parallelepiped π and define a new function F by

$$F(x, y, z) = \begin{cases} f(x, y, z), & \text{if } (x, y, z) \text{ is in } S \\ 0, & \text{if } (x, y, z) \text{ is in } \pi \text{ but not in } S. \end{cases}$$

We then define

$$\iiint_S f(x, y, z)\, dV = \iiint_\pi F(x, y, z)\, dV.$$ (6)

REMARK 1: If f is continuous over S and if S is a region of the type we will discuss below, then $\iiint_S f(x, y, z)\, dV$ will exist. The proof of this fact is beyond the scope of this text but can be found in any advanced calculus text.

REMARK 2: If $f \ge 0$ on S, then the triple integral $\iiint_S f(x, y, z)\, dV$ represents the "volume" in four-dimensional space $\mathbb{R}^4$ of the region bounded above by

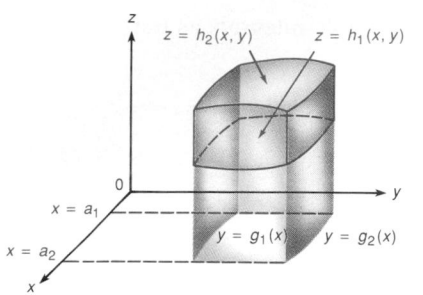

FIGURE 2
A REGION IN $\mathbb{R}^3$ OF THE FORM (7)

f and below by S. We cannot, of course, draw this volume, but otherwise the theory of volumes carries over to four (and more) dimensions.

Now let S take the form

$$S = \{(x, y, z): a_1 \leq x \leq a_2, g_1(x) \leq y \leq g_2(x), h_1(x, y) \leq z \leq h_2(x, y)\}. \quad (7)$$

What does such a solid look like? We first note that the equations $z = h_1(x, y)$ and $z = h_2(x, y)$ are the equations of surfaces in $\mathbb{R}^3$. The equations $y = g_1(x)$ and $y = g_2(x)$ are equations of cylinders in $\mathbb{R}^3$, and the equations $x = a_1$ and $x = a_2$ are equations of planes in $\mathbb{R}^3$. The solid S is sketched in Figure 2. We assume that g_1, g_2, h_1, and h_2 are continuous. If f is continuous, then $\iiint_S f(x, y, z)\, dV$ will exist and

---

THE TRIPLE INTEGRAL OVER A REGION S IN $\mathbb{R}^3$

$$\iiint_S f(x, y, z)\, dV = \int_{a_1}^{a_2} \int_{g_1(x)}^{g_2(x)} \int_{h_1(x,y)}^{h_2(x,y)} f(x, y, z)\, dz\, dy\, dx. \quad (8)$$

Note the similarity between equation (8) and equation (14.2.10).

EXAMPLE 2
EVALUATING A TRIPLE INTEGRAL
OVER A REGION S

Evaluate $\iiint_S 2x^3 y^2 z\, dV$, where S is the region

$$\{(x, y, z): 0 \leq x \leq 1, x^2 \leq y \leq x, x - y \leq z \leq x + y\}.$$

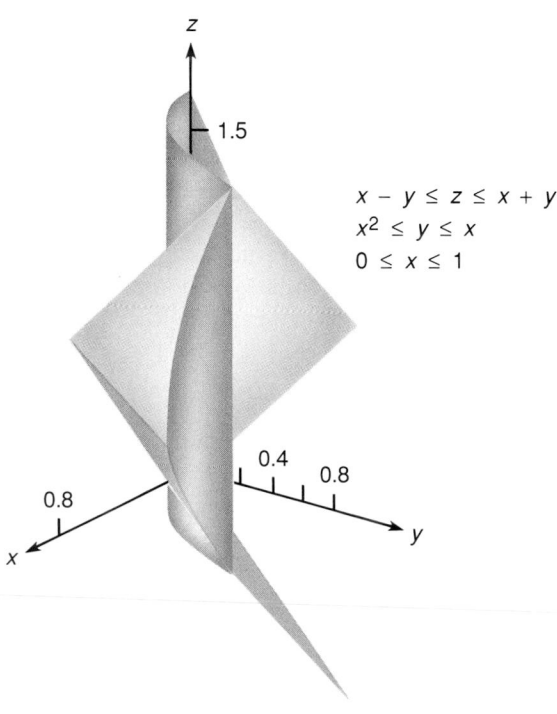

$x - y \leq z \leq x + y$
$x^2 \leq y \leq x$
$0 \leq x \leq 1$

FIGURE 3   COMPUTER-DRAWN SKETCH OF THE REGION $\{(x, y, z): 0 \leq x \leq 1,$
$x^2 \leq y \leq x, x - y \leq z \leq x + y\}$

SOLUTION: A computer-drawn sketch of the region S is given in Figure 3. We have

$$\iiint\limits_{S} 2x^3y^2z \, dV = \int_0^1 \int_{x^2}^x \int_{x-y}^{x+y} 2x^3y^2z \, dz \, dy \, dx$$

$$= \int_0^1 \int_{x^2}^x \left\{ x^3y^2z^2 \Big|_{x-y}^{x+y} \right\} dy \, dx$$

$$= \int_0^1 \int_{x^2}^x x^3y^2[(x+y)^2 - (x-y)^2] \, dy \, dx$$

$$= \int_0^1 \int_{x^2}^x 4x^4y^3 \, dy \, dx = \int_0^1 \left\{ x^4y^4 \Big|_{x^2}^x \right\} dx$$

$$= \int_0^1 (x^8 - x^{12}) \, dx = \frac{1}{9} - \frac{1}{13} = \frac{4}{117}.$$

Many of the applications we saw for the double integral can be extended to the triple integral. We present two of them below and one more in the problem sets.

VOLUME

Let the region S be defined by (7). Then, since $\Delta V = \Delta x \, \Delta y \, \Delta z$ represents the volume of a "box" in S, when we add up the volumes of these boxes and take a limit, we obtain the total volume of S. That is,

---

VOLUME AS A TRIPLE INTEGRAL

$$\text{volume of } S = \iiint\limits_{S} dV.$$

(9)

EXAMPLE 3
USING A TRIPLE INTEGRAL TO
COMPUTE VOLUME

Calculate the volume of the region of Example 2.

SOLUTION:

$$V = \int_0^1 \int_{x^2}^x \int_{x-y}^{x+y} dz \, dy \, dx = \int_0^1 \int_{x^2}^x \left\{ z \Big|_{x-y}^{x+y} \right\} dy \, dx$$

$$= \int_0^1 \int_{x^2}^x 2y \, dy \, dx = \int_0^1 \left\{ y^2 \Big|_{x^2}^x \right\} dx = \int_0^1 (x^2 - x^4) \, dx$$

$$= \frac{1}{3} - \frac{1}{5} = \frac{2}{15}.$$

EXAMPLE 4
USING A TRIPLE INTEGRAL TO
COMPUTE THE VOLUME OF A
TETRAHEDRON

Find the volume of the tetrahedron formed by the planes $x = 0$, $y = 0$, $z = 0$, and $x + (y/2) + (z/4) = 1$.

SOLUTION: The tetrahedron is sketched in Figure 4. We see that z ranges from 0 to the plane $x + (y/2) + (z/4) = 1$, or $z = 4[1 - x - (y/2)]$.

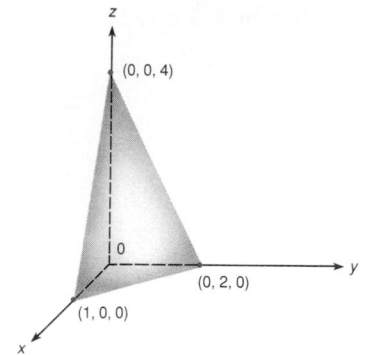

FIGURE 4
THE TETRAHEDRON WITH VERTICES AT
(0, 0, 0), (1, 0, 0), (0, 2, 0), AND (0, 0, 4)

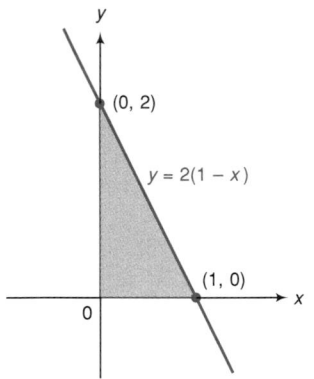

FIGURE 5
THE PROJECTION OF THE TETRAHEDRON
IN FIGURE 4 ONTO THE xy-PLANE

This last plane intersects the xy-plane in a line whose equation (obtained by setting $z = 0$) is given by $0 = 1 - x - (y/2)$ or $y = 2(1 - x)$, so that y ranges from 0 to $2(1 - x)$ (see Figure 5). Finally, this line intersects the x-axis at the point $(1, 0, 0)$, so that x ranges from 0 to 1, and we have

$$V = \int_0^1 \int_0^{2(1-x)} \int_0^{4(1-x-y/2)} dz\,dy\,dx$$

$$= \int_0^1 \int_0^{2(1-x)} 4\left(1 - x - \frac{y}{2}\right) dy\,dx = \int_0^1 -4\left(1 - x - \frac{y}{2}\right)^2 \Big|_0^{2(1-x)} dx$$

$$= 4 \int_0^1 (1 - x)^2\,dx = -\frac{4}{3}(1 - x)^3 \Big|_0^1 = \frac{4}{3}.$$

REMARK: It was not necessary to integrate in the order z, then y, then x. We could write, for example,

$$0 \le x \le 1 - \frac{y}{2} - \frac{z}{4}, \qquad 0 \le y \le 2 - \frac{z}{2}, \qquad 0 \le z \le 4$$

and could continue the problem by integrating in the order x, then y, then z.

DENSITY AND MASS IN $\mathbb{R}^3$

Let the function $\rho(x, y, z)$ denote the density (in kilograms per cubic meter, say) of a solid S in $\mathbb{R}^3$. Then for a "box" of sides Δx, Δy, and Δz, the approximate mass of the box will be equal to $\rho(x_i{}^*, y_j{}^*, z_k{}^*)\Delta x\,\Delta y\,\Delta z = \rho(x_i{}^*, y_j{}^*, z_k{}^*)\Delta V$ if Δx, Δy, and Δz are small. We then obtain

---

MASS AS A TRIPLE INTEGRAL

$$\text{total mass of } S = \mu(S) = \iiint_S \rho(x, y, z)\,dV. \tag{10}$$

EXAMPLE 5
USING A TRIPLE INTEGRAL TO
COMPUTE MASS

The density of the solid of Example 2 is given by $\rho(x, y, z) = x + 2y + 4z$ kg/m^3. Calculate the total mass of the solid.

SOLUTION:

$$\mu(S) = \int_0^1 \int_{x^2}^x \int_{x-y}^{x+y} (x + 2y + 4z)\,dz\,dy\,dx$$

$$= \int_0^1 \int_{x^2}^x \left\{ [(x + 2y)z + 2z^2] \Big|_{x-y}^{x+y} \right\} dy\,dx$$

$$= \int_0^1 \int_{x^2}^x (10xy + 4y^2)\,dy\,dx = \int_0^1 \left(5xy^2 + \frac{4y^3}{3}\right) \Big|_{x^2}^x dx$$

$$= \int_0^1 \left(5x^3 - 5x^5 + \frac{4}{3}x^3 - \frac{4}{3}x^6\right) dx$$

$$= \int_0^1 \left(\frac{19}{3}x^3 - 5x^5 - \frac{4}{3}x^6\right) dx = \frac{19}{12} - \frac{5}{6} - \frac{4}{21} = \frac{47}{84} \text{ kg.}$$

■ PROBLEMS 14.5

■ SELF-QUIZ

I. True–False: The volume of a region in $\mathbb{R}^3$ is the triple integral of the constant function 1 over that region.

II. Item _____ below is not equal to the other three items.

a. $\int_0^4 \int_2^5 \int_0^5 f(x, y, z)\, dy\, dz\, dx$

b. $\int_0^5 \int_2^5 \int_0^4 f(x, y, z)\, dx\, dz\, dy$

c. $\int_0^4 \int_0^5 \int_2^5 f(x, y, z)\, dx\, dy\, dz$

d. $\int_0^5 \int_0^4 \int_2^5 f(x, y, z)\, dz\, dx\, dy$

III. If we know that values of f are bounded above by 3 on the region $\Omega = \{(x, y, z): 0 \le x \le 2, 0 \le y \le 5, 1 \le z \le 3\}$, then we can infer that $\iiint_\Omega f(x, y, z)\, dV$ _____.

a. is positive

b. is negative

c. is less than or equal to $3 + 2 + 5 + 3$

d. is less than or equal to $3(2)(5)(3 - 1)$

IV. Suppose f is a function which is smooth throughout $\mathbb{R}^3$. True–False:

$$\int_1^2 \int_4^4 \int_0^5 f(x, y, z)\, dx\, dy\, dz = \int_1^2 \int_0^5 f(x, 4, z)\, dx\, dz.$$

■ DRILL

In Problems 1–8, evaluate the repeated triple integral.

1. $\int_0^1 \int_0^y \int_0^x y\, dz\, dx\, dy$

2. $\int_0^1 \int_0^y \int_z^y y\, dx\, dz\, dy$

3. $\int_0^? \int_{-z}^z \int_{y-z}^{y+z} 2xz\, dx\, dy\, dz$

4. $\int_{a_1}^{a_2} \int_{b_1}^{b_2} \int_{c_1}^{c_2} dy\, dx\, dz$

5. $\int_0^{\pi/2} \int_0^{\pi/2} \int_0^z \sin(x/z)\, dx\, dz\, dy$

6. $\int_1^2 \int_{1-y}^{1+y} \int_0^{yz} 6xyz\, dx\, dz\, dy$

7. $\int_0^1 \int_0^{\sqrt{1-x^2}} \int_0^x yz\, dz\, dy\, dx$

*8. $\int_0^1 \int_{-\sqrt{1-x^2}}^{\sqrt{1-x^2}} \int_{-\sqrt{1-x^2-y^2}}^{\sqrt{1-x^2-y^2}} z^2\, dz\, dy\, dx$

[Hint: Use polar coordinates.]

*9. Change the order of integration in Problem 1 and write the integral in the forms

a. $\int_?^? \int_?^? \int_?^? y\, dx\, dy\, dz$

b. $\int_?^? \int_?^? \int_?^? y\, dy\, dz\, dx$.

[Hint: Sketch the region in Problem 1 from the given limits and then find the new limits directly from that sketch.]

*10. Write the integral of Problem 2 in the forms

a. $\int_?^? \int_?^? \int_?^? 2xz\, dy\, dx\, dz$

b. $\int_?^? \int_?^? \int_?^? 2xz\, dz\, dy\, dx$.

*11. Write the integral of Problem 7 in the forms

a. $\int_?^? \int_?^? \int_?^? yz\, dy\, dz\, dx$

b. $\int_?^? \int_?^? \int_?^? yz\, dx\, dy\, dz$.

*12. Write the integral of Problem 8 in the form $\int_?^? \int_?^? \int_?^? z^2\, dx\, dz\, dy$.

13. Find $\iiint_S z\, dV$ where S is the region bounded by $x + y + z = 1$ and the coordinate planes.

14. Find $\iiint_S (x + y)\, dV$ where S is the region of Problem 13.

15. Find $\iiint_S (x + y + z)\, dV$ where S is the region of Problem 13.

16. Find $\iiint_S (x + y - 2z)\, dV$ where S is the region of Problem 13.

17. Compute $\iiint_S z^2\, dV$ where S is the region bounded by $x + 2y + 3z = 6$ and the coordinate planes.

18. Compute $\iiint_S (x - y)\, dV$ where S is the region of Problem 17.

■ APPLICATIONS

In Problems 19–26, find the volume of the given solid.

19. The tetrahedron with vertices at the points $(0, 0, 0)$, $(1, 0, 0)$, $(0, 1, 0)$, and $(0, 0, 1)$.

CONTINUED

20. The tetrahedron with vertices at the points $(0, 0, 0)$, $(a, 0, 0)$, $(0, b, 0)$, and $(0, 0, c)$.
21. The solid in the first octant bounded by the cylinder $x^2 + z^2 = 9$, the plane $x + y = 4$, and the three coordinate planes.
*22. The solid bounded by the planes $x - 2y + 4z = 4$, $-2x + 3y - z = 6$, $x = 0$, and $y = 0$.
23. The solid bounded above by the sphere $x^2 + y^2 + z^2 = 16$ and below by the plane $z = 2$.
24. The solid bounded by the parabolic cylinder $z = 5 - x^2$ and the planes $z = y$ and $z = 2y$ that lies in the half space $y \geq 0$.
25. The solid bounded by the ellipsoid $(x/a)^2 + (y/b)^2 + (z/c)^2 = 1$.

26. The solid bounded by the elliptic cylinder $9x^2 + y^2 = 9$ and the planes $z = 0$ and $x + y + 9z = 9$.

In Problems 27–30, find the mass of the given solid having the specified density.

27. The tetrahedron of Problem 19; $\rho(x, y, z) = x$.
28. The tetrahedron of Problem 20; $\rho(x, y, z) = x^2 + y^2 + z^2$.
29. The solid of Problem 21; $\rho(x, y, z) = z$.
30. The ellipsoid of Problem 25; $\rho(x, y, z) = \alpha x^2 + \beta y^2 + \gamma z^2$.

CENTER OF MASS

In $\mathbb{R}^3$, the first moments of a solid are considered with respect to the various coordinate planes. These moments are defined as weighted integrals of the signed-distances to the designated plane. For instance, the first moment with respect to the yz-plane is

$$M_{yz} = \iiint\limits_{\Omega} x\rho(x, y, z)\, dV.$$

Similarly, the moments with respect to the xz-plane and to the xy-plane are

$$M_{xz} = \iiint\limits_{\Omega} y\rho(x, y, z)\, dV$$

and

$$M_{xy} = \iiint\limits_{\Omega} z\rho(x, y, z)\, dV.$$

The **center of mass** of the region Ω is then given by

$$(\bar{x}, \bar{y}, \bar{z}) = \left(\frac{M_{yz}}{\mu}, \frac{M_{xz}}{\mu}, \frac{M_{xy}}{\mu}\right),$$

where $\mu = \iiint_{\Omega} \rho\, dV$ is the total mass. (If the density ρ is constant, then the center of mass is called the **centroid**.)

31. Find the centroid of the tetrahedron of Problem 20.
32. Find the centroid of the ellipsoid of Problem 25.
33. Find the center of mass for the tetrahedron of Problem 27.
34. Find the center of mass for the tetrahedron of Problem 28.
35. Find the center of mass for the solid of Problem 29.
36. Find the center of mass for the ellipsoid of Problem 30.
*37. The solid S lies in the first octant and is bounded by the planes $z = 0$, $y = 1$, $x = y$, and the hyperboloid $z = xy$. Its density is given by $\rho(x, y, z) = 1 + 2z$.
   a. Find the center of mass of S.
   b. Show that the center of mass lies inside S. [This is not obvious.]

ANSWERS TO SELF-QUIZ

I. True   II. c   III. d
IV. False (The left-hand side equals zero, the right-hand side need not.)

14.6 THE TRIPLE INTEGRAL IN CYLINDRICAL AND SPHERICAL COORDINATES

In this section, we show how triple integrals can be written by using cylindrical and spherical coordinates.

CYLINDRICAL COORDINATES

Recall from Section 11.9 that the cylindrical coordinates of a point in $\mathbb{R}^3$ are (r, θ, z), where r and θ are the polar coordinates of the projection of the point onto the xy-plane and z is the usual z-coordinate. In order to calculate an integral of a region given in cylindrical coordinates, we go through a procedure very similar to the one we used to write double integrals in polar coordinates in Section 14.4.

Consider the "cylindrical parallelepiped" given by

$$P_c = \{(r, \theta, z): r_1 \leq r \leq r_2; \theta_1 \leq \theta \leq \theta_2; z_1 \leq z \leq z_2\}.$$

This solid is sketched in Figure 1. If we partition the z-axis for z in $[z_1, z_2]$, we obtain "slices." For a fixed z the area of the face of a slice of π_c is, according to equation (14.4.5), given by

$$A_i = \int_{\theta_1}^{\theta_2} \int_{r_1}^{r_2} r\, dr\, d\theta. \tag{1}$$

The volume of a slice is, by (1), given by

$$V_i = A_i \Delta z = \left\{ \int_{\theta_1}^{\theta_2} \int_{r_1}^{r_2} r\, dr\, d\theta \right\} \Delta z.$$

Then adding these volumes and taking the limit as before, we obtain

$$\text{volume of } \pi_c = \int_{z_1}^{z_2} \int_{\theta_1}^{\theta_2} \int_{r_1}^{r_2} r\, dr\, d\theta\, dz. \tag{2}$$

In general, let the region S be given in cylindrical coordinates by

$$S = \{(r, \theta, z): \theta_1 \leq \theta \leq \theta_2, 0 \leq g_1(\theta) \leq r \leq g_2(\theta), h_1(r, \theta) \leq z \leq h_2(r, \theta)\}, \tag{3}$$

and let f be a function of r, θ, and z. Then the triple integral of f over S is given by

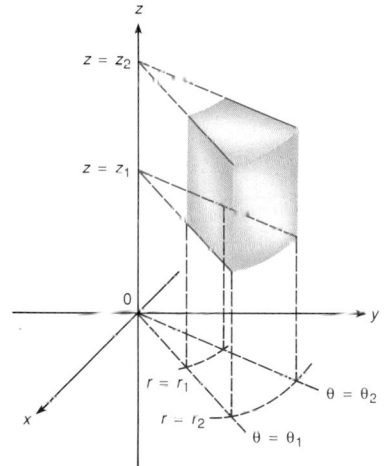

FIGURE 1
THE CYLINDRICAL PARALLELEPIPED BOUNDED BY THE PLANES $\theta = \theta_1$, $\theta = \theta_2$, $z = z_1$, $z = z_2$, AND THE CYLINDERS $r = r_1$, $r = r_2$

TRIPLE INTEGRAL IN CYLINDRICAL COORDINATES

$$\iiint\limits_S f = \int_{\theta_1}^{\theta_2} \int_{g_1(\theta)}^{g_2(\theta)} \int_{h_1(r,\theta)}^{h_2(r,\theta)} f(r, \theta, z)\, r\, dz\, dr\, d\theta. \tag{4}$$

NOTE: We can, of course, integrate in different orders. The order given in (4) is the most common one.

The formula for conversion from rectangular to cylindrical coordinates is virtually identical to formula (14.4.6). Let S be a region in $\mathbb{R}^3$. Since $x =$

$r \cos \theta$, $y = r \sin \theta$, and $z = z$, we have

CONVERTING A TRIPLE INTEGRAL FROM RECTANGULAR TO CYLINDRICAL COORDINATES

$$\iiint\limits_S f(x, y, z)\, dV = \iiint\limits_S f(r \cos \theta, r \sin \theta, z)r\, dz\, dr\, d\theta. \tag{5}$$

Given in rectangular   Given in cylindrical
coordinates   coordinates

REMARK: Again, do not forget the extra r when you convert to cylindrical coordinates.

EXAMPLE 1
COMPUTING MASS BY CONVERTING TO CYLINDRICAL COORDINATES

Find the mass of the solid bounded by the paraboloid $z = x^2 + y^2$ and the plane $z = 4$ if the density at any point is proportional to the distance from the point to the z-axis.

SOLUTION: The solid is sketched in Figure 2. The density is given by $\rho(x, y, z) = \alpha\sqrt{x^2 + y^2}$, where α is a constant of proportionality. Thus since the solid may be written as

$$S = \{(x, y, z): -2 \le x \le 2, -\sqrt{4 - x^2} \le y \le \sqrt{4 - x^2}, x^2 + y^2 \le z \le 4\},$$

we have

$$\mu = \iiint\limits_S \alpha\sqrt{x^2 + y^2}\, dV = \int_{-2}^{2} \int_{-\sqrt{4-x^2}}^{\sqrt{4-x^2}} \int_{x^2+y^2}^{4} \alpha\sqrt{x^2 + y^2}\, dz\, dy\, dx.$$

We write this expression in cylindrical coordinates, using the fact that $\alpha\sqrt{x^2 + y^2} = \alpha r$. We note that the largest value of r is 2, since at the "top" of the solid, $r^2 = x^2 + y^2 = 4$. Then

$$\mu = \int_0^{2\pi} \int_0^2 \int_{r^2}^4 (\alpha r)r\, dz\, dr\, d\theta = \int_0^{2\pi} \int_0^2 \alpha r^2(4 - r^2)\, dr\, d\theta$$

$$= \alpha \int_0^{2\pi} \left\{ \left(\frac{4r^3}{3} - \frac{r^5}{5} \right) \Bigg|_0^2 \right\} d\theta = \frac{64\alpha}{15} \int_0^{2\pi} d\theta = \frac{128}{15}\pi\alpha.$$

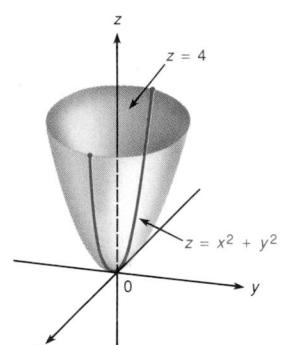

FIGURE 2
THE SOLID BOUNDED BY THE CIRCULAR PARABOLOID $z = x^2 + y^2$ AND THE PLANE $z = 4$

SPHERICAL COORDINATES

Recall from Section 11.9 that a point P in $\mathbb{R}^3$ can be written in the spherical coordinates (ρ, θ, φ), where $\rho \ge 0$, $0 \le \theta < 2\pi$, $0 \le \varphi \le \pi$. Here ρ is the distance between the point and the origin, θ is the same as in cylindrical coordinates, and φ is the angle between $\overrightarrow{OP}$ and the positive z-axis. Consider the "spherical parallelepiped"

$$P_S = \{(\rho, \theta, \varphi): \rho_1 \le \rho \le \rho_2, \theta_1 \le \theta \le \theta_2, \varphi_1 \le \varphi \le \varphi_2\}. \tag{6}$$

This solid is sketched in Figure 3. To approximate the volume of π_s, we partition the intervals $[\rho_1, \rho_2]$, $[\theta_1, \theta_2]$, and $[\varphi_1, \varphi_2]$. This partition gives

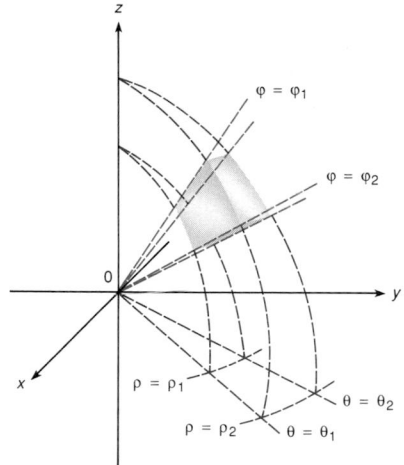

FIGURE 3
THE SPHERICAL PARALLELEPIPED
BOUNDED BY THE PLANES $\varphi = \varphi_1$, $\varphi = \varphi_2$,
$\theta = \theta_1$, $\theta = \theta_2$, AND THE SPHERES $\rho = \rho_1$,
$\rho = \rho_2$

us a number of "spherical boxes," one of which is sketched in Figure 4. The length of an arc of a circle is given by

$$L = r\theta, \tag{7}$$

where r is the radius of the circle and θ is the angle that "cuts off" the arc [see equation (8) in Appendix A1.1]. In Figure 4, one side of the spherical box is $\Delta\rho$. Since $\rho = \rho_i$ is the equation of a sphere, we find, from (7), that the length of a second side is $\rho_i \Delta\varphi$. Finally, the length of the third side is $\rho_i \sin \varphi_k \Delta\theta$ (see Figure 4). Thus, the volume of the box in Figure 4 is given, approximately, by

$$V_i \approx (\Delta\rho)(\rho_i \Delta\varphi)(\rho_i \sin \varphi_k \Delta\theta),$$

and using a familiar argument, we have

$$\text{volume of } P_S = \iiint_S \rho^2 \sin \varphi \, d\rho \, d\varphi \, d\theta = \int_{\theta_1}^{\theta_2} \int_{\varphi_1}^{\varphi_2} \int_{\rho_1}^{\rho_2} \rho^2 \sin \varphi \, d\rho \, d\varphi \, d\theta. \tag{8}$$

If f is a function of the variables ρ, θ, and φ, we have

$$\iiint_{P_S} f = \iiint_S f(\rho, \theta, \varphi)\rho^2 \sin \varphi \, d\rho \, d\varphi \, d\theta, \tag{9}$$

More generally, let the region S be defined in spherical coordinates by

$$S = \{(\rho, \theta, \varphi): \theta_1 \leq \theta \leq \theta_2, g_1(\theta) \leq \varphi \leq g_2(\theta), h_1(\theta, \varphi) \leq \rho \leq h_2(\theta, \varphi)\}. \tag{10}$$

Then

THE TRIPLE INTEGRAL IN SPHERICAL COORDINATES

$$\iiint_S f = \int_{\theta_1}^{\theta_2} \int_{g_1(\theta)}^{g_2(\theta)} \int_{h_1(\theta,\varphi)}^{h_2(\theta,\varphi)} f(\rho, \theta, \varphi)\rho^2 \sin \varphi \, d\rho \, d\varphi \, d\theta. \tag{11}$$

NOTE: As for cylindrical coordinates, we can integrate in different orders; the order given in (11) is the most common one.

Recall that to convert from rectangular to spherical coordinates, we have the formulas [see equations (11.9.5), (11.9.6), and (11.9.7)]

$$x = \rho \sin \varphi \cos \theta, \qquad y = \rho \sin \varphi \sin \theta, \qquad z = \rho \cos \varphi.$$

Thus, to convert from a triple integral in rectangular coordinates to a triple integral in spherical coordinates, we have

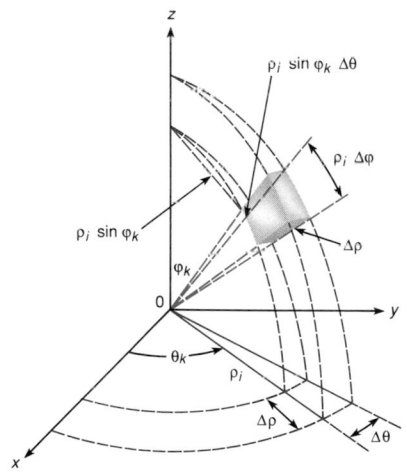

FIGURE 4
A SMALL SPHERICAL PARALLELEPIPED

**CONVERTING A TRIPLE INTEGRAL FROM
RECTANGULAR TO SPHERICAL COORDINATES**

$$\iiint_S f(x, y, z) \, dV \qquad \text{Given in rectangular coordinates}$$

$$= \iiint_S f(\rho \sin \varphi \cos \theta, \rho \sin \varphi \sin \theta, \rho \cos \varphi)\rho^2 \sin \varphi \, d\rho \, d\varphi \, d\theta.$$

Given in spherical coordinates $\tag{12}$

EXAMPLE 2
COMPUTING A MASS USING
SPHERICAL COORDINATES

Find the mass of the sphere $x^2 + y^2 + z^2 = a^2$ if its density at a point is proportional to the distance from the point to the origin.

SOLUTION: We have density $= \alpha\sqrt{x^2 + y^2 + z^2} = \alpha\rho$. Also, the sphere can be written (in spherical coordinates)

$$S = \{(\rho, \theta, \varphi): 0 \le \rho \le a, 0 \le \theta \le 2\pi, 0 \le \varphi \le \pi\}.$$

Thus,

$$\mu = \int_0^{2\pi} \int_0^{\pi} \int_0^a \alpha\rho(\rho^2 \sin \varphi)\,d\rho\,d\varphi\,d\theta = \frac{\alpha a^4}{4} \int_0^{2\pi} \int_0^{\pi} \sin \varphi\,d\varphi\,d\theta$$

$$= \frac{\alpha a^4}{4}(4\pi) = \pi a^4 \alpha.$$

■ PROBLEMS 14.6

■ SELF-QUIZ

I. True–False: The volume of a region is the integral of r over the region.

II. The volume of the cylinder $\{(r, \theta, z): 0 \le r \le 4, 0 \le \theta \le 2\pi, -4 \le z \le 4\}$ is _____.

a. $\int_0^4 \int_0^{2\pi} \int_{-4}^4 r\,dr\,d\theta\,dz$

b. $\int_{-4}^4 \int_0^{2\pi} \int_0^4 (1)r\,dr\,d\theta\,dz$

c. $\int_{-4}^4 \int_0^{2\pi} \int_0^4 r^2\,dr\,d\theta\,dz$

d. $\int_{-4}^4 \int_0^{2\pi} \int_0^4 r\,dx\,dy\,dz$

III. True–False: The volume of a region is the integral of $\rho^2 \sin \varphi$ over the region.

IV. The volume of the ball $\{(\rho, \theta, \varphi): 0 \le \rho \le 1, 0 \le \theta \le 2\pi, 0 \le \varphi \le \pi\}$ is _____.

a. $\int_0^1 \int_0^{2\pi} \int_0^{\pi} \rho^2 \sin \varphi\,d\rho\,d\theta\,d\varphi$

b. $\int_0^1 \int_0^{2\pi} \int_0^{\pi} \rho^2 \sin \varphi\,\rho^2 \sin \varphi\,d\rho\,d\theta\,d\varphi$

c. $\int_0^{\pi} \int_0^{2\pi} \int_0^1 \rho^2 \sin \varphi\,\rho^2 \sin \varphi\,d\rho\,d\theta\,d\varphi$

d. $\int_0^{\pi} \int_0^{2\pi} \int_0^1 (1)\rho^2 \sin \varphi\,d\rho\,d\theta\,d\varphi$

■ DRILL

Solve Problems 1–6 using cylindrical coordinates.

1. Find the volume of the region inside both the sphere $x^2 + y^2 + z^2 = 4$ and the cylinder $(x-1)^2 + y^2 = 1$.

2. Find the volume of the solid bounded above by the paraboloid $z = 4 - x^2 - y^2$ and below by the xy-plane.

3. Find the volume of the solid bounded by the plane $z = y$ and the paraboloid $z = x^2 + y^2$.

4. Find the volume of the solid bounded by the two cones $z^2 = x^2 + y^2$ and $z^2 = 16x^2 + 16y^2$ between $z = 0$ and $z = 2$.

*5. Find the volume of the solid bounded by the hyperboloid of two sheets $z^2 - x^2 - y^2 = 1$ and the cone $z^2 = x^2 + y^2$ for $0 \le z \le a$, $a \ge 1$.

6. Evaluate

$$\int_0^1 \int_0^{\sqrt{1-x^2}} \int_{\sqrt{x^2+y^2}}^{\sqrt{1-x^2-y^2}} z^3\,dz\,dy\,dx.$$

Solve Problems 7–14 by using spherical coordinates.

7. Evaluate

$$\int_0^1 \int_0^{\sqrt{1-x^2}} \int_0^{\sqrt{1-x^2-y^2}} \frac{1}{\sqrt{x^2 + y^2 + z^2}}\,dz\,dy\,dx.$$

8. Find the mass of the unit sphere if the density at any point is proportional to the distance to the boundary of the sphere.

*9. Find the volume of the solid inside the sphere $x^2 + y^2 + z^2 = 4$ and outside the cone $z^2 = x^2 + y^2$.

10. A solid fills the space between two concentric spheres centered at the origin. The radii of the spheres are a and b, where $0 < a < b$. Find the mass of the solid if the density at each point is inversely proportional to its distance from the origin.

CONTINUED

11. Find the volume of one of the smaller wedges cut from the unit sphere by two planes that meet at a diameter with an angle of $\pi/6$.

12. Find the mass of a wedge cut from a sphere of radius a by two planes that meet at a diameter with an angle of b radians if the density at any point on the wedge is proportional to the distance to that diameter.

13. Evaluate

$$\int_{-3}^{3} \int_{-\sqrt{9-x^2}}^{\sqrt{9-x^2}} \int_{-\sqrt{9-x^2-y^2}}^{\sqrt{9-x^2-y^2}} (x^2+y^2+z^2)^{3/2} \, dz \, dy \, dx.$$

14. Evaluate

$$\int_{0}^{a} \int_{0}^{\sqrt{a^2-x^2}} \int_{0}^{\sqrt{a^2-x^2-y^2}} \frac{z^3}{\sqrt{x^2+y^2}} \, dz \, dy \, dx.$$

■ **APPLICATIONS**

Problems 15–21 ask you to compute the centroid of a region or the center of mass of a solid object. Both of these were discussed immediately preceding Problem 14.5.31 on page 956; you might want to review that paragraph now before tackling the following problems.

15. Find the centroid of the region of Problem 1.

*16. Suppose the density of the region of Problem 1 is proportional to the square of the distance to the xy-plane and is measured in kg/m^3. Find the center of mass of the solid region.

17. Find the center of mass for the solid of Problem 2 if the density is proportional to the distance to the xy-plane.

18. Find the center of mass for the solid in Problem 4 if the density at any point is proportional to the distance to the z-axis.

19. Find the centroid of the "ice-cream cone" shaped region which is below the upper half of the sphere $x^2 + y^2 + z^2 = z$ and above the cone $z^2 = x^2 + y^2$.

20. Find the center of mass for the unit sphere if the density at a point is proportional to the distance from the origin.

21. Find the center of mass for the object in Problem 9 if the density at a point is proportional to the square of the distance from the region.

■ **SHOW/PROVE/DISPROVE**

*22. A "sphere" in $\mathbb{R}^4$ has the equation $x^2 + y^2 + z^2 + w^2 = a^2$.

   a. Explain why it is reasonable that the volume of this "sphere" be given by

$$\int_{-a}^{a} \int_{-\sqrt{a^2-x^2}}^{\sqrt{a^2-x^2}} \int_{-\sqrt{a^2-x^2-y^2}}^{\sqrt{a^2-x^2-y^2}} \int_{-\sqrt{a^2-x^2-y^2-z^2}}^{\sqrt{a^2-x^2-y^2-z^2}} dw \, dz \, dy \, dx$$
$$= 16 \int_{0}^{a} \int_{0}^{\sqrt{a^2-x^2}} \int_{0}^{\sqrt{a^2-x^2-y^2}} \int_{0}^{\sqrt{a^2-x^2-y^2-z^2}} dw \, dz \, dy \, dx.$$

   b. Using spherical coordinates, evaluate the integral established in part (a). [*Hint:* The first integral is easy and reduces the quadruple integral to a triple integral.]

■ **CHALLENGE**

23. Back in the days when you only knew how to work with functions of one variable, you computed volumes of solids of revolution in several ways; you may have been confused about when to use the "disk method" and when to use the "shell method." This problem offers you the chance to unify those methods.

Show that the "disk method" and the "shell method" only involve different repeated integrals which compute the value of a triple integral over the region in space of the function which has the constant value of 1.

Now that you also know about cylindrical and spherical coordinates and how to express a triple integral as repeated integrals in either system, you may want to include them in your discussion.

CONTINUED

■ **ANSWERS TO SELF-QUIZ**

  I. False (integrate 1 over the region.)   III. False (integrate 1 over the region.)
  II. b   IV. d

■ **PROBLEMS**   **CHAPTER 14 REVIEW**

■ **DRILL**

In Problems 1–12, evaluate the integral.

1. $\int_0^1 \int_0^2 x^2 y \, dx \, dy$ 2. $\int_0^1 \int_{x^2}^x xy^3 \, dy \, dx$

3. $\int_2^4 \int_{1+y}^{2+5y} (x - y^2) \, dx \, dy$

4. $\int_0^4 \int_{-\sqrt{16-x^2}}^{\sqrt{16-x^2}} 4y \, dy \, dx$

5. $\int_0^1 \int_0^y \int_0^x x^2 \, dz \, dx \, dy$

6. $\int_1^2 \int_{2-x}^{2+x} \int_0^{xz} 12xyz \, dy \, dz \, dx$

7. $\int_{-2}^2 \int_{-\sqrt{4-x^2}}^{\sqrt{4-x^2}} \int_{\sqrt{x^2+y^2}}^{\sqrt{4-x^2-y^2}} z^2 \, dz \, dy \, dx$

8. $\int_0^2 \int_0^{\sqrt{4-x^2}} \int_0^{\sqrt{4-x^2-y^2}} \frac{z^2}{\sqrt{x^2+y^2}} \, dz \, dy \, dx$

9. $\iint_\Omega (y - x^2) \, dA$, where $\Omega = \{(x, y): -3 \le x \le 3, 0 \le y \le 2\}$.

10. $\iint_\Omega (y - x^2) \, dA$, where Ω is the region in the first quadrant bounded by the x-axis, the y-axis, and the circle $x^2 + y^2 = 4$.

11. $\iint_\Omega (x + y^2) \, dA$, where Ω is the region in the first quadrant bounded by the x-axis, the y-axis, and the unit circle.

12. $\iint_\Omega (2x + y)e^{-(x+y)} \, dA$, where Ω is the first quadrant.

13. Change the order of integration of $\int_2^5 \int_1^x 3x^2 y \, dy \, dx$ and evaluate.

14. Change the order of integration of $\int_0^3 \int_0^{\sqrt{9-x^2}} (9 - y^2)^{3/2} \, dy \, dx$ and evaluate.

15. Change the order of integration of $\int_0^\infty \int_x^\infty f(x, y) \, dy \, dx$.

16. Change the order of integration in $\int_0^1 \int_0^y \int_0^x x^2 \, dz \, dx \, dy$ and write the integral in the following forms:

    a. $\int_?^? \int_?^? \int_?^? x^2 \, dx \, dy \, dz$

    b. $\int_?^? \int_?^? \int_?^? x^2 \, dy \, dz \, dx$

17. Evaluate $\iint_\Omega e^{-(x^2+y^2)} \, dA$ where Ω is the circle of radius 3 centered at the origin.

18. Find close lower and upper bounds for $\iint_\Omega e^{-(x^4+y^4)} \, dA$ where Ω is the same region as in the preceding exercise.

■ **APPLICATIONS**

19. Find the volume of the solid bounded by the plane $x + 2y + 3z = 6$ and the three coordinate planes.

20. Find the volume of the tetrahedron with vertices at (0, 0, 0), (2, 0, 0), (0, 1, 0), and (0, 0, 3).

21. Find the volume of the solid bounded by the cylinder $y^2 + z^2 = 4$ and the planes $x = 0$ and $x = 3$.

22. Find the volume enclosed by the ellipsoid $4x^2 + y^2 + 25z^2 = 100$.

23. Find the volume of the solid bounded by the paraboloid $x = y^2 + z^2$ and the plane $y = x - 2$.

24. Calculate the volume under the surface $z = r^5$ and over the circle of radius 6 centered at the origin.

25. Calculate the area of the region enclosed by the curve $r = 2(1 + \sin \theta)$.

26. Calculate the area of the region enclosed by the curve $r = 3(1 + \sin \theta)$ and outside the circle $r = 2$.

27. Find the volume of the solid bounded by the cone $z^2 = x^2 + y^2$ and the paraboloid $4z = x^2 + y^2$.

28. Find the volume between the planes $z = x + y$ and $z = 3x + 5y$ which lies over the region bounded by the curves $x = y^2$ and $x = y^3$.

29. Find the volume of the solid bounded by the parabolic cylinder $z = 4 - y^2$ and the planes $z = x$ and $z = 2x$ that lies in the half space $z \ge 0$.

30. Find the volume of the region inside both the sphere $x^2 + y^2 + z^2 = a^2$ and the cylinder $x^2 + [y - (a/2)]^2 = (a/2)^2$.

CONTINUED

31. Find the volume of the solid bounded by the plane $z = x$ and the paraboloid $z = x^2 + y^2$.

32. Find the volume of the solid inside the sphere $x^2 + y^2 + z^2 = 9$ and outside the cone $z^2 = x^2 + y^2$.

33. Find the volume of a wedge cut from a sphere of radius 2 by two planes that meet at a diameter with an angle of $\pi/3$.

34. Find the mass of the tetrahedron of Problem 20 if the density function is $\rho(x, y, z) = 2x + y^2$.

35. Find the centroid of the region bounded by the right-hand loop of the lemniscate $r^2 = 4 \cos 2\theta$.

36. Find the centroid of the region under the surface $z = \dfrac{y^4}{4} + \dfrac{1}{8y^2}$ and over the rectangle $\{(x, y):$ $0 \le x \le 3, 2 \le y \le 4\}$.

37. Find the centroid of the hemisphere $x^2 + y^2 + z^2 = 16, z \ge 0$.

38. Find the center of mass of the unit disk if the density at a point is proportional to the distance to the boundary of the disk.

39. Find the center of mass of the region $\Omega = \{(x, y): 0 \le x \le 3, 1 \le y \le 5\}$ if $\rho(x, y) = 3x^2 y$.

40. Find the center of mass of the triangular region bounded by the lines $y = x$, $y = 2 - x$, and the y-axis if $\rho(x, y) = 3xy + y^3$.

41. Find the center of mass of the solid bounded by the ellipsoid $4x^2 + y^2 + 25z^2 = 100$ if the density function is $\rho(x, y, z) = z^2$.

42. Suppose the density of the region of Problem 30 is proportional to the square of the distance to the xy-plane; find the center of mass of the region.

*43. Find the center of mass of the solid in Problem 33 if the density at a point is proportional to its distance to the origin.

▌ COMPUTER EXERCISES

1. Let D denote the disk of radius 1 with center at the origin. The disk is to be cut along a vertical line so that the center of mass of the right-hand piece lies halfway along a radius. That is, the disk is to be cut along a line $x = a$ so that the center of mass of the resulting piece is located at $(\frac{1}{2}, 0)$.
   a. Should a be chosen to be positive or negative? [*Hint:* Calculate the center of mass of the half disk. How does the center compare with $(\frac{1}{2}, 0)$?]

   b. Approximate a to two decimal places.

2. What value of a is required if the centroid is to be at $x = \frac{2}{3}$?

3. R is the region between the graphs of $\tan^{-1} x$ and $-\tan^{-1} x$ from $x = 0$ to $x = a$. Find a so that the center of mass of R is located at $(1, 0)$.

15 INTRODUCTION TO VECTOR ANALYSIS

15.1 VECTOR FIELDS

In Section 13.1, we discussed functions that assign a real number to each vector in a subset, D, of $\mathbb{R}^2$ or $\mathbb{R}^3$. More generally, suppose we assign a vector $\mathbf{F(p)}$ to each point $\mathbf{p}$ in a subset D of a set of vectors. We call this assignment a **vector field** on D and say that $\mathbf{F(p)}$ is a **vector-valued function** (or simply, **vector function**) with domain D. For most purposes, D will be a subset of $\mathbb{R}^2$ or $\mathbb{R}^3$.

Before defining a vector field, we need to define a region in $\mathbb{R}^2$ or $\mathbb{R}^3$.

DEFINITION **OPEN SET, CONNECTED SET, AND REGION**

A set Ω in $\mathbb{R}^2$ (or $\mathbb{R}^3$) is **open** if for every point $\mathbf{x} \in \Omega$, there is an open disk D (open sphere S) with center at $\mathbf{x}$ that is contained in Ω. An open set Ω in $\mathbb{R}^2$ (or $\mathbb{R}^3$) is **connected** if any two points in Ω can be joined by a piecewise smooth curve lying entirely in Ω. A **region** Ω in $\mathbb{R}^2$ or $\mathbb{R}^3$ is an open, connected set. ∎

These definitions are illustrated in Figure 1.

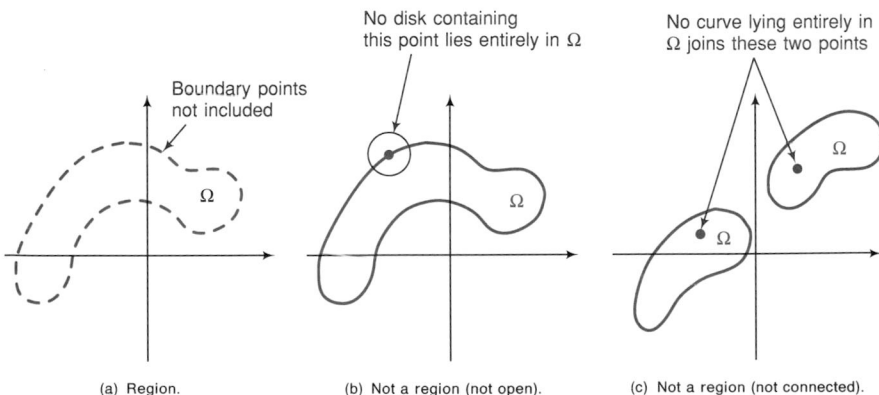

(a) Region.

(b) Not a region (not open).

(c) Not a region (not connected).

FIGURE 1 THREE SETS IN $\mathbb{R}^2$

DEFINITION **VECTOR FIELD IN $\mathbb{R}^2$**

Let Ω be a region in $\mathbb{R}^2$. Then $\mathbf{F}$ is a **vector field** in $\mathbb{R}^2$ if $\mathbf{F}$ assigns to every $\mathbf{x}$ in Ω a unique vector $\mathbf{F(x)}$ in $\mathbb{R}^2$. ∎

REMARK: Simply put, a vector field in $\mathbb{R}^2$ is a function whose domain is a region in $\mathbb{R}^2$ and whose range is a subset of $\mathbb{R}^2$. One vector field in $\mathbb{R}^2$ and

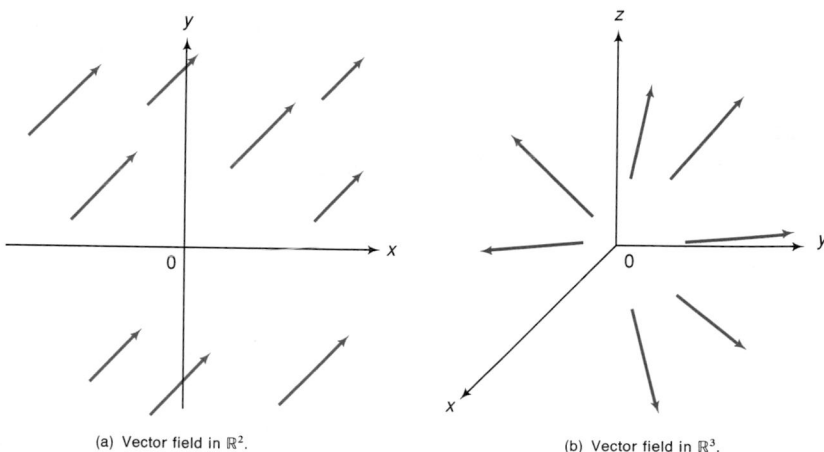

(a) Vector field in $\mathbb{R}^2$.

(b) Vector field in $\mathbb{R}^3$.

FIGURE 2 TWO VECTOR FIELDS

one in $\mathbb{R}^3$ are sketched in Figure 2. The meaning of this sketch is that to every point $\mathbf{x}$ in Ω a unique vector $\mathbf{F}(\mathbf{x})$ is assigned. That is, the function value $\mathbf{F}(\mathbf{x})$ is represented by an arrow with $\mathbf{x}$ at the "tail" of the arrow.

EXAMPLE 1
GRADIENT FIELD IN $\mathbb{R}^2$

Let $z = f(x, y)$ be a differentiable function. Then $\mathbf{F} = \nabla f = f_x \mathbf{i} + f_y \mathbf{j}$ is a vector field. It is called a **gradient vector field**.

The definition of a vector field can be extended, in an obvious way, to $\mathbb{R}^3$.

DEFINITION   **VECTOR FIELD IN $\mathbb{R}^3$**

Let S be a region in $\mathbb{R}^3$. Then $\mathbf{F}$ is a **vector field** in $\mathbb{R}^3$ if $\mathbf{F}$ assigns to each vector $\mathbf{x}$ in S a unique vector $\mathbf{F}(\mathbf{x})$ in $\mathbb{R}^3$. ∎

REMARK: A vector field in $\mathbb{R}^3$ is a function whose domain and range are subsets of $\mathbb{R}^3$.

A vector field in $\mathbb{R}^3$ is sketched in Figure 2(b).

EXAMPLE 2
GRADIENT FIELD IN $\mathbb{R}^3$

If $u = f(x, y, z)$ is differentiable, then $\mathbf{F} = \nabla f = f_x \mathbf{i} + f_y \mathbf{j} + f_z \mathbf{k}$ is a vector field. It is also called a **gradient vector field**.

EXAMPLE 3
GRAVITATIONAL FIELD

Let m_1 represent the mass of a (relatively) fixed object in space and let m_2 denote the mass of an object moving near the fixed object. Then the magnitude of the gravitational force between the objects is given by (see Example 2.2.5 on page 143)

$$|\mathbf{F}| = G\frac{m_1 m_2}{r^2}, \tag{1}$$

where r is the distance between the objects and G is a universal constant. If we assume that the first object is at the origin, then we may denote the position of the second object by $\mathbf{x}(t)$, and then, since $r = |\mathbf{x}|$, (1) can be written

$$|\mathbf{F}| = \frac{Gm_1m_2}{|\mathbf{x}|^2}. \tag{2}$$

Also, the force acts toward the origin, that is, in the direction opposite to that of the position vector $\mathbf{x}$. Therefore,

$$\text{direction of } \mathbf{F} = -\frac{\mathbf{x}}{|\mathbf{x}|}, \tag{3}$$

so from (2) and (3),

$$\mathbf{F}(t) = \frac{\alpha\mathbf{x}(t)}{|\mathbf{x}(t)|^3}, \tag{4}$$

where $\alpha = -Gm_1m_2$.

The vector field (4) is called a **gravitational field**. We can sketch this vector field without much difficulty because for every $\mathbf{x} \neq \mathbf{0} \in \mathbb{R}^3$, $\mathbf{F}(\mathbf{x})$ points toward the origin. The sketch appears in Figure 3.

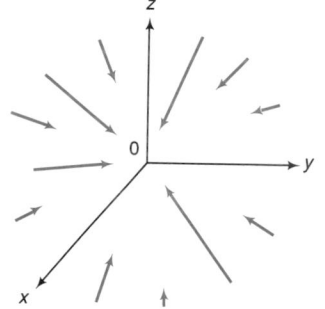

FIGURE 3
A GRAVITATIONAL FIELD

Vector fields arise in a great number of physical applications. There are, for example, mechanical force fields, magnetic fields, electric fields, velocity fields, and direction fields.

CONSERVATIVE VECTOR FIELDS

Suppose that $\mathbf{F} = \nabla f$ is a gradient field in $\mathbb{R}^2$. If $\mathbf{F} = -\nabla f$ for some function f, then $\mathbf{F}$ is said to be a **conservative vector field** and f is called a **potential function** for $\mathbf{F}$. The reason for this terminology will be made clear shortly. [If $\mathbf{F} = \nabla f$, then $\mathbf{F} = -\nabla(-f)$, so that the introduction of the minus sign does not cause any problem.]

Now let $\mathbf{g}(t) = x(t)\mathbf{i} + y(t)\mathbf{j}$ be a differentiable curve and suppose that a particle of mass m moves along it. Suppose further that the force acting on the particle at any time t is given by $\mathbf{F}(\mathbf{x}(t))$, where $\mathbf{F}$ is assumed to be a conservative vector field. By Newton's second law,

$$\mathbf{F}(\mathbf{x}(t)) = m\mathbf{a}(t) = m\mathbf{x}''(t).$$

But since $\mathbf{F}$ is conservative, $\mathbf{F}(\mathbf{x}) = -\nabla f(\mathbf{x})$ for some differentiable function f. Then we have

$$-\nabla f(\mathbf{x}(t)) = m\mathbf{x}''(t)$$

or

$$m\mathbf{x}'' + \nabla f(\mathbf{x}) = \mathbf{0}. \tag{5}$$

We now take the dot product of both sides of (3) with $\mathbf{x}'$ to obtain

$$m\mathbf{x}' \cdot \mathbf{x}'' + \nabla f(\mathbf{x}) \cdot \mathbf{x}' = 0. \tag{6}$$

But by the product rule,

$$\frac{d}{dx}|\mathbf{x}'(t)|^2 = \frac{d}{dt}(\mathbf{x}'(t) \cdot \mathbf{x}'(t)) = \mathbf{x}'(t) \cdot \mathbf{x}''(t) + \mathbf{x}''(t) \cdot \mathbf{x}'(t)$$
$$= 2\mathbf{x}'(t) \cdot \mathbf{x}''(t), \tag{7}$$

and by equation (13.6.3) on page 876,

$$\frac{d}{dt}f(\mathbf{x}(t)) = \nabla f(\mathbf{x}(t)) \cdot \mathbf{x}'(t). \tag{8}$$

Using (7) and (8) in (6), we obtain

$$\frac{d}{dt}\left[\frac{1}{2}m|\mathbf{x}'|^2 + f(\mathbf{x}(t))\right] = 0,$$

which implies that

$$\frac{1}{2}m|\mathbf{x}'|^2 + f(\mathbf{x}(t)) = C, \tag{9}$$

where C is a constant. This is one of the versions of the **law of conservation of energy**. The term $\frac{1}{2}m|\mathbf{x}'|^2 = \frac{1}{2}m|\mathbf{v}|^2$ is called the **kinetic energy** of the particle, and the term $f(\mathbf{x}(t))$ is called the **potential energy** of the particle. Equation (9) tells us simply that if the force function $\mathbf{F}$ is conservative, then the total energy of the system is constant and, moreover, the potential function f of $\mathbf{F}$ represents the potential energy of the system.

The **principle of the conservation of energy** states that energy may be transformed from one form to another but cannot be created or destroyed; that is, the total energy is constant. Thus it seems reasonable that force fields in classical physics are conservative (although since the work of Einstein, it has been found that energy can be transformed into mass and vice versa, so that there are forces that are not conservative). One example of a conservative force is given by the force of gravitational attraction. Let m_1 represent the mass of a (relatively) fixed object in space and let m_2 denote the mass of an object moving near the fixed object. Then, as in Example 3,

$$\mathbf{F}(t) = \frac{\alpha\mathbf{x}(t)}{|\mathbf{x}(t)|^3}$$

where $\alpha = -Gm_1m_2$. We now prove that $\mathbf{F}$ is conservative. To show this, we must come up with a function f such that $\mathbf{F} = -\nabla f$. Here we will pull f "out of a hat." In Section 15.3 we will show you how to construct such an f (if one exists). Let

$$f(\mathbf{x}) = \frac{\alpha}{|\mathbf{x}|}.$$

Then $f(\mathbf{x}) = \alpha/\sqrt{x^2 + y^2 + z^2}$ and

$$-\nabla f(\mathbf{x}) = -\alpha\left[\frac{-x}{(x^2 + y^2 + z^2)^{3/2}}\mathbf{i} + \frac{-y}{(x^2 + y^2 + z^2)^{3/2}}\mathbf{j}\right.$$
$$\left. + \frac{-z}{(x^2 + y^2 + z^2)^{3/2}}\mathbf{k}\right]$$
$$= \frac{\alpha}{|\mathbf{x}|^3}(x\mathbf{i} + y\mathbf{j} + z\mathbf{k}) = \frac{\alpha\mathbf{x}}{|\mathbf{x}|^3} = \mathbf{F}.$$

Thus **F** is conservative, so that, with respect to gravitational forces, the law of conservation of energy holds. The function f given above is called **gravitational potential**.

■ PROBLEMS 15.1

■ SELF-QUIZ

I. _____ is not a vector field in $\mathbb{R}^2$.
   a. $\{x\mathbf{i} + y\mathbf{j}: |y| \geq 1\}$ b. $3\mathbf{i} + 4\mathbf{j}$
   c. $y\mathbf{i} - x\mathbf{j}$ d. $xy - \mathbf{i} \cdot \mathbf{j}$
II. Suppose $f(x, y)$ is differentiable, then _____ is not a vector field in $\mathbb{R}^2$.

   a. $(f_x + f_y)\mathbf{i} + (f_x - f_y)\mathbf{j}$
   b. $f_y\mathbf{i} - f_x\mathbf{j}$
   c. $(f_x - f_y)(\mathbf{i} - \mathbf{j})$
   d. $(f_x + \mathbf{i}) - (f_y + \mathbf{j})$

■ DRILL

In Problems 1–20, compute the gradient vector field of the given function.

1. $f(x, y) = 1/\sqrt{x^2 + y^2}$ 2. $f(x, y) = \tan^{-1}(y/x)$
3. $f(x, y) = (x + y)^2$ 4. $f(x, y) = e^{\sqrt{xy}}$
5. $f(x, y) = \cos(x - y)$
6. $f(x, y) = \ln(2x - y + 1)$
7. $f(x, y) = y\tan(y - x)$
8. $f(x, y) = x^2 \sinh y$

9. $f(x, y) = \sec(x + 3y)$ 10. $f(x, y) = \dfrac{x - y}{x + y}$

11. $f(x, y) = \dfrac{x^2 - y^2}{x^2 + y^2}$ 12. $f(x, y) = \dfrac{e^{x^2} - e^{-y^2}}{3y}$

13. $f(x, y, z) = \sqrt{x^2 + y^2 + z^2}$
14. $f(x, y, z) = xyz$
15. $f(x, y, z) = \sin x \cos y \tan z$
16. $f(x, y, z) = \dfrac{x^2 - y^2 + z^2}{3xy}$
17. $f(x, y, z) = x \ln y - z \ln x$
18. $f(x, y, z) = xy^2 + y^2z^3$
19. $f(x, y, z) = (y - z)e^{x + 2y + 3z}$
20. $f(x, y, z) = \dfrac{x - z}{\sqrt{1 - y^2 + x^2}}$

■ APPLICATIONS

*21. Two wires, straight and infinite, pass through the points $(-1, 0)$ and $(1, 0)$. They are perpendicular to the xy-plane. If the wires are uniformly, but oppositely, charged with electricity, then they create an electric field that is the gradient of

$$F(x, y) = \ln\frac{\sqrt{(x - 1)^2 + y^2}}{\sqrt{(x + 1)^2 + y^2}}.$$

Compute and graph this vector field.

22. Verify that

$$\nabla \frac{1}{|\mathbf{x}|} = \frac{-1}{|\mathbf{x}|^2} \mathbf{u}$$

where $\mathbf{u}$ is the unit vector $\mathbf{x}/|\mathbf{x}|$.

23. Show that the force $\mathbf{F}(x, y) = y\mathbf{i} + x\mathbf{j}$ is conservative by finding a potential function for it.
24. Suppose that a moving particle is subjected to a force of constant magnitude that always points towards the origin. Show that this force is conservative. [*Hint:* If the path is given by $\mathbf{x}(t) = x(t)\mathbf{i} + y(t)\mathbf{j}$, find a unit vector that points toward the origin.]
*25. Show that the force $\mathbf{F}(\mathbf{x}) = -\alpha\mathbf{x}/|\mathbf{x}|^k$ is conservative if k is a positive integer.
26. Show that if $\mathbf{F}$ and $\mathbf{G}$ are conservative, then so is $\alpha\mathbf{F} + \beta\mathbf{G}$ where α and β are arbitrary constants.
27. Show that the force $\mathbf{F}(x, y) = y\mathbf{i} - x\mathbf{j}$ is *not* conservative. [*Hint:* Suppose there were an f such that $\mathbf{F} = -\nabla f$; then $-y = \dfrac{\partial f}{\partial x}$ and $x = \dfrac{\partial f}{\partial y}$; integrate to obtain a contradiction.]

■ ANSWERS TO SELF-QUIZ

I. a, d [(a): $\{(x, y): |y| \geq 1\}$ is neither open nor connected, (b) is a constant vector field.]

II. d

15.2 WORK AND LINE INTEGRALS

Recall from Section 11.2 that the work done by a force vector **F** in moving an object in the direction **d** is given by

$$W = \frac{\mathbf{F} \cdot \mathbf{d}}{|\mathbf{d}|} |\mathbf{d}| = \mathbf{F} \cdot \mathbf{d}. \qquad \text{equation (10) on page 735} \tag{1}$$

The component of **F** in the direction of motion is

$$\frac{\mathbf{F} \cdot \mathbf{d}}{|\mathbf{d}|}. \tag{2}$$

The vector **d** is called a **displacement vector**. That is, *the work done is the dot product of the force* **F** *and the displacement vector* **d**. Note that if **F** acts in the direction **d** and if φ denotes the angle (which is zero) between **F** and **d**, then $\mathbf{F} \cdot \mathbf{d} = |\mathbf{F}| |\mathbf{d}| \cos \varphi = |\mathbf{F}| |\mathbf{d}| \cos 0 = |\mathbf{F}| |\mathbf{d}|$, which is the scalar formula

$$W = Fd$$

(see p. 734).

EXAMPLE 1
COMPUTING THE WORK DONE
BY A FORCE

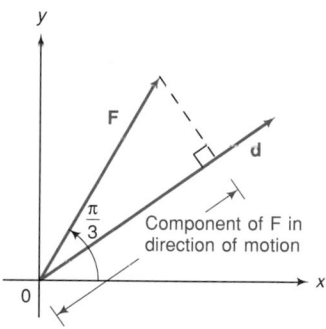

FIGURE 1
THE FORCE **F** ACTS IN THE DIRECTION **d**

A force of 4 N has the direction $\pi/3$. What is the work done in moving an object from the point $(1, 2)$ to the point $(5, 4)$, where distances are measured in meters?

SOLUTION: A unit vector with direction $\pi/3$ is given by $\mathbf{u} = (\cos \pi/3)\mathbf{i} + (\sin \pi/3)\mathbf{j} = (1/2)\mathbf{i} + (\sqrt{3}/2)\mathbf{j}$. Thus $\mathbf{F} = 4\mathbf{u} = 2\mathbf{i} + 2\sqrt{3}\mathbf{j}$. The displacement vector **d** is given by $(5 - 1)\mathbf{i} + (4 - 2)\mathbf{j} = 4\mathbf{i} + 2\mathbf{j}$. Thus,

$$W = \mathbf{F} \cdot \mathbf{d} = (2\mathbf{i} + 2\sqrt{3}\mathbf{j}) \cdot (4\mathbf{i} + 2\mathbf{j}) = (8 + 4\sqrt{3})$$
$$\approx 14.93 \text{ N·m} = 14.93 \text{ joules (J)}.$$

The component of **F** in the direction of motion is sketched in Figure 1.

We now calculate the work done when a particle moves along a curve C. In doing so, we will define an important concept in applied mathematics—the *line integral*.

Suppose that a curve in the plane is given parametrically by

$$C: \quad \mathbf{x}(t) = f(t)\mathbf{i} + g(t)\mathbf{j}, \qquad \text{or} \qquad \mathbf{x}(t) = x(t)\mathbf{i} + y(t)\mathbf{j}. \tag{3}$$

If a force is applied to a particle moving along C, then such a force will have magnitude and direction, so the force will be a vector function of the vector $\mathbf{x}(t)$. That is, the force will be a vector field. We write

$$\mathbf{F}(\mathbf{x}) = \mathbf{F}(x, y) = P(x, y)\mathbf{i} + Q(x, y)\mathbf{j} \tag{4}$$

where P and Q are scalar-valued functions. The problem is to determine the work done when a particle moves on C from a point $\mathbf{x}(a)$ to a point $\mathbf{x}(b)$ subject to the force $\mathbf{F}$ given by (4). We will assume in our discussion that the curve C is *smooth* or *piecewise smooth*. By that we mean that the functions $f(t)$ and $g(t)$ in (3) are continuously differentiable or that f' and g' exist and are piecewise continuous (see page 122 for the definition of piecewise continuity).

A typical curve C is sketched in Figure 2. Let

$$W(t) = \text{work done in moving from } \mathbf{x}(a) \text{ to } \mathbf{x}(t).$$

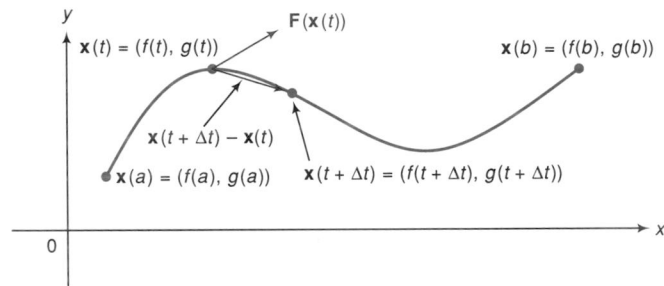

FIGURE 2 MOVING A PARTICLE ALONG A SMOOTH CURVE

The work done by the variable force $\mathbf{F}$ in moving a particle along C from $\mathbf{x}(t)$ to $\mathbf{x}(t + \Delta t)$ is, if Δt is small, approximately equal to $\mathbf{F}(\mathbf{x}(t)) \cdot [(\mathbf{x}(t + \Delta t) - \mathbf{x}(t)]$ because, if Δt is small, the part of the curve from $\mathbf{x}(t)$ to $\mathbf{x}(t + \Delta t)$ is "almost" a (straight) directed line segment.

Then the work done in moving from $\mathbf{x}(t)$ to $\mathbf{x}(t + \Delta t)$ is given by

$$W(t + \Delta t) - W(t).$$

Now if Δt is small, then the part of the curve between $\mathbf{x}(t)$ and $\mathbf{x}(t + \Delta t)$ is "close" to a straight line and so can be approximated by the vector

$$\mathbf{x}(t + \Delta t) - \mathbf{x}(t).$$

If Δt is small and if $\mathbf{F}(\mathbf{x})$ is continuous, then the force applied between $\mathbf{x}(t)$ and $\mathbf{x}(t + \Delta t)$ is approximately equal to $\mathbf{F}(\mathbf{x}(t))$. Thus by (1), if Δt is small, then

$$W(t + \Delta t) - W(t) \approx \mathbf{F}(\mathbf{x}(t)) \cdot [\mathbf{x}(t + \Delta t) - \mathbf{x}(t)]. \tag{5}$$

We divide both sides of (7) by Δt and take the limit as $\Delta t \to 0$ to obtain

$$W'(t) = \lim_{\Delta t \to 0} \frac{W(t + \Delta t) - W(t)}{\Delta t} = \lim_{\Delta t \to 0} \left\{ \mathbf{F}(\mathbf{x}(t)) \cdot \frac{[\mathbf{x}(t + \Delta t) - \mathbf{x}(t)]}{\Delta t} \right\}$$
$$= \mathbf{F}(\mathbf{x}(t)) \cdot \mathbf{x}'(t).$$

Also,

$$W(a) = 0 \quad \text{and} \quad W(b) = \text{total work done on the particle.} \tag{6}$$

Thus,

$$W = \text{total work done} = W(b) - W(a) = \int_a^b W'(t)\,dt, \tag{7}$$

or

WORK DONE BY A FORCE F IN MOVING A PARTICLE ALONG A CURVE x(t) FROM x(a) to x(b)

$$W = \int_a^b \mathbf{F}(\mathbf{x}(t)) \cdot \mathbf{x}'(t)\, dt. \tag{8}$$

We write equation (8) as

WORK WRITTEN AS A LINE INTEGRAL

$$W = \int_C \mathbf{F}(\mathbf{x}) \cdot \mathbf{dx}. \tag{9}$$

The symbol $\int_C$ is read "the integral along the curve C." The integral in (9) is called a *line integral of* $\mathbf{F}$ *over* C.

REMARK 1: If C lies along the x-axis, then C is given by $\mathbf{x}(t) = x(t)\mathbf{i} + 0\mathbf{j}^\dagger$ and $\mathbf{x}'(t) = x'(t)\mathbf{i}$, so that (10) becomes

$$\int_C \mathbf{F}(\mathbf{x}) \cdot \mathbf{dx} = \int_a^b \left[F(x(t))\mathbf{i} \cdot [x'(t)\mathbf{i}] \right] dt = \int_{x(a)}^{x(b)} F(x)\, dx, \tag{10}$$

which is our usual definite integral.

REMARK 2: Since $\mathbf{F}$ is given by (4), we can write equation (8) in a form more useful in computations:

ALTERNATIVE EQUATION FOR WORK

$$W = \int_a^b \left[P(x(t), y(t))x'(t) + Q(x(t), y(t))y'(t) \right] dt. \tag{11}$$

EXAMPLE 2
COMPUTING WORK

A particle is moving along the parabola $y = x^2$ subject to a force given by the vector field $2xy\mathbf{i} + (x^2 + y^2)\mathbf{j}$. How much work is done in moving from the point $(1, 1)$ to the point $(3, 9)$ if forces are measured in newtons and distances are measured in meters?

SOLUTION: The curve C is given parametrically by

$$\mathbf{x}(t) = t\mathbf{i} + t^2\mathbf{j} \qquad \text{where } 1 \le t \le 3.$$

We therefore have $f(t) = t$ and $g(t) = t^2$. Then $P = 2xy = 2t^3$ and $Q = x^2 + y^2 = t^2 + t^4$. Also, $f'(t) = 1$ and $g'(t) = 2t$, so by (11),

$$W = \int_1^3 \left[(2t^3)1 + (t^2 + t^4)(2t) \right] dt = \int_1^3 (4t^3 + 2t^5)\, dt = 322\frac{2}{3}\ \text{J.}$$

---

† We have substituted $x(t)$ for $f(t)$ and 0 for $g(t)$ here.

We used the notion of work to motivate the discussion of the line integral. We now give a general definition of the line integral in the plane.

DEFINITION   **LINE INTEGRAL IN THE PLANE**

Let P and Q be continuous on a region S containing the smooth (or piecewise smooth) curve C given by

$$C: \quad \mathbf{x}(t) = f(t)\mathbf{i} + g(t)\mathbf{j}, \qquad t \in [a, b].$$

Let the vector field $\mathbf{F}$ be given by

$$\mathbf{F}(x, y) = P(x, y)\mathbf{i} + Q(x, y)\mathbf{j}.$$

The **line integral** of $\mathbf{F}$ over C is given by

$$\int_C \mathbf{F}(\mathbf{x}) \cdot d\mathbf{x} = \int_a^b \left[P(f(t), g(t))f'(t) + Q(f(t), g(t))g'(t) \right] dt. \quad\blacksquare \quad \text{(12)}$$

REMARK: If C is piecewise smooth but not smooth, then C is made up of a number of "sections," each of which is smooth. Since C is continuous, these sections are joined. One typical piecewise smooth curve is sketched in Figure 3. If C is made up of the n smooth curves $C_1, C_2, \ldots, C_n$, then

$$\int_C \mathbf{F}(\mathbf{x}) \cdot d\mathbf{x} = \int_{C_1} \mathbf{F}(\mathbf{x}) \cdot d\mathbf{x} + \int_{C_2} \mathbf{F}(\mathbf{x}) \cdot d\mathbf{x} + \cdots + \int_{C_n} \mathbf{F}(\mathbf{x}) \cdot d\mathbf{x} \qquad \text{(13)}$$

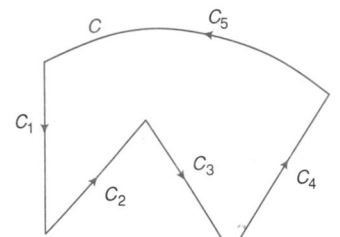

FIGURE 3
A PIECEWISE SMOOTH CURVE

EXAMPLE 3
COMPUTING A LINE INTEGRAL
OVER A RECTANGLE

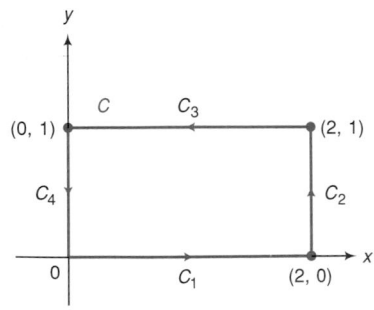

FIGURE 4
A RECTANGLE TRAVERSED IN THE
COUNTERCLOCKWISE DIRECTION

Calculate $\int_C \mathbf{F}(\mathbf{x}) \cdot d\mathbf{x}$, where $\mathbf{F}(x, y) = xy\mathbf{i} + ye^x\mathbf{j}$ and C is the rectangle joining the points $(0, 0)$, $(2, 0)$, $(2, 1)$, and $(0, 1)$ if C is traversed in the counterclockwise direction.

SOLUTION: The rectangle is sketched in Figure 4, and it is made up of four smooth curves (straight lines). We have

$$C_1: \quad \mathbf{x}(t) = t\mathbf{i}, \qquad\qquad 0 \le t \le 2 \qquad \mathbf{x}'(t) = \mathbf{i}$$

$$C_2: \quad \mathbf{x}(t) = 2\mathbf{i} + t\mathbf{j}, \qquad 0 \le t \le 1 \qquad \mathbf{x}'(t) = \mathbf{j}$$

$$C_3: \quad \mathbf{x}(t) = (2 - t)\mathbf{i} + \mathbf{j}, \qquad 0 \le t \le 2 \qquad \mathbf{x}'(t) = -\mathbf{i}$$

$$C_4: \quad \mathbf{x}(t) = (1 - t)\mathbf{j}, \qquad 0 \le t \le 1. \qquad \mathbf{x}'(t) = -\mathbf{j}$$

Note, for example, that on C_3, $t = 0$ corresponds to the point $2\mathbf{i} + \mathbf{j} = (2, 1)$ and $t = 2$ corresponds to $(2 - 2)\mathbf{i} + \mathbf{j} = (0, 1)$. Thus as t increases, we do move along C_3 in the direction indicated by the arrow in Figure 4. This illustrates why our parametrization of the rectangle is correct Now

$$\int_C \mathbf{F}(\mathbf{x}) \cdot d\mathbf{x} = \int_{C_1} \mathbf{F}(\mathbf{x}) \cdot d\mathbf{x} + \int_{C_2} \mathbf{F}(\mathbf{x}) \cdot d\mathbf{x} + \int_{C_3} \mathbf{F}(\mathbf{x}) \cdot d\mathbf{x}$$
$$+ \int_{C_4} \mathbf{F}(\mathbf{x}) \cdot d\mathbf{x}.$$

On C_1, $x = t$ and $y = 0$, so that $xy = 0$, $ye^x = 0$, and

$$\int_{C_1} \mathbf{F}(\mathbf{x}) \cdot d\mathbf{x} = 0.$$

On C_2, $x = 2$, $y = t$, $\mathbf{F(x)} = 2t\mathbf{i} + te^2\mathbf{j}$ and $\mathbf{x}'(t) = \mathbf{j}$, so that

$$\int_{C_2} \mathbf{F(x)} \cdot \mathbf{dx} = \int_0^1 te^2\,dt = \frac{e^2}{2}.$$

We leave it as an exercise (see Problems 7 and 8) to show that

$$\int_{C_3} \mathbf{F(x)} \cdot \mathbf{dx} = -2 \quad \text{and} \quad \int_{C_4} \mathbf{F(x)} \cdot \mathbf{dx} = -\frac{1}{2}.$$

Thus

$$\int_C \mathbf{F(x)} \cdot \mathbf{dx} = 0 + \frac{e^2}{2} - 2 - \frac{1}{2} = \frac{e^2}{2} - \frac{5}{2} \approx 1.2.$$

LINE INTEGRALS IN SPACE

The definition and theory of line integrals in $\mathbb{R}^3$ is very similar to the results in $\mathbb{R}^2$. We summarize some basic facts below.

Suppose that a curve C in space is given parametrically as

$$C. \quad \mathbf{x}(t) = f(t)\mathbf{i} + g(t)\mathbf{j} + h(t)\mathbf{k}, \quad \text{or} \quad \mathbf{x}(t) = x(t)\mathbf{i} + y(t)\mathbf{j} + z(t)\mathbf{k}. \quad (14)$$

As defined earlier, a **region** S in space is, as in $\mathbb{R}^2$, an open, connected set. We now define the line integral in space.

> **DEFINITION** **LINE INTEGRAL IN SPACE**
>
> Suppose that P, Q, and R are continuous on a region S in $\mathbb{R}^3$ containing the smooth or piecewise smooth curve C given by (14). Let the vector field $\mathbf{F}$ be given by
>
> $$\mathbf{F}(x, y, z) = P(x, y, z)\mathbf{i} + Q(x, y, z)\mathbf{j} + R(x, y, z)\mathbf{k}.$$
>
> Then the **line integral** of $\mathbf{F}$ over C is given by
>
> $$\int_C \mathbf{F(x)} \cdot \mathbf{dx} = \int_a^b \big[P(f(t), g(t), h(t))f'(t) + Q(f(t), g(t), h(t))g'(t)$$
> $$+ R(f(t), g(t), h(t))h'(t) \big]\,dt. \quad \blacksquare \qquad (15)$$

PROBLEMS 15.2

SELF-QUIZ

I. A force of 4 N acts in the direction of the positive x-axis to move a particle from $(2, 0)$ to $(5, 0)$ where the unit distance is meters. The work done is _____.
   a. $4(2 - 5) = -12$ N-m
   b. $4(5^2) - 4(2^2) = 84$ N-m

   c. $[4\mathbf{i} + 0\mathbf{j}] \cdot [(5 - 2)\mathbf{i} + 0\mathbf{j}] = 12$ N-m
   d. $[0\mathbf{i} + 4\mathbf{j}] \cdot [(5 - 2)\mathbf{i} + 0\mathbf{j}] = 0$ N-m

II. A force of _____ N acting in the direction of $\mathbf{i} + \mathbf{j}$ does 12 N-m work in moving a particle from $(2, 0)$ to $(5, 0)$ (unit distance is meters).
   a. 4   b. $4\sqrt{2}$ c. $4/\sqrt{2}$ d. 8

CONTINUED

III. $\int_0^4 [t\mathbf{i} - t\mathbf{j}] \cdot [\mathbf{i} + \mathbf{j}]\, dt = \int_C \mathbf{F(x)} \cdot \mathbf{dx}$ where $\mathbf{F}(x, y) = x\mathbf{i} - y\mathbf{j}$ and C is _____.
   a. the straight line segment from $(0, 0)$ to $(4, 4)$
   b. the straight line segment from $(4, 4)$ to $(0, 0)$
   c. the straight line segment from $(0, 0)$ to $(4, 0)$
   d. the parabola $y = x^2$ from $(0, 0)$ to $(4, 16)$
IV. If $\mathbf{F(x)} = \mathbf{x}$ and C is the straight line from $(1, 1)$ to $(4, 4)$, then $\int_C \mathbf{F(x)} \cdot \mathbf{dx} = $ _____.

   a. $\int_1^4 [t\mathbf{i} + t\mathbf{j}] \cdot [\mathbf{i} + \mathbf{j}]\, dt$
   b. $\int_0^3 [(1 + t)\mathbf{i} + (1 + t)\mathbf{j}] \cdot [\mathbf{i} + \mathbf{j}]\, dt$
   c. $\int_0^1 [(1 + 3u)\mathbf{i} + (1 + 3u)\mathbf{j}] \cdot [3\mathbf{i} + 3\mathbf{j}]\, du$
   d. $\int_0^1 [(1 + 3u)\mathbf{i} + (1 + 3u)\mathbf{j}] \cdot [\mathbf{i} + \mathbf{j}]\, du$

■ DRILL

In Problems 1–6, find the work done when the force with the given magnitude and direction moves an object from P to Q (all forces are in newtons and all distances are in meters).

1. $|\mathbf{F}| = 3$ N, $\theta = 0$, $P = (2, 3)$, $Q = (1, 7)$
2. $|\mathbf{F}| = 2$ N, $\theta = \pi/2$, $P = (5, 7)$, $Q = (1, 1)$
3. $|\mathbf{F}| = 6$ N, $\theta = \pi/4$, $P = (2, 3)$, $Q = (-1, 4)$
4. $|\mathbf{F}| = 4$ N, $\theta = \pi/6$, $P = (-1, 2)$, $Q = (3, 4)$
5. $|\mathbf{F}| = 4$ N, θ is in the direction of $2\mathbf{i} + 3\mathbf{j}$, $P = (2, 0)$, $Q = (-1, 3)$
6. $|\mathbf{F}| = 5$ N, θ is in the direction of $-3\mathbf{i} + 2\mathbf{j}$, $P = (1, 3)$, $Q = (4, -6)$

In Problems 7–26, calculate $\int_C \mathbf{F(x)} \cdot \mathbf{dx}$.

7. $\mathbf{F}(x, y) = xy\mathbf{i} + ye^x\mathbf{j}$; C is the curve $\mathbf{x}(t) = (2 - t)\mathbf{i} + \mathbf{j}$ for $0 \le t \le 2$ [*Note:* This is $\int_{C_3} \mathbf{F(x)} \cdot \mathbf{dx}$ for Example 3.]
8. $\mathbf{F}(x, y) = xy\mathbf{i} + ye^x\mathbf{j}$; C is the curve $\mathbf{x}(t) = (1 - t)\mathbf{j}$ for $0 \le t \le 1$ [*Note:* This is $\int_{C_4} \mathbf{F(x)} \cdot \mathbf{dx}$ for Example 3.]
9. $\mathbf{F}(x, y) = x^2\mathbf{i} + y^2\mathbf{j}$; C is the straight line segment from $(0, 0)$ to $(2, 4)$
10. $\mathbf{F}(x, y) = x^2\mathbf{i} + y^2\mathbf{j}$; C is the parabola $y = x^2$ from $(0, 0)$ to $(2, 4)$
11. $\mathbf{F}(x, y) = xy\mathbf{i} + (y - x)\mathbf{j}$; C is the line $y = 2x - 4$ from $(1, -2)$ to $(2, 0)$
12. $\mathbf{F}(x, y) = xy\mathbf{i} + (y - x)\mathbf{j}$; C is the curve $y = \sqrt{x}$ from $(0, 0)$ to $(1, 1)$

13. $\mathbf{F}(x, y) = xy\mathbf{i} + (y - x)\mathbf{j}$; C is the unit circle in the counterclockwise direction
14. $\mathbf{F}(x, y) = xy\mathbf{i} + (y - x)\mathbf{j}$; C is the triangle joining the points $(0, 0)$, $(0, 1)$, and $(1, 0)$ in the counterclockwise direction
15. $\mathbf{F}(x, y) = xy\mathbf{i} + (y - x)\mathbf{j}$; C is the triangle joining the points $(0, 0)$, $(1, 0)$, and $(1, 1)$ in the counterclockwise direction.
16. $\mathbf{F}(x, y) = e^x\mathbf{i} + e^y\mathbf{j}$; C is the curve of Problem 12
17. $\mathbf{F}(x, y) = (x^2 + 2y)\mathbf{i} - y^2\mathbf{j}$; C is the part of the ellipse $x^2 + 9y^2 = 9$ joining the points $(0, -1)$ and $(0, 1)$ in the clockwise direction
18. $\mathbf{F}(x, y) = (\cos x)\mathbf{i} - (\sin y)\mathbf{j}$; C is the curve of Problem 17
19. $\mathbf{F}(x, y) = e^{x+y}\mathbf{i} + e^{x-y}\mathbf{j}$; C is the curve of Problem 14
20. $\mathbf{F}(x, y) = e^{x+y}\mathbf{i} + e^{x-y}\mathbf{j}$; C is the curve of Problem 15
21. $\mathbf{F}(x, y) = (y/x^2)\mathbf{i} + (x/y^2)\mathbf{j}$; C is the straight line segment from $(2, 1)$ to $(4, 6)$
22. $\mathbf{F}(x, y) = (\ln x)\mathbf{i} + (\ln y)\mathbf{j}$; C is the curve $\mathbf{x}(t) = 2t\mathbf{i} + t^3\mathbf{j}$ for $1 \le t \le 4$
23. $\mathbf{F}(x, y, z) = x\mathbf{i} + y\mathbf{j} + z\mathbf{k}$; C is the curve $\mathbf{x}(t) = t\mathbf{i} + t^2\mathbf{j} + t^3\mathbf{k}$ from $(0, 0, 0)$ to $(1, 1, 1)$
24. $\mathbf{F}(x, y, z) = 2xz\mathbf{i} - xy\mathbf{j} + yz^2\mathbf{k}$; C is the curve of the preceding problem.
25. $\mathbf{F}(x, y, z) = x^2\mathbf{i} + y^2\mathbf{j} + z^2\mathbf{k}$; C is the helix $\mathbf{x}(t) = (\cos t)\mathbf{i} + (\sin t)\mathbf{j} + t\mathbf{k}$ from $(1, 0, 0)$ to $(0, 1, \pi/2)$
26. $\mathbf{F}(x, y, z) = yz\mathbf{i} + xz\mathbf{j} + xy\mathbf{k}$; C is the curve of the preceding problem

■ APPLICATIONS

In Problems 27–34, forces are given in newtons and distances are given in meters.

27. Calculate the work done when a force field $\mathbf{F}(x, y) = x^3\mathbf{i} + xy\mathbf{j}$ moves a particle from the point $(0, 1)$ to the point $(1, e^{\pi/2})$ along the curve $\mathbf{x}(t) = (\sin t)\mathbf{i} + e^t\mathbf{j}$.

28. Calculate the work done by the force field $\mathbf{F}(x, y) = 2xy\mathbf{i} + y^2\mathbf{j}$ when a particle is moved counterclockwise around the triangle with vertices $(0, 0)$, $(1, 0)$, and $(1, 1)$.

29. Calculate the work done when the force field $\mathbf{F}(x, y) = xy\mathbf{i} + (2x^3 - y)\mathbf{j}$ moves a particle

CONTINUED

around the unit circle in the counterclockwise direction.

30. What is the work done if the particle in the preceding problem is moved in the clockwise direction?

31. Calculate the work done by the force field $\mathbf{F}(x, y) = -xy^2\mathbf{i} + 2x\mathbf{j}$ when a particle is moved around the ellipse $(x/a)^2 + (y/b)^2 = 1$ in the counterclockwise direction.

32. Calculate the work done by the force field $\mathbf{F}(x, y) = 2x\mathbf{i} - xy^2\mathbf{j}$ when a particle is moved around the ellipse $(x/a)^2 + (y/b)^2 = 1$ in the counterclockwise direction.

33. Two electrical charges of like polarity (i.e., both positive or both negative) will repel each other. If a charge of α coulombs is placed at the origin and a charge of 1 coulomb of the same polarity

is at the point (x, y), then the force of repulsion is given by

$$\mathbf{F}(x, y) = \frac{\alpha x}{(x^2 + y^2)^{3/2}}\,\mathbf{i} + \frac{\alpha y}{(x^2 + y^2)^{3/2}}\,\mathbf{j}.$$

How much work is done by the force on the 1-coulomb charge as that charge moves on the straight line segment from $(1, 0)$ to $(3, -2)$?

34. How much work is done by the force on the charge in the preceding problem if the charge moves in the counterclockwise direction along the semicircle $x^2 + y^2 = a^2$, $y \geq 0$?

35. A particle moves along the elliptical helix $\mathbf{x}(t) = (\cos t)\mathbf{i} + 4(\sin t)\mathbf{j} + t\mathbf{k}$. It is subject to a force given by the vector field $x^2\mathbf{i} + y^2\mathbf{j} + 2xyz\mathbf{k}$. How much work is done in moving from the point $(1, 0, 0)$ to the point $(0, 4, \pi/2)$?

■ CHALLENGE

*36. Show that if $\mathbf{F}$ is continuous and if the curve C is smooth, then the value of $\int_C \mathbf{F}(\mathbf{x}) \cdot d\mathbf{x}$ does not depend on the parametrization of C. Note that the notion of a curve involves both a set of points and a direction. For instance, the straight line segment from $(0, 0)$ to $(1, 1)$ can be parametrized by $t\mathbf{i} + t\mathbf{j}$ for $0 \leq t \leq 1$ and by $\cos \theta\mathbf{i} + \cos \theta\mathbf{j}$ for $-\pi/2 \leq \theta \leq 0$ but $(1 - u^2)\mathbf{i} + (1 - u^2)\mathbf{j}$ for $0 \leq$

$u \leq 1$ goes in the wrong direction. There are many more parametrizations. You will prove that the value of the integral is independent of the one chosen. [Also note that you are *not* dealing with the question of whether a different path connecting the same endpoints, e.g., by way of $(1, 0)$, would yield the same value for the integral.]

■ ANSWERS TO SELF-QUIZ

I. c   II. b   III. a
IV. a, b, c (See Problem 36.)

15.3 EXACT VECTOR FIELDS AND INDEPENDENCE OF PATH

There are certain conditions under which the calculation of a line integral becomes very easy. We first illustrate what we have in mind.

EXAMPLE 1
COMPUTING A LINE INTEGRAL
OVER THREE DIFFERENT PATHS

Let $\mathbf{F}(x, y) = y\mathbf{i} + x\mathbf{j}$. Calculate $\int_C \mathbf{F}(\mathbf{x}) \cdot d\mathbf{x}$, where C is as follows:

(a) The straight line from $(0, 0)$ to $(1, 1)$.

(b) The parabola $y = x^2$ from $(0, 0)$ to $(1, 1)$.

(c) The curve $\mathbf{x}(t) = t^{3/2}\mathbf{i} + t^5\mathbf{j}$ from $(0, 0)$ to $(1, 1)$.

SOLUTION:

(a) C is given by $\mathbf{x}(t) = t\mathbf{i} + t\mathbf{j}$. Then $\mathbf{x}'(t) = \mathbf{i} + \mathbf{j}$ and $\mathbf{F} \cdot \mathbf{x}' = (t\mathbf{i} + t\mathbf{j}) \cdot (\mathbf{i} + \mathbf{j}) = 2t$, so

$$\int_C \mathbf{F}(\mathbf{x}) \cdot d\mathbf{x} = \int_0^1 2t\, dt = 1.$$

(b) C is given by $\mathbf{x}(t) = t\mathbf{i} + t^2\mathbf{j}$. Then $\mathbf{x}'(t) = \mathbf{i} + 2t\mathbf{j}$ and $\mathbf{F} \cdot \mathbf{x}' = (t^2\mathbf{i} + t\mathbf{j}) \cdot (\mathbf{i} + 2t\mathbf{j}) = 3t^2$, so

$$\int_C \mathbf{F}(\mathbf{x}) \cdot \mathbf{dx} = \int_0^1 3t^2 \, dt = 1.$$

(c) Here $\mathbf{x}'(t) = \frac{3}{2}\sqrt{t}\mathbf{i} + 5t^4\mathbf{j}$ and $\mathbf{F}(x, y) = t^5\mathbf{i} + t^{3/2}\mathbf{j}$, so $\mathbf{F} \cdot \mathbf{x}' = (t^5\mathbf{i} + t^{3/2}\mathbf{j}) \cdot (\frac{3}{2}\sqrt{t}\mathbf{i} + 5t^4\mathbf{j}) = \frac{3}{2}t^{11/2} + 5t^{11/2} = \frac{13}{2}t^{11/2}$. Then

$$\int_C \mathbf{F} \cdot \mathbf{x}' \, dt = \int_0^1 \frac{13}{2}t^{11/2} \, dt = 1.$$

INDEPENDENCE OF PATH

In Example 1, we saw that on three very different curves, we obtained the same answer as we moved between the points $(0, 0)$ and $(1, 1)$. In fact, as we will show in a moment, we will get the same answer if we integrate along any piecewise smooth curve C joining these two points. When this happens, we say that the line integral is **independent of the path**. A condition that ensures that a line integral is independent of the path over which it is integrated is given below.

> **THEOREM 1** **A FUNCTION IS A GRADIENT IF AND ONLY IF ITS LINE INTEGRAL IS INDEPENDENT OF THE PATH**
>
> Let $\mathbf{F}$ be continuous in a region Ω in $\mathbb{R}^2$. Then $\mathbf{F}$ is the gradient of a differentiable function f if and only if for any piecewise smooth curve C lying in Ω, the line integral $\int_C \mathbf{F}(\mathbf{x}) \cdot \mathbf{dx}$ takes the same value for any piecewise smooth curve C in Ω.

PROOF: We first assume that $\mathbf{F}(\mathbf{x}) = \nabla f$ for some differentiable function f. Recall the vector form of the chain rule:

$$\frac{d}{dt} f(\mathbf{x}(t)) = \nabla f(\mathbf{x}(t)) \cdot \mathbf{x}'(t) \tag{1}$$

[see equation (13.6.3) on page 876]. Now suppose that C is given by $\mathbf{x}(t)$: $a \le t \le b$, $\mathbf{x}(a) = \mathbf{x}_0$, and $\mathbf{x}(b) = \mathbf{x}_1$. We will assume that C is smooth. Otherwise, we could write the line integral in the form (15.2.13) and treat the integral over each smooth curve C_i separately. Using (1), we have

$$\int_C \mathbf{F}(\mathbf{x}) \cdot \mathbf{dx} = \int_a^b \mathbf{F}(\mathbf{x}(t)) \cdot \mathbf{x}'(t) \, dt \overset{\mathbf{F} = \nabla f}{=} \int_a^b \nabla f(\mathbf{x}(t)) \cdot \mathbf{x}'(t) \, dt$$

$$= \int_a^b \frac{d}{dt} f(\mathbf{x}(t)) \, dt = f(\mathbf{x}(b)) - f(\mathbf{x}(a))$$

$$= f(\mathbf{x}_1) - f(\mathbf{x}_0).$$

This proves that the line integral is independent of the path since $f(\mathbf{x}_1) - f(\mathbf{x}_0)$ does not depend on the particular curve chosen.

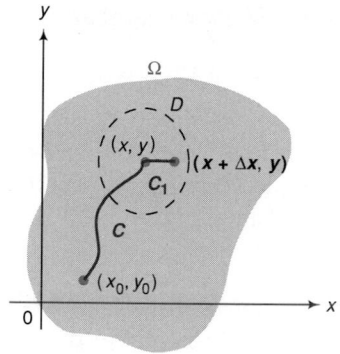

FIGURE 1
THE CURVE C JOINS (x_0, y_0) TO (x, y). THE HORIZONTAL LINE SEGMENT C_1 JOINS (x, y) TO $(x + \Delta x, y)$.

We now assume that $\int_C \mathbf{F}(\mathbf{x}) \cdot d\mathbf{x}$ is independent of the path and prove that $\mathbf{F} = \nabla f$ for some differentiable function f. Let $\mathbf{x}_0$ be a fixed point in Ω and let $\mathbf{x}$ be any other point in Ω. Since Ω is connected, there is at least one piecewise smooth path C joining $\mathbf{x}_0$ and $\mathbf{x}$, with C wholly contained in Ω. We define a function f by

$$f(\mathbf{x}) = \int_C \mathbf{F}(\mathbf{x}) \cdot d\mathbf{x}.$$

This function is well defined because, by hypothesis, $\int_C \mathbf{F}(\mathbf{x}) \cdot d\mathbf{x}$ is the same no matter what path is chosen between $\mathbf{x}_0$ and $\mathbf{x}$. Write $\mathbf{x}_0 = (x_0, y_0)$ and $\mathbf{x} = (x, y)$. Since Ω is open, there is an open disk D centered at (x, y) that is contained in Ω. Choose $\Delta x > 0$ such that $(x + \Delta x, y) \in D$, and let C_1 be the horizontal line segment joining (x, y) to $(x + \Delta x, y)$. The situation is depicted in Figure 1. We see that $C \cup C_1$ is a path joining (x_0, y_0) to $(x + \Delta x, y)$, so that

$$f(x + \Delta x, y) = \int_C \mathbf{F}(\mathbf{x}) \cdot d\mathbf{x} + \int_{C_1} \mathbf{F}(\mathbf{x}) \cdot d\mathbf{x}$$
$$= f(x, y) + \int_{C_1} \mathbf{F}(\mathbf{x}) \cdot d\mathbf{x}. \tag{2}$$

A parametrization for C_1 (which is a horizontal line) is

$$\mathbf{x}(t) = (x + t\,\Delta x, y), \qquad 0 \le t \le 1$$

and

$$\mathbf{x}'(t) = (\Delta x, 0).$$

Suppose that $\mathbf{F}(\mathbf{x}) = P(x, y)\mathbf{i} + Q(x, y)\mathbf{j}$. Then

$$\mathbf{F}(\mathbf{x}) \cdot d\mathbf{x} = P(x, y)\,\Delta x. \tag{3}$$

From (2) and (3) we compute

$$f(x + \Delta x, y) - f(x, y) = \int_{C_1} \mathbf{F}(\mathbf{x}) \cdot d\mathbf{x} = \int_0^1 P(x + t\,\Delta x, y)\,\Delta x\, dt,$$

so that

$$\frac{f(x + \Delta x, y) - f(x, y)}{\Delta x} = \int_0^1 P(x + t\,\Delta x, y)\, dt. \tag{4}$$

In (4), $P(x + t\,\Delta x, y)$ is a function of one variable (t), so we may apply the mean value theorem for integrals (Theorem 4.8.2) to see that there is a number $\bar{x}$ with $x < \bar{x} < x + \Delta x$ such that

$$\int_0^1 P(x + t\,\Delta x, y)\, dt = P(\bar{x}, y) \int_0^1 dt = P(\bar{x}, y).$$

Thus taking the limits as $\Delta x \to 0$ on both sides of (4), we have

$$\frac{\partial f(\mathbf{x})}{\partial x} = \lim_{\Delta x \to 0} \frac{f(x + \Delta x, y) - f(x, y)}{\Delta x}$$

$$P \text{ is continuous and } \bar{x} \to x \text{ as } \Delta x \to 0$$
$$\downarrow$$
$$= \lim_{\Delta x \to 0} P(\bar{x}, y) = P(x, y).$$

In a similar manner, we can show that $(\partial f / \partial y)(\mathbf{x}) = Q(x, y)$. This shows that $\mathbf{F} = \nabla f$, and the proof is complete. ∎

In proving this theorem, we also proved the following corollary.

COROLLARY 1

Suppose that $\mathbf{F}$ is continuous in a region Ω in $\mathbb{R}^2$ and $\mathbf{F} = \nabla f$ for some differentiable function f. Then for any piecewise smooth curve C in Ω starting at the point $\mathbf{x}_0$ and ending at the point $\mathbf{x}_1$,

$$\int_C \mathbf{F}(\mathbf{x}) \cdot d\mathbf{x} = f(\mathbf{x}_1) - f(\mathbf{x}_0). \tag{5}$$

That is, the value of the integral depends only on the endpoints of the path. ■

REMARK 1: This corollary is really the line integral analog of the fundamental theorem of calculus. It says that we can evaluate the line integral of a gradient field by evaluating at two points the function for which $\mathbf{F}$ is the gradient.

REMARK 2: In Theorem 1 it is important that $\mathbf{F}$ be continuous on Ω, not only on C. (See Problem 28.)

EXAMPLE 2
INTEGRATING A FUNCTION
THAT CAN BE WRITTEN AS A
GRADIENT

Since $y\mathbf{i} + x\mathbf{j} = \nabla(xy)$, we immediately find that for any curve C starting at $(0, 0)$ and ending at $(1, 1)$

$$\int_C \mathbf{F}(\mathbf{x}) \cdot d\mathbf{x} = xy\Big|_{(1,1)} - xy\Big|_{(0,0)} = 1 - 0 = 1.$$

We now state a general result that tells us whether $\mathbf{F}$ is a gradient of some function f. Half its proof is difficult and is suggested in Problem 15.4.24. The other half is suggested in Problem 27.

THEOREM 2 CONDITIONS FOR EXACTNESS IN $\mathbb{R}^2$

Let $\mathbf{F}(x, y) = P(x, y)\mathbf{i} + Q(x, y)\mathbf{j}$ and suppose that P, Q, $\partial P/\partial y$, and $\partial Q/\partial x$ are continuous in an open disk D centered at (x, y). Then, in D, $\mathbf{F}$ is the gradient of a function f if and only if

$$\frac{\partial P}{\partial y} = \frac{\partial Q}{\partial x}. \quad ■$$

DEFINITION EXACT VECTOR FIELD IN $\mathbb{R}^2$

We say that the vector field $P(x, y)\mathbf{i} + Q(x, y)\mathbf{j}$ is **exact** if $\mathbf{F}$ is the gradient of a function f. ■

COROLLARY 2

Let $\mathbf{F}(x, y) = P(x, y)\mathbf{i} + Q(x, y)\mathbf{j}$. Then if P, Q, $\partial P/\partial y$, and $\partial Q/\partial x$ are all continuous on an open disk D containing C and if $\partial P/\partial y = \partial Q/\partial x$, then $\int_C \mathbf{F}(\mathbf{x}) \cdot d\mathbf{x}$ is independent of the path. ■

EXAMPLE 3
INTEGRATING AN EXACT VECTOR
FIELD

Let $\mathbf{F}(x, y) = [4x^3y^3 + (1/x)]\mathbf{i} + [3x^4y^2 - (1/y)]\mathbf{j}$. Calculate $\int_C \mathbf{F}(\mathbf{x}) \cdot d\mathbf{x}$ for any smooth curve C, not crossing the x- or y-axis, starting at $(1, 1)$ and ending at $(2, 3)$.

SOLUTION: $P(x, y) = 4x^3y^3 + (1/x)$ and $Q(x, y) = 3x^4y^2 - (1/y)$, so

$$\frac{\partial P}{\partial y} = 12x^3y^2 = \frac{\partial Q}{\partial x}.$$

Thus, $\mathbf{F}$ is exact.
If $\nabla f = \mathbf{F}$, then $\partial f/\partial x = P$, so

$$f(x, y) = \int \left(4x^3y^3 + \frac{1}{x} \right) dx = x^4y^3 + \ln|x| + g(y).$$

Differentiating with respect to y, we have

$$Q = \frac{\partial f}{\partial y} = 3x^4y^2 + g'(y) = 3x^4y^2 - \frac{1}{y}.$$

Thus, $g'(y) = -1/y$, $g(y) = -\ln|y| + C$, and finally,

$$f(x, y) = x^4y^3 + \ln|x| - \ln|y| + C = x^4y^3 + \ln\left| \frac{x}{y} \right| + C.$$

We have shown that $\mathbf{F}$ is exact and that $\mathbf{F} = \nabla f$, where $f(x, y) = x^4y^3 + \ln|x/y|$. Thus $\int_C \mathbf{F}(\mathbf{x}) \cdot d\mathbf{x}$ is independent of the path, and

$$\int_C \mathbf{F}(\mathbf{x}) \cdot d\mathbf{x} = f(2, 3) - f(1, 1) = (16)(27) + \ln\left(\frac{2}{3} \right) - 1$$

$$= 431 + \ln\left(\frac{2}{3} \right).$$

EXAMPLE 4
INTEGRATING A VECTOR FIELD
THAT IS NOT EXACT OVER TWO
DIFFERENT PATHS

Let $\mathbf{F}(\mathbf{x}) = x\mathbf{i} + (x - y)\mathbf{j}$. Let C_1 be the part of the curve $y = x^2$ that connects $(0, 0)$ to $(1, 1)$, and let C_2 be the part of the curve $y = x^3$ that connects these two points. Compute $\int_{C_1} \mathbf{F}(\mathbf{x}) \cdot d\mathbf{x}$ and $\int_{C_2} \mathbf{F}(\mathbf{x}) \cdot d\mathbf{x}$.

SOLUTION: Along C_1, $\mathbf{x}(t) = t\mathbf{i} + t^2\mathbf{j}$ and $\mathbf{x}'(t) = \mathbf{i} + 2t\mathbf{j}$ for $0 \le t \le 1$, so that

$$\int_{C_1} \mathbf{F}(\mathbf{x}) \cdot d\mathbf{x} = \int_0^1 [t \cdot 1 + (t - t^2)2t] \, dt = \int_0^1 (t + 2t^2 - 2t^3) \, dt$$

$$= \left(\frac{t^2}{2} + \frac{2t^3}{3} - \frac{t^4}{2} \right)\Big|_0^1 = \frac{2}{3}.$$

Along C_2, $\mathbf{x}(t) = t\mathbf{i} + t^3\mathbf{j}$ and $\mathbf{x}' = \mathbf{i} + 3t^2\mathbf{j}$ for $0 \le t \le 1$, so

$$\int_{C_2} \mathbf{F}(\mathbf{x}) \cdot d\mathbf{x} = \int_0^1 [t \cdot 1 + (t - t^3)(3t^2)] \, dt = \int_0^1 (t + 3t^3 - 3t^5) \, dt$$

$$= \left(\frac{t^2}{2} + \frac{3}{4}t^4 - \frac{1}{2}t^6 \right)\Big|_0^1 = \frac{3}{4}.$$

We see that $\int_C \mathbf{F}(\mathbf{x}) \cdot d\mathbf{x}$ is *not* independent of the path. Note that here $\partial P/\partial y = 0$ and $\partial Q/\partial x = 1$, so $\mathbf{F}$ is *not* exact.

There is another important consequence of Theorem 1. Let C be a closed curve (i.e., $\mathbf{x}_0 = \mathbf{x}_1$).

THE LINE INTEGRAL OF AN EXACT VECTOR FIELD OVER A CLOSED CURVE IS ZERO

If $\mathbf{F}$ is continuous in a region Ω and if $\mathbf{F}$ is the gradient of a differentiable function f, then for any closed curve C lying in Ω,

$$\int_C \mathbf{F}(\mathbf{x}) \cdot d\mathbf{x} = f(\mathbf{x}_1) - f(\mathbf{x}_0) = 0. \tag{6}$$

INDEPENDENCE OF PATH IN $\mathbb{R}^3$

As in the plane, the computation of a line integral in space is not difficult if the integral is **independent of the path**. Theorem 3 generalizes Theorem 1. Its proof is essentially identical to the proof of that theorem.

THEOREM 3   CONDITIONS FOR EXACTNESS IN $\mathbb{R}^3$

Let $\mathbf{F}$ be continuous on a region S in $\mathbb{R}^3$. Then $\mathbf{F}$ is the gradient of a function f if and only if the following equivalent conditions hold:

(i) For any piecewise smooth curve C lying in S, the line integral

$$\int_C \mathbf{F}(\mathbf{x}) \cdot d\mathbf{x}$$

is independent of the path.

(ii) For any piecewise smooth curve C in S starting at the point $\mathbf{x}_0$ and ending at the point $\mathbf{x}_1$,

$$\int_C \mathbf{F}(\mathbf{x}) \cdot d\mathbf{x} = f(\mathbf{x}_1) - f(\mathbf{x}_0). \quad \blacksquare$$

In Theorem 2; we saw that $\mathbf{F}(x, y) = P(x, y)\mathbf{i} + Q(x, y)\mathbf{j}$ is the gradient of a function f if $\partial P/\partial y = \partial Q/\partial x$. If $\mathbf{F}(x, y, z) = P(x, y, z)\mathbf{i} + Q(x, y, z)\mathbf{j} + R(x, y, z)\mathbf{k}$, then there is a condition that can be used to check whether there is a differentiable function f such that $\mathbf{F} = \nabla f$. We give this result without proof.[†]

THEOREM 4   CONDITIONS FOR EXACTNESS IN $\mathbb{R}^3$

Let $\mathbf{F}(x, y, z) = P(x, y, z)\mathbf{i} + Q(x, y, z)\mathbf{j} + R(x, y, z)\mathbf{k}$ and suppose that $P, Q, R, \partial P/\partial y, \partial P/\partial z, \partial Q/\partial x, \partial Q/\partial z, \partial R/\partial x$, and $\partial R/\partial y$ are continuous in an open ball centered at (x, y, z).[‡] Then $\mathbf{F}$ is the gradient of a function f if and only if

$$\frac{\partial P}{\partial y} = \frac{\partial Q}{\partial x}, \qquad \frac{\partial R}{\partial x} = \frac{\partial P}{\partial z}, \qquad \frac{\partial Q}{\partial z} = \frac{\partial R}{\partial y}. \quad \blacksquare \tag{7}$$

[†] For a proof, see R. C. Buck and E. F. Buck, *Advanced Calculus*, 3rd ed. (New York: McGraw-Hill, 1978), pp. 497–498.

[‡] This theorem and Theorem 2 are also true in a more general region called a **simply connected region**. We define such a region on page 984.

DEFINITION   **EXACT VECTOR FIELD IN** $\mathbb{R}^3$

If **F** is the gradient of a differentiable function f, then **F** is said to be **exact**. ∎

EXAMPLE 5
INTEGRATING AN EXACT VECTOR
FIELD IN $\mathbb{R}^3$

Let $\mathbf{F}(x, y, z) = yz\mathbf{i} + xz\mathbf{j} + xy\mathbf{k}$. Verify that **F** is exact and compute $\int_C \mathbf{F}(\mathbf{x}) \cdot d\mathbf{x}$ where C is a smooth curve joining $(1, 1, 1)$ to $(2, -1, 3)$.

SOLUTION: We find that $\partial P/\partial y = z = \partial Q/\partial x$, $\partial R/\partial x = y = \partial P/\partial z$, and $\partial Q/\partial z = x = \partial R/\partial y$. Thus **F** is exact, and there is a differentiable function f such that $\mathbf{F} = \nabla f$, so that

$$f_x = yz, \qquad f_y = xz, \qquad f_z = xy.$$

Then

$$f(x, y, z) = \int f_x \, dx = \int yz \, dx = xyz + g(y, z)$$

for some differentiable function g of y and z only. (This means that g is a constant function *with respect to x*.) Hence,

$$f_y = xz + g_y(y, z) = xz,$$

so $g_y(y, z) = 0$ and $g(y, z) = h(z)$, where h is a differentiable function of z only. Then

$$f(x, y, z) = xyz + h(z)$$

and

$$f_z(x, y, z) = xy + h'(z) = xy,$$

so $h'(z) = 0$ and $h(z) = C$, a constant. Thus,

$$f(x, y, z) = xyz + C,$$

and it is easily verified that $\nabla f = \mathbf{F}$. Finally, we have

$$\int_C \mathbf{F}(\mathbf{x}) \cdot d\mathbf{x} = f(2, -1, 3) - f(1, 1, 1) = (2)(-1)(3) - 1 = -7.$$

There is another way to complete this problem. Since **F** is exact, the integral is independent of the path. The simplest path joining $(1, 1, 1)$ to $(2, -1, 3)$ is the straight line segment joining these two points. One parametrization of this line segment is (see Section 11.5)

$$x = 1 + t \qquad y = 1 - 2t \qquad z = 1 + 2t, \qquad 0 \le t \le 1.$$

Then

$$yz = (1 - 2t)(1 + 2t) = 1 - 4t^2$$

$$xz = (1 + t)(1 + 2t) = 1 + 3t + 2t^2$$

$$xy = (1 + t)(1 - 2t) = 1 - t - 2t^2$$

$$d\mathbf{x} = \mathbf{i} - 2\mathbf{j} + 2\mathbf{k}$$

and

$$\mathbf{F}(\mathbf{x}) \cdot d\mathbf{x} = 1 - 4t^2 - 2(1 + 3t + 2t^2) + 2(1 - t - 2t^2)$$
$$= 1 - 8t - 12t^2.$$

Thus,

$$\int_C \mathbf{F}(\mathbf{x}) \cdot d\mathbf{x} = \int_0^1 (1 - 8t - 12t^2)\, dt = (t - 4t^2 - 4t^3)\Big|_0^1$$
$$= 1 - 4 - 4 = -7.$$

Before leaving this section, we note that line integrals in space are often written in a different way. If $\mathbf{F}(x, y, z) = P(x, y, z)\mathbf{i} + Q(x, y, z)\mathbf{j} + R(x, y, z)\mathbf{k}$, and if we write $d\mathbf{x} = dx\mathbf{i} + dy\mathbf{j} + dz\mathbf{k}$, then

---

ALTERNATIVE WAY TO WRITE A LINE INTEGRAL IN SPACE

$$\int_C \mathbf{F}(\mathbf{x}) \cdot d\mathbf{x} = \int_C (P\, dx + Q\, dy + R\, dz) \tag{8}$$

Finally, we remind you of an important definition given in Section 15.1.

DEFINITION **CONSERVATIVE VECTOR FIELD AND SCALAR POTENTIAL**

Suppose that $\mathbf{F} = -\nabla f$ for some differential function f. Then $\mathbf{F}$ is said to be **conservative** and f is called a **scalar potential**. ∎

■ PROBLEMS 15.3

■ SELF-QUIZ

I. True–False: $0.333x\mathbf{i} - 0.667y\mathbf{j}$ is not exact because 0.333 is not exactly equal to $1/3$ and -0.667 is not exactly equal to $-2/3$.

II. True–False: $2y\mathbf{i} + 3x\mathbf{j}$ is exact because 2 and 3 are integers.

III. Let C be a smooth path connecting $(2, 3)$ to $(8, 7)$. Then $\int_C \nabla(xy) \cdot d\mathbf{x} = $ _____.
 a. $(8)(7) - (2)(3) = 50$
 b. $(8 - 2)(7 - 3) = 24$
 c. $(8)(2) - (7)(3) = -5$
 d. varies with the particular path chosen

IV. Let C be a smooth path connecting $(2, 3, 0)$ to $(8, 7, 0)$. Then $\int_C (y\mathbf{i} + x\mathbf{j} + 0\mathbf{k}) \cdot d\mathbf{x} = $ _____.
 a. $(8)(7) - (2)(3) = 50$

 b. $(8)(7)(0) - (2)(3)(0) = 0$
 c. $(8 - 2)(7 - 3)(0 - 0) = 0$
 d. varies with the particular path chosen

V. Suppose C is the square path from $(0, 0)$ to $(1, 0)$ to $(1, 1)$ to $(0, 1)$ and back to $(0, 0)$. Then $\int_C \nabla[\tan^{-1}(x - \ln|e^y + e^x|)] \cdot d\mathbf{x} = $ _____.
 a. 0
 b. $\coth(\ln\sqrt{5}) + (22/7)$
 c. $\pi/4 - e^2$
 d. $\sqrt{e} + \sqrt{\pi}$

VI. Suppose C is the square path from $(0, 0)$ to $(1, 0)$ to $(1, 1)$ to $(0, 1)$ and back to $(0, 0)$. Then $\int_C (y\mathbf{i} - x\mathbf{j}) \cdot d\mathbf{x} = $ _____.
 a. 0
 b. -1
 c. -2
 d. 2

■ DRILL

In Problems 1–16, test for exactness. If the given vector field $\mathbf{F}$ is exact, then find all functions f for which $\nabla f = \mathbf{F}$.

1. $\mathbf{F}(x, y) = 2xy\mathbf{i} + (x^2 + 1)\mathbf{j}$
2. $\mathbf{F}(x, y) = (4x^3 - ye^{xy})\mathbf{i} + (\tan y - xe^{xy})\mathbf{j}$
3. $\mathbf{F}(x, y) = (4x^2 - 4y^2)\mathbf{i} + (8xy - \ln y)\mathbf{j}$
4. $\mathbf{F}(x, y) = [x\cos(x + y) + \sin(x + y)]\mathbf{i}$
    $+ x\cos(x + y)\mathbf{j}$

5. $\mathbf{F}(x, y) = 2x(\cos y)\mathbf{i} + x^2(\sin y)\mathbf{j}$

6. $\mathbf{F}(x, y) = \left[\dfrac{\ln(\ln y)}{x} + \dfrac{2}{3}xy^3\right]\mathbf{i} + \left[\dfrac{\ln x}{y\ln y} + x^2y^2\right]\mathbf{j}$

7. $\mathbf{F}(x, y) = (x - y\cos x)\mathbf{i} - (\sin x)\mathbf{j}$
8. $\mathbf{F}(x, y) = e^{x^2y}\mathbf{i} + e^{x^2y}\mathbf{j}$
9. $\mathbf{F}(x, y) = (3x\ln x + x^5 - y)\mathbf{i} - x\mathbf{j}$

CONTINUED

10. $\mathbf{F}(x, y) = \left[\dfrac{1}{x^2} + y^2\right]\mathbf{i} + 2xy\mathbf{j}$

11. $\mathbf{F}(x, y) = (x^2 + y^2 + 1)\mathbf{i} - (xy + y)\mathbf{j}$

12. $\mathbf{F}(x, y) = \left[\dfrac{-1}{x^3} + 4x^3y\right]\mathbf{i} + [\sin y + \sqrt{y} + x^4]\mathbf{j}$

13. $\mathbf{F}(x, y, z) = \mathbf{i} + \mathbf{j} + \mathbf{k}$

14. $\mathbf{F}(x, y, z) = yz\mathbf{i} + xz\mathbf{j} + xy\mathbf{k}$

15. $\mathbf{F}(x, y, z) = [e^{yz} + y]\mathbf{i} + [xze^{yz} - x]\mathbf{j} + [xye^{yz} + 2z]\mathbf{k}$

16. $\mathbf{F}(x, y, z) = [2xy^3 + x + z]\mathbf{i} + [3x^2y^2 - y]\mathbf{j} + [x + \sin z]\mathbf{k}$

■ APPLICATIONS

In Problems 17–26, verify that $\mathbf{F}$ is exact, then use Corollary 1 to calculate $\int_C \mathbf{F}(\mathbf{x}) \cdot d\mathbf{x}$ where C is a smooth curve starting at $\mathbf{x}_0$ and ending at $\mathbf{x}_1$.

17. $\mathbf{F}(x, y) = 2xy\mathbf{i} + (x^2 + 1)\mathbf{j}$; $\mathbf{x}_0 = (0, 1)$, $\mathbf{x}_1 = (2, 3)$

18. $\mathbf{F}(x, y) = (4x^2 - 4y^2)\mathbf{i} + (\ln y - 8xy)\mathbf{j}$; $\mathbf{x}_0 = (-1, 1)$, $\mathbf{x}_1 = (4, e)$

19. $\mathbf{F}(x, y) = \lceil x\cos(x + y) + \sin(x + y)\rceil\mathbf{i} + [x\cos(x + y)]\mathbf{j}$; $\mathbf{x}_0 = (0, 0)$, $\mathbf{x}_1 = (\pi/6, \pi/3)$

20. $\mathbf{F}(x, y) = \left[\dfrac{1}{x^2} + y^2\right]\mathbf{i} + 2xy\mathbf{j}$; $\mathbf{x}_0 = (1, 4)$,
$\mathbf{x}_1 = (3, 2)$

21. $\mathbf{F}(x, y) = 2x(\cos y)\mathbf{i} - x^2(\sin y)\mathbf{j}$; $\mathbf{x}_0 = (0, \pi/2)$, $\mathbf{x}_1 = (\pi/2, 0)$

22. $\mathbf{F}(x, y) = [2xy^3 - 2]\mathbf{i} + [3x^2y^2 + \cos y]\mathbf{j}$; $\mathbf{x}_0 = (1, 0)$, $\mathbf{x}_1 = (0, -\pi)$

23. $\mathbf{F}(x, y) = e^y\mathbf{i} + xe^y\mathbf{j}$; $\mathbf{x}_0 = (0, 0)$, $\mathbf{x}_1 = (5, 7)$

24. $\mathbf{F}(x, y) = (\cosh x)(\cosh y)\mathbf{i} + (\sinh x)(\sinh y)\mathbf{j}$; $\mathbf{x}_0 = (0, 0)$, $\mathbf{x}_1 = (1, 2)$

25. $\mathbf{F}(x, y, z) = y^2z^4\mathbf{i} + 2xyz^4\mathbf{j} + 4xy^2z^3\mathbf{k}$; $\mathbf{x}_0 = (0, 0, 0)$, $\mathbf{x}_1 = (3, 2, 1)$

26. $\mathbf{F}(x, y, z) = [yz + 2]\mathbf{i} + \lceil xz - 3\rceil\mathbf{j} + [xy + 5]\mathbf{k}$; $\mathbf{x}_0 = (2, 1, 2)$, $\mathbf{x}_1 = (-1, 0, 4)$

■ SHOW/PROVE/DISPROVE

27. We omitted the proof of Theorem 2, but half of it is not difficult to prove. Show that if each of $P, Q, \dfrac{\partial P}{\partial y}$, and $\dfrac{\partial Q}{\partial x}$ are continuous in an open disk, and there is a smooth function f such that $P(x, y)\mathbf{i} + Q(x, y)\mathbf{j} = \nabla f(x, y)$ in that disk, then

$$\frac{\partial P}{\partial y} = \frac{\partial Q}{\partial x}.$$

[*Hint:* Use the equality of second mixed partials.]

*28. This problem gives an example that shows the warning of Remark 2 on page 978 is necessary. Let the force field $\mathbf{F}$ be given by

$$\mathbf{F}(x, y) = \frac{y}{x^2 + y^2}\mathbf{i} - \frac{x}{x^2 + y^2}\mathbf{j}.$$

  a. Show that $\mathbf{F}$ is exact.
  b. Let C be the unit circle traversed in the counterclockwise direction; calculate $\int_C \mathbf{F}(\mathbf{x}) \cdot d\mathbf{x}$.

  c. Explain why the integral computed in the preceding part is not equal to zero. Does your work on this problem contradict Theorem 1 or equation (6)?

29. Show that the work done by a conservative force field as it moves a particle completely around a closed path is zero, provided the path is contained in a region over which the force field is continuous.

*30. If a particle moves around a circle with constant angular speed ω, then the radial force pushing the particle away from the center is called **centrifugal force** and is given by $\mathbf{F}_C = \omega^2|\mathbf{x}|\mathbf{u}$ where $\mathbf{x}$ is the position of the particle and $\mathbf{u}$ is a unit vector pointing away from the center. The **potential**, f_C, of the centrifugal force is defined as the work done in moving the particle from $\mathbf{0}$ to $\mathbf{x}$. Assuming that $f_C(\mathbf{0}) = 0$, show that

$$f_C(\mathbf{x}) = \frac{1}{2}\omega^2|\mathbf{x}|^2.$$

■ ANSWERS TO SELF-QUIZ

  I. False ($0.333x\mathbf{i} - 0.667y\mathbf{j}$ is exact.)
  II. False ($2y\mathbf{i} + 3x\mathbf{j}$ is not exact.)

III. a   IV. a   V. a
VI. c [$y\mathbf{i} - x\mathbf{j}$ is not exact.]

15.4 GREEN'S THEOREM IN THE PLANE

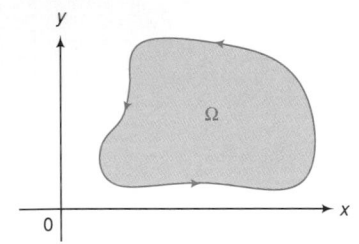

FIGURE 1
A TYPICAL REGION IN $\mathbb{R}^2$

In this section we state a result that gives an important relationship between line integrals and double integrals.

Let Ω be a region in the plane (a typical region is sketched in Figure 1). The curve (indicated by the arrows) that goes around the edge of Ω in the direction that keeps Ω on the left (the *counterclockwise* direction) is called the **boundary of** Ω and is denoted by $\partial\Omega$. Let

$$\mathbf{F}(x, y) = P(x, y)\mathbf{i} + Q(x, y)\mathbf{j}.$$

If the curve $\partial\Omega$ is given by

$$\partial\Omega: \quad \mathbf{x}(t) = x(t)\mathbf{i} + y(t)\mathbf{j},$$

we can write

$$\mathbf{F}(\mathbf{x}) \cdot \mathbf{dx} = P\,dx + Q\,dy. \tag{1}$$

We then denote the line integral of $\mathbf{F}$ around $\partial\Omega$ by

$$\oint_{\partial\Omega} P\,dx + Q\,dy. \tag{2}$$

IMPORTANT NOTATION: The symbol $\oint_{\partial\Omega}$ indicates that $\partial\Omega$ is a closed curve around which we integrate in the counterclockwise direction (the direction of the arrow).

DEFINITION   **SIMPLE CURVE**

A curve C is called **simple** if it does not cross itself. That is, suppose C is given by

$$C: \quad \mathbf{x}(t) = f(t)\mathbf{i} + g(t)\mathbf{j}, \qquad t \in [a, b].$$

Then C is simple if and only if $\mathbf{x}(t_1) \neq \mathbf{x}(t_2)$ whenever $t_1 \neq t_2$ (with the possible exception $t_1 = a$ and $t_2 = b$).   ∎

This notation is illustrated in Figure 2.

(a) Closed, simple.

(b) Not closed, simple.

(c) Not closed, not simple.

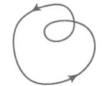

(d) Closed, not simple.

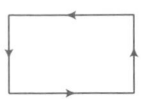

(e) Closed, simple.

FIGURE 2   FIVE CURVES

DEFINITION   **SIMPLY CONNECTED REGION**

A region Ω in the xy-plane is called **simply connected** if it has the following property: If C is a simple closed curve contained in Ω, then every point in the region enclosed by C is also in Ω.   ∎

Intuitively, a region is simply connected if it has no holes. We illustrate this definition in Figure 3.

(a) Simply connected.

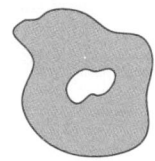
(b) Not simply connected (region has a hole).

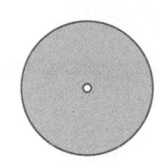
(c) Not simply connected (point is missing in this "punctured disk").

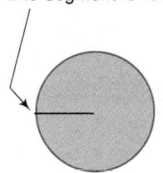
Every point on this line segment is removed

(d) Simply connected.

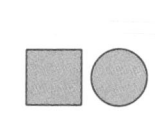
(e) Not a region (not connected).

FIGURE 3 FIVE SETS IN THE xy-PLANE

We are now ready to state the principal result of this section.

THEOREM 1 **GREEN'S THEOREM IN THE PLANE**

Let Ω be a simply connected region in the xy-plane bounded by a piecewise smooth curve $\partial\Omega$. Let P and Q be continuous with continuous first partials in an open disk containing Ω. Then

$$\oint_{\partial\Omega} P\,dx + Q\,dy = \iint_{\Omega} \left(\frac{\partial Q}{\partial x} - \frac{\partial P}{\partial y}\right) dx\,dy. \quad \blacksquare \tag{3}$$

REMARK: This theorem shows how the line integral of a function around the boundary of a region is related to a double integral over that region.

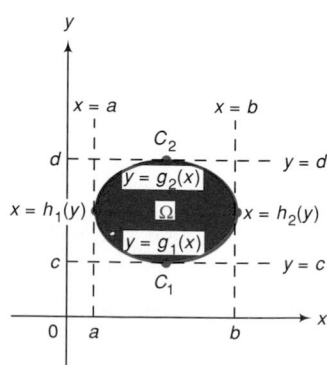

FIGURE 4
THE REGION Ω CAN BE WRITTEN

$$\{(x, y): a \le x \le b, g_1(x) \le y \le g_2(x)\}$$

AND

$$\{(x, y): c \le y \le d, h_1(y) \le x \le h_2(y)\}.$$

PARTIAL PROOF OF GREEN'S THEOREM We prove Green's theorem in the case in which Ω takes the simple form given in Figure 4. The region Ω can be written

$$\{(x, y): a \le x \le b, g_1(x) \le y \le g_2(x)\} \tag{4}$$

and

$$\{(x, y): c \le y \le d, h_1(y) \le x \le h_2(y)\}. \tag{5}$$

We first calculate

$$\iint_{\Omega} \frac{\partial P}{\partial y}\,dx\,dy = \int_a^b \left\{\int_{g_1(x)}^{g_2(x)} \frac{\partial P}{\partial y}\,dy\right\} dx = \int_a^b \left\{P(x, y)\Big|_{y=g_1(x)}^{y=g_2(x)}\right\} dx$$

$$= \int_a^b \left[P(x, g_2(x)) - P(x, g_1(x))\right] dx. \tag{6}$$

Now

$$\oint_{\partial\Omega} P\,dx = \int_{C_1} P\,dx + \int_{C_2} P\,dx, \tag{7}$$

where C_1 is the graph of $g_1(x)$ from $x = a$ to $x = b$ and C_2 is the graph of $g_2(x)$ from $x = b$ to $x = a$ (note the order). Since on C_1, $y = g_1(x)$, we have

$$\int_{C_1} P(x, y)\,dx = \int_a^b P(x, g_1(x))\,dx. \tag{8}$$

Similarly,

$$\int_{C_2} P(x, y)\, dx = \int_b^a P(x, g_2(x))\, dx = -\int_a^b P(x, g_2(x))\, dx. \tag{9}$$

Thus

$$\oint_{\partial \Omega} P\, dx = \int_a^b P(x, g_1(x))\, dx - \int_a^b P(x, g_2(x))\, dx$$

$$= -\int_a^b [P(x, g_2(x)) - P(x, g_1(x))]\, dx. \tag{10}$$

Comparing (6) and (10), we find that

$$\iint_\Omega -\frac{\partial P}{\partial y}\, dx\, dy = \oint_{\partial \Omega} P\, dx. \tag{11}$$

Similarly, using the representation (5), we have

$$\iint_\Omega \frac{\partial Q}{\partial x}\, dx\, dy = \int_c^d \left\{ \int_{h_1(y)}^{h_2(y)} \frac{\partial Q}{\partial x}\, dx \right\} dy$$

$$= \int_c^d [Q(h_2(y), y) - Q(h_1(y), y)]\, dy,$$

which by analogous reasoning yields the equation

$$\iint_\Omega \frac{\partial Q}{\partial x}\, dx\, dy = \oint_{\partial \Omega} Q\, dy. \tag{12}$$

Adding (11) and (12) completes the proof of theorem in the special case that Ω can be written in the form (4) and (5).[†] ∎

HISTORICAL NOTE: Green's theorem is named after George Green (1793–1841), a British mathematician and physicist who wrote an essay in 1828 on electricity and magnetism that contained this important theorem. Green was the self-educated son of a baker. His 1828 essay was published for private circulation. It was largely overlooked until it was rediscovered by Lord Kelvin in 1846. The theorem was independently discovered by the Russian mathematician Michel Ostrogradski (1801–1861), and to this day the theorem is known in Russia as *Ostrogradski's theorem.*

EXAMPLE 1
EVALUATING A LINE INTEGRAL
USING GREEN'S THEOREM

Evaluate $\oint_{\partial \Omega} xy\, dx + (x - y)\, dy$, where Ω is the rectangle $\{(x, y): 0 \le x \le 1, 1 \le y \le 3\}$.

SOLUTION: $P(x, y) = xy$, $Q(x, y) = x - y$, $\partial Q/\partial x = 1$, and $\partial P/\partial y = x$, so

$$\oint_{\partial \Omega} xy\, dx + (x - y)\, dy = \int_0^1 \int_1^3 (1 - x)\, dy\, dx = \int_0^1 \left\{ (1 - x)y \Big|_1^3 \right\} dx$$

$$= \int_0^1 2(1 - x)\, dx = 1.$$

[†] For a proof of Green's theorem for more general regions, see R. C. Buck and E. F. Buck, *Advanced Calculus,* 3rd ed. (New York: McGraw-Hill, 1978), p. 479.

EXAMPLE 2
EVALUATING A LINE INTEGRAL
USING GREEN'S THEOREM

Evaluate $\oint_C (x^3 + y^3)\,dx + (2y^3 - x^3)\,dy$, where C is the unit circle.

SOLUTION: We first note that $C = \partial\Omega$, where Ω is the unit disk. Next, we have $(\partial Q/\partial x) - (\partial P/\partial y) = \quad 3x^2 - 3y^2 = -3(x^2 + y^2)$. Thus,

$$\oint_C (x^3 + y^3)\,dx + (2y^3 - x^3)\,dy = -3 \iint_{\text{unit disk}} (x^2 + y^2)\,dx\,dy$$

Converting to polar coordinates
$$\downarrow$$
$$= -3 \int_0^{2\pi} \int_0^1 r^2 \cdot r\,dr\,d\theta$$

$$= -3 \int_0^{2\pi} \left\{ \frac{r^4}{4}\Big|_0^1 \right\} d\theta = -\frac{3}{4} \int_0^{2\pi} d\theta$$

$$= -\frac{3\pi}{2}.$$

Green's theorem can be useful for calculating area. Recall that

$$\text{area enclosed by } \Omega = \iint_\Omega dA. \tag{13}$$

But by Green's theorem,

FOUR INTEGRALS THAT REPRESENT AREA

$$\iint_\Omega dA = \oint_{\partial\Omega} x\,dy = \oint_{\partial\Omega} (-y)\,dx = \frac{1}{2} \oint_{\partial\Omega} [(-y)\,dx + x\,dy] \tag{14}$$

(explain why). Any of the line integrals in (14) can be used to calculate area.

EXAMPLE 3
CALCULATING THE AREA OF AN
ELLIPSE USING GREEN'S THEOREM

Use Green's theorem to calculate the area enclosed by the ellipse $(x^2/a^2) + (y^2/b^2) = 1$.

SOLUTION: The ellipse can be written parametrically as

$$\mathbf{x}(t) = a(\cos t)\mathbf{i} + b(\sin t)\mathbf{j}, \qquad 0 \le t \le 2\pi.$$

Then using the first line integral in (14), we obtain

$$A = \oint_{\partial\Omega} x\,dy = \int_0^{2\pi} (a \cos t) \frac{d}{dt}(b \sin t)\,dt$$

$$= \int_0^{2\pi} (a \cos t) b \cos t\,dt = ab \int_0^{2\pi} \cos^2 t\,dt$$

$$= \frac{ab}{2} \int_0^{2\pi} (1 + \cos 2t)\,dt = \pi ab.$$

Note how much easier this calculation is than the direct evaluation of $\iint_A dx\,dy$, where A denotes the area enclosed by the ellipse.

■ PROBLEMS 15.4

■ SELF-QUIZ

I. Suppose C is the square path from $(0, 0)$ to $(1, 0)$ to $(1, 1)$ to $(0, 1)$ and back to $(0, 0)$, then $\oint_C 3y\,dx - 2x\,dy = $ _____.

   a. $\int_0^1 \int_0^1 [(-2) - 3]\,dx\,dy$

   b. $\int_0^1 \int_0^1 [3 - (-2)]\,dx\,dy$

   c. $\int_0^3 \int_0^{-2} [(-2) - 3]\,dx\,dy$

   d. $\int_0^1 \int_0^1 [3y - 2x]\,dx\,dy$

II. Suppose D is the unit disk and C is the unit circle (traversed counterclockwise), then $\oint_C 3y^2\,dx - 2x\,dy = $ _____.

   a. $\iint_D [(-2x) - 3y^2]\,dA$

   b. $\iint_D [6y + (-2)]\,dA$

   c. $\iint_D [(-2) - 6y]\,dA$

   d. $\iint_D [y^3 - x^2]\,dA$

III. Suppose D is the unit disk and C is the unit circle (traversed counterclockwise), then $\iint_D 0\,dA \neq $ _____.

   a. $\oint_C x\,dx + y\,dy$ b. $\oint_C y\,dx + x\,dy$

   c. $\oint_C x\,dx - y\,dy$ d. $\oint_C y\,dx - x\,dy$

IV. Suppose C is the square path from $(0, 0)$ to $(1, 0)$ to $(1, 1)$ to $(0, 1)$ and back to $(0, 0)$, then $\int_0^1 \int_0^1 2x\,dA \neq $ _____.

   a. $\oint_C x^2\,dx$

   b. $\oint_C x^2\,dy$

   c. $\oint_C \sin x\,dx + x^2\,dy$

   d. $\oint_C \tan^{-1} x\,dx + (x^2 - \sin y)\,dy$

■ DRILL

In Problems 1–15, calculate the value of the line integral by means of Green's theorem.

1. $\oint_{\partial\Omega} 3y\,dx + 5x\,dy$;
   $\Omega = \{(x, y): 0 \leq x \leq 1, 0 \leq y \leq 1\}$

2. $\oint_{\partial\Omega} ay\,dx + bx\,dy$;
   $\Omega = \{(x, y): 0 \leq x \leq A, 0 \leq y \leq B\}$

3. $\oint_{\partial\Omega} e^x \cos y\,dx + e^x \sin y\,dy$; Ω is the region enclosed by the triangle with vertices at $(0, 0)$, $(1, 0)$, and $(0, 1)$.

4. The integral of Problem 3 where Ω is the region enclosed by the triangle with vertices at $(0, 0)$, $(1, 1)$, and $(1, 0)$.

5. The integral of Problem 3 where Ω is the region enclosed by the rectangle with vertices at $(0, 0)$, $(2, 0)$, $(2, 1)$ and $(0, 1)$.

6. $\oint_{\partial\Omega} 2xy\,dx + x^2\,dy$; Ω is the unit disk

7. $\oint_{\partial\Omega} (x^2 + y^2)\,dx - 2xy\,dy$; Ω is the unit disk

8. $\oint_{\partial\Omega} \frac{1}{y}\,dx + \frac{1}{x}\,dy$; Ω is the region bounded by the lines $y = 1$ and $x = 16$ and the curve $y = \sqrt{x}$

9. $\oint_{\partial\Omega} \cos y\,dx + \cos x\,dy$; Ω is the region enclosed by the rectangle $\{(x, y): 0 \leq x \leq \pi/4, 0 \leq y \leq \pi/3\}$

10. $\oint_{\partial\Omega} x^2 y\,dx - xy^2\,dy$; Ω is the disk $\{(x, y): x^2 + y^2 \leq 9\}$

11. $\oint_{\partial\Omega} y \ln x\,dy$;
    $\Omega = \{(x, y): 1 \leq y \leq 3, e^y \leq x \leq e^{y^3}\}$

12. $\oint_{\partial\Omega} \sqrt{1 + y^2}\,dx$;
    $\Omega = \{(x, y): -1 \leq y \leq 1, y^2 \leq x \leq 1\}$

13. $\oint_{\partial\Omega} ay\,dx + bx\,dy$; Ω is a region satisfying relations of types (4) and (5)

14. $\oint_{\partial\Omega} e^x \sin y\,dx + e^x \cos y\,dy$; Ω is the elliptical region $\{(x, y): (x/a)^2 + (y/b)^2 \leq 1\}$

15. $\oint_{\partial\Omega} \frac{-4x}{\sqrt{1 + y^2}}\,dx + \frac{2x^2 y}{(1 + y^2)^{3/2}}\,dy$;
    Ω is a region satisfying relations of types (4) and (5)

16. Use one of the line integrals in (14) to calculate the area of the circle $\mathbf{x}(t) = a(\cos t)\mathbf{i} + a(\sin t)\mathbf{j}$.

CONTINUED

■ APPLICATIONS

17. Use Green's theorem to calculate the area enclosed by the triangle with vertices at $(0, 0)$, $(5, 2)$, and $(-3, 8)$.

18. Use Green's theorem to calculate the area enclosed by the triangle with vertices at (a_1, b_1), (a_2, b_2), and (a_3, b_3), assuming that the three points are not collinear.

19. Use Green's theorem to calculate the area enclosed by the quadrilateral with vertices at $(0, 0)$, $(2, 1)$, $(-1, 3)$, and $(4, 4)$.

20. Use Green's theorem to calculate the area enclosed by the quadrilateral with vertices at (a_1, b_1), (a_2, b_2), (a_3, b_3), and (a_4, b_4), assuming that no three of these points are collinear and that no point is within the triangle whose vertices are the other three points.

■ SHOW/PROVE/DISPROVE

*21. Let $\partial\Omega$ be the ellipse satisfying $(x/a)^2 + (y/b)^2 = 1$; let $\mathbf{F}(x, y)$ be the vector field $-x\mathbf{i} - y\mathbf{j}$ that, at any point (x, y), points towards the origin. Show that $\oint_{\partial\Omega} \mathbf{F} \cdot \mathbf{T}\, ds = 0$, where $\mathbf{T}$ is the unit tangent vector to $\partial\Omega$ and ds is the differential of arc length (see page 813).

*22. Let Ω be the disk $\{(x, y): x^2 + y^2 \leq a^2\}$. Show that $\oint_{\partial\Omega} \alpha\sqrt{x^2 + y^2}\, dx + \beta\sqrt{x^2 + y^2}\, dy = 0$.

23. Let Ω be the same disk as in the preceding problem; suppose that g is continuously differentiable. Show that $\oint_{\partial\Omega} \alpha g(x^2 + y^2)\, dx + \beta g(x^2 + y^2)\, dy = 0$.

*24. Use Green's theorem to prove the "if" part of Theorem 15.3.2. Show that if each of P, Q, $\dfrac{\partial P}{\partial y}$, and $\dfrac{\partial Q}{\partial x}$ are continuous in an open disk and $\dfrac{\partial P}{\partial y} = \dfrac{\partial Q}{\partial x}$ then there is a smooth function f such that $P(x, y)\mathbf{i} + Q(x, y)\mathbf{j} = \nabla f(x, y)$ in that disk. [Hint: A line integral is independent of path in Ω if and only if the line integral along any closed curve in Ω is zero; use Theorem 15.3.1.]

■ ANSWERS TO SELF-QUIZ

I. a   II. c   III. d   IV. a

15.5 THE PARAMETRIC REPRESENTATION OF A SURFACE AND SURFACE AREA

In Section 12.1, we saw that a curve in $\mathbb{R}^2$ or $\mathbb{R}^3$ can be represented parametrically in terms of a single variable t. We have

$$\mathbf{f}(t) = x(t)\mathbf{i} + y(t)\mathbf{j}, \ a \leq t \leq b \qquad \text{or} \qquad \mathbf{f}(t) = x(t)\mathbf{i} + y(t)\mathbf{j} + z(t)\mathbf{k}, \ a \leq t \leq b$$

$$\text{in } \mathbb{R}^2 \qquad\qquad\qquad\qquad\qquad \text{in } \mathbb{R}^3$$

where every function of t is continuous. We can think of the function $\mathbf{f}$ in $\mathbb{R}^3$ as a continuous function that takes the interval $I = [a, b]$ and "deforms" it into a curve C in $\mathbb{R}^3$ (see Figure 1).

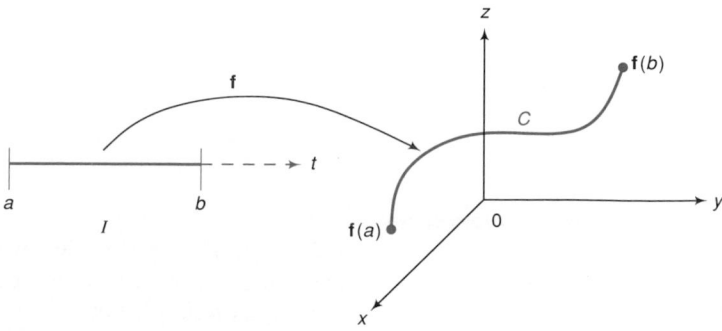

FIGURE 1 AS t VARIES OVER I, THE CURVE C IS TRACED OUT IN $\mathbb{R}^3$.

A surface in $\mathbb{R}^3$ can be represented parametrically in terms of two parameters.

PARAMETRIC SURFACE

Let $x = x(u, v)$, $y = y(u, v)$, and $z = z(u, v)$ be continuous. Then the set of points S given by

$$\mathbf{r}(u, v) = x(u, v)\mathbf{i} + y(u, v)\mathbf{j} + z(u, v)\mathbf{k} \tag{1}$$

where (u, v) ranges over a region Ω in $\mathbb{R}^2$ is called a **parametric surface**, or simply a **surface**, in $\mathbb{R}^3$. Equation (1) is a **parametric representation** of the surface. ∎

The function $\mathbf{r}$ in equation (1) is a function with domain Ω in $\mathbb{R}^2$ and range in $\mathbb{R}^3$. We can think of this function as deforming a set of points Ω in $\mathbb{R}^2$ into a set of points S in $\mathbb{R}^3$. This is illustrated in Figure 2.

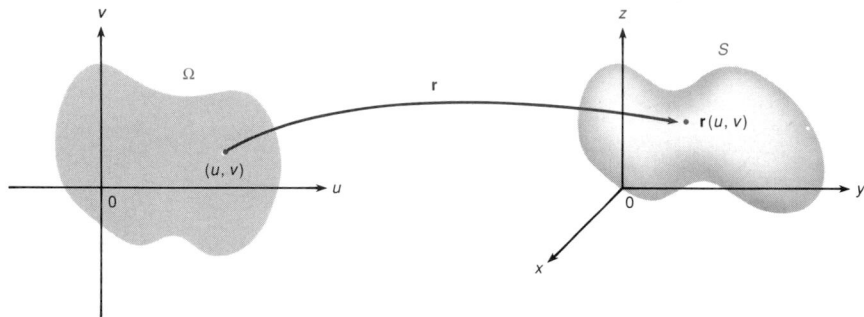

FIGURE 2 AS (u, v) VARIES OVER Ω, THE POINTS IN THE SURFACE S ARE OBTAINED.

It may be helpful to think of the function $\mathbf{r}$ as deforming a rectangle R into a "magic carpet," as in Figure 3.

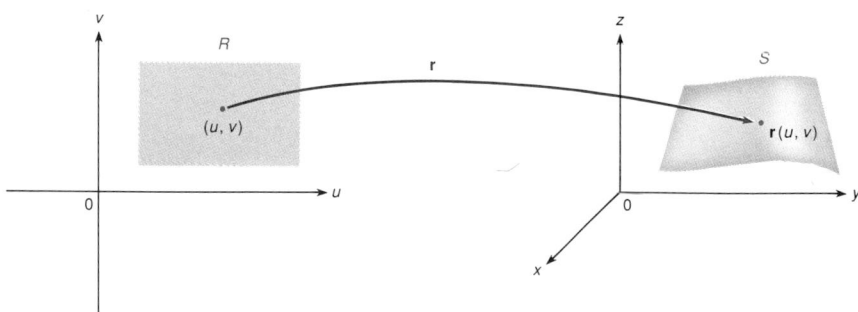

FIGURE 3 AS (u, v) VARIES OVER THE RECTANGLE R, THE POINTS IN A "MAGIC CARPET" ARE OBTAINED.

EXAMPLE 1
REPRESENTING THE SURFACE
$z = f(x, y)$ PARAMETRICALLY

In Section 13.1 (page 836), we said that the graph of the continuous function $z = f(x, y)$ is a surface in $\mathbb{R}^3$. We can write this surface in the form (1) as follows: Let

$$x(u, v) = u \qquad y(u, v) = v \qquad \text{and} \qquad z(u, v) = f(u, v). \tag{2}$$

Then

$$\mathbf{r}(u, v) = u\mathbf{i} + v\mathbf{j} + f(u, v)\mathbf{k}$$

yields the surface which is the graph of $z = f(x, y)$.

EXAMPLE 2
PARAMETRIC REPRESENTATION OF
A HYPERBOLIC PARABOLOID

Find a parametric representation of the hyperbolic paraboloid

$$z = \frac{x^2}{4} - \frac{y^2}{9}.$$

SOLUTION: Using (2), we set

$$x = u \qquad y = v \qquad \text{and} \qquad z = \frac{u^2}{4} - \frac{v^2}{9}.$$

Then the surface can be written

$$\mathbf{r}(u, v) = u\mathbf{i} + v\mathbf{j} + \left(\frac{u^2}{4} - \frac{v^2}{9}\right)\mathbf{k} \qquad \text{for } (u, v) \in \mathbb{R}^2.$$

EXAMPLE 3
PARAMETRIC REPRESENTATION OF A
PLANE THROUGH THE ORIGIN

Let Π be a plane passing through the origin. If $\mathbf{x}$ and $\mathbf{y}$ are two noncollinear vectors lying on Π, then every other vector $\mathbf{w}$ on Π can be written in the form

$$\mathbf{w} = \mathbf{r}(u, v) = u\mathbf{x} + v\mathbf{y}. \tag{3}$$

The representation (3) is called a **parametric representation of a plane**. You are asked to prove that (3) holds in Problems 40–42.

EXAMPLE 4
PARAMETRIC REPRESENTATION OF
THE PLANE $x + 2y + 3z = 0$

Find a parametric representation in the form (3) of the plane

$$\Pi: x + 2y + 3z = 0. \tag{4}$$

SOLUTION: From (4), we have

$$z = -\frac{1}{3}(x + 2y).$$

Then every vector on Π has the form

$$\left(x, y, -\frac{1}{3}(x + 2y)\right). \tag{5}$$

Two noncollinear vectors on Π are found by setting $x = 1$, $y = 0$, and $x = 0$, $y = 1$:

$$\mathbf{x} = \left(1, 0, -\frac{1}{3}\right) \qquad \text{and} \qquad \mathbf{y} = \left(0, 1, -\frac{2}{3}\right).$$

If u and v are real numbers, then

$$u\mathbf{x} + v\mathbf{y} = \left(u, v, -\frac{1}{3}(u + 2v)\right),$$

which is of the form (5), so $u\mathbf{x} + v\mathbf{y}$ is on Π. Conversely, if $\mathbf{w} = (\alpha, \beta, \gamma)$ is on Π, then $\gamma = -\frac{1}{3}(\alpha + 2\beta)$, so

$$\mathbf{w} = (\alpha, \beta, \gamma) = \left(\alpha, \beta, -\frac{1}{3}(\alpha + 2\beta) \right)$$

$$= \alpha\left(1, 0, -\frac{1}{3} \right) + \beta\left(0, 1, -\frac{2}{3} \right) = \alpha\mathbf{x} + \beta\mathbf{y}.$$

Thus Π consists precisely of those vectors that can be written in the form $u\mathbf{x} + v\mathbf{y}$.

EXAMPLE 5
PARAMETRIC REPRESENTATION OF A SPHERE

Recall that if $P = (x, y, z)$ is a point or vector in $\mathbb{R}^3$, then the Cartesian coordinates of P can be written in terms of the spherical coordinates of P as follows (see page 784):

$$x = \rho \sin \varphi \cos \theta$$

$$y = \rho \sin \varphi \sin \theta \qquad 0 \le \varphi \le \pi, 0 \le \theta \le 2\pi$$

$$z = \rho \cos \varphi$$

If $\rho = a$ is fixed, we have

$$x^2 + y^2 + z^2 = a^2. \qquad \text{Equation (8) on page 784}$$

This is the equation of a sphere centered at the origin with radius a. Thus the parametric representation of this sphere is

$$\mathbf{r}(\varphi, \theta) = a \sin \varphi \cos \theta\mathbf{i} + a \sin \varphi \sin \theta\mathbf{j} + a \cos \varphi\mathbf{k},$$

$$0 \le \varphi \le \pi, 0 \le \theta \le 2\pi \qquad \textbf{(6)}$$

Note that here the parameters φ and θ lie in the bounded rectangle given by (6).

EXAMPLE 6
PARAMETRIC REPRESENTATION OF A CYLINDER

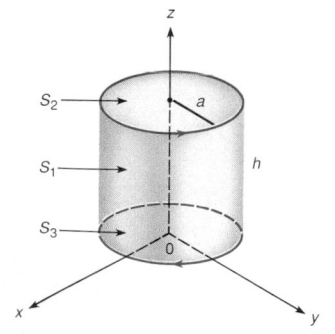

FIGURE 4
PARAMETRIZABLE SUBSURFACES OF A CYLINDER

A right circular cylinder with radius a and height b consists of three surfaces: the side S_1, the top S_2, and the bottom S_3.

Consider the right circular cylinder sketched in Figure 4. In order to write it in parametric form, it is convenient to parametrize the side, top, and bottom separately.

The side S_1 We use cylindrical coordinates with $r = a$, a constant (see equation (2) on page 782):

$$x = a \cos \theta, \qquad y = a \sin \theta, \qquad z = z,$$

so

$$\mathbf{r}(\theta, z) = a \cos \theta\mathbf{i} + a \sin \theta\mathbf{j} + z\mathbf{k}, \qquad 0 \le \theta \le 2\pi, 0 \le z \le h.$$

The top S_2 Here $z = h$ but both r and θ vary:

$$\mathbf{r}(r, \theta) = r \cos \theta\mathbf{i} + r \sin \theta\mathbf{j} + h\mathbf{k}, \qquad 0 \le \theta \le 2\pi, 0 \le r \le a$$

The bottom S_3 Now $z = 0$ and

$$\mathbf{r}(r, \theta) = r \cos \theta\mathbf{i} + r \sin \theta\mathbf{j} + 0\mathbf{k}, \qquad 0 \le \theta \le 2\pi, 0 \le r \le a.$$

REMARK: If it is convenient, as here, to divide the surface into different "subsurfaces" and parametrize each one separately, then each of the parametrizable subsurfaces is called a **surface patch**.

COMPUTING SURFACE AREA

Consider the surface

$$S: \mathbf{r}(u, v) = x(u, v)\mathbf{i} + y(u, v)\mathbf{j} + z(u, v)\mathbf{k}. \tag{7}$$

We assume that the six partial derivatives

$$\frac{\partial x}{\partial u}, \frac{\partial x}{\partial v}, \frac{\partial y}{\partial u}, \frac{\partial y}{\partial v}, \frac{\partial z}{\partial u}, \text{ and } \frac{\partial z}{\partial v} \tag{8}$$

exist and are continuous for (u, v) in a region Ω.

We assume further that Ω is a bounded region and, as in Section 14.1, we partition Ω into nm subrectangles, denoted by R_{ij}. A typical subrectangle is sketched in Figure 5(a). The subregion S_{ij} in Figure 5(b) is the image of R_{ij} under the function $\mathbf{r}$. If we know the area of each S_{ij}, we can compute the area of S by adding up the areas of the subsurfaces. If we can only approximate the areas of the subsurfaces, then we can obtain the area of S by adding the approximate values and taking a limit as Δu and Δv tend to zero. This gives us a double integral as in our calculation of volume in Section 14.1.

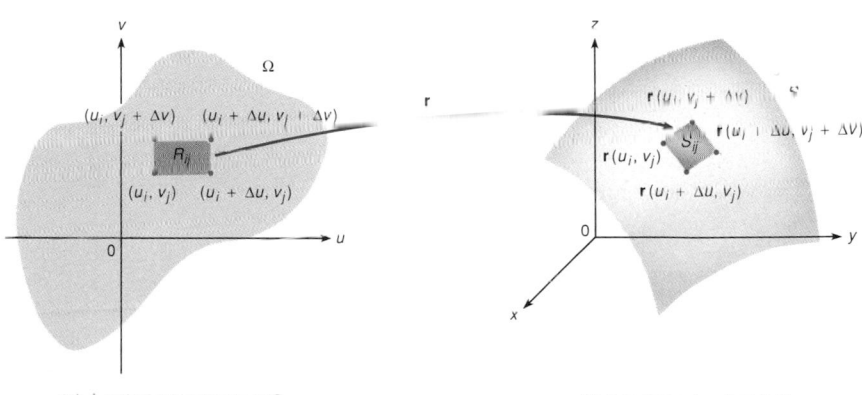

(a) A typical subrectangle in Ω.   (b) A typical subsurface in S.

FIGURE 5 THE SUBREGION S_{ij} IS THE IMAGE OF THE RECTANGLE R_{ij} UNDER THE FUNCTION $\mathbf{r}$.

$\mathbf{r}$ *takes the region Ω in $\mathbb{R}^2$ into the surface S in $\mathbb{R}^3$. $\mathbf{r}$ takes the subrectangle R_{ij} of Ω into the subsurface S_{ij} in S.*

Consider the point $\mathbf{r}(u_i, v_j)$ on the surface S. If v_j is held fixed and u is allowed to vary, then $\mathbf{r}(u, v_j)$ is a curve in $\mathbb{R}^3$ that passes through the point (u_i, v_j). A tangent vector to this curve is, from Chapter 12, given by the derivative of $\mathbf{r}$ with respect to u (since v is fixed). That is,

a tangent vector to the curve $\mathbf{r}(u, v_j)$ at the point (u_i, v_j)

$$= \mathbf{r}_u(u_i, v_j) = \frac{\partial x(u_i, v_j)}{\partial u}\mathbf{i} + \frac{\partial y(u_i, v_j)}{\partial u}\mathbf{j} + \frac{\partial z(u_i, v_j)}{\partial u}\mathbf{k}. \tag{9}$$

Similarly, if we hold $u = u_i$ fixed and we let v vary, we obtain

a tangent vector to the curve $\mathbf{r}(u_i, v)$ at the point (u_i, v_j)

$$= \mathbf{r}_v(u_i, v_j) = \frac{\partial x(u_i, v_j)}{\partial v}\mathbf{i} + \frac{\partial y(u_i, v_j)}{\partial v}\mathbf{j} + \frac{\partial z(u_i, v_j)}{\partial v}\mathbf{k}.$$

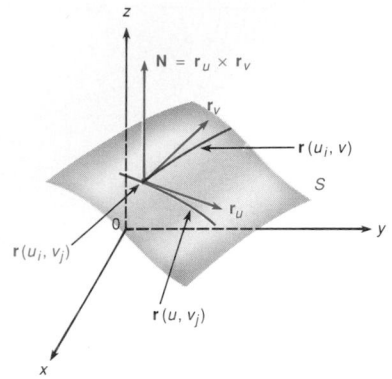

FIGURE 6
TWO TANGENT VECTORS TO THE SUR-
FACE S AT THE POINT $\mathbf{r}(u_i, v_j)$

$\mathbf{r}_u$ is tangent to the curve $\mathbf{r}(u, v_j)$; $\mathbf{r}_v$ is
tangent to the curve $\mathbf{r}(u_i, v)$ and $\mathbf{N} =
\mathbf{r}_u \times \mathbf{r}_v$ is orthogonal to both tangent
vectors.

We sketch these two tangent vectors in Figure 6. We now assume that

$$\mathbf{r}_u \times \mathbf{r}_v \neq \mathbf{0}.$$

This means that $\mathbf{r}_u$ and $\mathbf{r}_v$ determine a plane, called the **tangent plane** to the
surface S at the point $\mathbf{r}(u_i, v_j)$. A normal vector to this plane is given by (see
Section 11.7)

$$\mathbf{N} = \mathbf{r}_u(u_i, v_j) \times \mathbf{r}_v(u_i, v_j).$$

Suppose that Δu and Δv are small. We have

$$\frac{\mathbf{r}(u_i + \Delta u, v_j) - \mathbf{r}(u_i, v_j)}{\Delta u} = \frac{x(u_i + \Delta u, v_j) - x(u_i, v_j)}{\Delta u}\mathbf{i}$$

$$+ \frac{y(u_i + \Delta u, v_j) - y(u_i, v_j)}{\Delta u}\mathbf{j}$$

$$+ \frac{z(u_i + \Delta u, v_j) - z(u_i, v_j)}{\Delta u}\mathbf{k}.$$

But by the definition of $\dfrac{\partial x}{\partial u}$, if Δu is small, then

$$\frac{x(u_i + \Delta u, v_j) - x(u_i, v_j)}{\Delta u} \approx \frac{\partial x(u_i, v_j)}{\partial u}$$

or

$$x(u_i + \Delta u, v_j) - x(u_i, v_j) \approx \frac{\partial x(u_i, v_j)}{\partial u}\Delta u.$$

Then, using similar approximations for the two other terms, we obtain

$$\mathbf{r}(u_i + \Delta u, v_j) - \mathbf{r}(u_i, v_j) \approx \mathbf{r}_u(u_i, v_j)\Delta u.$$

Thus, one side of the subsurface S_{ij} is approximated by the vector from
$\mathbf{r}(u_i, v_j)$ to $\mathbf{r}(u_i + \Delta u, v_j)$, which in turn is approximated by the tangent vector
$\mathbf{r}_u(u_i, v_j)\Delta u$. Similarly, a second side of S_{ij} is approximated by the tangent
vector $\mathbf{r}_v(u_i, v_j)\Delta v$. Then, if Δu and Δv are small, we have

$$\text{area of } S_{ij} \approx \text{area of parallelogram with sides } \mathbf{r}_u\,\Delta u \text{ and } \mathbf{r}_v\,\Delta v$$

see page 762

$$\stackrel{\downarrow}{=} \left|\mathbf{r}_u\,\Delta u \times \mathbf{r}_v\,\Delta v\right|$$

or

$$\text{area of } S_{ij} \approx \left|\mathbf{r}_u(u_i, v_j) \times \mathbf{r}_v(u_i, v_j)\right|\Delta u\,\Delta v$$

(see Figure 7). Thus

$$\text{area of } S \approx \sum_{i=1}^{n}\sum_{j=1}^{n}\left|\mathbf{r}_u(u_i, v_j) \times \mathbf{r}_v(u_i, v_j)\right|\Delta u\,\Delta v$$

and, taking a limit as $\sqrt{\Delta u^2 + \Delta v^2} \to 0$, we obtain

$$\text{area of } S = \iint\limits_{\Omega}\left|\mathbf{r}_u(u, v) \times \mathbf{r}_v(u, v)\right|du\,dv. \tag{10}$$

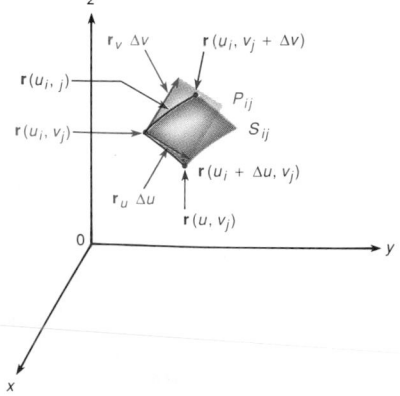

FIGURE 7
THE AREA OF S_{ij} IS APPROXIMATED BY
THE AREA OF THE PARALLELOGRAM WITH
SIDES $\mathbf{r}_u\,\Delta u$ AND $\mathbf{r}_v\,\Delta v$.

Before continuing, we stress that the computation that led to (10) did not prove anything. But it does make it plausible to use (10) as a definition of surface area. In order to do so, we must make two assumptions. First, we assume that $\mathbf{r}_u \times \mathbf{r}_v \neq \mathbf{0}$.

DEFINITION   SMOOTH SURFACE

The surface given by (7) is called **smooth** over the region Ω if

(i) all six partial derivatives (8) exist and are continuous for (u, v) in Ω, and

(ii) $\mathbf{N}(u, v) = \mathbf{r}_u(u, v) \times \mathbf{r}_v(u, v) \neq \mathbf{0}$ for (u, v) in Ω. ∎

We also require that $\mathbf{r}$ be **one-to-one** over Ω (so that we do not count some subsurfaces more than once). That is,

if $(u_1, v_1) \neq (u_2, v_2)$, then $\mathbf{r}(u_1, v_1) \neq \mathbf{r}(u_2, v_2)$.

DEFINITION   SURFACE AREA

Let $\mathbf{r}$ be a surface that is both smooth and one-to-one over a region Ω. Then the **surface area** σ of $\mathbf{r}$ over Ω is given by

$$\text{surface area} = \sigma = \iint_\Omega |\mathbf{N}(u, v)| \, du \, dv$$

$$= \iint_\Omega |\mathbf{r}_u(u, v) \times \mathbf{r}_v(u, v)| \, du \, dv. \quad \blacksquare \qquad (11)$$

EXAMPLE 7
THE SURFACE AREA OF A SPHERE

Compute the surface area of the sphere

$$x^2 + y^2 + z^2 = a^2.$$

SOLUTION: We write the sphere parametrically, using u and v instead of φ and θ (see Example 4):

$\mathbf{r}(u, v) = a \sin u \cos v \mathbf{i} + a \sin u \sin v \mathbf{j} + a \cos u \mathbf{k}$,
$0 \leq u \leq \pi, 0 \leq v \leq 2\pi$.

Then

$\mathbf{r}_u = a \cos u \cos v \mathbf{i} + a \cos u \sin v \mathbf{j} - a \sin u \mathbf{k}$

$\mathbf{r}_v = -a \sin u \sin v \mathbf{i} + a \sin u \cos v \mathbf{j} + 0 \mathbf{k}$

$$\mathbf{N}(u, v) = \mathbf{r}_u \times \mathbf{r}_v = \begin{vmatrix} \mathbf{i} & \mathbf{j} & \mathbf{k} \\ a \cos u \cos v & a \cos u \sin v & -a \sin u \\ -a \sin u \sin v & a \sin u \cos v & 0 \end{vmatrix}$$

$$= a^2 [\sin^2 u \cos v \mathbf{i} + \sin^2 u \sin v \mathbf{j} + \sin u \cos u \overbrace{(\sin^2 v + \cos^2 v)}^{=1} \mathbf{k}]$$

so

$$|\mathbf{N}(u, v)| = |\mathbf{r}_u \times \mathbf{r}_v|$$
$$= a^2[\sin^4 u \cos^2 v + \sin^4 u \sin^2 v + \sin^2 u \cos^2 u]^{1/2}$$
$$= a^2[\sin^4 u + \sin^2 u \cos^2 u]^{1/2}$$
$$= a^2[\sin^2 u(\sin^2 u + \cos^2 u)]^{1/2}$$
$$= a^2 \sin u.$$

Thus,

$$\sigma = \iint_{\Omega} a^2 \sin u \, du \, dv = a^2 \int_0^{2\pi} \int_0^{\pi} \sin u \, du \, dv = a^2 \int_0^{2\pi} \left(-\cos u \Big|_0^{\pi}\right) dv$$
$$= 2a^2 \int_0^{2\pi} dv = 2a^2(2\pi) = 4\pi a^2.$$

EXAMPLE 8
THE LATERAL SURFACE AREA OF A
RIGHT CIRCULAR CONE

Compute the lateral surface area of a right circular cone with radius a and height h.

SOLUTION: We place the cone with its vertex at the origin and its axis along the z-axis, as in Figure 8. By the **lateral** surface area, we mean the area of the side, not the circle on the top (which has area πa^2). In Figure 8, we drew the **slant height** $s = \sqrt{h^2 + a^2}$ and the vertex **semiangle** α. Consider the cross-section of the cone having a slant height of v. The radius of the circle is $v \sin \alpha$, so a parametrization of the cross-sectional circle is

$$v \sin \alpha(\cos u\mathbf{i} + \sin u\mathbf{j}), \qquad 0 \le u \le 2\pi.$$

The z-value corresponding to a slant height v is $v \cos \alpha$. Remember that α is fixed: $\alpha = \tan^{-1} \dfrac{a}{h}$. Then a parametrization of the cone is

$$\mathbf{r}(u, v) = v \cos u \sin \alpha\mathbf{i} + v \sin u \sin \alpha\mathbf{j} + v \cos \alpha\mathbf{k},$$
$$0 \le u \le 2\pi, 0 \le v \le s = \sqrt{a^2 + h^2}.$$

We compute

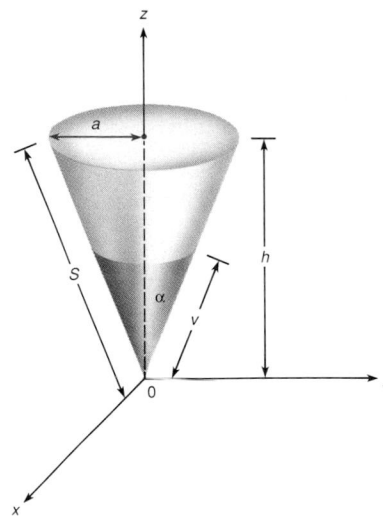

FIGURE 8
A RIGHT CIRCULAR CONE WITH RADIUS a
AND HEIGHT h

$$\mathbf{r}_u \times \mathbf{r}_v = \begin{vmatrix} \mathbf{i} & \mathbf{j} & \mathbf{k} \\ \dfrac{\partial x}{\partial u} & \dfrac{\partial y}{\partial u} & \dfrac{\partial z}{\partial u} \\ \dfrac{\partial x}{\partial v} & \dfrac{\partial y}{\partial v} & \dfrac{\partial z}{\partial v} \end{vmatrix}$$

$$= \begin{vmatrix} \mathbf{i} & \mathbf{j} & \mathbf{k} \\ -v \sin u \sin \alpha & v \cos u \sin \alpha & 0 \\ \cos u \sin \alpha & \sin u \sin \alpha & \cos \alpha \end{vmatrix}$$

$$= v \cos u \sin \alpha \cos \alpha\mathbf{i} + v \sin u \sin \alpha \cos \alpha\mathbf{j}$$
$$- v \sin^2 \alpha \underbrace{(\sin^2 u + \cos^2 u)}_{=1}\mathbf{k}$$

so

$$|\mathbf{r}_u \times \mathbf{r}_v| = [v^2 \sin^2 \alpha \cos^2 \alpha\overbrace{(\cos^2 u + \sin^2 u)}^{=1} + v^2 \sin^4 \alpha]^{1/2}$$
$$= [v^2 \sin^2 \alpha(\cos^2 \alpha + \sin^2 \alpha)]^{1/2} = v \sin \alpha.$$

Thus

$$\sigma = \text{lateral surface of the cone} = \int_0^s \int_0^{2\pi} v \sin \alpha \, du \, dv$$

$$= 2\pi \sin \alpha \int_0^s v \, dv = \pi s^2 \sin \alpha.$$

But, $\sin \alpha = \dfrac{h}{s}$. Thus

$$\sigma = \pi s^2 \left(\frac{h}{s} \right) = \pi h s = \pi h \sqrt{a^2 + h^2}.$$

SURFACE AREA OF THE SURFACE $z = f(x, y)$

If the surface is given in the form $z = f(x, y)$ for (x, y) in Ω, then, from Example 1, the surface is parametrized by

$$\mathbf{r}(u, v) = u\mathbf{i} + v\mathbf{j} + f(u, v)\mathbf{k}, \qquad (u, v) \in \Omega.$$

We compute

$$\mathbf{r}_u \times \mathbf{r}_v = \begin{vmatrix} \mathbf{i} & \mathbf{j} & \mathbf{k} \\ \dfrac{\partial x}{\partial u} & \dfrac{\partial y}{\partial u} & \dfrac{\partial z}{\partial u} \\ \dfrac{\partial x}{\partial v} & \dfrac{\partial y}{\partial v} & \dfrac{\partial z}{\partial v} \end{vmatrix} = \begin{vmatrix} \mathbf{i} & \mathbf{j} & \mathbf{k} \\ 1 & 0 & \dfrac{\partial f}{\partial u} \\ 0 & 1 & \dfrac{\partial f}{\partial v} \end{vmatrix} = -\frac{\partial f}{\partial u}\mathbf{i} - \frac{\partial f}{\partial v}\mathbf{j} + \mathbf{k}$$

so

$$|\mathbf{N}(u, v)| = |\mathbf{r}_u \times \mathbf{r}_v| = \sqrt{1 + \left(\frac{\partial f}{\partial u} \right)^2 + \left(\frac{\partial f}{\partial v} \right)^2}.$$

Reverting to the original parameters $x = u$ and $y = v$, we have

> **LATERAL SURFACE AREA OF THE SURFACE $z = f(x, y)$ OVER A REGION Ω**
>
> If $\dfrac{\partial f}{\partial x}$ and $\dfrac{\partial f}{\partial y}$ are continuous, then
>
> $$\text{surface area} = \sigma = \iint_\Omega \sqrt{1 + \left(\frac{\partial f}{\partial x} \right)^2 + \left(\frac{\partial f}{\partial y} \right)^2} \, dx \, dy. \qquad (12)$$

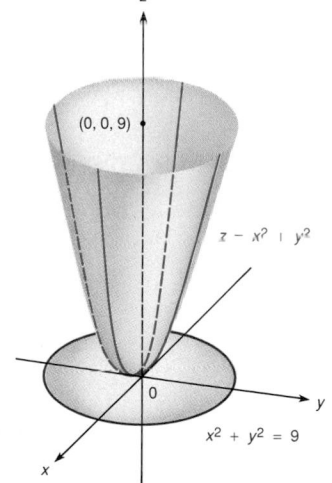

FIGURE 9
THE CIRCULAR PARABOLOID $z = x^2 + y^2$ FOR $0 \le z \le 9$

EXAMPLE 9
COMPUTING THE SURFACE AREA OF PART OF A CIRCULAR PARABOLOID

Find the lateral surface area of the circular paraboloid $z = x^2 + y^2$ between the xy-plane and the plane $z = 9$.

SOLUTION: The surface area requested is sketched in Figure 9. The region Ω is the disk $x^2 + y^2 \le 9$. We have

$$f_x = 2x \qquad \text{and} \qquad f_y = 2y,$$

so

$$\sigma = \iint_{\Omega} \sqrt{1 + 4x^2 + 4y^2} \, dA.$$

Clearly, this problem calls for the use of polar coordinates. We have

$$\sigma = \int_0^{2\pi} \int_0^3 \sqrt{1 + 4r^2} \, r \, dr \, d\theta = \int_0^{2\pi} \left\{ \frac{1}{12}(1 + 4r^2)^{3/2} \Big|_0^3 \right\} d\theta$$

$$= \frac{\pi}{6}(37^{3/2} - 1) \approx 117.3$$

EXAMPLE 10
OBTAINING A DOUBLE
INTEGRAL THAT REPRESENTS
SURFACE AREA

Calculate the area of the part of the surface $z = x^3 + y^4$ that lies over the square $\{(x, y): 0 \le x \le 1, 0 \le y \le 1\}$.

SOLUTION: $f_x = 3x^2$ and $f_y = 4y^3$, so

$$\sigma = \int_0^1 \int_0^1 \sqrt{1 + 9x^4 + 16y^6} \, dx \, dy.$$

However, this is as far as we can go unless we resort to numerical techniques to approximate this double integral. As with ordinary definite integrals, many double integrals cannot be integrated in terms of functions that we know. However, there are a great number of techniques for approximating a double integral numerically that parallel the techniques discussed in Section 7.9. A computer-generated sketch of the solid whose surface area we seek is given in Figure 10.

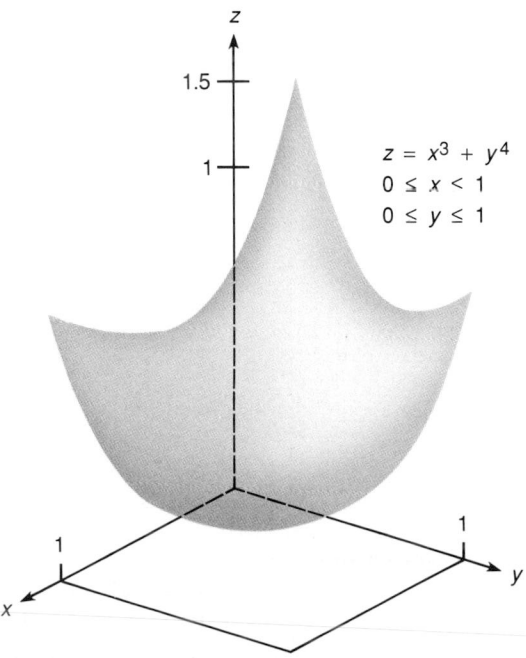

$z = x^3 + y^4$
$0 \le x < 1$
$0 \le y \le 1$

FIGURE 10   THE SOLID BELOW $z = x^3 + y^4$ AND OVER THE SQUARE $0 \le x \le 1, 0 \le y \le 1$

■ PROBLEMS 15.5

■ SELF-QUIZ

I. Which of the following (if any) are parametrizations of the plane $2x - y + 3z = 0$?
  a. $\mathbf{r}(u, v) = 2\mathbf{i} - \mathbf{j} + 3\mathbf{k}$
  b. $\mathbf{r}(u, v) = 2u\mathbf{i} - v\mathbf{j} + 3(u + v)\mathbf{k}$
  c. $\mathbf{r}(u, v) = u\mathbf{i} + v\mathbf{j} + \frac{1}{3}(v - 2u)\mathbf{k}$
  d. $\mathbf{r}(u, v) = u\mathbf{i} + (2u + 3v)\mathbf{j} + v\mathbf{k}$
  e. $\mathbf{w} = u(0, 3, 1) + v(1, 2, 0)$
  f. $\mathbf{w} = u(-1, 1, 1) + v(-2, -1, 1)$

II. The area of that part of the surface $z = 5 - x + y$ which lies over the region $\Omega = \{(x, y): 0 \le x \le 2, 0 \le y \le 3\}$ is computed by _____.
  a. $\int_0^3 \int_0^2 \sqrt{1 + (-1)^2 + (1)^2}\, dx\, dy$
  b. $\int_0^3 \int_0^2 \sqrt{1 + (-x)^2 + (y)^2}\, dx\, dy$
  c. $\int_0^2 \int_0^3 \sqrt{1 + (-x)^2 + (y)^2}\, dx\, dy$
  d. $\int_0^2 \int_0^3 \sqrt{1 - (x)^2 + (y)^2}\, dx\, dy$

III. If f is a constant function of (x, y) and if Ω is a simple region in the xy-plane, then the area on the surface $z = f(x, y)$ above Ω is _____ the area of the region Ω.
  a. greater than or equal to
  b. equal to
  c. less than
  d. unrelated to (i.e., can be larger than, or smaller than, or the same as)

IV. If f is a smooth function of (x, y) and if Ω is a simple region in the xy-plane, then the area on the surface $z = f(x, y)$ above Ω is _____ the area of the region Ω.
  a. greater than or equal to
  b. equal to
  c. less than
  d. unrelated to (i.e., can be larger than, smaller than, or the same as)

■ DRILL

In Problems 1–10, parametrize the given surface.

1. the plane $2x + 3y - z = 0$
2. the plane $x - 2y + 5z = 0$
3. the hemisphere $x^2 + y^2 + z^2 = 4$, $z \ge 0$
4. the sphere $(x - 1)^2 + (y + 1)^2 + (z - 2)^2 = 16$
5. the paraboloid $x^2 - y^2 - 4z = 0$; $0 \le x \le 1$, $2 \le y \le 3$
6. the cone $z = \sqrt{x^2 + y^2}$ lying above the disk $x^2 + y^2 \le 9$
7. the part of the plane $z = 4x - y$ that lies inside the cylinder $x^2 + y^2 = 1$
8. the part of the cylinder $x^2 + y^2 = 16$ that lies between the planes $z = 0$ and $z = y + 1$
9. the part of the unit sphere that lies above the cone $z = \sqrt{x^2 + y^2}$
*10. the **torus** T obtained by revolving a circle in the xz-plane with center at $(b, 0, 0)$ and radius $a < b$ about the z-axis (see Figure 6 on page 500 for a sketch of a torus).

In Problems 11–20, find the area of that part of the given surface which lies over the specified region.

11. $z = x + 2y$; $\Omega = \{(x, y): 0 \le x \le y, 0 \le y \le 2\}$

12. $z = 4x + 7y$; Ω is the region between $y = x^2$ and $y = x^5$
13. $z = ax + by$; Ω is upper half of unit circle
14. $z = y^2$; $\Omega = \{(x, y): 0 \le x \le 2, 0 \le y \le 4\}$
*15. $z = 3 + x^{2/3}$; $\Omega = \{(x, y): -1 \le x \le 1, 0 \le y \le 2\}$
16. $z = (x^4/4) + 1/(8x^2)$; $\Omega = \{(x, y): 1 \le x \le 2, 0 \le y \le 5\}$
17. $z = \frac{1}{3}(y^2 + 2)^{3/2}$; $\Omega = \{(x, y): -4 \le x \le 7, 0 \le y \le 3\}$
18. $z = 2\ln(1 + y)$; $\Omega = \{(x, y): 0 \le x \le 2, 0 \le y \le 1\}$
*19. $(z + 1)^2 = 4x^3$; $\Omega = \{(x, y): 0 \le x \le 1, 0 \le y \le 2\}$
*20. $y^2 + z^2 = 9$; $\Omega = \{(x, y): 0 \le x \le 1, 0 \le y \le 2\}$

In Problems 21–24, compute the surface area of each surface.

21. the surface in Problem 7.

22. the surface in Problem 6.

23. the surface in Problem 9.

*24. the torus in Problem 10.

■ APPLICATIONS

*25. Calculate the lateral surface area of the cylinder $y^{2/3} + z^{2/3} = 1$ for x in the interval $[0, 2]$.

26. Find the surface area of the hemisphere $x^2 + y^2 + z^2 = a^2$, $z \ge 0$.

CONTINUED

*27. Find the surface area of the part of the sphere $x^2 + y^2 + z^2 = a^2$ that is also inside the cylinder $x^2 + y^2 = ay$.

*28. Find the area of the surface in the first octant cut from the cylinder $x^2 + y^2 = 16$ by the plane $y = z$.

*29. Find the area of the portion of the sphere $x^2 + y^2 + z^2 = 16z$ lying within the circular paraboloid $z = x^2 + y^2$.

*30. Find the area of the surface cut from the hyperbolic paraboloid $4z = x^2 - y^2$ by the cylinder $x^2 + y^2 = 16$.

In Problems 31–34, find a repeated integral that represents the area of the given surface over the specified region. Do *not* try to evaluate your integral.

31. $z = x^3 + y^3$; Ω is the unit circle

32. $z = \ln(x + 2y)$; $\Omega = \{(x, y): 0 \le x \le 1, 0 \le y \le 4\}$

33. $z = \sqrt{1 + x + y}$; Ω is the triangle bounded by $y = x$, $y = 4 - x$, and the y-axis

34. $z = e^{x-y}$; Ω is the ellipse $4x^2 + 9y^2 = 36$.

*35. Find, but do not evaluate, a repeated integral that represents the surface area of the ellipsoid $(x/a)^2 + (y/b)^2 + (z/c)^2 = 1$.

36. A **spiral ramp** or **helicoid** is given parametrically by

$$\mathbf{r}(u, v) = v \cos u\,\mathbf{i} + v \sin u\,\mathbf{j} + u\mathbf{k},$$
$$0 \le u \le 2\pi, 0 \le v \le 1.$$

a. Sketch this surface.

b. Obtain its surface area.

37. Find a formula for the area of the triangle in $\mathbb{R}^3$ with vertices at $(a, 0, 0)$, $(0, b, 0)$, and $(0, 0, c)$.

■ **SHOW/PROVE/DISPROVE**

*38. Let $z = f(x, y)$ be the equation of a plane. Show that the area of the portion of this plane lying over the region Ω is

$$\sigma = \iint_\Omega \sec \gamma \, dA,$$

where γ is the angle between the normal vector $\mathbf{N}$ to the plane and the positive z-axis. Further, suppose that $f(x, y) = ax + by + c$; show, using the dot product, that

$$\cos \gamma = \frac{\mathbf{N} \cdot \mathbf{k}}{|\mathbf{N}|} = \frac{1}{\sqrt{1 + a^2 + b^2}}.$$

*39. Suppose g is a smooth function of one variable. The graph, in the xz-plane, of $z = g(x)$ for $a \le x \le b$ is an arc. By moving this arc parallel to the y-axis from $y = 0$ to $y = c$, we sweep out a patch on a surface in $\mathbb{R}^3$. Show that the surface area of this patch is equal to the arc length of the curve multiplied by the horizontal distance we move the curve parallel to the y-axis.

40. Let $\Pi: ax + by + cz = 0$ be an equation of a plane passing through the origin. Assume that $c \ne 0$.

a. Show that for (x, y, z) in Π, $z = -\dfrac{1}{c}(ax + by)$.

b. Show that $\mathbf{x} = (c, 0, -a)$ and $\mathbf{y} = (0, c, -b)$ are two noncollinear vectors in Π.

41. In Problem 40, show that $u\mathbf{x} + v\mathbf{y}$ is in Π for any (u, v) in $\mathbb{R}^2$.

42. In Problem 41, suppose that $\mathbf{w} = (\alpha, \beta, \gamma)$ is in Π. Find numbers u and v such that $\mathbf{w} = u\mathbf{x} + v\mathbf{y}$.

43. Suppose that the graph of $y = f(x) > 0$ for $a \le x \le b$ is revolved about the x-axis, as in Figure 11. Obtain a parametrization for the surface in Figure 11(b).

44. Show that the lateral surface area of the surface in Figure 11(b) is given by

$$\sigma = \text{lateral surface area}$$
$$= 2\pi \int_a^b f(x)\sqrt{1 + [f'(x)]^2}\, dx,$$

assuming that f is continuously differentiable.

45. Use the result of Problem 44 to compute the lateral surface area of a cone with radius a and height h. [*Hint*: Revolve a certain line about the x-axis.]

■ **ANSWERS TO SELF-QUIZ**

I. c, d, e, f II. a

III. b (since $\sqrt{1 + f_x^2 + f_y^2} = 1$)

IV. a (since $\sqrt{1 + f_x^2 + f_y^2} \ge 1$)

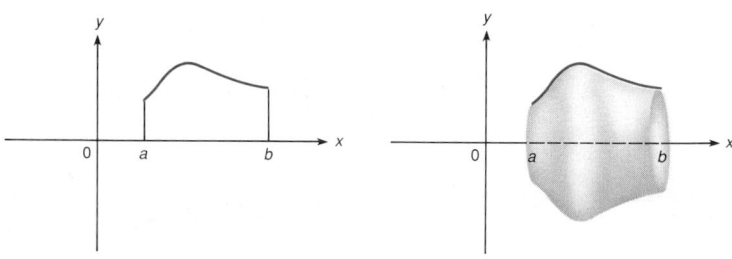

(a) When this curve is revolved about the x-axis . . .

(b) . . . this solid is obtained.

FIGURE 11

15.6 SURFACE INTEGRALS

In Section 15.4 we discussed Green's theorem, which gave a relationship between a line integral over a closed curve and a double integral over a region enclosed by that curve. In this section, we discuss the notion of a surface integral. This will enable us, in Section 15.8, to extend Green's theorem to integrals over regions in space.

Recall from Section 11.8 that a surface in space is defined as the set of points satisfying the equation

$$G(x, y, z) = 0, \tag{1}$$

where G is a continuous function defined on a region in $\mathbb{R}^3$. In this section we do not write the surface parametrically. Rather, we will only consider surfaces that can be written in the form

$$z = f(x, y) \tag{2}$$

for some function f. Note that (2) is a special case of (1), as can be seen by defining $G(x, y, z) = z - f(x, y)$ so that $z = f(x, y)$ is equivalent to $G(x, y, z) = 0$. We remind you of a definition given in Section 13.7.

DEFINITION   **SMOOTH SURFACE**

The surface $z = f(x, y)$ is called **smooth** at a point (x_0, y_0, z_0) if $\partial f/\partial x$ and $\partial f/\partial y$ are continuous at (x_0, y_0). If the surface is smooth at all points in the domain of f, we speak of it as a **smooth surface**. That is, if f is continuously differentiable, then the surface is smooth.[†] ∎

A surface integral is very much like a double integral. Suppose that the surface $z = f(x, y)$ is smooth for (x, y) in a bounded region Ω in the xy-plane. Since $f(x, y)$ is continuous on Ω, we find from the definition of a double

---

[†] This definition does not differ from the definition on page 995 because the added restriction $\mathbf{r}_u \times \mathbf{r}_v \neq \mathbf{0}$ is automatically satisfied in this case: $|\mathbf{r}_u \times \mathbf{r}_v| = \sqrt{1 + \left(\dfrac{\partial f}{\partial x}\right)^2 + \left(\dfrac{\partial f}{\partial y}\right)^2} \geq 1$.

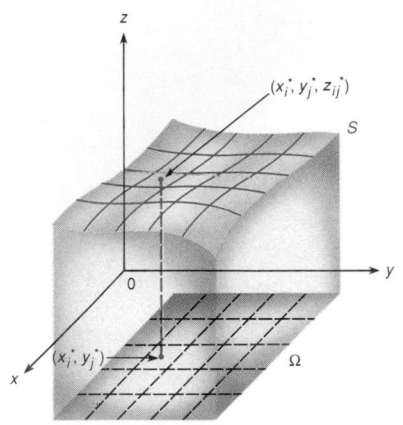

FIGURE 1
A PARTITION OF Ω INTO SUBRECTANGLES LEADS TO A PARTITION OF S INTO SUBSURFACES.

integral and Theorem 14.1.2, that

$$\iint_{\Omega} f(x, y)\, dA = \lim_{\Delta s \to 0} \sum_{i=1}^{n} \sum_{j=1}^{m} f(x_i^*, y_j^*)\, \Delta x\, \Delta y \tag{3}$$

exists, where $\Delta s = \sqrt{\Delta x^2 + \Delta y^2}$ and the limit is independent of the way in which the points (x_i^*, y_j^*) are chosen in the rectangle R_{ij}. We stress that in (3) the quantity $\Delta x \Delta y$ represents the *area* of the rectangle R_{ij}.

The double integral (3) is an integral over a region in the plane. A *surface integral*, which we will soon define, is an integral over a surface in space. Suppose we wish to integrate the function $F(x, y, z)$ over the surface S given by $z = f(x, y)$ where $(x, y) \in \Omega$ and, as before, Ω is a bounded region in the xy-plane. Such a surface is sketched in Figure 1.

We partition Ω into rectangles (and parts of rectangles) as before. This procedure provides a partition of S into mn "subsurfaces" S_{ij}, where

$$S_{ij} = \{(x, y, z): z = f(x, y) \text{ and } (x, y) \in R_{ij}\}. \tag{4}$$

We choose a point in each S_{ij}. Such a point will have the form (x_i^*, y_j^*, z_{ij}^*), where $z_{ij}^* = f(x_i^*, y_j^*)$ and $(x_i^*, y_j^*) \in R_{ij}$. We let $\Delta \sigma_{ij}$ denote the surface area of S_{ij}. This is analogous to the notation $\Delta x\, \Delta y$ as the area of the rectangle R_{ij}. Then we write the double sum

$$\sum_{i=1}^{n} \sum_{j=1}^{m} F(x_i^*, y_j^*, z_{ij}^*)\, \Delta \sigma_{ij}$$

and consider

$$\lim_{\Delta s \to 0} \sum_{i=1}^{n} \sum_{j=1}^{m} F(x_i^*, y_j^*, z_{ij}^*)\, \Delta \sigma_{ij}. \tag{5}$$

DEFINITION   **INTEGRAL OVER A SURFACE**

Suppose the limit in (5) exists and is independent of the way the surface S is partitioned and the way in which the points (x_i^*, y_i^*, z_{ij}^*) are chosen in S_{ij}. Then F is said to be **integrable** over S and the **surface integral** of F over S, denoted by $\iint_S F(x, y, z)\, d\sigma$, is given by

$$\iint_S F(x, y, z)\, d\sigma = \lim_{\Delta s \to 0} \sum_{i=1}^{n} \sum_{j=1}^{m} F(x_i^*, y_j^*, z_{ij}^*)\, \Delta \sigma_{ij}. \blacksquare \tag{6}$$

The key to evaluating the limit in (6) is to note that, from formula (12) on page 997,

$$\Delta \sigma_{ij} \approx \sqrt{f_x^2(x_i^*, y_j^*) + f_y^2(x_i^*, y_j^*) + 1}\, \Delta x\, \Delta y. \tag{7}$$

Then inserting (7) into the limit in (6) and noting that $z_{ij}^* = f(x_i^*, y_j^*)$, we have

$$\lim_{\Delta s \to 0} \sum_{i=1}^{n} \sum_{j=1}^{m} F(x_i^*, y_j^*, z_{ij}^*)\, \Delta \sigma_{ij}$$

$$= \lim_{\Delta s \to 0} \sum_{i=1}^{n} \sum_{j=1}^{m} F(x_i^*, y_j^*, f(x_i^*, y_j^*))\sqrt{f_x^2(x_i^*, y_j^*) + f_y^2(x_i^*, y_j^*) + 1}\, \Delta x\, \Delta y. \tag{8}$$

Now if $z = f(x, y)$ is a smooth surface over Ω, then f_x and f_y are continuous over Ω, so that $\sqrt{f_x^2 + f_y^2 + 1}$ is also continuous over Ω. Furthermore, if $F(x, y, z)$ is continuous for (x, y, z) on S, then by Theorem 14.1.2

$$\iint_\Omega F(x, y, f(x, y))\sqrt{f_x^2(x, y) + f_y^2(x, y) + 1}\, dA$$

exists and is equal to the right-hand limit in (8). We therefore have the following important result.

> **THEOREM 1** **WRITING A SURFACE INTEGRAL AS AN ORDINARY DOUBLE INTEGRAL**
>
> Let S: $z = f(x, y)$ be a smooth surface for (x, y) in the bounded region Ω in the xy-plane. Then if F is continuous on S, F is integrable over S, and
>
> $$\iint_S F(x, y, z)\, d\sigma = \iint_\Omega F(x, y, f(x, y))\sqrt{f_x^2(x, y) + f_y^2(x, y) + 1}\, dA. \quad \blacksquare \quad (9)$$

REMARK 1: Using (9), we can compute a surface integral over S by transforming it into an ordinary double integral over the region Ω that is the projection of S into the xy-plane.

REMARK 2: If $F(x, y, z) = 1$ in (9), then (9) reduces to

$$\sigma = \iint_S d\sigma = \iint_\Omega \sqrt{f_x^2(x, y) + f_y^2(x, y) + 1}\, dA. \qquad (10)$$

This is the formula for surface area given in equation (12) in the last section.

EXAMPLE 1
COMPUTING A SURFACE INTEGRAL

Compute $\iint_S (x^2 + y^2 + 3z^2)\, d\sigma$, where S is the part of the circular paraboloid $z = x^2 + y^2$ with $x^2 + y^2 \leq 9$.

SOLUTION: Here $f(x, y) = x^2 + y^2$, so that $f_x = 2x$, $f_y = 2y$, and $d\sigma = \sqrt{1 + 4x^2 + 4y^2}\, dx\, dy$. Thus,

$$I = \iint_S (x^2 + y^2 + 3z^2)\, d\sigma$$

$$= \iint_\Omega [x^2 + y^2 + 3(x^2 + y^2)^2]\sqrt{1 + 4x^2 + 4y^2}\, dA,$$

where Ω is the disk (in the xy-plane) $x^2 + y^2 \leq 9$. The problem is greatly simplified by the use of polar coordinates. We have, using $x^2 + y^2 = r^2$,

$$I = \int_0^{2\pi} \int_0^3 [r^2 + 3(r^2)^2]\sqrt{1 + 4r^2}\, r\, dr\, d\theta$$

$$= \int_0^{2\pi} \int_0^3 (r^3 + 3r^5)\sqrt{1 + 4r^2}\, dr\, d\theta$$

$$= 2\pi \int_0^3 (r^3 + 3r^5)\sqrt{1 + 4r^2}\, dr.$$

There are several ways to complete the evaulation of this integral. The easiest way is to make the substitution $u^2 = 1 + 4r^2$. The result is

$$I = 2\pi \left\{ 21(37)^{3/2} + \frac{1}{120}[1 - 55(37)^{5/2}] + \frac{1}{280}(37^{7/2} - 1) \right\} \approx 12{,}629.4.$$

EXAMPLE 2
COMPUTING A SURFACE
INTEGRAL

Evaluate $\iint_S (x + y + z)\, d\sigma$, where S is the part of the surface $z = x^3 + y^4$ lying over the square $\{(x, y): 0 \le x \le 1, 0 \le y \le 1\}$.

SOLUTION: We have $f_x = 3x^2$, $f_y = 4y^3$, and $d\sigma = \sqrt{1 + 9x^4 + 16y^6}\, dx\, dy$, so

$$\iint_S (x + y + z)\, d\sigma = \int_0^1 \int_0^1 (x + x^3 + y + y^4)\sqrt{1 + 9x^4 + 16y^6}\, dx\, dy.$$

This is as far as we can go because there is no way to compute this integral directly. The best we can do is to use numerical integration to approximate the answer.

REMARK: Example 2 is typical. Like the computations of arc length and surface area, the direct evaluation of a surface integral will often be impossible since it will involve integrals of functions for which antiderivatives cannot be found.

We can derive another way to represent a surface integral. First we need to define the orientation of a surface.

ORIENTATION OF A SURFACE

Consider the smooth surface $z = f(x, y)$. We can write this surface in two ways:

$$F(x, y, z) = f(x, y) - z = 0 \quad \text{and} \quad G(x, y, z) = z - f(x, y) = 0. \qquad \text{(11)}$$

From Section 13.7 (page 882), we know that the gradient vectors $\mathbf{\nabla}F$ and $\mathbf{\nabla}G$ are normal to the surface at every point on the surface. Then, if $\mathbf{\nabla}F$ and $\mathbf{\nabla}G \ne \mathbf{0}$,

$$\mathbf{n}_1 = \frac{\mathbf{\nabla}F}{|\mathbf{\nabla}F|} \quad \text{and} \quad \mathbf{n}_2 = \frac{\mathbf{\nabla}G}{|\mathbf{\nabla}G|} \qquad \text{(12)}$$

are unit normal vectors. In fact, it is evident from the way F and G are defined that $\mathbf{n}_2 = -\mathbf{n}_1$.

DEFINITION   **OUTWARD UNIT NORMAL VECTOR**

We choose one of these normal vectors, denote it by $\mathbf{n}$, and call it the **outward unit normal vector**. The direction of $\mathbf{n}$ is called the **positive normal direction** to the surface at a point. ■

Choosing **n** if S is a closed surface

REMARK: If S is a closed surface, such as the surface of a ball, then by convention we choose **n** so that it points away from the region bounded by the surface. This explains the use of the term *outward* unit normal vector. Put another way, if S is the boundary of a finite region Ω, then **n** points away from Ω.

Choosing **n** if S does not enclose any region

If S does not enclose any part of a region in space (for example, if S is part of a plane), then by convention we choose **n** so that its **k** component is positive. In this context **n** is called the **upper unit normal vector**.

DEFINITION **ORIENTABLE SURFACE**

Let P_0 be any point on the surface S. Let C be a closed curve on S that passes through P_0. Then S is said to be **orientable** if the outward unit normal vector at P_0 does not change direction as it is displaced continuously around C until it returns to P_0. ■

If a surface is orientable, then the choice of the outward unit normal vector **n** determines a positive direction on S by displacing **n** along curves lying in S.

All of the surfaces we have considered in this text are orientable. The most famous example of a nonorientable surface is given by the Möbius strip.[†] To construct a Möbius strip, take a long rectangular piece of paper, twist one of the ends once, and paste the ends together (as in Figure 2). In our terminology, the Möbius strip is not orientable because a normal vector moving once around the closed curve indicated by the dotted line will change direction.

In this text, we will write the surface $z = f(x, y)$ as $G(x, y, z) = z - f(x, y) = 0$. Then

$$\nabla G = G_x \mathbf{i} + G_y \mathbf{j} + G_z \mathbf{k} = -f_x \mathbf{i} - f_y \mathbf{j} + \mathbf{k}$$

and the outward (and upper) unit normal vector is

$$\mathbf{n} = \frac{\nabla G}{|\nabla G|} = \frac{-f_x \mathbf{i} - f_y \mathbf{j} + \mathbf{k}}{\sqrt{f_x{}^2 + f_y{}^2 + 1}}. \qquad \text{The outward unit normal vector to the surface } z = f(x, y) \qquad \textbf{(13)}$$

This vector determines our orientation (positive direction).

We now define the angle γ to be the acute angle between **n** and the positive z-axis. This angle is depicted in Figure 3. Since **k** is a unit vector having the direction of the positive z-axis, we have, from Theorem 11.2.2 on page 731,

$$\cos \gamma = \frac{\mathbf{n} \cdot \mathbf{k}}{|\mathbf{n}| \, |\mathbf{k}|} \overset{\substack{|\mathbf{n}| = |\mathbf{k}| = 1 \\ \downarrow}}{=} \frac{1}{\sqrt{f_x{}^2 + f_y{}^2 + 1}} \qquad \textbf{(14)}$$

and

$$\sec \gamma = \sqrt{f_x{}^2 + f_y{}^2 + 1}. \qquad \textbf{(15)}$$

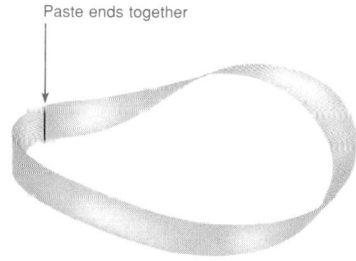

Paste ends together

FIGURE 2
A MÖBIUS STRIP

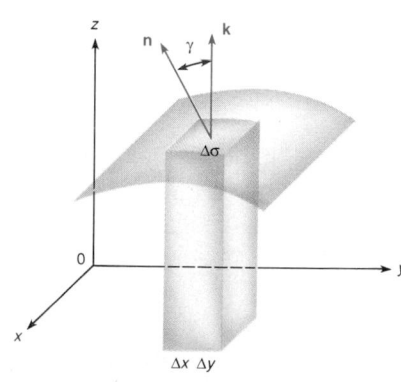

FIGURE 3
γ IS THE ANGLE BETWEEN THE OUTWARD UNIT NORMAL VECTOR **n** AND THE UNIT VECTOR **k**.

[†] Named after the German mathematician August Ferdinand Möbius (1790–1868). Möbius, who was a student of Gauss, did important work in geometry, mechanics, and number theory.

Hence, we have

A FORMULA FOR A SURFACE INTEGRAL

$$\iint_S F(x, y, z) \, d\sigma = \iint_\Omega F(x, y, f(x, y)) \sec \gamma(x, y) \, dx \, dy. \qquad (16)$$

EXAMPLE 3
COMPUTING SURFACE AREA
USING EQUATION (16)

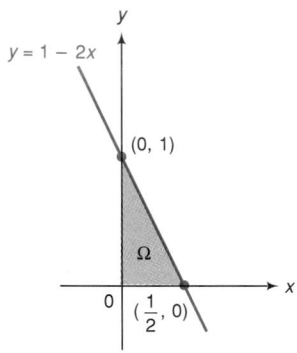

FIGURE 4
THE REGION BOUNDED BY THE COOR-
DINATE AXES AND THE LINE $y = 1 - 2x$

Compute $\iint_S (x + 2y + 3z) \, d\sigma$, where S is the part of the plane $2x - y + z = 3$, that lies above the triangular region Ω in the xy-plane bounded by the x- and y-axes and the line $y = 1 - 2x$.

SOLUTION: Here S is the plane $2x - y + z = 3$. From Section 11.7 we know that $\mathbf{N} = 2\mathbf{i} - \mathbf{j} + \mathbf{k}$ is normal to this surface, so $\mathbf{n} = \mathbf{N}/|\mathbf{N}| = (1/\sqrt{6})(2\mathbf{i} - \mathbf{j} + \mathbf{k})$ is the outward unit normal vector. Then $\cos \gamma = \mathbf{n} \cdot \mathbf{k} = 1/\sqrt{6}$ and $\sec \gamma = \sqrt{6}$, so from Figure 4,

$$\iint_S (x + 2y + 3z) \, d\sigma = \iint_\Omega [(x + 2y) + 3(3 - 2x + y)] \sqrt{6} \, dx \, dy,$$

$$= \sqrt{6} \int_0^{1/2} \int_0^{1-2x} (9 - 5x + 5y) \, dy \, dx$$

$$= \sqrt{6} \int_0^{1/2} \left(9y - 5xy + \frac{5y^2}{2} \right) \Big|_0^{1-2x} dx$$

$$= \sqrt{6} \int_0^{1/2} \left[9(1 - 2x) - 5x(1 - 2x) + \frac{5}{2}(1 - 2x)^2 \right] dx$$

$$= \sqrt{6} \int_0^{1/2} \left(20x^2 - 33x + \frac{23}{2} \right) dx = \frac{59}{24} \sqrt{6}.$$

■ PROBLEMS 15.6

■ SELF-QUIZ

I. The integral of $F(x, y, z) = 7 + z$ over the surface which is that patch of the plane $z = 3 + x + 5y$ lying above the unit disk $D = \{(x, y): x^2 + y^2 \leq 1\}$ is _____.

a. $\iint_D [3 + x + 5y] \, dA$

b. $\iint_D [7 + z] \, dA$

c. $\iint_D [7 + (3 + x + 5y)] \, dA$

d. $\iint_D [7 + (3 + x + 5y)] \sqrt{(1)^2 + (5)^2 + 1} \, dA$

II. Let S be the surface which is that patch of the plane $z = 3 + x + 5y$ lying above the unit disk D. The area of that surface is _____.

a. $\iint_S 1 \, d\sigma$

b. $\iint_S \sqrt{(1)^2 + (5)^2 + 1} \, d\sigma$

c. $\iint_S z \, d\sigma$

d. $\iint_D 1 \, dA$

e. $\iint_D \sqrt{(1)^2 + (5)^2 + 1} \, dA$

f. $\iint_D [3 + x + 5y] \, dA$

III. The integral of $F(x, y, z) = x^2 + y^2 - z$ over the surface S of the preceding problem is _____.

a. $\int_0^{2\pi} \int_0^1 \sqrt{(1)^2 + (5)^2 + 1}\, r\, dr\, d\theta$

b. $\int_0^{2\pi} \int_0^1 [r^2 - 3 - r(\cos\theta + 5\sin\theta)]r\, dr\, d\theta$

c. $\int_0^{2\pi} \int_0^1 [r^2 - 3 - r(\cos\theta + 5\sin\theta)] \times \sqrt{(1)^2 + (5)^2 + 1}\, r\, dr\, d\theta$

d. $\int_0^{2\pi} \int_0^1 [r^2 - 3 - r(\cos\theta + 5\sin\theta)] \times \sqrt{(2r\cos\theta)^2 + (2r\sin\theta)^2 + 1}\, r\, dr\, d\theta$

■ DRILL

In Problems 1–16, evaluate the given surface integral over the specified surface S.[†]

1. $\iint_S x\, d\sigma$, where $S = \{(x, y, z): z = x^2, 0 \le x \le 1, 0 \le y \le 2\}$
2. $\iint_S y\, d\sigma$, where S is as in Problem 1
*3. $\iint_S (x^2 - 2y^2)\, d\sigma$, where S is as in Problem 1
4. $\iint_S \sqrt{1 + 4z}\, d\sigma$, where S is as in Problem 1
*5. $\iint_S x\, d\sigma$, where S is the hemisphere $\{(x, y, z): x^2 + y^2 + z^2 = 4, x^2 + y^2 < 4, z > 0\}$
*6. $\iint_S xy\, d\sigma$, where S is as in Problem 5
7. $\iint_S (x + y)\, d\sigma$, where S is the planar patch $\{(x, y, z): x + 2y - 3z = 4, 0 \le x \le 1, 1 \le y \le 2\}$
8. $\iint_S yz\, d\sigma$, where S is as in Problem 7

9. $\iint_S z^2\, d\sigma$, where S is as in Problem 7
*10. $\iint_S (x^2 + y^2 + z^2)\, d\sigma$, where S is the part of the plane $x - y = 4$ that lies inside the cylinder $y^2 + z^2 = 4$
11. $\iint_S \cos z\, d\sigma$, where S is the planar patch $\{(x, y, z): 2x + 3y + z = 1, 0 \le x \le 1, -1 \le y \le 2\}$
*12. $\iint_S z\, d\sigma$, where S is the tetrahedron bounded by the coordinate planes and the plane $4x + 8y + 2z = 16$
13. $\iint_S |x|\, d\sigma$, where S is the hemisphere $\{(x, y, z): x^2 + y^2 + z^2 = 4, y^2 + z^2 \le 4, x \ge 0\}$
14. $\iint_S z\, d\sigma$, where S is the surface of Problem 13
15. $\iint_S z^2\, d\sigma$, where S is the hemisphere $\{(x, y, z): x^2 + y^2 + z^2 = 9, x^2 + z^2 \le 9, y \ge 0\}$
16. $\iint_S x^2\, d\sigma$, where S is the surface of Problem 15

■ APPLICATIONS

MASS: Suppose $z = f(x, y)$ for points (x, y, z) on a smooth thin metallic surface S. Also suppose $\rho(x, y, z)$ is the local density of that metallic surface. Then the **mass** of the surface is

$$\mu = \iint_S \rho(x, y, f(x, y))\, d\sigma.$$

Problem 33 asks you to work through a justification for this formula. For now, use this result to work Problems 17–20.

17. A metallic dome has the shape of a hemisphere centered at the origin with radius 4 m. Its area density at a point (x, y, z) in space is given by $\rho(x, y, z) = 25 - x^2 - y^2$ kg/m². Find the total mass of this dome.
18. Find the mass of a metallic sheet in the shape of the hemisphere

$$\{(x, y, z): x^2 + y^2 + z^2 = 9, x^2 + y^2 \le 9, z \ge 0\}$$

if its density at a point is proportional to the density from that point to the origin.
19. Find the mass of a triangular metallic sheet with corners at $(1, 0, 0)$, $(0, 1, 0)$, and $(0, 0, 1)$ if its density is a constant α. (Assume the units are meters and kg/m².)
20. Find the mass of the sheet of the preceding problem if its density is proportional to x^2.

FLUX: Immerse a porous surface in a moving fluid and think about the rate of flow through the surface. To be a bit more formal, let S be the surface and suppose $\mathbf{F}$ is the vector field which specifies how the fluid flows. In general, $\mathbf{F}$ is the product of a scalar density $\rho(x, y, z)$ and a vector velocity $\mathbf{v}(x, y, z)$. If $\mathbf{n} = \mathbf{n}(x, y, z)$ is the unit normal to the surface, then $\mathbf{F} \cdot \mathbf{n}$ is the component of $\mathbf{F}$ on $\mathbf{n}$ and it measures the local flow rate per unit of area; if we integrate $\mathbf{F} \cdot \mathbf{n}$ over the surface, then we obtain the total rate of

---

[†] If S does not enclose part of a region, then, to avoid ambiguity, choose $\mathbf{n}$ to be the upper unit normal vector. That is, choose $\mathbf{n}$ so that its $\mathbf{k}$-component is positive (if possible).

CONTINUED

flow through the whole surface:

$$\textbf{flux of F over S} = \iint\limits_{S} \textbf{F} \cdot \textbf{n}\, d\sigma.$$

Problem 34 includes an alternative expression for flux which may be easier to compute. In Problems 21–31, compute the flux $\iint_{S} \textbf{F} \cdot \textbf{n}\, d\sigma$ for the given surface S lying in the given vector field $\textbf{F}$.

21. $S = \{(x, y, z): z = xy, 0 \leq x \leq 1, 0 \leq y \leq 2\}$; $\textbf{F} = x^2 y \textbf{i} - z \textbf{j}$
22. $S = \{(x, y, z): z = 4 - x - y, x \geq 0, y \geq 0, z \geq 0\}$; $\textbf{F} = -3x \textbf{i} - y \textbf{j} + 3z \textbf{k}$
23. S is the unit sphere, $\textbf{F} = x \textbf{i} + y \textbf{j} + z \textbf{k}$
24. $S = \{(x, y, z): x^2 + y^2 + z^2 = 1, z \geq 0\}$; $\textbf{F} = x \textbf{i} + y \textbf{j} + z \textbf{k}$

25. $S = \{(x, y, z): x^2 + y^2 + z^2 = 1, y \geq 0\}$; $\textbf{F} = x \textbf{i} + y \textbf{j} + z \textbf{k}$
26. $S = \{(x, y, z): x^2 + y^2 + z^2 = 1, x \leq 0\}$; $\textbf{F} = x \textbf{i} + y \textbf{j} + z \textbf{k}$
27. $S = \{(x, y, z): z = \sqrt{x^2 + y^2}, x^2 + y^2 \leq 1\}$; $\textbf{F} = x \textbf{i} - y \textbf{j} + xy \textbf{k}$
28. $S = \{(x, y, z): x = \sqrt{y^2 + z^2}, y^2 + z^2 \leq 1\}$; $\textbf{F} = y \textbf{i} - z \textbf{j} + yz \textbf{k}$
29. $S = \{(x, y, z): z = \sqrt{x^2 + y^2}, x^2 + y^2 \leq 1, x \geq 0, y \geq 0\}$; $\textbf{F} = x^2 \textbf{i} + y^2 \textbf{j} + z \textbf{k}$
*30. S is region bounded by $y = 1$ and $y = \sqrt{x^2 + z^2}$; $x^2 + z^2 \leq 1$; $\textbf{F} = x \textbf{i} - z \textbf{j} + xz \textbf{k}$
*31. $S = \{(x, y, z): 0 \leq x \leq 1, 0 \leq y \leq 2, 0 \leq z \leq 3\}$; $\textbf{F} = x^2 y \textbf{i} - 2yz \textbf{j} + x^3 y^2 \textbf{k}$ [*Hint:* Compute the flux through each of the six faces separately; then add up those partial results.]

■ SHOW/PROVE/DISPROVE

32. Assuming all three of the following integrals exist, show that

$$\iint\limits_{S} [F(x, y, z) + G(x, y, z)]\, d\sigma$$

$$= \iint\limits_{S} F(x, y, z)\, d\sigma + \iint\limits_{S} G(x, y, z)\, d\sigma.$$

*33. Suppose $z = f(x, y)$ for points (x, y, z) on a smooth thin metallic surface S. Also suppose $\rho(x, y, z)$ is the local density of that metallic surface. Show that $\iint_{S} \rho(x, y, f(x, y))\, d\sigma$ is the limit of finite sums which approximate the total mass of the surface and comment on the reasonableness of defining that mass to be the value of the integral expression.

34. Suppose $S = \{(x, y, z): z = f(x, y), (x, y) \in \Omega\}$ is a smooth surface with upper unit normal $\textbf{n} = \textbf{n}(x, y, z)$. Also suppose $\textbf{F} = \textbf{F}(x, y, z) = P(x, y, z)\textbf{i} + Q(x, y, z)\textbf{j} + R(x, y, z)\textbf{k}$ is a vector field which is continuous on S. The paragraph preceding Problem 21 discusses the flux of $\textbf{F}$ over S and argues that $\iint_{S} \textbf{F} \cdot \textbf{n}\, d\sigma$ computes it. For this problem, show that

$$\iint\limits_{S} \textbf{F} \cdot \textbf{n}\, d\sigma = \iint\limits_{\Omega} [-Pf_x - Qf_y + R]\, dx\, dy.$$

35. Suppose S is the sphere $\{(x, y, z): x^2 + y^2 + z^2 = \alpha^2\}$ and $\textbf{F} = a \textbf{i} + b \textbf{j} + c \textbf{k}$ where a, b, c are constants. Show that $\iint_{S} \textbf{F} \cdot \textbf{n}\, d\sigma = 0$.
36. Let S be the hemisphere $\{(x, y, z): x^2 + y^2 + z^2 = a^2, x^2 + y^2 \leq a^2, z \geq 0\}$.
   a. Show that

$$\iint\limits_{S} (x^2 + y^2)\, d\sigma = \frac{4}{3} \pi a^4.$$

   b. Show that

$$\iint\limits_{S} x^2\, d\sigma = \iint\limits_{S} y^2\, d\sigma = \iint\limits_{S} z^2\, d\sigma$$

$$= \frac{1}{3} \iint\limits_{S} (x^2 + y^2 + z^2)\, d\sigma.$$

   c. Without performing any explicit integrations, explain why the last integral of part (b) equals $\frac{2}{3}\pi a^4$.
   d. Use part (c) to explain, without doing an integration, why $\iint_{S} (x^2 + y^2)\, d\sigma = \frac{4}{3}\pi a^4$.

■ ANSWERS TO SELF-QUIZ

I. d   II. $a = e = \sqrt{27}\pi$
III. c

CONTINUED

III. The integral of $F(x, y, z) = x^2 + y^2 - z$ over the surface S of the preceding problem is _____.

a. $\int_0^{2\pi} \int_0^1 \sqrt{(1)^2 + (5)^2 + 1}\, r\, dr\, d\theta$

b. $\int_0^{2\pi} \int_0^1 \left[r^2 - 3 - r(\cos\theta + 5\sin\theta)\right] r\, dr\, d\theta$

c. $\int_0^{2\pi} \int_0^1 \left[r^2 - 3 - r(\cos\theta + 5\sin\theta)\right] \times$ $\sqrt{(1)^2 + (5)^2 + 1}\, r\, dr\, d\theta$

d. $\int_0^{2\pi} \int_0^1 \left[r^2 - 3 - r(\cos\theta + 5\sin\theta)\right] \times$ $\sqrt{(2r\cos\theta)^2 + (2r\sin\theta)^2 + 1}\, r\, dr\, d\theta$

■ **DRILL**

In Problems 1–16, evaluate the given surface integral over the specified surface S.[†]

1. $\iint_S x\, d\sigma$, where $S = \{(x, y, z): z = x^2,\ 0 \le x \le 1,\ 0 \le y \le 2\}$

2. $\iint_S y\, d\sigma$, where S is as in Problem 1

*3. $\iint_S (x^2 - 2y^2)\, d\sigma$, where S is as in Problem 1

4. $\iint_S \sqrt{1 + 4z}\, d\sigma$, where S is as in Problem 1

*5. $\iint_S x\, d\sigma$, where S is the hemisphere $\{(x, y, z): x^2 + y^2 + z^2 = 4,\ x^2 + y^2 \le 4,\ z \ge 0\}$

*6. $\iint_S xy\, d\sigma$, where S is as in Problem 5

7. $\iint_S (x + y)\, d\sigma$, where S is the planar patch $\{(x, y, z): x + 2y - 3z = 4,\ 0 \le x \le 1,\ 1 \le y \le 2\}$

8. $\iint_S yz\, d\sigma$, where S is as in Problem 7

9. $\iint_S z^2\, d\sigma$, where S is as in Problem 7

*10. $\iint_S (x^2 + y^2 + z^2)\, d\sigma$, where S is the part of the plane $x - y = 4$ that lies inside the cylinder $y^2 + z^2 = 4$

11. $\iint_S \cos z\, d\sigma$, where S is the planar patch $\{(x, y, z): 2x + 3y + z = 1,\ 0 \le x \le 1,\ -1 \le y \le 2\}$

*12. $\iint_S z\, d\sigma$, where S is the tetrahedron bounded by the coordinate planes and the plane $4x + 8y + 2z = 16$

13. $\iint_S |x|\, d\sigma$, where S is the hemisphere $\{(x, y, z): x^2 + y^2 + z^2 = 4,\ y^2 + z^2 \le 4,\ x \ge 0\}$

14. $\iint_S z\, d\sigma$, where S is the surface of Problem 13

15. $\iint_S z^2\, d\sigma$, where S is the hemisphere $\{(x, y, z): x^2 + y^2 + z^2 = 9,\ x^2 + z^2 \le 9,\ y \ge 0\}$

16. $\iint_S x^2\, d\sigma$, where S is the surface of Problem 15

■ **APPLICATIONS**

MASS: Suppose $z = f(x, y)$ for points (x, y, z) on a smooth thin metallic surface S. Also suppose $\rho(x, y, z)$ is the local density of that metallic surface. Then the **mass** of the surface is

$$\mu = \iint_S \rho(x, y, f(x, y))\, d\sigma.$$

Problem 33 asks you to work through a justification for this formula. For now, use this result to work Problems 17–20.

17. A metallic dome has the shape of a hemisphere centered at the origin with radius 4 m. Its area density at a point (x, y, z) in space is given by $\rho(x, y, z) = 25 - x^2 - y^2$ kg/m². Find the total mass of this dome.

18. Find the mass of a metallic sheet in the shape of the hemisphere

$$\{(x, y, z): x^2 + y^2 + z^2 = 9,\ x^2 + y^2 \le 9,\ z \ge 0\}$$

if its density at a point is proportional to the density from that point to the origin.

19. Find the mass of a triangular metallic sheet with corners at $(1, 0, 0)$, $(0, 1, 0)$, and $(0, 0, 1)$ if its density is a constant α. (Assume the units are meters and kg/m².)

20. Find the mass of the sheet of the preceding problem if its density is proportional to x^2.

FLUX: Immerse a porous surface in a moving fluid and think about the rate of flow through the surface. To be a bit more formal, let S be the surface and suppose $\mathbf{F}$ is the vector field which specifies how the fluid flows. In general, $\mathbf{F}$ is the product of a scalar density $\rho(x, y, z)$ and a vector velocity $\mathbf{v}(x, y, z)$. If $\mathbf{n} = \mathbf{n}(x, y, z)$ is the unit normal to the surface, then $\mathbf{F} \cdot \mathbf{n}$ is the component of $\mathbf{F}$ on $\mathbf{n}$ and it measures the local flow rate per unit of area; if we integrate $\mathbf{F} \cdot \mathbf{n}$ over the surface, then we obtain the total rate of

[†] If S does not enclose part of a region, then, to avoid ambiguity, choose $\mathbf{n}$ to be the upper unit normal vector. That is, choose $\mathbf{n}$ so that its $\mathbf{k}$-component is positive (if possible).

CONTINUED

flow through the whole surface:

$$\textbf{flux of F over S} = \iint_S \mathbf{F} \cdot \mathbf{n}\, d\sigma.$$

Problem 34 includes an alternative expression for flux which may be easier to compute. In Problems 21–31, compute the flux $\iint_S \mathbf{F} \cdot \mathbf{n}\, d\sigma$ for the given surface S lying in the given vector field $\mathbf{F}$.

21. $S = \{(x, y, z): z = xy, 0 \le x \le 1, 0 \le y \le 2\}$; $\mathbf{F} = x^2 y\mathbf{i} - z\mathbf{j}$
22. $S = \{(x, y, z): z = 4 - x - y, x \ge 0, y \ge 0, z \ge 0\}$; $\mathbf{F} = -3x\mathbf{i} - y\mathbf{j} + 3z\mathbf{k}$
23. S is the unit sphere, $\mathbf{F} = x\mathbf{i} + y\mathbf{j} + z\mathbf{k}$
24. $S = \{(x, y, z): x^2 + y^2 + z^2 = 1, z \ge 0\}$; $\mathbf{F} = x\mathbf{i} + y\mathbf{j} + z\mathbf{k}$

25. $S = \{(x, y, z): x^2 + y^2 + z^2 = 1, y \ge 0\}$; $\mathbf{F} = x\mathbf{i} + y\mathbf{j} + z\mathbf{k}$
26. $S = \{(x, y, z): x^2 + y^2 + z^2 = 1, x \le 0\}$; $\mathbf{F} = x\mathbf{i} + y\mathbf{j} + z\mathbf{k}$
27. $S = \{(x, y, z): z = \sqrt{x^2 + y^2}, x^2 + y^2 \le 1\}$; $\mathbf{F} = x\mathbf{i} - y\mathbf{j} + xy\mathbf{k}$
28. $S = \{(x, y, z): x = \sqrt{y^2 + z^2}, y^2 + z^2 \le 1\}$; $\mathbf{F} = y\mathbf{i} - z\mathbf{j} + yz\mathbf{k}$
29. $S = \{(x, y, z): z = \sqrt{x^2 + y^2}, x^2 + y^2 \le 1, x \ge 0, y \ge 0\}$; $\mathbf{F} = x^2\mathbf{i} + y^2\mathbf{j} + z\mathbf{k}$
*30. S is region bounded by $y = 1$ and $y = \sqrt{x^2 + z^2}$; $x^2 + z^2 \le 1$; $\mathbf{F} = x\mathbf{i} - z\mathbf{j} + xz\mathbf{k}$
*31. $S = \{(x, y, z): 0 \le x \le 1, 0 \le y \le 2, 0 \le z \le 3\}$; $\mathbf{F} = x^2 y\mathbf{i} - 2yz\mathbf{j} + x^3 y^2\mathbf{k}$ [*Hint:* Compute the flux through each of the six faces separately; then add up those partial results.]

■ SHOW/PROVE/DISPROVE

32. Assuming all three of the following integrals exist, show that

$$\iint_S [F(x, y, z) + G(x, y, z)]\, d\sigma$$
$$= \iint_S F(x, y, z)\, d\sigma + \iint_S G(x, y, z)\, d\sigma.$$

*33. Suppose $z = f(x, y)$ for points (x, y, z) on a smooth thin metallic surface S. Also suppose $\rho(x, y, z)$ is the local density of that metallic surface. Show that $\iint_S \rho(x, y, f(x, y))\, d\sigma$ is the limit of finite sums which approximate the total mass of the surface and comment on the reasonableness of defining that mass to be the value of the integral expression.

34. Suppose $S = \{(x, y, z): z = f(x, y), (x, y) \in \Omega\}$ is a smooth surface with upper unit normal $\mathbf{n} = \mathbf{n}(x, y, z)$. Also suppose $\mathbf{F} = \mathbf{F}(x, y, z) = P(x, y, z)\mathbf{i} + Q(x, y, z)\mathbf{j} + R(x, y, z)\mathbf{k}$ is a vector field which is continuous on S. The paragraph preceding Problem 21 discusses the flux of $\mathbf{F}$ over S and argues that $\iint_S \mathbf{F} \cdot \mathbf{n}\, d\sigma$ computes it. For this problem, show that

$$\iint_S \mathbf{F} \cdot \mathbf{n}\, d\sigma = \iint_\Omega [-Pf_x - Qf_y + R]\, dx\, dy.$$

35. Suppose S is the sphere $\{(x, y, z): x^2 + y^2 + z^2 = \alpha^2\}$ and $\mathbf{F} = a\mathbf{i} + b\mathbf{j} + c\mathbf{k}$ where a, b, c are constants. Show that $\iint_S \mathbf{F} \cdot \mathbf{n}\, d\sigma = 0$.
36. Let S be the hemisphere $\{(x, y, z): x^2 + y^2 + z^2 = a^2, x^2 + y^2 \le a^2, z \ge 0\}$.
    a. Show that
    $$\iint_S (x^2 + y^2)\, d\sigma = \frac{4}{3}\pi a^4.$$
    b. Show that
    $$\iint_S x^2\, d\sigma = \iint_S y^2\, d\sigma = \iint_S z^2\, d\sigma$$
    $$= \frac{1}{3}\iint_S (x^2 + y^2 + z^2)\, d\sigma.$$
    c. Without performing any explicit integrations, explain why the last integral of part (b) equals $\frac{2}{3}\pi a^4$.
    d. Use part (c) to explain, without doing an integration, why $\iint_S (x^2 + y^2)\, d\sigma = \frac{4}{3}\pi a^4$.

■ ANSWERS TO SELF-QUIZ

I. d II. $a = e = \sqrt{27}\pi$
III. c

15.7 DIVERGENCE AND CURL

Let the function $F(x, y, x)$ be given. Then the gradient of F is given by

$$\nabla F = \frac{\partial F}{\partial x}\mathbf{i} + \frac{\partial F}{\partial y}\mathbf{j} + \frac{\partial F}{\partial z}\mathbf{k}. \tag{1}$$

We can think of the gradient as a function that takes a differentiable function of (x, y, z) into a vector field in $\mathbb{R}^3$. We write this function, symbolically, as

$$\nabla = \frac{\partial}{\partial x}\mathbf{i} + \frac{\partial}{\partial y}\mathbf{j} + \frac{\partial}{\partial z}\mathbf{k}. \tag{2}$$

The operator in (2) provides a useful device for writing things down. For example, (1) can be written as

$$\nabla F = \left(\frac{\partial}{\partial x}\mathbf{i} + \frac{\partial}{\partial y}\mathbf{j} + \frac{\partial}{\partial z}\mathbf{k}\right)F = \frac{\partial F}{\partial x}\mathbf{i} + \frac{\partial F}{\partial y}\mathbf{j} + \frac{\partial F}{\partial z}\mathbf{k}.$$

We now define the divergence and curl of a vector field $\mathbf{F}$ in $\mathbb{R}^3$ given by

$$\mathbf{F}(x, y, z) = P(x, y, z)\mathbf{i} + Q(x, y, z)\mathbf{j} + R(x, y, z)\mathbf{k}. \tag{3}$$

DEFINITION **DIVERGENCE AND CURL**

Let the vector field $\mathbf{F}$ be given by (3), where P, Q, and R are differentiable. Then the **divergence** of $\mathbf{F}$ (div $\mathbf{F}$) and **curl** of $\mathbf{F}$ (curl $\mathbf{F}$) are given by

$$\text{div } \mathbf{F} = \frac{\partial P}{\partial x} + \frac{\partial Q}{\partial y} + \frac{\partial R}{\partial z} \tag{4}$$

and

$$\text{curl } \mathbf{F} = \left(\frac{\partial R}{\partial y} - \frac{\partial Q}{\partial z}\right)\mathbf{i} + \left(\frac{\partial P}{\partial z} - \frac{\partial R}{\partial x}\right)\mathbf{j} + \left(\frac{\partial Q}{\partial x} - \frac{\partial P}{\partial y}\right)\mathbf{k}. \quad \blacksquare \tag{5}$$

NOTE: div $\mathbf{F}$ is a scalar function and curl $\mathbf{F}$ is a vector field.

Before giving examples of divergence and curl, we derive an easy way to remember how to compute them.

THEOREM 1

Let ∇ be given by (2) and let the differentiable vector field $\mathbf{F}$ be given by (3). Then,

(i) div $\mathbf{F} = \nabla \cdot \mathbf{F}$ $\tag{6}$

and

(ii) curl $\mathbf{F} = \nabla \times \mathbf{F}$. $\tag{7}$

PROOF:

(i) $\mathbf{V} \cdot \mathbf{F} = \left(\dfrac{\partial}{\partial x} \mathbf{i} + \dfrac{\partial}{\partial y} \mathbf{j} + \dfrac{\partial}{\partial z} \mathbf{k} \right) \cdot (P\mathbf{i} + Q\mathbf{j} + R\mathbf{k})$

$\qquad = \dfrac{\partial P}{\partial x} + \dfrac{\partial Q}{\partial y} + \dfrac{\partial R}{\partial z} = \text{div } \mathbf{F}.$

(ii) $\mathbf{V} \times \mathbf{F} = \begin{vmatrix} \mathbf{i} & \mathbf{j} & \mathbf{k} \\ \dfrac{\partial}{\partial x} & \dfrac{\partial}{\partial y} & \dfrac{\partial}{\partial z} \\ P & Q & R \end{vmatrix}$

$\qquad = \left(\dfrac{\partial R}{\partial y} - \dfrac{\partial Q}{\partial z} \right) \mathbf{i} + \left(\dfrac{\partial P}{\partial z} - \dfrac{\partial R}{\partial x} \right) \mathbf{j} + \left(\dfrac{\partial Q}{\partial x} - \dfrac{\partial P}{\partial y} \right) \mathbf{k}$

$\qquad = \text{curl } \mathbf{F}. \quad \blacksquare$

EXAMPLE 1
COMPUTING DIVERGENCE
AND CURL

Compute the divergence and curl of $\mathbf{F}(x, y, z) = xy\mathbf{i} + (z^2 - 2y)\mathbf{j} + \cos yz\mathbf{k}$.

SOLUTION:

$\text{div } \mathbf{F} = \dfrac{\partial}{\partial x} (xy) + \dfrac{\partial}{\partial y} (z^2 - 2y) + \dfrac{\partial}{\partial z} (\cos yz)$

$\qquad = y - 2 - y \sin yz$

and

$\text{curl } \mathbf{F} = \begin{vmatrix} \mathbf{i} & \mathbf{j} & \mathbf{k} \\ \dfrac{\partial}{\partial x} & \dfrac{\partial}{\partial y} & \dfrac{\partial}{\partial z} \\ xy & z^2 - 2y & \cos yz \end{vmatrix}$

$\qquad = \left[\dfrac{\partial}{\partial y} \cos yz - \dfrac{\partial}{\partial z} (z^2 - 2y) \right] \mathbf{i} + \left(\dfrac{\partial}{\partial z} xy - \dfrac{\partial}{\partial x} \cos yz \right) \mathbf{j}$

$\qquad + \left[\dfrac{\partial}{\partial x} (z^2 - 2y) - \dfrac{\partial}{\partial y} xy \right] \mathbf{k} = (-z \sin yz - 2z)\mathbf{i} - x\mathbf{k}.$

Recall from Theorem 15.3.4 that the vector field $\mathbf{F} = P\mathbf{i} + Q\mathbf{j} + R\mathbf{k}$ is the gradient of a function f if and only if

$$\dfrac{\partial P}{\partial y} = \dfrac{\partial Q}{\partial x}, \qquad \dfrac{\partial R}{\partial x} = \dfrac{\partial P}{\partial z}, \qquad \dfrac{\partial Q}{\partial z} = \dfrac{\partial R}{\partial y}. \tag{8}$$

Using (8) and (5), we obtain the following interesting result.

THEOREM 2

The differentiable vector field $\mathbf{F}$ is the gradient of a function f if and only if curl $\mathbf{F} = \mathbf{0}$. $\quad \blacksquare$

As we shall see, the curl of a vector field $\mathbf{F}$ represents the circulation per unit area at the point (x, y, z). If curl $\mathbf{F} = \mathbf{0}$ for every (x, y, z) in some region

W in $\mathbb{R}^3$, then the fluid flow $\mathbf{F}$ is called **irrotational**. The divergence of $\mathbf{F}$ at a point (x, y, z) represents the net rate of flow away from (x, y, z). If div $\mathbf{F} = 0$ for every (x, y, z) in W, then the flow $\mathbf{F}$ is called **incompressible** or **solenoidal**. Let us examine these ideas more closely.

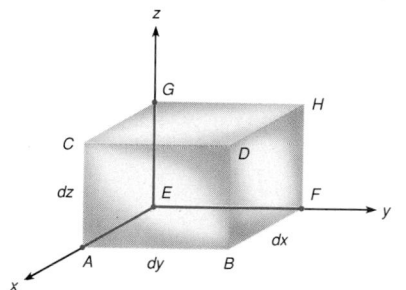

FIGURE 1
A BOX OF VOLUME $dx\,dy\,dz$

PHYSICAL INTERPRETATION OF DIVERGENCE AND CURL

Divergence Suppose that a fluid is flowing through the small volume $dx\,dy\,dz$ depicted in Figure 1. Let $\mathbf{v} = v_x\mathbf{i} + v_y\mathbf{j} + v_z\mathbf{k}$ and ρ denote, respectively, the velocity and the density of the fluid at the origin. Fluid is flowing in all directions. We consider the positive x-direction first. The fluid flowing into this volume per unit time through the face $EFGH$ is

$$\text{rate of flow in (face } EFGH) = \rho v_x\big|_{x=0}\, dy\,dz. \tag{9}$$

The components ρv_y and ρv_z of the flow are tangential to this face and contribute nothing to the flow through this face. The rate of flow out (in the positive direction) through the face $ABCD$ is

$$\text{rate of flow out (face } ABCD) = \rho v_x\big|_{x=dx}\, dy\,dz. \tag{10}$$

By the mean-value theorem, we have

$$\rho v_x\big|_{x=dx}\, dy\,dz - \rho v_x\big|_{x=0}\, dy\,dz = \left[\frac{\partial}{\partial x}(\rho v_x)\,dx\right] dy\,dz, \tag{11}$$

where the partial derivative is evaluated at some point where x takes a value in $(0, dx)$. Hence,

$$\begin{matrix}\text{net flow in}\\ \text{positive direction}\\ \text{at the point } (x, y, z)\end{matrix} \quad = \quad \text{flow out} - \text{flow in} = \frac{\partial}{\partial x}(\rho v_x)\,dx\,dy\,dz.$$

Similar results hold in the positive y- and z-directions and we have

$$\text{net flow per unit time} = \left[\frac{\partial}{\partial x}(\rho v_x) + \frac{\partial}{\partial y}(\rho v_y) + \frac{\partial}{\partial z}(\rho v_z)\right] dx\,dy\,dz$$

$$= \text{div}(\rho v)\,dx\,dy\,dz.$$

Therefore, the net flow of the compressible fluid from the volume element $dx\,dy\,dz$ per unit volume per unit time is $\text{div}(\rho v)$. This is why we call it *divergence*.

The divergence appears in a wide variety of physical problems, ranging from a probability current density in quantum mechanics to neutron leakage in a nuclear reactor.

Curl Consider the circulation of fluid around the loop in the xy-plane drawn in Figure 2. Let the velocity vector $\mathbf{v} = v_x\mathbf{i} + v_y\mathbf{j}$. Now

$$\text{circulation} = \text{flow along 1} + \text{flow along 2} + \text{flow along 3}$$

$$+ \text{flow along 4.}$$

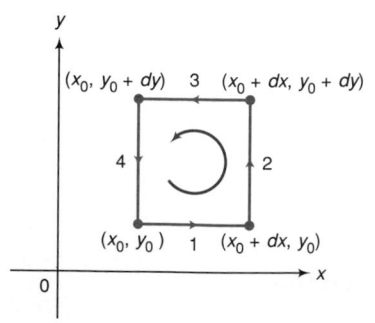

FIGURE 2
CIRCULATION AROUND A DIFFERENTIAL LOOP

Recall that directed distance is the integral of velocity. Thus, treating v_x and v_y as constants (since, dx and dy are small), we have

$$\text{flow along } 1 = \int_{x_0}^{x_0+dx} v_x(x_0, y_0)\, dx = v_x(x_0, y_0)\, dx$$

$$\text{flow along } 2 = \int_{y_0}^{y_0+dy} v_y(x_0 + dx, y_0)\, dy = v_y(x_0 + dx, y_0)\, dy$$

$$\text{flow along } 3 = \int_{x_0+dx}^{x_0} v_x(x_0, y_0 + dy)\, dx$$

$$= -\int_{x_0}^{x_0+dx} v_x(x_0, y_0 + dy)\, dx = -v_x(x_0, y_0 + dy)\, dx$$

$$\text{flow along } 4 = \int_{y_0+dy}^{y_0} v_y(x_0, y_0)\, dy = -v_y(x_0, y_0)\, dy.$$

Using Taylor expansions around (x_0, y_0), we have

$$v_y(x_0 + t, y_0) = v_y(x_0, y_0) + \frac{\partial v_y}{\partial x}(x_0, y_0)t + \text{higher-order terms}$$

and

$$v_y(x_0 + dx, y_0) = v_y(x_0, y_0) + \frac{\partial v_y}{\partial x}(x_0, y_0)\, dx + \text{higher-order terms},$$

so

$$\text{flow along } 2 \approx v_y(x_0, y_0)\, dy + \frac{\partial v_y}{\partial x}(x_0, y_0)\, dx\, dy.$$

Similarly,

$$v_x(x_0, y_0 + dy) = v_x(x_0, y_0) + \frac{\partial v_x}{\partial y}\, dy + \text{higher-order terms}$$

and

$$\text{flow along } 3 = -v_x(x_0, y_0)\, dx - \frac{\partial v_x}{\partial y}(x_0, y_0)\, dx\, dy.$$

Adding the circulations along 1, 2, 3, and 4 and ignoring the higher-order terms, we have

$$\text{circulation}_{1234} = \left(\frac{\partial v_y}{\partial x} - \frac{\partial v_x}{\partial y}\right) dx\, dy,$$

or, dividing by $dx\, dy$,

$$\text{circulation per unit area at } (x_0, y_0) = \text{curl } \mathbf{v}(x_0, y_0).$$

In principle, curl $\mathbf{v}$ at (x_0, y_0) could be determined by inserting a paddle wheel into the moving fluid at the point (x_0, y_0). The rotation of the paddle wheel would be a measure of the curl.

■ PROBLEMS 15.7

■ SELF-QUIZ

I. Let $\mathbf{F} = x^2 \cos y\mathbf{i} + 2x \sin y\mathbf{j} + 0\mathbf{k}$, then div $\mathbf{F} = $ _____ .
 a. 0   b. $4x \cos y$
 c. $(2 - x^2) \sin y$ d. $(2 - x^2) \sin y\mathbf{k}$

II. Let $\mathbf{G} = 2x \sin y\mathbf{i} + x^2 \cos y\mathbf{j} + 0\mathbf{k}$, then curl $\mathbf{G} = $ _____ .
 a. $0\mathbf{k} = \mathbf{0}$ b. $4x \cos y$
 c. $(2 - x^2) \sin y\mathbf{k}$ d. $4x \cos y\mathbf{k}$

III. True–False: If the smooth vector field $\mathbf{F}(x, y, z)$ is exact, then curl $\mathbf{F} = \mathbf{0}$.

IV. True–False: It is possible to produce an example of a smooth vector field $\mathbf{G}(x, y, z)$ which is conservative, but for which curl $\mathbf{G} \neq \mathbf{0}$.

■ DRILL

In Problems 1–10, compute the divergence and curl of the given vector field.

1. $x^2\mathbf{i} + y^2\mathbf{j} + z^2\mathbf{k}$
2. $(\sin y)\mathbf{i} + (\sin z)\mathbf{j} + (\sin x)\mathbf{k}$
3. $a\mathbf{i} + b\mathbf{j} + c\mathbf{k}$; a, b, c are constants
4. $\sqrt{1 + x^2 + y^2}\mathbf{i} + \sqrt{1 + x^2 + y^2}\mathbf{j} + z^4\mathbf{k}$
5. $xy\mathbf{i} + yz\mathbf{j} + xz\mathbf{k}$
6. $(y^2 + z^2)\mathbf{i} + (x^2 + z^2)\mathbf{j} + (x^2 + y^2)\mathbf{k}$
7. $e^{yz}\mathbf{i} + e^{xz}\mathbf{j} + e^{xy}\mathbf{k}$
8. $e^{xy}\mathbf{i} + e^{yz}\mathbf{j} + e^{zx}\mathbf{k}$
9. $\dfrac{x}{y}\mathbf{i} + \dfrac{y}{z}\mathbf{j} + \dfrac{z}{x}\mathbf{k}$
10. $\sqrt{y + z}\mathbf{i} + \sqrt{x + z}\mathbf{j} + \sqrt{x + y}\mathbf{k}$

In Problems 11–18, use the results of Problems 45 and 46 to calculate

 a. curl $\mathbf{F}$ b. $\oint_{\partial\Omega} \mathbf{F} \cdot \mathbf{T}\, ds$
 c. div $\mathbf{F}$ d. $\oint_{\partial\Omega} \mathbf{F} \cdot \mathbf{n}\, ds$

11. $\mathbf{F}(x, y) = ax\mathbf{i} + by\mathbf{j}$; $\Omega = \{(x, y): 0 \leq x \leq 1, 0 \leq y \leq 1\}$
12. $\mathbf{F}(x, y) = Ay\mathbf{i} + Bx\mathbf{j}$; Ω is the region of Problem 11
13. $\mathbf{F}(x, y) = x^2\mathbf{i} + y^2\mathbf{j}$; Ω is the region of Problem 11
14. $\mathbf{F}(x, y) = y^2\mathbf{i} + x^2\mathbf{j}$; Ω is the region of Problem 11
15. $\mathbf{F}(x, y) = x\mathbf{i} + y\mathbf{j}$; Ω is the unit disk
16. $\mathbf{F}(x, y) = y\mathbf{i} - x\mathbf{j}$; Ω is the unit disk
17. $\mathbf{F}(x, y) = y^3\mathbf{i} + x^3\mathbf{j}$, Ω is the unit disk
18. $\mathbf{F}(x, y) = xy\mathbf{i} + (y^2 - x^2)\mathbf{j}$; Ω is the region enclosed by the triangle with vertices at $(0, 0)$, $(1, 0)$, and $(0, 1)$

■ APPLICATIONS

LAPLACIAN: The **Laplacian** of a twice-differentiable scalar function $f = f(x, y, z)$, denoted by $\nabla^2 f$, is defined by

$$\text{Laplacian of } f = \nabla^2 f = \frac{\partial^2 f}{\partial x^2} + \frac{\partial^2 f}{\partial y^2} + \frac{\partial^2 f}{\partial z^2}.$$

A function that satisfies **Laplace's equation** $\nabla^2 f = 0$ is said to be a **harmonic** function.

In Problems 19–22, compute $\nabla^2 f$ and identify which functions are harmonic.

19. $f(x, y, z) = xyz$
20. $f(x, y, z) = x^2 + y^2 + z^2$
21. $f(x, y, z) = 2x^2 + 5y^2 + 3z^2$
22. $f(x, y, z) = 1/\sqrt{x^2 + y^2 + z^2}$

23. It is true, although we will not attempt to prove it, that if div $\mathbf{F} = 0$, then there is some vector field $\mathbf{G}$ such that $\mathbf{F} = \text{curl } \mathbf{G}$. Let $\mathbf{F} = 2x\mathbf{i} + y\mathbf{j} - 3z\mathbf{k}$.
 a. Verify that div $\mathbf{F} = 0$.
 b. Find a vector field $\mathbf{G}$ such that $\mathbf{F} = \text{curl } \mathbf{G}$.
24. Let $\mathbf{F} = 4xyz\mathbf{i} + 2x^2z\mathbf{j} + 2x^2y\mathbf{k}$.
 a. Verify that curl $\mathbf{F} = \mathbf{0}$.
 b. Find a function f such that $\mathbf{F} = \nabla f$.
25. Let $\mathbf{F}(x, y, z) = x^2\mathbf{i} + y^2\mathbf{j} + z^2\mathbf{k}$; verify that div$[\text{curl } \mathbf{F}] = 0$.
26. Let $f(x, y, z) = 1/\sqrt{x^2 + y^2 + z^2}$; verify that curl$[\text{grad } f] = \mathbf{0}$.
27. Verify that the vector flow $\mathbf{F}(x, y) = x\mathbf{i} + y\mathbf{j}$ is irrotational.
28. Is the vector flow $\mathbf{G}(x, y, z) = e^x \sin x\mathbf{i} + y^{5/2}\mathbf{j} + \tan^{-1} z\mathbf{k}$ irrotational?
29. Verify that the vector flow $\mathbf{F}(x, y) = y\sqrt{x^2 + y^2}\mathbf{i} - x\sqrt{x^2 + y^2}\mathbf{j}$ is incompressible.

CONTINUED

30. Is the vector flow $\mathbf{G}(x, y, z) = yz^2\mathbf{i} + zx^2\mathbf{j} + xy^2\mathbf{k}$ incompressible?

31. The electrostatic field at $\mathbf{x}$ caused by a point charge q located at the origin is $\mathbf{E}(\mathbf{x}) = \dfrac{q}{4\pi\epsilon_0|\mathbf{x}|^3}\,\mathbf{x}$.

Calculate div $\mathbf{E}$. What happens as $\mathbf{x}$ approaches the origin?

32. Verify that the gravitational force (see example 15.1.3) $\mathbf{F} = \dfrac{-Gm_1m_2}{|\mathbf{x}|^3}\,\mathbf{x}$ is irrotational.

■ **SHOW/PROVE/DISPROVE**

In Problems 33–41, assume that all given functions are smooth.

33. Show that $\operatorname{div}(\mathbf{F} + \mathbf{G}) = \operatorname{div}\mathbf{F} + \operatorname{div}\mathbf{G}$.
34. Show that $\operatorname{curl}(\mathbf{F} + \mathbf{G}) = \operatorname{curl}\mathbf{F} + \operatorname{curl}\mathbf{G}$.
35. If $f = f(x, y, z)$ is a scalar function, show that $\operatorname{div}(f\mathbf{G}) = f(\operatorname{div}\mathbf{G}) + (\nabla f) \cdot \mathbf{G}$.
36. If f is a scalar function, show that $\operatorname{curl}(f\mathbf{G}) = f(\operatorname{curl}\mathbf{G}) + (\nabla f) \times \mathbf{G}$.
37. If f is a smooth scalar function, show that $\operatorname{curl}(\nabla f) = \mathbf{0}$.
38. Show that $\operatorname{div}[\operatorname{curl}\mathbf{F}] = 0$.
39. Show that $\operatorname{div}(\mathbf{F} \times \mathbf{G}) = (\operatorname{curl}\mathbf{F}) \cdot \mathbf{G} - \mathbf{F} \cdot (\operatorname{curl}\mathbf{G})$.
40. Show that $\nabla^2 f = \operatorname{div}(\operatorname{grad} f)$.
41. The **Laplacian** of a vector field $\mathbf{F} = P\mathbf{i} + Q\mathbf{j} + R\mathbf{k}$ is given by $\nabla^2\mathbf{F} = (\nabla^2 P)\mathbf{i} + (\nabla^2 Q)\mathbf{j} + (\nabla^2 R)\mathbf{k}$. Show that $\operatorname{curl}(\operatorname{curl}\mathbf{F}) = \nabla\operatorname{div}\mathbf{F} - \nabla^2\mathbf{F}$.
42. Suppose $f(x)$ and $g(y)$ are smooth scalar functions. Show that $\mathbf{F}(x, y) = f(x)\mathbf{i} + g(y)\mathbf{j}$ is irrotational.
43. Suppose g is a smooth scalar function. Show that $\mathbf{F}(x, y) = -yg(x^2 + y^2)\mathbf{i} + xg(x^2 + y^2)\mathbf{j}$ is incompressible.
44. Let f, g, and h be differentiable functions of two variables. Show that the vector field $\mathbf{F}(x, y, z) = f(y, z)\mathbf{i} + g(x, z)\mathbf{j} + h(x, y)\mathbf{k}$ is incompressible.
*45. Suppose $\mathbf{F}(x, y)$ is smooth on the region Ω and $\mathbf{T}(s)$ is the unit tangent vector to $\partial\Omega$. Show that

Green's theorem implies

$$\oint_{\partial\Omega} \mathbf{F} \cdot \mathbf{T}\,ds = \iint_{\Omega}([\operatorname{curl}\mathbf{F}] \cdot \mathbf{k})\,dx\,dy.$$

*46. Suppose $\mathbf{F}(x, y)$ is smooth on the region Ω and $\mathbf{n}$ is the unit normal vector to $\partial\Omega$. Show that Green's theorem implies

$$\oint_{\partial\Omega} \mathbf{F} \cdot \mathbf{n}\,ds = \iint_{\Omega}\operatorname{div}\mathbf{F}\,dx\,dy.$$

47. Let

$$\mathbf{F}(x, y) = \frac{y}{x^2 + y^2}\,\mathbf{i} - \frac{x}{x^2 + y^2}\,\mathbf{j}.$$

   a. Show that $\operatorname{curl}\mathbf{F} = 0$.
   b. Show that $\oint_C \mathbf{F} \cdot \mathbf{T}\,ds \neq \mathbf{0}$ if C is the unit circle oriented counterclockwise.
   c. Explain why the results of parts (a) and (b) do not contradict Problem 45.

48. The velocity of a two-dimensional flow of liquid is given by $\mathbf{F} = u(x, y)\mathbf{i} + v(x, y)\mathbf{j}$. If the liquid is incompressible and the flow is irrotational, show that

$$\frac{\partial u}{\partial y} = \frac{\partial v}{\partial x} \quad \text{and} \quad \frac{\partial u}{\partial x} = -\frac{\partial v}{\partial y}.$$

(These are known as the **Cauchy-Riemann** equations.)

■ **ANSWERS TO SELF-QUIZ**

I. b   II. a   III. True
IV. False (Such an example cannot exist.)

15.8 STOKES'S THEOREM

Let $S: z = f(x, y)$ be a smooth surface for $(x, y) \in \Omega$, where Ω is bounded. We assume that the boundary of S, denoted by ∂S, is a piecewise smooth, simple closed curve in $\mathbb{R}^3$. The positive direction on ∂S corresponds to the positive direction of $\partial\Omega$, where $\partial\Omega$ is the projection of ∂S into the xy-plane. This orientation is illustrated in Figure 1.

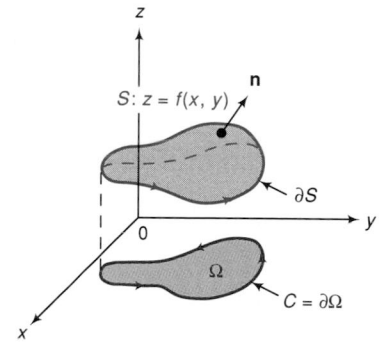

FIGURE 1
THE BOUNDARY OF Ω $(C = \partial\Omega)$ IS THE PROJECTION ONTO THE xy-PLANE OF THE BOUNDARY OF S.

We now state the second major result in vector calculus (the first was Green's theorem).

THEOREM 1   **STOKES'S THEOREM**[†]

Let $\mathbf{F}(x, y, z) = P(x, y, z)\mathbf{i} + Q(x, y, z)\mathbf{j} + R(x, y, z)\mathbf{k}$ be continuously differentiable on a bounded region W in space that contains the smooth surface S, and let ∂S be a piecewise smooth, simple closed curve traversed in the positive sense. Then

$$\oint_{\partial S} \mathbf{F} \cdot \mathbf{T}\, ds = \iint_S \text{curl }\mathbf{F} \cdot \mathbf{n}\, d\sigma \tag{1}$$

where $\mathbf{T}$ is the unit tangent vector to the curve ∂S at a point (x, y, z) and $\mathbf{n}$ is the outward unit normal vector to the surface S at the point (x, y, z). ∎

The proof of this theorem is difficult and is omitted.[‡]

REMARK 1:   If ∂S is given parametrically by $\mathbf{x}(t) = x(t)\mathbf{i} + y(t)\mathbf{j} + z(t)\mathbf{k}$, then by Theorem 12.5.3,

$$\mathbf{T} = \frac{d\mathbf{x}}{ds}, \qquad \text{or} \qquad d\mathbf{x} = \mathbf{T}\, ds \tag{2}$$

Thus Stokes's theorem can be written as

ALTERNATIVE FORM OF STOKES'S THEOREM

$$\oint_{\partial S} \mathbf{F} \cdot d\mathbf{x} = \iint_S \text{curl }\mathbf{F} \cdot \mathbf{n}\, d\sigma. \tag{3}$$

REMARK 2:   Using the notation given in equation (15.2.15) on page 973 we can also write Stokes's theorem as

---

SECOND ALTERNATIVE FORM OF STOKES'S THEOREM

$$\oint_{\partial S} P\, dx + Q\, dy + R\, dz = \iint_S \text{curl }\mathbf{F} \cdot \mathbf{n}\, d\sigma. \tag{4}$$

EXAMPLE 1
EVALUATING A LINE INTEGRAL
USING STOKES'S THEOREM

Use Stokes's theorem to evaluate $\oint_C \mathbf{F} \cdot d\mathbf{x}$, where $\mathbf{F}(x, y, z) = (z - 2y)\mathbf{i} + (3x - 4y)\mathbf{j} + (z + 3y)\mathbf{k}$ and C is the unit circle in the plane $z = 2$.

SOLUTION: C is given parametrically by $\mathbf{x}(t) = (\cos t)\mathbf{i} + (\sin t)\mathbf{j} + 2\mathbf{k}$. Clearly, C bounds the unit disk $x^2 + y^2 \le 1$, $z = 2$ in the plane $z = 2$.

[†] Named after the English mathematician and physicist Sir G. G. Stokes (1819–1903). Stokes is also known as one of the first to discuss the notion of uniform convergence (in 1848).

[‡] For proofs of Stokes's theorem and the divergence theorem, see *Multivariable Calculus, Linear Algebra, and Differential Equations*, 2nd ed. by S. I. Grossman (San Diego: Harcourt Brace Jovanovich, 1986), pp. 378, 381.

(It bounds many surfaces—for example, a hemisphere with this circle as its base. But the disk is the simplest one to use.) Then

$$\oint_C \mathbf{F} \cdot \mathbf{dx} = \iint_S \operatorname{curl} \mathbf{F} \cdot \mathbf{n} \, d\sigma.$$

We compute

$$\operatorname{curl} \mathbf{F} = \begin{vmatrix} \mathbf{i} & \mathbf{j} & \mathbf{k} \\ \dfrac{\partial}{\partial x} & \dfrac{\partial}{\partial y} & \dfrac{\partial}{\partial z} \\ z - 2y & 3x - 4y & z + 3y \end{vmatrix} = 3\mathbf{i} + \mathbf{j} + 5\mathbf{k},$$

and $\mathbf{n} = \mathbf{k}$ (since $\mathbf{k}$ is normal to any vector lying in a plane parallel to the xy-plane). Thus

$$\operatorname{curl} \mathbf{F} \cdot \mathbf{n} = 5,$$

so that

$$\oint_C \mathbf{F} \cdot \mathbf{dx} = 5 \iint_S d\sigma.$$

But $\iint_S d\sigma =$ the surface area of $S =$ the area of the unit disk $= \pi$. Thus,

$$\oint_C \mathbf{F} \cdot \mathbf{dx} = 5\pi.$$

EXAMPLE 2
EVALUATING A LINE INTEGRAL
AROUND A TRIANGLE USING
STOKES'S THEOREM

Compute $\oint_C \mathbf{F} \cdot \mathbf{dx}$, where $\mathbf{F}$ is as in Example 1 and C is the boundary of the triangle joining the points $(1, 0, 0)$, $(0, 1, 0)$, and $(0, 0, 1)$.

SOLUTION: We already have found that $\operatorname{curl} \mathbf{F} = 3\mathbf{i} + \mathbf{j} + 5\mathbf{k}$. The curve C lies in a plane. Two vectors on the plane are $\mathbf{i} - \mathbf{j}$ and $\mathbf{j} - \mathbf{k}$ (explain why). A normal vector to the plane is therefore given by

$$\mathbf{N} = (\mathbf{i} - \mathbf{j}) \times (\mathbf{j} - \mathbf{k}) = \begin{vmatrix} \mathbf{i} & \mathbf{j} & \mathbf{k} \\ 1 & -1 & 0 \\ 0 & 1 & -1 \end{vmatrix} = \mathbf{i} + \mathbf{j} + \mathbf{k},$$

so that $\mathbf{n} = (1/\sqrt{3})(\mathbf{i} + \mathbf{j} + \mathbf{k})$ is the required outward unit normal vector. Thus, $\operatorname{curl} \mathbf{F} \cdot \mathbf{n} = 9/\sqrt{3} = 3\sqrt{3}$, and we have

$$\oint_C \mathbf{F} \cdot \mathbf{dx} = 3\sqrt{3} \iint_S d\sigma = 3\sqrt{3} \times (\text{area of the triangle}).$$

The triangle is an equilateral triangle with the length, ℓ, of one side given by

$$\ell = \text{distance between } (1, 0, 0) \text{ and } (0, 1, 0) = \sqrt{2}$$

Then the area of the triangle is $\sqrt{3}/2$ (since the area of an equilateral triangle with side s is $(\sqrt{3}/4)s^2$). Thus,

$$\oint_C \mathbf{F} \cdot \mathbf{dx} = (3\sqrt{3})\left(\frac{\sqrt{3}}{2}\right) = \frac{9}{2}.$$

REMARK: In the preceding example, tangent vectors to the surface (a plane) are vectors lying on the plane. The vector $\mathbf{j} - \mathbf{k}$ is obtained from the vector $\mathbf{i} - \mathbf{j}$ by rotating $\mathbf{i} - \mathbf{j}$ 120° in the counterclockwise direction when viewed from the side of the plane that does not contain the origin. This procedure leads to $\mathbf{n} = (1/\sqrt{3})(\mathbf{i} + \mathbf{j} + \mathbf{k})$. If we chose the "other" counterclockwise direction, then we would obtain $\mathbf{n} = -(1/\sqrt{3})(\mathbf{i} + \mathbf{j} + \mathbf{k})$ and $\oint_C \mathbf{F} \cdot d\mathbf{x} = -\frac{9}{2}$. Which answer is correct? The first, because we choose $\mathbf{n}$ so that its $\mathbf{k}$-component is positive. If $\mathbf{n} = \dfrac{1}{\sqrt{3}}(\mathbf{i} + \mathbf{j} + \mathbf{k})$, then the $\mathbf{k}$-component is $\dfrac{1}{\sqrt{3}} > 0$. In the other case the $\mathbf{k}$-component is $-\dfrac{1}{\sqrt{3}} < 0$.

From Theorem 15.7.2, in a simply connected region, $\mathbf{F}$ is the gradient of a function f if and only if curl $\mathbf{F} = \mathbf{0}$. This result provides another proof of the fact that

$$\oint_C \mathbf{F} \cdot d\mathbf{x} = 0$$

if $\mathbf{F}$ is the gradient of a function f, since

$$\oint_C \mathbf{F} \cdot d\mathbf{x} = \iint_S \text{curl } \mathbf{F} \cdot \mathbf{n} \, d\sigma = 0,$$

where S is any smooth surface whose boundary is C.

We can combine several results to obtain the following theorem.

THEOREM 2   **FIVE EQUIVALENT CONDITIONS**

Let $\mathbf{F} = P(x, y, z)\mathbf{i} + Q(x, y, z)\mathbf{j} + R(x, y, z)\mathbf{k}$ be continuously differentiable on a simply connected region S in space. Then the following conditions are equivalent. That is, if one is true, all are true.

(i) $\mathbf{F}$ is the gradient of a differentiable function f.

(ii) $\dfrac{\partial P}{\partial y} = \dfrac{\partial Q}{\partial x}, \dfrac{\partial R}{\partial x} = \dfrac{\partial P}{\partial z}$, and $\dfrac{\partial Q}{\partial z} = \dfrac{\partial R}{\partial y}$.

(iii) $\int_C \mathbf{F} \cdot d\mathbf{x} = 0$ for every piecewise smooth, simple closed curve C lying in S.

(iv) $\int_C \mathbf{F} \cdot d\mathbf{x}$ is independent of path.

(v) Curl $\mathbf{F} = \mathbf{0}$. ∎

Many interesting physical results follow from Stokes's theorem, including the following.

AMPERE'S LAW

Suppose a steady current is flowing in a wire with an electric current density given by the vector field $\mathbf{i}$. It is well known in physics that such a current will set up a magnetic field, which is usually denoted by $\mathbf{B}$. In Figure 2 we see a collection of compass needles near a wire carrying (a) no current and

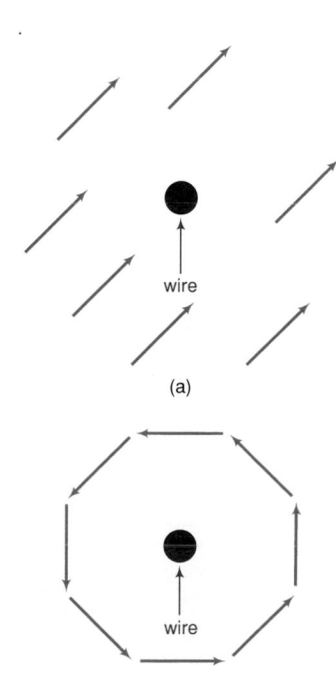

FIGURE 2
IN (a) THERE IS NO CURRENT; IN (b) THERE IS A STRONG CURRENT IN THE WIRE.

(b) a very strong current. A special case of one of the famous Maxwell equations[†] states that

$$\text{curl } \mathbf{B} = \mathbf{i}. \tag{5}$$

From this equation we can deduce an important physical law. Let S be a surface with smooth boundary ∂S. Then $\oint_{\partial S} \mathbf{B} \cdot \mathbf{T} \, ds$ is defined as the **circulation** of the magnetic field around ∂S. By Stokes's theorem

$$\oint_{\partial S} \mathbf{B} \cdot \mathbf{T} \, ds = \iint\limits_{S} \text{curl } \mathbf{B} \cdot \mathbf{n} \, d\sigma.$$

But by (5), $\iint_S \text{curl } \mathbf{B} \cdot \mathbf{n} \, d\sigma = \iint_S \mathbf{i} \cdot \mathbf{n} \, d\sigma$, so

$$\oint_{\partial S} \mathbf{B} \cdot \mathbf{T} \, ds = \iint\limits_{S} \mathbf{i} \cdot \mathbf{n} \, d\sigma. \tag{6}$$

In other words, (6) states that *the total current flowing through a surface S is equal to the circulation of the magnetic field induced by* $\mathbf{i}$ *around the boundary of S.* This important result is known as **Ampère's law.**[‡]

■ PROBLEMS 15.8

■ SELF-QUIZ

I. Let $\mathbf{F}(x, y, z) = x\mathbf{i} + x\mathbf{j} + x\mathbf{k}$ and let C be the square with vertices $(0, 0, 0), (1, 0, 0), (1, 1, 0)$, and $(0, 1, 0)$ traversed counterclockwise. Then $\oint_C \mathbf{F} \cdot \mathbf{dx} = $ _____.

  a. $\int_0^1 \int_0^1 -x \, dx \, dy$

  b. $\int_0^1 \int_0^1 x \, dx \, dy$

  c. $\int_0^1 \int_0^1 -1 \, dx \, dy$

  d. $\int_0^1 \int_0^1 1 \, dx \, dy$

II. Let $\mathbf{F}(x, y, z) = y^3\mathbf{i} + y^2\mathbf{j} + y\mathbf{k}$ and let C be the square with vertices $(0, 0, 0), (1, 0, 0), (1, 1, 0)$, and $(0, 1, 0)$ traversed counterclockwise. Then $\oint_C \mathbf{F} \cdot \mathbf{dx} = $ _____.

  a. $\int_0^1 \int_0^1 -3y^2 \, dx \, dy$

  b. $\int_0^1 \int_0^1 2y \, dx \, dy$

  c. $\int_0^1 \int_0^1 -1 \, dx \, dy$

  d. $\int_0^1 \int_0^1 0 \, dx \, dy$

III. Let $\mathbf{F}(x, y, z) = z^3\mathbf{i} + z^2\mathbf{j} + z\mathbf{k}$ and let C be the square with vertices $(0, 0, 5), (1, 0, 5), (1, 1, 5)$, and $(0, 1, 5)$ traversed counterclockwise. Then $\oint_C \mathbf{F} \cdot \mathbf{dx} = $ _____.

  a. $\int_0^1 \int_0^1 -3(5^2) \, dx \, dy$

  b. $\int_0^1 \int_0^1 (2z - 3z^2) \, dx \, dy$

  c. $\int_0^1 \int_0^1 -65 \, dx \, dy$

  d. $\int_0^1 \int_0^1 0 \, dx \, dy$

IV. Let $\mathbf{F}(x, y, z) = 3\mathbf{i} + 4\mathbf{j} + 5\mathbf{k}$ and let C be the square with vertices $(0, 0, 0), (0, 1, 0), (1, 1, 0)$, and $(1, 0, 0)$ traversed clockwise. Then $\oint_C \mathbf{F} \cdot \mathbf{dx} = $

_____.

  a. $\int_0^1 \int_0^1 3 \, dx \, dy$

  b. $\int_0^1 \int_0^1 0 \, dx \, dy$

  c. $\int_0^1 \int_0^1 -\sqrt{50} \, dx \, dy$

  d. $\int_0^1 \int_0^1 (-1/\sqrt{3}) \, dx \, dy$

[†] James Maxwell (1831–1879) was a British physicist. He formulated four equations, known as **Maxwell's equations**, which were supposed to explain all electromagnetic phenomena.

[‡] Named after André-Marie Ampère (1775–1836), a French physicist, mathematician, chemist, and philosopher. Ampère is credited with founding the science of electromagnetics (which he named electrodynamics). The unit of electric current is named after him.

■ **DRILL**

In Problems 1–8, evaluate the line integral by using Stokes's theorem. In each case S is the plane region enclosed by C.

1. $\oint_C \mathbf{F} \cdot d\mathbf{x}$, where $\mathbf{F}(x, y, z) = (x + y)\mathbf{i} + (z - 2x + y)\mathbf{j} + (y - z)\mathbf{k}$ and C is the unit circle in the plane $z = 5$.

2. $\oint_C \mathbf{F} \cdot d\mathbf{x}$, where $\mathbf{F} = a x\mathbf{i} + b y\mathbf{j} + c z\mathbf{k}$ and C is the unit circle in the plane $z = -7$.

3. $\oint_C \mathbf{F} \cdot d\mathbf{x}$, where $\mathbf{F}$ is as in Example 1 but C is the triangle joining the points $(2, 0, 0)$, $(0, 2, 0)$, and $(0, 0, 2)$.

4. $\oint_C \mathbf{F} \cdot d\mathbf{x}$, where $\mathbf{F}$ is as in Example 2 but C is the triangle with vertices $(d, 0, 0)$, $(0, d, 0)$, and $(0, 0, d)$.

*5. $\oint_C x^3 y^2\, dx + 2xyz^3\, dy + 3xy^2z^2\, dz$, where C is given parametrically by $\mathbf{x}(t) = 2\cos t\,\mathbf{i} + 3\mathbf{j} + 2\sin t\,\mathbf{k}$, $0 \le t \le 2\pi$. [*Hint:* C is a circle.]

6. $\oint_C \mathbf{F} \cdot d\mathbf{x}$, where $\mathbf{F} = 2y(x - z)\mathbf{i} + (x^2 + z^2)\mathbf{j} + y^3\mathbf{k}$ and C is the square $\{(x, y, z): 0 \le x \le 3, 0 \le y \le 3, z = 4\}$.

7. $\oint_C e^x\, dx + x\sin y\, dy + (y^2 - x^2)\, dz$, where C is the equilateral triangle formed by the intersection of the plane $x + y + z = 3$ with the three coordinate planes.

8. $\oint_C \mathbf{F} \cdot \mathbf{T}\, ds$, where $\mathbf{F} = -3y\mathbf{i} + 3x\mathbf{j} + \mathbf{k}$ and C is the circle $\{(x, y, z): x^2 + y^2 = 1, z = 3\}$.

In Problems 9–12, verify Stokes's theorem for the given S and $\mathbf{F}$.

9. $\mathbf{F} = z^2\mathbf{i} + x^2\mathbf{j} + y^2\mathbf{k}$; S is the part of the plane satisfying $x + y + z = 1$ which lies in the first octant (i.e., $x \ge 0$, $y \ge 0$, $z \ge 0$).

10. $\mathbf{F} = y^2\mathbf{i} + z^2\mathbf{j} + x^2\mathbf{k}$ and S is the part of the plane $x + 2y + 3z = 6$ lying in the first octant.

11. $\mathbf{F} = 2y\mathbf{i} + x^2\mathbf{j} + 3x\mathbf{k}$; S is the hemisphere $x^2 + y^2 + z^2 = 16$, $z > 0$.

12. $\mathbf{F} = y\mathbf{i} - x\mathbf{j} - z\mathbf{k}$; S is the circle $x^2 + y^2 \le 9$, $z = 2$.

■ **APPLICATIONS**

*13. Let $\mathbf{E}$ be an electric field and let $\mathbf{B}(t)$ be a magnetic field in space induced by $\mathbf{E}$. One of Maxwell's equations states that curl $\mathbf{E} = -\dfrac{\partial \mathbf{B}}{\partial t}$. Let S be a surface in space with smooth boundary C. We define

$$\text{voltage drop around } C = \oint_C \mathbf{E} \cdot d\mathbf{x}.$$

Prove **Faraday's law**, which states that the voltage drop around C is equal to the time rate of decrease of the magnetic flux through S. (Flux of a vector field over a surface is discussed preceding Problem 15.6.21.) [*Warning:* At one point in your work, it will be necessary to interchange the order of differentiation and integration. A proof that such manipulation is "legal" requires techniques from advanced calculus; for now, just assume that it can be done.]

■ **SHOW/PROVE/DISPROVE**

14. Show that Green's theorem is really a special case of Stokes's theorem.

15. Let S be a sphere. Use Stokes's theorem to show that $\iint_S [\text{curl } \mathbf{F}] \cdot \mathbf{n}\, d\sigma = 0$ if $\mathbf{F}$ is smooth over a region containing S.

■ *ANSWERS TO SELF-QUIZ*

I. d   II. a   III. d   IV. b

15.9 THE DIVERGENCE THEOREM

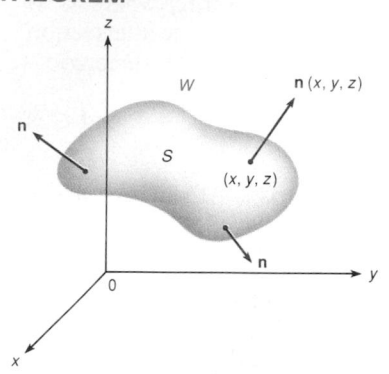

FIGURE 1
THE SURFACE *S* IS THE BOUNDARY OF THE SOLID *W*.

In this section we discuss the third major result in vector integral calculus: the **divergence theorem**.

Green's theorem gives us a relationship between a line integral in $\mathbb{R}^2$ around a closed curve and a double integral over the region in $\mathbb{R}^2$ enclosed by that curve. Stokes's theorem shows a relationship between a line integral in $\mathbb{R}^3$ around a closed curve and a surface integral over a surface that has the closed curve as a boundary. As we will see, the divergence theorem gives us a relationship between a surface integral over a closed surface and a triple integral over the solid bounded by that surface.

Let S be a surface that forms the complete boundary of a solid W in space. A typical region and its boundary are sketched in Figure 1. We will assume that S is smooth or piecewise smooth. This assumption ensures that $\mathbf{n}(x, y, z)$, the outward unit normal to S, is continuous or piecewise continuous as a function of (x, y, z).

THEOREM 1   THE DIVERGENCE THEOREM†

Let W be a solid in $\mathbb{R}^3$ totally bounded by the smooth or piecewise smooth surface S. Let $\mathbf{F}$ be a smooth vector field on W, and let $\mathbf{n}$ denote the outward unit normal to S. Then

$$\iint_S \mathbf{F} \cdot \mathbf{n} \, d\sigma = \iiint_W \operatorname{div} \mathbf{F} \, dv. \quad \blacksquare \tag{1}$$

REMARK: Just as Green's theorem transforms a line integral to an ordinary double integral, the divergence theorem transforms a surface integral to an ordinary triple integral.

The proof of the divergence theorem is omitted.

EXAMPLE 1
EVALUATING A SURFACE INTEGRAL USING THE DIVERGENCE THEOREM

Compute $\iint_S \mathbf{F} \cdot \mathbf{n} \, d\sigma$, where $\mathbf{F} = 2xy\mathbf{i} + 3y\mathbf{j} + 2z\mathbf{k}$ and S is the boundary of the solid bounded by the three coordinate planes and the plane $x + y + z = 1$.

SOLUTION: $\operatorname{div} \mathbf{F} = 2y + 3 + 2 = 2y + 5$, so

$$\iint_S \mathbf{F} \cdot \mathbf{n} \, d\sigma = \iiint_W (2y + 5) \, dx \, dy \, dz,$$

$$\int_0^1 \int_0^{1-y} \int_0^{1-x-y} (2y + 5) \, dz \, dx \, dy$$

$$= \int_0^1 \int_0^{1-y} (2y + 5)(1 - x - y) \, dx \, dy$$

$$= \int_0^1 \left[-(2y + 5) \frac{(1 - x - y)^2}{2} \Big|_{x=0}^{x=1-y} \right] dy$$

$$= \frac{1}{2} \int_0^1 (2y + 5)(1 - y)^2 \, dy = \frac{11}{12}.$$

---

† This theorem is also known as **Gauss's theorem**, named after the German mathematician Carl Friedrich Gauss (1777–1855). See the biographical sketch on the next page.

THE GREATEST MATHEMATICIAN of the nineteenth century, Carl Friedrich Gauss is considered one of the three greatest mathematicians of all time—the others being Archimedes and Newton.

Gauss was born in Brunswick, Germany, in 1777. His father, a hard-working laborer who was exceptionally stubborn and did not believe in formal education, did what he could to keep Gauss from appropriate schooling. Fortunately for Carl (and for mathematics), his mother, while uneducated herself, encouraged her son in his studies and took considerable pride in his achievements until her death at the age of 97.

Gauss was a child prodigy. At the age of three, he found an error in his father's bookkeeping. A famous story tells of Carl, age 10, as a student in the local Brunswick school. The teacher there was known to assign tasks to keep his pupils busy. One day he asked his students to add the numbers from 1 to 100. Almost at once, Carl placed his slate face down with the words, "There it is." Afterwards, the teacher found that Gauss was the only one with the correct answer, 5050. Gauss had noticed that the numbers could be arranged in 50 pairs, each with the sum $101(1 + 100, 2 + 99,$ and so on) and $50 \times 101 = 5050$. Later in life, Gauss joked that he could add before he could speak.

When Gauss was 15, the Duke of Brunswick noticed him and became his patron. The Duke helped him enter Brunswick College in 1795 and, three years later, to enter the university at Göttingen. Undecided between careers in mathematics and philosophy, Gauss chose mathematics after two remarkable discoveries. First, he invented the method of least squares a decade before the result was published by Legendre. Second, a month before his 19th birthday, he solved a problem whose solution had been sought for more than two thousand years. Gauss showed how to construct, using compass and ruler, a regular polygon with the number of sides not a multiple of 2, 3, or 5. On March 30, 1796, the day of this discovery, he began a diary, which contained as its first entry rules for construction of a 17-sided regular polygon. The diary, which contains 146 statements of results in only 19 pages, is one of the most important documents in the history of mathematics.

CARL FRIEDRICH GAUSS 1777–1855

After a short period at Göttingen, Gauss went to the University of Helmstädt and, in 1798 at the age of 20, wrote his now-famous doctoral dissertation. In it he gave the first mathematically rigorous proof of the fundamental theorem of algebra—that every polynomial of degree n, has, counting multiplicities, exactly n roots. Many mathematicians, including Euler, Newton, and Lagrange, had attempted to prove this result.

Gauss made a great number of discoveries in physics as well as in mathematics. For example, in 1801 he used a new procedure to calculate, from very little data, the orbit of the planetoid Ceres. In 1833, he invented the electromagnetic telegraph with his colleague Wilhelm Weber (1804–1891). However, while he did brilliant work in astronomy and electricity, it was Gauss's mathematical output that was astonishing. He made fundamental contributions to algebra and geometry.

In 1811, he discovered a result that led to the development of complex variable theory by Cauchy. He is encountered in courses in matrix theory in the Gauss-Jordan method of elimination. Students of numerical analysis study Gaussian quadrature—a technique for numerical integration.

Gauss became a professor of mathematics at Göttingen in 1807 and remained in that post until his death in 1855. Even after his death, his mathematical spirit remained to haunt nineteenth-century mathematicians. Often it turned out that an important new result was discovered earlier by Gauss and could be found in his published notes.

In his mathematical writings, Gauss was a perfectionist and is probably the last mathematician who knew everything in his subject. Claiming that a cathedral was not a cathedral until the last piece of scaffolding was removed, he endeavored to make each of his published works complete, concise, and polished. He used a seal that pictured a tree carrying only a few fruit together with the motto *pauca sed matura* (few, but ripe). But Gauss also believed that mathematics must reflect the real world. At his death, Gauss was honored by a commemorative medal on which was inscribed "George V. King of Hanover to the Prince of Mathematicians."

EXAMPLE 2
EVALUATING A SURFACE INTEGRAL USING THE DIVERGENCE THEOREM

Compute $\iint_S \mathbf{F} \cdot \mathbf{n} \, d\sigma$, where $\mathbf{F} = (5x + \sin y \tan z)\mathbf{i} + (y^2 - e^{x - 2\cos z^3})\mathbf{j} + (4xy)^{3/5}\mathbf{k}$ and S is the boundary of the solid bounded by the parabolic cylinder $y = 9 - x^2$, the plane $y + z = 1$, and the xy- and xz-planes.

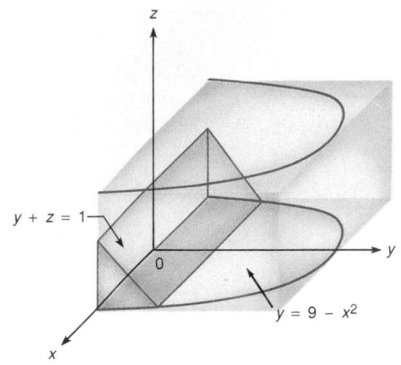

FIGURE 2
THE SOLID BOUNDED BY $y = 9 - x^2$
$y + z = 1$, AND THE xy- AND xz-PLANES

SOLUTION: The solid is sketched in Figure 2. It would be difficult to compute this surface integral directly, but with the divergence theorem it becomes relatively easy. Since div $\mathbf{F} = 5 + 2y$, we have

$$
\iint_S \mathbf{F} \cdot \mathbf{n} \, d\sigma = \iiint_W (5 + 2y) \, dx \, dy \, dz
$$

$$
= \int_0^1 \int_{-\sqrt{9-y}}^{\sqrt{9-y}} \int_0^{1-y} (5 + 2y) \, dz \, dx \, dy
$$

$$
= 2 \int_0^1 \int_0^{\sqrt{9-y}} (5 + 2y)(1 - y) \, dx \, dy
$$

$$
= 2 \int_0^1 \int_0^{\sqrt{9-y}} (5 - 3y - 2y^2) \, dx \, dy
$$

$$
= 2 \int_0^1 (5 - 3y - 2y^2) \sqrt{9 - y} \, dy
$$

$$
= \frac{56}{15} 8^{5/2} + \frac{64}{105} 8^{7/2} + 180 - \frac{180{,}792}{105} \approx 16.66.
$$

■ PROBLEMS 15.9

■ SELF-QUIZ

I. Let $\mathbf{F}(x, y, z) = x\mathbf{i} + y\mathbf{j} + z\mathbf{k}$ and let S be the surface of the rectangular solid with $(0, 0, 0)$ and $(1, 2, 5)$ at the ends of an interior diagonal. Then $\iint_S \mathbf{F} \cdot \mathbf{n} \, d\sigma = \underline{\qquad}$.

a. $\oint_C \mathbf{F} \cdot d\mathbf{x}$ where C is the rectangle with vertices $(0, 0, 0)$, $(1, 0, 5)$, $(1, 2, 5)$, $(0, 2, 5)$

b. $\int_0^1 \int_0^2 \int_0^5 [x + y + z] \, dx \, dy \, dz$

c. $\int_0^1 \int_0^1 \int_0^1 \sqrt{3} \, dx \, dy \, dz$

d. $\int_0^1 \int_0^2 \int_0^5 3 \, dx \, dy \, dz$

e. $\int_0^1 \int_0^2 \int_0^5 0 \, dx \, dy \, dz$

II. Let $\mathbf{F} = \mathbf{V}(xyz)$ and let S be the same surface as in the preceding problem. Then $\iint_S \mathbf{F} \cdot \mathbf{n} \, d\sigma = \underline{\qquad}$.

a. $\int_0^1 \int_0^2 \int_0^5 0 \, dx \, dy \, dz$

b. $\int_0^1 \int_0^2 \int_0^5 (yz + xz + xy) \, dx \, dy \, dz$

c. $\int_0^1 \int_0^2 \int_0^5 (3xyz) \, dx \, dy \, dz$

d. $\int_0^1 \int_0^2 \int_0^5 3 \, dx \, dy \, dz$

III. True–False: If $\mathbf{F}$ is a conservative vector field throughout a region containing the surface S, then $\iint_S \mathbf{F} \cdot \mathbf{n} \, d\sigma = 0$

■ DRILL

In Problems 1–15, evaluate the given surface integral by using the divergence theorem.

1. $\iint_S \mathbf{F} \cdot \mathbf{n} \, d\sigma$, where $\mathbf{F}(x, y, z) = x\mathbf{i} + y\mathbf{j} + z\mathbf{k}$ and S is the unit sphere.

2. $\iint_S \mathbf{F} \cdot \mathbf{n} \, d\sigma$, where $\mathbf{F} = x^2\mathbf{i} + y^2\mathbf{j} + z^2\mathbf{k}$ and S is the unit sphere.

3. $\iint_S \mathbf{F} \cdot \mathbf{n} \, d\sigma$, where $\mathbf{F} = x\mathbf{i} + y\mathbf{j} + z\mathbf{k}$ and S is the cylinder $x^2 + y^2 = 4, 0 \le z \le 3$.

4. $\iint_S \mathbf{F} \cdot \mathbf{n} \, d\sigma$, where $\mathbf{F} = y\mathbf{i} + z\mathbf{j} + x\mathbf{k}$ and S is as in the preceding problem.

5. $\iint_S \mathbf{F} \cdot \mathbf{n} \, d\sigma$, where $\mathbf{F} = (y^2 + z^2)^{3/2}\mathbf{i} + \sin[(x^2 - z^5)^{4/3}]\mathbf{j} + e^{x^2 - y^2}\mathbf{k}$ and S is the ellipsoid $(x/a)^2 + (y/b)^2 + (z/c)^2 = 1$.

6. $\iint_S \mathbf{F} \cdot \mathbf{n} \, d\sigma$, where $\mathbf{F} = x\mathbf{i} + y\mathbf{j} + z\mathbf{k}$ and S is the

surface of the unit cube $\{(x, y, z): 0 \le x \le 1, 0 \le y \le 1, 0 \le z \le 1\}$.

7. $\iint_S \mathbf{F} \cdot \mathbf{n} \, d\sigma$, where $\mathbf{F} = x^2\mathbf{i} + y^2\mathbf{j} - xy\mathbf{k}$ and S is as in Problem 6.

8. $\iint_S \mathbf{F} \cdot \mathbf{n} \, d\sigma$, where $\mathbf{F} = xyz\mathbf{i} + yz\mathbf{j} + z\mathbf{k}$ and S is as in Problem 6.

9. $\iint_S \mathbf{F} \cdot \mathbf{n} \, d\sigma$, where $\mathbf{F} = 2x\mathbf{i} + 3y\mathbf{j} + z\mathbf{k}$ and S is the boundary of the hemisphere $\{(x, y, z): x^2 + y^2 + z^2 \le 9, z \ge 0\}$.

10. $\iint_S \mathbf{F} \cdot \mathbf{n} \, d\sigma$, where $\mathbf{F} = x^2\mathbf{i} + y^2\mathbf{j} + z^2\mathbf{k}$ and S is as in the preceding problem.

11. $\iint_S \mathbf{F} \cdot \mathbf{n} \, d\sigma$, where $\mathbf{F} = xy\mathbf{i} + y^2\mathbf{j} + yz\mathbf{k}$ and S is the boundary of the tetrahedron with vertices at $(0, 0, 0)$, $(1, 0, 0)$, $(0, 1, 0)$, and $(0, 0, 1)$.

CONTINUED

12. $\iint_S \mathbf{F} \cdot \mathbf{n} \, d\sigma$, where $\mathbf{F} = y^2\mathbf{i} + x^2\mathbf{j} + z^2\mathbf{k}$ and S is as in the preceding problem.

13. $\iint_S \mathbf{F} \cdot \mathbf{n} \, d\sigma$, where $\mathbf{F} = x(1 - \sin y)\mathbf{i} + (y - \cos y)\mathbf{j} + z\mathbf{k}$ and S is as in Problem 11.

14. $\iint_S \mathbf{F} \cdot \mathbf{n} \, d\sigma$, where $\mathbf{F} = x\mathbf{i} + y\mathbf{j} + z\mathbf{k}$ and S is the surface of the region bounded by the parabolic

cylinder $z = 1 - y^2$, the plane $x + z = 2$, and the xy- and yz-planes.

15. $\iint_S \mathbf{F} \cdot \mathbf{n} \, d\sigma$, where $\mathbf{F} = (x^2 + e^{y \cos z})\mathbf{i} + (xy - \tan z^{1/3})\mathbf{j} + (x - y^{3/5})^{2/9}\mathbf{k}$ and S is as in the preceding problem.

■ **APPLICATIONS**

*16. One of Maxwell's equations states that an electric field $\mathbf{E}$ in space satisfies div $\mathbf{E} = (1/\epsilon_0)\rho$, where ϵ_0 is a constant and ρ is the charge density. Show that the flux of the displacement $\mathbf{D} = \epsilon_0\mathbf{E}$ across

a closed surface is equal to the charge q inside the surface. (This last result is known as **Gauss's law**.)

■ **SHOW/PROVE/DISPROVE**

17. Show that if $\mathbf{F}$ is twice continuously differentiable, then $\iint_S [\text{curl } \mathbf{F}] \cdot \mathbf{n} \, d\sigma = 0$ for any smooth closed surface S.

18. Show that if $\mathbf{F}$ is a constant vector field and if S is a smooth closed surface, then $\iint_S \mathbf{F} \cdot \mathbf{n} \, d\sigma = 0$.

19. Let $\mathbf{F} = x\mathbf{i} + y\mathbf{j} + z\mathbf{k}$ and suppose W is a solid with a smooth closed boundary S. Show that

volume of $W = \dfrac{1}{3} \iint_S \mathbf{F} \cdot \mathbf{n} \, d\sigma.$

*20. Suppose that a vector field $\mathbf{F}$ is tangent to a closed smooth surface S, where S is the boundary of a solid W. Show that $\iiint_W [\text{div } \mathbf{F}] \, dx \, dy \, dz = 0$.

■ **ANSWERS TO SELF-QUIZ**

I. d   II. a

III. False (e.g., let $\mathbf{F} = \mathbf{V}[x^2] = 2x\mathbf{i}$.)

15.10 CHANGING VARIABLES IN MULTIPLE INTEGRALS AND THE JACOBIAN

In many places in this and the last chapter we found it useful to evaluate a double integral by first converting to polar coordinates. In Section 14.4 we proved that [see equation (6) on page 946]

$$\iint_\Omega f(x, y) \, dx \, dy = \iint_\Omega f(\cos \theta, r \sin \theta) r \, dr \, d\theta. \tag{1}$$

In Section 14.6 we showed how to convert an ordinary triple integral from an integral in Cartesian coordinates to an integral in cylindrical or spherical coordinates.

In this section we show how, under certain conditions, it is possible to convert from one set of coordinates to another in double and triple integrals. Much of what we do is a generalization of the formula for changing variables in ordinary definite integrals. Let us recall that formula now. If $x = g(u)$ is a differentiable one-to-one function, then [see equation (7.4.1) on page 494]

$$\int_a^b f(x) \, dx = \int_c^d f(g(u))g'(u) \, du, \tag{2}$$

where $c = g^{-1}(a)$ and $d = g^{-1}(b)$.

Now suppose we wish to change the variables of integration in the double integral

$$\iint_\Omega f(x, y)\, dx\, dy. \tag{3}$$

We assume that the new variables are called u and v and that they are related to the old variables x and y by the relations

$$x = g(u, v) \qquad \text{and} \qquad y = h(u, v). \tag{4}$$

The functional relationship described by (4) is called a **mapping** from the uv-plane into the xy-plane. We will assume that there is a region Σ in the uv-plane that gets mapped onto Ω by the mapping described by (4) and that the mapping is one-to-one. That is:

(i) For every $(x, y) \in \Omega$, there is a $(u, v) \in \Sigma$ such that $x = g(u, v)$ and $y = h(u, v)$.
(ii) If $g(u_1, v_1) = g(u_2, v_2)$ and $h(u_1, v_1) = h(u_2, v_2)$, then $u_1 = u_2$ and $v_1 = v_2$.

Condition (ii) is the natural extension of the definition of one-to-one for functions of two variables. We illustrate what is going on in Figure 1.

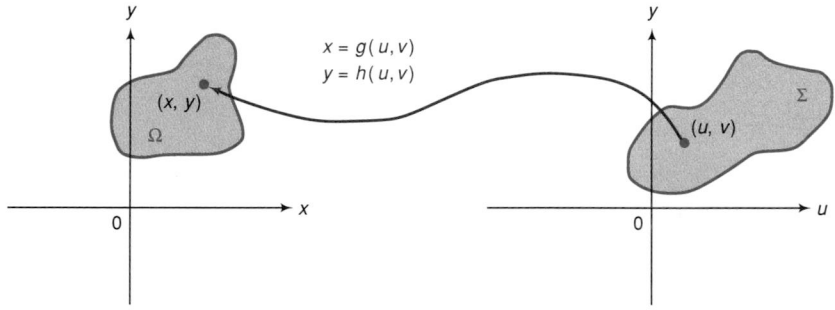

FIGURE 1   THE MAPPING TAKES A POINT (u, v) IN Σ (IN THE uv-PLANE) INTO A POINT (x, y) IN Ω (IN THE xy-PLANE).

EXAMPLE 1
FINDING THE IMAGES OF VERTICAL LINES UNDER A MAPPING

Let $x = u^2 - v^2$ and $y = 3uv$. We will compute the image of the vertical line $u = k$ in the uv-plane under the mapping. Substituting k for u in the equations given above, we have

$$x = k^2 - v^2 \qquad \text{and} \qquad y = 3kv.$$

Thus $v = y/3k$ and

$$x = k^2 - \frac{y^2}{9k^2}. \tag{5}$$

For every $k \neq 0$, equation (5) is the equation of a parabola in the xy-plane. Moreover, if $v = c$, a constant, then $x = u^2 - c^2$, $y = 3uc$, $u = y/3c$, and

$$x = \frac{y^2}{9c^2} - c^2, \tag{6}$$

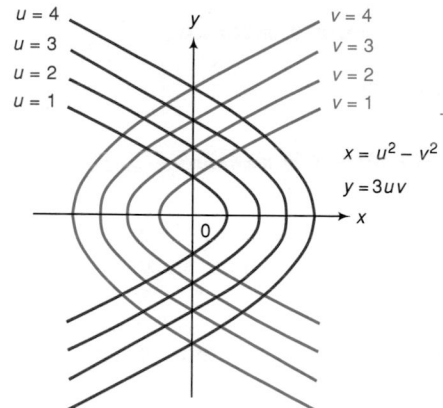

FIGURE 2
THE IMAGE OF $v = $ CONSTANT IS A PARA-
BOLA OPENING TO THE RIGHT IN THE
xy-PLANE. THE IMAGE OF $u = $ CONSTANT
IS A PARABOLA OPENING TO THE LEFT.

which is also a parabola for $c \neq 0$. Thus the functions given above map straight lines into parabolas. Some of these curves are sketched in Figure 2.

We now give an important definition.

DEFINITION   JACOBIAN IN $\mathbb{R}^2$

Let $x = g(u, v)$ and $y = h(u, v)$ be differentiable. Then the **Jacobian**[†] of x and y with respect to u and v, denoted by $\partial(x, y)/\partial(u, v)$, is given by

$$\frac{\partial(x, y)}{\partial(u, v)} = \frac{\partial x}{\partial u}\frac{\partial y}{\partial v} - \frac{\partial x}{\partial v}\frac{\partial y}{\partial u} = \begin{vmatrix} \dfrac{\partial x}{\partial u} & \dfrac{\partial x}{\partial v} \\ \dfrac{\partial y}{\partial u} & \dfrac{\partial y}{\partial v} \end{vmatrix}. \quad \blacksquare \qquad (7)$$

EXAMPLE 2
COMPUTING A JACOBIAN

If $x = u^2 - v^2$ and $y = 3uv$ as in Example 1, then $\partial x/\partial u = 2u$, $\partial x/\partial v = -2v$, $\partial y/\partial u = 3v$, and $\partial y/\partial v = 3u$, so

$$\begin{vmatrix} \dfrac{\partial x}{\partial u} & \dfrac{\partial x}{\partial v} \\ \dfrac{\partial y}{\partial u} & \dfrac{\partial y}{\partial v} \end{vmatrix} = \begin{vmatrix} 2u & -2v \\ 3v & 3u \end{vmatrix} = 6(u^2 + v^2).$$

EXAMPLE 3
COMPUTING THE JACOBIAN OF
THE POLAR COORDINATE
MAPPING

Let $x = r \cos \theta$ and $y = r \sin \theta$. Then

$$\frac{\partial x}{\partial r} = \cos \theta, \qquad \frac{\partial x}{\partial \theta} = -r \sin \theta, \qquad \frac{\partial y}{\partial r} = \sin \theta, \qquad \frac{\partial y}{\partial \theta} = r \cos \theta,$$

so

$$\frac{\partial(x, y)}{\partial(r, \theta)} = \begin{vmatrix} \cos \theta & -r \sin \theta \\ \sin \theta & r \cos \theta \end{vmatrix}$$
$$= r \cos^2 \theta + r \sin^2 \theta = r(\sin^2 \theta + \cos^2 \theta) = r.$$

To obtain our main result we need assumptions (i) and (ii) cited earlier. We need also to assume the following:

(iii) $C_1 = \partial\Omega$ and $C_2 = \partial\Sigma$ are simple closed curves, and as (u, v) moves once about C_2 in the positive direction, $(x, y) = (g(u, v), h(u, v))$ moves once around C_1 in the positive or negative direction.

(iv) All second-order partial derivatives are continuous.

---

† Named after the German mathematician Carl Gustav Jacob Jacobi (1804–1851). See the biographical sketch on page 1029.

THEOREM 1   **CHANGE OF VARIABLES IN A DOUBLE INTEGRAL**

If assumptions (i), (ii), (iii), and (iv) hold, then

$$\iint_{\Omega} f(x, y)\, dx\, dy = \pm \iint_{\Sigma} f[g(u, v), h(u, v)] \frac{\partial(x, y)}{\partial(u, v)}\, du\, dv. \tag{8}$$

The plus (minus) sign is taken if as (u, v) moves around C_2 in the positive direction $(x, y) = (g(u, v), h(u, v))$ moves around C_1 in the positive (negative) direction. ■

The proof, using Green's theorem, is not difficult but it is very long. For that reason, it is omitted.[†]

REMARK: Conditions (i) and (ii) given on page 1024 are often difficult to verify. It can be shown that these conditions hold if all partial derivatives are continuous in Ω and the Jacobian $\partial(x, y)/\partial(u, v)$ is not zero on Ω.

EXAMPLE 4
CHANGING TO POLAR
COORDINATES IN A DOUBLE
INTEGRAL

We can use Theorem 1 to obtain the polar-coordinate formula (1) very easily. For if $x = r \cos \theta$ and $y = r \sin \theta$, then $\partial(x, y)/\partial(r, \theta) = r$ by Example 3, and (1) follows immediately from (8).

EXAMPLE 5
CHANGING VARIABLES IN A
DOUBLE INTEGRAL

Let Ω be the region in the upper half of the xy-plane bounded by the parabolas $y^2 = 9 - 9x$ and $y^2 = 9 + 9x$ and by the x-axis. Compute $\iint_{\Omega} (x + y)\, dx\, dy$ by making the change of variables $x = u^2 - v^2, y = 3uv$.

SOLUTION: We saw in Example 1 that this mapping takes straight lines in the uv-plane into parabolas in the xy-plane. For example, if $y^2 = 9 - 9x$, then $y^2 = 9u^2v^2 = 9 - 9x$, or $u^2v^2 = 1 - x = 1 - u^2 + v^2$, or $u^2 + u^2v^2 = 1 + v^2 = u^2(1 + v^2)$, and $u^2 = 1$, so $u = \pm 1$. Similarly, if $y^2 = 9 + 9x$, then $v = \pm 1$. Since, in Ω, $y \geq 0$, we must have $uv \geq 0$, so that u and v have the same sign. We will choose $u = v = 1$ for reasons to be made clear shortly. Note also that if $u = 0$, then $y = 0$ and $x = -v^2 \leq 0$, so that the positive v-axis in the uv-plane is mapped into the negative x-axis in the xy-plane. Similarly, the positive u-axis in the uv-plane is mapped into the positive x-axis in the xy-plane. The situation is sketched in Figure 3. The reason we chose $u = v = 1$ is that moving around Σ in the positive direction corresponds to moving around Ω in the positive direction. [Try it: Take the path $v = 0$ (the u-axis) to $u = 1$ to $v = 1$ to $u = 0$.] Thus the integral around the "parabolic" region given in the problem can be reduced to an integral around a square. Also, as we computed in Example 2,

$$\frac{\partial(x, y)}{\partial(u, v)} = 6(u^2 + v^2).$$

[†] For a proof see Stanley I. Grossman, *Multivariable Calculus, Linear Algebra, and Differential Equations,* Second Edition (San Diego: Harcourt Brace Jovanovich, 1986), pp. 341–2.

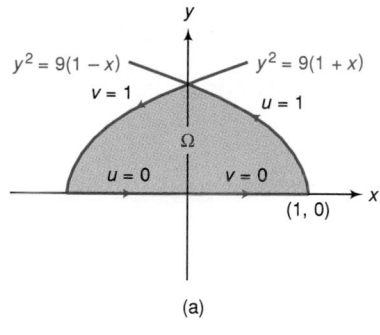

(a)

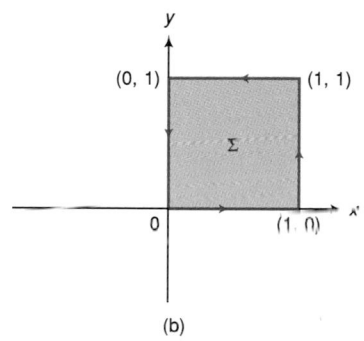

(b)

FIGURE 3
CHANGING VARIABLES IN A DOUBLE INTEGRAL

Under the mapping $x = u^2 - v^2$, $y = 3uv$, *the region bounded by two parabolas and the x-axis in (a) is mapped into the unit square in (b).*

Finally, $x + y = u^2 + 3uv - v^2$, so

$$\iint_\Omega (x + y)\, dx\, dy = 6 \int_0^1 \int_0^1 (u^2 + 3uv - v^2)(u^2 + v^2)\, du\, dv$$

$$= 6 \int_0^1 \int_0^1 (u^4 + 3u^3 v + 3uv^3 - v^4)\, du\, dv = \frac{9}{2}.$$

For triple integrals there is a result analogous to the one we have proven for double integrals. We will state this result without proof.
Let

$$x = g(u, v, w), \qquad y = h(u, v, w), \qquad z = j(u, v, w).$$

DEFINITION **JACOBIAN IN** $\mathbb{R}^3$

We define the **Jacobian** of the transformation from a region U in a *uvw*-space to a region W in *xyz*-space as

$$\text{Jacobian} = \frac{\partial(x, y, z)}{\partial(u, v, w)} = \begin{vmatrix} \dfrac{\partial x}{\partial u} & \dfrac{\partial x}{\partial v} & \dfrac{\partial x}{\partial w} \\ \dfrac{\partial y}{\partial u} & \dfrac{\partial y}{\partial v} & \dfrac{\partial y}{\partial w} \\ \dfrac{\partial z}{\partial u} & \dfrac{\partial z}{\partial v} & \dfrac{\partial z}{\partial w} \end{vmatrix}. \ \blacksquare$$

Then under hypotheses similar to the ones made in Theorem 1, we have the following theorem.

THEOREM 2 **CHANGE OF VARIABLES IN A TRIPLE INTEGRAL**

$$\iiint_W F(x, y, z)\, dx\, dy\, dz$$

$$= \pm \iiint_U F(g(u, v, w), h(u, v, w), j(u, v, w)) \frac{\partial(x, y, z)}{\partial(u, v, w)}\, du\, dv\, dw. \ \blacksquare$$

EXAMPLE 6
CHANGING TO CYLINDRICAL COORDINATES IN A TRIPLE INTEGRAL

Let $x = r \cos\theta$, $y = r \sin\theta$, and $z = z$. These are cylindrical coordinates. Then

$$\frac{\partial(x, y, z)}{\partial(r, \theta, z)} = \begin{vmatrix} \dfrac{\partial x}{\partial r} & \dfrac{\partial x}{\partial \theta} & \dfrac{\partial x}{\partial z} \\ \dfrac{\partial y}{\partial r} & \dfrac{\partial y}{\partial \theta} & \dfrac{\partial y}{\partial z} \\ \dfrac{\partial z}{\partial r} & \dfrac{\partial z}{\partial \theta} & \dfrac{\partial z}{\partial z} \end{vmatrix} = \begin{vmatrix} \cos\theta & -r\sin\theta & 0 \\ \sin\theta & r\cos\theta & 0 \\ 0 & 0 & 1 \end{vmatrix}$$

$$= r(\cos^2\theta + \sin^2\theta) = r,$$

so

$$\iiint\limits_{W} f(x, y, z)\, dx\, dy\, dz = \iiint\limits_{U} f(r \cos \theta, r \sin \theta, z) r\, dz\, dr\, d\theta.$$

This is equation (14.6.5) on page 958.

EXAMPLE 7
CHANGING TO SPHERICAL
COORDINATES IN A TRIPLE
INTEGRAL

Let $x = \rho \sin \varphi \cos \theta$, $y = \rho \sin \varphi \sin \theta$, and $z = \rho \cos \varphi$. These are spherical coordinates (see Section 11.9). Then

$$\frac{\partial(x, y, z)}{\partial(\rho, \varphi, \theta)} = \begin{vmatrix} \dfrac{\partial x}{\partial \rho} & \dfrac{\partial x}{\partial \varphi} & \dfrac{\partial x}{\partial \theta} \\[2mm] \dfrac{\partial y}{\partial \rho} & \dfrac{\partial y}{\partial \varphi} & \dfrac{\partial y}{\partial \theta} \\[2mm] \dfrac{\partial z}{\partial \rho} & \dfrac{\partial z}{\partial \varphi} & \dfrac{\partial z}{\partial \theta} \end{vmatrix}$$

$$= \begin{vmatrix} \sin \varphi \cos \theta & \rho \cos \varphi \cos \theta & -\rho \sin \varphi \sin \theta \\ \sin \varphi \sin \theta & \rho \cos \varphi \sin \theta & \rho \sin \varphi \cos \theta \\ \cos \varphi & -\rho \sin \varphi & 0 \end{vmatrix}$$

Expanding in the last row

$$= \cos \varphi \begin{vmatrix} \rho \cos \varphi \cos \theta & -\rho \sin \varphi \sin \theta \\ \rho \cos \varphi \sin \theta & \rho \sin \varphi \cos \theta \end{vmatrix}$$

$$+ \rho \sin \theta \begin{vmatrix} \sin \varphi \cos \theta & -\rho \sin \varphi \sin \theta \\ \sin \varphi \sin \theta & \rho \sin \varphi \cos \theta \end{vmatrix}$$

$$= \cos \varphi (\rho^2 \cos \varphi \sin \varphi \cos^2 \theta + \rho^2 \sin \varphi \cos \varphi \sin^2 \theta)$$
$$+ \rho \sin \varphi (\rho \sin^2 \varphi \cos^2 \theta + \rho \sin^2 \varphi \sin^2 \theta)$$
$$= \cos \varphi (\rho^2 \sin \varphi \cos \varphi) + \rho \sin \varphi (\rho \sin^2 \varphi)$$
$$= \rho^2 \sin \varphi (\cos^2 \varphi + \sin^2 \varphi) = \rho^2 \sin \varphi.$$

Thus

$$\iiint\limits_{W} f(x, y, z)\, dx\, dy\, dz$$

$$= \iiint\limits_{U} f(\rho \sin \varphi \cos \theta, \rho \sin \varphi \sin \theta, \rho \cos \theta) \rho^2 \sin \varphi\, d\rho\, d\varphi\, d\theta.$$

This is formula (14.6.12) on page 959.

THE SON OF A PROSPEROUS BANKER, Carl Gustav Jacob Jacobi was born in Potsdam, Germany, in 1804. He was educated at the University of Berlin, where he received his doctorate in 1825. In 1827, he was appointed Extraordinary Professor of Mathematics at the University of Königsberg. Jacobi taught at Königsberg until 1842, when he returned to Berlin under a pension from the Prussian government. He remained in Berlin until his death in 1851.

A prolific writer of mathematical treatises, Jacobi was best known in his time for his results in the theory of elliptic functions. Today, however, he is most remembered for his work on determinants. He was one of the two most creative developers of determinant theory, the other being Cauchy. In 1829, Jacobi published a paper on algebra that contained the notation for the Jacobian that we use today. In 1841 he published an extensive treatise titled *De determinantibus functionalibus*, which was devoted to results about the Jacobian. Jacobi showed the relationship between the Jacobian of functions of several variables and the derivative of a function of one variable. He also showed that n functions of n variables are linearly independent if and only if their Jacobian is not identically zero.

In addition to being a fine mathematician, Jacobi was considered the greatest teacher of mathematics of his generation. He inspired and influenced an astonishing number of students. To dissuade his students from mastering great amounts of mathematics before setting off to do their own research, Jacobi often remarked, "Your father would never have married, and you would not be born, if he had insisted on knowing all the girls in the world before marrying one."

Jacobi believed strongly in research in pure mathematics and frequently defended it against the claim that research should always be applicable to something. He once said, "The real end of science is the honor of the human mind."

CARL GUSTAV JACOB JACOBI

1804–1851

PROBLEMS 15.10

SELF-QUIZ

I. The Jacobian of the mapping $x = u + v$, $y = u - v$ is _____.
   a. 0   b. 1   c. 2
   d. -1 e. -2 f. $u^2 - v^2$

II. The Jacobian of the mapping $x = w$, $y = v$, $x = u$ is _____.
   a. 1   b. -1 c. uvw
   d. 0   e. 3   f. -3

DRILL

In Problems 1–20, compute the Jacobian of the given transformation.

1. $x = u + v$, $y = u - v$
2. $x = u^2 - v^2$, $y = u^2 + v^2$
3. $x = u^2 - v^2$, $y = 2uv$
4. $x = \sin u$, $y = \cos v$
5. $x = u + 3v - 1$, $y = 2u + 4v + 6$
6. $x = v - 2u$, $y = u + 2v$
7. $x = au + bv$, $y = bu - av$
8. $x = e^v$, $y = e^u$
9. $x = ue^v$, $y = ve^u$
*10. $x = u^v$, $y = v^u$

11. $x = \ln(u + v)$, $y = \ln uv$
12. $x = \tan u$, $y = \sec v$
13. $x = u \sec v$, $y = v \csc u$
14. $x = u \ln v$, $y = v \ln u$
15. $x = u + v + w$, $y = u - v - w$, $z = -u + v + w$
16. $x = au + bv + cw$, $y = au - bv - cw$, $z = -au + bv + cw$
17. $x = u^2 + v^2 + w^2$, $y = u + v + w$, $z = uvw$
18. $x = u \sin v$, $y = v \cos w$, $z = w \sin u$
19. $x = e^u$, $y = e^v$, $z = e^w$
20. $x = u \ln(v + w)$, $y = v \ln(u + w)$, $z = w \ln(u + v)$

CONTINUED

In Problems 21–25, transform the integral in (x, y) to an integral in (u, v) by using the given transformation. You need not evaluate the integral.

21. $\int_0^1 \int_y^1 xy\, dx\, dy$; $x = u - v$, $y = u + v$.

22. $\iint_\Omega e^{(x+y)/(x-y)}\, dx\, dy$, where Ω is the region in the first quadrant between the lines $x + y = 1$ and $x + y = 2$; $x = u + v$, $y = u - v$.

23. $\iint_\Omega y\, dx\, dy$, where Ω is the region

$7x^2 + 6\sqrt{3}x(y - 1) + 13(y - 1)^2 \leq 16$; use the transformation $x = \sqrt{3}u + (\frac{1}{2})v$, $y = 1 - u + (\sqrt{3}/2)v$.

24. $\int_0^1 \int_0^x (x^2 + y^2)\, dy\, dx$; $x = v$, $y = u$.

*25. $\iint_\Omega (y - x)\, dy\, dx$, Ω is the region bounded by $y = 2$, $y = x$, and $x = -y^2$; $x = v - u^2$, $y = u + v$.

■ APPLICATIONS

26. Let Ω be the region in the first quadrant of the xy-plane bounded by the hyperbolas $xy = 1$, $xy = 2$, and the lines $x = y$ and $x = 4y$. Compute $\iint_\Omega x^2 y^2\, dx\, dy$ by setting $x = u$ and $y = u/v$.

27. Let W be the solid enclosed by the ellipsoid $(x^2/a^2) + (y^2/b^2) + (z^2/c^2) = 1$. Then

$$\text{volume of } W = \iiint_W dx\, dy\, dz.$$

Compute this volume by making the transformation $x = au$, $y = bv$, $z = cw$. [*Hint:* In uvw-space you'll obtain a sphere.]

28. Compute $\iiint_W (xy + xz + yz)\, dx\, dy\, dz$, where W is the region of Problem 27.

■ ANSWERS TO SELF-QUIZ

I. e   II. b

■ PROBLEMS   CHAPTER 15 REVIEW

In Problems 1–6, compute the gradient vector field of the given function.

1. $f(x, y) = (x + y)^3$

2. $f(x, y) = \sin(x + 2y)$

3. $f(x, y) = \dfrac{x + y}{x - y}$

4. $f(x, y) = \sqrt{x/y}$

5. $f(x, y) = x^2 + y^2 + z^2$

6. $f(x, y) = xyz$

In Problems 7 and 8, compute and sketch the gradient field of the given function.

7. $f(x, y) = xy$

8. $f(x, y) = x^2 - y^2$

In Problems 9–14, calculate $\int_C \mathbf{F} \cdot d\mathbf{x}$.

9. $\mathbf{F}(x, y) = x^2\mathbf{i} + y^2\mathbf{j}$; C is the curve $y = x^{3/2}$ from $(0, 0)$ to $(1, 1)$.

10. $\mathbf{F}(x, y) = x^2 y\mathbf{i} - xy^2\mathbf{j}$; C is the unit circle in the counterclockwise direction.

11. $\mathbf{F}(x, y) = 3xy\mathbf{i} - y\mathbf{j}$; C is the triangle joining the points $(0, 0)$, $(1, 1)$, and $(0, 1)$ in the counterclockwise direction.

12. $\mathbf{F}(x, y) = e^y\mathbf{i} + e^x\mathbf{j}$; C is the curve from $(0, 0)$ to $(1, 1)$ and back to $(0, 0)$ which has the parametric description

$$\mathbf{x}(t) = \begin{cases} t\mathbf{i} + \sqrt{t}\mathbf{j} & \text{for } 0 \leq t \leq 1, \\ (2 - t)\mathbf{i} + (2 - t)^2\mathbf{j} & \text{for } 1 \leq t \leq 2. \end{cases}$$

13. $\mathbf{F}(x, y, z) = x\mathbf{i} + y\mathbf{j} + z\mathbf{k}$; C is the curve $\mathbf{x}(t) = t^3\mathbf{i} + t^2\mathbf{j} + t\mathbf{k}$ from $(0, 0, 0)$ to $(1, 1, 1)$.

14. $\mathbf{F}(x, y, z) = x^2\mathbf{i} + y^2\mathbf{j} + z^2\mathbf{k}$; C is the helix $\mathbf{x}(t) = (\sin t)\mathbf{i} + (\cos t)\mathbf{j} + 2t\mathbf{k}$ from $(0, 1, 0)$ to $(1, 0, \pi)$.

15. Calculate the work done when the force field $\mathbf{F}(x, y) = x^2 y\mathbf{i} + (y^3 + x^3)\mathbf{j}$ moves a particle around the unit circle in the counterclockwise direction.

16. Calculate the work done when the force field $\mathbf{F}(x, y) = 3(x - y)\mathbf{i} + x^5\mathbf{j}$ moves a particle around the triangle of Problem 11.

17. Verify that $\mathbf{F}(x, y) = 3x^2y^2\mathbf{i} + 2x^3y\mathbf{j}$ is exact, and calculate $\int_C \mathbf{F} \cdot \mathbf{dx}$, where C starts at $(1, 2)$ and ends at $(3, -1)$.

18. Verify that $\mathbf{F}(x, y) = e^{xy}(1 + xy)\mathbf{i} + x^2e^{xy}\mathbf{j}$ is exact, and calculate $\int_C \mathbf{F} \cdot \mathbf{dx}$, where C starts at $(1, 2)$ and ends at $(3, -1)$.

*19. Verify that $\mathbf{F}(x, y, z) = -(y/z)\mathbf{i} - (x/z)\mathbf{j} + (xy/z^2)\mathbf{k}$ is exact. Use that fact to evaluate $\int_C \mathbf{F} \cdot \mathbf{dx}$ where C is a piecewise smooth curve joining the points $(1, 1, 1)$ and $(2, -1, 3)$ and not crossing the xy-plane.

20. Evaluate $\oint_{\partial\Omega} 2y\, dx + 4x\, dy$, where $\Omega = \{(x, y): 0 \leq x \leq 2, 0 \leq y \leq 2\}$.

21. Evaluate $\oint_{\partial\Omega} x^2y\, dx + xy^2\, dy$, where Ω is the region enclosed by the triangle of Problem 11.

22. Evaluate $\oint_{\partial\Omega} (x^3 + y^3)\, dx + (x^2y^2)\, dy$, where Ω is the unit disk.

23. Evaluate $\oint_{\partial\Omega} \sqrt{1 + x^2}\, dy$, where $\Omega = \{(x, y): -1 \leq x \leq 1, x^2 \leq y \leq 1\}$.

24. Evaluate $\oint_{\partial\Omega} \dfrac{2x^2 + y^2 - xy}{\sqrt{x^2 + y^2}}\, dx + \dfrac{xy - x^2 - 2y^2}{\sqrt{x^2 + y^2}}\, dy$ where Ω is a region of the type shown in Figure 15.4.4 that does not contain the origin.

25. Let $\mathbf{F}(x, y) = xy^2\mathbf{i} + x^2y\mathbf{j}$ and let Ω denote the disk of radius 2 centered at $(0, 0)$. Calculate
   a. curl $\mathbf{F}$ b. $\oint_{\partial\Omega} \mathbf{F} \cdot \mathbf{T}\, ds$
   c. div $\mathbf{F}$ d. $\oint_{\partial\Omega} \mathbf{F} \cdot \mathbf{n}\, ds$

26. Let $\mathbf{F}(x, y) = y^2\mathbf{i} - x^2\mathbf{j}$ and let Ω be the triangle of Problem 11. Calculate the four items specified in the preceding problem.

27. Verify that the vector flow $\mathbf{F}(x, y) = (\cos x^2)\mathbf{i} + e^y\mathbf{j}$ is irrotational.

28. Verify that the vector flow $\mathbf{F}(x, y) = -(x^2y + y^3)\mathbf{i} + (x^3 + xy^2)\mathbf{j}$ is incompressible.

In Problems 29–31, parametrize the given surface.

29. the plane $2x - 3y + z = 0$
30. the hemisphere $x^2 + y^2 + z^2 = 25$, $z \geq 0$
31. the paraboloid $x^2 - y^2 - z = 0$; $0 \leq x \leq 2$; $2 \leq y \leq 4$

32. Compute the surface area of the surface in Problem 30.

33. Find an integral that gives the surface area of the surface in Problem 31. Do not evaluate it.

34. Compute the lateral surface area of a right circular cone with radius 2 and height 5.

35. Find the surface area of the hemisphere $x^2 + y^2 + z^2 = 16$, $x \leq 0$.

36. Find the surface area of the part of the surface $z = \dfrac{y^4}{4} + \dfrac{1}{8y^2}$ over the rectangle $0 \leq x \leq 3$, $2 \leq y \leq 4$.

In Problems 37–42, evaluate the integral of the given function over the specified surface.

37. $\iint_S y\, d\sigma$, where $S = \{(x, y, z): z = y^2, 0 \leq x \leq 2, 0 \leq y \leq 1\}$
38. $\iint_S (x^2 + y^2)\, d\sigma$, where S is as in the preceding problem
39. $\iint_S y^2\, d\sigma$, where S is the hemisphere $\{(x, y, z): x^2 + y^2 + z^2 = 9, x^2 + y^2 \leq 9, z \geq 0\}$
40. $\iint_S xz\, d\sigma$, where S is as in the preceding problem
41. $\iint_S y\, d\sigma$, where S is the surface of the tetrahedron bounded by the coordinate planes and the plane $x + y + z = 1$
42. $\iint_S (x^2 + y^2 + z^2)\, d\sigma$, where S is the part of the plane $y - x = 3$ that lies inside the cylinder $y^2 + z^2 = 9$

43. Find the mass of a triangular metallic sheet with corners at $(1, 0, 0)$, $(0, 1, 0)$, and $(0, 0, 1)$ if its density is proportional to y^2.

44. Find the mass of a metallic sheet in the shape of the hemisphere $x^2 + y^2 + z^2 = 1$, $x^2 + z^2 \leq 1$, $y \geq 0$ if its density is proportional to the distance from the y-axis.

In Problems 45–48, compute the flux $\iint_S \mathbf{F} \cdot \mathbf{n}\, d\sigma$ for the given surface immersed in the specified vector field.

45. $S = \{(x, y, z): z = 2xy, 0 \leq x \leq 1, 0 \leq y \leq 4\}$; $\mathbf{F} = xy^2\mathbf{i} - 2z\mathbf{j}$
46. $S = \{(x, y, z): x^2 + y^2 + z^2 = 1, z \geq 0\}$; $\mathbf{F} = x\mathbf{i} + 2y\mathbf{j} + 3z\mathbf{k}$
47. $S = \{(x, y, z): y = \sqrt{x^2 + z^2}, x^2 + z^2 \leq 4\}$; $\mathbf{F} = x\mathbf{i} - xz\mathbf{j} + 3z\mathbf{k}$
48. S is the unit sphere; $\mathbf{F} = x^2\mathbf{i} + y^2\mathbf{j} + z^2\mathbf{k}$

In Problems 49–54, compute the divergence and curl of the given vector field.

49. $\mathbf{F}(x, y, z) = x\mathbf{i} + y\mathbf{j} + z\mathbf{k}$
50. $\mathbf{F}(x, y, z) = (x - y)\mathbf{i} + (y - z)\mathbf{j} + (z - x)\mathbf{k}$
51. $\mathbf{F}(x, y, z) = yz\mathbf{i} + xz\mathbf{j} + xy\mathbf{k}$

CONTINUED

52. $\mathbf{F}(x, y, z) = (\ln x)\mathbf{i} + (\ln y)\mathbf{j} + (\ln z)\mathbf{k}$
53. $\mathbf{F}(x, y, z) = e^{yz}\mathbf{i} + e^{xz}\mathbf{j} + e^{xy}\mathbf{k}$
54. $\mathbf{F}(x, y, z) = (\cos y)\mathbf{i} + (\cos x)\mathbf{j} + (\cos z)\mathbf{k}$

In Problems 55–57, evaluate the line integral by using Stokes's theorem.

55. $\oint_C \mathbf{F} \cdot \mathbf{dx}$, where $\mathbf{F}(x, y, z) = (x + 2y)\mathbf{i} + (y - 3z)\mathbf{j} + (z - x)\mathbf{k}$ and C is the unit circle in the plane $z = 2$.
56. $\oint_C \mathbf{F} \cdot \mathbf{dx}$, where $\mathbf{F}(x, y, z) = x\mathbf{i} + y\mathbf{j} + z\mathbf{k}$ and C is the boundary of the triangle joining the points $(1, 0, 0)$, $(0, 1, 0)$, and $(0, 0, 1)$.
57. $\oint_C \mathbf{F} \cdot \mathbf{dx}$, where $\mathbf{F}(x, y, z) = y^2\mathbf{i} + x^2\mathbf{j} + z^2\mathbf{k}$ and C is the boundary of that part of the plane $x + y + z = 1$ which lies in the first octant.

58. Compute $\iint_S [\operatorname{curl} \mathbf{F}] \cdot \mathbf{n}\,d\sigma$, where S is the unit sphere and $\mathbf{F} = e^{xy}\mathbf{i} + \tan^{-1} z\mathbf{j} + (x + y + z)^{7/3}z^2\mathbf{k}$.

In Problems 59–64, evaluate the surface integral by using the divergence theorem.

59. $\iint_S \mathbf{F} \cdot \mathbf{n}\,d\sigma$, where $\mathbf{F} = x\mathbf{i} + 2y\mathbf{j} + 3z\mathbf{k}$ and S is the unit sphere.
60. $\iint_S \mathbf{F} \cdot \mathbf{n}\,d\sigma$, where $\mathbf{F} = ax\mathbf{i} + by\mathbf{j} + cz\mathbf{k}$ and S is the unit sphere.
61. $\iint_S \mathbf{F} \cdot \mathbf{n}\,d\sigma$, where $\mathbf{F} = x\mathbf{i} + 2y\mathbf{j} + 3z\mathbf{k}$ and S is the cylinder $x^2 + y^2 = 9$, $0 \le z \le 6$.

62. $\iint_S \mathbf{F} \cdot \mathbf{n}\,d\sigma$, where $\mathbf{F} = (y^3 - z)\mathbf{i} + x^2 e^z\mathbf{j} + (\sin xy)\mathbf{k}$ and S is the ellipsoid $(x/2)^2 + (y/4)^2 + (z/5)^2 = 1$.
63. $\iint_S \mathbf{F} \cdot \mathbf{n}\,d\sigma$, where $\mathbf{F} = x\mathbf{i} + 2y\mathbf{j} + 3z\mathbf{k}$ and S is the surface of the unit cube.
64. $\iint_S \mathbf{F} \cdot \mathbf{n}\,d\sigma$, where $\mathbf{F} = xz\mathbf{i} + xy\mathbf{j} + xyz\mathbf{k}$ and S is the unit cube in the first octant.

In Problems 65–72, compute the Jacobian of the given transformation.

65. $x = u + 2v$, $y = 2u - v$
66. $x = u^3 - v^3$, $y = u^3 + v^3$
67. $x = u \ln v$, $y = v \ln u$
68. $x = ve^u$, $y = ue^v$
69. $x = \dfrac{u}{v}$, $y = \dfrac{v}{u}$
70. $x = v \tan u$, $y = \tan uv$
71. $x = u + v + w$, $y = u - 2v + 3w$, $z = -2u + v - 5w$
72. $x = vw$, $y = uw$, $z = uv$

73. Transform the integral $\int_0^1 \int_x^1 xy\,dy\,dx$ by making the transformation $x = u + v$, $y = u - v$.
74. Transform $\iint_\Omega e^{(x-y)/(x+y)}\,dx\,dy$, where Ω is the region in the first quadrant between the lines $x + y = 2$ and $x + y = 3$, by making the transformation $u = x - y$ and $v = x + y$. Then evaluate the integral.

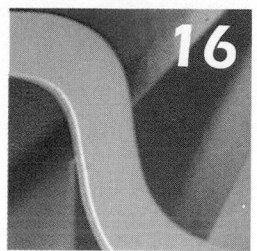

16 ORDINARY DIFFERENTIAL EQUATIONS

16.1 INTRODUCTION

Many of the basic laws of the physical sciences and, more recently, of the biological and social sciences are formulated in terms of mathematical relations involving certain known and unknown quantities and their derivatives. Such relations are called **differential equations**.

In Section 6.6 we studied the differential equation

$$\frac{dP}{dt} = \alpha P \tag{1}$$

with solution

$$P(t) = P(0)e^{\alpha t}. \tag{2}$$

Equation (2) is an equation of exponential growth. It describes a **continuous growth** process because the function $P(t)$, representing population, is continuous. The equation (1) is a **first-order** differential equation because the highest order derivative that appears is a first derivative.

In this chapter we discuss a number of first- and second-order differential equations and show how they can be used in a variety of important applications. After studying this chapter, you should begin to understand why differential equations comprise the most important part of that vast subject called *applied mathematics*.

16.2 FIRST-ORDER DIFFERENTIAL EQUATIONS: SEPARATION OF VARIABLES

A **first-order differential equation** involves a function of one independent variable and its first derivative. We will only consider those first-order equations that can be put in the form

$$\frac{dy}{dx} = f(x, y). \tag{1}$$

A **solution** to (1) is a differentiable function of x that satisfies the equation. Suppose that $f(x, y)$ can be written as

$$f(x, y) = g(x)h(y), \tag{2}$$

where g and h are each functions of only one variable. Then (1) can be written

$$\frac{1}{h(y)} \frac{dy}{dx} = g(x),$$

and integrating both sides with respect to x, we have

$$\int \frac{1}{h(y)} \, dy = \int \frac{1}{h(y)} \frac{dy}{dx} \, dx = \int g(x) \, dx + C. \tag{3}$$

16 ORDINARY DIFFERENTIAL EQUATIONS

The method of solution suggested in (3) is called the method of **separation of variables**. In general, if a differential equation can be written in the form $\dfrac{1}{h(y)} \, dy = g(x) \, dx$, then direct integration of both sides (if possible) will produce a set of solutions.

EXAMPLE 1
SOLVING A FIRST-ORDER DIFFERENTIAL EQUATION BY SEPARATING THE VARIABLES

Solve the differential equation $dy/dx = 4y$.

SOLUTION: We solved equations like this one in Section 6.6. We have $f(x, y) = 4y$, and we can write

$$\frac{dy}{y} = 4 \, dx, \qquad \int \frac{dy}{y} = \int 4 \, dx, \qquad \ln|y| = 4x + C$$

or

$$|y| = e^{4x + C} = e^{4x} e^C,$$

which can be written

$$y = k e^{4x},$$

where $k = \pm e^C$ can be any real number except zero. In addition, the constant function $y \equiv 0$ is also a solution.

As in Section 6.6, if we specify an **initial condition**, the infinite number of solutions reduces to one solution. For example, if we specify that $y(0) = 100$, then we obtain

$$y(x) = k e^{4x}$$

$$100 = y(0) = k e^0 = k$$

and the unique solution is

$$y(x) = 100 e^{4x}.$$

The differential equation $y' = f(x, y)$ together with the initial condition $y(x_0) = y_0$ is called an **initial-value problem**.

EXAMPLE 2
SOLVING A FIRST-ORDER INITIAL-VALUE PROBLEM BY SEPARATING THE VARIABLES

Solve the initial-value problem $dy/dx = y^2(1 + x^2)$, $y(0) = 1$.

SOLUTION: We have $f(x, y) = y^2(1 + x^2) = g(x)h(y)$, where $g(x) = 1 + x^2$ and $h(y) = y^2$. Then, successively,

$$\frac{dy}{y^2} = (1 + x^2) \, dx, \qquad \int \frac{dy}{y^2} = \int (1 + x^2) \, dx, \qquad -\frac{1}{y} = x + \frac{x^3}{3} + C,$$

or

$$y = -\frac{1}{x + (x^3/3) + C}.$$

For every number C this expression is a solution to the differential equation. Moreover, the constant function $y \equiv 0$ is also a solution. When

$x = 0$, $y = 1$, so

$$1 = y(0) = -\frac{1}{0 + 0 + C} = -\frac{1}{C},$$

implying that $C = -1$, and we obtain the unique solution to the initial-value problem:

$$y = -\frac{1}{x + (x^3/3) - 1}.$$

REMARK: **This solution (like any solution to a differential equation) can be checked by differentiation. You should *always* carry out this check.** To check, we have

$$\frac{dy}{dx} = \frac{1}{[x + (x^3/3) - 1]^2}(1 + x^2) = \left\{\frac{-1}{[x + (x^3/3) - 1]}\right\}^2 (1 + x^2) = y^2(1 + x^2).$$

Also, $y(0) = -1/(0 + 0 - 1) = 1$. Finally, observe that the solution is only defined as long as $x + x^3/3 - 1 \neq 0$. The only real zero of $x + x^3/3 - 1$ is, approximately, $x = 0.81773$. Thus the solution is only defined for $0 \leq x \leq 0.81773$.

EXAMPLE 3
LOGISTIC GROWTH

Let $P(t)$ denote the population of a species at time t. The **growth rate** of the population is defined as the growth in the population divided by the size of the population. Thus, for example, if the birth rate is 3.2 per 100 and the death rate is 1.8 per 100, then the growth rate is $3.2 - 1.8 = 1.4$ per $100 = 1.4/100 = 0.014$. We then write $dP/dt = 0.014P$.

Suppose that in a given population the average birth rate is a positive constant β. It is reasonable to assume that the average death rate is proportional to the number of individuals in the population. Greater populations mean greater crowding and more competition for food and territory. We call this constant of proportionality δ (which is greater than 0). Thus,

From the discussion above

$$\text{growth rate} = \frac{\text{growth in population}}{\text{population size}} = \frac{dP/dt}{P} = \beta - \delta P$$

or

$$\frac{dP}{dt} = P(\beta - \delta P). \tag{4}$$

This differential equation, together with the condition

$$P(0) = P_0 \qquad \text{(the initial population)}, \tag{5}$$

is an initial-value problem. To solve, we have

$$\frac{dP}{P(\beta - \delta P)} = dt,$$

or

$$\int \frac{dP}{P(\beta - \delta P)} = \int dt = t + C. \tag{6}$$

To calculate the integral on the left, we use partial fractions. We have (verify this)

$$\frac{1}{P(\beta - \delta P)} = \frac{1}{\beta P} + \frac{\delta}{\beta(\beta - \delta P)},$$

so

$$\int \frac{dP}{P(\beta - \delta P)} = \int \frac{1}{\beta}\frac{dP}{P} + \frac{\delta}{\beta}\int \frac{dP}{\beta - \delta P}$$

$$= \frac{1}{\beta}\ln P - \frac{1}{\beta}\ln(\beta - \delta P) = t + C,$$

or

$$\ln\left(\frac{P}{\beta - \delta P}\right) = \beta(t + C)$$

and

$$\frac{P}{\beta - \delta P} = C_1 e^{\beta t} \quad \text{where} \quad C_1 = e^{\beta C}. \tag{7}$$

Using the initial condition (5), we have

$$\frac{P_0}{\beta - \delta P_0} = C_1 e^0 = C_1,$$

so

$$C_1 = \frac{P_0}{\beta - \delta P_0}. \tag{8}$$

Finally, with the insertion of (8) into (7) it follows (after some algebra) that

---

SOLUTION TO THE LOGISTIC EQUATION

$$P(t) = \frac{\beta}{\delta + [(\beta/P_0) - \delta]e^{-\beta t}}. \tag{9}$$

Equation (4) is called the **logistic equation**, and we have shown that the solution to the logistic equation with initial population P_0 is given by (9). Sketches of the growth governed by the logistic equation are given in Figure 1.

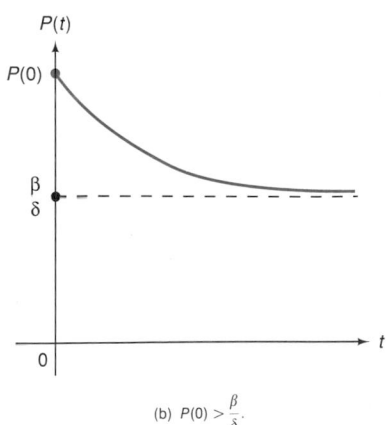

(a) $P(0) < \frac{\beta}{\delta}$.

(b) $P(0) > \frac{\beta}{\delta}$.

FIGURE 1

GRAPH OF $P(t) = \dfrac{\beta}{\delta + \left[\dfrac{\beta}{P(0)} - \delta\right]e^{-\beta t}}$

(LOGISTIC GROWTH)

We close this section by noting that most differential equations, even of the first order, cannot be solved by elementary methods (although there are other techniques of solution besides the technique of separation of variables). However, even when an equation cannot be solved in a "closed form" (i.e., by writing one variable in terms of another), numerical techniques usually exist for calculating a solution to as many decimal places of accuracy as needed.

In the next section we discuss a wide class of first-order equations for which closed-form solutions can be found.

A PERSPECTIVE: WHEN IS A DIFFERENTIAL EQUATION SEPARABLE?

In this section we saw how to solve a first-order differential equation when the equation was separable. However, it is not always clear that an equation is separable. For example, it is obvious that $f(x, y) = e^x \cos y$ is separable, but it is not so obvious that $f(x, y) = 2x^2 + y - x^2y + xy - 2x - 2$ is separable.[†] In this perspective, we give conditions that ensure that an equation is separable.[‡]

THEOREM 1

Suppose that $f(x, y) = g(x)h(y)$, where both g and h are differentiable. Then

$$f(x, y)f_{xy}(x, y) = f_x(x, y)f_y(x, y). \tag{10}$$

Here the subscripts denote partial derivatives.

PROOF: Observe that

$$f_x(x, y) = g'(x)h(y)$$

$$f_y(x, y) = g(x)h'(y)$$

$$f_{xy}(x, y) = g'(x)h'(y).$$

Hence,

$$f(x, y)f_{xy}(x, y) = g(x)h(y)g'(x)h'(y) = [g'(x)h(y)][g(x)h'(y)]$$
$$= f_x(x, y)f_y(x, y). \quad \blacksquare$$

It turns out that, under further conditions, if equation (10) holds, then $f(x, y)$ is separable. For what follows, D denotes an open disk in the xy-plane; that is, $D = \{(x, y): (x - a)^2 + (y - b)^2 < r^2\}$, where a, b, and r are real numbers and $r > 0$.

THEOREM 2   **CONDITIONS FOR SEPARABILITY**

Suppose that in D, f, f_x, f_y, and f_{xy} exist and are continuous, $f(x, y) \neq 0$, and equation (10) holds. Then there are continuously differentiable functions $g(x)$ and $h(y)$ such that, for every $(x, y) \in D$,

$$f(x, y) = g(x)h(y). \tag{11}$$

PROOF: Since $f(x, y) \neq 0$ and f is continuous on D, f has the same sign on D. Assume $f(x, y) > 0$ for $(x, y) \in D$. A similar proof

---

[†] $2x^2 + y - x^2y + xy - 2x - 2 = (1 + x - x^2)(y - 2)$.

[‡] The results here are based on the paper "When is an Ordinary Differential Equation Separable?" by David Scott in *American Mathematical Monthly* 92 (1985): 422–423.

works if $f(x, y) < 0$ if f is replaced by $-f$. Now, from the quotient rule of differentiation,

Equation (10)

$$\frac{\partial}{\partial y} \frac{f_x(x, y)}{f(x, y)} = \frac{f(x, y)f_{xy}(x, y) - f_x(x, y)f_y(x, y)}{f^2(x, y)} = 0.$$

When the partial derivative with respect to y of a function of x and y is zero, then that function must be a function of x only. Thus, there is a function $\alpha(x)$ such that

$$\frac{f_x(x, y)}{f(x, y)} = \alpha(x).$$

Also, since $f(x, y) > 0$, $\ln f(x, y)$ is defined and

$$\frac{\partial}{\partial x} \ln f(x, y) = \frac{f_x(x, y)}{f(x, y)} = \alpha(x).$$

The function $\alpha(x)$ is continuous in D because it is the quotient of continuous functions and the function in the denominator is non-zero. Let $\beta(x) = \int \alpha(x)\, dx$. Then

$$\ln f(x, y) = \int \left[\frac{\partial}{\partial x} \ln f(x, y) \right] dx = \int \alpha(x)\, dx = \beta(x) + \gamma(y),$$

where γ is a function of y only. (The partial derivative with respect to x of a function of y only is zero. Thus $\gamma(y)$ represents the most general constant of integration.) Finally, let $g(x) = e^{\beta(x)}$ and $h(y) = e^{\gamma(y)}$. Then

$$f(x, y) = e^{\ln f(x,y)} = e^{\beta(x) + \gamma(y)} = e^{\beta(x)}e^{\gamma(y)} = g(x)h(y). \quad \blacksquare$$

EXAMPLE 4
A SEPARABLE DIFFERENTIAL
EQUATION

Let $f(x, y) = 2x^2 + y - x^2y + xy - 2x - 2$. Then

$$f_x(x, y) = 4x - 2xy + y - 2$$

$$f_y(x, y) = 1 - x^2 + x$$

$$f_{xy}(x, y) = -2x + 1$$

$$f(x, y)f_{xy}(x, y) = (2x^2 + y - x^2y + xy - 2x - 2)(-2x + 1)$$
$$= -4x^3 - xy + 2x^3y - 3x^2y + 6x^2 + 2x + y - 2$$

and

$$f_x(x, y)f_y(x, y) = (4x - 2xy + y - 2)(1 - x^2 + x)$$
$$= 2x - xy + y - 2 - 4x^3 + 2x^3y - 3x^2y + 6x^2.$$

Since the last two expressions are equal, we conclude by Theorem 2 that $f(x, y)$ is separable (see the first footnote on page 1037).

EXAMPLE 5
A DIFFERENTIAL EQUATION THAT
IS NOT SEPARABLE

Let $f(x, y) = 1 + xy$. Then

$$f_x(x, y) = y$$

$$f_y(x, y) = x$$

$$f_{xy}(x, y) = 1$$

$$f(x, y)f_{xy}(x, y) = 1 + xy$$

and

$$f_x(x, y)f_y(x, y) = xy.$$

Since the last two expressions are unequal, we conclude that $f(x, y)$ is not separable.

■ PROBLEMS 16.2

■ SELF-QUIZ

I. $f(x) = e^{2x}$ is a solution of the differential equation _____.
   a. $y' + y = 0$ b. $y' - y = 0$
   c. $2y' - y = 0$ d. $y' - 2y = 0$
II. $g(x) = x^2$ is a solution of the differential equation _____.
   a. $y' - 2y = 0$ b. $2y' - y = 0$
   c. $xy' - 2y = 0$ d. $2xy' - y = 0$
III. _____ is a solution of the differential equation $xy' - y = 0$.

   a. $f(x) = 17$ b. $g(x) = x$
   c. $F(x) = x^2$ d. $G(x) = e^{1/x}$
IV. _____ is a solution of the differential equation $xy' - 3y = 0$.
   a. $f(x) = e^{3x}$ b. $g(x) = 13.807x$
   c. $F(x) = e^{2 \ln x}$ d. $G(x) = (-5x)^3$
V. _____ is the solution of the initial-value problem $y' + 2y = 0$, $y(0) = 4$.
   a. $f(x) = 4e^{2x}$ b. $g(x) = 4e^{-2x}$
   c. $F(x) = e^{-2x} + 3$ d. $G(x) = (e^{-x} + 1)^2$

■ DRILL

In Problems 1–12, solve the given differential equation by the method of separation of variables. If an initial condition is given, find the unique solution to that initial-value problem.

1. $\dfrac{dy}{dx} = -7x$

2. $\dfrac{dy}{dx} = e^{x+y}$; $y(0) = 2$

3. $\dfrac{dx}{dt} = \sin x \cos t$; $x(\pi/2) = 3$

4. $\dfrac{dy}{dx} = 1/y^3$

5. $x^3(y^2 - 1)\dfrac{dy}{dx} = (x + 3)y^5$

6. $\dfrac{dy}{dx} = 2x^2y^2$; $y(1) = 2$

7. $\dfrac{dx}{dt} = e^x \sin t$; $x(0) = 1$

8. $\dfrac{dy}{dx} = 1 + y^2$; $y(0) = 1$

9. $\dfrac{dx}{dt} = x(1 - \cos 2t)$; $x(0) = 1$

*10. $\dfrac{dx}{dt} = x^n t^m$; n, m integers
   [*Hint:* Consider the cases with $n = 1$ and $m = -1$ separately.]

11. $\dfrac{dy}{dx} = x^2(1 + y^2)$

12. $\dfrac{dx}{dt} = \dfrac{e^x t}{e^x + t^2 e^x}$

◼ APPLICATIONS

13. The population of Australia was 12,100,000 in 1968 and 13,268,000 in 1973. Assuming that Australia's growth rate is proportional to the size of its population, estimate that population in 1978, 1983, and 1988.

14. In the logistic equation of Example 3, assume that $\delta = 0.00005$, $\beta = 0.2$, and $P_0 = 20,000$. Calculate $P(10)$.

15. The population of a particular species of bacteria grows according to the logistic equation.
    a. Determine the **equilibrium population** P_e, given by $P_e = \lim_{t \to \infty} P(t)$.
    b. Let $Q = P/P_e$. Find a simple differential equation satisfied by Q.

16. Suppose $P(t)$ satisfies the logistic differential equation.
    a. What is the value of P at an inflection point of its graph? [*Hint:* Differentiate P' implicitly.]
    *b. What is the value of P at a point where $|P'|$ is maximal?

17. **Newton's law of cooling** states that the rate of change of the temperature difference between an object and its surrounding medium is proportional to the temperature difference. Suppose the air temperature is 30°C and an object with an initial temperature of 10°C warms to 14°C in one hour.
    a. What will its temperature be after two hours?
    b. How long will it take for the object to be warmed to 25°C?

*18. A remote lake is estimated to be able to support a maximum of 10,000 pike (that max is called its **carrying capacity**). Let $P(t)$ be the number of pike in the lake at a time t days after a moment when there are 2,000 pike in the lake. One plausible model for the net growth rate, P'/P, in the absence of any harvesting by fishermen, is that it is proportional to the unused capacity, $10,000 - P$. Suppose it takes 60 days for the number of pike to reach 3,000. Solve for P as an explicit function of t.

*19. An object of mass m that falls from rest, starting at a point near the Earth's surface, is subjected to two forces: a downward force mg and a resisting force proportional to the square of the velocity of the body. Thus,

$$F = ma = m\frac{dv}{dt} = mg - \alpha v^2,$$

where α is a constant of proportionality and the downward direction is treated as positive.
    a. Find $v(t)$ as a function of t. [*Hint:* Use the fact that $v(0) = 0$.]
    b. Verify that the velocity does not increase indefinitely but approaches the equilibrium value $\sqrt{mg/\alpha}$. (This value is called the **terminal velocity**.)

20. The economist Vilfredo Pareto (1848–1923) considered the consequences of a model that supposed the rate of decrease of the number, y, of people in a stable economy having an income of at least x dollars per year to be directly proportional to the number of such people and inversely proportional to their income. Obtain an expression (**Pareto's law**) for y in terms of x.

*21. Bacteria are supplied as food to a protozoan population at a constant rate μ. It is observed that the bacteria are consumed at a rate that is proportional to the square of their numbers. The concentration $c(t)$ of the bacteria therefore satisfies the differential equation $\dfrac{dc}{dt} = \mu - \lambda c^2$, where λ is a positive constant.
    a. Determine $c(t)$ in terms of $c(0)$.
    b. What is the equilibrium concentration of the bacteria?

22. In some chemical reactions certain products catalyze their own formation. If $x(t)$ is the amount of such a product at time t, a possible model for the reaction is given by the differential equation $\dfrac{dx}{dt} = \alpha \cdot (\beta - x)$, where α and β are positive constants. According to this model, the reaction is completed when $x = \beta$ since this condition indicates that one of the constituent chemicals has been depleted.
    a. Solve the equation in terms of the constants α, β, and the initial value $x(0)$.
    b. For $\alpha = 1$, $\beta = 200$, and $x(0) = 20$, draw a graph of $x(t)$ for $t > 0$.

*23. On a certain day it began to snow early in the morning and the snow continued to fall at a constant rate. The velocity at which a snowplow is able to clear a road is inversely proportional to the height of the accumulated snow. The snowplow started at 11 a.m. and cleared 4 mi of road by 2 p.m. By 5 p.m., it had cleared an additional 2 mi. When did it start snowing?

**24. A large open cistern has the shape of a hemisphere with radius 25 ft. This bowl has a circular

CONTINUED

hole of radius 1 ft in its bottom. By **Torricelli's law,**[†] water will flow out of the hole with the same speed it would attain in falling freely from the level of the water to that of the hole. If the cistern is filled with water, how long will it take for all that water to drain from the cistern?

■ CHALLENGE

**25. Suppose the inside of the cistern in the preceding problem is a surface of revolution. What shape should it be in order that water drains so that the level drops at a constant rate?[‡]

*26. In an isolated town with 1,000 people, the rate of spread of a rumor is proportional to the product of the number of residents who have already heard the rumor and the number who have not. Suppose a particular rumor started by 25 people at a party has spread in 2 hours so that 100 people have heard it by then. To the best extent possible, find the number of people who have heard the rumor as a function of time.

■ ANSWERS TO SELF-QUIZ

I. d   II. c   III. b
IV. d   V. b

[†] Evangelista Torricelli (1608–1647) was an Italian physicist.
[‡] You are being given the opportunity to rediscover a principle used by the ancient Egyptians in constructing water clocks to tell time.

16.3 FIRST-ORDER LINEAR DIFFERENTIAL EQUATIONS

A differential equation is said to be of **nth order** if it involves a function of one variable and some of its derivatives, and the highest-order derivative appearing is the **nth** derivative. An nth-order differential equation is called **linear** if it can be written in the form

$$\frac{d^n y}{dx^n} + a_{n-1}(x)\frac{d^{n-1}y}{dx^{n-1}} + \cdots + a_1(x)\frac{dy}{dx} + a_0(x)y = f(x). \tag{1}$$

For example, the most general first-order linear equation takes the form

$$\frac{dy}{dx} + a(x)y = f(x), \tag{2}$$

while a second-order linear equation can be written as

$$\frac{d^2 y}{dx^2} + a(x)\frac{dy}{dx} + b(x)y = f(x). \tag{3}$$

In this section, we will discuss first-order linear equations; we will discuss second-order linear differential equations in Sections 16.4 and 16.5. However, we begin by stating an important result that applies to linear differential equations of any order.

One nice fact about linear equations is given by the following theorem.[†]

[†] For a proof, see W. R. Derrick and S. I. Grossman, *Introduction to Differential Equations with Boundary Value Problems*, 3rd ed. (St. Paul: West, 1987), App. 3.

THEOREM 1   **EXISTENCE-UNIQUENESS THEOREM**

Let $x_0, y_0, y_1, \ldots, y_{n-1}$ be real numbers and let $a_0, a_1, \ldots, a_{n-1}$ be continuous. Then there is a unique solution $y = f(x)$ to the nth-order linear differential equation (1) that satisfies

$$f(x_0) = y_0, \qquad f'(x_0) = y_1, \quad \ldots, \quad f^{(n-1)}(x_0) = y_{n-1}. \qquad \text{(4)}$$

That is, if we specify n initial conditions, there exists a unique solution. ■

We say that the linear equation (1) is **homogeneous** if $f(x) = 0$ for every number x in the domain of f. Otherwise, we say that the equation is **nonhomogeneous**. If the functions $a_0(x), a_1(x), \ldots, a_{n-1}(x)$ are constant, then the equation is said to have **constant coefficients**. Otherwise, it is said to have **variable coefficients**. It turns out that we can solve by integration *all* first-order linear equations and all nth-order linear homogeneous and nonhomogeneous equations with constant coefficients. In this section, we show how solutions to linear equations in the form (2) can be explicitly calculated. We do this in three cases of increasing complexity.

Case 1: Constant coefficients, homogeneous   Then (2) can be written

$$\frac{dy}{dx} + ay = 0 \qquad \text{(5)}$$

where a is a constant, or

$$\frac{dy}{dx} = -ay.$$

Solutions (from Section 6.6) are given by

$$y = Ce^{-ax} \qquad \text{(6)}$$

for any constant C. Many examples of this type of equation were given in Section 6.6.

Case 2: Constant coefficients, nonhomogeneous   Then (2) can be written

$$\frac{dy}{dx} + ay = f(x), \qquad \text{(7)}$$

where a is a constant. It is now impossible to separate the variables. However, equation (7) can be solved by multiplying both sides of (7) by an **integrating factor**. We first note that

$$\frac{d}{dx}(e^{ax}y) = e^{ax}\frac{dy}{dx} + ae^{ax}y = e^{ax}\left(\frac{dy}{dx} + ay\right). \qquad \text{(8)}$$

Thus, if we multiply both sides of (7) by e^{ax}, we obtain

$$e^{ax}\left(\frac{dy}{dx} + ay\right) = e^{ax}f(x)$$

or using (8),

$$\frac{d}{dx}(e^{ax}y) = e^{ax}f(x)$$

and upon integration,

$$e^{ax}y = \int e^{ax}f(x)\,dx + C$$

where $\int e^{ax}f(x)\,dx$ denotes one particular antiderivative. This leads to the general solution

$$y = e^{-ax}\int e^{ax}f(x)\,dx + Ce^{-ax}. \tag{9}$$

The term e^{ax} is called an integrating factor *for (7) because it allows us, after multiplication, to solve the equation by integration.*

EXAMPLE 1
SOLVING A FIRST-ORDER LINEAR DIFFERENTIAL EQUATION

Find all solutions to

$$\frac{dy}{dx} + 3y = x. \tag{10}$$

SOLUTION: We multiply both sides of the equation by e^{3x}. Then

$$e^{3x}\left(\frac{dy}{dx} + 3y\right) = xe^{3x}, \quad\text{or}\quad \frac{d}{dx}(e^{3x}y) = xe^{3x},$$

and

$$e^{3x}y = \int xe^{3x}\,dx.$$

But setting $u = x$ and $dv = e^{3x}\,dx$, we find that $du = dx$, $v = \frac{1}{3}e^{3x}$, and

$$\int xe^{3x}\,dx = \frac{x}{3}e^{3x} - \frac{1}{3}\int e^{3x}\,dx = \frac{x}{3}e^{3x} - \frac{1}{9}e^{3x} + C,$$

so

$$e^{3x}y = \frac{x}{3}e^{3x} - \frac{1}{9}e^{3x} + C,$$

and

$$y = \frac{x}{3} - \frac{1}{9} + Ce^{-3x}.$$

This answer should be checked by differentiation.

Case 3: Variable coefficients, nonhomogeneous We first note the following facts, the first of which follows from the fundamental theorem of calculus (assuming that a is continuous):

(i) $\dfrac{d}{dx}\displaystyle\int a(x)\,dx = a(x).$ \hfill (11)

[This holds for *any* antiderivative $\int a(x)\,dx$.]

(ii) $\dfrac{d}{dx} e^{\int a(x)\,dx} = e^{\int a(x)\,dx}\dfrac{d}{dx}\displaystyle\int a(x)\,dx = a(x)e^{\int a(x)\,dx}.$ **(12)**

Now consider the equation (2),

$$\frac{dy}{dx} + a(x)y = f(x).$$

We multiply both sides by the integrating factor $e^{\int a(x)\,dx}$. Then we have

$$e^{\int a(x)\,dx}\frac{dy}{dx} + a(x)e^{\int a(x)\,dx}y = e^{\int a(x)\,dx}f(x).$$ **(13)**

Now we note that, from (12),

$$\frac{d}{dx}\,ye^{\int a(x)\,dx} = y\frac{d}{dx}\,e^{\int a(x)\,dx} + \frac{dy}{dx}\,e^{\int a(x)\,dx}$$

$$= a(x)e^{\int a(x)\,dx}y + \frac{dy}{dx}\,e^{\int a(x)\,dx}$$

$$= \text{the left-hand side of (13)}.$$

Thus, from (13),

$$\frac{d}{dx}\left[e^{\int a(x)\,dx}y\right] = e^{\int a(x)\,dx}f(x),$$

or integrating,

$$e^{\int a(x)\,dx}y = \int e^{\int a(x)\,dx}f(x)\,dx + C,$$

and

SOLUTION TO A FIRST-ORDER LINEAR EQUATION

$$y = e^{-\int a(x)\,dx}\int e^{\int a(x)\,dx}f(x)\,dx + Ce^{-\int a(x)\,dx}.$$ **(14)**

It is probably a waste of time to try to memorize the complicated-looking formula (14). Rather, it is important to remember that multiplication by the integrating factor $e^{\int a(x)\,dx}$ will always enable you to reduce the problem of solving a differential equation to the problem of calculating an integral.

EXAMPLE 2
SOLVING A FIRST-ORDER LINEAR
INITIAL-VALUE PROBLEM

(a) Find all solutions to the equation

$$\frac{dy}{dx} + \frac{4}{x}y = 3x^2.$$

(b) Find the unique solution that satisfies $y(1) = 2$.

SOLUTION:

(a) Here $a(x) = 4/x$, $\int a(x)\,dx = 4\ln x = \ln x^4$, and $e^{\int a(x)\,dx} = e^{\ln x^4} = x^4$ (since $e^{\ln u} = u$ for all $u > 0$). Thus, we can multiply both sides of the equation by the integrating factor x^4 to obtain

$$x^4 \frac{dy}{dx} + 4x^3 y = 3x^6.$$

But

$$\frac{d}{dx}(yx^4) = \frac{dy}{dx}(x^4) + y(4x^3),$$

and our equation has become

$$\frac{d}{dx}(yx^4) = 3x^6.$$

Then we integrate to find that

$$yx^4 = \frac{3x^7}{7} + C, \qquad \text{or} \qquad y = \frac{3x^3}{7} + \frac{C}{x^4}.$$

Note that $3x^3/7$ is one solution to the nonhomogeneous equation $(dy/dx) + (4/x)y = 3x^2$, while C/x^4 represents all solutions to the homogeneous equation $(dy/dx) + (4/x)y = 0$. This should be checked by differentiation.

(b) Inserting $x = 1$ and $y = 2$ into our solution we find that

$$2 = \frac{3}{7} + C, \qquad \text{or} \qquad C = \frac{11}{7}.$$

Thus, the unique solution to the initial-value problem is

$$y = \frac{1}{7}\left(3x^3 + \frac{11}{x^4}\right).$$

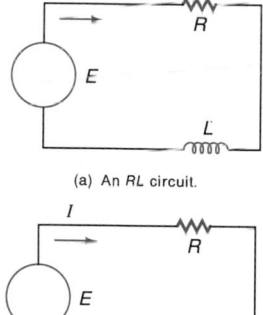

(a) An *RL* circuit.

(b) An *RC* circuit.

FIGURE 1
TWO ELECTRIC CIRCUITS

SIMPLE ELECTRIC CIRCUITS

As an application of the material in this section, we will consider simple electric circuits containing a resistor (R) and an inductor (L) or capacitor (C) in series with a source of electromotive force (emf). Such circuits are shown in Figures 1(a) and 1(b), and their action can be understood without any special knowledge of electricity.

(i) An electromotive force (emf) E (volts, V), usually a battery or generator, drives an electric charge Q (coulombs, C) and produces a current I (amperes, A). The current is defined as the rate of flow of the charge, and we can write

$$I = \frac{dQ}{dt}. \tag{15}$$

(ii) A resistor of resistance R (ohms, Ω) is a component of the circuit that opposes the current, dissipating the energy in the form of heat.

It produces a drop in voltage given by **Ohm's law**:

$$E_R = RI. \tag{16}$$

(iii) An inductor of inductance L (henrys, H) opposes any change in current by producing a voltage drop of

$$E_L = L\frac{dI}{dt}. \tag{17}$$

(iv) A capacitor of capacitance C (farads, F) stores charge. In so doing, it reduces the charge, causing a drop in the voltage of

$$E_C = \frac{Q}{C}. \tag{18}$$

The quantities R, L, and C are usually constants associated with the particular components in the circuit; E may be a constant or a function of time. The fundamental principle guiding such circuits is given by **Kirchhoff's voltage law**.

The algebraic sum of all voltage drops around a closed circuit is zero.

In the circuit of Figure 1(a), the resistor and the inductor cause voltage drops of E_R and E_L, respectively. The emf, however, provides a voltage of E (i.e., a voltage drop of $-E$). Thus, Kirchhoff's voltage law yields

$$E_R + E_L - E = 0.$$

Transposing E to the other side of the equation and using equations (16) and (17) to replace E_R and E_L, we have

$$L\frac{dI}{dt} + RI = E. \tag{19}$$

The following example illustrates the use of equation (19) in analyzing the circuit shown in Figure 1(a).

▦ EXAMPLE 3
FINDING THE CURRENT IN
AN RL CIRCUIT

An inductance of 2 H and a resistance of 10 Ω are connected in series with an emf of 100 V. If the current is zero when $t = 0$, what is the current at the end of 0.1 sec?

SOLUTION: Since $L = 2$, $R = 10$, and $E = 100$, equation (19) and the initial current yield the initial-value problem:

$$2\frac{dI}{dt} + 10I = 100, \qquad I(0) = 0. \tag{20}$$

Dividing both sides of (20) by 2, we note that the resulting linear first-order equation has e^{5t} as an integrating factor; that is,

$$\frac{d}{dt}(e^{5t}I) = e^{5t}\left(\frac{dI}{dt} + 5I\right) = 50e^{5t}. \tag{21}$$

Integrating both sides of equation (21), we get

$$e^{5t}I(t) = 10e^{5t} + C,$$

or

$$I(t) = 10 + Ce^{-5t}. \tag{22}$$

Setting $t = 0$ in (22) and using the initial condition $I(0) = 0$, we have

$$0 = I(0) = 10 + C, \tag{23}$$

which implies that $C = -10$. Substituting this value into (23), we obtain an equation for the current at all times t:

$$I(t) = 10(1 - e^{-5t}).$$

Thus, when $t = 0.1$, we have

$$I(0.1) = 10(1 - e^{-0.5}) \approx 3.93 \text{ A}.$$

For the circuit in Figure 1(b) we have $E_R + E_C - E = 0$, or

$$RI + \frac{Q}{C} = E.$$

Using the fact that $I = dQ/dt$, we obtain the linear first-order equation

$$R\frac{dQ}{dt} + \frac{Q}{C} = E. \tag{24}$$

The next example illustrates how to use equation (24).

▦ EXAMPLE 4
FINDING THE CHARGE AND
CURRENT IN AN RC CIRCUIT

A resistance of 2000 Ω and a capacitance of 5×10^{-6} F are connected in series with an emf of 100 V. Determine the charge and current at all times t and determine the current at $t = 0.1$ sec if $I(0) = 0.01$ A.

SOLUTION: Setting $R = 2000$, $C = 5 \times 10^{-6}$, and $E = 100$ in (24), we have

$$2000\frac{dQ}{dt} + \frac{1{,}000{,}000}{5}Q = 100,$$

or, dividing both sides by 2000,

$$\frac{dQ}{dt} + 100Q = \frac{1}{20}. \tag{25}$$

Then we can determine $Q(0)$ since

$$\frac{1}{20} = Q'(0) + 100Q(0) = I(0) + 100Q(0).$$

Thus

$$Q(0) = \frac{1}{100}\left[\frac{1}{20} - I(0)\right] = \frac{1}{100}\left[\frac{1}{20} - \frac{1}{100}\right]$$

$$= \frac{1}{100}\left(\frac{4}{100}\right) = 4 \times 10^{-4} \text{ C}.$$

Multiplying both sides of (25) by the integrating factor e^{100t}, we get

$$\frac{d}{dt}(e^{100t}Q) = \frac{e^{100t}}{20},$$

and integrating this equation yields

$$e^{100t}Q = \frac{e^{100t}}{2000} + k.$$

Dividing both sides by e^{100t} gives us

$$Q(t) = \frac{1}{2000} + ke^{-100t},$$

and setting $t = 0$, we find that $k = Q(0) - \frac{1}{2000} = 4 \times 10^{-4} - 5 \times 10^{-4} = -10^{-4}$. Thus the charge at all times t is

$$Q(t) = \frac{5 - e^{-100t}}{10^4},$$

and the current is

$$I(t) = Q'(t) = \frac{1}{100}e^{-100t}.$$

Thus, $I(0.1) = 10^{-2}e^{-10} \approx 4.54 \times 10^{-7}$ A.

■ PROBLEMS 16.3

■ SELF-QUIZ

I. The general solution to $\dfrac{dx}{dt} + 3x = 0$ is

_____.

  a. $y = e^{-3x} + C$ b. $y = Ce^{-3x}$

  c. $x = e^{-3t} + C$ d. $x = Ce^{-3t}$

II. e^{3x} is an integrating factor for _____.

  a. $\dfrac{dy}{dx} - 3y = x^2$

  b. $\dfrac{dy}{dx} + \dfrac{y}{3} = x - 3$

  c. $\dfrac{dy}{dx} + 3y = x + 4$

  d. $3\dfrac{dy}{dx} + y = \dfrac{x}{5}$

III. The general solution to $\dfrac{dy}{dx} + 3y = 9x$ is

_____.

  a. $y = e^{-3x} + C(3x - 1)$

  b. $y = (3x - 1) + Ce^{-3x}$

  c. $y = C(e^{-3x} + (3x - 1))$

  d. $y = e^{-3x} + (3x - 1) + C$

IV. _____ is an integrating factor for

$$\frac{dy}{dx} + 2xy = -2xe^{-x^2}.$$

  a. e^{-x^2} b. e^{x^2}

  c. $e^{x^2/2}$ d. $-2xe^{-x^2}$

V. $\sin x$ is an integrating factor for _____.

  a. $\dfrac{dy}{dx} + (\cot x)y = x$

  b. $\dfrac{dy}{dx} + (\cos x)y = -x$

  c. $\dfrac{dy}{dx} + y = -\cos x$

  d. $\dfrac{dy}{dx} + \dfrac{y}{x} = \cot x$

CONTINUED

VI. Suppose $f(x)$ is a solution of $\dfrac{dy}{dx} + y = 3$. Answer True or False to each of the following:

a. $\lim_{x \to \infty} f(x)$ does not exist.
b. $\lim_{x \to \infty} f(x) = 0$.
c. $\lim_{x \to \infty} f(x) = 3$.
d. $\lim_{x \to \infty} f'(x)$ does not exist.

e. $\lim_{x \to \infty} f'(x) = 3$.
f. $\lim_{x \to \infty} f'(x) = 0$.
g. If $f(0) > 3$, then f always decreases.
h. If $f(5) > \pi$, then f always decreases.
i. If $f(0) < \pi$, then $f'(x) > 0$ for all x.
j. If $f(-5) < 3$, then $f'(x) > 0$ for all x.

■ **DRILL**

In Problems 1–16, find all solutions to the given differential equation. If an initial condition is specified, then find the particular function which solves that initial-value problem.

1. $\dfrac{dy}{dx} = 4x$

2. $\dfrac{dy}{dx} + 22y = 0; \ y(1) = 2$

3. $\dfrac{dx}{dt} = x + 1; \ x(0) = 1$

4. $\dfrac{dy}{dx} + xy = 0; \ y(0) = 2$

5. $\dfrac{dx}{dt} + x = \sin t; \ x(0) = 1$

6. $\dfrac{dx}{dt} + x = \dfrac{1}{1 + e^{2t}}$

8. $\dfrac{dy}{dx} + (\tan x)y = 2x \sec x; \ y(\pi/4) = 1$

9. $\dfrac{dx}{dt} - ax = be^{at}; \ a, b$ constant

10. $\dfrac{dx}{dt} - ax = be^{ct}; \ x(0) = d; \ a, b, c, d$ are constant

11. $\dfrac{dy}{dx} = x + 2y \tan 2x$ 　　 12. $\dfrac{dy}{dx} = 2y + x^2 e^{2x}$

13. $(x^2 + 1)\dfrac{dy}{dx} + 2xy = x, \ y(0) = 1$

14. $\dfrac{dy}{dx} + \dfrac{y}{x^2} = \dfrac{3}{x^2}; \ y(1) = 2$

15. $\dfrac{dx}{dt} + \dfrac{4t}{t^2 + 1} x = 3t; \ x(0) = 4$

*7. $\dfrac{dy}{dx} - y \ln x = x^x$ 　　 16. $\dfrac{ds}{dt} + s \tan t = \cos t; \ s(\pi/3) = \tfrac{1}{2}$

■ **APPLICATIONS**

17. Intravenous infusion of glucose into the bloodstream of a patient is an important medical technique. To study this process, let $G(t)$ be the amount of glucose in the patient's bloodstream t minutes after the process begins. Suppose that glucose is infused into the bloodstream at the constant rate of k g/min. Also suppose that at the same time the glucose is converted and removed from the bloodstream at a rate proportional to the amount of glucose still present, with proportionality constant r.
    a. Write a first-order differential equation that is satisfied by $G(t)$.
    b. Assume that glucose is not present initially in the patient's bloodstream; solve the differential equation.
    c. Find $\lim_{t \to \infty} G(t)$.

18. An infectious disease is introduced to a large population. The proportion of people who have been exposed to the disease increases with time. Suppose that $P(t)$ is the proportion of people who have been exposed to the disease within t years of its introduction. If $P'(t) = (1 - P(t))/3$, and $P(0) = 0$, after how many years will the proportion have increased to 90%?

In Problems 19–20, assume that the RL circuit shown in Figure 1(a) has the given resistance, inductance, emf, and initial current. Find an expression for the current at all times t and calculate the current after 0.1 sec.

▦ 19. $R = 8$ ohms, $L = 1$ henry, $E = 6$ volts, $I(0) = 1$ ampere

CONTINUED

20. $R = 50$ ohms, $L = 2$ henrys, $E = 100$ volts, $I(0) = 0$ amperes

In Problems 21–22, use the given resistance, capacitance, emf, and initial charge on the capacitor in the RC circuit shown in Figure 1(b). Find an expression for the charge at all times t.

21. $R = 10$ ohms, $C = 0.001$ farad, $E = 10\cos 60t$ volts, $Q(0) = 0$ coulombs

22. $R = 1$ ohm, $C = 0.01$ farad, $E = \sin 60t$ volts, $Q(0) = 0$ coulombs

23. Solve the problem in Example 3 with an emf of $E = 100\sin 60t$ volts.

*24. Solve the problem in Example 4 with an emf of $E = 100\sin 120\pi t$ volts.

*25. An inductance of 1 H and a resistance of 2 Ω are connected in series with a battery of $6e^{-0.001t}$ volts. No current is flowing initially. When will the current measure 0.5 amperes?

26. A variable resistance $R = 1/(5 + t)$ ohms and a capacitance of 5×10^{-6} farads are connected in series with an emf of 100 volts. If $Q(0) = 0$ coulombs, what is the charge on the capacitor after one microsecond (10^{-6} sec)?

27. In the RC circuit (Figure 1(b)) with constant voltage E, how long will it take the current to decrease to one-half its original value?

28. Suppose that the voltage in an RC circuit is $E(t) = E_0 \cos \omega t$, where $2\pi/\omega$ is the period of the cycle. Assuming that the initial charge is zero, what are the charge and current as functions of R, C, ω, and t?

*29. Solve the equation

$$y - x \cdot \frac{dy}{dx} = \frac{dy}{dx} \cdot y^2 e^y$$

by reversing the roles of x and y (i.e., by considering x as a function of y). [Hint· Use the fact (from Section 6.1) that, under certain conditions, $dx/dy = 1/(dy/dx)$.]

30. Use the technique of Problem 29 to solve the differential equation

$$\frac{dy}{dx} = \frac{-1}{x - e^{-y}}.$$

■ SHOW/PROVE/DISPROVE

31. Suppose y_1 and y_2 are solutions of the nonhomogeneous linear differential equation $y' + a(x) \cdot y = f(x)$. Prove that $y_1 - y_2$ is then a solution of the homogeneous equation $y' + a(x) \cdot y = 0$.

32. Show that the current in Problem 28 consists of two parts: a **steady-state** term that has a period of $2\pi/\omega$ and a **transient** term that tends to zero as t increases.

33. In Problem 32, show that if R is small, then the transient term can be quite large for small values of t. (This is why fuses can blow when a switch is flipped.)

■ ANSWERS TO SELF-QUIZ

I. d	II. c	III. b	d. False	e. False	f. True
IV. b	V. a		g. True	h. True	i. False
VI. a. False	b. False	c. True	j. True		

16.4 SECOND-ORDER LINEAR, HOMOGENEOUS DIFFERENTIAL EQUATIONS WITH CONSTANT COEFFICIENTS

The most general second-order linear differential equation is

$$y''(x) + a(x)y'(x) + b(x)y(x) = f(x). \tag{1}$$

Unlike the analogous first-order equation (16.3.2), it is not possible, in general, to find solutions to (1) by integration. However, it is always possible to solve (1) if the functions $a(x)$ and $b(x)$ are constants. We will therefore consider the constant-coefficient equation

$$y''(x) + ay'(x) + by(x) = f(x) \tag{2}$$

and the related homogeneous equation

$$y''(x) + ay'(x) + by(x) = 0. \tag{3}$$

In this section, we will show how solutions to (3) can be found. In Section 16.5, we will deal with the nonhomogeneous equation (2) for a number of choices of the function f.

THEOREM 1 **PRINCIPLE OF SUPERPOSITION**

Let $y_1(x)$ and $y_2(x)$ be solutions to the homogeneous equation (3).

(i) For any constants c_1 and c_2, $c_1 y_1 + c_2 y_2$ is also a solution to (3).

(ii) Let $y(x)$ be any other solution to (3). If y_2 is not a constant multiple of y_1 and $y_1 \not\equiv 0$, then there exist constants k_1 and k_2 such that

$$y(x) = k_1 y_1(x) + k_2 y_2(x) \qquad (4)$$

for every x at which $y_1(x)$ and $y_2(x)$ are defined. ∎

REMARK 1: The fact in (i) is referred to as the **principle of superposition**. The expression $c_1 y_1 + c_2 y_2$ is called a **linear combination** of the functions y_1 and y_2. The principle of superposition states that any linear combination of solutions to (3) is again a solution to (3).

REMARK 2: Part (ii) tells us that if we know two solutions to (3) that are "independent" in the sense that one is not a constant multiple of the other,[†] then we know them all, for any other solution can be written as a linear combination of these two independent solutions.

The proof of the first part of the theorem is not difficult and is left as an exercise (see Problem 21). The proof of the second part of the theorem relies on the uniqueness theorem for initial-value problems (Theorem 16.3.1). However, the proof would take us too far afield and so is omitted.[‡]

We now turn to the problem of finding two independent solutions to (3). Recall that for the analogous first-order equation $y' + ay = 0$, the general solution is $y(x) = Ce^{-ax}$. It is then reasonable to "guess" that there may be a solution to (3) of the form $y(x) = e^{\lambda x}$ for some number λ. Setting $y(x) = e^{\lambda x}$ in (3), we obtain

$$y' = \lambda e^{\lambda x}, \qquad y'' = \lambda^2 e^{\lambda x},$$

and

$$y'' + ay' + by = (\lambda^2 + a\lambda + b)e^{\lambda x}.$$

We see that $y = e^{\lambda x}$ will be a solution to (3) if and only if

---

THE AUXILIARY EQUATION

$$\lambda^2 + a\lambda + b = 0. \qquad (5)$$

[†] y_2 is a constant multiple of y_1 if there exists a constant c such that $y_2(x) = cy_1(x)$ for every x for which $y_1(x)$ is defined.

[‡] For a proof, see W. Derrick and S. Grossman, *Introduction to Differential Equations with Boundary Value Problems*, 3rd ed. (St. Paul: West, 1987), Chapter 3.

Equation (5) is called the **auxiliary** or **characteristic equation** of the homogeneous differential equation (3). From the quadratic formula, equation (5) has the roots

$$\lambda_1 = \frac{-a + \sqrt{a^2 - 4b}}{2} \quad \text{and} \quad \lambda_2 = \frac{-a - \sqrt{a^2 - 4b}}{2}. \tag{6}$$

There are three possibilities: $a^2 - 4b > 0$, $a^2 - 4b = 0$, and $a^2 - 4b < 0$.

Case 1: Two Real Roots If $a^2 - 4b > 0$, then λ_1 and λ_2 are two distinct real numbers and $y_1(x) = e^{\lambda_1 x}$ and $y_2 = e^{\lambda_2 x}$ are independent solutions since, clearly, $e^{\lambda_2 x}$ is not a constant multiple of $e^{\lambda_1 x}$ if $\lambda_1 \neq \lambda_2$. The general solution to (3) is then given by

---

GENERAL SOLUTION IN CASE OF TWO REAL ROOTS

$$y(x) = c_1 e^{\lambda_1 x} + c_2 e^{\lambda_2 x} \tag{7}$$

---

where c_1 and c_2 denote arbitrary constants.

EXAMPLE 1
SOLVING A LINEAR,
HOMOGENEOUS SECOND-ORDER
DIFFERENTIAL EQUATION: TWO
REAL ROOTS

(a) Find the general solution to

$$y''(x) + 2y'(x) - 8y(x) = 0.$$

(b) Find the unique solution that satisfies $y(0) = 1$ and $y'(0) = 3$.

SOLUTION:

(a) The auxiliary equation is

$$\lambda^2 + 2\lambda - 8 = 0, \quad \text{or} \quad (\lambda + 4)(\lambda - 2) = 0,$$

with roots $\lambda = -4$ and $\lambda = 2$. Thus, two independent solutions are $y_1(x) = e^{-4x}$ and $y_2(x) = e^{2x}$, so the general solution is

$$y(x) = c_1 e^{-4x} + c_2 e^{2x}.$$

(b) $y'(x) = -4c_1 e^{-4x} + 2c_2 e^{2x}$, so

$$y(0) = c_1 + c_2 = 1$$

and

$$y'(0) = -4c_1 + 2c_2 = 3.$$

Multiplying the first equation by 4 and adding the two equations together, we find that

$$c_1 = -\frac{1}{6} \quad \text{and} \quad c_2 = \frac{7}{6},$$

so the solution to the initial-value problem is

$$y(x) = -\frac{1}{6}e^{-4x} + \frac{7}{6}e^{2x}.$$

This answer should be checked by differentiation.

Case 2: Real Repeated Root If $a^2 - 4b = 0$, then the roots in (6) are both equal to $-a/2$. One solution is, therefore,

$$y_1(x) = e^{-(a/2)x}. \tag{8}$$

Another, independent solution is given by

$$y_2(x) = xe^{-(a/2)x}. \tag{9}$$

We prove this by differentiation. We have

$$y_2'(x) = \left(1 - \frac{a}{2}x\right)e^{-(a/2)x}$$

and

$$y_2''(x) = \left(\frac{a^2}{4}x - a\right)e^{-(a/2)x},$$

so that

$$y'' + ay' + by = e^{-(a/2)x}\left[\left(\frac{a^2}{4}x - a\right) + a\left(1 - \frac{a}{2}x\right) + bx\right]$$

$$= e^{-(a/2)x}\left(-\frac{a^2}{4}x + bx\right).$$

But $a^2 - 4b = 0$, so $a^2/4 - b$ and

$$e^{-(a/2)x}\left(-\frac{a^2}{4}x + bx\right) = 0,$$

which completes the demonstration. Finally, in Case 2 we see that the general solution is given by

---

GENERAL SOLUTION IN CASE OF ONE REPEATED REAL ROOT

$$y(x) = c_1e^{-(a/2)x} + c_2xe^{-(a/2)x}. \tag{10}$$

EXAMPLE 2
SOLVING A LINEAR,
HOMOGENEOUS SECOND-ORDER
DIFFERENTIAL EQUATION: REAL
REPEATED ROOT

(a) Find the general solution of

$$y''(x) - 6y'(x) + 9y(x) = 0.$$

(b) Find the unique solution that satisfies $y(0) = 2$ and $y'(0) = 3$.

SOLUTION:

(a) Here $\lambda^2 - 6\lambda + 9 = 0 = (\lambda - 3)^2$, so the only root is $\lambda = 3$. Thus, the general solution is given [using (10)] by

$$y(x) = c_1e^{3x} + c_2xe^{3x}.$$

(b) We have $y'(x) = 3c_1e^{3x} + c_2(1 + 3x)e^{3x}$. Thus,

$$y(0) = c_1 = 2 \quad \text{and} \quad y'(0) = 3c_1 + c_2 = -3,$$

so that $c_1 = 2$, $c_2 = -9$, and the solution to the initial-value problem is

$$y(x) = 2e^{3x} - 9xe^{3x} = e^{3x}(2 - 9x).$$

Case 3: Two Complex Roots   If $a^2 - 4b < 0$, then the roots of (6) are not real numbers. Rather than digress with a discussion of complex numbers, we will simply tell you the two independent solutions. Let

$$\alpha = -\frac{a}{2} \quad \text{and} \quad \beta = \frac{\sqrt{4b - a^2}}{2}. \tag{11}$$

By assumption, $4b - a^2 > 0$, so that both α and β are real numbers. Then two independent solutions of (3) are given by

$$y_1(x) = e^{\alpha x} \cos \beta x \quad \text{and} \quad y_2(x) = e^{\alpha x} \sin \beta x, \tag{12}$$

and the general solution is

GENERAL SOLUTION IN CASE OF COMPLEX CONJUGATE ROOTS

$$y(x) = e^{\alpha x}(c_1 \cos \beta x + c_2 \sin \beta x). \tag{13}$$

We prove that $e^{\alpha x} \cos \beta x$ is a solution of $y'' + ay' + by = 0$. The proof for $e^{\alpha x} \sin \beta x$ is similar. If

$$y(x) = e^{\alpha x} \cos \beta x,$$

then

$$y'(x) = e^{\alpha x}(\alpha \cos \beta x - \beta \sin \beta x)$$

and

$$y''(x) = e^{\alpha x}[(\alpha^2 - \beta^2) \cos \beta x - 2\alpha\beta \sin \beta x].$$

We have

$$y'' + ay' + by = e^{\alpha x}[(\alpha^2 - \beta^2 + a\alpha + b) \cos \beta x - (2\alpha\beta + a\beta) \sin \beta x].$$

To show that $e^{\alpha x} \cos \beta x$ is a solution, we must show that

$$\alpha^2 - \beta^2 + a\alpha + b = 0 \quad \text{and} \quad 2\alpha\beta + a\beta = 0.$$

But from (11), $\alpha^2 = a^2/4$ and $\beta^2 = (4b - a^2)/4 = b - a^2/4$, so

$$\alpha^2 - \beta^2 + a\alpha + b = \frac{a^2}{4} - b + \frac{a^2}{4} - \frac{a^2}{2} + b = 0.$$

Similarly,

$$2\alpha\beta + a\beta = \beta(2\alpha + a) = \beta\left[2\left(-\frac{a}{2}\right) + a\right] = \beta \cdot 0 = 0.$$

EXAMPLE 3
THE HARMONIC OSCILLATOR

Consider a mass m attached to a spring resting on a horizontal friction-less surface. If the mass is pulled out beyond its equilibrium (resting) position, then the spring will generally exert a restoring force F, which is proportional to the distance the mass is pulled and is in the opposite

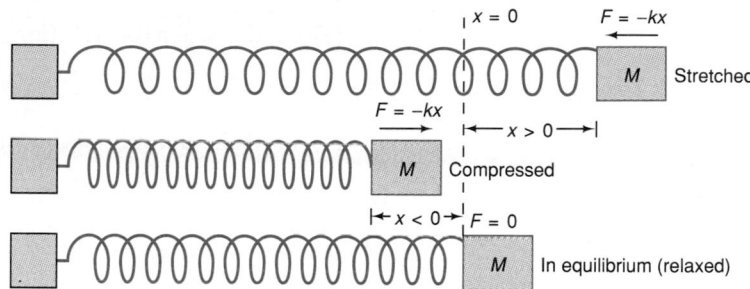

FIGURE 1   MOTION OF A VIBRATING SPRING

direction. This phenomenon can be expressed by

$$F - -kx,$$ (14)

where k is a positive number (called the **spring constant**). The minus sign indicates that the force acts in such a way as to bring the mass back to its equilibrium position. This is illustrated in Figure 1. The restoring force of the spring is the only force acting on the mass, if we ignore air resistance, since we have assumed that there is no friction. By Newton's second law of motion, the sum of the forces acting on the mass is given by $F = ma$, where a is the acceleration of the particle. But $a - d^2x/dt^2$ This leads to the equation $m\, d^2x/dt^2 = -kx$, or

$$m\frac{d^2x}{dt^2} + kx = 0.$$ (15)

Since k and m are positive numbers, we can write $k/m = \omega^2$, where ω is a positive number. Then if we divide (15) through by m, we obtain

$$\frac{d^2x}{dt^2} + \omega^2 x(t) = 0.$$ (16)

Equation (16) is called the **equation of the harmonic oscillator**.

The auxiliary equation of (16) is $\lambda^2 + \omega^2 = 0$. Here, $a = 0$, $b = \omega^2$, and $a^2 - 4b = -4\omega^2 < 0$. Thus, $\alpha = -a/2 = 0$ and $\beta = \sqrt{4b - a^2}/2 = \sqrt{4\omega^2}/2 = \omega$, so that, according to (13), the general solution is given by

$$x(t) = c_1 \cos \omega t + c_2 \sin \omega t.$$ (17)

Suppose that the initial position $(x(0))$ and the initial velocity $(x'(0))$ are known. Then, since $x'(t) = -\omega c_1 \sin \omega t + \omega c_2 \cos \omega t$, $\sin 0 = 0$, and $\cos 0 = 1$, we obtain

$$x(0) = c_1 \qquad \text{and} \qquad x'(0) = \omega c_2 \qquad \text{so } c_2 = x'(0)/\omega,$$

and the unique solution to (16) is

$$x(t) = x(0) \cos \omega t + \frac{x'(0)}{\omega} \sin \omega t.$$

We shall say more about vibratory motion in Section 16.6.

EXAMPLE 4
SOLVING A LINEAR,
HOMOGENEOUS SECOND-ORDER
DIFFERENTIAL EQUATION: TWO
COMPLEX ROOTS

(a) Find the general solution to

$$y'' + y' + y = 0.$$

(b) Find the unique solution that satisfies $y(0) = 1$ and $y'(0) = 3$.

SOLUTION:

(a) We have $\lambda^2 + \lambda + 1 = 0$ and $a^2 - 4b = 1 - 4 = -3$. Thus if we set

$$\alpha = -\frac{a}{2} = -\frac{1}{2} \quad \text{and} \quad \beta = \frac{\sqrt{4b - a^2}}{2} = \frac{\sqrt{4 - 1}}{2} = \frac{\sqrt{3}}{2},$$

wc obtain the general solution

$$y(x) = e^{-x/2}\left(c_1 \cos \frac{\sqrt{3}}{2} x + c_2 \sin \frac{\sqrt{3}}{2} x\right).$$

(b) We have

$$y'(x) = e^{-x/2}\left[\left(-\frac{1}{2}c_1 + \frac{\sqrt{3}}{2} c_2\right) \cos \frac{\sqrt{3}}{2} x \right.$$
$$\left. + \left(-\frac{\sqrt{3}}{2} c_1 - \frac{1}{2}c_2\right) \sin \frac{\sqrt{3}}{2} x\right].$$

Then

$$y(0) = c_1 = 1 \quad \text{and} \quad y'(0) = -\frac{1}{2}c_1 + \frac{\sqrt{3}}{2} c_2 = 3,$$

with solutions $c_1 = 1$ and $c_2 = 7/\sqrt{3}$. Thus, the solution to the initial-value problem is

$$y(x) = e^{-x/2}\left(\cos \frac{\sqrt{3}}{2} x + \frac{7}{\sqrt{3}} \sin \frac{\sqrt{3}}{2} x\right).$$

■ PROBLEMS 16.4

■ SELF-QUIZ

I. If a and b are constants, then $y'' - (a + b)y' + aby = 0$ is solved by $y = $ _____.
   a. e^{ax}
   b. e^{-ax}
   c. $e^{(a+b)x}$
   d. e^{abx}

II. $y'' - 5y' + 6y = 0$ is solved by $y = $ _____.
   a. $e^{2x} + e^{3x}$
   b. $-e^{-2x} - e^{-3x}$
   c. e^{-2x}
   d. $e^x + e^{-30x}$

III. e^{5x} does *not* solve _____.
   a. $y'' - (5 + k)y' + 5ky = 0$
   b. $y'' - 3y' - 10y = 0$

   c. $y'' + 8y' - 65y = 0$
   d. $y'' + 5y' + 50y = 0$

IV. $y'' + 4y = 0$ is *not* solved by _____.
   a. $\cos 2x$
   b. $3 \sin 2x - 5 \cos 2x$
   c. $6 \sin x - 10 \cos x$
   d. $\cos(2x + \pi/3)$

V. xe^x solves _____.
   a. $y'' - y = 0$
   b. $y'' - y/x = 0$
   c. $y'' - 2y' + y = 0$
   d. $y'' - 2xy' + x^2y = 0$

CONTINUED

■ **DRILL**

In Problems 1–20, find the general solution of each equation. Where initial conditions are specified, determine the particular solution that satisfies them.

1. $y'' - 4y = 0$
2. $y'' + y' - 3y = 0$; $y(0) = 0$, $y'(0) = 1$
3. $y'' + 2y' + 2y = 0$
4. $y'' - 3y' + 2y = 0$
5. $8y'' + 4y' + y = 0$; $y(0) = 0$, $y'(0) = 1$
6. $y'' + 8y' + 16y = 0$
7. $y'' + 5y' + 6y = 0$; $y(0) = 1$, $y'(0) = 2$
8. $y'' + y' + 7y = 0$

9. $4y'' + 20y' + 25y = 0$; $y(0) = 1$, $y'(0) = 2$
10. $y'' + y' + 2y = 0$
11. $y'' - y' - 6y = 0$; $y(0) = -1$, $y'(0) = 1$
12. $y'' - 5y' = 0$
13. $y'' + 4y = 0$; $y(\pi/4) = 1$, $y'(\pi/4) = 3$
14. $y'' - 10y' + 25y = 0$; $y(0) = 2$, $y'(0) = -1$
15. $y'' + 17y' = 0$; $y(0) = 1$, $y'(0) = 0$
16. $y'' + 2\pi y' + \pi^2 y = 0$; $y(0) = 1$, $y'(0) = 1/\pi$
17. $y'' - 13y' + 42y = 0$
18. $y'' + 4y' + 6y = 0$
19. $y'' = y$; $y(0) = 2$, $y'(0) = -3$
20. $y'' - y' + y = 0$; $y(0) = 3$, $y'(0) = 7$

■ **SHOW/PROVE/DISPROVE**

21. Prove Theorem 1 (i). [*Hint:* Let $y = c_1 y_1 + c_2 y_2$. Differentiate twice and show that y satisfies equation (3), using the fact that y_1 and y_2 satisfy equation (3).]

16.5 SECOND-ORDER NONHOMOGENEOUS DIFFERENTIAL EQUATIONS WITH CONSTANT COEFFICIENTS:
THE METHOD OF UNDETERMINED COEFFICIENTS

In this section, we present a method for finding a particular solution to the nonhomogeneous equation

$$y''(x) + ay'(x) + by(x) = f(x). \tag{1}$$

We need only find one solution to (1) since if y_p and y_q are solutions to (1), then $y_p - y_q$ is a solution to the homogeneous equation

$$y''(x) + ay'(x) + by(x) = 0. \tag{2}$$

To see this, we note that

$$(y_p - y_q)'' + a(y_p - y_q)' + b(y_p - y_q) = y_p'' - y_q'' + ay_p' - ay_q' + by_p - by_q$$
$$= (y_p'' + ay_p' + by_p) - (y_q'' + ay_q' + by_q)$$
$$= f - f = 0.$$

Thus, the general solution to (1) can be written as the sum of one particular solution y_p to (1) plus the general solution to the homogeneous equation (2).

Let $P_n(x)$ be a polynomial of degree n. We will show how to find a solution to (1) if $f(x)$ takes one of the following forms:

(i) $P_n(x)$ (3)

(ii) $P_n(x)e^{ax}$ (4)

(iii) $P_n(x)e^{ax} \sin bx$ or $P_n(x)e^{ax} \cos bx$ (5)

The technique we will use is to "guess" that there is a solution to (1) in the same basic "form" as $f(x)$ and then substitute this "guessed" solution into

(1) to determine the unknown coefficients. This technique is called the **method of undetermined coefficients** and is best illustrated with examples. There is a more general method for finding solutions to (1) for an arbitrary f, but that method, called **variation of constants**, is more complicated and is best left to a book on differential equations.

EXAMPLE 1
USING THE METHOD OF
UNDETERMINED COEFFICIENTS

Find the general solution to

$$y'' + y = x^2. \tag{6}$$

SOLUTION: We first solve the homogeneous equation $y'' + y = 0$. The general solution to this equation is (see Example 3 on page 1054)

$$c_1 \cos x + c_2 \sin x. \tag{7}$$

Next, we find a particular solution to the nonhomogeneous equation (6). Since x^2 is a polynomial of degree 2, we "guess" that there is a solution to (6) that is a polynomial of degree 2. Let

$$y_p(x) = ax^2 + bx + c.$$

Then

$$y_p' = 2ax + b, \qquad y_p'' = 2a,$$

and

$$y_p'' + y_p = ax^2 + bx + c + 2a.$$

If y_p is a solution to (6), then we must have

$$ax^2 + bx + c + 2a = x^2,$$

which implies that $a = 1$, $b = 0$, and $c + 2a = 0$, so that $a = 1$, $b = 0$, $c = -2$, and

$$y_p(x) = x^2 - 2.$$

You should verify that this is indeed a solution to (6). Since the general solution to $y'' + y = 0$ is

$$c_1 \cos x + c_2 \sin x,$$

the general solution to (6) is

$$y(x) = c_1 \cos x + c_2 \sin x + x^2 - 2. \tag{8}$$

EXAMPLE 2
USING THE METHOD OF
UNDETERMINED COEFFICIENTS

Find the general solution of

$$y'' - 4y' + 4y = xe^{3x}. \tag{9}$$

SOLUTION: We first solve the homogeneous equation

$$y'' - 4y' + 4y = 0.$$

The auxiliary equation is $\lambda^2 - 4\lambda + 4 = (\lambda - 2)^2 = 0$ with the single root $\lambda = 2$. Thus the general solution to the homogeneous equation is

$$c_1 e^{2x} + c_2 x e^{2x}. \tag{10}$$

Since x is a polynomial of degree 1, we seek a solution to (9) of the form

$$y_p(x) = (ax + b)e^{3x}.$$

Then

$$y_p' = (3ax + 3b + a)e^{3x} \quad \text{and} \quad y_p'' = (9ax + 9b + 6a)e^{3x},$$

so

$$y_p'' - 4y_p' + 4y_p = e^{3x}[9ax + 9b + 6a - 4(3ax + 3b + a) + 4(ax + b)]$$
$$= e^{3x}(ax + b + 2a) = xe^{3x}.$$

Thus

$$a = 1,$$
$$b + 2a = 0,$$

and $a = 1$ and $b = -2$, so that

$$y_p(x) = (x - 2)e^{3x}.$$

Finally, using the homogeneous solution (10), we find that the general solution to the nonhomogeneous equation (9) is

$$y(x) = c_1 e^{2x} + c_2 x e^{2x} + (x - 2)e^{3x}.$$

EXAMPLE 3
USING THE METHOD OF UNDETERMINED COEFFICIENTS WHEN THE NONHOMOGENEOUS TERM IS A SOLUTION TO THE HOMOGENEOUS EQUATION

Find a particular solution to

$$y'' - y = 3e^{-x}. \tag{11}$$

SOLUTION: We first note that the general solution to the related homogeneous equation

$$y'' - y = 0 \tag{12}$$

is $c_1 e^x + c_2 e^{-x}$ Thus $3e^{-x}$ is a solution to (12). If we try to determine a solution to (11) in the form $y_p(x) = ae^{-x}$, we find that

$$y_p' = -ae^{-x}, \quad y_p'' = ae^{-x},$$

and

$$y_p'' - y_p = ae^{-x} - ae^{-x} = 0.$$

But this should have been obvious anyway since ae^{-x} is a solution to (12) and, therefore, could not be a solution to (11). To find a solution to (11), we try the trick that worked for solving Case 2 of the homogeneous equation. We look for a solution to (12) in the form

$$y_p(x) = axe^{-x}.$$

Then

$$y_p'(x) = a(1 - x)e^{-x} \quad \text{and} \quad y_p''(x) = a(x - 2)e^{-x},$$

so

$$y_p'' - y_p = a(-2)e^{-x} = 3e^{-x}.$$

This implies that $a = -\frac{3}{2}$, so a particular solution is given by

$$y_p(x) = -\frac{3}{2}xe^{-x}.$$

We now describe in detail the method of undetermined coefficients to obtain a solution to the nonhomogeneous equation (1) when $f(x)$ is of the form (3), (4), or (5).

Case 1 No term in $y_p(x)$ (as given in Table 1) is a solution of the related homogeneous equation (2). Then a particular solution of (1) will have a form according to Table 1.

TABLE 1

$f(x)$	$y_p(x)$
$P_n(x)$	$a_0 + a_1x + a_2x^2 + \cdots + a_nx^n$
$P_n(x)e^{ax}$	$(a_0 + a_1x + a_2x^2 + \cdots + a_nx^n)e^{ax}$
$\left.\begin{array}{l} P_n(x)e^{ax}\sin bx \\ \quad\quad\text{or} \\ P_n(x)e^{ax}\cos bx \end{array}\right\}$	$\begin{array}{l}(a_0 + a_1x + a_2x^2 + \cdots + a_nx^n)e^{ax}\sin bx \\ \quad + (b_0 + b_1x + b_2x^2 + \cdots + b_nx^n)e^{ax}\cos bx\end{array}$

Case 2 If *any term* of $y_p(x)$ is a solution of (2), then multiply the appropriate function $y_p(x)$ of Case 1 by x^k, where k is the smallest integer having the property that no term of $x^k y_p(x)$ is a solution of (2).

EXAMPLE 4
MULTIPLYING BY x^2 TO FIND A
PARTICULAR SOLUTION

Find the general solution of

$$y'' - 4y' + 4y = e^{2x}.$$

SOLUTION: The homogeneous equation $y'' - 4y' + 4y = 0$ has the independent solutions e^{2x} and xe^{2x}. Thus, multiplying $f(x) = e^{2x}$ by x^2 (since, using Case 2, x^2e^{2x} is not a solution to the homogeneous equation but xe^{2x} is a solution), we look for a particular solution of the form $y_p = ax^2e^{2x}$. Then

$$y_p' = ae^{2x}(2x^2 + 2x)$$

and

$$y_p'' = ae^{2x}(4x^2 + 8x + 2),$$

so

$$y_p'' - 4y_p' + 4y_p = ae^{2x}(4x^2 + 8x + 2 - 8x^2 - 8x + 4x^2)$$
$$= 2ae^{2x} = e^{2x},$$

or $2a = 1$ and $a = \frac{1}{2}$. Thus $y_p = \frac{1}{2}x^2 e^{2x}$, and the general solution is

$$y(x) = c_1 e^{2x} + c_2 x e^{2x} + \frac{1}{2}x^2 e^{2x} = e^{2x}\left(c_1 + c_2 x + \frac{1}{2}x^2\right).$$

EXAMPLE 5
USING THE METHOD OF
UNDETERMINED COEFFICIENTS
WHEN ONE TERM OF THE
GUESSED SOLUTION IS A
SOLUTION TO THE
HOMOGENEOUS EQUATION

Find the general solution to

$$y'' + y = x \sin x.$$

SOLUTION: The guessed solution is

$$y_p = (Ax + B)\cos x + (Cx + D)\sin x.$$

Since $B \cos x + D \sin x$ solves $y'' + y = 0$, the modification in Case 2 is required. We therefore multiply by x and try a solution of the form

$$y_p = (Ax^2 + Bx)\cos x + (Cx^2 + Dx)\sin x.$$

Then

$$y_p' = [Cx^2 + (2A + D)x + B]\cos x$$
$$+ [-Ax^2 + (2C - B)x + D]\sin x,$$

$$y_p'' = [-Ax^2 + (4C - B)x + 2A + 2D]\cos x$$
$$+ [-Cx^2 - (4A + D)x + 2C - 2B]\sin x,$$

and

given
↓

$$y_p'' + y_p = [4Cx + 2A + 2D]\cos x + [-4Ax + 2C - 2B]\sin x = x\sin x.$$

This yields $A = -\frac{1}{4}$, $B = 0$, $C = 0$, $D = \frac{1}{4}$, and the particular solution

$$y_p = -\frac{1}{4}x^2\cos x + \frac{1}{4}x\sin x.$$

Thus the general solution is

$$y = c_1\cos x + c_2\sin x - \frac{1}{4}x^2\cos x + \frac{1}{4}x\sin x,$$

or, combining terms,

$$y = \left(c_1 - \frac{1}{4}x^2\right)\cos x + \left(c_2 + \frac{1}{4}x\right)\sin x.$$

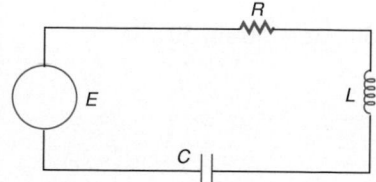

FIGURE 1
AN *RLC* CIRCUIT

MORE ON SIMPLE ELECTRIC CIRCUITS

We will make use of the concepts developed in Section 16.3 and the methods of this section to study a simple electric circuit containing a resistor, an inductor, and a capacitor in series with an electromotive force (Figure 1).

Suppose that R, L, C, and E are constants. Applying Kirchhoff's law, we obtain

$$L\frac{dI}{dt} + RI + \frac{Q}{C} = E. \tag{13}$$

Since $dQ/dt = I$, we differentiate (13) to get the second-order homogeneous differential equation

$$L\frac{d^2I}{dt^2} + R\frac{dI}{dt} + \frac{I}{C} = 0. \tag{14}$$

To solve this equation, we note that the auxiliary equation

$$\lambda^2 + \frac{R}{L}\lambda + \frac{1}{CL} = 0$$

has the following roots:

$$\lambda_1 = \frac{-R + \sqrt{R^2 - 4L/C}}{2L}, \qquad \lambda_2 = \frac{-R - \sqrt{R^2 - 4L/C}}{2L}. \tag{15}$$

Equation (14) may now be solved using the methods of Section 16.4.

EXAMPLE 6
FINDING CHARGE AND CURRENT
IN AN *RLC* CIRCUIT

Let $L = 1$ H, $R = 100\ \Omega$, $C = 10^{-4}$ F, and $E = 1000$ V in the circuit shown in Figure 1. Suppose that no charge is present and no current is flowing at time $t = 0$ when E is applied. Here $R^2 - 4L/C = 10{,}000 - 4 \times 10^4 = -30{,}000$, so we have Case 3 with $\alpha = -R/2L = -50$ and $\beta = \sqrt{30{,}000}/2 = 50\sqrt{3}$. Thus, the general solution to (14) is

$$I(t) = e^{-50t}(c_1 \cos 50\sqrt{3}t + c_2 \sin 50\sqrt{3}t).$$

Applying the initial condition $I(0) = 0$, we have $c_1 = 0$. Hence,

$$I(t) = c_2 e^{-50t} \sin 50\sqrt{3}t$$

and

$$I'(t) = c_2 e^{-50t}(50\sqrt{3} \cos 50\sqrt{3}t - 50 \sin 50\sqrt{3}t).$$

To establish the value of c_2, we must make use of equation (13) and the initial condition $Q(0) = 0$. Now

$$Q(t) = C\left(E - L\frac{dI}{dt} - RI\right)$$

$$= 10^{-4}[1000 - c_2 e^{-50t}(50\sqrt{3} \cos 50\sqrt{3}t - 50 \sin 50\sqrt{3}t + 100 \sin 50\sqrt{3}t)]$$

$$50 \cdot 10^{-4} = \frac{1}{200}$$

$$\Big\downarrow$$

$$= \frac{1}{10} - \frac{c_2}{200} e^{-50t}(\sin 50\sqrt{3}t + \sqrt{3} \cos 50\sqrt{3}t).$$

Thus,

$$Q(0) = \frac{1}{10} - \frac{c_2}{200}\sqrt{3} = 0 \qquad \text{so } c_2 = \frac{20}{\sqrt{3}}.$$

Finally,

$$Q(t) = \frac{1}{10} - \frac{1}{10\sqrt{3}} e^{-50t}(\sin 50\sqrt{3}t + \sqrt{3}\cos 50\sqrt{3}t)$$

and

$$I(t) = \frac{20}{\sqrt{3}} e^{-50t} \sin 50\sqrt{3}t.$$

From these equations we observe that the current will rapidly damp out and that the charge on the capacitor will rapidly approach its steady-state value of $\frac{1}{10}$ C.

■ PROBLEMS 16.5

■ SELF-QUIZ

I. _____ is *not* a solution of $y'' + y' = 1$.
   a. $e^{-x} + 38 + x$ b. $38e^{-x} \ 1 \mid x$
   c. $-5 + x - e^{-x}$ d. $-5x + 1 - 38e^{-x}$

II. The general solution of $y'' + y' = 1$ is _____.
   a. $C_1 e^{-x} + C_2 + x$
   b. $C_1 + C_2 x + e^{-x}$
   c. $C_1 x + C_2 e^{-x} + 1$
   d. $Ce^{-x} + C + x$

III. Suppose y_1 and y_2 are solutions of the nonhomogeneous linear differential equation $y'' - 7y' + \pi^2 y = \sin x^2$. Answer True or False to each of the following assertions.

   a. $y_1 + y_2$ solves that differential equation.
   b. $y_1 - y_2$ solves that differential equation.
   c. $y_1 - y_2$ solves $y'' - 7y' + \pi^2 y = 0$.
   d. $y_1 + y_2$ solves $y'' - 7y' + \pi^2 y = 0$.
   e. $y_2 - y_1$ solves $y'' - 7y' + \pi^2 y = 0$.
   f. $8y_2 - 7y_1$ solves $y'' - 7y' + \pi^2 y = 0$.

IV. The solution to the initial-value problem $y'' + 5y' + 6y = 30$, $y(0) = 2$, $y'(0) = 9$ is _____.
   a. $C_1 e^{-2x} + C_2 e^{-3x} + 5$
   b. $-e^{-2x} + 5$
   c. $-3e^{-3x} + 5$
   d. $15e^{-2x} - 13e^{-3x}$

■ DRILL

In Problems 1–16, find the general solution of each given differential equation. If initial conditions are specified, find the particular solution that satisfies them.

1. $y'' + y' = 4x^2$; $y(0) = 3$, $y'(0) = -1$
2. $y'' + y' - 6y = 10e^{-x}$; $y(0) = 0$, $y'(0) = 2$
3. $y'' + 3y' + 2y = 3e^{4x}$
4. $y'' + y' = x^2 - x^3 + 2$
5. $y'' + 6y = 4\cos x$
6. $y'' + y = 1 + x + x^2$
7. $y'' + y = \cos x$

8. $y'' + 16y = 8\cos 4x$; $y\left(\frac{\pi}{2}\right) = 0$, $y'\left(\frac{\pi}{2}\right) = 1$
9. $y'' - 4y' + 5y = \sin x$
10. $y'' - 9y' + 20y = 40x$; $y(0) = 2$, $y'(0) = 3$
11. $y'' - 2y' + y = -2e^{3x}$
12. $y'' - 6y' + 9y = 4xe^{3x}$; $y(0) = 1$, $y'(0) = 0$
13. $y'' - 2y' + y = 2e^x$
*14. $y'' + 4y' + 5y = \sinh 2x \sin x$
*15. $y'' + y' + y = e^{-x/2} \cos\left(\frac{\sqrt{3}}{2}x\right)$
16. $y'' - 2y' + 3y = e^x \sin\sqrt{2}x$

CONTINUED

■ **APPLICATIONS**

In Problems 17–20, use the result of Problem 25 to obtain a particular solution of the given equation.

17. $y'' + y = 3 - 4 \cos x$
18. $y'' - 2y' - 3y = e^{2x} + x - x^2$
19. $y'' + 9y = 2 \sin 3x + x^2$
20. $y'' + 6y' + 9y = xe^x + \cos x$

21. In Example 4, let $L = 10$ H, $R = 250 \, \Omega$, $C = 10^{-3}$ F, and $E = 900$ V. With the same assumptions, calculate the current and charge for all values of $t \geq 0$.

22. In the preceding problem, suppose instead that $E = 50 \cos 30t$ V. Find $Q(t)$ for $t \geq 0$.

In Problems 23 and 24, find the current in the RLC circuit of Figure 1. In each case, identify the steady-state component and the transient component.

*23. $L = 10$ H, $R = 40 \, \Omega$, $C = 0.025$ F, $E = 100 \cos 5t$ V
*24. $L = 5$ H, $R = 10 \, \Omega$, $C = 0.1$ F, $E = 25 \sin t$ V

■ **SHOW/PROVE/DISPROVE**

25. Show that if y_1 is a solution to $y'' + ay' + by = f_1(x)$ and if y_2 is a solution to $y'' + ay' + by = f_2(x)$, then $y_1 + y_2$ is a solution to $y'' + ay' + by = f_1(x) + f_2(x)$.

26. Let $p(x)$ be a polynomial of degree n. Show that for any constants a and $b \neq 0$, there is a solution to $y'' + ay + by = p(x)$ that is a polynomial of degree n.

■ **ANSWERS TO SELF-QUIZ**

I. d
II. a (Item (d) is the special case with $C_1 = C_2$.)

III. a. False   b. False   c. True
    d. False   e. True   f. False
IV. c

16.6 VIBRATORY MOTION (OPTIONAL)

Differential equations were used for the first time to describe the motion of particles. As a simple example, consider the motion of a mass m attached to a coiled spring, the upper end of which is securely fastened (see Figure 1).

We have denoted the equilibrium or rest position of the mass on the spring by the number 0. Suppose that the mass is given an initial displacement x_0 and an initial velocity v_0. We now describe the motion of the mass on the spring. By Hooke's law (see page 358), the magnitude of the force exerted on the mass by the spring is proportional to the distance the spring is displaced and is directed toward the equilibrium position. That is,

$$F = -kx, \tag{1}$$

where k is a constant of proportionality. But by Newton's second law,

$$F = ma = m \frac{d^2x}{dt^2}. \tag{2}$$

If we assume that there is no other force acting on the mass, then combining (1) and (2), we obtain

$$m \frac{d^2x}{dt^2} = -kx, \tag{3}$$

or

$$m \frac{d^2x}{dt^2} + kx = 0,$$

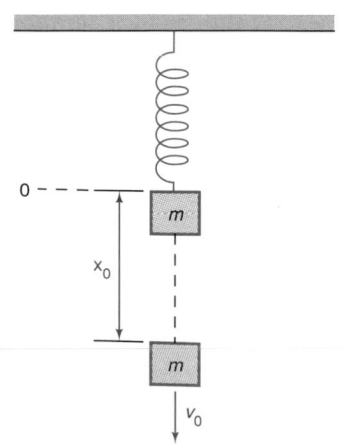

FIGURE 1
A MASS m IS ATTACHED TO A COILED SPRING.

which can be written

$$x'' + \frac{k}{m} x = 0. \tag{4}$$

In Example 16.4.3 (see page 1054) we found the unique solution (since $\omega = \sqrt{k/m}$):

EQUATION OF HARMONIC MOTION

$$x(t) = x_0 \cos \sqrt{\frac{k}{m}} t + v_0 \sqrt{\frac{m}{k}} \sin \sqrt{\frac{k}{m}} t. \tag{5}$$

We see from (5) that the motion of the mass is periodic with period

$$P = 2\pi \sqrt{\frac{m}{k}} \tag{6}$$

and **frequency** f (the frequency is the number of complete oscillations per unit time), given by

$$f = \frac{\sqrt{k/m}}{2\pi}. \tag{7}$$

It is not difficult to show (see Problem 10) that the maximum value taken by x, called the **amplitude** of the motion (and denoted by A), is given by

$$A = \sqrt{x_0{}^2 + v_0{}^2 \frac{m}{k}}. \tag{8}$$

Note that while the amplitude of the motion depends on the initial displacement and velocity, the period and frequency of the motion do not. These latter depend only on the spring constant k and the mass m.

EXAMPLE 1
DETERMINING AN EQUATION OF HARMONIC MOTION

A spring, fixed at its upper end, supports a 5-kg mass that stretches the spring 50 cm. Find the equation of motion of the mass if it is drawn to a position 20 cm below its equilibrium position and released.

SOLUTION: The force exerted by the mass on the spring is given by $F = mg = 5g$, where g is the acceleration due to gravity. Then by Hooke's law, we have

$$F = -kx, \qquad \text{or} \qquad 5g = -k(-50),$$

and

$$k = \frac{g}{10},$$

so

$$\sqrt{\frac{k}{m}} = \sqrt{\frac{g/10}{5}} = \sqrt{\frac{g}{50}}.$$

Since $x_0 = 20$ and $v_0 = 0$, we obtain, from (5),

$$x(t) = 20 \cos \sqrt{\frac{g}{50}}\, t.$$

Also, $P = 2\pi\sqrt{50/g}$, $f = (1/2\pi)\sqrt{g/50}$, and $A = 20$ cm. Note here that we must use the value $g = 981$ cm/sec^2 since our unit of length is the centimeter.

DAMPED HARMONIC MOTION

We now turn to a more practical (and complicated) situation. In the discussion above we made the assumption that there were no external forces acting on the mass. This assumption is, however, not very realistic. We must deal with other forces such as friction in the spring and air resistance, which tend to slow things down. Such forces are called **damping forces**. It is reasonable to assume that the magnitude of the damping force is proportional to the speed of the mass (e.g., the slower the motion, the smaller the air resistance). Thus to equation (3) we add the term $\mu\, dx/dt$, where μ, called the **damping constant**, depends on all the external factors. We then obtain the equation

$$m\frac{d^2x}{dt^2} = -kx - \mu\frac{dx}{dt},$$

or

$$x'' + \frac{\mu}{m}x' + \frac{k}{m}x = 0; \qquad x(0) = x_0, \qquad x'(0) = v_0. \tag{9}$$

The auxiliary equation for (9) is

$$\lambda^2 + \frac{\mu}{m}\lambda + \frac{k}{m} = 0, \tag{10}$$

and

$$\lambda = \frac{-\mu \pm \sqrt{\mu^2 - 4mk}}{2m}. \tag{11}$$

There are three cases to consider.

Case 1: $(\mu^2 > 4mk)$ Then both roots of (10) are negative and we obtain the general solution to (9):

$$x(t) = c_1 e^{\lambda_1 t} + c_2 e^{\lambda_2 t}, \tag{12}$$

where $\lambda_1 = (-\mu + \sqrt{\mu^2 - 4mk})/2m$ and $\lambda_2 = (-\mu - \sqrt{\mu^2 - 4mk})/2m$. Clearly, $x(t) \to 0$ as $t \to \infty$, and the mass quickly returns to its equilibrium position. (See Figure 2.)

Case 2: $(\mu^2 = 4mk)$ Then $\lambda_1 = \lambda_2 = -\mu/2m$ and

$$x(t) = c_1 e^{(-\mu/2m)t} + c_2 t e^{(-\mu/2m)t}, \tag{13}$$

and, again, $x(t) \to 0$ as $t \to \infty$.

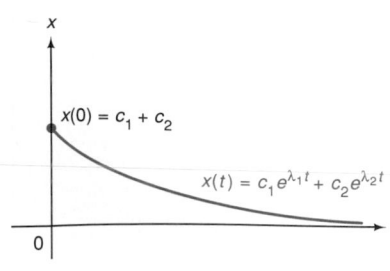

FIGURE 2
EXPONENTIAL DECAY: λ_1 AND λ_2 ARE NEGATIVE

Case 3: ($\mu^2 < 4mk$)   Then the general solution is

$$x(t) = e^{(-\mu/2m)t}\left(c_1 \cos \frac{\sqrt{4mk - \mu^2}}{2m} t + c_2 \sin \frac{\sqrt{4mk - \mu^2}}{2m} t\right). \tag{14}$$

This motion is more interesting and is called **damped harmonic motion**. The mass oscillates about its equilibrium position but with an ever-decreasing amplitude. The term $e^{(-\mu/2m)t}$ is called a **damping factor**. This motion is illustrated in Figure 3.

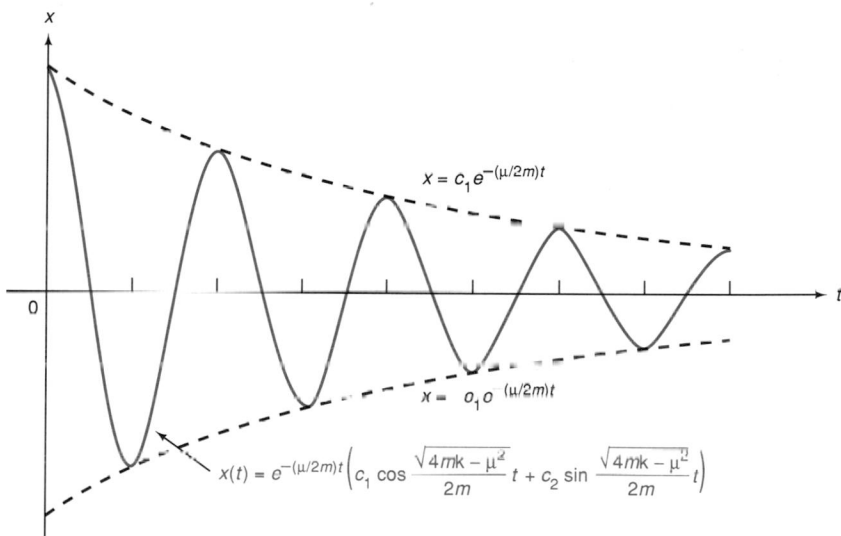

FIGURE 3   GRAPH OF DAMPED HARMONIC MOTION

EXAMPLE 2
DETERMINING AN EQUATION OF
DAMPED HARMONIC MOTION

In Example 1, assume that there are damping forces and that $\mu = 2$. Find the new equation of motion.

SOLUTION: We have $\mu^2 - 4mk = 4 - 4(5)(g/10) = 4 - 2g$, which is less than 0 (since $g = 981$). Thus since $\mu/2m = 2/10 = 0.2$, we obtain

$$x(t) = e^{-0.2t}\left(c_1 \cos \frac{\sqrt{2g - 4}}{10} t + c_2 \sin \frac{\sqrt{2g - 4}}{10} t\right).$$

Then

$$x'(t) = e^{-0.2t}\left[\left(c_2 \frac{\sqrt{2g - 4}}{10} - 0.2c_1\right) \cos \frac{\sqrt{2g - 4}}{10} t \right.$$
$$\left. + \left(-\frac{\sqrt{2g - 4}}{10} c_1 - 0.2c_2\right) \sin \frac{\sqrt{2g - 4}}{10} t\right].$$

Thus

$$20 = x(0) = c_1$$

and

$$0 = x'(0) = c_2 \frac{\sqrt{2g - 4}}{10} - 0.2c_1,$$

and we obtain

$$c_1 = 20, \qquad c_2 = \frac{40}{\sqrt{2g - 4}},$$

so

$$x(t) = e^{-0.2t}\left(20 \cos \frac{\sqrt{2g - 4}}{10}\, t + \frac{40}{\sqrt{2g - 4}} \sin \frac{\sqrt{2g - 4}}{10}\, t\right).$$

This motion is sketched in Figure 4.

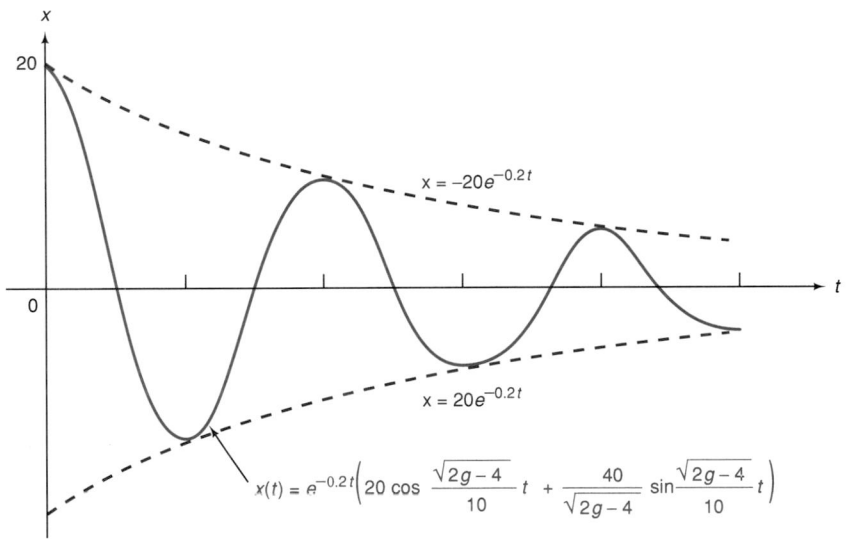

FIGURE 4   DAMPED HARMONIC MOTION

FORCED VIBRATIONS

We now turn to another situation. The motion of the mass considered in the three cases above is determined by the inherent forces of the spring-weight system and the natural forces acting on the system. Accordingly, the vibrations are called **free** or **natural vibrations**. We will now assume that the system is also subject to an external periodic force $\alpha \sin \omega t$, perhaps due to the motion of the object to which the upper end of the spring is attached (see Figure 1 again). In this case the mass will undergo **forced vibrations**.

Equation (9) may be replaced by the nonhomogeneous second-order differential equation

$$m \frac{d^2x}{dt^2} = -kx - \mu \frac{dx}{dt} + \alpha \sin \omega t,$$

which we write in the form

$$\frac{d^2x}{dt^2} + \frac{\mu}{m}\frac{dx}{dt} + \frac{k}{m}x = \frac{\alpha}{m}\sin \omega t. \tag{15}$$

By the method of undetermined coefficients, we know that $x(t)$ has a particular solution of the form

$$x_p(t) = A_1 \cos \omega t + A_2 \sin \omega t. \tag{16}$$

Substituting this function into (15) yields the simultaneous equations

$$\left(\frac{k}{m} - \omega^2\right)A_1 + \frac{\mu\omega}{m}A_2 = 0 \quad \text{and} \quad -\frac{\mu\omega}{m}A_1 + \left(\frac{k}{m} - \omega^2\right)A_2 = \frac{\alpha}{m}, \tag{17}$$

from which we obtain

$$A_1 = \frac{-\alpha\mu\omega}{(k - m\omega^2)^2 + (\mu\omega)^2} \quad \text{and} \quad A_2 = \frac{\alpha(k - m\omega^2)}{(k - m\omega^2)^2 + (\mu\omega)^2}. \tag{18}$$

To find the general solution of equation (15), we add the general solution of the homogeneous equation to the particular solution (16). The nature of the general solution again depends on the discriminant $\sqrt{\mu^2 - 4mk}$. If $\mu^2 \geq 4mk$, we are superimposing the periodic function (16) on either equation (12) or (13). Since (12) and (13) damp out as t increases, the motion for large t will be very close to (16). On the other hand, if $\mu^2 < 4mk$, we will suppose initially that $\mu = 0$. Then the general solution for $\sqrt{k/m} \neq \omega$ is

$$x(t) = c_1 \cos\sqrt{\frac{k}{m}}t + c_2 \sin\sqrt{\frac{k}{m}}t + \frac{\alpha}{m[(k/m) - \omega^2]}\sin \omega t. \tag{19}$$

Note that the effect of the sinusoidal force $\alpha \sin \omega t$ is merely to superimpose itself on the simple harmonic motion of the system. The period of this added motion is the same as that of the external force, with constant amplitude for each fixed ω, but the amplitude approaches infinity as ω approaches $\sqrt{k/m}$. If $\omega = \sqrt{k/m}$, then the particular solution must have the form (from Case 2 on page 1060)

$$x_p(t) = A_1 t \cos \omega t + A_2 t \sin \omega t. \tag{20}$$

Substituting (20) into (15) yields $A_1 = -\alpha/2\omega$ and $A_2 = 0$, so we have the general solution

$$x(t) = c_1 \cos \omega t + c_2 \sin \omega t - \frac{\alpha}{2\omega}t \cos \omega t. \tag{21}$$

We note that as t increases, the vibrations will get larger without bound. Of course, this situation cannot arise in practice, since some resistance to the motion will always be present ($\mu \neq 0$). However, it is evident that if the resistance is small and the external forcing motion agrees with the natural vibration, then the displacement x may become so large that the elastic limit of the spring will be exceeded and a permanent set or distortion will occur. If $\omega = \sqrt{k/m}$, we say the external force is in **resonance** with the vibrating mass. The phenomenon of resonance is of great importance in engineering. It is necessary in some cases, as with our spring problem, to avoid a resonant condition so that no undue stresses will occur; on the other hand, resonance is desirable in many radio circuit problems.

■ PROBLEMS 16.6

■ SELF-QUIZ

I. Which of the following equations are equations of damped harmonic motion

$$x'' + \frac{\mu}{m} x' + \frac{k}{m} x = 0; \quad x(0) = x_0, \quad x'(0) = v_0?$$

a. $\mu = 1, m = 1, k = 1$

b. $\mu = 2, m = 1, k = 1$
c. $\mu = 2, m = 1, k = 2$
d. $\mu = 2, m = 2, k = 1$
e. $\mu = 5, m = 2, k = 2$
f. $\mu = 5, m = 2, k = 3$
g. $\mu = 5, m = 3, k = 3$

■ APPLICATIONS

1. One end of a rubber band is fixed at a point A. A 2-kg mass, attached to the other end, stretches the rubber band vertically to the point B in such a way that the length AB is 32 cm greater than the natural length of the band. If the mass is further drawn to a position 16 cm below B and released, what will be its velocity (if we neglect resistance) as it passes the position B?

2. If in Problem 1 the mass is released at a position 16 cm above B, what will be its velocity as it passes 2 cm above B?

*3. A cylindrical block of wood of radius and height 1 ft and weighing 12.48 lb floats with its axis vertical in water (62.4 lb/ft³). If it is depressed so that the surface of the water is tangent to the block and is then released, what will be its period of vibration and equation of motion? Neglect resistance. [*Hint:* The upward force on the block is equal to the weight of the water displaced by the block.]

*4. A cubical block of wood, 1 ft on a side, is depressed so that its upper face lies along the surface of the water and is then released. The period of vibration is found to be 1 sec. Neglecting resistance, what is the weight of the block of wood (in pounds)?

5. A 10-g mass suspended from a spring vibrates freely, the resistance being numerically equal to half the velocity (in meters per second) at any instant. The period of the motion is 8 sec. What is the spring constant (in grams per second per second)?

6. A weight w (pounds) is suspended from a spring whose constant is 10 lb/ft. The motion of the weight is subject to a resistance (pounds) numerically equal to half the velocity (feet per second). If the motion is to have a 1-sec period, what are the possible values of w?

*7. A 50-g mass is hanging at rest on a spring that is stretched 10 cm by the weight. The upper end of the spring is given the periodic force $10 \sin 2t$ gram meters per second per second and the resistance has a magnitude $100\sqrt{17}$ (g/second) times the velocity (in meters per second). Find the equation of motion of the mass.

*8. Suppose in Problem 7 that the upper end of the spring is given, instead, the force $1 - e^{-\sqrt{g}t}$ gram meters per second per second. Find the displacement of the mass after 0.1 sec.

9. In Example 2, for what minimum damping constant would the mass fail to oscillate about its equilibrium?

■ SHOW/PROVE/DISPROVE

10. Let $x(t)$ be given by (5).
  a. Show that $x'(t) = 0$ when $\tan \sqrt{k/m}\, t = (v_0/x_0)\sqrt{m/k}$.
  b. Show that for t as given in (a),

$$\sin \sqrt{\frac{k}{m}} t = \frac{\pm v_0 \sqrt{m/k}}{\sqrt{x_0^2 + v_0^2(m/k)}} \quad \text{and}$$

$$\cos \sqrt{\frac{k}{m}} t = \frac{\pm x_0}{\sqrt{x_0^2 + v_0^2(m/k)}}.$$

  c. Show that the maximum value of $x(t)$ is $\sqrt{x_0^2 + v_0^2(m/k)}$.

▨ 16.7 NUMERICAL SOLUTIONS OF DIFFERENTIAL EQUATIONS: EULER'S METHOD

In this section, we present an elementary numerical method for "solving" differential equations. But before presenting this numerical technique, it is useful to discuss the situations in which numerical methods could or should be employed. Such methods are used frequently when other methods are not applicable. Even when other methods do apply, there may be an advantage in having a numerical solution; solutions in terms of more exotic special functions are sometimes difficult to interpret. There may also be computational advantages: the exact solution may be extremely tedious to obtain. Finally, there are situations for which we can solve a differential equation but for which the solution is given implicitly; in these cases a numerical method can tell us more about the solution.

On the other hand, care must always be exercised in using any numerical scheme, as the accuracy of the solution depends not only on the "correctness" of the numerical method being used but also on the precision of the device (hand calculator or computer) used for the computations.

We assume that the initial-value problem

$$\frac{dy}{dx} = f(x, y), \qquad y(x_0) = y_0 \tag{1}$$

has a unique solution $y(x)$. The technique we describe below approximates this solution $y(x)$ only at a finite number of points

$$x_0, \qquad x_1 = x_0 + h, \qquad x_2 = x_0 + 2h, \ldots, \qquad x_n = x_0 + nh,$$

where h is some (nonzero) real number. The method provides a value y_k that is an approximation of the exact value $y(x_k)$ for $k = 0, 1, \ldots, n$.

THE EULER METHOD[†]

This procedure is crude but very simple. The idea is to approximate $y(x_1) = y_1$ by assuming that $f(x, y)$ varies so little on the interval $x_0 \le x \le x_1$ that only a very small error is made by replacing it by the constant value $f(x_0, y_0)$. Integrating $dy/dx = f(x, y)$ from x_0 to x_1, we obtain

$$y_1 - y_0 = y(x_1) - y(x_0) = \int_{x_0}^{x_1} \left(\frac{dy}{dx}\right) dx$$

$$= \int_{x_0}^{x_1} f(x, y)\, dx \approx f(x_0, y_0)(x_1 - x_0).$$

Since $h = x_1 - x_0$,

$$y_1 = y_0 + hf(x_0, y_0). \tag{2}$$

[†] See the biographical sketch of Euler on page 34.

Repeating the process with (x_1, y_1) to obtain y_2, and so on, we obtain the equation

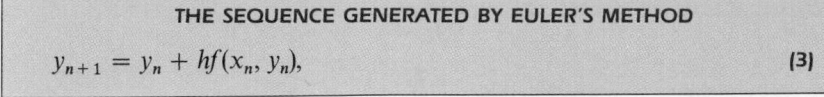

THE SEQUENCE GENERATED BY EULER'S METHOD

$$y_{n+1} = y_n + hf(x_n, y_n), \tag{3}$$

where $y_n = y(x_0 + nh), n = 0, 1, 2, \ldots$ We solve equation (3) iteratively—that is, by first finding y_1, then using it to find y_2, and so on.

The geometric meaning of equation (3) is illustrated in Figure 1. We are simply following the tangent to the solution curve passing through (x_n, y_n) for a small horizontal distance. Looking at Figure 1, where the smooth curve is the unknown exact solution to the initial-value problem (1), we see how equation (3) approximates the exact solution. Since $f(x_0, y_0)$ is the slope of the exact solution at (x_0, y_0), we follow this line to the point (x_1, y_1). Some solution to the differential equation passes through this point. We follow its tangent line at this point to reach (x_2, y_2), and so on. The differences Δ_k are errors of the kth stage in the process.

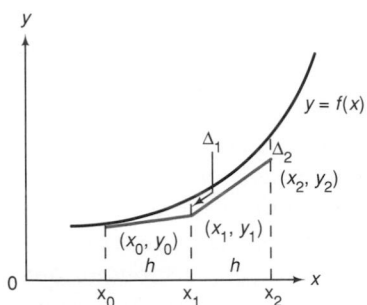

FIGURE 1
THE BLUE LINE IS THE EULER APPROXI-MATION TO THE SOLUTION CURVE.

📖 **EXAMPLE 1**
APPROXIMATING THE SOLUTION TO AN INITIAL-VALUE PROBLEM USING EULER'S METHOD

Use Euler's method to estimate $y(1)$ where $y(x)$ satisfies the initial-value problem

$$\frac{dy}{dx} = y, \qquad y(0) = 1.$$

SOLUTION: From Section 16.2, we know that the solution to this problem is $y(x) = e^x$. Thus $y(1) = e^1 = e \approx 2.71828$. Let us apply the Euler method with $h = \frac{1}{5}$ to this familiar problem to see if this answer is obtained.

We begin by dividing the interval $[0, 1]$ into five subintervals. Then $h = \frac{1}{5} = 0.2$ and $f(x_n, y_n) = y_n$. Thus we obtain, successively,

$$y_0 = y(0) = 1$$

$$y_1 = y_0 + 0.2y_0 = (1 + 0.2)y_0 = (1.2)y_0 = 1.2$$

$$y_2 = y_1 + 0.2y_1 = (1.2)y_1 = (1.2)^2$$

$$y_3 = y_2 + 0.2y_2 = (1.2)y_2 = (1.2)^3$$

$$y_4 = y_3 + 0.2y_3 = (1.2)y_3 = (1.2)^4$$

$$y_5 = y_4 + 0.2y_4 = (1.2)y_4 = (1.2)^5 \approx 2.48832.$$

Thus, with $h = 0.2$, the Euler method yields $y(1) \approx 2.48832$. Since we know that $y(1) = e \approx 2.71828$, our error at $x = 1$ is given by

$$\text{error} = 2.71828 - 2.48832 = 0.22996.$$

What happens if we double the number of subintervals? If $n = 10$, then $h = 0.1$, and, computing as before, we obtain $y(1) \approx y_{10} = (1.1)^{10} \approx 2.59374$. The error is given by

$$\text{error} = 2.71828 - 2.59374 = 0.12454.$$

In general, if $h = 1/n$, then

$$y_1 = y_0 + hy_0 = (1 + h)y_0 = 1 + h = 1 + \frac{1}{n}$$

$$y_2 = y_1\left(1 + \frac{1}{n}\right) = \left(1 + \frac{1}{n}\right)^2$$

$$y_3 = y_1\left(1 + \frac{1}{n}\right) = \left(1 + \frac{1}{n}\right)^3$$

$$\vdots$$

$$y_n = \left(1 + \frac{1}{n}\right)^n.$$

Thus,

$$y(1) \approx \left(1 + \frac{1}{n}\right)^n.$$

Different values of $(1 + 1/n)^n$ are given in Table 1 (to five decimal places of accuracy).

The numbers in Table 1 should not be surprising. In fact, from Section 6.2, (see equation (6.2.8) on page 402) the number e is given by

$$e = \lim_{n \to \infty}\left(1 + \frac{1}{n}\right)^n.$$

Thus we have shown that $e = \lim_{n \to \infty} y_n$, where y_n is the nth iterate in Euler's method with $h = 1/n$; that is, y_n approximates $y(1)$ with better and better accuracy as n increases.

TABLE 1

n	$\left(1 + \dfrac{1}{n}\right)^n$
1	2
2	2.25
5	2.48832
10	2.59374
100	2.70481
1,000	2.71692
10,000	2.71815
100,000	2.71827
1,000,000	2.71828

EXAMPLE 2
APPROXIMATING THE SOLUTION
TO AN INITIAL-VALUE PROBLEM
USING EULER'S METHOD

Find an approximate value for $y(1)$ if $y(x)$ satisfies the initial-value problem

$$\frac{dy}{dx} = y + x^2, \qquad y(0) = 1. \tag{4}$$

Use five subintervals in your approximation.

SOLUTION: Here $h = 1/n = \frac{1}{5} = 0.2$ and we wish to find $y(1)$ by approximating the solution at $x = 0.0, 0.2, 0.4, 0.6, 0.8$, and 1.0. We see that $f(x_n, y_n) = y_n + x_n^2$, and the Euler method [equation (3)] yields

$$y_{n+1} = y_n + h \cdot f(x_n, y_n) = y_n + h(y_n + x_n^2).$$

Since $y_0 = y(0) = 1$, we obtain

$$y_1 = y_0 + h(y_0 + x_0^2) = 1 + 0.2(1 + 0^2) = 1.2$$

$$y_2 = y_1 + h(y_1 + x_1^2) = 1.2 + 0.2[1.2 + (0.2)^2] = 1.448 \approx 1.45$$

$$y_3 = y_2 + h(y_2 + x_2^2) = 1.45 + 0.2[1.45 + (0.4)^2] \approx 1.77$$

$$y_4 = y_3 + h(y_3 + x_3^2) = 1.77 + 0.2[1.77 + (0.6)^2] \approx 2.20$$

$$y_5 = y_4 + h(y_4 + x_4^2) = 2.20 + 0.2[2.20 + (0.8)^2] \approx 2.77.$$

TABLE 2

x_n	y_n	$f(x_n, y_n) = y_n + x_n{}^2$	$y_{n+1} = y_n + hf(x_n, y_n)$
0.0	1.00	1.00	1.20
0.2	1.20	1.24	1.45
0.4	1.45	1.61	1.77
0.6	1.77	2.13	2.20
0.8	2.20	2.84	2.77
1.0	2.77		

We arrange our work as shown in Table 2. The value $y_5 = 2.77$, corresponding to $x_5 = 1.0$, is our approximate value for $y(1)$. Equation (4) has the exact solution $y = 3e^x - x^2 - 2x - 2$ (check this), so that $y(1) = 3e - 5 \approx 3.15$. Thus the Euler method estimate was off by about 12 percent.[†] This is not surprising, because we treated the derivative as a constant over intervals of length 0.2 units. The error that arises in this way is called **discretization error**, because the "discrete" function $f(x_n, y_n)$ is substituted for the "continuously valued" function $f(x, y)$. It is usually true that if we reduce the step size h, we can improve the accuracy of our answer, since then the "discretized" function $f(x_n, y_n)$ is closer to the true value of $f(x, y)$ over the interval $[0, 1]$. This is illustrated in Figure 2 with $h = 0.2$ and $h = 0.1$. Indeed, carrying out similar calculations with $h = 0.1$ yields an approximation of $y(1)$ of 3.07, which is a good deal more accurate (an error of less than 3 percent).[‡]

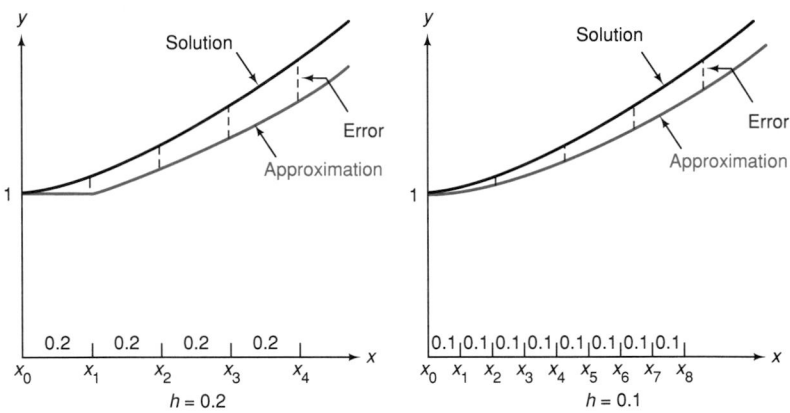

FIGURE 2   DISCRETIZATION ERROR DECREASES AS THE STEP SIZE h DECREASES

In general, reducing step size improves accuracy. However, we must attach a warning to this. Reducing step size increases the amount of work that must be done. Moreover, at every stage of the computation, **round-off errors** are introduced. For example, in our calculations with $h = 0.2$, we

[†] $y(1) - y_5 \approx 3.15 - 2.77 = 0.38$ and $0.38/3.15 \approx 0.12 = 12\%$.
[‡] $y(1) - y_{10} \approx 3.15 - 3.07 = 0.08$ and $0.08/3.15 \approx 0.0254 \approx 2\frac{1}{2}\%$.

rounded off the exact value 1.448 to the value 1.45 (correct to two decimal places). The rounded-off value was then used to calculate further values of y_n. It is not unusual for a computer solution of a more complicated differential equation to take several thousand individual computations, thus having several thousand round-off errors. In some problems, the accumulated round-off error can be so large that the resulting computed solution is sufficiently inaccurate to invalidate the result. Fortunately, this usually does not occur, since round-off errors can be positive or negative and tend to cancel one another out. The statement that reducing step size improves accuracy is made under the assumption (usually true) that the average of the round-off errors is zero. In any event, it should be clear that reducing the step size, thereby increasing the number of computations, is a procedure that should be carried out carefully. In general, each problem has an optimal step size, and a smaller than optimal step size yields a greater error due to accumulated round-off errors.

The material in this section is intended as a brief introduction to the numerical solution of differential equations. In real-life applications, Euler's method is rarely used because much more efficient methods are widely available. Most differential equations textbooks have descriptions of other, more effective methods.[†]

■ PROBLEMS 16.7

■ SELF QUIZ

I. The Euler method may produce an exact solution (i.e., with approximation error equal to zero) to the differential equation $y' = f(x, y)$ provided $f(x, y)$ is simple enough. For each of the following choices of the function f, answer (True–False) whether Euler's method will produce an exact solution.

a. $f(x, y) = 0$ for all x and y
b. $f(x, y) = -4$ for all x and y
c. $f(x, y) = x$ for all x and y
d. $f(x, y) = y$ for all x and y

■ DRILL

In Problems 1–24, use the Euler method to find approximately the solution of the given initial-value problem at the given x-value. Use the indicated step size h.

1. $y' = -x$, $y(0) = 1$, $h = 0.2$, $x = 1$
2. $y' = x$, $y(1) = 1.5$, $h = -0.2$, $x = 0$
3. $y' = -y$, $y(0) = 1$, $h = 0.2$, $x = 1$
4. $y' = -y$, $y(5) = 1$, $h = 0.2$, $x = 6$
5. $y' = x + y$, $y(0) = 1$, $h = 0.2$, $x = 1$
6. $y' = x - y$, $y(1) = 2$, $h = 0.4$, $x = 3$

7. $\dfrac{dy}{dx} = \dfrac{x - y}{x + y}$, $y(2) = 1$, $h = -0.2$, $x = 1$

8. $\dfrac{dy}{dx} = \dfrac{y}{x} + \left(\dfrac{y}{x}\right)^2$, $y(1) = 1$, $h = 0.2$, $x = 2$

9. $y' = x\sqrt{1 + y^2}$, $y(1) = 0$, $h = 0.4$, $x = 3$

10. $y' = x\sqrt{1 - y^2}$, $y(1) = 0$, $h = 0.125$, $x = 2$

11. $\dfrac{dy}{dx} = \dfrac{y}{x} - \dfrac{5}{2}x^2 y^3$, $y(1) = \dfrac{1}{\sqrt{2}}$, $h = 0.125$, $x = 2$

12. $\dfrac{dy}{dx} = \dfrac{-y}{x} + x^2 y^2$, $y(1) = \dfrac{2}{9}$, $h = \dfrac{1}{3}$, $x = 3$

13. $y' = ye^x$, $y(0) = 2$, $h = 0.2$, $x = 2$
14. $y' = xe^y$, $y(0) = 0$, $h = 0.1$, $x = 1$

[†] See, for example, W. R. Derrick and S. I. Grossman, *Introduction to Differential Equations With Boundary Value Problems*, 3rd ed. (St. Paul: West, 1987), Chapter 9.

CONTINUED

15. $y' = xy^2 + y^3$, $y(0) = 1$, $h = 0.02$, $x = 0.1$

16. $y' = \sqrt{y^2 - x^2}$, $y(0) = 1$, $h = 0.1$, $x = 1$

17. $y' = x + \sin(\pi y)$, $y(1) = 0$, $h = 0.2$, $x = 2$

18. $y' = x + \cos(\pi y)$, $y(0) = 0$, $h = 0.4$, $x = 2$

19. $y' = \sqrt{x^2 + y^2}$, $y(1) = 5$, $h = -0.2$, $x = 0$

20. $y' = \sqrt{x^2 + y^2}$, $y(0) = 1$, $h = 0.5$, $x = 5$

21. $y' = \sqrt{x + y^2}$, $y(0) = 1$, $h = 0.2$, $x = 1$

22. $y' = \sqrt{x + y^2}$, $y(1) = 2$, $h = -0.2$, $x = 0$

23. $y' = \cos(xy)$, $y(0) = 0$, $h = \pi/4$, $x = \pi$

24. $y' = \sin(xy)$, $y(0) = 1$, $h = \pi/4$, $x = 2\pi$

■ **CHALLENGE**

25. Discuss the relationship between the linearization of a function and Euler's method. Do we have two names for the same thing? Do we have two techniques exploiting the same geometric notion?

■ **ANSWERS TO SELF-QUIZ**

I. a. True   b. True
   c. False   d. False

■ **PROBLEMS**   **CHAPTER 16 REVIEW**

In Problems 1–28, find the general solution to the given differential equation. If initial conditions are specified, then find the unique solution to that initial-value problem.

1. $y' = 3x$

2. $y' = e^{x-y}$; $y(0) = 4$

3. $x' = e^x \cos t$; $x(0) = 3$

4. $x' = x^{13} t^{11}$

5. $y' + 3y = \cos x$; $y(0) = 1$

6. $\dfrac{dy}{dx} + 3y = \dfrac{1}{1 + e^{3x}}$

7. $x' = 3x + t^3 e^{3t}$; $x(1) = 2$

8. $y' + y \cot x = \sin x$; $y(\pi/6) = \frac{1}{2}$

9. $y'' - 5y' + 4y = 0$

10. $y'' - 9y' + 14y = 0$; $y(0) = 2$, $y'(0) = 1$

11. $y'' - 9y = 0$

12. $y'' + 9y = 0$

13. $y'' + 6y' + 9y = 0$

14. $y'' + 8y' + 16y = 0$; $y(0) = -1$, $y'(0) = 3$

15. $y'' - 2y' + 2y = 0$; $y(0) = 0$, $y'(0) = 1$

16. $y'' + 8y' = 0$; $y(0) = 2$, $y'(0) = -3$

17. $y'' + 4y = 2 \sin x$

18. $y'' + y' - 12y = 4e^{2x}$; $y(0) = 1$, $y'(0) = -1$

19. $y'' + y' + y = e^{-x/2} \sin(\sqrt{3}x/2)$

20. $y'' + 4y = 6x \cos 2x$; $y(0) = 1$, $y'(0) = 0$

21. $y'' + y = 0$; $y(0) = y'(0) = 0$

22. $y'' + y = x^3 - x$

23. $y'' - 2y' + y = e^{-x}$

24. $y'' - 8y' + 16y = e^{4x}$; $y(0) = 3$, $y'(0) = 1$

25. $y'' - y' - 6y = e^x \cos x$

26. $y'' - 2y' + 3y = e^x \cos \sqrt{2}x$

27. $y'' + y = x + e^x + \sin x$

28. $y'' + 16y = \cos 4x + x^2 - 3$

In Problems 29–34, use the Euler method with the specified value of h to obtain an approximate value for the solution of the given initial-value problem evaluated at the indicated value of x.

29. $\dfrac{dy}{dx} = \dfrac{e^x}{y}$; $y(0) = 2$. Find $y(3)$ with $h = \dfrac{1}{2}$.

30. $\dfrac{dy}{dx} = \dfrac{e^y}{x}$; $y(1) = 0$. Find $y\left(\dfrac{1}{2}\right)$ with $h = -0.1$.

31. $\dfrac{dy}{dx} = \dfrac{y}{\sqrt{1 + x^2}}$; $y(0) = 1$. Find $y(3)$ with $h = \dfrac{1}{2}$.

32. $xy \dfrac{dy}{dx} = y^2 - x^2$; $y(1) = 2$. Find $y(3)$ with $h = \dfrac{1}{2}$.

33. $\dfrac{dy}{dx} = y - xy^3$; $y(0) = 1$. Find $y(3)$ with $h = \dfrac{1}{2}$.

34. $\dfrac{dy}{dx} = \dfrac{2xy}{3x^2 - y^2}$; $y\left(\dfrac{-3}{8}\right) = \dfrac{-3}{4}$. Find $y\left(\dfrac{3}{8}\right)$ with $h = \dfrac{1}{8}$.

CONTINUED

35. a. Apply the Euler method with $h = 0.2$ to approximate $y(1)$ for the initial-value problem $y' = y^3$, $y(0) = 1$.
    b. Show that $y = (1, - 2x)^{-1/2}$ is a solution to the initial-value problem in part (a).
    c. Compare the results of parts (a) and (b); explain any apparent contradiction.

36. A spring, fixed at its upper end, supports a 10-kg mass that stretches the spring 60 cm. Find the equation of motion of the mass if it is drawn to a position 10 cm below its equilibrium position and released with an initial velocity of 5 cm/sec upward.

37. What are the period, frequency, and amplitude of the motion of the mass in Problem 36?

38. Find the equation of motion of the mass of Problem 36 if it is subjected to damping forces having the damping constant $\mu = 3$.

39. In Problem 38, for what minimum damping constant would the mass fail to oscillate about its equilibrium?

40. For what value of ω will the external force of $10 \sin \omega t$ newtons produce resonance in the spring of Problem 38?

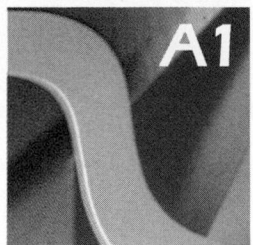

REVIEW OF TRIGONOMETRY

A1.1 ANGLES AND RADIAN MEASURE

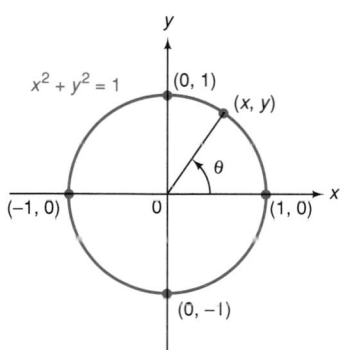

FIGURE 1
THE UNIT CIRCLE

We begin by drawing the unit circle, the circle with radius 1 centered at the origin (see Figure 1). **Angles** are measured starting at the positive x-axis. An angle is positive if it is measured in the counterclockwise direction, and it is negative if it is measured in the clockwise direction. We measure angles in degrees, using the fact that the circle contains 360°. Then we can describe any angle by comparison with the circle. Some angles are depicted in Figure 2. In Figure 2(e) we obtained the angle $-90°$ by moving in the negative (clockwise) direction. In Figure 2(f) we obtained an angle of 720° by moving around the circle twice in the counterclockwise direction.

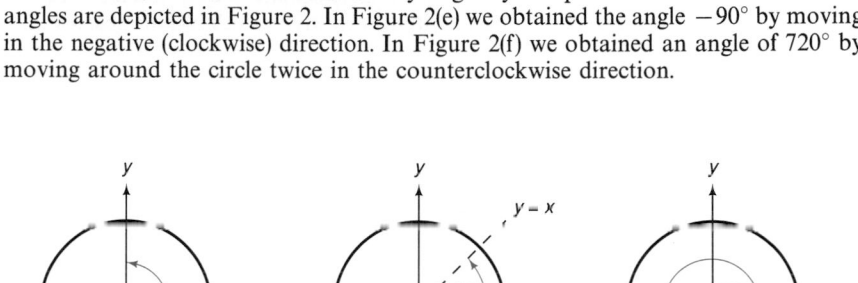

| (a) | (b) | (c) |

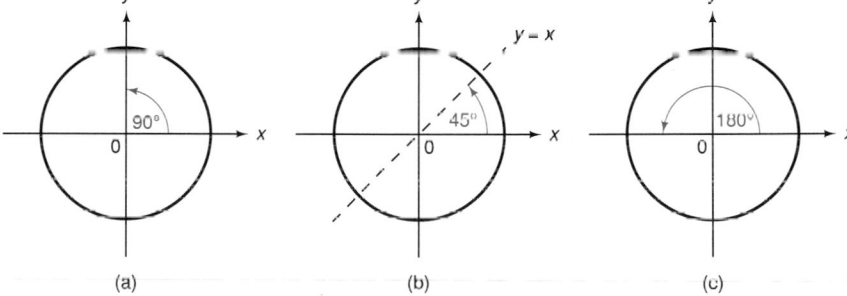

| (d) | (e) | (f) |

FIGURE 2   SIX ANGLES

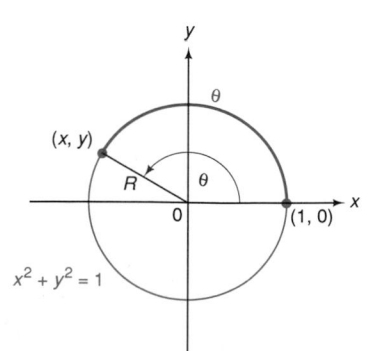

FIGURE 3
THE LENGTH OF THE ARC FROM (1, 0) TO (x, y) IS THE RADIAN MEASURE OF θ

Figure 2 illustrates the great advantage of using circles rather than triangles to measure angles. Any angle in a triangle must be between 0° and 180°. In a circle there is no such restriction.

There is another way to measure angles, which, in many instances, is more useful than measurement in degrees. Let R denote the radial line which makes an angle of θ with the positive x-axis. See Figure 3. Let (x, y) denote the point at which this radial line intersects the unit circle. Then the **radian measure** of the angle is the length of the arc of the unit circle from the point (1, 0) to the point (x, y).

Since the circumference of a circle is $2\pi r$, where r is its radius, the circumference of the unit circle is 2π. Thus,

$$360° = 2\pi \text{ radians.} \tag{1}$$

Then since $180° = \frac{1}{2}(360°)$, $180° = \frac{1}{2}(2\pi) = \pi$ radians. In general, θ (in degrees) is to $360°$ as θ (in radians) is to 2π radians. Thus $(\theta/360)(\text{degrees}) = (\theta/2\pi)(\text{radians})$, or

$$\theta(\text{degrees}) = \frac{180}{\pi} \theta(\text{radians}) \tag{2}$$

and

$$\theta(\text{radians}) = \frac{\pi}{180} \theta(\text{degrees}). \tag{3}$$

We can calculate that 1 radian $= 180/\pi \approx 57.3°$ and $1° = \pi/180 \approx 0.0175$ radians. Representative values of θ in degrees and radians are given in Table 1.

TABLE 1

θ (degrees)	0	90	180	270	360	45	30	60	−90	135	120	720
θ (radians)	0	$\frac{\pi}{2}$	π	$\frac{3\pi}{2}$	2π	$\frac{\pi}{4}$	$\frac{\pi}{6}$	$\frac{\pi}{3}$	$\frac{-\pi}{2}$	$\frac{3\pi}{4}$	$\frac{2\pi}{3}$	4π

The radian measure of an angle does not refer to "degrees"; it refers to distance measured along an arc of the unit circle. This is an advantage when discussing trigonometric functions that arise in applications having nothing at all to do with angles.

Let C_r denote the circle of radius r centered at the origin (see Figure 4). If $0P$ denotes a radial line as pictured in the figure, then $0P$ cuts an arc from C_r of length L. Let θ be the positive angle between $0P$ and the positive x-axis. If $\theta = 360°$, then $L = 2\pi r$. If $\theta = 180°$, then $L = \pi r$. In fact, it is evident from the figure that

$$\frac{\theta°}{360°} = \frac{L}{2\pi r}, \tag{4}$$

or

$$\theta° = 360° \frac{L}{2\pi r}. \tag{5}$$

If we measure θ in radians, then (4) becomes

$$\frac{\theta}{2\pi} = \frac{L}{2\pi r}, \tag{6}$$

or

$$\theta = \frac{2\pi L}{2\pi r} = \frac{L}{r}. \tag{7}$$

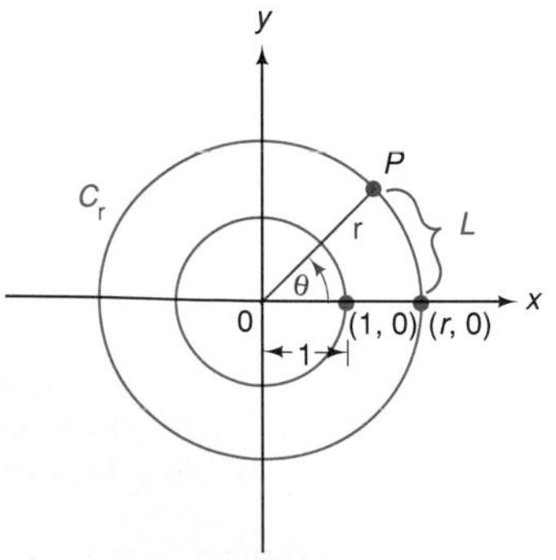

FIGURE 4
AN ANGLE OF θ RADIANS CUTS OFF AN ARC OF LENGTH $r\theta$ IN A CIRCLE OF RADIUS r.

Finally, rewriting (7), we obtain

$$L = r\theta. \tag{8}$$

That is, if θ is measured in radians, then the angle θ "cuts" from the circle of radius r centered at the origin an arc of length $r\theta$. Note that if $r = 1$, then (8) reduces to $L = \theta$, which is the definition of the radian measure of an angle.

EXAMPLE 1
FINDING THE ARC LENGTHS CUT FROM A CIRCLE

What is the length of an arc cut from the circle of radius 4 centered at the origin by an angle of **(a)** 45°, **(b)** 60°, **(c)** 270°?

SOLUTION: From (8) we find that $L = 4\theta$, where θ is the radian measure of the angle. We therefore have the following:

$$\textbf{(a)} \ L = 4 \cdot \frac{\pi}{4} = \pi \qquad \textbf{(b)} \ L = 4 \cdot \frac{\pi}{3} = \frac{4\pi}{3} \qquad \textbf{(c)} \ L = 4 \cdot \frac{3\pi}{2} = 6\pi$$

■ PROBLEMS A1.1

In Problems 1–6, convert from degrees to radians.

1. $\theta = 150°$ 2. $\theta = -45°$ 3. $\theta = 300°$
4. $\theta = 72°$ 5. $\theta = 144°$ 6. $\theta = 1080°$

In Problems 7–12, convert from radians to degrees.

7. $\pi/12$ 8. $7\pi/12$ 9. $\pi/8$
10. 3π 11. $-(\pi/3)$ 12. $5\pi/4$

13. Let C denote the circle of radius 2 centered at the origin. If a radial line cuts an arc of length π (starting at the point $(2, 0)$), what is the angle (in degrees) between this line and the positive x-axis?

14. If the radial line in Problem 13 makes an angle of 75° with the positive x-axis, what is the length of the arc it cuts from the circle?

A1.2 THE TRIGONOMETRIC FUNCTIONS AND BASIC IDENTITIES

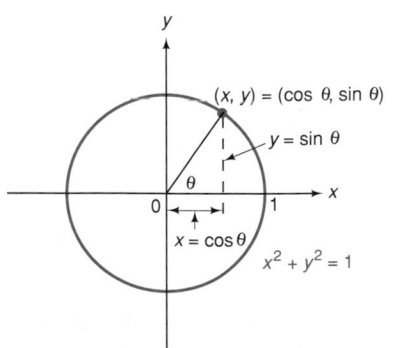

FIGURE 1
$\cos \theta$ AND $\sin \theta$ ARE THE COORDINATES OF THE POINT AT WHICH THE RADIAL LINE HITS THE UNIT CIRCLE.

We again begin with the unit circle (see Figure 1). An angle θ uniquely determines a point (x, y) where the radial line intersects the circle. We then define

$$\text{cosine } \theta = x \qquad \text{and} \qquad \text{sine } \theta = y. \tag{1}$$

These are the two basic trigonometric (or circular) functions, usually written $\cos \theta$ and $\sin \theta$. Since the equation of the circle is $x^2 + y^2 = 1$, we see that $\sin \theta$ and $\cos \theta$ satisfy the equation

$$\sin^2 \theta + \cos^2 \theta = 1. \tag{2}$$

We emphasize that $\cos \theta$ is the x-coordinate of the point (x, y) and $\sin \theta$ is the y-coordinate. As θ varies, $\cos \theta$ and $\sin \theta$ oscillate between $+1$ and -1. For example, if $\theta = 0$, then the radial line intersects the circle at the point $(1, 0)$, and we have $\cos 0 = 1$ and $\sin 0 = 0$. If $\theta = 90° = \pi/2$, then the radial line intersects the circle at the point $(0, 1)$, and we have $\cos 90° = \cos \pi/2 = 0$ and $\sin 90° = \sin \pi/2 = 1$. If $\theta = 45° = \pi/4$, then the radial line is the line $y = x$. Since $x^2 + y^2 = 1$ and $y = x$ at the point of intersection, we see that $x^2 + x^2 = 2x^2 = 1$, or $x = y = 1/\sqrt{2} = \sqrt{2}/2$. Thus, $\cos 45° = \cos \pi/4 = \sqrt{2}/2$ and $\sin 45° = \sin \pi/4 = \sqrt{2}/2$. It will be shown in Appendix 1.4 that $\cos 30° = \cos \pi/6 = \sqrt{3}/2$, $\sin 30° = \sin \pi/6 = \frac{1}{2}$, $\cos 60° = \cos \pi/3 = \frac{1}{2}$, and $\sin 60° = \sin \pi/3 = \sqrt{3}/2$. The most commonly used values of $\cos \theta$ and $\sin \theta$ are given in Table 1.

Some basic facts about the functions $\sin \theta$ and $\cos \theta$ can be derived by simply looking at the graph of the unit circle. First, we note that if we add 360° to the angle θ in Figure 1, then we end up with the same point (x, y) on the circle. Thus,

$$\cos(\theta + 360°) = \cos(\theta + 2\pi) = \cos \theta \tag{3}$$

and

$$\sin(\theta + 360°) = \sin(\theta + 2\pi) = \sin \theta.$$ **(4)**

TABLE 1

θ	0	$\dfrac{\pi}{6}$	$\dfrac{\pi}{4}$	$\dfrac{\pi}{3}$	$\dfrac{\pi}{2}$	π	$\dfrac{3\pi}{2}$	2π
$\cos \theta$	1	$\dfrac{\sqrt{3}}{2}$	$\dfrac{\sqrt{2}}{2}$	$\dfrac{1}{2}$	0	-1	0	1
$\sin \theta$	0	$\dfrac{1}{2}$	$\dfrac{\sqrt{2}}{2}$	$\dfrac{\sqrt{3}}{2}$	1	0	-1	0

In general, if α is the *smallest* positive number such that $f(x + \alpha) = f(x)$, we say that f is **periodic** of **period** α. Thus from (3) and (4) we see that the functions $\cos \theta$ and $\sin \theta$ are periodic of period 2π.

A glance at Figure 2 tells us the sign of the two basic functions. With all the information above we can draw a sketch of $y = \cos \theta$ and $y = \sin \theta$. These sketches are shown in Figures 3 and 4.

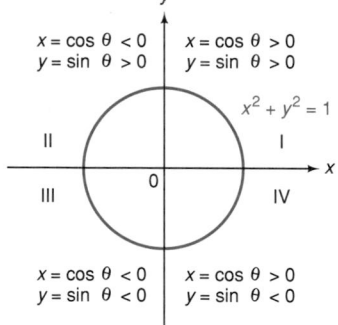

FIGURE 2
THE SIGNS OF $\cos \theta$ AND $\sin \theta$

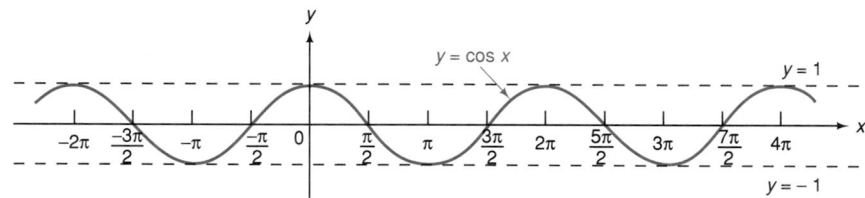

FIGURE 3   GRAPH OF $\cos x$

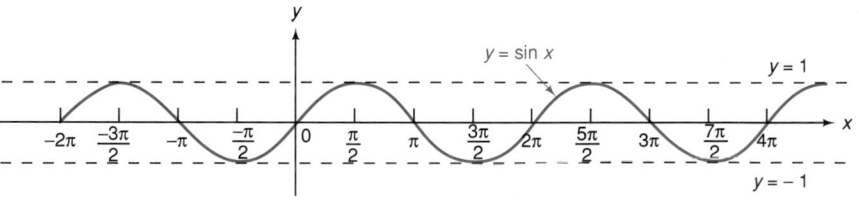

FIGURE 4   GRAPH OF $\sin x$

Now look at Figure 5. We see that in a comparison of θ and $-\theta$, the x-coordinates are the same while the y-coordinates have opposite signs. This suggests that

$$\cos(-\theta) = \cos \theta$$ **(5)**

and

$$\sin(-\theta) = -\sin \theta.$$ **(6)**

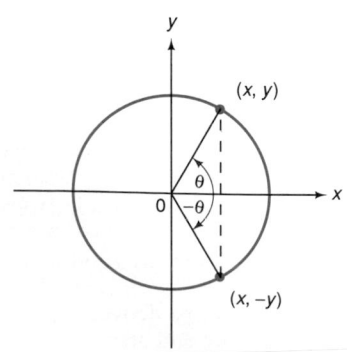

FIGURE 5
$\cos(-\theta) = \cos \theta$ AND $\sin(-\theta) = -\sin \theta$.

To obtain another identity, we add $180° = \pi$ to θ (Figure 6). Then the x- and y-coordinates of $\theta + \pi$ have signs opposite to those of the x- and y-coordinates of θ. Thus,

$$\cos(\theta + 180°) = \cos(\theta + \pi) = -\cos\theta \tag{7}$$

and

$$\sin(\theta + 180°) = \sin(\theta + \pi) = -\sin\theta. \tag{8}$$

Several other identities can be obtained by simply glancing at a graph of the unit circle.

We now obtain another identity that is very useful in computations.

THEOREM 1

$$\cos(\theta + \varphi) = \cos\theta\cos\varphi - \sin\theta\sin\varphi.$$

PROOF: We prove the theorem in the case θ and φ are between 0 and $\pi/2$. We will leave another case as a problem (see Problem 16). From Figure 7 we see that the arc $P_1 P_3$ has the same length as the arc $P_2 P_4$ (they are both equal to the radian measure of the angle $\theta + \varphi$). Then the distance from P_1 to P_3 is the same as the distance from P_2 to P_4. Using the distance formula (see equation (0.2.2)), we obtain

$$\overline{P_1 P_3}^{\dagger} = \sqrt{(c-a)^2 + (d+b)^2} = \sqrt{(e-1)^2 + f^2} = \overline{P_2 P_4}. \tag{9}$$

But

$$\cos\theta = a, \quad \cos\varphi = c, \quad \cos(\theta + \varphi) = e,$$
$$\sin\theta = b, \quad \sin\varphi = d, \quad \sin(\theta + \varphi) = f. \tag{10}$$

Then we square both sides of (9):

$$c^2 - 2ac + a^2 + d^2 + 2bd + b^2 = e^2 - 2e + 1 + f^2.$$

Since $a^2 + b^2 = c^2 + d^2 = e^2 + f^2 = 1$ (why?), we have

$$-2ac + 2bd + 2 = -2e + 2,$$

or

$$e = ac - bd. \tag{11}$$

Substituting (10) into (11) proves the theorem. ∎

There are many other identities that can be proved by using Theorem 1. We indicate in Table 2 some of the identities we will find useful in other parts of this text. The proofs of these identities are suggested in the problems. We use x and y instead of θ and φ in this table. These identities will be very useful to us when we discuss techniques of integration in Chapter 7.

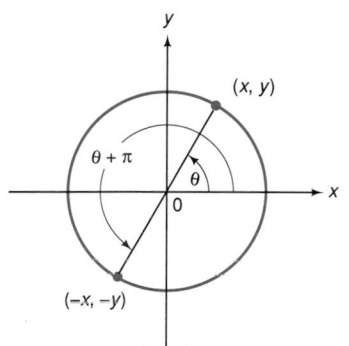

FIGURE 6
$\cos(\theta + \pi) = -\cos\theta$ AND
$\sin(\theta + \pi) = -\sin\theta$

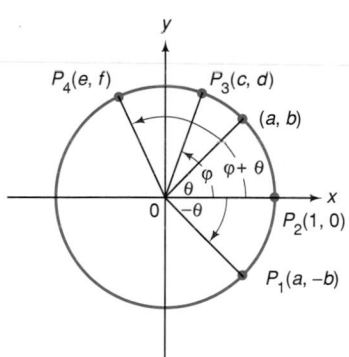

FIGURE 7
PROVING THAT
$\cos(\theta + \varphi) = \cos\theta\cos\varphi - \sin\theta\sin\varphi$

---

† The symbol $\overline{P_1 P_3}$ denotes the distance between the points P_1 and P_3.

TABLE 2 BASIC IDENTITIES INVOLVING cos x AND sin x

(i) $\sin^2 x + \cos^2 x = 1$

(ii) $\cos(-x) = \cos x$

(iii) $\sin(-x) = -\sin x$

(iv) $\cos(x + \pi) = -\cos x$

(v) $\sin(x + \pi) = -\sin x$

(vi) $\cos(\pi - x) = -\cos x$

(vii) $\sin(\pi - x) = \sin x$

(viii) $\cos\left(\dfrac{\pi}{2} + x\right) = -\sin x$

(ix) $\sin\left(\dfrac{\pi}{2} + x\right) = \cos x$

(x) $\cos\left(\dfrac{\pi}{2} - x\right) = \sin x$

(xi) $\sin\left(\dfrac{\pi}{2} - x\right) = \cos x$

(xii) $\cos(x + y) = \cos x \cos y - \sin x \sin y$

(xiii) $\sin(x + y) = \sin x \cos y + \cos x \sin y$

(xiv) $\cos(x - y) = \cos x \cos y + \sin x \sin y$

(xv) $\sin(x - y) = \sin x \cos y - \cos x \sin y$

(xvi) $\cos 2x = \cos^2 x - \sin^2 x = 2\cos^2 x - 1 = 1 - 2\sin^2 x$

(xvii) $\sin 2x = 2 \sin x \cos x$

(xviii) $\cos\dfrac{x}{2} = \pm\sqrt{\dfrac{1 + \cos x}{2}}$

(xix) $\sin\dfrac{x}{2} = \pm\sqrt{\dfrac{1 - \cos x}{2}}$ ⎱ The **half angle** formulas

(xx) $\cos^2 x = \dfrac{1 + \cos 2x}{2}$

(xxi) $\sin^2 x = \dfrac{1 - \cos 2x}{2}$

(xxii) $\cos x - \cos y = 2 \sin\dfrac{x + y}{2} \sin\dfrac{y - x}{2}$

(xxiii) $\sin x - \sin y = 2 \sin\dfrac{x - y}{2} \cos\dfrac{x + y}{2}$

■ PROBLEMS A1.2

In Problems 1–15, use the basic identities to calculate $\sin \theta$ and $\cos \theta$.

1. $\theta = 6\pi$

2. $\theta = -30°$

3. $\theta = 7\pi/6$

4. $\theta = 5\pi/6$

5. $\theta = 75°$

6. $\theta = 15°$

7. $\theta = 13\pi/12$

8. $\theta = -150°$

9. $\theta = -\pi/12$

10. $\theta = \pi/8$

11. $\theta = \pi/16$
    [*Hint:* Use the result of Problem 10.]

12. $\theta = 3\pi/8$

13. $\theta = 67\frac{1}{2}°$

14. $\theta = 7\pi/24$

15. $\theta = -7\frac{1}{2}°$

CONTINUED

16. Prove Theorem 1 in the case $0° < \theta < 90°$ and $90° < \varphi < 180°$.
17. Prove that $\cos[(\pi/2) + x] = -\sin x$.
18. Show that $\sin[(\pi/2) + x] = \cos x$. [*Hint:* Use Problem 17 to show that $\sin[x + (\pi/2)] = -\cos(x + \pi)$ and then use identity (iv).]
19. Use Problems 17 and 18 to show that $\sin(x + y) = \sin x \cos y + \cos x \sin y$. [*Hint:* Start with $\sin(x + y) = -\cos[(\pi/2) + x + y]$ and then apply Theorem 1.]
20. Prove identities (xiv) and (xv). [*Hint:* Use Theorem 1, Problem 19, and identities (ii) and (iii).]
21. Prove identities (vi), (vii), (x), and (xi). [*Hint:* Use Problem 20.]
22. Prove identities (xvi) and (xvii). [*Hint:* Use Theorem 1 and Problem 19.]
23. Prove that $\cos(x/2) = \pm\sqrt{(1 + \cos x)/2}$. [*Hint:* Use identity (xvi) to show that $\cos x = 2\cos^2(x/2) - 1$.]
24. Prove identity (xix). [*Hint:* Use identity (i) and Problem 23.]
25. Prove identities (xx) and (xxi). [*Hint:* Use identities (xviii) and (xix).]

26. Prove that $\cos x - \cos y = 2 \sin[(x + y)/2] \cdot \sin[(y - x)/2]$. [*Hint:* Expand the right side by using identities (xiii), (xv), and (xix).]
27. Prove identity (xxiii).
28. Graph the function $y = 3 \sin x$. The greatest value a periodic function takes is called the **amplitude** of the function. Show that in this case the amplitude is equal to 3.
29. Graph the function $y = -2 \cos x$. What is the amplitude?
30. Show that the function $y = \sin 2x$ is periodic of period π. Graph the function.
31. Show that the function $y = 4 \cos(x/2)$ is periodic of period 4π. What is its amplitude? Graph the curve.
32. Graph the curve $y = 3 \sin(x/3)$.
33. Graph the curve $y = \sin(x - 1)$. [*Hint:* See Section 0.5.] Show that its period is 2π.
34. Graph the curve $y = 2 \sin[(x/2) + 3]$. What is its period? What is its amplitude?
35. Graph the curve $y = 3 \cos(3x - \frac{1}{2})$. What is its period? What is its amplitude?

A1.3 OTHER TRIGONOMETRIC FUNCTIONS

In addition to the two functions we have already discussed, there are four other trigonometric functions, which can be defined in terms of $\sin x$ and $\cos x$:

(i) tangent $x = \tan x = \dfrac{\sin x}{\cos x}$ for $\cos x \neq 0$

(ii) cotangent $x = \cot x = \dfrac{\cos x}{\sin x} = \dfrac{1}{\tan x}$ for $\sin x \neq 0$

(iii) secant $x = \sec x = \dfrac{1}{\cos x}$ for $\cos x \neq 0$

(iv) cosecant $x = \csc x = \dfrac{1}{\sin x}$ for $\sin x \neq 0$

Each of these four functions grows without bound as x approaches certain values. When $x \to 0^+$, $\cot x$ and $\csc x$ approach $+\infty$; and when $x \to 0^-$, $\cot x$ and $\csc x$ approach $-\infty$. This is true because $\sin x > 0$ for x near 0 and positive, and $\sin x < 0$ for x near 0 and negative (see Figure 4 in Appendix A1.2). Also, $\cos x$ is near 1 for x near 0. Similarly, we obtain

$$\lim_{x \to \pi/2^-} \tan x = +\infty, \qquad \lim_{x \to \pi/2^+} \tan x = -\infty,$$

$$\lim_{x \to \pi/2^-} \sec x = +\infty, \qquad \lim_{x \to \pi/2^+} \sec x = -\infty.$$

These facts hold because $\cos x$ is positive for $x < \pi/2$ and x near $\pi/2$, and $\cos x$ is negative for $x > \pi/2$ and x near $\pi/2$. (See Figure 3 in Appendix A1.2.)

We also observe that

$$\tan 0 = \frac{\sin 0}{\cos 0} = \frac{0}{1} = 0 \qquad \text{and} \qquad \cot \frac{\pi}{2} = \frac{\cos(\pi/2)}{\sin(\pi/2)} = \frac{0}{1} = 0.$$

We note that since $-1 \leq \sin x \leq 1$ and $-1 \leq \cos x \leq 1$, we have $|\sec x| \geq 1$ and $|\csc x| \geq 1$. That is, $\sec x$ and $\csc x$ can never take values in the open interval $(-1, 1)$.

In addition,

$$\tan(x + \pi) = \frac{\sin(x + \pi)}{\cos(x + \pi)} = \frac{-\sin x}{-\cos x} = \frac{\sin x}{\cos x} = \tan x,$$

so that $\tan x$ is periodic of period π. Similarly, $\cot x$ is periodic of period π. Also,

$$\sec(x + 2\pi) = \frac{1}{\cos(x + 2\pi)} = \frac{1}{\cos x} = \sec x,$$

so that $\sec x$ (and $\csc x$) are periodic of period 2π.

TABLE 1

x	0	$\dfrac{\pi}{6}$	$\dfrac{\pi}{4}$	$\dfrac{\pi}{3}$	$\dfrac{\pi}{2}$	π	$\dfrac{3\pi}{2}$	2π
$\tan x$	0	$\dfrac{1}{\sqrt{3}}$	1	$\sqrt{3}$	undefined	0	undefined	0
$\cot x$	undefined	$\sqrt{3}$	1	$\dfrac{1}{\sqrt{3}}$	0	undefined	0	undefined
$\sec x$	1	$\dfrac{2}{\sqrt{3}}$	$\sqrt{2}$	2	undefined	-1	undefined	1
$\csc x$	undefined	2	$\sqrt{2}$	$\dfrac{2}{\sqrt{3}}$	1	undefined	-1	undefined

Values of $\tan x$, $\cot x$, $\sec x$, and $\csc x$ are given in Table 1. Putting this information all together and using our knowledge of the functions $\sin x$ and $\cos x$, we obtain the graphs given in Figure 1.

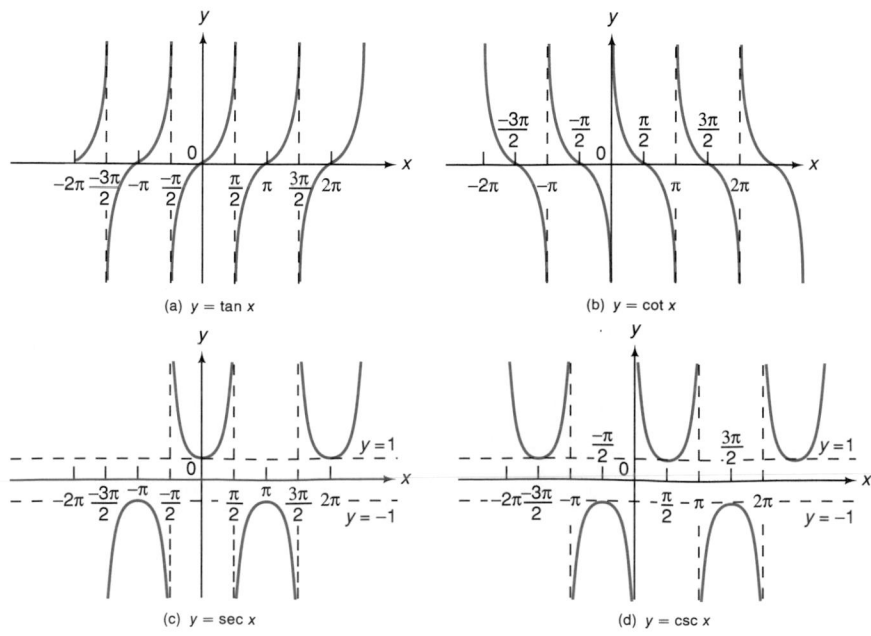

(a) $y = \tan x$

(b) $y = \cot x$

(c) $y = \sec x$

(d) $y = \csc x$

FIGURE 1   GRAPHS OF FOUR TRIGONOMETRIC FUNCTIONS

There are many identities involving the four functions introduced in this section. However, for our purposes there are two that will prove especially useful:

$$1 + \tan^2 x = \sec^2 x, \tag{1}$$

$$1 + \cot^2 x = \csc^2 x. \tag{2}$$

Both can be obtained by starting with the identity

$$\sin^2 x + \cos^2 x = 1$$

and then dividing both sides by $\cos^2 x$ to obtain (1) and by $\sin^2 x$ to obtain (2).

■ PROBLEMS A1.3

In Problems 1–15, calculate $\tan x$, $\cot x$, $\sec x$, and $\csc x$.

1. $x = 6\pi$
2. $x = -30°$
3. $x = 7\pi/6$
4. $x = 5\pi/6$
5. $x = 75°$
6. $x = 15°$
7. $x = 13\pi/12$
8. $x = -150°$
9. $x = -\pi/12$
10. $x = \pi/8$
11. $x = \pi/16$
12. $x = 3\pi/8$
13. $x = 67\frac{1}{2}°$
14. $x = 7\pi/24$
15. $x = -7\frac{1}{2}°$

In Problems 16–21, find the period of the given function and sketch the graph of the function.

16. $y = \tan 2x$
17. $y = 3 \sec x/3$
18. $y = 4 \cot(4x + 1)$
19. $y = 2 \csc 6x$
20. $y = -2 \tan\left(\dfrac{x}{5} + 1\right)$
21. $y = 8 \sec(5x + 5)$

22. Using the corresponding formulas for sine and cosine, show that

$$\tan(x - y) = \frac{\tan x - \tan y}{1 + \tan x \tan y}.$$

23. Show that

$$\tan(x + y) = \frac{\tan x + \tan y}{1 - \tan x \tan y}.$$

24. Show that

$$\tan \frac{x}{2} = \frac{1 - \cos x}{\sin x} = \frac{\sin x}{1 + \cos x}.$$

A1.4 TRIANGLES

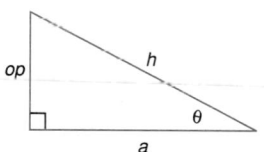

FIGURE 1
A RIGHT TRIANGLE

In many elementary courses in trigonometry the six trigonometric functions are introduced in terms of the ratios of sides of a right triangle. Consider the angle θ in the right triangle in Figure 1. The side opposite θ is labeled "op," the side adjacent to θ (which is not the hypotenuse) is labeled "a," and the hypotenuse (the side opposite the right angle) is labeled "h." Then we define

$$\sin \theta = \frac{\text{opposite}}{\text{hypotenuse}} = \frac{op}{h}, \qquad \cos \theta = \frac{\text{adjacent}}{\text{hypotenuse}} = \frac{a}{h},$$

$$\tan \theta = \frac{\text{opposite}}{\text{adjacent}} = \frac{op}{a}, \qquad \cot \theta = \frac{\text{adjacent}}{\text{opposite}} = \frac{a}{op},$$

$$\sec \theta = \frac{\text{hypotenuse}}{\text{adjacent}} = \frac{h}{a}, \qquad \csc \theta = \frac{\text{hypotenuse}}{\text{opposite}} = \frac{h}{op}.$$

Of course, these definitions are limited to angles between 0° and 90° (since the sum of the angles of a triangle is 180° and the right angle is 90°).

We now show that these "triangular" definitions give the same values as the "circular" definitions given earlier. It is only necessary to show this for the functions $\sin \theta$ and $\cos \theta$, since the other four functions are defined in terms of them. To verify this for $\sin \theta$ and $\cos \theta$, we place the triangle as in Figure 2. We then draw the circle with radius h that is centered at the origin and draw the unit circle. The triangles $0AB$ and $0CD$ are similar (since they have the same angles). Therefore the ratios of corresponding sides are equal. This fact tells us that

$$\frac{op}{h} = \frac{y}{1} \qquad \text{and} \qquad \frac{a}{h} = \frac{x}{1}.$$

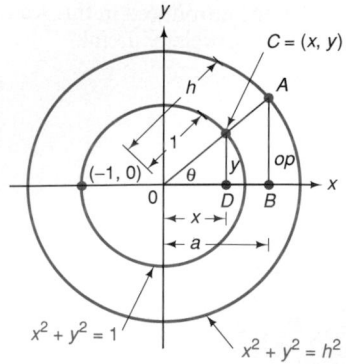

FIGURE 2

$\cos \theta = \dfrac{a}{h} = \dfrac{x}{1}$, $\sin \theta = \dfrac{op}{h} = \dfrac{y}{1}$ AND

$\tan \theta = \dfrac{op}{a} = \dfrac{y}{x}$.

But op/h is the triangular definition of $\sin \theta$, while $y/1 = y$ is the circular definition of $\sin \theta$. Thus the two definitions lead to the same function. In a similar fashion, we see that the two definitions of $\cos \theta$ lead to the same function.

Let L be any straight line. Its slope is the tangent of the angle θ that the line makes with the positive x-axis. To see this, we look at the line parallel to the given line that passes through the origin, and we draw the unit circle around that new line (see Figure 3). Then the slope of the new line (which is parallel to L) that contains the points $(0, 0)$ and $(\cos \theta, \sin \theta)$ is given by

$$m = \frac{\Delta y}{\Delta x} = \frac{\sin \theta - 0}{\cos \theta - 0} = \tan \theta.$$

Triangles are often useful for computations of values of trigonometric functions. For example, we can use a triangle to prove that $\sin 30° = \frac{1}{2}$. Look at the equilateral triangle in Figure 4. Let the sides of the triangle have lengths of 1 unit and let BD be the angle bisector of angle B, which is also the perpendicular bisector of side BD (this can be proven since the triangles ABD and DBC are congruent). The length of side BD is, from the Pythagorean theorem, equal to $\sqrt{3}/2$. Then

$$\sin 30° = \frac{op}{h} = \frac{\frac{1}{2}}{1} = \frac{1}{2}, \qquad \cos 30° = \frac{a}{h} = \frac{\sqrt{3}/2}{1} = \frac{\sqrt{3}}{2},$$

and so on. Another use of triangles is given in Example 1.

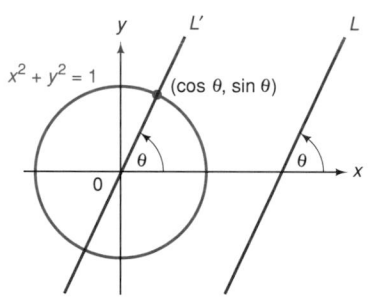

FIGURE 3
L AND L' ARE PARALLEL AND SO HAVE THE SAME SLOPE.

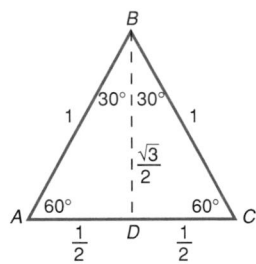

FIGURE 4
AN EQUILATERAL TRIANGLE

EXAMPLE 1
CALCULATING TRIGONOMETRIC VALUES USING A TRIANGLE

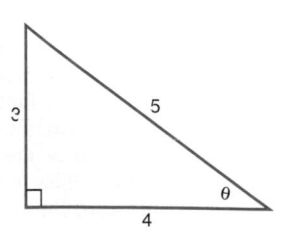

FIGURE 5
IN THIS TRIANGLE $\sin \theta = \frac{3}{5}$.

If $\sin \theta = \frac{3}{5}$, calculate $\cos \theta$, $\tan \theta$, $\cot \theta$, $\sec \theta$, and $\csc \theta$.

SOLUTION: We draw a triangle (see Figure 5). Since $\sin \theta = op/h = \frac{3}{5}$, we set $op = 3$ and $h = 5$. Then $a = \sqrt{5^2 - 3^2} = 4$, and from the triangle, $\cos \theta = \frac{4}{5}$, $\tan \theta = \frac{3}{4}$, $\cot \theta = \frac{4}{3}$, $\sec \theta = \frac{5}{4}$, and $\csc \theta = \frac{5}{3}$. There is another possible answer. The function $\sin \theta$ is positive in the second quadrant. In that quadrant $\cos \theta$ is negative, $\tan \theta$ is negative, $\cot \theta$ is negative, $\sec \theta$ is negative, and $\csc \theta$ is positive. Thus another possible set of answers is $\cos \theta = -\frac{4}{5}$, $\tan \theta = -\frac{3}{4}$, $\cot \theta = -\frac{4}{3}$, $\sec \theta = -\frac{5}{4}$, and $\csc \theta = \frac{5}{3}$. In order to get a unique answer, we must indicate the quadrant to which θ belongs.

The procedure outlined in Example 1 will be useful in Chapter 7. There are many other things that can be discussed by using triangles and are covered in any trigonometry course. Two interesting rules, the *law of cosines* and the *law of sines*, are stated in Problems 11 and 13.

■ PROBLEMS A1.4

In Problems 1–10, the value of one of the six trigonometric functions is given. Find the values of the other five functions in the indicated quadrant.

1. $\cos \theta = \frac{5}{11}$; first quadrant
2. $\tan \theta = 3$; first quadrant
3. $\sec \theta = 2$; fourth quadrant
4. $\cot \theta = -1$; second quadrant
5. $\csc \theta = 5$; second quadrant
6. $\sin \theta = -\frac{2}{3}$; third quadrant
7. $\sin \theta = -\frac{2}{3}$; fourth quadrant
8. $\sec \theta = -7$; second quadrant
9. $\tan \theta = 10$; third quadrant
10. $\cot \theta = 3$; first quadrant

*11. Let A, B, and C be the vertices of a triangle and let a, b, and c denote the corresponding opposite sides (see Figure 6(a)). Then the **law of cosines** states that

$$c^2 = a^2 + b^2 - 2ab \cos C.$$

Prove this result. [*Hint:* Place the triangle as in Figure 6(b). Use the circular definition of the functions $\sin x$ and $\cos x$ to show that the coordinates of the vertex A are $(b \cos C, b \sin C)$. Then use the distance formula to find $c = \overline{AB}$.]

12. The sides of a triangle are 2, 5, and 6. Use the law of cosines to calculate the angles of the triangle.

13. Let ABC be the vertices of a triangle. The **law of sines** states that

$$\frac{a}{\sin A} = \frac{b}{\sin B} = \frac{c}{\sin C}.$$

Prove this result. [*Hint:* Drop a perpendicular line from A to the side a (see Figure 7(a) or 7(b)). Show that, in either case,

$$\frac{h}{c} = \sin B \qquad \text{and} \qquad \frac{h}{b} = \sin C$$

(in Figure 7(b) we need the fact that $\sin(\pi - C) = \sin C$). Use these facts to complete the proof.]

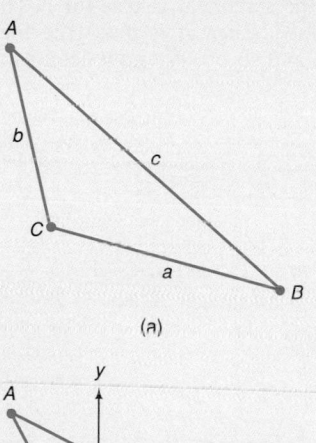

(a)

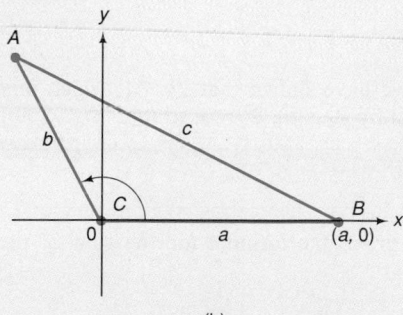

(b)

FIGURE 6

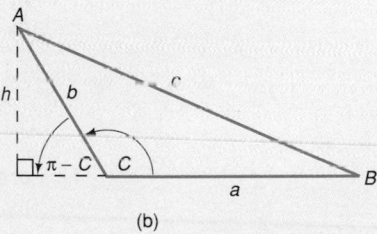

(a)

(b)

FIGURE 7

14. Two angles of a triangle are $23°$ and $85°$. The side between them has a length of 5. Use the law of sines to find the lengths of the other two sides.

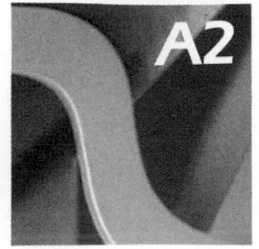

A2 MATHEMATICAL INDUCTION

Mathematical induction[†] is the name given to an elementary logical principle that can be used to prove a certain type of mathematical statement. Typically, we use mathematical induction to prove that a certain statement or equation holds for every positive integer. For example, we may need to prove that $2^n > n$ for all integers $n \geq 1$.

To do so, we proceed in two steps:

(i) We prove that the statement is true for some integer N (usually $N = 1$).

(ii) We *assume* that the statement is true for an integer k and then *prove* that it is true for the integer $k + 1$.

If we can complete these two steps, then we will have demonstrated the validity of the statement for *all* positive integers greater than or equal to N. To convince you of this fact, we reason as follows: Since the statement is true for N [by step (i)] it is true for the integer $N + 1$ [by step (ii)]. Then it is also true for the integer $(N + 1) + 1 = N + 2$ [again by step (ii)], and so on. We now demonstrate the procedure with some examples.

EXAMPLE 1
USING MATHEMATICAL INDUCTION TO PROVE AN INEQUALITY

Show that $2^n > n$ for all integers $n \geq 1$.

SOLUTION:

(i) If $n = 1$, then $2^n = 2^1 = 2 > 1 = n$, so $2^n > n$ for $n = 1$.

(ii) Assume that $2^k > k$, where $k \geq 1$ is an integer. Then

$$\text{Since } 2^k > k$$
$$\downarrow$$
$$2^{k+1} = 2 \cdot 2^k = 2^k + 2^k > k + k \geq k + 1.$$

This completes the proof since we have shown that $2^1 > 1$, which implies, by step (ii), that $2^2 > 2$, so that, again by step (ii), $2^3 > 3$, so that $2^4 > 4$, and so on.

EXAMPLE 2
A FORMULA FOR A SUM OF INTEGERS

Use mathematical induction to prove the formula for the sum of the first n positive integers:

$$1 + 2 + 3 + \cdots + n = \frac{n(n + 1)}{2}. \tag{1}$$

SOLUTION:

(i) If $n = 1$, then the sum of the first one integer is 1. But $(1)(1 + 1)/2 = 1$, so equation (1) holds in the case in which $n = 1$.

[†] This technique was first used in a mathematical proof by the great French mathematician Pierre de Fermat (1601–1665). See his biography on page 93.

(ii) Assume that (1) holds for $n = k$; that is,

$$1 + 2 + 3 + \cdots + k = \frac{k(k + 1)}{2}.$$

We must now show that it holds for $n = k + 1$. That is, we must show that

$$1 + 2 + 3 + \cdots + k + (k + 1) = \frac{(k + 1)(k + 2)}{2}.$$

But

$$1 + 2 + 3 + \cdots + k + (k + 1) = (1 + 2 + 3 + \cdots + k) + (k + 1)$$
$$= \frac{k(k + 1)}{2} + (k + 1)$$
$$= \frac{k(k + 1) + 2(k + 1)}{2} = \frac{(k + 1)(k + 2)}{2},$$

and the proof is complete.

You may wish to try a few examples to illustrate that formula (1) really works. For example,

$$1 + 2 + 3 + 4 + 5 + 6 + 7 + 8 + 9 + 10 = \frac{10(11)}{2} = 55.$$

EXAMPLE 3
A FORMULA FOR A SUM OF SQUARES

Use mathematical induction to prove the formula for the sum of the squares of the first n positive integers:

$$1^2 + 2^2 + 3^2 + \cdots + n^2 = \frac{n(n + 1)(2n + 1)}{6}. \tag{2}$$

SOLUTION:

(i) Since $1(1 + 1)(2 \cdot 1 + 1)/6 = 1 = 1^2$, equation (2) is valid for $n = 1$.
(ii) Suppose that equation (2) holds for $n = k$; that is, suppose that

$$1^2 + 2^2 + 3^2 + \cdots + k^2 = \frac{k(k + 1)(2k + 1)}{6}.$$

Then to prove that (2) is true for $n = k + 1$, we have

$$1^2 + 2^2 + 3^2 + \cdots + k^2 + (k + 1)^2 = \frac{k(k + 1)(2k + 1)}{6} + (k + 1)^2$$
$$= \frac{k(k + 1)(2k + 1) + 6(k + 1)^2}{6}$$
$$= \frac{k + 1}{6} \left[k(2k + 1) + 6(k + 1) \right]$$
$$= \frac{k + 1}{6} (2k^2 + 7k + 6)$$
$$= \frac{k + 1}{6} \left[(k + 2)(2k + 3) \right]$$
$$= \frac{(k + 1)(k + 2)[2(k + 1) + 1]}{6},$$

which is equation (2) for $n = k + 1$, and the proof is complete.

Again you may wish to experiment with this formula. For example,

$$1^2 + 2^2 + 3^2 + 4^2 + 5^2 + 6^2 + 7^2 = \frac{7(7+1)(2 \cdot 7 + 1)}{6}$$

$$= \frac{7 \cdot 8 \cdot 15}{6} = 140.$$

EXAMPLE 4
THE SUM OF A GEOMETRIC PROGRESSION

For $a \neq 1$, use mathematical induction to prove the formula for the sum of a geometric progression:

$$1 + a + a^2 + \cdots + a^n = \frac{1 - a^{n+1}}{1 - a}. \qquad (3)$$

SOLUTION:

(i) If $n = 0$, then

$$\frac{1 - a^{0+1}}{1 - a} = \frac{1 - a}{1 - a} = 1.$$

Thus equation (3) holds for $n = 0$. (We use $n = 0$ instead of $n = 1$ since $a^0 = 1$ is the first term.)

(ii) Assume that (3) holds for $n = k$; that is,

$$1 + a + a^2 + \cdots + a^k = \frac{1 - a^{k+1}}{1 - a}.$$

Then

$$1 + a + a^2 + \cdots + a^k + a^{k+1} = \frac{1 - a^{k+1}}{1 - a} + a^{k+1}$$

$$= \frac{1 - a^{k+1} + (1 - a)a^{k+1}}{1 - a} = \frac{1 - a^{k+2}}{1 - a},$$

so that equation (3) also holds for $n = k + 1$, and the proof is complete.

EXAMPLE 5
THE DERIVATIVE OF A SUM

Let $f_1, f_2, \ldots, f_n$ be differentiable functions. Use mathematical induction to prove that

$$\frac{d}{dx}(f_1 + f_2 + \cdots + f_n) = \frac{df_1}{dx} + \frac{df_2}{dx} + \cdots + \frac{df_n}{dx}. \qquad (4)$$

SOLUTION:

(i) For $n = 2$, equation (4) was demonstrated in the proof of Theorem 2.1.3.

(ii) Assume that equation (4) is valid for $n = k$; that is,

$$\frac{d}{dx}(f_1 + f_2 + \cdots + f_k) = \frac{df_1}{dx} + \frac{df_2}{dx} + \cdots + \frac{df_k}{dx}.$$

Let $g(x) = f_1(x) + f_2(x) + \cdots + f_k(x)$. Then

(by the case $n = 2$)

$$\frac{d}{dx}(f_1 + f_2 + \cdots + f_k + f_{k+1}) = \frac{d}{dx}(g + f_{k+1}) = \frac{dg}{dx} + \frac{df_{k+1}}{dx}$$

$$= \frac{d}{dx}(f_1 + f_2 + \cdots + f_k) + \frac{df_{k+1}}{dx} = \frac{df_1}{dx} + \frac{df_2}{dx} + \cdots + \frac{df_k}{dx} + \frac{df_{k+1}}{dx},$$

which is equation (4) in the case $n = k + 1$, and the theorem is proved.

■ PROBLEMS

1. Use mathematical induction to prove that the sum of the cubes of the first n positive integers is given by

$$1^3 + 2^3 + 3^3 + \cdots + n^3 = \frac{n^2(n+1)^2}{4}. \tag{5}$$

2. Let the functions $f_1, f_2, \ldots, f_n$ be integrable on $[0, 1]$. Show that $f_1 + f_2 + \cdots + f_n$ is integrable on $[0, 1]$ and that

$$\int_0^1 [f_1(x) + f_2(x) + \cdots + f_n(x)] \, dx$$
$$= \int_0^1 f_1(x) \, dx + \int_0^1 f_2(x) \, dx + \cdots + \int_0^1 f_n(x) \, dx.$$

3. Use mathematical induction to prove that the nth derivative of the nth-order polynomial

$$P_n(x) = x^n + a_{n-1}x^{n-1} + a_{n-2}x^{n-2} + \cdots + a_1 x^1 + a_0$$

is equal to $n!$ $[n! = n(n-1)(n-2) \cdots 3 \cdot 2 \cdot 1]$.

4. Show that if $a \neq 1$,

$$1 + 2a + 3a^2 + \cdots + na^{n-1} = \frac{1 - (n+1)a^n + na^{n+1}}{(1-a)^2}.$$

*5. Prove, using mathematical induction, that there are exactly 2^n subsets of a set containing n elements.

6. Use mathematical induction to prove that

$$\ln(a_1 a_2 a_3 \cdots a_n) = \ln a_1 + \ln a_2 + \cdots + \ln a_n,$$

if $a_k > 0$ for $k = 1, 2, \ldots, n$.

7. Let $\mathbf{u}, \mathbf{v}_1, \mathbf{v}_2, \ldots, \mathbf{v}_n$ be $n + 1$ vectors in $\mathbb{R}^2$. Prove that (see Section 11.2)

$$\mathbf{u} \cdot (\mathbf{v}_1 + \mathbf{v}_2 + \cdots + \mathbf{v}_n)$$
$$= \mathbf{u} \cdot \mathbf{v}_1 + \mathbf{u} \cdot \mathbf{v}_2 + \cdots + \mathbf{u} \cdot \mathbf{v}_n.$$

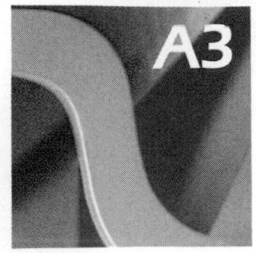

THE PROOFS OF SOME THEOREMS ON LIMITS, CONTINUITY, AND DIFFERENTIATION

In this appendix we will provide rigorous proofs of several theorems discussed in Chapters 1, 2, and 6. These proofs require a knowledge of the material in Section 1.8. In the proofs of the limit theorems, we will assume that x_0 and L are real numbers. The cases of infinite limits and limits at infinity will be discussed in the problem set.

THEOREM 1 (THEOREM 1.3.2 (i))

Let c be any real number and suppose that $\lim_{x \to x_0} f(x)$ exists. Then

$$\lim_{x \to x_0} cf(x) = c \lim_{x \to x_0} f(x).$$

PROOF: *Case (i):* $c = 0$. Then since $\lim_{x \to x_0} 0 = 0$ (see Problem 1), we have

$$\lim_{x \to x_0} cf(x) = \lim_{x \to x_0} 0 = 0 = 0 \cdot \lim_{x \to x_0} f(x).$$

Case (ii): $c \neq 0$. Let $\lim_{x \to x_0} f(x) = L$ and let $\epsilon > 0$ be given. Choose $\delta > 0$ so that if $0 < |x - x_0| < \delta$, then $|f(x) - L| < \epsilon/|c|$ [we can always do this since $\lim_{x \to x_0} f(x) = L$]. Then

$$|cf(x) - cL| = |c(f(x) - L)| = |c||f(x) - L| < |c| \frac{\epsilon}{|c|} = \epsilon$$

if $0 < |x - x_0| < \delta$, and the theorem is proved. ∎

THEOREM 2 (THEOREM 1.3.2 (ii))

If $\lim_{x \to x_0} f(x)$ and $\lim_{x \to x_0} g(x)$ both exist (and are finite), then

$$\lim_{x \to x_0} [f(x) + g(x)] = \lim_{x \to x_0} f(x) + \lim_{x \to x_0} g(x).$$

PROOF: Let $\lim_{x \to x_0} f(x) = L_1$ and $\lim_{x \to x_0} g(x) = L_2$. For a given $\epsilon > 0$, choose δ_1 such that if $0 < |x - x_0| < \delta_1$, then $|f(x) - L_1| < \epsilon/2$; and choose δ_2 such that if $0 < |x - x_0| < \delta_2$, then $|g(x) - L_2| < \epsilon/2$. Let δ be the smaller of δ_1 and δ_2 (denoted by $\delta = \min\{\delta_1, \delta_2\}$). Then if $0 < |x - x_0| < \delta$,

$$|(f(x) + g(x)) - (L_1 + L_2)| = |(f(x) - L_1) + (g(x) - L_2)|$$

$$\leq |f(x) - L_1| + |g(x) - L_2| = \frac{\epsilon}{2} + \frac{\epsilon}{2} = \epsilon,$$

and the theorem is proved. (The last step follows from the triangle inequality; see Section 0.1.) ∎

REMARK: It is easy to extend Theorem 2 to a finite sum. We have, if all indicated limits exist,

$$\lim_{x \to x_0} [f_1(x) + f_2(x) + \cdots + f_n(x)] = \lim_{x \to x_0} f_1(x) + \lim_{x \to x_0} f_2(x) + \cdots + \lim_{x \to x_0} f_n(x).$$

THEOREM 3   (THEOREM 1.3.2 (iii))

If $\lim_{x \to x_0} f(x)$ and $\lim_{x \to x_0} g(x)$ both exist, then

$$\lim_{x \to x_0} f(x) \cdot g(x) = \left[\lim_{x \to x_0} f(x) \right]\left[\lim_{x \to x_0} g(x) \right].$$

PROOF: Let $L_1 = \lim_{x \to x_0} f(x)$ and $L_2 = \lim_{x \to x_0} g(x)$. Let $\epsilon > 0$ be given. We choose four δ's as follows:

(i) Choose $\delta_1 > 0$ such that if $0 < |x - x_0| < \delta_1$, then $|f(x) - L_1| < 1$ and $|f(x)| < |L_1| + 1$.

(ii) Choose δ_2 such that if $0 < |x - x_0| < \delta_2$, then

$$|g(x) - L_2| < \frac{\epsilon}{2}\left(\frac{1}{|L_1| + 1}\right).$$

(iii) Choose $\delta_3 > 0$ such that if $0 < |x - x_0| < \delta_3$, then

$$|f(x) - L_1| < \frac{\epsilon}{2}\left(\frac{1}{|L_2| + 1}\right).$$

(iv) Choose $\delta = \min\{\delta_1, \delta_2, \delta_3\}$.

We now show that if $0 < |x - x_0| < \delta$, then $|f(x)g(x) - L_1 L_2| < \epsilon$. This step will complete the proof of the theorem. We have, for $0 < |x - x_0| < \delta$,

$$
\begin{aligned}
|f(x)g(x) - L_1 L_2| &= |f(x)g(x) - f(x)L_2 + f(x)L_2 - L_1 L_2| \\
&\leq |f(x)g(x) - f(x)L_2| + |f(x)L_2 - L_1 L_2| \\
&= |f(x)||g(x) - L_2| + |L_2||f(x) - L_1| \\
&< (|L_1| + 1)\left(\frac{\epsilon}{2}\right)\left(\frac{1}{|L_1| + 1}\right) + |L_2|\left(\frac{\epsilon}{2}\right)\left(\frac{1}{|L_2| + 1}\right) \\
&< \frac{\epsilon}{2} + \frac{\epsilon}{2} = \epsilon. \quad \blacksquare
\end{aligned}
$$

COROLLARY 1

If $\lim_{x \to x_0} f(x)$ exists and n is a positive integer, then

$$\lim_{x \to x_0} (f(x))^n = \left(\lim_{x \to x_0} f(x) \right)^n.$$

PROOF:

$$\lim_{x \to x_0} f^2(x) = \lim_{x \to x_0} f(x) \cdot \lim_{x \to x_0} f(x) = \left[\lim_{x \to x_0} f(x) \right]^2$$

For $n > 2$, simply use this argument as many times as necessary (i.e., use mathematical induction). ∎

COROLLARY 2 **(THEOREM 1.3.1)**

Let $P(x) = c_0 + c_1 x + c_2 x^2 + \cdots + c_n x^n$ be a polynomial, where c_0, c_1, $c_2, \ldots, c_n$ are real numbers. Then

$$\lim_{x \to x_0} P(x) = P(x_0) = c_0 + c_1 x_0 + c_2 x_0^2 + \cdots + c_n x_0^n.$$

PROOF: We use the fact that $\lim_{x \to x_0} c_0 = c_0$ and $\lim_{x \to x_0} x = x_0$ (see Problems 1 and 3). Then using Theorems 1, 2, and 3 and Corollary 1, we obtain

$$\lim_{x \to x_0} P(x) = \lim_{x \to x_0} (c_0 + c_1 x + c_2 x^2 + \cdots + c_n x^n)$$

$$= \lim_{x \to x_0} c_0 + c_1 \lim_{x \to x_0} x + c_2 \left(\lim_{x \to x_0} x \right)^2 + \cdots + c_n \left(\lim_{x \to x_0} x \right)^n$$

$$= c_0 + c_1 x_0 + c_2 x_0^2 + c_3 x_0^3 + \cdots + c_n x_0^n = P(x_0). \quad \blacksquare$$

THEOREM 4 **(THEOREM 1.3.2 (iv))**

If $\lim_{x \to x_0} f(x)$ and $\lim_{x \to x_0} g(x)$ both exist and $\lim_{x \to x_0} g(x) \neq 0$, then

$$\lim_{x \to x_0} \frac{f(x)}{g(x)} = \frac{\lim_{x \to x_0} f(x)}{\lim_{x \to x_0} g(x)}.$$

PROOF: As before, let $L_1 = \lim_{x \to x_0} f(x)$ and $L_2 = \lim_{x \to x_0} g(x)$. Since $L_2 \neq 0$, we show first that $\lim_{x \to x_0} 1/g(x) = 1/L_2$. We must show that for a given $\epsilon > 0$, there is a $\delta > 0$ such that when $|x - x_0| < \delta$,

$$\left| \frac{1}{g(x)} - \frac{1}{L_2} \right| < \epsilon. \tag{1}$$

But the inequality (1) is equivalent to

$$\left| \frac{L_2 - g(x)}{L_2 g(x)} \right| < \epsilon. \tag{2}$$

Select δ_1 such that $0 < |x - x_0| < \delta_1$ implies that $|g(x) - L_2| < |L_2|/2$. Then

$$|L_2| = |(L_2 - g(x)) + g(x)| \leq |L_2 - g(x)| + |g(x)| < \tfrac{1}{2}|L_2| + |g(x)|,$$

and so $|g(x)| > \tfrac{1}{2}|L_2|$. Then for $0 < |x - x_0| < \delta_1$,

$$|L_2 g(x)| = |L_2| |g(x)| > |L_2| \left| \frac{L_2}{2} \right| = \frac{L_2^2}{2},$$

or $1/|L_2 g(x)| < 2/L_2^2$. Similarly, there is a δ_2 such that if $0 < |x - x_0| < \delta_2$, then

$$|L_2 - g(x)| < \frac{L_2^2}{2} \epsilon.$$

Now choose $\delta = \min\{\delta_1, \delta_2\}$. Then for $0 < |x - x_0| < \delta$,

$$\left| \frac{L_2 - g(x)}{L_2 g(x)} \right| = \frac{1}{|L_2 g(x)|} \cdot |L_2 - g(x)| < \frac{2}{L_2^2} \cdot \frac{L_2^2}{2} \epsilon = \epsilon,$$

which shows that $\lim_{x \to x_0} 1/g(x) = 1/L_2$. Finally, we obtain from this result and Theorem 3

$$\lim_{x \to x_0} \frac{f(x)}{g(x)} = \lim_{x \to x_0} f(x) \cdot \lim_{x \to x_0} \frac{1}{g(x)} = L_1 \cdot \frac{1}{L_2} = \frac{L_1}{L_2},$$

and Theorem 4 is proved. ■

COROLLARY 3

Let $r(x) = p(x)/q(x)$ be a rational function (the quotient of two polynomials). Then if $q(x_0) \neq 0$,

$$\lim_{x \to x_0} r(x) = r(x_0) = \frac{p(x_0)}{q(x_0)}.$$

PROOF: From Theorem 4 and Corollary 2,

$$\lim_{x \to x_0} r(x) = \lim_{x \to x_0} \frac{p(x)}{q(x)} = \frac{\lim_{x \to x_0} p(x)}{\lim_{x \to x_0} q(x)} = \frac{p(x_0)}{q(x_0)} = r(x_0). \quad ■$$

THEOREM 5   (THEOREM 1.4.1: SQUEEZING THEOREM)

Suppose that $f(x) \leq g(x) \leq h(x)$ for x in a neighborhood of x_0 and $\lim_{x \to x_0} f(x) = \lim_{x \to x_0} h(x) = L$, where x_0 may be $+\infty$ or $-\infty$. Then

$$\lim_{x \to x_0} g(x) = L.$$

PROOF: We prove this theorem in the case that x_0 is finite. For the case $x_0 = +\infty$ or $-\infty$, see Problem 4. Let $\epsilon > 0$ be given. Choose δ_1 such that $|f(x) - L| < \epsilon$ if $0 < |x - x_0| < \delta_1$, and choose δ_2 so that $|h(x) - L| < \epsilon$ if $0 < |x - x_0| < \delta_2$. Let the inequalities $f(x) \leq g(x) \leq h(x)$ hold for x in some open interval (a, b) (recall the definition of a neighborhood). Then since x_0 is not one of the numbers a or b, there exists a δ_3 such that $a < x_0 - \delta_3 < x_0 < x_0 + \delta_3 < b$, and therefore, for $0 < |x - x_0| < \delta_3$, $f(x) \leq g(x) \leq h(x)$. Now let $\delta = \min\{\delta_1, \delta_2, \delta_3\}$. Then if $0 < |x - x_0| < \delta$,

$$g(x) - L < h(x) - L < \epsilon.$$

Since $|f(x) - L| < \epsilon$, we have $f(x) - L > -\epsilon$. Then

$$g(x) - L \geq f(x) - L > -\epsilon.$$

Putting this information all together, we obtain

$$|g(x) - L| < \epsilon,$$

and the squeezing theorem is proved. ■

We have continually talked about *the* limit. We have not yet proven that if a limit exists, then it is unique. We do so now.

THEOREM 6

If $\lim_{x \to x_0} f(x)$ exists, then it is unique.

PROOF: Suppose that $\lim_{x \to x_0} f(x) = L_1$ and $\lim_{x \to x_0} f(x) = L_2$. Let $\epsilon > 0$ be given. There are positive numbers δ_1 and δ_2 such that if $0 < |x - x_0| < \delta = \min\{\delta_1, \delta_2\}$, then $|f(x) - L_1| < \epsilon$ and $|f(x) - L_2| < \epsilon$.

Then

$$|L_1 - L_2| = |L_1 - f(x) + f(x) - L_2| \leq |L_1 - f(x)| + |f(x) - L_2|$$
$$= |f(x) - L_1| + |L_2 - f(x)| < \epsilon + \epsilon = 2\epsilon.$$

Thus $|L_1 - L_2| < 2\epsilon$ for *every* positive number ϵ. This can only happen if $L_1 = L_2$ and the theorem is proved. ∎

We now turn to an important limit theorem involving continuity. First, we recall our definition of continuity (see page 113).

DEFINITION 1   CONTINUITY

Let $f(x)$ be defined for every x in an open interval containing the number x_0. Then f is **continuous** at x_0 if all of the following three conditions hold:

(i) $f(x_0)$ exists (that is, x_0 is in the domain of f).
(ii) $\lim_{x \to x_0} f(x)$ exists.
(iii) $\lim_{x \to x_0} f(x) = f(x_0)$. ∎

Using our $\epsilon - \delta$ definition of a limit, we can give a second definition of continuity. This second definition will enable us to prove our important limit theorem.

DEFINITION 2

The function f is **continuous** at x_0 if f is defined in a neighborhood of x_0 and, for every $\epsilon > 0$, there is a $\delta > 0$ such that if $|x - x_0| < \delta$, then $|f(x) - f(x_0)| < \epsilon$. ∎

Before using these two definitions, we show that they are equivalent. That is, if f is continuous according to Definition 1, then it is continuous according to Definition 2, and vice versa.

THEOREM 7

Definitions 1 and 2 are equivalent.

PROOF:

(i) We first show that if a function is continuous according to Definition 1, then it is continuous according to Definition 2. Suppose that f is defined in a neighborhood of x_0 and that $\lim_{x \to x_0} f(x) = f(x_0)$. Let $\epsilon > 0$ be given. Then by the definition of limit, there is a δ such that if $0 < |x - x_0| < \delta$, then $|f(x) - f(x_0)| < \epsilon$. Further, $|f(x_0) - f(x_0)| = 0 < \epsilon$. Thus f is defined in a neighborhood of x_0, and for every $\epsilon > 0$ there is a $\delta > 0$ such that if $|x - x_0| < \delta$, then $|f(x) - f(x_0)| < \epsilon$. Hence f is continuous in the sense of Definition 2.

(ii) Conversely, if we assume that f is continuous at x_0 in the sense of Definition 2, then f is defined in a neighborhood of x_0, and for every $\epsilon > 0$ there is a $\delta > 0$ such that $|x - x_0| < \delta$ implies $|f(x) - f(x_0)| < \epsilon$. Then by the definition of limit, $\lim_{x \to x_0} f(x) = f(x_0)$, and this statement is Definition 1. ∎

We now state and prove our next limit theorem.

THEOREM 8 (THEOREM 1.7.2)

If f is continuous at a and if $\lim_{x \to x_0} g(x) = a$, then $\lim_{x \to x_0} f(g(x)) = f(a)$.

PROOF: Let $\epsilon > 0$ be given. We must show that there exists a $\delta > 0$ such that if $|x - x_0| < \delta$, then $|f(g(x)) - f(a)| < \epsilon$. We do this in two steps.

(i) Since f is continuous at a, by Definition 2, there is a $\delta_1 > 0$ such that if $|x - a| < \delta_1$, then $|f(x) - f(a)| < \epsilon$.

(ii) Since $\lim_{x \to x_0} g(x) = a$, there is a $\delta > 0$ such that if $0 < |x - x_0| < \delta$, then $|g(x) - a| < \delta_1$. But if $|g(x) - a| < \delta_1$, then, from (i), $|f(g(x)) - f(a)| < \epsilon$, and the theorem is proved. ∎

THEOREM 9 (THEOREM 6.1.3)

Let f be continuous on $[a, b]$ and let $c = f(a)$ and $d = f(b)$. If f has an inverse $g = f^{-1}$, then g is continuous on $[c, d]$ (or $[d, c]$ if $c > d$).

PROOF:

(i) We first show that f is monotone on $[a, b]$. Since f^{-1} exists, f is 1–1 on $[a, b]$. Suppose there are numbers x_1, x_2, and x_3 in $[a, b]$ such that $x_1 < x_2 < x_3$ and $f(x_1) < f(x_2)$ but $f(x_3) < f(x_2)$ (so that f is neither increasing nor decreasing). The situation is depicted in Figure 1(a).

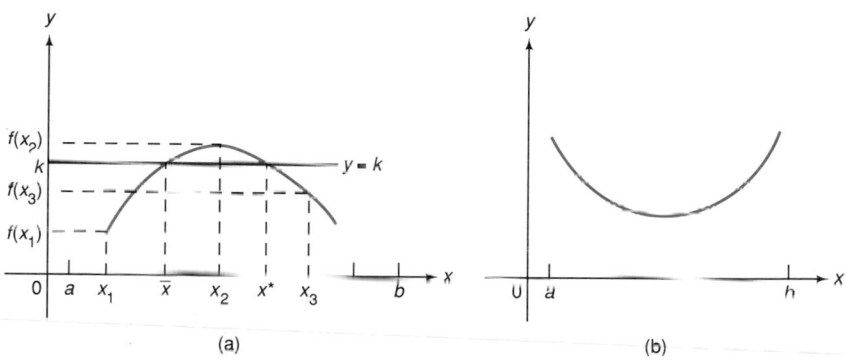

(a)   (b)

FIGURE 1 f IS NEITHER INCREASING NOR DECREASING IN (a, b).

Let k be a number between $f(x_3)$ and $f(x_2)$. By the intermediate-value theorem, there is a number $\bar{x}$ in (x_1, x_2) such that $f(\bar{x}) = k$. Similarly, there is a number x^* in (x_2, x_3) such that $f(x^*) = k$. Since $\bar{x} \neq x^*$ but $f(\bar{x}) = f(x^*) = k$, we have obtained a contradiction of the fact that f is 1–1. A similar argument rules out the graph in Figure 1(b). Thus we conclude that f is monotone.

(ii) We now show that $g = f^{-1}$ is monotone on $[c, d]$. Suppose that f is increasing (a similar argument works if f is decreasing), and suppose that $y_1, y_2 \in (c, d)$, with $y_1 < y_2$. There are numbers x_1 and x_2 in (a, b) such that $y_1 = f(x_1)$ and $y_2 = f(x_2)$. Suppose that $g(y_1) \geq g(y_2)$. Then since f is increasing,

$$f(\overbrace{g(y_1)}^{x_1}) \geq f(\overbrace{g(y_2)}^{x_2})$$

so that $y_1 = f(x_1) \geq f(x_2) = y_2$. This result contradicts the fact that $y_1 < y_2$. Thus g is also increasing. We have shown the following:

(a) If f is increasing, f^{-1} is increasing.

(b) If f is decreasing, f^{-1} is decreasing.

(iii) We show that $g = f^{-1}$ is continuous in $[c, d]$. That is, we must show that for every $\bar{y} \in [c, d]$,

$$\lim_{y \to \bar{y}} g(y) = g(\bar{y}), \tag{3}$$

where the limit in (3) is a one-sided limit if $\bar{y} = c$ or d. We will assume that $\bar{y} \in (c, d)$ and leave the endpoints as an exercise. Then (3) is equivalent to the following: For every $\epsilon > 0$, there is a number $\delta > 0$ such that

$$\text{if } |y - \bar{y}| < \delta, \qquad \text{then } |g(y) - g(\bar{y})| < \epsilon,$$

or equivalently,

$$\text{if } \bar{y} - \delta < y < \bar{y} + \delta, \qquad \text{then } g(\bar{y}) - \epsilon < g(y) < g(\bar{y}) + \epsilon. \tag{4}$$

Choose $\epsilon > 0$ and assume that $a < g(\bar{y}) - \epsilon < g(\bar{y}) + \epsilon < b$ (this is no restriction). Let $\bar{x} = g(\bar{y})$, so that $f(\bar{x}) = \bar{y}$. Let

$$\delta_1 = \bar{y} - f(\bar{x} - \epsilon), \qquad \delta_2 = f(\bar{x} + \epsilon) - \bar{y}, \qquad \delta = \min\{\delta_1, \delta_2\}. \tag{5}$$

Since f is increasing, δ_1, δ_2, and therefore δ are all positive. Suppose that $\bar{y} - \delta < y < \bar{y} + \delta$. Then from (5)

$$\bar{y} - \delta_1 \leq \bar{y} - \delta < y < \bar{y} + \delta \leq \bar{y} + \delta_2,$$

and

$$f(\bar{x} - \epsilon) < y < f(\bar{x} + \epsilon),$$

which implies, since g is increasing, that

$$g[f(\bar{x} - \epsilon)] < g(y) < g[f(\bar{x} + \epsilon)].$$

But since $g = f^{-1}$, we have

$$\bar{x} - \epsilon < g(y) < \bar{x} + \epsilon.$$

This completes the proof. ∎

Our last result is the intermediate-value theorem (Theorem 1.7.5 on page 119). In order to prove this theorem, we need to use the completeness axiom for real numbers discussed on page 118: Every nonempty set of numbers which is bounded above has a least upper bound (l.u.b.) and every nonempty set of numbers that is bounded below has a greatest lower bound (g.l.b.).

THEOREM 10   INTERMEDIATE VALUE THEOREM

Let f be continuous on $[a, b]$. Then if c is any number between $f(a)$ and $f(b)$, there is a number $\bar{x}$ in (a, b) such that $f(\bar{x}) = c$.

PROOF: Suppose that $f(a) < c < f(b)$. Let $S = \{x \in [a, b] : f(x) < c\}$. S is a nonempty set of real numbers ($a \in S$) that is bounded above by b. Thus S has an l.u.b. $\bar{x}$. Suppose $f(\bar{x}) < c$. Let $\epsilon = c - f(\bar{x})$. Then, since f is continuous, there is a $\delta > 0$ such that $|f(x) - f(\bar{x})| < \epsilon/2$ if $|x - \bar{x}| < \delta$. In particular $|f(\bar{x} + \delta) - f(\bar{x})| < \epsilon/2$ which means that $f(\bar{x} + \delta) < c$. Thus $\bar{x} + \delta \in S$. But this contradicts the fact that $\bar{x}$ is an upper bound for S. Thus $f(\bar{x}) \geq c$.

Now suppose that $f(\bar{x}) > c$. Then, by the same reasoning as before, $f(x) > c$ for $\bar{x} - \delta < x < \bar{x}$ for some $\delta > 0$. Thus $\bar{x} - \delta$ is an upper bound for S. But $\bar{x} - \delta < \bar{x}$ so $\bar{x}$ is not the *least* upper bound for S. This contradiction shows that $f(\bar{x}) \leq c$. Since $f(\bar{x}) \geq c$ also, we conclude that $f(\bar{x}) = c$. ∎

■ PROBLEMS

1. Prove from the definition of a limit that for any finite real number x_0, $\lim_{x \to x_0} c = c$ for any constant c.
2. Prove that $\lim_{x \to \infty} c = c$.
3. Prove that $\lim_{x \to x_0} x = x_0$ for every real number x_0.
4. Prove that if $f(x) \leq g(x) \leq h(x)$ for all x larger than some number N, and $\lim_{x \to \infty} f(x) = \lim_{x \to \infty} h(x) = L$, then $\lim_{x \to \infty} g(x) = L$. [*Hint:* Treat the cases $L < \infty$ and $L = \infty$ separately.]
5. Prove that $\lim_{x \to \infty} cf(x) = c \lim_{x \to \infty} f(x)$.
6. Prove that if $\lim_{x \to \infty} f(x)$ and $\lim_{x \to \infty} g(x)$ exist and are finite, then

$$\lim_{x \to \infty} [f(x) + g(x)] = \lim_{x \to \infty} f(x) + \lim_{x \to \infty} g(x).$$

*7. Give an example to show that the result of Problem 6 is false in general if the limits are not assumed to be finite. [*Hint:* Find $f(x)$ and $g(x)$ so that $\lim_{x \to \infty} f(x) = \lim_{x \to \infty} g(x) = \infty$ but that $f(x) - g(x)$ is a constant.]
8. Show that if $\lim_{x \to \infty} f(x)$ and $\lim_{x \to \infty} g(x)$ are finite, then

$$\lim_{x \to \infty} f(x)g(x) = \left[\lim_{x \to \infty} f(x) \right]\left[\lim_{x \to \infty} g(x) \right].$$

*9. Give an example to show that the result of Problem 8 is false in general if the limits are not finite. [*Hint:* Find $f(x)$ and $g(x)$ such that $\lim_{x \to \infty} f(x) = \infty$, $\lim_{x \to \infty} g(x) = 0$, and $f(x) \cdot g(x)$ is a constant. Note that the expression $0 \cdot \infty$ is not defined.]
10. Show that if $\lim_{x \to \infty} f(x)$ is finite and $\lim_{x \to \infty} g(x)$ is finite and nonzero, then

$$\lim_{x \to \infty} \frac{f(x)}{g(x)} = \frac{\lim_{x \to \infty} f(x)}{\lim_{x \to \infty} g(x)}.$$

*11. Give an example to show that the result of Problem 10 is false in general if the limits are not assumed to be finite.
12. Prove that if $\lim_{x \to \infty} f(x)$ exists, then it is unique.

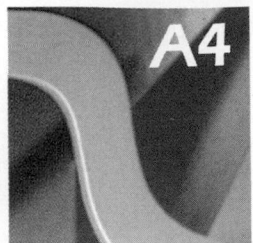

A4 DETERMINANTS

In several parts of this book we made use of determinants. In this appendix we show how determinants arise and how they can be computed.

We begin by considering the system of two linear equations in two unknowns:

$$a_{11}x_1 + a_{12}x_2 = b_1,$$
$$a_{21}x_1 + a_{22}x_2 = b_2. \tag{1}$$

For simplicity, we assume that the constants a_{11}, a_{12}, a_{21}, and a_{22} are all nonzero (otherwise, the system can be solved directly). To solve the system (1), we reduce it to one equation in one unknown. To accomplish this, we multiply the first equation by a_{22} and the second by a_{12} to obtain

$$a_{11}a_{22}x_1 + a_{22}a_{12}x_2 = a_{22}b_1,$$
$$a_{12}a_{21}x_1 + a_{22}a_{12}x_2 = a_{12}b_2. \tag{2}$$

Then subtracting the second equation from the first, we have

$$(a_{11}a_{22} - a_{12}a_{21})x_1 = a_{22}b_1 - a_{12}b_2. \tag{3}$$

Now we define the quantity

$$D = a_{11}a_{22} - a_{12}a_{21}. \tag{4}$$

If $D \neq 0$, then (3) yields

$$x_1 = \frac{a_{22}b_1 - a_{12}b_2}{D}, \tag{5}$$

and x_2 may be obtained by substituting this value of x_1 into either of the equations of (1). Thus if $D \neq 0$, *the system (1) has a unique solution.*

On the other hand, suppose that $D = 0$. Then $a_{11}a_{22} = a_{12}a_{21}$, and equation (3) leads to the equation

$$0 = a_{22}b_1 - a_{12}b_2.$$

Either this equation is true or it is false. If it is false (that is, if $a_{22}b_1 - a_{12}b_2 \neq 0$), then the system (1) has *no* solution. If the equation is true (that is, if $a_{22}b_1 - a_{12}b_2 = 0$), then the second equation of (2) is a multiple of the first and (1) consists essentially of only one equation. In this case, we may choose x_1 arbitrarily and calculate the corresponding value of x_2, which means that there are an *infinite* number of solutions. In sum, we have shown that *if $D = 0$, then the system (1) has either no solution or an infinite number of solutions.*

These facts are easily visualized geometrically by noting that (1) consists of the equations of two straight lines. A solution of the system is a point of intersection of the two lines. If the slopes are different, then $D \neq 0$ and the two lines intersect at a single point, which is the unique solution. It is easy to show (see Problem 21) that $D = 0$ if and only if the slopes of the two lines are the same. If $D = 0$, either we have two parallel lines and no solution, since the lines never intersect, or both equations yield the same line and every point on this line is a solution. These results are illustrated in Figure 1.

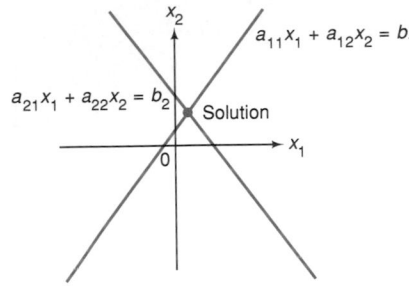

(a) Unique solution—lines intersect.

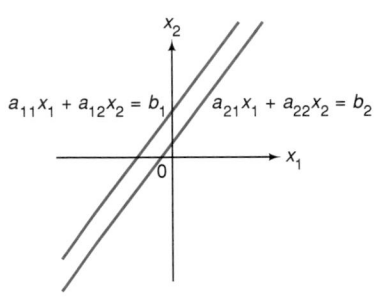

(b) No solution—lines are parallel.

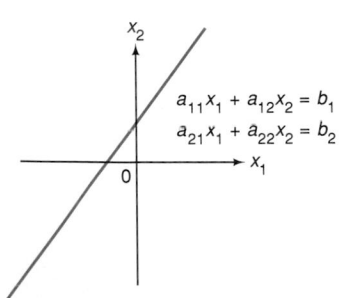

(c) Infinite number of solutions—lines coincide.

FIGURE 1
TWO SOLUTIONS IN TWO UNKNOWNS HAVE ONE SOLUTION, NO SOLUTION, OR AN INFINITE NUMBER OF SOLUTIONS.

EXAMPLE 1
DETERMINING WHETHER A
SYSTEM HAS A UNIQUE
SOLUTION BY LOOKING AT ITS
DETERMINANT

Consider the following systems of equations:

(i) $2x_1 + 3x_2 = 12$ (ii) $x_1 + 3x_2 = 3$ (iii) $x_1 + 3x_2 = 3$

$x_1 + x_2 = 5$ $3x_1 + 9x_2 = 8$ $3x_1 + 9x_2 = 9$

In system (i), $D = 2 \cdot 1 - 3 \cdot 1 = -1 \neq 0$, so there is a unique solution, which is easily found to be $x_1 = 3$, $x_2 = 2$. In system (ii), $D = 1 \cdot 9 - 3 \cdot 3 = 0$. Multiplying the first equation by 3 and then subtracting this from the second equation, we obtain the equation $0 = -1$, which is impossible. Thus, there is no solution. In (iii), $D = 1 \cdot 9 - 3 \cdot 3 = 0$. But now the second equation is simply three times the first equation. If x_2 is arbitrary, then $x_1 = 3 - 3x_2$, and there are an infinite number of solutions.

Returning again to the system (1), we define the **determinant of the system** as

$$D = a_{11}a_{22} - a_{12}a_{21}. \tag{6}$$

For convenience of notation, we denote the determinant by writing the coefficients of the system in a square array:

$$D = \begin{vmatrix} a_{11} & a_{12} \\ a_{21} & a_{22} \end{vmatrix} = a_{11}a_{22} - a_{12}a_{21}. \tag{7}$$

Therefore, a 2×2 determinant is the product of the two elements in the upper-left-to-lower-right diagonal minus the product of the other two elements.

We have proved the following theorem.

THEOREM 1

For the 2×2 system (1), there is a unique solution if and only if the determinant D is not equal to zero. If $D = 0$, then there is either no solution or an infinite number of solutions. ∎

Let us now consider the general system of n equations in n unknowns:

$$a_{11}x_1 + a_{12}x_2 + \cdots + a_{1n}x_n = b_1,$$
$$a_{21}x_1 + a_{22}x_2 + \cdots + a_{2n}x_n = b_2,$$
$$\vdots \qquad \vdots \qquad \qquad \vdots \quad \vdots \tag{8}$$
$$a_{n1}x_1 + a_{n2}x_2 + \cdots + a_{nn}x_n = b_n,$$

and define the determinant of such a system in order to obtain a theorem like the one above for $n \times n$ systems. We begin by defining the determinant of a 3×3 system:

$$D = \begin{vmatrix} a_{11} & a_{12} & a_{13} \\ a_{21} & a_{22} & a_{23} \\ a_{31} & a_{32} & a_{33} \end{vmatrix} = a_{11}\begin{vmatrix} a_{22} & a_{23} \\ a_{32} & a_{33} \end{vmatrix} - a_{12}\begin{vmatrix} a_{21} & a_{23} \\ a_{31} & a_{33} \end{vmatrix} + a_{13}\begin{vmatrix} a_{21} & a_{22} \\ a_{31} & a_{32} \end{vmatrix}. \tag{9}$$

We see that to calculate a 3×3 determinant, it is necessary to calculate three 2×2 determinants.

EXAMPLE 2
CALCULATING A 3×3
DETERMINANT

$$\begin{vmatrix} 3 & 5 & 2 \\ 4 & 2 & 3 \\ -1 & 2 & 4 \end{vmatrix} = 3\begin{vmatrix} 2 & 3 \\ 2 & 4 \end{vmatrix} - 5\begin{vmatrix} 4 & 3 \\ -1 & 4 \end{vmatrix} + 2\begin{vmatrix} 4 & 2 \\ -1 & 2 \end{vmatrix}$$

$$= 3 \cdot 2 - 5 \cdot 19 + 2 \cdot 10 = -69$$

The general definition of the determinant of the $n \times n$ system of equations (8) is simply an extension of this procedure:

$$D = \begin{vmatrix} a_{11} & a_{12} & \cdots & a_{1n} \\ a_{21} & a_{22} & \cdots & a_{2n} \\ \vdots & \vdots & & \vdots \\ a_{n1} & a_{n2} & \cdots & a_{nn} \end{vmatrix}$$
$$= a_{11}A_{11} - a_{12}A_{12} + \cdots + (-1)^{n+1}a_{1n}A_{1n}, \tag{10}$$

where A_{1j} is the $(n-1) \times (n-1)$ determinant obtained by crossing out the first row and jth column of the original $n \times n$ determinant. Thus an $n \times n$ determinant can be obtained by calculating n $(n-1) \times (n-1)$ determinants. Note that in definition (10) the signs alternate. The signs of the n^2 $(n-1) \times (n-1)$ determinants can easily be illustrated by the following schematic diagram:

$$\begin{vmatrix} + & - & + & - & + & - & \cdots \\ - & + & - & + & - & + & \cdots \\ + & - & + & - & + & - & \cdots \\ - & + & - & + & - & + & \cdots \\ + & - & + & - & + & - & \cdots \\ \vdots & \vdots & \vdots & \vdots & \vdots & \vdots & \ddots \end{vmatrix}$$

EXAMPLE 3
CALCULATING A 4×4 DETERMINANT

$$\begin{vmatrix} 1 & 3 & 5 & 2 \\ 0 & -1 & 3 & 4 \\ 2 & 1 & 9 & 6 \\ 3 & 2 & 4 & 8 \end{vmatrix} = 1\begin{vmatrix} -1 & 3 & 4 \\ 1 & 9 & 6 \\ 2 & 4 & 8 \end{vmatrix} - 3\begin{vmatrix} 0 & 3 & 4 \\ 2 & 9 & 6 \\ 3 & 4 & 8 \end{vmatrix}$$
$$+ 5\begin{vmatrix} 0 & -1 & 4 \\ 2 & 1 & 6 \\ 3 & 2 & 8 \end{vmatrix} - 2\begin{vmatrix} 0 & -1 & 3 \\ 2 & 1 & 9 \\ 3 & 2 & 4 \end{vmatrix}$$
$$= 1(-92) - 3(-70) + 5(2) - 2(-16) = 160$$

(The values in parentheses are obtained by calculating the four 3×3 determinants.)

The reason for considering determinants of systems of n equations in n unknowns is that Theorem 1 also holds for these systems (although this fact will not be proven here).[†]

THEOREM 2

For the system (8) there is a unique solution if and only if the determinant D, defined by (10), is not zero. If $D = 0$, then there is either no solution or an infinite number of solutions. ∎

It is clear that calculating determinants by formula (10) can be extremely tedious, especially if $n \geq 5$. For that reason, techniques are available for greatly simplifying these calculations. Some of these techniques are described in the theorems below.

[†] For proofs of theorems in this appendix, see S. I. Grossman, *Elementary Linear Algebra*, 4th ed. (Philadelphia, PA: Saunders, 1991), Chapter 2.

We begin with the result that states that the determinant can be obtained by expanding in any row.

THEOREM 3

For any i, $i = 1, 2, \ldots n$,

$$D = \begin{vmatrix} a_{11} & a_{12} & \cdots & a_{1n} \\ a_{21} & a_{22} & \cdots & a_{2n} \\ \vdots & \vdots & & \vdots \\ a_{n1} & a_{n2} & \cdots & a_{nn} \end{vmatrix}$$

$$= (-1)^{i+1} a_{i1} A_{i1} + (-1)^{i+2} a_{i2} A_{i2} + \cdots + (-1)^{i+n} a_{in} A_{in},$$

where A_{ij} is the $(n-1) \times (n-1)$ determinant obtained by crossing off the ith row and jth column of D. Notice that the signs in the expansion of a determinant alternate. (See the schematic diagram on page A-26.) ∎

EXAMPLE 4
CALCULATING A 3 × 3
DETERMINANT BY EXPANDING
IN THE SECOND ROW

Calculate $\begin{vmatrix} 3 & 5 & 2 \\ 4 & 2 & 3 \\ -1 & 2 & 4 \end{vmatrix}$ by expanding in the second row (see Example 2).

SOLUTION:

$$\begin{vmatrix} 3 & 5 & 2 \\ 4 & 2 & 3 \\ -1 & 2 & 4 \end{vmatrix} = (-1)^{2+1}(4) \begin{vmatrix} 5 & 2 \\ 2 & 4 \end{vmatrix} + (-1)^{2+2}(2) \begin{vmatrix} 3 & 2 \\ -1 & 4 \end{vmatrix}$$

$$+ (-1)^{2+3}(3) \begin{vmatrix} 3 & 5 \\ -1 & 2 \end{vmatrix}$$

$$= -4(16) + 2(14) - 3(11) = -69.$$

We remark that we can also get the same result by expanding in the third row.

THEOREM 4

For any j, $j = 1, 2, \ldots, n$,

$$D = \begin{vmatrix} a_{11} & a_{12} & \cdots & a_{1n} \\ a_{21} & a_{22} & \cdots & a_{2n} \\ \vdots & \vdots & & \vdots \\ a_{n1} & a_{n2} & \cdots & a_{nn} \end{vmatrix}$$

$$= (-1)^{1+j} a_{1j} A_{1j} + (-1)^{2+j} a_{2j} A_{2j} + \cdots + (-1)^{n+j} a_{nj} A_{nj},$$

where A_{ij} is as defined in Theorem 3. That is, we can obtain D by expanding in any column. ∎

EXAMPLE 5
EXPANDING IN THE THIRD
COLUMN

Calculate $D = \begin{vmatrix} 3 & 5 & 2 \\ 4 & 2 & 3 \\ -1 & 2 & 4 \end{vmatrix}$ by expanding in the third column.

SOLUTION:

$$\begin{vmatrix} 3 & 5 & 2 \\ 4 & 2 & 3 \\ -1 & 2 & 4 \end{vmatrix} = (-1)^{1+3}(2)\begin{vmatrix} 4 & 2 \\ -1 & 2 \end{vmatrix} + (-1)^{2+3}(3)\begin{vmatrix} 3 & 5 \\ -1 & 2 \end{vmatrix}$$

$$+ (-1)^{3+3}(4)\begin{vmatrix} 3 & 5 \\ 4 & 2 \end{vmatrix}$$

$$= 2(10) - 3(11) + 4(-14) = -69.$$

THEOREM 5

$$\text{Let } D = \begin{vmatrix} a_{11} & a_{12} & \cdots & a_{1n} \\ a_{21} & a_{22} & \cdots & a_{2n} \\ \vdots & \vdots & & \vdots \\ a_{n1} & a_{n2} & \cdots & a_{nn} \end{vmatrix}.$$

(i) If any row or column of D is zero, then $D = 0$.

(ii) If any row (column) is a multiple of any other row (column), then $D = 0$.

(iii) Interchanging any two rows (columns) of D has the effect of multiplying D by -1.

(iv) Multiplying a row (column) of D by a constant α has the effect of multiplying D by α.

(v) If any row (column) of D is multiplied by a constant and added to a different row (column) of D, then D is unchanged. ∎

EXAMPLE 6
ILLUSTRATING PROPERTY (ii)

$$\text{Calculate } D = \begin{vmatrix} 2 & 1 & 4 & 3 \\ 3 & 1 & -2 & -1 \\ 14 & -2 & 0 & 6 \\ 6 & 2 & -4 & -2 \end{vmatrix}.$$

SOLUTION: $D = 0$ according to (ii) since the fourth row is twice the second row. This can easily be verified.

The results in Theorem 5 can be used to simplify the calculation of determinants.

EXAMPLE 7
CALCULATING A 4 × 4
DETERMINANT

$$\text{Calculate } D = \begin{vmatrix} 1 & 3 & 5 & 2 \\ 0 & -1 & 3 & 4 \\ 2 & 1 & 9 & 6 \\ 3 & 2 & 4 & 8 \end{vmatrix}.$$

SOLUTION: This determinant was calculated in Example 3. The idea is to use Theorem 5 to make the evaluation of the determinant almost trivial. We begin by multiplying the first row by -2 and adding it to the third row. By (v), this manipulation will leave the determinant unchanged.

$$\begin{vmatrix} 1 & 3 & 5 & 2 \\ 0 & -1 & 3 & 4 \\ 2+(-2)1 & 1+(-2)3 & 9+(-2)5 & 6+(-2)2 \\ 3 & 2 & 4 & 8 \end{vmatrix} = \begin{vmatrix} 1 & 3 & 5 & 2 \\ 0 & -1 & 3 & 4 \\ 0 & -5 & -1 & 2 \\ 3 & 2 & 4 & 8 \end{vmatrix}$$

Now we multiply the first row by -3 and add it to the fourth row:

$$D = \begin{vmatrix} 1 & 3 & 5 & 2 \\ 0 & -1 & 3 & 4 \\ 0 & -5 & -1 & 2 \\ 0 & -7 & -11 & 2 \end{vmatrix}.$$

We now expand D by its first column:

$$D = 1 \begin{vmatrix} -1 & 3 & 4 \\ -5 & -1 & 2 \\ -7 & -11 & 2 \end{vmatrix} - 0 \begin{vmatrix} 3 & 5 & 2 \\ -5 & -1 & 2 \\ -7 & -11 & 2 \end{vmatrix} + 0 \begin{vmatrix} 3 & 5 & 2 \\ -1 & 3 & 4 \\ -7 & -11 & 2 \end{vmatrix}$$

$$+ 0 \begin{vmatrix} 3 & 5 & 2 \\ -1 & 3 & 4 \\ -5 & -1 & 2 \end{vmatrix} = \begin{vmatrix} -1 & 3 & 4 \\ -5 & 1 & 2 \\ -7 & -11 & 2 \end{vmatrix},$$

which is a 3×3 determinant. We can calculate it by expansion or we can reduce further. By Theorem 5, parts (iv) and (v),

$$\begin{vmatrix} -1 & 3 & 4 \\ -5 & -1 & 2 \\ -7 & -11 & 2 \end{vmatrix} = - \begin{vmatrix} 1 & -3 & -4 \\ -5 & -1 & 2 \\ -7 & -11 & 2 \end{vmatrix} = - \begin{vmatrix} 1 & -3 & -4 \\ 0 & -16 & -18 \\ 0 & -32 & -26 \end{vmatrix}$$

$$= - \begin{vmatrix} -16 & -18 \\ -32 & -26 \end{vmatrix}$$

$$= -[(-16)(-26) - (-18)(-32)] = 160.$$

In the second step we multiplied the first row by 5 and added it to the second, and multiplied the first row by 7 and added it to the third. Note how we were able to reduce the problem to the calculation of a single 2×2 determinant.

There is one further result about determinants that will be useful in Chapter 11.

THEOREM 6

Let $D = \begin{vmatrix} a_{11} & a_{12} & \cdots & a_{1n} \\ a_{21} & a_{22} & \cdots & a_{2n} \\ \vdots & \vdots & & \vdots \\ a_{i1} + b_{i1} & a_{i2} + b_{i2} & \cdots & a_{in} + b_{in} \\ \vdots & \vdots & & \vdots \\ a_{n1} & a_{n2} & \cdots & a_{nn} \end{vmatrix}$. Then

$$D = \begin{vmatrix} a_{11} & a_{12} & \cdots & a_{1n} \\ a_{21} & a_{22} & \cdots & a_{2n} \\ \vdots & \vdots & & \vdots \\ a_{i1} & a_{i2} & \cdots & a_{in} \\ \vdots & \vdots & & \vdots \\ a_{n1} & a_{n2} & \cdots & a_{nn} \end{vmatrix} + \begin{vmatrix} a_{11} & a_{12} & \cdots & a_{1n} \\ a_{21} & a_{22} & \cdots & a_{2n} \\ \vdots & \vdots & & \vdots \\ b_{i1} & b_{i2} & \cdots & b_{in} \\ \vdots & \vdots & & \vdots \\ a_{n1} & a_{n2} & \cdots & a_{nn} \end{vmatrix}. \tag{11}$$

EXAMPLE 8
ILLUSTRATING THEOREM 6

To illustrate Theorem 6, we note that

$$\begin{vmatrix} 2 & 1 & 4 \\ 3+5 & 2-3 & 1+2 \\ 0 & -4 & 2 \end{vmatrix} = \begin{vmatrix} 2 & 1 & 4 \\ 3 & 2 & 1 \\ 0 & -4 & 2 \end{vmatrix} + \begin{vmatrix} 2 & 1 & 4 \\ 5 & -3 & 2 \\ 0 & -4 & 2 \end{vmatrix}$$

$$= -38 - 86 = -124.$$

■ PROBLEMS

For each of the 2×2 systems in Problems 1–8, calculate the determinant D. If $D \neq 0$, find the unique solution. If $D = 0$, determine whether there is no solution or an infinite number of solutions.

1. $2x_1 + 4x_2 = 6$
   $x_1 + x_2 = 3$

2. $2x_1 + 4x_2 = 6$
   $x_1 + 2x_2 = 5$

3. $2x_1 + 4x_2 = 6$
   $x_1 + 2x_2 = 3$

4. $6x_1 - 3x_2 = 3$
   $-2x_1 + x_2 = -1$

5. $6x_1 - 3x_2 = 3$
   $-2x_1 + x_2 = 1$

6. $6x_1 - 3x_2 = 3$
   $-2x_1 + 2x_2 = -1$

7. $2x_1 + 5x_2 = 0$
   $3x_1 - 7x_2 = 0$

8. $2x_1 - 3x_2 = 0$
   $-4x_1 + 6x_2 = 0$

In Problems 9–20, calculate the determinant.

9. $\begin{vmatrix} 1 & 2 & 3 \\ 6 & -1 & 4 \\ 2 & 0 & 6 \end{vmatrix}$

10. $\begin{vmatrix} 4 & -1 & 0 \\ 2 & 1 & 7 \\ -2 & 3 & 4 \end{vmatrix}$

11. $\begin{vmatrix} 7 & 2 & 3 \\ 0 & 4 & 1 \\ 0 & 0 & 5 \end{vmatrix}$

12. $\begin{vmatrix} 1 & -1 & 4 \\ 3 & -2 & 1 \\ 5 & 1 & 7 \end{vmatrix}$

13. $\begin{vmatrix} 4 & 2 & 7 \\ 1 & 5 & 3 \\ -1 & 1 & 4 \end{vmatrix}$

14. $\begin{vmatrix} -1 & 0 & 4 \\ 7 & 3 & 2 \\ 4 & 1 & 5 \end{vmatrix}$

15. $\begin{vmatrix} 1 & 4 & 7 & 2 \\ 0 & 5 & 8 & 1 \\ 0 & 0 & -3 & 4 \\ 0 & 0 & 0 & 8 \end{vmatrix}$

16. $\begin{vmatrix} a_1 & a_2 & a_3 & a_4 \\ 0 & b_1 & b_2 & b_3 \\ 0 & 0 & c_1 & c_2 \\ 0 & 0 & 0 & d_1 \end{vmatrix}$

17. $\begin{vmatrix} 2 & 1 & 3 & 4 \\ 3 & -2 & 5 & 1 \\ 4 & 0 & 4 & 5 \\ 2 & 1 & 7 & -4 \end{vmatrix}$

18. $\begin{vmatrix} 1 & 3 & -1 & 7 \\ -2 & 5 & 2 & 8 \\ -3 & 7 & 3 & 3 \\ 5 & 0 & -5 & 11 \end{vmatrix}$

19. $\begin{vmatrix} 2 & 3 & 1 & 4 \\ 2 & 2 & 4 & 6 \\ 3 & -1 & -2 & 4 \\ 4 & 2 & -3 & -5 \end{vmatrix}$

20. $\begin{vmatrix} 1 & 0 & 2 & 3 & 1 \\ 0 & 4 & -1 & -2 & 3 \\ 2 & 1 & 0 & -1 & 1 \\ -3 & 2 & 2 & 0 & 5 \\ 0 & 3 & 6 & 1 & -3 \end{vmatrix}$

21. Show that two lines in system (1) have the same slope if and only if the determinant of the system is zero.

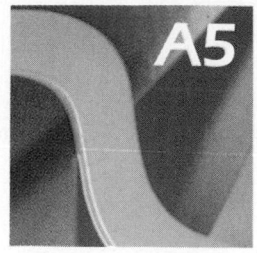

A5 COMPLEX NUMBERS

In algebra we encounter the problem of finding the roots of the polynomial

$$\lambda^2 + a\lambda + b = 0. \tag{1}$$

To find the roots, we use the quadratic formula to obtain

$$\lambda = \frac{-a \pm \sqrt{a^2 - 4b}}{2}. \tag{2}$$

If $a^2 - 4b > 0$, there are two real roots. If $a^2 - 4b = 0$, we obtain the single root (of multiplicity 2) $\lambda = -a/2$. To deal with the case $a^2 - 4b < 0$, we introduce the **imaginary number**[†]

$$i = \sqrt{-1}. \tag{3}$$

Then for $a^2 - 4b < 0$,

$$\sqrt{a^2 - 4b} = \sqrt{(4b - a^2)(-1)} = \sqrt{4b - a^2}\sqrt{-1} = \sqrt{4b - a^2}\,i,$$

and the two roots of (1) are given by

$$\lambda_1 = -\frac{a}{2} + \frac{\sqrt{4b - a^2}}{2}\,i \quad \text{and} \quad \lambda_2 = -\frac{a}{2} - \frac{\sqrt{4b - a^2}}{2}\,i.$$

EXAMPLE 1
SOLVING A QUADRATIC
EQUATION WITH COMPLEX
ROOTS

Find the roots of the quadratic equation $\lambda^2 + 2\lambda + 5 = 0$.

SOLUTION: We have $a = 2$, $b = 5$, and $a^2 - 4b = -16$. Thus $\sqrt{a^2 - 4b} = \sqrt{-16} = \sqrt{16}\sqrt{-1} = 4i$, and the roots are

$$\lambda_1 = \frac{-2 + 4i}{2} = -1 + 2i \quad \text{and} \quad \lambda_2 = -1 - 2i.$$

DEFINITION   **COMPLEX NUMBER**

A **complex number** is a number of the form $z = \alpha + i\beta$, $\tag{4}$

where α and β are real numbers. α is called the **real part** of z and is denoted by Re z. β is called the **imaginary part** of z and is denoted by Im z. Representation (4) is sometimes called the **Cartesian form** of the complex number z. ■

[†] You should not be troubled by the term "imaginary." It's just a name. The British mathematician Alfred North Whitehead, in the chapter on imaginary numbers in his *Introduction to Mathematics*, wrote:

At this point it may be useful to observe that a certain type of intellect is always worrying itself and others by discussion as to the applicability of technical terms. Are the incommensurable numbers properly called numbers? Are the positive and negative numbers really numbers? Are the imaginary numbers imaginary, and are they numbers?—are types of such futile questions. Now, it cannot be too clearly understood that, in science, technical terms are names arbitrarily assigned, like Christian names to children. There can be no question of the names being right or wrong. They may be judicious or injudicious; for they can sometimes be so arranged as to be easy to remember, or so as to suggest relevant and important ideas. But the essential principle involved was quite clearly enunciated in Wonderland to Alice by Humpty Dumpty, when he told her, apropos of his use of words, 'I pay them extra and make them mean what I like'. So we will not bother as to whether imaginary numbers are imaginary, or as to whether they are numbers, but will take the phrase as the arbitrary name of a certain mathematical idea, which we will now endeavour to make plain.

REMARK: If $\beta = 0$ in equation (4), then $z = \alpha$ is a real number. In this context we can regard the set of real numbers as a subset of the set of complex numbers.

EXAMPLE 2
THE REAL AND IMAGINARY PARTS
OF A COMPLEX NUMBER

In Example 1, Re $\lambda_1 = -1$ and Im $\lambda_1 = 2$.

We can add and multiply complex numbers by using the standard rules of algebra.

EXAMPLE 3
ADDING AND MULTIPLYING
COMPLEX NUMBERS

Let $z = 2 + 3i$ and $w = 5 - 4i$. Calculate **(a)** $z + w$, **(b)** $3w - 5z$, and **(c)** zw.

SOLUTION:

(a) $z + w = (2 + 3i) + (5 - 4i) = (2 + 5) + (3 - 4)i = 7 - i$.

(b) $3w = 3(5 - 4i) = 15 - 12i$, $5z = 10 + 15i$, and $3w - 5z = (15 - 12i) - (10 + 15i) = (15 - 10) + i(-12 - 15) = 5 - 27i$.

(c) $zw = (2 + 3i)(5 - 4i) = (2)(5) + 2(-4i) + (3i)(5) + (3i)(-4i) = 10 - 8i + 15i - 12i^2 = 10 + 7i + 12 = 22 + 7i$. Here we used the fact that $i^2 = -1$.

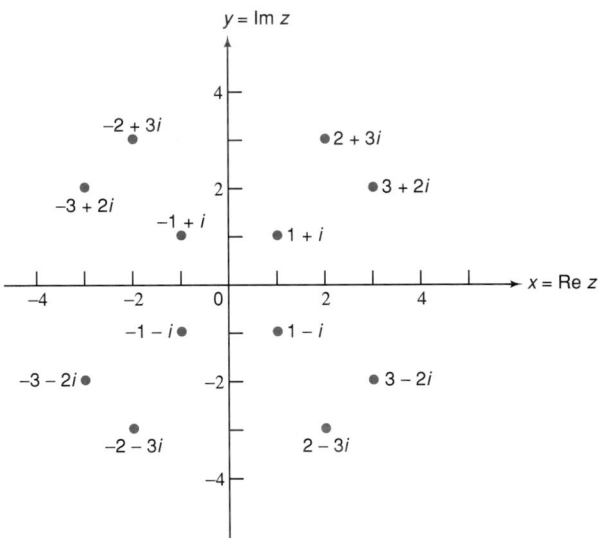

FIGURE 1   TWELVE COMPLEX NUMBERS

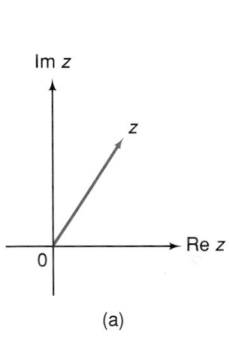

(a)

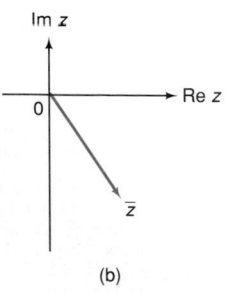

(b)

FIGURE 2
z AND ITS CONJUGATE $\bar{z}$

We can plot a complex number z in the xy-plane by plotting Re z along the x-axis and Im z along the y-axis. Thus each complex number can be thought of as a point in the xy-plane. With this representation the xy-plane is called the **complex plane.** Some representative points are plotted in Figure 1.

If $z = \alpha + i\beta$, then we define the **conjugate** of z, denoted by $\bar{z}$, by

$$\bar{z} = \alpha - i\beta. \tag{5}$$

Figure 2 depicts a representative value of z and $\bar{z}$.

EXAMPLE 4
COMPUTING FOUR COMPLEX
CONJUGATES

Compute the complex conjugates of **(a)** $1 + i$, **(b)** $3 - 4i$, **(c)** $-7 + 5i$, and **(d)** -3.

SOLUTION:

(a) $\overline{1 + i} = 1 - i$.   **(b)** $\overline{3 - 4i} = 3 + 4i$.
(c) $\overline{-7 + 5i} = -7 - 5i$.   **(d)** $\overline{-3} = -3$.

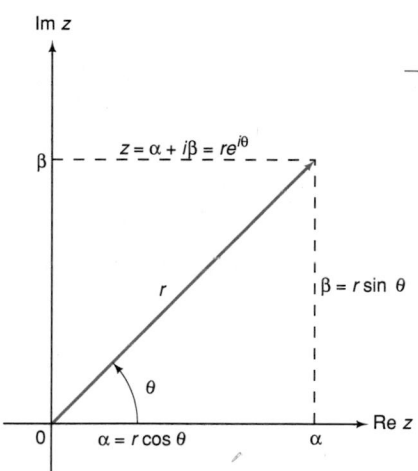

FIGURE 3
$r = |z| = \sqrt{\alpha^2 + \beta^2}$ AND $\tan \theta = \dfrac{\beta}{\alpha}$.

It is not difficult to show (see Problem 35) that

$$\bar{z} = z \qquad \text{if and only if } z \text{ is real.} \tag{6}$$

If $z = \beta i$ with β real, then z is said to be **pure imaginary**. We can then show (see Problem 36) that

$$\bar{z} = -z \qquad \text{if and only if } z \text{ is pure imaginary.} \tag{7}$$

Let $p_n(x) = a_0 + a_1 x + a_2 x^2 + \cdots + a_n x^n$ be a polynomial with real coefficients. Then it can be shown (see Problem 41) that the complex roots of the equation $p_n(x) = 0$ occur in complex conjugate pairs. That is, if z is a root of $p_n(x) = 0$, then so is $\bar{z}$. We saw this fact illustrated in Example 1 in the case in which $n = 2$.

For $z = \alpha + i\beta$ we define the **magnitude** of z, denoted by $|z|$, by

$$|z| = \sqrt{\alpha^2 + \beta^2}, \tag{8}$$

and we define the **argument** of z, denoted by arg z, as the angle θ between the line $0z$ and the positive x-axis. From Figure 3 we see that $r = |z|$ is the distance from z to the origin, and

$$\theta = \arg z = \tan^{-1} \frac{\beta}{\alpha}. \tag{9}$$

By convention, we always choose a value of $\tan^{-1} \beta/\alpha$ that lies in the interval

$$-\pi < \theta \le \pi. \tag{10}$$

From Figure 4 we see that

$$|\bar{z}| = |z| \tag{11}$$

and

$$\arg \bar{z} = -\arg z. \tag{12}$$

We can use $|z|$ and arg z to describe what is often a more convenient way to represent complex numbers.[†] From Figure 3 it is evident that if $z = \alpha + i\beta$, $r = |z|$, and $\theta = \arg z$, then

$$\alpha = r \cos \theta \qquad \text{and} \qquad \beta = r \sin \theta. \tag{13}$$

We will see at the end of this appendix that

$$e^{i\theta} = \cos \theta + i \sin \theta. \tag{14}$$

Since $\cos(-\theta) = \cos \theta$ and $\sin(-\theta) = -\sin \theta$, we also have

$$e^{-i\theta} = \cos(-\theta) + i \sin(-\theta) = \cos \theta - i \sin \theta. \tag{14'}$$

Formula (14) is called **Euler's formula**.[‡] Using Euler's formula and equation (13), we have

$$z = \alpha + i\beta = r \cos \theta + ir \sin \theta = r(\cos \theta + i \sin \theta),$$

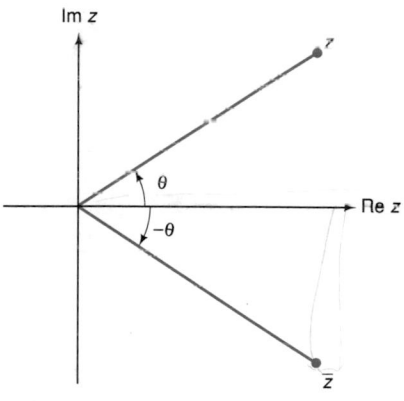

FIGURE 4
$z = |\bar{z}|$ AND arg $\bar{z} = -$arg z.

[†] If you have already studied polar coordinates in Chapter 8, you will find this new representation very familiar.

[‡] Named for the Swiss mathematician Leonhard Euler (1707–1783). See his biographical sketch on page 34.

or

$$z = re^{i\theta}.$$ (15)

Representation (15) is called the **polar form** of the complex number z.

EXAMPLE 5
THE POLAR FORMS OF SIX
COMPLEX NUMBERS

Determine the polar forms of the following complex numbers:

(a) 1 (b) -1 (c) i
(d) $1 + i$ (e) $-1 - \sqrt{3}$ (f) $-2 + 7i$

SOLUTION: The six points are plotted in Figure 5.

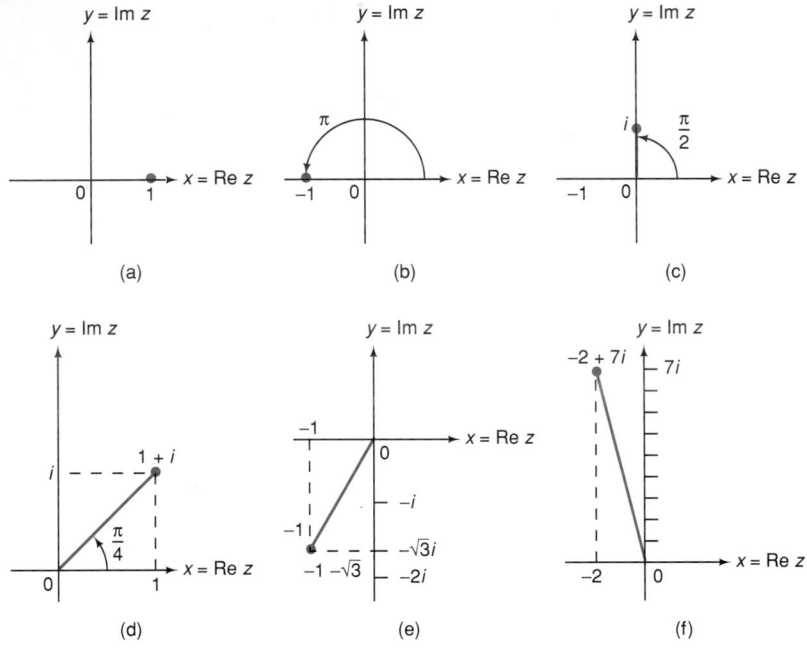

FIGURE 5 SIX COMPLEX NUMBERS

(a) From Figure 5(a) it is clear that $\arg 1 = 0$. Since $\operatorname{Re} 1 = 1$, we see that, in polar form,

$$1 = 1e^{i0} = 1e^{0} = 1.$$

(b) Since $\arg(-1) = \pi$ (Figure 5(b)) and $|-1| = 1$, we have

$$-1 = 1e^{\pi i} = e^{i\pi}.$$

(c) From Figure 5(c) we see that $\arg i = \pi/2$. Since $|i| = \sqrt{0^2 + 1^2} = 1$, it follows that

$$i = e^{i\pi/2}.$$

(d) $\arg(1 + i) = \tan^{-1} \frac{1}{1} = \pi/4$, and $|1 + i| = \sqrt{1^2 + 1^2} = \sqrt{2}$, so that

$$1 + i = \sqrt{2}\, e^{i\pi/4}.$$

(e) Here $\tan^{-1} \beta/\alpha = \tan^{-1} \sqrt{3} = \pi/3$. However, $\arg z$ is in the third quadrant, so $\theta = (\pi/3) + \pi = 4\pi/3$. Also, $|-1 - \sqrt{3}i| = \sqrt{1^2 + (\sqrt{3})^2} = \sqrt{1 + 3} = 2$, so that $-1 - \sqrt{3} = 2e^{4\pi i/3}$.

(f) To compute this complex number, we need a calculator. A calculator indicates that

$$\arg z = \tan^{-1}(-\tfrac{7}{2}) = \tan^{-1}(-3.5) \approx -1.2925.$$

But $\tan^{-1} x$ is defined as a number in the interval $(-\pi/2, \pi/2)$. Since from Figure 5(f) θ is in the second quadrant, we see that $\arg z = \tan^{-1}(-3.5) + \pi \approx 1.8491$. Next, we see that

$$|-2 + 7i| = \sqrt{(-2)^2 + 7^2} = \sqrt{53}.$$

Hence

$$-2 + 7i \approx \sqrt{53}\, e^{1.8491i}.$$

EXAMPLE 6
CONVERT FROM POLAR TO
CARTESIAN FORM

Convert the following complex numbers from polar to Cartesian form: **(a)** $2e^{i\pi/3}$, **(b)** $4e^{3\pi i/2}$.

SOLUTION:

(a) $e^{i\pi/3} = \cos \pi/3 + i \sin \pi/3 = \frac{1}{2} + (\sqrt{3}/2)i$. Thus $2e^{i\pi/3} = 1 + \sqrt{3}i$.

(b) $e^{3\pi i/2} = \cos 3\pi/2 + i \sin 3\pi/2 = 0 + i(-1) = -i$. Thus $4e^{3\pi i/2} = -4i$.

If $\theta = \arg z$, then by equation (12), $\arg \bar{z} = -\theta$. Thus since $|\bar{z}| = |z|$, we have the following:

If $z = re^{i\theta}$, then $\bar{z} = re^{-i\theta}$. $\hspace{2cm}$ **(16)**

Suppose we write a complex number in its polar form $z = re^{i\theta}$. Then

$$z^n = (re^{i\theta})^n = r^n(e^{i\theta})^n = r^n e^{in\theta} = r^n(\cos n\theta + i \sin n\theta). \hspace{1cm} \textbf{(17)}$$

Formula (17) is useful for a variety of computations. In particular, when $r = |z| = 1$, we obtain the **De Moivre formula**:[†]

$$(\cos \theta + i \sin \theta)^n = \cos n\theta + i \sin n\theta. \hspace{2cm} \textbf{(18)}$$

EXAMPLE 7
USING THE DE MOIVRE FORMULA

Compute $(1 + i)^5$.

SOLUTION: In Example 5(d) we showed that $1 + i = \sqrt{2}\, e^{\pi i/4}$. Then

$$(1 + i)^5 = (\sqrt{2}\, e^{\pi i/4})^5 = (\sqrt{2})^5 e^{5\pi i/4} = 4\sqrt{2}\left(\cos \frac{5\pi}{4} + i \sin \frac{5\pi}{4}\right)$$

$$= 4\sqrt{2}\left(-\frac{1}{\sqrt{2}} - \frac{1}{\sqrt{2}}i\right) = -4 - 4i.$$

This can be checked by direct calculation. If the direct calculation seems no more difficult, then try to compute $(1 + i)^{20}$ directly. Proceeding as above, we obtain

$$(1 + i)^{20} = (\sqrt{2})^{20} e^{20\pi i/4} = 2^{10}(\cos 5\pi + i \sin 5\pi)$$
$$= 2^{10}(-1 + 0) = -1024.$$

PROOF OF EULER'S FORMULA: We will show that

$$e^{i\theta} = \cos \theta + i \sin \theta \hspace{3cm} \textbf{(19)}$$

[†] Abraham De Moivre (1667–1754) was a French mathematician well known for his work in probability theory, infinite series, and trigonometry. He was so highly regarded that Newton often told those who came to him with questions on mathematics, "Go to M. De Moivre; he knows these things better than I do."

by using power series. If you have not covered Chapter 10, skip the proof. We have

$$e^x = 1 + x + \frac{x^2}{2!} + \frac{x^3}{3!} + \cdots, ^\dagger \tag{20}$$

$$\sin x = x - \frac{x^3}{3!} + \frac{x^5}{5!} - \cdots, \tag{21}$$

$$\cos x = 1 - \frac{x^2}{2!} + \frac{x^4}{4!} - \cdots. \tag{22}$$

Then

$$e^{i\theta} = 1 + (i\theta) + \frac{(i\theta)^2}{2!} + \frac{(i\theta)^3}{3!} + \frac{(i\theta)^4}{4!} + \frac{(i\theta)^5}{5!} + \cdots. \tag{23}$$

Now $i^2 = -1$, $i^3 = -i$, $i^4 = 1$, $i^5 = i$, and so on. Thus (23) can be written

$$e^{i\theta} = 1 + i\theta - \frac{\theta^2}{2!} - \frac{i\theta^3}{3!} + \frac{\theta^4}{4!} + \frac{i\theta^5}{5!} - \cdots$$

$$= \left(1 - \frac{\theta^2}{2!} + \frac{\theta^4}{4!} - \cdots\right) + i\left(\theta - \frac{\theta^3}{3!} + \frac{\theta^5}{5!} - \cdots\right)$$

$$= \cos \theta + i \sin \theta.$$

This completes the proof. ∎

■ PROBLEMS

In Problems 1–5, perform the indicated operation.

1. $(2 - 3i) + (7 - 4i)$
2. $3(4 + i) - 5(-3 + 6i)$
3. $(1 + i)(1 - i)$
4. $(2 - 3i)(4 + 7i)$
5. $(-3 + 2i)(7 + 3i)$

In Problems 6–15, convert the complex number to its polar form.

6. $5i$
7. $5 + 5i$
8. $-2 - 2i$
9. $3 - 3i$
10. $2 + 2\sqrt{3}i$
11. $3\sqrt{3} + 3i$
12. $1 - \sqrt{3}i$
13. $4\sqrt{3} - 4i$
14. $-6\sqrt{3} - 6i$
15. $-1 - \sqrt{3}i$

In Problems 16–25, convert from polar to Cartesian form.

16. $e^{3\pi i}$
17. $2e^{-7\pi i}$
18. $\frac{1}{2}e^{3\pi i/4}$
19. $\frac{1}{2}e^{-3\pi i/4}$
20. $6e^{\pi i/6}$
21. $4e^{5\pi i/6}$
22. $4e^{-5\pi i/6}$
23. $3e^{-2\pi i/3}$
24. $\sqrt{3}e^{23\pi i/4}$
25. e^i

In Problems 26–34, compute the conjugate of the given number.

26. $3 - 4i$
27. $4 + 6i$
28. $-3 + 8i$
29. $-7i$
30. 16
31. $2e^{\pi i/7}$
32. $4e^{3\pi i/5}$
33. $3e^{-4\pi i/11}$
34. $e^{0.012i}$

35. Show that $z = \alpha + i\beta$ is real if and only if $z = \bar{z}$. [Hint: if $z = \bar{z}$, show that $\beta = 0$.]
36. Show that $z = \alpha + i\beta$ is pure imaginary if and only if $z = -\bar{z}$. [Hint: if $z = -\bar{z}$, show that $\alpha = 0$.]
37. For any complex number z, show that $z\bar{z} = |z|^2$.
38. Show that the circle of radius 1 centered at the origin (the *unit circle*) is the set of points in the complex plane that satisfy $|z| = 1$.
39. For any complex number z_0 and real number a, describe $\{z : |z - z_0| = a\}$.
40. Describe $\{z : |z - z_0| \le a\}$, where z_0 and a are as in Problem 39.
*41. Let $p(\lambda) = \lambda^n + a_{n-1}\lambda^{n-1} + a_{n-2}\lambda^{n-2} + \cdots + a_1\lambda + a_0$ with $a_0, a_1, \ldots, a_{n-1}$ real numbers. Show that if $p(z) = 0$, then $p(\bar{z}) = 0$. That is, *the roots of polynomials with real coefficients occur in complex conjugate pairs*.
42. Derive expressions for $\cos 4\theta$ and $\sin 4\theta$ by comparing the De Moivre formula and the expansion of $(\cos \theta + i \sin \theta)^4$.
*43. Prove De Moivre's formula by mathematical induction. [Hint: Recall the trigonometric identities $\cos(x + y) = \cos x \cos y - \sin x \sin y$ and $\sin(x + y) = \sin x \cos y + \cos x \sin y$.]

$\dagger$ Although we will not prove it here, these series expansions are also valid when x is a complex number.

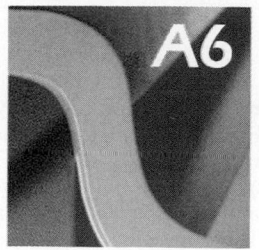

A6 GRAPHING USING A CALCULATOR

It has long been possible to generate graphs of a wide variety of functions on a computer. Recently, hand-held calculators with graphing capabilities have become available. In this appendix, we shall discuss some techniques that will help you use the graphing calculator more effectively. We shall also discuss some ways that graphing calculators can be used to solve algebraic problems like approximating the zeros of functions, finding where two curves intersect, and solving inequalities.

The graphing calculators currently available are produced by (in alphabetical order) Casio, Hewlett-Packard, Sharp, and Texas Instruments. This appendix is generic; that is, it is intended for use with any graphing calculator. All applications cited in this appendix can be carried out with any of the calculators currently available. Therefore, our discussion will focus on graphing techniques rather than on specific keystrokes; that is, we will not tell you which buttons to push. For that reason:

It is essential that you read the instruction manual that accompanies your graphing calculator before you read any further.

A6.1 OBTAINING A GRAPH ON A CALCULATOR

In order to obtain a graph on a calculator, two things must be done (not necessarily in the order given here).

A. Enter the Function to Be Graphed

Read your manual to learn the procedure that must be used to enter functions. There will be a special way to enter the function variable, which is most often denoted by X.

B. Determine the Range and Scale

You must tell the calculator the range of values over which you wish the function to be graphed and specify the scale on the x- and y-axes.

The range is given by entering the smallest and largest values to be taken for x and y. The scale is the length represented by each tick mark on an axis. If the x-scale is 2, for example, then the distance between two successive tick marks represents a length of 2 units. In this appendix, we use the following notation:

$x_{\min}$ = minimum value of x

$x_{\max}$ = maximum value of x

x_{scl} = the scale on the x-axis

$y_{\min}$ = minimum value of y

$y_{\max}$ = maximum value of y

y_{scl} = the scale on the y-axis

Most calculators have a key (sometimes labeled $\boxed{\text{RANGE}}$) that must be pressed in order to enter range and scale values.

NOTE: If you graph a function that you have entered but do not specify range and scale values, then one of three things will happen. First, the calculator may use the range and scale values that were entered for the previous graph that was sketched. Second, the calculator may use some "standard" built-in range and scale values. These are called "standard defaults." On one TI calculator, for example, the standard defaults for both the x- and y-axes are the intervals $[-10, 10]$ with a scale of 1. Third, if the calculator has built-in graphs, then preset ranges and scales will be used whenever one of these graphs is sketched. For example, on one Casio calculator the graph of $y = \ln x$ is built in. When this function is sketched, the calculator uses the following range and scale values:

$$x_{\min} = -1 \qquad y_{\min} = -1.6$$
$$x_{\max} = 8.4 \qquad y_{\max} = 2.368$$
$$x_{\text{scl}} = 2 \qquad y_{\text{scl}} = 1$$

Before pressing a graphing key, check the range and scale values. A number of things can happen if you do not enter these values yourself.

The hardest part about graphing on a calculator is choosing appropriate range values. Our first example illustrates why care in choosing these values is essential.

EXAMPLE 1

FINDING AN APPROPRIATE RANGE IN ORDER TO GENERATE A GRAPH

Sketch the graph of $y = -x^3 + 2x^2 + 5x - 6 = -(x + 2)(x - 1)(x - 3)$.

SOLUTION: Suppose that we do not notice that the cubic can be factored and we arbitrarily choose the following range and scale values:

Ranges: $-2 \le x \le 2; -3 \le y \le 3$

Scales: Each x-axis tick represents 1 unit, and each y-axis tick represents 2 units. That is,

$$x_{\min}: -2 \qquad y_{\min}: -3$$
$$x_{\max}: 2 \qquad y_{\max}: 3$$
$$x_{\text{scl}}: 1 \qquad y_{\text{scl}}: 2$$

After the function is entered, the graph in Figure 1 is obtained.

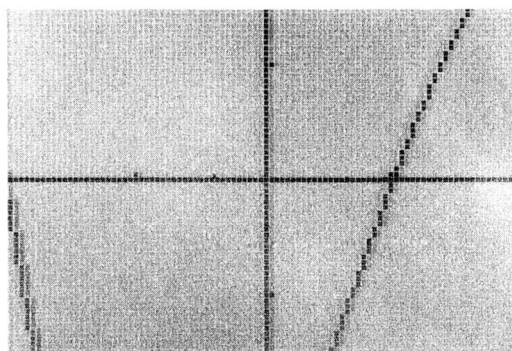

FIGURE 1   GRAPH OF $y = -x^3 + 2x^2 + 5x - 6$ for $-2 \le x \le 2, -3 \le y \le 3$

This graph is accurate but not very useful. We need to see what happens outside of our rather limited range. Let us greatly expand the ranges:

$$-20 \le x \le 20; \quad -200 \le y \le 200; \quad \text{with } x_{\text{scl}} = 1 \text{ and } y_{\text{scl}} = 20.$$

The graph now appears as in Figure 2.

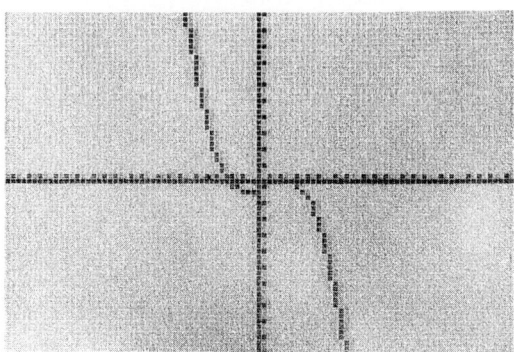

FIGURE 2   GRAPH OF $y = -x^3 + 2x^2 + 5x - 6$ FOR $-20 \leq x \leq 20$, $-20 \leq y \leq 200$

This graph looks different but is still not what we want. Its appearance should not be surprising. For $|x|$ large, $-x^3 + 2x^2 + 5x - 6 \approx -x^3$, so since we used a large range of x-values, we have obtained a graph that looks like the graph of $-x^3$.

We can get a much more revealing graph by thinking a bit before entering range values. If $f(x) = -x^3 + 2x^2 + 5x - 6$, then, for example, $f(-4) = 70$ and $f(4) = -18$. It is not hard to see that if $x < -4$, then $f(x) > 70$, and if $x > 4$, then $f(x) < -18$. Thus, most of the interesting behavior of this function occurs for $-4 \leq x \leq 4$ and $-18 \leq y \leq 70$. Setting these range values and letting $x_{\text{scl}} = 1$ and $y_{\text{scl}} = 2$, we obtain the graph in Figure 3.

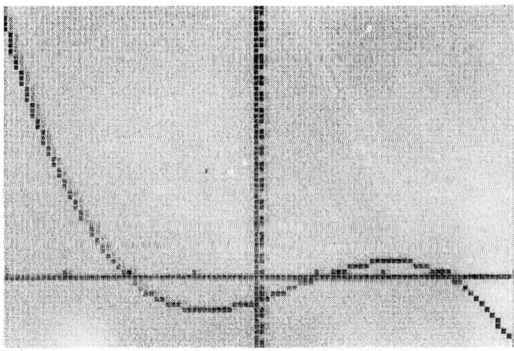

FIGURE 3   GRAPH OF $y = -x^3 + 2x^2 + 5x - 6$ FOR $-4 \leq x \leq 4$, $-18 \leq y \leq 70$

This is the type of graph we want. We can clearly see the zeros at -2, 1, and 3. Other interesting behavior, like intervals over which the function is increasing or decreasing, is plainly shown.

NOTE: On some calculators, there is a "zoom" or "factor" feature. This feature allows you to zoom in or zoom out by a factor you set. If you zoom out in Figure 1, for example, then you get a more global picture of the graph. If you zoom out too much, you might get a picture like the one in Figure 2. Then you could zoom in to obtain the more accurate graph in Figure 3. You can experiment with the zoom feature, zooming in and out, until you get a graph that looks accurate. You can save a lot of confusion, however, if you first think about what reasonable range values should be.

EXAMPLE 2

SKETCHING THE GRAPH OF A
RATIONAL FUNCTION HAVING
A VERTICAL ASYMPTOTE

Sketch the graph of $y = f(x) = \dfrac{x^2 - 2x + 5}{x + 2}$.

SOLUTION: We first note that the function is not defined at $x = -2$. As x gets close to -2 from either side, $|y|$ gets large. Also, $|y|$ gets large as $|x|$ gets large. To see this, we divide to obtain

$$\frac{x^2 - 2x + 5}{x + 2} = x - 4 + \frac{13}{x + 2}.$$

Since $\dfrac{13}{x + 2} \to 0$ as $x \to \pm\infty$, we see that, for x large,

$$\frac{x^2 - 2x + 5}{x + 2} \approx x - 4.$$

You must be very careful when you enter functions in your calculator. Without parentheses around the numerator and denominator of this function, you would not be graphing the function you want to graph. Specifically, if you enter

$$y = x^2 - 2x + 5 \div x + 2,$$

you will obtain the graph of

$$y = x^2 - 2x + \frac{5}{x} + 2.$$

This is very different from the correctly entered function

$$y = (x^2 - 2x + 5) \div (x + 2).$$

There are many ranges of values that will give us a suitable graph. Each one must include $x = -2$ and allow for reasonably large values for $|y|$. Here is one set of range values:

$$-12 \le x \le 10, \quad -25 \le y \le 15$$

We use scales of 1 on each axis.
The graph is given in Figure 4.

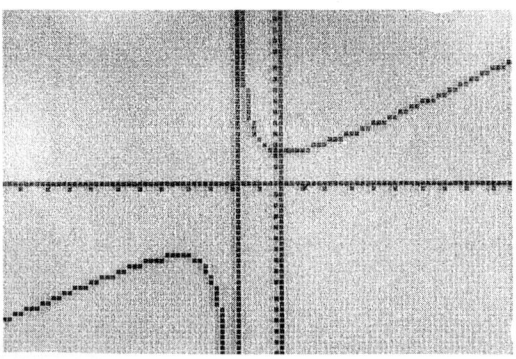
(a) Graph showing the vertical asymptote

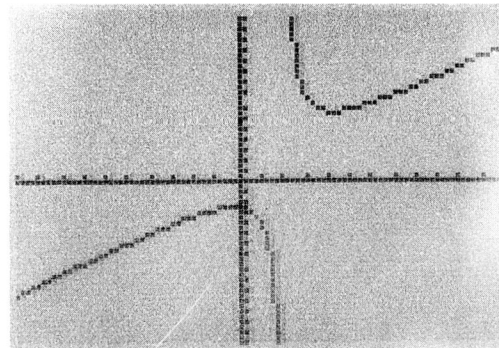
(b) Graph not showing the vertical asymptote

FIGURE 4   GRAPH OF $y = \dfrac{x^2 - 2x + 5}{x + 2}$ FOR $-12 \le x \le 10,\ -25 \le y \le 15$

If your calculator shows a vertical line at $x = -2$, it is because the calculator is attempting to connect the points on the graph. Thus, some calculators may show the vertical asymptote (Figure 4(a)) at $x = -2$ (see page 205 for a discussion of vertical asymptotes) and some may not (Figure 4(b)).[†]

EXAMPLE 3
SKETCHING THE GRAPH OF A RATIONAL FUNCTION WITH HORIZONTAL AND VERTICAL ASYMPTOTES

Sketch the graph of $y = \dfrac{2x^2 - 3x + 5}{x^2 - 1}$.

SOLUTION: We first note that $x^2 - 1 = 0$ when $x = 1$ and $x = -1$. Since the numerator is not zero at either of these values, the lines $x = 1$ and $x = -1$ are vertical asymptotes (see page 205). If we divide numerator and denominator by x^2, we obtain

$$\frac{2x^2 - 3x + 5}{x^2 - 1} = \frac{2 - \dfrac{3}{x} + \dfrac{5}{x^2}}{1 - \dfrac{1}{x^2}}.$$

The terms $\dfrac{3}{x}$, $\dfrac{5}{x^2}$, and $\dfrac{1}{x^2} \to 0$ as $x \to \pm\infty$, so $\dfrac{2x^2 - 3x + 5}{x^2 - 1} \to 2$ as $x \to \pm\infty$. Thus $y = 2$ is a horizontal asymptote to the graph (see page 205). This means that for x large in either direction, y is close to 2. On the other hand, $|y| \to \infty$ as x gets near 1 or -1 so we must allow for large values of y.

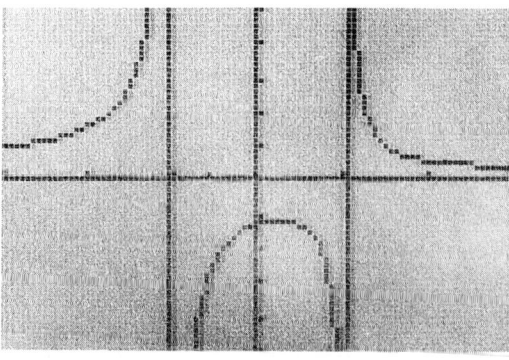

(a) Graph showing vertical asymptotes   (b) Graph not showing vertical asymptotes

FIGURE 5 GRAPH OF $y = \dfrac{2x^2 - 3x + 5}{x^2 - 1}$ FOR $-3 \le x \le 3$, $-20 \le y \le 20$, $x_{\text{scl}} = 1$, $y_{\text{scl}} = 5$

In Figure 5, we provide the graph for the values $-3 \le x \le 3$, $-20 \le y \le 20$.

This graph gives us an accurate picture. However, if we want to see what happens as $|x|$ gets larger (to see the approach to the asymptote $y = 2$), then we can increase the x-values and decrease the y-values.

In Figure 6, we give the graph for $-12 \le x \le 12$ and $-10 \le y \le 10$.

Now we see clearly that the graph approaches a horizontal line as $|x|$ gets large. However, we lose an accurate picture in the interval $-1 < x < 1$.

Which of the graphs in Figures 5 and 6 is better? The answer here is not clear. If you want to see what happens to the function as $|x|$ becomes large, then Figure 6 is better. However, Figure 5 provides a clearer picture of the shape of

[†] One TI calculator draws the asymptote if the range values for x are set to $[-12, 10]$ but does not draw it if the range values are set to $[-10, 10]$. Thus the appearance or nonappearance of asymptotes depends both on the calculator used and the range settings.

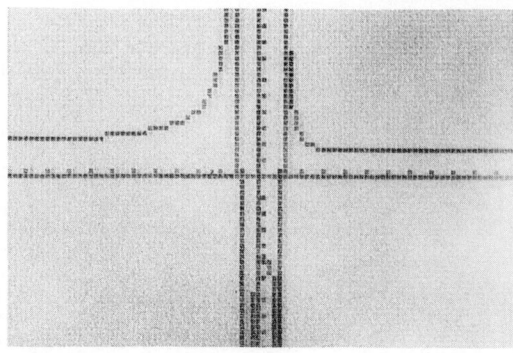

FIGURE 6   GRAPH OF $y = \dfrac{2x^2 - 3x + 5}{x^2 - 1}$ FOR $-12 \leq x \leq 12$, $-10 \leq y \leq 10$

the curve and is closer to the one that would appear in a textbook. A nice feature of a graphing calculator, especially one with a "zoom" feature, is that it allows you to experiment with different range values in order to obtain a graph that best suits your needs.

EXAMPLE 4
SKETCHING THE GRAPH OF
AN EXPONENTIAL FUNCTION

Sketch the graph of $y = f(x) = e^{x^2 + 1}$.

SOLUTION: $x^2 + 1 \geq 1$ so $e^{x^2 + 1} \geq e^1 = e \approx 2.718$. Thus the minimum value for y is 2.718. Also $f(-2) = f(2) = e^5 \approx 148$. One suitable set of ranges is

$$-2 \leq x \leq 2; \quad 0 \leq y \leq 20.$$

We allow the value $y = 0$ so that the graph will clearly exhibit the minimum value at $x = 0$. We use the scale values $x_{scl} = 0.25$ and $y_{scl} = 1$.
Consult your calculator manual to see how to enter the function $e^{x^2 + 1}$. When $e^{x^2 + 1}$ is entered, the graph in Figure 7 appears.

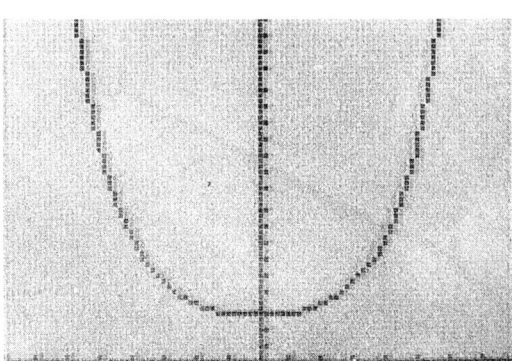

FIGURE 7   GRAPH OF $y = e^{x^2 + 1}$ FOR $-2 \leq x \leq 2$, $0 \leq y \leq 20$

EXAMPLE 5
SKETCHING THE GRAPH OF
A TRIGONOMETRIC FUNCTION

Sketch the graph of $y = f(x) = 4 \sin x + 2 \cos 2x$.

SOLUTION: Since $\sin x$ is periodic of period 2π and $\cos 2x$ is periodic of period π, $f(x)$ is periodic of period 2π (explain why) and we can obtain two complete cycles by letting the range of x include an interval having length of at least $4\pi \approx 12.6$. A suitable range is $-7 \leq x \leq 7$. Also, $|4 \sin x| \leq 4$ and $|2 \cos 2x| \leq 2$,

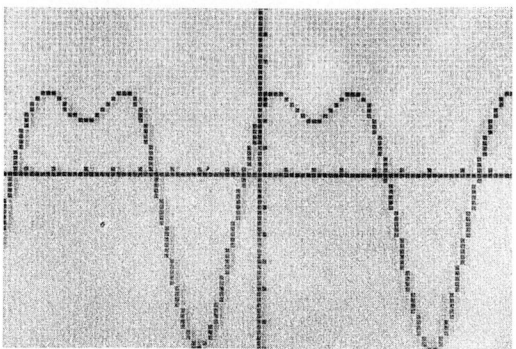

FIGURE 8   GRAPH OF $y = 4 \sin x + 2 \cos 2x$, $-7 \le x \le 7$, $-6 \le y \le 6$

so $|4 \sin x + 2 \cos 2x| \le 6$, and a suitable range for y is $-6 \le y \le 6$. Since we are dealing with a function of period 2π, a suitable x-scale is $\pi/4 \approx 0.785$. A suitable y-scale is 1. With these values, we obtain the sketch in Figure 8.

A6.2 OTHER USES OF CALCULATOR GRAPHING

There are many types of problems that can be solved on a calculator by using suitable graphs. Among these are (1) finding the zeros of a function and (2) finding points of intersections of graphs. The two problems are really the same because finding where $f(x) = g(x)$ is equivalent to finding the zeros of $f(x) - g(x)$. We will, however, treat these as distinct problems and give examples of each.

EXAMPLE 6
USING GRAPHS TO FIND THE ZEROS OF A POLYNOMIAL

Find all zeros of $f(x) = x^3 - 3x^2 + 7x - 8$.

SOLUTION: $f(0) = -8$, $f(x) < 0$ if $x < 0$, $f(1) = -3$, $f(2) = 2$, and $f(3) = 13$. Reasonable ranges are $-2 \le x \le 3$ and $-10 \le y \le 15$. Using a scale of 1 on each axis, we obtain the graph in Figure 9. (The numbers on the x-axis were added to make things clearer.)

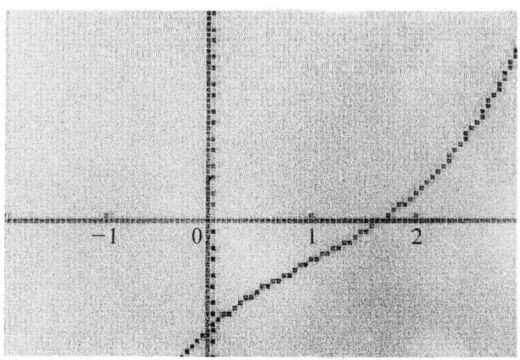

FIGURE 9   GRAPH OF $y = x^3 - 3x^2 + 7x - 8$ FOR $-2 \le x \le 3$, $-10 \le y \le 15$

The graph suggests that there is one and only one zero between 1 and 2. (We already knew that there was at least one zero because $f(1) < 0$ and $f(2) > 0$.)

There are several ways to approximate this zero more closely. If your calculator has a zoom feature, you can zoom in on the part of the graph where the curve crosses the x-axis and find the zero to at least 4 or 5 decimal place accuracy with little difficulty. Whether or not you have this feature, or any similar one, you can improve your estimate by changing the range values along the x-axis. This is what we do now.

Here are some new range and scale values:

$$1 \leq x \leq 2; \quad x_{\text{scl}} = 0.05; \quad -3 \leq y \leq 2; \quad y_{\text{scl}} = 0.1.$$

From the graph in Figure 10, we see that the graph crosses the x-axis in the interval $1.6 < x < 1.7$.

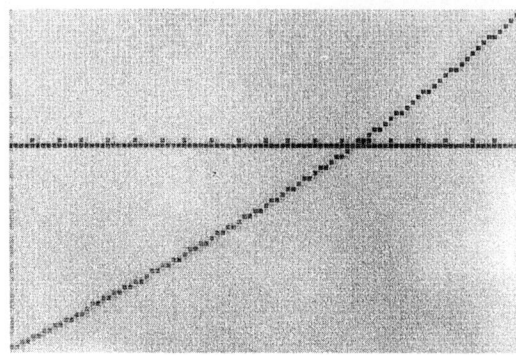

FIGURE 10   GRAPH OF $y = x^3 - 3x^2 + 7x - 8$ FOR $1 \leq x \leq 2$, $-3 \leq y \leq 2$, $x_{\text{scl}} = 0.05$, $y_{\text{scl}} = 0.1$

To get more precision, we can change range and scale again:

$$1.6 \leq x \leq 1.7; \quad x_{\text{scl}} = 0.01$$

$$-0.5 \leq y \leq 0.5; \quad y_{\text{scl}} = 0.1$$

Now we see, in Figure 11, that the zero falls between 1.67 and 1.68. We stop here because there are faster ways to get more precision if more precision is needed.

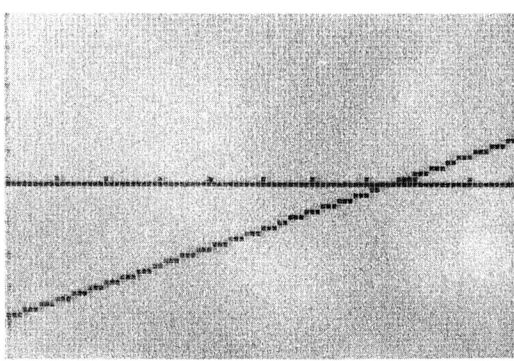

FIGURE 11   GRAPH OF $y = x^3 - 3x^2 + 7x - 8$ FOR $1.6 \leq x \leq 1.7$, $-0.5 \leq y \leq 0.5$, $x_{\text{scl}} = 0.01$, $y_{\text{scl}} = 0.1$

A NOTE ON ACCURACY: We say that a number x_A approximates a number x with *k decimal places of accuracy* if

$$|x - x_A| < \frac{1}{2} \times 10^{-k}.$$

For example, suppose we wish to approximate $\pi \approx 3.14159265359$.

A First Approximation

$$\pi \approx 3.1$$

Then

$$|\pi - 3.1| \approx 0.04159265359.$$

Now $10^{-1} = 0.1$ and $\frac{1}{2} \times 10^{-1} = 0.05$.

So

$$|\pi - 3.1| < 0.05 = \frac{1}{2} \times 10^{-1}.$$

So 3.1 approximates π with one decimal place accuracy.

A Second Approximation

$$\pi \approx 3.1415$$

Then

$$|\pi - 3.1415| \approx 0.00009265359.$$

Now

$$10^{-4} = 0.0001 \quad \text{and} \quad \frac{1}{2} \times 10^{-4} = 0.00005.$$

$$10^{-3} = 0.001 \quad \text{and} \quad \frac{1}{2} \times 10^{-3} = 0.0005.$$

So since $0.00005 < 0.00009265359 < 0.0005$, we see that 3.1415 approximates π with three (but not four) decimal place accuracy.

ANOTHER NOTE ON THE ZOOM FEATURE: On a calculator with a zoom feature, the zero in Example 5 can be approximated quicker than by rescaling manually. If we zoom in on the graph in Figure 9 by a scale of 10 in each axis, with the zooming centered at the place where the curve crosses the x-axis, then we obtain the graph in Figure 12.

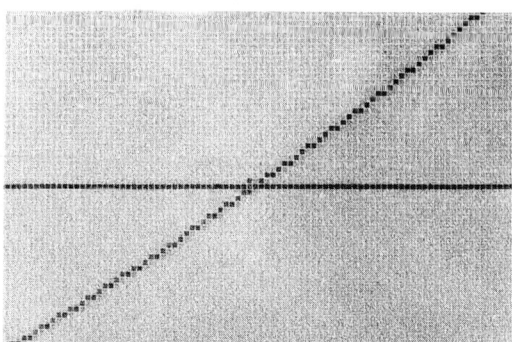

FIGURE 12   GRAPH OF $y = x^3 - 3x^2 + 7x - 8$ (zoomed in from Figure 9)

Using the trace feature on the calculator, we can move along the curve to locate two points near where the curve crosses the x-axis. One point should be to the left of the zero, and the other should be to the right. We can find, in Figure 12, that $1.65 <$ the zero < 1.69. If we zoom in again by a factor of 10, we get the graph in Figure 13.

Then we find that the zero is between $x = 1.671$ and $x = 1.674$. Thus we know that, to two decimal places, $x \approx 1.67$.

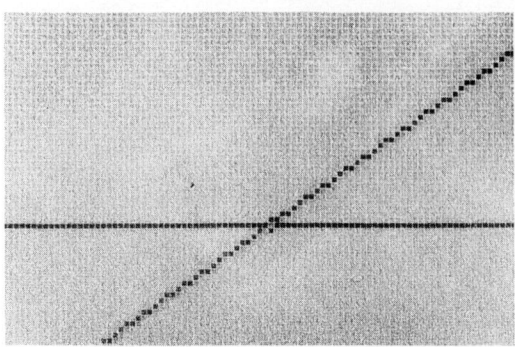

FIGURE 13 GRAPH OF $y = x^3 - 3x^2 + 7x - 8$ (zoomed in from Figure 12)

The reason that there are no tick marks in Figures 12 and 13 is that, in zooming in, the calculator we used did not change the scale values. Therefore, the scales stayed at 1 unit on each axis. Since the range of x-values is far less than 1 unit, no tick marks appear.

Alternatively, if your calculator has a zoom feature, it may also have a "box" feature. This allows you to draw a rectangle on the screen and then zoom in on the rectangle within. If you blow up a small rectangle around the zero, you can see more precisely where the curve crosses the x-axis. By doing this repeatedly, you can approximate the zero quickly and accurately.

We will say no more about zoom and box features in this appendix, but you should use them if they are available on your calculator.

EXAMPLE 7
FINDING THE ZEROS
OF A POLYNOMIAL

Find all real zeros of $p(x) = 2x^4 + 1.5x^3 - 9x^2 - x + 5$.

SOLUTION: We first observe that the $2x^4$ term dominates the other terms for x large. For example, if $x = 5$, then $2x^4 = 1250$ while $1.5x^3 - 9x^2 - x + 5 = -37.5$. The range values $-5 \le x \le 5$ include all the "interesting" behavior in the graph. If we set the y values at $-10 \le y \le 10$ with $x_{\text{scl}} = 1$ and $y_{\text{scl}} = 1$, we obtain the graph in Figure 14.

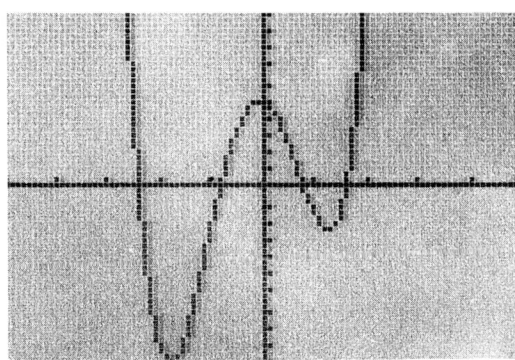

FIGURE 14 GRAPH OF $y = 2x^4 + 1.5x^3 - 9x^2 - x + 5$ FOR $-5 \le x \le 5$, $-10 \le y \le 10$

Evidently $p(x)$ has four zeros. There is one between -3 and -2, one between -1 and $-\frac{1}{2}$, one between $\frac{1}{2}$ and 1, and one between 1 and 2. We can obtain each zero by choosing x-values near the zero.

FIRST ZERO: We set $-3 \le x \le -2$, $-1 \le y \le 1$ and $x_{\text{scl}} = y_{\text{scl}} = 0.1$ to obtain the graph in Figure 15.

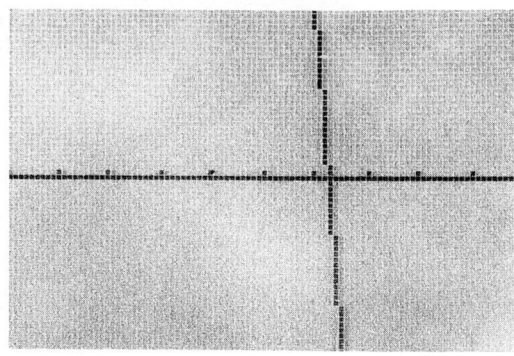

FIGURE 15 GRAPH OF $y = 2x^4 + 1.5x^3 - 9x^2 - x + 5$, $-3 \leq x \leq -2$, $-1 \leq y \leq 1$, $x_{scl} = 0.1$

The zero is between -2.4 and -2.3. Rescaling again for $-2.4 \leq x \leq -2.3$, $-1 \leq y \leq 1$, $x_{scl} = y_{scl} = 0.01$, we obtain the graph in Figure 16.

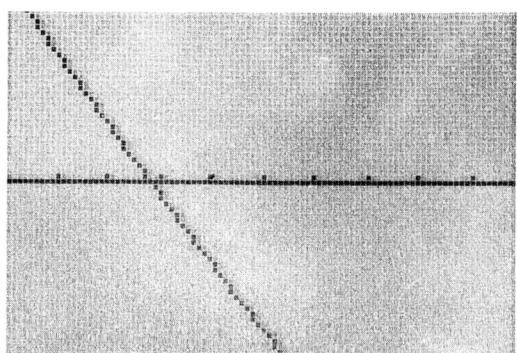

FIGURE 16 GRAPH OF $y = 2x^4 + 1.5x^3 - 9x^2 - x + 5$, $-2.4 \leq x \leq -2.3$, $x_{scl} = 0.01$

Now we see that the zero is between -2.38 and -2.37. Continuing in this manner, we find, to three decimal place accuracy, $x \approx -2.371$.

SECOND ZERO: Since the zero is between -1 and -0.5, we rescale with $-1.0 \leq x \leq -0.5$, $-1 \leq y \leq 1$, $x_{scl} = y_{scl} = 0.1$. The result is sketched in Figure 17.

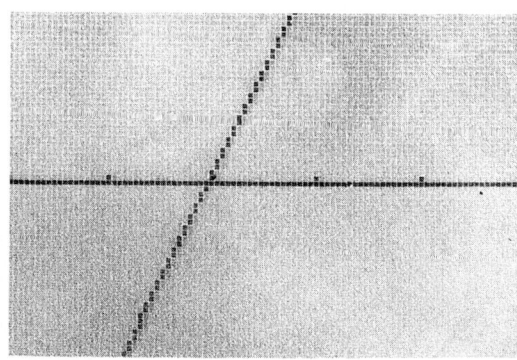

FIGURE 17 GRAPH OF $y = 2x^4 + 1.5x^3 - 9x^2 - x + 5$, $-1.0 \leq x \leq -0.5$

The zero is very close to -0.8. We can continue to find that, to three decimal places, $x \approx -0.807$.

THIRD ZERO: This zero is between $\frac{1}{2}$ and 1, so in Figure 18, we sketch the curve for $0.5 \leq x \leq 1.0$ and $x_{\text{scl}} = 0.1$.

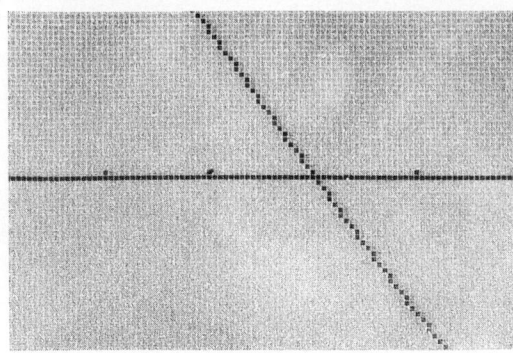

FIGURE 18   GRAPH OF $y = 2x^4 + 1.5x^3 - 9x^2 - x + 5$, $0.5 \leq x \leq 1.0$, $x_{\text{scl}} = 0.1$

The curve seems to cross the x-axis near $x = 0.8$. To three decimal places, $x = 0.803$.

FOURTH ZERO: Since this zero is between 1 and 2, we set $1 \leq x \leq 2$ with $x_{\text{scl}} = 0.1$ to obtain the graph in Figure 19.

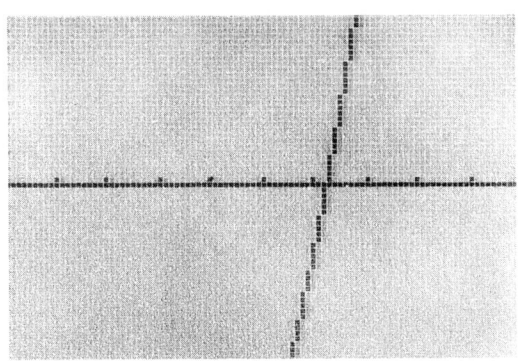

FIGURE 19   GRAPH OF $y = 2x^4 + 1.5x^3 - 9x^2 - x + 5$, $1 \leq x \leq 2$, $x_{\text{scl}} = 0.1$

The last zero is near 1.6. We find that, to three decimal places, $x \approx 1.626$.

Thus, to three decimal places, the four zeros of $p(x) = 2x^4 + 1.5x^3 - 9x^2 - x + 5$ are -2.371, -0.807, 0.803, and 1.626.

EXAMPLE 8
FINDING POINTS OF
INTERSECTION OF
TWO GRAPHS

Find, to two decimal place accuracy, all points of intersection of the graphs of $y = 2 - x$ and $y = \ln x$.

SOLUTION: $\ln x$ is defined only when $x > 0$ (see page 400). Also, if $x > 2$, $2 - x < 0$ and $\ln x > \ln 2 \approx 0.6931 > 0$, so there are no points of intersection for $x > 2$. We use the following ranges and scales on our initial graphs:

x_{min}: 0 y_{min}: -2

x_{max}: 2 y_{max}: 2

x_{scl}: 0.5 y_{scl}: 1

Depending on the calculator, we can sketch both curves on the same screen in one of two ways.

ON SOME CALCULATORS: You can enter several functions. When the "graph" key is pressed, the functions will be sketched, one after the other. (It is most common that four functions can be entered.)

ON OTHER CALCULATORS: Simply sketch one function and do not clear it. Then, sketch the second function. As long as the range and scale values are not changed, both functions will appear on the same graph.
The graphs of both functions are given in Figure 20.

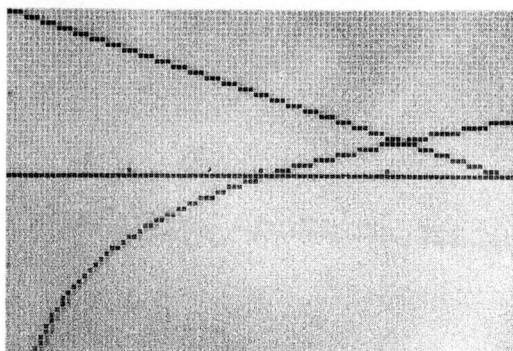

FIGURE 20 GRAPHS OF $y = 2 - x$ AND $y = \ln x$ FOR $0 < x \le 2$, $-2 \le y \le 2$, $x_{scl} = 0.5$, $y_{scl} = 1$

It appears that the graphs intersect between 1.4 and 1.6 and probably closer to 1.6.

In Figure 21, we draw the graphs for x in the interval $[1.5, 1.6]$:

x_{min}: 1.5 y_{min}: 0.4 (ln 1.5 ≈ 0.405)

x_{max}: 1.6 y_{max}: 0.5 (ln 1.6 ≈ 0.47)

x_{scl}: 0.01 y_{scl}: 0.01

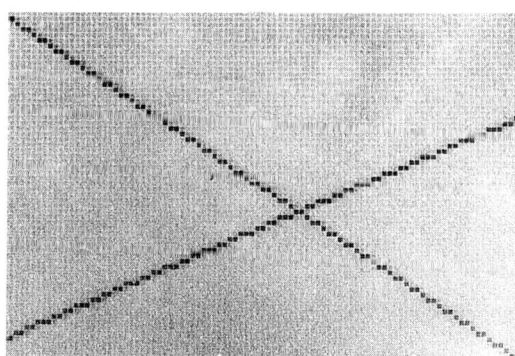

FIGURE 21 GRAPHS OF $y = 2 - x$ AND $y = \ln x$ FOR $1.5 \le x \le 1.6$, $0.4 \le y \le 0.5$, $x_{scl} = y_{scl} = 0.01$

The graphs seem to cross between 1.55 and 1.56. To be more precise, we sketch the graphs for x in the interval $[1.55, 1.56]$ (see Figure 22).

x_{min}: 1.55 y_{min}: 0.43 (ln 1.55 ≈ 0.438)

x_{max}: 1.56 y_{max}: 0.45 (ln 1.56 ≈ 0.445)

x_{scl}: 0.001 y_{scl}: 0.002

We see that the graphs intersect near the value 1.557. Therefore, to two decimal places, the graphs intersect at $x = 1.56$.

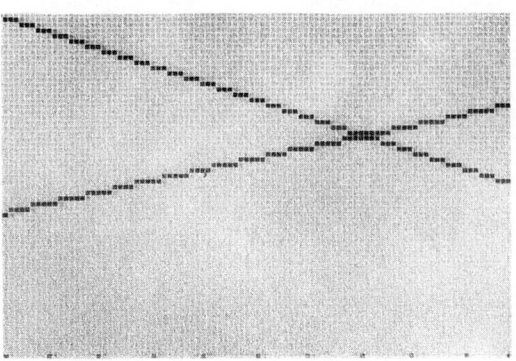

FIGURE 22   GRAPHS OF $y = 2 - x$ AND $y = \ln x$ FOR $1.55 \le x \le 1.56$, $0.43 \le y \le 0.45$, $x_{\text{scl}} = 0.001$, $y_{\text{scl}} = 0.002$

NOTE 1:   Using Newton's method (see Section 4.6), we can show that, to 9 decimal places, the curves intersect at $x = 1.557145599$.

NOTE 2:   This problem seemed more difficult than problems involving finding zeros because it is hard to see where the graphs intersect in relation to tick marks along the x-axis. This difficulty can be avoided in one of two ways: First, if your calculator has a trace function, and most do, then you can use it to find, approximately, the x- and y-coordinates of the point of intersection. If you do this, you don't have to worry about the tick marks.

Second, you can change the problem. If $2 - x = \ln x$, then

$$f(x) = 2 - x - \ln x = 0.$$

That is, you can change the problem into one of finding a zero of a function. This, as we have seen, is easier.

EXAMPLE 9
DETERMINING THE NUMBER OF
POINTS OF INTERSECTION
OF TWO GRAPHS

At how many points in the interval $[-2\pi, 2\pi]$ do the graphs of $y = \tan 2x$ and $y = \frac{1}{2}x$ intersect?

SOLUTION: The range of x-values is given as $-2\pi \le x \le 2\pi \approx -6.28 \le x \le 6.28$. Then since $y = \frac{1}{2}x$ for one of the graphs, we have $-\pi \le y \le \pi$ or $-3.14 \le y \le 3.14$. We use scales of $\frac{\pi}{4} \approx 0.785$ on the x-axis and 1 on the y-axis.

We can plot both curves on the same screen as in Example 8. When we do so, we obtain the graphs in Figure 23. In Figure 23(a), the calculator has sketched

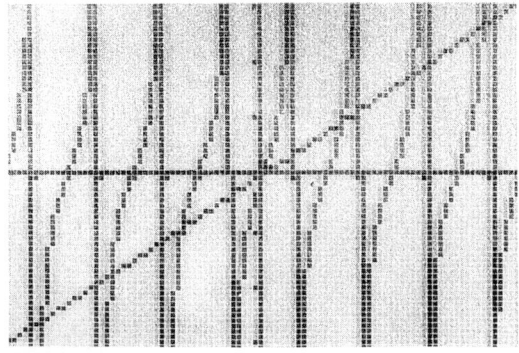

(a) Asymptotes sketched

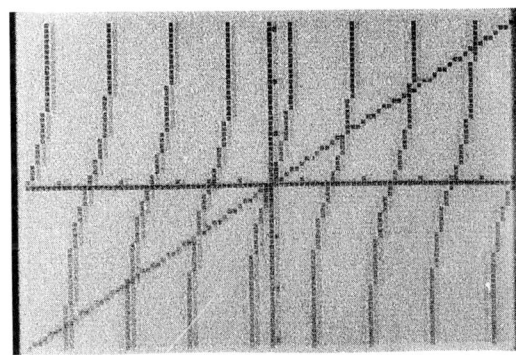

(b) Asymptotes not sketched

FIGURE 23   GRAPHS OF $y = \tan 2x$ AND $y = \frac{1}{2}x$ SHOWING SEVEN POINTS OF INTERSECTION IN THE INTERVAL $[-2\pi, 2\pi]$

the vertical asymptotes at $x = -\dfrac{7\pi}{4}, -\dfrac{5\pi}{4}, -\dfrac{3\pi}{4}, -\dfrac{\pi}{4}, \dfrac{\pi}{4}, \dfrac{3\pi}{4}$ and $\dfrac{5\pi}{4}$. In Figure 23(b), the calculator omitted the asymptotes. In either case, we see that there are seven points of intersection in the interval $[-2\pi, 2\pi]$.

EXAMPLE 10
USING GRAPHS TO SOLVE
A QUADRATIC INEQUALITY

Solve the quadratic inequality

$$4 + 4x - x^2 > 1 - 2x.$$

SOLUTION: We observe that

$$f(x) > g(x) \text{ is equivalent to } f(x) - g(x) > 0.$$

It is simpler to solve the equivalent inequality

$$(4 + 4x - x^2) - (1 - 2x) > 0$$

or

$$4 + 4x - x^2 - 1 + 2x > 0$$

or

$$3 + 6x - x^2 > 0.$$

Before continuing, we observe that $y = f(x) > 0$ for precisely those values of x at which the graph of f lies above the x-axis.

We therefore graph $f(x) = 3 + 6x - x^2$ and see where it lies above the x-axis.

We use the range and scale values:

$x_{\min}: -2 \qquad y_{\min}: -4$

$x_{\max}: 10 \qquad y_{\max}: 12$

$x_{\text{scl}}: 1 \qquad y_{\text{scl}}: 1$

The graph is given in Figure 24.

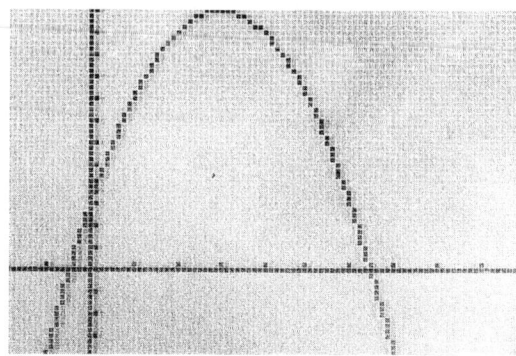

FIGURE 24 GRAPH OF $y = 3 + 6x - x^2$ FOR $-2 \le x \le 10,\ -4 \le y \le 12,$
$x_{\text{scl}} = y_{\text{scl}} = 1$

We see that $3 + 6x - x^2$ has two zeros, one between -1 and 0 and the other between 6 and 7. Moreover, $3 + 6x - x^2$ is above the x-axis between these two zeros.

We can find these zeros more precisely by using the method of Examples 6 and 7. To two decimal places, the zeros are -0.46 and 6.46. Thus the solution set of the inequality is, approximately,

$$-0.46 < x < 6.46.$$

We can check this answer by obtaining the points of intersection algebraically:

$$3 + 6x - x^2 = 0$$

$$x^2 - 6x - 3 = 0 \qquad \text{Multiply both sides by } -1$$

$$x = \frac{6 \pm \sqrt{36 - 4(1)(-3)}}{2} \qquad \text{Quadratic formula}$$

$$x = \frac{6 \pm \sqrt{48}}{2} = \frac{6 \pm 4\sqrt{3}}{2} = 3 \pm 2\sqrt{3}$$

The points of intersection are

$$x = 3 - 2\sqrt{3} \approx -0.464101615 \quad \text{and} \quad x \approx 3 + 2\sqrt{3} = 6.464101615.$$

This illustrates that the graphical technique is useful for obtaining an approximate solution to a problem. To obtain very accurate answers, other techniques may be more efficient.

A6.3 GRAPHING CONIC SECTIONS

As long as your calculator can plot two functions on the same screen, you can use it to obtain the graphs of conic sections. We illustrate this first with a simple example.

EXAMPLE 11
GRAPHING A CIRCLE

Graph the circle $x^2 + y^2 = 4$.

SOLUTION: This is the circle of radius 2 centered at the origin. We cannot graph this directly since we must enter a function in the form $y = f(x)$.[†]

$$y^2 = 4 - x^2$$

$$y = \pm \sqrt{4 - x^2} \tag{1}$$

Equation (1) is not the equation of a function. In fact, it is the equation of *two* functions:

$$y_1 = \sqrt{4 - x^2} \qquad \text{The upper half of the circle}$$

$$y_2 = -\sqrt{4 - x^2} \qquad \text{The lower half of the circle}$$

If we plot these two functions together, we obtain the entire circle. Using the range values $-2 \le x \le 2$ and $-2 \le y \le 2$ with $x_{\text{scl}} = y_{\text{scl}} = 1$, we obtain the graph in Figure 25.
   This figure appears as an ellipse since the built-in scales on the x- and y-axes are not the same. (The rectangular screen is longer along the x-axis than along the y-axis.) There are two ways to fix this. The easier way can be used if your

---

[†] Some calculators can sketch graphs given in parametric form. On such a calculator, we could graph the circle in another way. We will not discuss this method here, but we discuss how to obtain polar graphs using a parametric form beginning on page A-58.

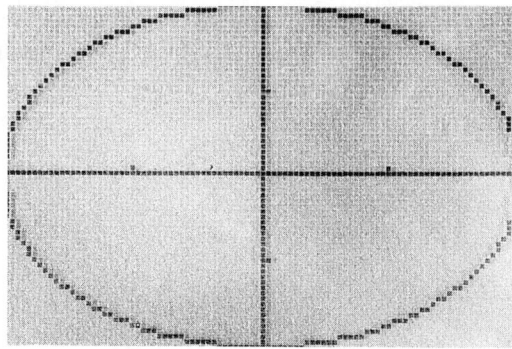

FIGURE 25 GRAPHS OF THE TWO FUNCTIONS $y_1 = \sqrt{4 - x^2}$ AND $y_2 = -\sqrt{4 - x^2}$, $-2 \le x \le 2$, $-2 \le y \le 2$

calculator has a function key labeled SQUARE. By using this function, the calculator will automatically rescale along the x-axis or the y-axis to make units along both axes have approximately the same length. (That is, a tick mark corresponding to a scale value of 1 will have approximately the same length on each axis.) When this is done, the graph appears as in Figure 26.

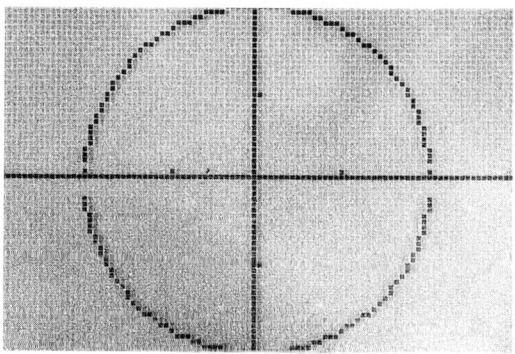

FIGURE 26 GRAPHS OF $y_1 = \sqrt{4 - x^2}$ AND $y_2 = -\sqrt{4 - x^2}$, RESCALED

Alternatively, rescale the screen yourself. In this case, the circle in Figure 26 is obtained if the x-range is expanded to $-3 < x \le 3$.

EXAMPLE 12
GRAPHING A HYPERBOLA

Graph the hyperbola

$$\frac{y^2}{9} - \frac{x^2}{16} = 1 \tag{2}$$

SOLUTION: We discuss hyperbolas in Section 8.3. To graph this hyperbola, we first solve equation (2) for y:

$$\frac{y^2}{9} = 1 + \frac{x^2}{16}$$

$$y^2 = 9 + \frac{9}{16}x^2$$

$$y = \pm \sqrt{9 + \frac{9}{16}x^2}$$

Since $x^2 \geq 0$, $\sqrt{9 + \frac{9}{16}x^2} \geq \sqrt{9} = 3$. Thus either $y \geq 3$ or $y \leq -3$. x can take on any real value. As in Example 11, we graph the two functions

$$y_1 = \sqrt{9 + \frac{9}{16}x^2} \quad \text{and} \quad y_2 = -\sqrt{9 + \frac{9}{16}x^2}$$

with range values, $-20 \leq x \leq 20$, $-6 \leq y \leq 6$, and $x_{\text{scl}} = y_{\text{scl}} = 1$. The graph is given in Figure 27.

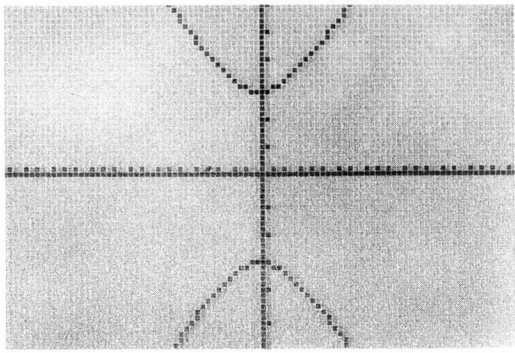

FIGURE 27 GRAPHS OF $y_1 = \sqrt{9 + \frac{9}{16}x^2}$ AND $y_2 = -\sqrt{9 + \frac{9}{16}x^2}$ FOR $-20 \leq x \leq 20$, $-6 \leq y \leq 6$

If we "square" the graph as in Example 11, we obtain the more accurate sketch in Figure 28.

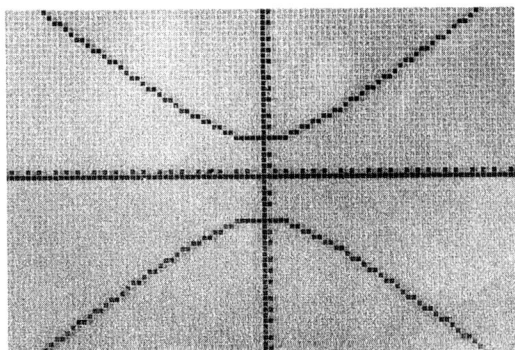

FIGURE 28 GRAPH OF THE HYPERBOLA RESCALED: $-20 \leq x \leq 20$, $-13.33 \leq y \leq 13.33$

EXAMPLE 13
GRAPHING AN ELLIPSE

Sketch the graph of the second-degree equation

$$9x^2 + 36x + 4y^2 - 8y + 4 = 0. \tag{3}$$

SOLUTION: We graphed this curve in Example 6 in Section 8.1 (see p. 545). To write equation (3) in a form that can be entered on a calculator, we first complete the squares.

$$9x^2 + 36x + 4y^2 - 8y + 4 = 0$$

$$9(x^2 + 4x) + 4(y^2 - 2y) = -4$$

$$9(x^2 + 4x + 4) - 9 \cdot 4 + 4(y^2 - 2y + 1) - 4 \cdot 1 = -4$$

$$9(x + 2)^2 - 36 + 4(y - 1)^2 - 4 = -4$$

$$9(x + 2)^2 + 4(y - 1)^2 = 36$$

$$4(y - 1)^2 = 36 - 9(x + 2)^2$$

$$(y - 1)^2 = 9 - \frac{9}{4}(x + 2)^2$$

$$y - 1 = \pm\sqrt{9 - \frac{9}{4}(x + 2)^2}$$

$$y = 1 \pm \sqrt{9 - \frac{9}{4}(x + 2)^2}$$

Before sketching these two curves, we observe that we must have

$$9 - \frac{9}{4}(x + 2)^2 \geq 0$$

$$\frac{9}{4}(x + 2)^2 \leq 9$$

$$(x + 2)^2 \leq 4$$

$$|x + 2| \leq 2$$

$$-2 \leq x + 2 \leq 2 \qquad \text{see page 6}$$

$$-4 \leq x \leq 0$$

This is our range of values for x.
  Also,

$$0 \leq (x + 2)^2 \leq 4$$

If $(x + 2)^2 = 4$, then $\sqrt{9 - \frac{9}{4}(x + 2)^2} = 0$.

If $(x + 2)^2 = 0$, then $\sqrt{9 - \frac{9}{4}(x + 2)^2} = \sqrt{9} = 3$.

Thus

$$0 \leq \sqrt{9 + \frac{9}{4}(x + 2)^2} \leq 3$$

so

$$1 \leq 1 + \sqrt{9 + \frac{9}{4}(x + 2)^2} \leq 4$$

and

$$-2 \leq 1 - \sqrt{9 + \frac{9}{4}(x + 2)^2} \leq 1.$$

Thus a suitable range of y-values is $-2 \leq y \leq 4$. Using these values, we obtain the ellipse in Figure 29.

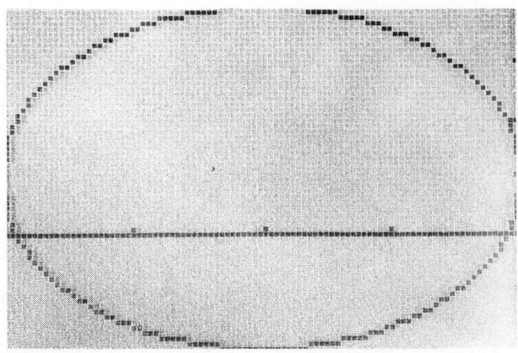

FIGURE 29 GRAPHS OF THE CURVES $y_1 = 1 + \sqrt{9 - \frac{9}{4}(x + 2)^2}$ AND
$y_2 = 1 - \sqrt{9 - \frac{9}{4}(x + 2)^2}$, $-4 \le x \le 0$, $-2 \le y \le 4$, $x_{\text{scl}} = y_{\text{scl}} = 1$

Finally, if we square to equalize the spacing along the axes, we obtain the curve in Figure 30.

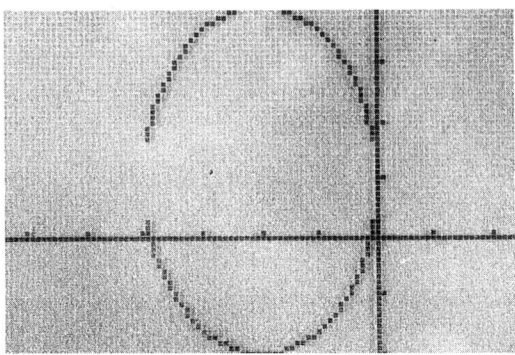

FIGURE 30 GRAPH OF THE ELLIPSE RESCALED: $-6.5 \le x \le 2.5$, $-2 \le y \le 4$

A6.4 GRAPHING CALCULATORS CANNOT GRAPH ALL FUNCTIONS ACCURATELY

We have seen that a remarkable amount of information can be obtained by graphing a function on a calculator. However, as our next example illustrates, graphing calculators do have their limitations.

EXAMPLE 14
A POLYNOMIAL WHOSE GRAPH CANNOT BE OBTAINED ACCURATELY ON A GRAPHING CALCULATOR

Graph the polynomial $p(x) = -\dfrac{x^5}{40} + x^4 + 3x^3 - 5x^2 - 10x + 4$, and find each of its zeros to one decimal place accuracy.

SOLUTION: We set our initial range and scale values as follows:

$$x_{\min} = -5 \qquad y_{\min} = -15$$
$$x_{\max} = 5 \qquad y_{\max} = 15$$
$$x_{\text{scl}} = 1 \qquad y_{\text{scl}} = 5$$

The graph in Figure 31 is obtained.

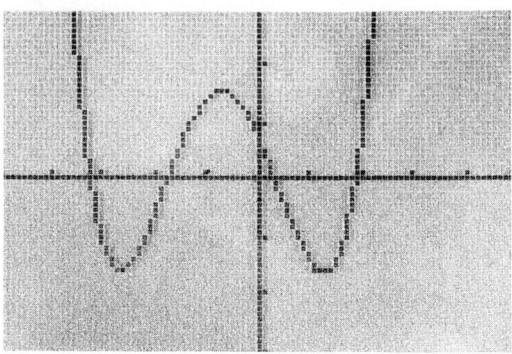

FIGURE 31 GRAPH OF $p(x) = -x^5/40 + x^4 + 3x^3 - 5x^2 - 10x + 4$ FOR $-5 \leq x \leq 5$, $-15 \leq y \leq 15$

We can see four zeros quite clearly. To one decimal place, they are -3.2, -1.7, 0.4, and 1.9.

But something is clearly wrong. If x is positive and large, then the term $-x^5/40$ will dominate the other terms. For example, if $x = 100$, then $-x^5/40 = -250,000,000$ while $x^4 + 3x^3 - 5x^2 - 10x + 4 = 102,949,004$ so $p(100) = -147,050,996$. We see that the graph of $p(x)$ must eventually turn downward and become (and stay) negative as x increases. Thus $p(x)$ has a fifth zero. To find it, we change our range and scale values as follows (after some trial and error):

$$x_{\min} = -5 \qquad y_{\min} = -100,000$$
$$x_{\max} = 45 \qquad y_{\max} = 350,000$$
$$x_{\text{scl}} - 5 \qquad y_{\text{scl}} = 100,000$$

We obtain the sketch in Figure 32.

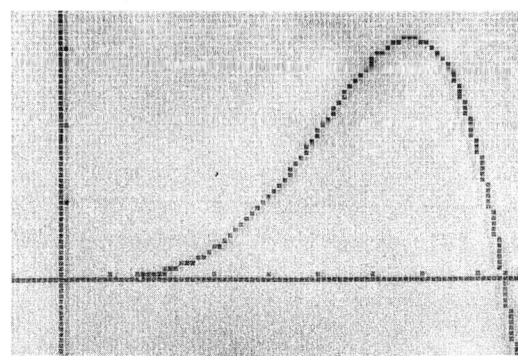

FIGURE 32 GRAPH OF $p(x) = -x^5/40 + x^4 + 3x^3 - 5x^2 - 10x + 4$ FOR $-5 \leq x \leq 45$, $-100,000 \leq y \leq 350,000$, $x_{\text{scl}} = 5$, $y_{\text{scl}} = 100,000$

From this figure, we can determine that $p(x)$ reaches a maximum height of approximately 312,000 when x is near 34. It then decreases rapidly and is zero at $x = 42.7$ (rounded to 1 decimal place).

However, in obtaining a graph that showed the behavior of $p(x)$ for x greater than 30, we completely lost the picture obtained in Figure 31. The reason for this should be clear. In the graph in Figure 31, $p(x)$ reaches an approximate minimum value of -8 at $x \approx -3$ and $x \approx 1.3$ and a maximum value of approximately 8 at $x \approx -0.7$. But the number 8 is negligible compared to the maximum height of 312,000 at $x \approx 34$. It is simply impossible to scale the x- and y-axes to show the behavior for $-4 \le x \le 4$ and $30 \le x \le 50$ simultaneously on your calculator screen. If we show one, we lose the other. In fact, an accurate sketch drawn *to scale* that portrayed the graph of $p(x)$ accurately would have to be drawn on a very large piece of paper indeed. For example, if we used a scale of 1 mm = 1 unit (this is very small—there are 25 mm in an inch), then we would need a positive y-axis that extended 312,000 mm = 312 m $\approx$ 1024 ft, which is almost a fifth of a mile! And if we use a scale much smaller than 1 mm per unit, then, as in Figure 32, we simply would not see how $p(x)$ behaved for $-4 \le x \le 4$. The graph in this interval would appear as part of the x-axis.

This example illustrates the fact that a bit of thought is needed when using even the best available graphing device. You should, among other things, think about what happens to the graph of a function as $|x|$ gets large. It may be that no range setting will give you an accurate picture.

A6.5 GRAPHING POLAR EQUATIONS

In Section 8.6, we discuss the graphs of polar equations. It is possible to graph polar equations on a calculator that can draw parametric graphs.

An equation is given *parametrically* if it is written in the form

$$x = f(t)$$
$$f = g(t) \qquad \text{for } a \le t \le b.$$

The number t is called a *parameter*. We will not discuss parametric equations in detail here. Rather, we will show how they can be used to obtain polar graphs.

Recall from equation (1) on page 578 that if the point (r, θ) is given in polar coordinates and (x, y) denotes the point in Cartesian coordinates, then

$$x = r \cos \theta \quad \text{and} \quad y = r \sin \theta. \tag{4}$$

Suppose we have the polar equation $r = f(\theta)$. Then, from (4), we obtain

$$x = r \cos \theta = f(\theta) \cos \theta$$

and

$$y = r \sin \theta = f(\theta) \sin \theta.$$

On calculators with parametric graphing capability, the parameter is often denoted by T. Then, after setting the mode to parametric graphing, we can graph $r = f(\theta)$ by graphing the functions

$$x = f(T) \cos T$$
$$y = f(T) \sin T. \tag{5}$$

In addition to x- and y-range and scale values, we also need to set $T_{\min}$, $T_{\max}$, and T_{step} values. The last setting tells the calculator how many points to plot. For example, if $T_{\min} = 0$, $T_{\max} = 1$, and $T_{\text{step}} = 0.1$, then the calculator will plot x- and y-values for $T = 0, 0.1, 0.2, \ldots, 0.9$, and 1.0. If $T_{\text{step}} = 0.2$, then it will plot values for $T = 0$, 0.2, 0.4, 0.6, 0.8, and 1.0.

If we are plotting a polar graph, then because $\cos \theta$ and $\sin \theta$ are periodic of period $2\pi \approx 6.28$, we will set $T_{\min} = 0$ and $T_{\max} = 6.28$ in our graphs. In addition, we will set $T_{\text{step}} = 0.1$. Finally, we set $x_{\min}$, $x_{\max}$, x_{scl}, $y_{\min}$, $y_{\max}$, and y_{scl} in the usual way.

In the three examples that follow, we will not discuss how to obtain range and scale values. These are obtained by the kind of thoughtful analysis that we have described for other graphing problems in this appendix.

EXAMPLE 15
GRAPHING A CIRCLE IN POLAR COORDINATES

Graph the circle $r = \cos \theta$.

SOLUTION: We can show that $r = \cos \theta$ is the equation of a circle because, from (4),

$$x = r \cos \theta = \cos \theta \, (\cos \theta) = \cos^2 \theta, \; y = r \sin \theta = (\cos \theta) \sin \theta$$

and

$$x^2 + y^2 = \cos^4 \theta + \cos^2 \theta \sin^2 \theta = \cos^2 \theta \, (\cos^2 \theta + \sin^2 \theta)$$

$$\underset{\underset{= \cos^2 \theta = x}{\downarrow}}{\cos^2 \theta + \sin^2 \theta = 1}$$

Then, completing the square,

$$x^2 - x - y^2 = 0$$

$$\left(x - \frac{1}{2}\right)^2 - \frac{1}{4} + y^2 = 0$$

$$\left(x - \frac{1}{2}\right)^2 + y^2 = \frac{1}{4}$$

This is the equation of a circle centered at $(\frac{1}{2}, 0)$ with radius $\frac{1}{2}$. To graph it, we set

$$T_{\min} = 0 \qquad x_{\min} = -1 \qquad y_{\min} = -1$$

$$T_{\max} = 6.28 \qquad x_{\max} = 1 \qquad y_{\max} = 1$$

$$T_{\text{step}} = 0.1 \qquad x_{\text{scl}} = 0.5 \qquad y_{\text{scl}} = 0.5$$

$$x(T) = \cos^2 T$$

$$y(T) = \cos T \sin T$$

The graph in Figure 33 is obtained.

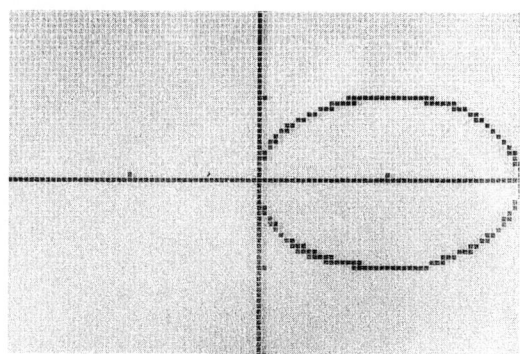

FIGURE 33   GRAPH OF $r = \cos \theta$, $-1 \le x \le 1$, $-1 \le y \le 1$

As in Example 11, our circle doesn't look very circular. But if we use the squaring function in the calculator, we obtain the graph in Figure 34.

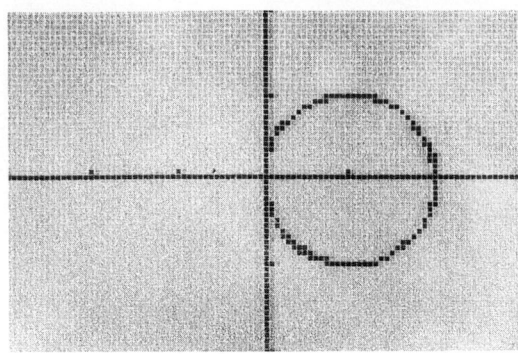

FIGURE 34 GRAPH OF $r = \cos \theta,\ -1.5 \le x \le 1.5,\ -1 \le y \le 1$

EXAMPLE 16
GRAPHING A LIMACON
WITH LOOP

Graph the polar equation $r = 1 + 2 \sin \theta$.

SOLUTION: We set the range and scale values as follows:

$$T_{\min} = 0 \qquad x_{\min} = -2 \qquad y_{\min} = -1$$
$$T_{\max} = 6.28 \qquad x_{\max} = 2 \qquad y_{\max} = 3$$
$$T_{\text{step}} = 0.1 \qquad x_{\text{scl}} = 1 \qquad y_{\text{scl}} = 1$$

Then we enter

$$x(T) = (1 + 2 \sin T) \cos T$$
$$y(T) = (1 + 2 \sin T) \sin T$$

The graph in Figure 35 is generated.

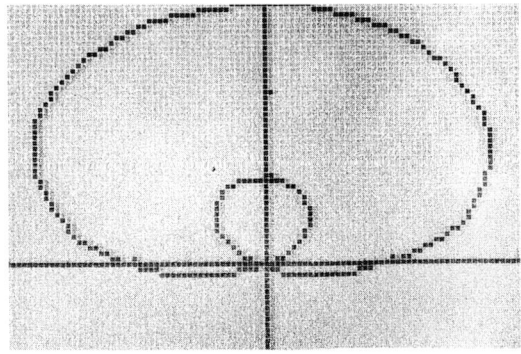

FIGURE 35 GRAPH OF $r = 1 + 2 \sin \theta,\ -2 \le x \le 2,\ -1 \le y \le 3$

We recognize this as the graph of a limaçon with loop (see page 586). We get the more accurate graph in Figure 36 by using the squaring function.

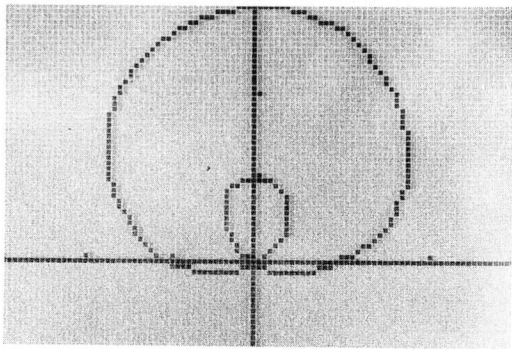

FIGURE 36   GRAPH OF $r = 1 + 2\cos\theta$, $-3 \le x \le 3$, $-1 \le y \le 3$

EXAMPLE 17
GRAPHING AN EIGHT-LEAFED
ROSE

Graph the polar equation $r = 3\cos 4\theta$.

SOLUTION: We set the following range and scale values:

$$T_{\min} = 0 \qquad x_{\min} = -3 \qquad y_{\min} = -3$$
$$T_{\max} = 6.28 \qquad x_{\max} = 3 \qquad y_{\max} = 3$$
$$T_{\text{step}} = 0.1 \qquad x_{\text{scl}} = 1 \qquad y_{\text{scl}} = 1$$

We then enter

$$x(T) - 3\cos 4T \cos T$$
$$y(T) = 3\cos 4T \sin T$$

to obtain the graph in Figure 37.

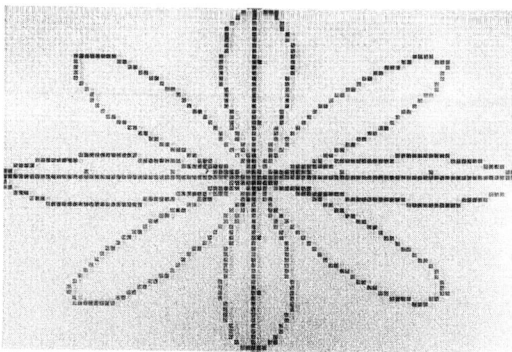

FIGURE 37   GRAPH OF $r = 3\cos 4\theta$, $-3 \le x \le 3$, $-3 \le y \le 3$

The curve in Figure 37 is called an **eight-leafed rose**. By using the squaring function, we obtain the more accurate graph in Figure 38.

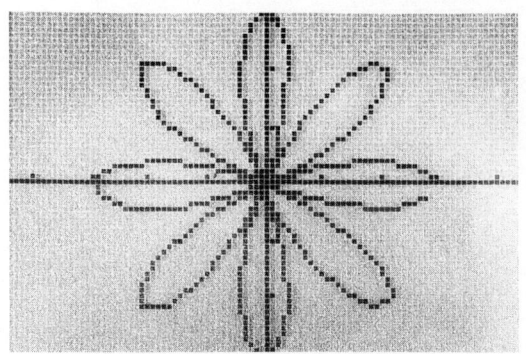

FIGURE 38 GRAPH OF $r = 3 \cos 4\theta$, $-4.5 \le x \le 4.5$, $-3 \le y \le 3$

■ PROBLEMS

In Problems 1–27 obtain sketches of each graph by first setting the appropriate calculator range and scale.

1. $y = x^2 + 2$ 2. $y = -x^2 + 4x + 2$
3. $y = x^3 - x$
4. $y = 2x^3 - 3x^2 + 4x - 5$
5. $y = -x^4 + 2x^2 - x + 3$
6. $y = x^5 - x^4 - x^3 + x^2 + x - 2$

7. $y = \dfrac{x}{x + 1}$ 8. $y = \dfrac{x^2 - 3}{x}$

9. $y = \dfrac{x + 2}{x - 4}$ 10. $y = \dfrac{x^3 - x^2 + 1}{x - 3}$

11. $y = \dfrac{x - 2}{x^2 - 1}$ 12. $y = e^{3x-2}$

13. $y = \ln(x^2 - 4)$
14. $y = 3 - 2\ln(1 - x) + 4e^x$

15. $y = \dfrac{e^{-1/x}}{x + 2}$ 16. $y = \dfrac{1 - \ln x}{x^2 + 4}$

17. $y = e^{x - \ln x}$ 18. $y = 1 - \sin x$
19. $y = 2\sin x - 5\cos 3x$
20. $y = \sin x + \sin 3x + \sin 5x$
21. $y = x\cos x$ 22. $y = e^{-x}\cos x$

23. $y = e^{2x}\sin 3x$ 24. $y = \tan^2 \dfrac{x}{3}$

25. $y = 3\sec 5x$ $\left[Hint: \sec x = \dfrac{1}{\cos x} \right]$

26. $y = \cot \dfrac{x}{2}$ 27. $y = \csc(x - 3)$

In Problems 28–37 approximate, to two decimal place accuracy, all zeros of each polynomial by sketching appropriate graphs. [*Note:* To be sure that you have two decimal place accuracy, you should approximate each zero to three decimal places.]

28. $x^2 - 3x - 10$ 29. $6x^2 + 5x - 56$
30. $x^3 + x^2 + 5$ 31. $x^3 - x^2 - 1$
32. $3x^5 + 2x^4 + x^2 + 3$ 33. $x^7 - x^2 - 12$
34. $6x^4 - 2x^3 + 3x^2 - 4x - 2$
35. $x^3 - 6x^2 - 15x + 4$
36. $x^3 + 14x^2 + 60x + 78$
37. $4x^4 - 4x^3 - 23x^2 + x + 10$

In Problems 38–43 find, by graphing, the number of solutions to each equation, and approximate each solution to two decimal place accuracy.

38. $\ln x = x - 4$ 39. $e^{-x} = \ln x$

40. $\dfrac{x^2 - 4}{x^2 - 9} = \dfrac{x}{4}$

41. $x^2 - 3x + 5 = \ln(x^4 + x + 5)$
42. $\sin 2x = 1 - x^2$ 43. $1 + 3\cos x = \sqrt{x + 5}$

In Problems 44–52 find an approximate solution to each inequality.

44. $2x + 3 > 5$
45. $|3x - 5| < 4$
   [*Hint:* Use the $\boxed{\text{Abs}}$ key or its equivalent on your calculator.]

46. $\left| \dfrac{x}{2} - 5 \right| < 8$

47. $x^2 - 2 > x + 5$ 48. $\dfrac{1}{x + 2} > \dfrac{x}{x + 1}$

49. $\dfrac{x}{x^2 - 3} < \dfrac{1}{2 - x}$

50. $2x^2 + 3x - 5 < 3 - 7x$
51. $x^3 - 2x^2 + 7x - 5 > 2x^2 - x + 4$
52. $3x^3 + 7 < 1 - 2x + x^4$

In Problems 53–67 sketch the graph of each conic section.

53. $x^2 + y^2 = 25$
54. $(x + 2)^2 + (y - 3)^2 = 12$

CONTINUED

55. $x^2 + 2x + y^2 - 4y = 20$

56. $2x^2 + 3y^2 = 12$

57. $7x^2 - 11y^2 = 47$ 58. $19y^2 - 43x^2 = 6$

59. $\dfrac{x^2}{37} + \dfrac{y^2}{121} = 1$ 60. $7x^2 - 3y = 4$

61. $3y^2 + 12x = 6$ 62. $\dfrac{x}{2} - \dfrac{y^2}{9} = 3$

63. $2x^2 + 7x + 5y^2 - 2y = 8$

64. $3x^2 + 6x + 9y - 7 = 0$

65. $2y^2 - 8y - 12x^2 + 12x = 16$

66. $17x^2 - 32x - 43y - 11y^2 = 33$

67. $\dfrac{1}{2}x^2 + 4x + \dfrac{1}{3}y^2 - 3y = 5$

In Problems 68–81 sketch the graph of each polar equation.

68. $r = 4$ 69. $r = 3 \sin \theta$

70. $r = 2 \cos \theta - 2 \sin \theta$ 71. $r = 1 - 4 \sin \theta$

72. $r = -2 - 5 \cos \theta$ 73. $r = 4 + 3 \sin \theta$

74. $r = 4 \cos 2\theta$ 75. $r = 2 \sin 3\theta$

76. $r = \theta$ 77. $r = \dfrac{\theta}{3}$

78. $r^2 = 4 \cos 2\theta$

[*Hint:* Graph $r = \sqrt{4 \cos 2\theta}$ and $r = -\sqrt{4 \cos 2\theta}$ separately.]

79. $r^2 = -9 \sin 2\theta$ 80. $r^2 = \theta$

81. $r = 1 + 2 \sec \theta$

TABLE OF INTEGRALS

All angles are measured in radians.

STANDARD FORMS

1. $\int a\,dx = ax + C$

2. $\int af(x)\,dx = a\int f(x)\,dx + C$

3. $\int u\,dv = uv - \int v\,du$ (integration by parts)

4. $\int u^n\,du = \dfrac{u^{n+1}}{n+1} + C, \quad n \neq -1$

5. $\int \dfrac{du}{u} = \ln u$ if $u > 0$ or $\ln(-u)$ if $u < 0$

$\quad\quad = \ln|u| + C$

6. $\int e^u\,du = e^u + C$

7. $\int a^u\,du = \int e^{u \ln a}\,du$

$\quad\quad = \dfrac{e^{u \ln a}}{\ln a} = \dfrac{a^u}{\ln a} + C, \quad a > 0, a \neq 1$

8. $\int \sin u\,du = -\cos u + C$

9. $\int \cos u\,du = \sin u + C$

10. $\int \tan u\,du = \ln|\sec u| = -\ln|\cos u| + C$

11. $\int \cot u\,du = \ln|\sin u| + C$

12. $\int \sec u\,du = \ln|\sec u + \tan u|$

$\quad\quad = \ln\left|\tan\left|\dfrac{u}{2} + \dfrac{\pi}{4}\right|\right| + C$

13. $\int \csc u\,du = \ln|\csc u - \cot u| = \ln\left|\tan\dfrac{u}{2}\right| + C$

14. $\int \sec^2 u\,du = \tan u + C$

15. $\int \csc^2 u\,du = -\cot u + C$

16. $\int \sec u \tan u\,du = \sec u + C$

17. $\int \csc u \cot u\,du = -\csc u + C$

18. $\int \dfrac{du}{u^2 + a^2} = \dfrac{1}{a}\tan^{-1}\dfrac{u}{a} + C$

19. $\int \dfrac{du}{u^2 - a^2} = \dfrac{1}{2a}\ln\left|\dfrac{u-a}{u+a}\right| + C$

$\quad\quad = -\dfrac{1}{a}\coth^{-1}\dfrac{u}{a} + C, \quad u^2 > a^2$

20. $\int \dfrac{du}{a^2 - u^2} = \dfrac{1}{2a}\ln\left|\dfrac{a+u}{a-u}\right| + C$

$\quad\quad = \dfrac{1}{a}\tanh^{-1}\dfrac{u}{a} + C, \quad u^2 < a^2$

21. $\int \dfrac{du}{\sqrt{a^2 - u^2}} = \sin^{-1}\dfrac{u}{|a|} + C$

22. $\int \dfrac{du}{\sqrt{u^2 + a^2}} = \ln(u + \sqrt{u^2 + a^2}) + C$

23. $\int \dfrac{du}{\sqrt{u^2 - a^2}} = \ln|u + \sqrt{u^2 - a^2}| + C$

24. $\int \dfrac{du}{u\sqrt{u^2 - a^2}} = \dfrac{1}{|a|}\sec^{-1}\left|\dfrac{u}{a}\right| + C$

25. $\int \dfrac{du}{u\sqrt{u^2 + a^2}} = -\dfrac{1}{a}\ln\left|\dfrac{a + \sqrt{u^2 + a^2}}{u}\right| + C$

26. $\int \dfrac{du}{u\sqrt{a^2 - u^2}} = -\dfrac{1}{a}\ln\left|\dfrac{a + \sqrt{a^2 - u^2}}{u}\right| + C$

INTEGRALS INVOLVING $au + b$

27. $\int \dfrac{du}{au + b} = \dfrac{1}{a}\ln|au + b| + C$

28. $\int \dfrac{u\,du}{au + b} = \dfrac{u}{a} - \dfrac{b}{a^2}\ln|au + b| + C$

29. $\int \dfrac{u^2\,du}{au + b} = \dfrac{(au+b)^2}{2a^3} - \dfrac{2b(au+b)}{a^3} + \dfrac{b^2}{a^3}\ln|au + b| + C$

30. $\int \dfrac{du}{u(au + b)} = \dfrac{1}{b}\ln\left|\dfrac{u}{au+b}\right| + C$

31. $\int \dfrac{du}{u^2(au + b)} = -\dfrac{1}{bu} + \dfrac{a}{b^2}\ln\left|\dfrac{au+b}{u}\right| + C$

32. $\int \dfrac{du}{(au + b)^2} = \dfrac{-1}{a(au+b)} + C$

33. $\int \dfrac{u\,du}{(au + b)^2} = \dfrac{b}{a^2(au+b)} + \dfrac{1}{a^2}\ln|au + b| + C$

34. $\int \dfrac{du}{u(au + b)^2} = \dfrac{1}{b(au+b)} + \dfrac{1}{b^2}\ln\left|\dfrac{u}{au+b}\right| + C$

35. $\int (au + b)^n\,du = \dfrac{(au+b)^{n+1}}{(n+1)a} + C, \quad n \neq -1$

36. $\int u(au+b)^n \, du = \dfrac{(au+b)^{n+2}}{(n+2)a^2} - \dfrac{b(au+b)^{n+1}}{(n+1)a^2} + C, \quad n \neq -1, -2$

37. $\int u^m (au+b)^n \, du = \begin{cases} \dfrac{u^{m+1}(au+b)^n}{m+n+1} + \dfrac{nb}{m+n+1} \int u^m (au+b)^{n-1} \, du \\[2mm] \dfrac{u^m (au+b)^{n+1}}{(m+n+1)a} - \dfrac{mb}{(m+n+1)a} \int u^{m-1}(au+b)^n \, du \\[2mm] \dfrac{-u^{m+1}(au+b)^{n+1}}{(n+1)b} + \dfrac{m+n+2}{(n+1)b} \int u^m (au+b)^{n+1} \, du \end{cases}$

INTEGRALS INVOLVING $\sqrt{au+b}$

38. $\int \dfrac{du}{\sqrt{au+b}} = \dfrac{2\sqrt{au+b}}{a} + C$

39. $\int \dfrac{u \, du}{\sqrt{au+b}} = \dfrac{2(au-2b)}{3a^2} \sqrt{au+b} + C$

40. $\int \dfrac{du}{u\sqrt{au+b}} = \begin{cases} \dfrac{1}{\sqrt{b}} \ln \left| \dfrac{\sqrt{au+b}-\sqrt{b}}{\sqrt{au+b}+\sqrt{b}} \right| + C, \quad b>0 \\[3mm] \dfrac{2}{\sqrt{-b}} \tan^{-1} \sqrt{\dfrac{au+b}{-b}} + C, \quad b<0 \end{cases}$

41. $\int \sqrt{au+b} \, du = \dfrac{2\sqrt{(au+b)^3}}{3a} + C$

42. $\int u\sqrt{au+b} \, du = \dfrac{2(3au-2b)}{15a^2} \sqrt{(au+b)^3} + C$

43. $\int \dfrac{\sqrt{au+b}}{u} \, du = 2\sqrt{au+b} + b \int \dfrac{du}{u\sqrt{au+b}}$ (See 40.)

INTEGRALS INVOLVING $u^2 + a^2$

44. $\int \dfrac{du}{u^2+a^2} = \dfrac{1}{a} \tan^{-1} \dfrac{u}{a} + C$

45. $\int \dfrac{u \, du}{u^2+a^2} = \dfrac{1}{2} \ln(u^2+a^2) + C$

46. $\int \dfrac{u^2 \, du}{u^2+a^2} = u - a \tan^{-1} \dfrac{u}{a} + C$

47. $\int \dfrac{du}{u(u^2+a^2)} = \dfrac{1}{2a^2} \ln\left(\dfrac{u^2}{u^2+a^2}\right) + C$

48. $\int \dfrac{du}{u^2(u^2+a^2)} = -\dfrac{1}{a^2 u} - \dfrac{1}{a^3} \tan^{-1} \dfrac{u}{a} + C$

49. $\int \dfrac{du}{(u^2+a^2)^n} = \dfrac{u}{2(n-1)a^2(u^2+a^2)^{n-1}} + \dfrac{2n-3}{(2n-2)a^2} \int \dfrac{du}{(u^2+a^2)^{n-1}}$

50. $\int \dfrac{u \, du}{(u^2+a^2)^n} = \dfrac{-1}{2(n-1)(u^2+a^2)^{n-1}} + C, \quad n \neq 1$

51. $\int \dfrac{du}{u(u^2+a^2)^n} = \dfrac{1}{2(n-1)a^2(u^2+a^2)^{n-1}} + \dfrac{1}{a^2} \int \dfrac{du}{u(u^2+a^2)^{n-1}}, \quad n \neq 1$

INTEGRALS INVOLVING $u^2 - a^2$, $u^2 > a^2$

52. $\int \dfrac{du}{u^2-a^2} = \dfrac{1}{2a} \ln\left|\dfrac{u-a}{u+a}\right| + C$

53. $\int \dfrac{u \, du}{u^2-a^2} = \dfrac{1}{2} \ln(u^2-a^2) + C$

54. $\int \dfrac{u^2 \, du}{u^2-a^2} = u + \dfrac{a}{2} \ln\left|\dfrac{u-a}{u+a}\right| + C$

55. $\int \dfrac{du}{u(u^2-a^2)} = \dfrac{1}{2a^2} \ln\left|\dfrac{u^2-a^2}{u^2}\right| + C$

56. $\int \dfrac{du}{u^2(u^2-a^2)} = \dfrac{1}{a^2 u} + \dfrac{1}{2a^3} \ln\left|\dfrac{u-a}{u+a}\right| + C$

57. $\int \dfrac{du}{(u^2-a^2)^2} = \dfrac{-u}{2a^2(u^2-a^2)} - \dfrac{1}{4a^3} \ln\left|\dfrac{u-a}{u+a}\right| + C$

58. $\int \dfrac{du}{(u^2-a^2)^n} = \dfrac{-u}{2(n-1)a^2(u^2-a^2)^{n-1}} - \dfrac{2n-3}{(2n-2)a^2} \int \dfrac{du}{(u^2-a^2)^{n-1}}$

59. $\int \dfrac{u \, du}{(u^2-a^2)^n} = \dfrac{-1}{2(n-1)(u^2-a^2)^{n-1}} + C$

60. $\int \dfrac{du}{u(u^2-a^2)^n} = \dfrac{-1}{2(n-1)a^2(u^2-a^2)^{n-1}} - \dfrac{1}{a^2} \int \dfrac{du}{u(u^2-a^2)^{n-1}}$

INTEGRALS INVOLVING $a^2 - u^2$, $< a^2$

61. $\displaystyle\int \frac{du}{a^2 - u^2} = \frac{1}{2a} \ln\left|\frac{a+u}{a-u}\right| + C = \frac{1}{a}\tanh^{-1}\frac{u}{a} + C$

62. $\displaystyle\int \frac{u\,du}{a^2 - u^2} = -\frac{1}{2}\ln|a^2 - u^2| + C$

63. $\displaystyle\int \frac{u^2 du}{a^2 - u^2} = -u + \frac{a}{2}\ln\left|\frac{a+u}{a-u}\right| + C$

64. $\displaystyle\int \frac{du}{u(a^2 - u^2)} = \frac{1}{2a^2}\ln\left|\frac{u^2}{a^2 - u^2}\right| + C$

65. $\displaystyle\int \frac{du}{(a^2 - u^2)^2} = \frac{u}{2a^2(a^2 - u^2)} + \frac{1}{4a^3}\ln\left|\frac{a+u}{a-u}\right| + C$

66. $\displaystyle\int \frac{u\,du}{(a^2 - u^2)^2} = \frac{1}{2(a^2 - u^2)} + C$

INTEGRALS INVOLVING $\sqrt{u^2 + a^2}$

67. $\displaystyle\int \frac{du}{\sqrt{u^2 + a^2}} = \ln(u + \sqrt{u^2 + a^2}) + C = \sinh^{-1}\frac{u}{|a|} + C$

68. $\displaystyle\int \frac{u\,du}{\sqrt{u^2 + a^2}} = \sqrt{u^2 + a^2} + C$

69. $\displaystyle\int \frac{u^2 du}{\sqrt{u^2 + a^2}} = \frac{u\sqrt{u^2 + a^2}}{2} - \frac{a^2}{2}\ln(u + \sqrt{u^2 + a^2}) + C$

70. $\displaystyle\int \frac{du}{u\sqrt{u^2 + a^2}} = -\frac{1}{a}\ln\left|\frac{a + \sqrt{u^2 + a^2}}{u}\right| + C$

71. $\displaystyle\int \sqrt{u^2 + a^2}\,du = \frac{u\sqrt{u^2 + a^2}}{2} + \frac{a^2}{2}\ln(u + \sqrt{u^2 + a^2}) + C$

72. $\displaystyle\int u\sqrt{u^2 + a^2}\,du = \frac{(u^2 + a^2)^{3/2}}{3} + C$

73. $\displaystyle\int u^2\sqrt{u^2 + a^2}\,du = \frac{u(u^2 + a^2)^{3/2}}{4} - \frac{a^2 u\sqrt{u^2 + a^2}}{8} - \frac{a^4}{8}\ln(u + \sqrt{u^2 + a^2}) + C$

74. $\displaystyle\int \frac{\sqrt{u^2 + a^2}}{u}\,du = \sqrt{u^2 + a^2} - a\ln\left|\frac{a + \sqrt{u^2 + a^2}}{u}\right| + C$

75. $\displaystyle\int \frac{\sqrt{u^2 + a^2}}{u^2}\,du = -\frac{\sqrt{u^2 + a^2}}{u} + \ln(u + \sqrt{u^2 + a^2}) + C$

INTEGRALS INVOLVING $\sqrt{u^2 - a^2}$

76. $\displaystyle\int \frac{du}{\sqrt{u^2 - a^2}} - \ln|u + \sqrt{u^2 - a^2}| + C$

77. $\displaystyle\int \frac{u\,du}{\sqrt{u^2 - a^2}} = \sqrt{u^2 - a^2} + C$

78. $\displaystyle\int \frac{u^2 du}{\sqrt{u^2 - a^2}} = \frac{u\sqrt{u^2 - a^2}}{2} + \frac{a^2}{2}\ln|u + \sqrt{u^2 - a^2}| + C$

79. $\displaystyle\int \frac{du}{u\sqrt{u^2 - a^2}} = \frac{1}{|a|}\sec^{-1}\left|\frac{u}{a}\right| + C$

80. $\displaystyle\int \sqrt{u^2 - a^2}\,du = \frac{u\sqrt{u^2 - a^2}}{2} - \frac{a^2}{2}\ln|u + \sqrt{u^2 - a^2}| + C$

81. $\displaystyle\int u\sqrt{u^2 - a^2}\,du = \frac{(u^2 - a^2)^{3/2}}{3} + C$

82. $\displaystyle\int u^2\sqrt{u^2 - a^2}\,du = \frac{u(u^2 - a^2)^{3/2}}{4} + \frac{a^2 u\sqrt{u^2 - a^2}}{8} - \frac{a^4}{8}\ln|u + \sqrt{u^2 - a^2}| + C$

83. $\displaystyle\int \frac{\sqrt{u^2 - a^2}}{u}\,du = \sqrt{u^2 - a^2} - |a|\sec^{-1}\left|\frac{u}{a}\right| + C$

84. $\displaystyle\int \frac{\sqrt{u^2 - a^2}}{u^2}\,du = -\frac{\sqrt{u^2 - a^2}}{u} + \ln|u + \sqrt{u^2 - a^2}| + C$

85. $\displaystyle\int \frac{du}{(u^2 - a^2)^{3/2}} = -\frac{u}{a^2\sqrt{u^2 - a^2}} + C$

INTEGRALS INVOLVING $\sqrt{a^2 - u^2}$

86. $\displaystyle\int \frac{du}{\sqrt{a^2 - u^2}} = \sin^{-1}\frac{u}{|a|} + C$

87. $\displaystyle\int \frac{u\,du}{\sqrt{a^2 - u^2}} = -\sqrt{a^2 - u^2} + C$

88. $\displaystyle\int \frac{u^2 du}{\sqrt{a^2 - u^2}} = -\frac{u\sqrt{a^2 - u^2}}{2} + \frac{a^2}{2}\sin^{-1}\frac{u}{|a|} + C$

89. $\displaystyle\int \frac{du}{u\sqrt{a^2 - u^2}} = -\frac{1}{a}\ln\left|\frac{a + \sqrt{a^2 - u^2}}{u}\right| + C$

90. $\displaystyle\int \frac{du}{u^2\sqrt{a^2 - u^2}} = -\frac{\sqrt{a^2 - u^2}}{a^2 u} + C$

91. $\displaystyle\int \sqrt{a^2 - u^2}\,du = \frac{u\sqrt{a^2 - u^2}}{2} + \frac{a^2}{2}\sin^{-1}\frac{u}{|a|} + C$

92. $\displaystyle\int u\sqrt{a^2 - u^2}\,du = -\frac{(a^2 - u^2)^{3/2}}{3} + C$

93. $\int u^2 \sqrt{a^2 - u^2} \, du = -\dfrac{u(a^2 - u^2)^{3/2}}{4} + \dfrac{a^2 u \sqrt{a^2 - u^2}}{8} + \dfrac{a^4}{8} \sin^{-1} \dfrac{u}{|a|} + C$

94. $\int \dfrac{\sqrt{a^2 - u^2}}{u} \, du = \sqrt{a^2 - u^2} - a \ln\left| \dfrac{a + \sqrt{a^2 - u^2}}{u} \right| + C$ **95.** $\int \dfrac{\sqrt{a^2 - u^2}}{u^2} \, du = -\dfrac{\sqrt{a^2 - u^2}}{u} - \sin^{-1} \dfrac{u}{|a|} + C$

INTEGRALS INVOLVING THE TRIGONOMETRIC FUNCTIONS

96. $\int \sin au \, du = -\dfrac{\cos au}{a} + C$ **97.** $\int u \sin au \, du = \dfrac{\sin au}{a^2} - \dfrac{u \cos au}{a} + C$

98. $\int u^2 \sin au \, du = \dfrac{2u}{a^2} \sin au + \left(\dfrac{2}{a^3} - \dfrac{u^2}{a} \right) \cos au + C$ **99.** $\int \dfrac{du}{\sin au} = \dfrac{1}{a} \ln(\csc au - \cot au) = \dfrac{1}{a} \ln\left| \tan \dfrac{au}{2} \right| + C$

100. $\int \sin^2 au \, du = \dfrac{u}{2} - \dfrac{\sin 2au}{4a} + C$ **101.** $\int u \sin^2 au \, du = \dfrac{u^2}{4} - \dfrac{u \sin 2au}{4a} - \dfrac{\cos 2au}{8a^2} + C$

102. $\int \dfrac{du}{\sin^2 au} = -\dfrac{1}{a} \cot au + C$

103. $\int \sin pu \sin qu \, du = \dfrac{\sin(p - q)u}{2(p - q)} - \dfrac{\sin(p + q)u}{2(p + q)} + C, \quad p \neq \pm q$

104. $\int \dfrac{du}{1 - \sin au} = \dfrac{1}{a} \tan\left(\dfrac{\pi}{4} + \dfrac{au}{2} \right) + C$

105. $\int \dfrac{u \, du}{1 - \sin au} = \dfrac{u}{a} \tan\left(\dfrac{\pi}{4} + \dfrac{au}{2} \right) + \dfrac{2}{a^2} \ln\left| \sin\left(\dfrac{\pi}{4} - \dfrac{au}{2} \right) \right| + C$

106. $\int \dfrac{du}{1 + \sin au} = -\dfrac{1}{a} \tan\left(\dfrac{\pi}{4} - \dfrac{au}{2} \right) + C$

107. $\int \dfrac{du}{p + q \sin au} = \begin{cases} \dfrac{2}{a\sqrt{p^2 - q^2}} \tan^{-1} \dfrac{p \tan \frac{1}{2}au + q}{\sqrt{p^2 - q^2}} + C, & |p| > |q| \\[3mm] \dfrac{1}{a\sqrt{q^2 - p^2}} \ln\left| \dfrac{p \tan \frac{1}{2}au + q - \sqrt{q^2 - p^2}}{p \tan \frac{1}{2}au + q + \sqrt{q^2 - p^2}} \right| + C, & |p| < |q| \end{cases}$

108. $\int u^m \sin au \, du = -\dfrac{u^m \cos au}{a} + \dfrac{m u^{m-1} \sin au}{a^2} - \dfrac{m(m-1)}{a^2} \int u^{m-2} \sin au \, du$

109. $\int \sin^n au \, du = -\dfrac{\sin^{n-1} au \cos au}{an} + \dfrac{n-1}{n} \int \sin^{n-2} au \, du$

110. $\int \dfrac{du}{\sin^n au} = \dfrac{-\cos au}{a(n-1) \sin^{n-1} au} + \dfrac{n-2}{n-1} \int \dfrac{du}{\sin^{n-2} au}, \quad n \neq 1$

111. $\int \cos au \, du = \dfrac{\sin au}{a} + C$ **112.** $\int u \cos au \, du = \dfrac{\cos au}{a^2} + \dfrac{u \sin au}{a} + C$

113. $\int u^2 \cos au \, du = \dfrac{2u}{a^2} \cos au + \left(\dfrac{u^2}{a} - \dfrac{2}{a^3} \right) \sin au + C$ **114.** $\int \dfrac{du}{\cos au} = \dfrac{1}{a} \ln|\sec au + \tan au| = \dfrac{1}{a} \ln\left| \tan\left(\dfrac{\pi}{4} + \dfrac{au}{2} \right) \right| + C$

115. $\int \cos^2 au \, du = \dfrac{u}{2} + \dfrac{\sin 2au}{4a} + C$ **116.** $\int u \cos^2 au \, du = \dfrac{u^2}{4} + \dfrac{u \sin 2au}{4a} + \dfrac{\cos 2au}{8a^2} + C$

117. $\int \dfrac{du}{\cos^2 au} = \dfrac{\tan au}{a} + C$

118. $\int \cos qu \cos pu \, du = \dfrac{\sin(q - p)u}{2(q - p)} + \dfrac{\sin(q + p)u}{2(q + p)} + C, \quad q \neq \pm p$

119. $\displaystyle\int \frac{du}{p + q \cos au} = \begin{cases} \dfrac{2}{a\sqrt{p^2 - q^2}} \tan^{-1}\left[\sqrt{(p - q)/(p + q)} \, \tan \frac{1}{2}au\right] + C, & |p| > |q| \\[2ex] \dfrac{1}{a\sqrt{q^2 - p^2}} \ln\left[\dfrac{\tan \frac{1}{2}au + \sqrt{(q + p)/(q - p)}}{\tan \frac{1}{2}au - \sqrt{(q + p)/(q - p)}}\right] + C, & |p| < |q| \end{cases}$

120. $\displaystyle\int u^m \cos au \, du = \frac{u^m \sin au}{a} + \frac{mu^{m-1}}{a^2} \cos au - \frac{m(m - 1)}{a^2} \int u^{m-2} \cos au \, du$

121. $\displaystyle\int \cos^n au \, du = \frac{\sin au \cos^{n-1} au}{an} + \frac{n - 1}{n} \int \cos^{n-2} au \, du$

122. $\displaystyle\int \frac{du}{\cos^n au} = \frac{\sin au}{a(n - 1) \cos^{n-1} au} + \frac{n - 2}{n - 1} \int \frac{du}{\cos^{n-2} au}$

123. $\displaystyle\int \sin au \cos au \, du = \frac{\sin^2 au}{2a} + C$

124. $\displaystyle\int \sin pu \cos qu \, du = -\frac{\cos(p - q)u}{2(p - q)} - \frac{\cos(p + q)u}{2(p + q)} + C, \quad p \neq \pm q$

125. $\displaystyle\int \sin^n au \cos au \, du = \frac{\sin^{n+1} au}{(n + 1)a} + C, \quad n \neq -1$

126. $\displaystyle\int \cos^n au \sin au \, du = -\frac{\cos^{n+1} au}{(n + 1)a} + C, \quad n \neq -1$

127. $\displaystyle\int \sin^2 au \cos^2 au \, du = \frac{u}{8} - \frac{\sin 4au}{32a} + C$

128. $\displaystyle\int \frac{du}{\sin au \cos au} = \frac{1}{a} \ln|\tan au| + C$

129. $\displaystyle\int \frac{du}{\cos au(1 \pm \sin au)} = \mp \frac{1}{2a(1 \pm \sin au)} + \frac{1}{2a} \ln\left|\tan\left(\frac{au}{2} + \frac{\pi}{4}\right)\right| + C$

130. $\displaystyle\int \frac{du}{\sin au(1 \pm \cos au)} = \pm \frac{1}{2a(1 \pm \cos au)} + \frac{1}{2a} \ln\left|\tan \frac{au}{2}\right| + C$

131. $\displaystyle\int \frac{du}{\sin au \pm \cos au} = \frac{1}{a\sqrt{2}} \ln\left|\tan\left(\frac{au}{2} \pm \frac{\pi}{8}\right)\right| + C$

132. $\displaystyle\int \frac{\sin au \, du}{\sin au \pm \cos au} = \frac{u}{2} \mp \frac{1}{2a} \ln|\sin au \pm \cos au| + C$

133. $\displaystyle\int \frac{\cos au \, du}{\sin au \pm \cos au} = \pm \left[\frac{u}{2} + \frac{1}{2a} \ln|\sin au \pm \cos au|\right] + C$

134. $\displaystyle\int \frac{\sin au \, du}{p + q \cos au} = -\frac{1}{aq} \ln|p + q \cos au| + C$

135. $\displaystyle\int \frac{\cos au \, du}{p + q \sin au} = \frac{1}{aq} \ln|p + q \sin au| + C$

136. $\displaystyle\int \sin^m au \cos^n au \, du = \begin{cases} -\dfrac{\sin^{m-1} au \cos^{n+1} au}{a(n + m)} + \dfrac{m - 1}{m + n} \int \sin^{m-2} au \cos^n au \, du, & m \neq -n \\[2ex] \dfrac{\sin^{m+1} au \cos^{n-1} au}{a(m + n)} + \dfrac{n - 1}{m + n} \int \sin^m au \cos^{n-2} au \, du, & m \neq -n \end{cases}$

137. $\displaystyle\int \tan au \, du = -\frac{1}{a} \ln|\cos au| = \frac{1}{a} \ln|\sec au| + C$

138. $\displaystyle\int \tan^2 au \, du = \frac{\tan au}{a} - u + C$

139. $\displaystyle\int \tan^n uu \sec^2 au \, du = \frac{\tan^{n+1} au}{(n + 1)a} + C, \quad n \neq -1$

140. $\displaystyle\int \tan^n au \, du = \frac{\tan^{n-1} au}{(n - 1)a} - \int \tan^{n-2} au \, du + C, \quad n \neq 1$

141. $\displaystyle\int \cot au \, du = \frac{1}{a} \ln|\sin au| + C$

142. $\displaystyle\int \cot^2 au \, du = -\frac{\cot au}{a} - u + C$

143. $\displaystyle\int \cot^n au \csc^2 au \, du = -\frac{\cot^{n+1} au}{(n + 1)a} + C, \quad n \neq -1$

144. $\displaystyle\int \cot^n au \, du = -\frac{\cot^{n-1} au}{(n - 1)a} - \int \cot^{n-2} au \, du, \quad n \neq 1$

145. $\displaystyle\int \sec au \, du = \frac{1}{a} \ln|\sec au + \tan au| = \frac{1}{a} \ln\left|\tan\left(\frac{au}{2} + \frac{\pi}{4}\right)\right| + C$

146. $\displaystyle\int \sec^2 au \, du = \frac{\tan au}{a} + C$

147. $\int \sec^3 au \, du = \dfrac{\sec au \tan au}{2a} + \dfrac{1}{2a} \ln|\sec au + \tan au| + C$

148. $\int \sec^n au \tan au \, du = \dfrac{\sec^n au}{na} + C$

149. $\int \sec^n au \, du = \dfrac{\sec^{n-2} au \tan au}{a(n-1)} + \dfrac{n-2}{n-1} \int \sec^{n-2} au \, du, \quad n \neq 1$

150. $\int \csc au \, du = \dfrac{1}{a} \ln|\csc au - \cot au| = \dfrac{1}{a} \ln\left|\tan \dfrac{au}{2}\right| + C$

151. $\int \csc^2 au \, du = -\dfrac{\cot au}{a} + C$ **152.** $\int \csc^n au \cot au \, du = -\dfrac{\csc^n au}{na} + C$

153. $\int \csc^n au \, du = -\dfrac{\csc^{n-2} au \cot au}{a(n-1)} + \dfrac{n-2}{n-1} \int \csc^{n-2} au \, du, \quad n \neq 1$

INTEGRALS INVOLVING INVERSE TRIGONOMETRIC FUNCTIONS

154. $\int \sin^{-1} \dfrac{u}{a} \, du = u \sin^{-1} \dfrac{u}{a} + \sqrt{a^2 - u^2} + C$ **155.** $\int u \sin^{-1} \dfrac{u}{a} \, du = \left(\dfrac{u^2}{2} - \dfrac{a^2}{4}\right) \sin^{-1} \dfrac{u}{a} + \dfrac{u\sqrt{a^2 - u^2}}{4} + C$

156. $\int \cos^{-1} \dfrac{u}{a} \, du = u \cos^{-1} \dfrac{u}{a} - \sqrt{a^2 - u^2} + C$ **157.** $\int u \cos^{-1} \dfrac{u}{a} \, du = \left(\dfrac{u^2}{2} - \dfrac{a^2}{4}\right) \cos^{-1} \dfrac{u}{a} - \dfrac{u\sqrt{a^2 - u^2}}{4} + C$

158. $\int \tan^{-1} \dfrac{u}{a} \, du = u \tan^{-1} \dfrac{u}{a} - \dfrac{a}{2} \ln(u^2 + a^2) + C$ **159.** $\int u \tan^{-1} \dfrac{u}{a} \, du = \dfrac{1}{2}(u^2 + a^2) \tan^{-1} \dfrac{u}{a} - \dfrac{au}{2} + C$

160. $\int u^m \sin^{-1} \dfrac{u}{a} \, du = \dfrac{u^{m+1}}{m+1} \sin^{-1} \dfrac{u}{a} - \dfrac{1}{m+1} \int \dfrac{u^{m+1}}{\sqrt{a^2 - u^2}} \, du$

161. $\int u^m \cos^{-1} \dfrac{u}{a} \, du = \dfrac{u^{m+1}}{m+1} \cos^{-1} \dfrac{u}{a} + \dfrac{1}{m+1} \int \dfrac{u^{m+1}}{\sqrt{a^2 - u^2}} \, du$

162. $\int u^m \tan^{-1} \dfrac{u}{a} \, du = \dfrac{u^{m+1}}{m+1} \tan^{-1} \dfrac{u}{a} - \dfrac{a}{m+1} \int \dfrac{u^{m+1}}{u^2 + a^2} \, du$

INTEGRALS INVOLVING e^{au}

163. $\int e^{au} \, du = \dfrac{e^{au}}{a} + C$ **164.** $\int u e^{au} \, du = \dfrac{e^{au}}{a}\left(u - \dfrac{1}{a}\right) + C$

165. $\int u^2 e^{au} \, du = \dfrac{e^{au}}{a}\left(u^2 - \dfrac{2u}{a} + \dfrac{2}{a^2}\right) + C$

166. $\int u^n e^{au} \, du = \dfrac{u^n e^{au}}{a} - \dfrac{n}{a} \int u^{n-1} e^{au} \, du$

$\qquad = \dfrac{e^{au}}{a}\left[u^n - \dfrac{nu^{n-1}}{a} + \dfrac{n(n-1)u^{n-2}}{a^2} - \cdots + \dfrac{(-1)^n n!}{a^n}\right]$ if n is a positive integer

167. $\int \dfrac{du}{p + q e^{au}} = \dfrac{u}{p} - \dfrac{1}{ap} \ln|p + q e^{au}| + C$ **168.** $\int e^{au} \sin bu \, du = \dfrac{e^{au}(a \sin bu - b \cos bu)}{a^2 + b^2} + C$

169. $\int e^{au} \cos bu \, du = \dfrac{e^{au}(a \cos bu + b \sin bu)}{a^2 + b^2} + C$

170. $\int u e^{au} \sin bu \, du = \dfrac{u e^{au}(a \sin bu - b \cos bu)}{a^2 + b^2} - \dfrac{e^{au}[(a^2 - b^2)\sin bu - 2ab \cos bu]}{(a^2 + b^2)^2} + C$

171. $\int u e^{au} \cos bu \, du = \dfrac{u e^{au}(a \cos bu + b \sin bu)}{a^2 + b^2} - \dfrac{e^{au}[(a^2 - b^2)\cos bu + 2ab \sin bu]}{(a^2 + b^2)^2} + C$

172. $\int e^{au} \sin^n bu \, du = \dfrac{e^{au} \sin^{n-1} bu}{a^2 + n^2 b^2} (a \sin bu - nb \cos bu) + \dfrac{n(n-1)b^2}{a^2 + n^2 b^2} \int e^{au} \sin^{n-2} bu \, du$

173. $\int e^{au} \cos^n bu \, du = \dfrac{e^{au} \cos^{n-1} bu}{a^2 + n^2 b^2} (a \cos bu + nb \sin bu) + \dfrac{n(n-1)b^2}{a^2 + n^2 b^2} \int e^{au} \cos^{n-2} bu \, du$

INTEGRALS INVOLVING ln u

174. $\int \ln u \, du = u \ln u - u + C$

175. $\int u \ln u \, du = \dfrac{u^2}{2} \left(\ln u - \dfrac{1}{2} \right) + C$

176. $\int u^m \ln u \, du = \dfrac{u^{m+1}}{m+1} \left(\ln u - \dfrac{1}{m+1} \right), \quad m \ne -1$

177. $\int \dfrac{\ln u}{u} \, du = \dfrac{1}{2} \ln^2 u + C$

178. $\int \dfrac{\ln^n u \, du}{u} = \dfrac{\ln^{n+1} u}{n+1} + C \quad \text{if } n \ne -1$

179. $\int \dfrac{du}{u \ln u} = \ln|\ln u| + C$

180. $\int \ln^n u \, du = u \ln^n u - n \int \ln^{n-1} u \, du + C, \quad n \ne -1$

181. $\int u^m \ln^n u \, du = \dfrac{u^{m+1} \ln^n u}{m+1} - \dfrac{n}{m+1} \int u^m \ln^{n-1} u \, du + C, \quad m, n \ne -1$

182. $\int \ln(u^2 + a^2) \, du = u \ln(u^2 + a^2) - 2u + 2a \tan^{-1} \dfrac{u}{a} + C$

183. $\int \ln|u^2 - a^2| \, du = u \ln|u^2 - a^2| - 2u + a \ln\left| \dfrac{u+a}{u-a} \right| + C$

INTEGRALS INVOLVING HYPERBOLIC FUNCTIONS

184. $\int \sinh au \, du = \dfrac{\cosh au}{a} + C$

185. $\int u \sinh au \, du = \dfrac{u \cosh au}{a} - \dfrac{\sinh au}{a^2} + C$

186. $\int \cosh au \, du = \dfrac{\sinh au}{a} + C$

187. $\int u \cosh au \, du = \dfrac{u \sinh au}{a} - \dfrac{\cosh au}{a^2} + C$

188. $\int \cosh^2 au \, du = \dfrac{u}{2} + \dfrac{\sinh au \cosh au}{2a} + C$

189. $\int \sinh^2 au \, du = \dfrac{\sinh au \cosh au}{2a} - \dfrac{u}{2} + C$

190. $\int \sinh^n au \, du = \dfrac{\sinh^{n-1} au \cosh au}{an} - \dfrac{n-1}{n} \int \sinh^{n-2} au \, du$

191. $\int \cosh^n au \, du = \dfrac{\cosh^{n-1} au \sinh au}{an} + \dfrac{n-1}{n} \int \cosh^{n-2} au \, du$

192. $\int \sinh au \cosh au \, du = \dfrac{\sinh^2 au}{2a} + C$

193. $\int \sinh pu \cosh qu \, du = \dfrac{\cosh(p+q)u}{2(p+q)} + \dfrac{\cosh(p-q)u}{2(p-q)} + C$

194. $\int \tanh au \, du - \dfrac{1}{a} \ln \cosh au + C$

195. $\int \tanh^2 au \, du = u - \dfrac{\tanh au}{a} + C$

196. $\int \tanh^n au \, du = \dfrac{-\tanh^{n-1} au}{a(n-1)} + \int \tanh^{n-2} au \, du$

197. $\int \coth au \, du = \dfrac{1}{a} \ln|\sinh au| + C$

198. $\int \coth^2 au \, du = u - \dfrac{\coth au}{a} + C$

199. $\int \operatorname{sech} au \, du = \dfrac{2}{a} \tan^{-1} e^{au} + C$

200. $\int \operatorname{sech}^2 au \, du = \dfrac{\tanh au}{a} + C$

201. $\int \operatorname{sech}^n au \, du = \dfrac{\operatorname{sech}^{n-2} au \tanh au}{a(n-1)} + \dfrac{n-2}{n-1} \int \operatorname{sech}^{n-2} au \, du$

202. $\int \operatorname{csch} au\, du = \frac{1}{a} \ln \left| \tanh \frac{au}{2} \right| + C$

203. $\int \operatorname{csch}^2 au\, du = -\frac{\coth au}{a} + C$

204. $\int \operatorname{sech} u \tanh u\, du = -\operatorname{sech} u + C$

205. $\int \operatorname{csch} u \coth u\, du = -\operatorname{csch} u + C$

SOME DEFINITE INTEGRALS

Unless otherwise stated, all letters stand for positive numbers.

206. $\int_0^\infty \frac{dx}{x^2 + a^2} = \frac{\pi}{2a}$

207. $\int_0^\infty \frac{x^{p-1}}{1 + x}\, dx = \frac{\pi}{\sin p\pi}$

208. $\int_0^a \frac{dx}{\sqrt{a^2 - x^2}} = \frac{\pi}{2}$

209. $\int_0^a \sqrt{a^2 - x^2}\, dx = \frac{\pi a^2}{4}$

210. $\int_0^\pi \sin mx \sin nx\, dx = \begin{cases} 0, & \text{if } m, n \text{ integers and } m \neq n \\ \dfrac{\pi}{2}, & \text{if } m, n \text{ integers and } m = n \end{cases}$

211. $\int_0^\pi \cos mx \cos nx\, dx = \begin{cases} 0, & \text{if } m, n \text{ integers and } m \neq n \\ \dfrac{\pi}{2}, & \text{if } m, n \text{ integers and } m = n \end{cases}$

212. $\int_0^\pi \sin mx \cos nx\, dx = \begin{cases} 0, & \text{if } m, n \text{ integers and } m + n \text{ is even} \\ \dfrac{2m}{(m^2 - n^2)}, & \text{if } m, n \text{ integers and } m + n \text{ is odd} \end{cases}$

213. $\int_0^{\pi/2} \sin^2 x\, dx = \int_0^{\pi/2} \cos^2 x\, dx = \frac{\pi}{4}$

214. $\int_0^\infty e^{-ax} \cos bx\, dx = \frac{a}{a^2 + b^2}$

215. $\int_0^\infty e^{-ax} \sin bx\, dx = \frac{b}{a^2 + b^2}$

216. $\int_0^\infty e^{-a^2 x^2}\, dx = \frac{\sqrt{\pi}}{2a}$

217. $\int_0^{\pi/2} \sin^{2m} x\, dx = \int_0^{\pi/2} \cos^{2m} x\, dx = \frac{1 \cdot 3 \cdot 5 \cdot \cdots \cdot (2m - 1)}{2 \cdot 4 \cdot 6 \cdot \cdots \cdot 2m} \frac{\pi}{2}, \quad m = 1, 2, 3, \ldots$

218. $\int_0^{\pi/2} \sin^{2m+1} x\, dx = \int_0^{\pi/2} \cos^{2m+1} x\, dx = \frac{2 \cdot 4 \cdot 6 \cdot \cdots \cdot 2m}{1 \cdot 3 \cdot 5 \cdot \cdots \cdot (2m + 1)}, \quad m = 1, 2, 3, \ldots$

219. $\int_0^\infty \frac{e^{-x}}{\sqrt{x}}\, dx = \sqrt{\pi}$

220. $\int_0^1 x^m (\ln x)^n\, dx = \frac{(-1)^n n!}{(m + 1)^{n+1}}, \quad m \neq -1$

ANSWERS TO ODD-NUMBERED PROBLEMS AND REVIEW EXERCISES

CHAPTER 0

Problems 0.1, page 8

1.

3. $(-\infty, 4)$ **5.** $(-\infty, 0)$ **7.** $[-\frac{1}{2}, 1]$ **9.** $(-1, 1)$

11. $(-\infty, -5] \cup [5, \infty)$ **13.** $\mathbb{R}$ **15.** $\emptyset$ (the empty set)   **17.** $(1, 3)$ **19.** $(-\infty, -7] \cup [1, \infty)$
21. $(-\frac{7}{2}, -\frac{1}{2})$ **23.** $(-\infty, 4] \cup [6, \infty)$ **25.** $(-\infty, -\frac{10}{3}) \cup (\frac{2}{3}, \infty)$ **27.** $(-\infty, \frac{4}{3}] \cup [\frac{8}{5}, \infty)$

29. $(-\infty, -\frac{53}{3}) \cup (\frac{19}{3}, \infty)$ **31.** $\left\{ x: x \le \dfrac{c-b}{a} \text{ or } x \ge -\dfrac{c+b}{a} \right\}$ **33.** $(0, \frac{1}{3})$ **35.** $(-1, 0) \cup (1, \infty)$

37. $\mathbb{R} - \{1\} = (-\infty, 1) \cup (1, \infty)$ **39. a.** $|x| \ge 3$ **b.** $|x - 1| < 3$ **c.** $|x - 3| < 7$ **d.** $|x - 8| > 3$ **e.** $|x - \frac{11}{2}| \ge \frac{7}{2}$
f. $|x - 5| < |x| [x > \frac{5}{2}]$ **49.** no, $[1.25^2, 1.35^2] = [1.5625, 1.8225]$

Problems 0.2, page 14

1. a. IV   **b.** II   **c.** III   **d.** IV   **e.** III   **f.** III   **3.** $\sqrt{122}$ **5.** $\sqrt{2}|b - a|$ **7.** $(x - 1)^2 + (y - 1)^2 = 2$
9. $(x + 1)^2 + (y - 4)^2 = 25$ **11.** $(x - 7)^2 + (y - 3)^2 = 80$ **13.** inside   **15. a.** $(\frac{7}{2}, \frac{17}{2})$ **b.** $(\frac{1}{2}, \frac{5}{2})$
17. $(x - 5)^2 + (y - 1)^2 = 34$ **21.** center $= (3, -2)$; radius $= 5$

Problems 0.3, page 21

1. -2 **3.** 1   **5.** 0   **7.** m is undefined

In answers 9–17, a point-slope equation is given first ("pinned" to the first point specified in the problem); then the slope-intercept form is given.

9. $y + 7 = 0(x - 4)$; $y = -7$ **11.** $y + 3 = -\frac{4}{3}(x - 7)$; $y = -\frac{4}{3}x + \frac{19}{3}$ **13.** $y - 2 = 2(x - 1)$; $y = 2x$
15. $y + 4 = \frac{11}{5}(x + 2)$; $y = \frac{11}{5}x + \frac{2}{5}$ **17.** $y - b = 0(x - a)$; $y = b$ **19.** $2x + 5y = 3$ **21.** $-3x + y = 1$

23. none (lines are parallel)   **25.** $(\frac{67}{45}, \frac{2}{15})$ **27.** $1/\sqrt{3}$ **29.** 1   **31.** $-\sqrt{3}$ **33.** $45° = \dfrac{\pi}{4}$ **35.** $150° = \dfrac{5\pi}{6}$

37. $63.435° \approx 1.107$ **39.** parallel (collinear)   **41.** parallel (collinear)   **43.** parallel   **45.** perpendicular
47. neither   **49.** $y = -x + \sqrt{2}$ **51.** $y = -\frac{1}{2}x + 2$ **53.** $1/\sqrt{13}$ **55.** $\sqrt{5}$

Problems 0.4, page 34

1. $f(0) = 0, f(5) = 25, f(v) = v^2, f(2 + h) = (2 + h)^2 = h^2 + 4h + 4$ **3.** $f(0) = 0, f(1) = 1, f(9) = 3, f(u^2) = |u|, f(z^4) = z^2$,
$f(4 + h) = \sqrt{4 + h}$ **5.** $f(0) = 0, f(2) = 16, f(-2) = 16, f(\sqrt{5}) = 25, f(s^{1/5}) = s^{4/5}$,
$f(s - 1) = (s - 1)^4 = s^4 - 4s^3 + 6s^2 - 4s + 1$ **7.** $g(0) = 1, g(-1) = 0, g(3) = 2, g(7) = 2\sqrt{2}, g(n^3 - 1) = n^{3/2}$,
$g(1/w) = \sqrt{\dfrac{1}{w} + 1}$ **9.** $h(0) = 1, h(2) = 7, h(\frac{1}{3}) = \frac{13}{9}, h(-\frac{1}{2}) = \frac{3}{4}, h(z^5) = 1 + z^5 + z^{10}, h\left(\dfrac{1}{u^2}\right) = 1 + \dfrac{1}{u^2} + \dfrac{1}{u^4}$ **11.** yes

13. no; because a has two images   **15.** yes

17. The function is not defined at w so it is a function but its domain is not D.   **19.** yes   **21.** yes   **23.** yes

25. no   **27.** yes   **29.** $\mathbb{R}; \mathbb{R}$ **31.** $\mathbb{R} - \{-1\}; \mathbb{R} - \{0\}$ **33.** $\mathbb{R}; (3, 4]$ **35.** $[1, \infty); [0, \infty)$

37. $\mathbb{R}; [0, \infty)$ **39.** $\mathbb{R} - \{-\frac{4}{3}\}; \mathbb{R} - \{\frac{2}{3}\}$ **41.** $(f + g)(x) = -2x - 5, \mathbb{R}; (f - g)(x) = 6x - 5, \mathbb{R};$

$(f \cdot g)(x) = -8x^2 + 20x, \mathbb{R}; (f/g)(x) = (2x - 5)/(-4x), \mathbb{R} - \{0\}$ **43.** $(f + g)(x) = \sqrt{x + 2} + \sqrt{2 - x}, [-2, 2];$

$(f - g)(x) = \sqrt{x + 2} - \sqrt{2 - x}, [-2, 2]; (f \cdot g)(x) = \sqrt{4 - x^2}, [-2, 2]; (f/g)(x) = \sqrt{(2 + x)/(2 - x)}, [-2, 2)$

45. $(f \circ g)(x) = 2x + 1, \mathbb{R}; (g \circ f)(x) = 2x + 2, \mathbb{R}$ **47.** $(f \circ g)(x) = 2 - |x + 1|, \mathbb{R}; (g \circ f)(x) = (3 - \sqrt{x})^2, [0, \infty)$

49. $(f \circ g)(x) = (x - 1)/(3x - 1), \mathbb{R} - \{0, \frac{1}{3}\}; (g \circ f)(x) = -2/x, \mathbb{R} - \{-2, 0\}$ **51.** $(f \circ g)(x) = 1, \mathbb{R} - \{0\};$

$(g \circ f)(x) = 1, \mathbb{R} - \{0\}$ **53.** $f(x) = x^2, g(x) = x - 1/x$ **55.** $f(x) = \sqrt{x}, g(x) = (x - 3)(x + 2)$

57. $f(x) = (1 - x)/(1 + x), g(x) = x^2$ **59.** $f(x) = x - 1/x = g(x)$ **61.** $f(2.34) = 12.4729, f(-1.89) = 25.913725,$

$f(10.6) = 115.186$ **63.** $h(38.2) \approx 0.0283114813, h(57.9) \approx 0.0181877369, h(238.4) \approx 0.0042477146$

65. $A = W(25 - W)$ with domain $= (0, 25)$ and range $= (0, \frac{625}{4}]$ **67.** $d(t) = \begin{cases} \sqrt{90^2 + (90 - 30t)^2} & 0 \le t < 3 \\ |180 - 30t| & 3 \le t \le 9 \\ \sqrt{90^2 + (30t - 270)^2} & 9 < t \le 12 \end{cases}$

69. *Hint:* Think about the difference in domains.   **73.** $g(x) = (x - b)/a$

Problems 0.5, page 44

1.

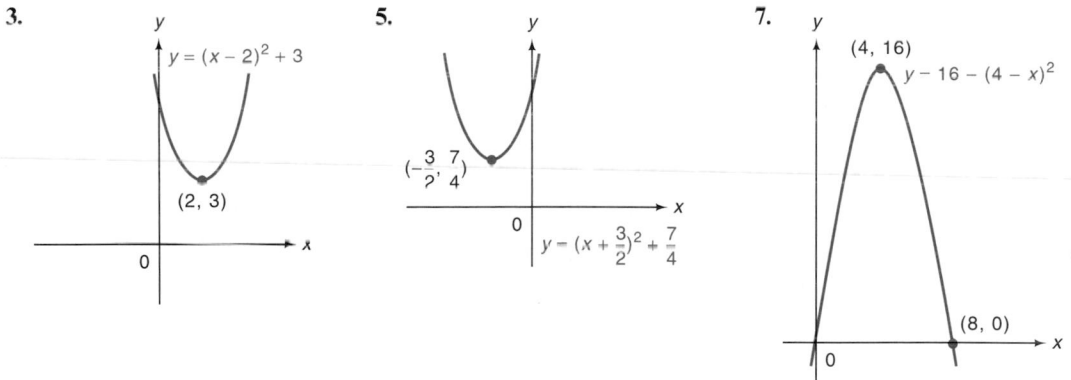

(a)   (b)   (c)   (U)

3.   **5.**   **7.**

9.

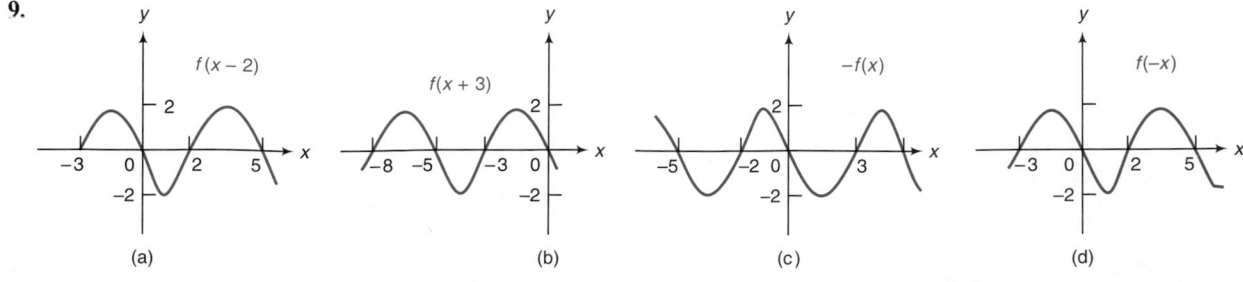

(a)   (b)   (c)   (d)

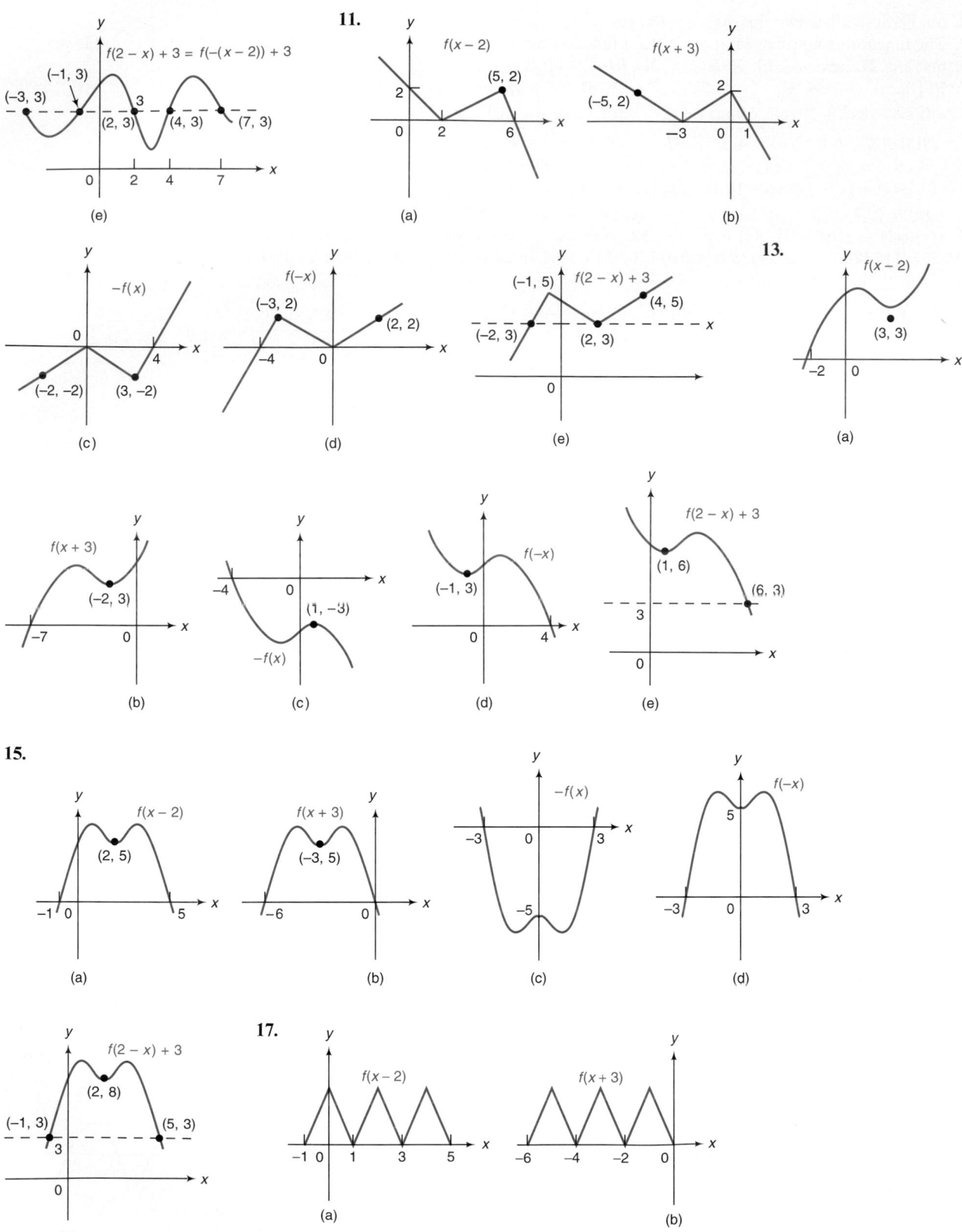

11.

(e) $f(2-x)+3 = f(-(x-2))+3$

(a) $f(x-2)$

(b) $f(x+3)$

(c) $-f(x)$

(d) $f(-x)$

(e) $f(2-x)+3$

13.

(a) $f(x-2)$

(b) $f(x+3)$

(c) $-f(x)$

(d) $f(-x)$

(e) $f(2-x)+3$

15.

(a) $f(x-2)$

(b) $f(x+3)$

(c) $-f(x)$

(d) $f(-x)$

(e) $f(2-x)+3$

17.

(a) $f(x-2)$

(b) $f(x+3)$

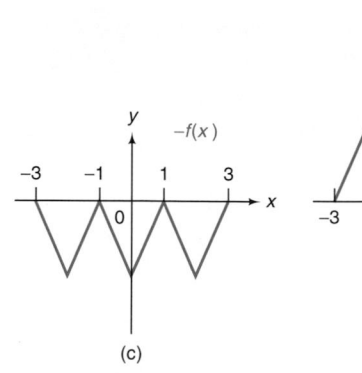

(c)

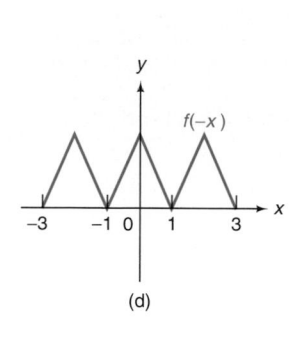

(d)

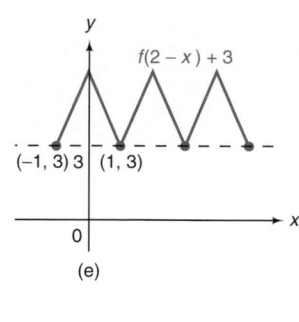

(e)

19. even   **21.** even   **23.** neither   **25.** odd   **27.** neither

29.

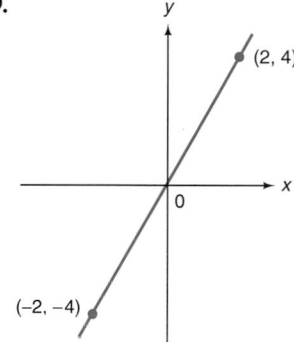

31.

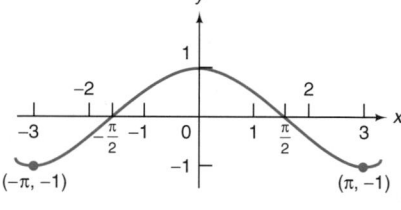

33.

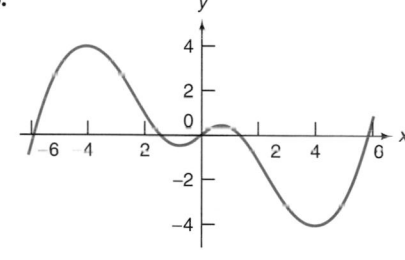

35.

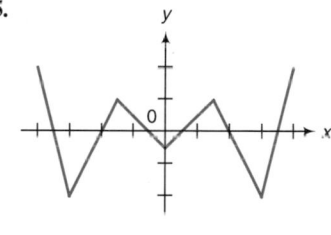

Computer Exercises, page 47

1. a. $(-\infty, -2] \cup [\frac{1}{2}, 5]$ **b.** $(-\infty, -2] \cup [\frac{1}{2}, 40]$ **c.** $(-\infty, -2) \cup (-2, \frac{1}{2}) \cup (4, \infty)$ **d.** $(-3, 5)$ **e.** $[-3, \frac{3}{2}]$

f. $\varnothing$ (the empty set)   **g.** $\{-2\} \cup [0, 2]$ **3.** $\left(-\infty, -\dfrac{\sqrt{5}+3}{2}\right) \cup \left(-\dfrac{\sqrt{13}+1}{2}, \dfrac{\sqrt{5}-3}{2}\right) \cup \left(\dfrac{\sqrt{13}-1}{2}, \infty\right)$

$\left[-\dfrac{\sqrt{5}+3}{2} \approx -2.618033989; \ -\dfrac{\sqrt{13}+1}{2} \approx -2.302775638; \dfrac{\sqrt{5}-3}{2} \approx -0.381966011\right]$

5. $\left(1 + \dfrac{42}{\sqrt{197}}, 2 - \dfrac{3}{\sqrt{197}}\right) \approx (3.992376099, 1.78625885)$ **7.** $d = 6$

CHAPTER 1

Problems 1.2, page 65

1. a.

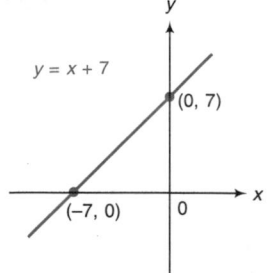

b. $f(1) = 8$, $f(1.5) = 8.5$, $f(1.9) = 8.9$, $f(1.99) = 8.99$

c. $f(3) = 10$, $f(2.5) = 9.5$, $f(2.1) = 9.1$, $f(2.01) = 9.01$ **d.** 9 **3. a.**

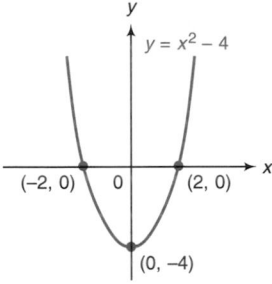

b. $f(0) = -4$, $f(0.5) = -3.75$; $f(0.9) = -3.19$, $f(0.99) = -3.0199$
c. $f(2) = 0$, $f(1.5) = -1.75$, $f(1.1) = -2.79$, $f(1.01) = -2.9799$ **d.** -3 **5. a.** division by zero is impossible
b. let $g(x) = x - 2$ **c.** -1 **7.** $\sqrt{x+1}$ is undefined for $x < -1$ so left-hand limit does not exist **9.** 19 **11.** $\frac{1}{2}$
13. 0 **15.** 3 **17.** limit does not exist (no left-hand limit) **19.** 0 **21.** 2 **23.** $\frac{1}{2}$ **25.** 0
27. 0 **29.** -6 **31.** 1 **33.** 0 **35.** 1 **37.** does not exist **39.** 3 **41.** 3 **43.** 1
45. 1 **47.** does not exist **49.** $\frac{1}{3}$ **51.** **b.** 0

53. a.

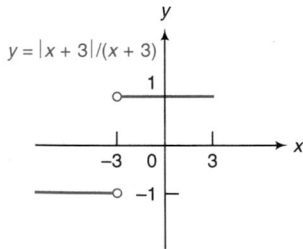

c. $1, -1$ **55. a.**

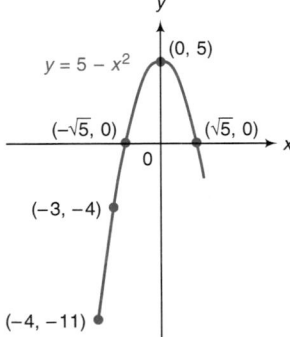

d. the slope of the secant line between $(-3, -4)$ and $(-3 - h, 5 - (-3 - h)^2)$
e. 6 **f.** 6

57. a. $f(-1) \approx 0.47380$, $f(-1.5) \approx 0.45182$, $f(-1.9) \approx 0.37170$, $f(-1.99) \approx 0.33771$, $f(-1.999) \approx 0.33378$
b. $f(-3)$ and $f(-2.5)$ are undefined, $f(-2.1) \approx 0.28051$, $f(-2.01) \approx 0.32882$, $f(-2.001) \approx 0.33289$
c. $\frac{1}{2}(f(-1.999) + f(-2.001) \approx 0.33333$ **d.** $f(-2) = \frac{2}{6} = \frac{1}{3}$ **59.** 0   **61.** 2   **63.** -1 **65.** $\frac{1}{4}$
67. -1 **69.** $\frac{1}{12}$ **71.** $\frac{1}{10}$ **73.** 500   **75.** 75   **77.** 80   **79.** $\frac{5}{3}$ **81.** -20 **83.** 2   **85.** 0
87. $\frac{3}{5}$ **89.** 20.08554   **91.** 1.58496   **93.** 1   **97.** both are 0   **99.** 1   **101.** 5   **103.** 5
105. $\frac{3}{2}$ **107.** -2 **109.** 1   **111.** -1 **113.** 0

Problems 1.3, page 73

1. 2   **3.** 3   **5.** 24   **7.** 243   **9.** $\frac{1}{19}$ **11.** $\frac{5}{4}$ **13.** 5   **15.** $-\frac{3}{2}$ **17.** 6   **19.** -4
21. 6   **23.** $\frac{1}{2}$ **25.** $\frac{1}{2}$ **27.** Let $f(x) = x - 7$ and $g(x) = 1/(x - 7)$
29. *Hint:* The assertion is true if "$<$" is replaced by "$\leq$".

Problems 1.4, page 81

1. ∞ **3.** does not exist   **5.** ∞ **7.** does not exist   **9.** 1   **11.** 0   **13.** -1 **15.** $\frac{2}{3}$
17. 0   **19.** $\frac{3}{7}$ **21.** $-\infty$ **23.** does not exist   **25.** -1 **27. b.** Pick x so that $x < -\sqrt{5000} \approx -70.71$.
29. $e^{-1} \approx 0.36788$ **31.** $e^0 = 1$ **33.** ∞ **35.** 0
37. *Hint:* Think about the difference between one-sided and two-sided limits.   **39.** It is false $[\text{let } f(x) = x]$

Problems 1.5, page 100

1. $f'(x) = 0$; $y = 4$ **3.** $f'(x) = -4$; $y = -4x + 6$ **5.** $f'(x) = 4x - 12$; $y = 4x - 14$
7. $f'(x) = -2x + 3$; $y = 3x + 5$ **9.** $f'(x) = 3x^2$; $y = 12x - 16$ **11.** $f'(x) = -1/(2\sqrt{x + 3})$; $y = -\frac{1}{6}x - 2$
13. $f'(x) = 4x^3$; $y = 32x - 48$ **19.** $x = \frac{1}{2}$ **21.** $a = \frac{3}{2}$ **23.** $f'(0) = -1$ **25. a.** $f(1.5) = -6.75$,
$f(2.1) = -13.23$, $f(1.99) = -11.8803$, $f(2.001) = -12.012003$ **b.** -10.5, -12.3, -11.97, -12.003; -12
c. $f'(x) = -6x$, $f'(2) = -12$ **d.** $y = -12x + 12$ **e.**

27. $-0.7071 \left(\text{actual value is } -\sin\frac{\pi}{4} = -\frac{\sqrt{2}}{2}\right)$ **29.** $2 \left(\text{actual value is } \sec^2\frac{\pi}{4} = 2\right)$ **31.** 1   **33.** $0.1 = \frac{1}{10}$

35. 1   **37.** 0.3679 (actual value is e^{-1})   **47.** $\dfrac{d}{dx}\ln x = \dfrac{1}{x}$ **49.** They are not equivalent.

Problems 1.6, page 111

1. 9   **3.** 65   **5.** $\frac{1}{4}$ **7.** $v(3) = 420$ ft/sec; $v(10) = 1400$ ft/sec   **9.** $P'(5) = 500$ indiv/hr; $P'(24) = 55{,}296$ indiv/hr;
no; at this rate the bacteria would soon take over the earth.   **11.** $V'(10) = 4\pi(10)^2 = 400\pi$ $\mu m^3/\mu m$
13. a. $V(S) = S^{3/2}/(6\sqrt{\pi})$ **b.** $V'(100) = (5/2\sqrt{\pi})$ $\mu m^3/\mu m^2$ **15. a.** $C'(q) = 6 - 0.02q + 0.03q^2$
b. Marginal cost is increasing (for $q > \frac{1}{3}$), so the answer is no.   **17.** $\frac{1}{4}$ kg/m

Problems 1.7, page 120

1. $(-\infty, \infty)$ **3.** $(-\infty, -2)$, $(-2, \infty)$ **5.** $(-\infty, -1)$, $(-1, 1)$, $(1, \infty)$ **7.** $(-\infty, \infty)$
9. $(n, n + 1)$ where n is an integer   **11.** yes; yes   **13.** no; no   **15.** no; no   **17.** yes; no   **19.** no; no
21. *Hint:* The intermediate value theorem applies here.

25.

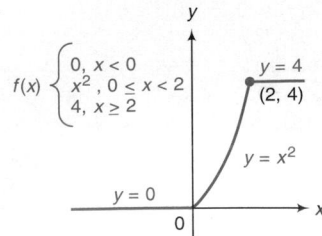

$f(x) \begin{cases} 0, & x < 0 \\ x^2, & 0 \le x < 2 \\ 4, & x \ge 2 \end{cases}$

$y = 4$ (2, 4)

$y = x^2$

$y = 0$

27. f is continuous at $x = 0$ and $x = 1$

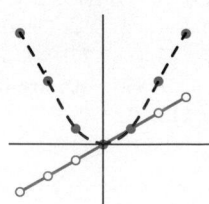

29. $\frac{1}{3}$ **31.** $(-1.5, -1)$; the zero ≈ -1.255 **33.** $(-1, -0.5)$; the zero ≈ -0.771 **35.** $(1.5, 2)$; the zero ≈ 1.764
37. $(1.8, 1.9)$; the zero ≈ 1.8955 **39.** $(4.4, 4.5)$; the zero ≈ 4.4934 **41. a.** division by zero is undefined   **b.** 2
c. $g(x) = x + 1$ **43.** $f(x) = (x-1)(x-2)(x-3)/(x-\alpha)$ has a removable discontinuity if α is 1, 2, or 3.
45. no on $[-3, 3]$; yes on $[-1, 1]$ **47.** jump discontinuities at $x = -1, 0, 1, 2$ **49.** jump discontinuity at -1
51. continuous   **53.** discontinuous at each integer, continuous everywhere else
55. $f(x) = \begin{cases} 1, & x \text{ an integer} \\ 0, & x \text{ not an integer} \end{cases}$, so piecewise continuous   **57.** $f(x) = \begin{cases} x - 1, & x \text{ an integer} \\ [x], & x \text{ not an integer} \end{cases}$
59. a. $f(x)g(x) = \begin{cases} (x+3)(x^2+6) & \text{if } x \le 1 \\ (2x+5)(5x^3-1) & \text{if } x > 1 \end{cases}$ **b.** $\lim_{x \to 1^-} f(x) = 4, \lim_{x \to 1^+} f(x) = 7$
c. $\lim_{x \to 1^-} g(x) = 7, \lim_{x \to 1^+} g(x) = 4$ **d.** $\lim_{x \to 1^-} f(x)g(x) = 28 = \lim_{x \to 1^+} f(x)g(x)$

Problems 1.8, page 129

1. $\frac{1}{70}, \frac{1}{700}$; $\delta = \epsilon/7$ **3.** $\frac{1}{50}, \frac{1}{500}$; $\delta = \epsilon/5$ **5.** $\frac{1}{70}, \frac{1}{700}$; $\delta = \min\{1, \epsilon/7\}$; a more precise choice is $\delta = \sqrt{9 + \epsilon} - 3$
7. $\frac{1}{11}, \frac{1}{110}$; $\delta = \min\{1, \epsilon/2\}$; a more precise choice is $\delta = (\sqrt{1 + 4\epsilon} - 1)/2$ **9. b.** $N \ge 10/\epsilon^2 = 100,000$
11. b. $N \ge 3 + 1/\epsilon = 103$ **23.** $\delta = \epsilon/2 = 0.005$ **25.** $\delta = \sqrt{\epsilon} \approx 0.32$ **27.** $\delta = \min\{1, 2\epsilon\} = 0.2$
29. $\delta = \min\{\frac{1}{2}, \frac{\epsilon}{2}\} = 0.05$

Problems 1 Review, page 131

1. $f(1) = 4, f(1.5) = 3.75, f(1.9) = 3.91, f(1.99) = 3.9901; f(3) = 6, f(2.5) = 4.75, f(2.1) = 4.11, f(2.01) = 4.0101;$
$\lim_{x \to 2}(x^2 - 3x + 6) = 4$ **3. a.** 0   **b.** -108 **c.** $\frac{76}{31}$ **d.** 0   **5. a.** -1 **b.** 4   **7. a.** does not exist   **b.** 5   **c.** $\frac{25}{7}$
d. 5   **e.** -6 **f.** 1024   **g.** $\frac{1}{3}$ **h.** c/f **9.** $\lim_{x \to 0^+} 1/x^3 = \infty$ but $\lim_{x \to 0^-} 1/x^3 = -\infty$; $\lim_{x \to 0^+} 1/x^4 = \infty = \lim_{x \to 0^-} 1/x^4$
11. a. 0   **b.** 0   **c.** 0   **d.** $\frac{1}{3}$ **e.** 0   **f.** ∞ **13. a.** positive at x_1, x_2, x_6; 0 at x_3, x_5, x_7; negative at x_4, x_8
b. positive at x_1, x_2, x_{10}; 0 at $x_3, x_4, x_5, x_7, x_8, x_9$; negative at x_6 **15.** $\lim_{x \to -1^-} dy/dx = -1 \ne 1 = \lim_{x \to -1^+} dy/dx$
17. 16π ft^3/ft   **19.** $-3, 2$ **21.** $\alpha = -15$; $\lim_{x \to 3} f(x) = 12$ **23.** $(-\infty, \infty)$
25. $(-\infty, -\sqrt{6}); (-\sqrt{6}, \sqrt{6}); (\sqrt{6}, \infty)$ **27.** $(-\infty, \infty)$ **29.** $(-\infty, -3), (-3, \infty)$; removable discontinuity at -3
31. $(-\infty, -1), (-1, \infty)$; removable discontinuity at -1

Computer Exercises, page 132

1. 2.72 (the limit is $e \approx 2.71828$)   **3.** The limit is b.   **5.** 1.39 (the limit is $\ln 4 \approx 1.38629$)   **7. d.** $\dfrac{d}{dx} \ln x = \dfrac{1}{x}$

CHAPTER 2

Problems 2.1, page 137

1. 0   **3.** $2x + a + b$ **5.** $-1 + 4t^3 - 7t^6$ **7.** $162w^5 - 15w^4 + 4$ **9.** $-36z^{11} + 36z^2$
11. $24s^7 - 48s^5 - 28s^3 + 4s$ **13.** $y = 4x - 3$ **15.** $y = 5x - 5$ **17.** $y = x + 1$ **19.** $y = 3x - 8$
21. $y = -x/3 + \frac{4}{3}$ **23.** $y = -x/3 - \frac{14}{3}$ **25.** x values are $(-1 \pm \sqrt{5})/2$
27. a. $3000 - 20\sqrt{t} - 30t + 30t^2$ **b.** $-10/\sqrt{t} - 30 + 60t$ organisms per hour   **c.** 205 organisms/hr; 326 organisms/hr
29. $\dfrac{d}{dr} \pi r^2 = 2\pi r$ **31.** $(1024, 32)$ **33.** $y = 6x - 6$; $y = -2x + 2$ **35.** $y = 2x$; $y = -6x$

Problems 2.2, page 144

1. $3x^2 - 10x - 9$ **3.** $-6/t^7$ **5.** $6x^2 + 2$ **7.** $-100t^{-101}$

9. $(1 + x + x^5)(-1 + 6x^5) + (2 - x + x^6)(1 + 5x^4) = 1 - 2x + 10x^4 + 7x^6 + 11x^{10}$ **11.** $-\frac{1}{2}x^{-3/2}$

13. $t^3\frac{1}{2}t^{-1/2} + (1 + \sqrt{t})3t^2 = 3t^2 + \frac{7}{2}t^{5/2}$ **15.** $1/[\sqrt{t}(1 - \sqrt{t})^2]$ **17.** $5v^4 + 7v^{5/2} - \frac{5}{2}v^{3/2} - 2$ **19.** $\frac{5}{2}x^{3/2}$

21. $-(9t^4 + 2)/[2t^{3/2}(t^4 + 2)^2]$ **23.** $(2x^{-3})(1 + x)(1 - x) + (1 - x^{-2})(1)(1 - x) + (1 - x^{-2})(1 + x)(-1) = 2/x^3 - 2x$

25. $(2x)(x^3 + 2)(x^4 + 3) + (x^2 + 1)(3x^2)(x^4 + 3) + (x^2 + 1)(x^3 + 2)(4x^3)$ **27.** $y = 28x - 20$ **29.** $y = (-7u + 11)/2$

31. $y = 2x - 1$ **33.** $y = 8x - 8$ **35.** $dR/dr = -4\alpha L/(0.2)^5 = -12{,}500\alpha L$ **37.** $(f^2 g' + f' g^2)/(f + g)^2$

Problems 2.3, page 153

1. $3(x + 1)^2$ **3.** $6(1 + x^6)^5 \cdot 6x^5$ **5.** $4(x^2 - x^3)^3(2x - 3x^2)$ **7.** $3(1 - x^2 + x^5)^2(-2x + 5x^4)$

9. $3(\sqrt{x} - x)^2[1/(2\sqrt{x}) - 1]$ **11.** $-4(y^2 + 3)^{-5} \cdot 2y$ **13.** $-6(t + 1)^2/(t - 1)^4$

15. $10x(x^2 + 2)^4(x^4 + 3)^3 + (x^2 + 2)^5 12x^3(x^4 + 3)^2$ **17.** $(-3t^2 + 2t - 4)/[\sqrt{t^2 + 1}(t + 2)^5]$

19. $\dfrac{[4x(x^2 + 1)(x^3 + 2)^3 + (x^2 + 1)^2 9x^2(x^3 + 2)^2]\sqrt{x^4 + 3} - (x^2 + 1)^2(x^3 + 2)^3(2x^3/\sqrt{x^4 + 3})}{x^4 + 3}$ **21.** $\dfrac{2x + x/\sqrt{1 + x^2}}{2\sqrt{x^2 + \sqrt{1 + x^2}}}$

23. $(-1)[(1 + x)^{-1} + (1 - x)^{-1}]^{-2}[(-1)(1 + x)^{-2} + (1 - x)^{-2}] = -x$ **25.** $2\,\text{Virgil}\,(2x + 1)$

27. $6\,\text{Homer}^5(x)\,\text{Virgil}\,(x)$ **29.** $\dfrac{1}{x}$ **31.** $\dfrac{1}{2x\sqrt{L(7x)}}$ **33.** $-\dfrac{24L^{11}\left(\dfrac{x + 1}{x - 1}\right)}{x^2 - 1}$ **35.** $\dfrac{1 - L(x)}{x^2}$ **37.** $4E(4x)$

39. $(3x^2 + 2)E(x^3 + 2x + 3)$ **41.** $\left(1 + \dfrac{x}{2}\right)E^{1/2}(x + 1)$ **43.** $\dfrac{E(L(x))}{x}$ **45.** $y = 180$ when $x = 4$

Problems 2.4, page 159

1. $\frac{2}{5}x^{-3/5} + \frac{2}{3}x^{-2/3}$ **3.** $\frac{10}{3}x(x^2 + 1)^{2/3}$ **5.** $\frac{3}{10}x^2(x^3 + 3)^{-9/10}$ **7.** $\frac{40}{7}t^9(t^{10} - 2)^{-3/7}$

9. $(u^2 + 1)(u^3 + 3u + 1)^{-2/3}$ **11.** $-\frac{28}{17}[(z^2 - 1)/(z^2 + 1)]^{-24/17}[z/(z^2 + 1)^2] = \dfrac{-28z}{17(z^2 - 1)^{24/17}(z^2 + 10)^{10/17}}$

13. $\frac{5}{3}r(r^2 + 1)^{-1/6}(r - 1)^{1/2} + \frac{1}{2}(r - 1)^{-1/2}(r^2 + 1)^{5/6} = (13r^2 - 10r + 3)/6(r^2 + 1)^{1/6}(r - 1)^{1/2}$ **15.** $3\sqrt{2}x^{\sqrt{2} - 1}$

17. $2\sqrt{3}t(t^2 + 1)^{\sqrt{3} - 1}$ **19.** $4(u^{\sqrt{2}} - u^{\sqrt{3}})^3(\sqrt{2}u^{\sqrt{2} - 1} - \sqrt{3}u^{\sqrt{3} - 1})$ **21.** $y = 2x$; $y = -\frac{1}{2}x + \frac{5}{2}$ **23.** $y = 0$; $x = 1$

25. a. $v_{rms} = 10\sqrt{3}/\sqrt{8.99} \approx 5.7767$ **b.** $dv_{rms}/d\rho = -\frac{1}{2}\sqrt{3P/\rho^3}$ **c.** $dv_{rms}/d\rho \approx -32.1285$

Problems 2.5, page 167

1. $3\cos 3x$ **3.** $\cos x - x \sin x$ **5.** $\cos x/2\sqrt{\sin x}$ **7.** $2\sin x \cos x$ **9.** $2\cos^2 x - 2\sin^2 x$ **11.** 0

13. $-\sin x$ **15.** $2\sec x \tan x/(1 - \sec x)^2$ **17.** $(3x^2 - 2)\cos(x^3 - 2x + 6)$ **19.** $2\tan x \sec^2 x$

21. $3\sec^3 x \tan x$ **23.** $-2x\csc x^2 \cot x^2$ **25.** $\frac{1}{2}$ **27.** 0 **29.** 16 **31.** $\frac{3}{4}$ **33.** 0 **35.** $\frac{4}{49}$

39. $\pi/180$

Problems 2.6, page 173

1. $-x^2/y^2$ **3.** $-\sqrt{y/x}$ **5.** $-(y/x)^{15/8}$ **7.** $2/[15(3xy + 1)^4] - y/x$ **9.** $-y/x$

11. $\dfrac{\frac{1}{2}(x + y)^{-1/2} - \frac{2}{3}x(x^2 + y)^{-2/3}}{\frac{1}{3}(x^2 + y)^{-2/3} - \frac{1}{2}(x + y)^{-1/2}}$ **13.** x/y **15.** y is not a function of x (since $\sin w$ can never equal 2)

17. $\cos x \cos y/(1 + \sin x \sin y)$ **19.** $(5x^4 - 2xy^2 - y)/(x + 2x^2 y)$ **21.** $-x/y$ **23.** $3x/5y$

25. vertical tangent at $(0, 1)$; horizontal tangent at $(1, 0)$

27. no vertical or horizontal tangents (on the curve $xy = 1$, $x \neq 0$ and $y \neq 0$)

29. no horizontal tangents; vertical tangents at $(-a, 0)$ and $(a, 0)$

31. horizontal tangents at $(\pi/2 + k\pi, (-1)^k)$ where k is an integer; no vertical tangents   **33.** $y = -x + 2$

Problems 2.7, page 176

1. $0; 0$ **3.** $8; 0$ **5.** $-\frac{1}{4}x^{-3/2}; \frac{3}{8}x^{-5/2}$ **7.** $6x^{-4}; -24x^{-5}$ **9.** $2a; 0$ **11.** $30(x+1)^{-7}; -210(x+1)^{-8}$
13. $-(1-x^2)^{-3/2}; -3x(1-x^2)^{-5/2}$ **15.** $-4x^2 \sin x^2 + 2 \cos x^2; -12x \sin x^2 - 8x^3 \cos x^2$ **17.** $2 \sec^2 x \tan x;$
$2 \sec^4 x + 4 \sec^2 x \tan^2 x$ **19.** $\csc^3 x + \csc x \cot^2 x = 2 \csc^3 x - \csc x; -6 \csc^3 x \cot x + \csc x \cot x$ **21.** 50
23. a. $x'(t) > 0$ if $0 < t < 2$ or $t > 8$ **b.** $t > 5$ **c.** $t > 8$ **d.** $2 < t < 5$
25. $\dfrac{1}{14}\left[\dfrac{1}{(x+7)^2} - \dfrac{1}{(x-7)^2}\right]; \dfrac{1}{14}\left[\dfrac{2}{(x-7)^3} - \dfrac{2}{(x+7)^3}\right]; \dfrac{1}{14}\left[\dfrac{6}{(x+7)^4} - \dfrac{6}{(x-7)^4}\right]$
27. $(uv)'' = u''v + 2u'v' + uv''; (uv)''' = u'''v + 3u''v' + 3u'v'' + uv'''$

Problems 2 Review, page 178

1. 3 **3.** $-(1+y)/(1+x)$ **5.** $6x\sqrt{x+1} + (3x^2+1)/(2\sqrt{x+1})$ **7.** $-(y/x)^{1/4}$
9. $\dfrac{1}{3}\left(\dfrac{x+2}{x-3}\right)^{-2/3}\left[\dfrac{-5}{(x-3)^2}\right]$ **11.** $\dfrac{-x^4 + 4x^3 - 4x^2 - 12x - 3}{(x^3 + 5x - 6)^2}$ **13.** $\dfrac{5x^2(x^3+4)^{6/7}}{2(x^3-3)^{1/6}} + \dfrac{18x^2(x^3-3)^{5/6}}{7(x^3+4)^{1/7}}$
15. $\dfrac{1 + \frac{1}{2}(x+y)^{-1/2}}{\frac{1}{3}y^{-2/3} - \frac{1}{2}(x+y)^{-1/2}}$ **17.** $3\sqrt{2}x^2(1+x^3)^{\sqrt{2}-1}$ **19.** $\dfrac{2x^2(x^5+3)^{3/4}(x^3-1)^{-1/3} - \frac{15}{4}x^4(x^3-1)^{2/3}(x^5+3)^{-1/4}}{(x^5+3)^{3/2}}$
21. $3x^2 \cos x^3$ **23.** $\tan^2 x \sec x + \sec^3 x$ **25.** $y = 2x + 2; y = -\frac{1}{2}x + \frac{9}{2}$ **27.** $y = -\frac{183}{8}x + \frac{327}{8};$
$y = \frac{8}{183}x + \frac{3286}{183}$ **29.** $y = -(3\sqrt{3}/4\pi)x + (\pi/4\sqrt{3}) + \frac{1}{2}; y = (4\pi/3\sqrt{3})x - (4\pi^3/27\sqrt{3}) + \frac{1}{2}$ **31.** $42x^5 - 210x^4 + 6x;$
$210x^4 - 840x^3 + 6$ **33.** $2(1+x)^{-3}; -6(1+x)^{-4}$ **35.** $(-48x^2 + 80)/(x^2+5)^3; (192x^3 - 960x)/(x^2+5)^4$
37. $(-x^2 \cos x + 2x \sin x + 2 \cos x)/x^3; (x^3 \sin x + 3x^2 \cos x - 6x \sin x - 6 \cos x)/x^4$

Computer Exercises, page 179

1. $\dfrac{d^n}{dx^n} \tan x$ can be written as a polynomial in $\tan x$ of degree $n+1$.
3. $f^{(n)}(0) = 0$ if n is odd; $f^{(n)}(0) = (-1)^{n/2}n!$ if n is even **5.** $(-0.450184, -0.232466)$

CHAPTER 3

Problems 3.1, page 183

1. $-\frac{10}{3}$ **3.** $\frac{8}{3}$ ft/sec **5.** 5 ft/sec **7.** $-11550/\sqrt{9000} \approx 121.75$ ft/sec
9. decreasing at $125/\sqrt{3625} \approx 2.08$ km/hr **11.** $1/400\pi \approx 0.0008$ m/min **13.** 50π cm/sec **15.** $-\frac{2}{9}$ cm/hr
17. decreasing at $\dfrac{3}{5}\dfrac{(0.01)(18)}{0.3} = 0.36$ m³/sec **19.** $\dfrac{dx}{dt} = \dfrac{dx}{d\theta}\dfrac{d\theta}{dt} = -13 \sin \theta \dfrac{d\theta}{dt} = -100\pi$ cm/min;

$\dfrac{dy}{dt} = 240\pi$ cm/min [*Note:* 1 rpm = 2π rad/min] **21. a.** $\sigma_z \approx 66,728$ lb/ft² **b.** $\dfrac{d\sigma_z}{dt} \approx -26,226$ lb/ft²/hr

23. $3; 75/\sqrt{17}$ **25.** $\dfrac{dH}{dt} = -\dfrac{1500}{99^{3/2}} \approx -1.52$ m/sec

Problems 3.2, page 196

1. a. increasing on $(-\frac{1}{2}, \infty)$ **b.** decreasing on $(-\infty, -\frac{1}{2})$ **c.** $-\frac{1}{2}$ **d.** y-intercept -30; x-intercepts $-6, 5$
e. global minimum of $-\frac{121}{4}$ at $x = -\frac{1}{2}$; no local maximum **f.**

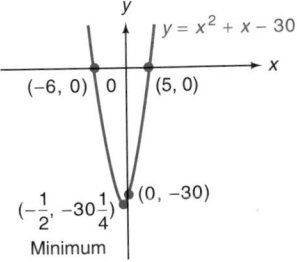

3. a. increasing on $(0, \infty)$ **b.** decreasing on $(-\infty, 0)$ **c.** 0 **d.** y-intercept -8; x-intercepts $\pm\sqrt{2}$
e. global minimum of -8 at $x = 0$; no local maximum **f.**

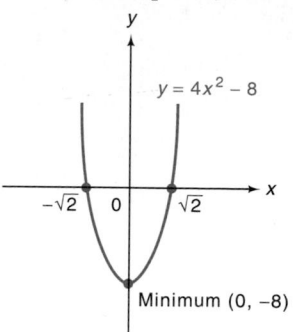

5. a. increasing on $(-1, \infty)$ **b.** decreasing on $(-\infty, -1)$ **c.** -1 **d.** y-intercept 6; no x-intercept
e. global minimum of 4 when $x = -1$; no local maximum **f.**

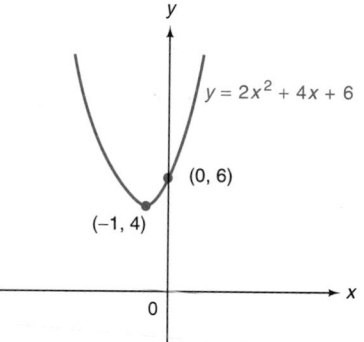

7. a. always increasing **b.** never decreasing **c.** no critical points **d.** y-intercept 0; x-intercept 0
e. no local minimum or local maximum **f.**

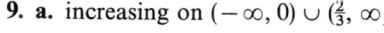

9. a. increasing on $(-\infty, 0) \cup (\frac{2}{3}, \infty)$

b. decreasing on $(0, \frac{2}{3})$ **c.** $0, \frac{2}{3}$ **d.** y-intercept 0; x-intercepts 0, 1 **e.** local maximum of 0 at $x = 0$;
local minimum of $-\frac{4}{27}$ at $x = \frac{2}{3}$ **f.**

11. a. increasing on $(-3, 0) \cup (3, \infty)$

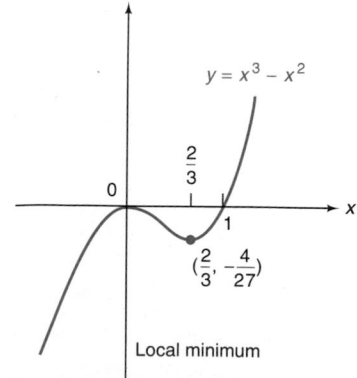

b. decreasing on $(-\infty, -3) \cup (0, 3)$ **c.** $-3, 0, 3$ **d.** y-intercept 0; x-intercepts $\pm 3\sqrt{2}$ **e.** local maximum of 0 at $x = 0$; local minimum of -81 at $x = -3$ and $x = 3$ **f.**   **13. a.** increasing on $(-\frac{7}{2}, \infty)$

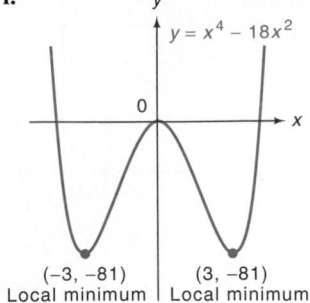

$y = x^4 - 18x^2$

$(-3, -81)$ Local minimum | $(3, -81)$ Local minimum

b. decreasing on $(-\infty, -\frac{7}{2})$ **c.** $-\frac{7}{2}$ **d.** y-intercept 7; x-intercept $-\frac{7}{2}$ **e.** no local maximum; global minimum of 0 at $x = -\frac{7}{2}$
f.   **15. a.** always increasing   **b.** never decreasing   **c.** -1 **d.** y-intercept 1; x-intercept -1

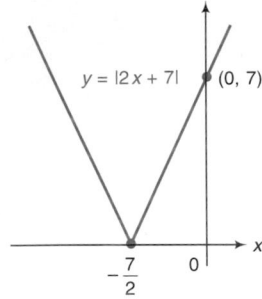

$y = |2x + 7|$ $(0, 7)$

$-\frac{7}{2}$ 0

e. no local maximum or minimum   **f.**

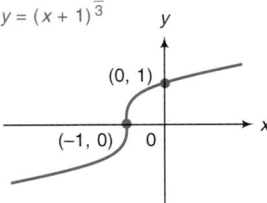

$y = (x + 1)^{\frac{1}{3}}$

$(0, 1)$

$(-1, 0)$ 0

17. a. increasing on $(-\pi/8 + k\pi/2, \pi/8 + k\pi/2)$ where k is an integer   **b.** decreasing on $(\pi/8 + k\pi/2, 3\pi/8 + k\pi/2)$
c. $\pi/8 + k\pi/4$ **d.** y-intercept 0; x-intercepts $k\pi/4$ **e.** local maximum of 1 at $\pi/8 + k\pi/2$ where k is an even integer;
local minimum of -1 at $\pi/8 + k\pi/2$ where k is an odd integer   **f.**

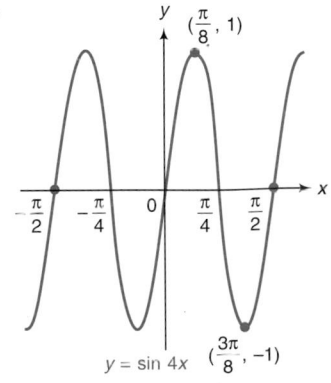

$(\frac{\pi}{8}, 1)$

$-\frac{\pi}{2}$ $-\frac{\pi}{4}$ 0 $\frac{\pi}{4}$ $\frac{\pi}{2}$

$y = \sin 4x$ $(\frac{3\pi}{8}, -1)$

19. c. y-intercept 1; x-intercept -1

d. vertical line $x = -1$ is tangent at $(-1, 0)$ because $f(x) \to 0$ and $f'(x) \to \infty$ as $x \to -1^+$

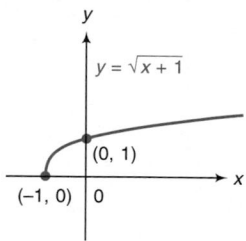

21. b. $\lim_{x \to \pm\infty} f(x) = 1$ **c.** $\lim_{x \to -1^-} f(x) = \infty$, $\lim_{x \to -1^+} f(x) = -\infty$

f.

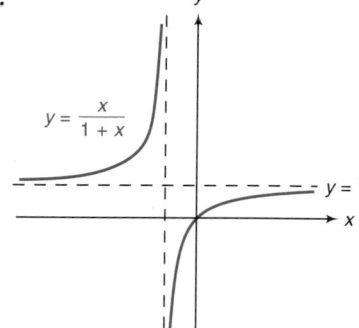

23.

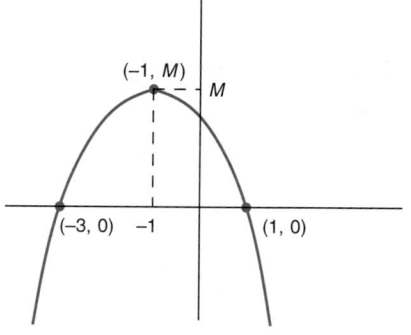

Problems 3.3, page 203

1.

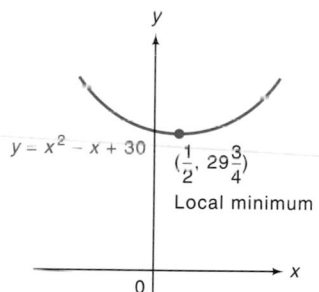

3.

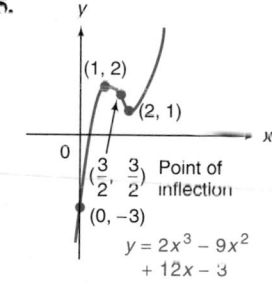

5.

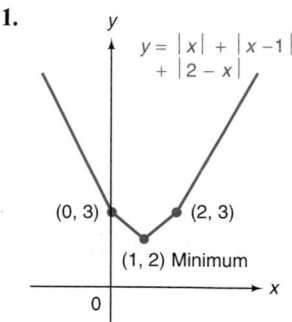

7.

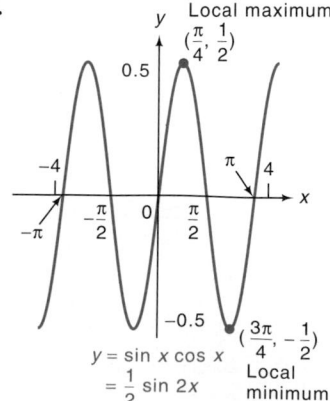

9.

11.

13.

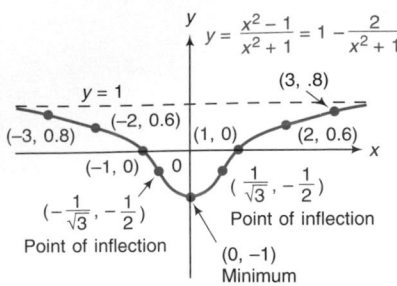

15.

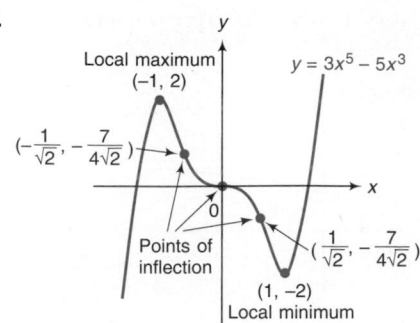

17.

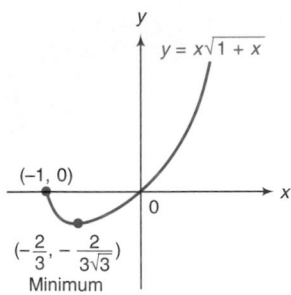

19.

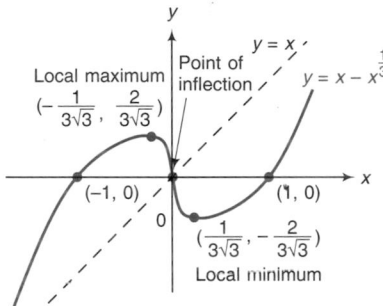

21.

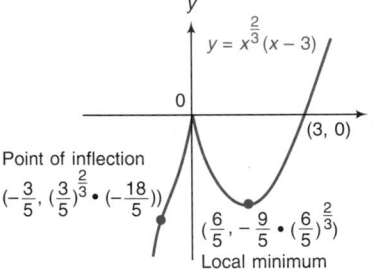

23.

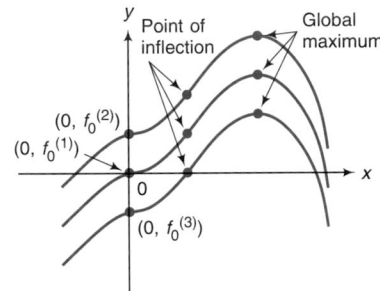

25. positive

Problems 3.4, page 211

1.

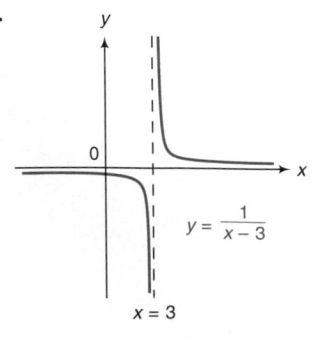

3.

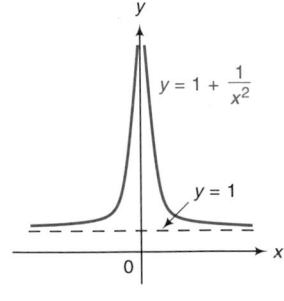

5.

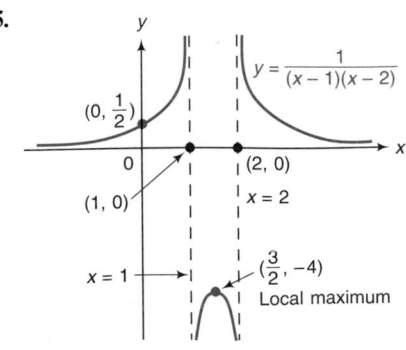

7.

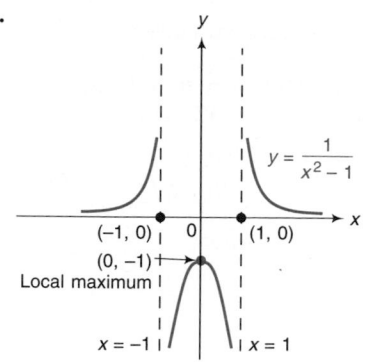

9.

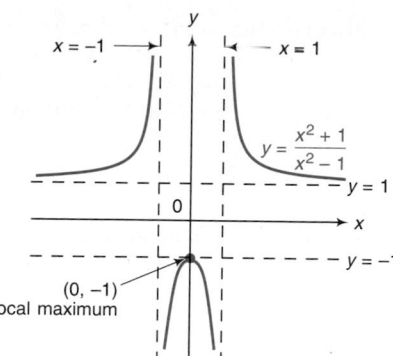

11.

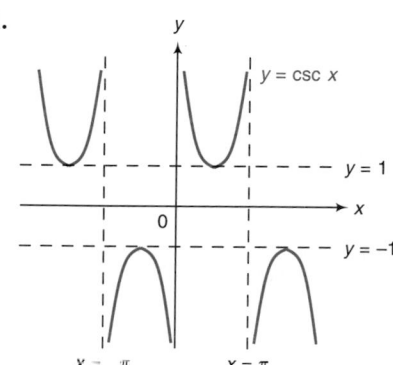

13.

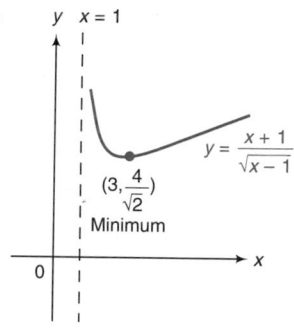

15.

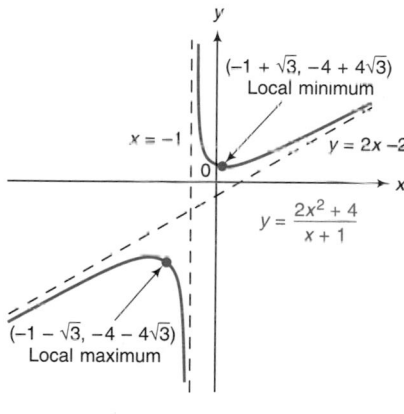

17.

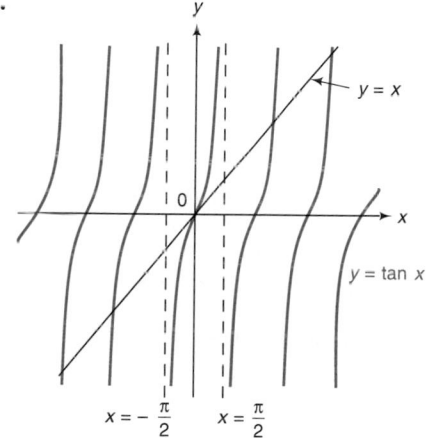

19. symmetric with respect to the y-axis because $(-x) \cdot \sin(-x) = x \cdot \sin x$.
21. *Hint:* $x + (\sin x)/x^2$ is a counterexample to the assertion.

Problems 3.5, page 223

1. minimum -870 at $x = -10$; maximum 890 at $x = 10$ **3.** minimum -1 at $x = -2$; maximum 2 at $x = 7$
5. minimum -150 at $x = 5$; maximum 106 at $x = -3$ **7.** minimum -2 at $x = -31$; maximum 2 at $x = 33$
9. minimum 0 at $x = 1$; maximum $\frac{4}{5}$ at $x = -3$ **11.** minimum 0 at $x = 0$ and $x = 2$; maximum 8 at $x = -2$
13. minimum $-\sqrt{2}$ at $x = 5\pi/4$; maximum $\sqrt{2}$ at $x = \pi/4$ **15.** minimum $\cos(\frac{5}{3}) \approx -0.096$ at $x = -2$;
maximum $\cos(0) = 1$ at $x = \frac{1}{2}$ **17.** minimum 0 at $x = 0$; maximum 16 at $x = -4$
19. $2 + 2y' = 0$ and $A' = y + xy' = 0$ imply $y' = -1$ and $x = y$; maximal A is $(1000/4)^2 = 62{,}500$ yd^2
21. Length = width = perimeter$/4 = \frac{300}{4} = 75$ m   **23.** $(\frac{76}{25}, -\frac{18}{25})$ **25. a.** 48 ft/sec   **b.** 1.5 sec   **c.** 36 ft

29. $h = 2r; r = \left(\dfrac{25}{\pi}\right)^{1/3} \approx 1.9965$ cm.   **31.** maximum $= \pi(35/2\pi)^2 \approx 97.5$ cm^2 when all wire used for circle;

minimum ≈ 42.88 cm^2 when radius of circle is $35/[2(\pi + 4)] \approx 2.45$ cm.   **33.** semicircle is $60/(4 + \pi)$ ft in diameter; rectangle is $30/(4 + \pi)$ ft high   **35.** point R should be 0.75 km from P on the shoreline to Q **37.** $\frac{40}{7}$ ft

39. $\pi/2$ (a right angle)   **41. b.** $\theta \approx 1.508$ radians $\approx 86.4°$ $(\cos \theta = \frac{1}{16})$ **43.** $\sin \theta_1 = \sin 38° \left(\dfrac{300{,}000}{220{,}400}\right)$; $\theta_1 \approx 56.93°$

49. If $D = (a_1 - x)^2 + (a_2 - x)^2 + \cdots + (a_n - x)^2$, then D is minimized when $x = (a_1 + a_2 + \cdots + a_n)/n$.

Problems 3.6, page 237

1. $[9, 10]$ **3.** 9.486832981 **5.** 2.236067977 **7.** $0.8074175964, 6.192582404$
9. 7.984792473 approximates the single real solution   **11. a.**   **b.** 0.7390851332

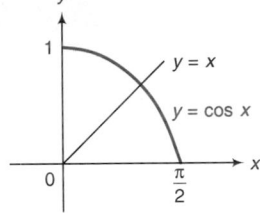

13. 1.036447266 **15.** $(x + \frac{5}{2})^2 + \frac{3}{4} > 0$; condition 1 is not satisfied
17. On any interval including $x_0 = 1$ and a zero of $f(x)$ the hypotheses of either theorem do not hold since $f'(x) = 0$ at at least one point in each such interval.

Problems 3.7, page 245

1. 0 **3.** $\frac{9}{4}$ **5.** $\sqrt{\frac{7}{3}}$ **7.** $(3 - 4\sqrt{3})/3 \approx -1.31$ **9.** $c \approx 0.69$ rad $(\sin c = 2/\pi)$

11. $c \approx 0.15$ rad $(\cos 2c = 3/\pi)$ **13.** $\epsilon_n = \left(\dfrac{1}{n}\right)^{1/(n-1)}$ **19. c.** $|x - 1|$ is not differentiable at $x = 1$.   **37.** $0 < b < \frac{1}{2}$

Problems 3.8, page 254

1. $1 + 6(x - 1); 1.6; [30(1.1^4)/2](1.1 - 1)^2 = 0.219615$ **3.** $2 + \frac{1}{4}(x - 4); 2.05; (\frac{1}{4}4^{-3/2}/2)(4.2 - 4)^2 = 0.000625$
5. $1 + 2(x - 0); 1.1; (4 \cdot 0.95^{-3}/2)(0.05 - 0)^2 \approx 0.0058$ [Without a calculator use $0.81^{-3/2}$.]   **7.** $1 + 0(x - 0) = 1; 1;$
$\frac{1}{2}(0.1 - 0)^2 = 0.005$ **9.** $0 + (-1)(x - \pi); 0.05; (|-\sin x|/2)((\pi - 0.05) - \pi)^2 \leq \frac{1}{2}(-0.05)^2 = 0.00125$ [On $(\pi - 0.05, \pi)$,

$|\sin x| \leq \left|\sin \dfrac{5\pi}{6}\right| = \dfrac{1}{2}$ so we can use $\frac{1}{2}$ instead of 1 to get the better estimate   error < 0.000625.]

15. a. $1 + 6(x - 1) + 15(x - 1)^2$ **b.** 1.75 **c.** $20(1.1)^3(0.1)^3 = 0.02662$ **17. a.** $2 + \frac{1}{4}(x - 4) - \frac{1}{64}(x - 4)^2$ **b.** 2.049375

c. $\dfrac{3}{8 \cdot 32}\left(\dfrac{0.2^3}{6}\right) = 0.000015625$ **19. a.** $1 + 2x + 2x^2$ **b.** 1.105 **c.** $\dfrac{12}{(0.95)^4}\left[\dfrac{(0.05)^3}{6}\right] \approx 0.000307$ **21. a.** $1 - x^2/2$

b. 0.995 **c.** $\dfrac{\sin 0.1}{6}(0.1)^3 \approx 0.0000166$ **23. a.** $-(x - \pi)$ **b.** 0.05 **c.** $\dfrac{0.05^3}{6} \approx 0.0000208$

Problems 3.9, page 260

11. $y = 3x^4 - 4x^3 - 12x^2 + 3$

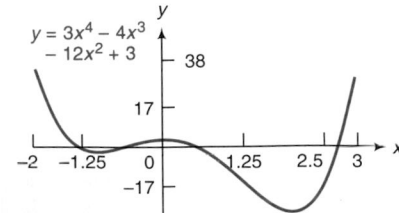

Problems 3 Review, page 261

1. $\sqrt{34}/2 \approx 2.915$ m/sec

3.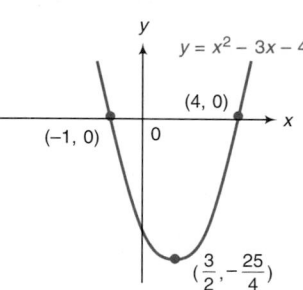

5.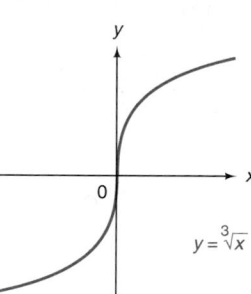

7.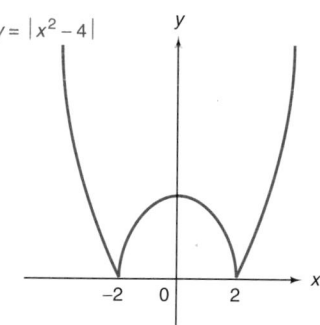

9.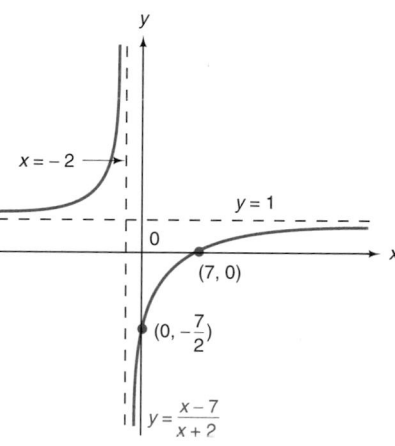

11.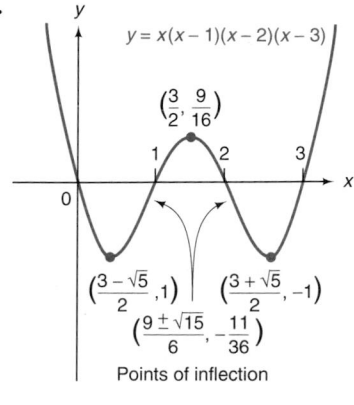

13.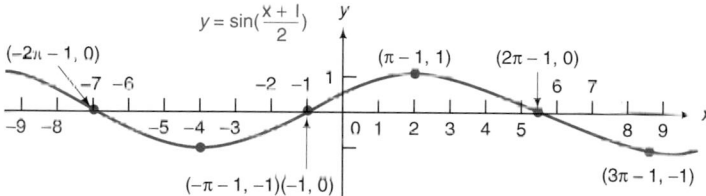

15.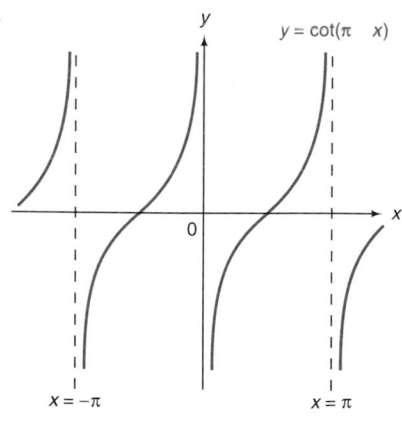

17.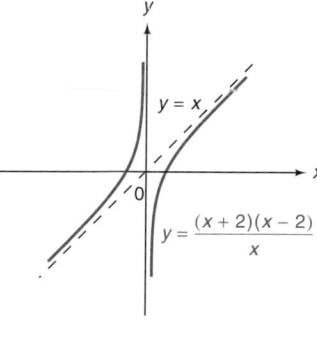

19. a. min is -10 at $x = 1$; max is 358 at $x = 5$ **b.** min is $-2^{1/3}$ at $x = 0$; max is $2^{1/3}$ at $x = 4$ **c.** min is $-\frac{2}{3}$ at $x = 1$; max is 0 at $x = -1$ **d.** min is $\frac{9}{5}$ at $x = -\frac{1}{5}$; function is unbounded above (i.e., no max) **21.** $40{,}000$ m^2 [use a square]

23. $r = \sqrt[3]{192} = 4\sqrt[3]{3}$; $h = \dfrac{128}{r^2} = \dfrac{8}{3}\sqrt[3]{3}$ **25.** 1.752171779 approximates the single real solution to 10 significant figures

27. $\left(\frac{1}{5}\right)^{1/4} \approx 0.6687$ **29.** $-\left(\frac{31}{5}\right)^{5/4} \approx -9.7834$ **31. a.** x **b.** 0.15 **c.** $\frac{\sin 0.15}{2}(0.15)^2 \approx 0.00168$

33. a. $\frac{1}{2} - \frac{1}{4}(x-2)$ **b.** 0.4875 **c.** $\frac{1}{4}\left(\frac{0.05^2}{2}\right) = 0.0003125$

Computer Exercises, page 263

1. a. $\sqrt{2} = 1.4142135623730950488016887242 10$ **b.** 6 iterations **3. a.** $\sqrt{703} = 26.514147167125704101832863574050$

b. 6 iterations **5.** *Hint:* $|x - \sqrt{a}| = \dfrac{|x^2 - a|}{|x + \sqrt{a}|}$ **7.** $(1.38027, 0.724495)$

9. base $10\frac{1}{2}$ ft, height $9\frac{1}{3}$ ft, bid \$2,567.565 **11. a.** The underwater portion of the pipeline tends to take a due west course.
b. $C'(0)$ is negative for all k. **13.** $x_1 = -1.96838523691318601326$
$x_2 = 0.77207792743967325093$
$x_3 = 4.925185020602793 22476$
Note: This computation was done on *Mathematica*, which gives the complex roots at the same time. These are
$x_4, x_5 = 0.13556114443535976879 \pm 0.61840355022780490593i$

CHAPTER 4

Problems 4.2, page 274

1. $-5x + C$ **3.** $5x^2 + C$ **5.** $x + x^2 + x^3 + x^4 + C$ **7.** $-\frac{1}{4}x^{-4} + C$ **9.** $14x^{1/2} + C$
11. $-\frac{289}{4}x^{4/17} + \frac{27}{5}x^{5/9} + C$ **13.** $-2\cos x - 3\sin x + C$ **15.** $2x^2 - 7\cos x + C$ **17.** $y = \frac{2}{3}x^3 + x^2 - \frac{28}{3}$
19. $y = \frac{3}{4}x^{4/3} + \frac{1}{2}x^2 - \frac{1}{2}x^{2/3} - \frac{35}{4}$ **21.** $y = \sin x + \frac{7}{2}$ **23.** $s(t) = (2.9t^2 + 0.2t + 25)$ meters **25.** $h = 2000t - 16t^2$
27. $\frac{3}{2}(100) + \frac{2}{3}(1000) - \frac{2}{5}(10)^{5/2} \approx 690.176$ kg **29.** \$120,890; \$2,302,223.33 **31.** unprofitable when $q > 3,333$

Problems 4.3, page 277

1. $\frac{15}{2}$ **3.** -21 **5.** $\frac{10973}{2520}$ **7.** $\sum_{j=0}^{4} 2^j$ **9.** $\sum_{j=2}^{7} \frac{j}{j+1} = \sum_{j=3}^{8} \frac{j-1}{j}$ **11.** $\sum_{j=1}^{8} (2j-1)(2j+1)$

13. $\sum_{j=0}^{5} (-2x)^j$ **15.** $\sum_{j=0}^{7} x^{3j}$ **17.** $\sum_{j=1}^{5} (j/5)^2$ **19.** $\sum_{j=0}^{4} 0.2\sin(0.1 + 0.2j)$ **21.** c **23.** c

27. $\sum_{k=1}^{n} k^3 = \left[\dfrac{n(n+1)}{2}\right]^2$ **29.** $\sum_{k=2}^{n} \dfrac{1}{k^2-1} = \dfrac{3}{4} - \dfrac{n+\frac{1}{2}}{n(n+1)} = \dfrac{(3n+2)(n-1)}{4n(n+1)}$

Problems 4.4, page 290

1. $s = \frac{155}{512} \approx 0.303$; $S = \frac{187}{512} \approx 0.365$ **5.** $28 < A < 84$; $42 < A < 70$; $49 < A < 63$; $A = 56$
7. $12\frac{3}{4} < A < 26\frac{1}{4}$; $16\frac{1}{8} < A < 22\frac{7}{8}$; $17\frac{13}{16} < A < 21\frac{3}{16}$; $A = 19\frac{1}{2}$

In the answers to Problems 9–19, computations used the left-hand endpoint of each subinterval.

9. a. $s_4 = \frac{7}{8}$ **b.** $s_8 = \frac{35}{32}$ **c.** $s_n = \dfrac{2}{n}\sum_{k=0}^{n-1}\dfrac{1}{2}\left(\dfrac{2k}{n}\right)^2 = \dfrac{4}{n^3}\sum_{k=0}^{n-1} k^2 = \dfrac{2(n-1)(2n-1)}{3n^2}$ **d.** $A = \frac{4}{3}$

11. a. $s_8 = 1 - (7 \cdot 15)/(6 \cdot 8^2) \approx 0.7266$ **b.** $s_{16} = 1 - (15 \cdot 31)/(6 \cdot 16^2) \approx 0.6973$

c. $s_n = \dfrac{1}{n}\sum_{k=0}^{n-1}\left[1 - \left(\dfrac{k}{n}\right)^2\right] = \left(\dfrac{1}{n}\sum_{k=0}^{n-1} 1\right) - \left(\dfrac{1}{n^3}\sum_{k=0}^{n-1} k^2\right) = 1 - \dfrac{(n-1)(2n-1)}{6n^2}$ **d.** $A = \frac{2}{3}$ **13. a.** $s_8 = (15 \cdot 55)/(6 \cdot 8^2) \approx 2.148$

b. $s_{16} = (31 \cdot 111)/(6 \cdot 16^2) \approx 2.240$ **c.** $s_n = \dfrac{1}{n}\sum_{k=0}^{n-1}\left(1 + \dfrac{k}{n}\right)^2 = 1 + \dfrac{n-1}{n} + \dfrac{(n-1)(2n-1)}{6n^2}$ **d.** $A = \frac{7}{3}$

15. a. $s_4 = (5^4 \cdot 3^2)/4^3 \approx 87.89$ **b.** $s_8 = (5^4 \cdot 7^2)/(4 \cdot 8^2) \approx 119.63$ **c.** $s_n = \dfrac{5}{n}\sum_{k=0}^{n-1}\left(\dfrac{5k}{n}\right)^3 = \left(\dfrac{5}{n}\right)^4\sum_{k=0}^{n-1} k^3 = \dfrac{625}{4}\left(\dfrac{n-1}{n}\right)^2$

d. $A = \dfrac{5^4}{4} = 156.25$ **17. a.** $s_6 = (5 \cdot 29)/(6 \cdot 36) \approx 0.6713$ **b.** $s_{12} = (11 \cdot 59)/(6 \cdot 144) \approx 0.7512$

c. $s_n = \dfrac{1}{n}\sum_{k=0}^{n-1}\left[\left(\dfrac{k}{n}\right)+\left(\dfrac{k}{n}\right)^2\right] = \dfrac{n-1}{2n}+\dfrac{(n-1)(2n-1)}{6n^2}$ d. $A = \frac{5}{6}$ 19. a. $s_4 = (7\cdot 11)/(2\cdot 4^2) = 2.40625$

b. $s_8 = (15\cdot 23)/(2\cdot 8^2) \approx 2.695$ c. $s_n = \dfrac{1}{n}\sum_{k=0}^{n-1}\left[1+2\left(\dfrac{k}{n}\right)+3\left(\dfrac{k}{n}\right)^2\right] = 1+\dfrac{n-1}{n}+\dfrac{(n-1)(2n-1)}{2n^2}$ d. $A = 3$

21. a. $a, a+\dfrac{b-a}{n}, a+2\left(\dfrac{b-a}{n}\right), \ldots, a+(n-1)\left(\dfrac{b-a}{n}\right), b$ b. $f(x_i^*) = \left[a+i\left(\dfrac{b-a}{n}\right)\right]^2$

c. $\sum_{i=1}^{n}\left[a^2+2a(b-a)\dfrac{i}{n}+(b-a)^2\dfrac{i^2}{n^2}\right]\left[\dfrac{b-a}{n}\right] = a^2(b-a)+\dfrac{a(b-a)^3(n+1)}{n}+\dfrac{(b-a)^2(n+1)(2n+1)}{6n^2}$ d. $A = \dfrac{b^3-a^3}{3}$

23. $s_{10} = 8.10656, s_{20} = 7.22666 \ [A = 6.4]$ 25. $s_{10} = 5.97, s_{20} = 5.8175 \ [A = \frac{17}{3} \approx 5.6667]$

27. $s_{10} \approx 18.115746; s_{20} = 17.728902 \ [A = \frac{52}{3} \approx 17.33333]$ 29. $s_{10} \approx 0.7865659, s_{20} \approx 0.7391767 \ [A = \ln 2 \approx 0.6931472]$

39. $A = \frac{1}{6}$ 41. $A = \frac{17}{12}$ 43. $A = \frac{2}{3}$ 45. $A = 2-\sqrt{2}$

47. a. S_n approximates the area of one-quarter of a circle with radius 1 and centered at the origin.   b. $\frac{1}{4}\pi 1^2 = \pi/4 \approx 0.785398$

Problems 4.5, Page 306

1. $12 \le \int \le 24$ 3. $0 \le \int \le 2$ 5. $\dfrac{99}{100} \le \int \le 99$ 7. $0 \le \int \le \dfrac{\pi}{2}$ 9. $(-5)^3 \le \int \le (-4)^3$ 11. 56

13. $37\frac{1}{2}$ 15. 10   17. $\frac{8}{3}$ 19. 0   21. $\frac{2}{3}$ 23. $\frac{25}{12}$ 25. $-\frac{1}{2}$ 27. 1   29. $(b^3-a^3)/3$

31. Use Theorem 7 and the fact that $s^{17} < s^{55/3}$ for $23 \le s \le 47$ 33. $\int_1^2 x\,dx$

35. *Additional hint:* $\sqrt{1+x^3}$ is an increasing function; thus $\sqrt{1+x^3} \le \sqrt{\frac{9}{8}}$ on $[0, \frac{1}{2}]$ and $\sqrt{1+x^3} \le \sqrt{2}$ on $[\frac{1}{2}, 1]$.

37. a. 128 ft/sec   b. 256 ft   c. 5 sec   43. a. and b. are false, c. is true.   45. $\frac{1}{2}(2b-[b]-1)[b]-\frac{1}{2}(2a-[a]-1)[a]$

Problems 4.6, page 317

1. $\frac{33}{5}$ 3. $\frac{26}{3}$ 5. $\frac{333}{4}$ 7. $2-\sqrt{2}$ 9. 0   11. $\frac{41}{6}$ 13. $\frac{1}{30}$ 15. 18   17. $\frac{80}{3}$ 19. 17

21. 36   23. $\frac{32}{3}$ 25. $(b-a)^3/6$ 27. $\frac{37}{12}$ 29. $\frac{1056}{5}$ 31. $2\sqrt{2}$ 33. $F'(x) = 1/(1+x^3); F'(2) = \frac{1}{9}$

35. $F'(x) = \sqrt{(x-1)/(x+1)}; F'(1) = 0$ 37. $F'(x) = 3/(1+3x); F'(1) = \frac{3}{4}$

39. Each integral equals 1. The integrals compute the areas of four congruent regions (the same region translated three times).   41. a. 10^6 m/sec^2 b. 55 m

43. 12,000 individuals added from the first day of the 9th week through the last day of the 24th week

45. $900   47. $f(a) = 0$ 49. C   51. $\frac{1}{3}$ 53. 2   55. $\frac{19}{3}$

57. $-400/\sqrt{800/g} = -80$ ft/sec (using $g = 32$ ft/sec^2)   59. $45

Problems 4.7, page 326

1. $3\,dx$ 3. $4x^3\,dx$ 5. $8x(1+x^2)^3\,dx$ 7. $(x/\sqrt{1+x^2})\,dx$ 9. $-(\sin\sqrt{x}/2\sqrt{x})\,dx$ 11. $\frac{1}{6}(1+x^2)^6 + C$

13. 20   15. $\frac{3}{2}$ 17. $\frac{74}{3}$ 19. $2(10^{3/2}-1)/27$ 21. $\frac{5}{72}(1+3x^4)^{6/5} + C$ 23. $\frac{2}{15}(s^5+5s)^{3/2} + C$

25. $\sqrt{20}-\sqrt{3}$ 27. $-(1+\sqrt{x})^{-2} + C$ 29. $-\frac{3}{16}[1+(1/v^2)]^{8/3} + C$ 31. $\frac{1}{3}(ax^2+2bx+c)^{3/2} + C$

33. $\frac{1}{3}(2^{3/2}-1)\alpha^3$ 35. $\dfrac{2}{nb}\sqrt{a+bs^n} + C$ 37. 0   39. $\frac{1}{2}$ 41. $-2\cos\sqrt{x} + C$ 43. $\frac{2}{3}(27-2\sqrt{2})$

45. $\frac{2}{3}(10^{3/2}-1)$ 47. $\frac{3}{2}$ 49. $\frac{2}{3}$ m   51. $8   53. $\frac{5}{38}$ kg/m

Problems 4.8, page 331

1. $5\cos(5x)$ 3. $-2\dfrac{1+2x}{\sqrt{1+(1+2x)^2}}$ 5. $(\cos x + \sin x - 10)\cos x \sin x$

Problems 4 Review, page 332

1. $f'(x) = x^3/\sqrt{x^2+17}; f'(8) = \frac{512}{9}$ 3. $F'(s) = 2(1+4s^2)^{2001}; F'(0) = 2$ 5. $\frac{1}{3}\,dx$ 7. $2x\,dx$

9. $\frac{1}{6}x^6 + C$ 11. $-\frac{32}{3}$ 13. $\frac{3}{4}$ 15. $\frac{4}{3}(t^3+7)^{1/4} + C$ 17. $\frac{37}{2}$ 19. $\frac{233}{6}$ 21. $\frac{9}{2}$ 23. $\frac{3}{10}$

25. a. 118.5 m   b. $\dfrac{118.5}{15} = 7.9$ m/sec

Computer Exercises, page 333

1. a. $\left[\dfrac{i-1}{n}, \dfrac{i}{n}\right]$ **b.** 0.553928, 0.593344, 0.597924   **c.** 0.638075, 0.610173, 0.606339   **e.** $0.597924 < A < 0.606339$, so the average of the numbers $= 0.602132$; this approximates the area with an error less than 0.0042.

3. The zeros of the polynomial $(p(x))$ are $\pm\sqrt{3}$ and $\sqrt[3]{2}$ so $A = \int_{-\sqrt{3}}^{\sqrt[3]{2}} p(x)\,dx - \int_{\sqrt[3]{2}}^{\sqrt{3}} p(x)\,dx \approx 14.50595612$

5. $\displaystyle\int \cos^2 x\,dx = \dfrac{x}{2} + \dfrac{\sin 2x}{4} + C = \dfrac{x}{2} + \dfrac{\sin x \cos x}{2} + C;$ $\displaystyle\int \cos^4 x\,dx = \dfrac{3x}{8} + \sin x\left(\dfrac{\cos^3 x}{4} + \dfrac{3}{8}\cos x\right) + C$

$\displaystyle\int \cos^6 x\,dx = \dfrac{5x}{16} + \sin x\left(\dfrac{\cos^5 x}{6} + \dfrac{5}{24}\cos^3 x + \dfrac{5}{16}\cos x\right) + C;$ $\displaystyle\int \cos^n x\,dx = ax + \sin x\, p(\cos x) + C$ where p is a polynomial of degree $n-1$.

CHAPTER 5

Problems 5.1, page 338

1. $\frac{1}{6}$ **3.** $\frac{343}{6}$ **5.** $\frac{125}{24}$ **7.** $\frac{1}{12} + \frac{34}{3} = \frac{137}{12}$ **9.** $\frac{32}{3}$ **11.** $\frac{126}{5} - 2\sqrt{5}$ **13.** $\frac{1}{2}$ (2 pieces)   **15.** 1

17. $\frac{125}{6}$ **19.** $\frac{27}{5}$ **21.** $\frac{128}{3}$ **23.** $\left(\dfrac{a}{12} + \dfrac{b}{2}\right) + \sqrt{a^2 + 4ab}\left(\dfrac{1}{12} + \dfrac{b}{3a}\right)$

$\left[\displaystyle\int_0^{x^*} (ax + b - ax^2)\,dx \text{ where } x^* = \dfrac{a + \sqrt{a^2 + 4ab}}{2a}\right]$ **25.** $\frac{7}{2}$ **27.** 49   **29.** $y = (1 - 2^{-2/3})a^2$

Problems 5.2, page 348

1. $8\pi/3$ **3.** $4\pi/5$ **5.** 16π **7.** 117π **9.** 16π **11.** 9π **13.** $\pi/4$ **15.** $3\pi/10$ **17.** $64\pi/15$
19. $4\pi r^3/3$ **21.** $2\pi^2 ar^2$ **23.** $68\pi/15$; $13\pi/6$ **25.** $\frac{112000}{3}\pi$ kg   **27.** $\frac{128}{3}$
29. $(\pi/3)(b_1 + b_2 + b_3)|(a_2 - a_1)(b_3 - b_1) - (b_2 - b_1)(a_3 - a_1)|$

Problems 5.3, page 354

1. $\displaystyle\int_0^1 \sqrt{1 + x^2}\,dx$ **3.** $\displaystyle\int_0^\pi \sqrt{1 + \sin^2 x}\,dx$ **5.** $4\sqrt{10}$ **7.** $(13^{3/2} - 8)/27$ **9.** $\frac{14}{3}$ **11.** $\frac{10}{3}$
13. $1 + 2/(3a^3)$ **15.** 6   **17.** 1.1 [The exact answer is $\frac{1}{2}(\sqrt{2} + \ln(1 + \sqrt{2})) \approx 1.147794$ (ln x is defined in Chapter 6)
19. 3.8 [To 6 decimal places the answer is 3.820198.]   **21. b.** Look at the definition of a *radian*.
23. a. Reflect one curve through the line $y = x$ to get the other.   **b.** $x = 2\sqrt{u}$

Problems 5.4, page 364

1. 120 ft-lb   **3.** 27 J   **5.** $\frac{4}{3}$ m   **7.** $(9.81)(1000)(25)(50)\pi$ J $\approx 3.85 \times 10^7$ J

9. $\approx 9.63 \times 10^6$ J (integrate from 0 to 5)   **11.** $(1000\pi)(9.81)\displaystyle\int_0^9 \left(3 - \dfrac{1}{3}x\right)^2 (17 - x)\,dx \approx 12{,}273{,}676$ J

13. 150 ft-lb $\left(= 1.5\displaystyle\int_0^{10} (15 - x)\,dx\right)$ **15.** 449,280 ft-lb   **17.** 45.38 sec (2 hp $= 2 \cdot 550$ ft-lb/sec)   **19.** 337.5 J
21. $\sqrt{140{,}000/3} \approx 216$ ft/sec ≈ 147 mi/hr   **23.** $v_0 = \sqrt{2gd} = \sqrt{2(3.92)(3{,}430{,}000)} \approx 5.186$ km/sec ≈ 3.222 mi/sec
25. KE $= mgx_0$; $v_{\text{final}} = \sqrt{2gx_0}$

Problems 5.5, page 368

1. 42 g·cm; 4.2 cm   **3.** 101 g·cm; $\frac{101}{18}$ cm   **5. a.** $\frac{1}{2}$ **b.** $\frac{1}{3}$ **c.** $\frac{2}{3}$ **7. a.** 1 **b.** 0 **c.** 0   **9. a.** $\frac{45}{4}$ **b.** $\frac{381}{7}$
c. $\frac{508}{105}$ **11.** It is true

Problems 5.6, page 376

1. $(-1.8, 3.4)$ **3.** $(\frac{101}{18}, \frac{3}{2})$ **5.** $(\frac{13}{24}, \frac{49}{24})$ **7.** $(\frac{4}{3}, \frac{4}{3})$ **9.** $(\frac{13}{20}, \frac{77}{50})$ **11.** $(4, \frac{1}{2})$ **13.** $(0, 1)$ **15.** $4\sqrt{2}\pi^2$

Problems 5.7, page 382

1. $72,000(9.81)\pi \approx 2,218,970$ N   **3.** (a) $P_A = 1000(13.6)(9.81)h = 133,416h$ **5.** 975 lb   **7.** 5297.4 N

9. $(1000)(9.81)\int_0^{25/4}\left(\frac{25}{4} - y\right)2\sqrt{25y}\,dy = 2,554,687.4$ N   **11.** $(1000)(9.81)\int_0^{25/4}\left(\frac{25}{4} - y\right)(10\sqrt{y})2\,dy \approx 5,109,378$ N

13. $3(1030)(9.81) = 30,312.9$ N   **15.** $(1000)(9.81)(3\sqrt{\frac{5}{2}})\int_0^1 (\frac{3}{2} - 3x/2)dx \approx 34,900$ N   **17.** $\frac{45}{4}(1000)(9.81) = 110,362.5$ N

Problems 5 Review, page 383

1. $\frac{233}{6}$ **3.** $\frac{9}{2}$ **5.** $\frac{9}{16}$ **7.** $2\sqrt{2}$ **9.** 56π **11.** 20π **13.** $\pi/7$ **15.** 80π **17.** 144   **19.** $5\sqrt{5}$
21. $\frac{2}{3}[2^{3/2} - 1]$ **23.** $\int_0^{\pi/2} \sqrt{1 + (x\cos x + \sin x)^2}\,dx$ **25.** 4.8 J   **27.** $\frac{11500}{3}$ J   **29.** -21 kg·m; $-\frac{7}{5}$
31. $\frac{9}{4}$ **33.** $(\frac{30}{13}, \frac{363}{130})$ **35.** $(-2, \frac{3}{5})$ **37.** 392,400 N   **39.** $(1000)(9.81)\int_0^2 \left[(3x^2 + 6x)/4\right]dx = 49,050$ N
41. (a) 3525.47 N   (b) 13,488.75 N   (c) 28,970.2 N

CHAPTER 6

Problems 6.1, page 394

1. $\mathbb{R}$; $x = (y - 5)/3$; $\dfrac{dx}{dy} = \dfrac{1}{3}$ **3.** $\mathbb{R} - \{0\}$; $x = 1/y$; $\dfrac{dx}{dy} = -x^2 = -1/y^2$

5. $\left[-\frac{3}{4}, \infty\right)$; $x = (y^2 - 3)/4$; $\dfrac{dx}{dy} = \dfrac{1}{2}\sqrt{4x + 3} = y/2$ **7.** $\mathbb{R}$; $x = \sqrt[3]{1 - y}$; $\dfrac{dx}{dy} = -\dfrac{1}{3x^2} = -\dfrac{1}{3}(1 - y)^{-2/3}$

9. $\mathbb{R} - \{0\}$; $x = \dfrac{1}{y - 1}$; $\dfrac{dx}{dy} = -x^2 = \dfrac{1}{(y - 1)^2}$ **11.** $\left(-\infty, \frac{5}{2}\right]$ and $\left[\frac{5}{2}, \infty\right)$;

$x_1 = (5 - \sqrt{1 + 4y})/2$, $x_2 = (5 + \sqrt{1 + 4y})/2$; $\dfrac{dx_1}{dy} = 1/(2x_1 - 5) = -1/\sqrt{1 + 4y}$, $\dfrac{dx_2}{dy} = 1/(2x_2 - 5) = 1/\sqrt{1 + 4y}$

13. $\mathbb{R}$; from the cubic formula (which few know but which you can find in the *CRC Handbook of Tables for Mathematics*)

$x = \sqrt[3]{\dfrac{y}{2} + \sqrt{\dfrac{y^2}{4} + \dfrac{1}{27}}} + \sqrt[3]{\dfrac{y}{2} - \sqrt{\dfrac{y^2}{4} + \dfrac{1}{27}}}$; $\dfrac{dx}{dy} = 1/(1 + 3x^2)$

15. $[0, \pi]$; $x = \cos^{-1} y$ (which we define formally in Section 6.8); $\dfrac{dx}{dy} = -1/\sin x = -1\sqrt{1 - y^2}$

17. $\left(-\dfrac{\pi}{2}, \dfrac{\pi}{2}\right)$; $x = \tan^{-1} y$; $\dfrac{dx}{dy} = \dfrac{1}{\sec^2 x} = \dfrac{1}{1 + \tan^2 x} = \dfrac{1}{1 + y^2}$

19.

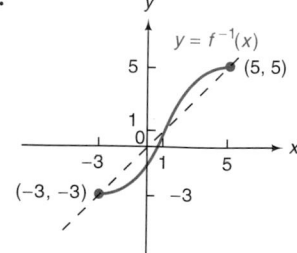

21.

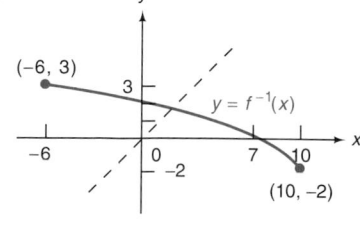

PROBLEMS 6.2, page 405

1. 2   **3.** -3 **5.** -4π **7.** 5   **9.** $-\frac{1}{2}$ **11.** -1 **13.** $\pm\sqrt{6}$ **15.** $\sqrt{2}$ **17.** e^π **19.** 4
21. $\frac{1}{2}$ **23.** $10^{10^{-23}}$ **25.** $\frac{27}{16}$ **27.** 24   **29.** $\sqrt[3]{25/2}$ **31.** 3   **33.** $\frac{7}{2}\log x$ **35.** $5\ln x - 18\ln(1 + x)$

37.

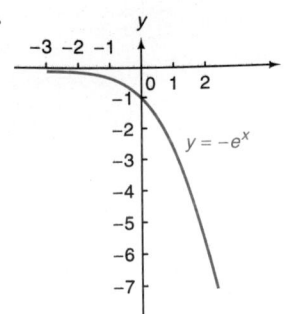

$y = -e^x$

39.

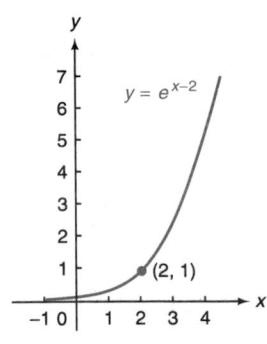

$y = e^{x-2}$ (2, 1)

41.

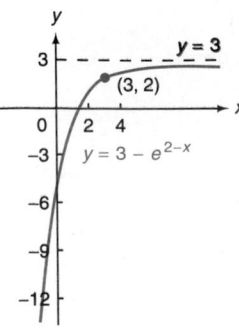

$y = 3$ (3, 2) $y = 3 - e^{2-x}$

43.

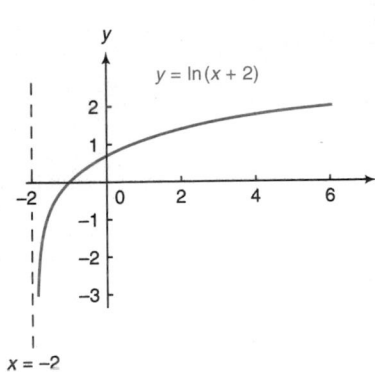

$y = \ln(x + 2)$ $x = -2$

45.

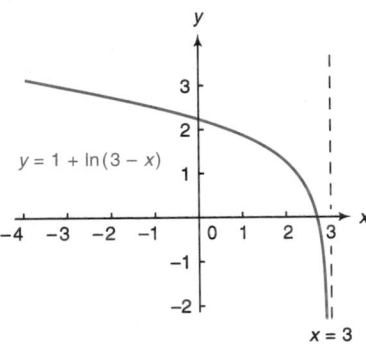

$y = 1 + \ln(3 - x)$ $x = 3$

47. a. $\text{pH} = -\log(4.2 \cdot 10^{-6}) = 6 - \log 4.2 \approx 5.37675.$ **b.** $\text{pH} = -\log(8 \cdot 10^{-6}) = 6 - \log 8 \approx 5.09691.$
c. $\text{pH} = -\log(0.6 \cdot 10^{-7}) = 7 - \log 0.6 \approx 7.22185.$ **49. (a)** $L_2/L_1 = 1 + (10 \log 2)/L_1$ if $I_2 = 2I_1$
(b) $I_2/I_1 = I_1/I_0 = I_1 \times 10^{12}$ if $L_2 = 2L_1$ **(c)** $I_2/I_1 = (I_1/I_0)^5 = I_1^5 \times 10^{60}$ if $L_2 = 6L_1$

Problems 6.3, page 417

1. $5 \ln|x|$ **3.** $\frac{7}{2} \ln x$ **5.** $\frac{1}{3} \ln(x^2 + 3) + \frac{1}{8} \ln(x^5 - 9)$ **7.** $(0, \infty)$; $1/x$ **9.** $(-\frac{1}{2}, \infty)$; $2/(1 + 2x)$
11. $(-\infty, -3) \cup (2, \infty)$; $1/(x - 2) + 1/(x + 3) = (2x + 1)/(x^2 + x - 6)$ **13.** $\mathbb{R} - \{-1, 0\}$; $1/(x + 1) - 1/x = -1/x(x + 1)$
15. $(2k\pi, (1 + 2k)\pi)$; $\cot x$ **17.** $(0, \infty)$; $\ln x + 1$ **19.** $\mathbb{R} - \{-1, 1\}$; $1/(x + 1) - 1/(x - 1) = -2/(x^2 - 1)$

21. $\sqrt[5]{\dfrac{x^3 - 3}{x^2 + 1}} \left(\dfrac{1}{5}\right) \left(\dfrac{3x^2}{x^3 - 3} - \dfrac{2x}{1 + x^2}\right)$ **23.** $\sqrt{x}\sqrt[3]{x + 2}\sqrt[5]{x - 1} \left[\dfrac{1}{2x} + \dfrac{1}{3(x + 2)} + \dfrac{1}{5(x - 1)}\right]$ **25.** $\ln(1 + x^2) + C$

27. $\frac{1}{3} \ln|7 + x^3| + C$ **29.** $\ln|x^2 - 1| + C$ **31.** $\dfrac{4}{x}$ **33.** $-2 \tan x$

35. $\ln 0.8 \approx -0.22314349$; $\ln 1.2 \approx 0.18232154$

37.

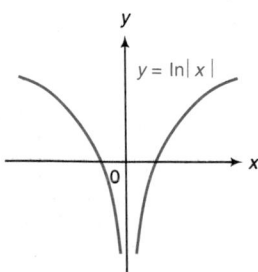

$y = \ln|x|$

39.

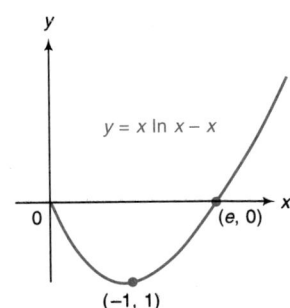

$y = x \ln x - x$ (e, 0) (−1, 1)

41. $\ln 125 = 3 \ln 5$

43. $\frac{1}{2}(\ln^2 4 - \ln^2 2) = \frac{3}{2} \ln^2 2$ **45.** $\ln x$

Problems 6.4, page 430

1. 5 **3.** -2.718 **5.** $4! = 24$ **7.** 3 **9.** $2 - \pi$ **11.** $2e^{2x+1}$ **13.** $\frac{1}{2}\sqrt{e^x} = \frac{1}{2}e^{x/2}$
15. $-e^{-x}(\cos x + \sin x)$ **17.** $(x + 1)e^x$ **19.** xe^x **21.** $(1/x) + 1$ **23.** $4 - e$ **25.** $\frac{1}{3}e^{3x} + C$
27. $\frac{1}{2}e^{2x+1} + C$ **29.** $-\cos e^x + C$ **31.** $-e^{1/x} + C$ **33.** $2e^2 - 2e$ **35.** $\frac{1}{8}(5 + e^{x^2})^4 + C$
37. a. 1.138828376 **b.** 0.690730947 **c.** $0.000000007; 0.000003383$
39.

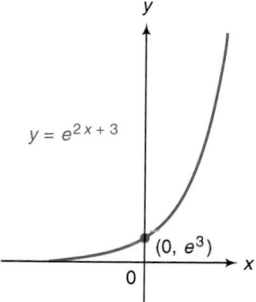

$y = e^{2x+3}$
$(0, e^3)$

41.

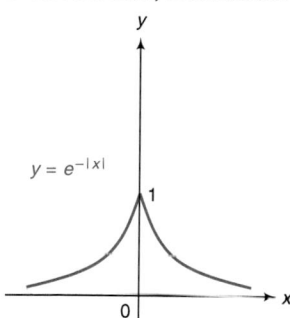

$y = e^{-|x|}$

43.

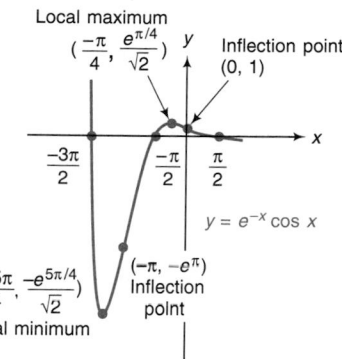

Local maximum
$\left(\frac{-\pi}{4}, \frac{e^{\pi/4}}{\sqrt{2}}\right)$ Inflection point $(0, 1)$
$\frac{-3\pi}{2}$ $\frac{-\pi}{2}$ $\frac{\pi}{2}$
$y = e^{-x}\cos x$
$\left(\frac{-5\pi}{4}, \frac{-e^{5\pi/4}}{\sqrt{2}}\right)$ $(-\pi, -e^\pi)$ Inflection point
Global minimum

45. $6 - e$ **47.** $(e^2 - e^{-2})/2$ **49.** $0.01 + 1596 \cdot e^{-200} \approx 0.01$ m/min^2 $[e^{-200} \approx 1.38 \times 10^{-87}]$

Problems 6.5, page 437

1. 4 **3.** -1 **5.** -5 **7.** $\frac{1}{2}$ **9.** $1 \cdot 2 \cdot 3 = 6$ **11.** $10^{10^{-23}}$ **13.** 4 **15.** 72
17. $\sqrt[3]{25/8} = 25^{1/3}/2$ **19.** $5^x \cdot \ln 5$ **21.** $3 \cdot 2^{3x} \cdot \ln 2$ **23.** $(\log e)/x$
25. $-2/[(3 - 2x)\ln \pi] = -2\log_\pi e/(3 - 2x)$ **27.** $(1 + \frac{1}{2}\ln x)x^{\sqrt{x}-1/2}$ **29.** $(\frac{1}{2})^x/\ln(\frac{1}{2}) + C$
31. $\pi^x/\ln \pi + C$ **33.** 2.292029674 **35.** -2.714065192 **37.** 0.605511561 **39.** 0.219054924
41. -2.965647273 **43.** $999^{1000} > 1000^{999}$ **45.**

$y = \pi^{x-1}$
$(0, \frac{1}{\pi})$

47.

$y = \log(x/10)$

49. a. $L_2/L_1 = 1 + (10\log 2)/L_1$ if $l_2 = 2l_1$ **b.** $l_2/l_1 = l_1/l_0 = 10^{12}l_1$ if $L_2 = 2L_1$ **c.** $l_2/l_1 = (l_1/l_0)^5 = 10^{60}l_1^5$ if $L_2 = 6L_1$
55. c. $e^x x^{-e}(1 - e/x)$ **57.** equality only when $x = y = \frac{1}{2}$

Problems 6.6, page 450

1. 62,500 after 20 days; 156,250 after 30 days **3. a.** 400 **b.** 10,780,782 **5. a.** 119°F
b. $\frac{1}{2}(\ln 0.2)/(\ln 0.7) \approx 2.256$ hr **7.** $5580(\ln 0.7)/(\ln 0.5) \approx 2871$ yr **9. a.** $2,800 **b.** $8,590.93 **c.** $8,739.13
d. $8,753.36 **11.** $(\ln 3)/15 \approx 0.07324 = 7.324\%$; $(\ln 2)/10 \approx 0.06931 = 6.931\%$
13. $\beta = (\frac{1}{1500})\ln(845.6/1013.25) \approx -1.205812 \cdot 10^{-4}$ **a.** 625.526 mbar **b.** 303.416 mbar **c.** 429.033 mbar
d. 348.632 mbar **e.** 57,396.3 m **15. a.** $25(0.6)^{24/10} \approx 7.3367$ kg **b.** $10(\ln 0.02)/(\ln 0.6) \approx 76.58$ hr **17.** 2036
19. a. 281,323,227 **b.** 2053 **21.** 10.83% **23.** 2.71% **25.** the president

Problems 6.7, page 454

1. $\frac{1}{2}\sin 2x + C$ **3.** $\frac{3}{4}$ **5.** $-e^{\cos x} + C$ **7.** $\ln(1 + \sin x) + C$ **9.** $\frac{3}{2}\ln 2$
11. $-\ln|\sin(2 - x)| + C = \ln|\csc(2 - x)| + C$ **13.** $\frac{3}{16}$ **15.** $\frac{1}{3}\sin^3 x + C$ **17.** $1/3\sqrt{2} = \sqrt{2}/6$
19. $\frac{3}{4}(\sin 2x)^{2/3} + C$ **21.** $\frac{1}{4}\sec 4x + C$ **23.** 1 **25.** $-\frac{1}{2}\csc(x^2) + C$ **27.** $2/\sqrt{\csc x} + C = 2\sqrt{\sin x} + C$

29. 3   **31.** $\ln|\tan x| + C$ **33.** $\frac{1}{2}(x - \sin x) + C$ **35.** $\pi/4$ **37.** $\ln|\sec x + \tan x| + C$
39. a. $\ln|1 + \sin x| + C$ **b.** $-\ln|\cos x| + C$ **41.** $\frac{1}{2}\ln(\frac{3}{2})$ **43.** 1   **45.** $3\pi/2$ **47. a.** $e^x(\sin x + \cos x)$
b. $e^x(\cos x - \sin x)$ **c.** $2e^x \cos x$ **d.** $2e^x \sin x$ **e.** $\frac{1}{2}e^x(\sin x - \cos x) + C$ **f.** $\frac{1}{2}e^x(\sin x + \cos x) + C$
49. The two indefinite integrals are essentially the same since $\sin^2 x$ and $-\cos^2 x$ differ only by 1.

Problems 6.8, page 462

1. $\pi/3$ **3.** $-\pi/6$ **5.** $\pi/6$ **7.** $-\pi/6$ **9.** $\frac{4}{5}$ **11.** $\frac{3}{4}$ **13.** $-5/\sqrt{26}$ **15.** $\sqrt{1 - x^2}$
17. $x/\sqrt{1 - x^2}$ **19.** $3/\sqrt{1 - 9x^2}$ **21.** $-1/\sqrt{1 - (x - 5)^2}$ **23.** $2/(4 + x^2)$ **25.** $1/(2\sqrt{x - x^2})$
27. $1 + x^2 > 1$ if $x \neq 0$; therefore, $\sin^{-1}(1 + x^2)$ is undefined except for $x = 0$ and no derivative exists.
29. $-3x^2/\sqrt{1 - (x^3 + 1)^2}$. This is only defined when $-1 < x < 0$.   **31.** $(\sin x)/\sqrt{1 - x^2} + (\sin^{-1} x) \cos x$
33. $2\sqrt{1 - x^2}$ **35.** $\tan^{-1} 2$ **37.** $\tan^{-1}(x - 1) + C$ **39.** $-\frac{1}{2}\tan^{-1}(\frac{1}{2}e^{-x}) + C$ **41.** $\tan^{-1}(\sin x) + C$
43. $\pi/6$ **45.** $\frac{1}{3}\sin^{-1}(3x) + C$ **47.** $\cos^{-1}(\frac{1}{2}\cos x) + C = -\sin^{-1}(\frac{1}{2}\cos x) + C_1$ **49.** $-\frac{1}{2}(\cos^{-1} x)^2 + C$
51. $\pi/4$ **53.** $\pi/6$ **55.** $-\frac{30}{136} \approx -0.22$ radians per minute   **57.** $3 \cdot 2 \cdot (\pi/9) = 2\pi/3 \approx 2.09$ km/sec
59. 0.894665817   **61.** -1.047197551 **63.** 0.667457216   **65.** 2.765209171   **67.** 1.570786327

73. It is true.   **77. b.** $\{x: |x| \leq 4\}$ is the domain of each   **c.** $g'(x) = \dfrac{2}{\sqrt{16 - x^2}}$;

$f'(x) = \dfrac{x}{|x|} \cdot \dfrac{2}{\sqrt{16 - x^2}} = \pm g'(x)[= g'(x)$ for $x > 0$ and $-g'(x)$ for $x < 0]$ **d.** $f(x) = g(x)$ for $0 \leq x \leq 4$

85. b. $A'(x) = B'(x)$ for $x \neq 0$ **c.** $A(x) - B(x) = \begin{cases} \pi & \text{if } x < 0 \\ \text{undefined} & \text{if } x = 0 \\ 0 & \text{if } x > 0 \end{cases}$

89. a. $\pi/3$ **b.** $2\pi/3$ **c.** $\pi/4$ **d.** $\pi/6$ **e.** 0   **f.** π **91.**

95. a. $\pi/6$ **b.** $-\pi/6$ **c.** $\pi/4$ **d.** $\pi/3$ **e.** $\pi/2$ **f.** $-\pi/2$ **97.** $\sec^{-1} x + \csc^{-1} x = \pi/2$ **99.** $-e^x/(1 + e^{2x})$

101. $\dfrac{4}{|4x + 2|\sqrt{(4x + 2)^2 - 1}}$ **103.** $-1/(2x\sqrt{x - 1})$, $x > 1$ **105.** $\dfrac{-\csc^2 x}{|\cot x|\sqrt{\cot^2 x - 1}}$
107. $\tan^{-1}(x - 3) + C = \cot^{-1}(3 - x) + C$ **109.** $\pi/6$
111. $|1/a| \sec^{-1}|x/a| + C = |1/a| \cos^{-1}|a/x| + C$

Problems 6.9, page 470

1. $\cosh x = \sqrt{29}/5$; $\tanh x = -2/\sqrt{29}$; $\coth x = -\sqrt{29}/2$; $\text{sech } x = 5/\sqrt{29}$; $\text{csch } x = -\frac{5}{2}$ **3.** $4\cosh(4x + 2)$

5. $\cosh x \cos(\sinh x)$ **7.** $-\text{sech}^2(1/x)/x^2$ **9.** $\dfrac{\text{sech}^2(\tan^{-1} x)}{1 + x^2}$

11. $\sin^{-1}(\cosh x)$ is defined only for $x = 0$ (since $\cosh x > 1$ otherwise); the derivative does not exist.
13. $\frac{1}{2}\cosh 2x + C$ **15.** $\frac{1}{2}\ln|1 + \sinh 2x| + C$ **17.** $2\tanh \sqrt{x} + C$ **19.** $\ln|\sinh x| + C$ **21.** $-\cosh(1/x) + C$
23. $2a^2 \sinh 1$

Problems 6.10, page 473

1. $\mathbb{R}; 3/\sqrt{(3x+2)^2+1}$ **3.** $1/[x(1-\ln^2 x)], 1/e < x < e$ **5.** $1/[2\sqrt{(x^2+1)}\sinh^{-1}x], x > 0$ **7.** $\mathbb{R}; x/\sqrt{x^2+1}$

9. $\dfrac{1}{2}\cosh^{-1}\left(\dfrac{x^2}{2}\right) + C = \dfrac{1}{2}\ln(x^2 + \sqrt{x^4-4}) + C_1$ **11.** $\dfrac{1}{4}\tanh^{-1}\left(\dfrac{x^2}{2}\right) + C = \dfrac{1}{8}\ln\left|\dfrac{2+x^2}{2-x^2}\right| + C$

13. $\sinh^{-1}(e^x) + C = \ln(e^x + \sqrt{e^{2x}+1}) + C$ **15.** $\sinh^{-1}(\sin x) + C = \ln(\sin x + \sqrt{1+\sin^2 x}) + C$

17. $\cosh^{-1}\left(\dfrac{\ln x}{5}\right) + C = \ln|\ln x + \sqrt{\ln^2 x - 25}| + C_1$

Problems 6 Review, page 475

1. 2   **3.** $\frac{1}{2}$ **5.** $10^{10^{-2}} = 10^{0.01}$ **7.** $\sqrt{51}/10$ **9.** $\sqrt{13}/3$ **11.** $\sqrt{40}/7$

13. y increases by $3\ln 2 = \ln 8 \approx 2.079$ **15.** $2x/(1+x^2)$ **17.** $2(x+1)e^{(x+1)^2}$

19. $[1 + 1/(x+3)]/[x + \ln(x+3)]$ **21.** $2^{x+5} \cdot \ln 2$ **23.** $-2x/(x^4+1)$ **25.** $-\csc\sqrt{x}\cot\sqrt{x}/(2\sqrt{x})$

27. $-\sec x$ **29.** $1/[3x^{2/3}(1+x^{2/3})]$ **31.** $-e^{-x}/\sqrt{1-e^{-2x}}$ **33.** $(2x+3)\cdot\cosh(x^2+3x)$

35. $-1/(2x\sqrt{1+x})$ **37.** $\frac{3}{2}\ln(1+x^2) + C$ **39.** $\frac{1}{2}e^{-1/x^2} + C$ **41.** $4^{\ln x}/\ln 4 + C$ **43.** $-5\ln(\sqrt{3}/2) \approx 0.72$

45. $\frac{3}{2}\tan^{-1}x^2 + C$ **47.** $\frac{1}{6}\sin^2 3x + C = -\frac{1}{6}\cos^2 3x + C_1$ **49.** $\ln|\sinh x| + C$ **51.** $\tanh(\ln x) + C$

53. $\cosh^{-1}(\frac{1}{2}\tan x) + C = \ln|\tan x + \sqrt{\tan^2 x - 4}| + C_1$ **55.** $[\frac{5}{2}, \infty]; x = \frac{1}{2}(5+y^2); \dfrac{dx}{dy} = y$

57. $(-\infty, \frac{3}{4}); x = \frac{1}{4}(3-e^y); \dfrac{dx}{dy} = -e^y/4$ **59.** $((k-\frac{1}{2})\pi, (k+\frac{1}{2})\pi); x = 2\tan^{-1}y; \dfrac{dx}{dy} = 2/(1+y^2)$

61. $250,000e^{0.6} \approx 455,530$ in 1980; $250,000e^{1.8} \approx 1,512,412$ in 2000   **63. a.** 54.85°C **b.** 67.57 min
65. 3.106 weeks; 6.213 weeks; 13.425 weeks   **67.** $16,103.24; $16,160.74
69.

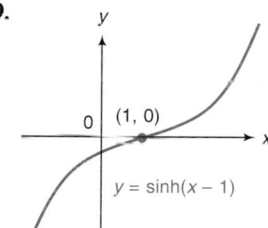
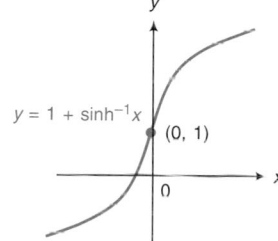

Computer Exercises, page 476

1. a. $-\ln|x-1| + \ln|x-2| - \ln\left|\dfrac{x-2}{x-1}\right| + C$ **b.** $\dfrac{1}{7}[\ln|x-4| - \ln|x+3|] = \dfrac{1}{7}\ln\left|\dfrac{x-4}{x+3}\right| + C$

c. $\dfrac{1}{3}[\ln|x-5| - \ln|x-2|] = \dfrac{1}{3}\ln\left|\dfrac{x-5}{x-2}\right| + C$

If $ax^2 + bx + c$ has zeros p and q, then $\displaystyle\int \dfrac{dx}{ax^2+bx+c} = A\ln|x-p| + B\ln|x-q| + C$, where A and B are constants.

3. Local maximum at (0.631939089, −0.728617821) Local minimum at (2.484341346, −1.730751533) Global maximum at (41.8076371, 10.21610066)   **5.** 5.03613 times a year. This is rounded to 5 times a year.

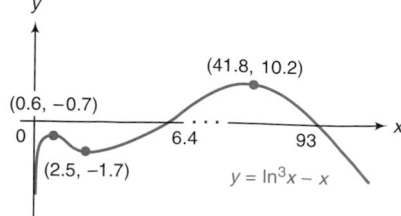

CHAPTER 7

Problems 7.2, page 485

1. $\frac{1}{3}xe^{3x} - \frac{1}{9}e^{3x} + C$ **3.** $-(x^2 + 2x + 2)e^{-x} + C$ **5.** $2\ln 2 - \frac{3}{4}$ **7.** $x\cosh x - \sinh x + C$

9. $\frac{16}{15}$ **11.** $\frac{3}{4}\sinh 2 - \frac{1}{2}\cosh 2$ **13.** $(x/2)[\sin(\ln x) + \cos(\ln x)] + C$ **15.** $\frac{1}{3}$

17. $\frac{1}{2}e^x(\cos x + \sin x) + C$ **19.** $-\dfrac{\sin^4 x \cos x}{5} - \dfrac{4\sin^2 x \cos x}{15} - \dfrac{8\cos x}{15} + C = -\cos x + \dfrac{2}{3}\cos^3 x - \dfrac{1}{5}\cos^5 x + C$

21. $\pi/4\sqrt{2} + 1 - 1/\sqrt{2}$ **23.** $\pi/4 - \frac{1}{2}\ln 2$ **25.** $\frac{1}{2}(1 + x^2)\tan^{-1} x - x/2 + C$

27. $e^{ax}(a\cos bx + b\sin bx)/(a^2 + b^2) + C$ **29.** $\frac{3}{5}\sinh 2x \sinh 3x - \frac{2}{5}\cosh 2x \cosh 3x + C$

31. $x\cos^{-1} x - \sqrt{1 - x^2} + \pi$ $[\sin(\cos^{-1} x) = \sqrt{1 - x^2}]$ **33.** $\pi^2/4$ **35.** $\pi(e - 2)$ **37.** $\pi/2$ **39.** $\pi(9\ln 3 - 4)$

41. $\pi(\pi/\sqrt{2} - 2)$ **43. a.**   **b.** $2 - 2/e$ **45.** $e^x(x^4 - 4x^3 + 12x^2 - 24x + 24)$

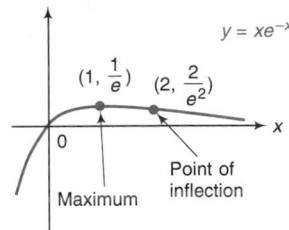

47. $\frac{1}{4}\tan x \sec^3 x + \frac{3}{8}\tan x \sec x + \frac{3}{8}\ln|\sec x + \tan x| + C$

51. The constants of integration for the two integrals are different.   **57. a.** $G_{n+1}(x) = -x^n e^{-x} + nG_n(x)$ if $n > 0$

b. $G_1(x) = 1 - e^{-x}$; $G_2(x) = 1 - (1 + x)e^{-x}$; $G_3(x) = 2 - (2 + 2x + x^2)e^{-x}$ **59. a.** $C_0(x) = x - \pi/2$; $C_1(x) = \ln|\sin x|$

b. $C_n(x) = -\dfrac{1}{n-1}\cot^{n-1}(x) - C_{n-2}(x)$ **c.** $C_2(x) = -\cot x - (x - \pi/2)$; $C_3(x) = -\frac{1}{2}\cot^2 x - \ln|\sin x|$

61. $J_n(x) = \dfrac{2}{2n - k + 2}x^{n-k+1}\sqrt{x^k + 1} - \dfrac{2(n - k + 1)}{2n - k + 2}J_{n-k}(x)$

Problems 7.3, page 490

1. $\frac{4}{15}$ **3.** $\frac{2}{3}\sin^{3/2} x - \frac{2}{7}\sin^{7/2} x + C$ **5.** $-\frac{1}{3}\cos^3 x + \frac{2}{5}\cos^5 x - \frac{1}{7}\cos^7 x + C$

7. $\frac{1}{3}\sin^3 x - \frac{2}{5}\sin^5 x + \frac{1}{7}\sin^7 x + C$ **9.** $\pi/8 + \frac{1}{4}$ **11.** $\frac{3}{8}x + \frac{1}{8}\sin 4x + \frac{1}{64}\sin 8x + C$

13. $\dfrac{x}{16} - \dfrac{\sin 4x}{64} - \dfrac{\sin^3 2x}{48} + C$ **15.** $\frac{1}{2} - \pi/8$ **17.** $\frac{4}{3}$ **19.** $\frac{1}{6}\tan^6 x + \frac{1}{4}\tan^4 x + C$ **21.** $\frac{7}{6}$

23. $\frac{1}{2}\cos x - \frac{1}{10}\cos 5x + C$ **25.** $\dfrac{-1}{\sqrt{2}}\cos\left(\dfrac{x}{\sqrt{2}}\right) - \dfrac{1}{3\sqrt{2}}\cos\left(\dfrac{3x}{\sqrt{2}}\right) + C$ **27.** $\frac{1}{180}\sin 90x + \frac{1}{220}\sin 110x + C$

29. $-\cot x - x + C$ **31.** $\frac{4}{3}$ **33.** $3^4/8 + 3^5/10 = \frac{1377}{40}$ **35.** $-\frac{1}{2}\csc x \cot x - \frac{1}{2}\ln|\csc x - \cot x| + C$

37. a. $\frac{1}{2}(\sin x + \cos x)^2 + C$ **b.** $\int \cos^2 x\,dx = \frac{1}{2}(x + \sin x \cos x) + C$; $\int \sin^2 x\,dx = \frac{1}{2}(x - \sin x \cos x) + C$
$= \sin x \cos x + C$

Problems 7.4, page 500

1. $\dfrac{\pi}{6}$ **3.** $-\sqrt{9 - x^2} + C$ **5.** $\pi/3 + \sqrt{3}/2$ **7.** $5(\sqrt{2} - 1)$ **9.** $\sqrt{x^2 - 4} + C$ **11.** $x/(36\sqrt{36 + x^2}) + C$

13. $\frac{1}{2}\ln(36 + x^2) + C$ **15.** $\pi + 9\sqrt{3}/4$ **17.** $2\sqrt{2} - 2\ln(1 + \sqrt{2})$ **19.** $\dfrac{x\sqrt{x^2 - 36}}{2} + 18\ln|x + \sqrt{x^2 - 36}| + C$

21. $\dfrac{1}{9}\left[\dfrac{4}{\sqrt{7}} - \dfrac{5}{4}\right]$ **23.** $\left|\dfrac{-1}{a}\right|\ln\left|\dfrac{a + \sqrt{x^2 + a^2}}{x}\right| + C$ **25.** $\frac{1}{2}(x\sqrt{a^2 + x^2} + a^2\ln|x + \sqrt{a^2 + x^2}|) + C$

27. $x - a\tan^{-1}(x/a) + C$ **29.** $\dfrac{-1}{a^4}\left[\dfrac{a^2}{2x^2} + \ln\left|\dfrac{x}{\sqrt{x^2 + a^2}}\right|\right] + C$ **31.** $\frac{1}{3}(a^2 - x^2)^{3/2} - a^2\sqrt{a^2 - x^2} + C$

33. $\sqrt{5x^2 + 9} - 3 \ln\left|\dfrac{\sqrt{5x^2 + 9} + 3}{\sqrt{5}x}\right| + C$ **35.** $-\dfrac{\sqrt{a^2 - x^2}}{2a^2x^2} - \dfrac{1}{2|a|^3} \ln\left|\dfrac{|a| + \sqrt{a^2 - x^2}}{x}\right| + C$

37. $\sqrt{9 - 5x^2} - 3 \ln|(3 + \sqrt{9 - 5x^2})/(\sqrt{5}x)| + C$ **39.** $\sqrt{5x^2 - 9} - 3 \cos^{-1}(3/|x|\sqrt{5}) + C$ **41.** $x/(a^2\sqrt{a^2 - x^2}) + C$

43. $\left(1 + \dfrac{a^2}{2x^2}\right)\sqrt{x^2 - a^2} - \dfrac{3}{2}|a| \sec^{-1}\left|\dfrac{x}{a}\right| + C$ **45.** $\frac{1}{5}(x^2 - 49)^{5/2} + \frac{49}{3}(x^2 - 49)^{3/2} + C$ **47.** $\pi^2/4$ **49.** $\pi/2$

51. $\dfrac{\pi}{2}(\sqrt{2} + \ln(\sqrt{2} + 1))$ **53. a.** the part of the unit circle lying in the first quadrant   **b.** $\pi/4$ **59.** $\frac{1}{2}$

Problems 7.5, page 504

1. $-\frac{1}{2}(x - 1)^{-2} - \frac{1}{3}(x - 1)^{-3} + C$ **3.** 578.7   **5.** $\frac{9}{140}$ **7.** $\frac{2}{5}(1 + x)^{5/2} - \frac{2}{3}(1 + x)^{3/2} + C$

9. $\frac{6}{5}(1 + x)^{5/2} - \frac{14}{3}(1 + x)^{3/2} + 8(1 + x)^{1/2} + C$ **11.** $-\sin^{-1}(e^{-x}) + C$

13. $\dfrac{-2(1 + \cos x)}{1 + \cos x + \sin x} + C = \dfrac{-2 \sin x}{1 - \cos x + \sin x} + C$ **15.** $(2/\sqrt{5}) \tan^{-1}(1/\sqrt{15})$ **17.** $-\sqrt{9 - x^2} + C$ (let $u = 9 - x^2$)

19. $\frac{1}{2} \ln(36 + x^2) + C$ (let $u = 36 + x^2$)   **21.** $(64 - 33\sqrt{3})/15$ (let $u = 4 - x^2$)

23. $-a^2\sqrt{a^2 - x^2} + \frac{1}{3}(a^2 - x^2)^{3/2} + C$ (let $u = a^2 - x^2$)   **25.** $\frac{1}{5}(x^2 - 49)^{5/2} + \frac{49}{3}(x^2 - 49)^{3/2} + C$ (let $u = x^2 - 49$)

27. $\pi(e - 2)$ **29.** $\pi(9 \ln 3 - 4)$ **31.** $\ln|\sec x + \tan x| + C$ **33.** $\ln|x + \sqrt{x^2 + 5}| + C$

Problems 7.6, page 510

1. $\frac{1}{2} \ln|2x - 5| + C$ **3.** $\frac{1}{3} \ln|3x + 11| + C$ **5.** $\frac{1}{3} \tan^{-1}(x/3) + C$ **7.** $\frac{1}{8} \ln|(x - 4)/(x + 4)| + C$

9. $x - 2 \ln|x + 1| + C$ **11.** $2x + 11 \ln|x - 4| + C$ **13.** $\frac{1}{6}x^3 - \frac{3}{4}x^2 + 4x - 12 \ln|x + 3| + C$

15. $(1/2\sqrt{3}) \tan^{-1}(x/2\sqrt{3}) + C$ **17.** $\frac{3}{2} \ln(x^2 + 9) + \frac{5}{3} \tan^{-1}(x/3) + C$ **19.** $\ln|x^2 - 4| - \dfrac{3}{4} \ln\left|\dfrac{x - 2}{x + 2}\right| + C$

21. $\frac{1}{2} \ln|x^2 - 6x + 16| + \dfrac{2}{\sqrt{7}} \tan^{-1}\left(\dfrac{x - 3}{\sqrt{7}}\right) + C$ **23.** $\frac{1}{2}x^2 - 2x + \frac{3}{2} \ln|x^2 + x + 1| + \dfrac{11}{\sqrt{3}} \tan^{-1}\left(\dfrac{2x + 1}{\sqrt{3}}\right) + C$

25. $2 - \ln 4$ **27.** $x - 4\sqrt{x + 2} + 4 \ln(1 + \sqrt{x + 2}) + C$ **29.** $(\cos x + 2)^{-1} + C$ **31.** $\frac{1}{4} - \ln \frac{5}{4}$

33. $12\left[\dfrac{x^{8/12}}{8} - \dfrac{x^{7/12}}{7} + \dfrac{x^{6/12}}{6} - \dfrac{x^{5/12}}{5} + \dfrac{x^{4/12}}{4} - \dfrac{x^{3/12}}{3} + \dfrac{x^{2/12}}{2} - x^{1/12} + \ln(1 + x^{1/12})\right] + C$

35. $6(\frac{1}{7}x^{7/6} - \frac{1}{5}x^{5/6} + \frac{1}{3}x^{3/6} - 5x^{1/6} + 5 \tan^{-1} x^{1/6}) + C$ **37.** $2 \tan^{-1}(2 + \tan(x/2)) + C$

Problems 7.7, page 517

1. $\frac{1}{3} \ln|(x - 4)/(x - 1)| + C$ **3.** $\frac{7}{5} \ln|x + 3| + \frac{8}{5} \ln|x - 7| + C$ **5.** $\dfrac{1}{b - a} \ln\left|\dfrac{x - b}{x - a}\right| + C$

7. $x + 4 \ln|x| - 2 \ln|x + 1| + C$ **9.** $\dfrac{x^3}{3} + x + \frac{1}{2} \ln\left|\dfrac{x - 1}{x + 1}\right| + C$ **11.** $\ln\dfrac{\sqrt{3}}{2} - \ln\dfrac{\sqrt{8}}{3} = \frac{3}{2} \ln 3 - \frac{5}{2} \ln 2$

13. $4 \ln|x - 1| - 14 \ln|x - 2| + 11 \ln|x - 3| + C$ **15.** $2 \ln|x| - \frac{3}{4} \ln|x + 1| - \frac{3}{2}(x - 1)^{-1} - \frac{5}{4} \ln|x - 1| + C$

17. $x^{-1} - \ln|x| + \ln|x - 1| + C$ **19.** $-x^{-1} - \tan^{-1} x + C$ **21.** $\dfrac{1}{4} \ln\left|\dfrac{x + 2}{x}\right| - \dfrac{1}{4x} - \dfrac{1}{4(x + 2)} + C$

23. $\dfrac{1}{4} \ln\left|\dfrac{x^2 - 1}{x^2 + 1}\right| + C$ **25.** $\frac{1}{5} \ln|x + 2| - \frac{1}{10} \ln(x^2 + 1) + \frac{2}{5} \tan^{-1} x + C$

27. $2 \ln(x^2 + 1) + 3 \tan^{-1} x - 2 \ln(x^2 + 2) - (3/\sqrt{2}) \tan^{-1}(x/\sqrt{2}) + C$ **29.** $5 \ln \frac{4}{3}$ **31.** $\tan^{-1} 2 - \pi/4$

33. $\dfrac{1}{2\sqrt{2}} \ln\left|\dfrac{\tan x - 1 + \sqrt{2}}{\tan x - 1 - \sqrt{2}}\right|$ **35.** $\frac{1}{4} \ln|x^4 - 2x^2| + C$ **37.** $\frac{4}{5} \ln|x| + \frac{7}{40} \ln|2x^4 + 5| + C$

Problems 7.8, page 521

1. $\frac{1}{3}\sqrt{16 + 3x^2} + C$
3. $\frac{1}{12} \ln\left|\frac{2x + 3}{2x - 3}\right| + C$ [Note: $|2x - 3| = |3 - 2x|$]
5. $\frac{x}{2} - \frac{3}{4} \ln|2x + 3| + C$

7. $\frac{1}{3} \ln\left|\frac{x}{3 + 2x}\right| + C$
9. $\frac{1}{5}(x - 1)(3 + 2x)^{3/2} + C$
11. $(\frac{1}{3}x^2 - \frac{2}{9}x + \frac{2}{27})e^{3x} + C$

13. $\left(\frac{-1}{7}\right)\left(\sin^2 x + \frac{2}{5}\right)\cos^5 x + C$
15. $\frac{1}{2}\cos x - \frac{1}{14}\cos 7x + C$
17. $\tan^3(x/3) - 3\tan(x/3) + x + C$

19. $\frac{1}{2}x^2 \tan^{-1} x^2 - \frac{1}{4}\ln(x^4 + 1) + C$
21. $\frac{1}{16}[(8x^2 - 1)\cos^{-1}(2x) - 2x\sqrt{1 - 4x^2}] + C$

23. $\frac{1}{2}x\sqrt{16 - 3x^2} + (8/\sqrt{3})\sin^{-1}(\sqrt{3}x/4) + C$
25. $\sqrt{10 + 2x^2} - \sqrt{10}\ln\left|\frac{\sqrt{10} + \sqrt{10 + 2x^2}}{\sqrt{2}x}\right| + C$

27. $\frac{1}{3}\cos^{-1}(3/\sqrt{2}x) + C = \frac{1}{3}\sec^{-1}(\sqrt{2}x/3) + C$
29. $\frac{-x}{32\sqrt{x^2 - 4}} + C$

31. $\frac{x}{4}\sqrt{2x^2 - 5} + \frac{5}{4\sqrt{2}}\ln|\sqrt{2}x + \sqrt{2x^2 - 5}| + C$
33. $\frac{1}{195}e^{4x^3}(4\cos 7x^3 + 7\sin 7x^3) + C$

Problems 7.9, page 532

In Problems 1–9, the answers are given in the order requested in the text. The last two values are the actual error in the trapezoidal and Simpson's approximations, respectively.

1. 0.5; 0.5; 0; 0; 0.5; 0; 0
3. $\frac{11}{32}$; $\frac{1}{3}$; $\frac{1}{96}$; 0; $\frac{1}{3}$; $-\frac{1}{96}$; 0
5. 1.727221905; 1.718318842; 0.0141577; 0.00005899; $e - 1 \approx 1.718281828$; -0.008940076; -0.000037013
7. 0.987115801; 1.000134585; 0.020186; 0.0002075; 1; 0.012884199; -0.000134585
9. 0.9956971321; 0.9999360657; 0.0117435512; 0.0003852527; 1; -0.0043028736; -0.000063934

In Problems 11–21 the trapezoidal approximation is given first.

11. 0.81944828; 0.83040900
13. 0.6590191268; 0.6593301635
15. 1.987795499; 1.994503740
17. 1.488736680; 1.493674110
19. 1.1123324; 1.111446
21. 0.98411199254; 0.9838189135
23. $|y''| = |-2 + 4x^2|e^{-x^2} \le 2$ on $[-1, 1]$ and $|y^{(4)}| = |12 - 48x^2 + 16x^4|e^{-x^2} \le 12$. Thus $|\epsilon_{10}^T| \le (2 \cdot 2^3)/(12 \cdot 10^2) \approx 0.01333$ and $|\epsilon_{10}^S| \le (12 \cdot 2^5)/(180 \cdot 10^4) \approx 0.00021333$.
25. $|y''| = |x(12 + 3x^3)/4(1 + x^3)^{3/2}| \le 1.468$ on $[0, 1]$ and $|y^{(4)}| = 9x^2(-80 + 56x^3 + x^6)/16(1 + x^3)^{7/2}| \le 7.02$. Thus $|\epsilon_{10}^T| \le 1.468/(12 \cdot 10^2) \approx 0.001223$ and $|\epsilon_{10}^S| \le 7.02/(180 \cdot 10^4) \approx 0.0000039$.
27. $|y''| = |e^x(1 + e^x)^{-2}| \le \frac{1}{4}$ on $[0, 1]$ and $|y^{(4)}| = e^x|(1 - 4e^x + e^{2x})(1 + e^x)^{-4}| \le 1/8$. Thus, $|\epsilon_8^T| \le 0.25/(12 \cdot 8^2) = 0.000326$ and $|\epsilon_8^S| \le 0.125/(180 \cdot 8^4) \approx 1.695 \cdot 10^{-7}$.
29. $|y^{(4)}| \le 3$ so we need $(3/\sqrt{2\pi})/(180n^4) \le 0.005$ (for half the integral—using the hint); $n = 2$ will do; we calculate 0.683058104 (the "true" value is 0.682689492 giving an error of 0.000368612)
31. a. On $[0, 50]$ we need $(3/\sqrt{2\pi})50^2/(180n^4) \le 0.05$ or $n \ge 82$; this leads to the approximation $2(0.4999994266) = 0.9999988532$. **b.** The limit of the integral from $-N$ to N as $N \to \infty$ is 1.
33. $|y''| = 2/x^3 \le 2$ on $[1, 2]$; need $2/(12n^2) \le 10^{-10}$, i.e., $2 \cdot 10^{10}/12 \le n^2$, which implies $n > 40{,}824.8$.
35. $|J_{1/2}''(x)| \le 8$ on $[\frac{1}{2}, 1]$. For the trapezoidal rule we need $8(\frac{1}{2})^3/(12 \cdot 0.01) \le n^2$ or $n > 2.89$. Using $n = 3$ yields 0.3095670957.

Problems 7 Review, page 535

1. $\frac{1}{2}\{\sqrt{5} + 4\ln[2/(1 + \sqrt{5})]\}$
3. $2\tan^{-1} x + x/(x^2 + 1) + C$
5. $\frac{2}{15}$
7. $-e^{-2x}[(x/2) + \frac{1}{4}] + C$
9. $\frac{1}{4}\sec^4 x - \sec^2 x - \ln|\cos x| + C$
11. $3\ln 2 - \frac{3}{2}$
13. $(-2/\sqrt{3}) + \sqrt{2} + \ln[(2 + \sqrt{3})/(\sqrt{2} + 1)]$
15. $\frac{1}{4}e^{2x}(\sin 2x - \cos 2x) + C$
17. $-\frac{1}{4}\cos^4 x + \frac{1}{6}\cos^6 x + C$
19. $-\frac{1}{5}\csc^5 x + \frac{1}{3}\csc^3 x + C$
21. $\frac{1}{3}x^3 + x + \ln|x/(x + 1)| + C$
23. $x\cosh x - \sinh x + C$
25. $-1/(x - 2) + C$
27. $\frac{1}{2}(\sin x + \frac{1}{13}\sin 13x) + C$
29. $-\frac{11}{14}\ln|x - 3| + \frac{2}{7}\ln|x + 4| + \frac{3}{2}\ln|x - 5| + C$
31. $\frac{1}{5}(x^2 - 4)^{5/2} + \frac{4}{3}(x^2 - 4)^{3/2} + C$ ($u = x^2 - 4$)
33. $\frac{1}{3}\tan^3 x + \tan x + C$
35. $x + 3\ln|x| - (3/x) - (1/2x^2) + C$
37. $5\pi/32$
39. $\ln(\sqrt{2} + 1)$
41. $\frac{1}{2\sqrt{6}}\ln\left|\frac{\tan(x/2) + 5 - \sqrt{24}}{\tan(x/2) + 5 + \sqrt{24}}\right| + C = \frac{1}{2\sqrt{6}}\ln\left|\frac{1 - \cos x + (5 - 2\sqrt{6})\sin x}{1 - \cos x + (5 + 2\sqrt{6})\sin x}\right| + C$

43. $-\frac{3}{2}(1 + x^4)^{-1/3} + C$ **45.** $\frac{1}{81}\left[-\frac{4}{7}(2 + 3x)^{7/2} + \frac{24}{5}(2 + 3x)^{5/2} - 10(2 + 3x)^{3/2} - 4(2 + 3x)^{1/2}\right] + C$ $(u = 2 + 3x)$

47. $\frac{9}{10}e^{-x}\left[-\cos(x/3) + \frac{1}{3}\sin(x/3)\right] + C$ **49.** $\frac{3}{8}\cosh x \sinh 3x - \frac{1}{8}\sinh x \cosh 3x + C$

51. $-(x^2 + 2x + 2)e^{-x} + C$ **53.** $\sin^2(\sqrt{x}) + C$ **55.** $\frac{1}{3}\ln|x/(3 + \sqrt{9 - 4x^2})| + C$ (let $u = 3/2x$)

57. $-(\sqrt{x^2 - 16}/x) + \ln|x + \sqrt{x^2 - 16}| + C$ **59.** $\frac{1}{7}\sin^{7/4}(2x^2) + C$ **61.** $\frac{16}{35}$ **63.** $\frac{1}{12}\sinh(6x) - (x/2) + C$

65. $-\frac{1}{39}e^{-3x^3}\left[3\sin(2x^3) + 2\cos(2x^3)\right] + C$ **67.** $\frac{1}{9}\left[\sec^2(3x) + 2\right]\tan(3x) + C$

69. $e^{2x}\left[\frac{3}{5}\cosh 3x - \frac{2}{5}\sinh 3x\right] + C = \frac{1}{10}e^{5x} + \frac{1}{2}e^{-x} + C$ **71.** $x - \frac{1}{3}\tanh(3x) + C$ **73.** $(1/\sqrt{7})\tan^{-1}(\sqrt{7}\tan x) + C$

75. $\left[-1/(1 + \tan x)\right] + C = \left[(\sin 2x)/(\cos 2x - \sin 2x - 1)\right] + C$ **77.** $(\frac{3}{16}x^2 - \frac{3}{128})\cos 4x + (\frac{1}{4}x^3 - \frac{3}{32}x)\sin 4x + C$

79. $\frac{1}{18}\tan^{-1}(x^6/3) + C$ **81.** $\frac{1}{12}[x^6\sqrt{9 + x^{12}} + 9\ln|x^6 + \sqrt{9 + x^{12}}|] + C$

83. $-\frac{2}{3}\ln|x - 1| + \frac{5}{6}\ln|x^2 + x + 1| + \sqrt{3}\tan^{-1}[(2x + 1)/\sqrt{3}] + C$ **85.** $[1/(x - 1)] + 2\ln|x - 1| + C$

87. $\frac{1}{8}\ln|(x + 4)/(x - 4)| + C$ **89.** $-\frac{1}{9}\sqrt{25 - 9x^2} + C$ **91.** $(x/2)\sqrt{16 + 9x^2} - \frac{8}{3}\ln|3x + \sqrt{16 + 9x^2}| + C$

93. $(\frac{1}{4}x^4 - \frac{1}{864})\cos^{-1}(3x) - (\frac{1}{48}x^3 + \frac{1}{288}x)\sqrt{1 - 9x^2} + C$ **95.** $\frac{1}{3}[x^3\tan^{-1}(x^3) - \frac{1}{2}\ln(1 + x^6)] + C$

97. $-(x - 4)^{-1} - 2(x - 4)^{-2} + C$ **99.** $\frac{1}{2}\ln|x^2 - 1| + C$ **101.** $-\frac{1}{6}\ln\left|\frac{x - 1}{x + 1}\right| + \frac{1}{12}\ln\left|\frac{x - 2}{x + 2}\right| + C$

103. $\frac{1}{2}x^2 - x + C$ **105.** $-\pi\ln 2$ **107.** $\frac{1}{16}$ **109.** $\pi(b - a)^2/8 =$ area of semicircle

111. 1.383213747 **113.** 0.9253959267 **115.** 1.0057092571

117. $|y''| = |3(2x + 3x^4)e^{x^3}| \le 15e$ on $[0, 1]$; need $15e/(12n^2) \le 0.01$ or $n \ge 18.4$.

119. $|y^{(4)}| < 120/x^6 \le 120$ on $[1, 2]$; we need $120/(180n^4) \le 0.0001$ or $n \ge 9.04$. Using Simpson's Rule with $n = 10$ subintervals yields the approximation 0.5000124699; the actual value of the integral is 0.5.

Computer Exercises, page 538

1. b. $f^{(4)}(x) = \dfrac{24(1 - 10x^2 + 5x^4)}{(1 + x^2)^5}$; on a graphing calculator the maximum is seen to be 24.

c. $\dfrac{24}{180n^4} < 10^{-11}$ yields $n > 339.8$, so 340 intervals are needed. **d.** $\pi \approx 3.14159265359$ **3.** $\ln 100 \approx 4.60517018599$

5. $\dfrac{16\sqrt{3}}{147}\tan^{-1}\left(\dfrac{2x + 1}{\sqrt{3}}\right) + \dfrac{3}{49}\ln(x^2 + x + 1) + \dfrac{1}{175}\ln|x + 3| - \dfrac{157}{1225}\ln|x - 2| - \dfrac{6}{35(x - 2)} + C$

CHAPTER 8

Problems 8.1, page 548

1. center: $(0, 0)$
foci: $(0, \pm 3)$
vertices: $(0, \pm 5)$
major axis: line segment between $(0, -5)$ and $(0, 5)$
minor axis: line segment between $(-4, 0)$ and $(4, 0)$

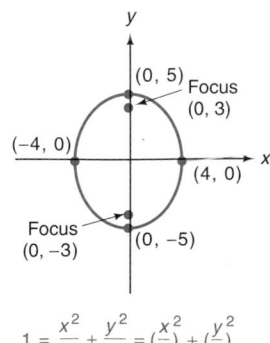

$$1 = \frac{x^2}{16} + \frac{y^2}{25} = \left(\frac{x}{4}\right)^2 + \left(\frac{y}{5}\right)^2$$

Note: In the following answers, $\overline{(a, b)(c, d)}$ denotes the line segment between (a, b) and (c, d).

3. center: $(0, 0)$
foci: $(0, \pm 2\sqrt{2})$
vertices: $(0, \pm 3)$
major axis: $\overline{(0, -3)(0, 3)}$
minor axis: $\overline{(-1, 0)(1, 0)}$

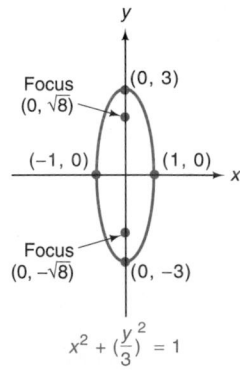

$$x^2 + \left(\frac{y}{3}\right)^2 = 1$$

5. center: $(0, 0)$
foci: $(\pm 2\sqrt{3}, 0)$
vertices: $(\pm 4, 0)$
major axis: $\overline{(-4, 0)(4, 0)}$
minor axis: $\overline{(0, -2)(0, 2)}$

$$\left(\frac{x}{4}\right)^2 + \left(\frac{y}{2}\right)^2 = 1$$

7. center: $(1, -3)$
foci: $(1, -6), (1, 0)$
vertices: $(1, -8), (1, 2)$
major axis: $\overline{(1, -8)(1, 2)}$
minor axis: $\overline{(-3, -3)(5, -3)}$

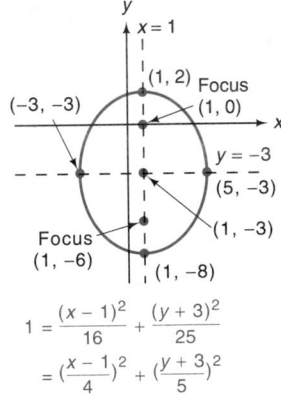

$$1 = \frac{(x-1)^2}{16} + \frac{(y+3)^2}{25}$$
$$= \left(\frac{x-1}{4}\right)^2 + \left(\frac{y+3}{5}\right)^2$$

9. The graph is a circle, centered at $(0, 0)$ with radius 1 (the unit circle).

11. center: $(0, 0)$
foci: $(\pm 3\sqrt{3}/2, 0)$
vertices: $(\pm 3, 0)$
major axis: $\overline{(-3, 0)(3, 0)}$
minor axis: $\overline{(0, -\frac{3}{2})(0, \frac{3}{2})}$

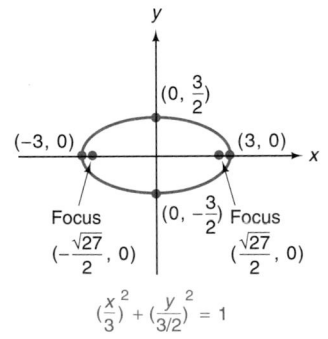

$$\left(\frac{x}{3}\right)^2 + \left(\frac{y}{3/2}\right)^2 = 1$$

13. center: $(-1, -3)$
foci: $(-1, -3 \pm 2\sqrt{3})$
vertices: $(-1, -7), (-1, 1)$
major axis: $\overline{(-1, -7)(-1, 1)}$
minor axis: $\overline{(-3, -3)(1, -3)}$

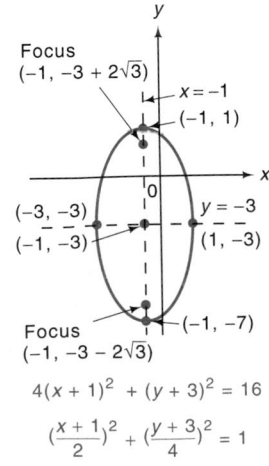

$$4(x+1)^2 + (y+3)^2 = 16$$
$$\left(\frac{x+1}{2}\right)^2 + \left(\frac{y+3}{4}\right)^2 = 1$$

15. center: $(-1, 3)$
  foci: $(-1, 3 \pm 2\sqrt{3})$
  vertices: $(-1, -1), (-1, 7)$
  major axis: $\overline{(-1, -1)(-1, 7)}$
  minor axis: $\overline{(-3, 3)(1, 3)}$

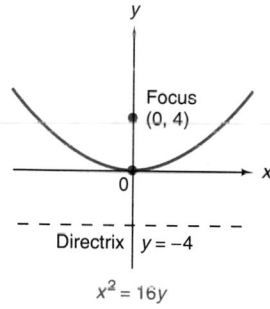

$$\left(\frac{x+1}{2}\right)^2 + \left(\frac{y-3}{2}\right)^2 = 1$$

17. center: $(-2, \frac{1}{4})$
  foci: $(-2 \pm \sqrt{\frac{325}{48}}, \frac{1}{4})$
  vertices: $(-2 \pm \sqrt{\frac{65}{6}}, \frac{1}{4})$
  major axis: $\overline{(-2 - \sqrt{\frac{65}{6}}, \frac{1}{4})(-2 + \sqrt{\frac{65}{6}}, \frac{1}{4})}$
  minor axis: $\overline{(-2, (1 - \sqrt{65})/4)(-2, (1 + \sqrt{65})/4)}$

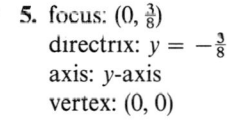

$$\left(\frac{x+2}{65/6}\right)^2 + \left(\frac{y-1/4}{65/16}\right)^2 = 1$$

19. $\frac{3}{5}$ **21.** $\sqrt{3}/2 \approx 0.87$ **23.** $\sqrt{\frac{5}{8}} \approx 0.79$ **25.** (g); vertices $(-1, 2)$ and $(7, 2)$ **27.** (d); vertices $(\pm\sqrt{21}, 0)$

29. (f); foci $(-2, -1 \pm 2\sqrt{3})$ **31.** (a); foci $(\pm\sqrt{7}, 0)$ **33.** $\dfrac{x^2}{9} + \dfrac{y^2}{25} = 1$ **35.** $y = -\dfrac{x}{3} + \dfrac{7}{3}$ **39.** 0.9679

41. 8.3 AU   **43.** Lanes 1 and 6 have $5\sqrt{7} \approx 13.2$ ft, lanes 2 and 5 have $10\sqrt{3} \approx 17.3$ ft, lanes 3 and 4 have $5\sqrt{15} \approx 19.4$ ft.
45. Let $D = C + A^2/4 + B^2/8$. Then (a) $D > 0$ (b) $D = 0$ (c) $D < 0$.

Problems 8.2, page 557

1. focus: $(0, 4)$
  directrix: $y = -4$
  axis: y-axis
  vertex: $(0, 0)$

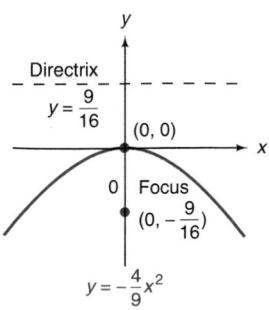

$x^2 = 16y$

3. focus: $(0, -4)$
  directrix: $y = 4$
  axis: y-axis
  vertex: $(0, 0)$

$x^2 = -16y$

5. focus: $(0, \frac{3}{8})$
  directrix: $y = -\frac{3}{8}$
  axis: y-axis
  vertex: $(0, 0)$

$y = \frac{2}{3}x^2$

7. focus: $(0, -\frac{9}{16})$
  directrix: $y = \frac{9}{16}$
  axis: y-axis
  vertex: $(0, 0)$

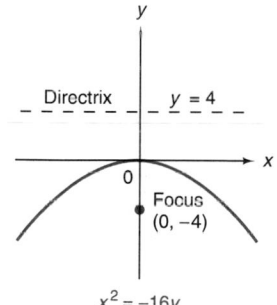

$y = -\frac{4}{9}x^2$

9. vertex: $(1, -3)$
  focus: $(1, -7)$
  directrix: $y = 1$
  axis: $x = 1$

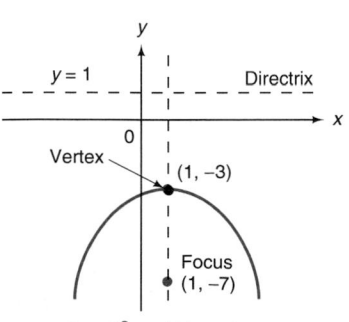

$(x - 1)^2 = -16(y + 3)$

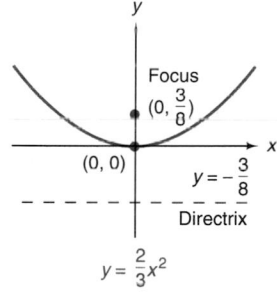

11. vertex: $(0, \frac{9}{4})$
focus: $(0, \frac{5}{4})$
directrix: $y = \frac{13}{4}$
axis: $x = 0$

13. vertex: $(-1, 0)$
focus: $(-1, -\frac{1}{4})$
directrix: $y = \frac{1}{4}$
axis: $x = -1$

15. vertex: $(-2, 4)$
focus: $(-2, \frac{15}{4})$
directrix: $y = \frac{17}{4}$
axis: $x = -2$

17. vertex: $(-2, -4)$
focus: $(-2, -\frac{15}{4})$
directrix: $y = -\frac{17}{4}$
axis: $x = -2$

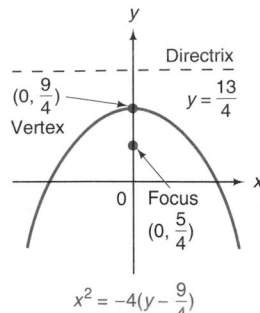

$x^2 = -4(y - \frac{9}{4})$

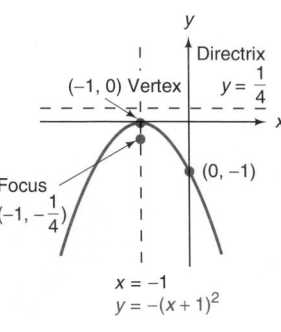

$y = -(x + 1)^2$

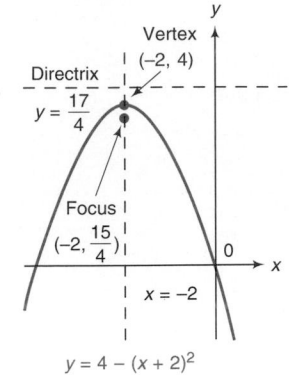

$y = 4 - (x + 2)^2$

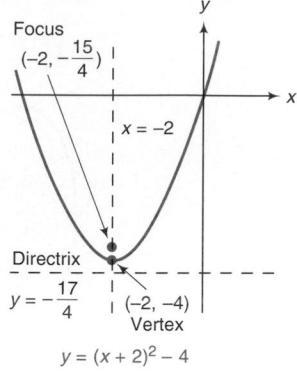

$y = (x + 2)^2 - 4$

19. (f)   **21.** (b)   **23.** (c)   **25.** (e)   **27.** $16y = x^2$ **29.** $16(y - 5) = (x + 2)^2$ **31.** $y = \frac{2}{3}x - 1$

33. If the asteroid's position on its orbit looks like

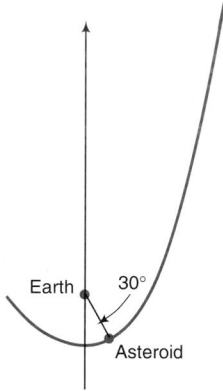

then its closest point to the earth is $\frac{150,000}{2}(1 + \sqrt{3}/2) \approx 139,952$ km. On the other hand, if the problem statement refers to a position like

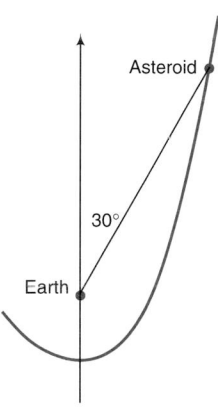

then the closest point is $\frac{150,000}{2}(1 - \sqrt{3}/2) \approx 10,048$ km.

35. 67.74 ft   **37b.** $(0, -y_0)$

Problems 8.3, page 568

1. center: $(0, 0)$
foci: $(\pm\sqrt{41}, 0)$
vertices: $(\pm 4, 0)$
transverse axis: $\overline{(-4, 0)(4, 0)}$
asymptotes: $y = \pm\frac{5}{4}x$

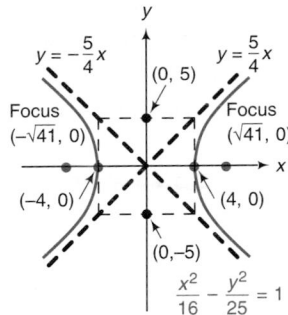

3. center: $(0, 0)$
foci: $(0, \pm\sqrt{41})$
vertices: $(0, \pm 5)$
transverse axis: $\overline{(0, -5)(0, 5)}$
asymptotes: $y = \pm\frac{5}{4}x \ (x = \pm\frac{4}{5}y)$

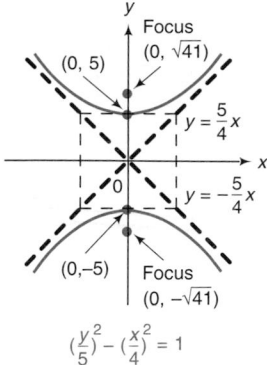

5. center: $(0, 0)$
foci: $(0, \pm\sqrt{2})$
vertices: $(0, \pm 1)$
transverse axis: $\overline{(0, -1)(0, 1)}$
asymptotes: $y = \pm x$

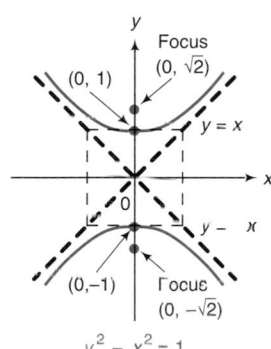

7. center: $(0, 0)$
foci: $(\pm 3\sqrt{5}/2, 0)$
vertices: $(\pm 3, 0)$
transverse axis: $\overline{(-3, 0)(3, 0)}$
asymptotes: $y = \pm x/2$

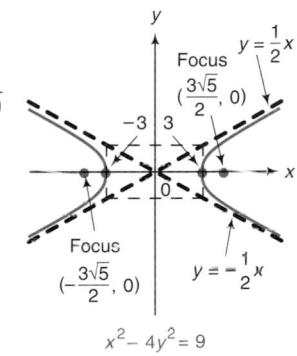

9. center: $(0, 0)$
foci: $(0, \pm 3\sqrt{5}/2)$
vertices: $(0, \pm 3)$
transverse axis: $\overline{(0, -3)(0, 3)}$
asymptotes: $y = \pm 2x$

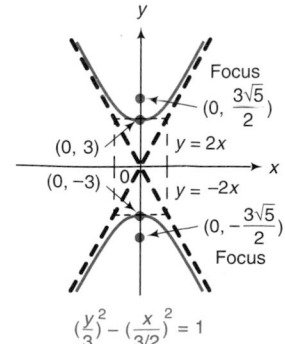

11. center: $(0, 0)$
foci: $(\pm\sqrt{\frac{10}{3}}, 0)$
vertices: $(\pm\sqrt{2}, 0)$
transverse axis: $\overline{(-\sqrt{2}, 0)(\sqrt{2}, 0)}$
asymptotes: $y = \pm\sqrt{\frac{2}{3}}x$

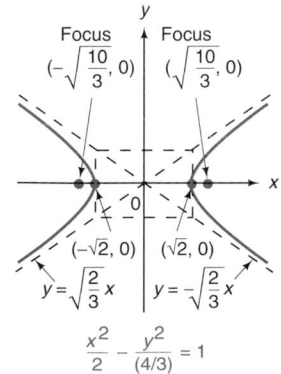

13. center: $(0, 0)$
  foci: $(0, \pm\sqrt{\frac{10}{3}})$
  vertices: $(0, \pm\sqrt{2})$
  transverse axis: $(0, -\sqrt{2})(0, \sqrt{2})$
  asymptotes: $y = \pm\sqrt{\frac{3}{2}}x$

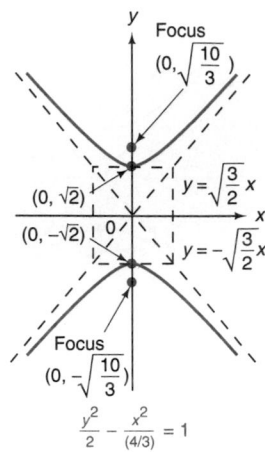

$$\frac{y^2}{2} - \frac{x^2}{(4/3)} = 1$$

15. center: $(1, -2)$
  foci: $(1 - \sqrt{5}, -2), (1 + \sqrt{5}, -2)$
  vertices: $(-1, -2), (3, -2)$
  transverse axis: $(-1, -2)(3, -2)$
  asymptotes: $y = \pm(\frac{1}{2})(x - 1) - 2$

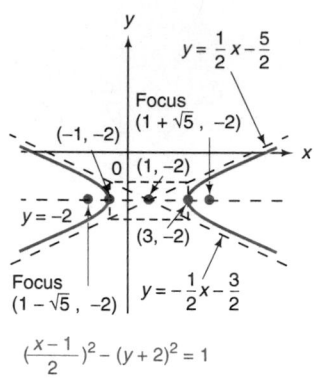

$$(\frac{x-1}{2})^2 - (y + 2)^2 = 1$$

17. center: $(-1, -3)$
  foci: $(-1 \pm 2\sqrt{5}, -3)$
  vertices: $(-3, -3), (1, -3)$
  transverse axis: $(-3, -3)(1, -3)$
  asymptotes: $y = \pm2(x + 1) - 3$

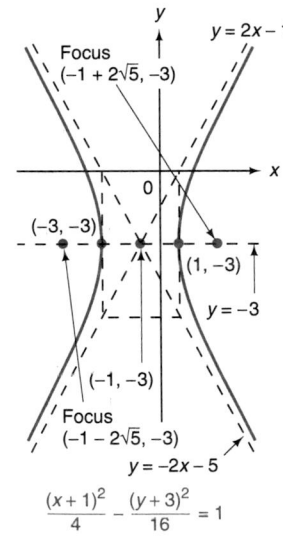

$$\frac{(x+1)^2}{4} - \frac{(y+3)^2}{16} = 1$$

19. center: $(4, ?)$
  foci: $(4 - \sqrt{\frac{325}{6}}, 2)(4 + \sqrt{\frac{325}{6}}, 2)$
  vertices: $(4 - \sqrt{\frac{65}{2}}, 2), (4 + \sqrt{\frac{65}{2}}, 2)$
  transverse axis: $(4 - \sqrt{\frac{65}{2}}, 2)(4 + \sqrt{\frac{65}{2}}, 2)$
  asymptotes: $y = \pm\sqrt{\frac{2}{3}}(x - 4) + 2$

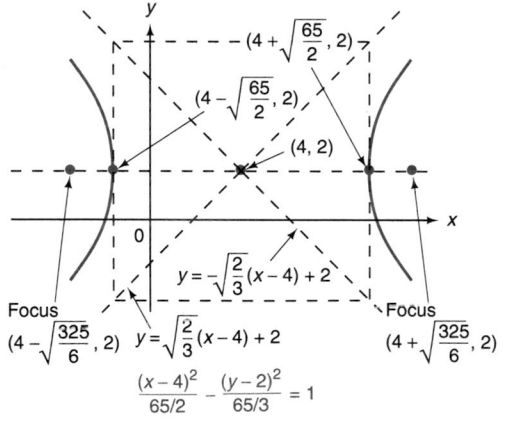

$$\frac{(x-4)^2}{65/2} - \frac{(y-2)^2}{65/3} = 1$$

21. $\dfrac{x^2}{9} - \dfrac{y^2}{7} = 1$ **23.** $\dfrac{y^2}{4} - \dfrac{x^2}{5} = 1$ **25.** $\dfrac{(x-2)^2}{4} - \dfrac{(y-1)^2}{5} = 1$ **27.** $\dfrac{x^2}{4} - \dfrac{y^2}{4} = 1$ **29.** $\dfrac{(x-3)^2}{4} - \dfrac{(y-1)^2}{16} = 1$

31. $\sqrt{41}/4$ **33.** $\sqrt{5}/2$ **35.** $\sqrt{41}/5$ **37.** $\sqrt{5}/2$ **39.** $\sqrt{\frac{5}{3}}$ **41.** $\left(\dfrac{x}{4}\right)^2 - \left(\dfrac{y}{3}\right)^2 = 1$ **43.** $\left(\dfrac{x}{2}\right)^2 - \left(\dfrac{y}{6}\right)^2 = 1$

45. $\left(\dfrac{x-4}{2}\right)^2 - \left(\dfrac{y+5}{4}\right)^2 = 1$ or $4x^2 - 32x - y^2 - 10y = -23$ **47.** $64x^2 - 260x + 300y - 120xy + 475 = 0$

49. $\dfrac{x^2}{(1533)^2} - \dfrac{y^2}{1,649,911} = 1$ where distances are measured in meters. $[1,649,911 = 2000^2 - 1533^2]$

55. $y = (\pm 1/\sqrt{3})(x+2) + 1$

Problems 8.4, page 575

1. ellipse; $\dfrac{(x')^2}{15} + \dfrac{(y')^2}{9} = 1$ **3.** $(\sqrt{3} - \frac{3}{2})x' - (1 + 3\sqrt{3}/2)y' = 6$

5. If $\theta = \pi/3$, the resulting equation is $3(x')^2 + 2\sqrt{3}x'y' + (y')^2 + 24x' - 24\sqrt{3}y' = 0$; if $\theta = 4\pi/3$, the equation becomes $3(x')^2 + 2\sqrt{3}x'y' + (y')^2 - 24x' + 24\sqrt{3}y' = 0$.

7. Rotate axes through $\theta = \frac{1}{2}\tan^{-1}\frac{4}{3} \approx 26.6°$ to obtain two parallel lines given by $5(x')^2 = 9$ or $x' = \pm 3/\sqrt{5}$

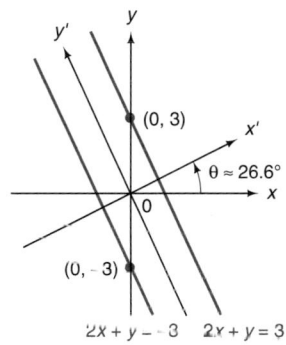

9. Rotate axes through $\theta = \frac{1}{2}\tan^{-1}(-\frac{2}{3}) \approx -16.8°$ (or $73.2°$) to obtain the hyperbola $[(\sqrt{13}+3)/10](x')^2 - [(\sqrt{13}-3)/10](y')^2 = 1$. Note that, since θ is in the fourth quadrant, $\sin\theta < 0$ and

$$\sin\theta = \sin\frac{1}{2}\tan^{-1}(-\tfrac{2}{3}) = -\sqrt{\frac{1 - \cos\tan^{-1}(-\frac{2}{3})}{2}} = -\sqrt{\frac{1 - 3/\sqrt{13}}{2}} = -\sqrt{\frac{\sqrt{13}-3}{2\sqrt{13}}}; \cos\theta = \sqrt{\frac{\sqrt{13}+3}{2\sqrt{13}}}.$$

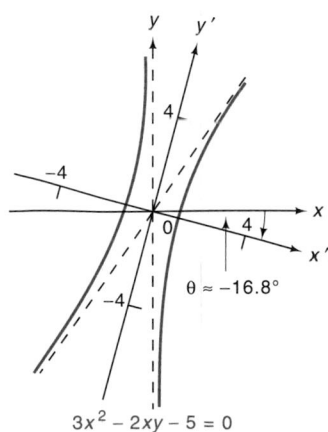

If $73.2°$ is used, the equation becomes $[(3 - \sqrt{13})/10](x')^2 + [(3 + \sqrt{13})/10](y')^2 = 1$.

11. Rotate axes through $\theta = \pi/4$ to obtain the hyperbola $[(x')^2/2a] - [(y')^2/2a] = 1$.

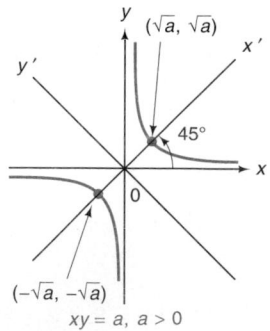

$xy = a, \ a > 0$

13. Rotate axes through $\theta = \frac{1}{2} \tan^{-1} \frac{4}{3} \approx 26.6°$ to obtain the parabola $[x' + (3/\sqrt{5})]^2 = (8/\sqrt{5})\left[y' + \left(\dfrac{9}{8\sqrt{5}}\right)\right]$.

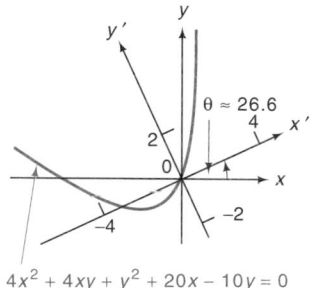

$4x^2 + 4xy + y^2 + 20x - 10y = 0$

15. Rotate axes through $\theta = \pi/8$ to obtain the ellipse $[(3 + \sqrt{2})/8](x')^2 + [(3 - \sqrt{2})/8](y')^2 = 1$.

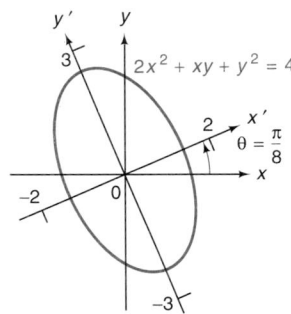

$2x^2 + xy + y^2 = 4$

17. Rotate axes through $\theta = \frac{1}{2} \tan^{-1} 3 \approx 35.8°$ to obtain the ellipse $[(4 - \sqrt{10})/36](x')^2 + [(4 + \sqrt{10})/36](y')^2 = 1$.

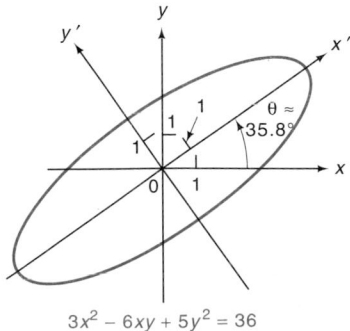

$3x^2 - 6xy + 5y^2 = 36$

19. Rotate axes through $\theta = \frac{1}{2}\tan^{-1}\frac{4}{3} \approx 26.6°$ to obtain the hyperbola $[4(y' + 2\sqrt{5})^2/35] - [(x' + \sqrt{5})^2/35] = 1$.

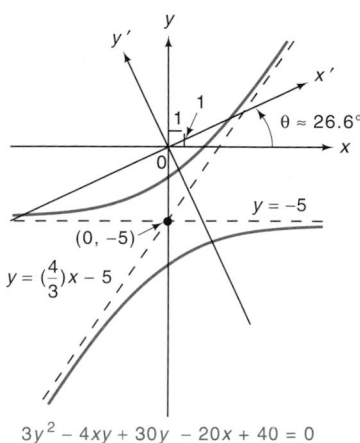

$3y^2 - 4xy + 30y - 20x + 40 = 0$

Problems 8.5, page 580

1. $(3, 0)$ **3.** $(2\sqrt{2}, 2\sqrt{2})$ **5.** $(-3\sqrt{3}, -3)$ **7.** $(-5, 0)$ **9.** $(\frac{5}{2}, -5\sqrt{3}/2)$ **11.** $(\sqrt{2}, -\sqrt{2})$

13. $(\sqrt{2}, -\sqrt{2})$ **15.** $(0, -1)$ **17.** $(0, -1)$ **19.** $(0, 0)$ **21.** $(3, 0)$ **23.** $(1, \pi/2)$ **25.** $(\sqrt{2}, \pi/4)$

27. $(3\sqrt{2}, 7\pi/4)$ **29.** $(4, \pi/3)$ **31.** $(10, 5\pi/3)$ **33.** $(2, \pi/6)$ **35.** $(8, 5\pi/6)$

Problems 8.6, page 588

1. circle centered at the origin with radius 5   **3.** straight line through origin with slope $-1/\sqrt{3}$

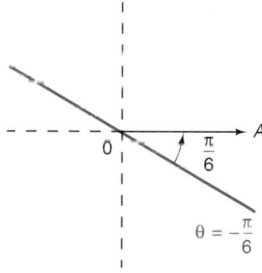

5. circle centered at $(0, \frac{5}{2})$ with radius $\frac{5}{2}$ **7.** circle centered at $(0, -\frac{5}{2})$ with radius $\frac{5}{2}$

9. two circles (called **osculating circles**) with centers at $(0, \pm\frac{1}{2})$, radii $= \frac{1}{2}$, symmetric about polar axis, pole, and $\theta = \pi/2$

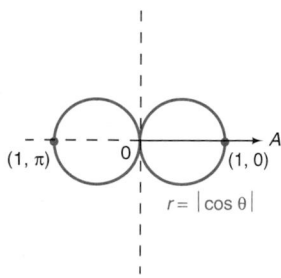

11. circle centered at $(\frac{5}{2}, \frac{5}{2})$ with radius $5/\sqrt{2}$ **13.** cardioid, symmetric about $\theta = \pi/2$

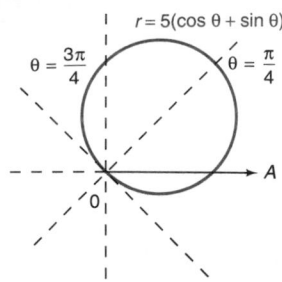

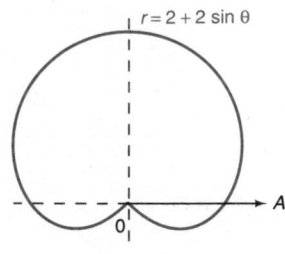

15. cardioid obtained by reflecting curve in Problem 13 about polar axis   **17.** same graph as in Problem 15
19. limaçon with loop, symmetric about $\theta = \pi/2$ **21.** limaçon with loop, symmetric about polar axis

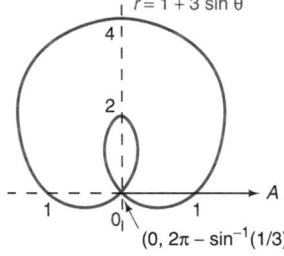

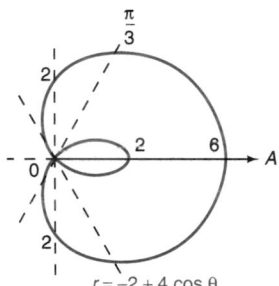

23. limaçon with loop, symmetric about polar axis   **25.** limaçon with loop, symmetric about $\theta = \pi/2$

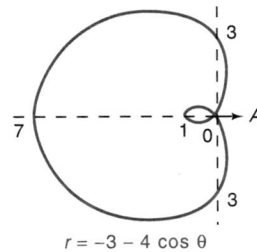

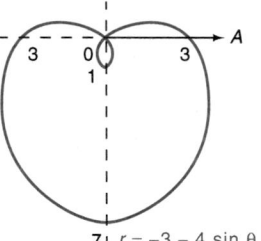

27. limaçon without loop, symmetric about polar axis   **29.** limaçon without loop, symmetric about $\theta = \pi/2$

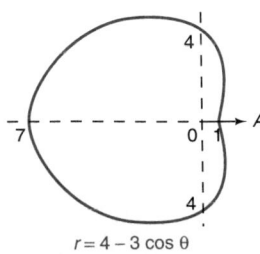

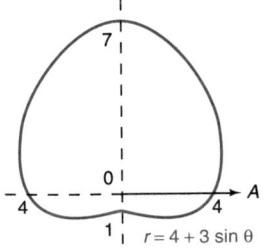

31. four-leafed rose, symmetric about polar axis, pole, and $\theta = \pi/2$ **33.** same graph as in Problem 31

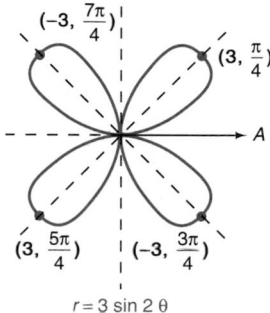

$r = 3 \sin 2\theta$

35. three-leafed rose, symmetric about $\theta = \pi/2$ **37.** this is the graph of Problem 35 reflected about polar axis

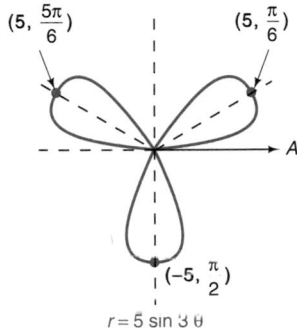

$r = 5 \sin 3\theta$

39. eight-leafed rose, symmetric about polar axis, pole, and $\theta = \pi/2$ **41.** Spiral of Archimedes, no symmetry

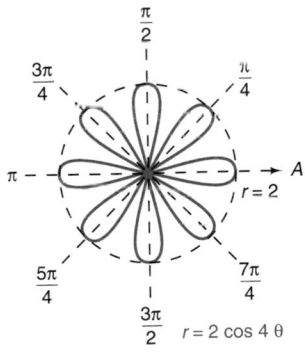

$r = 2 \cos 4\theta$

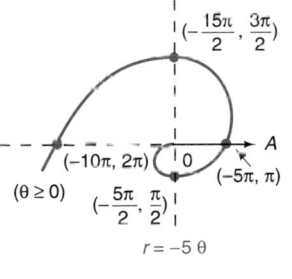

$r = -5\theta$

43. logarithmic spiral, no symmetry   **45.** lemniscate, symmetric about pole, $\theta = \dfrac{\pi}{4}$ and $\theta = \dfrac{3\pi}{4}$

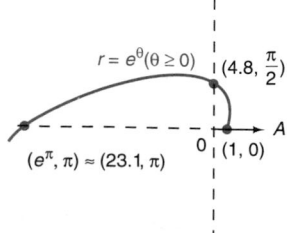

$r = e^{\theta} (\theta \geq 0)$

$(e^{\pi}, \pi) \approx (23.1, \pi)$

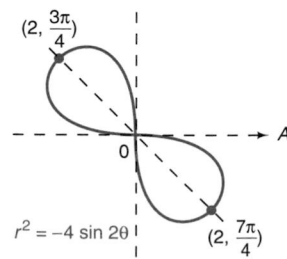

$r^2 = -4 \sin 2\theta$

47. lemniscate, symmetric about polar axis, pole, and $\theta = \pi/2$ **49.** cissoid, symmetric about polar axis

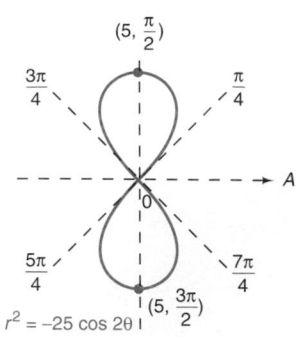

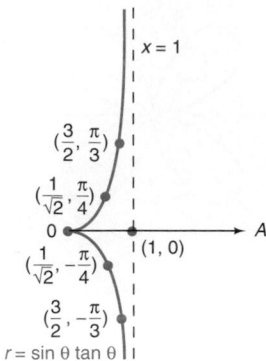

51. conchoid, symmetric about $\theta = \pi/2$ **53.** parabolic spiral

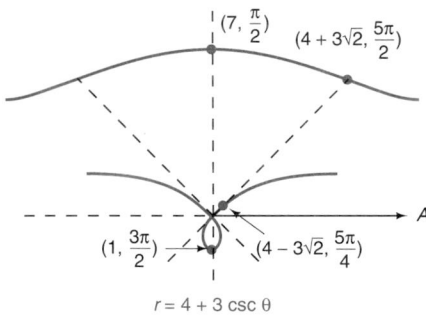

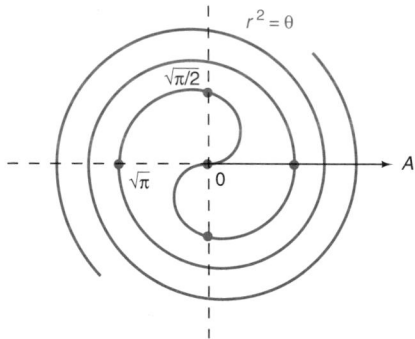

55. $r = -4 \cos \theta + 3 \sin \theta$

In Problems 57–75, the points of intersection are visible on the graphs.

57.

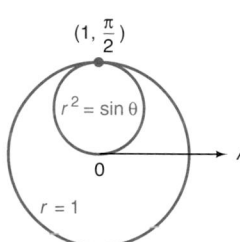

59. $r = \sin \theta$

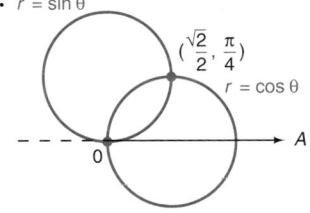

61.

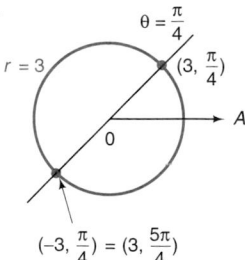

63. $r = -4 + 2 \sin \theta$

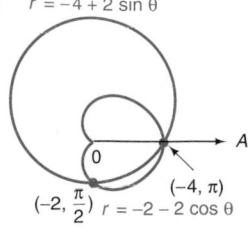

65.

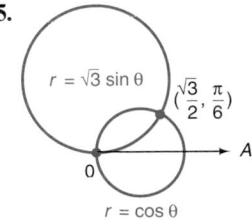

67. The graphs intersect at $(\frac{3}{2}, k\pi/6)$ for $k = 1, 2, 4, 5, 7, 8, 10, 11$.

69.

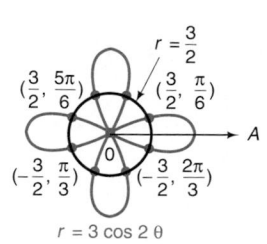

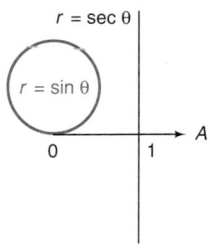

73. pole; $(1/\sqrt{2}, k\pi/8)$ for $k = 1, 3, 5, 7, 9, 11, 13, 15$

75. no common point

71. $r^2 = \theta, 0 \leq r < 2$

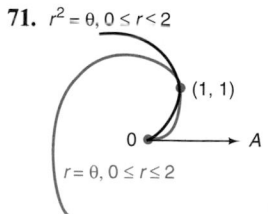

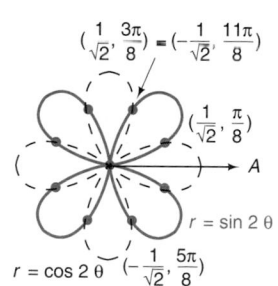

77. $(0, 0), (1, 0), (1 + 2/\sqrt{5}, \tan^{-1}2), (1 - 2/\sqrt{5}, \pi + \tan^{-1}2), (\frac{1}{5}, -\sin^{-1}(\frac{4}{5})) = (-\frac{1}{5}, \pi - \cos^{-1}(\frac{3}{5}))$
79. $(1, k\pi/8)$ for $k = 1, 3, 5, 7, 9, 11, 13, 15$ **81.** $(12, \pi/6), (12, 5\pi/6), (4, 7\pi/6), (4, 11\pi/6)$ **83.** $(-1, \pi) = (1, 0)$
85. $(0, 0), (1/2\sqrt{3}, \pi/6), (1/2\sqrt{3}, 11\pi/6)$ **87.** $(4k\pi, 0)$, for $k = 0, 1, 2, 3, \ldots$

Problems 8.7, page 592

1. $\pi^3/48$ **3.** π **5.** $\pi a^2/20$ **7.** $\pi^5/80$ **9.** $25\pi^9/18$
11. circle; center $(a/2, \pi/2)$, radius $a/2$, $A = \pi a^2/4$ **13.** circle; center $(a/\sqrt{2}, \pi/4)$, radius $a/\sqrt{2}$, $A = \pi a^2/2$

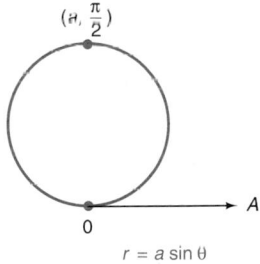

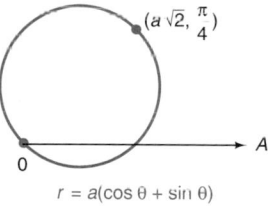

15. circle; center $(a/\sqrt{2}, 3\pi/4)$, radius $a/\sqrt{2}$, $A = \pi a^2/2$ **17.** limaçon without loop, symmetric about polar axis, $A = 11\pi$

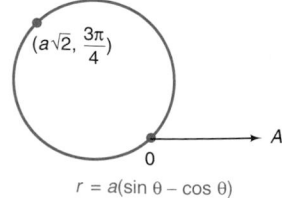

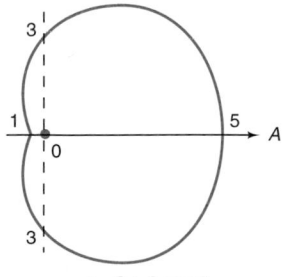

19. limaçon without loop, symmetric about polar axis, $A = 11\pi$

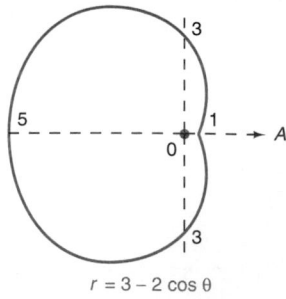

$r = 3 - 2\cos\theta$

21. four-leafed rose; $A = \pi a^2/2$

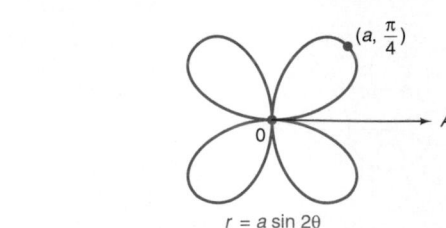

$r = a\sin 2\theta$

23. five-leafed rose; $A = \pi a^2/4$

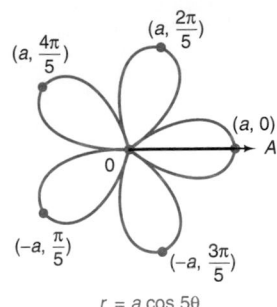

$r = a\cos 5\theta$

25. lemniscate; symmetric about pole, polar axis, and $\theta = \pi/2$; $A = a^2$

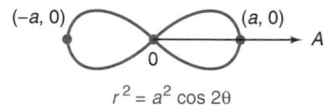

$r^2 = a^2\cos 2\theta$

27. $A = \dfrac{1}{2}\displaystyle\int_{2\pi/3}^{4\pi/3}(1 + 2\cos\theta)^2\,d\theta = \pi - 3\sqrt{3}/2$

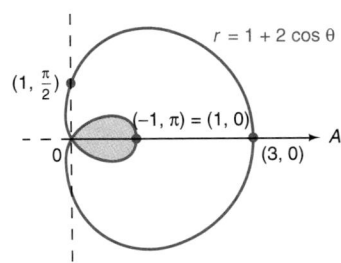

$r = 1 + 2\cos\theta$

29. $A = \displaystyle\int_{\cos^{-1}(-2/3)}^{\pi}(2 + 3\cos\theta)^2\,d\theta = \dfrac{17}{2}\pi - \dfrac{17}{2}\cos^{-1}\left(-\dfrac{2}{3}\right) - 3\sqrt{5} \approx 0.44$ **31.** $A = \pi + 3\sqrt{3}$

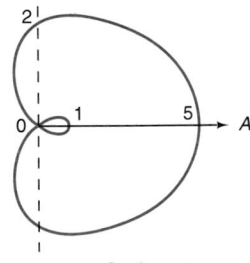

$r = 2 + 3\cos\theta$

33. outer area $= \displaystyle\int_{0}^{\cos^{-1}(-2/3)}(2 + 3\cos\theta)^2\,d\theta - \int_{\cos^{-1}(-2/3)}^{\pi}(2 + 3\cos\theta)^2\,d\theta = 17[\cos^{-1}(-\tfrac{2}{3}) - \pi/2] + 6\sqrt{5} \approx 25.82$

35. $a^2(\pi/4 + 2)$; $r = a(1 + \sin\theta)$

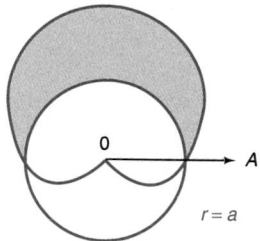

37. $\dfrac{1}{4}\left[9\tan^{-1}\left(\dfrac{4}{3}\right) - 12 + 16\tan^{-1}\left(\dfrac{3}{4}\right)\right] \approx 1.66$ (See the answer to Problem 41.)

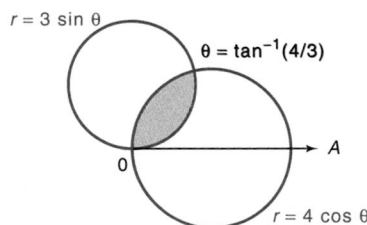

39. $A = \dfrac{\pi}{8} + 2 \cdot \dfrac{1}{2}\displaystyle\int_{\pi/4}^{\pi/3}(\cos 2\theta)^2\,d\theta + 2 \cdot \dfrac{1}{2}\displaystyle\int_{\pi/3}^{\pi/2}\cos^2\theta\,d\theta = \pi/4 - 3\sqrt{3}/16$

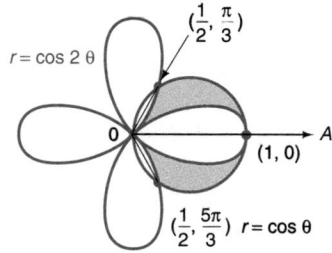

41. $A = \dfrac{1}{2}a^2\displaystyle\int_0^C\sin^2\theta\,d\theta + \dfrac{1}{2}b^2\displaystyle\int_C^{\pi/2}\cos^2\theta\,d\theta\left(\text{where } C = \tan^{-1}\dfrac{b}{a}\right) = \dfrac{1}{4}\left[a^2\tan^{-1}\left(\dfrac{b}{a}\right) - ab + b^2\tan^{-1}\left(\dfrac{a}{b}\right)\right].$
Note that $\dfrac{\pi}{2} - \tan^{-1}\dfrac{b}{a} = \cot^{-1}\dfrac{b}{a} = \tan^{-1}\dfrac{a}{b}$.

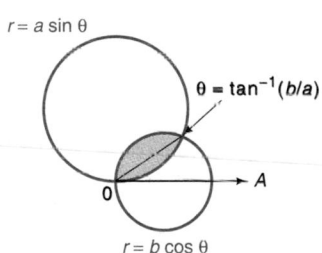

Problems 8 Review, page 593

1. Ellipse
center: $(0, 0)$
foci: $(0, \pm\sqrt{7})$
vertices: $(0, \pm 4)$
major axis: $\overline{(0, -4)(0, 4)}$
minor axis: $\overline{(-3, 0)(3, 0)}$
eccentricity: $\sqrt{7}/4$

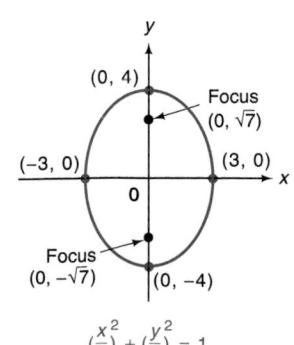

$\left(\dfrac{x}{3}\right)^2 + \left(\dfrac{y}{4}\right)^2 = 1$

3. Parabola
focus: $(0, \frac{9}{64})$
directrix: $y = -\frac{9}{64}$
axis: $x = 0$
vertex: $(0, 0)$

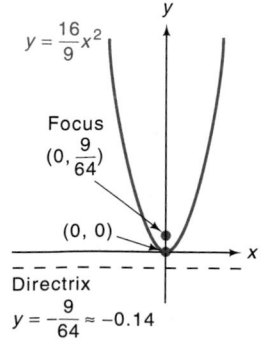

5. Hyperbola
center: $(0, 0)$
foci: $(0, \pm 5)$
vertices: $(0, \pm 3)$
eccentricity: $\frac{5}{3}$
transverse axis: $\overline{(0, -3)(0, 3)}$
asymptotes: $y = \pm\frac{3}{4}x$

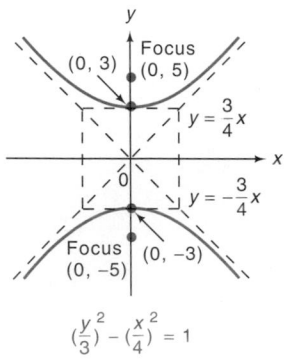

$$\left(\frac{y}{3}\right)^2 - \left(\frac{x}{4}\right)^2 = 1$$

7. Ellipse
center: $(1, -1)$
foci: $(1, -1-\sqrt{5})$, $(1, -1+\sqrt{5})$
vertices: $(1, -4)$, $(1, 2)$
major axis: $\overline{(1, -4)(1, 2)}$
minor axis: $\overline{(-1, -1)(3, -1)}$
eccentricity: $\sqrt{5}/3$

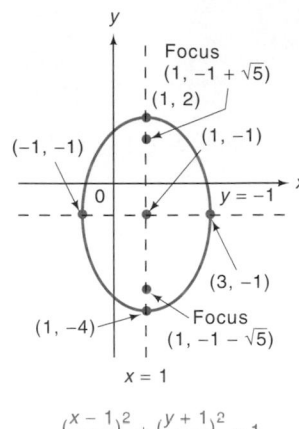

$$\left(\frac{x-1}{2}\right)^2 + \left(\frac{y+1}{3}\right)^2 = 1$$

9. single point $(-2, 5)$

11. two straight lines; i.e., a degenerate hyperbola

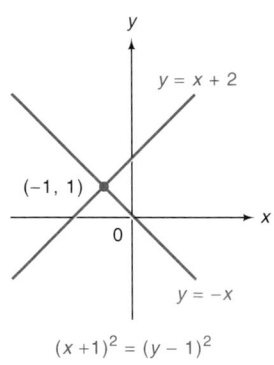

$$(x+1)^2 = (y-1)^2$$

13. Ellipse
center: $(-\frac{1}{2}, -4)$
foci: $(-\frac{1}{2}, -4 - 3/\sqrt{2})$, $(-\frac{1}{2}, -4 + 3/\sqrt{2})$
vertices: $(-\frac{1}{2}, -4 - 3\sqrt{2})$, $(-\frac{1}{2}, -4 + 3\sqrt{2})$
major axis: $\overline{(-\frac{1}{2}, -4 - 3\sqrt{2})(-\frac{1}{2}, -4 + 3\sqrt{2})}$
minor axis: $\overline{(-\frac{1}{2} - 3\sqrt{\frac{3}{2}}, -4)(-\frac{1}{2} + 3\sqrt{\frac{3}{2}}, -4)}$
eccentricity: $\frac{1}{2}$

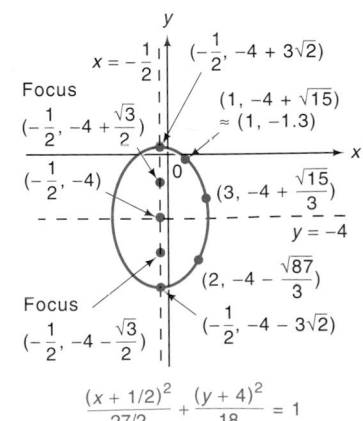

$$\frac{(x+1/2)^2}{27/2} + \frac{(y+4)^2}{18} = 1$$

15. unique ellipse: $(x/5)^2 + (y/4)^2 = 1$ **17.** $y^2/4 - x^2/5 = 1$

19. After the rotation, the curve satisfies the equation $5(x')^2 + 3(y')^2 = 20$; the curve is an ellipse. Referring to the (x', y') coordinates, we find

center: $(0, 0)$

foci: $(0, \pm 2\sqrt{\frac{2}{3}})$

vertices: $(0, \pm 2\sqrt{\frac{5}{3}})$

major axis: $\overline{(0, -2\sqrt{\frac{5}{3}})(0, 2\sqrt{\frac{5}{3}})}$

minor axis: $\overline{(-2, 0)(2, 0)}$

eccentricity: $\sqrt{\frac{2}{5}}$

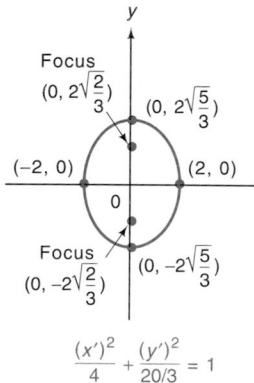

$$\frac{(x')^2}{4} + \frac{(y')^2}{20/3} = 1$$

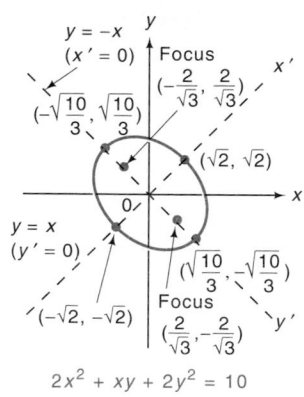

$$2x^2 + xy + 2y^2 = 10$$

21. $9 = x^2 + 4xy + 4y^2 = (x + 2y)^2$; the graph consists of the two parallel lines with equations $3 = x + 2y$ and $-3 = x + 2y$. If the coordinate axes are rotated through the angle $\frac{1}{2} \cot^{-1}((1 - 4)/4) \approx 63.43°$, the curve satisfies the equation $9 = 5(x')^2$ or $x' = \pm 3/\sqrt{5}$.

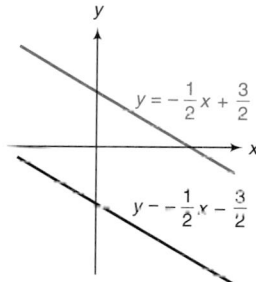

$$y = -\frac{1}{2}x + \frac{3}{2}$$

$$y = -\frac{1}{2}x - \frac{3}{2}$$

23. Rotate the axes through $\pi/4$ radians; the transformed equation is $-3 = xy = [(x')^2 - (y')^2]/2$ or $(y')^2/6 - (x')^2/6 = 1$.

	(x', y') coordinates	(x, y) coordinates
center:	$(0, 0)$	$(0, 0)$
foci:	$(0, \pm 2\sqrt{3})$	$(-\sqrt{6}, \sqrt{6}), (\sqrt{6}, -\sqrt{6})$
vertices:	$(0, \pm\sqrt{6})$	$(-\sqrt{3}, \sqrt{3}), (\sqrt{3}, -\sqrt{3})$
transverse axis:	$(0, -\sqrt{6}), (0, \sqrt{6})$	$(-\sqrt{3}, \sqrt{3}), (\sqrt{3}, -\sqrt{3})$
asymptotes:	$y' = \pm x'$	$x = 0, y = 0$
eccentricity:	$\sqrt{2}$	

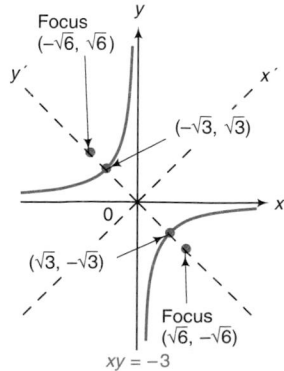

$$xy = -3$$

25. Rotate axes through $\theta = \frac{1}{2} \cot^{-1}((-2-4)/3) \approx 76.717°$; the curve satisfies the transformed equation $(3\sqrt{5}/2 + 1)(x')^2 - (3\sqrt{5}/2 - 1)(y')^2 = 5$; or, using numerical approximations, $(x')^2/1.148 - (y')^2/2.124 = 1$. At this stage, we can identify the curve as a hyperbola opening along the x'-axis.

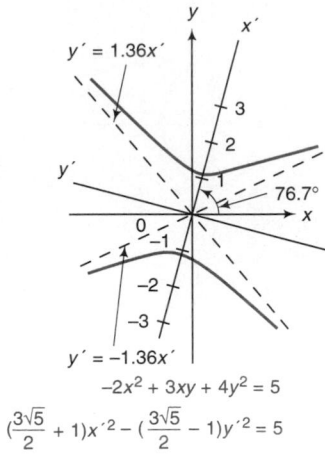

27. $(2, 0)$ **29.** $(0, -7)$ **31.** $(0, 1)$ **33.** $(2, 0)$ **35.** $(2, 11\pi/6)$ **37.** $(6\sqrt{2}, 5\pi/4)$
39. circle, radius 8, center at $(0, 0)$ **41.** circle; radius 1, center at $(1, 0)$
43. limaçon, symmetric about line $\theta = \pi/2$ **45.** four-leafed rose, symmetric about polar axis, pole, $\theta = \pi/2$

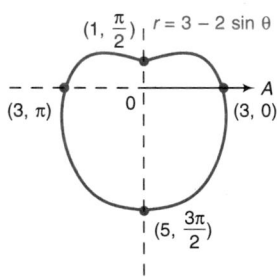

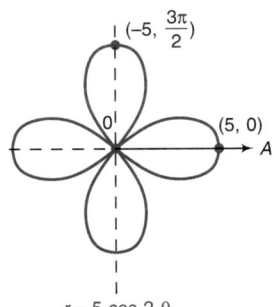

47. eight-leafed rose, symmetric about pole, polar axis, $\theta = \pi/2$ **49.** logarithmic spiral

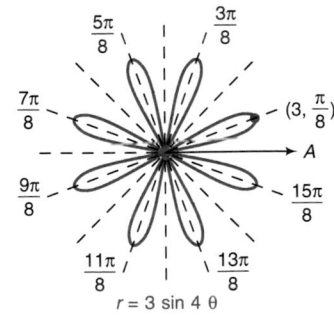

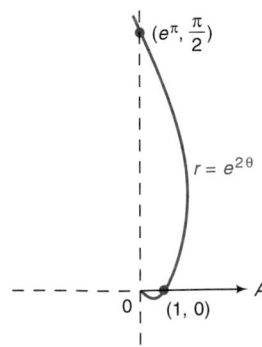

51. horizontal line $y = 4$ **53.** $(x - 2)^2 + (y - \frac{3}{2})^2 = (\frac{5}{2})^2$ **55.** $(1, 0)$ **57.** pole, $(1, \pi/2)$
59. pole, $(9, 9)$, and, in general, (θ, θ) where $\theta = \frac{9}{2} + \sqrt{(\frac{9}{2})^2 + 18k\pi}$, k is a positive integer
61. $(1, k\pi/12)$ for $k = 1, 5, 7, 11, 13, 17, 19, 23$ **63.** $\frac{1}{6}[(\pi/2)^3 - (\pi/6)^3] = 13\pi^3/648$ **65.** $\pi a^2/2$ **67.** 6π
69. $34[\pi/2 - \cos^{-1}(\frac{3}{4})] + 9\sqrt{7}$ **71.** $\pi/2 - 1$ **73.** $2\pi/5$

Computer Exercises, page 594

1. Rotate through an angle of $\cos^{-1}\sqrt{\dfrac{1+\dfrac{2}{\sqrt{13}}}{2}} \approx 28.15°$; obtain the ellipse $\dfrac{(x')^2}{\left(\dfrac{2}{4+\sqrt{13}}\right)} + \dfrac{(y')^2}{\left(\dfrac{2}{4-\sqrt{13}}\right)} = 1$.

3. The curves meet at $\theta \approx \pm 0.69572$; $A = \dfrac{1}{2}\displaystyle\int_{-0.69572}^{0.69572} \left[9\cos^2 t - (\sec t + 1)^2\right] dt \approx 2.29439$.

CHAPTER 9

Problems 9.1, page 600

1. $-\frac{3}{2}$ **3.** 0 **5.** 0 **7.** $\frac{3}{2}$ **9.** $1/(2\sqrt{x_0})$ if $x_0 > 0$; does not exist otherwise **11.** 0 **13.** $\frac{1}{2}$ **15.** $\frac{1}{3}$
17. 1 **19.** 0 **21.** 0 **23.** e^2 **25.** $\frac{2}{5}$ **27.** $\frac{1}{6}$ **29.** -2 **31.** 1 **33.** $f'(x)$ **39.** $f(a)$
41. $-1/(x-a) + C$ **43.** 2

Problems 9.2, page 606

1. $\frac{1}{2}$ **3.** 0 **5.** 0 **7.** 0 **9.** 1 **11.** 0 **13.** 1 **15.** 0 **17.** ∞ **19.** e **21.** $\frac{3}{2}$
23. 1 **25.** limit does not exist **27.** 1 **29.** ∞ **31.** $\frac{1}{2}$ **33. a.** e^{-k} **b.** ∞

Problems 9.3, page 615

1. diverges **3.** diverges **5.** $\frac{3}{2}$ **7.** $\frac{1}{2}$ **9.** 1 **11.** $\frac{1}{4}$ **13.** 0 **15.** diverges
17. $(1/|b|)[(\pi/2) - \tan^{-1}(a/|b|)]$ **19.** diverges **21.** diverges **23.** diverges **25.** diverges **27.** $\frac{1}{2}\ln\frac{11}{9}$
29. diverges **31.** $\frac{1}{2}$ **33.** $\pi/2$ **35.** diverges **37.** diverges **39.** diverges **41.** $\ln\sqrt{3}$
43. diverges **45.** -1 **47.** $\pi/2$ **49.** $1/\sqrt{x^3 + 1} + \sin x \le x^{-3/2}$ and $\int_4^\infty x^{-3/2}\, dx = 1$
51. If $x > 0$, then $x > \ln(1 + x) > 0$ and $1/\ln(1 + x) > 1/x$; now observe that $\int_1^\omega 1/x\, dx$ diverges.

53. $(\sin^2 x)/x^2 \le 1/x^2$ and $\int_1^\infty x^{-2}\, dx = 1$ **55.** $1 + \dfrac{1}{2} + \dfrac{1}{3} + \cdots + \dfrac{1}{N}$ [Make the substitution $u = 1 - x$] **57.** $1/s$

59. $1/(s-1)$ **61.** The value of each integral is 6. Because $u = x^{1/3}$ has an unbounded derivative as $x \to 0^+$, we actually only transform $\int_h^8 x^{-1/3}\, dx$ into $\int_{\sqrt[3]{h}}^2 3u\, du$ and we still need to calculate a limit as $h \to 0^+$.
73. $1/a$ **75.** $\sigma^2 = (b - a)^2/12$; $\sigma = |b - a|/2\sqrt{3}$

Problems 9.4, page 624

1. $\left(\dfrac{1}{\sqrt{2}}\right)\left[1 - \left(x - \dfrac{\pi}{4}\right) - \dfrac{(x-\frac{\pi}{4})^2}{2!} + \dfrac{(x-\frac{\pi}{4})^3}{3!} + \dfrac{(x-\frac{\pi}{4})^4}{4!} \overset{\omega}{\underset{c}{=}} \dfrac{(x-\frac{\pi}{4})^5}{5!} - \dfrac{(x-\frac{\pi}{4})^6}{6!}\right]$

3. $1 + \dfrac{(x-e)}{e} - \dfrac{(x-e)^2}{2e^2} + \dfrac{(x-e)^3}{3e^3} - \dfrac{(x-e)^4}{4e^4} + \dfrac{(x-e)^5}{5e^5}$ **5.** $1 - (x - 1) + (x - 1)^2 - (x - 1)^3 + (x - 1)^4$

7. $x + \frac{1}{3}x^3$ **9.** $(x - \pi) + \frac{1}{3}(x - \pi)^3$ **11.** $1 - x^2 + x^4$ **13.** $x + x^3/3!$ **15.** $-\frac{1}{2}[x - \pi/2]^2$

17. $\dfrac{1}{2} + \dfrac{1}{2^4}x + \dfrac{3}{2^7 \cdot 2!}x^2 + \dfrac{3 \cdot 5}{2^{10} \cdot 3!}x^3 + \dfrac{3 \cdot 5 \cdot 7}{2^{13} \cdot 4!}x^4 = \dfrac{1}{2} + \dfrac{1}{16}x + \dfrac{3}{256}x^2 + \dfrac{5}{2,048}x^3 + \dfrac{35}{65,536}x^4$

19. $1 + \beta x + \left(\dfrac{\beta^2}{2!}\right)x^2 + \left(\dfrac{\beta^3}{3!}\right)x^3 + \left(\dfrac{\beta^4}{4!}\right)x^4 + \left(\dfrac{\beta^5}{5!}\right)x^5 + \left(\dfrac{\beta^6}{6!}\right)x^6$ **21.** $x + x^3/3!$

23. $(a_0 + a_1 + a_2 + a_3) + (a_1 + 2a_2 + 3a_3)(x - 1) + (a_2 + 3a_3)(x - 1)^2 + a_3(x - 1)^3 = a_0 + a_1 x + a_2 x^2 + a_3 x^3$ **25.** x^2

Problems 9.5, page 632

1. $(\pi/4)^7/7! \approx 3.658 \times 10^{-5}$ **3.** $(\frac{105}{32})(\frac{1}{4})^{-9/2}(4 - 1)^5/5! = 3402$ (This error bound can be improved dramatically, to about 6.64, by considering subintervals $[\frac{1}{4}, 1]$ and $[1, 4]$ separately.) **5.** $|\beta|^5 \cdot \max\{e^0, e^\beta\}/5!$

7. $[16 \cdot 8 + 88 \cdot 4 + 16 \cdot 2] \cdot (\pi/4)^5/5! \approx 1.275$ **9.** $e^{1/9}\left(\dfrac{120}{3} + \dfrac{160}{27} + \dfrac{32}{243}\right)\left(\dfrac{(\frac{1}{3})^5}{5!}\right) \approx 0.001765$

11. $\dfrac{1}{2} + \dfrac{\sqrt{3}}{2}(0.2) - \dfrac{1}{2}\left(\dfrac{(0.2)^2}{2!}\right) - \dfrac{\sqrt{3}}{2}\left(\dfrac{(0.2)^3}{3!}\right) \approx 0.66205$ **13.** $\sum_{j=0}^{7}(1/j!) \approx 2.71825$

15. $\sin 33° = \sin\left(\dfrac{\pi}{6} + \dfrac{\pi}{60}\right); \dfrac{1}{2} + \left(\dfrac{\sqrt{3}}{2}\right)\left(\dfrac{\pi}{60}\right) - \dfrac{1}{2}\left(\dfrac{\pi}{60}\right)^2 \Big/ 2 \approx 0.54466 \left[\text{Actually } P_1\left(\dfrac{\pi}{60}\right) \text{ gives the desired accuracy}\right]$

17. $\int_0^{\pi/4}\left(x + \dfrac{x^3}{3}\right)dx = \left(\dfrac{\pi}{4}\right)^2 \Big/ 2 + \dfrac{(\pi/4)^4}{12} \approx 0.34013; \ |\text{max error}| \leq 512(\pi/4)^6/5! \approx 1.00145$

(The precise answer is $\frac{1}{2}\ln 2 \approx 0.34657$.)   **19.** $\int_0^{1/3}(1 + x^2 + x^4/2)\,dx = \frac{1}{3} + \frac{1}{3}(\frac{1}{3})^3 + \frac{1}{10}(\frac{1}{3})^5 \approx 0.34609;$

$|\text{error}| \leq \frac{1}{3} \cdot 0.001765 \approx 0.000588$ **21.** $\int_0^{1/5}(1 + x^3)\,dx = 0.2 + (0.2)^4/4 = 0.20040$

23. $\int_0^{\pi/6}\left(1 - \dfrac{x^4}{2}\right)dx = \dfrac{\pi}{6} - \dfrac{1}{10}\left(\dfrac{\pi}{6}\right)^5 \approx 0.519663$

25. a. $(1 + 0.03)^3 = 1 + 3 \cdot 0.03 + 3 \cdot 0.03^2 + 0.03^3 = 1 + 0.09 + 0.0027 + 0.000027 = 1.092727$
b. $(1 - 0.03)^4 = 1 + 4 \cdot (-0.03) + 6 \cdot (-0.03)^2 + 4 \cdot (-0.03)^3 + (-0.03)^4 =$
$1 - 0.12 + 0.0054 - 0.000108 + 0.00000081 = 0.88529281$ **c.** $(1 + 0.2)^4 = 1 + 4 \cdot 0.2 + 6 \cdot 0.2^2 + 4 \cdot 0.2^3 + 0.2^4 =$
$1 + 0.8 + 0.24 + 0.032 + 0.0016 = 2.0736$ **d.** $(1 - 0.2)^5 = 1 + 5 \cdot (-0.2) + 10 \cdot (-0.2)^2 + 10 \cdot (-0.2)^3$
$+ 5 \cdot (-0.2)^4 + (-0.2)^5 = 1 - 1.0 + 0.4 - 0.08 + 0.008 - 0.00032 = 0.32768$
27. $16\tan^{-1}(\frac{1}{5}) \approx 16[(\frac{1}{5}) - (\frac{1}{5})^3/3 + (\frac{1}{5})^5/5 - (\frac{1}{5})^7/7] \approx 3.158328076$ with an error bounded by $16(\frac{1}{5})^9/9 \approx 9.1 \times 10^{-7};$
$4\tan^{-1}(\frac{1}{239}) \approx 4 \cdot \frac{1}{239} \approx 0.01673640167$ with an error bounded by $4(\frac{1}{239})^3/3 \approx 9.8 \times 10^{-8}$. Thus,
$\pi \approx 3.158328076 - 0.01673640167 \approx 3.141591675$ with error bounded by 1.008×10^{-6}.

Problems 9 Review, page 634

1. $\frac{3}{2}$ **3.** does not exist   **5.** 0   **7.** ∞ **9.** 1   **11.** ∞ **13.** 1   **15.** $-\frac{1}{2}$ **17.** ∞ **19.** 0

21. $\frac{1}{3}$ **23.** $6/7^4$ **25.** $\frac{1}{4}\ln\frac{13}{12}$ **27.** diverges   **29.** $\frac{2}{3}$ **31.** diverges   **33.** $2\sqrt{2}$ **35.** diverges

37. diverges   **39.** diverges   **41.** $1 + x + x^2/2! + x^3/3!$ **43.** $\dfrac{1}{2} + \dfrac{\sqrt{3}}{2}\left(x - \dfrac{\pi}{6}\right) - \dfrac{1}{4}\left(x - \dfrac{\pi}{6}\right)^2 - \dfrac{\sqrt{3}}{12}\left(x - \dfrac{\pi}{6}\right)^3$

45. $-(x - \pi/2) - \frac{1}{3}(x - \pi/2)^3$ **47.** $3 + 2x - x^2 + x^3$ **49.** $(\pi/3)^6/6! \approx 0.00183$ **51.** $e/7! < 3/7! \approx 0.000595$
53. $\pi/3$ **55.** 2π **57.** $\int_0^{1/2}(1 - x^4/2)\,dx = \frac{159}{320} = 0.496875$ with error bounded by $0.5^9/4! \approx 8.14 \times 10^{-5}$
61. b. $(-1)^N N!$

Computer Exercises, page 636

1. b. $\displaystyle\int_1^{10} \approx 0.2864709075$

$\displaystyle\int_1^{100} \approx 0.2913939146$

$\displaystyle\int_1^{1000} \approx 0.2914412830$ **c.** error $< 5 \times 10^{-7}$ **3.** $\tan 0.1 \approx 0.1003346720885$

CHAPTER 10

Problems 10.1, page 643

1. $\frac{1}{3}, \frac{1}{9}, \frac{1}{27}, \frac{1}{81}, \frac{1}{243}$ **3.** $\frac{3}{4}, \frac{15}{16}, \frac{63}{64}, \frac{255}{256}, \frac{1023}{1024}$ **5.** $1, 0, -1, 0, 1$ **7.** $a_n = \dfrac{n}{n+1}$ **9.** $a_n = n \cdot 5^{n-1}$

11. 0   **13.** divergent   **15.** 0   **17.** 0   **19.** e^4 **21.** diverges to ∞ **23.** 0   **25.** 0   **27.** 1
29. $a_0 \in [0, 1]$ **37.** Disprove

Problems 10.2, page 650

1. $\frac{1}{2}$ **3.** 1   **5.** 1   **7.** 1   **9.** 2   **11.** $(\ln 3)/3 \approx 0.3662$
13. $\sin n\pi = 0$; the sequence is constant (the sequence is both monotone increasing and monotone decreasing)
15. strictly increasing   **17.** strictly decreasing   **19.** strictly increasing   **21.** strictly increasing
23. not monotonic   **25.** strictly increasing   **27.** not monotonic   **29. b.** 0.56714
31. 12.713 [*Note:* $x \approx 1.2959$ is another fixed point but fixed point iteration will not find it because $f'(x) > 1$ in $(0, 5)$]
33. a. $x = \pm \sqrt{2}$ **35. b.** 27

Problems 10.3, page 656

1. 364   **3.** $[1 - (-5)^6]/[1 - (-5)] = -2604$ **5.** $(0.3)^2[1 - (-0.3)^7]/1.3 \approx 0.0692$ **7.** $(b^{16} - 1)/(b^{16} + b^{14})$
9. $(1 - 16\sqrt{2})/(1 - \sqrt{2}) = 31 + 15\sqrt{2}$ **11.** $(-16)(\frac{1025}{5}) = -3280$ **13.** $\frac{4}{3}$ **15.** $\frac{3}{4}$ **17.** $\frac{10}{3}$ **19.** $\frac{1}{3}$
21. $n \geq 7$ **23.** $n \geq 459$

Problems 10.4, page 667

1. $\frac{1}{3}$ **3.** $\frac{35}{99}$ **5.** $\frac{71}{99}$ **7.** $\frac{501}{999} = \frac{167}{333}$ **9.** $\frac{11351}{99900}$ **11.** 10   **13.** $2 - 1 - \frac{1}{2} = \frac{1}{2}$ **15.** $\frac{3}{5}$
17. 125   **19.** $\frac{1}{2}$ **21.** 1   **23.** $2^3 \cdot (3 - 1 - \frac{2}{3}) = \frac{32}{3}$ **25.** $(5^{-2}/6^1)(\frac{5}{6})^4[1/(1 - 5/6)] = 5^2/6^4 = \frac{25}{1296}$
27. $\frac{3}{2} + 3 = \frac{9}{2}$ **29.** $\frac{15}{4} - \frac{28}{3} = -\frac{67}{12}$ **31.** $2 - \frac{7}{4} = \frac{1}{4}$ **33.** $1\frac{1}{11}$ hr $= 1:05\frac{5}{11}$ P.M. $\approx 1:05:27$ P.M.
35. $8 + 8 \cdot 2 \cdot \frac{2}{3} + 8 \cdot 2 \cdot (\frac{2}{3})^2 + 8 \cdot 2 \cdot (\frac{2}{3})^3 + \cdots = 40$ meters   **41.** Prove
45. $\sum_{n=0}^{\infty} a_n = a_1 + a_0 \sum_{k=1}^{\infty} (1/k!) = a_1 + a_0(e - 1)$; see Section 10.9

Problems 10.5, page 674

1. diverges   **3.** diverges   **5.** converges   **7.** diverges   **9.** converges   **11.** diverges   **13.** converges
15. converges   **17.** diverges   **19.** diverges   **21.** diverges   **23.** diverges   **25.** diverges   **27.** converges
29. converges   **31.** diverges   **33.** converges   **39.** diverges   **43. a.** ≈ 11.23 hours   **b.** ≈ 69.44 days
c. ≈ 28.22 years (using 1 year $= 365.25$ days)

Problems 10.6, page 679

1. diverges   **3.** converges   **5.** converges   **7.** diverges   **9.** converges   **11.** converges   **13.** diverges
15. converges   **17.** diverges   **19.** converges   **21.** converges   **23.** converges   **25.** converges

Problems 10.7, page 689

1. divergent   **3.** divergent   **5.** conditionally convergent   **7.** conditionally convergent
9. conditionally convergent   **11.** conditionally convergent   **13.** divergent   **15.** divergent
17. conditionally convergent   **19.** absolutely convergent   **21.** conditionally convergent
23. absolutely convergent   **25.** divergent   **27.** divergent   **29.** conditionally convergent
31. $\sum_{k=1}^{6} \frac{(-1)^{k+1}}{k!} \approx 0.63194$ **33.** $\sum_{k=1}^{9} \frac{(-1)^{k+1}}{k^4} \approx 0.947093$ **35.** $\sum_{k=1}^{5} \frac{(-1)^{k+1}}{k^k} \approx 0.783451$
37. $1 - \frac{1}{2} - \frac{1}{4} - \frac{1}{6} - \frac{1}{8} + \frac{1}{3} - \frac{1}{10} - \frac{1}{12} - \frac{1}{14} - \frac{1}{16}$ **39.** The most obvious example is to let $a_n = 1/n$.
45. The series is absolutely convergent; it, and any rearrangement, converges to $-\pi^2/12 \approx -0.822$.
49. *Hint:* If $\sum_{k=0}^{\infty} 2^k a_k$ converges, then $\sum_{k=0}^{\infty} a_k$ is absolutely convergent.

Problems 10.8, page 695

1. $6; (-6, 6)$ **3.** $3; (-4, 2)$ **5.** $\frac{1}{3}; (-\frac{1}{3}, \frac{1}{3})$ **7.** $1; [0, 2]$ **9.** $\infty; (-\infty, \infty)$ **11.** $1; (-1, 1)$
13. $\infty; (-\infty, \infty)$ **15.** $1; (-1, 1)$ **17.** $\infty; (-\infty, \infty)$ **19.** $\frac{1}{2}; [-\frac{1}{2}, \frac{1}{2}]$ **21.** $\frac{5}{2}; (-4, 1)$ **23.** $0; x = 0$
25. $1; (-1, 1)$ **27.** $0; x = -1$ **29.** $3; [-13, -7)$ **31.** $1; (-1, 1)$ **33.** $1; (-1, 1)$

Problems 10.9, page 703

1. $\sum_{k=0}^{\infty} (-1)^k x^{2k}$ **3.** $\sum_{k=0}^{\infty} \dfrac{(-1)^k x^{2k+1}}{2k+1}$ **5.** $\sum_{k=0}^{\infty} (-1)^k (x-1)^k$ for $0 < x < 2$

7. $T_{14} \approx 3.139220$ with an error of $|\pi - T_{14}| \leq 0.00445$. Note that by Theorem 10.7.4 on page 684,

$|\pi - T_n| \leq \dfrac{4}{(2n+1)(2n+3)}$. Alternatively, $\pi = 6 \tan^{-1} \dfrac{1}{\sqrt{3}} = S = 6 \sum_{k=0}^{\infty} \dfrac{(-1)^k}{(2k+1)} \left(\dfrac{1}{\sqrt{3}}\right)^{2k+1}$; then

$S_5 = 6\left(\dfrac{1}{\sqrt{3}} - \dfrac{1}{3(\sqrt{3})^3} + \dfrac{1}{5(\sqrt{3})^5} - \dfrac{1}{7(\sqrt{3})^7} + \dfrac{1}{9(\sqrt{3})^9} - \dfrac{1}{11(\sqrt{3})^{11}}\right) \approx 3.141309$; error < 0.00037.

9. $S_3 \approx 0.743$ with error ≤ 0.0046; $T_3 \approx 0.755$ with error ≤ 0.0096 **11.** $S_2 \approx 0.496884$ with error ≤ 0.000009; $T_3 \approx 0.496795$ with error ≤ 0.000005 **13.** $S_2 \approx 0.763542$ with error ≤ 0.000347; $T_2 \approx 0.7569$ with error ≤ 0.0068
15. $S_4 \approx 0.190031$ with error ≤ 0.0038; $T_3 \approx 0.1955$ with error ≤ 0.0074 **17.** 0.09871   **19.** 0.47725
23. $\ln 1.5 \approx 0.4055$; $\ln 0.5 \approx -0.6931$; $\ln 2 \approx 0.6931$

Problems 10.10, page 713

1. $\sum_{k=0}^{\infty} e(x-1)^k/k!$ **3.** $\dfrac{1}{\sqrt{2}}\left[1 - \left(x - \dfrac{\pi}{4}\right) - \dfrac{(x-\pi/4)^2}{2!} + \dfrac{(x-\pi/4)^3}{3!} + \dfrac{(x-\pi/4)^4}{4!} - \cdots\right]$ **5.** $\sum_{k=0}^{\infty} (\beta x)^k/k!$

7. $\sum_{k=0}^{\infty} x^{k+1}/k!$ **9.** $\sum_{k=0}^{\infty} (-1)^k x^{2k}/(2k+1)!$ **11.** $\sum_{k=1}^{\infty} (-1)^{k+1}(x-1)^{k+1}/k$; $(0, 2]$

13. $1 + 0 + (x - \pi/2)^2/2! + 0$; $(0, \pi)$ **15.** $2 + \dfrac{1}{4}(x-4) + \sum_{k=2}^{\infty} \dfrac{(-1)^{k+1} 1 \cdot 3 \cdot \cdots \cdot (2k-3)}{2^{3k-1}k!}(x-4)^k$; $R = 4$

17. $\sum_{k=0}^{\infty} \dfrac{(-1)^k x^{4k+2}}{(2k+1)!}$ **19.** $\sum_{k=0}^{\infty} (-1)^k x^{2k+1}/(2k+1)$; $R = 1$ **21.** $1 + \dfrac{1}{4}x - \dfrac{1}{4} \cdot \dfrac{3}{4} \cdot \dfrac{x^2}{2!} + \dfrac{1}{4} \cdot \dfrac{3}{4} \cdot \dfrac{7}{4} \cdot \dfrac{x^3}{3!} - \dfrac{1}{4} \cdot \dfrac{3}{4} \cdot \dfrac{7}{4} \cdot \dfrac{11}{4} \cdot \dfrac{x^4}{4!} + \cdots$

23. Using the first four terms we obtain the approximation 0.503807.   **25. a.** $\dfrac{2}{\sqrt{\pi}} \sum_{k=0}^{\infty} \dfrac{(-1)^k x^{2k+1}}{k!(2k+1)}$

b. $\text{erf}(1) \approx 0.842699$; $\text{erf}(0.5) \approx 0.520500$ **35. c.** $\text{Si}(x) = \sum_{k=0}^{\infty} \dfrac{(-1)^k x^{2k+1}}{(2k+1)!(2k+1)}$ **d.** $\text{Si}(1) \approx 0.946083$; $\text{Si}(0.5) \approx 0.493108$

Problems 10 Review, page 716

1. $-1, 0, \frac{1}{3}, \frac{1}{2}, \frac{3}{5}$ **3.** $a_n = (2n-1)/2^{n+2}$ **5.** converges to 0   **7.** converges to 0   **9.** converges to e^{-2}
11. unbounded, not monotonic   **13.** bounded below by $\frac{2}{3}$ and above by 1, strictly increasing
15. bounded above by 2 and below by 0, strictly decreasing   **17.** bounded below by $\frac{-6}{5}$ and above by 1, strictly increasing
19. $(4^{11} - 1)/3 - 5 = 1{,}398{,}096$ **21.** $\frac{27}{16} - \frac{8}{75} = \frac{1897}{1200} \approx 1.58$ **23.** $\frac{79}{99}$ **25.** converges   **27.** converges
29. diverges   **31.** diverges   **33.** converges   **35.** converges   **37.** conditionally convergent
39. conditionally convergent   **41.** absolutely convergent   **43.** conditionally convergent   **45.** divergent
47. divergent   **49.** $T_6 \approx 0.9021$ with error ≤ 0.00086; $S_{10} \approx 0.901116$ with error ≤ 0.000751
51. $9\frac{9}{11}$ hr $\approx 9{:}49{:}05$ P.M.   **53.** $3; (-3, 3)$ **55.** $1; [-1, 1]$ **57.** $\infty; (-\infty, \infty)$ **59.** $1; (\frac{2}{3}, \frac{8}{3})$

61. $1; (-1, 1)$ **63.** 0.461281 with error $\leq 9.04 \times 10^{-6}$ **65.** 0.014509 with error ≤ 0.0000488 **67.** $\sum_{k=0}^{\infty} \dfrac{x^{k+2}}{k!}$

69. $\dfrac{1}{2} + \sum_{k=0}^{\infty} \dfrac{(-1)^k x^{2k} 2^{2k-1}}{(2k)!} = \dfrac{1}{2}\left(1 + \sum_{k=0}^{\infty} \dfrac{(-1)^k 4^k x^{2k}}{(2k)!}\right)$

Computer Exercises, page 717

1. b. $\sum_{n=1}^{10} \dfrac{1}{n^2} = \dfrac{1968329}{1270080} \approx 1.54976\ 77311\ 66540\ 69035\ 02142.$ $\sum_{n=1}^{100} \dfrac{1}{n^2} \approx 1.63498\ 39001\ 84892\ 86507\ 71695.$

$\sum_{n=1}^{500} \dfrac{1}{n^2} \approx 1.64293\ 60655\ 14894\ 16980\ 27009.$ **c.** $\sum_{n=501}^{\infty} \dfrac{1}{n^2} < \int_{500}^{\infty} \dfrac{dx}{x^2} < \sum_{n=500}^{\infty} \dfrac{1}{n^2}.$

d. $\left| \sum_{n=1}^{\infty} \dfrac{1}{n^2} - \sum_{n=1}^{500} \dfrac{1}{n^2} \right| = \sum_{n=501}^{\infty} \dfrac{1}{n^2} < \int_{500}^{\infty} \dfrac{dx}{x^2} = \dfrac{1}{500} = 0.002.$

3. b. $\gamma_n \leq 1 - \ln 2 + \int_2^{\infty} \left(\dfrac{1}{x-1} - \dfrac{1}{x} \right) dx = 1$ **c.** $0.572256, 0.572617, 0.576716$

d. error $\leq \int_{1001}^{\infty} \left(\dfrac{1}{x-1} - \dfrac{1}{x} \right) dx = -\ln \dfrac{1000}{1001} \approx 0.000995$ **e.** error $\leq \int_{n+1}^{\infty} \left(\dfrac{1}{x-1} - \dfrac{1}{x} \right) dx = \ln \dfrac{n+1}{n} \leq 10^{-10}$;

therefore $n \geq \dfrac{1}{e^{10^{-10}} - 1} \approx 10^{10}$ **5. b.** i. $a \approx 0.5$ ii. $a \approx 0.77$ iii. $a \approx 0.93$

CHAPTER 11

Problems 11.1, page 728

1. $Q = (3, 3)$ **3.** $Q = (4, 4)$ **5.** $Q = (-2, -5)$ **7.** $|\mathbf{v}| = 4\sqrt{2}, \theta = \pi/4$ **9.** $|\mathbf{v}| = 4\sqrt{2}, \theta = 7\pi/4$
11. $|\mathbf{v}| = 2, \theta = \pi/6$ **13.** $|\mathbf{v}| = 2, \theta = 2\pi/3$ **15.** $|\mathbf{v}| = 2, \theta = 4\pi/3$
17. $|\mathbf{v}| = \sqrt{89}, \theta = \pi + \tan^{-1}(-\frac{8}{5}) \approx 2.13$ (in second quadrant)   **19.** $\mathbf{v} = \mathbf{j}$ **21.** $\mathbf{v} = -6\mathbf{i} + \mathbf{j}$ **23.** $\mathbf{v} = -9\mathbf{i} + 5\mathbf{j}$
25. $\mathbf{v} = -5\mathbf{i} + 5\mathbf{j}$ **27. a.** $(6, 9)$ **b.** $(-3, 7)$ **c.** $(-7, 1)$ **d.** $(39, -22)$ **29.** $(2/\sqrt{13})\mathbf{i} + (3/\sqrt{13})\mathbf{j}$ **31.** $\frac{3}{5}\mathbf{i} + \frac{4}{5}\mathbf{j}$
33. $-\frac{3}{5}\mathbf{i} + \frac{4}{5}\mathbf{j}$ **35.** $\sin \theta = -3/\sqrt{13}, \cos \theta = 2/\sqrt{13}$ **37.** $-(1/\sqrt{2})\mathbf{i} - (1/\sqrt{2})\mathbf{j}$ **39.** $(\frac{3}{5}, -\frac{4}{5})$ **41.** $\frac{3}{5}\mathbf{i} + \frac{4}{5}\mathbf{j}$
43. a. $(1/\sqrt{2})\mathbf{i} - (1/\sqrt{2})\mathbf{j}$ **b.** $(3/\sqrt{34})\mathbf{i} - (5/\sqrt{34})\mathbf{j}$ **c.** $(7/\sqrt{193})\mathbf{i} - (12/\sqrt{193})\mathbf{j}$ **d.** $-(2/\sqrt{53})\mathbf{i} + (7/\sqrt{53})\mathbf{j}$
45. $(3\sqrt{3}/2)\mathbf{i} + (\frac{3}{2})\mathbf{j}$ **47.** $-7\mathbf{i}$ **49.** $(1/\sqrt{2})\mathbf{i} + (1/\sqrt{2})\mathbf{j}$ **51.** $-8\mathbf{j}$

Problems 11.2, page 735

1. $0; 0$ **3.** $0; 0$ **5.** $20; \frac{20}{29}$ **7.** $-22; -22/(5\sqrt{53})$ **9.** $100; 20/\sqrt{481}$ **11.** parallel   **13.** neither
15. orthogonal   **17.** parallel   **19.** $\frac{4}{7}\mathbf{i} + \frac{3}{7}\mathbf{j}$ **21.** $\mathbf{0}$ **23.** $-\frac{2}{13}\mathbf{i} + \frac{3}{13}\mathbf{j}$ **25.** $\frac{14}{5}\mathbf{i} + \frac{28}{5}\mathbf{j}$ **27.** $-\frac{14}{5}\mathbf{i} + \frac{28}{5}\mathbf{j}$
29. $\dfrac{\alpha + \beta}{2}\mathbf{i} + \dfrac{\alpha + \beta}{2}\mathbf{j}$ **31.** $\dfrac{\alpha - \beta}{2}\mathbf{i} + \dfrac{\alpha - \beta}{2}\mathbf{j}$ **33. a.** $-\frac{3}{4}$ **b.** $\frac{4}{3}$ **c.** $(-96 - \sqrt{7500})/78 \approx -2.34$
d. $(-96 + \sqrt{7500})/78 \approx -0.12$ **35.** $52/(5\sqrt{113}) \approx 0.9783; 61/(\sqrt{34}\sqrt{113}) \approx 0.9841; -27/(5\sqrt{34}) \approx -0.9261$
37. $\text{Proj}_{\overrightarrow{PQ}} \overrightarrow{RS} = \dfrac{21}{13}\mathbf{i} + \dfrac{14}{13}\mathbf{j}; \text{Proj}_{\overrightarrow{RS}} \overrightarrow{PQ} = -\dfrac{7}{26}\mathbf{i} + \dfrac{35}{26}\mathbf{j}$ **39.** $\sqrt{5}$ **41.** $-2\mathbf{i} - 5\mathbf{j}$ N; $2\mathbf{i} + 5\mathbf{j}$ N
43. $-2\mathbf{i} - 3\mathbf{j}$ N; $2\mathbf{i} + 3\mathbf{j}$ N   **45.** $-2\sqrt{3}\mathbf{i} - 7\mathbf{j}$ lb; $2\sqrt{3}\mathbf{i} + 7\mathbf{j}$ lb   **47.** $(3/\sqrt{2})\mathbf{i} - [(3/\sqrt{2}) + 2]\mathbf{j}$ N;
$-(3/\sqrt{2})\mathbf{i} + [(3/\sqrt{2}) + 2]\mathbf{j}$ N   **49.** $(7\sqrt{2} - (7\sqrt{3}/2) - \frac{7}{2})(\mathbf{i} + \mathbf{j})$ N; $(-7\sqrt{2} + (7\sqrt{3}/2) + \frac{7}{2})(\mathbf{i} + \mathbf{j})$ N   **51.** -12 J
53. $(8\sqrt{3} + 4)$ J   **55.** $3\sqrt{2}$ J   **57.** $12/\sqrt{13}$ J   **59.** $500(\sin 20°/\sin 30°) \approx 342$ N
61. tugboat 1: $(500)(\cos 20°)(750) \approx 352{,}385$ J; tugboat 2: $(342)(\cos 30°)(750) \approx 221{,}149$ J   **63.** *Hint:* $\mathbf{u} \cdot \mathbf{v} = 0$.
65. *Hint:* Disprove.   **67.** $a_1 a_2 + b_1 b_2 > 0$

Problems 11.3, page 741

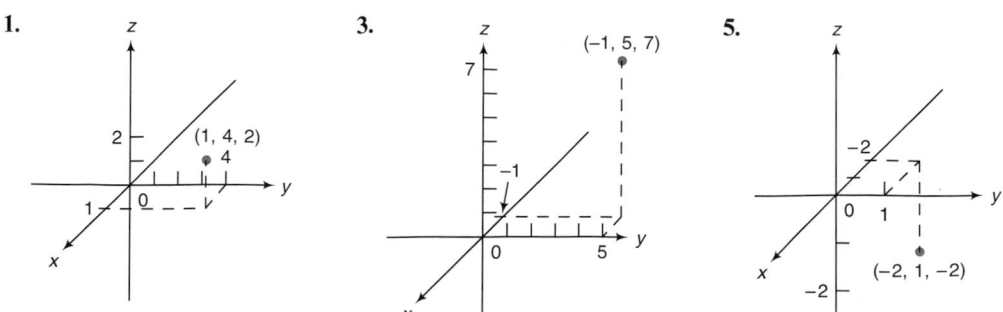

1.

3.

5.

7.

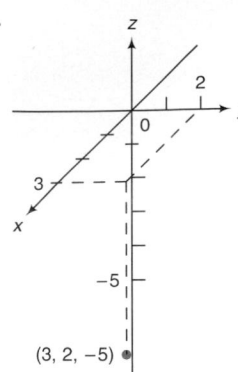

9.

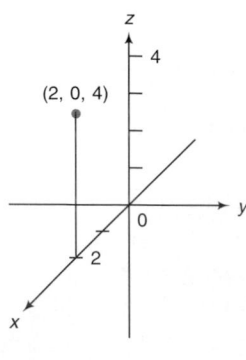

11.

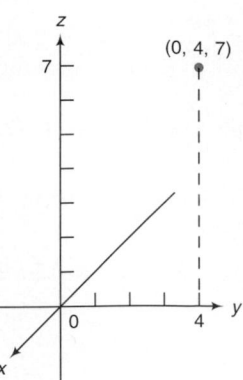

17. 2   **19.** 2   **21.** $6\sqrt{2}$ **23.** $\sqrt{6}$

13.

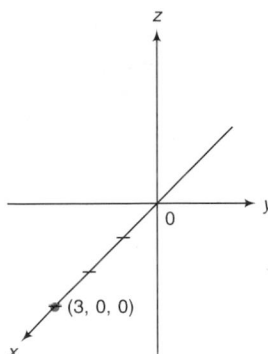

15.

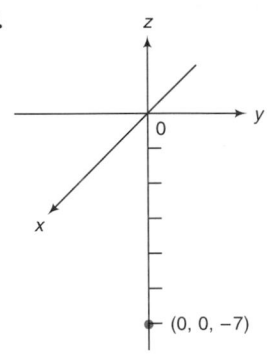

25. $\sqrt{118}$ **27.** $(x-2)^2 + (y+1)^2 + (z-4)^2 = 4$

29. *Hint:* $(3, 0, 1)$ is the midpoint of the segment between $(0, -4, 0)$ and $(6, 4, 2)$.   **31.** Center: $(2, 2, -4)$; radius: 4.

33. $(x-1)^2 + (y-2)^2 + (z+\frac{1}{2})^2 = 1$ **35.** $\frac{93}{4}$

Problems 11.4, page 749

1. $|\mathbf{v}| = 3$; 0, 1, 0   **3.** $|\mathbf{v}| = 14$; 0, 0, 1   **5.** $|\mathbf{v}| = \sqrt{17}$; $4/\sqrt{17}$, $-1/\sqrt{17}$, 0   **7.** $|\mathbf{v}| = \sqrt{13}$; $-2/\sqrt{13}$, $3/\sqrt{13}$, 0

9. $|\mathbf{v}| = \sqrt{3}$; $1/\sqrt{3}$, $-1/\sqrt{3}$, $1/\sqrt{3}$ **11.** $|\mathbf{v}| = \sqrt{3}$; $-1/\sqrt{3}$, $1/\sqrt{3}$, $1/\sqrt{3}$ **13.** $|\mathbf{v}| = \sqrt{3}$; $-1/\sqrt{3}$, $1/\sqrt{3}$, $-1/\sqrt{3}$

15. $|\mathbf{v}| = \sqrt{3}$; $-1/\sqrt{3}$, $-1/\sqrt{3}$, $-1/\sqrt{3}$ **17.** $|\mathbf{v}| = \sqrt{222}$; $-7/\sqrt{222}$, $2/\sqrt{222}$, $-13/\sqrt{222}$

19. $|\mathbf{v}| = \sqrt{82}$; $-3/\sqrt{82}$, $-3/\sqrt{82}$, $8/\sqrt{82}$ **21.** $-6\mathbf{j} + 9\mathbf{k}$ **23.** $-36\mathbf{i} + 54\mathbf{j} - 72\mathbf{k}$ **25.** $8\mathbf{i} - 14\mathbf{j} + 9\mathbf{k}$

27. $16\mathbf{i} + 29\mathbf{j} + 42\mathbf{k}$ **29.** $\sqrt{59}$ **31.** $\cos^{-1}(35/(\sqrt{29}\sqrt{59})) \approx 0.562 \approx 32°$ **33.** $\cos^{-1}(7/(\sqrt{38}\sqrt{50})) \approx 1.410 \approx 81°$

35. $\frac{25}{38}\mathbf{v} = -\frac{25}{19}\mathbf{i} - \frac{75}{38}\mathbf{j} + \frac{125}{38}\mathbf{k}$ **37.** $-\frac{1}{5}\mathbf{t} = -\frac{3}{5}\mathbf{i} - \frac{4}{5}\mathbf{j} - \mathbf{k}$ **39.** $\frac{35}{59}\mathbf{w} = \frac{35}{59}\mathbf{i} - \frac{245}{59}\mathbf{j} + \frac{105}{59}\mathbf{k}$ **41.** $\frac{1}{\sqrt{3}}\mathbf{i} + \frac{1}{\sqrt{3}}\mathbf{j} + \frac{1}{\sqrt{3}}\mathbf{k}$

43. $\frac{1}{\sqrt{26}}\mathbf{i} - \frac{3}{\sqrt{26}}\mathbf{j} + \frac{4}{\sqrt{26}}\mathbf{k}$ **45.** *Hint:* $\cos^2(\pi/6) + \cos^2(\pi/3) + \cos^2(\pi/4) = 1.5$.

47. $\{x, y, z: x + 3 = 0\}$. This is a plane (see Section 11.7).

49. $\sqrt{5610}/51 = \sqrt{\frac{110}{51}}$ (Note that the distance is given by $|\overrightarrow{QP} - \text{Proj}_{\overrightarrow{QR}} \overrightarrow{QP}|$; draw a picture.)   **51.** *Hint:* $\overrightarrow{PQ} \cdot \overrightarrow{PR} = 0$.

53. $\left(\frac{\pm 1}{\sqrt{419}}\right)(7\mathbf{i} - 17\mathbf{j} + 9\mathbf{k})$

Problems 11.5, page 754

1. $x\mathbf{i} + y\mathbf{j} + z\mathbf{k} = (2\mathbf{i} + \mathbf{j} + 3\mathbf{k}) + t(-\mathbf{i} + \mathbf{j} - 4\mathbf{k})$; $x = 2 - t$, $y = 1 + t$, $z = 3 - 4t$; $(x-2)/(-1) = y - 1 = (z-3)/(-4)$

3. $x\mathbf{i} + y\mathbf{j} + z\mathbf{k} = (\mathbf{i} + 3\mathbf{j} + 2\mathbf{k}) + t(\mathbf{i} + \mathbf{j} - 4\mathbf{k})$; $x = 1 + t$, $y = 3 + t$, $z = 2 - 4t$; $x - 1 = y - 3 = (z-2)/(-4)$

5. $x\mathbf{i} + y\mathbf{j} + z\mathbf{k} = (-4\mathbf{i} + \mathbf{j} + 3\mathbf{k}) + t(-\mathbf{j} - 2\mathbf{k})$; $x = -4$, $y = 1 - t$, $z = 3 - 2t$; $x = -4$ and $(y-1)/(-1) = (z-3)/(-2)$

7. $x\mathbf{i} + y\mathbf{j} + z\mathbf{k} = (\mathbf{i} + 2\mathbf{j} + 3\mathbf{k}) + t(2\mathbf{i} - 2\mathbf{k})$; $x = 1 + 2t$, $y = 2$, $z = 3 - 2t$; $y = 2$ and $(x-1)/2 = (z-3)/(-2)$

9. $x\mathbf{i} + y\mathbf{j} + z\mathbf{k} = (\mathbf{i} + 2\mathbf{j} + 4\mathbf{k}) + t(3\mathbf{k})$; $x = 1$, $y = 2$, $z = 4 + 3t$; $x = 1$ and $y = 2$

11. $x\mathbf{i} + y\mathbf{j} + z\mathbf{k} = (2\mathbf{i} + 2\mathbf{j} + \mathbf{k}) + t(2\mathbf{i} - \mathbf{j} - \mathbf{k})$; $x = 2 + 2t$, $y = 2 - t$, $z = 1 - t$; $(x - 2)/2 = (y - 2)/(-1) = (z - 1)/(-1)$

13. $x\mathbf{i} + y\mathbf{j} + z\mathbf{k} = (\mathbf{i} + 3\mathbf{k}) + t(\mathbf{i} - \mathbf{j})$; $x = 1 + t$, $y = -t$, $z = 3$; $x = -y$ or $x + y = 1$ and $z = 3$

15. $x\mathbf{i} + y\mathbf{j} + z\mathbf{k} = (-\mathbf{i} - 2\mathbf{j} + 5\mathbf{k}) + t(-3\mathbf{j} + 4\mathbf{k})$; $x = -1$, $y = -2 - 3t$, $z = 5 + 4t$; $x = -1$ and $\dfrac{y + 2}{-3} = \dfrac{z - 5}{4}$ or $4y + 3z = 7$

17. $x\mathbf{i} + y\mathbf{j} + z\mathbf{k} = (-\mathbf{i} - 3\mathbf{j} + \mathbf{k}) + t(-7\mathbf{j})$; $x = -1$, $y = -3 - 7t$, $z = 1$; $x = -1$ and $z = 1$

19. $x\mathbf{i} + y\mathbf{j} + z\mathbf{k} = (a\mathbf{i} + b\mathbf{j} + c\mathbf{k}) + t(d\mathbf{i} + e\mathbf{j})$; $x = a + dt$, $y = b + et$, $z = c$; $z = c$ and $(x - a)/d = (y - b)/e$

21. $x\mathbf{i} + y\mathbf{j} + z\mathbf{k} = (4\mathbf{i} + \mathbf{j} - 6\mathbf{k}) + t(3\mathbf{i} + 6\mathbf{j} + 2\mathbf{k})$; $x = 4 + 3t$, $y = 1 + 6t$, $z = -6 + 2t$; $(x - 4)/3 = (y - 1)/6 = (z + 6)/2$

23. The lines meet at $(2, -1, -3)$.   **25.** The lines meet at $(4, 3, 11)$.   **27.** no intersection   **29.** no intersection

33. $t = -(\overrightarrow{OP} \cdot \mathbf{v})/(\mathbf{v} \cdot \mathbf{v}) = -(\overrightarrow{OP} \cdot \mathbf{v})/|\mathbf{v}|^2$ **35.** $x = 2 + t$, $y = -3 + t$, $z = 1$; $x = 2 + 3t$, $y = -3$, $z = 1 - 4t$

Problems 11.6, page 765

1. $-6\mathbf{i} - 3\mathbf{j}$ **3.** $-\mathbf{i} - \mathbf{j} + \mathbf{k}$ **5.** $12\mathbf{i} + 8\mathbf{j} - 21\mathbf{k}$ **7.** $(bc - ad)\mathbf{j}$ **9.** $-5\mathbf{i} - \mathbf{j} + 7\mathbf{k}$ **11.** $\mathbf{0}$

13. $42\mathbf{i} + 6\mathbf{j}$ **15.** $-9\mathbf{i} + 39\mathbf{j} + 61\mathbf{k}$ **17.** $-4\mathbf{i} + 8\mathbf{k}$ **19.** $\mathbf{0}$ **21.** $\left(\dfrac{\pm 1}{\sqrt{181}}\right)(-9\mathbf{i} - 6\mathbf{j} + 8\mathbf{k})$

23. $\sqrt{30}/(\sqrt{6}\sqrt{29}) \approx 0.415$ **25.** $(x - 1)/2 = (y + 3)/(-26) = (z - 2)/(-22)$

27. $x = -2 + 13u$, $y = 3 + 22u$, $z = 4 - 8u$ **29.** $5\sqrt{5}$ **31.** $\sqrt{523}$ **33.** $\sqrt{a^2b^2 + b^2c^2 + c^2a^2}$ **35.** $\frac{1}{2}\sqrt{3778}$

37. $\frac{1}{2}\sqrt{3}$ **39.** 14 **41.** 23 **43.** $48/\sqrt{437}$ **45.** $23/\sqrt{27}$ **51.** $\frac{1}{6} \times$ volume of the parallelepiped

Problems 11.7, page 772

1. $x = 0$ (yz-plane)   **3.** $z = 0$ (xy-plane)   **5.** $x + z = 4$ **7.** $3x - y + 2z = 19$ **9.** $4x + y - 7z = -15$

11. $x + y + z = -6$ **13.** $20x + 13y - 3z = 58$ **15.** $x + y + z = 1$ **17.** coincident   **19.** orthogonal

21. orthogonal   **23.** coincident   **25.** $x = -1$ and $y + z = 2$ or $x = -1$, $y = t$, $z = 2 - t$

27. $x = t$, $y = -11 - 37t$, $z = -2 - 10t$ **29.** $\cos^{-1}(\frac{1}{3}) \approx 1.23 \approx 70.5°$ **31.** $\cos^{-1}(18/(\sqrt{26}\sqrt{69})) \approx 1.132 \approx 64.9°$

33. coplanar: $x - 22y - 17z = 0$ **35.** not coplanar   **37.** $33/\sqrt{59}$ **39.** $11/\sqrt{68}$

41.   **43.**   **45.**

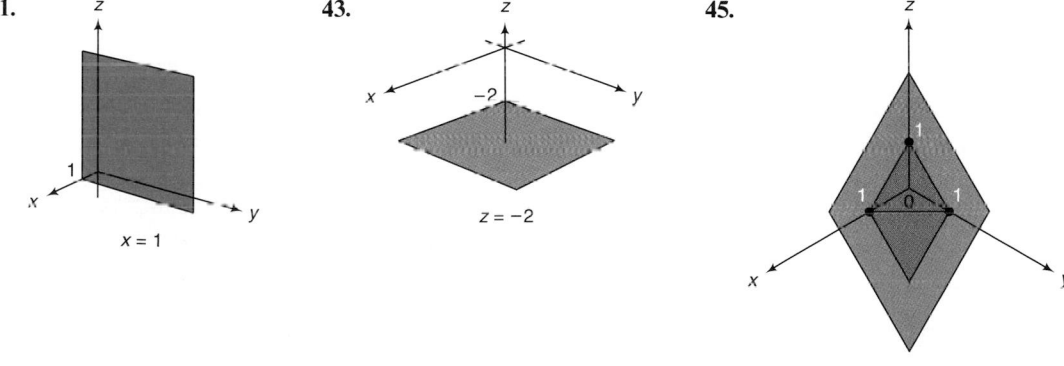

47.   **49.**

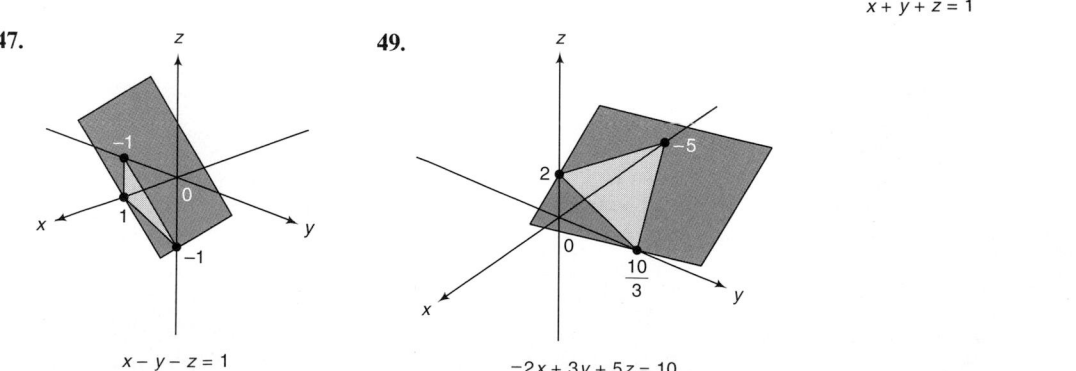

Problems 11.8, page 780

1. $y = \sin x$

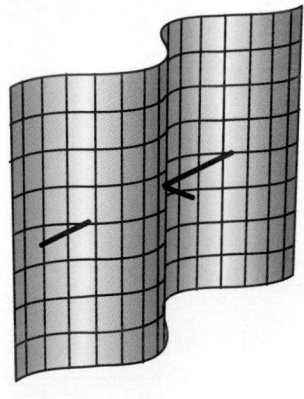

3. $y = \cos x$

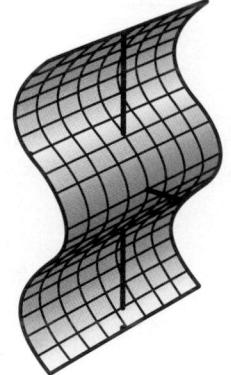

5. $z = x^3$

7. $|y| + |z - 5| = 1$

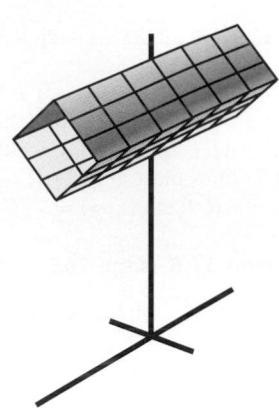

9. right circular cylinder, radius 2, centered on z-axis

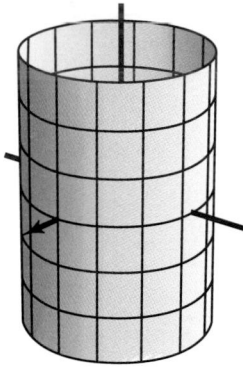

11. right circular cylinder, radius 2, centered on y-axis

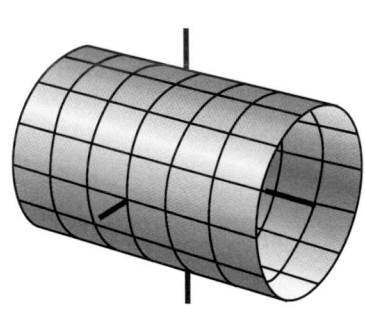

13. elliptic cylinder

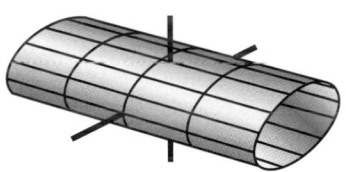

15. hyperbolic cylinder

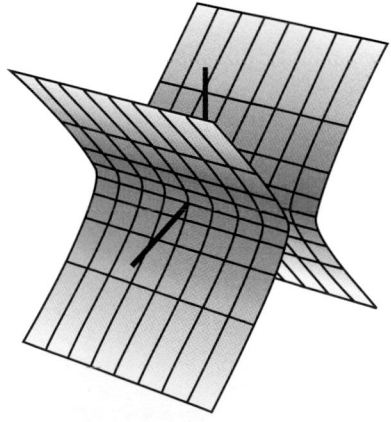

17. sphere of radius 1 centered at origin

19. hyperboloid of one sheet centered at $(0, -1, 0)$; cross-sections parallel to the xy-plane are ellipses

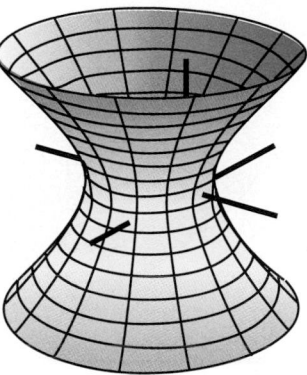

21. elliptic paraboloid; cross-sections parallel to the yz-plane are ellipses; defined for $x \le 4$

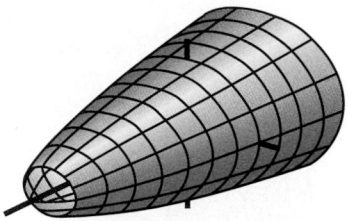

23. hyperboloid of two sheets centered at $(-2, -3, -4)$; cross-sections parallel to the xy-plane are ellipses; defined for $|z + 4| \ge 1$

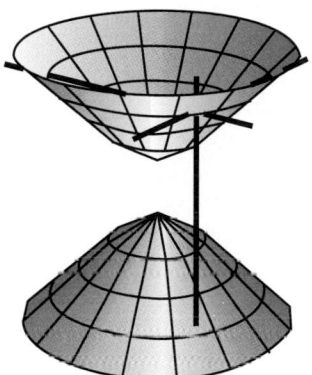

25. ellipsoid centered at $(0, 0, \frac{1}{2})$

27. hyperboloid of two sheets; cross-sections parallel to the yz-plane are ellipses; defined for $|x| \ge 2$

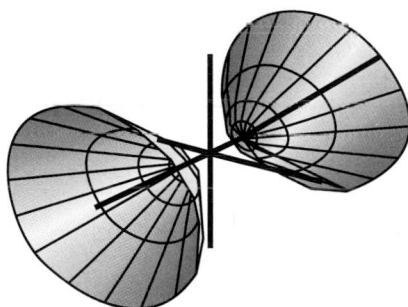

29. hyperboloid of two sheets; cross-sections parallel to the xz-plane are ellipses; defined for $|y| \ge 1$

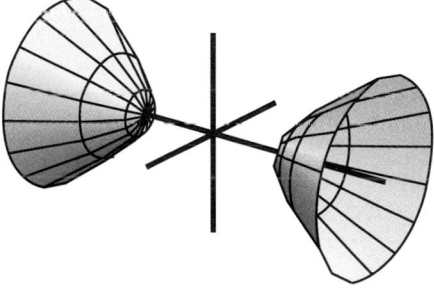

31. ellipsoid

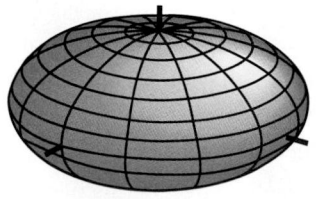

33. hyperbolic paraboloid; cross-sections parallel to the xy-plane are hyperbolas

35. elliptic (actually circular) paraboloid centered at $(\frac{1}{2}, -\frac{1}{2}, \frac{1}{2})$; cross-sections parallel to the xz-plane are ellipses

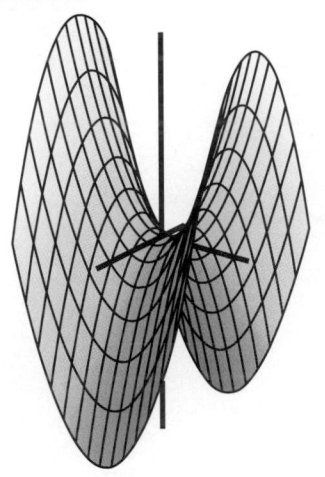

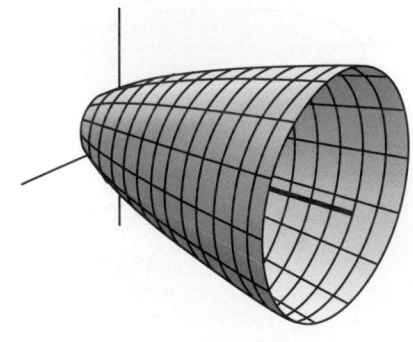

37. ellipsoid with equation $\dfrac{x^2 + z^2}{a^2} + \dfrac{y^2}{b^2} = 1$

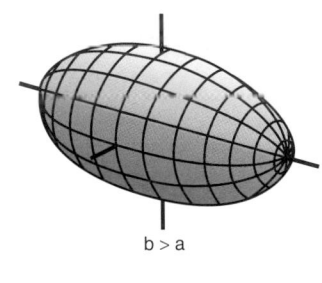

$b > a$

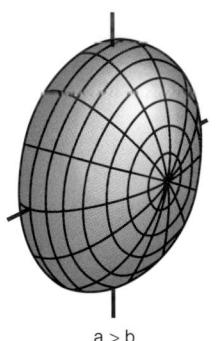

$a > b$

Problems 11.9, page 785

1. $(1, \sqrt{3}, 5)$ **3.** $(-4, 4\sqrt{3}, 1)$ **5.** $(-3/\sqrt{2}, 3/\sqrt{2}, 2)$ **7.** $(-10, 0, -3)$ **9.** $(-7/\sqrt{2}, -7/\sqrt{2}, 2)$

11. $(1, 0, 0)$ **13.** $(0, \theta, 1)$ where θ is arbitrary   **15.** $(\sqrt{2}, 3\pi/4, 4)$ **17.** $(4, 11\pi/6, 8)$ **19.** $(4, 7\pi/6, 1)$

21. $(\sqrt{3}, 0, 1)$ **23.** $(0, 3\sqrt{3}, 3)$ **25.** $(\frac{7}{2}, -\frac{7}{2}, -7/\sqrt{2})$ **27.** $(\sqrt{3}, 3, -2)$ **29.** $(5\sqrt{3}/4, -\frac{5}{4}, -5\sqrt{3}/2)$

31. $(\sqrt{2}, \pi/4, \pi/2)$ **33.** $(2, 7\pi/4, \pi/4)$ **35.** $(2\sqrt{2}, 5\pi/3, \pi/4)$

37. $(\sqrt{23}, \tan^{-1}(\sqrt{3}/2), \cos^{-1}(4/\sqrt{23})) \approx (4.796, 0.714, 0.584)\ [\approx (4.796, 41°, 33°)]$

39. $(\sqrt{23}, \pi + \tan^{-1}(2/\sqrt{3}), \cos^{-1}(-4/\sqrt{23})) \approx (4.796, 3.999, 2.557)\ [\approx (4.796, 229°, 147°)]$

[*Note:* $\cos\theta < 0$ and $\sin\theta < 0$ so θ is in the third quadrant.]   **41.** cylindrical: $r^2 + z^2 = 25$; spherical: $\rho = 5$

43. $x^2 + y^2 - 9y = 0$ **45.** cylindrical: $r^2 - z^2 = 1$; spherical: $\rho^2(\sin^2\varphi - \cos^2\varphi) = 1$ **47.** $z = x^2 + y^2$

49. $z\sqrt{x^2 + y^2} = 1$ **51.** $z = 2xy$ **53.** $\rho = 6\sin\varphi\sin\theta$

Problems 11 Review, page 786

1. a. $(10, 5)$ **b.** $(5, -3)$ **c.** $(-31, 12)$ **d.** $(0, 11)$ **3. a.** $(15, 0, 6)$ **b.** $(4, 3, -3)$ **c.** $(-8, -6, 6)$ **d.** $(0, -15, 23)$

5. $|\mathbf{v}| = 3\sqrt{2}, \theta = \pi/4$ **7.** $|\mathbf{u}| = 12\sqrt{2}, \theta = 5\pi/4$ **9.** $|\mathbf{w}| = 2, (\sqrt{3}/2, 1/2, 0)$ **11.** $|\mathbf{w}| = \sqrt{58}, (7/\sqrt{58}, -3/\sqrt{58}, 0)$

13. $|\mathbf{w}| = 5\sqrt{2}, (-3/5\sqrt{2}, 4/5\sqrt{2}, 1/\sqrt{2})$ **15.** $\sqrt{68}$ **17.** $\sqrt{216}$ **19.** $2\mathbf{i} + 2\mathbf{j}$ **21.** $4\mathbf{i} + 2\mathbf{j}$ **23.** $-4\mathbf{i} + 6\mathbf{j}$

25. $2\mathbf{i} - 4\mathbf{j} - 4\mathbf{k}$ **27.** $-(1/\sqrt{2})\mathbf{i} - (1/\sqrt{2})\mathbf{j}$ **29.** $-(10/\sqrt{149})\mathbf{i} + (7/\sqrt{149})\mathbf{j}$ **31.** $-(3/\sqrt{130})\mathbf{j} - (11/\sqrt{130})\mathbf{k}$

33. $-(1/\sqrt{14})\mathbf{i} + (2/\sqrt{14})\mathbf{j} + (3/\sqrt{14})\mathbf{k}$ **35.** $(7/\sqrt{53})\mathbf{i} - (2/\sqrt{53})\mathbf{j}$ **37.** $(1, \sqrt{3})$ **39.** $(-4, 0)$ **41.** $(3, 4, 0)$

43. $(1/\sqrt{2}, -1, -1/\sqrt{2})$ **45.** $-1; -1/\sqrt{10}$ **47.** $-22; -22/\sqrt{3965}$ **49.** $-2; -\frac{2}{5}$ **51.** $0; 0$ **53.** parallel
55. neither **57.** neither **59.** neither **61.** $7\mathbf{i} + 7\mathbf{j}$ **63.** $\frac{15}{13}\mathbf{i} + \frac{10}{13}\mathbf{j}$ **65.** $-\frac{90}{59}\mathbf{i} - \frac{210}{59}\mathbf{j} + \frac{30}{59}\mathbf{k}$
67. $\frac{26}{21}\mathbf{i} - \frac{52}{21}\mathbf{j} + \frac{13}{21}\mathbf{k}$ **69.** $-7\mathbf{i} - 7\mathbf{k}$ **71.** $-5\mathbf{i} - 41\mathbf{j} - 3\mathbf{k}$ **73.** $x\mathbf{i} + y\mathbf{j} + z\mathbf{k} = (3\mathbf{i} - \mathbf{j} + 4\mathbf{k}) + t(4\mathbf{i} - 7\mathbf{j} + 2\mathbf{k});$
$x = 3 + 4t, y = -1 - 7t, z = 4 + 2t; (x - 3)/4 = (y + 1)/(-7) = (z - 4)/2$
75. $x\mathbf{i} + y\mathbf{j} + z\mathbf{k} = (3\mathbf{i} + \mathbf{j} + 2\mathbf{k}) + t(3\mathbf{i} - \mathbf{j} - \mathbf{k}); x = 3 + 3t, y = 1 - t, z = 2 - t; (x - 3)/3 = (y - 1)/(-1) = (z - 2)/(-1)$
77. $y - z = 0$ **79.** $2y - 3z = -26$ **81.** $(3\sqrt{3}/2, \frac{3}{2}, -1)$ **83.** $(2\sqrt{2}, \pi/4, -4)$ **85.** $(3\sqrt{2}/4, 3\sqrt{6}/4, 3/\sqrt{2})$
87. $(2, 3\pi/4, 3\pi/4)$ **89.** cylindrical: $r^2 + z^2 = 5^2$; spherical: $\rho = 5$
91. cylindrical: $r^2(\cos^2\theta - \sin^2\theta) + z^2 = 1$; spherical: $\rho^2(\sin^2\varphi(\cos^2\theta - \sin^2\theta) + \cos^2\varphi) = 1$
93. plane parallel to z-axis which intersects xy-plane in the line $y = 3 - 5x$ **95.** parabolic cylinder

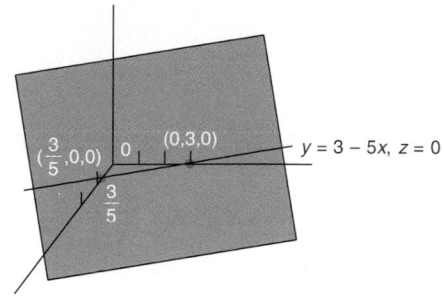
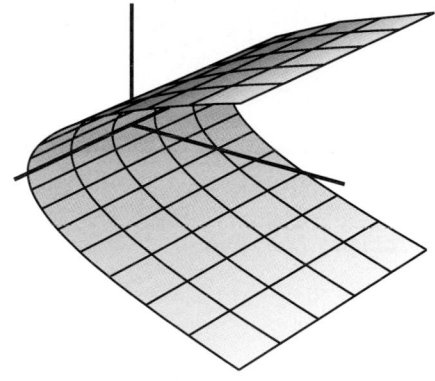

97. $\left(\frac{\pm 1}{\sqrt{2}}\right)(\mathbf{i} + \mathbf{j})$ **99. a.** 6 **b.** $-\frac{8}{3}$ **c.** $\frac{4}{5}$ **d.** $(192 \pm \sqrt{32448})/6$ **101.** $46x + 14y - 19z = -55$

103. $x = \frac{1}{2} - \frac{9}{2}t, y = \frac{7}{2} - \frac{11}{2}t, z = t$ **105.** $16/\sqrt{5}$ **107.** $5/\sqrt{6}$ **109.** $(x + 1)/14 = (y - 2)/(-26) = (z - 4)/(-11)$
111. the points are collinear **113.** $\sqrt{2065}$ **115.** right circular cylinder

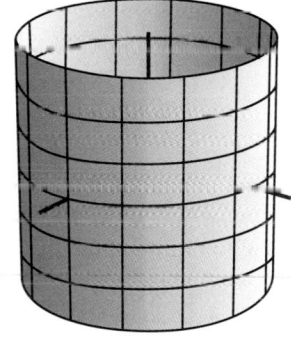

117. hyperboloid of one sheet; cross-sections parallel to xz-plane are ellipses

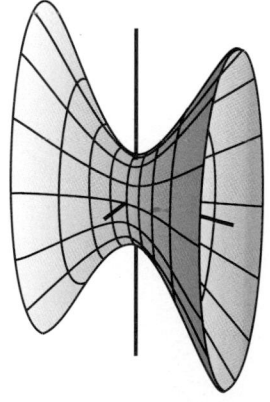

119. hyperboloid of two sheets; cross-sections parallel to xz-plane are ellipses; surface defined only for $|y| \geq \frac{5}{4}$

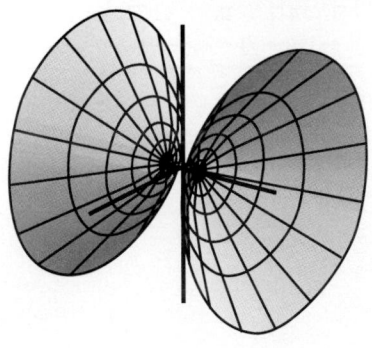

CHAPTER 12

Problems 12.1, page 793

1. $\mathbb{R} - \{0, 1\}$ **3.** $\mathbb{R} - \{-1, 1\}$ **5.** $(0, 1)$ **7.** $\mathbb{R} - \{k\pi/2: k \text{ is an integer}\}$
9. $x = y^2/4$ **11.** $x = y^{2/3}$ or $y = \pm x^{3/2}$ **13.** $y = 2x + 5$

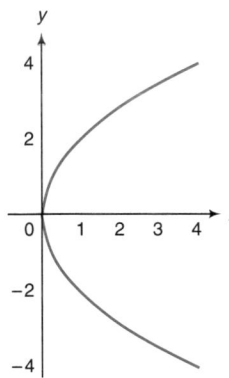

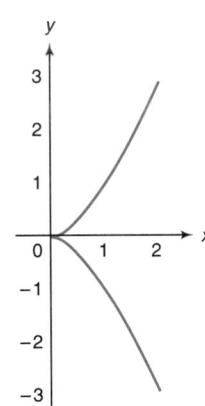

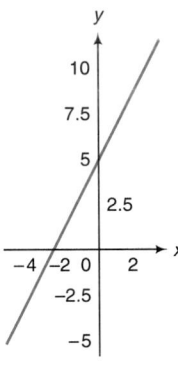

15. $x = y^2 + y + 1, y \geq 0$ **17.** $y = x^3 - 1$ **19.** $y = (\ln x)^2, x > 0$

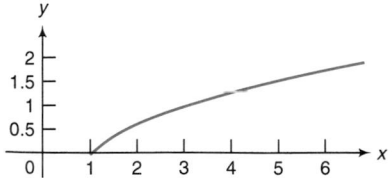

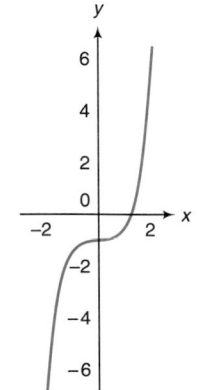

21. $x/y = \tan[(1/2)\ln(x^2 + y^2)]$. This is a logarithmic spiral. In polar coordinates, $r = e^{-\theta}$.　　**23.** $y = x^2$, $x > 0$

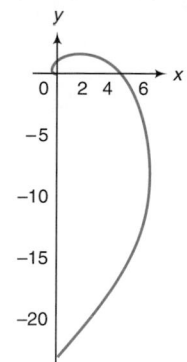

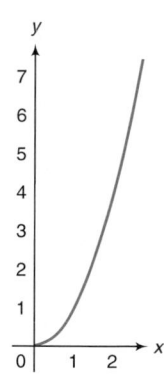

25. $z = x/2 = y/(-3)$　　　　**27.** $y = \cos x$ and $z = \sin x$ (a circular helix)　　**29.** $x = 2 - t$, $y = 4 + 2t$

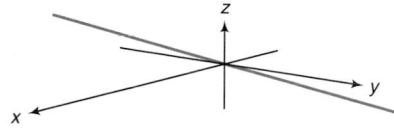

31. $x = 3 - 4t$, $y = 5 - 12t$　　**33.** $x = -2 + 6t$, $y = 3 + 4t$　　**35.** $x = 1 + t$, $y = 3 + t$, $z = 5 + t$
37. $x = -3 - 2t$, $y = 3t$, $z = 7 - 7t$　　**43. a.** $x = r(\alpha - \sin \alpha)$ and $y = r(1 - \cos \alpha)$

Problems 12.2, page 798

1. $\frac{4}{3}$　　**3.** -1　　**5.** $\sqrt{2}$　　**7.** $-\sqrt{3}/8$　　**9.** $-16/\pi^2$　　**11.** $y = \frac{2}{3}x + \frac{22}{3}$　　**13.** $y = -e^{-4}x + 2e^{-2}$
15. $y = \sqrt{2}$　　**17.** V: none; H: none　　**19.** V: none; H: $(0, 1)$　　**21.** V: $(\pm 2, 0)$, H: $(0, \pm 3)$
23. V: none $\left(\text{but } \dfrac{dy}{dx} \to \infty \text{ as } t \to -1^{+}\right)$; H: $(0, 0)$　　**25.** V: $((-1)^k, \cos(\frac{5}{3}(k + \frac{1}{2})\pi))$ for $k = 0, 1, 2, 3, 4, 5$;
H: $(\sin(\frac{3}{5}k\pi), (-1)^k)$ for $k = 0, 1, 2, 3, 4, 5, 6, 7, 8, 9$.　　**27.** $\sqrt{3}$　　**29.** $-\sqrt{3}$　　**31.** $5/\sqrt{3}$　　**33.** 2
35. $y = -x + 9$; $y = \frac{1}{3}x - \frac{17}{3}$　　**37.** $m = \sin \alpha/(1 - \cos \alpha)$

Problems 12.3, page 804

1. $\mathbf{i} - 5t^4\mathbf{j}$; $-20t^3\mathbf{j}$; $\mathbb{R}$　　**3.** $(2\cos 2t)\mathbf{i} - (3\sin 3t)\mathbf{j}$, $(-4\sin 2t)\mathbf{i} - (9\cos 3t)\mathbf{j}$; $\mathbb{R}$　　**5.** $t^{-1}\mathbf{i} + 3e^{3t}\mathbf{j}$; $-t^{-2}\mathbf{i} + 9e^{3t}\mathbf{j}$; $t > 0$
7. $(\sec^2 t)\mathbf{i} + (\sec t \tan t)\mathbf{j}$; $(2\sec^2 t \tan t)\mathbf{i} + (\sec^3 t + \sec t \tan^2 t)\mathbf{j}$; $t \neq \pi/2 + k\pi$, $k = 0, \pm 1, \pm 2, \ldots$
9. $-(\tan t)\mathbf{i} + (\cot t)\mathbf{j}$; $-(\sec^2 t)\mathbf{i} - (\csc^2 t)\mathbf{j}$; $\{t : \cos t > 0, \sin t > 0\} = \{t : 0 < t - 2k\pi < \pi/2\}$, $k = 0, \pm 1, \pm 2, \ldots$
11. $(1/\sqrt{13})(2\mathbf{i} + 3\mathbf{j})$　　**13.** $\mathbf{j}$　　**15.** $(1/\sqrt{2})(-\mathbf{i} + \mathbf{j})$　　**17.** $\mathbf{i}$　　**19.** $(1/\sqrt{97})(4\mathbf{i} - 9\mathbf{j})$　　**21.** $(1/\sqrt{14})(\mathbf{i} + 2\mathbf{j} + 3\mathbf{k})$
23. $(1/\sqrt{3})(\mathbf{i} + \mathbf{j} - \mathbf{k})$　　**25.** $(1/\sqrt{65})(-8\mathbf{i} + \mathbf{k})$　　**27.** $(-\frac{1}{2}\cos 2t + C_1)\mathbf{i} + (e^t + C_2)\mathbf{j}$　　**29.** $\frac{1}{2}(\mathbf{i} - \mathbf{j})$
31. $(t \ln t - t + C_1)\mathbf{i} + (te^t - e^t + C_2)\mathbf{j}$　　**33.** $\frac{8}{3}\mathbf{i} - 4\mathbf{j} + \frac{32}{5}\mathbf{k}$　　**35.** $(-\cos t^2 + C_1)\mathbf{i} + (\sin t^2 + C_2)\mathbf{j} + (e^{t^2} + C_3)\mathbf{k}$
37. $[(t^4/4) + 2]\mathbf{i} + [5 - (t^6/6)]\mathbf{j}$　　**39.** $(\sin t)\mathbf{i} - (\cos t)\mathbf{k}$　　**41.** $(1/\sqrt{a^2 + b^2})(-a\mathbf{i} + b\mathbf{j})$
43. $[1/(8 + 2\sqrt{2} - 2\sqrt{6})^{1/2}][(-1 - \sqrt{2})\mathbf{i} + (\sqrt{3} - \sqrt{2})\mathbf{j}] = \{1/[18 + (9/\sqrt{2})(1 - \sqrt{3})]^{1/2}\}[(-\frac{3}{2} - 3/\sqrt{2})\mathbf{i} + (3\sqrt{3}/2 - 3/\sqrt{2})\mathbf{j}]$
45. $\dfrac{845{,}000}{16.1} \approx 52{,}484.5$ ft ≈ 9.94 miles　　**47.** ≈ 209.9 m (after ≈ 2.42 sec)

Problems 12.4, page 810

1. $(2 + \sec^2 t)\mathbf{i} - [\sin t + \sec t \tan t]\mathbf{j}$　　**3.** 0　　**5.** $\left[\dfrac{1}{\sqrt{1 - t^2}} - \sin t\right]\mathbf{i} + \left[\cos t - \dfrac{1}{\sqrt{1 - t^2}}\right]\mathbf{j}$　　**7.** $5\mathbf{i} - 9t^2\mathbf{j} - 6t\mathbf{k}$

9. $(1/\sqrt{2})[(-\sin t)\mathbf{i} + (\cos t)\mathbf{j} + \mathbf{k}]$; $-\frac{1}{2}\mathbf{i} + \frac{1}{2}\mathbf{j} + (1/\sqrt{2})\mathbf{k}$　　**11.** $\dfrac{1}{\sqrt{10 + 7\cos^2 t}}[(-3\sin t)\mathbf{i} + (4\cos t)\mathbf{j} + \mathbf{k}]$; $(1/\sqrt{17})(4\mathbf{j} + \mathbf{k})$

13. $(1/\sqrt{1+4t^2})(\mathbf{j}+2t\mathbf{k})$; $(1/\sqrt{5})(\mathbf{j}+2\mathbf{k})$ **15.** $(1/\sqrt{6})[(\cos 2t - 2\sin 2t)\mathbf{i} + (\sin 2t + 2\cos 2t)\mathbf{j} + \mathbf{k}]$; $(1/\sqrt{6})(\mathbf{i}+2\mathbf{j}+\mathbf{k})$

17. $\mathbf{T}(t) = (-\sin 3t)\mathbf{i} + (\cos 3t)\mathbf{j}$; $\mathbf{n}(t) = (-\cos 3t)\mathbf{i} + (-\sin 3t)\mathbf{j}$; $\mathbf{T}(0) = \mathbf{j}$; $\mathbf{n}(0) = -\mathbf{i}$

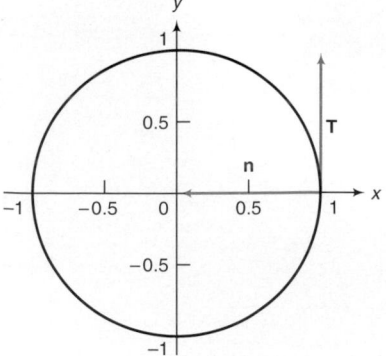

19. $\mathbf{T}(t) = (-\sin 4t)\mathbf{i} + (\cos 4t)\mathbf{j}$; $\mathbf{n}(t) = (-\cos 4t)\mathbf{i} + (-\sin 4t)\mathbf{j}$; $\mathbf{T}(\pi/4) = -\mathbf{j}$; $\mathbf{n}(\pi/4) = \mathbf{i}$

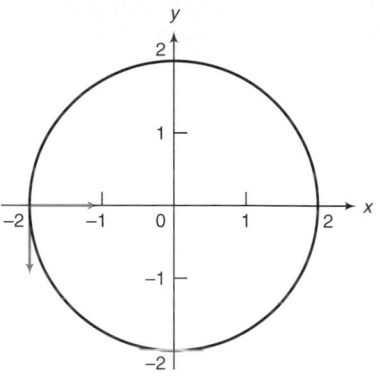

21. $\mathbf{T}(t) = (-\sin t)\mathbf{i} + (\cos t)\mathbf{j}$; $\mathbf{n}(t) = (-\cos t)\mathbf{i} + (-\sin t)\mathbf{j}$; $\mathbf{T}(\pi/4) = (1/\sqrt{2})(-\mathbf{i}+\mathbf{j})$; $\mathbf{n}(\pi/4) = (1/\sqrt{2})(-\mathbf{i}-\mathbf{j})$

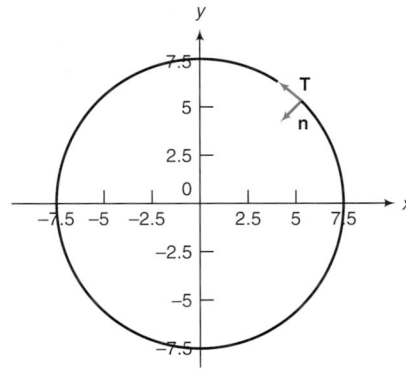

23. $\mathbf{T}(t) = \mathbf{T}(3) = (1/\sqrt{34})(3\mathbf{i}-5\mathbf{j})$; $\mathbf{n}(t) = \mathbf{n}(3) = \pm(1/\sqrt{34})(5\mathbf{i}+3\mathbf{j})$

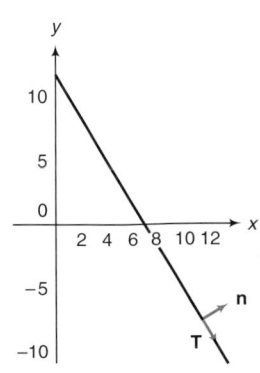

25. $\mathbf{T}(t) = \mathbf{T}(t_0) = (1/\sqrt{b^2 + d^2})(b\mathbf{i} + d\mathbf{j})$; $\mathbf{n}(t) = \mathbf{n}(t_0) = \pm(1/\sqrt{b^2 + d^2})(d\mathbf{i} - b\mathbf{j})$

27. $\mathbf{T}(t) = \sqrt{\dfrac{1 + \sin t}{2}}\,\mathbf{i} - \dfrac{\cos t}{\sqrt{2 + 2\sin t}}\,\mathbf{j}$; $\mathbf{n}(t) = \dfrac{\cos t}{\sqrt{2 + 2\sin t}}\,\mathbf{i} + \sqrt{\dfrac{1 + \sin t}{2}}\,\mathbf{j}$; $\mathbf{T}(\pi/2) = \mathbf{i}$; $\mathbf{n}(\pi/2) = \mathbf{j}$

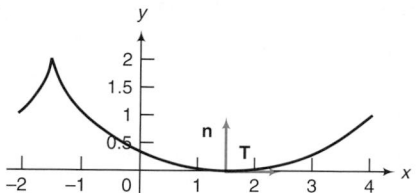

29. $\mathbf{T}(t)$ and $\mathbf{n}(t)$ are the same as for Problem 27; $\mathbf{T}\left(\dfrac{\pi}{4}\right) = \sqrt{\dfrac{\sqrt{2} + 1}{2\sqrt{2}}}\,\mathbf{i} - \dfrac{1}{\sqrt{4 + 2\sqrt{2}}}\,\mathbf{j} \approx 0.9239\mathbf{i} - 0.3827\mathbf{j}$

$\mathbf{n}\left(\dfrac{\pi}{4}\right) = \dfrac{1}{\sqrt{4 + 2\sqrt{2}}}\,\mathbf{i} + \sqrt{\dfrac{\sqrt{2} + 1}{2\sqrt{2}}}\,\mathbf{j} \approx 0.3827\mathbf{i} + 0.9239\mathbf{j}$

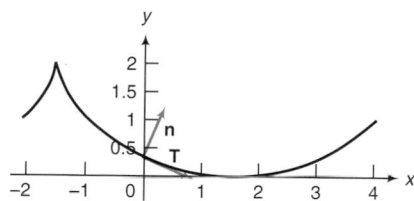

Problems 12.5, page 818

1. $\frac{2}{27}[13^{3/2} - 8]$ **3.** $8^{3/2} - 8$ **5.** $8|a|$ **7.** $\dfrac{1}{2}\left[\sqrt{2} - \left(\dfrac{\sqrt{6}}{5}\right) + \ln\left(\dfrac{\sqrt{5} + \sqrt{10}}{1 + \sqrt{6}}\right)\right] \approx 0.6861$

9. $\sqrt{2}(e^{\pi/2} - 1)$ **11.** $\sqrt{2}$ **13.** $|a|\pi/2$ **15.** $\frac{1}{3}[(4 + \pi^2)^{3/2} - 8]$ **17.** $8a$ **19.** $\sqrt{2}(e^3 - 1)$

21. $\mathbf{f} = 3\{[(s + 2)/2]^{3/2} - 1\}\mathbf{i} + 2\{[(s + 2)/2]^{3/2} - 1\}^{3/2}\mathbf{j}$ **23.** $\mathbf{f} = \frac{1}{27}\{[(27s + 8)^{2/3} - 4]^{3/2} + 27\}\mathbf{i} + \frac{1}{9}\{(27s + 8)^{2/3} - 13\}\mathbf{j}$

25. $\mathbf{f} = 3\cos(s/3)\mathbf{i} + 3\sin(s/3)\mathbf{j}$ **27.** $\mathbf{f} = a\cos(s/a)\mathbf{i} - a\sin(s/a)\mathbf{j}$ **29.** $\mathbf{f} = [a + b\cos(s/b)]\mathbf{i} + [c + b\sin(s/b)]\mathbf{j}$

31. $\mathbf{f} = a[1 - (2s/3a)]^{3/2}\mathbf{i} + a(2s/3a)^{3/2}\mathbf{j}$ **33.** $4\int_0^{\pi/2} \sqrt{a^2 \sin^2\theta + b^2\cos^2\theta}\,d\theta$

Problems 12.6, page 827

1. $\kappa = 2/5^{3/2}$; $\rho = 5^{3/2}/2$ **3.** $\kappa = \frac{1}{2}$; $\rho = 2$ **5.** $\kappa = \frac{4}{9}$; $\rho = \frac{9}{4}$ **7.** $\kappa = 24\sqrt{2}/125$; $\rho = 125/24\sqrt{2}$

9. $\kappa = \frac{1}{2}$; $\rho = 2$ **11.** $\kappa = 1/\sqrt{2}$; $\rho = \sqrt{2}$ **13.** $\kappa = 1/2\sqrt{2}$; $\rho = 2\sqrt{2}$ **15.** $\kappa = e/(e^2 + 1)^{3/2}$; $\rho = (e^2 + 1)^{3/2}/e$

17. $\kappa = 4/7^{3/2}$; $\rho = 7^{3/2}/4$ **19.** $\kappa = 2|a|/(1 + b^2)^{3/2}$; $\rho = (1 + b^2)^{3/2}/2|a|$ **21.** $\kappa = \rho = 1$ (compute limit as $x \to 1^-$)

23. $\kappa = 1/|a|$; $\rho = |a|$ **25.** $\kappa = 3/4|a|$; $\rho = 4|a|/3$ **27.** $\kappa = 3/2\sqrt{2}|a|$; $\rho = 2\sqrt{2}|a|/3$

29. $\mathbf{T} = (1/\sqrt{5})\mathbf{j} + (2/\sqrt{5})\mathbf{k}$; $\kappa = 2/5^{3/2}$; $\mathbf{n} = -(2/\sqrt{5})\mathbf{j} + (1/\sqrt{5})\mathbf{k}$; $\mathbf{B} = \mathbf{i}$

31. $\mathbf{T} = (1/\sqrt{a^2 + 1})[(a/\sqrt{2})\mathbf{i} - (a/\sqrt{2})\mathbf{j} + \mathbf{k}]$; $\kappa = a/(a^2 + 1)$; $\mathbf{n} = -(1/\sqrt{2})(\mathbf{i} + \mathbf{j})$; $\mathbf{B} = (1/\sqrt{a^2 + 1})[(1/\sqrt{2})\mathbf{i} - (1/\sqrt{2})\mathbf{j} - a\mathbf{k}]$

33. $\mathbf{T} = (1/\sqrt{b^2 + 1})(b\mathbf{j} + \mathbf{k})$; $\kappa = a/(b^2 + 1)$; $\mathbf{n} = -\mathbf{i}$; $\mathbf{B} = (1/\sqrt{b^2 + 1})(-\mathbf{j} + b\mathbf{k})$

35. $\mathbf{T} = (1/\sqrt{6})(\mathbf{i} + 2\mathbf{j} + \mathbf{k})$; $\kappa = \sqrt{5}/3$; $\mathbf{n} = (1/\sqrt{5})(-2\mathbf{i} + \mathbf{j})$; $\mathbf{B} = (1/\sqrt{30})(-\mathbf{i} - 2\mathbf{j} + 5\mathbf{k})$ **37.** $a_T = 0$, $a_n = 4$

39. $a_T = 4t/\sqrt{1 + 4t^2}$, $a_n = 2/\sqrt{1 + 4t^2}$ **41.** $a_T = \sin t\cos t/\sqrt{1 + \sin^2 t}$, $a_n = |\cos t|/\sqrt{1 + \sin^2 t}$

43. $a_T = (-e^{-2t} + e^{2t})/\sqrt{e^{-2t} + e^{2t}}$, $a_n = 2/\sqrt{e^{-2t} + e^{2t}}$ **45.** $\kappa_{max} = 2/(3\sqrt{3})$ at $x = 1/\sqrt{2}$ **47.** M^2

49. a. $[(10,000)(80,000)^2/(3600)^2] \cdot (1/\sqrt{2}) \approx 3,491,885.3$ N   **b.** $\sqrt{2.5\sqrt{2}(3600)^2(9.81)} \approx 21,201.4$ m/hr ≈ 21.2 km/hr $= 5.89$ m/sec

51. 0 **53.** $\dfrac{1}{2}$

Problems 12 Review, page 830

1. $y = 2x$ **3.** $x = (3 + y/2)^2$ **5.** $x^2 + y^2 = 1$ **7.** $x = y^3, x \geq 0$ **9.** 3; V: (0, 0), H: none
11. $\sqrt{3}$; V: $(\pm 1, 0)$, H: $(0, \pm 1)$ **13.** undefined; V: (1, 0), H: none   **15.** $-4/(3\sqrt{3})$; H: $(\pm 3, 0)$, V: $(0, \pm 4)$
17. $2i - 2tj$; $-2j$ **19.** $-5(\sin 5t)i + 2(\cos t)j$; $-25(\cos 5t)i - 2(\sin t)j$ **21.** $3t^2 i - 2tj + k$; $6ti - 2j$
23. $9i + \frac{243}{2}j$ **25.** $i + j + (\pi^2/8)k$ **27.** $4t - 3/2\sqrt{t}$ **29.** $4i - 3j + (3\cos t + 4\sin t)k$
31. $(1/\sqrt{41})(4i + 5j)$; $(1/\sqrt{41})(-5i + 4j)$ **33.** $(1/\sqrt{5})(2i + j)$; $(1/\sqrt{5})(-i + 2j)$ **35.** $\pi/3$ **37.** 16
39. $\sqrt{2}\pi/3$ **41.** $v = -\sqrt{3}i + j$, $|v| = 2$; $a = -2i - 2\sqrt{3}j$, $|a| = 4$
43. $v = (\ln 2 - 1)i + 2j$, $|v| = \sqrt{(\ln 2 - 1)^2 + 4}$; $a = (\ln^2 2 + 1)i$, $|a| = \ln^2 2 + 1$
45. $v = 2(\cosh 1)i + 4j$, $|v| = \sqrt{4\cosh^2 1 + 16}$; $a = 2(\sinh 1)i$, $|a| = 2\sinh 1$ **47.** $a_T = 0$, $a_n = 2$
49. $a_T = (6t + 12t^3)/\sqrt{t^2 + t^4}$, $a_n = 6t^2/\sqrt{t^2 + t^4}$ **51.** $\kappa = \rho = 1$ **53.** $\kappa = 36/(\frac{97}{2})^{3/2}$, $\rho = (\frac{97}{2})^{3/2}/36$
55. $\kappa = 4$, $\rho = 1/4$ **57.** $\kappa = e^{-1}/(1 + e^{-2})^{3/2}$, $\rho = e(1 + e^{-2})^{3/2}$ **59.** $\kappa = \frac{3}{4}$, $\rho = \frac{4}{3}$
61. $f = \frac{3}{4}[(2s + 1)^{2/3} - 1]i + [(2s + 1)^{2/3} - 1]^{3/2}j$ **63.** $f = 2\cos(s/2)i + 2\sin(s/2)j$
65. $T(2) = (1/\sqrt{5})(j + 2k)$; $\kappa(2) = 5^{-3/2}$; $n(2) = (1/\sqrt{5})(-2j + k)$; $B(2) = i$
67. $T(\pi/6) = (1/\sqrt{5})(-i + \sqrt{3}j + k)$; $\kappa(\pi/6) = 2/5$; $n(\pi/6) = \frac{1}{2}(-\sqrt{3}i - j)$; $B(\pi/6) = (1/2\sqrt{5})(i - \sqrt{3}j + 4k)$

Computer Exercises, page 831

1. Contestant 1: $\pi/\sqrt{g} \approx 3.14/\sqrt{g}$
    Contestant 2: $\approx 3.72/\sqrt{g}$
    Contestant 3: $\approx 3.28/\sqrt{g}$ (The pass of quickest decent is the brachistochrome.)

3. The curvature is $\dfrac{\sqrt{-32\cos^4 t - 4\cos^2 t(8\sin^2 t - 9) + 5}}{2(5 - 4\cos^2 t)^{3/2}}$. The critical points of the curvature occur
where $\sin t \cos t = 0$.

CHAPTER 13

Problems 13.1, page 839

1. $\mathbb{R}^2$; $[-1, 1]$ **3.** $\mathbb{R}^2$; $[0, \infty)$ **5.** $\{(x, y): y \neq 0\}$; $\mathbb{R}$ **7.** $\{(x, y): x^2 + 4y^2 \leq 1\}$; $[0,1]$
9. $\{(x, y): 1 + x^2 > y^2\}$; $\mathbb{R}$ **11.** $\{(x, y): x^2 \geq y^2 \text{ and } x \neq y\}$; $[0, \infty)$ **13.** $\{(x, y): -1 \leq x + y \leq 1\}$; $[-\pi/2, \pi/2]$
15. $\{(x, y): x \neq -y\}$; $\mathbb{R}$ **17.** $\mathbb{R}^3$; $\mathbb{R}$ **19.** $\mathbb{R}^3$; $(0, \infty)$ **21.** $\mathbb{R}^3$; $\mathbb{R}$ **23.** $\{(x, y, z): z \neq 0\}$; $\mathbb{R}$
25. $\{(x, y, x): y \neq 0\}$; $(-\pi/2, \pi/2)$ **27.** $\mathbb{R}^3$; $[-1, 1]$ **29.** $\{(x, y, z): x + y - z > 0\}$; $\mathbb{R}$
31. $\mathbb{R}^3 - \{(0, 0, 0)\}$; $(0, \infty)$ **33.** $\{(x, y, z): x^2 - y^2 - z^2 > 0\}$; $(0, \infty)$
35. $z = x - y^2$, parabolic cylinder   **37.** $y = x^2 + 4z^2$, elliptic paraboloid

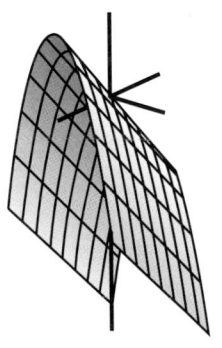

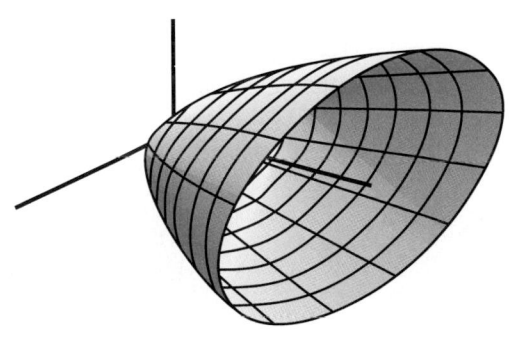

39. $z = x^2 - 4y^2$, hyperbolic paraboloid

41. $y = \sqrt{x^2 - 4z^2 + 4}$, hyperboloid of one sheet (right half)

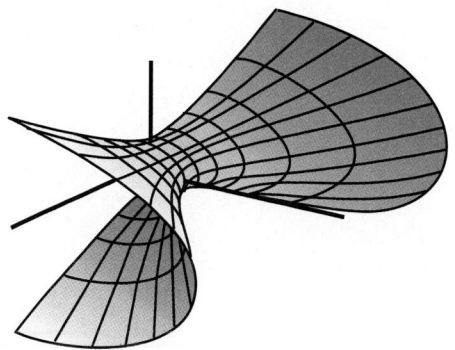

43. parallel straight lines (with slopes of -1) $y = -x + (z^2 - 1)$

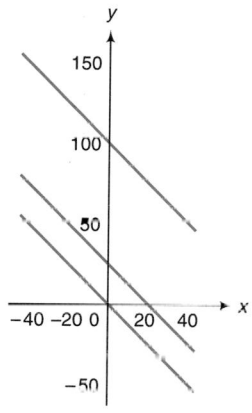

45. concentric ellipses (centered at the origin) with equations $x^2 + 4y^2 = 1 - z^2$ if $z < 1$; if $z = 1$, we obtain the single point $(0, 0, 0)$

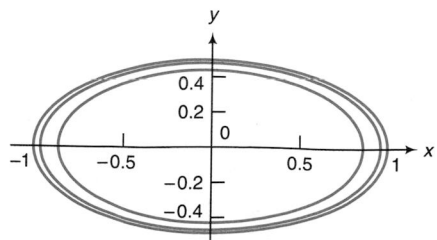

47. parallel straight lines: $y = x - \cos z$

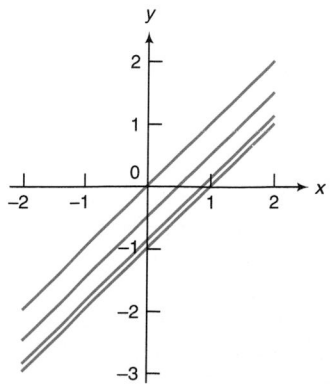

49. for each value of z we get a family of straight lines all of which have a slope of -1:
for $z = 0$ we obtain $y = -x + n\pi$ where n is an integer; for $z = 1$ we obtain $y = -x + (n + \frac{1}{4})\pi$;
for $z = -1$ we obtain $y = -x + (n - \frac{1}{4})\pi$; for $z = \sqrt{3}$ we obtain $y = -x + (n + \frac{1}{3})\pi$

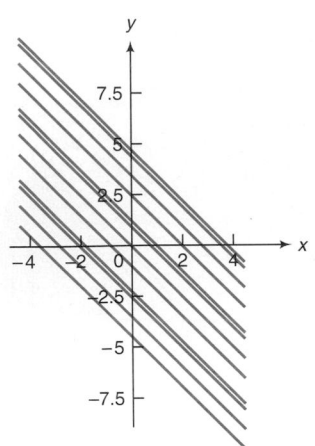

51. concentric ellipses $x^2/a^2 + y^2/b^2 = 1$ (centered at the origin) with $a = \sqrt{T - 20}$ and $b = \frac{1}{2}a$ for $T > 20$

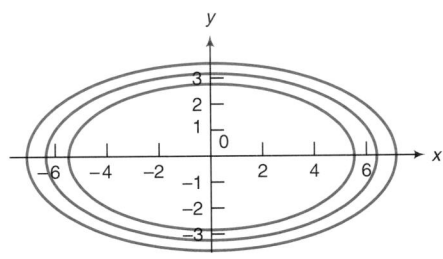

53. concentric ellipses (centered at the origin) with $a = \sqrt{(P - 100)/2}$ and $b = \sqrt{(P - 100)/3}$ for $P > 100$;
for $P = 100$ we obtain the single point $(0, 0, 0)$

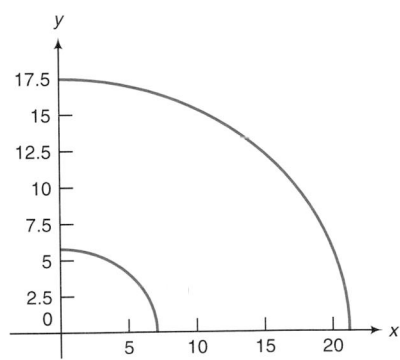

55. a. $25{,}000 \cdot 5^{1/3} \cdot 3^{2/3} \approx 88{,}922.3$ **b.** $F(2L, K/2) = 2^{-1/3}F(L, K) \approx 0.79\, F(L, K)$ (i.e., output decreases about 21%)
c. $F(L/2, 2K) = 2^{1/3}F(L, K) \approx 1.26F(L, K)$ (i.e., output increases about 26%)   **57.** $1984   **59.** f   **61.** a
63. b   **65.** g   **67.** l   **69.** i

Problems 13.2, page 850

1. (the circle itself is not included in this region)   **3.** (the sphere itself is not included in this region)

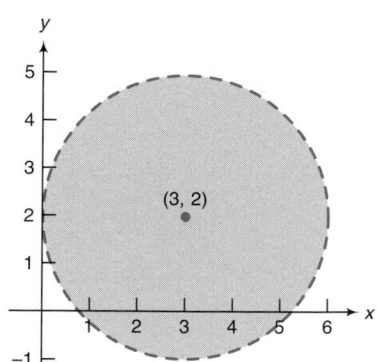

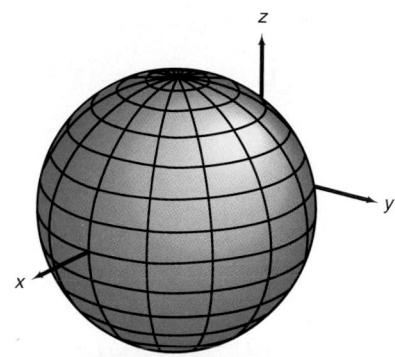

11. Approaching $(0, 0)$ along the line $y = -x$ we have $(x + y)/(x - y) = 0$ but along the line $y = 2x$ we find $(x + y)/(x - y) = -3$.
13. $xy/(x^2 + y^2) = m/(1 + m^2)$ along the line $y = mx$; $m/(1 + m^2)$ is not constant with respect to changes in m.
15. The function is undefined along the line $y = -x$; therefore the limit cannot exist because there is no neighborhood of the origin within which the function is defined.†
17. $(x^2 - 2y)/(y^2 + 2x) = (x - 2m)/(xm^2 + 2)$ along the line $y = mx$; $\lim_{x \to 0}(x - 2m)/(xm^2 + 2) = -m$
19. Approach $(0, 0, 0)$ within the plane $z = 0$ and use the answer to Problem 13.   **21.** -18 **23.** $-\frac{11}{13}$
25. $\ln(1 + e^3)$ **27.** 0   **29.** $\frac{20}{181}$ **31.** $\{(x, y): y < x\}$ **33.** $\{(x, y): x \neq y\}$ **35.** $\{(x, y): 3x + 2y + 6 > 0\}$
37. $\mathbb{R}^2$ **39.** $\{(x, y): (x/2)^2 + y^2 < 1\}$ **41.** $\{(x, y): x - y + 2z + 4 \neq 0\}$ **43.** $\{(x, y, z): -1 < x^2 - y < 1\}$
45. $g(x) = 2x$ **47.** $c = 0$ **49.** 0   **51.** 0

Problems 13.3, page 856

1. $2xy; x^2$ **3.** $3y^3e^{xy^3}; 9xy^2e^{xy^3}$ **5.** $4/y^5; -20x/y^6$ **7.** $3x^2y^5/(x^3y^5 - 2); 5x^3y^4/(x^3y^5 - 2)$
9. $\frac{4}{3}(x + 5y \sin x)^{1/3}(1 + 5y \cos x); \frac{4}{3}(x + 5y \sin x)^{1/3}(5 \sin x)$ **11.** $1/\sqrt{1 - (x - y)^2}; -1/\sqrt{1 - (x - y)^2}$
13. $z/(1 - x - 3yz^2); z^3/(1 - x - 3yz^2)$ **15.** $\cos(z - x)/[\cos(z - x) + y/z^2]; 1/z[\cos(z - x) + y/z^2]$ **17.** 3   **19.** 0
21. 1   **23.** $\frac{48}{169}$ **25.** $w_x = w_y = w_z = \frac{1}{2}(x + y + z)^{-1/2}$ **27.** $e^{x + 2y + 3z}; 2e^{x + 2y + 3z}; 3e^{x + 2y + 3z}$
29. $1/z; 1/z; -(x + y)/z^2$ **31.** $3x^2/(x^3 + y^2 + z); 2y/(x^3 + y^2 + z); 1/(x^3 + y^2 + z)$ **33.** 12   **35.** $\frac{1}{2}$ **37.** $\frac{1}{7}$
39. 16   **41.** $-e^6$ **43. a.** $x = 1, (y + 1)/1 = (z - 5)/(-12)$ **b.** $y = -1, (x - 1)/1 = (z - 5)/3$
45. $y = 1, (x - 1)/(-4) = z - 1$ **47.** $C_x = -50(2 + x)^{-2}, C_y = -250(3 + y)^{-3}$
51. $P_v = -10R293/2^2 = -732.5R$ [T is given in $^\circ K$] where $R \approx 8315$ joules per kg-mole-degree
53. $C_t = -5/(2 \cdot 10^{2/3}) \approx -0.5386$ **61. b.** no

Problems 13.4, page 863

1. $f_{xx} = 2y; f_{xy} = f_{yx} = 2x; f_{yy} = 0$ **3.** $f_{xx} = 3y^6e^{xy^3}; f_{xy} = f_{yx} = 9y^2(1 + xy^3)e^{xy^3}; f_{yy} = 9xy(2 + 3xy^3)e^{xy^3}$

5. $f_{xx} = 0; f_{xy} = f_{yx} = -20/y^6; f_{yy} = 120x/y^7$ **7.** $f_{xx} = \dfrac{-3xy^5(4 + x^3y^5)}{(x^3y^5 - 2)^2}; f_{xy} = f_{yx} = \dfrac{-30x^2y^4}{(x^3y^5 - 2)^2}; f_{yy} = \dfrac{-5x^3y^3(8 + x^3y^5)}{(x^3y^5 - 2)^2};$

9. $f_{xx} = \frac{4}{9}(1 + 5y \cos x)^2(x + 5y \sin x)^{-2/3} - \frac{20}{3}y \sin x(x + 5y \sin x)^{1/3};$
$f_{xy} = f_{yx} = \frac{20}{9} \sin x(1 + 5y \cos x)(x + 5y \sin x)^{-2/3} + \frac{20}{3} \cos x(x + 5y \sin x)^{1/3}; f_{yy} = \frac{100}{9} \sin^2 x(x + 5y \sin x)^{-2/3}$
11. $f_{xx} = (x - y)[1 - (x - y)^2]^{-3/2}; f_{xy} = f_{yx} = -(x - y)[1 - (x - y)^2]^{-3/2}; f_{yy} = (x - y)[1 - (x - y)^2]^{-3/2}$
13. $f_{xx} = 0; f_{yy} = 0; f_{zz} = 0; f_{yz} = f_{zy} = x; f_{xz} = f_{zx} = y; f_{xy} = f_{yx} = z$
15. $f_{xx} = 2y^3z^4; f_{yy} = 6x^2yz^4; f_{zz} = 12x^2y^3z^2; f_{yz} = f_{zy} = 12x^2y^2z^3; f_{xz} = f_{zx} = 8xy^3z^3; f_{xy} = f_{yx} = 6xy^2z^4$
17. $f_{xx} = 0; f_{yy} = 0; f_{zz} = 2(x + y)/z^3; f_{yz} = f_{zy} = -1/z^2; f_{xz} = f_{zx} = -1/z^2; f_{xy} = f_{yx} = 0$
19. $f_{xx} = 9y^2e^{3xy} \cos z; f_{yy} = 9x^2e^{3xy} \cos z; f_{zz} = -e^{3xy} \cos z; f_{yz} = f_{zy} = -3xe^{3xy} \sin z; f_{xz} = f_{zx} = -3ye^{3xy} \sin z;$
$f_{xy} = f_{yx} = 3(1 + 3xy)e^{3xy} \cos z$ **21.** $6y^2$ **23.** $24/(3x - 2y)^3$ **25.** 0

---

† This type of argument also works for Problems 11 and 53.

Problems 13.5, page 874

1. $2(x + y)\mathbf{i} + 2(x + y)\mathbf{j}$ **3.** $(e/2)\mathbf{i} + (e/2)\mathbf{j}$ **5.** $(x/\sqrt{x^2 + y^2})\mathbf{i} + (y/\sqrt{x^2 + y^2})\mathbf{j}$

7. $-y\sec^2(y - x)\mathbf{i} + [\tan(y - x) + y\sec^2(y - x)]\mathbf{j}$ **9.** $\sec 3 \tan 3\mathbf{i} + 3\sec 3 \tan 3\mathbf{j}$ **11.** $\dfrac{4xy^2}{(x^2 + y^2)^2}\mathbf{i} - \dfrac{4x^2y}{(x^2 + y^2)^2}\mathbf{j}$

13. $(3\sqrt{2}/4)\mathbf{i} - (\sqrt{6}/4)\mathbf{j} + \sqrt{2}\mathbf{k}$ **15.** $\frac{5}{6}\mathbf{i} - \frac{5}{12}\mathbf{j}$ **17.** $9\mathbf{i} + 6\mathbf{j} + 27\mathbf{k}$ **19.** $\mathbf{i} - \mathbf{k}$ **29. a.** $f_x(0, 0) = 0 = f_y(0, 0)$
31. $f(x, y) = \frac{1}{2}(x^2 + y^2) + C$

Problems 13.6, page 879

1. $3e^{3t}$ **3.** 1 **5.** $\dfrac{5(\cos 5t)(\cos 3t) + 3(\sin 5t)(\sin 3t)}{\sin^2 5t + \cos^2 3t}$ **7.** $2t$ **9.** $3e^{3t} + 4e^{4t} - 5e^{5t}$ **11.** $z_r = 2r, z_s = -2s$

13. $z_r = -e^{s-r}, z_s = e^{s-r}$ **15.** $z_r = 2r/s^2; z_s = -2r^2/s^3$ **17.** $w_r = s + 2, w_s = r$
19. $w_r = (-2r - s + t)/(s + 2t), w_s = (t - r)(2t - r)/(s + 2t)^2, w_t = (r + s)(2r + s)/(s + 2t)^2$
21. $w_s = \cosh(\sqrt{r + s} + 2\sqrt[3]{s - t} + 3/(r + t))[1/2\sqrt{r + s} + 2/3(s - t)^{2/3}]$
23. $w_r = 8r/(4r^2 + 5s^2 - t^2), w_s = 10s/(4r^2 + 5s^2 - t^2), w_t = -2t/(4r^2 + 5s^2 - t^2)$
25. $w_x = x/\sqrt{x^2 + y^2} - y^2/(x^2 + y^2)^{3/2}, w_y = y/\sqrt{x^2 + y^2} + xy/(x^2 + y^2)^{3/2}, w_z = 1$ **27.** $V_t = 7500\pi$ in^3/min; increasing
29. $P_t = (10R + 75)/1000$ N/cm^2/min; increasing **31.** $(2\pi - 38 - 2\sqrt{3})/\sqrt{676 - 240\sqrt{3}} \approx -2.18$ cm/sec; decreasing
33. $w_r = g_x \cos\theta + g_y \sin\theta, w_\theta = -g_x r \sin\theta + g_y r \cos\theta, w_t = g_z$
39. $z = g(y + ax)$ where g is an arbitrary smooth function of one variable

Problems 13.7, page 883

1. T: $y = 1$; N: $x = z = 0$ **3.** T: $(x - a)/a + (y - b)/b + (z - c)/c = 0$; N: $a(x - a) = b(y - b) = c(z - c)$
5. T: $(x - 4)/4 + (y - 1)/2 + (z - 9)/6 = 0$, or $3x + 6y + 2z = 36$; N: $4(x - 4) = 2(y - 1) = 6(z - 9)$
7. T: $2(x - 1) + (y - 2) + (z - 2) = 0$, or $2x + y + z = 6$; N: $(x - 1)/2 = y - 2 = z - 2$
9. T: $24(x - 3) - 2(y - 1) + 20(z + 2) = 0$, or $12x - y + 10z = 15$; N: $(x - 3)/24 = (y - 1)/(-2) = (z + 2)/20$
11. T: $(\pi/\sqrt{3})(y - 1) + \sqrt{3}(z - \pi/3) = 0$, or $\pi y + 3z = 2\pi$; N: $(\sqrt{3}/\pi)(y - 1) = (1/\sqrt{3})(z - \pi/3), x = \pi/2$
13. T: $(\ln 5)(x - 1) + (\ln 5)(y - 1) + (z - \ln 5) = 0$, or $\ln 5(x + y) + z = 3 \ln 5$; N: $(x - 1)/(\ln 5) = (y - 1)/(\ln 5) = z - \ln 5$
15. T: $z = x + 2y - 2$; N: $x\mathbf{i} + y\mathbf{j} + z\mathbf{k} = (\mathbf{i} + \mathbf{j} + \mathbf{k}) + t(-\mathbf{i} - 2\mathbf{j} + \mathbf{k})$
17. T: $z = 1$; N: $x\mathbf{i} + y\mathbf{j} + z\mathbf{k} = (\pi/8)\mathbf{i} + (\pi/20)\mathbf{j} + (1 + t)\mathbf{k}$
19. T: $x + y + 4z = -\pi$; N: $x\mathbf{i} + y\mathbf{j} + z\mathbf{k} = (-2\mathbf{i} + 2\mathbf{j} + (-\pi/4)\mathbf{k}) + t(\frac{1}{4}\mathbf{i} + \frac{1}{4}\mathbf{j} + \mathbf{k})$
21. T: $z = 2\sqrt{3}x - 2\sqrt{3}y + 2 - (2\sqrt{3}/3)\pi$; N: $x\mathbf{i} + y\mathbf{j} + z\mathbf{k} = ((\pi/2)\mathbf{i} + (\pi/6)\mathbf{j} + 2\mathbf{k}) + t(-2\sqrt{3}\mathbf{i} + 2\sqrt{3}\mathbf{j} + \mathbf{k})$
23. $5\mathbf{i} + \mathbf{j}$; $((x - 1)\mathbf{i} + (y - 5)\mathbf{j}) \cdot (5\mathbf{i} + \mathbf{j}) = 0$, or $5x + y = 10$
25. $-6\mathbf{i} + 8\mathbf{j}$; $((x - 4)\mathbf{i} + (y - 3)\mathbf{j}) \cdot (-6\mathbf{i} + 8\mathbf{j}) = 0$, or $-6x + 8y = 0$
27. $(\sqrt{2}/2)\mathbf{i} + (\sqrt{2}/4)\mathbf{j}$; $((x - \sqrt{2})\mathbf{i} + (y - 2\sqrt{2})\mathbf{j}) \cdot ((\sqrt{2}/2)\mathbf{i} + (\sqrt{2}/4)\mathbf{j}) = 0$, or $(\sqrt{2}/2)x + (\sqrt{2}/4)y = 2$
29. at $(0, 2, 4)$, $\cos\theta = -2/\sqrt{102}$; at $(1, 1, 2)$, $\cos\theta = 2/(3\sqrt{6})$

Problems 13.8, page 890

1. $9/\sqrt{10}$ **3.** $\sqrt{2}/7$ **5.** $-5/(4\sqrt{13})$ **7.** $(2e^2 + 3e)/\sqrt{2}$ **9.** $2\sqrt{3}$ **11.** $-17/\sqrt{30}$ **13.** $2e^{-3}/\sqrt{35}$
15. $-22/\sqrt{10}$ **17.** $10/\sqrt{42}$ **19.** $\sqrt{5}$ **21.** $1/\sqrt{2}$ **23.** $(1/\sqrt{6})(-\mathbf{i} + \mathbf{j} - 2\mathbf{k})$ **25. a.** $(0, 0, 0)$

b. $(1/\sqrt{14})(3\mathbf{i} - \mathbf{j} + 2\mathbf{k})$ **c.** $-(1/\sqrt{14})(3\mathbf{i} - \mathbf{j} + 2\mathbf{k})$, yes **27.** $\left|\dfrac{x}{a}\right|^{a^2} = \left|\dfrac{y}{b}\right|^{b^2}$

Problems 13.9, page 893

1. $y^3 \Delta x + 3xy^2 \Delta y$ **3.** $(x - y)^{-1/2}(x + y)^{-3/2}(y \Delta x - x \Delta y)$ **5.** $(2x + 3y)^{-1}(2 \Delta x + 3 \Delta y)$
7. $y^2z^5 \Delta x + 2xyz^5 \Delta y + 5xy^2z^4 \Delta z$ **9.** $\sinh(xy - z)(y \Delta x + x \Delta y - \Delta z)$ **11. a.** $x(\Delta y)^2 + 2y(\Delta x)(\Delta y) + (\Delta x)(\Delta y)^2$
b. -0.000309 **13. a.** $2000\pi \approx 6283.2$ cm^3 **b.** $19\pi \approx 59.7$ cm^3 **15. a.** $\frac{8}{3}\Omega$ **b.** $dR \approx 0.0305\ \Omega$

Problems 13.10, page 902

1. local minimum at $(0, 0)$ **3.** local minimum at $(-2, 1)$ **5.** local minimum at $(-2, 1)$
7. local minimum at $(\sqrt{5}, 0)$; local maximum at $(-\sqrt{5}, 0)$; $(-1, \pm 2)$ are saddle points
9. $(\frac{1}{2}, 1)$ and $(-\frac{1}{2}, -1)$ are saddle points; local minimum at $(-\frac{1}{2}, 1)$; local maximum at $(\frac{1}{2}, -1)$
11. local minimum at $(-2, -2)$ **13.** local maximum at $(\frac{4}{3}, \frac{4}{3})$; $(0, 0)$, $(0, 4)$ and $(4, 0)$ are saddle points
15. local minima at $(\pm 1, 1)$ **17.** $(0, 0)$ is a saddle point. (Note that $D = 0$ at the origin, but f can take positive and negative values in any neighborhood of the origin.)   **19.** $x = y = z = 50/3$; the maximum is $(50/3)^3$
21. Maximum product $\approx 36{,}168{,}981$ at $(\frac{25}{3}, \frac{50}{3}, 25)$.   **23.** $7/\sqrt{6}$ at $(\frac{13}{6}, \frac{4}{3}, \frac{5}{6})$ **25.** $x = y = \sqrt{\beta/3}$, $z = \frac{1}{2}\sqrt{\beta/3}$
27. Profit $P(x, y) = 40(8xy + 32x + 40y - 4x^2 - 6y^2) - 10x - 4y$ is maximized when $x = \frac{3501}{160} \approx 21.88 \approx 22$ and $y = 1433/80 \approx 17.91 \approx 18$. Maximum profit $\approx \$28{,}188.77$.   **29. a.** $P(a, b) = 150[300a/(a + 3) + 160d/(d + 5)] - a - d$
b. No   **31.** $y = \frac{30}{13} - \frac{27}{26}x$

Problems 13.11, page 910

1. $x = y = z = \frac{50}{3}$; the maximum is $(50/3)^3$ **3.** $x = \frac{25}{3}$, $y = \frac{50}{3}$, $z = 25$; the maximum $\approx 36{,}168{,}981$
5. $3/\sqrt{13}$ to the point $(\frac{7}{13}, \frac{17}{13})$ **7.** $5/\sqrt{3}$ to the point $(\frac{8}{3}, \frac{2}{3}, \frac{1}{3})$
9. $|d|/\sqrt{a^2 + b^2}$ to the point $(ad/\sqrt{a^2 + b^2}, bd/\sqrt{a^2 + b^2})$ **11.** $|ax_0 + by_0 - d|/(a^2 + b^2)$
13. $0 = 0^2 + 0^2$ is the minimum value; $x^2 + y^2$ is unbounded even when restricted to $x^3 + y^3 = 6xy$. [x or y can be negative]
15. $1 = (\pm 1)^2 + 0^2 + 0^2$ is the minimum value; $x^2 + y^2 + z^2$ is unbounded since y is unconstrained.
17. $\sqrt{3}$ is the maximum value and $-\sqrt{3}$ is the minimum value (at the points $(\pm 1)(1/\sqrt{3}, 1/\sqrt{3}, 1/\sqrt{3})$ respectively)
19. maximum is $\dfrac{|abc|}{3\sqrt{3}}$, at $\left(\dfrac{|a|}{\sqrt{3}}, \dfrac{|b|}{\sqrt{3}}, \dfrac{|c|}{\sqrt{3}}\right)$ and three other points; minimum is $\dfrac{-|abc|}{3\sqrt{3}}$ at
$\left(-\dfrac{|a|}{\sqrt{3}}, -\dfrac{|b|}{\sqrt{3}}, \dfrac{|c|}{\sqrt{3}}\right)$ and three other points   **21.** $\frac{121}{32}$ at $(\frac{5}{4}, \frac{5}{4}, -\frac{1}{2})$
23. $\min\{|a|, |b|, |c|\}$ is the minimum and $\max\{|a|, |b|, |c|\}$ is the maximum   **25.** $\sqrt{n}$ **27.** $4\sqrt{3}$
29. base is 2 m², height is 3 m   **31.** $x/y = \frac{1}{2}$ **33. a.** $L = \frac{1}{2}$, $K = \frac{150}{7}$ **b.** $250(\frac{1}{2})^{0.7}(\frac{150}{7})^{0.3} \approx 14{,}343$
c. marginal productivity of labor/marginal productivity of capital $= \frac{4}{7}$ **35. a.** $l = 150/\sqrt{1637}$, $h = 128/\sqrt{1637}$
b. $6584/\sqrt{1637}\,¢ \approx \$1.62$ **37.** 10 days in each city

Problems 13.12, page 915

1. b. $(x_1, y_1) = (-0.025, -0.075)$; $(x_2, y_2) = (-0.005573048, -0.0003463476)$ **c.** $(x_1, y_1) = (1.025, 1035)$,
$(x_2, y_2) = (1.00077655, 1.00096693)$ **3. a.** $(x_1, y_1) = (1.70833333, 0.875)$; $(x_2, y_2) = (1.69806217, 0.88330962)$
b. $(1.69806217, 0.88330962)$
5. i. a. $(x_1, y_1) = (-1.2, 1.3)$; $(x_2, y_2) = (-1.192857143, 1.278571423)$ **b.** $(-1.19287310, 1.27844411)$
    ii. a. $(x_1, y_1) = (2.925, -3.5125)$; $(x_2, y_2) = (2.923492752, -3.510016747)$ **b.** $(2.92349211, -3.51001556)$
7. i. a. $(x_1, y_1) = (2.596280088, 2.705142232)$; $(x_2, y_2) = (2.590876452, 2.692273937)$ **b.** $(2.59085858, 2.69222325)$
    ii. a. $(x_1, y_1) = (2.462540717, -0.7052117264)$; $(x_2, y_2) = (2.46272607, -0.6797490992)$
    b. $(2.4627339306, -0.6795568791)$
    iii. a. $(x_1, y_1) = (-0.0422535211, -0.492957746)$; $(x_2, y_2) = (-0.0491181274, -0.410806881)$
    b. $(-0.04932091, -0.40838904)$
    iv. a. $(x_1, y_1) = (-2.327133479, 2.15809628)$; $(x_2, y_2) = (-2.308094033, 2.110366365)$ **b.** $(-2.30789104, 2.10937054)$
    v. a. $(x_1, y_1) = (-2.415584416, -0.097402597)$; $(x_2, y_2) = (-2.410834885, -0.093801326)$
    b. $(-2.41082044, -0.09379760)$
    vi. a. $(x_1, y_1) = (-0.2821100917, 2.385321101)$; $(x_2, y_2) = (-0.2855592166, 2.380159497)$
    b. $(-0.2855601249, 2.3801497352)$
9. $(-0.090533054, -0.0998636709)$; $(1.2433857533, 0.0221338638)$; $(-0.2311447876, 1.225000791)$;
$(-1.1531141693, -0.2017059939)$; $(0.012485153, -1.1248379068)$; $(1.2912933376, 1.159132058)$;
$(-1.0604369649, 1.2569429491)$; $(1.1860751213, -1.180972109)$; $(-1.1980103894, -1.0558299813)$
13. a. $(-0.8471270884, 0.1528729116, 1.8471270884)$ **b.** $(1.1804604217, 2.1804604217, -0.1804604217)$

Problems 13 Review, page 917

1. $\{(x, y): |x| \geq |y|\}$; $[0, \infty)$ **3.** $\mathbb{R}^2$; $[-1, 1]$ **5.** $\{(x, y, z): x^2 + y^2 + z^2 > 1\}$; $(0, \infty)$
7. parallel straight lines with slopes of -1; $y = -x + (1 - z^2)$

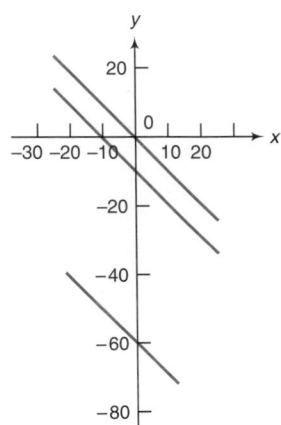

9. parallel straight lines with slopes of $\frac{1}{3}$; $x - 3y = e^z$ **11.**

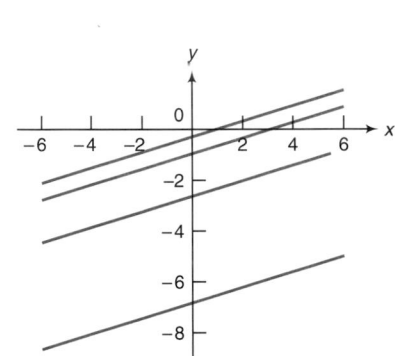

13. along $y = kx$ limit is $k/(k^2 - 1)$ **15.** since $0 \leq (x - y^2)^2 = x^2 - 2xy^2 + y^4$, $|2xy^2| \leq x^2 + y^4$ so that
$|4xy^3/(x^2 + y^4)| \leq |4xy^3/2xy^2| = 2|y|$ **17.** $-\frac{1}{4}$ **19.** $\{(x, y): 2x - 3y < 1\}$ **21.** $\{(x, y, z): x^2 - y^2 + z^2 < 1\}$
23. $f_x = -y/x^2$; $f_y = 1/x$ **25.** $f_x = -x/(x^2 - y^2)^{3/2}$; $f_y = y/(x^2 - y^2)^{3/2}$ **27.** $f_x = 1/(x - y + 4z)$;
$f_y = -1/(x - y + 4z)$; $f_z = 4/(x - y + 4z)$ **29.** $f_x = -2(y/x^3) \sinh(y/x^2)$; $f_y = (1/x^2) \sinh(y/x^2)$; $f_z = 0$
31. Let A denote $\frac{2}{3}(x^2y - y^3z^5 + x\sqrt{z})^{-1/3}$; then $f_x = A \cdot (2xy + \sqrt{z})$, $f_y = A \cdot (x^2 - 3y^2z^5)$, and $f_z = A \cdot (-5y^3z^4 + x/2\sqrt{z})$
33. $f_x = (y - z + 3w)/(y + 2w - x)^2$; $f_y = (-x + z - w)/(y + 2w - x)^2$; $f_z = -1/(y + 2w - x)$;
$f_w = (-3x + y + 2z)/(y + 2w - x)^2$ **35.** $f_{xx} = 0$; $f_{xy} = f_{yx} = 3y^2$; $f_{yy} = 6xy$ **37.** $f_{xx} = -y^2(x^2 - y^2)^{-3/2}$;
$f_{xy} = f_{yx} = xy(x^2 - y^2)^{-3/2}$; $f_{yy} = -x^2(x^2 - y^2)^{-3/2}$ **39.** Let $A = (2 - 3x + 4y - 7z)^{-2}$. Then $f_{xx} = -9A$, $f_{xy} = f_{yx} = 12A$,
$f_{yy} = -16A$, $f_{xz} = f_{zx} = -21A$, $f_{zz} = -49A$, $f_{yz} = f_{zy} = 28A$ **41.** 0   **43.** $2\mathbf{i} - 12\mathbf{j}$ **45.** $\frac{4}{25}\mathbf{i} - \frac{6}{25}\mathbf{j}$ **47.** $2\mathbf{i} + 6\mathbf{k}$
49. $-(a^2 + b^2 + c^2)^{-3/2}(a\mathbf{i} + b\mathbf{j} + c\mathbf{k})$ **51.** $2 \cos 2t$ **53.** $-2s^5/r^3$ **55.** $w_r = 5r^4s^2$, $w_s = 2r^5s$
57. T: $x + y + z = 3$; N: $x = y = z$ **59.** T: $3x - y + 5z = 15$; N: $(x + 1)/3 = (y - 2)/(-1) = (z - 4)/5$
61. T: $-3x + 6y - 2z = 18$; N: $(x + 2)/(-3) = (y - 1)/6 = (z + 3)/(-2)$ **63.** $-3/\sqrt{2}$ **65.** $-1/(2\sqrt{13})$
67. $(9/\sqrt{14})6^{-3/2}$ **69.** $3x^2y^2 \Delta x + 2x^3y \Delta y$ **71.** $\frac{1}{2}[(x + 1)(y - 1)]^{-1/2} \Delta x - \frac{1}{2}\sqrt{x + 1}\,(y - 1)^{-3/2} \Delta y$
73. $(x - y + 4z)^{-1}(\Delta x - \Delta y + 4\Delta z)$ **75.** local minimum at $(0, 0)$ **77.** local minimum at $(\sqrt{2}, 1/\sqrt{2})$;
$(\sqrt{2}, -1/\sqrt{2})$ and $(-\sqrt{2}, 1/\sqrt{2})$ are saddle points; local maximum at $(-\sqrt{2}, -1/\sqrt{2})$ **79.** local minima at $(2^{-1/5}, \pm 2^{3/10})$
81. $8/\sqrt{11}$ at $(\frac{14}{11}, -\frac{3}{11}, \frac{20}{11})$ **83.** $\frac{1}{2}(\frac{10}{3})^{3/2}$ [dimensions are $\sqrt{\frac{10}{3}} \times \sqrt{\frac{10}{3}} \times \frac{1}{2}\sqrt{\frac{10}{3}}$]   **85.** $86/31$ at $(\frac{44}{31}, \frac{1}{31}, -\frac{27}{31})$

Computer Exercises, page 919

1. a. (i) $f_x = \dfrac{y(x^4 + 4x^2y^2 - y^4)}{(x^2 + y^2)^2}$, $f_y = \dfrac{x(x^4 - 4x^2y^2 - y^4)}{(x^2 + y^2)^2}$, $(x, y) \neq (0, 0)$ (ii) $f_x = f_y = 0$ at $(0, 0)$

b. (i) $f_{xy} = f_{yx} = \dfrac{x^6 + 9x^4y^2 - 9x^2y^4 - y^6}{(x^2 + y^2)^3}$, $(x, y) \neq (0, 0)$ (ii) $f_{xy}(0, 0) = -1$, $f_{yx}(0, 0) = 1$

c. f_{xy} is not continuous at $(0, 0)$; for example $\lim_{x \to 0} f_{xy}(x, 0) = 1$ while $\lim_{y \to 0} f_{xy}(0, y) = -1$

d.

3. a. and **b.** local minimum at $(\frac{1}{2}, -\frac{5}{4})$; saddle points at $\left(\dfrac{3 - \sqrt{33}}{4}, \dfrac{7 - \sqrt{33}}{8}\right)$ and $\left(\dfrac{3 + \sqrt{33}}{4}, \dfrac{7 + \sqrt{33}}{8}\right)$

c.

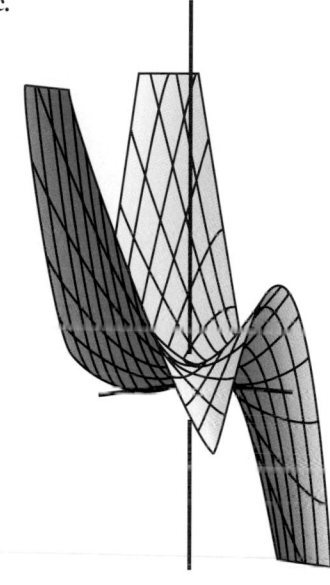

5. The bottom panel should be 1.64550 by 1.32817. The height is 0.457560. (The shorter of the dimensions of the bottom panel is the length of the left and right side panels.) The total cost of the box is $16.62.

CHAPTER 14

Problems 14.1, page 927

1. $\frac{45}{2}$ **3.** 0   **5.** 16   **7.** 39   **9.** 0   **11.** $\frac{15}{2}$ **13.** 10   **15.** 9   **17.** 41   **19.** 11
21. $0 \leq \iint \leq 6$ **23.** $-\sqrt{2}\pi/3 \leq \iint \leq \sqrt{2}\pi/3$ (These are crude bounds obtained from $|x - y| \leq \sqrt{2}$ and $1/(4 - x^2 - y^2) \leq \frac{1}{3}$ on the unit disk; a symmetry argument shows that $\iint = 0$.)   **25.** $0 \leq \iint \leq \frac{1}{4} \ln 2$
[The area of the triangle is $\frac{1}{4}$ and $x + y \leq 1$ over the triangle.]

Problems 14.2, page 938

1. $\frac{2}{3}$ **3.** $e^{-5} - e^{-1} - e^{-2} + e^2$ **5.** -31 **7.** $\frac{162}{5}$ **9.** $\frac{16}{3}$ **11.** $\frac{20}{3}$ **13.** $\frac{1}{2}(e^{19} - e^{17} - e^3 + e)$

15. $\frac{1}{3} + (\pi/16)$ **17.** $\int_0^{1/2} \int_x^{1-x} (x + 2y)\, dy\, dx = \frac{7}{24}$ **19.** $\int_0^{1/\sqrt{2}} \int_{x^2}^{1-x^2} (x^2 + y)\, dy\, dx = \dfrac{\sqrt{2}}{5}$

21. $\int_1^2 \int_x^2 \dfrac{y}{\sqrt{x^2 + y^2}}\, dy\, dx = \dfrac{1}{\sqrt{2}} - \dfrac{\sqrt{5}}{2} + 2\ln\left(\dfrac{2 + 2\sqrt{2}}{1 + \sqrt{5}}\right)$ **23.** 0 **25.** 0 **27.** $\frac{2}{15}$ **29.** 0 **31.** $-\frac{2}{5} + \frac{1}{2}\ln 5$

33. $\int_{-1}^3 \int_0^2 dy\, dx = 8$ **35.** $\int_1^2 \int_2^4 x^{-3}y^3\, dy\, dx + \int_2^4 \int_x^4 x^{-3}y^3\, dy\, dx = 27$

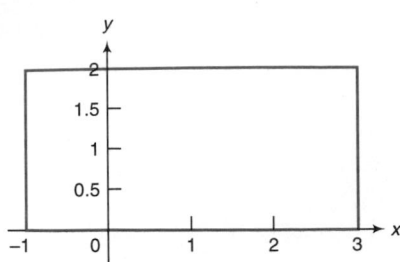
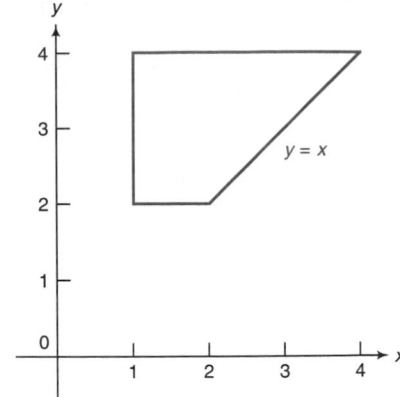

37. $\int_0^1 \int_0^y dx\, dy = \frac{1}{2}$ **39.** $\int_0^2 \int_0^{\sqrt{4-x^2}} (4 - x^2)^{3/2}\, dy\, dx = \dfrac{256}{15}$

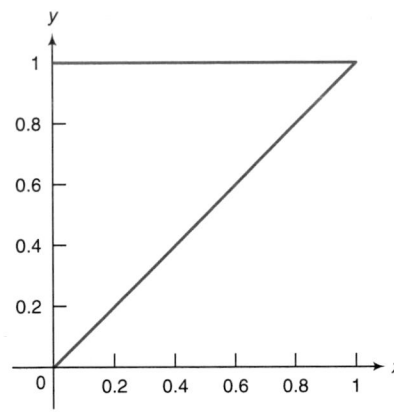
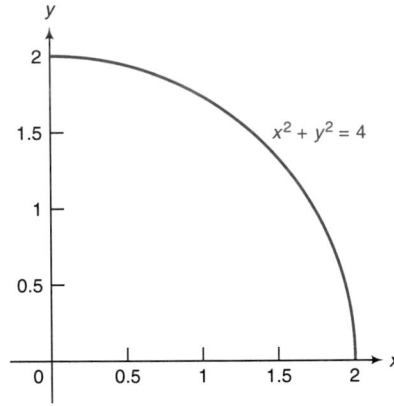

41. $\int_0^1 \int_0^{x^2} \sqrt{3 - x^3}\, dy\, dx = \frac{2}{9}(3^{3/2} - 2^{3/2})$ **43.** $\frac{9}{2}$ **45.** $\frac{128}{3}$ **47.** 48π **49.** 1 **51.** $\dfrac{\pi}{2}$

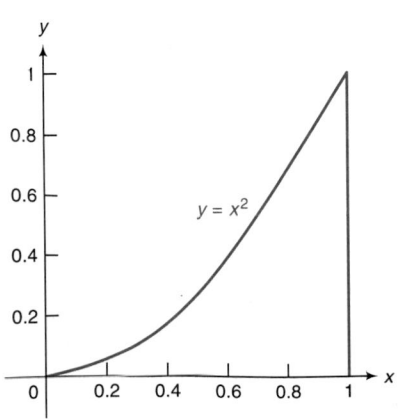

53. $\int_{-\sqrt{3}}^{-1}\int_{0}^{3-x^2} dy\,dx + \int_{-1}^{1}\int_{x^3+1}^{3-x^2} dy\,dx = 2\sqrt{3} + \frac{2}{3}$; also $\int_{-1}^{1}\int_{0}^{x^3+1} dy\,dx + \int_{1}^{\sqrt{3}}\int_{0}^{3-x^2} dy\,dx = 2\sqrt{3} - \frac{2}{3}$

55. 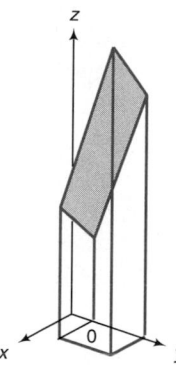 　　　　　　　　**59.** $\dfrac{\pi}{2}$

Problems 14.3, page 944

1. $\frac{16}{3}$; $(\frac{51}{32}, 0)$　　**3.** $(\sqrt{3}-1)/12$; $\left(\dfrac{1}{2} + \dfrac{\pi}{6} - \dfrac{\pi}{6(\sqrt{3}-1)}, \dfrac{1}{3} - \dfrac{\pi}{18(\sqrt{3}-1)}\right)$

5. $3e^3 - 3e + e^{-1} - e^{-3}$; $\left(\dfrac{10e^3 - 10e - 2e^{-1} + 2e^{-3}}{3e^3 - 3e + e^{-1} - e^{-3}}, \dfrac{6e^3 - 12e + 2e^{-1} - 4e^{-3}}{3e^3 - 3e + e^{-1} - e^{-3}}\right)$

7. $(\pi/16) + \frac{1}{3}$; $\left(\dfrac{62}{5(16 + 3\pi)}, \dfrac{16 + 15\pi}{5(16 + 3\pi)}\right)$　　**9.** $\frac{5}{24}$; $(\frac{11}{20}, \frac{1}{5})$　　**11.** $\frac{1}{2}$; $(1, 1)$　　**13.** $4\sqrt{2}\pi^2$

Problems 14.4, page 949

1. $2\pi a^{n+2}/(n+2)$　　**3.** 560π　　**5.** $3\pi/2$　　**7.** $2\pi + 3\sqrt{3}/2$　　**9.** 1　　**11.** $(a^2 + \frac{1}{2}b^2)\pi$　　**13.** $18\sqrt{3} - 4\pi$

15. $\frac{2}{3}\pi a^3(8 - 3\sqrt{3})$　　**17.** $\frac{512}{9}$　　**19.** $\frac{4}{3}\pi 8^{3/2}$　　**21.** $(\frac{1}{2}, 1)$　　**23.** $\left(\dfrac{b(b^2 + 4a^2)}{2(b^2 + 2a^2)}, 0\right)$

Problems 14.5, page 955

1. $\frac{1}{8}$　　**3.** 0　　**5.** $(\pi^3/16)(1 - \cos 1)$　　**7.** $\frac{1}{30}$　　**9. a.** $\int_0^1\int_z^1\int_z^y y\,dx\,dy\,dz$　　**b.** $\int_0^1\int_0^x\int_x^1 y\,dy\,dz\,dx$

11. a. $\int_0^1\int_0^x\int_0^{\sqrt{1-x^2}} yz\,dy\,dz\,dx$　　**b.** $\int_0^1\int_0^{\sqrt{1-z^2}}\int_z^{\sqrt{1-y^2}} yz\,dx\,dy\,dz$　　**13.** $\frac{1}{24}$　　**15.** $\frac{1}{8}$　　**17.** $\frac{12}{5}$　　**19.** $\frac{1}{6}$

21. $9(\pi - 1)$　　**23.** $40\pi/3$　　**25.** $\frac{4}{3}\pi abc$　　**27.** $\frac{1}{24}$　　**29.** $\frac{207}{8}$　　**31.** $(a/4, b/4, c/4)$　　**33.** $(\frac{2}{5}, \frac{1}{5}, \frac{1}{5})$

35. $(\frac{108}{115}, \frac{176}{115}, 18\pi/23 - \frac{72}{115})$　　**37. a.** $(\frac{258}{455}, \frac{372}{455}, \frac{7}{26})$

Problems 14.6, page 960

1. $\frac{16}{3}(\pi - \frac{4}{3})$　　**3.** $\pi/32$　　**5.** $\dfrac{\pi}{3} + (a - 1)\pi = (a - \frac{2}{3})\pi$　　**7.** $\pi/4$　　**9.** $32\pi(1 - 1/\sqrt{2})/3$　　**11.** $\pi/9$

13. 486π　　**15.** $(24/5(3\pi - 4), 0, 0)$　　**17.** $(0, 0, 2)$　　**19.** $(0, 0, \frac{7}{12})$　　**21.** $(0, 0, 0)$

Problems 14 Review, page 962

1. $\frac{4}{3}$　　**3.** $\frac{67}{3}$　　**5.** $\frac{1}{20}$　　**7.** 0　　**9.** -24　　**11.** $\frac{1}{3} + \pi/16$　　**13.** $\left(\int_1^2\int_2^5 + \int_2^5\int_y^5\right)3x^2y\,dx\,dy = \frac{4347}{5}$

15. $\int_0^\infty\int_0^y f(x, y)\,dx\,dy$　　**17.** $(1 - e^{-9})\pi$　　**19.** 6　　**21.** 12π　　**23.** $81\pi/32$　　**25.** 6π　　**27.** $32\pi/3$

29. $\frac{128}{15}$　　**31.** $\pi/32$　　**33.** $16\pi/9$　　**35.** $(\pi/2\sqrt{2}, 0)$　　**37.** $(0, 0, \frac{3}{2})$　　**39.** $(\frac{9}{4}, \frac{31}{9})$　　**41.** $(0, 0, 0)$

43. $\left(\dfrac{3\sqrt{3}}{5}, \dfrac{3}{5}, 0\right)$

CHAPTER 15

Problems 15.1, page 968

1. $(x^2 + y^2)^{-3/2}(-x\mathbf{i} - y\mathbf{j})$ **3.** $2(x + y)(\mathbf{i} + \mathbf{j})$ **5.** $\sin(x - y)(-\mathbf{i} + \mathbf{j})$
7. $-y\sec^2(y - x)\mathbf{i} + [\tan(y - x) + y\sec^2(y - x)]\mathbf{j}$ **9.** $\sec(x + 3y)\tan(x + 3y)(\mathbf{i} + 3\mathbf{j})$ **11.** $4xy(x^2 + y^2)^{-2}(y\mathbf{i} - x\mathbf{j})$
13. $(x^2 + y^2 + z^2)^{-1/2}(x\mathbf{i} + y\mathbf{j} + z\mathbf{k})$ **15.** $\cos x \cos y \tan z\mathbf{i} - \sin x \sin y \tan z\mathbf{j} + \sin x \cos y \sec^2 z\mathbf{k}$
17. $(\ln y - z/x)\mathbf{i} + (x/y)\mathbf{j} - \ln x\mathbf{k}$ **19.** $e^{x + 2y + 3z}[(y - z)\mathbf{i} + (1 + 2y - 2z)\mathbf{j} + (-1 + 3y - 3z)\mathbf{k}]$

21. $\nabla F(x, y) = \dfrac{2(x^2 - y^2 - 1)\mathbf{i} + 4xy\mathbf{j}}{[(x + 1)^2 + y^2][(x - 1)^2 + y^2]}$ **23.** $f(x, y) = -xy$

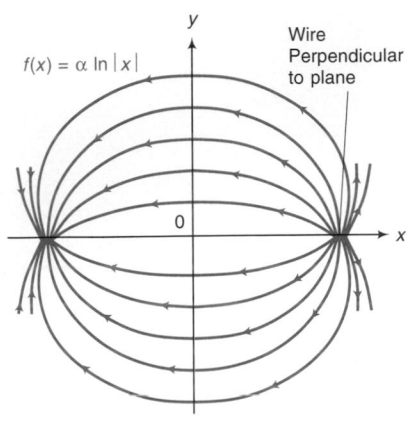

$f(x) = \alpha \ln|x|$ Wire Perpendicular to plane

25. If $k = 1$, let $f(\mathbf{x}) = \alpha|\mathbf{x}|$; if $k = 2$, $f(\mathbf{x}) = \alpha \ln|\mathbf{x}|$; if $k > 2$, then $f(\mathbf{x}) = [\alpha/(k - 2)]|\mathbf{x}|^{2 - k}$ is a potential function.

Problems 15.2, page 973

1. -3J **3.** $-6\sqrt{2}\text{J}$ **5.** $12\sqrt{13}\text{J}$ **7.** -2 **9.** 24 **11.** $-\frac{19}{3}$ **13.** $-\pi$ **15.** $-\frac{5}{6}$
17. $3\pi - \frac{2}{3}$ **19.** $\frac{1}{2}e + \frac{1}{2}e^{-1} - 2 = \cosh 1 - 2$ **21.** $\frac{5}{2}\ln 2 + \frac{2}{5}\ln 6 + \frac{1}{3}$ **23.** $\frac{3}{2}$ **25.** $\pi^3/24$
27. $\frac{9}{20} + (2e^{\pi}/5) \approx 9.706\text{ J}$ **29.** $\frac{3}{2}\pi\text{ J}$ **31.** $2ab\pi\text{ J}$ **33.** $\alpha(1 - 1/\sqrt{13})\text{ J}$ **35.** $(21 + \pi)\text{ J}$

Problems 15.3, page 982

1. $x^2y + y + C$ **3.** not exact   **5.** not exact   **7.** $\frac{1}{2}x^2 - y\sin x + C$ **9.** $\frac{3}{2}x^2\ln x - \frac{3}{4}x^2 + \frac{1}{6}x^6 - xy + C$

11. not exact   **13.** $x + y + z + C$ **15.** not exact   **17.** $\mathbf{F} = \nabla(x^2y + y)$; $\int_C \mathbf{F}(\mathbf{x}) \cdot d\mathbf{x} = 14$

19. $\mathbf{F} = \nabla[x\sin(x + y)]$; $\int_C \mathbf{F}(\mathbf{x}) \cdot d\mathbf{x} = \pi/6$ **21.** $\mathbf{F} = \nabla(x^2\cos y)$; $\int_C \mathbf{F}(\mathbf{x}) \cdot d\mathbf{x} = \pi^2/4$

23. $\mathbf{F} = \nabla(xe^y)$; $\int_C \mathbf{F}(\mathbf{x}) \cdot d\mathbf{x} = 5e^7$ **25.** $\mathbf{F} = \nabla(xy^2z^4)$; $\int_C \mathbf{F}(\mathbf{x}) \cdot d\mathbf{x} = 12$

Problems 15.4, page 988

1. 2 **3.** $\cos 1 - \sin 1 + e - 2$ **5.** $2(e^2 - 1)(1 - \cos 1)$ **7.** 0 **9.** $\pi(\frac{1}{8} - \frac{1}{3} + 1/3\sqrt{2}) = \dfrac{\pi}{24}(4\sqrt{2} - 5)$

11. $\frac{596}{15}$ **13.** $(b - a)(\text{area of }\Omega)$ **15.** 0 **17.** 23 **19.** 10

Problems 15.5, page 999

1. $\mathbf{r}(u, v) = u\mathbf{x} + v\mathbf{y}$ where $\mathbf{x} = \mathbf{i} + 2\mathbf{k}$ and $\mathbf{y} = \mathbf{j} + 3\mathbf{k}$
3. $\mathbf{r}(\theta, \varphi) = 2\cos\theta\sin\varphi\mathbf{i} + 2\sin\theta\sin\varphi\mathbf{j} + 2\cos\varphi\mathbf{k}$, $0 \leq \theta \leq 2\pi$, $0 \leq \varphi \leq \pi/2$

5. $\mathbf{r}(u, v) = u\mathbf{i} + v\mathbf{j} + \dfrac{u^2 - v^2}{4}\mathbf{k}$, $0 \leq u \leq 1$, $2 \leq v \leq 3$

7. $\mathbf{r}(r, \theta) = r \cos \theta \mathbf{i} + r \sin \theta \mathbf{j} + r(4 \cos \theta - \sin \theta)\mathbf{k}, \ 0 \le \theta \le 2\pi, \ 0 \le r \le 1$

9. $\mathbf{r}(\theta, \varphi) = \cos \theta \sin \varphi \mathbf{i} + \sin \theta \sin \varphi \mathbf{j} + \cos \varphi \mathbf{k}, \ 0 \le \theta \le 2\pi, \ 0 \le \varphi \le \pi/4$ **11.** $2\sqrt{6}$ **13.** $(\pi/2)\sqrt{1 + a^2 + b^2}$

15. $2\int_0^1 \int_0^2 \sqrt{1 + \frac{4}{9}x^{-2/3}} \, dy \, dx = \frac{4}{27}(13^{3/2} - 8)$ $\Big[$ Note that $\int_{-1}^1 \int_0^2 \sqrt{1 + \frac{4}{9}x^{-2/3}} \, dy \, dx$ is *wrong* since the integrand

is not defined at $x = 0$; evaluate the area as the sum of two improper integrals: $\int_{-1}^0 \int_0^2 + \int_0^1 \int_0^2.$ $\Big]$

17. 132   **19.** $\int_0^1 \int_0^2 \sqrt{1 + 9x} \, dy \, dx = \frac{4}{27}(10^{3/2} - 1)$ (This is the surface area for $z \ge -1$.)

21. $3\sqrt{2}\pi$ **23.** $(2 - \sqrt{2})\pi$ **25.** 12   **27.** $4a^2(\pi/2 - 1)$ **29.** 16π **31.** $\int_{-1}^1 \int_{-\sqrt{1-y^2}}^{\sqrt{1-y^2}} \sqrt{1 + 9x^4 + 9y^4} \, dx \, dy$

33. $\int_0^2 \int_x^{4-x} \sqrt{1 + \dfrac{1}{2(1 + x + y)}} \, dy \, dx = \left(\int_0^2 \int_0^y + \int_2^4 \int_0^{4-y} \right) \sqrt{1 + \dfrac{1}{2(1 + x + y)}} \, dx \, dy$

35. $2\int_{-a}^a \int_{-b\sqrt{1-x^2/a^2}}^{b\sqrt{1-x^2/a^2}} \sqrt{\dfrac{1 + \left(\dfrac{c^2}{a^2} - 1\right)\dfrac{x^2}{a^2} + \left(\dfrac{c^2}{b^2} - 1\right)\dfrac{y^2}{b^2}}{1 - \dfrac{x^2}{a^2} - \dfrac{y^2}{b^2}}} \, dy \, dx$ **37.** $\frac{1}{2}\sqrt{b^2c^2 + c^2a^2 + a^2b^2}$

43. $\mathbf{r}(u, v) = u\mathbf{i} + f(u) \cos v\mathbf{j} + f(u) \sin v\mathbf{k}, \ a \le u \le b, \ 0 \le v \le 2\pi$ **45.** $\pi a\sqrt{a^2 + h^2}$

Problems 15.6, page 1006

1. $(5^{3/2} - 1)/6$ **3.** $-\frac{101}{48}\sqrt{5} - \frac{131}{96} \ln(2 + \sqrt{5})$ **5.** 0   **7.** $2\sqrt{14}/3$ **9.** $2\sqrt{14}/81$
11. $(\sqrt{14}/6)(-\cos 7 + \cos 5 + \cos 2 - \cos 4)$ **13.** 8π **15.** 54π **17.** $536\pi/3$ **19.** $(\sqrt{3}/2)\alpha$ kg
21. $-\frac{2}{9}$ **23.** 4π **25.** 2π **27.** 0   **29.** $(\pi - 2)/6$ **31.** -12

Problems 15.7, page 1013

1. $\text{div } \mathbf{F} = 2(x + y + z); \ \text{curl } \mathbf{F} = \mathbf{0}$ **3.** $\text{div } \mathbf{F} = 0; \ \text{curl } \mathbf{F} = \mathbf{0}$ **5.** $\text{div } \mathbf{F} = x + y + z; \ \text{curl } \mathbf{F} = -y\mathbf{i} - z\mathbf{j} - x\mathbf{k}$
7. $\text{div } \mathbf{F} = 0; \ \text{curl } \mathbf{F} = x(e^{xy} - e^{xz})\mathbf{i} + y(e^{yz} - e^{yx})\mathbf{j} + z(e^{zx} - e^{zy})\mathbf{k}$

9. $\text{div } \mathbf{F} = \dfrac{1}{y} + \dfrac{1}{z} + \dfrac{1}{x}; \ \text{curl } \mathbf{F} = \left(\dfrac{y}{z^2}\right)\mathbf{i} + \left(\dfrac{z}{x^2}\right)\mathbf{j} + \left(\dfrac{x}{y^2}\right)\mathbf{k}$ **11. a.** 0   **b.** 0   **c.** $a \mid b$ **d.** $a + b$

13. a. 0   **b.** 0   **c.** $2x + 2y$ **d.** 2   **15. a.** 0   **b.** 0   **c.** 2   **d.** 2π **17. a.** $3(x^2 - y^2)\mathbf{k}$ **b.** 0   **c.** 0   **d.** 0
19. 0 (harmonic)   **21.** 20 (not harmonic)   **23.** $G = Cyz + (C - 3)xz + (C - 1)xy$ where C is an arbitrary constant
31. $\text{div } \mathbf{E} = 0$ if $|\mathbf{x}| \ne 0$ so the limit is 0   **47. c.** $\mathbf{F}$ is undefined at the origin so it is not smooth throughout the unit disk

Problems 15.8, page 1018

1. -3π **3.** 18   **5.** -108π **7.** $21 - \sin 3$ **9.** both integrals equal 1   **11.** both integrals equal -32π

Problems 15.9, page 1022

1. 4π **3.** 36π **5.** 0   **7.** 2   **9.** 108π **11.** $\frac{1}{6}$ **13.** $\frac{1}{2}$ **15.** $\frac{184}{35}$

Problems 15.10, page 1029

1. $[\partial(x, y)/\partial(u, v)] = -2$ **3.** $[\partial(x, y)/\partial(u, v)] = 4(u^2 + v^2)$
5. $[\partial(x, y)/\partial(u, v)] = -2$ **7.** $[\partial(x, y)/\partial(u, v)] = -(a^2 + b^2)$
9. $[\partial(x, y)/\partial(u, v)] = (1 - uv)e^{u+v}$ **11.** $[\partial(x, y)/\partial(u, v)] = [1/(u + v)][(1/v) - (1/u)]$
13. $[\partial(x, y)/\partial(u, v)] = \sec v \csc u(1 + uv \cot u \tan v)$ **15.** $[\partial(x, y, z)/\partial(u, v, w)] = 0$
17. $[\partial(x, y, z)/\partial(u, v, w)] = 2(u^2v - uv^2 + v^2w - vw^2 + w^2u - wu^2) = 2(u - v)(v - w)(u - w)$

19. $[\partial(x, y, z)/\partial(u, v, w)] = e^{u+v+w}$ **21.** $\int_0^1 \int_y^1 xy \, dx \, dy = \int_{-1/2}^0 \int_{-v}^{1+v} (u^2 - v^2)(2) \, du \, dv = \frac{1}{8}$

23. 2π **25.** $\frac{128}{15}$ **27.** $4\pi abc/3$

Problems 15 Review, page 1030

1. $3(x + y)^2(\mathbf{i} + \mathbf{j})$ **3.** $(x - y)^{-2}(-2y\mathbf{i} + 2x\mathbf{j})$ **5.** $2x\mathbf{i} + 2y\mathbf{j} + 2z\mathbf{k}$ **7.** $y\mathbf{i} + x\mathbf{j}$

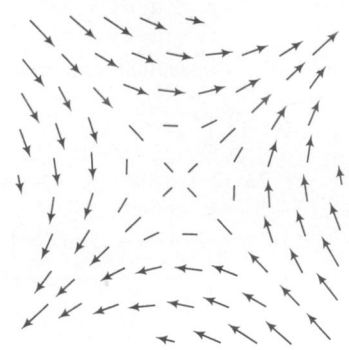

9. $\frac{2}{3}$ **11.** $-\frac{1}{2}$ **13.** $\frac{3}{2}$ **15.** $\pi/2$ **17.** 23   **19.** $\frac{5}{3}$ **21.** $\frac{1}{6}$ **23.** 0
25. a. 0   **b.** 0   **c.** $y^2 + x^2$ **d.** 8π **29.** $\mathbf{r}(u, v) = u\mathbf{x} + v\mathbf{y}$ where $\mathbf{x} = \mathbf{i} - 2\mathbf{k}$ and $\mathbf{y} = \mathbf{j} + 3\mathbf{k}$

31. $\mathbf{r}(u, v) = u\mathbf{i} + v\mathbf{j} + (u^2 - v^2)\mathbf{k}$, $0 \le u \le 2$, $2 \le v \le 4$ **33.** $\int_0^2 \int_2^4 \sqrt{1 + 4x^2 + 4y^2}\, dy\, dx$ **35.** 32π

37. $(5^{3/2} - 1)/6$ **39.** 54π **41.** $\sqrt{3}/6$ **43.** $c\sqrt{3}/12$ where c is the constant of proportionality
45. $-\frac{128}{3}$ **47.** $-32\pi/3$ **49.** div $\mathbf{F} = 3$; curl $\mathbf{F} = \mathbf{0}$ **51.** 0; $\mathbf{0}$
53. 0; curl $\mathbf{F} = x(e^{xy} - e^{xz})\mathbf{i} + y(e^{yz} - e^{yx})\mathbf{j} + z(e^{zx} - e^{zy})\mathbf{k}$ **55.** -2π **57.** 0   **59.** 8π **61.** 324π

63. 6   **65.** -5 **67.** $(\ln u)(\ln v) - 1$ **69.** 0   **71.** 3   **73.** $\int_{-1/2}^0 \int_{-v}^{1+v} 2(u^2 - v^2)\, du\, dv = \frac{1}{8}$

CHAPTER 16

Problems 16.2, page 1039

1. $(-\frac{7}{2})x^2 + C$ **3.** $\sin t + \ln|\csc x + \cot x| = 1 + \ln|\csc 3 + \cot 3|$ **5.** $y^{-4} - 2y^{-2} = -4x^{-1} - 6x^{-2} + C$
7. $x = -\ln|\cos t + e^{-1} - 1|$ **9.** $x = e^{t - (\sin 2t)/2}$ **11.** $y = \tan(\frac{1}{3}x^3 + C)$
13. 14,549,000 in 1978; 15,953,000 in 1983; 17,493,000 in 1988   **15. a.** β/δ **b.** $dQ/dt = \beta Q(1 - Q)$

17. a. $17.2°C$ **b.** $-(\ln 4)/(\ln 0.8) \approx 6.21$ hours   **19. a.** $v = \sqrt{\dfrac{mg}{\alpha}} \tanh\left(\sqrt{\dfrac{\alpha g}{m}}\, t\right)$

21. a. $c(t) = \sqrt{\dfrac{\mu}{\lambda}}\left(\dfrac{ke^{2\sqrt{\mu\lambda}t} - 1}{ke^{2\sqrt{\mu\lambda}t} + 1}\right)$, where $k = \dfrac{\sqrt{\mu} + c(0)\sqrt{\lambda}}{\sqrt{\mu} - c(0)\sqrt{\lambda}}$ **b.** $\sqrt{\mu/\lambda}$ **23.** 9:08:45.2 a.m.   **25.** $y = C \cdot x^4$

Problems 16.3, page 1048

1. $y = 2x^2 + C$ **3.** $x = 2e^t - 1$ **5.** $x = \frac{1}{2}(\sin t - \cos t) + \frac{3}{2}e^{-t}$ **7.** $y = x^x(1 + Ce^{-x})$ **9.** $x = (bt + C)e^{at}$
11. $y = \frac{1}{4} + \frac{1}{2}x \tan 2x + C \sec 2x$ **13.** $y = \frac{1}{2}[1 + 1/(1 + x^2)]$ **15.** $x = \frac{1}{2}(t^2 + 1) + \frac{7}{2}(t^2 + 1)^{-2}$

17. a. $\dfrac{dG}{dt} = k - rG$ **b.** $G(t) = \dfrac{k}{r} + \left[G(0) - \dfrac{k}{r}\right]e^{-rt} = \dfrac{k}{r}(1 - e^{-rt})$ **c.** $\lim_{t \to \infty} G(t) = \dfrac{k}{r}$

19. $I(t) = \frac{1}{4}(3 + e^{-8t})$; $I(0.1) \approx 0.862$ A   **21.** $Q(t) = \frac{1}{680}(3 \sin 60t + 5 \cos 60t - 5e^{-100t})$
23. $I(0.1) = \frac{1}{29}(2 \sin 6 - 24 \cos 6 + 24e^{-0.5}) \approx -0.3119$ A   **25.** $\approx 1000 \ln 6 \approx 1791.75$ sec $\approx \frac{1}{2}$ hr   **27.** $t = RC \ln 2$
29. $x = y(e^y + C)$

Problems 16.4, page 1056

1. $y = c_1 e^{2x} + c_2 e^{-2x}$ **3.** $y = c_1 e^{-x} \sin x + c_2 e^{-x} \cos x$ **5.** $y = 4e^{-x/4} \sin(x/4)$ **7.** $y = 5e^{-2x} - 4e^{-3x}$
9. $y = (1 + \frac{9}{2}x)e^{-5x/2}$ **11.** $y = -\frac{1}{5}e^{3x} - \frac{4}{5}e^{-2x}$ **13.** $y = \sin(2x) - \frac{3}{2}\cos(2x)$ **15.** $y = 1$
17. $y = c_1 e^{6x} + c_2 e^{7x}$ **19.** $y = -\frac{1}{2}e^x + \frac{5}{2}e^{-x}$

Problems 16.5, page 1063

1. $y = \frac{4}{3}x^3 - 4x^2 + 8x - 6 + 9e^{-x}$ **3.** $y = \frac{1}{10}e^{4x} + c_1 e^{-x} + c_2 e^{-2x}$ **5.** $y = \frac{4}{5}\cos x + c_1 \sin(\sqrt{6}x) + c_2 \cos(\sqrt{6}x)$

7. $y = (\frac{1}{2}x + c_1)\sin x + c_2 \cos x$ **9.** $y = \frac{1}{8}(\sin x + \cos x) + c_1 e^{2x}\sin x + c_2 e^{2x}\cos x$

11. $y = -\frac{1}{2}e^{3x} + (c_1 + c_2 x)e^x$ **13.** $y = (c_1 + c_2 x + x^2)e^x$ **15.** $y = \left[\left(c_1 + \frac{x}{\sqrt{3}}\right)\sin\left(\frac{\sqrt{3}x}{2}\right) + c_2 \cos\left(\frac{\sqrt{3}x}{2}\right)\right]e^{-x/2}$

17. $y = 3 - 2x\sin x$ **19.** $y = -\frac{1}{3}x\cos 3x + \frac{1}{9}x^2 - \frac{2}{81}$ **21.** $I(t) = 6(e^{-5t} - e^{-20t})$; $Q(t) = \frac{1}{10}(9 - 12e^{-5t} + 3e^{-20t})$

23. $I_{\text{steadystate}} = \frac{1050}{841}\sin 5t + \frac{1000}{841}\cos 5t$, $I_{\text{transient}} = (-\frac{1000}{841} + \frac{40}{29}t)e^{-2t}$

Problems 16.6, page 1070

1. $V = 4\sqrt{g/2}$ cm/sec (since $x = c_1 \cos\sqrt{(g/32)}t + c_2 \sin\sqrt{(g/32)}t$ with $x(0) = 16$ and $x'(0) = 0$)

3. period $2\pi/\sqrt{5\pi g}$ sec; $x(t) = [(1/5\pi) - 1]\cos\sqrt{5\pi g}t$ **5.** $k = (100\pi^2 + 1)/160$

7. $x(t) = e^{-\sqrt{17}t}(c_1 \cos 9t + c_2 \sin 9t) + [(47\sin 2t - 2\sqrt{17}\cos 2t)/22770]$ **9.** $\mu = \sqrt{2g} \approx 4.43$

Problems 16.7, page 1075

1. $y(1) \approx 0.60$ (no rounding)   **3.** $y(1) \approx 0.32768$ (no rounding)   **5.** $y(1) \approx 2.97664$ (no rounding)

7. $y(1) \approx 0.712$ **9.** $y(3) \approx 8.31$ **11.** $y(2) \approx 0.343$ **13.** $y(2) \approx 156.45$ [This is very inaccurate as $y(2) = 1190.6$.]

15. $y(0.1) \approx 1.1185$ **17.** $y(2) \approx 1.5198$ **19.** $y(0) \approx 1.6079$ **21.** $y(1) \approx 2.628$ **23.** $y(\pi) \approx 0.469$

Problems 16 Review, page 1076

1. $y = \frac{3}{2}x^2 + c$ **3.** $x = -\ln(e^{-3} - \sin t)$ **5.** $y = 0.7e^{-3x} + 0.3\cos x + 0.1\sin x$ **7.** $x = (2e^{-3} - \frac{1}{4} + \frac{1}{4}t^4)e^{3t}$

9. $y = c_1 e^x + c_2 e^{4x}$ **11.** $y = c_1 e^{3x} + c_2 e^{-3x}$ **13.** $y = (c_1 + c_2 x)e^{-3x}$ **15.** $y = e^x \sin x$

17. $y = \frac{2}{3}\sin x + c_1 \sin 2x + c_2 \cos 2x$ **19.** $y = [c_1 \sin(\sqrt{3}x/2) + (c_2 - x/\sqrt{3})\cos(\sqrt{3}x/2)]e^{-x/2}$

21. $y = 0$ (This is a harmonic oscillator that never gets going.)   **23.** $y = \frac{1}{4}e^{-x} + (c_1 + c_2 x)e^x$

25. $y = c_1 e^{3x} + c_2 e^{-2x} - \frac{7}{50}e^x \cos x + \frac{1}{50}e^x \sin x$ **27.** $y = x + \frac{1}{2}e^x - \frac{1}{2}x\cos x + c_1 \sin x + c_2 \cos x$

Note: In Problems 29–33, all computations were rounded to three decimal places at each step.

29. 6.076   **31.** 5.446   **33.** 0.625   **35. a.** 25.054   **37.** period $= 2\pi\sqrt{60/g}$; frequency $= (\sqrt{g/60}/2\pi)$,

amplitude $= \sqrt{10^2 + (-5)^2}\,10/(g/6) \approx 10.076$ cm   **39.** $\mu = \sqrt{4(10)(g/6)} \approx 80.87$

APPENDIX A1

Problems A1.1, page A-3

1. $5\pi/6$ **3.** $5\pi/3$ **5.** $4\pi/5$ **7.** $15°$ **9.** $22.5°$ **11.** $-60°$ **13.** $90°$

Problems A1.2, page A-6

In Problems 1–15 the value of $\sin\theta$ is given first.

1. 0, 1   **3.** $-\frac{1}{2}, -\sqrt{3}/2$ **5.** $(\sqrt{2}/4)(1 + \sqrt{3}), (\sqrt{2}/4)(\sqrt{3} - 1)$ **7.** $-\sqrt{2 - \sqrt{3}}/2, -\sqrt{2 + \sqrt{3}}/2$

9. $-\sqrt{2 - \sqrt{3}}/2, \sqrt{2 + \sqrt{3}}/2$ **11.** $\sqrt{2 - \sqrt{2 + \sqrt{2}}}/2, \sqrt{2 + \sqrt{2 + \sqrt{2}}}/2$ **13.** $\sqrt{2 + \sqrt{2}}/2, \sqrt{2 - \sqrt{2}}/2$

15. $-\sqrt{2 - \sqrt{2 + \sqrt{3}}}/2, \sqrt{2 + \sqrt{2 + \sqrt{3}}}/2$ **29.** amplitude $= 2$

31. amplitude $= 4$

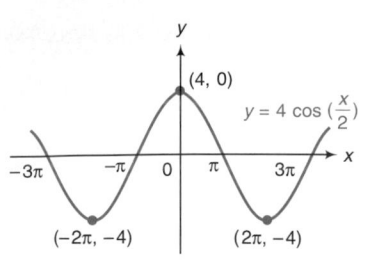

33.

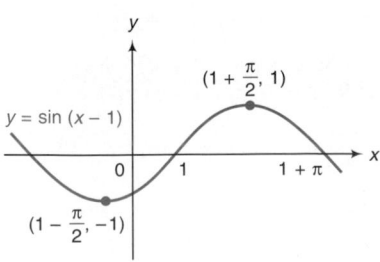

35. period $= 2\pi/3$, amplitude $= 3$

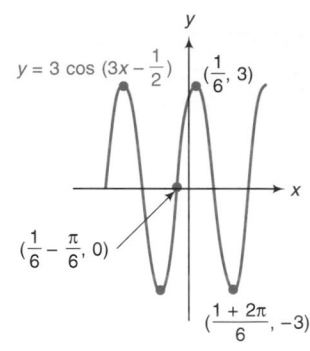

Problems A1.3, page A-9

In Problems 1–15 the answers are given in the order $\tan x$, $\cot x$, $\sec x$, $\csc x$.

1. 0, undefined, 1, undefined **3.** $1/\sqrt{3}$, $\sqrt{3}$, $-2/\sqrt{3}$, -2

5. $2 + \sqrt{3} = \sqrt{(2 + \sqrt{3})/(2 - \sqrt{3})}$, $2 - \sqrt{3}$, $2/\sqrt{2 - \sqrt{3}} = \sqrt{6} + \sqrt{2}$, $2/\sqrt{2 + \sqrt{3}} = \sqrt{6} - \sqrt{2}$

7. $2 - \sqrt{3}$, $2 + \sqrt{3}$, $-2/\sqrt{2 + \sqrt{3}} = \sqrt{2} - \sqrt{6}$, $-2/\sqrt{2 - \sqrt{3}} = -\sqrt{6} - \sqrt{2}$

9. $-2 + \sqrt{3}$, $-2 - \sqrt{3}$, $2/\sqrt{2 + \sqrt{3}}$, $-2/\sqrt{2 - \sqrt{3}}$

11. $\sqrt{2 - \sqrt{2 + \sqrt{2}}}/\sqrt{2 + \sqrt{2 + \sqrt{2}}}$, $\sqrt{2 + \sqrt{2 + \sqrt{2}}}/\sqrt{2 - \sqrt{2 + \sqrt{2}}}$, $2/\sqrt{2 + \sqrt{2 + \sqrt{2}}}$, $2/\sqrt{2 - \sqrt{2 + \sqrt{2}}}$

13. $\sqrt{2} + 1 = \sqrt{2 + \sqrt{2}}/\sqrt{2 - \sqrt{2}}$, $\sqrt{2} - 1$, $2/\sqrt{2 - \sqrt{2}}$, $2/\sqrt{2 + \sqrt{2}}$

15. $-\sqrt{2 - \sqrt{2 + \sqrt{3}}}/\sqrt{2 + \sqrt{2 + \sqrt{3}}}$, $-\sqrt{2 + \sqrt{2 + \sqrt{3}}}/\sqrt{2 - \sqrt{2 + \sqrt{3}}}$, $2/\sqrt{2 + \sqrt{2 + \sqrt{3}}}$, $-2/\sqrt{2 - \sqrt{2 + \sqrt{3}}}$

17. period $= 6\pi$ **19.** period $= \pi/3$

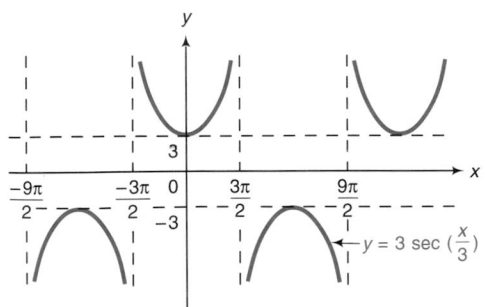

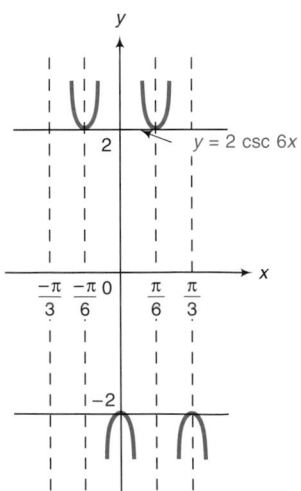

21. period $= 2\pi/5$

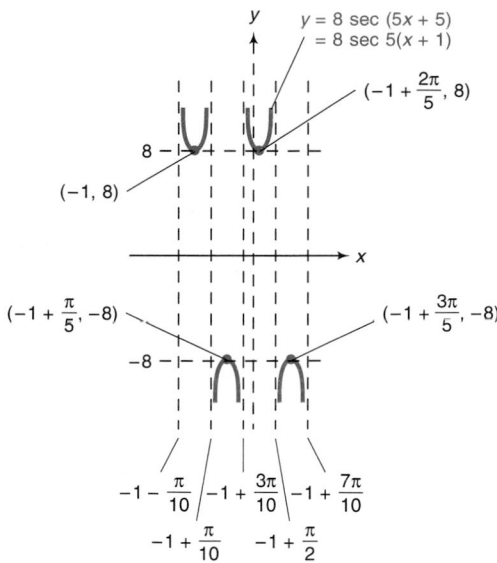

Problems A1.4, page A-10

1. $\sin \theta = \frac{4}{11}\sqrt{6}$, $\cos \theta = \frac{5}{11}$, $\tan \theta = 4\sqrt{6}/5$, $\csc \theta = 11/4\sqrt{6}$, $\sec \theta = \frac{11}{5}$, $\cot \theta = 5/4\sqrt{6}$
3. $\sin \theta = -\sqrt{3}/2$, $\cos \theta = \frac{1}{2}$, $\tan \theta = -\sqrt{3}$, $\csc \theta = -2/\sqrt{3}$, $\sec \theta = 2$, $\cot \theta = -1/\sqrt{3}$
5. $\sin \theta = \frac{1}{5}$, $\cos \theta = -2\sqrt{6}/5$, $\tan \theta = -1/2\sqrt{6}$, $\csc \theta = 5$, $\sec \theta = -5/2\sqrt{6}$, $\cot \theta = -2\sqrt{6}$
7. $\sin \theta = -\frac{2}{3}$, $\cos \theta = \sqrt{5}/3$, $\tan \theta = -2/\sqrt{5}$, $\csc \theta = -\frac{3}{2}$, $\sec \theta = 3/\sqrt{5}$, $\cot \theta = -\sqrt{5}/2$
9. $\sin \theta = 10/\sqrt{101}$, $\cos \theta = 1/\sqrt{101}$, $\tan \theta = 10$, $\csc \theta = -\sqrt{101}/10$, $\sec \theta = -\sqrt{101}$, $\cot \theta = \frac{1}{10}$

APPENDIX A2, page A-15

1. *Hint:* Verify the identity $[n^2(n+1)^2/4] + (n+1)^3 = (n+1)^2(n+2)^2/4$.   **3.** *Hint:* Apply the product rule to $xP_n(x) + b$.
5. *Hint:* Pick one element from the set and color it orange. Now count how many subsets contain this special orange element; count how many subsets do not contain it. Can there be any other subsets?   **7.** *Hint:* See Theorem 11.2.1(ii).

APPENDIX A3, page A-23

7. Let $f(x) = x + 1$ and $g(x) = x$ (more complicated examples can be found).   **9.** Let $f(x) = x$ and $g(x) = 1/x$.
11. Let $f(x) = g(x) = x$. Note that ∞/∞ is not defined.

APPENDIX A4, page A-30

1. $D = -2$; $x_1 = 3$, $x_2 = 0$ **3.** $D = 0$; infinite number—the entire line $x_1 + 2x_2 = 3$ **5.** $D = 0$; no solution
7. $D = -29$; $x_1 = x_2 = 0$ **9.** -56 **11.** 140 **13.** 96 **15.** -120 **17.** -132 **19.** -398

APPENDIX A5, page A-36

1. $9 - 7i$ **3.** 2 **5.** $-27 + 5i$ **7.** $5\sqrt{2}e^{\pi i/4}$ **9.** $3\sqrt{2}e^{-\pi i/4}$ **11.** $6e^{\pi i/6}$ **13.** $8e^{-\pi i/6}$
15. $2e^{-2\pi i/3}$ **17.** -2 **19.** $-1/(2\sqrt{2}) - (1/2\sqrt{2})i$ **21.** $-2\sqrt{3} + 2i$ **23.** $-\frac{3}{2} - (3\sqrt{3}/2)i$
25. $\cos 1 + i \sin 1 \approx 0.54 + 0.84i$ **27.** $4 - 6i$ **29.** $7i$ **31.** $2e^{-\pi i/7}$ **33.** $3e^{4\pi i/11}$
39. circle of radius a centered at z_0 if $a > 0$, single point z_0 if $a = 0$, empty set if $a < 0$

APPENDIX A6, page A-62

1. $y = x^2 + 2$

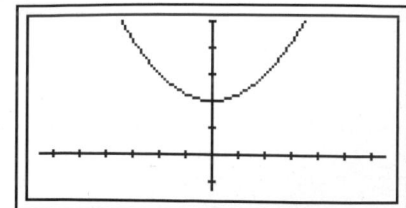

$-3 \leq x \leq 3 \qquad -1.3 \leq y \leq 5$
$x_{scl} = 0.5 \qquad y_{scl} = 1$

3. $y = x^3 - x$

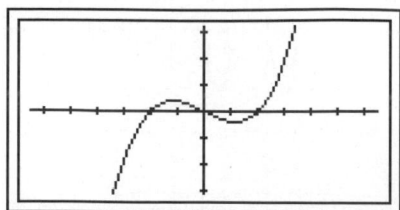

$-3 \leq x \leq 3 \qquad -3 \leq y \leq 3$
$x_{scl} = 0.5 \qquad y_{scl} = 1$

5. $y = -x^4 + 2x^2 - x + 3$

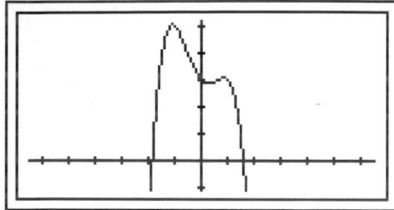

$-6.5 \leq x \leq 6.5 \qquad -1 \leq y \leq 5$
$x_{scl} = 1 \qquad y_{scl} = 1$

7. $y = \dfrac{x}{x + 1}$

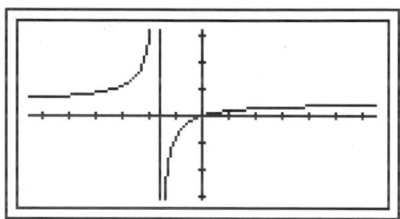

$-4 \leq x \leq 4 \qquad -6 \leq y \leq 6$
$x_{scl} = 0.6 \qquad y_{scl} = 2$

9. $y = \dfrac{x + 2}{x - 4}$

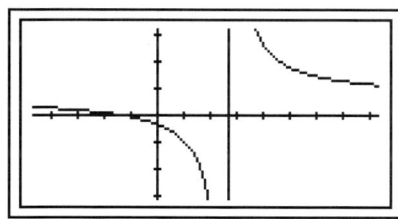

$-7 \leq x \leq 12.5 \qquad -4.7 \leq y \leq 4.8$
$x_{scl} = 1.5 \qquad y_{scl} = 1.5$

11. $y = \dfrac{x - 2}{x^2 - 1}$

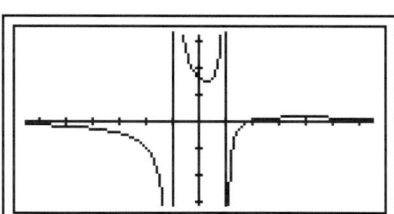

$-6.5 \leq x \leq 6.5 \qquad -4 \leq y \leq 4$
$x_{scl} = 1 \qquad y_{scl} = 1.2$

13. $y = \ln(x^2 - 4)$

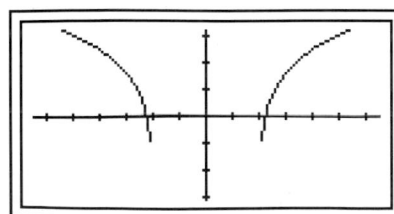

$-6.5 \leq x \leq 6.5 \qquad -3 \leq y \leq 3$
$x_{scl} = 1 \qquad y_{scl} = 1$

15. a. $y = \dfrac{e^{-1/x}}{x + 2}$

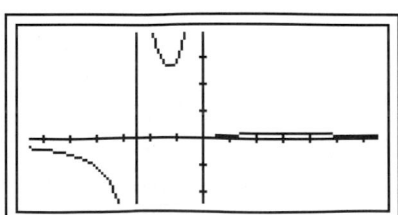

$-5 \leq x \leq 5 \qquad -2.5 \leq y \leq 4$
$x_{scl} = 0.8 \qquad y_{scl} = 1$

b. (enlarged view of first quadrant) $y = \dfrac{e^{-1/x}}{x+2}$

17. $e^{x-\ln x}$

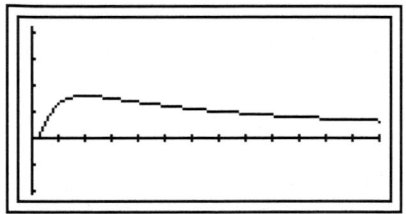

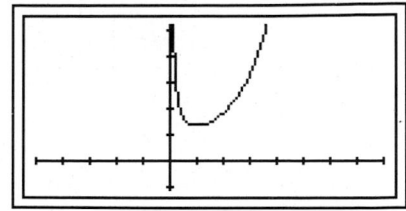

$0 \le x \le 13 \qquad -0.2 \le y \le 0.4$
$x_{\text{scl}} = 1 \qquad y_{\text{scl}} = 0.1$
19. $y = 2 \sin x - 5 \cos 3x$

$-5 \le x \le 8 \qquad -2 \le y \le 10$
$x_{\text{scl}} = 1 \qquad y_{\text{scl}} = 2$
21. a. $y = x \cos x$

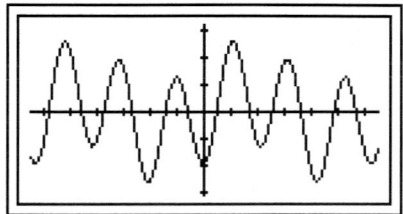

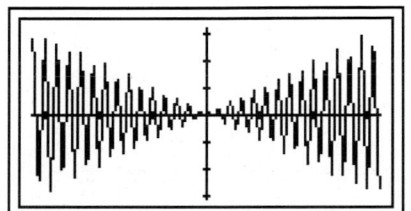

$-6.5 \le x \le 6.5 \qquad -8 \le y \le 8$
$x_{\text{scl}} = 2.6 \quad y_{\text{scl}} = 1$
b. (alternate view) $y - x \cos x$

$-500 \le x \le 500 \qquad -500 \le y \le 500$
$x_{\text{scl}} = 77 \qquad y_{\text{sol}} = 151$
23. a. $y - e^{2x} \sin 3x$

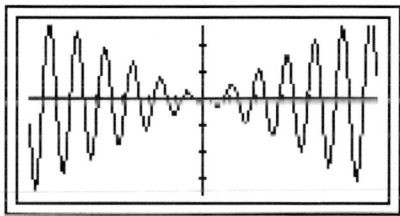

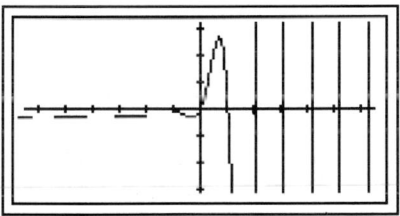

$-39 \le x \le 39 \qquad -35 \le y \le 35$
$x_{\text{scl}} = 6 \qquad y_{\text{scl}} = 11.5$
   b. (alternate view) $y = e^{2x} \sin 3x$

$-6.5 \le x \le 6.5 \qquad -4 \le y \le 4$
$x_{\text{scl}} = 1 \qquad y_{\text{scl}} = 1.3$
25. $y = 3 \sec 5x = 3/\cos 5x$

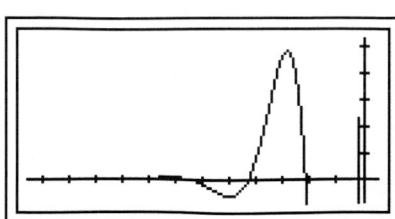

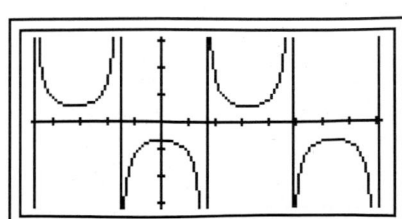

$-6 \le x \le 0.23 \qquad -0.01 \le y \le 0.055$
$x_{\text{scl}} = 0.5 \qquad y_{\text{scl}} = 0.01$

$-1.6 \le x \le 0.95 \qquad -15 \le y \le 15$
$x_{\text{scl}} = 0.2 \qquad y_{\text{scl}} = 4.5$

27. $y = \csc(x - 3) = 1/\sin(x - 3)$ **29.** $2.67, -3.50$ **31.** 1.47 **33.** 1.46 **35.** $-2.09, 0.24, 7.85$

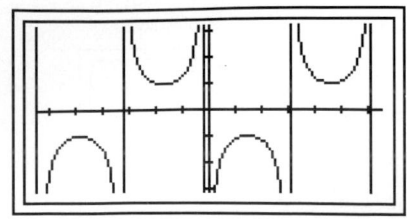

$-6.5 \le x \le 6.5$ $-3 \le y \le 3$
$x_{scl} = 1$ $y_{scl} = 1$
37. $-1.82, -0.72, 0.67, 2.87$ **39.** 1.31 **41.** $1.79, 2.62$ **43.** $-4.59, -1.25, 1.06, 5.56, 6.90$
45. $\{x: 0.33 < x < 3.00\}$ **47.** $\mathbb{R}$ **49.** $\{x: x < -1.73 \text{ or } -0.82 < x < 1.73 \text{ or } 1.82 < x < 2\}$ **51.** $\{x: x > 2.22\}$
53. $x^2 + y^2 = 25$

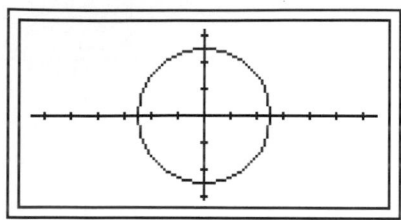

$-13 \le x \le 13$ $-6.2 \le y \le 6.4$
$x_{scl} = 2$ $y_{scl} = 2$
Note: The range values given above lead to a round circle on HP calculators. To get the same roundness on a TI or Casio calculator, it is necessary to use different values. $-9.6 \le x \le 9.4$ and $-6.4 \le y \le 6.2$ work on these machines. A similar comment applies to all the conic section graphs.
55. $x^2 + 2x + y^2 - 4y = 20$

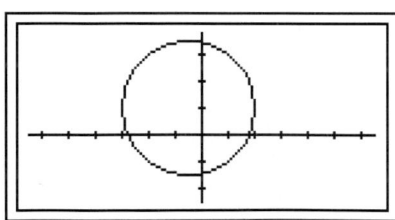

$-13 \le x \le 13$ $-5 \le y \le 7.6$
$x_{scl} = 2$ $y_{scl} = 2$
To get the same picture on a TI, the scale is $-14.4 \le x \le 14.1$ and $-9.6 \le y \le 9.3$.
57. $7x^2 - 11y^2 = 47$

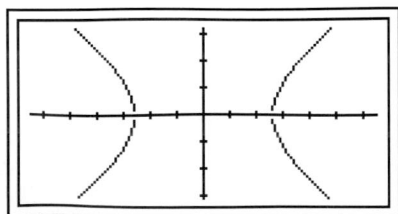

$-6.5 \le x \le 6.5$ $-3.1 \le y \le 3.2$
$x_{scl} = 1$ $y_{scl} = 1$
To get the same picture on a TI, the scale is $-4.8 \le x \le 4.7$ and $-3.2 \le y \le 3.1$.

59. $\dfrac{x^2}{37} + \dfrac{y^2}{121} = 1$

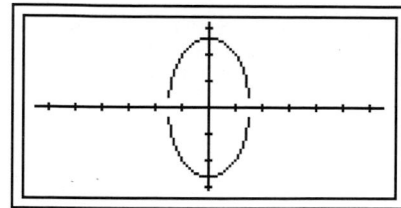

$-26 \le x \le 26 \qquad 13 \le y \le 13.5$
$x_{\text{scl}} = 4 \qquad y_{\text{scl}} = 4.2$
To get the same picture on a TI, the scale is $-19.2 \le x \le 18.8$ and $-12.8 \le y \le 12.4$.

61. $3y^2 + 12x = 6$

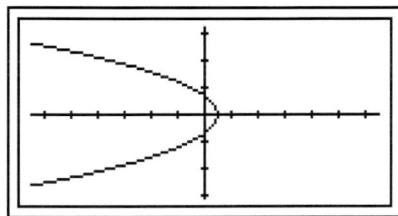

$-6.5 \le x \le 6.5 \qquad -6.2 \le y \le 6.4$
$x_{\text{scl}} = 1 \qquad y_{\text{scl}} = 2$
To get the same picture on a TI, the scale is $-9.6 \le x \le 9.4$ and $-6.4 \le y \le 6.2$.

63. $2x^2 + 7x + 5y^2 - 2y = 8$

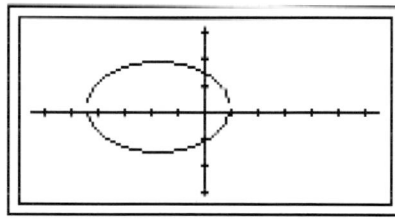

$-6.5 \le x \le 6.5 \qquad -3.1 \le y \le 3.2$
$x_{\text{scl}} = 1 \qquad y_{\text{scl}} = 1$
To get the same picture on a TI, the scale is $-6.8 \le x \le 2.7$ and $-3.2 \le y \le 3.1$.

65. $2y^2 - 8y - 12x^2 + 12x = 16$

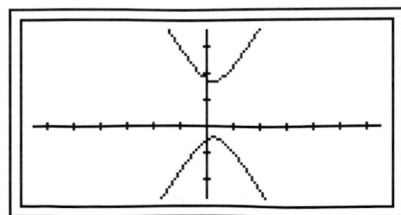

$-13 \le x \le 13 \qquad -8 \le y \le 11$
$x_{\text{scl}} = 2 \qquad y_{\text{scl}} = 3$
To get the same picture on a TI, the scale is $-14.4 \le x \le 14.1$ and $-7.6 \le y \le 11.3$.

67. $\frac{1}{2}x^2 + 4x + \frac{1}{3}y^2 - 3y = 5$

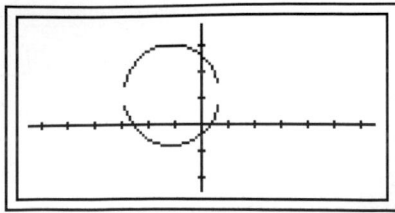

$-23 \le x \le 23 \qquad -10 \le y \le 15$
$x_{scl} = 3.5 \qquad y_{scl} = 4$
To get the same picture on a TI, the scale is $-19.2 \le x \le 18.8$ and $-9.8 \le y \le 15.4$.

69. $r = 3 \sin \theta$

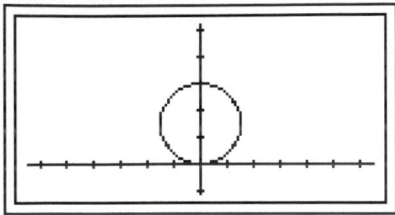

$-6.5 \le x \le 6.5 \qquad -1.1 \le y \le 5.2$
$x_{scl} = 1 \qquad y_{scl} = 1$
To get the same picture on a TI, the scale is $-4.8 \le x \le 4.7$ and $-1.2 \le y \le 5.1$.
In all the polar graphs except in Problem 77, we set the parameter range (T or θ) as $0 \le T \le 6.28$.

71. $r = 1 - 4 \sin \theta$

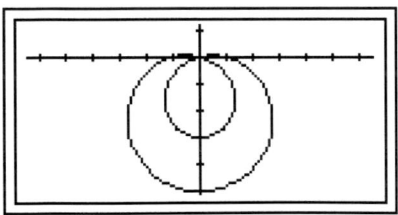

$-6.5 \le x \le 6.5 \qquad -5.1 \le y \le 1.2$
$x_{scl} = 1 \qquad y_{scl} = 1$
To get the same picture on a TI, the scale is $-4.8 \le x \le 4.7$ and $-5.2 \le y \le 1.1$.

73. $r = 4 + 3 \sin \theta$

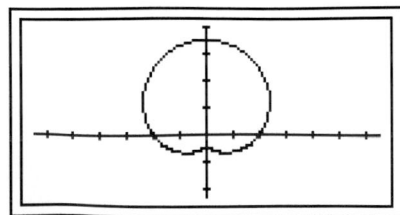

$-13 \le x \le 13 \qquad -4.6 \le y \le 8$
$x_{scl} = 2 \qquad y_{scl} = 2$
To get the same picture on a TI, the scale is $-14.4 \le x \le 14.1$ and $-9.6 \le y \le 9.3$.

75. $r = 2 \sin 3\theta$

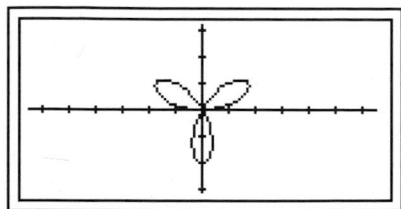

$-6.5 \leq x \leq 6.5 \qquad -3.1 \leq y \leq 3.2$
$x_{\text{scl}} = 1 \qquad y_{\text{scl}} = 1$
To get the same picture on a TI, the scale is $-4.8 \leq x \leq 4.7$ and $-3.2 \leq y \leq 3.2$.

77. $r = \dfrac{\theta}{3}$

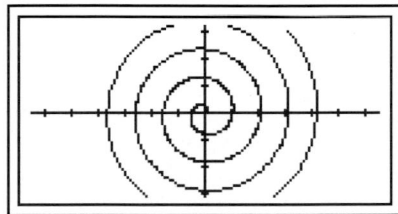

$-13 \leq x \leq 13 \qquad -6.2 \leq y \leq 6.4$
$x_{\text{scl}} = 2 \qquad y_{\text{scl}} - 2$
To get the same picture on a TI, the scale is $-9.6 \leq x \leq 9.4$ and $-6.4 \leq y \leq 6.2$. [Here we have $0 \leq T \leq 25.8$]

79. $r^2 = -9 \sin 2\theta$

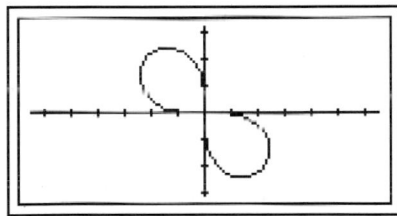

$-6.5 \leq x \leq 6.5 \qquad -3.1 \leq y \leq 3.2$
$x_{\text{scl}} = 1 \qquad y_{\text{scl}} = 1$
To get the same picture on a TI, the scale is $-9.6 \leq x \leq 9.4$ and $-6.4 \leq y \leq 6.2$.

81. $r = 1 + 2 \sec \theta$

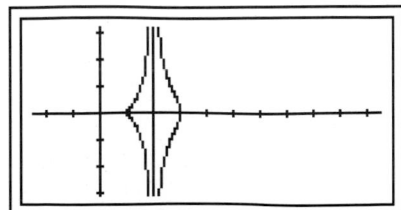

$-2.5 \leq x \leq 10.5 \qquad -9.3 \leq y \leq 9.6$
$x_{\text{scl}} = 1 \qquad y_{\text{scl}} = 3$
To get the same picture on a TI, the scale is $-1.8 \leq x \leq 7.8$ and $-9.3 \leq y \leq 9.6$.

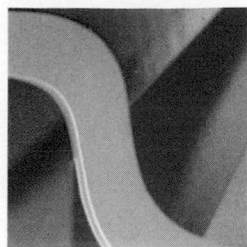

CREDITS

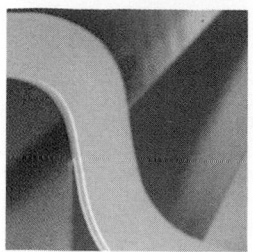

INDEX

ALGEBRAIC FORMULAS

$$|a + b| \leq |a| + |b| \qquad -|a| \leq a \leq |a|$$

$$ax^2 + bx + c = 0 \text{ if } x = \frac{-b + \sqrt{b^2 - 4ac}}{2a} \text{ or } x = \frac{-b - \sqrt{b^2 - 4ac}}{2a} \quad \text{(quadratic formula)}$$

$$n! = n(n - 1)(n - 2) \cdots (3)(2)(1) \quad \text{and} \quad 0! = 1$$

$$\binom{n}{k} = \frac{n!}{k!(n - k)!} = \frac{n(n - 1)(n - 2) \cdots (n - (k - 1))}{k!} \text{ for } n \geq k \geq 1 \quad \text{(binomial coefficient)}$$

TRIGONOMETRIC FORMULAS

$$\sin^2 x + \cos^2 x = 1 \qquad \sec^2 x = 1 + \tan^2 x \qquad \sin(-x) = -\sin x \qquad \cos(-x) = \cos x$$

$$\sin(x \pm y) = \sin x \cos y \pm \cos x \sin y \qquad \sin 2x = 2 \sin x \cos x \qquad \cos 2x = \cos^2 x - \sin^2 x$$

$$\cos(x \pm y) = \cos x \cos y \mp \sin x \sin y$$

$$\sin^2 x = \frac{1 - \cos 2x}{2} \qquad \cos^2 x = \frac{1 + \cos 2x}{2}$$

$$\tan(x \pm y) = \frac{\tan x \pm \tan y}{1 \mp \tan x \tan y}$$

$$\pi = 3.14159265358979323846 \ldots$$

EXPONENTIAL AND LOGARITHMIC FORMULAS

$$a^r = e^{r \ln a} \qquad \log_a bc = \log_a b + \log_a c$$

$$a^{b+c} = a^b a^c \qquad \log_a b^r = r \log_a b$$

$$(a^b)^c = a^{bc}$$

$$\log_a x = \frac{\ln x}{\ln a}$$

$$a^{-b} = \frac{1}{a^b}$$

$$e = 2.71828182845904523536 \ldots$$

VECTOR FORMULAS

$$|\overrightarrow{PQ}| = \sqrt{(x_1 - x_0)^2 + (y_1 - y_0)^2 + (z_1 - z_0)^2} \qquad \text{(distance formula)} \qquad \mathbf{T}(t) = \frac{\mathbf{r}'(t)}{|\mathbf{r}'(t)|} = \frac{d\mathbf{r}/dt}{|d\mathbf{r}/dt|}$$

$$|a_1\mathbf{i} + a_2\mathbf{j} + a_3\mathbf{k}| = \sqrt{a_1^2 + a_2^2 + a_3^2}$$

$$\mathbf{a} \cdot \mathbf{b} = a_1 b_1 + a_2 b_2 + a_3 b_3 = |\mathbf{a}|\,|\mathbf{b}| \cos \theta$$

$$\mathbf{N}(t) = \frac{\mathbf{T}'(t)}{|\mathbf{T}'(t)|} = \frac{d\mathbf{T}/dt}{|d\mathbf{T}/dt|}$$

$$\mathbf{a} \times \mathbf{b} = (a_2 b_3 - a_3 b_2)\mathbf{i} + (a_3 b_1 - a_1 b_3)\mathbf{j} + (a_1 b_2 - a_2 b_1)\mathbf{k}$$

$$|\mathbf{a} \times \mathbf{b}| = |\mathbf{a}|\,|\mathbf{b}| \sin \theta$$

$$k(t) = \frac{\mathbf{T}'(t)}{|\mathbf{r}'(t)|} = \frac{|\mathbf{v} \times \mathbf{a}|}{|\mathbf{v}|^3}$$

$$\mathbf{v} = \frac{d\mathbf{r}}{dt} \qquad \mathbf{a} = \frac{d\mathbf{v}}{dt}$$

$$|\mathbf{v}| = \frac{ds}{dt} = \sqrt{\left(\frac{dx}{dt}\right)^2 + \left(\frac{dy}{dt}\right)^2 + \left(\frac{dz}{dt}\right)^2}$$